SPIN PHYSICS

SPIN PHYSICS

18th International Spin Physics Symposium

Charlottesville, Virginia 6–11 October 2008

EDITORS

Donald G. Crabb
Donal B. Day
Simonetta Liuti
Xiaochao Zheng
University of Virginia
Charlottesville, VA

Matt Poelker
Jefferson Laboratory
Newport News, VA

Yelena Prok
Christopher Newport University
Newport News, VA

SPONSORING ORGANIZATIONS
University of Virginia
Brookhaven National Laboratory
European Journal of Physics
International Spin Committee
International Union of Pure and Applied Physics (IUPAP)
Jefferson Laboratory
Old Dominion University
RIKEN BNL

AMERICAN INSTITUTE OF PHYSICS

Melville, New York, 2009
AIP CONFERENCE PROCEEDINGS 1149

Editors

Donald G. Crabb
Donal B. Day
Simonetta Liuti
Xiaochao Zheng

Physics Department
University of Virginia
382 McCormick Rd.
Charlottesville, VA 22903

E-mail: dgc3q@virginia.edu
dbd@virginia.edu
sl4y@virginia.edu
xiaochao@jlab.org

Matt Poelker

Jefferson Laboratory
Postal Suite 500
m/s: ARC Bldg., Room 500-17
12050 Jefferson Avenue
Newport News, VA 23606

E-mail: poelker@jlab.org

Yelena Prok
Department of Physics
Computer Science and Engineering
Christopher Newport University
217-A Gosnold Hall
1 University Place
Newport News, VA 23606

E-mail: yprok@jlab.org

L.C. Catalog Card No. 2009905782

ISBN 978-0-7354-0686-5
ISSN 0094-243X

Printed in the United States of America

CONTENTS

PART ONE

SPIN 2008

PLENARY SESSION

PARALLEL SESSION CONTRIBUTIONS

Session A: Spin Beyond the Standard Model/Fundamental Symmetry and Spin

Conveners: D. Armstrong, Y. Prok, and X. Zheng

LONGITUDINAL SPIN CONTRIBUTIONS

Conveners: A. Deshpande, O. Rondon, and M. Vanderhaeghen

SPIN STRUCTURE OF THE NUCLEON—TRANSVERSE SPIN

Conveners: H. Avakian, A. Bacchetta, and J. P. Chen

SESSION D: SPIN STRUCTURE OF THE NUCLEON—COMBINED

Conveners: H. Avakian, J.-P. Chen, A. Bacchetta, M. Jones, and S. Liuti

GENERALIZED PARTON DISTRIBUTIONS AND NUCLEON FORM FACTORS

Conveners: M. Jones and S. Liuti

SESSION F: SPIN IN SOFT HADRONIC REACTIONS

Conveners: H. Gao and S. Kuhn

SPIN PHYSICS IN NUCLEI

Conveners: D. Day and H. WellerI

ACCELERATION, STORAGE, AND POLARIMETRY OF POLARIZED BEAMS (INCLUDE ANTI-PROTONS)

Conveners: M. Bai, S. Derbenev, E. Stephenson

POLARIZED SOURCES AND TARGETS CONTRIBUTIONS

Conveners: T. Averett, J. Clendenin, and M. Poelker

SESSION J: APPLICATION OF SPIN, ASSOCIATED TECHNOLOGIES & FUTURE FACILITIES

Convenors: T. Averett, G. Cates and B. Norum

PART TWO

Workshop on Polarized Electron Sources and Polarimeters—PESP 2008

Preface

The 18th International Symposium on Spin Physics (SPIN2008) was held on the Grounds of the University of Virginia in Charlottesville, Virginia, from October 6th – 11th, 2008. The symposium was the fifth to cover both Nuclear and High Energy Spin Physics after their merger at the 2000 Spin Symposium in Osaka, Japan. The previous Spin Symposia were SPIN2006 held in Kyoto, Japan, SPIN2004 in Trieste, Italy, and SPIN2002 in Brookhaven, USA. SPIN2008 followed the two year cycle of holding the Symposium in the USA, Europe and Asia. There were 282 registered attendees from 27 different countries. The participants ranged from students to senior physicists and some retirees, covering the gamut of theoretical and experimental Spin Physics.

The tradition of having a workshop on polarized electron sources connected with the Spin Symposium was reestablished. At SPIN2006 it was subsumed into the Symposium as a Parallel Session, but for SPIN2008 it was once again broken out as a separate workshop, PESP2008 (Workshop on Polarized Electron Sources and Polarimeters). It was held at Jefferson Laboratory from October 1st – 4th just before SPIN2008. PESP2008 was organized by Matt Poelker of Jefferson Laboratory who summarized it in a plenary talk. The talks presented at PESP2008 are included in these proceedings.

The scientific program reflected the current interests of the field with 195 talks given in Parallel Sessions and 33 Plenary Session presentations, including opening and closing session addresses. No Poster Session was included in the program thus every contribution was scheduled for an oral presentation. Included in each Parallel Session was at least one invited lead-off talk which was intended to provide some context to the rest of the papers in that session. A special session called the Directors' Session, was held on the evening of Friday October 11th , where Directors of laboratories from around the world, or their surrogates, presented their respective plans for Spin Physics.

Tributes to two colleagues, Oliver Overseth and Jacques Arvieux, both of whom had died since the last Symposium, were given by Homer Neal and Alan Krisch, acknowledging their pioneering roles in Spin Physics.

We are very grateful to our sponsors for providing financial and logistical support for the Symposium. These were: various units of the University of Virginia, Brookhaven National Laboratory, European Journal of Physics, International Spin Committee, IUPAP, Jefferson Laboratory, Old Dominion University and Riken BNL.

Finally we are extremely grateful to Michel Van Eersel *of Meetings etc* who by his management of accommodations, social program, transportation and similar duties for the Symposium lifted a great burden from the shoulders of the Local Organizing Committee, thereby allowing them to concentrate on organizing the Physics Program.

Donald Crabb
Donal Day
Simonetta Liuti
Matt Poelker
Yelena Prok
Xiaochao Zheng

International Spin Committee

K. Imai (Chair)	Kyoto
T. Roser (Past Chair)	Brookhaven
F. Bradamante	Trieste
E. D. Courant*	BNL
D. G. Crabb	Virginia
A. V. Efremov	JINR
G. Fidecaro*	CERN
H. Gao	Duke
W. Haeberli*	Wisconsin
K. Hatanaka	RCNP
A. D. Krisch	Michigan
G. Mallot	CERN
A. Masaike*	JSPS
R. G. Milner	MIT
C. Y. Prescott*	SLAC
F. Rathmann	COSY
H. Sakai	Tokyo
Yu. M. Shatunov	Novosibirsk
V. Soergel*	Heidelberg
E. Steffens	Erlangen
E. J. Stephenson	Indiana
N. E. Tyurin	IHEP
W. T. H. van Oers*	Manitoba

* Honorary Member

Local Organizing Committee

D. Crabb (Chair)	Virginia
H. Avakian	JLab
T. Averett	William and Mary
A. Bacchetta	JLab
M. Bai	Brookhaven
G. Cates	Virginia
J. P. Chen	JLab
J. Clendenin	SLAC
D. Day	Virginia
A. Deshpande	Stonybrook and RIKEN BNL
H. Gao	Duke
M. Jones	JLab
S. Kuhn	ODU
S. Liuti	Virginia
Z. Mezziani	Temple
B. Norum	Virginia
M. Poelker	JLab
Y. Prok	CNU
O. Rondon	Virginia
H. Weller	Duke
X. Zheng	Virginia/JLab

PART ONE
October 6 - 11, 2008
University of Virginia, Charlottesville, VA
SPIN 2008
The 18th International Spin Physics Symposium
International Advisory Committee:
K. Imai - Kyoto (Chair)
T. Roser - BNL (Past-Chair)
F. Bradamante - Trieste
E.D. Courant* - BNL
D.G. Crabb - Virginia
A.V. Efremov - JINR
G. Fidecaro* - CERN
H. Gao - Duke
G.Mallot - CERN
W. Haeberli* - Wisconsin
K. Hatanaka - RCNP
A.D. Krisch - Michigan
A. Masaike* - JSPS
R.G. Milner - MIT
C.Y. Prescott* - SLAC
F. Rathmann - COSY
H.Sakai - Tokyo
Yu. M. Shatunov - Novosibirsk
V.Soergel* - Heidelberg
E. Steffens - Erlangen
E.J. Stephenson - Indiana
N.E. Tyurin - IHEP
W.T.H. van Oers* - Manitoba
* Honorary Members
Local Organizing Committee:
D. Crabb - Virginia (Chair)
H. Avakian - Jefferson Lab
T. Averett - William & Mary
M. Bai - BNL
G. Cates - Virginia
J.P. Chen - Jefferson Lab
J. Clendenin - SLAC
D. Day - Virginia
A. Deshpande - SUNY Stony Brook & RBRC
H. Gao - Duke
M. Jones - Jefferson Lab
S. Kuhn - Old Dominion
S. Liuti - Virginia
Z. Meziani - Temple
B. Norum - Virginia
M. Poelker - Jefferson Lab
Y. Prok - Christopher Newport/Jefferson Lab
O. Rondon - Virginia
M. Vanderhaeghen - William &Mary/ Jefferson Lab
H. Weller - Duke
X. Zheng - Virginia
Hosted by: Physics Department, University of Virginia
Sponsored by: International Spin Committee
IUPAP
RIKEN BNL Research Center
University of Virginia
EPJ The European Physical Journal
Jefferson Lab
BROOKHAVEN NATIONAL LABORATORY
Photograph Copyright Edward Nemergut (photo.nemergut.com)
faculty.virginia.edu/spin2008

Introductory Remarks on Behalf of IUPAP

Willem T.H. van Oers

Department of Physics and Astronomy,
University of Manitoba, Winnipeg, MB, Canada R3T 2N2
and
TRIUMF, 4004 Wesbrook Mall, Vancouver, BC, Canada V6T 2A3

Abstract. Words of welcome expressed on behalf of IUPAP by the Vice-Chair of the IUPAP Commission on Nuclear Physics (C12) followed by a short description of IUPAP activities of relevance to the participants of the 18th International Symposium on Spin Physics."

Keywords: Particle Physics, Nuclear Physics, IUPAP, Conference Support
PACS: 10., 20.

It is a great pleasure to express words of welcome on behalf of the International Union of Pure and Applied Physics (IUPAP) at this 18th International Symposium on Spin Physics. These symposia have indeed a rich history and are the amalgamation of the International Symposia on Polarization Phenomena in Nuclear Physics and the International Conferences on High Energy Spin Physics. It is entirely satisfying to notice the large attendance to this 18th International Symposium with many younger scientists participating. On behalf of IUPAP and its Commission on Nuclear Physics I am wishing great success of what promises to be a very stimulating and exciting conference.

I would like to make use of the opportunity given to present a brief overview of IUPAP, its Commissions, in particular the Commission on Nuclear Physics (C12), and its Working Groups, in particular ICFA, the International Committee on Future Accelerators, and WG.9, the Working Group on International Cooperation in Nuclear Physics.

IUPAP - A BRIEF HISTORICAL OVERVIEW

The beginning was the formation of the International Research Council in 1919, largely through the efforts of representatives of the National Academy of Sciences, Washington, DC, USA, and of the Royal Society of London, United Kingdom. The objective of this body was the coordination of efforts in the different branches of science towards the formation of respective international unions.

The first General Assembly of the International Research Council took place in 1922 in Brussels, Belgium. The physicists present at the General Assembly decided that the formation of a Physics Union was indeed imperative. The physicists represented thirteen countries, in alphabetical order: Belgium, Canada, Denmark, France, Japan, the Netherlands, Norway, Poland, Spain, Switzerland, United Kingdom, United States of America, and the Union of South-Africa. A Steering Committee was formed with Sir William Bragg as President and Members: M. Brillouin, O.M. Corbino, M. Knudsen, M. Leblanc, R.A. Millikan, H. Nagoda, E. van Aubel, and H. Abraham; names that are encountered in first year physics text books.

CP1149, *Spin Physics, 18th International Spin Physics Symposium*
edited by D. G. Crabb, D. B. Day, S. Liuti, X. Zheng, M. Poelker, and Y. Prok

FIGURE 1. Sir William Bragg, President of the Steering Committee

The first General Assembly of the International Union of Pure and Applied Physics was held in Paris, France, in 1923. Three more countries joined the Union: Czechoslovakia, Italy, and Sweden. It confirmed the membership of the Executive of the Union with one replacement. And during the deliberations Madame Curie complained about the proliferation of scientific reports! IUPAP's mission is:

- to assist in the worldwide development of physics,
- to foster international cooperation in physics,
- to help in the application of physics toward solving problems of concern to humanity.

The activities of IUPAP can be listed as follows:

- sponsoring international meetings,
- encouraging research and education,
- fostering the free circulation of scientists (even today of great relevance !),
- promoting international agreements on symbols, units, and nomenclature,
- cooperating with other organizations on disciplinary and interdisciplinary problems,
- fostering communications and publications.

As already indicated above the membership of IUPAP was originally from the countries that had joined, but with the difficulty of representation from both the Republic of China (ROC) and the People's Republic of China, membership in IUPAP is now through identified physics communities in a geographical region. These identified physics communities form an "Adhering Body" with a National IUPAP Liaison Committee. Delegates appointed by the various National IUPAP Liaison Committees attend the IUPAP General Assembly. The number of delegates is determined by the number of (annually paid) shares in IUPAP.

The IUPAP Executive Council for 2006 - 2008 is as follows:

President:	Alan Astbury
Past President:	Yves Petroff
President Designate:	Sukekatsu Ushioda
Secretary-General:	Judy Franz
Associate Secretary-General:	Peter Melville
Vice Presidents (elected at large):	Jia-Er Chen, Carmen Cisneros, Bruce McKellar
Vice Presidents (of the Commissions)	Dan Dahlberg, Pavel Exner Pratibha Jolly, Leslie Pendrill, Annick Suzor-Weiner

The General Assembly is the governing body of IUPAP, which consists of the delegates appointed by the National IUPAP Liaison Committees, the Chairs or Secretaries of the nineteen Commissions, and the members of the Executive Council. The General Assembly is held every three years at a location that circulates the globe. The General Assembly elects the incoming Executive Council and the members of the Commissions. It also appoints the physics representatives to other international scientific bodies.

IUPAP COMMISSIONS

Currently there are nineteen IUPAP Commissions:

1 - Finance	11 - Particles and Fields
2 - SUNAMCO	12 - Nuclear Physics
3 - Statistical Physics	13 - Physics for Development
4 - Cosmic Rays	14 - Physics Education
5 - Low Temperature Physics	15 - Atomic, Molecular, and Optical Physics
6 - Biological Physics	16 - Plasma Physics
7 -	17 - Quantum Electronics
8 - Semiconductors	18 - Mathematical Physics
9 - Magnetism	19 - Astrophysics
10 - Structure and Dynamics of Condensed Matter	20 - Computational Physics

Among the activities of the IUPAP Commissions one can list:

- The organization of conferences in the given subdiscipline of physics.
- The awarding of prizes and medals.
- The establishment of authoritative websites for the particular subdiscipline.
  - with brief comprehensive reviews of the current activities,
  - with links to other authoritative sites, e.g., giving the latest accepted values of the fundamental constants, giving the compilation of atomic masses.
- The arrangement of representation on related scientific bodies, e.g., IUPAP Commission C2 on Symbols, Units, Nomenclature, Atomic Masses, and Fundamental Constants (SUNAMCO) has formal connections to Bureau International des Poids et Mesures (BIPM) [Sevres, France], to Comite Consultatif des Unites (CCU), International Organization for Standardization (ISO) [Geneva, Switzerland], Organisation

Internationale de Metrologie Legale (IOML) [Paris, France], International Union of Pure and Applied Chemistry (IUPAC).

- The publication of a Red Book on the standard usage of symbols and nomenclature.

The participants of this International Conference on Spin Physics are mainly represented by IUPAP Commissions C11 (Particles and Fields) and C12 (Nuclear Physics).

There exist also three affiliated Commissions: AC1 - International Commission for Optics, AC2 - International Commission on General Relativity and Gravitation, and AC3 - International Commission for Acoustics.

IUPAP WORKING GROUPS

At the present there are eight active IUPAP Working Groups. The Working Groups are formally established by the IUPAP General Assembly for three year renewable periods. The eight IUPAP Working Groups are:

- WG.1 - International Committee on Future Accelerators (ICFA).
- WG.2 - Communication in Physics.
- WG.3 - Facilities for Condensed Matter Physics with two subcommittees,
  - International Committee on the Future of Neutron Sources (ICFNS),
  - International Committee on High Magnetic Field Facilities (HMFF).
- WG.4 - Particle and Nuclear Astrophysics and Gravitation International Committee (PANAGIC).
- WG.5 - Women in Physics.
- WG.7 - International Committee on Ultrahigh Intensity Lasers.
- WG.8 - International Committee on Nanoscience.
- WG.9 - International Cooperation in Nuclear Physics (ICNP).

Of relevance for the delegates at this International Conference on Spin Physics are WG.1 (ICFA) and WG.9 (ICNP).

The International Committee on Future Accelerators (ICFA), established in 1976, has as mandate to promote international collaboration in all phases of the design, development, construction, and exploitation of very-high energy accelerators with chief use in high-energy physics. ICFA played a prominent role in the establishment of the Large Hadron Collider (LHC) at CERN and is responsible for the coordination of efforts worldwide towards the future International Linear Collider (ILC). ICFA also organizes meetings to formulate advice on future plans for regional facilities, to formulate advice on joint studies and uses, to organize workshops on the technology of very-high energy accelerator complexes and their international exploitation and to foster research and development of the necessary technology. The present Chair of ICFA is Albrecht Wagner from DESY and its Secretary is Roy Rubinstein from Fermilab [website: www.fnal.gov/directorate/icfa].

The Working Group on International Cooperation in Nuclear Physics (INPC) was established in 2003. Its mandate given is as follows:

1. To provide a description of the landscape of key issues in nuclear physics research for the next 10 to 20 years.

2. To produce and maintain a compendium of nuclear physics facilities existing or under development worldwide.
3. To establish a mapping of these facilities onto the scientic questions identified above.
4. To identify missing components that would have to be designed, developed, and constructed to provide an optimized, comprehensive network of international nuclear physics facilities.
5. To explore mechanisms and opportunities for enhancing international collaborations and cooperation in the nuclear sciences.
6. To identify those research and development projects that could benefit from international joint efforts.
7. To serve as a source of expert advice for governmental or intergovernmental organizations in connection with efforts to coordinate and promote nuclear science at the international level.
8. To serve as a forum for the discussion of future directions of nuclear science in the broadest sense.
9. To document the cross-disciplinary impact of nuclear physics and of the nuclear physics facilities and to identify mechanisms for fostering and expanding cross-disciplinary research.

To address the first four items above the IUPAP Working Group WG.9 published in July 2006 IUPAP Report 41 entitled "Research Facilities in Nuclear Physics" This report presented an overview of user facilities in nuclear physics worldwide as well as an outline of the five key questions that these and future nuclear physics facilities were intended to address. The report also provided a compilation consisting of short descriptions of some 91 nuclear physics facilities worldwide with a constituency of users. The electronic version of IUPAP Report 41 can be found by going to the IUPAP website: www.iupap.org under Working Groups and then WG.9, and clicking on its website. The report can be accessed as a pdf file. WG.9 has established a three year cycle for updating the electronic version of IUPAP Report 41. It will be soliciting updates of the descriptions of nuclear physics facilities in January 2009.

The IUPAP Working Group WG.9 (Chair and Secretary as well as the Chair of C12) was invited to join the Organization for Economic Co-operation and Development (OECD) Global Science Forum's Working Group on Nuclear Physics. The Global Science Forum (GSF) is a venue for consultations among senior science policy officials of OECD member and observer countries on matters relating to fundamental scientific research. The Global Science Forium's Working Group on Nuclear Physics presented its report in May 2008. The report described the field of nuclear physics, including the science opportunities, facilities, workforce, and investment worldwide. The scientific vision for the future was given in the form of five key research questions (as developed by various national and regional documents of the scientific communities and advisory committees) and the new capabilities that are planned to address these. Significantly, a global roadmap for large facilities was articulated, looking forward to the year 2020.

The IUPAP Working Group WG.9 has encouraged and initiated the formation of nuclear physics co-operation committees for Asia and for South-America. It has established several sub-committees to advice on specific questions. The present Chair of the IUPAP Working Group WG.9 is Anthony W. Thomas of Jefferson Laboratory and its Secretary is Willem T.H. van Oers at TRIUMF. The website is as given above.

I hope that in what I have described, I have given a glimpse of the many activities of IUPAP and its Commissions and Working Groups in particular as these pertain to nuclear physics and particle physics.

PLENARY SESSION

The Power of Spin: A scalpel-like probe of theoretical ideas

Elliot Leader

Imperial College London, Prince Consort Road, London SW7 2BW

Abstract. We survey briefly the important role that spin-dependent measurements have played, time and time again, in helping to shape our current understanding of particle physics.

Keywords: importance of spin-dependent measurements
PACS: 12.38.Bx, 12.38.-t, 13.15.+g, 13.60.Hb, 13.60.-r, 13.85.Ni, 13.88.+e, 14.20.Dh

INTRODUCTION

My aim is to remind you, or if there are any doubting Thomases, to convince them, that spin dependent measurements have a scalpel-like ability to probe a theory, which, for example, may have been able to fudge the results of ordinary cross-section measurements. Thus the path of spin is strewn with the wreckage of discarded theories. The positive aspect is that better (hopefully) theories arise from the ashes.

THE ANCIENT WORLD (PRE 1988)

I will remind you of two cases where spin played a crucial role in altering our viewpoint.

- Electroweak theory: When I was a student weak interactions were supposed to involve S-T coupling: 1 and $\frac{i}{2}[\gamma_\mu, \gamma_\nu]$. From the role of spin in comparing rates for $\pi \to e\bar{\nu}$ and $\pi \to \mu\bar{\nu}$ and from measurement of the helicity of the neutrino, we eventually learned that the coupling was V-A: $\gamma_\mu(1-\gamma_5)$. The unification of Weak and Electromagnetic interactions is one of the greatest achievements in Particle Physics. Without the discovery of the V-A structure this would have been impossible! The power of spin is vividly illustrated by experiments to determine the structure of the weak Hamiltonian. The most general form, allowing for parity violation, involves TEN coupling constants:

$$H = \sum_{i=1\ldots5} \left[C_i(\bar{\psi}_e\Gamma_i\psi_\mu)(\bar{\psi}_{\nu_\mu}\Gamma_i\psi_{\nu_e}) + C_i'(\bar{\psi}_e\Gamma_i\psi_\mu)(\bar{\psi}_{\nu_\mu}\Gamma_i\gamma_5\psi_{\nu_e})\right] \qquad (1)$$

  where the Γ_i are the usual five types of Dirac gamma matrix structures. The results of an experiment at SIN to test the V-A structure of the weak Hamiltonian [1] are shown in Fig. 1. With V-A all couplings should be zero except for the one labeled *LL* in the bottom line, which should have the value 1. The agreement is dramatic.
- Regge poles: In the 1960s-70s there was the dramatic discovery that diffraction peaks in $\frac{d\sigma}{dt}$ for elastic cross-sections

CP1149, *Spin Physics, 18th International Spin Physics Symposium*
edited by D. G. Crabb, D. B. Day, S. Liuti, X. Zheng, M. Poelker, and Y. Prok

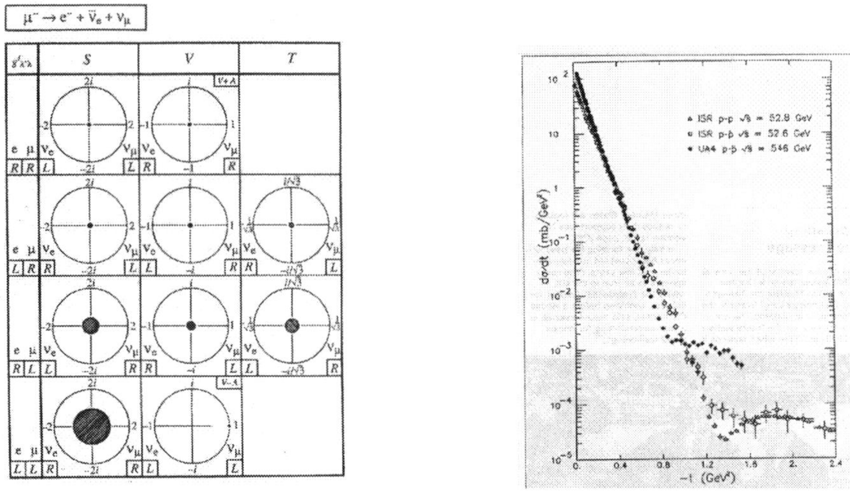

FIGURE 1. Left: V-A structure of the weak Hamiltonian. Right: Shrinking diffraction peaks

$$A(p_1) + B(p_2) \rightarrow A(p_3) + B(p_4) \qquad t = (p_1 - p_3)^2 \tag{2}$$

shrink as the energy increases ... a most non intuitive behavior. Fig. 1 shows CERN ISR and SPS UA(4) data. The theory of Complex Angular Momentum (in its simplest form, Regge poles), provided a beautiful explanation! There were many successes in relating different cross-sections, but problems arose with polarizations. Many questions remained unanswered, which could have been resolved by spin-dependent measurements. Our “Critical Tests for Regge Pole Theory" [2] , were, alas, never carried out. The deep level inclusion of spin in Regge theory provided a fantastically rich panoply of possibilities: conspiracies, evasions, daughters... most of which were never tested.

THE RENAISSANCE: THE EUROPEAN MUON COLLABORATION EXPERIMENT

A quick reminder about deep inelastic scattering in the parton model

The polarized cross-section is expressed in terms of two spin-dependent *structure functions*, $g_{1,2}$ (for a recent review and references see [3]). Our interest is in g_1:

$$g_1(x, Q^2) = \frac{1}{2} \sum_{flav} e_j^2 [\Delta q_j(x, Q^2) + \Delta \bar{q}_j(x, Q^2)] \tag{3}$$

The key ingredients here are the polarized quark densities. To appreciate the EMC experiment define

$$\Delta q = \int_0^1 dx \Delta q(x) \tag{4}$$

and the important flavor combinations :

$$a_3 = \Delta u + \Delta\bar{u} - \Delta d - \Delta\bar{d} = 1.267 \pm 0.0035 \tag{5}$$

$$a_8 = \Delta u + \Delta\bar{u} + \Delta d + \Delta\bar{d} - 2(\Delta s + \Delta\bar{s}) = 0.585 \pm 0.025 \tag{6}$$

$$\Delta\Sigma = \sum_f (\Delta q_f + \Delta\bar{q}_f) \tag{7}$$

where the values of $a_{3,8}$ are known from neutron and hyperon β-decay respectively. Note that

$$\Delta\Sigma = a_8 + 3(\Delta s + \Delta\bar{s}) \tag{8}$$

Ellis and Jaffe [4] had suggested it was safe to ignore $\Delta s + \Delta\bar{s}$ implying

$$\Delta\Sigma \simeq a_8 \simeq 0.59 \tag{9}$$

Now the EMC measurement [5] of

$$\Gamma_1^p = \int_0^1 dx g_1^p(x) = \frac{1}{12}\left[a_3 + \frac{1}{3}(a_8 + 4a_0)\right] \tag{10}$$

implied

$$a_0^{EMC} \simeq 0$$

But in the naive parton model $a_0 = \Delta\Sigma$, which is thus in gross contradiction with the Ellis-Jaffe result Eq. 9.

Moreover, since

$$\Delta\Sigma = 2\langle S_z^{quarks}\rangle \tag{11}$$

this seems to imply $\langle S_z^{quarks}\rangle \simeq 0$ and there appears to be "A crisis in the parton model: where, oh where, is the proton's spin?" [6].

Resolution (??) of the crisis; the anomalous gluon contribution

The Operator Product Expansion has no gluon operator contributing to the first moment of g_1, but a Feynman diagram approach yields the result:

$$a_0 = \Delta\Sigma - \frac{3\alpha_s(Q^2)}{2\pi}\Delta G(Q^2) \tag{12}$$

It was thus hoped that one could have a reasonable $\Delta\Sigma \simeq 0.6$ and still obtain a very small a_0. But even with present day estimates $a_0 \approx 0.2$ this requires

$$\Delta G \simeq 1.7 \text{ at } Q^2 = 1 GeV^2$$

Is this acceptable? What do we know about ΔG?

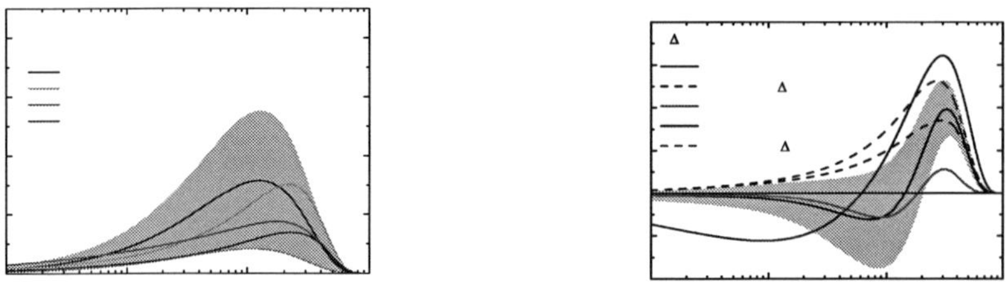

FIGURE 2. Polarized gluon density as in 2006 (left) and 2008 (right)

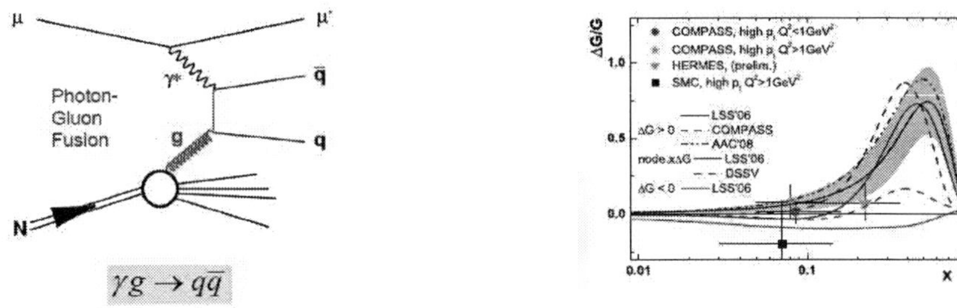

FIGURE 3. Left: Feynman diagram for $c\bar{c}$ production. Right: Direct measurements of ΔG compared with DIS results

THE PRESENT (NEW CHALLENGES)

Attempts to measure ΔG

There are three ways to access $\Delta G(x)$:

(i) In polarized DIS parametrize the polarized quark and gluon densities and fit data on $g_1(x,Q^2)$.The main role of the gluon is in the *evolution* with Q^2, but the range of Q^2 is very limited so the determination of $\Delta G(x)$ is imprecise. Fig. 2 shows the world results on $\Delta G(x)$ in 2006 and how this had changed by 2008.

Typically one has $\Delta G \approx 0.29 \pm 0.32$ ... much smaller than the desired 1.7 !

(ii) $c\bar{c}$ production in DIS. This requires a high energy lepton beam, ideal for COMPASS at CERN. Given that the nucleon has no *intrinsic* charm, the $c\bar{c}$ are produced via "gluon-photon fusion" (see Fig. 3).

Detecting *both* charmed particles would be an absolutely clean signal for the mechanism! But the intensity is too low—-a factor of 30 in rate is lost in detecting the second charmed meson—- so one relies on single charm production and also on back-to-back jets. Fig. 3 shows results from COMPASS, HERMES and SMC, suggesting a very small ΔG compatible with the DIS result quoted above.

(iii) A_{LL} with polarized protons: uniquely at RHIC.
There are several reactions e.g.

$$\vec{p}+\vec{p} \longrightarrow \pi^0 + X \text{ which needs Fragmentation Functions}$$

and

$$\vec{p}+\vec{p} \longrightarrow Jet + X$$

The dominant partonic reactions are

$$\vec{g}+\vec{g} \longrightarrow g+g : \text{dominates at smaller } p_T^2$$

$$\vec{g}+\vec{q} \longrightarrow g+q\text{: dominates at larger } p_T^2$$

As a test, it turns out that PQCD describes the cross-sections quite well.
The results are (see Fig. 4) that A_{LL} is small, consistent with zero gluon polarization! The spin crisis is still with us.

Transverse single-spin asymmetries

Consider hadronic reactions like

$$p^{\uparrow} + p \rightarrow \pi + X \tag{13}$$

where $p^{\uparrow}$ means a transversely polarized proton. The asymmetry under reversal of the direction of polarization is

$$A_N \equiv \frac{d\sigma^{\uparrow} - d\sigma^{\downarrow}}{d\sigma^{\uparrow} + d\sigma^{\downarrow}} \tag{14}$$

The partonic mechanism is shown in Fig. 4. In the simple collinear Parton Model $A_N \approx 0$. To get an idea of the size, at parton level

$$\hat{a}_N = \alpha_s \frac{m_q}{\sqrt{s}} f(\theta^*) \tag{15}$$

where $f(\theta^*)$ is of order 1. This gives asymmetries of a fraction of a percent. The data strongly contradict this, as shown in Fig. 5 [7]

This was a profound challenge to the theory. In fact it was a major problem that had been swept under the carpet for decades. The solution is to go beyond the simple parton model, by inventing new soft mechanisms which require the inclusion of the intrinsic transverse momentum $\boldsymbol{k}_T$ of the partons.
(i) In the *Sivers* mechanism the number density of quarks with momentum $x\boldsymbol{P}+\boldsymbol{k}_T$ depends on the polarization $\mathscr{P}$ of the parent hadron:

$$q(x,\boldsymbol{k}_T) = A + B_S\,\mathscr{P}\cdot(x\boldsymbol{P}\times\boldsymbol{k}_T)$$

But one can can show that this violates Parity and Time Reversal invariance *if*

$$hadron \longrightarrow quark + X$$

is treated as an independent reaction——-as it is in the parton model. To avoid this requires *initial* or *final* state interactions, thereby spoiling straightforward *universality* of the parton model.

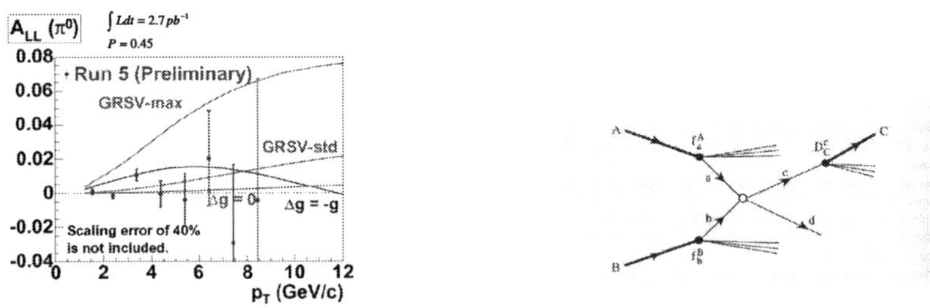

FIGURE 4. Left: PHENIX preliminary results on A_{LL} for π^0 production. Right: Partonic mechanism for hadronic $A+B\rightarrow C+X$

FIGURE 5. Left: E704 data on the transverse spin asymmetries for pions. Right: One photon diagram for $ep\rightarrow ep$

(ii) In the *Collins* mechanism, in the fragmentation of a quark of momentum $\boldsymbol{p}$ into a hadron, the hadron has intrinsic transverse momentum $\boldsymbol{k}_T$ relative to the quark. The number density of hadrons with momentum $\boldsymbol{P}_h = \frac{1}{z}\boldsymbol{p}+\boldsymbol{k}_T$ depends on the polarization $\mathscr{P}$ of the fragmenting quark.

$$D(z,\boldsymbol{P}_h) = A + B_C\,\mathscr{P}\cdot(\boldsymbol{p}\times\boldsymbol{P}_h)$$

Again, this vanishes if the fragmentation

$$q \to hadron + X$$

is treated as an independent reaction, as it is in the parton model. So again one loses straightforward universality. In summary, spin has caused a major rethink of the theory!

The biggest surprise of all: $ep \to ep$

The reaction is usually considered as a one photon exchange process (see Fig. 5) The photon-proton vertex is given by:

$$\bar{u}(p')[\gamma_\mu F_1^{EM}(Q^2) + \frac{i\sigma_{\mu\nu}q^\nu}{2M}\kappa F_2^{EM}(Q^2)]u(p) \tag{16}$$

where $q = p' - p$ κ = anomalous magnetic moment $Q^2 = -q^2$

$F_{1,2}$ are the Dirac and Pauli EM form factors. The Sachs form factors are more convenient: $G_E = F_1 - \kappa\tau F_2$ $G_M = F_1 + \kappa F_2$ where $\tau = \frac{Q^2}{4M^2}$. They are normalized to

$$G_E(0) = 1 \qquad G_M(0) = \text{total magnetic moment}(\mu) = 2.79 \tag{17}$$

The differential cross-section in the LAB is given by the *Rosenbluth* formula

$$\frac{d\sigma}{d\Omega} = (\frac{d\sigma}{d\Omega})'_{Mott}\left[\frac{G_E^2 + \tau G_M^2}{1+\tau} + 2\tau G_M^2 \tan^2(\theta/2)\right] \tag{18}$$

In the textbooks one is told that both G_E and G_M drop with increasing Q^2, and that experimentally

$$G_M(Q^2) \approx \mu G_E(Q^2)$$

as seen in Fig. 6.

However, a totally new kind of measurement, the polarization transfer to the proton from a longitudinally polarized electron colliding with an unpolarized target has produced startling results [8, 9, 10]! The *longitudinal* polarization of the recoil proton, in one-photon exchange, is given by

$$\mathscr{P}_L \propto \left[\frac{E+E'}{M}\right]\sqrt{\tau(1+\tau)}\, G_M^2 \tan^2(\theta/2) \tag{19}$$

The *transverse* (in scattering plane) polarization of the recoil proton is given by

$$\mathscr{P}_T \propto -2\sqrt{\tau(1+\tau)}\, G_E G_M \tan(\theta/2) \tag{20}$$

Fig. 6 compares $\frac{\mu G_E}{G_M}$ for the proton extracted from polarization and from Rosenbluth measurements. The disagreement implies that the assumed dynamical mechanism is incorrect.

This is one of the oldest, simplest reactions studied in particle physics. For decades

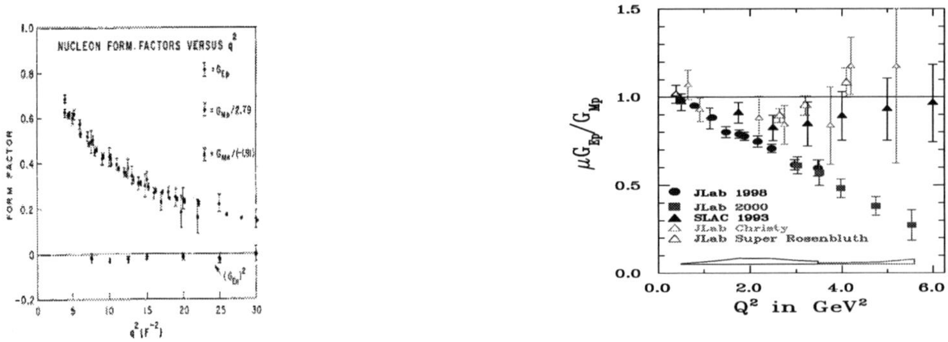

FIGURE 6. Left: Early data on the EM form factors of the nucleon. Right: Ratio of μG_E to G_M for protons, from Rosenbluth and from polarization measurements

it was believed to be totally understood. *Once again a spin-dependent measurement has upset the apple cart!!!*

CONCLUSION: A PUZZLE

Manifestly, spin-dependent measurements have played a crucial role in shaping theories of elementary particles. Why, then, has it *not* attracted a large following? Why is it *not* a subject of major interest to everyone in Particle Physics? There are possibly two reasons:

(i) Practical: Polarization measurements are notoriously difficult. Sources, acceleration, depolarizing resonances etc etc are a headache.

(ii) Pedagogical-psychological: Spin had a difficult birth! There was the conflict between the Stern-Gerlach experiment and the fine structure of hydrogen (the spin-orbit coupling). In most textbooks one finds words like:"Mysterious effects too complicated to explain in an undergraduate text". Here are some examples. The emphases are mine.

....

$$H_{magn} = \frac{1}{m^2c^2}\frac{1}{r}\frac{dV}{dr}\boldsymbol{L}.\boldsymbol{S} \qquad (18.100) \tag{21}$$

When the actual calculation is made with the **proper** Lorentz transformations for the fields, **it is found** that owing to purely kinematic effects we must add a term to the energy, which has the same form as (18.100) but a different coefficient Known as the Thomas term, this contribution to the hamiltonian is

$$H_{Thomas} = -\frac{1}{2}H_{magn} \qquad (18.101) \tag{22}$$

....

However, the kinematics used above is **nonrelativistic. Relativistically**, the electron also precesses about the nucleus (this is called the Thomas precession) with a certain frequency. The net upshot of this precession is that the magnetic field " seen" by the electron is only half as large as the one assumed in the derivation of equation (19.2), and therefore................

NB: Relativity produces a factor of 1/2 !!!!!!!!!!!!!

....

Note that a direct calculation of the spin-orbit coupling, with the **usual** formulations of special relativity , gives a value twice as large as (13.22), and therefore a fine-structure splitting twice too large. This is why Pauli, at the end of 1925, did not believe in the idea of spin, and called it "irrleher" (heresy) in a letter to Niels Bohr. However, in March 1926, L.H.Thomas remarked that the rest frame of of the electron is not an inertial frame, and that a correct calculation introduces a factor of 1/2 in the formula (the Thomas precession). This convinced Pauli of the validity of the spin-1/2 concept.

....

Thus spin was an unwanted child, its adolescence made even more difficult by the following. In the 1960s there was great excitement at the possibility of calculating strong interaction amplitudes reliably for the first time ever using the Mandelstam Representation. It gave rise to a massive effort to study the analytic properties of Feynman diagrams, a horribly complicated task, made much worse by the inclusion of spin. Hence the motto: *Spin is an inessential complication*...a dangerous generalization! Indeed, grossly incorrect as seen in Fig. 7 for the double spin asymmetry in elastic *pp* scattering.

Without doubt, spin-dependent measurements provide a very sharp and subtle probe of dynamical theories. They are scalpel-like in exposing theoretical weaknesses, far more so than are cross-section measurements. We have seen this on many occasions historically and I predict we will see it again often in the future.

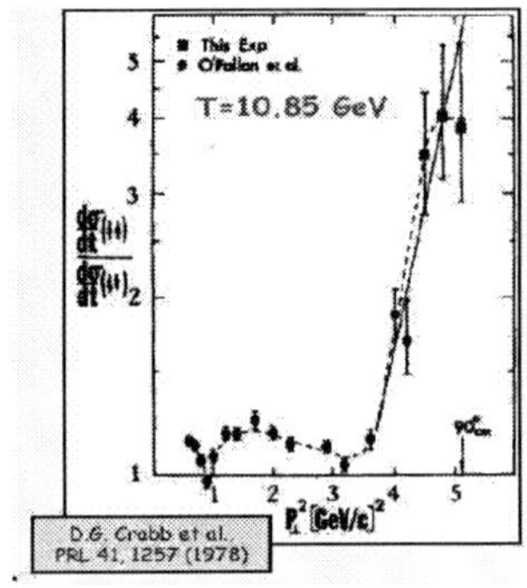

The greatest asymmetry in hadron physics ever seen by a human being

(Brodsky)

FIGURE 7. Asymmetries can be enormous!

ACKNOWLEDGMENTS

I am grateful to Umberto D'Alesio, Stan Brodsky, Alan Krisch and Dimitar Stamenov for help

REFERENCES

1. W. Fetscher, H. J. Gerber, and K. F. Johnson (1986), in *Heidelberg 1986, Proceedings, Weak and electromagnetic interactions in nuclei* 812-815. (see Conference Index).
2. E. Leader, and R. C. Slansky, *Phys. Rev.* **148**, 1491–1501 (1966).
3. S. E. Kuhn, J. P. Chen, and E. Leader (2008), to appear in Prog. Part. Nucl. Phys., `hep-ph/0812.3535`.
4. J. R. Ellis, and R. L. Jaffe, *Phys. Rev.* **D9**, 1444 (1974).
5. J. Ashman, et al., *Phys. Lett.* **B206**, 364 (1988).
6. E. Leader, and M. Anselmino, *Z. Phys.* **C41**, 239 (1988).
7. D. L. Adams, et al., *Phys. Lett.* **B264**, 462–466 (1991).
8. V. Punjabi, et al., *Phys. Rev.* **C71**, 055202 (2005), `nucl-ex/0501018`.
9. O. Gayou, et al., *Phys. Rev.* **C64**, 038202 (2001).
10. O. Gayou, et al., *Phys. Rev. Lett.* **88**, 092301 (2002), `nucl-ex/0111010`.

Spin Experiments - Technological Challenges and Selected Physics Highlights

K. Rith

DESY, Notkestrasse 85, 22603 Hamburg, Germany and University of Erlangen-Nürnberg, Physikalisches Institut, Erwin-Rommel-Str. 1, 91058 Erlangen, Germany

Abstract. Selected highlights in experimental spin physics are presented and dicussed.

INTRODUCTION

During the last two decades spin physics has made enormous progress. New high-precision experiments required substantially increased luminosities and beam and target polarizations. They profited very much from the many achievements in the technologies of polarized sources, beams and targets, which have been discussed in many contributions to this conference. In this contribution I will address shortly those achievements and then discuss a selection of highlights in experimental spin physics reflecting my personal taste. Due to lack of space I can only mention and cite a small fraction the experimental results and refer the reader to the corresponding contributions to this conference.

TECHNOLOGICAL ACHIEVEMENTS

Examples for the many technological achievements, which were the prerequisites for todays high-precision experiments are:

- strained-surface photocathode GaAs polarized electron sources, which nowadays routinely provide high current electron beams with polarizations $p_e \geq 0.8$, recently even around 0.9;
- high beam polarization in high-energy electron/positron storage rings. The degree of polarization depends critically on the machine energy and magnet alignment. At LEP the precise measurement of the beam energy by resonant depolarization of the transversely polarized beam [1] was crucial to reduce the uncertainty in the measurement of the Z^0 mass to its present low value of only 2.1 MeV. At HERA longitudinal polarization was provided for HERMES, H1 and ZEUS by systems of spin-rotators upstream and downstream of the experiments. Until the year 2000 polarization values of about 0.6 [2, 3] were routinely achieved;
- high flux polarized ion sources [4];

CP1149, *Spin Physics, 18th International Spin Physics Symposium*
edited by D. G. Crabb, D. B. Day, S. Liuti, X. Zheng, M. Poelker, and Y. Prok

- internal storage-cell polarized $\vec{H}$, $\vec{D}$, and $^3\vec{He}$ atomic gas targets [5] with a large fraction f of polarizable nucleons and a high degree of polarization ($f \cong 1$ and $p_T \geq 0.8$ for H and D);
- large polarized kryogenic targets [6];
- polarized proton beams in storage rings like the IUCF cooler ring, COSY or RHIC. The RHIC accelerator complex [7] contains a system of several full and partial Siberian Snakes both in the AGS and in RHIC to overcome depolarizing resonances and to preserve polarization during ramping of the beam energy. Average polarizations of about 0.6 have been achieved so far at a beam energy of 100 GeV for a running period of more than a month. In my view RHIC is a masterpiece of accelerator physics;
- fast and precise polarimeters.

Challenging projects for the next generation of accelerators and experiments are the development of

- high current polarized positron sources for the Next Linear Collider [8],
- polarized anti-proton beams for the anti-proton facility FAIR at GSI [9].

SELECTED PHYSICS HIGHLIGHTS

Highlight 1: A^f_{LR} from SLD A new era of experiments with polarized electron beams started with the development of electron sources with strained GaAs photo-cathode surfaces for the SLC which allowed to overcome the GaAs polarization limit of $p_{e,max} = 0.5$. After optimization of source laser wavelength a beam polarization of $p_e \cong 0.65$ was achieved in 1993 which was increased to $p_e \cong 0.80$ from 1994 onwards [10].

Only the electron beam was polarized. This allowed the measurement of the left-right asymmetry A^f_{LR}in the reaction $\vec{e}^-{}_{L,R}e^+ \longrightarrow f\bar{f}$, which is related to the weak vector and axial coupling of fermions of type f to the Z^0, $g_{V,f}$ and $g_{A,f}$.

Since longitudinal electron polarization was only available at SLC and not at LEP all the information on such left-right asymmetries listed in the Review of Particle Physics [11] is based on measurements at SLD. For $f \equiv e^-$ one obtains $g_{V,e-}/g_{A,e-} = 1 - 4sin^2\theta_W$, where θ_W is the Weinberg angle. These SLD measurements resulted in the statistically most precise single determination of $sin^2\theta_W$:

$$sin^2\theta_W(M_Z) = 0.23098 \pm 0.00026. \tag{1}$$

Highlight 2: PV in Moeller scattering Such polarized electron sources became standard equipment at essentially all electron facilities like AMPS and Bates (which meanwhile both were shut down), ELSA, JLAB, MAMI, JPARC, and others. Sophisticated photo-cathode surface structures (e.g. gradient-doped strained superlattice) were developed to overcome the charge limit and to achieve high polarization.

First measurements of parity violation (PV) in $\vec{e}^-{}_{L,R}e^- \longrightarrow e^-e^-$ were performed by E158 [12], the last fixed-target experiment at SLAC. A very high beam polarization of

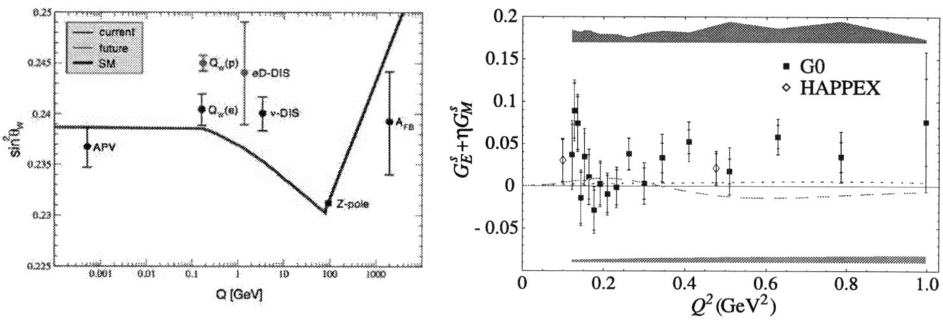

FIGURE 1. Left panel: $sin^2\theta_W$ as a function of momentum transfer Q (this figure is taken from the Review of Particle Physics [11]; Right panel: World data for $G_E^s + \eta G_M^s$ as a function of Q^2.

$p_e = 0.89 \pm 0.04$ was achieved. All beam-helicity related systematics had to be very precisely controlled, since the PV amplitude due to the electroweak interference of one-photon and Z^0 exchange was measured to be only

$$A_{PV} = A(Q^2,y)[1 - sin^2\theta_W^{eff}(Q)] = [-1.31 \pm 0.14(stat) \pm 0.10(sys)]10^{-7}, \quad (2)$$

corresponding to

$$sin^2\theta_W^{eff}(Q) = 0.2397 \pm 0.0010 \pm 0.0008. \quad (3)$$

The result is shown in the left panel of Fig. 1. It is 6.2σ away from the value at the Z^0 pole and demonstrates the running of the Weinberg-angle in perfect agreement with SM predictions.

Highlight 3: Parity violation in electron-nucleon scattering - Nucleon Strange Form Factors Parity violation in elastic electron-nucleon scattering has been measured with very high precision by several experiments (SAMPLE/BATES, G0/JLAB, HAPPEX/JLAB, A4/MAMI) with the aim to determine the contribution of strange quarks, G_E^s and G_M^s, to the electric and magnetic form factors of the nucleon. Very small asymmetries of O(10^6) have been obtained, from which the quantity $G_E^s + \eta G_M^s$, has been extracted, where η is a kinematical factor. Recent results for this quantity [13] are shown in the right panel of Fig.1 as a function of Q^2. For a recent discussion of the extraction of G_E^s and G_M^s see for example Refs. [14, 15].

Highlight 4: Total polarized CC electron-nucleon scattering cross section The two collider experiments H1 [16] and ZEUS [17] at HERA measured the polarization dependence of the cross-section for the charged-current (CC) interaction $\vec{e}_{L,R}p \longrightarrow \nu X$ for both incident electrons and positrons. The standard model (SM) predicts that the CC cross-section for right-handed electrons and for left-handed positrons vanishes and that the polarization dependence of the cross section is given by:

$$\sigma_{e\pm p}^{CC} = (1 \pm P_e)\sigma_{e\pm p}^{CC}(P_e = 0), \quad (4)$$

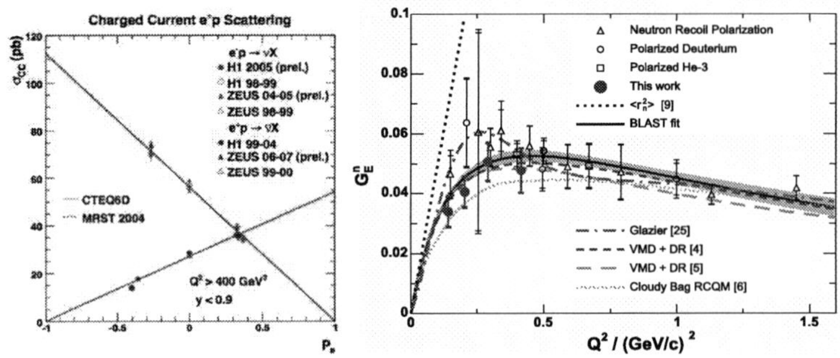

FIGURE 2. Left panel: ZEUS and H1 results for the deep-inelastic CC cross section for e^- and e^+ as a function of longitudinal beam polarization; Right panel: World data for G_E^n from double-polarization experiments as a function of Q^2 .

where the positive (negative) sign in the first bracket on the right-hand side of the equation holds for positrons (electrons) and $P_e = N_R - N_L/N_R + N_L$. The results of these measurements are shown in the left panel of Fig. 2 as a function of P_e.

This textbook plot demonstrates nicely the agreement with the SM. There is no sign of right-handed charged currents. Furthermore the data can be used to set a limit on the mass of a right-handed W: $m_{W,R} > 208\ GeV$ [16].

Highlight 5: Precise determination of G_E^n from double-spin experiments As already discussed in detail by E. Leader [18] recent measurements of the Q^2 dependence of the electric proton form factor, $G_E^p(Q^2)$, showed surprising differences between the Rosenbluth method, used for nearly half a century for the extraction of G_E^p and G_M^p, and the method based on the measurement of the polarization of the recoiling proton [19]. These differences might be related to contributions of two-photon exchange in the elastic scattering process [20]. The measurement of the small electric form factor of the neutron, G_E^n is rather difficult. Early measurements using the Rosenbluth method even resulted in negative values of $(G_E^n)^2$ [21].

Double-polarization experiments, where longitudinally polarized electrons are either scattered off a longitudinally polarized Deuterium or ^3He target or off an unpolarized liquid deuterium target and the polarization of the recoiling neutron is measured, offer a clean possibility to determine G_E^n as in this case it enters linearly in the asymmetry.

Recent results [22] of the Q^2 dependence of G_E^n, obtained by the double-polarization technique are shown in the right panel of Fig. 2.

Highlight 6: Asymmetries in polarized pp scattering at low energies The power of polarization experiments with internal polarized gas targets in storage rings is demonstrated by the measurements of the angular distributions of various spin-correlation parameters in elastic $\vec{p}\vec{p}$ scattering by the PINTEX collaboration at the Indiana cooler ring[23]. Data were taken for eight beam energies between 197 MeV and 445. Previously these distributions were essentially terra incognita while now the data points cover the angular range up to 45 degrees like a chain of pearls.

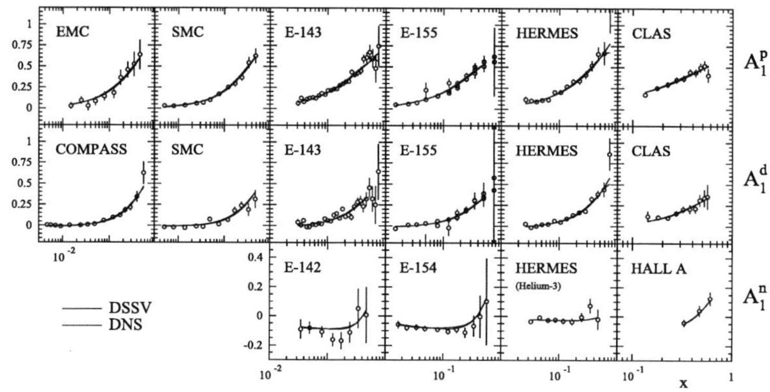

FIGURE 3. World data for the double-spin virtual-photon asymmetry $A_1(x)$ for proton, deuteron and neutron measured in polarized deep-inelastic charged-lepton nucleon scattering.

Highlight 7: Asymmetries in polarized DIS Most of the present experimental information about the spin structure of the nucleon is based on the measurement of double-spin asymmetries (DSA) in deep-inelastic scattering (DIS), of longitudinally polarized charged leptons from longitudinally polarized targets, $\vec{l}\vec{N} \longrightarrow l'X$. A detailed discussion of the underlying physics, the various experiments, the experimental results, and their theoretical interpretation can be found in the recent review by Burkardt, Miller and Nowak [24].

The experimental information from inclusive polarized DIS experiments is summarized in Fig. 3, where the virtual-photon asymmetry $A_1(x)$ for proton, deuteron and neutron is shown as a function of the Bjorken variable x [25]. From these asymmetries one obtains the spin-dependent structure function $g_1(x) = 1/2\sum_q e_q^2 q(x)$, where the sum runs over all quark and anti-quark flavors, e_q is the charge of the quark (anti-quark) in units of the elementary charge $\mid e \mid$ and $q(x)$ is the quark (anti-quark) helicity distribution function (DF) (for simplicity the additional dependence on the squared four-momentum transfer Q^2 is omitted here), and also the quantity $\Delta\Sigma$, which in the $\overline{MS}$-scheme can be interpreted as the fractional contribution of quark spins to the spin of the nucleon.

From their high statistics data taken with a polarized deuteron target, both HERMES and COMPASS have recently determined $\Delta\Sigma$ in NLO-QCD and the $\overline{MS}$-scheme. The HERMES result [26] $\Delta\Sigma = 0.330 \pm 0.025(stat) \pm \pm 0.030(sys)$ was obtained from the fractional first moment of $g_1^d(x)$ under the assumption that g_1^d vanishes for x-values below the x-range covered by the data. The COMPASS result [27] $\Delta\Sigma = 0.30 \pm 0.010(stat) \pm 0.020(syst)$ was obtained from a NLO-QCD fit to the data. The two results agree nicely within their uncertainties They have much higher precision than the original EMC result [28] $\Delta\Sigma = 0.12 \pm 0.09 \pm 0.14$ and we can safely conclude that quark spins contribute only about one third to the spin of the nucleon.

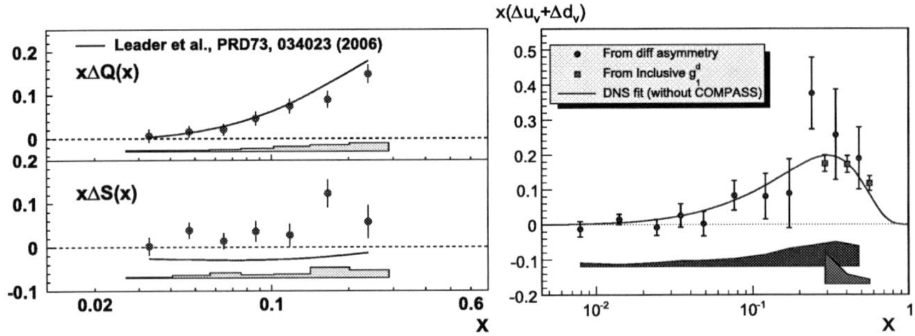

FIGURE 4. Left panel: HERMES results for the strange-quark helicity distribution; Right panel: COMPASS results for the helicity distribution of valence quarks.

Highlight 8: Quark helicity distributions Information about the flavor-separated quark (anti-quark) helicity DFs can be obtained from DSAs in semi-inclusive polarized DIS experiments, $\vec{l}\vec{N} \longrightarrow l'hX$ where in addition to the scattered lepton also (identified) hadrons h are detected. Such data have been published some time ago by HERMES [29]. $\Delta u(x)$ is positive, $\Delta d(x)$ is negative and the sea-quark helicity DFs are compatible with zero within their uncertainties. New results, which shed more light on the sea-quark DFs have been recently published by HERMES and COMPASS. They are presented in Fig. 4. The left panel shows the HERMES result [30] for the strange-quark helicity DF, $x\Delta s(x)$, obtained from the semi-inclusive asymmetry of the *sum* of positive and negative kaons. The first partial moment in the measured x-range [0.02,0.6] $\Delta s = 0.037 \pm 0.019 \pm 0.027$ is rather different from the result derived from the inclusive deuteron data within SU(3): $\Delta s = -0.85 \pm 0.019 \pm 0.027$. Probably this is due to a negative contribution from the unmeasured low-x region. The right panel shows the COMPASS result [31] for the helicity DF of valence quarks, $x(u_v + d_v)$, obtained from the semi-inclusive asymmetry of the *difference* of positive and negative pions and kaons, respectively. The first partial moment of this DF in the measured x-range [0.02,0.7] saturates at a value of $0.40 \pm 0.07 \pm 0.05$. From this result it is concluded that the assumption $\Delta\bar{u} = -\Delta\bar{d}$ for the helicity DFs of light sea-quarks agrees much better with the data than the assumption of a flavor symmetric polarized sea: $\Delta\bar{u} = \Delta\bar{d} = \Delta s = \Delta\bar{s}$.

Highlight 9: Gluon helicity distribution The gluon helicity DF $\Delta g(x_g, Q^2)$ can in principle be obtained from NLO-QCD fits to the double-spin asymmetries in inclusive polarized DIS. At present the $x - Q^2$ range of such data is not yet large enough to sufficiently constrain this important quantity. Data at higher values of Q^2 and at lower values of x are urgently needed, which could be obtained by a future polarized electron-proton collider like EIC.

More promising are at present measurements via processes where the gluon density enters the cross section directly. In lepton-nucleon scattering such a process is *photon-gluon fusion* (PGF) ($\gamma^* g \rightarrow q\bar{q}$). This process can be isolated either by studying reactions in which a charmed quark anti-quark pair is produced and the necessary hard scale is given by the mass of the c-quark, or alternatively by measuring a quark and an anti-quark jet or pairs of hadrons, respectively, or even single hadrons at large transverse momenta p_T relative to the direction of the virtual photon. Results for such measurements have been published by HERMES [32], SMC [33] and COMPASS [34].

DSAs in polarized proton-proton scattering are the second tool to gain direct information about $\Delta g(x_g)$. Several different reaction channels can be studied. Examples are: inclusive jet production, where the dominant partonic subprocesses are ($\vec{g}\,\vec{g} \to g\,g$) and ($\vec{g}\,\vec{q} \to g\,q$), inclusive pion production with the same subprocesses, direct photon production corresponding to ($\vec{g}\,\vec{q} \to \gamma\,q$) and heavy quark production corresponding to ($\vec{g}\,\vec{g} \to c\,\bar{c}, b\,\bar{b}$). These reactions involve a convolution of two parton distributions and the interpretation of the results is therefore always based on a comparison to a Monte Carlo simulation in NLO-QCD. Results for measurement of the longitudinal DSA for inclusive jet and π^0 production have been published by STAR [35] and PHENIX [36] and QCD fits in NLO have been performed to deduce from these results informations about $\Delta g(x)$ [37].

It is difficult to draw a firm conclusion from these results. A large positive gluon distribution is definitively excluded, the data favor a very small Δg, and $\Delta g(x_g)$ possibly has a node at x_g around 0.1. For further detailed discussions see the contributions by E. Leader [18] and E. Rondio [38].

Highlights 10 and 11: Transversity, Collins and Sivers Effect Measurements with transversely polarized targets allow to access transverse-momentum dependent DFs and fragmentation functions (FFs), like the chiral-odd transversity DF $\delta q\left(x, Q^2\right)$ [39] and Collins FF $H_1^\perp$ [40, 41], which is also odd under naive time reversal, and the time-reversal odd Sivers DF $f_{1T}^\perp$ [42, 43]. One especially interesting aspect of this DF is the possible relation to orbital angular momenta of quarks [44]. Very little experimental information about these quantities exists until now. In SIDIS they manifest themselves in SSAs in the distribution of hadrons in the azimuthal angles ϕ (ϕ_s) around the virtual photon direction between the lepton scattering plane and the hadron production plane (transverse component of the target spin vector). The Collins mechanism will cause a $\sin(\phi + \phi_S)$ moment proportional to a convolution of $\delta q(x)$ and $H_1^\perp(z)$, while the Sivers mechanism will cause a $\sin(\phi - \phi_S)$ moment proportional to the convolution of $f_{1T}^\perp(x)$ and the unpolarized FF $D(z)$.

Preliminary HERMES results [45] for the Collins and Sivers moments for charged pions and kaons, obtained from data taken with a transversely polarized proton target, are shown in Fig. 5. The measured Collins asymmetries (left panel) are small but different from zero providing evidence for the existence of both $\delta q(x)$ and $H_1^\perp(z)$. The large π^- moment indicates that the unfavored Collins FF has similar magnitude as the favored one, but opposite sign.

The π^+ and K^+ Sivers asymmetries (right panel) are significantly positive, providing the first evidence for a T-odd PDF appearing in leptoproduction. Consequently one has to conclude from this result that orbital angular momenta of quarks inside the nucleon are non-zero. At present it is, however, not jet possible to quantitatively relate the magnitude of this asymmetry to the fraction of nucleon spin which can be attributed to orbital angular momenta of quarks. The positive kaon amplitudes appear to be larger than the pion amplitudes, which might point to a large Sivers function for sea-quarks.

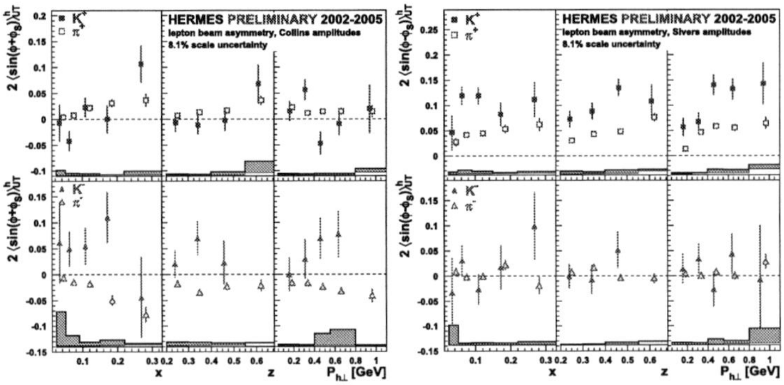

FIGURE 5. HERMES results for the "Collins moments" (left panel) and the "Sivers moments" (right panel) for charged pions and kaons obtained with a transversely polarised hydogen target.

The COMPASS results [46], obtained from data taken with a polarized deuteron ($\vec{Li}\vec{D}$) target, are compatible with zero within errors for both the Collins and the Sivers moments. Very likely this is caused by cancellations arising a deuteron target. Transversity and Sivers DFs for up-quarks seem to have about the same magnitude as those for down-quarks but opposite sign. Preliminary results from the COMPASS proton data indicate that the Collins moments are different from zero and are in agreement with the HERMES data while the Sivers moments are still compatible with zero. For a detailed discussion of these results and their interpretation see Ref. [47].

Large transverse SSAs or analyzing powers A_N, respectively, have been observed in pp reactions with transverse beam or target spins. For a review of the experimental results and their interpretation see Ref. [48]. Recent theoretical attempts for their interpretation include the role of partonic transverse momentum effects, transversity DF and Collins FF, the Sivers effect, and higher-twist effects arising from quark-gluon correlations [49].

Highlight 12: Hard exclusive processes - Determination of L_q and J_q The partonic structure of the nucleon has traditionally been described in terms of parton distribution functions. In recent theoretical developments these have been conceptually subsumed within the broader framework of Generalized Parton Distributions (GPDs), which also describe elastic form factors and hard exclusive reactions [50]. These distributions allow to obtain a 3-dimensional picture of the nucleon, correlating the longitudinal momentum fraction of partons with their transverse distance from the center of the nucleon. Furthermore moments of certain GPDs were found to relate directly to the total (including orbital) angular momentum carried by partons in the nucleon [51]. GPDs are accessible through hard exclusive processes, i.e. production of vector mesons, pseudoscalar mesons or real photons (Deeply Virtual Compton Scattering, DVCS), that involve at least two hard vertices, yet leave the target nucleon intact. They manifest themselves in various azimuthal angular asymmetries in the azimuthal angles ϕ and ϕ_s with respect to beam charge and beam and target polarization. Several pioneer measurements of various such asymmetries like the beam helicty asymmetry A_{LU}, the beam charge

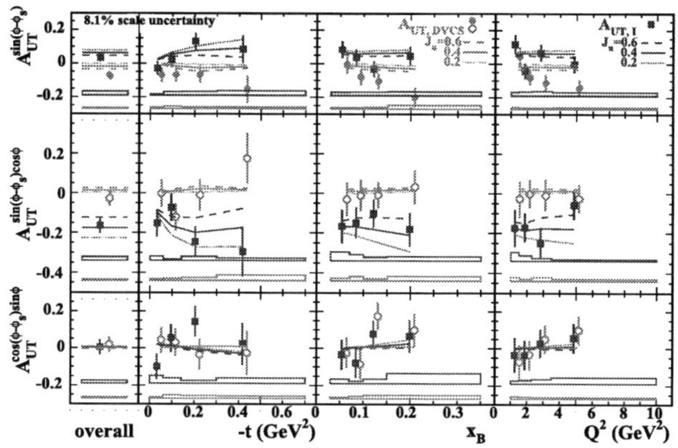

FIGURE 6. HERMES results for the asymmetry amplitudes in DVCS for the transversely polarized proton target.

asymmetry A_C, and the longitudinal (transverse) target spin asymmetry $A_{UL}(A_{UT})$ for DVCS or the transverse target asymmetry A_{UT} for ρ^0 and π^+ mesons have been performed by HERMES [52, 53] [53] and JLAB [54]. As an example various moments of the transverse target asymmetry A_{UT} are shown in Fig. 6 as a function of the kinematic variables $-t, x_B, Q^2$. The data show some (model-dependent) sensitivity to the total up-quark angular momentum J_u. The ultimate goal of such investigations is to constrain and to evaluate the up- and down-quark total (orbital) angular momentum contribution to the spin of the nucleon. HERMES has performed further measurements with a dedicated recoil detector and several 2nd generation experiments are presently being planned and prepared for future high precision measurements of this kind.

CONCLUSION

Spin experiments have produced exciting results which are often a challenge for theory. I am looking forward to see plenty even more challenging results from the next generations of high precision spin experiments.

REFERENCES

1. A. Blondel et al., Proceedings of the 12th Intrnational Symposium on High-Energy Spin Physics, 10-14 September 1996, Amsterdam, C.W. de Jager et al. (Eds.), 267-271 ((1997).
2. D. Barber et al., *Nucl. Instrum. Meth.* **A338**, 166-184 (1994).
3. D. Barber et al., *Phys. Lett.* **B343**, 436-443 (1995).
4. A. Zelensky, these Proceedings
5. W. Haeberli and E. Steffens, *Rep. Progr. Phys.* **66**, 1887-1935 (2003).
6. W. Meyer, these Proceedings
7. T. Roser, these Proceedings
8. J. Clarke, these Proceedings

9. E. Steffens, these Proceedings
10. R. Prepost, Proceedings of the 12th Intrnational Symposium on High-Energy Spin Physics, 10-14 September 1996, Amsterdam, C.W. de Jager et al. (Eds.), 127-135 ((1997).
11. C. Amsler et al., *Phys. Lett.* **B667**, 1(2008).
12. P. L. Anthony et al., *Phys. Rev. Lett.* **95**, 081601 (2005).
13. D. S. Armstrong et al., *Phys. Rev. Lett.* **95**, 0892001 (2005).
14. J. Liu et al., *Phys. Rev.* **C76**, 025202 (2007).
15. A. Acha et al., *Phys. Rev. Lett.* **98**, 032301 (2007).
16. A. Aktas et al., *Phys. Lett.* **B632**, 35 (2006).
17. S. Chekanov et al., *Eur. Phys. J.* **C42**, 1 (2005); DESY-08-177 (2008).
18. E. Leader, these Proceedings.
19. C. Hyde and K. de Jager, *Ann. Rev. Nucl. Part. Sci.* **54**, 217 (2004).
20. P. Guichon and M. Vanderhaeghen, *Phys. Rev. Lett.* **91**, 142303 (2003).
21. E. B. Hughes et al., *Phys. Rev.* **B139**, 458 (1965).
22. E. Geis et al., *Phys. Rev. Lett.* **101**, 042501 (2008).
23. F. Rathmann et al., *Phys. Rev.* **C58**, 658 (1998).
24. M. Burkardt, A. Miller, and W.-D. Nowak, arXiv:0812.1234 (2008).
25. W. Vogelsang, private communication.
26. A. Airapetian et al. (HERMES), *Phys. Rev.* **D75**, 012007 (2007).
27. V.Yu. Alexakhin et al. (COMPASS), *Phys. Lett.* **B647**, 8-17 (2007).
28. J. Ashman et al. (EMC), *Phys. Lett.* **B206**, 364 (1988).
29. A. Airapetian et al. (HERMES), *Phys. Rev.* **D71**, 012003 (2005).
30. A. Airapetian et al. (HERMES), *Phys. Lett.* **B666**, 466 (2008).
31. M. Alekseev et al. (COMPASS), *Phys. Lett.* **B660**, 458-465 (2008).
32. A. Airapetian et al. (HERMES), *Phys. Rev. Lett.* **84**, 2584-2588 (2000).
33. B. Adeva et al. (SMC), *Phys. Rev. D* **70**, 012002 (2004).
34. E.S. Ageev et al. (COMPASS), *Phys. Rev. D* **70**, 012002 (2004); *Phys. Lett. B* **633**, 25-32 (2006); M. Alekseev et al. (COMPASS), arXiv:0802.3023 (2008).
35. B.I. Abelev et al. (STAR), *Phys. Rev. Lett.* **97**, 252001 (2006); *Phys. Rev. Lett.* **100**, 232003 (2008).
36. S.S. Adler et al. (PHENIX), *Phys. Rev.* **D73**, 091102 (2006); A. Adare et al. (PHENIX), *Phys. Rev.* **D76**, 051106 (2007); arXiv:0810.0694(2008).
37. D. de Florian et al., *Phys. Rev. Lett.* **101**, 072001 (2008).
38. E. Rondio, these Proceedings.
39. V. Barone and P.G. Ratcliffe, *Transverse Spin Physics*, World Scientific, 2003.
40. J. Collins, *Nucl. Phys. B* **396**, 161-182 (1993).
41. J. Collins et al., *Nucl. Phys. B* **420**, 565-582 (1994).
42. P. J. Mulders and R. D. Tangerman, *Nucl.Phys. B* **461**, 197-237 (1996); Erratum *Nucl.Phys. B* **484**, 538-540 (1997).
43. D. W. Sivers, *Phys. Rev. D* **41**, 83-90 (1990); *Phys. Rev. D* **43**, 261-263 (1991).
44. M. Burkardt, *Phys. Rev. D* **66**, 114005 (2002), *Phys. Rev. D* **69**, 074032 (2004).
45. M. Diefenthaler (for the HERMES collaboration), Proceedings of the 15th International Workshop on Deep-Inelastic Scattering (DIS2007), 16-20 April 2007, 579-582 (2007).
46. V.Yu. Alexakhin et al. (COMPASS), *Phys. Rev. Lett.* **94**, 202002 (2005).
47. N. Makins, these Proceedings
48. C. Aidala, arXiv:0808.4139 (2008) and these Proceedings
49. J. Qiu and G. Sterman, *Phys. Rev.* **D59**, 014004 (1999); M. Anselmino et al., *Phys. Lett.* **B362**, 164 (1995); M. Anselmino et al., *Phys. Rev.* **D60**, 054027 (1999); Y. Koike, *AIP Conf. Proc.* **675**, 449 (2003).
50. C. Weiss, these Proceedings
51. X.D. Ji, *Phys. Rev. Lett.* **78**, 610 (1997).
52. A. Airapetian et al. (HERMES), *Phys. Rev. Lett.* **87**, 182001 (2001); *Phys. Rev.* **D75**, 011103 (2007).
53. A. Airapetian et al. (HERMES), *JHEP* **0806**, 66 (2008).
54. S. Stepanyan et al. (CLAS), *Phys. Rev. Lett.* **87**, 1820012 (2001); S. Chen et al. (CLAS), *Phys. Rev. Lett.* **97**, 072002 (2006); C. Munoz Camacho et al. (JLAB Hall A), *Phys. Rev. Lett.* **97**, 262002 (2006); F.X. Girod et al.(CLAS), *Phys. Rev. Lett.* **100**, 162002 (2008).

Exploring the Standard Model with the High Luminosity, Polarized Electron-Ion Collider

Richard G. Milner
Laboratory for Nuclear Science
Massachusetts Institute of Technology
Cambridge, MA 02139

Abstract. The Standard Model is only a few decades old and has been successfully confirmed by experiment, particularly at the high energy frontier. This will continue with renewed vigor at the LHC. However, many important elements of the Standard Model remain poorly understood. In particular, the exploration of the strong interaction theory Quantum Chromodynamics is in its infancy. How does the spin-½ of the proton arise from the fundamental quark and gluon constituents? Can we understand the new QCD world of virtual quarks and gluons in the nucleon? Using precision measurements can we test the limits of the Standard Model and look for new physics? To address these and other important questions, physicists have developed a concept for a new type of accelerator, namely a high luminosity, polarized electron-ion collider. Here the scientific motivation is summarized and the accelerator concepts are outlined.
Keywords: Quantum Chromodynamics, Nucleon spin, Saturation, DIS, Accelerators.
PACS: 12.38.-t, 24.85.+p, 29.20.db

INTRODUCTION

The Standard Model was developed in the closing decades of the twentieth century and has successfully been vindicated in many experimental tests. It describes the subatomic world via gauge theories of the electroweak and strong interactions. The Standard Model theory of the strong interaction is Quantum Chromodynamics (QCD) and describes hadrons as composed of highly relativistic light mass quarks bound via exchange of massless gluons. QCD has been most successful in describing high energy scattering experiments but exploration of its rich structure is in its infancy. QCD describes a world where, in the language of atomic physics, Lamb-shift physics dominates. Elementary properties of the proton and neutron, the building blocks of atomic nuclei, are not understood in terms of the quarks and gluons of QCD. In addition, the application of QCD to the lower energies typical of our physical world has been hindered by the inability to calculate the theory in the strong coupling limit. Numerical simulations on a discretized spacetime lattice, so-called lattice QCD, are becoming increasingly realistic using more powerful computers and more clever algorithms.

Physics theories are only believable to the extent that they describe experimental data. While the Standard Model has to date been successful in describing all experimental

CP1149, *Spin Physics, 18th International Spin Physics Symposium*
edited by D. G. Crabb, D. B. Day, S. Liuti, X. Zheng, M. Poelker, and Y. Prok

data, it is hard (at least for experimentalists) to imagine that it is the final word. Thus, the LHC will subject the Standard Model to experimental test in a completely new energy regime. While past experience indicates that the high energy frontier is a fertile area for discovery, precision experiments at modest energy have also proved to be very important, e.g. discovery of parity violation, and the first experimental observation of neutral currents. In considering the necessary new accelerator capabilities to explore the Standard Model, physicists have developed the concept of an electron-ion collider (EIC) [1] possessing high luminosity and having polarized beam capability. Over many decades, it has been established that lepton-hadron scattering is the technique of choice to cleanly study the strong interaction. The collider offers the possibility of a large range kinematic range in parton momentum x and momentum transfer Q^2, which are essential to deconvolute the effects of the virtual photon-quark interaction from those of hadron structure. The value of x determines whether valence or sea partons are probed and Q^2 determines the resolution of the scattering. EIC will probe down to values of x of 10^{-4} and will span a range in Q^2 from 1 to 1000 $(\text{GeV/c})^2$. The collider also provides a large range in center-of-mass energy which again extends the kinematic range. The polarized beam capability is demanded by the desire to understand how the spin-½ of the proton arises from the quark and gluon constituents. In Figure 1, EIC is shown [2] together with a representative example of existing machines on a plot of luminosity vs. center-of-mass energy. The electron-ion collider would provide a new capability far beyond that which is available at any present or planned accelerator.

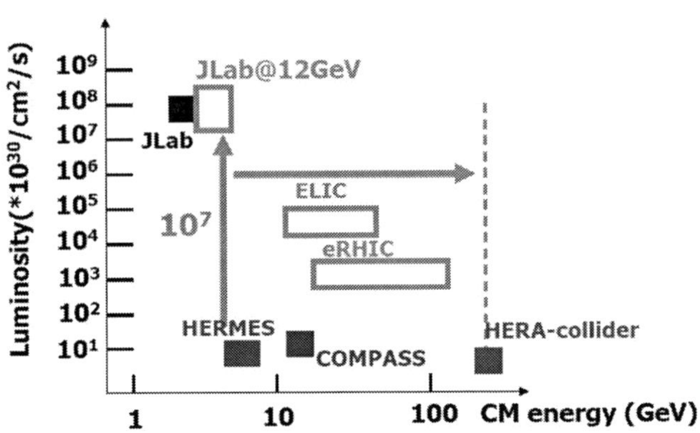

Figure 1. EIC (both JLAB/ELIC [3] and BNL/eRHIC [4]) together with a representative sample of existing capabilities on a plot of luminosity vs. center-of-mass energy.

In the last several years, interest in the electron-ion collider has increased beyond the U.S. with two such accelerators under consideration in Europe both for the LHC at CERN [5] and for the next generation nuclear physics accelerator FAIR at GSI [6].

SCIENTIFIC MOTIVATION

The physics motivation for EIC has been developed over about a decade [1]. Here I highlight three important areas of study which motivate EIC.

What is the origin of the spin of the nucleon?

A widely recognized early success of the quark model was its ability to explain the ratio of the magnetic moments of the proton and neutron. Since the discovery of deep inelastic lepton scattering, it has been a goal to understand how the spin of the nucleon arises dynamically from its quark and gluon constituents. In the valence region, the observed spin asymmetries are in agreement with the expectations of the quark model. However, to date the limited experimental determinations of the sea quark [7] and gluon polarizations [8] possible with existing facilities are all consistent with zero. Estimates of the contribution of the quark orbital angular momentum [9] using theoretical models as well as lattice QCD [10] suggest that the contribution here is also small. Further, the contribution of the gluon orbital angular momentum may not be accessible [11]. This present understanding of the spin structure of the proton is very unsatisfactory.

Further experiments will be carried out in the coming years using high energy colliding proton beams ($E_{CM} \leq 500$ GeV) at RHIC and low energy (~ 10 GeV) electron beams on fixed targets at Jefferson Lab. These are expected to yield more definitive information on the polarization of the sea quarks and gluons in a limited range of x and Q^2 as well as model dependent determinations of the valence quark orbital angular momentum.

It is clear that EIC would provide a dramatic advance beyond the present unsatisfactory situation with respect to the study of the spin structure of the nucleon. For example, the gluon polarization can be precisely extracted in three independent ways at EIC: using inclusive data with a QCD evolution analysis; using dijet production; and by means of charm production. Fig. 2 shows the projected uncertainties for the latter two techniques below [12]. EIC can provide access to all the constituents of the nucleon and precisely quantify their polarization over a large range in Q^2. The polarizations of the different flavors of sea quarks can be precisely determined using semi-inclusive deep inelastic scattering.

A fundamental test of the Standard Model using polarized deep inelastic scattering is the determination of the Bjorken sum rule [13]. This relates the measured proton and neutron inclusive structure functions to the axial vector coupling of nucleon β-decay. It has been experimentally validated to about ± 10% but EIC can push this to about ± 1% [1] with the control of systematic uncertainties, particularly due to knowledge of the beam polarizations, being a challenge. Testing the Standard Model via the Bjorken Sum Rule is analogous to testing the prediction for the value of the anomalous moment of the muon.

New more generalized families of parton distributions have been identified which hold great promise for carrying out a program of tomography on the nucleon [9]. In particular, they may be able to shed light on the contribution of orbital angular momentum to the spin of the proton. However, their unambiguous determination and their importance in understanding nucleon structure is undetermined at present. It is likely that their measurement at EIC will be carried out at lower center-of-mass energies and will demand higher collision luminosity. The high energies of EIC compared to planned fixed target experiments will be essential to definitively establish the significance of these new distributions and whether they can play a role in elucidating the spin structure of the proton.

The program of experiments on the spin structure of the nucleon can be very well carried out with an integrated luminosity of about 50 fb^{-1}, which is comfortably carried out with a luminosity of 10^{33} $cm^{-2}s^{-1}$ over about a decade.

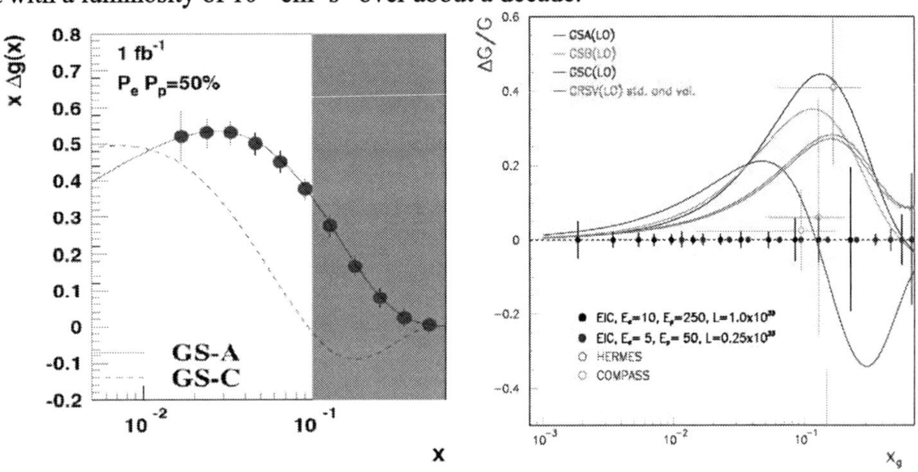

FIGURE 2. Left panel: expected statistical uncertainties of the extracted spin-dependent gluon distribution from measurements ep→ jet jet X at EIC, assuming 1 fb^{-1} luminosity. Right panel: projected uncertainties in the determination of the gluon polarization $\Delta g(x)/g(x)$ at EIC using the channel $D^0 \rightarrow K\pi^+$ in charm production. The integrated luminosity is 10 fb^{-1} for the 10 GeV electron on 250 GeV proton measurement, and 2.5 $fb^{-1}$for 5 GeV electrons on 50 GeV protons. The currently available COMPASS and HERMES data points are also shown.

Exploring the exotic world of low energy gluons

Our current understanding of hadron structure indicates that that the proton is overwhelmingly comprised of gluons at low $x \leq 0.01$. In nuclei, the regime of $x < 0.01$ is *terra incognita*. EIC offers the unprecedented opportunity to explore a new world dominated by the massless force carriers of the strong interaction. For example, QCD tells us that soft gluons can recombine into harder gluons – a completely new phenomenon.

Simple arguments in QCD suggest that at small x the field strengths of gluon fields should be the maximum possible found in nature, corresponding to a novel non-linear regime of the theory. The physics governing gluon interactions in this regime may be universal across hadrons and nuclei. Further, we believe that this non-linear regime is reached at larger values of x in heavy nuclei and its effects amplified by nuclear size. Studies of the properties of gluons and the accompanying sea quarks in this regime, across a wide range of nuclei, have the potential to fundamentally impact our understanding of QCD at high energies.

The glue dominated regime of nuclei will be explored by EIC by addressing the following questions:

- What is the momentum distribution of gluons in nuclei?
- What is the space-time distribution of gluons in nuclei?
- How do fast probes interact with an extended gluonic medium?
- Do strong color fields enhance the role of color neutral degrees of freedom in scattering off nuclei?

Fig. 3 indicates [12, 14] that EIC can carry out definitive measurements which will discriminate between different theoretical predictions.

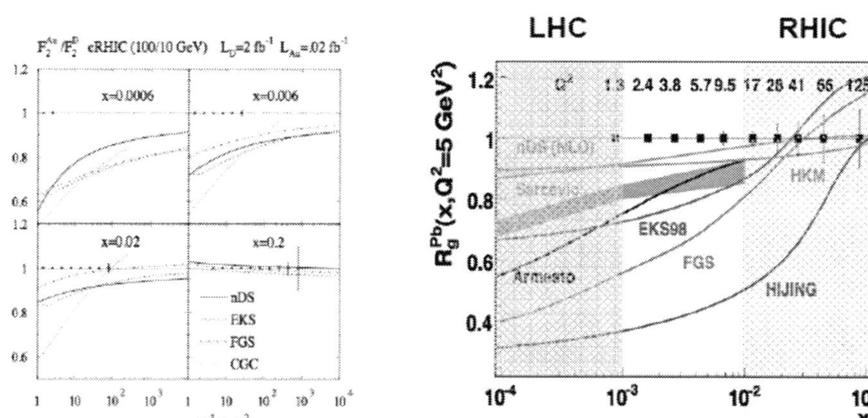

FIGURE 3. Left panel: statistical uncertainties of the ratio of F_2 (representing the sum of the quark and anti-quark distributions) in gold nuclei to deuterium. nDS, EKS, and FGS represent pQCD models with different shadowing; CGC is a saturation model prediction. Right panel: The ratio of gluon distributions in lead relative to deuterium for 10 fb^{-1} integrated luminosity. Additional shadowing parameterizations are shown. The shaded region is the kinematic domain for semi-hard processes in AA collisions at the LHC.

Tests of fundamental symmetries

Testing the validity and limits of the Standard Model is one of the most important research areas in physics. An important consequence of realizing a completely new accelerator is that it offers novel experiments to test the Standard Model. In this regard, EIC offers significant promise in testing the electroweak interaction in lepton-nucleon scattering. Further, the electroweak interaction can be used to test for violation of important symmetries in the standard model.

Precision measurement of parity violating spin-dependent electron (and positron) scattering from the nucleon directly determines one of the fundamental parameters of the Standard Model, namely the Weinberg angle θ_W. This would require a collision luminosity of about 10^{34} $cm^{-2}s^{-1}$ and requires precision beam polarimetery. Fig. 4 [15] shows that EIC could produce competitive measurements at a momentum transfer of 5-10 GeV/c.

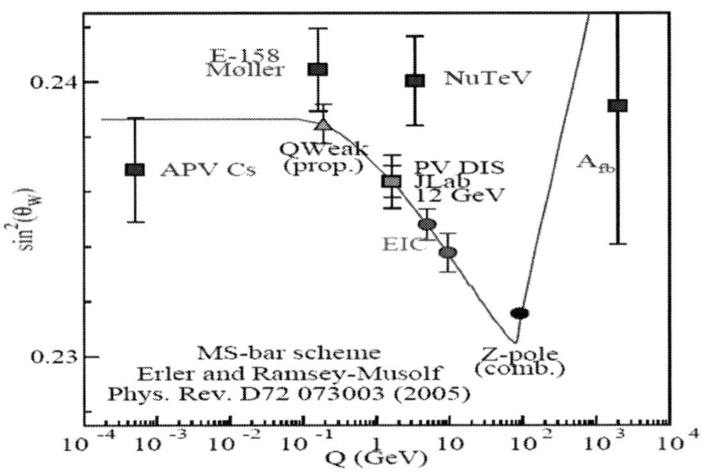

Figure 4. Projected statistical uncertainty from [15] for EIC measurement of the Weinberg angle assuming 500 fb^{-1} of integrated luminosity in 5 GeV electron scattering from 50 GeV deuterium.

Measurement of the electroweak charged current interaction [16] with electron and positron beams on proton and deuteron targets can provide important test of charge symmetry breaking [17]. The cross sections for these processes are low and this will also require higher collision luminosity.

ACCELERATOR CONCEPTS

The EIC machine design is driven by the requirements of the experiments. The comprehensive exploration of QCD demands a luminosity of at least 10^{33} $cm^{-2}s^{-1}$ with highly polarized e± and nucleon beams and the full range of nuclear beams from deuterium to uranium. Tests of the electroweak interaction demand a significantly higher luminosity of order 10^{34} $cm^{-2}s^{-1}$. The center of mass energy should span the onset of deep inelastic scattering (~ 10 GeV to ~ 100 GeV). The design of the EIC electron-ion interaction region is intimately connected with the design of the collider detector.

Accelerator physicists [3,4] have developed several concepts for EIC in the U.S. based around existing capabilities at BNL/RHIC and JLab/CEBAF. They fall into two distinct categories: a ring-ring machine and a linac-ring machine, where the electron beam is produced by a linac.

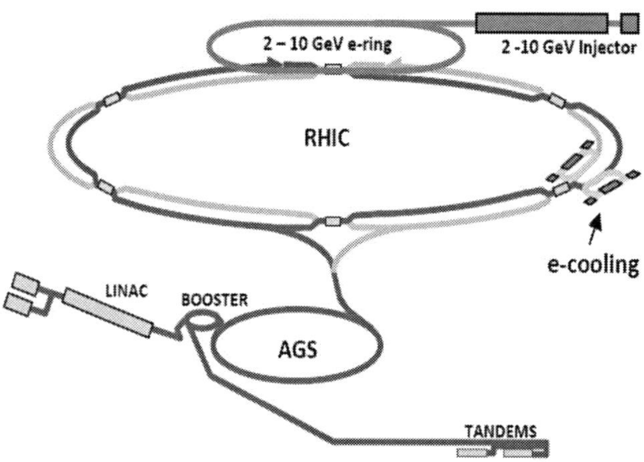

Figure 5. Schematic layout of the eRHIC ring-ring electron-ion collider [4].

Ring-Ring Concepts

The best understood EIC machine configuration is where a ~ 10 GeV electron storage ring is arranged to collide with the existing RHIC complex at BNL [4], as shown schematically in Fig. 5. This design takes advantage of the investment and capabilities of RHIC. Polarized positron beams can be straightforwardly produced. However, the present ring-ring machine design is limited in luminosity to ~ 3 x 10^{32} $cm^{-2}s^{-1}$. There are no significant R&D issues associated with this concept.

A more ambitious concept [3] is based around the existing CEBAF at Jefferson Lab and would require a new ion complex, as shown schematically in Fig. 6. This concept aims for a luminosity in excess of 10^{34} $cm^{-2}s^{-1}$ accomplished with a high collision frequency.

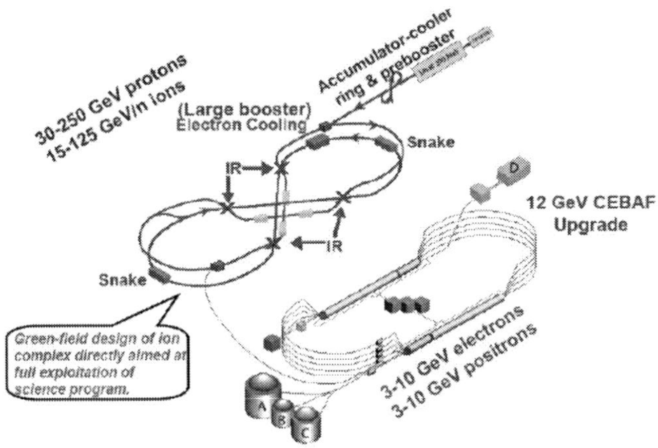

Figure 6. Schematic layout of the ELIC ring-ring electron-ion collider [3].

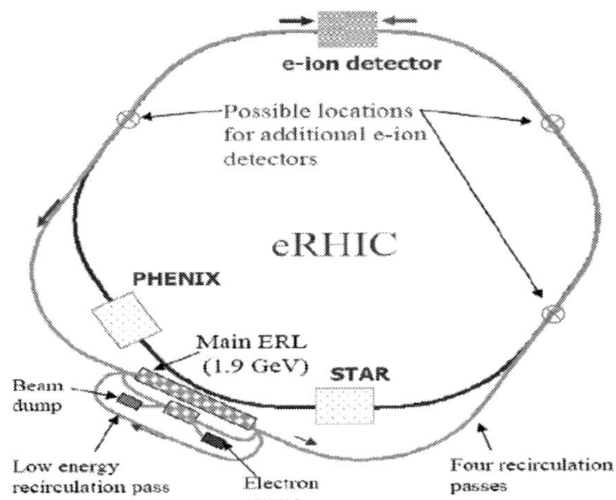

Figure 7. Schematic layout of the eRHIC linac-ring electron-ion collider [4].

Linac-Ring Concept

High energy (~10 GeV), high intensity (~100 mA) electron linear accelerators will require recovery of the energy due to the high power in the beam. There have been significant advances in this technology using superconducting RF. Thus, it is conceivable that the electron beam for EIC can be realized with an ERL. Indeed, the luminosity possible could substantially exceed that of the ring-ring design. Also, the linac-ring allows more space for the detectors in the interaction region. However, with the requirement that the electron beam be polarized, the intensity of the polarized electron source is much larger than exists with current technology.

Fig. 7 shows the layout for a linac-ring concept [4] using the RHIC complex. The beam is recirculated four times in the RHIC tunnel. Collision luminosities of $\sim 3 \times 10^{33}$ c^2 are attained with this design. A staged implementation of this concept is under development at present.

Accelerator R&D

R&D for the EIC machine is a major activity at present. A test ERL facility is under construction at BNL. Development of a high intensity polarized electron source is being carried out at MIT-Bates. Development of cooling techniques for the ion beam is a major focus of effort as this is essential to attain high luminosity. At Jefferson Lab, crab cavities are being studied as well as the ability to attain high collision frequencies.
At present the scientific reach and the cost of the different design concepts are being studied in detail. It is anticipated that within three years the most optimal design will be identified and costed.

DETECTOR CONSIDERATIONS

A successful science program at EIC will require a suite of optimized detectors. The conceptual design of these detectors is in progress. A solenoidal central detector will likely be required to detect the final state particles from $\sim 10^{\circ}$ to $\sim 170^{\circ}$ with respect to the incoming electron. In the forward and backward directions a dipole magnet looks like an attractive option. Indeed, a detailed design has been developed for this configuration [18]. Although the collision luminosity is high, the electroweak cross section is small so the trigger rate is not beyond state of the art. Particle identification is important as well as the ability to detect jets.

In the collider configuration many important experiments require the detection of final-state particles with momentum close to the beam momentum and at very forward angles. This requires specialized detectors located to the beamline.

Beam polarimetry is important for EIC as the precise knowledge of the beam polarization is essential for a number of important experiments. The electron and ion polarimeters must be designed as integral elements of the interaction region. Work on this has started [19].

ACKNOWLEDGEMENTS

The material here is based on the work of the Electron Ion Collider Collaboration (EICC). In particular, I would like to acknowledge my EICC Steering Committee colleagues A. Bruell, A. Deshpande, R. Ent, C. Hyde, P. Jacobs, T. Ulrich, R. Venugopalan and W. Vogelsang and my MIT colleagues D Hasell, R. Jaffe, J. Negele, B. Surrow, C. Tschalär, and F. Wang. Further information on the electron ion collider can be found at http://web.mit.edu/eicc/. The author's research is supported by the Office of Nuclear Physics of the U.S. Department of Energy under grant DE-FG02-94ER40818.

REFERENCES

1. Abhay Deshpande, Richard Milner, Raju Venugopalan and Werner Vogelsang, *Ann. Rev. Nucl. Part. Sc.* **55**, 165 (2005) and references therein.
2. D. Hasch, Invited Talk at EICC meeting, Hampton University,VA, June 2008.
3. *Zeroth Order Design Report for the Electron Light-Ion Collider at CEBAF*, prepared for the 2007 NSAC Long Range Plan, available at http://web.mit.edu/eicc/
4. *eRHIC Accelerator Position Paper*, prepared for the 2007 NSAC Long Range Plan, available at http://web.mit.edu/eicc/
5. S. Chattopadhay, Invited Talk at EICC meeting, Hampton University, VA, June 2008.
6. D. Von Harrach, Private communication.
7. A. Airapetian *et al.* (The HERMES Collaboration) Phys. Lett. B **666**, 446 (2008).
8. D. de Florian *et al.*, Phys. Rev. Lett. **101**, 072001 (2008).
9. *Exploring the 3D quark and gluon structure of the proton: electron scattering with present and future facilities*, a white paper prepared for the 2007 NSAC Long Range Plan, available at http://web.mit.edu/eicc/
10. Ph. Hägler *et al.*, Phys. Rev. D **77**, 094502 (2008).
11. R.L. Jaffe, private communication.
12. *A High Luminosity Electron-Ion Collider*, a white paper prepared for the 2007 NSAC Long Range Plan, 24th April 2007, available at http://web.mit.edu/eicc/
13. J.D. Bjorken, Phys. Rev. **148**, 1467 (1966).
14. *Physics Opportunities with e-A Collisions at an Electron-Ion Collider,* a white paper prepared for the 2007 NSAC Long Range Plan, 4thApril 2007, available at http://web.mit.edu/eicc/
15. R.J. Holt, Invited Talk at EICC meeting, Hampton University, VA, June 2008.
16. A.W. Thomas, Invited Talk at EICC meeting, Hampton University, VA, June 2008
17. Gerald A. Miller, Allena K. Opper, and Edward J. Stephenson, *Ann. Rev. Nucl. Part. Sc.* **56**, 253 (2006).
18. *Feasibility for a Detector for Forward Physics at eRHIC*, I. Abt, A. Caldwell, X. Liu, J. Sutiak, 29 July 2004, hep-ex/0407053
19. *Workshop on Precision Electron Beam Polarimetry for the Electron Ion Collider*, August 23-24 2007, University of Michigan, talks available at http://web.mit.edu/eicc/

Spin and Fundamental Symmetries*

William J. Marciano†

Brookhaven National Laboratory
Upton, New York 11973

Abstract. Two very different applications of spin for testing the Standard Model and uncovering "New Physics" are discussed. The first employs spin to monitor and measure (or bound) electromagnetic dipole moments. The second, more theoretical use, relates QCD corrections derived for spin structure functions and the Bjorken Sum Rule to hadronic effects in the CKM unitarity test $|V_{ud}|^2 + |V_{us}|^2 + |V_{ub}|^2 = 1$ and thereby reduces its uncertainty to $\pm 0.06\%$.

Keywords: Dipole Moments, CKM Unitarity
PACS: 13.40.Em, 12.15.Hh

INTRODUCTION

Particles, whether elementary or composite, belong to irreducible representations of the Poincaire group. As such, they are characterized by their Casimir invariants, mass and spin. Elementary spin 1/2 fermions provide the matter fields of the Standard Model, spin 1 gauge fields mediate strong and electroweak interactions and spin 0 Higgs scalars are responsible for electroweak symmetry breaking and mass generation.

Fermions and bosons nicely coexist and interact with one another in the Standard $SU(3)_c \times SU(2)_L \times U(1)_Y$ Model; but, their roles may be further intertwined if supersymmetry (SUSY) exists. That elegant symmetry rotates fermions and bosons into one another via generators that increase or decrease spin by 1/2 unit. Of course, if supersymmetry exists, it must be badly broken, since we don't observe any mass degenerate supersymmetry partners of the known elementary particles. Instead, supersymmetry and particle partners are assumed to be hiding at very high scales $\gtrsim 200$ GeV waiting to be uncovered. Of course, discovery of supersymmetry would dramatically alter our understanding of space-time and could point to superstrings as a deeper description of Nature.

So, spin is an important fundamental property of elementary particle physics and its underlying symmetries. However, it is not that feature that will directly concern me in this lecture. Instead, I consider some applications of spin for the purpose of studying the Standard Model and perhaps unveiling new interactions and their related symmetries, such as supersymmetry or general coordinate invariance of extra dimensions. Specifically, I consider two very different utilizations of spin, one direct and the other more indirect or theoretical.

---

* Plenary talk given at SPIN 2008, University of Virginia, Oct. 6–11, 2008.
† Work supported by DOE Grant DE-AC02-98CH10886

CP1149, *Spin Physics, 18th International Spin Physics Symposium*
edited by D. G. Crabb, D. B. Day, S. Liuti, X. Zheng, M. Poelker, and Y. Prok

I first consider electromagnetic dipole moments of spin 1/2 particles [1]. Those moments must be proportional to the particle's spin, since it is the only (pseudo) vector that characterizes the state. As such, the spin direction monitors electromagnetic dipole moments and can be used to measure their response to external electromagnetic fields. That technique has been used to precisely measure anomalous magnetic dipole moments and to search for electric dipole moments with high sensitivity. Currently, the measured [2] muon anomalous magnetic moment, $a_\mu \equiv (g_\mu - 2)/2$, deviates by about 3.7 sigma from the Standard Model prediction [1], perhaps indicating the presence of supersymmetry contributions [3]. However, recent experimentally based studies of hadronic contributions to a_μ have put theory into some turmoil [4]. I will review the current situation.

Electric dipole moments (edms) can more directly unveil "new physics". The Standard Model predicts those moments to be unobservably small; so, their observation would point to new interactions that violate CP (since edms violate P and T (CP) symmetries). Such a discovery could help explain the matter-antimatter asymmetry of our Universe, since its origin seems to require a new stronger source of CP violation. In this talk, I will emphasize the complementarity of neutron and proton edm searches.

Finally, I will give an indirect theoretical application [5] of spin or helicity in the study of QCD. Taking the chiral limit, $m_q \to 0$, one can relate the QCD corrections to spin structure functions studied in polarized eN deep-inelastic scattering to QCD corrections to super-allowed nuclear beta decays. That in turn improves the extraction of V_{ud} from experiment and the CKM unitarity test

$$|V_{ud}|^2 + |V_{us}|^2 + |V_{ub}|^2 = 1 \tag{1}$$

which can be used to constrain or search for "New Physics" effects.

ELECTROMAGNETIC DIPOLE MOMENTS

The matrix element of the electromagnetic current $J_\mu^{\rm em} = e\sum_f Q_f \bar{f}\gamma_\mu f$ between initial and final states of a spin 1/2 fermion is given by (with $q = p' - p$)

$$< f(p')|J_\mu^{\rm em}|f(p) >= \bar{u}_f(p')\Gamma_\mu u_f(p) \tag{2}$$

$$\Gamma_\mu = F_1(q^2)\gamma_\mu + iF_2(q^2)\sigma_{\mu\nu}q^\nu - F_3(q^2)\sigma_{\mu\nu}q^\nu\gamma_5 + \cdots \tag{3}$$

where the form factors $F_i(q^2)$ are real and denoted at $q^2 = 0$ by

$$F_1(0) = Q_f e = \text{electric charge} \tag{4}$$

$$F_2(0) = a_f Q_f \frac{e}{2m_f} = \text{anomalous magnetic moment} \tag{5}$$

$$F_3(0) = d_f Q_f = \text{electric dipole moment} \tag{6}$$

(For the neutron $Q_f = 0$, but one should use $Q_f = 1$ in eqs. (5) and (6).)

TABLE 1. Measured values of anomalous magnetic moments in units of $e/2m_f$. For nucleons, $e/2m_N$ with $m_N = (m_n + m_p)/2$ is traditionally used. Also given are bounds on their edms.

Fermion	$a_f(e/2m_f)$	$d_f(e-\text{cm})$
electron	0.00115965218073(28)	$< 2 \times 10^{-27}$
muon	0.00116592080(63)	$< 2 \times 10^{-19}$
proton	1.792158142(28)	$< 7.9 \times 10^{-25}$
neutron	-1.9130427(5)	$< 2.9 \times 10^{-26}$

The two real quantities F_2 and F_3 can be combined into a complex dipole, F_D, such that [1]

$$F_D(0) = F_2(0) + iF_3(0) = |F_D(0)|e^{\phi} \tag{7}$$

$$\tan\phi_f = d_f/a_f\frac{e}{2m_f} \tag{8}$$

where ϕ_f parametrizes the relative size of the CP violating edm relative to the anomalous magnetic moment. One can also define a "New Physics" (NP) phase, ϕ^{NP} that parametrizes the relative size of the NP contribution to a particle's edm as compared with its anomalous magnetic moment NP effect

$$\tan\phi_f^{\text{NP}} = \frac{d_f^{\text{NP}}}{a_f^{\text{NP}}\frac{e}{2m_f}} \tag{9}$$

In some theories ϕ_f^{NP} can be potentially large.

ANOMALOUS MAGNETIC MOMENTS

In table 1, I give the current experimental values of the electron, muon, proton and neutron anomalous magnetic moments [6, 7]. All values are precisely measured using the fact that the particle's magnetic moment is proportional to its spin; so, monitoring the spin precession of polarized particles in a magnetic field measures their magnetic moments.

The proton and neutron magnetic moments are dominated by hadronic effects that cannot be reliably computed; therefore, they cannot be used for "new physics" searches. The electron's anomalous magnetic moment is the most precisely measured and its Standard Model prediction is theoretically clean and computed to high order in $(\alpha/\pi)^n$. However, most "new physics" contributions to a_e scale as m_e^2/Λ_{NP}^2 where Λ_{NP} is the "new physics" scale. The smallness of m_e makes a_e not a good place for *NP* searches. Instead, the comparison of a_e^{exp} and a_e^{SM} is usually employed to obtain our most precise value of the fine structure constant [1]

$$\alpha^{-1}(a_e) = 137.035999084(51) \tag{10}$$

The muon's a_μ is $m_\mu^2/m_e^2 \simeq 43{,}000$ times more sensitive to a generic NP mass scale Λ_{NP} than the electron. By measuring a polarized muon's spin precision in a controlled storage ring with magic momentum $P_\mu \sim 3.1$ GeV, a_μ is directly determined with the E821 at Brookhaven result [2]

$$a_\mu^{\text{exp}} = 116592080(63) \times 10^{-11} \tag{11}$$

given in table 1. The possibility of moving the storage ring to a higher muon flux facility and improving a_μ^{exp} by about a factor of 4 is currently being explored.

For comparison, the SM prediction for a_μ is presently given by

$$a_\mu^{\text{SM}} = a_\mu^{\text{QED}} + a_\mu^{\text{EW}} + a_\mu^{\text{Hadronic}} \tag{12}$$

where [1]

$$a_\mu^{\text{QED}} = 116584718.1(2) \times 10^{-11} \tag{13a}$$

$$a_\mu^{\text{EW}} = 154(1)(2) \times 10^{-11} \tag{13b}$$

$$a_\mu^{\text{Hadronic}} = 6901(46)_{\text{VP}}(26)_{\text{LBL}} \times 10^{-11} \tag{13c}$$

with the hadronic loop uncertainty from vacuum polarization (VP) effects giving the dominant error.

The discrepancy between a_μ^{exp} and a_μ^{SM} (all errors added in quadrature)

$$a_\mu^{\text{exp}} - a_\mu^{\text{SM}} = 307(82) \times 10^{-11} \tag{14}$$

is currently 3.7 sigma. Its leading NP explanation is SUSY loops that can generically give a contribution of order the effect in eq. (14) for $m_{\text{SUSY}} \sim 200$–500 GeV, depending on the specific model's properties.

I should mention, three new important experimental results of relevance for a_μ^{SM} were announced at Tau 2008 [4]. The KLOE collaboration at Frascati gave an update of its radiative return $e^+e^- \to \pi^+\pi^-\gamma$ measurements at the ϕ resonance. They confirmed the hadronic contributions in eq. (13c). However, Belle confirmed earlier $\tau^+ \to \nu_\tau\pi^+\pi^0$ results which can be related to $e^+e^- \to \pi^+\pi^-$ in the $I = 1$ channel via isospin. The tau data suggests a reduction in eqs. (13c) and (14) by about 150×10^{-11}. That, in turn, would explain about half the current a_μ^{exp}–a_μ^{SM} discrepency. To confuse matters further, a preliminary BaBar study of the radiative return process, $e^+e^- \to \pi^+\pi^-\gamma$ seemed to confirm the tau results. However, the BaBar analysis is apparently being revised. Overall, I would characterize the hadronic vacuum polarization corrections to a_μ^{Hadronic} as in turmoil. Hopefully, continuing studies will clarify the situation.

ELECTRIC DIPOLE MOMENTS

Current bounds on various particle electric dipole moments (edms) are given in table 1. The electron and neutron constraints are the most stringent and they are expected to be further improved by several orders of magnitude over the next few years. Indeed, a major discovery may be on the horizon; one that may help explain the matter-antimatter asymmetry of the Universe, a core ingredient of our existence.

The proton edm bound has been recently improved [8] by a factor of 7, using indirect atomic ^{199}Hg edm constraints. Further improvement is expected; but, it is unlikely that d_p can become competitive with d_n in such an approach. For that to happen, a dedicated direct d_p measurement in a proton storage ring facility with 10^{-28} e-cm sensitivity appears necessary. Fortunately, such an approach now seems to be technically feasible; but, nonetheless extremely challenging [9].

Accepting the possibility of measuring or bounding both d_n and d_p with better than 10^{-28} e-cm sensitivity, it is natural to ask: What is their relationship? In the case of anomalous magnetic dipole moments

$$a_N^{(I=1)} = \frac{a_p - a_n}{2} \simeq 1.85 \tag{15a}$$

$$a_N^{(I=0)} = \frac{a_p + a_n}{2} \simeq -0.06 \tag{15b}$$

the isovector combination dominates by a factor of -30 over the isoscalar, *i.e.* $a_p \simeq -a_n$. That relationship is understood as a hadronic structure effect and is well described by the naive constituent quark model of nucleons.

Should one expect a similar $d_p \simeq -d_n$ isovector relationship between nucleon edms. In some "New Physics" scenarios such as θ induced edms, that appears to roughly be the case [10] (at least in a leading chiral log approximation [11]). However, in general, they are expected to be independent. In the complex formalism of eq. (8), there is no $\phi_p = \phi_n$ requirement. Indeed, to properly unfold a "New Physics" source of nucleon edms will certainly require at least the measurement of both d_n and d_p with similar sensitivity, since we must know both the isovector and isoscalar parts. In addition, similar sensitivity in studies of the deuteron edm as well as atomic edms would be very helpful additions [9].

CKM UNITARITY AND THE BJORKEN SUM RULE

One of the most elegant uses of spin occurs in studies of the Bjorken Sum Rule [12]. Employing deep-inelastic polarized electron scattering on polarized proton and neutron targets, one measures the isovector difference of their spin structure functions [13]

$$\Gamma_1^{p-n} \equiv \int_0^1 dx[g_1^p x - g_1^n(x)] \tag{16}$$

That quantity was first studied by Bjorken [12] in 1966. He found the remarkable result $\Gamma_1^{p-n} = \frac{1}{6} g_A$ where $g_A = G_A/G_V \simeq 1.27$. The sum rule connected a weak interaction parameter, g_A, measured in neutron β-decay with the purely QED spin derived quantity of eq. (16).

With the advent of QCD, scaling violations in the form of perturbative α_s corrections to the Bjorken Sum Rule were calculated [14]. (In fact, the first one loop effect actually predates QCD [15].) Today, due to the computational skills of a few talented theorists

[14], the QCD corrections are known in the massless u & d quark limit through 3 loop order

$$\Gamma_1^{p-n} = \frac{g_A}{6}\left[1 - \frac{\alpha_s(Q^2)_{\overline{MS}}}{\pi} - C_2\left(\frac{\alpha_s(Q^2)_{\overline{MS}}}{\pi}\right)^2 - C_3\left(\frac{\alpha_s(Q^2)_{\overline{MS}}}{\pi}\right)^3\right] \tag{17}$$

where

$$\begin{aligned} C_2 &= 4.583 - 0.333N_F \\ C_3 &= 41.440 - 7.607N_F + 0.177N_F^2 \end{aligned} \tag{18}$$

The prediction in eq. (17) has been validated experimentally at high Q^2. It also seems to work well down to low $Q^2 \simeq m_\rho^2$ without finding large higher twist effects [13].

Alberto Sirlin and I noticed [5], a few years ago, that the QCD corrections to the Bjorken Sum Rule and the QCD corrections to the short-distance electroweak radiative corrections to super-allowed, $0^+ \to 0^+$, nuclear β-decays were identical in the chiral limit, to all orders in perturbation theory [5]. Therefore, the QCD corrections in eq. (17) could be applied to Nuclear β-decay. Indeed, combining that perturbative improvement with a large N QCD, vector meson dominated, description of intermediate Q^2 loop momentum, we found that the theoretical uncertainty in the extraction of V_{ud} from experiment was reduced by about a factor of 2. Using our improved calculation, a recent survey of many super-allowed β-decays [16] recently found

$$|V_{ud}| = 0.97425(11)(19)_{\mathrm{RC}} = 0.97425(22) \tag{19}$$

The electroweak radiative corrections (RC) are still the largest uncertainty in eq. (19); however, without our improvements, one would have found $|V_{ud}| = 0.97418(42)$, a significantly larger error.

Combining the value of $|V_{ud}|$ in eq. (18) with the kaon average [17]

$$|V_{us}| = 0.2253(9) \tag{20}$$

and $|V_{ub}|^2 \simeq 1.5 \times 10^{-5}$ leads to

$$|V_{ud}|^2 + |V_{us}|^2 + |V_{ub}|^2 = 0.9999(6) \tag{21}$$

which is in outstanding agreement with unitarity. In fact, it confirms the Standard Model electroweak radiative corrections [18] ($\sim 3.6\%$) at about the 60 sigma level.

The agreement of eq. (21) with 1 can be used to constrain "New Physics" appendages to the Standard Model. I list a few examples, but direct the reader to [17] for details.

1. $BR(\mu^+ \to e^+ \bar{\nu}_e \nu_\mu) < 0.001$ — wrong neutrinos
2. $|V_{uD}| < 0.03$ — heavy quark mixing
3. $\Lambda_{\text{Compositeness}} > 30$ TeV
4. $m_{W^*} \gtrsim 2$–6 TeV — Kaluza-Klein Excitations
5. $m_{Z_\chi} \gtrsim 400$ GeV — SO(10) Z' Boson

6. $m_{H^\pm} \gtrsim 5.6 \tan\beta$ GeV Type II 2 Higgs Doublet

How much further can the unitary test in eq. (21) be pushed? Lattice gauge theory calculations of $f_+(0)$ [19] in $K_{\ell 3}$ decays and f_K/f_π [20] in $K_{\mu 2}$ and $\pi_{\mu 2}$ decays should reduce the $|V_{us}|$ error in eq. (20) to about ± 0.0005. Further improvement [5] in the radiative corrections to nuclear beta decay could reduce the error in eq. (19) to about ± 0.00015. Together they would improve the unitarity test to about ± 0.0004. To do much better would be difficult. Nevertheless, such a constraint will be extremely interesting to compare with anticipated discoveries made at the LHC.

REFERENCES

1. A. Czarnecki and W. Marciano, in Lepton Dipole Moments, to be published by World Scientific, B.L. Roberts and W. Marciano editors.
2. G.W. Bennett *et al.*, Phys. Rev. Lett. **92**, 161802 (2004).
3. D. Stockinger, J. Phys. **G34**, R45 (2007).
4. W. Marciano, in Tau 2008, Summary/Perspective Talk.
5. W. Marciano and A. Sirlin, Phys. Rev. Lett. **96**, 032002-1 (2006).
6. D. Hanneke, S. Fogwell and G. Gabrielse, Phys. Rev. Lett. **100**, 120801 (2008).
7. C. Amsler *et al.*, Phys. Lett. **B667**, 1 (2008).
8. W.C. Griffith *et al.*, arXiv 0901:2328 (unpublished).
9. Y. Semertzidis, private communication; Lect. Notes Phys. **741**:97 (2008).
10. W.J. Marciano, in CP Violation 1990, edited by S. Dawson and A. Soni (World Scientific, Singapore 1991).
11. R.J. Crewther *et al.*, Phys. Lett. **B88**, 123 (1979), *erratum* Phys. Lett. **B91**, 341 (1980).
12. J.D. Bjorken, Phys. Rev. **148**, 1467 (1966).
13. A. Deur *et al.*, Phys. Rev. Lett. **93**, 212001-1 (2004).
14. W. Bardeen, A. Buras, D. Duke and T. Muta, Phys. Rev. **D18**, 3998 (1978); S. Gorishnii and S. Larin, Phys. Lett. **B172**, 109 (1986); S. Larin and J. Vermaseren, Phys. Lett. **B259**, 345 (1991).
15. S. Adler and W. Tung, Phys. Rev. **D1**, 2846 (1970); M.A.B Beg, Phys. Rev. **D11**, 1165 (1975).
16. J.C. Hardy and I.S. Towner, arXiv:0812.1202vi (2008).
17. W. Marciano, "CKM Unitarity and New Physics".
18. A. Sirlin, Rev. Mod. Phys. **50**, 573 (1978); A. Czarnecki, W. Marciano and A. Sirlin, Phys. Rev. **D70**, 093006 (2004).
19. D.J. Antonio *et al.*, arXiv:hep-lat0702026.
20. E. Follana *et al.*, Phys. Rev. Lett. **100**, 062002 (2008); W. Marciano, Phys. Rev. Lett. **93**, 231803 (2004).

Review of Electric Dipole Moments of Fundamental Particles

Yannis K. Semertzidis

Physics Dept., Brookhaven National Laboratory, Upton, NY 11973-5000

Abstract. Electric dipole moments (EDM) experiments are in the research frontier of CP-violation beyond the standard model (SM). EDM experiments set the current limits on CP-violation beyond the SM and are most likely to be the first ones to discover if nature has indeed chosen that path.

INTRODUCTION

Fundamental particles have a well defined mass, charge and spin together with a well defined magnetic dipole moment (MDM). The magnitude of the MDM of a fundamental particle depends on the dynamics of its internal structure and quantum field fluctuations (QFF). The MDMs of leptons are dominated by QFF whereas those of hadrons are dominated by the dynamics of their internal structure. Therefore the study of the MDM of leptons is more likely to give information about the quantum fields, as it is the case for the muon and electron, whereas the study of the MDM of the proton and neutron is more likely to give information about the dynamics of the internal structure. Since QFF can be theoretically estimated with much higher accuracy than the effect of the internal dynamics of hadrons, leptonic MDM searches, like the muon g-2 experiment at BNL [1], have been used as sensitive probes to check the SM and physics beyond the SM.

The study of electric dipole moments on the other hand is different. Quantum mechanics dictates that the EDM vector like the MDM of a fundamental particle needs to be along the spin direction $\vec{d} = d\hat{\sigma}$, where $\hat{\sigma}$ is the unit vector along the particle spin. EDMs of fundamental particles then violate both P-parity (mirror reflection) symmetry and T (time reflection) symmetry as shown in equations (1) & (2). The EDM interaction energy (H) in an electric field (E) is

$$H = -\vec{d}\cdot\vec{E} = -d\hat{\sigma}\cdot\vec{E} \xrightarrow{T} -d(-\hat{\sigma})\cdot\vec{E} = \vec{d}\cdot\vec{E} \tag{1}$$

$$H = -\vec{d}\cdot\vec{E} = -d\hat{\sigma}\cdot\vec{E} \xrightarrow{P} -d\hat{\sigma}\cdot(-\vec{E}) = \vec{d}\cdot\vec{E} \tag{2}$$

and changes sign under P and T symmetries. Therefore either the EDM is zero or there is an interaction that allows P and T-symmetry violation.

It is only when the EDM vector is along the particle spin that it violates both P and T-symmetries. The C-symmetry (charge reflection) is conserved since in that case both vectors change sign and the total interaction energy keeps its original sign. T-violation,

CP1149, *Spin Physics, 18th International Spin Physics Symposium*
edited by D. G. Crabb, D. B. Day, S. Liuti, X. Zheng, M. Poelker, and Y. Prok

under the assumption that the combined CPT symmetry is conserved, means that there is CP-violation. The current status of CP-violating sources in the SM is:

- QED: No CP-violation is observed nor it is theoretically motivated.
- Weak Interactions: There is CP-violation discovered in both the kaon and B-system and it is accommodated by a CP-violating phase in the CKM matrix of the weak interactions.
- Strong interactions: No CP-violation has been observed but the QCD Lagrangian allows it! An irreducible QCD term that depends on an angular parameter theta-QCD ($\bar{\theta}$) with a possible value range of $[0,2\pi]$:

$$L_{CPV} = \bar{\theta}\frac{\alpha_s}{8\pi}G\tilde{G} \tag{3}$$

We can estimate $\bar{\theta}$ within an order of magnitude from the neutron EDM:

$$d_n(\bar{\theta}) \approx \bar{\theta}\frac{e}{m_n}\frac{m_*}{\Lambda_{QCD}} \approx \bar{\theta}\cdot(5\times 10^{-17})\,\mathrm{e\cdot cm} \tag{4}$$

with

$$m_* = \frac{m_u m_d}{m_u + m_d} \tag{5}$$

the reduced mass of the up and down quarks. Λ_{QCD} is the QCD scale and m_n the neutron mass. A more precise estimation [2, 3] yields

$$d_n(\bar{\theta}) \approx \bar{\theta}\cdot(3.6\times 10^{-16})\,\mathrm{e\cdot cm} \tag{6}$$

From the present neutron EDM limit [4] of $3\times 10^{-26}\,\mathrm{e\cdot cm}$ results a limit on theta-QCD: $\bar{\theta} \leq 10^{-10}$. The proton and deuteron EDM [5] at $10^{-29}\,\mathrm{e\cdot cm}$ could further push the limit to the $\bar{\theta} \leq 10^{-13}$ level.

The single, weak-interactions CP-violating phase contributes very little to EDM with contributions estimated to be several orders of magnitude below the current experimental limits. On the other hand EDMs of hadronic systems could just be due to a non-zero value of theta-QCD. Many models with physics beyond the SM also predict EDM values near the current experimental limits. Therefore experimental discovery of a non-zero EDM, possibly in the near future, would either mean the existence of physics beyond the SM or that theta-QCD is non-zero. EDMs of several systems need to be discovered in order to sort out the precise source of a non-zero CP-violating source [6].

The distinction between MDM and EDM is clear: even though the internal dynamics of hadrons is complicated and difficult to evaluate accurately, a non-zero EDM value will be already a revolutionary discovery. In simple systems, like the neutron, proton, and deuteron it will be possible to tell the CP-violating source if a non-zero EDM value is found in any of them with adequate accuracy [2, 3, 6].

CP-violation is one of the requirements for an initially symmetric Universe to develop to the one observed today [7]. The CKM CP-violating phase is very small and not able to produce the observed baryon-antibaryon asymmetry of our Universe (BAU) by about 10 orders of magnitude. A new, much stronger CP-violating source is needed to solve the BAU mystery and EDM experiments are well suited for this task.

Current beyond the SM models, including super-symmetry (SUSY), Multi-Higgs, Left-Right Symmetric, etc., all predict large EDM values accessible by the next generation EDM experiments. As an example the proton EDM in SUSY-type new physics models is given by

$$d_p \approx 10^{-24}\,\mathrm{e\cdot cm} \times \sin\delta \times \left(\frac{1\mathrm{TeV}}{M_{\mathrm{SUSY}}}\right)^2 \tag{7}$$

The proton at $10^{-29}\,\mathrm{e\cdot cm}$ has a physics reach of order 10^3 TeV or if there is new physics at the LHC scale it is sensitive to a CP-violating phase of $\delta \geq 10^{-5}$ rad, an unprecedented sensitivity level.

EXPERIMENTAL TECHNIQUES

Magnetic dipole moments precess in the presence of magnetic fields and electric dipole moments precess in the presence of electric fields. Since both MDMs and EDMs are along the spin direction the spin vector also undergoes the same precession:

$$\frac{d\vec{s}}{dt} = \vec{\mu} \times \vec{B} + \vec{d} \times \vec{E} \tag{8}$$

In order to probe the EDM of a particle one looks for the spin precession frequency change as the electric field direction changes. The spin precession rate for spin 1/2 and 1 particles is $ds/dt = 1/2\hbar\omega$ and $ds/dt = \hbar\omega$ respectively. As an example the spin of spin 1/2 particles will precess with frequency ω_1 and ω_2 when the E-field flips sign:

$$\hbar\omega_1 = 2\mu B + 2dE, \quad \hbar\omega_2 = 2\mu B - 2dE \tag{9}$$

and the EDM is then

$$d = \frac{\hbar(\omega_1 - \omega_2)}{4E} \tag{10}$$

For a neutron EDM of $d = 10^{-28}\,\mathrm{e\cdot cm}$ and $E = 100$ kV/cm, the difference in the angular frequencies of spin precession between the two E-field signs is 60 nrad/s.

The statistical sensitivity σ_d depends on the electric field strength, the total number of particles included in the measurement, the spin coherence time (SCT), or polarization lifetime (τ_p), of the particles measured and the total time (T) of the experiment:

$$\sigma_d \propto \frac{1}{\tau_p E}\frac{1}{\sqrt{N_{\mathrm{tot}}}} \tag{11}$$

if the measurement time is short compared to the machine re-filling time and

$$\sigma_d \propto \frac{1}{E}\frac{1}{\sqrt{N_{\mathrm{tot,c}}\tau_p T}} \tag{12}$$

if the measurement time is long compared to the machine re-filling time. N_{tot} is the total number of detected particles and $N_{\mathrm{tot,c}}$ is the total number of particles per cycle. Equation

(11) always holds for a single measurement, i.e. the error is inversely proportional to the SCT. However, for repeated measurements and depending on the relative values of the SCT and machine cycle time, the statistical error may become inversely proportional to the square root of SCT as shown in equation (12). This is so because when the spin coherence time is longer than the machine cycle, the total number of measurement cycles within a fixed total experimental time (T) becomes smaller and it reflects on the functional form of the statistical error reduction.

History and important steps in EDM experiments

The study of MDM of both charged and neutral particles can be done by placing the particles in high quality magnetic fields and measuring the spin precession rate with high accuracy. Since EDMs couple to electric fields, however, placing charged particles in strong electric fields is not as easy: the particle will feel a strong force and will leave the electric field region in a very short time making the measurement much harder to perform. This fact had a major influence in the history of the EDM field:

- 1950's, EDM searches started with N. Ramsey and Parcell [8] as a search of parity violation in nuclear interactions. They figured out that it was impossible to place a charged particle in an electric field region for a significant amount of time and instead they proceeded with the neutron.
- 1960's, Studying an electron as part of an atomic system was thought to be of no use. In order for the electron to stay in the atom and not move, the average electric field acting on it should be zero. It was wrongly concluded that the EDM effect would also be zero. Schiff looked for exceptions of this rule and was the first one to suggest that the nucleus size of the hydrogen atom would produce a very small, but non-zero EDM effect[9]. Non-the-less in the literature Schiff's name is associated with the EDM cancellation factor rather than the non-cancellation factor[1]. Salpeter [10] and later Sandars [11] suggested that in certain atoms, where relativistic effects are important, the average atomic EDM (d_a) effect is very strong even though the average E-field the electron feels is zero. In paramagnetic atoms the effect is larger than one according to
$$R = d_a/d_e = 10Z^3\alpha^2 \tag{13}$$
with Z the atomic number and α the fine structure constant, whereas in diamagnetic atoms, like ^{199}Hg the EDM effect is reduced by a large factor of order of $10^2 - 10^3$.
- 1970's, An indirect storage ring EDM method was applied and a limit on the muon EDM was obtained [12].
- 1980's, Theoretical studies on molecules with large enhancement factors were initiated.

---

[1] Normal Ramsey told the story that it was himself and Purcell that claimed that the EDM effect of an electron in an atom was not measurable since the total E-field applied on the electron is zero, but Schiff was the first one to suggest that there is a small but non-zero effect in the hydrogen atom.

- 1990's, First significant experimental attempts with molecules were not as fruitful as expected. A main obstacle is the difficulty in preparing large numbers of the interesting molecular state. The dedicated storage ring EDM method was first proposed [13].
- 2000's, Scientific approval of the deuteron EDM experiment at the $10^{-29}\,\mathrm{e \cdot cm}$ level using a dedicated storage ring method was obtained at BNL. Many electron EDM efforts under way using different molecules with large estimated enhancement factors. The neutron EDM experiment at the spallation neutron source (SNS) of Oak Ridge using a new method is making progress towards achieving its set goals. Two more neutron EDM experiments, one at ILL-Grenoble (France) and the other at PSI (Switzerland) are also closer to begin taking data towards a goal of about $10^{-28}\,\mathrm{e \cdot cm}$. Eric Cornell begun the EDM investigation of charged molecules.

Important stages for a successful EDM experiment include:

- Polarization: The state preparation with as high intensity beam and as high polarization possible are very important towards achieving a small statistical error.
- Interact with a large electric field: The electric field needs to be as high as possible and for as large time as possible. The SCT, i.e. the time where all the particle spins remain coherent is also a critical parameter in achieving a small statistical error.
- Analysis: A high efficiency and highly sensitive analyzer is again very important.
- Scientific interpretation of results: It is a lot easier to interpret the results for a simple system. Of course a non-zero EDM implying a new source of CP-violation would be great news anyway.

We can summarize the three main systems where sensitive EDM methods have been proposed and are in advanced stage:

- Neutron: Using ultra-cold neutrons (UCN) in a storage box where the kinetic energy of the neutrons is not high enough to overcome the potential barrier of the wall lattice. In the box there is a small B-field and a large E-field applied. Probe the frequency shift as the electric field flips sign. Sensitive systems, like ^{199}Hg atoms are used as co-magnetometers to track any changes in the magnetic field. The next generation neutron EDM experiments are planning to use SQUIDS as they are even more sensitive to small B-field changes.
- Atomic and Molecular Systems: Prepare states with large enhancement factors and probe the first order Stark effect change as the external E-field changes sign.
- Storage Ring EDM for charged particles: In the dedicated storage ring method the EDM precession causes the spin to precess in the vertical direction instead of the common g-2 in-plane precession.

Recent advances in the various methods:

- Neutron: Advances in reducing the effects from stray magnetic fields; Higher UCN intensities have been achieved.

- Atomic and Molecular Systems: The effective E-field is very large, in some systems is as much as several GV/m.
- Storage Ring EDM for charged particles: High intensity with high polarization beams are already available. High electric fields in the rest frame of the particles can be achieved. The spin precession techniques in storage rings are well understood.

Some of the current weaknesses of the various methods are:

- Neutron: The intensities are still small. They still have sensitivity to changing B-fields of order of pGauss. Motional B-fields (i.e. lab-frame E-fields appear as B-fields in the particle rest frame) are a serious systematic error because spin phase rotations don't cancel in three dimensions.
- Atomic and Molecular Systems: Low intensity of desired states. In some systems the physics interpretation is very difficult.
- Storage Ring EDM for charged particles: Sensitive to vertical electric fields. Some systematic errors are different from the g-2 experiment.

Neutron EDM experimental techniques

The neutron EDM experiments at ILL and PSI are using similar methods of probing the EDM signal. For neutrons with very low kinetic energy the energy barrier presented by the lattice atoms of polarized foils can be just right to reflect neutrons with one spin state whereas to allow them to go through with the opposite spin state. Thus very high neutron polarization can be obtained this way. Neutrons are guided through a specially coated waveguide to the storage box where there is a holding magnetic field keeping their spins aligned. A gate valve is closed after the neutron beam is injected to keep them from escaping.

The neutron EDM is probed using Ramsey's method of separated oscillatory fields: As the neutron spin is aligned along the direction of the holding B-field (of order of 2μT) an oscillating B(t) field is applied in a direction orthogonal to the holding field direction. The oscillation frequency is equal to the spin precession frequency due to the holding field. The oscillatory B-field is also called $\pi/2$ pulse since it causes the neutron spin to rotate by $\pi/2$ and become orthogonal to the holding field. The neutrons are then left to precess for a time that is comparable to their SCT in the presence of the holding B-field and a co-linear strong E-field. After that time (also called "free precession" time) another $\pi/2$ pulse is applied and the neutron spins acquire an additional $\pi/2$ phase and become oppositely polarized with their original spin direction. The gate valve is then opened and the neutrons go through an analyzing magnetized foil into a UCN detector. This process is repeated with the strong E-field flipping sign for every store and the numbers of detected neutrons are compared as a function of the E-field direction.

The UCN EDM experiment at SNS of Oak Ridge is planning to use a different method. It uses polarized neutrons mixed with polarized ^{3}He in a superfluid ^{4}He bath. It is taking advantage of the fact that the Landau-Feynman dispersion curve for ^{4}He excitations cross the dispersion curve for free neutrons at one point where both the momentum and energy are conserved and hence the interaction is resonant at E=1.03meV.

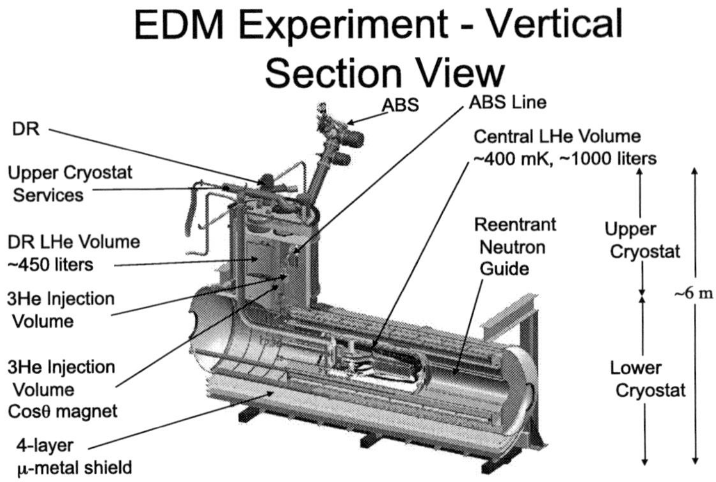

FIGURE 1. The current design of nEDM experimental apparatus at the SNS facility.

The neutrons dissipate their kinetic energy and fall into a lower energy state where there is no way of receiving back this energy since the dispersion curves don't cross again. This method, suggested by R. Golub and J.M. Pendlebury [14] has been already tried and showed great promise.

The neutron and ^{3}He gyromagnetic ratios (γ_n and γ_3 respectively) are the same to about 10% making the ^{3}He an excellent co-magnetometer. However, the ^{3}He turns out to be an excellent analyzer as well. The sensitivity of the ^{3}He atom to EDM is very small because the atomic electrons shield the nucleus from the external electric field very efficiently. Therefore any response of the neutron-^{3}He system to the external electric field would be a sign of a neutron EDM. The ^{3}He nucleus reacts with the neutron to give a tritium (t) plus a proton. The reaction rate depends on the relative direction of their spins. The cross section with their spin parallel is less than 10^2 barns and about 10^4 barns when their spins are opposite. The UCN loss rate is then

$$1 - \vec{p}_3 \cdot \vec{p}_n = 1 - p_3 p_n \cos\left[(\gamma_n - \gamma_3)B_0 + 2dE\right]t \quad (14)$$

The loss rate is monitored as a function of the E-field. Very small magnetic fields could still cause a problem unless the two gyromagnetic ratios were exactly the same. The SNS nEDM collaboration intends to use the spin dressing technique to effectively equalize their factors and further reduce this systematic error [15].

In the reaction

$$\mathrm{n} + {}^3\mathrm{He} \rightarrow \mathrm{p} + \mathrm{t} + 764\mathrm{keV} \quad (15)$$

the energetic particles produce excited state of He molecules (He_2^*) which decay in a few nsec and produce 80nm photons for which the superfluid ^4He is transparent. The photons are converted on the walls by a wave-shifter to visible light of 450nm that is detectable by a photo-multiplier. Approximately 1photon/keV deposited is expected. Figure (1) shows the current schematic of the experiment.

The experiment has received CD2 from DOE and is well advanced. Completion of the experiment and publication of the results is expected around 2020 with an nEDM sensitivity of a couple $\times 10^{-28}\,\mathrm{e\cdot cm}$.

Charged particle EDM experimental techniques

Electron EDM

Clearly the charge of free electron will get in the way if there is an attempt to place it in a high electric field region with the purpose of probing its EDM. It was argued that an atomic electron placed in the same electric field region will have to feel an average E-field equal to zero. Otherwise it would move. It was (incorrectly) concluded that the EDM effect would also be zero on average. Schiff suggested that the nuclear size plays a role and the atomic electron EDM effect is small but non-the-less non-zero. Later on Salpeter and later on Sandars suggested that due to relativistic effects the average atomic EDM for unpaired atomic $s_{1/2}$ and $p_{1/2}$ orbitals can be far from zero. An excellent description of the effect is given in reference [16] where it is shown that the average interaction energy is not zero because the EDM vector in the lab frame is velocity dependent. The enhancement factor works well in paramagnetic atoms, like ^{205}Tl. In diamagnetic atoms like ^{199}Hg the effect is opposite, only a small EDM effect survives and the atomic electrons effectively shield the nucleus.

The best electron EDM experimental limit comes from thallium [17] where the enhancement factor is -585. Currently there is a very large activity to take advantage of the much larger enhancement factors that exist in heavy molecules like PbO (Yale), ThO (Yale, Harvard), HfF^+ (NIST, NRC), PbF (Univ. of Oklahoma), YbF (Oxford) ,WC (Michigan), etc. Other experiments, using highly deformed nuclei are also under intense investigation at the TRIμP facility of KVI [18] and at Argonne National Laboratory [19]. Great progress has been made towards achieving capture of high intensity beams using magnetic optical traps (MOT).

Storage ring EDM: Proton, deuteron, and muon EDM experiments

A charged particle, such as the proton, will be lost quickly if it is placed in an electric field region. However, if it is placed in an appropriate storage ring the particle will be stored. In a storage ring with only a radial E-field the particle will feel a radial E-field force which is balanced by the centrifugal force. This method can be very effective if the spin of the particle keeps tangential to the orbit, along its momentum vector for the duration of the storage time as shown in figure 2. It turns out that particles with

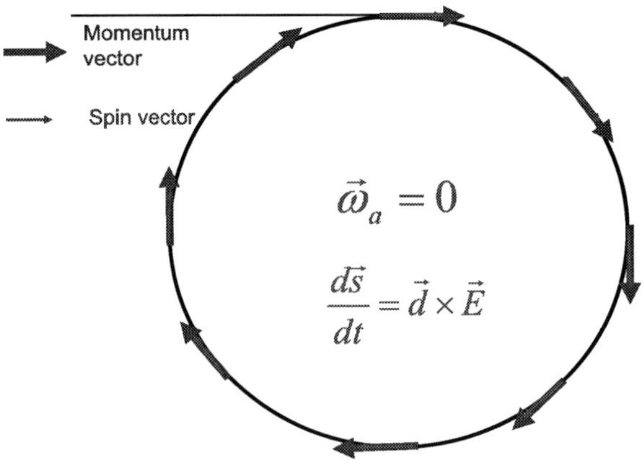

FIGURE 2. If the particle spin is kept along the momentum vector, i.e. the g-2 precession is equal to zero ($\vec{\omega}_a = 0$), the radial E-field will always be orthogonal to it for the duration of the storage time. The EDM signal will be a rise of the out of plane spin component as a function of time.

a positive anomalous magnetic moment value (defined as $a = (g-2)/2$), also known as g-2 value, can satisfy the required condition and their spins can always be aligned with the momentum vector in the presence of E-fields. The g-2 spin precession angular frequency ω_a, i.e. the relative spin precession compared to the momentum precession, in the presence of E-fields only is given by

$$\vec{\omega}_a = \frac{e}{m}\left[a - \left(\frac{m}{p}\right)^2\right]\vec{\beta} \times \vec{E} \tag{16}$$

which, for a positive a, becomes zero at the "magic" momentum of $p = m/\sqrt{a}$. Equation (16) shows that the g-2 cancellation does not depend on the E-field but only on the particle momentum. This, so called magic, momentum is 0.7GeV/c for the proton (*a=1.8*) and 3.1GeV/c for the muon (*a=0.0011*).

For any particle, larger E-field strengths give better EDM sensitivity with the additional benefit that the required storage ring size is smaller. For protons at the magic momentum and $E = 16.5$MV/m the ring radius needs to be 25m, a large but manageable size. The electric field that can be sustained between two plates depends on their distance. The theoretically possible E-field strength is actually much larger than has been obtained so far [20]. The prevailing model of E-field breakdown is that when a particle from one electrode gets released and flies to the opposite electrode, because of its charge, it acquires a kinetic energy equal to the voltage difference V between the two plates. This process will cause a breakdown when the kinetic energy will be enough to cause meltdown and a subsequent large pressure rise locally [21]. We can assume the charge of the dislodged material to be proportional to the electric field gradient $q=k\,E$. The breakdown energy is $W_{\rm br} = qV$ and

$$W_{\rm br} = qV = kE^2 d \Rightarrow E = const. \times \frac{1}{\sqrt{d}} \tag{17}$$

Therefore the electric field that can be sustained between two plates is inversely proportional to the distance between the plates. In addition, the E-field that can be sustained between two plates can be dramatically improved by treating the metal surfaces with high pressure water rinsing (HPR) [22] or with gas cluster ion beams [20]. Electric field gradients of better than 20MV/m are becoming practically attainable.

Protons. Polarized proton beams with small phase space emittance are already available at BNL, COSY, and other places in the world. A beam with polarization of 90% and intensity of 2×10^{10} with horizontal (95%, un-normalized) emittance of 3-mm-mrad, vertical emittance of 10-mm-mrad and rms $\delta p/p$ of 2.5×10^{-4} is readily available at BNL. The small emittance is needed to obtain a large SCT. It also helps it fit between the E-field plates. From equation (16) it is clear that the SCT of the beam will depend on the momentum spread. To further improve the SCT we need an RF-cavity to oscillate the particle momenta so that they all have the same average momentum value. The second order momentum effects plus the effects due to the betatron oscillations are eliminated using sextupole magnets around the ring lattice [23, 24, 5]. The ring lattice needs to be specially designed so that the sextupoles will eliminate all three sources of de-coherence. It is estimated [24, 5] that a SCT of order of 10^3s is achievable.

The spin vector of the particles is analyzed using elastic nuclear scattering between the proton beam and a solid ^{12}C target which serves as the limiting aperture in a specific location in the ring. Strip-lines located in a straight section of the ring will be used to add white noise and slowly increase the beam phase space and hence guide it to the limiting aperture ^{12}C solid target. Protons scattered from the target have a left-right angular distribution that strongly depends on the vertical component of their spin vector [5].

The statistical sensitivity of the experiment has been shown to be

$$\sigma_{d(p)} = \frac{3\hbar}{PAE\sqrt{N_{\rm tot,c}T_{tot}\tau_p}} \tag{18}$$

and holds for a beam extraction so that the number of counts are the same as a function of time. For SCT of $\tau_p = 10^3$ s, beam polarization $p = 0.9$, analyzing power of $A = 0.75$, a radial E-field, $E = 15$ MV/m, a total number of detected particles per cycle of $N_{\rm tot,c} = 2\times10^{10}\times f$, with $f = 0.01$ the detection efficiency, and $T_{\rm tot} = 10^7$s the total running time per year, we get an error of $\sigma_{d(p)} \approx 10^{-29}\,\mathrm{e}\cdot\mathrm{cm}$ /year.

Deuterons. The deuteron (-0.143), since its anomalous magnetic moment is negative, does not have a magic momentum. However, a combination of magnetic (B) and electric (E) fields can have the same result of freezing the horizontal spin precession along the momentum [13]. The g-2 precession, when the magnetic fields are included, is given by

$$\vec{\omega}_a = \frac{e}{m}\left[a\vec{B} + \left[a - \left(\frac{m}{p}\right)^2\right]\vec{\beta}\times\vec{E}\right] \tag{19}$$

In this case a radial E-field can be used to cancel the first term in the above equation (19), and it is found to be

$$E = \frac{aBc\beta\gamma^2}{1-a\beta^2\gamma^2} \approx aBc\beta\gamma^2 \tag{20}$$

with the approximation holding when $1-a\beta^2\gamma^2 \approx 1$.

Even though it is required to combine E and B-fields to cancel the horizontal spin precession locally (adding to the complication of the method), there is an advantage to the statistical power of the method. The rest frame E-field is dominated by the motional E-field, i.e. the lab-frame B-field which is partially converted to an E-field in the particle rest frame. The effective E-field, E^*, is then multiplied by an enhancement factor (*EF*), which can be estimated to be

$$EF = \frac{1}{a\gamma^2}(1+a) \tag{21}$$

For deuterons of 1GeV/c the absolute value of the enhancement factor is approximately 4.7, which is significant. Clearly the method is well suited for particles with small anomalous magnetic moment values. The statistical sensitivity of the deuterons (a spin 1 particle) is given by

$$\sigma_{d(d)} = \frac{6\hbar}{PAE^*\sqrt{N_{\text{tot,c}}T_{tot}\tau_p}} \tag{22}$$

which is also at the same level of $\sigma_{d(d)} \approx 10^{-29}\,\text{e}\cdot\text{cm}$ /year (the smaller asymmetry for 1GeV/c deuterons and the factor of two statistical loss because the deuterons have spin equal to one, are compensated for by the large enhancement factor).

The physics reach of the proton and deuteron EDM compared to other hadronic EDM experiments currently under investigation is given in table 1.

The storage ring EDM collaboration proposed the deuteron EDM experiment with sensitivity of $10^{-29}\,\text{e}\cdot\text{cm}$ to the BNL PAC and has received scientific approval. The collaboration is scheduled for a technical review in the fall of 2009 focused on the proton EDM (deemed to be cheaper and easier than the deuteron EDM experiment) and, if successful, will seek DOE funding after that.

Muons. The muons, where the first (indirect) storage ring EDM limit was obtained [12] have the advantage of being the only second generation particles whose EDM value have been probed with a significant sensitivity. The last CERN muon g-2 experiment and the BNL muon g-2 experiment were done at the muon magic momentum of 3.1GeV/c. If we ignore the effect of the E-field and include the EDM term we have for the total spin precession relative to momentum:

$$\vec{\omega}_{ae} = \frac{e}{m}\left[a\vec{B} + \frac{\eta}{2}\left(\vec{\beta}\times\vec{B}\right)\right] \tag{23}$$

TABLE 1. The current and future EDM limits in units of [e · cm] for some hadronic systems under active investigation. Each hadronic system has a different physics reach and to compare them we show their neutron equivalent as compiled by W. Marciano.

Hadronic System	Current limit	Future Limit	Neutron equivalent
neutron	$< 1.6 \times 10^{-26}$*	10^{-28}	10^{-28}
^{199}Hg atom	$< 3.1 \times 10^{-29}$†	$< 10^{-29}$	$10^{-26} - 10^{-27}$**
^{129}Xe atom	$< 6 \times 10^{-27}$	$10^{-30} - 10^{-33}$	$10^{-26} - 10^{-29}$‡
proton nucleus	$< 7.9 \times 10^{-25}$§	10^{-29}¶	$4 \times 10^{-29} - 2.5 \times 10^{-30}$
deuteron nucleus	N/A	10^{-29}‖	$3 \times 10^{-29} - 5 \times 10^{-31}$

* Limit from [4]
† Limit from [25]
** There is a loss of sensitivity due to the shielding of the nucleus by the atomic electrons.
† There is a loss of sensitivity due to the shielding of the nucleus by the atomic electrons.
§ Indirect limit from the ^{199}Hg experiment
¶ Direct limit at the end of the storage ring EDM experiment.
‖ Direct limit at the end of the storage ring EDM experiment.

where η plays the same role for EDM as the g-factor plays for MDM. $\eta = (m/e)(4dc/\hbar)$ for spin 1/2 articles and $\eta = (m/e)(2dc/\hbar)$ for spin 1 particles. Equation (23) implies that the spin precession plane is slightly tilted with respect to the horizontal plane. It also implies that the g-2 precession frequency will be slightly higher in the presence of a non-zero EDM term. Both those methods have been applied to set a limit on the muon EDM, indirectly from the g-2 experiment [26].

However, freezing the muon spin in a dedicated muon EDM experiment is still possible [13]. The method of freezing the muon spin can be either using only a radial E-field and a magic momentum similar to protons or using combined B and E-fields (and lower than the magic momentum) similar to the deuterons described above. For muons the ring would be much smaller in the combined B and E-fields case than in the magic momentum case. Muons decay with a P-violating weak process into an electron and two neutrinos. The energy of the electron in the muon rest frame is highest along the direction of the muon spin. If the muon spin precesses out of plane the decay electron energy will produce an up/down asymmetry in the detectors surrounding the ring. The statistical sensitivity is given by

$$\sigma_{d(\mu)} = \frac{\hbar}{2\gamma\tau\beta cBPA\sqrt{N_{\text{tot}}}} \approx \frac{\hbar a\gamma}{2\tau EPA\sqrt{N_{\text{tot}}}} \tag{24}$$

where τ is the smaller of the muon lifetime and the SCT. Equation (24) holds when the up/down asymmetry counts as a function of time are fit to a function with a slope and a dc offset.

The two disadvantages of the muon compared to the proton and deuteron is that its lifetime is too short, and they are produced from a secondary source, i.e. its phase space is inevitably too large if one is to collect a significant number of them. These are the reasons that one cannot get significantly better than 10^{-25} e · cm sensitivity for the muon EDM. Non-the-less, since muons are the only second generation particle whose EDM can be probed with high sensitivity, it is a worth-while experiment to pursue. Recently a dedicated muon EDM experiment has been proposed at PSI in Switzerland with a potential sensitivity at the $10^{-22}\,\mathrm{e\cdot cm}$ level [27].

CONCLUDING REMARKS

EDM efforts are on-going around the world and are pursued with a variety of methods. Most of them employ small groups but larger groups have also begun to appear. There is a renaissance in this field and it is more probable that the field will be even more successful and non-zero EDM values of fundamental particles will be discovered. If that turns out to be the case the EDM field will significantly contribute to the resolution of the BAU mystery.

ACKNOWLEDGMENTS

I wish to thank Gerco Onderwater for reading the document and for his many comments that improved the document. Many thanks to all those who provided me with the latest information regarding their EDM efforts. The EDM field is diverse and quite broad. It is impossible to cover it all in this document. The best thing that can happen to this field is that a non-zero EDM value is found in any system. Thanks to all for trying so hard.

REFERENCES

1. G.W. Bennett et al., Phys. Rev. **D73**, 072003 (2006).
2. I.B. Khriplovich, R.A. Korkin, Nucl. Phys. A **665**, 365 (2000); O. Lebedev et al., Phys. Rev. D **70**, 016003 (2004); M. Pospelov, A. Ritz, Ann. Phys. **318**, 119 (2005).
3. C.P. Liu and R.G.E. Timmermans, Phys. Rev. **C70**, 055501 (2004).
4. C.A. Baker et al., Phys. Rev. Lett. **97**, 131801 (2006).
5. Deuteron Storage Ring EDM Proposal to the BNL PAC, March 2008, available at `http://www.bnl.gov/edm/`.
6. W. Marciano, HEP seminar at BNL in 2008 on the theoretical aspects of deuteron, proton and neutron EDM.
7. A.D. Sakharov, JETP Lett., **5**, 24 (1967).
8. E.M. Parcel, and N.F. Ramsey, Phys. Rev. **78**, 807 (1950); J.H. Smith, E.M. Parcel, and N.F. Ramsey, Phys. Rev. **108**, 120 (1957) and references therein.
9. L.I. Schiff, Phys. Rev. **132**, 2194 (1963).
10. E.E. Salpeter, Phys. Rev. **112**, 1642 (1958).
11. P.G.H. Sandars, Phys. Lett. **14**, 194 (1965); Phys. Lett. **22**, 290 (1966).
12. J. Bailey et al., J. Phys. **G4**, 345 (1978).
13. Y.K. Semertzidis et al., AGS2000 proceedings, BNL 1996; Y.K. Semertzidis, Sandansky conference on "QED and the physics of vacuum", 1998; Y.K. Semertzidis et al., High Intensity Muon Sources (HIMUS99), Tsukuba, Japan Dec. 1999, hep-ph/0012087; F.J.M. Farley et al., Phys. Rev. Lett. **93**, 052001 (2004); Y.K. Semertzidis et al., AIP Conf. Proc. **698**, 200 (2004). More information on the storage ring EDM can be found at `http://www.bnl.gov/edm`.
14. R. Golub and J.M. Pendlebury Phys. Lett. **53A** (1975); Phys. Lett. **62A** (1977).
15. R. Golub and S.K. Lamoreaux, Phys. Rep. bf 237, 1 (1994).
16. E. Cummings et al., Am. J. Phys. **75**, 6 (2007).
17. B.C. Regan et al., Phys. Rev. Lett. **88**, 071805 (2002).

18. S. De et al., arXiv:0807.4100, 2008
19. J.R. Guest et al., Phys. Rev. Lett. **98**, 093001 (2007).
20. D.R. Swenson et al., Nucl. Instr. Meth. **B 261**, 630 (2007).
21. "High Voltage Technology", Edited by L.L. Alston, Oxford Univ. Press 1968, p. 82; L. Cranberg, J. of Appl. Phys. **23**, 518 (1952).
22. B.M. Dunham et al., Proceed. of PAC07, 1224 (2007).
23. I.B. Vasserman et al., Phys. Lett. **198**, 302 (1987); A.P. Lysenko et al., Particle accelerators **18**, 215 (1986).
24. Y. Orlov, Storage ring EDM note #61, 2001, unpublished.
25. W.C. Griffith et al., Phys. Rev. Lett. **102**, 101601 (2009).
26. G.W. Bennett et al., submitted to Phys. Rev. D and ArXiv:hep-ex/0811.1207v1, Nov. 2008.
27. A. Adelmann et al., ArXiv:hep-ex/0606034v2, Dec. 2008.

Parton Distributions in Impact Parameter Space

Matthias Burkardt

Department of Physics, New Mexico State University, Las Cruces, NM 88003, and Jefferson Lab, 12000 Jefferson Ave., Newport News, VA 23606.

Abstract. Parton distributions in impact parameter space, which are obtained by Fourier transforming GPDs, exhibit a significant deviation from axial symmetry when the target and/or quark is transversely polarized. In combination with the final state interactions, this transverse deformation provides a natural mechanism for naive-T odd transverse single-spin asymmetries in semi-inclusive DIS. The deformation can also be related to the transverse force acting on the active quark in polarized DIS at higher twist.

Keywords: GPDs, SSAs, higher twist
PACS: 13.30.Fz, 13.60.Hb

IMPACT PARAMETER DEPENDENT PDFS AND SSAS

The Fourier transform of the GPD $H_q(x,0,t)$ yields the distribution $q(x,\mathbf{b}_\perp)$ of unpolarized quarks, for an unpolarized target, in impact parameter space [1]

$$q(x,\mathbf{b}_\perp) = \int \frac{d^2\Delta_\perp}{(2\pi)^2} H_q(x,0,-\Delta_\perp^2)\, e^{-i\mathbf{b}_\perp\cdot\Delta_\perp}, \tag{1}$$

with $\Delta_\perp = \mathbf{p}'_\perp - \mathbf{p}_\perp$. For a transversely polarized target (e.g. polarized in the $+\hat{x}$-direction) the impact parameter dependent PDF $q_{+\hat{x}}(x,\mathbf{b}_\perp)$ is no longer axially symmetric and the transverse deformation is described by the gradient of the Fourier transform of the GPD $E_q(x,0,t)$ [2]

$$q_{+\hat{x}}(x,\mathbf{b}_\perp) = q(x,\mathbf{b}_\perp) - \frac{1}{2M}\frac{\partial}{\partial b_y}\int \frac{d^2\Delta_\perp}{(2\pi)^2} E_q(x,0,-\Delta_\perp^2)\, e^{-i\mathbf{b}_\perp\cdot\Delta_\perp} \tag{2}$$

$E_q(x,0,t)$ and hence the details of this deformation are not very well known, but its x-integral, the Pauli form factor F_2, is. This allows to relate the average transverse deformation resulting from Eq. (2) to the contribution from the corresponding quark flavor to the anomalous magnetic moment. This observation is important in understanding the sign of the Sivers function.

In a target that is polarized transversely (*e.g.* vertically), the quarks in the target nucleon can exhibit a (left/right) asymmetry of the distribution $f_{q/p^\uparrow}(x_B,\mathbf{k}_T)$ in their transverse momentum $\mathbf{k}_T$ [3, 4]

$$f_{q/p^\uparrow}(x_B,\mathbf{k}_T) = f_1^q(x_B,k_T^2) - f_{1T}^{\perp q}(x_B,k_T^2)\frac{(\hat{\mathbf{P}}\times\mathbf{k}_T)\cdot\mathbf{S}}{M}, \tag{3}$$

where $\mathbf{S}$ is the spin of the target nucleon and $\hat{\mathbf{P}}$ is a unit vector opposite to the direction of the virtual photon momentum. The fact that such a term may be present in (3) is known as

CP1149, *Spin Physics, 18th International Spin Physics Symposium*
edited by D. G. Crabb, D. B. Day, S. Liuti, X. Zheng, M. Poelker, and Y. Prok

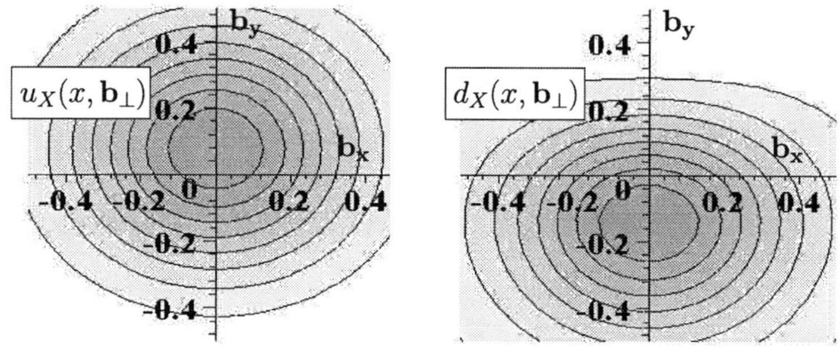

FIGURE 1. Distribution of the j^+ density for u and d quarks in the $\perp$ plane ($x_{Bj} = 0.3$ is fixed) for a proton that is polarized in the x direction in the model from Ref. [2]. For other values of x the distortion looks similar. The signs of the distortion are determined by the signs of the contribution from each quark flavor to the proton anomalous magnetic moment.

the Sivers effect and the function $f_{1T}^{\perp q}(x_{\mathrm{B}}, k_T^2)$ is known as the Sivers function. The latter vanishes in a naive parton picture since $(\hat{\mathbf{P}} \times \mathbf{k}_T) \cdot \mathbf{S}$ is odd under naive time reversal (a property known as naive-T-odd), where one merely reverses the direction of all momenta and spins without interchanging the initial and final states. The momentum fraction x, which is equal to x_{B} in DIS experiments, represents the longitudinal momentum of the quark *before* it absorbs the virtual photon, as it is determined solely from the kinematic properties of the virtual photon and the target nucleon. In contradistinction, the transverse momentum $\mathbf{k}_T$ is defined in terms of the kinematics of the final state and hence it represents the asymptotic transverse momentum of the active quark *after* it has left the target and before it fragments into hadrons. Thus the Sivers function for semi-inclusive DIS includes the final state interaction between struck quark and target remnant, and time reversal invariance no longer requires that it vanishes.

The significant distortion of parton distributions in impact parameter space (2) provides a natural mechanism for a Sivers effect. In semi-inclusive DIS, when the virtual photon strikes a u quark in a $\perp$ polarized proton, the u quark distribution is enhanced on the left side of the target (for a proton with spin pointing up when viewed from the virtual photon perspective). Although in general the final state interaction (FSI) is very complicated, we expect it to be on average attractive thus translating a position space distortion to the left into a momentum space asymmetry to the right and vice versa (Fig. 2) [5]. Since this picture is very intuitive, a few words of caution are in order. First of all, such a reasoning is strictly valid only in mean field models for the FSI as well as in simple spectator models [6]. Furthermore, even in such mean field or spectator models there is in general no one-to-one correspondence between quark distributions in impact parameter space and unintegrated parton densities (e.g. Sivers function) (for a recent overview, see Ref. [7]). While both are connected by an overarching Wigner distribution [8], they are not Fourier transforms of each other. Nevertheless, since the primordial momentum distribution of the quarks (without FSI) must be symmetric we find a qualitative connection between the primordial position space asymmetry and the momentum space asymmetry (with FSI). Another issue concerns the x-dependence of

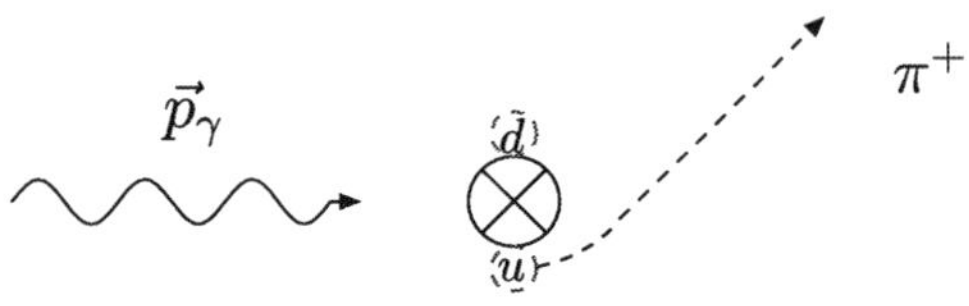

FIGURE 2. The transverse distortion of the parton cloud for a proton that is polarized into the plane, in combination with attractive FSI, gives rise to a Sivers effect for u (d) quarks with a $\perp$ momentum that is on the average up (down).

the Sivers function. The x-dependence of the position space asymmetry is described by the GPD $E(x,0,-\Delta_\perp^2)$. Therefore, within the above mechanism, the x dependence of the Sivers function should be related to the x dependence of $E(x,0,-\Delta_\perp^2)$. However, the x dependence of E is not known yet and we only know the Pauli form factor $F_2 = \int dx E$. Nevertheless, if one makes the additional assumption that E does not fluctuate as a function of x then the contribution from each quark flavor q to the anomalous magnetic moment κ determines the sign of $E^q(x,0,0)$ and hence of the Sivers function. With these assumptions, as well as the very plausible assumption that the FSI is on average attractive, one finds that $f_{1T}^{\perp u} < 0$, while $f_{1T}^{\perp d} > 0$. Both signs have been confirmed by a flavor analysis based on pions produced in a SIDIS experiment by the HERMES collaboration [9] and are consistent with a vanishing isoscalar Sivers function observed by COMPASS [10].

Evolution

At increasing Q^2, the nucleon is increasingly likely to be in a higher Fock component. In particular, quarks at low x are likely to have dressed themselves with perturbative gluons, which also changes the color of the quark. For example, a quark that was red before emitting the gluon may e.g. become green by emitting a red-antigreen gluon. As a result, that quark will no longer be attracted by the remaining spectators (which are still antired) but repelled (this repulsion is weaker than the attraction between red and antired and vanishes in the large N_C limit). Of course, that (now) green quark will still be attracted by the red-antigreen gluon. The main effect is that the 'chromodynamic lensing' mechanism described in the previous section essentially disappears after the active quark has radiated off a gluon (the perturbatively radiated off gluon will be at almost the same transverse position as the quark that radiated it off and hence the $\perp$ impulse excerted on the active quark will be small (and correlated with the quark spin and not the nucleon spin).

Of course, the active quark getting dressed by a gluon is not the only QCD correction. For example, the gluon that provides the FSI can also split, or the spectators can dress themselves with gluons, but neither one of these processes changes the momentum of the active quark. Therefore, while including the latter is essential for a proper determination of the Q^2 dependence of the running coupling, it is less important when considering the Q^2 evolution of the quark Sivers function at smaller values of x — a regime which is dominated by dressing of quark lines with gluons. As a result, at high Q^2 and small x one would not expect a significant Sivers effect. In particular, it is plausibele that COMPASS measurements of the Sivers function are already in that regime where the active quark is more likely than not to already have radiated off a gluon. Therefore, particularly at smaller x, the Sivers function at COMPASS i.e. a significant decrease of the Sivers function compared to what one would expect in the (x,Q^2) regime dominated by valence quarks.

TRANSVERSE FORCE ON QUARKS IN DIS

The chirally even spin-dependent twist-3 parton distribution $g_2(x) = g_T(x) - g_1(x)$ is defined as

$$\int \frac{d\lambda}{2\pi} e^{i\lambda x} \langle PS|\bar{\psi}(0)\gamma^\mu \gamma_5 \psi(\lambda n)|_{Q^2}|PS\rangle \\ = 2\left[g_1(x,Q^2)p^\mu(S\cdot n) + g_T(x,Q^2)S_\perp^\mu + M^2 g_3(x,Q^2)n^\mu(S\cdot n)\right].$$

neglecting m_q: $g_2(x) = g_2^{WW}(x) + \bar{g}_2(x)$, with $g_2^{WW}(x) = -g_1(x) + \int_x^1 \frac{dy}{y} g_1(y)$. $\bar{g}_2(x)$ involves quark-gluon correlations, e.g. [11, 12]

$$\int dx x^2 \bar{g}_2(x) = \frac{d_2}{6} \tag{4}$$

with

$$g\left\langle P,S\left|\bar{q}(0)G^{+y}(0)\gamma^+ q(0)\right|P,S\right\rangle = MP^+P^+S^x d_2 \tag{5}$$

At low Q^2, g_2 has the physical interpretation of a spin polarizability, which is why the matrix elements (note that $\sqrt{2}G^{+y} = B^x - E^y$)

$$\chi_E 2M^2\vec{S} = \langle P,S|q^\dagger \vec{\alpha}\times g\vec{E}q|P,S\rangle \qquad \chi_B 2M^2\vec{S} = \langle P,S|q^\dagger g\vec{B}q|P,S\rangle \tag{6}$$

are sometimes called spin polarizabilities or color electric and magnetic polarizabilities [13]. In the following we will discuss that at high Q^2 a better interpretation for these matrix elements is that of a 'force'.

As Qiu and Sterman have shown [14], the average transverse momentum of the ejected quark (here also averaged over the momentum fraction x carried by the active quark) in a SIDIS experiment can be represented by the matrix element

$$\langle k_\perp^y\rangle = -\frac{1}{2P^+}\left\langle P,S\left|\bar{q}(0)\int_0^\infty dx^- G^{+y}(x^+=0,x^-)\gamma^+ q(0)\right|P,S\right\rangle \tag{7}$$

which has a simple physical interpretation: the average transverse momentum is obtained by integrating the transverse component of the color Lorentz force along the trajectory of the active quark — which is an almost light-like trajectory along the $-\hat{z}$ direction, with $z = -t$: The $\hat{y}$-component of the Lorentz force acting on a particle moving, with (nearly) the speed of light $\vec{v} = (0,0,-1)$, along the $-\hat{z}$ direction reads

$$g\sqrt{2}G^{y+} = g\left(E^y + B^x\right) = g\left[\vec{E} + \vec{v}\times\vec{B}\right]^y. \tag{8}$$

We now rewrite Eq. (7) as an integral over time

$$\langle k_\perp^y \rangle = -\frac{\sqrt{2}}{2P^+}\langle P,S|\bar{q}(0)\int_0^\infty dt G^{+y}(t,z=-t)\gamma^+ q(0)|P,S\rangle \tag{9}$$

in which the physical interpretation of $-\frac{\sqrt{2}}{2P^+}\langle P,S|\bar{q}(0)G^{+y}(t,z=-t)\gamma^+ q(0)|P,S\rangle$ as being the averaged force acting on the struck quark at time t after being struck by the virtual photon becomes more apparent:

$$\begin{aligned} F^y(0) &\equiv -\frac{\sqrt{2}}{2P^+}\langle P,S|\bar{q}(0)G^{+y}(0)\gamma^+ q(0)|P,S\rangle \\ &= -\frac{1}{\sqrt{2}}MP^+S^x d_2 = -\frac{M^2}{2}d_2, \end{aligned} \tag{10}$$

where the last equality holds only in the rest frame ($p^+ = \frac{1}{\sqrt{2}}M$) and for $S^x = 1$, can be interpreted as the averaged transverse force acting on the active quark in the instant right after it has been struck by the virtual photon.

Lattice calculations of the twist-3 matrix element yield [15]

$$d_2^{(u)} = 0.010 \pm 0.012 \qquad\qquad d_2^{(d)} = -0.0056 \pm 0.0050 \tag{11}$$

renormalized at a scale of $Q^2 = 5\ \mathrm{GeV}^2$ for the smallest lattice spacing in Ref. [15]. Here the identity $M^2 \approx 5\mathrm{GeV/fm}$ is useful to better visualize the magnitude of the force.

$$F_{(u)} = -25 \pm 30\mathrm{MeV/fm} \qquad\qquad F_{(d)} = 14 \pm 13\mathrm{MeV/fm}. \tag{12}$$

In the chromodynamic lensing picture, one would have expected that $F_{(u)}$ and $F_{(d)}$ are of about the same magnitude and with opposite sign. The same holds in the large N_C limit. A vanishing Sivers effect for an isoscalar target would be more consistent with equal and opposite average forces. However, since the error bars for d_2 include only statistical errors, the lattice result may not be inconsistent with $d_2^{(d)} \sim -d_2^{(u)}$.

The average transverse momentum from the Sivers effect is obtained by integrating the transverse force to infinity (along a light-like trajectory) $\langle k^y \rangle = \int_0^\infty dt F^y(t)$. This motivates us to define an 'effective range'

$$R_{eff} \equiv \frac{\langle k^y \rangle}{F^y(0)}. \tag{13}$$

Note that R_{eff} depends on how rapidly the correlations fall off along a light-like direction and it may thus be larger than the (spacelike) radius of a hadron. Of cource, unless the functional form of the integrand is known, R_{eff} cannot really tell us about the range of the FSI, but if the integrand in (10) does not oscillate, (13) provides a reasonable estimate for the range over which the integrand in (10) is significantly nonzero.

Fits of the Sivers function to SIDIS data yield about $|\langle k^y\rangle| \sim 100$ MeV [17]. Together with the (average) value for $|d_2|$ from the lattice this translates into an effective range R_{eff} of several fm. It would be interesting to compare R_{eff} for different quark flavors and as a function of Q^2, but this requires more precise values for d_2 as well as the Sivers function.

Note that a complementary approach to the effective range was chosen in Ref. [18], where the twist-3 matrix element appearing in Eq. (10) was, due to the lack of lattice QCD results, estimated using QCD sum rule techniques. Moreover, the 'range' was taken as a model *input* parameter to estimate the magnitude of the Sivers function.

A measurement of the twist-4 contribution f_2 to polarized DIS allows determination of the expectation value of different Lorentz/Dirac components of the quark gluon correlator appearing in (5)

$$f_2 M^2 S^\mu = \frac{1}{2}\langle p,S|\bar{q}g\tilde{G}^{\mu\nu}\gamma_\nu q|p,S\rangle, \tag{14}$$

In combination with (5) this allows a decomposition of the force into electric and magnetic components using

$$F_E^y(0) = -\frac{M^2}{8}\chi_E \qquad\qquad F_B^y(0) = -\frac{M^2}{4}\chi_B \tag{15}$$

for a target nucleon polarized in the $+\hat{x}$ direction, where [13, 16]

$$\chi_E = \frac{2}{3}(2d_2 + f_2) \qquad\qquad \chi_M = \frac{1}{3}(4d_2 - f_2). \tag{16}$$

A relation similar to (10) can be derived for the x^2 moment of the twist-3 scalar PDF $e(x)$. For its interaction dependent twist-3 part $\bar{e}(x)$ one finds for an unpolarized target [19]

$$4MP^+P^+e_2 \quad = \quad g\langle p|\bar{q}\sigma^{+i}G^{+i}q|P\rangle, \tag{17}$$

where $e_2 \equiv \int_0^1 dx x^2 \bar{e}(x)$. The matrix element on the r.h.s. of Eq. (17) can be related to the average transverse force acting on a transversely polarized quark in an unpolarized target right after being struck by the virtual photon. Indeed, for the average transverse momentum in the $+\hat{y}$ direction, for a quark polarized in the $+\hat{x}$ direction, one finds

$$\langle k^y\rangle = \frac{1}{4P^+}\int_0^\infty dx^- g\langle p|\bar{q}(0)\sigma^{+y}G^{+y}(x^-)q(0)|p\rangle. \tag{18}$$

A comparison with Eq. (17) shows that the average transverse force at $t=0$ (right after being struck) on a quark polarized in the $+\hat{x}$ direction reads

$$F^y(0) = \frac{1}{2\sqrt{2}p^+}g\langle p|\bar{q}\sigma^{+y}G^{+y}q|p\rangle = \frac{1}{\sqrt{2}}MP^+S^x e_2 = \frac{M^2}{2}e_2, \tag{19}$$

where the last identify holds only in the rest frame of the target nucleon and for $S^x = 1$.

The impact parameter distribution for quarks polarized in the $+\hat{x}$ direction [20] was found to be shifted in the $+\hat{y}$ direction [21, 22]. Applying the chromodynamic lensing model implies a force in the negative $\hat{y}$ direction for these quarks and one thus expects $e_2 < 0$ for both u and d quarks. Furthermore, since $\kappa_\perp > \kappa$, one would expect that in a SIDIS experiment the $\perp$ force on a $\perp$ polarized quark in an unpolarized target on average to be larger than that on unpolarized quarks in a $\perp$ polarized target, and thus $|e_2| > |d_2|$.

Acknowledgements: I would like to thank A.Bacchetta, D. Boer, J.P. Chen, Y.Koike, and Z.-E. Mezziani for useful discussions. This work was supported by the DOE under grant numbers DE-FG03-95ER40965 and DE-AC05-06OR23177 (under which Jefferson Science Associates, LLC, operates Jefferson Lab).

REFERENCES

1. M.Burkardt, Phys. Rev. D **62**, 071503 (2000), Erratum-ibid. D **66**, 119903 (2002); M. Diehl, Eur. Phys. J. C **25**, 223 (2002); J.P. Ralston and B. Pire, Phys. Rev. D **66**, 111501 (2002).
2. M. Burkardt, Int. J. Mod. Phys. A **18**, 173 (2003).
3. D.W. Sivers, Phys. Rev. D **43**, 261 (1991).
4. A. Bacchetta et al., Phys. Rev. D **70**, 117504 (2004).
5. M. Burkardt, Phys. Rev. D **66**, 114005 (2002); Phys. Rev. D **69**, 057501 (2004).
6. S.J. Brodsky, D.S. Hwang, and I. Schmidt, Nucl. Phys. B **642**, 344 (2002); M. Burkardt and D.S. Hwang, Phys. Rev. **D69**, 074032 (2004); L.P. Gamberg et al., Phys. Rev. D **67**, 071504 (2003); D. Boer, S.J. Brodsky, and D.S. Hwang, Phys. Rev. D **67**, 054003 (2003); A. Bacchetta et al., Phys. Lett **B578**, 109 (2004); M. Radici et al., hep-ph/0708.0232; L.P. Gamberg et al. hep-ph/0708.0324; D. Boer et al., Phys. Rev. D **67**, 054003 (2003); L.P. Gamberg et al., Phys. Rev. D **67** (2003) 071504.
7. S. Meissner et al., Phys. Rev. D **76**, 034002 (2007)
8. A.V. Belitsky, X. Ji, and F. Yuan, Phys. Rev. D **69**, 074014 (2004).
9. A. Airapetian et al. (HERMES collaboration), Phys. Rev. Lett. **94**, 012002 (2005).
10. A. Martin (COMPASS collaboration), Czech. J. Phys. **56**, F33 (2006).
11. E. Shuryak and A.I. Vainshtein, Nucl. Phys. B **201**, 141 (1982).
12. R.L. Jaffe, Comm. Nucl. Part. Phys. **19**, 239 (1990).
13. B.W. Filippone and X. Ji, Adv. Nucl. Phys. **26**, 1 (2001).
14. J. Qiu and G. Sterman, Phys. Rev. Lett. **67**, 2264 (1991).
15. M. Göckeler et al., Phys. Rev. D **72**, 054507 (2005).
16. Z.-E. Mezziani et al., hep-ph/0404066.
17. M. Anselmino et al., hep-ph/0805.2677
18. A. Schäfer et al., Phys. Rev. D **47**, R1 (1993); E. Stein et al., hep-ph/9409212
19. Y. Koike and K. Tanaka, Phys. Rev. D **51**, 6125 (1995).
20. M. Diehl and P.Hägler, Eur. Phys. J. **C44**, 87 (2005).
21. M. Göckeler et al. (QCDSF collaboration), Phys. Rev. Lett. **98**, 222001 (2007); Ph. Hägler et al. (LHPC collaboration), hep-lat/0705.4295.
22. M. Burkardt and B. Hannafious, Phys. Lett. B **658**, 130 (2008).

Beyond the Standard Model: Future Prospects with Parity Violating Electron Scattering

Krishna S. Kumar

Department of Physics, University of Massachusetts, Amherst, MA 01003

Abstract. We discuss the status and prospects of an experimental program of parity-violating asymmetry measurements in the scattering of longitudinally polarized electrons off unpolarized fixed targets. In particular, we focus on those measurements where judicious choices of target and kinematics allows the theoretical predictions to be made purely in terms of fundamental electroweak couplings. If such asymmetries are measured with sufficient precision, they are sensitive to new physics at the TeV scale. The physics implications of recent results, experiments under construction and plans for the future are discussed.

Keywords: Parity Violation, Weak Neutral Current Interactions
PACS: 12.15.Mm Neutral Currents 11.30.Er C, P, T and other discrete symmetries

INTRODUCTION

Since the $SU(2)_L \times U(1)_Y$ gauge theory of electroweak interactions was established three decades ago as a cornerstone of the Standard Model, experiments have been searching for clues to address the model's many shortcomings. Compelling theoretical arguments point to pursuing measurements that study the TeV scale in two broad thrusts: high energy colliders and targeted complementary precision electroweak measurements at low energy. Important low energy measurements are being carried out both at accelerator as well as non-accelerator facilities. The most important accelerator-based low energy measurements focus on electroweak observables that can be calculated to high accuracy and aim to achieve sufficient precision so that indirect effects of new dynamics at the TeV scale might become manifest. In the following, we describe an important subset of such measurements which use the technique of parity-violating electrons scattering.

Nearly 50 years ago, soon after the discovery of parity violation in beta decay, Zel'dovich speculated that there might be an analogous parity violating neutral current interaction[1]. He noted that if such an interaction existed, then parity violation would be manifested in lepton-nucleon scattering due to the interference between the weak and electromagnetic amplitudes. He predicted that if one scatters longitudinally polarized electrons off unpolarized protons and flipped the sign of the beam polarization, the fractional difference in the cross-section would be:

$$A_{PV} \equiv \frac{\sigma_R - \sigma_L}{\sigma_R + \sigma_L} \simeq \frac{|A_Z|}{|A_\gamma|} \simeq \frac{G_F Q^2}{4\pi\alpha} \simeq 10^{-4} Q^2 \tag{1}$$

For typical fixed target experiments, A_{PV} ranges from roughly 10^{-4} to as small as 10^{-7}. In the mid-seventies, parity violation in deep inelastic electron nucleon scattering

CP1149, *Spin Physics, 18th International Spin Physics Symposium*
edited by D. G. Crabb, D. B. Day, S. Liuti, X. Zheng, M. Poelker, and Y. Prok

was first observed at SLAC[2], from which the electron-quark weak neutral current coupling could be extracted. The measurement was an important validation of the Standard Model, and the extracted value of the electroweak mixing angle $\sin^2\theta_W$ matched the corresponding value obtained from neutral current neutrino scattering experiments.

Over the past 20 years, the experimental techniques employed to measure these tiny left-right asymmetries have been steadily refined such that statistical errors approaching 0.01 parts per million (ppm) and systematic errors of a few parts per billion (ppb) are possible[3]. Depending on the choice of target and kinematic variables, this has facilitated measurements in several important physics topics, such as many-body nuclear physics, nucleon structure and searches for physis beyond the standard model at the TeV scale.

In this review, we focus on the last topic. Electroweak measurements continue to probe for the indirect effects of new physics at the TeV scale by making more and more precise measurements of electroweak parameters[4]. For example, weak neutral current interactions at $Q^2 \ll M_Z^2$ can probe for heavy Z′ bosons or leptoquarks whose effects might be highly suppressed in measurements on the Z pole. Since Z pole measurements are imaginary, there are no interference terms with new, real amplitudes. At low Q^2 on the other hand, interference effects might be measurable if sufficient accuracy is achieved[5]. The goal of low energy neutral current measurements is to reach a sensitivity on contact interactions at the 10 TeV level, similar sensitivity to that of the highest energy colliders.

Three measurement techniques have reached such sensitivity. The weak charge measurement in atomic Cesium[7], the NuTeV neutrino deep-inelastic scattering measurement[6] and the measurement of $A_{\rm PV}$ in electron-electron (Møller) scattering[8] at SLAC. These measurements are used to set important new constraints on new parity-violating contact interactions at the 10 TeV scale.

LEPTON-QUARK PHENOMENOLOGY

The interactions of the Z-boson and heavier particles can be approximated by four-fermion contact interactions. The parity-violating part of the electron-hadron interaction can then be given in terms of phenomenological couplings C_{ij}

$$\mathscr{L}^{PV} = \frac{G_F}{\sqrt{2}}[\bar{e}\gamma^\mu\gamma_5 e(C_{1u}\bar{u}\gamma_\mu u + C_{1d}\bar{d}\gamma_\mu d) + \bar{e}\gamma^\mu e(C_{2u}\bar{u}\gamma_\mu\gamma_5 u + C_{2d}\bar{d}\gamma_\mu\gamma_5 d)]$$

with additional terms as required for the heavy quarks. Here C_{1j} (C_{2j}) gives the vector (axial-vector) coupling to the jth quark. For the Standard Model,

$$C_{1u} = g_A^e g_V^u \approx -\frac{1}{2} + \frac{4}{3}\sin^2\theta_W \approx -0.19 \quad (2)$$

$$C_{1d} = g_A^e g_V^d \approx \frac{1}{2} - \frac{2}{3}\sin^2\theta_W \approx 0.34 \quad (3)$$

$$C_{2u} = g_V^e g_A^u \approx -\frac{1}{2} + 2\sin^2\theta_W \approx -0.030 \quad (4)$$

$$C_{2d} = g_V^e g_A^d \approx \frac{1}{2} - 2\sin^2\theta_W \approx 0.025 \quad (5)$$

The atomic Cesium weak charge measurement measures one combination of C_{1u} and C_{1d} precisely.

PARITY-VIOLATING ELASTIC ELECTRON-PROTON SCATTERING

At sufficiently forward angles and low Q^2, the hadronic structure undertainty in the WNC elastic electron-proton amplitude becomes small enough such that one can measure the underlying coherent $2u+d$ e-q amplitude combination to high precision, thus precisely constraining $2C_{1u}+C_{1d}$. This combination is proportional to $1-4\sin^2\theta_W$, so that a 4% measurement of A_{PV} would achieve a precision of $\delta(\sin^2\theta_W)=0.0007$. Such a measurement has been proposed by the Qweak collaboration in Hall C at Jefferson Lab. Construction is well under way and first commissioning is scheduled for 2010.

PARITY-VIOLATING DEEP INELASTIC SCATTERING

The upgrade of Jefferson Laboratory (Jlab) to 11 GeV incident energy will allow precision measurements in parity-violating deep inelastic scattering (PV DIS). While the two measurements will be carried out at 6 GeV [9] and 11 GeV [10], a dedicated spectrometer could for the first time allow the accumulation of high statistics at high $x\sim0.7$, where x is the fraction of the nucleon momentum carried by the struck quark. PV DIS provides access to novel aspects of nucleon structure, such as charge symmetry violation and investigation of higher-twist effects. Most importantly, PV DIS allows the isolation of the linear combination $2C_{2u}+C_{2d}$, which is difficult to measure using elastic scattering.

To comprehensively address these physics topics experimentally, a series of A_{PV} measurements with better than 1% accuracy are required for the x range from 0.3 to 0.7, with a lever arm of a factor of 2 in Q^2 while keeping $W^2_{\rm min}>4$ and $Q^2_{\rm min}>1$. With the upgrade of Jlab, high luminosity with a beam energy of 11 GeV becomes possible for the first time. However, to achieve sufficient statistics at the highest possible Q^2, a spectrometer with at least 50% acceptance in the azimuth is required.

It turns out that the central tracking magnetic field of one of the collider experiments BaBar, CDF or CLEO would be ideal for this purpose. A proposal was recently submitted to the JLab PAC for this comprehensive program and detailed simulations have been carried out for the case of using the BaBar magnet [11]. The proposal was conditionally approved and an engineering design of the detectors and support structures inside the solenoid will be carried out in 2009.

PARITY-VIOLATING MØLLER SCATTERING

The parity-violating asymmetry in electron-electron (Møller) scattering was measured for the first time by the E158 experiment at SLAC. This measurement established the running of $\sin^2\theta_W$ for the first time and provided important constraints on electron-

electron contact interactions at the TeV scale in a manner complementary to collider measurements.

With the JLab upgraded 11 GeV beam, it becomes feasible to design a new experiment that would improve on the E158 measurement by a factor of 5. The measurement takes advantage of the large beam current that JLab provides to compensate for the smaller beam energy compared to that for E158. A novel spectrometer system based on two toroids has been designed and a proposal was recently submitted to the JLab PAC [12]. That proposal was approved and an engineering design of the spectrometer toroids and R& D on detectors will begin in 2009.

The proposed JLab measurement would achieve a precision on the measurement of $\sin^2\theta_W$ that is comparable to the two best high energy collider measurements. This is especially important since these two measurements differ from each other by about 3 standard deviations. In addition, the proposed measurement would provide important complementary constraints on TeV scale contact interactions such as that produced by heavy Z' bosons and R-parity violating supersymmetric amplitudes [13].

ELECTROWEAK PHYSICS AT AN EIC

A logical next step in precision studies of the structure of the nucleon is being pursued via the design of a polarized lepton light ion collider (EIC). In such a machine, 5 to 20 GeV polarized electrons would collide with 100 to 250 GeV polarized ^1H, ^2H or ^3He. The luminosity that is envisioned for such a machine is sufficiently high ($10^{34}/cm^2/s$), that it becomes feasible to make precision measurements of DIS parity violating asymmetries.

Preliminary investigations of the physics potential have just been launched. Apart from being able to improve on the JLab 11 GeV PV DIS measurements, it becomes feasible for the first time to measure with sufficient precision the parity-violating asymmetry in DIS where the electron is unpolarized but the hadron is polarized. Such measurements would allow access to new spin structure functions $g_1^{\gamma Z}$ and $g_5^{\gamma Z}$ for the first time, which would provide new information on the helicity distributions of quarks and anti-quarks.

SUMMARY

Parity-violating electron scattering is a mature field and addresses fundamental questions in a variety of different topics. The E158 experiment at SLAC has produced the most precise measurement of the weak mixing angle at low energy. Future measurements at Jefferson Laboratory will lead to important new insights on the structure of the nucleon and yield new and more precise measurements of the weak mixing angle. In the far future, it is possible to probe the nucleon in novel ways at a polarized electron-polarized light ion collider.

ACKNOWLEDGMENTS

It is a pleasure to thank the organizers for a stimulating meeting. The contributions of all the experimental collaborations discussed in this review are acknowledged. This work is funded in part by US Department of Energy Grant No. DE-FG02-88R40415-A018.

REFERENCES

1. Zel'dovich,Ya.B., *J.Exptl.Theoret.Phys.* (U.S.S.R.), **36**, 1959, pp. 964-966.
2. Prescott,C.Y., et.al., *Phys.Lett.*, **B84**, 1979, 524.
3. Kumar,K.S. and Souder,P.A., *Prog.Part.Nucl.Phys.*, **45**, 2000, pp. S333-S395.
4. Ramsey-Musolf,M.J., *Phys.Rev.*, **C60**, 1999, 015501.
5. Kumar,K.S., et.al., *Mod.Phys.Lett.*, **A10**, 1995, 2979.
6. Zeller,G.P., et.al., *Phys.Rev.Lett.*, **88**, 2002, 091802.
7. Wood,C.S., et.al., *Science*, **275**, 1997, 1759.
8. Anthony,P.L., et.al, *Phys.Rev.Lett.*, **95**, 2005, 081601.
9. Jefferson Lab Experiment E05-007, X. Zheng, Contact Person
10. Jefferson Lab Proposal PR12-07-102, P. Reimer, Contact Person
11. Jefferson Lab Proposal PR-09-012, P.A. Souder, Contact Person
12. Jefferson Lab Proposal PR-09-005, K.S. Kumar, Contact Person
13. A. Kurylov, M.J. Ramsey-Musolf and S. Su, Phys. Rev. **D68**, (2003) 035008.

Spin beyond Standard Model: Theory

J. Erler

Departamento de Física Teórica, Instituto de Física, Universidad Autónoma de México, 04510 México D.F., México

Abstract. I use spin as a guide through the labyrinth of possibilities and ideas that go beyond the established understanding of the fundamental interactions.

Keywords: Higgs boson, technicolor, compositeness, extra gauge bosons, extra dimensions, SUSY.
PACS: 11.30.Cp,12.10.-g,12.15.-y,12.60.-i,14.80.Bn.

INTRODUCTION

What is required to define spin are Lorentz-covariant free particles and quantum mechanics. On the other hand, causality (from demanding a Lorentz invariant S-matrix) and the cluster decomposition principle (the requirement that distant experiments yield unrelated results which in relativistic theories naturally leads to the concept of quantum fields) are not necessary. Thus, spin can be considered even outside the framework of a quantum field theory (QFT) — but not conversely — and can serve as a very general organizing principle for the many types of physics beyond the Standard Model (SM) that have been suggested[1], including extra dimensions, strings, and M-theory.

The complete classification of unitary representations of the inhomogeneous Lorentz group according to Wigner [4] is recalled in Table 1. The little group is defined as the subgroup leaving some conveniently chosen "standard" four-momentum, k^μ, unchanged and gives rise to spin and helicity. Thus, for a massive particle the possible spin states are obtained from the $SO(3)$ Lie algebra and are consequently identical to those familiar from non-relativistic quantum mechanics. The little group for massless particles is the

TABLE 1. Wigner classification of unitary representations of the Poincaré group.

p^2	p_0	standard k^μ	little group	comment
>0	>0	$(M,0,0,0)$	$SO(3)$	massive particle
>0	<0	$(-M,0,0,0)$	$SO(3)$	$E<0$ (unphysical)
$=0$	>0	$(k,k,0,0)$	$ISO(2)$	massless particle
$=0$	$=0$	$(0,0,0,0)$	$SO(3,1)$	vacuum (no particles)
$=0$	<0	$(-k,k,0,0)$	$ISO(2)$	$E<0$ (unphysical)
<0	any	$(0,M,0,0)$	$SO(2,1)$	tachyon ($\lvert v\rvert > c$)

[1] For tests of the SM see the contributions by Bill Marciano [1] (muon anomalous magnetic moment and CKM unitarity), Yannis Semertzidis [2] (electric dipole moments and $\mu \rightarrow e$ conversion), and Krishna Kumar [3] (polarized electron scattering).

CP1149, *Spin Physics, 18th International Spin Physics Symposium*
edited by D. G. Crabb, D. B. Day, S. Liuti, X. Zheng, M. Poelker, and Y. Prok

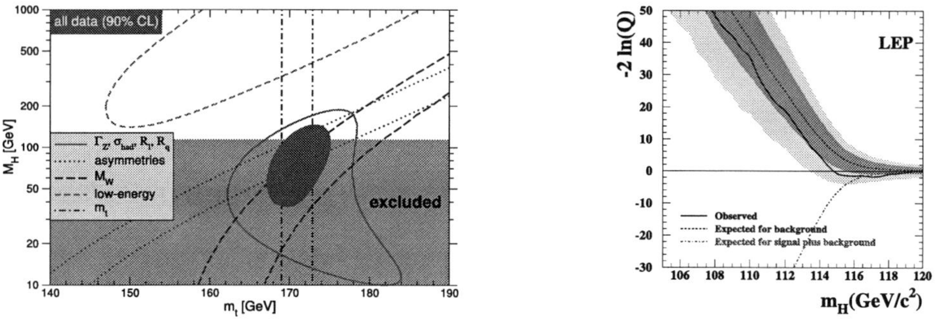

FIGURE 1. Left: 1 σ contours from various data sets in the M_H-m_t plane and 90% region allowed by all precision data. Right: Results from Higgs searches at LEP 2 [5]. The bands around the dotted line for the background expectation refer to the 1 and 2 σ regions. The solid line for the observation is below (above) the dotted line if an upward (downward) fluctuation is seen. A measure for the sensitivity to a given M_H is the distance between the dot-dashed line for the signal plus background hypothesis and the dotted line.

Euclidean group in two dimensions, $ISO(2)$ (generated by two translations and one rotation), and (being non-compact) permits, in general, a continuous parameter ("continuous spin"). One assumes (basically on phenomenological grounds) that physical states are non-trivially represented only with respect to the compact $SO(2) = U(1)$ subgroup, and identifies its "charges" with the helicity, h, of the massless particle. The trivial $U(1)$ Lie algebra would allow arbitrary values of h, but the topology of $SO(3,1) = SL(2,\mathbb{C})/Z_2$ can be shown to be that of the doubly connected space $\mathbb{R}^3 \times S^3/Z^2$, so that the requirement of a globally defined wave function restricts h to integer and half-integer values. Still, Lorentz invariance requires consideration of the entire $ISO(2)$ little group. I will return to this point in the discussion of states with $h \geq 1$.

The spin 0 category includes the (would-be) Nambu-Goldstone bosons of spontaneously broken continuous symmetries, such as Higgs bosons, familons (broken family symmetry), Majorons (broken lepton number conservation), or axions (broken Peccei-Quinn symmetries). Furthermore, there may be radions (graviscalar components of the metric tensor in models with extra spatial dimensions), dilatons (*e.g.*, in string theories), moduli (scalars with flat potential in supersymmetry), inflatons, scalar leptoquarks, and sfermions (scalar superpartners). Spin 1/2 states beyond the SM could be due to a fourth family, right-handed neutrinos, other exotic states (*e.g.*, those needed to cancel anomalies in models with new gauge symmetries), techniquarks from dynamical (strong) symmetry breaking models, or X-inos (fermionic superpartners other than gravitinos). Spin 1 states could be extra Z' or W' bosons or vector leptoquarks (*e.g.*, from Grand Unified Theories). Examples for spin 2 particles are the graviton (*predicted* by string and M-theory) and its Kaluza-Klein excitations (in models with extra dimensions). A massless particle with spin > 2 is not expected to give rise to a long-range force, but the towers of massive string excitations include states with arbitrarily high spins. On the other hand, a massless particle (or massive particle after spontaneous supersymmetry breaking) of spin 3/2 would uniquely point to the gravitino of local supersymmetry (supergravity).

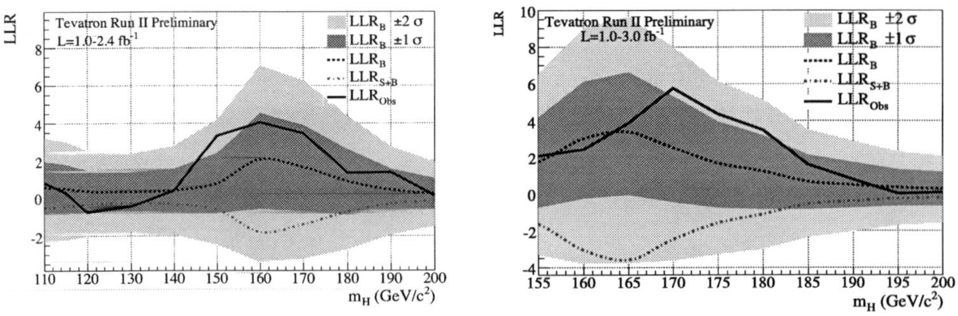

FIGURE 2. Results from Higgs searches at run II of the Tevatron [6]. The meaning of the lines and bands is as in the right panel of Figure 1. Notice, that the right-hand plot is from a slightly larger data set.

SPIN 0

The unique fundamental scalar within the SM is the yet to be discovered Higgs boson. Its mass, M_H, is constrained from direct (so far negative) searches and from indirect precision analyzes (quantum loop effects). The global fit to all indirect data currently yields $M_H = 89^{+29}_{-22}$ GeV, with a very reasonable goodness of the fit. The M_H constraints as functions of the top quark mass, m_t, are shown in Figure 1 for various data sets. As can be seen, these are consistent with each other with the exception of the low energy data which deviate mainly due to the result on deep inelastic neutrino scattering (NuTeV).

LEP 2 (Figure 1) excluded Higgs masses below 114.4 GeV (95% CL), and saw a small but by itself insignificant excess around $M_H = 117$ GeV. An upward fluctuation is also seen for similar M_H values at run II of the Tevatron (Figure 2). The best sensitivity is here for Higgs masses slightly above the threshold for decays into W pairs and thanks to a downward fluctuation values close to $M_H = 170$ GeV could already be ruled out. The combined result of all constraints (direct and indirect) is shown in Figure 3.

Even though the Higgs boson is a SM particle, it may provide clues about new physics. *E.g.*, if one requires that it remains perturbatively coupled up to the fundamental Planck scale, $\kappa_4 \approx 2.4 \times 10^{18}$ GeV, one obtains (in the absence of new physics) the upper limit, $M_H \lesssim 180$ GeV. Likewise, vacuum stability up to κ_4 implies $M_H \gtrsim 130$ GeV (if one allows a meta-stable vacuum this weakens to $M_H \gtrsim 115$ GeV). In general, one expects M_H to be of the form $M_H^2 = (M_H^0)^2 + c\frac{\alpha}{\pi}\kappa_4^2 = \mathscr{O}(G_F^{-1/2})$, so that $c = \mathscr{O}(1)$ would require a delicate cancellation of the tree and loop terms (the hierarchy problem[2]). The main mission of the LHC is then to find at least one Higgs boson and the new physics providing the solution to the hierarchy problem (ideally) also explaining (i) why the precision data appear to be consistent with the SM with so far no clear indication for new physics, (ii) the nature of the dark matter, and (iii) the baryon asymmetry of the universe.

---

[2] There is no hierarchy problem for fermions, because chiral symmetry protects massless fermions to acquire masses radiatively, and so the radiative corrections for massive fermions will be no larger than of the order of the chiral symmetry breaking tree mass terms themselves.

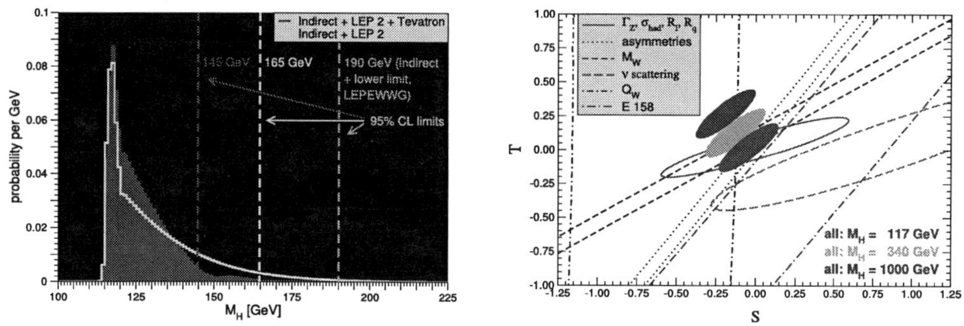

FIGURE 3. Left: Combination of direct and indirect constraints on M_H. The red (solid) histogram includes all data while the yellow line shows the result without the Tevatron constraints. The 95% upper limits are also shown, where LEP 2 obtained their 190 GeV upper limit by treating their lower bound as a step function. Right: 1 σ constraints on the parameters S and T (describing gauge boson self-energies) for various data sets and $M_H = 117$ GeV. The 90% allowed regions for all data and three reference values for M_H are also shown. In the SM, S = T = 0 (by definition) while in TC models typically S $\gtrsim$ 1.

SPIN 1/2

One way to address the hierarchy problem is to avoid fundamental scalars altogether. In the original *technicolor* (TC) idea [7, 8] (techni)fermions condense through non-Abelian gauge interactions, break electroweak symmetry (EWS), and generate the masses for the W and Z bosons. It is less straightforward to obtain the masses for the SM fermions and *extended technicolor* gauge interactions are needed. In general, this leads in turn to large flavor changing neutral current effects in conflict with observation. This can be cured by decelerating the running of the TC coupling (walking TC) so as to effectively decouple the extended TC gauge bosons. Another complication is the large m_t value and one considers models of top quark condensation (topcolor). Models in which the EWS breaks solely due to topcolor predict too large an m_t so that one arrives at hybrid models (topcolor assisted TC). One also has to introduce an extra $U(1)'$ gauge symmetry to prevent condensation of the much lighter bottom quarks. This is one of many examples in which a model of EWS breaking predicts a specific massive Z' boson to solve some model building problem. Thus, Z' diagnostics may be an additional analyzing tool for the new physics that breaks EWS. However, models of TC are generally in conflict with the S parameter (Figure 3). Incidentally, the S parameter also constrains a fourth fermion generation and rules it out if mass degenerate ($S = 2/3\pi$). The non-degenerate case is also disfavored and only marginally consistent for specific mass values.

Another way to avoid fundamental scalars is to assume that the Higgs field is *composite* [9]. A modern reincarnation of this idea is *Little Higgs Theory* [10] with a fundamental scale Λ of 5 to 10 TeV, and where the Higgs is lighter since it appears as a pseudo-Goldstone boson. In these models the quadratically divergent contributions to M_H are then postponed by one loop order (in some models by two orders). One can also assume that quarks and leptons are composite. This introduces effective contact interactions with an effective scale constrained by LEP 2 to satisfy $\Lambda \gtrsim \mathcal{O}(10\text{ TeV})$.

SPIN 1

In a QFT, a particle with $h = \pm 1$ can produce a long-range ($1/r^2$) force only if it is coupled to a conserved vector current. This can be traced to the ISO(2) little group mentioned in the introduction and implies the concept of gauge invariance [11]. Extra neutral gauge bosons (Z') can arise from extra $U(1)'$ gauge symmetries as contained, *e.g.*, in the E_6 unification group, $E_6 \to SO(10) \times U(1) \to SU(5) \times U(1)^2$, or in left-right symmetric models, $SU(2)_L \times SU(2)_R \times U(1) \to SU(2)_L \times U(1)^2$. The Z' could also be a Kaluza-Klein (KK) excitation (sequential Z'), the techni-ρ, or the topcolor-Z' mentioned above. The Z' mass, $M_{Z'}$, is constrained to be greater than 1305 GeV in the sequential case (DELPHI) and $M_{Z'} > 630$ to 891 GeV for E_6 scenarios (DØ), while electroweak (EW) data limit the mixing angle with the ordinary Z to < 0.01. Likewise, mass and mixing of an extra W' are limited, respectively, to > 1 TeV (DØ) and < 0.12 (OPAL).

SPIN 2

In a QFT, $1/r^2$-forces are produced by an $h = \pm 2$ particle only if it is coupled to the conserved energy-momentum tensor implying invariance under general coordinate transformations [11]. The ultraviolet completion of gravity requires its embedding into structures like *string* or *M-theory* whose growing number of known vacua led to the concept of the *string landscape* [12] and a revival of anthropic reasoning (the multiverse).

String [13] and M-theory [14] predict the existence of extra dimensions (EDs), some of which are conceivably much larger than κ_4^{-1}. One can consider [15] a factorized D-dimensional space, $\mathbb{R}^4 \times M^{D-4}$, where M is a *flat* space with volume $(2\pi R)^{D-4}$. The D-dimensional Planck scale, κ_D, is identified with the EW scale, and the hierarchy problem appears here as the puzzle why $R^{-1} = \kappa_D(\kappa_D/\kappa_4)^{2/(D-4)} \ll \kappa_D$ (the case $D = 4+1$ is excluded from precision studies of Newton's law). Collider processes such as $e^+e^- \to \gamma +$ missing energy require $\kappa_D > \mathcal{O}(1\text{ TeV})$, while astrophysical bounds, $\kappa_D \gtrsim \mathcal{O}(100\text{ TeV})$, can be avoided in models. An alternative is to introduce a *warped* 5-dimensional space [16] with metric, $ds^2 = e^{-yk}\eta_{\mu\nu}dx^\mu dx^\nu - dy^2$, where k is the anti-de-Sitter (AdS) curvature and $0 \leq y \leq \pi R$ the fifth coordinate. The EW scale is here exponentially suppressed ($\sim \kappa_4 e^{-\pi k R}$). Observables like the $\ell^+\ell^-$ or $\gamma\gamma$ invariant mass spectra can probe the interaction scale of KK gravitons with matter concluding $kR \gtrsim 11$.

In addition to the graviton one can also allow SM fields to propagate in the factorized EDs. *E.g.*, TeV-scale string compactification [17] with compactification radius $R \gtrsim (7\text{ TeV})^{-1}$ (from LEP 2) implies gauge fields propagate in the bulk. *Gauge-Higgs Unification* [18] protects the EW scale by defining the Higgs as a higher-dimensional gauge field component. *Universal Extra Dimensions* (UEDs) [19] allow all SM particles in the bulk. Grand Unified Theories (GUTs) in EDs with $R^{-1} \approx M_U$ (the gauge coupling unification scale) can solve many problems of conventional supersymmetric GUTs [20]. There are also models with SM fields propagating in warped EDs (but not the Higgs in order to protect the exponential suppression of the EW scale). These can be viewed as *dual* descriptions of walking TC by virtue of the AdS correspondence with conformal field theories [21]. No fully realistic models exist, but it is interesting that EWS breaking by boundary conditions of gauge fields in warped EDs yield *higgsless models* [22].

SPIN 3/2

In a QFT, an $h = \pm 3/2$ particle can produce a $1/r^2$-force only if it is coupled supersymmetrically. Conversely, supersymmetry [23] is the only possibility to extend the Poincaré algebra non-trivially and its non-renormalization theorems offer the most elegant solution to the hierarchy problem. Moreover, in models of supersymmetric unification a large m_t correctly predict EWS breaking (a fact that was known [24, 25] even before m_t was measured) and these are further supported by the approximate unification of gauge couplings at $M_U \sim 10^{-16}$ GeV $\lesssim \kappa_4$ in the minimal supersymmetric standard model (MSSM). Finally, the MSSM predicts $M_{h^0} \lesssim 135$ GeV for the lighter of the two CP-even Higgses, in remarkable agreement with Figure 3 (identifying h^0 with H). The additional assumption of R-parity conservation (sufficient to forbid proton decay by dimension 4 operators) implies that the lightest supersymmetric particle is stable offering an explanation for the observed dark matter (a similar statement applies to UED models).

ACKNOWLEDGMENTS

This work was supported by UNAM as DGAPA-PAPIIT project IN115207 and by CONACyT (México) as project 82291-F.

REFERENCES

1. W. J. Marciano, these proceedings.
2. Y. Semertzidis, these proceedings.
3. K. S. Kumar, these proceedings.
4. E. P. Wigner, *Annals Math.* **40**, 149–204 (1939) and *Nucl. Phys. Proc. Suppl.* **6**, 9–64 (1989).
5. ALEPH, DELPHI, L3, and OPAL Collaborations, and LEP Working Group for Higgs Boson Searches: R. Barate *et al.*, *Phys. Lett. B* **565**, 61–75 (2003).
6. CDF and DØ Collaborations and Tevatron New Phenomena and Higgs Working Group: G. Bernardi *et al.*, e-prints `arXiv:0804.3423` and `arXiv:0808.0534` [hep-ex].
7. S. Weinberg, *Phys. Rev.* D **19**, 1277–1280 (1979).
8. L. Susskind, *Phys. Rev.* D **20**, 2619–2625 (1979).
9. D. B. Kaplan, H. Georgi, and S. Dimopoulos, *Phys. Lett.* B **136**, 187–190 (1984).
10. N. Arkani-Hamed, A. G. Cohen, and H. Georgi, *Phys. Lett.* B **513**, 232–240 (2001).
11. S. Weinberg, *Phys. Rev.* **138**, B988–B1002 (1965).
12. L. Susskind, "The Anthropic Landscape of String Theory", e-print `arXiv:hep-th/0302219`.
13. M. B. Green, J. H. Schwarz, and E. Witten, *Superstring Theory* (Vols. 1 & 2), Cambridge, 1987.
14. P. Horava and E. Witten, *Nucl. Phys.* B **460**, 506–524 (1996).
15. N. Arkani-Hamed, S. Dimopoulos, and G. R. Dvali, *Phys. Lett.* B **429**, 263–272 (1998).
16. L. Randall and R. Sundrum, *Phys. Rev. Lett.* **83**, 3370–3373 (1999).
17. I. Antoniadis, *Phys. Lett.* B **246**, 377–384 (1990).
18. N. S. Manton, *Nucl. Phys.* B **158**, 141–153 (1979).
19. T. Appelquist, H. C. Cheng, and B. A. Dobrescu, *Phys. Rev.* D **64**, 035002 (2001).
20. L. J. Hall and Y. Nomura, *Phys. Rev.* D **66**, 075004 (2002).
21. J. M. Maldacena, *Adv. Theor. Math. Phys.* **2**, 231–252 (1998).
22. C. Csaki, C. Grojean, H. Murayama, L. Pilo, and J. Terning, *Phys. Rev.* D **69**, 055006 (2004).
23. J. Wess and B. Zumino, *Phys. Lett.* B **49**, 52–54 (1974).
24. L. Alvarez-Gaumé, M. Claudson, and M. B. Wise, *Nucl. Phys.* B **207**, 96–110 (1982).
25. L. E. Ibáñez and G. G. Ross, *Phys. Lett.* B **110**, 215–220 (1982).

Polarized Antiprotons – The Quest For A Missing Tool

Erhard Steffens

Physikalisches Institut, University of Erlangen-Nürnberg
Erwin-Rommel-Str. 1, D-91058 Erlangen, Germany
steffens@physik.uni-erlangen.de

Abstract. After termination of the LEAR facility in 1996 and the restriction of the CERN antiproton program to Trap experiments, stored antiprotons at low and medium energies are no longer available for experiments. FAIR at GSI (Darmstadt) will provide intense stored antiproton beams in less than a decade from now. This has renewed the interest in polarized antiprotons originally proposed for LEAR. In August 2007, an International Workshop was organized at the Cockcroft Institute (Daresbury) and methods to polarize stored antiprotons were discussed. In June 2008, a Heraeus Seminar at the Physikzentrum in Bad Honnef dealing with the same subject took place. The results of these workshops will be presented with some historical remarks and an account of the recent developments.

Keywords: Antiprotons, spin polarized beams, spin filtering.

PACS: 24.70.+s 25.43.+t 29.27.Hj

INTRODUCTION

The $\bar{p}p$ interaction has been studied extensively using secondary antiproton beams at BNL and KEK, and by utilizing stored antiprotons at LEAR/CERN (1983-1996). Phenomenologically, the $\bar{p}p$ interaction at low momenta can be described by two potentials, the elastic one, V_{el}, and the annihilation potential V_{ann}. The elastic $\bar{p}p$ potential is related to V_{el}^{pp} by the G-parity transformation. No analogue to V_{ann} exists in the pp channel. Experiments on the spin dependence in the $\bar{p}p$ system are restricted so far to single spin experiments, i.e. measurements of A_{0n00} or A_{000n} [1]. The study of the $\bar{p}p$ spin correlation in the entrance channel requires beams of polarized antiprotons. Attempts at LEAR to produce polarized $\bar{p}'s$ by double scattering were unsuccessful. The FILTEX proposal to polarize stored $\bar{p}'s$ through Spin Filtering by means of a polarized internal hydrogen gas target was submitted in 1985 [2]. The development of this new method took too long to become part of the LEAR program which was terminated in 1996. Since then, the CERN $\bar{p}$ program is restricted to experiments at very low energies which are conducted at the CERN Antiproton Decelerator (AD).

In 2007, the new international FAIR project initiated by GSI (Darmstadt) was officially launched. FAIR stands for 'Facility for Antiproton and Ion Research' [3]. The completion is scheduled for 2016. A substantial fraction of the program is devoted to antiproton physics with multi-GeV beams at the High Energy Storage Ring

CP1149, *Spin Physics, 18th International Spin Physics Symposium*
edited by D. G. Crabb, D. B. Day, S. Liuti, X. Zheng, M. Poelker, and Y. Prok

(HESR). A general consensus exists that a new powerful $\bar{p}$ source should be complemented by a polarized antiproton beam. The PAX collaboration [4] has proposed to polarize $\bar{p}'s$ in a small storage ring (APR) of suitable energy by means of Spin Filtering [4, 5]. As key experiment the study of the double polarized Drell-Yan process $p\uparrow\bar{p}\uparrow\rightarrow e^+e^-$ has been identified [6, 7].

Since the first workshop at Bodega Bay in 1985, organized by O. Chamberlain and A. Krisch [8], several methods to polarize $\bar{p}'s$ stored in a ring have been studied or proposed. In the following, previous attempts to develop an antiproton polarizer are discussed. In the wake of the PAX proposal, and 22 years after Bodega Bay, the next workshop on polarized $\bar{p}'s$ was conducted in 2007 at the Cockcroft Institute (Daresbury) [9], followed by a Heraeus Seminar at the Physikzentrum (Bad Honnef) in 2008 [10]. The results of the two workshops are summarized. These and recent results obtained at COSY by the ANKE and PAX collaboration on depolarization of protons by co-moving electrons [11] represent impressive progress towards polarized $\bar{p}'s$.

HISTORY OF POLARIZED ANTIPROTONS[1]

In the early 1980's, the CERN $Sp\bar{p}S$ collider came into operation leading to the discovery of the Z^0 and $W^{\pm}$ bosons. Its powerful $\bar{p}$ source enabled a low energy $\bar{p}$ program at LEAR, a cooler and stretcher ring. First ideas by Kilian and Möhl appeared to polarize antiprotons by Spin Filtering. This process was first proposed by Shapiro [12] in 1966 to polarize thermal neutrons by spin-dependent attenuation in a polarized hydrogen target. It should be noted that spin filtering of slow neutrons by means of a polarized ^{3}He gas cell is now a standard method in neutron physics [13].

At the Bodega Bay workshop in 1985 [8], eleven potential methods to polarize stored $\bar{p}'s$ were identified and discussed. Out of these eleven, two ideas were studied in depth afterwards: (i) The 'Spin Splitter' is based on repeated 'Stern-Gerlach kicks' resulting in a spatial separation of the two antiproton spin states. It has been discussed theoretically in a number of papers (e.g. [14]) and workshop contributions, but not yet been demonstrated experimentally. (ii) The Spin Filter proposal for LEAR [2] was actively pursued by the FILTEX collaboration. Polarization by Channeling through magnetized ferromagnetic foils was also discussed. Recently, the idea of Channeling by bent crystals as polarizer received some attention [9].

The Filtex Test Experiment at the TSR (Heidelberg)

The Filter Test was performed in the Heidelberg Test Storage Ring (TSR), a low energy ion storage ring with electron cooling and large acceptance [15]. The principle of the experiment is shown in Figure 1. Protons (or antiprotons in the final experiment) are stored in a ring with electron cooling. They traverse polarized hydrogen gas, confined in an open T-shaped storage cell fed by an Atomic Beam Source. The target polarization is oriented by means of a weak guide field vertically, parallel to the B-field in the ring dipoles, the stable spin direction. Low-ß quadrupoles

[1] for more details see reference 9, p. 1

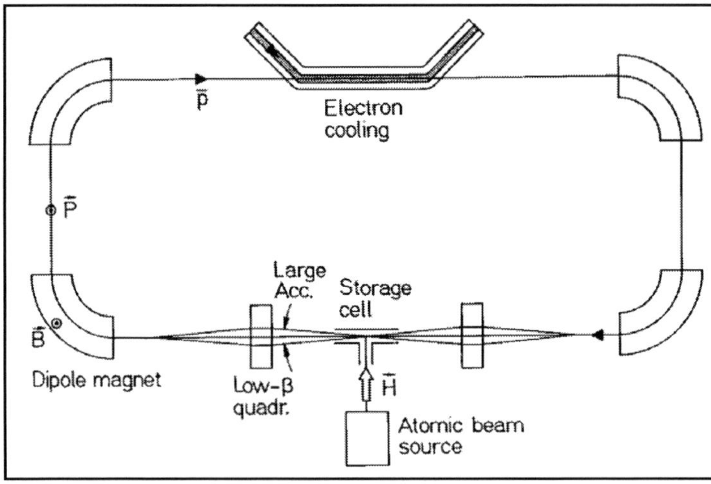

FIGURE 1. Principle of the Filter Experiment. See the text for an explanation.

at the target section provide a large angular acceptance. This way, the spin-independent losses by Coulomb scattering can be minimized. By measuring the left-right asymmetry of elastic protons, the beam polarization can be recorded continuously. The result of the test performed in 1992 is shown in Figure 2 [16].

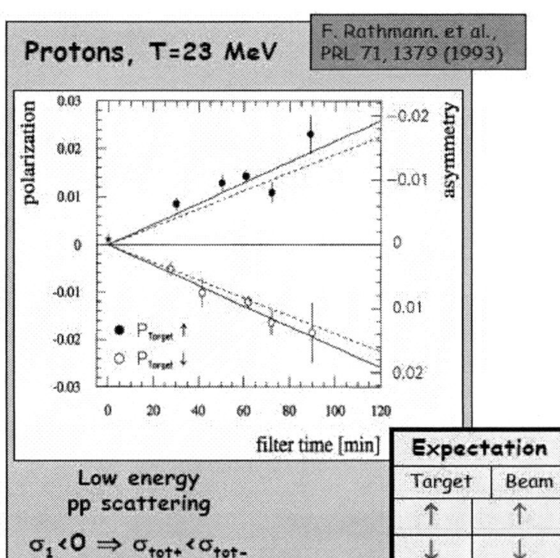

FIGURE 2. Result of the filter test for protons of 23 MeV. The beam polarization measured after a certain filter time is plotted for two orientations of the target polarization. The expected sign of the induced polarization is indicated in the insert and agrees with the data.

A clear build-up of polarization is visible, correlated with target polarization up or down. This result represents the first and only result on polarization build-up of a stored ion beam. The measured rate $dP/dt = (1.24 \pm 0.06)\cdot 10^{-2}$ /h was about half of the naive expectation [16] based on nuclear interaction. In 1994, Meyer [17] presented a quantitative explanation of the FILTEX result. Agreement with the data was obtained by including two more terms: 1. spin transfer from polarized target electrons, and 2. scattering of polarized target protons into the ring acceptance. As experimental opportunities were lacking, this explanation has not been questioned for a whole decade. With the advent of the FAIR project in 2003, antiproton polarization became a key issue. In 2005, the concept for the antiproton polarizer has been optimized [18] resulting in higher predicted $\overline{p}$ polarization. The electromagnetic spin-flip induced by the polarized target electrons calculated in [19], became the principal polarizing mechanism. According to the simulations, at an acceptance angle of 30 mrad the optimum beam energy was found around 40 MeV, resulting in a beam polarization in excess of 0.3 after two beam life times, both for protons or antiprotons.

In 2005, new theoretical activities by a Novosibirsk group (Milstein & Strakhovenko) and a Jülich group (Nikolaev and Pavlov) shed some doubt on the original Meyer-Horowitz approach. Both groups concluded that the build-up is entirely due to spin-dependent removal of beam particles by scattering to angles $\theta \geq \theta_{acc}$. In 2007, a new method was proposed by T. Walcher, H. Arenhövel et al. (Mainz) [20] with the promising feature of being virtually loss-free. The idea is to utilize spin transfer by **free** electrons to co-moving ions at very low relative velocity. The proposal is based on numerical calculations of this process by Arenhövel (cited in [20]) indicating that for attractive systems pe^- or $\overline{p}e^+$ the corresponding cross section may be as large as 10^{13} b = 10^{-11} cm^2, resulting in very short build-up times. Clearly, this contradiction needed to be resolved.

WORKSHOP AT THE COCKCROFT INSTITUTE

The workshop, entitled "Polarized Antiproton Beams – How?" was held at the Cockcroft Institute of Accelerator Science and Technology, Warrington, UK, from 29 to 31 August 2007 [9]. The workshop was attended by 28 participants from six countries. It was organized in topical sessions, followed by discussion periods and a workshop summary, presented by H.-O. Meyer (Indiana).

The workshop was opened with two papers by Buttimore and O'Brien (both Dublin) on the various processes which may be utilized in the polarizer, like channeling by bent crystals, polarization transfer and spin filtering with its evolution equation thus clarifying the different mechanisms. Nikolaev and Pavlov (Jülich) presented their study [21] on $\overline{p}$ spin filtering and the processes introduced by Meyer [17]. They concluded that due to the cancellation of terms the target electrons do not contribute to the polarization build-up. The resulting polarizing cross section is in fair agreement with the re-analyzed result of the FILTEX test. A similar study was presented by Strakhovenko (Novosibirsk) based on an earlier paper by the Novosibirsk group [22]. They also obtain reasonable agreement with the re-analyzed FILTEX data of 1992. By using the Paris potential, predictions for $\overline{p}$ spin filtering were made as well.

A number of contributions were related to proposals of the original Bodega Bay list. Barber (DESY and Cockcroft Institute) discussed the status of the various Spin Splitter proposals based on Stern-Gerlach (SG) forces. He stressed the importance to include all known processes that influence particle motion and stability in a storage ring. Otherwise the spatial separation estimated in, possibly over-simplified models is washed out by these effects. In his view, none of the papers claiming that polarization either by transverse or longitudinal SG forces can be produced, do meet these requirements.

FIGURE 3. Participants of the Workshop at the Cockrcoft Institute during a discussion session, lead by E. Leader (London), right.

The idea of Dynamical Nuclear Polarization (DNP) "in flight" was presented by Krisch (Michigan), based on his results obtained with Jeffries and Kleppner at Bodega Bay. A beam of polarized electrons $e^{\pm}$ overlaps with the stored $\overline{p}'s$ at equal velocity in a strong magnetic field. By means of microwave radiation, lepton polarization is transferred to antiprotons, similar to the original DNP in solids, proposed by Jeffries in the 1950's. According to their original estimates, the required $e^{\pm}$ densities are still far beyond present day capabilities. Unless a special trick is found to reduce the effective e-p radius, as already stated 22 years earlier, this proposal is far from realization.

As shown by Lehrach (Jülich), storage rings as the stage for all these proposed methods are in good shape to provide the necessary features. The role of resonances and their crossing is deeply understood. Stable longitudinal polarization can be maintained by means of a Siberian Snake. The beam life time, in particular in the presence of strong cooling, is still an issue. For filtering studies as planned at COSY very long life times in the order of 10^5 s are required. In such extreme situations, simulation studies both for the orbital and spin motion are of utmost importance.

Barber illustrated in his talk on spin tracking the methods which allow one in principle to include stochastic effects like *damping* of betatron oscillations by cooling, or *scattering*, i.e. intra-beam scattering, or scattering on the residual gas or target gas. These effects need to be taken into account if spin filtering over long storage times is to be simulated reliably.

Polarization by channeling was an issue as well at Bodega Bay, stimulated by the Kaminsky experiments on low energy deuteron beams using magnetized thin Ni crystals [23]. Later, the observed polarization was traced back to the formation of fast D atoms by pick-up of polarized electrons from the solid, resulting in nuclear polarization due to re-coupling of electron and deuteron spin. The recent channeling proposal was inspired by the amazing results obtained on deflection of very high energy ions by means of bent single crystals as pioneered by the Protvino laboratory. Fiorini (Ferrara) gave an overview of similar experiments at the CERN SPS [24], illustrating the capability of deflecting 400 GeV protons which follow the curvature of the bent Si single crystal. Ukhanov (Protvino) presented his proposal to utilize bent crystals as polarizer for energetic particle beams. It is well known that in elastic scattering to a particular side, e.g. to the left, the scattered particle has acquired some vector polarization perpendicular to the scattering plane. If many of these deflections *to the same side* occur as in a bent crystal, they should add up to some finite value. The total bending angle might be as large as 100 mrad or more resulting in a high number of scattering processes of the order 10^3 - 10^4. No measurements exist to date so that it is merely a matter of speculation which polarization to take per elementary scattering process. Experimental results of high energy elastic hadron scattering in the interference region were discussed by Buttimore, indicating rather unfavorable conditions for accumulating a significant polarization. Here experiments are urgently needed.

Two presentations dealt with electrons. Rathmann (Jülich) described an experiment foreseen for COSY to test the large cross sections for e-p polarization transfer at low relative velocity predicted by the Mainz group [20]. The idea is to store a polarized proton beam in COSY and detune the electron cooler periodically such that v_{rel} in the e-p system is in the range for which the huge cross sections should occur. This would then result in fast depolarization of the protons. The sensitivity of this method was estimated to be better than 10^{-5} of the predicted cross section. The experiment has been performed in the meantime and is summarized at the end of this report. The 2nd paper was also related to the Mainz proposal which requires a strong source for polarized positrons. Bailey (Cockcroft Inst., and Liverpool) described the concepts for sources of polarized positrons suitable for future e^+e^- colliders like ILC or CLIC. The technique of these sources would also be applicable within the new scheme to polarize $\overline{p}'s$ by lepton spin transfer.

The workshop was summarized by Meyer (Indiana) with exceptional clarity. He divided the proposed methods into two groups. The *three methods in limbo* were Channeling, Stern-Gerlach Separation and Dynamic Nuclear Polarization in Flight. Significant theoretical efforts have been spent on these methods. A final judgment depends on experimental tests which are lacking in all these cases. The other group of proposals deals with a *beam orbiting through a polarized target*. The latter may consist of atoms or electrons. Meyer pointed out that (at the time of the workshop) two contradicting explanations of the FILTEX result existed, one by the Jülich-Novosibirsk groups that spin transfer by electrons plays no role, and the other including this effect. The depolarization test proposed for COSY has the potential to discriminate between these models und is therefore highly desirable.

WE-HERAEUS SEMINAR AT BAD HONNEF

The 409th Wilhelm und Else Heraeus Seminar, entitled "Polarized Antiprotons", was held at the Physikzentrum of the DPG at Bad Honnef near Bonn from 22 to 25 June 2008, organized by P. Lenisa (Ferrara) and F. Rathmann (Jülich), see Figure 4. It was generously sponsored by the Wilhelm und Else Heraeus-Stiftung and attended by 57 participants. Out of 27 presentations, eight were about general spin physics theory and experiment and five on spin physics with polarized $\bar{p}'s$. Four presentations dealt with theoretical aspects of polarizing $\bar{p}'s$ and three with experimental ones. Six machine-related talks covered spin filtering and future machines for $\bar{p}\uparrow$ experiments. Available space does not permit to mention all speakers. The talks can be found at http://www.fe.infn.it/heraeus/talks/.

FIGURE 4. The organizers of the Heraeus Seminar, P. Lenisa and F. Rathmann, at the Physikzentrum during the workshop.

In the opening talk, Rith (Erlangen) gave an overview of experimental results in high energy spin physics. The lesson from the past is that new experiments are usually based on technological achievements. Polarized antiprotons have the potential of being a powerful new tool. In a number of talks, the status of experiments on nucleon spin structure like COMPASS or HERMES was discussed. Other topics were the potential of the Drell-Yan process for various measurements and its comparison to DIS, special aspects of $\bar{p}$ induced reactions and predictions from theory about spin observables, e.g. for $\bar{p}p$ interaction. The possibility to perform polarization experiments at PANDA detector at HESR were discussed by Maas (GSI and Mainz). An overview was given by Meyer (Indiana) on polarization experiments in storage rings, underlining the power of this technique.

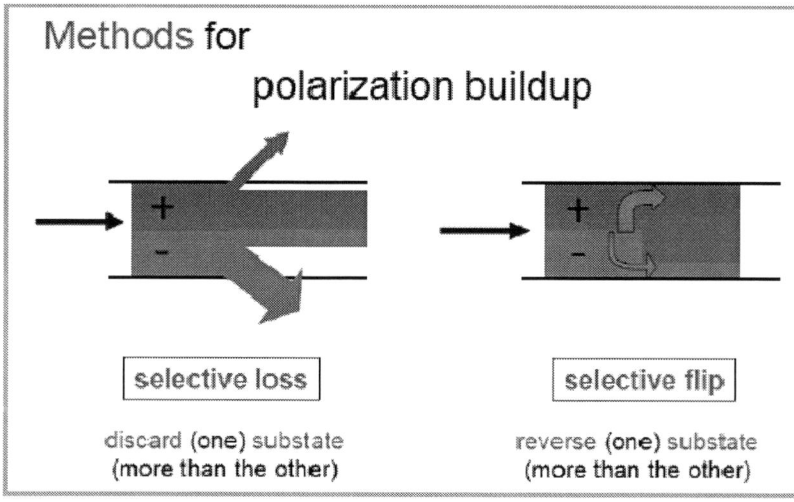

FIGURE 5. Classification of the two classes of mechanisms for a polarization build-up of a beam traversing a *polarizer*: selective loss (left) and selective spin flip (right). Taken from H.-O. Meyer's summary talk in [9].

The main topic at the Honnef meeting was what can be learned from the successful FILTEX test in 1992, and how the result can be understood quantitatively. The key question was how the polarization build-up works (see Fig. 5). The revised Mainz proposal was presented by Walcher (Mainz), based on the existence of electron spin transfer to the proton (or e^+ spin transfer to the antiproton). The alternative view was presented by Strakhovenko (Novosibirsk), concluding that this mechanism is by many orders of magnitude too low to contribute compared with selective removal. As mentioned at the Cockcroft workshop, a depolarization experiment was being prepared at COSY. The first result presented at the Heraeus Seminar by Oellers (Jülich) was eagerly awaited. It clearly indicated no significant depolarization of the stored protons by the electrons of the cooler which were periodically detuned to the right velocity difference predicted by the Mainz calculations. The sensitivity was such that a possible spin transfer cross section should be very small compared to the predicted 10^{13} barn. A more advanced status of the analysis has been presented at Spin2008 by Rathmann [11]. The conclusion has not been changed confirming that the result of the Novosibirsk-Jülich groups is correct.

This result has important implications. The possibility to utilize an electromagnetic and thus calculable process as polarizer does not exist. Spin filtering based on the spin dependence of the – partly unknown – $\overline{p}p$ interaction seems to be the only way. This makes pp filtering tests at COSY, and in particular $\overline{p}p$ filtering tests at the CERN AD indispensible mile stones on the way to polarized antiprotons!

In a number of presentations, the necessary ingredients for spin filter experiments were discussed:

1. Long polarization life time, which seems OK

2. Long beam life time, in particular at low energies, e.g. at COSY injection energy, where the observed times are shorter than the naive expectation based on chamber limits
3. Polarized H and D targets with high density and polarization; more density could be used if available
4. A target section with large acceptance angle to minimize the (spin-independent) Coulomb losses
5. A target cell with sufficient opening during beam injection; this is particularly demanding at the AD
6. Detector systems with large acceptance capable of standing the high rates involved
7. Siberian snakes to allow for longitudinal polarization at the target which gives a higher build-up rate

In his summary, Ströher (Jülich) presented the mile stones on the way towards polarized antiprotons at FAIR: (2010) measurement of proton spin filtering at COSY; (2011/12) measurement of antiproton spin filtering at AD; (2013) final design of antiproton polarizer, based on a preliminary design [5]. He reminded the community that NOW there is a unique chance. A clear roadmap and a strong commitment of the PAX collaboration exist. A NEW experimental result – sixteen years after the first one – has been achieved. Strong support of the community is needed in order to create a polarized antiproton facility at FAIR. Given the lively discussions at the Heraeus Seminar, and the many young and dedicated participants, the route to polarized antiprotons at FAIR seems accessible.

CONCLUSIONS

The lack of polarized antiprotons has been regarded as shortcoming of the LEAR program restricting the reach of what could be learnt about the spin dependence of the $\bar{p}p$ interaction. All attempts to produce polarized antiproton beams of significant quality were unsuccessful [25], or the results came too late to make an impact. The upcoming FAIR facility with its strong antiproton program has renewed the interest in antiproton polarization. This missing tool would enable exciting new physics as exemplified in the talk by Anselmino at this meeting [7] on the direct measurement of transversity of the nucleon in double-polarized Drell-Yan, or in a number of talks at the Heraeus Seminar [10], at this conference or elsewhere. Let us work together in order to add this missing, extremely powerful instrument to our tool box!

ACKNOWLEDGMENTS

I thank the german BMBF for financial support of the research grant 06ER144. Discussions with my colleagues, in particular with A. Kacharava, P. Lenisa, A. Nass, H.-O. Meyer, F. Rathmann, K. Rith and H. Ströher, are gratefully acknowledged.

REFERENCES

1. E. Klempt, F. Bradamante, A. Martin, and J.M. Richard, Phys. Rep. **368** (2002) 119.
2. E. Steffens et al., in: Proc. 3rd LEAR Workshop, Tignes, 1985, p.245; and CERN Proposal CERN/PSCC/85-80 (Nov. 5, 1985).
3. http://www.gsi.de/fair/index_e.html
4. http://www.fz-juelich.de/ikp/pax/
5. A. Garishvili et al: Design Study for an Antiproton Polarizer Ring (APR). Proceed. EPAC 2006, Edinburgh (2006), MOPCH083.
6. http://www.fz-juelich.de/ikp/pax/proposals/files/techproposal20060125.pdf
7. M. Anselmino: Transverse Spin Structure of the Nucleon. These proceedings.
8. Proc. of the Workshop on Polarized Antiprotons, Bodega Bay, CA, 1985. A.D. Krisch, A.M.T. Lin, O. Chamberlain (Edts.), AIP Conf. Proc. **145** (1986) 207.
9. Proc. of the Int. Workshop "Polarized Antiproton Beams – How?", Cockcroft Institute, Daresbury, UK 29 – 31 August 2007. S. Chattopadhyay, D.P. Barber, N. Buttimore, G. Court, E. Steffens (Edts.), AIP Conf. Proc. **1008** (2008).
10. http://www.fe.infn.it/heraeus/
11. F. Rathmann: Do unpolarized electrons affect the polarization of a stored proton beam? These proceedings.
12. F.L. Shapiro: Nuclear Structure Studies with Neutrons. Proc. Int. Conf. Study of Nucl. Structure with Neutrons, Antwerp 1965, North-Holland, Amsterdam (1966), p. 588.
13. T.R. Gentile et al, Physica **B 356** (2005) 96.
14. T.O. Niinikoski and R. Rossmanith, NIM **A 255** (1987) 460.
15. D. Habs et al, Nucl. Instr. Meth. **B 43** (1989) 390.
16. F. Rathmann et al., Phys. Rev. Lett. **71** (1993) 1379.
17. H.O. Meyer, Phys. Rev. **E 50** (1994) 1485.
18. F. Rathmann et al, Phys. Rev. Lett. **94** (2005) 014801.
19. C.J. Horowitz and H.O. Meyer, Phys. Rev. Lett. **72** (1984) 3982.
20. T. Walcher et al, Eur. Phys. J. **A 34** (2007) 447.
21. N. Nikolaev and F. Pavlov, arXiv:0701175v1 [hep-ph] (2005).
22. A.I. Milstein, V.M. Strakhovenko, Phys. Rev. **E 72** (2005) 066503.
23. M. Kaminsky, Phys. Rev. Lett. **23** (1969) 819.
24. W. Scandale et al., Phys. Rev. Lett. **98** (2007) 154801.
25. A. Martin et al., Nucl. Phys. **A 487** (1988) 563.

Nuclear Structure Studies with Polarized Radioactive Beams

K. Asahi,[a] H. Ueno,[b] K. Shimada,[c] T. Nagatomo,[b] A. Yoshimi,[b] D. Nagae,[d] D. Kameda,[b] M. Uchida,[a] T. Inoue,[a] M. Hatakeyama,[a] S. Kagami,[a] Y. Hasama,[a] K. Suzuki,[a] J. Murata,[e] H. Kawamura,[e,b] K. Narita[e] and M. Ishihara[b]

[a]*Department of Physics, Tokyo Institute of Technology, 2-12-1 Oh-okayama, Meguro-ku, Tokyo 152-8551, Japan*
[b]*RIKEN Nishina Center, RIKEN, 2-1 Hirosawa, Wako, Saitama 351-0198, Japan*
[c]*Cyclotron and Radioisotope Center, Tohoku University, 6-3 Aoba, Aramaki, Aoba-ku, Sendai, Miyagi, 980-8578, Japan*
[d]*Japan Atomic Energy Research Institute, Tokai-mura, Ibaraki 319-1195, Japan.*
[e]*Department of Physics, Rikkyo University, 3-34-1Nishi-Ikebukuro, Toshima-ku, Tokyo 171-8501, Japan*

Abstract. Nuclear moment studies on nuclei far from the β-stability line, performed by taking advantage of the projectile fragmentation reaction, are reviewed. A method to produce spin polarization/alignment in the projectile fragments was developed and utilized for the β-NMR/NQR measurements. Several new findings in nuclear structure physics were obtained from the nuclear moment experiments done with these techniques.

Keywords: Projectile fragmentation, spin polarization, nuclear moments, nuclear structure
PACS: 24.70+s, 29.38.Db, 21.10.Ky, 21.10.Hw

INTRODUCTION

After the spin-orbit force was introduced [1], regarding a nucleus as a collection of independent particles in an average field became a successful way of understanding major nuclear properties which are governed by a particular series of numbers - the magic numbers. Such properties result from the "bunching" character of the single-particle orbits, or the existence of nuclear shells. About half a century after, however, the present-day studies of nuclei in the region far from the β stability line reveal that the structure of nuclear shells may evolve [2]: As the N/Z ratio increases, the single particle orbits are considered to migrate upward or downward in energy depending on their $j = l \pm 1/2$ character, as is schematically depicted in Fig. 1. As a result the decisive role played by one orbit (the valence orbit) is taken over by another, and in some cases the nuclear deformation takes place even in a nucleus whose N or Z number is a (heretofore) magic number. In such a situation, the static nuclear moments, namely the magnetic dipole moments μ and electric quadrupole moments Q, serve as very useful indicators for alterations in nuclear structure.

CP1149, *Spin Physics, 18th International Spin Physics Symposium*
edited by D. G. Crabb, D. B. Day, S. Liuti, X. Zheng, M. Poelker, and Y. Prok

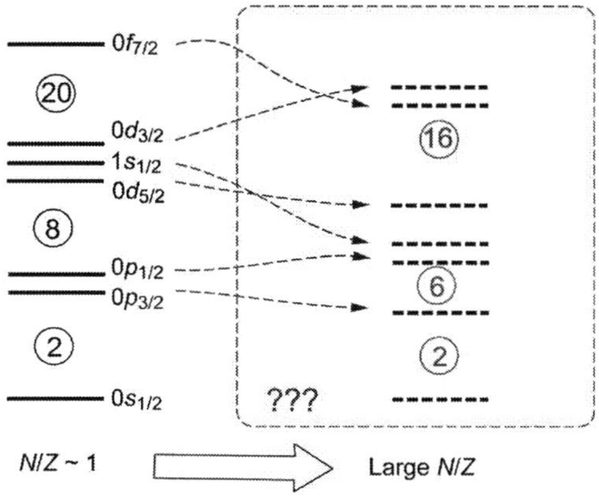

FIGURE 1. Schenatic illustration of the concept of altered shell structure, due to migration of the energies of the individual single-particle orbits.

Once the site of study moves from the region around the stable nuclei to those of exotic nuclei, however, the measurement of μ or Q becomes a nontrivial task, since such a measurement requires certainly high production yields and, moreover, the polarization or alignment of their spins. A breakthrough leading to the production of spin-polarized/aligned exotic nuclei was found in an experiment conducted by RIKEN-GANIL-Orsay collaboration [3] at the LISE spectrometer at GANIL: In this experiment, the nucleus produced from the projectile fragmentation reaction at intermediate energy proved to be spin-aligned when an appropriate selection in the outgoing momentum was made. Furthermore, a subsequent experiment carried out at RIKEN revealed that the projectile fragmentaion reaction also provides significant spin polarization in the fragment nuclei [4]. The observed behavior of polarization turned out to follow the prediction inferred from the result of the spin-alignment experiment above. After several experiments were devoted to the investigation of the mechanism and properties of the polarization/alignment at RIKEN [5,6] and GSI [7], and then a series of experiments to study static nuclear moments for exotic nuclei was initiated. We review results of those experiments performed until now using polarized/aligned projectile fragments.

SPIN POLARIZATION IN PROJECTILE FRAGMENTATION

The idea underlying the experiment [3] to look for spin orientation of nuclei produced in the fragmentation reaction was the following: The fragmentation reaction is known to show kinematical properties that are quite well explained by a simple model that some of the nucleons originally contained in the projectile nucleus (the participants) are removed "softly" so that the remaining nucleons (spectators) do not even "know" that their party has been abraded off, thus keeping their momenta unaltered. Accordingly the momentum $\boldsymbol{p}_{\mathrm{frag}}$ of the resulting fragment (a cluster of the spectator nucleons) is just written as $\boldsymbol{p}_{\mathrm{frag}} = \mathrm{M}\boldsymbol{v}_0 - \boldsymbol{p}_{\mathrm{part}}$ where $\boldsymbol{v}_0$ denotes the projectile velocity in the laboratory frame and $\boldsymbol{p}_{\rm part}$ the total momentum of the removed nucleons (participants) in the projectile-rest frame. Note that the removal takes place in a peripheral region of the projectile, and consequently the orbital angular momentum $\boldsymbol{L}_{\rm part}$ taken away by the participants is non-zero: $\boldsymbol{L}_{\rm part} = \boldsymbol{R}\times\boldsymbol{p}_{\rm part}$ where $\boldsymbol{R}$ denotes the average position of the participant nucleons at the moment of removal, measured from the center of the projectile. Thus, one expects that the angular momentum, or the nuclear spin, $\boldsymbol{L}_{\rm frag}$ of the fragment (assuming that the projectile had no spin, for sinplicity) is related to its linear momentum $\boldsymbol{p}_{\rm frag}$, as $\boldsymbol{L}_{\rm frag} = -\boldsymbol{L}_{\rm part} = \boldsymbol{R}\times(\boldsymbol{p}_{\rm frag} - \mathrm{M}\boldsymbol{v}_0)$. This situation is illustrated in Fig. 2(a).

Figure 2(b) presents the result obtained in the experiment in Ref. [3], where the spin alignment was measured for a nucleus ^{14}B produced in the fragmentation of a projectile ^{18}O on a Be target at E/A = 60 MeV/u. The obtained result clearly indicated that the produced fragment is in fact spin-aligned, and the behavior of the alignment is well compatible with the expected one.

The confirmed presence of the spin alignment in projectile fragmentation reaction stimulated us to look for spin *polarization* which should have more straightforward implication for the β-NMR method to measure the nuclear moments for exotic nuclei

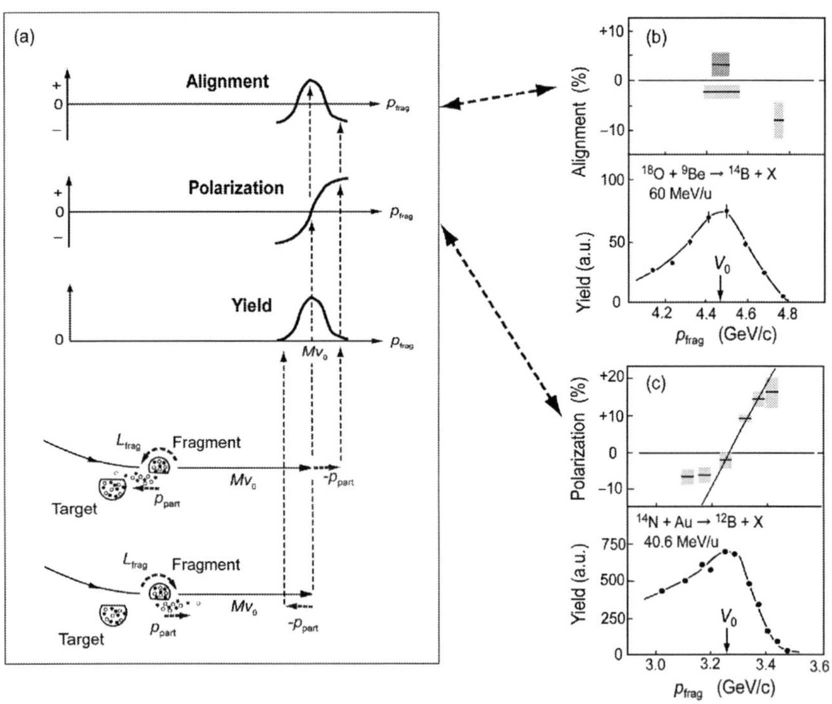

FIGURE 2. Spin polarization and alignment in projectile fragments. (a) The prediction from a simple kinematical model of projectile fragmentation reaction. (b) The result of the spin alignment measurement [3]. (c) The result of the spin polarization measurement [4]. For explanation, see text.

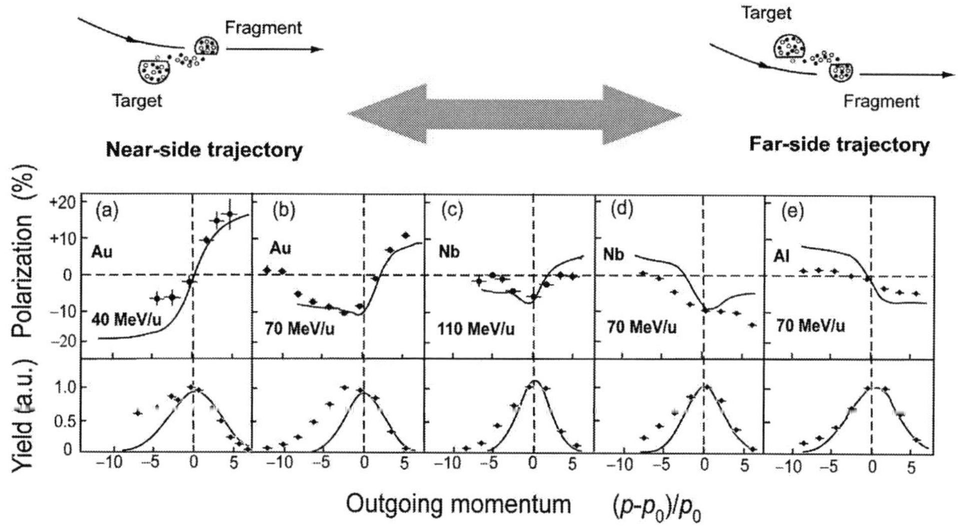

FIGURE 3. Behavior of the spin polarization of a fragment nucleus produced in the projectile fragmentation reaction with various targets and projectile energies. Here p is the outgoing fragment momentum and p_0 the momentum corresponding to the velocity of the projectile. Detailed conditions for the individual spectra are: (a) ^{14}N (39.4 MeV/u) + $^{197}Au \rightarrow {}^{12}B$ ($\theta_L = 5.0°$) + X, (b) ^{15}N (68.0 MeV/u) + $^{197}Au \rightarrow {}^{13}B$ ($\theta_L = 4.0°$) + X, (c) ^{15}N (109.6 MeV/u) + $^{197}Au \rightarrow {}^{13}B$ ($\theta_L = 2.0°$) + X, (d) ^{15}N (67.3 MeV/u) + $^{93}Nb \rightarrow {}^{13}B$ ($\theta_L = 2.5°$) + X, and (e) ^{15}N (68.0 MeV/u) + $^{27}Al \rightarrow {}^{12}B$ ($\theta_L = 1.0°$) + X. Also shown for comparison are the calculated yield spectra and $L_z/|\boldsymbol{L}|$ values (solid lines), where $\boldsymbol{L}$ is the angular momentum generated in the fragment nucleus by the fragmentation process. The $L_z/|\boldsymbol{L}|$ values were scaled by a factor 0.25 for all the reactions.

that have newly become accessible by use of the projectile fragmentation reaction. The result of a polarizationmeasurement on a ^{12}B fragment from the fragmentation of a ^{14}N projectile (40.6 MeV/u) on an ^{197}Au target [4], presented in Fig. 2(c), confirmed that the spin polarization is also available and is in fact compatible with the prediction by the model of Fig. 2(a). Figure 3 summarizes the results of developmental studies that followed [5,6], for the production of spin-polarized radioactive beams.The observed polarization with a heavy target ^{179}Au [Fig. 3(a)] takes negative values in the low momentum wing of the peak while it becomes positive in the high-momentum wing, The behavior of P reverses completely when a light target, ^{27}Al, is employed [Fig. 3(e)]. With a "medium weight" target, Nb, at the projectile energy of 110 MeV/u [Fig. 3(c)], an intermediate case occurs where the polarization takes a maximum at the peak momentum.

The reversal of the polarization pattern observed above is explained by the change in projectile trajectory from the near-side to far-side ones. In the cases of a Nb target [Fig. 3(c) and (d)] the polarization is no more a simple increasing or decreasing function of p, but shows a (negative) maximum at the yield peak. This behavior was explained as an effect of rescattering of the abraded nucleons [6]. It is also noted that raising the projectile energy [from (a) to (b), and from (d) to (c)] makes the pattern of

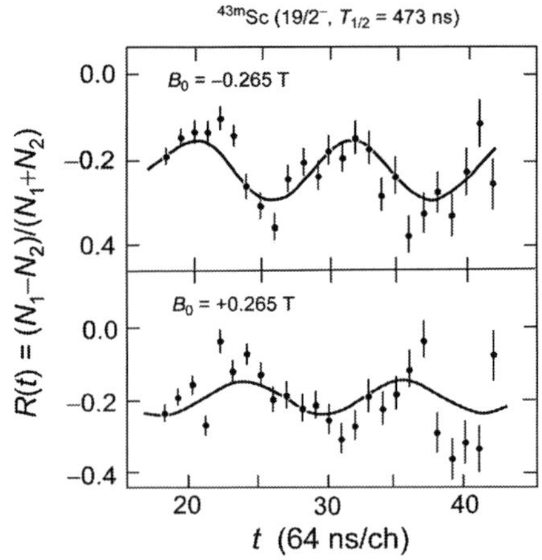

FIGURE 4. Spin rotation pattern function $R(t)$ for a ^{43m}Sc ($I^{\pi} = 19/2^-$, $T_{1/2} = 473$ ns) isomer from the fragmentation of 500 MeV/u ^{46}Ti projectiles at GSI, measured with a Pb stopper in a magnetic field of −0.265 T (upper) and +0.265 T (lower). In the definition of $R(t)$, N_1 and N_2 denote the gamma-ray count rates at Ge detectors located at +45° and −45°, respectively, from the fragment beam direction. Also shown are the line representing a least chisquared fit with a fixed period of 750 ns. The observed amplitude of $R(t)$ indicates that the ^{43m}Sc isomer produced in the center of the momentum distribution is spin aligned to about +35 %.

polarization to tend more to that of the far-side case. The spin alignment at a higher projectile energy 500 MeV/u was observed [7] by using a fragment separator FRS at GSI. In Fig. 4 the time-differential perturbed angular distribution (TDPAD) pattern $R(t)$ representing the spin rotation measured for a ^{43m}Sc (I^{π}= 19/2$^-$, $T_{1/2}$ = 473 ns) isomer from the fragmentation of 500 MeV/u ^{46}Ti projectiles at GSI is presented. The observed amplitude of $R(t)$ indicates that the ^{43m}Sc isomer produced in the center of the momentum distribution is spin aligned to about +35 %, suggesting that the projectile fragmentation reaction is useful also for isomer studies with TDPAD technique.

NUCLEAR MOMENT RESULTS AND THEIR IMPLICATIONS

Taking advantage of the results presented above, a variety of spin-polarized radioactive nuclear beams were produced and used for the nuclear moment measurements. In Fig. 5, species whose magnetic moment or quadrupole moment was determined with this technique at RIPS are indicated on the nuclear chart. Physics implications obtained from these studies may be categorized in three groups.

Looking for Modified Nuclear Properties for Increased *N*/*Z*

The first group of studies concerns with possible modification of basic nuclear properties in relation to increase of the *N*/*Z* ratio. The magnetic moment for the ground state of ^{14}B was determined [8] to be $|\mu(^{14}B)| = 1.185 \pm 0.005\mu_N$. With three protons in the p shell and one neutron in the sd shell, the $J^\pi = 2^-$ ground state of ^{14}B should be dominated by two configurations, $[\pi(p_{3/2})_{-1} \otimes \nu s_{1/2}]^{J=2}$ and $[\pi(p_{3/2})_{-1} \otimes \nu d_{5/2}]^{J=2}$. To leading order, therefore, the magnetic moment of ^{14}B is given by a weighted average of the moments μ_1 and μ_2 for these two configuration (note that the off-diagonal matrix element of M1 operator between these two configuration is zero) with weights being the probabilities $|c_1|^2$ and $|c_2|^2$ for the respective configurations, as

$$\mu(^{14}\mathrm{B}) = \frac{|c_1|^2 \mu_1 + |c_2|^2 \mu_2}{|c_1|^2 + |c_2|^2} \quad . \tag{1}$$

Since μ_1 and μ_2 are quite different from each other ($\mu_1 = +1.88\mu_N$ for the former and $\mu_2 = -0.98\mu_N$ for the latter, evaluated with the effective M1 operator used in literature [9]), the measurement of μ provides a sensitive measure for the relative importance between these two configurations. Although expression (1) holds only approximately

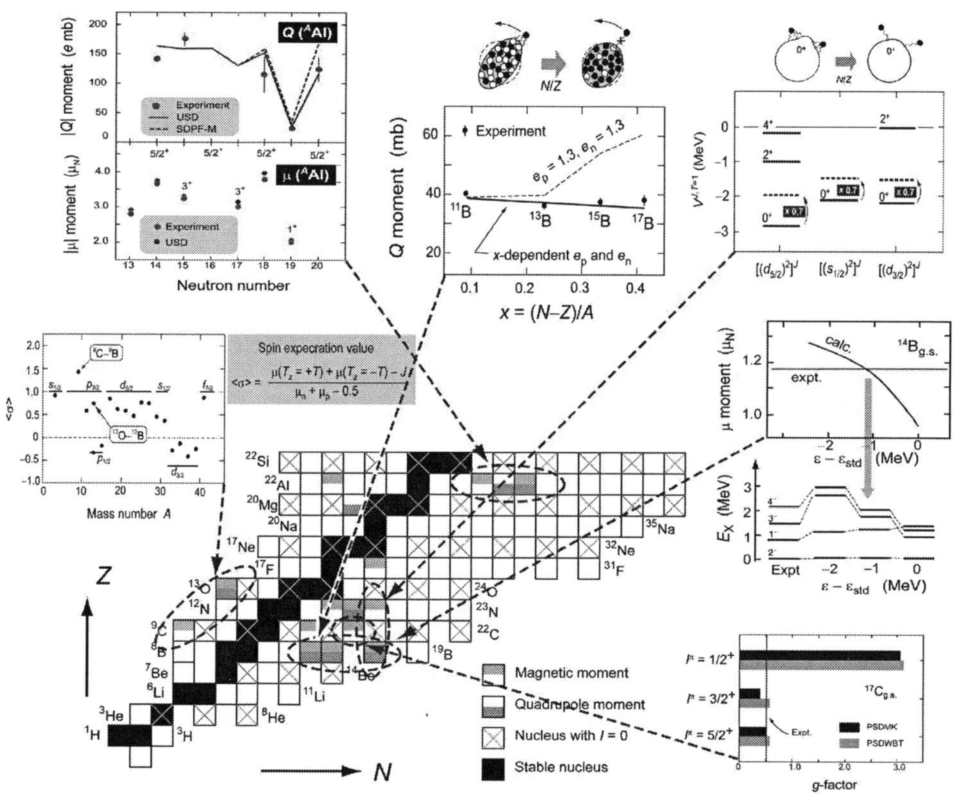

FIGURE 5. Nuclear species for which the magnetic moment or electric quadrupole moment has been determined at RIPS/RIKEN using spin polarized fragment beams from the projectile fragmentation reaction, and physics information derived from them. For explanation, see text.

(the two configurations above together share 87 % of the ^{14}B ground state in shell-model calculations with PSDMK interactions for example, with remaining 13 % probability being shared by many other minor configurations) the above statement would be unaltered. The measured μ turned out to be substantially larger than the shell-model calculation assuming the standard values of the $s_{1/2}$ and $p_{1/2}$ orbit single-particle energies [10], indicating that the ground state of ^{14}B contains more $[\pi(p_{3/2})_{-1}\otimes\nu s_{1/2}]^{J=2}$ component than usually considered. The enhanced contribution of configuration 1 relative to configuration 2 may occur either if the single-particle energy $\varepsilon_{s1/2}$ for the $s_{1/2}$ orbit is lower than usual, or if $\varepsilon_{d5/2}$ for the $d_{5/2}$ orbit is higher. In fact, good agreement between the calculation and experiment was obtained when the single-particle energy difference $\Delta\varepsilon = \varepsilon_{s1/2} - \varepsilon_{d5/2}$ between the $s_{1/2}$ and $p_{1/2}$ orbits are reduced by about 1 MeV from the value normally employed in the region of the β stability. The reduced Δε in the neutron-rich nucleus ^{14}B was also inferred from the position of the succeedingly found 1^+ state in a β-delayed neutron-γ spectroscopic experiment [11]. Systematic lowering of the $s_{1/2}$ orbit single-particle energy has been discussed in literature since long before [12].

The magnetic moment of ^{17}N is another example of measurements from which a modified feature of interactions in a neutron-rich nucleus is extracted [13]. The spin-parity $1/2^-$ of the ground state of ^{17}N strongly restricts the configurations that can enter: With one proton hole in the p shell and two neutrons in the $d_{5/2}$ -$s_{1/2}$ shell, this nucleus is allowed to contain only the configurations $\psi_0 = [\pi(p_{1/2})_{-1}\otimes\nu(d_{5/2}s_{1/2})^{Jn=0}]^{J=1/2}$ and $\psi_2 = [\pi(p_{3/2})_{-1}\otimes\nu(d_{5/2}s_{1/2})_2^{Jn=2}]^{J=1/2}$ are allowed (note that two neutrons within the $d_{5/2}$ and $s_{1/2}$ orbits cannot form $J_n = 1$ because of the symmetry). Since the $p_{3/2}$ proton has an order of magnitude larger μ value compared to the $p_{1/2}$ proton [$\mu_{s.p.}(p_{3/2}) = +3.80\mu_N$, $\mu_{s.p.}(p_{1/2}) = -0.276\mu_N$], the magnetic moment sensitively reflects the amount of the ψ_2 configuration relative to ψ_0. The measured $|\mu(^{17}N)|$ substantially exceeds $|\mu|$ for $p_{1/2}$ proton, and indicates excess contribution of $J_n = 2$ coupling compared to $J_n = 0$ coupling of neutrons in the ^{17}N nucleus. This may happen if the $J_n = 0$ pairing energy for neutrons, which is considered to have an extensive contribution from core-polarization effect [14], is weakened by 30 % in this neutron-rich nucleus. The magnetic moments [8,13] and binding energies of ^{17}B and ^{15}B also are explained with the hypothesis that the $J_n = 0$ pairing is weakened for neutrons by 30 % in neutron-rich nuclei.

The electric quadrupole moment is written as

$$Q = \langle I, M = +I | \sum e_k \hat{Q}_k | I, M = +I \rangle, \quad (2)$$

with $\hat{Q}_k \equiv \sqrt{16\pi/5} \cdot r_k^2 Y_{20}(\hat{r}_k)$ and the electric charge e_k for the k-th nucleon. Looking at this expression naively, it might appear to receive no contribution from neutrons. As a matter of fact, however, we need to use a model wave function $|\psi_{I,M=+I}\rangle$ in place of the true nuclear state $|I, M = +I\rangle$, and as a result we should take into consideration that the valence neutrons may polarize the core protons into configurations outside the model space. In such a case, e_k needs to be replaced by effective charges $e_n^{\text{eff}} = \delta e_n$ for neutron and $e_p^{\text{eff}} = e + \delta e_p$ for proton. Empirically the E2 polarization charges are known to take values $\delta e_n \approx 0.5e$ and $\delta e_p \approx 0.3e$ in light-mass nuclei close to the β stability line [15]. δe would be smaller than these, if the coupling of the valence nucleon with core is weaker than usual. The measured Q for ^{15}B and ^{17}B are found [16,17] to be very close to Q of the neutron-closed isotope ^{13}B, indicating that e_n^{eff} is quenched in ^{15}B and ^{17}B. Small experimental Q obtained for ^{18}N (Q of this nucleus is considered to be predominated by the neutron side because of a $p_{1/2}$ hole character of the proton side) indicates that e_n is significantly reduced also in this nucleus [18]. These results on Q-moments clearly demonstrate that the E2 effective charges are isospin-dependent. This has already been pointed out by the textbook of Bohr and Mottelson [19] considering the decoupling feature of loosely bound particles as well as modification of the high-frequency isoscalar and isovector quadrupole modes [17,20]. The reduced effective charge for neutron in a neutron-rich nucleus is also compatible with recent observation of hindered E2 transitions [21] from the first 2^+ excited states in ^{16,18}C.

Knowing about Inhabitants of Newly Accessible Regions

The ground-state spin-parity of ^{17}C ($T_{1/2}$ = 193 ms) has been studied using several theoretical and experimental approaches. According to shell model calculations, I^π = $1/2^+$, $3/2^+$ or $5/2^+$ are candidates for the ground state, but to choose one among them from theory is very difficult since the calculated energies for these three low-lying states are very close to each other. Experimentally, from the β-decay branching ratio, the $I^\pi = 5/2^+$ spin-parity assignment is excluded. Of the remaining two candidates, I^π = $1/2^+$ and $3/2^+$, the latter has been proposed from the β-delayed neutron spectroscopy and one-neutron removal reactions. Under such a circumstance, the measurement of the g-factor plays a decisive role because it is extremely sensitive to whether the spin-parity is $1/2^+$ or $3/2^+$. The g-factor of the ^{17}C ground state was measured to be $|g(^{17}C_{g.s.})| = 0.5054 \pm 0.0025$. The experimental $|g|$ is well reproduced by both calculations with the PSDMK and PSDWBT interactions for spin-parity $3/2^+$, while the calculated $|g|$'s for $I^\pi = 1/2^+$ are about 6 times larger. Thus, the ground-state spin-parity of ^{17}C was assigned to be $I^\pi = 3/2^+$, and the possibility of $I^\pi = 1/2^+$ was clearly excluded.

A new method of spin-parity assignment was developed that makes use of β-ray asymmetry measured in coincidence with delayed neutron in the β decay of a spin-polarized parent nucleus. Neutron-emitting excited states of ^{15}C were studied [23]. This technique has been utilized later in the study of excited states of ^{11}Be populated by the β decay of a laser polarized ^{11}Li at TRIUMF [24].

The magnetic moments of ^{9}C and ^{13}O were measured to extract the isoscaler magnetic moments for A = 9 and A = 13, T = 3/2 mirror pairs [25], and the spin expectation values $\langle\sigma\rangle$ deduced from them were used for a stringent test of nuclear structure theories.

Exploring a Domain of Anomaly

Neutron-rich isotopes of Ne, Na and Mg with neutron numbers $N \approx 20$ have been known to show anomalously large binding energies and lowered first 2^+ excited states

[26], and later experimental and theoretical works [27,28] revealed that the ground states in this region of isotopes were dominated by unusual nucleon configurations, with significant shape deformations. This region of neutron-rich nuclei around the magic number $N = 20$ is named the island of inversion. The island of inversion may extend beyond the elemental line of Mg, and may show a diffuse or clear-cut border, depending on what mechanism underlies the onset of such an anomaly. Investigations in neutron-rich isotopes of Al are in progress by making use of nuclear moments; the μ and Q moments as probes into nucleon configurations and nuclear shapes, respectively. Results obtained so far [29,30,31] for ^{30}Al, ^{31}Al, ^{32}Al, and ^{33}Al, suggest that the "northern" borderline of the island is likely to run between Mg and Al isotopes for $N = 19$ and 20, with a possible diffusion for the $N = 20$ case into ^{33}Al.

ACKNOWLEDGMENTS

The authors are grateful to staffs at the accelerators of RIKEN. They also like to express thanks to members of collaborations between RIKEN and several institutions in the World, namely GANIL, Orsay, GSI, MSU, Leuven, Sophia and T.I. Tech.

REFERENCES

1. M.G. Mayer and J.H.D. Jensen, *Elementar Theory of Nuclear Structure*, John Wiley and Sons, London, 1955.
2. T. Otsuka, T. Suzuki, R. Fujimoto, H. Grawe, and Y. Akaishi, *Phys. Rev. Lett.* **95**, 232502 (2005).
3. K. Asahi, M. Ishihara, T. Ichihara, M. Fukuda, T. Kubo, Y. Gono, A.C. Mueller, R. Anne, D. Bazin, D. Guillemaud-Mueller, R. Bimbot, W.-D. Schmidt-Ott, and J. Kasagi, *Phys. Rev.* C **43**, 456 (1991).
4. K. Asahi, M. Ishihara, N. Inabe, T. Ichihara, T. Kubo, M. Adachi, H. Takanashi, M. Kouguchi, M. Fukuda, D. Mikolas, D.J. Morrissey, D. Beaumel, T. Shimoda, H. Miyatake, and N. Takahashi, *Phys. Lett.* B **251**, 488 (1990).
5. K. Asahi, H. Okuno, H. Ueno, H. Sato, J. Kura, T. Kubo, T. Nakamura, N. Inabe, A. Yoshida, Y. Ohkubo, M. Adachi, T. Ichihara, M. Ishihara, T. Shimoda, H. Miyatake, N. Takahashi, D. Beaumel, D.J. Morrissey, and W.-D. Schmidt-Ott, *Hyperfine Interact.* **75**, 101 (1992).
6. H. Okuno, K. Asahi, H. Sato, H. Ueno, J. Kura, M. Adachi, T. Nakamura, T. Kubo, N. Inabe, A. Yoshida, T. Ichihara, Y. Kobayashi, Y. Ohkubo, M. Iwamoto, F. Ambe, T. Shimoda, H. Miyatake, N. Takahashi, J. Nakamura, D. Beaumel, D.J. Morrissey, W.-D. Schmidt-Ott, and M. Ishihara, *Phys. Lett.* B **335**, 29 (1994).
7. W.-D. Schmidt-Ott, K. Asahi, Y. Fujita, H. Geissel, K.-D. Gross, T. Hild, H. Irnich, M. Ishihara, K. Krumbholz, V. Kunze, A. Magel, F. Meissner, K. Muto, F. Nickel, H. Okuno, M. Pfutzner, C. Scheidenberger, K. Suzuki, M. Weber, C. Wennemann, *Z. Phys.* A **350**, 215 (1994).
8. H. Okuno, K. Asahi, H. Ueno, H. Izumi, H. Sato, M. Adachi, T. Nakamura, T. Kubo, N. Inabe, A. Yoshida, N. Fukunishi, T. Shimoda, H. Miyatake, N. Takahashi, W.-D. Schmidt-Ott, and M. Ishihara, *Phys. Lett.* B **354**, 41 (1995).
9. A. Arima, K. Shimizu, W. Bentz, and H. Hyuga, *Adv. Nucl. Phys.* **18**, 1 (1987); B.A. Brown, B.H. Wildenthal, *Nucl. Phys.* A **474**, 290 (1987); I.S. Towner, *Phys. Rep.* **155**, 263 (1987).
10. B.A. Brown, A. Etchegoyen, and B.D.M. Rae, Computer Code OXBASH, *MSU Cyclotron Laboratory Report,* No. **524** (1986).
11. N. Aoi, K. Yoneda, H. Miyatake, H. Ogawa, Y. Yamamoto, E. Ideguchi, T. Kishida, T. Nakamura, M. Notani, H. Sakurai, T. Teranishi, H. Wu, S. S. Yamamoto, Y. Watanabe, A. Yoshida, M. Ishihara, *Nucl. Phys.* A **616**, 181c (1997).
12. L. Talmi and L. Una, *Phys. Rev. Lett.* **4**, 469 (1960).
13. H. Ueno, K. Asahi, H. Izumi, K. Nagata, H. Ogawa, A. Yoshimi, H. Sato, M. Adachi, Y. Hori, K. Mochinaga, H. Okuno, N. Aoi, M. Ishihara, A. Yoshida, G. Liu, T. Kubo, N. Fukunishi, T. Shimoda,

H. Miyatake, M. Sasaki, T. Shirakura, N. Takahashi, S. Mitsuoka, and W.-D. Schmidt-Ott, *Phys. Rev.* C **53**, 2142 (1996).

14. G.E. Brown, *Unified Theory of Nuclear Models and Forces*, North-Holland, Amsterdam, 1971.
15. B.A. Brown and B.H. Wildenthal, *Ann. Rev. Nucl. Part. Sci.* **38**, 29 (1988).
16. H. Izumi, K. Asahi, H. Ueno, H. Okuno, H. Sato, K. Nagata, Y. Hori, M. Adachi, N. Aoi, A. Yoshida, G. Liu, N. Fukunishi, and M. Ishihara, *Phys. Lett.* B **366**, 51 (1996).
17. H. Ogawa, K. Asahi, K. Sakai, H. Ueno, T. Suzuki, H. Izumi, H. Miyoshi, N. Imai, Y.X. Watanabe, K. Yoneda, M. Nagakura, K. Yogo, A. Goto, T. Suga, T. Honda, A. Yoshimi, N. Fukuda, N. Aoi, Y. Kobayashi, W.-D. Schmidt-Ott, G. Neyens, S. Teughels, A. Yoshida, T. Kubo, and M. Ishihara, *Phys. Rev.* C **67**, 064308 (2004).
18. H. Ogawa, K. Asahi, K. Sakai, A. Yoshimi, M. Tsuda, Y. Uchiyama, T. Suzuki, K. Suzuki, N.Kurokawa, M. Adachi, H. Izumi, H. Ueno, T. Shimoda, S. Tanimoto, N. Takahashi, W.-D.Schmidt-Ott, M. Schaefer, S. Fukuda, A. Yoshida, M. Notani, T. Kubo, H. Okuno, H. Sato, N. Aoi, K. Yoneda, H. Iwasaki, N. Fukuda, N. Fukunishi, M. Ishihara, and H. Miyatake, *Phys. Lett.* B **451**, 11 (1999).
19. A. Bohr and B. R. Mottelson, *Nuclear Structure*, World Scientific, Singapore, 1998, Vol. II.
20. H. Sagawa and K. Asahi, *Phys. Rev.* C **63**, 064310 (2001).
21. H.J. Ong, N. Imai, D. Suzuki, H. Iwasaki, H. Sakurai, T. K. Onishi, M. K. Suzuki, S. Ota, S. Takeuchi, T. Nakao, Y. Togano, Y. Kondo, N. Aoi, H. Baba, S. Bishop, Y. Ichikawa, M. Ishihara, T. Kubo, K. Kurita, T. Motobayashi, T. Nakamura, T. Okumura, and Y. Yanagisawa, *Phys. Rev.* C **78**, 014308 (2008).
22. H. Ogawa, K. Asahi, H. Ueno, K. Sakai, H. Miyoshi, D. Kameda, T. Suzuki, H. Izumi, N. Imai, Y. Watanabe, K. Yoneda, N. Fukuda, H. Watanabe, A. Yoshimi, N. Aoi, M. Nagakura, T. Suga, K. Yogo, A. Goto, T. Honda, Y. Kobayashi, W.-D. Schimidt-Ott, G. Neyens, S. Teughels, A. Yoshida, T. Kubo, and M. Ishihara, *Eur. Phys. J.* A **13**, 81 (2002).
23. H. Miyatake, H. Ueno, Y. Yamamoto, N. Aoi, K. Asahi, E. Ideguchi, M. Ishihara, H. Izumi, T. Kishida, T. Kubo, S. Mitsuoka, Y. Mizoi, M. Notani, H. Ogawa, A. Ozawa, M. Sasaki, T. Shimoda, T. Shirakura, N. Takahashi, S. Tanimoto, and K. Yoneda, *Phys. Rev.* C **67**, 014306 (2003).
24. Y. Hirayama, T. Shimoda, H. Izumi, A. Hatakeyama, K.P. Jackson, C.D.P. Levy, H. Miyatake, M. Yagi, H. Yano, *Phys. Lett.* B **611**, 239 (2005).
25. K. Matsuta, M. Fukuda, M. Tanigaki, T. Minamisono, Y. Nojiri, M. Mihara, T. Onishi, T. Yamaguchi, A. Harada, M. Sasaki, T. Miyake, S. Fukuda, K. Yoshida, A. Ozawa, T. Kobayashi, I. Tanihata, J. R. Alonso, G. F. Krebs, T. J. M. Symons, Nucl. Phys. A 588, 153c (1995); K. Matsuta, T. Minamisono, M. Tanigaki, M. Fukuda, Y. Nojiri, M. Mihara, T. Onishi, T. Yamaguchi, A. Harada, M. Sasaki, T. Miyake, K. Minamisono, T. Fukao, K. Sato, Y. Matsumoto, T. Ohtsubo, S. Fukuda, S. Momota, K. Yoshida, A. Ozawa, T. Kobayashi, I. Tanihata, J. R. Alonso, G. F. Krebs and T. J. M. Symons, *Hyperfine Interact.* **97/98**, 519 (1996).
26. C. Thibault, R. Klapisch, C. Rigaud, A. M. Poskanzer, R. Prieels, L. Lessard, and W. Reisdorf, *Phys. Rev.* C 12, 644 (1975); C. Detrazt, M. Langevin, D. Guillemaud, M. Epherre, G. Audi, C. Thibault and, F. Touchard, *Nucl. Phys.* A **394**, 378 (1983).
27. M. Keim, U. Georg, A. Klein, R. Neugart, M. Neuroth, S. Wilbert, P. Lievens, L. Vermeeren4, B.A. Brown, and the ISOLDE Collaboration, *Eur. Phys. J.* A **8**, 31 (2000).
28. Y. Utsuno et al., T. Otsuka, T. Glasmacher, T. Mizusaki, and M. Honma, *Phys. Rev.* C **70**, 044307 (2004).
29. H. Ueno, H. Ueno, Y. Yamamoto, N. Aoi, K. Asahi, E. Ideguchi, M. Ishihara, H. Izumi, T. Kishida, T. Kubo, S. Mitsuoka, Y. Mizoi, M. Notani, H. Ogawa, A. Ozawa, M. Sasaki, T. Shimoda, T. Shirakura, N. Takahashi, S. Tanimoto, and K. Yoneda, *Phys. Lett.* B **615**, 186 (2005).
30. D. Kameda, H. Ueno, K. Asahi, M. Takemura, A. Yoshimi, T. Haseyama, M. Uchida, K. Shimada, D. Nagae, G. Kijima, T. Arai, K. Takase, S. Suda, T. Inoue, J. Murata, H. Kawamura, Y. Kobayashi, H. Watanabe, and M. Ishihara, *Phys. Lett.* B **647**, 93 (2007).
31. H. Ueno, D. Kameda, D. Nagae, M. Takemura, K. Asahi, K. Shimada, K. Takase, T. Sugimoto, T. Nagatomo, M. Uchida, T. Arai, T. Inoue, A. Yoshimi, T. Kawamura and K. Narita, *Eur. Phys. J. Special Topics* **150**, 185 (2007).

Polarized Solid Targets – Present and Future

W. Meyer

Institut für Experimentalphysik I, Ruhr-Universität Bochum, 44780 Bochum, Germany

Abstract. The development, in the early 1960s, of the dynamic nuclear polarization scheme in solid diamagnetic materials, doped with paramagnetic radicals, led to the use of solid polarized H and D targets in numerous nuclear and particle physics experiments. Since then steady progress has been made in all contributing sub-systems so that proton polarization values around 90 % and deuteron polarization values between 40 % and 80 % have been achieved in various set-ups. These polarization values have been measured with a relative accuracy of $\pm 5\,\%$ or better using nuclear magnetic resonance techniques. Many experiments with reasonably high luminosities have taken advantage of these developments and many more are being planned, especially with electromagnetic probes. Latest developments open up the possibility to use DNP hyperpolarized substances for medical applications.

INTRODUCTION

A detailed understanding of the structure and interaction of hadrons has been the main motivation for strong interaction physics for many decades. The advent of solid polarized targets in the early 1960s opened up a new era in the study of particle interactions.

Meanwhile quantum chromodynamics (QCD) has led to a theoretical description of the strong interaction in terms of the fundamental constituents, quarks and gluons.

The CERN, SLAC and DESY experiments – each with its own specific experimental power – have confirmed, with greatly improved precision, the original EMC result that quarks alone could not be responsible for the spin of the nucleon. In addition, the SMC and in particular the HERMES collaboration measured precisely the contributions to the nucleon spin of each valence quark flavour. Recently the HERMES experiment at DESY with its polarized internal nuclei targets and the COMPASS experiment at CERN, which uses like the original EMC experiment polarized muons and polarized solid targets, have started to investigate the gluon spin contribution to the nucleon spin. Together with the polarized beam experiments at the RHIC collider at the Brookhaven laboratory a full understanding of how spin $\frac{1}{2}$ quarks and spin 1 gluons together with their orbital angular momenta play one's part to give nucleons a spin of $\frac{1}{2}$ cannot be far off.

In the energy range up to several *GeV*, a rich program is in progress at the electron beam facilities at the University of Bonn, the University of Mainz and Thomas Jefferson National Accelerator Facility (TJNAF). From the list of proposed experiments at these facilities, it is clear that measurements involving spin degrees of freedom are becoming increasingly common, perhaps even prevalent. The common factor of all these experiments is that several ingredients contribute to the measured quantities and it is well known that spin observables often provide better discrimination separating one piece from another.

CP1149, *Spin Physics, 18th International Spin Physics Symposium*
edited by D. G. Crabb, D. B. Day, S. Liuti, X. Zheng, M. Poelker, and Y. Prok

A comprehensive review on polarized H, D and ^{3}He targets for particle physics experiments is given in ref. [1], as well as an overview of the relevant physics concepts for the dynamic nuclear polarization (DNP).

In the following, examples of state of-the-art-polarized solid targets are discussed.

Employing the DNP process new developments have been made with the focus of the use of hyperpolarized substances for medical applications.

GENERAL CONSIDERATIONS

Luminosity is one of the main quality factors in determining how fast an experiment can be done. The critical parameter is the acceptable beam current (i) due to the cryogenic requirements of the polarized solid targets and (ii) due to the radiation resistance of the polarization of the target materials. Solid H targets using ammonia as the hydrogen carrier operate with a density of about $10^{23} cm^{-2}$ at a beam current of $100nA$ or less yielding a luminosity of about $6 \cdot 10^{34} cm^{-2} s^{-1}$. Typical luminosities achieved with low intensity secondary or tertiary beams such as tagged photons or polarized muons are in the order of $10^{30} - 10^{32} cm^{-2} s^{-1}$. The quality of a polarization experiment is, however, not only given by the reaction counting rate, but is very much dependent on other factors like background from other sources and the polarization degree of the target material. In designing an experiment, one has to consider seriously these factors in addition to the luminosity.

A 'figure of merit' for polarized solid targets $FOM_{solid\ target}$, can then be defined

$$FOM_{solid\ target} = \rho \cdot \kappa \cdot P_t^2 \cdot f^2 \quad (1)$$

i.e. the larger $\rho \cdot \kappa \cdot P_t^2 \cdot f^2$ the shorter the running time required to achieve a chosen accuracy in the measured asymmetry observable. Here P_t is the measured target polarization degree and f is the dilution factor (ratio of polarizable nucleons to all nucleons in the target materials). ρ is the density of the target material and κ is the filling factor of the frozen material (normally beads or chips) in the target cell. A comparison of the $FOM_{solid\ target}$ for the favored proton and deuteron materials is given in table 1. For more details see ref. [1][2].

TABLE 1. $FOM_{solid\ target}$ values of the favored polarized solid target materials used in particle physcis experiments

	Butanol C_4H_9OH	Ammonia $^{14}NH_3$	D-Butanol C_4D_9OD	Lithium Deuteride ^6LiD
Target pol. P_t	0.90	0.90	0.8	0.5
Dilution factor f	0.135	0.175	0.238	0.5
Density ρ [kg/m^3]	990	870	1100	840
Filling factor κ	0.60	0.55	0.60	0.55
$FOM_{solid\ target} = \rho \cdot \kappa \cdot P_t^2 \cdot f^2$ [kg/m^3]	8.8	12.0	23.9	28.9

POLARIZED SOLID TARGETS

A polarized solid target can be assumed to be an ensemble of particles with spin, placed in a high magnetic field and cooled to very low temperature. The basic idea - to obtain a high polarization of nuclear spins - consists in using a microwave field, to transfer the polarization of electron spins to these nuclei. This process is called dynamic nuclear polarization (DNP). The main practical problem with DNP is finding a suitable combination of hydrogen - or deuterium - rich material and a paramagnetic dopant, i.e. material with an unpaired or quasi-free electron. Suitable means that the relaxation time of the electron spins is short (*ms*) and that of the nucleons (nuclei) is long (*min*), resulting in a high nucleon (nuclei) polarization.

Molecular hydrogen would be the ideal proton target material, except that at low temperature it devolves to the magnetically inert para-state, with spin zero and, hence, is unpolarizable. Various practical polarized solid target materials have been developed in which a diluted amount of paramagnetic centers has been introduced by chemical or radiation methods.

Polarized solid targets have some special advantages, such as:

- High density, which makes them unique for particle experiments with secondary beams e.g. (tagged) photon beams or muon beams.
- High construction flexibility, which enables an integration into almost any detector set-up.

A very important feature of polarized solid targets is the fact, that the DNP scheme works for any nucleus with spin. However, the efficiency of the DNP in the different target materials, i.e. to get highest nuclear polarization values, depends strongly on the nature and the interaction of the added paramagnetic centers in the respective solids.

In this context electron spin resonance (ESR) spectroscopy has been introduced as an important tool in polarized solid target material research, since it allows to study the paramagnetic centers, which are used for the DNP. In particular, the ESR linewidth of the paramagnetic centers is a very important parameter, especially concerning the deuterated target materials [3].

There are two operation modes depending on the intensity of the particle beams:

- Continuous mode, i.e. permanent microwave irradiation for DNP.
- Frozen spin mode, i.e. after the DNP process microwave irradiation is stopped.

In addition to the degree of polarization important criteria for the choice of a target material suitable for particle physics experiments are:

- High dilution factor f.
- Presence of other polarizable nuclei.
- Short polarization build-up time.
- Good resistance of the polarization against radiation damage with the possibility to repair the radiation damage.

REALIZATION OF POLARIZED SOLID TARGETS

Depending on the beam intensities there are two modes of operation for a polarized solid state target system. (Polarized) electron beams have to be operated at intensities of less than $100\,nA$ in order to handle cooling requirements and radiation damage of the target materials. In this case polarized targets with ^4He evaporative cooling are used, which have to work in a continuous mode, i.e. with permanent microwave irradiation to maintain the DNP process at $1\,K$. At this temperature the nucleus (nucleon) polarization relaxation time T_1 is relatively short. This continuous mode operation puts strong constraints on the design of polarizing magnets. Due to the field homogeneity requirements over the entire target volume, the large dimensions of the magnet coils limit the angles for the outgoing particles. This mode of operation has been used for the SLAC experiments E143, E155 and E155x and is presently used for experiments at TJNAF (see Figure 1(*l*)). Other electromagnetic probes such as tagged photon beams are intensity limited by their production techniques to about 10^8 particles/*s*. To obtain a reasonable counting rate, a wide opening angle with the ability to simultaneously measure a large kinematical range is needed. This can be achieved with the concept of the frozen spin target.

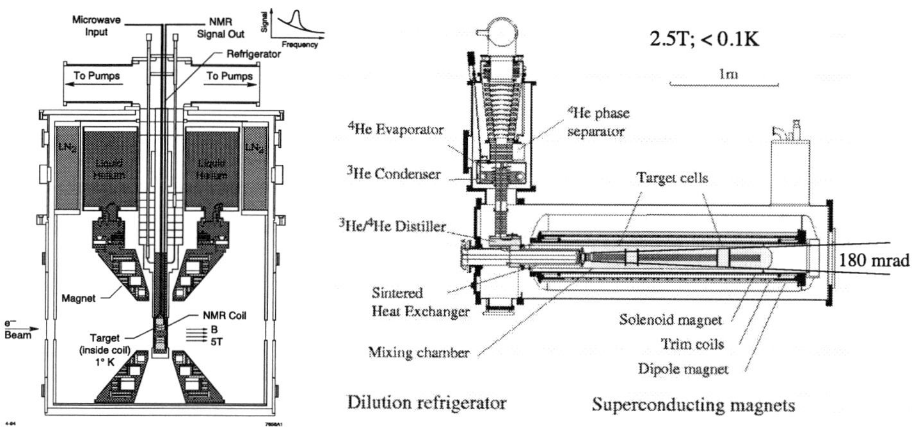

FIGURE 1. (*l*) $5T$ polarizing superconducting magnet with a vertically operated ^4He evaporation refrigerator [4]. (*r*) The CERN dilution $^3He/^4He$ dilution refrigerator with a superconducting $2.5T$ magnet and a $0.5T$ dipole magnet [5].

The operation of the frozen spin target is based on the experimental fact that the nucleon polarization relaxation time T_1 is a very strong function of the temperature and magnetic field. T_1 characterizes the polarization decay after switching off the DNP mechanism (microwaves). Typical values for T_1 are minutes at a temperature of $1\,K$ and days below $100\,mK$. The principle of the frozen spin target operation is to polarize the target material at a high field (e.g. $2.5\,T$) and in the temperature range $0.15-0.3\,K$. Once the target material is optimally polarized, the microwaves are switched off and the polarization is frozen at temperatures around $50\,mK$ (frozen spin mode). Due to the very long T_1 at these temperatures, it is possible to reduce the field to a value of $0.3-0.4\,T$ where the polarization decay is acceptable (holding field).

A very specific target design has been developed for the CERN polarized target experiments SMC and COMPASS. A huge $^3He/^4He$ dilution refrigerator ($T \leq 50mK$) is connected to a $2.5T$ superconducting solenoid which is surrounded by a dipole magnet (see Figure 1 (*r*)). With a beam intensity of $\approx 10^7 \mu^+/s$, which was more than four orders of magnitude less than at the SLAC experiments E155 and E155x, the luminosity was increased by having a target two orders of magnitude longer than at SLAC or TJNAF. In the CERN experiments it was not possible to flip the beam polarization (unless one changed to μ^- !) and with a polarization build-up time of hours, it was not very easy to change the target polarization frequently. So a strategy of rotating the solenoid field to the opposite direction was adopted. With $190\,GeV$ muons and a longitudinal field, the effect on the scattered particles was minimal. A dipole magnet was turned on at a strategic time so that the total magnetic field never went below $0.5\,T$ while the solenoid was crossing zero and the target was in the frozen spin mode.

Thus the CERN polarized target can be operated in the continuous mode at $2.5T$ as well as in a frozen spin mode $T \leq 50mK$ for horizontal or transverse polarization direction. Other systems using the frozen spin mode have been operated by SACLAY [6] and PSI [7] groups.

A new concept of a small superconducting holding magnet placed inside a $^3He/^4He$ dilution refrigerator has been developed in Bonn (see Figure 2) [8]. Thus the fringe field and its influence on nearby detector components is minimized and that makes it possible to operate polarized solid targets in a 4π-detector (e.g. DAPHNE in Mainz) [9].

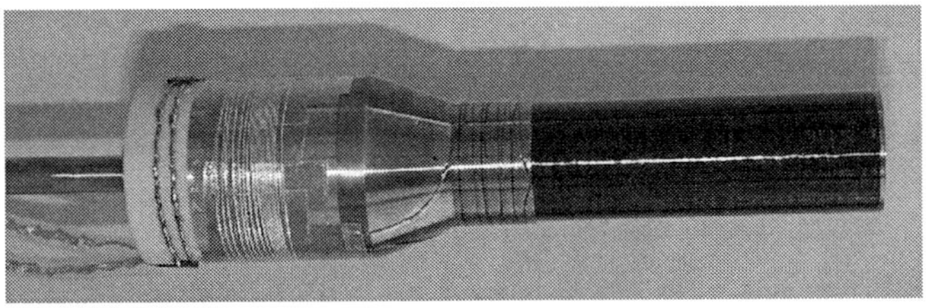

FIGURE 2. The Bonn superconducting holding magnet ($0.6T$) with a total thickness of ~1mm, placed inside a horizontal $^3He/^4He$ dilution refrigerator [8].

In the meantime this scheme is adopted from TJNAF and Mainz for the operation of their frozen spin targets in the CLAS detector and CRYSTAL BALL detector, respectively. At TJNAF further improvements of a polarized HD frozen spin target with its lengthy target material preparation method (brute force [10]) is considered to improve the signal to background ratio of the particle detection. More detailed information about the present polarized solid targets are given in Ref. [1][2].

PERSPECTIVES

Experiments with $^3He/^4He$ Dilution Refrigerators

The frozen spin target operation is a powerful tool for (double) polarization experiments at low intensity beams. However, the present polarized frozen spin target systems with their internal low mass holding coils allow – due to their horizontal geometry (see Figure 2) – only a polarization in the beam direction (z-direction). This will be changed by replacing the internal solenoidal holding coil by a saddle coil geometry which gives access to y- and x polarization directions. Such a coil is presently manufactured by the Bonn and TJNAF target groups.

Nevertheless the price for the good particle acceptance is a loss in beam time, which is needed to repolarize the target material. Furthermore a dedicated railway system to move the detector and external polarizing magnet for the polarization procedure is needed. In best cases a beam time efficiency of 80 % is attainable. The second restriction from the 'figure of merit' point of view is the decay of the polarization during the data taking. The overall target polarization of an experiment using a frozen spin target is given by the nucleon relaxation time and limited to approximately $0.8\,P_{\text{max}}$.

The goal for the future is to combine the advantages of the frozen spin technique with those of the 'continuons mode' operating target to a so called '4π continuons mode' polarized target. This new polarized target scheme leads to an improvement in the 'figure of merit' by a factor of 2 compared to the existing frozen spin target operation. The important points of such a target system are:

- large angular acceptance close to 4π
- high average polarization during the data taking (continuous DNP at high polarizing field and low temperature)
- high luminosities up to $10^{33} cm^{-2} s^{-1}$
- no moving system for the polarization process required
- good beam time efficiency (fast repolarization in case of radiation damage)

In practise, starting from the existing 'internal superconducting holding coil' a new coil capable of providing an increased field has to be implemented into the $^3He/^4He$ refrigerator as an internal polarizing magnet. It has to fulfill the requirements of the high homogeniety of the external polarization magnet and the low mass distribution of the internal holding coil to ensure a good detection probability for the outgoing particles. The most crucial point which has to be considered in the development of the internal polarizing magnet is the homogeneity of the small coil. Because of the minimized dimensions of the solenoid the magnetic volume is only a factor of about 20 larger than the target volume, which needs the good homogeneity of $\Delta B/B \leq 10^{-4}$. The investigation of this problem has led to a new approach for the design of this magnet, which relies of the geometrical shape of the winding [11].

As mentioned above, luminosities of up to $10^{33} cm^{-2} s^{-1}$ at high polarization can be achieved with this '4π continuons mode' polarized target i.e. beam intensities higher than 10^{10} particles/s upon some extended target area ($\sim cm^2$) can be accepted. However, at such levels of beam intensities the radiation resistance of the polarization due to material damage has to be considered. Then inorganic materials like ammonia (NH_3 and ND_3) or ^6LiD and ^7LiH are the target materials of choice.

DNP polarized target materials for medical applications

Magnetic Resonance Imaging (MRI) is a very successful imaging modality in terms of temporal and spatial resolution of morphological details. The sensitivity of NMR, imaging or spectroscopy, is strongly limited by the low polarization that can be reached at room temperature and any achievable field strength. In an attempt to improve this, DNP–hyperpolarization was invented.

Several imaging applications will benefit from the strong signal available from the ^{13}C agent, e.g. angiography and perfusion studies. Other applications have so far been impossible by MRI, e.g. studying the metabolic fate of a hyperpolarized agent during the first minute after injection. The ability to study metabolism of endogenous molecules could have great implications in e.g. oncology and cardiology. Conventional MRI contrast parameters have not been correlated with processes at the cellular level. The idea behind hyperpolarized ^{13}C-labelled contrast agents was to directly image by MRI an agent (e.g. endogenous molecule) that would have specificity to certain deceases.

An apparatus and method for producing strongly polarized molecules in solution has been developed [12]. The nuclear spins of a solid sample doped with a paramagnetic agent (Trityl or Tempo radicals) can be strongly polarized at low temperature ($\sim 1.2K$) and moderate magnetic field ($3.35T$) by exposing the sample to microwave irradiation ($94GHz$) close to the electron paramagnetic resonance. The solid sample is rapidly dissolved and an image or spectrum is acquired within a time of the order of the nuclear relaxation time T_1 (within minutes). By this method is has been possible to produce solutions of ^{13}C-labelled molecules with polarization of up to $\sim 40\%$ [13]. Just recently a ^{13}C-polarization of $\sim 80\%$ has been measured in trityl doped pyruvic acid by increasing the magnetic field up to $5.0T$ [14]. Polarization values between 5% and 10% are reported in Tempo doped ^6Li– or ^{15}N labelled molecules as well [15]. The progress in this field is demonstrated e.g. in ref. [16].

CONCLUSION

The subsequent developments of all aspects of a polarized solid state target system have opened up new experimental possibilities and allowed a choice of approaches to a particular physics problem. There are a variety of polarizable solid state target materials available with dilution factors varying between 0.13 and 0.5.

The highest polarization resistance against radiation damage is shown in materials which are prepared for DNP by irradiation. Polarizing them in ^4He evaporation refrigerators with their high cooling power at $1K$ beam intensities upon the targets up to $\approx 6 \cdot 10^{11}$ particles/s can be tolerated. Thus luminosities of higher than $10^{35}\,cm^{-2} \cdot s^{-1}$ can be achieved.

Polarized target experiments performed with low intensity beams such as polarized muons or (polarized) tagged photons have to rely on solid target materials. At intensities up to 10^8 particles/s the frozen spin mode operation of the target system is favorable in order to allow a particle detection over a large (4π) angular acceptance.

Maximum polarization values for protons $> 90\,\%$ and for deuterons between $40\,\%$

and 80 % are achieved. Polarization measurements are done by the NMR technique to a relative accuracy of $< \pm 3$ % for protons and $< \pm 5$ % for deuterons. Future developments as discussed will certainly improve the experimental situation for experiments with polarized solid targets in the next decade.

Latest developments open up the possibility to use DNP–hyperpolarized substances in the field of medicine. It is very likely that this spin–off activity will be an additional motor for further improvements of the polarized solid targets.

REFERENCES

1. St. Goertz, W. Meyer, G. Reicherz, Prog. Part. Nucl. Phys. **49**, (2002) 403.
2. W. Meyer, Proc. 11th Workshop on The Physics of Exited Nucleons, Bonn, Springer 2008.
3. J. Heckmann *et al.*, Phys. Rev. B **74**, (2006) 134418.
4. D.G. Crabb *et al.*, NIMA **356**, (1995) 9.
5. J. Kyynäräinen *et al.*, NIMA **356**, (1995) 47.
6. R. Bernard *et al.*, NIMA **249**, (1986) 176.
7. B. van den Brandt *et al.*, NIM A 356 (1995) 53.
8. H. Dutz *et al.*, NIMA **356**, (1995) 111.
9. H.-J. Arends *et al.*, Phys. Rev. Lett. **84** (2000) 5950.
10. A. Honig, Phys. Rev. Lett. **19** (1967) 1009.
11. A. Raccanelli, H. Dutz, R. Krause, Proc. of the 11th Workshop on Pol. Sources and Targets, Tokyo 2005, World Scientific (2007) 221.
12. Ardenkjaer-Larsen *et al.*, Proc. Math. Acad. Sci. USA (2003) 10158.
13. J. Wolber *et al.*, NIM A **526** (2004) 173.
14. J. Heckmann *et al.*, to be published.
15. F. Piegsa *et al.*, in these proceedings.
16. K. Golman *et al.*, Nature Medicine, Vol. 13, No 11, (2007).

Present Status And Future Prospects Of Polarized Ion Sources

A. Belov[a] and A. Zelenski[b]

[a]*INR, Russian Academy of Sciences, Moscow, Russia.*
[b]*Brookhaven National Laboratory, Upton, NY 11973, U.S.A.*

Abstract. Recent progress in polarized ion sources development is reviewed. New techniques for production of polarized H^- ion (proton), D^- (D^+) and $^3He^{++}$ ion beams are discussed. Feasibility studies of these techniques are in progress at BNL and other laboratories. Polarized deuteron beams will be required for the polarization program at the Dubna Nuclotron and at the deuteron Electric Dipole Moment experiment at BNL. Experiments with polarized $^3He^{++}$ ion beams are a part of the experimental program at the future Electron Ion Collider.

Keywords: Polarized Ion Sources.
PACS: 29.25.Lg

INTRODUCTION

Polarization is an intrinsic property of light (photons), electrons, protons, nuclear beams, and the study of polarization effects provides complimentary information on particle structure and their interactions. Collider experiments with polarized beams at RHIC and HERA (at HERA, the experimental program with polarized electron beam and polarized internal target, HERMES, has been completed) provide essential tests of QCD and Electroweak interaction. Polarization asymmetries and parity violation are strong signatures for the identification of the fundamental processes, which are otherwise inaccessible. Such experiments require the maximum available luminosity, and therefore polarization must be obtained as an extra beam quality without sacrificing intensity. This is already the case for electron accelerators. In a storage ring, an electron beam is self polarized by the Sokolov-Ternov effect. For linear accelerators, a great effort in polarized electron source development was finally rewarded by achievement of up to 90% polarization and high beam intensity which will be sufficient to run high-current accelerators at Jefferson Laboratory or the future International Linear Collider at maximum intensity with polarized beam.

There are a number of proposals to polarize the high-energy proton beam in a storage ring by the Stern-Gerlach effect (or antiprotons by the spin-filtering technique). But so far the only feasible option is to accelerate the polarized beam produced in the source and make sure that polarization will survive during acceleration and storage.

High intensity unpolarized H^- ion sources are presently a common choice for high-energy accelerators due to the advantage of stripping injection into the accelerator ring. Polarized deuteron beam will be required for the deuteron EDM (Electron Dipole

CP1149, *Spin Physics, 18th International Spin Physics Symposium*
edited by D. G. Crabb, D. B. Day, S. Liuti, X. Zheng, M. Poelker, and Y. Prok

Moment) experiment and is also planned for the Dubna "Nuclotron" facility. Experiments with accelerated polarized $^3He^{++}$ ion beams shall be a part of the program at future Electron Ion Collider.

This paper is based on materials presented at the 12th International Workshop on Polarized Sources, Targets and Polarimetry (PSTP 2007 at BNL) / 1 /.

POLARIZATION TECHNIQUES

Spin filtering techniques. The basic feature of these polarization techniques is an attenuation: scattering out, adsorption, defocusing, quenching of unwanted part of light, electron, atomic, proton, or nuclei beams having 'unwanted' direction of polarization. The adsorption of one component of linear polarization of light in some materials is called dichroism and is widely used in science, technology and everyday life (polarization sunglasses, photography). A strong spin dependence of thermal energy neutron beam capture in a polarized ^3He gas cell is often used for neutron beam polarization and polarization analysis. In Lamb-shift polarized sources the metastable hydrogen atoms in unwanted spin states are quenched by the "spin-filter" to the ground states and remaining polarized atoms can be produced by selective ionization from metastable states.

Selective focusing by sextupole separating magnets is used for hydrogen (deuterium) beam polarization by electron spin in Atomic Beam Sources (ABS) of polarized ions and polarized internal targets. In these sources the atomic hydrogen is produced by dissociation of hydrogen molecules in RF discharge. Hydrogen gas flows out the dissociator volume to vacuum forming gaseous jet. The atomic hydrogen beam is formed then from central part of the jet using skimmers and diaphragms. A typical velocity of atoms in the beam is about $(1\text{-}2)\cdot 10^5$ cm/s, which is achieved by cooling of the dissociator nozzle to a temperature of 30-80 K. The sextupole magnetic field acts on electron magnetic moment axially aligned with the field gradient towards magnet tips as a focusing lens. The other component having opposite electron spin direction is defocused. Then the electron polarization is transferred to the protons by means of radio-frequency transitions. The atomic beam of a selected spin-state is directed into an ionizer (or a storage cell). The polarized proton or H- ion beam can be produced by ionization of atomic beam in magnetic field of about 1.5 kG, which is sufficient to break electron-proton spin coupling while preserving proton polarization.

The ABS beam can be used as an internal target in accelerator-collider storage ring (polarized H-jet polarimeter at RHIC /2/), or for feeding storage cell type of internal targets. There is a proposal to use polarized atomic hydrogen target for antiproton polarization in the storage ring by filtering (removing) of unwanted states /3/.

Optical pumping. Electron or nuclear polarized atoms can be obtained in process of absorption of polarized photons from external source of polarized light and subsequent spontaneous emission of unpolarized photon (optical pumping). Angular momentum of atom electron shell is changed during the process and atom takes on electron polarization. The electron polarization is transferred to nuclear through spin-spin interaction if optical pumping takes place in sufficiently low magnetic field.

Population of spin states of atoms is changed during the optical pumping without filtering process due to absorption of external polarized photons which angular momentum is transmitted to atoms.

For optical pumping of rubidium atoms laser radiation with wavelength of 795 nm is used to excite transitions $5S_{1/2} \rightarrow 5P_{1/2}$. Polarized ^3He atoms are obtained by exciting transitions $2^3S_1 \rightarrow 2^3P_0$ (1083 nm). Unfortunately, direct optical pumping of hydrogen atoms is not used so far for polarized ion production due to absence of suitable light sources with wavelength of 121.5 nm (Lyman alpha radiation).

Polarization-transfer technique. In any type of polarized proton (H^- ion) source the first step is the generation of an electron-spin polarized atomic beam. The polarization is then transferred to the protons by hyperfine interaction and finally the beam is ionized. The difference is in the velocity of the atomic beam. It is comparatively easy to polarize a "slow" (thermal energy) beam by using separating magnets, as discussed above. The advantages of using "fast" (a few keV energy) beams are higher intensity and simple, more efficient ionization. In this technique optical pumping is used to get polarized alkali atoms. The electron–spin polarization of the "fast" H beam is produced either in a charge-exchange process, when primary protons capture polarized electrons from polarized atoms in a vapor cell, or in spin-exchange collisions /4/. This technique is called an "Optically-Pumped Polarized Ion Source" (OPPIS), although polarized electrons can also be captured from a ferromagnetic foil (as in the original Zavoiski proposal), or from hydrogen, or an alkali-metal atomic beam polarized by separating magnets.

ATOMIC BEAM SOURCE WITH RESONANT PLASMA IONIZER

A polarized ion source with resonant plasma ionizer has been developed at INR RAS /5/. A deuterium plasma injector is used in this source for production of polarized H^- ions. The injector generates plasma consisting mainly from D^+ and D^- ions.

The present version of the plasma injector is shown schematically in Figure 1. The plasma flux from the arc-discharge plasma source is enriched by negative ions in a surface-plasma converter. Positive ions are converted into neutral atoms with eV energy in collisions with a neutralizer internal surface. Polarized atomic hydrogen beam is injected into the plasma and polarized H^- ions are produced via reaction:

$$H^0\uparrow + D^- \Rightarrow H^-\uparrow + D^0$$

The plasma flux is guided to the internal surface of the neutralizer by the magnetic field created by the plasma coil (in longitudinal direction) and by the converter electromagnet (transversal direction). Plasma ions interact with the neutralizer surface and most of them are reflected as neutral hot atoms.

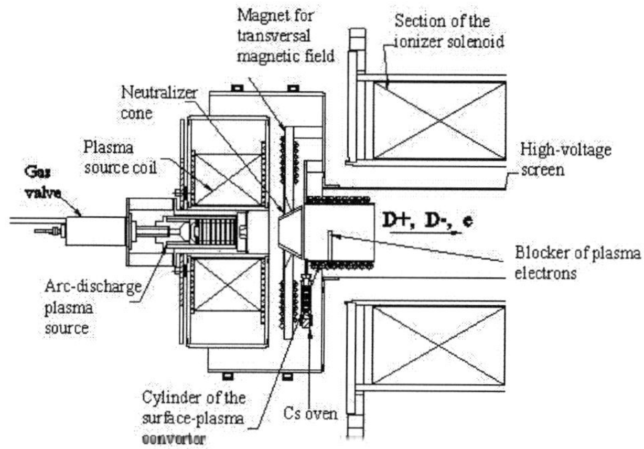

FIGURE 1 The injector of plasma consisting mainly from D^+ and D^- ions.

The reflected hot atoms hit the converter cylinder internal (molybdenum) surface where the atoms are converted partially into negative ions and injected into the ionization region along the fringing solenoid field lines.

Figure 2a shows the measured dependence of the H^- ion beam current from the unpolarized D^- ion current. In the Figure 2 the solid line is the result of calculation and the points are results of measurements of the polarized H^- ion beam current recorded by the Faraday cup downstream the bending magnet. Deviation of observed values for polarized H^- ion current from calculated curve at high D^- ion beam current (at high plasma density) is observed. The deviation can be explained by destruction of slow polarized H^- ions in plasma in collisions with plasma electrons and D^+ ions.

A polarized H^- ion beam with peak current of 4 mA has been obtained with D^- ion beam current of 62 mA. Respective oscillograms of polarized H^- and unpolarized D^- ion beam pulses are shown in Figure 2b.

The polarization of the H^- ion beam was measured to be 0.91 ± 0.03. The Lamb shift polarimeter has been used for the polarization measurements. Figure of merit of the polarized H^- ion beam produced $\mathbf{P^2I}$ has record value of 3.2 mA. Efficiency of direct conversion of polarized hydrogen atoms into polarized H^- ions reached of 12.5%. Polarized D^- ion beam with peak intensity up to 2 mA and polarization up to 90% from nominal of vector polarization ±1 and tensor polarization of +1,-2 have been obtained from the polarized ion source CIPIOS at IUCF /6/. The source was developed in a collaboration of IUCF and INR RAS. The source had a nearly resonant charge-exhange plasma ionizer and produced also polarized H^- ions (1.8 mA peak) and unpolarized H^- and D^- ion beams (40 mA and 30 mA respectively) with pulse duration of 300 μs and repetition rate of 2 Hz.

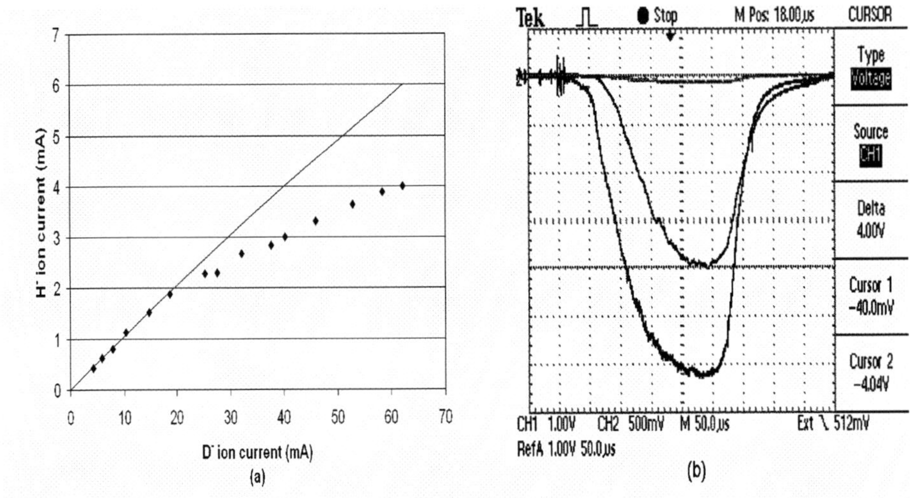

FIGURE 2. (a) Polarized H^- ion current vs unpolarized D^- ion current. Solid line is result of calculation using formula (5) with intensity of atomic hydrogen beam of $2\cdot10^{17}sec^{-1}$, points- results of the measurements. (b) Oscillograms of the polarized H^- ion pulse with atomic hydrogen beam „on" (middle trace) and „off" (upper trace) and unpolarized D^- ion pulse (lower trace).Vertical scale for the H^- ion pulse is 1 mA/div and it is 10 mA/div for the D^- ion pulse. Horizontal scale is 50 µs/div.

The normalized emittance of the polarized ion beams was measured to be 1.2 π mm mrad. A relatively small emittance was necessary due to acceptance limitation of the RFQ used (1 π mm mrad normalized). The emittance was minimized by a carefully designed focusing of an atomic hydrogen beam by permanent magnet sextupoles with magnetic field up to 1.4 T and by restriction of magnetic field in a charge-exchange region to value ≤1 kG.

PROJECT OF POLARIZED DEUTERON SOURCE FOR DUBNA NUCLOTRON ACCELERATOR

The program of polarization research at Joint Institute for Nuclear Research at Dubna, Russia is based on the acceleration of polarized deuteron beams up to 10 GeV energy in the NUCLOTRON. A high intensity pulsed source of polarized deuterons is required to achieve the number of accelerated deuterons of $\sim10^{10}$ deuterons/cycle (with the present one turn injection scheme). A new polarized deuteron source will be built in a collaboration of JINR and INR of RAS. Parts of the CIPIOS source from IUCF were delivered to JINR and are will be used for the Dubna polarized deuteron source /7/.

A project goal is a polarized deuteron beam from the source with a peak intensity of 10 mA and polarization of 90% from nominal vector polarization ±1 and tensor polarization +1,-2. A nearly resonant charge-exchange deuteron plasma ionizer with storage cell will be used to increase intensity of the polarized ion beam from the source, reduce the polarized ion beam emittance and the unpolarized proton current in the charge-exchange region, in comparison with a source without the storage cell, and respectively reduce background current of H_2^+ ions which will not be separated from polarized deuterons in the bending magnet.

POLARIZED SOURCE WITH CESIUM BEAM IONIZER AT COSY

The principle of the source is an ionization of pulsed polarized hydrogen or deuterium beams in collisions with a pulsed neutral Cesium beam having a kinetic energy of about 45 keV /8/.

In a charge exchange reaction taking place in a solenoid field, negatively charged hydrogen, or deuteron, ions are created and accelerated toward the extraction elements. Then the ions are bent magnetically by 90°, passed through a Wien-filter and enter the transporting source beam line that guides them into the cyclotron. The new record value of 50 μA and 90% polarization was reached during routine source operation in 2005. This exceeds the original design value of 30 μA. It was the result of the optimization of all source components /9/. A breakthrough decision was to develop a pulsed cesium beam matched to the short injection period of up to 20 ms for COSY, thus virtually eliminating the severe sputtering damage that had been an obstruction for reliable operation. Beam diagnostics, like beam scanner, faraday cups and viewer, for the cesium beam were added to successfully shape the transverse phase space for optimal overlap with the atomic hydrogen or deuterium beam.

OPTICALLY PUMPED POLARIZED H^- ION SOURCE AT RHIC

An example of a source based on polarization transfer technique is the OPPIS at RHIC /10/. In this source an ECR-type source produces a primary proton beam of 2.8 ~ 3.0 keV energy, which is converted to electron-spin polarized H atoms by electron pick-up in an optically pumped Rb vapor cell. Electrostatic deflection plates downstream of the polarized alkali remove any surviving H+ or other charged species.

The electron-polarized H beam then passes through a magnetic field reversal region, where the polarization is transferred to the nucleus, via hyperfine interaction (Sona transition). The polarized H atoms are then negatively ionized in a Na-jet vapor cell to form nuclear polarized H-ions (see Fig. 3). Alternatively, the H atoms can be ionized in a He gaseous cell to form polarized protons. This source is capable of producing in excess of 1.6 mA polarized H- ion current in dc operation. Detailed studies of polarization losses in the RHIC OPPIS and the source parameters optimization resulted in the OPPIS polarization increase to 86-90%. The numerical simulations of the Sona–transition efficiency were done using code which was developed at INR, Moscow /11/. This contributed to increasing polarization in the AGS and RHIC to 65 ~ 70% / 12/.

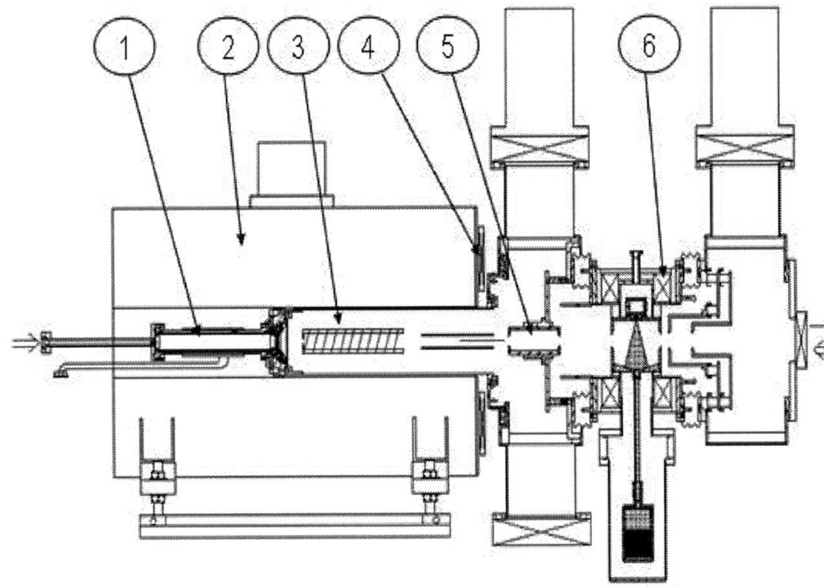

FIGURE 3. Layout of the RHIC OPPIS: 1) Electron Cyclotron Resonance primary proton source: 2). Super-conducting Solenoid; 3) Optically-Pumped Rubidium Cell; 4). Correction coil; 5). Sona-shield; 6). Na-jet ionizer.

OPPIS WITH THE FAST ATOMIC HYDROGEN BEAM SOURCE

The ECR proton source is operated in high magnetic field. It has low hydrogen gas consumption, which makes possible a dc OPPIS operation with intensity in excess of 1.0 mA. However, the ECR source has a comparatively low emission current density and high beam divergence. This limits further current increase and gives rise to inefficient use of the available laser power for optical pumping. In fact only about 15% of the electron-spin polarized hydrogen atoms produced in Rb cell is within the ionizer cell acceptance.

In pulsed operation, suitable for application at high-energy accelerators and colliders, the ECR source limitations can be overcome by using instead a high brightness proton source outside the magnetic field /13/. Following neutralization in hydrogen, the high brightness 3.0~4.0 keV atomic H beam is injected into a superconducting solenoid, where both a He ionizer cell and an optically-pumped Rb cell are situated in the same 25-30 kG solenoid field, which is required to preserve the electron-spin polarization. The injected H atoms are ionized in the He cell with 80% efficiency to form a low emittance intense proton beam, which enters the polarized Rb vapor cell. The protons pick up polarized electrons from the Rb atoms to become a beam of electron-spin polarized H atoms. A negative bias of about 1.0 kV applied to the He cell decelerates the proton beam produced in the cell. This allows energy separation of the polarized hydrogen atoms produced after lower energy proton neutralization in Rb vapor and residual hydrogen atoms of the primary beam. Atomic hydrogen beam current densities greater than 100 mA/cm^2 can be obtained at the Na jet ionizer location (about 180 cm from the source) by using a very high brightness fast atomic beam source developed at BINP, Novosibirsk /14/, and tested in experiments at TRIUMF, where more than 10 mA polarized H^- and 50 mA proton beam intensity was demonstrated /4/.

Higher polarization is also expected with the fast atomic beam source due to: a) elimination of neutralization in residual hydrogen - $\mathbf{B_{H2}} \sim 1$; b) better Sona-transition transition efficiency for the smaller ~ 1.5 cm diameter beam; c) use of higher ionizer field (up to 3.0 kG), while still keeping the beam emittance below 2.0 π mm·mrad, because of the smaller beam – 1.5 cm diameter. All these factors combined will further increase polarization in the pulsed OPPIS to ~ 90% and the source intensity to over 10 mA. The ECR-source replacement with an atomic hydrogen injector will provide the high intensity beam for polarized RHIC luminosity upgrade and for future eRHIC facilities.

PROPOSAL FOR POLARIZED $^3He^{++}$ SOURCE AT RHIC

Polarized beams of $^3He^{++}$ ions also contain the polarized neutron component (as a deuteron) and its magnetic moment is close to the proton magnetic moment value, therefore the AGS and RHIC Siberian snake should preserve polarization during acceleration /15/. In this case the difficulty is the polarized $^3He^{++}$ source. The proposed polarized $^3He^{++}$ acceleration in RHIC (and also for future RHIC upgrade to electron ion collider eRHIC) will require about $2 \cdot 10^{11}$ $^3He^{++}$ ions in the source pulse and about 10^{11} ions in the RHIC bunch. To deliver these ions in a 20 us pulse duration for injection to Booster, the source peak current has to be about 1000 uA, which is 10000 times higher than ever achieved in existed $^3He^{++}$ ion sources. A new technique has been proposed for production of high intensity $^3He^{++}$ ion beam. It is based on ionization of ^3He gas (polarized by metastability exchange technique) in the Electron Beam Ion Source (EBIS) /16/.

The highest 3He nuclear polarization close to 80% was achieved so far by the metastability exchange technique. In this method, 3He gas at typically 1 torr pressure is contained in a glass bulb and a weak RF discharge is maintained in the gas. Metastable atoms in the 2 3S1 state are produced in the discharge and may be polarized by means of optical pumping with circularly polarized (2 3S - 2 3P) 1083 nm light. The metastability is transmitted between atoms increasing the pumping rate. In the proposed technique, the polarized ^3He atoms consumption for injection to an ionizer is very small, of the order of 10^{13}-10^{14} He atoms sec^{-1} and higher polarization is expected.

An EBIS is under construction at BNL as an alternative to the Tandem heavy ion injector for RHIC. It is proposed to use the EBIS to produce 3He^{++} by ionization of the polarized 3He gas, which is fed from the polarizing cell. The ionization in the EBIS is produced in a 50 kG magnetic field, which preserves the nuclear 3He polarization while in the intermediate single-charged 3He$^+$ state. The ionization efficiency to the double-charged $3He^{++}$ will be close to 100% and the number of ions is limited to the maximum charge, which can be confined in the EBIS. From experiments with $Au32^{+}$ ion production, one expects about $2.5 \cdot 10^{11}$ $3He^{++}$ ions/pulse to be produced and extracted for subsequent acceleration and injection to RHIC.

After $^3He^{++}$ acceleration to a few MeV/nucleon He-D or He-Carbon collisions can be used for polarization measurements. The Lamb-shift polarimeter at the source energy of 10-20 keV can be used in the feasibility studies (similar to the OPPIS polarimeter). In this technique $^3He^{++}$ ions are partially converted to He^{+} (2S) – metastable ions in the alkali vapor cell. Then the hyperfine sublevel populations can be analyzed in the spin-filter device to extract the primary $^3He^{++}$ nuclear polarization.

A study of limitation on the maximum attainable nuclear polarization in the metastability exchange technique (at the very low polarized ^3He gas consumption rate) will be required to define the maximum attainable polarization. Possible depolarization effects during polarized ^3He gas injection to existing EBIS prototype and multi-step ionization process should be also studied. The expected $^3He^{++}$ ion beam intensity is in excess of $2 \cdot 10^{11}$ ions/pulse with polarization in excess of 70%.

PROPOSAL FOR POLARIZED SOURCE OF $^3He^{++}$ IONS BASED ON CHARGE-EXCHANGE IONIZATION

Polarized $^3He^{++}$ ions can be also produced using next nearly resonant charge-exchange reaction:

$$^3He^0\uparrow + {}^4He^{++} \rightarrow {}^3He^{++}\uparrow + {}^4He^0$$

Cross-section of this reaction at collision energy of 50 eV is $4 \cdot 10^{-16} cm^2$. A polarized $^3He^{++}$ ion source scheme using this reaction was proposed in ref. /17/, but was not checked experimentally yet. It was assumed in this scheme that polarized ^3He atoms will be produced by method of optical pumping with metastability exchange where nuclear polarization of ~ 80% is achieved. A cooled storage cell can be installed in a charge-exchange region. Exceptionally long relaxation time for polarized helium atoms will lead to small polarization losses and large polarized gaseous target thickness of $\sim 10^{15}$ cm^{-2}.

Unpolarized $^4He^{++}$ ions passing through the storage cell with thickness of $\sim 10^{15}$ cm^{-2} will undergo charge-exchange collisions and produce polarized $^3He^{++}$ ions with probability of ~0.4. Thus, polarized $^3He^{++}$ ion current will be comparable with unpolarized $^4He^{++}$ ion current transported through the cell. If $^4He^{++}$ ion current injected into the storage cell is ~10 mA then the expected polarized $^3He^{++}$ ion current is ~4 mA. It is important to note that the magnetic field in the charge-exchange region is necessary only for radial confinement of ions and can be at ~1 kG level, which should produce the polarized $^3He^{++}$ ion beam with relatively small emittance (~ 0.5 π mm mrad normalized).

SUMMARY

There are a number of proposals on polarization studies with polarized ion beams at new and existing accelerators and colliders, which will require high-intensity, high polarization proton, deuteron and $^3He^{++}$ ion beams. State-of-the-art atomic beam sources with resonant plasma ionizer and optically pumped polarized proton sources produce sufficient beam intensity (of a few mA H^- ion beam) for charging the high energy accelerators to full capacity (for colliders the intensity is limited by the beam-beam interaction). The proton polarization of about 90% has been achieved for the high intensity beams. The further increase to over 10 mA pulsed beam intensity has also been demonstrated and will be used at future Electron Ion Colliders. For the polarized $^3He^{++}$ source, feasibility studies of the proposals based on EBIS ionizer and resonant charge-exchange are required.

REFERENCES

1. Proceedings of 12th International Workshop on Polarized Sources Targets and Polarimetry, BNL 2007, AIP Conf.Proc. 980, 2008.
2. A. Zelenski et al., Nucl. Inst. & Meth. **A536**, 248 (2005)
3. A.Nass et al., PSTP 2007, AIP Conference Proceedings 980, p.153, ((2008).
4. A. Zelenski, Proc.SPIN2000, AIP Conference Proceedings 570, p.179, (2000).
5. Belov A., "Production of Polarized Ions with Nearly Resonant Charge-Exchange Collisions in Plasma", AIP Conference Proceedings 980, p.209, (2008).
6. V. P. Derenchuk, A. S. Belov, " Polarized D^- Operation and Development of the IUCF Ion Source CIPIOS", AIP Conference Proceedings 675, p. 887, 2003
7. N. N. Agapov, A. S. Belov, V. P. Derenchuk, V. V. Fimushkin, V. P. Vadeev, "IUCF Polarized Ion Source CIPIOS for JINR Accelerator Nuclotron", in *Proc. of 16th International Spin Physics Symposium (SPIN 2004)*, Trieste, Italy, 2004, World Scientific, eds. F. Brandamante, A. Bressan, A. Martin, 2005, p. 774.
8. W. Haeberli, NIM, **62**, 1968, pp. 355.
9. R.Gebel et al., PSTP 2007, AIP Conference Proceedings 980, p. 231, ((2008).
10. A. Zelenski et al., "OPPIS for RHIC Spin Physics", Rev.Sci.Instr., 73, p.888, (2002).
11. Antishev E., Belov A., "Study Of adiabaticity of Hydrogen Atoms Spin Motion in Changing Magnetic Fields", AIP Conf.Proc.**980**, p.263, (2008).
12. A.Zelenski et al., "The RHIC Optically-Pumped Polarized H- Ion Source", AIP Conf.Proc.980, p.221, (2008).
13. A.Zelenski et al., **NIM A245**, p.223, (1986).
14. V. Davydenko, **NIM A427**, p.230, (1999).
15. W.MacKay, "Prospects for Acceleration of D and He beams", AIP Conf.Proc.**980**, p.191, (2008).
16. Zelenski A., Alessi J., "Prospects on high intensity optically-pumped polarized H^-, D^- and $^3He^{++}$ions", ICFA Beam Dynamics Newsletter **No.30**, p.39, (2003).
17. A. S. Belov, Nucl. Instrum. Meth., **A402**, 205-207 (1998).

Polarization in Heavy Ion Physics

S.M. Troshin

Institute for High Energy Physics, Protvino, Moscow Region, 142281, Russia

Abstract. Role of polarization studies in heavy ion physics is discussed with emphasis on the search for quark-gluon plasma formation and studies of its dynamical properties.

Keywords: heavy-ion collisions, orbital angular momentum, polarization, directed flow
PACS: 12.38.Mh, 13.88.+e, 24.10.Ht, 24.70.+s

INTRODUCTION

The main goal of the experiments performed with heavy ion collisions is generation of the deconfined state of the QCD matter and studies of its properties. Existence of deconfined state was predicted prior to the QCD era (cf. e.g. [1] and references therein). Asymptotic freedom of QCD has led straightforwardly to the prediction that the deconfined state should be a weakly bounded state of quarks and gluons. Important tools in the studies of the nature of the new form of matter are the anisotropic flows which are the quantitative characteristics of the collective motion of the produced hadrons in the nuclear interactions. With their measurements one can obtain a valuable information on the early stages of reactions and observe signals of QGP formation. Despite that the polarization measurements in this field are complementary, they nevertheless are very useful tool for the searches for the deconfined state and study of its nature. Of course, in heavy–ion collisions spin states of final particles can only be measured.

POLARIZATION AND SEARCH FOR QUARK-GLUON PLASMA

Assuming different dynamics of hadron and nuclear collisions, i.e. that high densities of matter reached in nuclear collisions would lead to QGP formation, the idea to use vanishing polarization of Λ hyperons produced in nuclear collisions as a signal of QGP has been exploited in the early studies [2, 3, 4].

Indeed, in hadron interaction the experimental situation with hyperon polarization is widely known and stable for a long time. Polarization of Λ produced in the unpolarized inclusive pp–interactions is negative and energy independent. It increases linearly with x_F at large transverse momenta ($p_\perp \geq 1$ GeV/c), and for such values of transverse momenta is almost $p_\perp$-independent, this dependence is represented in Fig. 1. It should be noted that polarization was measured with respect to the production plane, i.e. the plane spanned over the vector of the initial momentum and the vector of the produced particle momentum. It was supposed that in the nuclear collisions the produced transient state has an isotropic distribution of parton momenta and therefore QGP lose any memory of the initial state momenta. The only existing vector is the vector of the final state

CP1149, *Spin Physics, 18th International Spin Physics Symposium*
edited by D. G. Crabb, D. B. Day, S. Liuti, X. Zheng, M. Poelker, and Y. Prok

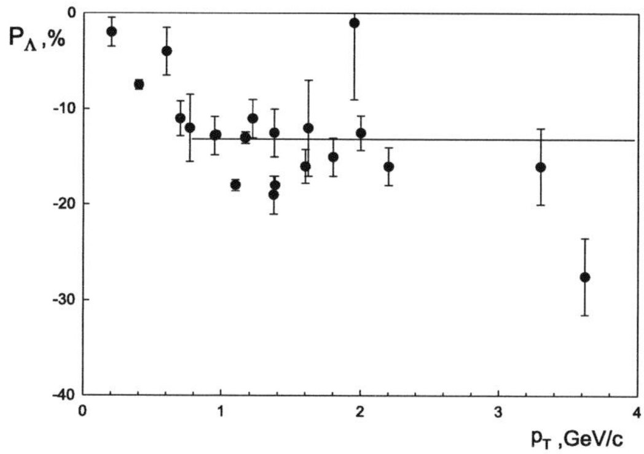

FIGURE 1. Transverse momentum dependence of polarization of Λ.

momentum. But due to parity conservation polarization along this direction should be zero. Thus, the absence of hyperon polarization w.r.t. production plane was predicted to be a signal of QGP formation. Vanishing hyperon polarization w.r.t. production plane in central nuclear collisions follows from various models also.

However, there are no experimental measurements at high energies of hyperon polarization w.r.t. **production** plane in nuclear collisions and this prediction is still a hypothesis.

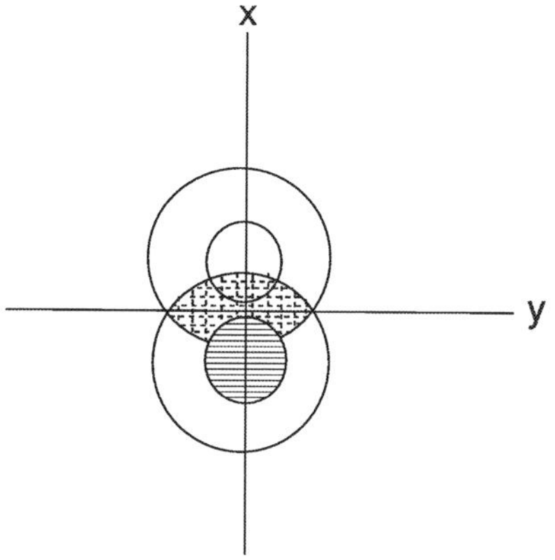

FIGURE 2. Reaction plane in hadron and nuclei collisions.

Here we turn to another aspect of the same problem. For this purpose it is important to make a remark that in the recent paper [5], which provides an emphasis on the historic aspects of the QGP searches, the conclusion made that the deconfined state of matter has being observed in hadronic reactions during a long time and it would be interesting to study collective properties of transient state in reactions with hadrons and nuclei simultaneously since the dynamics is qualitatively the same and differences are merely quantitative. Another array of theoretical predictions is related to prediction of significant polarization w.r.t. **reaction** plane in the non-central hadron and nuclear collisions dew to the presence of the large orbital angular momentum in such collisions. Reaction plane is the plane spanned over the vectors of the initial momentum $\vec{p}$ and the impact parameter of colliding particles or nuclei $\vec{b}$. This definition is illustrated in Fig. 2, where the normal to the reaction plane is directed along the y axis.

The determination of the reaction plane in the non-central hadronic collisions could be experimentally feasible with use of the standard procedure. The relationship of the impact parameter with the final state multiplicity is a useful tool in these studies similar to the studies of the nuclei interactions. The value of impact parameter can be determined through the centrality and then, e.g. global polarization, directed or elliptic flow can be analyzed by selecting events in a specific centrality ranges.

It is evident that orbital angular momentum can have very large values. It can be estimated as follows

$$L(s,b) \simeq \alpha b \frac{\sqrt{s}}{2} D_C(b), \tag{1}$$

where $D_C(b)$ describes distribution of matter in the overlap region (cf. Fig. 2). It should be noted that $L \to 0$ at $b \to \infty$ and $L = 0$ at $b = 0$ (Fig. 3). In nuclear collisions at RHIC

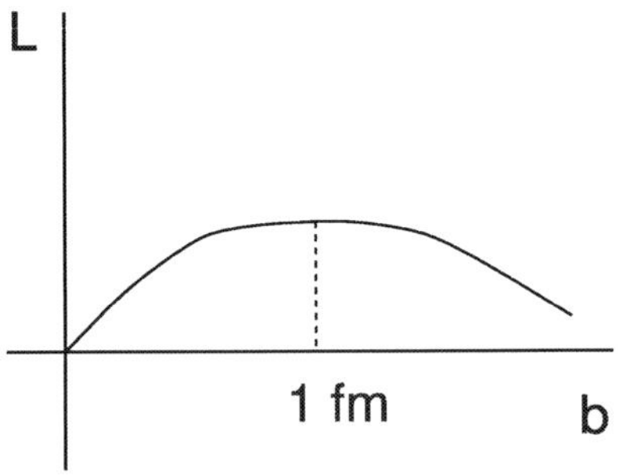

FIGURE 3. Qualitative dependence of the orbital angular momentum L on the impact parameter b.

and LHC average values of orbital angular momentum are of order ($\hbar = 1$) 10^5 (RHIC) and 10^7 (LHC)[1].

It is natural that the question arises: what are the observable effects of the large angular orbital momentum presented in the peripheral heavy-ion and hadron collisions? Carruthers supposed that orbital angular momentum would be released in the coherent burst of polarized vector mesons [6]. Yang and Chou pointed out to the strong necessity for the spins of the outgoing particles to line up parallel to each other in the transverse direction to the reaction plane [7], i.e. $(\sigma_1^T \cdot \sigma_2^T)_{average} > 0$ due to a net deficiency in the left and right-moving outgoing systems. The reaction plane is perpendicular to the orbital angular momentum direction. Thus, in a process of high energy hadron and nuclear collisions large initial orbital angular momentum can, in principle, be converted into the spin angular momentum of final particles resulting in their polarization relative to the reaction plane. We will try to connect this possibility with the nature of transient strong interaction matter.

[1] In nuclear collisions orbital angular momentum has a maximum at $b \simeq R_A$.

Weakly-coupled matter of QGP (parton model with final state interactions) does not allow coherent collective rotation of the system and therefore finite transverse gradient of the average longitudinal momentum per produced parton should exist in the overlap region. It is claimed that relative OAM in collision of partons will lead to global quark polarization due to spin-orbital coupling [8].

Significant (order of 0.3) polarization of hyperons relative to reaction plane was anticipated due to global polarization of quarks. Similar ideas were used for the hyperon polarization in hadron non-central collisions [9]. The idea to observe circularly polarized direct photons as a signal of quark polarization in the QGP was proposed in [10, 11].

The measurements of global polarization Λ and $\bar{\Lambda}$ were performed at RHIC (STAR Collaboration) in $Au+Au$ collisions at $\sqrt{s_{NN}} = 62.4$ and 200 GeV and upper limit $|P_{\Lambda,\bar{\Lambda}}| \leq 0.02$ has been obtained [12]. It should be noted that global spin alignment for ϕ and K^{*0} were also not observed for the different centralities [13]. Thus, if we will not consider scenario of complete dilution of polarization by hadronization mechanism, we should conclude that at the moment no experimental evidence exists for conversion of the orbital angular momentum into the spin angular momentum in nuclear collisions and this conclusion is correlated but not necessarily follow from the result on the strongly interacting nature of transient matter observed at RHIC.

It is well known that discovery of the deconfined state of matter has been announced by the four major experiments at RHIC Despite the highest values of energy and density have been reached, a genuine quark-gluon plasma QGP was not found. The deconfined state reveals the properties of the perfect liquid, being strongly interacting collective state and therefore it was labelled as sQGP. The question arise again: what are the experimental manifestations of the large orbital angular momentum could be in the case of strongly interacting transient matter?

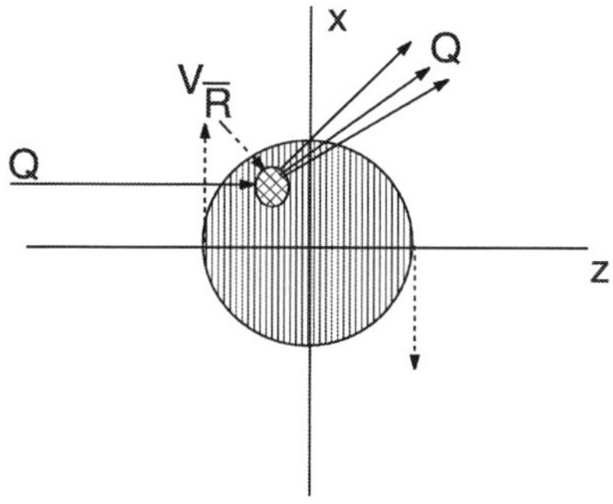

FIGURE 4. Interaction of the constituent quark with rotating quark-pion liquid.

First, observation of an ideal liquid as a transient state means that this state of matter has low viscosity, i.e. large interaction cross-section (estimates provide the value about 22 mb). Then we should conclude that in these circumstances the preexisting large orbital angular momentum will lead to rotation of the transient deconfined matter in the overlap region as a whole, i.e. all parts have the same angular velocity and the orbital angular momentum does not convert to the spin angular momentum. This rotation will lead to the directed flow v_1 as it was shown in [14]. Assumed there particle production mechanism at moderate transverse momenta is an excitation of a part of the rotating transient state of massive constituent quarks (interacting by pion exchanges) by the one of the valence constituent quarks with subsequent hadronization of the quark-pion liquid droplets (cf. Fig. 4). The directed flow v_1 in nuclei collisions as well as in hadron reactions depends on the rapidity difference $y - y_{beam}$ and not on the incident energy. The mechanism therefore can provide a qualitative explanation of the incident-energy scaling [15, 16, 17, 18] of v_1 observed at RHIC (Fig.5).

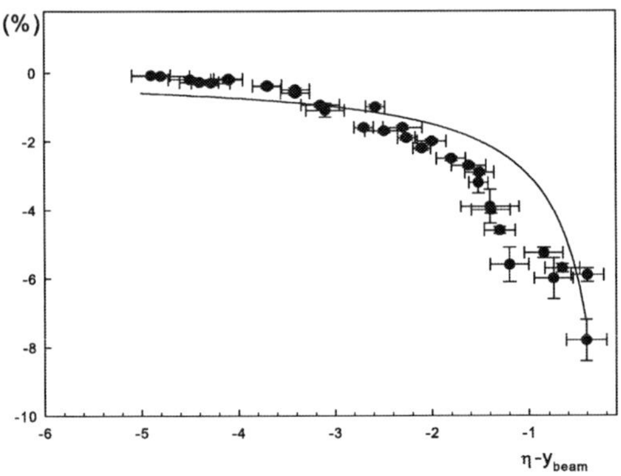

FIGURE 5. Dependence of directed flow on $\eta - y_{beam}$ in $Au+Au$ and $Cu+Cu$ collisions at 62.4 GeV and 200 GeV at RHIC.

Thus, the strongly interacting nature of transient state would lead to the non-zero directed flow v_1 and vanishing global polarization.

EXPECTATIONS FOR THE LHC ENERGIES

The evident question nowadays is about the nature of the matter which will be produced at the LHC.

If the vanishing directed flow v_1 would be observed then one could conclude on the weakly-coupled system, or genuine QGP. But then what about global polarization? Is this a signal of QGP formation? Not necessarily, and the reason is the existence of the reflective (antishadow) scattering at the LHC energies. This particular mechanism is not related to the question on what kind of deconfined matter was produced [19, 20]. The geometric picture at the energy which is beyond the black disc limit can be described as a scattering off the partially reflective and the partially absorptive disk surrounded by the black ring which turns out into a grey area at larger values of the impact parameter. The evolution with energy is characterized by increasing albedo due to the interrelated increase of reflection and decrease of absorption at small impact parameters. This mechanism results from unitarity saturation and leads to the peripheral production of secondary particles in the impact parameter space. The emerging global polarization of produced particles will appear then due to the imbalance of the orbital angular momentum in the initial and final states. Indeed, the enhancement of the peripheral particle production would destroy the balance between orbital angular momentum in the initial and final states; most of the particles in the final state would carry out significant orbital angular momentum. To compensate this orbital momentum the spins of the secondary particles should become lined up.

CONCLUSION

Concluding on the polarization studies in heavy ion physics, it is evident that there are more questions than answers in this field at the moment. Nevertheless, there are no doubts that polarization measurements combined with anisotropic flow measurements will be able to help in detecting QGP and revealing its dynamical properties more unambiguously. At RHIC there is another possibility of the directed flow measurements in the polarized proton collisions. As it was discussed above, the magnitudes of global polarization and directed flow are in a prompt relation with the nature of the transient matter, namely, depending on whether it is weakly or strongly coupled matter, these observables have different values. The role of polarization measurements in heavy ion physics certainly deserves further studies — both experimental and theoretical.

ACKNOWLEDGMENTS

Fruitful discussions with N. E. Tyurin and support are gratefully acknowledged. I am also grateful to D. G. Crabb for invitation and warm hospitality at the University of Virginia during Symposium SPIN 2008.

REFERENCES

1. L. McLerran, *J. Phys. G: Nucl. Part. Phys.* **35**, 104001 (2008).
2. P. Hoyer, *Phys. Lett. B* **187**, 162–166 (1987).
3. A. D. Panagiotou, *Phys. Rev. C* **33**, 1999–2002 (1986).
4. R. Stock et al., Proc. of the Conference on Quark Matter Formation in Heavy-Ion Collisions, Singapore 1982.
5. R. M. Weiner, *Int. J. Mod. Phys. E* **15**, 37-70 (2006).
6. P. Carruthers, *Nucl. Phys. A* **418**, 501C-523C (1984).
7. C. N. Yang, and T. T. Chou, *J. Phys. Soc. Japan, Suppl.* **55**, 53–52 (1986).
8. Z.-T. Liang, and X.-N. Wang, *Phys. Rev. Lett* **94** 102301 (2005).
9. S. A. Voloshin, *nucl-th/0410089v1* (2004).
10. A. Ipp, A Di Piazza, J. Evers, and C.H. Keitel, *arXiv: 0710.5700v1* (2007).
11. B. Betz, M. Gyulassy, G. Torrieri, *Phys. Rev. C* **76** 044901 (2007).
12. B.I. Abelev et al., *Phys. Rev. C* **76**, 024915 (2007).
13. I. Selyuzhenkov (for the STAR Collaboration) *J. Phys. G* **34**, S1099 (2007).
14. S. M. Troshin, and N. E. Tyurin, *Int. J. Mod. Phys. E* **17**, 1619–1633 (2008).
15. G. Wang (for the STAR Collaboration), *J. Phys. G* **34**, S1093 (2007).
16. B.B. Back et al., (PHOBOS Collaboration), *Phys. Rev. Lett.* **97**, 012301 (2006).
17. A.H. Tang (for the STAR collaboration), *J. Phys. G* **31**, S35 (2005).
18. J. Adams et al. (STAR Collaboration), *Phys. Rev. C* **73**, 034903 (2006).
19. S. M. Troshin, *Phys. Lett. B* **597**,391-393 (2004).
20. S. M. Troshin, and N. E. Tyurin, *Int. J. Mod. Phys. A* **22**, 4437–4449 (2007).

Spin in Hadron Reactions

Christine A. Aidala

University of Massachusetts, Amherst, MA 01003-9337, U.S.

Abstract. The Relativistic Heavy Ion Collider (RHIC) has brought the study of spin effects in hadronic collisions to a new energy regime. In conjunction with other experiments at facilities around the world, much can be learned from the high-energy polarized proton collisions RHIC provides, allowing the collider to serve as a powerful tool to continue to understand the rich subtleties and surprises of spin effects in QCD, some of which were originally discovered more than three decades ago.

Keywords: spin, polarized interactions, hadrons
PACS: 13.85.Ni, 13.88.+e

INTRODUCTION

Since the startling discovery of hyperon polarization in collisions of an unpolarized proton beam on a beryllium target in 1976, hadronic interactions have played an essential role in exploring spin effects in quantum chromodynamics (QCD). The Relativistic Heavy Ion Collider (RHIC), inaugurated in 2001 as a high-energy polarized proton collider, has taken the stage as the flagship facility for the exploration of spin in hadron reactions in recent years. Having achieved polarized proton collisions at center-of-mass energies as high as 200 GeV with approximately 60% polarization, and with polarized proton collisions at 500 GeV planned for 2009, RHIC offers a wealth of opportunities. The collider has already made a number of contributions toward unraveling the mysteries of spin effects in QCD, which can broadly be placed into three categories: hyperon polarization, the transverse spin structure of the nucleon and transverse-momentum-dependent parton distributions, and the helicity structure of the nucleon.

HYPERON POLARIZATION

More than three decades ago, an experiment at Fermilab was performed in which a 300-GeV beam of unpolarized protons was fired at a beryllium target, and lambda hyperons were measured in the final state. The self-analyzing decay of the lambda into a proton and a pion allowed determination of the polarization of the produced particles, and remarkably, transverse polarization of up to nearly 20% was observed, increasing with the transverse momentum (p_T) of the lambda up to the highest measured value of 1.6 GeV/c [1]. More than 30 years after this initial discovery, and despite plentiful experimental confirmations of the effect, the mechanism by which the hyperons acquire polarization remains very poorly understood.

With the availability of polarized beams, the polarization *transfer* to final-state hyperons can instead be measured, providing new pieces to help resolve the puzzle. This was

CP1149, *Spin Physics, 18th International Spin Physics Symposium*
edited by D. G. Crabb, D. B. Day, S. Liuti, X. Zheng, M. Poelker, and Y. Prok

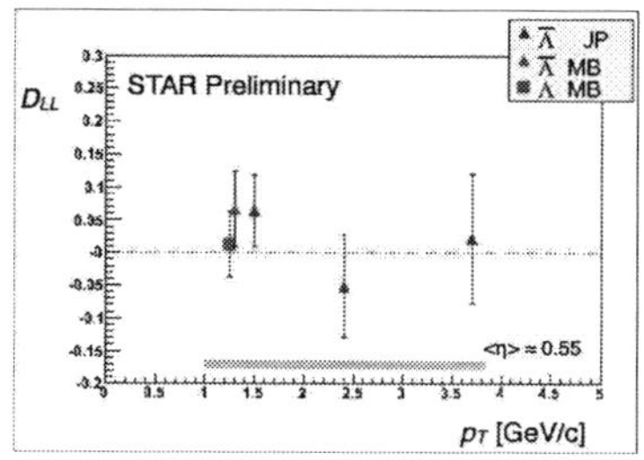

FIGURE 1. The spin transfer from a longitudinally polarized proton to a longitudinally polarized lambda or antilambda produced in proton-proton collisions at $\sqrt{s} = 200$ GeV, measured by STAR. "JP" and "MB" denote data taken with the STAR jet patch and minimum-bias triggers, respectively.

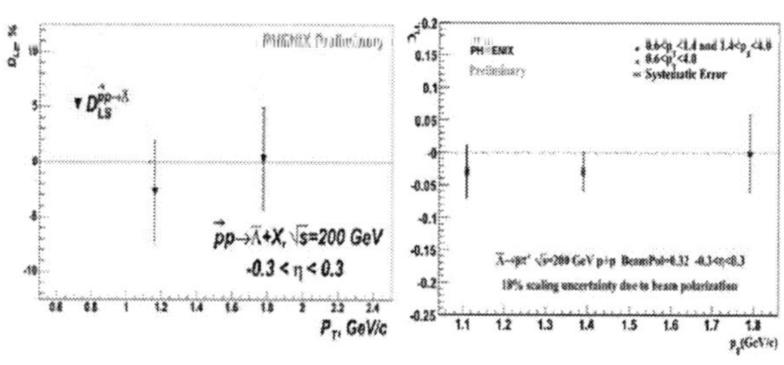

FIGURE 2. The spin transfer from a longitudinally polarized proton to a transversely (left) or longitudinally (right) polarized antilambda produced in proton-proton collisions at $\sqrt{s} = 200$ GeV, measured by PHENIX.

done in the mid-1990s by the E704 experiment at Fermilab, and a significant correlation between the transverse spin of the initial-state proton and the final-state lambda of up to $\sim 30\%$ was observed, rising again with p_T [2]. The polarized proton beams at RHIC allow similar measurements to be made. In Figs. 1-2 are shown spin-transfer measurements from initial-state, longitudinally polarized protons to final-state, transversely or longitudinally polarized (anti-)lambdas. The polarization transfer in both cases is consistent with zero for all measured p_T. Spin-transfer results with transverse polarization in the initial state are expected from RHIC in the future and will be important to compare to the E704 result performed at a center-of-mass energy one order of magnitude lower. While hyperon polarization remains poorly understood, new data combined with present work on transverse spin phenomena may eventually shed some light on the matter.

TRANSVERSE SPIN PHENOMENA

The study of transverse spin phenomena has witnessed a veritable explosion of activity over approximately the past seven years. For more detailed perspectives of the field than given here, see e.g. [3] and [4]. Surprisingly large single-spin asymmetries (SSA's) of magnitudes as large as 40-50% were initially observed in collisions of transversely polarized protons with other hadrons at the ZGS at Argonne National Laboratory [5,

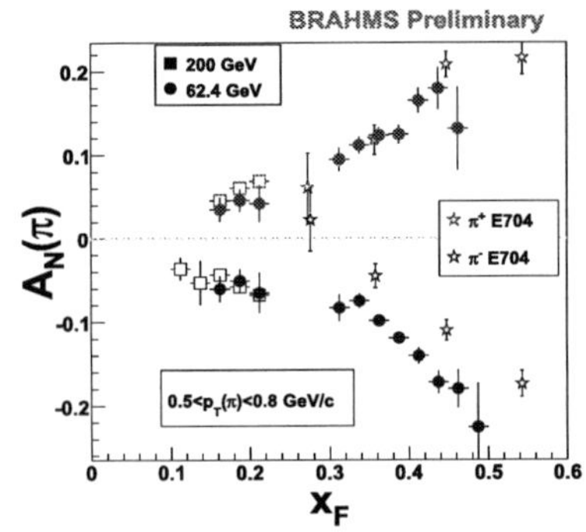

FIGURE 3. Charged pion asymmetries measured at 200 and 62.4 GeV by the BRAHMS experiment and at 19.4 GeV by the E704 experiment, shown for overlapping kinematic ranges.

6] as well as the PS at CERN [7] in the late 1970s and early 1980s, for center-of-mass energies between 5 and 10 GeV. It was these early hadronic data as well as preliminary hadronic data from Serpukhov, Protvino that inspired the groundbreaking paper by Dennis Sivers in 1989 [8], which proposed that these large asymmetries could be generated by a transverse-momentum-dependent partonic distribution (TMD), in which the transverse momentum of the partons is correlated with the spin of the proton.

Subsequent experimental data over a wide range of energies indicate that similar effects exist for particle production at center-of-mass energies of tens of GeV and even as high as 200 GeV at RHIC. In Fig. 3 are shown the transverse SSA's for charged pions measured at 19.4 [2], 62.4 [9], and 200 GeV. The higher-energy data allow perturbative QCD (pQCD) to be used in attempting to interpret the results. While much remains to be learned, exciting progress has been made in understanding the observed transverse single-spin asymmetries at high energies, and the wealth of experimental data now becoming available from RHIC and other experiments is helping to drive the field forward.

The complex task of understanding measurements performed in hadronic collisions is now being facilitated as relevant quantities are being studied in semi-inclusive deep-inelastic scattering (SIDIS) and e^+e^- annihilation. A definitively non-zero transverse-momentum-dependent Collins fragmentation function (FF) measured in e^+e^- annihilation was published by the BELLE experiment in 2005 [10]. The availability of the Collins FF makes it possible to extract the transversity distribution from asymmetry measurements in SIDIS and proton-proton collisions, and first extractions of transversity have been published [11]. First extractions of the Sivers TMD from SIDIS data have also been released [12, 13]. As constraints start to be provided on transversity and the various transverse-momentum-dependent distribution and fragmentation functions, more can in turn be learned from hadronic collision data.

The transversely polarized proton-proton data from RHIC have provided several surprises to the field, all of which remain unexplained. In 2002, a transverse SSA of approximately -10% was observed in the production of forward neutrons for $p+p$ collisions at 200 GeV [14], and a significant asymmetry has been seen to persist in

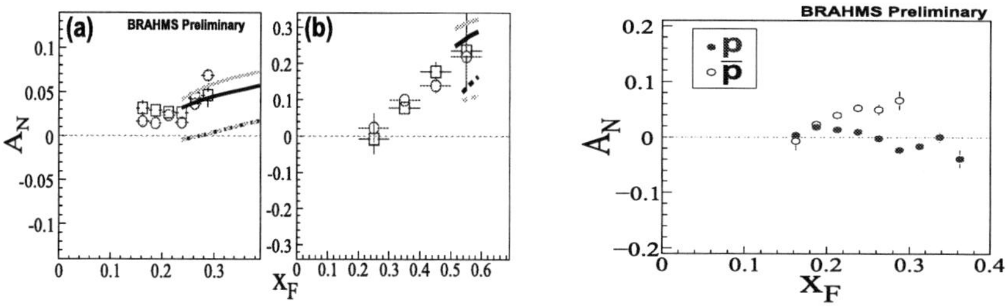

FIGURE 4. Transverse SSA's measured by BRAHMS. Kaons at 200 GeV (a) and 62.4 GeV [9] (b). Positive (negative) kaons are indicated by open circles (squares). Solid (dashed) curves are for positive (negative) kaon initial-state twist-three calculations with (black) and without (grey) sea-quark contributions. Protons and antiprotons at 200 GeV (right).

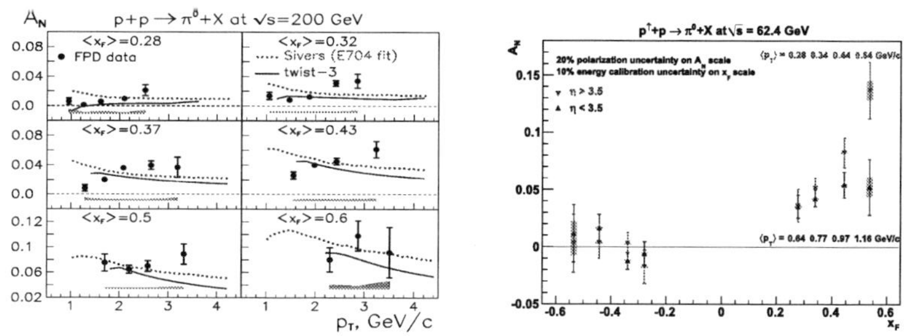

FIGURE 5. Left: p_T dependence of the transverse single-spin asymmetry of neutral pions produced at $\sqrt{s} = 200$ GeV, measured by STAR [15]. Right: Transverse single-spin asymmetry of neutral pions produced at $\sqrt{s} = 62.4$ GeV for two different pseudorapidity ranges, measured by PHENIX [16].

collisions up to 410 GeV. As can be seen in Fig. 4 (a, b), at both 200 and 62.4 GeV, the SSA's in forward charged kaon production are very similar for both kaon species. With the observed forward pion asymmetries understood as largely a valence quark effect, calculations of the SSA for negative kaons consistently underpredict the data. Even more startling is the significant non-zero antiproton asymmetry discovered by the BRAHMS experiment at 200 GeV, shown in Fig. 4 (right), while a SSA consistent with zero was observed for protons at both 200 and 62.4 GeV. Unfortunately no measurement of antiprotons is available at 62.4 GeV.

Theoretical work continues to attempt to unravel these mysteries, while the experimental data from RHIC become increasingly precise. For inclusive pion measurements, enough statistics are now available to allow studies of the asymmetry dependence on p_T and rapidity in addition to x_F, as shown for example in Fig. 5. As can be seen, the p_T dependence of the data shown disagrees with existing models. Close examination of the behavior of the SSA's as a function of various kinematic variables will be essential to constrain phenomenological models of the underlying physics generating the large asymmetries.

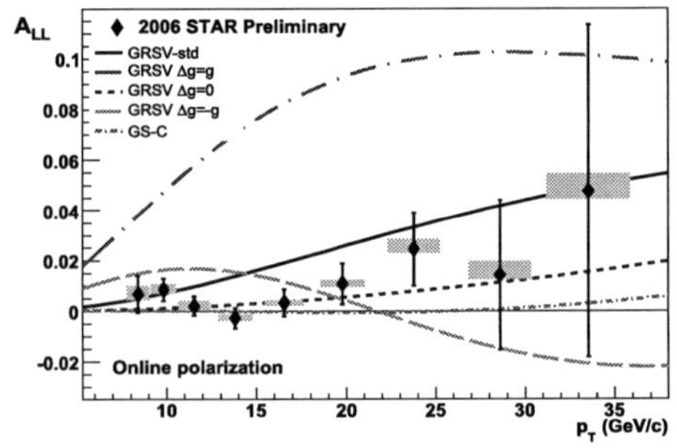

FIGURE 6. The double-helicity asymmetry in inclusive jet production in $p+p$ collisions at $\sqrt{s} = 200$ GeV [18], measured by STAR, compared to predictions based on several assumptions for ΔG within the GRSV parameterization [21].

HELICITY STRUCTURE OF THE NUCLEON

Since the so-called "proton spin crisis" was discovered by the EMC experiment more than 20 years ago [17], the quest has been on to understand the contributions of the spins of the gluons and the sea quarks to the total spin of the nucleon. RHIC, capable of polarized proton collisions for $50 < \sqrt{s} < 500$ GeV, is in a unique position to study both the gluons as well as the sea quarks. Performing hadronic collisions rather than deep-inelastic scattering, gluons contribute to the measured asymmetries at leading order. With polarized proton collisions at 500 GeV planned to start in 2009, the parity-violating longitudinal single-spin asymmetry in W boson production will allow flavor separation of the spin contributions of the sea quarks without reliance on fragmentation functions.

Several measurements have already been made at RHIC of double-helicity asymmetries for inclusive jet and hadron production, sensitive to ΔG, the gluon spin contribution to the spin of the proton. Jets represent the most abundant probe of ΔG available to STAR; PHENIX, with smaller acceptance and excellent electromagnetic calorimetry, has focused primarily on measurements of neutral pions. The latest results from both experiments at $\sqrt{s} = 200$ GeV can be seen in Figs. 6 [18] and 7 [19]. These measurements are sensitive only to a finite range in x_{gluon}. One way to extend the x range probed is by varying the center-of-mass energy. Measurements of neutral pions have already been made at 62.4 GeV [20], and 500 GeV collisions will begin this year.

Precise determination of parton distribution functions requires many experimental measurements covering a broad kinematic range. The proton-proton collision data from RHIC have already helped to constrain $\Delta g(x)$ [22], and further data will provide even stronger constraints over a wider kinematic range.

CONCLUSION

Hadronic reactions have a long history of seminal contributions to the study of spin effects in QCD. In particular in conjunction with knowledge gained from simpler ex-

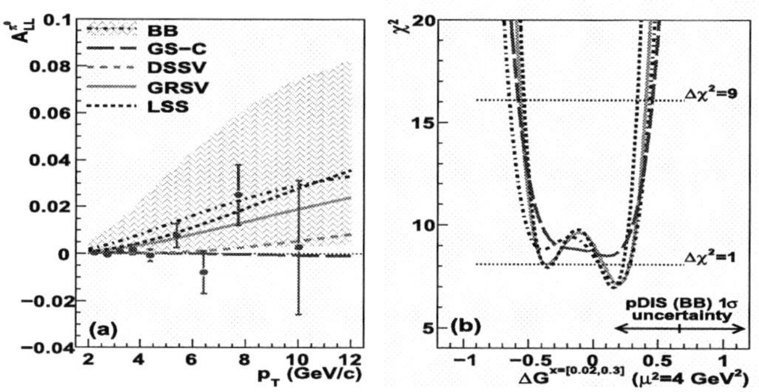

FIGURE 7. The double-helicity asymmetry in inclusive neutral pion production in $p+p$ collisions at $\sqrt{s} = 200$ GeV, measured by PHENIX, compared to expectations from several different parameterizations of the parton helicity distributions. See [19] for further details

perimental systems, much can be learned from the more complex interactions between hadrons. RHIC, as a high energy polarized proton collider, has opened up entirely new opportunities within the field; a wide variety of new results has already been released, and the program looks forward to many more years of data.

REFERENCES

1. G. Bunce, et al., *Phys. Rev. Lett.* **36**, 1113–1116 (1976).
2. A. Bravar, et al., *Phys. Rev. Lett.* **78**, 4003–4006 (1997).
3. M. Anselmino, these proceedings.
4. N. Makins, these proceedings.
5. R. D. Klem, et al., *Phys. Rev. Lett.* **36**, 929–931 (1976).
6. W. H. Dragoset, et al., *Phys. Rev.* **D18**, 3939–3954 (1978).
7. J. Antille, et al., *Phys. Lett.* **B94**, 523 (1980).
8. D. W. Sivers, *Phys. Rev.* **D41**, 83 (1990).
9. I. Arsene, et al., *Phys. Rev. Lett.* **101**, 042001 (2008).
10. K. Abe, et al., *Phys. Rev. Lett.* **96**, 232002 (2006).
11. M. Anselmino, et al., *Phys. Rev.* **D75**, 054032 (2007).
12. M. Anselmino, et al., *arXiv:0805.2677 [hep-ph]* (2008).
13. S. Arnold, A. V. Efremov, K. Goeke, M. Schlegel, and P. Schweitzer (2008).
14. A. Bazilevsky, et al., *Phys. Lett.* **B650**, 325–330 (2007).
15. B. I. Abelev, et al., *Phys. Rev. Lett.* **101**, 222001 (2008).
16. M. Chiu, *AIP Conf. Proc.* **915**, 539–542 (2007).
17. J. Ashman, et al., *Phys. Lett.* **B206**, 364 (1988).
18. M. Sarsour, these proceedings.
19. A. Adare, et al., *arXiv:0810.0694 [hep-ex]* (2008).
20. A. Adare, et al., *arXiv:0810.0701 [hep-ex]* (2008).
21. M. Glück, E. Reya, M. Stratmann, and W. Vogelsang, *Phys. Rev.* **D63**, 094005 (2001).
22. D. de Florian, R. Sassot, M. Stratmann, and W. Vogelsang, *Phys. Rev. Lett.* **101**, 072001 (2008).

The g_2 Structure Function

K. Slifer

University of New Hampshire

Abstract. We will discuss recent results for the spin structure functions, with an emphasis on g_2. High precision g_2 data allows for tests of the Burkhardt–Cottingham sum rule, and is needed to consistently evaluate higher twist effects.

Keywords: Spin Structure, Polarized Targets
PACS: 11.55.Hx,25.30.Bf,29.25.Pj,29.27.Hj

INTRODUCTION

Four independent structure functions are needed for a complete description of nucleon structure. All spin-dependent effects are contained in g_1 and g_2, while spin-independent effects are parameterized in F_1 and F_2. A simple physical interpretation is given in the impulse approximation of the parton model as distributions of quark momentum and spin in the nucleon:

$$\begin{aligned} F_1(x) &= \frac{1}{2}\sum e_i^2\,[q_i(x)+\bar{q}_i(x)] \\ F_2(x) &= 2xF_1(x) \\ g_1(x) &= \frac{1}{2}\sum e_i^2 \Delta q_i(x) \end{aligned} \tag{1}$$

Here the summation runs over all quark flavors.

Analysis of the spin structure functions (SSF) is typically facilitated via the Cornwall-Norton (CN) moments [1, 2]:

$$\begin{aligned} \Gamma_1^{(n)}(Q^2) &= \int_0^1 dx\, x^{n-1} g_1(x,Q^2) \\ \Gamma_2^{(n)}(Q^2) &= \int_0^1 dx\, x^{n-1} g_2(x,Q^2) \end{aligned} \tag{2}$$

where x is the Bjorken scaling variable, and Q^2 represents the positive definite four momentum transfer of the virtual photon, which mediates the interaction between the incident electron probe and the nucleon target. By convention, the superscript is usually dropped in the case of $n = 1$.

The Q^2 and $x-$dependence of the g_1 structure function has been measured [3, 4, 5, 6, 7, 8, 9, 10, 11, 12, 13] with impressive precision over a wide kinematic range. In this proceedings, we focus on the other structure function g_2, which has been historically neglected due to the technical difficulties of producing transversely polarized targets, and the lack of a simple interpretation of g_2 in the classic parton model.

CP1149, *Spin Physics, 18th International Spin Physics Symposium*
edited by D. G. Crabb, D. B. Day, S. Liuti, X. Zheng, M. Poelker, and Y. Prok

THE BURKHARDT-COTTINGHAM SUM RULE

The first CN moment of the g_2 structure function is predicted to vanish by the Burkhardt-Cottingham (BC) sum rule [14]:

$$\Gamma_2 = \int_0^1 dx\, g_2(x, Q^2) = 0 \qquad (3)$$

This sum rule arises from the unsubtracted dispersion relation for the spin-dependent virtual-virtual Compton scattering amplitude S_2. It is expected to be valid for any value of Q^2, although the reader is referred to [15] for a detailed discussion of scenarios which would lead to violations of the relation.

Fig. 1 displays existing world data for $\Gamma_2(Q^2)$ for the proton, neutron and ^{3}He. All available data are consistent with the nuclear sum rule for ^{3}He. Similarly, we find satisfaction of the neutron sum rule, within uncertainties, across several different experiments and a large range of Q^2. The picture is not so clear for the proton. The E155 collaboration found their data to be inconsistent with the proton BC sum rule at $Q^2 = 5$ GeV^2, and the only other data point is from the RSS collaboration at $Q^2 = 1.3$ GeV^2. Two upcoming JLab experiments will greatly expand our knowledge of g_2^p with measurements planned [16, 17] in the ranges $0.02 < Q^2 < 0.4$ and $2.5 < Q^2 < 6.5$ GeV^2.

The open symbols in Fig. 1 represent the experimentally measured data which typically covers the resonance region. To evaluate Eq. 3, the contributions from $x = 1$ and $x \to 0$ must be included. The nucleon elastic contribution at $x = 1$ can be easily evaluated using the form factor parametrizations of [22, 23], and the nuclear elastic contribution for ^{3}He is similarly determined from [24]. All form factor parameterizations are known to high precision over the relevant region.

To estimate the $x \to 0$ contribution for the g_2 integrals, the Wandzura-Wilczek [25] relation is used:

$$g_2^{WW}(x, Q^2) \equiv -g_1(x, Q^2) + \int_x^1 \frac{g_1(y, Q^2)}{y} dy \qquad (4)$$

This relation gives a prediction for the leading twist behaviour of g_2 entirely in terms of g_1. Using Eq. 4 for the low x contribution of the integral relies on the assumption that higher twist effects asymptotically vanish as $x \to 0$, which is, of course, untested due to the difficulty in measuring structure functions in this region. This contribution to the integral represents the largest contribution to the overall systematic uncertainty.

What can the BC sum rule tell us about low x?

Due to the extreme low counting rate, precise g_2 data at low Q^2 and low x are unlikely to be obtained anytime in the near future. However, if one were to accept the BC sum rule purely on theoretical grounds, it is possible to invert Eq. 3 to provide information on g_2 in this region. The high precision of the elastic form factors and the measured resonance region data allows us to determine the $x \to 0$ contribution to $\Gamma_2(Q^2)$ with relatively small uncertainty. Fig. 2 shows the resonance region (open symbols) and elastic contribution

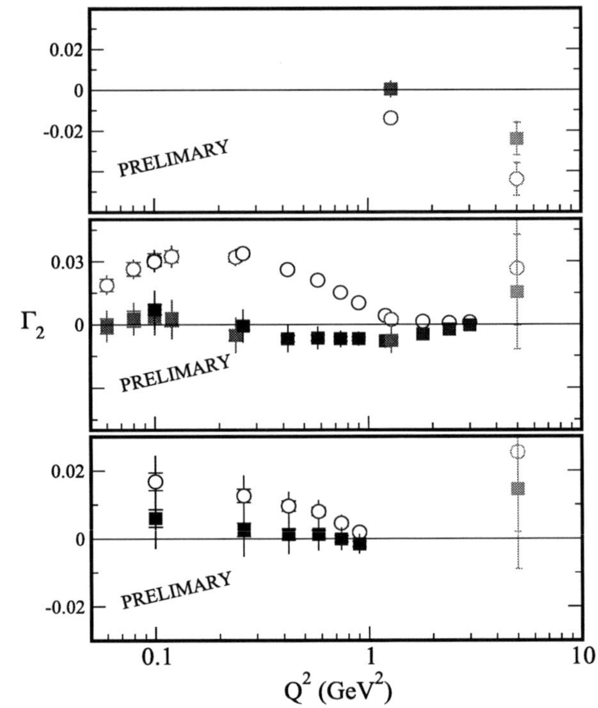

FIGURE 1. World data for $\Gamma_2(Q^2)$ for the proton (top), neutron (middle) and ^{3}He (bottom). Open symbols represent measured data (typically the resonance region), while the full symbols include an estimate of the unmeasured contributions to the integral. Inner (outer) error bars represent statistical (total) uncertainties. Brown: E155 collaboration [4]. Red: RSS [18, 19]. Black: E94010 [20, 11]. Green: E97110 [21] (Very Preliminary). Blue: E01012 [13] (Very Preliminary).

(solid line) to Eq. 3. The solid band is the $x \longrightarrow 0$ contribution inferred by assuming that Eq. 3 vanishes.

Violations of the BC Sum Rule?

As a necessary caveat to the preceding discussion, we stress that the experimental verification of the BC sum rule is still in question. In particular, Table 1 gives some perspective on the relative uncertainty of the determinations of Γ_2^n in the vicinity of $Q^2 \approx 1.0$ GeV^2. Satisfaction of the neutron BC sum rule is found only at the 2 to 2.5 σ level of the total uncertainty. As can be seen from the table, the uncertainty is typically dominated by systematic effects, while the statistical uncertainties are quite small. The dominant systematics of E94010 arose from the radiative corrections. Considering the significant improvements in knowledge of structure functions at low Q^2 that has occurred in recent years, a re-examination of this uncertainty is warranted in order to provide a conclusive

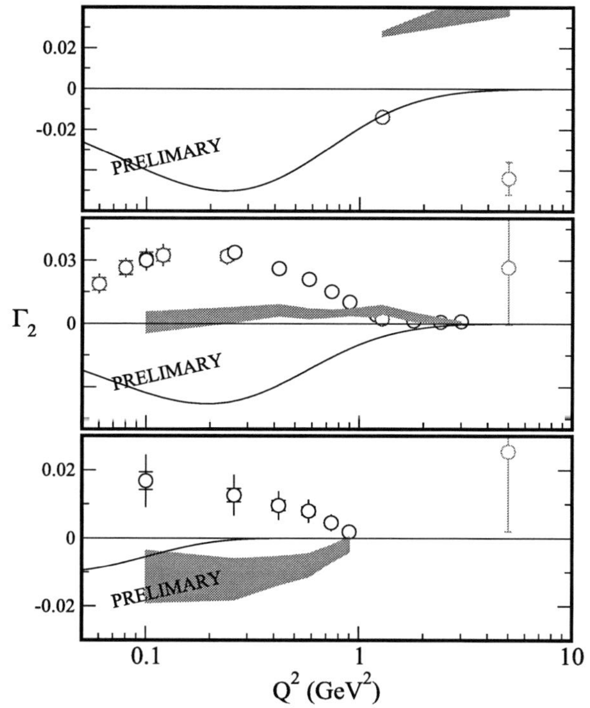

FIGURE 2. Full band represents an evaluation of $x \to 0$ contribution to $\Gamma_2(Q^2)$ assuming the validity of the BC sum rule. Proton (top), neutron (middle), and ^{3}He (bottom). Open symbols represent measured data (typically the resonance region). Inner (outer) error bars represent statistical (total) uncertainties. Brown: E155 collaboration [4]. Red: RSS [18, 19]. Black: E94010 [20, 11]. Green: E97110 [21] (Very Preliminary). Blue: E01012 [13] (Very Preliminary). Nucleon form factors from [22, 23]. Nuclear form factors from [24].

TABLE 1. Standard deviations of $\Gamma_2^n(Q^2)$ from zero. From E94010 [20] for $Q^2 < 1.0$ GeV^2 and E01012 [13] (Very Preliminary) for $Q^2 > 1.0$ GeV^2

Q^2(GeV^2)	Total Relative Uncertainty	Statistical Relative Uncertainty
0.74	1.7	6.6
0.90	2.0	7.9
1.20	2.4	2.9
1.80	1.9	2.2
2.40	1.8	2.2

test of the BC sum rule.

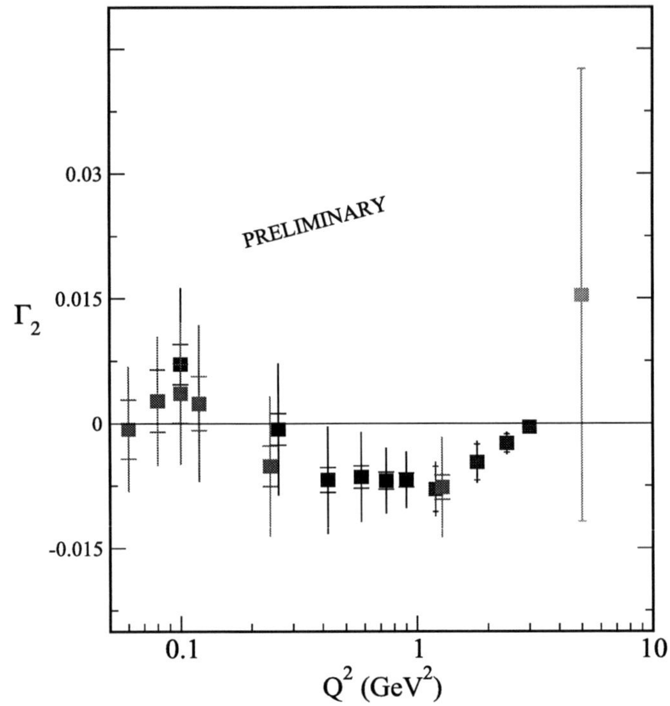

FIGURE 3. $\Gamma_2^n(Q^2)$. Full integral for neutron. Inner (outer) error bars represent statistical (total) uncertainties. Brown: E155 collaboration [4]. Red: RSS [18, 19]. Black: E94010 [20, 11]. Green: E97110 [21] (Very Preliminary). Blue: E01012 [13] (Very Preliminary).

HIGHER TWIST MEASUREMENTS

The Operator Product Expansion (OPE) [26, 27, 28] relates the structure functions to QCD quark and gluon matrix elements, which contain information on parton interactions. This effective theory of QCD is formulated in terms of the CN moments of Eq. 2. The OPE provides the following expansion of the lowest moment $\Gamma_1(Q^2)$:

$$\Gamma_1(Q^2) = \sum_{\tau=2,4,\ldots} \frac{\mu_\tau(Q^2)}{Q^{\tau-2}} \tag{5}$$

This expansion is performed in inverse powers of the *twist* τ. Successful predictions from the parton model map onto the leading twist of the OPE, while deviations from leading twist behaviour are known as higher twist effects. The parton model says nothing about the g_2 structure function, making it in an excellent quantity to study higher twist.

The coefficients μ_τ are identified with nucleon matrix elements of QCD operators of twist $\leq \tau$. The first coefficient beyond leading twist is μ_4, which can be decomposed as:

$$\mu_4 = \frac{1}{9}M^2\left(\tilde{a}_2 + 4\tilde{d}_2 + 4\tilde{f}_2\right) \tag{6}$$

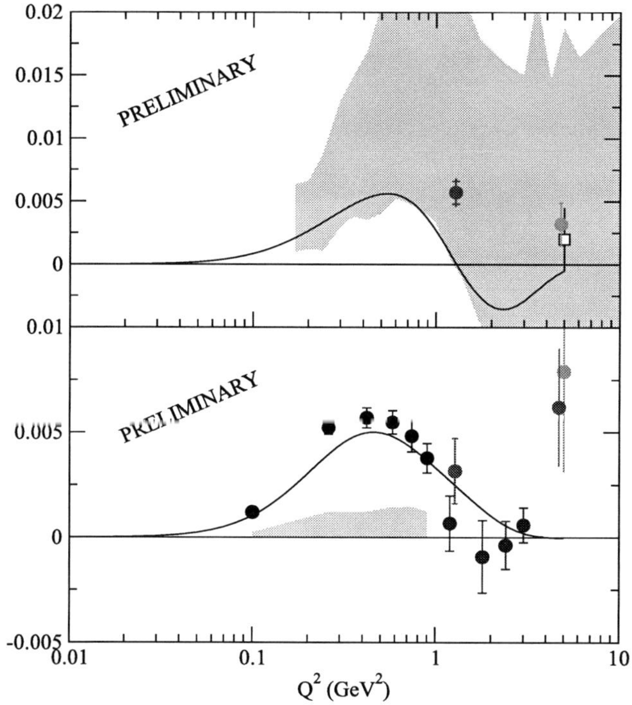

FIGURE 4. $I(Q^2)$ for proton (top) and neutron (bottom). Brown: E155 collaboration [4]. Black open square: QCDSF [29] Red: RSS [18, 19]. Black circle: E94010 [20]. Green: E97110 [21] (Very Preliminary). Blue: E01012 [13] (Very Preliminary). Magenta: E99117 [30]. Solid line: MAID model [31]. In the top panel, the filled band represents a determination of $I(Q^2)$ based on existing world data [10]. The large uncertainty reflects the lack of knowledge of g_2^p in this region.

This involves a combination of the leading twist, twist-3 and twist-4 matrix elements, respectively.

Higher Moments

The twist-2 term $\tilde{a}_2$ is purely kinematical in nature and can be evaluated [32] from the (target mass corrected) third moment of g_1:

$$\int_0^1 x^2 g_1(x,Q^2)dx = \frac{1}{2}\tilde{a}_2 + \mathscr{O}\left(\frac{M^2}{Q^2}\right) \tag{7}$$

Similarly, the third moment of g_1 provides information on $\tilde{d}_2$:

$$\int_0^1 x^2 g_2(x,Q^2)dx = \frac{1}{3}(\tilde{d}_2 - \tilde{a}_2) + \mathscr{O}\left(\frac{M^2}{Q^2}\right) \tag{8}$$

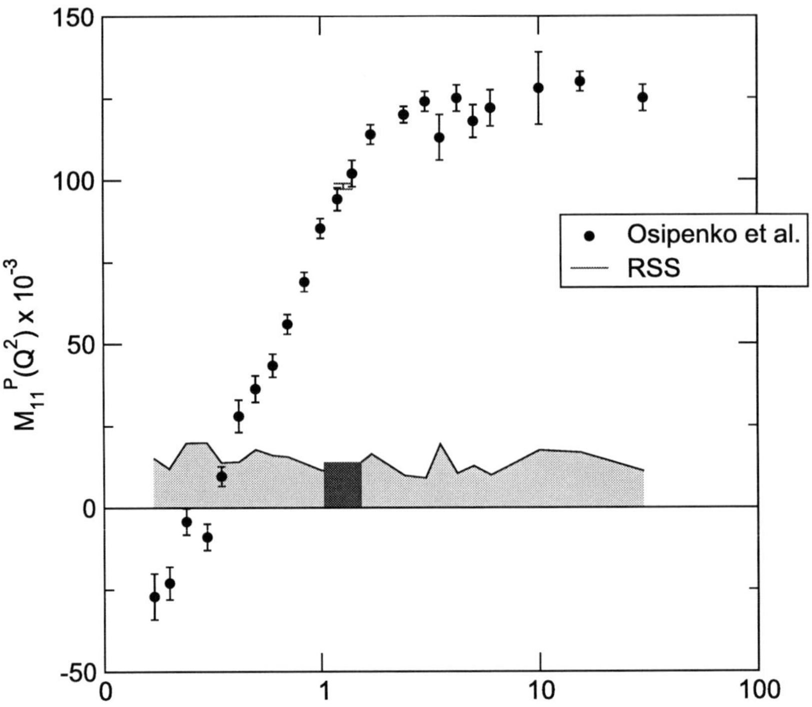

FIGURE 5. $M_1^1(Q^2)$ from [10] and RSS [18, 19] (Preliminary). Statistical errors are shown on the data points, while systematic errors are represented by the full bands on the horizontal axis.

Eqs. 7 and 8 are often combined to extract the twist-3 matrix element via:

$$I(Q^2) = \int_0^1 dx\, x^2 (2g_1 + 3g_2) \tag{9}$$

$$= \tilde{d}_2(Q^2) + \mathcal{O}\left(\frac{M^2}{Q^2}\right) \tag{10}$$

However, note that $I(Q^2)$ is not equivalent to the twist-3 matrix element $\tilde{d}_2$ in any kinematic region where M^2 is not negligible compared to Q^2 [32]. World data on $I(Q^2)$ is shown in Fig. 4. In the top panel, the filled band represents a determination of $I(Q^2)$ based on existing world data [10]. The large uncertainty reflects the lack of knowledge of g_2^p in this region.

TARGET MASS CORRECTIONS

As emphasized in [32], dynamical higher twists can be extracted from the measured SSFs by using the Nachtmann moments [33, 34], which utilize the variable $\xi = 2x/(1+$

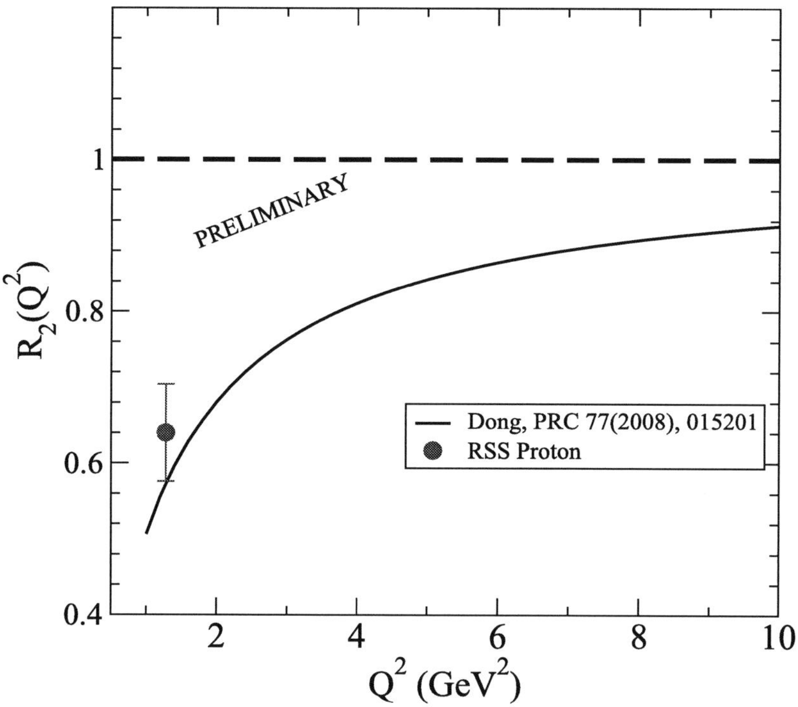

FIGURE 6. RSS (very preliminary) R_2 data compared to the prediction of Ref. [32]

$\sqrt{1+(2xM)^2/Q^2})$. The moments are defined:

$$M_1^n(Q^2) \equiv \frac{\tilde{a}_{n-1}}{2} \equiv \frac{1}{2}a_n E_1^n(Q^2,g) = \int_0^1 \frac{dx}{x^2}\xi^{n+1} \left[\left\{\frac{x}{\xi} - \frac{n^2}{(n+2)^2}\frac{M^2x\xi}{Q^2}\right\}g_1 - \frac{4n}{n+2}\frac{M^2x^2}{Q^2}g_2\right] \quad (11)$$

$$M_2^n(Q^2) \equiv \frac{\tilde{d}_{n-1}}{2} \equiv \frac{1}{2}d_n E_2^n(Q^2,g) = \int_0^1 \frac{dx}{x^2}\xi^{n+1} \left[\frac{x}{\xi}g_1 + \left\{\frac{n}{n-1}\frac{x^2}{\xi^2} - \frac{n}{n+1}\frac{M^2x^2}{Q^2}\right\}g_2\right] \quad (12)$$

In the limit $M^2/Q^2 \to 0$, (i.e. TMC vanish) the Nachtmann moments reduce to the more familiar CN moments:

$$M_1^1(Q^2) \rightarrow \Gamma_1(Q^2) \quad (13)$$

$$2M_2^3(Q^2) \rightarrow I(Q^2) \quad (14)$$

At finite Q^2, the ratio of Nachtmann to CN moments gives a quantitative measure of the TMC:

$$R_1(Q^2) = \frac{M_1^1(Q^2)}{\Gamma_1(Q^2)} \tag{15}$$

$$R_2(Q^2) = \frac{2M_2^3(Q^2)}{I(Q^2)} \tag{16}$$

The Nachtmann moments allow a clean separation of dynamical higher twists. In particular, Ref. [32] shows that M_2^3 gives access to the twist-3 matrix element that is accurate to $\mathscr{O}(M^8/Q^8)$.

A comparison of RSS data for M_1^1 to the analysis of Ref. [10] is shown in Fig. 5. Ref. [32] predicts that experimental determinations of $I(Q^2)$ significantly overestimate contributions to the twist-3 matrix element for $Q^2 < 10$ GeV^2. Fig. 6 provides support for this conclusion from preliminary RSS proton data.

CONCLUSION

We have presented an overview of existing tests of the Burkhardt–Cottingham sum rule. All neutron and ^{3}He data support the sum rule prediction within experimental uncertainties, although the agreement is only at the 2.5σ level for the neutron. Proton data is found to be lacking. We have also presented a Nachtmann moment analysis of existing proton g_2 data, which shows that CN moments can significantly overestimate determinations of the twist-3 matrix element.

ACKNOWLEDGMENTS

I would like to thank the spokesmen of RSS: Mark Jones and Oscar Rondon, E97110, Alexandre Deur, J.P. Chen and Franco Garibaldi, and E01012: Nilanga Liyanage, J.P Chen and Seonho Choi for kindly allowing me to present their preliminary data here. Many thanks also to Vince Sulkosky and Patricia Solvignon for providing the raw data. I would also like to thank Oscar Rondon for many instructive discussions on the topic material.

REFERENCES

1. J. M. Cornwall, and R. E. Norton, *Phys. Rev.* **177**, 2584–2586 (1969).
2. W. Melnitchouk, R. Ent, and C. Keppel, *Phys. Rept.* **406**, 127–301 (2005), hep-ph/0501217.
3. P. L. Anthony, et al., *Phys. Lett.* **B493**, 19–28 (2000), hep-ph/0007248.
4. P. L. Anthony, et al., *Phys. Lett.* **B553**, 18–24 (2003).
5. A. Airapetian, et al., *Phys. Rev.* **D75**, 012007 (2007), hep-ex/0609039.
6. M. Amarian, et al., *Phys. Rev. Lett.* **89**, 242301 (2002).
7. M. Amarian, et al., *Phys. Rev. Lett.* **93**, 152301 (2004).
8. Z. E. Meziani, et al., *Phys. Lett.* **B613**, 148–153 (2005), hep-ph/0404066.
9. M. Osipenko, et al., *Phys. Lett.* **B609**, 259–264 (2005), hep-ph/0404195.

10. M. Osipenko, et al., *Phys. Rev.* **D71**, 054007 (2005), hep-ph/0503018.
11. K. Slifer, et al., *Phys. Rev. Lett.* **101**, 022303 (2008), 0803.2267.
12. Y. Prok, et al. (2008), 0802.2232.
13. P. Solvignon, et al., *Phys. Rev. Lett.* **101**, 182502 (2008), 0803.3845.
14. H. Burkhardt, and W. N. Cottingham, *Annals Phys.* **56**, 453–463 (1970).
15. R. L. Jaffe, *Comments Nucl. Part. Phys.* **19**, 239 (1990).
16. A. Camsonne, J.P. Chen and K. Slifer, JLab Experiment E08-027. (????).
17. S. Choi, M. Jones, Z.-E. Meziani and O.A. Rondon, JLab Experiment E07-003. (????).
18. K. Slifer, et al. (2008), 0812.0031.
19. F. R. Wesselmann, et al., *Phys. Rev. Lett.* **98**, 132003 (2007), nucl-ex/0608003.
20. M. Amarian, et al., *Phys. Rev. Lett.* **92**, 022301 (2004), hep-ex/0310003.
21. A. Deur, J.P. Chen and F. Garibaldi, JLab Experiment E97-110. (????).
22. E. L. Lomon, *Phys. Rev.* **C66**, 045501 (2002), nucl-th/0203081.
23. P. Mergell, U. G. Meissner, and D. Drechsel, *Nucl. Phys.* **A596**, 367–396 (1996), hep-ph/9506375.
24. A. Amroun, et al., *Nucl. Phys.* **A579**, 596–626 (1994).
25. S. Wandzura, and F. Wilczek, *Phys. Lett.* **B72**, 195 (1977).
26. K. G. Wilson, *Phys. Rev.* **179**, 1499–1512 (1969).
27. J. Kodaira, Y. Yasui, and T. Uematsu, *Phys. Lett.* **B344**, 348–354 (1995), hep-ph/9408354.
28. B. Ehrnsperger, A. Schafer, and L. Mankiewicz, *Phys. Lett.* **B323**, 439–445 (1994), hep-ph/9311285.
29. M. Gockeler, et al., *Phys. Rev.* **D63**, 074506 (2001), hep-lat/0011091.
30. X. Zheng, et al., *Phys. Rev.* **C70**, 065207 (2004), nucl-ex/0405006.
31. D. Drechsel, O. Hanstein, S. S. Kamalov, and L. Tiator, *Nucl. Phys.* **A645**, 145–174 (1999), nucl-th/9807001.
32. Y. B. Dong, *Phys. Rev.* **C77**, 015201 (2008).
33. S. Matsuda, and T. Uematsu, *Nucl. Phys.* **B168**, 181 (1980).
34. A. Piccione, and G. Ridolfi, *Nucl. Phys.* **B513**, 301–316 (1998), hep-ph/9707478.

Gluon polarization in the nucleon

Ewa Rondio

A.Soltan Institute for Nuclear Studies, Poland
ewar@fuw.edu.pl

Abstract. Gluons can contribute to the nucleon spin and may provide the part missing when only quark spins are taken into account. The gluon helicity distribution can be obtained from QCD fits or from direct measurements in experiments selecting processes sensitive to gluons. Recently several important measurements were completed and others are in progress in fixed target lepton nucleon scattering as well as in collider experiments with polarized protons. Asymmetries are obtained in processes with significant contributions from gluons such as in lepton nucleon scattering the production of D mesons and production of hadrons with high transverse momentum, and in polarized pp scattering the production of jets , π^0 mesons and direct photons. These results are interpreted in terms of universal gluon helicity distribution, using assumptions and approximations specific for each analysis. The complementarity of the approaches is very important in controling systematic effects and in establishing the present conclusion and interpretation that the data show no signature of very high gluon polarization in the kinematic region of observations.

Keywords: spin, parton distributions, QCD fits, spin asymmetries, gluon polarization
PACS: 13.88.+2,14.70.Dj,13.60.-r,13.85.-t

Introduction

The understanding of contributions to spin of the nucleon has a long history. The spin structure of the nucleon has been studied in polarized **D**eep **I**nelastic lepton-nucleon **S**cattering **(DIS)** for quite a long time. The decomposition of the nucleon spin among quark spin, gluon spin and orbital angular momentum can be written as $\frac{1}{2} = \frac{1}{2}\Delta\Sigma + \Delta G + L_z$.

The experimental observation by the EMC [1], that a surprisingly small fraction of the nucleon spin is carried by quarks has had a major influence on subsequent spin physics programs. Many experiments were performed to confirm this result and check it on different targets. Other measurements are still in progress. Several theoretical ideas were proposed [2] to explain the EMC observation. They were based either on large negative sea quark polarization or on contributions from polarized gluons and orbital angular momenta of quarks and gluons. To check which of them is responsible for making up the nucleon spin one should determine a fraction of nucleon spin carried by gluons. Information about this quantity can be obtained from QCD fits to inclusive DIS data or from direct measurements of observables with direct sensitivity to gluons. The fits require assumptions on the shape at a selected fixed value of the initial scale Q_0^2 for each parton distribution that is to be extracted. The limited Q^2 range covered by the data present a futher complication in these analyses. Measurements in which the gluon is probed directly are thus essential for verification of the spin structure of the nucleon. The present generation of experiments, aiming for better precision on $\Delta G/G$ through

CP1149, *Spin Physics, 18th International Spin Physics Symposium*
edited by D. G. Crabb, D. B. Day, S. Liuti, X. Zheng, M. Poelker, and Y. Prok

processes directly sensitive to gluons, taking or analyzing data are: Hermas at DESY, Compass at CERN, Phenix and Star at BNL.

Gluon distributions from inclusive data

Gluons, unlike quarks, do not couple directly to virtual photon.

QCD fits to all inclusive measurements give information on the spin dependent parton distributions. The polarized quark distributions are determined, with best precision. However, information about the polarized gluon is also obtained from these fits since gluons contribute to the Q^2 evolution of spin dependent structure function g_1. Even though different mathematical approaches are used, the concept is to parameterize the parton distributions at an initial scale Q_0^2, to use DGLAP evolution equations to evaluate expected values of the g_1 structure function for (x,Q^2) of the measurements, and to adjust the parameters of the initial distributions until optimal agreement is obtained with the measured values of g_1.

From the time of the first measurements of the g_1 structure function several analyses of this kind were performed by experimental and theoretical groups. First attempts were done in the LO approximation for the evolution equations, but now all analyses use the NLO approximation. Data from all experiments on protons, deuterons and ^{3}He targets have been used. The history of such analyses started in the early 1990's. One of the first NLO analyses was performed by Ball, Forte and Ridolfi in 1994 [3] and subsequently by the SMC and E155 experimental groups [4]. The result of gluon polarization at Q^2=4 GeV^2 from some of the fits using the second generation of experiments are shown on the left panel in Fig.1. The flexibility of the assumed functional forms and number of parton distributions which can be fitted depends on the precision of available data. Very important for gluon determination is the lever arm in Q^2 covered by measurements. In early fits only 3 distributions were fitted (singlet and non-singlet quark and gluons) and

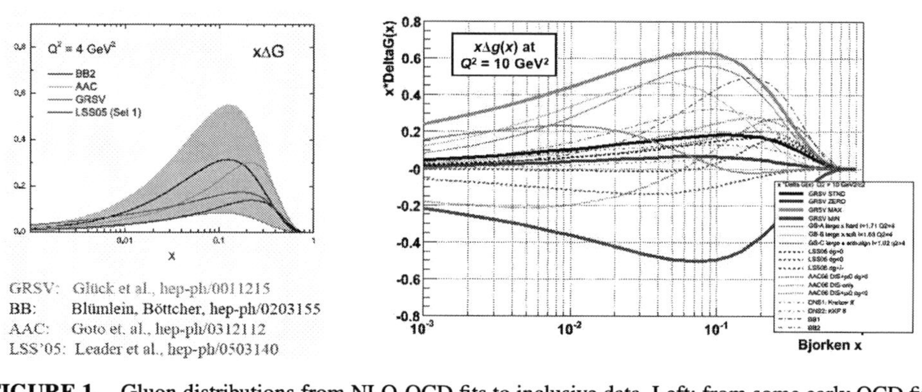

FIGURE 1. Gluon distributions from NLO-QCD fits to inclusive data. Left: from some early QCD fits to second generation of experiments measuring g_1 of p, d and ^3He. Right: compilation of most of presently existing fit results (from C.Gagliardi's talk at DIS2008).

number of parameters have to be restricted. With improving precision of measurements more flexibility in the fits was possible, but also it become clear that simple assumptions on functional forms do not allow good description of the better data. Very important constraint is provided by the precise measurements of g_1 at high x performed in J-Lab.

Present situation with fits to inclusive data

Reacently many other groups have done such analyses, incorporating more and more precise data. The new fits result in a much better determination of the quark distributions and allow even the separate extraction of distributions by flavor, but for the polarized gluon distribution the situation is not improving. Results on the polarized gluon distribution from a collection of such fits is shown in the right panel of Fig.1. The gluon polarization is still very weakly constrained by the data. Better precision of the data has allowed for more flexible functional forms for the gluon polarization as a function of x and it brought a wide spread of acceptable solutions including positive and negative distributions as well as distributions that change sign. Most recent analysis obtain a central value of the first moment of $|\Delta G|$ of the order of 0.2-0.3 at Q^2 around 3 GeV^2. The positive and negative solutions have similar χ^2, so there is no strong preference for the sign. Values of the first moment close to zero can also be obtained from the function changing sign. In present fits it is happening arond x=0.1 and $\Delta g(x, Q_0^2)$ is negative at small x. The differences in the polarized gluon distributions obtained by several groups originate in the assumptions made for the functional forms, in additional inputs, such as for example possible higher twist contributions, as well as in the inclusion of additional data sets. Many efforts are presently being undertaken to improve the situation.

Significant improvement of the gluon polarization determination from QCD fits to inclusive data, however, can be obtained only by enlarging the lever arm in Q^2 for QCD fit. To do this one would need data from a lepton-nucleon collider with polarized beams. Proposals for such machines exist [5], but the time scale is quite long. Therefore to improve our knowledge on gluon polarization we have to perform dedicated measurements more directly sensitive to this quantity.

Presently, efforts are made to get estimates on the gluon polarization from two kind of interactions: lepton-hadron scattering and polarized proton-proton collisions. In both cases one needs to select processes with direct contributions from hard scattering processes involving at least one gluon.

Direct measurements of the gluon polarization in DIS experiments

A process that involves gluons, required to measure the gluon polarization, is the so-called photon-gluon fusion (PGF). It occurs in a deep inelastic lepton-nucleon scattering together with other processes that are consider to be background. Diagrams of the PGF and of order processes contributing to the same and lower order are shown in Fig.2.

Two approaches are used to select PGF and reject background:

- the detection of at least one charmed hadron which can be produced in the PGF

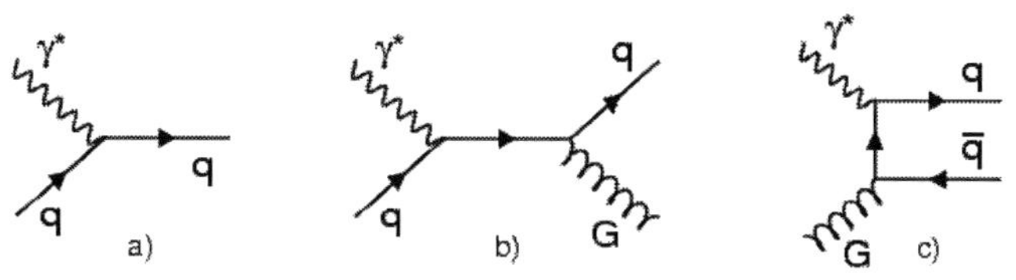

FIGURE 2. Diagrams of the lowest order processes contributing to DIS. The PGF process is shown in diagram c), the a) and b) show background processes for gluon polarization measurement.

process but not in the background processes in view of the exceedingly small intrinsic charm content of the nucleon. While this selection is "clean", it significantly reduces the data sample available for analysis in view of the smaller production cross section and the detection efficiency for charmed hadrons.

- the detection of pairs of hadrons (or jets) with high transverse momentum (p_T). This selects PGF pair production and reduces background from Fig.2a) since p_T in that process originates from the relatively small transverse quark momenta in the nucleon.

ΔG from charm production

This selection requires sufficiently high energy in the center of mass system, such that two heavy c quarks can be produced. It was used in the COMPASS experiment at CERN looking at the scattering of 160 GeV muons on polarized deuterons in a solid LiD target. The PGF process was selected by the detection of the D^0 charm meson via its decay to charged K and π, which has branching ratio of 3.8%. The vertex resolution is not sufficient to separate the D^0 production and decay vertices. Therefore the combinatorial background has to be reduced in analysis. Most important is particle identiÞcation of the decay kaon provided by RICH detector. In case of D^* meson an additional pion is produced in the decay and this provides a strong constraint on the difference of the invariant masses of $K\pi\pi$ and $K\pi$. Therefore, the analysis was pursued separately for the sample tagged as D^* and the sample containing other D^0 mesons. A simultaneous extraction of $\Delta G/G$ and the background asymmetry is performed using data from 2 or 3 target cells, which contain deuterons with alternating polarization directions. The events are weighted with their analyzing power to make best use of the available information. Based on a leading order QCD analysis an average gluon polarization

$$\frac{\Delta G}{G} = -049 \pm 0.27(stat) \pm 0.11(syst) \tag{1}$$

was obtained at a scale μ^2 of about 13 GeV^2 and an average gluon momentum fraction of 0.11 (see talk by F. Kunne). This result, still statistically limited, is shown in Fig.3. At the same time asymmetries of polarized to unpolarized cross sections for D^0 production were extracted in bins of D^0 transverse momentum and energy fraction (z) for possible use in the global QCD fits.

ΔG *from events with high* p_T *hadrons*

The selection of events with high p_T hadrons enhances the PGF fraction in the event sample. To extract $\Delta G/G$ one needs to know the contributions from other processes and this makes the analysis model-dependent. This kind of analyses were performed already before by the Hermes and SMC experiments and recently more precise results were presented by HERMES and COMPASS. Two models, based on Monte Carlo generators, were used depending on the kinematic region covered by the data: Pythia was used in the analyses by HERMES [7] and COMPASS [6] at low Q2, whereas for Q2 > 1 GeV2 Lepto was used in the COMPASS analysis (see talk by K.Kurek).

HERMES used several data samples, selecting single hadrons and hadron pairs, whereas COMPASS always used samples with 2 hadrons having p_T above a threshold. Fig.3 (left) summarizes the results, including the previous results and recent charm result. Fig.3 (right) shows the result of an alternative analysis by HERMES [7]. Here, events with high p_T hadrons were used to calculate the asymmetry and then several gluon polarization distributions were used in a MC program to calculate the same asymmetry. The two fitting best are shown with the systematic uncertainty. The difference between the curves can be used as an alternative estimate of the systematic uncertainty. The shaded area shows the x region covered by the data.

Measurements of the gluon polarization from the RHIC pp collider

Polarized proton collisions at RHIC allow the study of spin effects in the proton at very high center of mass energies and therefore in the region where perturbative QCD is applicable. The measured cross sections are well described in the NLO QCD framework over many orders of magnitude. This is illustrated, for selected processes with sensitivity

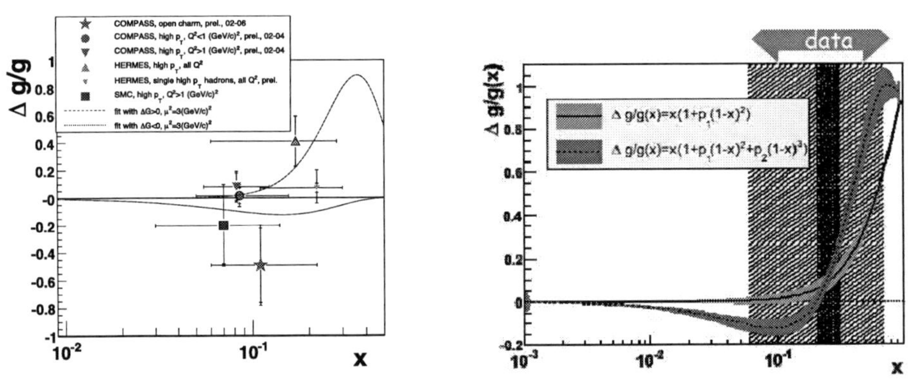

FIGURE 3. On the left - the average value of gluon polarization at the average gluon momentum fraction from analyses of samples with high p_T hadrons and the sample with charmed D_0 meson are shown. The right panel shows gluon polarization distributions which best reproduce the HERMES asymmetries for the sample of untagged high p_T hadrons.

to gluons, in Fig.4 where measurements by PHENIX [9] and STAR [10] experiments are compared with NLO calculations. The agreement is at the level of 20%, justifying the interpretation of the data within the NLO framework. Some efforts have been made to provide an alternative description [8] and the agreement seems to be at a similar level, but with a reduced scale uncertainty.

How to access gluons

In the proton proton interactions one has to consider the scattering of the two composite objects. The cross section therefore receives contributions from gluon-gluon, quark-gluon, and quark-quark scattering. The asymmetry is given by:

$$A_{LL} \approx \int dx_1 dx_2 a_{LL}^{gg} \frac{\Delta g}{g}\frac{\Delta g}{g} + a_{LL}^{qg}\frac{\Delta g}{g}\frac{\Delta q}{q} + a_{LL}^{qq}\frac{\Delta q}{q}\frac{\Delta q}{q} \qquad (2)$$

where the asymmetris a_{LL}'s are at the partonic level and are calculable in perturbative QCD.

From asymmetry to gluon polarization

The asymmetry of cross sections with different relative spin orientation of the colliding protons is the observable. To gain sensitivity to the polarized gluon distribution, reactions with significant gluon contributions to the cross section should be selected and the measurement should be done in the kinematic regions where the partonic asymmetry

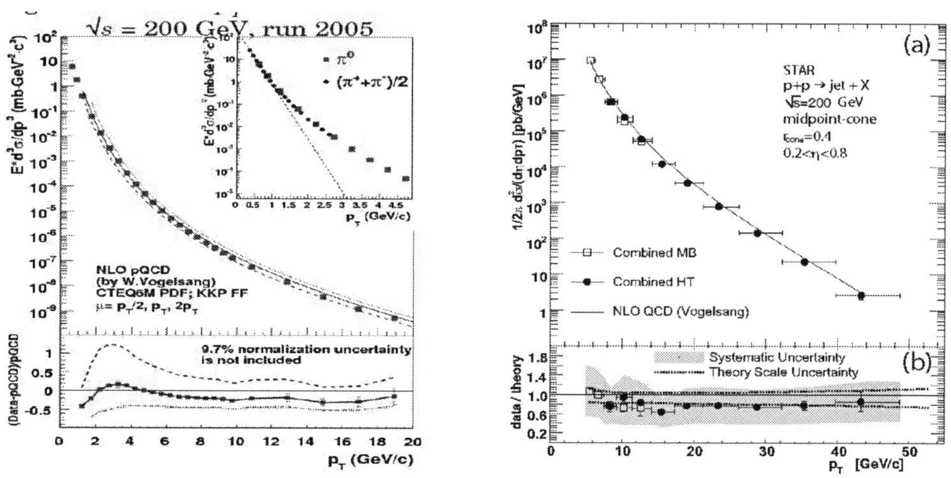

FIGURE 4. Cross section measurements compared with expectations from NLO-QCD: left - inclusive pion production as measured by PHENIX [9]; right - inclusive jet production as measured by STAR [10].

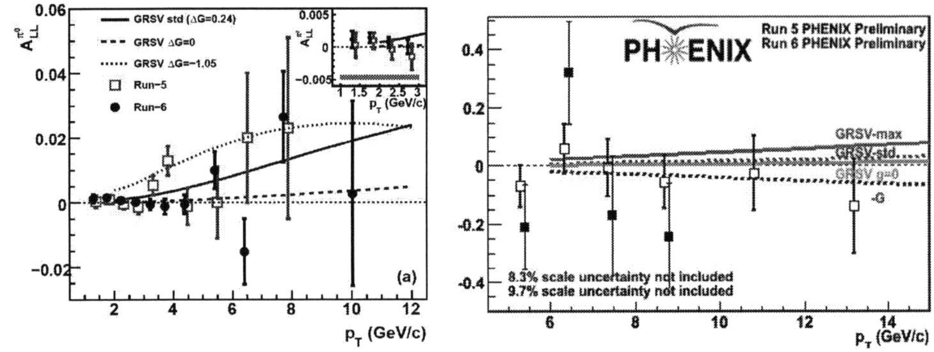

FIGURE 5. Double spin asymmetries: left: inclusive neutral pions (K. Boyle at this conf.); right: for events with direct photon production (R. Bennett at this conf.), measured by PHENIX

is not zero. If the selected process uses final state hadrons, such as charged or neutral pions, one needs to use externally measured and non-perturbative fragmentation functions, unlike in the case of jets. The observation of di-jet of photon-jet events makes it possible, in leading order, to determine the average partonic kinematics. Direct photon-jet production is a particularly attractive process, but it has a small cross section and very high demands on background rejection from π^0 production. The asymmetries for the observed processes are calculated in NLO perturbative QCD using different gluon polarization distributions and those resulting in best agreement with the experimental results are searched for. In this way several parametrizations can be eliminated.

Measurements are in progress for many processes. First results were obtained for inclusive pion production by PHENIX. Recent asymmetries for inclusive π^0 production are presented and compared with predictions on left part of Fig.5 (see talk by K. Boyle). Two sets of points show the first results and most recent data with significantly improved precision. The fractions of contributing hard processes as well as the momentum fraction carried by the gluons (x) in the hard scattering varies with p_T, therefore the asymmetries

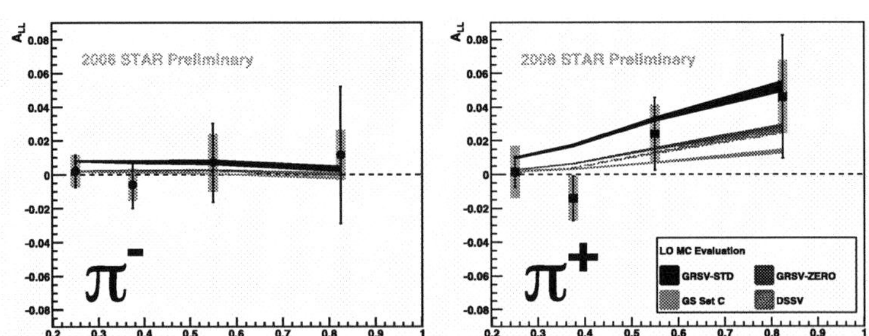

FIGURE 6. Preliminary results from STAR (A. Kocoloski at this conf.) on the asymmetries for charged pions in the hemisphere opposite to the triggering jet. The lines correspond to LO QCD predictions.

are extracted as a function of p_T. The right part of Fig.5 shows first preliminary result on direct photon production from PHENIX (see talk by R.Bennett).

An alternative approach is to study less inclusive processes, for example the production of identified particles within jets or the production of di-jets or di-hadrons. Fig.6 shows preliminary STAR results for negatively and positively charged pions which are observed opposite to the jet that triggered the read-out, so as to reduce biases caused by the trigger. The data are expressed as a function of the momentum ratio of the pion to the trigger jet, which is an approximation of fragmentation-z, and are shown together with LO predictions obtained from MC (see talk by A. Kocoloski). NLO predictions are not yet available. The π^+ asymmetries show significant sensitivity for high values of z, as expected.

The asymmetries for inclusive jet production from STAR [11] are shown in Fig.7, together with the corresponding confidence levels for different sets of polarized parton distributions. There is a group of parameterizations showing quite high confidence levels. Even more precise data have been shown at the conference (see talk of M. Sarsour). There are much more measurements from both experiments for selected processes, spe-

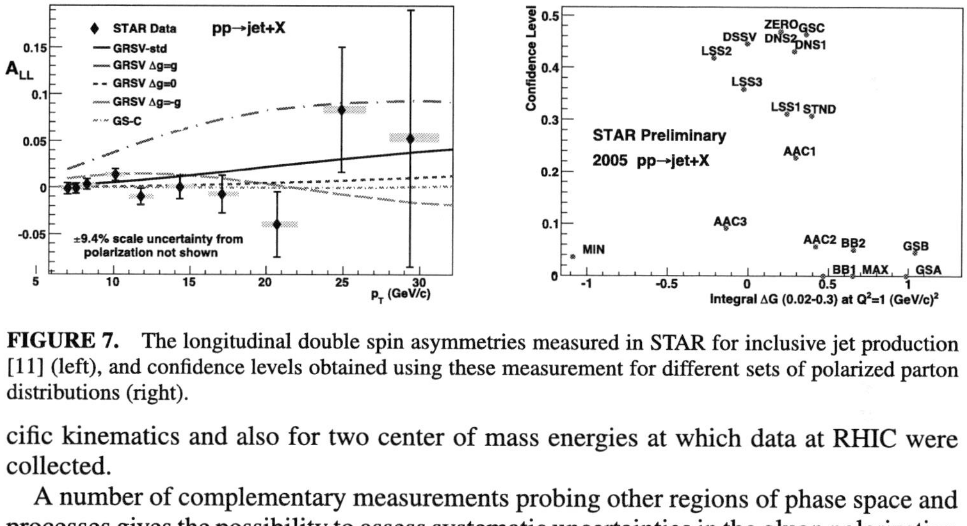

FIGURE 7. The longitudinal double spin asymmetries measured in STAR for inclusive jet production [11] (left), and confidence levels obtained using these measurement for different sets of polarized parton distributions (right).

cific kinematics and also for two center of mass energies at which data at RHIC were collected.

A number of complementary measurements probing other regions of phase space and processes gives the possibility to assess systematic uncertainties in the gluon polarization determination. Therefore it is very important to make such measurements and use them for testing the consistency of the obtained results. With increasing statistics, at RHIC, the experiments are approaching a phase when sensitive tests will become possible.

Fitting with all available data

The gluon polarization determination presented in the previous section is classifying rather than quantifying. Even with χ^2 and confidence levels for each set of parton distributions, the new data are used only passively. They do not influence the parton

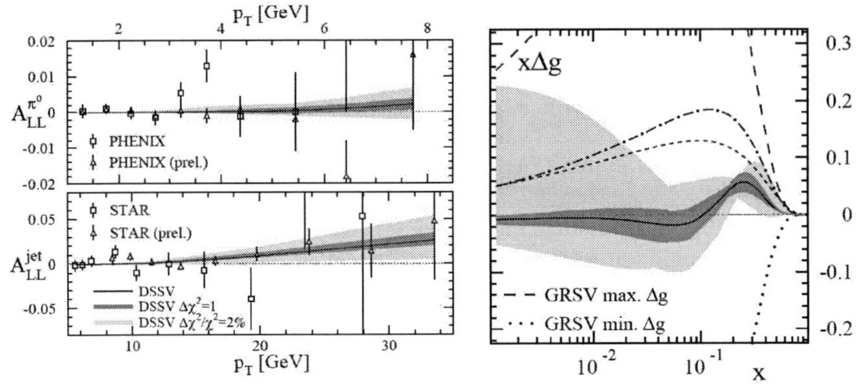

FIGURE 8. The asymmetries measured at RHIC that are used in the FSSV fit are shown on the left panel. On the right, the obtained gloun polarization distribution is shown. Note reduction of uncertainty obtained in the x region where pp data contribute. Both figures are from [12].

(mainly gluon) distribution functions. A better solution is using these new data in the fit on an equal footing with inclusive and also semi-inclusive DIS data. However, inclusion of the semi inclusive or exclusive processes requires additional information about fragmentation functions, bringing new source of systematic uncertainties.

This is by far easier saied than done, but recently this important step has been made [12]. In the simultaneous fit, all inclusie and most semi-inclusive DIS data are used together with neutral pion and jet asymmetries measured at RHIC. The evolution was done in Mellin space, to speed up calculations, parts of the expression were pre-computed on a grid in complex space and for specific observables sampling techniques were applied. The parton distributions were described by functions using 21 parameters. The total number of data points in the fit was 467 and about 10% originate from RHIC (shown in Fig.8 left). The quality of fit is very good giving $\chi^2/ndf = 0.86$, when statistical and systematic errors are used in a quadratic sum (see M. Stratmann talk). The central value for the first moment ΔG from this fit is -0.084 at a scale $Q^2 = 10$ GeV^2, however, the distribution is largely unconstrained for x below 0.001(Fig.8 right). Therefore, the authors have evaluated the contribution and uncertainty at the same scale over the restricted x-range from 0.001 to 1 and obtain a central value of 0.013 with asymmetric uncertainties smaller than 0.15 when χ^2 is allowed to vary by one unit. The authors comment that this is likely an underestimate of the true uncertainty and provide an alternative estimate using a relative variation in χ^2 of 2%. The conclusion that can be made from this fit, is that the gluon spin contribution is most probably small. A value close to zero, most likely in the range of ± 0.2 at $Q^2 = 1$ GeV^2, would seem reasonable.

Summary and outlook.

Despite considerable progress in understanding gluon polarization, one has to say that there is still a long way to go. Two ways of approaching ΔG are possible: one is the interpretation of data from individual experiments in terms of ΔG and the other is bringing more inputs to the global fits and providing theoretical background for interpretation. Both approaches are needed and should be viewed as complementary, rather than as competing. For example, the history of measurements of the strong coupling constant α_S, has repeatedly demonstrated the importance of cross checks. Comparison of individual results can bring best control of contributions to systematic errors. Measurements which can not be included in global fits, because there is insufficient theoretical insight to do it (yet), should nevertheless be encouraged and by comparing different such approaches one can get information about the importance of simplifications that are necessarily made in any of the analyses.

Quark spins contribute about a third of the proton spin and the gluon spin contribution is rather small. Nevertheless, several scenarios are still possible: gluon spins contribute 0.2–0.3, thus leaving little room for orbital momenta, or gluons are unpolarized and the still missing contribution must come from orbital angular momenta. If one takes suggestions from lattice calculations, that the quark contribution to the angular momentum is small, at face value, then gluon orbital angular momentum would be significant. The values of ΔG, as presently observed, show that the anomalous contribution to the axial charge a_0 is small and can not explain the "spin crisis". Spin measurements have a bright future, they are likely also to continue to bring surprises. Exciting progress has been made and the next years may well be equally exciting

Acknowledgments

The author would like to thank organizers for a very nice conference, stimulating a lot of interesting and useful discussion. Special thanks to A. Deshpande, K. Kurek and E. Sichtermann and many speakers from parallel sessions for discussions and help in material selection. The work was supported by Polish SPB 41/N-CERN/2007/0

REFERENCES

1. EMC Collaboration, J. Ashman *et al.,* Nucl. Phys. B **328** (1989) 1.
2. S.D Bass and A.W. Thomas, Prog. Part. Nucl. Phys **33** (1994) 449.
3. R. Ball, S. Forte, G. Ridolfi, Phys. Rev. Lett. B**378** (1996) 255.
4. SMC,B.Adeva *et. al.* Phys.Lett. B**412** (1997)414; E155,P.L.Anthony *et al.* Phys.Lett.B**493** (2000)19.
5. A. Deshpande, AIP Conf.Proc.792:1046-1051,2005 and Nucl.Phys.Proc.Suppl.105:178-184,2002.
6. COMPASS Collab., E. Ageev *et. al.,* Phys.Lett.B633:25-32,2006.
7. HERMES Collab., A. Airapetian *et al.*, JHEP (2008) 0806:017.
8. D. De Florian, W. Vogelsang and F. Wagner, Phys. Rev. **D76** (2007) 094021.
9. PHENIX Collab. A. Adare *et. al.,* Phys. Rev. **D76** (2007) 051106.
10. STAR Collab. B. Abelev *et. al.,* ,PRL **97** (2006) 252001.
11. M. Sarsour for STAR colab. at this conference and Phys. Rev. Lett 100(2008) 232003.
12. D. de Florian., R. Sassot, M. Stratmann, W. Vogelsang (2008) Phys. Rev. Lett. 101 (2008) 072001.

Generalized Parton Distributions: Status and Perspectives

C. Weiss

Theory Center, Jefferson Lab, Newport News, VA 23606, USA

Abstract. We summarize recent developments in understanding the concept of generalized parton distributions (GPDs), its relation to nucleon structure, and its application to high–Q^2 electroproduction processes. Following a brief review of QCD factorization and transverse nucleon structure, we discuss (a) new theoretical methods for the analysis of deeply–virtual Compton scattering (t–channel–based GPD parametrizations, dispersion relations); (b) the phenomenology of hard exclusive meson production (experimental tests of dominance of small–size configurations, model–independent comparative studies); (c) the role of GPDs in small–x physics and pp scattering (QCD dipole model, central exclusive diffraction). We emphasize the usefulness of the transverse spatial (or impact parameter) representation for both understanding the reaction mechanism in hard exclusive processes and visualizing the physical content of the GPDs.

INTRODUCTION

Generalized parton distributions (GPDs) have established themselves as a key concept in the study of hadron structure in QCD. On the experimental side, they enable a unified description of several high–Q^2 electroproduction processes (inclusive, exclusive) on the basis of QCD factorization. On the theoretical side, they provide us with information about the transverse spatial distribution of quarks/gluons and have given rise to new ideas of "quark/gluon imaging" of the nucleon. Studies of GPDs are the subject of completed and on-going experiments (HERA, HERMES, JLab 6 GeV) and are an essential part of the physics program of future facilities for ep/eA scattering (JLab 12 GeV, EIC). The purpose of this article is to summarize the status and future perspectives of this field, with emphasis on the interplay between experimental data and theoretical/phenomenological concepts; for more detailed reviews, see Ref. [1].

QCD FACTORIZATION IN HIGH–Q^2 ELECTROPRODUCTION

The basis for a partonic description of hadron structure is the idea of factorization in high–Q^2 electroproduction processes. At $Q^2 \gg R_{\text{had}}^{-2}$ the virtual photon has a transverse resolution much smaller than the typical hadronic size, and the scattering process takes place with a quasi–free parton in the nucleon. This is well–known in inclusive DIS (see Fig. 1a), where the cross section can be factorized into that of the parton–level process, involving transverse distances $\sim 1/Q$, and the distributions of partons in the nucleon (PDFs), governed by long–distance dynamics ($\sim R_{\text{had}}$). The latter are formally defined

CP1149, *Spin Physics, 18th International Spin Physics Symposium*
edited by D. G. Crabb, D. B. Day, S. Liuti, X. Zheng, M. Poelker, and Y. Prok

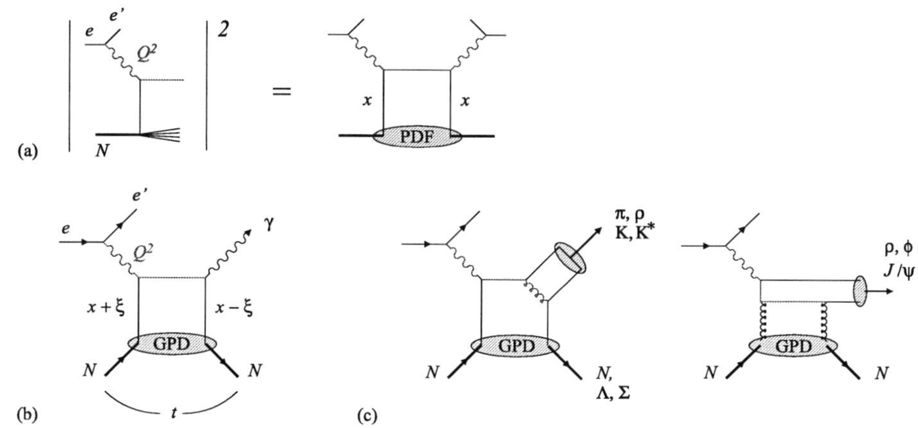

FIGURE 1. Factorization in high–Q^2 electroproduction. (a) Inclusive production $N(e,e')X$. (b, c) Exclusive production $N(e,e'M)N$ ($M = \gamma$, meson). The parameter ξ is related to the Bjorken scaling variable, $\xi = x_B/(2-x_B)$, and kinematically fixed (longitudinal momentum transfer to target).

as the expectation values of a certain quark bilinear operators in the nucleon state, $\langle N|\bar{\psi}(0)\ldots\psi(z)|N\rangle_{z^2=0}$, and can be interpreted in space–time as the densities of quarks with longitudinal momentum xP in a fast–moving nucleon (momentum $P \gg R_{\text{had}}^{-1}$).

The same reasoning applies to certain exclusive production processes at high Q^2, specifically deeply–virtual Compton scattering (DVCS) and exclusive meson production (see Fig. 1b and c), where the final–state photon (meson) is produced in the reaction with a single parton in the nucleon. The amplitudes can be factorized into the quark–level production amplitude and the GPDs, describing the coupling of the active parton to the nucleon. They depend on the momentum fractions of the initial and final parton, $x \pm \xi$ (-2ξ is the fractional longitudinal momentum transfer to the nucleon), as well as the transverse momentum transfer to the nucleon, $\mathbf{\Delta}_\perp$, or, alternatively, the invariant momentum transfer, t. The GPDs are formally defined as the transition matrix elements $\langle N'|\ldots|N\rangle$ of the quark (or gluon) bilinear operators and combine aspects of the parton densities with those of the nucleon elastic form factors; in fact, they contain both as limiting cases. QCD factorization guarantees that they are universal, process–independent characteristics of the nucleon, which can be probed using different partonic scattering processes (*e.g.* different meson production channels); this fact is essential not only for experimental studies of GPDs, but also for our ability to calculate them using non-perturbative QCD methods such as lattice simulations. (For a discussion of QCD radiative corrections and the scale dependence of GPDs, see Ref. [1].)

QCD factorization is formally justified is the $Q^2 \to \infty$ limit and assumes that the production process takes place predominantly over transverse distances $\sim 1/Q$. In processes at $Q^2 \sim$ few GeV^2 there can be significant contributions from finite–size configurations, leading to power ($1/Q^2$, higher twist) corrections to the amplitude, in particular in meson production (see below). Combining small–size and finite–size contributions in a consistent approach is an important problem for phenomenological studies.

GPDS AND NUCLEON STRUCTURE

The concept of GPDs has proved to be a most useful tool in the general effort to describe nucleon structure in terms of QCD degrees of freedom. The most interesting aspect is that the GPDs describe the transverse spatial distribution of quarks and gluons in the nucleon. This is seen most easily in the special case of zero longitudinal momentum transfer, $\xi = 0$, where the GPD represents the "transverse form factor" of the QCD operator measuring the density of partons with longitudinal momentum fraction x. Its Fourier transform $\boldsymbol{\Delta}_\perp \to \boldsymbol{b}$ describes the distribution of these partons with respect to their transverse displacement, $\boldsymbol{b}$, from the center–of–momentum of the nucleon (impact parameter representation, see Fig. 2a) [2]. It corresponds to a set of tomographic parton images of the nucleon at fixed longitudinal momentum fraction x (Fig. 2b). This representation provides a natural framework for visualizing the partonic structure and discussing its relation to long–distance hadronic dynamics. At $x \sim 0.3$ on "sees" the valence quarks (and gluons), which are distributed over transverse distances $b \ll 1$ fm. At $x < M_\pi / M_N = 0.15$ the pion cloud makes a distinct contribution to the partonic structure, extending up to transverse distances $b \sim M_\pi^{-1}$ [3]. At even smaller x the dominant partons are radiatively generated singlet quarks and gluons; the area of their transverse distribution $\langle b^2 \rangle$ grows logarithmically with $1/x$ (effective Regge slope).

The transverse coordinate representation also lends itself to the discussion of polarization effects [4, 5]. There are 4 helicity components of the quark GPDs, corresponding to the operators measuring the sum/difference of quark helicities (*cf.* unpolarized/polarized PDFs) and their nucleon helicity non-flip/flip matrix elements (*cf.* Dirac/Pauli and axial/pseudoscalar form factors). The Pauli form factor–type GPD (E) has an interesting interpretation in the transverse nucleon spin basis; it describes the distortion of the longitudinal motion of the quarks in a transversely polarized nucleon (Fig. 2c) [4]. Likewise, for a longitudinally polarized nucleon, one can form combinations of the unpolarized and polarized GPDs corresponding to definite quark helicity states ($H \pm \tilde{H}$), and study the spatial distribution of quarks polarized along or opposite to the nucleon spin.

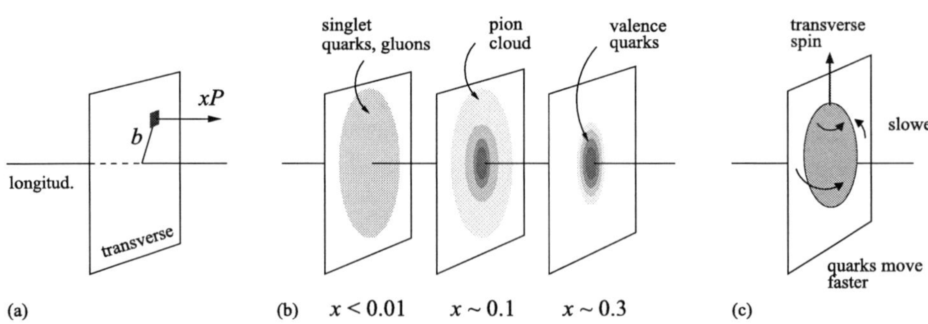

FIGURE 2. (a) Impact parameter representation of the GPDs ($\xi = 0$). (b) Tomographic parton images of the nucleon at fixed longitudinal momentum fraction x. (c) Distortion of the longitudinal motion of partons by transverse polarization of the nucleon.

Another reason for interest in the GPDs is that their moments are related to fundamental static properties of the nucleon. Integration over x with a weighting factor x^{n-1} projects out the spin–n component of the non–local quark bilinear operator. The spin–2 component coincides with the quark part of the QCD energy–momentum (EM) tensor, whose matrix elements cannot be measured directly with local electroweak probes. In particular, the EM tensor is related to the angular momentum operator; this is the basis of the Ji sum rule which expresses the total quark angular momentum in terms of the second moments of the unpolarized GPDs E and H [6, 7]. The components of the EM tensor also provide information about the QCD forces acting on a quark in the nucleon at rest (pressure, shear forces) [8]; these relations have recently been illustrated by a calculation within the chiral quark–soliton model of the nucleon [9].

Moments of the GPDs, as matrix elements of local operators, can also be calculated in lattice QCD [10, 11]. Present simulations are limited to $n \leq 4$, allowing for first conclusions about the correlation of x– and t–dependence in non-singlet GPDs (singlets require inclusion of disconnected diagrams). Accurate lattice results would have the potential to constrain future GPD parametrizations.

GPDS IN *EN* SCATTERING

Quantitative studies of GPDs draw from several sources of information. Important basic information comes from the parametrizations of PDFs and elastic nucleon form factors, which, respectively, constrain the GPDs in the zero momentum–transfer limit ($\xi = 0, t = 0$) and their first moments as functions of t. The "new" information about the correlation of the x, ξ and t–dependence, which *e.g.* determines the transverse profile of the nucleon and its change with x in Fig. 2b, comes from measurements of high–Q^2 exclusive processes, in particular DVCS and meson production (see Fig. 1b and c).

The analysis of high–Q^2 exclusive processes generally proceeds in two steps. One first investigates the reaction mechanism and tries to ascertain that a description based on short–distance dominance and QCD factorization is applicable; this is done by testing certain qualitative implications of the approach to the short–distance regime (Q^2–scaling, *etc.*) which do not depend on the specific form of the GPDs. One can then attempt to extract quantitative information about the GPDs by analyzing the kinematic dependences of the leading–twist (leading in $1/Q^2$) observables. The details of this program vary considerably between DVCS and meson production, because of the different complexity and reaction dynamics of the two processes.

Deeply–virtual Compton scattering. DVCS is generally regarded as the most promising channel for probing GPDs in the valence quark regime. In this process one expects short–distance dominance to set in already at $Q^2 \sim$ few GeV^2, based on the formal analogy of DVCS to inclusive DIS and the experience with other exclusive two–photon processes, such as $\gamma^*\gamma \to \pi^0$ studied in e^+e^- annihilation. In the measured $N(e, e'\gamma)N$ cross section the DVCS process interferes with the Bethe–Heitler (BH) process, whose amplitude is purely real and calculable in terms of the nucleon elastic form factors (see Fig. 3a); this circumstance amplifies the DVCS amplitude and allows one to measure it — and thus the GPDs — at the amplitude level, by isolating the interference term

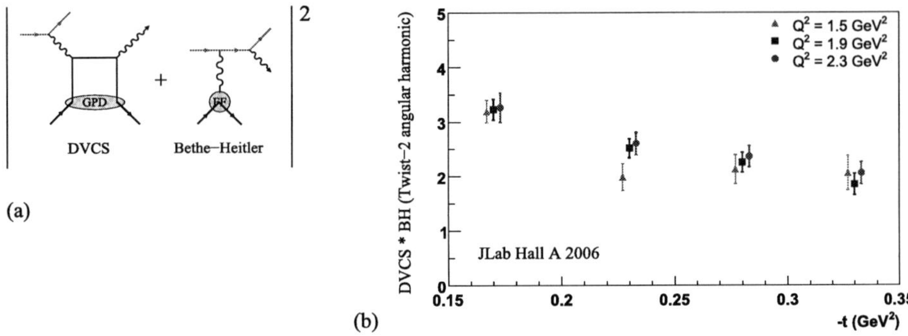

FIGURE 3. (a) Interference of DVCS and BH processes in $N(e,e'\gamma)N$. (b) The beam spin–dependent BH-DVCS interference cross section (Twist–2 angular harmonic) measured in the JLab Hall A experiment [12], as a function of t, for several values of Q^2.

through polarization observables.

Recent measurements of the beam spin–dependent BH–DVCS interference cross section in the JLab Hall A experiment [12] indicate an early approach to Q^2–scaling (see Fig. 3b), in accordance with theoretical expectations, suggesting that a description based on leading–twist QCD factorization should be applicable already at $Q^2 \sim$ few GeV^2. Present experimental efforts focus on separating the BH–DVCS interference and the DVCS^2 term in the beam spin–independent cross section, separating the nucleon helicity–components of the GPDs through measurement of target spin asymmetries, and measuring neutron GPDs through quasi–free neutron DVCS with nuclear targets [7].

The methods for analyzing DVCS observables at the leading–twist level and extracting information about GPDs are well developed and have been extensively discussed in the literature. Recently, two important new tools were added to the arsenal. One is GPD parametrizations based on the idea of a t–channel partial–wave expansion of the hard exclusive amplitude. In the so–called dual parametrization [13], inspired by the analogy with dual amplitudes in hadron–hadron scattering, the GPD *viz.* hard exclusive amplitude is represented as a formal series of t–channel resonance exchanges, which is then resummed and parametrized in terms of functions of the complexity of usual parton densities ("forward–like" functions; see Ref. [14] for a recent summary). In the approach of Ref. [15], based on the analogy with Regge theory, a similar representation is derived using complex angular momentum techniques. These t–channel representations have several advantages over the traditional double distribution (spectral) GPD parametrization [16]: (a) they diagonalize the QCD evolution equations for GPDs by using the conformal moments of the QCD operator; (b) they suggest a natural high–energy (small–ξ) expansion of the hard exclusive amplitudes. The representation of Ref. [15] has successfully been applied to fit the HERA DVCS data at NLO accuracy [17]; for a summary of the status of LO fits based on the dual parametrization see Ref. [18]. A simple GPD model based on the dual parameterization describes the bulk of the JLab spin–dependent cross section and asymmetry data [19]. However, as a general GPD parametrization the

t–channel representation is likely to be less useful in the large–x_B (large–ξ) region.

The other new tool are s–channel dispersion relations for the DVCS amplitude [20]; see also Ref. [17]. They allow one to restore the real part of the DVCS amplitude from the imaginary part, which is directly accessible in the spin–dependent BH–DVCS interference cross section, and an energy–independent subtraction constant. In the GPD description, the imaginary part is proportional to the GPD at the transition points $x = \pm\xi$, while the real part is given by a certain integral of the GPD over x. The existence of a dispersion relation thus implies that there is a "hidden" relation between the values of the GPDs at $x = \pm\xi$ and elsewhere; this property is in a subtle way related to the fact that moments of the GPD, as matrix elements of local spin–n operators, are required by Lorentz invariance to be polynomials of degree n in ξ ("polynomiality condition"). The subtraction constant turns out to be related to the D–term [21], a distinct component of the unpolarized GPDs required by polynomiality, which is located in the region $-\xi < x < \xi$ and drops out in the zero momentum–transfer limit ($\xi = 0$), and thus has no analog in the usual parton densities. If confirmed by further studies, these recent findings would have profound implications for the GPD analysis of DVCS data in the valence quark regime ($x_B > 0.1$). They would mean that in measurements at fixed Q^2 only the GPDs at $x = \pm\xi$ (as functions of ξ and t) and the D–term (as a function of t alone) are observable; whatever information from the GPDs at $x \neq \pm\xi$ enters in the amplitude could be reconstructed from the dispersion relation. The main task of experiments then would be to provide accurate and comprehensive data on the imaginary part of the DVCS amplitude through measurements of the spin–dependent interference cross section.

Exclusive meson production. In meson production the quantum numbers of the produced meson determine the spin (unpolarized/polarized), charge parity ($q \pm \bar{q}$), and flavor of the GPD, allowing one to probe the individual GPD components more selectively than in DVCS. However, because of the additional interaction required in the formation of the meson the partonic process is more complicated, and detailed study of the reaction mechanism is necessary before a factorized description can be applied.

QCD factorization for meson production relies on the fact that for $Q^2 \to \infty$ the meson is predominantly produced in a small–size configuration (transverse size $\sim 1/Q$), whose coupling to the target is weak ("color transparency") and can be computed perturbatively. The approach to the small–size regime with increasing Q^2 can be verified experimentally in a model–independent manner and constitutes a crucial test of the reaction mechanism. The t–slope of the differential cross section (more precisely, the $\Delta_\perp^2$ slope) measures the transverse area of the interaction region, reflecting the size of the target and that of the dominant configurations of the produced meson. As Q^2 increases and small–size meson configurations become more important, one expects the t–slope to decrease and eventually stabilize (see Fig. 4a). Exactly this behavior is seen in the HERA vector meson production data (see Fig. 4b), where the exponential t–slope of ρ^0 and ϕ production changes from the "soft" value of $B \sim 10\,\mathrm{GeV}^2$ at $Q^2 = 0$ to a value of $B \sim 5\,\mathrm{GeV}^2$ at $Q^2 \sim 20 - 30\,\mathrm{GeV}^2$; the latter value is close to the slope of J/ψ production, which is practically Q^2–independent because the J/ψ is produced in small–size configurations even at $Q^2 = 0$. These observations attest to the approach to the small–size regime in vector meson production at high Q^2. The limiting value of the t–slope for $Q^2 \to \infty$ then reflects the size of the target only and can be associated with

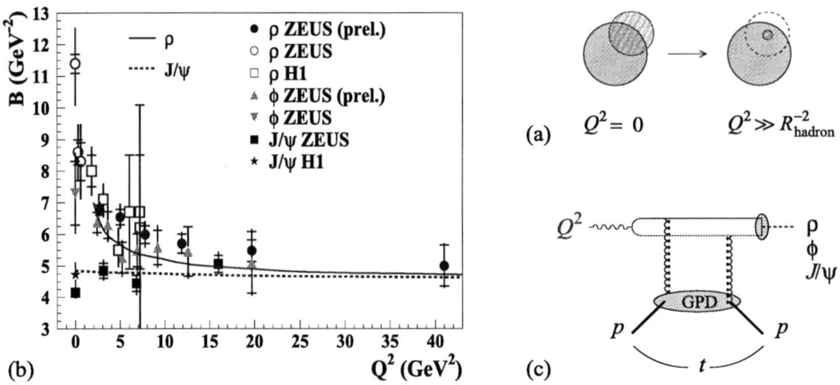

FIGURE 4. (a) Approach to the small–size regime in meson production (transverse view). (b) The exponential t–slope, B, for several vector meson electroproduction channels measured at HERA, as a function of Q^2 [22]. (c) In the small–size regime, the measured t–dependence can be identified with that of the gluon GPD.

the t–dependence of the GPD (see Fig. 4c).

In vector meson production at $Q^2 \sim$ few GeV^2 the data indicate substantial contributions from finite–size configurations in the produced meson. This is confirmed by theoretical calculations which account for the finite meson size by including the intrinsic quark transverse momentum in the partonic scattering process (higher–twist corrections) [23, 24, 25]. They describe well the high–energy data ($W > 10\,\text{GeV}$, HERA), where vector meson production couples mainly to the gluon GPD (see Fig. 4c) [23]. The model of Ref. [25] also reproduces the ϕ production cross section data at lower energies, including CLAS [26] (see Fig. 5a). Interestingly, none of the present calculations including quark GPDs can account for the fast rise of the ρ^0 production cross section below $W < 4\,\text{GeV}$, seen in the recent CLAS [27] and other data (see Fig. 5b). Whether this is due to shortcomings of the present quark GPD forms or inadequate modeling of higher–twist/finite–size corrections is presently under investigation. Model–independent tests of short–distance dominance through measurement of $t - (\Delta_\perp^2 -)$ slopes (see above), as well as comparison of vector meson channels involving gluon and quark GPDs in different proportion (ϕ: mostly gluons; ρ^0: gluons and singlet quarks; ρ^+: non-singlet quarks, no gluons) will help to distinguish between the two possibilities.

In the regime where small–size configurations dominate, meson production data can be used to extract information about the GPDs. The analysis of absolute cross sections is generally challenging, as the theoretical predictions depend on the detailed treatment of the hard scattering process (effective scale, higher–order QCD corrections; see Ref. [28] for a recent discussion) as well as on the GPD parametrization/model. One way of simplifying the analysis is to consider ratios of cross sections in which the uncertainties associated with the hard scattering process cancel, and one is left only with ratios of the the GPDs or their integrals. An example is the transverse target asymmetry in $\gamma_L^* p \to \rho^0 p$, which is sensitive to the Pauli form–factor type GPD E figuring in the angular momentum sum rule [1]. A similar asymmetry can be studied in $\gamma_L^* p \to K^{*+} \Lambda$ by

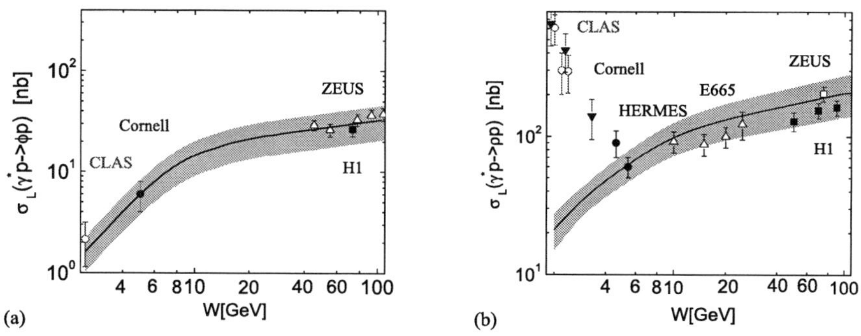

FIGURE 5. The longitudinal cross section for (a) exclusive ϕ, (b) exclusive ρ^0 production (adapted from Ref. [25]). The curve/error band show the GPD–based model calculation of Ref. [25].

measuring the recoil polarization of the produced Λ [29]; this process probes the $p \to \Lambda$ transition GPDs, which can be related to the usual flavor–diagonal GPDs in the proton using $SU(3)$ flavor symmetry (see Ref. [29] for other interesting "ratio observables").

Pseudoscalar meson production (π, η, K) in the small–size regime probes the polarized GPDs in the nucleon. A particular feature of π^+ (and to some extent also K^+) production is the existence of a "pole term" in the GPD, in which the QCD operator measuring the quark density is connected to the nucleon by t–channel $\pi^+(K^+)$ exchange; it gives a contribution to the amplitude proportional to the pion form factor, which is in fact the basis for measuring the latter in electroproduction experiments [30]. Better insight into the relation between the "pole" and the "non–pole" component of the GPD and their relative importance is necessary not only for improving the extraction of the pion form factor, but also for isolating the non-pole component related to the nucleon helicity structure. It could come *e.g.* from model–independent comparisons of π^+ (which has a pole) and π^0 (no pole) electroproduction data (see Ref. [29] for strange channels). Experimental studies of exclusive pseudoscalar meson production are challenging because the L/T virtual photon cross sections have to be separated by comparing data taken at different beam energies (Rosenbluth method). There are intriguing suggestions that σ_T in exclusive pion production above the resonance region could be described as the limit of semi-inclusive production via the fragmentation mechanism [31]; if confirmed, this could greatly aid the analysis of such processes.

GPDS IN SMALL–X PHYSICS AND PP SCATTERING

The notion of the transverse spatial distribution of partons conveyed by the GPDs has many important applications in small–x physics and high–energy pp collisions with hard processes. Measurements of the t–dependence of exclusive J/ψ photo/electroproduction at HERA (*cf.* Fig. 4b) and FNAL have provided a rather detailed picture of the transverse spatial distribution of gluons with $10^{-4} < x < 10^{-2}$ [22]. In particular, these experiments have shown that the nucleon's gluonic transverse size at $Q^2 \sim$ few GeV^2 is substantially

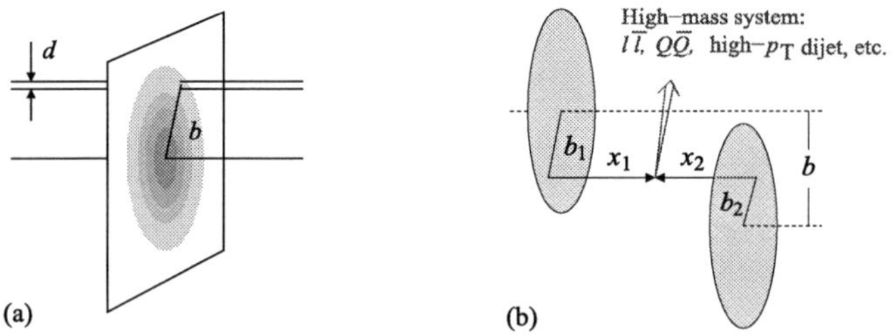

FIGURE 6. (a) The dipole model of high–energy scattering. The dipole–nucleon scattering amplitude at fixed impact parameter, b, is proportional to the local gluon density in the transverse plane. (b) Hard process in pp scattering.

smaller than its size in soft hadronic interactions at high energies, and increases less rapidly with energy (effective Regge slope $\alpha'_g \ll \alpha'_{\text{soft}}$). This information provides essential input for theoretical studies of high–energy scattering in the QCD dipole model (see Fig. 6a), where the dipole–nucleon interaction is governed by the local gluon density in the transverse plane, and for modeling the initial condition of non-linear QCD evolution equations describing the approach to the black–disk regime (unitarity limit).

In high–energy pp scattering with hard processes (see Fig. 6b), knowledge of the transverse spatial distribution of the partons in the colliding protons allows one to calculate the probability of the hard process as a function of the pp impact parameter, b, and thus describe in detail the spectator interactions in such processes, which are different from minimum bias events [32]. This is particularly important in hard diffractive pp scattering, where the spectator interactions determine the rapidity gap survival probability. By measuring the p_T dependence of central exclusive diffraction $pp \to p + H + p$ (H = high–mass system) one can even extract information about the transverse spatial distribution of gluons from the observed "diffraction pattern" [33].

FROM DENSITIES TO CORRELATIONS

A fast–moving hadron in QCD represents a complex many–body system, characterized by an equilibrium of creation and annihilation processes, capable of a wave function description as outlined by Gribov in the context of scalar field theory [34]. The GPDs summarize the single–particle structure of this many–body system, at a transverse resolution scale where one can unambiguously identify quasi–free partons (*i.e.*, integrated over transverse momenta up to a scale $\mu^2 \gg R_{\text{had}}^{-2}$). However, there is much more to be learned about this system than the single–particle densities. One example are the quantum fluctuations of the parton densities, which result from the quantum–mechanical superposition of configurations of different size, numbers of particles, *etc.* in the wave function, and can be revealed in hard diffractive ep scattering [35]. Another interesting property are multi–parton correlations, which are probed *e.g.* in pp scattering with multiple hard processes. There are indications of significant transverse correlations between partons, possibly related to the non–perturbative short–distance scale introduced by the spontaneous breaking of chiral symmetry in QCD (instanton size, "constituent quark" size) [36]. These correlations show up as higher–twist corrections in inclusive DIS [37] and likely play an essential role in the p_T/azimuthal dependence of semi–inclusive hadron production. Studying these correlations should be the next step in the exploration of the partonic structure of the nucleon.

Notice: Authored by Jefferson Science Associates, LLC under U.S. DOE Contract No. DE-AC05-06OR23177. The U.S. Government retains a non-exclusive, paid-up, irrevocable, world-wide license to publish or reproduce this manuscript for U.S. Government purposes.

REFERENCES

1. K. Gocke, M. V. Polyakov and M. Vanderhaeghen, Prog. Part. Nucl. Phys. **47**, 401 (2001). M. Diehl, Phys. Rept. **388**, 41 (2003). A. V. Belitsky and A. V. Radyushkin, Phys. Rept. **418**, 1 (2005).
2. M. Burkardt, Phys. Rev. D **62**, 071503 (2000) [Erratum-ibid. D **66**, 119903 (2002)]. P. V. Pobylitsa, Phys. Rev. D **66**, 094002 (2002). M. Diehl, Eur. Phys. J. C **25**, 223 (2002).
3. M. Strikman and C. Weiss, Phys. Rev. D **69**, 054012 (2004); arXiv:0811.3631 [hep-ph].
4. M. Burkardt, Int. J. Mod. Phys. A **18**, 173 (2003).
5. M. Diehl and Ph. Hagler, Eur. Phys. J. C **44**, 87 (2005).
6. X. D. Ji, Phys. Rev. Lett. **78**, 610 (1997); Phys. Rev. D **55**, 7114 (1997).
7. For a summary, see: E. Voutier, these proceedings, arXiv:0901.3016 [nucl-ex].
8. M. V. Polyakov, Phys. Lett. B **555**, 57 (2003).
9. K. Goeke *et al*, Phys. Rev. D **75**, 094021 (2007).
10. Ph. Hagler *et al.* [LHPC Collaborations], Phys. Rev. D **77**, 094502 (2008).
11. For a recent review, see: J. M. Zanotti, arXiv:0812.3845 [hep-lat].
12. C. Munoz Camacho *et al.* [JLAB Hall A & Hall A DVCS Coll.], Phys. Rev. Lett. **97**, 262002 (2006).
13. M. V. Polyakov and A. G. Shuvaev, arXiv:hep-ph/0207153.
14. K. M. Semenov-Tian-Shansky, Eur. Phys. J. A **36**, 303 (2008). M. V. Polyakov and K. M. Semenov-Tian-Shansky, arXiv:0811.2901 [hep-ph].
15. D. Mueller and A. Schafer, Nucl. Phys. B **739**, 1 (2006).
16. A. V. Radyushkin, Phys. Rev. D **56**, 5524 (1997); D **59**, 014030 (1999).
17. K. Kumericki, D. Mueller and K. Passek-Kumericki, Nucl. Phys. B **794**, 244 (2008).
18. V. Guzey and T. Teckentrup, Phys. Rev. D **79**, 017501 (2009).
19. M. V. Polyakov and M. Vanderhaeghen, arXiv:0803.1271 [hep-ph].
20. O. V. Teryaev, arXiv:hep-ph/0510031. I. V. Anikin and O. V. Teryaev, Phys. Rev. D **76**, 056007 (2007). M. Diehl and D. Y. Ivanov, Eur. Phys. J. C **52**, 919 (2007).
21. M. V. Polyakov and C. Weiss, Phys. Rev. D **60**, 114017 (1999).
22. L. Frankfurt, M. Strikman and C. Weiss, Ann. Rev. Nucl. Part. Sci. **55**, 403 (2005).
23. L. Frankfurt, W. Koepf and M. Strikman, Phys. Rev. D **54**, 3194 (1996).
24. M. Vanderhaeghen, P. A. M. Guichon and M. Guidal, Phys. Rev. D **60**, 094017 (1999).
25. S. V. Goloskokov and P. Kroll, Eur. Phys. J. C **42**, 281 (2005); C **50**, 829 (2007); C **53**, 367 (2008).
26. J. P. Santoro *et al.* [CLAS Collaboration], Phys. Rev. C **78**, 025210 (2008).
27. S. A. Morrow *et al.* [CLAS Collaboration], arXiv:0807.3834 [hep-ex].
28. M. Diehl and W. Kugler, Eur. Phys. J. C **52**, 933 (2007).
29. M. Strikman and C. Weiss, arXiv:0804.0456 [hep-ph].
30. T. Horn *et al.* [Jefferson Lab F(pi)-2 Collaboration], Phys. Rev. Lett. **97**, 192001 (2006). T. Horn *et al.*, arXiv:0707.1794 [nucl-ex]. H. P. Blok *et al.*, Phys. Rev. C **78**, 045202 (2008). G. M. Huber *et al.*, Phys. Rev. C **78**, 045203 (2008).
31. M. M. Kaskulov, K. Gallmeister and U. Mosel, Phys. Rev. D **78**, 114022 (2008).
32. L. Frankfurt, M. Strikman and C. Weiss, Phys. Rev. D **69**, 114010 (2004).
33. L. Frankfurt, C. E. Hyde, M. Strikman and C. Weiss, Phys. Rev. D **75**, 054009 (2007).
34. V. N. Gribov, arXiv:hep-ph/0006158.

35. L. Frankfurt, M. Strikman, D. Treleani and C. Weiss, Phys. Rev. Lett. **101**, 202003 (2008).
36. L. Frankfurt, M. Strikman and C. Weiss, Annalen Phys. **13**, 665 (2004).
37. A. V. Sidorov and C. Weiss, Phys. Rev. D **73**, 074016 (2006).

Generalized Parton Distributions, the Hunt for Quark Orbital Momenta

E. Voutier

Laboratoire de Physique Subatomique et de Cosmologie
IN2P3/CNRS - Université Joseph Fourier - INP
53 rue des Martyrs, 38026 Grenoble, France
E-mail address: voutier@lpsc.in2p3.fr

Abstract. The Generalized Parton Distributions (GPDs) are the appropriate framework for a universal description of the partonic structure of the nucleon. They characterize the dynamics of quarks and gluons inside the nucleon and consequently contain information about the spin of the nucleon. The current experimental knowledge about GPDs is rewieved with the emphasis on the determination of $E^q(Q^2,x,\xi,t)$, the least known and constrained GPD, of particular importance in the nucleon spin puzzle. The perspectives of this experimental program are also addressed.

INTRODUCTION

After several decades of intensive experimental efforts, the origins of the nucleon spin still keep a lot of secrets. While the recent measurements at CERN, DESY, JLab and SLAC have clearly established the quark helicity contributions to the nucleon spin [1], little is known about the gluon polarization contribution [2], and the quark and gluon orbital momenta contributions are essentially unknown. On the theoretical front, it is only recently that a comprehensive picture of the nucleon spin sum rules, enlightening the importance of the transversity distributions, was proposed [3]. Intuitively, these different distributions are the expression of a single unique reality, the dynamics of quarks and gluons constituting the nucleon, which should rely on more fundamental quantities. Generalized Parton Distributions (GPDs) [4, 5] provide a universal and powerful framework which unifies in the same formalism electromagnetic form factors, parton distributions, and the spin of the nucleon. GPDs encode the correlations between quarks, anti-quarks and gluons, and can be interpreted as the transverse distribution of partons carrying a certain longitudinal momentum fraction of the nucleon [6, 7], providing then a natural link with the transverse degrees of freedom. The GPD framework and its experimental knowledge is hereafter presented, particularly in the context of the nucleon spin puzzle.

GENERALIZED PARTON DISTRIBUTIONS

GPDs are universal non-perturbative objects entering the description of hard scattering processes. They are defined for each quark flavor and gluon, and correspond to the amplitude for removing a parton of longitudinal momentum fraction $x+\xi$ and restoring

CP1149, *Spin Physics, 18th International Spin Physics Symposium*
edited by D. G. Crabb, D. B. Day, S. Liuti, X. Zheng, M. Poelker, and Y. Prok

it with momentum fraction $x-\xi$ (fig. 1). In this process, the nucleon receives a four-momentum transfer $t=\Delta^2$ whose transverse component $\Delta_\perp$ is Fourier conjugate to the transverse position of partons. At the leading twist, the partonic structure of the nucleon [8, 9] is described by four quark helicity conserving and chiral even GPDs $(H^q,\widetilde{H}^q,E^q,\widetilde{E}^q)$ and four quark helicity flipping and chiral odd GPDs $(H^q_T,\widetilde{H}^q_T,E^q_T,\widetilde{E}^q_T)$, together with eight similar gluon GPDs. In the forward limit ($t\to 0,\xi\to 0$), the optical theorem links the H GPDs to the usual density, helicity, and tranversity distributions measured in deep inelastic scattering (DIS). The E GPDs, which involve a flip of the nucleon spin, do not have any DIS equivalent and consequently constitute a new piece of information about the nucleon structure. The first Mellin moments relate chiral even GPDs to form factors, as E^q with the Pauli electromagnetic form factor

$$\int_{-1}^{+1} dx\, E^q(Q^2,x,\xi,t) = F_2^q(t)\,, \tag{1}$$

and the second Mellin moments relate GPDs to the nucleon dynamics, particularly the total angular momentum carried by the partons, following Ji's sum rule [10]

$$J^q = \frac{1}{2}\Delta\Sigma + L^q = \frac{1}{2}\int_{-1}^{+1} dx\, x\left[H^q(Q^2,x,\xi,0)+E^q(Q^2,x,\xi,0)\right]. \tag{2}$$

$\Delta\Sigma$ being known from DIS, GPD measurements allow to access the contribution of the orbital momemtum to the nucleon spin. In this prospect, $E(Q^2,x,\xi,t)$ is of particular interest since it is not constrained by any DIS limit and remains essentially unknown. Similar relations have been proposed which relate chiral odd GPDs to the transverse spin-flavor dipole moment and the correlation between quark spin and angular momentum in an unpolarized nucleon [11].

DEEPLY VIRTUAL EXCLUSIVE SCATTERING

Deeply Virtual Compton Scattering (DVCS), corresponding to the absorption of a virtual photon by a quark followed quasi-instantaneously by the emission of a real photon, is the simplest reaction to access GPDs. In the Bjorken regime, $-t \ll Q^2$ and Q^2 much larger than the quark confinement scale, the leading contribution to the reaction amplitude is represented by the so-called handbag diagram (fig. 1) which figures the convolution of a known $\gamma^* q \to \gamma q$ hard scattering kernel with an unknown soft matrix element describing the partonic structure of the nucleon parametrized by GPDs [12, 13]. Consequently, GPDs (E^q) enter the reaction cross section through a Compton form factor ($\mathcal{E}$) which involves an integral over the intermediate quark propagators

$$\begin{aligned} \mathcal{E} &= \sum_q e_q^2\, \mathcal{P}\int_{-1}^{+1} dx \left(\frac{1}{x-\xi}+\frac{1}{x+\xi}\right) E^q(Q^2,x,\xi,t) \\ &\qquad - i\pi \sum_q e_q^2 \left[E^q(Q^2,\xi,\xi,t) - E^q(Q^2,-\xi,\xi,t)\right] \end{aligned} \tag{3}$$

e_q being the quark electric charge in units of the elementary charge. In addition to the DVCS amplitude, the cross section for electroproduction of photons gets contributions

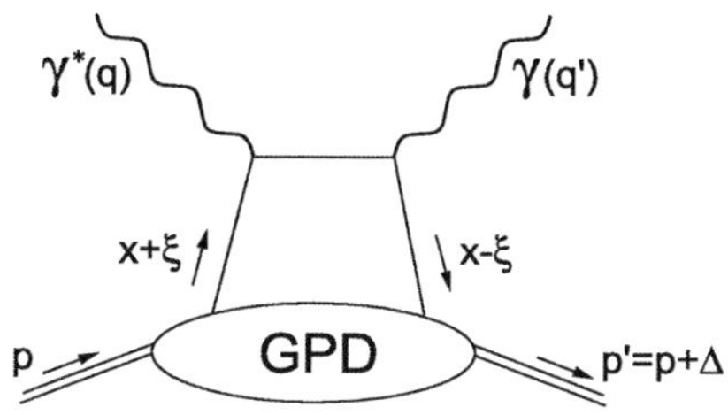

FIGURE 1. Lowest order (QCD) amplitude for the virtual Compton process. The four-momentum of the incident and scattered photon are q and q'. The four-momentum of the initial and final proton are p and p', with $\Delta=(p'-p)=(q-q')$. The DIS scaling variable is $x_B=Q^2/(2p\cdot q)$ and the DVCS scaling variable is $\xi=x_B/(2-x_B)$.

from the Bethe-Heitler (BH) process where the real photon is emitted by the initial or final lepton, leading to

$$\frac{d^5\sigma}{dQ^2dx_Bdtd\phi_e d\varphi} = \mathcal{T}_{BH}^2 + |\mathcal{T}_{DVCS}|^2 \mp 2\,\mathcal{T}_{BH}\Re e\{\mathcal{T}_{DVCS}\} \tag{4}$$

with ϕ_e the scattered electron azimuthal angle, and φ the out-of-plane angle between the leptonic and hadronic planes; the $+(-)$ sign of the interference amplitude stands for negative(positive) leptons. Though undistinguishable from DVCS, the BH amplitude is known and exactly calculable from the electromagnetic form factors. Beam charge and beam and/or target polarization observables can be advantageously used to select different contributions to the cross section. For instance, the polarized cross section difference for opposite beam helicities allows to isolate the imaginary part of the DVCS amplitude

$$\begin{aligned}\frac{d^5\Delta\sigma}{dQ^2dx_Bdtd\phi_e d\varphi} &= \frac{1}{2}\left[\frac{d^5\overrightarrow{\sigma}}{dQ^2dx_Bdtd\phi_e d\varphi} - \frac{d^5\overleftarrow{\sigma}}{dQ^2dx_Bdtd\phi_e d\varphi}\right] \\ &= \mathcal{T}_{BH}\Im m\{\mathcal{T}_{DVCS}\} + \Re e\{\mathcal{T}_{DVCS}\}\Im m\{\mathcal{T}_{DVCS}\}\end{aligned} \tag{5}$$

where $\Im m\{\mathcal{T}_{DVCS}\}$ appears now linearly and magnified by the BH amplitude. In practice, the measurements of the cross section and a selected set of single and double spin polarization observables allows to extract the real an imaginary parts of the Compton form factors, from which the GPDs can be deconvoluted [14].

Deeply Virtual Meson Production (DVMP), where the real photon is replaced by a meson, is another channel to access GPDs which provides in addition an elegant flavor decomposition. In this case, the factorization of the cross section applies only to longitudinal virtual photons and the GPDs entering the Compton form factors (eq. 3) are further convoluted with a meson distribution amplitude. The measurement of the angular distribution of the decay products of the vector mesons allows to extract the longitudinal part of the cross section and the longitudinal polarization of the vector mesons, assuming the s-channel helicity conservation. Other reaction mechanisms, like the 2-gluon exchange from a $q\bar{q}$ fluctuation of the virtual photon, may contribute to the production process. Similarly to DVCS, polarization observables help to single-out the pure handbag contributions.

EXPERIMENTAL STATUS

The pioneering studies of the electro-production of photons at DESY [15, 16] and JLab [17] did prove the existence of the DVCS mechanism by measuring sizeable beam spin asymmetries (BSA from the ratio of eq. 5 and eq. 4) in the valence region and significant cross sections in the gluon sector. Other limited studies showed the importance of the beam charge [18] and the target polarization observables [19]. The recent remarkable results of the starting dedicated DVCS experimental program of JLab is the strong indication for scaling in the valence region at Q^2 as low as 2 GeV^2, and the observation of an unexpected DVCS amplitude magnitude at JLab energies [20]. This early scaling is also supported by the φ angular dependence of the BSA measured at JLab with an unprecedented accuracy over a wide kinematic range [21]. In general, GPD based calculations provide a reasonable but unsatisfactory agreement with these data which turn out to be fairly reproduced by a more conventional Regge approach [22]. The significance of this duality has not yet been resolved.

In the meson sector, the experimental status with respect to GPDs remains controversial. Sizeable BSAs have been reported for exclusive neutral pion electro-production at JLab energies [23], which suggest that both longitudinal and transverse amplitudes contribute to the process. On the one hand, this forbids a direct GPD based interpretation of the data, and on the other hand, a Regge based approach fails to reproduce them. The longitudinal cross section for the electro-production of longitudinally polarized neutral rho was recently measured at JLab [24]. Standard GPD calculations, particularly successful at high energies, do not reproduce data in the valence region, while calculations based on hadronic degrees of freedom are in very good agreement over the complete energy range scanned by the world data. The observation that strongly modified GPDs allow for a partonic interpretation of these measurements raises the question whether current GPD parametrizations must be revisited or the existence of strong higher twist corrections percludes a GPD wise interpretation of DVMP in the valence region.

PERSPECTIVES

Transversely polarized proton targets are of primary importance for the determination of the quark orbital momenta, since they provide a direct access to the GPD E^q which is essentially parametrized in terms of the quark angular momenta [25]. In this respect, DVCS and DVMP transverse target asymmetries (TSA) are key observables. The first experimental indication of this sensitivity was reported by HERMES from DVCS TSA measurements off a proton target [26]. Within a simple twist-2 approach, the TSA data may be interpreted as a measurement of the GPDs combination $\approx t[F_2H - F_1E]/4M^2$. Relying on a given GPD model, the most probable parameters J_u and J_d, representing the u and d quarks angular momenta, can be extracted from the comparison between data and calculations, leading to the vertical band in fig. 2.

Another important observable connected to E^q is the DVCS polarized cross section difference (eq. 5) off a neutron target. Taking advantage of the smalness of the neutron Dirac electromagnetic form factor and of the cancellation between the u and d quarks

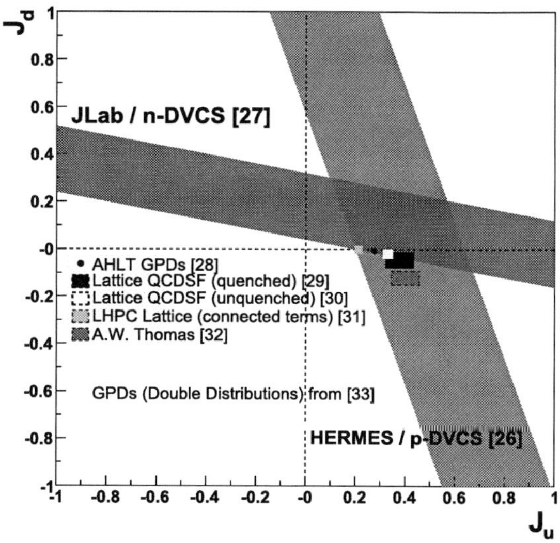

FIGURE 2. Model dependent experimental constraint on J_u and J_d quark angular momenta from the HERMES $\vec{p}$-DVCS [26] and the JLab n-DVCS [27] experiments. Different model calculations [28, 29, 30, 31, 32] are compared to an extrapolation of experimental data within the VGG double distribution description of GPDs [33].

in the GPD $\widetilde{H}$ at small t, this observable is, at leading twist, a measurement of the combination $tF_2E/4M^2$. The first data on this observable were obtained by the JLab n-DVCS experiment [27]. Similarly to TSA, the comparison with calculations for the same GPD model leads to the horizontal band in fig. 2.
In the present status of GPD models, the main message of this appealing picture is the experimental demonstration of the complementarity between transversely polarized proton and neutron targets in the search for quark orbital momenta. This fundamental feature is a direct consequence of isospin symmetry. Experimental advances in this quest are depending on the progress in the developments of high luminosity transversely polarized targets [34] and on the next generation of n-DVCS experiments for which transversely polarized targets may provide an interesting observable to extract E at very small t [35].

CONCLUSIONS

The exciting exploration of the GPDs world is progressing. A larger set of experimental data are expected from JLab 6 GeV and the HERMES device supplemented by a recoil detector. In a near future, the energy upgrade of JLab and the completion of a DVCS program at COMPASS will considerably enlarge the kinematic range of GPDs knowledge. Together with the evolution of lattice QCD calculations capabilities, one may reasonably expect to get a comprehensive and quantitative understanding of the nucleon structure, including the contribution of the quark orbital momenta to the nucleon spin, by the end of the next decade.

ACKNOWLEDGMENTS

I would like to thank F.-X. Girod, M. Guidal, N. d'Hose, M. Mazouz and A. Sandorfi for providing pictures and material for this presentation. This work was supported in part by the U.S. Department of Energy (DOE) contract DOE-AC05-06OR23177 under which the Jefferson Science Associates, LLC, operates the Thomas Jefferson National Accelerator Facility, the National Science Foundation, the French Atomic Energy Commission and National Center of Scientific Research, and the GDR n°3034 Physique du Nucléon.

REFERENCES

1. A. Airapetian et al., *Phys. Rev.* **D 75** (2007) 012007, and references therein.
2. F. Ellinghaus, *Contribution to these proceedings*, and references therein.
3. B.L. Bakker, E. Leader, T.L. Trueman, *Phys. Rev.* **D 70** (2004) 114001.
4. D. Müller, D. Robaschick, B. Geyer, F.M. Dittes, J. Hořejši, *Fortschr. Phys.* **42** (1994) 101.
5. A.V. Radyushkin, *Phys. Rev.* **D** 56 (1997) 5524.
6. M. Burkardt, *Phys. Rev.* **D** 62 (2000) 071503(R).
7. M. Diehl, *Eur. Phys. Jour.* **C** 25 (2002) 223.
8. M. Diehl, *Phys. Rep.* **388** (2003) 41.
9. A.V. Belitsky, A.V. Radyushkin, *Phys. Rep.* **418** (2005) 1.
10. X. Ji, *Phys. Rev. Lett.* **78** (1997) 610.
11. M. Burkardt, *Phys. Rev.* **D 72** (2005) 094020.
12. X. Ji, J. Osborne, *Phys. Rev.* **D 58** (1998) 094018.
13. J.C. Collins, A. Freund, *Phys. Rev.* **D 59** (1999) 074009.
14. M. Guidal, *Eur. Phys. J.* **A 37** (2008) 319.
15. A. Airapetian et al., *Phys. Rev. Lett.* **87** (2001) 182001.
16. C. Adloff et al., *Phys. Lett.* **B 517** (2001) 47.
17. S. Stepanyan et al., *Phys. Rev. Lett.* **87** (2001) 182002.
18. A. Airapetian et al., *Phys. Rev.* **D 75** (2007) 011103.
19. S. Chen et al., *Phys. Rev. Lett.* **97** (2006) 072002.
20. C. Muñoz Camacho, A. Camsonne, M. Mazouz, C. Ferdi, G. Gavalian, E. Kuchina et al., *Phys. Rev. Lett.* **97** (2006) 262002.
21. F.-X. Girod, R.A. Niyazov et al., *Phys. Rev. Lett.* **100** (2008) 162002.
22. J.-M. Laget et al., *Phys. Rev.* **C 76** (2007) 052201.
23. R. De Masi, M. Garçon, B. Zhao et al., *Phys. Rev.* **C 77** (2008) 042201(R).
24. S.A. Morrow, M. Guidal, M. Garçon, J.-M. Laget, E.S. Smith et al., *arXiv/hep-ex:* **0807.3834** (2008).
25. F. Ellinghaus, W.-D. Nowak, A.V. Vinnikov, Z. Ye, *Eur. Phys. J.* **C 46** (2006) 729.
26. A. Airapetian et al., *ArXiv/hep-ex:* **0802.2499** (2008).
27. M. Mazouz, A. Camsonne, C. Muñoz Camacho, C. Ferdi, G. Gavalian, E. Kuchina et al., *Phys. Rev. Lett.* **99** (2007) 242501.
28. S. Ahmad, H. Honkanen, S. Liuti, S.K. Taneja, *Phys. Rev.* **D 75** (2007) 094003.
29. M. Göckeler et al., *Phys. Rev. Lett.* **92** (2004) 042002.
30. G. Schierholz, Proc. of the *Workshop on Exclusive Reactions at High Momentum Transfer*, Edts. P. Stoler and A. Radyushkin, World Scientific (2008).
31. Ph. Hägler et al., *Phys. Rev.* **D 77** (2008) 094502.
32. A.W. Thomas, *Phys. Rev. Lett.* **101** (2008) 102003.

33. M. Vanderhaeghen, P.A.M. Guichon, M. Guidal, *Phys. Rev.* **D 60** (1999) 094017.
34. A. Sandorfi, Proc. of the *Workshop on Exclusive Reactions at High Momentum Transfer*, Edts. P. Stoler and A. Radyushkin, World Scientific (2008) 425.
35. E. Voutier, Proc. of the *XXVIIth International Workshop on Nuclear Theory*, Rila (Bulgaria), June 22-26, 2008; *arXiv/nucl-ex:* **0809.2670** (2008), and references therein.

SPIN@COSY: Spin-Manipulating Polarized Deuterons and Protons

M.A. Leonova[1], A.W. Chao[1,†], A.D. Krisch[1], V.S. Morozov[1], R.S. Raymond[1], D.W. Sivers[1], V.K. Wong[1]; A. Garishvili[2,‡], R. Gebel[2], A. Lehrach[2], B. Lorentz[2], R. Maier[2], D. Prasuhn[2], H. Stockhorst[2], D. Welsch[2]; F. Hinterberger[3], K. Ulbrich[3]; Ya.S. Derbenev[4]; A.M. Kondratenko[5]; E.J. Stephenson[6]

[1] *Spin Physics Center, University of Michigan, Ann Arbor, Michigan 48109-1040, USA*
[2] *Forschungszentrum Jülich, Institut für Kernphysik, Postfach 1913, D-52425 Jülich, Germany*
[3] *Helmholtz-Institut für Strahlen- und Kernphysik, Universität Bonn, D-53115 Bonn, Germany*
[4] *Thomas Jefferson National Accelerator Facility, Newport News, VA 23606, USA*
[5] *GOO Zaryad, Russkaya St. 41, Novosibirsk, 630058 Russia*
[6] *IUCF, Indiana University, Bloomington, Indiana 47408-0768 USA*

Abstract. We studied spin manipulation of 1.85 GeV/c polarized deuteron beam stored in COSY obtaining a spin-flip efficiency of 97 ± 1%. We first discovered experimentally and then explained theoretically interesting behavior of the deuteron tensor polarization. We, for the first time, studied systematically spin resonance strengths induced by rf dipoles and solenoids. We found huge disagreements between the strengths measured in controlled Froissart-Stora sweeps and the theoretical values calculated using the well-known formulae. These data instigated re-examination of these formulae. We tested Chao's proposed new matrix formalism for describing the spin dynamics due to a single spin resonance, which may be the first fundamental improvement of the Froissart-Stora equation in that it allows analytic calculation of the beam polarization's behavior inside a resonance. Our measurements of the deuteron's polarization near and inside the resonance agreed precisely with the Chao formalism's predicted oscillations. We tested Kondratenko's proposal to overcome depolarizing resonances by ramping through them with a crossing pattern, which should force the depolarizing contributions to cancel themselves. Our first test of this idea with 2.1 GeV/c protons was not conclusive but a later test with 1.85 GeV/c deuterons demonstrated a rather substantial reduction in the depolarization compared to the tune jump at the same rate.

Keywords: Spin resonance strength, Spin dynamics, Depolarization.
PACS: 29.27.Bd, 29.27.Hj, 41.75.Ak

INTRODUCTION

Precise polarized scattering experiments require frequent spin-direction reversals to reduce systematic errors. Moreover, one must efficiently overcome spin resonances to maintain the polarization. COSY is now the world's best facility to study the spin manipulations needed to maintain and flip 1 GeV – multi-TeV stored polarized beams.

---

† also at SLAC, 2575 Sand Hill Rd., Menlo Park, CA 94025.
‡ also at Erlangen-Nürnberg Univ., D-91058, Germany.

CP1149, *Spin Physics, 18th International Spin Physics Symposium*
edited by D. G. Crabb, D. B. Day, S. Liuti, X. Zheng, M. Poelker, and Y. Prok

In flat circular rings, each beam particle's spin precesses around the vertical fields of the ring's bending magnets. The spin tune ν_s (the number of spin precessions during one turn around the ring) is given by $\nu_s = G\gamma$, where G is the particle's gyromagnetic anomaly and γ is its Lorentz energy factor. The vertical polarization can be perturbed by any horizontal magnetic field. RF fields can induce an rf spin resonance when

$$f_r = f_c(k \pm \nu_s), \tag{1}$$

where f_r is the resonance frequency, f_c is the circulation frequency and k is an integer.

The deuteron's vector polarization [1] behaves in a similar way to the proton's (vector) polarization; ramping an rf magnet's frequency through f_r can flip each particle's spin. The modified [2] Froissart-Stora (FS) equation [3] relates the beam's initial polarization P_i to its final polarization P_f after crossing the resonance,

$$P_f = P_i\left\{(1+\hat{\eta})\exp\left[\frac{-(\pi\,\varepsilon_{FS}\,f_c)^2}{\Delta f/\Delta t}\right] - \hat{\eta}\right\}; \tag{2}$$

the parameter $\hat{\eta}$ is the limiting spin-flip efficiency, ε_{FS} is the spin resonance strength, Δf is the rf frequency's ramp range during the ramp time Δt.

Equation (2) is only valid if Δf is significantly larger than the spin resonance's width. Chao's new matrix formalism was proposed [4-5] to deal with conditions where the FS formula is not valid; the Chao formalism can be used to calculate the spin dynamics anywhere inside a piece-wise linear resonance crossing.

Several techniques are commonly used to overcome spin resonances. Siberian snakes are most useful at very high energies. In the 1 to 25 GeV region one uses harmonic correction of the imperfection resonances and fast crossing (FC) of the intrinsic resonances; these are less effective than Siberian snakes and often leave some depolarization at each of the many crossed resonances. Kondratenko proposed a new technique for overcoming depolarizing resonances of moderate strength [6]. It involves crossing each resonance using the Kondratenko Crossing (KC) pattern [7].

For a flat ring, with a short rf magnet causing the only spin perturbation, ε_{Bdl} due to an rf solenoid's longitudinal B or an rf dipole's transverse B should be given by [8-9]

$$\text{Solenoid:} \qquad \varepsilon_{Bdl} = \frac{1}{\pi 2\sqrt{2}}\frac{e(1+G)}{p}\int B_{rms}\,dl, \tag{3}$$

$$\text{Dipole:} \qquad \varepsilon_{Bdl} = \frac{1}{\pi 2\sqrt{2}}\frac{e(1+G\gamma)}{p}\int B_{rms}\,dl, \tag{4}$$

where e is the particle's charge, p is its momentum, and $\int B_{rms}dl$ is the rf magnet's rms magnetic field integral in its rest frame.

One can obtain the dependence of $\varepsilon_{FS}/\varepsilon_{Bdl}$ on the distance between the vertical betatron tune ν_y and the rf spin resonance's tune $\nu_r \equiv (k \pm f_r/f_c)$ where k is an integer; this dependence is hyperbolic [8, 10-11]

$$\varepsilon_{FS}/\varepsilon_{Bdl} = \left|A + \frac{B}{(\nu_y - \nu_r)}\right|. \tag{5}$$

The parameter B depends on details of the ring, but the parameter A should give the $\varepsilon_{FS}/\varepsilon_{Bdl}$ ratio when one is infinitely far from any intrinsic spin resonances.

EXPERIMENTAL RESULTS

Identical apparatus was used for both protons and deuterons, including the COSY storage ring, the EDDA detector, the low energy polarimeter, the electron cooler, the injector cyclotron, the polarized ion source, and the rf dipole or rf solenoid [5, 7, 10-14]. We started by experimentally finding the spin resonance's frequency, that had been approximately calculated using Eq. (1). To obtain the resonance strength ε_{FS} for each experimental setting, we first measured the final beam polarization P_f after passing through a spin resonance by ramping an rf magnet's frequency through a range Δf during a time Δt. The measured dependence of P_f on Δt was then fit to Eq. (2); this allowed us to obtain ε_{FS} as a fit parameter.

Spin-Flipping

We spin-flipped vertically polarized protons at COSY; after optimizing the crossing rate $\Delta f/\Delta t$ we measured spin-flip efficiency of 99.92 ± 0.04% [12] as shown in Fig. 1.

We studied the more complex spin manipulation of polarized deuteron beams stored in the IUCF Cooler Ring using an rf solenoid, and then in COSY using the Michigan-COSY ferrite rf dipole; the spin-flip efficiencies were 94.2 ± 0.3% [1] and 97 ± 1% [13], see Fig. 2, respectively. These polarized deuteron studies seem to be the first successful manipulation of the tensor polarization P_T of a stored beam of spin-1 bosons. These studies show striking and interesting behavior of P_T.

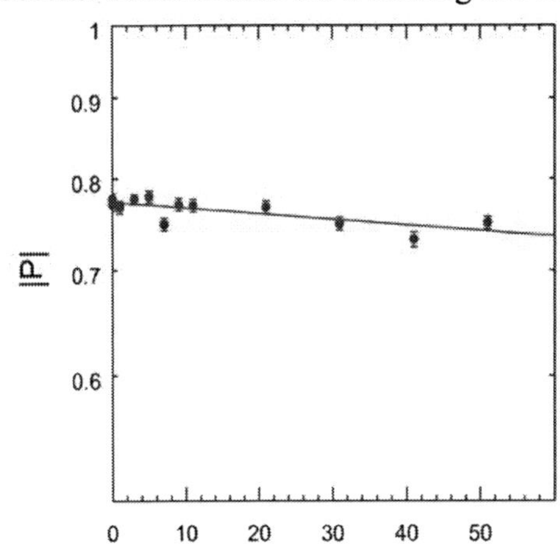

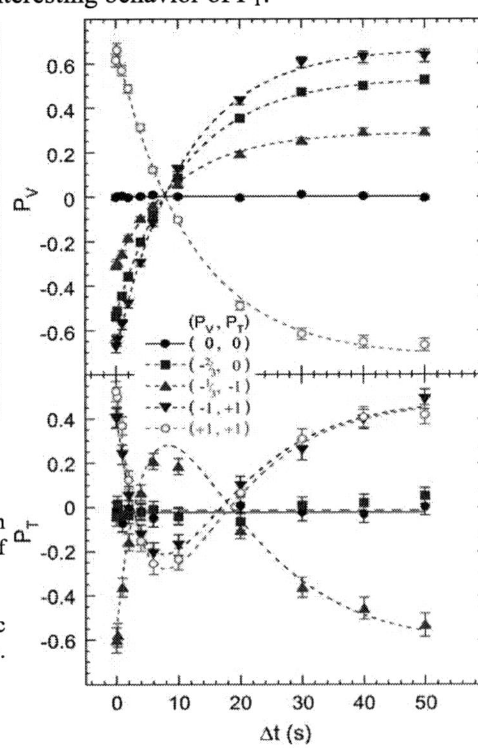

FIGURE 1. (LEFT) Measured 2.1 GeV/c proton polarization after n flips plotted vs. the number of flips.

FIGURE 2. (RIGHT) Measured 1.85 GeV/c deuteron vector and tensor polarizations plotted vs. the ramp time.

Chao Matrix Formalism

The Chao matrix formalism [4] allows analytic calculations of a beam's polarization behavior inside a spin resonance. We recently tested its prediction of polarization oscillations occurring in a stored beam of polarized particles near a spin resonance [5]. Using a 1.85 GeV/c polarized deuteron beam stored in COSY, we swept a new rf solenoid's frequency by 400 Hz during 100 ms, while varying the distance between the sweep's end frequency and the central frequency of an rf-induced spin resonance, as shown in Fig. 3. Our measurements of the deuteron's polarization near and inside the resonance agree with the Chao formalism's predicted oscillations, see Fig. 4.

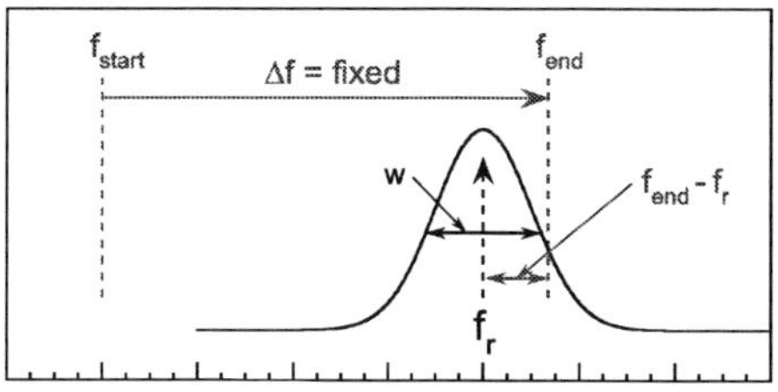

FIGURE 3. Schematic of Chao formalism test.

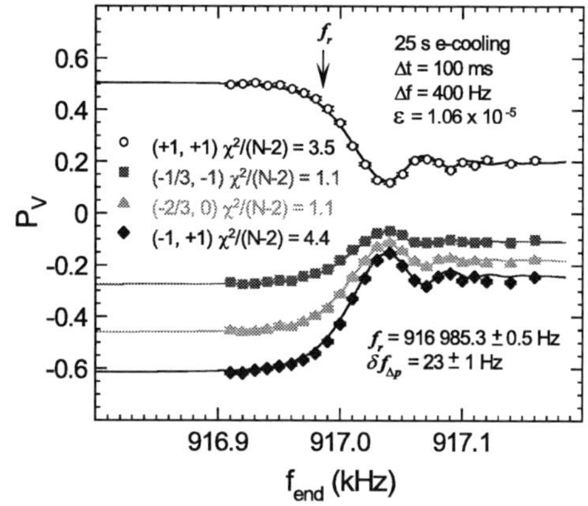

FIGURE 4. Measured 1.85 GeV/c deuteron vector polarizations plotted vs. the rf-solenoid sweep's end frequency f_{end} ; the electron cooling time was 25 s.

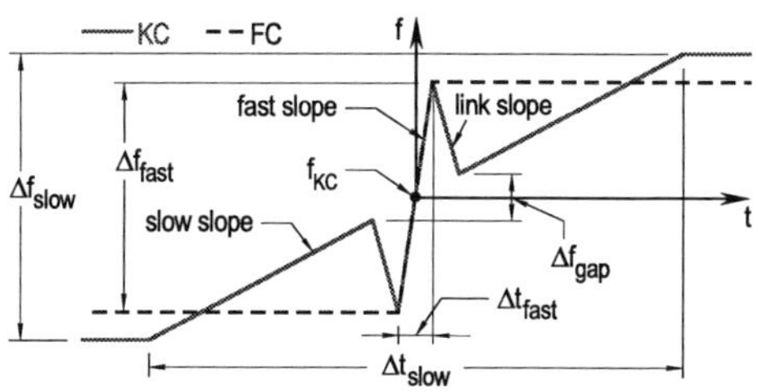

FIGURE 5. Kondratenko Crossing (KC) [solid line] and Fast Crossing (FC) [dashed line] patterns.

Kondratenko Crossing

We tested Kondratenko's Crossing proposal (KC) [6, 7] to avoid most polarization loss when crossing a spin resonance using stored 1.85 GeV/c vertically polarized deuterons. We crossed an rf depolarizing resonance by ramping an rf solenoid's frequency through KC pattern shown in Fig. 5. One can use the Chao matrix formalism to calculate analytically the polarization for this crossing pattern. The unbunched beam data agree with the predicted KC behavior. We varied different parameters of the KC pattern. With its optimal parameters, KC gave measured polarization losses of 3.3 ± 0.2% and 0.8 ± 0.3% with unbunched and bunched beams, respectively, while fast crossing (FC) at the same crossing rate, gave measured losses of 15.6 ± 0.2% and 15.0 ± 0.3%, respectively. This clearly shows KC's potential value. While the Chao formalism cannot yet calculate the KC behavior for bunched beams, the measured 19-fold polarization loss reduction (compared to FC at the same crossing rate) shows that Kondratenko Crossing may prove to be quite valuable for the bunched beams used in most accelerators.

RF Resonance Strength

We analyzed all available data on spin-flipping stored beams of protons, deuterons and electrons [10]. We calculated the rf-induced spin resonance strength ratios $\varepsilon_{FS}/\varepsilon_{Bdl}$, where ε_{FS} was obtained by fitting the measured polarization to Eq. (2); ε_{Bdl} was calculated using Eqs. (3) and (4). We found that $\varepsilon_{FS}/\varepsilon_{Bdl}$ was often 7 times lower than predicted for deuterons, and 12 – 170 times higher for protons. We systematically studied $\varepsilon_{FS}/\varepsilon_{Bdl}$ ratios with vertically polarized protons and deuterons in COSY.

With an rf dipole, we studied the dependence of $\varepsilon_{FS}/\varepsilon_{Bdl}$ on the beam size, momentum spread $\Delta p/p$, and distance to the nearest 1st-order intrinsic spin resonance for both protons [10] and deuterons [11], and on Δf for deuterons. We observed no dependence of $\varepsilon_{FS}/\varepsilon_{Bdl}$ on the beam's size or $\Delta p/p$ for either protons or deuterons, and no dependence of $\varepsilon_{FS}/\varepsilon_{Bdl}$ on Δf for deuterons. We varied the vertical betatron tune ν_y near a 1st-order intrinsic spin resonance and observed an enhancement of $\varepsilon_{FS}/\varepsilon_{Bdl}$ with a hyperbolic dependence on distance from the 1st-order intrinsic spin resonance for both protons and deuterons; the deuteron data are shown in Fig. 6. This can explain the enhancement for protons, but can not explain the deuteron's very small $\varepsilon_{FS}/\varepsilon_{Bdl}$.

Our recent experiment, using a new rf-solenoid magnet with stored polarized deuterons, measured $\varepsilon_{FS}/\varepsilon_{Bdl}$ of 1.02 ± 0.05 over a wide range of parameters [14]. As shown in Figs. 6 and 7, these new data agree precisely with Eq. (3), except for the data point centered on the intrinsic resonance; the rf solenoid's resonance strength has no dependence on ν_y. This is because, unlike an rf dipole, an rf solenoid does not excite coherent vertical beam oscillations. All available $\varepsilon_{FS}/\varepsilon_{Bdl}$ data are shown in Fig. 8.

These rf-solenoid data together with earlier rf-dipole results [11], which disagree with Eq. (4) by a factor of about 7, indicate that Eq. (3) for longitudinal rf fields is correct, while Eq. (4) for radial rf fields is incorrect. Efforts are being made to solve the problem of Eq. (4) [15-17]; apparently Eq. (4) must be replaced by more complex

FIGURE 6. (LEFT) Ratio $\varepsilon_{FS}/\varepsilon_{Bdl}$ for deuterons is plotted vs. vertical betatron tune ν_y. The long-dash line prediction is at 1. The medium-dash line is a fit of rf-dipole data to an asymmetric hyperbola [11]. The small-dash line is Kondratenko's parameter-free calculation [16]. The dotted-line fit to all rf-solenoid points, except the point exactly on the intrinsic resonance, gives $\varepsilon_{FS}/\varepsilon_{Bdl}$ of 1.02 ± 0.05.

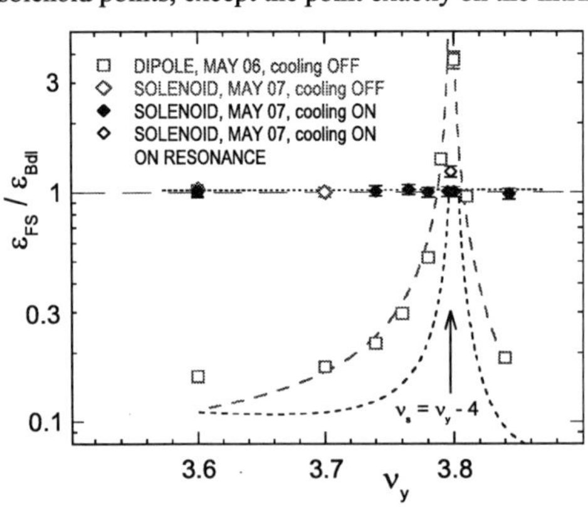

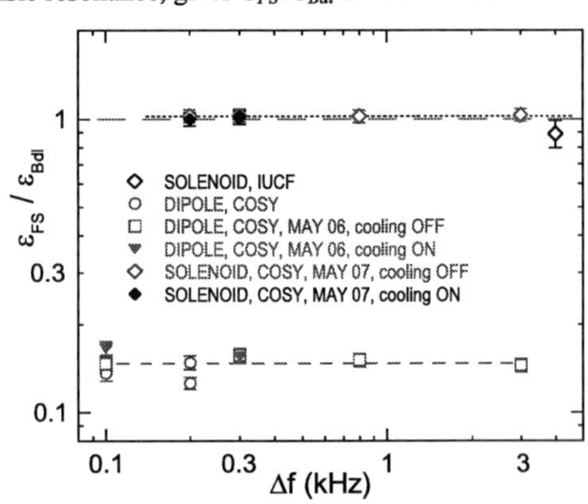

FIGURE 7. (RIGHT) Ratio $\varepsilon_{FS}/\varepsilon_{Bdl}$ for deuterons is plotted vs. the rf magnet's frequency sweep range Δf. The long-dash line prediction is at 1. The fit to all rf-dipole data (medium-dash line) gives $\varepsilon_{FS}/\varepsilon_{Bdl}$ of 0.15 ± 0.01. The fit to the COSY rf-solenoid data (dotted line) gives $\varepsilon_{FS}/\varepsilon_{Bdl}$ of 1.02 ± 0.05.

calculations, which depend on each ring's properties. One must properly include the modification of ε_{Bdl} due to beam's coherent oscillations caused by the rf dipole, which then interact with all the ring's radial and longitudinal magnetic fields [16].

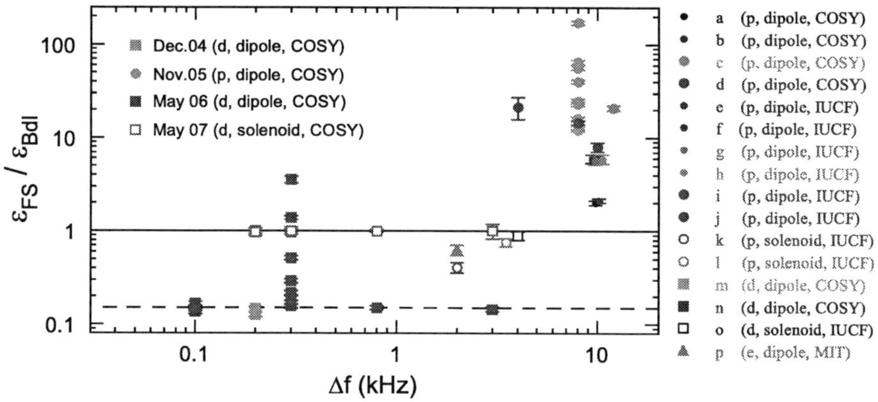

FIGURE 8. Ratio of ε_{FS} to ε_{Bdl} is plotted vs. rf magnet's frequency sweep range Δf.

ACKNOWLEDGMENTS

We thank the COSY staff for the successful operation of COSY, and E. D. Courant and Yu. F. Orlov for their help and advice. This research was supported by grants from the German BMBF Science Ministry, its JCHP-FFE program at COSY, the Helmholtz Association through funds provided to the virtual institute "Spin and strong QCD" (VH-VI-231), and the U.S. NSF.

REFERENCES

1. V. S. Morozov *et al.*, Phys. Rev. Lett. **91**, 214801 (2003).
2. V. S. Morozov *et al.*, Phys. Rev. ST Accel. Beams **4**, 104002 (2001).
3. M. Froissart and R. Stora, Nucl. Instr. Methods **7**, 297 (1960).
4. A. W. Chao, Phys. Rev. ST Accel. Beams **8**, 104001 (2005).
5. V. S. Morozov *et al.*, Phys. Rev. ST Accel. Beams **10**, 041001 (2007).
6. A. M. Kondratenko *et al.*, Physics of Particles and Nuclei Letters **1**, 255 (2004); SPIN@COSY Int. Rpt. (November 2006) http://spinbud.physics.lsa.umich.edu/OptKCpar.pdf.
7. V. S. Morozov *et al.*, submitted to Phys. Rev. Lett. (November 13, 2008).
8. S. Y. Lee, *Spin Dynamics and Snakes in Synchrotrons*, (World Scientific, Singapore, 1997), p. 79, Eq. (4.85); Phys. Rev. ST Accel. Beams **9**, 074001 (2006).
9. T. Roser, Handbook of Accelerator Physics and Engineering, (World Scientific, Singapore, 2002), p. 153, Eq. (7); M. Bai *et al.*, Phys. Rev. ST Accel. Beams **8**, 099001 (2005).
10. M. A. Leonova *et al.*, Phys. Rev. ST Accel. Beams **9**, 051001 (2006).
11. A. D. Krisch *et al.*, Phys. Rev. ST Accel. Beams **10**, 071001 (2007).
12. M. A. Leonova *et al.*, Phys. Rev. Lett. **93**, 224801 (2004).
13. V. S. Morozov *et al.*, Phys. Rev. ST Accel. Beams **8**, 099002 (2005).
14. For additional details and references see M. A. Leonova *et al.*, "Spin resonance strengths due to rf solenoids and dipoles for stored deuteron beams" submitted to Phys. Rev. Lett. (May 20, 2008).
15. A. M. Kondratenko, Proc. 9th International Spin Phys. Symp., Bonn, 1990, p. 140.
16. A. M. Kondratenko *et al.*, Physics of Particles and Nuclei Letters **6**, 528 (2008).
17. E. D. Courant, "Spin motion and resonances in accel. & storage rings", BNL Rpt. AP 292v2 (2008).

Summary Report of PESP2008 - The Workshop on Polarized Electron Sources and Polarimeters

M. Poelker

Thomas Jefferson national Accelerator Facility
12000 Jefferson Ave., Newport News, VA 23606

Abstract. The Workshop on Polarized Electron Sources and Polarimeters (PESP2008) was hosted by Jefferson Lab October 1-3, 2008, in association with SPIN2008, continuing a tradition since 1983. The workshop was well attended with 84 registrants, 38 oral presentations and 17 posters. The usual topics were covered: DC and RF photoguns, high voltage issues, photocathodes, drive lasers, vacuum and polarimetry. Detailed accounts of each contribution (oral and poster) appear within these Proceedings. This submission summarizes some of the highlights of the workshop.

Keywords: polarized electrons
PACS: 29.25.Bx; 29.27.Hj

EXISTING PROGRAMS, NEW FACILITIES AND INITIATIVES

Highly productive nuclear physics experimental programs are being conducted using spin-polarized electron sources at three facilities worldwide: ELSA at the University of Bonn, Germany, the Mainz Microtron at the University of Mainz, Germany and at CEBAF/Jefferson Lab, Newport News, Virginia. The GaAs photoguns at these facilities are the product of years of research and development – they are very reliable, with adequate operating lifetime, beam current capability and polarization (80% or higher).

Other groups with an extensive history of polarized electron source development continue to conduct polarized photogun research, although the experimental nuclear and high energy physics programs at MIT-Bates and SLAC have ended. The group at Nagoya University, Japan, has made noteworthy progress developing a 200kV polarized electron source for the International Linear Collider (ILC) and they have also constructed a novel low-voltage (20kV) polarized electron source with a back-illuminated photocathode for materials science and commercial applications. The source group at the MIT-Bates accelerator has teamed with Brookhaven National Lab (BNL) to develop a high current polarized electron source to meet the requirements of eRHIC - the BNL proposal for an Electron Ion Collider (EIC). And the source group at SLAC has been actively involved in the ILC program, designing (among other things) the electron-beam photoinjector and drive laser. The speakers from these institutions addressed the unique requirements of proposed new accelerators and the need for continued photogun R&D.

CP1149, *Spin Physics, 18th International Spin Physics Symposium*
edited by D. G. Crabb, D. B. Day, S. Liuti, X. Zheng, M. Poelker, and Y. Prok

It was gratifying to hear about the aspirations of polarized electron beam physics at the S-DALINAC, University of Darmstadt, Germany. The Darmstadt group has constructed a 100kV load locked polarized source and is working hard to install and commission this source at the accelerator during 2009.

PHOTOGUN DEVELOPMENT FOR LIGHT SOURCES: VERY HIGH VOLTAGE DC PHOTOGUNS

A number of accelerator-based light source programs use (or plan to use) DC high voltage photoguns to produce very bright electron beams, with small transverse and/or longitudinal emittance. Although these programs do not require polarized electrons, the photoguns at these facilities operate with GaAs photocathodes because this material provides very high quantum efficiency with readily available laser light. Extremely demanding emittance requirements necessitate extremely high bias voltage: typically -500kV or higher. A recurring theme of five presentations from photogun experts at the Jefferson Lab FEL, Cornell University, Daresbury Laboratory, and JAEA, related to high voltage problems, particularly field emission from components at high voltage and associated insulator damage that often leads to catastrophic vacuum loss. Eliminating field emission at gradients up to ~12 MV/m with bias voltage 500kV would be a major victory. Unfortunately, field emission-free operation for most labs seems limited to just ~ 7MV/m with associated maximum voltage of ~ 300kV. One topic of a lunchtime round table discussion was a desire for more robust insulators, particularly insulators with suitable bulk resistivity to "bleed off" accumulated charge stemming from field emission. Another approach uses segmented insulators incorporating metal shields to protect sensitive insulator regions from field emission. The possibility of using "inverted" insulators was discussed, where the insulator extends into the high voltage vacuum chamber eliminating the need for a metal cathode support structure that normally extends through the bore of a conventional insulator. The inverted insulator geometry has the advantage that electric field lines within the structure serve to direct field-emitted electrons toward the vacuum chamber wall at ground potential, rather than toward the insulator, preventing charge build up on the insulator and subsequent damage.

PHOTOGUN DEVELOPMENT FOR LIGHT SOURCES: WARM/COLD RF GUNS

The appealing feature of the RF gun rests in its ability to accelerate electrons to relativistic speed in a very short distance, thereby preserving good beam emittance. Beam energy from an RF gun would be in the ~ MeV range, and consequently an injector could be relatively compact, and with fewer components compared to injectors using DC photoguns. Indeed, numerous RF photoguns operate at laboratories worldwide however none produce polarized electrons and none operate CW. Prospects for producing polarized electrons from an RF gun seem distant mostly because GaAs (the only photocathode material providing polarized electrons to accelerators today) is a delicate material, not capable of withstanding the relatively harsh environment of an RF gun – from a vacuum and field emission point of view.

DULY Research has long advocated a novel warm RF gun design called the Plane Wave Transformer (PWT), a structure with relatively low-Q but an open geometry that supports lots of vacuum pumping in the vicinity of the photocathode compared to standard RF cavity designs, which would satisfy the UHV requirement of GaAs. Removing heat from the structure (and preserving good vacuum), especially when operating in CW mode, seems to be the dominant challenge. DULY has teamed with Fermi National Accelerator Lab to construct a PWT warm-RF gun with a GaAs photocathode for the ILC and experimental results could be forthcoming in 2009. DULY is also modeling a lower energy PWT photogun for CEBAF.

The superconducting RF gun has long been considered the "holy grail" of photogun R&D because it naturally provides superb vacuum with high accelerating gradient in CW mode, without the thermal heat-load problems and concomitant cooling requirements of copper structures. Rossendorf-ZFD has constructed their second SRF gun using a Cs2Te photocathode (unpolarized). They have successfully extracted 1uA of CW beam for hundreds of hours – a significant achievement – however a number of technological challenges remain, including: reliable photocathode exchange and elimination of particulate contamination within the RF structure and subsequent field emission. They plan to operate at significantly higher average current during 2009. BNL has been working with Advanced Energy Systems to construct a very high average current SRF photogun (unpolarized) for their energy recovery linac program and more recently, their partnership has expanded to include polarized e-beam SRF gun development.

In general, SRF gun programs are costly endeavors requiring significant engineering resources and generous construction schedules. These factors make it difficult for some labs to "get in the game". Fortunately, at least two groups are working on this project and positive results could have dramatic impact on design of future accelerators.

PHOTOCATHODES

It is now common for electron beam polarization at experimental facilities to exceed 80%, a monumental achievement stemming from the development of the strained-superlattice photocathode, a structure composed of multiple pairs of very thin layers of lattice-mismatched III-V semiconductor material, for example GaAs grown atop GaAsP. The strain induced within the GaAs crystal serves to break a valence-band energy-level degeneracy that limits polarization from bulk GaAs to < 50%. By growing very thin layers (~ few nanometers thick), one ensures that the strain does not relax and by growing multiple GaAs/GaAsP layer pairs, one ensures sufficiently high QE. There is one commercial supplier of strained-superlattice photocathode material - SVT Associates - and they are now developing alternatives to their GaAs/GaAsP superlattice structure using different III-V elements, with the goal of improving QE and minimizing/eliminating problematic behavior known as "surface charge limit", where QE falls as the power of the incident light is increased. Surface charge limit is a very serious problem that reduces the effective operating lifetime of the gun/photocathode, and likely represents one of the more serious challenges to new programs requiring high bunch charge and/or high average current. Surface charge limit is not well understood – e.g., onset, explanation of physical mechanism, prevention - and is one of the more interesting photogun research topics of the day.

The group at St. Petersburg State Polytechnic University, Russia has long been involved with development of high polarization photocathode material and they continue to model and grow new structures with success, both in terms of beam polarization and QE. A team of Japanese researchers led by T. Nakanishi of Nagoya University reported the development of a remarkable transmission photocathode, exhibiting excellent polarization and world-record beam brightness. The extracted beam is bright (i.e., the emittance small) because the transmission-style photocathode provides a means to illuminate the photocathode from behind using a short focal length lens. In this way, the drive laser beam can be made very small and consequently, the transverse dimensions of the extracted electron beam can be extremely small. The transmission photocathode is the central feature of a spin-polarized low-energy electron microscope (SPLEEM) they have built for material science research. Such a transmission photocathode could find application at accelerator-based physics research facilities, to improve beam quality/emittance and provide a means to simplify the vacuum beamline (no bend magnet required to introduce the drive laser beam into the photogun vacuum chamber, and no laser mirrors inside vacuum beamline downstream of the photogun).

LASERS

The telecommunications industry has made it possible to obtain inexpensive, high-power laser light at wavelengths ~1.06um and 1.56um, thanks to advances in Er and ErYb-doped fiber amplifier technology. This light can be converted to useful wavelengths for photoguns using conventional frequency conversion techniques. The noteworthy contribution to PESP2008 that illustrates the exciting potential of this approach was a description of the fiber-based drive laser developed at Cornell University for the 100mA energy recovery linac (ERL) program. A modelocked diode laser produces picosecond pulses at high pulse repetition rate (1.3GHz), which are delivered to high power fiber amplifiers, and then converted to green light using second harmonic generation to obtain over 15Watts average power. Such a laser system should be capable of producing the desired 100mA average beam current for the Cornell ERL, assuming the bulk GaAs photocathode inside the photogun can sustain reasonable QE (~ few %). A similar fiber-based laser system, albeit at lower average power (~ 2W), was developed at CEBAF/Jefferson Lab at 780nm wavelength suitable for high polarization photocathodes.

VACUUM

Vacuum technology and practice has always been a central issue for the polarized source community, because GaAs photocathodes can only survive within an ultrahigh vacuum environment. Initially, the challenge was to obtain suitably good vacuum to avoid chemical contamination of the photocathode surface. Now the challenge relates to making vacuum improvements that will lead to longer operating lifetime while delivering beam, especially at high current, since lifetime of modern DC high voltage photoguns is limited by ion-backbombardment, where the extracted electron beam ionizes residual gas within the cathode/anode gap. These ions are back-accelerated toward the photocathode surface, sputtering away the chemicals used to produce a negative electron affinity condition and/or damaging the photocathode crystal. The loose consensus within the polarized source community is that polarized photoguns today operate in the low 10^{-11} Torr range. To improve vacuum further, these questions (and likely others) must be addressed:

- where does the gas inside the photogun come from?
- how to limit this gas load?
- how best to pump the gas away?
- how to accurately measure pressure at lower and lower pressure scales?

C. Sinclair of Cornell University reported very low outgassing rates (~ 10^{-14} Torr L $s^{-1}cm^{-2}$) from thin-walled stainless steel baked at 400°C for an extended duration (~ 100 hours) [1]. Similar results were obtained by baking in air or under vacuum, and this low outgassing rate was maintained after venting the chamber to atmospheric pressure with subsequent bakeout to just 150°C. However, thick end flanges are problematic, exhibiting "typical" outgassing of 10^{-12} Torr L $s^{-1}cm^{-2}$, even when baked at 400°C, which can mask the low rate of the thin-walled chamber if flange material constitutes a significant portion of the total surface area of the vacuum chamber.

Non-evaporable getter pumps, first developed at CERN, were embraced by the polarized source community about 10 years ago. These passive pumps have very large hydrogen pump speed and are now considered indispensible, providing most of the pumping in every polarized electron source used at accelerators today. P. Manini of SAES Getters presented a detailed overview of NEG technology, including the relatively new technique of coating the interior vacuum chamber wall with a thin layer of NEG material. He was unaware of limitations these pumps might have in terms of achieving better vacuum (< 10^{-12}Torr).

SUMMARY

This year's PESP workshop marked the first time photogun experts from un-polarized beam programs were invited to attend. By all accounts, the expanded scope contributed to the success of PESP2008. It was clear that both gun communities (polarized and un-polarized) face many shared technological challenges and it was mutually beneficial to learn what each group is doing. For example, the programs to develop very high voltage DC photoguns for light sources will benefit new polarized-electron beam accelerator initiatives such as the ILC, CLIC and EIC, because these same photoguns, in principle, could generate polarized electron beams by simply changing the photocathode material and drive laser wavelength. Operation at considerably higher voltage compared to today's standard 100kV is appealing because photoguns at higher bias voltage produce a "stiffer" beam, suffering less space-charge induced emittance growth, especially when extracting high bunch charge which is a common attribute of electron beams at the new/proposed machines cited above. An ever-growing list of new workshops and conferences creates significant competition for experts and speakers - it might be difficult to gather the world's unpolarized-gun experts for each PESP workshop - but at a minimum, future PESP Organizers should invite gun experts from DC high voltage gun programs for light sources and CW-minded RF gun programs.

New accelerator proposals such as the ILC, CLIC, and EIC demand continued polarized photogun research. These machines are significantly different from today's machines. The polarized electron source community must continue to push the state of the art, even beyond the stated needs of host institutions, to eliminate potential objections to new facilities which might encounter criticism based on absence of required source technology. Some exciting R&D topics include very high average current operation, high bunch charge operation, beam handling at high average current and/or bunch charge, very high voltage operation, field emission suppression, surface charge limit reduction and extreme high vacuum.

REFERENCES

1. C. Park et al, JVST A **26**, 1166 (2008)

The Future of High Energy Polarized Proton Beams [1]

Thomas Roser

Brookhaven National Laboratory, Upton, New York 11793-5000, USA

Abstract. The acceleration and storage of high energy polarized proton beams has made tremendous progress over the last forty years challenging along the way the technologies, precision and the understanding of the beam dynamics of accelerators. This progress is most evident in that one can now contemplate high energy colliders with polarized beams and high luminosity at the same time. After a brief summary of the development and history of polarized proton beam acceleration the paper will focus on the highest energy polarized beams and the possibility of accelerating polarized beams to even higher energies in the future.

Keywords: Polarized beams
PACS: 29.27.Hj, 29.20.dk

SPIN DYNAMICS, RESONANCES AND SIBERIAN SNAKES

Accelerating polarized beams requires an understanding of both the orbital motion and spin motion. Whereas the effect of the spin on the orbit is negligible the effect of the orbit on the spin is usually very strong. The evolution of the spin direction of a beam of polarized protons in external magnetic fields such as exist in a circular accelerator is governed by the Thomas-BMT equation [1],

$$\frac{d\vec{P}}{dt} = -\left(\frac{e}{\gamma m}\right)\left[G\gamma\vec{B_\perp} + (1+G)\vec{B_\parallel}\right] \times \vec{P}$$

where the polarization vector P is expressed in the frame that moves with the particle. This simple precession equation is very similar to the Lorentz force equation which governs the evolution of the orbital motion in an external magnetic field:

$$\frac{d\vec{v}}{dt} = -\left(\frac{e}{\gamma m}\right)\left[\vec{B_\perp}\right] \times \vec{v}.$$

From comparing these two equations it can readily be seen that, in a pure vertical field, the spin rotates $G\gamma$ times faster than the orbital motion. Here $G = 1.7928$ is the anomalous magnetic moment of the proton and $\gamma = E/m$. In this case the factor $G\gamma$ then gives the number of full spin precessions for every full revolution, a number that is also called the spin tune ν_{sp}. At top RHIC energies this number reaches about 400.

[1] Work performed under the auspices of the United States Department of Energy, and with support of RIKEN (Japan)

CP1149, *Spin Physics, 18th International Spin Physics Symposium*
edited by D. G. Crabb, D. B. Day, S. Liuti, X. Zheng, M. Poelker, and Y. Prok

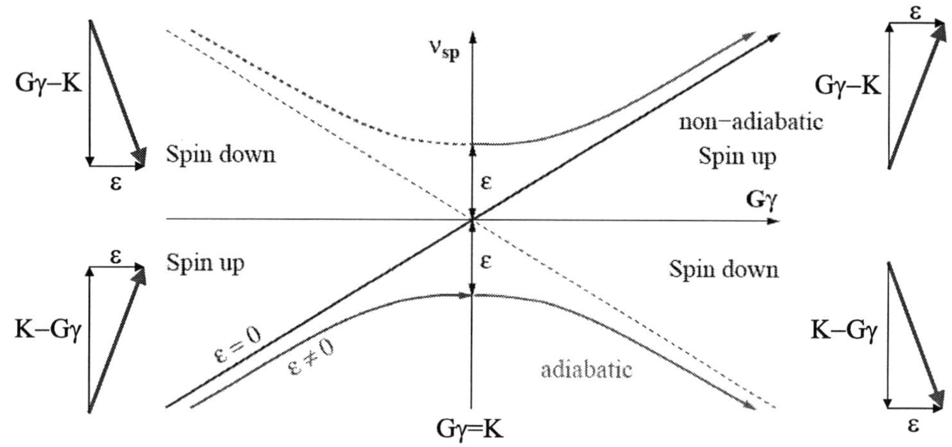

FIGURE 1. The evolution of the spin tune during the crossing of a resonance at $G\gamma = K$ with strength ε.

The Thomas-BMT equation also shows that at low energies ($\gamma \approx 1$) longitudinal fields $\overrightarrow{B_{\parallel}}$ can be quite effective in manipulating the spin motion, but at high energies transverse fields $\overrightarrow{B_{\perp}}$ need to be used to have any effect beyond the always present vertical holding field.

The acceleration of polarized beams in circular accelerators is complicated by the presence of numerous depolarizing spin resonances. During acceleration, a spin resonance is crossed whenever the spin precession frequency equals the frequency with which spin-perturbing magnetic fields are encountered. There are two main types of spin resonances corresponding to the possible sources of such fields: imperfection resonances, which are driven by magnet errors and misalignments, and intrinsic resonances, driven by the focusing fields.

The resonance conditions are usually expressed in terms of the spin tune ν_{sp}. For an ideal planar accelerator, where orbiting particles experience only the vertical guide field, the spin tune is equal to $G\gamma$, as stated earlier. The resonance condition for imperfection depolarizing resonances arise when $\nu_{sp} = G\gamma = n$, where n is an integer. Imperfection resonances are therefore separated by only $523\,MeV$ energy steps. The condition for intrinsic resonances is $\nu_{sp} = G\gamma = kP \pm \nu_y$, where k is an integer, ν_y is the vertical betatron tune and P is the superperiodicity. For example at the AGS, $P = 12$ and $\nu_y \approx 8.8$.

Close to a spin resonance the spin tune deviates away from its value of $G\gamma$ of the ideal flat machine. For a resonance with strength ε, which is the total spin rotation per radian of orbit deflection due to the resonance driving fields, the new spin tune is given by the equation

$$\cos(\pi\nu_{sp}) = \cos(\pi G\gamma)\cos(\pi\varepsilon)\,.$$

Fig. 1 shows the solutions of this equation with and without a resonance. A similar calculation can be done for the effective precession direction or, as it is often called, the stable spin direction or $\overrightarrow{n_0}$. The stable spin direction describes those polarization

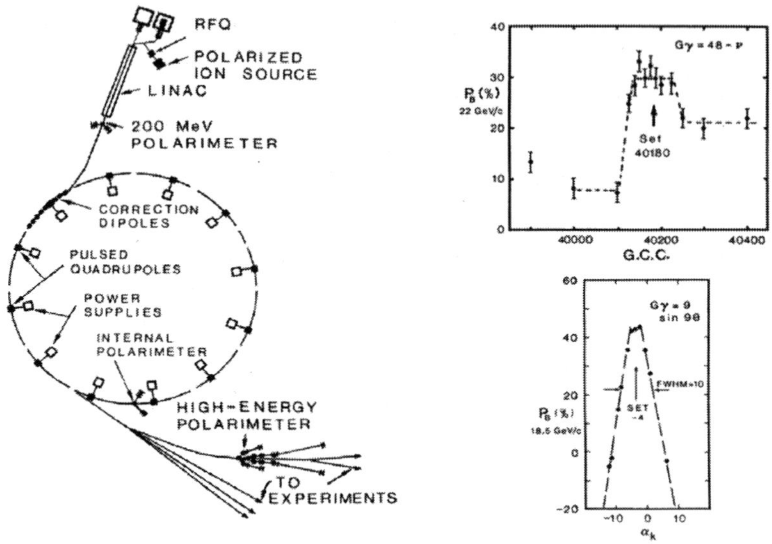

FIGURE 2. The first effort to accelerate polarized protons at the AGS used pused quadrupoles and correction dipoles, shown in the layout on the right to overcome depolarizeing resonances. Typical tuning graphs for the pulsed quadrupole timing (top) and correction dipole strengths (bottom) are shown on the right[4].

components that are repeated every turn. Note that both the stable spin direction and the spin tune are completely determined by the magnetic structure of the accelerator and the beam energy. The magnitude and sign of the beam polarization, however, depends on the beam polarization at injection and the history of the acceleration process.

The spin tune and stable spin direction calculations apply only to a time-independent static situation or if parameters are changed adiabatically. Far from the resonance the stable spin direction coincides with the main vertical magnetic field. Close to the resonance at $G\gamma = K$, the stable spin direction is perturbed away from the vertical direction by the resonance driving fields. When a polarized beam is accelerated through an isolated resonance at arbitrary speed, the final polarization can be calculated analytically [2] and is given by

$$P_f/P_i = 2e^{-\frac{\pi|\varepsilon|^2}{2\alpha}} - 1,$$

where P_i and P_f are the polarizations before and after the resonance crossing, respectively, and α is the change of $(K - \nu_{sp})$ per radian of the orbit deflection. When the beam is slowly ($\alpha \ll |\varepsilon|^2$) accelerated through the resonance, the spin vector will adiabatically follow the stable spin direction resulting in a spin flip as is indicated in Fig. 1. However, for a faster acceleration rate partial depolarization or partial spin flip will occur.

Traditionally, the intrinsic resonances were overcome by using a betatron tune jump, which makes α large, and the imperfection resonances were overcome with harmonic corrections of the vertical orbit to reduce the resonance strength ε. Both of these methods aim at making the resonance crossing non-adiabatic. They were used very successfully for the first two efforts of accelerating polarized proton beams to high energy. At the ANL ZGS polarized protons were accelerated to 12 GeV/c with a maximum polarization of 70% using tune jump quadrupoles[3]. Later, polarized protons were accelerated at the strong-focusing BNL AGS to 22 GeV/c. The intrinsic resonance strengths at the strong-focusing AGS were much stronger than at the ZGS requiring stronger and faster pulsed quadrupoles [4]. Fig. 2 shows the layout used at the AGS as well as tuning curves used to adjust the timing of the pulsed quadrupoles and the strengths of the correction dipoles. These resonance correction techniques require very accurate adjustments at every resonance crossing, which becomes very difficult and time consuming.

New techniques to cross both imperfection and intrinsic resonances adiabatically have been developed. The correction dipoles used to correct the imperfection resonance strength to zero were replaced by a localized spin rotator or 'partial Siberian snake', which makes all the imperfection resonance strengths large and causes complete adiabatic spin flip at every imperfection resonance [5]. The tune jump quadrupoles were initially replaced by a single rf dipole magnet, which increased the strength of the intrinsic resonances by driving large coherent betatron oscillations[6].

More recently two strong partial Siberian snakes were installed in the AGS that can overcome both imperfection and intrinsic resonances. With strong enough partial snakes a gap between the spin tune and an integer opens up that becomes large enough to place the fractional part of the betatron tune and therefore the intrinsic resonance inside this gap. Fig. 3 shows the measured asymmetry during acceleration in the AGS showing adiabatic spin flip at every integer value of $G\gamma$.

At higher energies a 'full Siberian snake' [7], which is a 180° spin rotator of the spin about a horizontal axis, will keep the stable spin direction unperturbed at all times as long as the spin rotation from the Siberian snake is much larger than the spin rotation due to the resonance driving fields. Therefore the beam polarization is preserved during acceleration. An alternative way to describe the effect of the Siberian snake comes from the observation that the spin tune with the snake is a half-integer and energy independent. Therefore, neither imperfection nor intrinsic resonance conditions can ever be met as long as the betatron tune is different from a half-integer.

Fig. 4 shows the lay-out of the Brookhaven accelerator complex highlighting the components required for polarized beam acceleration in the AGS and RHIC. The new 'Optically Pumped Polarized Ion Source'[8] is producing 10^{12} polarized protons per pulse. A single source pulse is captured into a single bunch, which is ample beam intensity to reach the nominal RHIC bunch intensity of 2×10^{11} polarized protons.

In the AGS two partial Siberian snakes are installed. One of them is an iron-based helical dipole[9] that rotates the spin by 11°. The other is a superconducting helical dipole that can reach a 3 Tesla field and a spin rotation of up to 45°[10]. Both helical dipoles have the same design with a variable pitch along the length of the magnet to minimize orbit excursions and also to fit into the 3 m available straight sections in

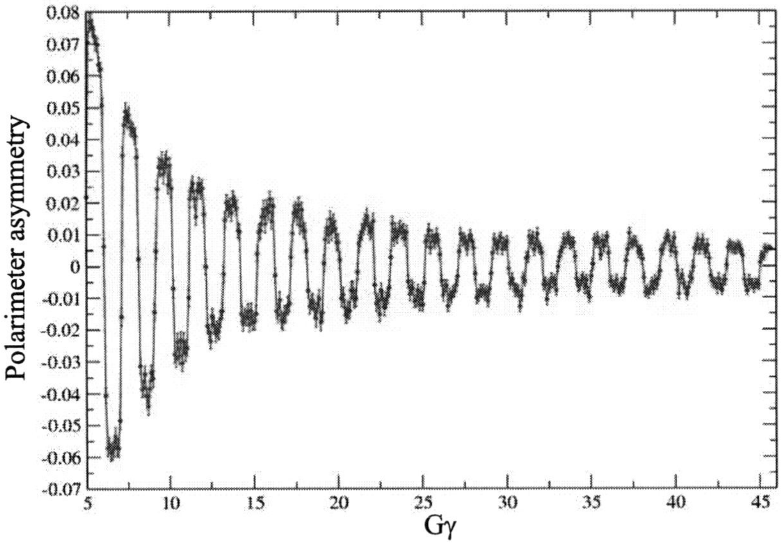

FIGURE 3. Left-right asymmetry measured by the AGS polarimeter during the the acceleration cycle. Note that the asymmetry flips sign at every integer value of $G\gamma$. The drop of the magnitude of the asymmetry during acceleration is mainly due to the decreasing analyzing power of the analyzing reaction of small-angle proton-carbon scattering.

the AGS. With the two partial snakes strategically placed with one third of the AGS ring between them all vertical spin resonances can be avoided up to the required RHIC transfer energy of about 25 GeV as long as the vertical betatron tune is placed at 8.98, very close to an integer[11]. This was achieved reliably over the whole acceleration cycle. With an 80% polarization from the source 65% polarization was reached at AGS extraction. The majority of the remaining polarization loss in the AGS comes from weak spin resonances driven by the horizontal motion of the beam.

The full Siberian snakes (two for each ring) and the spin rotators (four for each collider experiment) for RHIC each consist of four 2.4 m long, 4 T helical dipole magnet modules each having a full 360° helical twist. The two full snakes in RHIC both rotate the spin by 180° around an axis in the horizontal plane and are placed such that the beam deflection between the two snakes is exactly 180°, which guarantees that the spin tune is energy independent. The spin tune is then given by

$$\nu_{sp} = (\alpha_b - \alpha_a)/\pi.$$

Here α_a and α_b are the angles of the rotation axes of the two snakes relative to the beam direction. At RHIC these angles are $\pm 45°$ to achieve a half-integer spin tune. A spin tune of 1/2 reduces the number of snake resonances to a minimum giving more room for placing the betatron tune. With orthogonal snake rotation axes the spin tune is also independent of the betatron amplitude for a single spin resonance[12], which could also minimize the amplitude dependent spin tune shift in the presence of multiple resonances.

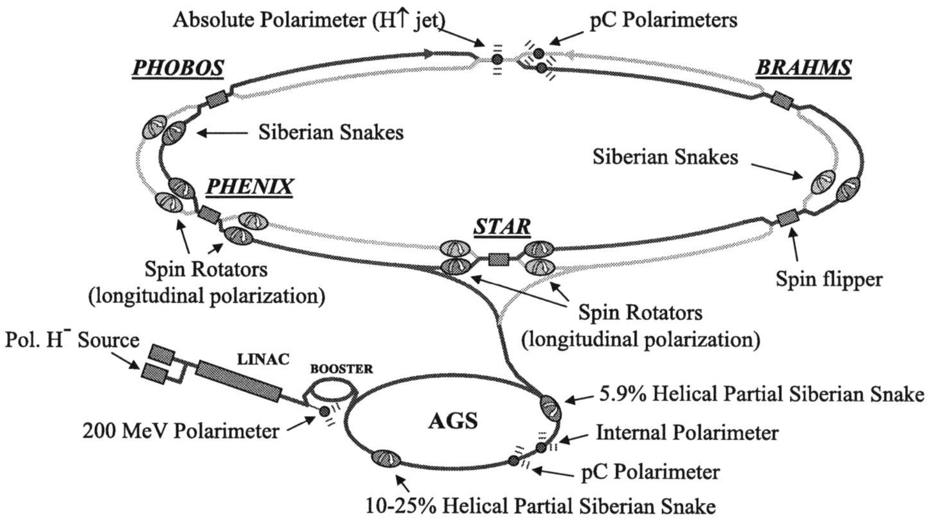

FIGURE 4. The RHIC accelerator complex with the elements required for the acceleration and collision of polarized protons highlighted.

Fig. 5 shows circulating beam current, luminosity and measured circulating beam polarization of a typical store during the 2006 run. A peak luminosity of about 35×10^{30} $\mathrm{cm}^{-2}\ \mathrm{s}^{-1}$ was reached. The beam polarization of up to 65% was calibrated at 100 GeV with an absolute polarimeter. To preserve beam polarization in RHIC during acceleration and storage the vertical betatron tune had to be controlled to better than 0.005 and the orbit had to be corrected to better than 1 mm rms to avoid depolarizing snake resonances[13].

More than 20 years after Y. Derbenev and A. Kodratenko[7] made their proposal to use local spin rotators to stabilize polarized beams in high energy rings, it has now been demonstrated that their concept is working flawlessly even in the presence of strong spin resonances at high energy.

A first successful test of polarization survival during acceleration to 250 GeV crossing three very strong spin resonances was performed[14]. A polarization of 45% was measured at 250 GeV using the polarimeter calibration from 100 GeV. It is not expected that the calibration would change significantly between these two energies, but a more accurate polarization will have to await a calibration measurement with the polarized gas jet performed at 250 GeV. Nevertheless, this preliminary result bodes well for a successful operation of RHIC with polarized 250 GeV proton beams producing collisions at $\sqrt{s} = 500$ GeV with a planned average store luminosity of up to $1.5 \times 10^{32}\ cm^{-2}\ s^{-1}$.

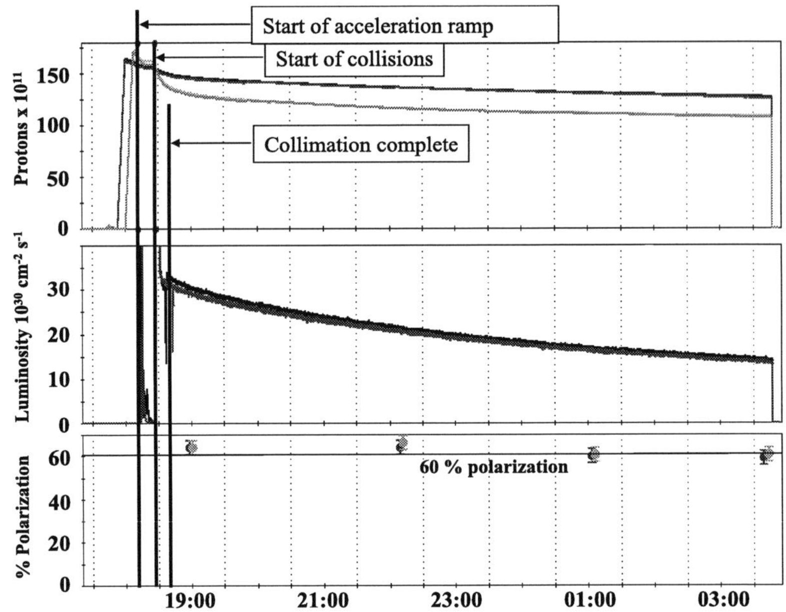

FIGURE 5. Circulating beam in the blue and yellow ring, luminosity at PHENIX (black) and STAR (red(grey)), as well as the measured circulating beam polarization in the blue and yellow RHIC ring (blue(dark) and yellow(light) lines and symbols, respectively) for one typical store.

ACCELERATING POLARIZED PROTONS TO EVEN HIGHER ENERGY

The strength of imperfection resonances approximately increases linearly with energy and the strength of intrinsic resonances increases with the square root of the beam energy. To first order the total spin rotation due to all full or partial Siberian snakes will have to be at least as large as the total spin rotation due to the resonance driving term. More realistically the snakes should at least provide twice the spin rotation of the resonances. This simple rule fits the experience of the AGS, where the partial snakes can overcome the typical AGS resonance strengths of 0.07 and at RHIC, where the two snakes can overcome resonance strengths up to about 0.5. It follows that at higher energy than RHIC multiple snake pairs will be needed. For example, acceleration of polarized protons in the 7 TeV LHC would require at least 16 full snakes or 8 snake pairs to cope with the expected resonance strengths of about 4.

For a ring with N snakes pairs the spin tune is energy independent and equal to 1/2 if

$$\sum_{i=1}^{N}\left(\theta_b^i-\theta_a^i\right)=\pi \text{ and } \sum_{\mathrm{i}=1}^{\mathrm{N}}\left(\alpha_{\mathrm{b}}^{\mathrm{i}}-\alpha_{\mathrm{a}}^{\mathrm{i}}\right)/\pi=1/2.$$

Here θ_a^i, θ_b^i, α_a^i, and α_b^i are the azimuthal locations and snake rotation angles of the

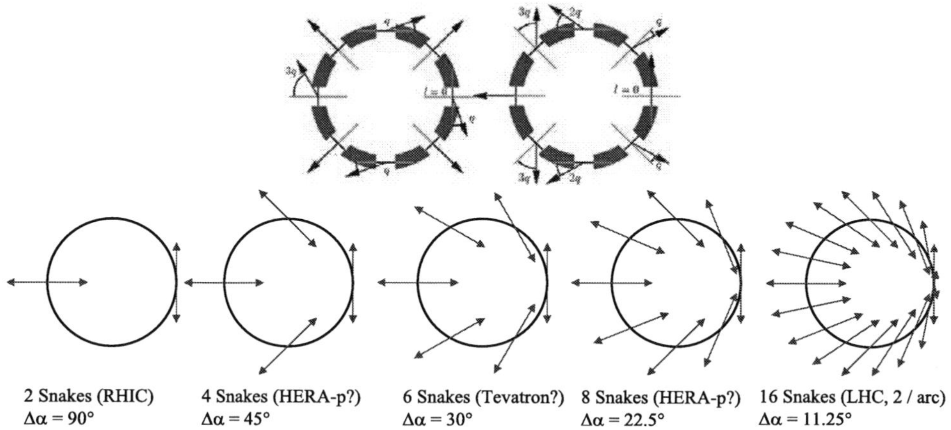

FIGURE 6. The top part shows two snake configurations showing the snake rotation axis for 8 equally spaced snakes[16]. The bottom part shows snake configurations of equally spaces snakes with rotation axis with steadily increasing angle[17]. Such configurations have been shown to have an emittance independent spin tune in a single resonance model

i^{th} snake pair, respectively. After satisfying the above conditions these parameters can then be chosen to maximize the beam polarization of particles with large betatron amplitude. This was first examined by K. Steffen[15]. Later, many configurations for 4 and 8 snakes were examined for optimum performance of the HERA-p machine by G. Hoffstaetter[16]. Two of the best configurations are shown in the top part of Fig. 6. S. Mane[17, 18] showed that snake configurations with rotation axis angle that increase by equal amounts around the ring preserve the property of the two snake configuration of RHIC that the spin tune is independent of the beam emittance for a single resonance and should therefore be a promising configuration with minimized amplitude dependent spin tune shift in the presence of multiple spin resonances. The bottom part of Fig. 6 shows such configurations up to 16 snakes, which would be applicable to LHC with one snake pair per arc.

The increasing strength of intrinsic resonances can be overcome with an increasing number of properly configured Siberian snakes. However, the random residual orbit distortions, after orbit corrections have been performed, drive imperfection resonances that are unaffected by Siberian snakes. S.Y. Lee and E.D. Courant[19] have shown that the maximum tolerable strength for such a resonance is about 0.05. For RHIC this corresponds to about $250\mu m$ rms residual orbit error, which is achievable. However, for LHC this translates to a very challenging $10\mu m$ rms residual orbit error.

Finally the strong focusing quadrupoles of the final focusing triplet of colliders will add spin precession for large amplitude particles leading to a reduced polarization at the interaction point. As Fig. 7 shows this can become a significant effect at high energy colliders.

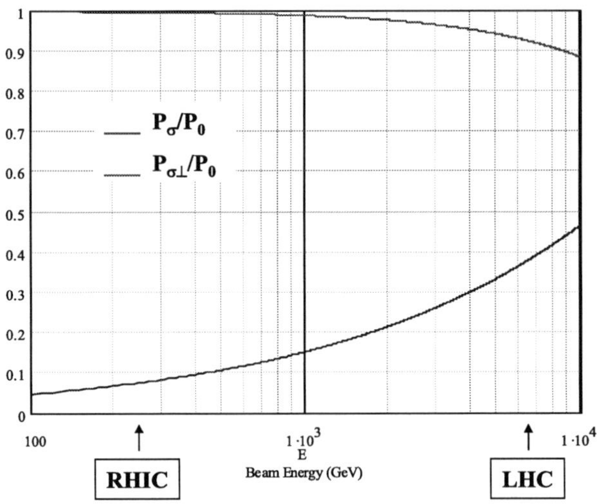

FIGURE 7. Polarization reduction (top) and transverse polarization (bottom) at the interation point at a high energy collider for a rms beam emittance of $3.5\mu m$ and a beta-star of 0.5 m.

POLARIMETRY, SPIN MANIPULATION AND SPIN FLIPPING

In addition to maintaining polarization, the accurate measurement of the beam polarization is of great importance. Very small angle elastic scattering in the Coulomb-Nuclear interference region offers the possibility for an analyzing reaction with a high figure-of-merit, which is not expected to be strongly energy dependent[21]. For polarized beam commissioning in the AGS and RHIC an ultra-thin carbon ribbon is used as an internal target, and the recoil carbon nuclei are detected to measure both vertical and radial polarization components. The detection of the recoil carbon with silicon detectors using both energy and time-of-flight information shows excellent particle identification. It was demonstrated that this polarimeter can be used to monitor polarization of high energy proton beams in an almost non-destructive manner and that the carbon fiber target could be scanned through the circulating beam to measure polarization profiles. A polarized atomic hydrogen jet was also installed in RHIC as an internal target for small angle proton-proton scattering, which allows the absolute calibration of the beam polarization to better than 5 %.

Artificially driven spin resonances can be used to control the polarization of the stored proton beam. Typically a horizontal magnetic dipole field that is modulated at or close to the spin tune frequency is used. These devices are called either rf or ac dipoles depending on the frequency of modulation. As described above an rf dipole was used in the AGS to overcome intrinsic resonances. By ramping the drive frequency through the resonance condition a full spin flip can be achieved[20]. For RHIC with a spin tune of 0.5 the two resonances that a simple rf dipole produces interfere during the resonance crossing and prevent a full spin flip. A new device that consists of two ac dipoles with a spin rotator in between can avoid this problem by producing only a single resonance by properly adjusting the relative amplitude and phase of the ac dipole excitations[22].

This device can also be used to measure the spin tune. By adiabatically exciting the spin flipper at a drive tune close to the spin tune the beam polarization will be tilted away from the otherwise stable vertical direction. If the strength of the spin flipper is larger than the spin tune spread the whole beam polarization will tilt in the same way and can be measured with the polarimeter. The ratio of the vertical and horizontal polarization component is then proportional to the difference between the drive tune and the spin tune.

ACKNOWLEDGMENT

The highly successful operation of the first high energy polarized collider was made possible by the excellent and dedicated RHIC design, construction, commissioning, and operations teams.

REFERENCES

1. L.H. Thomas, Phil. Mag. **3**, 1 (1927); V. Bargmann, L. Michel, V.L. Telegdi, Phys. Rev. Lett. **2**, 435 (1959).
2. M. Froissart and R. Stora, Nucl. Instr. and Meth. **1**, 297 (1960).
3. T. Khoe et al., Part. Accel. **6**, 213 (1975);
4. F.Z. Khiari, et al., Phys. Rev. D**39**, 45 (1989).
5. T. Roser, AIP Conf. Proc. No. 187, ed. K.J. Heller p.1442 (AIP, New York, 1988).
6. M. Bai et al., Phys. Rev. Lett. **80**, 4673 (1998)
7. Ya.S. Derbenev et al., Part. Accel. **8**, 115 (1978).
8. J. Alessi et al., "Polarized H- Source Performance during the 2003 RHIC Run", Proc. of PAC03, Portland, Oregon, p. 3282.
9. J. Takano et al., "Optimization of AGS Polarized Proton Operation with the Warm Helical Snake", Proc. of PAC05, Knoxville, Tennessee, p. 1003.
10. E. Willen et al., "Superconducting Helical Snake Magnet for the AGS", Proc. of PAC05, Knoxville, Tennessee, p. 2935.
11. H. Huang et al., Phys. Rev. Lett. **99**, 146802 (2007)
12. S.R. Mane, Nucl. Instr. and Meth. **A480**, 328 (2002)
13. M. Bai et al., Phys. Rev. Lett. **96**, 174801 (2006); M. Bai et al., "Observations of Snake Resonance in RHIC", Proceedings of PAC05, Knoxville, Tennessee, p. 2839.
14. M. Bai et al., Phys. Rev. Lett. 96, 174801 (2006); M. Bai et al., "Accelerating Polarized Protons to High Energy", Proc. of SPIN2006, to be published.
15. K. Steffen, AIP Conf. Proc. No. 187, ed. K.J. Heller p. 1093 (AIP, New York, 1988).
16. G.H. Hoffstaetter, Phys. Rev. ST Accel. Beams **7**, 121001 (2004)
17. S.R. Mane, Nucl. Instr. and Meth. **A587**, 188 (2008)
18. S.R. Mane, private communication.
19. S.Y. Lee and E.D. Courant, Phys. Rev. **D41**, 292 (1990).
20. D.D. Caussyn et al., Phys. Rev. Lett. **73**, 2857 (1994); B. B. Blinov et al., Phys. Rev. Lett. **81**, 2906 (1998); B. B. Blinov et al., Phys. Rev. Lett. **88**, 014801 (2002); V. S. Morozov et al., Phys. Rev. ST Accel. Beams **7**, 024002 (2004).
21. J. Tojo et al., Phys. Rev. Lett. **89**, 052302 (2002).
22. M. Bai and T. Roser, Phys. Rev. ST Accel. Beams **11**, 091001 (2008)

Nuclear Spin Physics via Polarization Measurements

Tomotsugu Wakasa

Department of Physics, Kyushu University, Higashi, Fukuoka 812-8581, Japan

Abstract. The pion is the most important ingredient for nuclear forces, and it leads to various phenomena such as the three nucleon forces, tensor correlations, and pion condensation. These very interesting phenomena have been extensively investigated by accumulating high-precision experimental data on various observables. Three topics are discussed by comparing these data with recent sophisticated calculations; three nucleon force effects in few nucleon systems, short-range and tensor correlations on spin and isospin responses, and pionic correlations at large momentum transfers.

Keywords: Few nucleon systems, spin-isospin responses, pionic correlations
PACS: 25.40.Kv, 24.70.+s, 25.10.+s, 25.40.Ep,

THREE NUCLEON FORCE EFFECTS IN FEW BODY SYSTEMS

Three Nucleon Forces and $T = 1/2$ Three Nucleon Systems

Modern two-nucleon forces (2NFs), such as CD Bonn [1], AV18 [2], and Nijmegen 93 [3], can describe two-nucleon scatterings and the deuteron property with high precision. However, they do not provide proper information on off-shell effects which appear in many nucleon systems. Indeed the properties of $A > 2$ nuclei could not be explained solely by these forces. Thus the three nucleon force (3NF) effect in these systems is one of the hottest topics in nuclear physics. The first indication of the 3NFs is the under-binding of ^{3}H and ^{3}He found by theoretical calculations. This under-binding of approximately 1 MeV can be resolved by introducing the 3NFs. Presently, 3NFs having 2π- and 3π-exchange diagrams are successfully used in *ab init.* calculations to describe the binding energies and levels of light nuclei up to $A \lesssim 12$ [4]. By introducing the 3NFs, these calculations can reproduce the experimental data reasonably well. In order to obtain further information on the dynamic properties of 3NFs, reaction data are important because energy and spin are controllable and measurable. However, the 3NF effects are expected to be small and thus easily masked by dominant 2NFs. Experimentally, the 3NF effect was first observed in proton-deuteron elastic scatterings at intermediate energies, where the 2NF contribution becomes small and the 3NF effects are expected to be visible at cross section minima at around $\theta_{\text{c.m.}} = 120°$ [5]. The study has been extended to wider range of energies covering 65–400 MeV$/A$, and various spin observables have been measured. Also other types of reactions such as the $p + d \rightarrow {}^3\text{He} + \gamma$ radiative capture and the breakup reactions, which arrow us to study in various kinematical conditions, have been extensively investigated. Some of recent experimental progress can be found in these proceedings [6].

CP1149, *Spin Physics, 18th International Spin Physics Symposium*
edited by D. G. Crabb, D. B. Day, S. Liuti, X. Zheng, M. Poelker, and Y. Prok

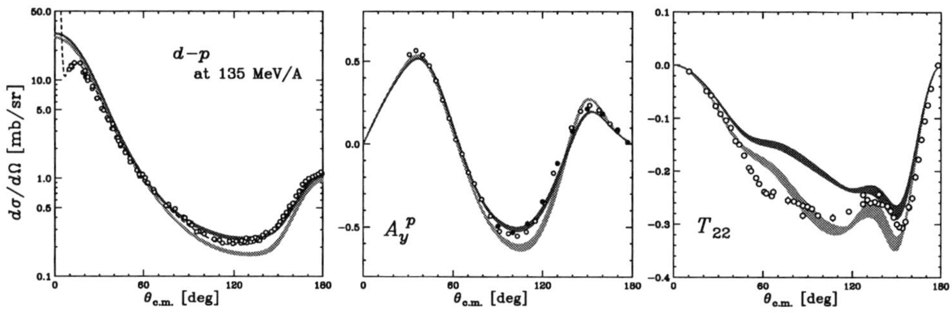

FIGURE 1. The cross section $d\sigma/d\Omega$ (left panel), vector analyzing power A_y (middle panel), and tensor analyzing power T_{22} (right panel) for *p-d* elastic scattering at $E_d = 135$ MeV$/A$. The light shaded bands contain the 2NF predictions, and the dark shaded bands contain the predictions including the 3NF effects. The dashed curve represents the Faddeev calculation with Coulomb force effects. See text for details.

Figure 1 shows the cross section $d\sigma/d\Omega$ and vector analyzing power A_y for *p-d* elastic scattering at $E_d = 135$ MeV$/A$ measured at RIKEN [7]. The light shaded bands are the Faddeev calculations with 2NF only. The bands represent the theoretical predictions based on the various modern 2NFs, which are very close to each other. A clear discrepancy at the cross section minima has been observed, and in addition, the A_y data are also not well-reproduced. The dark shaded bands show the calculations with the 3NF effects by using the 2π-exchange Tucson-Melbourne'(99) (TM'99) 3NF [8]. The 3NF effects can successfully remove the discrepancy between experiment and theory, and thus these data are considered as a clear signature of the 3NF effects. However, in the cross section $d\sigma/d\Omega$, there is still clear discrepancy at very forward angles. These calculations were performed without Coulomb forces because it is very difficult to treat the long-range Coulomb force in the Faddeev calculation. Thus the discrepancy at very forward angles has been considered to be due to Coulomb force effects. A recent progress has been the inclusion of Coulomb forces in Faddeev calculations. The conventional screening and renormalization technique for treating Coulomb forces has been refined and successfully employed for *p-d* scattering [9], and the result is shown by the dashed curve. The discrepancy at very forward angles is successfully removed, and it is found that the Coulomb force effects are limited in this region.

The cross section and vector analyzing power for *N-d* elastic scatterings at $E \lesssim 135$MeV$/A$ have been successfully reproduced by Faddeev calculations by considering 3NF effects. However, those at higher energies such as 250 MeV$/A$ [10, 11] and 400 MeV$/A$ [12] could not be reproduced especially at backward angles. The difference between the experimental data and theoretical calculations becomes significantly large at higher energies, by a factor of two for the cross section at 400 MeV$/A$. This energy dependence of the differences might suggest the importance of considering the relativistic effects because many Faddeev calculations have been performed in a non-relativistic framework. Recently relativistic calculations were performed by approximating the Lorentz-boosted two-nucleon potentials [13]. The effects are found to be limited in very backward angles with very small amplitude, and thus the differences could not be explained by the relativistic effects.

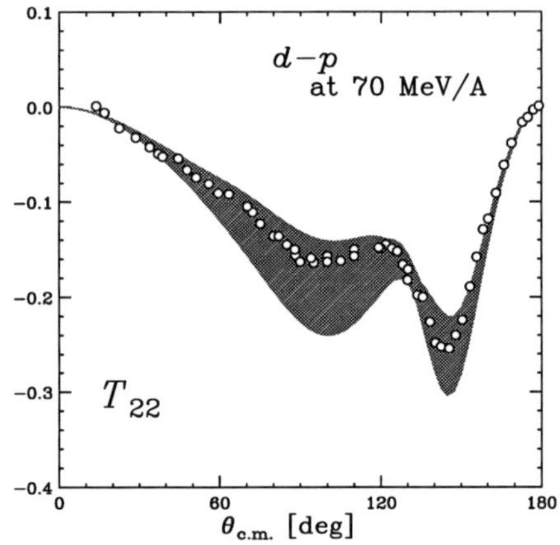

FIGURE 2. The tensor analyzing power T_{22} for *p-d* elastic scattering at $E_d = 70$ MeV/A. The shaded band is the Faddeev calculation based on the chiral effective field theory up to the next-to-next leading order. The band represents the theoretical uncertainty depending on the choice of the cut-off parameter.

Several possibilities have been proposed for better descriptions of data. One is the new type of 3NFs, such as π-ρ and ρ-ρ exchanges, or many Δ contributions with π-rings. The right panel of Fig. 1 shows the tensor analyzing power T_{22} for *p-d* elastic scattering at $E_d = 135$ MeV/A. The light and dark shaded bands represent the Faddeev calculations without and with the 3NF effects, respectively, and both calculations fail to reproduce the experimental data. Because the cross section is successfully reproduced, these disagreement suggest defects in the spin dependence of the present 3NFs. The short-range π-ρ and ρ-ρ 3NFs are expected to be important for resolving these defects. However, the 3NFs employed so far, such as TM'99-3NF, have been introduced more or less independently from 2NFs. The consistency among 2NFs and 3NFs is not clear and thus some consistent formalism is desired for further studies including the short-range 3NF effects. One of such theories is the Chiral Effective Field Theory (χ-EFT) [14]. In this theory, two-nucleon, three-nucleon, and many-nucleon forces can be systematically derived by expansion of the chiral Lagrangian. The net effect of 3NFs in the next leading-order (NLO) becomes zero at the non-relativistic limit. The non-vanishing 3NF appears in the next-to-next leading order (N^2LO). Figure 2 shows a comparison of the tensor analyzing power T_{22} for *p-d* elastic scattering at 70 MeV/A [7] with the Faddeev calculation using chiral effective field potential up to N^2LO [14]. The shaded band represents the theoretical uncertainty depending on the choice of the cut-off parameter. This calculation can reproduce the tensor analyzing power T_{22} within the theoretical uncertainty. Although the calculations are still limited to low energies of $\lesssim 100$ MeV/A, chiral effective field theory seems to be very important and useful to guide the theoretical refinement of 3NFs.

$T = 3/2$ Three Nucleon Systems

Experimental investigations of 3NFs have been extended besides the N-d systems. One is the ^{3}He(p,n) reaction to investigate the three proton system with isospin T = $3/2$. The search for resonances in the three-nucleon system with T = $3/2$ has a long and interesting story. Evidence for T = $3/2$ (three-proton) resonance was found in the ^{3}He(p,n) reaction at E_p = 48.8 MeV [15]. The maximum enhancement from the four-body phase space was observed at the excitation energy $E_x = 9 \pm 1$ MeV with a width $\Gamma = 10.5 \pm 1$ MeV in the three-proton system (where E_x = 0 at the three-proton rest mass energy). Similar enhancements have been observed for measurements at E_p = 24.9 MeV [16] and 197 MeV [17]. However, these studies have attributed the observed enhancements to a ^{1}S$_0$ two-proton final state interaction (FSI), rather than a three-proton resonance. Therefore, the interpretation of the enhancements in the cross-section remains controversial.

Recently three-neutron T = $3/2$ resonances have been studied in the framework of configuration-space Faddeev equations [18, 19]. Unfortunately, these studies have indicated the absence of three-neutron resonances. However, the calculations did not include 3NF effects. Thus, if there exists a three-nucleon resonance with T = $3/2$, the properties of this resonance would yield valuable information on the T = $3/2$ 3NFs [4], on which experimental data are scarce.

Spin-flip probabilities deduced from polarization transfer coefficients are sensitive to the spin–parity J^π of an excited state [20], and thus they are sensitive to the presence of a resonance that has a fixed J^π. Figure 3 shows the double-differential cross-section I and the ratio S_L/S_T of the spin-longitudinal spin-flip probability S_L to the spin-transverse spin-flip probability S_T for the ^{3}He(p,n) reaction at E_p = 346 MeV and $\theta_{\text{lab}} = 0°$ [21, 22]. The slope of the cross-section spectrum near the threshold is primarily determined by the phase space factor, and is consistent with that at E_p = 197 MeV [17]. Palarczyk *et al.* [17] reported phase-space calculations and showed that the increase in cross-section near the threshold is reproduced better by three-body than four-body phase space calculations. This means that the present ^{3}He(p,n) reaction can be described by $n(p,n)p$ quasi-elastic scattering on the neutron in ^{3}He, whereas the two protons in the relative ^{1}S$_0$ state act as a spectator. Thus, the measured spin-flip probabilities are expected to be well described by the corresponding free *NN* values if there is no resonance contribution.

The dashed curve in the lower panel of Fig. 3 represents the corresponding free *NN* values with the FA07 phase-shift solution [23]. Significant discrepancies are observed in S_L/S_T at $\omega_{\text{lab}} \sim 15$ MeV, which can be interpreted as evidence for three-proton resonance effects. The J^π of the resonance is estimated to be $1/2^-$ by comparing with distorted-wave impulse approximation (DWIA) predictions for S_L/S_T of the ^{3}He$(\vec{p},\vec{n})$ reaction.

The strength distribution of the resonance is deduced by assuming that the observed ^{3}He(p,n) cross-section comprises an incoherent sum of contributions from quasi-elastic scattering and the resonance with $J^\pi = 1/2^-$. The filled histogram in the upper panel of Fig. 3 represents the estimated contribution from the resonance with $J^\pi = 1/2^-$. By considering the contribution from $J^\pi = 1/2^-$, the S_L/S_T values are well reproduced as shown by the solid curve in the lower panel. The center ω_0 and the width Γ of the resonance are estimated to be $\omega_0 = 16 \pm 1$ MeV and $\Gamma = 11 \pm 3$ MeV, which is consistent

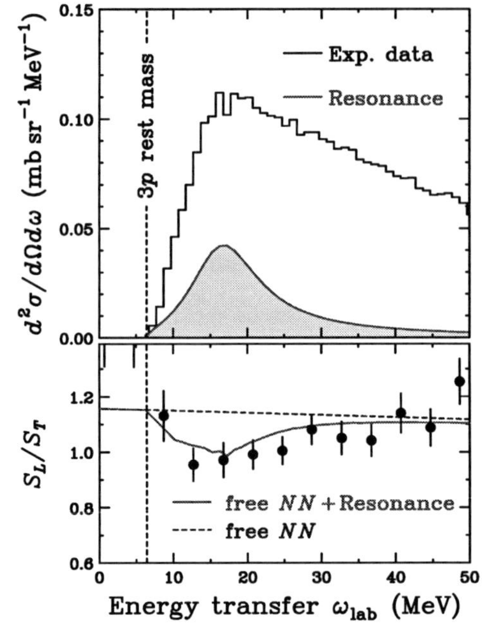

FIGURE 3. The double-differential cross section (upper panel) and the ratio of spin-flip probabilities S_L/S_T (lower panel) for the ^{3}He(p,n) reaction at $E_p = 346$ MeV and $\theta_{lab} = 0°$. The dashed curve is the corresponding free nucleon-nucleon values with the FA07 phase shift solution [23]. The filled histogram in the upper panel represents the estimated contribution from the $J^\pi = 1/2^-$ resonance. The solid curve in the lower panel shows the theoretical prediction including the resonance contribution.

with previous results obtained from the same reaction at E_p = 48.8 MeV. Further investigations are required for interpretation of the discrepancy in S_L/S_T, however, this data is challenging for the investigation of $T = 3/2$ 3NFs.

GAMOW-TELLER AND SPIN-DIPOLE STRENGTHS

Gamow-Teller Strength and Landau-Migdal Parameters

Recent high-quality (p,n) [24] and (n,p) [25] experiments at $E_p \simeq 300$ MeV have presented reliable information on the Gamow-Teller (GT) transition strength [26]. The total strength of the GT transition including the continuum is slightly quenched from its sum rule value $3(N-Z)$ [27]. A key concept to understand this phenomenon is the Landau-Migdal (LM) parameters, g'_{NN} and $g'_{N\Delta}$, which specify the particle-hole (ph) and Delta-hole (Δh) interactions. The LM parameters also specify the residual interaction at large momentum transfers. Thus the precise information on these parameters are very important to investigate the nuclear correlation effects, such as a precursor of pion condensation, at large momentum transfers.

The GT transition strengths, $R^{\pm}_{\mathrm{GT}}(\omega)$, are experimentally extracted from the $\Delta J^{\pi} = 1^{+}$ cross section, $d^2\sigma_{1^+}(q,\omega)/d\Omega d\omega$, determined by the multi-pole decomposition analysis (MDA) as [24]

$$\frac{d^2\sigma_{1^+}(q,\omega)}{d\Omega d\omega} = \hat{\sigma}_{\mathrm{GT}} F(q,\omega) R^{\pm}_{\mathrm{GT}}(\omega) \ , \tag{1}$$

with the GT unit cross section $\hat{\sigma}_{\mathrm{GT}}$ and the (q,ω) dependence factor $F(q,\omega)$.

The strength distribution of the GT^- transitions, namely, $R^{-}_{\mathrm{GT}}(\omega)$, from ^{90}Zr to ^{90}Nb was determined from the MDA of the (p,n) data [24, 28]. They cover not only the GT giant resonance (GTGR) region but also the continuum region up to $E_x = 50$ MeV. The GTGR peak position strongly depends on g'_{NN} in the random phase approximation (RPA) calculation, whereas the dependence on $g'_{N\Delta}$ is so weak. The RPA result with $g'_{NN} = 0.60 \pm 0.10$ [29] reproduces the experimental data reasonably well, which is consistent with the previous analyses [30, 31].

Recently, using the results of MDA applied to both the $^{90}\mathrm{Zr}(p,n)$ and $^{90}\mathrm{Zr}(n,p)$ data, Yako *et al.* [25] determined the GT quenching factor Q defined as

$$Q = \frac{S^{-}_{\mathrm{GT}}(\omega^{-}_{\mathrm{top}}) - S^{+}_{\mathrm{GT}}(\omega^{+}_{\mathrm{top}})}{3(N-Z)} \ , \tag{2}$$

with

$$S^{\pm}_{\mathrm{GT}}(\omega^{\pm}_{\mathrm{top}}) = \int^{\omega^{\pm}_{\mathrm{top}}} R^{\pm}_{\mathrm{GT}}(\omega) d\omega \ , \tag{3}$$

where $\omega^{\pm}_{\mathrm{top}}$ are the end energies of the integration of $R^{\pm}_{\mathrm{GT}}$. The $\omega^{-}_{\mathrm{top}}$ value is 57 MeV (E_x = 50 MeV of ^{90}Nb excitation) and the relevant $\omega^{+}_{\mathrm{top}}$ is chosen by taking into account for the Coulomb energy shift. The result is $Q = 0.86 \pm 0.07$ [25].

The quenching factor Q has been evaluated by the RPA calculations, and it is found that the results are almost completely determined by the $g'_{N\Delta}$ value. The $g'_{N\Delta}$ value is estimated to be 0.35 ± 0.16 [29], which coincides with the result of Arima *et al.* [32]. It should be noted that the $g'_{N\Delta}$ value is significantly smaller than the g'_{NN} value, and thus the universality ansatz, $g'_{NN} = g'_{N\Delta}$, does not hold.

Spin-Dipole Strength and Tensor Force Effects

The tensor force is a spin-dependent non-central interaction, and is a manifestation of meson exchange between nucleons. The tensor force effects in nuclear structures have been investigated over many decades. Recently the tensor force effects are predicted in the distribution of spin-dipole strengths. The spin-dipole has three different spin and parities, 0^-, 1^-, and 2^-, and the tensor force effects depend on spin and parities. Theoretical investigations have been extensively performed for a mass number A=12 system. Theoretically, the 1^- strength is expected to be fragmented due to the repulsive tensor force effect, whereas the 0^- strength is softened due to the attractive tensor force effect, and thus should be observed at $E_x \simeq 9$ MeV [33]. Experimentally, the 0^- state in $A = 12$ system was firstly observed by measuring the tensor analyzing powers for the

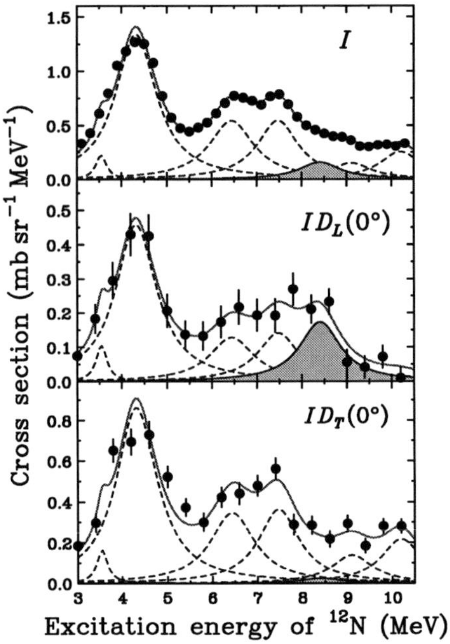

FIGURE 4. The double differential cross section I (upper panel), spin-longitudinal polarized cross section ID_L (middle panel), and spin-transverse polarized cross section ID_T (lower panel) for the $^{12}C(p,n)^{12}N$ reaction at E_p = 296 MeV and θ_{lab} = 0°. The curves represent the results of peak fitting. The contributions from the 0^- component at E_x = 8.4 MeV are shown by the filled histograms.

$^{12}C(\vec{d},{}^2He)^{12}B$ reaction, and the 0^- state in ^{12}B has been identified at E_x = 9.3 MeV [34, 35].

Later, the 0^- state in ^{12}N has been also identified by measuring a complete set of polarization transfer observables for the $^{12}C(\vec{p},\vec{n})^{12}N$ reaction at E_p = 296 MeV and θ_{lab} = 0° [36]. Figure 4 shows the experimental cross section I, spin-longitudinal polarized cross section ID_L, and spin-transverse polarized cross section ID_T in the spin-dipole region. Note that the 0^- and 1^- states would appear predominantly in the ID_L and ID_T spectra, respectively, whereas the 2^- state would appear in both spectra. The well-established 2^- state at $E_x \simeq 4$ MeV forms its peak in both ID_i spectra as expected. The curves are the results of peak fitting employing the parameters for excitation energies and widths in Ref. [37]. The spin-dipole resonance at $E_x \simeq 7$ MeV can be reproduced by two peaks at E_x = 6.4 and 7.5 MeV, and these peaks appear predominantly in both ID_i spectra. This result means that the spin-dipole resonance at $E_x \simeq 7$ MeV consists of mainly 2^- components and supports previous $^{12}C(\vec{d},{}^2He)^{12}B$ experiments [34, 35]. A large difference is observed at E_x = 8.4 MeV between the ID_i spectra; a clear peak appears in the ID_L spectrum but disappears in the ID_T spectrum, suggesting large contribution from the 0^- component. This result is consistent with the above-mentioned theoretical prediction in which the 0^- state is softened due to tensor correlations [33].

PIONIC CORRELATIONS IN SPIN AND ISOSPIN RESPONSES

The search for pionic enhancements in nuclei has a long and interesting history. These phenomena can be considered as a precursor of the pion condensation [38] that would be realized in neutron stars. Enhancements of the $M1$ cross section in proton inelastic scattering [39, 40, 41, 42] and of the ratio R_L/R_T, the spin-longitudinal (pionic) response function R_L to the spin-transverse (non-pionic) response function R_T, in the quasi-elastic scattering (QES) region [43, 44] were expected around a momentum transfer $q_{\mathrm{c.m.}} \simeq 1.7\,\mathrm{fm}^{-1}$. However, experimental data such as the differential cross section of the $M1$ transition ${}^{12}\mathrm{C}(p,p')^{12}\mathrm{C}(1^+,T=1)$ [45] and the ratio R_L/R_T extracted from ${}^{12}\mathrm{C}, {}^{40}\mathrm{Ca}(\vec{p},\vec{n})$ [46, 47] did not reveal any enhancements, and they were considered as evidences against the precursor phenomena of the pion condensation in normal nuclei. Several explanations exist to answer the question why no pionic enhancements were observed. For example, Bertsch, Frankfurt, and Strikman [48] suggest the modification of gluon properties in the nucleus that suppresses the pion field. Brown *et al.* [49] suggest the partial restoration of chiral invariance with density. However, we should note that the $M1$ cross section involves both the pionic and the non-pionic transitions, and similarly the non-pionic response R_T is equally important to determine the ratio R_L/R_T. Thus, in these indirect measurements, the pionic enhancement might be masked by the contribution from the non-pionic component. Recent analysis of the QES data [50] shows a pionic enhancement in the spin-longitudinal cross section that well represents the R_L, and suggest that the lack of enhancements of R_L/R_T is due to the non-pionic component.

In order to measure the pionic enhancement directly, it is desirable to investigate isovector $J^\pi = 0^-$, $0^\pm \to 0^\mp$ excitations because they carry the same quantum numbers as the pion and they are free from non-pionic contributions in the direct channel. Orihara *et al.* [51] measured the angular distribution of the ${}^{16}\mathrm{O}(p,n)^{16}\mathrm{F}(0^-)$ reaction at E_p = 35 MeV. They reported discrepancies between distorted wave Born approximation calculations and their data in the range of $q_{\mathrm{c.m.}}$ = 1.4–2.0 fm^{-1} that might be a signature of pionic enhancement. However, in both the ${}^{16}\mathrm{O}(p,p')^{16}\mathrm{O}(0^-,T=1)$ scattering at $E_p = 65$ MeV [52] and the ${}^{16}\mathrm{O}(p,n)^{16}\mathrm{F}(0^-)$ reaction at $E_p = 79$ MeV [53], such an enhancement was not observed. The differences in these (p,n) and (p,p') results might indicate contributions from complicated reaction mechanisms at these low incident energies.

The high-resolution West-South (WS) beam line [54] has been constructed at RCNP, Osaka University. This WS beam line enables us to perform very high-resolution measurements with less than 20 keV in full width at half maximum (FWHM) even at E_p = 400 MeV. At intermediate energies of 200–400 MeV, distortion effects become smallest, and thus the theoretical DWIA calculations should be reliable owing to the simple reaction mechanism. Figure 5 shows the measured data points and the calculated curves of the cross sections of the 0^-, $T=1$ transition in ${}^{16}\mathrm{O}(p,p')$ at $E_p = 295$ MeV [55] as a function of the momentum transfer $q_{\mathrm{c.m.}}$. The angular distribution was measured in the range of $q_{\mathrm{c.m.}} \simeq 1.0\,\mathrm{fm}^{-1}$ to $\simeq 2.1\,\mathrm{fm}^{-1}$ starting near the second maximum at $q_{\mathrm{c.m.}} \simeq 0.9\,\mathrm{fm}^{-1}$ and extending beyond the third maximum at $q_{\mathrm{c.m.}} \simeq 1.7\,\mathrm{fm}^{-1}$. The error bars of the data points are statistical uncertainties.

The dashed curve shows the DWIA calculation without pionic correlation. The data

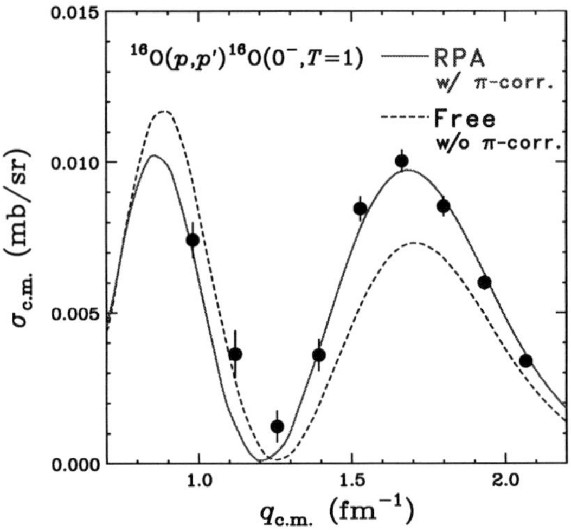

FIGURE 5. Comparison between experimental and theoretical cross section for the $^{16}O(p,p')^{16}O(0^-,T=1)$ scattering at $E_p = 295$ MeV. The dotted curve represents the DWIA result with the free response function. The solid curve denotes the DWIA result with the RPA response function with $g'_{NN} = 0.60$ and $g'_{N\Delta} = 0.35$.

show a clear enhancement from the calculation at large momentum transfers. Thus we performed DWIA calculations with the RPA response functions employing the $\pi+\rho+g'$ model interaction $V_{\rm eff}$ in order to investigate pionic correlation effects. The interaction $V_{\rm eff}$ is the sum of the one-π and one-ρ exchange interactions, and the LM interaction $V_{\rm LM}$ specified by the LM parameters. The solid curve in Fig. 5 shows the DWIA result with the RPA response function employing $g'_{NN} = 0.60$ and $g'_{N\Delta} = 0.35$ determined from the GT strength, This calculation reproduces the experimental data reasonably well. It shows large enhancement around $q_{\rm c.m.} \simeq 1.7\,{\rm fm}^{-1}$, reflecting the pionic enhancement. Around $q_{\rm c.m.} \simeq 1.7\,{\rm fm}^{-1}$, $V_{\rm eff}$ with these LM parameters is close to zero in the NN channel, but very attractive in the $N\Delta$ channel [50]. This attraction causes the pionic enhancement.

ACKNOWLEDGMENTS

I am grateful to K. Sekiguchi, K. Yako, Y. Maeda, M. Dozono, and E. Ihara for their valuable comments and help in preparing the presentation. I also acknowledge all collaborators of the relevant experiments performed at RCNP and RIKEN.

REFERENCES

1. R. Machleidt, *Phys. Rev. C* **63**, 024001 (2001).
2. R. B. Wiringa, V. G. J. Stoks, and R. Schiavilla, *Phys. Rev. C* **51**, 38 (1995).
3. V. G. J. Stoks, et al., *Phys. Rev. C* **49**, 2950 (1994).
4. S. C. Pieper, et al., *Phys. Rev. C* **64**, 014001 (2001).

5. H. Witała, et al., *Phys. Rev. Lett.* **81**, 1183 (1998).
6. N. Kalantar-Nayestanaki, *in Proceedings of the 18th International Spin Physics Symposium* (2009).
7. K. Sekiguchi, et al., *Phys. Rev. C* **65**, 034003 (2002).
8. S. A. Coon, and H. K. Han, *Few Body Syst.* **30**, 131 (2001).
9. A. Deltuva, A. C. Fonseca, and P. U. Sauer, *Phys. Rev. C* **71**, 054005 (2005).
10. K. Hatanaka, et al., *Phys. Rev. C* **66**, 044002 (2002).
11. Y. Maeda, et al., *Phys. Rev. C* **76**, 014004 (2007).
12. A. Tamii, et al., *AIP Conf. Proc.* **915**, 765 (2007).
13. H. Witała, et al., *Phys. Rev. C* **71**, 054001 (2005).
14. E. Epelbaum, *Prog. Part. Nucl. Phys.* **57**, 654 (2006).
15. L. E. Williams, et al., *Phys. Rev. Lett.* **23**, 1181 (1969).
16. A. D. Bacher, et al., *Phys. Lett. B* **29**, 573 (1969).
17. M. Palarczyk, et al., *Phys. Rev. C* **58**, 645 (1998).
18. H. Witała, and W. Glöckle, *Phys. Rev. C* **60**, 024002 (1999).
19. A. Hemmdan, W. Glöckle, and H. Kamada, *Phys. Rev. C* **66**, 054001 (2002).
20. J. M. Moss, *Phys. Rev. C* **26**, 727 (1982).
21. T. Wakasa, et al., *Phys. Rev. C* **77**, 054611 (2008).
22. E. Ihara, et al., *Phys. Rev. C* **78**, 024607 (2008).
23. R. A. Arndt, et al., *Phys. Rev. C* **76**, 025209 (2007).
24. T. Wakasa, et al., *Phys. Rev. C* **55**, 2909 (1997).
25. K. Yako, et al., *Phys. Lett. B* **615**, 193 (2005).
26. M. Ichimura, H. Sakai, and T. Wakasa, *Prog. Part. Nucl. Phys.* **56**, 446 (2006).
27. C. Gaarde, et al., *Nucl. Phys. A* **334**, 248 (1980).
28. T. Wakasa, et al., *J. Phys. Soc. Jpn.* **73**, 1611 (2004).
29. T. Wakasa, H. Sakai, and M. Ichimura, *Phys. Rev. C* **67**, 067303 (2005).
30. T. Suzuki, *Nucl. Phys. A* **379**, 110 (1981).
31. G. Bertsch, D. Cha, and H. Toki, *Phys. Rev. C* **24**, 533 (1981).
32. A. Arima, et al., *Phys. Lett. B* **499**, 104 (2001).
33. T. Suzuki, and H. Sagawa, *Nucl. Phys. A* **637**, 547 (1998).
34. H. Okamura, et al., *Phys. Rev. C* **66**, 054602 (2002).
35. M. A. de Huu, et al., *Phys. Lett. B* **649**, 25 (2007).
36. M. Dozono, et al., *J. Phys. Soc. Jpn.* **77**, 014201 (2008).
37. B. D. Anderson, et al., *Phys. Rev. C* **54**, 237 (1996).
38. A. B. Migdal, *Zh. Eksp. Teor. Fiz.* **61**, 2210 (1971).
39. H. Toki, and W. Weise, *Phys. Rev. Lett.* **42**, 1034 (1979).
40. J. Delorme, et al., *Phys. Lett. B* **89**, 327 (1980).
41. J. Delorme, A. Figureau, and N. Giraud, *Phys. Lett. B* **91**, 328 (1980).
42. H. Toki, and W. Weise, *Phys. Lett. B* **92**, 265 (1980).
43. W. M. Alberico, M. Ericson, and A. Molinari, *Phys. Lett. B* **92**, 153 (1980).
44. W. M. Alberico, M. Ericson, and A. Molinari, *Nucl. Phys. A* **379**, 429 (1982).
45. J. R. Comfort, et al., *Phys. Rev. C* **23**, 1858 (1981).
46. T. N. Taddeucci, et al., *Phys. Rev. Lett.* **73**, 3516 (1994).
47. T. Wakasa, et al., *Phys. Rev. C* **59**, 3177 (1999).
48. G. F. Bertsch, L. Frankfurt, and M. Strikman, *Science* **259**, 773 (1993).
49. G. E. Brown, et al., *Nucl. Phys. A* **593**, 295 (1995).
50. T. Wakasa, et al., *Phys. Rev. C* **69**, 054609 (2004).
51. H. Orihara, et al., *Phys. Rev. Lett.* **49**, 1318 (1982).
52. K. Hosono, et al., *Phys. Rev. C* **30**, 746 (1984).
53. R. Madey, et al., *Phys. Rev. C* **56**, 3210 (1997).
54. T. Wakasa, et al., *Nucl. Instrum. Methods Phys. Res. A* **482**, 79 (2002).
55. T. Wakasa, et al., *Phys. Lett. B* **632**, 485 (2006).

Project of the Nuclotron-based Ion Collider fAcility (NICA) at JINR, Dubna: Perspectives of heavy ion and spin physics

Richard Lednický[a,b]

[a]Joint Institute for Nuclear Research, RU-141980, Dubna, Moscow Region, Russia
[b]Institute of Physics ASCR, Na Slovance 2, CZ-18221, Prague, Czech Republic

Abstract. One of the main directions of the research program at the Joint Institute for Nuclear Research (JINR) in Dubna is the relativistic nuclear and spin physics. The present basic facility for this research is the 6 A GeV superconducting synchrotron – Nuclotron. In the coming years, the new JINR flagship program assumes the experimental study of hot and dense strongly interacting QCD matter and polarization phenomena at the new JINR facility. This goal is proposed to be reached by (i) development of the existing Nuclotron accelerator facility as a basis for generation of intense beams over atomic mass range from protons to uranium and light polarized ions, (ii) design and construction of the Nuclotron-based Heavy Ion Collider fAcility (NICA) with the maximum nucleon–nucleon center-of-mass energy of $\sqrt{s_{NN}}$ = 9 GeV and averaged luminosity 10^{27} cm^{-2} s^{-1} for U+U collisions, and (iii) design and construction of the Multipurpose Particle Detector (MPD) and Spin Physics Detector (SPD) at intersecting beams. Realization of the project will lead to unique conditions for research activity of the world community.

Keywords: Ion collider, phase transitions, spin physics
PACS: 13.88.+e, 25.75.-q, 29.20.db, 29.20.dk

SCIENTIFIC GOALS

The planned investigations at the Nuclotron-based Heavy Ion Collider Facility (NICA) [1-3] are relevant to understanding the evolution of the Early Universe after the Big Bang, formation of neutron stars, and the physics of heavy-ion collisions and spin phenomena.

The new JINR facility NICA will make it possible to study in-medium properties of hadrons and the nuclear matter equation of state, including a search for possible signatures of deconfinement and/or chiral symmetry restoration phase transitions and QCD critical endpoint in the region of $\sqrt{s_{NN}}$ = 4–9 GeV by means of careful scanning in beam energy, atomic number and centrality of the excitation functions. The first-stage measurements include multiplicities and global characteristics of identified hadrons, such as: fluctuations in multiplicity and transverse momenta; directed and elliptic flows for various hadrons; and femtoscopic momentum correlations. Electromagnetic probes (photons and dileptons) are supposed to be added at the second stage of the project.

CP1149, *Spin Physics, 18th International Spin Physics Symposium*
edited by D. G. Crabb, D. B. Day, S. Liuti, X. Zheng, M. Poelker, and Y. Prok

The beam energy of the NICA is very much lower than those of the RHIC and the LHC, but it sits right on top of the region where the baryon density at the freeze-out is expected to be the highest. In this energy range, the excited system occupies a maximal space–time volume in the mixed quark–hadron phase—the coexistence phase of hadron and quark–qluon matter similar to the water-vapor coexistence phase (see the schematic phase diagram in Fig. 1). Similar conditions will be achieved in collider experiments during low energy scan at RHIC in 2010 (however, with much lower luminosity) and, in the fixed target experiments at SPS CERN and the FAIR facility in Darmstadt (after the synchrotron SIS300 is put into operation in 2016).

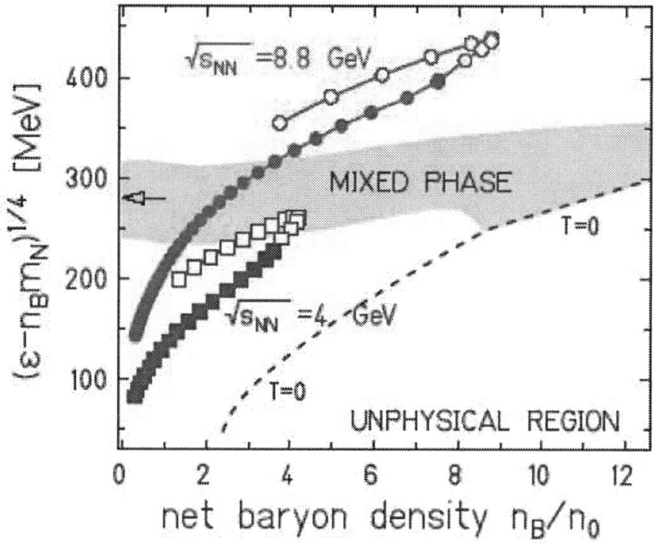

FIGURE 1. The phase diagram in terms of the reduced energy density vs the ratio of the net baryon density to the normal nuclear density [2]. The dynamical trajectories are shown for central Au+Au collisions at two extreme NICA energies $\sqrt{s_{NN}}$ = 8.8 and 4 GeV; the points correspond to the time steps of 0.3 and 0.5 fm/c, respectively.

Since 1987 the problem of the nucleon spin structure (so called spin crisis) remains to be one of the most intriguing puzzles of the high energy physics. While today a part of the spun-dependent Parton Distribution Functions (PDFs) can be considered as sufficiently well known, there is a number of PDFs which are either unknown or poorly known: these are distributions of longitudinally polarized light sea and strange quarks, and also gluons, as well as, distributions of both transversely polarized sea and valence quarks of all flavours; this is also a new class of PDFs, characterized by the non-trivial dependence on the transverse parton momentum, such as the Sivers and Boer-Mulders PDFs. Considerable interest is also devoted to the so called Generalized Parton Distributions (GPDs).

These open questions belong to the first priority tasks for the scientific research program to be realized at the second interaction point of the NICA facility with the polarized proton and deuteron beams up to the center-of-mass energy of ~20 GeV.

The high intensity and high polarization of the beams will provide unique possibilities for spin physics research (unavailable in existing and planned facilities), particularly for a study of Matveev-Muradyan-Tavkhelidze-Drell-Yan (MMT-DY) processes, not requiring the input from the (poorly known) fragmentation functions.

NICA GENERAL LAYOUT

The construction of the new facility is based on the existing buildings and infrastructure of the Synchrophasotron/Nuclotron of the JINR Veksler–Baldin Laboratory of High Energy Physics. The accelerator chain includes heavy-ion source, RFQ injector, linac, booster ring, Nuclotron and superconducting collider rings (see Fig. 2). The peak design kinetic energy of U^{92+} ions in the collider is 3.5 A GeV. Beam cooling and bunching systems are foreseen to achieve the average luminosity of 10^{27} cm^{-2} s^{-1} in the U+U collisions The maximum magnetic rigidity of the collider rings is chosen to be equal to the Nuclotron one in order to have a possibility to increase center-of-mass energy of the heavy ion collisions up to about $\sqrt{s_{NN}}$ = 11 GeV at the next stage of the project realization. The project design presumes the continuation of some of the fixed-target experiments, including those with polarized deuteron beam from the Nuclotron. The concept of the NICA project was first presented and discussed at the round table discussion in October 2006, the present project status is available in the Conceptual Design Report [2] and the Technical Design Report is supposed to be prepared during 2009.

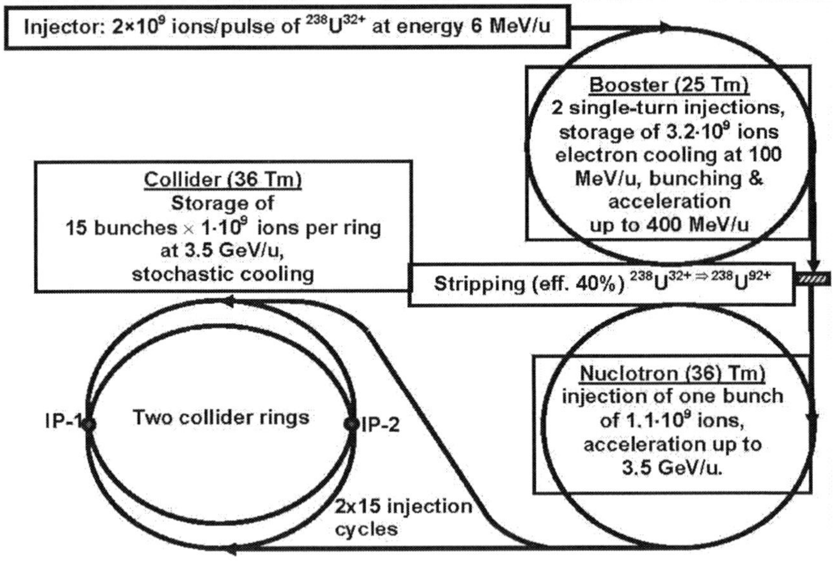

FIGURE 2. Schematic view of the NICA complex with the parameters for U-U collisions [2].

THE NUCLOTRON UPGRADE

The Nuclotron, a 6 A GeV synchrotron based on unique fast-cycling superferric magnets, was designed and constructed at JINR (1987–1992) and commissioned in March 1993. Ion beams up to iron and polarized deuterons have been accelerated and extracted from the accelerator. A view of the Nuclotron ring in the tunnel is shown in Fig. 3.

FIGURE 3. The Nuclotron in the tunnel.

The Nuclotron upgrade program in the NICA context was started in 2007 and contains the following main tasks: development of the heavy ion source ESIS based on the unique technology of highly charged state ion sources, presuming increase of the source magnetic field up to 6 Tesla; improvement of the vacuum conditions in the Nuclotron ring and linear injector; development of the Nuclotron power supply system in order to reach magnetic field in dipole magnets of 1.8−2 T; upgrade of the Nuclotron RF system; development of the extraction system; beam dynamics studies, minimization of the particle loss at all stages of the acceleration; design of a new injector. The Nuclotron upgrade is one of the key points in the realization of the NICA complex. The completion of the first stage work is planned by fall of 2009.

THE MULTIPURPOSE PARTICLE DETECTOR (MPD)

The experimental setup of the proposed MPD has to detect high-multiplicity events, perform particle identification and allow for accurate and controlled selection of the event centralities. The basic concept of the MPD is represented by a barrel and two endcaps located inside a magnetic field of about 0.5 T produced by a superconducting solenoid. The barrel part is a shell-like set of various detector systems surrounding the interaction point and aimed to reconstruct and identify both charged and neutral particles in the pseudorapidity region of $|\eta| <\mu$ 1.0. The two symmetric EndCup parts are designed to reconstruct and to measure the momenta of charged particles with higher pseudorapidity. The general design of the MPD is presented in Fig. 4. The option allowing to measure lepton pairs comprising a muon arm will be considered in the second stage of the NICA/MPD project. Some basic MPD parameters: the interaction rate of U+U collisions at a luminosity of 10^{27} cm^{-2} s^{-1} is 10 kHz (interaction rate of central collisions is of ~500 s−1); the accuracy of vertex reconstruction by means of silicon vertex detector is better than 0.2 mm; the TPC produces 30 hits on track and provides momentum measurement accuracy of ~1% in the momentum range of p = 0.2–2.0 GeV/c; the RPC TOF system has a resolution of ~100 ps and provides pion and kaon separation with a probability of 5% for p < 2 GeV/c.

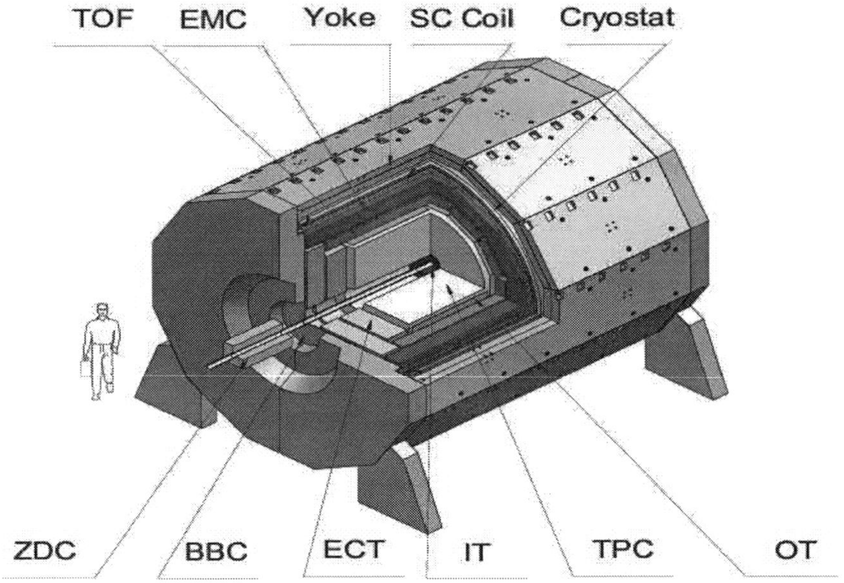

FIGURE 4. The MPD general view [3]. The subsystems are indicated: superconductor solenoid (SC Coil) and magnet yoke, inner trackerr (IT), straw-tube EndCup tracker (ECT), time-projection chamber (TPC), straw tube outer tracker (OT), time-of-flight stop counters (TOF), electromagnetic calorimeter (EMC), beam-beam counter (BBC), and zero degree calorimeter (ZDC).

Due to the high complexity of the search for the critical point and mixed phase in relativistiv heavy ion collisions and large uncertainties in the predicted signals, an accurate scanning of the considered phase diagram domain in the collision energy, impact parameter and system size is utterly needed. In this respect, it is important to provide a uniform detector acceptance over the whole energy range of interest. The operation in the collider mode, as proposed in the NICA/MPD project and in the low-energy RHIC program in 2010 (however, with the luminosity by several orders of magnitude lower than planned in the NICA project), naturally satisfies this demand and has an advantage as compared to the case of the fixed-target mode (SPS, FAIR).

THE SPIN PHYSICS DETECTOR (SPD)

The NICA facility will also give unique possibilities for spin physics, including proton and deuteron beams up to $\sqrt{s}$ of ~20 GeV, with both the transverse and longitudinal polarization higher than 50%, the luminosity higher than $10^{30}\,cm^{-2}\,s^{-1}$ and the polarimetry with ~3% accuracy. Nearly 4π Spin Physics Detetor (SPD) is under consideration for the second NICA interaction point.

The preliminary topics are: MMT-DY and J/ψ production processes with longitudinally and transversally polarized proton and deuteron beams for the extraction of unknown or poor known PDFs; spin effects in baryon, meson and photon production; spin effects in various exclusive reactions and diffractive processes; spin-dependent cross sections, helicity amplitudes and double spin asymmetries (Krisch effect) in elastic reactions; spectroscopy of quarkonia; polarimetry.

The one of conceptual SPD designs is shown in Fig. 5. Similar to the PAX setup [4], it assumes nearly 4π acceptance to enlarge the statistics of MMT-DY events, minimal radiation length to provide an effective detection of lepton pairs and a good angular resolution to allow for a measurement of azimuthal spin asymmetries in a wide kinematic region. The basic SPD parts are: a toroid magnet system with the integrated field of ~ 0.4T m, inner tracker (Silicon or MicroMega), main tracker (drift chambers or straw tubes), Cherenkov counter, electromagnetic calorimeter, trigger counters and EndCap detectors. Also considered is the possibility to detect the MMT-DY and J/ψ events using so-called beam-dump muon detector.

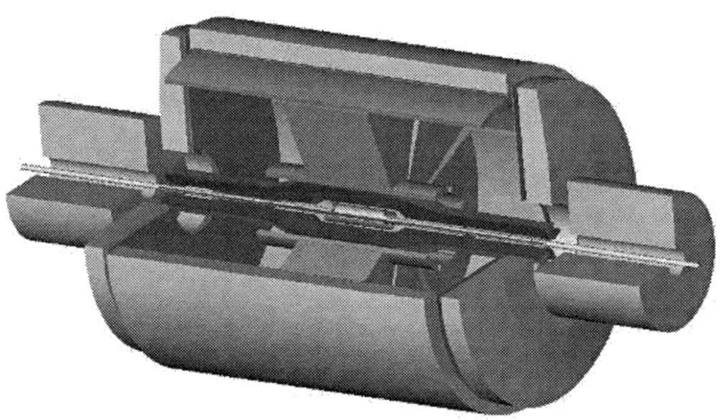

FIGURE 5. A conceptual SPD design.

The simulations of SPD performance for p+p collisions at $\sqrt{s}$ = 20 GeV with the luminosity of 10^{30} cm^{-2} s^{-1} shows that a statistics of ~5 (14) k MMT-DY events can be collected in a month of data taking with a cut of 2 (1.5) GeV on the lepton pair effective mass. Note that the 2 GeV cut, besides removing almost completely the background events, strongly reduces the signal. However, recent studies at COMPASS indicate that a dedicated detector and trigger system may allow for a softer cut and thus to substantially increase the statistics of MMT-DY events (see the numbers in brackets). One may see that a few years of data taking is thus required to access, with a good accuracy, a number of the yet practically unknown spin-dependent PDFs.

Note that the transversity, Boer-Mulders and Sivers PDFs can be accessed even with the help of single spin asymmetries, i.e. with much smaller errors than in the measurements of double spin asymmetries [5]. This possibility is of particular importance for future experiments at NICA, where the valence antiquarks will not be present in the initial state.

SUMMARY AND OUTLOOK

The strategic plan of JINR in high energy physics is targeting to the developments of home accelerator facility, NICA. It will make it possible to study very important unsolved problems of the physics of strongly interacting matter and spin phenomena in the forthcoming decade . The design and organizational work on this project started in 2006, recently resulting in a release of the NICA Conceptual Design Report and MPD Letter of Intent [1,2]. The Laboratory of High Energy Physics (LHEP) was founded (May 4, 2008) to concentrate efforts for realization of these plans in several stages: upgrade of the Nuclotron facility, preparation of the NICA technical design report, start of the prototyping of the NICA, MPD and SPD elements (2007–2009); design and construction of NICA, MPD and SPD (2009–2012); assembling (2011–2013); commissioning (2014). The collider rings will be operated without the beam acceleration in a symmetric mode. Further development of the facility presumes filling of the rings with different ions (asymmetric mode) and, possibly, a slow acceleration of the ions in the collider rings to the required energy. Worldwide cooperation is anticipated at R&D and construction stages of the work and, in the elaboration of corresponding scientific program.

ACKNOWLEDGMENTS

I am grateful to A. Nagaytsev and O. Shevchenko for the help in preparation of this report.

REFERENCES

1. A.N. Sissakian et al., *Part. Nucl. Lett.* **5**, 1 (2008); nucl-ex/0511018.
2 N. Agapov et al. (NICA Collaboration), "Design and Construction of Nuclotron-based Ion Collider fAcility (NICA). Conceptual Design Report", Dubna, 2008; www.nica.ru.
3 H.U. Abramyan et al. (MPD Collaboration), 'The MultiPurpose Detector (MPD) to study Heavy Ion Collisions at NICA, Letter of Intent, Dubna 2008, www.nica.ru.
4 P. Lenisa, F. Rathmann for the PAX Collaboration, hep-ex/0505054.
5. A.N. Sissakian et al., *Phys. Rev. D* **72**, 054027 (2005); *Eur. Phys. J. C* **46**, DOI:10.1140/epjc/s2006-0o2490-1 (2006); *Eur. Phys. J. C* **59,** DOI:10.1140/epjc/s10052-008-0806-0 (2008).

The Colorful Future of Spin

John P. Ralston

Department of Physics & Astronomy
The University of Kansas, Lawrence KS 66045 USA

Abstract. Activity and discoveries in spin physics of strong interactions has grown dramatically in recent years. What will we find in the future? I believe the future will be about the direct experimental exploration of confinement. While it is impossible to predict the "future spin" of the whole community, my personal vision indicates we will teach ourselves how to use new coordinates for the gauge sector. New coordinates will be more appropriate for the new questions being asked. Contrary to long-held beliefs, there does exist alternatives to *QCD*. Not only will the future be about discovering the micro-structure of hadrons, but also new fundamental questions of principle will be confronted.

Keywords: spin, qcd, strong interactions
PACS: 12.38.Aw, 12.38.Lg, 12.38.Qk, 13.88

For Once, And Thankfully, This is Not a Conference Summary

Skalak • Van Oers • Imai • Leader • Rith • Clarke • Milner • Marciano • Smertzidis • Burkardt • Anselmino • Kumar • Erler • Steffns • Asahi • Makins • Meyer • Saito • Zelenski • Troshin • Aildala • Slifer • Rondio • Weiss • Voutier • Leonova • Poelker • Maas • Vigdor • Montgomery • Tyurin • Roers • Wakasa • Lednicky • Wolf • Imai • Lorenzon • Poltoratska • Silva • Borodovitsyn • Bazilevsky • Makdisi • Zalenski • Ptitsyn • Chevtsov • Barber • Higinbotham • Montag • Luccio • MacKay • Huang • Onderwater • Lin • Luccio • Nass • Rathmann • Bai • Kondratenko • Courant • Morozov • Raymond • Balling • Bai • Ranjibar • Romalis • Wilson • West • Yin • Liu • Chevtsov • Eckardt • Goto • Engels • Saglam • He • Liyanage • Barish • Kurosawa • Tsutomu • Sowinski • Gardner • Fomin • Kohl • Paver • Quayle • Gericke • Sukhoruchkin • Pascke • Subedi • Shu • Yoshimi • Rivas • Kalantar • Dozono • Eibakyan • Furukawa • Silenko • Stibunov • Zetta • Uzikov • Neal • Dong • Sichtermann • Negrini • Belostotski • Goritshnig • Mochalov • Koike • Plyku • Kacharava • Finger • Brash • Kaiser • Kohl • Pentchev • Pierce • Aparajita • Zhimann • Girod • Mussgiler • Egiyan • Ahmad • Riordan • Weiss • Marukyan • Hafidi • Biselli • Goldstein • Mukherjee • Meissner • Schlegel • Zavada • Lin • Boyle • Hafidi • Seele • Fields • Ellighaus • Bennet • Datta • Hoffman • Kocoloski • Deur • De Nardo • Fantoni • Kao • Koretune • Bueltmann • El Alaouri • Liyanage • Sulkowsky • Strattmann • Myhrer • Kunne • Morreale • Nakano • Stamenov • Accardi • Muselmanbekov • Ramsey • Rubin • Honkanen • Saito • Dahiya • Prodhukin • Bosted • Pasquini • Conti • Aghasyan • Peng • Metz • Tanaka Zhou • Vogelsang • Chiu • Liu • Lu • Poljak • Drachenberg • Singh • Krimmer • Kim • Kovacs • Lolanoski • Zelenski • Pashke • Tanaka • Nass • Stancari • Wakui • Piegsa • Zhao • JNakajima • Urrego-Blanco • Thorn • Keith • Engels • Brown • Katich • Yuan • Joosten • Contalbrigo • Wollny • Giordano • Gamberg • Radici • Seidl • Wei • Koike • Kang • Bachetta • Jiang • Qiang • Qian • Gohn •

FIGURE 1. The speakers at the meeting I was sure to cite.

The conference organizers invited the talk recorded here, with the gracious suggestion it was **not** to be a comprehensive review of nearly 200 timely presentations. Instead, the organizers indicated they wanted *irresponsible visionary projections* about the future of spin physics. Evidently nobody on the organizing committee had volunteered.

CP1149, *Spin Physics, 18th International Spin Physics Symposium*
edited by D. G. Crabb, D. B. Day, S. Liuti, X. Zheng, M. Poelker, and Y. Prok

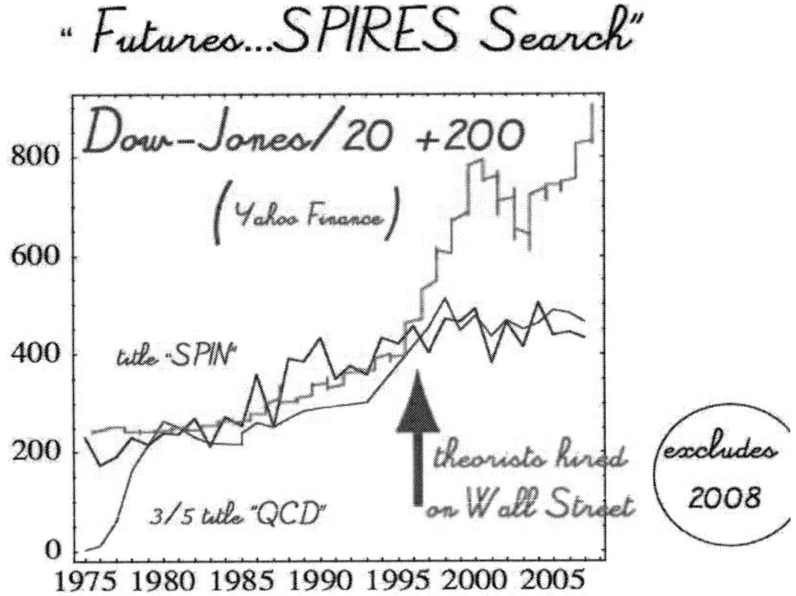

FIGURE 2. "Spin futures:" the number of papers per year with title *SPIN* and title *QCD* from SPIRES.

I felt it was important to give credit to those many concerned with the vision of future spin. Any omissions not cited in Fig. 1 receive my apologies. It is the community as a whole that has the vision and the energy the meeting was all about.

Bad Times Are Good for the Super Rich and the Super Creative

I collected data from SPIRES about the growth of spin physics over the past 30 years. Fig. 2 shows a chart of the investment history in "spin futures". The survey is unscientific but fun. The number of *papers per year* having the title "SPIN," and the number with the title "QCD" were complied. The SPIRES engine is intelligent and counts obscure variations of *QCD* such as "quantum chromodynamics." Spin and QCD have been strong and surprisingly comparable steady growth investments. Note that *QCD* rose from nothing and caught up with spin, not the other way around. Also shown is the Dow Jones average from *Yahoo Finance*. I used an offset and rescaling to convert Wall Street units to physics paper units. The other information on the chart probably relates to the entry of theorists into the financial markets. For decency's sake, I did not even collect the data from 2008. There can be no stronger argument for government support of basic research than keeping physicists out of finance.

Bad times can be good to stimulate the super-creativity of the spin community. *It seems to be time for the community to come together and define its own "spin-TEQ."*

The future will be a time when the priceless existing facilities of RHIC-SPIN and JLab will convince the world you've not done a single complete experiment until you've done all the spin experiments. It will be a time that mandates exceptional cooperation and great diversity in the use of every facility that has electricity. Whether or not the situation for funding improves, we hope the encouragement of creative new ideas and the magnificent international cooperation unique to spin physics will continue to flourish.

A Personal Vision

Although it is risky to project, I believe the future of spin physics of strong interactions was already verbalized by a politician. I think I heard Mr. Clinton say..."It's about Confinement, Stupid!" He was talking to his base. I maintain that *The future will be about the direct experimental exploration of confinement.* A few years ago[1] it was shown how to take photographs of the interiors of hadrons via *DVCS*. We discovered experiments could produce holographic resolution of three dimensional hadronic structure[1] . We worked on this because we cared: everyone wants to know *where is the color, spin and orbital angular momentum in every single hadron, from protons to nuclei ?* Some basic questions are:

- Do We Really Understand spin? Can we ever understand color?
- How does spin polarize color? What do we mean by "Color Stern-Gerlach?"
- What are the best way to explore the alternatives to *QCD*?

I believe we will make progress by developing new coordinates for the theory. Nature does not depend on your coordinates. The "gluon" is a particular set of coordinates for the gauge theory we want to describe, and it is not the best set for many purposes.

Understanding Spin

More than 25 years ago the Nobel Laureate C. N. Yang spoke on "THE Spin" at the 5th international Spin Symposium held at BNL. We all wondered whether the word THE was profound. Instead of Yang's paper getting a hundred citations he got one[3]. For a long time spin in *QCD* was taboo because of wrong folk-theorems. SPIN should not be taboo and is not that hard to understand. SPIN is the classical physics of polarization.

Quantum mechanics is defined by the Schroedinger equation. Focusing on one point in space, and a 2-component wave function ψ_j, the time evolution is defined by

$$i\dot{\psi}_j = H_{jk}\psi_k.$$

The solution is a transformation

$$\psi_j(t) = U_{jk}(t)\psi_k(0),$$

where $UU^\dagger = 1$. It is seldom explained that this kind of dynamics is old, and easy to understand. Consider[8] a set of real numbers q_i, p_i that are classical canonical variables, obeying

$$\dot{q}_i = \frac{\partial H}{\partial p_i}; \quad \dot{p}_i = -\frac{\partial H}{\partial q_i}.$$

Restrict attention to a generic linear theory, so that the Hamiltonian function $H(q_i, p_i)$ is bilinear:

$$H(q_i, p_i) = \frac{1}{2} p \cdot M^{-1} \cdot p + \frac{1}{2} q \cdot K \cdot q + q \cdot \Gamma \cdot p \tag{1}$$
$$\text{where } M = M^T; \; K = K^T, \; \Gamma = -\Gamma^T,$$

and super-T denotes the transpose. The "quantum map" is a simple assignment of a complex number to each canonical pair:

$$\psi_j = (q_j + i p_j)/\sqrt{2}$$

Hamilton's equations predict

$$\frac{\partial q_i}{\partial t} + i\frac{\partial p_i}{\partial t} = \frac{\partial H}{\partial p_i} - i\frac{\partial H}{\partial q_i} = -i\frac{\partial H}{\partial \psi_i^*}. \tag{2}$$

Since $H(q_i, p_i)$ was quadratic, the functional $H(\psi, \psi^*)$ is quadratic. If you want time evolution to be unitary, then ψ and ψ^* do not mix, which implies $H(\psi, \psi^*) \to \psi_i^* \hat{H}_{ij} \psi_j$, with $\hat{H}_{ij}$ a self-adjoint matrix of parameters. Eq. 2 becomes

$$i\dot{\psi}_i = \hat{H}_{ij}\psi_j,$$

showing the Schroedinger's equation is Hamilton's equations in Schroedinger's notation. The result extends to any dimension, including the infinite case. Statistical interpretations, and doctrine of measurement, are what is new. What we mean *particularly* by spin involves the spatial transformation properties of ψ_i. It is a nifty detail that spin 1/2 uses spinor representations, but all the dynamics is perfectly classical, canonical physics.

There is a group-theoretic basis for the quantum map[8]. Hamilton's equations for N degrees of freedom are invariant under $Sp(2N)$, the symplectic group. The group has an anti-symmetric metric J, and all $S \in Sp$ transformations preserve the metric via $S \cdot J \cdot S^T = J$. The symplectic group, with $N(2N+1)$ generators, has no immediate relation to keeping invariant $< \psi|\psi >= \sum_i (q_i^2 + p_i^2)/2$. That condition defines $SO(2N)$ with $N(N-1)/2$ generators. It is a fact that $U(N) = Sp(2N) \bigcap SO(2N)$ has only N^2 generators (Fig.3). Humans have historically thought that $U(N)$ defines the dynamical symmetries, but what it defines is a coordinate system. The quantum map diagonalizes J, so its eigenvalues $i = \sqrt{-1}$ show up in the $U(N)$-invariant Schroedinger form. This is good: yet all canonical coordinates are equally good.

It is an interesting diversion to ask what happened to $\hbar$, the operator structure, canonical quantization and so on. It is explored in Ref. [8], which also shows the relation to PT quantum mechanics[9]. The upshot is that once you are given a Hamiltonian theory you have everything else and quantization metaphors become obsolete. It is certainly obsolete to think about spin using little point-like gyroscopes that never did make sense.

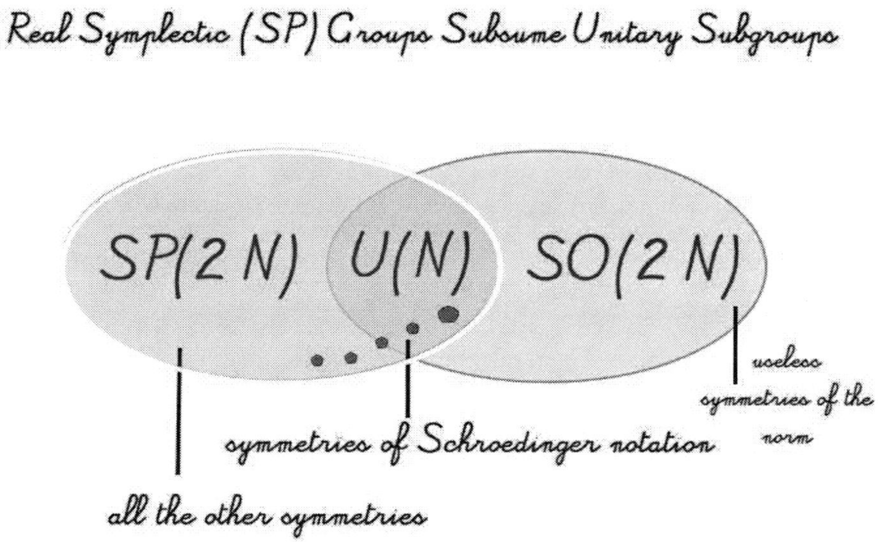

FIGURE 3. The relation of $Sp(2N)$, $U(N)$, and $SO(2N)$ groups. $Sp(2N)$ is the group of symmetries of the dynamics, and has more than twice as many generators as $U(N)$ traditionally invoked in quantum dynamics.

Can We Understand Color?

When color has been presented as an abstract direct product space of 3 complex dimensions it has always seemed difficult. However the quantum map tells us that the color degrees of freedom $Re(\psi_i) \rightarrow (q_1, q_2, q_3)$ are not inherently more complicated than a bouncing mass on a spring in three dimensions. It is interesting and important that Eq. 1 has mixed $q-p$ interactions that are not strictly Newtonian. Yet they are equivalent to an interaction with a background gauge field. That is, the symplectic dynamics of a gauge representation acts like it is immersed in a gauge field. If you pursue this you can show that $g = 2$ is the natural prediction for the g-factor of a spin 1/2 magnetic moment, without ever consulting relativity or Dirac.

It is sometimes forgotten that field theory has a Schroedinger representation with wave functions. The wave functionals tell you the amplitude to find a wave function, and here is where the quantum map is applied. There are so many degrees of freedom that sorting out the important ones has never been done. Indeed we ask experiments to do it because experiments can explore non-perturbative physics using coordinate-free methods every single time.

Correlations of Color and Spin

Any theorist can tell you that correlating color and spin runs into trouble with gauge invariance. Yet Nature does it, so there has to be a way.

The resolution comes at the level of beginning quantum mechanics. Suppose we have a field (wave function) with two indices $\psi_{a\alpha}$, where index a transforms on the color space and α on space-time. Any array with two indices can be diagonalized by finding suitable orthonormal frames on each space:

$$\psi_{a\alpha} = \sum_{M} e_a^M \Lambda_M \chi_\alpha^M.$$

The numbers Λ_M are joint invariants under left-right transformations. Write out the sum in coordinate free notation:

$$|\psi>= \Lambda_1 |e^{(1)}> |\chi^{(1)}> + \Lambda_2 |e^{(2)}> |\chi^{(2)}> + ...$$

The sum can only extend over the dimension of the smaller space, which is a maximum of 2 terms for spin 1/2.

This expression explains the Stern-Gerlach experiment. *Not all* of quantum measurement theory is "inexplicable principle", and some of it represents kinematic effects that are generic. There is always a strict correlation between any direct products of one attribute (polarization) and other (wave number, or position). There is also a strict correlation between color and polarization, by which an experiment "latching onto" one subspace (spin) will re-direct a unique subspace (color flow) and vice-versa. The spectacular observations from more than 30 years ago[4] of spin effects at high energies are color Stern-Gerlach experiments. Semi-inclusive *DIS* and transversity measures getting so much attention at Spin 2008 are also color Stern-Gerlach experiments. "Explanations" involving orbital angular momentum are appropriate, but might also be deemed "classifications" learned about channels contributing.

Theory still has trouble "predicting" the type and strength of the color-spin correlations that will be observed. At least we're no longer using leading twist perturbation theory or little gyroscopes to think about it.

Two Color Groups

Discussions of *QCD* talk about "the" group of local $SU(3)$ color transformations. It is good to think of this as the freedom to chose q_i, p_i phase spaces at each point in space, as in General Relativity.

However there are actually two $SU(3)$ groups waiting to be discussed. Let $Q_a^{(1)}$ be generators of gauge transformations of "type-1" on the matter fields, of "quarks". Let $Q_a^{(2)}$ be generators of gauge transformations of "type-2" on the gauge fields or "gluons." At equal time the theory is set up with commuting operators on each sector, and the transformations define an algebra of $SU_1(3) \times SU_2(3)$:

$$[Q_a^{(1)}(x), Q_b^{(1)}(x')] = ic_{abc} Q_c^{(1)}(x) \delta(x - x');$$

$$\begin{aligned}
[Q_a^{(2)}(x), Q_b^{(2)}(x')] &= ic_{abc}Q_c^{(2)}(x)\delta(x-x'); \\
[Q_a^{(1)}(x), Q_b^{(2)}(x')] &= 0.
\end{aligned}$$

The algebra transform each field by the rule it needs:

$$\begin{aligned}
[Q_a^{(1)}(x), \psi_b(x')] &= i\tau^a_{bc}\psi_c\delta(x-x'); \\
[Q_a^{(2)}(x), A_b(x')] &= ic_{abc}\delta A_c\delta(x-x') \\
&\textit{where} \\
A(x) &\rightarrow U(\theta(x))AU^\dagger(\theta(x)) - igU(\theta(x))\partial U^\dagger(\theta(x)). \quad (3)
\end{aligned}$$

The separate gauge *transformations* must be distinguished from the joint gauge *symmetry*. The $SU_{tot}(3)$ symmetry is found when the parameters (gauge angles) of the type-1 and type-2 groups are equal. It is naturally interpreted as one group transformation compensating another. Do we understand what that means?

There are nasty and nice ways to make sure gauge symmetry is good. The most nasty way starts with Lorentz invariant formalism, over-populates the gauge sector with non-existent fields, and then tries to get back to gauge invariance by imposing constraints and "ghosts". It has dominated perturbation theory, where a Lorentz covariant vector potential A is the tool of choice. The nicest way never creates a field that does not exist, and puts the gauge invariance into the physical states $|all>$ explicitly. It is the condition

$$(Q_a^{(1)}(x) + Q_a^{(2)}(x))|all> = 0. \quad (4)$$

Since $Q_a^{(1)} = \rho_a$ is the color charge density of the matter fields, and $Q_a^{(2)} = D \cdot E$ is the Gauss-Law generator of the gauge fields, the condition of Eq. 4 imposes Gauss' Law.

The problem of confinement is very close to the problem of solving Gauss' Law. If you think that confinement is "all states are color singlets" Eq. 4 is a proof. Unfortunately the proof is naive and allows color to cancel by spatially separated or extended means. Confinement needs all forms of color to stay inside the hadron. Asymptotic freedom tells us that the *QCD* Gauss' Law at short distances is not so different from the electrodynamic one. At long distances we wish we know what color was doing. And so the whole problem amounts to a few miserable Fermi of distance, which are just too many. Many distinguished workers have tried to solve the functional Gauss' Law equations for the *QCD* case, without ever producing a unique solution.

However when we view Gauss' Law as a transformation property the mystery switches. The color rotations on the quarks are being compensated by those by the gluons. There can't possibly be a unique solution to the transformation property, because any singlet function can multiply every expression and also be a good solution. After this, the rather shocking feature is that Gauss' Law requires the gauge sector to make "odd things" that transform like a quark: a color $\bar{3}$ is needed to compensate a color 3 and make an invariant. How is it possible to make a $\bar{3}$ from an octet gluon?

We call the "odd things" *dressing operators,* to express the fact a dressed quark has the fields attached to solve Gauss' Law. It is very important that all physical quarks are dressed quarks. There's no way to beat Gauss' Law. Therefore dressing operators really do exist.

Expressions for dressing operators using the octet gluons A may be the worst possible expression of dressing. The gluon is a perturbative coordinate designed for perturbative purposes, while gauge invariance is non-perturbative.

Gauge Links: The Handy Household Patch for Everyday Gauge Messes

Experimentalists are not always sympathetic to the distress caused by having quark fields separated by a non-zero space-time interval Δx. Yet almost every day a theorist somewhere on the planet finds a non-invariant gauge mess and needs an instant clean-up. Long ago Schwinger[10] solved 1+1 dimensional electrodynamics, descibing a narrow world with no room for the Gauss' Law flux except to make a phase:

$$e_{schwinger}(A) = e^{i\int_1^2 dx\cdot A}.$$

When the operator product expansion (*ope*) was later developed at leading twist, Schwinger's one-dimensional flux popped out as a one-dimensional (and generally unrecognized) quark dressing. The parton model including transverse coordinates has grown far beyond the straightjacket of the *ope* and no longer fits. Yet perturbation theory is saddled with a hidden coordinate dependence expressing everything in terms of the gluon field and gauge links. As if Nature cared about our perturbation theory!

Beyond leading twist, the deeply unphysical proposal to put three-dimensional Gauss' Law flux into a 1 dimensional tube continues as a patch for everyday gauge messes. The path dependence of the Schnger gauge-link is an obvious problem. If path dependence is then seen as profound, holonomy is mostly an awkward way to describe geometry that itself has no path dependence.

Dirac confronted gauge invariance differently[11]. For electrodynamics he made a functional $e_{dirac}(A)$ without path dependence that transforms like an electron. The formula is

$$e_{dirac}(A) = e^{ig\frac{\nabla\cdot A}{\nabla^2}}.$$

Only the longitudinal mode of a $U(1)$ gauge field changes in a gauge transformation:

$$\vec{A} = \vec{A}_L + \vec{A}_T;$$
$$\vec{A}_L = \nabla\frac{1}{\nabla^2}\nabla\cdot A \to \vec{A}_L + \vec{\nabla}\theta.$$

Dirac's dressing operator transforms

$$e_{dirac}(A) \to e_{dirac}(A)e^{-ig\frac{1}{\nabla^2}\nabla^2\theta} = e^{-ig\theta(x)}e_{dirac}(A).$$

There is nothing unsophisticated about the non-relativistic notation. In not introducing a zeroeth component ($A^0 = 0$) a Lorent-invariant theory can still be described, with the advantage of not mixing phony modes into the dynamics. Dirac only used the device to get rid of the longitudinal sector, so he could move on to free field theory of the transverse modes.

Dirac took care of $U(1)$ gauge invariance without the patch of path ordered exponentials. However the dressing fields are much more important than patches. If they are allowed their separate dynamical existence, it is possible to express the entire theory using them.

Consequences of Dressing

The non-local correlations of quarks fields ψ_a in parton distributions[6] become gauge invariant using the dressed operator $\psi^M = \bar{e}_a^M \psi_a$. Since we observe dressed correlations, the initial and final state (Sivers and Collins) correlations currently under-described with gauge links are going to measure dressing. Dressing is a dynamical and not a kinematic phenomenon. It can different in every different circumstance measured. Attempts to force notions of universality onto hadrons are conceptually flawed: Hadrons are incomplete to describe their own interactions[14]. As a consequence, I find[14] that the Soffer bound[5] on transversity[6] is not a valid fact of *QCD*. Current guesses about analytically continuing dressing operators are probably too naive to be true. When transversity is finally measured in Drell-Yan, I believe it will appear that factorization is violated. Drell-Yan lepton pair production with spin will be a watershed for Drell-Yan, which will never be a complete experiment until all spin dependent experiments are done.

Embedded Color Frames

Rather than express the dressing operators as non-local functions of the perturbative coordinate A, we got the idea to turn the tables and write A as a function of the dressing operators. It is basic "high-school" differential geometry that becomes really interesting when we remember that the dressing operators are already known to exist in Nature.

Let $e_a^M(x)$ be the Mth dressing field with color index a, that transforms like a 3-dimensional representation under the color group:

$$e_a^M(x) \rightarrow U_{ab}(x)e_b^M(x).$$

There will be many of these, labeled by index $M = 1...M_{max}$. Define combinations that are orthogonal:

$$e_a^M(x)\bar{e}_{bM}(x) = \delta_{ab}.$$

It is possible to use a Euclidean metric. It takes at least $M_{max} = 3$ solutions to do this: but we need $M_{max} > 3$ to get an interesting geometry. Otherwise e would be "pure gauge."

The geometrical interpretation of e_a^M is a *frame* of the space with labels a embedded in a big space of dimension M_{max}. Gauge symmetry is a local revision of the frame describing a manifold, without touching the manifold. From differential geometry, the formulas for locally gauge covariant curvatures are built up using an intermediate symbol, the *connection* $A_{ab}(e)$. The formula is

$$gA_{ab}(e) = -ie_a^M \partial \bar{e}_{bM}.$$

The spatial indices are suppressed. Then $A(e)$ automatically transforms like Eq. 3. It can be used everywhere in place of the ordinary vector potential. Very little literature has been found[12] on gauge frames. It seems new to recognize they have always existed, and represent gauge flux glued onto quarks.

It seems new to recognize that "no-go" theorems prohibiting gluon masses or productive alternatives to *QCD* are coordinate-dependent. They are written as theorems restricting Lagrangians $L = L(A, \partial A)$. The earliest such theorem was the great work of Utiyama shortly after Yang-Mills[13]. Although this was interpreted as requiring all Lagrangians to be local functions of A, color frames create a loophole.

If we consider gauge theories $L = L(e; \partial e; A(e), \partial A(e))$, then new terms of *lower dimension* are allowed. The gauge mass I have found[14] is written

$$L_2(e; \partial e; A(e), \partial A(e)) = \frac{m^2}{2}(\partial_\alpha e_a^M \partial^\alpha \bar{e}_{aM} - tr(A_\alpha A^\alpha),$$

where spatial indices α have been restored. I have proposed this term might exist as a fundamental parameter. It affects the infrared, while it is soft in the ultraviolet. It will be non-local when expressed ("solved") in terms of A, but why is that bad? Since the new term and other variations respect the gauge symmetry, and are not known to destroy the high-energy fit of conventional *QCD* to data, frame-based gauge theories might have a promising and colorful future.

ACKNOWLEDGMENTS

I thank all the organizers and especially Don Crabb for their good work. Research was supported in part under DOE Grant Number DE-FG02-04ER14308.

REFERENCES

1. J. P. Ralston and B. Pire, Phys. Rev. **D 66**, 111501 (2002).
2. C. N. Yang, in Proc. of 5th Int. Symp. on High Energy Spin Physics, Upton, N.Y.(1982).
3. B. Pire and J. P. Ralston, Phys. Rev. D **28**, 260 (1983).
4. G. Bunce *et al.*, Phys. Rev. Lett. **36**, 1113 (1976); K. J. Heller *et al.*, Phys. Rev. Lett. **41**, 607 (1978) [Erratum-ibid. **45**, 1043 (1980)].
5. J. Soffer, Phys. Rev. Lett. **74**, 1292 (1995)
6. J. P. Ralston and D. E. Soper, Nucl. Phys. B **152**, 109 (1979).
7. R. Utiyama, *Phys. Rev.* **101** , 1957 (1956).
8. J. P. Ralston, J. Phys. A: Math. Theor. **40**, 9883 (2007).
9. C. M. Bender and S. Boettcher, Phys. Rev. Lett. **80**, 5243 (1998) [arXiv:physics/9712001]; C. M. Bender, S. Boettcher and P. Meisinger, J. Math. Phys. **40**, 2201 (1999) [arXiv:quant-ph/9809072].
10. J. Schwinger, Phys. Rev. **82**, 664 (1951) .
11. P.A. M. Dirac, *The Physical Principles of Quantum Mechanics*, (Oxford 1958).
12. P. Haagensen, K. Johnson and C. S. Lam, Nuc. Phys. **B 477**, 273 (1996); M. Lavelle and D. McMullan, *Physics Reports*, **279** 1, 1997 and references therein; E. Bagan, M. Lavelle, and D. McMullan, *Phys.Rev.* **D57**, 4521(1998) ; *Mod.Phys.Lett* **A12** , 1815, (1997); Erratum-ibid.A12:2317, (1997).
13. R. Utiyama, Phys. Rev. 101, 1597 (1956).
14. J. P. Ralston, to appear in *Proceedings of Transversity '08* (World Scientific, Ferrara, 2008); arXiv:0810.0871 [hep-ph]. See also J. Phys. Conf. Ser. **69**, 012030 (2007).

PARALLEL SESSION CONTRIBUTIONS

Spin Identification of the Randall-Sundrum Graviton at the LHC

P. Osland*, A. A. Pankov†, A. V. Tsytrinov† and N. Paver**

*Department of Physics and Technology, University of Bergen, N-5020 Bergen, Norway
†The Abdus Salam ICTP Affiliated Centre, Technical University of Gomel, 246746 Gomel, Belarus
**University of Trieste and INFN-Trieste Section, 34100 Trieste, Italy

Abstract. Using as basic observable an angular-integrated asymmetry to be measured in Drell-Yan lepton-pair production at the LHC, we discuss the identification reach on the spin-2 of the lowest-lying Randall-Sundrum resonance predicted by gravity in one warped extra dimension, against the spin-1 and spin-0 hypotheses. Numerical results indicate that, depending on the graviton coupling strength to the standard model particles, such a spin-2 identification can extend up to mass scales of 1.0–1.6 TeV and 2.4–3.2 TeV for LHC integrated luminosities of 10 and 100 fb^{-1}, respectively.

Keywords: LHC, graviton, spin
PACS: 12.60.-i, 11.10.Kk, 12.60.Cn

INTRODUCTION

Heavy quantum states, with masses $M \gg M_{W,Z}$, are generally predicted by new physics (NP) models. If the masses are in the TeV range, such non-standard objects could be directly revealed as peaks, or resonances, in the cross sections for reactions among standard model (SM) particles at supercolliders.

The *discovery reach* represents the upper limit of the mass range where a peak can be observed experimentally. It depends, among other things, on the collider energy and the expected statistics, and determines an accessible region for the NP model parameters. However, once a peak is observed, the determination of the underlying model against others, potentially giving the same mass and number of events, is needed. Accordingly, for any model, the *identification reach* defines the upper limit of the mass range where the source of the peak can be determined or, equivalently, the competitor models can be excluded for all values of their parameters. Clearly, the identification reach is expected to select a subdomain of the model parameters accessible to discovery.

The resonance spin represents a powerful discriminating observable in this regard. Popular examples of NP scenarios that can produce (narrow) peaks in cross sections with the same mass and number of events, and can be discriminated by a spin analysis are: *i)* Models of gravity in extra spatial dimensions (spin-2); *ii)* Models with heavy neutral gauge bosons Z' (spin-1); and *iii)* SUSY models with R-parity breaking sneutrino couplings (spin-0). Here, we discuss the identification reach on the lowest-lying, spin-2, Randall-Sundrum (RS) graviton resonance [1], against *both* the spin-1 and spin-0 hypotheses, that can be obtained from the experimental study of the inclusive dilepton production process at LHC ($l = e, \mu$):

$$p + p \to l^+ l^- + X. \tag{1}$$

CP1149, *Spin Physics, 18th International Spin Physics Symposium*
edited by D. G. Crabb, D. B. Day, S. Liuti, X. Zheng, M. Poelker, and Y. Prok

While the total resonant cross section, integrated over the dilepton invariant mass under the peak at $M = M_G$, determines the number of events, hence the discovery reach, for the assessment of the spin-2 identification reach we adopt as basic observable a specific angular-integrated asymmetry, A_{CE}, at $M = M_G$ [2]. The angle is that between the final lepton and the initial quark or gluon in the dilepton center-of-mass frame. This asymmetry has the built-in feature of directly disentangling the spin-1 from other spin hypotheses [3, 4] and, being a ratio of cross sections, should be less affected by systematic uncertainties than other observables. Earlier attempts to discriminate spin-2 from spin-1, based on 'absolute' angular differential distributions, have been presented, *e.g.*, in Refs. [5, 6] and in the experimental Ref. [7].

NEW PHYSICS MODELS

RS model of gravity in extra dimensions

This scenario is a candidate solution to the gauge hierarchy problem, and its simplest version is based on one compactified 'warped' spatial extra dimension and a two-brane setup. The SM particles are localized to the TeV brane, gravity can propagate in the full 5-dimensional bulk and in particular, on the Planck brane, has an effective scale determined by $\overline{M}_{\text{Pl}} = 1/\sqrt{8\pi G_{\text{N}}} = 2.44 \times 10^{18}$ GeV. On the TeV brane the gravity scale Λ_π is suppressed by the exponential 'warp' factor, $\Lambda_\pi = \overline{M}_{\text{Pl}} \exp(-\pi k R_c)$, with $k \sim \overline{M}_{\text{Pl}}$ the 5-dimensional curvature and R_c the compactification radius. Boundary conditions at the branes determine a tower of spin-2 graviton resonances $G^{(n)}$ $(n \geq 1)$. Their predicted mass spectrum is $M_n = M_1 x_n / x_1$ with M_1 the lowest resonance mass and x_n the roots of the Bessel function $J_1(x_n) = 0$. Their TeV brane couplings to the SM particles are given by

$$-\mathscr{L}_{\text{TeV}} = \left[\frac{1}{\overline{M}_{\text{Pl}}} G^{(0)}_{\mu\nu}(x) + \frac{1}{\Lambda_\pi} \sum_{n=1}^{\infty} G^{(n)}_{\mu\nu}(x)\right] T^{\mu\nu}(x), \tag{2}$$

where $T^{\mu\nu}$ is the energy-momentum tensor and $G^{(0)}$ denotes the zero-mode, ordinary, graviton. For $kR_c \simeq 12$, Λ_π as well as the masses M_n (through the relation $M_1 = \Lambda_\pi k x_1 / \overline{M}_{\text{Pl}}$) are $\mathscr{O}(\text{TeV})$. This opens up the interesting possibility of observing gravity effects at colliders, in particular to reveal the graviton excitation exchange in the process (1). The contributing, tree level, partonic processes

$$q\bar{q} \to G \to l^+ l^- \qquad \text{and} \qquad gg \to G \to l^+ l^- \tag{3}$$

should yield cross-section peaks at the invariant dilepton mass $M = M_n$, with characteristic angular distributions [8].

The RS model thus depends on two independent parameters, that can be chosen as $M_G \equiv M_1$ and the universal 'coupling' $c = k/\overline{M}_{\text{Pl}}$ (in which case Λ_π is a derived parameter). Theoretically 'natural' limits are $0.01 \leq c \leq 0.1$ and $\Lambda_\pi < 10$ TeV [9]. Current 95% C.L. experimental lower limits on M_G, from the Tevatron collider, range from 300 GeV ($c = 0.01$) to 900 GeV ($c = 0.1$) [10]. One should notice that the unevenly spaced spectrum could be distinctive of the model by itself. However, in practice, due to the large masses involved, only the first RS resonance might be accessible at LHC, so that the spin-2 determination should be a necessary test of the RS model.

Heavy neutral gauge bosons
Turning to spin-1 exchanges in the process (1), Z's generally occur in electroweak models based on extended gauge symmetries. The leading-order partonic process, $q\bar{q} \to Z' \to l^+l^-$, should show up as a peak in the dilepton invariant mass distribution at $M = M_{Z'}$ with, in this case, the *same* angular distribution as the SM γ and Z exchanges. Besides the mass $M_{Z'}$, the model parameters are the vector and axial-vector Z' couplings to quarks and leptons. Popular scenarios are the cases where such couplings are specified theoretically: the list includes Z'_χ, Z'_ψ, Z'_η, $Z'_{\rm LR}$, $Z'_{\rm ALR}$ models, and the 'sequential' $Z'_{\rm SSM}$ model with the same couplings as the SM. Details can be found, *e.g.*, in the recent Ref. [11]. It turns out that, at the assumed LHC luminosity, the spin-2 RS resonance can be distinguished from the ALR and SSM spin-1 scenarios already at the level of signal events [2]. For the other Z' models, there are 'confusion regions' in the parameter spaces where spin-2 and spin-1 exchanges give rise to the same peaks and number of events, and therefore can be distinguished by a spin analysis only. Current experimental, model-dependent, lower limits on $M_{Z'}$, from the Tevatron collider, are in the range 500-900 GeV [12].

Sneutrino exchange
In SUSY theories with R-parity breaking, sparticles can be exchanged in the process (1) and appear as peaks in the dilepton invariant mass. This is the case of the spin-0 sneutrino formation by quark-antiquark annihilation, followed by leptonic decay [13]: $q\bar{q} \to \tilde{\nu} \to l^+l^-$. At the peak in the dilepton invariant mass, $M = M_{\tilde{\nu}}$, the spin-0 character implies a flat angular distribution. Basically, the cross section in this model depends on $M_{\tilde{\nu}}$ and on the product $X = (\lambda')^2 B_l$, where λ' is the R-parity breaking sneutrino coupling to $d\bar{d}$ and B_l the sneutrino leptonic branching ratio. Current constraints on X are very loose, and there exists an extended domain where $\tilde{\nu}$ production can mimic RS resonance formation (same mass and number of events under the peak), for details see [2].

$A_{\rm CE}$ ASYMMETRY AND SPIN-2 RS GRAVITON IDENTIFICATION

With $z = \cos\theta_{\rm cm}$ and $R = G, V, S$ denoting the spin-2, spin-1 and spin-0 hypotheses, respectively, we define the evenly integrated center-edge asymmetry:

$$A_{\rm CE}(M_R) = \frac{\sigma_{\rm CE}(R_{ll})}{\sigma(R_{ll})} \quad \text{with} \quad \sigma_{\rm CE}(R_{ll}) \equiv \left[\int_{-z^*}^{z^*} - \left(\int_{-z_{\rm cut}}^{-z^*} + \int_{z^*}^{z_{\rm cut}}\right)\right] \frac{d\sigma(R_{ll})}{dz}\, dz. \quad (4)$$

In (4): $0 < z^* < z_{\rm cut}$ is a priori free, and defines the separation between the "center" and the "edge" angular regions; $|z| < z_{\rm cut}$ accounts for detector angular acceptance; cross sections are integrated over the lepton-pair rapidity and over a bin in the lepton-pair invariant mass M centered at the peak $M = M_R$ and with size ΔM appropriate to account for the detector resolution, see, *e.g.*, Ref. [14]. To a very good approximation, the explicit z^* dependencies of $A_{\rm CE}$ are, for the three spin-hypotheses:

$$A_{\rm CE}^G = \varepsilon_q^{\rm SM} A_{\rm CE}^V + \varepsilon_q^G \left[2z^{*5} + \frac{5}{2} z^*(1 - z^{*2}) - 1\right] + \varepsilon_g^G \left[\frac{1}{2} z^*(5 - z^{*4}) - 1\right], \quad (5)$$

$$A_{\rm CE}^{V} \equiv A_{\rm CE}^{\rm SM} = \frac{1}{2} z^*(z^{*2}+3) - 1, \quad A_{\rm CE}^{S} = \varepsilon_q^{\rm SM} A_{\rm CE}^{V} + \varepsilon_q^{S}(2z^* - 1). \tag{6}$$

In (5), ε_q^G, ε_g^G and $\varepsilon_q^{\rm SM}$ are the fractions of resonant events for $q\bar{q}, gg \to G \to l^+l^-$ and SM background, respectively, with $\varepsilon_q^G + \varepsilon_g^G + \varepsilon_q^{\rm SM} = 1$. They are determined, as functions of M, by the overlaps of parton distribution functions, for which we choose the CTEQ6 ones [15]. Analogous definitions hold for Eq. (6). Strictly, Eqs. (5)-(6) are exact in the limit $z_{\rm cut} = 1$, whereas we will impose $z_{\rm cut} = 0.987$: the difference is numerically negligible at the 'optimal' values $z^* \simeq 0.5$ used in the subsequent analysis. The numerical results presented here are obtained from 'full' calculations with foreseen experimental cuts, such as lepton pseudorapidity $|\eta_l| < 2.5$ and transverse momenta $p_{T,l} > 20\,{\rm GeV}$. Also, a (perhaps optimistic) lepton identification efficiency of 90% has been assumed to evaluate the statistics.

One should notice, in Eq. (6), that $A_{\rm CE}^{V} \equiv A_{\rm CE}^{\rm SM}$ and, therefore, deviations of $A_{\rm CE}$ from the SM predictions definitely signal NP exchanges different from spin-1 models.

We now suppose that a peak is discovered in the dilepton mass distribution for the process (1) at $M = M_R$, and make the hypothesis that it is consistent with a spin-2 RS graviton resonance (in which case, M_R must be identified as M_G). To assess the level at which the spin-1 *and* spin-0 hypotheses can be excluded as competing sources of the peak with the same number of events, hence the spin-2 hypothesis being established, one can consider the deviations:

$$\Delta A_{\rm CE}^{V} = A_{\rm CE}^{G} - A_{\rm CE}^{V} \qquad \text{and} \qquad \Delta A_{\rm CE}^{S} = A_{\rm CE}^{G} - A_{\rm CE}^{S}. \tag{7}$$

As an example, Fig. 7 of Ref. [2] shows $A_{\rm CE}$ *vs* z^* for resonances with different spins, the same mass $M_R = 1.6$ TeV and the same number of events, $c = 0.01$ for the RS exchange, and LHC luminosity $\mathcal{L}_{\rm int} = 100\,{\rm fb}^{-1}$ (left panel). The right panel of that figure shows the corresponding deviations (7). The vertical bars attached to the dot-dashed line representing the RS model, give 2σ statistical uncertainties on the model itself. The figure suggests that, indeed, at the assumed LHC luminosity, the spin-2 RS graviton with $M_G = 1.6$ TeV and $c = 0.01$ can be discriminated from the other spin hypotheses by means of $A_{\rm CE}$ at $z^* \simeq 0.5$.

One can systematically generalize this example, and look at the domain in the RS parameter plane (M_G, c) where a peak can be *identified* as originating from spin-2 RS exchange, with the spin-1 and spin-0 hypotheses excluded. This domain will represent the searched for *identification reach* on the RS graviton resonance. For this purpose, one can adopt a simple-minded χ^2 criterion, where the χ^2 functions are defined as $\chi^2 = [\Delta A_{\rm CE}/\delta A_{\rm CE}]^2$, with $\Delta A_{\rm CE}$ given in (7), and $\delta A_{\rm CE}$ the corresponding expected statistical uncertainty pertaining to the RS model. The conditions $\chi^2 > \chi^2_{\rm C.L.}$ determine the ranges in (M_G, c) where the spin-0 and spin-1 hypotheses can be excluded to a given confidence level. The maximum sensitivity of $A_{\rm CE}$ to the spin-2 RS resonance parameters is generally achieved for $z^* = 0.5$.

Figure 1 shows the resulting allowed domain in the RS parameter plane for the spin-2 discrimination at 95% C.L. *vs.* the allowed domain for discovery at 5σ, for two values of LHC integrated luminosity, and the channels $l = e, \mu$ combined [2]. Theoretically suggested bounds on c and Λ_π are taken into account. The meaning of the lines in the two panels are as follows: the lowest RS resonance can be discovered if its representative

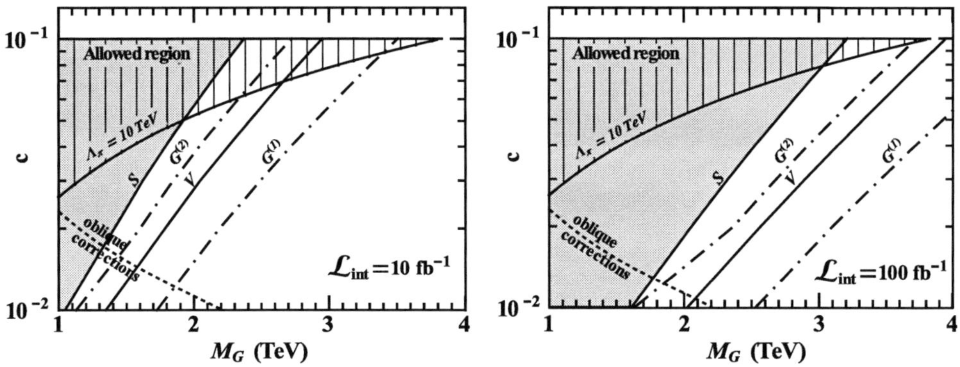

FIGURE 1. Discovery and Identification ranges as defined in the text.

point (M_G, c) lies to the left of the line "$G^{(1)}$"; in the region to the left of curve "V", of course included in the preceding one, the RS resonance spin-1 can be excluded, whereas the spin-0 hypothesis is still open; finally, if the observed resonance has the representative point in the domain to the left of the line "S", also the spin-0 hypothesis can be excluded. Accordingly, this is assumed to represent the searched for domain allowing spin-2 identification. Of course, these statements should be understood in a statistical sense, as specified by the confidence level.

As regards the mass ranges for discovery and identification of the RS graviton resonance, the results in Fig. 1 can be summarized as in Tab. 1. One should notice that, in Fig. 1, the line "S" always lies to the left of the line "V". This reflects the (general) fact that, as seen from Eqs. (5)–(7), $\Delta A^V_{\rm CE} > \Delta A^S_{\rm CE}$ for all z^*. Thus, if one were able to exclude spin-0, the exclusion of the whole class of spin-1 models would be automatically implied in a model-independent way. Conversely, the request of excluding the spin-0 hypothesis substantially reduces the extension of the parameter domain allowed by the weaker condition of only discriminating spin-2 from spin-1.

The somewhat 'low' values of M_G allowed to spin-0 exclusion suggest to look at the possibility of discovering the next RS resonance, $G^{(2)}$, in addition to $G \equiv G^{(1)}$, recalling that the ratio of masses is in the model a predicted number, $M_G/M^{(2)} = x_1/x_2$. Indeed, such 5σ discovery of $G^{(2)}$ turns out to be possible, with $\mathcal{L}_{\rm int} = 10\,{\rm fb}^{-1}$ for $M_G < 1.1$ TeV (2.7 TeV) at $c = 0.01$ ($c = 0.1$) and, with $\mathcal{L}_{\rm int} = 100\,{\rm fb}^{-1}$, for $M_G < 1.6$ TeV (3.7 TeV) at $c = 0.01$ ($c = 0.1$). Correspondingly, for a lowest resonance G in the (M_G, c) domain to the left of the line "$G^{(2)}$" in Fig. 1, also the higher graviton excitation with $n = 2$ can be discovered. One can see, therefore, that to the left of the line "S" the spin-2 of the lowest-lying RS resonance can be established and, in addition, also the characteristic RS mass spectrum can be tested by the discovery of the higher resonance, so that the model would be doubly clinched.

One can notice from Fig. 1 the dramatic role in RS graviton searches of the bound $\Lambda_\pi \leq 10\,$TeV, theoretically motivated by the need of not creating additional hierarchies in the model: taken literally, it would imply that, at the high luminosity of 100 fb^{-1}, the mass spectrum test should be feasible in the full discovery domain. However, in practice,

TABLE 1. Discovery and Identification reach [TeV]

	Discovery		Identification	
$\mathscr{L}_{\text{int}}$	$c = 0.01$	$c = 0.1$	$c = 0.01$	$c = 0.1$
10 fb^{-1}	1.7 TeV	3.5 TeV	1.1 TeV	2.4 TeV
100 fb^{-1}	2.5 TeV	4.6 TeV	1.6 TeV	3.2 TeV

such bound should be considered in a qualitative sense, as is the case for the indicative limits from the fit to oblique parameters, taken from [9, 16].

We may conclude by mentioning, as another RS resonance selective process, the inclusive diphoton production $p+p \to G \to \gamma\gamma + X$. Indeed, spin-1 could be excluded directly by $V \not\to \gamma\gamma$, leaving only the spin-2 and spin-0 hypotheses, and the RS model could be strongly tested by measurement of the ratio $Br(G \to \gamma\gamma)/Br(G \to l^+l^-)$ [17]. Currently, only the diphoton invariant mass distribution has been studied experimentally, but angular analysis should be possible, as mentioned in Refs. [5], notwithstanding the dominance of the partonic process $gg \to G \to \gamma\gamma$, strongly peaked near $z = \pm 1$ and potentially affected by the initial bremsstrahlung background. It could be interesting to attempt the application of the asymmetry A_{CE} to this process also.

REFERENCES

1. L. Randall and R. Sundrum, *Phys. Rev. Lett.* **83**, 3370 (1999); **83**, 4690 (1999).
2. P. Osland, A. A. Pankov, N. Paver and A. V. Tsytrinov, *Phys. Rev. D* **78**, 035008 (2008).
3. E. W. Dvergsnes, P. Osland, A. A. Pankov and N. Paver, *Phys. Rev. D* **69**, 115001 (2004).
4. P. Osland, A. A. Pankov and N. Paver, *Phys. Rev. D* **68**, 015007 (2003).
5. B. C. Allanach et al., *JHEP* **0009**, 019 (2000); *JHEP* **0212**, 039 (2002).
6. R. Cousins et al., *JHEP* **0511**, 046 (2005); D. Feldman, Z. Liu and P. Nath, *JHEP* **0611**, 007 (2006).
7. A. Abulencia et al. [CDF Collaboration], *Phys. Rev. Lett.* **95**, 252001 (2005).
8. T. Han, J. D. Lykken and R. J. Zhang, *Phys. Rev. D* **59**, 105006 (1999); G. F. Giudice, R. Rattazzi and J. D. Wells, *Nucl. Phys. B* **544**, 3 (1999).
9. H. Davoudiasl, J. L. Hewett and T. G. Rizzo, *Phys. Rev. Lett.* **84**, 2080 (2000); *Phys. Rev. D* **63**, 075004 (2001).
10. V. M. Abazov et al. [D0 Collaboration], *Phys. Rev. Lett.* **100**, 091802 (2008).
11. P. Langacker, arXiv:0801.1345 [hep-ph].
12. T. Aaltonen et al. [CDF Collaboration], *Phys. Rev. Lett.* **99**, 171802 (2007).
13. J. Kalinowski, R. Ruckl, H. Spiesberger and P. M. Zerwas, *Phys. Lett. B* **406**, 314 (1997); **414**, 297 (1997); T. G. Rizzo, *Phys. Rev. D* **59**, 113004 (1999).
14. ATLAS Collaboration, Reports No. CERN-LHCC-99-14, CERN-LHCC-99-15.
15. J. Pumplin et al., *JHEP* **0207**, 012 (2002).
16. T. Han, D. Marfatia and R.-J. Zhang, *Phys. Rev. D* **62**, 125018 (2000).
17. L. Randall and M. B. Wise, arXiv:0807.1746 [hep-ph].

Measuring the Properties of the Higgs Boson with the ATLAS Detector

W. Quayle on behalf of the ATLAS Collaboration

University of Wisconsin-Madison, 1150 University Ave., Madison, WI 53706, USA

Abstract. If an excess of events is discovered in the Higgs boson searches at LHC, it will be important to establish whether the excess is a Higgs boson or a different particle with similar decay modes. This talk reviews the measurements of the Higgs boson properties that are planned for the ATLAS detector at the LHC.

Keywords: Higgs
PACS: 14.80.Bn, 14.80.Cp

INTRODUCTION

One of the major goals of the LHC is to understand the mechanism of Electroweak symmetry breaking by searching for evidence of the Higgs boson [1, 2, 3]. The strongest such constraints presently come from searches at LEP, which has excluded a Standard Model Higgs boson with a mass below 114.4 GeV [4], and from Tevatron, which has ruled out the existence of a Standard Model Higgs boson with a mass of 170 GeV [5].

The ATLAS detector [6] at the LHC will either find evidence for a Higgs boson or place strong constraints on its production cross-section that will contradict the prediction of the Standard Model [7]. If a Higgs-like particle is found, its properties must be measured to establish that it is indeed a Higgs boson and not a different particle.

MASS, SPIN, AND CP PROPERTIES OF THE HIGGS BOSON

Observables such as the Higgs boson's branching ratios vary considerably as a function of mass; in order to interpret the other property measurements, one must measure the Higgs boson mass precisely. The Higgs boson mass determination at ATLAS is dominated by the $H \to \gamma\gamma$ and $H \to ZZ \to 4\ell$ channels.

The $H \to \gamma\gamma$ channel is sensitive in the mass range below about 150 GeV, and its mass resolution is 1.36 GeV for a Higgs boson with a mass of 120 GeV. The $H \to ZZ \to 4\ell$ channel is sensitive for a broad range of Higgs boson masses. Its resolution in the region below the ZZ threshold ranges from 1.8 GeV at $M_H = 120$ GeV to 2.7 GeV at $M_H = 180$ GeV. The resolution degrades slightly at higher mass, but for very heavy Higgs bosons the natural width is larger than the experimental resolution. The expected precision of a measurement of the natural width of the Higgs boson after collecting 300 fb^{-1} of integrated luminosity is better than 8% for masses above 270 GeV [8]. It is worth noting that the mass resolutions quoted here are taken from very recent studies, but other

CP1149, *Spin Physics, 18th International Spin Physics Symposium*
edited by D. G. Crabb, D. B. Day, S. Liuti, X. Zheng, M. Poelker, and Y. Prok

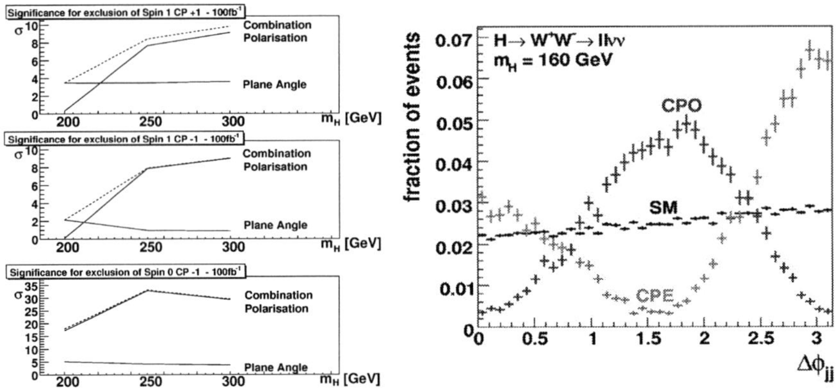

FIGURE 1. Left: the expected significance for the exclusion of several alternative spin/CP scenarios as a function of the Higgs boson mass, after 100 fb^{-1} of integrated luminosity. The top panel shows the result for a spin 1, CP +1 scenario; the middle panel shows spin 1, CP -1, and the bottom panel shows spin 0, CP -1. Right: the transverse opening angle between the two tag jets in the Weak Boson Fusion $H \to WW \to \ell\nu\ell\nu$ signal, for the Standard Model case (SM, black), a CP-odd scenario (CPO, red), and a CP-even scenario (CPE, green).

measurements, such as spin, CP, and coupling determinations, are based on older studies that have not yet been updated with the latest understanding of the detector.

A spin-1 boson cannot be produced by gluon fusion or decay to a photon pair [9], so if the decay $H \to \gamma\gamma$ is observed, then the spin of the Higgs boson cannot be 1. If the decay $H \to \gamma\gamma$ is not observed, it is still possible to extract information about the spin of the Higgs boson from angular correlations in the $H \to ZZ \to 4\ell$ channel and the angular distributions of the struck quarks in Higgs production by Weak Boson Fusion.

The Higgs boson spin and CP are extracted from $H \to 4\ell$ decays using three main observables: α, β, and R [10]. α and β are obtained by fitting $F(\phi) = 1 + \alpha\cos(\phi) + \beta\cos(2\phi)$ to the distribution of ϕ, the angle between the decay planes of the two Z bosons in the rest frame of the Higgs boson. R is given by $R = (L-T)/(L+T)$, where L and T are obtained by fitting $G(\theta) = T(1+\cos^2(\theta)) + L\ \sin^2(\theta)$ to the distribution of the angle θ between the Z boson decay products and the receding Higgs boson in the rest frame of the Z boson. The left plot in Fig. 1 shows the significance with which several alternative models can be ruled out with 100 fb^{-1} of integrated luminosity. The analysis of θ can rule out a non-Standard-Model Higgs boson with a mass larger than about 250 GeV; the analysis of ϕ through the coefficients α and β yields the strongest discriminating power for Higgs bosons with a mass around 200 GeV.

Searches for Higgs bosons via Weak Boson Fusion use several cuts on the two highest-P_T jets in the event ("tag" jets) to achieve a strong rejection against backgrounds [11]. The most sensitive search channels for this production mechanism are $H \to \tau\tau$ in the region around $M_H = 120$ GeV and $H \to WW \to \ell\nu\ell\nu$ in the region around $M_H = 160$ GeV. One can extract information about the tensor structure of the HVV vertex by analyzing the transverse angle $\Delta\phi_{jj}$ between the two leading jets [12]; the right plot in Fig. 1 shows the $\Delta\phi_{jj}$ distribution for $H \to WW$ signal in the Standard Model and

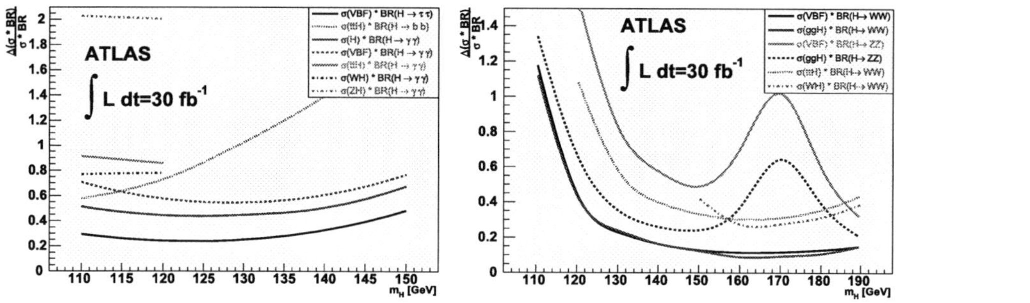

FIGURE 2. The expected relative error (after collecting 30 fb^{-1} of integrated luminosity) on the measurement of the rates for several channels. Left: $H \to b\bar{b}$ and $H \to \gamma\gamma$, for $110 < M_H < 150$ GeV. Right: $H \to WW$ and $H \to ZZ$, for $110 < M_H < 150$ GeV.

two alternate models. A χ^2 test is used to determine the dominant couplings. With 10 fb^{-1} of integrated luminosity, one would expect to exclude the CP-even (CP-odd) hypothesis with a confidence level corresponding to 5.4σ (4.6σ) using the $H \to WW$ channel if the Higgs boson mass is 160 GeV. Similarly, with 30 fb^{-1} of data, one would expect to exclude the CP-even (CP-odd) hypothesis with a confidence level corresponding to 2.5σ (2.0σ) using the $H \to \tau\tau$ channel if the Higgs boson mass is 120 GeV. After establishing that the dominant coupling is Standard Model-like, one can still test for a small CP-even anomalous coupling, since interference terms modify the $\Delta\phi_{jj}$ distribution. With 30 fb^{-1} of data, the expected error on the anomalous CP-even coupling is $\sigma(g_{5e}^{HZZ}) = 0.11$ if M_H is 160 GeV, or $\sigma(g_{5e}^{HZZ}) = 0.24$ if it is 120 GeV. These figures neglect systematic uncertainty in the background; adding a 10% background uncertainty leads to an additional error of about 0.02 in the measured coupling.

COUPLINGS TO GAUGE BOSONS AND FERMIONS

The most model-independent observable sensitive to the Higgs couplings is the total rate, $\sigma \times BR$, for each final state. In the case of $H \to WW$, the event selection assumes that the Higgs boson is CP-even and spin 0; the cut on the dilepton angle in the transverse plane exploits spin correlations to suppress the background. As shown in the last section, it is possible to verify this assumption through an analysis of $\Delta\phi_{jj}$ in Weak Boson Fusion events. Figure 2 shows the expected uncertainty on $\sigma \times BR$ for several channels [13]. For $M_H < 150$ GeV (left), the error in the rate ranges from 30% to 100%. For channels that can be seen in the mass range $110 < M_H < 190$ GeV, the error is smaller.

By assuming that there is only one Higgs boson, it is possible to extract a measurement of relative branching ratios such as Γ_Z/Γ_W. The $H \to WW$ channel is used as the normalization since it is measured with the smallest error. With 30 fb^{-1} of integrated luminosity, these ratios of partial widths can be measured with an accuracy better than 60% for $M_H > 120$ GeV, with a precision better than 35% for some masses.

Assuming that there are no extra particles or strong couplings to light fermions, it is possible to obtain estimates of the ratios of Higgs couplings. The cross-sections and

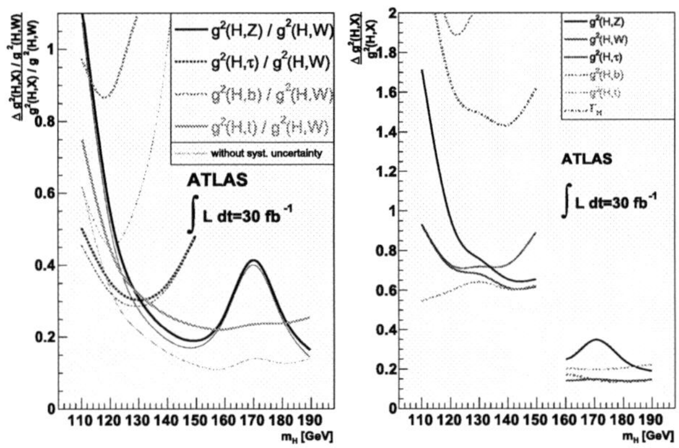

FIGURE 3. The expected error on the measurement of the relative (left) and absolute (right) couplings of the Higgs to gauge bosons and fermions after collecting 30 fb^{-1} of integrated luminosity. The discontinuity at $M_H = 150$ GeV is due to a change in the assumption about the sum of all branching ratios. For $M_H < 150$ GeV, decays to b and τ are included in the sum; for $M_H > 150$ GeV, they are not.

branching ratios are proportional to the squares of the couplings, and the proportionality constants are taken from a theoretical calculation. The left plot of Fig. 3 shows the expected relative error on the ratios of Higgs boson couplings after 30 fb^{-1} of integrated luminosity. The uncertainties range from about 20% up to 100%.

Finally, with the additional assumption that the sum of visible branching ratios is the same as in the Standard Model up to an error equal to the sum of undetected Standard Model branching ratios, the absolute couplings of the Higgs boson to the Z, the W, and the τ can be measured. The error is less than 100% for $120 < M_H < 200$ GeV, as shown in the right plot of Fig. 3.

REFERENCES

1. R. Brout and F. Englert, *Phys. Rev. Lett.* **13**, 321–323 (1964).
2. P. W. Higgs, *Phys. Lett.* **12**, 132-133 (1964).
3. P. W. Higgs, *Phys. Rev. Lett.* **13**, 508–509 (1964).
4. LEP Working Group for Higgs boson searches, *Phys. Lett.* **B565**, 61-75 (2003).
5. Tevatron New Phenomena Higgs Working Group, arXiv:0808.0534 [hep-ex] (2008).
6. ATLAS Collaboration, G. Aad et al., "The ATLAS Experiment at the Large Hadron Collider," JINST 3:S08003,2008.
7. ATLAS Collaboration, "Expected Performance of the ATLAS Experiment, Detector, Trigger and Physics," CERN-OPEN-2008-020, Geneva, 2008, to appear.
8. The ATLAS Collaboration, "ATLAS Detector and Physics Performance," CERN/LHCC 99-15 (1999).
9. C. N. Yang, *Phys. Rev.* **77**, 242-245 (1950).
10. C. P. Buszello et. al, *Eur. Phys. J.* **C32**, 209-219 (2004).
11. R. N. Cahn and S. Dawson, *Phys. Lett.* **B136**, 196 (1984).
12. C. Ruwiedel et al., *Eur. Phys. J.* **C51**, 285-414 (2007).
13. M. Dührssen, "Prospects for the Measurement of Higgs boson Coupling Parameters in the Mass Range from 110-190 GeV," ATLAS Note ATL-PHYS-2003-030 (2003).

NPDGamma: A Measurement of the Parity Violating Directional γ-Ray Asymmetry in Polarized Cold Neutron Capture on Hydrogen

Nadia Fomin
for the NPDGamma Collaboration

University of Tennessee, Knoxville, TN 37996, USA

Abstract. The NPDGamma experiment aims to measure the correlation between the neutron spin and the direction of the emitted photon in neutron-proton capture at low momentum transfer. An up-down parity violating asymmetry from this process can be related to the strength of the hadronic weak interaction between nucleons.

The first phase of the experiment was completed in 2006 at LANSCE. The methodology will be discussed and preliminary results will be presented. The next run will start in 2009 at the SNS at ORNL with many improvements that will yield a measurement with a projected statistical error of 1×10^{-8}, 20% of the predicted value for the asymmetry. This will allow the determination of the long range π contribution in the weak interaction between nucleons.

Keywords: parity violation, weak coupling constants, cold neutron capture

INTRODUCTION

The weak interaction between nucleons at low momentum is best accessed through measurements of small, parity-odd amplitudes. The weak interaction at low energies has traditionally been discussed in terms of the DDH Model [1], which uses the valence quarks to calculate effective parity violating meson-nucleon coupling directly from SM via 6 weak meson coupling constants. Experimental observables can be written as combinations of these couplings. In the context of modern effective field theory (EFT), the reaction is parametrized in terms of 5 low energy constants that correspond to 5 parity-odd S-P nucleon-nucleon amplitudes and an additional parameter that characterizes the long range weak pion-nucleon interaction [2]. In the case of NPDGamma, the observable is the up-down parity violating directional asymmetry, A_γ, of emitted gamma rays in polarized cold neutron capture on protons ($\vec{n} + p \rightarrow d + \gamma$). A_γ is sensitive primarily to the long-range pion-nucleon coupling constant ($A_\gamma \approx -0.11 f_\pi^1$) and is predicted to be -5×10^{-8} [1]. The result of the first phase of the NPDGamma experiment is in agreement with the only existing experimental result [3], and both are consistent with zero at the 2×10^{-7} level. The next phase of the NPDGamma experiment aims to make the first measurement of A_γ with enough sensitivity (1×10^{-8}) to test the prediction.

CP1149, *Spin Physics, 18th International Spin Physics Symposium*
edited by D. G. Crabb, D. B. Day, S. Liuti, X. Zheng, M. Poelker, and Y. Prok

EXPERIMENT

The first set of data for NPDGamma was taken at the Manual Lujan Jr. Neutron Scattering Center at LANSCE in 2006, where 800MeV proton pulses are incident on a Tungsten spallation target at 20Hz, and the vaporized neutrons from the spallation process are delivered to the experimental cave by a supermirror neutron guide after being cooled by a supercritical H_2 moderator [4]. The experiment was installed on flight path 12, which includes a beam chopper that was used to select neutrons with energies between 2meV and 15meV. Incoming neutrons were polarized using a ^{3}He spin filter cell [5]. A resonant RF spin rotator (RFSR)[6] was used to reverse (or not) neutron polarization for an individual pulse in order to construct a spin sequence ($\uparrow\downarrow\downarrow\uparrow\downarrow\uparrow\uparrow\downarrow$), chosen to minimize time-dependent effects, such as detector gain drifts. Finally, the neutrons were captured in a 16-liter liquid para-hydrogen target, surrounded by a 3π acceptance array of γ detectors. The detector array consists of 4 rings, each containing 12 CsI(Tl) detectors and was operated in current mode due to the high γ rate ($\approx$100MHz). The entire experiment [see Fig.1] was situated inside of a 10G holding field, used to preserve neutron polarization. The beam intensity and polarization were monitored with neutron transmission [5, 7] through ^{3}He ion chambers.

Great care was taken to align the detector array with the holding field. There are two parity-allowed left-right asymmetries (one comes from radiative capture [8] and the other from Mott-Schwinger scattering [9]) which can be mixed into the up-down asymmetry through detector misalignment.

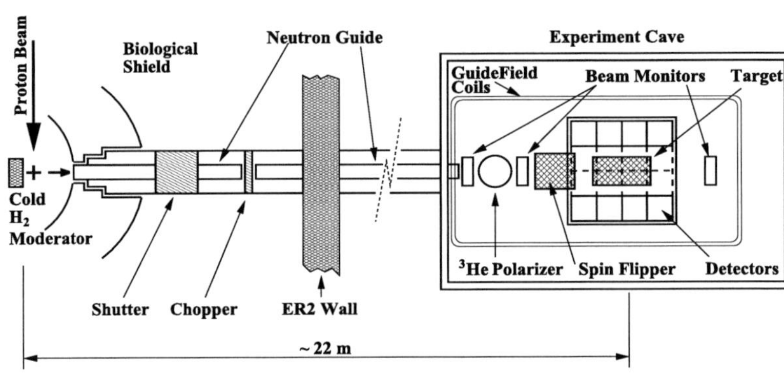

FIGURE 1. Top-down view of the experimental setup at LANSCE. The guide field coils produce a 10G vertical field.

ANALYSIS AND RESULTS

The analysis procedure is described in some detail in [10]. Almost 5000 good data runs of 8.5 minutes and 1250 spin sequences each were taken. The physics asymmetry is calculated for each detector pair in each ring separately, binned in 40ms time bins and by spin state of the neutron. For a given time bin and neutron spin state, the raw asymmetry is formed using detector yields Y_i,Y_j for a pair of detectors via:

$$A_{raw} = \frac{1}{2}\left[\frac{Y_i^{\uparrow} - Y_j^{\uparrow}}{Y_i^{\uparrow} + Y_j^{\uparrow}} + \frac{Y_j^{\downarrow} - Y_i^{\downarrow}}{Y_j^{\downarrow} + Y_i^{\downarrow}}\right] \tag{1}$$

The raw asymmetry, A_{raw}, contains contributions from the up-down, parity-violating asymmetry that we are interested in, A_{UD}, as well as the left-right, parity-conserving asymmetry, A_{LR}. The physics asymmetry is a combination of these and is related to the measured asymmetry for a detector pair in a given time bin via:

$$A_{UD}\cos\theta + A_{LR}\sin\theta = \frac{A_{raw} - A_g A_b - A_{noise}}{P_n \cdot \varepsilon_{pol} \cdot \varepsilon_{sf}}, \tag{2}$$

where A_g, A_b, A_{noise} are the gain, beam, and noise asymmetries, ε_{pol},ε_{sf} are the polarization and RSFR efficiencies, respectively, and finally, θ is the angle between the center of the detector and the guide field (see Fig. 2). Note that both physics asymmetries also contain background contributions, the largest of which was the result of neutron capture on aluminum windows of the target vessel ($\approx$25%, averaged over all 4 rings, but decreasing as one moves downstream). This was corrected for by using data taken on an empty target vessel in combination with Monte Carlo simulations of a full target vessel.

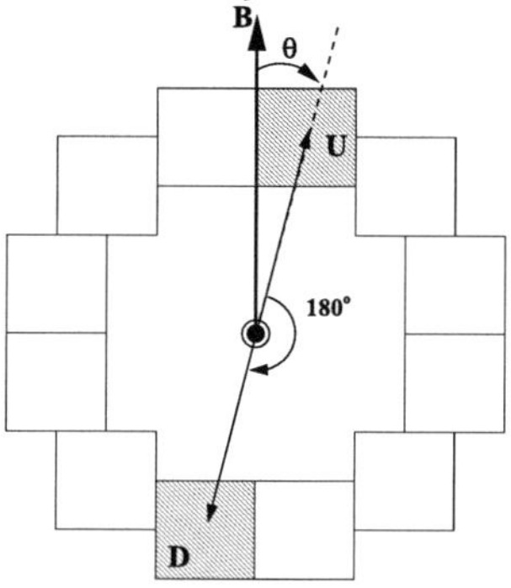

FIGURE 2. Geometry of a given detector ring. θ is defined as the angle between the center of the detector and the 10G magnetic holding field.

The time bins are then averaged over using energy-dependent and other correction factors (neutron polarization, RFSR efficiency, depolarization in the target) as weights. A physics asymmetry is extracted for each detector pair and a fit is performed for the asymmetries in each ring, as a function of the angle of the detector pair (see Fig. 3). The physics asymmetries from all the rings were then combined together for a final asymmetry using an error weighted mean. The preliminary analysis yields the following for the up-down and left-right asymmetries:

$$A_\gamma^{UD} = [-1.1 \pm 2.1\,(stat.) \pm 0.2\,(sys.)] \times 10^{-7}$$
$$A_\gamma^{LR} = [-1.9 \pm 2.0\,(stat.) \pm 0.2\,(sys.)] \times 10^{-7}.$$

Final analysis of the systematic effects is ongoing.

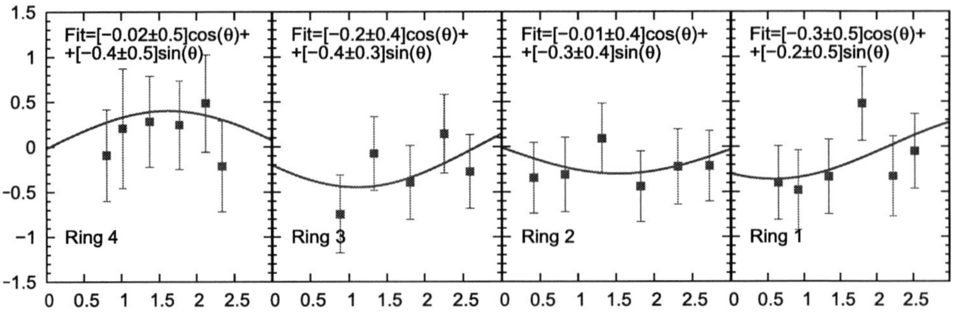

FIGURE 3. Measured asymmetries for 23 detector pairs (one pair in ring 3 was omitted due to frequent electronic problems) and corresponding fits for each detector ring.

NEXT PHASE AT THE SNS

The next phase of the experiment will be performed at Oak Ridge National Laboratory's new Spallation Neutron Source (SNS). The ramp-up of the SNS to full power (1.4MW) will be continuing during the running of NDPGamma, which will begin installation later in the year and will start data taking in early 2010. The high flux of the 60Hz pulsed neutrons from the SNS will allow the NPDGamma experiment to make a measurement sensitive enough to test the EFT prediction in 5000 hours. Thinner windows on the new target vessel will significantly decrease aluminum background contributions which were difficult to correct for in the previous analysis. Additionally, the ^{3}He polarizer will be replaced by a Supermirror Polarizer, giving an almost $4\times$ increase in the figure of merit. The installation of NPDGamma on the Fundamental Neutron Physics Beamline is currently under way.

REFERENCES

1. B. Desplanques, J. F. Donoghue, and B. R. Holstein, *Ann. Phys.* **124**, 449 (1980).
2. S.-L. Zhu, C. M. Maekawa, B. R. Holstein, M. J. Ramsey-Musolf, and U. van Kolck, *Nucl. Phys.* **A748**, 435–498 (2005).
3. J. F. Cavaignac, B. Vignon, and R. Wilson, *Phys. Lett.* **B67**, 148–150 (1977).
4. P.-N. Seo, et al., *Nucl. Instrum. Methods* **A517**, 285–294 (2003).
5. T. Chupp, et al., *Nucl. Instrum. and Methods* **A574**, 500–509 (2007).
6. P. N. Seo, et al., *Phys. Rev. ST Accel. Beams* **11**, 084701 (2008).
7. G. Jones, et al., *Nucl. Instrum. Methods* **A440**, 772–776 (2000).
8. A. Csoto, B. F. Gibson, and G. L. Payne, *Phys. Rev.* **C56**, 631–634 (1997).
9. M. T. Gericke, J. D. Bowman, and M. B. Johnson, *Phys. Rev.* **C78**, 044003 (2008).
10. M. Gericke, et al., *Nucl. Instr. Methods* (2008), proceedings prepared for the International Workshop on Particle Physics with Slow Neutrons, Grenoble, France.

The Time Reversal Experiment with Kaons (TREK) at J-PARC

Michael Kohl for the TREK collaboration

Hampton University, Hampton, VA 23668, USA, and Jefferson Lab, Newport News, VA 23606, USA

Abstract. The Time Reversal Experiment with Kaons (TREK) at J-PARC aims to find CP violation beyond the Standard Model in the semi-leptonic $K^+_{\mu 3}$ decay mode by measuring the T-violating transverse polarization P_T of outgoing muons. TREK makes use of the intense kaon beam at J-PARC stopped in a target and employs an optimized setup with excellent control of systematic uncertainties. The sensitivity at J-PARC is improved by a factor of 20 compared to the current uncertainty for P_T, well in the predicted range of various New Physics models. An overview of the planned experiment and current status will be presented.

Keywords: time reversal symmetry violation, transverse muon polarization, kaon decay
PACS: 11.30.Er, 12.60.Fr, 13.20.Eb

Despite its great successes in recent years culminating in the 2008 Nobel Prize in Physics, which was announced during this symposium, the Standard Model is widely believed to be incomplete as it leaves many fundamental questions unanswered such as the mechanism of baryogenesis and electroweak symmetry breaking. Experimentally, it is of greatest importance to find evidence for New Physics beyond the Standard Model in accessible observables. Electroweak theory allows to link T-odd observables (which change sign under time reversal transformation) to time reversal symmetry breaking, which can be interpreted as clear indication of New Physics. The CPT theorem then allows to connect T violation to CP violation. The transverse polarization P_T of muons in stopped K^+ decays in the $K^+_{\mu 3}$ mode $K^+ \to \mu^+ \pi^0 \nu_\mu$ is such a long-recognized example [1] and is depicted in Fig. 1.

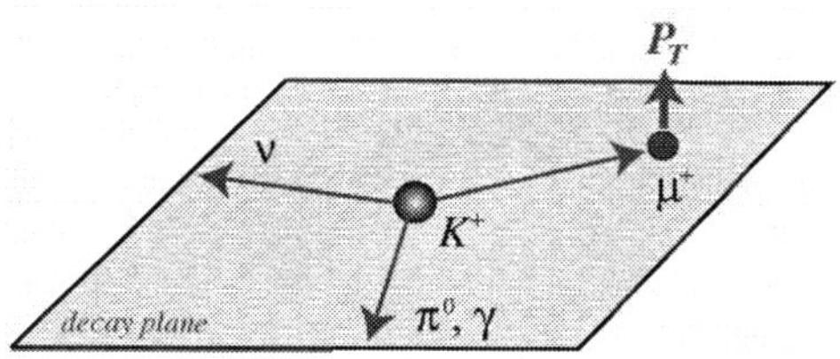

FIGURE 1. Transverse muon polarization P_T in $K^+_{\mu 3}$ decays at rest.

The Standard Model prediction for P_T is extremely small, of order 10^{-7}, arising from higher-order loop contributions. Final-state interaction in the semi-leptonic mode is precisely calculable and does not exceed 10^{-5} in the $K^+_{\mu 3}$ channel. On the other hand, models involving New Physics such as multi-Higgs doublet models, leptoquark models or supersymmetric models with R-parity breaking or s-quark mixing give rise to finite values of P_T ranging from 10^{-4} to 10^{-2}.

CP1149, *Spin Physics, 18th International Spin Physics Symposium*
edited by D. G. Crabb, D. B. Day, S. Liuti, X. Zheng, M. Poelker, and Y. Prok

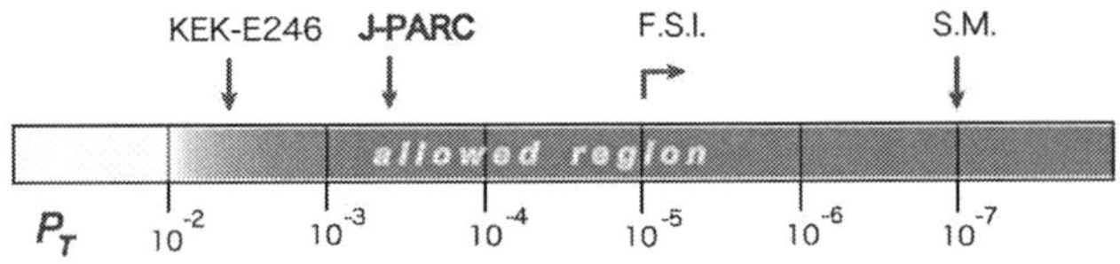

FIGURE 2. Sensitivity of P_T to New Physics. Also marked are the 90% confidence limits achieved by E-246 and projected for TREK.

The proposed Time Reversal Experiment with Kaons (TREK) at J-PARC [2] will be sensitive to P_T to the order 10^{-4}, thereby improving the current limit on P_T by a factor 20 and probing deeply inside the window of New Physics in this channel, as illustrated in Fig. 2.

Experimentally, P_T can be measured by polarimetry of muons after identifying the $K^+_{\mu 3}$ decay mode in stopped K^+ decays. The kinematics of this decay is completely determined by measuring energy and angle of the outgoing μ^+ (using a toroidal magnetic field and spatially sensitive tracking detectors) and the π^0 (using a large-acceptance CsI(Tl) calorimeter). In a second step, the transverse polarization of the outgoing muon is determined in a muon polarimeter based on the measured direction of emitted positrons in stopped $\mu^+ \to e^+ \nu_e \bar{\nu}_\mu$ decays.

The proposed experiment TREK at J-PARC reuses most of the existing setup of the predecessor experiment E-246 at KEK-PS shown in Fig. 3, which had resulted in the current upper limit of $P_T < 5 \cdot 10^{-3}$ (90% C.L.) [3]. The superconducting toroidal magnetic spectrometer has twelve sectors equally instrumented. The rotational symmetry of

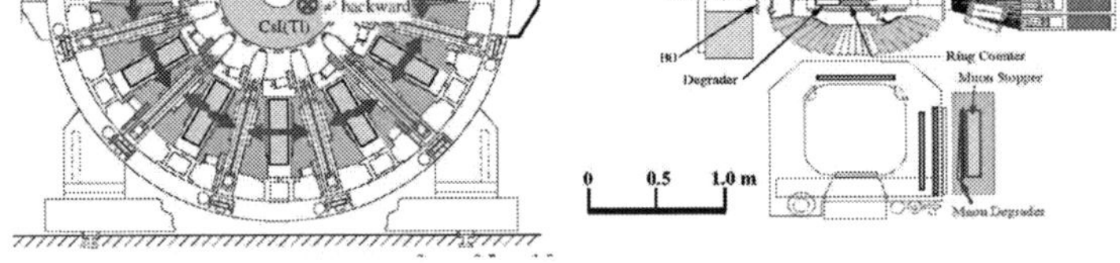

FIGURE 3. Schematic drawing of the E-246 apparatus to be reused by TREK in end view (left) and side view (right). The T-violating asymmetry is recorded from different positron count rates in clockwise and counterclockwise direction. Distinguishing between forward and backward pions corresponds to flipping of the decay plane, leading to a sign change of the asymmetry.

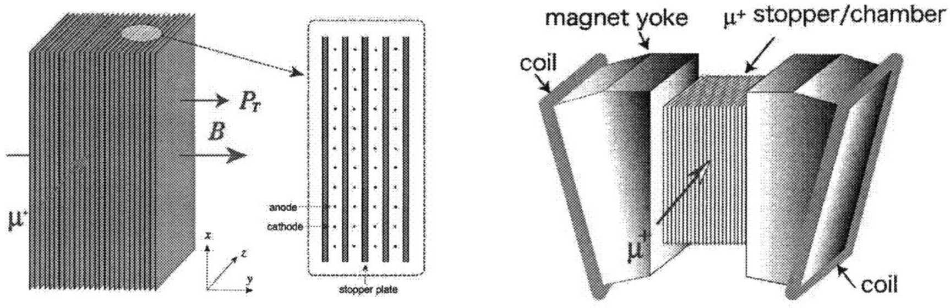

FIGURE 4. Schematic layout of the active muon polarimeter for TREK.

the arrangement helps reducing systematic uncertainties.

The previous E-246 system will be upgraded to meet the requirements of statistical and systematic uncertainties of TREK. The higher kaon beam intensity at J-PARC in combination with an increased acceptance of the upgraded setup with a newly designed active muon polarimeter will improve the statistical error of P_T by a factor 20 to 10^{-4} in less than one year of running time. While the previous passive polarimeter consisted of a stack of muon stoppers between simple plastic scintillators to count the clockwise and counterclockwise emitted positrons, the new active muon polarimeter (see Fig. 4) uses aluminum stopping plates parallel to the muon path, with the intermediate space between the plates designed as a wire chamber. This design results in a substantially larger solid angle acceptance to detect the decay positrons with good angular and certain energy resolution. In addition, a dedicated homogeneous holding field is used to ensure a polarization alignment relative to the field.

The higher count rates in TREK require a faster readout of the CsI(Tl) calorimeter, which will be based on avalanche photo diodes (APD) with new current amplifiers. The kaon stopping target has been redesigned with smaller diameter and higher segmentation by using a bundle of 432 scintillating fibers each 3 mm thick, resulting in 7.5 cm diameter. Promising tests of performance and radiation hardness have been conducted with multi-pixel photon counters (MPPC) as a novel fiber readout technology.

It has been demonstrated with detailed simulations that the systematic uncertainties of P_T can be reduced to $< 10^{-4}$. Alignments of tracking elements and the polarimeter will be calibrated with real tracks. The tracking capability for charged particles will be improved near the target region in order to identify and suppress backgrounds from charged pion tracks decaying in flight which resemble true $K^+_{\mu 3}$ events and which can cause spurious asymmetries. Identifying this class of events is only possible by adding near-target tracking elements in the field-free inner region of the toroid, which requires robust technology in the high-rate environment given. Therefore, the charged tracking upgrade will be based on the novel Gas Electron Multiplier (GEM) technology [4]. It is envisioned to construct a cylindrical GEM tracking detector (C0) surrounding the target assembly inside the CsI(Tl) calorimeter barrel, as well as additional planar GEM elements (C1) to cover the muon gaps outside the barrel in each of the twelve sectors. In

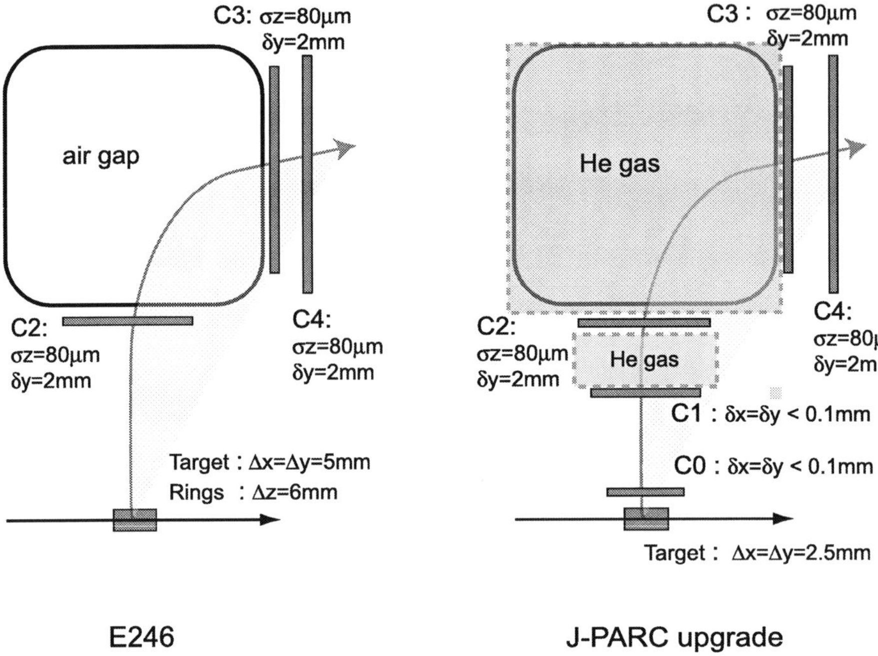

FIGURE 5. Proposed upgrade of the charged-particle tracking system.

addition, it is considered to fill previous large air gaps with He bags to reduce multiple scattering effects. The anticipated tracking upgrade for TREK is illustrated in Fig. 5.

In summary, a new generation of experimental search for T-violating transverse muon polarization in stopped-kaon decays has been proposed with TREK at J-PARC. The sensitivity of TREK will improve the current upper limit of P_T by at least a factor 20. The proposed experiment uses the previous E-246 apparatus after applying upgrades to various components.

This work has been supported in part by the US Department of Energy and National Science Foundation.

REFERENCES

1. J.J. Sakurai, Phys. Rev. **109**, 980 (1958).
2. K. Paton et al., TREK/E06 proposal for J-PARC (2006), `http://j-parc.jp/NuclPart/pac_0606/pdf/p06-Imazato_2.pdf`
3. M. Abe et al. [E246 collaboration], Phys. Rev. D **73**, 072005 (2006); Phys. Rev. Lett. **93**, 131601 (2004); Phys. Rev. Lett. **83**, 4253 (1999); J.A. Macdonald et al., Nucl. Instrum. and Meth. **A506**, 60 (2003).
4. F. Sauli, Nucl. Instr. and Meth. **A386**, 531 (1997); F. Simon et al., Nucl. Instr. and Meth. **A598**, 432 (2009).

The Q^p_{Weak} Experiment A Test for Physics Beyond the Standard Model via a Precision Measurement of the Proton Weak Charge

Michael T. Gericke (For the Qweak Collaboration)

University of Manitoba, Winnipeg R3T2N2, MB, Canada

Abstract. The Qweak collaboration is currently preparing a measurement of the proton's weak charge $Q^p_W = (1-4\sin^2\theta_W)$ at the Thomas Jefferson National Accelerator Facility (TJNAF). The experiment has been designed to produce a 4.1% measurement of the weak charge, via a 2.5% measurement of the parity violating asymmetry in the number of elastically scattered 1 GeV electrons from protons, at forward angles. At the proposed precision, the experiment would produce a 0.3% measurement of the weak mixing angle at a momentum transfer of ≈ 0.16 GeV, making it the most precise stand alone measurement of the weak mixing angle at low momentum transfer. In combination with other parity measurements, Qweak will also provide a high precision determination of the weak charges of the up and down quarks. Our proposed measurement of Q^p_W will be performed with significantly smaller statistical and systematic errors than existing low Q^2 data. Any significant deviation from the Standard Model prediction at low momentum transfer (Q^2) would be a signal of new physics, whereas agreement would place new and significant constraints on possible Standard Model extensions. The experiment is currently being prepared for installation, beginning in late 2009, and data taking beginning in mid 2010.

Keywords: Parity Violation, Electron Scattering, Weak Charge, Standard Model
PACS: <12.15.Mm, 12.60.Cn, 25.30.Bf, 11.30.Er>

INTRODUCTION AND PHYSICS

In the search for physics beyond the standard model, high energy colliders, such as the Large Hadron Collider (LHC) at CERN, search directly for signatures of new particles. Precision measurements at moderate energies, on the other hand, look for a discrepancy of suppressed, sensitive observables with their Standard Model predicted value and can reveal the signature of the new forms of matter found in high energy collider experiments [1, 2]. Results from the planned Q^p_W measurement at TJNAF, in conjunction with existing measurements of parity-violating electron scattering, will constrain the possibility of relevant physics beyond the Standard Model to the multi-TeV energy scale and beyond.

The Q^p_{Weak} experiment is currently being prepared for installation at TJNAF. The experiment will measure the parity violating asymmetry in the number of elastically scattered electrons ($E_{beam} = 1.165$ GeV) from protons at forward angles ($\langle\theta\rangle = 8°$), with an accuracy of 2.5%. In terms of the weak charges and nucleon form factors, the asymmetry is given by

CP1149, *Spin Physics, 18th International Spin Physics Symposium*
edited by D. G. Crabb, D. B. Day, S. Liuti, X. Zheng, M. Poelker, and Y. Prok

$$
\begin{aligned}
A_{LR}(\vec{e},p) &= k(A_{Q^p_W} + A_{H,V} + A_{H,A}) \\
A_{Q^p_W} &= Q^2 Q^p_W \\
A_{H,V} &= Q^n_W \frac{\varepsilon G_E^{p,\gamma} G_E^{n,\gamma} + \tau G_M^{p,\gamma} G_M^{n,\gamma}}{\varepsilon \left(G_E^{p,\gamma}\right)^2 + \tau \left(G_M^{p,\gamma}\right)^2} + Q^s_W \frac{\varepsilon G_E^{p,\gamma} G_E^{s} + \tau G_M^{p,\gamma} G_M^{s}}{\varepsilon \left(G_E^{p,\gamma}\right)^2 + \tau \left(G_M^{p,\gamma}\right)^2} \\
A_{H,A} &= Q^e_W \frac{\varepsilon' G_A^{p,Z} G_M^{p,\gamma}}{\varepsilon \left(G_E^{p,\gamma}\right)^2 + \tau \left(G_M^{p,\gamma}\right)^2},
\end{aligned}
$$

where $k \equiv -\frac{G_F}{4\pi\alpha\sqrt{2}}$, $\varepsilon \equiv \frac{1}{1+2(1+\tau)\tan^2\frac{\theta}{2}}$, $\varepsilon' \equiv \sqrt{\tau(1+\tau)(1-\varepsilon^2)}$, and $\tau \equiv -Q^2/4M_N^2$. The hadronic portion of the asymmetry has two terms, $(A_{H,V})$ and $(A_{H,A})$, corresponding to the vector and axial vector contributions. These are parameterized in terms of the electromagnetic form factors of the neutron and proton ($G_E^{p,\gamma}$, $G_E^{n,\gamma}$, $G_M^{p,\gamma}$, $G_M^{n,\gamma}$), as well as the strange and axial form factors (G_E^s, G_M^s, $G_A^{p,Z}$) respectively. The first term in the asymmetry is proportional to the weak charge of the proton $Q^p_W = \left(1 - 4\sin^2\theta_W\right)$. Both the asymmetry and the hadronic contribution decrease with lower momentum transfer and the experiment has been optimized for a maximal asymmetry, with a minimal contribution from hadronic constraints. The choice for the low momentum transfer ($Q^2 = 0.026$ GeV^2) for the Q^P_{Weak} experiment is based on this optimization. At this low Q^2, the hadronic contributions are well constrained by the wold data set [3, 4].

The Standard Model makes a firm prediction for Q^p_W, based on the running of the weak mixing angle, $\sin^2\theta_W$, from the Z^0 pole down to low energies, as shown in Fig. 1. The precise measurements near the Z^0 pole anchor the curve at one particular energy scale. The shape of the curve away from this point is a prediction of the Standard Model, and to test this prediction one needs precise off peak measurements. Beyond tree level, the weak charge of the proton can be written as [5] $Q^p_W = \rho_{PV}\left(1 - 4\kappa_{PV}\sin^2\theta_W\right) + \lambda_p = -2\left(2C_{1u} + C_{1d}\right)$, where $\rho_{PV} = 1 + \delta\rho^{SM} + \delta\rho^{SUSY}$, $\kappa_{PV} = 1 + \delta\kappa^{SM} + \delta\kappa^{SUSY}$, and $\lambda_p = 1 + \delta\lambda_p^{SM} + \delta\lambda_p^{SUSY}$ are parameters that contain both radiative corrections within the Standard Model, as well as contributions from possible new physics. The Standard Model radiative corrections combined in $\delta\kappa^{SM}(Q)$ include momentum transfer dependent terms from γ, Z^0 box and loop diagrams and collectively give rise to the scale dependence (running) of the weak mixing angle, as seen in Fig 1. The quark weak charges relevant for the Q^P_{Weak} experiment are $C_{1u} = \frac{1}{2}\rho'_{eq}\left(1 - \frac{8}{3}\kappa'_{eq}\sin^2\theta_W\right) + \lambda_{1u}$ and $C_{1d} = \frac{1}{2}\rho'_{eq}\left(-1 + \frac{4}{3}\kappa'_{eq}\sin^2\theta_W\right) + \lambda_{1d}$. Figure 1 shows the constraints on the quark weak charges, placed by the indicated completed measurements, together with the projected constraint added by the completed Q^P_{Weak} [4]. Combining the Q^p_W measurement with the previous experimental results will therefore lead to a significant improvement in the allowed range of values for C_{1u} and C_{1d}. Assuming the Standard Model holds, the resulting

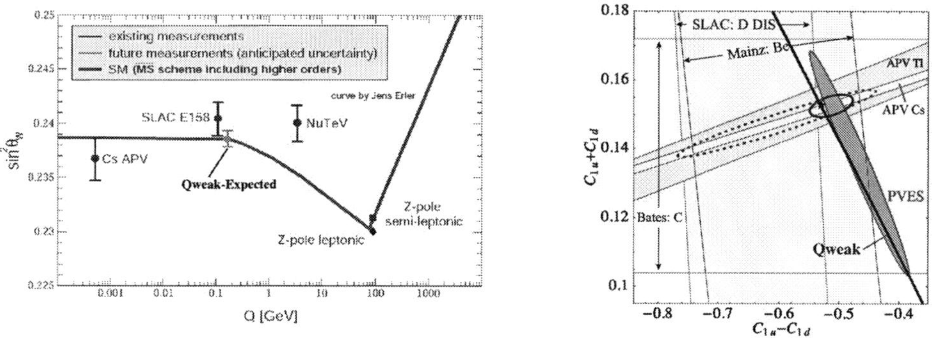

FIGURE 1. Left: Calculated running of the weak mixing angle in the Standard Model, as defined in the modified minimal subtraction scheme [6]. The uncertainty in the predicted running corresponds to the thickness of the blue curve. The black error bars show completed measurements [7, 8, 9, 10], while the grey error bar (with arbitrarily chosen vertical location) refers to the proposed 4% Qweak measurement. Right: Constraints on the neutral weak effective couplings [4]. The dotted contour displays the experimental limits (95% CL) reported in the PDG [10] together with the prediction of the Standard Model (black star). The filled ellipse denotes the current constraint provided by recent PVES scattering measurements on hydrogen, deuterium and helium targets (at 1σ), while the smaller solid contour (95% CL) indicates the full constraint obtained by combining all existing results. The solid blue line indicates the anticipated constraint from the planned Q_W^p measurement, assuming the SM value. All other experimental limits are at 1σ.

new limits on the values allowed for these fundamental constants will severely constrain relevant new physics to a mass scale for new weakly coupled physics of $\approx 2-6$ TeV.

EXPERIMENTAL OVERVIEW

The experimental layout is shown in Fig. 2. A 180 μA 85% longitudinally polarized beam of 1.165 GeV electrons is incident on a 2.5 kW LH_2 target. The scattered electrons are collimated to define an average scattering angle of $\theta = 8 \pm 3^\circ$ and a phi acceptance of 53% of 2π. An eight sector toroidal magnet focuses the scattered electrons onto a set of eight, two meter long quartz Čerenkov bars. Three sets of wire chambers are placed (two before the magnet at the collimators and one in front of the detectors) to allow for calibration mode (low beam current) Q^2 determination and background measurements. A 1 cm^2 scanning Čerenkov detector is placed behind the main Čerenkov detector to facilitate the Q^2 measurement at full current. A set of luminosity monitors are placed downstream of the main detectors. Møller and Compton polarimeters (not shown in figure) are located upstream of the target. At full current, the event rate per detector is about 800 MHz and the main detectors are therefore operated in current mode.

The main technical challenges for this experiment arise from the small expected asymmetry of approximately -0.3 ppm; we will measure this asymmetry to $\pm 2.1\%$ statistical and $\pm 1.3\%$ systematic errors. The experimental asymmetry must be corrected for inelastic and room background contributions as well as hadronic form factor uncertainties. Simulations indicate that the former will be small, the main contribution coming from

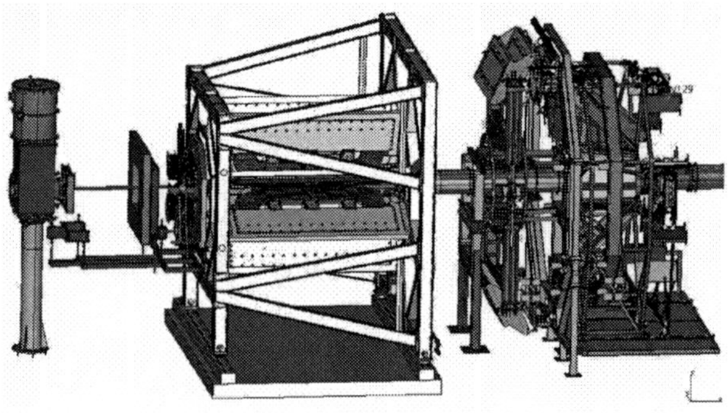

FIGURE 2. Engineering drawing of the experimental layout. The beam runs from left to right.

TABLE 1. Expected error budget.

Source of error	$\Delta A_{phys}/A_{phys}$	$\Delta Q_W^p/Q_W^p$
Counting statistics	2.1%	3.2%
Hadronic structure	-	1.5%
Beam polarimetry	1.0%	1.5%
Absolute Q^2	0.5%	1.5%
Backgrounds	0.5%	0.7%
Helicity correlated beam properties	0.5%	0.7%
Total	2.5%	4.1%

target walls, which can be measured and subtracted. The quadrature sum of systematic error contributions to Q_W^p, including the hadronic form factor uncertainty, is expected to be 2.6%. Experimental systematic errors are minimized by construction of a symmetric apparatus, optimization of the target design and shielding, utilization of feedback loops in the electron source to null out helicity correlated beam excursions and careful attention to beam polarimetry. A program of ancillary measurements will be carried out to determine the system response to helicity correlated beam properties and background terms. Table 1 summarizes the expected overall error budget for the experiment.

REFERENCES

1. J. Erler and M. J. Ramsey-Musolf. *Prog. Part. Nucl. Phys.*, **54**, 351 (2005).
2. M. J. Ramsey-Musolf and S. Su., arXiv: hep-ph/0612057 (2006).
3. J. Roche, R. D. Carlini, R.D. Young, and A.W. Thomas, *Phys. Rev. Lett.*, **97**, 102002 (2006).
4. R. D. Carlini, A.W. Thomas, R.D. Young, and J. Roche, *Phys. Rev. Lett.*, **99**, 122003 (2007).
5. Kurylov, Musolf, Su *Phys. Rev. D* **68**, 035008 (2003).
6. J. Erler and M. J. Ramsey-Musolf. *Phys. Rev.*, **D72**, 073003 (2005).
7. S. C. Bennett and C. E. Wieman. *Phys. Rev. Lett.*, **82**, 2484 (1999).
8. P. L. Anthony et al. *Phys. Rev. Lett.*, **95**, 081601 (2005).
9. G. P. Zeller et al. *Phys. Rev. Lett.*, **88**, 091802 (2002).
10. W. M. Yao et al. *J. Phys.*, **G33**, 1 (2006).

Parity-Violation in Deep Inelastic Scattering

K.D. Paschke

University of Virginia, Charlottesville, VA 22904, USA

Abstract. The completion of the planned 12 GeV upgrade at Jefferson Lab will open a new avenue for precision studies of the high-x structure of the nucleon through parity-violation in deeply-inelastic scattering (PV-DIS). PV-DIS would be a clean technique for studying quark-level charge symmetry violation, probing higher-twist effects and for a measurement of the structure function ratio d/u at high x. In addition to these topics in hadronic physics, these measurements would also provide access to a linear combination of the poorly measured axial electron-quark weak couplings C_{2u} and C_{2d}, and thus provide an important test of the electroweak Standard Model.

Keywords: parity, deep inelastic scattering, charge symmetry, parton distribution function, Standard Model
PACS: 13.60.-r, 11.30.Er, 13.60.Hb, 13.88.+e

INTRODUCTION

The measurement of parity-violation in deep-inelastic scattering (PV-DIS) provides both a powerful tool for studying partonic structure of nucleons and nuclei and a unique test of the Standard Model. In particular, this measurement would be sensitive to the poorly measured neutral-weak axial-quark to vector-electron coupling. These couplings are inaccessible in other measurements, leaving PV-DIS from the deuteron as a crucial component for any comprehensive search for physics beyond the Standard Model. PV-DIS will also have great value as a probe of the strong interaction, with sensitive to charge symmetry breaking in the deuteron and, for measurements of PV-DIS on hydrogen, the proton d/u ratio at high x.

A program for exploring these topics has been designed and proposed at Thomas Jefferson National Accelerator Facility [1]. A new spectrometer system is required to provide sufficiently high acceptance, resolution, and background suppression over a broad range of scattering angle. The physics program and experimental techniques are described below.

PARITY VIOLATION AND DEEP INELASTIC SCATTERING

The scattering of polarized electrons from unpolarized targets provides a clean window to study weak neutral current (WNC) interactions. At low values of Q^2, the parity-non-conserving WNC amplitude $\mathcal{M}_Z^{NC}$ is the leading term in the asymmetry:

$$A_{PV} = \frac{\sigma_R - \sigma_L}{\sigma_R + \sigma_L} \sim \frac{\mathcal{M}_Z^{NC}}{\mathcal{M}_\gamma} \sim \frac{G_F Q^2}{4\sqrt{2}\pi\alpha}(g_A^e g_V^T + \beta g_V^e g_A^T). \tag{1}$$

CP1149, *Spin Physics, 18th International Spin Physics Symposium*
edited by D. G. Crabb, D. B. Day, S. Liuti, X. Zheng, M. Poelker, and Y. Prok

In deep inelastic kinematics, the asymmetry for scattering polarized electrons from an unpolarized isoscalar target (such as deuterium) can be written

$$A_{PV}^{D}=\frac{3G_FQ^2}{2\sqrt{2}\pi\alpha}\left\{\frac{2C_{1u}-C_{1d}\left[1+R_s(x)\right]+Y(2C_{2u}-C_{2d})R_v}{5+R_s(x)}\right\} \quad \text{with} \tag{2}$$

$$R_s(x)=\frac{2s^{+}(x)}{u^{+}(x)+d^{+}(x)},\quad R_v(x)=\frac{u^{-}(x)+d^{-}(x)}{u^{+}(x)+d^{+}(x)},\quad \text{and } Y=\frac{1-(1-y)^2}{1+(1-y)^2-y^2\frac{R_{LT}}{1-R_{LT}}}.$$

Here the parton distribution functions appear as $q^{\pm}(x)=q(x)\pm\bar{q}(x)$ and $y=1-E'/E$. $R_{LT}=\sigma_L/\sigma_T\approx 0.2$ depends on x and Q^2. The C_{1q} represent the quark vector and electron axial couplings ($g_A^e g_V^q$), while the C_{2q} are the quark axial and electron vector coupling ($g_V^e g_A^q$). Above $x>0.4$, the sea quark contribution becomes negligible and the expression reduces to

$$A_{PV}^{D}=\frac{3G_FQ^2}{10\sqrt{2}\pi\alpha}\left\{2C_{1u}-C_{1d}+Y(2C_{2u}-C_{2d})\right\}. \tag{3}$$

This expression contains only combinations of the fundamental quark vector and axial couplings, and no longer carries any explicit dependence on the hadronic structure of the target.

One of the goals of a precision measurement of A_{PV}^{D} would be to test for parity-violating interactions beyond the Standard Model. As suggested by Eqn. 3, A_{PV}^{D} is sensitive to a linear combination of fundamental parameters, and a deviation from the asymmetry predicted under the Standard Model may be interpreted as evidence of a new interaction. However, this expression relies both on the assumption of charge symmetry and on the absence of significant higher-twist contributions. The validity of these assumptions must also be tested by a comprehensive program of PV-DIS measurements. The general strategy is to collect sufficient data for a precision measurement of A_{PV}^{D}, binned over a broad range of x from 0.3-0.75, with a range of about a factor of two in Q^2 at each $x<0.6$. There is a reasonable expectation that the contribution from charge symmetry violation and higher-twist will grow with increasing x, and along with the strong Q^2 signature of higher-twist. This behavior should allow these effects to be separately constrained in a global fit.

Charge symmetry violation (CSV) can be parameterized by two new PDFs: $\delta u\equiv u^p-d^n$ and $\delta d\equiv d^p-u^n$, where $u^{p(n)}$ and $d^{p(n)}$ are the up and down parton distribution functions in the proton (neutron). Non-zero values for δu and δd have been suggested in calculations considering QED effects in the Q^2 evolution [2, 3] or non-perturbative QCD effects [4, 5]. Corrections to A_{PV}^{D} due to CSV enter as the the ratio $R^{CSV}=(\delta u-\delta d)/(u+d)$. While δu and δd are small, the ratio R^{CSV} can be significant if the CSV PDFs drop more slowly than u and d with increasing x. There is presently no data which can directly test for parton-level CSV at the predicted levels, although these predictions of R^{CSV} would explain a significant fraction of the discrepancy between the NuTeV [6] result and the Standard Model [7].

Higher-twist contributions may also contribute, however most effects will contribute equally to the weak and electromagnetic amplitudes and therefore cancel in the asymmetry. Under the assumption of Conserved Vector Current, the one exception to this would

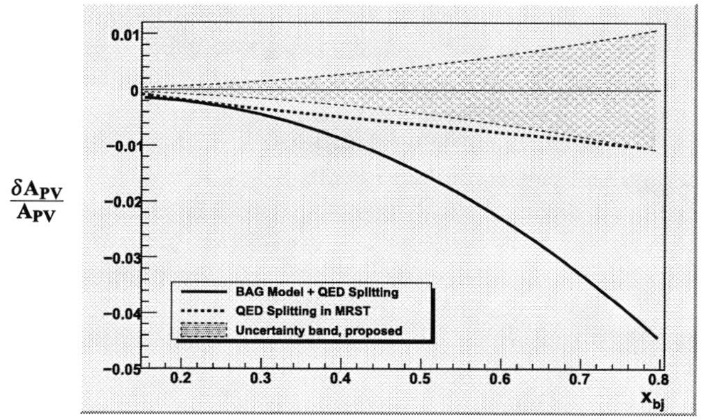

FIGURE 1. CSV predictions as a function of x. The vertical axis is the fractional change in A_{PV} due to CSV. The uncertainty band would result from the proposed study. The MRST results shown here account for QED splitting in the Q^2 evolution only, and do not include non-perturbative QCD effects [2].

be quark-quark correlations [8], which have not been previously isolated in studies of higher-twist effects. These effects might be apparent at higher x, corresponding the total momentum fraction carried by the correlated quarks together.

Existing spectrometers at JLab will be employed in an exploratory study of PV-DIS [9], but will lack the kinematic coverage necessary to separate these effects. With a new spectrometer system of the type described below, it will be feasible to collect a set of precise measurements of A_{PV}^D over a broad kinematic range. Sufficient range in x, Q^2, and Y can be obtained with a combination of measurements at beam energies of 11 GeV and 6.6 GeV. A global fit on this data could then simultaneously constrain a linear combination of fundamental electroweak parameters $M = 2C_{1u} - C_{2d} + \langle Y \rangle (2C_{2u} - C_{2d})$ along with parameterized contributions of R^{CSV} and higher-twist. For the envisioned measurement $\langle Y \rangle \approx 0.85$. The corresponding precision on the weak mixing angle $\sin^2\theta_W$ would approach $\delta(\sin^2\theta_W) \approx 0.0006$, comparable to any currently planned low-energy measurement from a hadronic target. The precision on R^{CSV} from this fit is illustrated in Fig. 1.

For PV-DIS from a proton target at high x, the asymmetry is approximately

$$A_{PV}^p = \frac{12C_{1u}u(x) - 6C_{1d}d(x)}{4u(x) + d(x)}. \tag{4}$$

A measurement of valence parton distribution functions d/u can be made, binned over a range of x, with a precision of around 2%. The highest bin would be at an average $x \approx 0.7$. This determination would not require any nuclear structure correction.

SOLENOIDAL LARGE INTENSITY DEVICE (SoLID)

The program outlined above requires measurements at high luminosity (540 pb^{-1} s^{-1}) and with large acceptance over a broad range of polar angles ($\sim 22° - 35°$) and scattered energies (1-5 GeV). A new spectrometer, SoLID, has been designed assuming the availability of a large superconducting solenoid, such as those used for the CLEO, BaBar, or D0 collider detectors. Collimation and instrumentation has been designed for a spectrometer system capable of making precision asymmetry measurements simultaneously throughout the broad kinematic range of interest.

The detector package will consist of tracking chambers for momentum and scattering angle determination and a gas Cerenkov detector to identify electrons and pions, triggered by an electromagnetic calorimeter. A series of 6 baffling collimators will block line-of-sight from the target to the detectors and intercept low momentum tracks which bend too quickly in the high axial magnetic field. The spectrometer design has been simulated under GEANT3, studying electromagnetic and pion backgrounds, acceptance, resolution and radiative losses. The BaBar solenoid was used as a model, with the field calculated using a redesigned yoke to reduce magnetic field in the region of the detectors. The trigger rate is expected to be on order of $\sim$220 kHz, including electromagnetic and pion backgrounds, which is manageable with the detector segmentation and pipeline electronics. GEM chambers will be required for tracking in areas of highest rate density, while straw tubes are planned for the larger tracking chambers. Acceptance, averaged around the azimuth, approaches 40% for the highest angle tracks. The kinematics of DIS events can be determined with sufficient precision, with $\sim$ 1.5 mrad resolution on scattering angle and 2-3% momentum resolution.

The program described above will also require electron beam polarimetry at the level of 0.4%. Compton polarimetry has been reported with a precision of 0.5% [10], and it may be that polarimetry using single-photon-detection mode will allow some further improvement. Moller polarimetry [11] using an ultra-cold trapped polarized atomic hydrogen target shows potential for similarly high precision. In combination, these promising techniques should enable robust and credible polarimetry at the level required for this ambitious program.

REFERENCES

1. P. Souder, et al. (2009), JLab PR12-09-012.
2. A. D. Martin, R. G. Roberts, W. J. Stirling, and R. S. Thorne, *Eur. Phys. J.* **C39**, 155–161 (2005), hep-ph/0411040.
3. M. Gluck, P. Jimenez-Delgado, and E. Reya, *Phys. Rev. Lett.* **95**, 022002 (2005), hep-ph/0503103.
4. E. Sather, *Phys. Lett.* **B274**, 433–438 (1992).
5. E. N. Rodionov, A. W. Thomas, and J. T. Londergan, *Mod. Phys. Lett.* **A9**, 1799–1806 (1994).
6. G. P. Zeller, et al., *Phys. Rev. Lett.* **88**, 091802 (2002), [Erratum-ibid. **90**, 239902 (2003)], hep-ex/0110059.
7. J. T. Londergan, D. P. Murdock, and A. W. Thomas, *Phys. Rev.* **D72**, 036010 (2005), hep-ph/0507029.
8. J. D. Bjorken, *Phys. Rev.* **D18**, 3239 (1978).
9. X. Zheng (2005), JLab E05-007.
10. K. Abe, et al., *Phys. Rev. Lett.* **84**, 5945–5949 (2000), hep-ex/0004026.
11. E. Chudakov, and V. Luppov, *Eur. Phys. J.* **A24S2**, 123–126 (2005).

$\vec{e}$-^{2}H Parity Violating Deep Inelastic Scattering at Jefferson Laboratory at 6 GeV

Ramesh R. Subedi*, Xiaoyan Deng*, Huaibo Ding†, Robert Michaels**, Paul E. Reimer‡, Diancheng Wang* and Xiaochao Zheng*

*University of Virginia, 382 McCormick Rd., Charlottesville, VA 22904, USA
†Tsinghua University, Beijing, 100084, P.R.China
**Thomas Jefferson National Accelerator Facility, Newport News, VA 23606, USA
‡Physics Division, Argonne National Laboratory, Argonne, IL 60439, USA

Abstract. The upcoming parity violating deep inelastic scattering (PVDIS) experiment E08-011 in Hall A at Thomas Jefferson National Accelerator Facility (Jefferson Lab) aims to measure a combination of the axial hadronic couplings of the electron with a factor of six improvement in precision over world data. Precise data for the couplings are essential to search for physics beyond the Standard Model. The experiment will measure a 10^{-4} asymmetry using polarized electron scattering from deuterium with a beam energy of 6 GeV. Recent progress on the preparation for this experiment, with an emphasis on the tests of the data acquisition system and the capability of measuring a small asymmetry, deadtime, and pileup effects are presented.

Keywords: parity, asymmetry, Standard Model
PACS: 21.10.Hw, 25.30.-c

INTRODUCTION

Parity violating electron scattering is one of the centerpieces of Jefferson Lab's physics programs. The goal of the PVDIS experiment (E08-011) is to measure precisely the asymmetry in parity violating deep inelastic scattering of a polarized electron on an unpolarized liquid deuterium target. This asymmetry is sensitive to the effective vector electron axial quark and axial electron vector quark couplings, the former of which is extremely poorly constrained by experimental data.

With σ_+ and σ_- as the left-handed and right-handed electron cross-sections, respectively, and Q^2 as the negative of the square of the four-momentum transfer, the parity violating asymmetry (A_{pv}) is given as

$$A_{pv} = \frac{\sigma_+ - \sigma_-}{\sigma_+ + \sigma_-} \cong Q^2 \; [100\,\mathrm{ppm/GeV^2}].$$

Taking effective coupling constants C_{iq} in the context of the Standard Model as:

$$C_{1u} = g^e{}_a\, g^u{}_v = -\frac{1}{2} + \frac{3}{4}\sin^2(\theta_w), \qquad C_{2u} = g^e{}_v\, g^u{}_a = -\frac{1}{2} + 2\sin^2(\theta_w),$$

$$C_{1d} = g^e{}_a\, g^d{}_v = \frac{1}{2} - \frac{2}{3}\sin^2(\theta_w), \qquad C_{2d} = g^e{}_v\, g^d{}_a = \frac{1}{2} - 2\sin^2(\theta_w),$$

CP1149, *Spin Physics, 18th International Spin Physics Symposium*
edited by D. G. Crabb, D. B. Day, S. Liuti, X. Zheng, M. Poelker, and Y. Prok

the important possible extraction of the experiment will be a high precision effective coupling constant combination $(2C_{2u}-C_{2d})$ from A_{pv}:

$$A_{pv}=(\frac{3G_FQ^2}{\pi\alpha^2\sqrt{2}})\ \frac{2C_{1u}[1+R_c(x)]-C_{1d}[1+R_s(x)]+Y(2C_{2u}-C_{2d})R_v(x)}{5+R_s(x)+4R_c(x)},$$

where G_F is the Fermi weak coupling constant, Y is a kinematic factor, R_s and R_v are sea- and valence-quark distribution functions (details in Ref. [1, 2]).

Another possible outcome of the experiment would be to extract an experimental result for the hadronic correction, $c(x)$, from higher-twist effects:

$$A_{pv}(x,Q^2)=A_{pv}(x)(1+c(x)/Q^2),$$

whose value is believed to be impossible to get from the first principles[3].

The data will be taken in Hall A using two standard spectrometer setups[4] at two Q^2 points of 1.10 (with incident electron energy $E=6.0$ GeV, scattered electron energy $E'=3.66$ GeV, Bjorken $x=0.25$, and scattered electron angle $\theta=12.9°$) and 1.90 (with $E=6.0$ GeV, $E'=2.63$ GeV, $x=0.30$, and $\theta=20.0°$) $(\mathrm{GeV}/c)^2$ using an 80% polarized electron beam of 85 μA on a 25-cm long deuterium target. The lowest expected physics asymmetry from the experiment is about 100 parts per million (ppm).

DEADTIME AND ASYMMETRY MEASUREMENTS

This experiment is introducing a scaler-based counting data acquisition (DAQ) system for the first time at Jefferson Lab. The new counting DAQ is essential, since the expected event rate of the experiment is about 0.5 MHz for each spectrometer and the background rate from pion electroproduction is high. The use of the conventional counting DAQ, which handles rates up to 4 kHz, is impossible. And the integrating DAQ used by previous parity violation experiments at JLab cannot reject pion backgrounds. The new DAQ will perform online particle identification and count electron events for a period of 30 msec helicity window in the scaler-readout. The scaler-readout is virtually deadtime-free, the only deadtime involved is from the electronic modules used.

The DAQ system has been assembled and a series of tests have been performed to determine its deadtime and to evaluate its performance in measuring small asymmetries. For the deadtime measurement, two different resolution times of the electronic modules were used: 20 ns ("narrow path") and 100 ns ("wide path"). Subsequent events falling within the resolution time of the previous events cannot be recorded by the DAQ, causing counting losses due to the deadtime.

Three methods were employed to measure the deadtime. The results are shown in Fig.1. In method-I, a pulser signal (called a tagger) with rate R_i was mixed (ORed) with a high rate random pulse ranging from a few Hz to a few MHz obtained from a photomultiplier tube (PMT). The mixed signal was then input to the PVDIS DAQ. The output trigger from the PVDIS DAQ was ANDed with the tagger giving R_o as the output. The fractional loss due to deadtime (D) was obtained as $D=1-(1-p)R_o/R_i$, where p is a correction factor for the pileup effects: when the tagger signal follows closely a random PMT signal, the DAQ output caused by the PMT signal would co-incide with the tagger

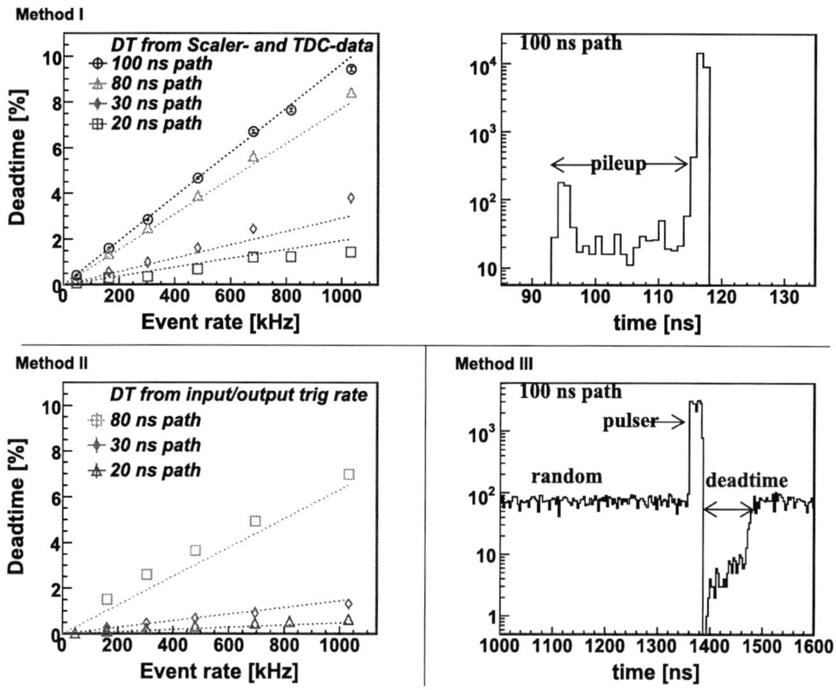

FIGURE 1. Results of deadtime measurements from three methods. Dotted lines show the expected fractional count loss due to deadtime as a function of the event rate. *Top-left:* results from method-I. *Top-right:* pileup appearing in the timing spectrum of the output trigger with respect to the reference tagger signal. *Bottom-left:* results from method-II. An intrinsic 15 ns deadtime in the input rate measurement was added to each point. *Bottom-right:* results from method-III.

signal, causing a false count that should have lost due to deadtimes. The pileup factor p was measured by the TDC spectrum as shown in the top-right pad in Fig.1. If there was no pileup, only the right-side peak would show up, hence the integral of the events other than this peak would correspond to the pileup amount. In method-II, the input trigger rate r_i and the output trigger rate r_o of the PVDIS DAQ were measured directly in the scaler module. The deadtime was obtained as $D = 1 - r_o/r_i$. In method-III, tagger signals are mixed with random PMT signals and then sent to a TDC. Following each tagger signal there exists a "dead-zone" region in the TDC spectrum during which the system does not respond to the random PMT signals, as shown in the bottom-right pad in Fig.1. This dead-zone region provides a measurement of the deadtime. The deadtime measured by these three methods were consistent with each other.

An artificial asymmetry signal, with a magnitude of a few ppm, was prepared using an asymmetry module locally made at Jefferson Lab. This asymmetry signal was mixed (ORed) with a high rate PMT signal and then sent as inputs to the PVDIS DAQ. Shown in Fig.2 are our test results from a few to a thousand ppm and the measured asymmetries agreed with the expected values within their statistical uncertainties.

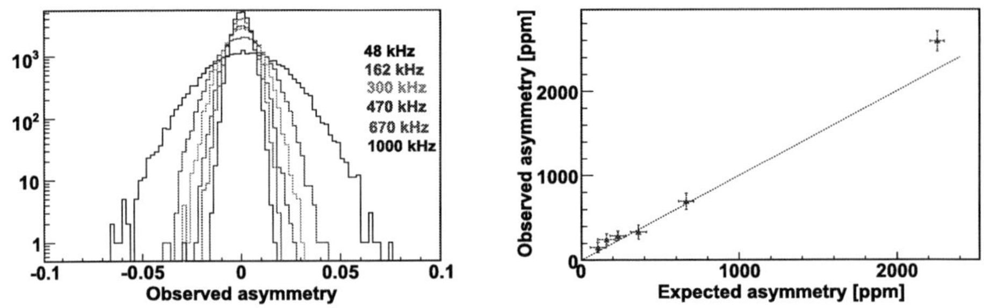

FIGURE 2. *Left*: width of asymmetry-histograms for different event rates. The rate of the innermost trace is the highest and goes decreasing as we march outwards. They agree with the expected width $\sqrt{\frac{1}{RT}}$ where R is the trigger rate and T is the integration time of a helicity window pair. *Right*: Expected and the observed asymmetries. The expected asymmetry of the experiment is in the region of $100-200$ ppm.

SUMMARY

Jefferson Lab experiment $E08-011$ will measure the parity violating asymmetry in deep inelastic scattering from a deuterium target. Significant efforts have been used to design, assemble and bench-test the fast-counting DAQ system needed by this experiment. Our tests provided the confidence that the different methods used for the deadtime determination and the technique for asymmetry measurement were working. The new DAQ is currently being tested by taking parasitic data with the electron beam.

ACKNOWLEDGMENTS

This work is supported in part by the U.S. National Science Foundation and the U.S. Department of Energy (DOE), under Contract No. DE-AC02-06CH1135.

Notice: Authored by Jefferson Science Associates, LLC under U.S. DOE Contract No. DE-AC05-06OR23177. The U.S. Government retains a non-exclusive, paid-up, irrevocable, world-wide license to publish or reproduce this manuscript for U.S. Government purposes.

REFERENCES

1. R. Michaels, P. Reimer, and X.-C. Zheng, *$\vec{e}$-^2H Parity Violating Deep Inelastic Scattering at CEBAF at 6 GeV*, Jefferson Lab Hall A Proposal E08-011.
2. X. Zheng, *Parity Violation in Deep Inelastic Scattering at JLab 6 GeV, in PAVI 2006 May 16-20: From Parity Violation to Hadronic Structure and more... (Part III) (Springer, Milos, Greece, 2006)*
3. K. S. Kumar, *Eur. Phys. J. A* **24**, s2, 191-195 (2005).
4. J. Alcorn, et al., *Nucl. Instr. and Meth.* **A 522**, 294 (2004).

Nuclear Spin Maser at Highly Stabilized Low Magnetic Field and Search for Atomic EDM

A. Yoshimi[a], K. Asahi[b], T. Inoue[b], M. Uchida[b], N. Hatakeyama[b], M. Tsuchiya[b] and S. Kagami[b]

[a]*RIKEN Nishina Center, Hirosawa 2-1, Wako, Saitama 351-0198, Japan.*

[b]*Department of Physics, Tokyo Institute of Technology, Oh-okayama 2-12-1, Meguro, Tokyo 152-8551, Japan*

Abstract. A nuclear spin maser is operated at a low static field through an active feedback scheme based on an optical nuclear spin detection and succeeding spin control by a transverse field application. The frequency stability of this optical-coupling spin maser is improved by installation of a low-noise current source for a solenoid magnet producing a static magnetic field in the maser operation. Experimental devices for application of the maser to EDM experiment are being developed.

Keywords: Electric dipole moment, time reversal, nuclear spin maser, spin polarization
PACS: 11.30.Er, 13.40.Em, 32.60.+i , 33.80.Be, 33.25.+k, 84.40.Ik

INTRODUCTION

Nuclear spin sometimes plays quite important roles in fundamental physics research. One of the most obvious examples would be found in the study of electric dipole moment (EDM) which, if non-vanishing, should constitute a clear evidence for violation of time-reversal (T) and space inversion (P) invariance [1]. In the EDM search experiments, EDM signal would arise as a frequency shift $\delta v = 4dE/h$ of the spin precession, when the direction of the electric field E is reversed along that of a static magnetic field. Here d indicates the non-zero EDM, and h the Plank constant. The experimental upper limit has been set $d_n < 2.9 \times 10^{-26}$ $e$cm for neutron EDM [2]. Therefore recent EDM experiments require the detection of a small frequency shift δv of the order of nHz. Recently searches for EDM in atomic or molecular system have been proposed and performed with various techniques such as optical pumping, trap, *etc.* These experiments have set the upper limit, for example, mercury atomic EDM $d(^{199}\text{Hg}) < 3.1 \times 10^{-29}$ $e$cm [3], and thallium atomic EDM $d(^{205}\text{Tl}) < 9.4 \times 10^{-25}$ $e$cm [4]. From these results for atomic EDM, more fundamental P,T-odd moments are derived, Schiff moment [5] $S(^{199}\text{Hg}) < 1.1 \times 10^{-12}$ $e\text{fm}^3$ and electron EDM $d_e < 1.6 \times 10^{-27}$ $e$cm. The microscopic calculations of the atomic EDM are continuously progressing [6-8]. These upper limits for the atomic and neutron EDM are discussed with some new CP violating mechanisms beyond the Standard Model, *e.g.,* Supersymmetry [9]. It should be noted the EDM predicted in physics beyond the Standard Model is larger by several

CP1149, *Spin Physics, 18th International Spin Physics Symposium*
edited by D. G. Crabb, D. B. Day, S. Liuti, X. Zheng, M. Poelker, and Y. Prok

orders than that predicted in the Standard Model, and are expected to be detectable in recent or near-future experiments.

The EDMs in diamagnetic atoms such as ^{199}Hg or ^{129}Xe are mainly induced by the Schiff moment, and these atomic EDM values are suppressed compared with neutron EDM. However, highly sensitive EDM measurement can be performed with the diamagnetic atoms. Noble gas atom ^{129}Xe with a nuclear spin 1/2 can have a long coherence time of the nuclear spin, and its spin polarization and relaxation have been extensively studied [10, 11]. We have developed a new type of the nuclear spin maser with the spin-polarized ^{129}Xe, in which the highly sensitive measurement of the spin precession can be performed at a low static magnetic field [12, 13].

R&D STATUS OF SPIN MASER AT LOW STATIC FIELD

The nuclear spin maser, where the nuclear spins continuously precess without losing their transverse polarization, can oscillate when an oscillating transverse magnetic field, whose phase is shifted by $\pi/2$ to the transverse polarization component and whose amplitude is proportional to that of the transverse polarization, is applied to the precessing spin system [12]. We have introduced the optical-detection of the ^{129}Xe nuclear precession which utilizes transverse spin-exchange from ^{129}Xe nucleus to Rb atom [12]. By feeding thus detected precession phase signal back to a coil to produce the oscillating magnetic field to operate the spin maser, the continuous nuclear precession of the ^{129}Xe can be realized. The spin maser with the optical-detection method can oscillate at low frequency of the spin precession (low static field), contrary to the conventional spin maser which oscillates due to the spin-coil coupling [14,15]. The operation of the spin maser at the low static field can be an advantageous tool for the EDM experiment, because fluctuations and drifts in the applied magnetic field would be smaller with lower field.

The experimental setup for the optical-coupling nuclear spin maser is described elsewhere in detail [12]. A typically observed maser signal is shown in Fig. 1 in which

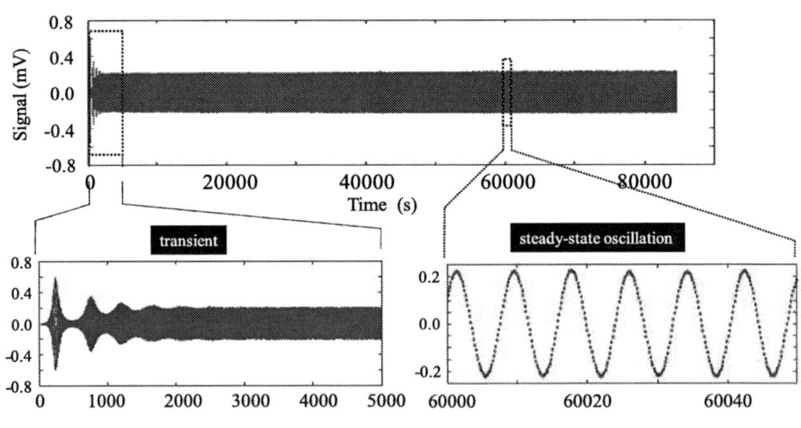

FIGURE 1. The observed spin maser signal.

the maser oscillation continues for one day. The signal was observed with a static magnetic field B_0 = 30.6 mG, corresponding to the precession frequency 36.0 Hz of ^{129}Xe nuclei. The signal is phase sensitively detected with lock-in devices, so that a beat signal is observed as shown in Fig. 1. A steady-state maser oscillation is observed after a transient phenomenon.

Frequency properties of the low-frequency spin maser were analyzed with two output signals V_x and V_y from the lock-in device whose phase is $\pi/2$ shifted each other. The frequency precision of the maser is deduced from a linear χ^2-fitting procedure for the phase data calculated from the V_x and V_y. The precession phase data for 3×10^4 s and the phase deviation are shown in Fig. 2. The frequency precision estimated from this procedure for typical 3×10^4 s data is 10 nHz. This precision corresponds to the EDM sensitivity of 1×10^{-27} $e$cm, assuming an applied electric field of 10 kV/cm. This precision of 10 nHz is improved by 2 orders from the previous measurement [16]. This improvement was done by introducing a newly fabricated current source for producing B_0 field with high stability of $\sim3\times10^{-6}$ during a 10^4 s, which is improved by 2 orders from a previous one [17].

We are now investigating long-term frequency stability of this spin maser. The frequency stability for the long term, over 10^4 s, is quite important for the EDM experiment. A fluctuation of temperature in some experimental devices including the current source and the glass cell was found to be one of main source of the frequency instability. Therefore a temperature feedback loop for the cell chamber is introducing, and a suppression of the temperature dependence of the current source is under preparation. We are also preparing to install a highly sensitive magnetometer with alkali vapor in order to monitor the fluctuation of the magnetic field. This magnetometer utilizes the nonlinear magneto-optical effect (NMOE) of Rb atoms, in which a narrow magnetic resonance width of 1 μG would be realized with an anti-relaxation coated glass cell. Recent studies on the NMOE magnetometer is expected to have a sensitivity of 10^{-12} G/$\sqrt{\text{Hz}}$ for small field region of 1μG [18], and 10^{-11} G/$\sqrt{\text{Hz}}$ for geophysical field range (< 500 mG) [19]. We are also fabricating field plates to apply the electric field for the EDM experiment.

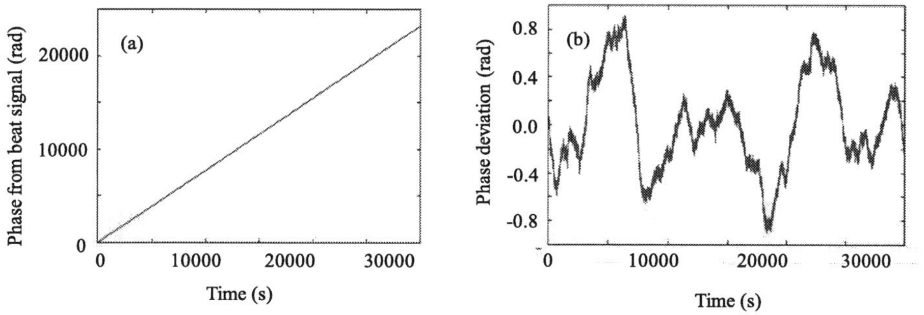

FIGURE 2. (a) Precession phase deduced from observed maser beat signal. (b) Phase deviation of maser signal from the linear-fitted function in (a).

SUMMARY

New scheme for the nuclear spin maser operation at low magnetic field of 30 mG, was developed to measure the precession frequency with high precision. The measured frequency precision was found to be 10 nHz with measurement time of 3×10^4 s, which was improved by 2 orders from the previous report with the newly fabricated current source with high stability. In order to proceed to the EDM search, further developments *e.g.*, temperature control system, precision atomic magnetometer, and electrodes for application of electric field are progressing.

ACKNOWLEDGMENTS

This work was supported by Grant-in-Aid for Scientific Research (B).

REFERENCES

1. I.B. Kriplovich and S.K. Lamoreaux, CP Violation without Strangeness, Springer-Verlag, Heidelberg, 1997.
2. C.A. Baker et al., *Phys. Rev. Lett.* **97**, 131801 (2006).
3. W.C. Griffith et al., *Phys. Rev. Lett.* **102**, 101601 (2009).
4. B.C. Regan et al., *Phys. Rev. Lett.* **88**, 071805 (2002).
5. L. I. Schiff, *Phys. Rev.* **132**, 2194 (1963).
6. J.S.M. Ginges and V.V. Flambaum, *Phys. Rep.* **397**, 63 (2004).
7. S. Oshima et al., *Phys. Rev. C* **75**, 035501 (2007).
8. C.-P. Liu et al., *Phys. Rev. C* **76**, 035503 (2007).
9. M. Pospelov and A. Ritz, *Ann. Phys.* **318**, 119 (2005).
10. W. Happer et al., *Phys. Rev. A* **29**, 3092 (1984).
11. G.D. Cates et al., *Phys. Rev. A* **45**, 4631 (1992).
12. A. Yoshimi et al., *Phys. Lett. A* **304**, 13 (2002).
13. A. Yoshimi et al., *Hyperfine Interact.* **159**, 401 (2004).
14. M.G. Richards et al., *J. Phys. B* **21**, 665 (1988).
15. T.E. Chupp et al., *Phys. Rev. Lett.* **72**, 2363 (1994).
16. K. Asahi et al., *AIP Conf. Proc.* **915**, 206 (2007).
17. A. Yoshimi et al., *Hyperfine Interact.* **181**, 111 (2008).
18. D. Budker et al., *Phys. Rev. A* **62**, 043403 (2000).
19. V. Acosta et al., *Phys. Rev. A* **73**, 053404 (2006).

On the Kinematics of the Centre of Charge of an Elementary Spinning Particle

Martín Rivas

Theoretical Physics Department, The University of the Basque Country, Apdo. 644, 48080 Bilbao, Spain

Abstract. In particle physics, most of the classical models consider that the centre of mass and centre of charge of an elementary particle, are the same point. This presumes some particular relationship between the charge and mass distribution, a feature which cannot be checked experimentally. If assumed different points, the centre of charge of an elementary spinning particle moves in a helical motion at the speed of light, and it thus satisfies, in general, a fourth order differential equation. This assumption of a separation betwen the centre of mass and centre of charge gives a contribution to the spin of the system and also justifies the existence of a magnetic moment produced by the relative motion of the centre of charge. This corresponds to an improved model of a charged elementary particle, than the point particle case.

Keywords: Classical Spin, Particle structure
PACS: 11.30.Ly, 11.10.Ef, 11.15.Kc

GEOMETRICAL ARGUMENTS

We analyze the most general differential equation satisfied by a point in three dimensional space, which is of fourth order. The free motion leads to consider that the center of charge is moving along a helix at a constant velocity and with a constant curvature and torsion. If this description has to be independent of any inertial observer, then necessarily the velocity has to be the speed of light.

A continuous and differentiable curve in three-dimensional space, $\mathbf{r}(s)$, is a regular curve if at any point s has a tangent vector $\dot{\mathbf{r}}(s) \neq 0$. It has associated three orthogonal unit vectors, $\mathbf{t}$, $\mathbf{n}$ and $\mathbf{b}$, called respectively the tangent, normal and binormal. If using the arc length s as the curve parameter, they satisfy the Frenet-Serret equations

$$\dot{\mathbf{t}} = \kappa\mathbf{n}, \quad \dot{\mathbf{n}} = -\kappa\mathbf{t} + \tau\mathbf{b}, \quad \dot{\mathbf{b}} = -\tau\mathbf{n},$$

where the overdot means the derivative with respect to s. In terms of the vector $\omega = \tau\mathbf{t} + \kappa\mathbf{b}$, known as Darboux vector, the Frenet-Serret equations can be rewritten as

$$\dot{\mathbf{t}} = \omega \times \mathbf{t}, \quad \dot{\mathbf{n}} = \omega \times \mathbf{n}, \quad \dot{\mathbf{b}} = \omega \times \mathbf{b},$$

so that, in units of arc length, Darboux vector represents the instantaneous angular velocity undergone by the local triad frame.

By changing the notation $\mathbf{r}^{(k)}(s) \equiv d^k\mathbf{r}/ds^k$, we have

$$\mathbf{r}^{(1)} = \mathbf{t}, \quad \mathbf{r}^{(2)} = \kappa\mathbf{n}, \quad \mathbf{r}^{(3)} = \dot{\kappa}\mathbf{n} + \kappa(-\kappa\mathbf{t} + \tau\mathbf{b})$$

CP1149, *Spin Physics, 18th International Spin Physics Symposium*
edited by D. G. Crabb, D. B. Day, S. Liuti, X. Zheng, M. Poelker, and Y. Prok

and this allows us to eliminate the three unit vectors **t**, **n** and **b**, in terms of the three derivatives $\mathbf{r}^{(k)}$, $k = 1,2,3$. If we replace them in the next order derivative $\mathbf{r}^{(4)}$, one obtains the most general differential equation satisfied by the point **r**, i.e., the fourth order differential system

$$\mathbf{r}^{(4)} - \frac{2\dot{\kappa}\tau + \dot{\tau}\kappa}{\kappa\tau}\mathbf{r}^{(3)} + \left(\kappa^2 + \tau^2 + \frac{\dot{\kappa}\dot{\tau} - \tau\ddot{\kappa}}{\kappa\tau} + \frac{2\dot{\kappa}^2}{\kappa^2}\right)\mathbf{r}^{(2)} + \frac{\kappa}{\tau}(\dot{\kappa}\tau - \dot{\tau}\kappa)\mathbf{r}^{(1)} = 0. \quad (1)$$

Let us consider that an elementary particle, is a localized mechanical system described by the evolution of a single point **r**, the centre of charge. We make the assumption that the centre of mass and the centre of charge are not necessarily the same point.

Let us assume now that the motion of the particle is free. This means that we cannot distinguish one point of the evolution from another, so that the motion has to be at a constant velocity such that the arc length $ds = |\mathbf{u}|dt$, where $\mathbf{u} = d\mathbf{r}/dt$ is the velocity of the charge, must be independent of the time t. At the same time, Darboux vector has to be also independent of time. The Frenet-Serret triad moves and rotates in a free motion with constant linear and angular velocities. The curvature and torsion are necessarily constants of the motion.

Thus $\dot{\kappa} = \dot{\tau} = 0$, and, in the free case, the equations (1) are reduced to

$$\mathbf{r}^{(4)} + \left(\kappa^2 + \tau^2\right)\mathbf{r}^{(2)} = \frac{d^2}{ds^2}\left(\mathbf{r}^{(2)} + \left(\kappa^2 + \tau^2\right)\mathbf{r}\right) = 0.$$

If the curvature and torsion are constant the curve is a helix, which can be factorized in terms of a central point

$$\mathbf{q} = \mathbf{r} + \frac{1}{\omega^2}\frac{d^2\mathbf{r}}{dt^2}, \quad \frac{d^2\mathbf{q}}{ds^2} = 0, \quad \omega^2 = \kappa^2 + \tau^2$$

which is moving in a straight trajectory, while the point **r** satisfies

$$\frac{d^2\mathbf{r}}{dt^2} + \omega^2(\mathbf{r} - \mathbf{q}) = 0,$$

an isotropic harmonic motion of frequency ω, around point **q**. The point **q** clearly represents the centre of mass position of the free particle.

The centre of charge of a free elementary particle is describing a helix at a constant velocity for any inertial observer. If we make a nonrelativistic analysis, the relationship of the velocity measurements among two arbitrary inertial observers O and O', is given by $\mathbf{u}' = R\mathbf{u} + \mathbf{v}$, where **v** is the constant velocity of O as measured by O' and the constant rotation matrix R is their relative orientation. Now,

$$u'^2 = u^2 + v^2 + 2\mathbf{v}\cdot R\mathbf{u}. \quad (2)$$

In a relativistic analysis

$$u'^2 = \frac{u^2 - c^2}{\gamma^2\left(1 + \mathbf{v}\cdot R\mathbf{u}/c^2\right)^2} + c^2. \quad (3)$$

Taking the time derivative of both expressions (2) and (3) if u' has to be also constant for observer O', we get that $\mathbf{v} \cdot R\dot{\mathbf{u}} = 0$, irrespective of $\mathbf{v}$ and of the rotation matrix R. This means that the vector $\mathbf{u}$ must be a constant vector. The centre of charge necessarily moves along a straight trajectory at a constant velocity, for every inertial observer, and the above general helix degenerates into a straight line and because the point is not accelerated $\mathbf{q} = \mathbf{r}$. This is the usual description of the spinless or pointlike free elementary charged particle, whose centre of charge and centre of mass are represented by the same point.

However, in the relativistic analysis, there is one alternative not included in the nonrelativistic approach. The possibility that the charge of an elementary particle will be moving at the speed of light and, in that case, $u = u' = c$, for any inertial observer. This means that the centre of the helix is always moving at a velocity $|d\mathbf{q}/dt| < c$, and, if it represents the centre of mass, this particle is a massive particle. In a variational description of this system the Lagrangian should depend up to the acceleration of the point $\mathbf{r}$ in order to obtain fourth order differential equations. This dependence on the acceleration will give a contribution to the spin of the particle and there is also another contribution from the rotation of the system, because the body frame rotates with angular velocity ω. The motion of the charge around the centre of mass produces the magnetic moment of the particle.

The velocity of the charge can never be reached by any inertial observer. Otherwise, the velocity of the charge will be zero at a certain time but different from zero for subsequent times, because the point is accelerated, which is contradictory with the requirement that the point describes a trajectory with a constant velocity.

In summary, there are only two possibilities for a free motion of the centre of charge of an elementary particle. One, the charge is moving along a straight line at any constant velocity, the centre of mass is the same point, the system has no magnetic moment and it can be described either in a relativistic or a non-relativistic framework. In the other, the particle has spin and magnetic moment, and the charge moves along a helix at the speed of light. The particle also has an internal frequency ω. Because all known elementary particles, quarks and leptons, are spin $1/2$ particles, and the charged ones have magnetic moment, we are left only with this last possibility if we want to give a more improved description of an elementary particle. Here, only the relativistic description is allowed.

This is consistent with Dirac's theory of the electron, because the eigenvalues of the components of Dirac's velocity operator are $\pm c$. This means that Dirac's spinor $\psi(t, \mathbf{r})$ is expressed in terms of the position of the charge $\mathbf{r}$, because the external fields $A_\mu(t, \mathbf{r})$ are defined and computed at this point.

This last possiblity is the description of the centre of charge of a relativistic spinning elementary particle obtained in the kinematical formalism [1], and which satisfies Dirac's equation when quantized.

In this formalism Dirac particles are localized and also orientable mechanical systems. By orientable we mean that we have to attach to the above point $\mathbf{r}$, a local cartesian frame to describe its spatial orientation. This frame could be the Frenet-Serret triad. The rotation of the frame will also contribute to the total spin of the particle. When quantizing the system, the spin $1/2$ is coming from the presence of the orientation variables. Otherwise, if there are no orientation variables, no spin $1/2$ structure is described when quantizing the system. This twofold structure of the classical spin has produced a pure kinematical interpretation of the gyromagnetic ratio [2]. The dependence of the Lagrangian on the acceleration is necessary for the particle to have magnetic moment and for the separation between the centre of mass and centre of charge.

SPIN STRUCTURE

The Lagrangians for decribing clasical elementary spinning particles [1] depend on the acceleration $\dot{\mathbf{u}}$ of the charge and on the angular velocity of the rotation of the body frame ω, $L(\dot{\mathbf{u}}, \omega)$, such that Noether's theorem leads to define the spin as

$$\mathbf{S} = \mathbf{u} \times \frac{\partial L}{\partial \dot{\mathbf{u}}} + \frac{\partial L}{\partial \omega}.$$

This spin represents the angular momentum of the particle with respect to the center of charge and not with respect to its center of mass. It satisfies the dynamical equation

$$\frac{d\mathbf{S}}{dt} = \mathbf{P} \times \mathbf{u},$$

irrespective of whether the particle is free or interacting. When quantizing the system it becomes Dirac's spin operator. This is important when analyzing the proton spin structure. By the spin of the proton we understand its angular momentum with respect to its center of mass. If the proton is a bound system of three quarks and each quark is a Dirac particle, it is impossible that the addition of the three Dirac spin operators of the quarks gave rise to the spin of the proton. The separation between the center of mass and center of charge of a Dirac particle is $R = \hbar/2mc$, half Compton's wavelength, and because of the binding energies the linear momenta of the quarks inside the proton have to be very high, of order of 320 MeV/c^2. This means that the angular momentum of each quark with respect to its center of mass is Dirac's spin operator plus the cross product of this linear momentum times the separation R.

If the three quarks are in a $L = 0$ orbital angular momentum state, what we have to add are the three spins of the quarks with respect to its own center of mass.

ACKNOWLEDGMENTS

This work was supported by The University of the Basque Country (Research Grant 9/UPV00172.310-14456/2002).

REFERENCES

1. M. Rivas, *Kinematical theory of spinning particles*, Kluwer, Dordrecht, 2001
   See also some other works related to this formalism through the webpage
   `<http://tp.lc.ehu.es/martin.htm>`
2. M. Rivas, J.M. Aguirregabiria and A. Hernández, *Phys. Lett. A* **257**, 21 (1999).

LONGITUDINAL SPIN CONTRIBUTIONS

Review of ΔG in DIS and pp

Frank Ellinghaus

University of Mainz, Staudinger Weg 7, 55128 Mainz, Germany
and
University of Colorado, Boulder, CO 80309, USA

Abstract. The polarized gluon distribution function $\Delta g(x)$ is presently studied at various deep inelastic scattering and pp experiments. The goal is to eventually determine the contribution of the gluon spin ΔG to the overall spin-1/2 of the proton, given that the quark spins are known to only contribute at the level of 30%. A review of the present status and prospects will be given.

Keywords: nucleon spin, gluon polarization, structure function, PDF, hadron electroproduction
PACS: 13.60.-r, 13.85.Ni, 13.88.+e, 14.20.Dh

In the most naive model, the so-called constituent quark model, in which the protons consist of three static quarks, two of the three spin 1/2 quarks will align their spins with the spin of the proton, the third one is anti-aligned, and *voila*, the three quarks account for the spin 1/2 of the proton. However, after pioneering experiments at SLAC starting in the mid-1970's it was the European Muon Collaboration (EMC) at CERN in the late 1980's that found the contribution of the quark spins to the spin of the nucleon to be small and actually consistent with zero within the uncertainties [1]. The most recent results by the HERMES Collaboration at DESY [2] and the COMPASS Collaboration at CERN [3] show that the quarks contribute about 30% to the spin of the nucleon, leading to our current understanding that a substantial fraction of the proton's spin has to arise from the quark orbital angular momentum and/or the gluons.

The possible contribution of the gluon spin $\Delta G(Q^2) = \int_0^1 dx \Delta g(x, Q^2)$ to the nucleon spin has first been studied in deep inelastic scattering (DIS). Here Q^2 and x denote the negative four-momentum squared of the virtual photon and the fraction of the momentum of the (fast) proton carried by the struck parton, respectively. The problem with studying gluons in DIS is that the electromagnetic probe, the photon, cannot directly couple to the gluon since gluons carry no electric charge. An indirect way of accessing the Parton Distribution Function (PDF) for polarized gluons, $\Delta g(x)$, is via next-to-leading order (NLO) fits to the polarized structure function $g_1(x, Q^2)$. The information on the polarized gluon PDF is contained in the dependence of g_1 on Q^2 at a given x, which is nearly constant. To be precise, it is the difference from the constant behavior, the so-called scaling violation, which contains the information. However, since all measurements of g_1 have been done at fixed-target experiments the range in Q^2 covered is very limited, leading to rather large uncertainties for $\Delta g(x)$, see, e.g., Ref.[4].

A more direct way to measure $\Delta g(x)$ at the DIS experiments is via a NLO process called photon-gluon fusion (PGF), in which the virtual photon and the gluon interact via the creation of a quark-antiquark pair as shown in the left panel of Fig. 1. A rather clean channel in order to tag the PGF process is the production of $c\bar{c}$ pairs that fragment into D

CP1149, *Spin Physics, 18th International Spin Physics Symposium*
edited by D. G. Crabb, D. B. Day, S. Liuti, X. Zheng, M. Poelker, and Y. Prok

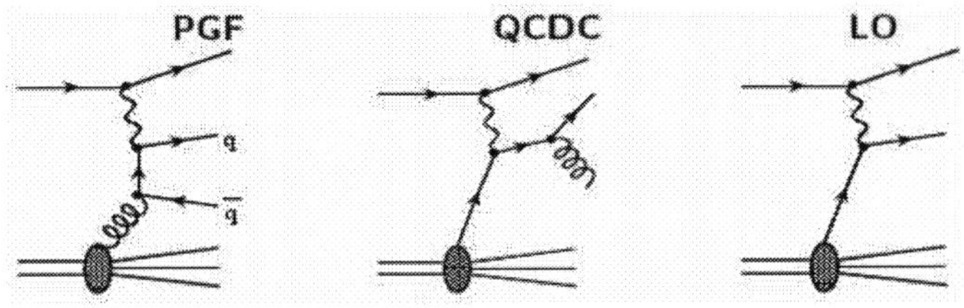

FIGURE 1. Feynman diagrams for the hard subprocesses Photon-Gluon Fusion (PGF), QCD Compton (QCDC), and Leading Order (LO) DIS.

mesons. This measurement was performed by the COMPASS Collaboration scattering a muon beam off deuterons. The μ^+ beam is longitudinally polarized with an average polarization of about -80% and the longitudinally polarized ^{6}LiD solid-state target has an average polarization of $\pm 50\%$ and a dilution factor of about 0.4. The target consists of two (now three) cells with alternating spin directions, which in addition are reversed every eight hours in order to minimize systematic uncertainties. Since the thick solid-state target does not allow for a vertex reconstruction with sufficient resolution in order to separate the D^0 production and decay vertices, the invariant mass distribution of the $D^0 \rightarrow K\pi$ decay modus (branching ratio 3.8%) is afflicted with a high combinatorial background. A cleaner but statistically even more limited sample of $D^*(2010)^{\pm}$ mesons identified through their decay into a D^0 meson and an additional "slow" $\pi^{\pm}$ (branching ratio about 68%) is also used. The D meson production is measured for the two cases where the spin of the muons and the deuterons is aligned and anti-aligned. The resulting yield difference divided by the sum of the yields, the so called asymmetry, is proportional to the asymmetry in the hard subprocess $\mu g \rightarrow \mu c\bar{c}$, the so-called analyzing power $\hat{a}_{LL}^{PGF}$, times $\Delta g(x)/G(x)$. The mass of the charm quark provides the hard scale and thus the analyzing power can be calculated in perturbative Quantum Chromodynamics (pQCD). However, since the analyzing power depends on the subprocess kinematics that are not fully accessible since only one D meson is required per event, the analyzing power cannot be calculated on an event-by-event basis in pQCD. Therefore the extraction of $\frac{\Delta g(x)}{G(x)}$ relies on an $\hat{a}_{LL}^{PGF}$ that is parameterized in terms of the measured kinematics using a Monte-Carlo in leading order QCD. The new preliminary result for this "open-charm" measurement based on the years 2002-2006 is $\frac{\Delta g}{G} = -0.49 \pm 0.27(stat.) \pm 0.11(sys.)$ at $x = 0.11^{+0.11}_{-0.05}$, as shown in Fig. 2. It is in agreement with the published one from the 2002-2004 data taking [5].

In order to increase the statistics for the PGF subprocess it is possible to take into account all hadrons or hadron pairs with a high transverse momentum p_T instead of just the D mesons. In this case the hard scale for the scattering process can be ensured by requiring $Q^2 > 1$ GeV^2. However, at least two additional subprocesses contribute in addition to PGF, the so-called QCD-Compton process ($\gamma^* q \rightarrow qg$) and the leading order DIS process ($\gamma^* q \rightarrow q$), as shown in Fig. 1. The fractional contributions and asymmetries of these background processes have to be taken into account in this analysis, leading in general to increased systematic uncertainties. A preliminary result from COMPASS based on data taken in the years 2002-2004 is available for the production of hadron pairs,

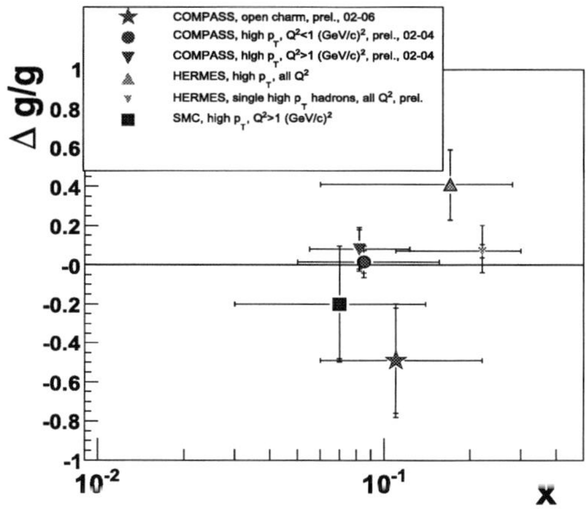

FIGURE 2. Results from various preliminary (see text) and the published HERMES[8] (stat. and sys. exp. uncertainties only) and SMC[6] extractions of $\frac{\Delta g}{G}$ as a function of x.

with $p_T > 0.7$ GeV for each hadron. The result ($\frac{\Delta g}{G} = 0.08 \pm 0.10(stat.) \pm 0.05(syst.)$ at $x = 0.082^{+0.041}_{-0.027}$) is shown in Fig. 2 together with an already published result on high-p_T hadron pair production from muon scattering at CERN by the SMC Collaboration [6].

A further increase in statistics can be gained by using data from all Q^2 values. This leads to a large additional contribution from the so-called resolved photon processes, i.e., partons from the resolved photon interact with either quarks or gluons from the nucleon. The former case contributes to the background, the latter case is actually sensitive to $\frac{\Delta g(x)}{G(x)}$ but involves the so far unmeasured polarized quark and gluon PDFs in the virtual photon. It is obvious that this analysis is even more difficult in terms of understanding the fractions of signal and background contribution as well as their different asymmetries. For the data with $Q^2 < 1$ GeV^2, $p_T > 0.7$ GeV for each hadron, and the quadratic sum of the two transverse momenta bigger than 2.5 GeV^2, the COMPASS Collaboration obtained a preliminary value of $\frac{\Delta g}{G} = 0.016 \pm 0.058(stat.) \pm 0.014(exp.syst.) \pm 0.052(MCsyst.) \pm 0.013(resolvedphoton)$ at $x = 0.095^{+0.08}_{-0.04}$ for the 2002-2004 data (see Fig. 2). See Ref. [7] for the published result based on the 2002 and 2003 and details on the analysis. The first attempt to extract $\frac{\Delta g(x)}{G(x)}$ using PGF has actually been made by the HERMES Collaboration, measuring high-p_T pairs at all Q^2 in the scattering of polarized positrons off polarized protons [8]. The result is shown in Fig. 2, however, the error bars do not include theoretical uncertainties. Finally, a preliminary result from the HERMES Collaboration using only single hadrons [9] at all Q^2 is also shown in the same figure.

A complementary approach is followed at the STAR and PHENIX experiments at RHIC (Relativistic Heavy Ion Collider) at BNL, where the gluon is studied in pp collisions. The advantages are that the involved energies are high and the gluon participates in leading order since the basic processes of interest are the scattering of quarks off

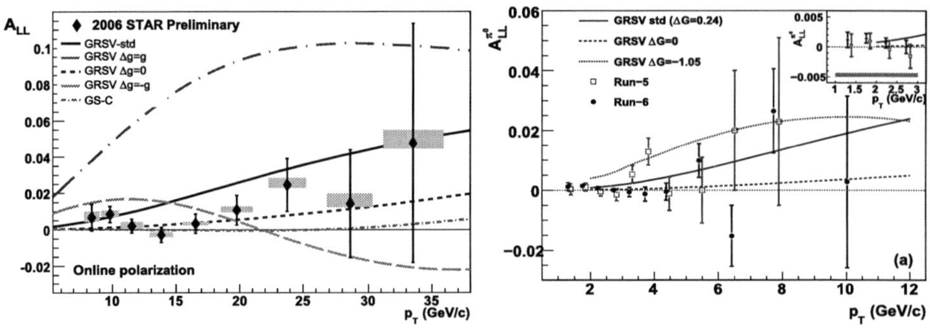

FIGURE 3. Double helicity asymmetry for mid-rapidity inclusive jet (left panel) production from the 2006 (preliminary, see Ref. [12] for final results from 2005 data) data and π^0 (right panel) production from the 2005 [11] and 2006 [13] data at $\sqrt{s} = 200$ GeV as a function of p_T. The results for the asymmetries are compared to pQCD calculations [14] based on different sets of polarized PDFs [15, 16].

gluons and of gluons off gluons. The so-called double helicity asymmetry

$$A_{LL} = \frac{\sigma^{++} - \sigma^{+-}}{\sigma^{++} + \sigma^{+-}} = \frac{\Delta\sigma}{\sigma}, \quad with \quad \Delta\sigma \propto \sum_{abc} \Delta f_a \otimes \Delta f_b \otimes \Delta\hat{\sigma}^{ab \to cX'} \otimes D_c^h, \qquad (1)$$

is measured, where the cross section σ^{++} (σ^{+-}) describes the reaction where both protons have the same (opposite) helicity. The spin–dependent term is given on the rhs of Eqn. 1, where Δf_a, Δf_b represent the spin–dependent PDFs for quarks (u,d,s) and gluons, which, from the NLO fits to g_1 as mentioned above, are well known for quarks but not well known for gluons. The fragmentation functions D_c^h represent the probability for a certain parton c to fragment into a certain hadron h, and thus they are not needed in the case of jet production. The spin–dependent hard scattering cross sections $\Delta\hat{\sigma}$ have to be calculated using perturbative QCD. Thus, the applicability of perturbative QCD in the kinematic regime of the A_{LL} measurements is a crucial prerequisite. This has been checked by comparing the results from NLO pQCD calculations using the well known unpolarized PDFs for quarks and gluons with measured unpolarized cross sections for, e.g., the inclusive production of jets [10] or neutral pions [11]. Good agreement has been found giving confidence in proceeding with measurements of the double helicity asymmetry (Eqn. 1).

The double helicity asymmetry as a function of the transverse momentum p_T for jet production at STAR and π^0 production at PHENIX are shown in Fig. 3. The results are compared to NLO pQCD calculations [14], using two sets of polarized PDFs. In addition to the GS-C PDFs [15], the GRSV PDFs [16] (GRSV-std) have been modified by assuming the polarized gluon PDF to be zero ($\Delta G = 0$) or negative ($\Delta G = -1.05$), and equal ($\Delta G = G$) or opposite ($\Delta G = -G$) to the unpolarized gluon PDF at the input scale. It is apparent from Fig. 3 that the maximum and minimum GRSV scenarios, which are not even shown anymore on the PHENIX plot, are ruled out by both experiments. While not excluded, GRSV-std is disfavored by the data as can be seen in Fig. 4, where for the STAR jet data (PHENIX π^0 data) the confidence level (χ^2 profile) is shown for several

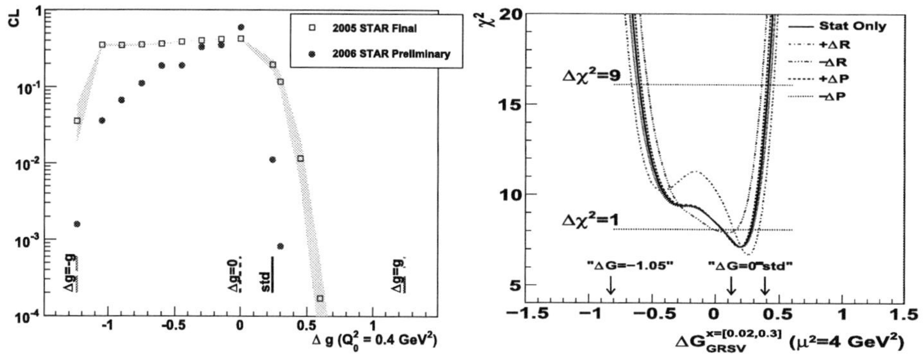

FIGURE 4. Confidence level of the STAR jet data (left panel) and χ^2 profile for the PHENIX π^0 data (right panel) resulting from a comparison to a calculation using the GRSV[16] set of PDFs with several values of ΔG at the input scale [14].

values of ΔG at the input scale. Note that no uncertainties have been calculated for the GRSV set of PDFs, and that they in general are rather large for the poorly constrained $\Delta g(x)$ from the DIS data as discussed above, hence the RHIC results remain consistent with the results from DIS.

Based on Fig. 4 a small or maybe negative contribution of the gluon spin ΔG to the nucleon spin appears to be most likely; however, this is a model dependent result. The present inclusive measurements cannot determine x but are sensitive to the gluon contribution integrated over the range $0.05 < x < 0.2$. Thus, in case the gluon contribution is large outside the presently accessible x–range it could still account for all the "missing" spin, even though the contribution in the limited accessible x range is rather small. An example is the predicted asymmetry based on the GS-C PDF also shown in the left panel of Fig. 3. It is consistent with the present data, but the GS-C PDF gives a large contribution of gluons to the nucleon spin due to the fact that it is large and positive at small x and becomes negative above $x \approx 0.1$. In other words, the gluon contribution is small in the measured region since it has a node, but the positive contribution at smaller x could account for all the spin due to the abundance of gluons at small x.

These features can also be seen in Fig. 5, which shows a recent NLO pQCD fit (DSSV [17]), including for the first time the *pp* data in addition to DIS and semi-inclusive DIS in the extraction of the polarized PDFs, a so-called global fit. In the x range accessible at RHIC the DSSV fit yields smaller values for $x\Delta g(x)$ than the earlier GRSV or DNS [18] fits with much reduced uncertainties. However, at smaller values of x the uncertainties are large and thus allow for a large $x\Delta g(x)$. A lower (higher) range in x can be probed running RHIC at higher (lower) energies. Results on π^0 production at PHENIX at $\sqrt{s} = 62.4$ GeV are already available [19], while running at $\sqrt{s} = 500$ GeV is foreseen to start in 2009. In addition, measurements at forward rapidity will allow access to a lower range in x. Also, with more data available in the future it is possible to advance from inclusive measurements, in which only one particle in the final state is detected, to measurements in which two or more leading particles or jets are detected in the final state, allowing for a much better handle on the x-dependence. Since in DIS the

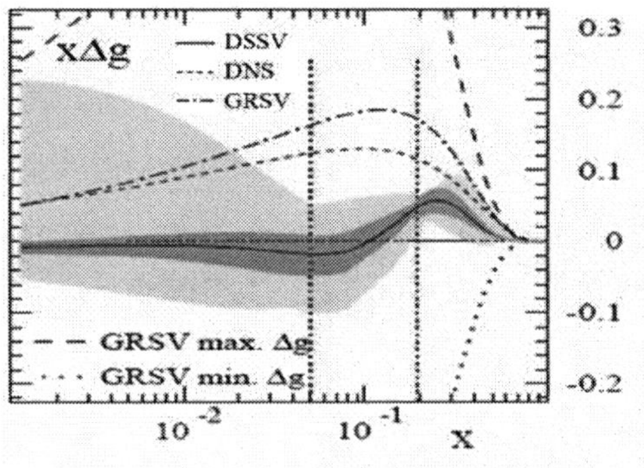

FIGURE 5. Polarized gluon PDF from a fit (DSSV [17]) that included for the first time the pp data (in addition to DIS and semi-inclusive DIS), compared to the earlier fits GRSV and DNS [18]. Alternative DSSV fits corresponding to $\Delta\chi^2 = 1$ ($\Delta\chi^2/\chi^2 = 2\%$) are given by the green/dark (yellow/light) bands. Figure taken from Ref. [17].

momentum fraction x can be measured directly, a precise measurement of the functional from of the polarized gluon PDF $\Delta g(x)$ will be possible at a proposed polarized electron-proton collider called EIC [20], which would be able to overcome the difficulties arising from the low energy scale and limited Q^2 range at fixed-target experiments. Further progress on the theoretical side can be achieved by including the open charm and/or other PGF related DIS asymmetry measurements into the global fit.

REFERENCES

1. J. Ashman *et al.*, *Phys. Lett. B* **206**, 364 (1988).
2. A. Airapetian *et al.*, *Phys. Rev. D* **75**, 012007 (2007).
3. V.Yu. Alexakhin *et al.*, *Phys. Lett. B* **647**, 8 (2007).
4. J. Blümlein, H. Böttcher, *Nucl. Phys. B* **636**, 225 (2002).
5. M. Alekseev et al., arXiv:0802.3023.
6. B. Adeva *et al.*, *Phys. Rev. D* **70**, 012002 (2004).
7. E.S. Ageev *et al.*, *Phys. Lett. B* **633**, 25 (2006).
8. A. Airapetian *et al.*, *Phys. Rev. Lett.* **84**, 2584 (2000).
9. P. Liebing for the HERMES Collaboration, Proceedings of the 17th Int. Spin Physics Symposium (SPIN06), Kyoto, Japan, 2-7 Oct 2006. AIP Conf.Proc. 915:331-334, (2007).
10. B.I. Abelev *et al.*, *Phys. Rev. Lett.* **97**, 252001 (2006).
11. A. Adare *et al.*, *Phys. Rev. D* **76**, 051106 (2007).
12. B.I. Abelev *et al.*, *Phys. Rev. Lett.* **100**, 232003 (2008).
13. A. Adare *et al.*, arXiv:0810.0694.
14. M. Stratmann and W. Vogelsang, private communication.
15. T. Gehrmann and W. J. Stirling, *Phys. Rev. D* **53**, 6100 (1996).
16. M. Glück, E. Reya, M. Stratmann, and W. Vogelsang, *Phys. Rev. D* **63**, 094005 (2001).
17. D. de Florian, R. Sassot, M. Stratmann, and W. Vogelsang, *Phys. Rev. Lett.* **101**, 072001 (2008).
18. D. de Florian, G.A. Navarro, and R. Sassot, *Phys. Rev. D* **71**, 094018 (2005).
19. A. Adare *et al.*, *Phys. Rev. D* **79**, 012003 (2009).
20. A. Deshpande, R. Milner, R. Venugopalan, and W. Vogelsang, *Ann. Rev. Nucl. Part. Sci.* **55**, 165 (2005).

Longitudinal Double Spin Asymmetry and Cross Section for Direct Photon Production Measured at Mid-rapidity in Polarized $\sqrt{s} = 200$ GeV pp Collisions at PHENIX

Robert Bennett for the PHENIX Collaboration

State University of New York at Stony Brook, Department of Physics and Astronomy, Stony Brook NY 11790, USA

Abstract. Direct photon production in pp collisions at RHIC is one of the important channels PHENIX will employ to determine the polarized gluon distribution ΔG. The direct photon A_{LL} is linear in ΔG, therefore sensitive to its sign and magnitude. To establish the applicability of perturbative Quantum Chromodynamics (pQCD) to this process, we present a comparison of the direct photon unpolarized cross section, with next-to-leading order pQCD calculations. We then evaluate the double helicity spin asymmetries, A_{LL}, from these data and compare with theoretical models. We present results and the current status the analysis of 2005 and 2006 data sets.

Keywords: Proton structure, Proton spin, direct photon, RHIC, PHENIX
PACS: 25.40.Ep, 14.20.Dh, 13.85Qk, 13.88+e

INTRODUCTION

Since the discovery, that quarks alone can not account for the total spin of the proton, by the EMC experiment [2], there has been a considerable effort put forth in nuclear physics to understand how the protons spin may be built up from it's constituent quarks and gluons. A major goal of the spin program at PHENIX is to measure the gluon contribution to the spin of the proton. Production of direct photons in $\sqrt{s} = 200$ GeV pp collisions at RHIC is one of the important channels PHENIX will employ to determine the polarized gluon distribution. The direct photon channel is a theoretically clean and well understood signal due to the small number of subprocesses which contribute to its production at leading order. With a proper choice of an isolation cut complications due to fragmentation functions may be removed. The direct photon measurement also benefits from the fact that the double longitudinal asymmetry show in equation 2 is linear in ΔG and therefor sensitive to both the sign and magnitude of ΔG. For these reasons the direct photon channel has been dubbed the *golden channel* for this measurement. To establish the theoretical understanding of this process, we compare the direct photon total cross section measured using our data, with the perturbative QCD calculations at next-to-leading order. This comparison shows good agreement between data and theoretical calculations over a broad range in photon transverse momentum (p_T).

CP1149, *Spin Physics, 18th International Spin Physics Symposium*
edited by D. G. Crabb, D. B. Day, S. Liuti, X. Zheng, M. Poelker, and Y. Prok

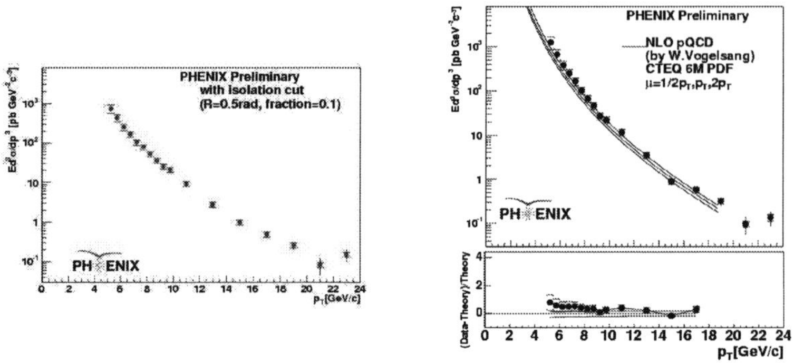

FIGURE 1. The cross section for direct photons was measured during the 2005 run using the isolation cut method, shown on the left[1]. It shows good agreement with both theory and an alternative method, where a simple statistical subtraction of the background was used, but the isolation cut was not applied.

DOUBLE LONGITUDINAL ASYMMETRY

The double longitudinal spin asymmetries, A_{LL}, is extracted from the 2005 data set based on a luminosity of 2.7 pb^{-1} and polarization 50% and the 2006 data set of $7.0pb^{-}1$ polarization of 57%. The form of A_{LL} is

$$A_{LL} = \frac{1}{|P_B||P_Y|}\frac{N_{++} - RN_{+-}}{N_{++} + RN_{+-}}, \quad R = \frac{N_{++}}{N_{+-}} \tag{1}$$

where $P_B(P_Y)$ is the polarization of the blue (yellow) beam and N_{++} (N_{+-}) are direct photon yields during like (unlike) sign helicity crossings. R is the relative luminosity, which is the ratio of the production rates. The asymmetry of the direct photon can then be related to gluon spin contribution in the following way

$$A_{LL} = \frac{\Delta G}{G} \otimes A_1^p \otimes \hat{a}_{LL}, \tag{2}$$

where A_1^p represents the quark spin contribution, previously measured in deep inelastic scattering experiments, and $\hat{a}_{LL}$ is the partonic asymmetry, which is calculable in the framework of pQCD.

DIRECT PHOTON MEASUREMENT

The extraction of direct photon A_{LL} relies on, first purifying the sample by considering only isolated photons as direct photon candidates. Photons originating from hadronic decays (π^0, η, ω...) will be found in close proximity to other jet material, as opposed to the direct photon which should remain isolated. A 0.5[rad] cone is defined around any candidate photon. A photon is considered isolated if the sum of the energy inside the cone is less then 10% of the energy of the candidate photon. This effectively removes photons originating from hadronic decays while leaving behind a cleaner direct photon sample. The cut is most efficient at high P_T .

Next, photons originating from hadronic decays (mostly $\pi^0 \rightarrow \gamma\gamma$) which pass the isolation cut are removed. These are hadronic decays in where the energy asymmetry of the photons is so great that the pair pass the isolation cut. These photons are identified by creating mass-pairs of photons which pass the cut. Any photon which forms a pair of 135 ± 30 MeV are simply removed from the isolated photon sample. We denote these photons by n_{π^0} . Finally we correct the asymmetry for hadronic decays in which one

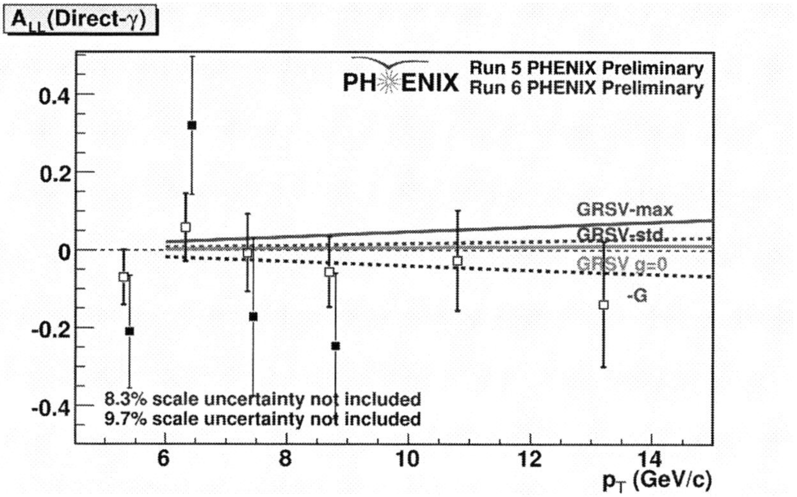

FIGURE 2. Direct Photon A_{LL} for the Run 5 and Run 6 data sets at PHENIX as a function of P_T with curves for GRSV[3] theoretical models

photon has missed the PHENIX acceptance. We estimate this from the total number of π^0's where both decay photons are in PHENIX acceptance, $N_{\pi^0}^{tagged}$. $N_{\pi^0}^{tagged}$ is then scaled by the probability that one photon will miss our acceptance ($R_{\pi^0}^{miss}$). $R_{\pi^0}^{miss}$ is estimated by a Monte Carlo simulation. We then assume other hadronic decays (η, ω...) will have similar behavior as the π^0 and their contribution can be calculated by comparing their branching ratios to that of the π^0. We denote all one-miss hadrons by, N^{bg}. The direct photon asymmetry and uncertainty are corrected by:

$$A_{LL}(direct-\gamma) = \frac{A_{LL}(iso-n_{\pi^0}) - rA_{LL}^{bg}}{1-r}, \; r = \frac{N_{bg}}{N_{iso-n_{\pi^0}}} \tag{3}$$

$$\Delta A_{LL}(sig)^2 = \frac{\Delta A_{LL}^2(iso-n_{\pi^0}) + r^2 \Delta A_{LL}^2(bg)}{(1-r)^2} \tag{4}$$

Here, $A_{LL}(iso-n_{\pi^0})$ is the asymmetry of the sample of isolated photons minus photons originating from hadronic decays that still pass the isolation cut. A_{LL}^{bg} is the asymmetry of the one-miss hadrons. r is the ratio of the number of one-miss hadrons to the number of isolated photons minus n_{π^0}. The uncertainly of each point, ΔA_{LL}, is then evaluated in the manner above.

RESULTS

Figure 2 shows the measured direct photon A_{LL} at PHENIX for the 2005 and 2006 RHIC polarized proton runs.

It is important to note that although we are statistically limited at the moment, we have shown that we have a consistent method for measuring the double longitudinal spin asymmetry and as gains are made in both luminosity and polarization this measurement will certainly complement the results of the neutral pion asymmetry analysis. Projections for the upcoming 2009 run with a luminosity of $27pb^{-1}$ and 70% polarization show that a significant measurement is within the near future.

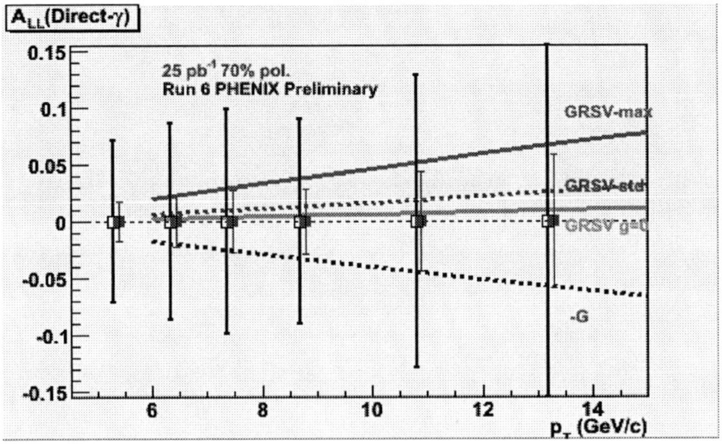

FIGURE 3. Projections for the upcoming 2009 run which expects a luminosity of $27pb^{-1}$ and 70%

REFERENCES

1. S.S. Adler *et al. Phys. Rev. Lett.* **98**, 012002 (2007).
2. J. Ashman *et al. Nucl. Phys.* **B328**, 1 (1989).
3. M. Glück, E.Reya, M. Stratmann and W. Vogelsang, *Phys. Rev. D* **53** (1996).

Double-Longitudinal Spin Asymmetry in Non-Identified Charged Hadron Production in pp Collisions at $\sqrt{s} = 62.4$ GeV at PHENIX

Amaresh Datta on behalf of PHENIX Collaboration

University of Massachusetts, Amherst, MA, 01003, USA
Brookehaven National Laboratory, Bldg 510C, Upton, NY, 11973, USA

Abstract. Charged hadron asymmetry measurements at PHENIX are interesting because of their possible sensitivity to both the sign and the magnitude of ΔG. Hadron production in pp scattering at $\sqrt{s} = 62.4$ GeV with transverse momentum < 5 GeV/c is dominated by quark-gluon scattering at leading order in pertubative QCD, providing sensitivity to the gluon helicity distribution in the proton. Within the PHENIX central arm [3] pseudo-rapidity coverage of $-0.35 < \eta < 0.35$, these measurements probe the scatering of partons with momentum fraction 0.05-0.2 [2]. The double longitudinal spin asymmetry of non-identified charged hadrons will provide new data points for extracting gluon polarization.

Keywords: double longitudinal asymmetry, spin, proton spin, gluon, ΔG, charge hadron, PHENIX
PACS: 13.85.Hd, 13.88.+e, 14.20.Dh, 25.40.Ep

INTRODUCTION

The gluon helicity distribution has been of interest to physicists trying to understand proton spin structure since the European Muon Collaboration constrained the contribution of quarks to proton spin. Later experiments (e.g. HERMES, SLAC, SMC) measured quark contribution to be 0.33 [4].

At the hard scale of interaction, the total cross-section (σ) of the polarised proton-proton scattering process can be factorized into distribution function (f_a, f_b) of partons, cross-section ($\widehat{\sigma}$) of partonic scattering and fragmentation ($D(z)_h$) of partons into final state particles (e.g. hadon h with momentun fraction z of the parent parton). The double longitudinal spin asymmetry, which is defined as:

$$A_{LL} = \frac{\sigma^{++} - \sigma^{+-}}{\sigma^{++} + \sigma^{+-}} \tag{1}$$

$$A_{LL} = \frac{1}{P_1 P_2} \frac{N^{++} - RN^{+-}}{N^{++} + RN^{+-}} \tag{2}$$

can be related to these quantities in the following way:

$$A_{LL} = \frac{\sum\limits_{a,b=q,\bar{q},g} \Delta f_a \otimes \Delta f_b \otimes \Delta\widehat{\sigma} \otimes D(z)_h}{\sum\limits_{a,b=q,\bar{q},g} f_a \otimes f_b \otimes \widehat{\sigma} \otimes D(z)_h} \tag{3}$$

(4)

CP1149, *Spin Physics, 18*th *International Spin Physics Symposium*
edited by D. G. Crabb, D. B. Day, S. Liuti, X. Zheng, M. Poelker, and Y. Prok

where $N^{++}(N^{+-})$ is the number of final state particles from same (opposite) helicity configuration of colliding protons, R is the Relative Luminosity of same and opposite helicity, P_1, P_2 are the beam polarizations and Δ indicates the difference of the quantity between same and opposite helicity configurations.

Contributions from different partonic subprocesses (quark-quark, quark-gluon, gluon-gluon) in the production of particles vary with transverse momentum (p_T) range of the final state particle. Theoretical calculations (Figure 1) lead us to believe that in $\sqrt{s} = 62.4$ GeV pp collisions, charged hadron production in the range of p_T $(0.5-4.5$ GeV/c) is dominated by quark-gluon processes, thereby rendering the measurement of asymmetries sensitive to gluon polarization (Δf_g or, simply, Δg) inside proton.

Another interesting aspect of charged hadron asymmetry is a possible signature of the sign of ΔG. A positive ΔG would be manifest in the ordering of asymmetries of pion production as:

$$A_{LL}(\pi^+) > A_{LL}(\pi^0) > A_{LL}(\pi^-) \tag{5}$$

whereas for a negative ΔG, the ordering will be the opposite way. As charged pions are the bigest contributors in the charged hadron samples analysed, we may hope to see the same signature in charged hadron asymmetries.

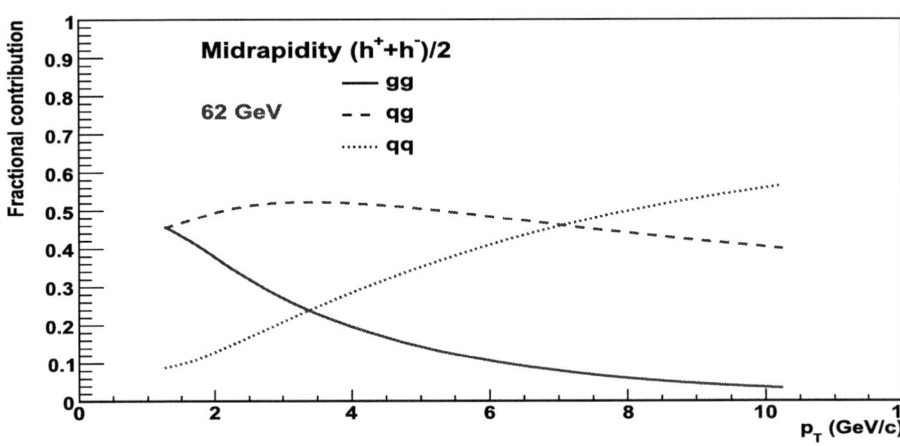

FIGURE 1. Contribution of subprocesses as function of transverse momentum (From W. Vogelsang).

MEASUREMENTS AND BACKGROUNDS

A total integrated luminosty of the 0.08 pb^{-1} of polarized pp collisions at $\sqrt{s} = 62.4$ GeV was recorded in one week at RHIC during 2006. About 40 nb^{-1} data triggered by requiring coincidence of hits on Beam Beam Counter [3] on both sides of nominal interaction region (PHENIX Minimum Bias trigger) were used for analysis. Average polarizations of the two beams were $\sim 48\%$ for the data set analysed. In the PHENIX

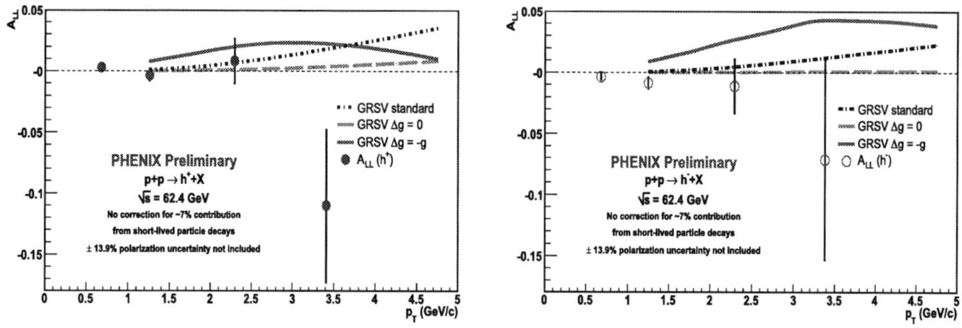

FIGURE 2. Preliminary results for A_{LL} of positive (left) and negative (right) hadrons in polarized proton proton collision at $\sqrt{s} = 62.4$ GeV.

detector system, charged tracks are reconstructed using Drift Chamber [3] and Pad Chambers [3]. High quality tracks with the additional requirement of hits on outer layer of Pad Chamber were selected for the analysis.

Electrons and positrons are some of the biggest cotributors to the background for the charged hadron candidates. However, Ring Imaging Cherenkov (RICH) [3] detector at PHENIX is fired by electrons and positrons with very low energy (17 MeV) as opposed to hadrons. Pions, the lightest of the hadron candidates, fire RICH at 4.7 GeV and heavier hadrons require even more energy. Electron and positron backgrounds were, therefore, removed efficiently by vetoing on RICH hits.

Another source of background is the decay products from weak decays of hadrons. Hadrons with typical $c\tau \sim 1 - 10$ m decay far from event vertex and the low momentum decay products often contribute to the background. We estimate the background by looking into the distribution of hits in outer detectors and correct the asymmetry measurement for the background asymmetry.

RESULTS

Preliminary results for the double spin asymmetries of charge-separated hadrons along with the corresponding predictions using GRSV framework [1] with three different initial ΔG assumptions are presented in Figure 2. Calculations for the predictions were done using DSS fragmentation and at factorization and renormalisation scale of $\mu = p_T$. We also show the comparison of charged hadron double spin asymmetries with neutral pion double spin asymmetry measured [5] at PHENIX at the same energy in Figure 3.

CONCLUSION

Longitudinal double spin asymmetry of charged hadron production in polarized proton proton collisions was found to be consistent with the small gluon polarization measured

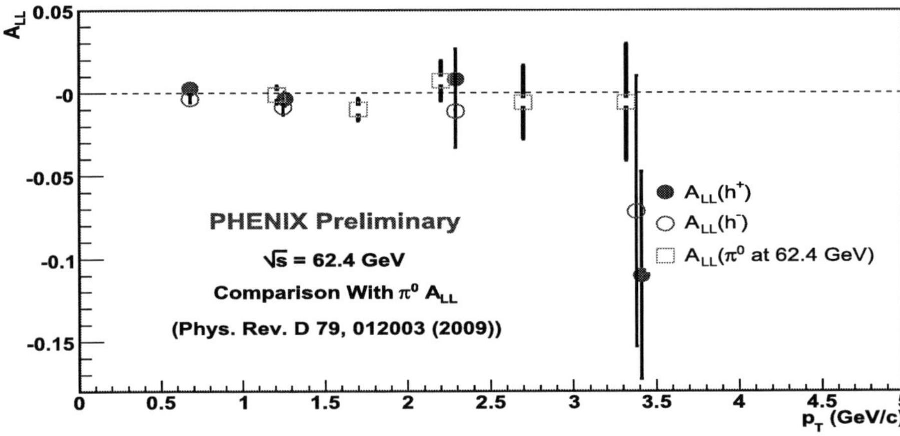

FIGURE 3. Comparison with neutral pion A_{LL} at $\sqrt{s} = 62.4$ GeV.

by other probes (e.g. neutral pions) in similar range of parton momentum fraction (x) at PHENIX [5] [6]. The results disfavour the models with significantly large positive or negative ΔG. Cross-section measurements of charged hadron production are in progress and will be compared with NLO and NLL calculations. The published result of neutral pion cross-section [5] from 2006 run at $\sqrt{s} = 62.4$ GeV already show good agreement with theoretical predictions [2] using GRSV framework and suggests that NLL corrections may be relevant at this energy. Although inconclusive in the present analysis, with enough statistics in future runs, the asymetry measurement of charged hadrons may be useful in indicating the sign of ΔG.

REFERENCES

1. M. Gluck, E. Reya, M. Stratmann, W. Vogelsang, *Phys. Rev. D* **63**, 094005 (2001)
2. D. de Florian, W. Vogelsang and F. Wagner, *Phys. Rev. D* **76**, 094021 (2007).
3. K. Adcox *et al.* (PHENIX Collaboration), *Nucl. Instr. and Meth. A* **499**, Pages 469-479, 489-507, 521-536 (2003).
4. A. Aipapetian *et al.* (HERMES Collaboration), *Phys. Rev. D* **75**, 012007 (2007).
5. A. Adare *et al.* (PHENIX Collaboration), *Phys. Rev. D* **79**, 012003 (2009).
6. A. Adare *et al.* (PHENIX Collaboration), arXiv:hep-ex/0810.0694 (Submitted to PRL).

Recent STAR Results from Neutral Pion Production in Polarized pp Collisions at $\sqrt{s} = 200$ GeV at RHIC

Alan Hoffman (for the STAR Collaboration)

Massachusetts Institute of Technology, 77 Massachusetts Ave. Cambridge, MA. 02139, USA

Abstract. A primary goal of the STAR spin physics program is measuring the polarized gluon distribution function, $\Delta g(x)$, accessible through the double longitudinal spin asymmetry, A_{LL}. The STAR detector is uniquely suited for measuring A_{LL} in a number of different final-state channels, such as inclusive jet and pion production, in polarized pp collisions. We present preliminary A_{LL} results from RHIC run 6 for neutral pion production at mid-rapidity. These A_{LL} measurements allow for discrimination among a number of different next-to-leading order (NLO) perturbative Quantum Chromodynamics (pQCD) predictions and provide meaningful constraints on $\Delta g(x)$.

Keywords: double longitudinal spin asymmetry, gluon, pion
PACS: 13.88+e, 13.85.Ni, 14.20.Dh

INTRODUCTION

A primary goal of the polarized pp program at the Relativistic Heavy Ion Collider (RHIC) is the measurement of the polarized gluon distribution function $\Delta g(x)$. RHIC, the world's only polarized proton collider, offers a unique environment to study gluon polarization as pp collisions provide access to gluons at leading order in the dominant subprocesses $gg \to gg$ and $qg \to qg$ [1]. Inclusive jet and pion final states have large production cross-sections at low and intermediate transverse momentum, p_T. This along with the large calculable hard partonic asymmetries at mid-rapidity make these final states suitable probes with which to study $\Delta g(x)$. Within pQCD, $\Delta g(x)$ is accessed through the observable A_{LL}, the double longitudinal spin asymmetry. Previous studies of the unpolarized π^0 cross section have validated the use of the next-to-leading order (NLO) perturbative Quantum Chromodynamics pQCD framework for studying inclusive π^0 production in pp collisions [2]. Neutral pions provide an independent measurement of A_{LL} with different experimental and theoretical biases, complimenting STAR's efforts with other final states in a global determination of $\Delta g(x)$.

DATA ANALYSIS AND RESULTS

Neutral pions are identified via the two-photon decay channel, which accounts for more than 98% of π^0 decays. Two primary sub-detectors, the barrel electromagnetic calorimeter (BEMC) and Shower Maximum Detector (SMD) are employed to reconstruct the decay photons from mid-rapidity π^0s. The BEMC, a lead scintillator sampling calorimeter[3], is segmented into 4800 towers, each one covering 0.05x0.05 in $\eta - \phi$

CP1149, *Spin Physics, 18th International Spin Physics Symposium*
edited by D. G. Crabb, D. B. Day, S. Liuti, X. Zheng, M. Poelker, and Y. Prok

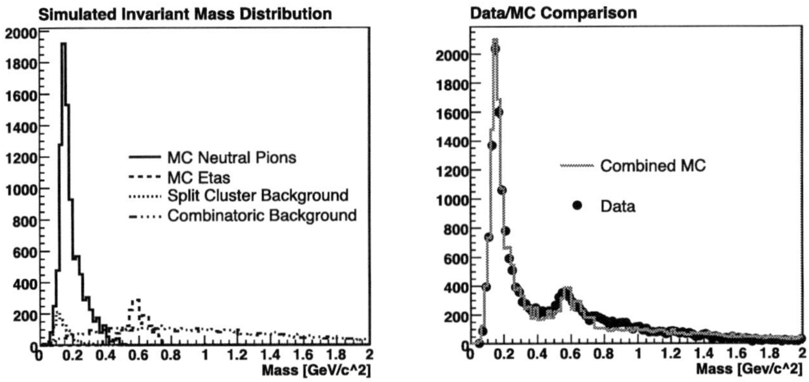

FIGURE 1. Simulated diphoton invariant mass distributions for various signal and background contributions (left) for candidates with 10.5 GeV/c $\leq p_T \leq$ 16.0 GeV/c. True (points) and Simulated (line) diphoton invariant mass distributions for candidates in the same p_T range (right).

space and with a depth of ~20 radiation lengths. In RHIC Run 6, for the first time, the full BEMC was installed and commissioned, providing pseudorapidity coverage from $-1 \leq \eta \leq 1$ and full azimuthal angle, ϕ coverage. For neutral pions with p_T above ~5 GeV/c, the opening angle between the decay photons is often smaller than the size of one tower. Thus the BEMC alone cannot resolve the individual decay photons; a detector with finer granularity is required. The SMD consists of two overlapping, perpendicular MWPCs that sit at a depth of $\sim 5X_0$ in each of the BEMC modules. The SMD wires, which are separated by 0.007 in $\Delta\eta$ (or $\Delta\phi$), provide the fine-grain position resolution necessary for reconstructing the two decay photons from high-$p_T \pi^0 s$.

Triggers are used to identify hard-scattering events and the high-p_T particles created therein. For Run 6, a new trigger, the Level-2 gamma trigger (L2γ), was designed and implemented specifically to identify events that produce high-p_T π^0s. This trigger selects events based on two energy criteria in the BEMC. A single tower must have an energy deposition greater than one threshold while the corresponding 3x3 tower patch (centered on the single high tower) must have an energy deposit greater than a second, higher threshold. This trigger algorithm better identifies events in which not all of the energy from a π^0 is deposited in one tower. The overall efficiency of this trigger stream is nearly twice that of a single high-tower trigger with an equivalent threshold. For Run 6 the tower/patch thresholds were set at 3.8 and 5.2 GeV respectively. Triggering on π^0 candidates themselves largely avoids introducing a systematic uncertainty from trigger reconstruction bias, which dominates the jet measurement uncertainty [4]. Along with the trigger requirement, a number of offline cuts are placed on the data to enhance purity. Decays with large a large photon energy asymmetry, $|\frac{E_{\gamma 1}-E_{\gamma 2}}{E_{\gamma 1}+E_{\gamma 2}}| > 0.8$, are excluded as are any BEMC energy depositions with charged tracks leading to them.

Figure 1(a) shows four simulated contributions to the invariant mass spectrum for candidates with 10.5 GeV/$c \leq p_T \leq$ 16.0 GeV/c. For the MC pion and eta distributions, single π^0 and η are generated in PYTHIA and reconstructed with a GEANT simulation

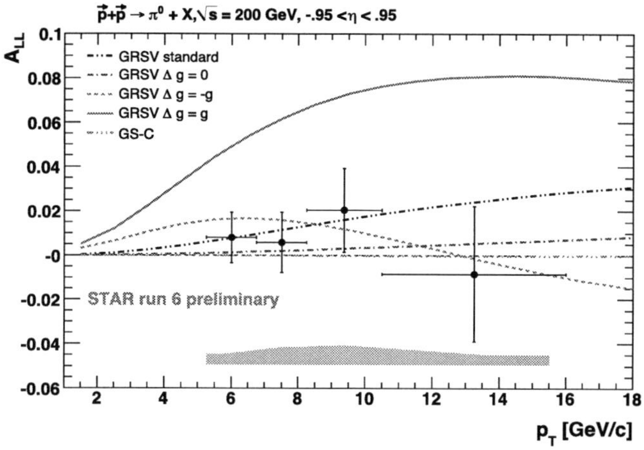

FIGURE 2. Double longitudinal spin asymmetry for inclusive π^0 production for $L2\gamma$ triggered events for RHIC run 6. A_{LL} is plotted along with theory curves form GRSV and DSSV. The error bars are purely statistical. Systematic uncertainties, excluding an overall uncertainty from beam polarization, are shown in the band below the data.

of STAR. The split cluster background is simulated by analyzing singly generated photons that, because of detector and algorithmic inefficiencies, are reconstructed as two separate photons. The combinatoric background is modeled using a mixed-event method in which photons from different events are combined into π^0 candidates. For more information on this procedure see Ref. [5]. Figure 1(b) shows a comparison of the diphoton invariant mass distribution for data and simulation. The combined MC distribution is the sum of the four individual distributions shown in 1(a).

The preliminary A_{LL} result presented here is calculated using a subset of the Run 6 data with an integrated luminosity for the L2γ trigger of $\sim 3.3 pb^{-1}$. Neutral pions are reconstructed at all azimuthal angles and in the range $-0.95 < \eta < 0.95$. The longitudinal double spin asymmetry is given by

$$A_{LL} = \frac{1}{P_1 P_2} \frac{(N^{++} - RN^{+-})}{(N^{++} + RN^{+-})} \tag{1}$$

Where $P_{1,2}$ are the two beam polarizations, $N^{++(+-)}$ are the π^0 yields for beams of equal (opposite) beam helicity configurations, respectively, and R is the relative luminosity for the two beam helicity configurations. The relative beam polarizations are measured in proton-carbon scattering using coulomb nucleon interference polarimeters and calibrated in pp elastic scattering using a hydrogen gas target of known polarization [6, 7]. The polarization values P_1 and P_2 are averaged over a beam fill and have values of $55\% - 60\%$. The relative luminosity value R is measured using the STAR beam-beam counters on a run-by-run basis.

Figure 2 shows A_{LL} for inclusive π^0 production for 5.2 GeV/$c \leq p_T^{\pi^0} \leq$16.0 GeV/c. Pions with p_T lower than 5.2 GeV/c fall below the trigger threshold and are excluded.

This measurement reaches to higher p_T values than any previous STAR or PHENIX π^0 asymmetry measurement [2, 8]. The measured values of A_{LL} are compared to a number of theoretical curves based on different assumptions about the gluon polarization [9, 10]. The curves are calculated using DSS fragmentation functions [10]. The systematic errors, which are dominated by background contamination effects, are shown in the gray band beneath the points. Other systematic effects include uncertainty in photon energy, non-longitudinal spin components, and errors in the relative luminosity measurements. Variations in π^0 selection criteria and studies of parity violating single-spin asymmetries show no evidence of systematic effects on A_{LL}. The four GRSV curves indicate predictions for A_{LL} based on the best fit to deep inelastic scattering (DIS) data (standard), extreme positive ($\Delta g = g$), extreme negative ($\Delta g = -g$), and vanishing ($\Delta g = 0$) polarization scenarios [9, 11]. The fifth curve indicates the asymmetry prediction based on the DSSV parameterization [12]. At the present level of precision, the data excludes the GRSV maximum gluon polarization scenario as in previous STAR and PHENIX measurements [4, 8, 13, 14, 15]. Note that this measurement samples only a limited range in parton momentum fraction: $0.03 \leq x_{bj} \leq 0.3$.

SUMMARY

We have reported preliminary A_{LL} measurement for inclusive π^0 production at mid-rapidity in polarized pp collisions $\sqrt{s} = 200$ GeV with $p_T^{\pi^0}$ between 5.2 and 16.0 GeV/c. This measurement, using $\sim 3.3pb^{-1}$ of triggered data from RHIC Run 6, represents a significant increase in precision over previous π^0 measurements at STAR. The data are consistent with predictions from the DSSV polarized gluon parameterization, as well the standard, maximum negative, and vanishing parameterizations at the starting scale within the GRSV framework. The GRSV-Max scenario is excluded.

REFERENCES

1. G. Bunce *et al.*, *Ann. Rev. Nucl. Part. Sci.* **50**, 525 (2000).
2. F. Simon *et. al.*, arXiv:hep-ex/0612004.
3. M. Beddo *et al.*, *Nucl. Instrum. Meth. A* **499**, 725 (2003).
4. M. Sarsour *et. al.*, *These proceedings.*
5. O. Grebenyuk, *Neutral Meson Production in d+Au and p+p Collisions at* $\sqrt(s_{NN}) = 200$ *GeV in STAR*, PhD dissertation, NIKEF and Utrecht University, (2007).
6. O. Jinnouchi *et. al.*, arXiv:nucl-ex/0412053.
7. H. Okada *et. al.*, arXiv:hep-ex/0601001.
8. A. Adare *et. al.*, *Phys. Rev. D* **76**, 051106 (2007).
9. M. Glück *et. al.*, *Phys. Rev. D* **63**, 094005 (2001).
10. D. de Florian *et. al. Phys. Rev. D* **75**, 114010 (2007).
11. W. Vogelsang, private communication (2006).
12. D. de Florain *et. al. Phys. Rev. Lett.* **101**, 072001 (2008).
13. B.I. Abelev *et. al.*, *Phys. Rev. Lett.* **100**, 232003 (2008).
14. B.I. Abelev *et. al. Phys. Rev. Lett.* **97**, 252001 (2008).
15. S.S. Adler *et. al. Phys. Rev. D* **73**, 091102(R) (2006).

Recent STAR Results from Charged Pion Production in Polarized pp Collisions at $\sqrt{s} = 200GeV$ at RHIC

Adam Kocoloski (for the STAR Collaboration)

Massachusetts Institute of Technology
77 Massachusetts Ave., Cambridge, MA 02139

Abstract. The STAR experiment at RHIC measures the longitudinal double-spin asymmetry A_{LL} for a variety of final states in collisions of longitudinally polarized protons to constrain the polarized gluon distribution in the proton. Asymmetries for mid-rapidity charged pion production benefit from large cross-sections and the excellent tracking and particle identification capabilities of the STAR Time Projection Chamber. This contribution presents a measurement of the ratio of cross sections for inclusive π^- and π^+ production using data collected in 2005, as well as a new measurement of A_{LL} for charged pions opposite a jet obtained from the 2006 RHIC run.

Keywords: gluon polarization, pion, jet, double-spin asymmetry
PACS: 13.88+e, 13.85.Ni, 13.87.Fh, 14.40.Aq

Collisions of polarized protons at the Relativistic Heavy Ion Collider (RHIC) offer a window into the spin composition of the proton through measurements of longitudinal double-spin asymmetries in a variety of final states [1], including charged pions where:

$$A_{LL} = \frac{\sum_{ABC} \Delta f_A \otimes \Delta f_B \otimes \hat{a}_{LL} \otimes \sigma_{AB \to CX} \otimes D_C^{\pi^{+(-)}}}{\sum_{ABC} f_A \otimes f_B \otimes \sigma_{AB \to CX} \otimes D_C^{\pi^{+(-)}}}. \quad (1)$$

Charged pion production at STAR [2, 3] is dominated by gg and qg scattering, providing sensitivity to the polarized gluon distribution $\Delta g(x, Q^2)$ over a restricted kinematic region of $0.03 < x < 0.3$. Furthermore, we know that a π^+ is more likely to have fragmented from a u quark ("favored" fragmentation) than a d quark ("disfavored" fragmentation). The converse is true for π^-, and the ratio of favored to disfavored fragmentation grows as the fraction of the parton momentum carried by the pion increases. This feature makes $A_{LL}(\pi^+)$ a particularly attractive measurement, since its analyzing power is magnified by the significant u quark polarization in the kinematic range where qg scattering dominates the pion production cross section.

One can write A_{LL} in terms of experimental quantities as

$$A_{LL} = \frac{1}{P_1 P_2} \frac{N^{++} - RN^{+-}}{N^{++} + RN^{++}}, \quad (2)$$

where $P_{1,2}$ are the polarizations of the colliding proton beams, N^{++} and N^{+-} are the identified particle yields when the proton helicities are aligned and anti-aligned, and R is the ratio of the beam luminosities in the two helicity configurations. The beam polarizations at RHIC are measured every few hours using a high-statistics Coulomb Nuclear

CP1149, *Spin Physics, 18th International Spin Physics Symposium*
edited by D. G. Crabb, D. B. Day, S. Liuti, X. Zheng, M. Poelker, and Y. Prok

Interference (CNI) polarimeter [4], and the analyzing power of the CNI polarimeter is normalized using a gas jet polarimeter [5]. The spin-dependent relative luminosities are measured at STAR using the Beam Beam Counters (BBCs), segmented scintillator annuli that provide full azimuthal coverage and span $3 < |\eta| < 5$ in rapidity . Coincident signals in the two BBCs define STAR's pp minimum-bias trigger condition; this approach samples 87% of the non-singly diffractive scattering cross section [6]. The BBCs can also be used to measure residual transverse beam polarization, which manifests as an azimuthal asymmetry in the scintillator tile counts [7].

The minimum-bias trigger condition is necessary but not sufficient to select events for this analysis. Also required is an electromagnetic energy deposit in the Barrel Electro-Magnetic Calorimeter (BEMC) which enhances STAR's sampling of hard scattering events. This jet patch (JP) trigger condition splits the BEMC into fixed patches with an extent of $\Delta\eta \times \Delta\phi = 1.0 \times 1.0$ and looks for summed energy above a threshold in at least one patch. In the 2005 RHIC run the JP trigger was implemented for $0 < \eta < 1$ with a primary threshold of 6.5 GeV. In 2006 the BEMC was fully commissioned and the JP trigger acceptance doubled to $|\eta| < 1.0$ with a threshold of 8.3 GeV. The BEMC has a depth of $\sim$1 hadronic interaction length, so charged pions often leave only a MIP signal of 264 MeV [8] in the calorimeter.

STAR reconstructs and identifies charged pions using a large Time Projection Chamber (TPC) situated inside a 0.5T magnetic field. The TPC can measure the transverse momenta of charged particles up to 20 GeV/c, and measurements of the energy loss along the track allow for particle identification across a wide range of momenta. The dE/dx and momentum of a track are nominally compared to a Bichsel parameterization of charged particle interactions in the TPC gas volume to obtain its PID; corrections to the Bichsel curves using particles identified via other means (electrons from the BEMC, pions and protons from strange decays) can improve the accuracy of dE/dx-based PID and allow for pion identification up to 15 GeV/c [9].

Figure 1 shows the ratio of cross sections for inclusive π^- and π^+ production as a function of p_T, using data taken by STAR during the 2005 run. Both cross sections are separately consistent with NLO predictions [10]. The ratio clearly diverges from unity at high transverse momenta, and this divergence is modeled well by NLO pQCD predictions incorporating modern fragmentation functions from the DSS [11] and AKK [12] collaborations.

The data sample obtained by STAR during the 2006 RHIC run offers a factor of 6 increase in the A_{LL} figure of merit ($P^4 * \mathscr{L}$) and a doubling of the jet patch trigger acceptance relative to 2005. However, the increased thresholds used in the 2006 version of the JP trigger lead to a significant fragmentation bias for jets that fire the trigger. Rather than fold this bias into a systematic uncertainty on A_{LL}, we restricted the 2006 charged pion A_{LL} to pions opposite a trigger jet ($\Delta\phi > 2.0$). Furthermore, we plot A_{LL} not against pion p_T but against the ratio of pion p_T and trigger jet p_T. This ratio forms a determination of the fragmentation variable z and, along with a restriction on the trigger jet p_T, allows the analysis to identify a region of phase space where the effect of $\Delta g(x)$ on A_{LL} is magnified by the large u quark polarization.

Events are selected for the A_{LL} analysis if they satisfy the BBC and JP trigger conditions outlined above. Jet reconstruction proceeds by clustering TPC tracks and BEMC towers using the midpoint-cone algorithm with a cone radius of 0.7. A jet is identified as

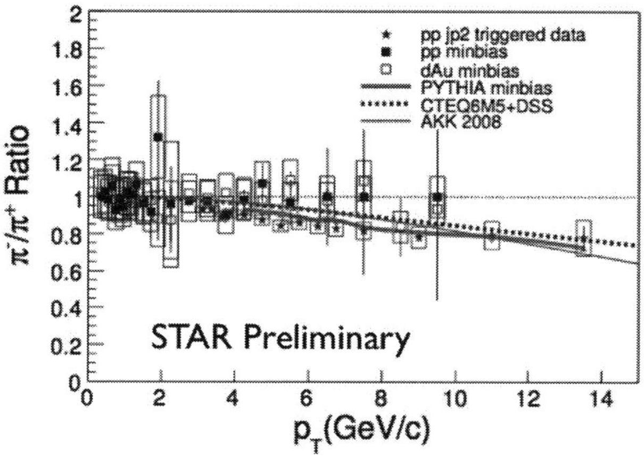

FIGURE 1. Ratio of invariant yields for charged pion production versus p_T. Statistical uncertainties are represented by vertical lines and systematics by the open boxes. The data are compared to predictions from PYTHIA [13] and NLO analyses incorporating modern fragmentation functions [11, 12].

a trigger jet if its axis points within 36° of the center of a JP above threshold. STAR also requires that the fraction of a jet's energy from the BEMC is less than 0.92 (to cut down on false jets formed out of beam background), and in this analysis we select jets with $9 < p_T < 25$ GeV/c to enhance our sample of qg scattering. Charged pions opposite the trigger jet are selected from a high-quality sample of TPC tracks with $p_T > 2$ GeV/c and $|\eta| < 1.0$, and are identified with a purity of better than 90% using dE/dx.

Figure 2 presents STAR's new preliminary results for charged pion A_{LL} opposite a trigger jet, plotted versus $z \equiv p_T(\pi)/p_T(jet)$. Full NLO predictions for this observable are in preparation [14]; in the interim, the data are compared to Monte Carlo evaluations of A_{LL} in which parton distribution functions [15, 16, 17] that have not been ruled out by earlier measurements are sampled at the kinematics specified by PYTHIA [13]. Point-to-point systematic uncertainties from a variety of sources are summed in quadrature and included as the gray bars.

The dominant systematic uncertainty in this analysis arises from the use of the JP trigger to select events. This trigger a) hardens the jet p_T spectrum in each z bin, and b) preferentially selects quark jets over gluon jets. We quantify a) by using Monte Carlo to determine the trigger efficiency for jets at a given p_T. In the range of jet p_T selected for this analysis we can parametrize the efficiency as

$$N_{jets,trigger}/N_{jets,total} = 1.149 - 0.2655 * p_T + 0.01857 * p_T^2 - 0.0003445 * p_T^3. \quad (3)$$

Theoretical predictions for this observable should incorporate the trigger efficiency. Future analyses with greater statistical precision may be able to avoid this requirement by binning in both jet p_T and z. For b) we investigate how our LO Monte Carlo evaluation of A_{LL} changes when we require the JP trigger condition. Specifically, we compare our MC asymmetries for the JP trigger with MC asymmetries that incorporate

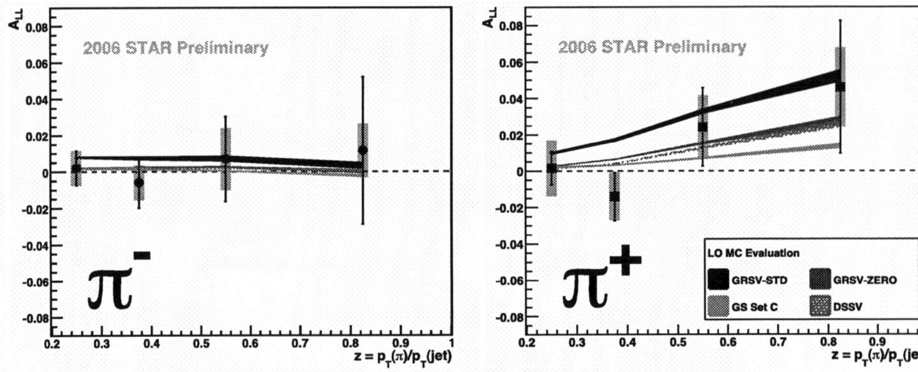

FIGURE 2. A_{LL} for charged pions opposite a trigger jet. The black error bars represent statistical uncertainties, and the gray bands total point-to-point systematics. A scale uncertainty of 8.3% from the beam polarization measurements is not included. The data are compared to leading order Monte Carlo evaluations of A_{LL} assuming various input distributions for $\Delta g(x)$ [15, 16, 17].

only the trigger *efficiency*. The difference between the two is a measure of how the trigger's preference for quark jets affects A_{LL}.

This contribution presented a ratio of the differential cross sections for inclusive charged pion production that shows a clear divergence from unity at high p_T, as well as a new measurement of A_{LL} for charged pion production opposite a trigger jet. A forthcoming NLO prediction for this observable will allow the measurement to be included in global analyses. Comparisons to LO MC evaluations exclude extreme scenarios for the gluon distribution function. Future, more precise measurements of $A_{LL}(\pi^+)$ have a great potential for providing a better understanding of the gluon polarization.

REFERENCES

1. G. Bunce, N. Saito, J. Soffer, and W. Vogelsang, *Ann. Rev. Nucl. Part. Sci.* **50**, 525–575 (2000).
2. K. H. Ackermann, et al., *Nucl. Instrum. Meth.* **A499**, 624–632 (2003).
3. J. Adams, et al., *Phys. Lett.* **B637**, 161–169 (2006).
4. O. Jinnouchi, et al. (2004), `nucl-ex/0412053`.
5. H. Okada, et al. (2006), `hep-ex/0601001`.
6. J. Adams, et al., *Phys. Rev. Lett.* **91**, 172302 (2003).
7. J. Kiryluk, et al. (2005), `hep-ex/0501072`.
8. T. M. Cormier, A. I. Pavlinov, M. V. Rykov, V. L. Rykov, and K. E. Shestermanov, *Nucl. Instrum. Meth.* **A483**, 734–746 (2002).
9. Y. Xu, et al. (2008), `0807.4303`.
10. Y. Xu, et al. (2009), `0901.0692`.
11. D. de Florian, R. Sassot, and M. Stratmann, *Phys. Rev.* **D75**, 114010 (2007).
12. S. Albino, B. A. Kniehl, and G. Kramer, *Nucl. Phys.* **B803**, 42–104 (2008).
13. T. Sjostrand, S. Mrenna, and P. Skands, *JHEP* **05**, 026 (2006).
14. D. de Florian, private communications (2008).
15. B. Jager, M. Stratmann, and W. Vogelsang, *Phys. Rev.* **D70**, 034010 (2004).
16. T. Gehrmann, and W. J. Stirling, *Phys. Rev.* **D53**, 6100–6109 (1996).
17. D. de Florian, R. Sassot, M. Stratmann, and W. Vogelsang, *Phys. Rev. Lett.* **101**, 072001 (2008).

The Strong Coupling Constant at Large Distances

A. Deur

Thomas Jefferson National Accelerator Facility, Newport News, VA 23606

Abstract. In this paper we discuss effective strong coupling constants. Those are well behaved in the low-Q^2 domain, contrarily to α_s from perturbative Quantum Chromodynamics (pQCD). We present an extraction of an effective strong coupling constant from Jefferson Lab polarized data at intermediate and low Q^2. We also show how these data, together with spin sum rules, allow us to obtain the effective coupling constant over the entire Q^2 range. We then discuss the relation between the experimentally extracted coupling constant and theoretical calculations at low Q^2. We conclude on the importance of such study for the application of the Anti-de-Sitter space/ Conformal Field Theory (AdS/CFT) correspondence to QCD.

Keywords: Strong coupling constant, QCD sum rules, non-perturbative, commensurate scale relations, Schwinger-Dyson, Lattice QCD, AdS/CFT
PACS: 12.38Qk,11.55Hx

In Quantum Chromodynamics (QCD), the magnitude of the strong force is given by the running coupling constant α_s. At large Q^2, in the perturbative QCD (pQCD) domain, α_s is well defined and is given by the series:

$$\mu\frac{\partial\alpha_s}{\partial\mu} = 2\beta(\alpha_s) = -\frac{\beta_0}{2\pi}\alpha_s^2 - \frac{\beta_1}{4\pi^2}\alpha_s^3 - \frac{\beta_2}{64\pi^3}\alpha_s^4 - ... \quad (1)$$

Where μ is the energy scale, to be identified to Q. The first terms of the β series are: $\beta_0 = 11 - \frac{2}{3}n$, with n the number of active quark flavors, $\beta_1 = 51 - \frac{19}{3}n$ and $\beta_2 = 2857 - \frac{5033}{9}n + \frac{325}{27}n^2$. The solution of the differential equation 1 is:

$$\alpha_s(\mu) = \frac{4\pi}{\beta_0 ln(\mu^2/\Lambda^2_{QCD})} \times \quad (2)$$

$$\left[1 - \frac{2\beta_1}{\beta_0^2}\frac{ln\left[ln(\mu^2/\Lambda^2_{QCD})\right]}{ln(\mu^2/\Lambda^2_{QCD})} + \frac{4\beta_1^2}{\beta_0^4 ln^2(\mu^2/\Lambda^2_{QCD})}\left(\left(ln\left[ln(\mu^2/\Lambda^2_{QCD})\right] - \frac{1}{2}\right)^2 + \frac{\beta_2\beta_0}{8\beta_1^2} - \frac{5}{4}\right)\right]$$

Equation 2 allows us to evolve the different experimental determinations of α_s to a conventional scale, typically $M^2_{z_0}$. The agreement between the α_s obtained from different observables demonstrates its universality and the validity of Eq. 1. One can obtain $\alpha_s(M^2_{z_0})$ with doubly polarized deep inelastic scattering (DIS) data and assuming the validity of the Bjorken sum rule [1]:

$$\Gamma_1^{p-n} = \int_0^1 (g_1^p - g_1^n)dx = \frac{g_a}{6}[1 - \frac{\alpha_s}{\pi} - 3.58\left(\frac{\alpha_s}{\pi}\right)^2 - 20.21\left(\frac{\alpha_s}{\pi}\right)^3 + ...] + O(\frac{1}{Q^2}), \quad (3)$$

where g_A is the well measured nucleon axial charge. Solving Eq. 3 using the experimental value of Γ_1^{p-n}, and then using Eq. 2 provides $\alpha_s(M^2_{z_0})$.

CP1149, *Spin Physics, 18th International Spin Physics Symposium*
edited by D. G. Crabb, D. B. Day, S. Liuti, X. Zheng, M. Poelker, and Y. Prok

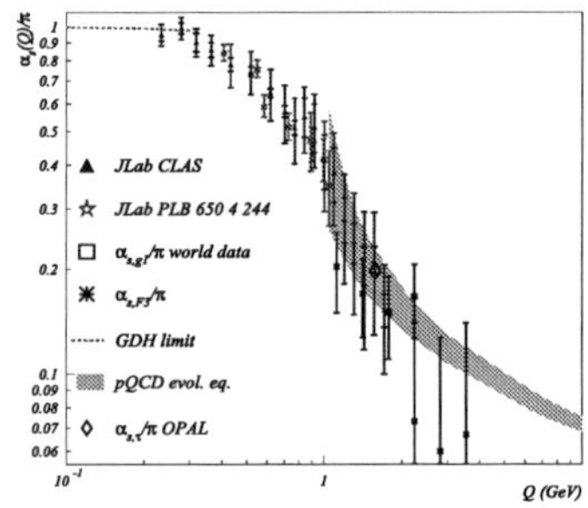

FIGURE 1. Value of $\alpha_{s,g_1}/\pi$ extracted from the world data on the Bjorken sum at $Q^2 = 5$ GeV^2 [4] and from JLab data [5] used to extract the Bjorken Sum [6]. Also shown are $\alpha_{s,\tau}$ extracted from the OPAL data on τ decay [2], and $\alpha_{s,GLS}$ extracted using the Gross-Llewellyn Smith sum rule [8] and its measurement by the CCFR collaboration [9]. The gray band indicates α_{s,g_1} extracted from the pQCD expression of the Bjorken sum at leading twist and third order in α_s (with α_s computed using Eq. 2). The values of $\alpha_{s,g_1}/\pi$ extracted using the Gerasimov-Drell-Hearn sum rule is given by the red dashed line.

Equation 2 leads to an infinite coupling at large distances, when Q^2 approaches Λ^2_{QCD}. This is not a conceptual problem since we are out of the validity domain of pQCD on which Eq. 2 is based. But since data show no sign of discontinuity or phase transition when crossing the intermediate Q^2 domain, one should be able to define an effective coupling α_s^{eff} at any Q^2 that matches α_s at large Q^2 but stays finite at small Q^2.

The Bjorken Sum Rule can be used to define α_s^{eff} at low Q^2. Defining α_s^{eff} from Eq. (3) truncated to first order: $\Gamma_1^{p-n} \equiv \frac{1}{6}(1 - \alpha_{s,g_1}/\pi)$, offers advantages. In particular, α_s^{eff} does not diverge near Λ_{QCD} and is renormalization scheme independent. However, α_s^{eff} becomes dependent on the choice of observable employed to define it. If Γ_1^{p-n} is used as the defining observable, the effective coupling is noted α_{s,g_1}. Relations, called *commensurate scale relations* [2], link the different effective couplings so in principle one effective coupling is enough to describe the strong force and the theory retains its predictive power.

The effective coupling definition in terms of pQCD evolution equations truncated to first order was proposed by Grunberg [3]. Following this definition, effective couplings have been extracted from different observables and have been compared to each other using the commensurate scale relations [7], see Fig. 1. There is good agreement between the effective couplings α_{s,g_1}, α_{s,F_3} and $\alpha_{s,\tau}$. The Gerasimov-Drell-Hearn (GDH) and Bjorken sum rules can be used to extract α_{s,g_1} at small and large Q^2 respectively [7]. This, together with the JLab data at intermediate Q^2, provides for the first time a coupling at any Q^2. A striking feature of Fig. 1 is that α_{s,g_1} becomes scale invariant at small Q^2. This was predicted by a number of calculations and it is known that color confinement leads to an infrared fixed point [10], but it is the first time it is seen experimentally.

A fit of the α_{s,g_1} has been performed with a form based on Eq. 2 at first order:

$$\alpha_{s,g_1}^{fit} = \gamma n(Q)/log(\frac{Q^2 + m_g^2(Q)}{\Lambda^2}) \tag{4}$$

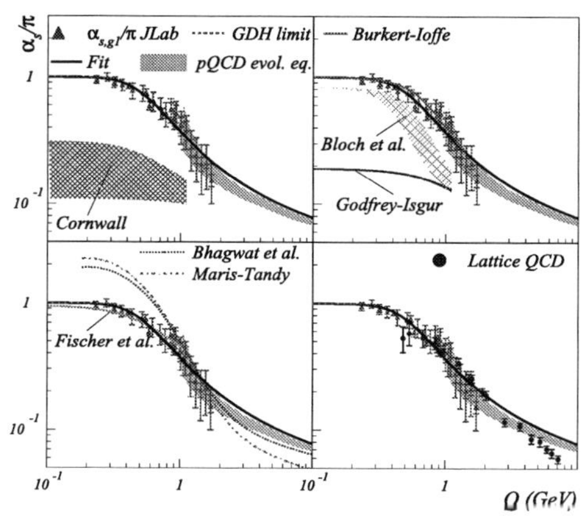

FIGURE 2. The effective coupling α_{s,g_1} extracted from JLab data, its fit, and its extraction using the Burkert and Ioffe [12] model to obtain Γ_1^{p-n}. The α_s calculations are: Top left: Schwinger-Dyson equations (Cornwall [11]); Top right: Schwinger-Dyson equations (Bloch) [13] and α_s used in a quark constituent model [14]; Bottom left: Schwinger-Dyson equations (Maris-Tandy [15]), Fischer, Alkofer, Reinhardt and Von Smekal [16] and Bhagwat et al. [17]; Bottom right: Lattice QCD [18].

where $\gamma = 4/\beta_0 = 12/(33-8)$, $n = \pi(1 + [\frac{\gamma}{log(m^2/\Lambda^2)(1+Q/\Lambda)-\gamma} + (bQ)^c]^{-1})$ and $m_g = (m/(1+(aQ)^d))$. The parameter values are: $\Lambda = 0.349 \pm 0.009$ GeV, $a = 3.008 \pm 0.081$ GeV^{-1}, $b = 1.425 \pm 0.032$ GeV^{-1}, $c = 0.908 \pm 0.025$, $m = 1.204 \pm 0.018$ GeV, $d = 0.840 \pm 0.051$. The fit is shown on Fig. 2 (plain black line). Eq. 4, used in $\Gamma_1^{p-n} \equiv \frac{1}{6}(1 - \alpha_{s,g_1}/\pi)$, can be employed to parametrize the generalized Bjorken and GDH sums.

In Figure 2, α_{s,g_1} is compared to theoretical results. There are several techniques used to predict α_s at small Q^2, e.g. lattice QCD, solving the Schwinger-Dyson equations, or choosing the coupling in a constituent quark model so that it reproduces hadron spectroscopy. However, the connection between these α_s is unclear, in part because of the different approximations used. In addition, the precise relation between α_{s,g_1} (or any effective coupling defined using [3] or [2]) and these computations is unknown. Nevertheless, one can still compare them to see if they share common features. The calculations and α_{s,g_1} present a similar behavior. Some calculations, in particular the lattice one, are in excellent agreement with α_{s,g_1}.

These works show that α_s is scale invariant (*conformal behavior*) at small and large Q^2 (but not in the transition region between the fundamental description of QCD in terms of quarks and gluons degrees of freedom and its effective one in terms of baryons and mesons). The scale invariance at large Q^2 is the well known asymptotic freedom. The conformal behavior at small Q^2 is essential to apply a property of *conformal field theories* (CFT) to the study of hadrons: the *Anti-de-Sitter space/Conformal Field Theory (AdS/CFT) correspondence* of Maldacena [19], that links a strongly coupled gauge field to weakly coupled superstrings states. Perturbative calculations are feasible in the weak coupling AdS theory. They are then projected on the AdS boundary, where they correspond to the calculations that would have been obtained with the strongly coupled CFT. This opens the possibility of analytic non-perturbative QCD calculations [20].

To sum up, thanks to the data on nucleon spin structure and to spin sum rules, an effective strong coupling can be extracted in any regime of QCD. The question of comparing it with theoretical calculations of α_s at low Q^2 is open, but such comparison exposes a similarity between these couplings. Apart for the parton-hadron transition region, the coupling shows that QCD is approximately a conformal theory. This is a necessary ingredient to the application of the AdS/CFT correspondence that may make analytical calculations possible in the non-perturbative domain of QCD.

ACKNOWLEDGMENTS

This work was done in collaboration with V. Burkert, J-P. Chen and W. Korsch. It is supported by the U.S. Department of Energy (DOE). The Jefferson Science Associates (JSA) operates the Thomas Jefferson National Accelerator Facility for the DOE under contract DE-AC05-84ER40150.

REFERENCES

1. J. D. Bjorken, *Phys. Rev.* **148**, 1467 (1966).
2. S. J. Brodsky and H. J. Lu, *Phys. Rev. D* **51**, 3652 (1995); S. J. Brodsky, G. T. Gabadadze, A. L. Kataev and H. J. Lu, *Phys. Lett. B* **372**, 133 (1996); See also S. J. Brodsky, hep-ph/0310289, S. J. Brodsky, S. Menke, C. Merino and J. Rathsman, *Phys. Rev. D* **67**, 055008 (2003).
3. G.Grunberg, *Phys. Lett. B* **95**, 70 (1980); *Phys. Rev. D* **29**, 2315 (1984); *Phys. Rev. D* **40**, 680 (1989).
4. K. Abe *et al.*, *Phys. Rev. Lett.* **79**, 26 (1997); P.L. Anthony *et al.*, *Phys. Lett. B* **493**, 19 (2000).
5. R. Fatemi *et al.*,*Phys. Rev. Lett.* **91**, 222002 (2003); J. Yun *et al.*,*Phys. Rev. C* **67**, 055204 (2003); V. Dharmawardane *et al.*, *Phys. Lett. B* **641**, 11 (2006); Y. Prok *et al.*, arXiv:0802.2232; M. Amarian *et al.*,*Phys. Rev. Lett.* **89**, 242301 (2002); M. Amarian *et al.*,*Phys. Rev. Lett.* **92**, 022301 (2004)
6. A. Deur *et al.*, *Phys. Rev. Lett.* **93**, 212001 (2004); A. Deur *et al.*, *Phys. Rev. D***78**, 032001 (2008).
7. A. Deur, V. Burkert, J.-P. Chen, W. Korsch, *Phys. Lett. B* **650**, 4 244 (2007); A. Deur, V. Burkert, J.-P. Chen, W. Korsch, *Phys. Lett. B* **665**, 349 (2008).
8. D. J. Gross and C. H. Llewellyn Smith, *Nucl. Phys. B* **14**, 337 (1969).
9. J. H. Kim *et al.*, *Phys. Rev. Lett.* **81**, 3595 (1998).
10. S. J. Brodsky, R. Shrock, *Phys. Lett. B* **666** 95 (2008); S. J. Brodsky, G. de Teramond, R. Shrock, arXiv:0807.2484
11. J. M. Cornwall, *Phys. Rev. D* **26**, 1453 (1982).
12. V. D. Burkert, B. L. Ioffe, *Phys. Lett. B* **296**, 223 (1992); *J. Exp. Theor. Phys.* **78**, 619 (1994).
13. J. C. R. Bloch, *Phys. Rev. D* **66**, 034032 (2002).
14. S. Godfrey and N. Isgur, *Phys. Rev. D* **32**, 189 (1985)
15. P. Maris and P. C. Tandy, *Phys. Rev. C* **60**, 055214 (1999).
16. C.S. Fischer, R. Alkofer, *Phys. Lett. B* **536**, 177 (2002); C.S. Fischer, R. Alkofer, H. Reinhardt, *Phys. Rev. D* **65** 125006 (2002); R. Alkofer, C.S. Fischer, L. Von Smekal, Acta Phys. Slov. **52** 191 (2002).
17. M.S. Bhagwat *et al.*, *Phys. Rev. C* **68**, 015203 (2003).
18. S. Furui and H. Nakajima, *Phys. Rev. D* **70**, 094504 (2004).
19. J. M. Maldacena, *Adv. Theor. Math. Phys.* 2 252 (1998); *Int. J. Theor. Phys.* **38**, 1113 (1999).
20. See for ex. J. Polchinski and M. J. Strassler, *Phys. Rev. Lett.* **88**, 031601 (2002); JHEP **0305**, 012 (2003); S. J. Brodsky and G. F. de Teramond, *Phys. Rev. Lett.* **94**, 201601 (2005); *Phys. Rev. Lett.* **96**, 201601 (2006). A. Karch, E. Katz, D. T. Son and M. A. Stephanov, *Phys. Rev. D* **74**, 015005 (2006).

Search for a 2-Photon Exchange in Inclusive DIS on a Transversely Polarized Hydrogen Target at HERMES

L. De Nardo*,†, A. Lopez-Ruiz** and A. Martinez de la Ossa (on behalf of the HERMES Collaboration)‡

*DESY, Notkestrasse 85, 22603 Hamburg, Germany
†Randall Laboratory of Physics, University of Michigan, Ann Arbor, Michgan 48109-1040, USA
**Department of Subatomic and Radiation Physics, University of Gent, 9000 Gent, Belgium
‡Nuclear Physics Laboratory, University of Colorado, Boulder, Colorado 80309-0390, USA

Abstract. Left-right single-spin asymmetries are measured in inclusive deep inelastic scattering at HERMES, with the goal of searching for a 2-photon exchange signal in the range $0.004 < x < 0.9$, $0.1 < Q^2 < 20$ GeV^2. In two separate regions, for $Q^2 > 1$ GeV^2 and $Q^2 < 1$ GeV^2, and for both electron and positron beams, the asymmetries are found to be consistent with zero showing that within the uncertainties no signal is detected.

Keywords: photon, DIS, inclusive
PACS: 13.60.Hb, 13.88.+e, 14.70.Bh

INTRODUCTION

In recent years the contribution of 2-photon-exchange to the cross section in electron-nucleon scattering has received increased attention. In elastic scattering it can be used to explain the different results obtained by the Rosenbluth and polarization transfer methods for the measurement of the electromagnetic proton form factors [1]. In inclusive Deep Inelastic Scattering (DIS) it is expected [2] to give rise to single-spin azimuthal asymmetries.

Little is known about the 2-photon exchange cross-section in DIS, other than the expected dependence on $\varepsilon_{\mu\nu\rho\sigma}S^\mu p^\nu k^\rho k'^\sigma \propto \vec{S}\cdot(\vec{k}\times\vec{k}')$, where k and k' are the initial and final lepton four-momenta, p is the initial four-momentum of the target nucleon, and S may be the polarization vector of either the lepton or the nucleon. The largest value of the cross section is thus obtained when the three vectors $\vec{S}$, $\vec{k}$ and $\vec{k}'$ are perpendicular to each other; so, for transverse target polarization, an asymmetry can be obtained between the number of leptons scattered to the left and those scattered to the right with respect to $\vec{S}$. Such an asymmetry is expected to be of the order of $\alpha_{em}M/Q$ [2], where M is the nucleon mass for transversely polarized target nucleons, while it is the lepton mass for the case of a transversely polarized lepton beam, thus showing that the effect is larger in the case of transversely polarized targets, where it is expected to be of the order of 10^{-2} for $Q^2 \simeq 1$ GeV^2.

CP1149, *Spin Physics, 18th International Spin Physics Symposium*
edited by D. G. Crabb, D. B. Day, S. Liuti, X. Zheng, M. Poelker, and Y. Prok

DATA ANALYSIS

The HERMES experiment at DESY has taken data with a transversely polarized proton target between 2002 and 2005 collecting a total of 3.5 M events with a positron beam in 2002 and 2004, and 6.1 M events with an electron beam in 2005, in the kinematic region $0.004 < x < 0.9$, $0.1 < Q^2 < 20$ GeV^2 and $W^2 > 3.24$ GeV^2, where, in the one-photon exchange approximation, $-Q^2$ is the squared four-momentum transfer, W is the invariant mass of the photon-nucleon system, and x is the Bjorken scaling variable, . For this analysis the Q^2 region was separated into a 'DIS' region ($Q^2 > 1$ GeV^2) and a 'low-Q^2' region ($Q^2 < 1$ GeV^2). Data were collected with the target polarized both in the 'up' ($P^\uparrow$) and 'down' ($P^\downarrow$) directions, with the polarization flipping every 90s to minimize systematic effects. The yields can be expressed as:

$$\begin{aligned} N_R^\uparrow &= (L^\uparrow \sigma_0 + L_P^\uparrow \Delta\sigma)\chi_R/2 \\ N_L^\uparrow &= (L^\uparrow \sigma_0 - L_P^\uparrow \Delta\sigma)\chi_L/2 \\ N_R^\downarrow &= (L^\downarrow \sigma_0 - L_P^\downarrow \Delta\sigma)\chi_R/2 \\ N_L^\downarrow &= (L^\downarrow \sigma_0 + L_P^\downarrow \Delta\sigma)\chi_L/2 \end{aligned} \tag{1}$$

where $L^{\uparrow,\downarrow}$ are the total luminosities in the two polarization states, $L_P^{\uparrow,\downarrow}$ are the polarization-weighted luminosities, $\chi_{L,R}$ are the left and right detector acceptances, and the 'true' asymmetry in the detector acceptance is $A_{true}^{meas} = \Delta\sigma/\sigma_0$. The measured asymmetry was calculated as:

$$A_N^{meas} = \left[\sqrt{\frac{N_L^\uparrow}{L_P^\uparrow}\frac{N_R^\downarrow}{L_P^\downarrow}} - \sqrt{\frac{N_R^\uparrow}{L_P^\uparrow}\frac{N_L^\downarrow}{L_P^\downarrow}} \right] \Big/ \left[\sqrt{\frac{N_L^\uparrow}{L^\uparrow}\frac{N_R^\downarrow}{L^\downarrow}} + \sqrt{\frac{N_R^\uparrow}{L^\uparrow}\frac{N_L^\downarrow}{L^\downarrow}} \right], \tag{2}$$

which has the advantage that effects coming from left and right acceptances cancel in the ratio [3], leaving no dependence on χ_L or χ_R. It can be approximated to $A_N^{meas} \simeq A_{true}^{meas}\left(1 + \frac{1}{2}\varepsilon_P^2\right)$ with $\varepsilon_P = (P^\uparrow - P^\downarrow)/(P^\uparrow + P^\downarrow)$. The measured asymmetry is equal to the true asymmetry if the two polarizations $P^\uparrow$ and $P^\downarrow$ are the same, which is true to a very good approximation, since they differ at most by 0.4%.

For a perfect detector, with a full 2π coverage in the azimuthal angle ϕ around the beam direction between the electron scattering plane and the upwards pointing direction of the target spin vector, and assuming that any underlying 2-photon asymmetry will behave in the simplest possible way, namely a *sinϕ* modulation which is also anticipated from the dependence $\vec{S}\cdot(\vec{k}\times\vec{k}')$, the expected measurement of a left-right asymmetry of amplitude $A^{\sin\phi}$ would lead to:

$$A_N^{2\pi} = \frac{\int_0^\pi d\phi\ \sigma_0\, A^{\sin\phi} sin(\phi)}{\int_0^\pi d\phi\ \sigma_0} = \frac{2}{\pi} A^{\sin\phi} = 0.64\, A^{\sin\phi}. \tag{3}$$

The HERMES detector is top-bottom symmetric [4], with a gap in the horizontal direction, and thus it does not cover the full 2π range in ϕ. In order to get an estimate of the *acceptance scaling-factor* because of this incomplete ϕ-coverage, an asymmetry was introduced in a Monte Carlo sample containing a full detector simulation and later

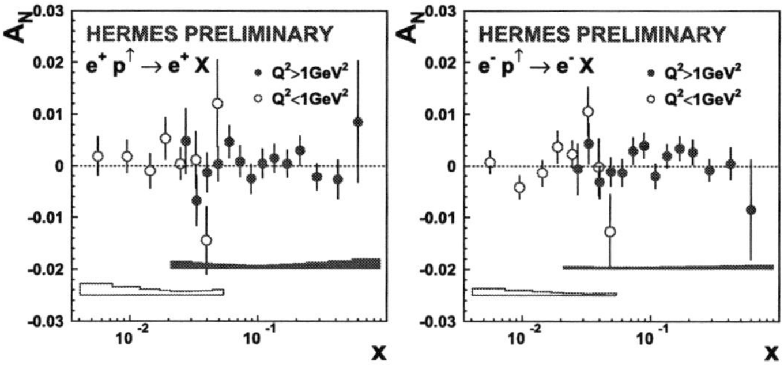

FIGURE 1. Normal asymmetry obtained from data taken with a positron beam (left) and with an electron beam (right). The open (closed) circles identify the data from $Q^2 < 1$ GeV^2 ($Q^2 > 1$ GeV^2).

extracted in the same way as done for real data. The ratio between the measured asymmetry A_N^{meas} and the input value $A^{\sin\phi}$ will be referred to as the acceptance scaling-factor χ: $A_N^{meas} = \chi A^{\sin\phi}$. The input amplitude $A^{\sin\phi}$ was varied between 0 and 0.5 and the stability of the resulting scaling factor was verified.

The final results for the measured asymmetries are shown in Figure 1 for positrons and electrons. In both cases the asymmetries are consistent with zero within uncertainties. The systematic bands include contributions from particle identification, trigger efficiencies, transverse magnet correction, target polarization and a Monte Carlo simulation used to estimate the effect of misalignment and beam curvature due to the transverse magnetic target field and beam shifts in position and angle. The acceptance factor χ is between 20% and 60% and is shown as a function of x in the middle panel of Figure 2. The results were not corrected for smearing and radiative effects, since at the moment there is no knowledge of the main background coming from the elastic 2-photon asymmetry. For this reason the estimated contribution coming from elastic events is shown as a panel in Figure 2. The overall asymmetries, obtained without binning in x and Q^2 but only separating the $0.004 < x < 0.9$ region into low and high Q^2, are shown in Table 1 together with statistical and systematic uncertainties. They are negligible within their uncertainties of the order of 10^{-3}. The average acceptance factors are 0.37 for the lower Q^2 region (58% of the value for full 2π acceptance) and 0.42 for the higher Q^2 region.

TABLE 1. Total asymmetry separated in the two Q^2 ranges under study.

beam	central $\times 10^{-3}$	stat $\times 10^{-3}$	total syst $\times 10^{-3}$
$Q^2 < 1$ GeV^2, $< x >= 0.02, < Q^2 >= 0.6$ GeV^2			
e^+	1.28	1.47	1.35
e^-	1.35	1.29	0.73
$Q^2 > 1$ GeV^2, $< x >= 0.14, < Q^2 >= 2.4$ GeV^2			
e^+	0.21	0.90	0.74
e^-	0.87	0.76	0.41

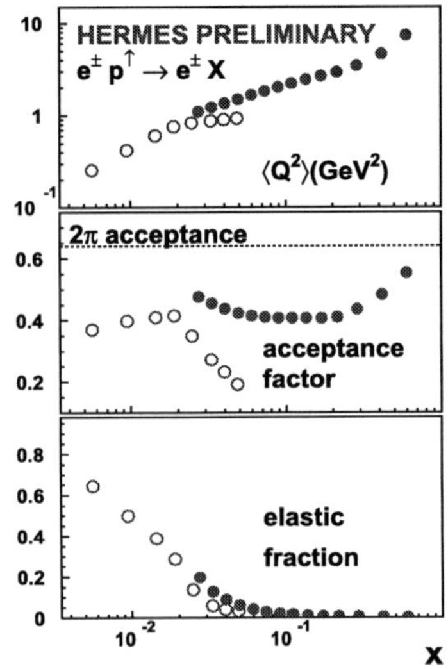

FIGURE 2. Top panel: Q^2 range of the data presented in Figure 1. Middle panel: acceptance factor under the assumption of a $\sin\phi$ asymmetry modulation. Bottom panel: percentage of elastic background events.

In conclusion, inclusive single spin asymmetries have been measured at HERMES on a transversely polarized proton target with the goal of searching for a signal coming from 2-photon exchange. No signal has been found within the uncertainties. At JLAB the proposed experiment JLAB-07-013 [5] will soon start taking data. Numerical estimates [6] for the kinematic range covered by this experiment result in expectations for the asymmetry of the order of 10^{-4}, the same order of magnitude of its anticipated precision.

ACKNOWLEDGMENTS

The authors would like to thank W.D. Nowak, K. Rith and P. Di Nezza for proof-reading the manuscript.

REFERENCES

1. A. V. Afanasev, et al., *Phys. Rev.* **D72**, 013008 (2005).
2. A. Metz, M. Schlegel, and K. Goeke, *Phys. Lett.* **B643**, 319–324 (2006).
3. H. Spinka, *RHIC Spin note 79* (1999).
4. K. Ackerstaff, et al., *Nucl. Instrum. Meth.* **A417**, 230 (1998).
5. X. Jiang, et al., "Target normal single spin asymmetry in inclusive DIS n(e,e') with a polarized ^{3}He target", *Jefferson Lab Hall A Experiment E-07-013* (2007).
6. A. V. Afanasev, M. Strikman, and C. Weiss, *Phys. Rev.* **D77**, 014028 (2008).

Nucleon Spin Polarizabilities in Heavy Baryon Chiral Perturbation Theory

Chung Wen Kao

Department of Physics, Chung Yuan Christian University, Chung-Li 32023, Taiwan

Abstract. We discuss the disagreement between the theoretical Heavy Baryon Chiral Perturbation Theory (HBχPT) predictions of the nucleon generalized forward spin polarizabilities and the experimental data of spin structure functions from Jefferson Laboratory.

Keywords: spin, polarizabilities, chiral symmetry, spin structure functions.
PACS: 11.55Hx, 13.60 Hb,14.20Dh

The spin-dependent doubly virtual Compton scattering amplitude in the forward direction (VVCS) are:

$$T(\nu, Q^2, \theta=0)^{spin} = i\vec{\sigma}\cdot(\vec{\varepsilon}_2^* \times \vec{\varepsilon}_1) g_{TT}(\nu, Q^2) - i\vec{\sigma}\cdot[(\vec{\varepsilon}_2^* - \vec{\varepsilon}_1)\times\hat{q}] g_{LT}(\nu, Q^2), \quad (1)$$

where ν is the virtual photon energy, Q^2 its virtuality, and $\hat{q}$ the unit vector along the virtual photon momentum in the Breit frame. The optical theorem relates the amplitudes above to the cross-section of inclusive electron-proton scattering and the spin structure functions g_1 and g_2. For the non-pole contributions to g_{TT} and g_{LT}, one can perform a Low energy expansion as follows:

$$\mathrm{Re}g_{TT}(\nu, Q^2) - \mathrm{Re}g_{TT}^{\mathrm{POLE}}(\nu, Q^2) = \left(\frac{2\alpha_{em}}{M_N^2}\right) I_A(Q^2)\nu + \gamma_0(Q^2)\nu^3 + \mathcal{O}(\nu^5), \quad (2)$$

$$\mathrm{Re}g_{LT}(\nu, Q^2) - \mathrm{Re}g_{LT}^{\mathrm{POLE}}(\nu, Q^2) = \left(\frac{2\alpha_{em}}{M_N^2}\right) Q I_3(Q^2) + Q\delta_{LT}(Q^2)\nu^2 + \mathcal{O}(\nu^4). \quad (3)$$

$\gamma_0(Q^2)$ and $\delta_{LT}(Q^2)$ are called generalized forward spin polarizabilities. Combining the low energy expansion and the dispersion relations, one acquires the sum rules as:

$$\gamma_0(Q^2) = \frac{\alpha_{em} 16 M_N^2}{Q^6} \int_0^{x_0} dx x^2 \left\{ g_1(x, Q^2) - \frac{4M_N^2}{Q^2} x^2 g_2(x, Q^2) \right\}, \quad (4)$$

and

$$\delta_{LT}(Q^2) = \frac{\alpha_{em} 16 M_N^2}{Q^6} \int_0^{x_0} dx x^2 \left\{ g_1(x, Q^2) + g_2(x, Q^2) \right\}. \quad (5)$$

The LHS of these sum rules have been calculated in Heavy Baryon Chiral Perturbation Theory (HBχPT) [1] and Relativistic Baryon Chiral Perturbation Theory (RBχPT) [2]. Up to $\mathcal{O}(p^4)$ the results of the calculations contain no unknown low energy constants (LECs). Namely the chiral dynamics should play a dominant role. Therefore these sum

CP1149, *Spin Physics, 18th International Spin Physics Symposium*
edited by D. G. Crabb, D. B. Day, S. Liuti, X. Zheng, M. Poelker, and Y. Prok

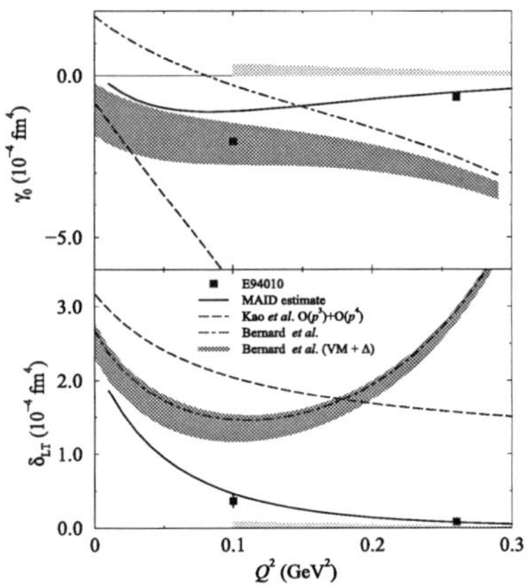

FIGURE 1. Generalized forward spin polarizabilities γ_0(upper panel) and δ_{LT}(lower panel) for the neutron. The dashed curve is the HBChPT result from [1]. The dot-dashed curves are RBχPT results from [2]. The dark band is RBChPT with $\Delta(1232)$ and the vector meson contributions. Solid curves are from the MAID model [4]. Solid squares are the experimental results with statistical uncertainties. The light bands are the systematic uncertainties.

rules become an excellent test ground for the χPT. For the purely transverse generalized forward spin polarizability γ_0, we find the expression at $\mathcal{O}(p^3)$ as :

$$\gamma_0^{\mathcal{O}(p^3)}(Q^2) = \frac{\alpha_{em} g_A^2}{(4\pi F_\pi)^2} \cdot \frac{4}{m_\pi^2} \left[\frac{1}{3} + \frac{1}{w^2} - \frac{4w^2+12}{3w^3\sqrt{w^2+4}} \sinh^{-1}[\frac{w}{2}] \right] . \tag{6}$$

Here, $w = \sqrt{Q^2/m_\pi^2}$ and the $\mathcal{O}(p^4)$ result is:

$$\begin{aligned} \gamma_0^{\mathcal{O}(p^4)}(Q^2) \;=\; & \frac{\alpha_{em} g_A^2}{(4\pi F_\pi)^2 M_N} \cdot \frac{\pi}{1152 m_\pi} \{ -576 w^2 + [(-1888 - 448\kappa_v) + (-416 - 128\kappa_s)\tau_3] \\ & + [(366 + 366\kappa_v) + (930 - 222\kappa_s)\tau_3] \cdot \frac{1}{w^2} \\ & + [(123 + 123\kappa_v) + (-2787 + 93\kappa_s)\tau_3] \cdot \frac{1}{w} \tan^{-1}[\frac{w}{2}] \\ & + [(2724 - 732\kappa_v) + (-1860 + 444\kappa_s)\tau_3] \cdot \frac{1}{w^3} \tan^{-1}[\frac{w}{2}] \\ & + [(272 + 272\kappa_v) + (16 + 16\kappa_s)\tau_3] \cdot \frac{1}{w^2+4} \\ & -192\tau_3 \cdot \frac{w^4 - 4w^2 - 24}{w^2+4} + 576 \cdot \frac{w^4 - 12}{w^4 + 4w^2} \} . \end{aligned} \tag{7}$$

Analogously, we find the expressions of δ_{LT} for the $\mathcal{O}(p^3)$ term as :

$$\delta_{LT}^{\mathcal{O}(p^3)}(Q^2) = \frac{\alpha_{em} g_A^2}{(4\pi F_\pi)^2} \cdot \frac{4}{m_\pi^2} \left[\frac{1}{3w\sqrt{w^2+4}} \sinh^{-1}[\frac{w}{2}] \right] , \tag{8}$$

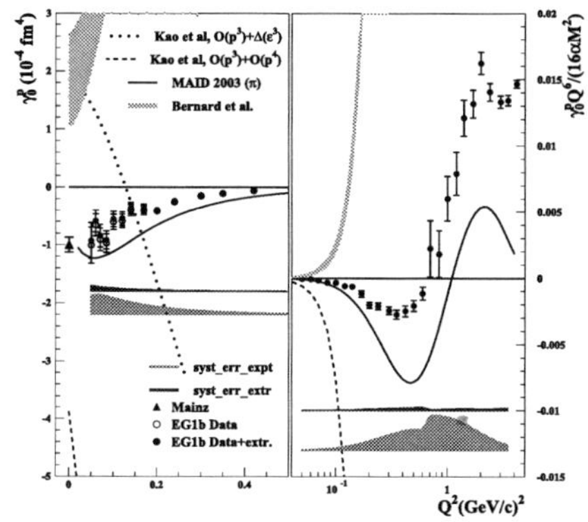

FIGURE 2. Generalized forward spin polarizabilities γ_0^p as a function of Q^2 for the full integral (closed circles), the measured portion of the integral (open circles). The triangle is the data at $Q^2 = 0$[5]. The data shown on the right are weighted by $Q^6/(16\alpha_{em}M_N^2)$. The systematic error on the measured (grey) and unmeasured (dark) contributions are indicated by bands. χPT calculations are shown along with the MAID model.

and for the $\mathcal{O}(p^4)$ correction as :

$$\begin{aligned}
\delta_{LT}^{\mathcal{O}(p^4)}(Q^2) &= \frac{\alpha_{em} g_A^2}{(4\pi F_\pi)^2 M_N} \cdot \frac{\pi}{192 m_\pi}\{(-48+24\kappa_v)+(-24+48\kappa_s)\tau_3 \\
&+[(-162+24\kappa_v)+(-18+24\kappa_s)\tau_3]\cdot\frac{1}{w^2} \\
&+[(-27-36\kappa_v)+(-27-12\kappa_s)\tau_3]\cdot\frac{1}{w}\tan^{-1}[\frac{w}{2}] \\
&+[(-252-48\kappa_v)+(36-48\kappa_s)\tau_3]\cdot\frac{1}{w^3}\tan^{-1}[\frac{w}{2}] \\
&+[12-(36+48\kappa_s)\tau_3]\cdot\frac{1}{4+w^2}+384\cdot\frac{3+w^2}{4w^2+w^4}\}. \qquad (9)
\end{aligned}$$

The data of γ_0^n and δ_{LT}^n from JLab [3] taken with the polarized ^{3}He target is shown in Fig. 1. The predictions of HBχPT fail to match both of them. On the other hand, the result of γ_0^n in RBχPT seems to be in good agreement with the data, but it also failed to give the correct values of δ_{LT}^n. It is remarkable that HBχPT gives the correct Q^2 dependence of δ_{LT}^n. On the contrary RBχPT predicted a rise when $Q^2 \geq 0.1$ GeV^2. The disagreement between χPT and the experimental data of δ_{LT}^n is a serious challenge for χPT since the resonance contribution to δ_{LT} has been calculated in [1] up to $\mathcal{O}(\varepsilon^3)$ and was found to be very small, hence this discrepancy cannot be explained by the resonance effects.

The new CLAS data on γ_0^p [6] further deepens this crisis. In Fig. 2 one finds that HBχPT again fails to match the CLAS data. Surprisingly, RBχPT is also in serious disagreement with the data, even for Q^2 as low as 0.05 GeV^2. Moreover, in Fig. 3 the results of $\gamma_0^p - \gamma_0^n$ and $\gamma_0^p + \gamma_0^n$ are presented by [7]. Since the $\Delta(1232)$ contribution is

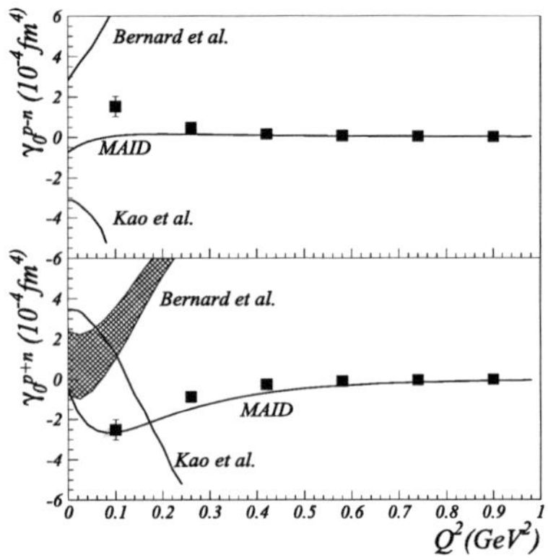

FIGURE 3. The isovector γ_0^p-γ_0^n (top) and isoscalar $\gamma_0^p + \gamma_0^n$ (bottom) generalized forward spin polarizabilities. The proton data are from CLAS [6] and the neutron data are from Hall A [3]. The curves are predictions of HBχPT [1]and relativistic BχPT [2] and the MAID model [4].

largely isoscalar such that isovector γ_0 should be dominated by the chiral physics. The CLAS data of $\gamma_0^p - \gamma_0^n$ at $Q^2 = 0.1$ GeV^2 is located between HBχPT and RBχPT curves. Both of the curves fail to reproduce the trend or magnitude of the data. Thjs implies that the current χPT calculations do not characterize the chiral physics as expected and casts doubt on the range of applicability of χPT. The conservative approach is to complete $\mathcal{O}(\varepsilon^4)$ calculation in HBχPT because the fourth order Δ loop effects, unlike the tree-level Δ diagrams, still would be significantly large for both isovector and isoscalar γ_0 and δ_{LT}. Although it is less likely because in general Δ loop contribution is usually quite tiny. After this, the next step is the calculation up to $\mathcal{O}(p^5)$ which contains the two-loop diagrams [8]. We also look forward to results from the forthcoming experiments at Jefferson Lab where spin structure functions will be measured down to $Q^2 = 0.01$ GeV^2. Those data will reveal more mystery of the low Q^2 spin properties of the nucleon.

ACKNOWLEDGMENTS

This work is support by NSC of Taiwan under grant No. NSC96-2112-M033-003-MY3.

REFERENCES

1. C. W. Kao, T. Spitzenberg and M. Vanderhaeghen, *Phys. Rev. D* **67**, 016001 (2003).
2. V. Bernard, T. R. Hemmert and U-G. Meissner, *Phys. Rev. D* **67**, 076008 (2003).
3. Duer et al., *Phys.Rev. Lett***93**, 212001(2004).
4. D. Drechsel, S. Kamalov, and L. Tiator, *Phys. Rev.* **D 63**, 114010 (2001).
5. H. Dutz et al., *Phys. Rev. Lett***91** , 192001(2003).
6. Y. Prok et al., *Phys. Lett. B* **672**, 12 (2009).
7. A. Duer et al., *Phys. Rev* **D 78**, 032001(2008).
8. C.W.Kao, *Int. Mod. Phys. A***21**,2027(2006).

New Angle on the Parton Distribution Functions: Self-Organizing Maps

H. Honkanen* and S. Liuti†

*Department of Physics and Astronomy, Iowa State University, Ames, IA 50011, USA
†Department of Physics, University of Virginia, P.O. Box 400714, Charlottesville, VA 22904-4714, USA

Abstract. Neural network (NN) algorithms have been recently applied to construct Parton Distribution Function (PDF) parametrizations, providing an alternative to standard global fitting procedures. Here we explore a novel technique using Self-Organizing Maps (SOMs). SOMs are a class of clustering algorithms based on competitive learning among spatially-ordered neurons. We train our SOMs with stochastically generated PDF samples. On every optimization iteration the PDFs are clustered on the SOM according to a user-defined feature and the most promising candidates are used as a seed for the subsequent iteration using the topology of the map to guide the PDF generating process. Our goal is a fitting procedure that, at variance with the standard neural network approaches, will allow for an increased control of the systematic bias by enabling user interaction in the various stages of the process.

Keywords: parton, distribution function, neural network
PACS: 13.60.Hb, 12.38.Bx, 84.35.+i

SELF-ORGANIZING MAPS

The new Parton Distribution Functions (PDF) fitting method we have recently proposed in Ref. [1] relies on the use of Self-Organizing Maps (SOMs). The aim of our approach is to both i) to be able to study the properties of the PDFs in a model independent way and yet ii) to be able to implement knowledge from the prior works on PDFs, and ultimately iii) to be able to guide the fitting procedure interactively with the help of the SOM properties.

The SOM [2] is a visualization algorithm that attempts to represent all the available observations with optimal accuracy using a restricted set of models. SOM consists of nodes, map cells, which are all assigned spatial coordinates, and the topology of the map is determined by a chosen distance metric $M_{\rm map}$. Each cell contains a map vector, that is isomorphic to the data samples used for training of the neural network. Our choice for the SOM shape is a simple 2-dimensional rectangular lattice with $L_1(x,y) = \sum_{i=1}^{2} |x_i - y_i|$ topology.

The implementation of SOMs proceeds in three stages: 1) initialization of the SOM, 2) training of the SOM and 3) associating the data samples with a trained map, i.e. clustering. For the details of the SOM implementation, see [1].

On a properly trained SOM, cells that are topologically close to each other have map vectors which are most similar to each other (according to a chosen similarity metric $M_{\rm data}$) compared to all the other map vectors. In the matching stage the actual data is matched against the map vectors of the trained map, and thus get distributed on the map

CP1149, *Spin Physics, 18th International Spin Physics Symposium*
edited by D. G. Crabb, D. B. Day, S. Liuti, X. Zheng, M. Poelker, and Y. Prok

according to the feature that was used as the similarity criterion. Clusters of similar data now emerge as a result of *unsupervised learning*. This local similarity property is the feature that makes SOM suitable for visualization purposes and facilitation of the user interaction with the data. Since each map vector now represents a class of similar objects, the SOM is an ideal tool for visualization of high-dimensional data by projecting it onto a low-dimensional map in clusters grouped by some desired similarity feature.

ENVPDF ALGORITHM

As a first step towards our goal, we give up, at variance with the "conventional" PDFs the functional form for the PDFs and rather rely on purely stochastical methods in generating the initial and training PDF samples. Our choice is a Genetic Algorithm - type analysis, in which our parameters are the values of PDFs at the initial scale for each flavor at each value of x where the experimental data exist. To obtain control over the shape of the PDFs we use some of the existing distributions (CTEQ6 [3], CTEQ5 [4], MRST02 [5], Alekhin [6] and GRV98 [7]) to establish an initial range, an envelope, within which we sample the candidate PDF values. For now we concentrate on DIS structure function data from H1 [8], BCDMS [9] and Zeus [10], which we use without additional kinematical cuts or normalization factors.

We generate the first set of sample PDFs by, for each flavour separately, selecting randomly either the range $[0.5,1]$, $[1.0,1.5]$ or $[0.75,1.25]$ times any of the initial PDF set. Next we generate values for each $x_{\rm data}$ using uniform distribution around the initial PDFs. Gaussian smoothing is applied to the resulting set of points, and the flavours combined to form a PDF set and scaled to conserve momentum, baryon number and charge.

We randomly initialize the $N \times N$ map with N^2 sample PDFs sets, such that each map vector consists of the PDF set itself, and of the observables $F_2^p(x,Q_0^2)$ derived from it. Then we train the map with $4N^2$ sample PDFs. We choose the similarity criterion to be the similarity of observables $F_2^p(x,Q^2)$ with $M_{\rm data} = L_1$. During the training all the observables (PDFs) of the map vectors get averaged with the observables (PDFs, flavor by flavor) matched within the neighbourhood. The resulting new averaged map vector PDFs are rescaled again to obey the sum rules. The map vector PDFs are evolved and the observables at every experimental data scale are computed and compared for similarity with the observables from the training PDFs.

For the second and the subsequent *iterations* we pick from the first trained SOM $2N$ best PDFs as the initial PDFs for a new map. These initial PDFs are introduced into the training set along with the sample PDFs, which are now constructed using each of the initial PDFs *in turn* as a center for a Gaussian random number generator, which assigns for *all* the flavours for each x a value around that *same* initial PDF such that $1-\sigma$ of the generator is given by the spread of the best PDFs in the topologically nearest neighbouring cells. In order to preserve the variety of the PDF shapes on the map, we also keep some copies of the first iteration generators in our generator mix. Since the best PDF candidates from the previous iteration are matched on this new map as unmodified, it is guaranteed that the χ^2/N as a function of the iteration either decreases or remains the same. We keep repeating the iterations until the χ^2/N saturates.

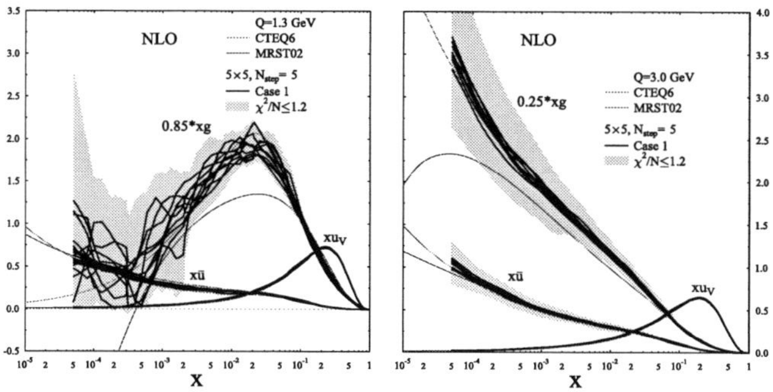

FIGURE 1. NLO ENVPDF best results and the $\chi^2/N \leq 1.2$ spreads of results from 10 separate runs.

The best χ^2/N values of the original initial PDFs (computed within our framework) are 1.67 for leading order (LO) (CTEQ6) and 1.89 for next-to-leading order (NLO) (MRST02). Our algorithm gives χ^2/N=1.04 (1.08) for LO (NLO) 5x5 map and χ^2/N=1.00 (1.07) for LO (NLO) 15x15 map.

Left panel of Fig. 1 presents the best NLO results, and the combined $\chi^2/N \leq 1.2$ spreads of the PDFs from any iteration, for 10 repeated 5×5, $N_{\text{step}} = 5$ runs at the initial scale. The average χ^2/N and the standard deviation σ for these runs are 1.122 and 0.029, corresponding to $\Delta\chi^2 \sim 20$. The right panel of the same Fig. 1 shows the 10 best result curves and the $\chi^2/N \leq 1.2$ spreads evolved up to $Q = 3.0$ GeV.

Clearly the seemingly large difference between the small-x gluon results at the initial scale gets smoothed out during the course of the Quantum Chromodynamics (QCD) evolution, and the initial scale wiggliness of the PDFs is thus mainly only a residual effect from our method of generating them and not linked to the over-training of the SOM. Therefore our simple method of producing the candidate PDFs by jittering random numbers inside a predetermined envelope is surprisingly stable when used together with a complicated PDF processing that SOMs provide. Remarkably then, even a single SOM run can provide a quick uncertainty estimate for a chosen $\Delta\chi^2$ without performing a separate error analysis.

FUTURE EXTENSION OF THE METHOD

ENVPDF algorithm is a relatively straightforward method of obtaining stochastically generated, parameter-free, PDFs, with an uncertainty estimate for a desired $\Delta\chi^2$, and the SOM method can be extended much further than that. What will ultimately sets the method apart from the standard global analyses or NNPDF [11] method are the clustering and visualization possibilities that it offers. Instead of setting $M_{\text{data}} = L_1$ and clustering according to the similarity of the observables, it is possible to set the clustering criteria to be anything that can be mathematically quantified. The stochastically generated PDFs with the desired feature can then be projected out from the SOM. Moreover, by combining the method with an interactive graphic user interface (GUI), it will be possible to change and control the shape and the width of the envelope as the minimization proceeds, and thus to guide the process by applying researcher's insight at various stages of the process. The uncertainty band produced by the SOM could further help the user to make decisions about the next steps of the minimization. With the GUI it would be e.g. possible to generate the sample PDFs by jittering a vector consisting of PDF parameters, instead of values of PDFs in each value of x of the data. This would lead to smooth, continuous type of solutions, either along the lines of global analyses, or NNPDFs using N SOMs for N Monte-Carlo sampled replicas of the data. For such a method, all the existing error estimates, besides an uncertainty band produced by the map, would be available as well. Since the solution would be required to stay within an envelope of selected width and shape, no restrictions for the parameters themselves would be required. The selection of the best PDF candidates for the subsequent iteration could then be made based on the user's preferences instead of solely based on the χ^2/N. That kind of method in turn could be extended to more complex cases, such as the Generalized Parton Distributions, where the experimental data are not numerous enough to allow for a model independent fitting, and the guidance and intuition of the user is therefore irreplaceable.

ACKNOWLEDGMENTS

We thank our computer science collaborators D. Brogan, J. Carnahan, Y. Loitiere and P. R. Reynolds. This work was financially supported by the US National Science Foundation grant no.0426971. HH was also supported by the U.S. Department of Energy, grant no. DE-FG02-87ER40371. SL is supported by the U.S. Department of Energy, grant no. DE-FG02-01ER41200.

REFERENCES

1. J. Carnahan, H. Honkanen, S. Liuti, Y. Loitiere and P. R. Reynolds, arXiv:0810.2598 [hep-ph].
2. T. Kohonen, *Self-organizing Maps*, Third Edition, Springer (2001); `http://www.cis.hut.fi/projects/somtoolbox`
3. J. Pumplin, D. R. Stump, J. Huston, H. L. Lai, P. Nadolsky and W. K. Tung, *JHEP* **0207**, 012 (2002). [arXiv:hep-ph/0201195].
4. H. L. Lai *et al.* [CTEQ Collaboration], *Eur. Phys. J. C* **12**, 375 (2000). [arXiv:hep-ph/9903282].
5. A. D. Martin, R. G. Roberts, W. J. Stirling and R. S. Thorne, *Eur. Phys. J. C* **23**, 73 (2002). [arXiv:hep-ph/0110215];
6. S. Alekhin, *Phys. Rev. D* **68**, 014002 (2003). [arXiv:hep-ph/0211096];
7. M. Gluck, E. Reya and A. Vogt, *Eur. Phys. J. C* **5**, 461 (1998). [arXiv:hep-ph/9806404].
8. C. Adloff *et al.* [H1 Collaboration], *Eur. Phys. J. C* **21**, 33 (2001). [arXiv:hep-ex/0012053].
9. A. C. Benvenuti *et al.* [BCDMS Collaboration], *Phys. Lett. B* **223**, 485 (1989); A. C. Benvenuti *et al.* [BCDMS Collaboration], *Phys. Lett. B* **237**, 592 (1990).
10. S. Chekanov *et al.* [ZEUS Collaboration], *Eur. Phys. J. C* **21**, 443 (2001). [arXiv:hep-ex/0105090].
11. J. Rojo *et al.* [NNPDF Collaboration], arXiv:0811.2288 [hep-ph]; R. D. Ball *et al.*, arXiv:0808.1231 [hep-ph].

Sum Rules for the Spin-Dependent Structure Functions from the Current Algebra on the Null-Plane

Susumu Koretune

Department of Physics, Shimane University,Matsue,Shimane,690-8504,Japan

Abstract. Spin dependent sum rules from the current algebra on the null-plane both from the current commutation and the current anti-commutation relations are given. By regularizing these sum rules, we explain that the Born term, resonances and the continuum are constrained severely by the sum rules. This fact explains why the sign change of the Gerasimov-Drell-Hearn (GDH) sum occurs in the very small Q^2. Further, combined with the regularized Burkhart and Cottingham(BC) sum rule, we find that g_2^{ep} has a similar behavior as g_1^{ep}. The method has been applied to the deuteron target.

Keywords: sum rule, polarized structure function
PACS: 24.85.+p.11.55.Hx,13.60.Hb,24.70.+s

INTRODUCTION

The current algebra based on the canonical quantization at equal-time gives us the very general constraint such as Adler-Weisberger sum rule and Adler sum rule. These sum rules are fundamental ingredients in Quantum Chromodynamics (QCD). This algebra was extended to the current algebra on the null-plane [1]. Apart from the technical points, the superiority of this algebra over the equal-time one is usually said to lie in the following points. (1) Corrections from the connected matrix element of the bilocal current appear in the sum rules where without the corrections considered to be peculiar. (2) We need not take a special frame such as the infinite-momentum frame and the sum rules hold in any frame. The null-plane formalism included further generalization to the current anti-commutation relation on the null-plane [2, 3]. We briefly explain these facts. Let us consider the Deser-Gilbert-Sudarshan (DGS) representation.[4]

$$\begin{aligned} C_{ab}(p\cdot q,q^2) &= \int d^4x \exp(iq\cdot x) < p|[J_a(x),J_b(0)]|p>_c \qquad (1)\\ &= \int d^4x \exp(iq\cdot x)\int_0^\infty d\lambda^2\int_{-1}^1 d\beta \exp(i\beta p\cdot x)h_{ab}(\lambda^2,\beta)i\Delta(x,\lambda^2)\\ &= (2\pi)\int_0^\infty d\lambda^2\int_{-1}^1 d\beta\,\delta((q+\beta p)^2-\lambda^2)\varepsilon(q^0+\beta p^0)h_{ab}(\lambda^2,\beta), \end{aligned}$$

$$\begin{aligned} C_{ab}(p\cdot q,q^2) &= \sum_n (2\pi)^4\delta^4(p+q-n)\langle p|J_a(0)|n\rangle\langle n|J_b(0)|p\rangle \qquad (2)\\ &- \sum_n (2\pi)^4\delta^4(p-q-n)\langle p|J_b(0)|n\rangle\langle n|J_a(0)|p\rangle. \end{aligned}$$

CP1149, *Spin Physics, 18th International Spin Physics Symposium*
edited by D. G. Crabb, D. B. Day, S. Liuti, X. Zheng, M. Poelker, and Y. Prok

In the rest frame, $p = (m, \vec{0})$, since the first term in Eq.(2) is restricted as $m + q^0 = n^0$, q^0 satisfies $q^0 \geqq M_s - m$, where M_s is the minimum mass of the s channel continuum. Similarly, since the second term is restricted as $m - q^0 = n^0$, q^0 satisfies $q^0 \leqq m - M_u$, where M_u is the minimum mass of the u channel continuum. Thus, if $m \leqq (M_s + M_u)/2$, the first and the second terms in Eq.(2) are disconnected.

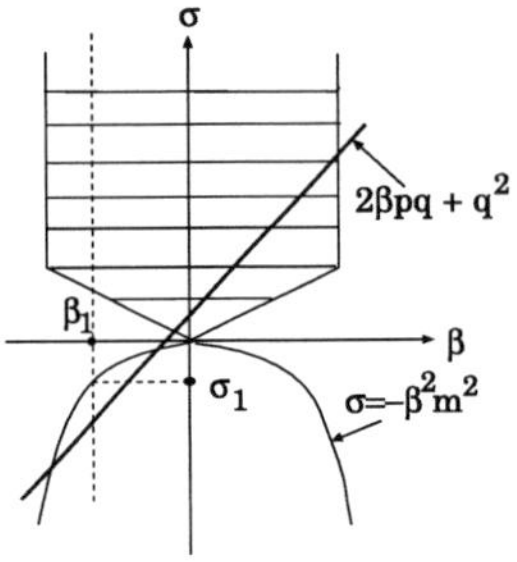

Now, in the DGS representation, $h_{ab}(\lambda^2, \beta)$ is not zero only in the shaded region. The integration path is $\sigma = 2\beta p \cdot q + q^2$, where $\sigma = \lambda^2 - \beta^2 m^2$. At the rest frame, we see that the point in the integration path where the sign changes by the factor $\varepsilon(p \cdot q + \beta m^2)$ lies always in the region $\sigma < -\beta^2 m^2$. In the s channel, since $p \cdot q > 0$, slope of the integration path is positive. Thus only the region $\varepsilon(p \cdot q + \beta m^2) = 1$ contributes,and hence, in the s channel,we obtain

$$(2\pi)\int_0^\infty d\lambda^2 \int_{-1}^1 d\beta\,\delta((q+\beta p)^2 - \lambda^2) h_{ab}(\lambda^2, \beta)\theta(q^0 + \beta p^0) = \sum_n (2\pi)^4 \delta^4(p+q-n)\langle p|J_a(0)|n\rangle\langle n|J_b(0)|p\rangle. \tag{3}$$

Similarly, in the u channel, we obtain

$$(2\pi)\int_0^\infty d\lambda^2 \int_{-1}^1 d\beta\,\delta((q+\beta p)^2 - \lambda^2) h_{ab}(\lambda^2, \beta)\theta(-(q^0 + \beta p^0)) = \sum_n (2\pi)^4 \delta^4(p-q-n)\langle p|J_b(0)|n\rangle\langle n|J_a(0)|p\rangle. \tag{4}$$

By combining these two relations we obtain

$$\begin{aligned} W_{ab}(p\cdot q, q^2) &= \int d^4x \exp(iq\cdot x) < p|\{J_a(x), J_b(0)\}|p>_c \\ &= \int d^4x \exp(iq\cdot x)\int_0^\infty d\lambda^2 \int_{-1}^1 d\beta \exp(i\beta p\cdot x) h_{ab}(\lambda^2,\beta)\Delta^{(1)}(x,\lambda^2) \\ &= (2\pi)\int_0^\infty d\lambda^2 \int_{-1}^1 d\beta\,\delta((q+\beta p)^2 - \lambda^2) h_{ab}(\lambda^2,\beta). \end{aligned} \tag{5}$$

The null-plane restriction of the current commutation relation or the anti-commutation relation can be obtained by the integration of C_{ab} and W_{ab} with respect to q^-. Now, if we take the scalar current $J_a(x) =: \phi^\dagger(x)\tau_a\phi(x) :$ with $[\phi^\dagger(x), \phi(0)]|_{x^+=0} = i\Delta(x)$, where $\Delta(x)$ at $x^+ = 0$ is $-\varepsilon(x^-)\delta(\vec{x}^\perp)/4$ being mass-independent, we can calculate the current commutation relation on the null-plane, and by substituting this result into the null-plane restriction of the DGS representation, we obtain information of the function h_{ab}. Using this information we obtain the connected matrix elements of the current anti-commutation relation on the null-plane since $\Delta^{(1)}(x)$ at $x^+ = 0$ is also mass-independent. A net result is that we simply change the $i\Delta(x)$ at $x^+ = 0$ to $\Delta^{(1)}(x)$ at $x^+ = 0$ which is given as $-\ln|x^-|\delta(\vec{x}^\perp)/2\pi$.

The condition necessary to get the expression on the null-plane can be studied by considering the integral of the $C_{ab}(p\cdot q,q^2)$ and $W_{ab}(p\cdot q,q^2)$ with respect to q^- such as

$$\lim_{\Lambda\to\infty}\int_{-\infty}^{\infty} dq^- \exp(-(q^-)^2/\Lambda^2)H_{ab}(p\cdot q,q^2), \tag{6}$$

where H_{ab} is C_{ab} or W_{ab}. This results in the condition to the h_{ab} and is known to be equivalent to the superconvergence relation which is required to get the fixed-mass sum rule, and in the null-plane formalism is equivalent to the interchange to set $q^+=0$ and $\nu = p\cdot q$ integration. The difference of the sum rule from the current commutation relation and those from the current anti-commutation relation lies in the following. Some of the current algebra sum rules are known to be divergent, and especially the sum rules which get corrections from the connected matrix element of the bilocal current need particular attention. In view of this fact it is important to regularize the sum rule to get physical information. Including such seemingly divergent sum rules, importance of the regularization of the sum rule is explained in the paper [5].

SPIN SUM RULES OF THE NUCLEON

The spin dependent part of the hadronic tensor is defined as

$$\begin{aligned} C_{ab}^{\mu\nu}|_{spin} &= \frac{1}{4\pi}\int d^4x \exp(iq\cdot x) < p,s|[J_a^\mu(x),J_b^\nu(0)]|p,s>_c|_{spin} \\ &= i\varepsilon^{\mu\nu\lambda\sigma}q_\lambda s_\sigma G_1^{ab} + i\varepsilon^{\mu\nu\lambda\sigma}q_\lambda(\nu s_\sigma - q\cdot s p_\sigma)G_2^{ab}. \end{aligned} \tag{7}$$

Under $q\to -q$, $a\leftrightarrow b$,and $\mu\leftrightarrow\nu$, spin structure function satisfy $G_1^{ab}(p\cdot q,q^2) = -G_1^{ba}(-p\cdot q,q^2)$, $G_2^{ab}(p\cdot q,q^2) = G_2^{ba}(-p\cdot q,q^2)$. From the current commutation relation on the null-plane for $\mu=+$ and $\nu=i$, the sum rules for the $g_1=\nu G_1$ and $g_2=\nu^2 G_2$ has been derived as[1]

$$\int_0^1 \frac{dx}{x} g_1^{[ab]}(x,Q^2) = -\frac{1}{16} f_{abc}\int_{-\infty}^{\infty} d\alpha\,\varepsilon(\alpha)[A_c^5(\alpha,0)+\alpha\bar{A}_c^5(\alpha,0)], \tag{8}$$

$$\int_0^1 \frac{dx}{x} g_2^{[ab]}(x,Q^2) = \frac{1}{16} f_{abc}\int_{-\infty}^{\infty} d\alpha\,\varepsilon(\alpha)\alpha\bar{A}_c^5(\alpha,0), \qquad \int_0^1 dx g_2^{(ab)}(x,Q^2) = 0, \tag{9}$$

where the connected matrix element of the anti-symmetric axial-vector bilocal current is defined as

$$< p,s|A_c^{5\beta}(x|0)|p,s>_c = s^\mu A_c^5(p\cdot x,x^2) + p^\mu(x\cdot s)\bar{A}_c^5(p\cdot x,x^2) + x^\mu(x\cdot s)\tilde{A}_c^5(p\cdot x,x^2), \tag{10}$$

where $g_i^{(ab)} = \frac{1}{2}(g_i^{ab}+g_i^{ba}), g_i^{[ab]} = \frac{1}{2i}(g_i^{ab}-g_i^{ba})$ for $i=1,2$. The sum rule for $g_1^{[ab]}$ by the equal-time formalism, where the right-hand side of it is zero, has been known as Bég sum rule [6]. This sum rule is very peculiar since it is violated even in the free model. In the null-plane formalism, this peculiarity disappears through the contribution of the bilocal current. The sum rule for $g_2^{(ab)}$ is derived by Burkhart and Cottingham(BC) [7], and also by Schwinger [8]. As Q^2 becomes large, Born term can be neglected and hence the sum rule becomes the one for the inelastic reaction.

The current anti-commutator on the null-plane is

$$<p,s|\{J_a^+(x),J_b^i(0)\}|p,s>_c|_{x^+=0} =<p,s|\{s^{+\beta i\alpha}\partial_\alpha[\Delta^{(1)}(x)G_{ab\beta}(x|0)] \\ -2g^{+\alpha}g^{i\beta}\partial_\alpha[\Delta^{(1)}(x)]G_{ab\beta}(x|0)-\varepsilon^{+i\alpha\beta}\partial_\alpha[\Delta^{(1)}(x)G^5_{ab\beta}(x|0)]\}|p,s>_c, \tag{11}$$

$$G^\beta_{ab}(x|0)=d_{abc}A^\beta_c(x|0)+f_{abc}S^\beta_c(x|0), \quad G^{5\beta}_{ab}(x|0)=d_{abc}S^{5\beta}_c(x|0)-f_{abc}A^{5\beta}_c(x|0), \tag{12}$$

where the matrix element of the symmetric axial-bilocal current $S^{5\beta}_c(x|0)$ is defined similarly as $A^{5\beta}_c(x|0)$. The fixed-mass sum rule can be derived as in the commutation relation, however, the crossing symmetry relation is opposite. Therefore we obtain the sum rule for the opposite combination with respect to a,b.

$$\int_0^1 \frac{dx}{x}g_1^{(ab)}(x,Q^2)=\frac{-1}{8\pi}d_{abc}\int_{-\infty}^{\infty}d\alpha\ln|\alpha|[S^5_c(\alpha,0)+\alpha\overline{S}^5_c(\alpha,0)], \tag{13}$$

$$\int_0^1 \frac{dx}{x}g_2^{(ab)}(x,Q^2)=\frac{1}{8\pi}d_{abc}\int_{-\infty}^{\infty}d\alpha\alpha\ln|\alpha|\overline{S}^5_c(\alpha,0), \qquad \int_0^1 dxg_2^{[ab]}(x,Q^2)=0. \tag{14}$$

Now, both in the commutator and the anti-commutator cases, the correction from the bilocal current is Q^2 independent. Therefore we have the equality in two different Q^2. However, if each term is divergent, this equality is useless. Because of this we first derive the sum rule in the non-forward direction, and noting that the correction from the bilocal current is Q^2 independent, we obtain the equality in two different Q^2 by assuming the moving Regge pole or cut,and we analytically continue it to the forward direction.[9] In this way we obtain the regularized sum rule in the *ep* reaction,[10]

$$\int_{x_c(Q)}^1 \frac{dx}{x}g_1^{ep}(x,Q^2)=B_1^{ep}(Q^2)-\frac{1}{8\pi^2\alpha_{em}}\int_{\nu_0}^{\nu_c}d\nu\{\sigma^{\gamma p}_{3/2}-\sigma^{\gamma p}_{1/2}\}+K_1^{ep}(E_c,Q^2), \tag{15}$$

where

$$B_1^{ep}(Q^2)=\frac{1}{2}\{F_1^p(0)(F_1^p(0)+F_2^p(0))-F_1^p(Q^2)(F_1^p(Q^2)+F_2^p(Q^2))\}, \tag{16}$$

$\nu_c=m_NE_c, \nu_c^Q=\nu_c+Q^2/2, x_c(Q)=Q^2/2\nu_c^Q$ with E the energy in the photoproduction at the laboratory frame and E_c is the cut-off energy.

$$K_1^{ep}(E_c,Q^2)=\frac{1}{8\pi^2\alpha_{em}}\int_{\nu_c}^{\infty}d\nu\{\sigma^{\gamma p}_{1/2}-\sigma^{\gamma p}_{3/2}\}-\int_{\nu_c^Q}^{\infty}\frac{d\nu}{\nu}g_1^p(x,Q^2). \tag{17}$$

It should be noted that the subtraction in K_1^{ep} should precede the integration. Now for a typical value such as E_c =2 GeV, the continuum contribution to $K_1^{ep}(E_c,Q^2)$ below Q^2 <1 GeV^2 is very small. Therefore the sum rule must be satisfied by resonances, Born term, and the nonresonant continuum below E_c = 2 GeV. Further, since Born

term changes rapidly in this region, the sum of the resonances and the nonresonant continuum must change rapidly to compensate it. Thus we see the sign change in the quantity $\int_{x_c(Q)}^{1} \frac{dx}{x} g_1^p(x,Q^2)$. This explains why the GDH sum $\Gamma^p(Q^2) = \int_0^1 dx g_1^p(x,Q^2)$ changes sign in the very small Q^2 region as experiment shows [11].
In case of g_2,[12] in the following relation

$$\triangle\sigma^{ab}(\nu,Q^2) = \sigma^{ab}_{3/2}(\nu,Q^2) - \sigma^{ab}_{1/2}(\nu,Q^2) = -\frac{8\pi^2\alpha_{em}}{K}\left(\frac{g_1^{ab}(x,Q^2)}{\nu} - \frac{m_N^2 Q^2 g_2^{ab}(x,Q^2)}{\nu^3}\right), \tag{18}$$

we take the derivative with respect to Q^2 and take the limit $Q^2 \to 0$

$$\frac{g_2^{ab}(x,0)}{\nu} = \frac{g_1^{ab}(x,0)}{2m_N^2} + \frac{\nu}{m_N^2}\frac{\partial g_1^{ab}(x,Q^2)}{\partial Q^2}\bigg|_{Q^2=0} + \frac{\nu^2}{8\pi^2 m_N^2 \alpha_{em}}\frac{\partial \triangle\sigma^{ab}(\nu,Q^2)}{\partial Q^2}\bigg|_{Q^2=0}. \tag{19}$$

where $K = (1 - \frac{Q^2}{2\nu})$, and the quantity related to $g_1(x,0)$ and $g_2(x,0)$ should be understood as $g_1(x,0) = \nu G_1(\nu,0)$, $g_2(x,0) = \nu^2 G_2(\nu,0)$, $\partial g_1(x,Q^2)/\partial Q^2|_{Q^2=0} = \partial \nu G_1(\nu,Q^2)/\partial Q^2|_{Q^2=0}$. Therefore $g_2^{ab}(x,0)/\nu$ is related to the experimentally measurable quantity, and we use the expression $g_2^{ab}(x,0)/\nu$ to mean this quantity. In this way we obtain the sum rule of g_2^{ab} similar to the sum rule of g_1^{ab}. However, the born term in this sum rule is proportional to Q^2, hence the behavior in the small Q^2 is masked by this factor. Then we notice that the expression divided by Q^2 is nothing but the one which appears in the BC sum rule. Thus we are led to consider the regularized BC sum rule. Using $G_2^{ep}(\nu,0)$ in Eq.(19), this sum rule is given as

$$\int_{\nu_0(Q)}^{\infty} d\nu G_2^{ep}(\nu,Q^2) - \int_{\nu_0}^{\infty} d\nu G_2^{ep}(\nu,0) = B_S^{ep}(Q^2), \tag{20}$$

$$B_S^{ep}(Q^2) = \frac{1}{4m_p^2+Q^2} G_M^p(Q^2)(G_M^p(Q^2) - G_E^p(Q^2)) - \frac{\mu_p(\mu_p-1)}{4m_p^2} \tag{21}$$

where the main contribution comes from the low energy and the small Q^2 region. Since Born term $B_S(Q^2)$ changes rapidly in this region, the left-hand side of this sum rule also changes rapidly. This is the same picture as in the sum rule for the g_1^{ep}. The Born terms $B_1^{ep}(Q^2)$ and $-B_S^{ep}(Q^2)$ take very similar values below Q^2 =1 GeV^2. Therefore the inelastic part of the corresponding regularized sum rule at moment $n = 0$ for the g_1^{ep} and that at the moment $n = 1$ for the g_2^{ep} take similar values in this region. The former quantity is related to the sign change of the GDH sum. Thus we observe such phenomena as rapid change of the elastic form factor being compensated by the rapid change of resonances and the non-resonant background which appears in the g_2^{ep}.

SUM RULES FOR THE DEUTERON TARGET

The method in the previous section can be applied to the deuteron target. The hadronic tensor is defined as in the paper [13]. In case of $\mu = \nu = +$, typical sum rules take the following form, where the elastic contributions are included.

Commutator	*Anti − commutator*
$\int_0^1 \frac{dx}{x} F_2^{[ab]}(x,Q^2) = \frac{1}{4} f_{abc}\Gamma_c$	$\int_0^1 \frac{dx}{x} F_2^{(ab)}(x,Q^2) = \frac{1}{4\pi} d_{abc} P \int_{-\infty}^{\infty} \frac{d\alpha}{\alpha} A_c(\alpha,0)$
$\int_0^1 \frac{dx}{x} b_2^{[ab]}(x,Q^2) = 0$	$\int_0^1 \frac{dx}{x} b_2^{(ab)}(x,Q^2) = \frac{1}{4\pi} d_{abc} \int_{-\alpha}^{\alpha} d\alpha \{\alpha A_c^P(\alpha,0) + 2\tilde{A}_c(\alpha,0)\}$

(22)

Other sum rules are for the quantity $xb_2^{ab} + 3xb_3^{ab} + 3xb_4^{ab}$. From the case $\mu = + \ \nu = i$, we obtain the sum rules for the structure functions g_1^{ab}/x , g_2^{ab}/x and g_2^{ab} , and for the structure functions $b_2^{ab} - 3b_4^{ab}$, b_2^{ab}/x^2, $b_2^{ab} + 2b_3^{ab} + b_4^{ab}$.
The sum rule in case of the current anti-commutator gets contribution from the connected matrix element of the bilocal current. Hence, a similar application as in the case of the nucleon target may be possible.

SUMMARY

Some sum rules from the current algebra on the null-plane depend on the connected matrix elements of the bilocal currents. In many cases, these sum rules may be divergent, however, we can regularize them. Further, as a connected matrix element between the stable hadron, we can generalize the current commutation relation on the null-plane to the current anti-commutation relations on the null-plane. Through this generalization, we can obtain the sum rules for the electromagnetic current which can be accessed experimentally far more easier than the neutrino reaction. In case of the spin sum rules of the nucleon target, we find that the Born term, resonances and the continuum are constrained by the sum rule. This fact explains why the sign change of the GDH sum occurs in the very small Q^2. Futher, combined with the regularized BC sum rule, we find that g_2^{ep} has a similar behavior as g_1^{ep}. The method has been applied to the deuteron target and new sum rules are found.

REFERENCES

1. D.A.Dicus,R.Jackiw, and V.L.Teplitze, *Phys. Rev.* **D4**, 1733 (1971).
2. S.B.Gerasimov and J.Moulin JINR pub. E2-7247;
   J.Moulin, JINR pub. E2-7638.
3. S.Koretune, *Phys. Rev.* **D21**,820 (1980); *Prog. Theor. Phys.* **72**, 821 (1984).
4. S.Deser,W.Gilbert,and E.C.G.Sudarshan, *Phys. Rev.* **115**, 731 (1959).
5. D.J.Broadhurst,J.F.Gunion, and R.L. Jaffe,*Ann. Phys.* , **81**, 88 (1973).
6. Bég, *Phys. Rev. Lett.* **17**, 333 (1966).
7. Burkhart and Cottingham, *Ann. Phys.* ,**56**, 543 (1970).
8. Schwinger, *Proc. Nat. Acad. Sci. USA* **72**, 1559 (1975).
9. S.P.Dealwis, *Nucl. Phys.* **B43**, 579 (1972).
10. S.Koretune, *Phys. Rev.* **C72**, 045205 (2005); **74**, 059901(E) (2006).
11. R.Fatemi et al . *Phys. Rev. Lett.* **91**, 222002 (2003).
12. S.Koretune and H.Kurokawa, *Phys. Rev.* **C75**, 025204 (2007).
13. P.Hoodbhoy,R.L.Jaffe and A.Manohar, *Nucl. Phys.* **B312**, 571 (1989).

Measurement of the Contribution of Strange Quarks to the Proton Spin

A. El Alaoui
(On Behalf of the HERMES Collaboration)

Laboratoire de Physique Subatomique et de Cosmologie
IN2P3/CNRS - Université Joseph Fourier - INP
53 rue des Martyrs 38026 Grenoble, France
e-mail address: ahmed@lpsc.in2p3.fr

Abstract. An analysis of HERMES data for charged-kaon production in deep inelastic scattering from a polarized deuteron target was performed to access the momentum and helicity distributions of strange quarks in the nucleon. This measurement involves the determination of the inclusive double-spin asymmetry and of the semi inclusive double-spin asymmetry for kaons. The shape of the momentum distribution for strange quarks is found to be softer than that of the average of the light sea anti-quarks. The strange helicity distribution is consistent with zero in the region $0.02 < x < 0.3$ and $Q^2 > 1$ GeV^2. Details of the analysis will be presented in this paper.

Keywords: Proton polarization, Strange quarks, semi inclusive deep inelastic scattering
PACS: 13.60.-r, 13.60.Hb, 13.60.Le, 13.88.+e, 14.20.Dh, 14.65.Bt

INTRODUCTION

The determination of the strange quarks distribution is of special interest as it allows to shed light on the nucleon structure in the sea sector. In current global Quantum Chromodynamics (QCD) analyses of unpolarized parton distribution functions (PDFs), the lack of information on $s(x)$ and $\bar{s}(x)$ is reflected by adopting the ansatz $s = \bar{s} = r(\bar{u}+\bar{d})/2$ [1] at some low factorization scale. The constant r is loosely constrained by data. On the other hand, QCD fit to the polarized structure function $g_1(x)$ extracted from polarized deep inelastic scattering (DIS), together with hyperon decay data, under the assumption of SU(3) flavor symmetry, have been vaguely used to access the quark helicity distributions. The most precise recent value [2] of the first moment of the strange helicity distribution ($\int \Delta s(x)\,dx$) is $-0.103 \pm 0.007(exp.) \pm 0.013(theo.) \pm 0.008(evol.)$. Another technique called 5-flavor decomposition based on polarized semi inclusive DIS (SIDIS) (where a hadron is detected in coincidence with the scattered lepton in the final state) allows a direct determination of the individual contribution of each quark to the proton spin. In this analysis $\int \Delta s(x)\,dx$ is found to be $0.028 \pm 0.033(stat.) \pm 0.009(syst.)$ [3]. In this paper a new extraction of the strange helicity distribution from polarized SIDIS on deuterium will be reported.

CP1149, *Spin Physics, 18th International Spin Physics Symposium*
edited by D. G. Crabb, D. B. Day, S. Liuti, X. Zheng, M. Poelker, and Y. Prok

HERMES SPECTROMETER

HERMES was a fixed target experiment located in the east part of the HERA collider at DESY. The self-polarized lepton beam interacted with a longitudinally polarized deuteron gas target. The beam polarization was measured continuously via Compton backscattering of circularly polarized laser beams. The target cell was fed by an atomic-beam source based on Stern-Gerlach separation with hyperfine transitions. Both the nuclear polarization of the atoms and the atomic fraction inside the target cell were continuously monitored. The beam and deuteron target polarizations were $55\% \pm 3\%$ and $85\% \pm 3.5\%$ respectively. The scattered lepton and the coincident hadrons were detected by the HERMES spectrometer [4] using a highly efficient detection system consisting of an electromagnetic calorimeter, a transition radiation detector, a preshower scintillator counter and a ring-imaging Čerenkov detector (RICH). The later plays a key role because it provides the kaon identification which is crucial for the present analysis. The measurement of the particle scattering angles and momenta was accomplished by a tracking system consisting of microstrip gas chambers and drift chambers.

STRANGE HELICITY EXTRACTION

Following the parton model, at leading order (LO), the semi-inclusive double-spin asymmetry for charged-kaon production can be written in terms of quark helicity distributions ($\Delta q(x,Q^2)$) and kaon fragmentation functions ($D_q^K(z,Q^2)$) as

$$A_1^K(x,Q^2,z) = \frac{\sum_q e_q^2 \Delta q(x,Q^2) D_q^K(z,Q^2)}{\sum_{q'} e_{q'}^2 q'(x,Q^2) D_{q'}^K(z,Q^2)}. \tag{1}$$

For a deuteron target, assuming isospin symmetry and charge conjugation invariance and by integrating over $z = E_h/\nu$, Eq. (1) becomes

$$A_{1,d}^K(x) = \frac{\Delta Q(x) \int \mathscr{D}_Q^K(z)dz + \Delta S(x) \int \mathscr{D}_S^K(z)dz}{Q(x) \int \mathscr{D}_Q^K(z)dz + S(x) \int \mathscr{D}_S^K(z)dz} \tag{2}$$

where $Q(x) = u(x) + \bar{u}(x) + d(x) + \bar{d}(x)$, $S(x) = s(x) + \bar{s}(x)$, $\int \mathscr{D}_Q^K(z)dz = 4\int D_u^K(z)dz + \int D_d^K(z)dz$ and $\int \mathscr{D}_S^K(z)dz = 2\int D_s^K(z)dz$. The Q^2 dependence is omitted for simplicity. The corresponding inclusive double spin asymmetry is given by

$$A_{1,d}(x) = \frac{5\Delta Q(x) + 2\Delta S(x)}{Q(x) + S(x)} \tag{3}$$

The data used for the asymmetry extraction are subject to the following kinematic cuts $Q^2 > 1$ GeV^2, $W^2 > 10$ GeV^2, $y = \nu/E < 0.85$ where W^2 is the photon-nucleon invariant mass. In addition, the kaons detected in coincidence with the scattered lepton should fulfill the following criteria: $0.2 < z < 0.8$ and $x_F \simeq 2p_L/W > 0.1$. p_L is the kaon longitudinal momentum with respect to the direction of the virtual photon in the photon-nucleon center of mass frame. The recorded kaon yields were corrected for RICH detection inefficiencies through an event weighting procedure. Other effects which account for detector resolution, QED radiative effects and detector acceptance were simulated and then used by a technique [2] that unfolds kinematic migration of events. Figure 1 shows the measured asymmetries.

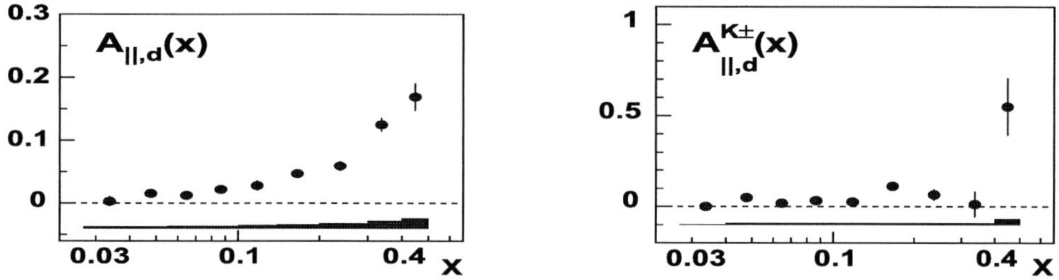

FIGURE 1. Lepton-nucleon polarized cross section asymmetries $A_{\|,d}(x)$ for inclusive DIS and $A^{K}_{\|,d}(x)$ for semi inclusive DIS off a deuteron target as function of Bjorken x, for identified charged kaons. The error bars are statistical, and the bands at the bottom represent the systematic uncertainties.

The non-strange $(\Delta Q(x))$ and the strange $(\Delta S(x))$ helicity distributions are obtained by combining Eqs. (2) and (3) in a matrix form:

$$\begin{pmatrix} A_{1,d}(x) \\ A^{K}_{1,d}(x) \end{pmatrix} = \begin{pmatrix} P_Q(x) & P_S(x) \\ P^{K}_{Q}(x) & P^{K}_{S}(x) \end{pmatrix} \begin{pmatrix} \Delta Q(x)/Q(x) \\ \Delta S(x)/S(x) \end{pmatrix} \tag{4}$$

where $$P_Q(x) = \frac{5Q(x)}{5Q(x)+2S(x)}, \quad P^{K}_{Q}(x) = \frac{Q(x)\int \mathscr{D}^{K}_{Q}(z)dz}{Q(x)\int \mathscr{D}^{K}_{Q}(z)dz + S(x)\int \mathscr{D}^{K}_{S}(z)dz}, \tag{5}$$

and $$P_S(x) = \frac{2S(x)}{5Q(x)+2S(x)}, \quad P^{K}_{S}(x) = \frac{S(x)\int \mathscr{D}^{K}_{S}(z)dz}{Q(x)\int \mathscr{D}^{K}_{Q}(z)dz + S(x)\int \mathscr{D}^{K}_{S}(z)dz}. \tag{6}$$

In order to get the non-strange and the strange helicity distributions, we need to determine $Q(x)$, $S(x)$, $\int \mathscr{D}^{K}_{Q}(z)dz$ and $\int \mathscr{D}^{K}_{S}(z)dz$. $Q(x)$ is taken from the CTEQ6L parameterization [1] and $\int \mathscr{D}^{K}_{S}(z)dz$ is taken from the most recent analysis on fragmentation functions [5]. $S(x)$ and $\int \mathscr{D}^{K}_{Q}(z)dz$ are to be extracted from the charged-kaon multiplicities which, for a deuteron target and at LO, are expressed by

$$\frac{d^2N^K(x)/dxdQ^2}{d^2N^{DIS}(x)/dxdQ^2} = \frac{Q(x)\int \mathscr{D}^{K}_{Q}(z)dz + S(x)\int \mathscr{D}^{K}_{S}(z)dz}{5Q(x)+2S(x)}. \tag{7}$$

By neglecting the term 2S(x) compared to 5Q(x) we get

$$S(x)\int \mathscr{D}^{K}_{S}(z)dz = Q(x)\left[5\frac{d^2N^K(x)}{d^2N^{DIS}(x)} - \int \mathscr{D}^{K}_{Q}(z)dz\right]. \tag{8}$$

At $x > 0.15$, S(x) is assumed to be negligible so that the non-strange fragmentation function can be obtained from a linear fit to the charged-kaon multiplicities which is constant. A value of $\int \mathscr{D}^{K}_{Q}(z)dz = 0.398 \pm 0.010$ was found which is in good agreement with the most recent value 0.435 ± 0.044 from [5] at $Q^2 = 2.5$ GeV^2. The value

0.398 ± 0.010 is then used in Eq. (8) together with values of $Q(x)$ from CTEQ6L and the measured charged-kaon multiplicities to pin down the product $S(x) \int \mathscr{D}_S^K(z)dz$. The shape of $xS(x)$ at $Q_0^2 = 2.5\, GeV^2$ was obtained from the product $S(x) \int \mathscr{D}_S^K(z)dz$ by dividing by $\int \mathscr{D}_Q^S(z)dz = 1.27 \pm 0.13$ taken from [5]. Figure 2 (left) shows the strange quarks distribution $xS(x)$ compared to the same quantity from CTEQ6L and to the sum of light antiquarks from CTEQ6L. The shape of $S(x)$ is substantially different from the averaged light sea quarks. The non-strange and strange helicity distribution can now be extracted from Eq. (4). The result is depicted in Fig. 2 (right). The systematic uncertainties include contributions from the beam and target polarizations, the neglection of the transverse spin structure function $g_2(x)$, and RICH kaon identification. The first moment of the helicity distributions in the measured region are presented

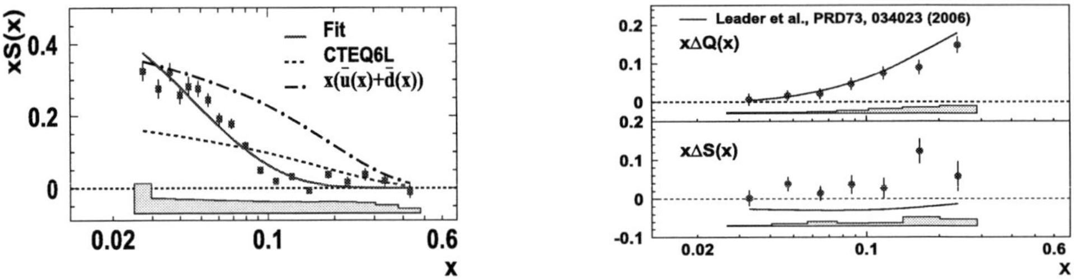

FIGURE 2. (Left) The strange parton distribution $xS(x)$ from the measured charged-kaon multiplicities on deuteron. (Right) The non-strange and strange quark helicity distributions evolved to $Q_0^2 = 2.5\, GeV^2$. The error bars are statistical, and the bands at the bottom represent the systematic uncertainties

in Table 1. The non-strange first moment is in good agreement with the one obtained from HERMES $g_1(x)$ data ($0.381 \pm 0.010(stat.) \pm 0.027(syst.)$). The strange first moment is found to be slightly positive and not in serious disagreement with the value of $-0.0435 \pm 0.010(stat.) \pm 0.004(syst.)$ reported by the same HERMES $g_1(x)$ data. The first moment of the octet combination $\Delta q_8(x) = \Delta Q(x) - 2\Delta S(x)$ given in Table 1 disagrees with the axial charge $a_8 = \Delta q_8 = 0.586 \pm 0.031$ extracted from hyperon decay constants using SU(3) symmetry. Possible explanation for this disagreement can be attributed to the violation of SU(3) symmetry or to the missing octet strength for $x < 0.02$.

TABLE 1. First moments of various helicity distributions in the Bjorken x range 0.02-0.30 at a scale of $Q_0^2 = 2.5\, GeV^2$

$\int \Delta Q(x)\, dx$	$0.359 \pm 0.026(stat.) \pm 0.018(syst.)$
$\int \Delta S(x)\, dx$	$0.037 \pm 0.019(stat.) \pm 0.027(syst.)$
$\int \Delta q_8(x)\, dx$	$0.285 \pm 0.046(stat.) \pm 0.057(syst.)$

REFERENCES

1. J. Pumplin, *et al.*, *J. High Energy Phys.*, **0207**, 12 (2002).
2. A. Airapetian, *et al.*, HERMES Collaboration, *Phys. Rev. D* **71**, 012003 (2005).
3. A. Airapetian, *et al.*, HERMES Collaboration, *Phys. Rev. D* **75**, 012007 (2007).
4. K. Ackerstaff, *et al.*, HERMES Collaboration, *Nucl. Instrum. Methods A* **417**, 230 (1998).
5. D. de Florian, *et al.*, *Phys. Rev. D* **75**, 114010 (2007).

Spin Structure of the Deuteron - New Results from CLAS

Nevzat Guler (for CLAS Collaboration)

Old Dominion University, Department of Physics, Norfolk, VA

Abstract. Double spin asymmetries for the proton and the deuteron have been measured in the EG1B experiment using the CLAS detector at Jefferson Lab. Longitudinally polarized electrons at energies 1.6, 2.5, 4.2 and 5.75 GeV were scattered from longitudinally polarized NH_3 and ND_3 targets. The double spin asymmetry $A_{\|}$ for the proton and the deuteron has been extracted from these data as a function of W and Q^2 with unprecedented precision. The virtual photon asymmetry A_1 and the spin structure function g_1 can be calculated from these measurements by using parametrization to the world data for the virtual photon asymmetry A_2 and the unpolarized structure functions F_1 and R. The large kinematic coverage of the experiment (0.05 GeV^2 < Q^2 < 5.0 GeV^2 and 1.08 GeV < W < 3.0 GeV) helps us to better understand the spin structure of the nucleon, especially in the transition region between hadronic and quark-gluon degrees of freedom. The results on A_1, g_1 and Γ_1 using the entire data set for the deuteron data are presented.

Keywords: Spin Structure, Nucleon Spin, Deuteron Spin
PACS: 13.60.Hb

INTRODUCTION

The spin structure of the nucleon is still of growing interest among the scientific community. Double polarized inclusive electron scattering remains one of the most powerful methods used to extract the spin structure of the nucleon from experimental observables. In particular, data at high x and moderate Q^2 can improve our knowledge of nuclear resonance structure and the transition from hadronic to quark degrees of freedom, as well as higher twist matrix elements. The most extensive data set on the deuteron (and the proton) in this kinematic region has become available, on which we report here.

We measured the double spin asymmetry $A_{\|} = (\sigma^+ - \sigma^-)/(\sigma^+ + \sigma^-)$ where $\sigma^+(\sigma^-)$ is the cross section for beam and target polarization antiparallel (parallel) to each other. This asymmetry is directly related to the virtual photon asymmetries A_1 and A_2 as $A_{\|} = D(A_1 + \eta A_2)$ where $D = (1 - E'\varepsilon/E)/(1 + \varepsilon R)$ is the polarization transfer of the virtual photon, $\eta = Q\varepsilon/(E - E'\varepsilon)$, ε is the virtual photon polarization parameter and $R = \sigma_L/\sigma_T$ is the ratio of longitudinal to transverse cross section. A_1 measures the asymmetry of virtual photon absorption leading to final state helicity 1/2 minus 3/2 and contains information on the quark helicity in a polarized nucleon in the scaling region. A_2 is the longitudinal-transverse interference asymmetry and was parametrized for our extraction of A_1. The virtual photon asymmetries are connected to the spin structure function g_1 of the nucleon by:

$$g_1(x, Q^2) = \frac{F_1(x, Q^2)}{1 + \gamma^2}(A_1 + \gamma A_2) \tag{1}$$

CP1149, *Spin Physics, 18th International Spin Physics Symposium*
edited by D. G. Crabb, D. B. Day, S. Liuti, X. Zheng, M. Poelker, and Y. Prok

where F_1 is the unpolarized structure function, $\gamma^2 = Q^2/\nu^2$ in which ν represents the energy transfer of the virtual photon. We used world data parametrization for F_1 and A_2. The first moment of g_1 is also a particularly interesting quantity to calculate:

$$\Gamma_1(Q^2) = \int_0^1 dx g_1(x, Q^2) \tag{2}$$

As Q^2 approaches infinity, the first moment is closely related to the spin carried by the quarks and is constrained by the Bjorken Sum Rule [1]. As Q^2 approaches zero, at the real photon point, it is constrained by the GDH sum rule [2]. The dramatic change of sign of Γ_1 from DIS-regime to the value at the real photon point draws our attention to the Q^2 evolution of the first moment of g_1. In the scaling region, we can compare Γ_1 with theoretical predictions from the Operator Product Expansion (OPE) while at small Q^2 we can test the Chiral Perturbation Theory (ChPT) [5].

THE EG1B EXPERIMENT AND RESULTS

During the EG1B experiment, longitudinally polarized electrons at energies of 1.6, 2.5, 4.2 and 5.75 GeV were scattered from longitudinally polarized NH_3 and ND_3 targets. The double spin asymmetry $A_{\|}$ for the proton and the deuteron has been extracted from the data as a function of W and Q^2 with unprecedented precision. The range of Q^2 covered was 0.05 to 4.5 GeV^2. Electron scattering events with final hadronic state masses up to $W = 3$ GeV were observed.

The experiment has been carried out at Jefferson Lab's Hall B by using the CEBAF Large Acceptance Spectrometer (CLAS). During the experiment, dynamically polarized NH_3 and ND_3 targets were used. The NH_3 polarization was 65-75% while the ND_3 polarization was about 25-35%. The targets were kept in a 1 K liquid helium bath. Also ^{12}C, ^{15}N and empty targets were used to measure the background contribution to the polarized events. The raw count rate asymmetry was extracted first, $A_{raw} = (N^+ - N^-)/(N^+ + N^-)$, where $N^-(N^+)$ is the scattered electron count normalized to the live time gated beam charge when electron and target spins are parallel (antiparallel). A_{raw} was measured in small Q^2 and W bins in order to determine its kinematical evolution. Later, corrections were applied to account for background contributions, radiative effects and total beam and target polarization. The results of these corrections yield the longitudinal double spin asymmetry: $A_{\|} = A_{raw}/(F_D P_b P_t f_{RC}) + A_{RC}$, where F_D is the dilution factor to correct for unpolarized background scattering, f_{RC} and A_{RC} are terms for radiative corrections and $P_b P_t$ is the beam times target polarization. The virtual photon asymmetry A_1 and the spin structure function g_1 were calculated from these measurements by using the world data parametrization for the virtual photon asymmetry A_2 and the unpolarized structure functions F_1 and R.

Preliminary results are shown in Fig. 1 for the structure function g_1 on the deuteron versus the Bjorken scaling variable x for selected kinematic bins. Data for the proton also covers the same kinematical region with even better statistical errors. The structure function g_1 is strongly affected by the resonance contribution at small four-momentum transfers. The most prominent contribution comes from the $\Delta(1232)$ resonance.

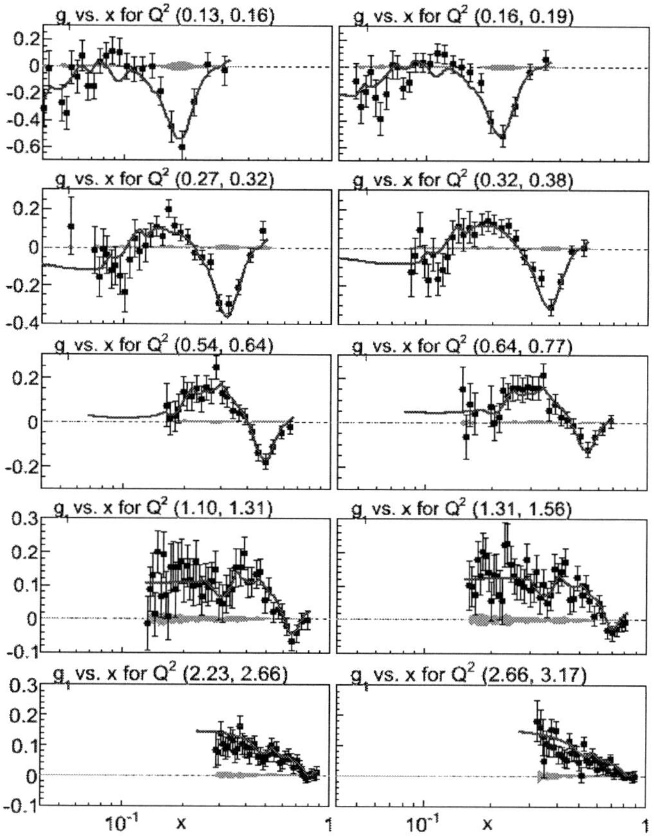

FIGURE 1. Preliminary results on the g_1 structure function of the deuteron are shown versus the Bjorken x (in log scale) for various Q^2 bins (in GeV^2). The plots shown above represent 40% of the Q^2 bins available from the data. Systematic errors are shown as bands at the zero line. Parametrized models are also shown as lines on the plots. We are currently improving our parametrization by using available world data. g_1 for the proton is also available from the EG1B data with even better statistical errors.

Figure 2 shows the preliminary results for Γ_1 on the deuteron and some of the world data currently available. Other phenomenological and theoretical predictions are also drawn on the figure for comparison [3] [4] [5]. We observe that the Burkert-Ioffe [3] parametrization describes the data well. The chiral perturbation models only follow data below $Q^2 \approx 0.06$ GeV^2. The contribution of the EG1B experiment to the world knowledge on the Q^2 evolution of Γ_1 is significant. The recent EG4 experiment will map the small Q^2 region for a better comparison with predictions of ChPT calculations. The results shown here represent the full statistics of the EG1B data. 1.6 and 5.7 GeV data from the experiment have been analyzed before and the results were published [6] [7]. The current analysis includes all beam energies (1.6. 2.5, 4.2, 5.7 GeV). With the full statistics of the experiment, the error bars especially in the medium Q^2 region are

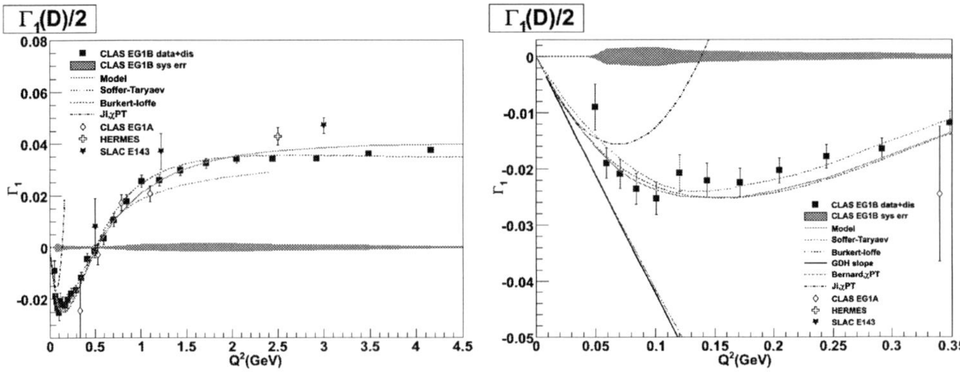

FIGURE 2. Preliminary results are shown for Γ_1 of the deuteron (per nucleon) as a function of Q^2. Some world data available from different experiments are also plotted. The left plot is for the full Q^2 range covered by the EG1B experiment. The right plot shows the behavior at small Q^2. The shaded band around the zero line represents the systematic error. Phenomenological predictions (Burkert-Ioffe, Soffer-Teryaev, ChPT) and the GDH slope are also shown.

reduced significantly. This will enable us to get more precise information on quark-hadron duality in the polarized structure functions and to extract higher-twist matrix elements from the moments of g_1.

CONCLUSION AND OUTLOOK

The EG1 data set has been reanalyzed with full statistics. While 1.6 and 5.7 GeV data have been available before, the addition of the remaining data sets to the analysis extends our kinematic coverage and reduces the errors on the measured quantities significantly. It is observed that the g_1 structure function is strongly affected by the resonance contributions at small Q^2. The EG1B contribution to Γ_1 in the medium Q^2 region is improved. The agreement between the results of the old analysis and the new one is a satisfying confirmation of the validity of our analysis methods. Both the proton and the deuteron data were analyzed in parallel. Our future goal is to extract the spin structure functions of the neutron by using the available data from the EG1B experiment.

REFERENCES

1. J.D. Bjorken *et al.*, *Phys. Rev.* **148**, 1467 (1966).
2. S.D. Drell and A.C. Hearn, *Phys. Rev. Lett.* **16**, 908 (1966).
3. V.D. Burkert and B.L. Ioffe, *Phys. Lett. B* **296**, 223 (1992).
4. J. Soffer and O.V. Teryaev, *Phys. Lett. B* **545**, 323 (2002).
5. X. Ji *et al.*, *Phys. Lett. B* **472**, 1 (2000).
6. K.V. Dharmawardane *et al.*, *Phys. Lett. B* **641**, 11 (2006).
7. Y. Prok *et al.*, *Phys. Lett. B* **672**, 12 (2009).

The Neutron and ^{3}He Spin Structure Functions at Low Q^2

Vincent A. Sulkosky

12000 Jefferson Avenue, Suite #4, Newport News, VA 23606

Abstract. Experiment E97-110 was performed at the Thomas Jefferson National Accelerator Facility in Hall A to provide a precise measurement of the moments of the neutron and ^{3}He spin structure functions. A longitudinally-polarized electron beam was scattered from a longitudinally or transversely polarized ^{3}He target. The extended Gerasimov-Drell-Hearn integral and other moments of the neutron and ^{3}He spin structure functions were extracted at very low momentum transfers ($0.02 < Q^2 < 0.3$ [GeV$/c]^2$). These data allow us to make a benchmark check of Chiral Perturbation Theory calculations in a region where they are expected to be valid. In these proceedings, the experimental details are discussed and preliminary results on the moments of the spin structure functions are presented.

Keywords: neutron,spin structure,chiral perturbation theory
PACS: 14.20.Dh

INTRODUCTION

Over the past few decades, the study of nucleon spin structure has been an active field, both experimentally and theoretically [1, 2]. Spin-dependent sum rules such as the Gerasimov-Drell-Hearn (GDH) [3] sum rule are important tools available to study nucleon spin structure. The GDH sum rule was originally derived for real photon absorption but was extended to virtual photon absorption [4]. The extension by Ji and Osborne [5] forms a Q^2-dependent sum rule. The relationship between the forward Compton scattering amplitudes and the spin-dependent structure functions $g_1(x,Q^2)$ and $g_2(x,Q^2)$ serves as a natural extension of the sum rule to virtual-photon scattering. This unique relation, valid at any four-momentum transfer (Q^2), can be used to make comparisons between theoretical calculations and experimental data, where they are available. This comparison allows an examination of the transition between the non-perturbative region and the perturbative region of Quantum Chromodynamics (QCD). In particular, low-energy effective field theories such as Chiral Perturbation Theory provide calculations for sum rules in the non-perturbative regime of QCD.

THE EXPERIMENT

Experiment E97-110 [6] acquired data at Jefferson Lab in Hall A [7]. The data were collected using seven incident electron beam energies and two angles to obtain the necessary kinematic coverage to extract the moments at constant Q^2. The experiment covered an invariant mass range including the elastic, quasi-elastic and resonance regions up to

CP1149, *Spin Physics, 18th International Spin Physics Symposium*
edited by D. G. Crabb, D. B. Day, S. Liuti, X. Zheng, M. Poelker, and Y. Prok

an invariant mass $W = 2.5$ GeV. The scattered electrons were detected by one of the Hall A High Resolution Spectrometers [7] with the addition of a horizontally-bending dipole magnet [8] to achieve scattering angles of 6° and 9°; without this magnet the minimum spectrometer angle is 12.5°. For this experiment, we used a polarized ^{3}He target as an effective polarized neutron target.

A longitudinally polarized electron beam was scattered from a longitudinally or transversely (in-plane) polarized ^{3}He target [7] to measure the inclusive reaction $^3\vec{\mathrm{He}}(\vec{\mathrm{e}},\mathrm{e}')$. The electron beam was used with a current of up to 10 μA and had an average polarization of $\sim$ 75%. The ^{3}He nuclei were polarized by spin-exchange optical pumping with rubidium atoms and were oriented in either the longitudinal or transverse direction (with respect to the beam direction) with the aid of two sets of Helmholtz coils. Two independent polarimetries [7] were used to monitor the polarization: NMR and EPR. The average target polarization was $\sim$ 40%.

By measuring both target polarization orientations, the spin structure functions $g_1(x,Q^2)$ and $g_2(x,Q^2)$ were extracted from the measured polarized cross-section differences without the need for models. The first moments of the ^{3}He polarized structure functions were then formed. The first moment of the g_1 structure function is given by

$$\overline{\Gamma}_1(Q^2) = \int_0^{x_0} g_1\left(x,Q^2\right) dx, \tag{1}$$

where x_0 is the inelastic threshold. This integral is related to the GDH sum rule as $Q^2 \to 0$ [9]. The Burkhardt-Cottingham (BC) sum rule [10], which is uniquely zero for any Q^2, is related to the first moment of g_2:

$$\Gamma_2(Q^2) = \int_0^1 g_2\left(x,Q^2\right) dx = 0. \tag{2}$$

The neutron moments were extracted from the measured ^{3}He quantities by using the prescription described in [12]. At $Q^2 \leq 0.1$ [GeV/c]^2, the systematic uncertainty due to this extraction was estimated to be 10%. In addition, we have used the parameterization in [11] to account for the unmeasured DIS contribution.

PRELIMINARY RESULTS

The left-hand side of Fig. 1 shows the preliminary results for the integral of $g_1\left(x,Q^2\right)$ compared to theoretical calculations and previous data [13, 14, 15, 16]. The solid triangles represent the new data, and the inverted triangles show the results from a previous neutron experiment [16] in Hall A for $0.1 \leq Q^2 \leq 0.9$ [GeV/c]^2. The error bars on the points include both systematic and statistical uncertainties. For the final results, we expect to significantly improve the (largely dominant) systematic uncertainties. The preliminary results agree well with the higher Q^2 data and also with the two model calculations [17, 18]. As $Q^2 \to 0$, the results indicate a smooth transition toward zero. Near $Q^2 = 0$, two calculations from Chiral Perturbation Theory are available [19, 20]. One of the calculations includes an estimate of the Δ-resonance contribution and is

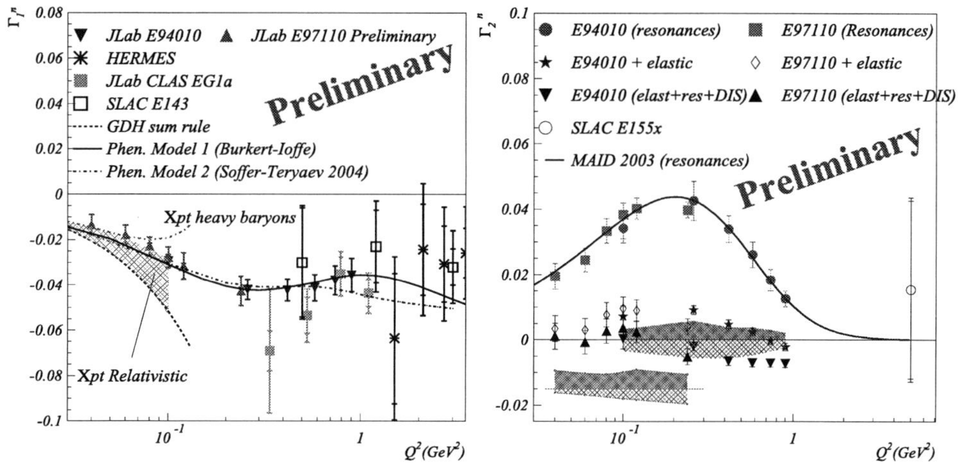

FIGURE 1. Preliminary neutron Γ_1 (left) and Γ_2 (right) results (see text for details)

represented by the band at low Q^2. The band is the result of the uncertainties on the parameterization of the Δ resonance. The data agree with the relativistic baryon calculation [19] within the experimental uncertainties, and the three lowest-Q^2 points agree well with the heavy baryon calculation [20].

The integral Γ_2^n is shown in the right-hand side of Fig. 1 for the measured region (solid squares). The open diamonds represent the integral after inclusion of the elastic contribution. The solid triangles represent the full integral with an estimate of the DIS contribution assuming $g_2 = g_2^{WW}$, where g_2^{WW} is the twist-2 part of g_2 derived by Wandzura and Wilczek [21]. The preliminary data agree well with the earlier neutron results [16], and the resonance data follow the MAID model [22]. The band centered near -0.015 represents the systematic errors. The darker band shows the total systematic uncertainty, while the lighter band shows the error for the unmeasured low-x part. The results, including the elastic and DIS contributions, are consistent with the BC sum rule over the measured Q^2 region. Also on the plot is a high-Q^2 point from the SLAC E155x collaboration [23], which is also consistent with zero but with a large uncertainty.

SUMMARY

Experiment E97-110 was conducted to provide precise spin structure data for the neutron and ^{3}He at low Q^2 between 0.02 and 0.3 [GeV/c]^2. From the measured structure functions, the generalized GDH integral and other moments of the spin structure functions are extracted. At low Q^2, these quantities provide an important benchmark check of Chiral Perturbation Theory calculations. Here, the preliminary results on the integrals of $g_1\left(x,Q^2\right)$ and $g_2\left(x,Q^2\right)$ have been presented. The $\overline{\Gamma}_1(Q^2)$ integral shows good agreement with published data at higher Q^2 and the relativistic baryon calculation, whereas only the three lowest-Q^2 points agree with the heavy baryon calculation. The integral Γ_2^n is consistent with the BC sum rule prediction. Systematic checks are ongoing, and final neutron results are expected soon. In the near future, the other moments of the spin structure functions, including the generalized forward spin polarizabilities [24] will also be extracted.

ACKNOWLEDGMENTS

This work was supported by the U.S. Department of Energy contract DE-AC05-84ER40150, Modification No. M175, under which the Southeastern Universities Research Association operates the Thomas Jefferson National Accelerator Facility.

REFERENCES

1. B. W. Filippone and X. Ji, *Adv. Nucl. Phys.* **26**, 1 (2001).
2. S. E. Kuhn, J.-P. Chen and E. Leader, arXiv:08123535 (2008).
3. S. B. Gerasimov, *Sov. J. Nucl. Phys.* **2**, 430 (1966);
   S. D. Drell and A. C. Hearn, *Phys. Rev. Lett.* **16**, 908 (1966).
4. M. Anselmino, B. L. Ioffe and E. Leader, *Sov. J. Nucl. Phys.* **49**, 136 (1989).
5. X. Ji and J. Osborne, *J. Phys. G* **27**, 127 (2001).
6. For details see V. Sulkosky's thesis at `http://hallaweb.jlab.org/experiment/E97-110/thesis.html`.
7. J. Alcorn *et al.*, *Nucl. Instrum. Meth. A* **522**, 294 (2004).
8. P. Brindza *et al.*, IEEE Transactions on Applied Superconductivity **11**, 1594 (2001).
9. D. Drechsel and L. Tiator, *Annu. Rev. Nucl. Part. Sci.* **54**, 69 (2004).
10. H. Burkhardt and W. N. Cottingham, *Ann. Phys.* **56**, 453 (1970).
11. N. Bianchi and E. Thomas, *Nucl. Phys. B* (*Proc. Suppl.*) **82**, 256 (2000).
12. C. Ciofi degli Atti and S. Scopetta, *Nucl. Phys. B* **404**, 223 (1997).
13. P. L. Anthony *et al.*, *Phys. Rev. D* **54**, 6620 (1996);
    K. Abe *et al.*, *Phys. Rev. D* **58**, 112003 (1998);
    K. Abe *et al.*, *Phys. Rev. Lett.* **79**, 26 (1997).
14. A. Airapetian *et al.*, *Eur. Phys. J. C* **26**, 527 (2003).
15. J. Yun *et al.*, *Phys. Rev. C* **67**, 055204 (2003).
16. M. Amarian *et al.*, *Phys. Rev. Lett.* **89**, 242301 (2002);
    M. Amarian *et al.*, *Phys. Rev. Lett.* **92**, 022301 (2004);
    M. Amarian *et al.*, *Phys. Rev. Lett.* **93**, 152301 (2004).
17. V. D. Burkert and B. L. Ioffe, *Phys. Lett. B* **296**, 223 (1992).
18. J. Soffer and O. V. Teryaev, *Phys. Rev. D* **70**, 116004 (2004).
19. V. Bernard, T. R. Hemmert and U.-G. Meissner, *Phys. Lett. B* **545**, 105 (2002);
    V. Bernard, T. R. Hemmert and U.-G. Meissner, *Phys. Rev. D* **67**, 076008 (2003).
20. X. Ji, C. Kao and J. Osborne, *Phys. Lett. B* **472**, 1(2000).
21. S. Wandzura and F. Wilczek, *Phys. Lett. B* **72**, 195 (1977).
22. D. Drechsel, S. S. Kamalov and L. Tiator, *Phys. Rev. D* **63**, 114010 (2001).
23. P. L. Anthony *et al.*, *Phys. Lett. B* **553**, 18 (2003).
24. J.-P. Chen, A. Deur and Z.-E. Meziani, *Mod. Phys. Lett. A* **20**, 2745 (2005).

Global QCD Analysis of Polarized Parton Densities

Marco Stratmann

Institut für Theoretische Physik, Universität Regensburg, 93040 Regensburg, Germany
Institut für Theoretische Physik und Astrophysik, Universität Würzburg, 97074 Würzburg, Germany
Radiation Laboratory, RIKEN, 2-1 Hirosawa, Wako, Saitama 351-0198, Japan

Abstract. We focus on some highlights of a recent, first global Quantum Chromodynamics (QCD) analysis of the helicity parton distributions of the nucleon, mainly the evidence for a rather small gluon polarization over a limited region of momentum fraction and for interesting flavor patterns in the polarized sea. It is examined how the various sets of data obtained in inclusive and semi-inclusive deep inelastic scattering and polarized proton-proton collisions help to constrain different aspects of the quark, antiquark, and gluon helicity distributions. Uncertainty estimates are performed using both the robust Lagrange multiplier technique and the standard Hessian approach.

Keywords: Parton Densities, Spin Structure, Global QCD Analysis, PDF Uncertainties
PACS: 13.88.+e, 12.38.Bx, 13.60.Hb, 13.85.Ni

INTRODUCTION

The structure of a nucleon in a helicity eigenstate is foremost described by the (anti)quark and gluon helicity parton distribution functions (PDFs), defined by

$$\Delta f(x,Q^2) \equiv f^+(x,Q^2) - f^-(x,Q^2). \tag{1}$$

Here, $f^+(x,Q^2)$ $[f^-(x,Q^2)]$ denotes the distribution of a parton of type f with positive [negative] helicity in a nucleon with positive helicity, having light-cone momentum fraction x of the nucleon momentum and being probed at a hard scale Q. The integral $\Delta f^1(Q^2) \equiv \int_0^1 \Delta f(x,Q^2)dx$ measures the spin contribution of parton f to the proton spin, which is one reason why there are world-wide efforts to extract the $\Delta f(x,Q^2)$ from experimental data [1].

The only way to effectively deconvolute the experimental information on PDFs, which in its raw form is smeared over x, summed over many different partonic subprocesses, and is taken at different Q for each data point, is a "global Quantum Chromodynamics (QCD) analysis". It treats all available probes simultaneously to extract the set of universal PDFs that yields the optimum theoretical description of the combined data. A first such global QCD analysis of the nucleon spin structure at next-to-leading (NLO) accuracy was completed only very recently [2] based on the world-data on polarized deep inelastic scattering (DIS), semi-inclusive DIS (SIDIS), and proton-proton (pp) collisions at BNL-RHIC. The data sets are complementary in the sense that they probe different aspects of the helicity dependent PDFs. Fully inclusive DIS data from many different experiments are pivotal in precisely determining the sum of quarks and antiquarks, SIDIS

CP1149, *Spin Physics, 18th International Spin Physics Symposium*
edited by D. G. Crabb, D. B. Day, S. Liuti, X. Zheng, M. Poelker, and Y. Prok

data help to tell different quark flavors and quark and antiquarks apart, and RHIC pp data give a first direct constraint on the gluon polarization in a certain range of x.

The best set of parameters in the multi-dimensional space defining the PDFs is determined by minimizing an effective χ^2 function that combines all phenomenological inputs of the analysis as a quantitative measure of the goodness of the global fit. Next, a proper assessment of PDF uncertainties and how they propagate to other observables of interest is required, which is a rather formidable task in practice. The specific challenge of a global QCD analysis is to incorporate a large body of data from many experiments with diverse characteristics and errors. The complications are compounded by uncertainties inherent to the theoretical framework used to describe the data which are notoriously difficult to quantify, like from the choice of the factorization scale or the functional form used to parameterize the PDFs.

The standard Hessian method to explore the dependence of χ^2 near its global minimum is based on the not necessarily adequate assumption of a quadratic behavior in the displacements of all parameters from their best fit values [3]. The much more robust method of Lagrange multipliers is free of such assumptions. Here, the idea is to explore directly how the fit to data deteriorates if one enforces certain values for an observable away from its best fit value [3]. The Lagrange multiplier technique provides necessary benchmarks for testing the accuracy and applicability of the Hessian method.

A global fit and the determination of PDF uncertainties requires an extensive number of calculations and minimizations of the effective χ^2 function to explore the very complex and entangled sensitivity of the data to variations in the parameters describing the PDFs. This calls for new and more efficient calculational tools to include all observables consistently at NLO accuracy without resorting to potentially unreliable approximations. We have devised a novel Monte-Carlo sampling technique [4] which supplements in an ideal way analysis tools based on the Mellin transform technique proposed in [5] and speeds up all relevant NLO calculations to a level that they can be incorporated in the global analysis. The new computational strategy is completely general and applies also to less inclusive observables like, e.g., two-particle correlations in hadronic collisions.

RESULTS OF THE GLOBAL QCD ANALYSIS

In the DSSV analysis [2] we propose the following versatile parametric form

$$x\Delta f(x,Q_0^2) = N_f x^{\alpha_f}(1-x)^{\beta_f}(1+\gamma_f\sqrt{x}+\eta_f x)\,, \tag{2}$$

with independent parameters for $\Delta u+\Delta\bar{u}$, $\Delta d+\Delta\bar{d}$, $\Delta\bar{u}$, $\Delta\bar{d}$, $\Delta\bar{s}=\Delta s$, and Δg at the input scale $Q_0 = 1\,\mathrm{GeV}$ for the Q^2 evolution. The ansatz (2) is flexible enough to accommodate all hard scattering data within their range of uncertainties, allows the PDFs to develop nodes, and to deviate from an SU(3) flavor symmetric sea.

Rather than imposing SU(2) and SU(3) flavor symmetry *assumptions* we allow for variations in our fit, i.e.,

$$\Delta\mathscr{U} - \Delta\mathscr{D} = (F+D)[1+\varepsilon_{\mathrm{SU(2)}}]\,, \tag{3}$$

$$\Delta\mathscr{U} + \Delta\mathscr{D} - 2\Delta\mathscr{S} = (3F-D)[1+\varepsilon_{\mathrm{SU(3)}}]\,, \tag{4}$$

where $\Delta\mathscr{F} \equiv \left[\Delta f^1 + \Delta \bar{f}^1\right](Q_0^2)$. In practice, we trade the input parameters $N_{u+\bar{u}}$ and $N_{d+\bar{d}}$ in Eq. (2) for $\varepsilon_{\text{SU}(2,3)}$ and fit the latter using $F+D = 1.269 \pm 0.003$ and $3F-D = 0.586 \pm 0.031$, including the experimental bounds in the χ^2 minimization. For the optimum DSSV fit we find only very small violations of the canonical constraints:

$$\varepsilon_{\text{SU}(2)} = 0.0011\,,\ \varepsilon_{\text{SU}(3)} = -0.0035\,. \tag{5}$$

Since the currently available data do not fully constrain the entire x dependence of Δf imposed in Eq. (2), we are forced to make some assumptions on the parameter space. For the sea quark and gluon densities we set $\gamma_f = 0$ in Eq. (2). This only marginally limits the freedom in the functional form and still allows nodes. In addition, the small x behavior, i.e., α_f in (2), of $\Delta u + \Delta\bar{u}$ and $\Delta d + \Delta\bar{d}$ is tied to that of $\Delta\bar{u}$ and $\Delta\bar{d}$, respectively, which is very reasonable as sea quarks dominate in this region. Since the SIDIS data are not yet sufficient to distinguish Δs from $\Delta\bar{s}$, we assume $\Delta s = \Delta\bar{s}$. The parameter $\alpha_{\bar{s}}$ is always close to $\alpha_{\bar{d}}$, so we set them equal. In general, the $x \to 1$ behavior of all PDFs is rather unconstrained, as is the case even for current sets of unpolarized PDFs. To avoid problems with the positivity constraint $|d\Delta\sigma| \leq d\sigma$, we make sure that $(1-x)^{\beta_f}$ for all Δf vanishes faster than for some typical reference set of unpolarized PDFs, for which we take the NLO $\overline{\text{MS}}$ set of MRST [6].

The total χ^2 of the best fit is 392.6 for 467 data points used in our NLO global analysis [2]. All data sets can be very satisfactorily described by a *universal* set of polarized PDFs which validates the notion of factorization assumed in pQCD calculations. We have explored to no avail the possibility of normalization shifts for individual data set to further improve the quality of the global fit. We make use of the uncertainty estimates for fragmentation functions (FFs) from the DSS analysis [7] and propagate them to all spin asymmetries with detected hadrons analyzed in our global fit. The estimates are included as a theoretical error in the χ^2 computation. Since FFs are fairly well determined [7] this amounts to additional uncertainties of a few percent at most.

An example of the excellent theoretical description of the data within the DSSV global analysis is given in Fig. 1 where we compare to the SIDIS world-data. The fit of inclusive DIS data is not significantly modified as compared to all previous NLO analyses. Apart from purely kinematic corrections to relate A_1 and the fitted structure function ratio g_1/F_1 in DIS, no genuine higher twist corrections are needed, and both g_1 and F_1 are computed with leading twist, NLO QCD expressions. Nevertheless, power corrections may be individually significant in g_1 and F_1 at high x and low Q^2. Comparisons to RHIC pp data for single-inclusive pion and jet production can be found in [2].

The results in Fig. 1 also demonstrate the significance of using the recent DSS FFs [7] in our global analysis. We recall that DSS is the only set of FFs which gives a satisfactory description of all corresponding unpolarized multiplicities or cross sections at the same kinematics relevant for the fitted spin asymmetries. This is of particular importance for the description of the SIDIS data with identified kaons, which in turn has a crucial impact on the obtained strangeness density Δs as we shall see next.

Figure 2 shows the individual polarized PDFs at $Q^2 = 10\,\text{GeV}^2$ as obtained in the DSSV global fit. Also shown are uncertainty estimates based on Lagrange multipliers and an improved, iterative Hessian method [3], both corresponding to an increase in χ^2 by one unit. We note that due to the complex nature of a global QCD analysis, the

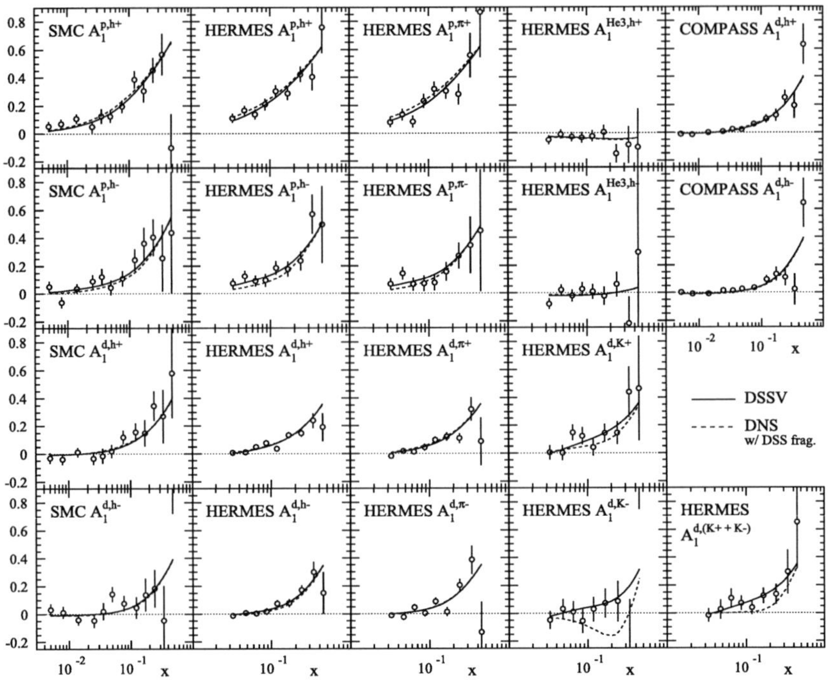

FIGURE 1. Comparison of the SIDIS spin asymmetries included in the global analysis with the DSSV best fit (solid lines). Also shown (dashed lines) are the results obtained with the old DNS fit [8].

statistical significance of a certain increase $\Delta\chi^2$ is debatable and requires careful studies. As can be seen, both methods to estimate PDF uncertainties in Fig. 2 agree fairly well for $\Delta\chi^2 = 1$ and can be used simultaneously. We have verified that this is the case also for other observables like truncated first moments or spin asymmetries [4]. For a larger increase $\Delta\chi^2$, the Hessian method becomes unreliable as non-quadratic terms in the functional dependence of χ^2 on the fit parameters are more pronounced.

Figure 2 already reveals the main results of the DSSV analysis: the gluon distribution $\Delta g(x)$ is rather small with a possible node in the x range covered by data; the sea polarizations clearly deviate from SU(3) symmetry with $\Delta\bar{u}(x) - \Delta\bar{d}(x) > 0$; $\Delta\bar{s}(x) > 0$ for $x \gtrsim 0.02$ contrary to previous fits. To understand these finding better and to identify the role of the individual data sets in constraining the PDFs, we analyze the partial contributions to $\Delta\chi^2$ from the different observables. More specifically, we apply the Lagrange multiplier technique to truncated first moments of PDFs, defined as

$$\Delta f^{1,[x_{\min}\to x_{\max}]}(Q^2) \equiv \int_{x_{\min}}^{x_{\max}} \Delta f(x,Q^2)dx. \qquad (6)$$

Such studies also test the overall consistency of the analysis and the data sets used.

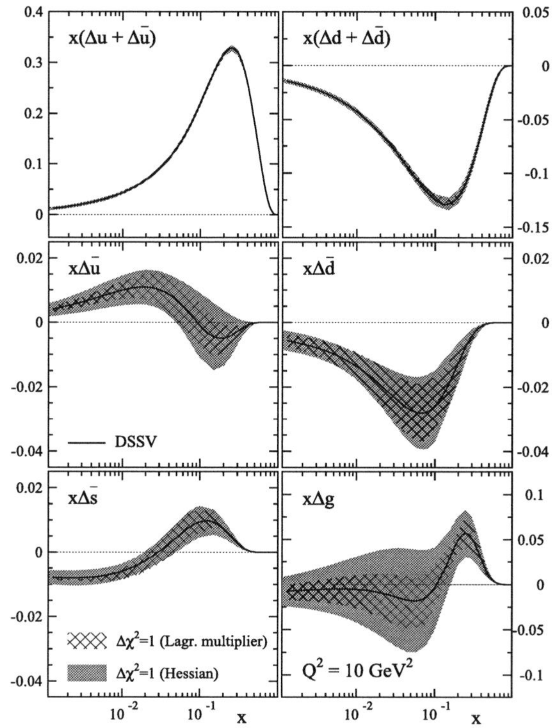

FIGURE 2. Helicity-dependent PDFs from the DSSV analysis and uncertainty estimates obtained with the Lagrange multiplier technique (hatched bands) and the Hessian method (solid bands).

The small gluon polarization is mainly driven by the combined RHIC *pp* data from PHENIX and STAR, which put a strong constraint on the size of Δg in the range $0.05 \lesssim x \lesssim 0.2$ as can be seen from the variations of $\Delta g^{1,[0.05\to 0.2]}$ in Fig. 3. We also observe a nice synergy of the different data sets in determining Δg. The resulting truncated moment is compatible with zero, with uncertainties depending on the tolerated $\Delta\chi^2$, see [2, 4]. The small x region remains still largely unconstrained, making statements about the entire first moment Δg^1, which enters the proton spin sum rule, not yet possible.

The amount of strange quark polarization has been a focus of interest since the very beginning of the so-called spin crisis, with a sizable negative Δs as a possible remedy. Indeed, fits to only DIS data have always preferred a sizable negative Δs, presumably related to symmetry assumptions for the sea polarizations and to a too restricted functional form, not allowing any nodes. At variance with these results, we find a positive polarization at large x. As can be inferred from Fig. 3, the shape of $\Delta\chi^2$ for $\Delta s^{1,[0.001\to 1]}$ is dominated by SIDIS data and here mainly by measurements with an identified kaon. We emphasize again the importance of reliable FFs when analyzing

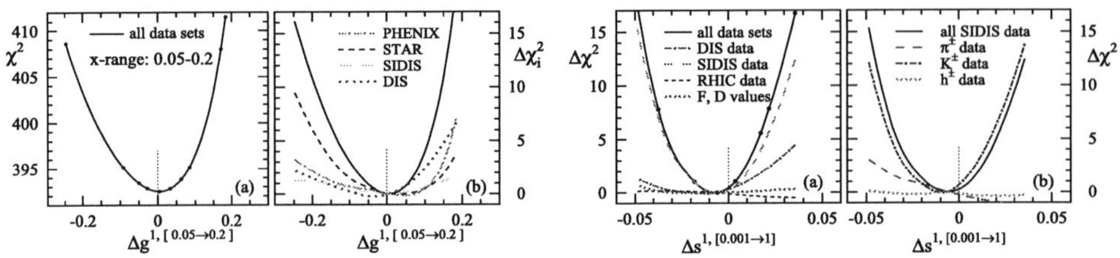

FIGURE 3. Partial contributions to $\Delta\chi^2$ of the different data sets for variations of $\Delta g^{1,[0.05\to0.2]}$ (left panel) and $\Delta s^{1,[0.001\to1]}$ (right panel) obtained with the Lagrange multiplier technique.

SIDIS data. Since we find that the SU(2) and SU(3) breaking parameters are small, see Eq. (5), the full first moment Δs^1 must be negative, which explains why $\Delta s(x)$ in Fig. 2 turns negative at small x, gaining most of its area Δs^1 there. The novel behavior of $\Delta s(x)$ also leaves its imprint on the quark singlet $\Delta\Sigma$, with a significantly higher value for $\Delta\Sigma^{1,[0.001\to1]} = 0.366$ than in previous fits of only DIS data. From Fig. 3 one can easily read off the uncertainties for the truncated moments for different $\Delta\chi^2$ values tolerated. More details and results can be found in Ref. [4].

The DSSV analysis and the estimates of PDF uncertainties will be continuously refined and updated, and new data sets can be straightforwardly included at NLO accuracy once they become available. In the near future, one can expect important new results from SIDIS measurements at COMPASS, for single-inclusive spin asymmetries with considerably improved statistics from RHIC, and for two-jet correlations from STAR. The latter will allow to map $\Delta g(x)$ much more precisely. The W boson program at RHIC will provide a crucial independent determination of Δu, $\Delta\bar{u}$, Δd, and $\Delta\bar{d}$ in the future.

ACKNOWLEDGMENTS

I am grateful to Daniel de Florian, Rodolfo Sassot, and Werner Vogelsang for a fruitful collaboration on the results presented here. Partial financial support from the organizers of SPIN'08 is greatly appreciated.

REFERENCES

1. See, e.g., M. Stratmann and W. Vogelsang, *J. Phys. Conf. Ser.* **69**, 012035 (2007) and reference therein.
2. D. de Florian, R. Sassot, M. Stratmann, and W. Vogelsang, *Phys. Rev. Lett.* **101**, 072001 (2008).
3. See, e.g., J. Pumplin *et al.*, *Phys. Rev. D* **65**, 014011 (2001); **65**, 014012 (2001); **65**, 014013 (2001).
4. D. de Florian, R. Sassot, M. Stratmann, and W. Vogelsang, to appear.
5. W. Vogelsang and M. Stratmann, *Phys. Rev. D* **64**, 114007 (2001).
6. A.D. Martin, R.G. Roberts, W.J. Stirling, and R.S. Thorne, *Eur. Phys. J. C* **28**, 455 (2003).
7. D. de Florian, R. Sassot, and M. Stratmann, *Phys. Rev. D* **75**, 114010 (2007); **76**, 074033 (2007).
8. D. de Florian, G.A. Navarro, and R. Sassot, *Phys. Rev. D* **D71**, 094018 (2005).

Measurement of the Gluon Polarization $\Delta g/g$ from Open Charm at COMPASS

Fabienne Kunne[a], on behalf of the COMPASS collaboration

[a]*IRFU SPhN, CEA Saclay, F91191 Gif/Yvette, France*

Abstract. We have measured the gluon polarization in the nucleon by detecting charm production via D^0 meson decay to charged K and π in polarized muon scattering off a longitudinally polarized deuteron target. The dominant process for charm production is the photon gluon fusion into a charm anti-charm quark pair. By using all deuteron statistics from COMPASS accumulated between 2002 and 2006, we extract double spin asymmetries in bins of the transverse momentum and the energy of the D^0 meson and we perform a leading order analysis of the data to extract the gluon polarization $<\Delta g/g>=-0.49 \pm 0.27$ (stat) ± 0.11 (syst) at a Quantum Chromodynamics (QCD) scale $\mu^2= 13$ GeV^2 and at a gluon momentum fraction $<x>=0.11$.

Keywords: Spin of the nucleon; Gluon Polarization
PACS: 13.60Hb

INTRODUCTION

The main objective of the COMPASS muon physics program is the study of the nucleon spin, and in particular the measurement of the gluon polarization ΔG and of the quark spin distributions, longitudinal and transverse. The nucleon spin can be decomposed into the contributions from the quarks, the gluons and the orbital angular momentum as $\frac{1}{2} = \frac{1}{2}\Delta\Sigma + \Delta G + L_z$. Predictions from the naive quark parton model as well as from a Quantum Chromodynamics (QCD) approach (neglecting strange quark polarization) lead to a large polarization of the quarks, *i.e.* $\Delta\Sigma$=0.60. On the contrary, measurements from the last decades show that a_0, the singlet axial matrix element related to $\Delta\Sigma$, is small, ~0.30. In fact, in some QCD schemes, $a_0= \Delta\Sigma - (3\alpha_s/2\pi)\,\Delta G$. Thus a large value of ΔG could help restoring $\Delta\Sigma$ ~0.6. This fact, together with the ½ nucleon spin sum rule, has motivated the efforts for direct measurements of ΔG.

THE MEASUREMENT OF $\Delta G/G$

At COMPASS, the direct measurement of the gluon polarization is performed via the double spin asymmetry of photon gluon fusion (PGF) events (Fig.1). PGF events are searched for in two distinct channels: the *'open charm'* channel, discussed here, where the outgoing quarks are charmed quarks (charm-anti-charm pairs), and the *'high p_T hadron pair'* channel, where the two outgoing quarks can be light quarks and

CP1149, *Spin Physics, 18th International Spin Physics Symposium*
edited by D. G. Crabb, D. B. Day, S. Liuti, X. Zheng, M. Poelker, and Y. Prok

hadronize with high transverse momentum p_T. The second channel ('*high* p_T') has high statistics but suffers from physical background and requires dedicated Monte Carlo study to estimate the contributions from the competing processes[1].

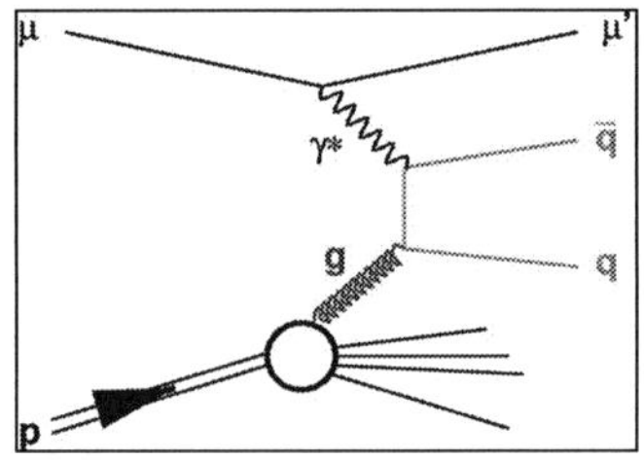

FIGURE 1: Photon Gluon Fusion process.

On the contrary, the *open charm* channel provides a clean but rare signature of the PGF process since there is basically no contamination from other physics channels. In particular there is no intrinsic charm in the nucleon in this kinematical domain. The hard QCD scale is set by the mass of the charm quark. The presence of an outgoing charm quark is signed by the production of a charmed D^0 meson decaying into a $K\pi$ pair. However, the combinatorial background and the limited statistics due to the small charm cross section and the low branching ratio of D^0 decay, makes the experiment difficult. The analysis is therefore based on a weighted approach in order to minimize the statistical error. The double spin asymmetries between PGF cross sections with opposite target polarizations are related to the gluon polarization $\Delta g/g$ through the PGF analyzing power a_{LL}^{PGF} .

The 160 GeV polarized muon beam from CERN, with $2\text{x}10^8$ muons per 4.8s spill and a polarization of 80% is scattered on a longitudinally polarized deuteron target. The target material, ^{6}LiD, has been chosen for its high dilution factor (fraction of polarized nucleons ~ 50%). To reduce false asymmetries, two 60 cm long target cells are polarized in opposite directions to about 50%. The COMPASS spectrometer[2] is a two stage spectrometer for large and small angle tracking, with precise momentum measurement and particle identification including π, K, p separation between 10 and 50 GeV thanks to a RICH detector. Data were taken during four years between 2002 and 2006. Before the 2006 run, several upgrades were performed. A new and larger solenoid magnet was installed, which increased the acceptance from 70 to 180 mrad. A three-cell target replaced the two-cell one, reducing the risk of false asymmetries. Tracking devices were added to cope with the enlarged acceptance. A major upgrade of the RICH detector (photon detectors with multi-anode PMTs in the central area, and a faster read-out system of the chambers at the periphery) improved its performance.

Since the polarized target is thick, the vertex of the D^0 (or anti-D^0) meson cannot be reconstructed, and the event selection is done by calculating the $K\pi$ invariant mass. D^0s are also identified by tagging D^* mesons which decay into D^0 and a soft π. The small mass difference between the D^* and the D^0 provides a strict and very effective selection cut which drastically reduces the background. Figure 2 shows the invariant mass distribution of the $K\pi$ pair for the untagged D^0s *(left)*, and for the D^0s tagged by a D^* *(right)*.

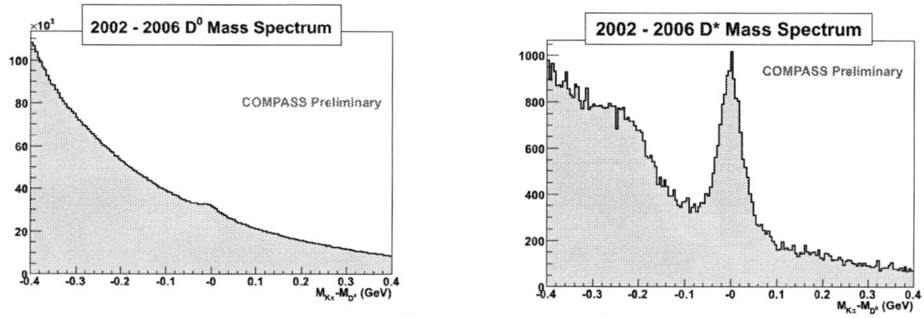

FIGURE 2. Invariant mass spectra for D^0 events *(left)* and "D^0 tagged by D^*" events *(right)*. The additional bump due to D^0 decaying into $K\,\pi\,\pi^0$(right) is not yet used in the present analysis.

The total statistics amounts to ~37400 untagged and ~8700 tagged D^0s events. Both samples of events are independent and contribute about the same weight to the final result, the first one thanks to a higher statistics, and the second to the higher signal to background ratio *S/(S+B)*.

At LO, the experimental spin asymmetry is related to $\Delta g/g$ through the analyzing power a_{LL}^{PGF}, the signal strength $R_{PGF}=S/(S+B)$, various dilution factors (beam and target polarizations, fraction of polarizable nucleons) and the background asymmetry:

$$A_{\exp} = P_B P_T\, f \left[R_{PGF}\, a_{LL}^{PGF} \frac{\Delta g}{g} + (1-R_{PGF}) A_{bkg} \right]$$

In order to account for the large variations of both a_{LL}^{PGF} (varying between -0.6 and +1) and $R_{PGF}=S/(S+B)$ in the phase-space, events are weighted in the analysis by those factors. The gain in statistics from the *S/(S+B)* weighting alone is +45% for D^0 which has the worse *S/(S+B)* ratio, and 15% for D^* tagged events.

RESULTS

The preliminary result for the gluon polarization extracted at LO from the charm channel is $<\Delta g/g>=-0.49 \pm 0.27$ (stat) ± 0.11 (syst) at a QCD scale $\mu^2= 13$ GeV^2 and at an average gluon momentum fraction $<x>=0.11$. In Fig.3 this result is shown together with the two preliminary '*high* p_T' points from COMPASS[1] and previous '*high* p_T' result from SMC[4] and HERMES[5]. Note that the μ^2 QCD scale is not identical for all data. The 3 curves are QCD next-to-leading order (NLO) fits to g_1 from GRSV giving integrals of $\Delta G= 0.2$, 0.6 and 2.4 respectively, at $\mu^2 =3$ (GeV/c)^2. For this assumed shape of $\Delta g(x)$, COMPASS data exclude large values for the integral of ΔG. Note that recent RHIC Spin results on A_{LL} (π^0 or jet production) also favor a small ΔG.

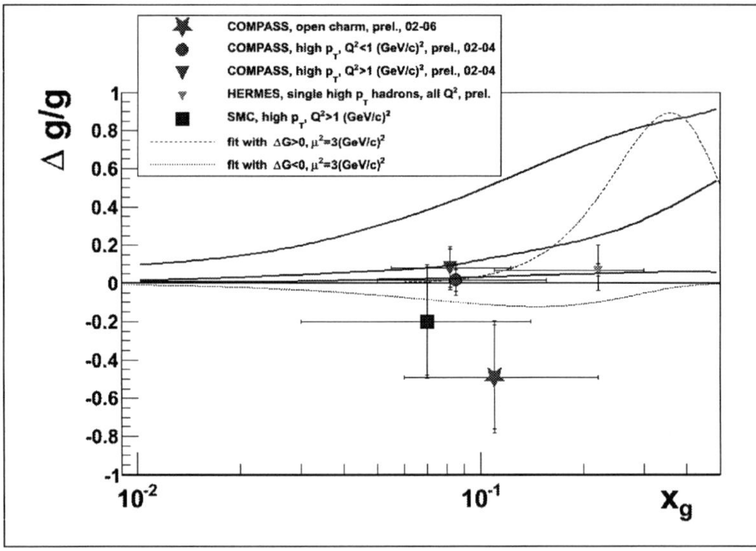

FIGURE 3. Direct measurements of $<\Delta g/g(x)>$ from COMPASS open charm (prel.) [3], high p_T pairs from COMPASS[1], SMC[4] and HERMES[5]. Curves are *g1* NLO fits from GRSVand COMPASS[6]

SUMMARY AND CONCLUSION

We have measured double spin asymmetries for charm D^0 meson production in the scattering of 160 GeV polarized muons off polarized deuterons. A significant improvement in statistics is obtained since our previous release, thanks to the addition of 2006 data (increased acceptance and improved PID) and to the use of *S/B* weighting and improved reconstruction. The model independent asymmetries, extracted in (p_T,E_D) bins, have also been produced and may be used in global NLO QCD analyses to constrain the values of $\Delta g(x)$. A direct extraction of $\Delta g/g$ using a LO approach, at $<x_g> = 0.11$, leads to a negative value, but still less than 2σ away from zero. More statistics will come by adding 2007 proton data and events from $D^0 \rightarrow K\pi\pi^0$ decay.

REFERENCES

1. E.Ageev et al., *Phys. Lett. B* **633** 25 (2006). K.Kurek, theese proceedings, *Δg/g high p_T*
2. P. Abbon *et. al., Nucl. Instr. Meth. A* **577,** 455 (2007).
3. F.Robinet, Ph.D. Thesis, Paris XI, September 2008; and proceedings of DIS2008
4. Adeva et al. *Phys. Rev. D* ***70,*** 012002 (2004).
5. A.Airapetian *et al., Phys. Rev. D* **75,** 012007 (2007) .
6. V. Alexakhin *et al., Phys. Lett. B* **647,** 8 (2007).

The Spin Content of the Proton

F. Myhrer

Dept. Physics & Astronomy, Univ. South Carolina, Columbia, SC 29208

Abstract. Considerable progress has been made in our knowledge of the spin distribution within the proton. The recently measured limits on polarized gluons in the proton suggest polarized gluons contribute very modestly to the proton spin. We will show that a modern, relativistic and chirally symmetric description of the nucleon structure naturally explains the current proton spin data. Most of the "missing" spin is carried by confined quark and antiquarks' angular momentum.

Keywords: Confined relativistic interacting quarks, cloudy bag model
PACS: 14.20.Dh,12.39.Ba

Introduction

During the past two decades there has been a concerted level of activity mapping the distribution of spin (and angular momentum) onto the quarks and gluons that compose the nucleon. This effort was sparked by the discovery by the European Muon Collaboration (EMC) of a proton "spin crisis" [1]. EMC observed that the valence quarks carried only a very small fraction Σ of the proton spin [1]. The published measurement of the fraction $\Sigma \simeq 14 \pm 9 \pm 21\%$, indicated that it could possibly be equal to zero whereas the Ellis-Jaffe sum rule, based on the non-relativistic quark model (NRQM), predicts that $\Sigma = 1$. The unexpectedly small EMC value for Σ generated a tremendous level of theoretical and experimental activity. Theoretically several well-known aspects of hadron structure were explored [2, 3, 4] but none could generate such a small value for Σ. It was however quickly realized that the famous $U(1)$ axial anomaly could strongly influence the value of Σ and that the proton might contain a large quantity of polarized glue, see e.g., Refs. [5, 6, 7, 8] for a mathematically elegant formulation of this possible contribution to the proton spin. In addition, and in contrast to the NRQM treatment, since the u and d quarks in the proton behave relativistically, their angular momenta will contribute to the spin content of the proton. Schematically the proton spin content can be written as

$$\frac{1}{2} = \frac{1}{2}\Sigma + \Delta_{Glue} + L_z$$

The very recent experimental measurements at CERN, DESY, JLAB, RHIC and SLAC have shown that Δ_{Glue} gives (at best) a small contribution to the proton spin [9]. Furthermore, the accuracy of the measured Σ-value has increased and we now know that the sum of the quark helicities in the proton is about 1/3,

$$\Sigma = 0.33 \pm 0.03 \pm 0.05\,,$$

CP1149, *Spin Physics, 18th International Spin Physics Symposium*
edited by D. G. Crabb, D. B. Day, S. Liuti, X. Zheng, M. Poelker, and Y. Prok

which is considerably higher than the initial EMC suggestion. That polarized glue is not the explanation for the spin problem leads us to focus again on the suggestions which were based on physics that is more familiar to those modeling non-perturbative Quantum Chromodynamics (QCD). In particular, we [10] suggest that most of the missing spin of the proton must be carried as orbital angular momentum by the quarks and anti-quarks.

Before explaining the key physics which, when combined, appears to provide a natural explanation of Σ, in more detail, see e.g. Ref. [10], we present a quick summary of these phenomenologically well-established physics "factors". They are:

- The relativistic motion of the confined valence quarks
- The virtual excitation of anti-quarks in low-lying p-states generated by the one-gluon-exchange hyperfine interaction – in nuclear physics terms this is called "an exchange current correction"
- The pion cloud of the nucleon's "quark core".

These three pieces of physics, tested in many independent ways (see below), all have the effect of converting quark spin to orbital angular momentum. As explained in more detail below (and in a recent publication [10]) the first reduces the spin by about one third, the second yields a reduction by an amount of order 0.15, and the third gives a multiplicative reduction by a factor of order 0.7. Combining these physics effects reduces the fraction of the proton spin carried by its quarks to about one third, i.e. $\Sigma = (1-0.35-0.15)\cdot 0.7 = 0.35$. One important recent new observation, based on a chiral analysis of data from lattice QCD [11], suggests very strongly that the pion cloud contributes very little to the Δ-N mass splitting. This new finding provides the justification for combining the corrections to the spin sum arising from one-gluon-exchange to that from the pion cloud. We now present some details of these three major reduction factors, which lead to the small value of Σ.

Relativistic valence quark motion

This effect on Σ was understood at the time of the EMC discovery. A spin-up, very light quark in an s-state, moving in a confined space, has a lower Dirac component in p-wave. The angular momentum coupling is such that for this component the spin is preferably down and reduces the "spin content" of the valence quarks. In the bag model, for example, where the massless quark's ground state energy equals $\Omega/R \simeq 2.043/R$, the reduction factor $B = \Omega/3(\Omega-1) \simeq 0.65$. The same factor reduces the value of g_A from 5/3 to $\simeq 1.09$ in a bag model and this value changes little if one uses typical light quark current masses. (In the discussion section we will briefly indicate how our model leads to a realistic g_A value $\simeq 1.27$.) The quark energy, Ω/R is determined by the bag confinement condition that the quark current out of the spherical bag cavity of radius R is zero, i.e. in Dirac's notation $\hat{r}\cdot\vec{j} = i\hat{r}\cdot\psi^\dagger\vec{\alpha}\psi = 0$ for $r = R$. Even in more modern relativistic models, where quark confinement is simulated by forbidding on-shell propagation through proper-time regularization, the reduction factor is very similar – e.g., in Ref. [12] $\Delta u + \Delta d$ is 0.67. The relativistic motion transfers roughly 35% of the nucleon spin from quark spin to valence quark orbital angular momentum.

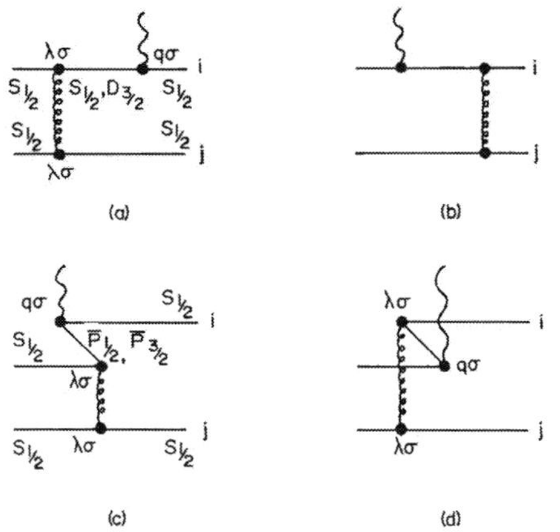

FIGURE 1. Illustration of the quark-quark hyperfine contributions which involve excited intermediate quark states. In the figures the external probe (top vertical wavy line) couples to the i'th quark which interacts with the second j'th quark via the effective confined gluon exchange. The intermediate quark propagator is evaluated as a sum over confined quark modes. Figs. (a) and (b) show the three-quark intermediate states, and (c) and (d) the one anti-quark and four quarks intermediate states. The mode sum converges rapidly and the lowest anti-quark $P_{1/2}$ and $P_{3/2}$ modes dominate the mode-sum [14].

The one-gluon-exchange hyperfine interaction

It is well established that the spin-spin interaction between quarks in a baryon, arising from the exchange of a single gluon, explains a major part of the mass difference between the octet and decuplet baryons – e.g., the nucleon-Δ mass difference [13]. This spin-spin interaction must therefore also play a role when an external probe interacts with the three-quark baryon state. In the context of spin sum rules, the probe couples to the all possible axial currents in the nucleon. *That is, the probe not only senses a single quark current but a two-quark current as well.* The latter has an intermediate confined-quark propagator connecting the vertex of the probe and the spin-spin interaction between two quarks, and is similar to the exchange-current corrections which are well known in nuclear physics. In the case of the two-quark current, investigated in detail in Ref. [14] using the MIT bag model, the confined quark propagator was written as a sum over quark eigenmodes and the dominant contributions were found to come from the intermediate p-wave anti-quark states. The primary focus of Ref. [14] was however the one-gluon-exchange corrections to the magnetic moments and semi-leptonic decays of the baryon octet (see below).

Myhrer and Thomas [2] realized the importance of this correction to the flavor singlet axial charge and hence to the proton spin, finding that it reduced the fraction of the spin of the nucleon carried by quarks, calculated in the naive bag model by 0.15, i.e., $\Sigma \rightarrow \Sigma - 3G$ [2]. The correction term, G, is proportional to α_s times certain bag model matrix elements [14], where α_s is determined by the "bare" nucleon-Δ mass difference. Again, the spin lost by the quarks is compensated by orbital angular momentum of the quarks and anti-quarks (predominantly $\bar{u}$ in the p-wave).

The pion cloud

The pion cloud is an effective description of the quark-antiquark excitations which are required by the spontaneous breaking of chiral symmetry in QCD. In fact, describing a physical nucleon as having a pion cloud which interacts with the valence quarks of the quark core (the "bare" nucleon), in a manner dictated by the requirements of chiral symmetry, has been very successful in describing the properties of the nucleon [15, 16, 17]. The cloudy bag model (CBM) [15, 16] reflects this nucleon description where the nucleon consists of a bare nucleon, $|N>$, with a probability $Z \sim 1 - P_{N\pi} - P_{\Delta\pi} \sim 0.7$, in addition to being described as a nucleon and a pion and a Δ and a pion, with probabilities $P_{N\pi} \sim 0.20 - 0.25$ and $P_{\Delta\pi} \sim 0.05 - 0.10$, respectively. The phenomenological constraints on these probabilities were discussed, e.g. [18, 19].

The pion cloud effect on Σ was investigated early by Schreiber and Thomas [4], who wrote the corrections to the spin sum-rules for the proton and neutron explicitly in terms of the probabilities above. To summarize Ref. [4]: the pion cloud correction to the flavor singlet combination modifies the proton spin in the following manner:

$$\Sigma \to \left(Z - \frac{1}{3} P_{N\pi} + \frac{5}{3} P_{\Delta\pi} \right) \Sigma. \tag{1}$$

From the point of view of the spin problem, the critical feature of the pion cloud is that the coupling of the spin of the nucleon to the orbital angular momentum of the pion in the $N\pi$ Fock state favors a spin down nucleon and a pion with +1 unit of orbital angular momentum. This too has the effect of replacing quark spin by quark and anti-quark orbital angular momentum. Note that in the $\Delta\pi$ Fock component the spin of the baryon tends to point up (and the pion angular momentum down), thus enhancing the quark spin. Nevertheless, the wave function renormalization factor, Z, dominates, yielding a reduction by a factor between 0.7 and 0.8 for the range of probabilities quoted above.

Other "spin observables" affected by these corrections

Some of the quark hyperfine interaction (OGE) and the pion cloud corrections are illustrated in the following two Tables. A brief summary of these corrections are:

The effective OGE (i) This correction is vital in the understanding of the measured strength of $\Sigma^- \to n + e^- + \bar{\nu}_e$, see Table 2.
(ii) Essential to explain the magnetic moment inequality $|\mu_\Lambda| < |\mu_{\Xi^-}|$, Table 1,
(iii) OGE introduces configuration mixing in baryon-octet ground states which can affect strongly the radiative decay of excited baryons to the ground state baryons.

Pion cloud (i) Crucial component of the neutron charge distribution.
(ii) Provides a large iso-vector contribution to the nucleon magnetic moment.
(iii) Gives the leading non-analytic chiral-loop corrections to nucleon observables.

TABLE 1. The baryon magnetic moments from the valence quark, the pion cloud and the OGE contributions as evaluated in [14].

Baryon	Quark	Pion	OGE	Mag mom	PDG − 2002
μ_p	μ_q	$\delta\mu_\pi$	0	2.79	+2.79
μ_n	$-\frac{2}{3}\mu_q$	$-\delta\mu_\pi$	$\frac{2}{3}G'$	−1.92	−1.91
μ_{Σ^+}	$\frac{8}{9}\mu_q+\frac{1}{9}\mu_s$	$\frac{1}{2}\delta\mu_\pi^*$	0	2.46	$+2.458\pm0.010$
μ_{Σ^-}	$-\frac{4}{9}\mu_q+\frac{1}{9}\mu_s$	$-\frac{1}{2}\delta\mu_\pi^*$	$-\frac{2}{3}G'$	−1.20	-1.160 ± 0.025
μ_{Ξ^0}	$-\frac{2}{9}\mu_q-\frac{4}{9}\mu_s$	$\simeq 0$	$\frac{2}{3}G'$	−1.20	-1.250 ± 0.014
μ_{Ξ^-}	$\frac{1}{9}\mu_q-\frac{4}{9}\mu_s$	$\simeq 0$	$-\frac{2}{3}G'$	−0.73	-0.6507 ± 0.0025
μ_Λ	$-\frac{1}{3}\mu_s$	0	$\frac{1}{3}G'$	−0.61	−0.61

TABLE 2. The semi-leptonic decays of some baryons showing only the valence quarks and OGE contributions [14]. Other corrections are implicit.

$B' \to B$	Quark + OGE	PDG − 2002
$n \to p$	$\frac{5}{3}B+G \simeq 1.25$	1.27
$\Sigma^- \to n$	$-\frac{1}{3}B-2G \simeq -0.34$	−0.34
$\Lambda \to p$	$B \simeq 0.72$	0.72
$\Xi^- \to \Lambda$	$\frac{1}{3}B-G \simeq 0.19$	0.25 ± 0.05

Epilogue

When the pion cloud or gluon exchange corrections were first discussed, each one alone did not yield a correction large enough to resolve the "spin crisis". Furthermore, since the pion might contribute a substantial fraction of the observed mass-splitting between the N and Δ, to combine these two corrections would reduce the strength of OGE. However, in the last few years the chiral analysis of quenched and full lattice QCD calculations for the N and Δ masses as a function of quark mass [11], concluded that pion effects likely contribute 50 MeV or less of the 300 MeV $N-\Delta$ mass difference. We can therefore without too large an error combine the one-gluon-exchange and pion cloud corrections in the quark spin sum. This combined correction will give a Σ between 0.35 ($P_{N\pi}=0.25, P_{\Delta\pi}=0.05$) and 0.40 ($P_{N\pi}=0.20, P_{\Delta\pi}=0.10$) in excellent agreement with the modern data. We note that g_A is reproduced by the same corrections affecting the Σ. Relativity reduces the value of g_A from 5/3 to 1.09 and OGE and the pion cloud as well as the center-of-mass corrections will increase the g_A value from 1.09 to 1.27. As seen in Table 1 these corrections are crucial in order to reproduce the baryon magnetic moments, i.e. the pion isovector cloud is an important correction to the nucleon magnetic moments and the OGE restores the ratio $\mu_p/\mu_n \simeq -3/2$![14].

We have used a model of confined quarks to compute the matrix elements of the axial current to find a Σ value relevant in the limit $Q^2 \to \infty$. Our result, $\Sigma \in (0.35, 0.40)$, agrees very well with the experimental value. The flavor singlet spin operator however has a non-zero anomalous dimension, γ, and the observable Σ should be renormalization group independent and gauge-invariant as defined by Larin *et al.* [20]. Motivated by the observation that a valence dominated quark model can only match experiment for parton distribution functions at a low scale, e.g. [21], our value of the quark spin would need to be multiplied by a non-perturbative factor involving the QCD β–function and γ. This evolution factor is truly non-perturbative and its three-loop perturbation theory evaluation by Larin *et al.* [20] is at best semi-quantitative.Nevertheless, it is rigorously less than unity and at three-loops gave a value of order 0.6–0.8 [22]. Multiplying the quark spin obtained above by this factor gives $\Sigma \in (0.21, 0.32)$, in excellent agreement with the current experimental value.

In conclusion, the impressive experimental progress aimed at resolving the spin problem has established that the quarks carry about 1/3 of proton's spin and that the gluonic contribution appears to be too small to account for the difference. Instead, well known nucleon structure features like the pion cloud, the quarks' hyperfine interaction, and the relativistic motion of the confined quarks, appear to explain the modern value of Σ. These new insights make us conclude that the missing spin should be accounted for by the orbital angular momentum of the quarks and anti-quarks – the latter associated with the pion cloud of the nucleon and the p-wave anti-quarks excited by the one-gluon-exchange hyperfine interaction. Exploring the angular momentum carried by quarks and anti-quarks is a major focus of the scientific program of the 12 GeV Upgrade at Jefferson Lab., and is a promising way to test the model ideas present here.

This work was supported in part by the NSF grant PHY-0758114.

REFERENCES

1. J. Ashman *et al.* [European Muon Collaboration], *Phys. Lett. B* **206**, 364 (1988).
2. F. Myhrer and A.W Thomas, *Phys. Rev. D* **38**, 1633 (1988)
3. H. Høgaasen and F. Myhrer, *Phys. Lett. B* **214**, 123 (1988)
4. A.W. Schreiber and A.W. Thomas, *Phys. Lett. B* **215**, 141 (1988)
5. A.V. Efremov and O.V. Teryaeb, Dubna report *JINR-E2-88-287* (1988)
6. E. Leader and M. Anselmino, *Z. Phys. C* **41**, 239 (1988); *AIP Conf. Proc.* **187**, 764 (1989).
7. G. Altarelli and G.G. Ross, *Phys. Lett. B* **212**, 391 (1988)
8. R.D. Carlitz, J.C. Collins and A.H. Mueller, *Phys. Lett. B* **214**, 229 (1988)
9. See these proceedings for COMPASS, HERMES, PHENIX and STAR collab. contributions.
10. F. Myhrer and A. W. Thomas, *Phys. Lett. B* **663**, 302 (2008);
    A.W. Thomas, preprint arXiv:0803.2775 [hep-ph].
11. R. D. Young *et al.* Phys. Rev. D **66**, 094507 (2002); *Nucl. Phys. Proc. Suppl.* **109A**, 55 (2002).
12. I.C. Cloet, W. Bentz and A.W. Thomas, *Phys. Lett. B* **659**, 214 (2008)
13. A. De Rujula *et al. Phys. Rev. D* **12**, 147 (1975); A. Chodos *et al. Phys. Rev. D* **10**, 2599 (1974).
14. H. Høgaasen and F. Myhrer, *Phys. Rev. D* **37**, 1950 (1988).
15. S. Theberge, A. W. Thomas and G. A. Miller, *Phys. Rev. D* **22**, 2838 (1980)
16. A. W. Thomas, *Adv. Nucl. Phys.* **13**, 1 (1984).
17. F. Myhrer, G.E. Brown and Z. Xu, *Nucl. Phys. A* **362**, 317 (1981);
18. J. Speth and A. W. Thomas, *Adv. Nucl. Phys.* **24**, 83 (1997).
19. W. Melnitchouk, J. Speth and A. W. Thomas, *Phys. Rev. D* **59**, 014033 (1999)
20. S. A. Larin, *Phys. Lett. B* **334**, 192 (1994); S. A. Larin *et al. Phys. Lett. B* **404** , 153 (1997).
21. M. Gluck and E. Reya, *Z. Phys. C* **43**, 679 (1989)
22. H. Høgaasen and F. Myhrer, *Z. Phys. C* **68**, 625 (1995)[arXiv:hep-ph/9501414].

Gluon Polarisation in the Nucleon from High Transverse Momentum Hadron Pairs at COMPASS

Krzysztof Kurek

A.Soltan Institute for Nuclear Studies, Hoza 69,00-681, Poland
kurek@fuw.edu.pl
On behalf of the COMPASS Collaboration

Abstract. The main goal of the COMPASS spin physics programme is the measurement of the helicity contribution of the gluons, $\Delta G/G$, to the nucleon spin. It is determined from the longitudinal spin asymmetry in the scattering of 160 GeV polarised muons off a polarised LiD target. The new analysis and new COMPASS $\Delta G/G$ result obtained from high transverse momentum hadron pairs for $Q^2 > 1$ (GeV/c)^2 is presented. The weighted method based on a Neural Network approach is discussed. The result of $\Delta G/G = 0.08 \pm 0.10(stat) \pm 0.05(syst)$ disfavors a large positive contribution to the nucleon spin for a gluon momentum fraction around 0.08 and agrees with the previously obtained result from the analysis for the small Q^2 kinematical region ($Q^2 < 1$ (GeV/c)^2).

Keywords: spin, parton distributions, QCD fits, gluon polarization
PACS: 13.88.+2,14.70.Dj,13.60.-r,13.85.-t

Introduction

The spin structure of the nucleon has been studied in polarized deep inelastic scattering (DIS) for a long time. The EMC spin asymmetry measurement [1] and the naive interpretation of the results following the Ellis-Jaffe sum rule [2] have introduced the so-called "spin crisis" to Particle Physics: quarks carry very small fraction of the nucleon's helicity. The next experiments at CERN, DESY and SLAC confirmed that quarks are only responsible for roughly $1/3$ of nucleon's helicity. The quark helicity distributions $\Delta q_i(x,Q^2)$ are related to vector-axial quark current which is not conserved due to the Adler-Bell-Jackiw anomaly. This fact allows to explain the spin crisis by changing the interpretation of the measurement: instead of quark spin contents $\Delta\Sigma = \int_0^1 \sum_{i=1}^{n_f} \Delta q_i(x,Q^2)dx$, the combination $\Delta\Sigma - \frac{3\alpha_s}{2\pi}\Delta G$ is measured, where ΔG is a gluon polarization inside the nucleon. The spin crisis and the violation of the Ellis-Jaffe sum rule can be then avoided if ΔG is large enough. To verify this fact one should determine fraction of the nucleon spin carried by gluons. In this paper a new result for a direct measurement of gluon polarization $\Delta G/G$ in the DIS kinematical region ($Q^2 > 1$ (GeV/c)^2) obtained from high transverse momentum hadron pairs is presented. The data sets were collected in years 2002-2004. The COMPASS experiment is using a 160 GeV polarized muon beam from the SPS at CERN scattered off a polarized ^{6}LiD target (see [3]). In leading order Quantum Chromodynamics (QCD) approximation the only subprocess which probes gluons inside nucleon is Photon-Gluon Fusion (PGF).

CP1149, *Spin Physics, 18th International Spin Physics Symposium*
edited by D. G. Crabb, D. B. Day, S. Liuti, X. Zheng, M. Poelker, and Y. Prok

There are two ways allowing direct access to gluon polarisation via the PGF subprocess available in the COMPASS experiment: the open-charm channel where the events with reconstructed D^0 mesons are used and the production of two hadrons with relatively high-p_T in the final state. The estimation of the gluon polarisation in the open-charm channel is much less Monte-Carlo (MC) dependent than in the two high-p_T hadrons method, where the complicated background requires very good MC description of the data. On the other hand the statistical precision in high-p_T hadrons method is much higher than in the open charm channel. The new gluon polarisation result from open-charm channel from the COMPASS experiment can be also found in these proceedings [4].

ΔG from events with high-p_T hadrons

The idea that the selection of events with high-p_T hadrons enhances the fraction of PGF in the sample has been first time discussed in [5] and then revised in [6]. An estimation of $\Delta G/G$ needs to know contributions from other processes and this makes the analysis model dependent. This kind of analysis has been already performed before by the HERMES and the SMC experiments. Depending on the kinematic region covered by the data, two models based on Monte Carlo generators were used: PYTHIA at low Q^2 (HERMES [7] and COMPASS [8]), and LEPTO for $Q^2 > 1$ (GeV/c)^2 (SMC [9]). The new COMPASS result for large Q^2 is presented in this article.

The helicity asymmetry for two high-p_T hadrons can be expressed as follows:

$$A_{LL}^{2h} = R_{PGF} a_{LL} \frac{\Delta G}{G} + A_{Bkg.}, \tag{1}$$

where a_{LL} and R_{PGF} are the analyzing power and the fraction of the PGF subprocess (estimated from MC) and A_{Bkg} denotes the asymmetry from different background processes which contribute to the observed two hadron final state. For large Q^2 these background processes are LO DIS and QCD-Compton. Taking into account that the inclusive asymmetry can also be decomposed in the same way the final formula for the gluon polarization can be schematically written as follows:

$$\Delta G/G \quad = \quad \frac{A_{LL}^{2h} + A^{corr}}{\beta}, \tag{2}$$

where A^{corr} is a combination of the inclusive asymmetry A_1 with some coefficients which together with β depend on the analyzing powers and the fractions of the background processes in the high-p_T as well as in the inclusive sample. The situation becomes more complicated in the case of low Q^2 region where additional background processes have to be taken into account [8]. To increase the statistical precision of the gluon polarisation measurement a weighted method of asymmetry extraction was used. The weight is $fDP_b\beta$, where f and D are the dilution and the depolarization factors, respectively, P_b is the beam polarisation. The weight has to be known on the event by event basis. Only f, D and P_b are calculated from data; β has to be obtained from MC.

A Neural Network (NN) trained on MC samples was used for the parametrization of the quantities which define β (fractions and a_{LL} for different sub-processes). As an input to the NN training for the inclusive sample, x_{Bj} and Q^2 were selected, while for the high-p_T sample in addition transverse and longitudinal momenta of the two hadrons were used. This method depends largely on MC. Good data description with the MC as well as good NN parameterizations are "key points" of this analysis.

Data selection, MC and NN parameterizations

The data used in the analysis were collected in 2002-2004 years. An incoming, a scattered muon and an interaction vertex in the target were required for each event. The kinematic cuts $0.1 < y < 0.9$ and $Q^2 > 1$ (GeV/c)^2 were used. The latter cut ensures that the scale of the hard process is high enough and perturbative QCD can be used. At least two charged hadrons in the interaction vertex with transverse momenta greater than 0.7 GeV/c each are required. Their Feynmann x_F as well as z have to be larger than 0. In addition the sum of the energy fraction of the two hadrons $z_1 + z_2 < 0.95$ is required. Finally the invariant mass of the two hadrons should be greater than 1.5 GeV/c^2. The total number of selected events is about half-million. The LEPTO generator and the full simulation of the COMPASS spectrometer were used. To improve the data and MC agreement some LEPTO parameters were tuned (k_T and parameters of fragmentation). Standard LEPTO tuning was used for systematic studies. In Fig. 1 examples of the data and MC comparison are shown for the transverse momenta of the hadrons. In the right plot the comparison between standard and tuned LEPTO is shown for $\sum p_T^2$. We observe good agreement between data and MC, also the necessity for LEPTO tuning is clearly visible. Also the NN parameterizations of the variables used in the weight were compared with original MC samples and good agreement was found.

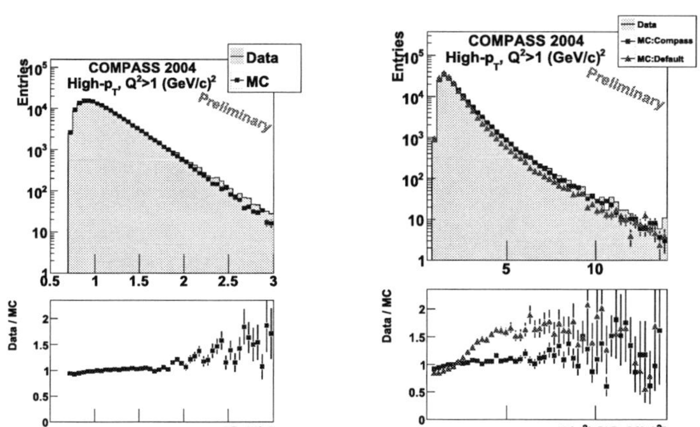

FIGURE 1. Data and MC comparisons for the transverse momenta of the hadrons. Leading hadrons are shown on the left. On the right $\sum p_T^2$ is shown for different LEPTO tunings.

Results and systematic studies

The main contribution to the systematic error comes from the MC. Other contributions like false asymmetries, uncertainty of the beam and target polarisation, radiative corrections, inclusive asymmetries and NN parameterizations were found to be much smaller. To estimate the systematic error coming from the MC four MC samples were used: tuned and standard LEPTO with parton shower ON/OFF. In addition, three different extractions of $\Delta G/G$ with each of the MC samples have been done; the first one with data and MC as they are, the second one using NN trained on selected events from regions of phase space where data and MC agree very well and in the last one MC events were re-weighted to have ratio data/MC equal to 1. Finally 12 values of $\Delta G/G$ have been obtained. From their distribution the MC systematic error were estimated to be about 0.04.

The preliminary result for $\Delta G/G$ from high transverse momentum hadron pairs for large Q^2 ($Q^2 > 1$ (GeV/c)^2) is $\Delta G/G = 0.08 \pm 0.10 \pm 0.05$. The average x_G was estimated to be 0.08 and the average scale of the process, μ^2, is about 3 (GeV/c)^2. The new result is also in very good agreement with $\Delta G/G$ obtained for the low Q^2 high-p_T analysis of 2002-2004 data: $\Delta G/G = 0.016 \pm 0.058(stat.) \pm 0.055(syst.)$ at $x_G \simeq 0.085^{+0.07}_{-0.035}$ and scale $\mu^2 \simeq 3$ (GeV/c)^2.

Summary

The preliminary result for $\Delta G/G$ for high transverse momentum hadron pairs for DIS domain ($Q^2 > 1$ (GeV/c)^2) has been presented. The result is compatible with zero as well as with other direct measurements of gluon polarisation from COMPASS, HERMES and SMC.

ACKNOWLEDGMENTS

Work supported by Ministry of Science and Higher Education grant 41/N-CERN/2007/0

REFERENCES

1. J.Ashman *et al., Nucl. Phys. B* **328**, 1 (1989).
2. J.Ellis and R.L.Jaffe, *Phys. Rev. D* , 1444 (1974); *Phys. Rev. D* **10**, 1669 (1974).
3. P. Abbon *et al., Nucl. Instr. Meth. A* **577**, 455 (2007).
4. F. Kunne for COMPASS Collab. at this conference.
5. R.D.Carlitz, J.C.Collins, A.H.Mueller, *Phys. Lett. B* **214**, 229 (1988).
6. A.Bravar, D.von Harrach, A.Kotzinian, *Phys. Lett. B* **421**, 349 (1998).
7. HERMES Collab. at DIS2008 Conference.
8. E. Ageev *et. al.,,Phys. Lett. B* **633**, 25 (2006).
9. B.Adeva *et al., Phys. Rev. D* **70**, 012002 (2004).

Accessing the Sign and Magnitude of Δg via high p_T, $A_{LL}^{\pi^\pm}$ in Polarized p+p Collisions at PHENIX

Astrid Morreale on behalf of the PHENIX Collaboraton

University of California, Riverside, CA, 92521, USA

Abstract. The double-helicity asymmetries (A_{LL}) of π^+ and π^- production in polarized proton proton collisions are specially interesting probes of the gluon's polarization. Relative differences among A_{LL} of positive, neutral, and negative pions at high transverse momentum are sensitive to the sign and of magnitude of Δg. Quark-gluon (qg) scattering starts to dominate mid-rapidity pion production at RHIC at transverse momenta above 5GeV/c. In this kinematic region the favored and unfavored fragmentation functions for each pion species are sensitive to both the gluon and the quark distributions, with different flavors having different weights for each pion species. Charged pion asymmetry measurements will be an important component in future global analyses, aiming to determine the gluon polarization over a wide range in x, the most recent results using polarized longitudinal data at $\sqrt{s}$ =200GeV, are presented.

INTRODUCTION

Measurements of $\pi^\pm$ A_{LL} are known to be an important process in the Δg global analysis that aims to disentangle all of the partonic contributions to the proton spin puzzle.[6] With polarized proton-proton collisions and measurements of pions at PHENIX central rapidities, spin assymetries are valuable tools used to access information about the gluon's polarization. Under factorization, a differential cross section can be written as the convolution of a parton density function (pdf) and a hard scattering process.This, along with universality of pdf's and fragmentation functions (FF, $D_{h},$) allows for separation of long and short distance(large momentum transfer interaction) contributions. In practice, what is measured are the asymmetries: the ratio of the polarized to unpolarized cross sections (Eq.1). Asymmetries give an elegant way of accessing parton information by counting observed particle yields in different helicity states of incident protons (++, −−, versus, +−, −+) normalized by the polarization in each beam ($P_{B,Y}$):

$$A_{LL}^{\pi} = \frac{\sum_{a,b,c=q,\bar{q},g} \Delta f_a \otimes \Delta f_b \otimes \Delta\hat{\sigma} \otimes D_{\pi/c}}{\sum_{a,b,c=q,\bar{q},g} f_a \otimes f_b \otimes \hat{\sigma} \otimes D_{\pi/c}} = \frac{\sigma_{++} - \sigma_{+-}}{\sigma_{++} + \sigma_{+-}},$$

$$A_{LL}^{\pi} = \frac{1}{P_B P_Y} \frac{N^{++} - RN^{+-}}{N^{++} + RN^{+-}}, \qquad R(luminosity) = \frac{L_{++}}{L_{+-}}. \tag{1}$$

CP1149, *Spin Physics, 18th International Spin Physics Symposium*
edited by D. G. Crabb, D. B. Day, S. Liuti, X. Zheng, M. Poelker, and Y. Prok

π MESONS

As pions form an isospin triplet, asymmetry measurements from all three pion species (for $5 < p_T < 12$ GeV/c) are particularly sensitive to the sign of Δg as qg interactions dominate pion production in this p_T range. Neutral pions have been already measured and provide some constraints to the magnitude of the gluon's polarization.[2, 3] Charged pions are a model independent probe which are sensitive to both the sign and magnitude of Δg as preferential fragmentation of up quarks (u) to π^+, and down quarks(d) to π^-, leads to the dominance of u-g, and d-g contributions. This dominance of u or d combined with the different signs of their polarized distributions translates into asymmetry differences for the different species π^+ , π^0 and π^- that depend on the sign of Δg. For example, a positive Δg could be indicated by an order of π asymmetries, i.e: $A_{LL}(\pi^+) > A_{LL}(\pi^0) > A_{LL}(\pi^-)$, and viceversa for a negative contribution.

PHENIX Detection of $\pi^{\pm}$

An important characteristic of the PHENIX detector at RHIC his the fine-grained calorimetry-100 times finer than previous collider detectors. The resolution of the electromagnetic calorimeter (EMCal)is $\delta\eta * \delta\phi = 0.01 * 0.01$ [4], allowing for high precision cluster identification. A high charged p_T signal is selected by requiring an energy cluster in the EMCal in coincidence with the collision trigger. The charged particle which triggered the event is then associated to a high quality track in the drift chamber; only these charged triggered particles are then selected. The $\pi^{\pm}$ signal is selected with pion Čerenkov light emision thresholds in the ring imaging Čerenkov detector (RICH). Electromagnetic shower-shape cuts in the EMCal clusters aid in the purification of the sample. An additional $\frac{E}{p}$ cut is also used, where E is the energy deposited in the EMCal (one interaction length) and p is the momentum measured using the drift chambers. The main background for this measurement comes from electron conversions which proceed far from the collision vertex and will produce light in the Čerenkov counter with high efficiency, albeit with very low energy. Background is removed by comparing a model of the charged track with the measured track signal. The background's vertex will show high deviation of the comparisong of model and track. The estimated remaining background in the sample is less than 2% in the highest p_T bin (7-10 GeV/c), and less than 1% in the lower p_T bins(5-7GeV/c) for both pion charges.

RESULTS

The data used in this measurement consist of 56,081(47,662) $\pi^+(\pi^-)$ identified with the PHENIX detector using 2006 longitudinally polarized data. The approximate integrated luminosity analyzed was $5.8pb^{-1}$ which is more than twice the luminosity and more than 3 times the figure of merit ($Luminosity * Polarization^4$) of the previous reported measurement [1]. The measured charge separated pion double spin assymetries are shown in Figure 1. The measurements are compared with previous charged pion results

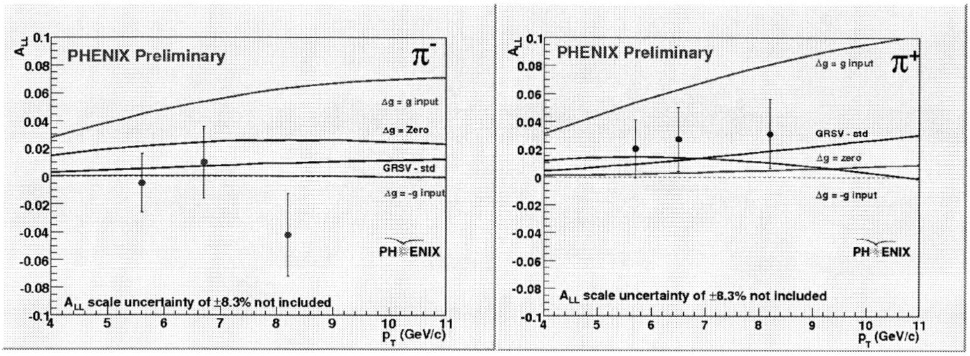

FIGURE 1. Measured of A_{LL} of π^-(left) and π^+(right)

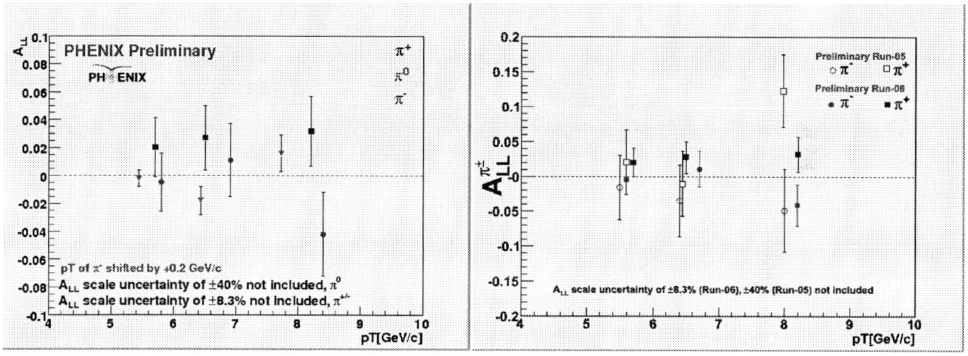

FIGURE 2. Comparison of charged and neutral assymetries at high pT(left), and previous reported measurements[1]

and with neutral pion assymetries, see figure 2. Figure 3 shows a comparison of these results to new GRSV parametrisations. These new parametrisations are calculated using the de Florian-Sassot-Stratmann (DSS) fragmentation functions ($D_{\pi^\pm}$) , which use the most updated $e^\pm$ and RHIC data. These calculations contain the best information on fragmentation functions current available [5], unlike the previously used KKP curves, also distinguish between positive and negative charges.[Fig. 3]

CONCLUSIONS

The measured $\pi^\pm$ A_{LL} (2006) have been compared with a new set of theoretical calculations. These for the first time use charge separated $\pi^\pm$ data for $D_{\pi^\pm}$ extraction. While the result has statistical improvement from the 2005 measurement, further statistics along with a cross section measurement would be useful in interpreting the results and comparing with pQCD predictions. The projections of charged pion sensitivities for future

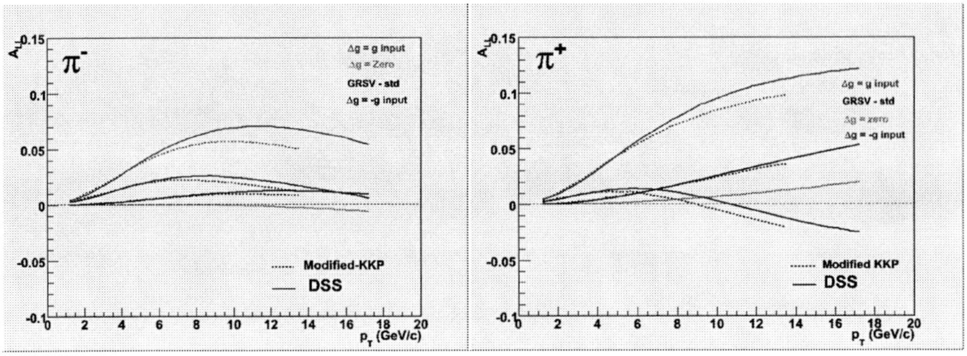

FIGURE 3. Comparison of GRSV curves using KKP and DSS fragmentation functions[5]

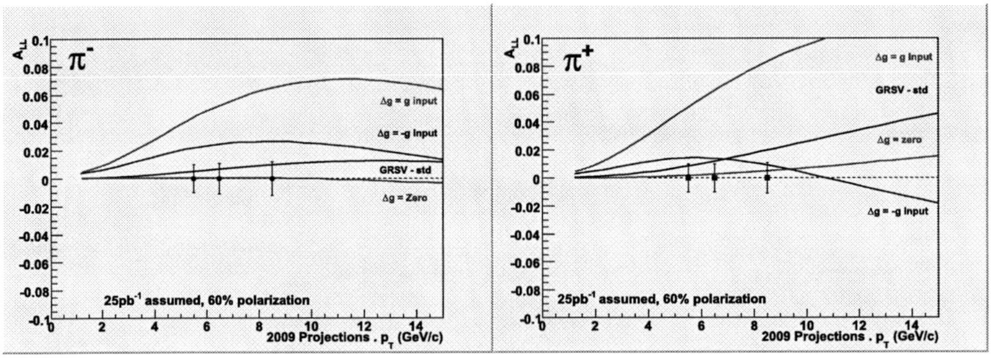

FIGURE 4. Charged pion sensitivity projections

RHIC longitudinally polarized runs are found in figure 4. The analyzing power of $\pi^{\pm}$ asymmetries are expected to be major players in future global analyses of Δg and are thus an important part of the 200GeV longitudinal program at RHIC.

REFERENCES

1. A. Morreale, *17th International Spin Physics Symposium* , AIP Conference Proceedings, American Institute of Physics, New York, Volume 915, pp. 359-362, 2007.
2. K. Boyle, *17th International Spin Physics Symposium*, AIP Conference Proceedings, American Institute of Physics, New York, Volume 915, pp. 335-338, 2007.
3. K. Boyle, *Physical Review D*, Volume 76,American Physical Society,New York, 2007, pp. 051106.
4. K. Adcox et al, *Nuclear Instruments and Methods A*, Volume 499, pp. 489-507, 2003.
5. D. de Florian, R. Sassot and M. Stratmann *Physical Review D*, Volume 75, American Physical Society, New York, 2007 pp. 114010. e-Print: hep-ph/0703242.
6. S. D Bass *The Spin Structure of the Proton*, World Scientific Publishing, New Jersey, 2008, pp 152.

Determination of Δg from Inclusive π^0 Measurement at RHIC-PHENIX

Kenichi Nakano on behalf of the PHENIX Collaboration

RIKEN, 2-1, Hirosawa, Wako-shi, Saitama-ken 351-0198, Japan

Abstract. One of the goals of the PHENIX experiment is to determine the contribution of the gluon spin (ΔG) to the proton spin using longitudinally polarized proton-proton collisions. Double spin asymmetries (A_{LL}) of π^0 production at mid-rapidity were measured at $\sqrt{s} = 200$ and 62.4 GeV with data taken in 2005 and 2006. The comparison with the next-to-leading order (NLO) perturbative Quantum Chromodynamics (pQCD) calculations imposed the limit as $\int_{0.02}^{0.3} dx \Delta g_{GRSV}(x) = 0.2 \pm 0.1$ (stat.) ± 0.1 (sys.) or ${}^{+0.2}_{-0.8}$ (3σ-level stat.) at $Q^2 = 4$ GeV^2.

Keywords: Proton Spin Structure, Polarized Gluon Distribution Function
PACS: 14.20.Dh, 13.88+e, 13.85.Hd

INTRODUCTION

It was a large impact on hadron physics when in 1988 the EMC experiment revealed the contribution of the quark spin to the proton spin ($\frac{1}{2}\Delta\Sigma$) to be only $(12 \pm 9 \pm 14)$%[1]. After that, many deep inelastic scattering experiments have measured $\frac{1}{2}\Delta\Sigma$ more precisely. Recent analyses[2][3] reported that $\frac{1}{2}\Delta\Sigma \sim 0.3$ with an error of $\sim$0.03, which is 30% of the proton spin. The remaining components must be carried by the gluon spin (ΔG) and the orbital angular momenta of quarks and gluons ($L_q + L_g$); $\frac{1}{2} = \frac{1}{2}\Delta\Sigma + \Delta G + L_q + L_g$.

One of the goals of the PHENIX experiment at RHIC is to measure Δg with longitudinally-polarized proton-proton collisions. It is suited for Δg measurement because gluon-involved scatterings, such as $qg \to qg$, dominate in the cross section. Δg is evaluated by measuring double helicity asymmetries (A_{LL}) of reactions such as jet, π^0 or direct photon production processes; $A_{LL} = \frac{\sigma_{++} - \sigma_{+-}}{\sigma_{++} + \sigma_{+-}}$. In A_{LL} measurements most systematic errors cancel out and thus a high precision can be achieved.

The main channel of the A_{LL} measurements at the PHENIX experiment is the inclusive measurement of π^0's from jet fragmentation at mid-rapidity. The final results of the PHENIX π^0 A_{LL} measurement at $\sqrt{s} = 200$ and 62.4 GeV with data taken in 2005 and 2006 have been released recently[4][5]. This paper presents these results.

EXPERIMENTAL SETUP

Longitudinally polarized proton-proton collisions were provided by RHIC at BNL. Table 1 summarizes the status of collision data. The lower $\sqrt{s}$ is more sensitive to higher x region.

Figure 1 shows the cross section of the PHENIX detector. The measurement of π^0

CP1149, *Spin Physics, 18th International Spin Physics Symposium*
edited by D. G. Crabb, D. B. Day, S. Liuti, X. Zheng, M. Poelker, and Y. Prok

TABLE 1. The status of longitudinally-polarized proton-proton collisions.

Year	$\sqrt{s}$ (GeV)	Delivered Luminosity, L (pb^{-1})	Polarization, P (%)	Figure of Merit, P^4L ($10^{-3}\ \mathrm{pb}^{-1}$)
2005	200	3.4	49	196
2006	200	7.5	60	972
	62.4	0.1	48	5.3

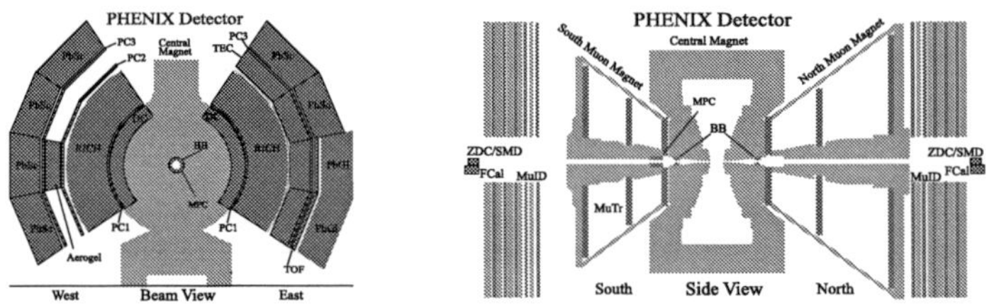

FIGURE 1. The cross section of the PHENIX detector.

used the PHENIX Central (East and West) Arms, which consist of two arms that are positioned almost back to back. Each arm covers the pseudo-rapidity region $|\eta| < 0.35$ and 90 degrees in azimuthal angle. The electromagnetic calorimeter (EMCal) was used to detect photons from π^0 decay. It has a fine segmentation, 0.01×0.01 rad/segment, and thus a photon pair from π^0 can be separated up to a π^0 energy of 20 GeV. Background particles were excluded by an electromagnetic-shower-shape cut, a veto against matching charged tracks and a time-of-flight cut.

The luminosity and the polarization direction of two colliding beams were measured with the Beam-Beam Counter (BBC) and the Zero-Degree Calorimeter (ZDC) placed near the beam pipe. High-p_T events were accumulated with the coincidence of a minimum bias trigger made by BBC and a high-energy-photon trigger made by EMCal. The efficiency of the latter trigger is 50% at $E \sim 1.4$ GeV and 100% at $E \gtrsim 2.5$ GeV.

MEASUREMENT OF π^0 A_{LL}

The method of measuring π^0 A_{LL} at PHENIX has been developed since 2003 [6][7] and was followed in this analysis.

The left panel of Fig. 2 shows an example of photon-pair invariant mass distribution. The fraction of combinatorial background under the π^0 mass peak, $w_{BG} \equiv N_{BG}/N_{all}$, is about 20% at $p_T^{\pi^0} = 2$ GeV/c, 8% at 5 and 5% at 10 in case of $\sqrt{s} = 200$ GeV.

A_{LL} can be expressed with measured quantities as

$$A_{LL} = \frac{1}{|P_B||P_Y|}\frac{N_{++} - RN_{+-}}{N_{++} + RN_{+-}}. \tag{1}$$

where N_{++} and N_{+-} are the number of measured π^0's with same and opposite beam helicities, respectively; $R \equiv L_{++}/L_{+-}$ is the relative luminosity; and P_B and P_Y are beam polarizations. We estimated that the error of $P_B \cdot P_Y$ is 10% and the error of

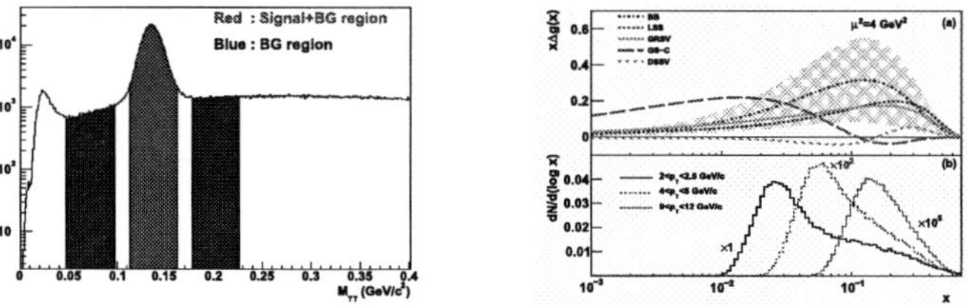

FIGURE 2. **Left**: An example of photon-pair invariant mass distribution. In A_{LL} measurement, the mass range under the π^0 peak and the two side bands were used in evaluating the yields of signal+background and background, respectively. **Right**: **(a)** $x\Delta g(x)$ in various parameterizations; **(b)** x distributions of gluons that originate π^0 in one of three p_T ranges at $\sqrt{s} = 200$ GeV.

A_{LL} due to R is 2×10^{-4} in 2005 and 7×10^{-4} in 2006. The measured A_{LL} has been corrected for combinatorial background as $A_{LL}^{\pi 0} = \frac{A_{LL}^{\pi 0+BG} - w_{BG} A_{LL}^{BG}}{1 - w_{BG}}$, where $A_{LL}^{\pi 0+BG}$ and A_{LL}^{BG} were measured with the yields in the red and blue regions in the left panel of Fig. 2, respectively.

This measurement probes only a part of the x range since p_T and acceptance in measurement are limited. It has been evaluated with a Monte-Carlo calculation using NLO pQCD as an input. The right panel of Fig. 2 Right (b) shows x distributions of gluons that originate π^0 in some p_T ranges. The range of gluon x to which this measurement is sensitive was estimated to be $0.02 < x < 0.3$ at $\sqrt{s} = 200$ GeV and $0.06 < x < 0.4$ at $\sqrt{s} = 62.4$ GeV.

RESULTS FROM 2005 AND 2006 DATA

Prior to the A_{LL} measurement, the cross section of π^0 production has been measured to confirm the applicability of NLO pQCD and the smallness of the contribution of soft-scattering events at low p_T. Figure 3 shows π^0 A_{LL} measured at two $\sqrt{s}$ for the 2005 and 2006 data. The calculations based on the GRSV parameterization[8] are also drawn. In the result at $\sqrt{s} = 200$ GeV, the size of the statistical error approaches that of the relative luminosity error at low p_T bins. In the result at $\sqrt{s} = 62.4$ GeV, the difference between the NLO and NLL calculations is not pronounced in small-Δg cases, where the NLL correction is important in cross section.

Since the measured π^0 A_{LL} at $\sqrt{s} = 200$ GeV has the best statistical accuracy at present, the constraint of this data on Δg was evaluated by deriving the χ^2 with calculations, which are based on the three models shown in Fig. 3 Center and also other models that have a middle Δg value. Figure 3 Right shows the χ^2 as a function of Δg of the models. The experimental errors due to the relative luminosity and the beam polarization were represented by red and blue lines, respectively. The former is rather large, typically ± 0.1. By taking a Δg value where χ^2 increases by 1 or 9, we imposed the limit on Δg at $Q^2 = 4$ GeV^2 as

$$\Delta g_{GRSV}^{[0.02,\,0.3]} \equiv \int_{0.02}^{0.3} dx \Delta g_{GRSV}(x) = 0.2 \pm 0.1 \ (1\sigma \text{ limit}) = 0.2^{+0.2}_{-0.8} \ (3\sigma \text{ limit}). \quad (2)$$

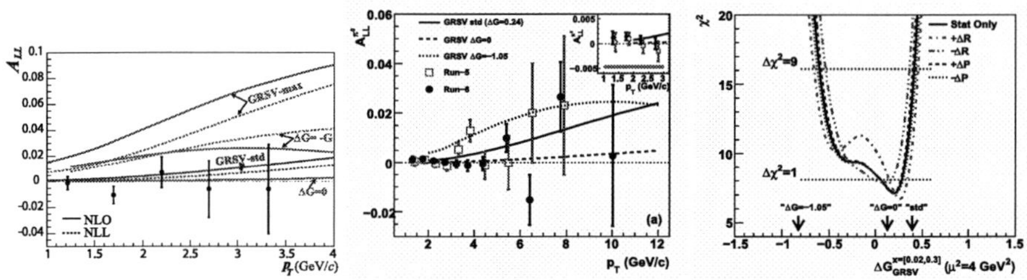

FIGURE 3. Double helicity asymmetry (A_{LL}) in π^0 production at $\sqrt{s}$ = 62.4 GeV **(Left)** and 200 GeV **(Center)**. **Right**: χ^2 between measured and calculated A_{LL}s as a function of $\int_{0.02}^{0.3} dx\Delta G_{GRSV}(x,\mu^2 = 4\text{GeV}^2)$.

Two uncertainties in the theoretical calculation, the shape of $\Delta g(x)$ and the theoretical scale μ, have been checked by trying various parameterizations and scales in the A_{LL} calculation. The 3 σ limit is consistent between various parameterizations within ± 0.1. The variation of the 3 σ limit due to different scales ($\mu = p_T$, $p_T/2$ and $2p_T$) is ± 0.1 at the upper limit and ± 0.2 at the lower limit.

CONCLUSION

The PHENIX experiment at RHIC is measuring the gluon polarization Δg in the proton with longitudinally-polarized proton-proton collisions. The measurements of inclusive π^0 at mid-rapidity have been done with high-statistics data taken in 2005 and 2006 at $\sqrt{s}$ = 200 GeV and 62.4 GeV. The π^0 A_{LL} at $\sqrt{s}$ = 200 GeV made the constraint as $\Delta g_{GRSV}^{[0.02,\ 0.3]} = 0.2 \pm 0.1$ (1σ limit) $= 0.2^{+0.2}_{-0.8}$ (3σ limit) at $Q^2 = 4$ GeV^2 with a ± 0.1 relative luminosity error. The comparison with various parameterizations indicates that the data can determine the size of Δg regardless of the shape.

ACKNOWLEDGMENTS

A part of data production and analysis have been done at RIKEN CCJ.

REFERENCES

1. J. Ashman *et al.* [European Muon Collaboration], *Phys. Lett. B* **206**, 364 (1988).
2. A. Airapetian *et al.* [HERMES Collaboration], *Phys. Rev. D* **75**, 012007 (2007).
3. D. de Florian, R. Sassot, M. Stratmann and W. Vogelsang, *Phys. Rev. Lett.* **101**, 072001 (2008).
4. A. Adare *et al.* [PHENIX Collaboration], arXiv:0810.0694 [hep-ex].
5. A. Adare *et al.* [PHENIX Collaboration], arXiv:0810.0701 [hep-ex].
6. A. Adare *et al.* [PHENIX Collaboration], *Phys. Rev. D* **76**, 051106 (2007).
7. S. S. Adler *et al.* [PHENIX Collaboration], *Phys. Rev. D* **73**, 091102 (2006).
8. M. Gluck, E. Reya, M. Stratmann and W. Vogelsang, *Phys. Rev. D* **63**, 094005 (2001).

Status of Longitudinal Polarized Parton Densities and Higher Twist

E. Leader*, A.V. Sidorov,† and D.B. Stamenov**

*Imperial College London, London WC1E 7HX, England
†Bogoliubov Theoretical Laboratory, Joint Institute for Nuclear Research, 141980 Dubna, Russia
**Institute for Nuclear Research and Nuclear Energy, Bulgarian Academy of Science, Sofia 1784, Bulgaria

Abstract. The present status of the longitudinal polarized parton densities (PDFs) and the contribution of their first moments to the nucleon spin is discussed. Special attention is paid to the role of higher twist effects in determining the PDFs and to the polarized strange quark and gluon densities, which are still not well determined from the present data.

Keywords: QCD, spin structure functions, polarized parton densities, higher twist
PACS: 13.60.Hb, 12.38-t, 14.20.Dh

INTRODUCTION AND METHODS OF ANALYSIS

Our present knowledge about the nucleon spin structure comes mainly from polarized inclusive and semi-inclusive deep inelastic scattering (DIS) experiments at SLAC, CERN, DESY and JLab, polarized proton-proton collisions at RHIC and polarized photoproduction experiments. One of the important and best studied aspects of this knowledge is the determination of the longitudinal polarized parton densities in Quantum Chromodynamics (QCD) and their first moments, which correspond to the spins carried by the quarks and gluons in the nucleon.

There are some peculiarities which make QCD analysis of the data more complicated and difficult than that in the unpolarized case:

i) A half of the present polarized DIS data are at moderate Q^2 and W^2 ($Q^2 \sim 1-4$ GeV^2, 4 GeV$^2 < \mathrm{W}^2 < 10$ GeV^2), or in the so-called preasymptotic region. So, in contrast to the unpolarized case, the $1/Q^2$ terms (*kinematic* - γ^2 factor, target mass corrections, and *dynamic* - higher twist corrections to the spin structure function g_1) have to be accounted for in the analysis in order to determine *correctly* the polarized PDFs.

ii) Due to the lack of charged current neutrino data, only the sum $(\Delta q + \Delta\bar{q})$ can be determined from inclusive DIS and information from semi-inclusive DIS (SIDIS) is needed for the flavor decomposition of the polarized sea ($\Delta\bar{u}$, $\Delta\bar{d}$, $\Delta\bar{s}$). Such an analysis, however, requires knowledge of the fragmentation functions which are not well known at present. Note that in the unpolarized case, SIDIS processes have never been used for a flavor separation of the parton densities.

The best manner to determine the polarized PDFs is to perform a QCD fit to the data on g_1/F_1, which can be obtained if both $A_{||}$ and $A_{\perp}$ asymmetries are measured. In the case if only $A_{||}$ is measured, the quantity $A_{||}/D(1+\gamma^2)$ is a good approximation

CP1149, *Spin Physics, 18th International Spin Physics Symposium*
edited by D. G. Crabb, D. B. Day, S. Liuti, X. Zheng, M. Poelker, and Y. Prok

of g_1/F_1, where D is the depolarization factor and $\gamma^2 = 4M^2x^2/Q^2$. The data on the photon-nucleon asymmetry A_1 are not suitable for the determination of PDFs because the structure function g_2 is not well known in QCD and the approximation $(A_1)^{theor}(= g_1/F_1 - \gamma^2 g_2/F_1) \approx (g_1/F_1)^{theor}$ in the preasymptotic region used by some of the groups is *not* reasonable.

In QCD, one can split g_1 and F_1 into leading (LT) and higher twist (HT) pieces

$$g_1 = (g_1)_{\rm LT} + (g_1)_{\rm HT}, \quad F_1 = (F_1)_{\rm LT} + (F_1)_{\rm HT}. \tag{1}$$

Then, approximately

$$\frac{g_1}{F_1} \approx \frac{(g_1)_{\rm LT}}{(F_1)_{\rm LT}}[1 + \frac{(g_1)_{\rm HT}}{(g_1)_{\rm LT}} - \frac{(F_1)_{\rm HT}}{(F_1)_{\rm LT}}]. \tag{2}$$

Note that LT pieces of g_1 and F_1 are expressed by the polarized and unpolarized PDFs, respectively.

There are mainly two methods to fit the data - taking or NOT taking into account the HT corrections to g_1. According to the first [1], the data on g_1/F_1 have been fitted taking into account the contribution of the first term $h(x)/Q^2$ in $(g_1)_{\rm HT}$ and using the experimental data for the unpolarized structure function F_1

$$\left[\frac{g_1(x,Q^2)}{F_1(x,Q^2)}\right]_{exp} \Leftrightarrow \frac{g_1(x,Q^2)_{\rm LT} + h(x)/Q^2}{F_1(x,Q^2)_{exp}}. \tag{3}$$

According to the second approach [2] only the LT expression in (1) have been used in the fit to the g_1/F_1 data

$$\left[\frac{g_1(x,Q^2)}{F_1(x,Q^2)}\right]_{exp} \Leftrightarrow \frac{g_1(x,Q^2)_{\rm LT}}{F_1(x,Q^2)_{\rm LT}}. \tag{4}$$

It is obvious that the two methods are equivalent in a pure DIS region where HT can been ignored. To be equivalent in the *preasymptotic* region requires a cancellation between the ratios $(g_1)_{\rm HT}/(g_1)_{\rm LT}$ and $(F_1)_{\rm HT}/(F_1)_{\rm LT}$ in (2). Figure 1 (left), based on our recent results on g_1 [3] and the unpolarized results of [4], demonstrates the validity of this for $x \geq 0.15$, but clearly indicates that ignoring HT terms in the ratio g_1/F_1 below $x = 0.15$, as was done by groups using the second method or some its modifications [5, 6], is incorrect. Note that in this case the HT effects are absorbed in the extracted PDFs which differ from those determined in the presence of HT (see [3] for more details).

POLARIZED PARTON DENSITIES AND HIGHER TWIST

Let us start our discussion with the impact of the COMPASS [6] and CLAS [7] inclusive DIS data on the *longitudinal* polarized PDFs and higher twist effects. The precise CLAS data on g_1^p/F_1^p and g_1^d/F_1^d are data at low $Q^2(1-4\ \text{GeV}^2)$ while COMPASS has presented data on A_1^d mainly at large Q^2 and very small x $(0.004 \leq x \leq 0.02)$, the *only*

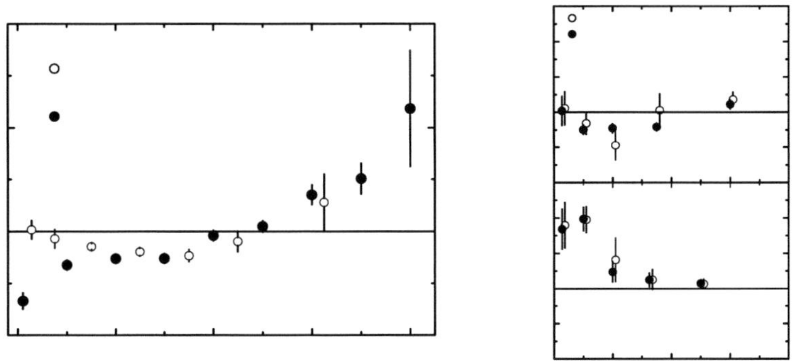

FIGURE 1. Comparison of HT terms in g_1 and F_1 (left). HT effects for protons and neutrons (right).

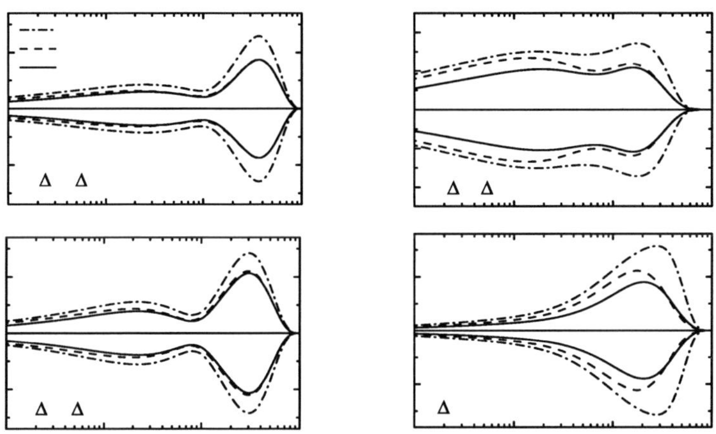

FIGURE 2. Impact of CLAS and COMPASS data on the uncertainties for NLO($\overline{\text{MS}}$) polarized PDFs.

precise data at such small x. Compared to the HT values obtained in the LSS'05 analysis [8], the uncertainties in the HT values at each x in the CLAS region ($x \sim 0.1 - 0.6$) are significantly reduced by the CLAS data, as seen in Fig. 1 (right). As expected, the central values of the polarized PDFs are practically *not* affected by the CLAS data. This is a consequence of the fact that at low Q^2 the deviation from logarithmic in Q^2 pQCD behaviour of g_1 is accounted for by the higher twist term in g_1. The accuracy of the determination of polarized PDFs, however, is essentially improved. This improvement is illustrated in Fig. 2 (dashed curves). It is a consequence of the much better determination of higher twist contribution in g_1, as discussed above. The impact of COMPASS data on the PDFs uncertainties is also shown in Fig. 2 (the solid curves). As seen, they help

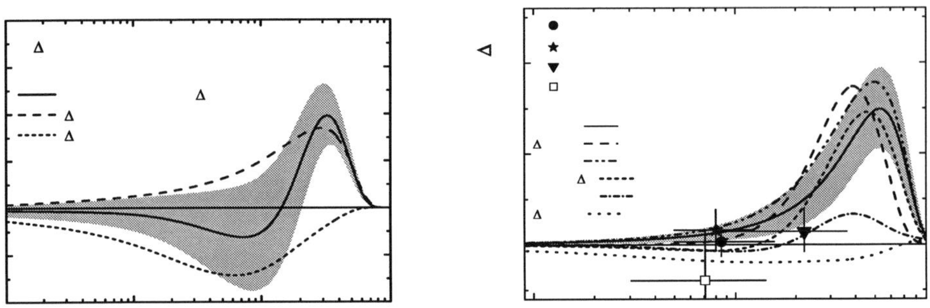

FIGURE 3. The three different LSS'06 solutions for $\Delta G(x)$ (left). Comparison with directly measured $\Delta G(x)/G(x)$ (right).

to improve in addition the accuracy of the determination of the gluon and strange sea quark polarized densities at $x < 0.2$ and $x < 0.1$, respectively. Another important effect of COMPASS data (see [3] and slides of Stamenov's talk at this Symposium [9]) is that $\Delta s(x)$ and respectively, $\Delta\Sigma(x)$, are less negative at x smaller than 0.1.

The present *inclusive* DIS data cannot distinguish between positive, negative and sign-changing $\Delta G(x)$ (see Fig. 3 (left)). In all the cases the magnitude of ΔG is small: $|\Delta G| < 0.4$ and the corresponding polarized quark densities are very close to each other. COMPASS analysis finds also a negative $\Delta G(x)$, but has some peculiarities which suggest it is not physical. Also, we have found that if in the input gluon parametrization a term $(1+\gamma x)$ is added, where γ is a free parameter to be determined from the fit to the data, a negative ΔG solution is impossible. So, although the negative ΔG cannot be formally ruled out by the data, it seems to be *not* reasonable.

A primary goal of the RHIC spin program is to determine the gluon polarization. There are two sources of information: $\vec{p}+\vec{p} \rightarrow \pi^0(\pi^{+/-})+X$, $\vec{p}+\vec{p} \rightarrow jet+X$. We will mention here the combined QCD analysis of the inclusive DIS and RHIC π^0-production data [10] which has been recently performed by AAC [11]. Two solutions for $\Delta G(x)$, positive and node, have been found with values for the first moments at $Q^2 = 1\ GeV^2 : \Delta G = 0.40$ +/- 0.28 and -0.12 +/- 1.78, respectively. Note that for the node $\Delta G(x)$ solution the positivity condition $\Delta G(x) \leq G(x)$ is broken for $x \geq 0.4$, which makes it questionable. For more details about the possibilities of the RHIC data to constrain the gluon polarization ΔG see the talks presented by STAR and PHENIX Collaborations at this Symposium [9]. The main conclusion is that the magnitude of ΔG is smaller than 0.4, but the present RHIC data cannot determine the form of the gluon density $\Delta G(x)$.

In Figure 3 (right) we compare the polarized gluon densities obtained by different QCD analyses (divided by the MRST'02 version of the unpolarized gluon density $G(x)$) with directly measured values of $\Delta G(x)/G(x)$ at $\mu^2 = 3\ GeV^2$. One can see from Fig. 3(right) that the most precise values of $\Delta G/G$, the COMPASS ones [12], are well consistent with most of the polarized gluon densities determined in the recent QCD analysis, and conclude that the present high p_t measurements cannot distinguish between

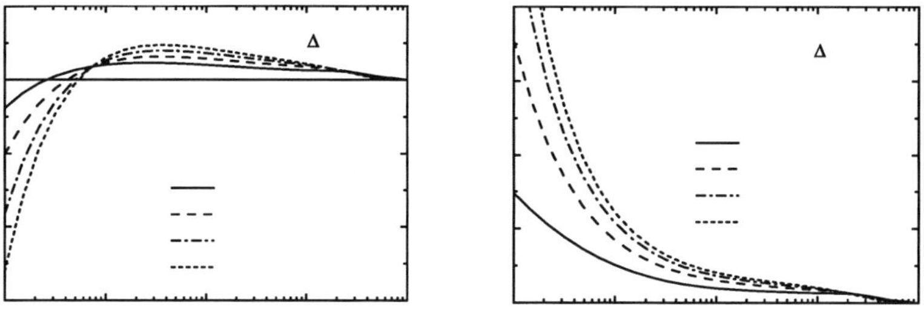

FIGURE 4. g_1^p at different Q^2 calculated using the LSS'06 PDFs corresponding to $\Delta G > 0$ and $\Delta G < 0$.

the different scenarios for $\Delta G(x)$.

It seems clear that present day data cannot distinguish between the three scenarios for $\Delta G(x)$. A clean distinction would be possible in the Electron-Ion Collider where very precise measurements of $g_1^p(x, Q^2)$ at very small x ($x \geq 0.00075$) could be performed. Fig. 4 shows $g_1^p(x, Q^2)$, calculated using the LSS'06 PDFs corresponding to positive and negative $\Delta G(x)$. There is a dramatic difference at small x.

Very recently the DSSV group has presented results on polarized PDFs [13] obtained from the first global analysis of the polarized DIS, SIDIS and RHIC polarized pp scattering data. Due to the SIDIS data a flavor decomposition of the polarized sea is achieved. This analysis yields changing in sign $\Delta\bar{s}(x, Q^2)$: positive for $x > 0.03$ and negative for small x (see Fig. 5). Its first moment is negative (fixed in practice by the SU(3) symmetric value of a_8) and almost identical with that obtained in the inclusive DIS analysis. It was shown in the talk by R. Windmolders at this Symposium [9] that the determination of $\Delta\bar{s}(x)$ from SIDIS strongly depends on the fragmentation functions (FFs) and that for such an unexpected behavior of $\Delta\bar{s}(x)$ the new FFs [14] are crucial. So, the model independent extraction of FFs is very important. The NLO($\overline{MS}$) PDFs determined from analyses of different sets of data: LSS'06 (inclusive DIS data), AAC'08 (inclusive DIS and RHIC π^0-production data), DSSV (DIS, SIDIS and RHIC data) are compared in Fig. 5. Note that in the LSS analysis the HT corrections have been taken into account while AAC'08 and DSSV have used the second method described above (without account for HT terms). Although the first moments are almost identical, the quark densities themselves are different, especially $\Delta\bar{s}(x)$. Note that all analyses of the inclusive DIS data yield $\Delta\bar{s}(x)$ negative. So, the DSSV result on $\Delta\bar{s}(x)$ is a big challenge for our understanding of spin properties of the nucleon. We would like to emphasize also that the very accurate DIS data in the preasymptotic region require a more precise confrontation of QCD to the data, as was already mentioned in the Introduction. Otherwise, the results on PDFs will differ. A comment on an incorrect confrontation of QCD to the inclusive DIS data in the DSSV analysis is presented in Stamenov's talk [9] at this Meeting.

Let us finally discuss the present status of the proton spin sum rule. Using the values for $\Delta\Sigma(Q^2)$ and $\Delta G(Q^2)$ at $Q^2 = 4\ GeV^2$ obtained in LSS'06 analysis [3] one can find

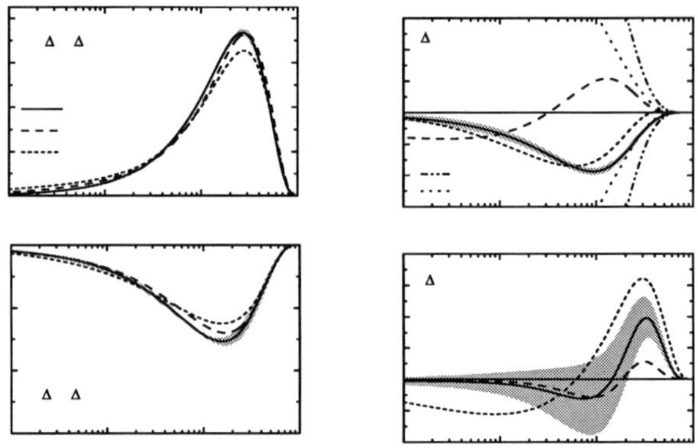

FIGURE 5. Comparison between LSS'06, AAC'08 and DSSV NLO PDFs in $(\overline{\text{MS}})$ scheme.

for the spin of the proton (the numbers in brackets correspond to node ΔG):

$$S_z = \frac{1}{2} = \frac{1}{2}\Delta\Sigma(Q^2) + \Delta G(Q^2) + L_z(Q^2) = 0.55(0.15) \pm 0.25(0.49) + L_z(Q^2). \quad (5)$$

Although the central values of parton contribution are very different in the two cases, in view of the big uncertainty in (5) coming mainly from the gluons, one cannot make a definite conclusion about the quark-gluon contribution in the spin of the nucleon.

ACKNOWLEDGMENTS

This research was supported by the RFBR grants (No 07-02-01046, 08-01-00686) and by the Bulgarian National Science Foundation under Contract DO 02-288.

REFERENCES

1. E. Leader, A.V. Sidorov, and D.B. Stamenov, *Phys. Rev. D* **67**, 074017 (2003).
2. M. Glück, E. Reya, M. Stratmann, and W. Vogelsang, *Phys. Rev. D* **63**, 094005 (2001).
3. E. Leader, A.V. Sidorov, and D.B. Stamenov, *Phys. Rev. D* **75**, 074027 (2007).
4. S.I. Alekhin, *Phys. Rev. D* **68**, 014002 (2003).
5. J. Blumlein, H. Bottcher, *Nucl. Phys.* **B 636**, 225 (2002); AAC, M. Hirai et al., *Phys. Rev.* **D 69**, 054021 (2004).
6. V.Y. Alexakhin *et al.* (COMPASS Collaboration), *Phys. Lett. B* **647**, 8 (2007).
7. K.V. Dharmwardane *et al.* (CLAS Collaboration), *Phys. Lett. B* **641**, 11 (2006).
8. E. Leader, A.V. Sidorov, and D.B. Stamenov, *Phys. Rev. D* **73**, 034023 (2006).
9. Parallel Longitudinal Spin Session, http://spreadsheets.google.com/pub?key=pqhuBYtzsrrUuuVad8k4TSA.
10. A. Adare *et al.* (PHENIX Collaboration), *Phys. Rev. D* **76**, 051106 (2007).
11. M. Hirai and S. Kumano, arXiv:0808.0413 [hep-ph].
12. See the talk presented by K.Kurek at this Symposium.
13. D. de Florian, R. Sassot, M. Stratmann, and W. Vogelsang, Phys. Rev. Lett. **101**, 072001 (2008).
14. D. de Florian, R. Sassot, and M. Stratmann, *Phys. Rev. D* **75**, 114010 (2007); D **76**, 074033 (207).

Target Mass Corrections and Beyond

Alberto Accardi

Hampton University, Hampton, VA 23668, USA and Jefferson Lab, Newport News, VA 23606, USA

Abstract. I examine the uncertainty of perturbative Quantum Chromodynamics (QCD) factorization for (un)polarized hadron structure functions in deep inelastic scattering at a large value of the Bjorken variable x_B. The focus will be on Target Mass Corrections and Jet Mass Corrections in the collinear factorization framework.

Keywords: polarized DIS, target mass corrections, collinear factorization, OPE
PACS: 12.38.Bx, 13.60.Hb

Precise parton distribution functions (PDFs) at large parton fractional momentum x are vital for understanding the nonperturbative structure of the nucleon, and the effects of color confinement on its partonic constituents. For instance, the d/u quark distribution ratio near $x = 1$ is very sensitive to the nature of the quark-quark forces in the nucleon, as are parton-parton correlations in the nucleon wave function. Moreover, the magnitude and sign of the ratios of spin-polarized to spin-averaged PDFs, $\Delta u/u$ and particularly $\Delta d/d$, in the limit $x \to 1$ reflect the nonperturbative quark-gluon dynamics in the nucleon, and can shed light on the origin of the nucleon's spin. Improved PDFs at large x will be crucial at the upcoming 12 GeV Jefferson Lab Upgrade for the planning and analysis of a host of experiments, as well as at upcoming neutrino oscillation experiments such as MINOS and MINERνA at Fermilab to control and reduce systematic errors to a minimum. They will also directly impact experiments at hadron colliders such as the Large Hadron Collider at CERN, by facilitating accurate extraction of new physics signals as excess compared to benchmark perturbative Quantum Chromodynamic (pQCD) calculations, and by better constraining PDFs at lower x and larger Q^2 via pQCD evolution.

Parton distribution functions can be extracted from experimental data through global QCD fits, which combine data from many different processes and observables, and analyze them by means of pQCD calculations. Currently, the unpolarized parton distributions are well-determined for $x \lesssim 0.5$ for quarks and $x \lesssim 0.3$ for gluons. To better constrain them at large x it is necessary to study hard scattering processes near kinematic thresholds, such as Deep Inelastic Scattering (DIS) at large Bjorken invariant x_B and low 4-momentum transfer squared Q^2, or Drell-Yan (DY) lepton pair production at large Feynman variable x_F. The large-x and low-Q^2 region is even more important in the case of parton helicity and transversity distributions, which measure the parton's spin distribution and correlation to its orbital angular momentum, since much of the presently available experimental data lie in this kinematic domain.

Theoretical improvements in calculating leading-twist perturbative cross sections are needed in order to confidently access large-x / low-Q^2 PDFs. In this talk, I use the collinear factorization framework to formulate and quantify corrections to DIS structure

CP1149, *Spin Physics, 18th International Spin Physics Symposium*
edited by D. G. Crabb, D. B. Day, S. Liuti, X. Zheng, M. Poelker, and Y. Prok

functions arising from target and jet mass corrections [1, 2]. Threshold resummation, higher twists, quark-hadron duality in the pre-asymptotic domain, and nuclear effects such as binding, Fermi motion and nucleon off-shellness are left for future efforts.

Target mass corrections (TMC) for spin-averaged nucleon structure functions were first considered by Georgi & Politzer within the Operator Product Expansion (OPE) [3, 4]. For spin-dependent scattering they were evaluated within the same OPE formalism in Refs. [4, 5]. One of the limitations of the OPE formulation of TMCs is the so-called "threshold problem", in which the target mass corrected structure functions remain nonzero at $x_B \geq 1$. This arises from the failure to consistently incorporate the elastic threshold in moments of structure functions at finite Q^2, resulting in nonuniformity of the $Q^2 \to \infty$ and $n \to \infty$ limits, where n is the rank of the moment. As a consequence, after performing an inverse Mellin transform on the moments, the structure functions acquire incorrect support at large x_B [6]. Various prescriptions have been considered to tame the unphysical behavior as $x_B \to 1$. However, these approaches are not unique and sometimes introduce additional complications.

An alternative approach, which avoids the threshold ambiguities from the outset, involves formulating TMCs directly in momentum space using the collinear factorization (CF) formalism [1]. A further advantage of this formalism is that it can be readily extended to processes such as semi-inclusive DIS, where the OPE is not available, and indeed to any other hard scattering process. Additional corrections to structure functions at large x_B, such as discussed in the introduction, can also be naturally incorporated together with TMCs. At any order in the strong coupling constant α_s, and considering only massless partons, the CF target-mass corrected transverse and longitudinal structure functions, F_T and F_L, read

$$F_{T,L}(x_B, Q^2, M^2) = \sum_f \int_{\xi}^{\xi/x_B} \frac{dx}{x} h_{f|T,L}\left(\frac{\xi}{x}, Q^2\right) \varphi_f(x, Q^2) \,, \tag{1}$$

where f is the parton flavor, M the target's mass, ξ the Nachtmann scaling variable,

$$\xi = \frac{2x_B}{1+\sqrt{1+\gamma^2}} \qquad \gamma^2 = 4x_B^2 M^2/Q^2 \,, \tag{2}$$

h_f the perturbatively calculable coefficients, and φ_f the unpolarized PDFs. The polarized g_1 structure function reads

$$g_1(x_B, Q^2, M^2) = \frac{1}{1+\gamma^2} \sum_f \int_{\xi}^{\xi/x_B} \frac{dx}{x} g_{1,f}\left(\frac{\xi}{x}, Q^2\right) \Delta\varphi_f(x, Q^2) \,, \tag{3}$$

where $g_{1,f}$ are the perturbatively calculable coefficients, and $\Delta\varphi_f$ are the parton helicity distributions. The upper limit of integration $x \leq \xi/x_B$ is imposed by combining four-momentum and baryon number conservation in the kinematics of the handbag diagram. As a result, the structure functions are zero when $x_B \geq 1$, which solves the threshold problem. Note also that contrary to previous claims [7], $F_{T,L}(x_B) \neq F_{T,L}^{(0)}(\xi)$, and $g_1(x_B) \neq (1+\gamma^2)^{-1} g_1^{(0)}(\xi)$, where the superscript (0) indicates "massless" structure functions computed with $M = 0$, and the upper limit on the x integration is $x \leq 1$.

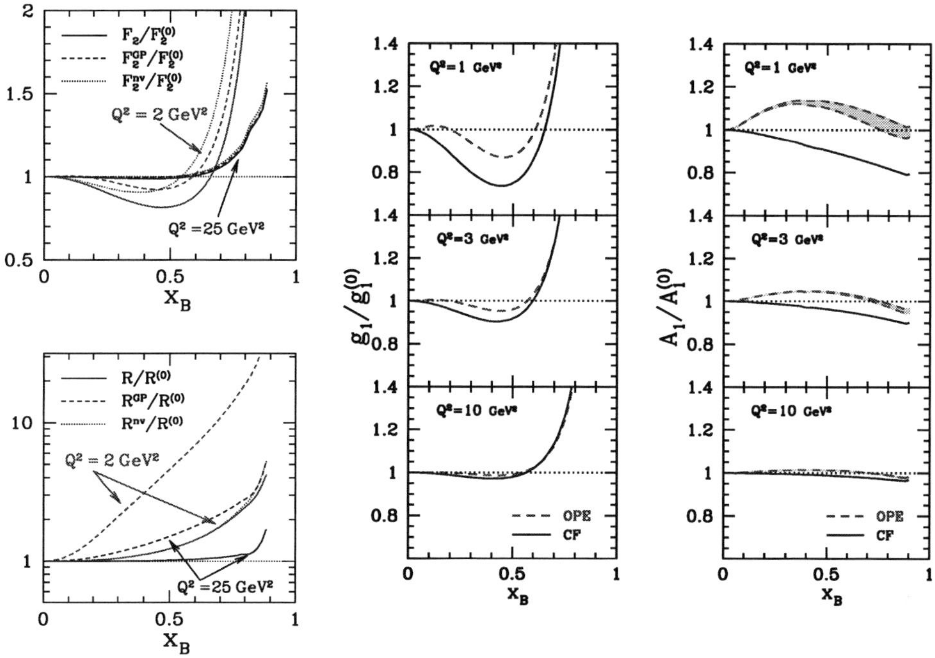

FIGURE 1. *Square plots:* Comparison of prescriptions for NLO target mass corrections to F_2 and $R = F_L/F_1$, computed with MRST2002 parton distributions [1]. (The "GP" curves are computed in OPE; the "nv" curves are computed using 1 as upper bound for the x integration in Eq. (1). *Vertical plots:* Same for the LO g_1 structure function (left) and A_1 polarization asymmetry (right) [2]. For A_1, the shaded band for the OPE result indicates the uncertainty due to the use of the Wandzura-Wilczek relation (lower bound) or the identity $g_1 + g_2 = 0$ (upper bound) in deriving A_1.

An estimate of TMC in collinear factorization for polarized and unpolarized structure functions, and for the $A_1 = (1+\gamma^2)g_1/F_1$ asymmetry is presented in Fig. 1, and compared to the TMC computed in OPE. The numerical results for the target mass corrections to the F_2 and g_1 structure functions are qualitatively similar in the CF and OPE computations. At leading order (LO), the differences of up to 30% for low Q^2 values disappear for $Q^2 \gtrsim 10$ GeV^2. At next-to-leading order (NLO), the differences are generally larger than at LO for intermediate x_B values. In all cases the TMCs remain significant at $x_B \gtrsim 0.7$ even for $Q^2 > 10$ GeV^2, and need to be taken into account when analyzing large-x_B data. In the longitudinal F_L structure function the CF and OPE corrections are larger by themselves, and differ by more than factors of several at large and not-so-large x_B, even at large Q^2 values usually considered TMC-free. The choice of TMC scheme will affect, e.g., the determination of the gluon PDF, to which F_L is much more sensitive than F_2, and is not a negligible matter. Since the TMCs are qualitatively similar for the g_1 and F_1 structure functions, they largely cancel in the A_1 asymmetry, although the sign of the correction is opposite in the CF and OPE approaches over most of the range of x_B, but can be as large as 20% at low Q^2 , disappearing at $Q^2 \gtrsim 10$ GeV^2. The commonly used approximation relating A_1 directly to the longitudinal asymmetry $A_\|$ can be shown to be equivalent to assuming $A_1 \approx A_1^{(0)}$ [2]. Figure 1 shows that it breaks down at low Q^2: an accurate measurement of g_1, e.g., at JLab, requires both $A_\|$ and $A_\perp$ asymmetries.

Even after solving the threshold problem as described, the structure functions remain unphysically positive as $x_B \to 1$ because the assumption that at LO the final state current jet is made of a single massless quark makes the LO perturbative coefficients proportional to $\delta(x - \xi/x_B)$. Such an assumption is clearly unphysical because the quark has to hadronize due to color confinement, so that the current jet will have an invariant mass m_j^2. Using this mass in the kinematics of the LO handbag diagram, and assuming that the invariant jet mass has a probability distribution J_m, one obtains

$$F_T^{JMC}(x_B, Q^2, M^2) = \int_0^{\frac{1-x_B}{x_B}Q^2} dm_j^2 J_m(m_j^2) F_T^{(0)}\left(\xi(1 + m_j^2/Q^2), Q^2\right). \qquad (4)$$

If the "jet function" $J_m(m_j^2)$ is a sufficiently smooth function of m_j^2, one obtains $F_T^{JMC}(x_B, Q^2, M^2) \to 0$ as $x_B \to 1$, as it should be. The jet mass corrections (JMC) so introduced are of order $O(m_j^2/Q^2)$, and it is easy to see that in the limit $Q^2 \gg \langle m_j^2 \rangle$ one recovers the structure functions with TMC only. It turns out that the ansatz (4) is theoretically correct, and that the jet function is the spectral function, J_2, of a vacuum quark propagator,

$$\int_0^\infty dm_j^2 J_2(m_j^2) 2\pi\delta(l^2 - m_j^2)\,\theta(l^0) = \frac{1}{4l^-}\int d^4z e^{iz\cdot l}\mathrm{Tr}\left[\gamma^- \langle 0|\overline{\psi}(z)\psi(0)|0\rangle\right], \qquad (5)$$

additionally smeared by soft momentum exchanges with the target jet [1]. A toy model estimate of these Jet Mass Corrections shows that they may in fact be comparable in size to the TMC. However their phenomenology is just at its beginning.

ACKNOWLEDGMENTS

This work was supported by the DOE contract No. DE-AC05-06OR23177, under which Jefferson Science Associates, LLC operates Jefferson Lab, and NSF award No. 0653508.

REFERENCES

1. A. Accardi and J. W. Qiu, JHEP **0807** (2008) 090.
2. A. Accardi and W. Melnitchouk, *Phys. Lett. B* **670**, 114 (2008).
3. H. Georgi and H. D. Politzer, *Phys. Rev. D* **14**, 1829 (1976); A. De Rujula, H. Georgi and H. D. Politzer, *Annals Phys.* **103**, 315 (1977).
4. J. Blumlein and A. Tkabladze, *Nucl. Phys. B* **553**, 427 (1999).
5. S. Matsuda and T. Uematsu, *Nucl. Phys. B* **168**, 181 (1980); A. Piccione and G. Ridolfi, *Nucl. Phys. B* **513**, 301 (1998); J. Blumlein and A. Tkabladze, *Nucl. Phys. B* **553**, 427 (1999).
6. K. Bitar, P. W. Johnson and W. K. Tung, *Phys. Lett. B* **83**, 114 (1979); P. W. Johnson and W. K. Tung, Print-79-1018 (Illinois Tech) *Contribution to Neutrino '79*, Bergen, Norway, June 18-22, 1979. For a review, see I. Schienbein *et al.*, *J. Phys. G* **35**, 053101 (2008).
7. M. A. G. Aivazis, F. I. Olness and W. K. Tung, *Phys. Rev. D* **50**, 3085 (1994); S. Kretzer and M. H. Reno, *Phys. Rev. D* **69**, 034002 (2004).

Phenomenological Determination of the Orbital Angular Momentum

Gordon P. Ramsey

Physics Department, Loyola University, Chicago, IL 60626
and High Energy Physics Division, Argonne National Lab, Argonne, IL 60439

Abstract. Measurements involving the gluon spin, $\Delta G(x,t)$ and the corresponding asymmetry, $A(x,t) = \Delta G(x,t)/G(x,t)$ play an important role in quantitative understanding of proton structure. We have modeled the asymmetry perturbatively and calculated model corrections to obtain information about non-perturbative spin-orbit effects. These models are consistent with existing COMPASS and HERMES data on the gluon asymmetry. The $J_z = \frac{1}{2}$ sum rule is used to generate values of orbital angular momentum at LO and NLO. For models consistent with data, the orbital angular momentum is small. Our studies specify accuracy that future measurements should achieve to constrain theoretical models for nucleon structure.

Keywords: Proton structure, gluon asymmetry, orbital angular momentum
PACS: 13.88.+e, 14.20.Dh

One of the goals of high energy spin physics studies is to investigate the role of quark and gluon constituents in contributing to the spin of nucleons. Recent deep inelastic scattering (DIS) experiments [1] have significantly lowered the measurement errors of the quark spin contribution ($\Delta\Sigma$) to the proton. The COMPASS collaboration analysis quotes a result

$$< \Delta\Sigma > = 0.30 \pm 0.01(\text{stat}) \pm 0.02(\text{evol}), \qquad \text{all data} \tag{1}$$

while the HERMES collaboration analysis quotes a result

$$< \Delta\Sigma > = 0.330 \pm 0.025(\text{exp}) \pm 0.011(\text{th}) \pm 0.028(\text{evol}), \qquad \text{all data.} \tag{2}$$

Several experimental programs have devised strategies aimed at providing a measure of the proton's spin weighted gluon density, $\Delta G(x,t)$ and the corresponding gluon asymmetry $A(x,t) \equiv \Delta G(x,t)/G(x,t)$, where x is the Bjorken scaling variable and $t \equiv \log[\alpha_s(Q_0^2)]/\log[\alpha_s(Q^2)]$ is the Q^2 evolution variable. The relation between the various components of proton spin is summarized by the $J_z = \frac{1}{2}$ sum rule:

$$J_z = \frac{1}{2} \equiv \frac{1}{2} < \Delta\Sigma(t) > + < \Delta G(t) > + L_z(t), \tag{3}$$

where $< \Delta\Sigma(t) >$ and $< \Delta G(t) >$ are the projections of the spin carried by all quarks and the gluons on the z-axis, respectively. The last term, L_z is the net z-component of the orbital angular momentum (OAM) of the constituents. We do not attempt to identify independent flavor components of L_z within the sum rule.

CP1149, *Spin Physics, 18th International Spin Physics Symposium*
edited by D. G. Crabb, D. B. Day, S. Liuti, X. Zheng, M. Poelker, and Y. Prok

The purpose of this work is to model ΔG perturbatively and use equation (3) to extract information on the OAM of the proton constituents. We write the polarized gluon asymmetry using the decomposition

$$A(x,t) \equiv \Delta G(x,t)/G(x,t) = A_0^{\varepsilon}(x) + \Delta g_{\varepsilon}(x)/G(x,t), \tag{4}$$

where we assume that $A_0^{\varepsilon}(x)$ is calculable in perturbative Quantum Chromodynamics (pQCD) by the definition

$$A_0^{\varepsilon}(x) \equiv \left[\left(\frac{\partial \Delta G(x,t)}{\partial t}\right) / \left(\frac{\partial G(x,t)}{\partial t}\right) \right]. \tag{5}$$

The small correction, $\Delta g_{\varepsilon}(x)$, is necessary due to slight differences in the evolutions of $G(x,t)$ and $\Delta G(x,t)$ at leading order (LO). The "gluon excess" $\Delta g_{\varepsilon}(x)$ is not gauge-invariant, but can be interpreted as a small correction to account for any differences in the perturbatively calculated gluon polarization and the experimentally measured values. This function can be modeled to generate a suitable prediction for $\Delta G(x,t)$ and, in turn, lend insight into the OAM of the proton constituents.

To establish an initial reference point, practical constraints for the asymmetry at LO should include:

- strong positivity: $|A(x,t_0)| \leq 1$ for all $0 \leq x \leq 1$, and
- endpoint values: $A(0,t_0) = 0$ and $A(1,t_0) = 1$.

In kinematic regions where the DGLAP (Dokshitzer-Gribov-Lipatov-Altarelli-Parisi) evolution equations are valid, equation (5) allows one to generate $A_0^{\varepsilon}(x)$ by

$$A_0^{\varepsilon} = \left[\frac{\Delta P_{Gq} \otimes \Delta q + \Delta P_{GG} \otimes (\Delta G)}{P_{Gq} \otimes q + P_{GG} \otimes G} \right], \tag{6}$$

where ΔG is replaced by

$$\Delta G(x,t) = A_0^{\varepsilon}(x) \cdot G(x,t) + \Delta g_{\varepsilon}(x). \tag{7}$$

The calculated asymmetries are determined by testing parameterizations of Δg_{ε} that together satisfy the constraints listed above. For each of these parameterizations, we start with an A_0^{ε} in the form

$$A_0^{\varepsilon} \equiv Ax^{\alpha} - (B-1)x^{\beta} + (B-A)x^{\beta+1}, \tag{8}$$

which automatically satisfies the endpoint constraints. The specific asymmetries were calculated using the CTEQ5 and MRST 2001 unpolarized parton distributions and the polarized distributions of reference [3]. Details of the calculation can be found in reference [4]. Only parameterizations of Δg_{ε} that conform to the condition $| < \Delta g_{\varepsilon} > | \leq 0.25$ satisfy the positivity constraint. Resulting parameterizations are then compared to existing data to narrow the possibilities for both ΔG and the asymmetry.

Using the current data for $\Delta\Sigma$ as input, we combine equations (1), (2) and (3) to write

$$L_z(t) + < \Delta G(t) > \approx \frac{1}{2}(1 - < \Delta\Sigma >) \approx 0.35 \pm 0.02. \tag{9}$$

TABLE 1. Gluon Asymmetry Parameters at $Q_0^2 = 1$ GeV^2

Δg_ε	$< \Delta g_\varepsilon >$	A_0^ε	$< \Delta G >$	L_z
$-2(1-x)^7$	-0.25	$1.75x^{1.1} - 1.5x^{2.1} + 0.75x^{3.1}$	0.18	0.17
$-90x^2(1-x)^7$	-0.25	$3.5x^{1.3} - 4.5x^{2.2} + 2x^{3.2}$	0.25	0.10
$-9x(1-x)^7$	-0.125	$3.25x^{1.4} - 3.75x^{2.2} + 1.5x^{3.2}$	0.25	0.10
0	0	$3x^{1.5} - 3x^{2.2} + x^{3.2}$	0.28	0.07
$x(x-0.25)(1-x)^5$	0	$1.75x^{1.9} - 5x^{2.4} + 4.25x^{3.4}$	-0.01	0.36

in a chiral factorization prescription. The viable asymmetries that agree to within 1σ of the existing data from COMPASS and HERMES are in lines one through three of Table 1 [2]. These give integrals $< \Delta G >$ ranging from about 0.18 to 0.25, implying that the gluon polarization is likely not large at these small $Q_0^2 = 1$ GeV^2 values. Those agreeing to within 2σ of the data are in the remaining lines of Table 1 and have integrals $< \Delta G >$ between $-0.01 \rightarrow 0.28 \pm 0.04$. All of the parameterizations that best agree with data have $\Delta g_\varepsilon \leq 0$.

Defining $< \Delta G > \equiv < \Delta G(t_0) >$, the parameterizations of Δg_ε allowed by the "practical" constraints give the bounds

$$-0.09 \quad \leq < \Delta G > \pm 0.04 \leq 0.59 \quad \text{and} \tag{10}$$
$$-0.16 \quad \leq < L_z > \pm 0.04 \leq 0.44. \tag{11}$$

The range only weakly depends on the input distributions. The ranges of $< \Delta G >$ and $< L_z >$ appear to be numerically limited by the constraints. When we require consistency of our asymmetry with existing experimental data (less the negative value given by the COMPASS charm production data) to within 1σ our approach gives the constraints

$$0.18 \quad \leq < \Delta G > \pm 0.04 \leq 0.25 \quad \text{and} \tag{12}$$
$$0.10 \quad \leq < L_z > \pm 0.04 \leq 0.17. \tag{13}$$

Requiring consistency of the asymmetry at a level of 2 σ gives the bounds

$$-0.01 \quad \leq < \Delta G > \pm 0.04 \leq 0.36 \quad \text{and} \tag{14}$$
$$0.07 \quad \leq < L_z > \pm 0.04 \leq 0.36. \tag{15}$$

This includes a significant portion of the ranges compatible with data at both the 1σ and 2σ levels. If the charm data from COMPASS is taken into account: $A = -0.49 \pm 0.27 \pm 0.11$ at $x = 0.11$, then the constraints become

$$-0.09 \quad \leq < \Delta G > \pm 0.04 \leq 0.25 \quad \text{and} \tag{16}$$
$$0.10 \quad \leq < L_z > \pm 0.04 \leq 0.44. \tag{17}$$

The results for the multiplicity of scenarios we have shown are not significantly different from each other. However, we do not rule out the necessity of altering this approach to consider more flexible models if data change significantly. We consider it likely that additional accurate data for the gluon asymmetry and ΔG over a wider kinematic range will allow us to further restrict our range of consistency for both ΔG and $A(x,t)$. Two immediate experimental possibilities are

- (1) the measurement of $\Delta G/G$ in a wider kinematic range of x_{Bj} at RHIC and
- (2) extraction of ΔG by prompt photon production or jet production in polarized pp collisions (RHIC).

We have proposed a method to calculate the gluon spin asymmetry $\frac{\Delta G}{G}$ independent of a priori assumptions about its shape of size. This is complementary to polarized data fitting schemes that exist in the literature. [5] These are presently somewhat limited by the existing data. This approach and the $J_z = \frac{1}{2}$ sum rule have allowed us to extract information about the OAM of the proton constituents. The allowable range for ΔG and L_z is restricted by physical constraints. Existing experimental measurements, within reported errors, suggest that the allowed values of $\Delta G(x,t)$ and $A(x,t)$ are considerably narrowed. This indicates that having more data available on ΔG and $A(x,t)$ over a wider kinematic range can further constrain these values.

ACKNOWLEDGMENTS

This work is supported by the U.S. Department of Energy, Division of High Energy Physics, Contract DE-AC02-06CH11357. ANL-HEP-CP-09-13

REFERENCES

1. F. Bradamante, *Phys. Lett.* **B612**, 154, (2005); *Prog. Part. Nucl. Phys.* **61**, 229, (2008); A. Airpetian, *et al.*,, *Phys. Lett.* **84**, 2584 (2000).
2. PHENIX Collaboration, *Phys. Rev.* **B76**, 051106, (2007) and arXiv:0802.3023 [hep-ex] (COMPASS).
3. L. E. Gordon, M. Goshtasbpour and G. P. Ramsey, *Phys. Rev.* **D58**, 094017 (1998) and hep-ph/9803351.
4. G. Ramsey, SPIN 2007 Proceedings (http://theor.jinr.ru/ spin), arXiv:0801.1128 [hep-ph].
5. M. Hirai and S. Kumano, arXiv 0808.0413 (hep-ph) and KEK-TH-1208 and M. Hirai, S. Kumano and N. Saito, hep-ph/0601087 (NSPIN).

The *New* $\Delta q(x)$ at HERMES

J. G. Rubin (on behalf of the HERMES Collaboration)

University of Illinois Department of Physics, 1110 W. Green St. Urbana, IL 61801, USA

Abstract. Since the HERMES Collaboration published its 5-flavor $\Delta q(x)$ extraction in 2005 [1], several aspects of the analysis and its related longitudinal double-spin asymmetries have been revisited. Work has been done to improve statistical uncertainties, more precisely determine previously overestimated systematic uncertainties, and to extract new related quantities. This presentation deals with two features of this new analysis. First, the $p_{h_\perp}$ dependence of the asymmetry A_1 is presented. No $p_{h_\perp}$ dependence is observed within the precision of the available statistics. Second, a novel approach for estimating the systematic uncertainty contribution to the extracted $\Delta q(x)$ values from uncertainty in the tune of the fragmentation model has been developed and is discussed.

Keywords: Polarized DIS, SIDIS, Transvese Momentum, Fragmentation Model Uncertainty
PACS: 13.50.-r, 13.88.+e, 14.20.Dh, 25.20.-x, 14.65.Bt

INTRODUCTION

The HERMES longitudinally polarized data set was acquired during the years 1996 through 2000 with 27.6 GeV HERA beam electrons (or positrons) incident upon the HERMES internal gas target containing either hydrogen or deuterium. The beam had an average polarization of $\sim 53\%$ and the target of $\sim 82\%$. Through the application of the inclusive kinematic cuts on Q^2 and W^2 given in Table 1, deep inelastic scattering (DIS) events are selected. While this is adequate for constructing inclusive asymmetries,

$$A_1(x) \equiv \frac{\sigma_{1/2} - \sigma_{3/2}}{\sigma_{1/2} + \sigma_{1/2}}(x) = \frac{g_1}{F_1}(x) = \frac{1}{2F_1(x)} \sum_q e_q^2 \Delta q(x), \tag{1}$$

more is required to disentangle the individual quark flavor contributions to $g_1(x)$.

In order to extract the quark helicity densities, the quark contributions must be unraveled. To accomplish this, semi-inclusive asymmetries are constructed as the final state hadron variety is expected to bare some correlation with the struck quark in the scattering process. If one rewrites Eq. 1 with an identified hadron, h, in the final state,

$$A_1^h(x,z,p_{h_\perp}) = \frac{\sum_q e_q^2 \Delta q(x) D_q^h(z,p_{h_\perp})}{\sum_{q'} e_{q'}^2 q'(x) D_{q'}^h(z,p_{h_\perp})} = \sum_q P_q^h(x,z,p_{h_\perp}) \frac{\Delta q(x)}{q(x)}, \tag{2}$$

where the *purity*, P_q^h, represents the likelihood that an observed hadron of species h was produced by striking a quark of flavor q. In practice, these purities are computed using JETSET [2], a phenomenological Monte Carlo fragmentation model.

While in prior analyses [1] semi-inclusive asymmetries were binned purely in terms of x, part of this new work has been to explore the benefits of including the new kinematic

CP1149, *Spin Physics, 18th International Spin Physics Symposium*
edited by D. G. Crabb, D. B. Day, S. Liuti, X. Zheng, M. Poelker, and Y. Prok

TABLE 1. Relevant kinematic variables and SIDIS requirements.

Variable	Description	Requirement
x	Parton Light Cone Momentum Fraction	
Q^2	Negative Squared Virtual Photon Momentum	$Q^2 > 1$ GeV^2
W^2	Final State Invariant Mass Squared	$W^2 > 10$ GeV^2
$p_{h_\perp}$	Hadron Transverse Momentum w.r.t. Photon	
z_h	Ratio of Hadron Energy to that of the Photon	$0.2 > z_h > 0.8$

dimensions z and $p_{h_\perp}$. In addition to potentially improving statistical leverage in the $\Delta q(x)$ extraction, this has given us access to new observables like $A_1^h(p_{h_\perp})$.

HADRON TRANSVERSE MOMENTUM DEPENDENCE OF A_1^h

One potentially interesting and largely unexplored feature of A_1^h is its dependence on $p_{h_\perp}$. In its simplest form, the $p_{h_\perp}$ dependence is understood to depend on a combination of the fragmentation-related and intrinsic parton transverse momentum. While it will likely be a complex task to extract the underlying distribution and fragmentation functions, theoretical work on this topic has already begun [3]. By placing constraints on certain hadron transverse momentum sources, the authors of [3] have produced $A_1(p_{h_\perp})$ calculations for various values of the widths of the underlying distributions. These calculations have been made at typical kinematics of HERMES and CLAS who have also recently begun to show preliminary results at conferences [4].

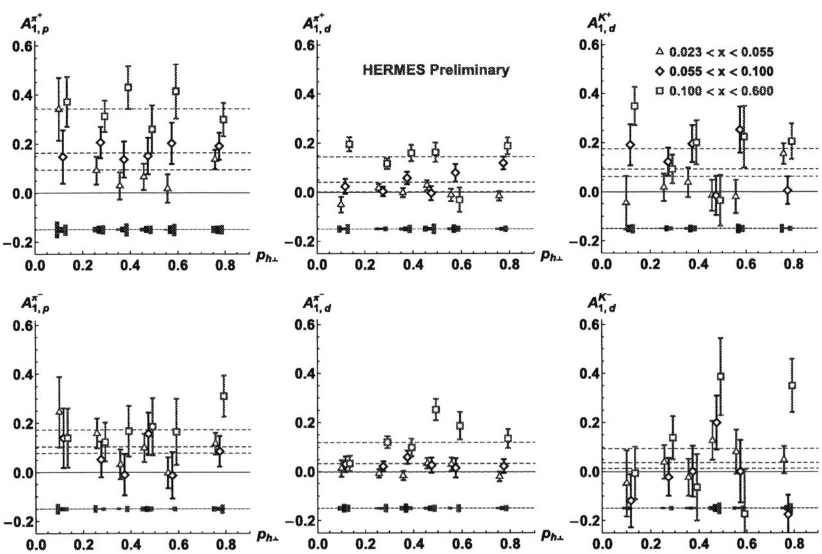

FIGURE 1. $A_1^h(p_{h_\perp})$ in three different x bins. The dotted lines represent the best fit of the functional form $A_1^h(x) = C_1^h + C_2^h x$. The small bands beneath the points are total systematic uncertainties.

TABLE 2. Agreement of $p_{h_\perp}$-independent and dependant fits with data.

	$p \to \pi^+$	$p \to \pi^-$	$d \to \pi^+$	$d \to \pi^-$	$d \to K^+$	$d \to K^-$
$\chi^{2\ (NDF=16)}_{C_1^h + C_2^h x}$	11.8	13.7	42.9	25.8	32.0	29.3
$\chi^{2\ (NDF=15)}_{C_1^h + C_2^h x + C_3^h p_{h_\perp}}$	11.7	13.3	41.1	25.2	32.0	27.4

The HERMES result for $A_1(p_{h_\perp})$ can be seen in Fig. 1. Because A_1^h depends strongly on x which is to some degree correlated with $p_{h_\perp}$, it would be misleading to present $A_1(p_{h_\perp})$ integrated over all x or without a correction of some sort. We have chosen to present these data in three distinct x ranges. This greatly reduces differences in average x across the range of $p_{h_\perp}$ for each x-bin.

In order to make a quantitative statement about any potential $p_{h_\perp}$ dependence, the data points have been fitted with two simple functions, one with $p_{h_\perp}$ dependence, $A_1^h(p_{h_\perp}) = C_1^h + C_2^h x + C_3^h p_{h_\perp}$, and one without, $A_1^h(p_{h_\perp}) = C_1^h + C_2^h x$. The latter is overlaid in Figure 1. The total χ^2 for each fit is presented in Table 2. Within the statistical precision available, no evidence for $p_{h_\perp}$ dependence is observed.

FRAGMENTATION MODEL UNCERTAINTIES

Another feature of this reanalysis has been to revisit certain sources of systematic uncertainty. In the published $\Delta q(x)$ result [1], the dominant source of systematic uncertainty in many kinematic bins was uncertainty in the tune of the fragmentation Monte Carlo used to produce the purity matrices of Eq. 2.

The difficulty with precisely determining this uncertainty stems from 1) the large number of parameters (>10) guiding JETSET, 2) the computational effort required to test each point in the input parameter space, and 3) the lack of a convenient analytic form for the model making more conventional error propagation difficult. In order to tune the model to the HERMES kinematic regime, different combinations of these input parameters are used to produce Monte Carlo multiplicities which are compared to HERMES data multiplicities. By minimizing the Monte Carlo to data multiplicity χ^2, a best tune is found.

In the past, this uncertainty was estimated by comparing $\Delta q(x)$ extracted using the best fit purities with $\Delta q(x)$ extracted using purities from two tunes known to represent fragmentation at HERMES kinematics poorly. For lack of a more rigorous method and adequate computational power, this was expected to produce a conservative estimate. A further potential problem was that without an exhaustive scan of the input parameter space in the vicinity of the tune, there was a *possibility* that correlations between parameters could produce a trough in the χ^2 surface – an ambiguity in the best tune that could have a significant impact on the final extracted $\Delta q(x)$ result.

Both of these issues have been addressed by the new uncertainty estimation method depicted schematically in Fig. 2. The procedure works as follow:

1. Generate a set of randomly selected tunes, produce Monte Carlo multiplicities for each, and compute χ^2 for each parameter space point by comparing Monte Carlo

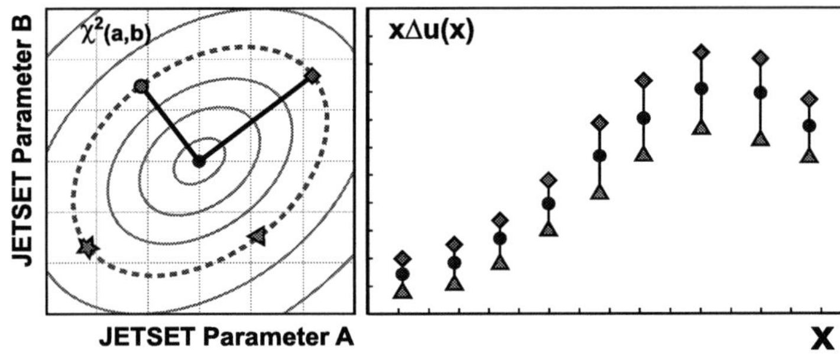

FIGURE 2. (Left) A schematic depiction of a contour plot of a quadratic polynomial fit to the χ^2 surface, associated uncorrelated parameter directions, and their intersections with the 1σ contour (dotted). (Right) Schematic representation of $\Delta q(x)$ extracted with purities at different parameter space points in order to sweep out the final fragmentation-related error band.

with unpolarized data multiplicities.

2. Fit the χ^2 surface with a quadratic polynomial assuming smoothness around the best tune and find uncorrelated *Hessian* parameter directions by diagonalizing the matrix of second derivatives of the fit function.
3. Compute the intersections of the Hessian vectors with a contour of fixed height above χ^2_{min} representing the 1σ uncertainty. This height is a function of both the numbers of parameters involved and the known imperfection in the model ($\chi^2_{min} \neq N\text{param}$).
4. Extract $\Delta q(x)$ using purities generated at each of the 1σ Hessian intersections to sweep out the uncertainty band corresponding to the tune uncertainty.

The result of this procedure is a set of rigorously determined systematic uncertainties that are indeed significantly smaller than those from the conservative estimate produced previously. Additionally, by mapping parameter correlations, the possibility of a misleading and dangerous χ^2_{min} degeneracy is excluded.

CONCLUSIONS

We have presented here only two of several new features which are part of the reanalysis of $\Delta q(x)$ at HERMES. In addition to presenting new observables, like $A_1(p_{h_\perp})$, statistical and systematic uncertainties on $\Delta q(x)$ have been revisited and in most cases improved.

REFERENCES

1. A. Airapetian, et al., *Phys. Rev.* **D71**, 012003 (2005), hep-ex/0407032.
2. T. Sjostrand, Pythia 5.7 and jetset 7.4 physics and manual (1995), hep-ph/9508391.
3. M. Anselmino, A. Efremov, A. Kotzinian, and B. Parsamyan, *Phys. Rev.* **D74**, 074015 (2006).
4. H. Avakian, "Studies on transverse spin effects at Jlab," in *2nd Workshop On QCD Structure of the Nucleon*, 2006.

Semi-leptonic Octet Baryon Weak Axial-Vector Form Factors in the Chiral Constituent Quark Model

Harleen Dahiya*, Neetika Sharma*, P.K. Chatley* and Manmohan Gupta†

*Department of Physics, Dr. B.R. Ambedkar National Institute of Technology, Jalandhar, Punjab-144 011, India
†Department of Physics, Panjab University, Chandigarh-160 014, India

Abstract. The weak vector and axial-vector form factors have been investigated for the semi-leptonic octet baryon decays in the chiral constituent quark model with configuration mixing (χCQM$_{\rm config}$). The effects of SU(3) symmetry breaking and configuration mixing have also been investigated and the results are not only in good agreement with existing experimental data but also show improvement over other phenomenological models.

Keywords: Chiral constituent quark model, weak form factors
PACS: 14.20.Dh, 12.39.Fe, 13.30.Ce

The measurements of the polarized structure functions in the deep inelastic scattering (DIS) experiments indicate that the valence quarks of the proton carry only about 30% of its spin and also establish the asymmetry of the quark distribution functions [1]. This is referred to as the "proton spin problem" and it relates the DIS measurements to the weak vector and axial-vector form factors ($f_{i=1,2,3}(Q^2)$ and $g_{i=1,2,3}(Q^2)$) of the semi-leptonic baryon decays [2]. The data to study the form factors in the early eighties was analyzed under the assumptions of exact SU(3) symmetry [3]. However, the experiments performed recently [4] are more precise and indicate that SU(3) symmetry breaking effects are important. The broader question of SU(3) symmetry breaking has been recently discussed by several authors in various phenomenological models [5] however, the results of all these models are not in agreement with each other on the magnitude as well as sign of these form factors. Pending further experiments, it is of interest to estimate the role of SU(3) symmetry breaking in the weak form factors.

Chiral constituent quark model with configuration (χCQM$_{\rm config}$) [6, 7, 8] can yield an adequate description of the "quark sea" generation through the chiral fluctuations and has been successful in explaining various general features of the quark flavor and spin distribution functions and baryon magnetic moments. It, therefore, becomes desirable to carry out a detailed phenomenological analysis within the framework of χCQM$_{\rm config}$ to calculate the axial-vector form factors [9]. The purpose of the present communication is to carry out a detailed analysis of the weak vector and axial-vector form factors at low energies for the baryon semi-leptonic decays within the framework of χCQM$_{\rm config}$ and to understand the role of SU(3) symmetry breaking and constituent quark masses.

In the χCQM formalism [7], the fluctuation process is $q^{\pm} \rightarrow {\rm GB} + q^{'\mp} \rightarrow (q\bar{q}^{'}) + q^{'\mp}$, where GB represents the Goldstone boson and $q\bar{q}^{'} + q^{'}$ constitute the "quark sea". The

CP1149, *Spin Physics, 18th International Spin Physics Symposium*
edited by D. G. Crabb, D. B. Day, S. Liuti, X. Zheng, M. Poelker, and Y. Prok

effective Lagrangian describing interaction between quarks and a nonet of GBs can be expressed as $\mathscr{L} = g_8\bar{q}(\Phi)q$, where g_8 is the coupling constants and Φ is the GB field. The SU(3) symmetry breaking is introduced by considering $m_s > m_{u,d}$ as well as by considering the masses of GBs to be nondegenerate ($M_{K,\eta} > M_\pi$). The nucleon wavefunction with configuration mixing can be expressed as $|B\rangle = \cos\phi|56,0^+\rangle_{N=0} + \sin\phi|70,0^+\rangle_{N=2}$, where ϕ represents the mixing angle. The spin structure of a nucleon is defined as $\hat{B} \equiv \langle B|N|B\rangle$ [7], where N is the number operator $N = n_{u^+}u^+ + n_{u^-}u^- + n_{d^+}d^+ + n_{d^-}d^- + n_{s^+}s^+ + n_{s^-}s^-$, $n_{q^\pm}$ being the number of $q^\pm$ quarks.

The transition matrix element for the semi-leptonic hadronic decay process $B_i \to B_f + l + \bar{\nu_l}$ is given as $M = \frac{G_F}{\sqrt{2}}(V_{q_iq_f})\langle B_f|J_h^\mu|B_i\rangle(\bar{u_l}(p_l)\gamma_\mu(1-\gamma_5)u_\nu(p_\nu))$. The weak hadronic current can be expressed in terms of the vector and axial-vector current as $\langle B_f(p_f)|J_h^\mu|B_i(p_i)\rangle = \langle B_f(p_f)|J_V^\mu - J_A^\mu|B_i(p_i)\rangle$ and the respective matrix elements for the vector and axial-vector current are given as

$$\bar{u}_f(p_f)\left(f_1(Q^2)\gamma^\mu - i\frac{f_2(Q^2)}{M_i+M_f}\sigma^{\mu\nu}q_\nu + \frac{f_3(Q^2)}{M_i+M_f}q^\nu\right)u_i(p_i), \tag{1}$$

$$\bar{u}_f(p_f)\left(g_1(Q^2)\gamma^\mu\gamma^5 - i\frac{g_2(Q^2)}{M_i+M_f}\sigma^{\mu\nu}q_\nu\gamma^5 + \frac{g_3(Q^2)}{M_i+M_f}q^\nu\gamma^5\right)u_i(p_i), \tag{2}$$

where M_i (M_f) and $u_i(p_i)$ ($\bar{u}_f(p_f)$) are the masses and Dirac spinors of the initial (final) baryon states, respectively. The four momenta transfer is given as $Q^2 = -q^2$, where $q \equiv p_i - p_f$. The functions $f_i(Q^2)$ and $g_i(Q^2)$ ($i = 1,2,3$) are the dimensionless vector and axial-vector form factors required to be real by $G-$parity. These equations can be solved using Gordon equality, in the Breit-frame (Lorentz frame), where $p_i = -p_f = \frac{1}{2}q$ in the nonrelativistic limit ($q^2 \ll M_i^2, M_f^2$) , giving the vector and axial-vector form factors in terms of the Sachs form factors at $Q^2 \approx 0$.

Further, the linear part of form factors can be solved in the quark sector and the form factors for the quark currents can be used to obtain the corresponding vector and axial-vector form factors for the baryons which are expressed as

$$f_1 = f_1(0), \quad f_2 = \left(\frac{\Sigma M}{\Sigma m}\frac{G_A}{G_V} - 1\right)f_1(0), \quad f_3 = \frac{\Sigma M}{\Sigma m}\left(E\frac{G_A}{G_V} - \varepsilon\right)f_1(0), \tag{3}$$

$$g_1 = g_1(0), \; g_2 = \left(\frac{\Sigma M}{\Sigma m}\varepsilon - \frac{1}{2}(1+\frac{\Sigma M^2}{\Sigma m^2})E\right)g_1(0), \; g_3 = \left(\frac{1}{2}(1-\frac{\Sigma M^2}{\Sigma m^2}) + \frac{\Sigma M^2}{\Sigma m^2}g_3^q\right)g_1(0), \tag{4}$$

where the higher order terms involving $E \equiv \frac{\Delta M}{\Sigma M}$ and $\varepsilon \equiv \frac{\Delta m}{\Sigma m}$ have been neglected. The ratio of g_1 and f_1 is the non-singlet combination of the quark spin polarizations given as $\Delta_3 = \frac{G_A}{G_V} = \frac{g_1(0)}{f_1(0)}$. The hyperon decays considered in the present work are $n \to p$, $\Sigma^\mp \to \Lambda$, $\Sigma^- \to \Sigma^0$ and $\Xi^- \to \Xi^0$ corresponding to the strangeness conserving decays and $\Sigma^- \to n$, $\Xi^- \to \Sigma^0$, $\Xi^- \to \Lambda$, $\Lambda \to p$ and $\Xi^0 \to \Sigma^+$ corresponding to the strangeness changing decays.

The χCQM$_{\text{config}}$ involves five parameters, four of these a, $a\alpha^2$, $a\beta^2$, $a\zeta^2$ representing respectively the probabilities of fluctuations to pions, K, η, η', following the hierarchy

TABLE 1. Weak vector and axial-vector form factors for the semi-leptonic octet baryons decays in the χCQM$_{\text{config}}$.

Decay	f_1	f_2	f_3	g_1	g_2	g_3
$n \to pe^-\bar{\nu}$	1.00	2.612	0.003	1.270	−0.004	−232.9
$\Sigma^- \to \Sigma^0 e^-\bar{\nu}$	1.414	1.033	0.005	0.676	−0.010	−201.3
$\Sigma^- \to \Lambda e^-\bar{\nu}$	0	2.265	0.080	0.646	−0.152	−271.4
$\Sigma^+ \to \Lambda e^-\bar{\nu}$	0	2.257	0.072	0.646	−0.136	−245.9
$\Xi^- \to \Xi^0 e^-\bar{\nu}$	−1.00	2.253	0.003	0.314	−0.007	113.8
$\Sigma^- \to ne^-\bar{\nu}$	−1.0	1.813	0.616	0.314	0.017	−9.2
$\Xi^- \to \Sigma^0 e^-\bar{\nu}$	0.707	2.029	−0.291	0.898	0.310	−29.1
$\Xi^- \to \Lambda e^-\bar{\nu}$	1.225	−0.450	−0.658	0.262	0.047	−8.9
$\Lambda \to pe^-\bar{\nu}$	−1.225	−1.037	0.415	−0.909	−0.170	20.7
$\Xi^0 \to \Sigma^+ e^-\bar{\nu}$	1.0	2.854	−0.414	1.27	0.446	−40.7

$a > \alpha > \beta > \zeta$, while the fifth representing the mixing angle. The mixing angle ϕ is fixed from the consideration of neutron charge radius, whereas the other parameters have been fitted using Δu, Δ_3 [10] as well as $\bar{u} - \bar{d}$, $\bar{u}/\bar{d}$ [1] leading to $a = 0.12$, $\zeta = -0.15$, $\alpha = \beta = 0.45$ as the best fit values.

In Table 1, we have given the individual values of vector and axial-vector form factors in the χCQM$_{\text{config}}$. Some important conclusions can be drawn regarding the dependence of these form factors on configuration mixing and SU(3) symmetry breaking. The ratio $\frac{g_1}{f_1}$ ratio depends on configuration mixing as well as SU(3) symmetry breaking. The other form factors f_2, f_3, g_2 and g_3 are found to be dependent on the constituent quark masses only. It can be clearly seen from the results that the contributions of second class currents f_3 and g_2 are very small for the same isospin multiplets. This is because of the small mass difference between the initial and final decay particles.

In Table 2, we have presented the values of $\frac{g_1}{f_1} = \frac{G_A}{G_V}$ at $Q^2 = 0$. The χCQM$_{\text{config}}$ is able to give a fairly good account for most of the decays where the experimental data is available for example, in the case of $\frac{G_A}{G_V}^{\Sigma^- \to n}$, $\frac{G_A}{G_V}^{\Xi^- \to \Lambda}$, $\frac{G_A}{G_V}^{\Lambda \to p}$ and $\frac{G_A}{G_V}^{\Xi^0 \to \Sigma^+}$. The SU(3) symmetry breaking as well as configuration mixing effects have been incorporated in the calculations and it clearly leads to a better agreement with data. Therefore, both are simultaneously important to obtain the desired agreement. A refinement in the case of the measurements with the assumption of SU(3) symmetry breaking would have important implications for the basic tenets of χCQM as well as SU(3) symmetry breaking.

In conclusion, we would like to state that at the leading order constituent quarks and the weakly interacting Goldstone bosons constitute the appropriate degrees of freedom in the nonperturbative regime of QCD in the χCQM$_{\text{config}}$ and SU(3) symmetry breaking is the key in understanding the spin content and the weak axial-vector form factors of the semi-leptonic octet baryon decays.

TABLE 2. Predictions for the the axial-vector form factors G_A/G_V in χCQM$_{\rm config}$.

Decay	Data [10]	NQM	χCQM$_{\rm config}$ with SU(3) symmetry	χCQM$_{\rm config}$ with SU(3) symmetry breaking
$\frac{G_A}{G_V}^{n\to p}$	1.2695 ± 0.0029	1.67	0.95	1.27
$\frac{G_A}{G_V}^{\Sigma^-\to\Sigma^0}$	–	0.67	0.39	0.48
$\frac{G_A}{G_V}^{\Sigma^-\to\Lambda}$	$\frac{f_1}{g_1} = 0.01 \pm 0.1$	0.82	0.45	0.65
$\frac{G_A}{G_V}^{\Sigma^+\to\Lambda}$	–	0.82	0.45	0.65
$\frac{G_A}{G_V}^{\Xi^-\to\Xi^0}$	–	−0.33	−0.16	−0.31
$\frac{G_A}{G_V}^{\Sigma^-\to n}$	-0.340 ± 0.017	−0.33	−0.16	−0.31
$\frac{G_A}{G_V}^{\Xi^-\to\Sigma^0}$	–	1.67	0.95	1.27
$\frac{G_A}{G_V}^{\Xi^-\to\Lambda}$	0.25 ± 0.05	0.33	0.21	0.21
$\frac{G_A}{G_V}^{\Lambda\to p}$	0.718 ± 0.015	1.00	0.58	0.74
$\frac{G_A}{G_V}^{\Xi^0\to\Sigma^+}$	1.21 ± 0.05	1.67	0.95	1.27

ACKNOWLEDGMENTS

H.D. would like to thank CCSTDS, Government of India and the organizers of SPIN2008, University of Virginia, for financial support.

REFERENCES

1. European Muon Collaboration, J. Ashman *et al.*, *Nucl. Phys. B* **328**, 1 (1989); E866/NuSea Collaboration, R. S. Towell *et al.*, *ibid.* **64**, 052002 (2001).
2. Lepton and Quarks, L.B. Okun, Elsevier Science Publishers (1984); F.E. Close, *The Nucleon Spin Crisis Bible*, RAL-93-034, 1993 (unpublished); F.E. Close and R.G. Roberts, *Phys. Lett. B* **316**, 165 (1993).
3. M. Bourquin *et al.*, *Z. Phys.* **21**, 17 (1983); *ibid* **21**, 27 (1983).
4. S.Y. Hsueh *et al.*, *Phys. Rev. D* **38**, 2056 (1988); A. Alavi-Harati *et al.*, KTeV Collab., *Phys. Rev. Lett.* **87**, 132001 (2001); J.R. Batley *et al* [NA48/1 Collabotration] *Phys. Lett. B* **645**, 36 (2007).
5. V. Mateu, A. Pich, JHEP **0510**, 041 (2005); R. Flores-Mendieta and C.P. Hofmann, *Phys. Rev. D* **74**, 094001 (2006); A. Lacour, B. Kubis and U.G. Meissner, JHEP **0710**, 083 (2007); T. Ledwig, A. Silva, H.-Ch. Kim, K. Goeke, JHEP **0807**, 132 (2008)
6. S. Weinberg, *Physica A* **96**, 327 (1979); A. Manohar and H. Georgi, *Nucl. Phys. B* **234**, 189 (1984).
7. E.J. Eichten, I. Hinchliffe and C. Quigg, *Phys. Rev. D* **45**, 2269 (1992); T.P. Cheng and Ling Fong Li, *Phys. Rev. Lett.* **80**, 2789 (1998); J. Linde, T. Ohlsson and H. Snellman, *Phys. Rev. D* **57**, 452 (1998).
8. H. Dahiya and M. Gupta, *Phys. Rev. D* **64**, 014013 (2001); *ibid.* **67**, 074001 (2003); *ibid. Phys. Rev. D* **66**, 051501(R) (2002), *ibid.* **67**, 114015 (2003).
9. T. Ohlsson and H. Snellman, *Eur. Phys. J. C* **6**, 285 (1999); *ibid.* **7**, 501 (1999); *ibid.* **C 12**, 271 (2000).
10. C. Amsler *et al.*, *Phys. Lett. B* **667**, 1 (2008).

Global QCD Analysis of Fragmentation and Parton Distribution Functions

Yoshiyuki Miyachi, Yoshimitsu Imazu, YuKi Kobayashi and Toshi-Aki Shibata

Department of Physics, Tokyo Institute of Technology, O-okayama 2-12-1, Meguro, Tokyo, 152-8551, Japan

Abstract. A new Quantum Chromodynamics (QCD) analysis framework of fragmentation and parton distribution functions has been developed by the TTAP (Team Towards global Analysis of the Proton and partons) collaboration. The Mellin transformation technique used in the analysis framework improved the computing speed, which allows us to perform further detailed analysis.

Keywords: parton distribution function, fragmentation function, global QCD analysis
PACS: 13.60.Hb

GLOBAL QCD ANALYSIS FRAMEWORK

We, (Team Towards global Analysis of the Proton and partons; TTAP), have developed a global Quantum Chromodynamical (QCD) analysis framework for fragmentation functions (FF) and parton distribution functions (PDF) [1]. In our earlier framework, the Dokshitzer-Gribov-Lipatov-Altarelli-Parisi (DGLAP) equation was numerically computed in the (x, Q^2) or (z, Q^2) space, which is called the "brute force" method. The brute force method is rather simple, but it requires intensive computing power in terms of memory and computing time. We had analyzed FF in the original framework with hadron production cross-section in single inclusive annihilation (SIA), and polarized PDF with double spin asymmetries in inclusive deep inelastic scattering (DIS) and semi-inclusive deep elastic scattering (SIDIS). Impact of the FF uncertainty on the polarized PDF analysis had also been studied [1]. However, due to a limit of the computing resources, it was not efficient to perform further detailed analysis, such as uncertainly correlation among the different hadron fragmentation functions, implementation of other hard scattering processes, and so on.

"Mellin Transform" technique has been intensively developed in the QCD analyses, i.e., [2], which simplifies a convolution integral in the analysis to a simple multiplication in moment space. We have developed an independent analysis framework based on the Mellin technique. For the new analysis framework, C++ was chosen instead of FORTRAN used before. It helps to keep the framework structure equivalent for both FF and PDF. Handling of physical observables, such as hadron production cross section in SIA, structure functions, F_1, g_1, and hadron multiplicity in SIDIS becomes more transparent, thus implementation of new scattering process in the framework takes less efforts. The SIDIS observables depend on x and z, so that "Mellin inversion" is extended to two dimensional transformation on the both variables.

CP1149, *Spin Physics, 18th International Spin Physics Symposium*
edited by D. G. Crabb, D. B. Day, S. Liuti, X. Zheng, M. Poelker, and Y. Prok

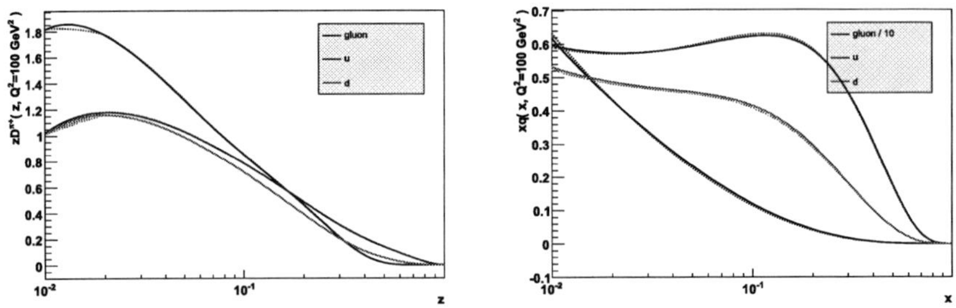

FIGURE 1. Kretzer π fragmentation function [5] and MRST1998 parton distribution function [6] calculated in the QCD analysis framework at $Q^2 = 100$ GeV^2 are shown left and right, respectively, with the solid lines. The corresponding grid data are also shown with the dashed lines.

It has become more important to determine uncertainty on the extracted parameters. However, non-trivial correlation among parameters sometime prevents us from determination of the error matrix of the parameters accurately. The advanced methods have been proposed to estimate uncertainty of parameters and uncertainty of the derived observables [3, 4]. These methods require iteration of fit with various conditions, which may be similar to "multi-dimensional parameter scan". In the previous framework, the calculation speed was the bottleneck, so that it was not realistic to perform such detailed error analysis, especially on SIDIS processes.

FRAGMENTATION AND PARTON DISTRIBUTION FUNCTIONS

There are several parameterizations available which provide the grid information in addition to the parameter itself. The present available grid information is compiled at `http://durpdg.dur.ac.uk/hepdata/`. Most of the parametrization are utilized with functional form which can analytically transform to the Mellin space, i.e., $f(x) = x^{\alpha} \cdot (1-x)^{\beta} \rightarrow f(N) = B(\beta+1, \alpha+N)$, where B is Beta function. Adopting such parameterizations, one can test if the developed QCD analysis framework works properly or not.

In Fig. 1, fragmentation and parton distribution functions calculated in the developed framework are shown. Kretzer pion parametrization [5], and MRST1998 unpolarized parton distribution parametrization [6] are chosen, and the functions are calculated at $Q^2 = 100$ GeV^2, with the corresponding values form the grid information. The framework reproduces almost identical results to the grid results.

HADRON PRODUCTION IN SINGLE INCLUSIVE ANNIHILATION

Using the fragmentation functions, one can calculate hadron production cross section in SIA. Under the same analysis condition in [5]: the functional form, the constraint on

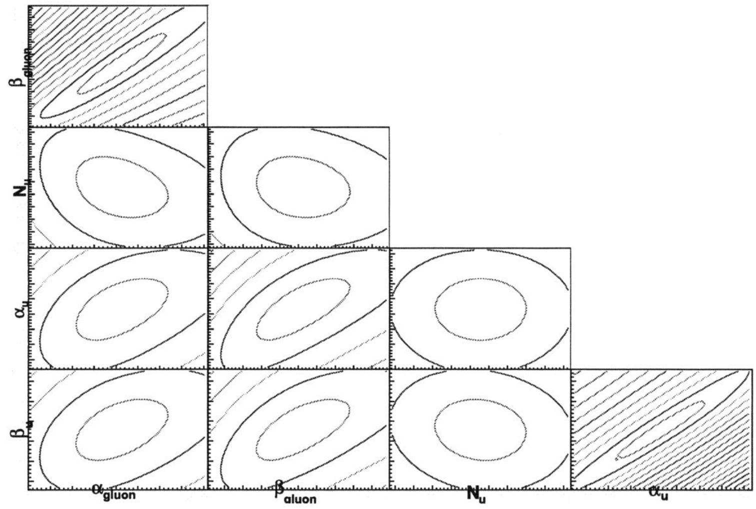

FIGURE 2. χ^2 contours obtained from parameter scan around the best fit value for π fragmentation function under the fitting condition used in [5] are shown for the gluon and u-quark parameters, α_g^π, β_g^π, N_u^π, α_u^π, and β_u^π. Range of the vertical or horizontal axis is corresponding to the one standard deviation of the fitted parameter. The contour levels are defined as $\Delta\chi^2 = (0.5 * i)^2$, where $i = 1, 2, 3, \ldots$.

parameters, the value of Λ_{QCD}, π fragmentation functions were determined by fitting the experimental data using the MINUIT library. The obtained parameters are consistent with the results in [5]. In Fig. 2, χ^2 contour for the gluon and u-quark parameters are shown. In each plot, the parameters were varied within one standard deviation around the best values, and the fitting procedure keeping the parameters fixed was repeated about 1000 times to map the entire χ^2 contour. Under the condition [5], there are 11 free parameters in the fit. In order to cover the possible parameter combination, about 60 thousand fits are required. In the new framework it took about one week. In the previous framework, the same procedure may have taken about 8 months. Strong correlations were observed between the parameter α and β for both gluon and u-quark, which may indicate need of additional parameters to control intermediate z region, as suggested in [7].

OTHER HARD SCATTERING PROCESSES

We have implemented unpolarized and polarized nucleon structure functions in DIS and SIDIS at NLO, Drell-Yan cross-section at LO, and hadron multiplicity in SIDIS. It is also planned to implement various cross sections in $p-p$ scattering.

In Fig. 3, charged pion multiplicity calculated at NLO are shown with the preliminary results from the HERMES experiment [8]. For PDF and FF functions, the MRST2001 [9] and DSS [7] parameterizations were taken. The framework reproduced the experimental results.

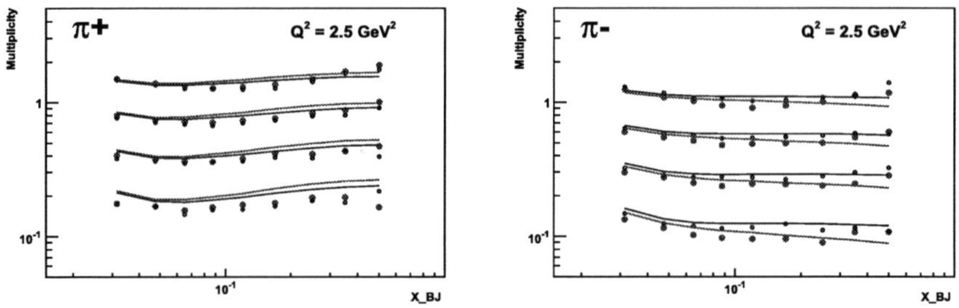

FIGURE 3. Multiplicities for $\pi^{+,-}$ in SIDIS at $Q^2 = 2.5$ GeV^2 calculated in the framework using MRST2001 [9] and DSS [7] parametrization are shown with HERMES preliminary results [8].

SUMMARY

The new analysis framework based on the Mellin technique has been developed. It was tested using the available parametrization and the corresponding grid information. We have confirmed that Q^2 evolution for FF and PDF and Mellin inversion are calculated properly. Several physics observables: hadron production cross section in SIA, unpolarized and polarized structure functions in DIS and SIDIS were implemented. The fit to the available SIA data demonstrated that the framework has the computing performance adequate for the complicated scattering processes and the further detailed uncertainty analysis.

ACKNOWLEDGMENTS

Work supported by Research Fellowships of the Japan Society for the Promotion of Science for Young Scientists.

REFERENCES

1. Y. Imazu, T.-A. Shibata, and Y. Miyachi, *AIP Conf. Proc.* **915**, 408–411 (2007).
2. A. Vogt, *Comput. Phys. Commun.* **170**, 65–92 (2005), hep-ph/0408244.
3. D. Stump, et al., *JHEP* **10**, 046 (2003), hep-ph/0303013.
4. D. de Florian, G. A. Navarro, and R. Sassot, *Phys. Rev.* **D71**, 094018 (2005), hep-ph/0504155.
5. S. Kretzer, *Phys. Rev.* **D62**, 054001 (2000), hep-ph/0003177.
6. A. D. Martin, R. G. Roberts, W. J. Stirling, and R. S. Thorne, *Eur. Phys. J.* **C4**, 463–496 (1998), hep-ph/9803445.
7. D. de Florian, R. Sassot, and M. Stratmann, *Phys. Rev.* **D75**, 114010 (2007), hep-ph/0703242.
8. A. Hillenbrand (2005), DESY-THESIS-2005-035.
9. A. D. Martin, R. G. Roberts, W. J. Stirling, and R. S. Thorne, *Eur. Phys. J.* **C28**, 455–473 (2003), hep-ph/0211080.

The Double Spin Asymmetry in Exclusive π^+ Electroproduction with CLAS

Joshua Pierce

University of Virginia, Charlottesville Virginia
Current Address: Oak Ridge National Laboratory, Oak Ridge Tennessee

Abstract. The eg1b run at Jefferson Lab used the CLAS detector in Hall B along with a 1.6 GeV - 5.6 GeV polarized electron beam and polarized nuclear targets (composed of NH_3 and ND_3) to measure single and double spin asymmetries. This analysis is of the double and single spin asymmetries in the exclusive electro-production of positive pions ($ep \to e\pi^+n$). The asymmetries were measured as a function of W, Q^2, $\cos\theta^*$, and ϕ^*. The value of these asymmetries can be used to help determine the spin structure of the resonances in the baryon resonance region. Preliminary results for the asymmetries are shown.

Keywords: Resonances, Asymmetries
PACS: 13.60.Hb

INTRODUCTION

Exclusive pion production is a useful tool for analyzing the nucleon resonances because of the large branching ratio of many of the resonances into the $N\pi$ channel, for example the $P_{11}(1440)$, or Roper resonance, $D_{13}(1520)$ and the $F_{15}(1680)$. This analysis is of the spin dependence of the $ep \to e\pi^+N$ reaction using data taken with CLAS [1] during the eg1b run period at Jefferson Lab. A longitudinally polarized electron beam and a longitudinally polarized ammonia target were used. The spin dependence is explored by measuring double and single spin asymmetries. Some results for the double spin and target spin asymmetries are shown.

SPIN DEPENDENCE OF THE CROSS-SECTION

When both electron and the proton are spin polarized, the virtual photon cross-section for exclusive π^+ production can be written in terms of the polarized response functions, R [2]. These are functions of three independent variables, W, Q^2 and $\cos\theta^*$ (the angle between the pion momentum and the direction of the virtual photon in the center of mass of the pion-neutron system). These equations are in the reference frame where $\hat{z}$ is along the virtual photon 3-momentum $\vec{q}$, and $\hat{y}$ is normal to the hadronic scattering plane.

CP1149, *Spin Physics, 18th International Spin Physics Symposium*
edited by D. G. Crabb, D. B. Day, S. Liuti, X. Zheng, M. Poelker, and Y. Prok

$$\begin{aligned} \frac{d\sigma_v}{d\Omega_\pi} &= \frac{|\vec{q}|}{q_\gamma^{CM}}[R_T + P_y R_T^y + \varepsilon_L(R_L + P_y R_L^y) \\ &+ \sqrt{2\varepsilon_L(1+\varepsilon)}((R_{LT} + P_y R_{LT}^y)\cos\phi^* + (P_x R_{LT}^x + P_z R_{LT}^z)\sin\phi^*) \\ &+ \varepsilon((R_{TT} + P_y R_{TT}^y)\cos 2\phi^* + (P_x R_{TT}^x + P_z R_{TT}^z)\sin 2\phi^*) \\ &+ h\sqrt{2\varepsilon_L(1-\varepsilon)}((R_{LT'} + P_y R_{LT'}^y)\sin\phi^* + (P_x R_{LT'}^x + P_z R_{LT'}^z)\cos\phi^*) \\ &+ h\sqrt{1-\varepsilon^2}(P_x R_{TT'}^x + P_z R_{TT'}^z)] \end{aligned} \quad (1)$$

where h is the helicity of the electron, ε is the polarization of the virtual photon, and P_x,P_y,P_z are the components of the target polarization vector.

The target polarization can be re-written in terms of ϕ^*, which is the angle between the hadronic and leptonic interaction planes, θ_γ, which is the production angle of the virtual photon, and P_T, the polarization as measured in the lab.

$$\begin{aligned} P_x &= P_T \sin\theta_\gamma \cos\phi^* \\ P_y &= -P_T \sin\theta_\gamma \sin\phi^* \\ P_z &= P_T \cos\theta_\gamma \end{aligned} \quad (2)$$

The cross-section can be written in terms of a polarization independent part, a part that depends only on the beam polarization (h or P_B), a part that depends only on the target polarization P_T, and a part that depends on both the beam and target polarization.

$$\sigma = \sigma_0 + P_B\sigma_e + P_T\sigma_t - P_B P_T \sigma_{et} \quad (3)$$

Since both single spin dependent parts of the cross section are non zero, separating them from the double spin dependent part of the cross-section requires that data be taken in all four possible combinations of flipped target and beam polarizations.

EXPERIMENT

The eg1b run used the CEBAF polarized electron beam and a polarized nuclear target [3]. The target consisted of an $^{15}NH_3$ sample, polarized using the dynamic nuclear polarization technique. The data were taken using the CLAS detector in experimental Hall B at Jefferson Lab.

Events were selected by detecting the scattered electron and the produced π^+, solving for the missing mass of the undetected particle, and applying a cut on this mass around the mass of the neutron.

Equations 4 and 5 show the relation between the double spin and target spin asymmetries, respectively and the measured counts in each polarization state. The N' terms represent the charge normalized counts for each combination of beam and target polarization. The product of the beam and target polarizations was determined by measuring the asymmetry for elastic scattering on the proton and comparing that to the known asymmetry. The beam polarization is known independently from runs using a Moeller polarimeter, meaning that the two can be separated.

$$A_{et} = \frac{\sigma_{et}}{\sigma_0} = \frac{1}{P_B^+ P_T^-} \frac{(N'_{-+} - N'_{++}) + r_B(N'_{+-} - N'_{--})}{(N'_{++} + N'_{-+}) + r_T(N'_{+-} + N'_{--})} \tag{4}$$

$$A_t = \frac{\sigma_t}{\sigma_0} = \frac{1}{P_T^-} \frac{(N'_{++} + N'_{-+}) - (N'_{+-} + N'_{--})}{(N'_{++} + N'_{-+}) + r_T(N'_{+-} + N'_{--})} \tag{5}$$

Since the target polarization is only reversed once per beam energy setting, and the positive and negative target polarization are not generally equal, it is necessary that the polarizations in the different beam and target configurations be normalized to each other. This is done in Eqns. 4 and 5 with the terms r_B and r_T which are the ratios of the beam and target polarizations, respectively for the periods when the target had negative polarization and the target had positive polarization.

Contributions from the ^{15}N and other materials, such as the liquid He surrounding the target and the target window material, are accounted for in each kinematic bin by scaling up the missing mass spectrum obtained from dedicated ^{12}C runs to approximate the unpolarized part of the $^{15}NH_3$ spectrum. This contribution is then subtracted as a background in each polarization state.

RESULTS

The data have been analyzed and the double spin asymmetries have been extracted. The asymmetry was extracted as an independent function of the four kinematic variables W, Q^2, $\cos\theta^*$, and ϕ^*, as previously defined. In order to display the results, the asymmetry is averaged over one or more kinematic variables, weighted by the error. The advantage of this in comparison to integrating the counts over a large kinematic range, or over one variable entirely, is that it limits the effects of acceptance on the results.

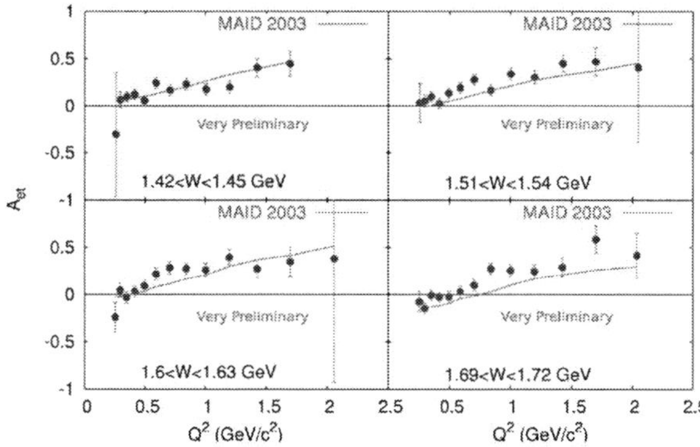

FIGURE 1. A_{et} as a function of Q^2 for fixed values of W, averaged over $\cos\theta^*$ and ϕ^*. The predictions from MAID2003 is shown for comparison.

All of the results shown here are from the 4.2 GeV beam energy run. In all of the figures, comparisons are shown to predictions from MAID2003 [4]. The values for

MAID2003 were generated over the same four dimensional space as the asymmetries were measured in, and then averaged together with the same weight as the asymmetry. Statistical error associated with the background subtraction and the measurement of the product of beam and target polarization are included in the statistical error bars shown, although systematic errors are not.

The target spin asymmetry has also been extracted for all of the data. Like the double spin asymmetry, the target spin asymmetry was measured as a function of all four kinematic variables W, Q^2, $\cos\theta^*$, and ϕ^* independently, and that is the format in which the data is most useful, but one or more kinematic variables are still generally averaged over to display the result. Since the ϕ^* dependence of the target spin asymmetry is such that the asymmetry is always zero if averaged over this variable, the results are shown as a function of ϕ^*.

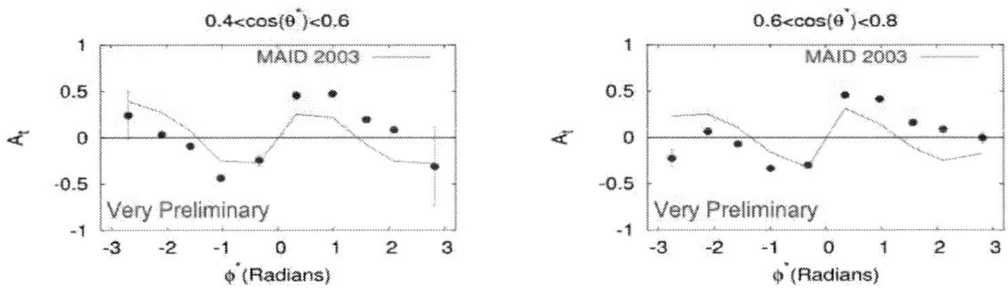

FIGURE 2. A_t as a function of ϕ^* for fixed values of $\cos\theta^*$, averaged over Q^2 and W. The predictions from MAID2003 is shown for comparison.

CONCLUSIONS AND OUTLOOK

Analysis of the asymmetry from the other energy settings of the experiment is proceeding, which will provide greater kinematic coverage. The high statistics and high polarizations of the eg1b run allow for the data be analyzed for a variety other asymmetries including inclusive asymmetries on both the proton and deuteron (which was the primary motivation for the run), as well as other exclusive and semi-inclusive reactions.

The large amount of data collected in this experiment will enable us to significantly increase our knowledge of the spin structure of the resonances. The addition of this data set should help in the development and enhancement of models of the multi-pole terms associated with the resonances.

REFERENCES

1. B. A. Mecking *et al.*, *Nucl. Instrum. Meth.A* **503**, 513 (2003).
2. A. S. Raskin and T. W. Donnelly, *Annals of Phys.* **191**, 78 (1989).
3. C. D. Keith *et al.*, *Nucl. Instrum. Meth. A* **501**, 327 (2003).
4. D. Drechsel *et al.*, *Nucl. Phys. A* **645**, 145 (1999).

Longitudinally Polarized Hadroproduction of Heavy Quarks

Johann Riedl*, Andreas Schäfer* and Marco Stratmann*,†

*Institut für Theoretische Physik, Universität Regensburg, 93040 Regensburg, Germany
†Institut für Theoretische Physik und Astrophysik, Universität Würzburg,
97074 Würzburg, Germany

Abstract. Measurements of double-spin asymmetries for single-inclusive heavy flavor production and for heavy quark correlations are expected to provide valuable information about the spin structure of the nucleon, in particular, the elusive gluon polarization. We discuss progress towards a versatile parton-level Monte-Carlo code for the production and decay of heavy quarks in longitudinally polarized proton-proton collisions at next-to-leading order accuracy of Quantum Chromodynamics (QCD). First phenomenological studies are presented for BNL-RHIC.

Keywords: Heavy Quarks, Perturbative QCD, Spin Structure
PACS: 12.38.Bx,13.88.+e

INTRODUCTION

Heavy flavors are an important tool to probe different aspects of Quantum Chromodynamics (QCD), ranging from heavy quark parton distribution functions (PDFs) and the hadronization of heavy quarks into heavy mesons to the dynamics of quantum chromodynamics (QCD) hard-scattering.

Modern theoretical calculations of heavy quark production rates aim at comparisons with data as close as possible to the observational level [1]. This involves modeling the heavy quark-to-heavy meson transition and often their subsequent semi-leptonic decay into the experimentally observed leptons. In addition, next-to-leading order (NLO) accuracy for the hard-scattering cross sections is the state-of-the-art throughout and allows to quantify theoretical ambiguities from the renormalization and factorization procedure. If necessary, these fixed-order calculations can be further improved by systematically resumming large corrections to the perturbative series to all orders. In this way, earlier discrepancies between theory and data, most famously for bottom production at the TeVatron, have been reduced to a point that no longer appears significant [2].

NLO calculations for longitudinally polarized hadro- and photoproduction of heavy flavors are more recent achievements and have concentrated so far on single-inclusive spectra [3] . With sufficient statistics collected at BNL-RHIC, measurements of double-spin asymmetries are expected to provide valuable new information about the spin structure of the nucleon, in particular, the polarized gluon distribution $\Delta g(x)$ and the dynamics of spin-dependent QCD hard-scattering beyond pion and jet production explored so far. The latter put a strong constraint on the size of $\Delta g(x)$ in the range $0.05 \lesssim x \lesssim 0.2$ as was revealed in a recent global QCD analysis of polarized PDFs [4].

CP1149, *Spin Physics, 18th International Spin Physics Symposium*
edited by D. G. Crabb, D. B. Day, S. Liuti, X. Zheng, M. Poelker, and Y. Prok

In this talk, we present the recent theoretical progress towards a flexible parton-level Monte-Carlo code for heavy quark production in polarized pp collisions at NLO accuracy. Contrary to the largely analytical calculations available so far [3], our code allows to implement experimental cuts on the final-state, includes the hadronization of heavy quarks into heavy mesons and their semi-leptonic decays, and to study heavy quark correlations, which are of great phenomenological interest at RHIC. It is planned to further extend the code to account also for photoproduction to analyze the existing COMPASS measurements consistently at NLO.

TECHNICAL ASPECTS AND PHENOMENOLOGICAL RESULTS

The polarized cross section for the hadroproduction of heavy flavors can be schematically written in factorized form

$$d\Delta\sigma = \sum_{ab} \Delta f_a \otimes \Delta f_b \otimes d\Delta\hat{\sigma}_{ab\to QX} \otimes D(Q \to H_Q)[\otimes f(H_Q \to l)] \quad (1)$$

as a convolution (denoted by $\otimes$) of the helicity PDFs $\Delta f_{a,b}$, the perturbatively calculable hard partonic cross sections $d\Delta\hat{\sigma}_{ab\to QX}$, and a phenomenological function $D(Q \to H_Q)$ describing the hadronization of the heavy quark Q into a heavy meson H_Q. Depending on the experimental setup, the subsequent decay of H_Q needs to be modeled as well, for instance, by a function $f(H_Q \to l)$ parameterizing the semi-leptonic decay spectrum of the meson H_Q into the detected lepton l. A generalization of Eq. (1) to $Q\bar{Q}$ correlations is straightforward, including the decays into the experimentally relevant signatures, like, e.g., muon-electron coincidences.

Our versatile parton-level Monte-Carlo code can account for all these observables at NLO accuracy and, furthermore, allows to straightforwardly implement experimental cuts as all phase-space integrations of the hard-scattering matrix elements are performed numerically, using the so-called subtraction method to deal with singularities in intermediate steps of the calculation. All technical details are very similar to the corresponding unpolarized calculation which can be found in Ref. [5].

At BNL-RHIC, heavy quark production can be studied in polarized pp collisions with center-of-mass energies $\sqrt{S}$ of up to 500 GeV. At present, both experiments, PHENIX and STAR, detect heavy flavors through their decay leptons and thus cannot distinguish produced charm from bottom quarks, which need to be summed also in theoretical calculations. PHENIX and STAR can detect electrons at central rapidities, $|y| \leq 0.35$ and $|y| \leq 1$, respectively. In addition, PHENIX is also sensitive to muons in the range $1.2 \leq |y| \leq 2.4$, which allows to study electron-muon correlations stemming from decays of the the produced $Q\bar{Q}$ pair.

Here, we present a first preliminary study of single-inclusive electron spectra from heavy quark production at $\sqrt{S} = 200$ GeV, following closely a corresponding study in the unpolarized case [1]. The fragmentation function $D(Q \to H_Q)$ is taken from Ref. [6], and the decay spectrum $f(H_Q \to l)$ is extracted from CLEO and BaBar data [7].

Figure 1 shows the polarized cross sections at the D and B meson level for three different sets of helicity PDFs. The DSSV set [4] was obtained from a recent global QCD analysis including all data sensitive to spin-dependent PDFs. The polarized gluon

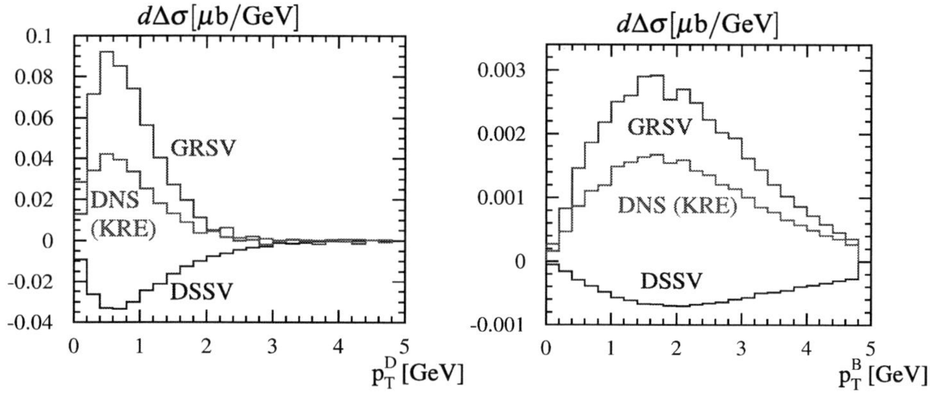

FIGURE 1. The polarized cross sections for charm (left panel) and bottom (right panel) production, differential in the transverse-momentum of the heavy D and B meson, respectively. Results are shown for three different sets of helicity PDFs.

density $\Delta g(x)$ was found to be very small in the range $0.05 \lesssim x \lesssim 0.2$ with a possible node. The other two older sets, GRSV and DNS(KRE) [8], are characterized by a moderate, somewhat larger and positive definite $\Delta g(x)$. As can be seen from Fig. 1, using DSSV in Eq. (1) yields a negative $d\Delta\sigma$ since the annihilation of light quarks, $q\bar{q} \to Q\bar{Q}$, dominates for which $\Delta\hat{\sigma}_{q\bar{q}} = -\hat{\sigma}_{q\bar{q}}$. For sets with larger $\Delta g(x)$, the gluon-gluon fusion subprocess starts to dominate and leads to positive cross sections, the size depending on the amount of gluon polarization. In general, all $\Delta g(x)$ compatible with present RHIC data for pion and jet yields predict also very small cross sections and double-spin asymmetries for heavy quark production. Nevertheless, their measurement would add valuable new information as the underlying dynamics of the QCD hard-scattering is quite different for pion and jet production on the one hand and heavy flavor production on the other.

Figure 2 includes the semi-leptonic decay of the heavy meson to the observed electron for three channels, $D \to e$, $B \to e$, and the cascade $B \to D \to e$. The lower right corner shows the phenomenologically relevant double-spin asymmetry for all three decay contributions summed up, as they cannot be distinguished in experiment. The rapidity of the electron is limited to $|y| < 0.5$. Much more detailed phenomenological studies of heavy flavor production at RHIC, including the semi-leptonic decay into muons as well as electron-muon and muon-muon correlations can be found in an upcoming publication [9].

ACKNOWLEDGMENTS

We are grateful to R. Vogt, M. Liu, and P. Kurnadi for useful discussions. J.R. is supported by a grant of the "Cusanuswerk", Bonn, Germany. This work is supported in part by the BMBF, Germany.

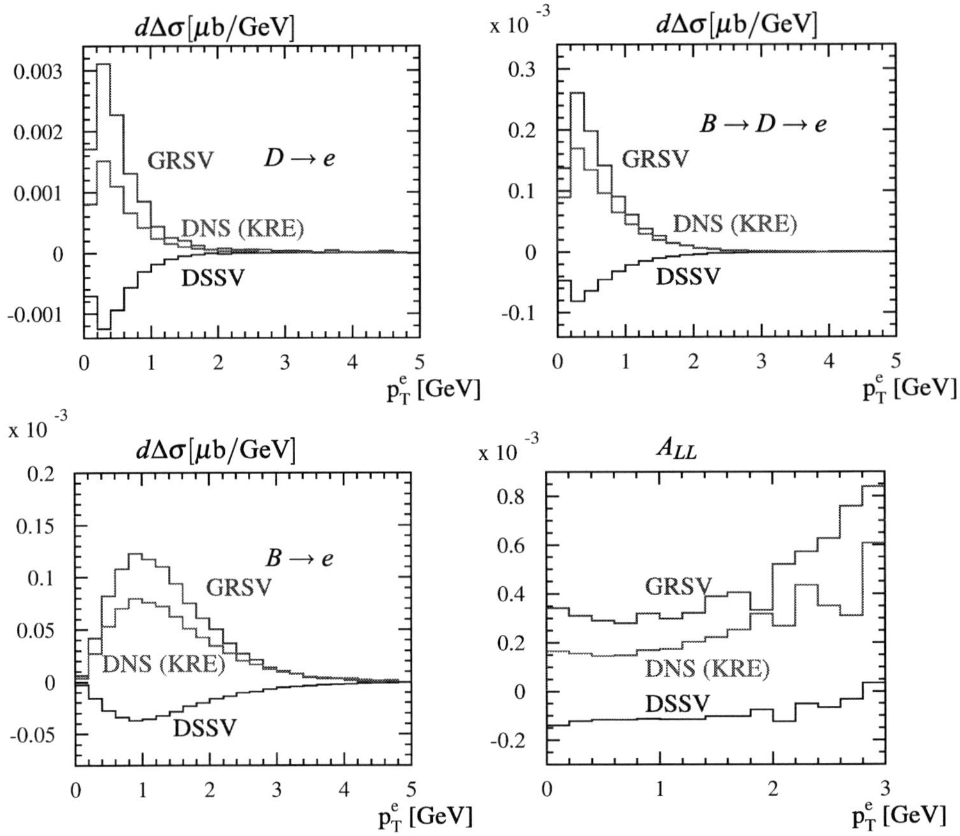

FIGURE 2. Different contributions to the transverse momentum spectrum of the decay electron and the resulting double-spin asymmetry A_{LL} (lower right corner).

REFERENCES

1. See, e.g., M. Cacciari, P. Nason, and R. Vogt, *Phys. Rev. Lett.* **95**, 122001 (2005).
2. See, e.g., M. Cacciari and P. Nason, *Phys. Rev. Lett.* **89**, 122003 (2002).
3. I. Bojak and M. Stratmann, *Phys. Lett. B* **433**, 411 (1998); *Nucl. Phys. B* **540**, 345 (1999); *Phys. Rev. D* **67**, 034010 (2003); A. P. Contogouris *et al.*, *Phys. Rev. D* **62**, 114509 (2000); J. Riedl, A. Schäfer, and M. Stratmann, *Eur. Phys. J. C* **52**, 987 (2007).
4. D. de Florian, R. Sassot, M. Stratmann, and W. Vogelsang, *Phys. Rev. Lett.* **101**, 072001 (2008).
5. M. L. Mangano, P. Nason, and G. Ridolfi, *Nucl. Phys. B* **373**, 295 (1992).
6. V. G. Kartvelishvili, A. K. Likhoded and V. A. Petrov, *Phys. Lett. B* **78**, 615 (1978).
7. B. Aubert *et al.* [BABAR Collab.], *Phys. Rev. D* **69**, 111104 (2004); A. H. Mahmood *et al.* [CLEO Collab.], *Phys. Rev. D* **70**, 032003 (2004); J. Yelton, CLEO talk 04-42; J. Wiss, in Proc. of "4th Flavor Physics and CP Violation Conference (FPCP 2006)", Vancouver, 2006, `hep-ex/0605030`.
8. M. Glück, E. Reya, M. Stratmann, and W. Vogelsang, *Phys. Rev. D* **63**, 094005 (2001); D. de Florian, G. A. Navarro, and R. Sassot, *Phys. Rev. D* **71**, 094018 (2005).
9. J. Riedl, A. Schäfer, and M. Stratmann, to appear.

The Strange Quark Polarisation from Charged Kaon Production on Deuterons

R.Windmolders (On behalf of the COMPASS collaboration)

Universität Bonn, Physikalisches Institut, 53115 Bonn, Germany

Abstract. The strange quark helicity distribution $\Delta s(x)$ is derived at LO from the semi-inclusive and inclusive spin asymmetries measured by the COMPASS experiment at CERN. The significance of the results is found to depend critically on the ratio of the $\bar{s}$ and u quark fragmentation functions into kaons $\int D_{\bar{s}}^{K+}(z)dz / \int D_{u}^{K+}(z)dz$.

Keywords: Deep inelastic scattering; Polarised parton densities
PACS: 13.60.Hb; 13.60.Le; 13.88.+e

In contrast to other sea quarks, the strange quark contribution to the nucleon spin is accessible in inclusive lepton-nucleon scattering experiments. Already twenty years ago the EMC [1] has found its first moment $\Delta s + \Delta\bar{s}$ to be negative under the assumption of $SU(3)_F$ symmetry in hyperon β decays and this result has been confirmed with improved precision by recent measurements performed by HERMES [2] ($\Delta s + \Delta\bar{s} = -0.103 \pm 0.007(\text{exp.}) \pm 0.013(\text{theor.}) \pm 0.008(\text{evol.})$) and by COMPASS [3]:

$$\Delta s + \Delta\bar{s} = -0.09 \pm 0.01(\text{stat.}) \pm 0.02(\text{syst.}). \tag{1}$$

Inclusive experiments provide a direct evaluation of the first moment of $(\Delta s + \Delta\bar{s})$ only. However the distribution $\Delta s(x, Q^2)$ can be obtained from semi-inclusive channels in which interactions of strange quarks are favored, such as charged kaon production. These measurements require final state particle identification and became only feasible in recent experiments [4, 5].

In this paper we present a leading order (LO) evaluation of the polarised parton distributions (PDFs) $\Delta u_v + \Delta d_v$, $\Delta\bar{u} + \Delta\bar{d}$ and $\Delta s (= \Delta\bar{s})$ derived from the inclusive and semi-inclusive spin asymmetries measured by the COMPASS experiment. The data sample covers the years of 2002-2004 and 2006, and corresponds to the full statistics collected on the longitudinally polarised ^{6}LiD target exposed to the 160 GeV muon beam at CERN. The averaged beam and target polarisations were -0.80 and 0.50, respectively. Each selected event contains a beam muon with energy $140 < E_\mu < 180$ GeV, a scattered muon and, for the semi-inclusive samples, at least one charged hadron originating from the interaction vertex. The deep inelastic scattering (DIS) conditions $Q^2 \geq 1(\text{GeV}/c)^2$ and $0.1 \leq y \leq 0.9$ are required in the muon kinematics. The hadron is required to carry a fractional energy $0.20 < z < 0.85$ and must be identified as a pion or a kaon by the RICH detector, which limits its momentum to the range $10 < p < 50$ GeV/c. The total number of identified hadrons is $23 \cdot 10^6$ π^+, $21 \cdot 10^6$ π^-, $4.8 \cdot 10^6$ K^+ and $3.3 \cdot 10^6$ K^-. The purity of the selected samples is above 95 % for the pions and between 80 and 90 % for the kaons.

CP1149, *Spin Physics, 18th International Spin Physics Symposium*
edited by D. G. Crabb, D. B. Day, S. Liuti, X. Zheng, M. Poelker, and Y. Prok

The inclusive and semi-inclusive asymmetries are shown in Fig. 1 as a function of the scaling variable x in a range limited by the kinematical threshold $Q^2 \geq 1(\mathrm{GeV}/c)^2$, which implies $x > 0.004$, and a value where sea quarks distributions become insignificant ($x < 0.3$).

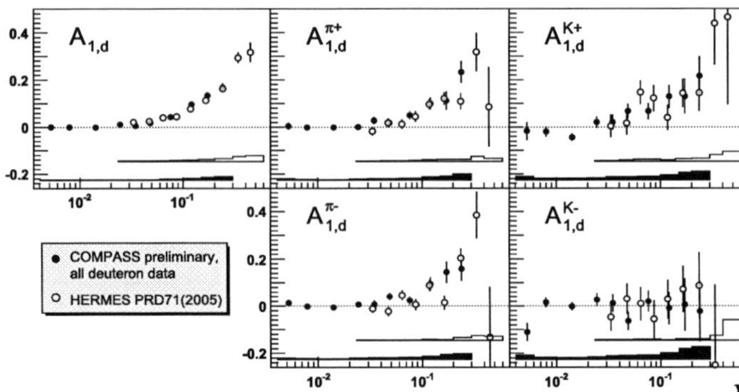

FIGURE 1. Spin asymmetries of COMPASS compared to the HERMES results [4]. The bands show the systematic uncertainties for the two sets of data.

The correction applied to account for the contamination of the pion and kaon samples has only a small effect on the semi-inclusive asymmetries. The HERMES data [4], shown on the same plots, are in good agreement with the present ones in the region of overlap and their quoted errors are generally also comparable. The COMPASS systematic error includes an overall scale uncertainty due to the uncertainties on the beam and target polarisation, on the dilution factor and on the depolarisation factor which amount to 8 % when added in quadrature. In addition, each point has an uncertainty due to possible apparatus induced asymmetries which are estimated to be less than 0.4 σ(stat.).

The PDFs have been evaluated at $Q_0^2 = 3(\mathrm{GeV}/c)^2$ in each bin of x by a least square fit on the asymmetries assumed to be independent of Q^2 (Fig. 2). In this calculation, unpolarised PDFs are taken from the MRST parameterisation [6] and fragmentation functions (FFs) from the recent DSS parameterisation [7], both at LO. In order to test the dependence of the polarised PDFs on the FFs, we also show the values obtained with the EMC FFs [8]. In contrast to other parameterisations which are derived from global fits, the latter ones have been extracted from the EMC data only, so that only the u quark fragmentation could be measured. Therefore, in addition to the usual assumptions made to reduce the number of FFs, it was also assumed that $D_{\bar{s}}^{K+} = D_u^{\pi+}$.

The comparison of the two sets of values shows that the valence and non-strange sea distributions depend very little on the choice of FFs while $\Delta s(x)$ and its statistical error increase by a factor of 2-3 when the DSS FFs are replaced by the EMC ones. The valence distribution is in good agreement with the DNS parameterisation [9] and with a previous evaluation based on the difference asymmetry for non-identified hadrons $A^{h^+-h^-}$ [3]: its first moment truncated to the measured range is $0.28 \pm 0.06(\mathrm{stat.}) \pm 0.03(\mathrm{syst.})$, vs. $0.26 \pm 0.07 \pm 0.04$ at $Q^2 = 10$ $(\mathrm{GeV}/c)^2$ in Ref. [3]. The non-strange sea distribution is consistent with zero for all values of x. The strange quark distribution obtained with the DSS FFs is also consistent with zero and its truncated moment is $-0.01 \pm 0.01 \pm 0.01$

(vs. $-0.05 \pm 0.03 \pm 0.01$ with the EMC FFs).

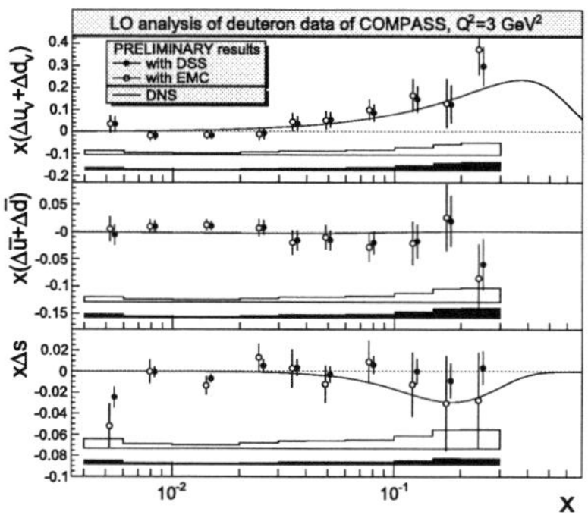

FIGURE 2. The quark helicity distributions evaluated at $Q_0^2 = 3(\mathrm{GeV}/c)^2$ as a function of x for two sets of fragmentation functions (DSS [7] and EMC[8]). The bands show the corresponding systematic uncertainties. The curves represent the LO DNS parameterisation of polarised PDFs [9].

The dependence of $\Delta s(x)$ on the FFs can be further explored by introducing the charged kaon asymmetry

$$A^{K^++K^-} = \left[\sigma^{K^+} A^{K^+} + \sigma^{K^-} A^{K^-}\right] / \left[\sigma^{K^+} + \sigma^{K^-}\right]. \tag{2}$$

These asymmetries are found to be very stable with respect to changes in the cross-section ratio $\sigma^{K^-}/\sigma^{K^+}$. At LO this ratio only depends on unpolarised PDFs and on the ratios of unfavoured to favoured and strange to favoured FFs: $R_{UF} = \int D_d^{K^+}(z)dz / \int D_u^{K^+}(z)dz$ and $R_{SF} = \int D_{\bar{s}}^{K^+}(z)dz / \int D_u^{K^+}(z)dz$, which are equal to 0.14 and 6.6 for the DSS FFs at $Q^2 = 3(\mathrm{GeV}/c)^2$ (0.35 and 3.4 for the EMC FFs). The values shown in Fig. (3, left) have been obtained with the MRST PDFs and the DSS FFs. As for the K^+ and K^- asymmetries, they are in very good agreement with the HERMES values of Ref. [4].

The strange quark polarisation is related to $A^{K^++K^-}$ and to the inclusive asymmetry A_1^d by the linear relation

$$\frac{\Delta s}{s} = A_1^d + (A^{K^++K^-} - A_1^d)\frac{Q/s + \alpha}{\alpha - 0.8} \tag{3}$$

where $Q = u + \bar{u} + d + \bar{d}$ is the non-strange quark density and $\alpha = (2R_{UF} + 2R_{SF})/(3R_{UF} + 2)$. As expected, the use of the above formula with the $A^{K^++K^-}$ values of Fig. 3 leads to values of Δs practically equal to those of Fig. 2, with slightly larger statistical errors. Several other interesting features can be derived from Eq.(3). In the special case where $A^{K^++K^-} = A_1^d$, Δs becomes insensitive to the FFs and its first moment is very small and positive. Otherwise the main dependence on the FFs is due to R_{SF} which appears only in the numerator of α, and its effect is amplified by the large values of the ratio Q/s. At low x where $A_1^d \approx 0$, negative values of $A^{K^++K^-}$ imply $\Delta s < 0$. The COMPASS values of $A^{K^++K^-}$ at low x provide at least a hint that Δs may become negative in the previously unmeasured region $x < 0.02$, as predicted in the recent DSSV fit [10].

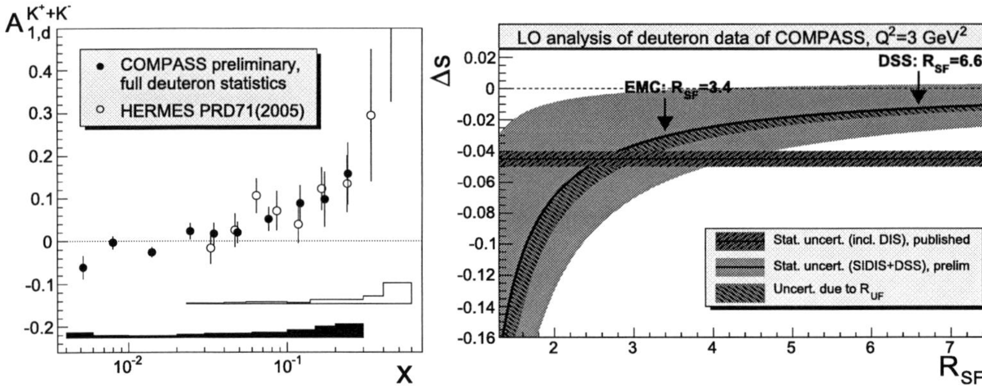

FIGURE 3. *Left:* Charged kaon asymmetries obtained with cross-section weights from MRST PDF's and DSS FF's. *Right:* Integral of Δs over the measured range of x, as a function of the ratio R_{SF} for R_{UF} fixed at the DSS value of 0.14 (thick solid curve). The light-gray area shows the statistical uncertainty and the hatched band inside of it shows the effect of increasing R_{UF} to 0.35 (EMC value). The horizontal band represents the full moment of Δs derived from the COMPASS value of Γ_1^N (Eq. 1).

Figure (3, right) shows the variation of the first moment of Δs truncated to the measured region as a function of R_{SF}. For $R_{SF} \gtrsim 5$, we observe that the values are close to zero and larger than the full moment derived from the inclusive analysis (Eq.1). In particular this is the case for the DSS FFs where $R_{SF} = 6.6$. The difference never exceeds two standard deviations so that no firm conclusions can be drawn from the COMPASS data alone but, as shown on Fig. (3, left) the HERMES data lead to a similar result. In contrast, if $R_{SF} \lesssim 4$, $A^{K^+ + K^-}$ becomes less and less sensitive to Δs because $D_{\bar{s}}^{K^+}$ is small.
The conclusions on Δs obtained in the present analysis are thus conditional on the ratio R_{SF} which will later be derived from the K^+ and K^- rates observed in the COMPASS data.

REFERENCES

1. EMC, J. Ashman, *et al., Phys. Lett. B* **206**, 364-370 (1988).
2. HERMES Collaboration, A. Airapetian, *et al., Phys. Rev. D* **75**, 012007, 1-48 (2007).
3. COMPASS Collaboration, M. Alekseev, *et al., Phys. Lett. B* **660**, 458-465 (2008).
4. HERMES Collaboration, A. Airapetian, *et al., Phys. Rev. D* **71**, 012003, 1-36 (2005).
5. HERMES Collaboration, A. Airapetian, *et al., arXiv:0803.2993v1* (hep-ex),20 Mar 2008.
6. A. D. Martin, W. J. Sterling, R. S. Thorne, *Phys. Lett. B* **636**, 259-264 (2006).
7. D. de Florian, *et al., Phys. Rev. D* **75**, 114010, 1-26 (2007).
8. EMC, M. Arneodo, *et al., Nucl. Phys. B* **321**, 541-560 (1989).
9. D. de Florian, G. A. Navarro, R. Sassot, *Phys. Rev. D* **71**, 094018, 1-12 (2005).
10. D. de Florian, R. Sassot, M. Stratmann, W. Vogelsang, *Phys. Rev. Lett.* **101**, 072001, 1-4 (2008).

A Model Independent Approach for Determining the Fragmentation Functions

Ekaterina Christova* and Elliot Leader†

*Institute for Nuclear research and Nuclear Energy, Sofia, Bulgaria
†Imperial College, London, United Kingdom

Abstract. We show that the difference cross sections in unpolarized semi-inclusive deep inelastic scattering (SIDIS) $e+N \to e+h+X$ and pp hadron production $p+p \to h+X$ determine independently in a model independent way, in any order in Quantum Chromodynamics (QCD), the two FFs: $D_u^{h-\bar{h}}$ and $D_d^{h-\bar{h}}$, $h=\pi^\pm, K^\pm$ or a sum over charged hadrons. If both $K^\pm$ and K_s^0 are measured, then $e^+e^- \to K+X$, $e+N \to e+K+X$ and $p+p \to K+X$ present independent measurements of just one FF: $D_{u-d}^{K^++K^-}$. The above results allow to test the existing parameterizations, obtained with various different assumptions about the FFs, and to test the Q^2 evolution and factorization.

Keywords: fragmentation functions
PACS: 13.87.Fh

INTRODUCTION

Extracting information about the polarized Parton Distribution Functions (PDFs) from semi-inclusive deep inelastic scattering (SIDIS) revealed the importance for a precise knowledge not only of the PDFs, but of the Fragmentation Functions (FFs) as well.

There are three types of processes that give information about the FFs:

$$e^+ + e^- \to h + X \tag{1}$$
$$l + N \to l + h + X, \qquad N = p, d \tag{2}$$
$$p + p \to h + X. \tag{3}$$

where h is any hadron. Schematically, following the factorization theorem, their cross sections can be written in the form:

$$d\sigma^h(z) \simeq \sum_c \hat{\sigma}^c_{e^+e^-} \otimes D_c^h \tag{4}$$
$$d\sigma_N^h(x,z) \simeq \sum_{q,c} f_q(x) \otimes \hat{\sigma}^c_{lq} \otimes D_c^h \tag{5}$$
$$d\sigma_{pp}^h(x,z) \simeq \sum_{a,b,c} f_a \otimes f_b \otimes \hat{\sigma}^c_{ab} \otimes D_c^h \tag{6}$$

where $\hat{\sigma}^c$ are the corresponding, calculable in perturbative Quantum Chromodynamics (QCD), parton-level cross sections for producing a parton c, f_a are the unpolarized PDFs and D_c^h are the FFs for a parton c to turn into a hadron h. Process (1) measures directly the FFs and provides the most precise information. However it determines

CP1149, *Spin Physics, 18th International Spin Physics Symposium*
edited by D. G. Crabb, D. B. Day, S. Liuti, X. Zheng, M. Poelker, and Y. Prok

only the combination $D_q^h + D_{\bar{q}}^h = D_q^{h+\bar{h}}$. The other two processes, (2) and (3), allow the determination of D_q^h and $D_{\bar{q}}^h$ separately, but they involve too many FFs to determine them uniquely. Analysing the data, usually different assumptions are made. At present, several sets of FFs exist and share the common features, such as 1) nearly all of them impose different theoretical relations among the FFs, and 2) there is a significant disagreement between the various analyses for some of the FFs.

In this talk we show that if instead of $d\sigma_N^h$ and $d\sigma_{pp}^h$, the key experimental ingredients are the difference cross sections for producing hadrons and producing their antiparticles i.e. data on $d\sigma_N^{h-\bar{h}} \equiv d\sigma_N^h - d\sigma_N^{\bar{h}}$ and $d\sigma_{pp}^{h-\bar{h}} \equiv d\sigma_{pp}^h - d\sigma_{pp}^{\bar{h}}$, one obtains without any assumptions, using only C-invariance of strong interaction, information about the non-singlet (NS) combinations $D_q^{h-\bar{h}}$ of the FFs. This talk is a summary of the results presented in [1].

THE DIFFERENCE CROSS SECTIONS

From C-invariance of strong interactions it follows

$$D_g^{h^+-h^-} = 0, \qquad D_q^{h^+-h^-} = -D_{\bar{q}}^{h^+-h^-}, \tag{7}$$

which eliminates $D_g^{h-\bar{h}}$ and $D_{\bar{q}}^{h-\bar{h}}$ in the difference cross sections. This implies that in any order of QCD $d\sigma_p^{h^+-h^-}$ and $d\sigma_{pp}^{h^+-h^-}$ are expressed solely in terms of NS combinations of the FFs. In NLO we have

$$d\sigma_p^{h^+-h^-}(x,z,Q^2) = \frac{1}{9}\left[4u_V \otimes D_u + d_V \otimes D_d + s_V \otimes D_s\right]^{h^+-h^-} \otimes \left(1 + \frac{\alpha_s}{2\pi} C_{qq}\right) \tag{8}$$

$$d\sigma_d^{h^+-h^-}(x,z,Q^2) = \frac{1}{9}\left[(u_V + d_V) \otimes (4D_u + D_d) + 2s_V \otimes D_s\right]^{h^+-h^-} \otimes \left(1 + \frac{\alpha_s}{2\pi} C_{qq}\right) \tag{9}$$

$$\begin{aligned} E^h \frac{d\sigma_{pp}^{h^+-h^-}}{d^3P^h} &= \frac{1}{\pi} \int dx_a \int dx_b \int \frac{dz}{z} \times \\ &\times \sum_{q=u,d,s} \left[L_q(x_b,t,u) q_V(x_a) + L_q(x_a,u,t) q_V(x_b)\right] D_q^{h^+-h^-}(z) \end{aligned} \tag{10}$$

where u_V and d_V are the valence quarks PDFs and $s_V = s - \bar{s}$. L_q are functions of the *known* quark densities $\tilde{q} \equiv q + \bar{q}$ and partonic cross sections:

$$L_u(x,t,u) = \tilde{u}(x)\, d\hat{\Sigma}(s,t,u) + [\tilde{d}(x) + \tilde{s}(x)]\, d\hat{\sigma}_{qq'}^{qX}(s,t,u) + g(x)\, d\hat{\sigma}_{qg}^{(q-\bar{q})X}(s,t,u)$$

$$L_d(x,t,u) = \tilde{d}(x)\, d\hat{\Sigma}(s,t,u) + [\tilde{u}(x) + \tilde{s}(x)]\, d\hat{\sigma}_{qq'}^{qX}(s,t,u) + g(x)\, d\hat{\sigma}_{qg}^{(q-\bar{q})X}(s,t,u)$$

$$L_s(x,t,u) = \tilde{s}(x)\, d\hat{\Sigma}(s,t,u) + [\tilde{u}(x) + \tilde{d}(x)]\, d\hat{\sigma}_{qq'}^{qX}(s,t,u) + g(x)\, d\hat{\sigma}_{qg}^{(q-\bar{q})X}(s,t,u)$$

$$d\hat{\Sigma} \equiv \left[d\hat{\sigma}_{qq}^{qX}(s,t,u) + \frac{1}{2} d\hat{\sigma}_{q\bar{q}}^{(q-\bar{q})X}(s,t,u)\right] \qquad \tilde{q} \equiv q + \bar{q} \tag{11}$$

Here x, z, Q^2 are the usual deep inelastic kinematic variables and s,t,u are the Mandelstein variables for $d\hat{\sigma}_{ab}^{cX}$. In LO $s+t+u=0$ which implies that one of the integrations in

(10) can be immediately done. In addition, (11) implies that $\sigma_{pp}^{h-\bar{h}}$ is much simpler than the corresponding process σ_{pp}^{h}, mainly – in LO it involves 4, and in NLO it is 6 partonic cross sections that enter $\sigma_{pp}^{h-\bar{h}}$, while it is 8 in LO and 20 in NLO that enter σ_{pp}^{h}.

Common for the cross sections (8) - (10) is that they all have the same structure 1) only NS of the FFs, $D_q^{h-\bar{h}}$, enter and 2) the FFs enter multiplied by $q_V = q - \bar{q}$, i.e. in the combination $q_V D^{h-\bar{h}}$. This implies that the contributions of D_u^h and D_d^h are enhanced by the large valence quark densities, while D_s^h is suppressed by the small factor $(s-\bar{s})$. Recently a strong bound on $(s-\bar{s})$ was obtained from neutrino experiments, $|s-\bar{s}| \leq 0.025$ [2]. This implies that the large uncertainties in D_s^h should not affect strongly the results for $\sigma_N^{h^+-h^-}$, and we expect the contribution of $(s-\bar{s})D_s^{h^+-h^-}$ to be within the experimental error and to be negligible. Thus, *ep*, *ed* and *pp* semi-inclusive difference cross sections provide three independent measurements for the two unknown FFs: $D_u^{h^+-h^-}$ and $D_d^{h^+-h^-}$. Note that the SIDIS *ep* and *ed* cross sections involve only u_V and d_V, which are the best known parton densities, with 2%-3% accuracy at $x \simeq 0.7$.

THE $\pi^\pm$ AND $K^\pm$ CROSS SECTIONS

Further information can be obtained specifying the final hadrons.

1) If $h = \pi^\pm$ the difference cross sections will determine without any assumptions $D_u^{\pi^+-\pi^-}$ and $D_d^{\pi^+-\pi^-}$, which would allow to test the usually made assumption

$$D_u^{\pi^+-\pi^-} = -D_d^{\pi^+-\pi^-}. \tag{12}$$

In [3] it was suggested, for the first time, that this relation might be violated up to 10 %.

2) If $h = K^\pm$, the difference cross sections will determine without any assumptions $D_u^{K^+-K^-}$ and $D_d^{K^+-K^-}$, which would allow to test the usually made assumption

$$D_d^{K^+-K^-} = 0. \tag{13}$$

One can formulate the above results also like this: If the relations $D_u^{\pi^+-\pi^-} = -D_d^{\pi^+-\pi^-}$ or $D_d^{K^+-K^-} = 0$ hold, then Eqs. (8), (9) and (10) for $h = \pi^\pm$ or $h = K^\pm$ are expressed solely in terms of *one* FF $D_u^{\pi^+-\pi^-}$ or $D_u^{K^+-K^-}$, respectively, and look particularly simple.

THE COMBINED $K^\pm$ AND K_s^0 CROSS SECTIONS

If in addition to the charged $K^\pm$ also neutral kaons $K_s^0 = (K^0 + \bar{K}^0)/\sqrt{2}$ are measured, no new FFs are introduced into the cross-sections. This is a consequence of SU(2) invariance that relates $D_q^{K^0}$ to $D_q^{K^++K^-}$. Then we show [1] that the combination

$$\sigma^{\mathscr{K}} \equiv \sigma^{K^+} + \sigma^{K^-} - 2\sigma^{K_s^0} \tag{14}$$

formed from the three types of semi-inclusive processes ($K = K^{\pm}, K_s^0$):

$$e^+ + e^- \to K + X, \tag{15}$$
$$e + N \to e + K + X, \qquad N = p, d, \tag{16}$$
$$p + p \to K + X, \tag{17}$$

always measures only one NS combination of FFs, namely $(D_u - D_d)^{K^+ + K^-}$. This result relies only on SU(2) invariance for the kaons and does not involve *any* assumptions about PDFs or FFs, it holds in any order in QCD. As $\sigma_{e^+e^-}^{\mathscr{K}}, \sigma_N^{\mathscr{K}}$ and $\sigma_{pp}^{\mathscr{K}}$ are model independent, it would be interesting to compare the resulting FF to the existing parameterizations from e^+e^- data, obtained with various assumptions. In addition, comparing the FF obtained in $\sigma_{e^+e^-}^{\mathscr{K}}$ at $Q^2 \simeq m_Z^2$, to those from $\sigma_N^{\mathscr{K}}$ at quite low Q^2 would allow to test the Q^2-evolution and provide a test of universality of the FFs. The three processes were considered separately and the expressions for the cross sections can be found in [1].

CONCLUSIONS

The discussed cross section differences require precise data on $h^{\pm}$ production.

We are informed that such accurate data on $\pi^{\pm}$ and $K^{\pm}$ production for SIDIS is feasible in HERMES [4], and for pp collisions – in BRAHMS [5] and STAR [6].

The discussed approach, considering differences of cross sections for produced hadrons and anti-hadrons and based only on charge conjugation invariance, was first worked out for polarized SIDIS [7] for extracting in a model independent way the polarized valence-quark distributions. In this paper we apply it for obtaining a model independent information about NS combinations of FFs. Analogously it should be applicable also to other cross section differences, for example the single and double spin asymmetries measured by CLAS at JLab [8] or the azimuthal SIDIS asymmetries measured by COMPASS at CERN [9].

ACKNOWLEDGMENTS

The paper is supported by 288/2008 Grant of Bulgarian National Science Foundation.

REFERENCES

1. E. Christova and E. Leader, *Eur. Phys. J. C* **51**, 825 (2007), arXiv:0809.0191(hep-ph), to appear in *Phys. Rev. D*
2. C. Bourrely, J. Soffer and F. Buccella *Phys. Lett. B* **648**, 39 (2007).
3. D. de Florian, R. Sassot and M.Stratmann, *Phys. Rev. D* **75**, 114010 (2007); **76**, 074033 (2007).
4. A. Airapetian et al, (HERMES collaboration), *Eur. Phys. J. C* **21**, 599 (2001), Achim Hillenbrand, Ph.D. theses, DESY-2005
5. BRAHMS collaboration, I. Arsene et al., hep-ex/0701041
6. STAR collaboration, B.I. Abelev et al., nucl-ex/0607033
7. E. Christova and E. Leader, *Nucl. Phys. B* **607**, 369 (2001).
8. H. Avakian et al. for the CLAS collaboration, nucl-ex/0509032
9. W. Kafer for the COMPASS collaboration, arXiv:0808.0114 (hep/ex)

Status and Prospects of Di-jet Production in High-energy Polarized Proton-Proton Collisions at RHIC at $\sqrt{s} = 200\,\text{GeV}$

Bernd Surrow (For the STAR Collaboration)

Massachusetts Institute of Technology
77 Massachusetts Avenue, Cambridge, MA 02139, USA

Abstract. The STAR experiment at the Relativistic Heavy Ion Collider (RHIC) at Brookhaven National Laboratory (BNL) is carrying out a spin physics program colliding transversely or longitudinally polarized proton beams at $\sqrt{s} = 200 - 500$ GeV to gain deeper insight into the spin structure and dynamics of the proton. These studies provide fundamental insight into Quantum Chromodynamics (QCD). One of the main objectives is the determination of the polarized gluon distribution function, Δg, through the measurement of the longitudinal double-spin asymmetry, A_{LL}, for various processes. Inclusive hadron and jet production from polarized pp collision data collected so far at $\sqrt{s} = 200\,\text{GeV}$ using the STAR detector at RHIC have placed important constraints on Δg. Di-jet production provides direct access to the initial parton kinematics at leading order (LO) QCD and thus provides sensitivity to the Bjorken-x dependence of Δg. The status of the mid-rapidity di-jet cross section analysis from the 2005 RHIC run and the longitudinal double-spin asymmetry at mid-rapidity for the 2006 data sample are discussed in these proceedings. Projections on future di-jet measurements at STAR are provided.

Keywords: BNL, RHIC, STAR, pp collisions, QCD, proton spin, A_{LL}, gluon polarization, global analysis
PACS: 12.38.-t, 13.85.Ni, 13.87.-a, 13.87.Fh, 13.88.+e, 14.70.Dj

INTRODUCTION

The longitudinal STAR spin physics program profits enormously from the unique capabilities of the STAR experiment for measuring large acceptance jet production, identified hadron production and photon production. Constraining the polarized gluon distribution function, Δg, through inclusive measurements has been, so far, the prime focus of the physics analysis program of the Run 3/4 [1], Run 5 [2] and Run 6 [3] data samples. The recent STAR inclusive jet [2, 3] results along with the PHENIX neutral pion results [4] have been used for the first time to constrain Δg in a next-to-leading order (NLO) global analysis along with semi-inclusive and inclusive deep inelastic scattering (DIS) data [5]. The RHIC data sets have been shown to provide strong constraints on Δg for $0.05 < x < 0.2$ in this analysis. Inclusive measurements, such as inclusive jet production, integrate over a fairly large x region for a given jet transverse momentum region. While those measurements provide a strong constraint on the value of Δg integrated over a range in x, those measurements do not permit a direct sensitivity to the actual x dependence. This motivates the need for correlation measurements in polarized proton-proton collisions.

CP1149, *Spin Physics, 18th International Spin Physics Symposium*
edited by D. G. Crabb, D. B. Day, S. Liuti, X. Zheng, M. Poelker, and Y. Prok

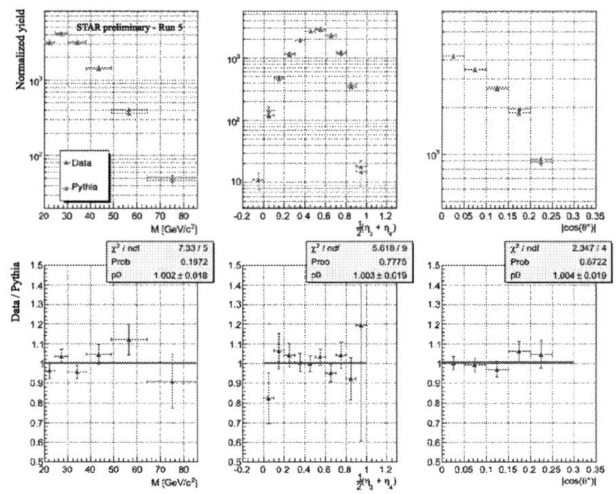

FIGURE 1. *Comparison of data (2005 RHIC run) and a PYTHIA-MC sample for three di-jet variables, the invariant mass (M), the mean of the di-jet pseudo-rapidities* ($1/2(\eta_3+\eta 4)$) *and the cosine of the center-of-mass scattering angle* ($\cos\theta^*$)*.*

STATUS AND PROSPECTS OF DI-JET PRODUCTION

Correlation measurements such as those for di-jet production allow for a better constraint of the partonic kinematics and thus the shape of Δg. At LO, the di-jet invariant mass, M, is proportional to the product of the x values of the partons, $M=\sqrt{s}\sqrt{x_1x_2}$, whereas the pseudorapidity sum of the final-state jets, $\eta_3+\eta_4$, is proportional to the logarithm of the ratio of the x values, $\eta_3+\eta_4=\ln(x_1/x_2)$. Photon-jet coincidence measurements are expected to provide a theoretically clean way to extract Δg [6]. A LO extraction of Δg alone would allow a model-independent way to constrain the x dependence, which would be an important contribution, without making an a priori assumption on the functional form of Δg as is currently required in a global analysis. This has been shown for photon-jet events in simulations [6]. The feasibility for a LO extraction of Δg as a function of x for the case of di-jet production still has to be demonstrated. Measurements at both $\sqrt{s}=200\,\text{GeV}$ and $\sqrt{s}=500\,\text{GeV}$ are preferred to maximize the kinematic reach in x and possibly provide a means to observe effects of scaling violations at fixed x by measuring different p_T values. The wide acceptance of the STAR experiment permits reconstruction of di-jet events with different topological configurations, i.e. different η_3/η_4 combinations, ranging from symmetric ($x_1=x_2$) partonic collisions to asymmetric ($x_1<x_2$ or $x_1>x_2$) partonic collisions. This, together with the variation of the center-of-mass energy, constrains Δg over a wide range in x of approximately $\sim 2\cdot 10^{-3}<x<0.3$ for di-jet and photon-jet events. The NLO framework for correlation measurements does exist and therefore those measurements can be used in a global analysis [7].

STAR reconstructs jets with the midpoint cone algorithm using clusters of charged track momenta measured with the STAR Time Projection Chamber (TPC) and tower energy deposits in the STAR Barrel Electromagnetic Calorimeter (BEMC) within a cone

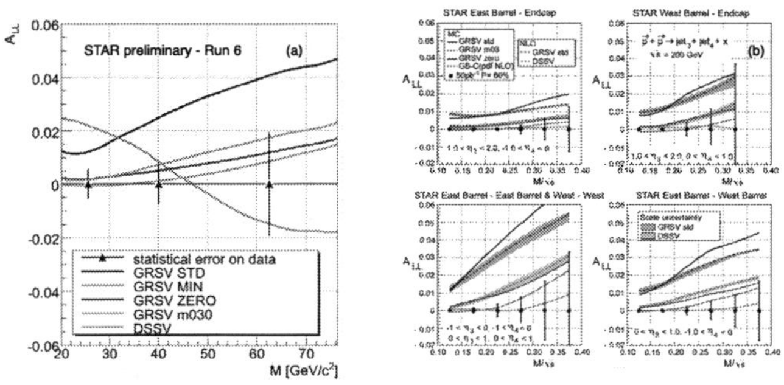

FIGURE 2. *(a) Statistical uncertainty of the longitudinal double-spin asymmetry, A_{LL}, as a function of the di-jet invariant mass, M, for the 2006 RHIC data sample. (b) Statistical precision of the longitudinal double-spin asymmetry, A_{LL}, for di-jet production as function of the ratio $M/\sqrt{s}$ for different topological combinations of the STAR BEMC and the STAR EEMC acceptance region.*

radius of $R \equiv \sqrt{\Delta\eta^2 + \Delta\phi^2}$ [3]. The jet axis was required to be within a fiducial range of $-0.05 < \eta_{\mathrm{JET}} < 0.95\,(-0.7 < \eta_{\mathrm{Detector}} < 0.9)$ for the 2005 (2006) data sample with a cone radius of $R = 0.4\,(0.7)$. The dominant fraction of di-jet events in both the 2005 and 2006 data samples are based on a jet patch (JP) trigger that required a minimum energy deposition for a group of towers over a region of $\Delta\eta \times \Delta\phi = 1.0 \times 1.0$. This trigger was taken in coincidence with a minimum-bias condition using the STAR Beam-Beam Counter (BBC). The di-jet analysis status presented below is based on an integrated luminosity of approximately $2\,\mathrm{pb}^{-1}$ and $5\mathrm{pb}^{-1}$ for the 2005 and 2006 data samples.

Figure 1 shows a comparison of data based on the 2005 RHIC run and a PYTHIA-MC sample for three di-jet variables, the invariant mass (M), the mean of the di-jet pseudorapidities ($1/2(\eta_3 + \eta 4)$) and the cosine of the center-of-mass scattering angle ($\cos\theta^*$). The top panels show the actual yield for each variable. The bottom panels display ratios of the data and MC histograms to quantify the comparison of data and MC distributions. The relative normalization has been fixed by the invariant mass distribution and then applied to both the $1/2(\eta_3 + \eta_4)$ and $\cos\theta^*$ distributions, i.e. the data/MC comparison reflects only one normalization factor. The comparison is carried out for asymmetric transverse momentum cuts on both jets using min($(p_T) \geq$ 7 GeV/c and max(p_T) $\geq$ 10 GeV/c where min(p_T) (max(p_T)) refers to the p_T of the jet in the di-jet pair with the smaller (larger) jet p_T. Such an asymmetric requirement on the final-state jet transverse momenta has been suggested as a means to keep NLO calculations under control since soft gluon emissions in a back-to-back jet configuration are not taken into account and would require resummations. The shape for each di-jet variable in data and MC is in good agreement. This agreement between data and MC is not dependent on the asymmetric jet transverse momentum requirement and even holds for symmetric cuts. A direct comparison of the actual di-jet cross section to NLO calculations requires further studies such as the completion of a full evaluation of hadronization and underlying event effects.

Figure 2 a) shows the statistical precision of the longitudinal double-spin asymmetry, A_{LL}, as a function of the di-jet invariant mass, M. These uncertainties, extracted from the 2006 data sample, are compared to a LO MC evaluation of A_{LL} computed with a PYTHIA MC sample using different event weights to account for different polarized gluon distribution functions of GRSV [8] and DSSV [5] similar to the ones discussed in [3]. The size of the statistical uncertainty at the highest invariant mass bin is at the level of the difference between GRSV-STD and DSSV.

Figure 2 b) shows the statistical precision of the longitudinal double-spin asymmetry, A_{LL}, for di-jet production as a function of $M/\sqrt{s}$ for different topological combinations of the STAR BEMC and the STAR Endcap Electromagnetic Calorimeter (EEMC) acceptance regions. At LO, the ratio $M/\sqrt{s}$ is equal to $\sqrt{x_1 x_2}$. Taking into account the different η ranges covered, and equivalently, the different $\cos\theta^*$ regions being probed, each panel represents a different range in x_1/x_2. At LO, $\cos\theta^*$ amounts to $\tanh\left(\frac{\eta_3-\eta_4}{2}\right)$. The upper left panel effectively probes asymmetric partonic collisions where predominantly a low-x gluon collides with a high-x quark at large invariant masses. The effective variation of x_1 and x_2 amounts to $0.2 < x_1 < 0.6$ and $0.07 < x_2 < 0.2$. In contrast, a kinematic region of larger x values in Δg is probed at predominantly symmetric partonic collisions such as the one shown in the lower right panel. The effective variation of x_1 and x_2 is roughly equal and given by the horizontal axis of the lower right panel. The projected uncertainties are shown for a luminosity of $50\,\mathrm{pb}^{-1}$ and a beam polarization of 60%. Those projected uncertainties are compared to a LO evaluation of A_{LL} and a full NLO A_{LL} calculation. Scale uncertainties are shown as a shaded band for DSSV and GRSV-STD reflecting a variation of the invariant mass M as a hard scale of $2M$ and $0.5M$. Asymmetric cuts are imposed for the LO MC and the NLO determination of $\min(p_T) \geq 7$ GeV/c and $\max(p_T) \geq 10$ GeV/c. The result of a LO MC evaluation using GS-C [9] for Δg is also shown. This particular choice of Δg is reflected in a large positive gluon polarization at low x, a node around $x \sim 0.1$ and a negative gluon polarization at large x at the initial scale of $4\,\mathrm{GeV}^2$. GS-C is still consistent with the current inclusive jet results [3]. A cone radius of $R = 0.7$ has been used. Good agreement is found between a LO MC evaluation of A_{LL} and a full NLO calculation. Scale uncertainties are found to be small in comparison to the variation of the chosen polarized gluon distribution functions, in particular, at large values of M. The projected uncertainties are those requested in the STAR Beam Use Proposal for the upcoming RHIC data taking period at $\sqrt{s} = 200\,\mathrm{GeV}$ in 2009. Di-jet production will play a critical role in the future, deepening our understanding of Δg, in particular, by constraining its shape.

REFERENCES

1. B. I. Abelev, et al., *Phys. Rev. Lett.* **97**, 252001 (2006).
2. B. I. Abelev, et al., *Phys. Rev. Lett.* **100**, 232003 (2008).
3. M. Sarsour, *(STAR Collaboration), These proceedings,* (2008).
4. A. Adare, *(PHENIX Collaboration), Submitted to Phys. Rev. Lett., hep-ex/0810.0694,* (2008).
5. D. de Florian, R. Sassot, M. Stratmann, and W. Vogelsang, *Phys. Rev. Lett.* **101**, 072001 (2008).
6. L. Bland, *Workshop at EPIC 1999, Bloomington, IN, hep-ex/9907058,* (1999).
7. D. de Florian, S. Frixione, A. Signer, and W. Vogelsang, *Nucl. Phys.* **B539**, 455–476 (1999).
8. M. Glueck, E. Reya, M. Stratmann, and W. Vogelsang, *Phys. Rev.* **D63**, 094005 (2001).
9. T. Gehrmann, and W. J. Stirling, *Phys. Rev.* **D53**, 6100–6109 (1996).

Constraints on ΔG through Longitudinal Double Spin Asymmetry Measurements of Inclusive Jet Production in Polarized pp Collisions at 200 GeV

M. Sarsour for the STAR Collaboration

Cyclotron Institute, Texas A&M University, College Station, TX 77843

Abstract. We report measurements of the longitudinal double spin asymmetry for inclusive jet production using polarized pp collisions at $\sqrt{s} = 200$ GeV at RHIC. Results from the 2005 and 2006 runs are presented. These results set substantial new constraints on the polarized gluon distribution in the proton over the kinematic range $0.02 < x < 0.3$, when compared to next-to-leading order global analyses of deep inelastic scattering (DIS) data. The first measurement of the transverse single-spin asymmetry for inclusive jets at mid-rapidity is also presented.

Keywords: gluon polarization, jets, longitudinal double spin asymmetry, STAR
PACS: 13.88.+e, 14.20.Dh, 13.87.Ce

INTRODUCTION

Polarized pp collisions at the Relativistic Heavy Ion Collider (RHIC) provide a very suitable environment, rich with strongly interacting probes, to constrain ΔG, the integral of the gluon polarized distribution function, $\Delta g(x, Q^2)$, evaluated at the input scale $Q_0^2 = 0.4$ GeV^2, directly and more precisely than previously attained. There are many processes where the gluon participates directly, and the high center of mass energy, $\sqrt{s} = 200$ GeV, and high transverse momentum, p_T, make next-to-leading order (NLO) perturbative Quantum Chromodynamics (pQCD) analysis more reliable. Indeed, several unpolarized pp cross sections for reactions sensitive to gluons have already been measured at RHIC and are described well by pQCD predictions [1]. The spin program of the Solenoid Tracker at RHIC (STAR) experiment [2] aims, in the short term, to utilize these advantages to constrain ΔG over the kinematic range $0.02 < x < 0.3$. Coincidence measurements can provide the x dependence of ΔG, but as RHIC was still developing higher luminosity and polarization, the abundant channels for inclusive jet and pion production were exploited. These proceedings report on the inclusive jet production asymmetry, which is an excellent channel to study the gluon polarization due to its large cross section and relative independence from fragmentation functions. The longitudinal double-spin asymmetry, A_{LL}, for inclusive jet production is defined as,

$$A_{LL} = \frac{\sigma^{++} - \sigma^{+-}}{\sigma^{++} + \sigma^{+-}} \tag{1}$$

where σ^{++} (σ^{+-}) is the inclusive jet cross section when the colliding proton beams have equal (opposite) helicities. The inclusive measurements from RHIC provide significant constraints on ΔG in the accessible kinematic range.

CP1149, *Spin Physics, 18th International Spin Physics Symposium*
edited by D. G. Crabb, D. B. Day, S. Liuti, X. Zheng, M. Poelker, and Y. Prok

EXPERIMENT AND ANALYSIS

The STAR detector subsystems [2] used in this measurement include the Time Projection Chamber (TPC), the Barrel (BEMC) and EndCap (EEMC) electromagnetic calorimeters and the Beam-Beam Counters (BBC). The TPC with pseudorapidity $|\eta| < 1.3$ and full azimuthal coverage is used to determine the momentum of charged particles. The BEMC and EEMC are lead-scintillator sampling calorimeters and have full azimuthal coverage spanning the pseudorapidities $|\eta| < 1.0$ and $1.08 < \eta < 2.0$, respectively. The calorimeters provide triggering and detection of photons and electrons. In 2005 only the west side of the BEMC, $0.0 < \eta < 1.0$, was active and the EEMC information was not included in the analysis. The BBC is mounted around the beam at longitudinal positions $z = \pm 370$ cm, with full azimuthal coverage and pseudorapidity $3.3 < |\eta| < 5.0$. It was used for triggering, luminosity measurement, and local polarimetry to determine non-longitudinal polarization components.

Jets were reconstructed using the midpoint cone algorithm [3] which clusters TPC charged track momenta and BEMC (and EEMC in 2006) tower energy deposits within a cone radius $R = \sqrt{\Delta\eta^2 + \Delta\phi^2}$. Only reconstructed TPC tracks and BEMC energy deposits above 0.2 GeV were used by the algorithm. The energy seed threshold was 0.5 GeV and jets were merged if more than 50% of their energy was of common origin. A minimum jet $p_T = 5$ GeV/c was also required. The jet axis was required to be within a fiducial range $0.2 < \eta < 0.8$ $(-0.7 < \eta < 0.9)$ for 2005 (2006) data combined with a cone radius of $R = 0.4(0.7)$ to reduce the edge effects of the calorimeter acceptance. In addition, the BBC time information was used to select events with a reconstructed vertex on the beam axis within approximately $\pm$ 60 cm from the TPC center to ensure a uniform tracking efficiency. Data were collected both with a high tower trigger (HT) that required a minimum energy deposition in a $\Delta\eta \times \Delta\phi = 0.05 \times 0.05$ tower and with a jet patch trigger (JP) that required a minimum energy deposition over a $\Delta\eta \times \Delta\phi = 1 \times 1$ region, in coincidence with the minimum bias trigger condition. JP triggers were dominant in both 2005 and 2006, and in this presentation we show only JP triggers from 2006 data.

RESULTS AND SUMMARY

In 2005 we had an order of magnitude increase in the figure of merit over our previous measurements [1] with coverage in jet p_T up to 30 GeV/c. This allowed us to perform a quantitative comparison of our measured results [4] to global fits of polarized DIS data within the GRSV framework [5] for various fixed values of ΔG [6]. These comparisons showed that the global fits that predict very large ΔG values and large A_{LL} are excluded [4]. However, some fits like GS-C [7] which has a large positive gluon polarization at low x, a node near $x \sim 0.1$, and a negative gluon polarization at large x, are still consistent with the data.

In 2006, we performed the first measurements of the transverse single-spin asymmetry, A_N, for inclusive jet production at mid-rapidity. Figure 1 shows A_N for four different pseudorapidity ranges covering the calorimeter acceptance. A_N is consistent with zero within the statistical uncertainties.

The left panel of Fig. 2 shows preliminary 2006 A_{LL} versus jet p_T corrected for

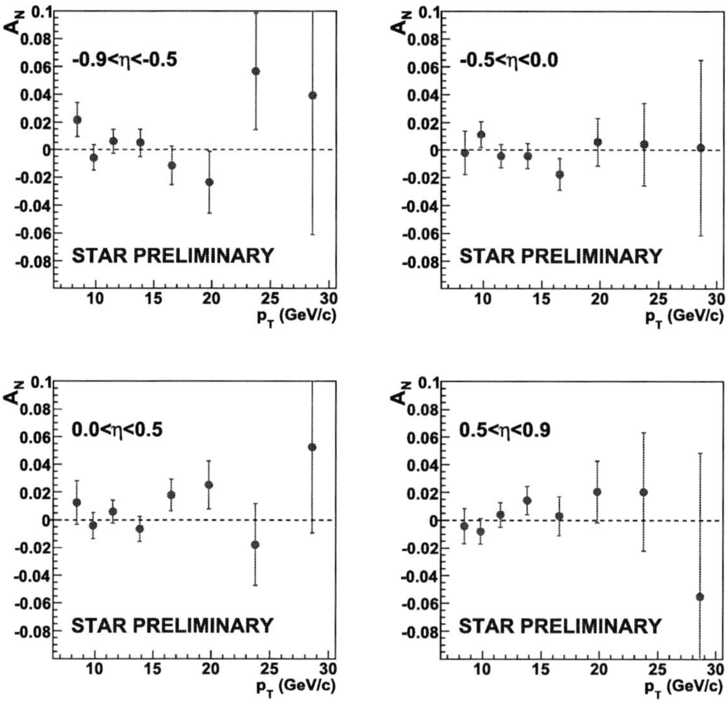

FIGURE 1. Preliminary 2006 A_N for inclusive jet production at $\sqrt{s} = 200$ GeV versus jet p_T for four different η bins.

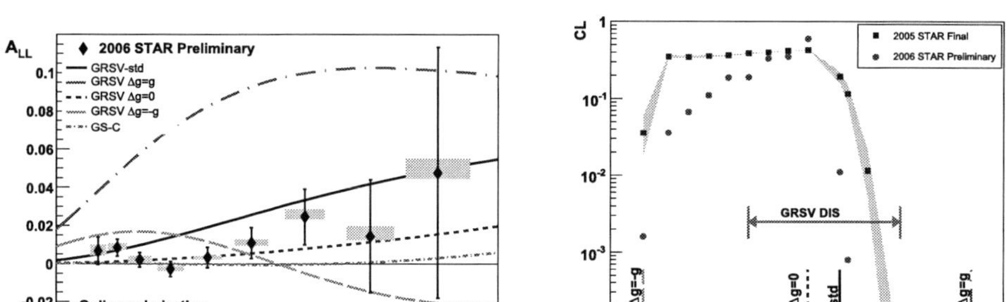

FIGURE 2. Left panel: Preliminary 2006 A_{LL} for inclusive jet production at $\sqrt{s} = 200$ GeV versus jet p_T, where the error bars are statistical and the gray bands indicate the systematic uncertainties. Right panel: Confidence level calculations of various predictions in the GRSV framework representing the A_{LL} results from 2005 and 2006.

detector response. The points with error bars are the data with statistical uncertainties. As for the gray bands, the height indicates the systematic uncertainties on A_{LL} while the width indicates the systematic uncertainties on p_T. The left panel of Fig. 2 also shows NLO pQCD calculations which incorporate different scenarios for $\Delta g(x)$, including the best global fit to the inclusive DIS data (std) [5]. The other three curves span the full physically allowed region from maximally positive gluon polarization ($\Delta g(x) = g(x)$) at the input scale to maximally negative gluon polarization ($\Delta g(x) = -g(x)$) and passing through zero gluon polarization ($\Delta g(x) = 0$). In addition, it includes a prediction derived from GS-C which has a node near $x \sim 0.1$ as described earlier. The statistical uncertainties in the 2006 A_{LL} measurements at high p_T are a factor of 3 to 4 smaller than they were in the 2005 data [4], which leads to significantly more stringent constraints on gluon polarization models as illustrated in the right panel of Fig. 2. It shows the confidence levels of global analysis predictions within the framework of GRSV [5, 6] representing the A_{LL} results from both runs 2005 and 2006.

These inclusive jet asymmetries were included in a recent "global" NLO analysis, DSSV [8], which is the first analysis to include inclusive DIS, semi-inclusive DIS, and RHIC pp collision data together. The result of this analysis showed that $\Delta g(x, Q^2)$ is small in the accessible range of momentum fraction, with a possible node in the distribution near $x \sim 0.1$, basically evolving away at higher scales with the opposite phase from GS-C. A study of the DSSV χ^2 profile and the partial contributions $\Delta\chi^2$ of the individual data sets shows that these STAR data provide the strongest limits on negative gluon polarization over the range $0.05 < x < 0.2$ and also contribute significantly to the limits on positive gluon polarization over the same range.

In summary, we reported on A_{LL} measurements from inclusive jet data from 2005 and 2006. These results provide significant constraints on the gluon spin contribution to the proton spin when compared to NLO pQCD calculations. They play a significant role in the first global NLO analysis to consider DIS, SIDIS and RHIC data together. In addition, we reported a preliminary A_N measurement for inclusive jets at mid-rapidity. It is statistically consistent with zero.

REFERENCES

1. B. I. Abelev et al. (STAR collaboration), *Phys. Rev. Lett.* **97**, 252001 (2006).
2. K. H. Ackermann et al. (STAR Collaboration), *Nucl. Instrum. Meth. A* **499**, 624 (2003).
3. G. C. Blazey et al., arXiv:*hep-ex*/0005012.
4. B. I. Abelev et al. (STAR collaboration), *Phys. Rev. Lett.* **100**, 232003 (2008).
5. M. Glück, E. Reya, M. Stratmann and W. Vogelsang, *Phys. Rev. D* **63**, 094005 (2001).
6. M. Stratmann and W. Vogelsang (private communication), 2007.
7. T. Gehrmann and W. J. Stirling, *Phys. Rev. D* **53**, 6100 (1996).
8. D. de Florian, R. Sassot, M. Stratmann, and W. Vogelsang, *Phys. Rev. Lett.* **101**, 072001 (2008).

Longitudinal Double Spin Asymmetries in Heavy Flavor Production in Polarized $p+p$ Collisions at $\sqrt{s} = 200$ GeV

Xiaorong Wang for the PHENIX Collaboration

New Mexico State University, Las Cruces, NM 88003

Abstract. One of the main goals of the RHIC spin program is to determine the contribution of gluons to the proton spin. At RHIC energies, heavy quark (charm and beauty) production is expected to be dominated by gluon-gluon interactions, so a measurement of the double longitudinal spin asymmetry A_{LL} in heavy quark production in polarized $p+p$ collisions directly probes the polarized gluon distribution inside the proton. We present the latest results on an A_{LL} measurement for J/ψ and open charm production measured by the PHENIX detector.

Keywords: Proton Structure, Heavy Flavor, Gluon Polarization
PACS: 14.20.Dh

INTRODUCTION

The quark contribution to the proton spin was found to be small by the European Muon Collaboration (EMC) with polarized lepton-nucleon deep inelastic scattering (DIS) [1]. Since then, a large number of DIS experiments have established that only $10 \sim 30\%$ of the proton spin is carried by the quarks and anti-quarks, which is smaller than the naive quark parton model expectation. The rest of the spin must be coming from the gluon spin and the parton orbital angular momentum. Since gluons are neutral and do not couple to virtual photons in leading order, the polarized gluon distribution, $\Delta g(x)$, is not well constrained by the DIS experiments. With RHIC, we can collide quarks and gluons at high energy. Furthermore, heavy quark production in polarized $p+p$ collisions at RHIC energy is dominated by the gluon-gluon interaction, thus providing direct access to the (polarized) gluon distribution in the proton. The double longitudinal spin asymmetry in heavy quark(onium) production can be written as

$$A_{LL}^{Q\bar{Q}} \sim \frac{\Delta g(x_1)}{g(x_1)} \times \frac{\Delta g(x_2)}{g(x_2)} \times \hat{a}_{LL}^{Q\bar{Q}} \tag{1}$$

where $g(x)$ ($\Delta g(x)$) is the unpolarized (polarized) gluon distribution function and $\hat{a}_{LL}^{Q\bar{Q}}$ is the double spin asymmetry for the partonic subprocess $g+g \rightarrow Q\bar{Q}(Q = c$ or $b)$, and has been estimated to be $\sim O(0.1)$ in pQCD calculations [2, 3, 4].

CP1149, *Spin Physics, 18th International Spin Physics Symposium*
edited by D. G. Crabb, D. B. Day, S. Liuti, X. Zheng, M. Poelker, and Y. Prok

MEASUREMENT AND ASYMMETRY CALCULATION

In this report we present the latest results for the longitudinal double spin asymmetry in J/Ψ and open heavy flavor production in polarized $p+p$ collisions at RHIC. Preliminary results for A_{LL} for J/ψ production were presented at SPIN 2006 [5]. The PHENIX muon spectrometers have measured single muons from heavy flavor semi-leptonic decays at forward rapidities ($1.2 < |\eta| < 2.4$). The recorded luminosity within the PHENIX acceptance was 2.23 pb^{-1} for the run in 2005 and 3.19 pb^{-1} in 2006.

Until silicon Vertex Detector (VTX) and Forward Vertex detector (FVTX) upgrades are complete, the PHENIX muon arms do not have the ability to measure displaced vertices. As a result, an open heavy flavor measurement in the muon arms is limited to an indirect statistical measurement. Extraction of signal muons in this analysis is predicated upon successful estimation and statistical subtraction of all background sources which are defined to be tracks not resulting from the semi-leptonic decay of heavy flavor mesons. The tracks penetrating deeply into the muon identifier last gap along beam direction consist of muons from heavy flavor decay and muons from light hadrons decay (K and π). The tracks stopping earlier in the muon identifiers are mostly hadrons based on our simulations [6].

Our strategy to extract the muon from heavy flavor decay asymmetry is to study the inclusive tracks which penetrated deeply in the muon identifier, while using hadrons which stopped earlier to estimate the background asymmetry.

The inclusive asymmetry $A_{LL}^{incl(hadron)}$ is given as

$$A_{LL}^{incl(hadron)} = \frac{1}{P_B \cdot P_Y} \frac{N^{++} - R \cdot N^{+-}}{N^{++} + R \cdot N^{+-}}, \tag{2}$$

where P_B and P_Y are the polarization for blue and yellow beams respectively. N^{++} (N^{+-}) is the number of muons from the same (opposite) helicity beam collisions, and $R = L^{++}/L^{+-}$ is the relative luminosity measured with Beam Beam Counter (BBC) and Zero Degree Calorimeter (ZDC).

A_{LL} of muon from heavy flavor decay vs p_T: The A_{LL} can be derived from

$$A_{LL}^{signal} = \frac{A_{LL}^{incl} - r \cdot A_{LL}^{BG}}{1-r}, \tag{3}$$

$$\delta A_{LL}^{signal} = \frac{\sqrt{(\delta A_{LL}^{incl})^2 + r^2 \cdot (\delta A_{LL}^{BG})^2}}{1-r}, \tag{4}$$

where r is the background fraction and A_{LL}^{BG} is the background asymmetry.

Table 1 lists the background fraction we calculate from the signal to background ratio from the $p+p$ run in year 2005 single muon analysis [7].The systematic errors introduced by background study was estimated. We use the upper and lower limits to estimate the error introduced from the background fraction study.

Figure 1 shows the double spin asymmetry results for muons from heavy flavor decays from the data for the runs in 2005 and 2006.

TABLE 1. Signal to background ratio and background fraction (r).

$p_T(GeV)$	S/B ratio	r	e_r
1-5	0.408	71.25%	$^{+7.18\%}_{-6.46\%}$
1.75-2.0	0.420	70.4567.98%	$^{+6.52\%}_{-5.50\%}$
2.5-3.0	0.492	67.01%	$^{+6.87\%}_{-5.74\%}$
3.0-3.85	0.470	68.05%	$^{+7.75\%}_{-6.53\%}$
3.85-5.0	0.336	75.07%	$^{+7.95\%}_{-7.16\%}$

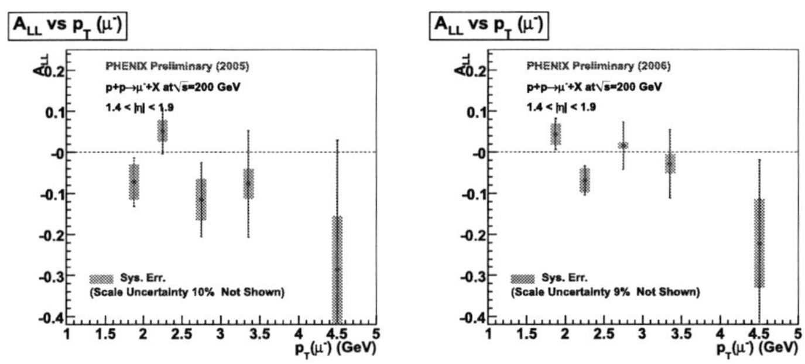

FIGURE 1. Double spin asymmetry A_{LL} for μ^- from heavy flavor decay as a function of p_T from $p+p$ runs in 2005 and 2006, with averaged polarization 50% and 60%, respectively.

A_L of muon from heavy flavor decay vs η: Single spin asymmetries are studied for transverse momenta $1.75 < p_T < 5.0$ GeV. The averaged asymmetries in the forward (average $\eta = 1.65$) and backward (average $\eta = -1.65$) directions are shown in figure 2.

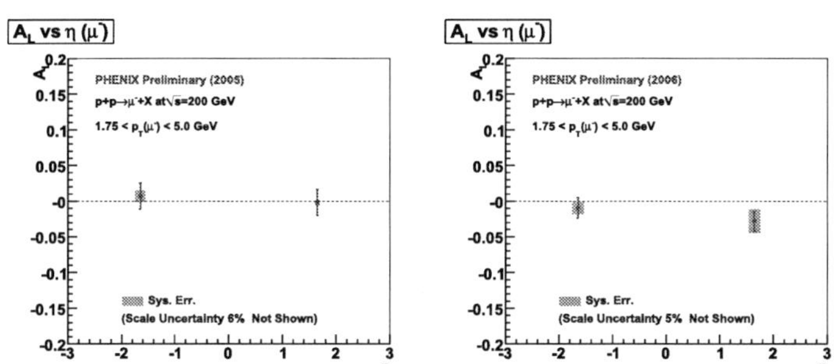

FIGURE 2. Single spin asymmetry A_L for μ^- from heavy flavor decay as a function of p_T from $p+p$ runs in 2005 and 2006, with averaged polarization 50% and 60%, respectively.

The systematic uncertainty has been checked with a bunch shuffling technique with randomly assigned polarization direction for bunches in each fill. The bunch-to-bunch and fill-to-fill systematic errors are much smaller than the statistical errors for this data set.

SUMMARY AND OUTLOOK

The longitudinal double spin asymmetry A_{LL} for J/Ψ and open heavy flavor have been measured for the first time at forward rapidity and $\sqrt{s}$ = 200 GeV with the PHENIX muon spectrometers. No nonzero A_{LL} has been observed at the current precision. Our results are limited by statistics. These results can not improve upon the constrains already placed on gluon polarization by other data.

The VTX and FVTX upgrades have been approved for construction. They will extend our coverage in gluon momentum fraction x up to 0.03 and down to 10^{-3}. Muon from heavy flavor decays originate from a vertex that is displaced from the original event vertex. These vertex detectors will allow us to identify displaced vertices by measuring the distance-of-closest-approach of the track to the collision axis. The capability of displace vertex measurement will improve our heavy flavor signal to hadronic background ratio by about a factor of 10. With expected recorded luminosity of 83 pb^{-1}, our double spin asymmetry measurements will allow us to constrain the gluon polarization. The projected sensitivity plot is shown in Figure 3

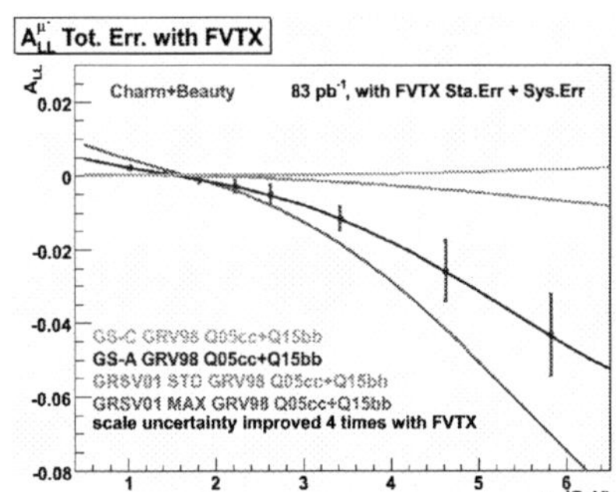

FIGURE 3. Double spin asymmetry A_{LL} for μ^- from heavy decay as a function of p_T from $p+p$ runs with FVTX detector detector with recording luminosity 83 pb^{-1}.

REFERENCES

1. J. Ashman *et al.*, *Phys. Lett.* **B 206**, 364 (1988); *Nucl. Phys.* **B328**, 1 (1989).
2. I. Bojak and M. Stratmann, *Phys. Rev. D* **67**, 034010 (2003).
3. A. Tkabladze and O. Teryaev, *Phys. Rev. D* **56**, 7331 (1997).
4. M. Klasen *et al.*, *Phys. Rev. D* **68**, 034017 (2003).
5. M. Liu, AIP conference proceeding vol. **915**, 494 (2006).
6. PHENIX collaboration, *Phys. Rev. C* **71**, 034908 (2005).
7. D. Hornback, *J. Phys. G* **35**, 104113 (2008).

Longitudinal Double-spin Asymmetries for π^0 Production at Forward Rapidity in *pp* Collisions at $\sqrt{s}$ = 200 GeV at STAR

Scott W. Wissink, for the STAR Collaboration

Department of Physics and Cyclotron Facility, Indiana University, Bloomington, IN 47408, USA

Abstract. Measurements of the double-spin asymmetry A_{LL} for inclusive π^0 production in hard *pp* collisions can provide important constraints on gluonic spin contributions to the proton. The STAR Endcap Electromagnetic Calorimeter is well-suited for these studies, with full azimuthal coverage for pseudorapidities $1.086 < \eta < 2$, good depth segmentation, and excellent transverse granularity. Details of the π^0 reconstruction algorithms we employed are described. Preliminary results from the 2006 data set for A_{LL} for π^0's at forward rapidities are presented and compared to several NLO pQCD calculations that assume different values for the integral ΔG.

Keywords: pions, gluons, double-spin asymmetry, forward rapidity
PACS: 13.88.+e, 13.85.Ni, 14.20.Dh

INTRODUCTION

The RHIC facility at Brookhaven National Laboratory is unique, in that one can study hard partonic scattering in polarized *pp* collisions, using the polarization of one parton to probe helicity preferences of another. To date, the strongest constraints on ΔG, the contribution of the gluonic spin to that of the proton, from RHIC have come from the double-spin asymmetry A_{LL} for inclusive jets at STAR [1] and for π^0's at PHENIX [2] near mid-rapidity. To confirm these results, and to test assumptions made by current theoretical models, it is essential that similar measurements be performed for different reactions and over a broad range of kinematics. Determining A_{LL} at forward rapidities probes new regions in energy and p_T, and hence x_g, and is driven by alternate mixtures of competing partonic subprocesses. Inclusive π^0 production is a high-rate process that fulfills these requirements; calculations [3] suggest that for π^0's emitted at $\eta \approx 1.5$, the sensitivity to ΔG will be greatest at intermediate p_T's (5-15 GeV/*c*), where polarized quarks serve as efficient analyzers of the struck gluon's helicity. Calculations carried out in next-to-leading order (NLO) perturbative Quantum Chromodynamics (pQCD), combined with current models of parton distribution functions and fragmentation functions, allow us to place statistically meaningful constraints on ΔG.

EXPERIMENTAL DETAILS

At STAR, neutral pions of pseudorapidity $1.1<\eta<2$ are reconstructed by detecting two decay photons in the Endcap e.m. calorimeter (EEMC) [4]. The EEMC is a Pb/scint. sampling device, providing full azimuthal coverage over this η range via 720 optically

CP1149, *Spin Physics, 18th International Spin Physics Symposium*
edited by D. G. Crabb, D. B. Day, S. Liuti, X. Zheng, M. Poelker, and Y. Prok

isolated projective towers (~22 X_0). The first two, and last, of the 24 scintillator layers are read out individually for pre- and post-shower information, respectively. At ~5 X_0 is a shower-maximum detector (SMD), divided into 12 overlapping 30° sections, each consisting of two orthogonal (*u* and *v*) planes of 288 scintillating strips. These planes provide the 1-mm transverse resolution required to locate the shower position in space.

For this analysis, based on data taken during Run 6, only events that satisfied both a single tower and surrounding 'patch' energy thresholds at the trigger level were used, thereby selecting events with large energy deposition concentrated in a relatively small ($\eta \times \phi \approx 0.2 \times 0.3$) area. This enhanced our ability to trigger directly on π^0's, reducing bias in our event selection. Our reconstruction philosophy was simple: maximize the π^0 yield by considering all combinations of all photon (e.m. shower) candidates, while keeping all threshold values low, even as our 'background' increased. To estimate the level of background, we determined the invariant mass for each photon pair:

$$m_{\gamma\gamma}{}^2 = 2E_1E_2(1-\cos\varphi_{\gamma\gamma}) \tag{1}$$

where the sum E_1+E_2 is measured by the towers, the opening angle is determined by the SMD, and the energy sharing (E_1/E_2) by an algorithm involving both detectors.

A total of 2.2M events, from 347 runs distributed among 41 RHIC fills, were used in the final analysis [5]. Just over 200k pions were reconstructed, yielding about 10% efficiency per triggered event. While most features of the π^0 peak in the invariant mass spectrum were well reproduced in simulation (details below), the background beneath the peak exhibited a yield at low mass substantially larger than we had expected based on EEMC 'slow simulator' results. This discrepancy—which completely dominates our combinatoric contribution—was eventually traced to real strip-to-strip fluctuations in the SMD response that are not present in our simulations. While we investigate this issue further, we have in parallel adopted a phenomenological approach, and extract our (spin-sorted) pion yields by simply fitting the background underlying the peak in our invariant mass distributions for different p_T bins.

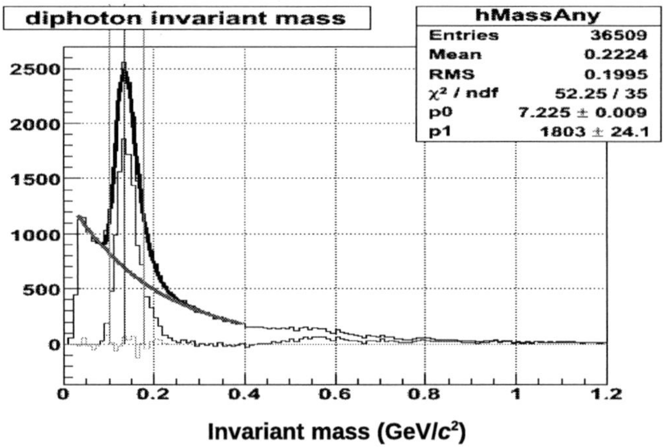

FIGURE 1. Invariant mass distribution for all π^0 candidates with $6 < p_T < 7$ GeV/*c*. Heavier lines show features of the fit, while lighter lines are derived from the data, and are explained in the text. Final pion yields are found by summing the light blue curve between the two green (vertical) lines.

A typical invariant mass distribution is presented in Fig.1. The heavy black curve is a fit to the data for $0.05 < m_{\gamma\gamma} < 0.40$ GeV, and consists of a skewed Gaussian peak plus a purely exponential (*i.e.*, two-parameter) background:

$$y(x) = A_1 \exp[-\alpha x] + A_2 \exp\left[-\tfrac{1}{2}\{[x-\mu]/\sigma[1+k(x-\mu)\}^2\right] \qquad (2)$$

The pink curve is the fitted background component, while the thin blue line shows the pion spectrum (data – fitted background). The vertical red and green lines indicate the true π^0 mass and our summing limits, respectively. In our analysis, the pion yields are not derived from the Gaussian fit parameters, but are determined by summing all counts in the blue histogram between the two green lines. To reduce our sensitivity to potential spin-dependent fitting errors, we proceed by first summing the mass spectra for each p_T bin over all four spin states, and performing a single fit to the function in Eq. 2. The spin-dependent yields are then extracted by fitting each of the four spectra separately, but allowing only the amplitudes A_1 and A_2 to vary. The skewness factor k was held fixed at 6.1 for all p_T bins; changing this by ±20% had essentially no effect.

RESULTS AND DISCUSSION

Several crucial 'sanity checks' were performed, to test the integrity and the robustness of our fitting results. The extracted centroids for the pion peak (μ in Eq. 2) were quite stable, and increased monotonically by ~3 MeV for p_T = 5 to 15 GeV. Absolute peak widths agreed well with simulation, as did their p_T dependence. To test for luminosity errors, the normalized, spin-sorted yields were used to extract longitudinal single-spin asymmetries A_L for each beam; these parity-violating observables were consistent with zero, with statistical errors of ±0.003 when averaged over the entire p_T range. Finally, in order to estimate the sensitivity of our final values for A_{LL} to (somewhat) arbitrary choices made during our fitting procedures, we carried out three analyses: vary the fit range (nominally 50-400 MeV) by ±10 MeV at each end; vary the analytic form of the background function, adding an additional parameter that kept normalized χ^2 constant; and for each p_T bin, hold μ and σ fixed during spin-sorted fits, but displaced from their 'best' values by twice their error. The resultant changes in A_{LL} were added linearly (in case of correlations) to form the dominant contribution to our systematic error budget.

Preliminary results [5] for the longitudinal double-spin asymmetry A_{LL} for inclusive π^0 production, integrated over the EEMC acceptance, are shown in Fig. 2 below. The vertical error bars indicate statistical errors only, which include contributions from background subtraction, while the horizontal 'errors' indicate the range of p_T for each point. Systematic errors, represented by the height of the tan band near the bottom of the plot, are dominated by variations in extracted values for A_{LL} when the fit function, its range, or the selection of parameters held fixed are changed, as discussed above.

The four curves shown are from the GRSV group [6], and are derived from fits to the 2001 DIS database. The black curve represents a best fit, while the red, green, and blue curves assume gluonic spins of maximum, minimum (*i.e.*, oppositely directed), and zero contribution to the proton spin, respectively. Our results strongly disfavor scenarios in which the gluon spins are fully aligned with that of the proton (red curve), consistent with other RHIC analyses [1,2]. Discrimination among the remaining curves, though, will clearly require data of higher statistical precision than has been achieved thus far, especially at the higher p_T values.

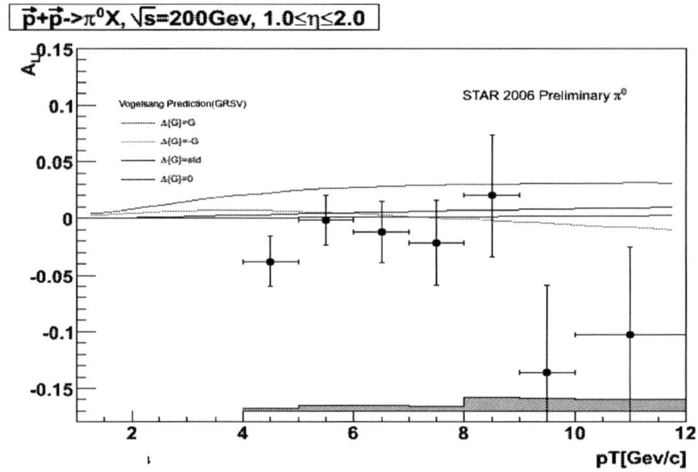

FIGURE 2. STAR preliminary values for A_{LL} for inclusive π^0 production at $1.1<\eta<2$. The error bars indicate statistical error only; systematic error is indicated by the height of the tan band at the bottom. The curves shown are from Ref. [6], and are explained in detail in the text.

CONCLUSIONS AND OUTLOOK

We present here the first measurements of the double-spin observable A_{LL} for inclusive π^0 production in the forward η region, thus extending the kinematic reach of the STAR spin program to new and lower values of x_{gluon}. Despite significant backgrounds, yield extraction techniques developed for this work appear robust, and associated systematic errors are small and under control. Limited statistics, and the reduction by a factor of 2-3 in the theoretically predicted magnitude for A_{LL} in this regime (compared to mid-rapidity), have diminished our sensitivity to gluonic polarization and compromised the ability of this data set, by itself, to place substantially tighter constraints on ΔG. When included as part of new global analyses [7], however, the trend toward negative values exhibited by our data in this regime could prove to have an important impact.

This work was supported in part by the US National Science Foundation.

REFERENCES

1. B.I. Abelev *et al.* (STAR Collaboration), *Phys. Rev. Lett.* **100**, 232003 (2008).
2. A. Adare *et al.* (PHENIX Collaboration), *Phys. Rev. D* **76**, 051106 (2007).
3. V. Guzey, M. Strikman, and W. Vogelsang, *Phys. Lett. B*, **603**, 173 (2004).
4. C.E. Allgower *et al., Nucl. Instrum. Meth.* A **499**, 740 (2003).
5. W. He, "Double Spin Asymmetry in Inclusive π^0 Production for Longitudinally Polarized proton-proton Collisions at $\sqrt{s}$ = 200 GeV with the Endcap Electromagnetic Calorimeter at STAR," Ph.D. Thesis, Indiana University, 2008.
6. M. Glueck, E. Reya, M. Stratmann and W. Vogelsang, *Phys. Rev. D* **63**, 094005 (2001); and M. Stratmann and W. Vogelsang, private communication (2005).
7. D. de Florian *et al., Phys. Rev. Lett.* **101**, 072001 (2008).

SPIN STRUCTURE OF THE NUCLEON—TRANSVERSE SPIN

Theoretical Overview on Recent Developments in Transverse Spin Physics

Feng Yuan

RIKEN/BNL Research Center, Building 510A, Brookhaven National Laboratory, Upton, NY 11973
Nuclear Science Division, Lawrence Berkeley National Laboratory, Berkeley, CA 94720, USA[1]

Abstract. Transverse-spin physics has been very active and rapidly developing in the last few years. In this talk, I will briefly summarize recent theoretical developments, focusing on the associated QCD dynamics in transverse spin physics.

Keywords: Single transverse-spin asymmetries
PACS: 12.38.Bx, 12.39.St, 13.85.Qk

There have been strong experimental interests on transverse spin physics around the world, from the deep inelastic scattering experiments such as the HERMES collaboration at DESY, SMC at CERN, and Hall A and CLAS at JLab, the proton-proton collider experiment from RHIC at Brookhaven, and the very relevant e^+e^- annihilation experiment from BELLE at KEK. One of the major goals in transverse spin physics is to study the quark transversity distribution, the last unknown leading-twist quark distribution in nucleon. Besides the quark transversity distribution, the transverse spin physics also opened a new window to explore the partonic structure of nucleon, the so-called transverse momentum dependent (TMD) parton distributions. TMD parton distribution is an extension to the usual Feynman parton distributions. These distributions allow us to study the three-dimension picture of partons inside the nucleon, and they are also closely related to the generalized parton distributions and the parton orbital angular momenta. Especially, the single transverse spin asymmetry (SSA) phenomena in high energy hadronic processes have attracted many theoretical and experimental investigations. The SSA is defined as the asymmetry when one of the hadrons' transverse spin is flipped, $A_N \sim (d\sigma(S_\perp) - d\sigma(-S_\perp))/(d\sigma(S_\perp) - d\sigma(-S_\perp))$. It has been a great theoretical challenge in the understanding of these phenomena. This is because the leading partonic contribution to the SSA vanish in the leading order, whereas the experimental observation show that these SSAs are in tens of percentage in the forward scattering of the polarized nucleon.

Recent theoretical developments have made great progress in the exploration of the underlying physics for the single spin phenomena. It is impossible to cover all these exciting physics in this short talk. Rather, I would like to focus on one important subject,

---

[1] This work was supported in part by the U.S. Department of Energy under contract DE-AC02-05CH11231. We are grateful to RIKEN, Brookhaven National Laboratory and the U.S. Department of Energy (contract number DE-AC02-98CH10886) for providing the facilities essential for the completion of this work.

CP1149, *Spin Physics, 18th International Spin Physics Symposium*
edited by D. G. Crabb, D. B. Day, S. Liuti, X. Zheng, M. Poelker, and Y. Prok

i.e., the nontrivial QCD dynamics associated with transverse spin physics: the QCD factorization, the universality of the parton distributions and fragmentation functions, and their scale evolutions.

Among those TMD parton distributions and fragmentation functions, two functions have been mostly discussed: the Sivers quark distribution and the Collins fragmentation function. The Sivers quark distribution represents a distribution of unpolarized quarks in a transversely polarized nucleon, through a correlation between the quark's transverse momentum and the nucleon polarization vector. The Collins function represents a correlation between the transverse spin of the fragmenting quark and the transverse momentum of the hadron relative to the "jet axis" in the fragmentation process. Although they both belong to the so-called "naive-time-reversal-odd" functions, they do have different universality properties. For the quark Sivers function, because of the initial/final state interaction difference, they differ by signs for the SIDIS and Drell-Yan processes [1, 2, 3, 4]. On the other hand, there have been several studies showing that the Collins function is universal between different processes, primarily in the SIDIS and e^+e^- annihilation [5, 6, 7, 8], and recently in pp collisions [9]. In the following, I will take the example of the Collins contribution to the azimuthal asymmetric distribution of hadrons inside a high energy jet in the transversely polarized pp collision to demonstrate this universality property,

$$p(P_A,S_\perp)+p(P_B)\rightarrow jet(P_J)+X\rightarrow H(P_h)+X\ , \tag{1}$$

where a transversely polarized proton with momentum P_A scatters on another proton with momentum P_B, and produces a jet with momentum P_J. The three momenta of P_A, P_B and P_J form the so-called reaction plane. Inside the produced jet, the hadrons are distributed around the jet axis, where we define transverse momentum P_{hT} relative to the jet axis. The correlation between P_{hT} and the polarization vector $S_\perp$ introduces the Collins contribution to the single spin asymmetry in this process.

We need to generate a phase from the scattering amplitudes to have a non-vanishing SSA. If the phase comes from the vertex associated with the fragmenting quark and the final state hadron, or from the dressed quark propagator, it is easy to argue the universality of the Collins function between this process and the SIDIS/e^+e^- process, because they are the same. The main issue of the universality discussion concerns the extra gluon exchange contribution between the spectator of the fragmentation process and hard partonic part. In Fig. 2, we have shown all these interactions for a particular partonic channel $qq'\rightarrow qq'$ contribution, including the gluon attachments to the incident quarks (a,c), and final state balancing quark (d) and the internal gluon propagator (b). The contributing phases of the diagrams in Fig. 2 come from the cuts through the internal propagators in the partonic scattering amplitudes. In Fig. 2, we labeled these cut-poles by short bars in the diagrams. From the calculations, we will find that all these poles come from a cut through the exchanged gluon and the fragmenting quark in each diagram, and all other contributions either vanish or cancel out each other. For example, in Fig. 2(d), we show two additional cuts, which contribute however opposite to each other and cancel out completely. Therefore, by using the Ward identity at this particular order, the final results for all these diagrams will sum up together into a factorized form, where the cross section is written as the hard partonic cross section for $q(S_\perp)q'\rightarrow q(s_\perp)q'$ subprocess

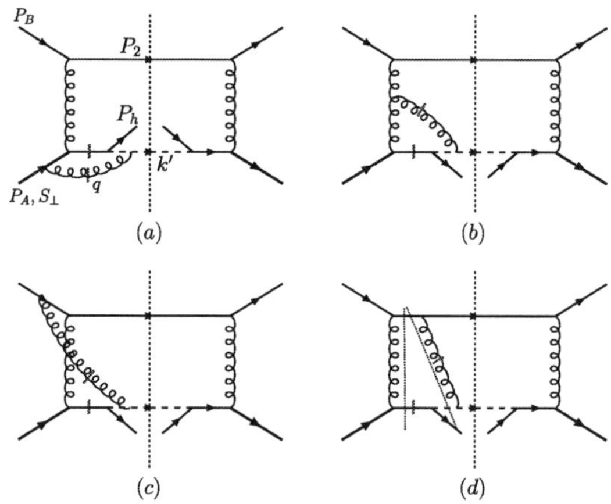

FIGURE 1. *Gluon exchange diagrams contributions to the Collins asymmetry in pp collisions. The short bars indicate the pole contributions to the phase needed for a non-vanishing SSA. The additional two cuts in (d) cancel out each other.*

multiplied by a Collins fragmentation function. The exchanged gluon in Fig. 2 is now attaching to a gauge link from the fragmentation function definition. Similar calculations can be performed for the other two processes SIDIS and e^+e^- annihilation, and the same Collins function will be observed. This argument can also be extended to two-gluon exchange diagrams [9].

The key steps in the above derivation are the eikonal approximation and the Ward identity. The eikonal approximation is valid when we calculate the leading power contributions in the limit of $P_{hT} \ll P_J$. The Ward identity ensure that when we sum up the diagrams with all possible gluon attachments we shall get the eikonal propagator from the gauge link in the definition of the fragmentation function. The most important point to apply the Ward identity in the above analysis is that the eikonal propagator does not contribute to the phase needed to generate a nonzero SSA.

This observation is very different from the SSAs associated with the parton distributions, where the eikonal propagators from the gauge link in the parton distribution definition play very important role [1, 2, 3, 4]. It is the pole of these eikonal propagators that contribute to the phase needed for a nonzero SSA associated with the naive-time-reversal-odd parton distributions, which also predicts a sign difference for the quark Sivers function between the SIDIS and Drell-Yan processes. More complicated results have been found for the SSAs in the hadronic dijet-correlation [10, 11], where a normal TMD factorization breaks down [12]. The reason is that the eikonal propagators from the initial and final state interactions in dijet-correlation process do contribute poles in the cross section [11, 12]. Because of this, the Ward identity is not applicable, and the standard TMD factorization breaks down, although a modified factorization may be valid if we modify the definition of the TMD parton distributions to take into account all the initial and final state interaction effects [10].

In particular, there is a sign change between the SSAs in SIDIS and Drell-Yan processes [1, 2],

$$\text{Sivers SSA}|_{\text{DY}} = -\text{Sivers SSA}|_{\text{DIS}} . \tag{2}$$

This nontrivial result of the opposite signs between the above two processes will still hold when gluon radiation contributions are taken into account, where the large transverse momentum Sivers function is generated from the twist-three quark-gluon correlation function [13]. It is of crucial to test this nontrivial QCD predictions by comparing the SSAs in these two processes. The Sivers single spin asymmetry in SIDIS process has been observed by the HERMES collaboration, and the planned Drell-Yan measurement at RHIC and other facility will test this prediction.

Another interesting probe for the initial/final state interaction effects is the SSA in heavy quark and antiquark production in hadronic process. Because the heavy quark and antiquark can be detected by their decay products, their SSAs can be measured separately. The heavy quark and antiquark produced in short distance partonic processes will experience different final state interactions with the nucleon spectator due to their different color charges, and therefore the SSAs for heavy quark and antiquark will be different. Detailed calculations show that the difference could be as large as a factor of 3 if the quark-antiquark channel contribution dominates [14]. Certainly, heavy quark production in high energy hadronic process is dominated by gluon-gluon fusion contributions. Therefore, the single spin asymmetry from gluonic sector will be important at RHIC energy. This part of contribution has recently been studied in the twist-three approach [15]. The twist-three three-gluon correlation functions contain two independent functions because of different color factors: one with f_{abc} and one with d_{abc} where a,b,c are the color indices for the three gluons [16]. Both correlations and the quark-gluon correlation function mentioned above will contribute to heavy quark single spin asymmetry in pp collisions. However, at RHIC energy kinematics, the quark-gluon correlation functions alone generate only a very small asymmetry for open charm production. Therefore, the observation of any significant single-spin asymmetry would be a clear indication of the presence of three-gluon correlations inside a polarized proton [15].

Most recently, there has been very exciting progress in studying the scale evolution equations for the quark-gluon and three-gluon correlation functions and their implications to the energy dependence of the relevant SSA observables [17, 18, 19]. General structure of the evolution equations for the twist-three quark-gluon correlation functions has been known in the literature [20]. However, the correlation functions responsible to the single transverse spin asymmetries are special projections of the general twist-three quark-gluon correlations, and their evolutions are not directly available from the already known results [20]. Earlier attempts [21] have been made to derive the evolution equations for the correlation functions, but were not complete. On the other hand, from the large transverse momentum quark Sivers function calculated in [13], we would already obtain the evolution equation for $T_F(x)$ (which is the transverse momentum moment of the quark Sivers function), since the collinear divergence in that calculation will lead to the splitting function of $T_F(x)$. This splitting function was confirmed by a complete calculation of next-to-leading order QCD correction to the transverse-momentum weighted spin asymmetry in Drell-Yan lepton pair production [17] and the derivations of the scale evolution equations directly [18, 19]. In particular, the scale evolution for the quark-gluon correlation function $T_F(x)$ is found to be,

$$\frac{\partial}{\partial \ln \mu^2} T_F(x_B, \mu^2) = \frac{\alpha_s}{2\pi} \int \frac{dx}{x} \left[C_F \left\{ \frac{1+z^2}{(1-z)_+} + \frac{3}{2} \delta(1-z) \right\} T_F(x,x) + \frac{C_A}{2} \left\{ \frac{1+z}{1-z} T_F(xz,x) - \frac{1+z^2}{1-z} T_F(x,x) + \tilde{T}_F(xz,x) \right\} \right] , \quad (3)$$

where $z = x_B/x$.

Furthermore, the NLO perturbative-QCD correction to the transverse momentum weighted single spin asymmetry in Drell-Yan lepton pair production in hadronic collisions has engaged the transverse spin physics to a more solid theoretical ground [17]. It has been shown that the collinear divergences can be absorbed into the NLO twist-three quark-gluon correlation function of the transversely polarized nucleon and the unpolarized quark distribution of the unpolarized nucleon. This calculation suggests that a general factorization formula exists for the transverse momentum weighted spin-dependent cross section in the Drell-Yan process, in extension of the general factorization arguments given in [22].

One important feature of this result is its behavior near "partonic threshold", that is in the large-z limit of the integrand, corresponding to $\hat{s} \sim Q^2$, when the initial partons have "just enough" energy to produce the virtual photon. Setting the scale $\mu = Q$, we have the following structure of the NLO correction in this case:

$$\frac{d\langle q_\perp \Delta\sigma(S_\perp)\rangle}{dQ^2} = \sigma_0 \frac{\alpha_s}{2\pi} \int \frac{dx}{x} \frac{dx'}{x'} T_F(x,x;\mu^2) \bar{q}(x';\mu^2) \left[4C_F(1+z^2) \left(\frac{\ln(1-z)}{1-z} \right)_+ \right] , \quad (4)$$

where we only keep the "double-logarithmic" term which dominates near threshold in the $\overline{\text{MS}}$ scheme. The structure of this expression is identical to that for the spin-averaged $q_\perp$-integrated NLO cross section near threshold,

$$\frac{d\sigma}{dQ^2} = \sigma_0 \frac{\alpha_s}{2\pi} \int \frac{dx}{x} \frac{dx'}{x'} q(x;\mu^2) \bar{q}(x';\mu^2) \left[4C_F(1+z^2) \left(\frac{\ln(1-z)}{1-z} \right)_+ \right] . \quad (5)$$

This means that the soft gluon contribution is spin-independent. It contributes in the same way to the spin-averaged and single-spin-dependent cross sections, and will lead to the same soft-gluon threshold resummation effects to these cross sections, at least at the leading double logarithmic level. This observation is very similar to that made for the transverse momentum resummation in the Drell-Yan process [23]. This will likely have the phenomenological consequence that the single-spin asymmetry for the Drell-Yan process will be quite stable under NLO corrections, in particular when $\tau = Q^2/s$ is large.

In summary, transverse spin physics has attracted much attention and has been rapidly developed in the last few years. In this overview, I could not cover all these important developments, rather I emphasized a few examples, including the universality of the parton distribution and fragmentation functions, and QCD evolution and next-to-leading order corrections to the relevant observables. Fortunately, there are many talks on the transverse spin physics in this conference, and I believe that they will present more comprehensive reviews on these exciting developments in this physics. This has shown that the transverse spin physics is playing a very important role in the strong interaction physics for hadronic spin physics. We will learn more about QCD dynamics and nucleon structure from these studies.

REFERENCES

1. S. J. Brodsky, D. S. Hwang and I. Schmidt, Phys. Lett. B **530**, 99 (2002); Nucl. Phys. B **642**, 344 (2002).
2. J. C. Collins, Phys. Lett. B **536**, 43 (2002).
3. X. Ji and F. Yuan, Phys. Lett. B **543**, 66 (2002); A. V. Belitsky, X. Ji and F. Yuan, Nucl. Phys. B **656**, 165 (2003).
4. D. Boer, P. J. Mulders and F. Pijlman, Nucl. Phys. B **667**, 201 (2003).
5. A. Metz, Phys. Lett. B **549**, 139 (2002).
6. J. C. Collins and A. Metz, Phys. Rev. Lett. **93**, 252001 (2004).
7. L. P. Gamberg, A. Mukherjee and P. J. Mulders, Phys. Rev. D **77**, 114026 (2008).
8. S. Meissner and A. Metz, arXiv:0812.3783 [hep-ph].
9. F. Yuan, Phys. Rev. Lett. **100**, 032003 (2008); Phys. Rev. D **77**, 074019 (2008).
10. C. J. Bomhof, P. J. Mulders and F. Pijlman, Phys. Lett. B **596**, 277 (2004); Eur. Phys. J. C **47**, 147 (2006); JHEP **0702**, 029 (2007); A. Bacchetta, C. J. Bomhof, P. J. Mulders and F. Pijlman, Phys. Rev. D **72**, 034030 (2005); C. J. Bomhof and P. J. Mulders, arXiv:0709.1390 [hep-ph].
11. J. W. Qiu, W. Vogelsang and F. Yuan, Phys. Lett. B **650**, 373 (2007); Phys. Rev. D **76**, 074029 (2007); W. Vogelsang and F. Yuan, Phys. Rev. D **76**, 094013 (2007).
12. J. Collins and J. W. Qiu, Phys. Rev. D **75**, 114014 (2007); J. Collins, arXiv:0708.4410 [hep-ph].
13. X. Ji, J. W. Qiu, W. Vogelsang and F. Yuan, Phys. Rev. Lett. **97**, 082002 (2006); Phys. Rev. D **73**, 094017 (2006); Phys. Lett. B **638**, 178 (2006).
14. F. Yuan and J. Zhou, arXiv:0806.1932 [hep-ph].
15. Z. B. Kang and J. W. Qiu, Phys. Rev. D **78**, 034005 (2008); Z. B. Kang, J. W. Qiu, W. Vogelsang and F. Yuan, Phys. Rev. D **78**, 114013 (2008).
16. X. D. Ji, Phys. Lett. B **289**, 137 (1992).
17. W. Vogelsang, F. Yuan, to be submitted.
18. Z. B. Kang and J. W. Qiu, arXiv:0811.3101 [hep-ph].
19. J. Zhou, F. Yuan and Z. T. Liang, arXiv:0812.4484 [hep-ph].
20. See, for example: I. I. Balitsky and V. M. Braun, Nucl. Phys. B **311**, 541 (1989); P. G. Ratcliffe, Nucl. Phys. B **264**, 493 (1986); X. D. Ji and C. h. Chou, Phys. Rev. D **42**, 3637 (1990); Y. Koike and K. Tanaka, Phys. Rev. D **51**, 6125 (1995); J. Kodaira, Y. Yasui, K. Tanaka and T. Uematsu, Phys. Lett. B **387**, 855 (1996); A. V. Belitsky and D. Mueller, Nucl. Phys. B **503**, 279 (1997); A. V. Belitsky, Nucl. Phys. B **558**, 259 (1999); A. V. Belitsky, Nucl. Phys. B **574**, 407 (2000); A. V. Belitsky, X. D. Ji, W. Lu and J. Osborne, Phys. Rev. D **63**, 094012 (2001). I. I. Balitsky, V. M. Braun, Y. Koike and K. Tanaka, Phys. Rev. Lett. **77**, 3078 (1996).
21. A. A. Henneman, D. Boer and P. J. Mulders, Nucl. Phys. B **620**, 331 (2002).
22. J. W. Qiu and G. Sterman, Nucl. Phys. B **353**, 105 (1991); Nucl. Phys. B **353**, 137 (1991).
23. A. Idilbi, X. d. Ji, J. P. Ma and F. Yuan, Phys. Rev. D **70**, 074021 (2004).

Measurement of TMDs in Semi-Inclusive DIS

Rainer Joosten

HISKP, University Bonn, Nußallee 14-16, 53115 Bonn, Germany

Abstract. Over the past decades, the colinear parton model has been very successful in describing the structure of nucleons. However, in the past years it became evident, that to fully describe the structure of the nucleon this picture has to be extended taking into account the transverse momentum of the partons. This gives rise to a new set of distribution and fragmentation functions entering the cross sections, namely the transverse momentum dependent (TMD) functions.

The measurement of these functions in semi-inclusive deep inelastic scattering (SIDIS) in the framework of the COMPASS and HERMES experiments is presented in this contribution.

Keywords: Semi-inclusive DIS, transverse single-spin asymmetries, transverse momentum
PACS: 13.60.Hb, 13.60.-r, 13.88.+e, 14.20.Dh, 14.65.-q

Deep Inelastic Scattering (**DIS**), where pointlike high energy leptons scatter off nucleon targets, is an excellent tool to investigate the structure of the nucleon. One of the most successful concepts to describe the results of DIS experiments is the *quark parton model*, where a fast moving nucleon is considered as a jet of quasi-free nearly massless partons moving almost colinear with respect to each other and the incoming lepton. In this picture, DIS can be considered as quasi-elastic scattering of leptons off one of these pointlike partons. Thus the cross section is considered to be the convolution of the elementary elastic lepton-parton scattering cross section calculable in QED with the unknown distributions of these partons, described by *Parton Distribution Functions* (**PDF**s). In the case of semi-inclusive DIS (**SIDIS**), where additionally a produced hadron is detected, the DIS cross section is convoluted with another function, describing the probability for quarks of given flavour q to fragment into certain hadrons, the *Parton Fragmentation Functions* (**PFFs**). Therefore, the cross section is given by

$$d\sigma^{l+N\rightarrow l'+X} = \mathrm{PDF} \otimes d\sigma^{l+q\rightarrow l'+q'} \otimes \mathrm{PFF} \tag{1}$$

A general scheme of all leading and higher-twist parton distribution functions[1] shows that to fully specify the quark structure of the nucleon at twist-two level, only three PDFs are required in the strictly colinear parton model. The unpolarised PDF $f_1^q(x)$ describes the probability to find a quark that carries the fraction x of the nucleon momentum. The *helicity* distribution function $g_{1L}^q(x)$ describes the probability to find a parton polarised in the same direction as the nucleon for a longitudinally polarised nucleon, while the *transversity* distribution function $h_1^q(x)$ describes the same for a *transversely* polarised nucleon.

Releasing the strict colinearity assumption, also the parton transverse momentum has to be taken into account, as it can lead to corrections to the dominant effects as well as to the rise of new effects. It can enter the PDFs as the transverse momentum k_T of the quarks in the struck nucleon and the PFFs as the transverse momentum p_T of the fragmenting quark or P_T^h of the produced hadron. Additionally it can be a source of

CP1149, *Spin Physics, 18th International Spin Physics Symposium*
edited by D. G. Crabb, D. B. Day, S. Liuti, X. Zheng, M. Poelker, and Y. Prok

kinematical effects. This gives rise to a completely new set of k_T and P_T^h dependent distribution and fragmentation functions entering the SIDIS cross section.

The full cross-section in one photon exchange (equation 2) has been calculated at leading and first subleading twist[2], introducing 18 structure functions $F_{XY,Z}^{moment}$. Here the superscript, if any, identifies an azimuthal asymmetry created by the given structure function, while the subscript indicates the polarisation of the beam (λ), the target (P_L,P_T) and the virtual photon. These functions in general depend on x, the squared four momentum transfer Q^2, the fraction z of the transfered energy carried by the hadron and P_T^h. It was shown [3] that they can be factorized into a hard part as well as TMD parton distribution and fragmentation functions, where the parton distribution functions depend on x and k_T while the fragmentation functions depend on z and P_T^h.

$$
\begin{aligned}
\frac{d\sigma}{d\phi_h} &\propto \frac{1}{Q^2}\Big\{F_{UU,T}+F_{UU,L}+F_{UU}^{\cos\phi_h}\cos\phi_h+F_{UU}^{\cos(2\phi_h)}\cos(2\phi_h)+\lambda F_{LU}^{\sin\phi_h}\sin\phi_h \\
&+ P_L\Big[F_{UL}^{\sin\phi_h}\sin\phi_h+F_{UL}^{\sin(2\phi_h)}\sin(2\phi_h)\Big]+P_L\lambda\Big[F_{LL}+F_{LL}^{\cos\phi_h}\cos\phi_h\Big] \\
&+ |P_T|\Big[(F_{UT,T}^{\sin(\phi_h-\phi_S)}+F_{UT,L}^{\sin(\phi_h-\phi_S)})\sin(\phi_h-\phi_S)+F_{UT}^{\sin(3\phi_h-\phi_S)}\sin(3\phi_h-\phi_S) \qquad (2)\\
&+ F_{UT}^{\sin(\phi_h+\phi_S)}\sin(\phi_h+\phi_S)+F_{UT}^{\sin\phi_S}\sin\phi_S+F_{UT}^{\sin(2\phi_h-\phi_S)}\sin(2\phi_h-\phi_S)\Big] \\
&+ |P_T|\lambda\Big[F_{LT}^{\cos(\phi_h-\phi_S)}\cos(\phi_h-\phi_S)+F_{LT}^{\cos\phi_S}\cos\phi_S+F_{LT}^{\cos(2\phi_h-\phi_S)}\cos(2\phi_h-\phi_S)\Big]\Big\}
\end{aligned}
$$

Considering the production of single unpolarised hadrons only, the cross section contains two *TMD* fragmentation functions, the spin independent fragmentation function $D_1^h(z,P_T^{h2})$ and the *Collins function*[4] $H_1^{\perp h}(z,P_T^{h2})$, describing the fragmentation of transversely polarised quarks into unpolarised hadrons. The nucleon can then be characterized by the 8 *TMD* distribution functions given in table 1, depending on the polarisation of the quarks inside the nucleon and the target nucleon polarisation. Here, the distribution function $h_{1T}(x,k_T^2)$ describes the distribution of transversely polarised quarks along the transverse polarisation of the nucleon, while $h_{1T}^{\perp}(x,k_T^2)$, descibes the quark transverse polarisation along an intrinsic transverse quark momentum inside a transversely polarised nucleon. The *transversity distribution* $h_1(x)$ is a result of combining these two distribution functions and integrating them over the transverse momentum k_T. The *Boer-Mulders function*[5] $h_1^{\perp}(x,k_T^2)$ descibes for an unpolarised nucleon the quark transverse polarisation along the normal to the plane defined by an intrinsic transverse quark momentum and the nucleon momentum.

In *SIDIS* all PDFs are accessible as the structure functions are the source of azimuthal modulations of the cross section, depending on different combinations of two angles: the target spin ϕ_S azimuthal angle and the azimuthal angle of the produced hadron ϕ_h with respect to the virtual photon direction. The Collins fragmentation function is independently accessible from electron-positron collider experiments [6].

As all these functions do depend on the individual quark flavors, it is important to perform the measurement on different target materials. Ideally, proton and neutron targets should be used, as the proton is dominated by the u-quark, while the neutron gives access to the d-quark contribution. As there are no pure neutron targets, materials have to be choosen which do have a neutron contribution.

TABLE 1. Transverse momentum dependent distribution functions for different nucleon (columns) and quark (rows) polarisation

q/N	U	L	T
U	$f_1(x,k_T^2)$		$f_{1T}^{\perp}(x,k_T^2)$
L		$g_{1L}(x,k_T^2)$	$g_{1T}(x,k_T^2)$
T	$h_1^{\perp}(x,k_T^2)$	$h_{1L}^{\perp}(x,k_T^2)$	$h_{1T}(x,k_T^2)$, $h_{1T}^{\perp}(x,k_T^2)$

EXPERIMENTAL RESULTS

In this article only the two European SIDIS experiments, HERMES at DESY in Hamburg and COMPASS at CERN in Geneva, will be discussed. The focus will be laid on the unpolarised cross-sections, namely the Cahn effect and the Boer-Mulders effect, as well as the on the *Pretzelosity* PDF $h_{1T}^{\perp}(x,k_T^2)$. Detailed results from the COMPASS and HERMES experiments on transversity and on the Sivers function can be found elsewhere in this proceeding [7][8].

The **HERMES** experiment used the 27.6 GeV positron beam of the HERA collider and an internal gas target. Both hydrogen and deuterium targets have been used, the later however not in transverse target polarisation. The HERMES detector [9] is a forward angle magnetic spectrometer whose geometric acceptance was confined to two regions in scattering angle, arranged symmetrically above and below the beam pipe covering the ranges $\pm(40-140)$ mrad in the vertical component of the scattering angle and ± 170 mrad in the horizontal component. Thus, only particles produced with a polar angle greater than 40 mrad with respect to the beam axis are visible. Charged pions are identified using a dual-radiator ring-imaging Cerenkov detector.

The **COMPASS** experiment [10] is a fixed target experiment set up at CERN, using both muon and hadron beams. For SIDIS measurements a -80% longitudinally polarised 160 GeV/c μ^+ beam was scattered off solid polarised targets, consisting of several cylindrical cells along the beam direction. As target materials ^6LiD and NH_3 were used, giving access to the properties of the deuteron and proton repectively. The target polarisation direction could be set either parallel or orthogonal to the beam direction and is periodically reversed to reduce systematic effects due to the different acceptances of the cells. Particle tracking and identification are performed in a two-stage spectrometer, covering a wide kinematical range for the full apperture of the polarised target of ± 180 mrad. A Ring Imaging Cherenkov detector (RICH-1) covering the complete spectrometer acceptance and two hadron calorimeters provide particle identification.

In 1978 Cahn [11] showed, that for unpolarised lepton nucleon scattering with intrinsic transverse parton momentum $k_T \neq 0$ leading order *QED* results in azimuthal asymmetries in the hard partonic cross-section, depending on the parton azimuthal angle ϕ_q:

$$\frac{d\sigma}{d\phi_q} \propto x^2(1+(1-y)^2)\left(1+2\frac{k_T}{Q}D_{\cos\phi_q}(y)\cos\phi_q+(\frac{k_T}{Q})^2 D_{\cos 2\phi_q}(y)\cos 2\phi_q\right) \quad (3)$$

leading after fragmentation to contributions in the $\cos\phi_h$ and $\cos 2\phi_h$ moments. As the $cos 2\phi_q$ term is kinematically suppressed compared to the $\cos\phi_q$ term, this *Cahn effect* is expected to be dominant for the $\cos\phi_h$ moment.

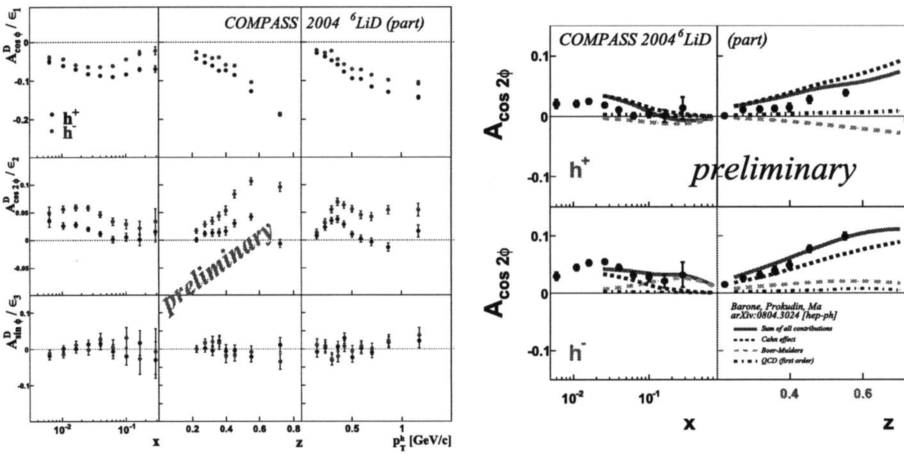

FIGURE 1. **Left:** Unpolarised amplitudes extracted from the COMPASS experiment for positive hadrons (black points) and negative hadrons (red points) *vs.* x and z. **Right:** $cos2\phi_h$ amplitudes extracted from the COMPASS experiment for positive hadrons (top) and negative hadrons (bottom) *vs.* x and z. The red (solid) curve shows predictions from [12], which are the sum of the Cahn effect (blue dotted line), the Boer-Mulders effect (green dashed line) and for the pQCD contribution (black dash–dotted line).

A further contribution for unpolarised beam and target comes from the Boer-Mulders PDF $h_1^\perp(x,k_T^2)$, convoluted with the Collins FF $H_1^{\perp h}(z,P_T^{h2})$. A model calculation [12] shows that the Boer-Mulders contribution to the $cos2\phi_h$ modulation might be of similar magnitude than the contribution from the Cahn effect. It may also contribute to the $cos\phi_h$ asymmetry [5], but the size of this effect is so far unknown and no predictions exist.

Perturbative QCD (pQCD) introduces a third ϕ_h dependence giving rise to azimuthal asymmetries of the form $sin\phi_h$ depending on the beam polarisation λ. However, they are expected to be small for COMPASS and HERMES kinematics [13].

Fig. 1(left) shows the first extraction of all three modulations for charge separated hadrons vs. x, z and P_T performed by COMPASS on a deuteron target [14]. While the $sin\phi_h$ asymmetries are consistent with zero, large asymmetries of up to 40 % are found for the $cos\phi_h$ modulations, and the $cos2\phi_h$ asymmetries are in the order of 5 %. However, there are significant differences in both modulations when comparing positive and negative hadrons. In case of the $cos\phi_h$ modulations this is surprising, as the Cahn effect which is expected to be the predominant source of this modulations is charge independent. Possibly this is a hint of a charge dependent Boer-Mulders contribution in this modulation. For the $cos2\phi_h$ modulation, the charge dependence can be a result of the Boer-Mulders effect, which is expected to change sign with the hadron charge. Fig. 1(right) shows a model-calculation[12] compared to the COMPASS $cos2\phi_h$ results. As can be seen, only the contribution of the Boer-Mulders function changes significantly with the hadron charge. The COMPASS measurement agrees qualitatively well with the predictions, giving a strong hint for a non zero Boer-Mulders function.

Also HERMES presented data on unpolarised cross sections at this conference [15]. The charge separated $cos\phi_h$ and $cos2\phi_h$ asymmetries have been extracted from proton

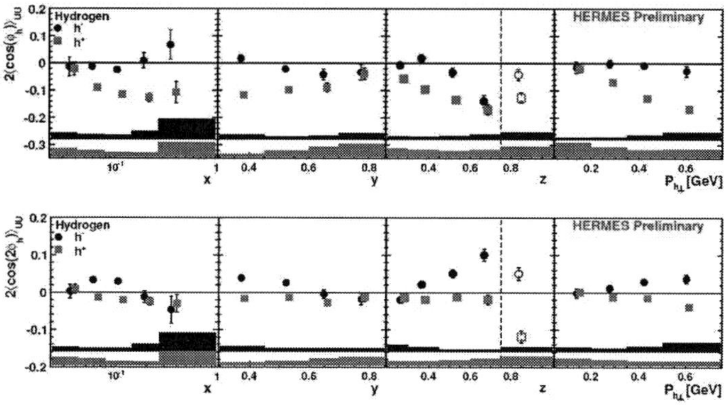

FIGURE 2. $cos\phi_h$ and $cos2\phi_h$ amplitudes extracted from the HERMES experiment for positive hadrons (blue squares) and negative hadrons (black circles) *vs.* x, y, z and P_T for a hydrogen target.

and deuteron targets. Figure 2 shows the results for the hydrogen target. Negative $cos\phi_h$ moments are extracted for positive and negative hadrons, with a larger absolute value for the positive ones. This is qualitatively in agreement with the COMPASS data, however, the amplitudes are significantly smaller for the HERMES results, possibly due to the different kinematic regime. The results for the $cos2\phi_h$ moments are slightly negative for the positive hadrons and positive for the negative hadrons. Also here, the charge dependence can be interpreted as an evidence of a non-zero Boer-Mulders function. Another important result is, that no significant differences for the amplitudes extracted from proton and deuteron targets could be observed. This implies that the Boer-Mulders function of u- and d-quarks might be of the similar size and sign.

Lately, the PDF $h_{1T}^{\perp}(x,k_T^2)$ has been intensely discussed. $h_{1T}^{\perp}(x,k_T^2)$ is chirally odd and can be accessed in combination with the Collins FF, where it gives rise to an azimuthal single spin asymmetry proportional to $sin(3\phi_h - \phi_s)$. As it requires nucleon wave functions components with two units of orbital momentum difference [16], it can be associated to the deviation of the nucleon shape from a sphere [17], hence the name *Pretzelosity*. In [18] it is observed that the difference between helicity and transversity distributions, which is often related to 'relativistic effects' in the nucleon, is given by the pretzelosity distribution. This relation can not be strictly valid in QCD, as all PDFs are independent, but it has to be understood under which conditions this simple relation can help to estimate the size of pretzelosity in terms of transversity and helicity.

So far, only COMPASS has released data on this asymmetry on a deuteron target [19]. Figure 3 shows the results together with a theoretical limit from [18], which takes only the positivity limit $2|h_{1T}^{\perp}(x)| \leq f_1(x) - g_1(x)$ into account. As can be seen, no significant asymmetry could be observed. Either this is due to the predicted suppression of $h_{1T}^{\perp}(x,k_T^2)$ for small and large values of x [16] or the contributions of u- and d-quarks cancel, as it is the case for the Sivers PDF and the Collins effect.

HERMES also included the $sin(3\phi_h - \phi_s)$ azimuthal modulation on a transversely polarised proton target [20] into their fits. However, no data were shown yet. To gain insight into this PDF data on the proton target are urgently needed. Currently, the

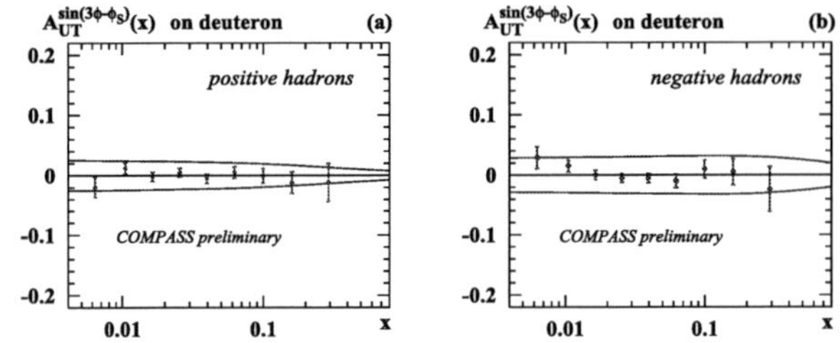

FIGURE 3. $h_{1T}^{\perp}(x,k_T^2)$ PDF as extracted from the COMPASS experiment on a deuteron target. The curve shows theoretical limits from [18]

COMPASS data from the 2007 proton run are being analysed and also the HERMES collaboration has plans to determine these asymmetries.

CONCLUSION

Research in the field of transverse momentum dependent distribution and fragmentation functions has been very active over the last years on both the experimental and theoretical side. Important results have be extracted from data taken by a new generation of SIDIS experiments, so that for the first time the existence of the transversity distribution function as well as the Sivers function could be proved and their size extracted. Data on semi-inclusive unpolarised scattering also give strong hints on the existence of the Boer-Mulders function. Moreover, the full set of TMD functions is accessible in the data which have been taken on proton and deuteron targets. Therefore, the ongoing full analysis will allow for the first time for a full extraction of flavour dependent PDFs.

REFERENCES

1. R.L. Jaffe and X. Ji, **PRL 67** (1991),552
2. A. Bacchetta, M. Diehl, K. Goeke, A. Metz, P. J. Mulders and M. Schlegel, **JHEP 0702** (2007), 093
3. X. Ji, J.P. Ma, F. Yuan, **PLB 597** (2004), 299
4. J. Collins, bf NPB 396 (1993), 161
5. D. Boer and P.J. Mulders, Phys. Rev. D **57** (1998) 5780
6. A. Vossen for the BELLE collaboration, this proceeding
7. H. Wollny for the COMPASS collaboration, this proceeding, hep-ex/0902.0519
8. M. Contalbrigo and R. Lamb for the HERMES collaboration, this proceeding, hep-ex/0901.2438
9. K. Ackerstaff et al. [HERMES Collaboration], Nucl. Instrum. Meth. A **417** (1998) 230
10. P. Abbon et al. [COMPASS Collaboration], Nucl. Instrum. Meth. A **577** (2007) 455
11. R.N. Cahn, **PLB 78** (1978), 269
12. V. Barone, A. Prokudin, B.Q. Ma, Phys. Rev. D **78** (2008) 045022
13. A. Daleo, D. de Florian and R. Sassot, Phys. Rev. D **71** (2005) 023013
14. W. Käfer, Proceeding of *Transversity 2008*, Ferrara 2008, hep-ex/0808.0114
15. F. Giordano for the HERMES collaboration, this proceeding
16. M. Burkhardt, hep-ph/0709.2966
17. G.A. Miller, Phys. Rev. C **76** (2007) 065209
18. H. Avakian, V.A. Efremov, P. Schweitzer, F. Yuan, hep-ph/0808.3982
19. A. Kotzinian for the COMPASS Collaboration, hep-ex/0705.2402
20. M. Diefenthaler [HERMES Collaboration], hep-ex/0706.2242 and hep-ex/0612010

Transverse Spin Physics at HERMES

Contalbrigo Marco
(On behalf of HERMES Collaboration)

INFN Sezione di Ferrara

Abstract. HERMES results on azimuthal single-spin asymmetries in semi-inclusive leptoproduction of pions and charged kaons from a transversely polarised hydrogen target are presented. Preliminary results for both Collins and Sivers Fourier amplitudes are extracted with a much higher statistical significance than the evidence firstly published by HERMES for charged pions in 2005. The first evidence for a correlation between the transverse target polarization and the azimuthal orientation of the plane containing a pair of produced pions is also observed. It is expected to be related to the product of the transversity and an as-yet unmeasured dihadron fragmentation function.

Keywords: Nucleon Structure, Spin, Transversity, Transverse Spin Effects, TMDs
PACS: 13.60.-r, 13.88.+e, 14.20.Dh, 14.65.-q

A complete description of the quark structure of the nucleon, after integrating over the quark transverse momentum, requires at leading twist the unpolarised distribution $q(x,Q^2)$, the helicity distribution $\Delta q(x,Q^2)$ and a third quark distribution, named transversity $\delta q(x,Q^2)$ [1]. The latter reflects the difference in probabilities to find quarks in a transversely polarised nucleon with their spin aligned or anti-aligned to the spin of the nucleon. Since the strong interaction preserves chirality, transversity cannot be measured in inclusive processes due to its chiral-odd nature, but only in a process in which it is combined with another chiral-odd object. The most direct approach is to measure transversity from double transverse-spin asymmetries in polarised Drell-Yan processes [2] where it appears twice. In single spin cross section asymmetries (SSA) in semi-inclusive deep inelastic electroproduction (SIDIS) of single hadrons or hadron pairs on a transversely polarised nucleon target, transversity enters the cross-section combined with novel chiral-odd fragmentation functions (FFs).

In the semi-inclusive production of pion pairs, transversity couples with the as-yet unknown dihadron fragmentation function describing the transfer of the transverse polarization of the fragmenting quark to the relative orbital angular momentum of the produced pion pair. Since this mechanism does not require transverse momentum of the pair, the collinear approach holds with simplified factorization and evolution issues. The dihadron SSA amplitudes, measured to be non-zero for the first time at HERMES [3], are shown in Figure 1 to be positive over the entire range of the relevant kinematical variables. The $M_{\pi\pi}$ dependence of the measured amplitude shows no sign change at the ρ^0 mass, contrary to the predictions in [4] and in qualitative agreement with [5].

In single hadron semi-inclusive production, the Collins fragmentation function $H_1^\perp(z)$ describes a spin-momentum correlation where the transverse polarisation of the struck quark influences the momentum component of the produced hadron transverse to the virtual photon, thereby leading to an azimuthal dependence of the cross-section. In the single hadron case, azimuthal spin asymmetries can be also generated by the T-odd

CP1149, *Spin Physics, 18th International Spin Physics Symposium*
edited by D. G. Crabb, D. B. Day, S. Liuti, X. Zheng, M. Poelker, and Y. Prok

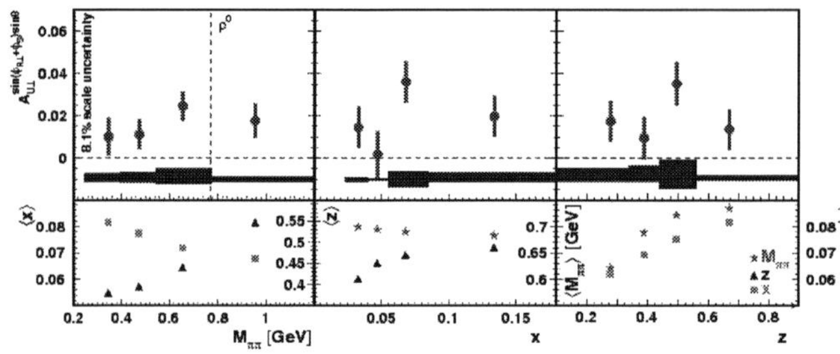

FIGURE 1. On the top the amplitude of the dihadron asymmetry as a function of the invariant mass $M_{\pi\pi}$ of the pair, x and z. On the bottom the average values of the variables that were integrated over. The error bars (shaded bands) show the statistical (systematic) uncertainty. The 8.1% scale uncertainty arises from the uncertainty on the target polarisation.

Sivers distribution function $f_1^{\perp}(x,k_T)$ that appears in the cross-section together with the unpolarised FF $D(z)$ [6]. The Sivers distribution describes a correlation between the transverse polarisation of the target nucleon and the intrinsic transverse momentum k_T of the struck quark. Its existence requires non-zero orbital angular momentum of quarks. The Collins and Sivers functions [7] are examples of transverse momentum dependent distributions (TMDs), which are under intense investigation since they provide new insight into the nucleon intrinsic dynamics. In SIDIS off a transversely polarised nucleon target, they can be separated thanks to their different azimuthal dependences with respect of the axis of target polarisation [6]. The first measurement of non-zero Collins and Sivers amplitudes has been published by HERMES [8]. The first results were for charge pions; they were based on a subsample of the transverse-target data and presented as a function of x, z and transverse hadron momentum $P_{h\perp}$.

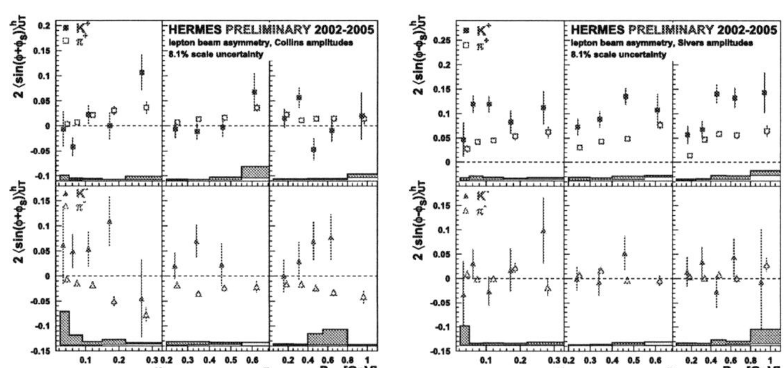

FIGURE 2. Collins (left) and Sivers (right) amplitudes for charged pions and kaons as a function of x, z and transverse hadron momentum $P_{h\perp}$. Error bars represent the statistical uncertainties, while the shaded bands represent the systematic uncertainty due to acceptance effects, instrumental smearing and QED radiation, hadron misidentification as well as possible contributions from $\cos\phi$ amplitudes of the unpolarised cross-section. A common 8.1% scale uncertainty is due to the uncertainty on the target polarisation.

The preliminary result for charged pions based on the full data-set is presented in Figure 2. The Collins amplitude is measured to be significantly positive for π^+ and negative for π^-. These results confirm the previously published ones [8] and demonstrate that both the transversity and Collins functions are non-zero. The large absolute magnitude of the π^- asymmetry is surprising, suggesting a substantial disfavored Collins function (i.e. u-quark fragmenting into π^-) with a sign opposite to that of the favored function (i.e. u-quark into π^+). This observation is consistent with the measurements at BELLE [9] and COMPASS [10], provided that transversity for u and d quarks has opposite sign resembling the helicities densities [11]. The Sivers amplitude is significantly positive for π^+, providing an indication of non-zero orbital angular momentum of quarks in the nucleon and the first evidence for a T-odd distribution function, while it is consistent with zero for π^-. The naively T-odd Sivers function, as extracted in SIDIS reactions, can in future be compared to its appearance in polarised Drell-Yan processes, where it is predicted to have opposite sign due to its fundamental time reversal invariance in QCD [12, 2]. The Collins and Sivers amplitudes were also extracted for π^0 mesons: the measured values are intermediate between those of π^+ and π^- in agreement with the expectation based on the isospin symmetry of the π-meson triplet.

For the first time, the transverse SSA amplitudes have been extracted also for charged kaons: the results are shown in Figure 2. While there is no reason to expect a similar amplitude for K^- and π^-, due to their different quark content, the u-quark dominance in SIDIS would suggest a similar amplitude for K^+ and π^+. Contrary to this expectation, the observed amplitude of of the Sivers moment for K^+ is larger than that for π^+ in the region $x \approx 0.1$. This suggests that the sea quarks may provide an important contribution to the Sivers function and thus may carry significant orbital angular momentum in the nucleon [11]. Indeed, a recent parametrization of fragmentation functions, based on world data on SIDIS and hadron reaction experiments, in addition to the usual e^+e^- collider data, favors an enhanced role of the $\bar{s}$ fragmentation into K^+ [13].

The amplitudes of the transverse target SSA in the difference between π^+ and π^- cross sections are shown in Figure 3. Such observables are interesting since the contribution from exclusive ρ^0 mesons cancels out. Moreover in their interpretation only target valence quark matter within some reasonable assumptions of symmetry in fragmentation and on transverse momentum dependences.

The interpretation of the transverse SSA amplitudes for single hadron production is

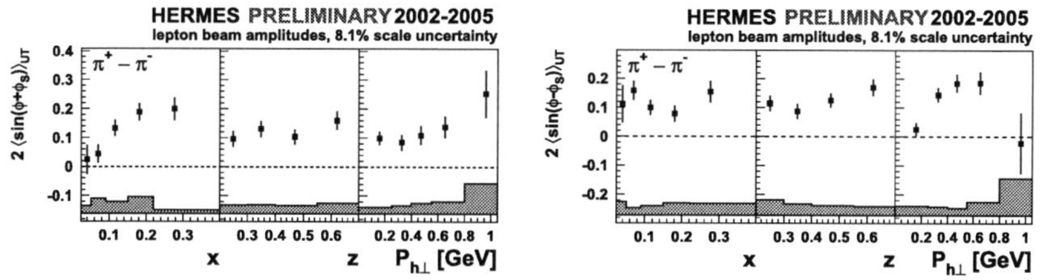

FIGURE 3. Collins (left) and Sivers (right) amplitudes for charged-pion cross-section difference as a function of x, z and transverse hadron momentum $P_{h\perp}$. Error bars and shaded bands are coded as in Figure 2.

complicated since the related cross section is described in terms of quark distribution and fragmentation functions unintegrated over transverse momentum. The dependence of these functions on transverse momentum is currently only poorly known and even less is known about their evolution properties (which are important when trying to explain the SIDIS results with the fragmentation functions measured at the higher scales of the e^+e^- collider machines). Thanks to the high statistics accumulated, the Collins and Sivers amplitudes of charged pions could be extracted in a multi dimensional analysis. At HERMES, binning in x means simultaneously binning in Q^2 since a strong correlation between x and Q^2 is introduced by the detector acceptance. The results for binning in x (Q^2) and $P_{h\perp}$ is shown in Figure 4. Multidimensional extractions provide the most complete experimental information that can be provided to phenomenological analyses, in trying to disentangle all the various contributions to the polarised cross section.

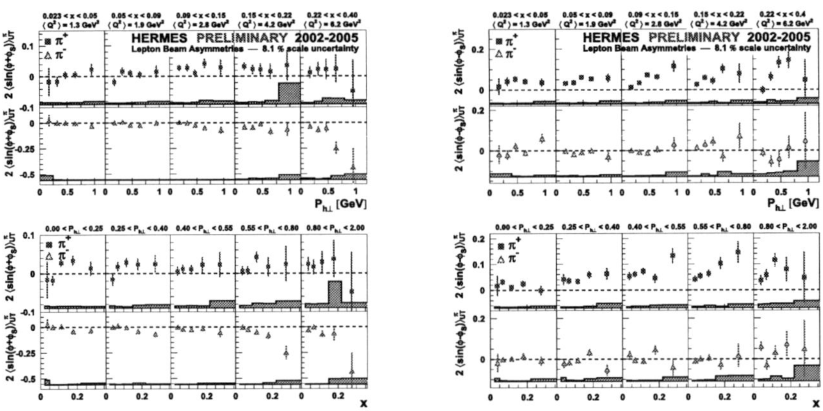

FIGURE 4. Collins (left) and Sivers (right) amplitudes for charged pions as a function of x and transverse hadron momentum $P_{h\perp}$ (2-dimensional analysis). Error bars and shaded bands are coded as in Figure 2.

REFERENCES

1. M. Burkardt, C. A. Miller and W.-D. Nowak, *e-Print* arXiv:0812.2208, submitted to *Rep. Prog. Phys.*
2. PAX Collab. (V. Barone *et al.*), *e-Print* hep-ex/0505054.
3. HERMES Collab. (A. Airapetian *et al.*), *JHEP* **06**, 017 (2008).
4. R.L. Jaffe *et al. Phys. Rev. Lett.* **80**, 1166 (1998).
5. A. Bacchetta and M. Radici, *Phys. Rev. D* **74**, 114007 (2006); J. She *et al. Phys. Rev. D* **77** 014035 (2008).
6. A. Bacchetta, M. Diehl, K. Goeke, A. Metz, P.J. Mulders and M. Schlegel, *JHEP* **0702**, 093 (2007).
7. J.C. Collins, *Nucl. Phys. B* **396**, 161 (1993); D.W. Sivers, *Phys. Rev. D* **41**, 83 (1990).
8. HERMES Collab. (A. Airapetian *et al.*), *Phys. Rev. Lett.* **94**, 012002 (2005).
9. BELLE Collab. (K. Abe *et al.*), *Phys. Rev. Lett.* **96**, 232002 (2006).
10. COMPASS Collab. (E.S. Ageev *et al.*), *Nulc. Phys. B* **765**, 31 (2007).
11. M. Anselmino, M. Boglione, U. D'Alesio, A. Kotzinian, S. Melis, F. Murgia, A. Prokudin and C. Turk, *e-Print* arXiv:0807.0173 for Collins and *e-Print* arXiv:0807.0166 for Sivers mechanism.
12. J.C. Collins, *Phys. Lett. B* **536**, 43 (2002).
13. D. de Florian, R. Sassot, M. Stratmann, *Phys. Rev. D* **75**, 114010 (2007).

Transverse Spin Effects at COMPASS

H. Wollny on behalf of the COMPASS collaboration

Physikalisches Institut der Albert-Ludwigs-Universität Freiburg
Hermann-Herder Str. 3, 79104 Freiburg, Germany

Abstract. The measurement of transverse spin effects in semi-inclusive deep-inelastic scattering (SIDIS) is an important part of the COMPASS physics program. In the years 2002-2004 data was taken by scattering a 160 GeV/c muon beam off a transversely polarized deuteron target. In 2007, additional data was collected on a transversely polarized proton target. New preliminary results for the Collins and Sivers asymmetries from the analysis of the proton data are presented.

Keywords: Semi-Inclusive Deep Inelastic Scattering; SIDIS; Transversely polarised proton target; Collins asymmetry; Sivers asymmetry; COMPASS

INTRODUCTION

Single spin asymmetries in semi-inclusive deep-inelastic scattering (SIDIS) off transversely polarized nucleon targets have been under intense experimental investigation over the past few years [1, 2, 3]. They provide new insights into QCD and the nucleon structure. For instance, they allow the determination of the third yet-unknown leading-twist quark distribution function $\Delta_T q(x)$, the transversity distribution [4, 5]. Additionally, they give insight into the parton transverse momentum distribution and angular momentum.

THE COMPASS EXPERIMENT

COMPASS is a fixed target experiment at the CERN SPS accelerator with a wide physics program focused on the nucleon spin structure and on hadron spectroscopy. COMPASS investigates transversity and the transverse momentum structure of the nucleon in SIDIS. A 160 GeV/c muon beam is scattered off a transversely polarized hydrogen or deuterium target. The scatterd muon and the produced hadrons are detected in a wide-acceptance two-stage detector with excellent particle identification capabilities [6]. In the years 2002, 2003 and 2004, data was collected on a transversely polarized ^6LiD target [7]. In 2007 data were taken with a transversely polarized NH_3 target. In the following we will focus on the new results of the data taken with the proton target.

THE COLLINS ASYMMETRY

The chiral-odd transversity distribution $\Delta_T q(x)$ can be measured in combination with the chiral-odd Collins fragmentation function $\Delta_0^T D_q^h(x)$ in SIDIS. According to Collins the fragmentation of a transversely polarized quark into an unpolarized hadron leads

CP1149, *Spin Physics, 18th International Spin Physics Symposium*
edited by D. G. Crabb, D. B. Day, S. Liuti, X. Zheng, M. Poelker, and Y. Prok

to a sine modulation in $\Phi_C = \phi_h + \phi_S - \pi$ in the SIDIS cross section [5]. Here ϕ_h is the azimuthal angle of the hadron with respect to the scattering plane and ϕ_S is the azimuthal angle between the spin of the initial quark and the scattering plane [8]. The number of produced hadrons N can be written as:

$$N(\Phi_C) = N_0 \cdot (1 + f \cdot P_t \cdot D_{nn} \cdot A_{Coll} \cdot \sin\Phi_C) \tag{1}$$

in which N_0 is the average hadron yield, f the fraction of polarized material in the target, P_t the target polarization and $D_{nn} = (1-y)/(1-y+y^2/2)$ the depolarization factor.

THE SIVERS ASYMMETRY

Another source of azimuthal asymmetry is related to the Sivers effect [9]. The Sivers parton distribution function $\Delta_0^T q(x, \mathbf{k_T})$ describes the correlation of quarks with intrinsic transverse momentum $\mathbf{k_T}$ in a transversely polarized nucleon. The Sivers effect leads to an azimuthal modulation of the number of produced hadrons

$$N(\Phi_S) = N_0 \cdot (1 + f \cdot P_t \cdot A_{Siv} \cdot \sin\Phi_S)\,, \tag{2}$$

with $\Phi_S = \phi_h - \phi_S$.

Since the Collins and Sivers asymmetries are othogonal to each other, both asymmetries can be determined independently from the same dataset.

DATA SAMPLE AND EVENT SELECTION

In 2007 COMPASS took data with a transversely polarized proton target (NH_3). The polarization of the material is $\sim 90\%$ with a dilution factor f of 0.15. The target consists of three cells in a row, where the two outer cells are polarized in one direction and the middle cell is polarized oppositely. To reduce systematics the polarization was reversed every 4-5 days. The new solenoid magnet installed in 2005 increased the angular acceptance up to 180 mrad.
The quality and the stability of the data was carefully checked. For the results presented here, about 20 % of the collected data have been analyzed.

To select DIS events, kinematic cuts on the squared four momentum transfer $Q^2 > 1\,(\mathrm{GeV}/c)^2$, the fractional energy transfer of the muon $0.1 < y < 0.9$ and the hadronic invariant mass $W > 5\,\mathrm{GeV}/c^2$ were applied. The hadron sample consists of all charged hadrons originating from the reaction vertex with $p_T^h > 0.1\,\mathrm{GeV}/c$ and $z > 0.2$. To reject muons an energy deposit of $\geq 5\,\mathrm{GeV}$ in the hadronic calorimeters was demanded.

To extract the asymmetries a maximum likelihood method binned in ϕ_h and ϕ_S was used. In each bin the product of the detector acceptance and the expression for the transverse polarization dependent part of the SIDIS cross section [10] was fitted. To disentangle acceptance and spin dependent modulations two samples of one period with opposite polarization and two samples of a consecutive period with reversed target polarization were coupled. To fix the acceptance it was assumed that in each bin and each sample the change of acceptance between the two consecutive periods is described

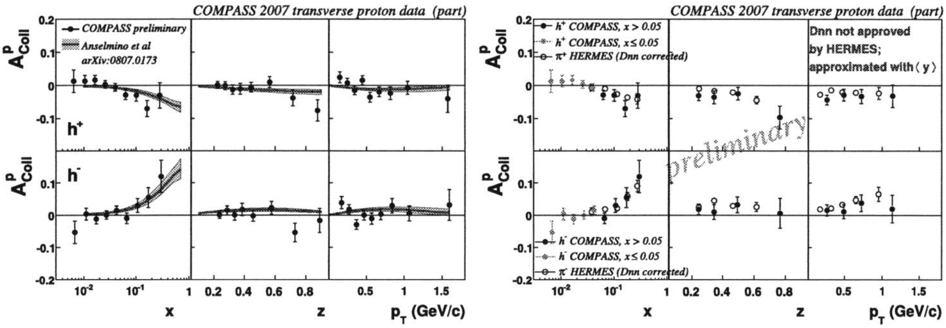

FIGURE 1. Collins asymmetry. Left: Comparison with recent predictions of Anselmino et al. Right: Comparison with results of HERMES. Only statistical errors are shown. The systematical error is 30 % of the statistical error

by one constant. The results have been checked by several other methods described in [1].

RESULTS

The Collins and Sivers asymmetries were evaluated as a function of x, z, and p_T^h respectively integrating over the remaining two variables.

In the left part of Fig. 1 the results for the Collins asymmetry are shown. For $x > 0.05$ the measured asymmetries are non-zero and negative in sign for positive hadrons and positive in sign for negative hadrons. For $x < 0.05$ they are small and compatible with zero. One can see the excellent agreement with the recent prediction of [11]. The right part of Fig. 1 shows the comparison between COMPASS and HERMES [12]. For comparing the results of the two experiments as a function of z and p_T a cut on $x > 0.05$ on the COMPASS data was applied and the HERMES values were corrected with $-1/D_{nn}$. The minus sign takes into account the different definitions of Φ_C and the depolarisation factor $D_{nn}(y)$ takes care of the different y-domains of the two experiments. For HERMES the D_{nn} values for each bin have been approximated with the corresponding mean values of y taken from [13]. The results of COMPASS and HERMES are in reasonable good agreement.

In the left part of Fig. 2 the results for the Sivers asymmetry are shown. The asymmetry for both charges is small and compatible with zero within the statistical error. The comparison with the recent prediction of [14] shows a good agreement for negative hadrons whereas it is at variance for the positive hadrons. The right part of Fig. 2 shows the comparison with the results of HERMES [12]. For comparing the asymmetries as a function of z and p_T, again a cut on $x > 0.05$ was applied. The agreement for the negative hadrons is very good whereas the result for positive hadrons is at variance with HERMES. This also explains the disagreement with the prediction, since it is also based on the results of HERMES.

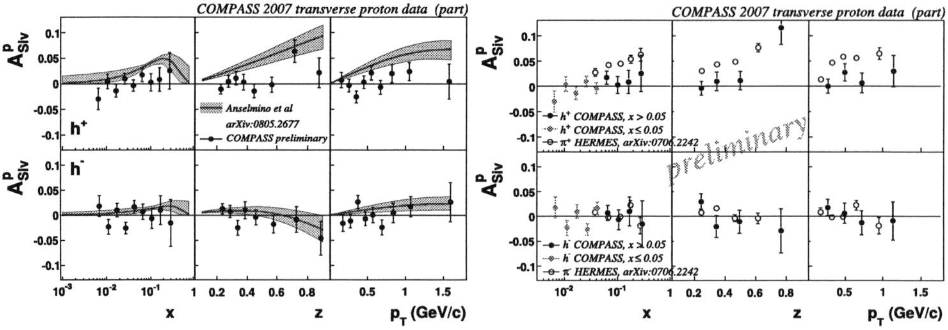

FIGURE 2. Sivers asymmetry. Left: Comparison with recent predictions of Anselmino et al. Right: Comparison with results of HERMES. Only statistical errors are shown. The systematical error is 50 % of the statistical error

SUMMARY

First preliminary results for Collins and Sivers asymmetries measured at COMPASS in SIDIS on a transversely polarized proton target have been presented. The measured Collins asymmetry is non-zero for $x > 0.05$ and of opposite sign for positive and negative hadrons. For both charges the results agree very well within present statistics with recent predictions of Anselmino et al. and with the results of the HERMES group. The Sivers asymmetry for negative and positive hadrons is small and compatible with zero within the statistical precision. Whereas the result for negative hadrons is in good agreement with recent predictions of Anselmino et al. and the results of the HERMES group, the result for positive hadrons is at variance.

REFERENCES

1. E. Ageev, et al., *Phys. Lett. B* **612**, 154 (2005).
2. V. Alexakhin, et al., *Nucl. Phys. B* **94**, 205 (2007).
3. A. Airpetian, et al., *Phys. Rev. Lett* **94**, 235 (2005).
4. X. Artru, and J. Collins, *Z. Phys. C* **69**, 277 (1996).
5. J. Collins, et al., *Nucl. Phys. B* **420**, 565 (1994).
6. P. Abbon, et al., *NIM A* **577**, 455–518 (2007).
7. R. Joosten, et al., *Measurement of Transverse Momentum Dependent Parton Distributions and Fragmentation Functions in Semi-Inclusive Deep Inelastic Scattering, in this proceedings* (2008).
8. X. Artru, *hep-ph/0207309v2* (2002).
9. D. Sivers, *Phys. Rev. D* **41**, 83 (1991).
10. A. Bacchetta, et al., *hep-ph/0611265v2* (2006).
11. M. Anselmino, et al., *arXiv:0807.0173* (2008).
12. M. Diefentaler, et al., *arXiv:0706.2242* p. 579 (2007).
13. U. Elschenbroich, *Eur. Phys. J. C* **94**, 125–250 (2007).
14. M. Anselmino, et al., *arXiv:0805.2677* (2008).

Measurement of azimuthal asymmetries of the unpolarized cross section at HERMES

Francesca Giordano* and Rebecca Lamb (on behalf of the HERMES Collaboration)†

*INFN & Università degli studi di Ferrara, giordano@fe.infn.it
†University of Illinois, rlamb2@illinois.edu

Abstract. A multi-dimensional $(x, y, z, P_{h\perp})$ extraction of $\cos\phi_h$ and $\cos 2\phi_h$ azimuthal asymmetries of unpolarized Semi-Inclusive Deep Inelastic Scattering at HERMES is discussed. The use of data taken with hydrogen and deuterium targets and the separation of positive and negative hadrons allow to access flavor-dependent information about quark intrinsic transverse momenta and spin-orbit correlations. This flavor sensitivity allows for a discrimination between theoretical models in the HERMES kinematic regime.

Keywords: Semi-Inclusive DIS; Azimuthal asymmetries; Intrinsic transverse momentum and spin.
PACS: 13.88.+e, 13.60.-r

INTRODUCTION

In Deep Inelastic Scattering (DIS), the structure of the nucleon is probed by the interaction of a high energy lepton and a target nucleon, via the exchange of a virtual boson. If at least one of the produced hadrons is detected in coincidence with the scattered lepton, the reaction is called Semi-Inclusive Deep Inelastic Scattering (SIDIS):

$$l(\mathbf{k}) + N(\mathbf{P}) \rightarrow l'(\mathbf{k}') + h(\mathbf{P}_h) + X(\mathbf{P}_X), \tag{1}$$

where l (l') is the incident (scattered) beam lepton, N is the target nucleon, h the detected hadron and X the target remnant. The quantities in parentheses are the corresponding four-momenta.

If the cross section is unintegrated over the hadron momentum component transverse to the virtual photon direction $P_{h\perp}$ (Fig. 1), an azimuthal dependence around the outgoing hadron direction exists [1]:

$$\begin{aligned}\frac{d\sigma}{dx\,dy\,dz\,dP_{h\perp}^2\,d\phi_h} = \frac{\alpha^2}{xyQ^2}(1+\frac{\gamma^2}{2x})\{A(y)\,F_{UU,T} + B(y)\,F_{UU,L} + \\ C(y)\,\cos\phi_h F_{UU}^{\cos\phi_h} + B(y)\,\cos 2\phi_h F_{UU}^{\cos 2\phi_h}\},\end{aligned} \tag{2}$$

where ϕ_h is the azimuthal angle of the hadron plane around the virtual-photon direction (Fig. 1). Here Q^2 and y are respectively the negative squared four-momentum and the fraction of the lepton's energy transferred to the virtual photon, x is the Bjorken scaling variable and z is the virtual photon's fractional energy transferred to the produced hadron. For the structure functions F, the subscript UU denotes Unpolarized beam and

CP1149, *Spin Physics, 18th International Spin Physics Symposium*
edited by D. G. Crabb, D. B. Day, S. Liuti, X. Zheng, M. Poelker, and Y. Prok

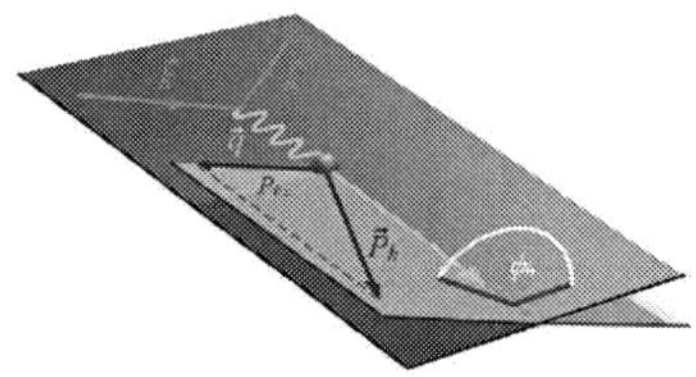

FIGURE 1. Definition of the azimuthal angle ϕ_h between the scattering plane (grey) and the hadron production plane (yellow).

Unpolarized target, T (L) indicates the Transverse (Longitudinal) polarization of the virtual photon, α is the electromagnetic coupling constant, $\gamma = 2Mx/Q$ with M the target mass, $A(y) \approx (1-y+1/2y^2)$, $B(y) \approx (1-y)$ and $C(y) \approx (2-y)\sqrt{1-y}$.

Two mechanisms are expected to give important contributions to the azimuthal dependence of the unpolarized cross section in the hadron transverse momentum range accessed at HERMES:

- the Cahn effect, a pure kinematic effect, generated by the non-zero intrinsic transverse motion of quarks, already pointed out by R. Cahn in 1978 [2];
- the Boer-Mulders effect, introduced more recently (1997) by D. Boer and P. J. Mulders [3]; this mechanism originates from a coupling between quark transverse momentum and quark transverse spin.

The HERMES experiment

The fixed target HERMES experiment was operated for more than 10 years until 2007 at the electron-positron storage ring of HERA at DESY. The HERMES spectrometer [4] was a forward angle instrument consisting of two symmetric halves (top, bottom) above and below the horizontal plane. It was characterized by very high efficiency ($98 \div 99\%$) in electron-hadron separation, provided by a transition radiation detector, a preshower scintillation counter and an electromagnetic calorimeter. In addition, a dual-radiator Ring-Imaging CHerenkov detector (RICH) provided good hadron-type identification for momenta above 2 GeV/c.

MULTI-DIMENSIONAL UNFOLDING AND RESULTS

The cross section azimuthal modulations can be measured via the $\langle \cos n\phi_h \rangle$-moments:

$$\langle \cos n\phi_h \rangle = \frac{\int \cos n\phi_h \, d^5\sigma}{\int d^5\sigma}, \tag{3}$$

where $n = 1, 2$ and $\int d^5\sigma$ stands for $\int dx\, dy\, dz\, dP_{h\perp}^2\, d\phi_h \frac{d^5\sigma}{dx\, dy\, dz\, dP_{h\perp}^2\, d\phi_h}$.

The extraction of these cosine moments from data is challenging because they couple to a number of azimuthal modulations that are due to experimental sources, *e.g.* detector

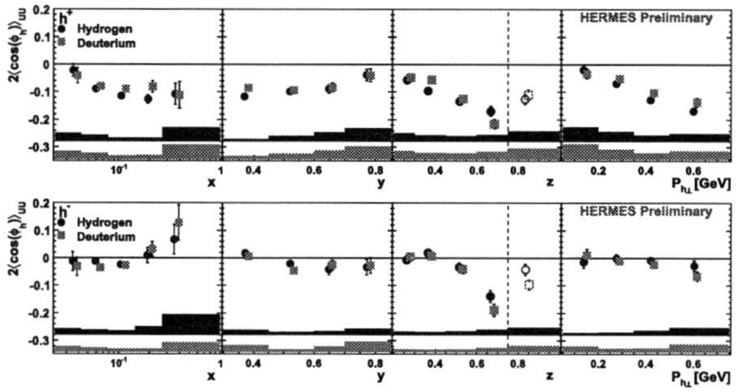

FIGURE 2. The $\cos\phi_h$ moments for positive (upper panel) and negative (lower panel) hadrons, extracted from hydrogen (circles) and deuterium (squares) data, shown as projection versus the kinematic variables x, y, z and $P_{h\perp}$.

geometrical acceptance and higher-order QED effects (radiative effects). Moreover, typically the event sample is binned only in one variable (1-dimensional analysis), and integrated over the full range of all the other ones, while the structure functions F used in equation 2 and the instrumental contributions depend on all the kinematic variables x, y, z and $P_{h\perp}$ simultaneously.

Therefore, in order to determine the cosine moments corrected for radiative and detector smearing, an unfolding procedure [6] was used, in which the event sample is binned simultaneously in all the relevant variables (*multi-dimensional analysis*[1]).

The unfolding algorithm is based on the relation between the unknown distribution of *Born* yields $B(j)$ and the distribution of *measured* yields $X(i)$:

$$X(i) = \sum_{j=1}^{n_b} S(i,j)B(j) + \beta(i). \tag{4}$$

where n_b is the total number of bins and $\beta(i)$ is a vector that contains the events smeared into the measured sample from outside the acceptance. The *Smearing matrix* $S(i,j)$ describes the probability that an event originating from the *Born bin* j, corresponding to the original kinematics (free from experimental distortions), is actually observed in the *measured bin* i. Both the background $\beta(i)$ and the smearing matrix $S(i,j)$ are determined by a detailed Monte Carlo simulation of the experimental apparatus.

Assuming a non-singular $S(i,j)$ matrix one obtains:

$$B(j) = \sum_{i=1}^{n_b} S^{-1}(j,i)\left[X(i) - \beta(i)\right]. \tag{5}$$

The extraction of cosine moments from the *Born* yields $B(j)$ can be performed by linear regression that takes into account the correlations between bins introduced by the smearing. In this way one pair of moments can be obtained in each kinematic bin ($\langle\cos\phi_h\rangle$, $\langle\cos 2\phi_h\rangle$), which represents results that are fully differential in all variables.

[1] For a more detailed discussion about 1- and multi-dimensional analysis see [5].

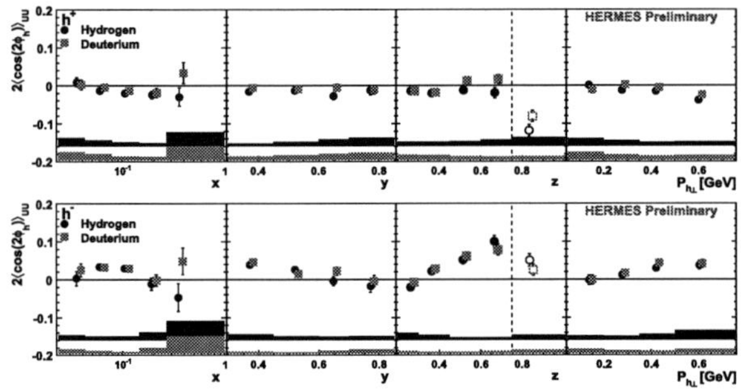

FIGURE 3. The $\cos 2\phi_h$ moments for positive (upper panel) and negative (lower panel) hadrons, extracted from hydrogen (circles) and deuterium (squares) data, shown as projection versus the kinematic variables x, y, z and $P_{h\perp}$.

The dependence of a moment on a single variable can be obtained by projecting the fully differential result onto the variable under study by weighting the moment in each bin k with the corresponding unpolarized 4π cross section $\sigma_k^{4\pi}$ (defined by a Monte Carlo), for instance in case of the ℓ-th x-bin: $\langle\cos n\phi_h\rangle(x_\ell) = \sum_k \sigma_k^{4\pi}\langle\cos n\phi_h\rangle_k / \sum_k \sigma_k^{4\pi}$, where k runs over all the n_b bins corresponding to x_ℓ.

The $\cos\phi_h$ moments from hydrogen and deuterium data are shown in figure 2 as projections versus the relevant kinematic variables for both hadron charges. Both hydrogen and deuterium data show similar behavior: the $\langle\cos\phi_h\rangle$ moments are found to be sizable and negative for positive hadrons. The signal increases with $P_{h\perp}$ and with the hadron energy fraction z, except in the very high z range, where the partonic interpretation of the cross section is no longer valid[2]. The signal for the negative hadrons is significantly lower, but the dependence versus z and $P_{h\perp}$ exhibits similar features.

Figure 3 shows the $\cos 2\phi_h$ moments that are found to be slightly negative for positive hadrons, and slightly positive for negative hadrons. Different results for positive and negative hadrons are not unexpected because both experimental evidence and theoretical models predict opposite Boer-Mulders contributions for differently charged hadrons.

REFERENCES

1. A. Bacchetta et al., *JHEP* **0702**, 093 (2007).
2. R. N. Cahn, *Phys. Lett. B* **78**, 269 (1978), R. N. Cahn, *Phys. Rev. D* **40**, 3107 (1989).
3. D. Boer and P. J. Mulders, *Phys. Rev. D* **57**, 5780 (1998).
4. K. Ackerstaff et al. (HERMES collaboration), *Nucl. Instr. Meth. A* **417**, 230 (1998).
5. F. Giordano, Proceedings of Transversity 2008 workshop, May 28-31, 2008, Ferrara, Italy, to be published by World scientific.
6. G. Cowan, *Statistical Data Analysis*, Clarendon Press, Oxford, 1998.

[2] The highest z-bin is plotted for completeness but it is not used for projecting moments onto the single variables.

Universality and Gluonic Pole Matrix Elements

Leonard Gamberg*, Asmita Mukherjee† and Piet Mulders**

*Penn State University Berks, Department of Physics, Reading,Pennsylvania 19610-USA
†Indian Institute of Technology, Department of Physics, Powai, Mumbai 400076-India
**Vrije University, Department of Physics and Astronomy, NL-1081 HV Amsterdam, the Netherlands

Abstract. We investigate the spectral properties of quark-gluon-quark and quark-quark-gluon correlators in order to study gluonic pole matrix elements. Such matrix elements appear both for distribution functions such as the Sivers function, and fragmentation functions such as the Collins function. We find that for a large class of spectator models, the contribution of the gluonic pole matrix element for fragmentation functions vanishes. This result is important in the study of universality for fragmentation functions.

Keywords: Transverse Momentum, Gluonic Poles, Fragmentation Functions
PACS: 13.87.Fh, 13.88.+e

INTRODUCTION

In high-energy scattering processes the structure of hadrons is accounted for using quark and gluon correlators; forward matrix elements of non-local quark and gluon operators between hadronic states. Making an expansion in the (inverse) hard scale, the leading dynamical effects come from two-field configurations at two light-like separated points, which are easily interpreted as parton densities or parton decay functions [1, 2]. These parton distribution functions depend on the momentum fraction x relating the parton momentum $k = xP$ to the hadron momentum P, or the fragmentation functions of partons into hadrons depending on the momentum fraction z, relating the parton momentum k to the hadron momentum $P = zk$. At sub-leading order in the hard scale or when explicitly measuring transverse momenta, other matrix elements become important such as the three-parton correlators containing parton fields at three different space-time points with light-like separations and two-parton correlators with also transverse separation (light-front correlations). These latter (light-front) correlators are described in terms of transverse momentum dependent (TMD) distribution and fragmentation functions, which are sensitive to the intrinsic transverse momenta of partons in hadrons, $k = xP + k_T$ in a frame in which the hadron does not have transverse momentum ($P_T = 0$) or for fragmentation $k = \frac{1}{z}P + k_T$. In this case one often refers to the hadron transverse momentum $P_\perp = -zk_T$ (in a frame in which the parton does not have a transverse momentum ($k_\perp = 0$)).

Here, we investigate multi-parton correlators with one additional gluon in which the zero-momentum limit is explored [3, 4]. These so-called gluonic pole matrix elements or Qiu-Sterman matrix elements, have opposite time-reversal (T) behavior as compared to the matrix elements without the gluon. Such matrix elements involving time-reversal odd (T-odd) operator combinations are of interest because they are essential for understand-

CP1149, *Spin Physics, 18th International Spin Physics Symposium*
edited by D. G. Crabb, D. B. Day, S. Liuti, X. Zheng, M. Poelker, and Y. Prok

ing single spin asymmetries in high energy scattering processes e.g. semi-inclusive deep inelastic scattering (SIDIS) and Drell-Yan scattering. In the collinear case T-symmetry can be used as a constraint on the parton correlators, limiting the distribution functions (DFs) to T-even ones. This constraint does not apply for the fragmentation correlator because the final state hadron is part of a jet and as such is not a plane wave, allowing both T-even and T-odd fragmentation functions (FFs).

Including transverse momentum dependence, both the distribution and fragmentation correlators (Φ and Δ) are no longer constrained by T-symmetry. The reason is that the appropriate color gauge invariant operators in the correlator, are not T-invariant. The T-odd operator structure can be traced back to the color gauge link that necessarily appears in correlators to render them color gauge-invariant. But the operator structure of the correlator is also a consequence of the necessary resummation of all contributions that arise from collinear gluon polarizations, i.e. those along the hadron momentum. How this resummation takes effect is a matter of calculation [5]. The result is a process dependence in the path in the gauge link. After azimuthal weighting of cross sections one simply finds that the T-odd features originating from the gauge link lead to specific factors with which the T-odd functions appear in observables. For DFs this provides a mechanism leading to T-odd functions, such as the Sivers function [6]. Comparing T-odd effects in DFs in semi-inclusive deep inelastic scattering (SIDIS) and the Drell-Yan process one finds a relative minus sign [7, 8]. Similarly, comparing T-odd effects in FFs in SIDIS and electron-positron annihilation one also finds a relative minus sign, at least for the T-odd effect originating from the operator structure (gauge link) [9]. But, for FFs there are now in principle two mechanisms leading to T-odd functions [9]. However, the two mechanisms can be distinguished. The effect coming from the hadron-jet final state not being a plane wave will not lead to process dependent factors, whereas the effect coming from the operator structure of gauge link would result in process dependence.

In order to understand the basic features of these matrix elements we perform a spectral analysis by modeling the distribution and fragmentation functions under *reasonable* assumptions [10]. In particular we consider the differences between distribution and fragmentation functions using a spectral analysis while restricting the momentum dependence and asymptotic behavior of the vertices. In this context, the relevant gluonic pole matrix elements that we study [10] are $\Phi_G(k,k-k_1)$ and $\Delta_G(k,k-k_1)$ shown in Figs. 1 and 2. Of these matrix elements only the dependence on the collinear components x and x_1 in the expansion of the momenta are needed (note, the gluon momentum is parameterized as $k_1 = [k_1^-, x_1, k_{1T}]$ in these figures). We find that while both $\Phi_G(x,x-x_1)$ and $\Delta_G(x,x-x_1)$ are nonzero, taking the limit $x_1 \to 0$, $\Phi_G(x,x)$ remains non-zero, while $\Delta_G(x,x)$ vanishes. The vanishing of the T-odd gluonic pole matrix elements is important in the study of universality of TMD PDFs and TMD FFs.

QUARK-GLUON CORRELATORS AND GLUONIC POLES

In our analysis we express the quark-quark correlator in terms of the collinear and transverse components of the quark momentum, $k = xP + \sigma n + k_T$, where the Sudakov vector n is an arbitrary light-like four-vector $n^2 = 0$ that has non-zero overlap $P \cdot n$ with the hadron's momentum P, and $k^- \sim \sigma$ (which is suppressed w/r to the hard scale) is

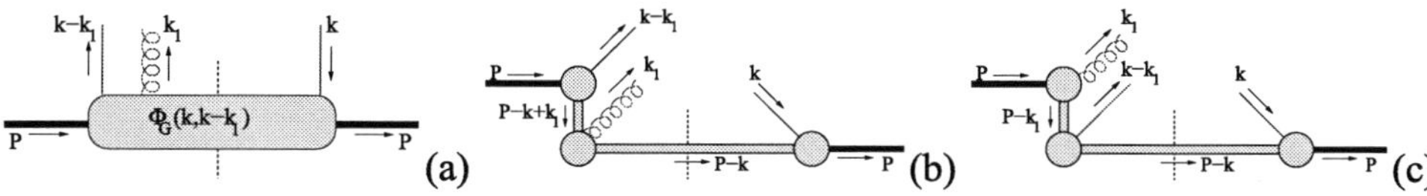

FIGURE 1. The graphical representation of the quark-quark-gluon correlator Φ_G for the case of distributions including a gluon with momentum k_1 (a), and the possible intermediate states (b) and (c) in a spectator model description. Conjugate contribution to (b) and (c) not shown.

given by,

$$\Phi_{ij}^{[\mathscr{U}]}(x,k_T) = \int \frac{d(\xi\cdot P)\,d^2\xi_T}{(2\pi)^3}\, e^{ik\cdot\xi} \langle P|\,\overline{\psi}_j(0)\,\mathscr{U}_{[0;\xi]}\,\psi_i(\xi)\,|P\rangle\big|_{\rm LF}\,. \tag{1}$$

LF ($\xi\cdot n = 0$) designates the light-front. The *gauge link* is the path-ordered exponential, $\mathscr{U}_{[\eta;\xi]} = \mathscr{P}\exp\left[-ig\int_C ds{\cdot}A^a(s)\,t^a\right]$ along the integration path C with endpoints at η and ξ. Its presence in the hadronic matrix element is required by gauge-invariance and the integration path C in the gauge link is process-dependent. This is due to the observation that the operator structure of the correlator is also a consequence of the necessary resummation of all contributions that arise from collinear gluon polarizations, i.e. those along the hadron momentum [5]. Collinear quark distribution functions are obtained from the TMD correlator after integration over p_T,

$$\Phi(x) \;=\; \int d^2k_T\, \Phi^{[\mathscr{U}]}(x,k_T) = \int \frac{d(\xi\cdot P)}{2\pi}\, e^{ix\xi\cdot P} \langle P|\,\overline{\psi}(0)\,\mathscr{U}^n_{[0;\xi]}\,\psi(\xi)\,|P\rangle\big|_{\rm LC}\,. \tag{2}$$

The non-locality is restricted to the light-cone (LC: $\xi\cdot n = \xi_T = 0$) and the gauge link is unique, being the straight-line path along n. In azimuthal asymmetries one needs the transverse moments contained in the correlator

$$\Phi_{\partial}^{\alpha\,[\mathscr{U}]}(x) = \int d^2k_T\; k_T^\alpha\, \Phi^{[\mathscr{U}]}(x,k_T)\,. \tag{3}$$

The TMD correlator, expanded in distribution functions depending on x and k_T^2 contains T-even and T-odd functions, since the correlator is not T-invariant. This property is attributed to the gauge link, that depending on the process, accounts for specific initial and/or final state interactions depending on the color flow in the process. For the collinear case, the link structure is unique in the case of integration over k_T (Eq. 2).

For the collinear weighted case, the transverse moments in Eq. (3) retains a nontrivial link-dependence that prohibits the use of T-invariance as a constraint. It is possible however, to decompose the weighted quark correlators as

$$\Phi_{\partial}^{\alpha\,[\mathscr{U}]}(x) = \tilde{\Phi}_{\partial}^{\alpha}(x) + C_G^{[\mathscr{U}]}\,\pi\Phi_G^\alpha(x,x)\,, \tag{4}$$

with calculable process-dependent gluonic pole factors $C_G^{[\mathscr{U}]}$ and process (link) independent correlators $\tilde{\Phi}_\partial$ and Φ_G. The correlator $\tilde{\Phi}_\partial$ contains the T-even operator combination, while Φ_G contains the T-odd operator combination. The latter is precisely the soft limit $x_1 \to 0$ of a quark-gluon correlator

$$\Phi_G^\alpha(x,x-x_1) = n_\mu \int \frac{d(\xi\cdot P)}{2\pi}\frac{d(\eta\cdot P)}{2\pi}\, e^{ix_1(\eta\cdot P)} e^{i(x-x_1)(\xi\cdot P)} \times \langle P|\,\overline{\psi}(0)\,U^n_{[0;\eta]}\,gG^{\mu\alpha}(\eta)\,U^n_{[\eta;\xi]}\,\psi(\xi)\,|P\rangle\,\big]_{\rm LC}\,. \tag{5}$$

The universal T-odd distribution functions in the parameterization of $\Phi_G(x,x)$ appear in T-odd observables such as single spin asymmetries with the specific gluonic pole factors from Eq. 4.

The situation for fragmentation functions is different. The TMD fragmentation correlator depending on the collinear and transverse components of the quark momentum, $k = \frac{1}{z}P + k_T + \sigma n$, is given by [9, 5]

$$\Delta_{ij}^{[\mathscr{U}]}(z,k_T) = \sum_X \int \frac{d(\xi\cdot P_h)\,d^2\xi_T}{(2\pi)^3}\, e^{ik\cdot\xi}\,\langle 0|\mathscr{U}_{[0,\xi]}\psi_i(\xi)|P,X\rangle\langle P,X|\overline{\psi}_j(0)|0\rangle|_{LF}\,. \tag{6}$$

The collinear, k_T-integrated correlator $\Delta(z) = \int d^2k_T \Delta^{[\mathscr{U}]}(z,k_T)$ only contains a T-even operator combination. Nevertheless one could in principle have T-even and T-odd fragmentation functions depending on z since the hadronic state $|P,X\rangle$ is an out-state, which is not T-invariant. In the transverse moments obtained after k_T-weighting,

$$\Delta_\partial^{\alpha\,[\mathscr{U}]}(z) = \int d^2k_T\; k_T^\alpha \Delta^{[\mathscr{U}]}(z,k_T) = \tilde{\Delta}_\partial^\alpha\left(\frac{1}{z}\right) + C_G^{[\mathscr{U}]}\,\pi\Delta_G^\alpha\left(\frac{1}{z},\frac{1}{z}\right), \tag{7}$$

the two link independent correlators $\tilde{\Delta}_\partial$ and Δ_G contain again a T-even and T-odd operator combination, respectively. The gluonic pole correlator is again the soft limit, $z_1^{-1} = x_1 \to 0$, of the quark-gluon correlator

$$\Delta_{Gij}^\alpha(x,x-x_1) = \sum_X \int \frac{d(\xi\cdot P)}{2\pi}\frac{d(\eta\cdot P)}{2\pi}\, e^{ix_1(\eta\cdot P)} e^{i(x-x_1)(\xi\cdot P)} \times \langle 0|\mathscr{U}^n_{[0,\eta]}\,gG^{n\alpha}(\eta)\,\mathscr{U}^n_{[\eta,\xi]}\,\psi_i(\xi)|P,X\rangle\langle P,X|\overline{\psi}_j(0)|0\rangle\Big|_{LC}\,. \tag{8}$$

As stated above, because of the appearance of hadronic states $|P,X\rangle$, each of correlators in Eq. 7 contains in principle T-even and T-odd functions. Rather than having a doubling of T-odd functions, we will show in a spectator model approach that $\Delta_G(x,x) = 0$, which implies that T-odd fragmentation functions in the transverse moments only come from $\tilde{\Delta}_\partial$, which appear with a universal strength (no gluonic pole factors). We demonstrate this starting with the collinear quark-gluon correlators, Eqs. 5 and 8 rather than the model approaches [11, 12, 13, 14, 15, 16] that looked at the transverse momentum dependent quark correlators in Eqs. 1 and 6.

GLUONIC POLES IN THE SPECTATOR MODEL

As mentioned above the T-odd gluonic pole matrix elements arise in the decomposition of the transverse weighted quark correlators and are precisely the soft limits ($k_1 \to 0$

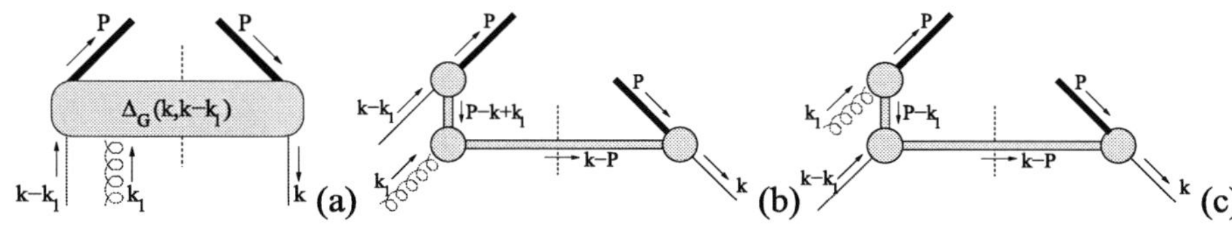

FIGURE 2. The graphical representation of the quark-quark-gluon correlator Δ_G in the case of fragmentation including a gluon with momentum k_1 (a) and the possible intermediate states (b) in a spectator model description. Conjugate contribution to (b) and (c) not shown.

or x_1, and z_1^{-1} or $x_1 \to 0$) of the quark-gluon-quark matrix elements [9] Eqs. 4 and 7 (see Figs. 1 and 2). The process-dependent gluonic pole factors $C_G^{[\mathscr{U}]}$ are calculable and the process (link) independent correlators $\tilde{\Phi}_\partial$ and $\tilde{\Delta}_\partial$ contains the T-even operator combination while Φ_G and Δ_G contain the T-odd operator combination. Turning to the analysis of the quark-gluon-quark and gluon-quark-quark matrix elements within a spectator model we consider the distribution and fragmentation correlators where the target and decay remnant in the distribution and fragmentation correlators respectively are represented by a spectator of mass M_s. These correlators are depicted in Figs. 1 (b) and (c) and Figs. 2 (b) and (c). We parameterize the gluon momentum as $k_1 = [k_1^-, x_1, k_{1T}]$ where $k_1^- = k_1 \cdot P - \frac{1}{2} x_1 M^2$ is the first component to be integrated over [10]. The result for the quark-gluon-quark correlator $\Phi_G(x, x-x_1, k_T, k_T - k_{1T})$

$$
\begin{aligned}
&\sim \frac{1}{(k^2-m^2)} \left\{ \int dk_1^- \, \frac{F_1\left(k_1^-, x, x_1, k_T, k_{1T}\right)}{(k_1^2 - m_1^2 + i\varepsilon)((k-k_1)^2 - m^2 + i\varepsilon)((P-k+k_1)^2 - M_{s1}^2 + i\varepsilon)} \right. \\
&\quad \left. + \int dk_1^- \, \frac{F_2\left(k_1^-, x, x_1, k_T, k_{1T}\right)}{(k_1^2 - m_1^2 + i\varepsilon)((k-k_1)^2 - m^2 + i\varepsilon)((P-k_1)^2 - M_{s2}^2 + i\varepsilon)} \right\} \\
&= \left\{ \int \frac{dk_1^-}{(\mu^2 - k_T^2)} \, \frac{(1-x)F_1\left(k_1^-, x, x_1, k_T^2, k_{1T}^2\right)}{(x_1 k_1^- - A_1 + i\varepsilon)((x_1 - x)k_1^- - A_2 + i\varepsilon)((1-x+x_1)k_1^- - B_1 + i\varepsilon)} \right. \\
&\quad \left. + \int \frac{dk_1^-}{(\mu^2 - k_T^2)} \, \frac{(1-x)F_2\left(k_1^-, x, x_1, k_T^2, k_{1T}^2\right)}{(x_1 k_1^- - A_1 + i\varepsilon)((x_1 - x)k_1^- - A_2 + i\varepsilon)((x_1 - 1)k_1^- - B_2 + i\varepsilon)} \right\},
\end{aligned}
$$

where, $F_i\left(k_1^-, x, x_1, k_T^2, k_{1T}^2\right)$ contains the numerators of propagators and/or traces of them in the presence of Dirac Gamma matrices, as well as the vertex form factors (see for example [17]) and where $\{A, B\}_i(\{m_i^2\}, \{k_{iT}^2\}, \{x_i\})$ are functions of masses and transverse quark momentum. Assuming that the numerator does not grow with k_1^- one can easily perform the k_1^- integrations assuming that the F_i are independent of k_1^-. Taking the limit $x_1 \to 0$ of the basic result for the quark-gluon correlators $\Phi_G(x, x - x_1, k_T, k_T - k_{1T})$ we obtain the gluonic pole correlators, for distribution functions ($0 \leq x \leq 1$) (see [10] for details)

$$
\Phi_G(x,x) \;=\; -\int d^2k_T \, d^2k_{1T} \, \frac{(1-x)\,F_1(x,0,k_T,k_{1T})\,\theta(1-x)}{(\mu^2 - k_T^2)\,(xB_1 + (1-x)A_2)\,A_1}, \tag{9}
$$

where $\mu^2(x) = xM_s^2 + (1-x)m^2 - x(1-x)M^2$, and for fragmentation functions

$$\Delta_G(x,x) = 0, \qquad (10)$$

where, $x = 1/z \geq 1$. This result depends on the assumption that the numerator has sufficient convergence in k_1^- otherwise one does not get the required $x_1\,\theta(x_1)$ behavior in the calculation [10]. In models, terms proportional to $k_1^- \sim k_1 \cdot P$ may easily arise from numerators of fermionic propagators [18] which in turn may easily be suppressed by form factors at the vertices. To prove a proper behavior within QCD one would need to study the fully unintegrated correlators such as e.g. in Ref. [19] and show that they fall off sufficiently fast as a function of $k_1 \cdot P$.

While our analysis is not yet the full proof that gluonic pole matrix elements vanish in the case of fragmentation, it is a step towards such a proof and the possible direction to obtain such a proof by considering the appropriate color gauge-invariant soft matrix elements. Such a proof is important as it eliminates a whole class of matrix elements parameterized in terms of T-odd fragmentation functions besides the T-odd fragmentation functions in the parameterization of the two-parton correlators (see [20] for a new model independent analysis).

ACKNOWLEDGMENTS

L.G. thanks the organizers of SPIN 2008 for the invitation to present this work. L.G. acknowledges support from U.S. Department of Energy under contract DE-FG02-07ER41460.

REFERENCES

1. J. C. Collins, and D. E. Soper, *Nucl. Phys.* **B194**, 445 (1982).
2. R. L. Jaffe, and X.-D. Ji, *Nucl. Phys.* **B375**, 527–560 (1992).
3. A. V. Efremov, and O. V. Teryaev, *Sov. J. Nucl. Phys.* **36**, 140 (1982).
4. J.-W. Qiu, and G. Sterman, *Phys. Rev. Lett.* **67**, 2264–2267 (1991).
5. C. J. Bomhof, and P. J. Mulders, *JHEP* **02**, 029 (2007).
6. D. W. Sivers, *Phys. Rev.* **D41**, 83 (1990).
7. J. C. Collins, *Phys. Lett.* **B536**, 43–48 (2002).
8. S. J. Brodsky, D. S. Hwang, and I. Schmidt, *Nucl. Phys.* **B642**, 344–356 (2002).
9. D. Boer, P. J. Mulders, and F. Pijlman, *Nucl. Phys.* **B667**, 201–241 (2003).
10. L. P. Gamberg, A. Mukherjee, and P. J. Mulders, *Phys. Rev.* **D77**, 114026 (2008).
11. X.-d. Ji, and F. Yuan, *Phys. Lett.* **B543**, 66–72 (2002).
12. A. Metz, *Phys. Lett.* **B549**, 139–145 (2002).
13. L. P. Gamberg, G. R. Goldstein, and K. A. Oganessyan, *Phys. Rev.* **D67**, 071504 (2003).
14. A. Bacchetta, A. Schaefer, and J.-J. Yang, *Phys. Lett.* **B578**, 109–118 (2004).
15. A. Bacchetta, L. P. Gamberg, G. R. Goldstein, and A. Mukherjee, *Phys. Lett.* **B659**, 234–243 (2008).
16. L. P. Gamberg, G. R. Goldstein, and M. Schlegel, *Phys. Rev.* **D77**, 094016 (2008).
17. R. Jakob, P. J. Mulders, and J. Rodrigues, *Nucl. Phys.* **A626**, 937–965 (1997).
18. L. P. Gamberg, D. S. Hwang, A. Metz, and M. Schlegel, *Phys. Lett.* **B639**, 508–512 (2006).
19. J. C. Collins, T. C. Rogers, and A. M. Stasto, *Phys. Rev.* **D77**, 085009 (2008).
20. S. Meissner, and A. Metz (2008), arXiv:0812.3783[hep-ph].

Measurement of Transverse Spin Dependent Fragmentation Functions at Belle

Anselm Vossen*, Matthias Grosse Perdekamp*, Martin Leitgab*, Ralf Seidl†, Akio Ogawa** and Kieran Boyle‡

*University of Illinois at Urbana Champaign
†RBRC (RIKEN BNL Research Center)
**BNL/RBRC
‡RBRC

Abstract. The topic of this article is the measurement of the transverse spin dependent Collins and dihadron interference fragmentation functions at the Belle experiment. Recent results and projections for future measurements will be presented.

PACS numbers: 13.88.+e, 13.66.-a, 14.65.-q, 14.20.c

Keywords: Fragmentation, Collins Effect, Interference Fragmentation Function, Transversity, Belle

INTRODUCTION

Transverse spin dependent fragmentation functions can serve as spin analyzers for the fragmenting quark. This is possible because they exhibit an azimuthal dependence relative to the spin of the fragmenting quark. Since they are chiral odd, they can be used to measure the chiral odd transversity distribution function δq of a parton q in the nucleon. Together with the unpolarized and the helicity distribution function of the quarks it completely describes the spin structure of the nucleon at leading order in a collinear picture. A review can be found in [1]. Currently the measurement of δq together with Collins or dihadron interference fragmentation functions seems to be most promising. Results from the HERMES and COMPASS experiments at DESY and CERN, respectively indicate that the product of transversity with either Collins or dihadron interference fragmentation functions leads to a non-vanishing effect [2][3][4][5]. Together with the results recently published by the Belle collaboration, they made the first global analysis of Collins fragmentation function and the transversity distribution in the nucleon possible [6]. This paper is organized as follows. First the Belle experiment is introduced. Then the properties of the Collins fragmentation function are recalled and recent results are presented. In the following section the same is done for the dihadron interference fragmentation function. However, since the analysis is still ongoing, only projections are shown. First results and prospect in the measurement of transversity with IFFs at PHENIX are given. Finally, an outlook is provided at the future plans for the measurement of fragmentation functions at Belle.

CP1149, *Spin Physics, 18th International Spin Physics Symposium*
edited by D. G. Crabb, D. B. Day, S. Liuti, X. Zheng, M. Poelker, and Y. Prok

THE BELLE EXPERIMENT

The Belle experiment is situated at the KEKB accelerator in Japan, which is a b-factory using asymmetric annihilations of electron and positrons with energies of about 8 GeV and 3.5 GeV respectively which collide under an angle $\theta = 20$ mrad. It is described in [7]. Most data is taken on the $\Upsilon(4S)$ resonance which decays into B meson pairs at a center of mass energy of $\sqrt{s} = 10.58$ GeV. But for background estimation around 10% of the data is taken off-resonance at $\sqrt{s} = 10.52$ GeV. At this energy the continuum production of light and charm quarks (u,d,s.c) dominates. Presently more than 700 fb^{-1} on-resonance and 60 fb^{-1} off-resonance data have been collected. The Belle detector exhibits the typical structure of a high energy physics experiment with the exception that it is desgned to be symmetric in the center of mass frame of the virtual photon. This leads to an asymmmetry in the lab frame. Tracking is done by a drift chamber which also provides dE/dx measurements for particle identification. In the barrel region, which is used for this analysis, further PID capabilities come from aerogel Cherenkov counters, time of flight counters and multiple layers of iron and resistive plate chambers to distinguish K_Ls and μ from other particles. The detectors are within a 1.5 T solenoid field which also contains a $CSI(Ti)$ electromagnetic calorimeter containing about 16 radiation lengths. Altogether, the Belle detector has a large acceptance and good tracking and particle identification capabilities.

Collins Asymmetries in unpolarized e^+e^- Collisions

In 1993 J.C Collins [9] proposed the now so called Collins effect in quark fragmentation as a possible analyzer sensitive to the transverse spin of the fragmenting quark. He pointed out that due to final state effects the fragmentation of a transversely polarized quark in an unpolarized hadron can have an azimuthal dependence proportional to $(\mathbf{k} \times \mathbf{P}_{hT}) \cdot \mathbf{s}_q$. Here $\mathbf{k}$ is the momentum of the fragmenting quark, $\mathbf{P}_{hT}$ the transversal momentum of the produced hadron with respect to the quark axis and $\mathbf{s}_q$ the spin vector of the quark. In a model by X. Artru this asymmetry has a descriptive interpreation in the Lund string fragmentation model. It can come about if a quark-antiquark pair is produced from the vacuum with the quantum numbers of the vacuum [8]. If the transverse direction is choosen as the quantization axis, this means orbital angular momentum $L_y = 1$ and $s_y = -1$. Together with the fragmenting quark with spin $s_y = \frac{1}{2}$ the antiquark forms a pseudoscalar hadron, for example a pion. Due to the conservation of orbital angular momentum the momentum distribution of the produced hadron has a left-right asymmetry. In general, the unpolarized fragmentation function of quark q into hadron h will be called $D_1^{q,h}(z)$ and the Collins fragmentation function, which contains the transverse spin dependent part of the fragmentation, $H_1^{\perp q,h}(z, P_{hT}^2)$. Here the dependence on quark transverse momentum of $D_1^{q,h}$ was neglected.

Because in unpolarized $e^+ - e^-$ annihilation the spin directions of the produced $q - \bar{q}$ pair is unknown, any spin dependent angular modulation will average out. However, it is known that the spins of the produced quark pair have to be antiparallel and the transverse spin component follows a $\sin^2\theta$ distribution where θ is the angle of quark and electron axis in the center of mass system. Therefore the azimuthal momentum distribution of hadrons originating from the produced quark has to be correlated with the one of hadrons originating from the produced antiquark. It follows that the angular momentum dependence of such a hadron pair is of the form $\cos(\phi_1+\phi_2)$ [10]. The azimuthal angles ϕ_i, $i \in \{1,2\}$ are measured around the $q-\overline{q}$ axis $\hat{\mathbf{n}}$ between the scattering plane and the plane which contains the hadron momentum vectors $P_{h\,i}$ in a system in which the quarks are back to back:

$$\phi_{12} = \mathrm{sgn}[\hat{\mathbf{n}}\cdot\{(\hat{\mathbf{z}}\times\hat{\mathbf{n}})\times(\hat{\mathbf{n}}\times\mathbf{P}_{h\;1,2})\}]\times\arccos\left(\frac{\hat{\mathbf{z}}\times\hat{\mathbf{n}}}{|\hat{\mathbf{z}}\times\hat{\mathbf{n}}}\cdot\frac{\hat{\mathbf{n}}\times\mathbf{P}_{h\;1,2}}{|\hat{\mathbf{n}}\times\mathbf{P}_{h\;1,2}|}\right\} \quad (1)$$

The ^ signifies unit vectors. Including the previously introduced θ dependence the cross section for the production of two hadrons can be written as

$$\frac{(e^+e^-\to h_1h_2X)}{d\Omega dz_1dz_2d\phi_1d\phi_2} = \sum_{q\overline{q}}\frac{3\alpha^2}{Q^2}\frac{e_q^2}{4}z_1^2z_2^2\Big\{(1+\cos^2\theta)D_1^{q,[0]}(z_1)\overline{D_1^{q,[0]}}(z_2) + \sin^2\theta\cos(\phi_1+\phi_2)H_1^{\perp,[1],q}(z_1)\overline{H_1^{\perp,[1],q}}(z_2)\Big\}. \quad (2)$$

The fragmentation functions appear as moments of the absolute value of transverse momenta: $F^{[n]}(z)=\int d|\mathbf{k}_T|^2[\frac{|\mathbf{k}_T|}{M_h}]^nF(z,\mathbf{k}_T^2)$. If the dependence on the thrust axis which is used to approximate the $q-q$ axis is integrated out, a dependence of the cross section on the angle ϕ_0 remains, which is the azimuthal angle between the planes spanned by one hadron momentum and the lepton momenta and the transverse momentum of the second hadron with respect to the first hadron momentum. This angle is given by

$$\phi_0 = \mathrm{sgn}[\mathbf{P}_{h2}\cdot\{(\hat{\mathbf{z}}\times\mathbf{P}_{h2})\times(\mathbf{P}_{h2}\times\mathbf{P}_{h1})\}]\times\arccos\left(\frac{\hat{\mathbf{z}}\times\mathbf{P}_{h2}}{|\hat{\mathbf{z}}\times\mathbf{P}_{h2}|}\cdot\frac{\mathbf{P}_{h2}\times\mathbf{P}_{h1}}{|\mathbf{P}_{h2}\times\mathbf{P}_{h1}|}\right) \quad (3)$$

and the cross section differential in ϕ_0 becomes proportional to $\cos(2\phi_0)$ [11]. Because it does not depend on the thrust axis anymore, transverse momentum dependence is also integrated over and instead of the product of fragmentation functions one measures the convolution of fragmentation functions.

In the following, the extracted asymmetries A_{12} and A_0 refer to the first $\cos(\phi_1+\phi_2)$ and $\cos(2\phi_0)$ moments of the cross section which can be estimated from the count rates.

The Collins asymmetry was extracted for charged pion pairs from a dataset containing 547 fb^{-1}, 55 fb^{-1} of which was measured off-resonance at a CM energy of 10.52 GeV and 492 fb^{-1} on the $\Upsilon(4S)$ resonance at 10.58 GeV. Since the B mesons are produced and decay nearly at rest, a cut of T>0.8 on the thrust value removes 98% percent of B meson pairs, leaving light and charm quarks whose fragmentation leads to a two jet topology. Contributions to the measured asymmetries from fragmenting charm quarks are corrected for using azimuthal asymmetries measured from a charm enhanced data sample. To minimize the contributions from decaying τ leptons, a cut on the visible Energy of $E_{\text{vis}} > 7\ GeV$ is applied. Only charged tracks lying in the barrel region $-0.6 < cos(\theta_{\text{Lab}}) < 0.9$ with θ_{Lab} being the polar angle in the lab system, are selected. Furthermore there are cuts on the vertex position which has to have a radial distance smaller than 2 cm and a distance in beam direction of smaller than 4 cm from the $e^+ - e^-$ interaction and $Q_T < 3.5\ GeV$. From the remaining data samples hadrons identified as pions are selected with likelihood ratios based on energy loss in the CDC, number of Cherenkov photons measured in the ACC and time of flight information. The number of misidentified pions is below 10% in all z_1, z_2 bins. The minimum fractional energy is $z_{1,2} > 0.2$. The asymmetries A_{12} and A_0 are then extracted from pion pairs where the hadrons are required to lie in different hemispheres with respect to the plane normal to the thrust axis $\hat{\mathbf{n}}$. When selecting pion pairs, there are three different possible combinations for which asymmetries A_{12} and A_0 can be extracted. Unlike (A_i^U) and like sign (A_i^L) pairs and charged pion combinations of any sign (A_i^C). Each asymmetry contains different combinations of favored and unfavored fragmentation functions, where favored functions describe the fragmentation of a quark into a hadron containing a valence quark of the same flavor. In the following strange and heavy quarks are ignored and the notation $D_q^{\text{fav/dis}}$, $H_q^{\perp,\text{fav/dis}}$ is used for favored and disfavored unpolarized and Collins fragmentation function respectively. Then the different asymmetries correspond to the following combination of fragmentation functions:

$$A^U \propto \frac{H_u^{\perp,\text{fav}} \times H_{\bar{u}}^{\perp,\text{fav}} + H_u^{\perp,\text{dis}} \times H_{\bar{u}}^{\perp,\text{dis}}}{D_u^{\text{fav}} \times D_u^{\text{fav}} + D_u^{\text{dis}} \times D_u^{\text{dis}}} \qquad A^L \propto \frac{H_u^{\perp,\text{fav}} \times H_{\bar{u}}^{\perp,\text{dis}} + H_u^{\perp,\text{dis}} \times H_{\bar{u}}^{\perp,\text{fav}}}{D_u^{\text{fav}} \times D_u^{\text{dis}} + D_u^{\text{dis}} \times D_u^{\text{fav}}} \tag{4}$$

$$A^C \propto \frac{\left(H_u^{\perp,\text{fav}} + H_u^{\perp,\text{dis}}\right)\left(H_{\bar{u}}^{\perp,\text{fav}} + H_{\bar{u}}^{\perp,\text{dis}}\right)}{\left(D_u^{\text{fav}} + D_u^{\text{dis}}\right)\left(D_u^{\text{fav}} + D_u^{\text{dis}}\right)} \tag{5}$$

In order to minimize systematic effects due to QCD radiation and acceptance effects, ratios or differences of the $A_i^{U,L,C}$ can be build. Both give similar results and here the ratios $A_i^{UC} = A_i^U / A_i^C$ and $A_i^{UL} = A_i^U / A_i^L$ are given for which acceptance effects cancel and contributions from QCD radiation which are independent on the electric charges cancel to first order. Systematic studies were done into effects due to beam polarization, additional moments in the fits, differences between on and off resonance data, the correction for charm contributions with MC and charm enhanced data samples and uncertainties due to the particle identification. Details can be found in [13]. As can be seen in fig. 1 large asymmetries of the order of 10 % can be observed at sufficiently high z_1, z_2.

INTERFERENCE FRAGMENTATION FUNCTION

The two hadron interference fragmentation function (IFF) is another probe for transversely polarized quarks. Like the Collins fragmentation function it is chiral odd and thus provides an independent way to access the transversity distribution. The additional

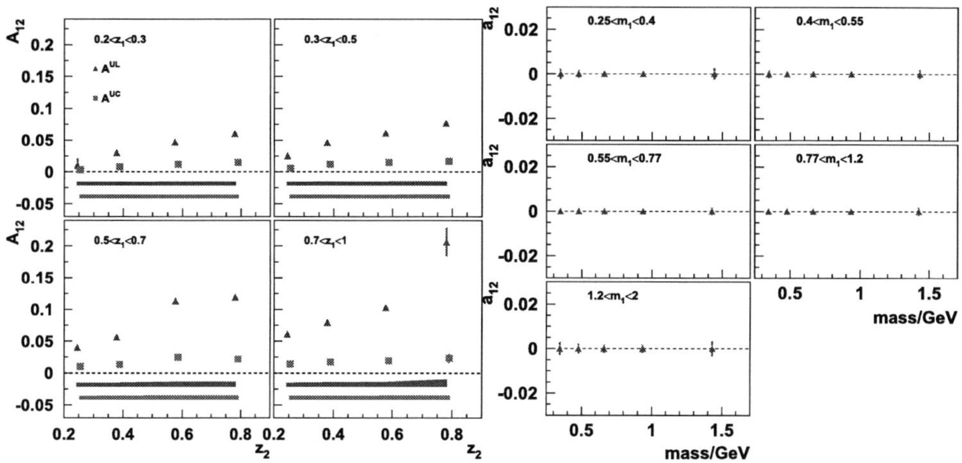

FIGURE 1. Belle results. On the left results for the Collins asymmetries A_{12} extracted from $547 fb^{-1}$. On the right projections for the IFF precision from $55 fb^{-1}$ of data.

hadron measured in the final state provides allows to construct an asymmetry that does not vanish if transverse momentum is integrated over. Therefore colinear factorization can be used and there is no dependence on the model of the k_T distribution. However this comes at the cost of an additional dependence on the invariant mass of the hadron pair. There exist various predictions for the two hadron IFF at COMPASS, HERMES and Belle kinematics [12].

The observable for the two hadron IFF is an angular dependence of the count rates similar to the Collins asymmetry. But instead of using the azimuthal angle between one hadron momentum and the plane spanned by $e^+ - e^-$ axis and the thrust axis (A_{12}) or the other hadron momentum (A_0), the azimuthal angle between the vector $\mathbf{R} = \mathbf{P}_{h1} - \mathbf{P}_{h2}$ is used. Here $\mathbf{P}_{hi}$ are the two hadron momenta measured in the final state in one hemisphere.

The analysis for the two hadron IFF for pions at Belle is ongoing. The cuts are similar to the ones used in the Collins asymmetry. However, no double ratios were used since observed false asymmetry are significantly lower than the statistical error. Fig. 1 shows the predicted sensitivity for π^+/π^- pairs from the 55 fb^{-1} off-resonance dataset used for the extraction of the Collins asymmetries. It is well below the sensitivity needed to measure the asymmetries predicted e.g. in [12].

There is an effort underway to extract transversity at the PHENIX experiment at BNL [14]. The two hadron IFF channel was choosen due to the capabilities of the detector. First results were presented in [14] which are shown in fig. 2. Phenix is still analyzing data taken in 2008 and will continue to measure in the future. Together with the Belle results the shown sensitivities promise interesting results in the near future.

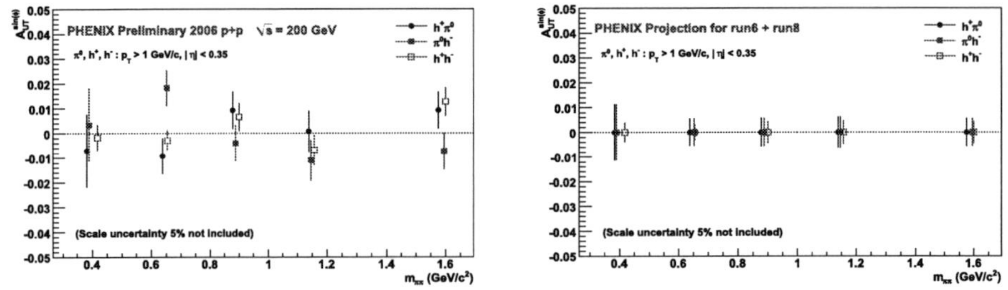

FIGURE 2. Phenix results for IFF asymmetries. On the left asymmetries extracted from $2.7pb^{-1}$ of data taken in 2006. On the right expected statistical uncertainties in combination with $5.2pb^{-1}$ of data taken in 2008.

OUTLOOK

The Collins function for charged pions has been measured. We are currently working on the extraction of the two hadron IFF and plan to release results for charged and neutral pions in the near future. Thereafter it is planned to extend the analysis to charged kaons. At the same time there are efforts to extract unpolarized fragmentation functions with very high precision from the Belle data that are needed for the extraction of spin dependent fragmentation functions and polarized quark distribution functions from the measured asymmetries.

REFERENCES

1. V. Barone, P.G. Ratcliffe, Transverse Spin Physics, World Scientific, River Edge USA , 2003
2. A. Airapetian et. al. [HERMES Collaboration], Phys. Rev. Lett. **94** 012002 (2005)
3. M. Diefenthaler et. al. [HERMES Collaboration] arXiv:0706.2242 [hep-ex], Proceedings of the 15th International Workshop on Deep Inelastic Scattering (DIS 2007), Munich, Germany (2007)
4. E.S. Ageev et. al. [COMPASS Collaboration], Nucl. Phys. B **765** 31 (2007)
5. S. Levorato et. al. [COMPASS Collaboration], arXiv:0808.0086 [hep-ex], Proceedings of Transversity 2008, Ferrara, Italy (2008)
6. M. Anselmino, M. Boglione, U. D'Alesio, A. Kotzinian, F. Murgia, A. Prokudin and C. Turk, Phys. Rev. D **75** 054032 (2007)
7. A. Abashian et. al. [Belle Collaboration], Nucl. Inst. Meth. A **479** 117 (2002)
8. X. Artru and M. Mekhfi, Transversely polarized parton densities, their evolution and their measurement, Z. Phys., **C45**:669, (1990)
9. J. C. Collins, Nucl. Phys. B**396** 161 (1993)
10. D. Boer, Ph.D. Thesis, RX1630, Vrije U., Amsterdam, 1998
11. D. Boer, R. Jakob and P.J. Mulders, Nucl. Phys. B **504** 345 (1997)
12. A. Bacchetta, F. A. Ceccopieri, A. Mukherjee, M. Radici, JLAB-THY-08-917
13. R. Seidl et. al. [Belle Collaboration], Phys. Rev. **D78** 032011 (2008),0805.2975
14. R. Yang [PHENIX Collaboration], PKU-RBRC Workshop on Transverse Spin Physics, Beijing 2008 http://rchep.pku.edu.cn/workshop/0806/20080627-iff.pdf

Transverse Single Spin Asymmetry in Heavy Flavor Production in Polarized p+p Collisions at PHENIX

Han Liu for the PHENIX Collaboration

Los Alamos National Laboratory, Los Alamos, NM 87545, USA

Abstract. At RHIC energy, heavy flavor production is dominated by gluon-gluon fusion, so the Collins effect has minimum impact on transverse single spin asymmetry (A_N) as the gluon's transversity is zero. The measurement of A_N in heavy flavor production is sensitive to the gluon Sivers distribution which is related to the orbital angular momentum of gluons inside the polarized protons. The most recent results of A_N in J/Ψ and open heavy flavor production at forward rapidity are presented.

Keywords: Proton structure, transverse spin
PACS: 14.20.Dh

INTRODUCTION

The transverse spin structure of the proton is poorly understood. The measurement of transverse single spin asymmetries gives us an opportunity to probe the quark and gluon structure of transversely polarized nucleons. Large transverse single spin asymmetries were first observed in hadronic collisions at the ZGS at Argonne National Laboratory [1], as well as at the PS at CERN [2] for center of mass energies between 5 to 10 GeV. Similar effects have been found to persist for particles produced at large x_F at center of mass energies at 20 GeV[3], 62 GeV[4] and 200 GeV[5], although they were expected to vanish in pQCD calculations at the leading order[6].

A number of pQCD based models have been developed to explain this phenomenon. Among them are the Sivers effect [7], the Collins effect [8], and the higher twist effects [9, 10]. In addition, single spin azimuthal asymmetries have been observed recently in semi-inclusive deep-inelastic scattering experiments at HERMES experiment at DESY[11] and in e^+e^- annihilation at the BELLE experiment at KEKB[12], which provided evidences that Sivers and Collins effects are real physical mechanisms. A unified picture of the Sivers mechanism and the higher twist mechanism has been shown by Ji and his collaborators, where they found although the two mechanisms each have their own domain of validity, they describe the same physics in the kinematic region where they overlap[13].

At RHIC energy, heavy flavor (heavy quarkonium and open heavy flavor) production is dominated by gluon gluon interaction. Any large transverse single spin asymmetry observed in heavy flavor production cannot originate from the Collins effect because the gluon's transversity is zero. Therefore, the production of heavy flavor particles in transversely polarized p+p collisions at the PHENIX experiment offers a good opportunity to

CP1149, *Spin Physics, 18th International Spin Physics Symposium*
edited by D. G. Crabb, D. B. Day, S. Liuti, X. Zheng, M. Poelker, and Y. Prok

study the gluon's Sivers function.

MEASUREMENT AND ASYMMETRY CALCULATION

During 2006 RHIC run, the PHENIX experiment has collected 2.7 pb^{-1} data with averaged beam polarization about 50% for both beams in transversely polarized p+p collisions at $\sqrt{s}$=200GeV. Muons from J/Ψ and heavy quark decays have been measured with muon spectrometers in full azimuthal coverage at forward rapidities ($1.2 < |\eta| < 2.4$). Muons are identified through the use of Iarocci tubes interleaved between steel absorber plate and momentum of muons are determined using cathode strip chambers inside a magnetic field.

The J/Ψs have been measured through their decays in the $\mu^+\mu^-$ channel. At least two tracks in the muon identifier (MuID) of appropriate absorber depth are required. The backgrounds are estimated by using the like (unlike) sign dimuon pairs within 2.0 GeV $< M_{\mu\mu} <$ 2.5 GeV, and like sign dimuon pairs within 2.8 GeV$< M_{\mu\mu} <$ 3.5 GeV.

The high p_T single muons from semi-leptonic decay of open heavy hadrons are measured through the statistical subtraction of background sources, with the remaining yield attributed to open heavy flavor decays. The dominate sources of backgrounds are: (1) muons from light hadron decay before reaching the pre Muon Tracker (MuTr) absorber; (2) punch through hadrons which penetrate the 1.5m of steel absorber to reach the deepest layer of MuID; (3) background tracks which are dominated by hadron which decay into a muon after reaching the MuTr. The detail of how to estimate the background contributions can be found in [14].

In this analysis, the transverse single spin left-right asymmetry A_N is calculated by:

$$A_N = \frac{1}{P_b}\frac{\sigma^{\uparrow} - \sigma^{\downarrow}}{\sigma^{\uparrow} + \sigma^{\downarrow}} = \frac{1}{P_b}\frac{\sqrt{N_L^{\uparrow}\cdot N_R^{\downarrow}} - \sqrt{N_L^{\downarrow}\cdot N_R^{\uparrow}}}{\sqrt{N_L^{\uparrow}\cdot N_R^{\downarrow}} + \sqrt{N_L^{\downarrow}\cdot N_R^{\uparrow}}}, \quad (1)$$

where P_b is the beam polarization from the CNI and jet polarimeter analysis, $\sigma^{\uparrow}(\sigma^{\downarrow})$ is the production cross section from crossings with polarized bunch spin up(down), $N^{\uparrow}(N^{\downarrow})$ is the particle yield in the left and right direction relative to the transversely polarized projectile direction.

The measured A_N^{Incl} is corrected for the contribution from background by using

$$A_N^{Phys} = \frac{A_N^{incl} - r\cdot A_N^{BG}}{1-r}, \quad (2)$$

$$\delta A_N^{Phys} = \frac{\sqrt{(\delta A_N^{incl})^2 + r^2\cdot(\delta A_N^{BG})^2}}{1-r}, \quad (3)$$

where $r = N^{BG}/N^{Incl}$ is the background fraction, and A_N^{BG} is the background asymmetry.

The systematic uncertainty has been checked with bunch shuffling technique with randomly assigned polarization direction for bunches in each fill. The bunch-to-bunch and fill-to-fill systematic errors are much smaller than the statistical errors.

RESULTS AND DISCUSSION

The details of the J/Ψ production mechanism and hadronization still remain an open question. Many theoretical attempts have been made to describe the J/Ψ production, however, currently no model can reproduce both the observed cross sections and the polarization[15]. It is hoped that new results from polarized p+p collisions could shed new light on our understandings of the J/Ψ production mechanisms.

The transverse single spin asymmetry in heavy quarkonium production has been studied in the framework of non-relativistic QCD (NRQCD), where the heavy quark pairs are produced at short distance in a color-singlet or color-octet configurations[17]. The asymmetry is predicted to vanish if J/Ψs are produced through the color-singlet channel but survive in the color-octet channel in ep scattering. However, in pp scattering, the asymmetry vanishes if J/Ψs are produce through color-octet channel but survives in the color-singlet channel. Therefore, the transverse single spin asymmetry in J/Ψ production shall allow us to gain new information on the production mechanism for the heavy quarkonium production. A preliminary result of A_N in J/Ψ production has been reported in [16]. The result is statistically limited and is consistent with model calculation assuming the gluon Sivers function is $0.5x(1-x)$ times the unpolarized gluon distribution and 30% of J/Ψ comes from feed down of χ_c [18]. We expect to double the statistics with the full data sample from the year 2006 and 2008.

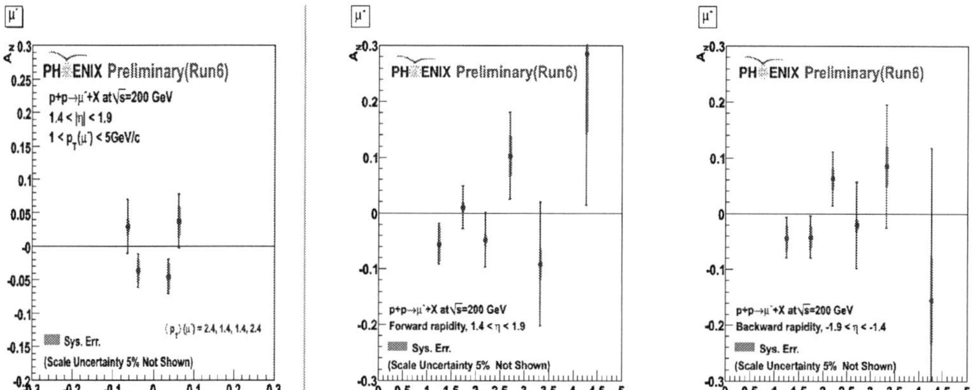

FIGURE 1. Transverse single spin asymmetry of prompt single muons (mostly from charm and beauty decays) at forward rapidities at $\sqrt{s}$=200GeV. A_N vs. x_F (left), and vs. p_T in the forward (middle) and backward rapidities (right).

At RHIC energy, transverse single spin asymmetry in open charm production was suggested to probe gluon Sivers function [19] and the trigluon correlation functions in the nucleon [20]. The first measurement of A_N of single muons from open heavy flavor decay has been made as a function of x_F, and p_T in the forward and backward rapidities (Figure 1). The large uncertainty of background contribution dominates the systematic uncertainty. A direct comparison between the prediction for D mesons and muons from open heavy flavor (open charm and beauty) decay will be available soon. However, the observed asymmetries in single muon production is significantly smaller than the model prediction for D mesons in the same kinematics region (x_F and p_T) [19].

SUMMARY

The transverse single spin asymmetries for J/Ψ and open heavy flavor have been measured with the PHENIX muon spectrometers at $\sqrt{s}$ =200GeV. The J/Ψ A_N provides a new probe to study the J/Ψ production mechanism. Heavy quark A_N can put a strong constrain on the gluon Sivers function, and help us to access the trigluon correlation functions in the nucleon.

The PHENIX Forward Silicon Vertex Detector (FVTX) upgrade will allow us directly measure prompt muons from heavy quark decays with much reduced systematic uncertainty. In addition, with much improved dimuon mass resolution, we can measure A_N in other channes, like Ψ' and Drell-Yan $\rightarrow \mu^+\mu^-$. It is expected that the FVTX will significantly enhance the physics capability in both longitudinal and transverse spin programs.

REFERENCES

1. R. D. Klem *et al.*, Phys. Rev. Lett. **36**, 929 (1976); W. H. Dragoset *et al.*, Phys. Rev. D **18**, 3939 (1978).
2. J. Antille *et al.*, Phys. Lett. B **94**, 523 (1980).
3. D.L. Adams *et al.*, Phys. Lett. B **264**, 462 (1991); A. Bravar *et al.*, Phys. Rev. Lett. **77**, 2626 (1996).
4. I. Arsene *et al.*, Phys. Rev. Lett. **101**, 042001 (2008).
5. J. Adams *et al.*, Phys. Rev. Lett. **92**, 171801 (2004).
6. G. L. Kane *et al.*, Phys. Rev. Lett. **41**, 1689 (1978).
7. D. Sivers, Phys. Rev. D **41**, 83 (1990).
8. J. Collins, Nucl. Phys. B **396**, 161 (1993).
9. J.W. Qiu *et al.*, Phys. Rev. D **59**, 014004 (1998).
10. Y. Kanazawa *et al.*, Phys. Lett. B **478**, 121 (2000).
11. A. Airapetian *et al.*, Phys. Rev. Lett. **84**, 4047 (2000); Phys. Rev. Lett. **94**, 012002 (2005).
12. R. Seidl *et al.*, Phys. Rev. Lett. **96**, 232002 (2006).
13. X.D. Ji *et al.*, Phys. Rev. Lett. **97**, 082002 (2006).
14. S.S. Alder *et al.*, Phys. Rev. D **76**, 092002 (2007); D. Hornback, J. Phys. G: Nucl. Part. Phys. **35**, 104113 (2008).
15. F. Abe *et al.*, Phy. Rev. Lett **69** 3704 (1992); S. Abachi *et al.*, Phys. Lett. B **370**, 239 (1996); B. Abbot *et al.*, Phys. Rev. Lett **82**, 35 (1999); F. Abe *et al.*, Phys. Rev. D **66**, 092001 (2002); D. Acosta *et al.*, Phys. Rev. D **71**, 032001 (2005); R. Vogt, Phys. Rep. **10**, 197 (1999); M. H. Schub *et al.*, Phys. Rev. D **52**, 1307 (1995).
16. H. Liu, *17th International Spin Physics Symposium*, AIP Conference Proceedings, American Institute of Physics, New York, Vol. **915**, 551 (2007).
17. F. Yuan, Phys. Rev. D **78**, 014024 (2008).
18. F. Yuan, private communication.
19. M. Anselmino *et al.*, Phys. Rev. D **70**, 074025 (2004).
20. Z.B. Kang *et al.*, Phys. Rev. D **78**, 114013 (2008).

Trigluon correlations and single transverse spin asymmetry in open charm production

Zhong-Bo Kang and Jian-Wei Qiu

Department of Physics and Astronomy, Iowa State University, Ames, IA 50011, USA

Abstract. We study the single transverse-spin asymmetry for open charm production in the semi-inclusive lepton-hadron deep inelastic scattering (SIDIS) and pp collision. Within collinear factorization approach, we find that the asymmetry is sensitive to the twist-3 trigluon correlation functions in the proton. With a simple model for the trigluon correlation functions, we estimate the asymmetry in SIDIS for both COMPASS and eRHIC kinematics, as well as in pp collision at RHIC energy. We discuss the possibilities of extracting the trigluon correlation functions in these experiments.

Keywords: Single transverse spin asymmetry, trigluon correlations, open charm production
PACS: 12.38.Bx, 12.39.St, 13.88.+e, 14.40.Lb

INTRODUCTION

Large single transverse-spin asymmetries (SSAs) have been consistently observed in various experiments at different collision energies. From the parity and time-reversal invariance of the strong interaction dynamics, the measured large asymmetries in high energy collisions should be directly connected to the transverse motion of partons inside a polarized hadron. Understanding the QCD dynamics behind the measured asymmetries should have the profound impact on our knowledge of strong interaction and hadron structure.

Within the QCD collinear factorization formalism for the cross sections with a large momentum transfer, the observed phenomenon of SSAs are attributed to three-parton correlations inside a polarized hadron. For the leading order gluonic pole contribution to the SSAs [1, 2], the relevant partonic correlations are given by the twist-3 quark-gluon correlation function

$$T_{q,F}(x,x) = \int \frac{dy_1^- dy_2^-}{4\pi} e^{ixP^+y_1^-} \langle P,s_T|\bar{\psi}_q(0)\gamma^+ \left[\epsilon^{s_T\sigma n\bar{n}} F_\sigma{}^+(y_2^-)\right] \psi_q(y_1^-)|P,s_T\rangle, \quad (1)$$

and by a twist-3 trigluon correlation function [3, 4, 5],

$$T_G(x,x) = \int \frac{dy_1^- dy_2^-}{2\pi} e^{ixP^+y_1^-} \frac{1}{xP^+} \langle P,s_T|F^+{}_\alpha(0) \left[\epsilon^{s_T\sigma n\bar{n}} F_\sigma{}^+(y_2^-)\right] F^{\alpha+}(y_1^-)|P,s_T\rangle. \quad (2)$$

There is a quark-gluon correlation function, $T_{q,F}$, for each quark (anti-quark) flavor q ($\bar{q}$), and there are two independent trigluon correlation functions, $T_G^{(f)}(x,x)$ and $T_G^{(d)}(x,x)$, because of the fact that the color of the three gluon field strengths in Eq. (2) can be neutralized by contracting with either the antisymmetric if^{ABC} or the symmetric d^{ABC} tensors with color indices, A, B, and C [3, 4, 5]. For the high order gluonic pole

CP1149, *Spin Physics, 18th International Spin Physics Symposium*
edited by D. G. Crabb, D. B. Day, S. Liuti, X. Zheng, M. Poelker, and Y. Prok

contribution as well as the fermionic pole contribution [1, 2], another set of quark-gluon and trigluon correlation functions is also relevant [6].

In this talk, we present the calculation of the leading gluonic pole contribution to the SSA of open charm production in both the SIDIS and pp collisions. We explore the role of the trigluon correlation functions in generating the SSA of D (or $\bar{D}$) mesons.

SINGLE TRANSVERSE-SPIN ASYMMETRY IN SIDIS AND pp COLLISION

We first study the SSAs in SIDIS, $e(\ell)+p(P,s_\perp)\to e(\ell')+h(P_h)+X$, with $s_\perp$ the transverse spin vector of the initial hadron and h the observed D or $\bar{D}$ meson . We work in the approximation of one-photon exchange, and define the virtual photon momentum $q=\ell-\ell'$ and its invariant mass $Q^2=-q^2$. We adopt the usual SIDIS variables:

$$S_{ep}=(P+\ell)^2, \qquad x_B=\frac{Q^2}{2P\cdot q}, \qquad y=\frac{P\cdot q}{P\cdot\ell}=\frac{Q^2}{x_B S_{ep}}, \qquad z_h=\frac{P\cdot P_h}{P\cdot q}. \tag{3}$$

The single transverse-spin asymmetry is defined as

$$A_N=\frac{\sigma(s_\perp)-\sigma(-s_\perp)}{\sigma(s_\perp)+\sigma(-s_\perp)}=\frac{d\Delta\sigma(s_\perp)}{dx_B dy dz_h dP_{h\perp}^2 d\phi}\Bigg/\frac{d\sigma}{dx_B dy dz_h dP_{h\perp}^2 d\phi}, \tag{4}$$

where ϕ is the azimuthal angle between the hadron plane defined by initial- and final-hadron and the lepton plane defined by the initial- and final-lepton.

Within the collinear factorization formalism, the leading order contribution for open charm production comes from the photon-gluon fusion subprocess at the partonic level. The single transverse-spin-dependent cross section for D meson production can be written as

$$\begin{aligned}\frac{d\Delta\sigma(s_\perp)}{dx_B dy dz_h dP_{h\perp}^2 d\phi} &= \sigma_0\int_{x_{min}}^1 dx\int\frac{dz}{z}D(z)\,\delta\left(\frac{P_{h\perp}^2}{z_h^2}-\frac{(1-\hat{x})(1-\hat{z})}{\hat{x}\hat{z}}Q^2+\hat{z}^2m_c^2\right)\\ &\times\left(\frac{1}{4}\right)\left[\varepsilon^{P_h s_\perp n\bar{n}}\left(\frac{\sqrt{4\pi\alpha_s}}{z\hat{t}}\right)\left(1+\frac{\hat{t}}{\hat{u}}\right)\right]\\ &\times\sum_{j=f,d}\sum_{i=1}^4\mathscr{A}_i\left[-x\frac{d}{dx}\left(\frac{T_G^{(j)}(x,x)}{x}\right)\hat{W}_i+\left(\frac{T_G^{(j)}(x,x)}{x}\right)\hat{N}_i\right],\end{aligned} \tag{5}$$

where $\sigma_0=e_c^2\alpha_{em}^2\alpha_s y/(8\pi z_h^2 Q^2)$, $\hat{x}=x_B/x$, $\hat{z}=z_h/z$, and e_c and m_c are the fractional charge and mass of the charm quark, respectively. The spin-averaged cross section $d\sigma$, and the functions $\mathscr{A}_i$, $\hat{W}_i$ and $\hat{N}_i$ are given in Ref. [4]. From Eq. (5), we find that the SSA for D meson production is proportional to $T_G^{(f)}+T_G^{(d)}$. We also calculate the SSA for $\bar{D}$ meson production, and find that the SSA for $\bar{D}$ has the same functional form as that for the D meson produciton except the sum of the trigluon correlation functions, $T_G^{(f)}+T_G^{(d)}$, is replaced by $T_G^{(f)}-T_G^{(d)}$. That is, the D and $\bar{D}$ meson production should have the same SSA if $T_G^{(d)} = 0$, but, with an opposite sign if $T_G^{(f)} = 0$. If both trigluon correlations vanish, there should be no significant SSA for open charm production in SIDIS.

The fully differential cross sections can be decomposed in terms of the independent angular distributions as follows,

$$\frac{d\sigma}{dx_B dy dz_h dP_{h\perp}^2 d\phi} = \sigma_0^U + \sigma_1^U \cos\phi + \sigma_2^U \cos 2\phi,$$
$$\frac{d\Delta\sigma}{dx_B dy dz_h dP_{h\perp}^2 d\phi} = \sin\phi_s \left(\Delta\sigma_0 + \Delta\sigma_1 \cos\phi + \Delta\sigma_2 \cos 2\phi\right). \quad (6)$$

In order to present the SSA and its angular dependence on the ϕ, we introduce the ϕ-integrated single spin azimuthal asymmetries: $\langle 1 \rangle = \Delta\sigma_0/\sigma_0^U$, $\langle \cos\phi \rangle = \Delta\sigma_1/2\sigma_0^U$, $\langle \cos 2\phi \rangle = \Delta\sigma_2/2\sigma_0^U$ [4]. To estimate these SSAs, we adopt the model for trigluon correlation functions $T_G^{(f,d)}$ introduced in Ref. [4]. In Fig. 1 we plot the SSAs (only the contribution from $T_G^{(f)}$) as a function of z_h for the COMPASS (left) and eRHIC (right) kinematics. The SSA, $\langle 1 \rangle$, is of the order of 10% (5%), which could be measurable at the COMPASS (eRHIC) experiment. If the contribution from $T_G^{(d)}$ is also included, the SSAs for D meson will be doubled if $T_G^{(d)} = T_G^{(f)}$ and vanish if $T_G^{(d)} = -T_G^{(f)}$. However, the situation will be opposite for $\bar{D}$ meson. So the difference between the SSAs of D and $\bar{D}$ production can uniquely separate contributions from these two trigluon correlation functions, unless both vanish.

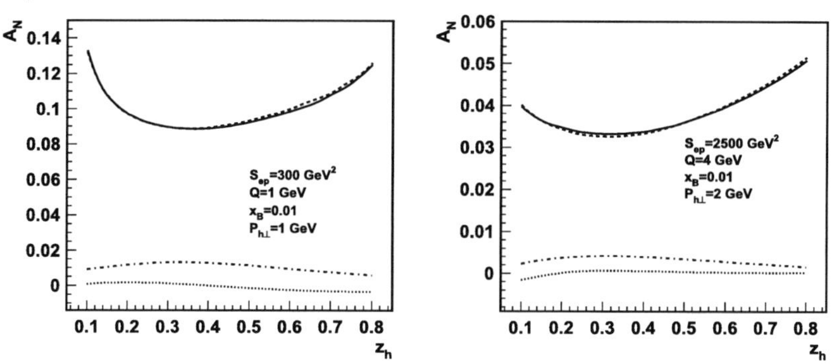

FIGURE 1. The SSAs for D^0 production in SIDIS for COMPASS (left) and eRHIC (right) kinematics. The curves are: solid-$\langle 1 \rangle$, dashed-$\langle 1 \rangle$ with derivative-term only, dot-dashed-$\langle \cos\phi \rangle$, and dotted-$\langle \cos 2\phi \rangle$.

The SSAs for open charm production in hadronic pp collisions can also be calculated similarly. Different from SIDIS process, both quark-gluon correlation function $T_{q,F}(x,x)$ and trigluon correlation functions $T_G^{(f,d)}(x,x)$ will contribute to the SSAs at the leading order. Following the same feature as in SIDIS, the SSAs for D meson ($\bar{D}$ meson) will depend on the sum (difference) of two trigluon correlation functions. In Fig. 2, we show the $P_{h\perp}$-dependence of the SSAs for both D and $\bar{D}$ mesons at RHIC kinematics. The different curves correspond to different assumptions on the two trigluon correlation functions [5]. One can clearly see that the quark-gluon correlation functions alone (dashed lines) generate only a very small asymmetry. Therefore, the observation of any sizable SSAs would be a clear evidence of nonvanishing trigluon correlations inside a polarized proton [5]. Due to the different dependence on the trigluon correlation functions, the asymmetries and the difference in the asymmetries of D and $\bar{D}$ meson production could be powerful observables for extracting both trigluon correlation functions inside a transversely polarized proton.

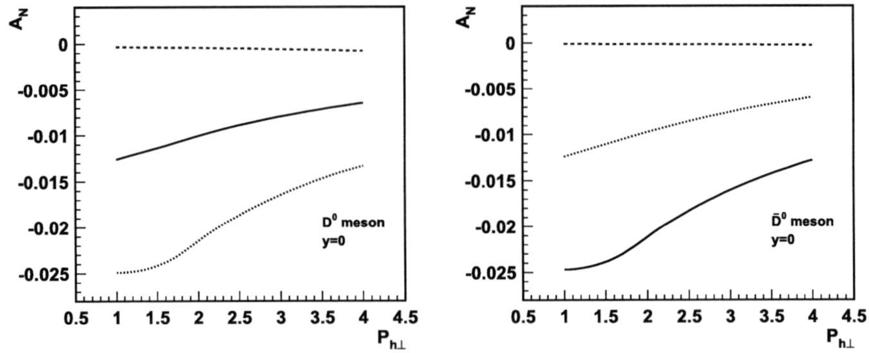

FIGURE 2. The SSA as a function of $P_{h\perp}$ for D^0 (left) and $\bar{D}^0$ mesons (right) at mid-rapidity, $y = 0$, and $\sqrt{s} = 200$ GeV. The curves are: solid ($T_G^{(d)} = T_G^{(f)}$), dashed ($T_G^{(f,d)} = 0$), dotted ($T_G^{(d)} = -T_G^{(f)}$).

In summary, we have studied the single transverse-spin asymmetries of D and $\bar{D}$ meson production in both SIDIS and pp collisions. We found that two trigluon correlation functions play important but very different role in generating the SSAs for D and $\bar{D}$ mesons. With data on SSAs for open charm production becoming available from RHIC [7] and perhaps in the future in SIDIS, we will be able to extract the trigluon correlation functions and to learn for the first time about the dynamics of quantum correlations of gluons inside a polarized proton.

Acknowledgements We thank Werner Vogelsang and Feng Yuan for discussions and collaborations. We also thank Ming Liu and Han Liu for correspondence on experimental measurements of SSAs in open charm production at RHIC. This work was supported in part by the U. S. Department of Energy under Grant No. DE-FG02-87ER40371.

REFERENCES

1. A. V. Efremov and O. V. Teryaev, Sov. J. Nucl. Phys. **36**, 140 (1982) [Yad. Fiz. **36**, 242 (1982)]; A. V. Efremov and O. V. Teryaev, Phys. Lett. B **150**, 383 (1985).
2. J. W. Qiu and G. Sterman, Phys. Rev. Lett. **67**, 2264 (1991); Nucl. Phys. B **378**, 52 (1992); Phys. Rev. D **59**, 014004 (1999).
3. X. D. Ji, Phys. Lett. B **289**, 137 (1992).
4. Z. B. Kang and J. W. Qiu, Phys. Rev. D **78**, 034005 (2008) [arXiv:0806.1970 [hep-ph]].
5. Z. B. Kang, J. W. Qiu, W. Vogelsang and F. Yuan, Phys. Rev. D **78**, 114013 (2008) [arXiv:0810.3333 [hep-ph]].
6. Z. B. Kang and J. W. Qiu, arXiv:0811.3101 [hep-ph]; Y. Koike, W. Vogelsang and F. Yuan, Phys. Lett. B **659**, 878 (2008) [arXiv:0711.0636 [hep-ph]]; and references therein.
7. H. Liu [PHENIX Collaboration], talk presented at "The 18th International Symposium on Spin Physics (Spin 2008)", Charlottesville, Virginia, Oct. 2008.

What can we learn from TMD measurements?[1]

Alessandro Bacchetta

Theory Center, Jefferson Lab, 12000 Jefferson Ave, Newport News, VA 23606, USA

Abstract. Transverse-momentum-dependent parton distribution and fragmentation functions describe the partonic structure of the nucleon in a three-dimensional momentum space. They are subjects of flourishing theoretical and experimental activity. They provide novel and intriguing information on hadronic structure, including evidence of the presence of partonic orbital angular momentum.

Keywords: parton distribution functions, semi-inclusive DIS, transverse momentum
PACS: 12.38.-t, 13.60.-r, 13.88.+e

TMDs is an acronym for Transverse Momentum Distributions or Transverse Momentum Dependent parton distribution functions, also called unintegrated parton distribution functions. Most of our knowledge of the inner structure of nucleons is encoded in parton distribution functions (PDFs). We introduce them to describe hard scattering processes involving nucleons. The presence of a hard probe in these processes — e.g., in DIS the photon with virtuality Q^2 — identifies a longitudinal direction, and a plane perpendicular to that, the transverse plane. Intuitively, standard collinear PDFs describe the probability to find in a fast-moving nucleon a parton with a specific fraction of the nucleon's longitudinal momentum. TMDs describe also the probability that the parton has a specific transverse momentum. They are therefore a natural extension of standard PDFs from one to three dimensions in momentum space.

Although useful from the intuition point of view, the probabilistic interpretation of PDFs and TMDs has some technical problems and is not strictly needed [1]. What is essential is that PDFs and TMDs can be defined in a formally clear way, through the application of factorization theorems. They reveal crucial aspects of the dynamics of confined partons, they can be extracted from experimental data, and allow us to make prediction for hard-scattering experiments involving nucleons. In this sense, the information contained in TMDs is as important as that contained in standard PDFs.

The main difference between collinear PDFs and TMDs is that the latter do not appear in totally inclusive processes. For instance, they do not appear in totally inclusive DIS, but they are needed when semi-inclusive DIS is studied and the transverse momentum of one outgoing hadron, $P_{h\perp}$, is measured. They are necessary to describe Drell–Yan processes when the transverse momentum of the virtual photon, q_T, is measured.

Factorization for processes involving TMDs has been worked out explicitly at leading twist (twist 2) and one-loop order and argued to hold at all orders [2, 3]. For instance, in

[1] Authored by Jefferson Science Associates, LLC under U.S. DOE Contract No. DE-AC05-06OR23177. The U.S. Government retains a non-exclusive, paid-up, irrevocable, world-wide license to publish or reproduce this manuscript for U.S. Government purposes.

CP1149, *Spin Physics, 18th International Spin Physics Symposium*
edited by D. G. Crabb, D. B. Day, S. Liuti, X. Zheng, M. Poelker, and Y. Prok
2009 American Institute of Physics 978-0-7354-0686-5/09/$25.00

unpolarized semi-inclusive DIS we can measure the structure function $F_{UU,T}$, which in the region $P_{h\perp}^2 \ll Q^2$ can be expressed as [4]

$$F_{UU,T} = \left|H\left(x\zeta^{1/2}, z^{-1}\zeta_h^{1/2}, \mu_F\right)\right|^2 \sum_a x e_a^2 \int d^2 p_T\, d^2 k_T\, d^2 l_T \\ \times \delta^{(2)}\left(p_T - k_T + l_T - P_{h\perp}/z\right) f_1^a(x, p_T^2; \zeta, \mu_F)\, D_1^a(z, k_T^2; \zeta_h, \mu_F)\, U(l_T^2; \mu_F) . \quad (1)$$

Apart from the transverse-momentum-dependent PDFs and fragmentation functions, the formula contains the soft factor U, a nonperturbative and process-independent object.

For the specific case of unpolarized observables integrated over the azimuthal angle of the measured transverse momentum, the analysis is usually performed in b-space in the Collins–Soper–Sterman framework [5]. The region of $P_{h\perp}^2 \gg M^2$, or $b^2 \ll 1/M^2$, can be calculated perturbatively, but when $P_{h\perp}^2 \approx M^2$ a nonperturbative component has to be introduced and its parameters must be fitted to experimental data. This component is usually assumed to be a flavor-independent Gaussian [6].

At present, especially for azimuthally-dependent structure functions, phenomenological analyses are often carried out using the tree-level approximated expression

$$F_{UU,T} = \sum_a x e_a^2 \int d^2 p_T\, d^2 k_T\, \delta^{(2)}\left(p_T - k_T - P_{h\perp}/z\right) f_1^a(x, p_T^2)\, D_1^a(z, k_T^2) . \quad (2)$$

Also in this case, the transverse-momentum dependence of the partonic functions is assumed to be a flavor-independent Gaussian [7]. The tree-level approximation and the Gaussian assumption are known to be inadequate at $P_{h\perp}^2 \gg M^2$, but they could still effectively describe the physics at $P_{h\perp}^2 \approx M^2$. Especially for low-energy experiments, this is where the bulk of the data is.

The definition of quark TMDs is [1, 3] (taking the example of the fully unpolarized distribution of a quark with flavor a)

$$f_1^a(x, p_T^2; \zeta, \mu_F) = \int \frac{d\xi^- d^2\xi_T}{(2\pi)^3}\, e^{ip\cdot\xi} \langle P|\bar{\psi}^a(0)\, \mathcal{L}^{v\dagger}_{(\pm\infty,0)}\, \gamma^+ \mathcal{L}^{v}_{(\pm\infty,\xi)}\, \psi^a(\xi)|P\rangle \bigg|_{\xi^+=0} . \quad (3)$$

The Wilson lines, $\mathcal{L}$, guarantee the gauge invariance of the TMDs. They depend on the gauge vector v and contain also components at infinity running in the transverse direction. A remarkable property of TMDs is that the detailed shape of the Wilson line is process-dependent. This immediately leads to the conclusion that TMDs are not universal. However, the situation is not hopeless and the predictive power of TMD factorization is not completely destroyed, for the following reasons

- For transverse-momentum-dependent fragmentation functions, the shape of the Wilson line appears to have no influence on physical observables [8].
- In SIDIS and Drell–Yan, the difference between the Wilson line consists in a simple direction reversal and leads to calculable effects, namely a simple sign reversal of all T-odd TMDs [9].
- In hadron-hadron collisions to hadrons, standard universality cannot be applied. It is however conceivable that only a manageable number of TMDs with distinct Wilson lines are needed, preserving part of the predictive power of the formalism [10].

quark pol.

nucleon pol.	U	L	T	
U	$\boldsymbol{f_1}$		$h_1^\perp$	
L		$\boldsymbol{g_1}$	$h_{1L}^\perp$	
T	$f_{1T}^\perp$	g_{1T}	$\boldsymbol{h_1}$	$h_{1T}^\perp$

TABLE 1. Twist-2 transverse-momentum-dependent distribution functions. The U,L,T correspond to unpolarized, longitudinally polarized and transversely polarized nucleons (rows) and quarks (columns). Functions in boldface survive transverse momentum integration. Functions in gray cells are T-odd.

- If we consider specific transverse-momentum-weighted observables instead of unintegrated observables, it should be possible to obtain factorized expressions in terms of transverse moments of TMDs multiplied by calculable, process-dependent factors [11].

Our understanding of TMDs and their extraction from data has made giant steps in the last years, thanks to new theoretical ideas and experimental measurements. In the near future, more experimental data are expected from HERMES, COMPASS, BELLE and JLab.

When the spin of the nucleon and that of the quark are taken into account, eight twist-2 functions can be introduced. They are listed in Tab. 1. As with collinear PDFs, extracting TMDs calls for global fits to semi-inclusive DIS, Drell–Yan, and e^+e^--annihilation data. Care has to be taken when considering the peculiar universality properties of TMDs. At the moment, we have some information only about the two functions in the first column of the table: f_1 (unpolarized function) and $f_{1T}^\perp$ (Sivers function).

Ultimately, the knowledge of TMDs should allow us to build tomographic images of the inner structure of the nucleon in momentum space, complementary to the impact-parameter space tomography that can be achieved by studying generalized parton distribution functions (GPDs). An example of tomographical images of the nucleon based on a model calculation of TMDs [12] is presented in Fig. 1.

TMDs measurements should allow us to address some intriguing questions, e.g.,

- Are there differences between the TMDs of different quark flavors (and of gluons)? We know that collinear PDFs are different, not only in normalization, but also in shape. We can expect that also the transverse momentum distribution is different. See Ref. [13] for an example of an experimental analysis of this issue.
- How does the transverse momentum dependence change with x? Such a dependence has already been introduced to describe data at low x [6].
- Does the transverse momentum dependence of fragmentation functions change for different quark flavors and different produced hadrons?
- Are there reasons to abandon a Gaussian ansatz? We know that this assumption fails at high transverse momentum, but there are no compelling reasons to take a Gaussian shape even for the low-transverse-momentum, nonperturbative region.

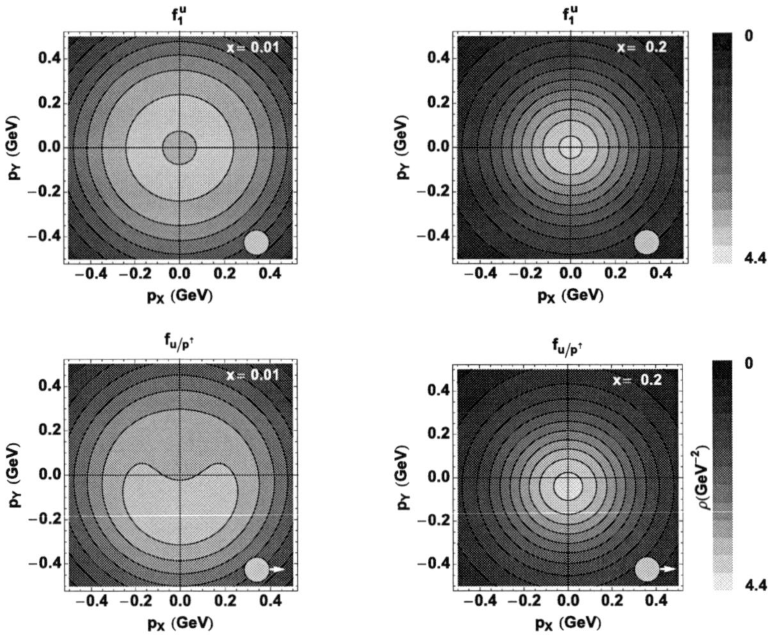

FIGURE 1. Momentum-space tomographic "images" of the up quarks in a nucleon obtained from a model calculation of TMDs [12]. The circle with the arrow indicates the nucleon and its spin orientation. The distortion in the lower panels is due to the Sivers function. In the future, it should be possible to reconstruct these images from experimental data.

The last item of the list is connected also to another fundamental issue that makes TMDs interesting, i.e., the observation of partonic orbital angular momentum. In non-relativistic quantum mechanics, it is well known that wavefunctions with orbital angular momentum vanish at zero momentum. This is a general statement independent of the specific potential in which the wavefunction is computed. This feature is reflected also in TMDs: contributions from partons with nonzero angular momentum have to vanish at zero transverse momentum (and therefore cannot be described by a simple Gaussian). In general, a downturn of a TMD going to zero transverse momentum signals the presence of nonzero orbital angular momentum. While this effects could barely be visible in unpolarized TMDs, certain combinations of polarized TMDs could filter out more clearly the configurations with nonzero orbital angular momentum. Fig. (2) shows an example of this phenomenon, using a model calculation for illustration purposes.

Apart from the details of their shape, all the TMDs that are not boldface in Tab. 1 vanish in the absence of orbital angular momentum due to angular momentum conservation. Measuring any one of them to be nonzero is already an unmistakable indication of the presence of partonic orbital angular momentum. We know already from other sources (in particular the measurement of nucleons' anomalous magnetic moments) that partonic orbital angular momentum is not zero, however TMDs have the advantage that

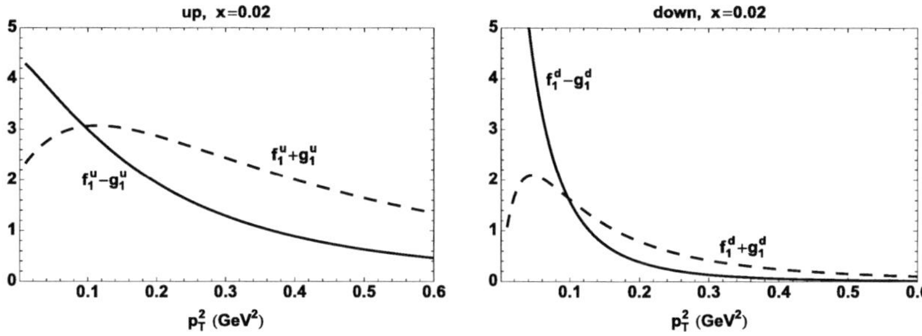

FIGURE 2. An illustration of how the presence of orbital angular momentum can influence the shape of TMDs. The model calculation shows different combinations of the f_1 and g_1 TMDs for u and d quark at $x = 0.02$. The downturns for $p_T^2 \to 0$ are due to the presence of orbital angular momentum.

they can be flavor-separated and that they are x dependent. Thus, they allow us to say if orbital angular momentum is present for each quark flavor and for gluons, and at each value of x.

If stating that a fraction of partons have nonzero orbital angular momentum is relatively simple, it is not easy to make a quantitative estimate of the net partonic orbital angular momentum using TMDs. Any statement in this direction is bound to be model-dependent. Generally speaking, TMDs have to be computed in a model and the parameters of the model have to be fixed to reproduce the TMDs extracted from data. Then, the total orbital angular momentum can be computed in the model. Unfortunately, it is possible that two models describe the data equally well, but give two different values for the total orbital angular momentum.

As an example of a procedure of this kind, let us take the measurement of the Sivers function. We know that the proper way to measure the quark total angular momentum is by measuring the combination [14]

$$J^a = \int_0^1 dx\, x \left(H^a(x,0,0) + E^a(x,0,0)\right), \tag{4}$$

where the definition of the generalized parton distribution E in terms of light-cone wavefunctions is

$$\begin{aligned} E(x,0,0) = \lim_{q_T \to 0} \Bigg(-\frac{1}{q_x - i q_y} \int \frac{d^2 p_T}{16\pi^2} \Big[& \psi_+^{+*}(x,p_T)\, \psi_+^-(x, p_T + (1-x) q_T) \\ & + \psi_-^{+*}(x,p_T)\, \psi_-^-(x, p_T + (1-x) q_T) \Big] \Bigg). \end{aligned} \tag{5}$$

On the other hand, the definition of the Sivers function in terms of light-cone wavefunctions can be written as

$$f_{1T}^{\perp}(x,p_T) = \frac{1}{16\pi^3} \mathrm{Im}\Big[\psi_+^{+*}(x,p_T)\, \psi_+^-(x,p_T) + \psi_-^{+*}(x,p_T)\, \psi_-^-(x,p_T) \Big]. \tag{6}$$

In spite of the similarities between the two expressions and the fact that the same light-cone wavefunctions are involved, in general there is no straightforward connection between the Sivers function and the GPD E [15]. Nevertheless, in a certain class of spectator models it turns out that

$$f_{1T}^{\perp a}(x) = -L(x)\, E^a(x,0,0). \qquad (7)$$

Exploiting this very simple relation and using for illustration purposes the results of the Sivers function fit from Ref. [16] we obtain

$$\frac{E^a(x,0,0)}{E^u(x,0,0)} = \frac{f_{1T}^{\perp a}(x)}{f_{1T}^{\perp u}(x)} = \frac{A_a}{A_u}\frac{f_1^a(x)}{f_1^u(x)}, \qquad (8)$$

where

$$A_d/A_u = -1.8, \quad A_{\bar{u}}/A_u = -1.1, \quad A_{\bar{d}}/A_u = 1.3, \quad A_s/A_u = -A_{\bar{s}}/A_u = -4.8. \qquad (9)$$

Although assumption-based, the above analysis shows that the measurement of the Sivers function can be used to give interesting constraints on the GPD E and ultimately on the amount of total orbital angular momentum for each flavor.

In summary, TMDs open new dimensions in the exploration of the partonic structure of the nucleon. They require challenging extensions of the standard formalism used for collinear parton distribution functions, leading us to a deeper understanding of QCD. Among other things, they give evidence of the presence of partonic orbital angular momentum and, with model assumptions, they can help constraining its size.

REFERENCES

1. J. C. Collins, *Acta Phys. Polon.* **B34**, 3103 (2003).
2. J. C. Collins, and D. E. Soper, *Nucl. Phys.* **B193**, 381 (1981).
3. X. Ji, J.-P. Ma, and F. Yuan, *Phys. Rev.* **D71**, 034005 (2005).
4. A. Bacchetta, D. Boer, M. Diehl, and P. J. Mulders, *JHEP* **08**, 023 (2008).
5. J. C. Collins, D. E. Soper, and G. Sterman, *Nucl. Phys.* **B250**, 199 (1985).
6. F. Landry, R. Brock, P. M. Nadolsky, and C. P. Yuan, *Phys. Rev.* **D67**, 073016 (2003).
7. U. D'Alesio, and F. Murgia, *Phys. Rev.* **D70**, 074009 (2004) .
8. J. C. Collins, A. Metz, *Phys. Rev. Lett.* **93**, 252001 (2004); F. Yuan, *Phys. Rev.* **D77**, 074019 (2008).
9. J. C. Collins, *Phys. Lett.* **B536**, 43–48 (2002).
10. J. Collins, and J.-W. Qiu, *Phys. Rev.* **D75**, 114014 (2007); W. Vogelsang, and F. Yuan, *Phys. Rev.* **D76**, 094013 (2007).
11. A. Bacchetta, C. J. Bomhof, P. J. Mulders, and F. Pijlman, *Phys. Rev.* **D72**, 034030 (2005); C. J. Bomhof, and P. J. Mulders, *JHEP* **02**, 029 (2007); A. Bacchetta, C. Bomhof, U. D'Alesio, P. J. Mulders, and F. Murgia, *Phys. Rev. Lett.* **99**, 212002 (2007).
12. A. Bacchetta, F. Conti, and M. Radici, *Phys. Rev.* **D78**, 074010 (2008).
13. H. Mkrtchyan, et al., *Phys. Lett.* **B665**, 20–25 (2008).
14. X. Ji, *Phys. Rev. Lett.* **78**, 610–613 (1997).
15. S. Meissner, A. Metz, and K. Goeke, *Phys. Rev.* **D76**, 034002 (2007).
16. S. Arnold, A. V. Efremov, K. Goeke, M. Schlegel, and P. Schweitzer (2008), arXiv:0805.2137.

Measurement of Single Target-Spin Asymmetry in Semi-Inclusive Pion Electroproduction on a Transversely Polarized ^{3}He Target

Yi Qiang[1]

Duke University, Durham, NC 27705

Abstract. We recently measured the neutron target single spin asymmetry in the semi-inclusive deep inelastic $^3\overrightarrow{He}(e,e'\pi^{\pm})X$ reactions with a transversely polarized ^{3}He target. The experiment was performed in Hall A at Jefferson Lab from October 2008 to February 2009 using a 40-cm polarized ^{3}He target. The pions were detected in the left high-resolution spectrometer in coincidence with the scattered electrons detected by the BigBite spectrometer. The kinematic coverage focuses on the valence quark region, $x = 0.13 - 0.41$, at $Q^2 = 1.31 - 3.10$ (GeV$/c)^2$. With good particle identifications using a RICH detector and an Aerogel threshold Cherenkov counter, data on the kaons were obtained at the same time. The data from this experiment, when combined with the world data, will provide constraints on the transversity and Sivers distributions on both u-quark and d-quark in the valence quark region.

Keywords: Transversity, TMD, neutron
PACS: 14.20.Dh,

INTRODUCTION

The unpolarized parton distribution function (PDF), $f_1(x,Q^2)$, has been extracted with excellent precision over a large range of the longitudinal momentum fraction x from deep inelastic scattering (DIS), Drell-Yan and other processes after several decades of experimental and theoretical efforts. Motivated by the original "spin crisis" from the EMC experiment in the 1980s, the longitudinal polarized parton distribution function, $g_1(x,Q^2)$, has been determined with significantly improved precision over a large region of x and Q^2 from polarized DIS experiments in the last two decades. What remains least known among the three leading twist parton distribution functions is the transversity function, $h_1(x,Q^2)$. It was first discussed by Ralston and Soper in 1979 [1] and later by Jaffe and Ji [2] in 1991. In the quark-parton models, the nucleon transversity distribution describes the net quark transverse polarization in a transversely polarized nucleon. In the non-relativistic limit, the transversity distribution function is the same as $g_1(x,Q^2)$. Therefore, the transversity distribution function probes the relativistic nature of the quarks inside the nucleon.

Besides these three parton distributions (f_1, g_1 and h_1), there are five more transverse momentum dependent distribution functions (TMDs) at leading twist [3, 4]. TMDs give the description of the parton distribution beyond the collinear approximation. They

[1] for Jefferson Lab Hall A and E06-010 Collaborations

CP1149, *Spin Physics, 18th International Spin Physics Symposium*
edited by D. G. Crabb, D. B. Day, S. Liuti, X. Zheng, M. Poelker, and Y. Prok

depend not only on x, but also on the transverse momentum (k_T) of the quark.

Because of the chiral-odd nature of transversity, its experimental determination requires processes where an additional chiral-odd function is involved. Double polarized Drell-Yan processes which involves two transversity distributions and single target spin azimuthal asymmetries from semi-inclusive DIS pion electro-production in which transversity is convoluted with a chiral-odd fragmentation function, called the Collins function [5] are two methods to measure transversity.

Considering the differential cross section in a single spin semi-inclusive DIS (SIDIS) $(e,e'h)$ reaction with a transversely polarized target, the transversity function convoluted with Collins fragmentation function gives rise to an azimuthal asymmetry in $sin(\phi_h + \phi_S)$, where azimuthal angles of both the hadron (ϕ_h) and the target spin (ϕ_S) are about the virtual photon axis and relative to the lepton scattering plane. Other than this, three more leading twist TMDs contribute to the azimuthal asymmetry.

- the Boer-Mulder function [4] has an azimuthal angular dependence in $sin(2\phi_h)$ due to the transversely polarized quarks inside an unpolarized nucleon;
- the Sivers function [6] refers to the azimuthal asymmetry of $sin(\phi_h - \phi_S)$ due to the correlation between the nucleon's transverse spin and the quark's transverse momentum, which involves the quark orbital motion;
- the Pretzelosity, with a $sin(3\phi_h - \phi_S)$ asymmetry in SIDIS, is the difference between the helicity and transversity distribution and measures the relativistic effects in the nucleon [7].

The orthogonal nature of these four azimuthal asymmetries gives them an unambiguous separation in SIDIS.

Recently, the HERMES Collaboration [8] and the COMPASS Collaboration [9, 10, 11] reported results on single-spin asymmetries from SIDIS on transversely polarized proton and deuteron targets. The asymmetry related to the Collins fragmentation function and the quark transversity distribution was also measured by the BELLE Collaboration [12] in the $e^+e^- \to h_1h_2X$ reaction. Currently, the world data from different collaborations on the proton Collins asymmetry are in good agreement. However, the preliminary Sivers asymmetry of positively charged hadrons on proton target from COMPASS [11] is not consistent with the HERMES result [8].

THE EXPERIMENT

The goal of the reported experiment, E06-010 [13], is to get the first direct data of the neutron single spin asymmetry in the SIDIS $\overrightarrow{^3He}(e,e'\pi^{\pm})X$ reactions on a transversely polarized ^{3}He target. It was performed in Jefferson Lab Hall A using a 6 GeV electron beam.

The experiment used the Hall A left side High-Resolution-Spectrometer (HRS_L) at 16° as the hadron arm. The HRS_L was equipped with standard vertical drift chambers (VDCs) and two scintillator planes to provide tracking and triggers. An Aerogel threshold Cherenkov counter ($n = 1.015$) and a ring imaging Cherenkov detector ($n = 1.28$) provided hadron identification (π, K and p) while a short gas Cherenkov detector and two layers of lead glass were installed for e/π discrimination. Therefore, the asymmetries in the kaon electro-production were measured as well during this experiment.

The large acceptance BigBite spectrometer was put at 30° beam-right as the electron arm. Three wire chambers including 18 wire planes in total were installed to provide precise tracking of charged particles. The combination of the pre-shower and shower lead glass detectors were used to identify electrons. In addition to these, a newly assembled gas Cherenkov detector was also installed to give extra electron PID information.

The kinematics were selected to cover the valence quark region and the highest possible invariance mass of the whole hadronic system with a 6 GeV beam, $0.135 < x < 0.405$ and $2.33 < W < 3.05$ GeV/c^2, corresponding to $1.31 < Q^2 < 3.10$ (GeV/c)^2. It was also chosen to detect the leading fragmentation pion which carries $z \approx 0.5$ of the energy transfer to favor the current fragmentation region. The value of hadronic missing mass was chosen to be as high as possible with a cut of $W' > 1.5$ GeV/c^2 to avoid contributions from resonance production channels. Therefore, the hadron arm momentum setting was fixed at $p = 2.4$ GeV/c. Because of the large momentum bite of the BigBite spectrometer ($\sim$ 64 msr), only one BigBite setting is needed to cover the whole kinematics range.

Three 40-cm Rb-K hybrid high pressure ^{3}He cells were used throughout the experiment. With optical pumping using traditional diode laser arrays, these newly developed hybrid cells have already been demonstrated to have significantly shorter spin-up times and higher polarization compared to old Rb only cells (6 hours and 55% compared to 20 hours and 45%). To have even more efficient optical pumping, 3 COMET narrow band lasers were used for the first time in Hall A. As a result, the spin up time and maximum polarization were improved to 4 hours and 70%, repectively. The whole target system was also redesigned to allow optical pumping in all three directions: longitudinal, transverse and vertical respected to the electron scattering plane. During the production of this experiment, the beam time was evenly distributed into the four configurations: transverse and vertical pumpings combined with negative and positive hadron charges. The target spin was also flipped every 20 minutes using adiabatic fast passage (AFP) to have the whole 2π coverage of ϕ_S and reduce systematic errors. Thanks to the very short spin up time, the polarization drop caused by the AFP loss was minimized and the in-beam average target polarization throughout the experiment was about 65% (compared to 42% in the original proposal)!

SUMMARY

Started from Oct 22, 2008, the first experiment on the SIDIS neutron single spin asymmetry using a transversely polarized ^{3}He ran smoothly through Feb 5, 2009 with a 15 μA beam in Jefferson Lab Hall A. The total statistics has reached the proposed goal. The data will allow us to extract the three azimuthal asymmetries from neutron related to Transversity, Sivers and Pretzelosity distribution functions. The projected statistics of the Collins and Sivers asymmetries are shown in Figure 1. When combined with world data, the results will provide important constraints on the TMDs on both u, d and s quarks in the valence quark region.

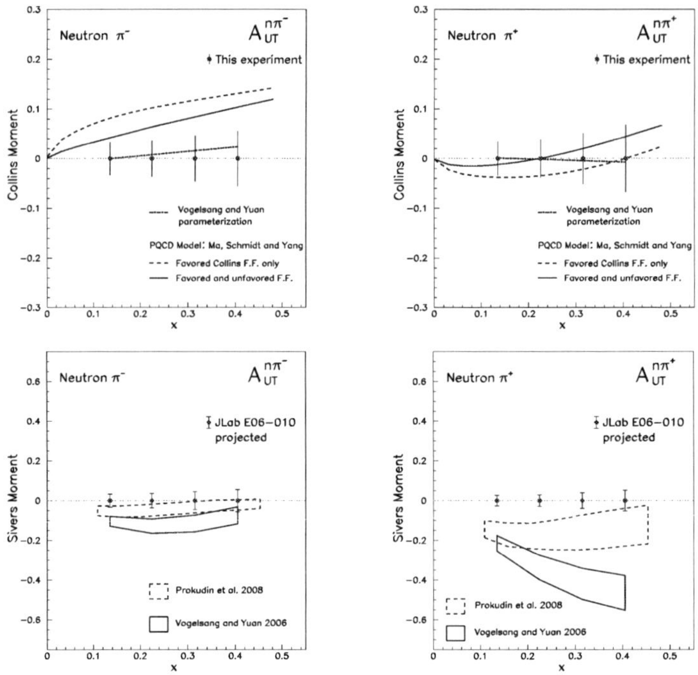

FIGURE 1. Projected statistics on Collins and Sivers asymmetries

ACKNOWLEDGMENTS

This work was supported by the US Department of Energy DE-FG02-03ER41231, under which the Jefferson Science Associates operates the Thomas Jefferson National Accelerator Facility.

REFERENCES

1. J. P. Ralston, and D. E. Soper, *Nucl. Phys.* **B152**, 109 (1979).
2. R. L. Jaffe, and X. Ji, *Phys. Rev. Lett.* **67**, 552 (1991).
3. P. J. Mulders, and R. D. Tangerman, *Nucl. Phys.* **B461**, 197 (1996).
4. D. Boer, and P. J. Mulders, *Phys. Rev.* **D57**, 5780 (1998).
5. J. C. Collins, *Nucl. Phys.* **B396**, 161 (1993).
6. D. Sivers, *Phys. Rev.* **D43**, 261 (1991).
7. H. Avakian, A. V. Efremov, P. Schweitzer, and F. Yuan, *Phys. Rev.* **D78**, 114024 (2008).
8. the HERMES Collaboration, *Phys. Rev. Lett* **94**, 012002 (2005).
9. the COMPASS Collaboration, *Phys. Rev. Lett.* **94**, 202002 (2005).
10. the COMPASS Collaboration, *Nucl. Phys.* **B765**, 31 (2007).
11. the COMPASS Collaboration, Proceeding for transversity 2008 workshop (2008).
12. the BELLE Collaboration, *Phys. Rev. Lett.* **96**, 232002 (2006).
13. Jlab experiment e06-010 (2006), spokespersons: X. Jiang, J. P. Chen, E. Cisbani, H. Gao and J. C. Peng.

Study of semi-inclusive deep inelastic $(e,e'\pi^{\pm})$ production with a polarized ^{3}He target

X. Qian[1], H. Gao[1], J.-P. Chen[2], E. Chudakov[2]

[1]*Duke University and the Triangle Universities Nuclear Laboratory, Durham, NC*
[2]*Jefferson Lab, Newport News, VA*

Abstract. The 12 GeV upgrade at Jefferson Lab will open a new era of precision measurement of the Transverse Momentum Dependent parton distribution functions (TMDs) through the Semi-Inclusive Deep-Inelastic Scattering (SIDIS). In this talk, we discussed the projected results on measurements of TMDs using a newly proposed large acceptance solenoid detector together with the standard Hall A polarized ^{3}He target. The high luminosity and the large acceptance detector will allow for a precise 3-D (x, z and P_T) mapping of azimuthal asymmetries. Furthermore, the full 2π azimuthal angular coverage for the solenoid detector is essential to control the relevant systematic uncertainties in extracting the azimuthal asymmetries. With a 40 cm long transversely polarized ^{3}He target, a total of 60 days of beam time at a beam current of 15 μA, we can determine tensor charge of d quark to 10% in a model independent way. The extracted Sivers asymmetry will provide important information to understand the correlation between the quark orbital angular momentum and the nucleon spin.

Keywords: Single transverse-spin asymmetries
PACS: 12.38.Bx, 12.39.St, 13.85.Qk

At leading twist if we integrate over the transverse momenta of quarks, there are three quark distribution functions: the unpolarized parton distribution f_1, the longitudinal polarized parton distribution g_1, and the quark transversity distribution, h_1. The transversity function, a chirally odd quark distribution function, is the least known among the three leading twist parton distribution functions. It was discussed for the first time by Hidaka, Monsay and Sivers [1] in 1978, by Ralston and Soper [2] in 1979, and later by Jaffe and Ji in early 1990s [3]. The lowest moment of h_1^q measures the "tensor charge", which is a fundamental property of the nucleon, analogous to the axial charge, and has been calculated from lattice QCD [4] and various models [5, 6, 7, 8].

Besides the three parton distributions which survive the integration over the quark transverse momentum (f_1, g_1 and h_1), there are five more transverse momentum dependent distribution functions (TMDs) at leading twist [9, 10]. TMDs give the description of the parton distribution beyond the collinear approximation. They depend not only on the longitudinal momentum fraction x, but also on the transverse momentum (k_T). Both the Sivers function $f_{1T}^{\perp}$ and the Boer Mulders function $h_1^{\perp}$ require wave function components with nonzero orbital angular momentum and thus provide information about the correlation between the quark orbital angular momentum (OAM) and the nucleon/quark spin, respectively. Furthermore, they are T-odd functions which rely on the final state interactions (FSI) experienced by the active quark in a SIDIS experiment as both functions vanish without FSI. Combining the wealth of information from all these TMDs could thus be invaluable for disentangling the spin orbit correlations in the nucleon wave func-

CP1149, *Spin Physics, 18th International Spin Physics Symposium*
edited by D. G. Crabb, D. B. Day, S. Liuti, X. Zheng, M. Poelker, and Y. Prok

tion, thus providing important information about the quark orbital angular momentum. Together with Generalized Parton distributions (or Impact Parameter Dependent distributions), which describe the probability of finding a parton with certain longitudinal momentum fraction and at certain transverse position b, TMDs give a complementary 3-D description of the nucleon structure. All eight TMDs can be accessed through the SIDIS [11].

The measurements of single target spin azimuthal asymmetry have been performed by pioneer works in HERMES [12, 13], COMPASS [14, 15, 16] and JLab [17, 18]. The Collins asymmetry is sizeable for proton and it has opposite sign for π^+ and π^-. The Collins asymmetry for deuteron is consistent with zero from COMPASS. HERMES shows a non-zero Sivers asymmetry for π^+ on proton, while the π^- on proton and the asymmetries from deuteron (COMPASS) are consistent with zero. The measurements of TMDs lead to very active theoretical and experimental studies.

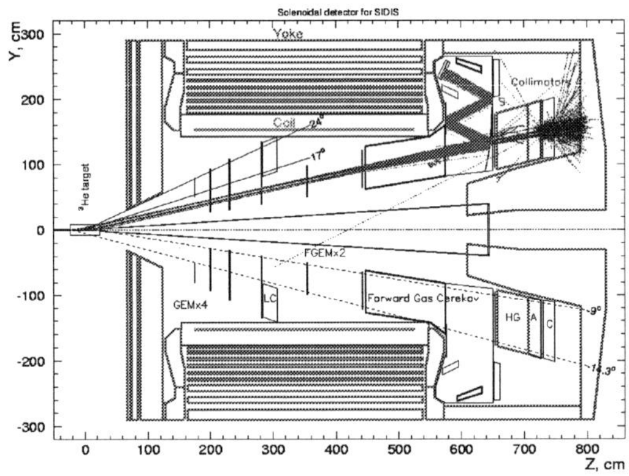

FIGURE 1. The experimental layout of the SoLID.

The aforementioned studies play important roles in exploring the TMDs of nucleon, in particular for transversity and Sivers distribution functions. They established the existence and some important features of the transversity, Sivers distribution and the Collins fragmentation function for both proton and neutron. However, compared with the unpolarized and longitudinal polarized parton distribution, the knowledge on the transversity is rather limited. We propose to use a large acceptance solenoid detector together with a high luminosity polarized ^3He target to measure the SSAs on neutron. The solenoid spectrometer is being proposed for a 11 GeV PVDIS experiment [19] and 11 GeV TRANSVERSITY experiment [20]. The experimental setup can be found in Fig. 1. The polar angle coverage for the forward angle detector (for both electron and pion detection) is from 8.5° to 16° and the momentum coverage is from 1.0 GeV/c to 7.0 GeV/c. The total solid angle is about 95 msr for this momentum coverage. GEM detectors will be used as tracking detectors. A combination of an electromagnetic calorimeter, gas Čerenkov counters and an aerogel Čerenkov counter will be used for particle identifications. PMTs of the calorimeter and Čerenkov counters will be placed in a low magnetic field region outside the magnetic coils. The polar angle coverage for the large angle detector is from 16° to 25°. They are mainly used for the electron detections in a momentum range of 4.0-6.0 GeV/c. The total solid angle is about 280 msr for this momentum range.

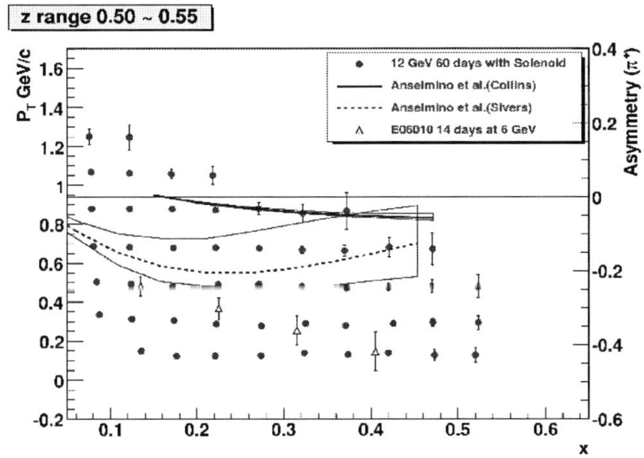

FIGURE 2. The projections at $0.5 < z < 0.55$ bin. The left axis shows the transverse momentum for each bin. The right axis shows the size of asymmetry. In addition, we also plotted the projections of the ongoing E06-010 experiment as the open triangle points.

With a 15 uA electron beam on a 40 cm long, 10 amgs transversely polarized ^{3}He target for 60 days, the projected results at $0.55 > z > 0.5$, are shown in Fig. 2 together with theoretical predictions from Anselmino *et al.*[21] and Vogelsang and Yuan [22] for both Collins and Sivers asymmetry. There are eight z bins in total and only one of them is shown in here. The precision measurement will allow for a determination of the tensor charge of d quark to 10% in a model independent way and providing important informations on Sivers and Pretzelosity asymmetries. The double spin asymmetries with a longitudinal polarized beam will allow for a parasitic measurement of the A_{LT} which can be used to access another leading twist distribution, g_{1T}. With a longitudinal polarized target, we can also access A_{LL} and A_{UL} which are related to g_{1L} and $h_{1L}^{\perp}$.

The proposed measurements with the solenoid detector will provide a unprecedented precision 3-D (x, z and P_T) mapping of SSA. It will also allow a complete factorization studies with x, z and P_T dependence. The results from this proposed measurement, when combined with future proton data, can extract various TMDs for both u, d and sea quarks. The combined analysis will lead to a 10% determination of the tensor charge. Furthermore, information on Sivers/Pretzelosity TMDs will reveal important correlations between the spin and the orbital angular momentum.

This work is supported in part by the U.S. Department of Energy and the National Science Foundation under contract number DE-FG02-03ER41231, and other DOE and NSF grants.

REFERENCES

1. K. Hidaka, E. Monsay and D. Sivers, Phys. Rev. **D19**, 1503 (1979).
2. J.P. Ralston and D.E. Soper, Nucl. Phys. **B152**, 109 (1979).
3. R.L. Jaffe and X. Ji, Phys. Rev. Lett. **67**, 552 (1991).
4. M. Gockeler *et al.*, Phys. Lett. **B627**, 113 (2005).
5. I. C. Cloet, W. Bentz and A. W. Thomas, Phys. Lett. **B659**, 214 (2008).
6. M. Wakamatsu, Phys. Lett. **B653**, 398 (2007).
7. H.-X. He and X. Ji, Phys. Rev. **D52**, 2960 (1995).
8. B. Pasquini, M. Pincetti, and S. Boffi, Phys. Rev. **D72**, 094029 (2005).
9. P. J. Mulders and R. D. Tangerman, *Nucl. Phys.* **B461** 197 (1996).
10. D. Boer, P.J. Mulders, Phys. Rev. **D57**, 5780 (1998).
11. A. Bacchetta, *et al.*, JHEP, 0702, 093 (2007), hep-ph/0611265.
12. A. Airapetian *et al.*, (HERMES), Phys. Rev. Lett. **84**, 4047 (2002); Phys. Rev. D **64**, 097101 (2001); Phys. Lett. **B562**, 182 (2003).
13. A. Airapetian *et al.* (HERMES), Phys. Rev. Lett. **94**, 012002 (2005).
14. The COMPASS Collaboration, Phys. Rev. Lett. **94**, 202002 (2005).
15. E.S. Ageev *et al.*, the COMPASS Collaboration, Nucl. Phys. **B765**, 31 (2007), hep-exp/0610068.
16. S. Levorato, for the COMPASS Collaboration, arXiv:0808.0086.
17. H. Avakian *et al.*, for the CLAS Collaboration, Phys. Rev. **D69**, 112004 (2004), hep-ex/0301005.
18. JLab Experiment E06-010, Spokespersons: J. P. Chen, E. Cisbani, H. Gao, X. Jiang and J. C. Peng.
19. JLab 12 GeV PVDIS proposal PR-09-012.
20. JLab 12 GeV TRANSVERSITY proposal PR-09-014.
21. M. Anselmino and A. Prokudin, private communications. Predictions are based on the extractions of Ref. e-Print: arXiv:0807.0173.
22. W. Vogelsang and F. Yuan, private communications.

Beam spin asymmetries from semi-inclusive pion electroproduction in deep inelastic scattering

W. Gohn*, H. Avakian†, K. Joo* and M. Ungaro*

*University of Connecticut, Storrs, CT
†Jefferson Lab, Newport News, VA

Abstract. Measurements of beam-spin asymmetries from semi-inclusive pion electro-production in deep inelastic scattering (DIS) have been performed. Data were taken with the CEBAF Large Acceptance Spectrometer (CLAS) at Jefferson Lab using a 5.7 GeV longitudinally polarized electron beam and an unpolarized liquid hydrogen target. All three pion channels (π^+, π^0 and π^-) were measured simultaneously over a large range of kinematics at the intermediate Q^2 range ($Q^2 \approx 1$ - 3 GeV^2). We present our preliminary studies of single-spin azimuthal asymmetries of the two charged pion channels as functions of z and P_T. This new high statistical data will provide important means of studying physics underlying the spin structure of the nucleon.

Keywords: SIDIS,beam-spin asymmetry
PACS: 14.20.Dh, 13.85.Ni

INTRODUCTION

Beam-spin asymmetries (BSA) from pion electroproduction in semi-inclusive deep-inelastic scattering (SIDIS) can provide important information pertaining to the spin structure of the proton. Using a polarized electron beam and an unpolarized target in the CLAS detector, the $\sin\phi$ moment can be measured via its direct relationship to the BSAs. This analysis is being performed for each of the three pion channels, π^+, π^0, and π^- with high statistics in each.

Assuming single photon exchange, the single pion electroproduction cross section can be expressed in a model-independent way by a set of structure functions. The cross section as a function of the azimuthal angle ϕ between the lepton scattering plane and hadron production plane [1] can be written as:

$$d\sigma = d\sigma_0(1 + A_{UU}^{\cos\phi}\cos\phi + A_{UU}^{cos2\phi}\cos 2\phi + \lambda_e A_{LU}^{\sin\phi}\sin\phi) \tag{1}$$

with beam helicity $\lambda_e = \pm 1$. A_{UU} and A_{LU} are ratios of corresponding structure functions for an unpolarized target with unpolarized or polarized lepton beams, respectively. We measure the z and P_T dependence of $A_{LU}^{\sin\phi}$ by fitting the BSA, which is calculated by:

$$BSA = \frac{1}{P_e}\frac{N^+ - N^-}{N^+ + N^-} \tag{2}$$

where P_e is the average beam polarization and $N^{+(-)}$ is the number of positive (negative) helicity events. Relevant physical quantities can be extracted by writing the $\sin\phi$ moment in terms of structure functions as: $A_{LU}^{\sin\phi} = F_{LU}^{\sin\phi}/F_{UU,T}$.

CP1149, *Spin Physics, 18th International Spin Physics Symposium*
edited by D. G. Crabb, D. B. Day, S. Liuti, X. Zheng, M. Poelker, and Y. Prok

$F_{LU}^{\sin\phi}$ is higher-twist by nature [2]. Presently, it is not known whether factorization proven for leading-twist observables in SIDIS [3, 4] holds also at subleading twist. Assuming factorization, the sub-leading observables provide unique insights into the quark-gluon-quark dynamics. Recently such higher-twist effects have been partly related to the average transverse force acting on quarks just after the interaction with the virtual photon in the collision [5]. Specifically, the LU term is due to the imaginary part of the interference between longitudinal and transverse photon amplitudes. The structure function $F_{LU}^{\sin\phi}$ is given in a tree-level formalism in terms of twist-3 TMD's by [6]:

$$F_{LU}^{\sin\phi} = \sqrt{2\varepsilon(1+\varepsilon)}\frac{2M}{Q}[-\frac{\hat{\mathbf{h}}\cdot\mathbf{k_T}}{M_h}(xeH_1^{\perp}+\frac{M_h}{M}f_1\frac{G^{\perp}}{z})+\frac{\hat{\mathbf{h}}\cdot\mathbf{p_T}}{M}(xg^{\perp}D_1+\frac{M_h}{M}h_1^{\perp}\frac{E}{z})] \quad (3)$$

In the above definition, e is a twist-3 parton distribution function (PDF), $H_1^{\perp}$ is the naive time-reversal odd Collins fragmentation function, $f_1(x)$ is a PDF, $g^{\perp}$ is the T-odd distribution function, which describes longitudinally polarized quarks in an unpolarized proton, D_1 is a fragmentation function, $h_1^{\perp}$ is the leading-order transverse momentum dependent (TMD) parton distribution, and E and $G^{\perp}$ are twist-3 fragmentation functions that respectively describe fragmentation of transversely and longitudinally polarized quarks to hadrons.

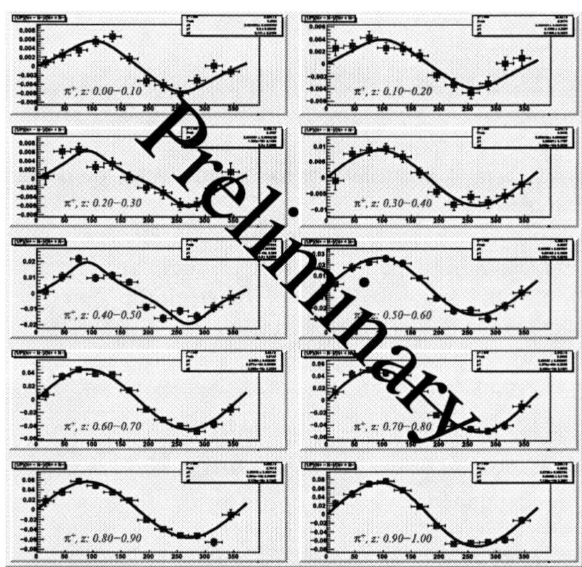

FIGURE 1. *Preliminary:* BSA as a function of ϕ. $A_{LU}^{\sin\phi}$ is determined for π^+ as a function of z by fitting the BSA with the function $\frac{A\sin\phi}{1+B\cos\phi+C\cos2\phi}$ in each z bin. The $\sin\phi$ moment is then given by the A coefficient from the fit. Error bars are statistical only.

EXPERIMENTAL PROCEDURES

The CEBAF Large Acceptance Spectrometer (CLAS) is used to study three semi-inclusive processes, $ep \to e\pi^+X$, $ep \to e\pi^-X$, and $ep \to e\pi^0X$, each with high statistics. An event is only considered as a semi-inclusive candidate if $W > 2$ GeV and $Q^2 > 1GeV^2$, where $Q^2 = -q^2$ and $W^2 = (q+P)^2$. Here $q = k - k'$ and P is the four-vector of the targeted proton. The four-vectors k and k' represent the incident and scattered electrons respectively. Other important kinematic variables are P_T, the transverse component of the hadron's momentum in the center-of-mass frame, and $z = \frac{E_h}{\nu}$, where E_h is the energy of the scattered hadron and $\nu = E - E'$, the energy lost by the scattered electron.

The experiment ran during the e1f run period between April and July of 2003. A polarized electron beam with an energy of 5.496 GeV and average polarization of 75.1% was scattered off of an unpolarized liquid hydrogen target. CLAS uses a toroidal magnetic field produced by six superconducting coils to affect the trajectory of charged particles. The current in the coils was reduced to 60% of its nominal value to increase acceptance of pion events of both charges. CLAS contains thirty-four layer drift

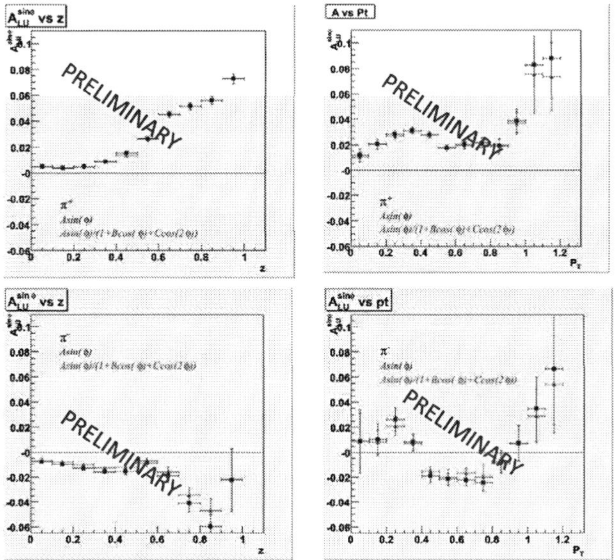

FIGURE 2. Preliminary results of $A_{LU}^{\sin\phi}$. The two plots on top show $A_{LU}^{\sin\phi}$ (square markers) for π^+ as a function of z and P_T respectively. The two plots below show the same for π^-. A comparison is made to fits with the function $A\sin\phi$ (trianglular markers). π^0 results are forthcoming, but similar CLAS results can be seen in the talk by M. Aghasyan *et al.* in these proceedings. Error bars show statistical error only.

chambers to provide information such as a particle's momentum and charge. Electron identification is carried out by an electromagnetic calorimeter and Cerenkov counter, and particle timing information comes from a time-of-flight detector. This timing combined with the particle's charge and momentum are used to determine the particle's mass, and hence identify which type of particle is detected. π^0's were detected by measuring two photons and reconstructing their invariant mass. Using these methods, approximately 6.4 million π^+, 1.4 million π^-, and 3.2 million π^0 events were recorded.

PRELIMINARY RESULTS

$A_{LU}^{\sin\phi}$ is obtained by fitting the data with the function $\frac{A\sin\phi}{1+B\cos\phi+C\cos2\phi}$. The data is first divided into ten bins in z and twelve bins in P_T, and each bin is fit independently. An example of this fitting procedure is shown in Fig. 1.

The fits in Fig.1 are repeated for each pion channel, preliminary results of which are shown in Fig. 2. Error bars are purely statistical, but many previous studies in CLAS suggest that the systematic error will be small. The current results are in good agreement with previous π^+ results from CLAS [7] as well as with previous results from all three pion channels from HERMES [8]. The P_T fits include a cut around the SIDIS region of $0.4 < z < 0.7$. The region of $P_T > 1$, where the statistical errors are high, will be explored extensively with CLAS12.

CONCLUSIONS

Beam spin asymmetries are being measured simultaneously and with high statistics for each of the three semi-inclusive pion channels in CLAS. These measurements will provide access to important information about quark-gluon-quark correlations in the proton, as well as standing as an important precursor to research that will take place with CLAS12 after the Jefferson Lab upgrade.

REFERENCES

1. A. Bacchetta *et al.*, arXiv:hep-ph/0410050v2 (2004)
2. J. Levelt and P.J. Mulders, Phys.Lett.B338 357 (1994)
3. X. Ji *et al.*, Phys.Lett.B610 247 (2005)
4. J.C. Collins *et al.*, arXiv:hep-ph/0409313v1 (2004)
5. M. Burkardt, 0807.2599 (2008)
6. A. Bacchetta, M. Diehl, K. Goeke, P. Mulders, and M. Schlegel, hep-ph/0611265v2, (2007)
7. CLAS collaboration (H. Avakian *et al*)., Phys.Rev.C69:042201 (2004)
8. HERMES collaboration (A. Airapetian *et al*), hep-ex/0612059v1 (2006)

Transversity and Collins Fragmentation Functions: Towards a New Global Analysis

M. Anselmino*, M. Boglione*, U. D'Alesio†,**, A. Kotzinian‡, S. Melis*, F. Murgia** and A. Prokudin*

*Dipartimento di Fisica Teorica, Università di Torino and INFN, Sezione di Torino, Via P. Giuria 1, I-10125 Torino, Italy
†Dipartimento di Fisica, Università di Cagliari, Cittadella Universitaria di Monserrato, I-09042 Monserrato (CA), Italy
**INFN, Sezione di Cagliari, C.P. 170, I-09042 Monserrato (CA), Italy
‡CEA-Saclay, IRFU/Service de Physique Nucléaire, 91191 Gif-sur-Yvette, France; Yerevan Physics Institute, 375036 Yerevan, Armenia; JINR, 141980 Dubna, Russia

Abstract. We present an update of a previous global analysis of the experimental data on azimuthal asymmetries in semi-inclusive deep inelastic scattering (SIDIS), from the HERMES and COMPASS Collaborations, and in $e^+e^- \to h_1 h_2 X$ processes, from the Belle Collaboration. Compared to the first extraction, a more precise determination of the Collins fragmentation function and the transversity distribution function for u and d quarks is obtained.

Keywords: Transversity, Double transverse-spin asymmetries
PACS: 12.38.Bx, 12.39.St, 13.85.Qk

The study of the nucleon spin structure has recently made remarkable progress. Our understanding of the unpolarized quark distribution, $q(x)$, and the helicity distribution, $\Delta q(x)$, is now quite accurate.

A full knowledge of the nucleon quark structure in the collinear, $\boldsymbol{k}_\perp$ integrated, configuration cannot be reached without information on the third twist-two parton distribution function: the transversity distribution, $\Delta_T q(x)$ (also denoted as $h_1(x)$). Despite its fundamental importance [1] and the intense theoretical work of the last decade [2], this function has only very recently been accessed experimentally.

The only way to access this chiral-odd distribution is by coupling it to another chiral-odd quantity. To such a purpose one can look for a chiral-odd partner either in the initial or the final state. In the first case the most promising approach, and the cleanest one from the theoretical point of view, is the study of the double transverse spin asymmetry, A_{TT}, in Drell-Yan processes.

Meanwhile, the most accessible and fruitful channel is the azimuthal asymmetry $A_{UT}^{\sin(\phi_h+\phi_S)}$ in SIDIS processes, namely $\ell p^\uparrow \to \ell h X$, involving the convolution of the transversity distribution with the Collins fragmentation function $\Delta^N D_{h/q^\uparrow}(z, p_\perp)$ or $H_1^{\perp q}(z, p_\perp)$ [3]: the spin and transverse momentum dependent (TMD) function parameterizing a left-right asymmetry in the fragmentation of a transversely polarized quark

---

[1] Presenting author.

CP1149, *Spin Physics, 18th International Spin Physics Symposium*
edited by D. G. Crabb, D. B. Day, S. Liuti, X. Zheng, M. Poelker, and Y. Prok

into an unpolarized hadron. This possibility has been and still is under active investigation by the HERMES, COMPASS and JLab Collaborations.

A crucial breakthrough has been achieved thanks to the independent measurement of the Collins function (or rather, of the convolution of two Collins functions), in $e^+e^- \to h_1h_2X$ unpolarized processes by the Belle Collaboration at KEK [4]. By combining the SIDIS data from HERMES [5] and COMPASS [6] with the Belle data a global fit leading to the first extraction of the transversity distribution and the Collins fragmentation functions was performed in Ref. [7].

Recently, much higher statistics data on these spin azimuthal asymmetries have become available [8], [9], [10]. Therefore, we reconsider here our previous analysis and study the impact of these new data on the extraction of both the transversity and the Collins functions.

We recall here the main steps necessary to calculate the azimuthal asymmetries in SIDIS and in e^+e^- annihilation, addressing to Ref. [7] for details. This approach is based on an extension of the ordinary collinear factorization theorems with inclusion of a new class of spin and TMD distributions [11, 12, 13, 14]. For such processes TMD factorization has been proven [15, 16, 17] in the regime of low observed transverse momenta (compared to the large scale in the processes). Another important result, crucial for this analysis, is the universality of the Collins function entering SIDIS and e^+e^- processes, as discussed in Refs. [18, 19]. We will restrict ourselves to leading order expressions neglecting the unknown soft factor coming from gluon resummation [15, 16, 17], responsible for potential Sudakov suppression [20].

Let us consider the single spin asymmetry for the process $\ell p^\uparrow \to \ell' hX$:

$$A_{UT}^{\sin(\phi_h+\phi_S)} \propto \frac{\sum_q e_q^2 \Delta_T q(x,k_\perp) \otimes \Delta^N D_{h/q^\uparrow}(z,p_\perp)}{\sum_q e_q^2 f_{q/p}(x,k_\perp) \otimes D_{h/q}(z,p_\perp)}\,, \tag{1}$$

where $\otimes$ stands for a convolution on the transverse momenta (see Eq. (4) of Ref. [7] for full details and related comments).

This analysis can be further simplified by working at $\mathcal{O}(k_\perp/Q)$ and adopting a Gaussian and factorized parameterization of TMDs. In particular for the unpolarized parton distribution (PDF) and fragmentation (FF) functions we use:

$$f_{q/p}(x,k_\perp) = f_{q/p}(x)\,\frac{e^{-k_\perp^2/\langle k_\perp^2\rangle}}{\pi\langle k_\perp^2\rangle}\;;\quad D_{h/q}(z,p_\perp) = D_{h/q}(z)\,\frac{e^{-p_\perp^2/\langle p_\perp^2\rangle}}{\pi\langle p_\perp^2\rangle}\,, \tag{2}$$

with $\langle k_\perp^2\rangle$ and $\langle p_\perp^2\rangle$ fixed to the values found in Ref. [21] by analyzing unpolarized SIDIS: $\langle k_\perp^2\rangle = 0.25\ \mathrm{GeV}^2$, $\langle p_\perp^2\rangle = 0.20\ \mathrm{GeV}^2$.

The GRV98LO PDF set [22] and the DSS fragmentation function set [23] are respectively used for the integrated parton distribution and fragmentation functions $f_{q/p}(x)$ and $D_{h/q}(z)$.

For the transversity distribution, $\Delta_T q(x,k_\perp)$, and the Collins FF, $\Delta^N D_{h/q^\uparrow}(z,p_\perp)$, we adopt the following parameterizations [7]:

$$\Delta_T q(x,k_\perp) = \frac{1}{2}\mathcal{N}_q^T(x)\,[f_{q/p}(x)+\Delta q(x)]\frac{e^{-k_\perp^2/\langle k_\perp^2\rangle_T}}{\pi\langle k_\perp^2\rangle_T}, \tag{3}$$

$$\Delta^N D_{h/q^\uparrow}(z,p_\perp) = 2\mathcal{N}_q^C(z)\,D_{h/q}(z)h(p_\perp)\frac{e^{-p_\perp^2/\langle p_\perp^2\rangle}}{\pi\langle p_\perp^2\rangle}, \tag{4}$$

with

$$\mathcal{N}_q^T(x) = N_q^T\,x^\alpha(1-x)^\beta\,\frac{(\alpha+\beta)^{(\alpha+\beta)}}{\alpha^\alpha\beta^\beta} \tag{5}$$

$$\mathcal{N}_q^C(z) = N_q^C\,z^\gamma(1-z)^\delta\,\frac{(\gamma+\delta)^{(\gamma+\delta)}}{\gamma^\gamma\delta^\delta}\ ;\ h(p_\perp) = \sqrt{2e}\,\frac{p_\perp}{M_h}\,e^{-p_\perp^2/M_h^2}, \tag{6}$$

and $-1 \le N_q^T \le 1$, $-1 \le N_q^C \le 1$. We assume $\langle k_\perp^2\rangle_T = \langle k_\perp^2\rangle$. The helicity distributions $\Delta q(x)$ are consistently taken from Ref. [24]. Notice that with these choices both the transversity and Collins function automatically obey their proper positivity bound.

When data or phenomenological information at different Q^2 values are considered, we take into account, at leading order (LO), the QCD evolution of the transversity distribution. For the Collins FF, $\Delta^N D_{h/q^\uparrow}$, as its scale dependence is unknown, we tentatively assume the same Q^2 evolution as for the unpolarized FF, $D_{h/q}$.

Two methods have been adopted in the experimental analysis performed by Belle. These can be schematically described as (for details and definitions see, e.g., Refs. [25, 7, 10]):
i) the "$\cos(\varphi_1+\varphi_2)$ method" in the Collins-Soper frame where the jet thrust axis is used as the $\hat{z}$ direction and the $e^+e^- \to q\bar{q}$ scattering defines the $\widehat{xz}$ plane;
ii) the "$\cos(2\varphi_0)$ method", using the Gottfried-Jackson c.m. frame where one of the produced hadrons (h_2) identifies the $\hat{z}$ direction and the $\widehat{xz}$ plane is determined by the lepton and the h_2 directions.

For the $\cos(\varphi_1+\varphi_2)$ method the cross section for the process $e^+e^- \to h_1h_2X$ reads:

$$\frac{d\sigma^{e^+e^-\to h_1h_2X}}{dz_1\,dz_2\,d\cos\theta\,d(\varphi_1+\varphi_2)} = \frac{3\alpha^2}{4s}\sum_q e_q^2\Big\{(1+\cos^2\theta)\,D_{h_1/q}(z_1)\,D_{h_2/\bar{q}}(z_2)$$
$$+\frac{\sin^2\theta}{4}\cos(\varphi_1+\varphi_2)\,\Delta^N D_{h_1/q^\uparrow}(z_1)\,\Delta^N D_{h_2/\bar{q}^\uparrow}(z_2)\Big\},$$

where θ is the angle between the lepton direction and the thrust axis and $\Delta^N D_{h/q^\uparrow}(z) \equiv \int d^2p_\perp \Delta^N D_{h/q^\uparrow}(z,p_\perp)$.

A combined fit of SIDIS asymmetries together with $e^+e^- \to h_1h_2X$ data allows the simultaneous extraction of the transversity distribution and the Collins fragmentation functions. We assume flavour independent values of α and β (neglecting transversity distributions of sea quarks) and, similarly, we assume that γ and δ are the same for

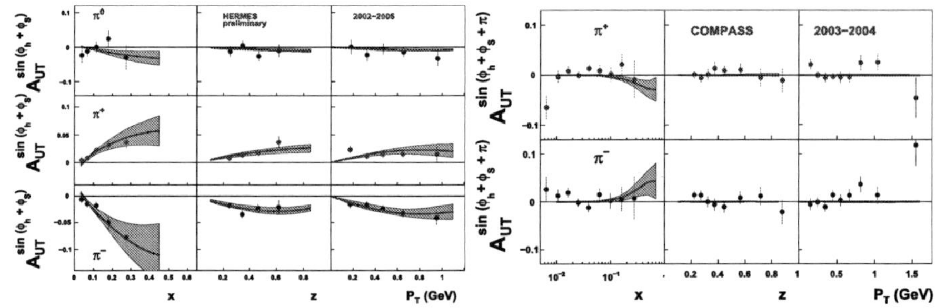

FIGURE 1. Fit of HERMES [8] (left panel) and COMPASS [9] data (right panel). The shaded area corresponds to the statistical uncertainty in the parameter values, for details see Appendix A of Ref. [27].

favoured and unfavoured Collins fragmentation functions; we then remain with a total number of 9 parameters.

New preliminary data from COMPASS operating on a transversely polarized hydrogen target have recently been released [26]: these are not included in the fit, but compared with our predictions.

The two sets of Belle data, coming from two analyses of the same experimental events, are not independent. Therefore we include only one set of data in the fit, either A_0 or A_{12} data. In this analysis we report the results obtained by using A_{12} data, the $\cos(\varphi_1 + \varphi_2)$ method. The consequences of fitting A_0 instead of A_{12} are presently under investigation.

In Fig. 1 we show the best fit to the HERMES [8] and COMPASS [9] data, respectively. Fig. 2 (left panel) shows the fit to the Belle A_{12} asymmetry, whereas in the right panel our corresponding predictions for the A_0 asymmetry are compared with data [10].

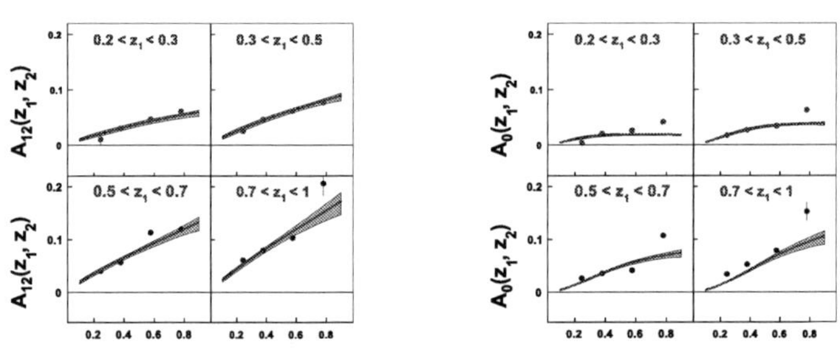

FIGURE 2. Fit of the Belle [10] data on the A_{12} (left panel) asymmetry (the $\cos(\varphi_1 + \varphi_2)$ method) and the corresponding estimate for the A_0 (right panel) asymmetry (the $\cos(2\,\varphi_0)$ method).

Table 1 collects the parameters of our best fit to the new data sets [8, 9, 10], while in Fig. 3 we show our updated transversity distribution and Collins fragmentation functions together with the uncertainty bands of our previous extraction [7]. We can definitely say that the two extractions are compatible with each other, with the new error bands being strongly reduced.

TABLE 1. Best values of the free parameters for the u and d transversity distribution functions and for the favoured and unfavoured Collins fragmentation functions. We obtain a $\chi^2/\text{d.o.f.} = 1.3$.

$N_u^T = 0.64 \pm 0.34$	$N_d^T = -1.00 \pm 0.02$	$\alpha = 0.73 \pm 0.51$	$\beta = 0.84 \pm 2.30$
$N_{fav}^C = 0.44 \pm 0.07$	$N_{unf}^C = -1.00 \pm 0.06$	$\gamma = 0.96 \pm 0.08$	$\delta = 0.01 \pm 0.05$
$M_h^2 = 0.91 \pm 0.52\ \text{GeV}^2$			

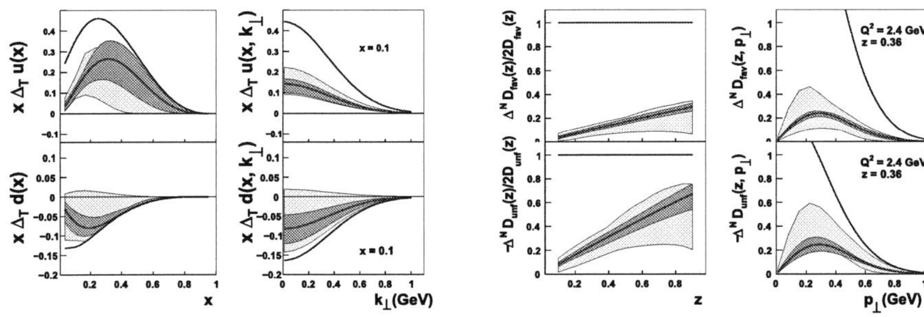

FIGURE 3. Left panel: The transversity distribution functions for u and d flavours as determined by our global fit, at $Q^2 = 2.4\ \text{GeV}^2$; we also show the Soffer bound [28] (highest or lowest lines) and the (wider) uncertainty bands of our previous extraction [7]. Right panel: Favoured and unfavoured Collins fragmentation functions as determined by our global fit, at $Q^2 = 2.4\ \text{GeV}^2$; we also show the positivity bound and the (wider) uncertainty bands as obtained in Ref. [7].

Another interesting quantity, related to the first x-moment of the transversity distribution, is the tensor charge:

$$\delta q = \int_0^1 dx\,(\Delta_T q - \Delta_T \bar{q}) = \int_0^1 dx\,\Delta_T q \tag{7}$$

where the last equality is valid only for zero antiquark transversity, as assumed in our approach. From our analysis we get, at $Q^2 = 0.8\ \text{GeV}^2$,

$$\delta u = 0.54^{+0.09}_{-0.22} \qquad \delta d = -0.23^{+0.09}_{-0.16}. \tag{8}$$

Such values compare reasonably well with various model predictions [29, 30, 31, 32, 33] for tensor charges which span the ranges $0.5 \leq \delta u \leq 1.5$ and $-0.5 \leq \delta d \leq 0.5$.

In Fig. 4 we show our predictions for the asymmetry $A_{UT}^{\sin(\phi_h+\phi_S)}$ at Jlab and COMPASS compared with available preliminary data [26]. The agreement is very good.

We acknowledge the support of the European Community - Research Infrastructure Activity under the FP6 "Structuring the European Research Area" program (HadronPhysics, RII3-CT-2004-506078). M.A., M.B., and A.P. acknowledge partial support by MIUR under PRIN 2006 and by the Helmholtz Association through funds provided to the virtual institute "Spin and strong QCD"(VH-VI-231).

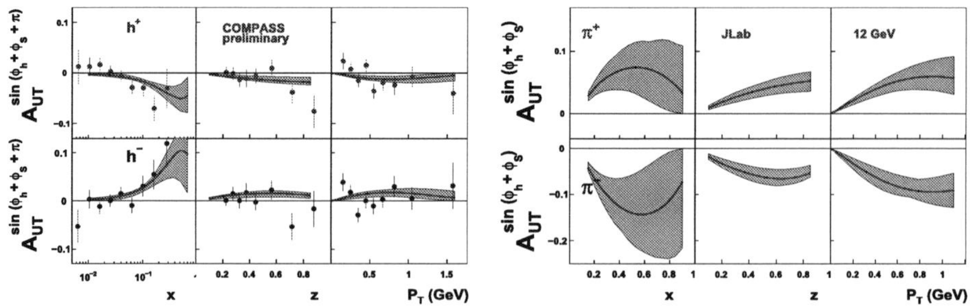

FIGURE 4. Left panel: Predictions for the single spin asymmetry $A_{UT}^{\sin(\phi_h+\phi_S+\pi)}$ compared to preliminary data by the COMPASS experiment operating with a transversely polarized hydrogen target [26]. Right panel: Estimates of the single spin asymmetry $A_{UT}^{\sin(\phi_h+\phi_S)}$ for JLab operating with proton target.

REFERENCES

1. J.P. Ralston and D.E. Soper, Nucl. Phys. B152 109 (1979)
2. V. Barone, A. Drago and P.G. Ratcliffe, Phys. Rep. 359 1, (2002), hep-ph/0104283.
3. J.C. Collins, Nucl. Phys. B396 161, hep-ph/9208213 (1993) .
4. Belle, K. Abe et al., Phys. Rev. Lett. 96 232002, (2006) hep-ex/0507063
5. HERMES, A. Airapetian et al., Phys. Rev. Lett. 94 012002, (2005) hep-ex/0408013.
6. COMPASS, E.S. Ageev et al., Nucl. Phys. B765 31, (2007) hep-ex/0610068 .
7. M. Anselmino et al., Phys. Rev. D75 054032 (2007) , hep-ph/0701006.
8. HERMES, M. Diefenthaler, (2007), arXiv:0706.2242 [hep-ex].
9. COMPASS, M. Alekseev et al., (2008), arXiv:0802.2160 [hep-ex].
10. Belle, R. Seidl et al., Phys. Rev. D78 032011 (2007) , arXiv:0805.2975 [hep-ex].
11. P. Mulders and R. Tangerman, Nucl. Phys. B461 197 (1996), hep-ph/9510301.
12. D. Boer and P.J. Mulders, Phys. Rev. D57 5780 (1998) , hep-ph/9711485.
13. A. Bacchetta et al., JHEP 02 093 (2007), hep-ph/0611265.
14. U. D'Alesio and F. Murgia, Prog. Part. Nucl. Phys. 61 394 (2008), arXiv:0712.4328 [hep-ph].
15. J.C. Collins and D.E. Soper, Nucl. Phys. B193 381 (1981).
16. X.-d. Ji, J.-p. Ma and F. Yuan, Phys. Lett. B597 299 (2004) , hep-ph/0405085.
17. X.-d. Ji, J.-p. Ma and F. Yuan, Phys. Rev. D71 034005, (2005) hep-ph/0404183.
18. A. Metz, Phys. Lett. B549 139 (2002) .
19. J.C. Collins and A. Metz, Phys. Rev. Lett. 93 252001 (2004) , hep-ph/0408249.
20. D. Boer, Nucl. Phys. B603 195 (2001) , hep-ph/0102071.
21. M. Anselmino et al., Phys. Rev. D71 074006 (2005), hep-ph/0501196.
22. M. Glück, E. Reya and A. Vogt, Eur. Phys. J. C5 461 (1998), hep-ph/9806404.
23. D. de Florian, R. Sassot and M. Stratmann, Phys. Rev. D75 114010 (2007) , hep-ph/0703242.
24. M. Glück et al., Phys. Rev. D63 094005 (2001), hep-ph/0011215.
25. D. Boer, R. Jakob and P.J. Mulders, Nucl. Phys. B504 345, (1997) hep-ph/9702281.
26. COMPASS, S. Levorato, (2008), arXiv:0808.0086 [hep-ex].
27. M. Anselmino et al., Eur. Phys. J. A39 89, (2009) arXiv:0805.2677 [hep-ph].
28. J. Soffer, Phys. Rev. Lett. 74 1292 (1995) , hep-ph/9409254.
29. I.C. Cloet, W. Bentz and A.W. Thomas, Phys. Lett. B659 214, (2008) arXiv:0708.3246 [hep-ph].
30. M. Wakamatsu, Phys. Lett. B653 398,(2007) arXiv:0705.2917 [hep-ph].
31. QCDSF, M. Gockeler et al., Phys. Lett. B627 113, (2005) hep-lat/0507001.
32. B. Pasquini, M. Pincetti and S. Boffi, Phys. Rev. D **72**, 094029 (2005) [arXiv:hep-ph/0510376].
33. H.-x. He and X.-d. Ji, Phys. Rev. D52 2960, (1995) hep-ph/9412235.

TMDs and Azimuthal Spin Asymmetries in a Light-Cone Quark Model

B. Pasquini*, S. Boffi*, A. V. Efremov† and P. Schweitzer**

* *Dipartimento di Fisica Nucleare e Teorica, Università degli Studi di Pavia, and Istituto Nazionale di Fisica Nucleare, Sezione di Pavia, I-27100 Pavia, Italy*
† *Joint Institute for Nuclear Research, Dubna, 141980 Russia*
***Department of Physics, University of Connecticut, Storrs, CT 06269, USA*

Abstract. The main properties of the leading-twist transverse momentum dependent parton distributions in a light-cone constituent quark model of the nucleon are reviewed, with focus on the role of the spin-spin and spin-orbit correlations of quarks. Results for azimuthal single spin asymmetries in semi-inclusive deep inelastic scattering are also discussed.

Keywords: transverse momentum dependent parton distributions, azimuthal spin asymmetries
PACS: 13.88.+e, 12.39.-x,13.85.Ni,13.60.-r

TMDS AND THE LIGHT-CONE CQM

In recent years much work has been devoted to study semi-inclusive deep inelastic scattering (SIDIS), Drell-Yan dilepton production and hadron production in e^+e^- annihilation as powerful tools to understand the nucleon structure. According to the factorization theorem, the physical observables of such processes can be expressed as convolution of hard partonic scattering cross sections, parton distribution functions (PDFs) and/or fragmentation functions (FFs) [1–6]. With respect to the usual inclusive deep inelastic scattering (DIS) where PDFs only depend on the longitudinal momentum fraction carried by the parton, now PDFs, as well as FFs, also depend on the transverse momentum. At leading twist there are eight transverse momentum dependent PDFs (TMDs) [3, 4], three of them surviving when integrated over the transverse momentum and giving rise to the familiar parton density, helicity and transversity distributions. Data [7–16] on SIDIS are available and many more are expected to come in future, giving first insights on the TMDs [17–24]. However, model calculations [25–34] play an important role for unraveling the information on the quark dynamics encoded in these novel functions. In this contribution we will review the results for the TMDs in a light-cone constituent quark model (CQM) which was successfully applied also for the calculation of the electroweak properties of the nucleon [35] and generalized parton distributions [36].
A convenient way to describe parton distributions is to use the representation in terms of overlaps of light-cone wave functions (LCWFs), which are the probability amplitudes to find a given N-parton configuration in the Fock-space expansion of the hadron state. This representation becomes useful in phenomenological applications where one can reasonably truncate the expansion of the hadron state to the Fock components with a few partons. In our approach, we consider the minimum Fock sector with just three-valence quarks. This truncation allows to describe the parton distributions in those kinematical

CP1149, *Spin Physics, 18th International Spin Physics Symposium*
edited by D. G. Crabb, D. B. Day, S. Liuti, X. Zheng, M. Poelker, and Y. Prok

region where the valence degrees of freedom are effective, while the contributions from sea quarks and gluons are suppressed. The three-quark component of the LCWF, keeping the full transverse momentum dependence of the partons, can be classified in a model independent way in terms of six independent light-cone amplitudes [37], which serve to parametrize the contribution from the four different orbital angular momentum components L_z compatible with total angular momentum conservation, i.e. $L_z = 0, \pm 1, 2$. In Ref. [27], these six amplitudes have been explicitly derived in a light-cone CQM, considering the relativistic spin dynamics arising from the boost of instant-form wave function into the light-cone. The instant-form wave function is constructed as a product of a momentum wave function which is in a pure S-wave state and invariant under permutations, and a spin-isospin wave function determined by SU(6) symmetry. The corresponding solution in light-cone dynamics is obtained through the unitary Melosh rotations acting on the spin of the individual quarks. The relativistic effects of the Melosh rotations are evident in the presence of spin-flip terms generating non-zero orbital angular momentum components which fit the model-independent classification of the three-quark LCWF [27, 37]. The explicit expressions of these light-cone amplitudes can be found in Ref. [27], while the corresponding results for the time-even TMDs are

$$
\begin{aligned}
f_1^a(x,p_T) &= N^a \int d[X]\; \delta(x-x_3)\delta(\mathbf{p}_T-\mathbf{p}_{\perp 3})\; |\psi(\{x_i,\mathbf{p}_{\perp i}\})|^2, \\
g_{1L}^a(x,p_T) &= P^a \int d[X]\; \delta(x-x_3)\delta(\mathbf{p}_T-\mathbf{p}_{\perp 3})\; \frac{(m+xM_0)^2-\mathbf{p}_T^2}{(m+xM_0)^2+\mathbf{p}_T^2}\; |\psi(\{x_i,\mathbf{p}_{\perp i}\})|^2, \\
g_{1T}^a(x,p_T) &= P^a \int d[X]\; \delta(x-x_3)\delta(\mathbf{p}_T-\mathbf{p}_{\perp 3})\; \frac{2M(m+xM_0)}{(m+xM_0)^2+\mathbf{p}_T^2}\; |\psi(\{x_i,\mathbf{p}_{\perp i}\})|^2, \\
h_{1L}^{\perp a}(x,p_T) &= -P^a \int d[X]\; \delta(x-x_3)\delta(\mathbf{p}_T-\mathbf{p}_{\perp 3})\; \frac{2M(m+xM_0)}{(m+xM_0)^2+\mathbf{p}_T^2}\; |\psi(\{x_i,\mathbf{p}_{\perp i}\})|^2, \\
h_{1T}^{\perp a}(x,p_T) &= -P^a \int d[X]\; \delta(x-x_3)\delta(\mathbf{p}_T-\mathbf{p}_{\perp 3})\; \frac{2M^2}{(m+xM_0)^2+\mathbf{p}_T^2}\; |\psi(\{x_i,\mathbf{p}_{\perp i}\})|^2, \\
h_1^a(x,p_T) &= P^a \int d[X]\delta(x-x_3)\delta(\mathbf{p}_T-\mathbf{p}_{\perp 3})\frac{(m+xM_0)^2}{(m+xM_0)^2+\mathbf{p}_T^2}|\psi(\{x_i,\mathbf{p}_{\perp i}\})|^2, \quad (1)
\end{aligned}
$$

where the integration measure is defined as in Ref. [27], M_0 is the mass of the non-interacting three-quark system, and the flavor dependence is given by the factors $N^u = 2$, $N^d = 1$, and $P^u = 4/3$, $P^d = -1/3$, as dictated by SU(6) symmetry. A further consequence of the assumed SU(6) symmetry is the factorization in Eqs. (1) of the momentum-dependent wave function $\psi(\{x_i,\mathbf{p}_{\perp i}\})$ from the spin-dependent factor arising from the Melosh rotations. As a result one finds the following relations

$$
2h_1^a(x,p_T) = g_{1L}^a(x,p_T) + \frac{P^a}{N^a} f_1^a(x,p_T), \qquad h_{1L}^{\perp a}(x,p_T) = -g_{1T}^a(x,p_T), \quad (2)
$$

$$
\frac{P^a}{N^a} f_1^a(x,p_T) = h_1^a(x,p_T) - \frac{p_T^2}{2M^2} h_{1T}^{\perp a}(x,p_T). \quad (3)
$$

These relations are common to several quark model calculations [25–27, 34] though not all [28]. The common feature of such models is that gluon degrees of freedom

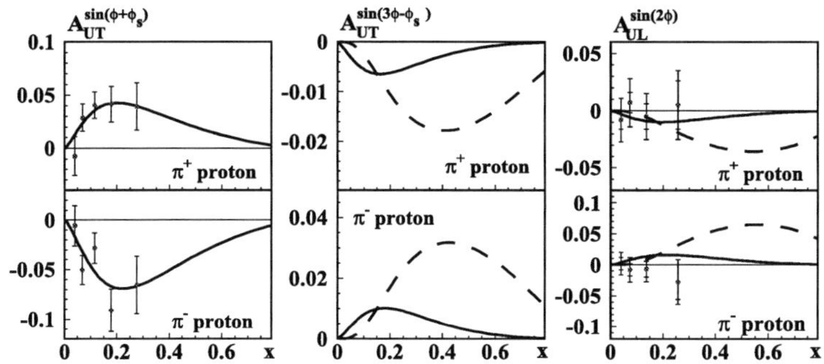

FIGURE 1. The SSAs $A_{UT}^{\sin(\phi+\phi_S)}$ (left column), $A_{UT}^{\sin(3\phi-\phi_S)}$ (middle column), and $A_{UL}^{\sin(2\phi)}$ (right column) in DIS production of charged pions off proton target, as function of x. The solid curves show the results on the basis of the light-cone CQM evolved to the scale $Q^2 = 2.5$ GeV^2, and the dashed curves correspond to the predictions at the low scale of the model. The experimental data are from Refs. [8, 9].

are neglected. On the other side, the recent model calculation of Ref. [34] found the interesting result that SU(6) symmetry is not a necessary condition for the relation (3).

RESULTS FOR AZIMUTHAL SSAS

The results in Eqs. (1) are general and can be applied to any CQM adopting the appropriate nucleon wave function. In the following we will take the momentum wavefunction from Schlumpf [38]. In Fig. 1 we show the results for the single spin asymmetries (SSAs) with unpolarized (U) beam and transversely (T) polarized proton target in SIDIS of positive (upper panels) and negative (lower panels) pions. The asymmetries $A_{UT}^{\sin(\phi+\phi_S)}$ (left column), $A_{UT}^{\sin(3\phi-\phi_S)}$ (middle column), and $A_{UL}^{\sin(2\phi)}$ (right column) are due to the Collins function and to the three chirally-odd TMDs h_1, $h_{1T}^{\perp}$, and $h_{1L}^{\perp}$, respectively. For the Collins function we use the results extracted in [20]. In the denominator of the asymmetries we take f_1 from [39] and the unpolarized FF from [40], both valid at the scale $Q^2 = 2.5$ GeV^2. The model results for h_1 evolved from the low hadronic scale of the model to $Q^2 = 2.5$ GeV^2 ideally describe the HERMES data [8] for $A_{UT}^{\sin(\phi+\phi_S)}$. This is in line with the favourable comparison between our model predictions [33] and the phenomenological extraction of the transversity and the tensor charges in Ref. [21]. In the case of $A_{UL}^{\sin(2\phi)}$ and $A_{UT}^{\sin(3\phi-\phi_S)}$ we compare the results obtained using the TMDs at the scale of the model (dashed curves) and the TMDs evolved at leading order to $Q^2 = 2.5$ GeV^2 (solid curves) using the evolution pattern of the transversity. Although this is not the correct evolution pattern, it may give us a rough insight on the possible size of effects due to evolution (for a more detailed discussion we refer to [42]). In the case of $A_{UT}^{\sin(3\phi-\phi_S)}$, the evolution effects give smaller asymmetries in absolute value and shift the peak at lower x values. Measurements in range $0.1 \lesssim x \lesssim 0.6$ are planned with the CLAS 12 GeV upgrade [41] and will be able to discriminate between the two scenarios. In the region $x \lesssim 0.2$, there exist also preliminary deuteron target data [16] which are compatible, within error bars, with the model predictions both at the hadronic and the evolved scale. Similar conclusions can be drawn also in the case of $A_{UL}^{\sin(2\phi)}$, where we compare our results with HERMES data [9].

REFERENCES

1. J. C. Collins, Acta Phys. Polon. B **34**, 3103 (2003).
2. J. C. Collins, Nucl. Phys. B **396**, 161 (1993).
3. P. J. Mulders and R. D. Tangerman, Nucl. Phys. B **461**, 197 (1996) and **484**, 538E (1997).
4. D. Boer and P. J. Mulders, Phys. Rev. D **57**, 5780 (1998).
5. S. J. Brodsky, D. S. Hwang and I. Schmidt, Phys. Lett. B **530**, 99; Nucl. Phys. B **642**, 344 (2002).
6. J. C. Collins, Phys. Lett. B **536**, 43 (2002).
7. M. Arneodo *et al.* [European Muon Collaboration], Z. Phys. C **34**, 277 (1987).
8. M. Diefenthaler, AIP Conf. Proc. **792**, 933 (2005); L. Pappalardo, Eur. Phys. J. A **38**, 145 (2008).
9. A. Airapetian *et al.* [HERMES], Phys. Rev. Lett. **84**, 4047 (2000).
10. A. Airapetian *et al.* [HERMES Collaboration], Phys. Rev. D **64**, 097101 (2001).
11. A. Airapetian *et al.* [HERMES], Phys. Lett. B **562**, 182 (2003).
12. A. Airapetian *et al.* [HERMES Collaboration], Phys. Lett. B **648** (2007) 164.
13. H. Avakian *et al.* [CLAS], Phys. Rev. D **69**, 112004 (2004); AIP Conf. Proc. **792**, 945 (2005).
14. A. Airapetian *et al.* [HERMES], Phys. Rev. Lett. **94**, 012002 (2005).
15. V. Alexakhin *et al.* [COMPASS], Phys. Rev. Lett. **94**, 202002 (2005); Nucl. Phys. B **765**, 31 (2007).
16. A. Kotzinian [on behalf of the COMPASS collaboration], arXiv:0705.2402 [hep-ex].
17. A. V. Efremov *et al.*, Phys. Rev. D **67**, 114014 (2003); Eur. Phys. J. C **32**, 337 (2003).
18. A. V. Efremov *et al.*, Phys. Lett. B **612**, 233 (2005).
19. W. Vogelsang and F. Yuan, Phys. Rev. D **72**, 054028 (2005).
20. A. V. Efremov, K. Goeke and P. Schweitzer, Phys. Rev. D **73**, 094025 (2006).
21. M. Anselmino *et al.* Phys. Rev. D **75**, 054032 (2007); arXiv:0812.4366 [hep-ph].
22. S. Arnold *et al.*, arXiv:0805.2137 [hep-ph].
23. M. Anselmino *et al.*, arXiv:0805.2677 [hep-ph].
24. B. Zhang, Z. Lu, B. Q. Ma and I. Schmidt, Phys. Rev. D **77**, 054011 (2008).
25. R. Jakob, P. J. Mulders and J. Rodrigues, Nucl. Phys. A **626**, 937 (1997).
26. H. Avakian, A. V. Efremov, P. Schweitzer and F. Yuan, arXiv:0805.3355 [hep-ph].
27. B. Pasquini, S. Cazzaniga and S. Boffi , Phys. Rev. D**70**, 034025 (2008).
28. A. Bacchetta, F. Conti and M. Radici, Phys. Rev. D**78**, 074010 (2008).
29. S. Meissner, A. Metz and K. Goeke, Phys. Rev. D **76**, 034002 (2007).
30. F. Yuan, Phys. Lett. B **575**, 45 (2003).
31. L. P. Gamberg, G. R. Goldstein and M. Schlegel, arXiv:0708.2580; Phys. Rev. D**77**, 094016 (2008).
32. P. Schweitzer *et al.*, Phys. Rev. D **64**, 034013; K. Goeke *et al.*, Acta Phys. Polon. B **32**, 1201 (2001).
33. B. Pasquini *et al.*,Phys. Rev. D **72**, 094029 (2005); Phys. Rev. D **76**, 034020 (2007).
34. A. V. Efremov, P. Schweitzer,O. V. Teryaev and P. Závada, arXiv:0812.3246 [hep-ph].
35. B. Pasquini, and S. Boffi , Phys. Rev. D **76**, 074011 (2007).
36. S. Boffi and B. Pasquini, Riv. Nuovo Cim. **30**, 387 (2007); S. Boffi , B. Pasquini, M. Traini, Nucl. Phys. B **649** 243, (2003).
37. X. Ji, J.-P. Ma and F. Yuan, Nucl. Phys. B **652**, 383 (2003). Eur. Phys. J. C **33**, 75 (2004);
38. F. Schlumpf, doctoral thesis, University of Zurich, 1992; hep-ph/9211255.
39. M. Glück, E. Reya and A. Vogt, Eur. Phys. J. C **5**, 461 (1998) [arXiv:hep-ph/9806404].
40. S. Kretzer, Phys. Rev. D **62** (2000) 054001 [arXiv:hep-ph/0003177].
41. H. Avakian, *et al.*, JLab LOI 12-06-108 (2008).
42. S. Boffi , A.V. Efremov, B. Pasquini, and P. Schweitzer, in preparation.

Twist-3 Single-Spin Asymmetry for SIDIS and its Azimuthal Structure

Yuji Koike* and Kazuhiro Tanaka†

*Department of Physics, Niigata University, Ikarashi, Niigata 950-2181, Japan
†Department of Physics, Juntendo University, Inba-gun, Chiba 270-1695, Japan

Abstract. We derive the complete twist-3 single-spin-dependent cross section for semi-inclusive DIS, $ep^{\uparrow} \to e\pi X$, associated with the complete set of the twist-3 quark-gluon correlation functions in the transversely polarized nucleon, extending our previous study. The cross section consists of five independent structure functions with different azimuthal dependences, consistently with the transverse-momentum-dependent (TMD) factorization approach in the low q_T region. Correspondence with the inclusive DIS limit and comparison with the TMD approach are briefly discussed.

Keywords: Single spin asymmetry, Twist-3, Azimuthal asymmetry in semi-inclusive DIS
PACS: 12.38.Bx, 13.88.+e, 13.60.Le

In our recent paper [1], we have presented the twist-3 single-spin-dependent cross section for the large-q_T pion production in semi-inclusive DIS (SIDIS), $ep^{\uparrow} \to e\pi X$, in the framework of the collinear factorization, in particular, the cross section associated with the twist-3 quark-gluon correlation functions. There we have also clarified the gauge-invariance and the factorization property of the corresponding twist-3 cross section. This report update the result, supplying the calculation not presented in [1].

For the kinematic variables of SIDIS, $e(\ell) + p(p, S_\perp) \to e(\ell') + \pi(P_h) + X$, we use $S_{ep} = (\ell + p)^2$, $Q^2 = -q^2 = -(\ell - \ell')^2$, $x_{bj} = \frac{Q^2}{2p\cdot q}$, $z_f = \frac{p\cdot P_h}{p\cdot q}$ and $q_T = \sqrt{-q_t^2}$ with $q_t^\mu = q^\mu - \left(\frac{P_h\cdot q}{p\cdot P_h}\right) p^\mu - \left(\frac{p\cdot q}{p\cdot P_h}\right) P_h^\mu$. We work in a frame where the momenta of the virtual photon and the initial nucleon are collinear, and use the lepton plane as a reference plane to define the azimuthal angles ϕ_h and ϕ_S of the hadron plane and the transverse spin vector $S_\perp^\mu$, respectively. In this frame the magnitude of the transverse momentum of the final pion is given by $P_{h\perp} = z_f q_T$. With this notation, one can present the azimuthal dependence of the twist-3 single-spin-dependent cross section in the following form [2]:

$$\frac{d^5\sigma^{\rm tw3}}{[d\omega]} = \sin(\phi_h - \phi_S)(\Delta\sigma_1 + \Delta\sigma_2 \cos(\phi_h) + \Delta\sigma_3 \cos(2\phi_h)) + \cos(\phi_h - \phi_S)(\Delta\sigma_4 \sin(\phi_h) + \Delta\sigma_5 \sin(2\phi_h)), \quad (1)$$

with the differential element $[d\omega] \equiv dx_{bj}dQ^2dz_f dq_T^2 d\phi_h$. In [1], we have presented the result only for $\Delta\sigma_{1,2,3}$. In addition, some diagrams that produce the "soft-fermion-pole (SFP)" contribution to $\Delta\sigma_i$ were missing as pointed out in [3]. In this report we will present the main features of the full result, taking these points into account.

In the twist-3 mechanism for SSA, quark-gluon correlation functions in the transversely polarized nucleon play the central role: There exist two independent twist-3 distribution functions $\{G_F(x_1,x_2), \widetilde{G}_F(x_1,x_2)\}$ defined from the Fourier transform of the

CP1149, *Spin Physics, 18th International Spin Physics Symposium*
edited by D. G. Crabb, D. B. Day, S. Liuti, X. Zheng, M. Poelker, and Y. Prok

correlation function $\langle pS_\perp|\bar{\psi}(0)F^{\alpha+}(\xi n)\psi(\lambda n)|pS_\perp\rangle$ on the light cone ($n^2=0$, $F^{\alpha\beta}$ the gluon's field strength), where $x_{1,2}$ denote the momentum fractions associated with the quark fields $\psi, \bar{\psi}$. $G_F(x_1,x_2)$ is symmetric, while $\widetilde{G}_F(x_1,x_2)$ is anti-symmetric, under $x_1 \leftrightarrow x_2$. By replacing the above nonlocal operator $\bar{\psi}F^{\alpha+}\psi$ by its charge conjugate, one can define the twist-3 distribution functions for the "anti-quark" flavour [2]. Also, replacing $F^{\alpha+}$ by the covariant derivative D^α, one can define other twist-3 distributions; the relation between $\{G_F, \widetilde{G}_F\}$ and those other twist-3 distributions was also clarified in [4]. In the twist-3 mechanism, the single-spin-dependent cross section occurs as pole contributions from an internal propagator in the partonic hard part. For SIDIS, three kinds of poles contribute, which are classified as soft-gluon-pole (SGP), hard-pole (HP) and SFP. Below we discuss the characteristic features of each contribution from these poles.

1. SGP contribution

The SGP contribution occurs from diagrams shown in Figs. 8 and 10 of [1]. Those diagrams have a propagator pole at $x_1=x_2$, and its evaluation gives rise to SGP contribution with $G_F(x,x)$ only, since $\widetilde{G}_F(x,x)=0$ due to the symmetry property. One can directly calculate the contribution to $\Delta\sigma_{4,5}$ from these diagrams. Alternatively, one can use the fact that the SGP cross section for SIDIS can be derived directly from the twist-2 unpolarized cross section formula as proved in [5]. This way, we reported the full SGP result in the erratum of [5] as

$$\frac{d^5\sigma^{\rm tw3,SGP}}{[d\omega]} = \frac{\alpha_{em}^2\alpha_s}{8\pi x_{bj}^2 S_{ep}^2 Q^2}\frac{\pi M_N}{C_F}\sum_q e_q^2 \sum_{j=q,g}\mathscr{C}_j\int\frac{dz}{z}\int\frac{dx}{x}D_j(z)\left[\frac{q_T}{Q^2}\sin(\phi_h-\phi_S)\right.$$
$$\times\sum_{k=1}^4\mathscr{A}_k\left\{\frac{\widehat{x}}{1-\widehat{z}}\widehat{\sigma}_k^{jq}x\frac{dG_F^q(x,x)}{dx}+\left(\frac{1}{\widehat{z}}Q^2\frac{\partial\widehat{\sigma}_k^{jq}}{\partial q_T^2}-\frac{\widehat{x}}{1-\widehat{z}}\frac{\partial(\widehat{x}\widehat{\sigma}_k^{jq})}{\partial\widehat{x}}\right)G_F^q(x,x)\right\}$$
$$\left.-\frac{\cos(\phi_h-\phi_S)}{\widehat{z}q_T}\left(\frac{1}{2}\mathscr{A}_8\widehat{\sigma}_3^{jq}+\mathscr{A}_9\widehat{\sigma}_4^{jq}\right)G_F^q(x,x)\right]\delta\left(\frac{q_T^2}{Q^2}-\left(\frac{1}{\widehat{x}}-1\right)\left(\frac{1}{\widehat{z}}-1\right)\right), \quad (2)$$

where the sum over q runs over all quark and anti-quark flavours and $\widehat{\sigma}_k^{jq}$ ($k=1,\cdots,4$) are the partonic hard cross section for the twist-2 unpolarized cross section as listed, for example, in [1]. $\mathscr{C}_q=-1/(2N)$ and $\mathscr{C}_g=N/2$ are the color factors, and $\widehat{x}=x_{bj}/x$, $\widehat{z}=z_f/z$. The factors $\mathscr{A}_k$ ($k=1,\cdots,4,8,9$) are defined as $\mathscr{A}_1=1+\cosh^2\psi$, $\mathscr{A}_2=-2$, $\mathscr{A}_3=-\cos\phi_h\sinh 2\psi$, $\mathscr{A}_4=\cos 2\phi_h\sinh^2\psi$, $\mathscr{A}_8=-\sin\phi_h\sinh 2\psi$, $\mathscr{A}_9=\sin 2\phi_h\sinh^2\psi$ with $\cosh\psi=2x_{bj}S_{ep}/Q^2-1$. From (2), one sees that the SGP contributions to $\Delta\sigma_4$ and $\Delta\sigma_5$ are related to those for $\Delta\sigma_2$ and $\Delta\sigma_3$, respectively, although $\Delta\sigma_{4,5}$ do not receive "derivative" contribution of the SGP function, $dG_F^q(x,x)/dx$, unlike the $\Delta\sigma_{2,3}$ terms.

2. HP contribution

Diagrams for the HP contributions are shown in Fig. 2 of [1], whose internal propagator has a pole at $x_1=x_{bj}$. Evaluating it, both $G_F(x_{bj},x)$ and $\widetilde{G}_F(x_{bj},x)$ contribute. By the direct calculation of the $\mathscr{V}_{8,9}$ components in the expansion of the hadronic tensor from these diagrams (see Eq.(50) of [1]), one can obtain the HP contribution to $\Delta\sigma_{4,5}$ in (1). It turned out that both G_F and $\widetilde{G}_F$ contribution satisfy the relation,

$$\Delta\sigma_4^{\rm HP}=\Delta\sigma_2^{\rm HP}, \qquad \Delta\sigma_5^{\rm HP}=\Delta\sigma_3^{\rm HP}. \quad (3)$$

We do not know if this relation holds or not in the higher-order calculation.

3. SFP contribution

The SFP contributions, from the propagator pole at $x_1 = 0$, exist only in the anti-quark fragmentation and gluon fragmentation channels for the "quark distribution", $G_F(0,x)$ and $\widetilde{G}_F(0,x)$ with $x > 0$. The diagrams for those are shown in Fig. 1. In the quark fragmentation channel with the same $G_F(0,x)$, $\widetilde{G}_F(0,x)$, the diagrams in Fig. 6 of [1] and those in Fig. 1 of [3] cancel with each other and thus there is no SFP contribution as was shown in [3]. (By reversing the arrows of the fermion lines in Fig. 1, one obtains the SFP contributions in the quark/gluon fragmentation channels, but with the "anti-quark SFP functions", $G_F(0,x)$, $\widetilde{G}_F(0,x)$ with $x < 0$.) This combination of the twist-3 distributions and the fragmentation functions is a feature peculiar to the SFP contribution.

In Fig. 1, the left and right diagrams for anti-quark and gluon fragmentation channels, respectively, are different only in the position of the final-state cut, and thus the partonic hard part for the SFP cross sections differ only in the overall sign for the two channels.

By collecting all the pole contributions associated with G_F and $\widetilde{G}_F$, one obtains the complete cross section formula in the form of (1) [2]. To make connection with the transverse-momentum-dependent (TMD) factorization approach, we recast (1) into

$$\frac{d^5\sigma^{\rm tw3}}{[d\omega]} = \sin(\phi_h-\phi_S)\,F^{\sin(\phi_h-\phi_S)} + \sin(2\phi_h-\phi_S)\,F^{\sin(2\phi_h-\phi_S)} + \sin(\phi_S)\,F^{\sin(\phi_S)} + \sin(3\phi_h-\phi_S)\,F^{\sin(3\phi_h-\phi_S)} + \sin(\phi_h+\phi_S)\,F^{\sin(\phi_h+\phi_S)}, \qquad (4)$$

with the obvious relations $F^{\sin(\phi_h-\phi_S)} = \Delta\sigma_1$, $F^{\sin(2\phi_h-\phi_S)} = (\Delta\sigma_2+\Delta\sigma_4)/2$, $F^{\sin(\phi_S)} = (-\Delta\sigma_2+\Delta\sigma_4)/2$, $F^{\sin(3\phi_h-\phi_S)} = (\Delta\sigma_3+\Delta\sigma_5)/2$, $F^{\sin(\phi_h+\phi_S)} = (-\Delta\sigma_3+\Delta\sigma_5)/2$. The five types of azimuthal dependence appearing in (4) are the same as what was derived by the TMD factorization approach [6]. We note, however, that our formula which is valid at $q_T \sim Q \gg \Lambda_{\rm QCD}$ contains only two independent (twist-3) distributions of nucleon, while the TMD formula contains more number of (TMD) distributions of nucleon.

It is instructive to consider the limit of inclusive DIS from our result. As is well known [7], SSA vanishes in the inclusive DIS. In our formula this limit corresponds to setting formally all the twist-2 fragmentation functions equal to one, and to integrating over ϕ_h, q_T and z_f. Integration over ϕ_h keeps only the $F^{\sin(\phi_S)}$-term in (4). Since the SFP partonic hard cross sections have opposite signs with the same magnitude between the anti-quark and gluon fragmentation channels, no SFP contribution survives. In addition, owing to the relation (3), there is no HP contribution to $F^{\sin(\phi_S)}$ from the beginning. The remaining SGP contribution also proves to vanish after integration over q_T^2 [5]. This way our formula is consistent with the fact that there is no SSA in inclusive DIS.

Finally we mention briefly the connection and consistency of the present result with that of the TMD factorization approach in the intermediate region of q_T, $\Lambda_{\rm QCD} \ll q_T \ll Q$. For $F^{\sin(\phi_h-\phi_S)}$ in (4), it's been shown that the two frameworks give identical result [8, 3] as in Drell-Yan process [9]. To make connection with the TMD factorization approach, it is necessary to look at the $q_T \to 0$ limit of the other structures in (4). This limit was studied in [10], using the result presented in [1, 4]. With our present update, we obtain the small-q_T behavior of the structure functions in (4) as $F^{\sin(\phi_h-\phi_S)} \sim 1/q_T^3$,

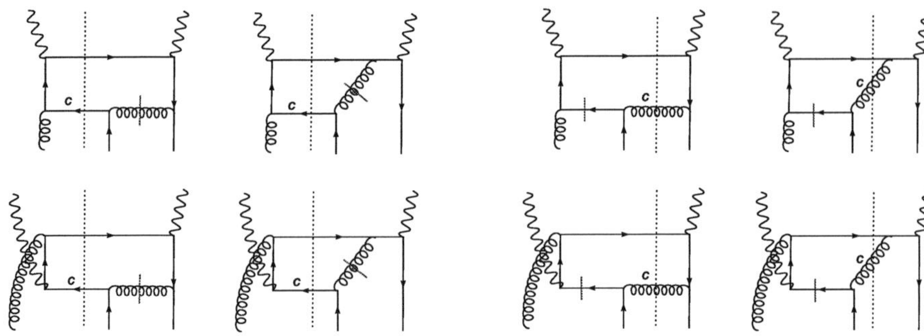

FIGURE 1. Diagrams for the SFP contribution to SSA in SIDIS $ep^{\uparrow} \to e\pi X$. Left four diagrams are for the anti-quark fragmentation channel, while right four are for the gluon fragmentation channel. The parton labeled by "c" fragments into the pion, and the bar denotes the pole contribution of the propagator.

$F^{\sin(2\phi_h-\phi_S)} \sim 1/q_T^2$, $F^{\sin(\phi_S)} \sim O(1)$, $F^{\sin(3\phi_h-\phi_S)} \sim 1/q_T$, $F^{\sin(\phi_h+\phi_S)} \sim 1/q_T$. Here the behavior of $F^{\sin(\phi_S)}$ is consistent with the limit of inclusive DIS mentioned above. We also note that the SFP contributions are always more power-suppressed compared to the SGP and HP contributions in all structure functions. In [4], the contribution from the twist-3 fragmentation function of the pion, $\widehat{E}_F(z_1,z_2)$, was calculated, keeping only the derivative term of the SGP contribution. This contribution gives the small-q_T behavior as $F^{\sin(\phi_S)}(\widehat{E}_F) \sim 1/q_T^2$, $F^{\sin(\phi_h+\phi_S)}(\widehat{E}_F) \sim 1/q_T^3$ and $F^{\sin(\phi_h-\phi_S)}(\widehat{E}_F) \sim 1/q_T$, with no contribution to the other two azimuthal structures. The first relation does not contradict with the inclusive-DIS limit, since $\widehat{E}_F \to 0$ in this limit. The leading term with $F^{\sin(\phi_h+\phi_S)}(\widehat{E}_F) \sim 1/q_T^3$ may be connected to the Collins effect in SIDIS, for which one needs more study.

ACKNOWLEDGMENTS

We thank F. Yuan for a useful comment. The work of Y.K. is supported in part by Uchida Energy Science Promotion Foundation. The work of K.T. is supported by the Grant-in-Aid for Scientific Research No. B-19340063.

REFERENCES

1. H. Eguchi, Y. Koike and K. Tanaka, Nucl. Phys. **B763** 193 (2007).
2. Y. Koike and K. Tanaka, in preparation.
3. Y. Koike, W. Vogelsang, F. Yuan, Phys. Lett. **B659** 878 (2008).
4. H. Eguchi, Y. Koike and K. Tanaka, Nucl. Phys. **B752** 1 (2006).
5. Y. Koike and K. Tanaka, Phys. Lett. B **646**, 232 (2007) [Erratum-ibid. B **668**, 458 (2008)]; See also Y. Koike and K. Tanaka, Phys. Rev. D **76**, 011502 (2007), for other processes.
6. A. Bacchetta, M. Diehl, K. Goeke, A. Metz, P. J. Mulders and M. Schlegel, JHEP **0702**, 093 (2007).
7. N. Christ and T.D. Lee, Phys. Rev. **143**, 1310 (1966).
8. X.D. Ji, J.W. Qiu, W. Vogelsang, F. Yuan, Phys. Lett. **B638** 178 (2006).
9. X.D. Ji, J.W. Qiu, W. Vogelsang, F. Yuan, Phys. Rev. **D73** 094017 (2006).
10. A. Bacchetta, D. Boer, M. Diehl and P. J. Mulders, JHEP **0808**, 023 (2008).

Beam-Spin Asymmetry Measurement at CLAS

M. Aghasyan for the CLAS collaboration

12000 Jefferson Ave., Newport News VA 23606.

Abstract. Beam Single Spin Asymmetries in single neutral semi-inclusive pion electroproduction off an unpolarized hydrogen target in the deep inelastic scattering regime ($Q^2 > 1GeV^2, W^2 > 4GeV^2$) have been measured using a polarized electron beam of 5.776 GeV with the CEBAF Large Acceptance Spectrometer at the Thomas Jefferson National Accelerator Facility (JLab). The measured kinematical dependences are compared with published data and existing theoretical predictions.

Keywords: spin, transverse momentum distributions, single spin asymmetries
PACS: 13.60.-r; 13.87.Fh; 13.88.+e; 14.20.Dh; 24.85.+p

In the recent years it became clear that understanding of the orbital motion and in particular the role of partonic initial and final state interactions of quarks is crucial in construction of a more complete picture of the nucleon in terms of elementary quarks and gluons. Parton distribution functions have been generalized to contain information not only on the longitudinal but also on the transverse distributions of partons in a fast moving hadron. Intense investigation of Transverse Momentum Distributions (TMDs) of partons both from the experimental and theoretical side, indicate that nontrivial QCD-dynamics inside hadrons, is much richer than what one can learn from collinear PDFs.

Two fundamental mechanisms have been identified leading to Single Spin Asymmetries SSAs in hard processes; the Sivers mechanism [1, 2, 3, 4, 5], which generates an asymmetry in the distribution of quarks due to orbital motion of partons, and the Collins mechanism [4, 6], which generates an asymmetry during the hadronization of quarks. TMDs also contain unique information on the role of initial and final state interactions of active partons in hard scattering processes [3, 4, 5]. Many experiments are currently trying to pin down various TMD effects through semi-inclusive deep-inelastic scattering (in experiments such as HERMES at DESY [7, 8, 9, 10], COMPASS at CERN [11], CLAS and Hall-A at Jefferson Lab [12, 13]) polarized proton-proton collisions (PHENIX, STAR and BRAHMS at RHIC) [14, 15, 16] and electron-positron annihilation (Belle at KEK) [17].

The cross section for single pion production by longitudinally polarized leptons scattering from unpolarized protons may be written in terms of a set of structure functions. Its helicity (λ_e) dependent part (σ_{LU}) [6, 18] arises from the anti-symmetric part of the hadronic tensor:

$$\frac{d\sigma_{LU}}{dx_B dy\, dz d^2 P_\perp} \propto \lambda_e \sqrt{y^2+\gamma^2}\sqrt{1-y-\frac{1}{4}\gamma^2}\sin\phi\, F_{LU}. \tag{1}$$

The subscripts in σ_{LU} specify the beam and target polarizations, respectively (L stands for longitudinally polarized and U for unpolarized). The azimuthal angle ϕ is the angle

CP1149, *Spin Physics, 18th International Spin Physics Symposium*
edited by D. G. Crabb, D. B. Day, S. Liuti, X. Zheng, M. Poelker, and Y. Prok

between leptonic and hadronic planes and kinematic variables x_B, y, and z are defined as: $x_B = Q^2/2(P_1q)$, $y = (P_1q)/(P_1k_1)$, $z = (P_1P)/(P_1q)$, where $Q^2 = -q^2$, $q = k_1 - k_2$ is the 4-momentum of the virtual photon, P_1 and P are the momenta of the target and the observed final-state hadron, and $\gamma^2 = 4M^2x_B^2y^2/Q^2$.

The structure function F_{LU} arises due to the interference of the longitudinal and transverse photon contributions. In the partonic description of SIDIS, structure functions factorize into different cotributions appearing as convolutions of distribution and fragmentation functions depending on the scaling variables x and z, respectively [6, 19].

The beam-spin asymmetries in single-pion production off the unpolarized target are higher-twist by their nature [20, 21]. The higher twist observables are important for understanding the long-range quark-gluon dynamics and they may also be accessible as leading contributions through the measurements of certain asymmetries [22, 23, 24, 18] and in particular beam SSAs. Recently, higher twist effects in SIDIS were interpreted in terms of average transverse force acting on the active quarks in the instant after being struck by the virtual photon [25].

Different contributions to the beam SSAs, discussed so far, provide information on different leading and sub-leading parton distribution and fragmentation functions, related both to Collins and Sivers production mechanisms. Large beam SSAs were predicted [21] based on the Sivers mechanism as the dynamical origin of the observable. Within this framework, the asymmetry generated at the distribution level, is given by either the convolution of the T-odd parton distribution $h_1^{\perp}$ with the twist-3 fragmentation function $E(x)$ [22], or the convolution of the twist-3 T-odd distribution function $g^{\perp}$ with unpolarized fragmentation function D_1[26]. Parton model analyzes [27, 20, 21, 26] of sub-leading single-spin asymmetries observed at HERMES [28] and CLAS [12] lead to introduction of the full set of twist-3 T-odd distribution functions [29, 30].

HERMES [9] and Belle [17] measurements indicate that favored and unfavored Collins functions are roughly equal and have opposite signs. Since the π^0 fragmentation function (FF) is the sum of π^+ and π^- FFs, its favored and unfavored FFs will be equal and the Collins FF describing the fragmentation of polarized quarks will be strongly suppressed. The contributions related to Sivers mechanism can thus be studied without a significant background from the Collins mechanism.

In this contribution we present measurement of beam-spin asymmetries in single-pion semi-inclusive leptoproduction using a 5.77 GeV electron beam and the CEBAF Large Acceptance Spectrometer (CLAS) [31] at JLab. Scattering of longitudinally polarized electrons off a liquid-hydrogen target was studied over a wide kinematic region. The beam polarization, frequently measured with a Møller polarimeter, was on average 0.80 ± 0.03. Beam helicity was flipped every 30 msec to minimize the helicity correlated systematics. The scattered electrons were detected in CLAS. Electron candidates were selected by a hardware trigger using a coincidence between the gas Cherenkov counters and the lead-scintillator electromagnetic calorimeters. The π^0 events were identified by the invariant mass of two high energy photons measured using the CLAS electromagnetic calorimeter (EC) and the Inner Calorimeter (IC)[32]. Gaussian plus linear fit for all bins, and 3 σ cut around π^0 peak have been used to select events with π^0 for given bin (see Fig.1).

The total number of electron-π^0 coincidences in the DIS range ($Q^2 > 1$ GeV^2, $W^2 > 4$

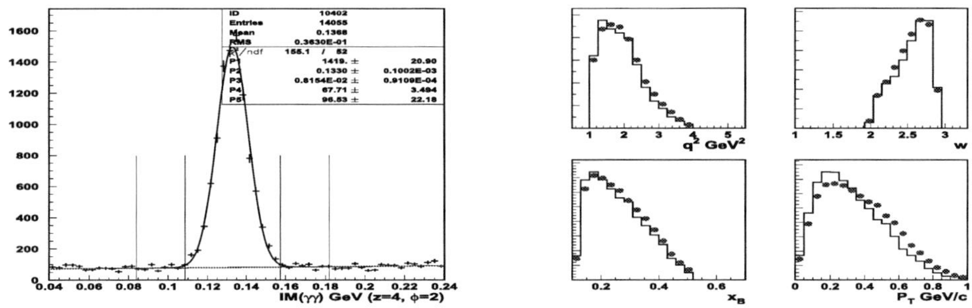

FIGURE 1. Example of fitting of the invariant mass $IM(\gamma\gamma)$ in one of the bins (left). MC (black) and data (red) after the cuts (right).

GeV^2) selected by quality, vertex, acceptance, fiducial, and kinematic cuts was $\approx 4 \times 10^5$. Events with low missing mass of the $e\pi^0$ system ($M_X < 1.1$)were discarded to reduce contributions from exclusive processes.

The semi-inclusive production of pions in e1-dvcs experiment was simulated using the LUND MC generator [33] as input for GEANT simulation and events were reconstructed using the same reconstruction program as for data. The reconstructed MC sample, which provides a pretty good description of kinematical distributions (see Fig.1) was used to define the kinematical cuts, extract reconstruction efficiencies and test the procedure for asymmetry extraction.

The measured beam SSA for π^0 appeared to be comparable with π^+ beam SSA from older CLAS data sets [34] both in the sign and value (Fig. 2), indicating that contribution from Collins mechanism could not be the dominating one. The asymmetry shows no significant x-dependence within the measured range. There is an indication that decrease of A_{LU} at large P_T, expected from perturbative QCD, starts already at $P_T \sim 0.6$. Observed large BSA for π^0 indicates that major contribution in pion SSAs may be due to the Sivers mechanism.

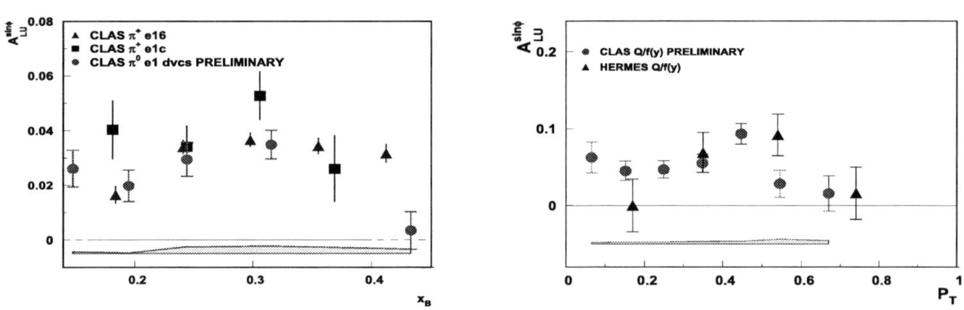

FIGURE 2. The π^0 beam SSA A_{LU} vs x_B for π^+ from e1c (CLAS@4.3 GeV) [12], e16 (CLAS@5.7 GeV) and π^0 from e1dvcs (CLAS@5.8 GeV) datasets (left). Beam SSA A_{LU} multiplied by the kinematical factor $< Q > /f(y)$ vs P_T from CLAS compared with HERMES data [10] (right).

A similar measurement has been performed by the HERMES collaboration [10] at higher beam energy (27.6 GeV). After taking into account the kinematical factor in the expression of the beam spin dependent contribution, CLAS and HERMES measurements were found to be consistent (see Fig.2), indicating that at energies as low as 4-6 GeV the behavior of spin asymmetries is consistent with higher energy measurements. CLAS data provides significant improvement of beam SSA measurement in the kinematical region where two data sets overlap, and extends beam SSA measurement to large x_B region not accessible by HERMES, providing a very significant improvement in statistical errors and an important input for studies of higher twist effects. Despite the fact that partonic formalism is much better suited for higher energy reactions, there is a reasonable agreement in size and behavior of beam SSAs measured in a wide energy range.

REFERENCES

1. D. W. Sivers, *Phys. Rev.* **D43**, 261–263 (1991).
2. M. Anselmino, and F. Murgia, *Phys. Lett.* **B442**, 470–478 (1998).
3. S. J. Brodsky, D. S. Hwang, and I. Schmidt, *Phys. Lett.* **B530**, 99–107 (2002).
4. J. C. Collins, *Phys. Lett.* **B536**, 43–48 (2002).
5. X. Ji, and F. Yuan, *Phys. Lett.* **B543**, 66–72 (2002).
6. P. J. Mulders, and R. D. Tangerman, *Nucl. Phys.* **B461**, 197–237 (1996).
7. A. Airapetian, et al., *Phys. Rev. Lett.* **84**, 4047–4051 (2000).
8. A. Airapetian, et al., *Phys. Rev.* **D64**, 097101 (2001).
9. A. Airapetian, et al., *Phys. Rev. Lett.* **94**, 012002 (2005).
10. A. Airapetian, et al., *Phys. Lett.* **B648**, 164–170 (2007).
11. V. Y. Alexakhin, et al., *Phys. Rev. Lett.* **94**, 202002 (2005).
12. H. Avakian, et al., *Phys. Rev.* **D69**, 112004 (2004).
13. H. Avakian, P. Bosted, V. Burkert, and L. Elouadrhiri, *AIP Conf. Proc.* **792**, 945–948 (2005).
14. J. Adams, et al., *Phys. Rev. Lett.* **92**, 171801 (2004).
15. M. Chiu, *AIP Conf. Proc.* **915**, 539–542 (2007).
16. I. Arsene, et al., *Phys. Rev. Lett.* **101**, 042001 (2008).
17. K. Abe, et al., *Phys. Rev. Lett.* **96**, 232002 (2006).
18. J. Levelt, and P. J. Mulders, *Phys. Lett.* **B338**, 357–362 (1994).
19. X. Ji, J. Ma, and F. Yuan, *Phys. Rev.* **D71**, 034005 (2005).
20. A. Afanasev, and C. E. Carlson (2003).
21. F. Yuan, *Phys. Lett.* **B589**, 28–34 (2004).
22. R. L. Jaffe, and X.-D. Ji, *Nucl. Phys.* **B375**, 527–560 (1992).
23. R. D. Tangerman, and P. J. Mulders, *Phys. Rev.* **D51**, 3357–3372 (1995).
24. A. Kotzinian, *Nucl. Phys.* **B441**, 234–248 (1995).
25. M. Burkardt, *hep-ph* **0807.2599** (2008).
26. A. Metz, and M. Schlegel, *Eur. Phys. J.* **A22**, 489–494 (2004).
27. A. V. Efremov, K. Goeke, and P. Schweitzer, *Nucl. Phys.* **A711**, 84–88 (2002).
28. A. Airapetian, et al., *Phys. Rev.* **D71**, 012003 (2005).
29. J. C. Collins, and A. Metz, *Phys. Rev. Lett.* **93**, 252001 (2004).
30. D. Boer, P. J. Mulders, and F. Pijlman, *Nucl. Phys.* **B667**, 201–241 (2003).
31. B. A. Mecking, et al., *Nucl. Instrum. Meth.* **A503**, 513–553 (2003).
32. F. X. Girod, et al., *Phys. Rev. Lett.* **100**, 162002 (2008).
33. L. Mankiewicz, A. Schafer, and M. Veltri, *Comput. Phys. Commun.* **71**, 305–318 (1992).
34. H. Avakian, and L. Elouadrhiri, *High Energy Spin Physics Proc., Editors A.V. Efremov and O.V. Teryaev* p. 239 (2003).

Drell-Yan Process and Nucleon Spin

J.C. Peng

University of Illinois at Urbana-Champaign, Urbana, IL 61801

Abstract. The role of the Drell-Yan process in studying nucleon spin is discussed. Recent results from the Fermilab E866 experiment on the angular distributions of Drell-Yan dimuons in $p+p$ and $p+d$ at 800 GeV/c are presented. These data are compared with the pion-induced Drell-Yan data, and with models which attribute the $\cos 2\phi$ azimuthal distribution to the presence of the transverse-momentum-dependent Boer-Mulders structure function $h_1^\perp$. Constraints on the magnitude of the sea-quark $h_1^\perp$ structure functions are obtained. Future prospects for studying the TMDs with Drell-Yan experiments at Fermilab and J-PARC are also discussed.

Keywords: Drell-Yan, azimuthal asymmetry, Boer-Mulders functions
PACS: 13.85.Qk, 14.20.Dh, 24.85.+p, 13.88.+e

INTRODUCTION

The Drell-Yan process [1], in which a charged lepton pair is produced in a hadron-hadron interaction via the electromagnetic $q\bar{q} \rightarrow l^+l^-$ process, has provided unique information on parton distributions. In particular, the Drell-Yan process has been used to determine the antiquark contents of nucleons and nuclei [2], as well as the quark distributions of pions, kaons, and antiprotons [3]. Such information is difficult, if not impossible, to obtain from DIS experiments. As the Drell-Yan process can be well described by next-to-leading order QCD calculations [4], a firm theoretical framework exists for utilizing the Drell-Yan process to extract the parton distributions.

The study of the transverse momentum dependent (TMD) parton distributions of the nucleon has received much attention in recent years as it provides new perspectives on the hadron structure and QCD [5]. These novel TMDs can be extracted from semi-inclusive deep-inelastic (SIDIS) scattering experiments. Recent measurements of the SIDIS by the HERMES [6] and COMPASS [7] collaborations have shown clear evidence for the existence of the T-odd Sivers functions. These data also allow the first determination [8] of the magnitude and flavor structure of the Sivers functions and the nucleon transversity distributions.

The TMD and transversity parton distributions can also be probed in Drell-Yan experiments. As pointed out [9] long time ago, the double transverse spin asymmetry in polarized Drell-Yan, A_{TT}, is proportional to the product of transversity distributions, $h_1(x_q)h_1(x_{\bar{q}})$. The single transverse spin asymmetry, A_N, is sensitive to the Sivers function [10], $f_{1T}^\perp(x)$ of the polarized proton (beam or target). Even unpolarized Drell-Yan experiments can be used to probe the TMD distribution function, since the $\cos 2\phi$ azimuthal angular dependence is proportional to the product of two Boer-Mulders functions [11], $h_1^\perp(x_1)\bar{h}_1^\perp(x_2)$. A unique feature of the Drell-Yan process is that, unlike the SIDIS, no fragmentation functions are involved. Therefore, the Drell-Yan process pro-

CP1149, *Spin Physics, 18th International Spin Physics Symposium*
edited by D. G. Crabb, D. B. Day, S. Liuti, X. Zheng, M. Poelker, and Y. Prok

vides an entirely independent technique for measuring the TMD functions. Furthermore, the proton-induced Drell-Yan process is sensitive to the sea-quark TMDs and can lead to flavor separation of TMDs when combined with the SIDIS data. Finally, the intriguing prediction [12] that the T-odd TMDs extracted from DIS will have a sign-change for the Drell-Yan process remains to be tested experimentally.

In this article, we first present the measurement of Drell-Yan azimuthal angular distributions in the Fermilab experiment E866. Implication of the data on the TMD Boer-Mulders function is discussed. Prospects for future Drell-Yan experiments for studying TMD and transversity distributions at Fermilab and J-PARC will also be discussed.

DRELL-YAN ANGULAR DISTRIBUTIONS AND BOER-MULDERS FUNCTIONS

Despite the success of perturbative QCD in describing the Drell-Yan cross sections, it remains a challenge to understand the angular distributions of the Drell-Yan process. Assuming dominance of the single-photon process, a general expression for the Drell-Yan angular distribution is [13]

$$\frac{d\sigma}{d\Omega} \propto 1+\lambda\cos^2\theta+\mu\sin 2\theta\cos\phi+\frac{\nu}{2}\sin^2\theta\cos 2\phi, \quad (1)$$

where θ and ϕ denote the polar and azimuthal angle, respectively, of the l^+ in the dilepton rest frame. In the "naive" Drell-Yan model, where the transverse momentum of the quark is ignored and no gluon emission is considered, $\lambda = 1$ and $\mu = \nu = 0$ are obtained. QCD effects [14] and non-zero intrinsic transverse momentum of the quarks [15] can both lead to $\lambda \neq 1$ and $\mu, \nu \neq 0$. However, λ and ν should still satisfy the relation $1-\lambda = 2\nu$ [13]. This so-called Lam-Tung relation, obtained as a consequence of the spin-1/2 nature of the quarks, is analogous to the Callan-Gross relation [16] in deep-inelastic scattering.

The first measurement of the Drell-Yan angular distribution was performed by the NA10 Collaboration for $\pi^- + W$ with the highest statistics at 194 GeV/c [17, 18]. The $\cos 2\phi$ angular dependences showed a sizable ν, increasing with dimuon transverse momentum (p_T) and reaching a value of ≈ 0.3 at $p_T = 2.5$ GeV/c. The Fermilab E615 Collaboration subsequently performed a measurement of $\pi^- + W$ Drell-Yan production at 252 GeV/c with broad coverage in the decay angle θ [19]. The E615 data showed that the Lam-Tung relation, $2\nu = 1-\lambda$, is clearly violated.

Boer pointed out [20] that the $\cos 2\phi$ angular dependences observed in NA10 and E615 could be due to the k_T-dependent parton distribution function $h_1^\perp$. This so-called Boer-Mulders function [11] is an example of a novel type of k_T-dependent parton distribution function, and it characterizes the correlation of a quark's transverse spin and its transverse momentum, k_T, in an unpolarized nucleon. The Boer-Mulders function is the chiral-odd analog of the Sivers function and owes its existence to the presence of initial/final state interactions [21]. While the Sivers function is beginning to be quantitatively determined from the SIDIS experiments, very little is known about the Boer-Mulders function so far.

To shed additional light on the origins of the NA10 and E615 Drell-Yan angular distributions, we recently analyzed $p+p$ and $p+d$ Drell-Yan angular distribution data at 800 GeV/c from Fermilab E866. There had been no report on the azimuthal angular distributions for proton-induced Drell-Yan – all measurements were for polar angular distributions [2, 22, 23]. Proton-induced Drell-Yan data provide a test of theoretical models. For example, the $\cos 2\phi$ dependence is expected to be much reduced in proton-induced Drell-Yan if the underlying mechanism involves the Boer-Mulders functions. This is due to the expectation that the Boer-Mulders functions are small for the sea-quarks [24]. However, if the QCD vacuum effect [25] is the origin of the $\cos 2\phi$ angular dependence, then the azimuthal behavior of proton-induced Drell-Yan should be similar to that of pion-induced Drell-Yan.

The measurement of the $\cos 2\phi$ dependence for the $p+d$ Drell-Yan cross sections has already been reported [26]. As shown in Fig. 1, significantly smaller (but non-zero) $\cos 2\phi$ azimuthal angular dependence was observed in the $p+d$ reaction. While pion-induced Drell-Yan process is dominated by annihilation between a valence antiquark in the pion and a valence quark in the nucleon, proton-induced Drell-Yan process involves a valence quark annihilating with a sea antiquark in the nucleon. Therefore, the $p+d$ result suggests [26] that the Boer-Mulders functions for sea antiquarks are significantly smaller than those for valence quarks.

A recent analysis [27] of the $p+d$ $\cos 2\phi$ data showed that the sea-quark Boer-Mulders functions are indeed smaller by a factor ~ 5 than the valence-quark Boer-Mulders functions. This analysis also indicated that the E866 $p+d$ data are consistent with the u and d Boer-Mulders functions having the same signs, as predicted by various models. However, the $p+d$ data alone can not provide an unambiguous determination of the flavor dependence of the Boer-Mulders functions. Moreover, it was recently pointed out [28, 29] that QCD processes would lead to sizeable $\cos 2\phi$ effect which has not been taken into account in the extractions [20, 27, 30] of Boer-Mulders functions from the Drell-Yan data. The $p+p$ Drell-Yan data should provide further constraints on the flavor dependence of the Boer-Mulders functions [30]. It is also interesting to compare the $p+p$ and $p+d$ data with the prediction of QCD.

Figure 2 shows the parameter ν vs. p_T for the $p+p$ and $p+d$ Drell-Yan data. The solid curves are calculations [27, 30] for $p+p$ and $p+d$ using parametrizations based on a fit to the $p+d$ Drell-Yan data. The predicted larger values of ν for $p+p$ compared to $p+d$ in the region of $p_T \sim 1.5$ GeV/c are not observed. Furthermore, the shape of the predicted p_T dependence differs from that of the data. This strongly suggests that there could be other mechanisms contributing to the $\cos 2\phi$ azimuthal angular dependence at large p_T region. In a recent paper [29], the QCD contribution to the $\cos 2\phi$ azimuthal angular dependence is given as

$$\nu = \frac{Q_\perp^2/Q^2}{1+\frac{3}{2}Q_\perp^2/Q^2}, \tag{2}$$

where $Q_\perp$ is the dimuon transverse momentum. The predicted QCD contribution is shown as the dotted curve is Fig. 2. Although neither the prediction based on Boer-Mulders functions nor the QCD prediction could describe the entire range of p_T, it is evident that the QCD contribution becomes important at high p_T and the Boer-Mulders

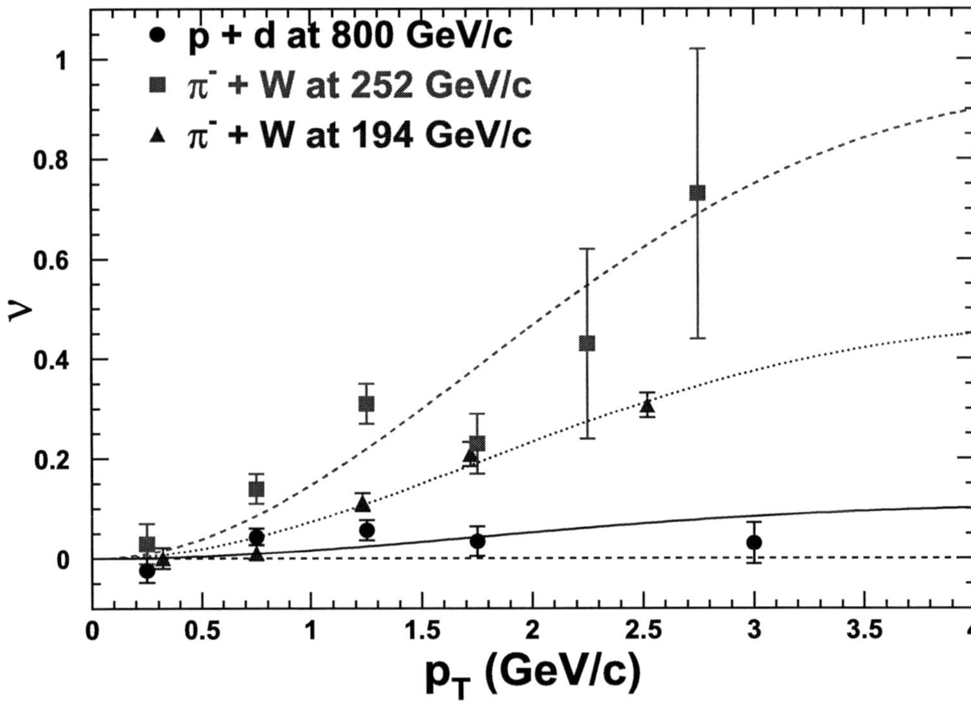

FIGURE 1. Parameter ν vs. p_T in the Collins-Soper frame for three Drell-Yan measurements. Fits to the data using an empirical expression [20] are also shown.

functions contribute at lower p_T. An analysis combining both effects is required before a reliable extraction of the Boer-Mulders functions could be obtained.

FUTURE PROSPECTS AT FERMILAB AND J-PARC

Future fixed-target dimuon experiments have been proposed at the 120 GeV Fermilab Main Injector (FMI) and the 50 GeV J-PARC facilities. The Fermilab proposal [31], E906, has been approved and is expected to start data-taking around 2011. Two dimuon proposals (P04 [32] and P24 [33]) have also been submitted to the J-PARC for approval. The lower beam energies at FMI and J-PARC present opportunities for extending the Drell-Yan measurements to larger x ($x > 0.25$). For given values of x_1 and x_2, the Drell-Yan cross section is proportional to $1/s$, hence a gain of ~ 16 times in the Drell-Yan cross sections can be obtained at the J-PARC energy of 50 GeV. This would extend the measurement of Boer-Mulders functions to larger x. In addition, the nuclear dependence of the Boer-Mulders functions can also be measured.

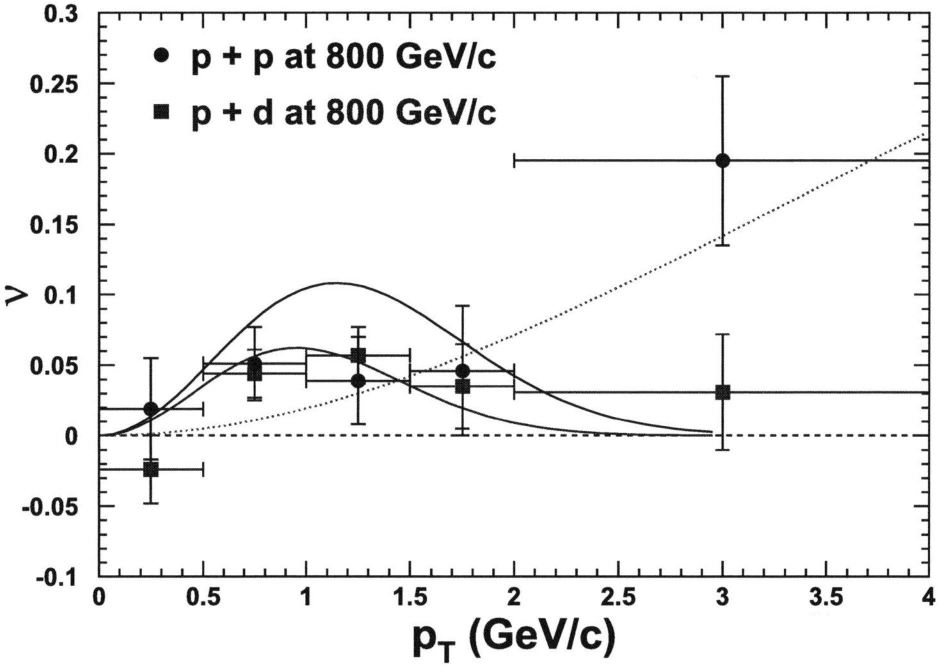

FIGURE 2. Parameter ν vs. p_T for the $p+p$ and $p+d$ Drell-Yan data. The solid curves are calculations [27] for $p+p$ and $p+d$ using parametrizations based on a fit to the $p+d$ Drell-Yan data. The dotted curve is the contribition from QCD process (Eq. 2).

With the possibility to accelerate polarized proton beams at J-PARC, the spin structure of the proton can also be investigated with the proposed dimuon experiments. In particular, polarzied Drell-Yan process with polarized beam and/or polarized target at J-PARC would allow a unique program on spin physics complementary to polarized DIS experiments and the RHIC-Spin program. Specific physics topics include the measurements of T-odd Boer-Mulders distribution function in unpolarized Drell-Yan, the extraction of T-odd Sivers distribution functions in singly transversely polarized Drell-Yan, the helicity distribution of antiqaurks in doubly longitudinally polarized Drell-Yan, and the transversity distribution in doubly transversely polarized Drell-Yan. It is worth noting that polarized Drell-Yan is one of the major physics program at the GSI Polarized Antiproton Experiment (PAX). The RHIC-Spin program will likely provide the first results on polarized Drell-Yan. However, the high luminosity and the broad kinematic coverage for the large-x region at J-PARC would allow some unique measurements to be performed in the J-PARC dimuon experiments.

ACKNOWLEDGMENTS

I am very grateful to Lingyan Zhu, Paul Reimer, and my collaborators on the E866 experiment at Fermilab. Illuminating comments from Matthias Burkardt, Feng Yuan, Bo-Qiang Ma, Werner Vogelsang, and Jianwei Qiu are greatly appreciated. I would also like to acknowledge the collaboration with Shin'ya Sawada and Yuji Goto on the J-PARC dimuon proposals.

REFERENCES

1. S.D. Drell and T.M. Yan, Phys. Rev. Lett. **25**, 316 (1970); Ann. Phys. (NY) **66**, 578 (1971).
2. P.L. McGaughey, J.M. Moss, and J.C. Peng, Annu. Rev. Nucl. Part. Sci. **49**, 217 (1999).
3. I.R. Kenyon, Rep. Prog. Phys. **45**, 1261 (1982); K. Freudenreich, Int. J. Mod. Phys. **A5**, 3643 (1990).
4. W.J. Stirling and M.R. Whalley, J. Phys. **G19**, D1 (1993).
5. V. Barone, A. Drago, and P. G. Ratcliffe, Phys. Rep. **359**, 1 (2002).
6. HERMES Collaboration, A. Airapetian *et al.*, Phys. Rev. Lett. **94**, 012002 (2005); M. Diefenthaler, arXiv: 0706.2242.
7. COMPASS Collaboration, V. Yu. Alexakhin *et al.*, Phys. Rev. Lett. **94**, 202002 (2005); M. Alekseev *et al.*, arXiv: 0802.2160.
8. W. Vogelsang and F. Yuan, Phys. Rev. D **72**, 054028 (2005); M. Anselmino *et al.*, Phys. Rev. D **72**, 094007 (2005); M. Anselmino *et al.*, arXiv:0807.0166.
9. J.P. Ralston and D.E. Soper, Nucl. Phys. B **152**, 109 (1979).
10. D. Sivers, Phys. Rev. D **41**, 83 (1990).
11. D. Boer and P.J. Mulders, Phys. Rev. D **57**, 5780 (1998).
12. J.C. Collins, Phys. Lett. B **536**, 43 (2002).
13. C.S. Lam and W.K. Tung, Phys. Rev. D **18**, 2447 (1978).
14. P. Chiappetta and M. LeBellac, Z. Phys. C **32**, 521 (1986).
15. J. Cleymans and M. Kuroda, Phys. Lett. B **105**, 68 (1981).
16. C.G. Callan and D. J. Gross, Phys. Rev. Lett. **22**, 156 (1969).
17. NA10 Collaboration, S. Falciano *et al.*, Z. Phys. C **31**, 513 (1986).
18. NA10 Collaboration, M. Guanziroli *et al.*, Z. Phys. C **37**, 545 (1988).
19. E615 Collaboration, J.S. Conway *et al.*, Phys. Rev. D **39**, 92 (1989); J.G. Heinrich *et al.*, Phys. Rev. D **44**, 1909 (1991).
20. D. Boer, Phys. Rev. D **60**, 014012 (1999).
21. D. Boer, S.J. Brodsky, and D.S. Hwang, Phys. Rev. D **67**, 054003 (2003).
22. E866 Collaboration, T.H. Chang *et al.*, Phys. Rev. Lett. **91**, 211801 (2003).
23. E866 Collaboration, C.N. Brown *et al.*, Phys. Rev. Lett. **86**, 2529 (2001).
24. Z. Lu, B.-Q. Ma, and I. Schmidt, Phys. Rev. D **75**, 014026 (2007).
25. A. Brandenburg, O. Nachtmann, and E. Mirkes, Z. Phys. C **60**, 697 (1993).
26. E866 Collaboration, L.Y. Zhu *et al.*, Phys. Rev. Lett. **99**, 082301 (2007).
27. B. Zhang, Z. Lu, B.-Q. Ma, and I. Schmidt, Phys. Rev. D **77**, 054011 (2008).
28. D. Boer and W. Vogelsang, Phys. Rev. D **74**, 014004 (2006).
29. E.L. Berger, J.-W. Qiu, and R.A. Rodriguez-Pedraza, Phys. Lett. B **656**, 74 (2007); Phys. Rev. D **76**, 074006 (2007).
30. B. Zhang, Z. Lu, B.-Q. Ma, and I. Schmidt, arXiv: 0807.0503.
31. http://www.phy.anl.gov/mep/drell-yan; D. Geesaman, P. Reimer, *et al.*, Fermilab Proposal P906 (1999).
32. http://j-parc.jp/NuclPart/pac_0606/pdf/p04-Peng.pdf; J. Peng, S. Sawada, *et al.*, J-PARC Proposal P04 (2006).
33. http://j-parc.jp/NuclPart/pac_0801/pdf/Goto.pdf; Y. Goto, H. Sato, *et al.*, J-PARC Proposal P24 (2007).

Exploring TMDs through Drell-Yan processes

Simone Arnold*, Andreas Metz† and Marc Schlegel**

*Institut für Theoretische Physik II, Ruhr-Universität Bochum, 44780 Bochum, Germany
†Department of Physics, Temple University, Philadelphia, PA 19122-6082, USA
**Theory Center, Jefferson Lab, 12000 Jefferson Avenue, Newport News, VA 23606, USA

Abstract. We present a comprehensive formalism for dilepton production from the collision of two polarized spin-$\frac{1}{2}$ hadrons by identifying the general angular distribution of the cross section in combination with a complete set of structure functions. The various structure functions are computed in the parton model approximation where we mainly consider the case when the transverse momentum of the dilepton pair is much smaller than its invariant mass. In this kinematical region dilepton production can be described in terms of transverse momentum dependent parton distributions.

Keywords: Dilepton Production, Polarization Observables, Parton Model
PACS: 12.38.Bx, 12.39.St, 13.85.Qk

INTRODUCTION

During the past decades dilepton production in high-energy collisions (the so-called Drell-Yan process [1, 2]) has played an important role in order to pin down parton distributions (PDFs) of hadrons. Though experimentally the Drell-Yan process is quite challenging, from a theoretical point of view it is the cleanest hard hadron hadron scattering process. The fact that no hadron is detected in the final state simplifies the proof of factorization in comparison to hadron hadron collisions with hadronic final states. This important point is one of the main reasons for the continued interest in the Drell-Yan reaction.

Currently, six programs for future Drell-Yan measurements are pursued (at COMPASS (CERN), FAIR (GSI), IHEP (Protvino), JINR (Dubna), J-PARC (KEK), and RHIC). While past measurements exclusively considered the unpolarized cross section, all the future programs are also aiming at polarization observables. Including polarization of the incoming hadrons opens up a variety of new opportunities for studying the strong interaction in both the perturbative and the non-perturbative regime. One major aim of the scheduled polarized Drell-Yan measurements is to explore transverse momentum dependent parton distributions (TMDs). The TMDs not only depend on the longitudinal momentum of a parton inside a hadron, but also on its (intrinsic) transverse momentum and, in general, describe the strength of various intriguing spin-spin or spin-orbit correlations of the parton-hadron system

In order to analyze upcoming polarized Drell-Yan data it is necessary to have a general and concise formalism at hand. Such a framework was provided in Ref. [3] and is summarized in this contribution. We present a decomposition of the process into a complete set of structure functions, and also express — for low transverse momenta q_T of the dilepton pair — the leading structure functions through TMDs. Our work completes previous studies [4, 5, 6, 7] on the polarized Drell-Yan process.

CP1149, *Spin Physics, 18th International Spin Physics Symposium*
edited by D. G. Crabb, D. B. Day, S. Liuti, X. Zheng, M. Poelker, and Y. Prok

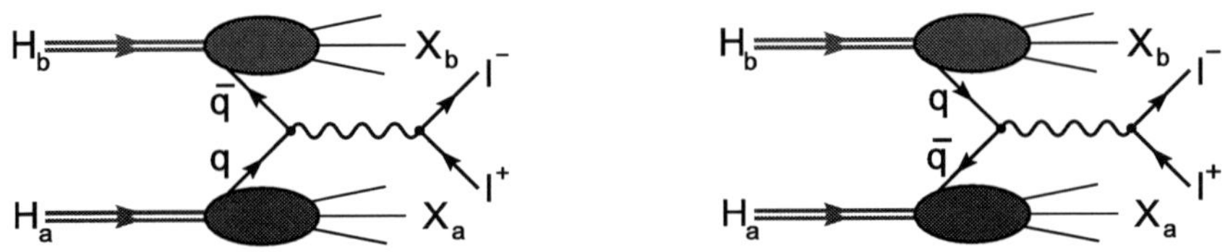

FIGURE 1. Amplitude for dilepton production in parton model approximation. Both diagrams have to be taken into account. The spectator systems X_a and X_b of the two hadrons are not detected.

CROSS SECTION AND HADRONIC TENSOR

To be now specific we consider the dilepton production

$$H_a(P_a,S_a)+H_b(P_b,S_b)\to l^-(l,\lambda)+l^+(l',\lambda')+X, \tag{1}$$

with (P_a,S_a) and (P_b,S_b) denoting the 4-momenta and the spin vectors of the incoming hadrons. One has $P_a^2=M_a^2$, $P_a\cdot S_a=0$, $S_a^2=-1$, and corresponding relations for the second hadron. The mass of the leptons in the final state is neglected, and we will sum over the lepton helicities λ, λ'. At large invariant mass q of the dilepton pair the process (1) can be described in the Drell-Yan model [1, 2], which corresponds to the parton model approximation (see also Fig. 1).

In the one-photon exchange approximation the (frame-independent) cross section of the Drell-Yan process is given by

$$\frac{l^0 l'^0\, d\sigma}{d^3\vec{l}\, d^3\vec{l}'}=\frac{\alpha_{em}^2}{F\, q^4}L_{\mu\nu}W^{\mu\nu}, \tag{2}$$

where $F\approx 2s$ represents the flux of the incoming hadrons and $q=l+l'$ the 4-momentum of the virtual photon. In Eq. (2), $L^{\mu\nu}$ denotes the spin-averaged leptonic tensor, while

$$W^{\mu\nu}(P_a,S_a;P_b,S_b;q)=\int\frac{d^4x}{(2\pi)^4}\,e^{iq\cdot x}\,\langle P_a,S_a;P_b,S_b\,|\,J_{em}^{\mu}(0)J_{em}^{\nu}(x)\,|\,P_a,S_a;P_b,S_b\rangle \tag{3}$$

is the hadronic tensor, which is determined by the electromagnetic current operator J_{em}^{μ}. The tensor $W^{\mu\nu}$ *a priori* is unknown and contains the information on the hadron structure. It has to fulfill certain constraints due to electromagnetic gauge invariance, parity, and hermiticity [6, 3]. By using these constraints, in Ref. [3] the hadronic tensor for two polarized spin-$\frac{1}{2}$ particles in the initial state has been decomposed into a set of 48 basis tensors multiplied by scalar functions (structure functions). The total hadronic tensor can be split up into the unpolarized, single polarized (for hadron H_a and hadron H_b), and double polarized tensor according to

$$W^{\mu\nu}=W_u^{\mu\nu}+W_a^{\mu\nu}+W_b^{\mu\nu}+W_{ab}^{\mu\nu}. \tag{4}$$

Here we merely have to consider the symmetric part of $W^{\mu\nu}$ because the spin-averaged leptonic tensor is symmetric under the exchange $\mu\leftrightarrow\nu$.

In the following we discuss in some detail the construction of the unpolarized hadronic tensor. The treatment of the polarized case essentially follows the same pattern. Since $W_u^{\mu\nu}$ depends on the 4-vectors q^μ, P_a^μ, and P_b^μ one can immediately write down the tensor basis

$$h_{u,1}^{\mu\nu},\ldots,h_{u,7}^{\mu\nu} = \Big\{g^{\mu\nu}\,,\, q^\mu q^\nu\,,\, P_a^\mu P_a^\nu\,,\, P_b^\mu P_b^\nu\,,\, q^\mu P_a^\nu + q^\nu P_a^\mu\,,\, q^\mu P_b^\nu + q^\nu P_b^\mu\,,\\ P_a^\mu P_b^\nu + P_a^\nu P_b^\mu\Big\}. \tag{5}$$

The expressions in (5) constitute a complete list of basis tensors being in accordance with the parity constraint. Therefore one can write in a first step

$$W_u^{\mu\nu} = \sum_{i=1}^{7} h_{u,i}^{\mu\nu}\,\tilde{V}_{u,i}\,, \tag{6}$$

where the structure functions $\tilde{V}_{u,i}$ depend on the invariants $P_a\cdot q$, $P_b\cdot q$, and q^2. The hermiticity constraint implies that the structure functions are real.

In order to implement gauge invariance we make use of a very elegant method proposed in Ref. [8] which is based on projection operators. We define

$$P^{\mu\nu} = g^{\mu\nu} - \frac{q^\mu q^\nu}{q^2}\,, \tag{7}$$

and let this projection operator act on the basis tensors in (5) according to $P^\mu{}_\rho\, h_{u,i}^{\rho\sigma}\, P_\sigma{}^\nu$. Doing so one ends up with four nonzero tensors which are gauge invariant by construction because of the property $q_\mu P^{\mu\nu} = P^{\mu\nu} q_\nu = 0$. This already provides the final form of the unpolarized hadronic tensor:

$$W_u^{\mu\nu} = \sum_{i=1}^{4} t_{u,i}^{\mu\nu}\, V_{u,i}\,, \tag{8}$$

with the four structure functions $V_{u,i}$, and the tensor basis

$$t_{u,1}^{\mu\nu},\ldots,t_{u,4}^{\mu\nu} = \left\{g^{\mu\nu} - \frac{q^\mu q^\nu}{q^2}\,,\, \tilde{P}_a^\mu \tilde{P}_a^\nu\,,\, \tilde{P}_b^\mu \tilde{P}_b^\nu\,,\, \tilde{P}_a^\mu \tilde{P}_b^\nu + \tilde{P}_a^\nu \tilde{P}_b^\mu\right\}. \tag{9}$$

In Eq. (9) we make use of the vectors

$$\tilde{P}_a^\mu = P_a^\mu - \frac{P_a\cdot q\, q^\mu}{q^2}\,, \qquad \tilde{P}_b^\mu = P_b^\mu - \frac{P_b\cdot q\, q^\mu}{q^2}\,. \tag{10}$$

These vectors vanish upon contraction with q which ensures electromagnetic gauge invariance of $W_u^{\mu\nu}$. It is needless to say that the tensor in (8) is frame-independent, and the contraction with the leptonic tensor can be performed in any frame.

To construct the polarization dependent parts of the hadronic tensor also the spin vectors S_a and S_b have to be included in the parameterization. One can then proceed along the lines discussed for the unpolarized case. Special care, however, has to be taken to remove redundant tensors. (See [3] for more details.) While $W_a^{\mu\nu}$ and $W_b^{\mu\nu}$ contain eight basis tensors, $W_{ab}^{\mu\nu}$ is given by 28 structures leading to a total of 48 independent tensors for $W^{\mu\nu}$. In Ref. [3] a complete list of these tensors was derived for the first time.

ANGULAR DISTRIBUTION OF THE CROSS SECTION

By means of the general form of the hadronic tensor one can now write down the full angular distribution of the Drell-Yan cross section. This is most conveniently done in a dilepton rest frame like the Collins-Soper frame [9] or the Gottfried-Jackson frame [10]. In such a frame the orientation of the dilepton pair is entirely specified by two angles $(\theta\,,\,\phi)$. One finds the following general form of the cross section [3]:

$$
\begin{aligned}
\frac{d\sigma}{d^4q\,d\Omega} = \frac{\alpha_{em}^2}{F\,q^2}\times \\
& \Big\{\Big((1+\cos^2\theta)\,F_{UU}^1 + (1-\cos^2\theta)\,F_{UU}^2 + \sin 2\theta\cos\phi\,F_{UU}^{\cos\phi} + \sin^2\theta\cos 2\phi\,F_{UU}^{\cos 2\phi}\Big) \\
& + S_{aL}\Big(\sin 2\theta\sin\phi\,F_{LU}^{\sin\phi} + \sin^2\theta\sin 2\phi\,F_{LU}^{\sin 2\phi}\Big) \\
& + |\vec{S}_{aT}|\Big[\sin\phi_a\Big((1+\cos^2\theta)\,F_{TU}^1 + (1-\cos^2\theta)\,F_{TU}^2 + \sin 2\theta\cos\phi\,F_{TU}^{\cos\phi} \\
& \qquad + \sin^2\theta\cos 2\phi\,F_{TU}^{\cos 2\phi}\Big) \\
& \qquad + \cos\phi_a\Big(\sin 2\theta\sin\phi\,F_{TU}^{\sin\phi} + \sin^2\theta\sin 2\phi\,F_{TU}^{\sin 2\phi}\Big)\Big] \\
& + S_{aL}\,S_{bL}\Big((1+\cos^2\theta)\,F_{LL}^1 + (1-\cos^2\theta)\,F_{LL}^2 + \sin 2\theta\cos\phi\,F_{LL}^{\cos\phi} \\
& \qquad + \sin^2\theta\cos 2\phi\,F_{LL}^{\cos 2\phi}\Big) + \ldots\Big\}\,. \qquad (11)
\end{aligned}
$$

In Eq. (11) we have limited ourselves to the unpolarized case, single polarization of one hadron, as well as double longitudinal polarization. The object ϕ_a is the azimuthal angle of the transverse component $\vec{S}_{aT}$ of the spin vector for hadron H_a. In total one finds 48 independent angular structures which are multiplied by the structure functions F_i. Of course there exist relations between the structure functions in (11) and the V_i discussed in the previous section which are, however, not needed here. We emphasize that the angular distribution in (11) holds for any dilepton rest frame, but the numerical values of the structure functions change when going from one frame to another.

Concerning the angular dependence of the unpolarized cross section the so-called Lam-Tung relation [11, 12]

$$
F_{UU}^2 = 2\,F_{UU}^{\cos 2\phi} + \mathcal{O}(\alpha_s^2)\,, \qquad (12)
$$

has attracted a lot of attention in the past. This relation is exact if one computes the process to $\mathcal{O}(\alpha_s)$ in collinear perturbative QCD. On the other hand, data for $\pi^- N \rightarrow \mu^-\mu^+ X$ taken at CERN [13, 14] and at Fermilab [15] are in disagreement with this relation. In Ref. [7] it was pointed out that intrinsic transverse motion of initial state partons might be responsible for the observed violation of the Lam-Tung relation. In the following section we will briefly return to this point. It is quite interesting that one has a corresponding relation for double longitudinal polarization [16, 17],

$$
F_{LL}^2 = 2\,F_{LL}^{\cos 2\phi} + \mathcal{O}(\alpha_s^2)\,. \qquad (13)
$$

Unfortunately, experimental data on the structure functions in (13) do not yet exist.

PARTON MODEL APPROXIMATION AND TMDS

In this section we are dealing with the parton model description of the Drell-Yan process, i.e., of the structure functions in (11). Up to this point we didn't specify the external kinematics of the reaction. In the following we will now consider the kinematical regime where the transverse momentum q_T of the dilepton pair is of the order of a typical hadronic mass scale which means, in particular, that it is much smaller than the hard scale q. This is the region where TMDs enter the description of the Drell-Yan process in a natural way [4, 6, 7, 3]. Our treatment is restricted to leading twist, i.e., to the leading order of an expansion in powers of $1/q$. Mainly because of the potential problems of subleading twist TMD-factorization pointed out in Refs. [18, 19] we refrain here from including the twist-3 case.

At leading twist one has eight TMDs [20, 21, 22, 23] for quarks, where six of them are T-even and two are T-odd (the Sivers function $f_{1T}^{\perp}$ [24] and the Boer-Mulders function $h_1^{\perp}$ [21]). In general, the latter can be responsible for single spin asymmetries in hard semi-inclusive reactions. A complete list of the structure functions F_i in Eq. (11) expressed in terms of these leading TMDs can be found in [3]. For brevity we give here just a very symbolic notation for the various polarized cross sections [3]:

$$\begin{aligned}
\sigma_{UU} &\propto f_1 f_1 + \cos 2\phi \, h_1^{\perp} h_1^{\perp} \,, \\
\sigma_{LU} &\propto \sin 2\phi \, h_{1L}^{\perp} h_1^{\perp} \,, \\
\sigma_{TU} &\propto f_{1T}^{\perp} f_1 + \sin 2\phi \, h_1 h_1^{\perp} + \sin 2\phi \, h_{1T}^{\perp} h_1^{\perp} \,, \\
\sigma_{LL} &\propto g_{1L} g_{1L} + \cos 2\phi \, h_{1L}^{\perp} h_{1L}^{\perp} \,, \\
\sigma_{TL} &\propto g_{1T} g_{1L} + \cos 2\phi \, h_1 h_{1L}^{\perp} + \cos 2\phi \, h_{1T}^{\perp} h_{1L}^{\perp} \,, \\
\sigma_{TT} &\propto f_{1T} f_{1T} + g_{1T} g_{1T} + \cos 2\phi \, h_1 h_1 + \cos 2\phi \, h_1 h_{1T}^{\perp} + \cos 2\phi \, h_{1T}^{\perp} h_{1T}^{\perp} \,. \qquad (14)
\end{aligned}$$

In (14) we have omitted any dependence on the azimuthal angles ϕ_a and ϕ_b. We have also suppressed the dependence of the TMDs on the longitudinal as well as the transverse quark momenta. In addition, we left out a convolution for the transverse parton momenta. At leading twist there exist 16 independent structure functions [3]. Each of them contains two TMDs — one associated with hadron H_a, the other one with hadron H_b. The so-called chiral-even TMDs (denoted by the letter f or g) appear exactly twice, while each chiral-odd TMD (denoted by h) appears four times. Therefore, not only a complete experiment for mapping out the leading twist quark TMDs is possible, but rather one also can perform cross checks. In this respect the Drell-Yan process has an advantage over semi-inclusive deep-inelastic scattering with a polarized incoming lepton and target for which each TMD appears just once [23]. We note in passing that all the chiral-odd TMDs drop out when integrating upon the azimuthal angle ϕ of the dilepton pair.

According to (14) the Boer-Mulders TMD $h_1^{\perp}$ gives rise to a $\cos 2\phi$ modulation of the unpolarized cross section. It may therefore be responsible for the observed violation of the Lam-Tung relation discussed in the previous section [7]. However, more experimental and theoretical studies are needed before one can unambiguously conclude that a possible breakdown of the Lam-Tung relation can be traced back to transverse parton motion. We also point out that the TMD $h_{1L}^{\perp}$ can generate a corresponding $\cos 2\phi$ dependence for double longitudinal polarization of the incoming hadrons.

SUMMARY

We have presented a general formalism for dilepton production from the collision of two polarized spin-$\frac{1}{2}$ hadrons. To this end we have identified a full set of structure functions, and supplemented our model-independent treatment by a parton model calculation [3]. Our work completes previous studies in the literature [4, 5, 6, 7], and can serve as a basis for analyzes of future polarized Drell-Yan data. Polarized Drell-Yan is an almost ideal reaction for mapping out TMDs, which contain a large amount of information on the partonic structure of hadrons. Last but not least, by (transversely) polarizing one of the hadrons one can check the predicted sign reversal of T-odd TMDs [25]. This sign change is at the heart of our present understanding of single spin asymmetries in QCD.

ACKNOWLEDGMENTS

This work has partially been supported by the Verbundforschung "Hadronen und Kerne" of the BMBF and by the Deutsche Forschungsgemeinschaft (DFG).

REFERENCES

1. S.D. Drell and T.M. Yan, Phys. Rev. Lett. **25**, 316 (1970) [Erratum-ibid. **25**, 902 (1970)].
2. S.D. Drell and T.M. Yan, Annals Phys. **66**, 578 (1971) [Annals Phys. **281**, 450 (2000)].
3. S. Arnold, A. Metz, and M. Schlegel, arXiv:0809.2262 [hep-ph].
4. J.P. Ralston and D.E. Soper, Nucl. Phys. **B152**, 109 (1979).
5. B. Pire and J.P. Ralston, Phys. Rev. D **28**, 260 (1983).
6. R.D. Tangerman and P.J. Mulders, Phys. Rev. D **51**, 3357 (1995).
7. D. Boer, Phys. Rev. D **60**, 014012 (1999).
8. W.A. Bardeen and W.K. Tung, Phys. Rev. **173**, 1423 (1968).
9. J.C. Collins and D.E. Soper, Phys. Rev. D **16**, 2219 (1977).
10. K. Gottfried and J.D. Jackson, Nuovo Cim. **33**, 309 (1964).
11. C.S. Lam and W.K. Tung, Phys. Rev. D **18**, 2447 (1978); Phys. Rev. D **21**, 2712 (1980).
12. J.C. Collins, Phys. Rev. Lett. **42**, 291 (1979).
13. S. Falciano *et al.* [NA10 Collaboration], Z. Phys. C **31**, 513 (1986).
14. M. Guanziroli *et al.* [NA10 Collaboration], Z. Phys. C **37**, 545 (1988).
15. J.S. Conway *et al.* [E615 Collaboration], Phys. Rev. D **39**, 92 (1989).
16. E.L. Berger, J.W. Qiu, and R.A. Rodriguez-Pedraza, Phys. Rev. D **76**, 074006 (2007).
17. W. Vogelsang, private communication, 2008.
18. L.P. Gamberg, D.S. Hwang, A. Metz, and M. Schlegel, Phys. Lett. B **639**, 508 (2006).
19. A. Bacchetta, D. Boer, M. Diehl, and P. J. Mulders, JHEP **0808**, 023 (2008).
20. P.J. Mulders and R.D. Tangerman, Nucl. Phys. **B461**, 197 (1996) [Erratum-ibid. **B484**, 538 (1997)].
21. D. Boer and P.J. Mulders, Phys. Rev. D **57**, 5780 (1998).
22. K. Goeke, A. Metz, and M. Schlegel, Phys. Lett. B **618**, 90 (2005).
23. A. Bacchetta, M. Diehl, K. Goeke, A. Metz, P. Mulders, and M. Schlegel, JHEP **0702**, 093 (2007).
24. D.W. Sivers, Phys. Rev. D **41**, 83 (1990); Phys. Rev. D **43**, 261 (1991).
25. J.C. Collins, Phys. Lett. B **536**, 43 (2002).

Double-Spin Asymmetry A_{LT} for Polarized Drell-Yan Process in $p\bar{p}$ Collisions: Wandzura-Wilczek Contribution

Yuji Koike*, Kazuhiro Tanaka† and Shinsuke Yoshida*

*Department of Physics, Niigata University, Ikarashi, Niigata 950-2181, Japan
†Department of Physics, Juntendo University, Inba, Chiba 270-1695, Japan

Abstract. The longitudinal-transverse spin asymmetry A_{LT} in the polarized Drell-Yan process depends on twist-3 quark distributions of nucleon. In addition to the contributions associated with the twist-3 operators, these distributions contain the "Wandzura-Wilczek (WW)" part, which is determined by a certain integral of twist-2 distributions. The recently obtained empirical information on the transversity allows a realistic estimate of the WW contribution to A_{LT} for the polarized $p\bar{p}$ collisions. Our results indicate that rather large A_{LT} ($\sim$ 10%) can be observed in the proposed spin experiments at GSI, with novel pattern as a function of dilepton mass compared with A_{TT} and A_{LL}.

Keywords: Drell-Yan process, Antiprotons, Longitudinal-transverse spin asymmetry, Twist-3
PACS: 12.38.-t,13.85.Qk,13.88.+e

The proposed polarization experiments with antiprotons at GSI provide promising opportunities to probe chiral-odd spin-dependent parton distributions h_1, h_L [1]. The double transverse-spin asymmetry for the Drell-Yan (DY) process in collisions of transversely polarized protons and antiprotons, $p^\uparrow \bar{p}^\uparrow \to l^+ l^- X$, is given as ($d\omega \equiv dQ^2 dx_F d\Omega$)

$$A_{TT} = \frac{d\sigma^{\uparrow\uparrow}/d\omega - d\sigma^{\uparrow\downarrow}/d\omega}{d\sigma^{\uparrow\uparrow}/d\omega + d\sigma^{\uparrow\downarrow}/d\omega} = \hat{a}_{TT}(\theta,\phi)\frac{\sum_a e_a^2 h_1^a(x_1,Q^2) h_1^a(x_2,Q^2)}{\sum_a e_a^2 f_1^a(x_1,Q^2) f_1^a(x_2,Q^2)}, \tag{1}$$

at the leading order (LO) QCD, for the dilepton production with the invariant mass Q and the direction given by the angle $\Omega = (\theta,\phi)$. h_1^a and f_1^a denote the transversity and unpolarized quark-distributions inside proton, and the summation is over all quark and anti-quark flavors with e_a the corresponding electric charge. The scaling variables $x_{1,2}$ represent the momentum fractions associated with the partons annihilating via the DY mechanism, such that $Q^2 = (x_1 P_1 + x_2 P_2)^2 = x_1 x_2 s$ and $x_F = x_1 - x_2$, where $s = (P_1 + P_2)^2$ is the CM energy squared of $p^\uparrow \bar{p}^\uparrow$. $\hat{a}_{TT}(\theta,\phi)$ represents the asymmetry in the parton level. In particular, moderate energies at GSI, $30 \lesssim s \lesssim 200$ GeV^2, allow us to measure (1) for $0.2 \lesssim Q/\sqrt{s} \lesssim 0.7$, and probe $h_1(x_{1,2},Q^2)$ in the "valence region". It turns out [2] that A_{TT} at GSI is quite stable when including the QCD (resummation and fixed-order) corrections, and that the recent empirical information on h_1 [3] leads to large value of (1) at GSI. We note that the double longitudinal-spin asymmetry A_{LL} for $p^\rightarrow \bar{p}^\rightarrow \to l^+ l^- X$ is given by (1) with the replacement $h_1 \to g_1$, $\hat{a}_{TT} \to \hat{a}_{LL}$, where g_1 is the chiral-even, helicity quark-distribution and $\hat{a}_{LL}$ is the corresponding partonic asymmetry.

It should not be overlooked that the double-spin longitudinal-transverse asymmetry $A_{LT} = [d\sigma^{\rightarrow\uparrow}/d\omega - d\sigma^{\rightarrow\downarrow}/d\omega]/[d\sigma^{\rightarrow\uparrow}d\omega + d\sigma^{\rightarrow\downarrow}/d\omega]$ is also readily accessible in the DY experiments at GSI: A_{LT} plays a distinguished role in spin physics, allowing us

CP1149, *Spin Physics, 18th International Spin Physics Symposium*
edited by D. G. Crabb, D. B. Day, S. Liuti, X. Zheng, M. Poelker, and Y. Prok

to access the twist-3 spin-dependent parton distributions as leading effects (see [1]),

$$A_{LT} = \frac{M}{Q}\hat{a}_{LT}(\theta,\phi)\frac{\sum_a e_a^2\left[g_1^a(x_1,Q^2)x_2g_T^a(x_2,Q^2)+x_1h_L^a(x_1,Q^2)h_1^a(x_2,Q^2)\right]}{\sum_a e_a^2 f_1^a(x_1,Q^2)f_1^a(x_2,Q^2)}, \quad (2)$$

with the proton mass M, although suppressed by M/Q compared to A_{TT}, A_{LL} that receive contribution only from twist-2 distributions. g_T^a and h_L^a denote the twist-3 distributions associated with the transversely and longitudinally polarized proton, respectively; the chiral-even distribution g_T^a is also accessible by the longitudinal-transverse asymmetry associated with the structure function g_2 in the polarized DIS [4], but the chiral-odd h_L^a is not accessible by inclusive DIS. We assess the potential of A_{LT} at GSI experiments [1], extending the previous study on A_{LT} in pp-collision at RHIC energy [5].

The distributions h_L and g_T contain the piece expressed by twist-2 matrix element as

$$h_L^a(x,Q^2) = 2x\int_x^1 dy\frac{h_1^a(y,Q^2)}{y^2}+\cdots, \qquad g_T^a(x,Q^2) = \int_x^1 dy\frac{g_1^a(y,Q^2)}{y}+\cdots, \quad (3)$$

where the ellipses stand for "genuine twist-3" contributions given as matrix element of the twist-3 operators; those twist-3 operators can be reexpressed as quark-gluon-quark three-body correlation operators on the lightcone, using the QCD equations of motion (see [6, 7]). We call the twist-2 component, shown explicitly in (3), the "Wandzura-Wilczek (WW)" part. Because the operators with different "geometric twist" do not mix with each other under QCD evolution with Q^2, both x- and Q^2-dependences of the WW part are determined solely by those of the corresponding twist-2 distributions.

We work at LO QCD with (2), which provides sufficient accuracy for our first estimate of A_{LT} at GSI. We may anticipate that QCD corrections to A_{LT} at GSI kinematics could be small, similarly to the case for A_{TT} mentioned above. We also employ the "WW approximation" to A_{LT}, i.e., evaluate (2) using (3) with the genuine twist-3 contributions omitted: the WW approximation for g_T of (3) is supported by the data on the structure function g_2 from the polarized DIS experiments [4], as well as from lattice QCD simulation [8]. Estimates from nucleon models, combined with the QCD evolution which is different between the relevant twist-2 [9] and twist-3 operators [6, 7] such that the latter is more strongly suppressed for high Q^2 than the former, also suggest that the WW part of (3) dominates h_L and g_T for $Q^2 \gg 1$ GeV^2 [5, 10].

We use the LO GRV98 [11] and GRSV2000 ("standard scenario") [12] distributions for the unpolarized and longitudinally-polarized quark distributions f_1 and g_1, respectively. For the LO transversity distribution h_1, we are guided by the recent information from the LO global fit [3]: we find that a useful estimate can be obtained by assuming the relation, $h_1^a(x,\mu^2) = g_1^a(x,\mu^2)$, at a low scale μ ($\mu^2 = 0.26$ GeV^2 using the GRSV2000 g_1); its QCD evolution from μ^2 to Q^2 is controlled by the LO DGLAP kernel [9] for h_1. Comparing the resulting transversity [1] with h_1 from the LO global fit [3], our h_1 lies slightly outside the error band of the fit, for the valence region $0.2 \lesssim x \lesssim 0.7$ relevant for the GSI kinematics. Thus our h_1 will provide a realistic estimate of the upper bound of the relevant asymmetries using (1)-(3), implied by the present empirical uncertainty in transversity. Similar choice for h_1 has been adopted to calculate QCD corrections for A_{TT} at GSI [2]. In the valence region relevant for the GSI kinematics, the dominance

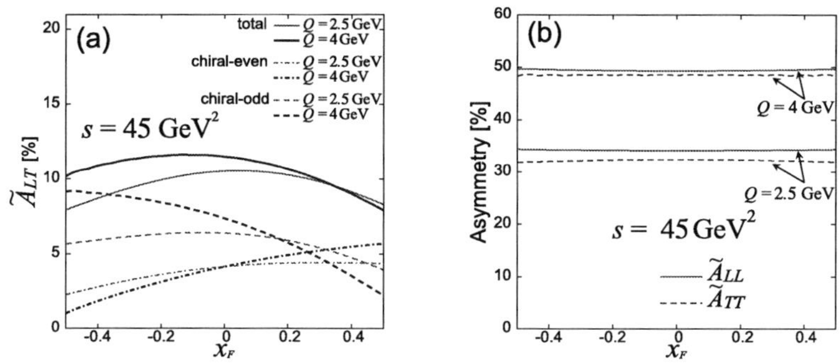

FIGURE 1. (a) $\widetilde{A}_{LT}$ and (b) $\widetilde{A}_{LL}$ and $\widetilde{A}_{TT}$ as a function of x_F for $Q = 2.5$ and 4 GeV at $s = 45$ GeV^2.

of the u-quark contribution is observed for all twist-2 quark-distributions, $(h_1^u)^2 \gg (h_1^a)^2$ with $a = \bar{u}, d, \bar{d}, \ldots$ [1, 3], and similarly for g_1 and f_1 [12, 11]. Calculating (3) using our twist-2 distributions, we find the similar u-quark dominance for h_L and g_T in the WW approximation [1]. Therefore, from (2) at GSI kinematics,

$$\widetilde{A}_{LT} \equiv \frac{A_{LT}}{(M/Q)\hat{a}_{LT}(\theta,\phi)} \simeq \frac{g_1^u(x_1,Q^2)x_2\int_{x_2}^1 dy \frac{g_1^u(y,Q^2)}{y} + 2x_1^2\int_{x_1}^1 dy \frac{h_1^u(y,Q^2)}{y^2} h_1^u(x_2,Q^2)}{f_1^u(x_1,Q^2)f_1^u(x_2,Q^2)} + \cdots, \tag{4}$$

where the ellipses denote the contributions associated with the genuine twist-3 operators. Thus the WW approximation to A_{LT} at GSI is directly related to the behavior of h_1^u.

We present the results [1] for the "reduced asymmetries", $\widetilde{A}_{LT}$, defined in (4), and $\widetilde{A}_{YY} \equiv A_{YY}/\hat{a}_{YY}$ $(Y = L, T)$. Figure 1(a) shows $\widetilde{A}_{LT}$, along with the separated contributions from the chiral-even and -odd distributions corresponding to the first and second terms in the numerator in (2). The value of $\widetilde{A}_{LT}$ is significant. Its x_F-behavior is in strong contrast to the flat and symmetric behavior of $\widetilde{A}_{LL}$, $\widetilde{A}_{TT}$ at the same kinematics, shown in Fig. 1(b). $\widetilde{A}_{LT}$ is smaller than $\widetilde{A}_{TT}$, $\widetilde{A}_{LL}$ by the additional $x_{1,2}$ factor in (2) compared with (1), and also by the behavior of h_L and g_T where the integral for the WW part, with the factors $1/y^2$, $1/y$ in (3), shifts the peak of the distributions to lower x with the suppressed peak-height, compared with the corresponding twist-2 distributions [1]. Indeed, for pp collisions where the sea-quark region is probed, these effects, in particular the additional $x_{1,2}$ factor, lead to $\widetilde{A}_{LT}$ much smaller than the corresponding $\widetilde{A}_{LL}$, $\widetilde{A}_{TT}$ [5].

Figure 2 shows the relevant asymmetries with $x_1 = x_2 = Q/\sqrt{s}$ as a function of Q. The Q dependence of $\widetilde{A}_{TT}$ as well as $\widetilde{A}_{LL}$ directly reflects the x dependence of the corresponding distributions, because the u-quark dominance in (1) implies that $\widetilde{A}_{TT}$ is controlled by the ratio $h_1^u(x,Q^2)/f_1^u(x,Q^2)$ and the scale dependence in this ratio turns out to almost cancel between the numerator and denominator in the valence region relevant at GSI [2]. With GRV and GRSV parameterizations, the ratio $h_1^u(x, 1\text{GeV}^2)/f_1^u(x, 1\text{GeV}^2)$, as well as $g_1^u(x, 1\text{GeV}^2)/f_1^u(x, 1\text{GeV}^2)$, is actually an increasing function of x, leading to the Q-dependence in Fig. 2(b) through $x = Q/\sqrt{s}$. For $\widetilde{A}_{LT}$, however, the cancellation of the

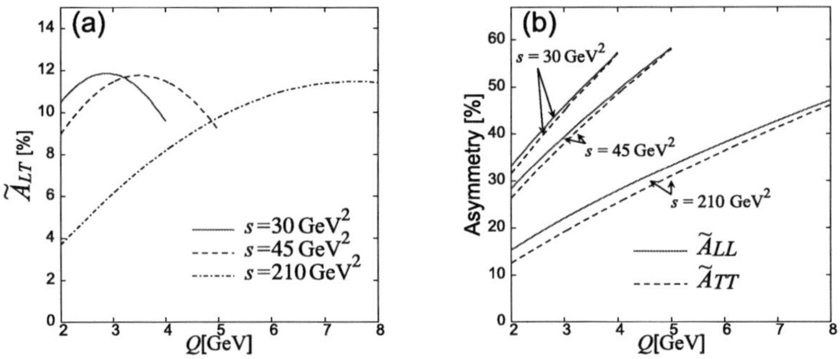

FIGURE 2. (a) $\widetilde{A}_{LT}$ and (b) $\widetilde{A}_{LL}$ and $\widetilde{A}_{TT}$ at $x_F = 0$ as a function of Q for $s = 30$, 45 and 210 GeV^2.

scale dependence between the numerator and denominator in (4) is less complete due to the additional y-integral for the WW part, which, combined with the additional factor x_1 or x_2 $(= Q/\sqrt{s})$, results in the novel Q-dependence in Fig. 2(a).

We have presented a first realistic estimate of the WW contribution to A_{LT} at GSI. Our results indicate that rather large $A_{LT} \sim 10\%$ can be observed. If the strong deviation from our results were observed in the future GSI measurements of A_{LT}, this would provide an indication of large twist-3 effect, associated with the quark-gluon-quark correlation inside nucleon, in particular, in the chiral-odd spin structure. Such data can be analyzed utilizing the QCD evolution of the three-body operators for large number of colors [7], which allows us to avoid the notorious complication in the evolution at higher twist.

ACKNOWLEDGMENTS

The work of Y.K. is supported by Uchida Energy Science Promotion Foundation. The work of K.T. is supported by the Grant-in-Aid for Scientific Research No. B-19340063.

REFERENCES

1. Y. Koike, K. Tanaka and S. Yoshida, Phys. Lett. **B668**, 286 (2008).
2. H. Kawamura, J. Kodaira and K. Tanaka, Phys. Lett. **B662**, 139 (2008).
3. M. Anselmino *et al.*, Phys. Rev. **D75**, 054032 (2007); arXiv:0812.4366 [hep-ph].
4. P.L. Anthony et al. [SLAC E155 Collaboration], Phys. Lett. **B553**, 18 (2003).
5. Y. Kanazawa, Y. Koike and N. Nishiyama, Phys. Lett. **B430**, 195 (1998).
6. Y. Koike and K. Tanaka, Phys. Rev. **D51**, 6125 (1995); J. Kodaira, Y. Yasui, K. Tanaka and T. Uematsu, Phys. Lett. **B387**, 855 (1996); J. Kodaira and K. Tanaka, Prog. Theor. Phys. **101**, 191 (1999).
7. A. Ali, V.M. Braun and G. Hiller, Phys. Lett. **B266**, 117 (1991); I.I. Balitsky, V.M. Braun, Y. Koike and K. Tanaka, Phys. Rev. Lett. **77**, 3078 (1996).
8. H. W. Lin, T. Blum, S. Ohta, S. Sasaki and T. Yamazaki, Phys. Rev. **D78**, 014505 (2008).
9. X. Artru and M. Mekhfi, Z. Phys. **C45**, 669 (1990); see also A. Hayashigaki, Y. Kanazawa and Y. Koike, Phys. Rev. **D56**, 7350 (1997); W. Vogelsang, Phys. Rev. **D57**, 1886 (1998).
10. Y. Kanazawa and Y. Koike, Phys. Lett. **B403**, 357 (1997).
11. M. Glück, E. Reya and A. Vogt, Eur. Phys. J. **C5**, 461 (1998).
12. M. Glück, E. Reya, M. Stratmann, and W. Vogelsang, Phys. Rev. **D63**, 094005 (2001).

Double Transverse-Spin Asymmetries for Small-Q_T Drell-Yan Pair Production in $p\bar{p}$ Collisions

Hiroyuki Kawamura* and Kazuhiro Tanaka†

*Department of Mathematical Sciences, University of Liverpool, Liverpool L69 3BX, UK
†Department of Physics, Juntendo University, Inba, Chiba 270-1695, Japan

Abstract. We discuss the double-spin asymmetries in transversely polarized Drell-Yan process, calculating all-order gluon resummation corrections up to the next-to-leading logarithmic accuracy. This resummation is relevant when the transverse-momentum Q_T of the produced lepton pair is small, and reproduces the (fixed-order) next-to-leading QCD corrections upon integrating over Q_T. The resummation corrections in $p\bar{p}$-collision behave differently compared with pp-collision cases, and are small at the kinematics in the proposed GSI experiments. This fact allows us to predict large value of the double-spin asymmetries at GSI, using recent empirical information on the transversity.

Keywords: Transversity, Drell-Yan process, Antiprotons, Soft gluon resummation
PACS: 12.38.-t,12.38.Cy,13.85.Qk,13.88.+e

The double-spin asymmetry in Drell-Yan (DY) process with transversely-polarized protons and antiprotons, $p^{\uparrow}\bar{p}^{\uparrow} \to l^+l^-X$, for azimuthal angle ϕ of a lepton measured in the rest frame of the dilepton l^+l^- with invariant mass Q and rapidity y, is given by

$$A_{TT} = \frac{d\sigma^{\uparrow\uparrow}/d\omega - d\sigma^{\uparrow\downarrow}/d\omega}{d\sigma^{\uparrow\uparrow}/d\omega + d\sigma^{\uparrow\downarrow}/d\omega} \equiv \frac{\Delta_T d\sigma/d\omega}{d\sigma/d\omega} = \frac{\cos(2\phi)}{2} \frac{\sum_q e_q^2 \delta q(x_1,Q^2)\delta q(x_2,Q^2) + ..}{\sum_q e_q^2 q(x_1,Q^2) q(x_2,Q^2) + ..} \quad (1)$$

where $d\omega \equiv dQ^2 dy d\phi$, the summation is over all quark and antiquark flavors with $\delta q(x,Q^2)$ and $q(x,Q^2)$ being the transversity and unpolarized quark-distributions inside a proton, and the ellipses stand for the corrections of next-to-leading order (NLO) and higher in QCD perturbation theory. The scaling variables $x_{1,2}$ represent the momentum fractions associated with the partons annihilating via the DY mechanism, such that $Q^2 = (x_1P_1 + x_2P_2)^2 = x_1x_2S$ and $y = (1/2)\ln(x_1/x_2)$, where $S = (P_1+P_2)^2$ is the CM energy squared of $p^{\uparrow}\bar{p}^{\uparrow}$. In the proposed polarization experiments at GSI [1], moderate energies, $30 \lesssim S \lesssim 200$ GeV^2, allow us to measure (1) for $0.2 \lesssim Q/\sqrt{S} \lesssim 0.7$, and probe the product of the two quark-transversities in the "valence region". Recently, QCD corrections in (1) at GSI kinematics have been studied: the NLO ($O(\alpha_s)$) corrections [2] as well as the higher order ones in the framework of threshold resummation [3] are rather small, so that the LO value of A_{TT}, which turns out to be large, is rather robust.

When the transverse momentum Q_T of the final l^+l^- is also observed, we obtain the new double-spin asymmetry as the ratio of the Q_T-differential cross sections, $\mathscr{A}_{TT}(Q_T) \equiv [\Delta_T d\sigma/d\omega dQ_T]/[d\sigma/d\omega dQ_T]$. The bulk of l^+l^- pair is produced at small $Q_T \ll Q$, where the cross sections $(\Delta_T)d\sigma/d\omega dQ_T$ receive large perturbative corrections with logarithms $\ln(Q^2/Q_T^2)$ multiplying α_s at each order, by the recoil from gluon

CP1149, *Spin Physics, 18th International Spin Physics Symposium*
edited by D. G. Crabb, D. B. Day, S. Liuti, X. Zheng, M. Poelker, and Y. Prok

radiations, and those have to be treated in an all-order resummation in QCD [4]. The corresponding "Q_T-resummation" has formally some resemblance to the threshold resummation [3], but embodies contributions from different "edge region" of phase space. The Q_T-resummation for $\Delta_T d\sigma/d\omega dQ_T$ has been derived recently [5], summing the corresponding large logarithms up to next-to-leading logarithmic (NLL) accuracy. Combined with that for $d\sigma/d\omega dQ_T$ [4], we get [6] ($b_0 \equiv 2e^{-\gamma_E}$ with γ_E the Euler constant)

$$\mathscr{A}_{TT}(Q_T) = \frac{\cos(2\phi)}{2} \frac{\int d^2b\, e^{i\mathbf{b}\cdot\mathbf{Q}_T} e^{S(b,Q)} \sum_q e_q^2 \delta q(x_1, b_0^2/b^2) \delta q(x_2, b_0^2/b^2) + \cdots}{\int d^2b\, e^{i\mathbf{b}\cdot\mathbf{Q}_T} e^{S(b,Q)} \sum_q e_q^2 q(x_1, b_0^2/b^2) q(x_2, b_0^2/b^2) + \cdots}, \quad (2)$$

where the numerator and denominator are, respectively, reorganized in the impact parameter b space in terms of the Sudakov factor $e^{S(b,Q)}$ resumming soft and flavor-conserving collinear radiation, while the ellipses involve the remaining contributions of the $O(\alpha_s)$ collinear radiation, which can be absorbed into the exhibited terms as $\delta q \to \Delta_T C_{qq} \otimes \delta q$, $q \to C_{qq} \otimes q + C_{qg} \otimes g$ using the corresponding coefficient functions $(\Delta_T)C_{ij}$; note that there is no "chiral-odd" gluon distribution to participate in the numerator of (2) as well as (1). Using *universal* Sudakov exponent $S(b,Q)$ with the first nonleading anomalous dimensions in (2), the first three towers of large logarithmic contributions to the cross sections, $\alpha_s^n \ln^m(Q^2/Q_T^2)/Q_T^2$ $(m = 2n-1, 2n-2, 2n-3)$, are resummed to all orders in α_s, yielding the NLL resummation. In addition to these resummed components relevant for small Q_T, the ellipses in (2) also involve the other terms of the fixed-order α_s, which treat the LO processes in the large Q_T region, so that (2) is the ratio of the NLL+LO polarized and unpolarized cross sections. We include a Gaussian smearing as usually as $S(b,Q) \to S(b,Q) - g_{NP}b^2$, corresponding to intrinsic transverse momentum of partons inside proton. The integrations of the NLL+LO cross sections $\Delta_T d\sigma/d\omega dQ_T$ and $d\sigma/d\omega dQ_T$ over Q_T coincide, respectively, with the (fixed-order) NLO cross sections $\Delta_T d\sigma/d\omega$ and $d\sigma/d\omega$, associated with A_{TT} of (1) [5, 6].

The resummation makes $1/b \sim Q_T$ the relevant scale in (2), in contrast to Q in (1). Figure 1 shows [6] the numerical evaluation of (2), as well as of $\Delta_T d\sigma/d\omega dQ_T$ associated with its numerator, at GSI kinematics using the NLO transversities that saturate the Soffer bound, $2\delta q(x,\mu^2) \leq q(x,\mu^2) + \Delta q(x,\mu^2)$, at a low scale μ with Δq the helicity distribution. The NLL resummed component dominates $\Delta_T d\sigma/d\omega dQ_T$ in small and moderate Q_T region, and similarly for $d\sigma/d\omega dQ_T$ reflecting universality of the large Sudakov effects, which leads to almost constant $\mathscr{A}_{TT}(Q_T)$, in particular, with even flatter behavior than the corresponding asymmetry [5] for the pp-collision case. Remarkably, $\mathscr{A}_{TT}(Q_T)$ at NLL+LO has almost the same value as that at LL; this is in contrast to the pp case where the resummation at higher level enhances the asymmetry [5]. We note that $\mathscr{A}_{TT}(Q_T)$ at LL is given by (2) omitting all nonleading corrections, i.e., omitting the ellipses, replacing $S(b,Q)$ by that at the LL level, and replacing the scale of the parton distributions as $b_0^2/b^2 \to Q^2$, so that the result coincides with A_{TT} at LO (see (1)). Therefore, at GSI, both $\mathscr{A}_{TT}(Q_T)$ and A_{TT} are quite stable when including the QCD (resummation and fixed-order) corrections, with $\mathscr{A}_{TT}(Q_T) \simeq \mathscr{A}_{TT}(0)$, and

$$\mathscr{A}_{TT}(Q_T) \simeq A_{TT}. \quad (3)$$

To clarify the relevant mechanism, $\mathscr{A}_{TT}(Q_T) \simeq \mathscr{A}_{TT}(0)$ allows us to consider the $Q_T \to 0$ limit of (2): for $Q_T \approx 0$, the b integral is controlled by a saddle point $b = b_{SP}$,

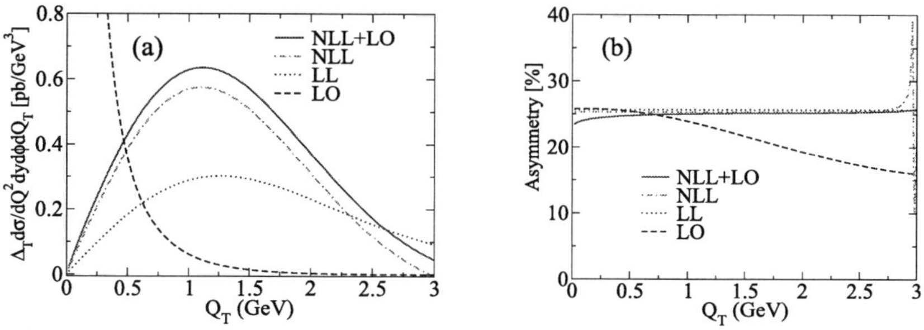

FIGURE 1. (a) $\Delta_T d\sigma/d\omega dQ_T$ (b) $\mathscr{A}_{TT}(Q_T)$ with GSI kinematics, $S = 210$ GeV^2, $Q = 4$ GeV, $y = \phi = 0$, and with $g_{NP} = 0.5$ GeV^2, using the NLO transversity distributions that correspond to the Soffer bound.

which has the same value between the numerator and denominator in (2) such that [5, 6]

$$\mathscr{A}_{TT}(0) \simeq \frac{\cos(2\phi)}{2} \frac{\sum_q e_q^2 \delta q(x_1, b_0^2/b_{SP}^2) \delta q(x_2, b_0^2/b_{SP}^2)}{\sum_q e_q^2 q(x_1, b_0^2/b_{SP}^2) q(x_2, b_0^2/b_{SP}^2)}, \tag{4}$$

omitting the small corrections from the LO components involved in the ellipses in (2). This saddle-point evaluation is exact at NLL accuracy; in particular, the $O(\alpha_s)$ contributions from the coefficients $(\Delta_T)C_{ij}$, e.g., $C_{qg} \otimes g$ associated with gluon distribution, completely decouple as $Q_T \to 0$ (see [4, 5]). The simple form of (4) is reminiscent of A_{TT} of (1) at LO, but is different from the latter in the unconventional scale b_0^2/b_{SP}^2; the actual position of the saddle point implies $b_0/b_{SP} \simeq 1$ GeV, irrespective of the values of Q and g_{NP} [5, 6]. In the valence region relevant for GSI kinematics, the u-quark contribution dominates in (4) and (1), so that these asymmetries are controlled by the ratio, $\delta u(x_{1,2}, \mu^2)/u(x_{1,2}, \mu^2)$ with $\mu^2 = b_0^2/b_{SP}^2$ and Q^2, respectively. Actually the scale dependence in this ratio almost cancels between the numerator and denominator as $\delta u(x, b_0^2/b_{SP}^2)/u(x, b_0^2/b_{SP}^2) \simeq \delta u(x, Q^2)/u(x, Q^2)$ (see Fig. 3 in [6]); this implies (3) at GSI. Note that this is not the case for pp collisions because of very different behavior of the sea-quark components under the evolution between transversity and unpolarized distributions [5]; indeed $\mathscr{A}_{TT}(Q_T) > A_{TT}$ at RHIC and J-PARC [5]. A similar logic applied to (2) also explains why $\mathscr{A}_{TT}(Q_T)$ at GSI are flatter than in pp collisions.

Another consequence of the similar logic is that $\delta u(x_{1,2}, 1\ \text{GeV}^2)/u(x_{1,2}, 1\ \text{GeV}^2)$ as a function of $x_{1,2}$ directly determines the Q- as well as S-dependence of the value of (3) at GSI, through $x_{1,2} = (Q/\sqrt{S})e^{\pm y}$. In Fig. 2, using the NLO transversity distributions corresponding to the Soffer bound, the symbols "$\triangle$" plot $\mathscr{A}_{TT}(Q_T \simeq 1\ \text{GeV})$ of (2) at NLL+LO [6]. The dashed curve draws the result using (4); this simple formula indeed works well. Also plotted by the two-dot-dashed curve is A_{TT} of (1) at LO with the transversities corresponding to the Soffer bound at LO level, to demonstrate (3). The Q- and S-dependence of these results reflect that the ratio $\delta u(x, 1\ \text{GeV}^2)/u(x, 1\ \text{GeV}^2)$ is an increasing function of x. These results using the Soffer bound show the "maximally possible" asymmetry, i.e., optimistic estimate. A more realistic estimate of (2) (with $Q_T \simeq 1$ GeV) and (4) is shown [6] in Fig. 2 by the symbols "∇" and the dot-dashed curve, respectively, with the NLO transversity distributions assuming $\delta q(x, \mu^2) = \Delta q(x, \mu^2)$ at

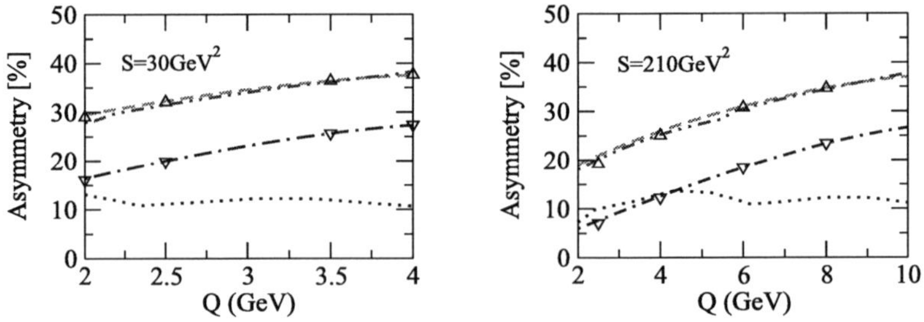

FIGURE 2. The double transverse-spin asymmetries at GSI as functions of Q with $y = \phi = 0$.

a low scale μ, as suggested by nucleon models and favored by the results of empirical fit for transversity [7]. The new estimate gives smaller asymmetries compared with the Soffer bound results, but still yields rather large asymmetries [6]. Based on (3), these results in Fig. 2 may be considered as estimate of A_{TT} of (1).

At present, empirical information of transversity is based on the LO global fit, using the semi-inclusive DIS data and assuming that the antiquark transversities vanish, $\delta\bar{q}(x) = 0$, so that the corresponding LO parameterization is available only for u and d quarks [7]. Fortunately, however, the dominance of the u-quark contribution in the GSI kinematics allows quantitative evaluation of A_{TT} at LO using only this empirical information [6]: the upper limit of the error band for the u- and d-quark transversities obtained by the global fit [7] yields the "upper bound" of A_{TT} shown by the dotted curve in Fig. 2; using (3), this result may be considered also as estimate of $\mathscr{A}_{TT}(Q_T)$. In the small Q region, our full NLL+LO result of $\mathscr{A}_{TT}(Q_T \simeq 1\text{ GeV})$, shown by "$\nabla$", can be consistent with estimate using the empirical LO transversity, but these results have rather different behavior for increasing Q, because the u-quark transversity for the former lies slightly outside the error band of the global fit for $x \gtrsim 0.3$ [6]. Thus, in the large asymmetries to be observed at GSI, the behavior of $\mathscr{A}_{TT}(Q_T)$, A_{TT} as functions of Q will allow us to determine the detailed shape of transversity distributions.

ACKNOWLEDGMENTS

This work was supported by the Grant-in-Aid for Scientific Research No. B-19340063.

REFERENCES

1. V. Barone *et al.* [PAX Collaboration], arXiv:hep-ex/0505054.
2. V. Barone, A. Caferella, C. Coriano, M. Guzzi and P. G. Ratcliffe, Phys. Lett. **B639**, 483 (2006).
3. H. Shimizu, G. Sterman, W. Vogelsang and H. Yokoya, Phys. Rev. **D71**, 114007 (2005).
4. J. C. Collins, D. E. Soper and G. Sterman, Nucl. Phys. **B250**, 199 (1985).
5. H. Kawamura, J. Kodaira, H. Shimizu and K. Tanaka, Prog. Theor. Phys. **115**, 667 (2006); H. Kawamura, J. Kodaira and K. Tanaka, Nucl. Phys. **B777**, 203 (2007); H. Kawamura, J. Kodaira and K. Tanaka, Prog. Theor. Phys. **118**, 581 (2007).
6. H. Kawamura, J. Kodaira and K. Tanaka, Phys. Lett. **B662**, 139 (2008).
7. M. Anselmino *et al.*, Phys. Rev. **D75**, 054032 (2007); arXiv:0812.4366 [hep-ph].

Drell-Yan Lepton Pair Azimuthal Asymmetry in Hadronic Processes

Jian Zhou*, Feng Yuan † and Zuo-Tang Liang**

*Nuclear Science Division, Lawrence Berkeley National Laboratory, Berkeley, CA 94720
†Nuclear Science Division, Lawrence Berkeley National Laboratory, Berkeley, CA 94720 and RIKEN BNL Research Center, Building 510A, Brookhaven National Laboratory, Upton, NY 11973
**School of Physics, Shandong University, Jinan, Shandong 250100, China

Abstract. We study the Drell-Yan lepton pair azimuthal asymmetry ($\cos 2\phi$) in hadronic scattering process at moderate transverse momentum region, taking into account the contributions from the twist-three quark-gluon correlations from the unpolarized hadrons. This contribution is found to dominate this asymmetry, and is not power suppressed by $q_\perp/Q$ where $q_\perp$ and Q are the transverse momentum and invariant mass of the lepton pair. Accordingly, the Lam-Tung relation will be violated at this momentum region.

Keywords: Azimuthal asymmetry, Drell-Yan process, Twist-three approach
PACS: 12.38.Bx , 13.88.+e

1. Introduction. Drell-Yan lepton pair production in hadronic scattering process [1] has been playing very important role in studying nucleon structure and QCD dynamics [2], and as an important complementary to the deep inelastic scattering studies [3]. More over, the later development on the angular distribution of lepton pair has laid ground for parton model and QCD dynamics studies [4, 5, 6]. The lepton pair production in hadronic scattering,

$$H_1 + H_2 \to \gamma^* + X \to \ell^+\ell^- + X \,, \tag{1}$$

comes from the virtual photon decays. In the leading order, virtual photon is produced through quark-antiquark annihilation process, $q\bar{q} \to \gamma^*$ in the parton picture [1]. In the rest frame of the lepton pair, we can define two angles [4]: one is the polar angle θ between one lepton momentum and the hadron; the azimuthal angle ϕ is defined as the angle between the hadronic plane and the lepton plane. Here and in the following discussions, we follow the Collins-Soper frame to define these angles. Our results can be translated to other frames too. The general expression for the lepton pair angular distribution can be written as [4],

$$\frac{dN}{d\Omega} = (1+\cos^2\theta) + A_0\left(\frac{1}{2} - \frac{3}{2}\cos^2\theta\right) + A_1 \sin 2\theta \cos\phi + \frac{A_2}{2}\sin^2\theta\cos 2\phi \,. \tag{2}$$

There exists an interesting relation among these coefficients, so-called Lam-Tung relation [5]: $2\nu - (1-\lambda) = 0$, where $\lambda = (2-3A_0)/(2+A_0)$ and $\nu = 2A_2/(2+A_0)$. At the leading power,this relation is obviously valid because $\lambda = 1$ and $\nu = 0$. Finite transverse momentum of the lepton pair ($q_\perp$) can be generated from gluon radiation from the Leading partonic process, for example, through the quark-antiquark annihilation channel $q\bar{q} \to \gamma^* g$. leading order perturbative QCD calculation has shown that the Lam-Tung

CP1149, *Spin Physics, 18th International Spin Physics Symposium*
edited by D. G. Crabb, D. B. Day, S. Liuti, X. Zheng, M. Poelker, and Y. Prok

relation is still valid for any value of $q_\perp$ from this contribution [6]. Higher order perturbative corrections to this relation has been calculated in the literature [7], where it was found that the violation is numerically very small. These studies have motivated many experimental investigations of the lepton pair angular distributions in hadronic scattering [8]. In particular, the π induced fixed target experiment found large $\cos 2\phi$ azimuthal asymmetry, which is difficult to understand [8, 9]. The intrinsic transverse momentum effects have been proposed to explain these effects, which however is limited to small transverse momentum region whereas the experimental data are in the moderate transverse momentum region [10].

Here, we study the lepton pair azimuthal asymmetry, in particular for the A_2 coefficient in Eq. (2) from different perspective. We are interested in the $\cos 2\phi$ asymmetry at the moderate transverse momentum region $\Lambda_{QCD} \ll q_\perp \ll Q$. At this region, there are two large momentum scales $q_\perp$ and Q. The gluon radiation contribution to A_2 from Eq. (3) is power suppressed by $q_\perp^2/Q^2$. However, from the following calculations we find that there exits contribution from the twist-three quark-gluon correlation functions from both incident hadrons,which is not power suppressed by $q_\perp^2/Q^2$.

2. Twist-three times twist-three contributions to the lepton pair azimuthal asymmetry. From the general analysis of twist-three functions [13], we find the only twist-three function is $T_F^{(\sigma)}(x,y)$ which is equivalent to $E(x,y)$ studied in the literature [13, 14]. It is defined as

$$T_F^{(\sigma)}(x_1,x_2) = \int \frac{dy_1^- dy_2^-}{4\pi} e^{iy_2^-(x_2-x_1)P^+ + iy_1^- x_1 P^+} \langle P|\overline{\psi}(0^-)\sigma^{+\mu} g F^+{}_\mu(y_2^-)\psi(y_1^-)|P\rangle \ , \quad (3)$$

where μ is a transverse index, the sums over color and spin indices are implicit, $|P\rangle$ denotes the unpolarized hadron state with momentum $P = (P^+, 0^-, 0_\perp)$ and $P^\pm = (P^0 \pm P^z)/\sqrt{2}$, ψ is the quark field, and $F_{+\mu}$ the gluon field tensor, and the gauge link has been suppressed. This correlation is a chiral-odd function, and can generate transverse polarized Hyperon production in unpolarized hadronic collisions [15]. Because of this chirality property, we have to introduce two correlation functions from both incoming hadrons. They will couple to produce Drell-Yan lepton pair, and contribute to the $\cos 2\phi$ azimuthal asymmetry.

To calculate its contribution, we follow the procedure outlined in [16], and recent developments for the similar calculations [17, 18, 19]. In the collinear factorization framework, the general factorization formula for the contributions from the above correlation functions from both hadrons can be written as,

$$\frac{d\sigma}{d^4q d\Omega} = \sum_q \int \frac{dx dx'}{x} \frac{dz dz'}{z} T_{F,q}^{(\sigma)}(x,x') T_{F,\bar{q}}^{(\sigma)}(z,z') \mathscr{H}(x,x';z,z';Q^2,q_\perp) \ , \quad (4)$$

where $T_{F,q}^{(\sigma)}(x,x')$ is the correlation function associated with the quark from hadron H_1 and $T_{F,\bar{q}}^{(\sigma)}(z,z')$ for the antiquark from hadron H_2, $\mathscr{H}$ is the hard part and can be calculated from perturbative partonic process. As we mentioned above, we are interested in the cross section contributions in the moderate transverse momentum region $q_\perp \ll Q$. In carrying out these calculations, we will utilize the power counting method and only keep the leading power contributions and neglect all higher power corrections in terms of $q_\perp/Q$. In the limit of $q_\perp \ll Q$, the final result is

$$\begin{aligned}\frac{d\sigma}{d^4q d\Omega} &= \sigma_0 \sin^2\theta\cos(2\phi)\frac{2}{q_\perp^4}\frac{\alpha_s}{2\pi^2}\sum_q e_q^2 \\ &\int\frac{dx}{x}\frac{dz}{z}\left\{AT_{F,\bar{q}}^{(\sigma)}(z,z)\delta(\hat{\xi}-1)+\bar{A}T_{F,q}^{(\sigma)}(x,x)\delta(\xi-1)\right. \\ &\left.+2C_F\delta(\xi-1)\delta(\hat{\xi}-1)T_{F,q}^{(\sigma)}(x,x)T_{F,\bar{q}}^{(\sigma)}(z,z)\ln\frac{Q^2}{q_\perp^2}\right\}, \quad (5)\end{aligned}$$

where $\sigma_0 = \alpha_{em}^2/6SQ^2$, $\xi = x_0/x$, $\hat{\xi} = z_0/z$ with $x_0 = \frac{Q}{\sqrt{S}}e^y$ and $z_0 = \frac{Q}{\sqrt{S}}e^{-y}$, and y is the rapidity of the lepton pair in the center of mass frame of incoming two hadrons and S is the hadronic center of mass energy square. The coefficient A is defined as

$$A = \frac{1}{2N_c}\left\{\left[x\frac{\partial}{\partial x}T_{F,q}^{(\sigma)}(x,x)\right]2\xi + T_{F,q}^{(\sigma)}(x,x)\frac{2\xi(\xi-2)}{(1-\xi)_+}\right\} + \frac{C_A}{2}T_{F,q}^{(\sigma)}(x,x_0)\frac{2}{(1-\xi)_+}, \quad (6)$$

and similar expression holds for $\bar{A}$ in the above equation. From the above equations, we can see that this contribution to A_2 coefficient is not power suppressed by $q_\perp^2/Q^2$ at moderate transverse momentum region. It is the same order as the $(1+\cos^2\theta)$ term in this power counting. Of course, it is suppressed by $\Lambda^2/q_\perp^2$ because of the higher-twist nature. This can also be seen from the above expression.

More importantly, the above result can also be reproduced by a transverse momentum dependent (TMD) factorization formalism [21, 22, 20, 23, 24] with the so-called Boer-Mulders function from both hadrons [25] at large transverse momentum which has been calculated in [15]. This demonstrates that in the intermediate transverse momentum region, for this part contribution, the twist-three times twist-three collinear factorization approach and TMD factorization approach are consistent for the $\cos(2\phi)$ azimuthal asymmetry in unpolarized Drell-Yan process. This is a nontrivial demonstration, because it goes beyond previous examples studied in the literature [19] where only one twist-three effect was considered. Because of this consistency, the energy evolution equation (the Collins-Soper evolution equation) can be derived for this contribution, and the soft gluon resummation can be accordingly performed. This will significantly change the relative sizes of this contribution and the contribution from Eq. (3) to the $\cos 2\phi$ asymmetry.

3. Conclusion. In summary, we have investigated the higher-twist effects to the Drell-Yan lepton pair angular distributions in hadron-hadron scattering process. We found that the $\cos 2\phi$ azimuthal asymmetry contribution is not power suppressed by $q_\perp^2/Q^2$, rather by $\Lambda^2/q_\perp^2$ at the moderate transverse momentum. We further argue that this part of contribution will not be affected by the soft gluon resummation effects, and the Lam-Tung relation will be modified at small and moderate transverse momentum region. It will be interested to compare our predictions with the experimental data [8] and check the phenomenological importance of our results. It will also be interested to extend to other processes like the $\cos 2\phi$ asymmetry in semi-inclusive hadron production in deep inelastic scattering and back-to-back two hadron production in e^+e^- annihilation processes where the similar effects shall play very important roles.

We thank Jianwei Qiu and Werner Vogelsang for interesting discussions. This work was supported in part by the U.S. Department of Energy under contract DE-AC02-05CH11231 and the National Natural Science Foundation of China under the approval No. 10525523. J.Z. is partially supported by China Scholarship Council.

REFERENCES

1. S. D. Drell and T. M. Yan, Phys. Rev. Lett. **25**, 316 (1970) [Erratum-ibid. **25**, 902 (1970)].
2. R. Brock *et al.* [CTEQ Collaboration], Rev. Mod. Phys. **67**, 157 (1995).
3. E. D. Bloom *et al.*, Phys. Rev. Lett. **23**, 930 (1969); M. Breidenbach *et al.*, Phys. Rev. Lett. **23**, 935 (1969).
4. J. C. Collins and D. E. Soper, Phys. Rev. D **16**, 2219 (1977).
5. C. S. Lam and W. K. Tung, Phys. Rev. D **18**, 2447 (1978); Phys. Lett. B **80**, 228 (1979); Phys. Rev. D **21**, 2712 (1980).
6. J. C. Collins, Phys. Rev. Lett. **42**, 291 (1979).
7. E. Mirkes and J. Ohnemus, Phys. Rev. D **51**, 4891 (1995) [arXiv:hep-ph/9412289].
8. see for example, J. S. Conway *et al.*, Phys. Rev. D **39**, 92 (1989); L. Y. Zhu *et al.* [FNAL-E866/NuSea Collaboration], Phys. Rev. Lett. **99**, 082301 (2007); arXiv:0811.4589 [nucl-ex].
9. P. Chiappetta and M. Le Bellac, Z. Phys. C **32**, 521 (1986).
10. D. Boer, Phys. Rev. D **60**, 014012 (1999).
11. A. Idilbi, X. d. Ji, J. P. Ma and F. Yuan, Phys. Rev. D **70**, 074021 (2004).
12. V. Braun, R. J. Fries, N. Mahnke and E. Stein, Nucl. Phys. B **589**, 381 (2000) [Erratum-ibid. B **607**, 433 (2001)].
13. R. L. Jaffe and X. D. Ji, Nucl. Phys. B **375**, 527 (1992).
14. J. Zhou, F. Yuan and Z. T. Liang, arXiv:0812.4484 [hep-ph].
15. J. Zhou, F. Yuan and Z. T. Liang, Phys. Rev. D **78**, 114008 (2008).
16. J.W. Qiu and G. Sterman, Phys. Rev. Lett. **67**, 2264 (1991); Nucl. Phys. B **378**, 52 (1992); Phys. Rev. D **59**, 014004 (1998).
17. C. Kouvaris, J. W. Qiu, W. Vogelsang and F. Yuan, Phys. Rev. D **74**, 114013 (2006).
18. H. Eguchi, Y. Koike and K. Tanaka, Nucl. Phys. B **752**, 1 (2006); Nucl. Phys. B **763**, 198 (2007).
19. X. Ji, J. W. Qiu, W. Vogelsang and F. Yuan, Phys. Rev. Lett. **97**, 082002 (2006); Phys. Rev. D **73**, 094017 (2006); Phys. Lett. B **638**, 178 (2006).
20. X. Ji and F. Yuan, Phys. Lett. B **543**, 66 (2002); A. V. Belitsky, X. Ji and F. Yuan, Nucl. Phys. B **656**, 165 (2003).
21. S. J. Brodsky, D. S. Hwang and I. Schmidt, Phys. Lett. B **530**, 99 (2002); Nucl. Phys. B **642**, 344 (2002).
22. J. C. Collins, Phys. Lett. B **536**, 43 (2002).
23. D. Boer, P. J. Mulders and F. Pijlman, Nucl. Phys. B **667**, 201 (2003).
24. X. Ji, J. P. Ma and F. Yuan, Phys. Rev. D **71**, 034005 (2005); Phys. Lett. B **597**, 299 (2004).
25. D. Boer and P. J. Mulders, Phys. Rev. D **57**, 5780 (1998); D. Boer, S. J. Brodsky and D. S. Hwang, Phys. Rev. D **67**, 054003 (2003). A. Metz, P. Schweitzer and T. Teckentrup, arXiv:0810.5212 [hep-ph].

QCD Corrections in Transversely Polarized Scattering

Werner Vogelsang

Physics Department, Brookhaven National Laboratory, Upton, NY 11973

Abstract. We discuss two recent calculations of higher-order QCD corrections to scattering of transversely polarized hadrons.

Keywords: Nucleon spin structure, Soft-gluon resummation, Single-spin asymmetry
PACS: 12.38Bx, 13.88.+e, 13.60.-r, 13.85.-t

INTRODUCTION

A basic concept underlying much of the theoretical description of high-energy hadronic scacttering is the factorization theorem, which states that large momentum-transfer reactions may be factorized into long-distance pieces that contain information on the structure of the nucleon in terms of its parton densities, and parts that are short-distance and describe the hard interactions of the partons. Two crucial points are that on the one hand the long-distance contributions are universal, i.e., they are the same in any inelastic reaction under consideration, and that on the other hand the short-distance pieces depend only on the large scales related to the large momentum transfer in the overall reaction and, therefore, may be evaluated using QCD perturbation theory.

The lowest order for the latter can generally only serve to give a rough description of the reaction under study. It merely captures the main features, but does not usually provide a quantitative understanding. The first-order ("next-to-leading order" (NLO)) corrections are generally indispensable in order to arrive at a firmer theoretical prediction for hadronic cross sections, and in some cases even an all-order resummation of large perturbative corrections is needed. In the present paper we will discuss two calculations [1, 2] of higher-order QCD corrections to transversely polarized scattering.

THRESHOLD RESUMMATION FOR $p^\uparrow p^\uparrow \to \pi X$

In spite of extensive studies in recent years, the partonic structure of spin-$1/2$ nucleons is not yet completely known. Among the leading-twist collinear parton distributions, the unpolarized (f), longitudinally polarized (Δf), and transversely polarized (δf) densities, there is so far very little information about the latter. These "transversity" distributions δf are defined as the differences of probabilities to find a parton of flavor f at scale μ and light-cone momentum fraction x with its spin aligned ($\uparrow\uparrow$) or anti-aligned ($\uparrow\downarrow$) with that of the transversely polarized nucleon:

$$\delta f(x,\mu) \equiv f_{\uparrow\uparrow}(x,\mu) - f_{\uparrow\downarrow}(x,\mu) \,. \tag{1}$$

CP1149, *Spin Physics, 18th International Spin Physics Symposium*
edited by D. G. Crabb, D. B. Day, S. Liuti, X. Zheng, M. Poelker, and Y. Prok

Unlike the longitudinally polarized distribution functions which can be measured directly in deep-inelastic scattering (DIS), transversity is not accessible in inclusive DIS due to its chiral-odd nature. It will be highly desirable in the future to have direct probes of transversity. These are available in double-transverse spin asymmetries in hadronic collisions,

$$A_{TT} \equiv \frac{\frac{1}{2}[d\sigma(\uparrow\uparrow) - d\sigma(\uparrow\downarrow)]}{\frac{1}{2}[d\sigma(\uparrow\uparrow) + d\sigma(\uparrow\downarrow)]} \equiv \frac{d\delta\sigma}{d\sigma}, \tag{2}$$

where the arrows denote the transverse polarization of the scattering hadrons. A program of polarized pp collisions is now well underway at the BNL Relativistic Heavy Ion Collider (RHIC), and measurements of A_{TT} for various reactions should be feasible with sufficient beam-time for transverse polarization [3]. In the more distant future, there is also hope to have transversely polarized $\bar{p}p$ collisions at the GSI-FAIR facility, where there are plans to have an asymmetric polarized $\bar{p}p$ collider. Likewise, transversely polarized pp collisions could become a possibility at the J-PARC facility.

We will consider single-inclusive pion production in transversely polarized hadronic collisions, $p^{\uparrow}p^{\uparrow} \to \pi X$, $\bar{p}^{\uparrow}p^{\uparrow} \to \pi X$, where the pion has large transverse momentum p_T. Compared to the Drell-Yan process, which is usually considered the "golden channel" for transversity measurements in hadronic scattering, the spin asymmetry for hadron production is typically considerably smaller, mostly due to the large contribution from gluonic scattering in the denominator of the asymmetry [4, 5, 6]. On the other hand, pions are very copiously produced in hadronic scattering, resulting in much smaller statistical uncertainties, and A_{TT} in single-inclusive hadron production could thus conceivably present a viable alternative to Drell-Yan at RHIC as well as at GSI-FAIR and J-PARC [7].

For energies typically available in fixed-target experiments, previous work [8] has shown that the NLO framework is no longer sufficient but that all-order resummations of large logarithmic corrections are needed. Here the value of $x_T \equiv 2p_T/\sqrt{S}$, with $\sqrt{S}$ the center-of-mass (c.m.) energy of the collision, is generally quite large. It turns out that the partonic hard-scattering cross sections relevant for $pp \to \pi X$ are then largely probed in the "threshold"-regime, where the initial partons have just enough energy to produce the high-transverse momentum parton that subsequently fragments into the hadron, and its recoiling counterpart. Relatively little phase space is then available for additional radiation of partons. In particular, gluon radiation is inhibited and mostly constrained to the emission of soft and/or collinear gluons. The cancellation of infrared singularities between real and virtual diagrams then leaves behind large double- and single-logarithmic corrections to the partonic cross sections. These logarithms appear for the first time at NLO, where they arise as terms of the form $\alpha_S \ln^2(1-\hat{x}_T^2)$ in the rapidity-integrated cross section, where $\hat{x}_T \equiv 2\hat{p}_T/\sqrt{\hat{s}}$ with $\hat{p}_T$ the transverse momentum of the produced parton and $\hat{s}$ the c.m. energy of the initial partons. At yet higher (kth) order of perturbation theory, the double-logarithms are of the form $\alpha_S^k \ln^{2k}(1-\hat{x}_T^2)$. When the threshold regime dominates, it is essential to take into account the large logarithms to all orders in the strong coupling α_S, a technique known as "threshold resummation" [9, 10]. In our earlier work, we examined the effects of threshold resummation on the single-inclusive hadron cross section and on the double-longitudinal spin asymmetry A_{LL} [8]. We found very significant enhancements of the cross section by resummation, which in fact lead to a relatively good agreement between resummed theory and experimental data. We also found a moderate decrease of the resummed A_{LL}, with respect to NLO. We concluded that threshold resummation is an essential part of the theoretical description in the fixed-target kinematic regime. Its effects at higher energies (such as at RHIC) are much smaller, even though it has to be said that one is typically much further away from the threshold regime here, so that the applicability of threshold resummation is not entirely clear.

Here we apply threshold resummation to the case of transverse polarization of the scattering partons. As we discussed in [8], the resummed partonic cross section for each subprocess can be written in Mellin moment space in the rather simple form

$$\frac{d\delta\hat{\sigma}^{(\mathrm{res})}_{ab\to cd}(N)}{d\phi} = \delta C_{ab\to cd}\,\Delta^a_N\Delta^b_N\Delta^c_N J^d_N\left[\sum_I \delta G^I_{ab\to cd}\,\Delta^{(\mathrm{int})ab\to cd}_{IN}\right]\frac{d\delta\hat{\sigma}^{(\mathrm{Born})}_{ab\to cd}(N)}{d\phi}, \quad (3)$$

where $\delta\hat{\sigma}^{(\mathrm{Born})}_{ab\to cd}(N)$ denotes the moments of the LO term in the perturbative expansion for a given partonic process. Each of the functions J^d_N, Δ^i_N, $\Delta^{(\mathrm{int})ab\to cd}_{IN}$ is an exponential and embodies part of the resummation. $\Delta^{a,b}_N$ represent the effects of soft-gluon radiation collinear to initial partons a,b, and similarly for Δ^c_N for the final-state fragmenting parton c. The function J^d_N embodies collinear, soft or hard, emission by the "non-observed" recoiling parton d. Large-angle soft-gluon emission is accounted for by the factors $\Delta^{(\mathrm{int})ab\to cd}_{IN}$, which depend on the color configuration I of the participating partons. Finally, the coefficients $\delta C_{ab\to cd}$ contain $N-$independent hard contributions arising from one-loop virtual corrections. When performing the inverse Mellin transform, we use the *Minimal Prescription* proposed in Ref. [11] to treat the singularity of the perturbative strong coupling constant in the resummed exponent.

We first present our results [1] for $\bar{p}p$ collisions at $\sqrt{S}=14.5$ GeV, corresponding to the GSI-FAIR project. At this relatively modest energy one expects the perturbative threshold corrections that we resum here to be particularly important. To match the experimental conditions realistically, we have to take into account the rapidity range that would be covered by the experiments. We assume this range to be $-1<\eta_{\mathrm{lab}}<2.5$, where η_{lab} is the pseudorapidity of the pion in the laboratory frame. We count positive rapidity in the forward direction of the antiproton. The results for the NLL resummed cross sections integrated over the range $-1<\eta_{\mathrm{lab}}<2.5$ are shown in Fig. 1. We also present in the figure the uncertainties in the prediction resulting from variation of the scale μ in the range $p_T\le\mu\le 4p_T$. One can see that the scale uncertainty remains rather large even after resummation.

We now turn to $p^\uparrow p^\uparrow\to\pi X$ at $\sqrt{S}=62.4$ GeV and at $\sqrt{S}=10$ GeV, as relevant for experiments at RHIC (here, PHENIX) and J-PARC. In order to allow comparison of our theoretical predictions with future data we consider the regions of pseudo-rapidity as $\eta_{cm}>0$ and $|\eta|<0.38$ for J-PARC and RHIC (PHENIX), respectively, over which we again integrate. The first choice is based on the assumption of a forward spectrometer geometry with 200 mrad acceptance, similar to the one used by the COMPASS experiment at CERN. The results for the two cases are presented in Fig. 2 which shows the NLL spin-averaged and spin-dependent cross sections. As before we have taken into account the theoretical uncertainties in the factorization and renormalization scales by

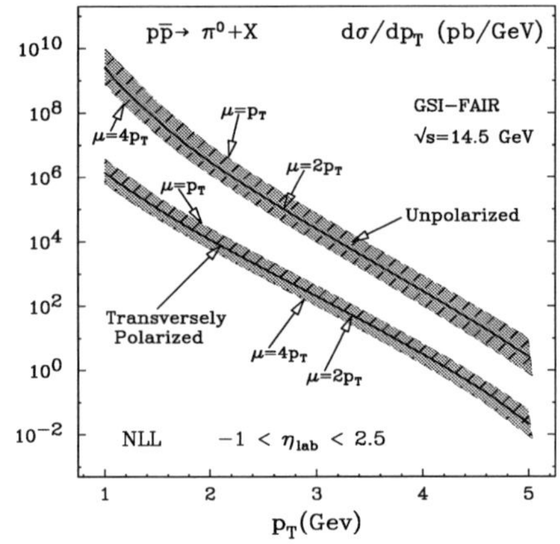

FIGURE 1. *NLL resummed cross section for unpolarized and transversely polarized* $\bar{p}p \rightarrow \pi^0 X$ *at* $\sqrt{S} = 14.5$ *GeV for* $-1 < \eta_{lab} < 2.5$ *[1]. The shaded bands represent the changes of the results if the factorization/renormalization scale is varied in the range* $p_T \leq \mu \leq 4p_T$. *The solid line corresponds to* $\mu = 2p_T$.

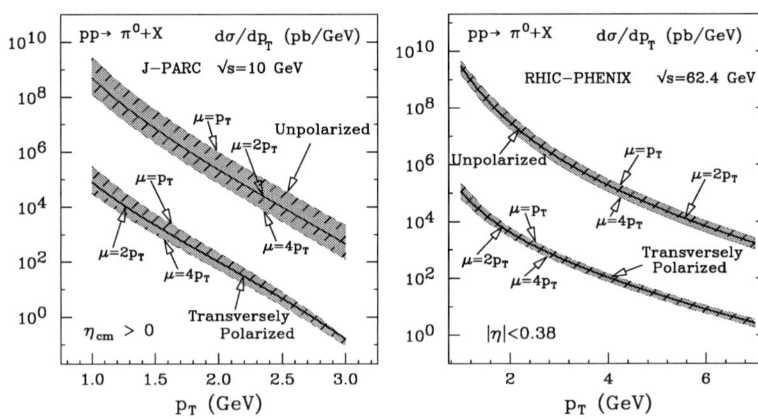

FIGURE 2. *NLL resummed cross section for unpolarized and transversely polarized* $pp \rightarrow \pi^0 X$ *at fixed-target* $\sqrt{S} = 10$ *GeV for* $\eta_{cm} > 0$ *(left) and at* 62.4 *GeV for* $|\eta| < 0.38$ *(right) [1]. The shaded bands represent the changes of the results if the factorization/renormalization scale is varied in the range* $p_T \leq \mu \leq 4p_T$. *The solid line corresponds to* $\mu = 2p_T$.

varying $\mu = \zeta p_T$, with $\zeta = 1,2,4$. It is worth noticing that after resummation the scale dependence is considerably reduced in the case of RHIC. As can be expected, the effects of resummation are much larger at $\sqrt{S} = 10$ GeV than at $\sqrt{S} = 62.4$ GeV. It is also interesting to notice that the K-factor is again much larger for the spin-averaged cross section than for the spin-dependent case, especially at $\sqrt{S} = 10$ GeV. This immediately implies that the spin asymmetry A_{TT}^{π} will be strongly reduced when going from NLO to the NLL resummed case.

NLO CALCULATION OF THE SINGLE TRANSVERSE SPIN ASYMMETRY IN THE DRELL-YAN PROCESS

Single transverse spin asymmetries (SSAs) in high energy hadronic reactions continue to attract much theoretical and experimental interest. They are defined as differences of cross sections when one of the initial hadrons' transverse spin is flipped, divided by the sum: $A_N \sim (d\sigma(S_\perp) - d\sigma(-S_\perp))/(d\sigma(S_\perp) + d\sigma(-S_\perp))$. The theoretical description of SSAs has presented challenges, since the leading collinear partonic contribution to the asymmetries vanishes. Over the past few years, there have been a number of theoretical developments that have led to much progress in the exploration of the underlying physics for single spin asymmetry phenomena. These developments mainly follow two lines: the transverse momentum dependent (TMD) approach, which uses parton distributions that depend on partonic transverse momentum [12], and the twist-three quark-gluon correlation function approach [13, 14]. More recently, it has been found in some cases that the two approaches are closely related and describe the same physics [15].

So far, phenomenological applications of the approaches have been limited to the "bare" parton model, that is, to the zeroth order of perturbation theory without any QCD corrections, as the latter were generally not available. This situation was remedied very recently when the leading-order (LO) kernels for the scale evolution of the relevant twist-three correlation functions were derived [16]. Here we take a further step [2] by calculating next-to-leading order (NLO) QCD corrections for a particular observable, the SSA in Drell-Yan lepton pair production weighted by the transverse momentum of the virtual photon:

$$\langle q_\perp \Delta\sigma(S_\perp)\rangle \equiv \int d^2q_\perp |q_\perp| \sin\phi \frac{d\Delta\sigma(S_\perp)}{d^2q_\perp} \propto \int \frac{dx_1}{x_1}\frac{dx_2}{x_2}\frac{dx'}{x'} T_F^q(x_1,x_2)\bar{q}(x')H(x_1,x_2;x'), \quad (4)$$

where T_F^q denotes a "Qiu-Sterman" matrix element [13], and H is the hard part that we are calculating to NLO.

The NLO real-gluon radiation diagrams have already been studied in [15], and the results can be carried over to our present calculation with relatively little effort. Some of the virtual diagrams are shown in Fig. 3. Their result is in fact proportional to that for the NLO virtual corrections for the ordinary spin-averaged Drell-Yan cross section. We have verified that the soft divergence in real-gluon radiation is canceled by that in the virtual diagrams, so that we are left with only collinear divergences, which can be absorbed into the parton distribution and/or the twist-three correlation function, where they give rise to the scale evolution of the distributions. We find the following evolution equation for the "diagonal" twist-three quark-gluon correlation function at $x_1 = x_2 = x$:

$$\begin{aligned}\frac{\partial}{\partial \ln\mu^2} T_F(x,x;\mu^2) &= \frac{\alpha_s(\mu^2)}{2\pi}\int \frac{dx'}{x'}\left\{ C_F T_F(x',x';\mu^2)\left(\frac{1+z^2}{1-z}\right)_+ \right. \quad (5)\\ &\left. + \frac{N_c}{2}\left[T_F(x',x';\mu^2)z - (T_F(x',x';\mu^2) - T_F(x',x'z;\mu^2))\frac{1+z}{1-z}\right]\right\}.\end{aligned}$$

This evolution equation is consistent with the results derived recently by different methods [16]. The full NLO results may be found in [2].

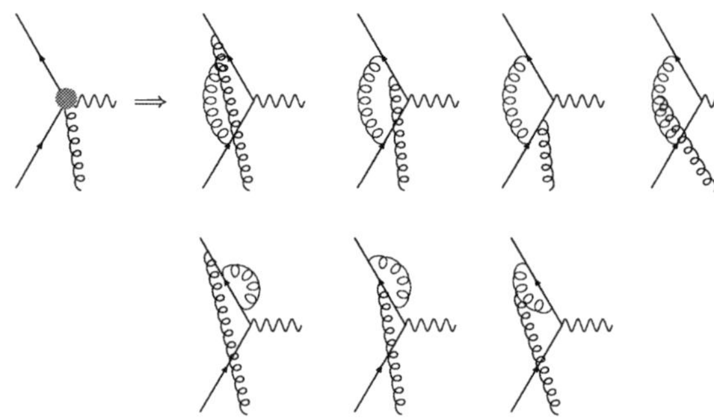

FIGURE 3. *Corrections to the quark-antiquark-photon vertex with gluon attachment.*

ACKNOWLEDGMENTS

I am grateful to D. de Florian, F. Wagner, and F. Yuan for collaboration on the topics presented here. I thank the U.S. Department of Energy (contract number DE-AC02-98CH10886) for providing the facilities essential for the completion of this work.

REFERENCES

1. D. de Florian, W. Vogelsang and F. Wagner, *Phys. Rev. D* **78**, 074025 (2008).
2. F. Yuan, W. Vogelsang, *in preparation.*
3. G. Bunce, N. Saito, J. Soffer and W. Vogelsang, *Ann. Rev. Nucl. Part. Sci.* **50**, 525 (2000); G. Bunce *et al.*, http://spin.riken.bnl.gov/rsc/report/spinplan_2008/spinplan08.pdf, *Plans for the RHIC Spin Physics Program.*
4. X. Artru and M. Mekhfi, *Z. Phys. C* **45**, 669 (1990).
5. X. D. Ji, *Phys. Lett. B* **284**, 137 (1992).
6. R. L. Jaffe and N. Saito, *Phys. Lett. B* **382**, 165 (1996); J. Soffer, M. Stratmann and W. Vogelsang, *Phys. Rev. D* **65**, 114024 (2002).
7. A. Mukherjee, M. Stratmann and W. Vogelsang, *Phys. Rev. D* **72**, 034011 (2005).
8. D. de Florian and W. Vogelsang, *Phys. Rev. D* **71**, 114004 (2005); *Phys. Rev. D* **76**, 094021 (2007).
9. G. Sterman, *Nucl. Phys. B* **281**, 310 (1987); S. Catani and L. Trentadue, *Nucl. Phys. B* **327**, 323 (1989); S. Catani, M. L. Mangano and P. Nason, *JHEP* **9807**, 024 (1998); E. Laenen, G. Oderda and G. Sterman, *Phys. Lett. B* **438**, 173 (1998); N. Kidonakis and G. Sterman, *Nucl. Phys. B* **505**, 321 (1997); N. Kidonakis, G. Oderda and G. Sterman, *Nucl. Phys. B* **525**, 299 (1998); *Nucl. Phys. B* **531**, 365 (1998); R. Bonciani, S. Catani, M. L. Mangano and P. Nason, *Phys. Lett. B* **575**, 268 (2003).
10. N. Kidonakis, G. Oderda and G. Sterman, *Nucl. Phys. B* **525**, 299 (1998); *Nucl. Phys. B* **531**, 365 (1998); N. Kidonakis and J. F. Owens, *Phys. Rev. D* **63**, 054019 (2001).
11. S. Catani, M. L. Mangano, P. Nason and L. Trentadue, *Nucl. Phys. B* **478**, 273 (1996).
12. D. W. Sivers, *Phys. Rev. D* **43**, 261 (1991); for review, see: V. Barone, A. Drago and P. G. Ratcliffe, *Phys. Rept.* **359**, 1 (2002).
13. J.W. Qiu and G. Sterman, *Phys. Rev. D* **59**, 014004 (1998).
14. H. Eguchi, Y. Koike and K. Tanaka, *Nucl. Phys. B* **763**, 198 (2007).
15. X. Ji, J. W. Qiu, W. Vogelsang and F. Yuan, *Phys. Rev. Lett.* **97**, 082002 (2006); *Phys. Rev. D* **73**, 094017 (2006); *Phys. Lett. B* **638**, 178 (2006).
16. Z. B. Kang and J. W. Qiu, *Phys. Rev. D* **79**, 016003 (2009); J. Zhou, F. Yuan and Z. T. Liang, arXiv:0812.4484 [hep-ph].

Azimuthal angle dependence of di-jet production in unpolarized hadron scattering

Zhun Lu and Ivan Schmidt

Departamento de Física, Universidad Técnica Federico Santa María, Valparaíso, Chile and Center of Subatomic Physics, Valparaíso, Chile

Abstract. We study the azimuthal asymmetry of back-to-back di-jet production in unpolarized hadron scattering, arising from the product of two Boer-Mulders functions, which describe the transverse spin distribution of quarks inside an unpolarized hadron. We find that there is a $\cos\delta\phi$ angular dependence of the di-jet, with $\delta\phi$ the difference of the azimuthal angle of tow jets respectively. In the case of $J_q + J_q$ production, we find that there is a color factor enhancement in the gluonic cross-section due to the multiple initial-/final-state interactions, compared with the result from the standard generalized parton model. We estimate the $\cos\delta\phi$ asymmetry of the total di-jet production at RHIC, showing that the color factor enhancement in the azimuthal asymmetric cross section of $J_q + J_q$ production will reverse the sign of the asymmetry.

Keywords: azimuthal asymmetry, gauge-link, transverse momentum, Boer-Mulders function
PACS: 12.38.Bx, 13.85.Ni, 13.88.+e

INTRODUCTION

The role of transverse momentum dependent (TMD) distributions [1, 2] in high energy physics has received a great deal of attention recently. This major interest in TMD distributions relies on the belief that they are responsible for various azimuthal asymmetries observed in different single particle inclusive processes involving at least two hadrons. It is found that leading-twist time-reversal-odd (T-odd) distributions play an essential role in these asymmetries. A key ingredient of T-odd distributions is the path-ordered exponential contained in their definition, the so called Wilson line or gauge-link [3, 4, 5] due to initial-state or final-state interactions [6], which also plays key role in the TMD Factorization in SIDIS, Drell-Yan and e^+e^- annihilation processes [7, 8], within the allowance of a sign reversal for T-odd functions in different processes. Unlike in SIDIS and Drell-Yan processes, in hadron-hadron collisions both initial- and final-state interactions contribute to the process resulting more complicated gauge-link for TMD distribution functions in hadron-hadron collisions than in SIDIS or Drell-Yan processes. Studies [9] show that in hadron-hadron collisions the form of the gauge-link depends even on the hard partonic process. In Refs. [10, 11], the authors show that in second order of gluon exchange from initial- and final-state interaction, the standard TMD factorization is violated in back-to-back hadron (jet) pair production, i.e., the TMD distributions need to be modified in a process-dependent way.

Here, we will study the azimuthal angle dependence of back-to-back di-jet produced in unpolarized hadron-hadron scattering which has interesting phenomenologies. We will show that there is a $\cos\delta\phi$ angular dependence of the di-jet due to the product of two Boer-Mulders function, with $\delta\phi = \phi_1 - \phi_2$ the difference of the azimuthal

CP1149, *Spin Physics, 18th International Spin Physics Symposium*
edited by D. G. Crabb, D. B. Day, S. Liuti, X. Zheng, M. Poelker, and Y. Prok

angle of tow jets respectively. Similar result has been obtained in photon-jet production in unpolarized hadron scattering [12]. Further more we will investigate the effect of multiple initial- and final-state interactions on different subprocesses which can contribute to di-jet production in unpolarized hadron scattering, We found that in the $\cos\delta\phi$ asymmetry of $J_q + J_q$ production, there is a color factor enhancement with size $-(7N^2+9)/(N^2-1) = -9$ in the gluonic cross-section, compared with the result from the standard generalized parton model. Therefore a quantitative study of the effect of multiple initial- and final-state interaction on unpolarized hadron scattering will shed light on the QCD dynamics that is present in azimuthal asymmetries.

ANGULAR DEPENDENCE OF DI-JET PRODUCTION IN UNPOLARIZED HADRON SCATTERING

The process we study is $h_A(P_A) + h_B(P_B) \longrightarrow J_1(P_1) + J_2(P_2) + X$, where h_A and h_B are unpolarized. We are interested in the kinematical region where the transverse momenta of the di-jet, $P_{1\perp}$ and $P_{2\perp}$, have similar size and are almost back to back $\boldsymbol{P}_{1\perp} \approx -\boldsymbol{P}_{2\perp}, |\boldsymbol{P}_{1\perp}| \approx |\boldsymbol{P}_{2\perp}| \approx P_\perp$, and where the total transverse momentum of the di-jet $\boldsymbol{q}_\perp = \boldsymbol{P}_{1\perp} + \boldsymbol{P}_{2\perp}$ is much smaller than $\boldsymbol{P}_\perp$: $q_\perp = |\boldsymbol{q}_\perp| \ll P_\perp$. Therefore the process is sensitive to the intrinsic transverse momenta of the partons inside the hadron. We label the azimuthal angles of $\boldsymbol{P}_{1\perp}$ and $P_{2\perp}$ as ϕ_1 and ϕ_2 respectively, and define $\delta\phi = \pi - (\phi_2 - \phi_1)$ to describe the imbalance of the di-jet from a back to back configuration.

The amplitude square of the di-jet production can be expressed as a convolution of the parton-parton to di-jet hard amplitudes and the correlation functions [13]:

$$
\begin{aligned}
|\mathscr{M}|^2 &= \int dx_1 d^2\boldsymbol{k}_{1\perp} dx_2 d^2\boldsymbol{k}_{2\perp} (2\pi)^4 \delta^4(k_1 + k_2 - P_1 - P_2) \\
&\times \ \mathrm{Tr}\{\Phi(x_1, \boldsymbol{k}_{1\perp}^2)\Phi(x_2, \boldsymbol{k}_{2\perp}^2) H(k_1, k_2, P_1, P_2) H^\star(k_1, k_2, P_1, P_2)\}. \qquad (1)
\end{aligned}
$$

The correlation matrix element (soft part) for the unpolarized hadron can be simply decomposed as [1, 2]

$$
\Phi^{[\mathscr{U}]}(P, x, \boldsymbol{k}_\perp^2) = \frac{1}{2}\left[f_1^{[\mathscr{U}]} v(x, \boldsymbol{k}_\perp^2) \not{P} + \frac{h_1^{\perp[\mathscr{U}]}(x, \boldsymbol{k}_\perp^2)}{M} k_\perp^\mu P^\nu \sigma_{\mu\nu} \right], \qquad (2)
$$

where $h_1^\perp(x, \boldsymbol{k}_\perp^2)$ is the Boer-Mulders function describing the transverse spin distribution of a quark in an unpolarized hadron [2], and $\mathscr{U}$ is the gauge-link which depend on the hard partonic process in di-jet production.

We found [14] that there is a $\cos\delta\phi$ asymmetric part in the cross-section due to the production of two Boer-Muders functions. We then can define a weighted $\cos\delta\phi$ asymmetry in $h_A + h_B \longrightarrow J_q + J_q + X$ subprocess as

$$
R_{qq} = \frac{d\sigma^{\mathscr{W}}[J_q J_q]/dP_\perp^2 d\eta_1 d\eta_2}{d\sigma^O[J_q J_q]/dP_\perp^2 d\eta_1 d\eta_2}
$$

$$= \frac{4(7N^2+9)\sum_q h_1^{\perp(1)q}(x_1)h_1^{\perp(1)q}(x_2)}{\left\{\frac{N^3-N}{2}\left(\frac{1+(1-y)^2}{y^2}+\frac{1+y^2}{(1-y)^2}\right)-(N^2-1)\frac{1}{y(1-y)}\right\}\sum_q f_1^q(x_1)f_1^q(x_2)}, \tag{3}$$

where $\mathscr{W} = \cos\delta\phi$. The main result given by this asymmetry is that there is an additional factor $-\frac{7N^2+9}{N^2-1}$, apart from the one obtained in the standard generalized parton model. This factor can be obtained by firstly multiplying the gluonic pole strengths contributed by the process dependent gauge-link appearing in $h_1^{\perp[\mathscr{U}]}$, and then summing over the color flows of the partonic process: $\frac{2N^2}{N^2-1}\times 1\times 1-\frac{N^2+1}{N^2-1}\times 3\times 3 = -\frac{7N^2+9}{N^2-1}$. The size of this gluonic pole factor is -9 for $N=3$, which is quite large and will greatly enhance the asymmetry. Moreover, comparing to the result from generalized parton model, the sign of the asymmetry is reversed by this gluonic pole factor.

In the same way we can give the $\cos\delta\phi$ asymmetry of $J_q+J_{\bar{q}}$ subprocess production:

$$\begin{aligned} R_{q\bar{q}} &= \frac{d\sigma^{\mathscr{W}}[J_qJ_{\bar{q}}]/dP_\perp^2 d\eta_1 d\eta_2}{d\sigma^{O}[J_qJ_{\bar{q}}]/dP_\perp^2 d\eta_1 d\eta_2} \\ &= \frac{A(y)\sum_q\left(h_1^{\perp,(1),q}(x_1)\,h_1^{\perp,(1),\bar{q}}(x_2)+(x_1\leftrightarrow x_2)\right)}{B(y)f_1^g(x_1)f_1^g(x_2)+C(y)\sum_q f_1^q(x_1)f_1^{\bar{q}}(x_2)+(x_1\leftrightarrow x_2)}, \end{aligned} \tag{4}$$

where the function $A(y)$, $B(y)$ and $C(y)$ are given in Ref. [14]. Unlike J_q+J_q production, in this case there is no color factor enhancement for the azimuthal angle asymmetric cross-section.

In a practical measurement of di-jet production, the flavor of the jets usually is not measured. Therefore we calculate the $\cos\delta\phi$ asymmetry of any di-jet production R_{JJ}, which is measurable and defined as

$$R_{JJ} = \frac{\sum_{a,b=q,\bar{q},g} d\sigma^{\mathscr{W}}[J_aJ_b]/dP_\perp^2 d\eta_1 d\eta_2}{\sum_{a,b=q,\bar{q},g} d\sigma^{O}[J_aJ_b]/dP_\perp^2 d\eta_1 d\eta_2}. \tag{5}$$

We assume that the partonic processes that contribute to the azimuthal angular dependence of di-jet production are $qq\to qq$ and $q\bar{q}\to q\bar{q}$. In the calculation we apply two sets of Boer-Mulders functions. Set I is to saturate the positivity bound, Set II is the parametrization given in Ref. [15]. In Fig. 1a we show by thin lines the asymmetry R_{JJ} at RHIC for $P_\perp = 20$ GeV vs y, from the two sets of Boer-Mulders functions . As a comparison, we also show by thick lines the result of R_{JJ} from the conventional generalized parton model (GPM) calculation where gluonic pole contributions are not considered. In Fig. 1b, we show the asymmetry R_{JJ} at RHIC for $P_\perp = 10$ GeV (thick lines) and $P_\perp = 20$ GeV (thin lines) vs $\eta_1+\eta_2$. In the calculations we restrict $-1<\eta_{1/2}<2$ to consider the pseudorapidity acceptance of jet at RHIC. The curves given in Fig. 1a indicate a sign reversal of the asymmetries R_{JJ} between the results from the two different approaches. This is due to the additional factor -9 in the azimuthal angular asymmetric part of $h_A+h_B\to J_q+J_q+X$, coming from the multiple initial- and final-state interactions. We remind that a similar sign reversal also happens in the SSA in hadronic photon-jet production [16], which is due to a similar effect. Therefore the measurement of the $\cos\delta\phi$

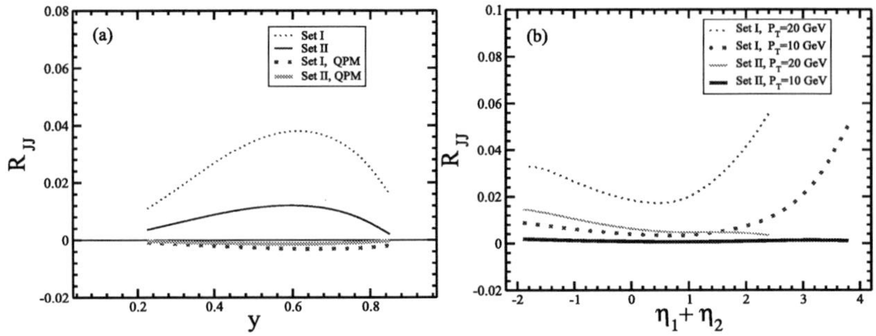

FIGURE 1. (a), The y-dependent $\cos\delta\phi$ asymmetry R_{JJ} in di-jet production at RHIC for $P_\perp = 20$ GeV. (b), The $\cos\delta\phi$ asymmetry R_{JJ} vs $\eta_1+\eta_2$ in di-jet production at RHIC for $P_\perp = 10$ GeV (thick lines) and $P_\perp = 20$ GeV (thin lines) from Boer-Mulders functions of Set I (dotted lines) and II (solid lines).

asymmetry R_{JJ} in unpolarized di-jet production at RHIC, especially the enhancement of the size and the sign reverse of the asymmetry, will provide further tests on the role of initial- and final-state interactions as QCD dynamics of T-odd distributions in hadron scattering.

Acknowledgements We acknowledge the support by the PBCT project ACT-028 "Center of Subatomic Physics".

REFERENCES

1. P.J. Mulders and R.D. Tangerman, Nucl. Phys. **B461**, 197 (1996).
2. D. Boer and P.J. Mulders, Phys. Rev. **D57**, 5780 (1998).
3. J.C. Collins, Phys. Lett. **B536**, 43 (2002).
4. X. Ji and F. Yuan, Phys. Lett. **B543**, 66 (2002); A.V. Belitsky, X. Ji, and F. Yuan, Nucl. Phys. **B656**, 165 (2003).
5. D. Boer, P.J. Mulders, and F. Pijlman, Nucl. Phys. **B667**, 201 (2003).
6. S.J. Brodsky, D.S. Hwang, and I. Schmidt, Phys. Lett. B **530**, 99 (2002); Nucl. Phys. **B642**, 344 (2002).
7. X. Ji, J.P. Ma, and F. Yuan, Phys. Lett. **B597**, 299 (2004).
8. J.C. Collins and A. Metz ,Phys. Rev. Lett. **93**, 252001 (2004).
9. C. J. Bomhof, P. J. Mulders, and F. Pijlman, Eur. Phys. J. C **47**, 147 (2006).
10. John Collins, arXiv:0708.4410 [hep-ph].
11. W. Vogelsang and F. Yuan, Phys. Rev. D **76**, 094013 (2007).
12. D. Boer, P.J. Mulders, and Pisano, Phys. Lett. **B660**, 360 (2008)
13. A. Bacchetta, C. J. Bomhof, P. J. Mulders, and F. Pijlman, Phys. Rev. D **72**, 034030 (2005).
14. Z. Lu and I. Schmidt, Phys. Rev. D **78**, 034041 (2008).
15. B. Zhang, Z. Lu, B.-Q. Ma, and I. Schmidt, Phys. Rev. D **77**, 054011 (2008).
16. A. Bacchetta, C. Bomhof, U. D'Alesio, P.J. Mulders, and F. Murgia, Phys. Rev. Lett. **99**, 212002 (2007).

π^0 Transverse Single-Spin Asymmetries at $\eta = 4.1$ in $p+p$ Collisions at $\sqrt{s} = 200$ GeV

J. L. Drachenberg[1], for the STAR Collaboration

[1] *Cyclotron Institute, Texas A&M University, College Station, TX 77843*

Abstract. Large transverse single-spin asymmetries, A_N, have been observed in forward hadron production at RHIC. STAR has reported precision measurements of the x_F-p_T dependence of A_N for forward π^0 production. Contrary to expectation, the asymmetry does not fall with p_T. However, questions remain about the trend at the lower end of the studied range, $p_T \sim 1$ GeV/c. Analysis of data from the 2008 RHIC run has extended the π^0 data in the $x_F > 0.4$, low p_T region. Preliminary results are reported.

Keywords: Transverse Single-spin Asymmetries
PACS: 13.88.+e

INTRODUCTION

In 2006, STAR measured transverse single-spin asymmetries, A_N, for forward rapidity π^0 production in $p+p$ collisions at $\sqrt{s} = 200$ GeV to $p_T \sim 4$ GeV [1]. Results showed large A_N at high-x_F as seen in previous experiments at lower center-of-mass energies, such as Fermilab E704 [2, 3]. In contrast to the E704 kinematics, pQCD predictions provide a reasonable description of the observed cross-sections for forward π^0 production at $\sqrt{s} = 200$ GeV [4]. Model calculations based on the Sivers [5] and Collins [6] effects provide fair descriptions of the x_F-dependence of the measured asymmetries. The Sivers model relates the asymmetry to correlations involving initial-state parton transverse momentum [7] while the Collins model relates the asymmetry to final-state transverse momentum dependent fragmentation [8]. However, the observed asymmetries do not provide an indication of decreasing A_N with increasing p_T, as predicted by theory [5, 6]. Isolating the origin or, in the case of a combination of effects, disentangling the origins of the observed asymmetries would provide crucial information about proton spin structure.

The previous STAR measurements [1] provide a hint of an initial decrease of A_N with p_T for $x_F > 0.4$. The data obtained in this low-p_T, high-x_F region were recorded during 2003 and 2005, and are limited by low statistics and low polarization. For the 2008 RHIC run, STAR extended measurements to low-p_T, high-x_F kinematics to further investigate this region with better statistics and polarization.

FORWARD PION DETECTOR

To obtain the measurements for the 2008 run, data were taken with the STAR Forward Pion Detector (FPD). The FPD is an electromagnetic calorimeter, consisting of two

CP1149, *Spin Physics, 18th International Spin Physics Symposium*
edited by D. G. Crabb, D. B. Day, S. Liuti, X. Zheng, M. Poelker, and Y. Prok

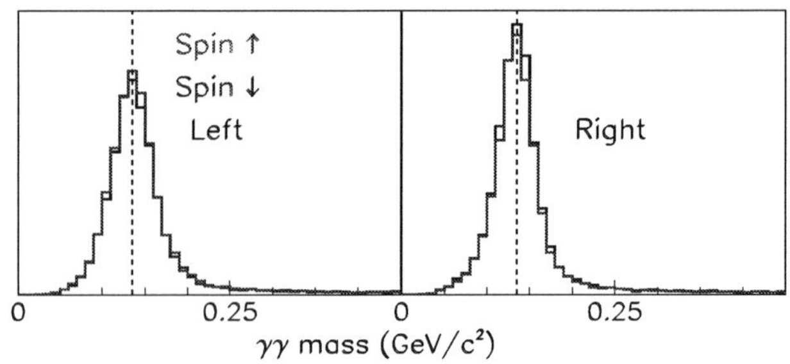

FIGURE 1. Spin-sorted di-photon mass peaks. The dashed lines mark the known mass value of the π^0.

modules, left and right of the beamline on the east side of STAR. Each module consists of a 7×7 array of PbGl calorimeters positioned 809 cm from the interaction point [9]. Each PbGl calorimeter cell measures $3.8 \times 3.8 \times 45$ cm^3. To access the desired low-p_T, high-x_F kinematics, the modules were positioned at $< \eta > \sim -4.1$.

EVENT CUTS AND SYSTEMATICS

Data were taken from $p_\uparrow + p \rightarrow \pi^0 + X$ yielding $L_{int} \sim 0.5$ pb^{-1} with beam polarization, $P_{beam} \sim 0.44$. Events were required to pass a hardware trigger that required a nominal summed energy of 25 GeV in a single FPD module. Software cuts were applied for di-photon events, requiring energy sharing $z_{\gamma\gamma} < 0.7$ and 0.07 GeV/c^2 $< m_{\gamma\gamma} <$ 0.3 GeV/c^2, and a software total energy threshold of $E_{total} > 25$ GeV was applied. The events were required to fall within a fiducial volume of one-half the width of a PbGl cell from the edge of each module. Calibration procedures involved analyzing raw ADC's channel-by-channel and run-by-run for pedestal shifts. For gain calibration, π^0 mass peaks were reconstructed channel-by-channel and corrected in an iterative process until convergence to the known value was achieved. Energy-dependent corrections were evaluated by fitting the π^0 peak in energy bins, then obtaining a linear fit of the ratio of the known value over the fitted centroids vs. energy. Finally, run-by-run summed-mass values are used to correct for run-dependent gain shifts. The result is agreement between the di-photon peaks in each module and the known value of the π^0 for polarization spin-up or spin-down, Fig. 1.

Systmatic errors are dominated by gain calibration issues and yields under the pion mass peak. Gain calibration issues arise because of limited statistics in the PbGl cells that are farthest from the beam in each module. To correct for yields under the pion peak, varying mass cuts were studied. The total systematic uncertainty was obtained by combining the effects in quadrature.

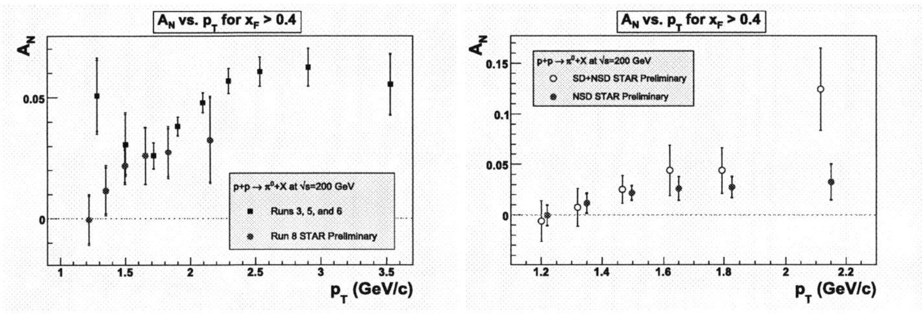

FIGURE 2. π^0 transverse single-spin asymmetries (A_N) versus transverse momentum (p_T). For the left-hand figure and NSD sample (Right) inner error bars are statistical errors only, and outer error bars are statistical plus systematic uncertainties. Points for the mixed event sample (Right) include only statistical uncertainties and have been offset to the left by 0.03 GeV/c.

RUN-8 ASYMMETRIES

Asymmetries shown have been calculated using a cross-ratio method,

$$A_N = \frac{1}{P_{beam}} \times \frac{\sqrt{N_{L\uparrow} \times N_{R\downarrow}} - \sqrt{N_{L\downarrow} \times N_{R\uparrow}}}{\sqrt{N_{L\uparrow} \times N_{R\downarrow}} + \sqrt{N_{L\downarrow} \times N_{R\uparrow}}} \tag{1}$$

where $N_{L(R)\uparrow(\downarrow)}$ is the number of events in the left (right) module with spin up (down). Results for $x_F < 0$ are consistent with zero, so we show data only for $x_F > 0$. Results are shown for two different software conditions on the STAR Beam-Beam Counter (BBC). First, a software trigger requiring a coincidence between BBC-east (BBCE) and BBC-west (BBCW) is required, producing a nearly pure non-singly-diffractive (NSD) event sample. These data are shown in comparison to previously published data from RHIC runs 3, 5, and 6 [1]. A second event sample is analyzed requiring a hit in BBCE, the same side of the collision as the FPD, and no hit in BBCW. This results in a mixture of NSD and singly-diffractive (SD) events. Results from analysis of the mixed sample are shown in comparison with the nearly-pure NSD event sample. Statistical errors are calculated as $\Delta A_N = 1/(P_{beam} \times \sqrt{N_{L\uparrow} + N_{L\downarrow} + N_{R\uparrow} + N_{R\downarrow}})$ and scaled to account for remaining acceptance asymmetries, which are less than 20%. For the nearly-pure NSD event sample, figures show inner error bars for statistical errors and outer error bars for systematic errors combined in quadrature with statistical errors. Errors for the mixed event sample are statistical only. The systematic uncertainties for the mixed event sample are expected to be less than the statistical errors.

RESULTS

Results from RHIC run-8 are mostly consistent with results from previous runs, Fig. 2 (Left). However, in the low-p_T region where previous data were dominated by RHIC run 3 with limited statistics, the run-8 data show A_N continuing to decrease. Likewise,

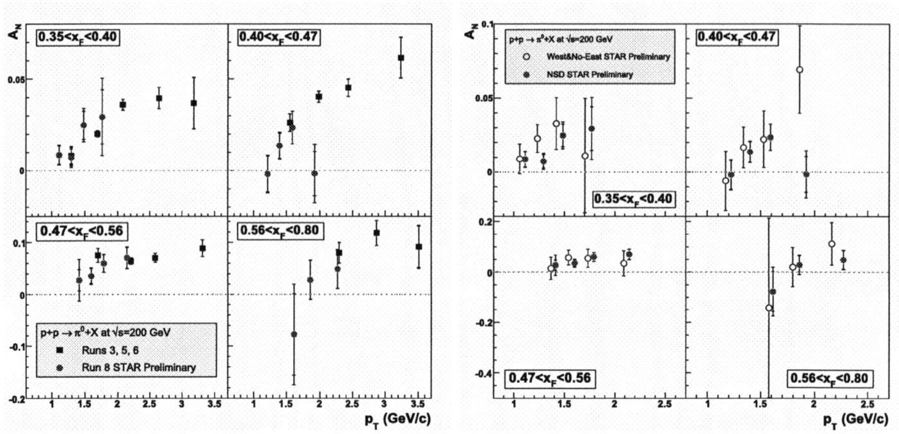

FIGURE 3. π^0 transverse single-spin asymmetries (A_N) versus transverse momentum (p_T) in x_F bins. For the left-hand figure and NSD sample (Right) inner error bars are statistical errors only, and outer error bars are statistical plus systematic uncertainties. Points for the mixed event sample (Right) include only statistical uncertainties and have been offset to the left by 0.06 GeV/c.

observing the p_T-dependence at fixed x_F, Fig. 3 (Left), run-8 results show general agreement with previous results in overlapping regions and smooth fall-off toward lower-p_T.

The results from the mixed event sample show general agreement with the nearly-pure NSD sample in terms of p_T, Figs. 2 and 3 (Right).

SUMMARY

We have shown STAR measurements of transverse single-spin asymmetries, A_N, from $p_\uparrow + p \to \pi^0 + X$ at $\sqrt{s} = 200$ GeV from the 2008 RHIC run. The data show general consistency with previous measurements and show A_N continuing to decrease at low-p_T. We have also shown results from a mixed event sample combining singly-diffractive and non-singly-diffractive events. Results from the mixed event sample are consistent with results from the nearly-pure non-singly-diffractive event sample.

REFERENCES

1. B. I. Abelev *et al.*, *Phys. Rev. Lett.* **101**, 222001 (2008).
2. D. L. Adams *et al.*, *Phys. Lett. B* **261**, 201 (1991).
3. D. L. Adams *et al.*, *Phys. Lett. B* **264**, 462 (1991).
4. J. Adams *et al.*, *Phys. Rev. Lett.* **97**, 152302 (2006).
5. U. D'Alesio, F. Murgia, *Phys. Rev. D* **70**, 074009 (2004).
6. F. Yuan, arXiv:hep-ex/0804.3047v2.
7. D. Sivers, *Phys. Rev. D* **41**, 83 (1990).
8. J. Collins, *Nucl. Phys. B* **396**, 161 (1993).
9. L. C. Bland, arXiv:hep-ex/0602012.

Spin-dependent Forward Particle Correlations in p+p Collisions at $\sqrt{s}$ = 200 GeV

Nikola Poljak[a] for the STAR collaboration

[a]*University of Zagreb, Croatia*

Abstract. The STAR collaboration has reported precision measurements of the transverse single spin asymmetries for the production of the π^0 mesons from polarized proton collisions at $\sqrt{s}$ = 200 GeV. These measurements were obtained using modular forward detectors. The Forward Meson Spectrometer (FMS), covering a region of $2.5 < \eta < 4.0$, and its engineering prototype, provide increased acceptance, as needed for spin-dependent correlation studies that could disentangle contributions to the forward π^0 asymmetries. We report on the status of the analysis.

Keywords: p p, colliding beams, polarized beam, hadroproduction, small-angle, STAR, single spin asymmetry, transverse momentum, x_F-dependence, electromagnetic calorimeter, 200 GeV
PACS: 13.88.+e; 13.85.Ni; 12.38.Qk

INTRODUCTION

For a particle produced in a collision of transversely polarized protons, the analyzing power (A_N) is defined as the difference of spin-up and spin-down cross sections divided by their sum. A_N is just one of many examples of transverse single spin asymmetries. A_N is expected to be near zero in a leading-twist and collinear perturbative QCD (pQCD) description of particle production. Cross sections for pions produced with large Feynman-x ($2p_L/\sqrt{s}$) and moderate p_T in p+p collisions are found to be in agreement with next-to-leading order pQCD calculations at $\sqrt{s}$=200 GeV [1,2]. Precision measurements of the asymmetry as a function of x_F and p_T were reported, showing large A_N at large x_F [3]. Also, it was shown that the x_F dependence matches the Sivers effect [4,5] expectations qualitatively. The p_T dependence at fixed x_F is not consistent with expectations of p-QCD based calculations.

Since these reports, there was a significant number of developments both in theory and experiment. New phenomenological analyses within a generalized parton model can now explain Sivers moments in semi-inclusive deep inelastic scattering and many features of $p\uparrow+p \rightarrow \pi+X$ [6]. Previous expectations that the Collins effect [7] is suppressed in $p\uparrow+p \rightarrow \pi+X$ were found incorrect due to a sign error [8]. COMPASS analyzed 20 percent of their transversely polarized proton data and reported non-zero Collins moments and Sivers moments compatible with zero, although the expected Sivers moments are small in the (x,Q^2) range of their experiment. To conclude, the need still remains to separate Collins and Sivers effects in $p\uparrow+p \rightarrow \pi+X$.

The Sivers effect manifests itself as an asymmetry in the forward jet or gamma production while the Collins as an asymmetry in the forward jet fragmentation. To

CP1149, *Spin Physics, 18th International Spin Physics Symposium*
edited by D. G. Crabb, D. B. Day, S. Liuti, X. Zheng, M. Poelker, and Y. Prok

distinguish between the two one has to look beyond inclusive π events. Since the Collins effect is an azimuthal modulation of hadrons around the thrust axis of an outgoing quark, integrating over the azimuthal angle cancels it, leaving only the Sivers effect. This is one possible path to distinguishing the two effects.

EXPERIMENTAL SETUP

The detectors used for the measurements done at STAR in the previous runs measured the inclusive π cross section as well as the single beam spin asymmetry for their inclusive production. These detectors, called Forward Pion Detector (FPD), were modular electromagnetic detectors, covering regions of $\langle\eta\rangle = 3.3$, 3.7 or 4.0. The Forward Meson Spectrometer (FMS) now provides STAR with full coverage of the azimuth in the range $2.5 \le \eta \le 4.0$. The forward acceptance of STAR has been increased 20x by the FMS. Studies of inclusive production in p+p and d(p)+Au collisions over a broad range of x_F and p_T for mesons that decay electromagnetically to observed photons, electrons and positrons are now enabled. Also possible are particle correlation studies within a single forward jet and correlation studies of fragments from jet pairs.

The FMS is built from lead-glass detectors of two types: 788 cells of square cross section 5.8 cm on a side and 18 radiation lengths, and 476 cells of square cross section 3.8 cm on a side and 18 radiation lengths. The FMS spans a 2x2 m^2 area perpendicular to the beam and its front face is 706 cm from the interaction point.

The methodologies that were developed to calibrate the FPD have been adapted to calibrate the FMS. Individual detector conversion gains are iteratively adjusted to give the Gaussian centroid at the π mass in di-photon invariant mass distributions associated with each detector. Events used for calibration were required to satisfy a hardware-level individual detector threshold condition and have at least two photons reconstructed within a fiducial volume of the calorimeter. The two highest energy photons in an event are used to compute the mass. The energy sharing, defined as $|E_1 - E_2|/(E_1+E_2)$, must be < 0.7 for the small cells. In addition to adjustments of linear calibration factors for each detector, energy dependent corrections are found to be required. The centroid of the π peak in the mass distributions is found to increase with leading photon energy for both the data and full GEANT simulations. For the latter, it is found that when energy dependent corrections are applied to give the correct centroid for the pion peak in di-photon invariant mass distributions binned in leading photon energy, the reconstructed di-photon energy, on average, agrees with the generated pion energy.

The calibrations proved to be stable at the level of a few percent, as determined by analysis of the time dependence of the mass peak centroid. Another tool checking the calibration stability was a light-emitting diode (LED) monitoring system placed on the front face of the detector. This system had the capability of flashing a number of LED signals at a given time so a response from the detector could be read out.

Full PYTHIA/GEANT simulations have been made with adequate statistics to study neutral pion production to moderate x_F at large p_T. Energy-binned mass distributions from data reconstructions are in reasonable agreement with similar distributions from reconstruction of simulation data.

GOALS AND FIRST RESULTS

It is anticipated that the FMS will provide new information about transverse single spin asymmetries. To make a point of contact with previous work, a first analysis of the x_F dependence of the A_N for the $p\uparrow + p \rightarrow \pi + X$ from the FMS was performed.

To verify patterns of polarization signs for the colliding beam bunches, analysis of concurrently measured spin asymmetries from the STAR beam-beam counter (BBC) was performed. It was previously known that the azimuthal distribution of multiplicity in the BBC has a significant analyzing power [9], making it useful as a colliding beam polarimeter. Spin asymmetries from the BBC, averaged over all bunches, and bunch-by-bunch spin asymmetries from the BBC for each fill, confirm the veracity of the spin directions of the polarized colliding beams for each fill.

In the analysis of $p\uparrow + p \rightarrow \pi + X$, we divided the FMS into octants for inclusive π spin sorting. A cross ratio analysis of the resulting distributions from opposite octants was then made. Analyzing powers comparable to prior measurements are found, from analysis of 75% of the 2008 run data, as shown in Fig. 1(b). Quantitative comparison of this data with earlier measurements requires binning in both x_F and p_T. The dependence of the asymmetry on $\langle\cos\varphi\rangle$, where $\langle\cos\varphi\rangle$ is determined from data, is well represented by a linear function, as expected. The zero crossing in Fig. 1. (a) is shifted from $\langle\cos\varphi\rangle$ at the level of 2σ. Statistical errors are included in the result, with systematical errors still being evaluated. A first estimate based on significance distributions [3] yields $\sigma_{tot.} \leq 1.2\ \sigma_{stat.}$ The figures presented here are a different presentation of the same results shown at the SPIN-2008 symposium.

"Jet-like" events were also looked at. "Jet-like" clusters are formed in an event by considering energy depositions > 0.4 GeV in all cells of the FMS. A cluster consists of N cells, where N is the maximal subset of cells that are found to be within a cone of radius 0.5 in the η-φ space. The x_F and p_T for the cluster are given by the vector sum of momenta from each cell, assuming the energy deposition is from photons originating from the collision vertex. Clusters with at least 15 cells with deposited energy > 0.4 GeV having cluster p_T > 1 GeV/c and x_F >0.2 are required in the analysis. A requirement that the cluster centroid is within the calorimeter volume by at least two cell widths is also imposed.

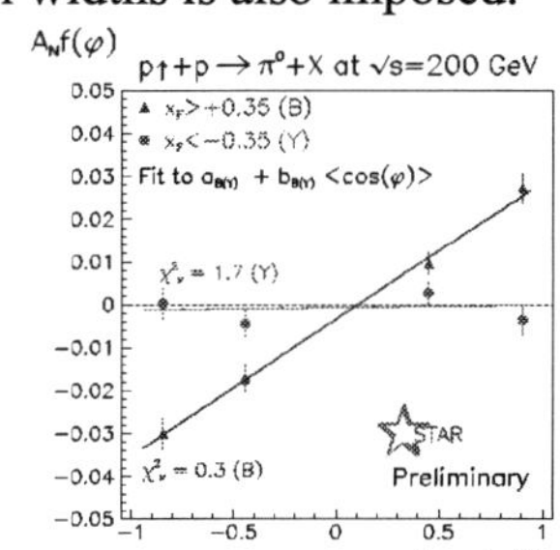

FIGURE 1. (a) A_N versus $\langle\cos\varphi\rangle$, as determined from data. The positive x_F fit gives a φ-independent fit parameter that is non-zero at the level of 2σ, possibly due to a statistical fluctuation.

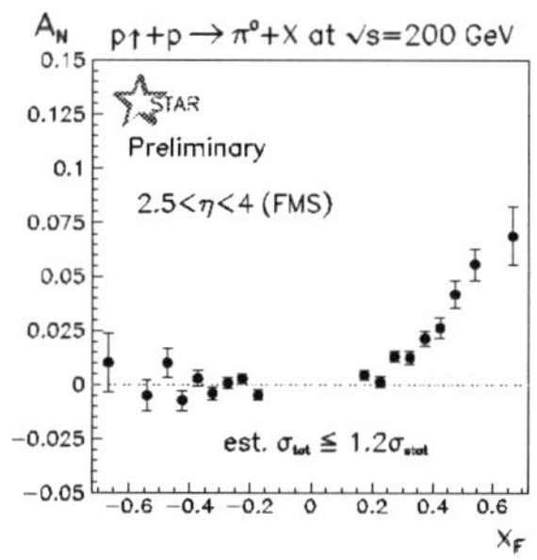

(b)Variation of A_N as a function of x_F integrated over the FMS acceptance. The negative x_F values are consistent with zero. The indicated uncertainties are statistical.

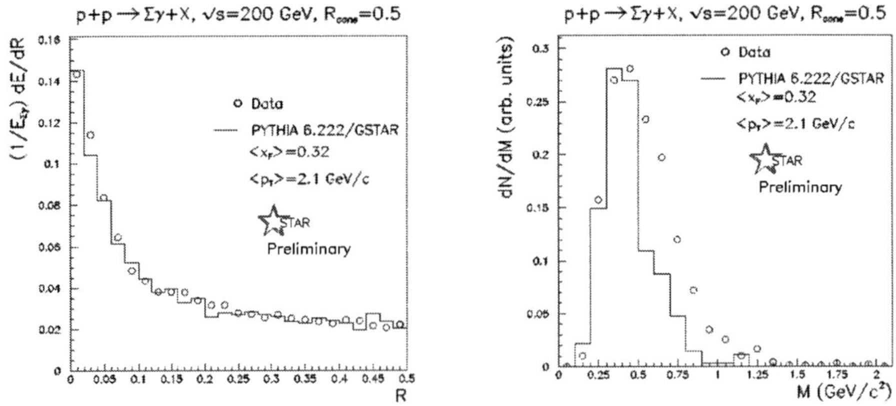

FIGURE 2. (a) "Jet-shape" distribution compared in data and simulation. The image shows the deposition of energy within the jet as a function of distance from the jet axis.

(b) Invariant mass distribution of "jet-like" clusters. Fair agreement is found between data and simulations, although the mass distribution is broader in data than from PYTHIA/GSTAR simulations.

The clustering analysis is applied to both data and PYTHIA/GEANT simulations. Good agreement for the jet-shape is found, and fair agreement for the jet-mass is found, as shown in Fig. 2.

CONCLUSIONS AND OUTLOOK

The FMS is complete, in place, commissioned and operated in the 2008 run. It has 20 times more acceptance than the previous modular detectors. The reconstruction and calibration procedures were successfully ported from the FPD to the FMS. The calibration is mostly complete and shows good agreement with the simulated sample of events. The inclusive pion $A_N(x_F)$ is comparable to prior FPD precision measurements and the analysis of "jet-like" events is under way.

We aim to finish the analysis of "jet-like" events and extend the inclusive pion $A_N(p_T)$ measurements. Also, we want to determine A_N for the final state that contains π pairs as well as for states with heavier mesons. In the future runs, the goal is to go beyond π detection to direct photons+jet final state A_N.

REFERENCES

1. J. Adams et al., *Physical Review Letters* **97**, 152302 (2006)
2. I. Arsene et al., *Physical Review Letters* **98**, 252001 (2007)
3. B. I. Abelev et al., *Physical Review Letters* **101**, 222001 (2008)
4. D. Sivers, *Physical Review D* **41**, 83-90 (1990)
5. D. Sivers, *Physical Review D* **43**, 261-263 (1991)
6. M. Boglione, U. D'Alesio and F. Murgia, *Physical Review D* **77**, 051502 (2008)
7. J. C. Collins, *Nuclear Physics B* **396**, 161-182 (1993)
8. M. Anselmino et al., *Physical Review D* **71**, 014002 (2005)
9. J. Kiryluk et al., *hep-ex 0501072*, published in Spin 2004 Conference Proceedings, Trieste, Italy

SPIN STRUCTURE OF THE NUCLEON—COMBINED

Exclusive Meson Electroproduction: GPDs, Regge and Dispersion Relations

Gary R. Goldstein* and Simonetta Liuti†

*Tufts University, Department of Physics and Astronomy, Medford, MA 02155, USA
†University of Virginia, Department of Physics, Charlottesville, VA 22901, USA

Abstract. Exclusive pi0 electroproduction from nucleons at large photon virtuality can be described in terms of Generalized Parton Distributions, particularly the chiral odd subset. Chiral odd GPDs are related to transversity and can be accessed experimentally for various special choices of observables. This is accomplished by choosing C-parity odd and chiral odd combinations of t-channel exchange quantum numbers. These GPDs are calculated in a spectator model, and are constrained by boundary functions. Alternatively, the production amplitudes correspond to C-odd Regge exchanges with final state interactions. The helicity structure of the virtual photoproduction amplitudes provides relations between the partonic description via GPDs and the hadronic, Regge description of C-odd processes, all via quark helicity flip. GPDs, in general, are analytic functions of energy variables. Their integrals over unobservable parton momenta thereby satisfy Dispersion Relations (DR). Using DR's could allow the real part of the amplitudes, the integrated GPDs, to be extracted from the more easily measured imaginary parts. However, at non-zero momentum transfer DRs require integration over unphysical regions of the variables. We show that the relevant unphysical region of the non-forward DRs is considerable. This will vitiate the efforts to avoid the actual measurement of the real parts more directly.

Keywords: exclusive, electroproduction, GPD, Regge, dispersion relation
PACS: 11.55.Fv, 12.40.Nn, 13.60.-r, 13.88.+e

1. EXCLUSIVE π^0 ELECTROPRODUCTION

Data are accumulating on the cross section and asymmetries in exclusive π^0 electroproduction at intermediate energies ν, low to high photon virtuality Q^2 and small momentum transfer t. This is the kinematical region in which both Regge and GPD descriptions overlap. Characteristic of this reaction is that it has chiral odd and C-parity odd t-channel quantum numbers. In the Regge picture the dominant exchanges will be neutral vector mesons (ρ^0, ω with $J^{PC} = 1^{--}$) and neutral axial vectors (b_1^0, h_1 with 1^{+-}) along with rescattering corrections. In the factorized GPD picture the chiral odd GPDs (soft physics) will be singled out for either pairs of quantum numbers and for both longitudinal and transverse photons (for the hard processes in the upper part of the handbag). The quantum numbers of this reaction lead to the possibility of measuring the *transversity* properties of the nucleon.

A successful Regge cut model was developed to fit photoproduction data many years ago [1]. That model essentially involves as input the vector and axial vector meson trajectories that factorize into couplings to the on-shell $\gamma + \pi^0$ vertex and the nucleon vertex. The cuts or absorptive corrections destroy that factorization, but fill in the small t and $t \approx -0.5$ amplitude zeroes. To connect to electroproduction, the upper vertex factor must acquire Q^2 dependence. This is accomplished by replacing the elementary,

CP1149, *Spin Physics, 18th International Spin Physics Symposium*
edited by D. G. Crabb, D. B. Day, S. Liuti, X. Zheng, M. Poelker, and Y. Prok

t-dependent couplings with Q^2 dependent transition form factors. In this Regge picture the factorization for the longitudinal virtual photon is not different from the transverse photon, except for the additional power of Q^2 for the longitudinal case. This is in contrast to the proofs of factorization for the longitudinal case in the GPD picture [2]. With our form factor approach to the upper vertex (including Sudakov factors to soften the endpoint singularities) we anticipate a similar factorization for the transverse case.

The Regge picture is implemented by singling out the 6 independent helicity amplitudes and noting that at large s and small $|t|$ the leading natural parity and unnatural parity Regge poles contribute to opposite sums and differences of pairs of helicity amplitudes. This gives rise to striking behavior for certain observables.

Now the crucial connection to the 8 GPDs that enter the partonic description of electroproduction is through the helicity decomposition [3], where, for example, one of the chiral even helicity amplitudes is given by

$$A_{++,++}(X,\xi,t) = \frac{\sqrt{1-\xi^2}}{2}(H^q + \tilde{H}^q - \frac{\xi^2}{1-\xi^2}(E^q + \tilde{E}^q)),$$

while one of the chiral odd amplitudes is given by

$$A_{++,--}(X,\xi,t) = \sqrt{1-\xi^2}(H_T^q + \frac{t_0-t}{4M^2}\tilde{H}_T^q - \frac{\xi}{1-\xi^2}(\xi E_T^q + \tilde{E}_T^q)).$$

There are relations to PDFs, $H^q(X,0,0) = f_1^q(X), \tilde{H}^q(X,0,0) = g_1^q(X), H_T^q(X,0,0) = h_1^q(X)$. The first moments of these are the charge, the axial charge and the tensor charge, for each flavor q, respectively. Further, the first moments of $E(X,0,0)$ and $2\tilde{H}_T^q(X,0,0) + E_T^q(X,0,0)$ are the anomalous moments κ^q, κ_T^q, with the latter defined by Burkardt [4].

Chiral even GPDs have been modeled in a thorough analysis [5], based on diquark spectators and Regge behavior at small X, and consistent with constraints from PDFs, form factors and lattice calculations. That analysis is used to obtain chiral odd GPDs via a multiplicative factor that fits the phenomenological $h_1(x)$ [6]. With that *ansatz* the observables can be determined in parallel with the Regge predictions. In Fig. 1 the beam asymmetry is shown for both models and compared to recent data [7]. It is remarkable that these model predictions are not fit to that data, but determined by other constraints. Figure 2 shows a range of predictions from the chiral odd GPD model for the target asymmetry, A_{UT}, depending on the choice of δu. This indicates that measurements will pin down values of the tensor charge. A thorough discussion of this work is given in Ref. [8].

2. GPDS AND DISPERSION RELATIONS - LIMITATIONS

Deeply Virtual Compton Scattering (DVCS) amplitudes satisfy Lorentz covariance and analyticity in energy at fixed angles. It has long been known that analyticity of two body amplitudes leads to dispersion relations (DR). Using Mandelstam variables s and t, or $\nu = (s-u)/4M$ and t, we introduce a generic scattering amplitude $T(\nu,t,Q^2)$ to represent any of the invariant amplitudes in DVCS. For this discussion we can take all particles spinless. Then this amplitude would have its square magnitude proportional to

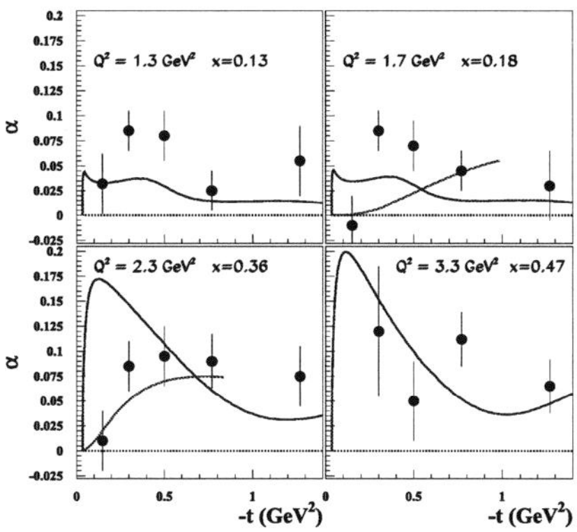

FIGURE 1. Beam asymmetry α in Regge (blue) and GPD pictures. Data are from Ref. [7].

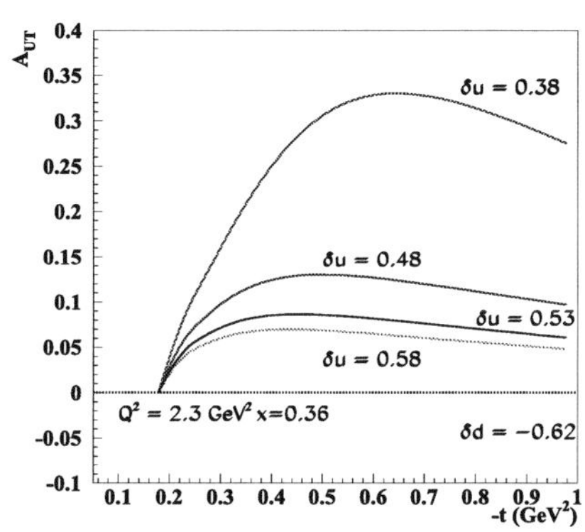

FIGURE 2. GPD predictions for transverse spin asymmetry, A_{UT}, for varying tensor charges.

$d\sigma/dt$ at fixed Q^2, t. The dispersion relation would be

$$\Re\mathrm{e}T^-(\nu,t,Q^2) = \frac{2\nu}{\pi}\int_{\nu_{threshold}}^{\infty} d\nu' \frac{\Im\mathrm{m}T^-(\nu',t,Q^2)}{\nu'^2-\nu^2}, \tag{1}$$

for an amplitude that is odd under the crossing symmetry $\nu \to -\nu$ (and thereby will avoid an extra subtraction for convergence) or

$$\Re\mathrm{e}T^+(\nu,t,Q^2) = \frac{\nu^2}{\pi}\int_{\nu_{threshold}}^{\infty} d\nu' \frac{\Im\mathrm{m}T^+(\nu',t,Q^2)}{\nu'^2(\nu'^2-\nu^2)} + \Delta, \tag{2}$$

for an amplitude that is even under the crossing symmetry. The latter has a subtraction Δ to assure convergence in ν'.

Where are the poles and branch cuts of $T^{\pm}$ as a function of ν'? Figure 3 shows the lowest hadronic *continuum* intermediate states, $\pi+N$. It is also clear that there will be a nucleon pole below the process threshold. The imaginary part of the amplitude is the discontinuity across the branch cut, beginning at $s=(M+m_\pi)^2$ and the corresponding value of $\nu=\nu_{continuum}$. As ν increases the next branch cut begins at the $2\pi+N$ threshold, and so on. There is a problem here, however. For non-forward values of t, the physically accessible threshold can begin at higher values of ν than $\nu_{continuum}$. But the DR requires an integrand defined down to the continuum. This was recognized to be a problem after non-forward DRs were first introduced and then studied carefully by Lehmann [9]. In the low energy region a partial wave decomposition allows for analytic continuation (via the "Lehmann ellipse"). In the high energy region the partial wave approach is of no use, so a model of some kind must be used to complete the DR integration. We will show some examples below.

The important reason for considering these issues is that the possibility has been raised that DRs can be applied to the partonic, factorized form of the amplitudes [10] - the handbag picture with GPDs of Fig. 3. An example is the spin independent $H_f(x,\xi,t)$, for flavor f, which is related to the amplitude for DVCS [11] $T^{\mu\nu}(\nu,Q^2,t)$ through

$$T^{\mu\nu}(\nu,Q^2,t)=\frac{1}{2}g^{\mu\nu}\bar{u}(p')\hat{n}u(p)\sum_{flavors}e_f^2\mathcal{H}_f(\xi,t), \tag{3}$$

where

$$\mathcal{H}_f(\xi,t)=\int_{-1}^{+1}dx\frac{H_f(x,\xi,t)}{x-\xi+i\varepsilon}, \tag{4}$$

from which it follows (via $1/(x-\xi+i\varepsilon)=P.V./(x-\xi)+\delta(x-\xi)/\pi$) that

$$\Im m(\mathcal{H}_f(\xi,t))=H(\xi,\xi,t) \text{ and } \Re e(\mathcal{H}_f(\xi,t))=\frac{1}{\pi}P.V.\int_{-1}^{+1}dx\frac{H_f(x,\xi,t)}{x-\xi}. \tag{5}$$

With the relations among the independent variables

$$x_{Bj} = \frac{Q^2}{2M\nu_{Lab}},\ X=\frac{k\cdot P}{q\cdot P'},\ \zeta=\frac{q\cdot P'}{q\cdot P} \tag{6}$$

$$\xi = \frac{\zeta}{(2-\zeta)}=\frac{Q^2}{4M\nu},\ x=\frac{(2X-\zeta)}{(2-\zeta)} \tag{7}$$

the DR in Eq. (2) appear quite similar to Eq. (5);

$$\Re e(\mathcal{H}_f(\xi,t))=\frac{1}{\pi}P.V.\int_{-1}^{+1}dx\frac{H_f(x,\xi,t)}{x-\xi}=\frac{1}{\pi}P.V.\int_{-1}^{+1}dx\frac{H_f(x,x,t)}{x-\xi}+\Delta(\xi). \tag{8}$$

It is this relation that led to the conclusion that using the imaginary part of $\mathcal{H}$ is sufficient to determine the entire Compton Form Factor $\mathcal{H}$. However, there are two possible problems. Consider Fig. 3 where we see that the handbag factorized form of the amplitude does not have physical, on-shell observable intermediate states. It is not clear where the unitarity cut should begin in evaluating the DR for the GPD. Perhaps

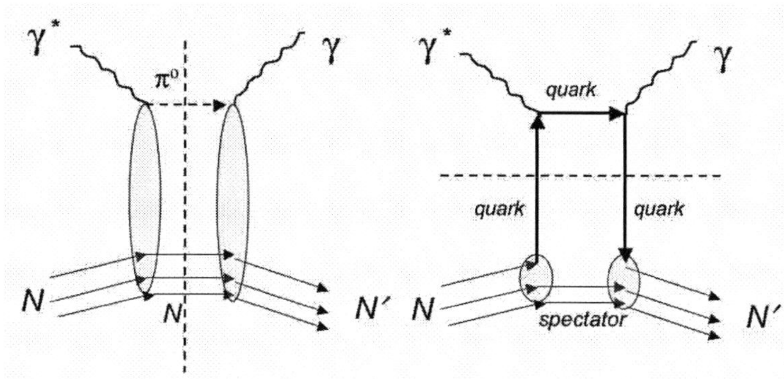

FIGURE 3. Left: hadronic s-channel unitarity cut for DVCS; Right: partonic "handbag" factorized diagram.

the relevant thresholds would be the jets associated with the partons, as suggested in Ref. [12]. But the threshold of such a branch cut is ill-defined. Putting that question aside for later exploration, we are left with the problem of the integration through an unphysical region between the continuum and the physical threshold for different values of t and Q^2.

In Fig. 4 we show the mismatch between the physical threshold of ν for different t values when, for exmple $Q^2 = 2$ GeV^2. This means that at $Q^2 = 2$ GeV^2 there are no measurements for $\Im m\mathcal{H}(\xi,t)$ for $\xi > \xi(\nu_{threshold})$. To complete the integration for the real part, though, requires continuing to $\xi = 1$. How can this continuation be achieved? An example of the continuation for DVCS is a Regge pole model, continued down to $\nu = 0$,

$$H^R(\nu,Q^2,t) = \beta(Q^2,t)(1 - e^{-i\pi\alpha(t)})(\nu/\nu_0)^{\alpha(t)}, \tag{9}$$

from which it follows that $\Re eH^R(\nu,Q^2,t) = \tan(\pi\alpha(t)/2)\Im mH^R(\nu,Q^2,t)$ exactly. The DR relation is

$$\Re eH^R(\nu,Q^2,t) = \frac{2\nu}{\pi}\int_{\nu_{threshold}}^{\infty} d\nu' \frac{\Im mH(\nu',Q^2,t)}{\nu'^2 - \nu^2}. \tag{10}$$

The comparison of the actual real part with the result of the dispersion integration is shown in Fig. 4 for a few values of Q^2. In terms of the variable ξ it is clear that the dispersion integral only converges to the real part at very small ξ or large ν (we have taken the odd amplitude here for simplicity).

We have checked this comparison with a simple, spinless diquark spectator model also. The model satisfies polynomiality and should thus satisfy the DR [10]. Again we find that the convergence for non-zero Q^2 is slow because of the threshold mismatch. A more relevant model is the GPD parameterization of AHLT [5]. We are in the process of checking the DR's for this determination of GPD's. In conclusion, DRs appear to be a parsimonious determination of the full analyticity of integrated GPDs. In practice, however, the unphysical region involved in implementing the integration requires model dependent interpolation.

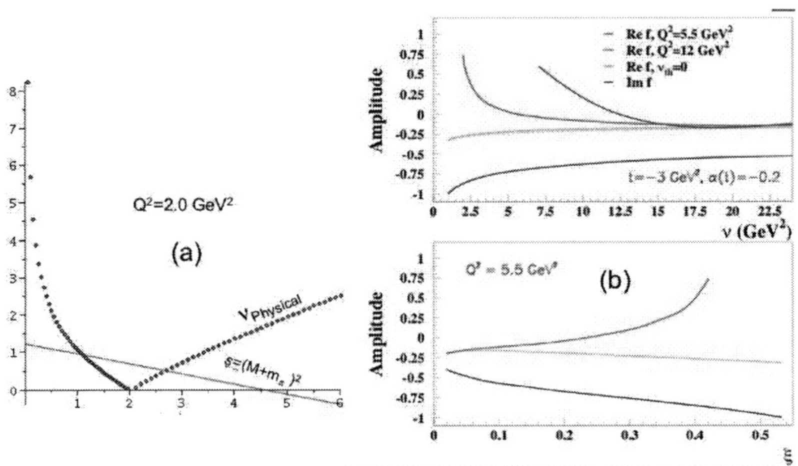

FIGURE 4. (a) Values of $\nu_{physical}$ vs. $-t$ for $Q^2 = 2$ GeV^2. The solid line is the DR lower limit for $s = (M + m_\pi)^2$. For $-1.2 > t > -2.7$ GeV^2 there is no gap.(b) Dispersion relation for real part of Regge amplitude for $\nu_{threshold} = 0$ (third line down) which is exact, and for two other thresholds corresponding to different Q^2 values. The input imaginary part is the lower curve of both graphs.

ACKNOWLEDGMENTS

We thank the organizers of Spin 2008. For initial work we thank L. Gamberg and for many helpful comments and questions, C. Weiss, J. Ralston, P. Stoler, H. Avakian, S. Ahmad. This work is supported by the U.S. Department of Energy grants no. DE-FG02-01ER4120 (S.L), and no. DE-FG02-92ER40702 (G.R.G.).

REFERENCES

1. G. R. Goldstein and J. F. Owens, *Phys. Rev. D* **7** (1973) 865.
2. J.C. Collins, L. Frankfurt and M. Strikman, *Phys. Rev. D* **56**, 2982 (1997).
3. M. Diehl, *Eur. Phys. J. C* **19**, 485 (2001).
4. M. Burkardt, *Int. J. Mod. Phys. A* **18**, 173 (2003); *Phys. Rev. D* **62**, 071503 (2000); [Erratum-*ibid.* **66**, 119903 (2002)].
5. S. Ahmad, H. Honkanen, S. Liuti and S. K. Taneja, *Phys. Rev. D* **75**, 094003 (2007); arXiv:0708.0268.
6. M. Anselmino, M. Boglione, U. D'Alesio, A. Kotzinian, F. Murgia, A. Prokudin and C. Turk, *Phys. Rev. D* **75**, 054032 (2007).
7. R. De Masi et al., *Phys. Rev. C* **77**, 042201 (2008).
8. S. Ahmad, G.R. Goldstein and S. Liuti, arXiv:0805.3568v2; to be published, *Phys. Rev. D.*
9. H. Lehmann, *Nuovo Cim.* **10**, 579 (1958).
10. I.V. Anikin and O.V. Teryaev, *Phys. Rev. D* **76**, 056007 (2007).
11. M. Diehl and D.Yu Ivanov, *Eur. Phys. J. C* **52**, 919 (2007).
12. J.C. Collins, T.C. Rogers and A.M. Stasto, *Phys. Rev. D* **77**, 085009 (2008).

Chiral Odd Generalized Parton Distributions in Position Space

A. Mukherjee*, D. Chakrabarti† and R. Manohar*

*Department of Physics, Indian Institute of Technology, Powai, Mumbai 400076, India
†Department of Physics, Swansea University, Singleton Park, Swansea, SA2 8PP, U.K.

Abstract. We report on a calculation of the chiral odd generalized parton distributions (GPDs) for non-zero skewness ζ in transverse and longitudinal position spaces by taking Fourier transforms with respect to the transverse and longitudinal momentum transfers respectively using overlaps of light-front wave functions (LFWFs).

Keywords: DVCS, generalized parton distributions, chiral odd
PACS: 12.38.Bx, 12.38.Aw, 13.88.+e

INTRODUCTION

At leading twist, there are three forward parton distributions (PDFs), namely, the unpolarized, helicity and transversity distributions. Similarly, three leading twist generalized quark distributions can be defined, which in the forward limit reduce to these three forward PDFs. The third one is chiral odd and is called the generalized transversity distribution F_T. This is defined as the off-forward matrix element of the bilocal tensor charge operator. It is parametrized in terms of four generalized parton distributions (GPDs), namely H_T, $\tilde{H}_T$, E_T and $\tilde{E}_T$ in the most general way [1, 2, 3]. The chiral-odd GPDs affect the transversely polarized quark distribution both in unpolarized and in transversely polarized nucleon in various ways. A relation for the transverse total angular momentum of the quarks has been proposed in [3], in analogy with Ji's relation, which involves a combination of second moments of H_T, E_T and $\tilde{H}_T$ in the forward limit.

In a previous work we have investigated the chiral odd GPDs for a simple spin-1/2 composite particle for skewness $\zeta = 0$ in impact parameter space [4]. In this work, using an overlap formula in the terms of the light-front wave functions (LFWFs) both in the DGLAP ($n \to n$) and ERBL ($n+1 \to n-1$) regions [5], we investigate them in a simple model, namely for the quantum fluctuations of a lepton in QED at one-loop order [6]. We generalize this analysis by assigning a mass M to the external electrons and a different mass m to the internal electron lines and a mass λ to the internal photon lines with $M < m + \lambda$ for stability. In effect, we shall represent a spin-$\frac{1}{2}$ system as a composite of a spin-$\frac{1}{2}$ fermion and a spin-1 vector boson [7, 8, 9, 10, 11]. This field-theory-inspired model has the correct correlation between the Fock components of the state as governed by the light-front eigenvalue equation, something that is extremely difficult to achieve in phenomenological models. GPDs in this model satisfy general properties like polynomiality and positivity. So it is interesting to investigate the general properties of GPDs in this model. By taking Fourier transforms with respect to $\Delta_\perp$, we

CP1149, *Spin Physics, 18th International Spin Physics Symposium*
edited by D. G. Crabb, D. B. Day, S. Liuti, X. Zheng, M. Poelker, and Y. Prok

express the GPDs in transverse position space and by taking Fourier transforms with respect to ζ we express them in longitudinal position space.

Chiral Odd GPDs using Overlap Formula

We use the parametrization of [3] for the chiral odd GPDs. Following [6, 7], we take a simple composite spin $1/2$ state, namely an electron in QED at one loop to investigate the GPDs. The light-front Fock state wavefunctions corresponding to the quantum fluctuations of a physical electron can be systematically evaluated in QED perturbation theory. The state is expanded in Fock space and there are contributions from $|e^-\gamma\rangle$ and $|e^-e^-e^+\rangle$, in addition to renormalizing the one-electron state. Both the two- and three-particle Fock state components are given in [12]. In the domain $\zeta < x < 1$, there are diagonal $2 \to 2$ overlaps. Using the overlap formula in [5] we calculate the chiral odd GPDs in this kinematic region. They are given by

$$E_T(x,\zeta,t) = -\frac{e^2}{8\pi^3}\frac{2M\pi}{1-\zeta}(M-\frac{m}{x})x(1-x)I_3, \tag{1}$$

$$\begin{aligned}\tilde{E}_T(x,\zeta,t) &= \frac{e^2}{8\pi^3}\frac{M\pi}{1-\zeta}\Big[-(1-x)\Big\{(M-\frac{m}{x})x+(M-\frac{m}{x'})x'\Big\}I_1 \\ &\quad +(M-\frac{m}{x})x(1-x)I_2\Big],\end{aligned} \tag{2}$$

$$\begin{aligned}H_T(x,\zeta,t) &= \frac{e^2}{8\pi^3}\frac{\pi}{2}\Big[\frac{x+x'}{2(1-x)}\ln(\frac{\Lambda^4}{DD'})+\Big\{\frac{x+x'}{2(1-x)}B(x,\zeta)+\frac{\zeta M}{1-\zeta}(M-\frac{m}{x})x(1-x)\Big\}I_2 \\ &\quad -\frac{\zeta M}{1-\zeta}\Big\{(M-\frac{m}{x})x(1-\zeta)+(M-\frac{m}{x'})x'\Big\}(1-x')I_1\Big],\end{aligned} \tag{3}$$

where

$$I_1 = \int_0^1 dy\frac{1-y}{Q(y)}, \tag{4}$$

$$I_2 = \int_0^1 dy\frac{1}{Q(y)}, \tag{5}$$

$$I_3 = \int_0^1 dy\frac{y}{Q(y)} \tag{6}$$

with $Q(y) = y(1-y)(1-x')^2\Delta_\perp^2 - y(M^2x(1-x)-m^2(1-x)-\lambda^2x)-(1-y)(M^2x'(1-x')-m^2(1-x')-\lambda^2x')$, $D = M^2x(1-x)-m^2(1-x)-\lambda^2x$ and $D' = M^2x'(1-x')-m^2(1-x')-\lambda^2x'$. In order to regulate the ultraviolet divergences we use a cutoff Λ on the transverse momentum $k^\perp$. $\tilde{H}_T(x,\zeta,t)$ is zero in this model. In Fig. 1 we have shown the chiral odd GPDs as functions of ζ for fixed t and different values of x.

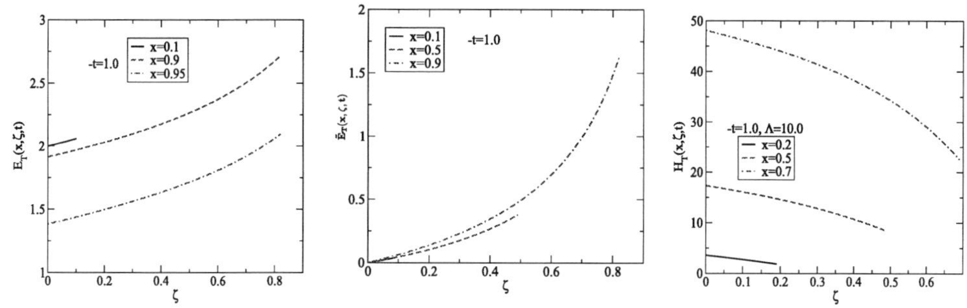

FIGURE 1. (Color online) Chiral-odd GPDs vs. ζ for fixed t in MeV^2 and different values of x.

Introducing the Fourier conjugate $b_\perp$ (impact parameter) of the transverse momentum transfer $\Delta_\perp$, the GPDs can be expressed in impact parameter space. Like the chiral even counterparts, chiral odd GPDs as well have interesting interpretations in impact parameter space [3]. In most experiments ζ is nonzero, and it is of interest to investigate the chiral odd GPDs in $b_\perp$ space with nonzero ζ. The probability interpretation is no longer possible as now the transverse positions of the initial and final protons are different as there is a finite momentum transfer in the longitudinal direction. The GPDs in impact parameter space probe partons at transverse position $|b_\perp|$ with the initial and final proton shifted by an amount of order $\zeta|b_\perp|$. Note that this is independent of x and even when GPDs are integrated over x in an amplitude, this information is still there [13]. Thus the chiral odd GPDs in impact parameter space gives the spin orbit correlations of partons in protons with their centers shifted with respect to each other. Taking the Fourier transform with respect to the transverse momentum transfer $\Delta_\perp$ we get the GPDs in the transverse impact parameter space:

$$\begin{aligned}\mathscr{E}_T(x,\zeta,b_\perp) &= \frac{1}{(2\pi)^2}\int d^2\Delta_\perp e^{-i\Delta_\perp\cdot b_\perp}E_T(x,\zeta,t)\\ &= \frac{1}{2\pi}\int \Delta d\Delta J_0(\Delta b)E_T(x,\zeta,t),\end{aligned} \tag{7}$$

where $\Delta=|\Delta_\perp|$ and $b=|b_\perp|$. The other impact parameter dependent GPDs $\tilde{\mathscr{E}}_T(x,\zeta,b_\perp)$ and $\mathscr{H}_T(x,\zeta,b_\perp)$ can also be defined in the same way. It is to be noted that $\mathscr{H}_T(x,\zeta,b_\perp)$ for a free Dirac particle is expected to be a delta function; the smearing in $|b_\perp|$ space is due to the spin correlation in the two-particle LFWFs. Figure 2 shows the plots of the above three functions for fixed ζ and different values of x. For given ζ, the peak of $\mathscr{H}_T(x,\zeta,b_\perp)$ as well as $\tilde{\mathscr{E}}_T(x,\zeta,b_\perp)$ increases with increase of x, however for $\mathscr{E}_T(x,\zeta,b_\perp)$ it decreases.

In [14], a phase space distribution of quarks and gluons in the proton is given in terms of the quantum mechanical Wigner distribution $W(\vec{r},\vec{p})$, in the rest frame of the proton, which are functions of three position and three momentum coordinates. Wigner distributions are not accessible in experiment. However, if one integrates two momentum components one gets a reduced Wigner distribution $W_\Gamma(\vec{r},x)$ which is related to the

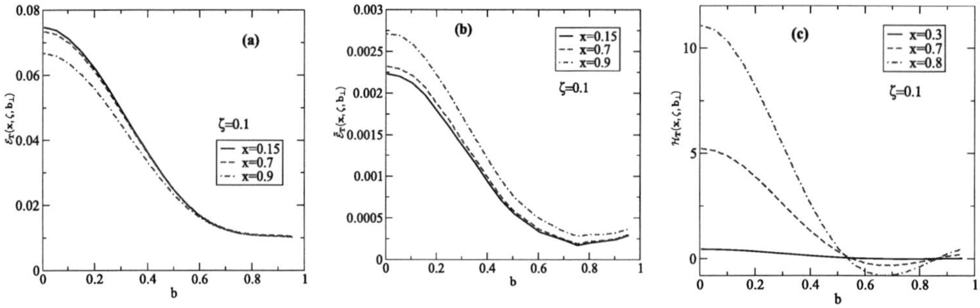

FIGURE 2. (Color online) Fourier spectrum of the chiral-odd GPDs vs. $|b_\perp|$ for fixed ζ and different values of x

GPDs by a Fourier transform. For a given x, this gives a 3D position space picture of the partons inside the proton. If the probing wavelength is comparable to or smaller than the Compton wavelength $\frac{1}{M}$, where M is the mass of the proton, electron-positron pairs will be created. As a result, the static size of the system cannot be probed to a precision better than $\frac{1}{M}$ in relativistic quantum theory. However, in light-front theory, transverse boosts are Galilean boosts which do not involve dynamics. So one can still express the GPDs in transverse position or impact parameter space and this picture is not spoiled by relativistic corrections. However, rotation involves dynamics here and rotational symmetry is lost. In [7], a longitudinal boost invariant impact parameter σ has been introduced which is conjugate to the longitudinal momentum transfer ζ. It was shown that the DVCS amplitude expressed in terms of the variables $\sigma, b_\perp$ show diffraction pattern analogous to diffractive scattering of a wave in optics where the distribution in σ measures the physical size of the scattering center in a 1-D system. In analogy with optics, it was concluded that the finite size of the ζ integration of the Fourier transforms acts as a slit of finite width and produces the diffraction pattern. We define a boost invariant impact parameter conjugate to the longitudinal momentum transfer as $\sigma = \frac{1}{2}b^- P^+$ [7]. The chiral odd GPD E_T in longitudinal position space is given by

$$\begin{aligned} \mathcal{E}_T(x,\sigma,t) &= \frac{1}{2\pi}\int_0^{\zeta_f} d\zeta e^{i\frac{1}{2}P^+\zeta b^-} E_T(x,\zeta,t) \\ &= \frac{1}{2\pi}\int_0^{\zeta_f} d\zeta e^{i\sigma\zeta} E_T(x,\zeta,t). \end{aligned} \tag{8}$$

Similarly one can obtain $\mathcal{H}_T(x,\sigma,t)$ and $\tilde{\mathcal{E}}_T(x,\sigma,t)$ as well. Figure 3 shows the plots of the Fourier spectrum chiral odd GPDs in longitudinal position space as a function of σ for fixed $x = 0.5$ and different values of t. Both $\mathcal{E}_T(x,\sigma,t)$ and $\mathcal{H}_T(x,\sigma,t)$ show diffraction pattern as observed for the DVCS amplitude in [7]; the minima occur at the sames values of σ in both cases. However $\tilde{\mathcal{E}}_T(x,\sigma,t)$ does not show diffraction pattern. This is due to the distinctively different behavior of $\tilde{E}_T(x,\zeta,t)$ with ζ compared to that of $E_T(x,\zeta,t)$ and $H_T(x,\zeta,t)$. $\tilde{E}_T(x,\zeta,t)$ rises smoothly from zero and has no flat plateau in ζ and thus does not exhibit any diffraction pattern when Fourier transformed with respect to ζ. The position of first minima in Fig.3 is determined by ζ_f. For $-t = 5.0$ and 1.0, $\zeta_f \approx x = 0.5$ and thus the first minimum appears at the same position while for $-t = 0.1$, $\zeta_f = \zeta_{max} \approx 0.45$ and the minimum appears slightly shifted. This is analogous to the single slit optical diffraction pattern. ζ_f here plays the role of the slit width. Since the positions of the minima (measured from the center of the diffraction pattern) are inversely proportional to the slit width, the minima move away from the center as the slit width (i.e., ζ_f) decreases. The optical analogy of the diffraction pattern in σ space has been discussed in [7] in the context of DVCS amplitudes.

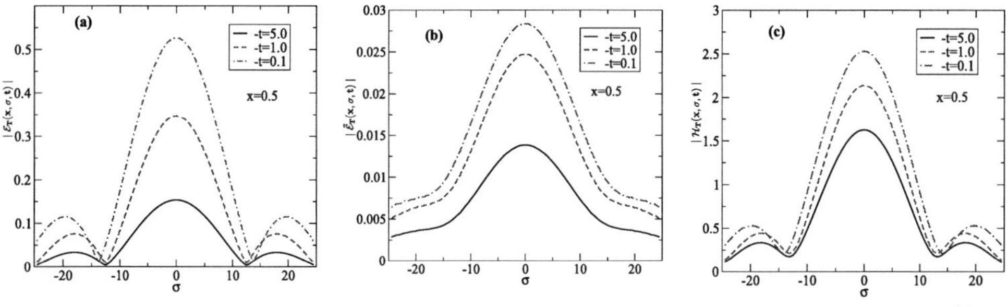

FIGURE 3. (Color online) Fourier spectrum of the chiral-odd GPDs vs. σ for fixed x and different values of $-t$ in MeV^2.

CONCLUSION

In this work, we studied the chiral-odd GPDs in transverse and longitudinal position space. Working in light-front gauge, we used overlap formulas for the chiral odd GPDs in terms of proton light-front wave functions in the DGLAP region. We used a self consistent relativistic two-body model, namely the quantum fluctuation of an electron at one loop in QED. We used its most general form [6], where we have a different mass for the external electron and different masses for the internal electron and photon. The impact parameter space representations are obtained by taking Fourier transforms of the GPDs with respect to the transverse momentum transfer. When ζ is non-zero, the initial and final proton are displaced in the impact parameter space relative to each other by an amount proportional to ζ. As this is the region probed by most experiments, it is of interest to investigate this. By taking a Fourier transform with respect to ζ we presented the GPDs in the boost invariant longitudinal position space variable σ. H_T and E_T show diffraction pattern in σ space.

ACKNOWLEDGMENTS

AM thanks the organizers of Spin 2008 for invitation and support and MHRD for support.

REFERENCES

1. M. Diehl, *Eur. Phys. J.* **C 19**, 485 (2001).
2. M. Diehl and P. Hagler, *Eur. Phys. J.* **C 44**, 87 (2005).
3. M. Burkardt, *Phys. Rev.* **D 72**, 094020 (2005).
4. H. Dahiya, A. Mukherjee, *Phys. Rev.* **D 77**, 045032 (2008).
5. D. Chakrabarti, R. Manohar, A. Mukherjee; arXiv:0811.0521 [hep-ph].
6. S. J. Brodsky and S. D. Drell, *Phys. Rev.* **D 22**, 2236 (1980).
7. S. J. Brodsky, D. Chakrabarti, A. Harindranath, A. Mukherjee and J. P. Vary, *Phys. Lett.* **B 641**, 440 (2006); *Phys. Rev.* **D 75**, 014003 (2007).
8. A. Harindranath, R. Kundu, W. M. Zhang, *Phys. Rev.* **D 59**, 094013 (1999); A. Harindranath, A. Mukherjee, R. Ratabole, *Phys. Lett.* **B 476**, 471 (2000); *Phys. Rev.* **D 63**, 045006 (2001).
9. D. Chakrabarti and A. Mukherjee, *Phys. Rev.* **D 71**, 014038 (2005).
10. D. Chakrabarti, A. Mukherjee, *Phys. Rev.* **D 72**, 034013 (2005).
11. A. Mukherjee and M. Vanderhaeghen, *Phys. Lett.* **B 542**, 245 (2002); *Phys. Rev.* **D 67**, 085020 (2003).
12. S. J. Brodsky, M. Diehl, D. S. Hwang, *Nucl. Phys.* **B 596**, 99 (2001); M. Diehl, T. Feldmann, R. Jacob, P. Kroll, *Nucl. Phys.* **B 596**, 33 (2001), Erratum-*ibid.* **605**, 647 (2001).
13. M. Diehl, *Eur. Phys. J.* **C 25**, 223 (2002), Erratum-*ibid.* **C 31**, 277 (2003).
14. X. Ji, *Phy. Rev. Lett.* **91**, 062001 (2003); A. Belitsky, X. Ji, F. Yuan, *Phys. Rev.* **D 69** 074014 (2006).

Properties and Applications of GTMDs

Stephan Meißner*, Andreas Metz† and Marc Schlegel**

*Institut für Theoretische Physik II, Ruhr-Universität Bochum, 44780 Bochum, Germany
†Department of Physics, Temple University, Philadelphia, PA 19122, USA
**Theory Center, Jefferson Lab, 12000 Jefferson Avenue, Newport News, VA 23606, USA

Abstract. We present first results from our analysis of the most general quark-quark correlator of the nucleon, which can be parameterized in terms of so-called generalized transverse momentum dependent parton distributions. These results include the first complete parameterization of the nucleon GPDs and TMDs as well as new results on possible nontrivial relations between them.

Keywords: Deep Inelastic Scattering, Spin and Polarization Effects, Parton Model
PACS: 12.38.Bx, 13.85.Ni, 13.85.Qk

INTRODUCTION AND DEFINITIONS

Parton distributions play an important role in the QCD description of hard scattering processes. In particular, generalized parton distributions (GPDs) and transverse momentum dependent parton distributions (TMDs) have extensively been studied during the last decade. We focus here on the most general form of quark-quark correlation functions, the so-called generalized transverse momentum dependent parton distributions (GTMDs). These GTMDs are connected to the so-called Wigner distributions of the hadron-parton system [1] and contain the GPDs and TMDs in certain limits. Recently a first complete analysis of these interesting objects has been performed, that was, however, restricted to the special case of spin-0 hadrons [2]. In this short note we will extend the analysis to the case of spin-1/2 hadrons and give a brief overview of the main applications for nucleon GTMDs.

We start by recalling some definitions, that will be needed later on. First of all, we use the common definition of the nucleon GPDs by the correlator

$$F_{\lambda\lambda'}^{[\Gamma]}(x,\xi,\vec{\Delta}_T) = \frac{1}{2}\int \frac{dz^-}{2\pi}\, e^{ik\cdot z}\, \langle p',\lambda' |\, \bar{\psi}(-\tfrac{1}{2}z)\,\Gamma\, \mathscr{W}_{\rm GPD}\, \psi(\tfrac{1}{2}z)\, |p,\lambda\rangle \big|_{z^+=\vec{z}_T=0}\,. \quad (1)$$

For $\Gamma = \gamma^+$ this correlator is parameterized in terms of two GPDs,

$$F_{\lambda\lambda'}^{[\gamma^+]}(x,\xi,\vec{\Delta}_T) = \frac{1}{2P^+}\,\bar{u}(p',\lambda')\left[\gamma^+ H(x,\xi,t) + \frac{i\sigma^{+\mu}\Delta_\mu}{2M}\, E(x,\xi,t)\right] u(p,\lambda)\,, \quad (2)$$

where $P = (p+p')/2$ denotes the average nucleon momentum and $\Delta = p' - p$ the nucleon momentum transfer. The GPDs depend on three kinematical variables: the longitudinal momentum fraction $x = k^+/P^+$ of the parton, the skewness parameter $\xi = -\Delta^+/(2P^+)$, and the momentum transfer $t = \Delta^2$. For the nucleon TMDs which depend on the intrinsic transverse momentum $\vec{k}_T$ of the parton we also use the common

CP1149, *Spin Physics, 18th International Spin Physics Symposium*
edited by D. G. Crabb, D. B. Day, S. Liuti, X. Zheng, M. Poelker, and Y. Prok

definition by the correlator

$$\Phi^{[\Gamma]}_{\lambda\lambda'}(x,\vec{k}_T) = \frac{1}{2}\int \frac{dz^-}{2\pi}\,\frac{d^2\vec{z}_T}{(2\pi)^2}\,e^{ik\cdot z}\,\langle P,\lambda'|\,\bar{\psi}(-\tfrac{1}{2}z)\,\Gamma\,\mathscr{W}_{\text{TMD}}\,\psi(\tfrac{1}{2}z)\,|P,\lambda\rangle\big|_{z^+=0}\,. \tag{3}$$

Taking again $\Gamma = \gamma^+$ and introducing the nucleon spin vector S, which may be expressed through certain combinations of the helicities λ and λ', one obtains two TMDs,

$$\Phi^{[\gamma^+]}(x,\vec{k}_T;S) = f_1(x,\vec{k}_T^2) - \frac{\varepsilon_T^{ij}k_T^i S_T^j}{M}\,f_{1T}^{\perp}(x,\vec{k}_T^2)\,. \tag{4}$$

GTMDS OF THE NUCLEON AND THEIR APPLICATIONS

By examining the correlators of GPDs (1) and TMDs (3) one immediately finds that they both look very similar. Therefore, it is evident to ask, whether it is possible to study a more general correlator, that contains both in some limits. Of course this correlator is given by the canonical extension of the correlators (1) and (3) to the most general quark-quark correlator of the nucleon

$$W^{[\Gamma]}_{\lambda\lambda'}(x,\xi,\vec{k}_T,\vec{\Delta}_T) = \frac{1}{2}\int \frac{dz^-}{2\pi}\,\frac{d^2\vec{z}_T}{(2\pi)^2}\,e^{ik\cdot z}\,\langle p',\lambda'|\,\bar{\psi}(-\tfrac{1}{2}z)\,\Gamma\,\mathscr{W}_{\text{GTMD}}\,\psi(\tfrac{1}{2}z)\,|p,\lambda\rangle\big|_{z^+=0}\,, \tag{5}$$

which contains the other two correlators in the limits

$$F^{[\Gamma]}_{\lambda\lambda'}(x,\xi,\vec{\Delta}_T) = \int d^2\vec{k}_T\,W^{[\Gamma]}_{\lambda\lambda'}(x,\xi,\vec{k}_T,\vec{\Delta}_T)\,, \tag{6}$$

$$\Phi^{[\Gamma]}_{\lambda\lambda'}(x,\vec{k}_T) = W^{[\Gamma]}_{\lambda\lambda'}(x,0,\vec{k}_T,0)\,. \tag{7}$$

The parton distributions, which are defined by this correlator, are the so-called generalized transverse momentum dependent parton distribution (GTMDs). These GTMDs may be of importance for the phenomenology of hard exclusive reactions. It is already known, for instance, that gluon GTMDs enter the description of diffractive vector meson [3] and Higgs production [4] even at leading twist.

The parameterization of the correlator (5) can be performed by applying parity, hermiticity, and time-reversal. Analogously to Ref. [2], where the case of a spin-0 hadron has been discussed, this yields some constraints. It follows for example that the parity constraint reduces the number of independent terms in the correlator (5) significantly. Taking again $\Gamma = \gamma^+$ one obtains four complex valued GTMDs,

$$W^{[\gamma^+]}_{\lambda\lambda'}(x,\xi,\vec{k}_T,\vec{\Delta}_T) = \frac{1}{2M}\bar{u}(p',\lambda')\left[F_{1,1} + \frac{i\sigma^{i+}k_T^i}{P^+}F_{1,2} + \frac{i\sigma^{i+}\Delta_T^i}{P^+}F_{1,3} + \frac{i\sigma^{ij}k_T^i\Delta_T^j}{M^2}F_{1,4}\right]u(p,\lambda)\,, \tag{8}$$

with $F_{1,n} = F_{1,n}(x,\xi,\vec{k}_T^2,\vec{k}_T\cdot\vec{\Delta}_T,\vec{\Delta}_T^2)$. In addition, the hermiticity constraint implies

$$F^*_{1,n}(x,\xi,\vec{k}_T^2,\vec{k}_T\cdot\vec{\Delta}_T,\vec{\Delta}_T^2) = \pm F_{1,n}(x,-\xi,\vec{k}_T^2,-\vec{k}_T\cdot\vec{\Delta}_T,\vec{\Delta}_T^2)\,, \tag{9}$$

where the plus sign holds for $n = 1,3,4$ and the minus sign for $n = 2$. Finally, the time-reversal constraint allows one to split each GTMD into two real valued functions,

$$F_{1,n}(x,\xi,\vec{k}_T^2,\vec{k}_T\cdot\vec{\Delta}_T,\vec{\Delta}_T^2) = F_{1,n}^e(x,\xi,\vec{k}_T^2,\vec{k}_T\cdot\vec{\Delta}_T,\vec{\Delta}_T^2) + iF_{1,n}^o(x,\xi,\vec{k}_T^2,\vec{k}_T\cdot\vec{\Delta}_T,\vec{\Delta}_T^2)\,, \tag{10}$$

the so-called T-even and T-odd part of the GTMD, which have a different behavior under time-reversal. While the T-even part $F_{1,n}^e$ remains the same under time-reversal the T-odd part $F_{1,n}^o$ changes its sign.

As the correlator of GTMDs (5) contains the correlators of GPDs (1) and TMDs (3) in the limits (6) and (7), GTMDs can be considered as the mother distributions of GPDs and TMDs. Therefore, one should be able to obtain the parameterization of GPDs in Eq. (2) and of TMDs in Eq. (4) directly from the parameterization of the GTMDs in Eq. (8). This is indeed possible and one finds for the GPDs

$$H(x,\xi,t) = \int d^2\vec{k}_T\left[F_{1,1}^e + 2\xi^2\left(\frac{\vec{k}_T\cdot\vec{\Delta}_T}{\vec{\Delta}_T^2}F_{1,2}^e + F_{1,3}^e\right)\right], \tag{11}$$

$$E(x,\xi,t) = \int d^2\vec{k}_T\left[-F_{1,1}^e + 2(1-\xi^2)\left(\frac{\vec{k}_T\cdot\vec{\Delta}_T}{\vec{\Delta}_T^2}F_{1,2}^e + F_{1,3}^e\right)\right], \tag{12}$$

with $F_{1,n}^e = F_{1,n}^e(x,\xi,\vec{k}_T^2,\vec{k}_T\cdot\vec{\Delta}_T,\vec{\Delta}_T^2)$ and for the TMDs

$$f_1(x,\vec{k}_T^2) = F_{1,1}^e(x,0,\vec{k}_T^2,0,0)\,, \tag{13}$$

$$f_{1T}^{\perp}(x,\vec{k}_T^2) = -F_{1,2}^o(x,0,\vec{k}_T^2,0,0)\,. \tag{14}$$

Consequently, a complete parameterization of all GTMDs immediately yields a complete parameterization of all GPDs and TMDs. We have performed such a parameterization of the correlator (5) to all twists and found a total of 64 complex valued GTMDs for the nucleon, which can be split into 64 real valued T-even and 64 real valued T-odd parts. In addition we also studied the limits of GPDs and TMDs and found that in both cases 32 real valued functions survive (see Tab. 1 for an overview). To our knowledge, this result is new for GPDs, as so far no complete parameterization of the correlator of GPDs (1) has been performed to all twists. For the TMDs our result is in perfect agreement with earlier works [5].

TABLE 1. Number of independent GTMDs, GPDs, and TMDs to all twists.

		GTMDs		**GPDs**		**TMDs**	
twist:	Γ	T-even	T-odd	T-even	T-odd	T-even	T-odd
2:	γ^+ / $\gamma^+\gamma_5$ / $i\sigma^{i+}$	4 / 4 / 8	4 / 4 / 8	2 / 2 / 4	0 / 0 / 0	1 / 2 / 3	1 / 0 / 1
3:	1 / γ_5 / γ^i	4 / 4 / 8	4 / 4 / 8	2 / 2 / 4	0 / 0 / 0	1 / 0 / 1	1 / 2 / 3
	$\gamma^i\gamma_5$ / $i\sigma^{ij}$ / $i\sigma^{+-}$	8 / 4 / 4	8 / 4 / 4	4 / 2 / 2	0 / 0 / 0	3 / 2 / 1	1 / 0 / 1
4:	γ^- / $\gamma^-\gamma_5$ / $i\sigma^{i-}$	4 / 4 / 8	4 / 4 / 8	2 / 2 / 4	0 / 0 / 0	1 / 2 / 3	1 / 0 / 1

Another interesting application for the GTMDs is the test of possible nontrivial relations between GPDs and TMDs. Such relations can be established in model calculations (see Ref. [6] for a review), but so far no model-independent proof has been found. The most prominent example of a possible nontrivial relation between GPDs and TMDs is probably the relation between the GPD E and the Sivers function $f_{1T}^{\perp}$, which reads

$$\int d^2\vec{b}_T\, \vec{\mathscr{I}}(x,\vec{b}_T)\,\frac{\varepsilon_T^{ij} b_T^i S_T^j}{M}\,\mathscr{E}'(x,\vec{b}_T^2) = -\int d^2\vec{k}_T\, \vec{k}_T\,\frac{\varepsilon_T^{ij} k_T^i S_T^j}{M}\, f_{1T}^{\perp}(x,\vec{k}_T^2) \qquad (15)$$

in impact parameter space [7]. Here $\mathscr{E}$ denotes the Fourier-transformed GPD E and $\vec{\mathscr{I}}$ the so-called lensing function. However, from Eqs. (12) and (14) it is obvious that the nontrivial relation (15) cannot hold in general as the involved GPD and TMD have different mother distributions. The same applies to all other nontrivial relations discussed in Ref. [6], in particular, also to the one between the GPD $\tilde{H}_T$ and the pretzelocity distribution $h_{1T}^{\perp}$.

SUMMARY

In summary we have performed the first complete parameterization of the most general quark-quark correlator of the nucleon in terms of GTMDs, which can be considered as the mother distributions of GPDs and TMDs. As a by-product we also obtained the first complete parameterization of the nucleon GPDs and TMDs to all twists. This allowed us to perform a model-independent test of possible nontrivial relations between GPDs and TMDs, which can be established in model calculations. Our results indicate, that no such nontrivial relations exist as the involved GPD and TMD have different mother distributions.

ACKNOWLEDGMENTS

This work has partially been supported by the Verbundforschung "Hadronen und Kerne" of the BMBF and by the Deutsche Forschungsgemeinschaft (DFG).

REFERENCES

1. A. V. Belitsky, X.-D. Ji, and F. Yuan, *Phys. Rev.* D **69**, 074014 (2004).
2. S. Meissner, A. Metz, M. Schlegel, and K. Goeke, *JHEP* **08**, 038 (2008).
3. A. D. Martin, M. G. Ryskin, and T. Teubner, *Phys. Rev.* D **62**, 014022 (2000).
4. V. A. Khoze, A. D. Martin, and M. G. Ryskin, *Eur. Phys. J.* C **14**, 525 (2000).
5. K. Goeke, A. Metz, and M. Schlegel, *Phys. Lett.* B **618**, 90 (2005).
6. S. Meissner, A. Metz, and K. Goeke, *Phys. Rev.* D **76**, 034002 (2007).
7. M. Burkardt and D. S. Hwang, *Phys. Rev.* D **69**, 074032 (2004).

Two-Photon Exchange in (Semi-)Inclusive DIS

M. Schlegel* and A. Metz†

*Theory Center, Jefferson Lab, Newport News, VA 23606, USA
†Department of Physics, Temple University, Philadelphia, PA 19122, USA

Abstract. In this note we consider effects of a Two-Photon Exchange (TPE) in inclusive DIS and semi-inclusive DIS (SIDIS). In particular, transverse single spin asymmetries are generated in inclusive DIS if more than one photon is exchanged between the lepton and the hadron. We briefly summarize the TPE in DIS in the parton model and extend our approach to SIDIS, where a new leading twist $\sin(2\phi)$ contribution to the longitudinal beam spin asymmetry shows up. Possible TPE effects for the Sivers and the Collins asymmetries in SIDIS are power-suppressed.

Keywords: Two Photon Exchange, Inclusive DIS, SIDIS
PACS: 13.60.Hb

TWO-PHOTON EXCHANGE IN INCLUSIVE DIS

It was already argued in the 60s that, due to time-reversal invariance, single spin asymmetries (SSAs) are forbidden in inclusive deep-inelastic lepton scattering off nucleons if one considers only a One-Photon Exchange (OPE) between the lepton and the nucleon (Christ-Lee theorem [1]). Early experiments confirmed this statement and found transverse single target spin asymmetries in DIS being consistent with zero at the percent level [2, 3]. Measurements of this observable have been repeated very recently at HERMES with higher precision (of the order 10^{-3}), and the effect is still compatible with zero [4]. An ongoing Hall A experiment at Jefferson Lab even plans to measure the target SSA in DIS with an accuracy of the order 10^{-4} [5].

If one considers DIS beyond the OPE approximation, transverse SSAs may well exist [1, 6, 7]. TPE can generate a nontrivial phase in the amplitude of the process — a necessary condition for a SSA — and therefore provide a nonzero single spin correlation of the form $\varepsilon^{\mu\nu\rho\sigma}S_\mu P_\nu l_\rho l'_\sigma$ (cf. Fig. 1). The transverse SSAs generated by TPE were calculated in the naive parton model in Ref. [6] by assuming that both photons couple to the same quark (Fig. 1). The observables were considered in the Bjorken limit where the momentum transfer to the nucleon, $q^2 = (l-l')^2 \equiv -Q^2$, is large, and power corrections in $1/Q$ were neglected. In the following we will briefly discuss this calculation.

Both the transverse target and lepton beam SSA receive their leading contributions in α — the electromagnetic fine structure constant — from an interference of OPE DIS amplitudes and TPE amplitudes (Fig. 1). In the parton model for DIS one considers the hard electromagnetic interaction of a quasi-free parton with the lepton and treats the other partons as spectators (impulse approximation). We assume that, for the particular observables we are interested in here, the dominant contribution (in the sense of a twist expansion) of TPE is picked out if one works with a single active parton. This approximation is represented in Fig. 1. The two-photon box diagram is the only QED

CP1149, *Spin Physics, 18th International Spin Physics Symposium*
edited by D. G. Crabb, D. B. Day, S. Liuti, X. Zheng, M. Poelker, and Y. Prok

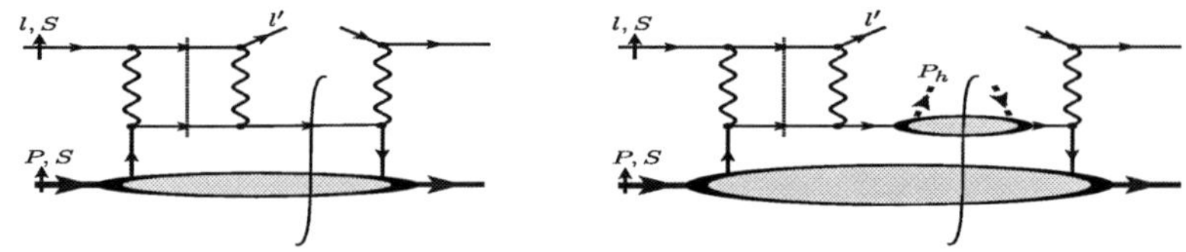

FIGURE 1. Two-Photon Exchange in the parton model. Merely the coupling of the two photons to the same quark is considered. Also the complex conjugate diagrams (not shown here) have to be taken into account. **Left Panel**: TPE in inclusive DIS. **Right Panel**: TPE in semi-inclusive DIS.

diagram in the parton model which carries an imaginary part. Other radiative corrections of the same order in α such as bremsstrahlung effects are purely real and hence cannot give rise to transverse SSAs.

The result for the transverse lepton beam SSA, calculated from the left graph in Fig. 1 (and its complex conjugate counterpart), reads [6]

$$A_{TU}(x_B, y, \phi_s) = \alpha \frac{m_l}{2Q} |\vec{S}_T| \sin(\phi_s) \frac{y^2\sqrt{1-y}}{1-y+\frac{1}{2}y^2} \frac{\sum_q e_q^3 f_1(x_B)}{\sum_q e_q^2 f_1(x_B)}, \tag{1}$$

where $x_B = Q^2/(2P\cdot q)$ and $y = (P\cdot q)/(P\cdot l)$ denote the common DIS variables, e_q the quark charge, $\vec{S}_T$ the transverse polarization vector of the lepton, and ϕ_s the angle of this vector with respect to the lepton plane. The ordinary unpolarized parton distribution of a quark flavor q is represented by f_1^q. We point out that divergent terms, which appear at intermediate steps of the calculation and can be regulated by a photon mass, cancel in the final result (1). Since the asymmetry (1) is not only proportional to $\alpha \simeq 1/137$ but also to the lepton mass m_l one can expect rather small effects. In fact, by assuming u-quark dominance in a proton target and an electron beam one readily estimates this asymmetry to be of the order 10^{-6}. Asymmetries roughly two-hundred times larger can be generated by a muon beam. One might also speculate about enhanced results from effects beyond the naive parton model. In Refs. [8, 9, 10, 11] (double) logarithms of the type $\log(Q^2/m_l^2)$ were advocated in connection with the transverse beam SSA in elastic lepton-nucleon scattering. Such logarithmic terms might also increase the beam SSA in DIS considered here. However, further work is required in order to decide whether and how precisely such effects show up for the DIS case.

We now turn to the parton model result for the transverse target SSA in DIS. The contribution of the left diagram in Fig. 1 (and its complex conjugate counterpart) reads [6]

$$A_{UT}(x_B, y, \phi_s) = \alpha \frac{x_B M}{2Q} \frac{y(1-y)\sqrt{1-y}}{1-y+\frac{1}{2}y^2} |\vec{S}_T| \sin(\phi_s) \left(\ln \frac{Q^2}{\lambda^2} + \text{finite} \right) \frac{\sum_q e_q^3 g_T^q(x_B)}{\sum_q e_q^2 f_1^q(x_B)}, \tag{2}$$

where M denotes the nucleon mass, and g_T a higher twist parton distribution. Note that a small photon mass λ has been introduced in order to get a finite result. The divergence for $\lambda \to 0$ signals a violation of electromagnetic gauge invariance in the parton model (for the particular observable A_{UT}). It was suggested that gauge invariance might be restored by considering contributions beyond the naive parton model such as multiparton correlations [6]. Since the SSA (2) is a twist-3 observable one indeed expects contributions from quark-gluon correlations. Whether the inclusion of such terms provides a finite result remains to be seen. Another candidate for the restoration of electromagnetic gauge invariance are multiquark correlations, i.e. configurations where the two photons couple to two different quarks.

Focusing on the finite contributions in (2) one expects the asymmetry to be at most of the order 10^{-3} (for $x \simeq 0.1$). (A very rough estimate of A_{UT}, where one just considers the suppression due to α, gives an effect of the order 10^{-2}.) However, because of the divergence, a reliable numerical result can hardly be extracted from Eq. (2). Additional contributions beyond the naive parton model not only are supposed to cancel the divergence but also to yield finite terms which have to be taken into account as well. It is also worthwhile to mention that a quark-mass effect to A_{UT} — a particular contribution beyond the parton model — was studied in detail in Ref. [7]. Such an effect is proportional to the quark transversity distribution and leads, according to [7], to an asymmetry of the order 10^{-4}. In addition, we note that a diquark spectator model calculation for the neutron also predicts $A_{UT} \simeq 10^{-4}$ [12].

TWO-PHOTON EXCHANGE IN SEMI-INCLUSIVE DIS

It is straightforward to extend the naive parton model calculation of Ref. [6] to SIDIS in the kinematical regime where the transverse momentum $P_{h\perp}$ of the produced hadron is much smaller than the momentum transferred to the nucleon, $P_{h\perp} \ll Q$. For this kinematics, observables can be described in the parton model in terms of transverse momentum dependent (TMD) parton distributions and fragmentation functions. Like in inclusive DIS, TPE in general also generates SSAs in SIDIS. In the following we discuss results for the SSAs calculated in the naive parton model. The contribution of the right graph in Fig. 1 reads for a longitudinally polarized lepton beam and a produced pion,

$$A_{LU}^{\sin(2\phi)} = \alpha \frac{y\left(1+\frac{2-y}{1-y}\ln y\right)}{1-y+\frac{1}{2}y^2} \sin(2\phi) \frac{\sum_q e_q^3 \mathscr{C}\left[\frac{2(\vec{h}\cdot\vec{k}_T)(\vec{h}\cdot\vec{p}_T)-\vec{k}_T\cdot\vec{p}_T}{2Mm_\pi} h_1^{\perp q} H_1^{\perp q}\right]}{\sum_q e_q^2 \mathscr{C}\left[f_1^q D_1^q\right]}. \tag{3}$$

The angle ϕ represents the angle between the lepton and the production plane (in the conventions of Ref. [13]), while D_1 denotes the unpolarized TMD fragmentation function, $h_1^\perp$ the Boer-Mulders function [14], and $H_1^\perp$ the Collins function [15]. We also use the unit-vector $\vec{h} \equiv -\vec{P}_{h\perp}/|\vec{P}_{h\perp}|$. The symbol $\mathscr{C}[\ldots]$ denotes a convolution in the transverse momentum space (w being some kinematical prefactor),

$$\mathscr{C}[wfD] = \int d^2k_T d^2p_T\, \delta^{(2)}(\vec{k}_T - \vec{p}_T - \vec{P}_{h\perp}/z_h)\, w(\vec{k}_T,\vec{p}_T) f^q(x_B,\vec{k}_T^2) D^q(z_h,\vec{p}_T^2). \tag{4}$$

Both TMD functions in Eq. (3) are so-called (naive) time-reversal odd entities for which initial/final state interactions play an important role. We assume that the interplay between initial/final state interactions and TPE does not spoil the QCD-factorization in terms of TMD correlation functions.

Like Eq. (1), the SSA in (3) is free of divergences. Furthermore, it is a leading twist $\sin(2\phi)$ modulation of the beam spin asymmetry that is absent in a model-independent decomposition of the SIDIS cross section into structure functions assuming OPE [16, 17, 13]. This means that the asymmetry (3) is generated by TPE only, and possible $\sin(2\phi)$ modulations from QCD radiative corrections in the OPE approximation are ruled out. A similar effect has also been discussed in [18] for TPE in electro-excitation of a Δ resonance. Since Eq. (3) is proportional to α and since one expects the Boer-Mulders effect in unpolarized SIDIS from a $\cos(2\phi)$ modulation to be of the order of a few percent (see, e.g., the model calculation [19]) one can roughly estimate the asymmetry in (3) to be of the order $10^{-4} - 10^{-3}$. At this point one might again speculate about possible logarithmic enhancements as discussed above for the transverse beam SSA in inclusive DIS.

We also calculated the target asymmetries A_{UL} and A_{UT}, where we again encounter divergences like in the case of the transverse target SSA in DIS. The comments on this feature which we made above apply here, too. Interestingly, in the case of the target SSAs, TPE provides only twist-3 contributions to azimuthal modulations like $\sin\phi$ for A_{UL}, as well as $\sin\phi_s$ and $\sin(2\phi - \phi_s)$ for A_{UT}. Such twist-3 modulations are already present in the OPE approximation [13]. In particular, we find that for A_{UT} TPE corrections to the modulations $\sin(\phi - \phi_s)$ and $\sin(\phi + \phi_s)$ which are typically analyzed in terms of the (leading twist) Sivers and Collins effect, respectively, appear at most at the level of twist-4.

Notice: Authored by Jefferson Science Associates, LLC under U.S. DOE Contract No. DE-AC05-06OR23177. The U.S. Government retains a non-exclusive, paid-up, irrevocable, world-wide license to publish or reproduce this manuscript for U.S. Government purposes.

REFERENCES

1. N. Christ, and T. D. Lee, *Phys. Rev.* **143**, 1310–1321 (1966).
2. J. R. Chen, et al., *Phys. Rev. Lett.* **21**, 1279–1282 (1968).
3. S. Rock, et al., *Phys. Rev. Lett.* **24**, 748–752 (1970).
4. L. DeNardo, *these proceedings* (2008).
5. X. Jiang, et al., *Jefferson Lab Hall A Experiment E-07-013* (2007).
6. A. Metz, M. Schlegel, and K. Goeke, *Phys. Lett.* **B643**, 319–324 (2006).
7. A. Afanasev, M. Strikman, and C. Weiss, *Phys. Rev.* **D77**, 014028 (2008).
8. A. V. Afanasev, and N. P. Merenkov, *Phys. Rev.* **D70**, 073002 (2004).
9. A. V. Afanasev, and N. P. Merenkov, *Phys. Lett.* **B599**, 48 (2004).
10. D. Borisyuk, and A. Kobushkin, *Phys. Rev.* **C73**, 045210 (2006).
11. M. Gorchtein, *Phys. Rev.* **C73**, 055201 (2006).
12. A. Afanasev, M. Schlegel, and C. Weiss, *in preparation* (2009).
13. A. Bacchetta, M. Diehl, K. Goeke, A. Metz, P. Mulders, and M. Schlegel, *JHEP* **02**, 093 (2007).
14. D. Boer, and P. J. Mulders, *Phys. Rev.* **D57**, 5780–5786 (1998).
15. J. C. Collins, *Nucl. Phys.* **B396**, 161–182 (1993).
16. A. Kotzinian, *Nucl. Phys.* **B441**, 234–248 (1995).
17. M. Diehl, and S. Sapeta, *Eur. Phys. J.* **C41**, 515–533 (2005).
18. V. Pascalutsa, C. E. Carlson, and M. Vanderhaeghen, *Phys. Rev. Lett.* **96**, 012301 (2006).
19. L. P. Gamberg, G. R. Goldstein, and M. Schlegel, *Phys. Rev.* **D77**, 094016 (2008).

The Pretzelosity Distribution Function and Intrinsic Motion of the Constituents in Nucleon

A. V. Efremov*, P. Schweitzer†, O. V. Teryaev* and P. Závada**

*Bogoliubov Laboratory of Theoretical Physics, JINR, 141980 Dubna, Russia
†Department of Physics, University of Connecticut, Storrs, CT 06269, USA
**Institute of Physics, Academy of Sciences of the Czech Republic, Na Slovance 2, CZ-182 21 Prague 8, Czech Republic

Abstract. The pretzelosity distribution function $h_{1T}^{\perp}$ is studied in a covariant quark-parton model which describes the structure of the nucleon in terms of 3D quark intrinsic motion. This relativistic model framework supports the relation between helicity, transversity and pretzelosity observed in other relativistic models *without* assuming SU(6) spin-flavor symmetry. Numerical results and predictions for SIDIS experiments are presented.

Keywords: transverse momentum dependent distribution function, single spin asymmetry (SSA)
PACS: 13.88.+e, 13.85.Ni, 13.60.-r, 13.85.Qk

1. INTRODUCTION

Transverse parton momentum dependent parton distribution (TMDs) and fragmentation functions [1–6] offer the access to novel information on the nucleon structure [7]. TMDs can be accessed in processes like semi-inclusive deep-inelastic lepton nucleon scattering (SIDIS) [8]. Data on such reactions [9–14] provide first insights [18–24]. However, model calculations play an important role for the understanding of the novel functions [25–37].

An important question in this context is whether it is possible to relate unknown TMDs with possibly better known ones. Such relations cannot be exact, since all TMDs are independent. Approximations motivated partly by data were discussed in [38]. The ideal playground to motivate and test any such relations among TMDs are models.

An interesting relation between pretzelosity $h_{1T}^{\perp q}$, transversity h_1^q and helicity g_1^q was observed in bag model [26]. The name pretzelosity reflects that this function 'measures' an appropriately defined deviation of the nucleon from spherical shape which could look like a pretzel [7]. This relation holds also in the spectator model [25], and was subsequently confirmed in the constituent quark model [27] but not in the model of [28].

The purpose of this work is to study pretzelosity, and its possible relations, in the covariant model of the nucleon of Ref. [30]. In this model the intrinsic motion of partons inside the nucleon is described in terms of a covariant momentum distribution. The model was applied to the study of unpolarized and polarized structure functions accessible in DIS $f_1^a(x)$, $g_1^a(x)$ and $g_T^a(x)$ [30, 31] and extended to compute transversity $h_1^a(x)$ [32]. In this work we will generalize the approach to the description of TMDs, focusing on chiral-odd TMDs accessible with transverse nucleon polarization.

CP1149, *Spin Physics, 18th International Spin Physics Symposium*
edited by D. G. Crabb, D. B. Day, S. Liuti, X. Zheng, M. Poelker, and Y. Prok

2. CHIRAL-ODD TMDS WITH TRANSVERSE POLARIZATION

We focus on chiral-odd TMDs in a nucleon polarized transversely, e.g. in SIDIS, with respect to the hard virtual photon $q^\mu = (q^0, |\vec{q}|, 0, 0)$. The light-front quark-correlator with the process-dependent Wilson-link $\mathcal{W}$ [6] where $z^\pm = (z^0 \pm z^1)/\sqrt{2}$ etc.,

$$\phi(x, \vec{p}_T, \vec{S}_T)_{ij} = \int \frac{\mathrm{d}z^- \mathrm{d}^2\vec{z}_T}{(2\pi)^3}\, e^{ipz}\, \langle P, \vec{S}_T | \bar{\psi}_j(0)\, \mathcal{W}(0, z, \text{path})\, \psi_i(z) | P, \vec{S}_T \rangle \Big|_{z^+=0,\, p^+=xP^+}, \tag{1}$$

allows to define (3 out of the 4) chiral-odd TMDs in the nucleon as follows

$$\frac{1}{2}\,\mathrm{tr}\Big[i\sigma^{j+}\gamma_5\, \phi(x, \vec{p}_T, \vec{S}_T) \Big] \;=\; S_T^j h_1 + \frac{(p_T^j p_T^k - \frac{1}{2}\vec{p}_T^{\,2}\delta^{jk}) S_T^k}{M_N^2} h_{1T}^\perp + \frac{\varepsilon^{jk} p_T^k}{M_N} h_1^\perp, \tag{2}$$

where $\varepsilon^{32} = -\varepsilon^{23} = 1$ and zero else. The only structure surviving the $\vec{p}_T$-integration in (1) is transversity $h_1^a(x)$. Nucleon polarizations and Dirac-structures other than that in Eqs. (1, 2) lead to further leading- and subleading-twist TMDs [3, 4, 16].

3. THE COVARIANT MODEL OF THE NUCLEON AND TMDS

The starting point for the calculation of the chiral-even functions accessible in DIS, $f_1^a(x)$, $g_1^a(x)$, $g_T^a(x)$, is the hadronic tensor [30, 31]. In the model it is assumed that DIS can be described as the incoherent sum of the scattering of electrons off non-interacting quarks, whose momentum distributions inside the nucleon are given in terms of the scalar functions: $G = G^\uparrow + G^\downarrow$ for unpolarized and $H = G^\uparrow - G^\downarrow$ for polarized quarks.

$G^i(pP/M)$ denotes the distribution of quarks of some (not indicated) flavour that are polarized parallel (antiparallel) $\uparrow(\downarrow)$ to the i-axis, where p is the quark momentum and M the nucleon mass. Though all expressions can be formulated in a manifestly covariant way, it is convenient to work in the nucleon rest-frame, where the G^i become functions of $p^0 = \sqrt{\vec{p}^{\,2} + m^2}$ with m the quark mass, and the distributions are rotationally symmetric.

The chiral-odd $h_1^q(x)$ cannot be accessed in DIS through the hadronic tensor. However, for theoretical purposes one may consider the auxiliary process described by the interference of a vector and a scalar current, described on the quark level by $T_\alpha^q = \varepsilon_{\alpha\beta\lambda\nu} p^\beta q^\lambda w^\nu$ where w^ν is the quark polarization vector. The nucleon current follows from convoluting T_α^q with the momentum distribution of polarized quarks $H(p^0)$ and reads

$$T_\alpha(x) = \frac{1}{2Pq} \varepsilon_{\alpha\beta\lambda\nu} q^\lambda \int \frac{\mathrm{d}^3 p}{p^0} H(p^0)\, \delta\!\left(\frac{p^0 - p^1}{M} - x\right) p^\beta w^\nu . \tag{3}$$

The auxiliary current is related to transversity as

$$2M\, T_\alpha(x)\, \varepsilon^{\alpha j} = S_T^j h_1^q(x) . \tag{4}$$

Before attempting to extend the approach to TMDs, let us stress that the QCD definition of TMDs includes a Wilson line absent in our model with no gauge boson degrees of freedom. In such an approach time-reversal (T) odd TMDs, such as the Boer-Mulders function $h_1^\perp$ in (2), vanish [5, 6].

Now we turn to the question how to extend the approach to describe of TMDs, focusing here on chiral-odd ones in a transversely polarized nucleon. For that we observe that the expression for the auxiliary current (3) is of the type: $T_\alpha(x) = \int d^2p_T T_\alpha(x,\vec{p}_T)$. In the following we explore the consequences of what happens if one *does not integrate out* transverse momenta in this expression.

With S^μ denoting the nucleon polarization vector (here $S^\mu = (0,0,\vec{S}_T)$ with $|\vec{S}_T| = 1$) the most general expression [31] for the covariant quark polarization vector w^μ reads

$$w^\mu = -\frac{pS}{pP+mM}P^\mu + S^\mu - \frac{M}{m}\frac{pS}{pP+mM}p^\mu \,. \tag{5}$$

From (5) we obtain for the 'unintegrated' auxiliary current contracted with $\varepsilon^{\alpha j}$ the result

$$2M\,T_\alpha(x,\vec{p}_T)\,\varepsilon^{\alpha j} = \int\frac{dp^1}{p^0}\,H(p^0)\,\delta\left(\frac{p^0-p^1}{M}-x\right)\left\{S_T^j(p^0-p^1) - p_T^j\,\frac{\vec{S}_T\vec{p}_T}{p^0+m}\right\}. \tag{6}$$

By comparing to (2) we read off the following results:

$$h_1^q(x,p_T) \;=\; \int\frac{dp^1}{p^0}\,H(p^0)\,\delta\left(\frac{p^0-p^1}{M}-x\right)\left[p^0-p^1-\frac{\vec{p}_T^{\,2}}{2(p^0+m)}\right], \tag{7}$$

$$h_{1T}^{\perp q}(x,p_T) \;=\; \int\frac{dp^1}{p^0}\,H(p^0)\,\delta\left(\frac{p^0-p^1}{M}-x\right)\left[-\frac{M^2}{p^0+m}\right], \tag{8}$$

and $h_1^{\perp q}(x,p_T) = 0$. Several comments are in order. First, in our approach the vanishing of the T-odd $h_1^{\perp q}$ is consistent. Second, integrating in Eq. (7) over $\vec{p}_T$ yields the model expression for $h_1^q(x) \equiv \delta q(x)$ from [32]. Third, $h_{1T}^{\perp q} \neq 0$ implies non-sphericity in the nucleon in the sense of [7] inspite of a spherically symmetric $H(p_0)$. Forth, adding $h_1^q(x)$ and $h_{1T}^{\perp(1)q}(x) = \int d^2p_T\frac{\vec{p}_T^{\,2}}{2M^2}h_{1T}^{\perp q}(x,p_T)$ yields the model expression for $g_1^q(x) \equiv \Delta q(x)$ derived in [31], i.e. we recover the remarkable relation [26]:

$$g_1^q(x) - h_1^q(x) = h_{1T}^{\perp(1)q}(x)\,. \tag{9}$$

This relation is satisfied in several [25–27] though not all [28] quark models. Remarkably, it follows in our approach *without* assuming SU(6) spin-flavour symmetry of the nucleon wave function as was done in [25–27]. This is an important observation: SU(6) is not a necessary condition for the relation (9) to be satisfied in a *quark model*. What is a necessary condition is the absence of gluon degrees of freedom [29].

Finally, we remark that in the chiral limit $m \to 0$ it is possible to relate the transverse moment of pretzelosity to the twist-3 parton distribution function $g_T^q(x)$ [31] as follows

$$h_{1T}^{\perp(1)q}(x) + g_T^q(x) = \mathcal{O}\left(\frac{m}{M}\right). \tag{10}$$

Since in the model the WW-relation holds, $g_T^q(x) = \int_x^1 dy\,g_1^q(y)/y + \mathcal{O}(\frac{m}{M})$ [31], this offers a possibility to *estimate* pretzelosity numerically in the model framework.

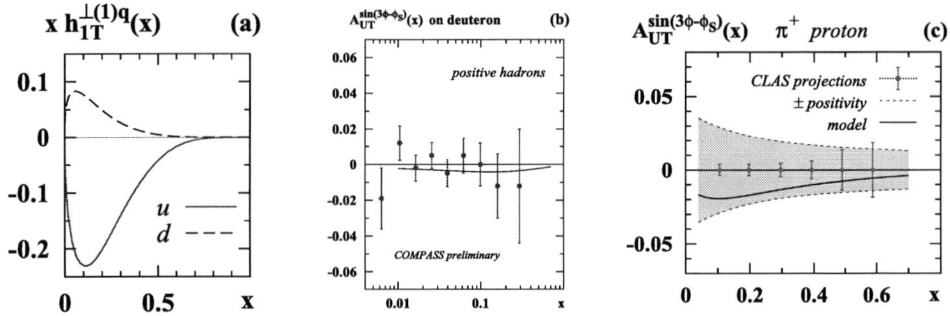

FIGURE 1. (a) $x h_{1T}^{\perp(1)q}(x)$ vs. x from the present approach. (b) $A_{UT}^{\sin(3\phi-\phi_S)}$ as function of x computed with pretzelosity from Fig. 1a for positive hadrons from a deuterium target at COMPASS in comparison to preliminary data [14]. (c) The same as Fig. 1b but for π^+ from proton target at CLAS. The shaded area is the region allowed by positivity [26], the error projections are from [40].

4. RESULTS AND PHENOMENOLOGY

We estimate $h_{1T}^{\perp(1)q}(x)$ in our approach using (10) and the WW-approximation for $g_T^q(x)$ with $g_1^q(x)$ at a scale of $2.5\,\mathrm{GeV}^2$ from [39]. We obtain the results shown in Fig. 1a.

The azimuthal SSA from transversely polarized targets, $A_{UT}^{\sin(3\phi-\phi_S)} \propto \sum_a e_a^2 h_{1T}^{\perp(1)a} H_1^{\perp a}$, allows to access pretzelosity in SIDIS due to the Collins effect [2], see [3, 26] for details. We use the information on $H_1^\perp$ from [20–22]. Figure 1b shows that the model results for the SSA are compatible with preliminary COMPASS deuteron target data [14]. Figure 1c shows estimates for the SSA in the kinematics of the CLAS 12 GeV beam experiment. The error projections from [40] included in the plot demonstrate that CLAS will be able to measure effects of pretzelosity of the size predicted by the model.

5. CONCLUSIONS

A generalization of the covariant model [30–32] to the description of TMDs was suggested, and applied to compute the pretzelosity distribution function $h_{1T}^\perp$. In particular, it was shown that the relation between helicity, transversity and pretzelosity [26] is satisfied in this model — remarkably, without assuming SU(6) symmetry.

A numerical estimate of $h_{1T}^{\perp(1)q}$ was presented, and used to compute $A_{UT}^{\sin(3\phi-\phi_S)}$, the leading-twist SSA in SIDIS due to Collins effect and pretetzelosity. The model results are compatible with the preliminary deuteron target data from COMPASS [14]. Predictions of this observable in the kinematics of the CLAS experiment with upgraded 12 GeV beam suggest that information on pretzelosity is accessible at Jefferson Lab [40].

ACKNOWLEDGMENTS

A. E. and O. T. are supported by the Grants RFBR 06-02-16215 and 07-02-91557, RF MSE RNP.2.2.2.2.6546 (MIREA) and by the Heisenberg-Landau and (also P.Z.) Votruba-Blokhitsev Programs of JINR. P. Z. is supported by the project AV0Z10100502 of the Academy of Sciences of the Czech Republic.

REFERENCES

1. J. C. Collins, *Acta Phys. Polon. B* **34**, 3103 (2003) [arXiv:hep-ph/0304122].
2. J. C. Collins, *Nucl. Phys. B* **396**, 161 (1993) [arXiv:hep-ph/9208213].
3. P. J. Mulders and R. D. Tangerman, *Nucl. Phys. B* **461** (1996) 197 and **484** (1997) 538E.
4. D. Boer and P. J. Mulders, *Phys. Rev. D* **57**, 5780 (1998) [arXiv:hep-ph/9711485].
5. S. J. Brodsky, D. S. Hwang and I. Schmidt, *Phys. Lett. B* **530**, 99; *Nucl. Phys. B* **642**, 344 (2002).
6. J. C. Collins, *Phys. Lett. B* **536**, 43 (2002) [arXiv:hep-ph/0204004].
7. G. A. Miller, arXiv:0708.2297 [nucl-th]. M. Burkardt, arXiv:0709.2966 [hep-ph].
8. X. D. Ji, J. P. Ma and F. Yuan, *Phys. Rev. D* **71**, 034005 (2005); *Phys. Lett. B* **597**, 299 (2004). J. C. Collins and A. Metz, *Phys. Rev. Lett.* **93**, 252001 (2004) [arXiv:hep-ph/0408249].
9. M. Arneodo et al. [European Muon Collaboration], *Z. Phys. C* **34** (1987) 277.
10. A. Airapetian et al. [HERMES], *Phys. Rev. Lett.* **84**, 4047 (2000); *Phys. Rev. D* **64**, 097101 (2001); *Phys. Lett. B* **562**, 182 (2003); *Phys. Lett. B* **622**, 14 (2005); *Phys. Lett. B* **648** (2007) 164.
11. H. Avakian et al. [CLAS], *Phys. Rev. D* **69**, 112004 (2004); *AIP Conf. Proc.* **792** (2005) 945.
12. A. Airapetian et al. [HERMES], *Phys. Rev. Lett.* **94**, 012002 (2005) [arXiv:hep-ex/0408013].
13. V. Alexakhin et al. [COMPASS], *Phys. Rev. Lett.* **94**, 202002 (2005); *Nucl. Phys. B* **765** (2007) 31.
14. A. Kotzinian [on behalf of the COMPASS collaboration], arXiv:0705.2402 [hep-ex].
15. K. Abe et al. [Belle Collaboration], *Phys. Rev. Lett.* **96**, 232002 (2006); arXiv:hep-ex/0607014; arXiv:0805.2975 [hep-ex].
16. A. Bacchetta, M. Diehl, K. Goeke, A. Metz, P. J. Mulders and M. Schlegel, *JHEP* **0702** (2007) 093.
17. A. Bacchetta, M. Boglione, A. Henneman and P. J. Mulders, *Phys. Rev. Lett.* **85**, 712 (2000).
18. A. V. Efremov et al., *Phys. Rev. D* **67**, 114014 (2003) *Eur. Phys. J. C* **32** (2003) 337.
19. A. V. Efremov et al., *Phys. Lett. B* **612**, 233 (2005) [arXiv:hep-ph/0412353].
20. W. Vogelsang and F. Yuan, *Phys. Rev. D* **72** (2005) 054028 [arXiv:hep-ph/0507266].
21. A. V. Efremov, K. Goeke and P. Schweitzer, *Phys. Rev. D* **73**, 094025 (2006).
22. M. Anselmino et al., *Phys. Rev. D* **75**, 054032 (2007) [arXiv:hep-ph/0701006].
23. S. Arnold et al., arXiv:0805.2137 [hep-ph]. M. Anselmino et al., arXiv:0805.2677 [hep-ph].
24. B. Zhang, Z. Lu, B. Q. Ma and I. Schmidt, *Phys. Rev. D* **77**, 054011 (2008).
25. R. Jakob, P. J. Mulders and J. Rodrigues, *Nucl. Phys. A* **626**, 937 (1997) [arXiv:hep-ph/9704335].
26. H. Avakian, A. V. Efremov, P. Schweitzer and F. Yuan, arXiv:0805.3355 [hep-ph].
27. B. Pasquini, S. Cazzaniga and S. Boffi, arXiv:0806.2298 [hep-ph].
28. A. Bacchetta, F. Conti and M. Radici, arXiv:0807.0323 [hep-ph].
29. S. Meissner, A. Metz and K. Goeke, *Phys. Rev. D* **76**, 034002 (2007) [arXiv:hep-ph/0703176].
30. P. Závada, *Phys. Rev. D* **55**, 4290 (1997) [arXiv:hep-ph/9609372].
31. P. Závada, *Phys. Rev. D* **65**, 054040 (2002); *ibid.* **67**, 014019 (2003); *Eur. Phys. J. C* **52**, 121 (2007).
32. A. V. Efremov, O. V. Teryaev and P. Závada, *Phys. Rev. D* **70**, 054018 (2004).
33. P. Závada, hep-ph/0609027; hep-ph/0511142; hep-ph/0412206; hep-ph/0408182.
34. P. Schweitzer et al., *Phys. Rev. D* **64**, 034013. K. Goeke et al., *Acta Phys. Polon. B* **32**, 1201 (2001).
35. F. Yuan, *Phys. Lett. B* **575**, 45 (2003) [arXiv:hep-ph/0308157].
36. L. P. Gamberg, G. R. Goldstein and M. Schlegel, arXiv:0708.2580, arXiv:0708.0324.
37. B. Pasquini et al.,*Phys. Rev. D* **72**, 094029 (2005); *Phys. Rev. D* **76**, 034020 (2007).
38. H. Avakian et al., *Phys. Rev. D* **77**, 014023 (2008). A. Metz, P. Schweitzer and T. Teckentrup, arXiv:0810.5212 [hep-ph].
39. M. Glück, E. Reya, M. Stratmann and W. Vogelsang, *Phys. Rev. D* **63**, 094005 (2001).
40. H. Avakian et al., JLab LOI 12-06-108 (2008).

A Review of Nucleon Spin Calculations in Lattice QCD

Huey-Wen Lin

Jefferson Laboratory, 12000 Jefferson Avenue, Newport News, VA 23606, USA

Abstract. We review recent progress on lattice calculations of nucleon spin structure, including the parton distribution functions, form factors, generalization parton distributions, and recent developments in lattice techniques.

Keywords: Lattice QCD, nucleon spin structures, form factors, GPDs
PACS: 12.38.-t 12.38.Gc 13.40.Gp 14.20.Dh 13.60.Fz

INTRODUCTION

Quantum chromodynamics (QCD) has been successful in describing many properties of the strong interaction. In the weak-coupling regime, we can rely on perturbation theory to work out the path integral which describes physical observables of interest. However, for long distances perturbative QCD no longer converges. Instead, we use a discretization of space and time in a finite volume to calculate these quantities from first principles numerically; such research forms the regime of lattice QCD[1].

To keep the systematic error due to discretization under control, one follows Symanzik improvement order by order in terms of the ultraviolet cutoff (a) for both the action and operators. However, the breaking of continuous (Euclidean) SO(4) symmetry allows many new degrees of freedom, leading to various lattice actions that return to the same continuum action once the symmetry is restored. Thus, there exist many gauge and fermion actions for us to choose from. Today, most gauge actions used are $O(a^2)$-improved and leave small discretization effects ($O(a^3\Lambda_{\rm QCD}^3)$) due to gauge choices. On the other hand, most fermion actions are only $O(a)$-improved and have systematic errors of $O(a^2\Lambda_{\rm QCD}^2)$ that become dominant. For this reason, lattice calculations are generally distinguished according to the fermion action used. Differences among the actions are benign once all systematics are included, and the choice of fermion action is constrained by limits of computational and human power and by the main physics focus. The commonly used actions are: domain-wall fermions (DWF), overlap fermions, Wilson/clover fermions, twisted-Wilson fermions and staggered fermions.

Since the real world is effectively continuous and infinitely large, we will have to take limits of $a \to 0$ and $V \to \infty$ to eliminate the artifacts introduced in a discretized finite box. With the most state-of-the-art supercomputer, we are close but yet to simulate at the physical pion mass. Using calculations at multiple heavier pion masses, which are affordable for available computational resources, we can apply chiral perturbation theory to extrapolate quantities of interest to the physical limit. A recent work by the BMW collaboration[2] calculating multiple lattice spacings, volumes and pion masses as light

CP1149, *Spin Physics, 18th International Spin Physics Symposium*
edited by D. G. Crabb, D. B. Day, S. Liuti, X. Zheng, M. Poelker, and Y. Prok

as 180 MeV provided an excellent demonstration of how ground-state hadron masses with fully understood and controlled systematics are consistent with experiment. Such calculations with multiple pion masses also help to determine the low-energy constants of chiral effective theory.

A typical nucleon interpolating field used in the lattice calculation is $\chi_N = \sum_{\vec{x},a,b,c} e^{i\vec{p}\cdot\vec{x}} \varepsilon^{abc} \left[u_a^T C\gamma_5 d_b\right] u_c$, and the nucleon two- and three-point Green functions are obtained from

$$\Gamma^{(2)}(t_{\rm src},t) = \langle \chi_N(t)\chi_N^\dagger(t_{\rm src})\rangle \quad (1)$$

$$\Gamma^{(3)}(t_{\rm src},t,t_{\rm snk}) = \langle \chi_N(t_{\rm snk},\vec{p}_{\rm snk})\,\mathscr{O}(t,\vec{q})\,\chi_N^\dagger(t_{\rm src},\vec{p}_{\rm src})\rangle, \quad (2)$$

where $\mathscr{O}$ is the operator of interest. For the vector (axial) current, the operator is $\mathscr{O} = \overline{\psi}\gamma_\mu(\gamma_5)\psi$. For the structure functions, the operators are

$$\begin{aligned}
\langle x^n\rangle_q: \quad & \mathscr{O}^q_{\mu_1\ldots\mu_n} &=& \; i^{n-1}\overline{\psi}\gamma_{\{\mu_1}\overleftrightarrow{D}_{\mu_2}\cdots\overleftrightarrow{D}_{\mu_n\}}\psi \\
\langle x^n\rangle_{\Delta q}: \quad & \mathscr{O}^{5q}_{\mu_1\ldots\mu_n} &=& \; i^{n-1}\overline{\psi}\gamma_{\{\mu_1}\gamma_5\overleftrightarrow{D}_{\mu_2}\cdots\overleftrightarrow{D}_{\mu_n\}}\psi \\
\langle x^n\rangle_{\delta q}: \quad & \mathscr{O}^{\sigma q}_{\mu\mu_1\ldots\mu_n} &=& \; i^{n-1}\overline{\psi}\gamma_5\sigma_{\mu\{\mu_1}\overleftrightarrow{D}_{\mu_2}\cdots\overleftrightarrow{D}_{\mu_n\}}\psi,
\end{aligned}$$

where $\overleftrightarrow{D} = \frac{1}{2}(\overrightarrow{D} - \overleftarrow{D})$ is the difference between forward and backward covariant derivatives. We calculate only the "connected" diagrams, which means the inserted quark current is contracted with the valence quarks in the baryon interpolating fields, as in the majority of lattice three-point calculations. However, due to isospin symmetry, isovector quantities have a cancellation that removes the unknown disconnected piece. (Disconnected diagrams are notoriously difficult to calculate directly on the lattice. They require that expensive fermion-matrix inversion be applied to numerous source vectors. Much effort has been devoted to solving this difficulty in the near future with new techniques.) For more details, please refer to a selection of plenary talks: Refs. [3] and references within.

LATTICE QCD REVEALS THE STRUCTURE OF THE NUCLEON

The nucleon axial charge is well measured in neutron β-decay experiments, so it is a natural candidate for demonstrating how well lattice QCD can be extrapolated to the physical pion point. The isovector axial charges g_A are defined as the zero-momentum-transfer limits of the isovector axial form factors. Results from various collaborations are shown on the left-hand side of Fig. 1. We show a small-scale–expansion fit (gray band) to the 2+1-flavor (DWF sea quarks) RBC data[5] (blue filled circles). The results are consistent with the LHPC data[6] (DWF valence with staggered sea); RBC's 2f and 0f results are consistent with QCDSF's 2f Clover and 0f overlap fermion numbers[7] respectively. The lowest–pion-mass values among RBC's 2+1f and 2f results may suffer from sizable finite-volume systematic errors; larger-volume calculations should be carried out to confirm these suspicions.

Results for the zeroth moment of transversity $\langle 1\rangle_{\delta q}$ are given on the right-hand side of Fig. 1. We observe rather weak dependence (roughly linear) on the quark mass, which remains consistent for calculations with the same number of sea-quark flavors. The chiral

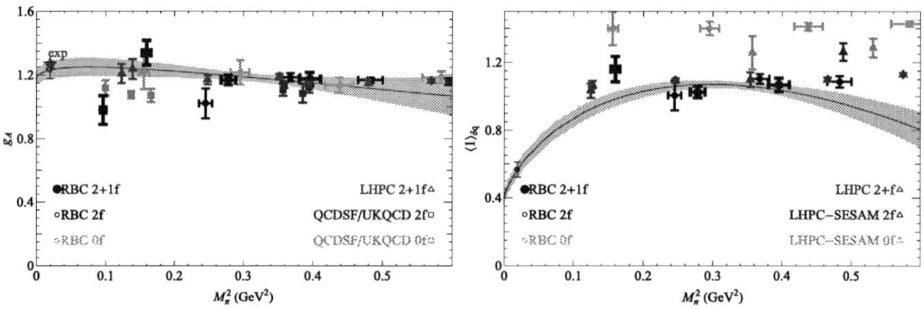

FIGURE 1. Left: Renormalized axial charge versus pseudoscalar mass squared from various lattice groups. Right: Zeroth moment of transversity from different lattice groups. The bands in both plots are chiral extrapolations fit to the RBC 2+1f data.

extrapolation form[4] is applied to the RBC 2+1f data[5], yielding 0.56(4) at the physical pion mass. However, this is close to what has been found by LHPC with mixed action[6]: about 0.7. These extrapolated values are significantly smaller than those found at the simulated pion masses, which are near 1. We urgently need data at the lightest pion mass to confirm the rapidly decreasing behavior predicted by the chiral effective theory.

Figure 2 shows the latest calculations of the first moments of the quark momentum fraction (left) and helicity (right) distributions. Here again the 0f, 2f and 2+1f results are compared between different collaborations with different choices of fermion actions and are seen to be consistent among calculations with the same number of sea-quark flavors. The chiral extrapolation[4] is performed using RBC's 2+1f data, which gives 0.133(13) and 0.203(23) for the first moments of the quark momentum fraction and helicity distributions respectively, consistent with experiment. We see strong curvature due to the chiral form; more light-pion points should be taken to reduce extrapolation uncertainties. These extrapolation numbers are also consistent with the LHPC's mixed-action calculation[6].

Higher moments of the isovector distributions are calculated by LHPC (SCRI, SESAM) with 0f and 2f Wilson and clover fermions[6] and by QCDSF with 0f clover fermions at multiple lattice spacings[7]. Consistent results are seen among different groups: The second and third moments are about 25% and 10% of the first moments respectively. However, for moments with $n \geq 4$, divergences occur involving lower-dimension operators at finite a, which limits the number of moments accessible to lattice QCD.

The twist-3 first moment of the polarized structure function d_n is another interesting feature to consider. It is related to the polarized structure functions g_1 and g_2 and the Wandzura-Wilczek relation[8]. The lowest moment d_1 from RBC's 2+1f data extrapolated to the physical pion mass is consistent with zero $d_1^{\mathrm{bare}} = -0.002(2)$. Combined with the small value of d_2 found by QCDSF's 2f calculation[7], we conclude that the Wandzura-Wilczek relation between the moments of g_1 and g_2, which asserts vanishing d_n, is at least approximately true.

Studying the momentum-transfer (Q^2) dependence of the elastic electromagnetic form factors is important in understanding the structure of hadrons at different scales. There

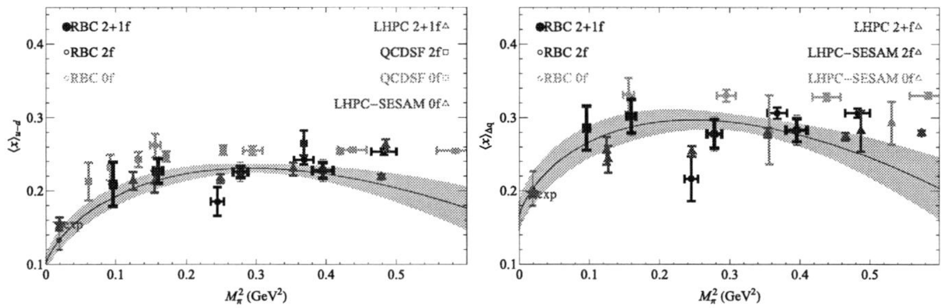

FIGURE 2. Global comparison of the first moments of the quark momentum fraction (left) and helicity (right) distributions in terms of M_π^2 and their chiral extrapolations. The bands in both plots are chiral extrapolations fit to the RBC 2+1f data.

have been many experimental studies of these form factors on the nucleon. A recent such experiment, the Jefferson Lab double-polarization experiment (with both a polarized target and longitudinally polarized beam) revealed a non-trivial momentum dependence for the ratio G_E^p/G_M^p. This contradicts results from the Rosenbluth separation method, which suggested $\mu_p G_E^p/G_M^p \approx 1$. The contradiction has been attributed to systematic errors due to two-photon exchange that contaminate the Rosenbluth separation method more than the double-polarization. (For details and further references, see the recent review articles: Refs. [9].) Lattice calculations can make valuable contributions to the study of nucleon form factors, since they allow access to both the pion-mass and momentum dependence of such form factors. Recently, the limitations of the largest-available Q^2 (in terms of the quality of the signal-to-noise ratios) has been overcome[10]. An exploratory study using clover fermions extends the range of momentum transfer to 6 GeV^2, as shown in Fig. 3. The Q^2 dependence of the neutron has exceeded the range of the current existing data. Such calculations will provide interesting comparisons for data collected after the future 12-GeV upgrade at Jefferson Lab.

The generalized parton distributions (GPDs) have been calculated by LHPC (2+1f mixed action, $M_\pi \sim$ 350–760 MeV)[6] and QCDSF (2f clover action, $M_\pi \sim$ 340–950 MeV)[7]. One of the main topics of physics interest derived from GPDs is to study the origins of the nucleon's spin. The quark spin ($\Delta\Sigma^q$) and total quark contribution (J^q) to the angular momentum can be connected to GPDs via[11]

$$\frac{1}{2}\Delta\Sigma^q = \frac{1}{2}\tilde{A}_{10}^q(0); \; J^q = \frac{1}{2}[A_{20}^q(0) + B_{20}^q(0)], \tag{3}$$

where $\tilde{A}_{n0}^q$ and $\{A,B\}_{n0}^q$ are the polarized and unpolarized generalized form factors. The left-hand side of Fig. 4 shows the quark spin and orbital angular momentum (L^q) calculated by LHPC and QCDSF; they are consistent. One found the total d quark angular momentum and the angular momentum from the sum of u and d quarks to be consistent with zero. This is because L^d and L^u are of the same magnitude but have opposite signs. A similar relation holds for $\Delta\Sigma^d$ and L^d. Note that in both of the calculations only connected diagrams are included.

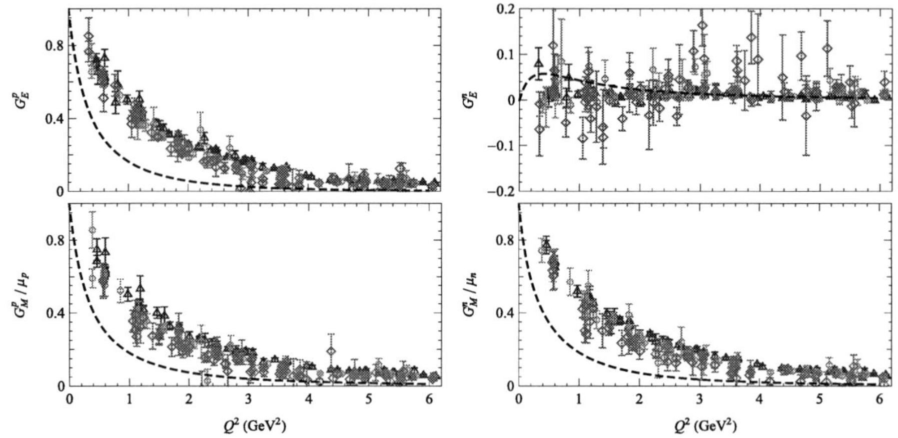

FIGURE 3. Nucleon form factors with pion masses of 480 (triangles), 720 (circles) and 1080 (diamonds) MeV. The dashed lines are plotted using experimental form-factor fit parameters[9].

For the transverse structure of the proton, QCDSF calculates the lowest two moments of the transverse spin densities of quarks in the nucleon[7]. An exploratory attempt has been made to calculate transverse-momentum distributions (TMDs) using lattice link products to approximate the Wilson line in light-cone frame in spatial coordinates and Fourier transforming into momentum-space to map the density distributions[6]; this calculation used 2+1f mixed action (DWF on staggered, $M_\pi \sim 500$ MeV).

Another new development during the last year is progress on disconnected diagrams. An indirect way of calculating the strangeness form factors via charge symmetry in 2+1f lattices is shown on the right-hand side of Fig. 4. The direct approach has also made great progress. For example, the strange-quark distribution $\langle x \rangle_s$ has been calculated by the χQCD collaboration on 2+1f lattices[12]. The same group has also calculated the gluon momentum fraction $\langle x \rangle_g$ with improved signal.

This is an exciting era for the use of lattice QCD in nuclear physics: there have been huge leaps due to increasing computational resources worldwide and improved algorithms, allowing continual improvement in lighter pion masses, larger volumes and finer lattice spacings. Various groups have demonstrated universality with the consistent results coming from independent calculations using different lattice actions. By reproducing well measured experimental values, we solidify confidence in lattice predictions of quantities that have not or cannot be measured by experiment. There are many different aspects of hadron structure that one can do with lattice QCD; only a few examples have been presented here.

In the near future, pion masses around 200 MeV or lighter with multiple volumes and lattice spacings will become commonplace. There will be less need to depend on chiral perturbation theory for extrapolations, and once the physical pion mass becomes accessible, we can check its correctness. Full-contraction calculations (including disconnected diagrams) in all matrix elements, form factors and GPDs will lead to precision

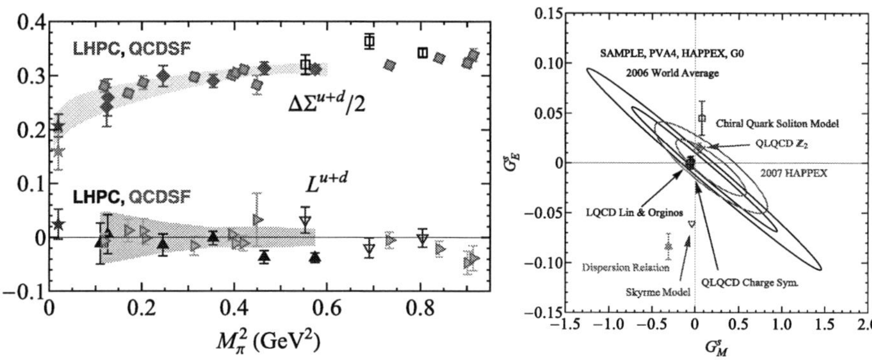

FIGURE 4. Left: The quark-component contributions to quark spin and orbital angular momentum in the spin of the nucleon from LHPC (squares, diamonds and upward triangles) and QCDSF (tilted squares and rightward triangles). The filled/open symbols represent dynamical/quenched data. The bands are chiral extrapolations based on LPHC's dynamical data. Right: Experimental results (ellipses) and various theoretical calculations (points) of G_M^s-G_E^s

calculations for individual quark components or individual proton and neutron quantities. Further improvements in methodology and expanding computational resources will allow direct calculations of the gluon content of the nucleon. The future is unlimited with lattice QCD.

Acknowledgements: Authored by Jefferson Science Associates, LLC under U.S. DOE Contract No. DE-AC05-06OR23177. The U.S. Government retains a non-exclusive, paid-up, irrevocable, world-wide license to publish or reproduce this manuscript for U.S. Government purposes.

REFERENCES

1. T. DeGrand, and C. E. Detar, new Jersey, USA: World Scientific (2006) 345 p.
2. S. Durr, et al., *Science* **322**, 1224–1227 (2008).
3. J. M. Zanotti, *PoS* **LAT2008**, 007 (2008); P. Hagler, *PoS* **LAT2007**, 013 (2007); K. Orginos, *PoS* **LAT2006**, 018 (2006); H.-W. Lin (2008), arXiv:0812.0411[hep-lat].
4. W. Detmold et al., *Phys. Rev. Lett.* **87**, 172001 (2001); D. Arndt et al., *Nucl. Phys.* **A697**, 429–439 (2002); J.-W. Chen et al., *Nucl. Phys.* **A707**, 452–468 (2002); W. Detmold et al., *Phys. Rev.* **D66**, 054501 (2002).
5. H.-W. Lin (2007), 0707.3844; H.-W. Lin et al., *Phys. Rev.* **D78**, 014505 (2008); T. Yamazaki et al., *Phys. Rev. Lett.* **100**, 171602 (2008); K. Orginos et al., *Phys. Rev.* **D73**, 094503 (2006).
6. R. G. Edwards et al., (2006), hep-lat/0610007; D. Dolgov et al., *Phys. Rev.* **D66**, 034506 (2002); P. Hagler, et al., *Phys. Rev.* **D77**, 094502 (2008); B. U. Musch, et al. (2008), 0811.1536.
7. M. Ohtani, *PoS* **LAT2007**, 158 (2007); M. Gockeler et al., *Phys. Rev.* **D71**, 114511 (2005); M. Gockeler et al., *Phys. Rev.* **D63**, 074506 (2001); M. Gockeler et al., *Phys. Rev. Lett.* **98**, 222001 (2007).
8. S. Wandzura, and F. Wilczek, *Phys. Lett.* **B72**, 195 (1977).
9. J. Arrington et al., *Phys. Rev.* **C76**, 035205 (2007); C. F. Perdrisat et al., *Prog. Part. Nucl. Phys.* **59**, 694–764 (2007); J. Arrington et al., *J. Phys.* **G34**, S23–S52 (2007); J. J. Kelly, *Phys. Rev.* **C70**, 068202 (2004).
10. H.-W. Lin et al., *Phys. Rev.* **D78**, 114508 (2008); H.-W. Lin et al., (2008), 0810.5141.
11. X.-D. Ji, *Phys. Rev. Lett.* **78**, 610–613 (1997), hep-ph/9603249.
12. M. Deka et al. (2008), 0811.1779; T. Doi et al. (2008), 0810.2482.

Recent Results and Future Prospects for the PHENIX Spin Program

Kieran Boyle and for the PHENIX Collaboration

RIKEN BNL Research Center, Brookhaven National Lab P.O. Box 5000 Upton, NY 11973, USA

Abstract. The PHENIX spin program studies the spin structure of the proton through longitudinally and transversely polarized proton collisions at the Relativistic Heavy Ion Collider (RHIC) at Brookhaven National Laboratory. The PHENIX detector is well suited to measure many final state particles, including π^0, $\pi^{\pm}$, direct photons and muons. Recent π^0 $A_{\rm LL}$ results have been shown to significantly constrain the gluon polarization. Large transversely polarized data sets from 2006 and 2008 will allow better understanding of the proton's transverse spin structure. We review recent results, show new results and discuss future prospects.

Keywords: Proton structure, Proton spin
PACS: 14.20.Dh, 13.85.Ni, 13.88.+e

INTRODUCTION

The PHENIX spin program at the Relativistic Heavy Ion Collider (RHIC) examines both the longitudinal and transverse spin structure of the proton in polarized proton collisions. Current measurements with longitudinally polarized protons at $\sqrt{s} = 200$ GeV have been shown by DSSV to significantly constrain ΔG, the integral of the gluon helicity distribution, $\Delta g(x)$. Future measurements at 500 GeV will allow flavor separation of the quark and anti-quark helicity distributions, $\Delta q(x)$ and $\Delta \bar{q}(x)$. Measurements using transversely polarized protons access the quark transversity (δq) and the quark and gluon Siver's distribution ($f_{1T}^{\perp}$). Increasing luminosity (L) and polarization (P) over the past several years (see Tables 1 and 2) are necessary to achieve the precise measurements expected at RHIC.

The PHENIX detector combines a high rate data acquisition system with specialized triggers to sample mid-rapidity ($|\eta| < 0.35$) π^0, η and γ via electromagnetic calorimetry. Charged pions at mid-rapidity are detected with the drift chamber and ring imaging Čerenkov detector. Muon are identified with the MuonID and Muon Tracker in the rapidity range $1.2 < \eta < 2.4$.

CONSTRAINING THE GLUON HELICITY DISTRIBUTION

The gluon helicity, ΔG, can be measured in $p+p$ collisions via the double helicity asymmetry ($A_{\rm LL}$) defined as

$$A_{\rm LL} = \frac{1}{\langle P_B P_Y \rangle} \frac{N_{++} - RN_{+-}}{N_{++} + RN_{+-}}, \quad R = \frac{L_{++}}{L_{+-}} \tag{1}$$

CP1149, *Spin Physics, 18th International Spin Physics Symposium*
edited by D. G. Crabb, D. B. Day, S. Liuti, X. Zheng, M. Poelker, and Y. Prok

TABLE 1. PHENIX longitudinally polarized runs.

Year	$\sqrt{s}$	Recorded Luminosity	Polarization	Figure of Merit (P^4L)
2003 (Run3)	200	0.35 pb^{-1}	27 %	1.5 nb^{-1}
2004 (Run4)	200	0.12 pb^{-1}	40 %	3.3 nb^{-1}
2005 (Run5)	200	3.40 pb^{-1}	49 %	200 nb^{-1}
2006 (Run6)	200	7.50 pb^{-1}	57 %	790 nb^{-1}
2006 (Run6)	62.4	0.04 pb^{-1}	48 %	2.1 nb^{-1}

TABLE 2. PHENIX transversely polarized runs.

Year	$\sqrt{s}$	Recorded Luminosity	Polarization	Figure of Merit (P^2L)
2001 (Run1)	200	0.15 pb^{-1}	15 %	3.4 nb^{-1}
2005 (Run5)	200	0.16 pb^{-1}	47 %	38 nb^{-1}
2006 (Run6)	200	2.70 pb^{-1}	51 %	700 nb^{-1}
2006 (Run6)	62.4	0.02 pb^{-1}	48 %	4.6 nb^{-1}
2008 (Run8)	200	5.20 pb^{-1}	46 %	1100 nb^{-1}

with $P_{B,Y}$ the beam polarizations, N the particle yield in same ('++') or opposite ('+−') helicity collisions, L the luminosity and R the relative luminosity. In Run6 (Run5), the p_T independent uncertainty in A_{LL} from relative luminosity was 7×10^{-4} (2×10^{-4}).

In order to interpret measured spin asymmetries in terms of the proton spin structure, we rely on next to leading order (NLO) pertabative QCD (pQCD). NLO pQCD has been shown to describe cross section measurements from PHENIX for π^0 at $\sqrt{s} = 200$ GeV [1] and 62.4 GeV [1] and direct photon [2], indicating that the extraction of ΔG has a strong theoretical foundation.

Results for $A_{LL}^{\pi^0}$ [1] are plotted in Fig. 1 with NLO pQCD calculations based on different input of $\Delta g(x)$ from fits to polarized DIS data. The resent global fit, DSSV, is also shown. The p_T bins cover broad overlapping distributions in x, the gluon momentum fraction. The $\sqrt{s} = 200$ GeV data set covers roughly $0.02 < x < 0.3$. The $\sqrt{s} = 62.4$

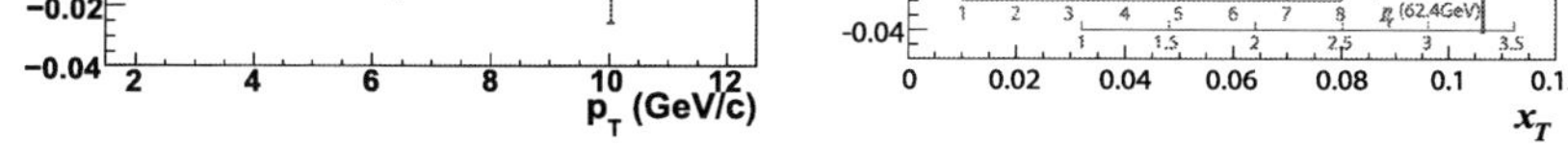

FIGURE 1. Left: $A_{LL}^{\pi^0}$ vs. p_T at $\sqrt{s} = 200$ GeV . Right: $A_{LL}^{\pi^0}$ vs. x_T at $\sqrt{s} = 62.4$ GeV compared with 2005 result at 200 GeV. The data are compared with expectations based on different gluon helicity distributions.

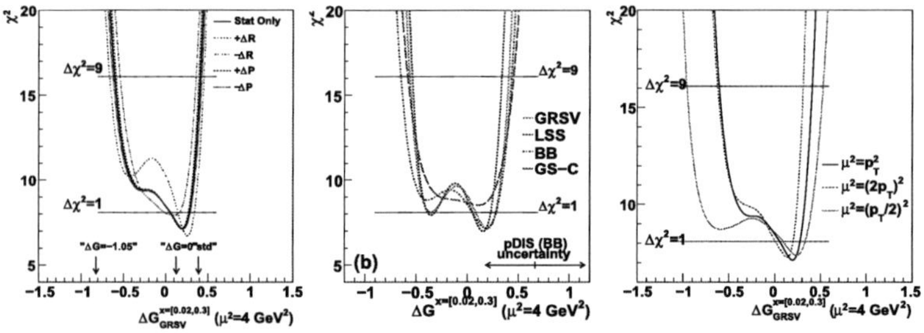

FIGURE 2. χ^2 profile as a function of $\Delta G^{[0.02,0.3]}$ considering variation of (left) experimental statistical and systematic uncertainties from relative luminosity and polarization, (center) the parameterization of $\Delta g(x)$ and (right) the theoretical scale μ in the NLO pQCD calculation.

GeV data [1] cover a higher x range, as can be seen in Fig. 1 from the different x_T ranges, where $x_T = p_\mathrm{T}/(\sqrt{s}/2)$ is proportional to the peak of the x distribution. Future measurements at $\sqrt{s} = 500$ GeV can access lower x.

By varying ΔG in a given pDIS fit result, we can create a family of A_LL expectations for different initial values of ΔG. Calculating the χ^2 for each expectation with our data, χ^2 profiles as a function of ΔG can be generated (see Fig. 2). The impact of experimental uncertainties results is an uncertainty in ΔG of 0.1 at 1σ, while varying the parameterization among a representative set of fits or the choice of theoretical scale μ does lead to variations in the constraint. At 3σ ($\Delta\chi^2 = 9$) however, we find a consistent constraint of $-0.7 < \Delta G < 0.5$ in the sampled x range ($0.02 < x < 0.3$) for all the different fits [1].

Figure 3 shows some of the other channels measured at PHENIX that are sensitive to ΔG. Charged pion [3] and η [4] at $\sqrt{s} = 200$ GeV and charged hadrons at $\sqrt{s} = 62.4$ GeV [5] sample ΔG similarly to π^0. However, the fragmentation functions in each case weight ΔG differently. For example, favored quark fragmentation functions in pion production (up to π^+, down to π^-), coupled with the different quark polarizations for up and down quarks, can lead to an ordering of the charged and neutral pion asymmetries. Other particles measurable at PHENIX, such as direct photon [2] and heavy flavor [6] have significantly different production mechanism, and so offer different methods for access ΔG.

Understanding the Orbital Angular Momentum and Related Distributions

While current uncertainties still allow for the missing portion of the proton's spin to be carried by the gluon helicity, it is important to understand the contribution of orbital angular momentum (OAM). Incite into the quark and gluon OAM can be gained from understanding of k_T dependent distributions (Siver's effect). Measurements of the azimuthal asymmetry in transversely polarized single spin collisions are sensitive to both the gluon and quark Siver's distribution. Figure 4 shows two such measurements

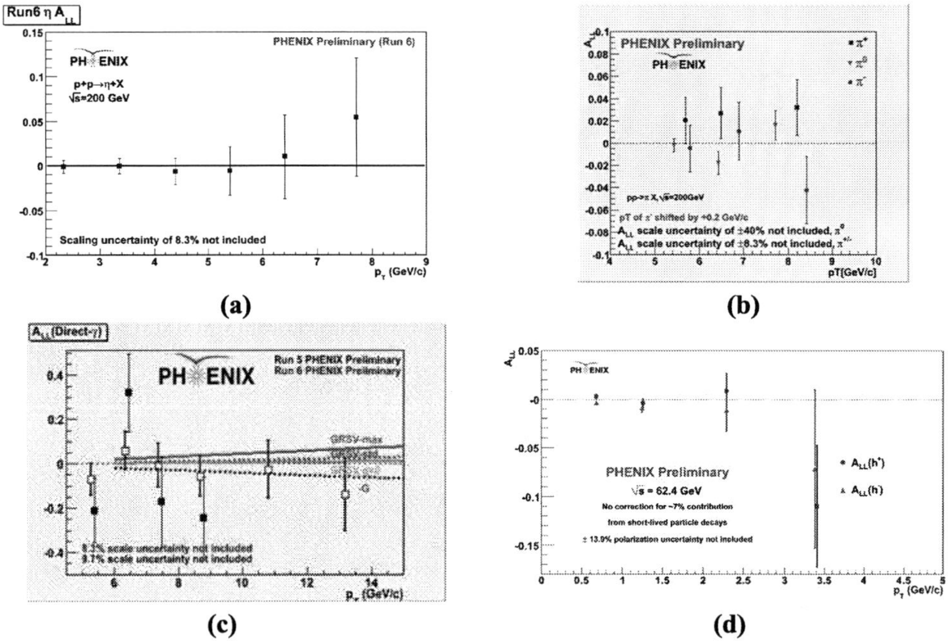

FIGURE 3. $A_{\rm LL}$ vs. $p_{\rm T}$ for (a) η, (b) charged (and neutral) pions and (c) direct photons at $\sqrt{s} = 200$ GeV and for (d) unidentified charged hadrons at $\sqrt{s} = 62.4$ GeV.

in the forward direction for π^0 [7] and muons [8]. The sizable asymmetries in forward π^0 production, measured with a forward calorimeter at $3.1 < \eta < 3.65$, are indicative of a nonzero quark Siver's function. Muons from charm production are expected to be sensitive to the gluon Siver's effect.

In the case of longitudinal spin, the helicity dependent jet k_T distribution may be sensitive to the transverse distribution of quarks, and so the quark OAM. The PHENIX measurement [9] of this quantity is shown in Fig. 4.

Another method for accessing the Siver's function is to measure the spin dependent deviation from back to back jets. Due to the limited acceptance in PHENIX, dihadron pairs are used. By measuring the asymmetry when the pair transverse momentum is perpendicular or parallel to the spin axis, the same effect could be observed. Figure 5 shows the results in the case of triggered π^0 and associated hadrons [10].

The transversity distribution is the equivalent of the helicity distribution, but for transversely polarized protons. The quark transversity distribution is T-odd, and so must be coupled with a second T-odd distribution, such as the Interference Fragmentation Function (IFF), which is being measured at BELLE. At PHENIX, measurements of the asymmetry in dihadron production from the IFF (Fig. 5) will help determine the transversity distribution [11].

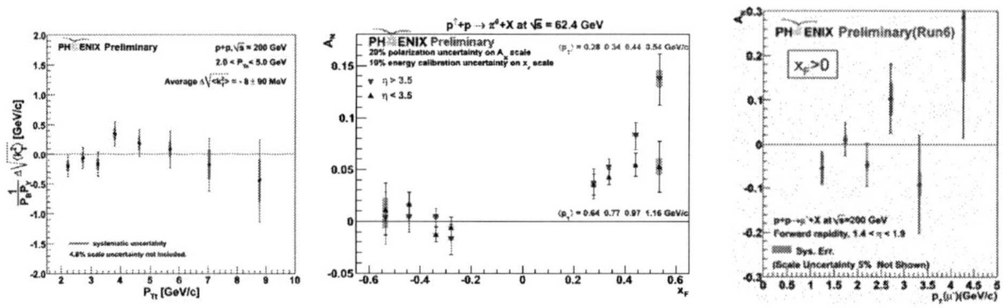

FIGURE 4. (Left) Helicity dependent k_T asymmetry at $\sqrt{s} = 200$ GeV. Azimuthal single spin asymmetry (center) for π^0 as a function of x_F at $\sqrt{s} = 62.4$ GeV and (right) for muons as a function of p_T at $\sqrt{s} = 200$ GeV.

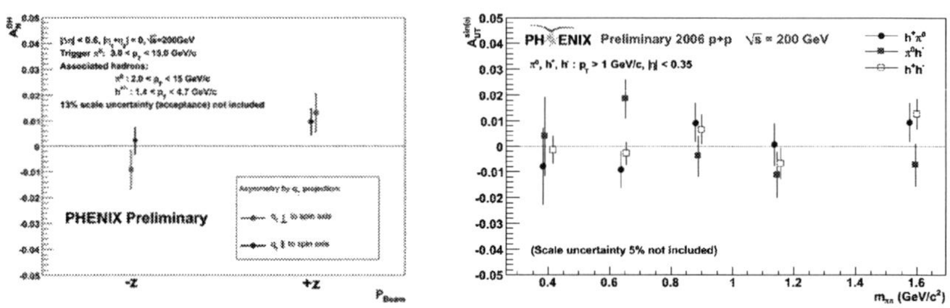

FIGURE 5. (Left) Back to back dihadron asymmetry. (Right) Interference fragmentation function asymmetry for π^0-hadron pairs as a function of the pair mass.

Future Prospects

$\sqrt{s} = 500$ GeV running at RHIC will begin soon, both allowing access to ΔG at lower x and offering a constraint on the flavor dependent sea quark helicity distributions via the parity violating asymmetry, A_L, in W boson production. To be able to trigger on high p_{T} muons from W decays, new hardware is required. Resistive Plate Chambers (RPC) will be installed in the current muon arms [12], along with additional triggering electronics in the Muon Tracker [13], to ensure that PHENIX can cleanly trigger on these high p_{T} muons. The large asymmetries expected based on several different pDIS fits are shown in Fig. 6, along with the expected uncertainties.

As heavy flavor and direct photon can access ΔG in unique ways, a vertex detector (VTX) will be added [14]. This detector will offer good separation of bottom and charm quark from the different decay lengths. The larger coverage ($\Delta\phi = 2\pi$, $|\eta| < 1.2$) will allow gamma jet measurements, with the photon detected by the current central calorimeter. Another forward calorimeter is proposed [15], which would extend the gamma jet measurement to include asymmetric partonic collisions, greatly increasing the rapidity coverage of the measurement.

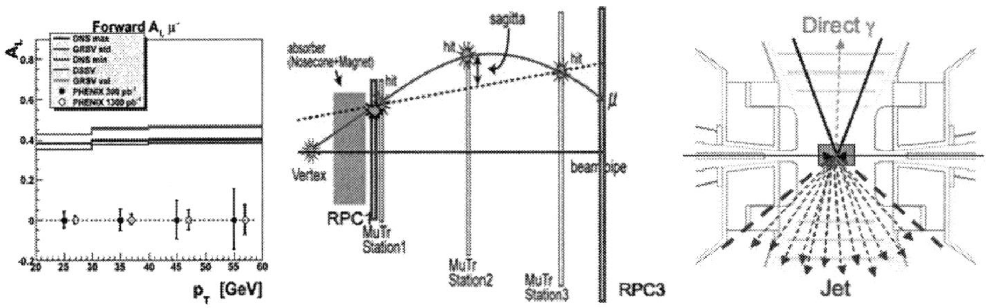

FIGURE 6. (Left) Forward muon A_L expectations. (Center) Additional tracking planes added to high p_T muon trigger. (Right) Idea of gamma-jet measurement with the VTX.

Conclusion

The PHENIX experiment is making important measurements of the proton spin structure, and will continue to do so in the future. π^0 double helicity asymmetry significantly constrain ΔG in the measured range $0.02 < x < 0.3$. Charged pions, η, direct photon and heavy flavor A_LL measurement will add to this constraint, as will measurements at different $\sqrt{s}$. Flavor separated sea quark and antiquark helicity will be measured in the near future from W boson production at $\sqrt{s} = 500$ GeV. Transversity and the Siver's function will be accessed from transverse spin asymmetries.

At Spin2008, we have reported Run5 and Run6 A_LL measurements for neutral and charged pions, direct photons, η, and heavy flavor at $\sqrt{s} = 200$ GeV and π^0 and charged hadrons at $\sqrt{s} = 62.4$ GeV. We have also reported on measurements of muon and π^0 A_N, Run6 dihadron IFF asymmetry, Run5 and Run6 longitudinally polarized k_T asymmetries and Run6 dihadron Siver's asymmetries. Please refer to PHENIX contributions to this volume [1-15] for details on each topic.

REFERENCES

1. A. Adare et al., arXiv:0810.0694 [hep-ex]. A. Adare et al., *Phys. Rev. D* **79**, 012003 (2009). K. Nakano, these proceedings.
2. R. Bennett, these proceedings.
3. A. Morreale, these proceedings.
4. J. Seele, these proceedings.
5. A. Datta, these proceedings.
6. X. Wang, these proceedings.
7. M. Chiu, these proceedings.
8. H. Liu, these proceedings.
9. D. Fields, these proceedings.
10. F. Wei, these proceedings.
11. A. Vossen, these proceedings.
12. M. Tsutomu, these proceedings.
13. X. He, these proceedings.
14. M. Kurosawa, these proceedings.
15. K. Barish, these proceedings.

Spin-Orbit Correlations in Kaon SIDIS at JLab

H. Avakian* and K. Hafidi†

*12000 Jefferson Avenue, Newport News, VA 23606, USA
†Argonne National Lab, Argonne, IL 60439, USA

Abstract. We present studies of the transverse momentum dependence of valence quark transverse and longitudinal spin distributions through measurements of spin-azimuthal asymmetries in Semi-Inclusive Deep Inelastic Scattering (SIDIS) of kaons using the upgraded JLab 11 GeV polarized electron beam and the CLAS12 detector with longitudinally polarized proton and deuteron targets. Measurements with kaons are complementary to measurements with pions and will provide additional information on the Collins fragmentation mechanism. The P_T dependence of the double spin asymmetry provides additional information on the flavor and polarization dependence of transverse-momentum-dependent helicity distributions of quarks, which provides complementary to the Single-Spin Asymmetry (SSA) measurements access to spin-orbit correlations. The large acceptance of CLAS12 would allow simultaneous detection of scattered electrons and leading hadrons, providing information on the struck quark's flavor and transverse momentum.

Keywords: spin, transverse momentum distributions, single spin asymmetries, kaons
PACS: 13.60.-r; 13.87.Fh; 13.88.+e; 14.20.Dh; 24.85.+p

Correlations of spin and transverse momentum of quarks are by now universally recognized as essential ingredients of the structure of hadrons. They are described by a number of Transverse Momentum dependent Distribution functions (TMDs) [1, 2] which give rise to various observables in hard hadronic processes [3]. Large Single Spin Asymmetries (SSAs) have been among the most difficult phenomena to understand from first principles in QCD and can be also related to TMDs. Two fundamental mechanisms have been identified leading to SSAs in hard processes; the Sivers mechanism [4, 5, 6, 7, 8], which generates an asymmetry in the distribution of quarks due to orbital motion of partons, and the Collins mechanism [1, 7], which generates an asymmetry during the hadronization of quarks.

Significant progress has been made recently in understanding the role of partonic initial and final state interactions [6, 7, 8]. The interaction between the active parton in the hadron and the spectators was included in gauge-invariant TMD distributions [6, 7, 8, 9, 10]. Furthermore, QCD factorization for SIDIS scattering at low transverse momentum in the current-fragmentation region has been established [11, 12], providing a rigorous basis to study the TMD parton distributions from SIDIS data using different spin-dependent and independent observables.

The TMDs $f_{1T}^{\perp}$ (chiral-even) and $h_1^{\perp}$ (chiral-odd) known as the Sivers and Boer-Mulders functions [4, 5, 7, 8, 9, 13], describe unpolarized quarks in the transversely polarized nucleon and transversely polarized quarks in the unpolarized nucleon, respectively. They require both orbital angular momentum, as well as non-trivial phases from the final state interaction, that survive in the Bjorken limit. The T-even counterpart of the Boer-Mulders function, $h_{1L}^{\perp}$, first introduced by Ralston and Soper [14], describes the correlations of the transverse spin of quarks in a longitudinally polarized nucleon

CP1149, *Spin Physics, 18th International Spin Physics Symposium*
edited by D. G. Crabb, D. B. Day, S. Liuti, X. Zheng, M. Poelker, and Y. Prok

and their transverse momentum.

Similar correlations arise in the hadronization process. One particular case is the Collins T-odd fragmentation function $H_1^\perp$ [15] describing fragmentation of transversely polarized quarks into unpolarized hadrons. The Collins function is one of the most fundamental quantities accessible in the hard fragmentation processes. It is of essential importance for spin physics because it works as an analyzer of the spin of the quark, but it is also interesting on its own because it allows the exploration of spin and orbital degrees of freedom of the QCD vacuum. It is universal [16, 12, 17], once measured in e^+e^- annihilation, it can be used in SIDIS and pp collisions, and vice versa. For kaons in particular, the u-goes-to-kaon Collins fragmentation function is sensitive to the strange vacuum, whereas the s-goes-to-kaon Collins function is sensitive to the strangeness spin distributions.

Many experiments worldwide are currently trying to pin down various TMD effects through SIDIS (in experiments such as HERMES at DESY [18, 19, 20], COMPASS at CERN [21], CLAS and Hall-A at JLab [22, 23]), polarized pp collisions (PHENIX, STAR and BRAHMS at RHIC) [24, 25, 26], and e^+e^- annihilation (Belle at KEK) [27]. Measurements by the HERMES Collaboration for the first time directly indicated significant azimuthal moments generated both by Collins ($\sigma_{UT}^{\sin(\phi+\phi_S)}$) and Sivers ($\sigma_{UT}^{\sin(\phi-\phi_S)}$) effects [2]. Significant differences were observed between charged pion and kaon, both for Sivers and Collins effects. Currently no satisfactory explanation exists for these observed differences.

So far no experimental information is available about the Collins fragmentation function for kaons. Recent direct calculations of the kaon Collins function [28] indicates that it may be comparable with the pion Collins function. HERMES measurements of the Collins asymmetry for pions and kaons, though, with large uncertainties, indicate that the differences may be significant.

The distribution function giving rise to SSAs for a longitudinally polarized target, $h_{1L}^\perp$, describes transversely polarized quarks in the longitudinally polarized nucleon. The physics of σ_{UL} (cross section for an unpolarized beam and a longitudinally polarized target), which involves the Collins fragmentation function $H_1^\perp$ and the Ralston-Soper-Mulders-Tangerman distribution function $h_{1L}^\perp$, was first discussed by Kotzinian and Mulders in 1996 [1, 29, 30]. Measurements of the $\sin 2\phi$ SSA [30] probe the Collins effect with no contamination from other mechanisms. A recent measurement of the $\sin 2\phi$ moment of σ_{UL} by HERMES [18] is consistent with zero. A measurably large asymmetry has been predicted only at large x ($x > 0.2$), a region well-covered by JLab [31].

Simultaneous measurements of the Kotzinian-Mulders asymmetry for pions and kaons will provide an independent measurement of ratios of Collins functions for pions and kaons, providing complementary measurements to e^+e^- annihilation. That measurement will also provide a check of the chiral limit, where that ratio is expected to be unity. Measurements of single and double spin asymmetries in SIDIS with longitudinally polarized targets (Fig. 1) will provide information on transverse momentum dependence of polarized and unpolarized partonic distributions, as well as fragmentation functions for kaons, complementary to transverse target (Fig. 2) measurements and future measurements in e^+e^- by BELLE. Projections for kaons were performed assuming two sectors of the CLAS12 detector being covered by a proximity focusing RICH detector similar to the one operating in Hall A at JLab [32, 33].

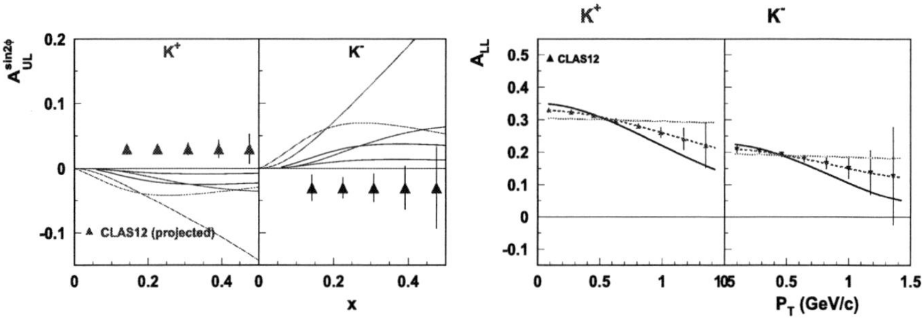

FIGURE 1. The SSA A_{UL} (left) and double spin asymmetry A_{LL} (right) for the NH_3-target as a function of the x_B and the transverse momentum of hadrons, P_T, averaged over the range $0.4 < z < 0.7$. The projected statistical errors are shown for 80 days of running with the hydrogen target. Curves are calculated using a constituent quark model for $h_{1L}^{\perp}$ [34] for different Collins functions. For A_{LL} calculations, different k_T widths for helicity distributions were used [35].

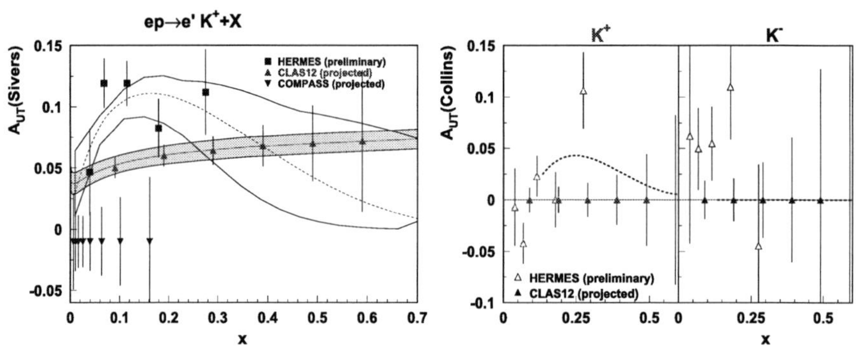

FIGURE 2. Expected Sivers SSA $A_{UT}^{\sin(\phi-\phi_S)}$ (left) and Collins SSA $A_{UT}^{\sin(\phi+\phi_S)}$ (right) as a function of x for K^+. The projected statistical errors are shown for 80 days of running with the hydrogen target. The squares represent preliminary results from HERMES [36]. The colored band [37] uses Kretzer fragmentation functions and the GRV98 PDF set, and the white band [38] uses DSS fragmentation functions and the GRV98 set with an eight parameter fit. The prediction for the Collins asymmetry is from Ref. [35].

The JLab 12-GeV upgrade will provide the unique combination of wide kinematic coverage, high beam intensity (luminosity), high energy, high polarization, and advanced detection capabilities necessary to study the transverse momentum dependence and spin correlations in double-polarized semi-inclusive processes both in the target and current fragmentation regions for variety of hadron species. Measured single and double spin asymmetries for all pions and kaons over a large range of kinematic variables (x_B, Q^2, z, $P_\perp$, and ϕ), combined with measurements with unpolarized targets, will provide detailed information on the flavor and polarization dependence of the transverse momentum distributions of quarks in the valence region.

ACKNOWLEDGMENTS

Jefferson Science Associates (JSA) operates the Thomas Jefferson National Accelerator Facility for the U. S. Department of Energy under contract DE-AC05-060R23177. This work was also supported by the U.S. Department of Energy, Office of Nuclear Physics, under Contract No. DE-AC02-06CH11357.

REFERENCES

1. P. J. Mulders, and R. D. Tangerman, *Nucl. Phys.* **B461**, 197–237 (1996).
2. A. Bacchetta, et al., *JHEP* **02**, 093 (2007).
3. V. Barone, A. Drago, and P. G. Ratcliffe, *Phys. Rept.* **359**, 1–168 (2002).
4. D. W. Sivers, *Phys. Rev.* **D43**, 261–263 (1991).
5. M. Anselmino, and F. Murgia, *Phys. Lett.* **B442**, 470–478 (1998).
6. S. J. Brodsky, D. S. Hwang, and I. Schmidt, *Phys. Lett.* **B530**, 99–107 (2002).
7. J. C. Collins, *Phys. Lett.* **B536**, 43–48 (2002).
8. X. Ji, and F. Yuan, *Phys. Lett.* **B543**, 66–72 (2002).
9. A. V. Belitsky, X. Ji, and F. Yuan, *Nucl. Phys.* **B656**, 165–198 (2003).
10. D. Boer, P. J. Mulders, and F. Pijlman, *Nucl. Phys.* **B667**, 201–241 (2003).
11. X. Ji, J. Ma, and F. Yuan, *Phys. Rev.* **D71**, 034005 (2005).
12. J. C. Collins, and A. Metz, *Phys. Rev. Lett.* **93**, 252001 (2004).
13. S. J. Brodsky, D. S. Hwang, and I. Schmidt, *Nucl. Phys.* **B642**, 344–356 (2002).
14. J. P. Ralston, and D. E. Soper, *Nucl. Phys.* **B152**, 109 (1979).
15. J. C. Collins, *Nucl. Phys.* **B396**, 161–182 (1993).
16. A. Metz, *Phys. Lett.* **B549**, 139–145 (2002).
17. F. Yuan, *Phys. Rev.* **D77**, 074019 (2008).
18. A. Airapetian, et al., *Phys. Rev. Lett.* **84**, 4047–4051 (2000).
19. A. Airapetian, et al., *Phys. Rev. Lett.* **94**, 012002 (2005).
20. A. Airapetian, et al., *Phys. Lett.* **B648**, 164–170 (2007).
21. V. Y. Alexakhin, et al., *Phys. Rev. Lett.* **94**, 202002 (2005).
22. H. Avakian, et al., *Phys. Rev.* **D69**, 112004 (2004).
23. H. Avakian, P. Bosted, V. Burkert, and L. Elouadrhiri, *AIP Conf. Proc.* **792**, 945–948 (2005).
24. J. Adams, et al., *Phys. Rev. Lett.* **92**, 171801 (2004).
25. M. Chiu, *AIP Conf. Proc.* **915**, 539–542 (2007).
26. I. Arsene, et al., *Phys. Rev. Lett.* **101**, 042001 (2008).
27. K. Abe, et al., *Phys. Rev. Lett.* **96**, 232002 (2006).
28. A. Bacchetta, L. P. Gamberg, G. R. Goldstein, and A. Mukherjee, *Phys. Lett.* **B659**, 234–243 (2008).
29. A. Kotzinian, *Nucl. Phys.* **B441**, 234–248 (1995).
30. A. M. Kotzinian, and P. J. Mulders, *Phys. Rev.* **D54**, 1229–1232 (1996).
31. A. V. Efremov, K. Goeke, and P. Schweitzer, *Phys. Rev.* **D67**, 114014 (2003).
32. F. Garibaldi, et al., *Nucl. Instrum. Meth.* **A502**, 117–122 (2003).
33. M. Iodice, et al., *Nucl. Instrum. Meth.* **A553**, 231–236 (2005).
34. B. Pasquini, S. Cazzaniga, and S. Boffi, *Phys. Rev.* **D78**, 034025 (2008).
35. M. Anselmino, A. Efremov, A. Kotzinian, and B. Parsamyan, *Phys. Rev.* **D74**, 074015 (2006).
36. M. Diefenthaler, *Proceedings of the XV International Workshop on Deep Inelastic Scattering* **Munich,Germany, April 16th - April 20th** (2008).
37. S. Arnold, A. V. Efremov, K. Goeke, M. Schlegel, and P. Schweitzer, *hep-ph/0805.2137* (2008).
38. M. Anselmino, et al., *hep-ph/0805.2677* (2008).

What Can We Learn From η Production In Proton-Proton Collisions?

Joseph Seele (for the PHENIX Collaboration)

Massachusetts Institute of Technology, 77 Massachusetts Ave., Cambridge, MA 02139, USA
and
University of Colorado at Boulder, 390 UCB, Boulder, CO 80309, USA

Abstract. Measurements of high p_T mid-rapidity η production in polarized and unpolarized proton-proton collisions provide information about a number of quantities that are currently not well constrained by other measurements. Specifically, the double helicity asymmetry in high p_T mid-rapidity η meson production is sensitive to the gluon helicity distribution in a polarized proton. In order to relate the double helicity asymmetry for an inclusive single particle final state to a polarized parton distribution, the fragmentation functions must be known. Despite the relatively large production rate of the η meson and its ease of detection, an extraction of the η fragmentation functions has only been performed recently, aided by a measurement of the cross section for high p_T mid-rapidity η meson production in proton-proton collisions.

Measurements of the double helicity asymmetry and cross section for high p_T mid-rapidity η meson production at PHENIX as well as a first extraction of the η fragmentation functions from world data will be reviewed. Lastly, an estimate of the gluon contribution to the proton spin is calculated using these results.

Keywords: Double helicity asymmetry, Gluon, ΔG, Proton Spin Structure, η meson, PHENIX
PACS: 13.85 Ni, 13.88 +e, 14.20 Nh, 14.40 Aq

INTRODUCTION

The division of the proton's spin among its quarks and gluons is a fundamental question in modern nuclear and particle physics. The PHENIX experiment at the Relativistic Heavy Ion Collider (RHIC) measures asymmetries in polarized proton-proton collisions at $\sqrt{s} = 200$ GeV, where s is the Mandelstam variable for the two incoming protons, to garner information about the polarizations of the quarks and gluons.

Each measured final state is produced from a different mix of partonic subprocesses. The different subprocesses give different sensitivities to the possible initial states. Specifically, the η meson should have an increased sensitivity, relative to π^0 measurements, to the strange quarks and gluons. To that end, measurements of high p_T mid-rapidity η meson production in proton-proton collisions yield information about the spin structure of the proton that is complementary to existing measurements [1, 2, 3].

In order to access information about the polarized parton distribution functions of a proton a double helicity asymmetry is measured. In factorized pQCD, a double helicity asymmetry is written as

$$A_{LL} = \frac{\sum_{abc} \Delta f_a \otimes \Delta f_b \otimes \Delta \hat{\sigma} \otimes D_c}{\sum_{def} f_d \otimes f_e \otimes \hat{\sigma} \otimes D_f}, \tag{1}$$

CP1149, *Spin Physics, 18th International Spin Physics Symposium*
edited by D. G. Crabb, D. B. Day, S. Liuti, X. Zheng, M. Poelker, and Y. Prok

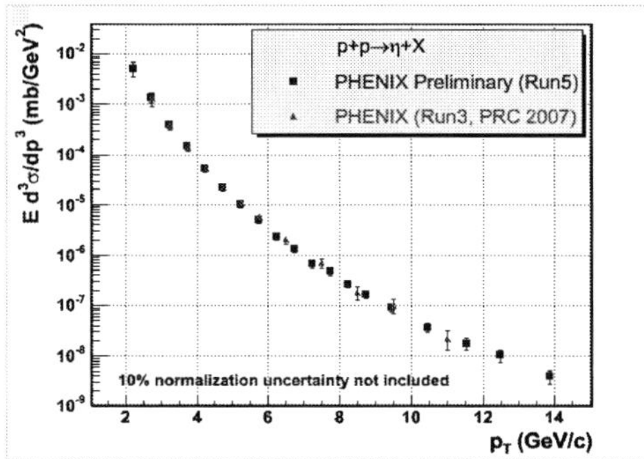

FIGURE 1. The run5 and run3 cross sections [4] for high p_T mid-rapidity η meson production in proton-proton collisions at $\sqrt{s} = 200$ GeV.

where the sums are over the different partons, Δf is a polarized parton distribution, f is an unpolarized parton distribution, $\Delta\hat{\sigma}$ is the polarized hard partonic cross section difference, $\hat{\sigma}$ is the unpolarized hard partonic cross section, and D is a fragmentation function. From this formula it can be seen that in order to obtain information regarding the polarized parton distribution functions from a double helicity asymmetry the fragmentation functions must be known. The fragmentation functions can be measured in $e^+ + e^-$ annihilation or extracted from single inclusive particle production in proton-proton collisions. In factorized pQCD, a cross section for single inclusive particle production in proton-proton collisions can be written as

$$\sigma = \sum_{abc} f_a \otimes f_b \otimes \hat{\sigma} \otimes D_c, \tag{2}$$

where the quantities were the same as they were in the formula for the double helicity asymmetry. With the η fragmentation functions, measurements of the double helicity asymmetry can be used to constrain the contribution of the spins of the gluons to the spin of the proton.

MEASUREMENTS AND EXTRACTIONS

At PHENIX, the primary method of reconstructing the η meson is through its two photon decay channel which has a branching ratio of approximately 0.39 [5]. The two photons are measured using the PHENIX electromagnetic calorimeter (EMCal) which has a spatial resolution of $\Delta\eta \times \Delta\phi(rad) = 0.01 \times 0.01$ [6] where η is the pseudorapidity with respect to the proton beams and ϕ is the polar angle with respect to the proton beams. There is also a set of detectors used for charged particle identification, tracking and subsequent veto [7].

PHENIX has measured both the cross section and the double helicity asymmetry for high p_T mid-rapidity η meson production in proton-proton collisions at $\sqrt{s} = 200$ GeV [4]. The measured cross sections are shown in Fig. 1 and the measured double helicity asymmetries are shown in Fig. 2. The cross section will be used in a global

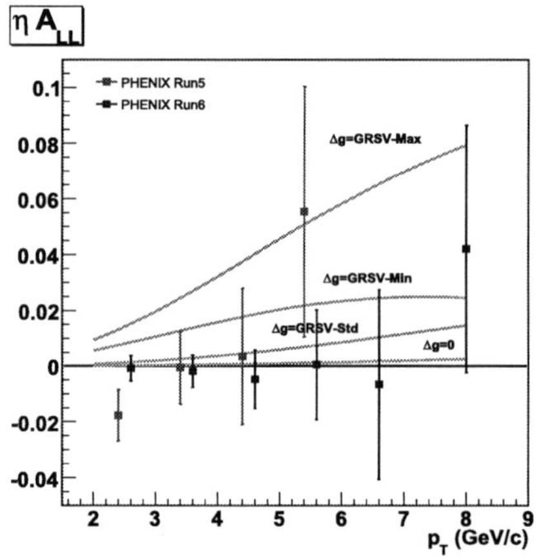

FIGURE 2. The run5 and run6 double helicity asymmetries for high-p_T mid-rapidity η meson production in proton-proton collisions at $\sqrt{s} = 200$ GeV. The scaling uncertainties due to the polarization uncertainty are not shown (run5 = 9.4% and run6 = 8.3%).

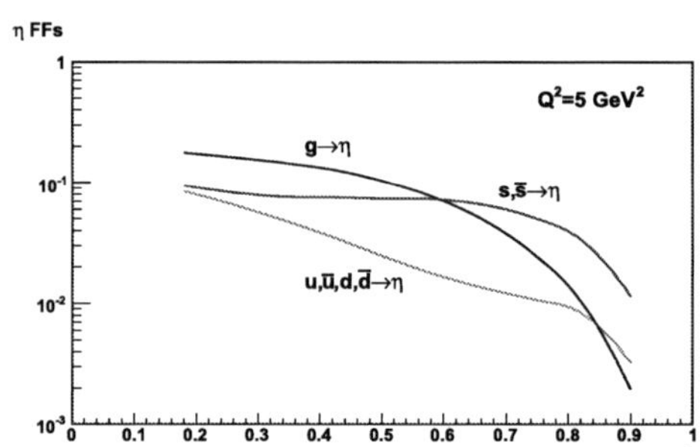

FIGURE 3. The η fragmentation functions as a function of z at a scale of $Q^2 = 5$ GeV^2. The light quarks, strange quark and gluon fragmentation functions are shown.

extraction of the η fragmentation functions which will then be used with the double helicity asymmetry to constrain the polarized gluon distribution in the proton.

As was mentioned previously, $e^+ + e^-$ annihilation data and proton-proton data can be used to extract a fragmentation function. There was no previous extraction of the η fragmentation function at next-to-leading order (NLO). A first extraction using the world data on η production was performed by Christine Aidala, Joseph Seele, and Marco Stratmann using the method outlined in [8]. Some preliminary results can be seen in Fig. 3. These results assume a strangeness suppression mechanism similar to what is commonly assumed for the kaon fragmentation functions.

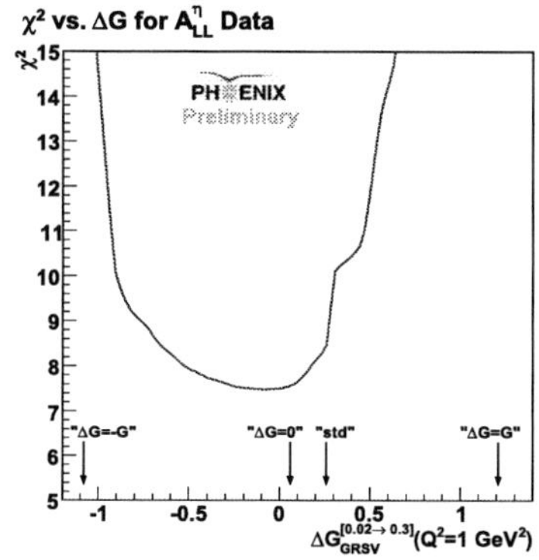

FIGURE 4. The χ^2 values for double helicity asymmetries and the model calculation by Vogelsang [9] as a function of the truncated moment of the polarized gluon distribution at a scale of $Q^2 = 1$ GeV^2.

DISCUSSION

With the fragmentation functions, the double helicity asymmetry can be used to provide a constraint on the polarized parton distribution functions and more specifically the polarized gluon distribution. In order to provide a model estimate of the preferred polarized gluon distribution, a series of constrained global fits were performed by Werner Vogelsang [9] where the shape and integral of the polarized gluon distribution were fixed at the input scale. The χ^2 between the measured run5 and run6 double helicity asymmetries in high p_T mid-rapidity η production and the calculated double helicity asymmetries from the constrained fits is shown in Fig. 4 where it can be seen that the η double helicity asymmetry prefers a small positive or negative polarized gluon distribution in this model.

REFERENCES

1. S. S. Adler, et al., *Phys. Rev.* **D73**, 091102 (2006), hep-ex/0602004.
2. A. Adare, et al., *Phys. Rev.* **D76**, 051106 (2007), 0704.3599.
3. B. I. Abelev, et al., *Phys. Rev. Lett.* **100**, 232003 (2008), 0710.2048.
4. S. S. Adler, et al., *Phys. Rev.* **C75**, 024909 (2007), nucl-ex/0611006.
5. W. M. Yao, et al., *J. Phys.* **G33** (2006).
6. L. Aphecetche, et al., *Nucl. Instrum. Meth.* **A499** (2003).
7. K. Adcox, et al., *Nucl. Instrum. Meth.* **A499** (2003).
8. D. de Florian, R. Sassot, and M. Stratmann, *Phys. Rev.* **D76**, 074033 (2007), 0707.1506.
9. W. Vogelsang, *private communication* (2008).

Helicity Dependence of Jet Properties from Di-hadrons in Longitudinally Polarized $p+p$ Collisions at $\sqrt{s} = 200$ GeV

Douglas E. Fields

University of New Mexico, 800 Yale NE MSC07-4220 Albuquerque, NM 87131, USA

Abstract. We examine the double-longitudinal difference of the RMS partonic transverse momentum $\sqrt{\langle k_T^2 \rangle}$ derived from di-hadron azimuthal correlations in $\vec{p}+\vec{p}$ collisions at $\sqrt{s} = 200$ GeV in PHENIX data from the 2005 and 2006 RHIC runs. The motivation is to explore the possible contribution of proton-spin-correlated partonic transverse momentum (orbital angular momentum) to the measured inclusive partonic $\sqrt{\langle k_T^2 \rangle}$ as described in [1]. We follow the analysis procedure described in [2] that relates the width of the away-side peak in the azimuthal two-particle distributions to the transverse momentum of the di-jet. The difference, $\Delta\sqrt{\langle k_T^2 \rangle}$ between like- and unlike-helicity collisions is then extracted. When averaged over the triggered particle transverse momentum, $\left\langle \Delta\sqrt{\langle k_T^2 \rangle} \right\rangle_{p_{Tt}} = -8 \pm 90$ MeV/c.

Keywords: Intrinsic, Transverse, Momentum, Orbital, Partons
PACS: 13.85.Fb,13.87.Fh,14.20.Dh

Being the lowest energy baryon state, one would, with good foundation, expect the quarks which make up the proton to be in an *S*-state. The naïve expectation then, is that there is no orbital angular momentum contribution to the proton spin. However, measurements of the quark spin contribution to the proton spin, ($\Delta\Sigma$), have determined that only a small fraction of the proton spin is due to quark polarization [3, 4, 5]. Recent measurements of ΔG, the gluon polarization inside the proton, are still statistically limited but have excluded large values of gluon polarization [6, 7], and the most recent global study indicates a vanishing gluon polarization together with a small quark polarization [8]. Forthcoming data from RHIC should place tighter constraints on ΔG and shed new light on the spin-puzzle, however, progress in the quark and gluon helicity distributions has rekindled interest in orbital angular momentum. In fact, it has been shown in a model independent way that the proton anomalous magnetic moment *requires* orbital angular momentum of the quarks, although the *net* orbital angular momentum may be zero [9].

The history of the theoretical interest in orbital angular momentum inside hadrons can be traced to a paper by Chou and Yang in 1976 [10] describing hadronic matter current inside a polarized hadron. Later, Meng Ta-chung et al. [1] proposed two experiments to access hadronic matter currents, one in semi-inclusive deep inelastic scattering of unpolarized leptons on transversely polarized protons, and the second in the measurement of the net transverse momentum of Drell-Yan pairs in collisions of longitudinally polarized protons. The latter lays the theoretical basis for this analysis, and is premised on the

CP1149, *Spin Physics, 18th International Spin Physics Symposium*
edited by D. G. Crabb, D. B. Day, S. Liuti, X. Zheng, M. Poelker, and Y. Prok

idea that if a part of the transverse momentum of the partons in the initial state is correlated to the (longitudinal) spin direction, as it would be in the case of orbital angular momentum, then hard collisions involving these circulating partons would lead to final states with more or less average transverse momentum k_T, depending upon the relative orientation of the spin directions and the impact parameter of the collision.

Since, at present, there is no good experimental tool to determine the collision impact parameter in $p+p$ collisions, one should ask if the effect remains when the impact parameter is undetermined. In [1], with a rather simple picture of the transverse spatial and momentum distributions, it was found that a net overall effect still exists when integrated over the impact parameter.

Here, we propose to probe the spin-correlated transverse momentum of partons within longitudinally polarized protons involved in hard collisions leading to jet-like events at the PHENIX experiment at RHIC. However, in PHENIX, due to our limited acceptance (in the central region, $\Delta\phi = \pi \times 2$ and $|\eta| < 0.35$ [11]), we cannot reconstruct the true jet kinematics to access the jet transverse momentum. An alternative method has been developed [2] that examines the di-hadron azimuthal angle correlation to extract the average pair transverse momentum on a statistical basis. The extracted average transverse momentum $\sqrt{\langle k_T^2 \rangle}$ for two subsets of the data, like-helicity collisions and unlike-helicity collisions, can then be compared as a measure of the helicity dependence of the interacting parton transverse motion.

Since, in contrast with the Drell-Yan experiment preoposed in [1], we deal here with hadronic final states, there could in principle be spin-dependent contributions to the measured k_T which are not related to the initial partonic transverse momentum. In a factorization ansatz, one can separately examine initial-state, hard scattering and fragmentation sources of a measured partonic k_T to conjecture that

$$\langle k_T^2 \rangle = \langle k_T^2 \rangle^{init} + \langle k_T^2 \rangle^{hard} + \langle k_T^2 \rangle^{frag} \tag{1}$$

With this factorization conjecture, one might attempt to understand the helicity dependence of each term. Then, we can propose that:

$$\Delta \langle k_T^2 \rangle^{init} \neq 0 \tag{2}$$

as is postulated in [1], stating that there may be a non-zero helicity difference in the combined k_T of the initial partons. The hard scattering k_T difference takes into account any helicity dependence of the induced transverse momentum from, e.g., an unequal probability for three-jet events. This contribution should be calculable to leading logarithm, and data on the double longitudinal asymmetry (A_{LL}) of neutral pion production [12] might place limits on the contributions from this term. Finally, any difference in the k_T of the final-state fragmentation could only contribute at higher twist and is thus suppressed. Since the fragmentation transverse momentum $\langle k_T^2 \rangle^{frag} \equiv j_T$ is also measured in this analysis, this can be confirmed experimentally.

In this analysis we used PHENIX high-p_T photon triggered data (trigger efficiency $\sim$ 5% at transverse momentum $p_T(\pi^0) \sim 1$ GeV/c and $\sim 90\%$ for $p_T(\pi^0) > 3.5$ GeV/c) from RHIC Runs 5 and 6 at $\sqrt{s} = 200$ GeV, as has previously been published for the PHENIX π^0 cross-section asymmetry (A_{LL}) analysis.

The azimuthal correlation function is obtained by measuring the distribution of the azimuthal angle difference, $\Delta\phi = \phi_t - \phi_a$, between a π^0 (triggered particle) and a charged hadron (associated particle). The data is analyzed in eight bins of π^0 transverse momentum from $2.0GeV/c < p_{Tt}(\pi^0) < 10.0GeV/c$ and the associated charged hadron transverse momentum bin is fixed to $2.0GeV/c < p_{Ta}(h^{\pm}) < 5.0GeV/c$ throughout this analysis. Whenever a π^0 is found in the event, the *real* $(dN_{real}/d\Delta\phi)$ and *mixed* $(dN_{mix}/d\Delta\phi)$ distributions are accumulated. The mixed event distribution is applied as a correction factor to account for the limited PHENIX acceptance. Mixed events are obtained by pairing a π^0 taken from a di-hadron event with a charged hadron taken from a different event, randomly selected from a minimum bias data set (no high-p_T photon required).

The jet fragmentation transverse momentum $\sqrt{\langle j_T^2 \rangle}$ and the partonic transverse momentum $\sqrt{\langle k_T^2 \rangle}$ are related to the widths of the two peaks in the correlation function, as explained in [2].

The data was analyzed in eight bins of p_{Tt} from $2.0GeV/c < p_{Tt}(\pi^0) < 10.0GeV/c$. $\sqrt{\langle j_T^2 \rangle}$ and $\sqrt{\langle k_T^2 \rangle}$ are calculated from the σ_{near} and $\sqrt{\langle p_{out}^2 \rangle}$ values obtained from the fit, using the procedure developed in [2].

The process of extracting $\sqrt{\langle j_T^2 \rangle}$ and $\sqrt{\langle k_T^2 \rangle}$ was repeated using two subsets of the data corresponding to collisions involving like- and unlike-helicity proton bunches at RHIC. Since any spin dependent effects should scale with the polarization of each beam, all helicity asymmetries are scaled by $\frac{1}{P_B P_Y}$, where P_B and P_Y are the run-averaged beam polarizations for the two colliding beams, ("blue" and "yellow") respectively, and are taken to be $P_B = 0.50$ and $P_Y = 0.49$ in Run 5 and $P_B = 0.56$ and $P_Y = 0.57$ in Run 6. Uncertainties on the polarizations were propagated together for the two data sets, resulting in a 4.8% scale uncertainty in our asymmetries.

The helicity difference for $\sqrt{\langle k_T^2 \rangle}$ (averaged over the Run 5 and Run 6 data sets) is shown in Fig. 1. Assuming no p_{Tt} dependence, the bin-averaged difference $\Delta\sqrt{\langle k_T^2 \rangle} = -37 \pm 88(stat) \pm 14(syst)$ MeV/c. Taken together with the dominance of the $g-g$ initial state indicated in PYTHIA and our model assumptions, this could qualitatively indicate a small gluon orbital angular momentum integrated over the kinematic region $0.002 < x_F \sim \frac{2p_T}{\sqrt{s}} < 0.1$.

REFERENCES

1. T.-C. Meng, J.-C. Pan, Q.-B. Xie, and W. Zhu, *Phys. Rev.* **D40**, 769 (1989).
2. S. S. Adler, et al., *Phys. Rev.* **D74**, 072002 (2006), hep-ex/0605039.
3. J. Ashman, et al., *Nucl. Phys.* **B328**, 1 (1989).
4. A. Airapetian, et al., *Phys. Rev.* **D75**, 012007 (2007), hep-ex/0609039.
5. V. Y. Alexakhin, et al., *Phys. Lett.* **B647**, 8–17 (2007), hep-ex/0609038.
6. B. I. Abelev, et al., *Phys. Rev. Lett.* **100**, 232003 (2008), 0710.2048.
7. PHENIX (2008), 0810.0694.
8. D. de Florian, R. Sassot, M. Stratmann, and W. Vogelsang, *Phys. Rev. Lett.* **101**, 072001 (2008), 0804.0422.

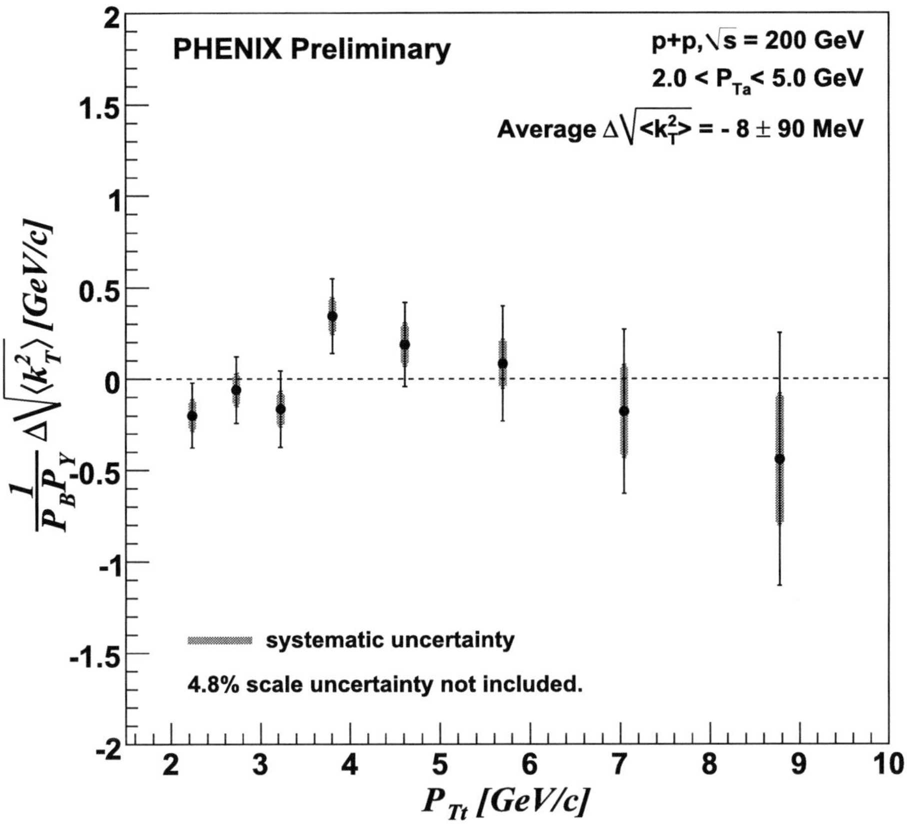

FIGURE 1. Difference in $\sqrt{\langle k_T^2 \rangle}$ for like- minus unlike-helicity combinations. A systematic uncertainty on the vertical scale due to uncertainty in the beam polarizations (4.8%) is not shown. However, this uncertatinty only affects the relative vertical scale.

9. M. Burkardt, and G. Schnell, *Phys. Rev.* **D74**, 013002 (2006), hep-ph/0510249.
10. T. T. Chou, and C. N. Yang, *Nucl. Phys.* **B107**, 1 (1976).
11. K. Adcox, et al., *Nucl. Instrum. Meth.* **A499**, 469–479 (2003).
12. A. Adare, et al., *Phys. Rev.* **D76**, 051106 (2007), arXiv:0704.3599[hep-ex].

GENERALIZED PARTON DISTRIBUTIONS AND NUCLEON FORM FACTORS

Future GPD Measurements

Ralf Kaiser

Department of Physics and Astronomy, University of Glasgow, Glasgow G12 8QQ, United Kingdom

Abstract. Generalised Parton Distributions (GPDs) have grown into one of the main topics in hadron physics. They are playing a central role in the physics of the JLab 12 GeV upgrade as well as in the future physics programme of the COMPASS experiment at CERN. This paper explores the future of GPD measurements in the short, medium and long term. The short term includes the analysis of already existing data from HERMES and JLab and planned measurements at JLab before the 12 GeV upgrade. In the medium term this concerns the JLab programme after the upgrade, measurements at COMPASS and at PANDA/FAIR. The EIC project or possible alternatives form the long term perspective. The main focus of the considerations lies on DVCS measurements and related experiments.

Keywords: Generalised Parton Distributions, Hard Exclusive Reactions, Deeply Virtual Compton Scattering
PACS: 12.38Aw, 12.38Gc, 12.38Qk, 12.39St, 13.40Gp, 13.60Fz, 13.60Hb, 13.88.+e, 24.85.+p

INTRODUCTION

Generalised Parton Distributions (GPDs) are the key to the present and future stages in the exploration of the structure of hadrons. Technological progress in accelerator performance and detector design over the last years and progress in the theory of hadron structure have put a 3-dimensional picture of the proton within reach. The description of hadron structure in terms of GPDs allows an approach that will simultaneously reveal the distributions of partons inside hadrons in one dimension of longitudinal momentum space and two dimensions of transverse impact parameter space [1, 2].

A wide variety of data can be used to constrain GPD models, from form factor measurements to hard exclusive vector meson production. The most crucial role, however, is currently played by asymmetries measured in Deeply Virtual Compton Scattering (DVCS) experiments. The first DVCS asymmetries were measured in 2001 by HERMES[3] and CLAS[4]. These measurements were followed by absolute cross section measurements by H1 and ZEUS[5] and further asymmetry data from HERMES[6], CLAS[7] and Hall A[8]. In the meantime HERMES has completed its final run in 2006/7, collecting as much DVCS data as in the decade before. These data are currently being analysed.

JLab has a very active ongoing DVCS programme, at CLAS as well as at Hall A, and GPD measurements are the very core of the 12 GeV physics programme. The COMPASS collaboration at CERN is about to submit a Letter of Intent for a future GPD programme starting in 2011. In 2015 they will be joined by the new PANDA experiment at FAIR/GSI. This rough time line of GPD experiments is illustrated in Figure 1. It is obvious from this figure that there will be a succession of experiments in Europe, with

CP1149, *Spin Physics, 18th International Spin Physics Symposium*
edited by D. G. Crabb, D. B. Day, S. Liuti, X. Zheng, M. Poelker, and Y. Prok

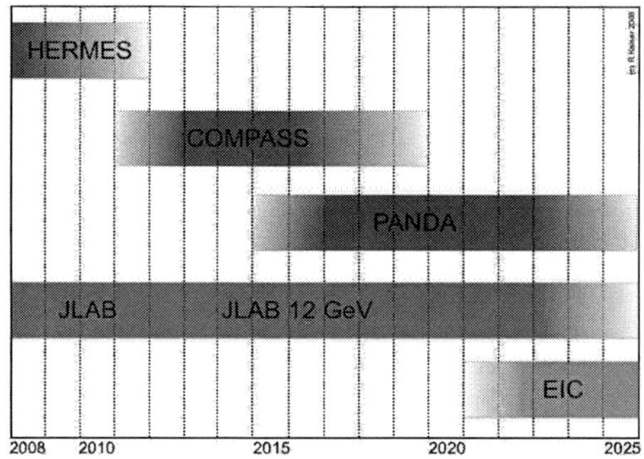

FIGURE 1. Rough time lines for the main existing or approved European (blue) and North-American (red) experiments and facilities for GPD measurements, plus the future EIC project (green).

some overlap, while the US will have one continuous programme at a single facility for about the next 10-15 years. In both cases this leads to the question of how GPD measurements will develop past 2020. The answer lies in the Electron Ion Collider (EIC) project, in one or several of its incarnations.

In the following sections some of the most recent DVCS results will be highlighted, followed by projections for the medium term future. Finally we will return to the EIC and more closely explore the different options.

RESULTS AND PROJECTIONS FOR DVCS MEASUREMENTS

HERMES

The most recently published DVCS results from the HERMES experiment concern the data collected with a transversely polarised hydrogen target and the HERA electron/positron beam over the years 2002-2005. The integrated luminosities for the electron and positron samples are about 100 pb^{-1} and 70 pb^{-1}, respectively. A maximum likelihood fit method was used to simultaneously extract the transverse target spin asymmetry (TTSA) and the beam charge asymmetry (A_C, BCA) from the data. Figure 2 shows the extracted values for the beam charge asymmetry.

The results of the TTSA measurement were used together with GPD model calculations to obtain model-dependent limits on the total angular momenta J_u and J_d of u- and d-quarks.The resulting bands are shown in Figure 2, together with a corresponding band from neutron DVCS results from JLab Hall A and lattice QCD results. Figure 2 is probably the best example to highlight the potential of DVCS measurements as well as the problems they are currently still facing. It illustrates very nicely what a future figure will look like that actually pinpoints J_u and J_d. At the same time the figure itself is deeply

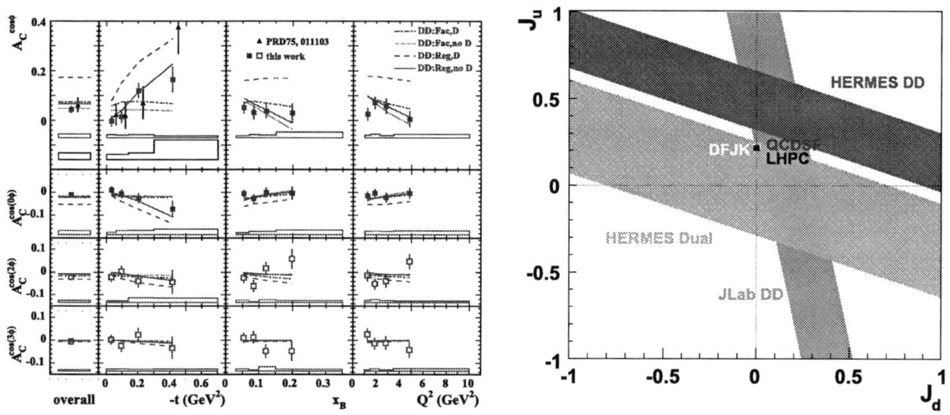

FIGURE 2. Left: Asymmetry amplitudes describing the dependence of the DVCS-BH interference term on the beam charge (A_C). Right: Model-dependent constraints on u-quark total angular momentum J_u vs d-quark total angular momentum J_d, obtained by comparing DVCS experimental results and theoretical calculations. For details see [9].

flawed: The model and the code [10] that are the basis for the HERMES and JLab double distribution bands (labeled DD) do not describe all HERMES DVCS data correctly. The dual model [11] used for the second HERMES band describes the HERMES data very well, but an error has been found in the derivation and an erratum published [12]. An updated fit is not yet available, i.e. the band in the current figure is simply wrong. The impressively small points from lattice QCD calculations are shown with statistical errors only, i.e. without systematic errors, and therefore appear to be more accurate than they actually are.

CLAS and CLAS12

GPD measurements in general and DVCS measurements in particular are at the very heart of the 12 GeV upgrade physics programme at JLab. This is true for both Hall A and CLAS; however here I will focus on CLAS. The first CLAS results on DVCS asymmetries were published in the same issue of PRL as the first HERMES results [4, 3]. Since then DVCS measurements at JLab have been refined and in particular massively increased in statistics 3.

CLAS experiment E08-021, scheduled for 2011 still with the 6 GeV beam, will measure the transverse target spin asymmetry (TTSA) in DVCS with improved precision (at somewhat different kinematics) compared to the recent HERMES results. It stands to hope that theoretical GPD models and fits will have improved by then and that limits on J_U and J_d (see figure 2) can be realised with much better precision.

DVCS results at JLab after the 12 GeV upgrade will be limited by systematics and GPD models, not by the statistics of the measurement. At the latest at that point a model-dependent extraction of the GPD H will be possible (see Figure 4).

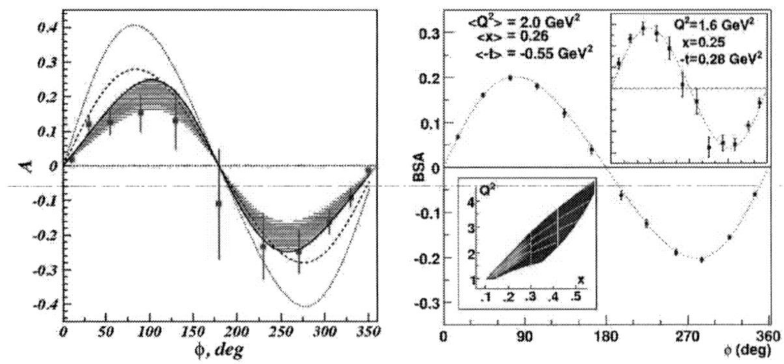

FIGURE 3. Left: CLAS first measurement of the beam spin asymmetry in DVCS [4]; Right: two (x,Q^2)-bins of the most recent CLAS DVCS BSA paper [7]

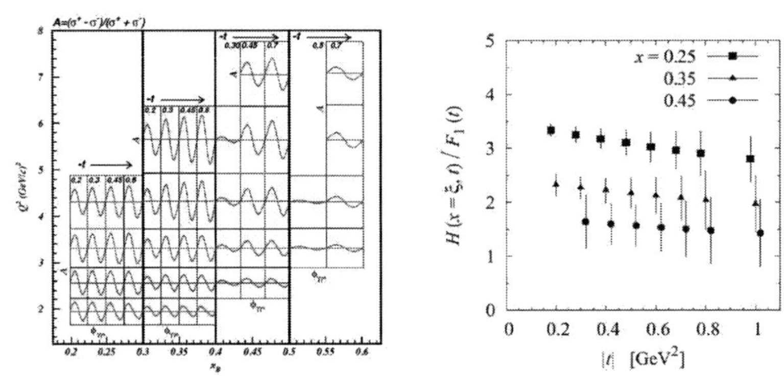

FIGURE 4. Left:Projection for DVCS BSA measurement with CLAS12. Right: Projection for extracted GPD H [13]

COMPASS

For the last few years the COMPASS experiment has carried out a systematic R&D programme into a 4 m long recoil detector for a 2.5 m long liquid hydrogen target as the key hardware component for a DVCS programme. A full-length detector prototype has been tested successfully and a smaller, but otherwise identical, 1 m long recoil detector is currently being operated as part of the COMPASS hadron spectroscopy programme.

Due to the nature of the muon beam COMPASS will measure a combined beam spin and charge asymmetry at higher Q^2 and lower x than the JLab measurements. The results will therefore be complimentary to the JLab results. Figure 5 shows projections for the beam spin and charge asymmetry for 6 kinematic bins.

The COMPASS experiment is preparing a Letter of Intent for a GPD physics programme that will be submitted to CERN in the near future.

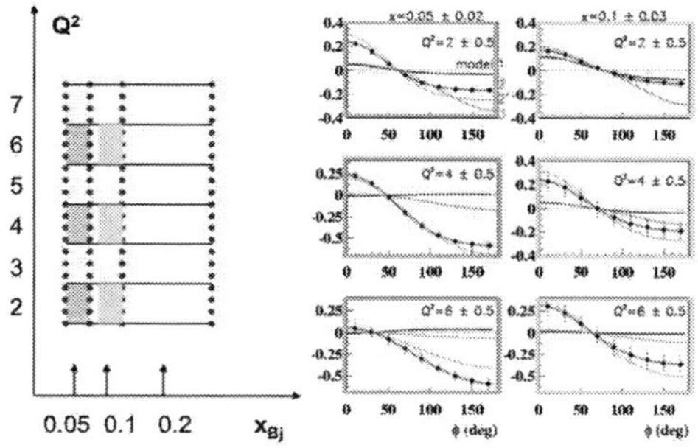

FIGURE 5. Projections for BS&CA asymmetries in DVCS measured by COMPASS.

PANDA

The PANDA experiment at FAIR will feature a 15 GeV anti-proton beam and an internal hydrogen pellet target. The $p\bar{p}$ reactions that will be studied include $p\bar{p} \rightarrow e^- e^+ \gamma$, the crossed channel reaction to DVCS. The same factorisation proof as for DVCS unfortunately does not hold, so that the reaction cannot be understood in terms of GPDs, but in terms of transition distribution amplitudes (TDAs). First calculations of TDA measurement cross sections show that the measurements will be viable at PANDA [15].

THE FAR FUTURE: ELECTRON-ION COLLIDER

If the current and medium-term GPD experiments are collected in a plot of luminosity vs center-of-mass energy (see Figure 6), it becomes apparent that the European facilities are located at lower luminosities, but higher energy, and the North-American facilities the other way around at high luminosities, but lower energies.

The obvious step for a future facility will be to combine the strengths of both approaches, i.e. the high c.m. energy and high luminosities. Additionally, there should be an overlap with CLAS or at least HERMES kinematics as well as with H1/Zeus kinematics. The resulting requirements for such an Electron-Ion Collider are a luminosity of at least 10^{33} cm^{-2}s^{-1} and a center-of-mass energy of 10 - 1000 GeV. There are currently four different proposals: eRHIC at BNL, ELIC at JLab, LHeC at CERN and the most recent one, ENC at FAIR/GSI.

It seems difficult to cover the entire desired kinematic range with a single machines; the best option could therefore in fact be two facilities that complement one another, one in Europe and one in North-America.

All of this will only be successful in the combination of experiments, lattice calculations and GPD model fits to the data. One of the most important tasks now is the improvement of GPD models and GPD based fits to the data.

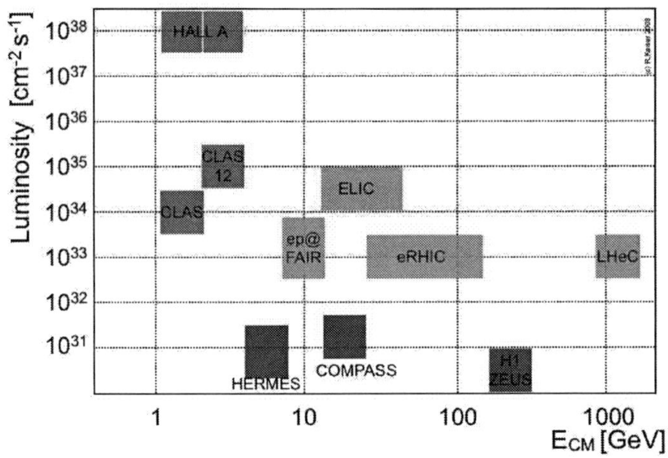

FIGURE 6. Current experiments and future approved projects for GPD measurements in North-America (red) and Europe (blue), as well as different Electron-Ion Collider projects (green).

ACKNOWLEDGMENTS

I would like to thank N.D'Hose and A.Sandorfi for the provision of some of the slides and collaborators and colleagues from the HERMES, PANDA, COMPASS and CLAS collaborations for discussions.

REFERENCES

1. M. Burkhardt, *Phys. Rev. D* **62** (2000) 071503 [Erratum ibid. 66 (2002) 119903]
2. M. Diehl, *Eur. Phys. J. C* **25** (2002) 223 [Erratum ibid. 31 (2003) 277]
3. HERMES Collaboration (A. Airapetian et al.), *Phys. Rev. Lett.* **87** (2001)
4. CLAS Collaboration (S. Stepanyan et al.), *Phys. Rev. Lett.* **87** (2001) 182002
5. H1 Collaboration (A. Aktas et al.), *Eur. Phys. J. C* **44** (2005) 1-11
6. HERMES Collaboration (A. Airapetian et al.), *Phys. Rev. D* **75** (2007) 011103
7. CLAS Collaboration (F.X. Girod et al.), *Phys. Rev. Lett.* **100** (2008) 162002
8. Hall A Collaboration (M. Mazouz et al.), *Phys. Rev. Lett.* **99** (2007) 242501
9. HERMES Collaboration (A. Airapetian et al.), *JHEP* **06** (2008) 066
10. M. Vanderhaeghen, P.A.M. Guichon, M. Guidal, *Phys. Rev. D* **60** (1999) 094017
11. V. Guzey, T. Teckentrup, *Phys. Rev. D* **74** (2006) 054027
12. V. Guzey, T. Teckentrup, *Phys.Rev.D* **79** (2009) 017501
13. CLAS Collaboration, Hall B 12 GeV Upgrade pCDR, http://www.jlab.org/Hall-B/clas12/Documentation/
14. N. D'Hose, private communication
15. B. Pire, L. Szymanowski, *Phys.Lett.B* **622** (2005) 83, J.P. Lansberg et al. *Nucl.Phys.* **A782** (2007) 16-23

Nucleon Form Factors from BLAST

Michael Kohl for the BLAST collaboration

Hampton University, Hampton, VA 23668, USA, and Jefferson Lab, Newport News, VA 23606, USA

Abstract. The BLAST (Bates Large Acceptance Spectrometer Toroid) experiment has been carried out at the MIT-Bates Linear Accelerator Center to study spin-dependent electron scattering from protons and deuterons with small systematic uncertainties. The experiment used a longitudinally polarized, intense electron beam stored in the Bates South Hall Ring in combination with isotopically pure, highly-polarized internal targets of polarized hydrogen and vector- and tensor-polarized deuterium from an atomic beam source. The BLAST data have been used to extract precise results for the elastic form factor ratios G_E/G_M of the proton and the neutron at low momentum transfer.

Keywords: polarization, electron scattering, elastic form factor
PACS: 13.40.Gp, 13.88.+e, 14.20.Dh, 25.30.Bf

Measurements of the elastic form factors of the proton and neutron and their theoretical descriptions have long been and continue to be a focus of modern nuclear physics. The BLAST (Bates Large Acceptance Spectrometer Toroid) experiment at the MIT-Bates Linear Acelerator Center was designed specifically to fully exploit polarization degrees of freedom in electron scattering in a systematic study of elastic nucleon and deuteron form factors, quasielastic deuteron electrodisintegration, and inelastic electron scattering in the Δ resonance region. The experimental configuration was the first to use an electron storage ring with longitudinally polarized beam and internal polarized hydrogen and vector- and tensor-polarized deuterium target supplied by an atomic beam source (ABS) in combination with a large-acceptance magnetic spectrometer.

Precise single and double polarization observables in inclusive scattering (e,e') along with the exclusive channels (e,e'p), (e,e'n), (e,e'd) und (e,e'π) were measured simultaneously with the BLAST detector, in order to extract e.g. the proton electric to magnetic form factor ratio G_E^p/G_M^p, the form factors of the neutron G_E^n and G_M^n, electromagnetic structure observables of the deuteron such as the tensor analyzing power T_{20} or the structure functions of the quasielastic deuteron breakup as well as pion electroproduction in the Δ resonance region.

The BLAST detector is a toroidal spectrometer with eight normal-conducting magnetic coils producing a maximum field of 0.38 T. The two opposite in-plane sectors were instrumented symmetrically with detectors to track charged particles and neutrons. The angle acceptance covered scattering angles between 20° and 80° as well as up to $\pm 15°$ out of plane. The South Hall Ring (SHR) was periodically filled with $\sim$ 200 mA of polarized electrons at 850 MeV. The beam polarization P_b of $\sim$66% was kept longitudinal at the interaction point with the help of spin rotators (Siberian Snake) in the opposite straight section of the SHR and was monitored continually with a Compton backscattering polarimeter. The ABS provided hydrogen or deuterium atoms in separate nuclear spin states with a proton polarization $P_t \sim 83\%$ and a deuteron vector (tensor) polarization of up to 89% (68%). The target spin direction was aligned with Helmholtz coils and

CP1149, *Spin Physics, 18th International Spin Physics Symposium*
edited by D. G. Crabb, D. B. Day, S. Liuti, X. Zheng, M. Poelker, and Y. Prok

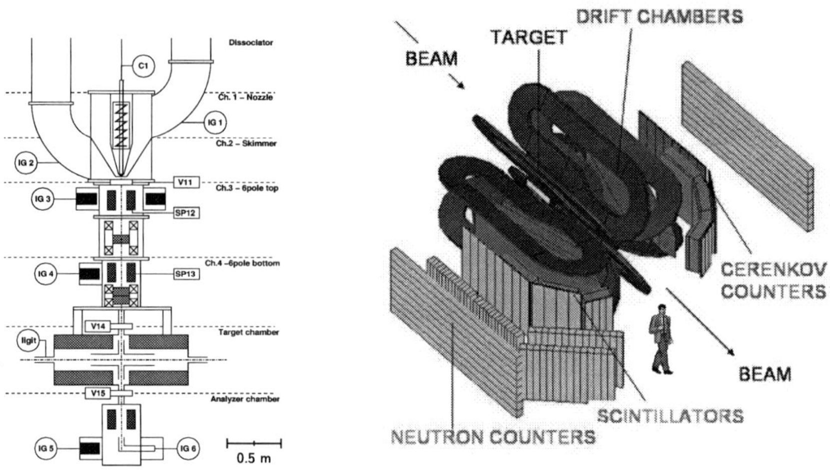

FIGURE 1. Schematic view of the Atomic Beam Source (left) and the BLAST detector (right).

carefully mapped, resulting in average values of 48.0° for the ^{1}H(e,e'p) and to 31.6° and 46.3° for the ^{2}H(e,e'n) event samples, relative to the beam direction into the left sector of BLAST. The ABS and the BLAST detector are shown schematically in Fig. 1. Details of the BLAST experimental setup can be found in [1].

With longitudinally polarized electron beam and polarized free-nucleon target, the spin-dependence of the elastic scattering cross section leads to a beam-target double asymmetry that is sensitive to the ratio of electric and magnetic form factors $G_E(Q^2)$ and $G_M(Q^2)$, viz.

$$A = -P_b P_t \frac{v_z \cos\theta^* + v_x \sin\theta^* \cos\phi^* G_E/G_M}{(\tau + \varepsilon(G_E/G_M)^2)\ /\ [\varepsilon(1+\tau)]}, \tag{1}$$

where θ^* and ϕ^* are the polar and azimuthal angles of the target polarization relative to the three-momentum transfer direction and $\tau = Q^2/(4M^2)$ with the nucleon mass M. The quantity $\varepsilon = [1 + 2(1+\tau)\tan^2(\theta_e/2)]^{-1}$ is known as the virtual photon polarization, where θ_e is the electron scattering angle, and $v_z = 2\tau\tan(\theta_e/2)\sqrt{1/(1+\tau)+\tan^2(\theta_e/2)}$, $v_x = 2\tan(\theta_e/2)\sqrt{\tau/(1+\tau)}$ are kinematic factors. Note that with the normalization $G_M(0) = \mu$, the nucleon magnetic moment, the "perpendicular" term $\propto \sin\theta^*$ alters its sign between protons ($\mu_p \simeq 2.79$) and neutrons ($\mu_n \simeq -1.91$). Also, with $G_E^n << G_E^p$, the magnitude of this term is quite different for protons and neutrons.

The elastic beam-target asymmetry has been measured with polarized protons for Q^2 between 0.15 and 0.65 (GeV/c)^2 in eight bins, each for two spin angle configurations simultaneously using the left-right symmetry of the setup. The kinematics of BLAST is such that the asymmetry with electrons scattered into the left (right) sector corresponds to $\theta^* \sim 90°$ ($\sim 0°$). The coincident elastic events were selected based on redundant kinematic information. The two asymmetry measurements in each Q^2 bin allows one to

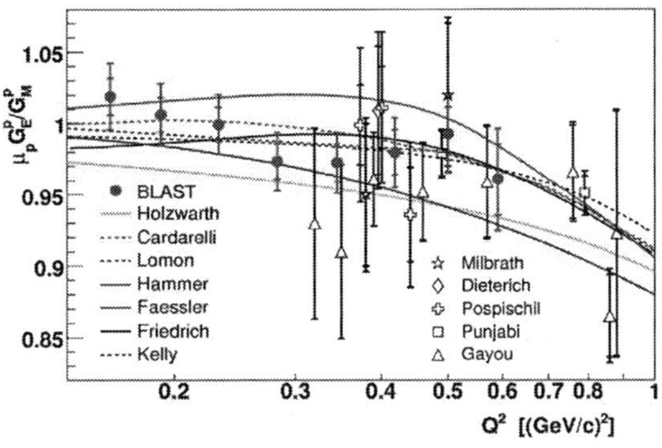

FIGURE 2. Results of $\mu_p G_E^p/G_M^p$ shown with the world polarized data [3] and several models [4, 5, 6, 7, 8, 9, 10].

extract both P_bP_t and $\mu_p G_E^p/G_M^p$ for each bin. However, since P_bP_t did not exhibit any Q^2 dependence, as expected, only one value for P_bP_t was extracted in a fit procedure together with $\mu_p G_E^p/G_M^p$ in all bins, thereby maximizing the information used toward extracting the form factor ratio. The BLAST results [2] are shown in Fig. 2 along with previous recoil polarization experiments [3] and various models [4, 5, 6, 7, 8, 9, 10]. In the Q^2 region below 0.3 (GeV/c)^2 the BLAST data mark the first measurements of $\mu_p G_E^p/G_M^p$ from a double polarization observable. In the overlap region with previous experiments the precision of the available data has been significantly improved.

Due to the lack of free neutron targets, any information on neutrons must be extracted from light nuclei such as the deuteron or ^{3}He. Instead of elastic electron-neutron scattering one has to consider e.g. quasielastic scattering in the ^{2}H(e,e'n) reaction. For nucleons in the S state, the nucleon spin is aligned with that of the vector-polarized deuteron. Nuclear corrections have to be applied for final state interactions (FSI), meson exchange currents (MEC), isobar configurations (IC) and relativistic corrections (RC), for which the electrodisintegration model of [11] has been used. The neutron elastic form factor ratio $\mu_n G_E^n/G_M^n$ has been determined with BLAST for Q^2 between 0.10 and 0.55 (GeV/c)^2 applying the model framework in a Montecarlo simulation of the experiment. The size of FSI correction at low Q^2 is substantial and its validity has been independently verified with the simultaneously measured tensor asymmetry, which is dominated by FSI and in good agreement with the data [12]. The neutron electric form factor was extracted assuming a parameterized form for G_M^n taken from [9]. The results are shown in Fig. 3. The new data from BLAST [12] significantly improve the data quality at low Q^2 and remove the large model uncertainty of G_E^n when extracted from elastic electron-deuteron scattering. The data agree very well with recent VMD fits [6, 18] and show no evidence for any narrow structure in Q^2.

Further analyses of BLAST data are in progress and will yield further form factor results on G_E^p/G_M^p from $^2\vec{\text{H}}(\vec{\text{e}},\text{e'p})$, and on G_M^n from the $^2\vec{\text{H}}(\vec{\text{e}},\text{e'})$ inclusive channel.

This work has been supported in part by the US Department of Energy and National Science Foundation.

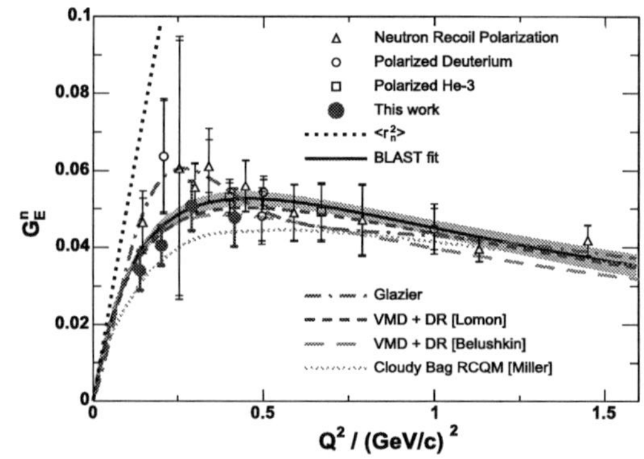

FIGURE 3. Results on G_E^n from BLAST [12] shown along with the world data from neutron recoil polarization experiments with unpolarized ^{2}H target [13] (open triangles) and experiments with polarized ^{2}H (open circles = Ref. [14]) and ^{3}He targets [15] (open squares), and with various model curves [16, 17, 9, 6, 18, 19].

REFERENCES

1. D. Hasell *et al.*, accepted by Nucl. Instrum and Meth.
2. C.B. Crawford *et al.*, Phys. Rev. Lett. **98**, 052301 (2007).
3. B. Milbrath *et al.*, Phys. Rev. Lett. **80**, 452 (1998); Erratum-ibid. Phys. Rev. Lett. **82**, 2221(E) (1999); S. Dieterich *et al.*, Phys. Lett. **B500**, 47 (2001); T. Pospischil *et al.*, Eur. Phys. J. **A12**, 125 (2001); M. Jones *et al.*, Phys. Rev. Lett. **84**, 1398 (2000); V. Punjabi *et al.*, Phys. Rev. C **71**, 055202 (2005); Erratum-ibid. Phys. Rev. C **71**, 069902(E) (2005); O. Gayou *et al.*, Phys. Rev. C **64**, 038202 (2001).
4. G. Holzwarth, Z. Phys. **A356**, 339 (1996).
5. F. Cardarelli and S. Simula, Phys. Rev. **C62**, 065201 (2000); S. Simula, Proc. of Workshop on the Physics of Excited Nucleons, World Scientific, 2001, p. 135.
6. E.L. Lomon, Phys. Rev. C **64** 035204 (2001); Phys. Rev. C **66**, 045501 (2002); arXiv:nucl-th/0609020v2.
7. H.-W. Hammer and Ulf-G. Meissner, Eur. Phys. J. **A20**, 469 (2004).
8. A. Faessler, T. Gutsche, V.E. Lyubovitskij, K. Pumsa-ard, Phys. Rev. D **73**, 114021 (2006).
9. J. Friedrich and Th. Walcher, Eur. Phys. J. A **17**, 607 (2003).
10. J.J. Kelly, Phys. Rev. C **70**, 068202 (2004).
11. H. Arenhövel, W. Leidemann, and E.L. Tomusiak, Eur. Phys. J. A **23**, 147 (2005); Phys. Rev. C **46**, 455 (1992); Z. Phys. A **331**, 123 (1988); Erratum Z. Phys. A **334**, 363 (1989).
12. E. Geis *et al.*, Phys. Rev. Lett. **101**, 042501 (2008).
13. B. Plaster *et al.*, Phys. Rev. C **73**, 025205 (2006) superseding R. Madey *et al.*, Phys. Rev. Lett. **91**, 122002 (2003); D.I. Glazier *et al.*, Eur. Phys. J. A **24**, 101 (2005); M. Ostrick *et al.*, Phys. Rev. Lett. **83**, 276 (1999); C. Herberg *et al.*, Eur. Phys. J. A **5**, 131 (1999); T. Eden *et al.*, Phys. Rev. C **50**, R1749 (1994).
14. G. Warren *et al.*, Phys. Rev. Lett. **92**, 042301 (2004); H. Zhu *et al.*, Phys. Rev. Lett. **87**, 081801 (2001); I. Passchier *et al.*, Phys. Rev. Lett. **82**, 4988 (1999).
15. J. Bermuth *et al.*, Phys. Lett. **B564**, 199 (2003) superseding D. Rohe *et al.*, Phys. Rev. Lett. **83**, 4257 (1999); J. Golak *et al.*, Phys. Rev. C **63**, 034006 (2001) applying FSI corrections to J. Becker *et al.*, Eur. Phys. J. A **6**, 4988 (1999).
16. S. Kopecky *et al.*, Phys. Rev. Lett. **74**, 2427 (1995); Phys. Rev. C **56**, 2229 (1997).
17. D.I. Glazier *et al.*, Eur. Phys. J. A **24**, 101 (2005).
18. M.A. Belushkin, H.-W. Hammer, and Ulf-G. Meißner, Phys. Rev. C **75**, 035202 (2007).
19. G.A. Miller, Phys. Rev. C **66**, 032201(R) (2002).

Experimental Validation of the Born Approximation in Proton Form Factor Measurements With Polarization Method

Lubomir Pentchev

The College of William and Mary, Williamsburg VA, USA
for the GEP-2Gamma collaboration

Keywords: Electromagnetic form factors; Polarizations in interactions and scattering
PACS: 13.40.Gp;13.88.+e

The discrepancy between the proton elastic form factor ratio data obtained at JLab using the polarization method [1, 2] and the previous measurements [3, 4] using the Rosenbluth method, questioned the validity of the lepton scattering technique to probe the nucleon structure. Both methods are based on the one-photon exchange mechanism (Born approximation) for electron-hadron scattering. Using this well established formalism, the measured cross sections or the polarization observables are directly related to the electromagnetic form factors. In this contribution the first experiment [5] aimed to study effects beyond the Born approximation at a percent level using the polarization method will be presented. The proton form factor ratio as extracted from the ratio of the proton recoil polarization components, was measured at a fixed Q^2 of 2.5 GeV^2 over a wide kinematic range. In the one-photon exchange process, the form factors depend only on Q^2 and any deviation from constant in this experiment would indicate a mechanism beyond the Born approximation. In addition, the proton longitudinal polarization was measured separately and can be compared independently to the Born approximation prediction.

In the general case, neglecting only the electron mass, elastic ep scattering is described by three complex amplitudes [6]: $\tilde{G}_M$, $\tilde{G}_E$, and $\tilde{F}_3$, the first two chosen as generalizations of the Sachs magnetic and electric form factors, G_M and G_E, and the last one, $\tilde{F}_3$, vanishing in the case of the Born approximation. While the Sachs form factors depend only on Q^2, in the general case the amplitudes depend also on a second variable, $\varepsilon = [1+2(1+\tau)\tan^2\frac{\theta_e}{2}]^{-1}$, where θ_e is the lab electron angle, and $\tau = Q^2/4M_p^2$. The reduced cross section σ_{red}, and the proton polarization transfer components P_t and P_l, can be expressed [6] as

$$\sigma_{red}/G_M^2 = 1+\frac{\varepsilon R^2}{\tau}+f_\sigma\left(Re(\tilde{G}_M),Re(\tilde{G}_E),Re(\tilde{F}_3)\right), \text{ where } R = G_E/G_M \quad (1)$$

$$P_t = -\sqrt{\frac{2\varepsilon(1-\varepsilon)}{\tau}}\frac{G_M^2 R}{\sigma_{red}}+f_t\left(Re(\tilde{G}_M),Re(\tilde{G}_E),Re(\tilde{F}_3)\right) \quad (2)$$

$$P_l = \sqrt{(1-\varepsilon^2)}\frac{G_M^2}{\sigma_{red}}+f_l\left(Re(\tilde{G}_M),Re(\tilde{G}_E),Re(\tilde{F}_3)\right), \quad (3)$$

CP1149, *Spin Physics, 18th International Spin Physics Symposium*
edited by D. G. Crabb, D. B. Day, S. Liuti, X. Zheng, M. Poelker, and Y. Prok

where the first terms represent the Born approximation, and the functions f_σ, f_t, and f_l depend on the real parts of the three amplitudes, Q^2, and ε.

In Born approximation, in the Rosenbluth method, the form factor ratio is obtained from the slope of the reduced cross-section as function of ε (Eq. 1). The radiative corrections [7] are an important part of the Rosenbluth procedure. Using the polarization technique, one can calculate (Eqs. 2,3) the form factor ratio: $R = -P_t/P_l\sqrt{\tau(1+\varepsilon)/2\varepsilon}$ from the ratio of the polarizations P_t/P_l. The polarization method doesn't require absolute cross section measurements, change of beam energy or detector angle, eliminating important sources of systematic uncertainties. Radiative corrections have been shown to be very small for polarization observables [8].

A possible explanation of the proton form factor discrepancy is the two-hard-photon exchange mechanism that is not taken into account in the standard radiative corrections [7]. In contrast to the polarization method, even small, ε-dependent correction to the cross-section, could change the Rosenbluth result significantly.

In [9] the two-photon exchange contributions are calculated at the partonic level assuming factorization of the soft nucleon-quark part, and the hard electron-quark interaction where two-photon exchange takes place via box diagram. The quarks are real and the box diagram is calculated exactly, while different Generalized Parton Distribution (GPD) models are used to calculate the nucleon-quark amplitudes, giving similar results. The GPD approach is valid at either high Q^2, or large ε.

In another approach [10, 11, 12], the two-photon exchange effects are calculated at the hadronic level. Similar two-photon exchange box diagrams are used where the quarks are replaced by nucleons. In the calculations [10] only nucleon intermediate states are taken into account. The inclusion of Δ in ref. [11] reduces the two-photon exchange correction. The contribution of higher nucleon resonances has smaller effect [12]. For the nucleon vertices in the loop integral, on-shell nucleon form-factors are used. However, it was shown [13] that using the off-shell ones makes almost no difference.

The two theoretical results [9, 10] for the corrections of the cross section and form factor ratio $\mu G_E/G_M$ as measured using polarization technique for $Q^2 = 2.5$ GeV^2, are shown together in Fig. 1 (right plot). Both theories describe the cross section data, however their predictions for the form factor ratio correction calculated from the polarization components have opposite signs. Although for Q^2 of 2.64 GeV^2 both theories can resolve the discrepancy, this is not the case for higher Q^2; at $Q^2 \sim 6$ GeV^2 the calculations explain only about half of the discrepancy.

The two-photon exchange mechanism is a possible but not yet sufficient explanation of the discrepancy. In [14] the discrepancy is explained by higher order radiative corrections that change the slope of the Rosenbluth plot.

The JLab experiment E04-019 (called GEP-2G) was performed in Hall C using a longitudinally polarized electron beam from the CEBAF machine. The proton from a elastic *ep* scattering was registered in the High Momentum Spectrometer (HMS). A new double polarimeter placed in the focal plane was used to measure the proton polarization. The elastic electron was registered in the newly constructed big (1744 channels) lead-glass calorimeter (BigCal) matching the acceptance of the HMS.

The key idea of the experiment, is to study the ε dependence of the recoil proton polarization at a fixed value of Q^2, and therefore at a fixed proton momentum. It means, first, that the analyzing power , A_y, of the reaction used to measure the proton polarization, is the same for all the kinematics. This makes it possible for the first time to measure separately the longitudinal polarization, P_l, with a point to point uncertainty less than one percent. Would the proton momenta be different the precision would be dominated by the uncertainty in the absolute value of the analyzing power which is of the order of 10%. Second, since the setting of the HMS is fixed, the transport of the spin from the focal plane were it was measured, back to the target is the same for all the data points. This is a significant advantage in the measurement of the polarization ratio P_t/P_l compared the standard polarization experiments, where the error is dominated by the spin transport (since the analyzing power cancels out in the ratio).

For the preliminary results presented here, only a fraction of the statistics was used. Tight cuts were applied to separate the elastic events so that the contribution from the inelastic reactions is reduced down to a fraction of a percent. In addition, the momentum acceptance of the HMS for the high ε points was restricted to coincide with the acceptance of the lowest ε point which makes it easier to compare the results at the different kinematics without applying additional corrections. No radiative corrections were applied.

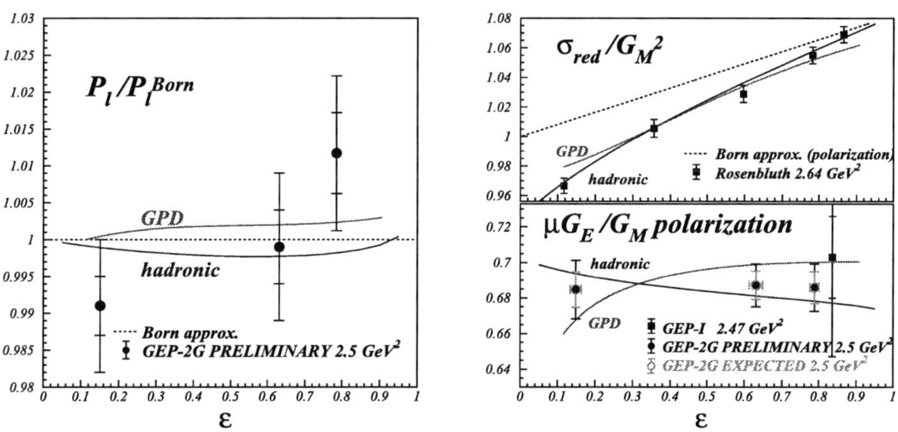

FIGURE 1. Left:Preliminary longitudinal polarization from this experiment normalized to the Born approximation (statistical (inner) and p.t.p. systematical (outer) errors shown) as a function of ε compared to the GPD [9] and hadronic [10] calculations. The mean value of all data points is normalized to one. Right Top: Normalized reduced cross-section as measured in [4] (data points) compared to the predicted from the the polarization data [2] Born approximation (dashed line), and to the two-photon exchange corrections [9] (GPD), and [10] (hadronic). Right Bottom: Form factor ratio as measured in GEP-I [2] and present GEP-2G experiment (solid symbols are preliminary results, open symbols represent expected final errors) compared to the same [9] and [10] theoretical predictions, separately normalized to the data (see text).

Figure 1 shows the preliminary results of the presented experiment. For the ratio of the measured longitudinal polarization to the Born approximation P_l/P_l^{Born} (left plot), the systematic errors are dominated by the point to point uncertainty in the beam polarization measurements estimated to be less than 0.5%. Both hadronic [10] and GPD

[9] calculations predict very small deviations from Born approximation. The data points are consistent with a constant. The data points for the form factor ratio (right bottom panel) as extracted from the polarization ratio, lay almost on the same constant line thus showing no deviations from the Born approximation at a percent level. **This supports the validity of the polarization method in measuring the proton form factor ratio.** On the same plot the two theoretical predictions [10, 9] are drawn. These are corrections to the Born approximation. That is why when compared to the data, different values of the Born form-factor ratio were used in the two theoretical curves to fit the data points. Even though the results favor the hadronic calculations, within the present statistical errors the GPD prediction is not excluded. On the same plot the expected final errors are shown. They are small enough allowing to ultimately distinguish between the two theories.

The precise measurements in this experiment of two polarization observables at Q^2 of 2.5 GeV^2, combined with the cross-section data [4] at a slightly higher value of 2.64 GeV^2, allow to address [15] the problem of the reconstruction of the real part of the three elastic amplitudes using exclusively the data. The two polarization quantities of the present measurements and the e^+p/e^-p cross-section ratio are sensitive to different amplitude combinations (Eqs. 1,2,3) and therefore, complementary in investigating the effects beyond the one-photon exchange approximation.

I thank my colleague C.F. Perdrisat for the important discussions of the contribution; the work was supported by U.S. NSF grant, PHY-0753777. Authored by Jefferson Science Associates, LLC under U.S. DOE Contract DE-AC05-06OR23177.

REFERENCES

1. M. K. Jones *et al.*, Phys. Rev. Lett. **84** 1398 (2000); O. Gayou *et al.*, Phys. Rev. Lett. **88** 092301 (2002).
2. V.Punjabi, C.F.Perdrisat et al. Phys. Rev. C71 055202 (2005); Erratum-ibid. C71 069902 (2005).
3. L. Andivahis *et al.*, Phys. Rev. D **50** 5491 (1994); M.E.Christy *et al.*, Phys. Rev. **C70** 015206 (2004); J.Arrington, Phys. Rev C **69**, 022201 (2004).
4. I.A. Quatan *et al.*, Phys. Rev. Lett. **94** 142301 (2005).
5. R.Gilman, L.Pentchev, C.F.Perdrisat, and R.Suleiman, "Measurements of the Two-photon Exchange Contribution in *ep* Elastic Scattering using Recoil Polarization", Jefferson Lab experiment 04-019 (2004).
6. P.A.M. Guichon and M. Vanderhaeghen, Phys. Rev. Lett. **91** 142303 (2003).
7. L.W. Mo and Y.S. Tsai, Rev. Mod. Phys. **41** 205 (1969); Y.S. Tsai, Phys. Rev. **122** 1898 (1961).
8. A.V. Afanansev *et al.*, Phys Lett. B **514**, 269 (2001).
9. Y.-C.Chen, A.Afanasev, S.J.Brodsky, C.E.Carlson, and M.Vanderhaeghen, Phys. Rev. Lett. **93** 122301 (2004); A.Afanasev, S.J.Brodsky, C.E.Carlson, Y.-C.Chen,and M.Vanderhaeghen, Phys.Rev. D72 013008 (2005).
10. P.G. Blunden, W.Melnitchouk and J.A.Tjon, Phys. Rev. Lett. **91** 142304 (2003).
11. S.Kondratyuk, P.G. Blunden, W.Melnitchouk and J.A.Tjon, Phys. Rev. Lett. **95** 172503 (2005).
12. S.Kondratyuk and P.G. Blunden, Phys. Rev. **C75** 038201 (2007) .
13. D.Borisyuk and A.Kobushkin, arXiv:0804.4128.
14. Yu.M.Bystritskiy, E.A.Kuraev and E.Tomasi-Gustafsson, Phys.Rev. C75 015207 (2007).
15. L.Pentchev, Proceedings of the Sixth International Conference on Perspectives in Hadronic Physics, (invited talk) 12- 16 May 2008, Trieste, Italy, ed. by S.Boffi, C.Ciofi degli Atti, M.Giannini, D.Treleani, pp 357-364, AIP Conference Proceedings 1056 (2008).

Magnetic Moments of Baryons and Exotics in Quasi Particle Diquark Model

A.Bhattacharya, A. Sagari, B.Chakrabarti, S.Mani

Department of Physics
Jadavpur University
Calcutta 700032, India.

Abstract. A quasi particle model for the diquark is suggested in an analogy with the quasi particle model in the crystal lattice. The magnetic moments of the exotics and the non exotic members of the baryon antidecuplet have been estimated considering the diquark-diquark- antiquark picture for the baryons. The masses of the baryon antidecuplet have also been estimated. The results are found to be in agreement with the existing theoretical and experimental results.

Keywords: Baryon magnetic moments, exotic baryon states
PACS: 14.20-c; 14.65-q; 14.20Dh; 12.39Pn

INTRODUCTION

The magnetic moment is an important quantum number for studying the internal structure of a particle. At the theoretical level the baryon antidecuplet has been represented as the third rotational state excitation in the context of the soliton model. The soliton picture of baryons suggested the existence of a baryon antidecuplet with spin $\frac{1}{2}$ as rotational excitation in addition to low lying $SU_{fl}(3)$ octet and decuplet [1]. It is interesting to identify the non exotic members of the baryon anti decuplet as the baryon resonances.The magnetic moments of the antidecuplet baryons have been investigated by a number of authors considering pentaquark configuration [2,3]. In the present work we have studied the magnetic moment and mass of the baryon antidecuplet considering the diquark-diquark-anti quark picture describing a diquark as a quasi particle. The spin contribution of the diquarks to the $\triangle$ - N has been computed attributing the mass difference as mass difference of scalar-axial vector diquarks.

QUASI PARTICLE PICTURE OF DIQUARK

The role of diquark in baryon spectroscopy have been discussed by number of authors [4]. We have described a diquark in which two quarks bound together forming a quasi particle in an analogy with an electron in crystal lattice [5]. The quasi particles are particle like entity arising in some system of interacting particles. They are low lying excited state possessing the energy very close to the ground state and the properties of the system can be obtained by investigating the properties of the quasi particles to a considerable extent. Such states are often observed in the usual condense matter physics. A crystal electron is subjected to two type of forces namely the effect of crystal field

CP1149, *Spin Physics, 18th International Spin Physics Symposium*
edited by D. G. Crabb, D. B. Day, S. Liuti, X. Zheng, M. Poelker, and Y. Prok

(∇V) and an external force (F) which accelerates the electron [9]. Under the influence of the two forces, an electron in a crystal behaves like a quasi particle whose effective mass m^* reflects the inertia of electrons which are already in a crystal field and the bare electrons (with normal mass m) are affected by the lattice force -∇V, Hence the ratio of the normal mass (m) to the effective mass (m^*) can expressed as:[5]

$$m/m^* = 1 - \frac{1}{F}[\frac{\delta \overline{V}}{\delta x}] \tag{1}$$

We have proposed a similar type of picture for diquark $[ud]_0$ as a quasi particle inside a baryon. It is now well understood experimentally that the hadrons are complicated system and complete description includes with virtual sea, gluons, $\bar{s}$ in addition to the valance quarks. So situation may be resembled to the situation of an electron in the crystal lattice. We assume that the diquark is a fundamental constituent and an independent body which is under the influence of two type of forces inside the hadron which simulates the many body interaction of the complicated structure of the hadrons. One is due to the background meson cloud which is represented by potential $\overline{V}$ =-(2/3)α_s/r and resembles the crystal field on a crystal electron. For the external force we have considered an average harmonic oscillator type of force which is of confinement type. The potential may be expressed as;

$$V_{ij} = -\frac{\alpha}{r} + (F_i F_j)(-\frac{1}{2}Kr^2) \tag{2}$$

where the coupling constant α = $\frac{2}{3}\alpha_s$; $F_i F_j$ = -$\frac{2}{3}$ for qq interaction [12]and K is the strength parameter.The ratio of the constituent mass and the the effective mass of the diquark can be estimated by using the same formalism as in equation (3) in quasi particle approximation and can be expressed as:

$$\frac{m}{m^*} = 1 + \frac{\alpha}{2ar^3} \tag{3}$$

where a=K/3,m is constituent mass and m*is the effective mass of the diquark, r-radius parameter of the diquark. To calculate the effective mass of the diquark we need the radius parameter r of the diquark which is not exacly known. We have considered the radius of the $[ud]_0$ diquark as the effective radius of the π meson as r_{ud} = 5.38 GeV^{-1} [6]whereas the radius of the $[us]_0$ diquark is computed as r_{us}=6.06 GeV^{-1} fitting the experimental mass of the Ξ with two $[us]_0$ diquark and one $\bar{u}$ configuration. With m_u= m_d = 360 MeV,m_s = 540 MeV,α = $\frac{2}{3}\alpha_s$ = .393 as α_s = 0.58, K= 241.5 MeVfm^{-2} [7] the diquarks masses have been estimated as $[m_{ud}]$= 506 MeV and $[m_{us}]$ =700 Mev. It may be mentioned that in QCD both gluon exchange interaction and instanton induced interaction favour spin singlet and colour antisymmetric diquark combination and predicts the diquark mass as 420 MeV. Magnetic Moment of multiquark states:-

Considering diquark as the boson like particle with spin zero as in JW model [3], the magnetic moment for pentaquark system with two scalar diquark and one antiquark system can be written as [2],

$$\overrightarrow{\mu} = \overrightarrow{l_1}\mu_1 + \overrightarrow{l_2}\mu_2 + g_3 \frac{\overrightarrow{1}}{2}\mu_3 \tag{4}$$

As diquarks are considered to be spin zero particle the magnetic moment arises due to the relative momentum of the diquarks and the spin of the antiquark. Considering the wave function for diquarks by Liu et al [2]we have estimated the magnetic moment of the baryon antidecuplet members and displayed in Table-1 along with the results from JW model[3]with two sets of diquark mass [3,8].The magnetic moment is very sensitive to diquark mass.Our results show agreement with the set-2 [3].

ANTIDECUPLET MASS

To estimate the mass of the antidecuplet we have considered the pentaquark configuration as in [3]. The diquark in the present model has been considered to be independent body behaving as quasi particle analogous to the many body systems which resembles the low lying excitation and as indicated earlier they are non interacting entities. Therefore as in the case of ideal gas of quasi particle , their energies may be assumed to be additive [9] and their effective mass characterises the dynamic properties of the quasi particle. Hence the mass of the pentaquarks can be represented as:

$$M_x = 2M_D + M_{\overline{q}} \tag{5}$$

where M_D is the mass of the diquark. The results are displayed in Table 2 along with results of Jaffe et al [3].Recently we have suggested diquark picture of nucleon (proton) considering it as diquark-quark system [10]. In the framework of the diquark picture the Δ-N mass difference may be attributed to the spin contribution of the scalar and axial vector diquark . The spin contribution runs as[11]:

$$E_{SD} = \frac{8}{9}\frac{\alpha_s}{m_1 m_2}\vec{s_1}.\vec{s_2}|\psi(0)^2| \tag{6}$$

with radius parameter r_{ud} for axial vector as 0.8fm from Nagata et al [12] the mass difference between the axial vector and scalar diquark and their spin contribution is obtained as:

$$M_\Delta - M_N = 288MeV \tag{7}$$

The experimental prediction is M_Δ -M_N $\sim$ 300MeV

CONCLUSIONS

The diquark has been described as a fundamental building block of hadrons, and as a quasi particle which embodies the interaction of the background medium. These are low lying excited states behaving as independent particles. The situation is somewhat like the phonon particle we come across in the context of superfluidity in usual superfluid like He-II as suggested by Landau. Each phonon creates an environment around itself so that it behaves like a free particle. It may be suggested that the superfluid nature of the vacuum may favour the formation of the diquark and its quasi particle behaviour. The results obtained are in good agreement with the current experimental results. However it

TABLE 1. AntiDecuplet Magnetic Moment

Y,I,I_3	*SetI*	*SetII*	*SetIII*
$2,0,0$	0.08	0.29	0.219
$1,\frac{1}{2},\frac{1}{2}$	0.037	0.22	0.155
$1,\frac{1}{2},\frac{-1}{2}$	0.12	0.22	0.195
$0,1,1$	-0.009	0.14	0.085
$0,1,0$	0.06	0.12	0.1004
$0,1,-1$	0.13	0.09	0.116
$-1,\frac{3}{2},\frac{3}{2}$	-0.06	0.06	0.008
$-1,\frac{3}{2},\frac{1}{2}$	0	0	0
$-1,\frac{3}{2},\frac{-1}{2}$	0.06	-0.06	-0.008
$-1,\frac{3}{2},\frac{-3}{2}$	0.12	-0.12	-0.017

TABLE 2. AntiDecuplet Mass

Particle	$JW[3]$	*OurResults*
θ^+	1540	1552
N	1440	1372
Σ	1660	1566
Λ	1600	1566
N_s	1700	1746
Σ_s	1890	1940

should be mentioned that accurate measurements of the properties of baryon resonances, and consequently a full compariaon with our model are not complete.

Acknowlwdgement: The authors are thankful to the Department of Science and Technology, India for financial assistance.

REFERENCES

1. G.S.Adkins et al., Nucl.Phys. **B 228**, 552 (1983); D.Diakonov et al., Z.Phys.**A 359**, 305 (1997).
2. Y.R.Liu et al;hep-ph/0312074 and references therein; M.V.Polyakov et al.,Eur.Phys.J.**A 18** 691 (2003).
3. R.L.Jaffe and F.Wilczek.,Phys.Rev.Lett **91**, 232003 (2003).
4. A.S.Castro et al.,Z.Phys.**C 57**, 315 (1993); M.Anselmino et al.,Rev.Mod.Phys**65** 1199 (1993); K.Nagata et al.,Prog.Theor.Phys.**111** 857 (2004); E.V.Shuryak.,Nucl.Phys.**B 50** 93 (1982); G. R. Goldstein and J. Maharana, Nuovo Cim. A **59**, 393 (1980); T.Schafer et al.,Rev.Mod.Phys **70** 323 (1998).
5. A.Haug: Theoretical Solid State Physics,p-100,Pergamon,Elmsford (1972).
6. S.N.Banerjee et al.,Int.J.Mod.Phys. **A2**, 1829 (1987).
7. W. Lucha et al ., Phys.Rep.**200**,168 (1991); R.K.Bhaduri et al.; Phys.Rev. Lett.**44)**1369 (1980).
8. M. Karliner and H.J.Lipkin., hep-ph/0307243.
9. L.D.Landau and E.M.Lifshitz, Statistical Physics, vol-5, 3rd Ed.Part-1, p-216.
10. A.Bhattacharya et al.,Int.J.Theor.Phys.DOI 10.1007/s10773-008-9684-3.
11. A.K. Rai et al;Nucl.Phys. **A 782** 406C (2007).
12. K.Nagata et al., annurep/2006/sec 2/nagata.pdf.

Experimental overview of Generalized Parton Distribution results from HERMES

B. Zihlmann

Jefferson Lab, 12000 Jefferson Avenue, Newport News, VA 23606
ON BEHALF OF THE HERMES COLLABORATION

Abstract. Over the course of more than a decade the HERMES experiment has accumulated a wealth of data with electron and positron beams on various gaseous targets from Hydrogen up to Xenon. In addition, the beams and targets can be polarized. This data set is viewed in the context of Generalized Parton Distributions, a theoretical formalism with an explicit three dimensional view of the structure of the nucleon. It provides a link between experimental observables and the total angular momentum of the quarks in the nucleon.

Keywords: GPD, DVCS, Nucleon structure
PACS: 21.10.Hw, 25.30.Rw, 25.30.Hm

INTRODUCTION

Hard exclusive reactions in lepton scattering can be expressed in terms of Generalized Parton Distributions (GPDs) [1, 2, 3]. GPDs are the probability amplitude for emitting a quark with longitudinal momentum fraction $x+\xi$ and absorbing a quark with longitudinal momentum fraction $x-\xi$ (fig. 1). The parton distribution functions (PDFs), known from inclusive and semi-inclusive deep-inelastic scattering, and the nucleon elastic form factors (FFs) are special limiting cases and moments of GPDs, respectively [2]. FFs are derived from elastic scattering experiments and describe the transverse distribution of the partons inside the nucleon, while PDFs describe the longitudinal momentum distribution of these partons. The combination of these two aspects of the nucleon structure is contained in the GPDs thereby providing a three dimensional picture of the nucleon [4, 5, 6]. In addition, a relation between the second moment of the GPDs H and E and the total angular momentum of the quarks in the nucleon is provided by the Ji relation [2]. The GPD $H^q(x,\xi,t)$ represents the unpolarized nucleon helicity non-flip GPD, while $E^q(x,\xi,t)$ represents the unpolarized nucleon helicity flip GPD. The Ji relation sparked a growing interest in measuring hard exclusive processes in lepton scattering. In this context the HERMES experiment [7] is well suited to measure azimuthal asymmetries of such reactions with its large acceptance using polarized electron or positron beams in combination with a longitudinal or transversely polarized Hydrogen target or with unpolarized nuclear targets like Deuterium, Nitrogen, Neon, Krypton and Xenon.

CP1149, *Spin Physics, 18th International Spin Physics Symposium*
edited by D. G. Crabb, D. B. Day, S. Liuti, X. Zheng, M. Poelker, and Y. Prok

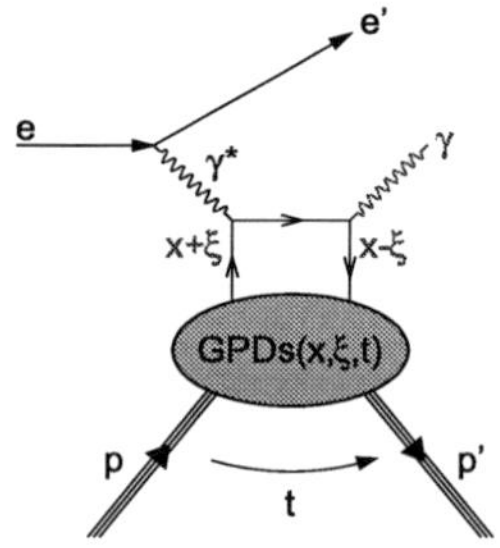

FIGURE 1. Handbag diagram of the DVCS process.

DEEPLY VIRTUAL COMPTON SCATTERING

Deeply Virtual Compton Scattering (DVCS) is the theoretically least complicated exclusive reaction to access GPDs. DVCS is the process where the virtual photon is absorbed by the struck quark which subsequently radiates a real hard photon before recombining with the spectator quarks to form the final state nucleon, as is depicted in the handbag diagram of figure 1. Because this reaction has the same final state as the Bethe-Heitler (BH) process where the photon is radiated by the lepton beam particle rather then the struck parton (see figure 1), the cross section

$$d\sigma \propto |\mathcal{T}_{BH}|^2 + |\mathcal{T}_{DVCS}|^2 + (\mathcal{T}^*_{BH}\mathcal{T}_{DVCS} + \mathcal{T}^*_{DVCS}\mathcal{T}_{BH}) \tag{1}$$

involves an interference term $\mathcal{I} = \mathcal{T}^*_{BH}\mathcal{T}_{DVCS} + \mathcal{T}^*_{DVCS}\mathcal{T}_{BH}$ that allows direct access to the DVCS amplitude and the phase. The individual contributions to the cross section can be expressed in terms of series of Fourier components and is explicitly given in equation 2 up to twist-3 as

$$\begin{aligned}
|\mathcal{T}_{BH}|^2 &\propto c_0^{BH} + \sum_{n=1}^{2} c_n^{BH}\cos(n\phi) + s_1^{BH}\sin(\phi) \\
|\mathcal{T}_{DVCS}|^2 &\propto c_0^{DVCS} + \sum_{n=1}^{2} c_n^{DVCS}\cos(n\phi) + \lambda s_1^{DVCS}\sin(\phi) \\
\mathcal{I} &\propto \pm\left(c_0^I + \sum_{n=1}^{3} c_n^I\cos(n\phi) + \lambda\sum_{n=1}^{2} s_n^I\sin(n\phi)\right)
\end{aligned} \tag{2}$$

for an unpolarized nucleon target where ϕ is the azimuthal angle between the lepton scattering plane and the photon production plane as defined by the virtual and real photon in DVCS. The term λ indicates the beam polarization and the term $\pm$ indicates the beam charge for electron or positron beams.

Azimuthal Asymmetries

To gain knowledge about the GPD H, one can use data taken with and unpolarized target and form azimuthal asymmetries with respect to either the beam charge or the beam polarization. This projects out different Fourier components of the interference term that depend on the real or imaginary part of the Compton Form Factor $\mathcal{H}$ [8][9]. The CFF $\mathcal{H}$ is a convolution of the GPD H and the hard scattering kernel. Explicitly, the beam charge asymmetry (BCA) exhibits a $\cos(\phi)$ modulation of the azimuthal asymmetry proportional to the real part of $\mathcal{H}$, while the beam spin asymmetry (BSA) exhibits a $\sin(\phi)$ modulation of the azimuthal asymmetry and is proportional to the imaginary part of the $\mathcal{H}$. In a combined analysis of data with polarized electron and positron beams the ratio of the beam-polarized cross section to the unpolarized cross section can be written as a sum of asymmetries

$$\frac{\sigma_{LU}}{\sigma_{UU}} = 1 + e_l \cdot A_C(\phi) + P_b \cdot A_{LU}^{DVCS}(\phi) + e_l \cdot P_b \cdot A_{LU}^{I}(\phi). \tag{3}$$

The first letter in the subscript of σ_{XY} refers to the beam polarization state while the second letter refers to the target polarization state using U for unpolarized and L for longitudinally polarized. The sign of the lepton beam charge is represented by e_l and the beam polarization by P_b. The asymmetries are calculated from the data as

$$\begin{aligned}
A_C(\phi) &= \frac{Y^{+\rightarrow} - Y^{-\rightarrow} + Y^{+\leftarrow} - Y^{-\leftarrow}}{Y^{+\rightarrow} + Y^{-\rightarrow} + Y^{+\leftarrow} + Y^{-\leftarrow}} \\
A_{LU}^{DVCS}(\phi) &= \frac{Y^{+\rightarrow} + Y^{-\rightarrow} - Y^{+\leftarrow} - Y^{-\leftarrow}}{Y^{+\rightarrow} + Y^{-\rightarrow} + Y^{+\leftarrow} + Y^{-\leftarrow}} \\
A_{LU}^{\mathcal{I}}(\phi) &= \frac{Y^{+\rightarrow} - Y^{-\rightarrow} - Y^{+\leftarrow} + Y^{-\leftarrow}}{Y^{+\rightarrow} + Y^{-\rightarrow} + Y^{+\leftarrow} + Y^{-\leftarrow}}
\end{aligned} \tag{4}$$

with $Y^{+(-)\rightarrow(\leftarrow)}$ being the yield of DVCS events with positron (electron) beam polarized parallel (anti-parallel) to the beam direction. Using the maximum likelihood method in combination with equations 2-4, allows for a simultaneous extraction of all the contributing moments. The resulting moments of the beam charge asymmetry are shown in figure 2. The amplitudes have been extracted separately as a function of the momentum transfer to the nucleon $-t$, the Bjorken scaling variable x_B and the negative square of the 4-momentum of the virtual photon Q^2, respectively. The main contribution to the BCA is the $\cos(\phi)$ term. The accuracy of the measurement allows for a rejection of certain GPD models and their parameters. In particular, the usage of a D-term in the VGG model [10, 12] is disfavored (the t-dependence in the VGG calculation shown here is modeled in a Regge-motivated ansatz), as well as a factorized ansatz for the t-dependence in both the VGG model and the dual model [11]. In the case of the BSA (not shown) the VGG model overshoots the data for the $\sin(\phi)$ moment of the interference term by almost a factor of two and this is the case regardless of the anstaz being taken for the t-dependence. A similar behavior is expected from the Dual model [14]. These examples show that the HERMES results on DVCS provide valuable information to reject GPD models and can serve as guidance in future model developments.

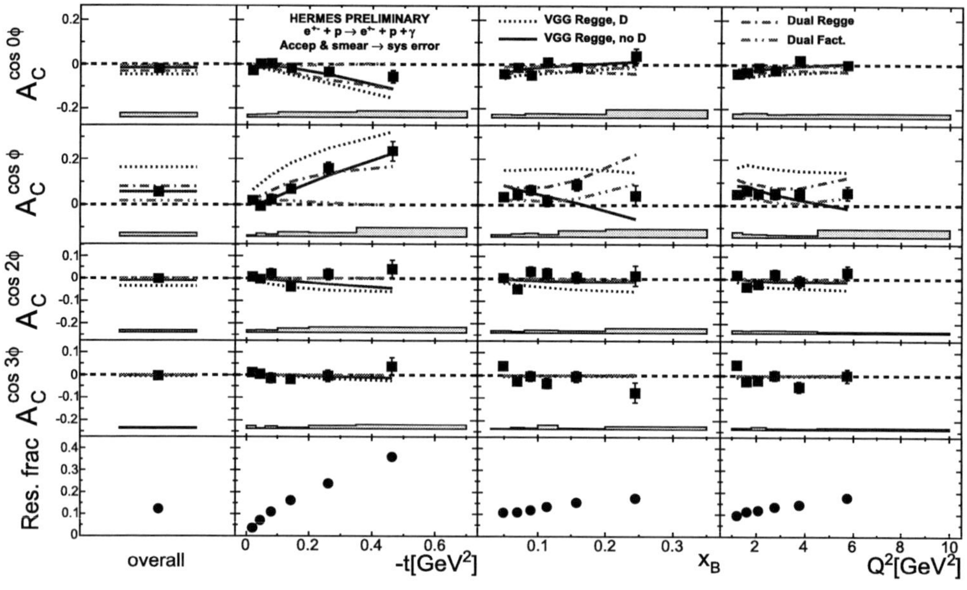

FIGURE 2. Moments of the beam charge asymmetry obtained from a maximum likelihood fit to the data.

GPD E

All observables discussed in the previous section are sensitive to the GPD H. In order to extract information about the total angular momentum of quarks in the nucleon using the Ji relation [2] it is necessary to measure observables that are sensitive to the GPD E. One of the rare observables in the HERMES regime, where the contribution from the GPD E is not kinematically suppressed, is the transverse target spin asymmetry (TTSA). This observable requires a transversely polarized nucleon target. The HERMES experiment used such a target between the years 2002 and 2006 in combination with polarized electron and positron beams. Similar to the case with longitudinal beam polarization as given by equation 3, the cross section with respect to the transverse target polarization can be written as

$$\frac{\sigma_{UT}}{\sigma_{UU}} = 1 + e_l \cdot A_C(\phi, \phi_S) + P_T \cdot A_{UT}^{DVCS}(\phi, \phi_S) + e_l \cdot P_T \cdot A_{UT}^{I}(\phi, \phi_S), \tag{5}$$

where the subscript T refers to a transversely polarized target. As a consequence of the transverse target polarization an additional azimuthal angle ϕ_S is introduced, that defines the angle between the lepton scattering plane and the polarization vector of the target nucleon. Similarly to equation 4 three asymmetries can be calculated. Equation 2 for the Fourier terms needs to be extended to include transverse target contributions [13]. In particular, the interference term will lead to a $\sin(\phi - \phi_S) \cdot \cos(\phi)$ modulation of the cross

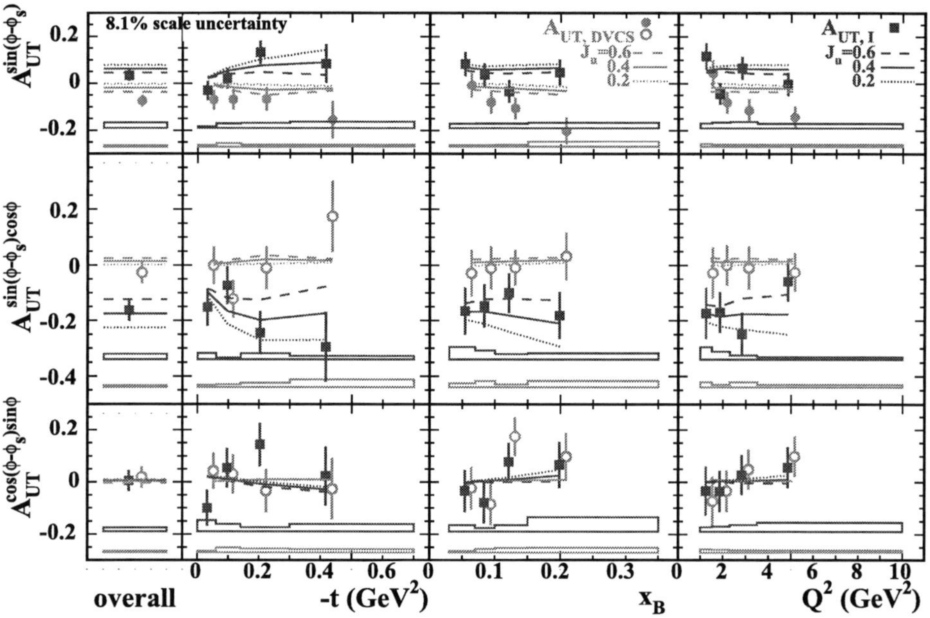

FIGURE 3. Moments of TTSA for the DVCS and interference term obtained from a maximum likelihood fit to the data.

section that can be projected out by the TTSA and is proportional to $Im[F_2\mathcal{H} - F_1\mathcal{E}]$. F_1 and F_2 are the Dirac and Pauli form factors and $\mathcal{E}$ is the CFF of E. The full transverse data set is analyzed using the maximum likelihood method in the same way as shown above. The resulting ϕ and ϕ_S dependent modulations from the DVCS and interference terms are shown in figure 3. As in the previous case, these amplitudes have been extracted as a function of $-t$, x_B and Q^2. The squares in figure 3 represent the moments from the interference term, while the circles represent those from the pure DVCS term. The filled symbols are related to twist-2 contributions while the open circles are related to twist-3 terms. The most important moment related to the interference term, is the $\sin(\phi - \phi_S) \cdot \cos(\phi)$ moment and it is significantly different from zero. The three different curves, solid, dotted and dashed, represent VGG-model calculations using different values for the total angular momentum of the u-quark J_u, while leaving the d-quark contribution at zero. Such comparisons give an impression to which degree the current data set can help to understand the contribution from orbital angular momenta of quarks to the spin of nucleon.

Deeply Virtual Compton Scattering on Nuclei

The HERMES experiment collected also data with a deuterium target and other heavier nuclei: He, N, Ne, Kr, Xe. At small values of $-t$ coherent processes dominate the cross section, while at larger values the incoherent scattering on protons and neutrons dominates. Comparing results on nuclei to those on hydrogen allows to investigate nuclear dependencies. Data at low $-t$ is selected in such a way that the mean value of Q^2 and x_B of each data sample on all different targets is approximately the same. This approach however results in different combinations of coherent to incoherent contributions in the data sample for different nuclear targets. Therefore the term "coherent enriched" has been created to emphasize the fact that the data sample is a mixture of coherent and incoherent processes and its ratio depends on the nuclear target while keeping the average kinematic variables fixed. The resulting ratios for the data samples of the incoherent enriched part to that of hydrogen are consistent with unity while for the coherent enriched part the ratios are systematically larger than unity for all nuclear targets.The fit value found for the coherent enriched sample is consistent with a very basic prediction of 5/3 for spin-0 and spin-1/2 targets, based on the ratio of the involved valence-quark charges squared [15], as well as with calculations done specifically for neon and krypton [16].

EXCLUSIVE MESON PRODUCTION

Requiring a meson instead of a hard photon in the final state provides an independent process to study GPDs. In the case of exclusive pseudo-scalar meson production (π^+, π^o) observables are sensitive to the polarized GPDs $\tilde{H}$ and $\tilde{E}$, while in the case of vector meson production (ρ, ϕ, ω) observables are sensitive to the GPDs H and E. These processes are somewhat more complicated than DVCS because the vertex of the outgoing meson involves a soft part as the struck quark has to fragment into a meson. In addition factorization is proven to hold only for longitudinal virtual photons, contrary to DVCS where it is also proven for transverse photons. In the case of the ρ^o an L/T separation is feasible when assuming s-channel helicity conservation and separating longitudinal and transverse ρ^o-mesons by their angular decay distributions.

REFERENCES

1. D. Müller, et al., *Fortschr. Phys.*, **42** 101 (1994).
2. X. Ji, *Rev. Letters*, **78** 610 (1997); *Phys. Rev.*, **D55** 7114 (1997).
3. A. V. Radyushkin, *Phys. Lett*, **B380** 417 (1996); *Phys. Rev.*, **D56** 5524 (1997).
4. M. Burkardt, *Phys. Rev.*, **D62** 071503 (2000); Erratumž013ibid. **D66** 119903 (2002).
5. A. V. Belitsky and D. Müller, *Nucl. Phys.*, **A711** 118 (2002).
6. P. Ralston and B. Pire, *Phys. Rev.*, **D66** 111501 (2002).
7. K. Ackerstaff et al., *Nucl. Instr. and Meth.* **A417** 230 (1998).
8. M. Diehl et al., *Phys. Lett.* **B411** 193 (1997).
9. A. V. Belitsky, D. Müller and A. Kirchner, *Nucl. Phys.* **B629** 323 (2002).
10. M. Vanderhaeghen, P. A. M. Guichon and M. Guidal, *Phys. Rev.* **D60** 094017 (1999).
11. V. Guzey and T. Teckentrup, *Phys. Rev.* **D74** 054027 (2006).
12. K. Goeke, M. V. Polyakov and M. Vanderhaeghen, *Prog. Part. Nucl. Phys.* **47** 401 (2001).
13. M. Diehl, *Phys. Rept.* **388** 41 (2003).
14. V. Guzey and T. Teckentrup, *hep-ph* 0810.3899
15. A. Kirchner and D. Müller, *Eur. Phys. J.* **C32** 347 (2003).
16. V. Guzey and M. Strikman, *Phys. Rev.* **C68** 015204 (2003).

Status and Prospects of the HERMES Recoil Detector

Andreas Mussgiller
(On Behalf of the HERMES Collaboration)

Deutsches Electronen-Synchrotron DESY, 22607 Hamburg, Germany

Abstract. Hard exclusive processes provide access to generalized parton distributions (GPDs), which extend our description of the nucleon structure beyond the standard parton distributions. The Deeply Virtual Compton Scattering (DVCS) process provides the theoretically cleanest access to the GPDs. For the final two years of data taking, a recoil detector was installed at HERMES for the purpose of improving the ability to measure hard-exclusive processes. In addition the recoil detector allows one to measure the individual background contributions which can be used to refine previously published results on DVCS. The progress of the ongoing data analysis is presented.

Keywords: DVCS, Recoil Detector, GPDs, Exclusive reactions
PACS: 13.60.-r, 13.60.Fz, 14.20.Dh

INTRODUCTION

The formalism of generalized parton distributions (GPDs) allows a complete description of nucleon structure. In different limiting cases the GPDs incorporate the well-known nucleon form factors determined from elastic scattering as well as parton distributions functions (PDFs) determined from measurements of inclusive and semi-inclusive deep inelastic lepton-nucleon scattering (DIS and SIDIS, respectively). Interest in the GPD framework has also strengthened because of the fact that GPDs encode the individually unknown total angular momentum of quarks and gluons within the nucleon (J_q and J_g, respectively). With the knowledge of the quarks' spin contribution to the spin of the nucleon, knowledge of the GPDs also allows, in principle, access to the orbital angular momentum of quarks (L_q) [1, 2].

The theoretically cleanest process to constrain GPDs is Deeply-Virtual Compton Scattering (DVCS), in which a highly virtual photon (emitted by the incoming beam lepton) is absorbed by a parton of the target nucleon and produces a single real photon in the final state along with the recoiling nucleon in its ground state.

The DVCS process has the same final state as the Bethe-Heitler (BH) process, in which a real photon is radiated by either the incident or the scattered lepton. At HERMES kinematics the BH process is the dominant contribution to the exclusive single photon production cross section. However, the DVCS amplitude can be accessed via DVCS-BH interference by measuring various cross section asymmetries and their dependence on the azimuthal angle ϕ, which is defined as the angle between the lepton scattering plane and the photon production plane.

CP1149, *Spin Physics, 18th International Spin Physics Symposium*
edited by D. G. Crabb, D. B. Day, S. Liuti, X. Zheng, M. Poelker, and Y. Prok

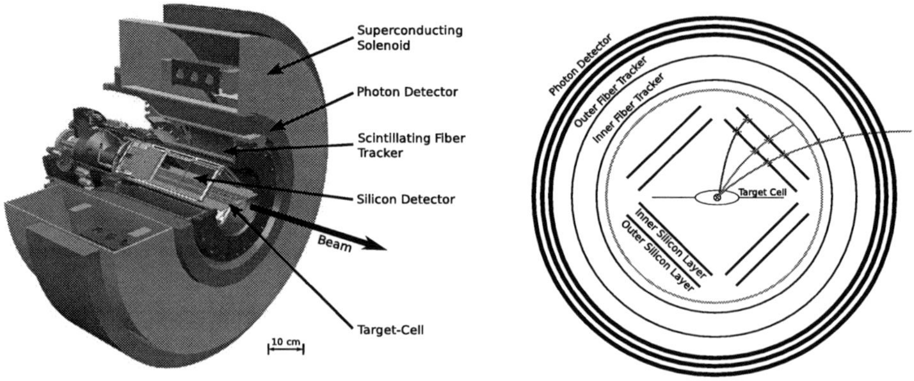

FIGURE 1. Left: A 3D CAD drawing of the recoil detector. Right: Schematic view of the detector in beam direction with three tracks representing the various track categories.

THE HERMES EXPERIMENT

HERMES [3] was a fixed target experiment that used the 27.6 GeV electron and positron beam provided by HERA at DESY in Hamburg. For the data collected up to 2005, exclusive DVCS events were selected by identifying the scattered beam lepton and the produced real photon in the HERMES spectrometer. The recoiling proton remained undetected and therefore a cut on the missing mass was applied. For the last two years of data taking (2006 and 2007) a new detection system was installed in the target region in order to detect the recoiling proton from exclusive reactions and hence to improve the selection of those events.

THE RECOIL DETECTOR

The HERMES recoil detector consisted of three sub-detector components (see left panel of Fig. 1). The innermost detector was a silicon detector (SSD) made up of two layers of double-sided sensors arranged around the target cell inside the HERA beam vacuum. The second component was a scintillating fiber tracker (SFT) that consisted of two barrels each with two layers of fibers parallel to the beam and two layers of fibers with a 10° stereo angle. The outermost detector component was a photon detector (PD) made up of 3 layers each with a tungsten and a scintillator sub-layer. The whole detection system was contained inside the 1 Tesla field of a superconducting solenoid.

Both the SSD and the SFT provided energy deposit and position information which is used in the track and momentum reconstruction. The right panel of Fig. 1 shows a schematic view of the recoil detector in beam direction with three tracks representing the variuos track categories. Starting from a momentum of about $125\,\mathrm{MeV/c}$ protons punched through the inner SSD layer and got stopped in the outer SSD layer as indicated by the left track in Fig. 1 (blue). In this case the momentum is simply calculated from the sum of both energy deposits. At $p \approx 145\,\mathrm{MeV/c}$ protons also punched through the

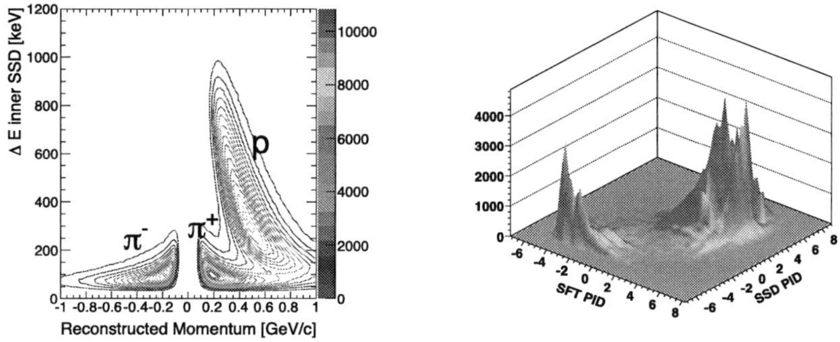

FIGURE 2. Left: Energy deposit in the inner SSD layer vs. reconstructed momentum. Right: Combined PID value for the SSD vs. combined PID value for the SFT.

outer SSD layer, but did not have enough energy to reach the SFT layers (indicated by the middle track in the figure; magenta). Here the momentum is reconstructed by both SSD energy deposits and the dE/dx behaviour for protons in silicon. For even higher momenta ($p > 200\,\mathrm{MeV/c}$) the additional position information from the SFT layers is used to reconstruct the momentum via the bending in the magnetic field (indicated by the right track in the figure; green).

PARTICLE IDENTIFICATION

For tracks reconstructed using both information from the SSD and the SFT, the particle identity can be obtained by means of the energy deposits in the detection layers and parent distributions. The left panel of Fig. 2 shows the energy deposit in the inner SSD layer vs. the reconstructed momentum. A clear separation of the proton and π^+ bands is visible which can be improved by combining the PID information from all detection layers. Here PID is defined as

$$PID = log_{10} \frac{P_p(dE,p)}{P_{\pi^+}(dE,p)} \tag{1}$$

with $P_p(dE,p)$ and $P_{\pi^+}(dE,p)$ being the probabilities of the particle being a proton or a π^+, respectively. The right panel of Fig. 2 shows the combined PID values for the SSD vs. the PID value for the SFT. The two peaks correspond to protons (positive PID values) and π^+ (negative PID values).

FIRST DVCS RELATED RESULTS

Based on the existing HERMES data productions including data from the recoil detector a first DVCS analysis has been performed. The starting point for this analysis was the requirement of having exactly one reconstructed photon and the scattered beam lepton reconstructed in the HERMES spectrometer [4]. These events are denoted as DVCS

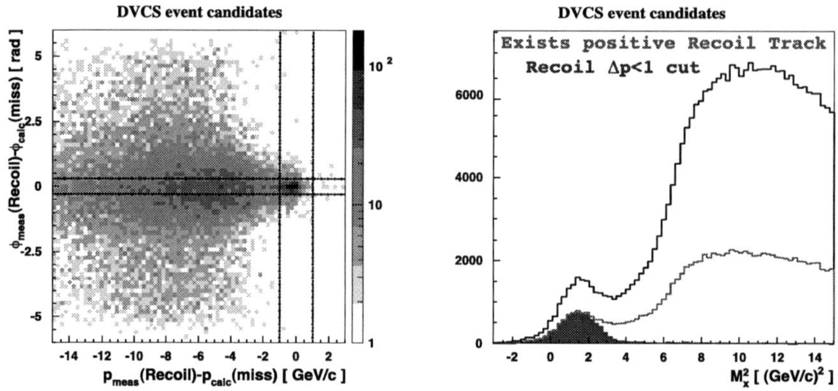

FIGURE 3. Left: Difference between the calculated and measured ϕ angles ($\Delta\phi$) vs. the difference between the calculated and measured recoiling proton momentum (Δp). Right: Squared missing mass distribution for DVCS candidate events (black), candidate events with a positively charge track in the recoil detector (red) and events with $|\Delta p| < 1\,\mathrm{GeV/c}$ (shaded blue peak).

candidate events from which the kinematics of the recoiling proton was calculated. The left panel of Fig. 3 shows the difference between the calculated ϕ and the measured ϕ angles ($\Delta\phi$) vs. the difference between the calculated and reconstructed momenta (Δp) of the recoiling proton. The lines in the figure represent potential cuts to select the recoiling proton. The corresponding squared missing mass distribution is shown in the right panel of the figure. The black curve shows the distribution for all DVCS candidate events, whereas the red curve shows the distribution for the candidate events with the additional requirement of having a positively charged track reconstructed in the recoil detector. The blue area represents events with an additional cut of $|\Delta p| < 1\,\mathrm{GeV/c}$.

SUMMARY AND OUTLOOK

All sub-detectors of the HERMES recoil detector have been calibrated and the efficiencies for protons in the individual layers is above 99 %. A first version of the particle identification is available and shows good performance in the proton/π^+ separation.

The first DVCS analysis using recoil detector data looks promising. Up to now PID information has not been utilized which will be one of the next steps to improve the event selection. In addition, single hits in the inner SSD layer will be included in the analysis to increase statistics and reach even lower recoiling proton momenta.

REFERENCES

1. X. Ji, *Phys. Rev. Lett.* **78**, 610 (1997).
2. X. Ji, *Phys. Rev.* **D55**, 7114 (1997).
3. HERMES Coll., K. Ackerstaff *et. al*, *Nucl. Instr. and Meth.* **A417**, 230 (1998).
4. A. Airapetian *et. al*, *Phys. Rev.* **D75**, 011103 (2007), `hep-ex/0605108`.

DVCS with a Longitudinally Polarized Target with CLAS12

Hovanes Egiyan*, Angela Biselli† and the CLAS Collaboration**

*University of New Hampshire, New Hampshire, 03824, USA
†Fairfield University, Connecticut, 06824, USA
**Newport News, Virginia, 23606, USA

Abstract. We describe a planned experiment to measure the longitudinal target-spin asymmetry in the DVCS process using CLAS12 detector. While the beam-spin asymmetry is more sensitive to H GPD, measurements of the target-spin asymmetry allow us to access the $\widetilde{H}$ GPD. These data will extend the Q^2 range accessable by the existing CLAS data and the scheduled 6 GeV CLAS DVCS experiments to $Q^2 = 8$ GeV^2.

Keywords: DVCS, polarized, target, GPD, Compton, formfactor, asymmetry
PACS: 13.60.Fz, 14.20.Dh, 13.88.+e, 13.40.Gp

INTRODUCTION

The recently introduced Generalized Parton Distributions (GPD) formalism [1, 2, 3, 4] provides us with a unifying framework to describe the nucleon structure. These functions contain the usual form factors and parton distributions, but in addition they include correlations between states of different longitudinal momentum and transverse momentum dependence. GPDs can therefore give three-dimensional pictures of the nucleon, providing information such as the transverse spatial distribution as a function of the longitudinal momentum fraction of the quarks.

Deeply Virtual Compton Scattering (DVCS) is the simplest and, therefore, the cleanest process which can be described in terms of GPDs. It was shown [1, 2, 5, 3, 4] that the DVCS reaction $\gamma^* p \to \gamma p$, in the Bjorken limit, can be factorized into a hard scattering kernel and a non-perturbative part (see Fig.1 left), described at twist-2 level by four GPDs E, H, $\widetilde{E}$ and $\widetilde{H}$. This factorization is valid when the virtuality of the incoming photon is large, but the momentum transfer to the nucleon is small ($-t << Q^2$). The GPDs are functions of three variables x, ξ and t, where x characterizes the average light-cone momentum fraction of the struck quark in the loop, ξ is the longitudinal momentum fraction of the transfer to the proton $\Delta = p - p'$, and $t = \Delta^2$ is the standard transfer between the virtual and real photons. Recent results of DVCS Beam Spin Asymmetries (BSA) from Jefferson Lab and HERMES [6, 7, 8, 9, 10] suggested that the GPD framework can be applied to DVCS experiments at the electron beam energies as low as 6 GeV.

In the forward limit ($\xi = t = 0$), the GPDs H and $\widetilde{H}$ reduce to the usual parton distributions $q(x)$ and $\Delta q(x)$:

$$H^q(x, \xi = 0, t = 0) \quad = \quad q(x)\,, \tag{1}$$

CP1149, *Spin Physics, 18th International Spin Physics Symposium*
edited by D. G. Crabb, D. B. Day, S. Liuti, X. Zheng, M. Poelker, and Y. Prok

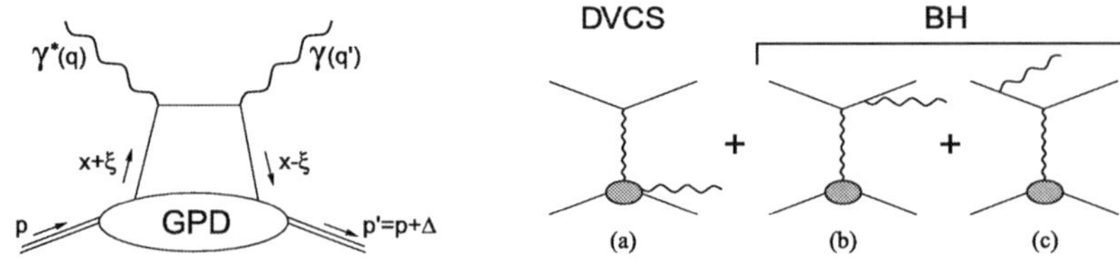

FIGURE 1. Handbag diagram to the DVCS process (left); diagrams contributing to the photon electroproduction process(right).

$$\widetilde{H}^q(x,\xi=0,t=0) \quad = \quad \Delta q(x)\,, \tag{2}$$

where q stands for the quark flavor. The first moments of the GPDs and the elastic form factors are related by the following relations:

$$\int_{-1}^{+1} dx H^q(x,\xi,t) \quad = \quad F_1^q(t)\,, \quad \int_{-1}^{+1} dx E^q(x,\xi,t) = F_2^q(t)\,, \tag{3}$$

$$\int_{-1}^{+1} dx \widetilde{H}^q(x,\xi,t) \quad = \quad g_A^q(t)\,, \quad \int_{-1}^{+1} dx \widetilde{E}^q(x,\xi,t) = h_A^q(t)\,, \tag{4}$$

where F_1^q and F_2^q are the Dirac and Pauli form factors, g_A^q is the axial form factor and h_A^q is the induced pseudoscalar form factor. The second moment of the GPDs and the total angular momentum carried by quarks are related via Ji's sum rule:

$$\sum_q \frac{1}{2}\int_{-1}^{+1} dx x[H^q(x,\xi,t=0) + E^q(x,\xi,t=0)] = J_q\,. \tag{5}$$

Since $\Delta\Sigma$ is constrained in DIS experiments, if one makes enough measurements to extract the second moments of the GPDs, the sum rule will determine the quark orbital momentum contribution to the nucleon spin.

The $ep \to ep\gamma$ process can either occur by radiation along one of the electron lines (Bethe-Heitler or BH) or by emission of a real photon by the nucleon (DVCS) as shown in right panel of Fig. 1. The total amplitude $\mathscr{T}$ is the coherent sum of the BH and DVCS amplitudes:

$$|\mathscr{T}|^2 \quad = \quad |\mathscr{T}_{BH}|^2 + |\mathscr{T}_{DVCS}|^2 + \mathscr{I} \tag{6}$$

$$\mathscr{I} \quad = \quad \mathscr{T}_{DVCS}^*\mathscr{T}_{BH} + \mathscr{T}_{DVCS}\mathscr{T}_{BH}^*\,, \tag{7}$$

where $\mathscr{T}_{DVCS}$ and $\mathscr{T}_{BH}$ are the amplitudes for the DVCS and Bethe-Heitler processes, and $\mathscr{I}$ denotes the interference between these amplitudes. The $|\mathscr{T}_{DVCS}|^2$ term is predicted to be smaller than other contributions, and therefore we do not keep it in this discussion. On the other hand, it may be possible to make empirical estimates of the relative magnitude of the $|\mathscr{T}_{DVCS}|^2$ term using future accurate data. Neglecting the $|\mathscr{T}_{DVCS}|^2$ term, the longitudinal target-spin asymmetry can be expressed in terms of Compton formfactors (CFFs) as:

$$A_{LU} = \frac{d\sigma^{\rightarrow} - d\sigma^{\leftarrow}}{\sigma^{\rightarrow} + d\sigma^{\leftarrow}} = \frac{2\cdot \mathscr{T}_{BH}\cdot Im(\mathscr{T}_{DVCS})}{|T_{BH}|^2 + 2\cdot \mathscr{T}_{BH}\cdot Re(\mathscr{T}_{DVCS})} =$$
$$= \left\{F_1\widetilde{\mathscr{H}} + \frac{x_B}{2-x_B}(F_1+F_2)\left(\mathscr{H} + \frac{x_B}{2}\mathscr{E}\right) - \frac{x_B}{2-x_B}\left(\frac{x_B}{2}F_1 + \frac{t}{4M^2}F_2\right)\widetilde{\mathscr{E}}\right\} \quad (8)$$

We can see from Eq. 8 that the main contributions to A_{LU} come from $\widetilde{\mathscr{H}}$ and $\mathscr{H}$, whereas the $\widetilde{\mathscr{E}}$ contribution is suppressed. Clearly we need at least another measured observable, for instance the beam-spin asymmetry, to be able to separate the $\widetilde{\mathscr{H}}$ and $\mathscr{H}$ terms. The extraction of GPDs will be done in the same manner as Parton Distribution Functions are extracted from structure function measurements, by means of a global fit of data using GPD parametrization.

EXPERIMENTAL DATA

The proposed experiments will use the standard CLAS12 setup. Six symmetrically arranged superconducting coils will provide a toroidal magnetic field for the spectrometer. Each of the six sectors of CLAS12 will be instrumented with identical detector packages, consisting of drift chambers for tracking, Cherenkov counters and time-of-flight systems for particle identification, electromagnetic shower and pre-shower calorimeters for triggering and neutral particle detection. A central detector system, consisting of a silicon vertex detector and a time-of-flight system will provide us with a detection and identification capabilities for slower backward-going particles. A special lead-glass calorimeter will be installed downstream of the target to detect the photons at small angles. The target will be polarized via the method of Dynamic Nuclear Polarization (DNP) which is a well established technique that has been used extensively in nuclear and particle physics experiments.

The experiments will run for 120 days at the luminosity of $L = 2\times 10^{35}$ cm^{-2}s^{-1} using 11 GeV electron beam provided by the upgraded accelerator of Jefferson Laboratory. The statistical uncertainties of the asymmetry measurements are expected to be between 2% and 10%, depending on the value of x_B and Q^2, because the number of expected events at low x_B is much larger than the number of events at higher $x_B > 0.4$. An example of the projected precision for the target-spin asymmetries assuming 80% target polarization is shown in Fig. 2. These differences between the blue dashed curve and the red solid curve clearly demonstrate the sensitivity of the target-spin asymmetry to the $\widetilde{H}$ GPD.

SUMMARY

The expected DVCS data from the experiment with a longitudinally polarized target and CLAS12 detector will extend the Q^2 range of the existing CLAS data and upcoming dedicated 6 GeV experiment to $Q^2 = 8$ GeV^2, at the same time providing a

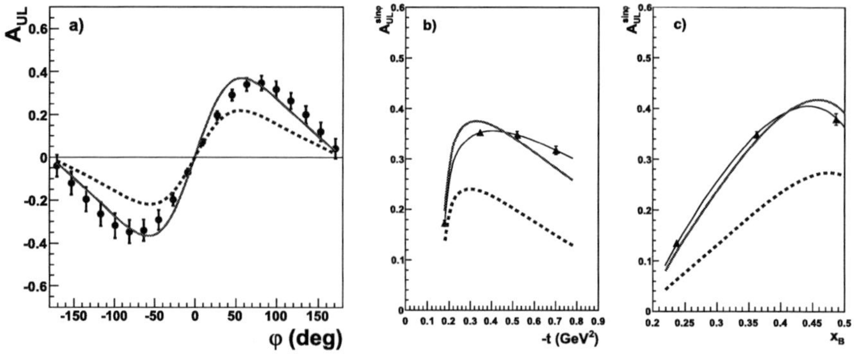

FIGURE 2. Target spin asymmetry versus φ for $Q^2 = 4.1$ GeV^2, $x_B = 0.36$ and $-t = 0.52$ GeV^2 (a). The black points show the values from [11] using CTEQ6 PDFs with the estimated errors from the proposed measurement. The red solid curve is using MRST02 PDFs with $E = \widetilde{E} = 0$, and for the blue dashed curve is $\widetilde{H}$ is also set to zero. $\sin\varphi$ moments of the target spin asymmetry versus $-t$ at $Q^2 = 4.1$ GeV^2 and $x_B = 0.36$ (b), and versus x_B at $Q^2 = 4.1$ GeV^2 and $-t = 0.52$ GeV^2 (c). The projected error bars represent the statistical uncertainties only.

large angular coverage and exceptional statistical accuracy. Along with the measurements of the beam-spin asymmetry, transversely polarized target-spin asymmetry and the helicity-dependent cross section, these data will provide a solid basis for a global fit of parametrized GPDs. With such a wide t and Q^2 coverage, it will be possible to study the onset of twist-3 (and higher) effects which enter the beam-spin asymmetries and target-spin asymmetries with at least an additional power of $\sqrt{-t}/Q$.

ACKNOWLEDGMENTS

This work was supported by the U.S. Department of Energy and the National Science Foundation, the French Commissariat a l'Energie Atomique, the Italian Istituto Nazionale di Fisica Nucleare, and the Korean Science and Engineering Foundation.

REFERENCES

1. X.-D. Ji, *Phys. Rev.* **D55**, 7114–7125 (1997), hep-ph/9609381.
2. X.-D. Ji, *Phys. Rev. Lett.* **78**, 610–613 (1997), hep-ph/9603249.
3. A. V. Radyushkin, *Phys. Lett.* **B380**, 417–425 (1996), hep-ph/9604317.
4. A. V. Radyushkin, *Phys. Rev.* **D56**, 5524–5557 (1997), hep-ph/9704207.
5. X.-D. Ji, W. Melnitchouk, and X. Song, *Phys. Rev.* **D56**, 5511–5523 (1997), hep-ph/9702379.
6. C. Munoz Camacho, et al., *Phys. Rev. Lett.* **97**, 262002 (2006), nucl-ex/0607029.
7. F. X. Girod, et al., *Phys. Rev. Lett.* **100**, 162002 (2008), 0711.4805.
8. A. Airapetian, et al., *Phys. Rev. Lett.* **87**, 182001 (2001), hep-ex/0106068.
9. S. Stepanyan, et al., *Phys. Rev. Lett.* **87**, 182002 (2001), hep-ex/0107043.
10. S. Chen, et al., *Phys. Rev. Lett.* **97**, 072002 (2006), hep-ex/0605012.
11. M. Vanderhaeghen, P. A. M. Guichon, and M. Guidal, *Phys. Rev.* **D60**, 094017 (1999), hep-ph/9905372.

DVCS with longitudinally polarized target using CLAS at 6 GeV

Angela Biselli
for the CLAS collaboration

Fairfield University, Fairfield, CT 06824

Abstract. Deeply Virtual Compton Scattering (DVCS) is one of the simplest processes that can be described in terms of Generalized Parton Distributions (GPDs). The target single-spin asymmetry (target SSA) in the reaction $e\vec{p} \to ep\gamma$ is directly proportional to the imaginary part of the DVCS amplitude, and gives access to a combination of GPDs namely $\tilde{H}$, H, and E. This asymmetry will be measured in a dedicated experiment at Jefferson Lab using the CEBAF 6-GeV polarized electron beam, a polarized solid-state $^{14}NH_3$ target, and the CEBAF Large Acceptance Spectrometer (CLAS) together with the Inner Calorimeter (IC). The expected asymmetry from leading-order calculations is in the range of 20% to 40%, depending on the kinematics and on the GPD model used. The DVCS amplitude will be mapped out in the Q^2 region from 1 to 4 GeV^2, x_B from 0.15 to 0.55 and $-t$ from 0.1 to 2 GeV^2 providing new constraints on the GPDs.

Keywords: DVCS
PACS: 13.60.Fz, 13.60.-r, 13.88.+e, 14.20.Dh, 24.85.+p

MOTIVATION

The Generalized Parton Distributions formalism allows to describe the nucleon in terms of fully correlated quark distributions in both coordinate and momentum space. These distributions contain crucial information such as the angular momentum distribution of quarks in the nucleon [1, 2, 3].

The cleanest way to access GPDs is through the Deeply Virtual Compton Scattering, the electro-production of a photon off one of the quarks in the proton. As shown in Fig. 1, this exclusive process can be factorized into a hard scattering part and a nucleon structure part parameterized via GPDs.

The soft process is described by 4 GPDs: H, $\tilde{H}$, E, and $\tilde{E}$, which depend on the momentum fraction of the struck quark, x, the longitudinal momentum fraction transferred to the proton, $\xi = x_B/(2-x_B)$, and the momentum transfer. t, between virtual and real photons. GPDs provide a link between the charge density measured with elastic scattering (form factors) and longitudinal parton momentum fraction measured in Deep Inelastic Scattering (DIS). More precisely, in the DIS limit of $t = \xi = 0$, H and $\tilde{H}$ become the unpolarized and polarized parton distributions, $q(x)$ and $\Delta q(x)$, respectively, while the first moments of the GPDs correspond to the form factors:

$$\int_{-1}^{1} H^q(x,\xi,t)dx = F_1^q(t), \qquad \int_{-1}^{1} E^q(x,\xi,t)dx = F_2^q(t), \qquad (1)$$

CP1149, *Spin Physics, 18th International Spin Physics Symposium*
edited by D. G. Crabb, D. B. Day, S. Liuti, X. Zheng, M. Poelker, and Y. Prok

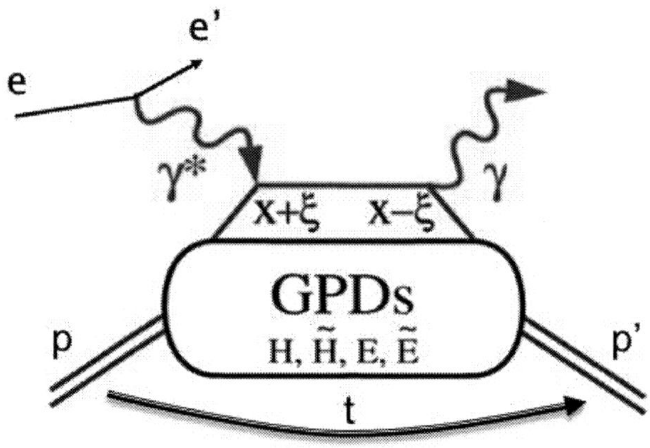

FIGURE 1. Handbag diagram for the DVCS process.

$$\int_{-1}^{1} \tilde{H}^q(x,\xi,t)dx = G_A^q(t), \qquad \int_{-1}^{1} \tilde{E}^q(x,\xi,t)dx = G_P^q(t), \quad \forall\xi. \tag{2}$$

From the experimental point of view, the DVCS cannot be disentangled from Bethe Heitler process, where the final-state photon is emitted by either the incoming or the outgoing electron,

$$\frac{d^4\sigma}{dQ^2 dx_B dt d\phi} \propto |T_{\mathrm{DVCS}} + T_{\mathrm{BH}}|^2. \tag{3}$$

The Bether Heitler cross section can dominate over the DVCS amplitude depending on the kinematics and the beam energy, however the DVCS signal can be enhanced measuring asymmetries which are proportional to the interference between the two processes.

$$A = \frac{\sigma^+ - \sigma^-}{\sigma^+ + \sigma^-} \propto \frac{I}{|T_{\mathrm{DVCS}}|^2 + |T_{\mathrm{BH}}|^2 + I}, \tag{4}$$

where $I = T_{\mathrm{DVCS}} T^*_{\mathrm{BH}} + T^*_{\mathrm{DVCS}} T_{\mathrm{BH}}$. Depending on the beam and target polarization configuration, one can extract different asymmetries which have different sensitivity to GPDs. For instance, the beam-spin asymmetry A_{LU}, measured in the dedicated CLAS experiment [4, 5], can be expressed as [6]

$$A_{\mathrm{LU}}(\phi) \propto \Im\mathrm{m}\left\{ F_1 \mathcal{H} + \frac{x_{\mathrm{B}}}{2 - x_{\mathrm{B}}}(F_1 + F_2)(\widetilde{\mathcal{H}} - \frac{\Delta^2}{4M^2} F_2 \mathcal{E}) + ... \right\} \sin\phi, \tag{5}$$

where $\Im\mathrm{m}\mathcal{H} \propto H(x = \pm\xi)$ etc, is sensitive to H, $\tilde{H}$, and E, particularly to H in the CLAS kinematic. The longitudinal target-spin asymmetry A_{UL}

$$A_{\mathrm{UL}}(\phi) \propto \Im\mathrm{m}\left\{ F_1 \widetilde{\mathcal{H}} + \frac{x_{\mathrm{B}}}{2 - x_{\mathrm{B}}}(F_1 + F_2)(\mathcal{H} + \frac{x_{\mathrm{B}}}{2} \mathcal{E}) + ... \right\} \sin\phi, \tag{6}$$

is equally sensitive to H and $\tilde{H}$. Furthermore the transversely-polarized target measurement provides additional sensitivity to E. A measurement of the target asymmetry is crucial to gain knowledge on GPDs, in particular $\tilde{H}$, to constrain GPDs models and ultimately understand the structure of the nucleon.

THE DVCS EXPERIMENT WITH A LONGITUDINALLY POLARIZED TARGET AND THE CLAS DETECTOR

The first dedicated DVCS experiment with the longitudinally polarized target is scheduled to run at Jefferson Lab at the beginning of 2009 [7]. The Continuous Electron Beam accelerator Facility (CEBAF) delivers an electron beam with an energy up to 6 GeV to three experimental halls. Hall B hosts the CEBAF Large Acceptance Spectrometer (CLAS), which allows multiple-particle identification with a nearly 2π acceptance. The beam is scattered off a longitudinally polarized solid, 1.5-cm-long, ammonia target ($^{14}NH_3$), polarized up to 70% via Dynamic Nuclear Polarization [8]. Two Helmholtz coils around the target provide the uniform ($\Delta B/B = 10^{-4}$) holding field of 5 T for the polarization and at the same time focus the low-energy Moller electrons toward the beam line. The trigger is a signal in the Cherenkov counters and the electromagnetic calorimeters, which allow the identification of electrons. Protons, deflected by a superconducting toroidal field, are detected by three layers of drift chambers, for momentum identification, and scintillators paddles, for time-of-flight identification. Neutral particles are detected by the electromagnetic calorimeter from 17° to 43°, and by the Inner Calorimeter (IC) from 4-15°. The IC was build for the dedicated DVCS experiment [4] in order to detect the photons of the DVCS reaction which are emitted mostly at low angles. The IC consists of 424 lead tungstate ($PbWO_4$) crystals, 16 cm in length (18 radiation lengths) with an energy and angle resolution of 4.5% and 4 mrad, respectively. The large acceptance of the CLAS detector in conjunction with the IC allows to completely detect the three final state particles from the reaction ep $\rightarrow$ epγ over a large kinematic range. The complete detection of the final state allows to apply kinematic cuts that drastically reduce the nuclear background from ^{14}N. Contamination from π^0 is corrected by measuring the epπ^0 asymmetry and by estimating the amount of contamination using Monte Carlo simulations. The accessible range in Q^2 is from 1 to 4 GeV^2, and in x_B from 0.15 to 0.55.

OUTLOOK

The longitudinal target single spin asymmetry has been measured for the first time, both at HERMES [9] and at CLAS [10]. The CLAS measurement was an outcome of the eg1 experiment, which was run in 2000 with beam energies from 1.6 to 5.7 GeV, with the primary goal to measure the spin structure function g_1 of the proton and the neutron. With a factor 10 increase in running time and a factor of 2 improvement in acceptance, due to the addition of the IC, the proposed experiment is expected to have statistical errors 4 times smaller and systematic errors 2 times smaller than the previous CLAS measurement. The increased statistics will allow for the first time to map out the Q^2,

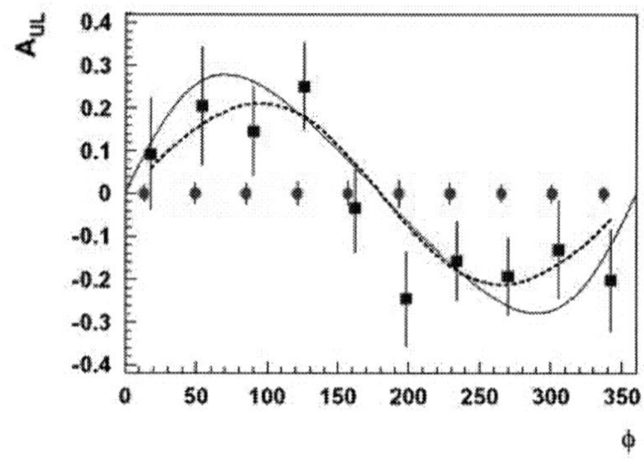

FIGURE 2. The ϕ dependence of the target single spin asymmetry. The square points are the result in [10], the circle points illustrate the expected statistical accuracy for the proposed experiment.

x_B, and t dependence of the target single-spin asymmetry over a large kinematic range, providing new constraints for the GPDs. Figure 2 shows the comparison between the CLAS results by Chen *et al.* and the expected results from the proposed experiment. Finally, since the experiment will be run using a beam of longitudinally polarized electrons (with polarization up to 80%), the double spin asymmetry will be extracted, providing additional information on the real part of the DVCS amplitude.

ACKNOWLEDGMENTS

We would like to thank the staff of the Accelerator and Physics Divisions at the Jefferson Laboratory for their outstanding efforts to provide us with the high quality beam and the facilities for the data analysis. This work was supported by the U.S. Department of Energy and the National Science Foundation, the French Commissariat a l'Energie Atomique, the Italian Istituto Nazionale di Fisica Nucleare, and the Korean Science and Engineering Foundation. The Southeastern Universities Research Association (SURA) operates the Thomas Jefferson National Accelerator Facility for the United States Department of Energy under Contract No. DE-AC05-84ER40150.

REFERENCES

1. D. Mueller, D. Robaschik, B. Geyer, F. M. Dittes and J. Horejsi, Fortsch. Phys. **42**, 101 (1994).
2. X. D. Ji, Phys. Rev. Lett. **78**, 610 (1997).
3. A. V. Radyushkin, Phys. Rev. D **56**, 5524 (1997).
4. V.D. Burkert, L. Elouadrhiri, M. Garçon, S. Stepanyan *et al.*, CEBAF experiment 01-113.
5. F. X. Girod *et al.* [CLAS Collaboration], Phys. Rev. Lett. **100**, 162002 (2008).
6. A. V. Belitsky, D. Mueller and A. Kirchner, Nucl. Phys. B **629**, 323 (2002).
7. A. Biselli, L. Elouadrhiri, K. Joo, S. Niccolai *et al.*, CEBAF experiment 05-114.
8. C. D. Keith *et al.*, Nucl. Instrum. Meth. A **501**, 327 (2003).
9. M. Kopytin HERMES Coll., talk at the DIS2005 Workshop, Madison (2005).
10. S. Chen *et al.* [CLAS Collaboration], Phys. Rev. Lett. **97**, 072002 (2006).

Measurements of the Electric Form Factor of the Neutron at High Momentum Transfer

Seamus Riordan

University of Virginia
CEBAF Center F206, Jefferson Laboratory
12000 Jefferson Ave., Newport News, VA 23606

Abstract. The electric and magnetic form factors of the nucleon provide experimental access to the underlying charge and magnetic moment distributions of quarks. We have measured the electric form factor of the neutron at four kinematic points between 1.2 and 3.5 GeV^2 in Hall A at Jefferson Lab. This more than doubles the momentum transfer region for which this quantity has previously been measured, providing new information on the structure of the neutron. Preliminary results for G_E^n at $Q^2 = 1.7, 2.5$, and 3.5 GeV^2 were presented and were compared with QCD-based models and phenomenological approaches.

Keywords: neutron, form factors, electric form factor, sachs
PACS: 13.40.Gp

INTRODUCTION

The electromagnetic current of a spin 1/2 target with structure interacting with a single virtual photon can we described using two form factors, the Sachs electric and magnetic form factors, G_E and G_M,

$$J^\mu = e\bar{u}(p')\left[\gamma^\mu \frac{G_E(Q^2)+\tau G_M(Q^2)}{1+\tau} + \frac{i\sigma^{\mu\nu}}{2M} q_\nu \frac{G_M(Q^2)-G_E(Q^2)}{1+\tau}\right]u(p)$$

where p and p' are the final and initial momenta of the nucleon, respectively, q^μ is the four momentum of the virtual photon, Q^2 is the four-momentum transfer, τ is $Q^2/4M^2$, and M is the mass of the target. These form factors carry the physical interpretation of being related to the Fourier transforms of the electric charge distribution, for G_E, and magnetic moment distribution, for G_M distributions in the Briet frame.

Form factors are important to our understanding of QCD because they provide experimental access to the underlying structure of the valence and sea quarks in the nucleon. Measurements of these form factors provide important testing grounds for QCD and QCD-inspired models and can provide insight into the dominant mechanisms of nucleon structure. In 1989, measurements of the electric form factor of the proton using techniques involving spin-observables yielded surprising results that were in severe disagreement with previous Rosenbluth separation techniques. Standard Rosenbluth separation had measured G_E^p to be consistent with the so-called "dipole parameterization", but polarization observables showed a reduction in G_E^p as one went to higher Q^2 [1]. One of the mechanisms for this reduction was suggested to be due to quark orbital angular momentum in the nucleon [2].

CP1149, *Spin Physics, 18th International Spin Physics Symposium*
edited by D. G. Crabb, D. B. Day, S. Liuti, X. Zheng, M. Poelker, and Y. Prok

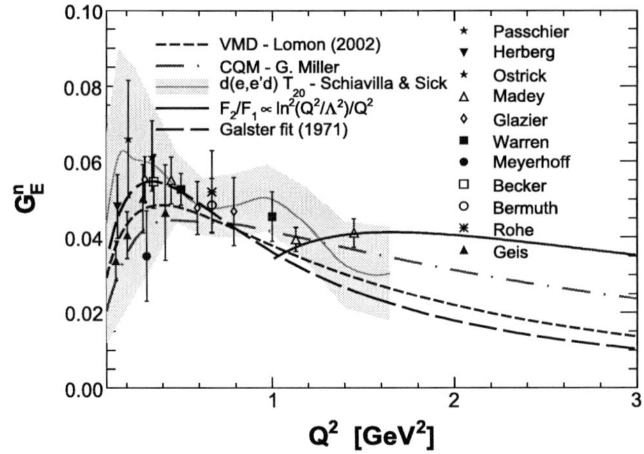

FIGURE 1. Previous measurements of G_E^n.

While the Sachs form factors of the proton have presently been relatively well measured up to and beyond $Q^2 = 10$ GeV^2, precise data on the form factors of the neutron only range up to $2 \sim 4$ GeV^2. In particular, the electric form factor, G_E^n has the smallest Q^2 coverage of all the form factors, Fig. 1. This is due to two reasons: first, the neutron is an overall neutral particle, making the value of G_E^n small compared to G_M^n. This is a hindrance when attempting to employ techniques such as Rosenbluth separation, where the cross section is dominated by G_M^n at high Q^2. Second, there are no sufficient free-neutron targets, requiring the neutron to be bound in a nucleus when studied, typically ^{2}H or ^{3}He.

JLAB HALL A EXPERIMENT E02-013

In spring 2006, we measured the electric form factor of the neutron, G_E^n, at $Q^2 = 1.4, 1.7, 2.5$, and 3.5 GeV^2 through the reaction $^3\overrightarrow{\text{He}}(e,e'n)pp$. This experiment more than doubles the previously measured Q^2 range. In this reaction we are sensitive to the form factor ratio G_E^n/G_M^n, which is related to the helicity-dependent cross section asymmetry

$$\begin{aligned} A_{\text{phys}} &= \frac{\sigma_+ - \sigma_-}{\sigma_+ + \sigma_-} \\ &= -\frac{2\sqrt{\tau(\tau+1)}\tan(\theta_e/2)G_E/G_M \sin\theta^* \cos\phi^*}{(G_E/G_M)^2 + (\tau + 2\tau(1+\tau)\tan^2(\theta_e/2))} \\ &\quad -\frac{2\tau\sqrt{1+\tau+(1+\tau)^2\tan^2(\theta_e/2)}\tan(\theta_e/2)\cos\theta^*}{(G_E/G_M)^2 + (\tau + 2\tau(1+\tau)\tan^2(\theta_e/2))}, \end{aligned}$$

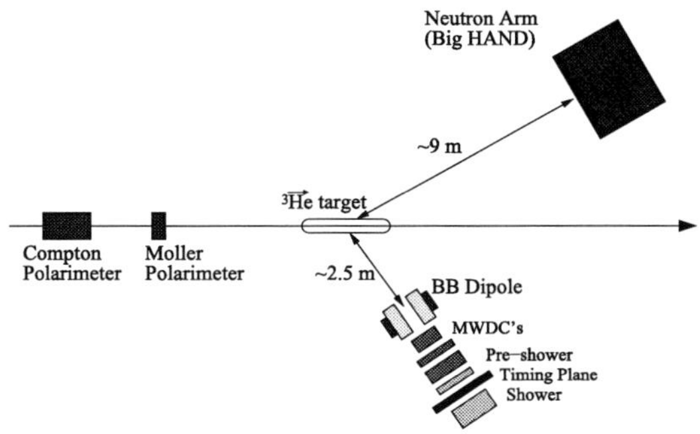

FIGURE 2. The detector configuration used in E02-013 (not to scale).

where $\sigma_{+/-}$ is the measured cross section for two electron beam helicities, θ_e is the scattering angle of the electron, and θ^* and $\phi*$ are the polar and azimuthal angles of the momentum transfer with respect to the target polarization and electron scattering plane.

A diagram of our experimental setup can be found in Fig. 2. A polarized ^{3}He target acts as our effective polarized neutron target. This target was polarized using a hybrid spin-exchange optical pumping technique, where both Rb and K vapors are used to efficiently transfer angular momentum from circularly polarized photons to the ^{3}He nucleus. From this target we were able to achieve just below 50% nucleus polarization during a majority of data taking.

To detect the scattered electron, we used the BigBite spectrometer, a large angular and momentum acceptance spectrometer consisting of a large dipole magnet of field integral approximately 1 T m with a newly constructed detector stack. A set of multiple wire drift chambers were used to perform high resolution hit based tracking. This spectrometer provided an overall momentum resolution of approximately $\sigma_{(\delta p/p)} =$ 1.0% while operating in a high rate environment of luminosity 5×10^{36} Hz/cm^2.

The recoiling nucleon was detected by a large array of scintillators that formed the neutron arm. It consists of seven layers, each containing $40 \sim 50$ scintillator bars. Sandwiched between these layers are large plates of iron and lead to enhance the nucleon interaction probability. In front of these layers are two additional layers of scintillator called the veto layers which are used to provide basic nucleon charge identification. The nucleon three-momentum was reconstructed through hit position and time of flight.

Analysis

Analysis of the data is underway. Quasielastic events are extracted from the data by selecting events with small missing parallel and perpendicular momentum and with invariant mass near the nucleon mass. This selection significantly eliminates accidental and inelastic background events as well as suppresses final state interactions.

The asymmetry is corrected for the degree of polarization of the target and polarization of the beam, both periodically measured during the experiment, as well as dilutions due to various background contributions, which are evaluated from the data. By choosing our cuts conservatively, corrections due to inelastic contributions are unnecessary, as supported by Monte Carlo simulations.

PRELIMINARY RESULTS DISCUSSION

Currently, we have performed a preliminary analysis on Q^2 points 1.7, 2.5, and 3.5 GeV^2. The remaining analysis requires the development of a complete Monte Carlo simulation of the experiment for confirmation of the interpretation of detector responses, primarily in the neutron arm. Also remaining are final calculations of final state interaction contributions, which are currently underway.

Our present results are consistent with the Galster parameterization [3], proposed in 1971. They do not show scaling from logarithmic pQCD predictions [4] which was shown to have set in for the proton at surprisingly low Q^2. Our results are also consistent with a calculation performed at Argonne [5, 6].

Finally, our results can be used in the direct interpretations of the transverse charge distributions for an unpolarized and transversely polarized nucleon in the infinite momentum frame using an impact parameter representation [7], [8].

REFERENCES

1. O. Gayou *et al.*, Phys. Rev. Lett. **88**, 092301 (2002).
2. G. Miller, Phys. Rev. C **66**, 032201 (2002).
3. S. Galster *et al.*, Nucl. Phys. B **32**, 221 - 237 (1971).
4. A. V. Belitsky, X. Ji, and F. Yuan, Phys. Rev. Lett. **91**, 092003 (2003).
5. M. S. Bhagwat *et al.*, arXiv:nucl-th/0610080
6. I. Cloet *et al.*, arXiv:nucl-th/0804.3118
7. G. Miller, Phys. Rev. C **78**, 032201 (2008).
8. C. Carlson and M. Vanderhaeghen, Phys. Rev. Lett. **100**, 032004 (2008).

Azimuthal asymmetries in DVCS on unpolarized hydrogen and deuterium targets

H. Marukyan
(on behalf of the HERMES Collaboration)

Yerevan Physics Institute, Alikhanian Brs. St. 2, Yerevan, 375036, Armenia

Abstract. We report the latest results of azimuthal asymmetries in the DVCS process measured at the HERMES experiment on unpolarized hydrogen and deuterium targets. Exploiting the ability of HERA to provide lepton beams with both charges and spin orientations, it is possible to extract simultaneously asymmetry amplitudes attributed to the pure DVCS process and to the interference between the Bethe-Heitler and DVCS processes. The results are compared to the theoretical calculations.

Keywords: DVCS, proton, deuteron
PACS: 13.60.-r 24.85.+p 13.60.Fz 14.20.Dh

INTRODUCTION

Recent interest in hard exclusive processes has its origin in the theoretical description of the Deeply-Virtual Compton Scattering (DVCS) process in terms of the Generalized Parton Distribution (GPD) formalism [1, 2]. The GPD formalism incorporates the well-known nucleon Form Factors (FFs) and parton momentum distributions functions (PDFs) in a 3-dimensional picture of the nucleon. Interest in the GPD framework has also been intensified by the fact that the unknown total angular momentum components of the quarks and gluons are encoded in the GPDs [3].

DVCS MEASUREMENT AT HERMES

One of the theoretically cleanest channels to probe GPDs is DVCS in which a single real photon is produced in the final state along with the recoiling nucleon (nucleus) intact. Measurement of the DVCS process is intrinsically mixed with that of the Bethe-Heitler (BH) process in which a real photon is radiated by the lepton rather than the quark. The cross section for the leptoproduction of real photons is therefore proportional to the square of the sum of the amplitudes [4]

$$\frac{\mathrm{d}\sigma}{\mathrm{d}x_A\,\mathrm{d}Q^2\,\mathrm{d}|t|\,\mathrm{d}\phi} = \frac{x_A y^2}{32\,(2\pi)^4 Q^4} \frac{|\tau_{BH}|^2 + |\tau_{DVCS}|^2 + \tau_{BH}^* \tau_{DVCS} + \tau_{DVCS}^* \tau_{BH}}{\sqrt{1+\varepsilon^2}}, \tag{1}$$

where $-Q^2$ represents the virtual photon four-momentum squared, y the fraction of the incident lepton energy carried by the virtual photon in the target rest frame, t the Mandelstam variable, x_A is related to the Bjorken scaling variable and $\varepsilon \equiv 2x_A M_A/Q$.

CP1149, *Spin Physics, 18th International Spin Physics Symposium*
edited by D. G. Crabb, D. B. Day, S. Liuti, X. Zheng, M. Poelker, and Y. Prok

The azimuthal angle ϕ is defined as the angle between the lepton scattering plane and the photon production plane spanned by the virtual and real photons.

The cross section for a longitudinally polarized lepton beam scattered off an unpolarized proton/deuteron target can be given by

$$\sigma_{LU}(P_l, e_l, \phi) = \sigma_{UU}(\phi) \cdot \{1 + P_l A_{LU}^{DVCS}(\phi) + e_l A_C(\phi) + e_l P_l A_{LU}^{I}(\phi)\}, \quad (2)$$

where e_l and P_l are the lepton beam charge and polarization. The beam spin asymmetries (BSA) A_{LU}^{DVCS} and A_{LU}^{I} are related to certain Compton Form Factors in a Fourier series of expansion in ϕ [5] of the pure DVCS squared and interference term, respectively. The beam charge asymmetry (BCA) A_C is related only to the even Fourier harmonics of the interference term. The various amplitudes of these asymmetries have been extracted simultaneously using combined data taken with both charges and spin orientations of the beam. The following functional expansion for these asymmetries has been used in a Maximum Likelihood method fit, which is described in details in Ref. [6],

$$A_C = \sum_{n=0}^{3} A_C^{\cos(n\phi)} \cos(n\phi), \quad A_{LU}^{I(DVCS)} = \sum_{n=1}^{3} A_{LU,I(DVCS)}^{\sin(n\phi)} \sin(n\phi). \quad (3)$$

In order to extract these asymmetries from the data, events were selected when they contained one photon and one scattered lepton track. Photons were identified by the energy deposition in the calorimeter and preshower counter in the absence of a track. The requirements imposed on the lepton kinematics were $Q^2 > 1$ GeV^2 and $W > 3$ GeV. The opening angle $\Theta_{\gamma^*\gamma}$ between the virtual and real photon was limited to a range of 5 to 45 mrad.

Since the recoiling nucleon (or nucleus) was not detected, the exclusive DVCS events were selected imposing an additional constraint on the missing mass M_x. The exclusive region was defined as $-1.5 < M_x < 1.7$ GeV (negative M_x values are obtained as a consequence of the finite energy resolution of the spectrometer).

RESULTS

The results of certain asymmetry amplitudes from Eq. 3 are shown in Figs. 1 and 2 as a function of $-t$, x_B and Q^2 in four bins as well as for the 'overall' kinematics (left column). The errors bars (bands) represent the statistical (systematic) uncertainties. The systematic uncertainties include the contributions from the effects of acceptance, smearing, bin width and the alignment of the detectors with respect to the beam direction and are estimated from a Monte Carlo simulation. The last rows on these figures show the fractions of coherent and associated (nucleon excites to resonances in the final states) processes. The latter remains the part of the signal, while the results have been corrected for the semi-inclusive background. The two hatched bands in these figures are the theoretical predictions, based on a model, developed in Refs. [7, 8], and carried out with 'factorized' and 'Regge' ansätzs for modeling GPDs.

A sizable magnitude of first cosine and sine amplitudes have been found for the BCA and BSA ($A_{LU,I}^{\sin\phi}$), respectively. The higher amplitudes which are related to the twist-three term and gluonic GPDs are compatible with zero.

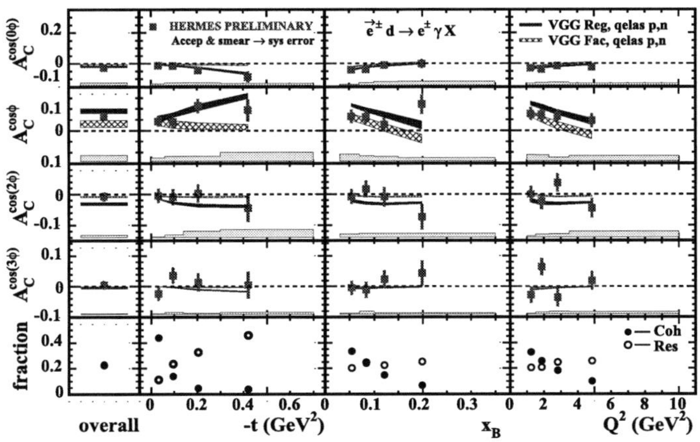

FIGURE 1. The BCA amplitudes for the hard exclusive electroproduction of photons from deuterium target at the HERMES experiment. See text for details.

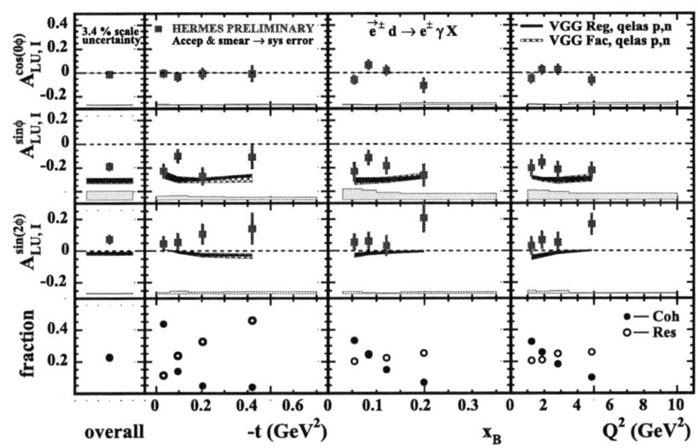

FIGURE 2. The BSA amplitudes $A_{LU,I}^{\sin(n\phi)}$ for the hard exclusive electroproduction of photons from deuterium target at the HERMES experiment. See text for details.

Figures 3 and 4 show the comparison of BCA and BSA (related only to the interference term) amplitudes, respectively, extracted from the deuterium and hydrogen data at HERMES. The deuteron and proton results appear to be consistent in most kinematic regions. For both targets, the BCA amplitudes become sizable only at higher values of $-t$. The effects from the coherent scattering on the deuteron could be expected in the first $-t$ bin, but are not apparent. The difference in BCA amplitudes $A_C^{\cos\phi}$ at large $-t$ values can be explained by the contribution from incoherent scattering on the neutron inside the deuteron. The BSA $A_{LU,DVCS}^{\sin(n\phi)}$ amplitudes are found to be compatible with zero (not shown here).

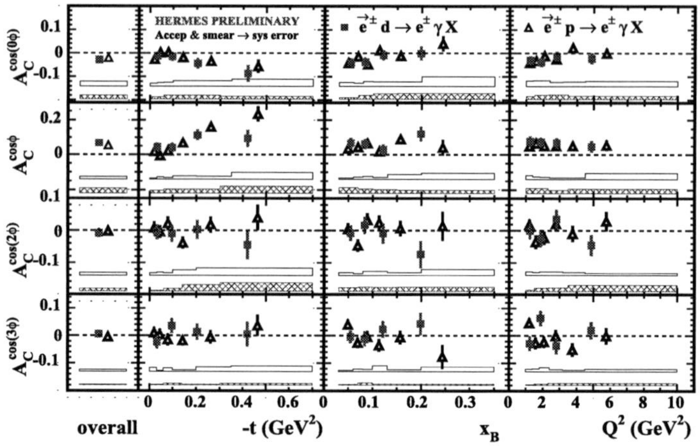

FIGURE 3. The BCA amplitudes for the hard exclusive electroproduction of photons from hydrogen and deuterium targets.

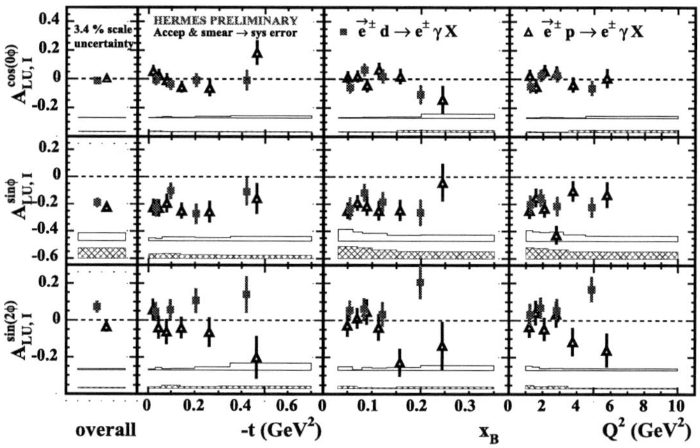

FIGURE 4. The BSA amplitudes $A_{LU,I}^{\sin(n\phi)}$ for the hard exclusive electroproduction of photons from hydrogen and deuterium targets.

REFERENCES

1. F.M. Dittes et al., *Phys. Lett.* **B209**, 325 (1988).
2. D. Müller et al., *Fortsch. Phys.* **42**, 101 (1994).
3. X. Ji, *Phys. Rev. Lett.* **78**, 610 (1997).
4. M. Diehl et al., *Phys. Lett.* **B411**, 193 (1997).
5. A. V. Belitsky, D. Müller and A. Kirchner, *Nucl. Phys.* **B269**, 323 (2002).
6. HERMES coll., A. Airapetian et al., JHEP, 06:066, 0802.2499 (2008).
7. M. Vanderhaeghen, P.A.M. Guichon and M. Guidal, *Phys. Rev.* **D60**, 094017 (1999).
8. K. Goeke, M.V. Polyakov and M. Vanderhaeghen, *Prog. Part. Nucl. Phys.* **47**, 401 (2001).

The Hybrid Generalized Parton Distributions Parametrization

S. Ahmad*, H. Honkanen†, S. Liuti** and S. K. Taneja‡

*University of Wisconsin-Whitewater, 800 West Main St., Whitewater, Wisconsin 53190, USA.
†Iowa State University, Ames, IA 50011, USA.
**University of Virginia, 382 McCormick Road, Charlottesville, Virginia 22904, USA.
‡Stony Brook, NY, USA

Abstract. We construct the ERBL region of the unpolarized Generalized Parton Distributions using the phenomenological constraints obtained from recent measurements of the $ep \to e'\gamma p$ reaction, from the nucleon form factor data, and from deep inelastic inclusive measurements.

Keywords: Deeply Virtual Compton Scattering, Generalized Parton Distributions
PACS: 12.38.-t,12.38.Cy,13.85.Qk,13.88.+e

THE ERBL REGION

The ERBL [1] region of the Generalized Parton Distributions (GPDs) is defined as the region where the returning quark in the factorized Deeply Virtual Compton Scattering (DVCS) amplitude diagram has negative longitudinal momentum, therefore describing the process of a quark antiquark pair emerging from the nucleon. We construct the ERBL region of the unpolarized Generalized Parton Distributions using the phenomenological constraints obtained from recent data on $ep \to e'\gamma p$, including the Bethe-Heitler (BH) process, DVCS, and BH-DVCS interference term. These constraints represent important additions to the set of constraints used in previous parametrizations of Generalized Parton Distributions (GPDs) including data on the nucleon electromagnetic form factors,inclusive scattering data [2, 3, 4, 5], and lattice QCD evaluations of GPD Mellin moments [5] (The latter are, however, known to introduce a somewhat large model dependence, see discussion in [5]). This provides a phenomenological determinations, of GPDs *i.e.* relying entirely on experimental data, at values of the skweness paramter, $\zeta > 0$. We used experimental data from Jefferson Lab [6] given at the scales $Q^2 = 1.9, 2.3$ GeV^2, separately for the imaginary part of the DVCS amplitude which gives information on GPDs at $X = \zeta$, and for the real part that provides insight into the ERBL region, corresponding to $0 < X < \zeta$. Notice that the data from Hermes at low Bjorken x [7] were not included in our analysis which as we will see is mainly centered in the valence region. We reiterate that given the way that GPDs enter the amplitude for DVCS and other exclusive processes observables, namely through a convolution over the variable X, a *direct* determination of these quantities from experiment cannot be obtained. A necessary step is therefore to devise theoretically motivated parametrizations that although relying on a set of assumptions, are sufficiently flexible to allow for meaningful fits of both the DVCS and the inclusive data. In this paper we propose a parametrization based on such criteria. We describe the sets of constraints available in particular from the new

CP1149, *Spin Physics, 18th International Spin Physics Symposium*
edited by D. G. Crabb, D. B. Day, S. Liuti, X. Zheng, M. Poelker, and Y. Prok

data, and we give the parametrizations for our unpolarized GPDs.

PHENOMENOLOGICAL CONSTRUCTION

The high precision measurements from Jefferson Lab [6] provide both the real and imaginary parts of the Bethe-Heitler BH-DVCS interference term at $Q^2 \approx 2$ GeV^2, and $x_{Bj} = 0.36$, showing the dominance of the twist-2 contribution. This, in turn, can be written in terms of GPDs as:

$$\mathscr{C}^{\mathscr{I}} = F_1(t)\mathscr{H}(\zeta,t) + \frac{\zeta}{2-\zeta} G_M(t)\widetilde{\mathscr{H}}(\zeta,t) - \frac{t}{4M^2} F_2(t)\mathscr{E}(\zeta,t), \quad (1)$$

with:

$$\mathscr{F}(\zeta,t) = \Re e\mathscr{F}(\zeta,t) + i\Im m\mathscr{F}(\zeta,t) \quad (2)$$

and:

$$\Re e\mathscr{F}(\zeta,t) = \mathscr{P}\int_0^1 dX \left(\frac{1}{X-\zeta} + \frac{1}{X}\right) F^q(X,\zeta,t). \quad (3a)$$

$$\Im m\mathscr{F}(\zeta,t) = -\pi \sum_q e_q^2 \left[F^q(\zeta,\zeta,t) + F^{\bar{q}}(\zeta,\zeta,t)\right] \quad (3b)$$

In Eqs.(2,3a,3b), $\mathscr{F} \equiv \{\mathscr{H}, \mathscr{E}, \widetilde{\mathscr{H}}\}$, and $F^q \equiv \{H^q, E^q, \widetilde{H^q}\}$, H^q and E^q being the unpolarized GPDs and $\widetilde{H}^q$ the polarized one; $q = u,d,s$. Finally, $F_{1(2)}$ are the Dirac (Pauli) form factors, $G_M = F_1 + F_2$ (for details on the photon electroproduction cross section and the harmonic decomposition of its (BH)^2, (DVCS)^2, and BH-DVCS interference terms see Ref.[8]). In previous work [4, 5] we presented our parametrization of unpolarized GPDs in the flavor Non Singlet (NS) sector. The parametrization is valid at both zero and non-zero skewness, and it uses constraints from both the nucleon form factors and the parton distribution functions,$q(X)$, or the forward limit of H^q, $q(X) \equiv H^q(X,0,0)$. At $\zeta = 0$ these constraints determine the quantities $H^q(X,0,t)$ and $E^q(X,0,t)$, accurately, for the mere fact that the functional dependence on ζ needs not to be determined. The parametrization provided in [4] offers the additional advantage that it can be extended smoothly to $\zeta \neq 0$ in the $X > \zeta$ region – the so-called DGLAP [9] region – by a simple kinematical shift in the quarks longitudinal momenta [5]. In Ref.[5] we predicted the values of the imaginary part of the amplitudes in Eq.(1). Predictions for both the proton and neutron data are in excellent agreement wih the data, as well as the values for the total orbital angular momentum extracted [5].

We specify the parametrization's form:

$$H_{DGLAP}(X,\zeta,t) = G(X,\zeta,t)R(X,\zeta,t) \quad (4)$$

(a similar form is obtained for $E(X,\zeta,t)$), where $R(X,\zeta,t)$ is a Regge motivated term that describes the low X and t behaviors, while the contribution of $G(X,\zeta,t)$, obtained using a spectator model, is centered at intermediate/large values of X. In order to describe the $X < \zeta$ – ERBL [1] – region one needs additional constraints since here

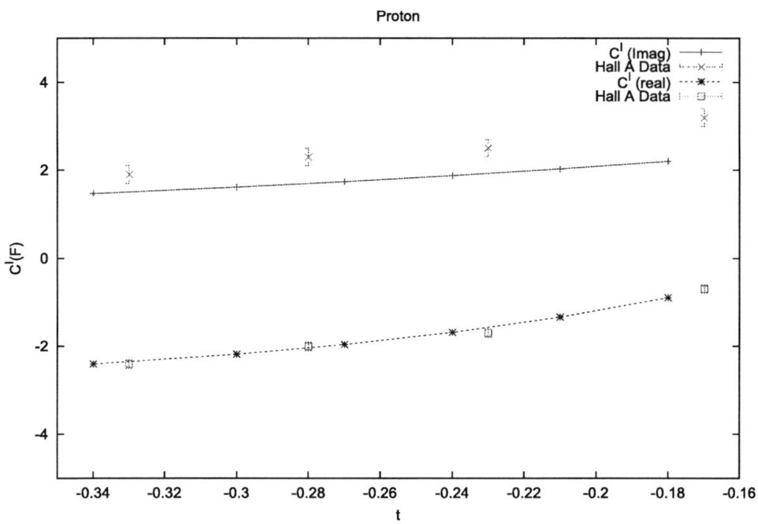

FIGURE 1. Bethe-Heitler BH-DVCS Interference term for Proton, real and imaginary parts.

the interpretation in terms of quark and gluon structure is not so obvious. It is in fact becoming increasingly manifest that in order to extract and interpret the amplitudes from a variety of hard exclusive experiments including DVCS in the ERBL region, it is fundamental to understand and connect the partonic structure of t-channel exchanges to previous information from Regge phenomenology [10, 11]. Here there is dominance of the scattering from a $q\overline{q}$ pair emerging from the initial nucleon in these processes.

We present preliminary results obtained with a fit function:

$$F(X,\zeta) = X^{\alpha}(a+bX) + \frac{cX(X-\zeta)}{d} \tag{5}$$

The fit parameters are a, b, c and d. This provides a negative dip in the GPD approaching $x = \zeta$, thus giving the right t dependence for the DVCS data. This is perhaps similar to a dip in ERBL region found in our previous study using the Lattice Data for the moments for GPDs [5]. This function satisfies the polynomiality condition up to ξ^2 terms.

The form factor constraint is imposed by:

$$\int_0^{\zeta} dX H^q_{ERBL}(X,\zeta,t) = F_1^q(t) - \int_{\zeta}^1 dX H^q_{DGLAP}(X,\zeta,t) \tag{6a}$$

$$\int_0^{\zeta} dX E^q_{ERBL}(X,\zeta,t) = F_2^q(t) - \int_{\zeta}^1 dX E^q_{DGLAP}(X,\zeta,t) \tag{6b}$$

with $H_{DGLAP}(X,\zeta,t)$ and $E_{DGLAP}(X,\zeta,t)$ determined from our earlier parametrization [4].

Fig. 1 shows our fit to the Hall A data for the proton. The amount of polynomiality violation for the higher moments is shown in Fig. 2.

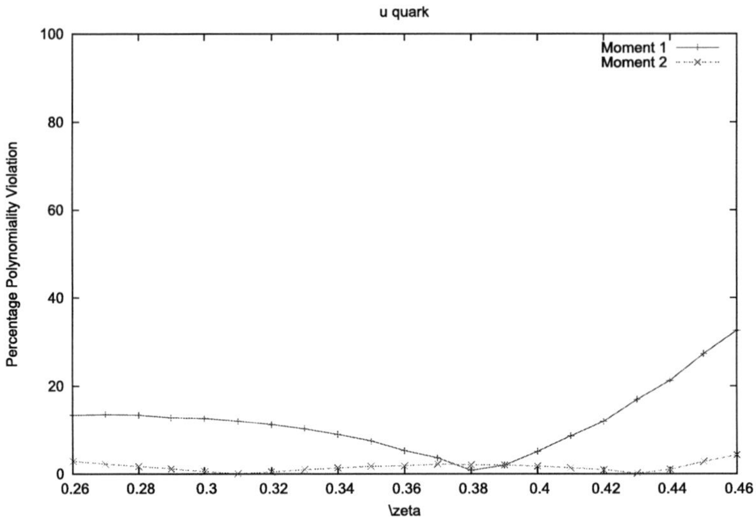

FIGURE 2. Percentage of polynomiality violation for higher moments.

CONCLUSIONS

We use Lattice data on the Moments of Generalized Parton Distributions as well as the experimental data available to construct a parametrization of GPDs. It is a physically motivated approach aimed at a model independent extraction of GPDs from experimental data.

REFERENCES

1. A.V. Efremov and A.V. Radyushkin, Theor. Math. Phys. 42, 97 (1980); Phys. Lett. 94B, 245 (1980); S.J. Brodsky and G.P. Lepage, 87B, 359 (1979).
2. M. Diehl, T. Feldmann, R. Jakob and P. Kroll, Eur. Phys. J. C **39**, 1 (2005).
3. M. Guidal, M. V. Polyakov, A. V. Radyushkin and M. Vanderhaeghen, Phys. Rev. D **72**, 054013 (2005).
4. S. Ahmad, H. Honkanen, S. Liuti and S.K. Taneja, Phys. Rev. D **75**, 094003 (2007).
5. S. Ahmad, H. Honkanen, S. Liuti and S.K. Taneja, EPJC, *in press*; arXiv:0708.0268 [hep-ph]
6. C. Munoz Camacho et al., Phys.Rev.Lett.97:262002,2006.
7. A. Airapetian *et al.* [HERMES Collaboration], Phys. Rev. Lett. **87**, 182001 (2001).
8. M. Diehl, Phys. Rept. **388**, 41 (2003).
9. V.N. Gribov and L.N. Lipatov, Sov. Phys. J. 15, 438 (1972); 675 (1972); Yu.L. Dokshitzer, Sov. Phys. JETP 46, 641 (1977); G. Altarelli and G. Parisi, Nucl. Phys. B126, 298 (1977)
10. A. P. Szczepaniak and J. T. Londergan, Phys. Lett. B **643**, 17 (2006)
11. V. Guzey and M. V. Polyakov, Eur. Phys. J. C **46**, 151 (2006) Phys. Rev. Lett. **97**, 262002 (2006)

SPIN IN SOFT HADRONIC REACTIONS

Physics Program at COSY-Jülich with Polarized Hadronic Probes

Andro Kacharava

Institut für Kernphysik, Forschungszentrum Jülich and Jülich Center for Hadron Physics Leo-Brandt-Str. 1, 52425 Jülich, Germany

Abstract. Hadron physics aims at a fundamental understanding of all particles and their interactions that are subject to the strong force. Experiments using hadronic probes could contribute to shed light on open questions on the structure of hadrons and their interaction as well as the symmetries of nature. The COoler SYnchrotron COSY at the Forschungszentrum Jülich accelerates protons and deuterons with momenta up to 3.7 GeV/c. The availability of both an electron cooler as well as a stochastic beam cooling system allows for precision measurements, using polarized proton and deuteron beams in combination with polarized Hydrogen or Deuterium targets.

This contribution summarizes the ongoing physics program at the COSY facility using ANKE, WASA and TOF detector systems with polarized hadronic probes, highlighting recent results and outlining the new developments.

Keywords: Hadron physics, polarized proton and deuteron beams, COSY-Jülcih
PACS: 21.30.Fe, 25.40.-h, 29.20.-c, 29.25.Pj, 29.27.Hj, 29.40.-n

INTRODUCTION

Hadron physics aims at a fundamental understanding of all bound and resonant systems and their interactions that are subject to the strong force. Many aspects of the theory of the strong interactions (QCD) are not understood yet. It is one of the great challenges of contemporary physics to understand this key feature of QCD, the confinement of quarks and gluons. As a consequence of confinement information about quarks and gluons can only be obtained through investigations of hadronic systems – mesons, baryons, and possibly exotic states – their production, decays and interactions. Intriguing tests of QCD are possible with high precision experiments eventually involving polarization.

This contribution gives an overview of the ongoing physics program at the COSY facility with the emphasis on the polarization experiments using ANKE, WASA and TOF installations, as well as highlighting a few selective results.

EXPERIMENTAL TOOLS

The COoler SYynchrotron COSY–Jülich stores and accelerates both unpolarized and polarized proton and deuteron beams injected from the injector cyclotron JULIC after preacceleration up to momenta of 0.3 GeV/c for protons and 0.55 GeV/c for deuterons [1]. COSY subsequently accelerates those particles up to 3.7 GeV/c. It exploits electron cooling at injection energy and stochastic cooling between 0.8 GeV/c and the maximum momentum to provide precision beams for both internal and external tar-

CP1149, *Spin Physics, 18th International Spin Physics Symposium*
edited by D. G. Crabb, D. B. Day, S. Liuti, X. Zheng, M. Poelker, and Y. Prok

gets. To preserve polarization during acceleration of polarized protons, well-established methods are being used. A fast tune jumping system has been developed to overcome intrinsic resonances. Polarization across imperfect resonances is preserved by the excitation of the vertical orbit using correcting dipoles to induce total spin flips. The polarization of the circulating beam in COSY can be monitored continuously during acceleration with the internal EDDA detector. The achieved polarization for protons is higher than 75% up to final momentum. Vector and tensor polarized deuterons are also routinely accelerated in COSY with polarizations up to 60% [2].

The Jülich Center for Hadron Physics (JCHP), recently established in Forschungszentrum Jülich (FZJ) [3], operates several experimental facilities for in-beam and external beam experiments, in cooperation with large international collaborations.

- **ANKE** (**A**pparatus for studies of **N**ucleon and **K**aon **E**jectiles), is a large acceptance forward magnetic spectrometer at an internal target station in the COSY ring [4]. The central dipole is movable to adjust the momentum of the detected particles independent of the beam momentum. Using deuterium cluster targets, reactions on the neuteron are tagged by detecting the low energy recoil proton in silicon strip detectors in vacuum next to the target. A Polarized Internal Target (PIT) system comprising with: an Atomic Beam Source (ABS), a Lamb–Shift Polarimeter (LSP), and a Storage Cell (SC) is in use at ANKE since 2005 [5].
- **TOF** (**T**ime–**O**f–**F**light), a non–magnetic spectrometer combining excellent tracking capability with large acceptance and full azimuthal symmetry allowing to measure complete Dalitz plots [6]. TOF is optimized for final states with strangeness. With the new straw tube tracking system, TOF will have a significantly improved mass resolution and reconstruction efficiency.
- **WASA** (**W**ide **A**ngle **S**hower **A**pparatus), an internal 4π spectrometer with large solid angle acceptance, is operated at the internal COSY beam [7]. WASA comprises an electromagnetic calorimeter, a very thin superconducting solenoid, inner and forward tracking detectors and a frozen–pellet target. Charge and neutral reaction products and their decay products can be measured exclusively.

In addition, the unique COSY capabilities are used by the **SPIN@COSY**- and the **dEDM**-collaborations to investigate spin-manipulations [8] and to prepare a dedicated EDM-storage ring experiment [9]. The **PAX** collaboration [10] is addressing the spin–filtering studies by accelerated proton beam and polarized targets.

EXPERIMENTAL PROGRAM AT COSY

Overview

The hadron physics program at COSY can be summarized to be focused around the issues of spectroscopy, spin, and symmetry. Under the headline spectroscopy the foremost research object is the nucleon an its mutual interactions as well as its excited states and possible exotics. The role and manifestation of the strange quark is an issue studied in associated production of hyperons and strange mesons. The symmetries expected in strong interactions, such as chiral symmetry and isospin, and their breaking have come into the focus of investigations. The medium modifications in the form of final state interactions or possible bound states are also being investigated. With proton and deuteron beams, isospin and polarization of beam and target are used as tools. COSY research also provides a window into the future **FAIR** project [11] (Facility for Antiproton and Ion Research to be built at GSI, Darmstadt) with studies on spin manipulation and polarization build–up of protons (antiprotons) in polarized targets (PAX experiment [10]).

In the following chapters a few selective results will be presented.

NN–Interaction

An understanding of the *NN* interaction is fundamental to the whole of nuclear and hadronic physics. The database on proton–proton elastic scattering is enormous and the wealth of spin–dependent quantities measured by the EDDA collaboration [12] has allowed the extraction of *NN* phase shifts in the isospin $I = 1$ channel up to a beam energy of at least 2 GeV. The situation is far less advanced for the isoscalar channel where the much poorer neutron–proton data only permit the $I = 0$ phase shifts to be evaluated up to at most 1.3 GeV but with significant ambiguity above about 800 MeV.

The ANKE collaboration has embarked on a systematic program to measure the differential cross section and analyzing powers of the small momentum transfer deuteron charge–exchange $dp \to \{pp\}n$ breakup reaction up to the maximum energy at COSY of 1.15 GeV per nucleon, with the aim of deducing information on the *np* amplitudes. Higher energies per nucleon will be achieved through the use of a proton beam and deuterium target. Spin correlations will also be studied with a polarized beam and target. However, for these to be valid objectives, the methodology has to be checked in a region where the *np* amplitudes are reasonably well known.

The charge–exchange amplitude of the elementary $np \to pn$ scattering may be written in terms of five scalar amplitudes in the c.m. system as:

$$f_{np} = \alpha(q) + i\gamma(q)(\vec{\sigma}_1 + \vec{\sigma}_2)\cdot\vec{n} + \beta(q)(\vec{\sigma}_1\cdot\mathbf{n})(\vec{\sigma}_2\cdot\vec{n})$$
$$+\delta(q)(\vec{\sigma}_1\cdot\vec{m})(\vec{\sigma}_2\cdot\vec{m}) + \varepsilon(q)(\vec{\sigma}_1\cdot\vec{l})(\vec{\sigma}_2\cdot\vec{l}),$$

where α is the spin–independent amplitude between the initial neutron and final proton, γ is a spin–orbit contribution, and β, δ, and ε are the spin–spin terms [13]. Impulse approximation applied to $dp \to \{pp\}_{^1S_0}n$ reaction leads to predictions for the differential cross section and deuteron spherical analyzing powers in such a form that allows in the $^1S_0$ limit to extract values of $|\beta(q)|^2 + |\gamma(q)|^2$, $|\delta(q)|^2$, and $|\varepsilon(q)|^2$ for small values of the momentum transfer $\vec{q}$ between the initial proton and final neutron. The variation of the cross section with momentum transfer can be found in Fig. 1 for $E_{pp} < 3$ MeV [13]. The impulse approximation describes well the dependence on this variable up to $q =$ 140 MeV/*c*. Our experimental values of the two tensor analyzing powers are shown in Fig. 1 as a function of the momentum transfer.

Although all the experimental data agree with the impulse approximation model one might invert the question. How well can one determine the amplitudes if there were no information available from the *np* phase shifts? Although the data reported here were

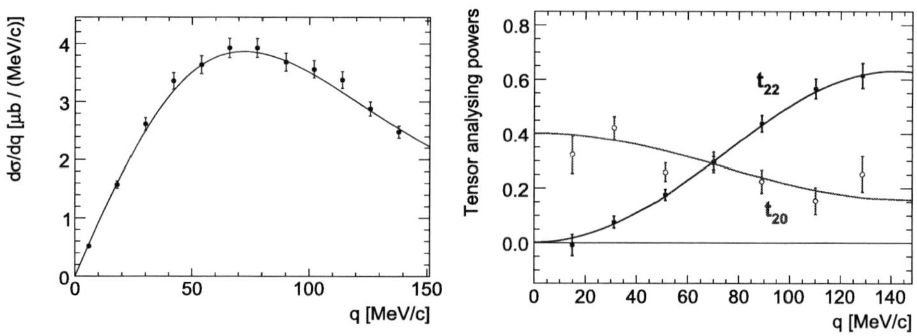

FIGURE 1. Left panel: Unpolarized differential cross section at $T_d = 1170$ MeV for the $dp \to \{pp\}n$ reaction of $E_{pp} < 3$ MeV. Right panel: Spherical tensor analyzing powers t_{20} (open symbols) and t_{22} (closed) for the $\vec{d}p \to \{pp\}n$ reaction at $T_d/2 = 585$ MeV for $E_{pp} < 1$ MeV. The solid curves are the impulse approximation predictions.

obtained over a short run, these are already sufficient to determine quite well the ratio of the $|\varepsilon(0)|/|\beta(0)|$ in the forward direction. The value obtained at the origin gives $t_{20} = 0.37 \pm 0.02$, where the error is purely statistical. The uncertainty introduced by the beam polarization would, however, contribute less than ± 0.01 to this. Since there is little or no dilution of the analyzing power by *p*-waves at $q = 0$, this result translates into an amplitude ratio of $|\varepsilon(0)|/|\beta(0)| = 0.61 \pm 0.03$, to be compared with SAID [14] value of 0.58. The proof–of–principle achieved in [13] for the method suggests that measurements at higher energies will provide useful information in regions where the existing *np* database is far less reliable.

Few Nucleon Systems with Diproton Final States

Properties of few nucleon systems have long been a subject of intensive theoretical and experimental study. A particularly important role in this study has been played by processes with the deuteron, a bound state of proton and neutron. In this respect, a particular interest arises for the investigation of the diproton, an unbound system of two protons with small excitation energy. The diproton, being the isospin partner of the deuteron, possesses different quantum numbers. Selection of the excitation energy of the proton pair $E_{pp} < 3$ MeV ensures dominance of the $^1S_0$ state of the diproton, which simplifies significantly the theoretical interpretation. The reactions with formation of a diproton involve transitions in the *NN* system, different from the case of the deuteron, and, in particular, the role of the Δ isobar is expected to be much suppressed because the *S*-wave ΔN intermediate state is forbidden. The diproton program at ANKE involves several experiments:

- Study of mesonless deuteron breakup $pd \to \{pp\}_s n$ in collinear kinematics: (i) at low momentum transfers (charge–exchange reaction) [13], and (ii) at high momentum transfers [15, 16].

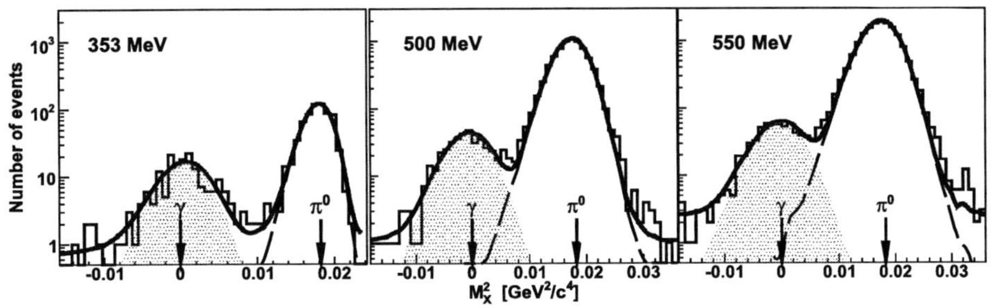

FIGURE 2. Missing-mass-squared for the $pp \to ppX$ reaction with $E_{pp} < 3$ MeV. The shaded area corresponds to the predicted γ peak, the dashed line to the π^0, and the solid to the sum of these and a straight line background.

- Study of meson production in $pN \to \{pp\}_s X$, where X can be : (i) a single pion [17], and (ii) the $(\pi\pi)$ system used to study ABC effect in $pp \to \{pp\}_s(2\pi)$ [18].
- Study of the inverse diproton photodisintegration $pp \to \{pp\}_s\gamma$.

The fundamental reaction $pp \to \{pp\}_s\gamma$ has been observed for proton beam energies of $T_p = 353, 500$, and 550 MeV [19]. This is equivalent to photodisintegration of a free $^1S_0$ diproton for photon energies $E_\gamma \approx T_p/2$. The events from this process produce a peak at $M_x^2 \approx 0$ in the experimental missing-mass spectra (Fig. 2).

The differential cross sections measured for c.m. angles $\theta_{pp}^{cm} < 20°$ exhibit a steep increase with angle that is compatible with E1 and E2 multipole contributions [19]. The ratio of the measured cross sections to those of $np \to \gamma d$ is on the $10^{-3} - 10^{-2}$ level. The increase of the $pp \to \{pp\}_s\gamma$ cross section with T_p might reflect the influence of the $\Delta(1232)$ excitation. Further information on this process will be obtained within the planned measurement of the $pp \to \{pp\}_s\pi^0$ and the $pn \to \{pp\}_s\pi^-$ process with a polarized proton beam at ANKE [20].

Single pion production and the charge–exchange breakup reaction are the key double polarization experiments in the physics program at ANKE [21].

η–He Interaction and η–Mass Measurement

New and very precise data on the $dp \to {}^3\text{He}\,\eta$ reaction near threshold [22, 23], taken at COSY, confirm the energy dependence of the total cross section found in earlier experiments, but with much finer steps in energy over an extended range. The observed variation of the total cross section and the angular distributions near threshold means that the both, the phase and magnitude of the *S*–wave amplitude, falls very rapidly with the η centre–of–mass momentum [24]. Such a behaviour is expected from a quasi–bound or virtual η^3He state. The interpretation can be investigated further through the measurements of $dp \to {}^3\text{He}\,\eta$ analyzing powers and spin correlations, which will allow us to pursue the investigation without some of the simplifying assumptions which have been made in the current analysis. Such an experiment has been conducted at ANKE [25] and the analysis is in progress.

Using the experience gained in this investigation, it has been proposed to determine the mass of the η–meson in a precision measurement at COSY in a new and innovative way [26]. The experiment made use of a polarized deuteron beam incident on an unpolarized cluster jet target inside the COSY–ring. The reaction to be exploited was $\vec{d}p \rightarrow ^3\mathrm{He}\eta$, which has been identified by ^{3}He detection in the magnetic spectrometer ANKE and a missing–mass analysis. The techniques which have been employed are: (i) ramping of the deuteron beam energy across the η production threshold; (ii) ^{3}He identification with full acceptance over the full energy range; (iii) determination of the excess energy from kinematics ('radius method') [26]; and (iv) determination of the absolute deuteron beam momentum ('spin resonance method') [27]. All of these steps have been tested in previous experiments at COSY and have been shown to work. Taking the uncertainty limits, which have already been achieved, we expect the final uncertainty of the η–mass to be less than 50 keV/c^2, *i.e.* equivalent to or better than the best existing measurements.

Strangeness Production: The $\Lambda - \mathrm{N}$ Scattering Length

For systems with strangeness, there are still many open questions and it is not even clear if the kaon is more appropriately treated as a heavy or a light particle. To improve further our understanding of the dynamics of systems containing strangeness, better data are needed. The insights to be gained are relevant, not only for few–body physics, but also for the formation of hypernuclei, and might even be of significance for the structure of neutron stars. The hyperon–nucleon scattering lengths are obvious quantities of interest in this context.

The Jülich (IKP) theory group has developed a method to enable one to deduce a scattering length directly from data on a production reaction, such as $pp \rightarrow pK^+\Lambda$, in terms of an integral over the invariant Λp mass (m_X) distribution [28]. Using this method it can be seen that the inclusive Saclay $pp \rightarrow K^+X$ data [29], which had a mass resolution of 4 MeV, allow the extraction of a scattering length with an experimental uncertainty of only 0.2 fm. However, the actual value of the scattering length obtained in this way is not meaningful, since it represents the incoherent sum of the $^3S_1$ and the $^1S_0$ Λp final states with unknown relative weights. It is important to try to separate them.

The COSY-TOF detector has been designed particularly for the investigation of final states with strangeness. The detector combines excellent tracking capability with large acceptance and full azimuthal symmetry [6]. The tracking information, in particular close to the interaction point, allows to identify the production of strange hadrons with almost no background, based on the detection of the displaced decay vertices of Λ hyperons ($\Lambda \rightarrow p\pi^-$) and K_S mesons ($K_S \rightarrow \pi^+\pi^-$).

The available experimental data on Λp elastic scattering and on the Λp final state interaction in production reactions only allow to deduce the spin-averaged value of the Λp scattering length, with yet large uncertainty. COSY-TOF plans to measure the spin-triplet Λp scattering length by analysing the $\vec{p}p \rightarrow \Lambda pK^+$ reaction around 90° K^+ polar angle in the center-of-mass frame. The goal of the measurement is to achieve an accuracy below 0.3 fm, which is also the theoretical uncertainty of the method [28] to deduce the scattering length from production reactions.

The Λn triplet final state could be isolated unambiguously by measuring the unpolarized K^+ spectrum in the forward direction and this weighted with the incident spin correlation, obtained using a transversally polarized beam and target. This we will achieve by looking at the $\vec{p}\vec{d} \to p_{sp}K^+X$ reaction using the ANKE installation [21].

Symmetry and Symmetry Breaking

Symmetry and symmetry breaking patterns as well as the structure of hadrons are the primary focus of the WASA-at-COSY collaboration. One of the key experiments is a measurement of $\vec{d}d \to {}^4\mathrm{He}\,\pi^0$ to extract the P-wave contribution to the Charge Symmetry (CS) breaking amplitude [7]. In preparation of data taking with a polarized deuteron beam, the reaction has been measured with an unpolarized beam at a momentum of p_d = 1.2 GeV/c (T_d =351 MeV) to establish the so far unknown cross section, which can only be estimated based on data available at much lower excess energy and to demonstrate the separation of the helium isotopes after installation of the new forward tracking hodoscope [31]. Data analysis is presently in progress, and from first checks of the data a total statistics of 500 to 1000 events is expected for the CS breaking reaction.

Spin–Filtering Studies at COSY

An entirely new chapter in studies of hadron physics might unfold with the advent of a polarized antiproton beam provided an efficient method for polarizing antiprotons can be demonstrated. Polarized antiproton beams could be possibly employed e.g. at FAIR [11], and at the European Organization for Nuclear Research (CERN). There are major technological challenges to be met in order to produce beams of polarized antiprotons. Foremost is the question of how to polarize antiprotons effectively. The only viable method demonstrated so far that could yield a beam of polarized antiprotons is spin–filtering of a stored beam by selective loss and/or selective spin–flip through repetitive interaction with a polarized internal target. (The proof-of-principle experiment (FILTEX) has been carried out with protons at TSR, Heidelberg [32].) Although the FILTEX experiment clearly demonstrated that the spin–filtering technique works, a unique interpretation of the result is not available yet. In order to make spin–filtering a practicable method for polarizing antiprotons, preparatory experiments with protons and antiprotons are required. In order to accomplish this goal, a proper experimental setup and a theoretical structure have to be implemented to support a series of dedicated experimental investigations, which has recently started at COSY [33].

SUMMARY

COSY in combination with its complementary experimental facilities and polarized hadronic beams and targets provide unique opportunities to investigate a variety of aspects in hadron physics. The experience gained in undertaking polarization measurements will be put to good use for future developments at FAIR.

ACKNOWLEDGEMENTS

I would like to acknowledge the help and cooperation of all of my colleagues at the Institute for Nuclear Physics at the Research Center Jülich (Germany) as well as those within the collaborations that come to COSY to perform experiments.

REFERENCES

1. R. Maier, *Nucl. Instr. Meth. A* **390**, 1 (1997).
2. D. Chiladze, *Phys. Rev. ST–AB* **9**, 050101 (2006).
3. See http://www.fz-juelich.de/ikp/jchp/en/research.shtml
4. S. Barsov et al., *Nucl. Instr. Meth. A* **462**, 364 (2001).
5. See contribution from Ralf Engels to this proceedings.
6. A. Böhm et al., *Nucl. Instr. Meth. A* **443**, 238 (2000).
7. WASA-at-COSY Collaboration, proposal for *WASA at COSY*, nucl-ex/0411038.
8. SPIN@COSY Collaboration, http://www.fz-juelich.de/ikp/spin@cosy/
9. COSY proposal #176, http://www.fz-juelich.de/ikp/publications/List-of-all-COSY-Proposals.shtml
10. P. Lenisa and F. Rathmann for the PAX Collaboration, http://www.fz-juelich.de/ikp/pax/
11. See http://www.gsi.de/fair/index-e.html
12. M. Altmeier et al., *Phys. Rev. Lett.* **85**, 1819 (2000); *Eur. Phys. J. A* **23**, 351 (2005); F. Bauer et al., *Phys. Rev. C* **71**, 054002 (2005).
13. D. Chiladze et al., nucl-ex/0811.3288 (2008).
14. R.A. Arndt, et al., *Phys. Rev. C* **48**, 1926 (1993); http://gwdac.phys.gwu.edu/analysis/pd-analysis.html
15. V. Komarov et al., *Phys. Lett. B* **553**, 179 (2003).
16. S. Yaschenko et al., *Phys. Rev. Lett.* **94**, 072304 (2005).
17. V. Kurbatov et al., *Phys. Lett. B* **661**, 22 (2008).
18. S. Dymov et al., nucl-ex/0902.0715 (2009).
19. V. Komarov et al., *Phys. Rev. Lett.* **101**, 102501 (2008).
20. COSY proposals: #158, #192 (2008), http://www.fz-juelich.de/ikp/anke/en/proposals.shtml.
21. COSY proposal #152 (2005), *Spin Physics From COSY To FAIR*, nucl-ex/0511028.
22. T. Mersmann et al., *Phys. Rev. Lett.* **98**, 242301 (2007).
23. J. Smyrski et al., *Phys. Lett. B* **649**, 258 (2007).
24. C. Wilkin et al., *Phys. Lett. B* **654**, 92 (2007).
25. COSY Proposal #157 (2006), http://www.fz-juelich.de/ikp/anke/en/proposals.shtml
26. COSY Proposal #187 (2007), http://www.fz-juelich.de/ikp/anke/en/proposals.shtml
27. See contribution from Maria Leonova to this proceedings.
28. A. Gasparyan et al., *Phys. Rev. C* **69**, 034006 (2004).
29. R. Siebert et al., *Nucl. Phys. A* **567**, 819 (1994).
30. COSY proposal #178 (2007), *Strangeness Physics at COSY–TOF*.
31. COSY proposal #173 (2007), http://www.fz-juelich.de/ikp/wasa/physics.shtml
32. F. Rathmann et al., *Phys. Rev. Lett.* **71**, 1379 (1993).
33. See contributions from Frank Rathmann and Alexander Nass to this proceedings.

Spin Physics Program At IHEP-Protvino

V.V Mochalov[a], A.N. Vasiliev[a], N.A. Bazhanov[b], N.I.Belikov[a], A.A.Belyaev[c], N.S. Borisov[b], V.N. Grishin[a], A.M. Davidenko[a], A.A.Derevschikov[a], V.G. Kolomiets[b], V.I. Kravtsov[a], A.B. Lazarev[b], A.A.Lukhanin[c], Y.A. Matulenko[a], Y.M. Melnik[a], A.P. Meschanin[a], N.G.Minaev[a], D.A. Morozov[a], A.B. Neganov[b], L.V. Nogach[a], S.B.Nurushev[a], Yu.A. Plis[b], A.F. Prudkoglyad[a], A.V. Ryazantsev[a], P.A.Semenov[a], O.N. Shchevelev[b], L.F. Soloviev[a], Y.A. Usov[b], A.E.Yakutin[a]

[a] *IHEP,1Pobedy str., Protvino, Moscow reg, 142281, Protvino, Russia*
[b]*JINR, Joliot-Curie 6, 141980 Dubna, Moscow region, Russia*
[b]*NSC KIPT, 1, Akademicheskaya St., Kharkov, 61108, Ukraine*

Abstract. A new single-spin asymmetry A_N measurement has been carried out at IHEP at 50 GeV. A_N was found to be $-(6.2\pm1.5)\%$ in the polarized target fragmentation region at $-0.6<x_F<-0.25$. The result proves that asymmetry does not depend on beam energy. A review of polarization results from Protvino as well as a proposal of the new experiment SPASCHARM are also presented.

Keywords: Polarization, single-spin asymmetry, inclusive meson production.
PACS: 13.88.+e, 13.85.Ni

INTRODUCTION

The year 2008 is a specific one for IHEP-Protvino spin physics program. The polarization measurements in the hadron-hadron elastic scattering started at IHEP 40 years ago. The JINR-IHEP polarized target was installed at Protvino 30 years ago to measure single-spin asymmetry in exclusive and inclusive reactions of neutral mesons. Polarized beam from Λ-decay was created for the FODS experiment 15 years ago to measure asymmetry in inclusive charge particle production.

We are presenting the overview of spin physics program at IHEP including single-spin asymmetry measurements in the polarized target fragmentation region and the proposal of new polarized SPASCHARM experiment.

ASYMMETRY IN EXCLUSIVE REACTIONS

Asymmetry was measured in the reactions $\pi^- p_\uparrow \rightarrow \pi^0(\eta,\eta'(958),\omega(783),f2(1270))n$ at 40 GeV. Significant effects were found in the all reactions (Fig. 1 – Fig. 4). Polarization (asymmetry) has the following features:

- polarization has a minimum when differential cross-section changes it's slope;

CP1149, *Spin Physics, 18th International Spin Physics Symposium*
edited by D. G. Crabb, D. B. Day, S. Liuti, X. Zheng, M. Poelker, and Y. Prok

- there are oscillations in polarization behavior;
- there is an indication that asymmetry is bigger for heavier particles.

All these asymmetries were measured more than 30 years ago, nevertheless there are no theoretical models describing the data.

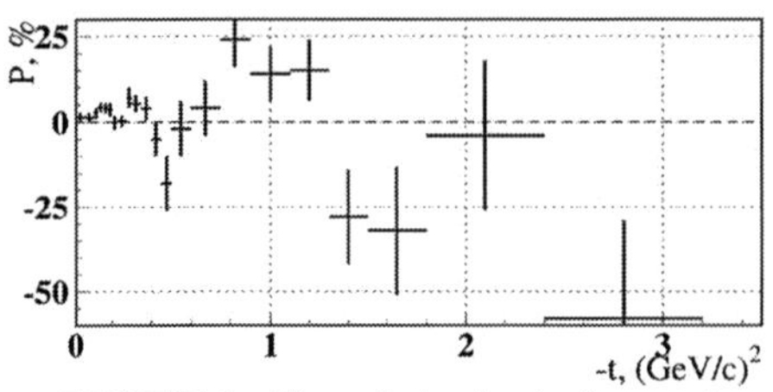

FIGURE 1. The polarization in the reaction $\pi^- p_\uparrow \to \pi^0 n$ [1]

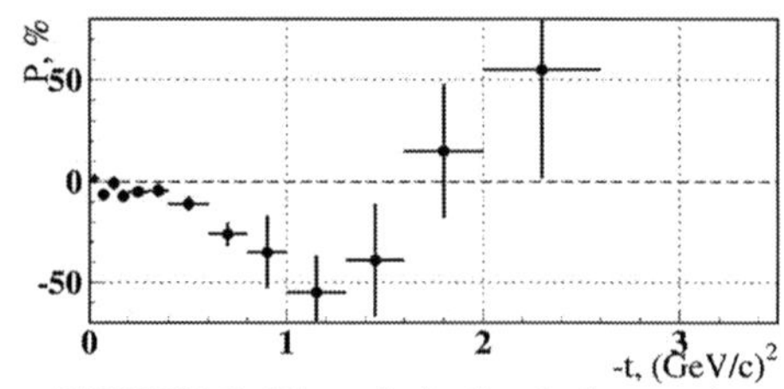

FIGURE 2. The polarization in the reaction $\pi^- p_\uparrow \to \eta n$ [2].

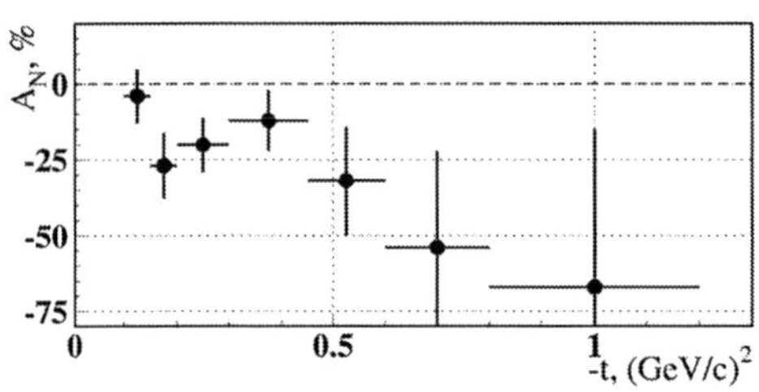

FIGURE 3. Asymmetry in the reaction $\pi^- p_\uparrow \to \omega(783) n$ [3]. ω was registered via $\pi\gamma$ decay (branching only 8.9%).

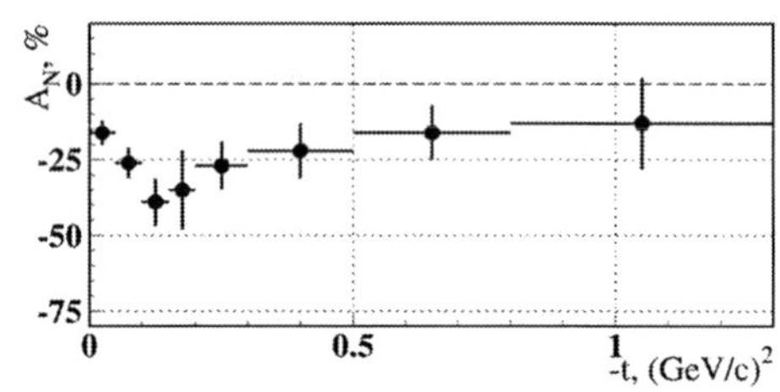

FIGURE 4. Asymmetry in the reaction $\pi^- p_\uparrow \to f_2(1270) n$ [4].

ASYMMETRY IN INCLUSIVE REACTIONS

The polarization study at IHEP was continued by inclusive program. Asymmetry in the reaction $\pi^- d_\uparrow \to \pi^0 X$ (Fig. 5) is zero at small p_T, and surprisingly achieves (16±3)% at $1.0<p_T<1.8$ GeV/c (systematic error 4%) in the unpolarized beam fragmentation region at $0.7<x_F<0.9$ [5]. A_N in inclusive π^0-production at large p_T at forward production and polarization in the charge exchange reaction $\pi^- p_\uparrow \to \pi^0 n$ are similar in the same kinematic region.

Asymmetry in the reaction $\pi^- p_\uparrow \to \pi^0 X$ (Fig. 6) in the polarized target fragmentation region also is zero at $|x_F| \approx 0$, increases by magnitude with $|x_F|$ increased and equals to -(6.9±2.8) at $x_F<-0.3$ [6]. In the both reaction the central region corresponds to small transverse momentum p_T. Nevertheless asymmetry in the reaction $\pi^- N(p,d)_\uparrow \to \pi^0 X$ (Fig. 7) increases by magnitude with p_T increased and equals to -(39.±7) at $p_T>2$ GeV/c [7]. The asymmetry does not depend on nuclei flavor.

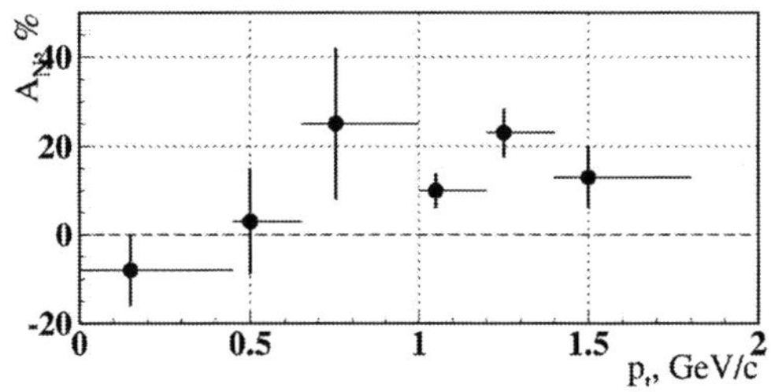

FIGURE 5. Asymmetry in the reaction $\pi^- d_\uparrow \to \pi^0 X$ at $0.7 < x_F < 0.9$ [5] in beam fragmentation region.

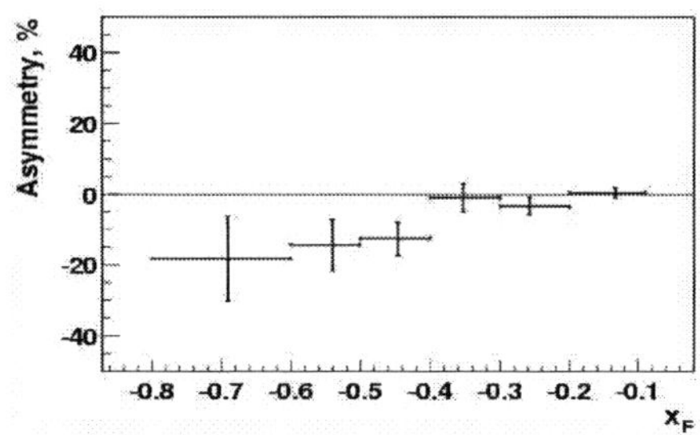

FIGURE 6. Asymmetry in the reaction $\pi^- p_\uparrow \to \pi^0 X$ in polarized target fragmentation region [6].

At the same time asymmetry in the reaction $pp_\uparrow \to \pi^0 X$ in the central region [8] is zero in the whole interval of p_T measured (Fig. 8) in agreement with E-704 data [9]. We may conclude that in the central region asymmetry depends on beam particle flavor. The difference may be explained by contribution of valence *u*-quark-antiquark combination due to π^- quark content.

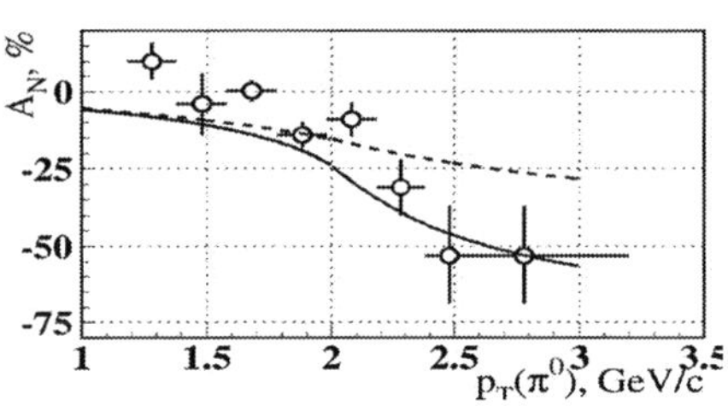

FIGURE 7. Asymmetry in the reaction $\pi^- N(p,d)_\uparrow \to \pi^0 X$ in central region [7]

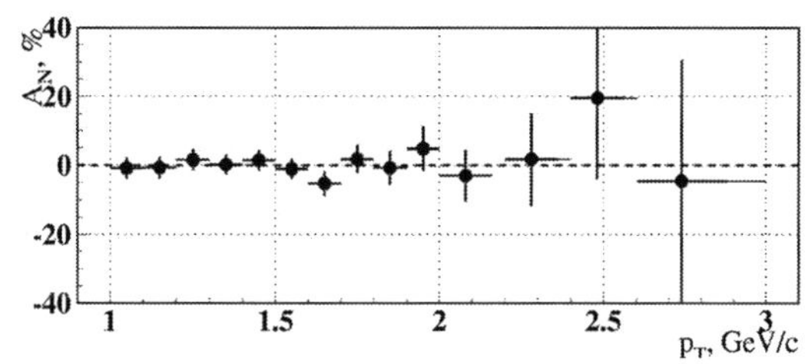

FIGURE 8. Asymmetry in the reaction $pp_\uparrow \to \pi^0 X$ in central region [8]

ASYMMETRY IN THE REACTION $PP_\uparrow \to \Pi^0 X$ IN THE TARGET FRAGMENTATOIN REGION

We are presenting a new result of asymmetry measurement in the reaction $pp_\uparrow \to \pi^0 X$ in the polarized target fragmentation region at 50 GeV. Two sets of data (2005 and 2007) are being used for analysis. The experiment was carried out at PROZA-2M setup (Fig. 9).

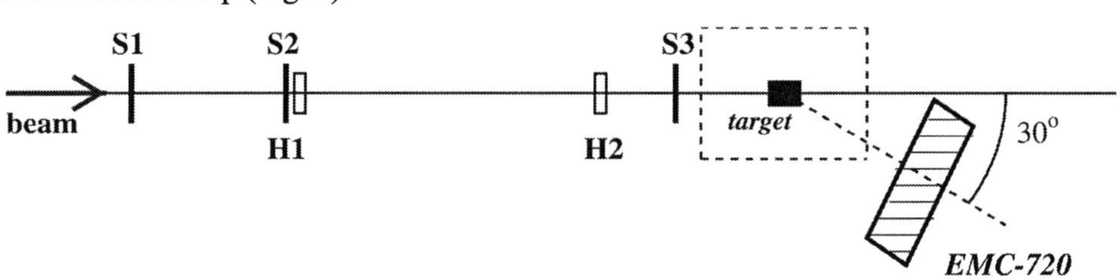

FIGURE 9. Experimental setup PROZA-2M. S1-S3 – trigger scintillation counters, H1-H2 – beam hodoscopes, EMC-720 – electromagnetic calorimeter from lead-glass counters.

A proton beam with intensity up to $7 \cdot 10^6$ *p*/cycle was extracted from U70 vacuum chamber using a bent crystal. Beam momentum spread ΔP/P was less than 0.013% (Fig. 10), beam size σ_x=2.7 mm, σ_y=3.7 mm.

The measurements in the backward hemisphere are very difficult, due to small acceptance and dependence of gamma-reconstruction on angle. Special procedure [10] allowed us to reconstruct π^0-mesons with a good efficiency (Fig. 11).

TABLE 1. A_N in the reaction $pp_\uparrow \to \pi^0 X$ on dependence of x_F.

x_F	-(0.13-11)	-(0.17-13)	-(0.21-17)	-(0.25-21)	-(0.35-25)	-(0.45-35)	-(0.60-45)
A_N,%	-(0.8±2.2)	-(2.0±1.8)	-(1.5±1.9)	-(2.3±2.1)	-(5.8±1.8)	-(6.8±4.1)	-(11.0±5.8)

Asymmetry dependence on x_F is presented in Table 1 and Fig. 12. Our result of A_N measurement in the inclusive π^0 production at polarized particle fragmentation region at $0.25 < x_F < 0.6$ at 50 GeV (6.2±1.5)% is in a very good agreement with A_N in the reaction $\pi^- p_\uparrow \to \pi^0 X$ at 40 GeV (6.9±2.8)% and with E704 data at 200 GeV (6.3±0.7)%.

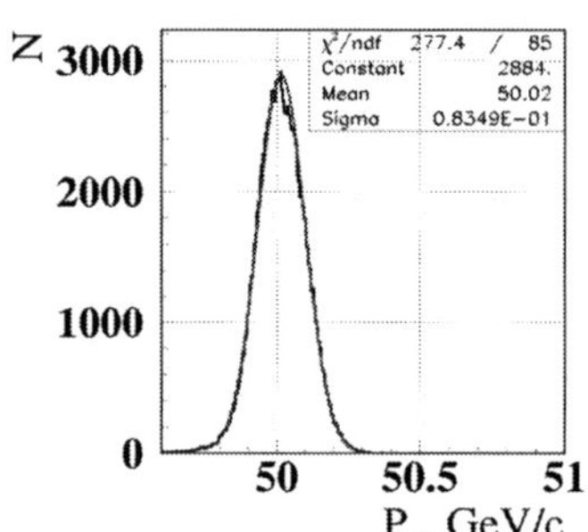

FIGURE 10. Experimental beam momentum distribution.

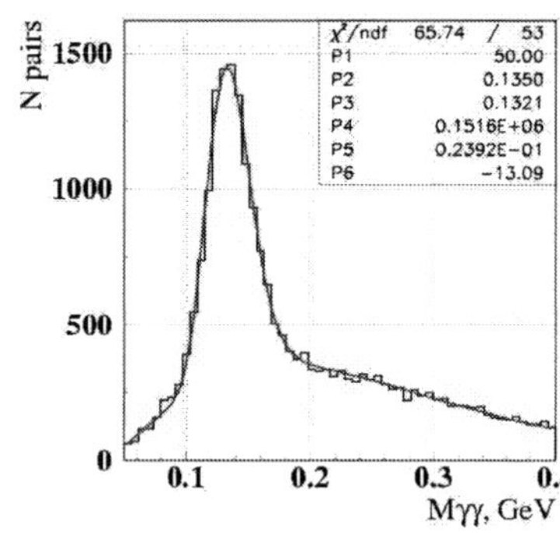

FIGURE 11. Experimental mass spectra of 2γ-events.

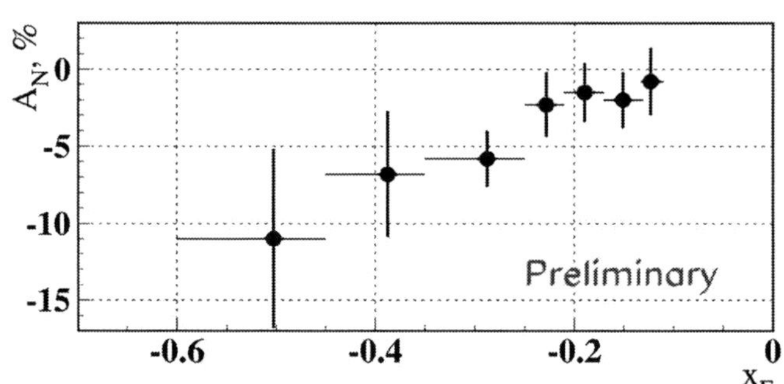

FIGURE 12. A_N in the reaction $pp_\uparrow \to \pi^0 X$ on dependence of x_F at 50 GeV.

ASYMMETRY MEASUREMENTS AT FODS-2 SETUP

Inclusive production of charged particles at IHEP is studied using FODS-2 setup. The results are presented in Fig. 13. FODS-2 experimental setup uses polarized beam from Λ-decays and allows to measure inclusive asymmetry of charged particles at different angles. Beam polarization is 40%. Asymmetry presented was measured for $\pi^{\pm}$, $K^{\pm}$, $p^{\pm}$ at two different kinematic ranges:

1. $|x_F| \le 0.2$, $0.5 \le p_T \le 4$ GeV/c, $\theta \approx 86^o$
2. $-0.1 \le x_F \le 0.7$, $0.3 \le p_T \le 2.0$ GeV/c, $\theta \approx 48^o$.

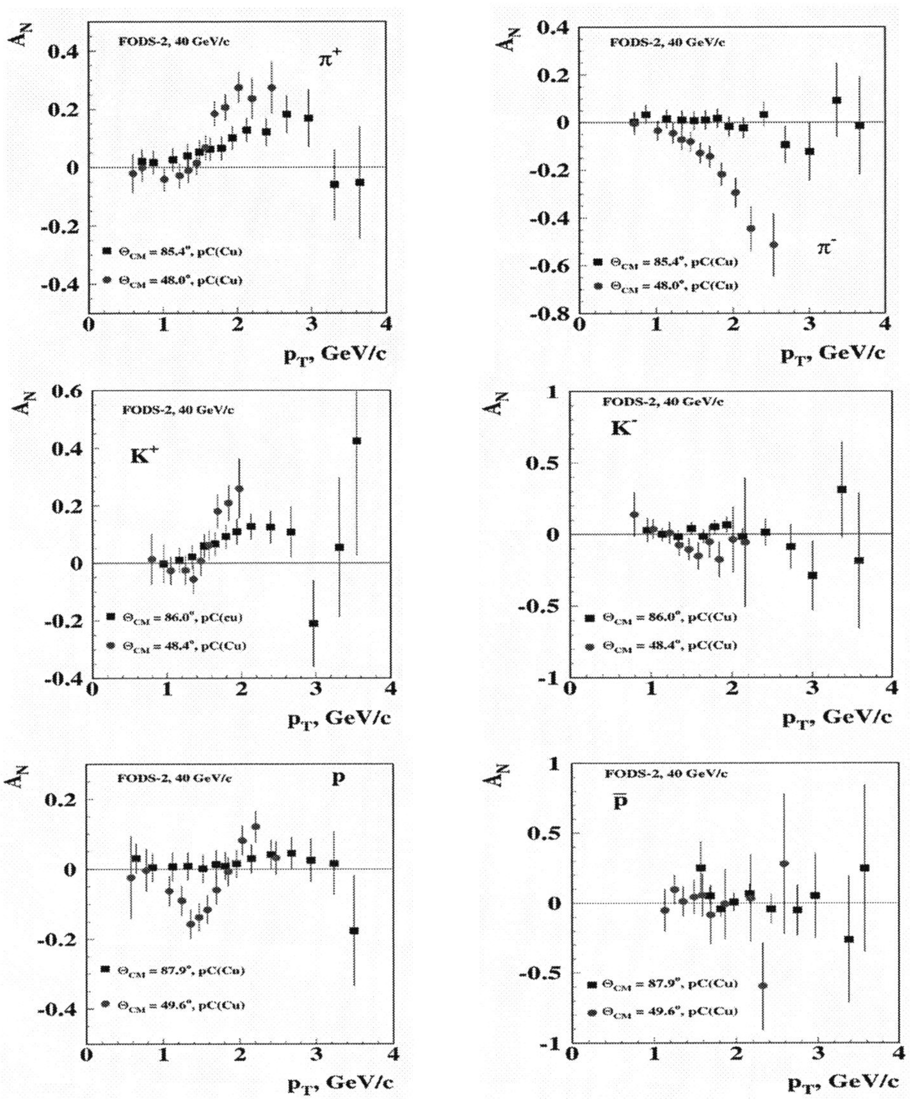

FIGURE 13. A_N in different inclusive reactions at the FODS-2 setup using polarized beam.

The detail description of FODS-2 results is presented elsewhere [11].

NEW SPIN PROGRAM AT IHEP

New polarization spin program SPASCHARM is under preparation now. Final goal of the project is to measure SPin ASymmetries in CHARMonium production. The second stage of the experiment is devoted to the measurements of double-spin asymmetry A_{LL} using both longitudinally polarized proton beam and target in charmonium production to study gluon polarization $\Delta G/G(x)$ at large x_B (around 0.3). The experimental setup layout is presented in Fig. 14. The detail description of the proposal was presented at XII Advanced Research Workshop on SPIN Physics [12].

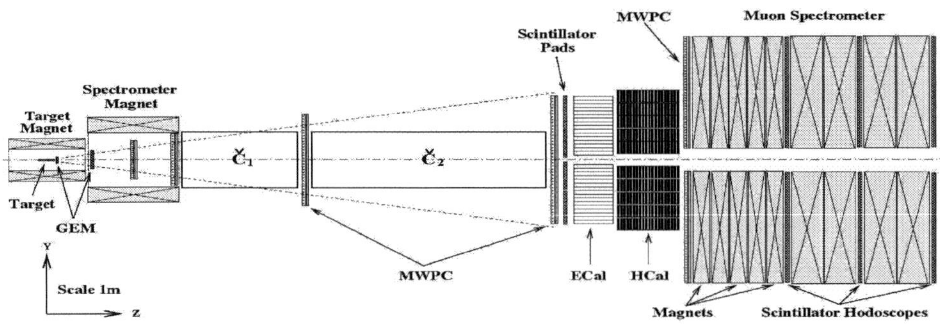

FIGURE 14. Experimental Setup SPASCHARM.

The main goal of the first stage of the new project is a precise study of single-spin asymmetry in miscellaneous inclusive and exclusive production of the light (*u, d, s* – quarks) resonances in the unpolarized beam fragmentation region. New experimental setup is under construction to start the experiment in 2011.

New experimental setup will allow us to measure resonances in both charge and neutral modes. We will be able to measure asymmetry in the exclusive charge exchange reactions $\pi^- p_\uparrow \rightarrow \omega(782)n$ and $\pi^- p_\uparrow \rightarrow \eta'(958)n$ with statistics 20 times more than measured previously and with statistic errors less than 2%. The statistics for the reaction $\pi^- p\uparrow \rightarrow f_2(1270)$ n will be three times higher, and statistic errors of 5% to be achieved.

Statistic errors of asymmetry measurement in the reactions $\pi^- p_\uparrow \rightarrow \omega(782)X$, ρX, $\eta'(958)X$ are 0.3–3% for different kinematic intervals. Simulated three-pion mass spectrum is presented in Fig. 15. K^+K^- mass spectrum is shown in Fig. 16.

Accuracy of the asymmetry measurement in the reaction $\pi^- p\uparrow \rightarrow f_2(1270)$ X is expected to be at the level of 0.1–1% (see Table 2). The errors are calculated for two different setup configurations: with two Cherenkov counters (7 m) as shown on Fig.14 and without second Cherenkov counter. In the last case electromagnetic calorimeter will be placed at 4 m downstream the target.

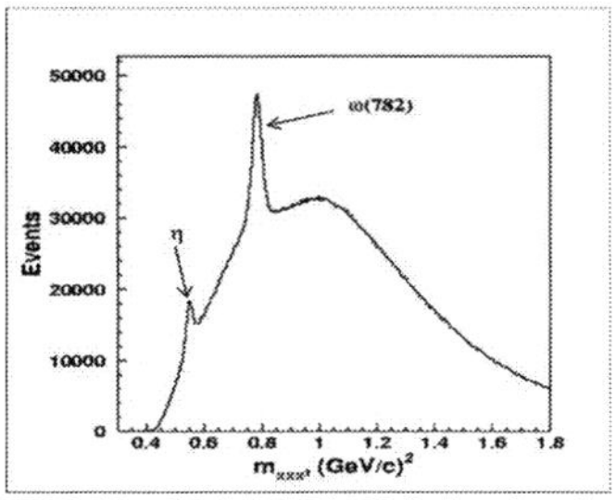

FIGURE 15. SPASCHARM simulated $\pi^+\pi^-\pi^0$ invariant mass spectrum.

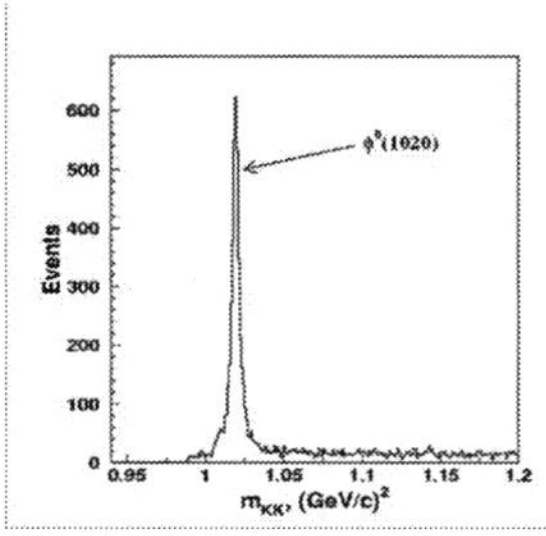

FIGURE 16. SPASCHARM simulated K^+K^- invariant mass spectrum.

TABLE 2. A_N accuracy (%) in the reaction $\pi^-p_\uparrow \to f_2X$ at 34 GeV for two different positions of electromagnetic calorimeter (summary error is written in **bold**).

p_T,GeV/c / x_F	0.5 – 0.6		0.6 – 0.7		0.7 – 0.8		0.8 – 0.9		0.9 – 1.0	
0.5-1.0 "4 m"	0.2		0.3		0.3		0.3		0.8	
"7 m"	0.2	**0.1**	0.3	**0.2**	0.3	**0.2**	0.4	**0.2**	1.0	**0.6**
1.0-1.5 "4 m"	0.3		0.3		0.3		0.4		1.7	
"7 m"	0.4	**0.2**	0.4	**0.2**	0.5	**0.3**	0.6	**0.3**	1.7	**1.2**
1.5-2.0 "4 m"	0.4		0.5		0.6		1.0			
"7 m"	0.6	**0.3**	0.7	**0.4**	0.8	**0.5**	1.0	**0.7**		
2.0-2.5 "4 m"	0.7		0.8		1.1		5.1			
"7 m"	1.4	**0.6**	1.3	**0.7**	1.7	**0.9**	7.0	**4.1**		

The simulation was done for all detectors to satisfy the requirements of the second stage of the experiment, i.e. the possibility to disentangle between three different states of χ_c states. The accuracy of momentum reconstruction should be at the level of 0.4% at 10 GeV while electromagnetic calorimeter (ECal) energy resolution should be not worse than 2.5%/√E (Fig.17).

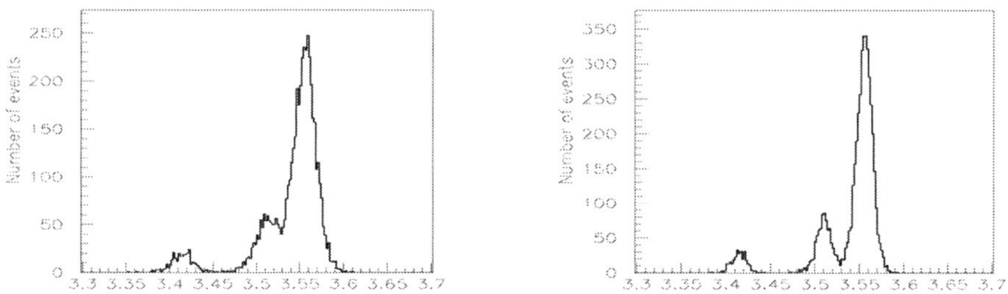

μ^+ and μ^- reconstruction from J/ψ decay, Δp/p= 0,003 at 10 GeV. 1C fit is not applied for J/ψ mass. ECal accuracy σ(E)/E = 12,0%/√E.

μ^+ and μ^- reconstruction from J/ψ decay Δp/p= 0,004 at 10 GeV. 1C fit is applied for J/ψ mass. ECal accuracy σ(E)/E = 2,5%/√E.

FIGURE 17. Reconstruction of χ_0(3410), χ_1(3510) and χ_2 (3555) masses at different conditions.

The separation between three χ_c states is required due to their different production mechanism. By measuring the cross-section and asymmetries of χ_c states as well as J/ψ production, one can understand the mechanism of charmonium production and gluon contribution to the proton spin (at least at $x_B \approx 0.3$, where it's contribution is expected to have maximal value).

Polarized beam is required to measure double-spin effects. We are discussing to possibilities to obtain polarized beam at the SPASCHARM polarized target:

- to produce polarize beam locally from Λ decay;
- to accelerate polarized proton beam in frame of new Super-ferric accelerator ring at IHEP.

SUMMARY

Essential spin effects have been found in both inclusive and exclusive reactions at IHEP. Single-spin asymmetry in many cases is NOT explained by a theory.

Asymmetry of inclusive pion production in the polarized projectile fragmentation region does NOT depend on energy (similar to hyperon polarization).

Intermediate (40-70 GeV) beam energy is a perfect tool to study polarization effects in miscellaneous reactions. High precision is getting more important than high energy.

The new experiment to study precisely spin effects in miscellaneous reactions is under preparation.

ACKNOWLEDGMENTS

The work was supported by State Atomic Energy Corporation "Rosatom" and RFBR grants 06-02-16119, 08-02-90455 and 09-02-00198.

We thank the FODS collaboration for the contribution of their result in our report.

REFERENCES

1. V.D. Apokin et al., *Sov. J. Nucl. Phys.* **45**, 840 (1987).
2. V.D. Apokin et al., *Z. Phys.* **C35**, 173 (1987).
3. I.A. Avvakumov et al. *Yad. Fiz.* **42,** 1146 (1985).
4. V.D. Apokin et al. *Yad. Fiz.* **47**, 727 (1988).
5. V.D. Apokin et al. *SPIN-91 (Protvino) Proc.*, 288, (1992).
6. A.N. Vasiliev et al, *Phys. At. Nucl.*, **67,** 1495 (2004).
7. V.D. Apokin et al, *Phys. Lett. B* **243**, 461 (1990).
8. A.N. Vasiliev et al, *Phys. At. Nucl.*, **67,** 1487 (2004).
9. D. L. Adams et al., *Z. Phys. C* **56**, 181 (1992).
10. A. Vasilev et al, *Instrum. Exp. Tech.* **50**, 458 (2007).
11. V.V. Abramov et al., e-Print: hep-ex/0511036 (2005).
12. A.N. Vasiliev, *Proc XII Advanced Research Workshop on SPIN Physics (DSPIN-07)*, 368 (2008).

Planned Spin Studies at the Large Hadron Collider

Homer A. Neal

Department of Physics, University of Michigan, Ann Arbor, Michigan 48109, USA

Abstract. This note summarizes how quantum spin can elucidate the physics of new energy regimes and presents a selection of planned studies at the LHC in which spin is expected to have special relevance.

Keywords: polarization, spin, hyperons, SUSY, LHC
PACS: 13.85.Ni; 13.88.+e;14.80.Ly

INTRODUCTION

Spin continues to be a very mysterious quantum entity. It divides fundamental particles into two distinct categories with strikingly different behavior, it underlies much of chemistry, and it may be the key to linking particle masses in theories such as SUSY. It is responsible for many technological advances, ranging from medical imaging to atomic clocks that make GPS systems possible. Yet, perhaps more than with any other such entity in quantum mechanics, it eludes our ability to acquire a satisfactory explanation for its existence.

Spin effects can be studied in many ways, including, for example, observing its role in defining nuclear and atomic energy states. Spin can also be studied in particle scattering experiments as one seeks to understand how it determines scattering probabilities, and by examining how the effects of particle scattering itself affects the spin directions of the participating particles.

If one examines the preponderance of theoretical predictions of the role of spin in particle interactions over past decades one sees a persistent tendency that suggests that spin effects should disappear as we move to the "next energy regime" – whatever that regime might be. These predictions have uniformly been incorrect. One only has to note the very large polarizations seen in inclusive hyperon production at several hundred GeV/c to be reminded that the spin mystery survives. This dilemma gives rise to the question as to what should be expected as we enter the energy domain of the LHC. Accordingly, following a brief historical detour, this note will focus on plans for pursing spin studies, or exploiting spin properties, at energies anticipated at the Large Hadron Collider.

CP1149, *Spin Physics, 18th International Spin Physics Symposium*
edited by D. G. Crabb, D. B. Day, S. Liuti, X. Zheng, M. Poelker, and Y. Prok

SPIN AS A TOOL FOR INVESTIGATING PARTICLE STRUCTURE

Quark compositeness will clearly be an area of inquiry in LHC experiments. For this reason it is useful to inquire if spin studies played any role in our last major leap in divining the structure of matter-- the discovery that protons had something inside of them which we now call quarks. To address this question, the reader's attention is directed to Fig. 1.

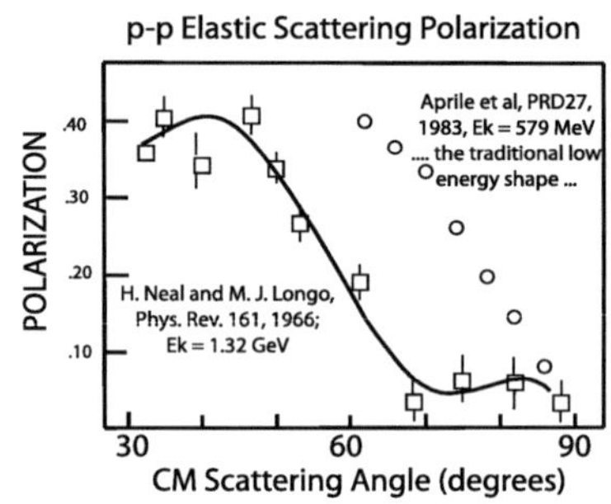

FIGURE 1. The first observation of a double-zero in the p-p elastic polarization, a feature that now seems to persist into the colliding energy realm beyond a hundred GeV [1].

The open squares in this figure represent the measured polarization in p-p elastic scattering [H. Neal and M. Longo, Phys. Rev. 161, 1966] as a function of center-of-mass scattering angle, with the incident proton beam having a kinetic energy of 1.32 GeV, near the highest energy available at the time. The circles represent measurements at a lower energy by Aprile et al [2] at .579 GeV, which is included to remind the reader of the traditional shape of p-p polarization at low energies. Particle symmetry requires the polarization to be zero at ninety degrees. An immediate question arises as to why the higher energy data departs from the smooth approach to zero at ninety degrees seen in the lower energy data, and why, instead, does it seem to develop a double-zero. It has been argued in subsequent publications [3,4] that this double-zero behavior is a direct result of there being at least two objects inside the proton, and that the polarization structure just represented the transition in the number of scatters taking place from one *q-q* scattering to two. Indeed, further analysis at higher energies shows evidence for three quarks using this argument. Moreover, in essentially every measurement in subsequent decades, the *p-p* elastic system has exhibited the behavior of having a double-zero near the t value of .7 $(\mathrm{GeV}/c)^2$, as is the case above. The results in Fig. 1 predate the formal announcement of the discovery of quarks and reminds us that structure in polarization should be kept in mind as we go to the new energy regime of the LHC. Indeed, polarization may have revealed quarks before any other measurements did.

EXAMPLES OF SPIN MEASUREMENTS PLANNED AT THE LHC

Numerous experiments at the LHC will either study spin effects directly, or use known spin properties to look for new physics, such as SUSY. Space constraints here prevent a full review of these projects, so only a few examples will be presented. One to be mentioned is a planned study of the inclusive polarization in Λ_b production, in an attempt to better understand the large polarization seen at lower energies in inclusive Λ production and to be alert for new physics in this domain. Another topic presented below is an idea for exploiting spin properties to confirm that any putative SUSY candidates indeed are due to SUSY and not extra-dimensions or some other mechanism. Included in this section also is a discussion of how the J^{CP} of the Higgs might be determined through an analysis of the angular correlations of its decay products..

Λ_b Polarization

Plans have been developed by a Michigan-Lancaster-Technion group [5] for measuring the spin parameters of the Λ_b hyperon. The decay mode

$$\Lambda_b \to J/\psi(\mu^+\mu^-)\Lambda(p\pi)$$

is employed to extract the Λ_b signal from what is expected to be a low background environment, given that the final state has four charged particles and a displaced secondary vertex. The polarization and parity violating parameter α_b will be determined from the relevant angular correlations between the final state particles. The team expects to accumulate approximately 13000 Λ_b events (and a similar number of anti- Λ_b) with an integrated luminosity of 30 fb^{-1} . This estimate is based on the latest reconstruction software and trigger simulation for the ATLAS experiment.

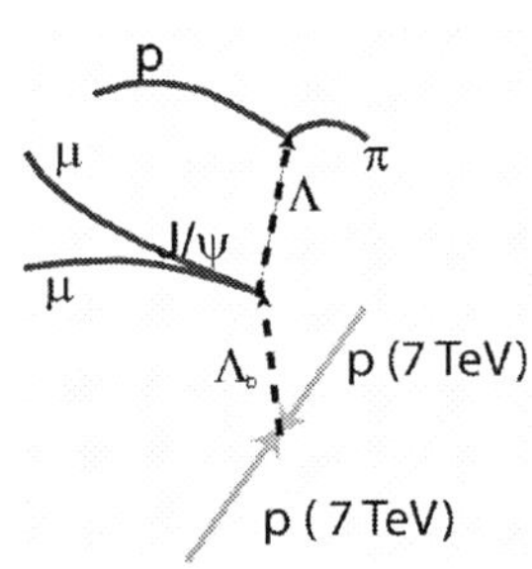

FIGURE 2. Illustration of the production and decay of the Λ_b hyperon in the channel to be studied. The correlation of the angles between the various final state particles suffice to determine both the polarization of the Λ_b and its parity violating parameter α.

The Λ_b is the lightest baryon containing a b quark, and since its discovery in 1991 by the UA1 Collaboration [6] it has created a great deal of interest. Besides the so-called Λ_b lifetime puzzle [7], the Λ_b has been the subject of various theoretical studies ranging from proposed tests of *CP* violation [8], *T* violation tests and new physics studies [9], measurement of top quark spin correlation functions [10] and the extraction of the weak phase of the CKM matrix [11]. Specific physics interest in the parity violating α_b parameter derives from its ability to serve as a test for various heavy quark factorization models and perturbative QCD.

Λ_b studies are also of interest because of the continuing mystery of why hyperons have consistently displayed large polarizations even when produced at energies as high as several hundred GeV and at large p_T where most models predict zero polarization. An additional unexplained phenomena is why the Λ polarization at fixed Feynman x appears to plateau near $p_t \sim 1$ GeV/c, as shown in Fig. 3. It is not known if these large polarization effects can be explained by some not yet understood aspects of existing physics, or if they point to new physics altogether.

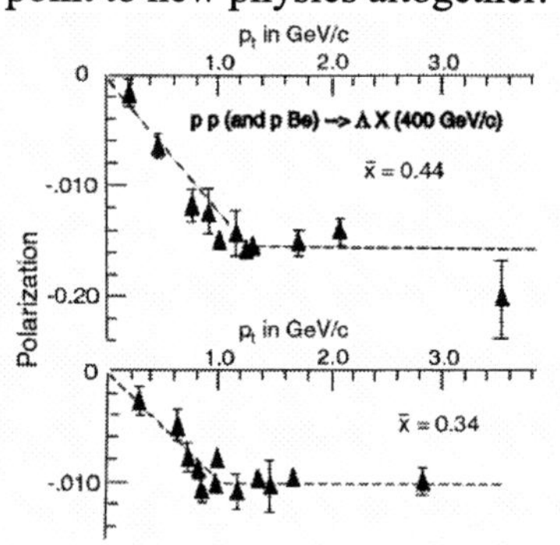

FIGURE 3. Compilation of Λ inclusive polarization in bins of Feynman x, illustrating a significant unexplained plateau near $p_t \sim 1$ GeV/c. These results are not inconsistent with a single quark scattering achieving almost a 100% saturated polarization (see Neal and Burelo [12]).

Λ_b polarization studies hold the possibility of illuminating just how polarized b quarks are produced [13] and, indeed, they may have relevance to how fermions are produced in all pp induced processes.

The group planning these measurements examined the primary technical challenges of measuring the Λ_b polarization in ATLAS by simulating large samples of Λ_b baryons with various known polarizations, allowing them to decay in the detector using model predicted amplitudes, and then reconstructing these events using standard ATLAS packages. These samples permitted the testing of our ability to reconstruct events and to confirm that we could precisely recover the input polarization and the decay amplitudes. The extensive simulations permitted various polarization extraction methods to be tested and also allowed the assessment of the impact of detector acceptance and resolution effects.

Correlation of Decay Planes of the Higgs Boson

The decay planes of the two Z bosons from the Higgs decay ($H \rightarrow ZZ$) are expected to be correlated. This correlation is largely attributed to the transversely polarized Z bosons.

The main utility of such a correlation, whose strength of course depends on the ultimate Higgs mass, is that it allows a direct determination of the J^{CP} of particles presumed to be the Higgs. Observation of particles that seemed to be Higgs but which would not conform to the Standard Model J^{CP} would clearly be a development of considerable interest.

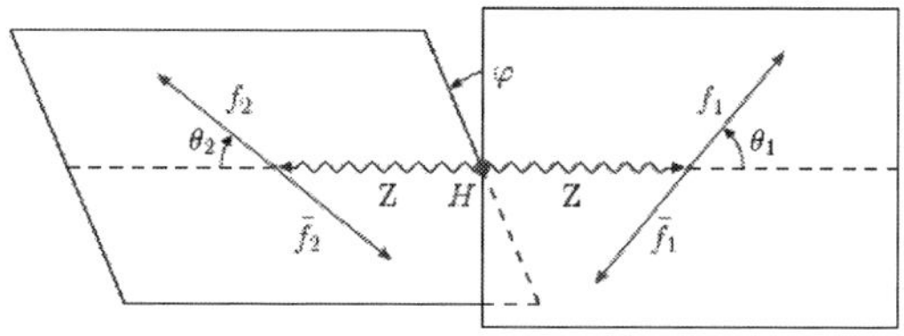

FIGURE 4. Illustration of the plane angle ϕ whose distribution is sensitive to the parent Higgs J^{CP} [14].

Neutralino Spin Measurements in ATLAS

If SUSY-like particles are observed at the LHC there will remain the challenge of making sure those particles are truly from SUSY mechanisms and not, for example, due to extra dimensions. An ATLAS group [15] has proposed an interesting test to discriminate against competing interpretations of SUSY-like events.

If one presumes that the second lightest neutralino has spin ½ and decays through the following chain

$$\tilde{q}_L \rightarrow \tilde{\chi}_2^0 q \rightarrow \tilde{l}^{\pm} l^{\mp} q \rightarrow l^+ l^- q \tilde{\chi}_1^0$$

there are certain observable consequences. It is shown by the authors that, since the relative rate of squarks to anti-squarks is expected to be large at LHC energies, there could be a significant charge asymmetry in the effective mass distributions between the final state quark and the "near" final state positive lepton when compared to the effective mass distribution formed with a negatively charged "near" lepton. This asymmetry is driven by the assumption that the second lightest neutralino has spin ½ and the angular distribution for its decay is thus not isotropic. The authors expect that, if the"SU3" point is appropriate, the two charge distributions can, with a high probability, be excluded from being equal with less than 10 fb^{-1} of data.

Other Spin Studies Planned

Many other studies are planned at the LHC where spin, in one of its roles, uniquely enters the proposed physics quest. Certain Higgs searches will use τ polarization to enhance the signal/background ratio in the search [16], both ATLAS and CMS will use the ZZ decay plane angles to discriminate between SM and non-SM Higgs [17], the W polarization in top decays will be used to test the SM [18], and correlations between the spins of top and anti-top quarks produced in semi-leptonic decays will used for an additional test of the SM [19]. Studies in ALICE will test the possible color flow dependence of single spin asymmetries and test the universality of traverse momentum dependent functions [20]. Plans have been developed in ATLAS [21] to make a careful study of the polarization of $J/\psi \rightarrow \mu\,\mu$ to test the "Color Octet" model.

SUMMARY

An exciting period of research awaits us at the LHC. Whether overtly, or through its usual subtle way, spin will continue to help us understand other mysteries in high energy physics, even while it continues to be mysterious itself.

ACKNOWLEDGEMENTS

With my longstanding ties to Michigan, the place where the discoverers of electron spin and proton spin spent most of their professional careers, and the place where the first g-2 measurement was made by Dick Crane, and where Oliver Overseth, who was a leader in the first measurement of the Λ parity violating parameter, both spent most of their lives, I want to note with sadness the recent passing of the last two individuals. They contributed much to the field's current understanding of spin, and were valued colleagues to those of us at Michigan and to many of you attending this impressive conference. They will be missed.

I want to acknowledge the contributions by Natasha Panikashvili, Eduard Burelo, Michael Longo, Alan Krisch, Shlomit Tarem, and Maria Smizanska. I also want to especially recognize Meghan Shanks for her diligent efforts to help identify the various spin-related projects planned for the LHC.

REFERENCES

1. H. A. Neal and M.J. Longo, *Phys. Rev.* **161**, 1374 (1967).
2. E. Aprile et al., *Phys. Rev. D* **27**, 2600 (1983).
3. H. A. Neal, H. B. Nielsen, *Phys. Lett. B* **51**, 79 (1974).
4. H. A. Neal, J. B. Kuah and H. B. Nielsen, *Phys. Lett. B* **439**, 407 (1998).
5. M. Biglietti; E. De La Cruz Burelo, C. Ferretti, H. A. Neal, N. Panikashvili, M. Smizanska, S. Tarem, *Nucl. Phys. B, Proc. Suppl.* **156**, 151 (2006).
6. C. Albajar et al., *Phys. Lett. B* **243**, 540 (1991).
7. F. Gabbiani et al., *Phys. Rev. D* **68**, 114006 (2003).
8. I. Dunietz, *Z. Phys. C* **56**, 129 (1992).
9. C. Q. Geng et al., *Phys. Rev. D* **65**, 091502 (2002).
10. C. A. Nelson, *Eur. Phys. J. C* **19**, 323 (2001).
11. A. K. Giri et al., *Phys. Rev. D* **65,** 073029 (2002).
12. H. A. Neal, E. De La Cruz Burelo, *AIP Conf. Proc.* **915**, 449 (2007).
13. B. Andersson, G. Gustafson, and G. Ingelman, *Phys. Lett. B* **85**, 417 (1979).
14. C. Ruwiedel [ATLAS and CMS Collaborations] arXiv: 0710.1954 [hep-ph] (2007).
15. M. Biglietti et al., Neutralino spin measurement with ATLAS detector at LHC, ATL-PHYS-CONF-2007-020, 05 December 2007.
16. M. Guchait and D. P. Roy, arXiv: 0808.0438 [hep-ph] (2008).

17. C. Ruwiedel [ATLAS and CMS Collaborations] arXiv: 0710.1954 [hep-ph] (2007).
18. F. Hubaut, E. Monnier, P. Pralavorio, V. Simak and K. Smolek, "ATLAS sensitivity to top quark and *W* boson polarization" SN-ATLAS-2005-052, 2005.
19. M. Baarmand, A. Mermerkay and L. Vodoniyangv, "Measurement of Spin Correlation in top Quark Pair Production in Semi-Leptonic Final State." 9 June, 2008. CMS note 2006/111.
20. D. Boer, C. J. Bomhof, D. S. Hwang and P.J. Mulders, *Phys. Lett. B* **659**, 127 (2008).
21. E. Etzion et al., "Quarkonium Production and Polarization with Early Data at ATLAS." ATL-SLIDE-2008-066 , 24 June, 2008.

Polarization Measurements of Hyperons and Vector Mesons in Heavy Ion Collisions at STAR

Xin Dong (for the STAR Collaboration)

Lawrence Berkeley National Lab
MS70R0319, One Cyclotron Road, Berkeley, CA 94720, USA

Abstract. Particles produced in non-central relativistic heavy ion collisions could become globally polarized due to the initial large orbital angular momentum and QCD spin-orbital coupling. Measurements of the polarization w.r.t. various kinematic planes may give insight into the mechanisms by which the polarization arises. We report the polarization measurements of $\Lambda(\bar{\Lambda})$, and spin alignment of K^* and ϕ w.r.t. the *reaction plane* in Au+Au collisions at RHIC. In addition the spin alignment of the K^* and ϕ can be extracted w.r.t. a second plane, the *production plane*, which gives insight into the possible size of polarization phenomena in the production dynamics of vector mesons, is also reported.

Keywords: Heavy Ion Collisions, Global Polarization, Spin Alignment, Hyperons, Vector Mesons
PACS: 24.70.+s, 14.20.Jn, 25.75.Ld, 25.75.Nq

The system produced in noncentral relativistic heavy ion collisions possesses large orbital angular momentum. Due to the spin-orbit coupling of QCD, this orbital motion may result in a net polarization of the produced particles along the direction of the initial angular momentum, that is, perpendicular to the reaction plane, yielding a global hadronic polarization in the final state [1, 2]. The magnitude and the transverse-momentum (p_T) dependence of the global polarization are expected to be sensitive to different hadronization scenarios [3]. Therefore, the study of this global polarization, as well as the polarization of final-state hadrons with respect to (w.r.t.) the particle's production plane may provide valuable insights into the evolution of the system, the hadronization mechanism, and the origin of hadronic spin preferences.

We report the global polarization (reaction plane) measurement for the $\Lambda(\bar{\Lambda})$ hyperons produced at midrapidity in Au + Au collisions at $\sqrt{s_{NN}} = 200$ GeV and 62.4 GeV [4], as well as the spin alignment measurements for the K^{*0} and ϕ vector mesons w.r.t. both the reaction plane and the production plane in Au + Au collisions at $\sqrt{s_{NN}} = 200$ GeV [5]. The data sets used were Au + Au minimum bias collision samples taken with the STAR detector at RHIC. The main subsystem used for these analyses was the Time Projection Chamber (TPC), located in a 0.5 T solenoidal magnet, which provided precision tracking and particle identification for charged hadrons [6].

The Λ hyperons were reconstructed from their weak decay topology, $\Lambda \to p\pi^-$ [7]. The K^{*0} and ϕ mesons were recontructed through their respective hadronic decay channels, $K^{*0} \to K^+\pi^-$ and $\phi \to K^+K^-$ [8, 9]. The decay pions, kaons and protons were selected based on the tracking and the ionization energy loss (dE/dx) measured in the TPC. For the Λ, the background contribution was estimated to be less than 8%. For the K^{*0} and ϕ, the combinatorial background was reconstructed using the mixed-event technique [8]. The signal-to-background ratio is $\sim$1/1000 for K^{*0} and $\sim$1/25 for ϕ in

CP1149, *Spin Physics, 18th International Spin Physics Symposium*
edited by D. G. Crabb, D. B. Day, S. Liuti, X. Zheng, M. Poelker, and Y. Prok

20-60% Au + Au collisions.

The global polarization of hyperons can be determined from the angular distribution of hyperon decay products relative to the system orbital momentum **L** - the direction perpendicular to the reaction plane.

$$\frac{dN}{d\cos\theta^*} \sim 1 + \alpha_H P_H \cos\theta^*$$

where P_H is the hyperon global polarization, α_H is the hyperon decay parameter, and θ^* is the angle between the three-momentum of the decay daughter in the hyperon rest frame and the system orbital momentum **L**.

The analysis technique is similar to that used in directed flow measurements in heavy ion collisions. The reaction plane angle was estimated via the standard event plane flow vector method [10]. The first-order event plane vector was determined from the forward TPCs, spanning a pseudorapidity region $2.7 < |\eta| < 3.9$ [11]. The direction of the system angular momentum is fixed by the convention that spectator neutrons are deflected along the direction of the impact parameter, and thus their directed flow is positive for a positive pseudorapidity value. The detector acceptance effect was included in the final systematic errors.

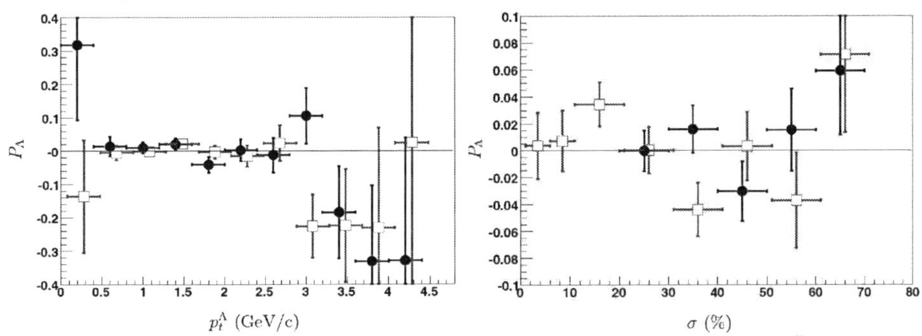

FIGURE 1. Global polarization of Λ hyperons in Au + Au collisions at $\sqrt{s_{NN}} = 200$ GeV (filled circles) and 62.4 GeV (open squares) as a function of p_T on the left and collisions centrality given as a fraction of the total inelastic hadronic cross section (0 means most central) on the right.

Figure 1 presents the results of global polarization for Λ hyperons in Au + Au collisions at $\sqrt{s_{NN}} = 200$ and 62.4 GeV as a function of p_T on the left and collisions centrality on the right. Results for $\bar{\Lambda}$ hyperons are similar. A constant fit to the data points shows P_Λ is consistent with 0 to within a statistical error of ~0.01. The result illustrates the magnitude of the Λ global polarization is very small, and there is no significant dependence on either p_T or the collision centrality.

The main contributions to the systematic errors include the reaction plane uncertainty (30%), the Σ^0 feed-down (30%), detector acceptance effects (20%), multistrange feed-downs (15%) and the background in reconstructing Λ candidates (8%). Considering all the systematic and statistical uncertainties, our results suggest that the global $\Lambda(\bar{\Lambda})$ polarizations are ≤ 0.02 in magnitude.

Spin matrix element ρ_{00} for vector mesons can be detemined from the angular distribution of the decay products

$$\frac{dN}{d\cos\theta^*} \sim (1-\rho_{00}) + (3\rho_{00}-1)\cos^2\theta^*$$

where θ^* is the angle between the polarization direction $\hat{n}$ and the momentum of the decay daughter in the vector meson rest frame [3]. The alignment can be extracted w.r.t. two different planes. In the case of a global spin alignment measurement, $\hat{n}$ is along the orbital angular momentum of the colliding system **L**, which is determined by the reaction plane, requiring only the second-order term for vector mesons. For the production plane measurement, $\hat{n}$ lies along the normal to the production plane, which is determined by the momentum of the vector meson and of the colliding beams.

In this analysis, invariant mass distributions were obtained for K^{*0} and ϕ for each $\cos\theta^*$ and p_T interval. The raw K^{*0} and ϕ yields in each of these distributions were obtained by subtracting the corresponding combinatorial backgrounds and fitting the remaining distributions with a Breit-Wigner function plus a polynomial curve to describe the residual background. The raw yields were then corrected for detection efficiency and acceptance determined from Monte Carlo GEANT simulations. The corrected distributions were fit with the above equation to extract the ρ_{00}. The deviation from $1/3$ for the ρ_{00} for these strong decays indicates a net spin alignment.

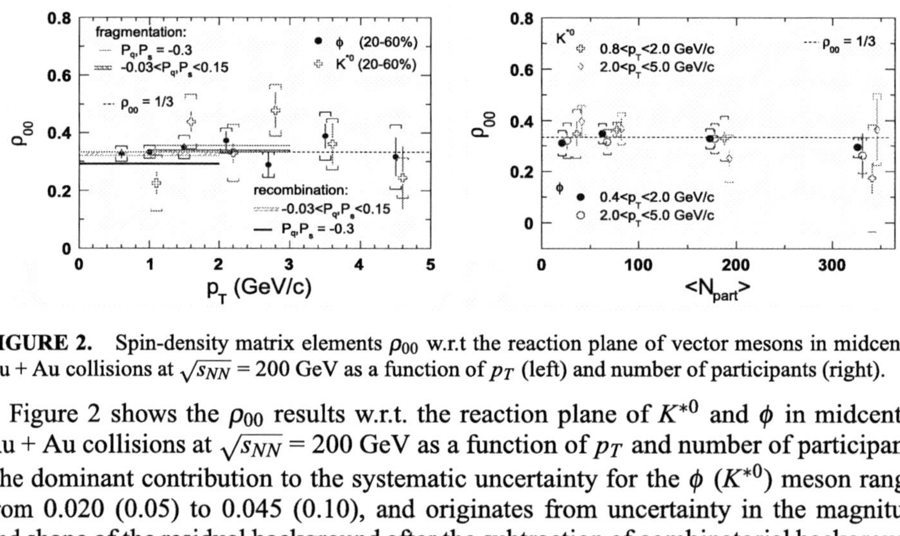

FIGURE 2. Spin-density matrix elements ρ_{00} w.r.t the reaction plane of vector mesons in midcentral Au + Au collisions at $\sqrt{s_{NN}} = 200$ GeV as a function of p_T (left) and number of participants (right).

Figure 2 shows the ρ_{00} results w.r.t. the reaction plane of K^{*0} and ϕ in midcentral Au + Au collisions at $\sqrt{s_{NN}} = 200$ GeV as a function of p_T and number of participants. The dominant contribution to the systematic uncertainty for the ϕ (K^{*0}) meson ranges from 0.020 (0.05) to 0.045 (0.10), and originates from uncertainty in the magnitude and shape of the residual background after the subtraction of combinatorial background. Other point-to-point systematic uncertainties include particle identification, fitting range in extracting the raw yields and particle tracking quality. The contributions from elliptic flow effects and the event plane resolution are found to be negligible. The average ρ_{00} is $0.32 \pm 0.04 \pm 0.09$ for the K^{*0} ($0.8 < p_T < 5.0$ GeV/c) and $0.34 \pm 0.02 \pm 0.03$ for the ϕ ($0.4 < p_T < 5.0$ GeV/c), which is consistent with the unploarized expectation ρ_{00} = 1/3. The data points show no significant dependence on p_T and collision centrality within the current precision.

Figure 3 presents the K^{*0} and ϕ spin alignment measurements w.r.t. the production plane in midcentral Au + Au collisions at $\sqrt{s_{NN}} = 200$ GeV together with the ϕ meson results in $p+p$ collisions at the same incident energy. The Au + Au data for ρ_{00} are consistent with 1/3 to within 2 times the total uncertainties, although the central values tend to increase with decreasing p_T for $p_T < 2.0$ GeV/c. The $p+p$ results are consistent with the Au + Au results and with 1/3. No conclusive evidence is found for large polarization phenomena in the production dynamics of vector mesons in the covered kinematic region within the precision of current measurements.

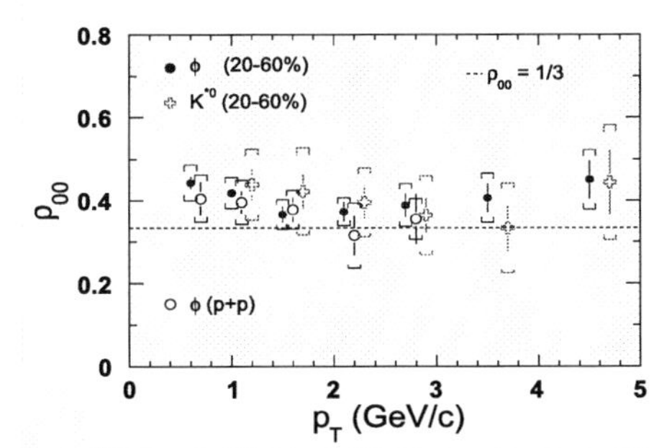

FIGURE 3. Spin-density matrix elements ρ_{00} w.r.t the production plane of vector mesons in midcentral Au + Au collision and $p+p$ collisions at $\sqrt{s_{NN}} = 200$ GeV as a function of p_T.

In summary, we measured the global polarization for Λ hyperons and the spin alignment for K^{*0} and ϕ vector mesons at midrapidity at RHIC. The size of the $\Lambda(\Lambda)$ global polarization is smaller than 0.02. The results on ρ_{00} for K^{*0} and ϕ w.r.t. the reaction plane are consistent with the unpolarized expectation of 1/3. Hence, no evidence is found for a strong global polarization despite the large orbital angular momentum in noncentral Au + Au collisions at RHIC. The results of ρ_{00} for K^{*0} and ϕ w.r.t. the production plane are found to be less than 2 standard deviations above 1/3 in Au + Au collisions and are consistent with the results in $p+p$ collisions at the same collision energy, suggesting vector mesons in the measured p_T region at mid-rapidity carry no significant polarization through production dynamics.

REFERENCES

1. Z. Liang, and X. Wang, *Phys. Rev. Lett.* **94**, 102301 (2005).
2. Z. Liang, and X. Wang, *Phys. Rev. Lett.* **96**, 039901(E) (2006).
3. Z. Liang, and X. Wang, *Phys. Lett. B* **629**, 20 (2005).
4. B. Abelev, et al., *Phys. Rev. C* **76**, 024915 (2007).
5. B. Abelev, et al., *Phys. Rev. C* **77**, 061902(R) (2008).
6. M. Anderson, et al., *Nucl. Instr. Meth. A* **499**, 659 (2003).
7. C. Adler, et al., *Phys. Rev. Lett.* **89**, 132301 (2002).
8. J. Adams, et al., *Phys. Rev. C* **71**, 064902 (2005).
9. J. Adams, et al., *Phys. Lett. B* **612**, 181 (2005).
10. A. Poskanzer, and S. Voloshin, *Phys. Rev. C* **58**, 1671 (1998).
11. K. Ackermann, et al., *Nucl. Instr. Meth. A* **499**, 713 (2003).

Λ Polarization with a Transversely Polarized Proton Target at the COMPASS Experiment

Teresa Negrini on behalf of the COMPASS collaboration

Helmholtz-Institut für Strahlen- und Kernphysik der Universität Bonn
Nußallee 14-16, 53115 Bonn, Germany

Abstract. The transverse polarization of Λ and $\overline{\Lambda}$ hyperons is investigated at the COMPASS experiment at CERN. In 2007 a 160 GeV/c longitudinally polarized muon beam and a transversely polarized NH_3 target were employed to study events in semi-inclusive deep-inelastic scattering. Preliminary results for the Λ and Λ polarizations extracted from this new data set are presented as a function of x_{Bj} and z. The polarizations of Λ and $\overline{\Lambda}$ are compatible with zero within their error bars and no dependence on x_{Bj} or z is observed.

Keywords: Transversely polarized Λ particles; transversely polarized proton target; transversely polarized fragmentation function; transversity function; Λ and $\overline{\Lambda}$ hyperons; COMPASS
PACS: 13.88.+e - 13.60.Rj - 13.87.Fh - 14.20.Jn

INTRODUCTION

The measurement of the so-called transversity distribution functions $\Delta_T q(x_{Bj})$ in the nucleon is an important part of the physics program of the COMPASS experiment[1] at CERN. These distribution functions, being chiral-odd objects, are not accessible in inclusive deep-inelastic scattering (DIS), but require the presence of another chiral-odd object. At COMPASS, $\Delta_T q(x_{Bj})$ can be measured in semi-inclusive deep-inelastic lepton-nucleon scattering (SIDIS), where this additional object is provided by the fragmentation functions $\Delta_T D_q^{\Lambda}(z)$. A promising channel for the measurement of transversity in SIDIS is the spin transfer to Λ hyperons[2]. When a lepton from the incoming beam interacts with one of the quarks of a transversely polarized nucleon, a scattered quark may leave the nucleon in a polarization state which is determined by its transverse spin distribution function inside the nucleon. The struck quark from a DIS reaction hadronizes with a certain probability into a Λ hyperon and transfers a fraction of its polarization to the Λ. Thus the measurement of the spin transfer to the Λ hyperon provides an insight of $\Delta_T q(x_{Bj})$. The polarization of the Λ is expressed by[3]

$$P_{\Lambda}(x_{Bj},z) \quad = \quad f P_T D(y) \frac{\sum_q e_q^2 \Delta_T q(x_{Bj}) \Delta_T D_q^{\Lambda}(z)}{\sum_q e_q^2 q(x_{Bj}) D_q^{\Lambda}(z)}, \tag{1}$$

where f is the dilution factor and P_T is the target polarization. The factor $D(y) = 2(1-y)/(1+(1-y)^2)$ is the spin transfer coefficient. Here, y is the fraction of the incoming lepton energy carried by the exchanged virtual photon. It appears that $\Delta_T q(x_{Bj})$ is coupled to $\Delta_T D_q^{\Lambda}(z)$, having different factorizations of x_{Bj} and z, respectively. By measuring $P_{\Lambda}(x_{Bj},z)$ in different bins of x_{Bj} and z, it is aimed to obtain knowledge about

CP1149, *Spin Physics, 18th International Spin Physics Symposium*
edited by D. G. Crabb, D. B. Day, S. Liuti, X. Zheng, M. Poelker, and Y. Prok

$\Delta_T q(x_{Bj})$ by modeling $\Delta_T D_q^\Lambda(z)$, or to gain insight on both $\Delta_T q(x_{Bj})$ and $\Delta_T D_q^\Lambda(z)$ in different regions of x_{Bj} and z[2].

EXTRACTION OF Λ POLARIZATION

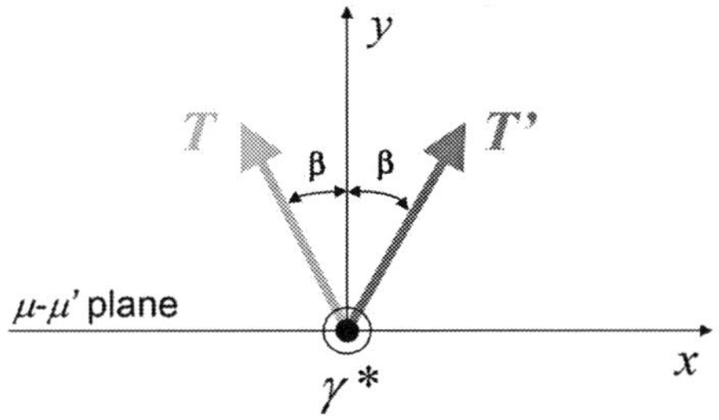

FIGURE 1. Definition of the reference frame: The initial (T) and final (T') transverse quark spin-polarization vectors are shown with respect to the μ - μ' scattering plane [4].

In Fig. 1 the reference system of the Λ production is shown. Here, the virtual photon γ^* moves along the positive z-axis. The transversely polarized target fixes the initial transverse quark-spin polarization vector T. Symmetric to the T-vector with respect to the normal of the μ - μ' scattering plane is the final transverse quark spin-polarization vector T'. The Λ polarization P_T^Λ is measured with respect to its reference axis T'. P_T^Λ is accessible through the angular distribution of the parity violating weak decay in the Λ rest frame by

$$\frac{dN}{d\cos\theta_{T'}} = \frac{N}{2}(1+\alpha P_T^\Lambda \cos\theta_{T'}), \tag{2}$$

where N is the final number of produced Λ hyperons, $\theta_{T'}$ is the decay angle of the proton with respect to the T'-axis, $\alpha = \pm 0.642 \pm 0.013$ is the analyzing power of the parity violating Λ decay. The final number of Λ hyperons in each $cos\theta_{T'}$ bin is determined by a fit of invariant Λ mass distributions. Since P_T^Λ can not be directly extracted from the experimental data due to apparatus acceptance effects, symmetry relations of the apparatus are combined to cancel any terms related to these effects. Two data subsets taken under exactly the same experimental conditions but with opposite target cell polarizations are combined by a geometric mean method in such a way that only the term proportional to the Λ polarization remains. In this analysis, the $cos\theta_{T'}$ bins with two target spin orientations and two data taking periods are combined to a counting rate asymmetry $\varepsilon_T(\theta_{T'}) = \alpha P_T^\Lambda \cos\theta_{T'}$ [5]. The Λ polarization is extracted from the slope of this distribution.

EVENT SELECTION

This analysis is based on a fraction of the data sample taken during the 2007 data recording with the proton target polarization oriented transversely with respect to the incoming muon beam at the COMPASS experiment. DIS events are selected at $Q^2 > 1\ (\mathrm{GeV}/c)^2$ and $0.1 < y < 0.9$. A reconstructed incoming muon μ and a scattered muon μ' are required along with a primary vertex position within the NH_3 target material. Λ particles are searched by looking for secondary vertices (i.e. vertices without an incoming particle) with two tracks of oppositely charged hadrons originating from it. A cut to the fiducial volume of both target cells has been applied in order to ensure an equal beam flux in the target cells. To reduce the background of photon conversion and false combinatorial track association, a cut on the collinearity angle is applied: The angle between the reconstructed Λ momentum and the vector between the primary and decay vertex must be smaller than 10 mrad. The contamination from electron positron pairs from photon conversion is reduced by requiring a minimal transverse momentum $p_T > 23$ MeV/c of the individual hadrons with respect to the reconstructed V^0 momentum. Since the probability of fake secondary vertices is higher close to the primary vertex, a Λ decay length cut has been optimized to require a decay length 7 times larger than its error. The last step to reduce the remaining background is to employ information from the RICH detector. Since the momentum threshold of 22 GeV/c to identify a proton is too high and thus limiting the available statistics significantly, a veto condition is used instead. The veto condition is based on a likelihood method and rejects $e^+(e^-)$, $\pi^+(\pi^-)$ and $K^+(K^-)$ particles from the proton (anti-proton) candidates in the Λ ($\overline{\Lambda}$) decay. The background is significantly reduced by the RICH cut by a factor of 5 whereas the number of particles in the Λ signal stays unchanged. The resulting invariant mass distributions are fitted with a Gaussian peak superimposed by a 3rd degree polynomial for background parametrization. Since the background is strongly reduced by using the RICH for particle discrimination, the error of the fit decreases significantly.

RESULTS & DISCUSSION

The polarization of the Λ and $\overline{\Lambda}$ hyperons is shown in Fig. 2 as a function of x_{Bj} and z. The polarizations of both Λ and $\overline{\Lambda}$ show no significant deviation from zero in the whole explored range. The shaded bands show the size of the corresponding systematic errors, which originate from the uncertainty of the background subtraction, the variation of selection cuts, and the uncertainty of the target polarization. The systematic errors have been estimated by calculating the pulls distributions. For the Λ, the systematic error is below 60% of the statistical error, for the $\overline{\Lambda}$ the systematic error is below 40% of the statistical one. The polarization of the spinless K^0s is extracted to be close to zero. Therefore it is assumed that the derivation of the Λ and $\overline{\Lambda}$ polarizations is free of biases. The observed P_T^{Λ} in the DIS region might indicate that transversity distributions are absent for the Λ production in the investigated kinematic range. Recent publications from Belle, Hermes and COMPASS prove the existence of opposite transversity distributions for u and d quarks[6]. The transverse Λ and $\overline{\Lambda}$ polarizations extracted from COMPASS data collected on a deuteron target[7] show no polarization which could be attributed to the cancellation of the u and d quark contributions. However, since u quarks dominate in the proton, one might expect a different transverse Λ polarization in the proton target data than in the deuteron target data. In the low range of x_{Bj} the transversity distribution function is known to be zero, however the limited statistics especially in the high range of x_{Bj} needs to be improved because it might be suspected that the transversity distribution function is different there. Another factor to be taken into consideration is the fact that the spin transfer from quark to Λ hyperon is proportional to the product of $\Delta_T q(x_{Bj})$ and $\Delta_T D_q^{\Lambda}(z)$. If $\Delta_T D_q^{\Lambda}(z)$ is too small, even a larger $\Delta_T q(x_{Bj})$ will be difficult to detect. The analysis of the full 2007 data sample will allow to extend the statistics in the full range of both x_{Bj} and z, thus gaining more insight about $\Delta_T q(x_{Bj})$ and $\Delta_T D_q^{\Lambda}(z)$.

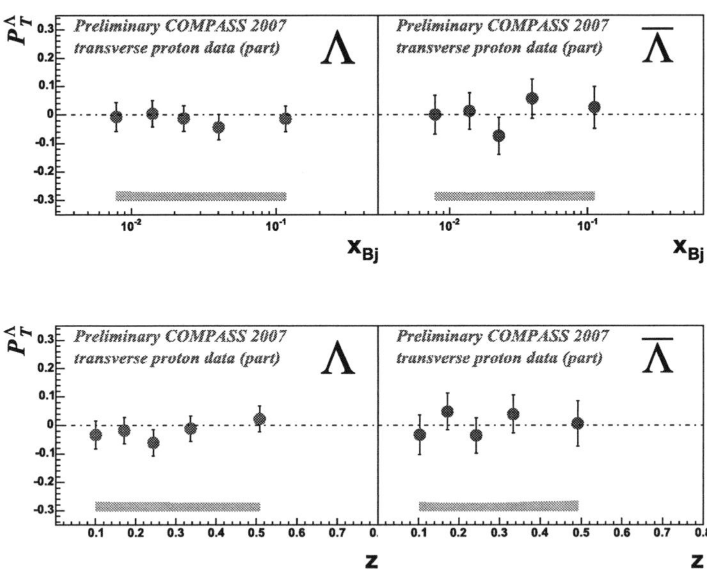

FIGURE 2. Transverse Λ and $\overline{\Lambda}$ polarizations as a function of x_{Bj} and z for $Q^2 > 1$ (GeV/c)^2 in the 2007 transverse proton data. The lower band shows the upper limit of the systematic error.

REFERENCES

1. P. Abbon et al., *Nucl. Inst. Meth.* **A 577**, 455 (2007).
2. M. Anselmino, *"Transversity and Lambda polarization", Proceeding of the Workshop on Future Physics at COMPASS,* Sept. 26-27, 2002, CERN.
3. M. Anselmino et al., *Phys. Lett.* **B 481**, 253 (2000).
4. M. Anselmino & F. Murgia, *Phys. Lett.* **B 483**, 74 (2000).
5. A. Ferrero, *"Lambda asymmetries",Proceeding of Transversity 2005*, Sept. 7-10, 2005, Como, Italy.
6. M. Anselmino et al., *"Transversity and Collins functions from SIDIS and e^+e^- data"*, arXiv:hep-ph/0701006v3.
7. A. Ferrero, *"Measurement of transverse Λ and $\overline{\Lambda}$ polarization at COMPASS", Proceeding of SPIN06,* Oct. 2-7. 2006, Kyoto, Japan.

Longitudinal Spin Transfer of Λ and $\bar{\Lambda}$ Hyperons in Polarized Proton-Proton Collisions at $\sqrt{s} = 200\,\mathrm{GeV}$

Ernst Sichtermann for the STAR Collaboration

Nuclear Science Division, Lawrence Berkeley National Laboratory, Berkeley, CA 94720, USA

Abstract. The longitudinal spin transfer D_{LL} of Λ and $\bar{\Lambda}$ hyperons produced with large transverse momenta p_T in polarized proton-proton collisions is sensitive to the helicity distribution functions of strange quarks and anti-quarks, and to polarized fragmentation functions. This contribution reports a measurement of the longitudinal spin transfer D_{LL} in the inclusive production of Λ and $\bar{\Lambda}$ hyperons at central rapidities in polarized proton-proton collisions at a center of mass energy of $\sqrt{s} = 200\,\mathrm{GeV}$. The data were collected with the STAR experiment at RHIC and correspond to a $\approx 3\,\mathrm{pb}^{-1}$ luminosity and a $\approx 50\%$ beam polarization.

Keywords: Spin transfer, Lambda polarization
PACS: 13.88.+e,13.85.Ni,13.87.Fh,14.20.Jn

INTRODUCTION

The combination of data on the spin dependent structure function g_1 in polarized deep-inelastic lepton-nucleon scattering and the couplings in weak hyperon decay lead to the puzzling notion that the sum of strange quark and anti-quark spins preferentially point opposite to the proton spin [1]. Semi-inclusive deep-inelastic scattering measurements form a complementary path to the (strange) polarization of the nucleon sea, but currently provide no evidence for such polarization [2, 3]. The semi-inclusive data, however, cover a smaller kinematic range than the inclusive data.

In the hard scattering of protons, the longitudinal spin transfer D_{LL} of Λ and $\bar{\Lambda}$ hyperons produced with large transverse momenta p_T is expected to be sensitive to the helicity distribution functions of strange quarks and strange anti-quarks, and to polarized fragmentation functions [4]. D_{LL} quantifies the transfer of longitudinal polarization from the polarized beam proton to the Λ in singly polarized proton-proton collisions,

$$D_{LL} \equiv \frac{\sigma(pp_+ \to \Lambda_+ X) - \sigma(pp_+ \to \Lambda_- X)}{\sigma(pp_+ \to \Lambda_+ X) + \sigma(pp_+ \to \Lambda_- X)}, \tag{1}$$

and similarly for $\bar{\Lambda}$. Here, σ denotes the inclusive production cross section and the subscripts $+$ and $-$ denote particle helicity. In a proton-proton collision experiment with one fully polarized beam, D_{LL} is the Λ ($\bar{\Lambda}$) polarization along its moving direction. Its precise measurement may thus shed light on the above puzzle. This contribution reports on measurements of D_{LL} in the inclusive production of Λ and $\bar{\Lambda}$ hyperons in polarized proton-proton collisions at a center of mass energy of $\sqrt{s} = 200\,\mathrm{GeV}$. These data cover p_T up to 4 GeV with 10% precision, thus far limited by statistics. Prior STAR

CP1149, *Spin Physics, 18th International Spin Physics Symposium*
edited by D. G. Crabb, D. B. Day, S. Liuti, X. Zheng, M. Poelker, and Y. Prok

measurements have determined that the inclusive $\Lambda + \bar{\Lambda}$ spin averaged cross section at $\sqrt{s} = 200\,\text{GeV}$ is in reasonable agreement with perturbative QCD expectations [5].

MEASUREMENT

The present data were collected with the STAR experiment [6] at RHIC in the year 2005 and correspond to a $\sim 3\,\text{pb}^{-1}$ luminosity and a $\sim 50\%$ average longitudinal beam polarization. RHIC beam bunches with net positive and negative proton helicities were collided at the STAR interaction region in four different combinations, $++$, $+-$, $-+$, and $--$ and STAR events were sorted accordingly.

The events used in this analysis were recorded if they satisfied a minimum bias (MB) trigger condition, defined by coincident signals from Beam-Beam Counters around the beam pipe at either side of the STAR interaction region. About 3×10^6 of such beam collision events were analyzed. In addition, about $3\,\text{pb}^{-1}$ was sampled with a jet patch (JP) trigger condition, imposing the minimum bias condition and a threshold of about 6.5 GeV on the transverse energy deposited in at least one patch of towers in the Barrel Electromagnetic Calorimeter. This trigger is the same as in the 2005 STAR jet spin-asymmetry measurement [8].

Charged particle tracks were reconstructed with the Time-Projection Chamber (TPC) in the 0.5 T STAR magnetic field for pseudo-rapidities $|\eta| \leq\sim 1.3$ and all azimuthal angles ϕ. Particles were identified using the agreement of the measured and expected ionization loss dE/dx in the TPC where possible. The Λ and $\bar{\Lambda}$ candidates were reconstructed by searching the events for the parity-violating hadronic decays $\Lambda \to p\pi^-$ and $\bar{\Lambda} \to \bar{p}\pi^-$, which have a branching ratio of 63.9(5)% [7]. Topological selections similar to those in Ref. [5] were applied to reduce backgrounds. Candidate events were retained only if at least one $p\pi^-$ or $\bar{p}\pi^+$ track pair was found and its reconstructed invariant mass $1.105 < M < 1.125\,\text{GeV}$ was in agreement with the $\Lambda(\bar{\Lambda})$ mass $m_{\Lambda,\bar{\Lambda}} = 1.115683(6)\,\text{GeV}$ [7].

The parity-violating hadronic decays $\Lambda \to p\pi^-$ and $\bar{\Lambda} \to \bar{p}\pi^-$ are self spin-analyzing. The Λ ($\bar{\Lambda}$) polarization can thus be determined from the angular distribution of the decay (anti-)proton in the Λ ($\bar{\Lambda}$) rest frame, which is usually described in terms of the rest-frame angle θ between the (anti-)proton momentum and the Λ ($\bar{\Lambda}$) polarization vector. Figure 1 shows a) the distribution of M versus $\cos\theta$ for Λ candidates from MB data after topological selections and b) the projection onto M after $\cos\theta$ was restricted to $\cos\theta < -0.2$ to eliminate a $K^0_s \to \pi^+\pi^-$ background originating from misidentified pions. In total about 30×10^3 Λ and 24×10^3 $\bar{\Lambda}$ candidates were retained from MB data. Their average p_T is about 1.3 GeV and $|x_F| \approx 8 \times 10^{-3}$. The candidate sample obtained from JP triggered data was about 80% larger.

The spin asymmetry D_{LL} was extracted in small intervals in $\cos\theta$ as follows,

$$D_{LL} = \frac{1}{\alpha P_b \langle\cos\theta\rangle} \frac{N^+ - N^-}{N^+ + N^-}, \tag{2}$$

where $\alpha = +(-)0.642(13)$ is the $\Lambda(\bar{\Lambda})$ decay parameter [7], P_b is the measured proton beam polarization at the STAR interaction region [9, 10, 11], and the yields N^+

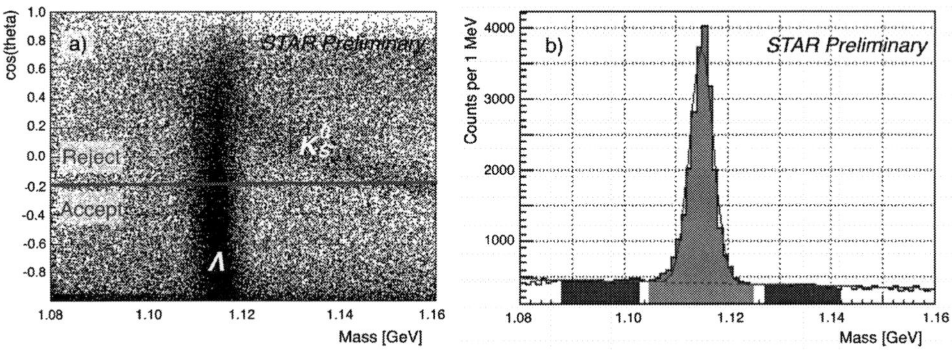

FIGURE 1. a) The invariant mass distribution of $p\pi^-$ track pairs from MB data versus $\cos\theta$ after topological selections and b) the invariant mass distribution of the Λ candidates with the additional constraint $\cos\theta < -0.2$ used to eliminate K_s^0 background. The shaded area with $1.105 < M < 1.125$ GeV below the peak was used to select signal candidates and the shaded intervals away from the peak were used in assessing residual background.

and N^- are obtained from the spin-sorted $\Lambda(\bar{\Lambda})$ candidate yields n^{++}, n^{+-}, n^{-+}, and n^{--} weighted by measured ratios of luminosities L [12]. Specifically, $N^+ = n^{++} \cdot L^{--}/L^{++} + n^{+-} \cdot L^{--}/L^{+-}$ and $N^- = n^{-+} \cdot L^{--}/L^{-+} + n^{--}$. Spin-sorted yields from intervals in M above and below $m_{\Lambda,\bar{\Lambda}}$, as indicated in Fig. 1b), were used to correct for residual background contributions. The analysis method makes use of QCD symmetries in the $\Lambda(\bar{\Lambda})$ production process to form a ratio in which the detector acceptance (largely) cancels [13].

RESULTS AND SUMMARY

Figure 2a) shows preliminary results for the Λ and $\bar{\Lambda}$ D_{LL} from MB data for positive and negative η. Positive η is taken along the momentum direction of the polarized beam. The data are statistics limited. Systematic uncertainty at the level of 0.01 is estimated from uncertainties in the measured ratios of luminosities, residual transverse beam polarization components at the STAR interaction region, and uncertainties in α and the beam polarization. At the present level of precision, the Λ and $\bar{\Lambda}$ D_{LL} are found consistent with each other and consistent for both η values with no spin transfer effects.

The Λ and $\bar{\Lambda}$ results from MB data are compared in Fig. 2b) with $\bar{\Lambda}$ results obtained from JP data for positive η. As expected the JP data extend to considerably larger values of p_T up to 4 GeV. In this region the inclusive $\Lambda + \bar{\Lambda}$ spin averaged yield is in reasonable agreement with perturbative QCD expectations [5] for suitable choices of the spin averaged fragmentation function. The systematic uncertainties in the JP D_{LL} measurement are estimated to be somewhat larger than those in the MB measurement, mostly because of possible biases introduced by the JP trigger requirement. These effects were studied through Monte Carlo simulation. The data on D_{LL} remain statistics limited and to within the present uncertainties of about 10% no evidence is found for p_T

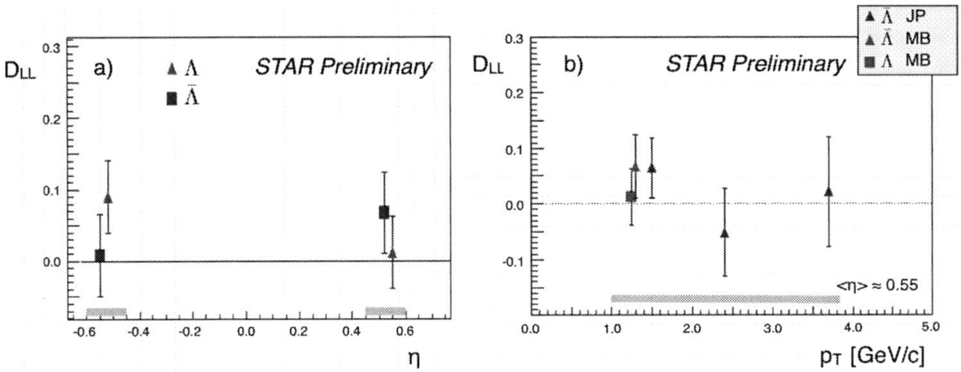

FIGURE 2. Left: preliminary results for the longitudinal spin transfer D_{LL} of inclusive Λ and $\bar{\Lambda}$ hyperons in proton-proton collisions at $\sqrt{s} = 200\,\mathrm{GeV}$ versus pseudo-rapidity η from data obtained with a beam-collision trigger. Positive η is taken along the direction of the polarized beam. Right: preliminary results for D_{LL} for Λ and $\bar{\Lambda}$ at positive η compared to preliminary results obtained for the $\bar{\Lambda}$ with a jet trigger. The uncertainties on the data points are statistical and the shaded bands indicate the estimated size of the total systematic uncertainty, not including the scale uncertainty introduced by the beam polarization measurement.

dependence.

In summary, the present measurements of D_{LL} form a proof-of-principle that pave the way for future measurements with better precision, which may provide new constraints on the polarized (anti-)strange sea quark polarization in the polarized nucleon and polarized fragmentation functions.

REFERENCES

1. J. Ashman et al. [European Muon Collaboration], *Nucl. Phys. B* **328**, 1 (1989).
2. A. Airapetian et al. [HERMES Collaboration], *Phys. Rev. D* **71**, 012003 (2005).
3. R. Windmolders for the COMPASS Collaboration, *these proceedings.*
4. Y. Chen et al., *Phys. Rev. D* **78**, 054007 (2008), *and references therein.*
5. B. I. Abelev et al. [STAR Collaboration], *Phys. Rev. C* **75**, 064901 (2007).
6. Special Issue on RHIC and Its Detectors, edited by M. Harrison, T. Ludlam, and S. Ozaki, *Nucl. Instrum. Methods Phys. Res., Sect. A* **499**, Nos.2-3 (2003).
7. C. Amsler et al. [Particle Data Group], *Phys. Lett. B* **667**, 1 (2008).
8. B. I. Abelev et al. [STAR Collaboration], *Phys. Rev. Lett.* **100**, 232003 (2008).
9. O. Jinnouchi et al., arXiv:nucl-ex/0412053.
10. H. Okada et al., arXiv:hep-ex/0601001.
11. J. Kiryluk et al. [STAR Collaboration], arXiv:hep-ex/0501072.
12. J. Kiryluk et al. [STAR Collaboration], *AIP Conf. Proc.* **675** (2003) 424.
13. Q.-H. Xu et al. [STAR Collaboration], *AIP Conf. Proc.* **842**, 71 (2006).

Polarization Effects in Lambda and Anti-Lambda Production at HERMES

S.L. Belostotski on behalf of the HERMES collaboration[a]

[a]*Petersburg Nuclear Physics institute, PNPI 188300 Gatchina, Lenigrad district, Russia*

Abstract. Study of the Λ and Λ hyperon production in the HERMES experiment is discussed. The data were collected using a 27.6 GeV longitudinally polarized positron (or electron) beam of the HERA accelerator and a polarized or unpolarized target. The transverse polarizations (perpendicular to the production plane) of the Λ and Λ hyperons have been measured. The spin transfer K_{LL} from the longitudinally polarized target has been found for inclusively detected Λ and Λ events. Longitudinal spin transfer D_{LL} from the beam to the Λ hyperon in the deep inelastic scattering regime has been analyzed. The D_{LL} for Λ has been measured for the first time.

Keywords: lambda, polarization.
PACS: 13.88.+e,13.60.-r, 13.60.Rj.

INTRODUCTION

Given the nontrivial spin structure of the proton, it is of interest to consider the spin structure of other baryons. In this respect the Λ hyperon, as the lightest strange baryon of the SU(3) spin-1/2 octet, is particularly interesting. The HERMES experiment [1] has studied Λ and Λ electro and photo-production using polarized positron (electron) beam and polarized or unpolarized targets. In these experiments, the most important observable is polarization of the produced hyperon which is accessed through the parity violating weak decay channel $\Lambda^0 \rightarrow p\pi^- \ \left(\bar{\Lambda}^0 \rightarrow \bar{p}\pi^+\right)$.

The results on transverse ("spontaneous") Λ (Λ) polarization in quasi-real photoproduction and on longitudinal spin transfer $D_{LL'}$ to the Λ from the beam in semi-inclusive Deep Inelastic Scattering (DIS) based on a limited number of events corresponding to HERMES RUN I (the years 1996-2000) have already been published by the HERMES collaboration [2-4]. Recently, HERMES has reported on measurement of the spin transfer K_{LL} to the Λ (Λ) hyperon from the longitudinally polarized target [5]. Here we present the data on D_{LL} obtained in HERMES RUN I and RUN II during the years 1996-2000 and 2002-2005, respectively, analyzed jointly. A recently developed method of D_{LL} extraction which does not require exact balance of the positive and negative helicity in the primary beam has been used. In the previous analysis, only unpolarized target data was analyzed. In this analysis, polarized target data has also been included under condition that the frequently flipped target polarization is well helicity balanced which corresponds to average zero polarization.

CP1149, *Spin Physics, 18th International Spin Physics Symposium*
edited by D. G. Crabb, D. B. Day, S. Liuti, X. Zheng, M. Poelker, and Y. Prok

Finally, the gain in statistics of a factor of 4 has been obtained which allows one to better study D_{LL} kinematical distributions and to extract for the first time D_{LL} for Λ.

LONGITUDINAL SPIN TRANSFER D_{LL}

The longitudinal spin transfer $D_{LL'}$ to the Λ hyperon was studied by the LEP experiments OPAL and ALEPH at an energy corresponding to the Z_0 pole [6,7]. In these experiments the hyperons were predominantly produced *via* the strange quark anti-quark decay channel in which the primary strange quarks initiating the hadronization process were negatively polarized at the level of 91%. The OPAL and ALEPH (LEP) data show a significant polarization transfer from the $s(\bar{s})$ quark to the Λ (Λ) hyperon, thus confirming that the *s* quark dominates the spin of Λ hyperon. In contrast to the LEP experiments, the production of Λ hyperons in DIS is originated predominately from the struck *u* or *d* quark emitted from the target nucleon, the contribution from the strange quark being on average small. The spin transfer coefficient $D_{LL'}$ describes the probability that the polarization of the struck quark is transferred to the Λ hyperon along the secondary quantization axis L' in the Λ rest frame. In this analysis, the quantization axis L' is chosen along the direction of the Λ momentum (as it is not affected by the relativistic transformation to the Λ rest frame). Another possible choice of the spin transfer direction is $L' = L$, i.e. along the spin of the virtual photon. Note that for HERMES kinematics the angle between the virtual photon momentum (spin) and Λ momentum in the Λ rest frame is not small, i.e., unlike the case of very high energies, *L* and *L'* are not collinear. In principle, the spin transfer may have two components in the Λ-production plane: longitudinal one, along the virtual photon spin (*L*), and transverse one, perpendicular to this direction, and L' axis (direction of the spin transfer) does not necessarily aligned with the Λ momentum. An analysis with the two components of the spin transfer reconstructed independently is now in progress, while here in this report it is *assumed* that L' in the Λ rest frame is along the direction of Λ momentum.

The $D_{LL'}$ has been extracted using a data sample consisting of 26714 Λ events and 3610 Λ events with the restriction Q^2 > 0.8 GeV applied, where Q^2 is the negative four-momentum transfer squared. In order to cancel the spectrometer acceptance effect, the $D_{LL'}$ has been determined by combining the two data sets measured with opposite beam polarizations [8]. It is important that in this approach the extracted $D_{LL'}$ does not contain an uncertainty coming from the Monte Carlo simulation of the spectrometer acceptance.

As the HERMES spectrometer is a forward detector, the Feynman variable x_F for Λ production is mostly positive ranging from -0.25 to 1. The $x_F > 0$ requirement cuts off only a small fraction of events. The spin-transfer coefficient for Λ averaged within full x_F range is found to be $D_{LL'} = 0.102 \pm 0.056_{stat} \pm 0.02_{syst}$. With the requirement $x_F > 0$, $D_{LL'} = 0.082 \pm 0.061_{stat} \pm 0.02_{syst}$. For the Λ in full x_F range it is found $D_{LL'} = -0.040 \pm 0.112_{stat} \pm 0.03_{syst}$. As Λ cannot be produced from a target remnant mechanism, in this case the cut on x_F has not been applied. The HERMES results on $D_{LL'}$ versus x_F for Λ are presented in Fig. 1 (left panel) together with the results of NOMAD[9] and COMPASS[10] experiments at CERN and E665[11] at FNAL. Two points for the Λ and Q^2 dependence for the Λ are shown in Fig. 1 (right panel). Most of NOMAD data have been obtained at negative x_F. For positive x_F, all existing to date experimental results indicate that the $D_{LL'}$ is compatible with zero within the statistical uncertainty about ± 0.10. In principle, the observation of a small (or equal to zero) value of $D_{LL'}$ points to the dominance of scattering from u or d quarks whose polarization within the Λ hyperons is small. Note that recent lattice-QCD calculations find small light quark polarizations in the Λ hyperon: $\Delta u^{\Lambda} = \Delta d^{\Lambda} = -0.02 \pm 0.04$ and $\Delta s^{\Lambda} = 0.68 \pm 0.04$ [12].

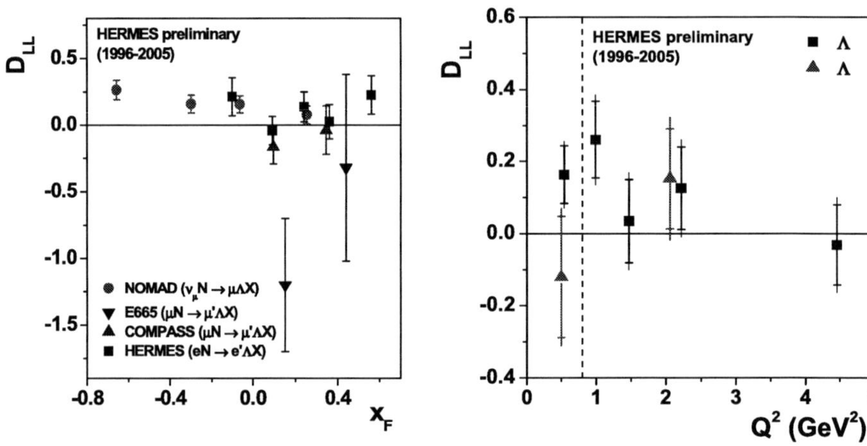

FIGURE 1. Longitudinal spin transfer from the beam to the Λ versus x_F (left panel) together with available world data. Longitudinal spin transfer from the beam to the Λ and Λ versus Q^2 (right panel). The vertical dashed line is positioned at $Q^2 = 0.8$ GeV^2. Two points for Λ are averaged results for $Q^2 < 0.8$ GeV^2 and $Q^2 > 0.8$ GeV^2 domains, respectively.

On the other hand, the condition $\Delta u^{\Lambda} = \Delta d^{\Lambda} \equiv 0$ does not necessarily mean that the $D_{LL'}$ vanishes everywhere. Thus, authors [13] have employed a phenomenological model to explore the dependences of $D_{LL'}$ on kinematical variables. They predict increasing positive values of the spin transfer with $D_{LL'} \to 1$ at $x_F \to 1$ $(z = E^{\Lambda}/\nu \to 1)$.

Unfortunately, interpretation of the spin transfer coefficient is not straightforward. For a Λ produced in DIS regime via a current quark q the "partial" spin-transfer from a quark q can be expressed as $D_{LL'}^{q\Lambda} = \frac{\Delta F^{q\Lambda}}{F^{q\Lambda}}$, where $\Delta F^{q\Lambda}$ and $F^{q\Lambda}$ is spin-dependent and spin-independent $q \to \Lambda$ fragmentation function, respectively. If a fractional contribution from a quark q (purity) is ω_q $(\sum_q \omega_q = 1)$ the total spin transfer is given by $D_{LL'} = \sum_q \omega_q D_{LL'}^{q\Lambda}$. Taking into account u-dominance, a naïve expectation of $D_{LL'}$ in DIS is $D_{LL'} \approx D_{LL'}^{u\Lambda} = \frac{\Delta F^{u\Lambda}}{F^{u\Lambda}} \approx \frac{\Delta u^{\Lambda}}{u^{\Lambda}}$. One should remember, however, that at the moderate energy of the HERMES experiment the distinction between the current and target fragmentation domain is not very sharp and Λ fragmentation may be affected by the target remnant mechanism even at positive x_F. Further, the fraction of Λ hyperons produced via heavier hyperon decays is significant: only about 40% of the Λ hyperons are produced directly from the string fragmentation. The contribution from heavier hyperons Σ^0, Σ^* or Ξ, which may decay into the polarized Λ hyperons and thus modify the $D_{LL'}$, must be taken into account.

ACKNOWLEDGMENTS

I gratefully acknowledge Yu. Naryshkin and D. Veretennikov for their essential contribution to the analysis of the presented experimental data.

REFERENCES

1. HERMES Collaboration, K. Ackerstaff et al., *Nucl. Instrum. Methods A* **417**, 230 (1998).
2. HERMES Collaboration, A. Airapetian et al., *Phys. Rev. D* **76**, 092008 (2007).
3. HERMES Collaboration, A. Airapetian et al., *Phys. Rev. D* **74**, 072004 (2006).
4. HERMES Collaboration, A. Airapetian et al., *Phys. Rev. D* **64**, 112005 (2001).
5. D. Veretennikov in *Proceedings of the XVI International Workshop on Deep-Inelastic Scattering and Related Topics (spin physics),* London, England, April 7-11, (2008).
6. OPAL Collaboration, K. Ackerstaff et al., *Eur. Phys. J. C* **2**, 49 (1998).
7. ALEPH Collaboration, D. Buskulic et al., *Phys. Lett. B* **374**, 319 (1996).
8. S. Belostotski in *Proceedings of the IX-Workshop on High Energy Spin Physics (SPIN-01),* Dubna2001, preprint LINR E1,2-2002-103, 192 (2002).
9. NOMAD Collaboration, P. Astier et al., *Nucl. Phys. B* **588**, 3 (2000).
10. COMPASS Collaboration, M. Sapozhnikov (private communication).
11. E665 Collaboration, M. R. Adams et al., *Eur. Thys. J. C* **17**, 263 (2000).
12. OSDSF Collaboration, M. Göckeler et al., *Phys. Rev. Lett. B* **545**, 112 (2002).
13. B.-Q. Ma, I. Schmidt, J. Soffer, and J.-J. Yang, *Phys. Rev. D* **65**, 034004 (2002).

$p\overline{p} \to \Lambda_c\overline{\Lambda}_c$ within a Handbag Picture – Cross Section and Spin Observables

A.T. Goritschnig*, P. Kroll† and W. Schweiger*

*Institut für Physik, Karl-Franzens-Universität Graz, Austria
†Fachbereich Physik, Universität Wuppertal, Germany

Abstract. We study the process $p\overline{p} \to \Lambda_c\overline{\Lambda}_c$ within the generalized parton picture. Our starting point is the double handbag diagram which factorizes into soft generalized parton distributions and a hard subprocess amplitude for $u\overline{u} \to c\overline{c}$. Our cross-section predictions may become interesting in view of forthcoming experiments at FAIR in Darmstadt.

Keywords: generalized parton distributions, hard hadronic reactions, exclusive Λ_c production
PACS: 12.38.Bx, 24.85.+p, 25.43.+t

INTRODUCTION

For exclusive hadronic reactions which require the production of heavy quark-antiquark pairs QCD perturbation theory has a good chance to provide a substantial part of the process amplitude already at moderately large energies. The reasons are that the mass of the heavy quark can serve as a hard scale and that certain reaction mechanisms can immediately be ruled out, since the heavy-flavor content of the quark sea in the proton is small.

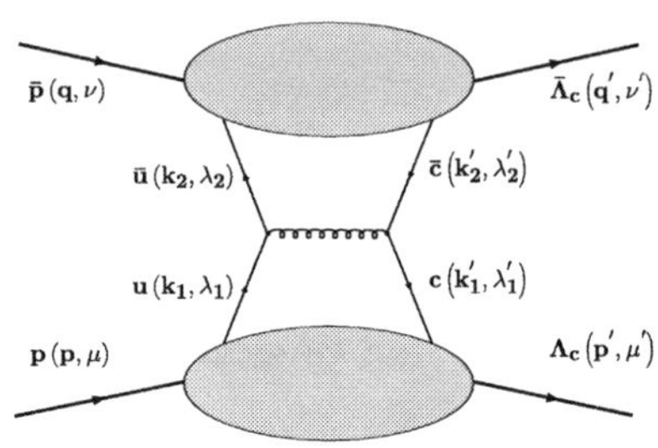

FIGURE 1. Double-handbag contribution to $p\overline{p} \to \Lambda_c\overline{\Lambda}_c$

For the process $p\overline{p} \to \Lambda_c\overline{\Lambda}_c$ the reaction mechanism which we assume to be dominant in the forward hemisphere for energies well above production threshold is depicted in Fig. 1. We analyze it in terms of generalized parton distributions [1], [2].

If the dynamical mechanism underlying $p\overline{p} \to \Lambda_c\overline{\Lambda}_c$ scattering is given by Fig. 1, it can be shown [3] – in analogy to Compton scattering [4], [5] – that the corresponding scattering amplitude factorizes into soft hadronic matrix elements for the $p \to \Lambda_c$ (and $\overline{p} \to \overline{\Lambda}_c$) transition and a hard subprocess amplitude for $u\overline{u} \to c\overline{c}$. The subprocess am-

CP1149, *Spin Physics, 18th International Spin Physics Symposium*
edited by D. G. Crabb, D. B. Day, S. Liuti, X. Zheng, M. Poelker, and Y. Prok

plitude can be calculated perturbatively, whereas the hadronic matrix elements describe the non-perturbative transition of the (anti-)proton to the (anti-)Λ_c by emission of a $(\bar{u})$ u quark and reabsorption of a $(\bar{c})$ c quark. It is just the hadronic matrix elements which give rise to generalized parton distribution functions (GPDs) and to form factors (FFs), which are essentially $1/x$ moments of the GPDs [1], [2] after decomposing them into their covariant structures. At this point eight FFs have to be introduced. If it is assumed that contributions from non-zero orbital angular momenta play a minor role in the proton and Λ_c wave functions, only three FFs remain, which do not necessarily require orbital angular momenta different from zero. In the covariant decomposition of the hadronic matrix elements they show up as coefficients of γ^+, $\gamma^+\gamma^5$ and $i\sigma^{+j}$, respectively. Taking only into account these dominant form factors we end up with comparably simple expressions for the $p\bar{p} \to \Lambda_c\bar{\Lambda}_c$ helicity amplitudes. See Eqs. (3.11)-(3.15) in Ref. [6]. A more detailed account of these results and their derivation will be given in a forthcoming publication [3].

MODELLING THE GPDS

In the forward hemisphere, well above production threshold, we have to consider only the DGLAP region where, according to Ref. [7], the GPDs can be constructed as overlap of light cone wave functions. This is due to the kinematical requirement for the production of a heavy $c\bar{c}$ pair in the hard subprocess, in which the virtuality of the gluon has to be at least $4m_c^2$. For the Λ_c it is certainly a good approximation to consider only its valence Fock state. For the proton higher Fock states may be important, but the required overlap with the Λ_c projects out only appropriate spin-flavor combinations of its valence Fock state. In addition, heavy quark effective theory (HQET) implies that, up to corrections in the inverse of the heavy quark mass, the light degrees of freedom in the Λ_c couple to spin and isospin zero so that the c quark carries the helicity of the Λ_c [8]. In this case the three form factors which remain, if non-zero angular momentum contributions in the wave functions are neglected, become even identical.

Since the light quarks in the Λ_c act only as spectators it is tempting to replace them by a single particle, namely a (scalar, isoscalar) diquark, and to model both, the Λ_c and the proton, as quark-diquark bound states. Only the quark becomes active and undergoes a hard scattering. The quantum numbers of the spectators are subsumed in the diquark. In Ref. [9] $p\bar{p} \to \Lambda_c\bar{\Lambda}_c$ has already been studied within a quark-diquark model, but without using the general framework of GPDs. For our present investigation we take a quark-diquark wave function for the Λ_c which is similar to the one used for a calculation of heavy-baryon transition form factors [10]:

$$|\Lambda_c^+ : \pm\rangle \sim |c_\pm S_{[ud]}\rangle \quad \text{with} \quad \Psi_\Lambda(x, k_\perp) = N_\Lambda\,(1-x)\, e^{-\frac{a_\Lambda^2}{x(1-x)}k_\perp^2}\, e^{-a_\Lambda^2 M^2 \frac{(x-x_0)^2}{x(1-x)}}\,. \tag{1}$$

The momentum fraction carried by the heavy quark is denoted by x, its transverse momentum component by $k_\perp$. The relevant quark-diquark wave function of the proton, on the other hand, is chosen similar to the one in Ref. [9]:

$$|p : \pm\rangle \sim |u_\pm S_{[ud]}\rangle \quad \text{with} \quad \Psi_p(x, k_\perp) = N_p\,(1-x)\, e^{-\frac{a_p^2}{x(1-x)}k_\perp^2}\,. \tag{2}$$

These are pure s-wave wave functions with the helicity of the active quark being identical with the helicity of the baryon. Thus, according to the foregoing remarks, only three form factors are non-zero and they become even identical. For the model wave functions (1) and (2) the overlap integral can then be done analytically and the form factor can immediately be evaluated from the GPD.

NUMERICAL RESULTS AND CONCLUSIONS

For our numerical calculations the transverse size parameters $a_\Lambda = a_p = 1.1\ \text{GeV}^{-1}$ are chosen in such a way that they provide a value of ≈ 280 MeV for the mean intrinsic transverse momentum $< k_\perp^2 >^{1/2}$ of a quark inside the proton. The wave-function normalizations are fixed such that the probabilities to find the quark-diquark states in a proton and a Λ_c are 0.5 and 0.9, respectively.

The integrated cross section for $p\bar{p} \to \Lambda_c\bar{\Lambda}_c$ is plotted in Fig. 2. It is of the order of nb, which is still in the range of high precision experiments and comparable in magnitude with the result in Ref. [9]. Shown is a band which indicates the uncertainty in the integrated cross section caused by varying the transverse size parameters a_Λ and a_p from $0.75\ \text{GeV}^{-1}$ to $1.1\ \text{GeV}^{-1}$. If the probabilities to find the quark-diquark states in a proton and a Λ_c are scaled by factors π_p and π_Λ, respectively, the cross section has to be multiplied by $(\pi_p \pi_\Lambda)^2$.

Single spin observables are zero since all our amplitudes are real. But we have various non-vanishing spin-spin correlation functions. Generically a spin-spin correlation function is defined as

$$\mathcal{O}_{ij} = \frac{\sigma(ij) + \sigma(-i-j) - \sigma(i-j) - \sigma(-ij)}{\sigma(ij) + \sigma(-i-j) + \sigma(i-j) + \sigma(-ij)}, \tag{3}$$

with $\sigma(ij)$, $i,j = L,N,S$ denoting cross sections for the case that two of the particles participating in the scattering process are polarized into longitudinal, normal (to the scattering plane) or sideways direction, respectively ($L \perp N \perp S$). Such spin-spin correlation functions are, e.g., A_{ij} (initial spin correlations), C_{ij} (final spin correlations), D_{ij} (polarization transfer from p to Λ_c), K_{ij} (polarization correlation between $\bar{p}$ and $\bar{\Lambda}_c$).

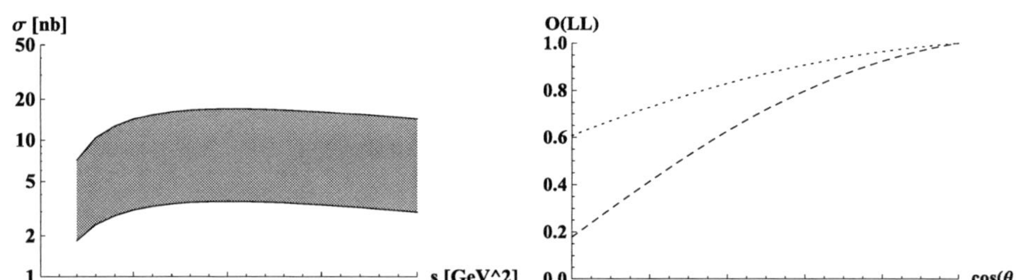

FIGURE 2. *Left:* The $p\bar{p} \to \Lambda_c\bar{\Lambda}_c$ integrated cross section σ versus Mandelstam s for a simple quark-diquark model. *Right:* The corresponding spin-spin correlation functions $-C_{LL}$ and $D_{LL} = -K_{LL}$ (dashed and dotted lines, respectively).

Because we have only one GPD in our model the spin-spin correlation functions of the process $p\bar{p} \rightarrow \Lambda_c\bar{\Lambda}_c$ are equal to the ones of the subprocess $u\bar{u} \rightarrow c\bar{c}$, since the form factor appears as a prefactor in the cross sections and cancels when calculating the spin correlators. The longitudinal spin-correlation functions $-C_{LL}$ and $D_{LL} = -K_{LL}$ are shown in Fig. 2. Since helicity is conserved if the gluon couples to a light quark, proton and antiproton are forced to have opposite helicities and hence $A_{LL} = -1$.

Non-trivial spin-effects for $p\bar{p} \rightarrow \Lambda_c\bar{\Lambda}_c$, which differ from those of the $u\bar{u} \rightarrow c\bar{c}$ subprocess, can, e.g., be achieved by taking more sophisticated wave functions for the proton and the Λ_c. One could, e.g., think of an admixture to the Λ_c wave function in which the helicity of the c quark is opposite to the Λ_c helicity. This would lead to three different form factors, which do not cancel anymore in the spin-spin correlation functions.

Since the constraints posed by HQET (in the limit $m_c \rightarrow \infty$) are already implemented in the simple quark diquark model, we do not expect that the use of three-quark wave functions for the proton and the Λ_c will change the results dramatically. Details of the wave functions will most likely show up in spin observables. Existing models for the three-quark light-cone wave function of the proton [11], [12] get some support from their success in phenomenological applications and from lattice QCD [13], but much less is known about the Λ_c wave function. The sensitivity of unpolarized and polarized cross sections on the model for the Λ_c wave function is presently under investigation [3]. It would be interesting to see experimentally, whether polarization phenomena in $p\bar{p} \rightarrow \Lambda_c\bar{\Lambda}_c$ are governed by the c-quark polarization, as in our simple quark-diquark model, or whether other wave function components will be necessary.

ACKNOWLEDGMENTS

A.T.G. acknowledges the support of the "Fonds zur Förderung der wissenschaftlichen Forschung in Österreich" (project DK W1203-N08).

REFERENCES

1. A. V. Radyushkin, *Phys. Rev. D* **56**, 5524 (1997).
2. X. D. Ji, *Phys. Rev. D* **55**, 7114 (1997).
3. A.T. Goritschnig, P. Kroll and W. Schweiger, in preparation.
4. M. Diehl, T. Feldmann, R. Jakob and P. Kroll, *Eur. Phys. J. C* **8**, 409 (1999).
5. M. Diehl, T. Feldmann, R. Jakob and P. Kroll, *Phys. Lett. B* **460**, 204 (1999).
6. A. T. Goritschnig, P. Kroll and W. Schweiger, *PoS* **(LC2008)**, 021 (2008) [arXiv:0811.2661 [hep-ph]].
7. M. Diehl, T. Feldmann, R. Jakob and P. Kroll, *Nucl. Phys. B* **596**, 33 (2001) [Erratum-*ibid.* B **605**, 647 (2001)].
8. N. Isgur and M. B. Wise, *Nucl. Phys. B* **348**, 276 (1991).
9. P. Kroll, B. Quadder and W. Schweiger, *Nucl. Phys. B* **316**, 373 (1989).
10. J. G. Körner and P. Kroll, *Z. Phys. C* **57**, 383 (1993).
11. J. Bolz and P. Kroll, *Z. Phys. A* **356**, 327 (1996).
12. J. Bartels and L. Motyka, arXiv:0711.2196 [hep-ph].
13. M. Göckeler *et al.*, *Phys. Rev. Lett.* **101**, 112002 (2008).

Soft-Fermion-Pole Mechanism to Single Spin Asymmetry in Hadronic Pion Production

Yuji Koike and Tetsuya Tomita

Department of Physics, Niigata University, Ikarashi, Niigata 950-2181, Japan

Abstract. Single spin asymmetry (SSA) is a twist-3 observable in the collinear factorization approach. We present a twist-3 single-spin-dependent cross section formula for the pion production in pp-collision, $p^{\uparrow}p \to \pi X$, relevant to RHIC experiments. In particular, we calculate the soft-fermion-pole (SFP) contribution to the cross section from the quark-gluon correlation functions. We show that its effect can be as large as the soft-gluon-pole (SGP) contribution owing to the large SFP partonic hard cross section, even though the derivative of the SFP function does not participate in the cross section.

Keywords: Single spin asymmetry, Twist-3, Soft-fermion-pole
PACS: 12.38.Bx, 13.88.+e, 13.85.Ni

The observed large single spin asymmetry (SSA) in $p^{\uparrow}p \to hX$ ($h = \pi, K$, *etc.*) can not be explained by the usual parton model and perturbative QCD. In the framework of collinear factorization, SSA is a twist-3 observable and can be described in terms of the twist-3 quark-gluon-correlation correlation functions in the nucleon and/or pion [1]-[9]. Among various components of the twist-3 cross sections, a large contribution is expected to come from the twist-3 distribution functions in the transversely polarized nucleon. (See [4], however.) There are two independent twist-3 distributions $G_F^a(x_1,x_2)$ and $\widetilde{G}_F^a(x_1,x_2)$ which are defined from a light-cone correlation function in the nucleon $\sim \langle \bar{\psi}^a F^{\alpha+} \psi^a \rangle$ where ψ^a is the quark field of flavor a and $F^{\alpha+}$ is the gluon's field strength. For the explicit definition of $\{G_F^a(x_1,x_2), \widetilde{G}_F^a(x_1,x_2)\}$ and the relation between different expressions for the twist-3 distribution functions, see [5, 7]. Due to the naively T-odd nature of SSA, the relevant cross section occurs as a pole contribution of an internal propagator in the partonic hard part. For $p^{\uparrow}p \to \pi X$, two kinds of poles contribute: soft-gluon-pole (SGP) leading to $x_1 = x_2$ and soft-fermion-pole (SFP) leading to $x_i = 0$ ($i = 1$ or 2). Thus the contribution to the cross section from these functions can be schematically written as

$$\begin{aligned} \Delta\sigma^{\rm tw3} &= \left(G_F(x,x) - x\frac{dG_F(x,x)}{dx} \right) \otimes f(x') \otimes D(z) \otimes \hat{\sigma}_{\rm SGP} \\ &+ G_F(0,x) \otimes f(x') \otimes D(z) \otimes \hat{\sigma}_{\rm SFP} + \widetilde{G}_F(0,x) \otimes f(x') \otimes D(z) \otimes \hat{\sigma}'_{\rm SFP}, \quad (1) \end{aligned}$$

where we have used the fact that the SGP contribution appears in the combination of $G_F(x,x) - x\frac{dG_F(x,x)}{dx}$ [6, 9]. Note only $G_F(x,x)$ contributes through the SGP, since $\widetilde{G}_F(x,x) = 0$ due to the anti-symmetric nature of $\widetilde{G}_F(x_1,x_2)$ under $x_1 \leftrightarrow x_2$. The origin of this combination and the connection of $\hat{\sigma}_{\rm SGP}$ to the twist-2 cross section was clarified in [8, 9]. Previous analyses [2, 6] focussed on the SGP contribution (first term of (1)),

CP1149, *Spin Physics, 18th International Spin Physics Symposium*
edited by D. G. Crabb, D. B. Day, S. Liuti, X. Zheng, M. Poelker, and Y. Prok

assuming that this one is a dominant contribution. However, there is no clue that the SFP functions $G_F(0,x)$ and $\widetilde{G}_F(0,x)$ themselves are small, and its importance depends on the behavior of the partonic hard cross sections $\hat{\sigma}_{\rm SFP}$ and $\hat{\sigma}'_{\rm SFP}$.

The purpose of this report is to present the SFP cross section for $p^\uparrow p \to \pi X$ and to study its impact compared with the SGP contribution, assuming that the SFP functions $G_F(0,x)$ and $\widetilde{G}_F(0,x)$ have the same order of magnitude as the SGP function $G_F(x,x)$.

For the calculation of the SFP contribution to $p^\uparrow(p,S_\perp) + p(p') \to \pi(P_h) + X$, we followed the formalism in [7]. With the Mandelstam variables for this process, $S = (p+p')^2$, $T = (p-P_h)^2$ and $U = (p'-P_h)^2$, the SFP contribution can be written as [10]:

$$
\begin{aligned}
E_h\frac{d^3\Delta\sigma^{\rm SFP}}{dP_h^3} &= \frac{\alpha_s^2}{S}\frac{M_N\pi}{2}\,\varepsilon^{-+P_hS_\perp}\sum_{a,b,c}\int_{z_{min}}^1\frac{dz}{z^3}\int_{x'_{min}}^1\frac{dx'}{x'}\int\frac{dx}{x}\frac{1}{x'S+T/z} \\
&\quad\times\delta\left(x-\frac{-x'U/z}{x'S+T/z}\right)\left[\sum_{a,b,c}\left(G_F^a(0,x)+\widetilde{G}_F^a(0,x)\right)\right. \\
&\times\left\{q^b(x')\left(D^c(z)\hat{\sigma}_{ab\to c}+D^{\bar{c}}(z)\hat{\sigma}_{ab\to\bar{c}}\right)+q^{\bar{b}}(x')\left(D^c(z)\hat{\sigma}_{a\bar{b}\to c}+D^{\bar{c}}(z)\hat{\sigma}_{a\bar{b}\to\bar{c}}\right)\right\} \\
&\quad+\sum_{a,b}\left(G_F^a(0,x)+\widetilde{G}_F^a(0,x)\right)\left(q^b(x')D^g(z)\hat{\sigma}_{ab\to g}+q^{\bar{b}}(x')D^g(z)\hat{\sigma}_{a\bar{b}\to g}\right) \\
&\quad+\sum_{a,c}\left(G_F^a(0,x)+\widetilde{G}_F^a(0,x)\right)G(x')\left(D^c(z)\hat{\sigma}_{ag\to c}+D^{\bar{c}}(z)\hat{\sigma}_{ag\to\bar{c}}\right) \\
&\quad\left.+\sum_{a}\left(G_F^a(0,x)+\widetilde{G}_F^a(0,x)\right)G(x')D^g(z)\hat{\sigma}_{ag\to g}\right], \qquad (2)
\end{aligned}
$$

where the summation $\sum_{a,b,c}$ implies that the sum of a is over all quark and anti-quark flavors ($a = u,d,s,\bar{u},\bar{d},\bar{s},\cdots$) and the sum of b and c is restricted onto quark flavors when a is a quark and onto anti-quark flavors when a is an anti-quark. (For an anti-quark b, $\bar{b}$ denotes quark flavor.) The integration region in the convolution formula is specified by $x'_{min} = \frac{-T/z}{S+U/z}$ and $z_{min} = -\frac{(T+U)}{S}$. $G(x')$ represents the gluon distribution and $D^g(z)$ is the gluon fragmentation function for the pion. $\hat{\sigma}_{ab\to c}$, *etc.* represents partonic hard cross sections where c is the flavor of the parton fragmenting into the pion. They are functions of the Mandelstam variables in the parton level; $\hat{s} = (xp+x'p')^2 = xx'S$, $\hat{t} = (xp-P_h/z)^2 = \frac{x}{z}T$ and $\hat{u} = (x'p'-P_h/z)^2 = \frac{x'}{z}U$ and are given as follows ($N=3$ is the number of color):

$$
\begin{aligned}
\hat{\sigma}_{ab\to c} &= \frac{-(N^2\hat{s}+2\hat{t})(\hat{s}^2+\hat{u}^2)}{N^2\hat{t}^3\hat{u}}\delta_{ac}+\frac{-(N^2\hat{t}+\hat{u}-\hat{s})\hat{s}}{N^3\hat{t}\hat{u}^2}\delta_{ab}\delta_{ac}, \\
\hat{\sigma}_{ab\to\bar{c}} &= 0, \\
\hat{\sigma}_{a\bar{b}\to c} &= \frac{(N^2\hat{u}+2\hat{t})(\hat{s}^2+\hat{u}^2)}{N^2\hat{t}^3\hat{u}}\delta_{ac}+\frac{(N^2\hat{u}+2\hat{s})(\hat{t}^2+\hat{u}^2)}{N^2\hat{s}^2\hat{t}\hat{u}}\delta_{ab}-\frac{(N^2-1)\hat{u}^2}{N^3\hat{s}\hat{t}^2}\delta_{ab}\delta_{ac}, \\
\hat{\sigma}_{a\bar{b}\to\bar{c}} &= \frac{-(N^2\hat{t}+2\hat{s})(\hat{t}^2+\hat{u}^2)}{N^2\hat{s}^2\hat{t}\hat{u}}\delta_{ab}+\frac{-N^2\hat{s}+\hat{t}-\hat{u}}{N^3\hat{u}^2}\delta_{ab}\delta_{ac}, \qquad (3)
\end{aligned}
$$

$$\hat{\sigma}_{ab\to g} = \frac{(N^2\hat{s}+2\hat{t})(\hat{s}^2+\hat{u}^2)}{N^2\hat{t}^3\hat{u}} + \frac{-1}{N^3\hat{s}\hat{t}\hat{u}^2}\left(N^2(\hat{s}^3+3\hat{s}^2\hat{u}-2\hat{u}^3)+\hat{s}^3-\hat{s}^2\hat{u}\right)\delta_{ab},$$

$$\hat{\sigma}_{a\bar{b}\to g} = \frac{-(N^2\hat{u}+2\hat{t})(\hat{s}^2+\hat{u}^2)}{N^2\hat{t}^3\hat{u}} + \left\{\frac{1}{N^3}\left(\frac{\hat{u}}{\hat{s}\hat{t}}+\frac{1}{\hat{u}}\right)+\frac{1}{N}\left(\frac{\hat{s}^2+\hat{s}\hat{t}+\hat{t}^2}{\hat{s}\hat{u}^2}-\frac{\hat{u}}{\hat{t}^2}\right)+\frac{N(\hat{u}^3-\hat{t}^3)(\hat{t}^2+\hat{u}^2)}{\hat{s}^2\hat{t}^2\hat{u}^2}\right\}\delta_{ab}, \quad (4)$$

$$\hat{\sigma}_{ag\to c} = \left\{\frac{N^2(\hat{s}^3-\hat{u}^3)(\hat{s}^2+\hat{u}^2)}{(N^2-1)\hat{s}\hat{t}^3\hat{u}^2} + \frac{\hat{s}\hat{u}(\hat{s}^2+\hat{s}\hat{u}-\hat{u}^2)-N^2(\hat{s}^4+\hat{s}^3\hat{u}+\hat{s}^2\hat{u}^2+\hat{s}\hat{u}^3+\hat{u}^4)}{N^2(N^2-1)\hat{s}\hat{t}^2\hat{u}^2}\right\}\delta_{ac} + \frac{(N^2\hat{u}+2\hat{s})(\hat{t}^2+\hat{u}^2)}{N(N^2-1)\hat{s}^2\hat{t}\hat{u}},$$

$$\hat{\sigma}_{ag\to\bar{c}} = \frac{\hat{s}+2\hat{t}-N^2\hat{s}}{N^2(N^2-1)\hat{u}^2}\delta_{ac} + \frac{-(N^2\hat{t}+2\hat{s})(\hat{t}^2+\hat{u}^2)}{N(N^2-1)\hat{s}^2\hat{t}\hat{u}}, \quad (5)$$

$$\hat{\sigma}_{ag\to g} = \frac{-N^2}{(N^2-1)\hat{s}^2\hat{t}^3\hat{u}^2}\left(4\hat{s}^6+11\hat{s}^5\hat{t}+19\hat{s}^4\hat{t}^2+22\hat{s}^3\hat{t}^3+19\hat{s}^2\hat{t}^4+11\hat{s}\hat{t}^5+4\hat{t}^6\right) + \frac{1}{N^2(N^2-1)\hat{s}\hat{t}^2\hat{u}^2}\left\{-\hat{s}\hat{t}\hat{u}^2+N^2(\hat{s}^4+\hat{s}^3\hat{t}+2\hat{s}^2\hat{t}^2+\hat{s}\hat{t}^3+\hat{t}^4)\right\}. \quad (6)$$

A remarkable feature of (2) is that the partonic hard cross sections for $G_F(0,x)$ and $\widetilde{G}_F(0,x)$ are the same, even though each diagram gives different contributions for the two functions [10]. Accordingly, they appear in the combination of $G_F^a(0,x)+\widetilde{G}_F^a(0,x)$ in (2). We also note that some terms in the hard cross section (the first terms in $\hat{\sigma}_{ab\to c}$, $\hat{\sigma}_{ab\to g}$, $\hat{\sigma}_{ag\to c}$, $\hat{\sigma}_{ag\to g}$, *etc.*) accompany the large color factors compared to the SGP cross section derived in [6], and show the steep rising behavior in the forward direction (*i.e.* large S/T). This suggests that the SFP contribution gives rise to the non-negligible contribution to the asymmetry.

To see the impact of the SFP contribution, we have performed a numerical calculation of A_N for $p^\uparrow p\to\pi X$, assuming that the SFP function is of the same order of magnitude as the SGP function. Kouvaris *et al.* [6] have parametrized the SGP function $G_F^a(x,x)$ so that the SGP contribution reproduces the A_N data obtained from RHIC and FNAL data. Their analysis shows that both data are reasonably well reproduced. In particular, they found that the derivative term in (1) brings dominant contribution compared to the nonderivative contribution. Here we adopt Fit(I) of [6] and assume $G_F^a(0,x)+\widetilde{G}_F^a(0,x)=G_F^a(x,x)$ $(a=u,d)$. Figure 1 shows A_N for the pion at $\sqrt{S}=200$ GeV and $P_{hT}=1.5$ GeV with and without SFP contribution. As is seen from the figure that the SFP contribution brings large effect in the positive x_F region, while its effect is negligible in the negative x_F region. From this figure it is clear that the SFP contribution can affect A_N significantly even though it does not receive enhancement by the derivative unlike SGP contribution, unless the SFP function itself is small.

To summarize, we have calculated the SFP contribution to the cross section for $p^\uparrow p\to hX$ associated with the twist-3 quark-gluon correlation functions in the polarized

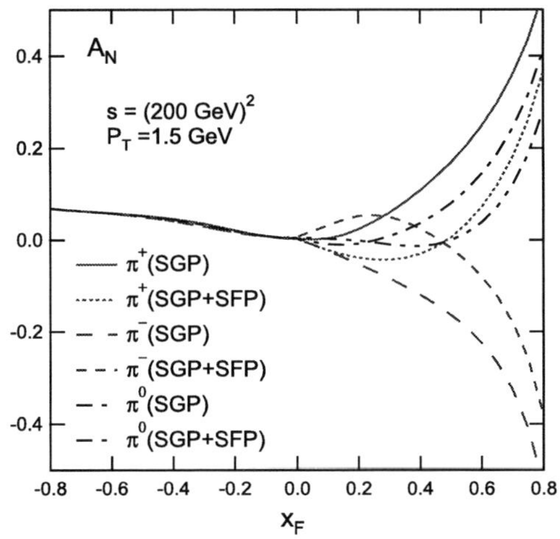

FIGURE 1. A_N for $p^{\uparrow}p \to \pi X$ at $\sqrt{S} = 200$ GeV and $P_{hT} = 1.5$ GeV. Solid, long-dashed, and dash-dot lines are, respectively, A_N for π^+, π^- and π^0 obtained with only the SGP contribution. Dotted, short-dashed, and dash-double-dot lines are, respectively, A_N for π^+, π^- and π^0 obtained with both SGP and SFP contributions.

nucleon, and have shown that its effect is significant and should be included in the analysis of A_N. For a more complete analysis, one needs to include the contribution from the triple-gluon twist-3 distribution function as well. Recent study [11] shows that A_N for the open charm production has a potential to determine the function. We hope that the global analysis of all those data including all the QCD effects will clarify the origin of observed SSA.

ACKNOWLEDGMENTS

The work of Y.K. is supported in part by Uchida Energy Science Promotion Foundation.

REFERENCES

1. J. Qiu and G. Sterman, *Nucl. Phys.* **B378**, 52 (1992).
2. J. Qiu and G. Sterman, *Phys. Rev. D* **59**, 014004 (1999).
3. Y. Kanazawa and Y. Koike, *Phys. Lett.* **B478**, 121 (2000); Phys. Lett. **B490**, 99 (2000).
4. Y. Koike, *AIP Conf. Proc.* **675**, 449 (2003); [arXiv:hep-ph/0210396]; *Nucl. Phys.* **A721**, 364 (2003).
5. H. Eguchi, Y. Koike and K. Tanaka, *Nucl. Phys.* **B752**, 1 (2006).
6. C. Kouvaris, J. W. Qiu, W. Vogelsang and F. Yuan, *Phys. Rev. D* **74**, 114013 (2006).
7. H. Eguchi, Y. Koike and K. Tanaka, *Nucl. Phys.* **B763**, 193 (2007).
8. Y. Koike and K. Tanaka, *Phys. Lett. B* **646**, 232 (2007) [Erratum-*ibid.* **668**, 458 (2008)].
9. Y. Koike and K. Tanaka, *Phys. Rev. D* **76**, 011502 (2007).
10. Y. Koike and T. Tomita, in preparation.
11. Z. B. Kang, J. W. Qiu, W. Vogelsang and F. Yuan, *Phys. Rev. D* **78**, 114013 (2008).

Polarized Proton-Proton Elastic Scattering at RHIC

D. Plyku (*for the STAR Collaboration*)

Old Dominion University, Physics Department, Norfolk, VA 23529, USA

Abstract. The Relativistic Heavy Ion Collider provides colliding polarized proton beams of 25 to 250 GeV/c momentum. With silicon strip detectors installed in Roman Pots at about 58 m on either side of the interaction point (IP) of the STAR detector, elastic scattering will be studied at a center of mass (cms) energy of 200 GeV and four momentum transfer squared range of 0.003 GeV^2/c^2 ≤ $|t|$ ≤ 0.03 GeV^2/c^2, with 100% acceptance and full ϕ coverage. Measurements of single and double spin asymmetries will be carried out with special beam optics. An upgrade with additional Roman pots closer to the IP will increase the kinematic range to $|t|$ ≤ 1.5 GeV^2/c^2 at $\sqrt{s}$ = 500 GeV. Simulation studies of the detector acceptance for 200 GeV cms energy will be given.

Keywords: Elastic scattering. Polarization. Roman pots.

INTRODUCTION AND MOTIVATION

The Relativistic Heavy Ion Collider (RHIC) has the unique capability of accelerating and colliding polarized protons with a high average polarization of 0.7 for each beam and high luminosity reaching $L = 2\cdot 10^{32}$ cm^{-2}sec^{-1}, at a wide and previously unexplored cms energy range of 50 GeV ≤ $\sqrt{s}$ ≤ 500 GeV. This enables the study of both unpolarized and polarized proton-proton (*pp*) scattering processes at high cms energies. The pp2pp experiment, now combined with the STAR experiment at RHIC, is designed to study these processes in *pp* elastic and diffractive scattering. Silicon strip detectors installed in Roman pots and positioned on both sides of the interaction point (IP) at STAR are being used to detect scattered protons in the forward direction [1, 2]. Diffractive scattering can also be studied by using the STAR detector to measure central production [3].

The pp2pp experiment at RHIC, designed to study elastic scattering at low four-momentum transfer squared range of 0.003 GeV^2/c^2 ≤ $|t|$ ≤ 0.03 GeV^2/c^2, had a successful period of data taking in 2002, performing the first measurement of elastic scattering at $\sqrt{s}$ = 200 GeV. During the second proton run at RHIC in 2003, pp2pp measured for the first time single and double spin transverse asymmetries at $\sqrt{s}$ = 200 GeV. Future experimental goals include improving the measurements of forward scattering and the spin dependent parameters at $\sqrt{s}$ = 200 GeV, and also performing these experiments at $\sqrt{s}$ = 500 GeV.

CP1149, *Spin Physics, 18th International Spin Physics Symposium*
edited by D. G. Crabb, D. B. Day, S. Liuti, X. Zheng, M. Poelker, and Y. Prok

Elastic Proton-Proton Scattering at Small $|t|$

High energy elastic scattering at small $|t|$ cannot be described by pQCD since the coupling constant (α_s) is large, thus preventing the use of perturbative techniques in QCD. As described by Regge theory, high energy elastic scattering is dominated by the exchange of a color singlet object, the Pomeron, with quantum numbers $J^{PC} = 0^{++}$. In pQCD, a Pomeron can be described as a color singlet combination of two-gluons. Pomeron exchange models have been successful in describing the slowly rising cross section of pp and $p\bar{p}$ scattering.

In order to describe elastic pp scattering at small $|t|$, contributions from both the Coulomb, dominant at $-t$ below 10^{-3} GeV$^2/c^2$, and the strong hadronic interactions need to be considered. The region where the two exchanges are of comparable magnitude is called the Coulomb-Nuclear Interference (CNI) region. The forward scattering parameters, which include the total scattering cross section σ_{tot}, the ratio of the real to the imaginary part of the forward scattering (t = 0) amplitude, ρ, and the nuclear slope parameter B, depend on the cms energy $\sqrt{s}$. At high values of $|t|$, around 1.4 GeV$^2/c^2$, the differential elastic cross section shows a diffractive minimum (dip structure) and then it continues to decrease slowly.

The Coulomb amplitude is calculable, thus a measurement of the elastic differential count rate, dN/dt, in the region below $-t = 5\cdot10^{-4}$ GeV$^2/c^2$, yields the luminosity L, which is necessary to obtain the total cross section. The ρ-parameter can be measured in the CNI region and the real part of the forward scattering amplitude can be obtained by using the measured total cross section. In the nuclear region, the nuclear slope parameter B can be measured. In 2003, the pp2pp experiment accomplished the first measurement of the nuclear slope parameter B at 200 GeV [1].

Spin Dependence of Elastic Scattering

Polarization effects of pp elastic scattering will also be studied in this experiment. Five independent helicity amplitudes are required to describe pp elastic scattering [4]. There is a measurable asymmetry in pp scattering in the CNI region, near the point of maximum interference of the two amplitudes. In this range of t, the single spin asymmetry A_N, originates mainly from the interference between electromagnetic spin-flip and hadronic nonflip amplitudes [4]. However, the measurement of A_N in the CNI region is a sensitive probe of the hadronic spin-flip amplitude, leading to another possible interference term between the electromagnetic non-flip and hadronic spin-flip amplitudes [5]. The maximum value of the asymmetry $A_{\max} \cong 0.04$ for 200 GeV is at $-t \cong 0.0024$ GeV$^2/c^2$. The analyzing power, A_N, was measured by the pp2pp experiment in 2003, at 200 GeV [6].

EXPERIMENTAL SETUP

Since the scattered protons scatter at very small angles they stay within the beam pipe of the accelerator. Thus, to be able to detect the scattered protons, the detectors need to be moved close to the beam for data taking. This is achieved by using insertion devices in the shape of cylindrical vessels, called Roman pots (RP), which house the detectors. The Roman pots can be moved close to the beam, while the detectors inside the pot remain isolated from the beam vacuum. Silicon strip detectors, with 100 μm strip pitch, are used for position measurements [2]. Two Roman pot stations, one horizontal and one vertical and with two pots per station, are installed in the outgoing beam on either side of the IR at STAR. The position of the detectors along the beam line is chosen to be where the scattered protons are well separated from the beam protons.

The scattered particles pass through bending and focusing magnets and follow trajectories determined by the accelerator magnets until they reach the detectors, which measure the particle's coordinates in the transverse plane. The position, x_D, and angle, θ_x^D, of the scattered particles at the detection point are related to the corresponding values at the intersection point (* = at IP) by the beam transport equations

$$x_D = a_{11} \cdot x^* + a_{12} \cdot \theta_x^*$$
$$\theta_x^D = a_{21} \cdot x^* + a_{22} \cdot \theta_x^*$$

where the coefficients are the beam transport matrix elements and are determined by the accelerator optics. Similar equations but with different coefficients exist for the y-coordinate. The optimum condition of the experiment is to minimize the dependence of the measured coordinates on the unknown collision vertex at the IP. This requires using special accelerator optics and is achieved by minimizing a_{11} and maximizing a_{12} in the beam transport equations. With this condition satisfied the coordinate measured at the detection point is directly proportional to the scattering angle at the IP. The angle θ_x^D needs to be measured to eliminate the dependence on the position at the IP, x^*. Having two sets of Roman pots allows the measurement of the local angle, which is equal to θ_x^D, since there are no focusing magnets between the Roman pot stations.

Experimental Plan and Simulation Results

The experimental plan has two phases. The acceptance of the detectors that will be used in Phase I and II of the experiment was studied for two different cms energies, 200 and 500 GeV, by using the simulator program called Hector [7].

In Phase I we are planning to use the existing Roman pots installed at about 58 m on both sides of the STAR IP to study elastic scattering. The plots below show the coordinates of the scattered particles in the transverse plane, Fig. 1. (a), and the normalized number of detected particles as a function of $|t|$, Fig. 1 (b). Since there are two RP stations, one horizontal and one vertical, there is a region in $|t|$ where the acceptance is constant and unity, Fig. 1. Beyond this region the acceptance falls of as t approaches the maximum value that can be reached, since the acceptance is limited by the magnets' apertures. The minimum value in $|t|$ that can be reached is related to the minimum distance of approach to the beam, which depends on the beam size at the detection point and the dead space of the detector package. The range in $|t|$ with 100 % acceptance and full coverage in azimuthal angle ϕ, that can be accessed with detectors positioned at about 58 m away from the IP was determined to be 0.003 GeV^2/c^2 < $|t|$ <

0.03 GeV^2/c^2 for $\sqrt{s}$ = 200 GeV, which is close to the CNI region where the measurement of ρ-parameter and the single spin asymmetry can be performed, Fig. 1 (b). Given the luminosity $2{\cdot}10^{29}$ cm^{-2} sec^{-1} and for the amount of data that is expected to be taken during four days of data taking in 2009, the estimated error on the slope parameter is $\Delta B = 0.31$ $(\text{GeV}/c)^{-2}$ and on the ratio of the real to the imaginary part $\Delta\rho$ = 0.01 [8]. With polarization 0.5 and 2.3 mb cross section, the number of events within our acceptance is calculated to be $40{\cdot}10^6$ events and the errors on the spin asymmetries are estimated as $\Delta A_N = 0.0017$ and $\Delta A_{NN} = \Delta A_{SS} = 0.003$ [8]. Data taking with this experimental setup requires special running conditions with special proton beam optics, in order to reach small $|t|$ values. With existing power supplies, this special accelerator optics is determined to be at $\beta^* = 21$ m for $\sqrt{s} = 200$ GeV.

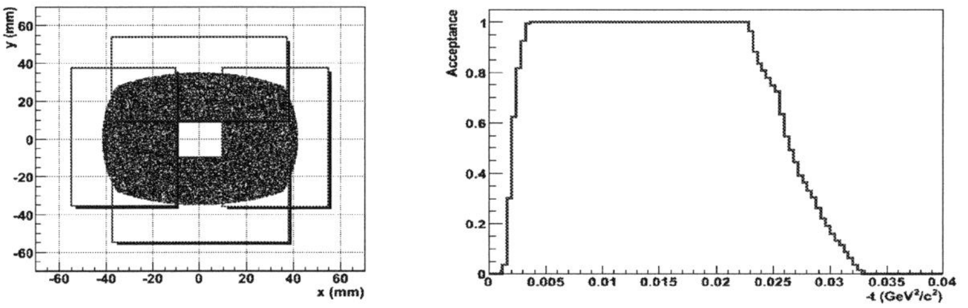

(a) (b)

FIGURE 1. (a) Transverse Positions of scattered protons and detector planes (red lines) at 58 m from the IR; (b) Acceptance as a function of t for Phase I at $\sqrt{s} = 200$ GeV.

In Phase II an additional set of Roman pots will be installed at about 18 m away from the IP, between the DX-D0 dipole magnets. The advantage for this is to reach higher values in $|t|$ and to be able to take data with the standard proton beam optics, taking data concurrent with STAR. The maximum t that can be reached with 100 GeV/c protons in this region is extended to 0.246 GeV^2/c^2, allowing the measurement of the slope parameter B and with 250 GeV/c protons to 1.54 GeV^2/c^2, allowing the study of the diffraction dip region.

To sum up, by using the forward detectors we can measure elastic scattering and its spin dependence in the forward direction and integration with the STAR detector allows measurements in the central region. The experimental setup has been successfully integrated with STAR and is ready for data taking.

REFERENCES

1. S. Bültmann et al., *Phys. Lett. B* **579,** 245 (2004).
2. S. Bültmann et al., *Nucl. Instr Meth. A* **535,** 415 (2004).
3. A. Bravar, W. Guryn, S.R.Klein, D. Milstead, B. Surrow, *J. Phys.G* **28,** 2885 (2002).
4. N. H. Buttimore et al., *Phys. Rev. D* **59,** 114010 (1999).
5. B. Z. Kopeliovich and B.G. Zakharov, *Phys. Lett. B* **266,** 156 (1989).
6. S. Bültmann et al., *Phys. Lett. B* **632,** 167 (2006).
7. J. de Favereau, X. Rouby, K. Piotrzkowski, Hector, a fast simulator for the transport of particles in beamlines, arXiv: 0707.1198 [physics.acc-ph].
8. W. Guryn, *for the STAR Collaboration*, arXiv: 0808.3961 [nucl-ex].

SPIN PHYSICS IN NUCLEI

What Have We Learned About Three-nucleon Systems at Intermediate Energies?

N. Kalantar-Nayestanaki

KVI, University of Groningen, Zernikelaan 25, 9747 AA, Groningen, The Netherlands

Abstract. In the past few years, a program has been set up at KVI to study the effects of three-nucleon forces in various reactions. Several observables in elastic and break-up channels in proton-deuteron scattering have been measured and compared with state-of-the-art Faddeev calculations using modern two-body forces and a number of three-body forces. For the elastic scattering process, the data-base is now large enough to perform a partial wave analysis to provide an unambiguous description of nuclear forces. For the break-up reaction, data are emerging at intermediate energies revealing interesting aspects of nuclear forces.

Keywords: nuclear forces, NN interaction, 3N forces, proton-deuteron elastic scattering, scaling
PACS: 13.75.Cs, 21.30.-x, 21.45.+v, 25.10.+s

In the last decade, several semi-phenomenological two-nucleon models, have become available, which describe two-nucleon scattering observables accurately. Furthermore, due to the availability of fast computers, Faddeev-type three-nucleon calculations are now-a-days routinely performed. The use of modern two-nucleon potentials, to describe three-nucleon scattering observables, leads to various degrees of agreement between the calculations and the experimental data, depending on the observable being studied and the incident beam energy. At low incident beam energies, up to ≈ 30 MeV, the differential cross sections in nucleon-deuteron scattering are described generally well using solely two-nucleon potentials, with small changes when the three-nucleon force is included. In contrast, efforts to describe the analyzing power have failed and the inclusion of three-nucleon forces (3NFs) into the calculations has not remedied the discrepancy, leading to the well-known A_y-puzzle. At intermediate and higher energies, calculations based on two-nucleon forces generally fail to describe many experimental observables [1, 2, 3, 4, 5, 6].

In order to have a comprehensive model of the three-body forces, one should not only study spin-averaged cross section but also observables which involve spin such as analyzing power, spin-transfer coefficients, and spin-correlation coefficients for a large energy and angular range. Only with such a reservoir of data, which has now emerged from various laboratories for the elastic channel, one can understand in detail the interplay between the two and three-body forces. The understanding of the nature of forces is strongly coupled to the theoretical efforts which attempt to generate nuclear forces [7, 8, 9] in a consistent manner such as those in the effective field theory [10, 11, 12].

The Dutch-Polish-Japanese collaboration has set up a program to study many of the observables for various reactions of interest systematically. These include elastic proton-deuteron scattering [3, 4, 5], proton-deuteron break-up reaction [13] and proton-deuteron

CP1149, *Spin Physics, 18th International Spin Physics Symposium*
edited by D. G. Crabb, D. B. Day, S. Liuti, X. Zheng, M. Poelker, and Y. Prok

capture reaction [14]. In this article, the results of the elastic-scattering cross sections which have been measured in the past are revisited from a different point of view. In addition, preliminary results for a sample of data from the proton-deuteron break-up reaction will be presented.

PROTON-DEUTERON ELASTIC SCATTERING

About a decade ago, it was proposed by Witała et al. and Nemoto et al. to search for the effects of three-body forces in the minimum of the proton-deuteron elastic scattering cross section [15, 16]. At KVI, a number of measurements were carried out to study this effect as a function of the beam energy. Measurements were done with six incident energies, namely 108, 120, 135, 150, 170 and 190 MeV. The results of these measurements have already been published in Refs. [4]. At 135 MeV, a discrepancy has arisen between the KVI data and those obtained at RIKEN [17]. To resolve this disagreement, a new measurement has been performed at KVI with a different detection system. The results of the new measurement [18] show that both experiments performed in the past, and in particular the RIKEN data, had an absolute normalization problem. In addition, the angular distributions of the cross sections turned out to be different in shape in both experiments. In order to do a global analysis for the elastic channel in proton-deuteron scattering, a partial-wave analysis (PWA) is, therefore, called for. Within the framework of the PWA, problems with the shapes and normalization factors of various experiments could be examined more thoroughly. Presently, the task of performing a PWA seems to be not feasible, though. Instead, we have carried out a systematic study of elastic cross sections as a function of momentum transfer.

The cross section of the elastic proton-deuteron scattering process as a function of angle shows that scattering at small angles (low momentum transfers) is dominated by the direct term in the two-nucleon potential which should then be rather well described by the calculations using only two-nucleon forces. At the backward angles (large momentum transfers), the exchange term in the two-nucleon potential is responsible for the increase of the cross sections. In these two regions, the effect of the three-body forces are predicted to be small. Finally, in the intermediate angles where the cross section has a minimum, one would expect the effects of three-nucleon forces to be largest as was shown in Refs. [15, 16]. The cross section can be expressed as a multiplication of the square of the invariant matrix element with a phase space factor which has a large variation with the incident beam energy. The dynamical part of the cross section, i.e. the square of the invariant matrix element, can be obtained and studied by taking out the trivial phase-space dependence by multiplying the cross section with $p^2 s$ where p is the momentum of the incident beam and s is the total energy available in the center of mass. This is done in Fig. 1 for a large number of bombarding energies between 70 and 250 MeV where the square of the invariant matrix element is plotted against the Mandelstam variable, $t = |p_1 - p_3|^2$, where p_1 and p_3 are the four momenta of the incoming and the outgoing particles. The variable t corresponds to the relativistic momentum transfer in the reaction. The calculations are all done by the Hannover-Lisabon group and include the Coulomb effect [19].

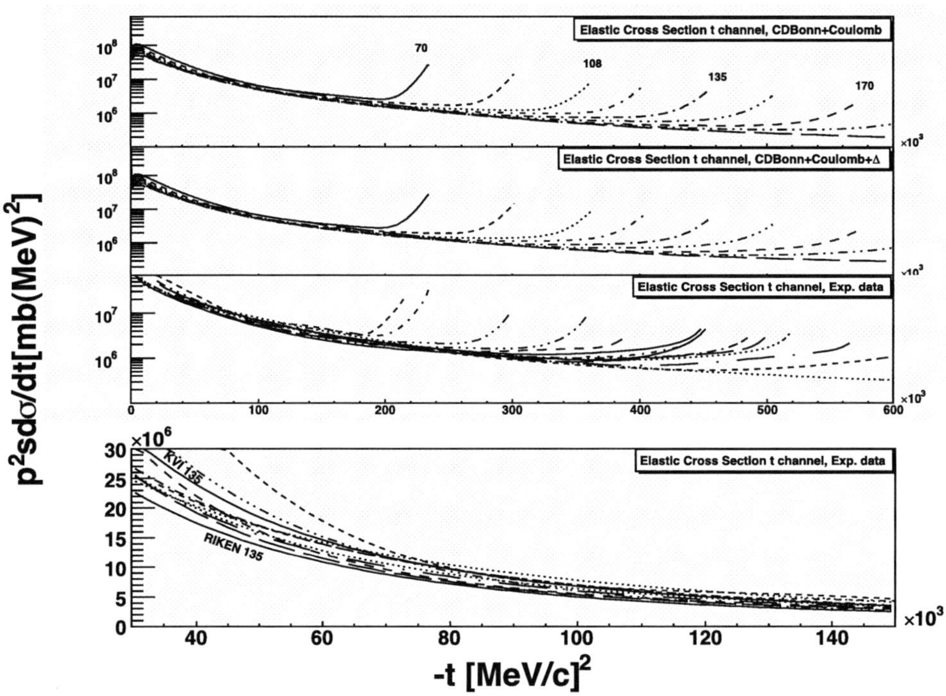

FIGURE 1. The elastic invariant matrix element is shown as a function of t, for a number of incident beam energies between 70 and 250 MeV. In the top panel the results of the calculations using the CD-Bonn potential are shown, while in the second panel from the top, the effect of three-nucleon force has also been added. All calculations include the Coulomb force. The third panel shows the same but for the experimental data. The lowest panel zooms into a small area at low momentum transfers.

There are several features which are very clear: at low values of t, one observes scaling for the square of the invariant matrix element at all incident energies shown in the figure. This implies that the dynamical part of the cross section predominantly depends on momentum transfer and has a very small energy dependence. It is exactly in this region where one could expect all the experiments to agree with each other (modulo a small energy dependence). For the intermediate values of t, one enters the region of the cross section minimum. In this region, the results of the calculations including the three-nucleon force show indeed that their effect is as large as 50% in the minimum. However, the energy dependence of the effect is again very small in the energy range shown (see the second panel from the top).

The third panel from the top in Fig. 1 shows the experimentally-obtained square of the invariant matrix element. Each line corresponds to a fit through the data for a given energy. The spread in the band at lower values of t represents the uncertainty in the normalization of various data sets. The width of this band is generally explained by the systematic uncertainties quoted by various experiments. Incidentally, the two recent data sets at 135 MeV which disagree with each other reside on the two sides of the band (the

KVI data above and the RIKEN data below the band). For more details, see the lowest panel in Fig. 1.

Similarly, one can study the dependence on the Mandelstam variable, $u = |p_1 - p_4|^2$ for which p_4 is the momentum of the recoil particle. This variable represents the relativistic exchange momentum transfer. The conclusions are again the same: there is the same sort of scaling as in Fig. 1 but then at backward angles. One should note that this simple "scaling" breaks down for energies below 100 MeV (see the curve for 70 MeV in Fig. 1). The origin of this may be due to the fact that the de Broglie wavelength is getting so large that the coherence effects cannot be ignored anymore.

PROTON-DEUTERON BREAK-UP CHANNEL

The break-up channel offers a very rich kinematics as it involves three particles in the final state. Kinematically, one would then need to measure five out of the nine available kinematical variables to define an event completely. For this purpose, a detection system was designed which has an almost complete coverage of the phase space of the break-up reaction. This detector which is called Big Instrument for Nuclear-polarization Analysis (BINA) has been operational since 2005. BINA is a setup with a nearly 4π geometrical acceptance and has been used in various few-nucleon scattering experiments to measure the scattering angles and energies of protons and deuterons with the possibility for particle identification. The target used for these studies is liquid hydrogen or deuterium [20].

For the proton-deuteron break-up reaction, three measurements have been performed at beam energies of 100 (deuteron), 135 (proton) and 190 (proton) MeV with BINA. Many kinematical configurations have been measured and analyzed. Here, only a very small sample of the preliminary analyzing-power results are shown for the experiment done at an incident proton beam energy of 190 MeV. The results shown in Fig. 2 are for configurations where the two protons scatter to forward angles of $(25^\circ, 25^\circ)$ (left panels) and $(25^\circ, 20^\circ)$ (right panels). The azimuthal opening angles between the two protons vary between 180° (top panels) to 20° (bottom panels). The results of various calculations are also shown. As can be seen from the top panels, various models agree with each other and also with experimental data for co-planar geometries. However, as one decreases the co-planarity angle from 180° to 20°, serious disagreements between various calculations set in and more importantly, they all disagree with the experimental data. Adding a 3NF to two-nucleon potential increases the disagreement even further. Even the effect of relativity [21] goes in the wrong direction. This problem needs to be further investigated.

CONCLUSIONS

To understand the properties of the three-nucleon forces, a large data-base including spectra of light nuclei and scattering data in three-body systems with all possible observables measured is mandatory. For the cross section of the elastic proton-deuteron scattering, a large data-base has been obtained which asks for a partial wave analysis for this channel. The results show unambiguously that two-nucleon potentials are not enough to describe the bulk of data. However, none of the three-nucleon potentials available in

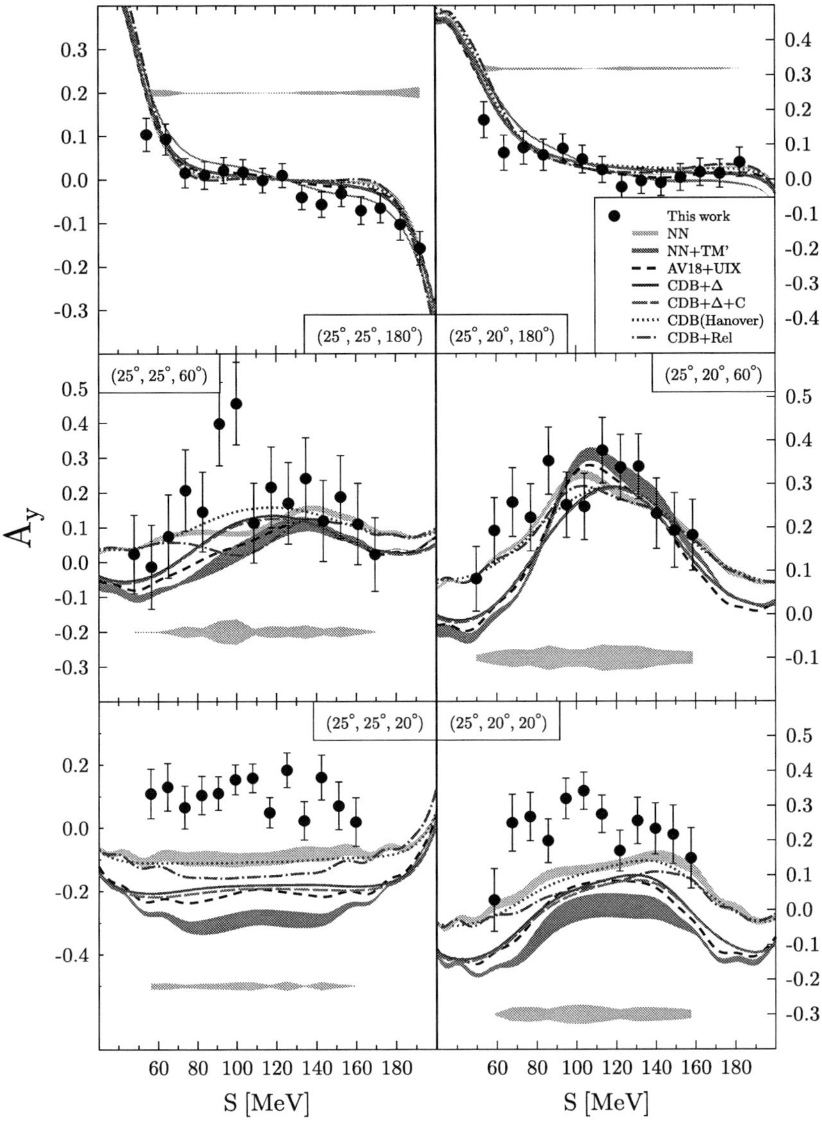

FIGURE 2. The preliminary results for analyzing powers of proton-deuteron break-up reaction at $E_p = 190$ MeV as a function of the kinematical variable, S, for various polar-angle combinations of the two outgoing protons and the opening azimuthal angle between them as marked in the figure by $(\theta_1, \theta_2, \phi_{12})$. The calculations are from the Hannover-Lisabon group (CDB, CDB+Δ, and CBD+Δ+Coulomb) and Bochum-Cracow group (NN, NN+TM', AV18+UIX, and CDB+Relativity). The systematic errors are shown by the light gray (cyan) band in the panels.

the literature is at a stage of describing all the data presented here and elsewhere consistently. For the break-up channel, results show clearly that the nucleon-nucleon interaction alone is not enough to describe the data obtained at many configuations. However, the existing three-nucleon forces and also the model which includes some relativistic effects are far from being able to describe the analyzing-power data, specially at small co-planarity angles. For these kinematics, they even make the disagreement larger. The self-consistent treatment of the Δ in the Hanover approach which does a relatively good job in describing the analyzing powers in the elastic proton-deuteron scattering at energies between 100 and 200 MeV, performs less satisfactorily for vector analyzing powers in the break-up channel. The advantage of this calculation is the inclusion of Coulomb force for the first time which is shown to be rather small in the elastic channel except at small scattering angles. Note that the Coulomb effects are rather sizable for some configurations in the break-up channel. Preliminary cross-section data for the break-up channel also show large disagreements between the data and the calculations but then for large co-planarity angles.

It is clear that one needs to go beyond the nucleon-nucleon interaction to understand the dynamics of the three-body system. However, the characteristics of the three-nucleon forces, such as the central part and the spin part of the force, are not yet under control to be able to make high-precision predictions for scattering observables. The illusive nature of three-body forces might be exposed completely when the results of calculations based on effective field theory become available for these intermediate energies.

I would like to thank my graduate students Karsten Ermisch, A. Ramazani-Moghadam-Aranai, and H. Mardanpour whose Ph.D. thesis works were presented in this presentation. A. Deltuva from the Hannover-Lisabon group and H. Witała from the Bochum-Cracow group should be thanked for providing the calculations shown in the figures.

REFERENCES

1. N. Sakamoto et al., Phys. Lett. B367, 60 (1996).
2. H. Sakai et al., Phys. Rev. Lett. 84, 5288 (2000).
3. R. Bieber et al., Phys. Rev. Lett. 84, 606 (2000).
4. K. Ermisch et al., Phys. Rev. Lett. 86, 5862 (2001); Phys. Rev. C68, 051001 (2003); Phys. Rev. C71, 064004 (2005).
5. H. Mardanpour et al., Eur. Phys. J. **A31**, 383 (2007).
6. K. Sekiguchi et al., Phys. Rev. C65, 034003 (2002); Phys. Rev. C70, 014001 (2004).
7. S.A. Coon et al., Phys. Rev. C48, 2559 (1993); Few Body Syst. 30, 131 (2001).
8. B.S. Pudliner et al., Phys. Rev. C56, 1720 (1997).
9. S.C. Pieper et al., Phys. Rev. C64, 014001 (2001).
10. U. van Kolck, Phys. Rev. C49, 2932 (1994).
11. E. Epelbaum et al., Phys. Rev. C66, 064001 (2002).
12. N. Kalantar-Nayestanaki and E. Epelbaum, Nuclear Physics News Inter. 17, No. 3, 22 (2007) and nucl-th/0703089.
13. St. Kistryn et al., Phys. Rev. C68, 054004 (2003); Phys. Rev. C72, 044006 (2005); Phys. Lett. B641, 23 (2006).
14. A. Mehmandoost-Khajeh-Dad et al., Phys. Lett. B617, 18 (2005).
15. H. Witała et al., Phys. Rev. Lett. 81, 1183 (1998).
16. S. Nemoto et al., Phys. Rev. C58, 2599 (1998).
17. K. Sekiguchi et al., Phys. Rev. Lett. **95**, 162301 (2005).
18. A. Ramazani-Moghaddam-Arani et al., Phys. Rec. C78, 014006 (2008).

19. A. Deltuva et al., Phys. Rev. C68, 024005 (2003); Phys. Rev. C71, 054005 (2005); A. Deltuva et al., Phys. Rev. C72, 054004 (2005).
20. N. Kalantar-Nayestanaki et al., Nucl. Inst. and Meth. in Phys. Res. A417, 215 (1998).
21. H. Witała et al., Phys. Lett. B634, 374 (2006).

Hadron Formation in Semi-Inclusive DIS on Nuclei at HERMES

G. Elbakyan

on behalf of the Hermes Collaboration
Alikhanian Br. st. 2, 0036 Yerevan, Armenia

Abstract. The influence of the nuclear environment on the hadronization of charged pions, kaons and (anti)protons in semi-inclusive deep inelastic scattering has been studied by the HERMES experiment at DESY using a 27.6 GeV positron beam. Identified hadron multiplicities have been measured for helium, neon, krypton and xenon relative to that of deuterium as a function of ν, z and p_T^2. A large increase in the multiplicity ratios with p_T^2 was observed, clarifying the role of FSI in the nuclear transverse momentum broadening. A clear effect of braodening from measurement of the difference in the width of the p_T^2 distribution for hadrons different types from heavy nuclei and deuterium been also obtained. This effect remains constant as a function of z, but vanishes as z goes to 1 due to energy conservation and increases as a function of Q^2. In addition, double-hadron production in the nuclear medium was investigated. Nuclear effects on the double hadron-production ratio provided an additional tool for studying modifications of hadronization in nuclear matter.

Keywords: Hadronization, nuclear broadening, double-hadron attenuation
PACS: 43.35.Ei, 78.60.Mq

INTRODUCTION

The study of hadron production in deep inelastic scattering (DIS) offers the possibility to investigate quark propagation and hadronization. Quark propagation in the nuclear medium involves competing process such as hadronization of quarks, quark energy loss through multiple scattering, and gluon radiation. The relatively clean nuclear environment of lepton-induced reactions allows the investigation of the space-time evolution of the hadronization process already in the early stage of hadron production.

ATTENUATION OF IDENTIFIED HADRONS

The experimental results are presented in terms of the multiplicity ratio R_A^h. This is the ratio of the number of hadrons produced per DIS event for a nuclear target with mass number A to that from a deuterium (D) target. It is defined as

$$R_A^h(\nu,z,Q^2,p_T^2) = \left(\frac{N_h^{SIDIS}(\nu,Q^2,z,p_T^2)}{N_e^{DIS}(\nu,Q^2)}\right)_A \Bigg/ \left(\frac{N_h^{SIDIS}(\nu,Q^2,z,p_T^2)}{N_e^{DIS}(\nu,Q^2)}\right)_D, \quad (1)$$

where z, Q^2 and p_T^2 represent the fraction of the virtual photon energy ν transferred to the hadron, the virtuality of the photon, and the square of the hadron momentum component transverse to the virtual photon direction, respectively. $N_h^{SIDIS}(\nu,z,p_T^2)$ is the number of

CP1149, *Spin Physics, 18th International Spin Physics Symposium*
edited by D. G. Crabb, D. B. Day, S. Liuti, X. Zheng, M. Poelker, and Y. Prok

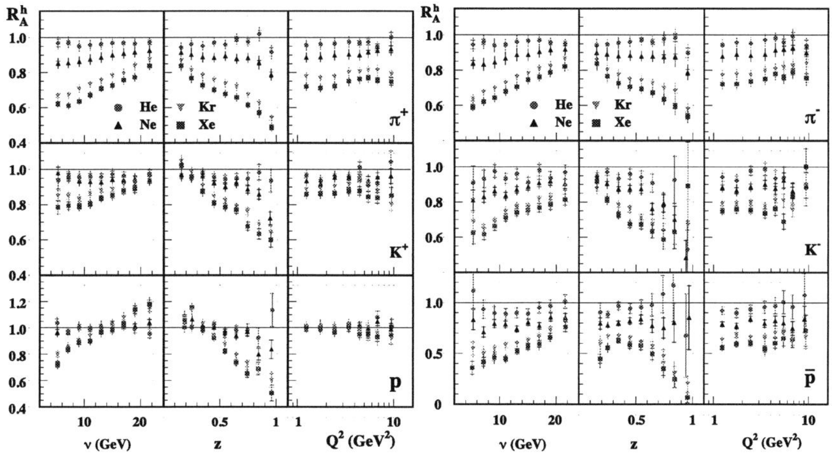

FIGURE 1. The multiplicity ratio R_A^h as a function of ν, z, and Q^2 [2]. The inner error bars represent the statistical uncertainty, while the outer ones show the total uncertainty.

semi-inclusive (SIDIS) hadrons (pions, kaons or (anti)protons) in a given (ν,z,p_T^2)-bin, and $N_e^{DIS}(\nu,Q^2)$ the number of inclusive DIS positrons in the same (ν,Q^2)-bin. The multiplicity ratio was determined as a function of ν, z and p_T^2, while integrating over all other kinematic variables. Nuclear attenuation is defined as 1-R_A^h.

The multiplicity ratio R_A^h has been measured at HERMES using D, He, Ne, Kr and Xe gas targets internal to the 27.6 GeV HERA positron storage ring, identifying both the scattered positron and the produced hadrons in the HERMES spectrometer[1].

The basic feature of the data presented on Fig.1 is the decrease of R_A^h with increasing value of the mass number A of the nucleus. Qualitatively this is understood as being due to increased partonic (quark energy loss) or hadronic (absorption) effects. Furthermore, there is a large similarity between the data for π^+, π^-, and a clear difference between those for K^+ and K^-, and those for p and $\overline{p}$. The data show that R_A^h is reduced at low ν and high z. The reduction is stronger for heavy nuclei. This clearly shows the influence of the nuclear environment on hadron formation. The difference for the kaons may be partially due to differences in the hadron-nucleon interaction cross section (σ_{hN}) for K^+ and K^-. The results for protons cannot really be related to those for any of the other particles. Because protons are already present in a nucleus, an appreciable fraction of them may not come from hadronization especially at low z. Furthermore, in final-state interactions (FSI) they generally are not absorbed, but give rise to more nucleons (both protons and neutrons), thus possibly even increasing R_A^h at lower z. Antiprotons feature a rather strong attenuation, which might be attributed to the relatively large $\overline{p}N$ cross section.

TRANSVERSE MOMENTUM BROADENING

R_A^h as a function of p_T^2 is shown for $\pi^{\pm}$, $K^{\pm}$, p and $\overline{p}$ in Fig. 2. An enhancement of the number of hadrons at high p_T^2 with atomic mass number A is observed. The p_T^2 spectra are determined mainly by the type and charge of the hadrons, and depend only weakly on the target mass. Thus, the observed enhancements in the pion and kaon spectra are more suppressed than in the proton spectra. Such an effect was first observed by EMC [3] for all charged hadrons taken together, while has been measured for hadrons by HERMES. The phenomenon is also known from heavy-ion collisions, where it is referred to as the Cronin effect [4]. Compared to experiments with heavy ions, the use of a lepton probe has the advantage that initial-state interactions do not play a role, except for shadowing effects, which are small in the x_{Bj} range of the present experiment. The observed rise of R_A^h at high p_t^2 is attributed to a broadening of the p_t^2 distribution. In principle this can result from partonic rescattering as well as from hadronic final-state interactions.

We report also the results of the measurement of the so-called p_t-broadening ($\Delta\!< p_t^2 >_A^h$) for several different hadron types, as function of the relevant kinematic variables. The p_t-broadening is defined to be the difference of the average squared transverse momentum of the detected hadron of type h produced on a nuclear target with mass A and that from a D target. As a function of ν the broadening is constant, reaches zero for z going to one and increases linearly with Q^2 (Fig. 3).
These measurements will provide theoretical insights on parton propagation and energy loss in cold nuclear matter as well as on the space time development of the hadronization process.

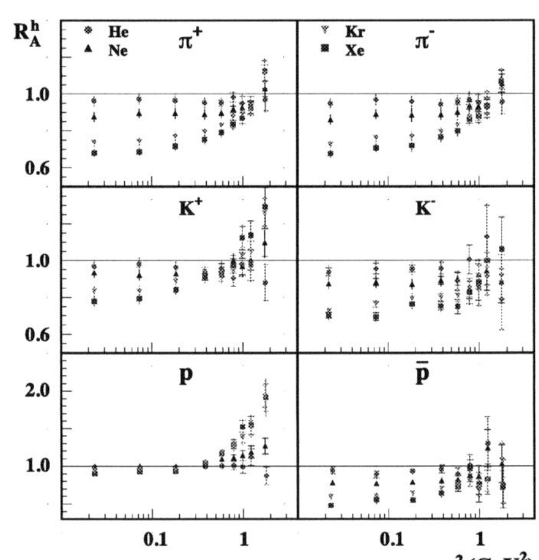

FIGURE 2. R_A^h for positively (left panel) and negatively (right panel) charged hadrons as a function of p_T^2. The data from Ref. [2]

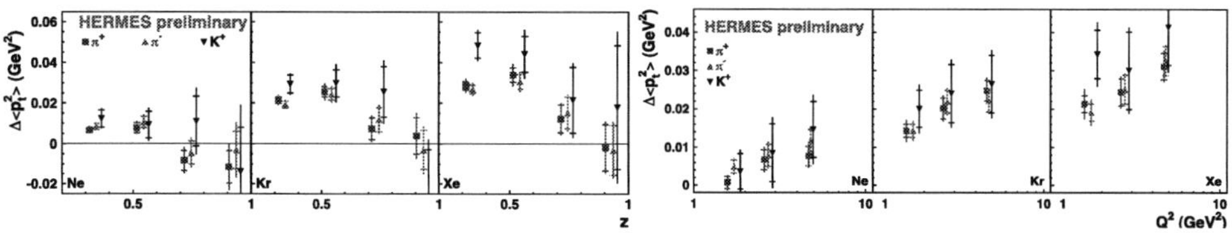

FIGURE 3. z (left) and Q^2 (right) dependence of p_t broadening for different hadron types produced on Ne, Kr and Xe targets.

DOUBLE-HADRON ATTENUATION

Double-hadron leptoproduction offers an additional tool to study hadronization. The experimental results for N, Kr and Xe relative to D are presented in terms of the ratio

$$R_{2h}(z_2) = \left(\frac{dN^{z_1>0.5}(z_2)/dz_2}{N^{z_1}}\right)_A \Bigg/ \left(\frac{dN^{z_1>0.5}(z_2)/dz_2}{N^{z_1}}\right)_D, \tag{2}$$

where z_1 and z_2 correspond to the leading (largest z) and sub-leading (second largest z) hadrons, respectively, $dN^{z_1>0.5}$ is the number of events with at least two detected hadrons in a bin of width dz_2 at z_2 with $z_1 > 0.5$, and $N^{z_1>0.5}$ is the number of events with at least one detected hadron with $z_1 > 0.5$. All charged hadrons and neutral pions are considered.

The HERMES data [5] clearly show that nuclear effects in the double-hadron ratio are smaller than for single-hadron attenuation measured under the same kinematic condition. Fig. 3-left (upper panel) shows calculations by means of a BUU transport model

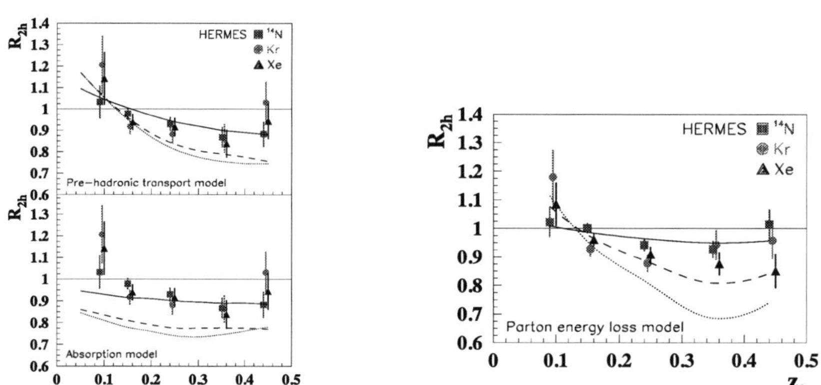

FIGURE 4. The ratio R_{2h} as function of z_2 with $z_1 > 0.5$. In the upper left panel the curves (solid for ^{14}N, dashed for Kr, dotted for Xe) are calculated within a BUU transport model[6]. In the bottom left panel with calculations that assume only absorption for the three nuclei [6]. The right panel the curves are from the parton energy loss model descibed in Refs [7, 8] (same line types in the all plots). The data from Ref. [5]

[6]. Although the general trend of the data is reproduced, the model prediction is not supported by the data. Fig. 3-left (bottom panel) shows the same data compared to a calculation with a purely absorptive treatment of the interaction of the pre-hadronic or the final hadronic states. This assumption [6] is ruled out by the data. The curve in Fig. 3-right represent the model [7, 8] in which modifications of the fragmentation functions arises from parton energy loss. Contrary to naïve expectations, this model predicts a significant A-dependence, in conflict with data. Thus, the double-hadron observables in SIDIS provide, as did the jet correlation measurements in heavy-ion collisions, new information for differentiating between the models of hadronization in nuclei that are indistinguishable in single-hadron measurements.

REFERENCES

1. HERMES Collab. (K. Ackerstaff *et al.*), *Nucl. Instr. and Meth. A* **417**, 230 (1998).
2. A. Airapetian *et al.*, *Nucl Phys B* **780**, 1 (2007).
3. J. Ashman *et al.* [EMC], Z. Phys. C52 (1991) 1.
4. J.W. Cronin *et al.*, Phys. Rev. D11 (1975) 3105.
5. A. Airapetian *et al.*, *Phys. Rev. Lett.* **96**, 162301 (2006).
6. T. Falter *et al.*, *Phys.Lett. B* **594**, 61 (2004); *Phys.Rev. C* **70**, 054609 (2004); K. Gallmeister and W. Cassing, *Nucl.Psys. A* **748**, 241 (2005).
7. X.N. Wang, *Psys. Lett. B* **579**, 299 (2004).
8. A. Majumder, *Eur. Phys. J. C* **43**, 259 (2005).

Study of $^{12}C(\vec{p},\vec{n})^{12}N(g.s.;1^+)$ at 296 MeV and Nuclear Correlation Effects

M. Dozono*, T. Wakasa*, E. Ihara*, S. Asaji*, K. Fujita*, K. Hatanaka†, T. Ishida**, T. Kaneda†, H. Matsubara†, Y. Nagasue*, T. Noro*, Y. Sakemi**, Y. Shimizu‡, H. Takeda*, Y. Tameshige†, A. Tamii† and Y. Yamada*

*Department of Physics, Kyushu University, Fukuoka 812-8581, Japan
†Reseach Center for Nuclear Physics, Osaka University, Osaka 567-0047, Japan
**Cyclotron and Radioisotope Center, Tohoku University, Sendai 980-8578, Japan
‡Center for Nuclear Study, The University of Tokyo, Tokyo 113-0033, Japan

Abstract.
We present measurements of the cross section and a complete set of polarization observables for the Gamow-Teller $^{12}C(\vec{p},\vec{n})^{12}N(g.s.;1^+)$ reaction at an incident proton energy of 296 MeV. The complete set of polarization observables is used to separate the cross section into spin-longitudinal and spin-transverse components. From the comparison with the distorted wave impulse approximation calculations employing random phase approximation response functions, we found that our data showed the existence of the pionic and rho-mesonic correlations in nuclei.

Keywords: complete set of polarization transfer observables, nuclear correlation effects
PACS: 24.70.+s, 25.40.Kv, 27.20.+n

INTRODUCTION

Nuclear correlation effects in spin-isospin modes [1] have been of considerable concern in nuclear physics. At large momentum transfers of $q \simeq$ 1–3 fm^{-1}, Alberico *et al.* [2] made a very interesting prediction for the quasielastic region. They pointed out that the spin-longitudinal (pionice) response function should be enhanced by the attractive spin-longitudinal interaction with respect to the free response function, while the spin-transverse (rho-mesonic) response function should be quenched by the repulsive spin-transverse interaction. In particular, the enhancement of the spin-longitudinal mode has aroused much interest in connection with the precursor phenomena of the pion condensation [2, 3, 4, 5, 6]. Some experiments [7, 8, 9, 10] have been made in order to assess these nuclear correlation effects. In the measurement of quasielastic (p,n) reactions [11], the enhancement of the spin-longitudinal mode has been observed, which supports the existence of pionic correlations. However, the enhancement of the spin-transverse mode has been also observed, which contradicts the theoretical prediction.

Recent experimental progresses such as the development of high efficiency neutron polarimeters have allowed us to measure a complete set of polarization observables D_{ij} for the $^{12}C(\vec{p},\vec{n})^{12}N(g.s.;1^+)$ reaction at large momentum transfers up to $q \simeq 2$ fm^{-1}. This Gamow-Teller (GT) transition is the isobaric analog to the M1 excitation of the $^{12}C(p,p')^{12}C^*(1^+,T=1)$ scattering in which possible evidence of the precursor was proposed for the first time [3, 4]. Thus it is very interesting to study nuclear correlation

CP1149, *Spin Physics, 18th International Spin Physics Symposium*
edited by D. G. Crabb, D. B. Day, S. Liuti, X. Zheng, M. Poelker, and Y. Prok

effects in this reaction by separating the cross section into the spin-longitudinal and spin-transverse components with D_{ij}.

In this paper, we present the measurement of the cross section and a complete set of D_{ij} for the $^{12}C(\vec{p},\vec{n})^{12}N(g.s.;1^+)$ reaction. The proton beam energy is 296 MeV where distortion effects become minimum and GT transitions are predominantly excited. Therefore we can examine nuclear correlation effects reliably by a comparison between the experimental data and distorted wave impulse approximation (DWIA) calculations employing random phase approximation (RPA) response functions.

EXPERIMENTAL METHOD

The experiment was performed using the Neutron Time-Of-Flight (NTOF) facility [12] and the neutron detector/polarimeter NPOL3 [13] at the Research Center for Nuclear Physics (RCNP), Osaka University. The NTOF facility consists of a beam swinger magnet, a Neutron Spin Rotation (NSR) magnet, and a 100–m Time-Of-Flight (TOF) tunnel. The typical proton beam current and polarization were 500 nA and 0.70, respectively. In the beam swinger magnet, the beam bombarded a self-supporting ^{nat}C (98.9% ^{12}C) target with a thickness of 89 mg/cm^2. Neutrons from the target passed through the NSR magnet and were measured by the NPOL3 system in the 100-m TOF tunnel with a resolution of about 500 keV FWHM. The neutron detection efficiency of NPOL3 was determined by normalizing to the $^7Li(p,n)^7Be(g.s.+0.43\ MeV)$ reaction at 0°, whose cross section is known for $T_p = 80$–795 MeV [14]. The obtained detection efficiency was 0.025 ± 0.002, where uncertainty mainly comes from the uncertainty of the cross section for $^7Li(p,n)^7Be(g.s.+0.43\ MeV)$ at 0°. The neutron polarimetry of NPOL3 was calibrated by using polarized neutrons from the $^{12}C(\vec{p},\vec{n})^{12}N(g.s.,1^+)$ reaction at 0° [13]. The resulting effective analyzing power $A_{y;\text{eff}}$ of NPOL3 was $0.151 \pm 0.07 \pm 0.04$, where the first and second uncertainties are statistical and systematic, respectively.

RESULTS AND DISCUSSION

Figure 1 shows the cross section for the $^{12}C(p,n)^{12}N(g.s.,1^+)$ raction at $T_p = 296$ MeV as a function of momentum transfer q. We can obtain the smooth angular distribution over a wide range of $q = 0.1$–$2.2\ fm^{-1}$. The open circles are the data for the $^{12}C(p,p')^{12}C^*(1^+,T=1)$ scattering at $T_p = 295$ MeV [15]. These data are shown by multiplying a factor of two because of the difference of the isospin Clebsh-Gordan (CG) coefficients between the $^{12}C(p,n)$ reaction and the $^{12}C(p,p')$ scattering. Both data sets are consistent with each other taking into account the statistical and systematic uncertainties.

We deduce the spin-longitudinal (ID_q) and spin-transverse (ID_p and ID_n) cross sections with D_{ij} following the procedure in Ref. [16]. The results are shown in Fig. 2. The error bars represent the statistical uncertainties of the data. The systematic uncertainties are, for example, about 30% at $q \simeq 1.6\ fm^{-1}$. We compared the experimental results with DWIA calculations using the computer code CRDW [16] in order to investigate nuclear correlation effects. The dashed curves represent the DWIA results with the free

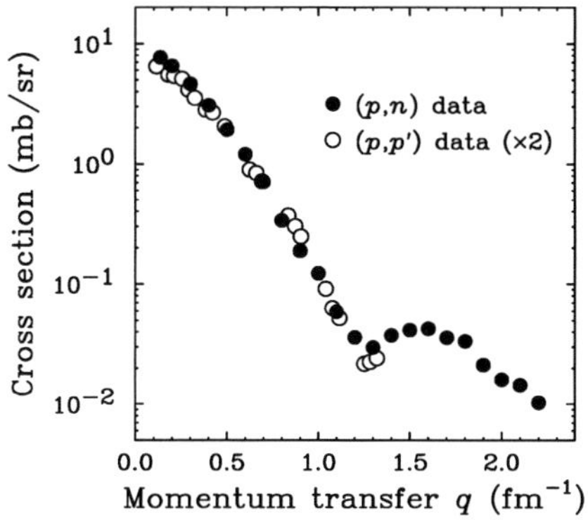

FIGURE 1. Measured cross section for $^{12}C(p,n)^{12}N(\text{g.s.}, 1^+)$ at $T_p = 296$ MeV (filled circles). The open circles are (p, p') results [15], multiplied by a factor of two, as explained in the text.

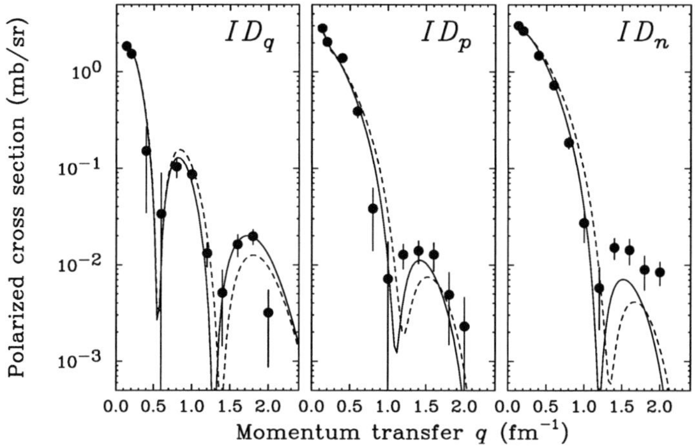

FIGURE 2. Obtained spin-longitudinal (ID_q) and two spin-transverse (ID_p and ID_n) cross sections for $^{12}C(\vec{p},\vec{n})^{12}N(\text{g.s.}; 1^+)$ at 296 MeV. The solid and dashed curves are the DWIA results with RPA and free response functions, respectively.

response function (= without nuclear correlations). These calculations reproduce the experimental data reasonably well for small momentum transfers of $q < 0.5$ fm^{-1}. However, the calculations shift the momentum transfer dependences to higher momentum transfer and significantly underestimate the cross section at large momentum transfers of $q \simeq 1.6$ fm^{-1}. We also performed DWIA calculations with RPA response functions (= with nuclear correlations) employing the $\pi + \rho + g'$ model interaction V_{eff}. The interaction V_{eff} is the sum of the pion and rho-meson exchange interactions, and the Landau-

Migdal interaction specified by the Landau-Migdal parameters, g'_{NN}, $g'_{N\Delta}$ and $g'_{\Delta\Delta}$. The solid curves represent the DWIA results with standard values of $g'_{NN} = 0.65$, $g'_{N\Delta} = 0.35$ and $g'_{\Delta\Delta} = 0.5$. For all three ID_i, their momentum transfer dependences are shifted to lower momentum transfer and those around $q = 1.6\ \mathrm{fm}^{-1}$ are enhanced by inducing the RPA correlations, which improves the agreement with the data. Thus our data support the existence of pionic and rho-mesonic correlations in nuclei.

ACKNOWLEDGMENTS

We are grateful to the RCNP cyclotron crew for providing a good quality beam for our experiments. We also thank H. Tanabe for his help during the experiments. This work was supported in part by the Grants-in-Aid for Scientific Research Nos. 14702005 and 16654064 of the Ministry of Education, Culture, Sports, Science, and Technology of Japan.

REFERENCES

1. M. Ichimura, H. Sakai, and T. Wakasa, *Prog. Part. Nucl. Phys.* **55**, 446 (2006).
2. W. M. Alberico, M. Ericson, and A. Molinali, *Nucl. Phys.* **A379**, 429 (1982).
3. M. Ericson, and J. Delorme, *Phys. Lett.* **76B**, 182 (1978).
4. H. Toki, and W. Weise, *Phys. Rev. Lett.* **42**, 1034 (1979).
5. W. M. Alberico, M. Ericson, and A. Molinari, *Phys. Lett.* **92B**, 153 (1980).
6. H. Toki, and W. Weise, *Phys. Lett.* **92B**, 265 (1980).
7. T. A. Carey, et al., *Phys. Rev. Lett.* **53**, 144 (1984).
8. T. N. Taddeucci, et al., *Phys. Rev. Lett.* **73**, 3516 (1994).
9. T. Wakasa, et al., *Phys. Rev. C* **59**, 3177 (1999).
10. C. Hautala, et al., *Phys. Rev. C* **65**, 034612 (2002).
11. T. Wakasa, et al., *Phys. Rev. C* **69**, 054609 (2004).
12. H. Sakai, et al., *Nucl. Instrum. Methods Phys. Res. A* **369**, 120 (1996).
13. T. Wakasa, et al., *Nucl. Instrum. Methods Phys. Res. A* **547**, 569 (2005).
14. T. N. Taddeucci, et al., *Phys. Rev. C* **41**, 2548 (1990).
15. A. Tamii, et al., *Nucl. Phys.* **A788**, 53c (2007).
16. K. Kawahigashi, K. Nishida, A. Itabashi, and M. Ichimura, *Phys. Rev. C* **63**, 044609 (2001).

The Tensor Analysing Power Component T_{21} of Exclusive π^--meson Photoproduction on Deuteron in the Resonance Region.

V.N. Stibunov*, L.M. Barkov†, V.F. Dmitriev†, A.Yu. Loginov*, S.I. Mishnev†, D.M. Nikolenko†, A.V. Osipov*, I.A. Rachek†, R.Sh. Sadykov†, Yu.V. Shestakov†, A.A. Sidorov*, D.K. Toporkov† and S.A. Zevakov†

**Nuclear Physics Institute at Tomsk Polytechnical University, 634050, Tomsk, Russia*
†Budker Institute of Nuclear Physics, 630090, Novosibirsk, Russia

Abstract.
The experimental data about the tensor analyzing power component T_{21} of the $\pi-$meson photoproduction on the polarized deuterons are presented here along with the theoretical calculation. The experiment has been performed at the 2 GeV Novosibirsk VEPP-3 electron storage ring using an internal polarized deuterium target by detecting the two protons in coincidence.

Keywords: tensor polarized deuteron, storage ring, tensor target asymmetry
PACS: 13.60.Le, 24.70.+s, 25.20.Lj

INTRODUCTION

The spin-dependent observables in π-meson photoproduction measured on the tensor polarized deuteron offer unique opportunities to test models of the deuteron structure and the nucleon-nucleon interaction at short distances. The tensor analyzing power components of these reactions are expected to be essential sensitive to small but dynamically important amplitudes which usually disappear in spin-independent observables.

The first measurement of tensor polarization observables in π-meson photoproduction on deuteron was carried out in our experiment at VEPP–3 electron storage ring using an internal gas target [1]. In this experiment the tensor analyzing power components T_{20} and T_{22} of the exclusive π^--meson photoproduction on deuteron were measured. New measurement of the tensor asymmetry in exclusive π^--meson photoproduction on deuteron is performed at the large polar angles of the outgoing protons with momenta above $340 MeV/c$ [2]. In order to obtain a high statistical accuracy in this experiment we have used both the cooling storage cell target and the novel high-intensity cryogenic atomic beam source (ABS), we have modified the magnetic optics of the VEPP-3 [3].

EXPERIMENTAL SETUP

The measurement has been performed simultaneously with the elastic electron-deuteron scattering experiment [4]. Novel ABS fed the storage cell with the tensor polarized

CP1149, *Spin Physics, 18th International Spin Physics Symposium*
edited by D. G. Crabb, D. B. Day, S. Liuti, X. Zheng, M. Poelker, and Y. Prok

deuterium. The flux of the polarized deuterium atoms is found to be $8.2 \times 10^{16} at/s$. The ABS delivered two polarized states of the deuterium atoms, one with P_{zz}^{+}=1 and other with P_{zz}^{-}=-2. The switching between the polarization settings was done every 30 seconds. The vector polarization always was equal to zero. The polarization of the deuterium beam monitored by the Breit-Rabi polarimeter. In order to increase a target density the cell was cooled by a liquid nitrogen. The inner surface of the cell was coated with drifilm to inhibit atom depolarization. The target polarization has been determined by a polarimeter (LQP) which measured the target asymmetry in elastic ed-scattering at the small transfered momentum ~ 1.6 fm^{-1} [5]. The polarimeter operated simultaneously with the main detector. Tensor polarization averaged over the whole of data taking run was found to be $P_{zz}^{+} = 0.397 \pm 0.013[stat.] \pm 0.018[syst.]$. The magnetic induction vector $\vec{B}$ was in the median plane of the proton detectors and its polar angle was $\theta_{\vec{B}} \approx 120°$. The particle detector system composed of two identical detectors [2] [4]. Each detector consists of electron calorimeter on one side of the beam and a hadron scintillator hodoscope on the other side. The hadron hodoscopes were used to detect two protons in coincidence. The drift chambers were positioned in front of detector arms to registrate particle tracks. Each hadron hodoscope included three plastic scintillator counters with the thicknesses 20, 126 and 120 mm. The angular acceptance of each hadron arm is $\theta_{1,2} = 55° \ldots 88°$ and $\Delta\phi_{1,2} = 60°$. The range of the measured proton kinetic energy is $50 - 200\ MeV$.

RESULTS AND DISCUSSIONS

The measured four-momenta $(E_i(i=1,2), \vec{p}_i(p_i, \theta_i, \phi_i))$ of the emitted protons define all the remaining kinematics: photon energy and four - momentum of the undetected pion. The incoming photon beam defines the z - axis while the slow protons define the xz - reaction plane. We restrict ourselves to coplanar geometry of the final particles. If we assumed the photon energy is known, the cross-section may be written as follows:

$$\frac{d\sigma}{dp_1 d\Omega_1 d\Omega_2} = \frac{d\sigma^0}{dp_1 d\Omega_1 d\Omega_2}\left\{1 - \frac{\sqrt{3}}{2} P_z \sin\theta_B \sin\varphi_B T_{11} + \frac{P_{zz}}{\sqrt{2}}\left[\frac{3\cos^2\theta_B - 1}{2} T_{20} + \right.\right.$$
$$\left.\left. + \sqrt{\frac{3}{8}}\sin 2\theta_B \cos\varphi_B T_{21} + \sqrt{\frac{3}{8}}\sin^2\theta_B \cos 2\varphi_B T_{22}\right]\right\}, \quad (1)$$

where $\frac{d\sigma^0}{dp_1 d\Omega_1 d\Omega_2}$ – an unpolarized cross-section; P_z (P_{zz}) – the degree of the target vector (tensor) polarization; θ_B, φ_B – determine the magnetic hold field direction. The whole tensor target asymmetry is defined as:

$$A^t = \sqrt{2}\frac{(N^+ - N^-)}{(N^- P_{zz}^+ - N^+ P_{zz}^-)}, \quad (2)$$

where N^+, N^- are the yields of pp-events obtained with the positive and negative polarized target. As the slow proton was detected by the both proton detectors the

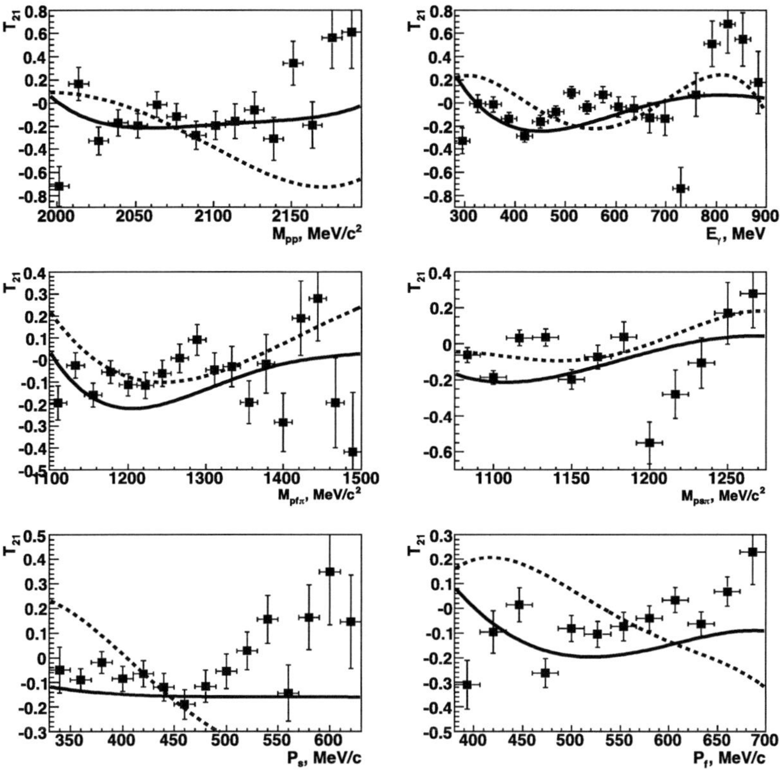

FIGURE 1. Experimental and theoretical dependences of the tensor analyzing power component T_{21} on the various kinematic variables: solid line – calculations with total amplitude, dashed – PWIA. Vertical bars are statistical errors.

experimental data were divided into two parts whith $\varphi_B = \pm 30°$ and $\varphi_B = 180° \pm 30°$. Then the T_{21} component of the tensor analyzing power is defined from equation (1) as:

$$T_{21} = \frac{k_\phi}{2d_{21}} \left(A^t_{\varphi_B=0} - A^t_{\varphi_B=\pi} \right), \tag{3}$$

where $k_\varphi = 1.047$ – correction for the azimuthal acceptance, $d_{21} = \sqrt{3/8} sin 2\theta_{\vec{B}}$.

The collected data files were subjected to programmed processing in a few consecutive stages. The reconstruction of the particle tracks and the hodoscope responses, calculation of the kinematic variables and the selection of the *pp*- events were performed. In this way we got data points containing information about momenta of all particles and another kinematic quantity that is useful in describing the reaction. The tensor target asymmetry T_{21} can be presented as a function of the single variable. The reaction amplitude is calculated in a diagrammatic approach [6]. We take into account the contributions those come from diagrams corresponding to the spectator model (PWIA-amplitude) and diagrams corresponding to πN-and NN- rescattering (total amplitude). We have used the deuteron wave functions of the Bonn NN-potential.

In order to compare the experimental results with theoretical predictions we have simulated the exclusive π^- meson photoproduction on polarized deuteron by Monte Carlo (MC) code. For each generated data point we have obtained all kinematic variables, cross section and the polarization observables, corresponding to experimental data point. The number of events on a phase space point neighbourhood and polarized cross-section in this point are proportional each other. On this base we have performed an event weighted averaging of the theory over a phase-space of each experimental bin. So, we have obtained the possibility to present experimental and theoretical results by one way.

The tensor analyzing power component T_{21} as a function of the different kinematic variables(the photon energy E_γ, the fast and slow proton momentum p_f, p_s, and pp, $p_f\pi, p_s\pi$- invariant masses) is shown on Fig.1. We selected the events where the final particle momenta are coplanar restricting the differences of the azimuth angles of the protons and pion withing 12°. The theoretical predictions are shown by solid and dashed curves.

It has been found that theoretical predictions, which correspond to total amplitude, are in the qualitative agreement with experimental data. However, dependence of the T_{21}-component from the invariant masses of the final systems demonstrate the noticeable discrepancies between the experiment and theory not only in the wide, comparable to the width of meson-nucleon resonances, but also in very narrow mass intervals. However, in dependences T_{21} - component on the invariant masses of the final systems the noticeable discrepancies between the experimental and theory not only in the wide, comparable to the width of meson-nucleon resonances, but also in very narrow mass intervals are observed. Similar result was obtained for the tensor target asymmetry component a_{20} [1, 7].

This work was supported in part by the RFBR under grant 08-02-00624-a,

REFERENCES

1. A.Yu. Loginov et al. *JETF Letters*, **v.67**, N 10,(1998), pp. 770-775.
2. V.N. Stibunov et al., *SPIN 2004 Proceed. of the 16 Internat. Spin Phycics Symp.*, edited by K. Aulenbucher et al., WSP, Singapore (2005), pp. 593-596.
3. M.V. Dyug et al. *Nucl. Instrum. Methods*, **A495**, 8 (2002).
4. D.M. Nikolenko et al., *Phys. Rev. Lett.*, **90**, N 7, 072501 (2003).
5. M.V. Dyug et al. *Nucl. Instr. and Meth.*, **A536**, 344 (2005).
6. A.Yu. Loginov, A.A. Sidorov and V.N. Stibunov, *Phys. of Atomic Nuclei* **v.63**, (2000), pp. 459-462.
7. V.N. Stibunov et al., *Few-Body System Suppl.*, **10**, 507 (1999), pp. 507-510.

^{117}In and ^{118}Sn Homologous State Identification via the ^{120}Sn $(\vec{p},\alpha)^{117}$In and ^{121}Sb$(\vec{p},\alpha)^{118}$Sn Reactions

L. Zetta[a], P. Guazzoni[a], M. Cavallaro[b], T. Faestermann[c], G. Graw[d], R. Hertenberger[d], H.-F. Wirth[d], and M. Jaskóła[e]

[a]*Dipartimento di Fisica dell'Università degli Studi, and I.N.F.N., I-20133 Milano, Italy*
[b]*Dipartimento di Fisica e Astronomia and I.N.F.N., I-95025 Catania, Italy*
[c]*Physik Department TUM, D-85748 Garching, Germany*
[d]*Sektion Physik der Universität, D-85748 Garching, Germany*
[e]*Soltan Institute for Nuclear Studies, 00-681 Warsaw, Poland*

Abstract. To investigate the spectator role of the $1g_{5/2}$ unpaired proton outside the Z=50 closed shell, the ^{120}Sn$(\vec{p},\alpha)^{117}$In and ^{121}Sb$(\vec{p},\alpha)^{118}$Sn reactions have been measured in high resolution experiments at 23 MeV incident proton energy of the Munich MP Tandem accelerator, using the Stern-Gerlach type polarized hydrogen ion source and the Q3D magnetic spectrograph.

Keywords: Spin in Nuclei, Polarized proton beam, Transfer reactions, Homologous states, Cross section and Asymmetry measurement.
PACS: 21.10.Hw, 24.70.+s, 25.40.Hs, 27.60.+j.

INTRODUCTION

The (p,α) reaction on nuclei around singly or doubly magic shells displays properties that make it a useful spectroscopic tool for supplementing level structure information obtained by other charged-particle reactions. Of particular interest is the occurrence of the corresponding homologous states in neighboring systems, which imply the persistence of preferred structures that survive after the addition of one particle that behaves as spectator, as we already tested in Z=40, Z=50 and Z=82 regions [1,2,3,4,5]. In order to further investigate the spectator role of the $1g_{5/2}$ unpaired proton outside the Z=50 closed shell, and to test the validity of the homology concept in the ^{121}Sb nucleus, the reactions ^{120}Sn$(\vec{p},\alpha)^{117}$In and ^{121}Sb$(\vec{p},\alpha)^{118}$Sn have been studied.

THE EXPERIMENT AND THE RESULTS

The angular distributions of cross sections and analyzing powers of the triton pickup reactions ^{120}Sn$(\vec{p},\alpha)^{117}$In and ^{121}Sb$(\vec{p},\alpha)^{118}$Sn have been measured in high resolution experiments, performed with the 23 MeV proton beam of the Munich MP Tandem accelerator, using the Stern-Gerlach source for intense, bright beams of

CP1149, *Spin Physics, 18th International Spin Physics Symposium*
edited by D. G. Crabb, D. B. Day, S. Liuti, X. Zheng, M. Poelker, and Y. Prok

negatively polarized hydrogen ions [6]. This source with electron-cyclotron resonance (ECR) ionizer and Cesium vapour target for charge exchange is able to produce proton beams with intensity up to 1.5 μA on the target. The reaction products were analyzed with the Q3D magnetic spectrograph and detected in the light ion focal plane detector with cathode strip readout [7]. The beam current in the present experiment was up to 700 nA, to save the Sn targets, and the beam polarization ~75%. High resolving power of the spectrometer, favorable peak to background ratio, and large solid angle define the good quality of the spectra.

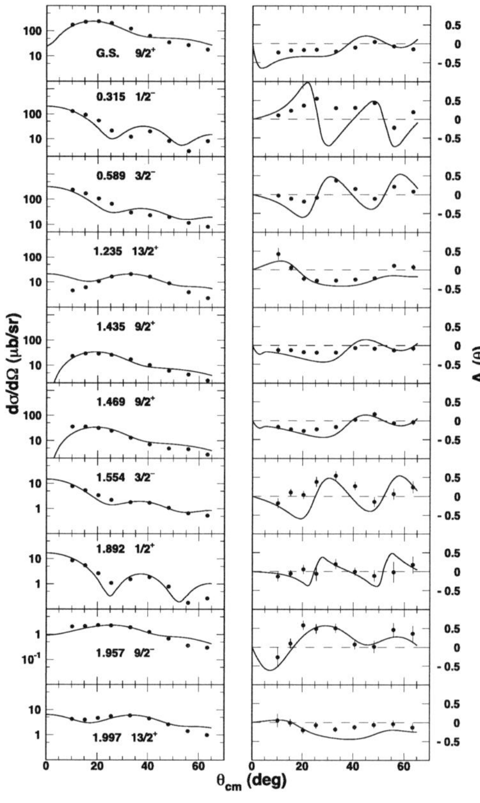

FIGURE 1. Angular distributions of cross sections and asymmetries for some of ^{117}In states. Dots refer to the experimental values and solid lines to the cluster DWBA calculations.

Transitions to final states of ^{117}In up to an excitation energy of 2000 keV and ^{118}Sn up to 5100 keV were studied.

A DWBA analysis of angular distributions of cross sections and asymmetries for the $^{120}\mathrm{Sn}(\vec{p},\alpha)^{117}\mathrm{In}$ reaction has been carried out in finite range approximation assuming a semi-microscopic triton cluster pickup mechanism. The calculations have been carried out with the code TWOFNR [8] using a Gaussian proton-triton interaction potential.

Fig. 1 shows the comparison between experimental and calculated σ(θ) and $A_y(\theta)$ for some ^{117}In states.

For this even-even target nucleus only one transferred orbital and total angular momentum, ***l*** and ***j***, contribute to the excitation of a given final state, greatly simplifying the theoretical calculations.

The dependence of σ(θ) and $A_y(\theta)$ of the emitted particles on the transferred total angular momentum ***J*** is of the greatest importance for identifying the spin and parity of the levels excited in a nuclear reaction. In particular for $(\vec{p},\alpha)$ reactions, as a general rule, the mean behavior is strictly J^{π} dependent, as shown in Fig.2 for an hypothetical level at E_x=1500 keV.

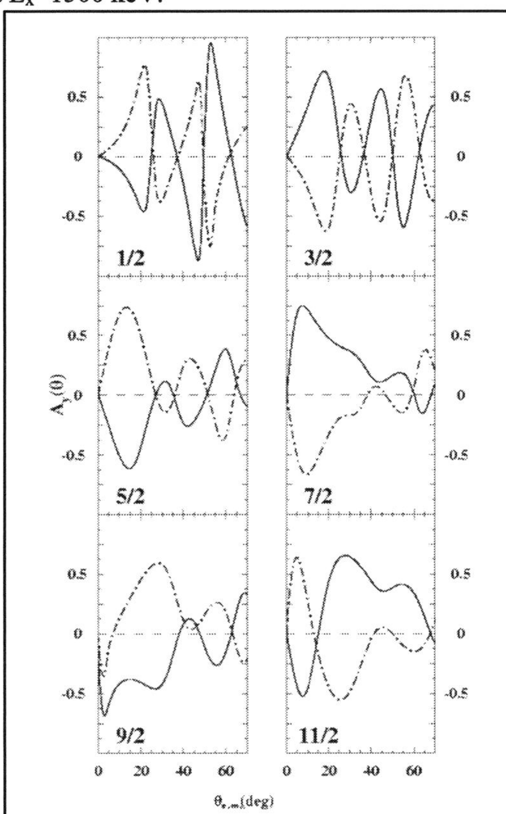

FIGURE 2. Dependence of $A_y(\theta)$ of the emitted particles on the transferred J^{π}. Solid lines refer to positive parity states, while dot-dashed lines to the negative ones.

In the case of an odd-mass target nucleus as ^{121}Sb and for transitions to states with spin values different from 0, several ***l*** and ***j*** may contribute to each transition, which have to be added incoherently in a DWBA calculation. Due to the high number of three-particle configurations involved, the analysis may be quite complicated and spectroscopic information less accurate and more difficult to deduce. On the contrary, in the case of homologous states, populated in the reaction on a near-magic target nucleus having one proton outside a completely filled magic shell, one has unique ***l***-transfer, that is given by the transition to the corresponding parent state.

Our aim in studying the $^{120}Sn(\vec{p},\alpha)^{117}In$ reaction is to obtain accurate measurements of angular distributions of cross sections and asymmetries for comparison with those measured for the $^{121}Sb(\vec{p},\alpha)^{118}Sn$ reaction. In this case one expects that the proton outside the magic proton shell acts as spectator so that only rather high-energy states of ^{118}Sn are excited. These homologous states arise from the coupling of the $1g_{5/2}$ unpaired proton with the configurations excited in ^{117}In. The homology between the low lying states of ^{117}In and the higher energy states of ^{118}Sn allows us to attribute spin and parity to these latter states, fulfilling the expectation of the weak coupling model.

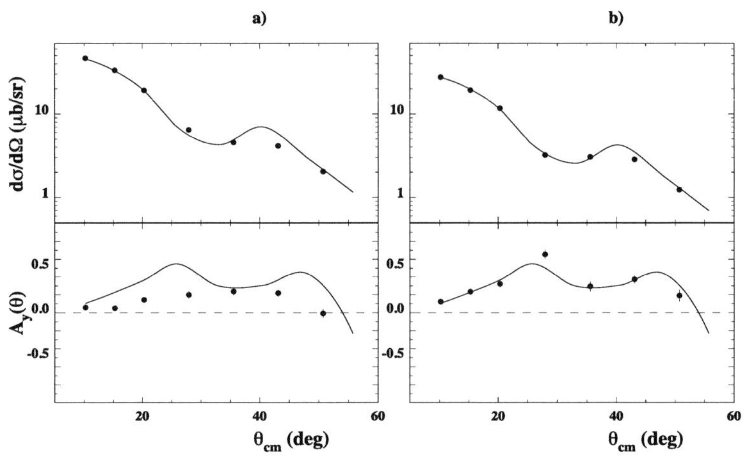

FIGURE 3. Angular distributions of cross sections and asymmetries for the ^{118}Sn 4.045 MeV 3^- (a) and 4.121 Mev 2^- (b) daughter states, homologous to the 0.315 MeV $1/2^-$ ^{117}In parent state. The dots refer to the experimental values of the daughter states and the solid lines to the experimental values of the parent state.

The identification of homologous states in ^{118}Sn is in progress. In Fig. 3 the angular distributions of cross sections and asymmetries for the ^{118}Sn 4.045 MeV 3^- and 4.121 MeV 2^- states homologous to the 0.315 MeV $1/2^-$ ^{117}In parent state are reported.

REFERENCES

1. P. Guazzoni et al., *Z. Phys.* A 356 (1997) 381
2. P. Guazzoni et al., *Eur. Phys. J.* A1 (1998) 365
3. P. Guazzoni et al., *Phys. Rev.* C72 (2005) 044604
4. P. Guazzoni et al., *Phys.Rev.* C49 (1994) 2784
5. J.N. Gu et al., *Phys.Rev.* C55 (1997) 2395
6. R. Hertenberger et al., *Nucl. Instrum. Methods Phys. Res.* A536 (2005) 266.
7. H.-F. Wirth et al., *Jahresbericth 2000*, Beschleunigerlaboratorium München, p. 71 (2001).
8. M. Igarashi, *Computer Code TWOFNR* (1977), unpublished.

ACCELERATION, STORAGE AND POLARIMETRY OF POLARIZED BEAMS (INCLUDE ANTIPROTONS)

Precision Electron Beam Polarimetry

W. Lorenzon

Randall Laboratory of Physics, University of Michigan, Ann Arbor, Michigan 48109-1040, USA

Abstract. Measurements at future electron accelerators will require unprecedented precision in the determination of electron beam polarization. Existing and novel electron beam polarization measurement schemes are examined, and ideas are being presented of how to to achieve sub-1% precision in polarization determination.

INTRODUCTION

In order to reach the physics goals at future electron/positron accelerators such as the Electron Ion Collider (EIC) or the International Linear Collider (ILC), strong requirements are placed on precision polarimetry for electrons and positrons. The current design of the EIC project foresees collisions of 3-20 GeV longitudinally polarized electrons on 30-250 GeV protons or 50-100 GeV/u heavy ions (such as gold) with bunch separations of 3-35 ns. It is anticipated that the electron beam polarization is 70% or better and that it needs to be measured with high precision ($\lesssim$1% systematic uncertainty). The current design of the ILC project foresees collisions of 45.6-500 GeV longitudinally polarized electrons and positrons. It is anticipated that the electron (positron) beam polarization is 80% (50%) or better and that it needs to be measured with unprecedented precision ($\lesssim$0.25% systematic uncertainty). This is a factor of two better than ever achieved before, and thus raises the question whether this is a realistic goal. Before trying to address this question, it is advantageous to take a step back and consider what has previously been achieved in precision beam polarimetry.

There are many polarimeters that have been in use, are in use, or are planned at various laboratories. Table 1 shows an overview of polarimeters that are or have been in use, and their precision [1]. The systematic uncertainties in beam polarization measurements for Compton polarimeters are reported to be in the 0.5-2% range, but they can get larger as measurements get pushed to lower beam energies ($E_b \lesssim 1.0$ GeV). For Møller scattering the systematic uncertainties are typically 2-3%, and may approach 1% or below at high magnetic fields.

Although the precision of the JLab polarimeters, listed in Table 1, do not reach the precision required by the EIC nor the ILC, they still serve a valuable lesson in understanding the ultimate precision reachable in polarization measurements. The systematic uncertainties of the various polarimeters in the three experimental halls at JLab were each evaluated individually. Since it was possible to compare the polarization of the five polarimters with a special arrangement of the CEBAF accelerator, the "Spin Dance" Experiment was performed in July 2000 [2]. In this experiment, a multi-hall cross-normalization of the relative analyzing power of the five JLab electron polarime-

CP1149, *Spin Physics, 18th International Spin Physics Symposium*
edited by D. G. Crabb, D. B. Day, S. Liuti, X. Zheng, M. Poelker, and Y. Prok

TABLE 1. Overview of polarimeters that are or have been in use, and their precision

Laboratory	Polarimeter	Relative precision	Dominant systematic uncertainty
JLab	5 MeV Mott	~1%	Sherman function
	Hall A Møller	~2-3%	target polarization
	Hall B Møller	1.6% (→ 2-3%)*	target polarization, Levchuk effect
	Hall C Møller	0.5% (→ 1.3%)†	target polarization, Levchuk effect, high current extrapolation
	Hall A Compton	1% (@ > 3 GeV)	detector acceptance + response
HERA	LPol Compton	1.6%	analyzing power
	TPol Compton	3.1%	focus correction + analyzing power
	Cavity LPol Compton	?	still unknown
MIT-Bates	Mott	~3%	Sherman function + detector response
	Transmission	>4%	analyzing power
	Compton	4%	analyzing power
SLAC	Compton	0.5%	analyzing power

* 1.6% is quoted by Hall B. 2% or even larger might be more realistic.

† 1.3% is best quoted value in an experiment. 0.5%, as quoted by Hall C polarimeter group, seems possible.

ters was performed. The purpose of this comparison between the Mott, Compton, and Møller polarimeters was to reveal possible differences between the polarimeters that are systematic in nature and have not previously been accounted for. The results are displayed in Fig. 1 with the solid symbols. There is significant discrepancy between the polarimeters, even if the systematic uncertainties are included.

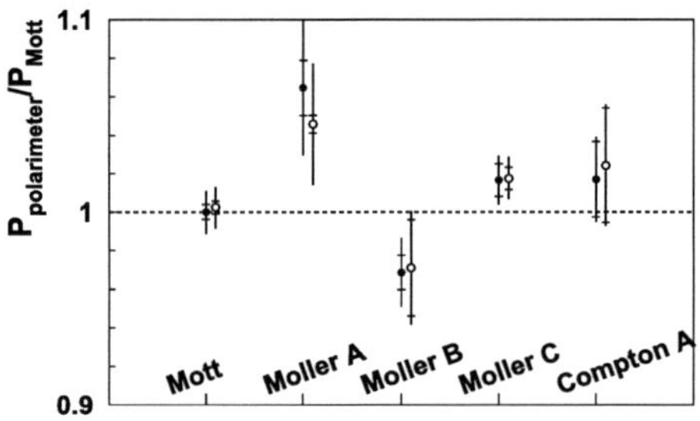

FIGURE 1. Relative analyzing power for the five JLab electron beam polarimeters, normalized to the Mott polarimeter for comparison. The solid symbols are the results for the entire data set. The open symbols represent the results for the data set limited to be within 25% of the maximum measured polarization.

Since the Hall A and B Møller polarimeters may have systematic effects that depend on the transverse components of the electron beam polarization, which are large when the longitudinal components are small, the data shown in open symbols have been restricted to be within 25% of the maximum polarization value. These results indicate that the horizontal component of polarization may be an important source of systematic effects for the Hall A Møller polarimeter. For the reduced data set, the discrepancy among the five polarimeters becomes less significant. As a result of the Spin Dance Experiment, the Hall A Møller polarimeter is implementing a Hall C style target to be able to isolate instrumental from target polarization effects.

LESSONS LEARNED

Many lessons have been learned from these earlier polarization measurements. Experience at many laboratories has taught us that it is imperative to include polarization diagnostics and monitoring capabilities in the design of the beam lattice. It is important to ensure that the beam polarization can be measured continuously during data taking to minimize systematic uncertainties associated with the beam polarization, such as drifts or luminosity related variations in polarization. If at all possible, the beam polarization should be measured at the IP, or as close to the IP as possible. The laser and beam polarizations have to be flipped at intervals that are short compared to any drifts in polarization. The cross-comparison of the analyzing power of various polarimeters at JLab has shown that providing or even proving precision at the 1% level is challenging. Since the requirements at future electron/positron accelerators are even stricter, it is absolutely crucial that multiple devices are employed for testing the systematic uncertainties of each polarimeter. There has to be at least one technique that can measure the absolute polarization of the beam, while others can do relative measurements.

Further, Compton scattering appears to be the ideal process for measuring the polarization of high energy, high beam current electron (positron) beams. Compton scattering is a pure QED process where no atomic or nuclear corrections have to be applied, and where radiative correction uncertainties are at the 0.1% level [3]. In Compton scattering, the target (i.e. laser beam) is effectively 100% polarized while in Møller scattering the target (i.e. polarized electrons in iron foils) is only about 8% polarized. Furthermore, Compton scattering measurements are non-invasive, while Møller scattering measurements are destructive due the need of using iron or iron-alloy foils. Compton scattering is ideal for high beam currents, while Møller scattering measurements suffer from beam induced foil heating effects at beam currents above a few μA.

To achieve sub-1% precision in the electron (positron) beam polarization determination, all these consideration have to be taken into account, and if possible, new and innovative ideas have to be employed.

NEW IDEAS FOR FUTURE ELECTRON ACCELERATORS

Most of the major disadvantages of Møller scattering might be overcome with a new idea that employs polarized atomic hydrogen in an ultra-cold magnetic trap [4]. It is argued that at 300 mK, the electrons of the hydrogen atoms are brute-force polarized to 100% within a factor of 10^{-3}, and a polarization measurement with a statistical uncertainty of 1% can be achieved in 10 min with a beam current of 100 μA and a target density of 3×10^{15} cm^{-3}. Employing an atomic hydrogen target has the advantage that it is non-invasive and can be used to continuously measure the beam polarization, and that it may provide a systematic uncertainty below 0.5%. It has the disadvantage, though, that the target is very complex, and that gas heating effects by radiation grow with the beam intensity squared. This might be a serious limitation for the high currents envisioned for the EIC or ILC. One solution around the gas heating effect might be to consider a hydrogen jet target instead. Further studies are underway to explore this interesting idea.

New developments in laser technology might give a big boost to Compton scattering based polarization measurements, where it has been necessary to build either delicate laser cavity lasers or use high power pulsed lasers to get Compton rates that allow polarization measurements within reasonable time scales. This technology is being borrowed from fiber based drive lasers at electron sources that provide very high power, and use gain switching, as compared to mode locking which is sensitive to mode lock problems. The advantages are that they can be phase locked to the actual beam of the accelerator, therefore providing a nearly 100% duty cycle. In addition, fiber lasers can be easily accessed since they are external to the beam line vacuum system (unlike cavity lasers). They further provide excellent stability, low maintenance, and straightforward implementation. Efforts are underway to build a Compton polarimeter using the fiber lasers for a new Hall C Compton polarimeter.

There is general agreement that detection of Compton electrons is the best tool for high precision polarimetry. Since the analyzing power depends strongly on the momentum of the Compton electrons, Compton electrons are typically analyzed by fitting the asymmetry shape over parts or the entire available momentum range. Alternatively, the Compton edge (which corresponds to the minimum energy of the back-scattered Compton electrons), can be used to determine the electron beam polarization. These methods however depend strongly on the response function of the detector, which must be calibrated and monitored carefully. A new idea to do a zero-crossing Compton electron analysis is suggested. It relies on the well-defined energies of the zero crossing of the asymmetry (corresponding to 90° scattering in the electron rest frame) and of the Compton edge. This analysis is based on a linear fit of the zero crossing of the Compton asymmetry, and an integration of the asymmetry spectrum from that point to the Compton edge, instead of a fit to the spectrum shape between those points. It has the advantage that no absolute energy response calibration of the detector is necessary, and that the corrections due to finite detector position and energy resolutions are small ($\ll 1\%$).

Polarimetry at the EIC

A rough idea for a polarimeter suitable to withstand the high luminosity at EIC is shown in Fig. 2. A chicane is introduced to separate Compton polarimetry from the electron beam. This introduces ample space for Compton photon detection, and a convenient way to separate the Compton electrons from the beam electrons in one of the soft bending magnets of the chicane. The chicane helps to minimize background rates: minimizing bremsstrahlung background requires to have a short section of beam line, while minimizing synchrotron background requires a chicane with soft bends. The Compton photon detector can be operated in counting and integrating mode. A pair spectrometer consisting of a variable converter (to select the appropriate rate of e^+e^- pairs in counting mode), a dipole magnet (to separate the pair-produced electrons and positrons), and position sensitive detectors could be configured for photon counting. Downstream of the pair spectrometer a radiation hard and fast (< 35 ns) position sensitive sampling calorimeter could be used to operate in integrating mode. If appropriate photomultiplier tubes are used, the sampling calorimeter can be operated in counting and integrating mode.

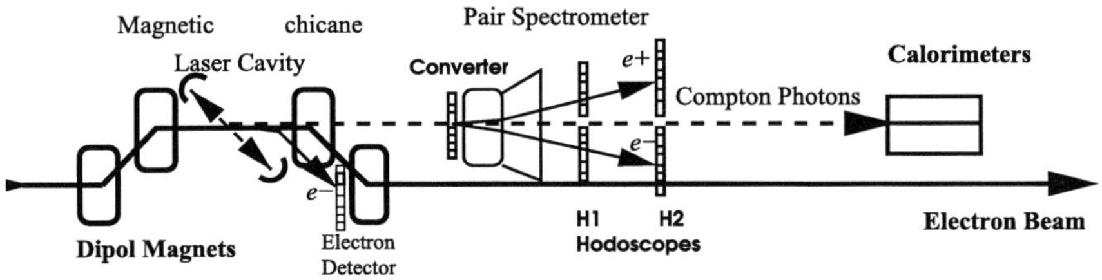

FIGURE 2. Schematic view of a possible Compton polarimeter for the EIC.

Polarimetry at the ILC

The current scheme for polarimetry at the ILC proposes three ways to measure polarization: an upstream Compton polarimeter (1.8 km from IP) shown in Fig. 3, a downstream Compton polarimeter (150 m from IP) and the $e^+e^- \to W^+W^-$ process which is measured at the IP and which is very sensitive to electron and positron polarization (at the 0.1% level). There is however some significant complication in this scheme. The polarization at the IP is the luminosity-weighted polarization which is not identical to the beam polarization at the upstream polarimeter. Furthermore, although the downstream polarimeter measures the luminosity-weighted polarization, there are depolarization and spin transport effects that are estimated to be at the 0.1-0.4% levels. Those effects are relatively large and of the same magnitude as the required accuracy, which makes it imperative that the uncertainties in these effects are well understood and kept small.

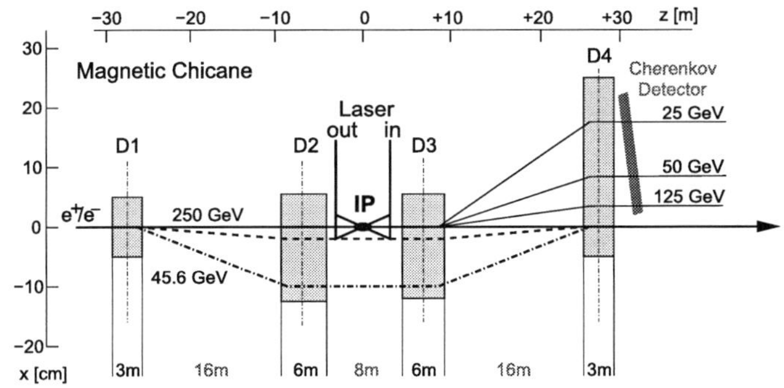

FIGURE 3. Schematic view of a possible upstream Compton polarimeter for the ILC.

In order to make a convincing case for a 0.25% precision in electron and positron polarimetry, all three techniques are needed. It is necessary to determine the most accurate polarization value for each polarimeter separately. It is then important to make sure that all depolarization and spin transport effects are understood. Only at this point the values for the up- and downstream polarimeters should be compared. If everything is understood properly, they should agree. The final calibration of the absolute polarization scale should be performed with the with $e^+e^- \to W^+W^-$ process. Only if all three measurements are in agreement, good confidence can be gained that the required precision has been achieved.

Summary

In summary, it appears that electron beam polarimetry can be measured with high precision. There are no serious obstacles to achieve a precision of 1%. It is imperative though to include polarimetry in the beam lattice and the interaction region design. And it is crucial to use multiple devices and even techniques to control the systematic uncertainties at the sub-1% level.

Reaching a precision of 0.25% presents a big challenge, but there are no apparent show stoppers. The large beam energy at the ILC is crucial, since it provides analyzing powers in Compton scattering that approach unity. Reaching a precision of 0.1% in contrast enters entirely new territory. This is well beyond traditional polarimetry and requires to utilize specific physics processes that are very sensitive to polarization.

There are many issues that require careful scrutiny. The beam crossing frequency is proposed to be somewhere between 3-35 ns at the EIC. This is very different from the crossing frequencies at RHIC (106 ns) and HERA (96 ns). HERA has demonstrated that beam-beam induced depolarization becomes important at high luminosities. An entirely new concept for the EIC is crab-crossing of bunches. What effect will it have on the beam polarization, and how can these effects be measured? Finally, depolarization and spin transport effects need to be understood to high accuracy at the ILC.

To tackle these issues, it will require a group of experienced physicists who can build on their experience, and at the same time are open to new developments in the field. With much work already done, extensive modeling will be necessary to optimize the design (i.e. analyzing power calibration, modeling of beam depolarization and spin transport, etc.) for polarimetry at a future electron/positron accelerator.

ACKNOWLEDGMENTS

The author's research is supported in part by the U.S. National Science Foundation, Intermediate Energy Nuclear Science Division under grant No. PHY-0555423. The author thanks Eugene Chudakov, Wouter Deconinck, Dave Gaskell, Avetik Hayrapetyan, Jenny List, Jeff Martin, and Kent Paschke for many fruitful discussions and for crucial contributions to this paper.

REFERENCES

1. Lorenzon W., "Precision Electron Beam Polarimetry", AIP Conference Proceedings No. **980**, ed. A. Kponou, Y. Makdisi, A. Zelenski p. 407 (2008).
2. Grames, J. M. et al., *Phys. Rev. ST Accel. Beams* **4**, 042802 (2004).
3. Swartz M.L., et al., Phys. Rev. **D58**, 014010 (1998).
4. Chudakov, E. A., and Luppov, V. G., *IEEE Trans. Nucl. Sci.* **51**, 1533 (2004).

Polarized Electrons in Darmstadt: Recent Developments[1]

Y. Poltoratska*, R. Barday*, U. Bonnes*, M. Brunken*, C. Eckardt*, R. Eichhorn*, J. Enders*, C. Heßler*, C. Ingenhaag*, W. F .O. Müller†, M. Platz*, M. Roth*, B. Steiner†, M. Wagner* and T. Weiland†

*Institut für Kernphysik, Technische Universität Darmstadt, Germany
†Institut für Theorie Elektromagnetischer Felder, Technische Universität Darmstadt, Germany

Abstract. The injector of the superconducting Darmstadt electron linear accelerator S-DALINAC will soon be extended by a source of polarized electrons. A test set-up of the future source consists of a 100 keV GaAs polarized gun and associated beamline including an assembly for polarization manipulation and determination. We give a short overview of the existing and planned polarimeters at the different accelerator sites, introduce briefly the future implementation of the polarized source at the S-DALINAC and list experiments that are planned with polarized electron and photon beams.

Keywords: S-DALINAC, polarized electron source, polarimetry
PACS: 29.20 Ex, 29.27.Hj, 79.60.-i

INTRODUCTION

The superconducting Darmstadt linear accelerator S-DALINAC [1] is a recirculating electron accelerator dedicated to nuclear physics experiments. It provides electron beams in an energy range of a few MeV up to 130 MeV with an average current of up to 60 μA. Due to the usage of superconducting accelerating structures, a 3 GHz continuous-wave (cw) time structure of the electron beam is possible. It is generated by chopping and bunching a DC electron beam currently produced at a thermionic gun. To complement the present experimental program at the S-DALINAC with electron and photon scattering experiments that are sensitive to polarization observables, a new injector for polarized electrons, the S-DALINAC Polarized Injector SPIN, has been developed [2]. The new injector consists of an electron source that produces longitudinally polarized electrons, a Wien filter and Mott polarimeter for spin manipulation and polarization measurement, beam steering and diagnostic elements, and of a laser system for the production of polarized light. To test the electron source independently from the operation of the S-DALINAC, a standalone offline-teststand has been constructed. Electron beams with degree of polarization of 86% were produced using strained superlattice GaAs photocathode. Details of the teststand construction can be found in Ref. [3]. In this contribution, we summarize current status at the polarized injector and give a brief overview on the present and future polarimetry at the S-DALINAC.

---

[1] Work supported by Deutsche Forschungsgemeinschaft through SFB 634.

CP1149, *Spin Physics, 18th International Spin Physics Symposium*
edited by D. G. Crabb, D. B. Day, S. Liuti, X. Zheng, M. Poelker, and Y. Prok

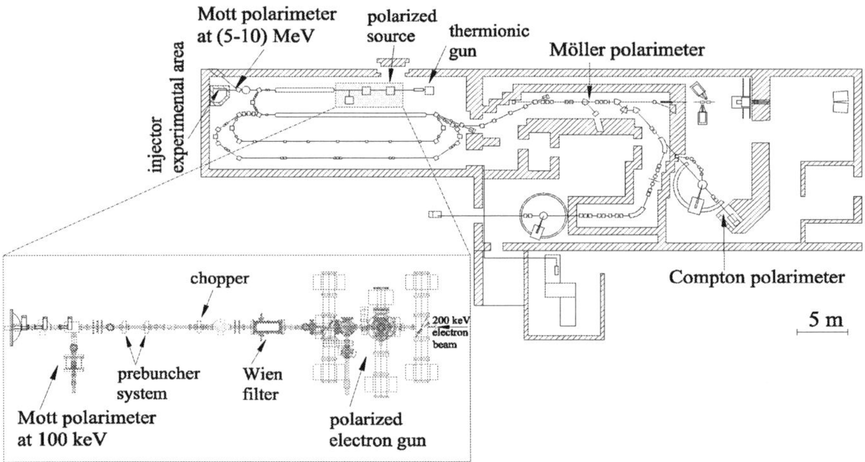

FIGURE 1. Schematic view of the S-DALINAC with new polarized injector. The locations of the polarimeters are indicated.

S-DALINAC AND THE SOURCE OF POLARIZED ELECTRONS

A schematic drawing of the S-DALINAC with the planned new injector layout and its experimental areas is shown in Fig. 1. Installation of the polarized electron source is currently foreseen between the unpolarized thermionic gun and the first superconducting accelerating structure. To separate the ultra-high vacuum part of the polarized gun ($\sim 10^{-11}$ mbar) from the rest of the S-DALINAC ($\sim 10^{-8}$ mbar) a differential pumping stage needs to be installed. A newly developed chopper and prebuncher system will be installed and used both for the polarized and the unpolarized beam. For that purpose, the energy of the thermionic gun will be reduced from 250 kV to 200 kV. The polarized source setup was built as compact as possible (Fig. 2) to accommodate the spatial constraints in the accelerator hall. It is planned to operate the polarized electron source at the S-DALINAC in pulsed as well as in cw mode. For this purpose in addition to the existing diode laser system [5] the Ti:Sapphire system with pulses of some 100 fs to some ps will be installed. The pulsed operation allows time-of-flight experiments to be carried out and/or background events to be removed.

POLARIMETRY

For the qualitative analysis of the planned photon and electron-induced experiments with polarized beams, the degree of polarization needs to be known. Typical set-ups - that are also under developments for the S-DALINAC - use electron-nucleus (Mott) and electron-electron (Møller) scattering as well as the attenuation of polarized photons due to photon-electron (Compton) scattering. The polarimeters will be installed at different experimental sites. To determine the polarization directly after the emission the 100 keV Mott polarimeter [4] was constructed. It is currently operated at the teststand

FIGURE 2. Photograph of the electron source with the part of the corresponding beam line.

beam line and will be transfered to the S-DALINAC with the source. To support experiments at the injector experimental area of the S-DALINAC one more Mott polarimeter for 5 – 10 MeV electrons is being developed. For the electrons with energies between 30 and 130 MeV, Møller scattering is well suited. However, the low and varying energy range of the electron beam requires a large acceptance and yet compact electron spectrometer. Electron-optical design work is underway. To monitor beam polarization during the electron scattering experiments, a Compton transmission polarimeter is foreseen. Polarimetry at the S-DALINAC is described in references [6, 7].

EXPERIMENTS WITH POLARIZED ELECTRONS AND PHOTONS

Longitudinally polarized electron beams at the S-DALINAC will make polarized electron scattering experiments and experiments with circularly polarized photons possible. The scattering of polarized electrons allow one to measure the fifth structure function in the break-up of nuclei at low momentum transfer. This structure function is sensitive to the final state interaction. When applied to light nuclei, effects of three-body force may become visible [8]. Polarized gamma-ray beams will be produced at the S-DALINAC by bremsstrahlung. Planned experiments include the search for parity-violating effects in photon scattering and photo-induced fission. Examples for future experiments with polarized beams at the S-DALINAC are discussed elsewhere Ref. [9].

SUMMARY AND OUTLOOK

A source of polarized electrons was developed for the S-DALINAC. It has been set up, characterized, and operated at a teststand. Installation at the S-DALINAC is currently scheduled for the second half of 2009. In addition to the available 100 keV Mott polarimeter, polarization measurement will be provided near the experimental sites. Experiments with polarized electrons and photons at the S-DALINAC may commence as early as 2010.

ACKNOWLEDGMENTS

We would like to thank K. Aulenbacher for helpful collaboration and many useful discussions. Yuliya Poltoratska thanks the SPIN 2008 organizing committee for financial support. This work is supported by Deutsche Forschungsgemeinschaft through SFB 634.

REFERENCES

1. A. Richter "Operational experience at the S-DALINAC," EPAC'96, Sitges, 1996 p. 110
2. C. Heßler "Konzeption, Aufbau und Test einer Quelle spinpolarisierter Elektronen für den supraleitenden Darmstädter Elektronenlinearbeschleuniger S-DALINAC", Doctoral Thesis D17, TU Darmstadt, 2008
3. Y. Poltoratska et al. "Status report of the new Darmstadt polarized electron injector", these proceedings
4. Y. Poltoratska, " Design and setup of a compact Mott polarimeter for the future S-DALINAC polarized injector", Diploma thesis, Kharkiv Karazin National University and TU Darmstadt, 2005, unpublished.
5. C. Heßler, "Status of the polarized-electron gun at the S-DALINAC", EPAC'06, Edinburgh, 2006, p. 3173.
6. C. Eckardt et al., "Polarimetry at the S-DALINAC", these proceedings.
7. R. Barday et al., "Electron beam polarimetry at S-DALINAC", EPAC'08, Genoa, 2008 p. 1476.
8. J. Golak, private communication, 2004
9. C. Eckardt et al., "Polarized electrons for experiments at low momentum transfer SPIN@S-DALINAC", these proceedings.

Towards High Precision Deuteron Polarimetry

M. da Silva e Silva for the dEDM collaboration

KVI, University of Groningen, Groningen, The Netherlands

Abstract. A finite electric dipole moment (EDM) in any fundamental system would constitute a signal for new physics. The deuteron presents itself as an optimal candidate both experimentally and theoretically. A new storage ring technique is being developed for which a small change in the vertical polarization would be a signal of a non-zero EDM. A novel polarimeter concept is under investigation. Besides being highly efficient, this polarimeter should continuously monitor the beam polarization, guaranteeing optimal sensitivity. Detailed studies on systematic error control, in addition to the measurement of cross sections and analyzing powers, were carried out at KVI-Groningen in The Netherlands. Measurements were conducted at COSY-Jülich in Germany yielding high efficiencies. The (statistics limited) ability to track changes in polarization at the level of a few hundred parts-per-million has been demonstrated. Further studies and developments to meet the final goal of sub-part-per-million sensitivity are in progress.

Keywords: Polarimetry; Deuteron; Storage Rings; Systematic errors
PACS: 95.75.Hi; 20.20.db

INTRODUCTION

The quest for physics beyond the Standard Model constitutes one of the main endeavours in physics research. Among the different approaches is the search for EDMs. The existence of a non-vanishing EDM violates both Parity and Time Reversal symmetries. Assuming CPT invariance, the violation of Time Reversal symmetry is equivalent to the violation of the Charge-Parity symmetry, the latest believed to be a necessity for the matter-antimatter inequality in the Universe. Several existing Standard Model Extensions predict EDMs larger than those of the Standard Model (where they are immeasurably small and below experimental sensitivity) and closer to experimental limits nowadays. As such, EDM experiments have become sensitive probes for New Physics. The deuteron is a promising candidate particle to search for an EDM. Besides being theoretically attractive [1], its small anomalous magnetic moment makes it a suitable candidate for the use of the frozen spin method, described elsewhere in these proceedings [2, 3].

DEDM METHOD

The frozen spin method, like other EDM experiments, comprises three stages. The first step is the *Preparation of a spin polarized beam.* An intense deuteron bunch (10^{12} particles), with a momentum of 1 GeV/c and a high vector polarization, is injected into the storage ring where it will *Interact with the deuteron-frame electric field* generated by passage through the magnetic field of the ring bending magnets. A charged particle moving in a magnetic field experiences a motional electric field in its rest frame. The interaction between this electric field and the EDM causes the spin to precess. The

CP1149, *Spin Physics, 18th International Spin Physics Symposium*
edited by D. G. Crabb, D. B. Day, S. Liuti, X. Zheng, M. Poelker, and Y. Prok

resulting change in the vertical polarization component with time signals an EDM. This signal has to be monitored continuously and with high efficiency through *Polarimetry*, the final step. The reachable statistical precision, σ_{EDM}, is given by

$$\sigma_{EDM} \simeq \frac{4\hbar}{\sqrt{\tau T} E_{eff} P_0 A \sqrt{\eta N}}. \quad (1)$$

The initial polarization P_0, the number of particles N and the polarization lifetime τ are determined in the first stage of the experiment and the effective electric field E_{eff} in the second. The beam lifetime T, the analyzing power A and the efficiency η are predominantly polarimetry dependent. Thus, optimization of the sensitivity largely depends on how well the polarimeter works. This will constitute the main subject of this article.

POLARIMETER CONCEPT

The design of a sensitive polarimeter is, at this stage, focused on the optimization of T, A and η. At the same time, robustness against systematic errors is considered. The concept under development relies on the strong spin dependence of large angle *d-C* scattering. The multiple scattering cross section is many orders of magnitude larger than the nuclear scattering cross section and usually dominates the removal of particles from the storage ring. The probability for a nuclear reaction can be enhanced by increasing the target thickness to several centimeters, thus enhancing the efficiency. Inserting such a target directly into the beam would lead to an unacceptably small T. Therefore, the beam is slowly extracted onto the thick target via a heating mechanism such as a very thin gas target (through Coulomb scattering) or a weak excitation of beam resonances (through a "noisy" kicker). In this way, the thick target becomes the defining aperture for the storage ring. The size and shape of the aperture, in combination with the extraction mechanism, will define T. Once the deuterons hit the target, the efficiency η will be determined by its geometry and composition. The latter, together with the detector acceptance, will also determine A. After scattering from the target, the particles are detected by a fast, segmented detector so that the scattering angles can be reconstructed. The precise configuration and composition of the target and detection system will determine the sensitivity (the ability to detect small signals) and selectivity (the ability to avoid false signals).

SENSITIVITY

For the optimization of the sensitivity, we concentrate on η and A and assume that T can be tuned independently with a suitably chosen extraction mechanism. Tuning parameters include the target material and geometry, as well as the scattering angle acceptance of the detection system. Carbon was found to be the ideal target material. It has a fairly large nuclear elastic scattering cross section, allowing for a high scattering probability (few %/cm). The cross section and analyzing power should be averaged over the deuteron energy. For optimum efficiency, the target thickness must be a significant fraction of the deuteron range. As a consequence, the deuteron energy varies by several 10's of percent. This is particularly important for the average analyzing power. Like for other low-Z targets, the (vector-)analyzing power for *d-C* scattering varies only slowly with scattering angle and energy, contrary to that for high-Z materials. This allows for a detection system with a large scattering angle acceptance and thus higher η, while maintaining a high average analyzing power. The EDM experiment requires the polarimeter to operate at extremely high incoming particle rates. This implies that only a minimum of information can be retained, and that sophisticated analysis, such as energy reconstruction and particle identification, may be impossible. It might be that only the rate of incoming (charged) particles will be available. Hence, the cross section σ and analyzing power of all reactions with charged particles in the final state should be considered. This includes inelastic deuteron scattering to discrete ($d+^{12}C \rightarrow d+^{12}C^{*}$) and continuum states ($d+^{12}C \rightarrow d+X$), as well as transfer-reactions ($d+^{12}C \rightarrow p+^{13}C$) and deuteron break-up ($d+C \rightarrow p+n+X$). As data are scarce in the literature, a comprehensive set of data was collected at KVI for incoming deuteron energies of 76 and 113 MeV. Future data gathering at higher deuteron energies is foreseen at COSY. Maximizing the figure of merit, given by the average σA^2, yielded an optimum operating point at a target thickness of about 5 cm for a scattering angle range of 5 to 20°. To confirm the findings of the models, experiments were conducted at COSY using the EDDA detector. Two runs were performed with different configurations. The first one made use of a deuteron momentum of 1.2 GeV/*c*, a scattering angle coverage between 11 and 72° and only the first layer of EDDA bar detectors. In this configuration, the detector has a large energy acceptance with a low energy threshold. As a consequence, a large fraction of break-up protons are counted in addition to the scattered deuterons. This run yielded $A \sim 0.048$ and $\eta \sim 1.5 \times 10^{-2}$. In the second run, the deuteron momentum was slightly lowered to 0.97 GeV/*c*, such that elastically scattered deuterons would stop in the second layer of EDDA, composed of ring detectors. The requirement of a hit in the second layer of detectors raised the energy threshold of the detector, simulating the effect of placing absorbers in the setup. This run yielded $A \sim 0.22$ and $\eta \sim 10^{-3}$. Because the count rate went down without significantly affecting the figure of merit compared to that from the first run, this configuration is preferred. Both runs used a thick tube-shaped carbon target of 1.5 cm length that was limitted by the available space at the EDDA detector. An additional factor of 2 to 3 in sensitivity is expected for longer targets.

SELECTIVITY

The configuration that yields the best sensitivity is not necessarily the one that yields the smallest systematic error. The latter is strongly dependent on the precise method used to analyze the data. We considered two analysis methods based on: (1) the traditional Left-Right asymmetry, $\varepsilon_{LR} = \frac{L-R}{L+R}$; and (2) the Cross Ratio Asymmetry, $\varepsilon_{CR} = \frac{1-r}{1+r}$, where $r = \sqrt{\frac{L^-R^+}{L^+R^-}}$ (+ and − stand for plus and minus vector states). For the given geometry, models were constructed to predict the effect of beam and detector imperfections. The main sources for systematic errors were found to be changes in position and angle of either the beam, or the detector, or both. Furthermore, the deuteron being a spin-1 particle implies that there are tensor effects to be taken into consideration, in particular the T_{21} analyzing power combined with a rotation in the azimuthal angle, ϕ which can mimic an EDM asymmetry. Experimental verification has been performed at KVI and at COSY. At KVI, a detailed study was performed with the IBP (In Beam Polarimeter) detector [4] for a single pass situation. Calibrated beam shifts were introduced in the system to understand the effect of such errors. After a first analysis, the Cross Ratio method, known for its robustness against changes in intensity and solid angle, still presented clear second order effects in the asymmetry as a function of beam shifts. Preliminary analysis also shows that these errors match those expected from our model. This study was then continued at COSY, so far at an exploratory level, to understand the behavior of the system in a storage ring environment. Confirmation of the actual behavior with simulations and a more robust analysis method are under development. Detailed understanding of the polarimeter response to the error drivers is the main challenge in the design of the polarimeter.

POLARIMETER DEMONSTRATION

A demonstration of continuous polarization monitoring was performed at COSY. In an environment similar to what expected for a true EDM experiment, controlled changes in the polarization were induced by ramping an RF solenoid through the $1-G\gamma$ depolarizing resonance, causing the polarized states to flip the sign of their polarization.

CONCLUSION

A novel, highly efficient, polarimeter concept is under development for the dEDM search. The optimal operating point of the polarimeter has been established. Statistical and systematic studies have been performed at KVI and COSY. Measurements at COSY yielded $A \sim 0.22$ and $\eta \sim \times 10^{-3}$ and are, at this moment, statistics limited. Studies at KVI showed unambiguous second order effects left over in the Cross Ratio method and more robust techniques are being devised. A demonstration of continuous polarization monitoring has been performed at COSY.

ACKNOWLEDGMENTS

This research is in part funded through an Innovational Research Grant of The Netherlands Organization for Scientific Research (NWO).

REFERENCES

1. C. P. Liu, and R. G. E. Timmermans, *Phys. Rev.* **C70**, 055501 (2004).
2. Y. K. Semertzidis, The EDM experiment with polarized deuteron beams in a storage ring.
3. C. J. G. Onderwater, Storage rings for EDM searches.
4. R. Bieber, et al., *Nucl. Instrum. Phys.* **A457**, 12–21 (2001).

Precision Measurements of the Proton Beam Absolute Polarization With RHIC Polarimeters

A. Bazilevsky*, I. Alekseev†, A. Bravar**, G. Bunce*, K. Boyle‡, C.M. Camacho§, V. Dharmawardane¶, R. Gill*, H. Huang*, H. Liu¶, Y. Makdisi*, B. Morozov*, I. Nakagawa‖,‡, H. Okada*, D. Svirida†, M. Sivertz* and A. Zelenski*

*Brookhaven National Laboratory, Upton, NY 11973, USA
†Institute for Theoretical and Experimental Physics (ITEP), 117259 Moscow, Russia
**University of Geneva, 1205 Geneva, Switzerland
‡RIKEN-BNL Research Center, Upton, NY 11973, USA
§Los Alamos National Laboratory, Los Alamos, NM 87545, USA
¶New Mexico State University, Las Cruces, NM 88003, USA
‖RIKEN, 2-1 Hirosawa Wako, Saitama 351-0198, Japan

Abstract. We discuss the beam polarization measurement strategy at RHIC with achieved relative precision of better than 5%.

Keywords: Elastic scattering, spin, coulomb nuclear interference, polarized proton
PACS: 13.88.+e, 13.85.Dz, 29.27.Pj, 29.27.Hj

INTRODUCTION

Polarimetry is one of the crucial part of the RHIC Spin Program [1]. It provides precise measurements of beam polarizations which is a necessary component of all high precision spin dependent measurements at RHIC.

Two types of polarimeters are used at RHIC, which are based on small angle elastic scattering, with sensitivity to the proton beam polarization coming from interference between electromagnetic and hadronic amplitudes (Coulomb-Nuclear Interference (CNI) region).

One type of polarimeter (pC) uses an ultra-thin (10–20 μg/cm^2) carbon ribbon target, and provides fast relative polarization measurements with a few percent statistical uncertainty in 20–30 sec. The other type polarimeter (H-Jet) uses a polarized hydrogen gas target and utilizes the analyzing power in proton-proton elastic scattering in the CNI region. It accumulates data over the entire fill and provides absolute polarization measurements, with 10% statistical uncertainty in a fill (6–8 hours). H-Jet data accumulated over several fills are used for the absolute calibration of the pC polarimeters.

H-JET POLARIMETER

The polarized hydrogen jet target polarimeter locates at one of the collision points in the RHIC accelerator and measures polarization of one of the two RHIC beams at a time. In

CP1149, *Spin Physics, 18th International Spin Physics Symposium*
edited by D. G. Crabb, D. B. Day, S. Liuti, X. Zheng, M. Poelker, and Y. Prok

2008 the the H-Jet polarimeter was tested running both beams separated vertically, and both incident on the jet. In future RHIC runs, it will allow the H-Jet to simultaneously monitor the polarization of both beams in all RHIC physics stores.

H-Jet polarimeter consists of a polarized atomic jet target with a Breit-Rabi polarimeter to measure target polarization, and a set of detectors to identify the elastically scattered recoil proton from beam-target interaction, on the right and on the left relative to vertically polarized beam or target. The details of H-Jet setup and Data Acquisition system are described elsewhere [2].

The elastically scattered protons are identified with small background contamination ($\sim 4\%$) using the kinematical correlation between recoil proton kinetic energy and time of flight, and kinetic energy and scattering angle. Then asymmetry between scattering on the left and on the right are measured relative to beam polarization direction or target polarization direction. Since beam polarization direction alternates every bunch (separated by 106 ns) or pair of bunches, and target polarization flips every $\sim$ 10 min, we simultaneously collect data for all beam-target spin configurations: $N_{\uparrow}^{L}$, $N_{\downarrow}^{L}$, $N_{\uparrow}^{R}$ and $N_{\downarrow}^{L}$, where subscript denotes the beam or target polarization direction, and superscript relates to left or right detector, defined relative to beam direction. Using so called square-root formula:

$$\varepsilon = \frac{\sqrt{N_{\uparrow}^{L} \cdot N_{\downarrow}^{R}} - \sqrt{N_{\uparrow}^{R} \cdot N_{\downarrow}^{L}}}{\sqrt{N_{\uparrow}^{L} \cdot N_{\downarrow}^{R}} + \sqrt{N_{\uparrow}^{R} \cdot N_{\downarrow}^{L}}}, \tag{1}$$

we cancel effects from left-right detector acceptance asymmetry and spin up-down beam luminosity asymmetry, when measuring raw asymmetry relative to beam polarization, ε_{beam}, or relative to target polarization ε_{target}.

Beam polarization P_{beam} is then defined from the measured asymmetries ε_{beam} and ε_{target} and known target polarization P_{target}:

$$P_{beam} = -\frac{\varepsilon_{beam}}{\varepsilon_{target}} \cdot P_{target} \tag{2}$$

Target polarization was monitored during the run and showed a stable polarization value of 92.4% after correction for $\sim 3\%$ molecular hydrogen contamination, which diluted jet polarization.

The asymmetry measurements were performed for the recoil proton kinetic energy range $T_R = 1-4$ MeV, in which the proton-proton elastic scattering analyzing power is slightly T_R dependent and on the average is $\langle A_N^{pp} \rangle \sim 4\%$.

The accumulated statistics in H-Jet in the first long polarized proton RHIC run in 2006 provided 2.4% uncertainty for the pC polarimeter absolute normalization for each of 100 GeV beams. Other uncertainties came from the molecular hydrogen background, $\sim 2\%$, and from background in the identification of elastically scattered recoil protons, $\sim 1.5\%$. The latter background didn't show any beam or target polarization dependence, so that it diluted both ε_{beam} and ε_{target} in the same way, and according to Eq. (2) didn't affect beam polarization measurements.

PROTON-CARBON POLARIMETER

Both RHIC rings are equipped with pC polarimeters, which provide fast relative measurements of beam polarizations, which subsequently is normalized to the absolute polarization measurements performed by H-Jet polarimeter. Repeated measurements during a fill also provide a handle on the beam polarization vs time as well as measurements of beam polarization profile in the transverse plane. The latter directly enters the average beam polarization observed in collisions at the interaction regions of the RHIC experiments.

The pC polarimeters consist of a carbon target and six silicon strip detectors mounted in a vacuum chamber at 45, 90 and 135 degrees relative to vertical axis in both left and right sides with respect to the beam direction. It allows to monitor not only vertical component of beam polarization through left-right asymmetry, but also horizontal component through up-down asymmetry using 45 and 135 degree detectors. The details of pC polarimeter setup are described elsewhere [3].

pC polarimeters utilize the analyzing power for the elastic polarized proton-carbon scattering in the CNI region. For the the recoil carbon kinetic energy range T_R =400–900 keV used for the asymmetry measurements the average analyzing power was $\langle A_N^{pC} \rangle \sim$ 1.2%. Within this energy range, A_N is T_R dependent and drops from $\sim$ 1.5% at 400 keV to $\sim$ 0.5% at 900 keV.

The identification of recoil carbon is performed through the correlation between recoil carbon kinetic energy and time of flight measured by silicon detectors. The recoil carbon energy includes corrections due to energy losses [4]. After applying 3σ mass cut corresponding to carbon mass, where σ is reconstructed carbon mass peak width, the background under carbon mass peak was estimated on the level of 1% [3].

Scanning carbon ribbon target across the beam allows to measure beam intensity and polarization profiles in both vertical and horizontal directions in transverse plane. For non-flat polarization profile the beam polarization will be seen differently by H-Jet polarimeter, pC polarimeter (in fixed target measurement mode) and when colliding beams in the interaction regions of RHIC experiments. Due to that the jet target in H-Jet polarimeter is much wider than beam width, the polarization profile is weighted with beam intensity profile in the average beam polarization. For colliding beams, the polarization profile is weighted with a product of two beam intensity profiles in transverse plane. The polarization measurement in pC polarimeter will depend on the positioning of the carbon target, and on whether horizontal or vertical target is used for the measurements (in case beam horizontal and vertical profiles are different).

Assuming gaussian profiles ($I(x,y)$ for intensity and $P(x,y)$ for polarization), the average beam polarization seen by H-Jet polarimeter can be expressed:

$$\langle P \rangle_{HJet} = \frac{\iint P(x,y) I(x,y) dxdy}{\iint I(x,y) dxdy} = \frac{P_{max}}{\sqrt{(1+R_X) \cdot (1+R_Y)}}, \tag{3}$$

by pC polarimeter for the case with vertical target positioned at beam maximum intensity (and polarization):

$$\langle P \rangle_{pC} = \frac{\int P(y) I(y) dy}{\int I(y) dy} = \frac{P_{max}}{\sqrt{(1+R_Y)}}, \tag{4}$$

and by experiments in two beam collision, for $I_{1,2}$ relating to intensity profiles for two beams respectively:

$$\langle P\rangle_{Exp} = \frac{\iint P(x,y)I_1(x,y)I_2(x,y)dxdy}{\iint I_1(x,y)I_2(x,y)dxdy} = \frac{P_{max}}{\sqrt{(1+\frac{1}{2}R_X)\cdot(1+\frac{1}{2}R_Y)}}, \tag{5}$$

with P_{max} the polarization at beam maximum intensity and polarization in transverse plane, R_X and R_Y the squared ratio of the intensity profile width and polarization profile width $(\sigma_I/\sigma_P)^2$, for X and Y projections respectively.

These relations between average polarizations are taken into account when normalizing pC measurements to H-Jet absolute polarization measurements and when providing polarization values for RHIC experiments. For example for the typical at RHIC values of $R \sim 0.1-0.2$, beam polarization seen by H-Jet is lower than polarization seen by experiments in beam collisions by $\sim 5-10\%$.

The main sources of systematic uncertainties in polarization measurements by pC polarimeter comes from the drift of the recoil carbon energy correction (2.5% in Run6), and from corrections due to polarization profile, particularly in runs when one or both transverse projections of the profile are not measured regularly (2% in Run6).

Combining these uncertainties with uncertainties from polarization normalization using H-Jet, the final systematic uncertainties for the fill by fill measurements by pC polarimeter were 4.7% and 4.8% for two RHIC beams and 8.3% for a product of two beam polarizations in 2006 for 100 GeV beam running.

SUMMARY

RHIC polarimeters have provided RHIC experiments with store by store beam polarization measurements with relative accuracy better than 5%. The pC polarimeter upgrade in under way for the 2009 RHIC run, which will further improve polarization measurements at RHIC particularly in high event rate environment anticipated in future high luminosity RHIC runs [5, 6].

REFERENCES

1. G. Bunce *et al.*, Ann. Rev. Nucl. Part. Sci. **50**, 525 (2000).
2. H. Okada et al., Proc. of the 12th International Workshop on Polarized Ion Sources, Targets and Polarimetry, Upton, New York, Sep 10-14, 2007, p. 370.
3. I. Nakagawa et al., Proc. of the 12th International Workshop on Polarized Ion Sources, Targets and Polarimetry, Upton, New York, Sep 10-14, 2007, p. 380.
4. I. Nakagawa et al., Proc. of the 17th International Spin Physics Symposium, Kyoto, Japan, Oct 2-7, 2006, p. 912.
5. A. Zelenski et al., these proceedings.
6. Y. Makdisi et al., these proceedings.

Towards Improved Proton Polarimetry at RHIC

Y. Makdisi, G. Atoian, A. Bazilevski, G. Bunce, R. Gill, H. Huang, B. Morozov, M. Sivertz, A. Zelenski

Brookhaven National Laboratory, Upton, NY 11973

Abstract. The RHIC polarized proton collider polarimeters employ the analyzing power in p-Carbon elastic scattering in the Coulomb Nuclear Interference region to measure the proton beam polarization. These are calibrated by the polarized hydrogen jet target that measures the absolute beam polarization utilizing the pp elastic scattering process in the CNI region. This paper describes the status of these polarimeters and the efforts employed to improve the performance in terms of better resolution, rate handling capability, and reduced systematic uncertainties.

Keywords: Silicon detectors, polarimeters, beam polarization.
PACS: 13.88.+e, 13.85.Dz, 29.27.Pj, 29.27.Hj t

INTRODUCTION

The RHIC polarimeters consist of a vacuum vessel in which several thin carbon targets are mounted on a special target drive operated by a motor to select horizontal or vertical targets for scanning across the beam to measure both the beam polarization and beam polarization profile. The targets are surrounded by three pairs of silicon detectors that are mounted at approximately 18 cm away with one pair at 90 degrees in the horizontal plane and the other two at + /- 45 degrees above and below. Figure 1.

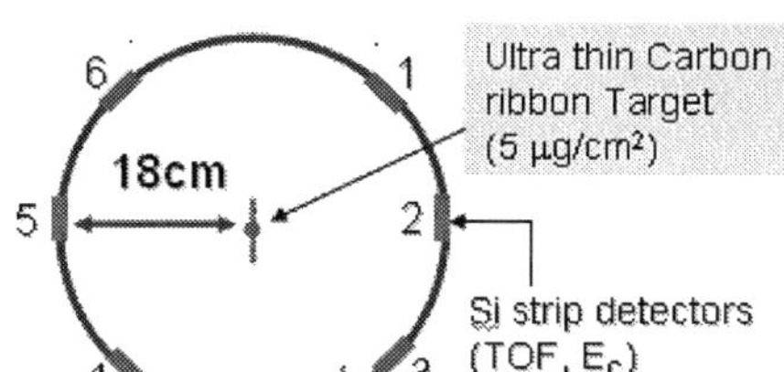

FIGURE 1. A beam view of the silicon detectors in the RHIC polarimeters

The silicon detectors, manufactured by the BNL instrumentation division, comprise 12 strips each and independently measure the scattered recoil carbon energy and time of arrival. The left-right scattering asymmetry determines the degree of the beam polarization. The large p-carbon elastic scattering cross section provides a fast measurement with 2% statistical accuracy in less than one minute. The polarimeter upgrades are discussed in this symposium by Zelenski[1]. The polarized jet target is located within 100 meters from the polarimeters and alternately measures the absolute

CP1149, *Spin Physics, 18th International Spin Physics Symposium*
edited by D. G. Crabb, D. B. Day, S. Liuti, X. Zheng, M. Poelker, and Y. Prok

beam polarization to better than 10% in an 8-hour store. It takes a few stores to calibrate each polarimeter to an accuracy of 5%. The status of the polarimetry analyses is described in this symposium by S. Bazilevsky[2].

The polarimeters silicon detectors are calibrated using americium sources that emit alphas with energy of 5.5 MeV. This is large compared to the range of the carbon recoils that range from 300 keV to about 1 MeV. While this gives a starting point, it renders the polarimeter silicon detectors energy calibration quite an involved process that relies on fitting the data to the time energy relation as well as data in the energy deposition of carbon in silicon[3]. Our experience has shown that this process results in a measured effective dead layer in the silicon 3 times larger than the physical one. Another issue facing the polarimeters is increasing rate as the RHIC bunch intensity moves upwards.

The polarized jet silicon energy calibration utilizes both americium as well as gadolinium sources, the latter emits a 3.2 MeV alpha particle as such the energies do overlap those of the recoil protons of interest and range from 1 to 5 MeV rendering a more comparable calibration. The jet measurement on the other hand is limited by the unpolarized molecular hydrogen background that at this stage is estimated at 3% and contributes a large systematic accuracy of the order of 2%. In what follows we will describe the efforts to overcome some of these difficulties

THE SILICON ENERGY AND RATE RESPONSE

At BNL we are fortunate to have the Tandem complex which serves as the HI injector for RHIC but with a versatile capability to accelerate various ion species at prescribed energies and intensities. Our plan calls to use a carbon beam from the Tandem to systematically scan energies of interest with varying intensities up to $4.10^6/cm^2$ to study the BNL manufactured silicon detectors response:

- The beam energy will span from 0.3 – 5 MeV wider than the recoil carbon range of 0.3 – 1 MeV as we need to understand detector response to the alpha energy from the americium source.
- Vary the beam carbon charge of +1, +2, and +3 to see if the detectors response reflects any differences or assess how soon charge equilibration occurs.
- Study potential non-linearity to the energy response which once determined will then be added to the calibration process.
- Check for energy resolution vs. beam energy
- Use alpha sources impinging forward and backward to determine the effective silicon thickness and assess potential degradation
- A recoil carbon suffers some energy loss as it traverses the target, we will use a foil to simulate the polarimeter carbon target in these tests.
- The tests will utilize the same readout preamplifiers, shapers, and waveform digitizers currently employed by the polarimeters to test the response to increased rate. We will also use alternate electronics to assess differences if any.

NEW POLARIMETER DETECTORS AND ELECTRONICS R&D

We have procured from Hamamatsu several types of silicon photodiode detectors of several thicknesses a 5μm S9724-005 (1x1cm2) detector, a 300 μm S3590-19 (1x1 cm2) detector and a strip-array: S4114-35 without window-35 strips (4.4x 9) mm2 in an effort to attain better energy resolution as well as to develop some understanding of the detector sensitivity to the large number of prompt particles that stream though concurrently when the detectors are responding to the recoil carbon scatters. Again we use the Tandem carbon beams for these tests:

- Test new photodiode detectors and array strip detector under similar conditions to the BNL silicon detectors. We anticipate a factor of four in energy resolution which will allow us to go to lower recoil carbon energies (lower t) and thus higher analyzing power.
- Test a duel-silicon detector system with
    - a thin 5 μm followed by the 300 μm detectors to assess the charge equilibration process.
    - Identify the carbon charge with the thin detector at 10 MeV. This then drives the energy loss (dead layer) determination from the earlier data
    - An added bonus, the thin detector could provide a trigger as it is likely to be blind to minimum ionizing prompts, so less rate dependent.
    - Test the existing charge amplifier with a lower shaping time.
    - Test a new low capacitance cable between the detector and preamp
- Test a current amplifier concept. While this is better for high capacitance detectors (thin detectors) and high rate environment, the associated noise levels are likely to be higher and the question is are they at acceptable levels?
- For high rate we will study reduced silicon volume current by reducing the detector area and thickness.

OPERATIONAL IN SITU TESTS WITH RHIC BEAM

The upgraded polarimeter vessels (see Zelenski) allow for dual polarimeter set ups in each ring. In the Blue ring we will install a pair of photodiode strip detectors at 90 degrees in the horizontal plane instrumented with the current readout system. With a vertical carbon target, this polarimeter response will be compared and contrasted with the two 45 degree BNL silicon polarimeters looking at the same target during the normal data taking. Similarly one of the Yellow ring polarimeters will utilize the 90 degree slots to install the Hamamatsu single photodiode detectors (two to a side) to carry out in situ comparisons. Dedicated studies will compare the responses to varying the number of bunches and bunch intensities. We are also equipped with two target thicknesses to study rate dependences. Another aspect is to assess any changes the detectors suffer due to radiation damage in a prolonged run.

The new polarimeters will have an added feature a scintillation counter and phototube mounted on the outside a thin 2mm flange to detect the arrival time of prompt particles. This is an attempt to better define t0 and better decouple the time and energy measurements.

LOOKING AHEAD WITH THE JET TARGET

In Run-8 the jet experienced an increase in the background under the recoil elastic signal. We attribute this to having removed a collimator system is an effort to gain vertical acceptance. Towards the end of the run, we tested the idea of having the two RHIC beams vertically separated by 5 mm simultaneously impinging on the jet target. We accumulated data in this fashion for three days. This was quite successful in that the data analysis did not reveal any more background under the elastic signal compared to running with one beam on target.

Buoyed with this success and in preparation for Run-9, we re-installed the jet collimators, re-aligned the jet with respect to the RHIC beam line, and along with an overhaul of the RHIC beam position monitors we expect to run with simultaneous measurement of both beams. We will also try to push for a smaller vertical beam separation consistent with acceptable beam-beam conditions to increase our vertical acceptance. The benefit is continuous and full monitoring and calibration of both RHIC polarimeters and doubling our statistics.

The effort to better monitor and understand the molecular hydrogen jet component rests on a collaboration with our Russian colleagues for another attempt to quantify this background. In addition, we will continue the R&D on the in situ luminescence monitors which we hope will attain better sensitivity to molecular hydrogen.

SUMMARY

We described the ongoing work to improve our understanding of the energy responses, calibration, and rate capabilities of the current RHIC polarimeters. R&D is underway to test an improved set of silicon detectors that will provide better energy resolution, rate capabilities, and allow access to higher analyzing powers. Work continues to improve the polarized jet target calibration of the RHIC polarimeters. Of note is the maturing data analysis process and the improved ability to provide the RHIC experiments with results in a timely fashion.

ACKNOWLEDGMENTS

We are grateful for significant technical help and support from D. Steski, G. Mahler, J. Ritter, T. Curcio, D. Lehn, and T. Russo and the BNL Tandem operations crew. Thanks to H. Okada and V. Dharmawardane for diligent data analyses of Run-8.

This work is performed under Brookhaven Science Associates, LLC, under contract No. DE-AC02-98CH10886 with the U.S. Department of Energy. Funding is also provided from the RIKEN BNL Research Center.

REFERENCES

1. A. Zelenski et al., These proceedings.
2. A. Bazilevsky et al., These proceedings.
3. I. Nakagawa et al., Proc. Of the 12th International Workshop on Polarized Ion Sources, Targets, and Polarimetry, Upton, NY, Sept. 10-14, 2007, p. 370.

The RHIC p-Carbon CNI Polarimeter Upgrade For The Beam Polarization And Intensity Measurements

A. Zelenski, A. Bazilevski, G. Bunce, R. Gill, H. Huang, G. Mahler, Y. Makdisi, B. Morozov, S.Nemesure, T. Russo, D. Steski, M. Sivertz

Brookhaven National Laboratory, Upton, NY 11973 U.S.

Abstract. Proton polarization measurements in the AGS and RHIC (Relativistic Heavy Ion Collider at the beam energies 24-250 GeV) are based on proton-carbon and proton-proton elastic scattering in the Coulomb Nuclear Interference (CNI) region. Polarimeter operation in the scanning mode also gives polarization profile and beam intensity profile (beam emittance) measurements. Bunch by bunch emittance measurement is a very powerful tool for machine setup. Presently, the polarization and beam intensity profile measurements (in both vertical and horizontal planes) are restricted by the long target switching time and possible target destruction during this complicated motion. The RHIC polarimeters were operated near the limit of the counting rate for present silicon strip detectors. The ongoing polarimeter upgrade for the 2009 run will address all these problems. The upgrade should allow significant reduction of the polarization measurement errors by making feasible the complete polarization measurements, which includes polarization profiles in both the horizontal and vertical planes.

Keywords: Polarimeters
PACS: 29.27.Hg

INTRODUCTION

The RHIC is the first collider where the "Siberian Snake" technique was very successfully implemented to avoid resonance depolarization and polarized proton beams were accelerated to energy 250 GeV[1]. Precise control of the machine parameters is required to prevent residual depolarisation. Therefore, fast polarization measurements are essential for machine set-up and data quality monitoring during the physics stores. The RHIC polarimetry is based on proton-Carbon elastic scattering with low momentum transfer (CNI region) and measurement of the asymmetry in recoil carbon nuclei production described in detail in [2-4]. This process has a large cross-section and analyzing power of a few percents, and is expected to have weak energy dependence in the energy range of a 24-250 GeV. The absolute beam polarization at 100 GeV beam energy was measured with a polarized H-jet polarimeter, which is also based on proton-proton elastic scattering in the CNI region [4]. The simultaneous measurements in pC and H-jet polarimeters provide the calibration for pC analyzing power. The fast pC polarimeter measures possible polarization losses during the energy ramp and possible polarization decay during the RHIC store.

CP1149, *Spin Physics, 18th International Spin Physics Symposium*
edited by D. G. Crabb, D. B. Day, S. Liuti, X. Zheng, M. Poelker, and Y. Prok

High intensity recoil nuclei from the scattering of the circulating proton beam in the thin carbon target is efficiently utilized in the silicon strip detectors and data acquisition system, which is capable of processing an event rate up to a few million/second. This makes it possible for fast, practically non-destructive polarization measurements. The polarization measurement on the beam energy ramp was implemented in the AGS and RHIC, providing locations of polarization losses. Polarimeter operation in the scanning mode also gives polarization profile and beam profile (including bunch by bunch) measurements.

In previous polarized runs at RHIC, the polarization and beam intensity profile measurements (in both vertical and horizontal planes) were restricted by the long target switching time and possible target destruction during this complicated motion. The RHIC polarimeters were operated near the limit of the counting rate for the present silicon strip detectors, but the development of new detectors was very limited by the absence of development time in the polarized RHIC run. The ongoing polarimeter upgrade for 2009 run will address all these problems. New polarimeter vacuum chambers will be installed in the RHIC ring. Two identical target motion mechanisms and detector assemblies will be installed in each chamber (each ring). One polarimeter can be used for the vertical polarization and intensity profile measurements and the second can be used for the horizontal profile measurements. It will minimize number of target motions, speed up the measurements and reduce the risk of target damage. An additional set of detector ports will facilitate new detector development, while keeping the established technique intact. Two sets of targets will extend the period between target replacement. The upgrade should allow significant reduction of the polarization measurement errors by making feasible the "complete" polarization measurements, which includes polarization profiles in both horizontal and vertical planes.

POLARIMETER UPGRADE

Two identical polarimeter vacuum chambers are located in the warm RHIC sections, which are separated for the two circulating beams and have the separate vacuum systems. The new vacuum chambers were expanded to host the second set of target drives and detectors, however, due to vacuum pump ports integration in the chamber the total polarimeter length is not changed. Each chamber will be pumped by an ion and a NEG (Non-Evaporable Getter) cartridge pump. The pumps are isolated by the gate valves, which utilize the full pumping speed capacity and reduce the pumping time after the polarimeter maintenance. The initial pumping time from atmospheric pressure is long to reduce the risk of damage to the fragile target strips by the air-flow in the chambers. Two vacuum ports for fore-pumping located at the chambers' periphery will allow to reduce this time due to reduced directional air-flow in the chambers.

The use of thin targets in the polarimeter is necessary to reduce multiple scattering for recoil carbon ions and also keep the event rate within the detectors' and DAQ capabilities. The carbon ribbons used in the polarimeters contain about 10^{15} carbon atoms/cm of target length (25 mm). The high intensity circulating beam knocks out about 10^7-10^8 carbon nuclei, which causes the eventual target destruction. It is demonstrated that targets survive in the RHIC beam for 100-200 measurements at the full beam intensity, which corresponds to a target lifetime of one-to-two weeks. Six target strips are attached to each target ladder (horizontal and vertical) to extend the time between maintenance periods. Two identical target motion mechanisms and detector assemblies will be installed in each chamber during the upgrade. It will allow doubling the number of targets. In addition, one polarimeter in the tandem will be used for the vertical polarization and intensity profile measurements and the second will be used for the horizontal profile measurements. It will minimize the number of target motions, speed up the measurements and reduce the risk of the target losses.

The combination of linear and rotational motion of the target mechanism provides the target positioning, replacement and scanning mode operations. The use of precision (low backlash) linear and rotary mechanisms driven by step-motors (and high-resolution encoders measurements) allows accurate target positioning control limited only by the target straightness.

Silicon-strip detectors are used in the polarimeters since they allow measurements of energy and arrival time of Carbon recoil nuclei of a few hundreds keV energy. After the recent progress of increasing the RHIC bunch intensity (up to $2 \cdot 10^{11}$ proton/bunch) the polarimeters were operated near the limit of the counting rate for present silicon strip detectors even with thinnest practically usable targets. The development of new detectors was very limited due to the absence of development time in the polarized RHIC runs. After the upgrade, the second set of detector ports will facilitate new detector development, while keeping the established technique intact. The status of the new detector development is presented in a paper by Y. Makdisi et al. at this Conference [5]. The multiplexing of the detector signals after preamplifiers, and step-motors for target drives will allow using the existing DAQ system and position control electronics, since only one polarization measurement can be done at a time.

POLARIZATION AND BEAM INTENSITY PROFILE MEASUREMENTS

The carbon target width is much smaller than the beam size therefore intensity and polarization profile can be measured by the target scan. Part of these profiles is produced in AGS (which was measured by the similar technique directly in AGS). There are also some possible polarization profile generation in RHIC at the ramp and during the store. The effect depends on machine parameters and can be different for the two rings and for the horizontal and vertical planes. The RHIC experiments are sensitive to the convolution of the two colliding beams polarization and polarization profiles therefore all polarization measurements for the physics stores were done in the scanning mode and polarization profiles were measured and correction to the polarization values as seen by the experiments were implemented. The limited knowledge of the polarization profiles (so far most of the measurements were done by scanning in one horizontal or vertical plane due to long target switching time) contributed the largest systematic error to this analysis [4]. After the upgrade, the complete measurements, which includes vertical and horizontal polarization profiles will be the main operational mode. In a scanning mode of polarimeter operation the counting rate dependence on the target position can be used for the beam intensity profile (emittance) measurements in addition to polarization measurements.

With WFDs, every elastic event has a time stamp attached and each event is sorted by the bunch crossing. With the high event rate, a large statistical sample can be accumulated in a short time (of a few hundred milliseconds for the target scan through the beam), which were used for beam profile (emittance) bunch-by-bunch measurements. These measurements have negligible effect to the beam losses and provide valuable tools for accelerator set-up and monitoring. The absolute accuracy of these emittance measurements is limited by the target straightness and the knowledge of the beta function of the beam at the polarimeters. The better quality measurements were obtained with the double thickness polarimeter targets (one of these targets was attached to each of multiple target holders) and these measurements were used for the Ionization Profile Monitors calibrations. After the upgrade the number of vertical targets will be doubled and number of horizontal targets will be tripled (12 targets instead of 4). This will allow sparing (from longer beam exposure during polarization measurements) better quality targets for beam emittance studies and calibration only and extend time between target replacements.

SUMMARY

The pC CNI polarimeters at AGS and RHIC provide fast, practically non-destructive polarization and intensity profile measurements. The polarimeter upgrade is under way for the 2009 polarized Run. The upgrade should allow significant reduction of the polarization measurement errors by making feasible the complete polarization measurements, which includes polarization profile scans in both the horizontal and vertical planes. In addition accurate beam intensity profiles measurements (beam emittances including bunch by bunch emittances) will provide complimentary diagnostics for machine set-up and operation.

ACKNOWLEDGEMENTS

We acknowledge the very important contribution of the RIKEN-BNL Research Centre to this project. We are thankful to the Collider Accelerator Department staff: S. Bugros, T. Cursio, D. Gassner, D. Lehn, M. Mapes, R. Todd, J. Tuozzolo and many other people for the effort to complete the polarimeter upgrade project on time for the polarized 2009 Run.

REFERENCES

[1] T. Roser, AIP Conf. Proc. 980, p.15, (2008).
[2] I. Nakagava et al., AIP Conf. Proc. 980, p.380, (2008).
[3] A. Bazilevsky et al., "Precision Measurements of the Absolute Proton Beam Polarization with RHIC Polarimeters", this conference.
[4] A. Bazilevsky et al., Proc. of EPAC2008.
[5] Y. Makdisi et al., "Towards Improved Polarimetry at RHIC", this conference.

Polarized Beams at EICs

V. Ptitsyn

Brookhaven National Laboratory, Collider-Accelerator Department, Upton, NY 11973, USA

Abstract. Designs of future electron-ion colliders (EIC) are presently under development at BNL and TJNAF. Polarized beams of electrons, protons and light ions are essential feature of those designs. The talk reviews approaches and techniques used in the EIC designs for delivering high-energy polarized beams and describes existing challenges.

Keywords: electron-ion collider; beam polarization; electron spin; proton spin; figure-8 ring.
PACS: 29.20.db; 29.20.Ej; 29.27.Hj.

INTRODUCTION

Designs of future electron-ion colliders are presently under development in several laboratories over the world [1,2]. The work on the eRHIC machine at BNL [3,4] and the ELIC machine at TJNAF [5,6] has started at the beginning of this decade and in this paper we will concentrate on the aspects of the polarized beam technology at those two machines. Polarized beam of electrons (positrons), protons and light ions are considered as essential requirements for those colliders, with the major scientific goal of resolving the puzzle of nucleon spin. Typical required level of the beam polarization is at 70% and more. Most of the planned physics experiments ask for the longitudinal polarization of both electron and proton beams at the collision points. More recently the work also has began on electron-ion collider designs in CERN and GSI laboratories [7], where the application of the polarized beams is also under the consideration.

Approaches accepted for the design of eRHIC and ELIC machines are quite different (Figure 1). For eRHIC, a so-called "linac-ring" scheme is presently considered as the baseline design scheme. The electron beam after the acceleration in the energy recovery linac (ERL) collides with the protons circulating in the RHIC ring in one or more interaction points. The ELIC collider uses the "ring-ring" scheme, when both electrons and protons circulate in storage rings (figure-8 shaped), which intersect each other at four locations. Because of this difference in the design approaches the issues and challenges related with the delivery of highly polarized electron beams are very different. But, surprisingly, quite different are also the ways of providing high energy polarized proton beams. Some major design choices were affected by considerations for polarized beams. For instance, the unusual figure-8 shape of ELIC storage rings was chosen to simplify tasks of polarization preservation and polarization orientation control for proton (and ion) beams. And one of the factors, which led to the choice of the "linac-ring" approach for eRHIC, was the polarization transparency in all energy range for the electron beam.

CP1149, *Spin Physics, 18th International Spin Physics Symposium*
edited by D. G. Crabb, D. B. Day, S. Liuti, X. Zheng, M. Poelker, and Y. Prok

In later sections we will review and compare main technological aspects related with the beam polarization in eRHIC and ELIC machine designs. Table 1 lists some selected machine parameters relevant for polarized beams.

TABLE 1. Some of the beam parameters of eRHIC and ELIC machines.

	ELIC		eRHIC	
	p	e	p	e
Energy, GeV	250	10	250	10
Bunch intensity, 10^{10}	0.27	0.69	20	12
Average beam current, A	0.22	0.55	0.42	0.26
Norm. rms emittance, mm mrad	0.7/0.03	51/2	1	77
Polarization, %	>70	80	70	>80
Peak luminosity, 10^{33} cm^{-2}s^{-1}	29		2.6	
Energy range, GeV	30-250	3-10	25-250	3-20
Polarized ion species	p,d, ^{3}He		p, ^{3}He	

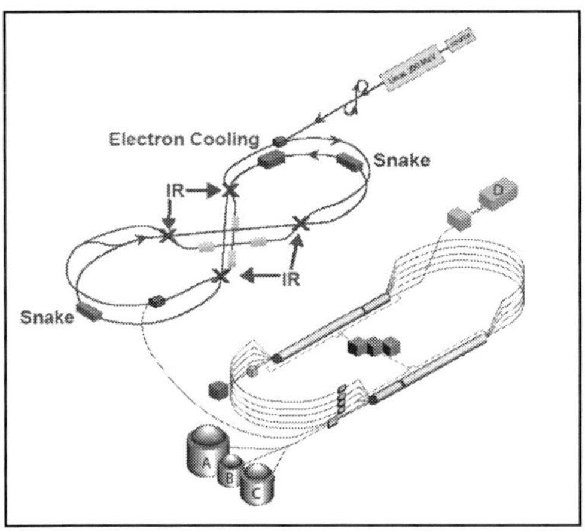

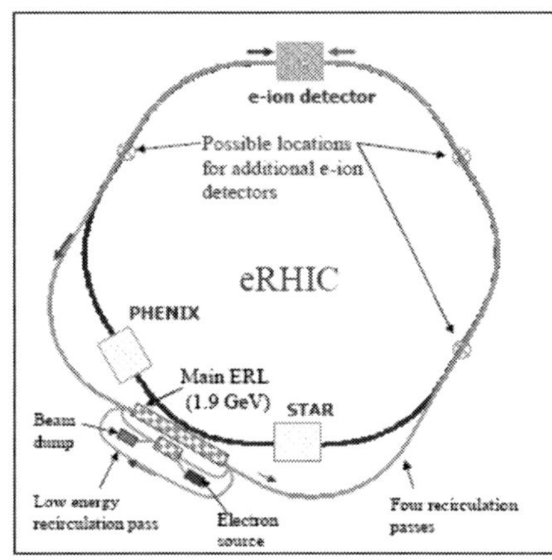

FIGURE 1. The present design layouts of the electron-ion colliders ELIC (left) and eRHIC (right).

POLARIZED ELECTRONS

Since the design approaches for the electron accelerator part of the colliders vary significantly, the issues and challenges related with the delivery of highly polarized electron beam are also quite different. For eRHIC the biggest challenge presents the development of a polarized electron source able to produce very large average electron beam current (50-260 mA). In ELIC the major challenge is related with overcoming depolarization effects caused by the synchrotron radiation in the presence of the complex spin rotator scheme.

In order to achieve the luminosity level of 10^{33} cm^{-2}s^{-1} in eRHIC the average electron current going through an interaction point has to exceed 100 mA (260 mA current is assumed presently as a nominal design value). The electron beam current is formed by bunches following with the 14 MHz repetition frequency. Since the electron beam is guided to the collision point directly from the ERL, the high average current, as well as the time structure of the current, have to be obtained directly from a polarized electron source. The presently studied approach to the extraction of the high polarized current from the electron source is based on the standard strained GaAs cathode technology and consists in the enlargement of the cathode area to a few square centimeters maintaining at the same time the current density output approximately at 50 mA/cm^2 level. Major issues, which have to be resolved, are related with the cathode lifetime, limited by the ion bombardment, and with the development of the efficient cooling of the cathode in order to remove the considerable heat load generated by the absorbed laser light. R&D work, which addresses the design issues of a high current source and aims to create a gun prototype, is underway at MIT-Bates [8]. Recent studies have confirmed benefits of a ring-like cathode shape from the point of view of the cathode lifetime. Relaxed transverse emittance requirements for the eRHIC electron beam are somewhat helpful for the design of the high intensity source. Initial stage of eRHIC, called MEeIC (the medium energy electron-ion collider), asks for 50 mA polarized electron current. Higher luminosity stages may consider increasing the average electron current by combining the output from several polarized electron guns. The efforts are also underway to develop an efficient cooling technique for high energy protons. If such technique would be available the requirement for the polarized electron current could be decreased proportionally to the reduction of the transverse proton beam emittance. R&D for the coherent electron cooling [9], which may be the effective cooling method, has started at BNL.

Aside from the high polarized electron current challenge, the linac-ring scheme used in the eRHIC design presents several advantages for the electron beam polarization. Without depolarizing resonances, essential in storage rings, the high beam polarization become available at any energy from the operating energy range (3-20 GeV) of the machine. The control of the spin orientation at the collision point does not require the application of strong field spin rotators around the interaction points. Instead the proper spin orientation can be managed at low beam energy, right after the polarized source, using a Wien filter and a solenoid magnet. Alternative technique involves a proper selection of accelerating gains in the main and pre-accelerator linacs of eRHIC which also produces the longitudinal spin orientation at the collision point at any beam energy for a certain geometric arrangement of the linacs and the recirculation arcs [10]. The additional important advantage of the eRHIC "linac-ring" scheme is that detector magnets do not introduce serious depolarization problems as in the case of a storage ring. This leads to a simpler interaction region design.

In ELIC the electron beam will be stored in a figure-8 shaped storage ring [5]. The acceleration will be done by upgraded CEBAF recirculation linacs. The requirements on the current of the polarized electron source in such scheme are at the level of the present day technology. To provide non-zero level of the equilibrium polarization in such scheme, the spin direction should be reversed between two halves of the figure 8. The electron storage ring is not flat, the vertical orbit excursions are done in the interaction regions in order to intersect the proton ring. To produce the longitudinal spin orientation in the collision points a complex scheme of spin rotation is used, which combines the rotation by solenoid spin rotators with the rotation by horizontal and vertical dipole field of main bending magnets [11]. Inside the rotator insertions, where the spin orientation is longitudinal, additional solenoids are used for the spin tune control. The effect on the electron polarization from the field of detector magnets has to be also taken into the consideration. The spin matching in such the complicated spin rotation geometry presents quite interesting and challenging task. The purpose of the spin matching is to prevent the leakage of $\gamma d\mathbf{n}/d\gamma$ function to the arcs, which otherwise will enhance the beam depolarization, reduce the depolarization time and decrease the achievable equilibrium polarization level [12].

The important advantage of the ELIC "ring-ring" scheme is the possibility to have polarized positrons. The positron beam can be self-polarized in the storage ring of ELIC on reasonable time scale ($\tau_{pol} \sim 2h$ at 7 GeV).

POLARIZED PROTONS AND IONS

Both, eRHIC and ELIC machines, aim to accelerate polarized proton beam to very high energies (250 GeV in both machines). The preservation of the polarization through all acceleration process requires application of special state-of-art techniques and hardware. Existing RHIC accelerator complex already has demonstrated the ability to accelerate the polarized protons to 250 GeV [13]. First physics run with polarized proton collisions at 250 GeV happens in early 2009. Several previous RHIC runs used the polarized protons with the energy up to 100 GeV, and the beam polarization up to 65% has been reached at that energy. Highly polarized beam (H^-) is produced in the high intensity optically-pumped polarized ion source [14]. The preservation of the beam polarization throughout the acceleration process is provided by using a number of dedicated hardware, which includes the resonance harmonic correction in the Booster, partial helical snakes in AGS and full Siberian Snakes in RHIC rings [15]. To control the polarization orientation at the collision points, the spin rotators are applied. Both the spin rotators and the Siberian snakes in RHIC are based on the helical magnets to minimize the orbit excursion in those devices [16]. For polarization measurements p-C CNI polarimeter and H-jet polarimeter are used [17]. The acceleration of polarized protons in eRHIC won't require any special additional hardware.

However, for the experiments in eRHIC the polarized $^3He^{+2}$ ions will be also needed in the addition to the polarized protons. The EBIS ion source, which is presently under construction, will provide a capability for the polarized ^3He production. The spin motion of $^3He^{+2}$ ions is more sensitive to the magnetic field than that of protons (G=-4.7, $G\gamma$ =744 for ^3He versus G=1.79, $G\gamma$ =450 for protons). That would allow the usage of the same helical snake and rotator magnets in both RHIC and AGS for the preservation and control of the ion spin [18]. On the other side, the enhanced spin sensitivity leads to the increased strength of depolarizing resonances. Additional studies, including spin dynamics simulations, are needed to verify the efficiency of the polarization preservation of $^3He^{+2}$ ions up to the top energy. The studies should also derive the tolerances for the orbit, betatron tune and spin tune control required for the polarization preservation.

In the ELIC design the special figure-8 form of the booster and main rings is used to provide the polarization preservation during the acceleration process and to simplify the control of the polarization orientation [5]. The figure-8 shape presents the alternative way, to the usage of the Siberian snakes, for the elimination of the dependence of the spin tune on the beam energy, thus avoiding the spin resonance crossing during the acceleration. While the value of the spin tune in RHIC, with Siberian Snakes, is ½, the figure-8 scheme holds the spin tune value near 0. The exact value of the spin tune is defined by the field strength of a stabilizing solenoid (Figure 2). The longitudinal field of the solenoid has to be strong enough to overcome the effect of vertical orbit errors, quite similar to the application of a partial snake. An additional advantage given by the figure-8 scheme is that the placement of the stabilizing solenoid in an interaction region straight automatically provides the longitudinal spin orientation throughout this straight section (thus through the two detector locations). To achieve the longitudinal polarization in another straight interaction region section the present design suggests using two full longitudinal axis snakes in the arc symmetry points. The radial polarization direction at the collision point can be also achieved by the arrangement of a proper vertical orbit excursion in the arcs. Besides the polarized protons, the ELIC design allows for the polarization preservation at the acceleration of polarized deuterons and ^{3}He ions.

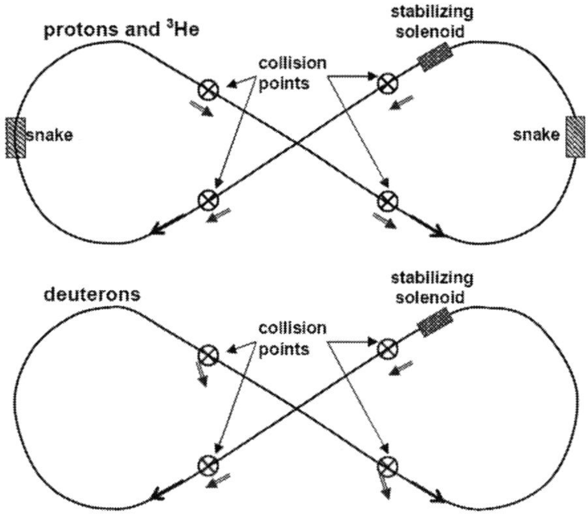

FIGURE 2. The schematics (from the ELIC ZDR) of the beam polarization control in the ELIC proton (ion) ring. For deuterons (due to small G) the snakes can not be used.

SUMMARY

The polarized beams of electrons (positrons), protons and light ions are the essential part of the designs of future electron-ion colliders eRHIC and ELIC.

Challenges for the delivery of the high polarization electron beam include the development of a high intensity polarized electron source for eRHIC and the spin matching of the complex spin rotator scheme to minimize depolarization effects in ELIC.

For the acceleration of polarized protons and ^3He ions in eRHIC the presently existing state-of-the-art hardware and tools of the RHIC accelerator complex will be used. In ELIC design, the figure-8 shape ring has been proposed to provide the polarization preservation of protons and light ions during the acceleration and to simplify the control of the spin orientation at the collision points.

ACKNOWLEDGMENTS

My acknowledgements to M.Bai, D.P.Barber, P.Chevtsov, Y.S.Derbenev, V.N.Litvinenko, W.W.MacKay, T.Roser and E.Tsentalovich who have been working on the beam polarization issues of ELIC and eRHIC designs and provided materials for this report.

REFERENCES

1. R. Milner, "Future electron ion/proton collider", these Proceedings.
2. V. Ptitsyn, "From HERA to Future Electron-Ion Colliders", Proceed. of 22nd Particle Accelerator Conference, WEZAKI02, (2007).
3. "eRHIC ZDR", BNL CA-D Note 142, (2004).
4. "eRHIC Accelerator Position Paper" for NSAC Long Range Plan, (2007), http://web.mit.edu/eicc/DOCUMENTS/Accelerator_Position_Paper-200701.pdf
5. ELIC ZDR (Draft), http://casa.jlab.org/research/elic/elic_report.shtml
6. S.A. Bogacz et al., "Advances on ELIC Design Studies", Proceed. of 11th European Particle Accelerator Conference, WEPP049, (2008).
7. Materials of 1st ECFA-CERN LHeC Workshop, (2008).
8. E. Tsentalovich, Proceed. of Workshop on Sources of Polarized Electrons and High Brightness Electron Beams (2008).
9. V. N. Litvinenko and Y. S. Derbenev, "FELs and High-Energy Electron Cooling", Proceed. 29th Int. FEL Conference, TUCAU01, (2007).
10. V.N. Litvinenko (private communications).
11. P. Chevtsov, "Electron Spin Matching for ELIC", these Proceedings.
12. Y.S. Derbenev, A.M. Kondratenko, *Journal of Experimental and Theoretical Physics (in Russian)*, v.64, 1918, (1973).
13. M.Bai et al., "Accelerating polarized protons to 250 GeV", Proceed. of 22nd Particle Accelerator Conference, TUODKI04, (2007).
14. A. Zelenski, "The RHIC polarized ion source", these Proceedings.
15. M. Bai, "Accelerating Polarized Protons to High Energy", Proceed, of 7th Intern. Spin Physics Symposium, (2006).
16. V. Ptitsyn and Yu.M. Shatunov, "Helical spin rotators and snakes", *Nuclear Instruments and Methods*, A398, p.126, (1997).
17. A. Bazilevski, "Precision Measurements of the Absolute Proton Beam Polarization with RHIC Polarimeters", these Proceedings.
18. W.W. MacKay and M. Bai, the section "Beam Polarization Issues" in eRHIC ZDR, BNL CA-D Note 142, (2004).

Electron Spin Rotation And Matching Scheme For ELIC, A High-Luminosity Ring-Ring Electron-Ion Collider

A. Bogacz, P. Chevtsov, Y. Derbenev, G. Krafft, Y. Zhang

Jefferson Lab, Newport News, VA 23606, USA

Abstract. A unique design feature of a polarized Electron-Ion Collider (ELIC) based on CEBAF is its Figure-8 shaped storage rings for both electrons and ions, which significantly simplifies beam polarization maintenance and manipulation. The CEBAF accelerator is used as a full energy injector of polarized electron beams into the electron storage ring. While electron polarization is maintained vertical in arcs of the ring, a stable longitudinal spin at four collision points is achieved through vertical crossing bending magnets, solenoid spin rotators, and horizontal orbit bends. Spin matching technique needs to be implemented in order to enhance quantum self-polarization and minimize depolarization effects. In this paper, we also discuss several important issues related to the use of positron beams, radiative polarization and quantum depolarization effects, as well as spin in ELIC.

Keywords: Electron-Ion Collider, Spin dynamics, Spin rotators.
PACS: 29.27. eg, 29.27.Hj

INTRODUCTION

ELIC is proposed as a future development of the Continuous Electron Beam Accelerator Facility (CEBAF) at Jefferson Lab [1, 2], beyond the 12 GeV Upgrade. The CEBAF accelerator with its polarized electron source will serve as a full energy injector into an electron storage ring providing required electron beam current, energy, and polarization. The electron and ion storage rings, which will be major additions to CEBAF, are designed as Figure-8 shaped double rings of 2.1 km total length sharing the same tunnel, with the electron ring above the ion ring. These rings consist of two identical arcs connected by two 330 m long crossing straight beam lines. The choice of the Figure-8 shape helps to eliminate any issues with spin maintenance and manipulation for all ion species. The electron complex of ELIC will deliver high current (up to 3 A) electron beams of the energy in the range of 3 to 9 GeV with longitudinal polarization at the Interaction Points (IPs) of not less than 70%. This complex has two major facilities: the existing CEBAF machine upgraded to 12 GeV and a new electron storage ring. CEBAF consists of two superconducting recirculating linacs operating at the 1.5 GHz RF frequency. Longitudinally polarized electrons are generated by a polarized DC photo-injector and then accelerated to the desired energy of 3 to 9 GeV in the CEBAF. After that, they are injected into the electron storage ring with vertical polarization in arcs and accumulated there until their average current reaches a desired value. The ELIC IPs will accommodate up to four detectors simultaneously. After beam stacking and accumulation is complete, both storage rings

CP1149, *Spin Physics, 18th International Spin Physics Symposium*
edited by D. G. Crabb, D. B. Day, S. Liuti, X. Zheng, M. Poelker, and Y. Prok

are switched to the collider mode, in which electrons are bent vertically with a crossing angle of about 0.022 radians to collide with ions.

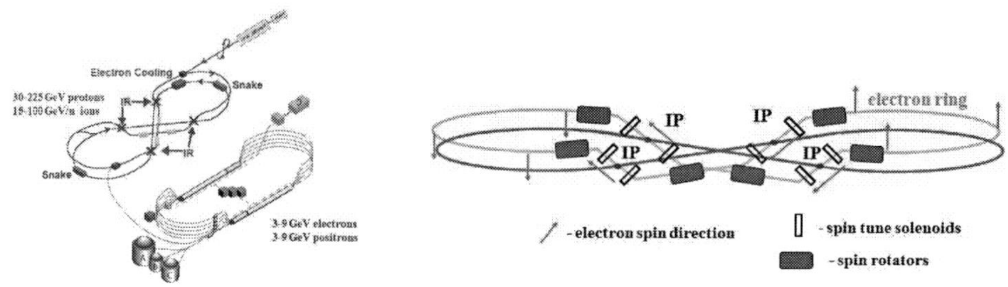

FIGURE 1. ELIC design overview and electron spin dynamics in the collider.

ELECTRON SPIN ROTATORS IN ELIC

The precession of the spin vector $\vec{S}$ of a relativistic electron with the mass m_e and electric charge e in external electro-magnetic fields is given by the Thomas-BMT equation [3]. In storage rings, where the electric fields are parallel to the particle momentum $\vec{P}$, this equation can be written as

$$\frac{d\vec{S}}{dt} = -\frac{e}{\gamma m_e}[(1+a\gamma)\vec{B}_\perp + (1+a)\vec{B}_\parallel]\times\vec{S} \qquad (1)$$

where $\vec{B}_\perp$ and $\vec{B}_\parallel$ are the magnetic fields perpendicular and parallel to the particle momentum respectively, γ is a relativistic factor, $a = 0.0011596$ is the anomalous magnetic moment of the electron. Eq.1 has the next two consequences, which are very important for our work.

In a solenoid with the length l and magnetic field $\vec{B}$ parallel to the particle momentum $\vec{P}$, the electron spin precesses around the magnetic field with respect to the momentum vector by the angle

$$\theta_{sol} = (1+a)\frac{Bl}{B_0\rho} \approx \frac{Bl}{B_0\rho} \qquad (2)$$

In a bending magnet with the magnetic field transverse to the particle momentum this precession angle $\theta_{bend} = a\gamma\theta_d$, where θ_d is the electron deflection angle. The value $a\gamma = \frac{E(GeV)}{0.44065}$ is known as the electron spin tune.

From the above equations follows that, for example,

a) to rotate spin of a 9 GeV electron by the angle $\pi/2$, a solenoid is required to have the total strength Bl equal to about 15π (T m);

b) if the electron deflection angle in a bending magnet at 9 GeV equals to 0.022, then the spin of this electron rotates by about 0.45 radians or 25.8 arc degrees;

c) in order to be longitudinal at the exit from a dipole magnet with a bending angle α, the electron spin vector $\vec{S}$ has to have the angle $a\gamma\alpha$ to the beam trajectory as it enters this magnet (see Fig. 2).

As it was mentioned above, polarized electrons from the CEBAF accelerator are injected into the Figure-8 ELIC electron ring with vertical polarization. However, nuclear physics experiments usually require longitudinal beam polarization at the IPs. Special devices are needed to rotate spin from vertical to longitudinal right before the IR and then back to vertical just after it. Such devices are called spin rotators.

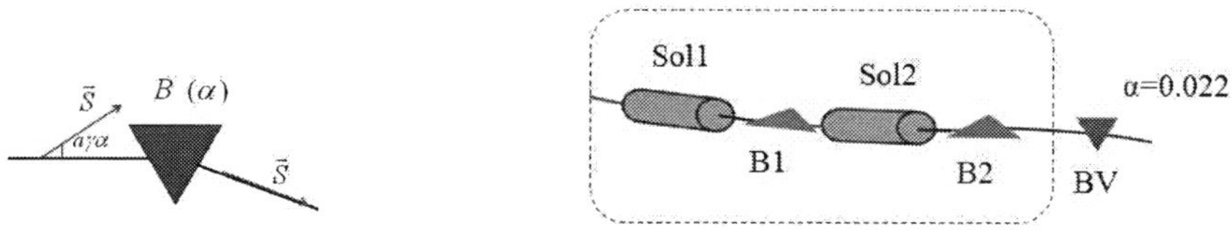

FIGURE 2. Electron spin rotation in a dipole magnet B with a bending angle α (left picture) and the electron spin rotator scheme for the Arc-to-IP spin transformation (right picture).

A special spin rotation scheme has been developed to transform the electron spin in ELIC from vertical in the arcs to longitudinal in all IPs. It consists of a combination of superconducting solenoids with horizontal arc and vertical crossing bends, which does not change the beam orbit at any available electron beam energy (3-9 GeV). One possible implementation of such a scheme is shown in Fig. 2. The last two arc dipoles (B1 and B2) and a vertical crossing bend (BV) interleave with two solenoids (Sol1 and Sol2). Consider this implementation in more detail. If we denote the spin rotation angles in the elements represented in Fig. 2 as φ_1 (Sol1), φ_2 (Sol2), $a\gamma\alpha_1$ (B1), $a\gamma\alpha_2$ (B2), and $a\gamma\alpha$ (BV) with the usual arrangement of positive and negative rotations, then it is not difficult to show that to make the required spin transition from vertical at the entrance to Sol1 to longitudinal after BV, the spin rotation angles in solenoids must be the solutions of the system of the next two equations:

$$\begin{aligned} \sin\varphi_1 &= \cos(\gamma a\alpha)\frac{\cos(a\gamma\alpha_2)}{\sin(a\gamma\alpha_1)} \\ \sin(\gamma a\alpha) &= -\sin\varphi_1\sin\varphi_2\cos(a\gamma\alpha_1) + \cos\varphi_1\cos\varphi_2 \end{aligned} \tag{3}$$

First of all, we note that the second equation in Eq.3 has the solution φ_2 for any values of the other rotation angles. If we now choose B1 to be two times stronger than B2 ($\alpha_1 = 2\alpha_2$), then the first equation in Eq.3 can be written as:

$$\sin\varphi_1 = \cos(\gamma a\alpha)\frac{1}{2\sin(a\gamma\alpha_2)} \tag{4}$$

The solution φ_1 of Eq.4 exists for any values of α if, for example,

$$\frac{\pi}{6} \le a\gamma\alpha_2 \le \frac{\pi}{2} \quad \text{or} \quad -\frac{\pi}{2} \le a\gamma\alpha_2 \le -\frac{\pi}{6} \tag{5}$$

depending on whether α_2 is positive (a counter-clockwise rotation) or negative (a clockwise rotation). This gives us the ratio of $a\gamma\alpha_2$ values in both expressions of Eq.3

equal to the ratio of the planned ELIC electron beam energies (3-9 GeV). If we now choose B2 rotating electron spin by the angle $\pi/2$ ($-\pi/2$) at 9 GeV, then our spin rotator will effectively work for all available beam energies. At that, the bending angle of B2 is equal to about 0.077 radians or 4.4 arc degrees. We note that in the designed spin rotators only one solenoid can be used at both the lowest (3 GeV, Sol1) and the highest (9 GeV, Sol2) beam energies. We also note that in addition to all spin rotators, eight solenoids are placed in straight sections of the ELIC before and after all IPs (see Fig. 2). They rotate spin around the longitudinal direction (this obviously doesn't affect the equilibrium spin, which is longitudinal there) forming an ELIC electron spin tuning set (ESTS).

ELIC ELECTRON RING OPTICS DESIGN

The spin behavior of the electrons that have the equilibrium energy and travel in spin rotators along the reference trajectory is similar to the dynamics of the equilibrium polarization vector $\vec{n}$, which is a periodical solution of the Thomas-BMT equation. The electrons with a non-equilibrium energy experience betatron oscillations. They pass through the rotator optical components in the fields, which are different from the fields acting on equilibrium particles. As a result, their spins acquire additional components to $\vec{n}$. The spread in electron spin directions can appear in rotators, which after a multiple passage through these rotators can lead to the beam depolarization. To avoid depolarization, the optics must be arranged in the way that perturbations caused by the elements in the whole system compensate each other. This is the main idea of spin matching [4, 5]. A great effort has been made to ensure the "spin transparency" [4] and spin matching in the ELIC design. The ELIC ESTS solenoids are used to make the machine spin tune equal to ½. As a result, most of the spin perturbations, which otherwise would add up over many revolutions compensate each other in consecutive revolutions [6]. The most critical elements of the ELIC electron ring, which can reduce of the electron polarization level, are vertical bends causing vertical dispersion and thus exiting vertical betatron oscillations as well as solenoids introducing horizontal and vertical (X-Y) beam coupling. In order to maintain an uncoupled optics and avoid generation of vertical beam instabilities, the ELIC spin rotators described above were modified by introducing additional quadrupoles into their structure. Without influencing the equilibrium spin direction, these quadrupoles effectively cancel X-Y beam coupling. Following the original approach of V. Litvinenko and A. Zholents [7], each spin rotating solenoid in the ELIC project is replaced by two solenoids of equal strength and a set of quadrupoles inserted between them. To minimize optics and orbit distortions, each spin rotator transfer matrix ($T_x = -T_y$) is made equal to the matrix of the drift L occupied by the insertion. The work on the design of ELIC optics integrating spin manipulating elements is in progress.

SOME ELIC SPIN PREPARATION AND MANIPULATION ISSUES

The uniqueness of the ELIC electron complex makes it possible to consider a variety of options for the spin preparation and manipulation. Normally, the CEBAF accelerator will inject polarized electrons in the ELIC electron ring with polarization, which is vertical and anti-parallel to the magnetic guiding fields in arcs. In this case, highly polarized electrons almost immediately become available for nuclear physics experiments and the Sokolov-Ternov mechanism will support the existing beam polarization. With the addition of a positron source to the CEBAF injector, any of two ELIC rings can accumulate positrons directly from CEBAF and then collide self-polarized positron beams with electrons or ions. To reduce self-polarization time, positrons can be injected and stored in the ring at the highest available energy (about 9 GeV), which can later be ramped to the value required by experiments. To reduce systematic errors, some nuclear physics experiments require frequent electron spin direction reversals or spin flipping. In ELIC, this can be done by RF dipoles inducing spin resonance frequencies [8] or with the use of a special electron bunch train structure provided by the CEBAF polarized source. This structure will contain electron bunch trains with the opposite direction of polarization at the required frequency. When the structure significantly degrades due to the Sokolov-Ternov effect, the electron beam can be dumped from the system and the ring can easily be filled with "fresh" electron bunch trains.

CONCLUSION

Special spin rotators, which do not change the beam orbit in the ELIC electron ring have been designed. These rotators and spin stabilizing solenoids placed in the beam interaction regions build up a spin manipulation system, which can effectively function at all available electron beam energies, from 3 to 9 GeV. The work on the electron ring optics integrating spin rotators and compensating their effects is almost finished. Preliminary theoretical estimates show that the designed electron polarization level for the ELIC electron ring is feasible.

ACKNOWLEDGMENTS

Authored by Jefferson Science Associates, LLC under U.S. DOE Contract No. DE-AC05-06OR23177. The U.S. Government retains a non-exclusive, paid-up, irrevocable, world-wide license to publish or reproduce this manuscript for U.S. Government purposes.

REFERENCES

1. A. Bogacz et al. "Design Studies of a High-Luminosity Electron-Ion Collider at CEBAF", PAC-2007, p. 1935-1937.
2. A. Bogacz et al. "Advances on ELIC Design Studies", EPAC-2008.
3. V. Bargmann, L. Michel, V. Telegdi, *Physical Review Letters*, May 15, 1959, p. 435-436.

4. D. Barber et al. “A Solenoid Spin Rotator For Large Electron Storage Rings”, DESY 84-102, October 1984.
5. D. Shatilov, A. Zholents “Spin Matching Conditions for a Storage Ring with Solenoid Rotators”, EPAC-1988, p. 917-919.
6. Y. Derbenev, A. Kondratenko, *Particle Accelerators*, Vol.8, 115 (1978).
7. A. Zholents and V. Litvinenko, DESY Internal Report, DESY, L-Trans 289.
8. V. Morozov et al. Phys. *Rev. ST Acc. Beams*, 4, 2001.

The Invariant Polarisation–Tensor Field For Spin-1 Particles In Storage Rings

D.P. Barber *,† and M. Vogt*

*Deutsches Elektronen–Synchrotron, DESY, Notkestrasse 85, 22607 Hamburg, Germany
†Cockcroft Institute, Daresbury Science and Innovation Campus and University of Liverpool, UK

Abstract. We complement the concept of the invariant spin field (ISF) in storage rings by defining the invariant polarisation–tensor field (ITF) for spin-1 particles and we suggest how to calculate it by stroboscopic averaging or directly from the invariant spin field. The ITF the ISF are used to construct equilibrium spin density–matrix fields, and thereby offer a clean framework for describing equilibrium ensembles of spin-1 particles such as deuterons in storage rings.

Keywords: Spin polarisation, deuterons, density matrix, storage rings, simulation
PACS: 29.27.Hj, 29.20.Db, 02.70.-c, 03.65.-w

INTRODUCTION

Spin motion for particles moving in electric and magnetic fields is governed by the T–BMT equation [1]. This describes the rate of precession of the rest–frame, pure–state, spin expectation value $\vec{S}$ ("the spin") of a particle. In circular particle accelerators and storage rings we take the distance around the ring as the independent variable [2]. Then, at the position s along the design orbit and the point u in the 6–dimensional phase space, we write the T–BMT equation as $d\vec{S}/ds = \vec{\Omega}(u;s) \times \vec{S}$ where the vector $\vec{\Omega}(u;s)$ describing the precession axis and the rate of precession, depends on the electric and magnetic fields in the laboratory, and on the reference energy of the ring. The motion of a particle is governed by the Lorentz force [1]. Thus both the motion of the particle and the motion of the spin expectation value can be treated classically. Nevertheless, we still need to look at the quantum mechanics. For this we exploit the spin density matrix.

The spin state of an ensemble of spin-1/2 fermions is completely specified by means of the 2×2 density matrix $\rho = \frac{1}{2}\{I + \vec{P} \cdot \vec{\sigma}\}$ where $\vec{P}$ is the vector polarisation, I is the unit matrix, the

$$\sigma_1 = \begin{pmatrix} 0 & -i \\ i & 0 \end{pmatrix}, \quad \sigma_2 = \begin{pmatrix} 1 & 0 \\ 0 & -1 \end{pmatrix}, \quad \sigma_3 = \begin{pmatrix} 0 & 1 \\ 1 & 0 \end{pmatrix},$$

are the Pauli spin matrices [3] and $\vec{\sigma}$ is the corresponding matrix-valued 3-vector. The vector polarisation is $\langle\vec{\sigma}\rangle = \vec{P} = Tr(\rho\vec{\sigma})$ where $\langle\rangle$ denotes a mixed-state expectation value.

Spin-1 particles such as deuterons need a 3×3 density matrix and we write it as [3]

$$\rho = \frac{1}{3}\left\{ I + \frac{3}{2}\vec{P} \cdot \vec{\mathfrak{J}} + \sqrt{\frac{3}{2}} \sum_{i,j} T_{ij} \left(\mathfrak{J}_i\mathfrak{J}_j + \mathfrak{J}_j\mathfrak{J}_i\right) \right\},$$

CP1149, *Spin Physics, 18th International Spin Physics Symposium*
edited by D. G. Crabb, D. B. Day, S. Liuti, X. Zheng, M. Poelker, and Y. Prok

where the three matrices $\mathfrak{J}$

$$\mathfrak{J}_1 = \frac{1}{\sqrt{2}}\begin{pmatrix} 0 & -i & 0 \\ i & 0 & -i \\ 0 & i & 0 \end{pmatrix}, \quad \mathfrak{J}_2 = \begin{pmatrix} 1 & 0 & 0 \\ 0 & 0 & 0 \\ 0 & 0 & -1 \end{pmatrix}, \quad \mathfrak{J}_3 = \frac{1}{\sqrt{2}}\begin{pmatrix} 0 & 1 & 0 \\ 1 & 0 & 1 \\ 0 & 1 & 0 \end{pmatrix},$$

representing the normalised spin operators $\hat{s}_i \ \ (i = 1,2,3)$, are the analogues for spin-1 of the Pauli matrices and where $\vec{\mathfrak{J}}$ is the corresponding matrix–valued 3–vector. The rank–2, 3×3, real, symmetric, traceless, Cartesian tensor T has just five independent components. With the three components of the vector polarisation there is a total of eight parameters for defining ρ as required for a Hermitian matrix with a definite trace (unity). The degree of tensor polarisation $\mathfrak{T} \equiv \sqrt{\sum_{i,j} T_{ij}^2} = \sqrt{Tr(T^2)}$ is at most unity [3].

In accelerator physics, we define a density matrix for each point u and s. Then at each u and s we write the density matrix for spin-1/2 fermions in terms of the local vector polarisation $\vec{P}_{\rm loc}(u;s)$. For spin-1 particles we need the local polarisation tensor $T_{\rm loc}(u;s)$ too. If the vector polarisation for spin-1/2 fermions is periodic from turn to turn, the density matrix can be described in terms of the vector $\hat{n}$ of the Invariant Spin Field (ISF) and the value of the equilibrium vector polarisation. Is there an analogous object for the polarisation tensor? This paper answers this in the affirmative by describing the new concepts of the Invariant Tensor Field (ITF) and the equilibrium tensor polarisation. Together with the ISF and the equilibrium vector polarisation they provide a complete description of the equilibrium spin state of a beam of spin-1 particles such as deuterons in a storage ring.

This paper provides just a brief summary. More information can be found in [4].

THE INVARIANT SPIN AND TENSOR FIELDS

Since the T–BMT equation is linear in $\vec{S}$ and since the particles at (u,s) all see the same $\vec{\Omega}(u;s)$, $\vec{P}_{\rm loc}(u(s);s)$ obeys the T–BMT equation along trajectories. Furthermore, $P_{\rm loc} \equiv |\vec{P}_{\rm loc}|$ is constant along a trajectory. For a storage ring at fixed energy, $\vec{\Omega}$ is 1–turn periodic in s at a fixed position in phase space u so that $\vec{\Omega}(u;s) = \vec{\Omega}(u;s+C)$ where C is the circumference. This opens the possibility that the polarisation is the same from turn to turn in the sense that $\vec{P}_{\rm loc}(u;s+C) = \vec{P}_{\rm loc}(u;s)$. We denote such a $\vec{P}_{\rm loc}$ by $\vec{P}_{\rm eq}$ since, if the phase space density has the same periodicity, the polarisation of the whole beam is 1-turn periodic, i.e. in equilibrium. Since $\vec{P}_{\rm eq}$ also obeys the T–BMT equation we then have $\vec{P}_{\rm eq}(M(u;s+C,s);s+C) = \vec{P}_{\rm eq}(M(u;s+C,s);s) = R(u;s+C,s)\vec{P}_{\rm eq}(u;s)$. where $M(u;s+C,s)$ is the new position in phase space after one turn starting at u and s, $R(u;s+C,s)$ is the corresponding 3×3 spin transfer matrix representing the solution to the T–BMT equation for one turn from s to $s+C$ and where here, $\vec{P}_{\rm eq}$ is represented by a column matrix of its components.

These relations motivate the introduction of a vector *field* $\hat{n}(u;s)$ of unit length, obeying similar constraints, namely $\hat{n}(M(u;s+C,s);s+C) = \hat{n}(M(u;s+C,s);s) = R(u;s+C,s)\hat{n}(u;s)$ where $\hat{n}$ is represented by a column matrix of its components. Since the vector field $\hat{n}(u;s)$ is invariant from turn to turn and independent of the real state of a beam it is called the invariant spin field (ISF) [5, 6]. We define $\hat{n}_0(s) \equiv \hat{n}(0;s)$.

We assume that u can be parametrised in terms of three pairs of action–angle variables $(J_i, \phi_i,\ i = 1,2,3)$ which we abbreviate by (J, ϕ). The actions J are constants of motion. Thus the orbital phase space is partitioned into disjoint tori characterised by unique sets J. We also assume that the orbital motion is nonresonant so that, in time, a trajectory could cover its torus. Then from the definition of $\vec{P}_{\rm eq}$, $P_{\rm eq} \equiv |\vec{P}_{\rm eq}|$ must be the same at all points ϕ on a torus but can depend on the J. Away from spin-orbit resonances and apart from the sign, $\hat{n}(J,\phi;s)$ is a unique function of ϕ and s, 2π–periodic in ϕ [5]. So we may write $\vec{P}_{\rm eq}(J,\phi;s) = P_{\rm eq}(J)\,\hat{n}(J,\phi;s)$, bearing in mind that $|\hat{n}| = 1$. If we also require that the particles on a torus are distributed uniformly in ϕ, we find that the polarisation for the torus, defined as the average $P_{\rm eq}(J)/(2\pi)^3 \int_0^{2\pi}\int_0^{2\pi}\int_0^{2\pi} \hat{n}(J,\phi;s)d\phi$, is invariant from turn to turn, i.e., in equilibrium. The key aspects of the ISF are summarised in [5]. The existence of a 1–turn periodic $\vec{P}_{\rm eq}$ and the corresponding ISF implies the existence of an equilibrium spin density–matrix field (ESDeMF), $\rho^{\rm eq}$, obeying the periodicity condition $\rho^{\rm eq}(u;s+C) = \rho^{\rm eq}(u;s)$ and which we can write in obvious notation as:

$$\rho^{\rm eq}_{P_{\rm eq}(J)}(J,\phi;s) = \frac{1}{2}\{I + P_{\rm eq}(J)\,\hat{n}(J,\phi;s)\cdot\vec{\sigma}\}\,.$$

We now extend the concept of invariance to the polarisation tensor by defining the Invariant Tensor Field $T^{\rm I}$ by the periodicity condition $T^{\rm I}(M(u;s+C,s);s+C) = T^{\rm I}(M(u;s+C,s);s) = R(u;s+C,s)\,T^{\rm I}(u;s)R^{\rm T}(u;s+C,s)$ where we have used the transformation rule [7] of a Cartesian tensor. We normalise so that $\sqrt{Tr\,([T^{\rm I}]^2)} = 1$.

Then we can construct an ESDeMF on a torus J for spin-1 particles too:

$$\rho^{\rm eq}_{[P_{\rm eq},\mathfrak{T}_{\rm eq}](J)}(J,\phi;s) = \frac{1}{3}\left\{I + \frac{3}{2}P_{\rm eq}(J)\,\hat{n}\cdot\vec{\mathfrak{J}} + \sqrt{\frac{3}{2}}\mathfrak{T}_{\rm eq}(J)\sum_{i,j} T^{\rm I}_{ij}\,(\mathfrak{J}_i\mathfrak{J}_j + \mathfrak{J}_j\mathfrak{J}_i)\right\}\,,$$

where $\mathfrak{T}_{\rm eq}(J)$ is the degree of equilibrium tensor polarisation on the torus. The ISF and ITF provide "scaffolding" on which to "hang" equilibrium spin distributions for spin-1 particles and the ESDeMF then depends on just two scalar parameters, $P_{\rm eq}(J)$ and $\mathfrak{T}_{\rm eq}(J)$.

The most general, *model–independent* way, to construct the ISF is by so–called stroboscopic averaging [8, 9, 10, 11]. This just requires a spin–orbit tracking code such as SPRINT [10] or EpsSLICK [12] which delivers 3×3 spin transport matrices along particle trajectories. As explained in [8, 9, 10, 11], with stroboscopic averaging the ISF, $\hat{n}(u_0;s_0)$, at the starting positions u_0 and $s = s_0$ can be found in terms of multi–turn spin transfer matrices by taking the average

$$\vec{f}_N(u_0;s_0) \equiv \frac{1}{N+1}\sum_{k=0}^{N} R(u(s_0 - kC);s_0,s_0 - kC))\hat{n}_0(s_0)\,,$$

for very large N and normalising this to unity: $\hat{n}(u_0;s_0) = \vec{f}_N(u_0;s_0)/|\vec{f}_N(u_0;s_0)|$. In this expression we have used notation similar to that in [8, eq.22] and chosen the "seed" spin field to be $\hat{n}_0$, although more general choices could be used [11].

The ITF can also be found by stroboscopic averaging:

$$g_N(u_0;s_0) \equiv \frac{1}{N+1}\sum_{k=0}^{N} R(u(s_0-kC);s_0,s_0-kC)\; T(s_0)\; R^{\mathrm{T}}(u(s_0-kC);s_0,s_0-kC)$$

where N is very large and $T(s_0)$ is a fixed 3×3 symmetric matrix with zero trace. The ITF is obtained as $T^{\mathrm{I}}(u_0;s_0) \equiv g_N/\sqrt{Tr(g_N^2)}$.

Given the above periodicity condition defining the ITF it is tempting to try to express the ITF in terms of the ISF. An ansatz in terms of $\hat{n}$ fulfilling this condition is:

$$T^{\mathrm{I}} = \pm\sqrt{\frac{3}{2}}\left\{\hat{n}\hat{n}^{\mathrm{T}} - \frac{1}{3}I\right\}.$$

Moreover, using the relation $T_{ij} = \frac{1}{2}\sqrt{\frac{3}{2}}\left\{\langle\hat{s}_i\hat{s}_j + \hat{s}_j\hat{s}_i\rangle - \frac{4}{3}\delta_{ij}\right\}$ [3], we find that this form for T^{I} is precisely that required if $\rho^{\mathrm{eq}}(J,\phi;s)$ is to be able to represent an ensemble of spins in the eigenstate of $\hat{n}\cdot\vec{\mathfrak{J}}$ whose eigenvalue is +1, -1 or 0. This is also the form obtained numerically by stroboscopic averaging. If this ansatz is unique the equilibrium spin-1 state on a torus J is then completely defined just by the vector field $\hat{n}(J,\phi;s)$ and the scalars $P_{\mathrm{eq}}(J)$ and $\mathfrak{T}_{\mathrm{eq}}(J)$:

$$\rho^{\mathrm{eq}} = \left\{\frac{I}{3} + \frac{1}{2}P_{\mathrm{eq}}\,\hat{n}\cdot\vec{\mathfrak{J}} + \mathfrak{T}_{\mathrm{eq}}\sum_{i,j}(\hat{n}_i\hat{n}_j - \frac{1}{3}\delta_{ij})\frac{(\mathfrak{J}_i\mathfrak{J}_j + \mathfrak{J}_j\mathfrak{J}_i)}{2}\right\}.$$

For example, for the eigenvalues ± 1 of $\hat{n}\cdot\vec{\mathfrak{J}}$ we have $P_{\mathrm{eq}} = \pm 1$ and $\mathfrak{T}_{\mathrm{eq}} = 1/2$. If the eigenvalue is 0, $P_{\mathrm{eq}} = 0$ and $\mathfrak{T}_{\mathrm{eq}} = -1$.

We thank J.A. Ellison, K. Heinemann, A. Kling and E. Leader for useful discussions.

REFERENCES

1. J.D. Jackson, *"Classical Electrodynamics"*, 3rd edition, Wiley (1998).
2. D.P. Barber, K. Heinemann and G. Ripken, Z. f. Physik C **64** (1994) 117.
3. E. Leader, *"Spin in particle physics"*, Cambridge Univ. Press (2001).
4. D.P. Barber and L. Malysheva, Cockcroft Report Cockcroft-08-01 (2008).
5. D.P.Barber, J.A. Ellison and K. Heinemann, Phys. Rev. ST Accel. Beams **7**, 124002 (2004).
6. Ya.S. Derbenev and A.M. Kondratenko, Sov. Phys. JETP **37** (1973) 968.
7. J.S. Bell, *"Polarised particles for accelerator physicists"*, CERN Report 75-11 (1975).
8. K. Heinemann and G.H. Hoffstaetter, Phys.Rev. E **54**(4) (1996) 4240.
9. G.H. Hoffstaetter, *"High Energy Polarised Proton Beams: a Modern View"*, Springer Tract in Modern Physics, Vol 218 (2006).
10. M. Vogt, PhD Thesis, University of Hamburg, Germany, DESY-THESIS-2000-054 (2000).
11. J.A. Ellison and K. Heinemann, Physica D **234**(2) (2007) 131.
12. D.P. Barber, DESY Report DESY 09-15 (2009).
13. D.P. Barber et al., five articles in proceedings of ICFA workshop "Quantum Aspects of Beam Physics", Monterey, U.S.A., 1998, edited by P. Chen, World Scientific (1999).

Electron Spin Precession at CEBAF

D. W. Higinbotham

Jefferson Lab, 12000 Jefferson Ave., Newport News, VA 23606

Abstract.
The nuclear physics experiments at the Thomas Jefferson National Accelerator Facility often require longitudinally polarized electrons to be simultaneously delivered to three experimental halls. The degree of longitudinal polarization to each hall varies as function of the accelerator settings, making it challenging in certain situations to deliver a high degree of longitudinal polarization to all the halls simultaneously. Normally, the degree of longitudinal polarization the halls receive is optimized by changing the initial spin direction at the beginning of the machine with a Wien filter. Herein, it is shown that it is possible to further improve the degree of longitudinal polarization for multiple experimental halls by redistributing the energy gain of the CEBAF linacs while keeping the total energy gain fixed.

Keywords: Polarized Electron, Spin Precession
PACS: 29.20.Ej, 29.27.Eg, 29.27.Hj

INTRODUCTION

Longitudinally polarized electron beams are produced at Jefferson Lab with a strained superlattice GaAs wafer and a DC photogun [1, 2]. These polarized electrons are required for a large fraction of the Jefferson Lab nuclear physics experiments; but the spin precession from the injector to the experimental halls is complicated by the fact that the accelerator runs with different total energies, different number of passes around the machine for different halls, and even with different bend angles into the various halls. In the past, once a beam energy was established, the only parameter that could be changed to manipulate the spin was the Wien angle at the beginning of the machine [3, 4]. If only one hall required polarization, by adjusting the Wien angle, that hall could receive the full injected longitudinal polarization. The challenge is to provide the highest possible longitudinal polarization simultaneously to multiple halls.

SPIN PRECESSION

In general, the spin precession, $\Delta\phi$, of an electron of fixed energy as it bends through a series of dipole magnets is given by:

$$\Delta\phi = \frac{\mathrm{E}}{440.65\mathrm{MeV}} \times \Delta\theta \tag{1}$$

where E is the energy of the electron and $\Delta\theta$ is the bend angle [5]. A simple example of CEBAF spin precession for first pass beam being hypothetically delivered to all three halls is shown in Fig. 1. In this example, the spin direction starts at the injector along the

CP1149, *Spin Physics, 18th International Spin Physics Symposium*
edited by D. G. Crabb, D. B. Day, S. Liuti, X. Zheng, M. Poelker, and Y. Prok

velocity direction ends in the three end stations with angles of $2\phi+\beta$, 2ϕ, and $2\phi-\beta$ with respect to the velocity direction. For even this simple calculation, two different energies need to be taken into account as the electrons increase in energy for each pass through a linac. In the example, the spin precesses by ϕ to the midpoint of the arc and by then another ϕ as it completes its bend around the arc all with an energy $E_{north\ linac}$; but after passing through the south linac the precession needs to be calculated of an energy of $E_{north\ linac}+E_{south\ linac}$ and a bend angle of 37°, 0°, -37° bend for Hall A, B, or C, respectively. Since Jefferson Lab is a race track accelerator, the energies continue to increase for higher passes. Nevertheless, a simple series of precession calculations are sufficient to calculate the total precession to any given Hall and at given pass. In fact, this relation between the beam energy and the amount of precession has been used to help determine the beam energy at Jefferson Lab [6].

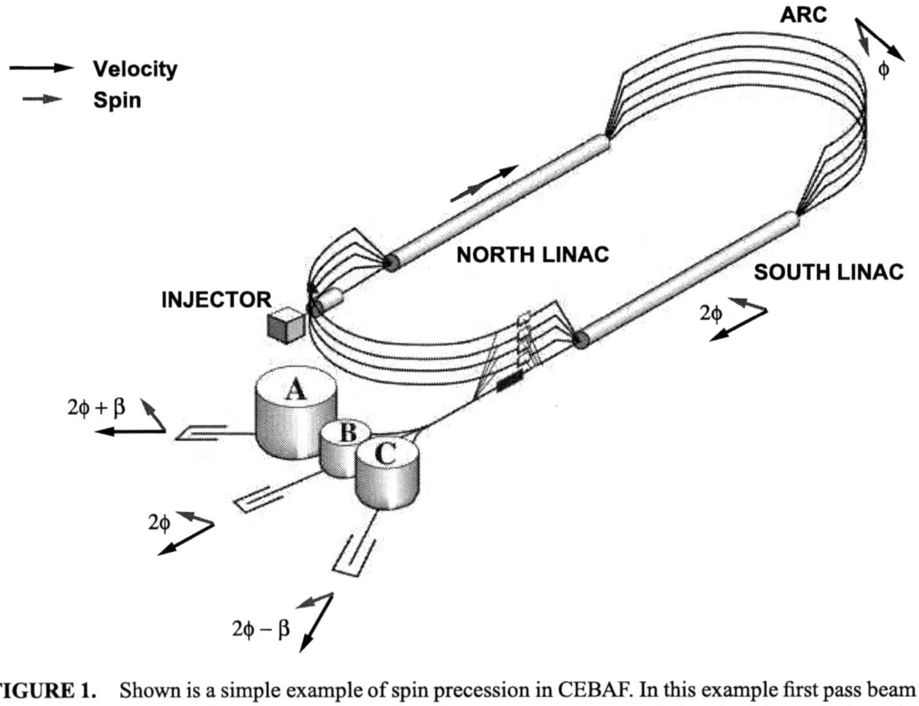

FIGURE 1. Shown is a simple example of spin precession in CEBAF. In this example first pass beam is being sent to all three halls. The velocity and spin directions start out parallel at the injector and continue parallel through the north linac. As the beam passes through the arc, spin precession causes the two vectors to start to point in different directions. By the time the beam has arrived at the three different end stations, the angle between the velocity and spin has taken the values of $2\phi+\beta$, 2ϕ, and $2\phi-\beta$, for Hall A, B, and C, respectively.

LINAC ENERGY IMBALANCE

In the summer of 2008, the CEBAF accelerator was running with only Hall C requiring high longitudinal polarization. The results of Moller measurements done in the halls during this time along with curves showing the expected degree of longitudinal polarization as a function of Wien angle is shown in Fig. 2. For spin experiments that use only beam polarization, like the one that was running in Hall C, the figure of merit of the experiment is proportional to P^2, the polarization square.

During the second half of this same run period, both Hall A & C required high polarization while Hall B required that the beam energy stay fixed. In the past, the Wien angle was the only parameter to change; but this would have only provided a P^2 of only 0.8 to Halls A & C and thus effectively caused a 10% reduction in beam time compared to the P^2 of 0.9 they were expecting. By running with imbalanced energy gains in the north and south linacs, the accelerator was able to precess the spin in such a may as to provide a P^2 of 0.9 to both Halls A & C while still maintaining the same total energy. The effect on the spin precession of the imbalance is shown in Fig. 3 and summarized in Table 1.

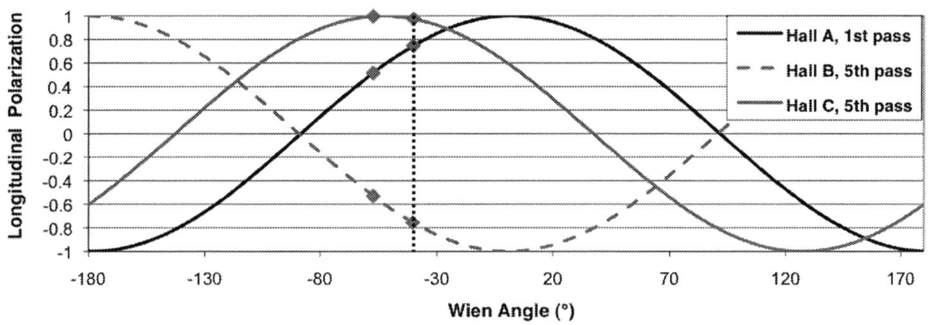

FIGURE 2. Shown is the degree of longitudinal polarization to the end stations for CEBAF setup with the injector set to 63.5 MeV and each of the linacs set to a gain of 565 MeV per pass. The points represent normalized Moller measurements done by the various Halls at different Wien settings. During this run period, in the summer of 2008, Hall C was the only experiment running which required the highest polarization. The dotted line indicates the Wien angle setting of -40° for this run period.

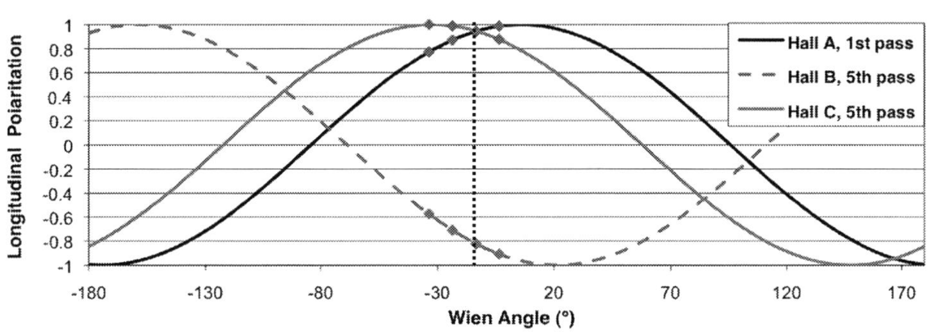

FIGURE 3. By reducing the north linac by 10 MeV and increasing the south linac by 10 MeV, we were able to deliver high polarization to both Hall's A & C for the second half of the summer 2008 run period while not changing the beam energy Hall B was receiving. The points represent normalized Moller measurements done by the various Halls at different Wien settings. The dotted line indicates the Wien angle setting of -13.5° that was used after the linacs were imbalanced.

TABLE 1. Shown are the conditions before and after imbalance the CEBAF linacs. With Halls A and C both doing polarization transfer experiments, being able to keep both Halls at a polarization squared, P^2, of 0.9 prevented an effective 10% reduction in beam time.

	Injector	**North Linac**	**South Linac**	P^2
Balanced	63.5 MeV	565 MeV	565 MeV	0.8
Unbalanced	63.5 MeV	555 MeV	575 MeV	0.9

Now that this method of optimizing the spin precession for multiple halls by imbalance the energy gain of the accelerator's north and south linacs has been demonstrated, Jefferson Lab's accelerator division is determining its potential for future experiments. With two degrees of freedom, the Wien angle and linac imbalance, the physics experiments should receive, for the same beam energy, higher longitudinal polarization then was previously achievable.

ACKNOWLEDGMENTS

Special thanks to my summer student Marie-Isabelle Holdrinet for her work on this topic and to the Jefferson Science Associates initiatives fund that make her stay at Jefferson Lab possible. And thanks Arne Freyberger and the Jefferson Lab accelerator scientists for their collaboration and for turning the idea of running the machine imbalanced into a reality.

REFERENCES

1. C. Hernandez-Garcia, M. L. Stutzman, and P. G. O'Shea, *Phys. Today* **61N2**, 44–49 (2008).
2. C. K. Sinclair, et al., *Phys. Rev. ST Accel. Beams* **10**, 023501 (2007).
3. C. W. Leemann, D. R. Douglas, and G. A. Krafft, *Ann. Rev. Nucl. Part. Sci.* **51**, 413–450 (2001).
4. S. R. Mane, Y. M. Shatunov, and K. Yokoya, *Reports on Progress in Physics* **68**, 1997–2265 (2005).
5. B. W. Montague, *Phys. Rept.* **113**, 1–96 (1984).
6. J. M. Grames, et al., *Phys. Rev. ST Accel. Beams* **7**, 042802 (2004).

RHIC Polarized Proton Performance in Run-8

C. Montag, M. Bai, W. W. MacKay, T. Roser, N. Abreu, L. Ahrens, D. Barton, J. Beebe-Wang, M. Blaskiewicz, J. M. Brennan, K. A. Brown, D. Bruno, G. Bunce, R. Calaga, P. Cameron, R. Connolly, T. D'Ottavio, A. Drees, A. V. Fedotov, W. Fischer, G. Ganetis, C. Gardner, J. Glenn, T. Hayes, H. Huang, P. Ingrassia, D. A. Kayran, J. Kewisch, R. C. Lee, F. Lin, V. N. Litvinenko, A. U. Luccio, Y. Luo, Y. Makdisi, N. Malitsky, G. Marr, A. Marusic, R. Michnoff, J. Morris, B. Oerter, F. Pilat, P. Pile, G. Robert-Demolaize, T. Russo, T. Satogata, C. Schultheiss, M. Sivertz, K. Smith, S. Tepikian, D. Trbojevic, N. Tsoupas, J. Tuozzolo, A. Zaltsman, A. Zelenski, K. Zeno, S. Y. Zhang

Brookhaven National Laboratory, Upton, NY 11973

Abstract. During Run-8, the Relativistic Heavy Ion Collider (RHIC) provided collisions of spin-polarized proton beams at two interaction regions. Physics data were taken with vertical orientation of the beam polarization, which in the "Yellow" RHIC ring was significantly lower than in previous years. We present recent developments and improvements as well as the luminosity and polarization performance achieved during Run-8, and we discuss possible causes of the not as high as previously achieved polarization performance of the "Yellow" ring.

Keywords: RHIC performance
PACS: 29.20.-c, 29.20.Dh, 29.27.Hj

INTRODUCTION

The RHIC polarized proton run in fiscal year 2008 (Run-8) lasted for six weeks, from January 28 to March 10, following a long deuteron-gold run [1]. This period includes the necessary start-up of the machine as well as the actual physics run. The main goal of this short run was to provide proton-proton collisions for comparison purposes with deuteron-gold data, and to develop higher luminosity for future runs.

MACHINE CONFIGURATION

During Run-6 luminosity performance with polarized protons was limited by beam-beam effects [2]. To overcome this limitation, a new near-integer working point was proposed and extensively studied in both simulations and experiments; however, the latter had been limited to studies at injection energy [3]. Simulations indicated that the dynamic aperture at the near-integer working point was significantly larger than at the Run-6 fractional tunes of (.685/.695). Therefore the initial set-up of RHIC was performed with the "Blue" ring at store fractional tunes of (.96/.95), while the working point of the "Yellow" ring remained at (.695/.685).

CP1149, *Spin Physics, 18th International Spin Physics Symposium*
edited by D. G. Crabb, D. B. Day, S. Liuti, X. Zheng, M. Poelker, and Y. Prok

TABLE 1. RHIC parameters during the polarized protons physics run.

Beam energy [GeV]	100
No. of bunches/beam	109
No. of protons/bunch [10^{11}]	1.5
95% norm. emittance [μm]	20 – 25
L_{peak} [10^{30} cm^{-2}sec^{-1}]	35
$L_{store\ avg.}$ [10^{30} cm^{-2}sec^{-1}]	23
β^* [m]	1.0
No. of collision points	2
Spin orientation at PHENIX	vertical
Spin orientation at STAR	vertical

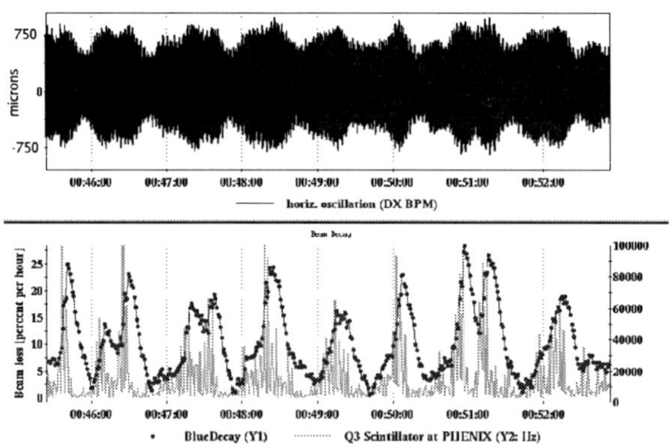

FIGURE 1. Horizontal oscillation of the "Blue" beam (black, top graph), together with the hourly beam loss rate (blue curve in lower graph, in percent) and the beam-induced background at PHENIX (green bars).

When first collisions were provided in this configuration [4], detector backgrounds from the "Blue" beam were found to be unacceptably high, and in near-perfect correlation to the fluctuating amplitude of the 10 Hz orbit oscillations [5], which had been increased by a factor five due to the near-integer working point as shown in Figure 1. The near-integer working point therefore had to be abandoned, and RHIC was set up in the same configuration as during Run-6. Table 1 lists the machine configuration and parameters used during the physics run.

PERFORMANCE

During the Run-8 physics run, working points of both beams were placed below the diagonal, around (.695/.685). The necessary tune separation to avoid coherent beam-beam effects was provided by separating the tunes along the $Q_x = Q_y$ diagonal, rather than by mirroring them across this diagonal. In this configuration, no obvious beam-beam limit was observed, in contrast to the Run-6 experience with working points mirrored across the diagonal. Bunch intensities of $1.6 \cdot 10^{11}$ protons/bunch were routinely injected and accelerated; the maximum bunch intensity reached $1.8 \cdot 10^{11}$ protons/bunch. Typical peak luminosities at the beginning of stores were approximately $30 \cdot 10^{30}\,\mathrm{cm}^{-2}\mathrm{sec}^{-1}$, with an average store luminosity of $23 \cdot 10^{30}\,\mathrm{cm}^{-2}\mathrm{sec}^{-1}$ (Figure 2). This resulted in a delivered integrated luminosity of approximately $18\,\mathrm{pb}^{-1}$, close to the maximum projected for the run.

Polarization in both rings was significantly lower than during Run-6, with roughly 50 percent in the "Blue" and 40 percent in the "Yellow" ring, compared to approximately 60 percent in both rings during Run-6. This low polarization can be attributed to several reasons. First, the beam polarization at the source reached only 80 percent, vs. 85 percent in Run-6. Furthermore, the AGS was set up in a new mode with injection "on-the-fly" [6]. While this was expected to provide a higher polarization due to a faster ramp rate across the lowest-energy depolarizing resonance, it resulted in a stronger intensity-dependence of the polarization at extraction, effectively lowering the polarization at intensities required for RHIC. For the final week of the physics run, the Run-6 set-up for the AGS was therefore re-loaded.

During the entire run the polarization transmission in the "Yellow" ring was significantly lower than in the "Blue" ring. This was most likely caused by the non-parallel "Yellow" beam orbits through the two Siberian snakes, which are located on opposite sides of the ring. This resulted in a deviation of the spin tune from its ideal value of $\nu_s = 1/2$, causing the depolarizing resonances to split into pairs. This splitting effectively reduced the available resonance-free tune space between depolarizing resonances.

Due to a broken beam position monitor in the snake area of the "Yellow" ring, orbit information were not readily available to confirm this theory. However, survey data taken during the shutdown after Run-8 (Figure 3) show that in the "Yellow" ring the two snakes are aligned with a net angle error of $-0.11\,\mathrm{mrad}$, while the corresponding angle in the "Blue" ring is only $-0.03\,\mathrm{mrad}$. These angles translate into spin tune errors that appear to be consistent with direct measurements of the spin tune at injection. In preparation for future runs, the alignment in the snake region has now been corrected, and the BPM has been repaired.

MACHINE DEVELOPMENT

To develop the machine towards higher luminosities in future runs, the β-functions at the two experiments were reduced from the operational $\beta^* = 1.0\,\mathrm{m}$ to $0.68\,\mathrm{m}$ during several dedicated development periods. Up to 56 bunches in both rings were accelerated and brought into collision. The luminosity scaled well with the reduced β^*, and after collimation detector background rates were at similar levels as in the regular configuration. Beam lifetimes at store were acceptable after correcting the nonlinear chromaticity.

SUMMARY

The polarized proton operations during Run-8 was successful, with weekly integrated luminosities reaching levels similar to those in Run-6. However, polarization levels in

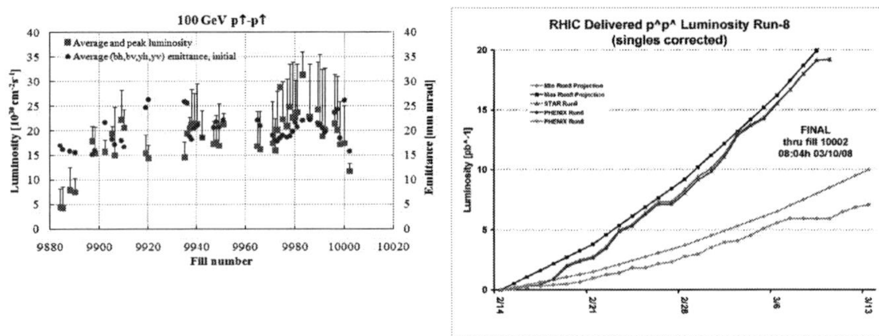

FIGURE 2. Left: Peak (vertical bars) and average (red squares) store luminosity vs. store number during the polarized proton Run-8. The average emittance, as calculated from measured luminosity and beam intensities for round beams, is indicated by blue dots. Right: Integrated delivered luminosity for PHENIX (blue) and STAR (red) during Run-8. For comparison, the integrated delivered luminosity during Run-6 is included (green line). The two smooth lines indicate the minimum and maximum projections for Run-8.

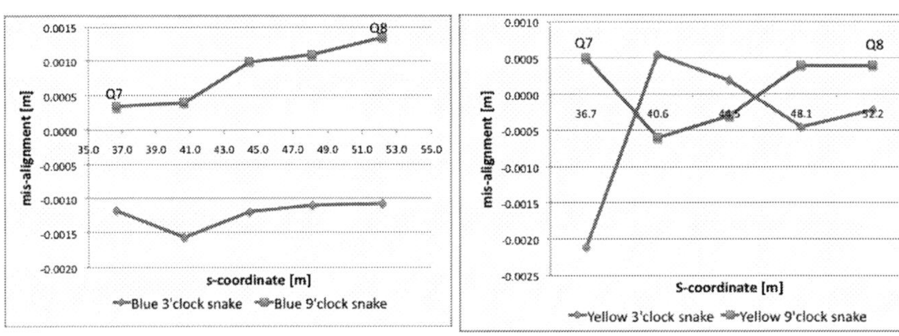

FIGURE 3. Measured horizontal magnet positions in the snake regions of the "Blue" (left) and "Yellow" (right) RHIC rings.

both rings were lower than in Run-6 for several reasons. The reduced β^*, which was developed but not used operationally during Run-8, is expected to significantly boost the luminosity in future runs.

REFERENCES

1. C. J. Gardner et al., "Setup and Performance of RHIC for the 2008 Run with Deuteron and Gold Collisions", Proc. EPAC 2008
2. V. Ptitsyn et al., "RHIC performance as polarized protons collider in Run-6", Proc. EPAC 2006
3. C. Montag et al., "A Near-Integer Working Point for Polarized Protons in the Relativistic Heavy Ion Collider", Proc. PAC 2007
4. C. Montag et al., "Operational Experience with a Near-integer Working Point at RHIC", Proc. EPAC 2008
5. C. Montag et al., Nucl. Instr. Meth. A 564, 26 - 31 (2006)
6. H. Huang et al., "AGS Polarized Proton Operation in Run 8", Proc. EPAC 2008

SPINK, A Thin Elements Spin Tracking Code [1]

Alfredo U. Luccio

Brookhaven National Laboratory, Upton, NY

Abstract. *spink* is a spin tracking code for spin polarized particles. The code tracks both trajectories in 3D and spin. It works using thick element modeling from *MAD* and thin element modeling based on the BMT equation to track spin. The code is written in Fortran and typically runs on a Linux platform, either sequentially or MPI-parallel.

Keywords: Polarized Beams
PACS: M 29-27.Hj

INTRODUCTION

Our purpose was to develop a code that tracks orbit and spin through the dynamically varying lattice of a real accelerator described in all its details, and that is fast enough to track millions of turns of a synchrotron or storage ring in a manageable run time.

For machine lattice descriptor and linear orbit tracking we use *MAD*[1], because it is widely used for accelerator work since it calculates twiss functions and several other machine parameters, it has a convenient way to deal with magnet errors and generates symplectic first order orbit transfer maps to propagate 6 canonical phase space coordinates through 'thick' machine elements. For higher order orbit tracking *MAD* is notoriously inadequate and we tend to use in *spink* other map generators, like *UAL*[2] or *COSY-infinity*[3]. For *spink*, the output of *MAD* is transformed by a pre-processor to a lattice descriptor file.

For spin tracking we track 3 coordinates representing the Cartesian components of an unitary spin vector (tensor polarization tracking is being implemented). Spin tracking is done using a 'thin' kick formalism that calculates the spin rotation in the magnetic field at the particle instantaneous position. We track one or several particles sequentially, averaging the final results, or "trivially" in parallel using the MPI library[4]. We follow the 'accelerator' coordinate convention with $\hat{x}$ radial, $\hat{y}$ vertical, and $\hat{z}$ longitudinal.

To reproduce and simulate 'control room' environment, many varying quantities, like acceleration rate and tune path, are given as input through tables, which are used to dynamically define and update the accelerator lattice read at start from *MAD*.

---

[1] Work supported by the US Department of Energy

CP1149, *Spin Physics, 18th International Spin Physics Symposium*
edited by D. G. Crabb, D. B. Day, S. Liuti, X. Zheng, M. Poelker, and Y. Prok

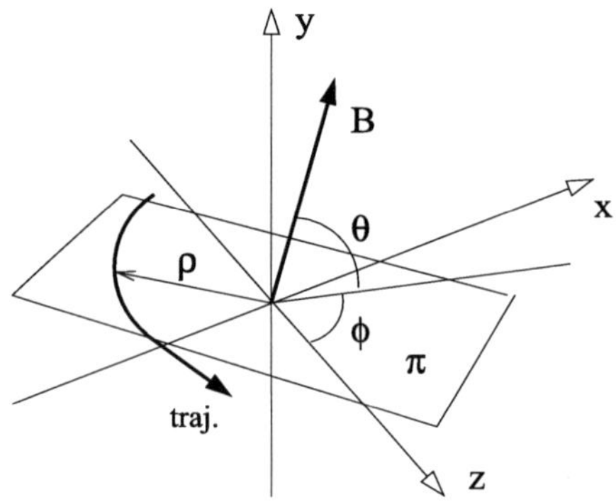

FIGURE 1. Plane π containing the particle orbit in an accelerator magnet

THOMAS-BMT EQUATION, FOR THIN SPIN KICK

Spin propagation is calculated from the Thomas-BMT equation[5]

$$\left\{ \frac{d\mathbf{S}}{dt} = \frac{q}{\gamma m} \mathbf{S} \times \mathbf{F}, \quad with \quad \mathbf{F} = (1+G\gamma)\mathbf{B}_{\perp} + (1+G)\mathbf{B}_{\parallel} \right. \tag{1}$$

using the magnetic field components transverse and longitudinal to the velocity of the particle, respectively. G is the gyromagnetic ratio, γ the energy or Lorentz factor.

A sketch of the formalism used in *spink* is the following. In a thin element write the BMT Eq. in matrix form, with a spin rotation (spin angle kick) μ

$$\mathbf{S} = \mathscr{M}\, \mathbf{S}_0, \quad \mu = \omega(\mathbf{P})\, \delta s. \tag{2}$$

where $\mathbf{P}$ is the phase space coordinate of the particle and δs the trajectory path length through the element

Spin rotation is completely defined by 3 angles: the angle μ of spin rotation and the angles ϕ, longitude and θ, latitude that define the orientation of the axis of rotation, coincident with the local magnetic field as seen by the particle at $\mathbf{P}$, see Fig.1. The angles are a function of the particle coordinate $\mathbf{P}$, calculated by orbit tracking.

By integration of a system of three differential equations, equivalent to the vector Equation(1), we obtain an explicit representation of the 3×3 matrix $\mathscr{M}$

$$\mathscr{M} = \begin{pmatrix} 1-(A_y^2+A_z^2)c & A_xA_yc+A_zs & A_xA_zc-A_ys \\ A_xA_yc-A_zs & 1-(A_x^2+A_z^2)c & A_yA_zc+A_xs \\ A_xA_zc+A_ys & A_yA_zc-A_xs & 1-(A_x^2+A_y^2)c \end{pmatrix} \tag{3}$$

with

$$c = 1-\cos\mu, \quad s = \sin\mu; \quad A_x = \tfrac{1}{\omega}a_x, \quad A_y = \tfrac{1}{\omega}a_y, \quad A_z = \tfrac{1}{\omega}a_z \tag{4}$$

and

$$a_x = f_x - \tfrac{1}{\rho_x}, \quad a_y = f_y - \tfrac{1}{\rho_y}, \quad a_z = f_z, \quad \omega^2 = a_x^2 + a_y^2 + a_z^2 \tag{5}$$

where ρ_x and ρ_y are the local radii of curvature of the orbit in the reference frame, in the horizontal and vertical plane, respectively, and $\mathbf{f} = \frac{1}{B\rho}\mathbf{F}$, with $B\rho$ the rigidity of the particle.

In the code we propagate the matrix $\mathscr{M}$ along the tracking, in order to calculate spin tune, as it will be discussed next, and the stable spin axis. For this, we can either: (1) propagate the matrix itself (9 numbers), or (2) propagate the 3 angles and reconstruct the matrices at each step. The second strategy is equivalent to use uniquaternions and may save running time[2].

SPIN TUNE AND OTHER SUBJECTS

In *spink* we track working particles relative to a reference particle. The distribution of spin tune, *i.e.* the number of spin revolution per turn, for each particle in a bunch is a function of the initial particle position **P** in phase space. Spin tune is calculated from the eigenvalues of the cumulative spin matrix at the end of one spin closed turn, defined as the turn where the particle is back to its starting position in phase space, within a prescribed accuracy. This happens after a substantial number of machine revolutions. Spin stable axis components are also calculated from the same one-turn spin matrix.

About the strength of the spin kick in the various elements in the accelerators, Eq.(1) shows that in an accelerator element where the field is transverse to the trajectory of the working particle (bends, quadrupoles), the spin kick to lowest order in the orbit coordinates should be proportional to $(1+G\gamma)$, while in elements where the field is parallel (solenoids) it should be proportional to $(1+G)$, *i.e.* independent of the energy of the particle. *Spink* automatically takes care of the expansion.

As the tracking shows, the result is not always this, particularly in the main bends of a circular accelerator, where the horizontal radius of curvature that appears in Eq.(5) has always the same sign. In a bend, the net result is that to lowest order the "1" in $1+G\gamma$ is canceled along the trajectory and the spin kick becomes proportional to $G\gamma$ with a spin rotation simply $G\gamma\theta$, with θ the bend angle [3].

In quadrupoles, the curvature of the trajectory, directed in any direction in space, is continuously changing from one quad to the next, in sign and amplitude, with the net result that the average spin kick coefficient remains $1+G\gamma$. An exception may be an RF Dipole with an horizontal field that modulates the vertical betatron motion and that may produce a constant sign of the vertical curvature in vicinity of a resonance. Also see [7].

Siberian snakes break the periodic condition producing spin resonances and polarization losses. These devices rotate the spin in a prescribed way, and are treated in *spink* in three ways: or as synthetic, represented for spin tracking by the three spin rotation angles, table driven, or analytic. In a Synthetic Snake, like in other elements where the spin rotation and axis orientation are prescribed (*e.g* spin rotators), the matrix $\mathscr{M}$ simply contains μ and the three quantities $A_x = \cos\theta\sin\phi$, $A_y = \sin\theta$, $A_z = \cos\theta\cos\phi$ i.e. the components of the spin rotation axis.

---

[2] Des Barber, private communication

[3] This result is discussed, e.g. in S .Y. Lee book[6].

About Eq.(2) for spin rotation in a thin element, δs is the physical length of a machine element, but a better expression should contain the length of the orbit through the element. Also, for complicated devices, like the Alternating Gradient combined functions dipoles of the AGS, a good practice is to divide the element in multiple sections, each represented by a spin kick.

NOTES AND ACKNOWLEDGMENTS

spink has a user's manual[8] and a graphical window driven GUI to help prepare its input.

The code has been around for several years. We extensively use it in Brookhaven for polarized proton studies at the AGS and RHIC. *spink* has been used in a few laboratories, notably Jülich in Germany (polarized deuterons) and KEK-JPARC in Japan. Other codes address some of the same issues, such as *DEPOL* by E.Courant[9], *SNAKE* by J. Buon[10], *SPRINT* by D.Barber and M.Vogt[11], *ASPIRRIN* by V.Ptitsyn and Y.Shatunov[12].

Many colleagues contributed to *spink*, directly or indirectly: Mei Bai, Haixin Huang, Fanglei Lin, Waldo MacKay, Nick Tsoupas, at Brookhaven, Andreas Lehrach at Jülich and Meiqin Xiao at RIKEN. Constant support by Thomas Roser has been important for the success of the code.

Vahid Ranjbar and Dan Abell at TechX have done much work on the code on a DOE grant, in order to consolidate its results, plus adding new features.

REFERENCES

1. H.Grote, and F.Ch.Iselin, The mad program, vers.8.19, Tech. Rep. CERN/SL/90-13, European Organization for Nuclear Research, Geneva, CH (1996).
2. N.Malitsky, and R.Talman, Ual user guide, Tech. Rep. 71010-2003, Brookhaven National Laboratory. Upton, NY (2002).
3. K.Makino, and M.Berz, *Nuclear Inst. Meth.* p. 346 (2003).
4. A. W.Gropp, E.Lusk, *Using MPI*, MIT Press, 1998.
5. D.J.Jackson, *Classical Electrodynamics*, Wiley, New York, 1975, second edition.
6. S.Y.Lee, *Spin Dynamics and Snakes in Synchrotrons*, World Scientific. Singapore, 1997.
7. E.D.Courant, Spin motion and resonances in accelerators and storage rings (preprint), Tech. Rep. C-A/AP/293, Brookhaven National Laboratory. Upton, NY, Upton, NY (2007).
8. A.U.Luccio, Spink user's manual. version v.2.21-beta, Tech. Rep. C-A/AP/283, Brookhaven National Laboratory. Upton, NY (2007).
9. E.D.Courant, and R.D.Ruth, The acceleration of polarized protons in circular accelerators, Tech. Rep. BNL 51270, Brookhaven National Laboratory. Upton, NY (September 12, 1980).
10. J.Buon, "Depolarization effects in a ring equipped with snakes," in *AIP Conference Proceedings Vol. 145*, July 1986, pp. 164–169.
11. D. Barber, and M. Vogt, The amplitude dependent spin tune and the invariant spin field in high energy proton accelerators (EPAC 1998, p.1362).
12. E. V.Ptitsyn, and Yu.M.Shatunov, "Spin-Orbital Function Formalism and ASPIRRIN Code," in *Proc 15th Intern.Spin Physics Symposium, SPIN2004, AIP Conference Proceedings*, 2003, pp. 761–765.

Effect of Various Errors on the Spin Tune and Stable Spin Axis[1]

W. W. MacKay

Brookhaven National Laboratory, Upton, NY 11973, USA.

Abstract. Even though RHIC has two full Siberian snakes in each ring, there are various perturbations to the ideal case including orbit errors at the snakes, experiment solenoids, injection bumps, and interlaced horizontal-vertical bumps at the hydrogen jet polarimeter. These errors can cause depolarization by shifting the spin tune and tilting the stable spin direction away from vertical. Tilting of the stable spin axis can enhance horizontal depolarizing resonances. This paper presents preliminary results for some of these error scenarios, as well as their impact on the stable spin directions at STAR and PHENIX.

Keywords: Polarized ion beams, RHIC, spin, polarization
PACS: 29.27.Hj, 29.20.db, 29.20.dk, 13.40.Em

In order to vary the polarization direction for collisions, RHIC has four spin rotators around the STAR experiment and another four around the PHENIX experiment [1]. This arrangement allows independent control of the polarization axis for either beam and at either experiment. Each rotator consists of four helical dipoles with the inner and outer pairs each connected in series to separate power supplies [1]. In each ring, the rotators on either side of the experiment are powered with opposite currents to create a spin transparent insertion for each of the two IR's. Table 1 shows typical operating currents for longitudinal polarization at both experiments during the 2006 run [2] adjusted to minimize the transverse components of polarization with the local polarimeters at the experiments [3]. The predicted settings are from tracking through an analytic model of helical dipoles. The deviations in the settings for the Blue rotators around PHENIX are much larger than one would expect from magnet construction errors and measurement errors of transverse polarization components.

TABLE 1. Typical rotator magnet currents for longitudinal polarization at 100 GeV for STAR and PHENIX.

	PHENIX		STAR		
	Blue	Yellow	Blue	Yellow	Prediction
$I_{\rm out}$ [A]	214	221	211	215	221
$I_{\rm in}$ [A]	204	184	184	187	184

[1] Work supported by Brookhaven Science Associates, LLC under Contract No. DE-AC02-98CH10886 with the U.S. Department of Energy.

CP1149, *Spin Physics, 18th International Spin Physics Symposium*
edited by D. G. Crabb, D. B. Day, S. Liuti, X. Zheng, M. Poelker, and Y. Prok

In order to understand discrepancies in the measured direction of the transverse polarization components for different running scenarios, I have resurrected code from an old online model called `bl` [4]. Spin tracking has been added in the new code `blspin` to explore the closed-orbit spin axis $\hat{n}_0$ and spin tune ν_{sp} with various contributions including solenoids, bumps, orbit errors from rotators, snakes and quadrupole misalignments as well as dipole field errors. At present the program only deals with linear optics. The algorithms for spin motion were derived from the "standard theory" as presented in Ref. [5]. The approximations made for spin ignore the longitudinal field components of dipole fringes and the radial field components of solenoids. For RHIC at 100 GeV, these fringe field effects are small, although I plan to add these in the near future.

The twist angle of the helical dipoles as constructed for the snakes and rotators is systematically about $\Delta\alpha = 0.4^\circ$ shy of a full 360° rotation. For a single helical dipole with field B_0 and helical twist wavelength λ, the resulting vertical orbit deflection is approximately [6]

$$\Delta y' \approx -\frac{qB_0}{p}\frac{\lambda}{2\pi}\sin\Delta\alpha, \tag{1}$$

and is independent of the handedness of the helix. While being well within the design tolerances, for the rotators this does impart significant vertical deflections which must be locally corrected. Fig. 1 shows a simulated closed orbit with rotator steering errors corrected by local bumps in addition to closed separation bumps at four IR's.

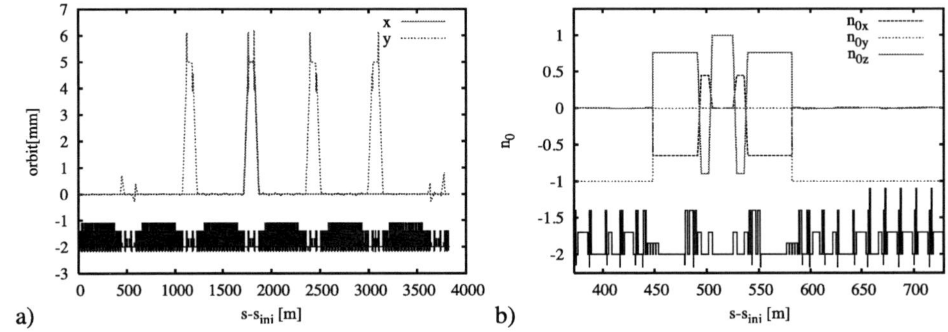

FIGURE 1. a) Closed orbit with vertical orbit errors from snakes and rotators and vertical separation bumps at four IR's. The errors from the rotators are corrected locally, but the smaller errors from the snakes remain uncompensated. The PHENIX and STAR interaction points are located at 515 m and 3710 m, respectively. A horizontal bump at the hydrogen jet is also shown at 1793 m. The snakes are located at 1034 and 2952 m. b) Closed-orbit spin components through the PHENIX IR. The PHENIX rotators are shown centered about positions 449 and 582 m, and the STAR rotators are similarly placed around the STAR IR. (Rotators and snakes are treated as thin elements in the model.)

Horizontal orbit errors through the snakes can cause tilts of the stable spin direction and spin-tune shifts [1] $|\Delta\nu_{sp}| = |\nu_{sp} - 0.5| \approx 2G\gamma|\Delta x'|/\pi$, where $\Delta x' = x'_2 - x'_1$ is the deviation from parallel of the horizontal orbit at the snakes on opposite sides of the ring. For example, an error $\Delta x' = 0.1$ mr at the snakes would shift the spin tune by 0.012. Orbit errors through the rotators can also cause sizable shifts.

TABLE 2. Calculated spin tune and transverse components of $\vec{n}$ with various perturbations. The column headings are as follows. Rot: polarization direction at [PHENIX, STAR] – v for vertical and l for longitudinal; Sol for PHENIX and STAR solenoids – on or off; Vsep: vertical separation bumps at IR's 2, 4, 10, and 12 (not STAR and PHENIX); Hsep(Hjet): horizontal separation bump at IR 12; Misalign: random misalignment of arc quadrupoles; ν_{sp}: spin tune; n_y, n_x: vertical and radial components of $\hat{n}_0$ at CNI polarimeter. All calculations are for 100 GeV with $\beta^* = 1$ m at STAR and PHENIX and 10 m at the other four IR's.

	Rot	Sol	Vsep	Hsep(Hjet)	Misalign	ν_{sp}	n_y	n_x
1	v,v	off	off	off	no	0.500004	0.999999	−0.000097
2	v,v	on	off	off	no	0.500011	0.999859	0.006818
3	v,v	off	5 mm	off	no	0.500005	0.999795	−0.000243
4	v,v	on	5 mm	off	no	0.500012	0.999387	0.006673
5	v,v	on	5 mm	5 mm	no	0.499974	0.999398	0.002919
6	l,l	off	5 mm	off	no	0.499447	0.999816	0.001904
7	l,l	on	5 mm	off	no	0.496647	0.999817	0.001829
8	l,l	on	5 mm	5 mm	no	0.496684	0.999817	−0.001600
9	l,l	on	5 mm	5 mm	no	0.496684	0.999817	−0.001600
10	l,l	on	5 mm	5 mm	yes	0.493656	0.999979	−0.006179

Table 2 shows calculated spin tune, and transverse spin components at the location of the p+C CNI polarimeter (at 1864 m) for various combinations of bumps, solenoids, rotator settings, and quadrupole misalignment. The first row shows a slight tune shift caused by the small vertical errors and path-length increase from the snakes. (The dispersion function is different at the two snakes.) The effect on spin tune from the experiment solenoids is negligible unless the rotators are turned on; compare rows 1–4 with 6 and 7. Vertical separation at IR's without rotators does not affect the spin, since the four horizontal bends for each IR crossing (two D0 and two DX magnets) have no quadrupoles between them. The last line of Table 2 shows an example of random misalignments of the arc quadrupoles (from Q8 to Q8). The random misalignments were generated with a flat distribution between ± 0.3 mm in both planes. This particular set of misalignments resulted in the rms orbit deviations (Fig. 2a) in the arcs of $\sigma_x = 0.57$ mm and $\sigma_y = 0.42$ mm. The transverse components of $\hat{n}$ for this last case were $(n_x, n_y) = (-0.0823, 0.0023)$ for PHENIX and $(-0.0404, -0.0104)$ for STAR. For the last row, $\vec{n}_0$ was tilted away from the longitudinal by about 5° at PHENIX.

At 100 GeV the combined horizontal and vertical separation bumps with respective amplitudes x and y at IR 12 with the Hjet polarimeter do not make a significant shift of ν_{sp}, but at 250 GeV (see Fig. 2b) the effect, proportional to xy^2, is larger.

Basically any spin rotation not about the vertical direction can, in principle, cause a distortion of spin tune and direction. Since rotations are noncommutative, the effects can add nonlinearly as demonstrated by combinations of rotators and solenoids and by combined horizontal and vertical separation bumps. Vertical orbit errors around the ring can cause tilts and spin-tune shift, and if the dice roll badly, then combined horizontal-vertical orbit errors can compound this. For other random quadrupole misalignments with $\sigma_{x,y} \sim 0.5$ mm, in the arcs, I have calculated $\Delta\nu_{sp} > 0.01$.

Preliminary results using the new spin tracking code `blspin` show tilts similar to what has been reported at times by the p+C CNI, STAR, and PHENIX polarimeters.

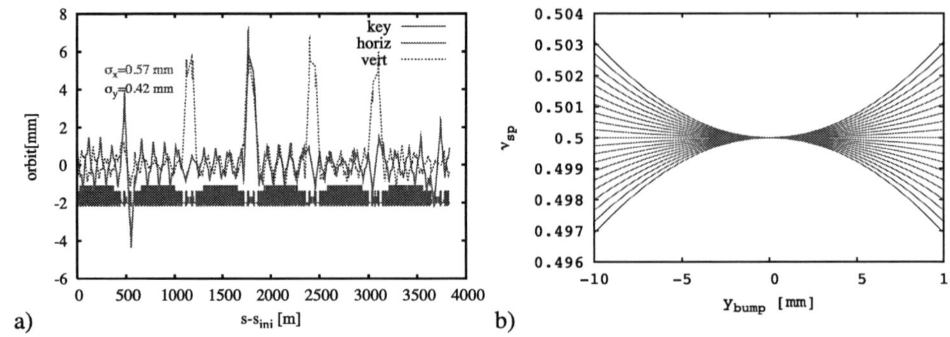

FIGURE 2. a) Closed orbit for conditions of last line of Table 2 (at 100 GeV). Notice the separation bumps at the 10, 12 (Hjet), 2, and 4 o'clock IR bumps. b) Effect on spin tune caused by combined horizontal and vertical bumps at the Hjet as calculated for 250 GeV with ideal snakes and no rotators or errors. The curves are plotted for 1 mm steps in the horizontal bump from -10 to 10 mm.

Since the original `bl` was written as an online code to simulate orbit steering, it can quickly simulate different alignment and bump configurations.

The setting of the Blue rotators at PHENIX relative to the predicted values corresponds to a vertical deflection of about 7–10° away from the longitudinal. The simulated tilt at PHENIX for conditions in Fig. 2a was a horizontal deflection; however, so far I have not tried to simulate the actual orbit errors at snakes and rotators. By examining actual orbit data taken during longitudinal pp running, and trying to match the orbit slopes measured at the snakes and rotators, it may be possible to explain this discrepancy.

Future improvements to `blspin` will add contributions to spin precession from the longitudinal fringe contributions of bend magnets and the radial fringe fields of the solenoids. Higher order multipoles will be added for both orbit and spin calculations, in particular sextupoles. Calculation of the invariant spin field $\hat{n}$ for phase space points other than the closed orbit will be explored. I eventually plan to develop a version with spin-1 spinors for deuterons.

I would like to thank D. Barber, A. U. Luccio, S. Mane, C. Montag, V. Ptitsyn, and T. Roser for fruitful discussions. During the workshop, Des Barber and others pointed out the need for inclusion of spin precession from the longitudinal component of the dipole end fields in dealing with low energy deuterons.

REFERENCES

1. I. Alekseev et al., *Configuration Manual Polarized Proton Collider at RHIC*, BNL, (2006). URL:`http://www.rhichome.bnl.gov/RHIC/Spin/design/CMan/CMan.pdf`
2. V. Ptitsyn et al., "RHIC Performance as Polarized Proton Collider in Run-6", PAC06 (2006).
3. W. W. MacKay et al., "Commissioning Spin Rotators in RHIC", PAC03, 1697 (2003).
4. W. W. MacKay, "An Interactive Beam Line Simulator Module for RHIC", PAC97, 2541 (1997).
5. S. R. Mane, Yu. M. Shatunov, and K. Yokoya, *Rep. Prog. Phys.*, **68**, 1997 (2005).
6. M. J. Syphers, "Field Quality Issues for RHIC Helical Dipole Magnets", (1995). URL: `http://www.rhichome.bnl.gov/RHIC/Spin/spinnotes/sn015.pdf`

Fighting the Residual Polarization Loss in the AGS

H. Huang, L.A. Ahrens, M. Bai, K. Brown, C. Gardner, J.W. Glenn, F. Lin, A.U. Luccio, W.W. MacKay, T. Roser, S. Tepikian, N. Tsoupas, K. Yip, A. Zelenski and K. Zeno

Brookhaven National Laboratory, Upton, NY 11973, USA

Abstract. A dual partial snake scheme has been used for AGS polarized proton operation for several years. It has provided polarized proton beams with 1.5×10^{11} protons per bunch and 65% polarization for the RHIC spin program. There is still residual polarization loss due to both snake resonances and horizontal resonances as shown in the data. Several schemes were tested or proposed in the AGS to mitigate the loss, such as putting horizontal tune into the spin tune gap, injection into a accelerating bucket, and tune jump across the horizontal resonances. This paper presents the experiment and simulation results and analyses.

Keywords: polarized protons, acceleration, spin
PACS: 41.75.Ak, 07.77.+p, 29.27.Bd, 29.27.Hj

INTRODUCTION

AGS has been running as RHIC polarized proton injector with dual partial snakes [1] since 2006. It has delivered 65% polarization with 1.5×10^{11}/bunch intensity with 82% as input polarization [2]. The focus in past two years was to overcome the residual polarization losses.

Imperfection and vertical intrinsic resonances have been taken care of by the dual partial snakes, except a few vertical intrinsic resonances near injection as vertical tunes are outside the spin tune gap in the early part of the energy ramp. Besides the spin mismatch at injection and extraction, and the partial snake resonances associated with vertical betatron motion, there are additional causes for polarization loss. In the presence of a partial snake, the stable spin direction is not purely vertical. For the horizontal component of polarization, the vertical magnetic field can drive spin resonances. Therefore, the perturbing fields that rotate the spin away from the stable spin direction have vertical as well as horizontal components. Particles undergoing horizontal betatron oscillations encounter vertical field deviations at the horizontal oscillation frequency. As a result, resonances are driven by the horizontal betatron oscillations, and will occur whenever the spin tune satisfies $G\gamma = k \pm \nu_x$, where $G = (g-2)/2 = 1.7928$ is the gyromagnetic anomaly of the proton, and γ is the Lorentz factor, k is integer and ν_x is the horizontal tune [3].

A series spin tracking study was performed to find out the residual polarization loss. The remaining polarization losses seen in the spin tracking can be divided into three categories [4]:

1. the loss due to horizontal intrinsic resonances along the ramp, which accounts for

CP1149, *Spin Physics, 18th International Spin Physics Symposium*
edited by D. G. Crabb, D. B. Day, S. Liuti, X. Zheng, M. Poelker, and Y. Prok

about 10% loss;
2. the loss associated with snake resonances near the strongest resonance $36+\nu_y$;
3. the polarization loss resulted from vertical intrinsic resonances near injection due to slow ramp rate and large lattice distortion.

POLARIZATION LOSS NEAR INJECTION

Normally, the vertical intrinsic resonances near injection are very weak. However, since the superperiod $P = 1$ with insertion of the two partial snakes and large lattice distortion associated with them, the effect of these resonances is not negligible especially when the acceleration rate is very slow at the beginning of the energy ramp.

To speed up the acceleration, the idea of injecting proton beam into the accelerating bucket to maintain a high acceleration rate in the early part is proposed. The tuning of injection is harder for this scheme since the RF bucket is smaller and synchronization between the AGS Booster and AGS is critical. However, in the new injection scheme, we encountered the transverse emittance growth problem.

The large lattice distortion causes injection mismatch which could be a source of emittance growth. Since the beam is relatively bright, the space charge related emittance growth is also a concern. The emittance growth issue has to be solved before we can use the injection-on-the-fly.

The vertical tune outside the spin tune gap is another source of polarization loss. Simulation shows that the vertical tune can be raised inside the spin tune gap near injection (from 8.88 to >8.95) with price of modest increased beta function near injection (shown in Fig. 1). This optics will be tested in the coming run.

MEASURES TO REDUCE EMITTANCE GROWTH

The emittances measured in the AGS are larger than the ones from linac. If we can find ways to reduce the emittance growth, it is going to affect polarization positively. Linac beam transfer line upgrade is underway. This will reduce emittance growth in the linac. The stripping foil in the linac to booster transfer line will be thinner and smaller. In addition, the beta function at the foil location will be reduced, too. All these will reduce the emittance growth there. A beam study will be carried out in the Booster to AGS transfer line (BtA) to verify model of BtA line with beam and achieve a better match at AGS injection. This would also reduce AGS emittances. The AGS horizontal emittances is expected to be reduced by about 30% after all these efforts. The observed emittance dependence on intensity hints some space charge or instability related to brightness of the beam. Change h=12 to h=6 for AGS will reduce the peak current and consequently reduce the emittance growth.

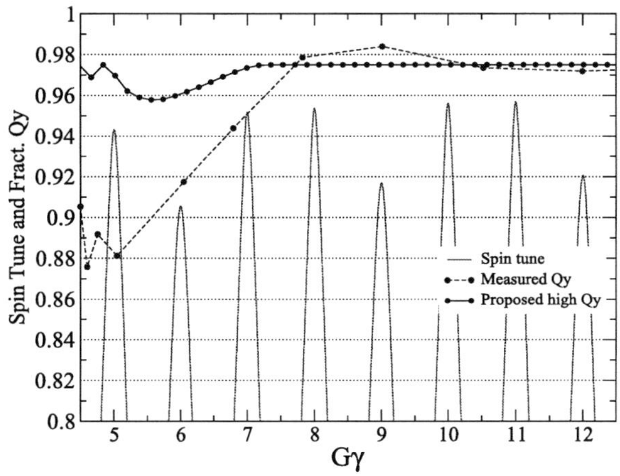

FIGURE 1. The solid curve is spin tune. The dashed line and dots are the measured vertical tune through the energy ramp. The solid line and dots are the proposed new vertical tune along early part of the ramp. It will merge to the blue trace at higher energies.

TUNE JUMP FOR HORIZONTAL RESONANCES

To overcome the horizontal depolarizing resonances, two quadrupoles which can ramp (or jump) very fast (ramp horizontal tune by 0.04 in $100\mu s$) are going to be installed in the AGS. The idea is to cross the resonances effectively faster by jumping the horizontal tune (see Fig.2). The timing of the jump is critical: it has to start slightly before the resonance. These resonances can be divided into two categories: one with $N+\nu_x$; one with $N-\nu_x$. The tune has to be jumped up for the $N-\nu_x$ resonances and to be jumped down for the $N+\nu_x$ resonances. The change of ν_x is 0.04 in 100 μs. This increases the crossing speed by about 4 times. For practical operation, we need to maintain the horizontal tune and radius constant throughout the ramp. Given the existence of the polarization profile due to horizontal resonances, the polarization measured at the center (peak position for polarization and beam intensity) and as a whole beam will be different. As a example, benefit on polarization transfer efficiency for a beam with horizontal 15π is given in Table 1:

TABLE 1. The benefit of horizontal tune jumps

Crossing speed	$P_f/P_{i(peak)}$	$P_f/P_{i(whole)}$
regular speed	0.912	0.832
Double speed	0.955	0.912
4X speed	0.977	0.955

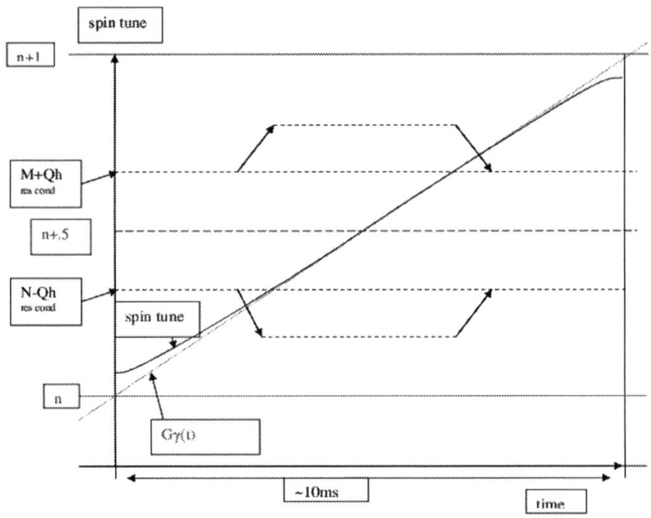

FIGURE 2. The sketch of horizontal tune jump.

CONCLUSIONS

Acceleration of 1.5×10^{11} protons to 23 GeV with 65% polarization was achieved in the AGS using dual helical partial snakes. The dual partial snake scheme avoids depolarization from both imperfection and intrinsic spin resonances in medium energy accelerators and also maintains good matching to the vertical polarization in the injection and extraction beam lines. The residual polarization losses are caused by horizontal resonances, snake resonances, and vertical intrinsic resonances near injection. For the future operation, emittance growth will be reduced and the polarization loss will mitigated accordingly. A new lattice with vertical tune in the spin tune gap near injection will be tested. In addition, a pair of tune jump quads will be used for overcoming the horizontal depolarizing resonances.

ACKNOWLEDGMENTS

This work was performed under contract No. DE-AC02-98CH1-886 with the auspices of the DoE of United States, and with support of RIKEN(Japan)

REFERENCES

1. T. Roser, *et al.*, Proc. of SPIN2004 (2005), p. 687.
2. H. Huang, *et al.*, Phys. Rev. Lett. **99**, 184501 (2007).
3. F. Lin, *et al*, PR. STAB **10**, 044001(2007).
4. F. Lin, *et al*, these proceedings.

Do Unpolarized Electrons Affect the Polarization of a Stored Beam?[1]

Frank Rathmann

Institut für Kernphysik and Jülich Center for Hadron Physics, Forschungszentrum Jülich, Germany

Abstract. We present a short overview of the PAX physics case for polarized antiprotons. In order to progress towards a stored polarized antiproton beam, it is crucial to understand the interaction of polarized protons with unpolarized electrons. Therefore investigations that address in particular the contributions of electrons to the polarization buildup of a stored proton beam are presented here in more detail. The measurement of the depolarizing $\vec{p}e$ cross section settled a long–standing controversy about the role of electrons in the polarization buildup of a stored beam by spin–filtering. Instead of studying the buildup of polarization in an initially unpolarized beam, here the inverse situation was investigated by observation of the depolarization of an initially polarized beam. For the first time, electrons in the electron cooler have been used as a target to study their depolarizing effect on a 49.3 MeV proton beam orbiting in COSY. The foreseen spin–filtering experiments at COSY–Jülich and at the AD of CERN are briefly discussed as well.

Keywords: Polarization, antiprotons, *ep* interaction, polarized targets
PACS: 24.70.+s, 25.43.+t, 13.88.+e

THE PAX EXPERIMENT

The QCD physics potential of experiments with high energy polarized antiprotons is enormous, but up to now high–luminosity experiments have been impossible. The situation would change dramatically with the production of stored polarized antiproton beams, and the subsequent realization of a double–polarized high–luminosity antiproton–proton collider. The list of fundamental physics issues to be addressed with such a collider includes the measurement of transversity, the quark transverse polarization inside a transversely polarized proton, which constitutes the last missing leading twist piece of the QCD description of the partonic structure of the nucleon. The transversity can be directly accessed only via double–polarized antiproton–proton Drell–Yan production. It should be noted, that without a measurement of the transversity, the spin tomography of the proton will remain incomplete. Other items of interest are the measurement of the phases of the timelike form factors of the proton, and double–polarized hard antiproton–proton scattering.

The PAX collaboration (**P**olarized **A**ntiproton e**X**periments) has formulated its ambitious physics program [1] and a Technical Proposal [2] has recently been submitted for the new Facility for Antiproton and Ion Research (FAIR) to be built at GSI in Darmstadt, Germany. The uniqueness and the strong scientific merits that would become available with the advent of stored beams of polarized antiprotons have been well received [3],

[1] for the ANKE and PAX collaborations

CP1149, *Spin Physics, 18th International Spin Physics Symposium*
edited by D. G. Crabb, D. B. Day, S. Liuti, X. Zheng, M. Poelker, and Y. Prok

and there is now an urgency to convincingly demonstrate experimentally that a high degree of antiproton beam polarization can be reached. The suggested collider aims at luminosities in excess of 10^{31} cm^{-2}s^{-1}. An integral part of such a machine is a dedicated large–acceptance Antiproton Polarizer Ring (APR) [4], for which a basic design has been developed [5].

TOWARDS POLARIZED ANTIPROTONS

For more than two decades, physicists have tried to produce beams of polarized antiprotons [6], generally without success. Conventional methods like atomic beam sources, appropriate for the production of polarized protons and heavy ions cannot be applied, since antiprotons annihilate with matter. Polarized antiprotons have been produced from the decay in flight of $\bar{\Lambda}$ hyperons at Fermilab. The intensities achieved with antiproton polarizations $P > 0.35$ never exceeded $1.5 \cdot 10^5$ s^{-1} [7]. Scattering of antiprotons off a liquid hydrogen target could yield polarizations of $P \approx 0.2$, with beam intensities of up to $2 \cdot 10^3$ s^{-1} [8]. Small angle $\bar{p}C$ scattering has been studied at LEAR, but the obsertved antiproton polarizations are negligibly small [9, 10]. Unfortunately, all the above mentioned approaches do not allow efficient accumulation in a storage ring, which would greatly enhance the luminosity. Spin splitting using the Stern–Gerlach separation of the given magnetic substates in a stored antiproton beam was proposed in 1985 [11]. Although the theoretical understanding has much improved since then [12], spin splitting using a stored beam has yet to be observed experimentally. In contrast to that, a convincing proof of the spin–filtering principle was provided by the FILTEX experiment at the TSR–ring in Heidelberg [13].

DO UNPOLARIZED ELECTRONS AFFECT THE POLARIZATION OF A STORED BEAM?

The original idea of PAX to use polarized electrons to produce a polarized beam of antiprotons [4] has triggered further theoretical work on the subject, which led to a new suggestion by a group from Mainz [14, 15] to use co–moving electrons (or positrons) at slightly different velocities than the orbiting protons (or antiprotons) as a means to polarize the stored beam. The cross section for $e\vec{p}$ spin–flip predicted by the Mainz group in a *numerical* calculation is as large as about $2 \cdot 10^{13}$ barn, if the relative velocities between proton (antiproton) and electron (positron) are adjusted to $v/c \approx 0.002$. At the same time, *analytical* predictions for the same quantity by a group from Novosibirsk [16] range well below a mbarn. Thus prior to the experiment, the two theoretical estimates differed by about 16 orders of magnitude. It should be also emphasized, that the practical use of the $\vec{e}p$ or $\vec{e}^+\bar{p}$ processes to polarize anything is excluded if the spin–flip cross sections are smaller than about 10^7 barn.

In order to provide an experimental answer for this puzzle, the ANKE and PAX collaborations joined forces at COSY and mounted an experiment, where for the first time, electrons in the electron cooler have been used as a target, and in which the effect of electrons on the polarization of a 49.3 MeV proton beam orbiting in COSY was

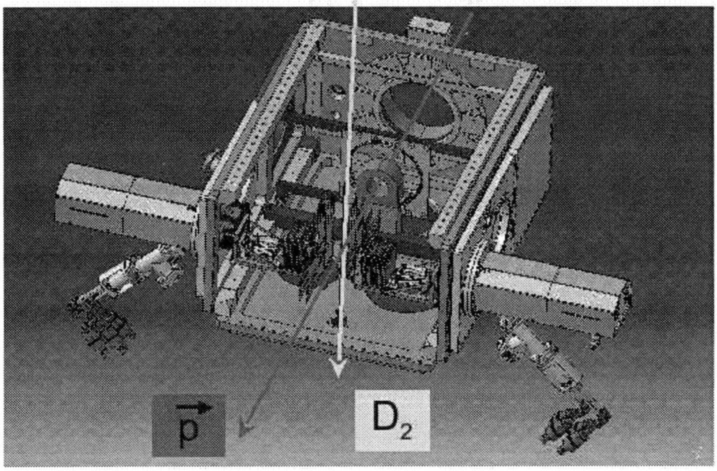

FIGURE 1. Using events from elastic $\vec{p}d$ scattering, two Silicon detector telescopes mounted near the deuterium beam (D_2) of the ANKE cluster jet target measure the change of the proton ($\vec{p}$) beam polarization induced by the depolarizing $e\vec{p}$ spin–flips in the COSY electron cooler.

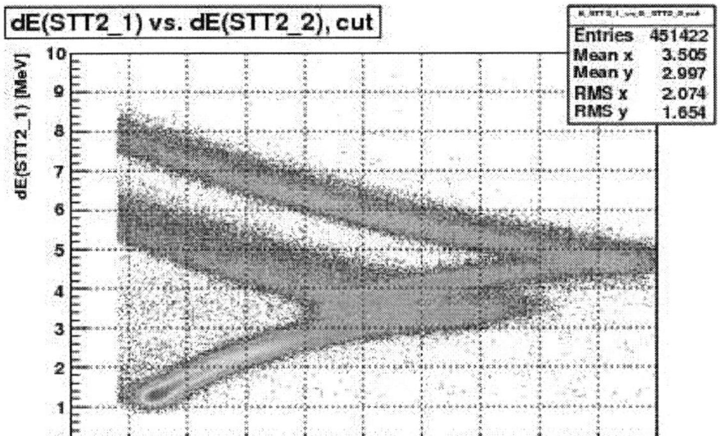

FIGURE 2. Identification of *pd* elastic scattering in the detector system.

determined. Instead of studying the buildup of polarization in an initially unpolarized beam, here the inverse situation was investigated by observation of the depolarization of an initially polarized beam. The proton beam polarization has been measured by making use of the analyzing power of $\vec{p}d$ elastic scattering on a deuterium cluster jet target. The experimental setup of the polarimeter consisting of two telescopes, each of them containing two 300 μm thick layers of Silicon, is depicted in Fig. 1.

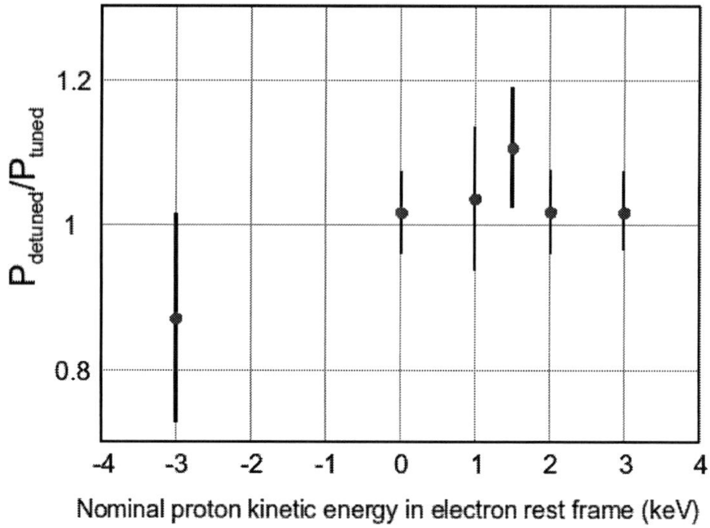

FIGURE 3. Ratio of the measured polarizations $\frac{P_{\text{detuned}}}{P_{\text{tuned}}}$ as function of the proton kinetic energy in the electron rest frame.

The energy deposit in the first detector layer is plotted as a function of the energy deposit in the second layer in Fig. 2. The upper band corresponds to deuterons and the lower one to protons from *pd* elastic scattering.

The depolarizing cross section is determined from the ratio of the measured beam polarizations P_{detuned} and P_{tuned} (Fig. 3). These polarizations correspond to well–defined changes of the electron velocity with respect to the protons, which were achieved by detuning the accelerating voltage in the electron cooler by a specific amount.

The depolarizing cross section is plotted in Fig. 4 as a function of the magnitude of the electron velocity in the proton rest frame. The measurement shows that the predictions by the Mainz group were too large by at least six orders of magnitude. It should be noted that very recently, the Mainz group has submitted two errata to their original publications stating that because of numerical problems in the calculation their theoretical estimates were too large by about 15 orders of magnitude.

SPIN–FILTERING EXPERIMENTS AT COSY AND AD

The depolarization study constitutes the first step of investigations at COSY, shedding light on the *ep* spin–flip cross sections when the target electrons are unpolarized. The experimental finding rules out the practical use of polarized leptons to polarize a beam of antiprotons with present–day technologies. This leaves us with the only proven method to polarize a stored beam in situ, namely *spin filtering* by the strong interaction. At present, we are lacking a complete quantitative understanding of all underlying processes, therefore the PAX collaboration is aiming at high–precision polarization

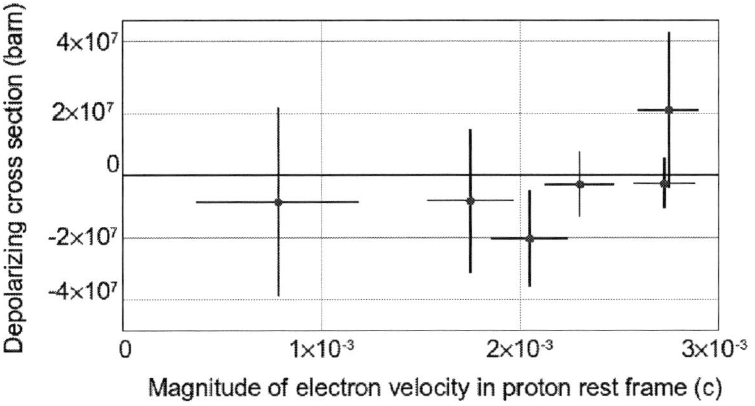

FIGURE 4. The depolarizing cross section is plotted as a function of the magnitude of the electron velocity in the proton rest frame, indicating an upper limit of a few 10^7 b. Prior to the experiment, the two theoretical estimates of this cross section differed by about 16 orders of magnitude at a relative velocity of $v/c = 2 \cdot 10^{-3}$ [15, 16]. The experiment ruled out the higher estimate.

buildup studies with transverse and longitudinal polarization using stored protons in COSY [17]. The buildup process itself can be studied in detail, because in this situation the spin–dependence of the *pp* interaction is completely known. The polarized internal target required for these investigations was previously used at the HERMES experiment at HERA/DESY. Including the target polarimeter, it has meanwhile been relocated to Jülich [18], where it is presently set up to be installed together with a large–acceptance detector system for the determination of target and beam polarizations in a dedicated low–β section at COSY.

In contrast to the *pp* system, the experimental basis for predicting the polarization buildup in a stored antiproton beam by spin filtering is practically non–existent. Therefore, it is of high priority to perform, subsequently to the COSY experiments, a series of dedicated spin–filtering experiments using stored antiprotons. The AD–ring at CERN is a unique facility at which stored antiprotons in the appropriate energy range are available with characteristics that meet the requirements for the first–ever antiproton polarization buildup studies. At the same time, the equipment required for the dedicated spin–filtering experiments at the AD, i.e. the polarized internal target and the new low–β section, an efficient polarimeter to determine target and beam polarizations, and a Siberian snake to maintain the longitudinal beam polarization, must be extensively commissioned and tested at COSY, prior to the experimental investigations at CERN. Already in 2005, the PAX collaboration suggested in a Letter–of–Intent [19] to the SPS committee of CERN to study the polarization buildup via spin filtering of stored antiprotons by multiple passage through a polarized internal hydrogen gas target, because only through these investigations, one can obtain direct access to the spin dependence of the total $\bar{p}p$ cross sections. Apart from the obvious interest for the general theory of $\bar{p}p$ interactions, the knowledge of these cross sections is necessary for the interpretation of unexpected features of the $\bar{p}p$, and other antibaryon–baryon pairs, contained in final states in J/Ψ and B–decays. Once these experiments have provided an experimental data base, the design of a dedicated APR can be targeted, as the next major milestone.

REFERENCES

1. F. Rathmann [PAX Collaboration], *Proceedings of the 17th International Spin Physics Symposium*, Kyoto, Japan, 2006, Eds. Kenichi Imai, Tetsuya Murakami, Naohito Saito, Kiyoshi Tanida, AIP Conf. Proc. **915**, 924 (2007).
2. Technical Proposal for *Antiproton–Proton Scattering Experiments with Polarization*, PAX Collaboration, spokespersons: P. Lenisa (Ferrara University, Italy) and F. Rathmann (Forschungszentrum Jülich, Germany), available from `http://arXiv:hep-ex/0505054` (2005). An update of this proposal can be found at the PAX website `http://www.fz-juelich.de/ikp/pax`.
3. The reports from the various committees can be found at the PAX collaboration website at `http://www.fz-juelich.de/ikp/pax`.
4. F. Rathmann *et al.*, Phys. Rev. Lett. **94**, 014801 (2005).
5. A. Garishvili *et al.*, *Proc. of the EPAC 2006*, Edinbough, Scotland, 2006, Eds. C. Biscari, H. Owen, Ch. Petit-Jean-Genaz, J. Poole, and J. Thomason, ISBN 92-9083-278-9 and ISBN 978-92-9083-278-2. Available from `http://accelconf.web.cern.ch/AccelConf/e06/` MOPCH083, p. 223.
6. *Proc. of the Workshop on Polarized Antiprotons*, Bodega Bay, CA, 1985, Eds. A. D. Krisch, A. M. T. Lin, and O. Chamberlain, AIP Conf. Proc. **145** (AIP, New York, 1986).
7. D.P. Grosnick *et al.*, Nucl. Instrum. Methods **A 290**, 269 (1990).
8. H. Spinka *et al.*, *Proc. of the 8th Int. Symp. on Polarization Phenomena in Nuclear Physics*, Bloomington, Indiana, 1994, Eds. E.J. Stephenson and S.E. Vigdor, AIP Conf. Proc. **339** (AIP, Woodbury, NY, 1995), p. 713.
9. A. Martin *et al.*, Nucl. Phys. A 487, 563 (1988).
10. R. Birsa *et al.*, Phys. Lett. B **155**, 437 (1985).
11. T. O. Niinikoski and R. Rossmanith, Nucl. Instrum. Methods **A 255**, 460 (1987).
12. P. Cameron *et al.*, *Proc. of the 15th Int. Spin Physics Symp.*, Upton, New York, 2002, Eds. Y. I. Makdisi, A. U. Luccio, and W. W. MacKay, AIP Conf. Proc. **675** (AIP, Melville, NY, 2003), p. 781.
13. F. Rathmann *et al.*, Phys. Rev. Lett. **71**, 1379 (1993).
14. H. Arenhövel, Eur. Phys. J. A, **34**, 303 (2007).
15. T. Walcher *et al.*, Eur. Phys. J. A, **34**, 447 (2007).
16. I. Milstein, S.G. Salnikov, and V.M. Strakhovenko, Nucl. Instrum. Methods **B 266**, 3453 (2008).
17. Letter–of–Intent for Spin Filtering Studies at COSY, PAX Collaboration, spokespersons: P. Lenisa (Ferrara University, Italy) and F. Rathmann (Forschungszentrum Jülich, Germany), available from the PAX website `http://www.fz-juelich.de/ikp/pax/`.
18. A. Nass *et al.* [PAX Collaboration], *Proceedings of the 17th International Spin Physics Symposium*, Kyoto, Japan, 2006, Eds. Kenichi Imai, Tetsuya Murakami, Naohito Saito, Kiyoshi Tanida, AIP Conf. Proc. **915**, 1002 (2007).
19. Letter–of–Intent for *Measurement of the Spin–Dependence of the $\bar{p}p$ Interaction at the AD–Ring*, PAX Collaboration, spokespersons: P. Lenisa (Ferrara University, Italy) and F. Rathmann (Forschungszentrum Jülich, Germany), available from `http://arXiv:hep-ex/0512021`, (2005).

Study By Spin Tracking of A Storage Ring For Deuteron Electric Dipole Moment

F.Lin, N.D.Malitsky, A.U.Luccio, W.M.Morse, Y.K.Semertzidis*, C.J.G.Onderwater† and Y.F.Orlov**

*Brookhaven National Laboratory, Upton, NY
†University of Groningen, NL-9747AA Groningen, the Netherlands
**Cornell University, Ithaca, NY

Abstract. Spin tracking of polarized deuterons for a proposed experiment to measure a possible Electric Dipole Moment (EDM) of the deuteron was done by using the codes UAL and SPINK. In the experiment the direction of spin polarization will be frozen using crossed electric and magnetic fields. Systematics, in particular the effects of non-linearities of the lattice on a beam with finite emittance and energy spread, have been extensively simulated and the effect of sextuple corrections to increase the spin coherence time has been studied.

Keywords: deuteron Electric Dipole Moment, spin polarization, spin coherence time
PACS: M 29-27.Hj

INTRODUCTION

The discovery of an intrinsic electric dipole moment (EDM) of the deuteron will directly indicate violation of time and parity symmetry. A new proposed approach to EDM studies is to employ a polarized deuteron beam in a storage ring, where the external magnetic and electric fields will act on the EDM signal and modify the observed spin precession.

This novel technique appears to be capable of reaching a sensitivity of 10^{29} e.cm [1], providing a powerful probe in new physics. If CPT symmetry is obeyed, the violation of time reversal would imply CP violation, which is necessary to explain the surplus of matter over anti-matter that we observe in the present-day universe. EDM search with the limit of sensitivity down by one to two orders of magnitude with respect to other proposed methods will allow super-symmetry theories to predict a missing source of CP-violation.

STORAGE RING EXPERIMENT

In a storage ring the spin precession due to the interaction of the magnetic dipole moment with magnetic and electric fields is described, in the particle rest frame, by the Thomas-BMT equation [2] :

$$\frac{d\vec{S}}{dt} = \frac{e}{m}\vec{S} \times [(a+\frac{1}{\gamma})\vec{B} - a\frac{\gamma}{\gamma+1}(\vec{\beta}\cdot\vec{B})\vec{\beta} - (\frac{g}{2} - \frac{\gamma}{\gamma+1})\frac{\vec{\beta}\times\vec{E}}{c}], \tag{1}$$

CP1149, *Spin Physics, 18th International Spin Physics Symposium*
edited by D. G. Crabb, D. B. Day, S. Liuti, X. Zheng, M. Poelker, and Y. Prok

where $a = (g-2)/2$ is the anomalous magnetic moment, $\vec{B}$ and $\vec{E}$ stand for the magnetic field and electric field, respectively. The spin precession due to the electric dipole moment is given by

$$\frac{d\vec{S}}{dt} = d[\vec{S} \times (c\vec{\beta} \times \vec{B} + \vec{E})]. \tag{2}$$

In a conventional storage ring with vertical magnetic field $\vec{\beta} \cdot \vec{B} = 0$, the precession frequency of spin direction is $\vec{\omega} = \vec{\omega}_a + \vec{\omega_{edm}}$. $\vec{\omega}_a = \frac{e}{m} a\vec{B}$ is proportional to the magnetic momentum, and $\vec{\omega_{edm}} = d\frac{c}{\hbar}\vec{\beta} \times \vec{B}$ would produce an EDM signal that will increase the vertical polarization. The change in the vertical polarization with time is

$$\Delta P_v(t) = P\frac{\omega_{edm}}{\Omega}\sin(\Omega t + \theta_0), \quad \Omega = \sqrt{\omega_{edm}^2 + \omega_a^2}, \tag{3}$$

where P is the polarization of the beam. Thus minimizing ω_a is the way to maximize the contribution to ΔP_v due to the EDM signal. This can be realized by adding a radial electric field of magnitude $E_r = (aB_v c\beta\gamma^2)/(1 - a\beta^2\gamma^2)$ to the vertical dipole magnetic field of the bending magnet (BE section) to cancel or minimize the spin precession frequency ω_a (frozen spin method, or FSM) [3, 4].

Ideally, the FSM can produce $\omega_a = 0$. However in a real beam with finite emittance and momentum spread, the horizontal and vertical betatron oscillation and off-momentum spread will modify ω_a. Any linear terms of these effects will average to zero over time, but not the nonlinear, in particular quadratic, terms. Our goal of prolonging the spin coherence time as much as possible requires to cancel the second order perturbation in ω_a. Since sextupoles produce quadratic fields, an arrangement of sextupoles in the ring lattice can conceptually cure quadratic effect.

SIMULATION

Parameters of an EDM deuteron storage ring (dEDM) based on FSM and with provisions to cancel second order perturbations of $\Delta\omega_a$ are shown in Table 1. This ring has 16 BE sections, 2 straight sections, 36 quadrupoles and three sets of sextupoles. Since the correction sextupoles are most effective at large beam transverse displacement, they are accordingly located along the lattice where the beta Twiss functions have large values and dispersion is large.

Numerical simulations by spin tracking to investigate spin coherence time were carried out on a multi-particle beam with Gaussian distribution in phase space. In order to separately study the effects on spin coherence time due to horizontal, vertical oscillation and off-momentum, spin trackings were done on only one dimension of motion. Subsequently, the spin trackings were done by combining multi dimensions. The code SPINK treats the spin as a 3-dimensional real vector of norm 1. The system of coordinate is x radial, y vertical, and z longitudinal. In all cases, trackings start with the initial spin vectors of particles pointing to the direction z, of longitudinally forward as required by the experiment.

To study the quadratic perturbation of the spin coherence time, a beam is used in the simulation with 95% normalized emittance in horizontal and vertical plane of $3\mu m$ and

TABLE 1. Deuteron EDM ring parameters

Deuteron Momentum	1.0 GeV/c	Rigidity (B-E/β)R	3.336 Tm
Magnetic field Bv	0.482 T	Radial electric field E0	12.0 MV/m
Length of BE section	3.3 m	BE section radius R	8.406 m
Length of orbit L	85.408 m	Horizontal tune	4.477
Vertical tune	3.469	$\beta_{x,max}$	12.5 m
$\beta_{y,max}$	16.0 m	Dispersion maximum	2.92 m
Momentum compaction factor α	0.149	Focu. quads gradient in bending section, l=0.15m	7.564 T/m
Defocu. quads gradient in bending section, l=0.15m	-6.593 T/m	quads gradient in straight section, l=0.375	12.079 T/m

$5\mu m$, respectively, and a full momentum spread $\Delta p/p$ is 10^{-3}. To limit computer run time, each tracking was performed in about 1.5 seconds, corresponding to 1 million turns in the ring.

Figure 1 and Fig.2 show the spin motion as only a function of time considering transverse motion: separated and combined horizontal and vertical motion. The three components of the spin have been averaged over all particles. The quantity $P = \sqrt{<Sx>^2 + <Sy>^2 + <Sz>^2}$ vs. time is used to measure the spin coherence time. The figures show how the action of the sextupoles in the lattice produce longer spin coherence time.

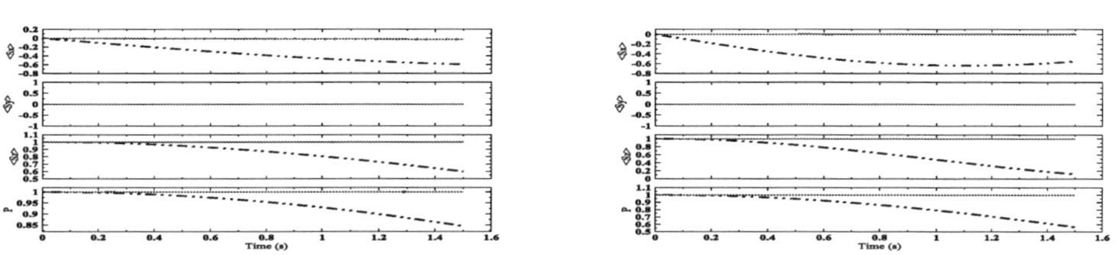

FIGURE 1. Spin motion as a function of time. The back dash lines are for no sextupoles. Left: Horizontal betatron motion only. The red solid lines for one set of sextupoles locating at large β_x positions. Right: Vertical betatron motion only. The red solid lines for one set of sextupoles locating at large β_y positions.

Since the horizontal and vertical motion are mostly separated, the combination of these two different sets of sextupoles can well correct the quadratic perturbation due to the transverse motion as shown in the Fig.2 (left). The polarization can still maintain a value of 92% after 1.5s with an initial given 100%. Not only the transverse motion, but also the distributed dispersion can introduce a perturbation in spin coherence time due to the momentum spread because of the coupling between the horizontal and longitudinal motion. As shown in Fig.2 (right), the spin coherence time can be promoted from 92.6% to 98.4% with an additional set of sextupoles locating at large dispersion positions, where those sextupoles can be fully utilized for this purpose. The coupling of horizontal

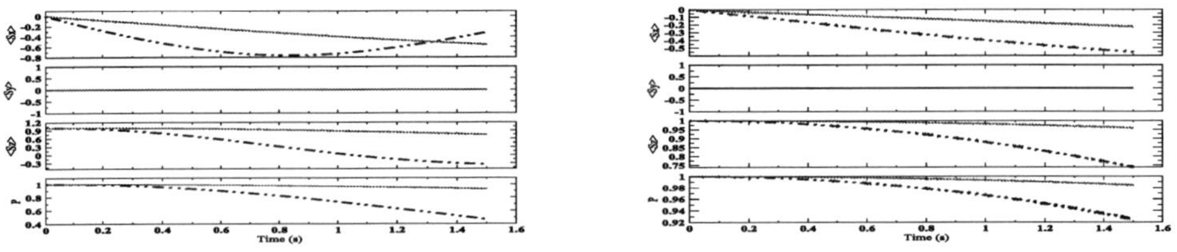

FIGURE 2. Spin motion as a function of time with both horizontal and vertical betatron motion. The back dash lines are for no sextupoles. Left: The red solid lines for two sets of sextupoles locating at large β_x and β_y positions respectively. Right: The red solid lines for three sets of sextupoles locating at large β_x, β_y and dispersion positions respectively.

and longitudinal motion complicates the correction of the quadratic perturbation in spin precession frequency. One solution to simplify this problem is to separate the minimum horizontal beta function and dispersion function. This leads to the least effect to the horizontal motion when one set of sextupoles are located at large dispersion for the correction of longitudinal momentum spread. In reality, the dispersion is almost proportional to the horizontal beta function in a dipole field. The only possible locations are in the straight section. This mechanism is under development by revising the strength of quadrupoles in the ring lattice.

CONCLUSION

The frozen spin method (FSM) has been employed to prolong the spin coherence time in a proposed experiment to measure a deuteron Electric Dipole Moment (dEDM). In this paper, several preliminary spin trackings of polarized deuterons are carried out to investigate the spin coherence time using a new designed dEDM ring lattice, which includes a radial electric field inside each dipole magnet and distributed sextupole magnets along the ring. Those sextupoles contribute to the cancellation of quadratic perturbation in $\Delta\omega_a$ to maintain the spin coherence time as shown in these simulation results.

REFERENCES

1. "AGS Proposal: Search for a permanent electric dipole moment of the deuteron nucleus at the 10^{29} e.cm level." Spokesperson Y.K.Semertzidis (2008)
2. L.H.Thomas, Phil. Mag. 3, 1 (1927) ; V.Bargmann, L.Michel, V.L.Telegdi, Phys. Rev. Lett. 2, 435 (1956).
3. F.J.M. Farley et al., Phys. Rev. Lett. 93, 052001 (2004)
4. D.F. Nelson, A.A. Schupp, R.W. Pidd, and H.R. Crane, Phys. Rev. Lett. 2, 492 (1959)
5. Y.F. Orlov, "Principal Scheme of a Deuteron EDM Ring with a Long Spin Coherence Time", Muon EDM Note No.61 (2004)

Spin Filtering Studies at COSY and AD

Alexander Nass for the PAX-Collaboration

Phys. Institut der Universität Erlangen Nürnberg, E.-Rommel-Str. 1, 91058 Erlangen, Germany

Abstract. The high physics potential of experiments with stored high-energy polarized antiprotons led to the proposal of PAX (Polarized Antiproton eXperiment) [1] for the High Energy Storage Ring (HESR) of the FAIR at GSI (Darmstadt/Germany). It is proposed to polarize a stored antiproton beam by means of spin filtering with a polarized H (D) gas target. The feasibility of spin filtering has been demonstrated in the FILTEX experiment. The current interpretation foresees a self-cancellation of the electron contribution to the filtering process and only the hadronic contribution is effective. Several experimental studies with protons (at COSY/Jülich) as well as antiprotons (at AD/CERN) will be carried out to test the principle and measure $\bar{p}\vec{p}$ and $\bar{p}\vec{d}$ cross sections. A polarized internal gas target (PIT) with surrounding Silicon detectors immersed into a low-β section has to be set up.

Keywords: Polarized Targets, Elastic Proton scattering, Antiproton-induced reactions
PACS: 29.25.Pj, 25.40.Cm, 25.43.+t

PRINCIPLE OF SPIN FILTERING

Several methods to polarize antiprotons were reviewed on workhops held in Bodega Bay, 1985 [2], Daresbury, UK, 2007 [3] and the Heräus Workshop in Bad Honnef, Germany, 2008 [4]. The only successfully tested method is spin filtering. It is based on the effect of selective removal of (anti)protons of a stored beam by a polarized target. The total cross section

$$\sigma_{tot} = \sigma_0 + \sigma_\perp \vec{P} \cdot \vec{Q} + \sigma_\parallel (\vec{P} \cdot \vec{k})(\vec{Q} \cdot \vec{k}) \tag{1}$$

consists of a transverse and longitudinal part, where $\vec{P}$ is the proton beam polarization, $\vec{Q}$ the target polarization and $\vec{k}$ the proton beam direction. For initially equally populated states $\uparrow$ ($m = +\frac{1}{2}$) and $\downarrow$ ($m = -\frac{1}{2}$) the total cross sections for the transverse and longitudinal cases are

$$\sigma^{\perp}_{tot\pm} = \sigma_0 \pm \sigma_\perp \vec{Q} \quad \text{and} \quad \sigma^{\parallel}_{tot\pm} = \sigma_0 \pm (\sigma_\perp + \sigma_\parallel) \cdot \vec{Q}\,, \tag{2}$$

respectively. Therefore an initially unpolarized (anti)proton beam will become polarized.

Experiments at the Test Storage Ring (TSR) at Heidelberg in 1993 (Fig. 1) showed that the spin-filtering technique works in principle [5]. A polarization buildup was observed (Fig. 1, right panel) with a cross section is $\sigma_\perp = 72.5 \pm 5.8$ mb. The current interpretation [6, 7] shows that by considering only elastic pp-scattering, a total cross section of $\sigma_\perp = 85.6$ mb is obtained.

Further experimental tests are necessary to fully understand the polarization buildup process. Depolarization experiments were already carried out at COSY [8]. Spin-filtering experiments [9] will be carried out at COSY with protons, followed by spin-filtering experiments with antiprotons at the Antiproton Decelerator Ring (AD/CERN) [10].

CP1149, *Spin Physics, 18th International Spin Physics Symposium*
edited by D. G. Crabb, D. B. Day, S. Liuti, X. Zheng, M. Poelker, and Y. Prok

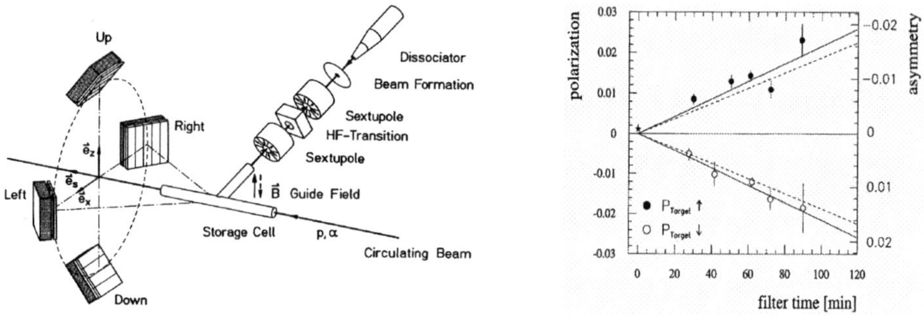

FIGURE 1. The setup of the test experiment at TSR and the results [5].

Required for the spin-filtering experiments is a highly polarized internal gas target with areal densities of up to 10^{14}atoms/cm^2 using a storage cell. A low-β section is necessary to pass the stored (anti)proton beam through the storage cell and to reduce the Coulomb losses in order to achieve long storage times of several hours. It is expected that nuclear polarized deuterium could be equally effective for spin filtering as hydrogen. Therefore the target should run with hydrogen and deuterium with nuclear polarization in variable target holding fields. Because no analyzing power measurement for $\bar{p}\vec{d}$ scattering exists at this energy range, the target gas has to be exchanged with hydrogen in order to measure the antiproton beam polarization. For longitudinal spin filtering a Siberian snake has to be implemented in order to preserve the longitudinal polarization of the beam.

EXPERIMENTAL SETUP

The setup for the spin-filtering experiments (Fig. 2) consists of an Atomic Beam Source (ABS) to produce polarized target gas, a target chamber with storage cell and a detector system to detect forward and recoil (anti)protons. A so-called Breit-Rabi Polarimeter (BRP) is used to measure the polarization of the target gas. A low-β section consisting of 4(6) magnets is necessary for the measurements at COSY(AD). This is designed for the interaction point TP 1 at COSY, and for one of the straight sections at AD. For longitudinal spin filtering at COSY, the solenoids of WASA and the electron cooler, located in the opposite straight section, can be used as a Siberian snake, but at the AD, an additional snake is required.

The former HERMES-ABS was set up in Jülich with a modified vacuum system, mounted on a new support. The cryogenic pumps were replaced by turbo molecular pumps and an oil-free forevacuum system. It was completely recabled to allow a fast assembly and disassembly at COSY and AD. The control system was renewed to allow for a full remote control via computer. The vacuum system with the microwave dissociator is operating well. The achieved pressures show that the new vacuum system is functioning as anticipated. None of the large turbo pumps run into their compression limit. After construction of a new analysis chamber with QMS and a calibrated compression tube,

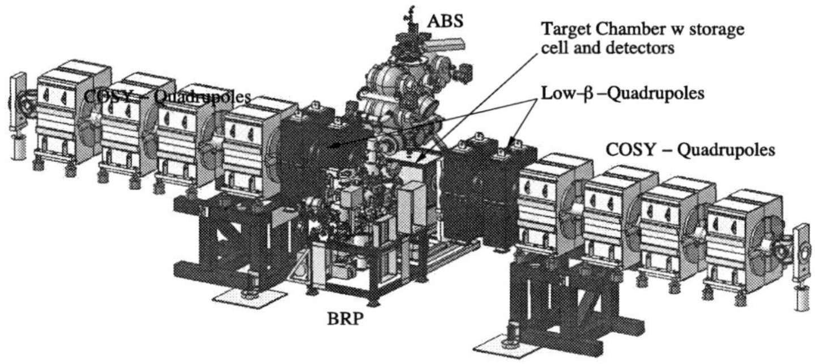

FIGURE 2. Overview over the setup for the spin filtering experiments.

the first intensity measurements were carried out. The measurements showed intensities of up to $6 \cdot 10^{16}$atoms/s for hydrogen in two hyperfine states. The new and unique feature of the present setup is that the ABS will be able to produce nuclear polarized hydrogen **or** deuterium beams in short sequence (5 min) without mechanical changes of components.

Since spin filtering requires areal densities of up to 10^{14}atoms/cm^2, the use of a storage cell is mandatory. The present cell design consists of 5 μm Teflon walls supported by an aluminum frame. Thin walls allow low energy recoil particles to pass through and be detected by the Silicon Tracking Telescopes (STT's). Teflon also suppresses depolarization and recombination of the target gas inside the cell. The cell has to be openable to provide enough space for the beam during injection at AD. The cell will be closed after the beam is decelerated and cooled. Subsequently the target gas is injected. Longitudinal and transverse weak holding field coils, added to the outside of the target chamber, provide the quantization axis for the polarized atoms.

The BRP is necessary to determine the nuclear polarization of deuterium at the AD because of the unknown analyzing power for $\bar{p}\vec{d}$ scattering. The former HERMES-BRP was rebuilt on a new support stucture with modifications due to the new configuration with the ABS. Tracking calculations led to a modified sextupole magnet configuration to adjust for the higher temperature of 300 K of the effusive hydrogen/deuterium beam out of the uncooled storage cell; at HERMES the cell was at 100 K. In addition a new strong field transition "Dual Cavity" was designed in order to induce transitions between hyperfine levels of hydrogen **or** deuterium.

The beam polarization will be measured by using pp (p$\bar{p}$)-elastic scattering. To this aim a detector system consisting of 12 STT's will be implemented around the target cell. The telescopes will detect both the low energy recoil particles (< 8 MeV) as well as the forward scattered particle with a large angular coverage and high resolution. In addition the proton polarization of the target can be measured with the initially unpolarized (anti)proton beam. This allows for the calibration of the BRP.

The vacuum system of the target section comprises three turbo molecular pumps, backed with smaller turbo molecular pumps and a dry forevacuum pump. This will ensure that most of the target gas exiting the storage cell is pumped away in the target

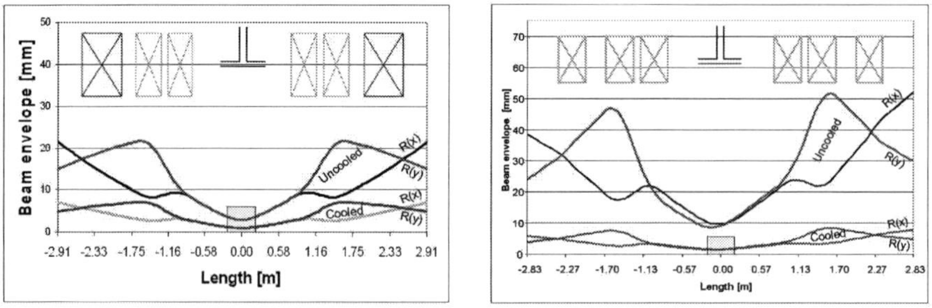

FIGURE 3. The beam size along the low-β-section at COSY (left) and AD (right).

chamber. Flow limiters will be installed between the target chamber and the magnet chambers in order to reduce the gas load.

The low-β-section will consist of 4(6) normal conducting quadrupole magnets. They will be implemented into the COSY (AD) lattice prior to the installation of the target. Calculations show the dependence of the beam envelope along the target section (Fig. 3).

PLANNED MEASUREMENTS

A first set of spin filtering measurements is planned to be carried out at COSY/Jülich with an initially unpolarized proton beam and a nuclear polarized target in a weak holding field. It will determine the polarization buildup effect [9] and comission the experiment for the AD. A second set of measurements is planned at AD/CERN. These measurements will provide data to estimate the achievable polarization in spin filtering with an antiproton beam [10].

REFERENCES

1. *Antiproton-Proton Scattering Experiments with Polarization*, Technical Proposal for the HESR at FAIR, Jülich (2005), e-Print Archive: hep-ex/0505054.
2. A. D. Krisch et al. (Eds.), *Polarized antiprotons*, AIP Conf. Proceedings **145** (1986).
3. D. P. Barber et al. (Eds.), *Polarized Antiproton Beams - How?*, AIP Conf. Proceedings **1008** (2008).
4. *Polarized Antiprotons*, WE-Heräus Seminar, Bad Honnef, Germany (2008), talks available under *http://www.fe.infn.it/heraeus/index.html*
5. F. Rathmann et al., *Phys. Rev. Lett.* **71**, 1379 (1993).
6. A.I. Milstein and V.M. Strakhovenko, *Phys. Rev. E* **72**, 066503 (2005).
7. N. N. Nikolaev and F. Pavlov *Spin Filtering of Stored (Anti)Protons: from FILTEX to COSY to AD to FAIR* (2007), e-Print Archive: hep-ph/0701175
8. F. Rathmann et al., *Do unpolarized electrons affect the polarization of a stored proton beam?*, published in these proceedings (2009).
9. *Spin-Filtering Studies at COSY*, Letter of Intent, Jülich (2006), available at http://www.fz-juelich.de/ikp/pax/.
10. *Measurement of the Spin-Dependence of the* $\bar{p}p$ *Interaction at the AD-Ring*, Letter of Intent, Jülich (2005), available at http://www.fz-juelich.de/ikp/pax/.

Simulation of Spin Interference and Spin Echo

M. Bai[a], A. Chao[b], T. Roser[a]

[a]*Brookhaven National Laboratory, Upton, NY 11973*
[b]*Physics Department, Stanford University, Stanford, CA 94305-4060*

Abstract. Successively jumping across a depolarization resonance twice produces interesting spin dynamics effects. The dynamics involves an interplay among four parameters of depolarization resonance strength ε, the jump size in spin tune, the time duration between the two jumps and the spin tune spread of the beam. By adjusting these parameters, the beam polarization after the second jump exhibits a wealth of effects of constructive interference, destructive interferences and spin echo.
A simulation code is written to study these effects. Subtleties such as control of numerical noise and how to prepare the beam polarization condition at the launch are resolved. The simulation results are compared with analysis [Chao/Courant, PRST-AB, 2007]. Experiments are proposed for AGS and RHIC.

Keywords: spin echo, resonance jump, rf spin rotator
PACS: 29.20.db

INTRODUCTION

The effect of crossing a single isolated depolarizing resonance[1] on the beam polarization is well described by the Froissart-Stora formula[2] as a function of the speed of resonance crossing and the strength of the spin resonance. However, this model only works well if the spin tune linearly ramps from a distance far away from the resonance to the other side, which is also far away. In reality, a spin resonance can be crossed with more complicated patterns. Sometimes, a single resonance can also be crossed multiple times. Lately, Chao developed a matrix formalism by solving the equation of spinor $\vec{\psi}$ motion at the neighborhood of a spin resonance to extend the spin motion beyond the Froissart-Stora formula regime[1].

The application of the matrix formalism to multiple spin resonance crossings revealed very interesting behaviors of the spin motion[1,2,3]. In the example of a spin resonance being jumped twice, the final polarization after the second resonance crossing not only depends on resonance strength and the size of the resonance jump, it also depends on the time between the two resonance crossings. When certain conditions are met, the effect of the two resonance crossings can either destructively cancel each other, a destructive case, or constructively enhance each other, a constructive case [3]. Eq. 1 shows the condition for the two cases of interference[1]:

$$2\pi(n_2 - n_1) = \begin{cases} 2k\pi\dfrac{1}{\Omega} & \text{destructive case} \\ (k\pi + \dfrac{\pi}{2})\dfrac{1}{\Omega} & \text{constructive case} \end{cases} \qquad (1)$$

CP1149, *Spin Physics, 18th International Spin Physics Symposium*
edited by D. G. Crabb, D. B. Day, S. Liuti, X. Zheng, M. Poelker, and Y. Prok

where $\Omega = \sqrt{A^2 + |\varepsilon|^2}$, k is an integer. ε is the spin resonance strength[2]. $A = Q_{res} - Q_s$ is the distance between the spin tune and spin resonance tune. The spin resonance was first jumped from A to –A at turn n_1 and then was jumped from –A to A at n_2 turn.

For a beam of particles with a spin tune spread, the coherent spin motion in the case when the difference of spin phase between particles is small, first begin to oscillate after each spin resonance crossing, but is soon damped when the spin phase difference between different particles grows large enough. If this phase smearing is perfect, the spin motion during the subsequent spin resonance should be independent of the previous resonance crossing. However, analysis[3] shows that with a finite spin tune spread, a short burst of spin coherent motion then peaks up at the $(n_2\text{-}n_1)^{th}$ turn after the second resonance jump if $n_2 - n_1 << \frac{\sqrt{2}\Omega}{2\pi|A|\sigma_{Q_s}}$ [3] where σ_{Q_s} is the rms spin tune spread.

This is the spin echo effect. The peak of the vertical projection of the spin echo signal is $P_y = \frac{2A^2|\varepsilon|^4}{\Omega^5}$, and a maximum spin echo can be achieve by choosing $|A| = \frac{|\varepsilon|}{2}$.

SIMULATION OF SPIN ECHO

The tracking of the spin motion was carried out by multiplying the 2x2 one turn spin transfer matrix for each orbital revolution. In the absence of the spin resonance, the one turn spin transfer matrix is described as $\vec{\psi}_{n+1} = e^{-\frac{i}{2}G\gamma 2\pi}\vec{\psi}_n$ where $\vec{\psi}_{n+1}$ and $\vec{\psi}_n$ are the spinors at the $(n+1)^{th}$ and n^{th} orbital revolution, respectively. The spin resonance was simulated using a localized rf spin rotator[4] which kicks the spin vector by an angle of $\tilde{\phi}_{res}$ around an axis which rotates in the horizontal plane at a frequency of Q_{res} times of orbital revolution frequency. Q_{res} is the tune of the spin resonance, and the spin resonance strength is $\varepsilon = \frac{\tilde{\phi}_{res}}{2\pi}$.

In the simulation, the beam energy was kept constant and the rf spin rotator tune was jumped cross the spin tune, i.e. $Q_s = G\gamma$, for each spin resonance crossing. For all the simulations, the rf spin rotator was first adiabatically ramped from zero to its maximum strength with its tune at a distance A below the spin tune, i.e. $A = Q_s - Q_{res} > 0$. Its tune then jumped to –A in one turn at the n_1^{th} turn and then at the n_2^{th} turn, the rf spin rotator tune was then jumped back to A in a single turn.

The spin echo simulation was done with 1201 particles with Gaussian distributed momentum spread that corresponds to a spin tune spread of 0.000022. The resonance strength amplitude as well as the resonance jump size were chosen to maximize the spin echo signal, i.e. A=0.002, or from 0.599 to 0.601, and a resonance strength of $\varepsilon = 0.002$.There were a total of 40,000 orbital turns between the two resonance jumps.

Fig. 2 shows the simulation result where an echo signal peaked up at 40,000 turns after the second resonance jump and the size of the echo signal is 0.57 as expected.

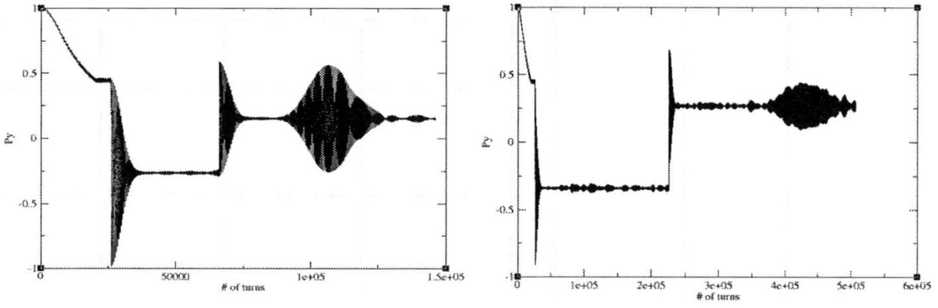

(a) (b)

FIGURE 1. The two plots show the vertical component of average spin of 1201 particles as a function of orbital turns. Both simulations were done in similar conditions, namely, the spin resonance was first ramped from 0 to 0.002 in the first 20000 turns with its tune set below spin tune by 0.001. This tune then jumped to 0.001 above the spin tune at 25000th turn. The spin resonance tune jumped back to 0.001 below the spin tune after the second jump. The left plot corresponds to an interval between the two resonance jumps of 40000 turns. The echo signal then peaked up at 40000 turns after the second spin resonance jump as expected. The right plot shows the tracking when the two resonance jumps were spaced by 300,000 turns. The amplitude of the echo signal was significantly reduced.

However, it was found in simulation that the size of the echo signal can be significantly reduced when there is more number of turns between the two jumps. As shown in Fig. 1(b) when the time between the two jumps went up to 300,000 turns, the spin echo was attenuated. This spin echo smearing effect turned out was due to the high order momentum. The linear dependence of Ω as shown in the second term in Eq. 2 introduces an extra term in the size of the spin echo amplitude.

$$\Omega = \sqrt{\left(A + \sigma_{Q_s}\right)^2 + |\varepsilon|^2} \approx \sqrt{A^2 + |\varepsilon|^2} + \frac{A\sigma_{Q_s}}{\sqrt{A^2 + |\varepsilon|^2}} + \frac{|\varepsilon|^2\left(\sigma_{Q_s}\right)^2}{2(A^2 + |\varepsilon|^2)^{\frac{3}{2}}} \qquad (2)$$

Possible Experiment of Observing Spin Echo

Since spin echo normally is expected to be short lived, one of the challenges in experimentally observing spin echo is how to obtain statistically meaningful beam polarization measurements with traditional polarimeters. The BNL AGS could be an ideal machine for the spin echo experiment because the analyzing power of the AGS CNI polarimeter[5] at its injection energy ($G\gamma = 4.6$) is very large. However, the AGS rf dipole can only reach an maximum strength of 20 Gauss-m that corresponds to a resonance strength of ~0.0001. Hence, in order to observe the maximum echo signal, the spin resonance jump size has to be 0.00005 and the momentum spread of the beam has to be about $4.6\text{x}10^{-6}$ which is very difficult to achieve in the AGS.

RHIC on the other hand has a couple of advantages for observing spin echo. First, because of the dual snake configuration[6], the spin tune is independent of beam energy and the spin tune spread only comes from higher order effects, which can be controlled with local orbit and optics at the snakes. Another advantage of RHIC is that a spin flipper made of two rf dipoles and four DC dipoles to simulate the rf spin rotator is under construction and scheduled to be commissioned in 2009.

In order to confirm that the spin motion of two resonance jumps in the presence of dual snakes is similar to the spin motion without snakes, two full snakes 180° apart from each other were added in the simulation. In this configuration, the spin tune is determined by the angle between the axes of the two snakes around which spin vector rotates. With the snake axis perpendicular to each other, the spin tune is ½. Similar to the simulation in Fig. 1, a total of 1201 particles were tracked with an rms spin tune spread of $4x10^{-5}$. The Gaussian distributed spin tune spread was generated by a spread of the angle between the axis of the two snakes. For the center particle, the axes of the two snakes are perpendicular. The spin resonance excited by the rf spin rotator was first ramped from zero to 0.00092 with its tune 0.00046 below the spin tune. Each resonance jump consisted of the rf spin rotator tune jumping in 200 turns from 0.49956 to 0.50048. The two resonance jumps were spaced by 70,000 turns and the spin echo peaked up at 70,000 turns after the second resonance jump, as expected.

FIGURE 2. This plot shows the simulation result of the spin motion after jumping across a spin resonance twice in the presence of two snakes, separated apart from each other by half of the ring. The plot shows the vertical component of the average spin of 1200 particles as a function of number of orbital turns. The two resonance jumps were spaced by 70,000 turns and spin echo signal peaked at 70,000 turns after the second resonance jump.

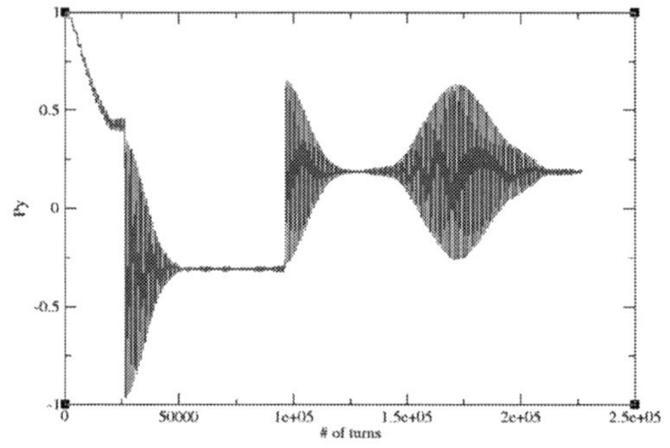

Conclusion

The numerical simulations confirm the spin echo effect of a dual spin resonance jump. The simulation also revealed the smearing effect of the spin echo signal.

We also conclude that it is extremely difficult to observe spin echo at the AGS since it has only a single ac dipole. Hence, we plan to observe the spin echo at RHIC with its new design of rf spin rotator[4]. The simulation also confirmed the spin echo effect in the presence of dual snakes.

REFERENCES

1. A. Chao, Phys. Rev. ST Accel. Beams 8, 014001 (2005). V.S. Morozov, et al., Phys. Rev. Lett. 100, 054801 (2008).
2. M. Froissart and R. Stora, Nucl. Instrum. Methods, 7, 297 (1960).
3. A. Chao and E. Courant, Phys. Rev. ST Accel. Beams 10, 014001 (2007).
4. M. Bai and T. Roser, Phys. Rev. ST Accel. Beams 10, 014001 (2008).
5. J. Tojo et al., Phys. Rev. Lett. 89, 052302 (2002)
6. K. Brown and et al, Polarized Proton Collider at RHIC Design Manual, 1998

Multiple Spin Resonance Crossing in Accelerators

A.M. Kondratenko*, M.A. Kondratenko* and Yu. N. Filatov†

*GOO Zaryad, Russkaya St. 41, Novosibirsk, 630058 Russia
†JINR, Joliot-Curie St. 6, Dubna, Moscow region, 141980 Russia

Abstract. In this paper the method of beam depolarization compensation during spin resonance crossing is developed. This method is based on the restoration of the spin adiabatic invariant after the spin resonance crossing. The case when the restoration of spin adiabatic invariant occurs due to multiple crossing of the resonance effective area is considered. At the same time it is necessary to provide the certain difference of spin phases between the moments of crossing of the resonance effective area. The conditions of beam polarization preservation when the resonance effective area is repeatedly crossed are obtained. Numerical examples are presented.

Keywords: polarized beam acceleration, spin resonance crossing, spin adiabatic invariant
PACS: 29.27.Bd, 29.27.Hj, 41.45.Ak

INTRODUCTION

Anomalous magnetic moment of particles leads to growth of spin precession with energy at polarized beam acceleration in synchrotrons. This leads to crossing the numerous spin resonances and the depolarization of the beam. In high energy rings (RHIC of BNL) spin resonances are overcome by introduction of helical type of strong dipole magnets (so-called Siberian Snakes). However, in medium and low energy rings (AGS, Nuclotron, U-70 and other machines) this technique is impractical due to large orbit excursion by snakes with energy. Use of partial snakes at acceleration in AGS reduces effects of spin resonances, but a significant depolarization still exists. We will describe a new technique of crossing spin resonances in low and medium energy synchrotrons (boosters) advantageous compared to the well-known methods of fast or adiabatic crossing. This technique has been offered in papers [1-3]. Here the method consists of a multiple crossing of a resonance with an intermediate rate while taking into account an interference of spin precession between the succeeding crossings. The conditions for transparent crossing in result are formulated. This technique also can be applied to organizing flipping of the spin in storage rings for experimental needs.

For the analysis of spin resonance crossings the representation based on the concept of spin adiabatic invariant is useful [4-8]. The similar parameter of spin adiabatic invariant (as well as invariant field) is used in other papers [9, 10, 11, etc.]

At stationary case during the particles motion in cyclic accelerator there is the exactly defining direction of precession axis $\vec{n}(A_j, \Psi_j, \theta)$ for each path of particle. This direction $\vec{n}$ is the function of amplitudes A_j, phases of betatron and synchrotron oscillations Ψ_j and azimuth of particle θ. The precession axis $\vec{n}$ is the periodic function of all this phases. The generalized frequency of spin precession $\nu(A_j)$ (spin tune) is determined as

CP1149, *Spin Physics, 18th International Spin Physics Symposium*
edited by D. G. Crabb, D. B. Day, S. Liuti, X. Zheng, M. Poelker, and Y. Prok

the frequency of spin vector rotation around the precession axis $\vec{n}$. For a quantum particle the precession axis $\vec{n}$ is the spin axis of quantization. The spin projection $J = \vec{s}\cdot\vec{n}$ to the precession axis is the spin adiabatic invariant (SAI). In particular, SAI can have discrete meanings $\pm\hbar/2$ for a quantum particle with spin $1/2$.The spin projection J to the axis $\vec{n}$ remains constant and stable in the case of slow deviations of the spin detuning ε_k and the resonance strength w_k.

The degree of beam polarization is determined by average value of SAI. The beam polarization will be directed along the precession axis $\vec{n}$. Thus, average value of SAI determines the degree of a beam polarization. The beam polarization will be directed along the precession axis $\vec{n}$.

For the isolated spin resonance the value of J is kept, if [4]:

$$\left| d\vec{h}_k/d\theta \right| \ll h_k^2 = \varepsilon_k^2 + w_k^2 . \tag{1}$$

The beam polarization degree can deviate only in areas where the condition (1) is violated. This condition (1) is violated in the effective resonance area and the spin precession essentially differs from a stationary one.

The problem of the spin resonance with constant speed crossing has been solved in the paper of M. Froissart, R. Stora [14]. The well-known F-S formula connects asymptotic values of vertical spin projection before and after crossing of resonance. The connection of spin adiabatic invariants before and after the resonance crossing has been represented in paper [4].

The solution is simple in adiabatic limit $\varepsilon_k' \to 0$. Spin adiabatic invariant is kept, and the precession phase is determined by the spin tune $\nu = \sqrt{\varepsilon_k^2 + w_k^2}$:

$$J = const\,, \qquad \Psi = \int_0^\theta h_k d\theta\,, \qquad h_k = \sqrt{\varepsilon_k^2 + w_k^2}$$

The deviation of spinor is described by a diagonal matrix of rotation around the precession axis $\vec{n}$:

$$\chi(\theta) = M_z(\theta)\chi(0)\,, \qquad M_z(\Psi) = \exp\left(-\frac{i\Psi}{2}\sigma_z\right), \tag{2}$$

where $\vec{\sigma} = (\sigma_x, \sigma_y, \sigma_z)$ are Pauli matrices.

The spin motion is more complicated in the effective resonance area. The connection of spinors before and after resonance crossing can be obtained without solving the equation of spin motion in the effective resonance area by the use of the WKB (Wentzel-Kramers-Brillouin) approximation. The connection of spinors is as following [4]:

$$\chi(\theta_f) = M\chi(\theta_i)\,, \quad M = M_z(\Psi_f)M_y(2\alpha)M_z(\Psi_i)\,, \quad M_y(\alpha) = \exp\left(-\frac{i\alpha}{2}\sigma_y\right), \tag{3}$$

$$\text{where} \quad \Psi_i = \int_{\theta_i}^0 h_k d\theta + \varphi\,, \quad \Psi_f = \int_0^{\theta_f} h_k d\theta + \varphi\,, \quad \sin(\alpha) = \exp\left(-\frac{\pi w_k^2}{4|\varepsilon_k'|}\right)$$

The resonance crossing occurs at the moment $\theta = 0$. Here φ is an additional shift of a rotation phase after resonance crossing [4].

Thus, influence of a resonance can be replaced by an additional abrupt rotation around the axis $\vec{n}$ at an angle 2α and an abrupt shift of a rotation phase around the precession axis at an angle 2φ.

Very important circumstance is that characteristic time of a spin invariant J and vertical spin projection S_z deviation strongly differ from each other after resonance crossing. The amplitude of invariant J oscillations decreases exponentially as a function of time τ, the amplitude of vertical projection S_z oscillations decreases only in inverse proportion of time ($\sim \tau^{-1}$).

It should be noted that it is impossible to keep completely a beam polarization at single crossing with constant speed.

TWO TIMES CROSSING OF THE SPIN RESONANCE

Unlike single resonance crossing, double crossing of a resonance allows to exclude completely a beam depolarization. Fig. 1 shows the deviation of resonance detuning ε_k versus time θ with double crossing of a resonance at the moments of time $\theta = a$ and $\theta = b$ (resonance crossing shape).

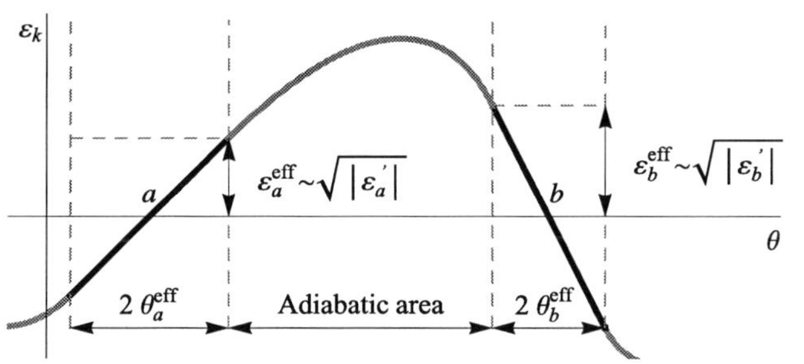

FIGURE 1. The resonance crossing shape at two-times crossing of the resonance.

For the reason above-stated the spinor transformation matrix can be written as

$$M(\theta_i,\theta_f) = M_z(\Psi_f)M_z(-2\beta)M_z(\Psi_{ab})M_y(2\alpha)M_z(\Psi_i)\,.$$

Parameters α and β are determined by crossing speed in points a and b accordingly:

$$\sin(\alpha) = \exp\left(-\frac{\pi}{4}\frac{w_k^2}{|\varepsilon_a'|}\right), \qquad \sin(\beta) = \exp\left(-\frac{\pi}{4}\frac{w_k^2}{|\varepsilon_b'|}\right)$$

and phases $\Psi_{ab} = \int_a^b h_k d\theta + \varphi_a + \varphi_b$, $\Psi_i = \int_{\theta_i}^a h_k d\theta + \varphi_a$, $\Psi_f = \int_b^{\theta_f} h_k d\theta + \varphi_b$,

designate precession phases of corresponding intervals with accounting of resonance phase shifts φ_a and φ_b at the moments of crossing $\theta = a$ and $\theta = b$.

Equating matrix elements M_{11} and M_{12} to zero, we receive, that for preservation of polarization it is necessary to satisfy following conditions ($m = 1,2,\ldots$):

a) case of SAI keeps sign $(M_{12} = 0, J_f = J_i)$: $\quad \alpha = \beta, \quad \Psi_{ab} = 2\pi m \qquad (4)$

b) case of SAI-flip $(M_{11} = 0, J_f = -J_i)$: $\quad \alpha = \pi/2 - \beta, \quad \Psi_{ab} = 2\pi m + \pi \qquad (5)$

The fulfilling of conditions (4), (5) is necessary to preserve of beam polarization at double crossing of a resonance area.

In a real situation the complete compensation of depolarization doesn't occur because of the deviations of crossing parameters from the values of the given conditions (4) or (5). For example, such deviations are caused by resonance detuning ε_k spread, resonance strength w_k spread or spreads of crossing speeds ε'_a and ε'_b. In our case, when adiabatic area between the moments of a resonance crossings exists and the number m is great enough, first of all it is necessary to take into account the deviation of a phase from the value of the given conditions (4) or (5). Thus the degree of depolarization is determined by a simple formula:

$$D = 1 - |J_f| = 2\sin^2\frac{\Delta\Psi_{ab}}{2}\sin^2 2\alpha .$$

Fig. 2 shows the SAI dependences on normalized time $\tau = w_k\theta$ at the parameters: $\alpha = \pi/2 - \beta = \pi/4$, $m = 4$ for the left graph and $\alpha = \beta = \pi/6$, $m = 4$ for the right graph. The moments of crossing of the resonance point are $\tau_a = w_k a$ and $\tau_b = w_k b$.

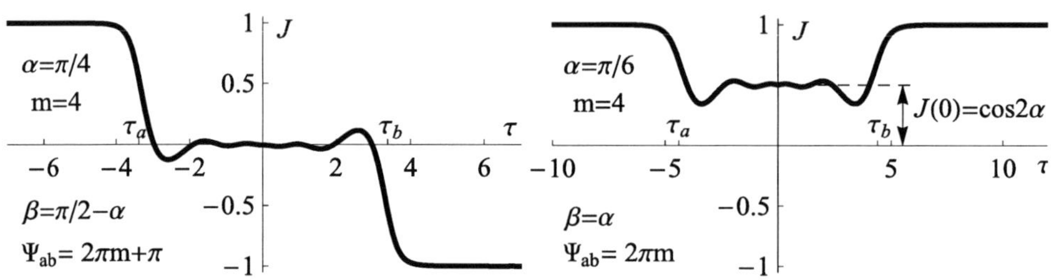

FIGURE 2. The graph on the left side shows the SAI dependence versus normalized time in the case of SAI-flip. The graph on the right side shows the SAI dependence versus normalized time in the case of SAI keeps sign.

It should be noted the result received by A. Chao studying "the spin echo". As it is pointed out in works [12, 13], in model of two abrupt "jumps" of resonance detuning under the condition $h_k(b-a) = 2\pi m$ the spin echo disappears. This condition coincides with the condition of polarization restoration (4) when the total resonant shift of precession phase after two jumps of resonance detuning is equal to zero: $\varphi_a + \varphi_b = 0$.

The condition of polarization restoration (5) with SAI-flip in model of two abrupt "jumps" of resonance detuning is not possible to fulfill.

THREE TIMES CROSSING OF THE SPIN RESONANCE

The case when the effective resonance area is crossed three times is actual for beam accelerations. In this case it is also possible to compensate depolarizing effects completely by means of organization of an appropriate interference between the moments of resonance crossings. Resonance crossing occurs at the moments of time $\theta = a$, $\theta = b$ and $\theta = c$.

There are more wide possibilities to keep polarization, than for the case of double crossing of resonance. Therefore, it is reasonable to find the solutions less sensitive to the deviation of phases Ψ_{ab} and Ψ_{bc}, than for the case of double crossing, when the degree of depolarization will be proportional not to the square of the phases deviation, but to the fourth power, i.e. $D \sim (\Delta\Psi)^4$.

One can fulfill the conditions of compensation with weak dependence on spread of resonance detuning and spread of resonance strength at the following relations of crossing parameters

$$\frac{1}{\sqrt{|\varepsilon_b'|}} = \frac{1}{\sqrt{|\varepsilon_a'|}} + \frac{1}{\sqrt{|\varepsilon_c'|}}, \quad b = \sqrt{\frac{|\varepsilon_b'|}{|\varepsilon_a'|}}\,a + \sqrt{\frac{|\varepsilon_b'|}{|\varepsilon_c'|}}\,c, \quad \Psi_{ab} = 2\pi m_1, \quad \Psi_{bc} = 2\pi m_2.$$

In particular, for symmetric crossing at $\varepsilon_a' = \varepsilon_c'$ and $\Psi_{ab} = \Psi_{bc}$ one obtains, that speeds of a resonance crossing in points a, b, and c should be connected among themselves by the relations

$$|\varepsilon_a'| = |\varepsilon_c'| = 4\,|\varepsilon_b'|, \tag{6}$$

and phases and should be multiple 2π:

$$\int_a^b h_k\,d\theta + \frac{\pi}{2} = \int_b^c h_k\,d\theta + \frac{\pi}{2} = 2\pi m. \tag{7}$$

In last formula with considered accuracy the phase shifts are equal to $\varphi_a \approx \varphi_b \approx \pi/4$.

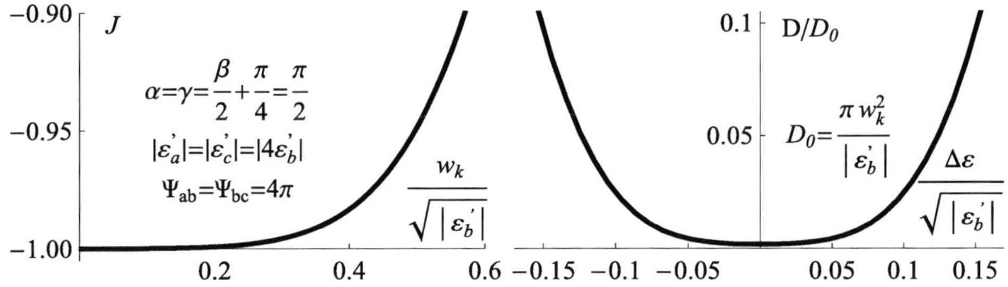

FIGURE 3. The graph on the left side shows the SAI dependence J vs resonance strength. The graph on the right side shows the dependence of degree of depolarization D vs the resonance detuning spread.

Fig. 3 shows the example of SAI dependence versus normalized resonance strength and the degree of depolarization versus resonance detuning spread for the symmetric three-times crossing, satisfying to conditions (6), (7).

We can see, that even at $\Psi_{ab} = \Psi_{bc} = 4\pi$, the SAI restoration (the SAI-flip) occurs to high accuracy. Thus, the considered model allows to simplify significantly the analysis of multiple crossing of a spin resonance at condition (1) of existence of adiabatic areas between the moments of crossings.

SUMMARY

In order to diminish the depolarization effect of the spin resonances in the process of acceleration, various known techniques are employed [4, 8, 13-20, etc]. The given method allows to exclude completely a beam depolarization at crossing of the spin-resonances.

Because of beam parameters spread, conditions of SAI restoration for all particles can not be fulfilled exactly. The chosen scheme of resonance shape (or resonance strength) should satisfy to the admissible level of loss of the polarization. It is especially important at the case of incoherent resonance crossing, when there exists a wide resonance strength spread.

Comparing the considered method of beam depolarization compensation with the method of betatron tune jump for the incoherent resonance, it is necessary to note less demands for the resonance shape. So, to diminish the degree of depolarization in N times, for the considered method it is required to provide amplitude of detuning deviation to be $\sqrt{N}$ times smaller, and time detuning deviation as much greater. Thus, in the considered method considerably smaller speed of resonance crossing is required (approximately $q \cdot N$ times less, where q is a number about of unit) for compensation of beam depolarization.

It is possible to introduce the "inset" of additional magnetic fields to the accelerator which will deviate only the spin tune [18]. In comparison with a method of betatron tune jump such insert does not excite betatron motion in the accelerator. Moreover, such insert is universal at crossing both coherent and incoherent spin resonances.

ACKNOWLEDGMENTS

We would like to thank Ya.S. Derbenev, A.D. Krisch, E.D.Courant, V.S. Morozov, M.A. Leonova, A.W. Chao and to other participants of a teleconference SPIN@COSY for the useful discussions of some questions, considered in this paper.

REFERENCES

1. A.M. Kondratenko, M.A. Kondratenko, Yu.N. Filatov, Pepan Letters, 1(5), p. 266, (2004).
2. A.M. Kondratenko, M.A. Kondratenko, Yu.N. Filatov, AIP Proc., Japan, v. 915, pp. 874, (2007).
3. A.M. Kondratenko, M.A. Kondratenko, Yu.N. Filatov, Baldin ISHEPP Proceedings, (2006).
4. Ya.S. Derbenev, A.M. Kondratenko, A.N. Skrinsky, Sov. Phys.-JETP, v. 33, p. 658, (1971).
5. Ya.S. Derbenev, A.M. Kondratenko, Sov.Phys.-JETP, v. 35, p. 1230 (1972).
6. Ya.S. Derbenev, A.M. Kondratenko, Sov.Phys.-JETP, v. 37. p. 968 (1973).
7. Ya.S. Derbenev, A.M. Kondratenko, Sov.Phys.Dokl. v. 19, p. 438 (1975).
8. Ya.S. Derbenev, A.M. Kondratenko, Sov.Phys.Dokl. v. 20, p. 562 (1976).
9. D.P. Barber, M. Vogt, AIP Proceedings, Kyoto, Japan, v. 915, (2007).
10. D.P. Barber, R. Jaganathan and M. Vogt, AIP proceedings v. 675 (2003).
11. G. Hoffstaetter, M. Vogt, Phys. Rev. E 70, 056501, (2004).
12. A.W. Chao, AIP Proceedings,Kyoto, Japan, v. 915, (2007).
13. A.W. Chao, Phys. Rev. ST Accel. Beams v. 8, 104001, (2005).
14. M. Froissart, R. Stora, Nuclear Instruments and Methods, v. 7, 3, p.p. 297-305, (1960).
15. T. Knoe et. al., Part. Accel., v. 6, p. 213, (1975).
16. E.D. Courant and R.D. Ruth, BNL 51270, (1980).
17. I.G. Ratner, BNL, AIP Conference Proceedings, 95, v. 28, p. 412, (1982).
18. N.I. Golubeva, A.M. Kondratenko, Yu.N. Filatov, "Deuteron-93" proc., Dubna, p. 374, (1994).
19. F.Z. Khiari, et al, Phys. Rev., D 39, p.p. 45-85, (1989).
20. M. Bai, L. Ahrens, and T. Roser, AIP Proc., Melville, New York, v. 667, p.p. 15-29 (2003).
21. A. Lehrach et al, AIP Proc., Melville, New York, v. 667, p.p. 30-39 (2003).

Spin Resonance Strength Calculations

E. D. Courant

Brookhaven National Laboratory, Upton NY 11973

Abstract. In calculating the strengths of depolarizing resonances it may be convenient to reformulate the equations of spin motion in a coordinate system based on the actual trajectory of the particle, as introduced by Kondratenko, rather than the conventional one based on a reference orbit. It is shown that resonance strengths calculated by the conventional and the revised formalisms are identical. Resonances induced by radiofrequency dipoles or solenoids are also treated; with rf dipoles it is essential to consider not only the direct effect of the dipole but also the contribution from oscillations induced by it.

Keywords: Polarized beams, spin resonances
PACS: 29.27.Hj

INTRODUCTION

The dynamics of particle spin in accelerators has been explored extensively [1]. In the process of acceleration beams tend to be depolarized by resonances with perturbing fields. But in a few recent papers [2] the experimental results have led to apparent disagreements between the resonance strengths inferred from the experimental results and theoretical values. Therefore it is felt to be desirable to review and reexamine the theory of depolarizing resonance strengths, especially in cases where an RF dipole or solenoid is present.

The Thomas-BMT equation for the behavior of spin in magnetic fields, as formulated by Froissart and Stora [3,4], may be written

$$\frac{d\vec{S}}{dt} = \frac{q}{m\gamma}\vec{S}\times[(1+G\gamma)\vec{B_\perp} + (1+G)\vec{B_{||}}] \tag{1}$$

where $\vec{B_{||}} = (\hat{v}\cdot\vec{B})\hat{v}$ and $\vec{B}_\perp = \vec{B} - \vec{B_{||}} = (\hat{v}\times \vec{B})\times\hat{v}$ are the longitudinal and transverse parts of the magnetic field $\vec{B}$, $\hat{v}$ being the unit vector in the direction of the particle velocity, q and m are its charge and mass, and γ is the Lorentz energy factor.

The conventional analysis employs a Frenet-Serret coordinate system based on a closed *reference orbit* as we consider particles whose motion takes place near (though not exactly on) that orbit. We assume the reference orbit is in a plane and has a circumference $2\pi R$. We define the coordinates to be: s = the distance along the reference orbit from an origin point (arbitrarily chosen) on the reference orbit to the

CP1149, *Spin Physics, 18th International Spin Physics Symposium*
edited by D. G. Crabb, D. B. Day, S. Liuti, X. Zheng, M. Poelker, and Y. Prok

point on the reference orbit closest to the particle; x and z = the horizontal and vertical components of the vector from that point to the particle.

Using s instead of the time t as the independent variable, expressing the components of the magnetic field in terms of the excursions of the particle, and using the spinor-SU2 formalism for analysis of rotation dynamics (as invented by Hamilton 150 years ago), (1) becomes

$$\Phi' = -\frac{i}{2}\begin{pmatrix} -G\gamma / R & \zeta e^{-i\chi} \\ \zeta^* e^{i\chi} & G\gamma / R \end{pmatrix}\Phi \tag{2}$$

where the prime denotes differentiation by s; the complex 2-component spinor Φ describes the components S_1, S_2, S_3 of the spin in the x, s and z directions via $S_1 = \langle \Phi^\dagger \sigma_1 \Phi \rangle$; $S_2 = \langle \Phi^\dagger \sigma_2 \Phi \rangle$; $S_3 = \langle \Phi^\dagger \sigma_3 \Phi \rangle$; $\chi = G\gamma(\Theta - \theta)$, $\Theta = \int_0^s ds'/\rho - x'$ is the turning angle, $\theta = s/R$ is the azimuth along the reference orbit, and the depolarizing term ζ equals

$$\zeta = -(1+G\gamma)z'' - i[G(\gamma-1)z'/\rho - (1+G)z(1/\rho)' + (1+G)B_{sol}/B\rho] \tag{3}$$

The diagonal elements of the matrix in (2) mean that the basic precession frequency about the vertical (spin tune) is $G\gamma$, and the off-diagonal terms produce changes (depolarization) in the spin. The depolarizing term $\zeta e^{-i\chi}$ is a combination of oscillations at various frequencies:

$$\zeta e^{-i\chi} = \frac{1}{R}\sum_r \varepsilon_r e^{-i\kappa_r \theta} \tag{4}$$

and ε_r is the strength of the depolarizing resonance which occurs at the energy where the spin tune $G\gamma$ equals κ_r. Here κ_r, the r-th resonance value of $G\gamma$, may be imperfection resonances, κ_r= any integer k; "intrinsic" resonances (due to vertical betatron oscillations) $\kappa_r = kP \pm \nu_z$, where ν_z is the vertical betatron tune, P is the periodicity of the magnet structure, k is any integer; broken periodicity resonances $\kappa_r = k \pm \nu_z$ which occur when the structure periodicity P is inexact; RF resonances $\kappa_r = k \pm \omega_{rf}/\omega_{orbit}$ induced by radiofrequency dipoles and/or solenoids placed somewhere on the orbit. When the orbit is the closed orbit produced by imperfections, without betatron oscillations, (4) is a straightforward Fourier series and resonances occur when $G\gamma$ = an integer k; the resonance strengths ε_k can be calculated by Fourier analysis of $\zeta e^{-i\chi}$ as defined in (3):

$$\varepsilon_k = \frac{1}{2\pi}\int_0^{2\pi R} \zeta e^{-i(k\theta+\chi)} ds = \frac{1}{2\pi}\int_0^{2\pi R} \zeta e^{-iG\gamma\Theta} ds \tag{5}$$

If betatron oscillations are present we have intrinsic and/or broken periodicity resonances $\kappa_r = k \pm \nu_z$; we consider z to be the trajectory of betatron oscillations of frequency ν_z. The frequencies κ_i in (4) are not multiples of one frequency; therefore (4) is no longer a simple Fourier series but a general combination of oscillations (an *almost periodic* function), and Fourier analysis (5) does not apply directly. But since $z = z_+(\theta)e^{i\nu_z\theta} + z_-(\theta)e^{-i\nu_z\theta}$, where $z_+(\theta)$ and $z_-(\theta) = z_+^*(\theta)$ are periodic functions, we can separate ζ into parts containing the factors $e^{i\nu_z\theta}$ and $e^{-i\nu_z\theta}$, which we call $\zeta_+ + \zeta_-$; $\zeta_+ e^{-i\nu_z\theta}$ and $\zeta_- e^{i\nu_z\theta}$ are periodic functions, and we perform Fourier analysis for each part, finding, for $G\gamma = \kappa = k \pm \nu_z$,

$$\varepsilon_\kappa = \frac{1}{2\pi}\int_0^{2\pi R} \zeta_\pm e^{-i(\kappa\theta+\chi)} ds = \frac{1}{2\pi}\int_0^{2\pi R} \zeta_\pm e^{-iG\gamma\Theta} ds \tag{6}$$

To evaluate the resonance strengths in any particular case it is necessary to express the integration factors ζ, ζ_+, ζ_- in (5) and (6) in terms of the components of the lattice structure, as is done in [5].

TRAJECTORY-BASED COORDINATE SYSTEM

The particles do not necessarily travel on the reference orbit, therefore the spin components in the directions $\hat{e}_x, \hat{e}_s, \hat{e}_z$ are not exactly in the directions transverse and longitudinal to the motion of a particle. Since the dynamical equation (1) shows that the transverse and longitudinal spin components behave differently, it may be desirable to formulate equations of motion that maintain this distinction, i.e. describe, for a given particle, the behavior of the components of spin parallel and perpendicular to the direction of motion of that particle, rather than the components parallel and perpendicular to the reference orbit.

In several papers Kondratenko [6] and Sivers [7] introduce a "natural" or "local" reference frame based on the actual trajectory of the particle. In their formalism the basis vector $\hat{u}_2$ is taken to be the unit vector $\hat{v}$ in the direction of the instantaneous particle velocity, and the other two are in the local radial and vertical direction orthogonal to $\hat{v}$ and to each other. In what follows, we ignore second and higher terms in the excursions x and z of the particle from the reference orbit. In this coordinate system the depolarizing term ζ in the spinor equation (2) becomes

$$\zeta = -G\gamma z'' + i[Gz'/\rho + (1+G)z(1/\rho)' - (1+G)B_{sol}/B\rho] \tag{7}$$

Note that the leading term $-G\gamma z''$ in (7) is different from the leading term $-(1+G\gamma)z''$ in (3), which led some people, including the present author [8], to the erroneous conjecture that resonance strengths in the two frames differ by the factor $G\gamma/(1+G\gamma)$. Kondratenko [6] shows that this conjecture is incorrect and that the resonance strength is the same in both systems. His derivation, in our terminology, is: The difference between the resonance strength calculated in the two different frames is (using a superscript r for the reference-orbit based system, and t for the trajectory-based system)

$$\varepsilon_\kappa^{(r)} - \varepsilon_\kappa^{(t)} = \frac{1}{2\pi}\int_0^{2\pi R} \left(\zeta^{(r)} - \zeta^{(t)}\right) e^{iG\gamma\Theta} ds \tag{8}$$

The integrand of (8) is

$$(\zeta^{(r)} - \zeta^{(t)})e^{iG\gamma\Theta} = -[z'' + iG\gamma\Theta' z']e^{iG\gamma\Theta} = -\frac{d}{ds}(z' e^{iG\gamma\Theta}) \qquad (9)$$

which is a perfect derivative. The integral of a perfect derivative over a period averages to zero. Therefore $\varepsilon_K{}^{(t)} = \varepsilon_K{}^{(r)}$: the resonance strength is independent of the frame in which it is calculated.

It follows that in any particular case where the strengths of resonances are to be calculated one may select the trajectory- based formalism or the reference-orbit formalism, whichever is more convenient. The algorithms given in [5] for calculation of resonance strengths for a given magnet structure, leading to the computer program DEPOL, remain valid.

RESONANCE STRENGTH WITH RF SOLENOIDS OR DIPOLES

RF solenoids or dipoles may be inserted in an accelerator or storage ring in order to deliberately excite spin resonances, either for the purpose of reversing (flipping) the spin, or to enhance intrinsic resonances to a strength where the spin reverses completely rather than partially.

The strength of these induced resonances may be calculated by using the results of the previous sections. RF solenoids or dipoles may be inserted in an accelerator or storage ring in order to deliberately excite spin resonances, either for the purpose of reversing (flipping) the spin, or to enhance intrinsic resonances to a strength where the spin reverses completely rather than partially.

The strength of these induced resonances may be calculated by using the results of the previous sections.

RF Solenoids

First; consider a radiofrequency solenoid with field $B_{sol} = \sqrt{2}B_{rms}\cos\omega_{rf}t$ .at one point in the ring, which we designate as θ=0. It rotates the spin by an angle

$$\Delta_s \cos\omega_{rf}t;\ \Delta_s = (1+G)\frac{\int\sqrt{2}B_{rms}ds}{B\rho} \qquad (10)$$

once per revolution, and does not affect the orbit. (The integral over B_{rms} extends over the length of the rf magnet). Therefore we may set $z = z' = 0$, and (8) simplifies to

$$\zeta = -i(1+G)B_{sol} / B\rho = -i\Delta_s\delta_p(\theta)\cos\nu_{rf}\theta \qquad (11)$$

where $\delta_p(\theta)$ is the periodic delta function and $\nu_{rf} = \omega_{rf} / \omega_{orbit}$ is the rf frequency normalized to the revolution frequency. Since $\cos\nu_{rf}\theta = (e^{i\nu_{rf}\theta} + e^{-i\nu_{rf}\theta})/2$ we can, as before, divide ζ into $\zeta_+ + \zeta_-$ and obtain, for $G\gamma=\kappa=k\pm\nu_z$,

$$\varepsilon_K = \frac{1}{2\pi}\int_0^{2\pi R}\zeta_\pm e^{iG\gamma\Theta}ds = -\frac{i(1+G)}{4\pi}\frac{\int\sqrt{2}B_{rms}ds}{B\rho} \qquad (12)$$

in agreement with [9] and [10]. Thus there are two resonances of equal strength and equal phase in each interval of $G\gamma$ between one integer and the next . If ν_{rf} is exactly a half integer these coalesce into a single resonance of twice the strength. Note that, since

Δ_s contains the factor $1/B\rho$ these resonances become weak at high energy; therefore rf solenoids are primarily useful for low-energy rings such as IUCF or COSY.

RF Dipoles

An alternative is to use rf dipoles with transverse fields. We assume that we have a radiofrequency dipole with radial horizontal field $\sqrt{2}B_{rms}\cos\omega_{rf}t$ at one point in the ring, which we designate as θ=0. The beam deflection it produces is

$$\Delta\cos\omega_{rf}t; \quad \Delta = \frac{\int\sqrt{2}B_{rms}ds}{B\rho} \tag{13}$$

The spin rotation associated with this deflection is, according to Eq. (1), just $(1+G\gamma)$ times the deflection, and this leads to the naïve equation

$$\varepsilon = (1+G\gamma)\frac{\Delta}{4\pi} = (1+G\gamma)\frac{\int\sqrt{2}B_{rms}ds}{4\pi B\rho} \tag{14}$$

for the resonance strength at $G\gamma = k \pm \nu_{rf}$ [9,10

the

effect

must also be considered, as recognized in the text of [10] and by other authors [11].

The equation of orbit motion for the forced oscillations is

$$\frac{d^2z}{ds^2} + K(s)z = \frac{\Delta}{R}\delta_p(\theta)\cos\nu_{rf}\theta \tag{15}$$

where $K(s)$ is the focusing gradient function, $\theta = s/R$ is the normalized azimuth, $\delta_p(\theta)$ is the periodic delta function, and $\nu_{rf} = \omega_{rf}/\omega_{orbit}$. We assume the solution of the homogeneous equation corresponding to (15) is known; the vertical tune is ν_z and the orbit functions are $\beta_z(s),\ \alpha_z(s)$. The solution of (15) is

$$z = \frac{\nu_z\beta_z\Delta}{4\pi}\sum_{k=-\infty}^{\infty}\left(\frac{e^{i(k+\nu_{rf})\varphi}}{\nu_z^2-(k+\nu_{rf})^2} + \frac{e^{i(k-\nu_{rf})\varphi}}{\nu_z^2-(k-\nu_{rf})^2}\right) \tag{16}$$

The equations of spin dynamics expressed in terms of the trajectory excursion $z(s)$ remain valid . As in the discussion leading to Eq. (6) the depolarizing term is $\zeta = \zeta_+ + \zeta_-$, where, in the trajectory-based frame, with $\kappa_\pm = k \pm \nu_{rf}$,

$$\zeta_\pm = \frac{\nu_z\Delta}{4\pi}\sum_{k=-\infty}^{\infty}\frac{e^{i\kappa_\pm\varphi}}{\nu_z^2-\kappa_\pm^2}\left[\frac{1+G\gamma}{\beta_z}\left(K\beta_z^2 - 1 + \frac{\kappa_\pm^2}{\nu_z^2}\right) + G\frac{\gamma-1}{\rho}\left(i\alpha_z + \frac{\kappa_\pm}{\nu_z}\right) + i(1+G)\beta_z\left(\frac{1}{\rho}\right)'\right] \tag{17}$$

The resonance strengths are, analogously with (6), for $G\gamma = \kappa = k \pm \nu_{rf}$,

$$\varepsilon_\kappa = \frac{1}{2\pi}\int_0^{2\pi R}\zeta_\pm e^{i\kappa\varphi}ds = \frac{\nu_z}{2\pi}\int_0^{2\pi}\beta_z\zeta_\pm e^{i\kappa\varphi}d\varphi \tag{18}$$

In the simple case of uniform focusing,

$$\beta_z = R/\nu_z;\ \alpha_z = 0;\ K\beta_z^2 = 1;\ \rho = R;\ \varphi = \theta, (1/\rho)' = 0$$

and (17) becomes

$$\zeta_\pm = \frac{\Delta}{4\pi R}\sum_{k=-\infty}^{\infty}\frac{e^{i\kappa_\pm\varphi}}{\nu_z^2 - \kappa_\pm^2}\left[(1+G\gamma)\kappa_\pm^2 + G(\gamma-1)\kappa_\pm\right] \quad (19)$$

For each k only a single term of the infinite series contributes to (20): For $\kappa = G\gamma = n + \nu_{rf}$ (n any integer) this is the term $k = -n$, $\kappa_- = -n - \nu_{rf} = -G\gamma$ in ζ_-, obtaining

$$\zeta_{\kappa_-} = \frac{\Delta}{4\pi R}\frac{e^{-iG\gamma\varphi}}{\nu_z^2 - (G\gamma)^2}\left[(1+G\gamma)(G\gamma)^2 - G(\gamma-1)G\gamma\right]$$

$$\varepsilon_\kappa = \frac{R}{2\pi}\int_0^{2\pi}\zeta_{\kappa-}e^{iG\gamma\varphi}d\varphi = \frac{\int\sqrt{2}B_{rms}ds}{4\pi B\rho}\frac{(G\gamma)^3 + G^2\gamma}{\nu_z^2 - (G\gamma)^2} \quad (20)$$

and the same expression for $\kappa = G\gamma = n - \nu_{rf}$. Kondratenko, Kondratenko and Filatov [12] have found the same result for the case of uniform focusing.

ACKNOWLEDGMENTS

This work had its origin in a series of telephone conferences led by A. D. Krisch. The author has benefited from discussions with participants in these conferences and others, in particular A. W. Chao, H. Huang, A. M. Kondratenko, M. A. Leonova, A. U. Luccio, W. MacKay, S. R. Mane, V. A. Morozov, Y. F. Orlov, T. Roser, and D. W. Sivers.

REFERENCES

1. See, for example, S. Y. Lee, Spin Dynamics and Snakes in Synchrotrons (World Scientific, 1997).
2. A. D. Krisch et al., Phys. Rev. ST AB **10**, 071001 (2007); M. A. Leonova et al., Phys. Rev. ST AB **9**, 051001 (2006).
3. L. H. Thomas, Phil. Mag., Series 7, vol. 3, #13 (1927); V Bargmann, L Michel, V Telegdi, Phs. Reve. Lett. **2**, 435 (1959).
4. M. Froissart and R. Stora, Nucl. Inst. Meth. **7**, 297 (1960).
5. E D Courant and R D Ruth, BNL Report BNL-51270 (1980).
6. A. M. Kondratenko, contributions to Michigan-COSY teleconferences 6/21/07, 7/24/07 and 9/13/07; Sov.Phys. JETP, **33** 658 (1971).
7. D W Sivers et al, contributions to Michigan-COSY teleconferences 3/20/07, 4/19/07 and 8/9/07.
8. E. D. Courant, BNL Report C-A/AP/292 (2007).
9. Reference [1], chapters 2 and 3; T. Roser in Handbook of Accelerator Physics and Engineering, A. W. Chao and M. Tigner (editors), (World Scientific, 1998), Sections 2.7.4, 2.7.5.
10. M. Bai, W. MacKay, T. Roser, Phys. Rev. ST AB **8**, 099001 (2005).
11. S. Y. Lee, Phys. Rev. ST AB **9**, 074001 (2006).
12. A. M. Kondratenko, M. A. Kondratenko, Y. N. Filatov, contribution to Michigan-COSY teleconference 12/11/07.

Experimental Verification of Predicted Oscillations near a Spin Resonance

V.S. Morozov[1], A.W. Chao[1,†], A.D. Krisch[1], M.A. Leonova[1], R.S. Raymond[1], D.W. Sivers[1], V.K. Wong[1], A. Garishvili[2,‡], R. Gebel[2], A. Lehrach[2], B. Lorentz[2], R. Maier[2], D. Prasuhn[2], H. Stockhorst[2], D. Welsch[2], F. Hinterberger[3], K. Ulbrich[3], A. Schnase[4], E.J. Stephenson[5], N.P.M. Brantjes[6], C.J.G. Onderwater[6], and M. da Silva[6]

[1] *Spin Physics Center, University of Michigan, Ann Arbor, Michigan 48109-1040, USA*
[2] *Forschungszentrum Jülich, Institut für Kernphysik, Postfach 1913, D-52425 Jülich, Germany*
[3] *Helmholtz-Institut für Strahlen- und Kernphysik, Universität Bonn, D-53115 Bonn, Germany*
[4] *JAEA/J-PARC, Ring RF Group, Shirakata-Shirane 2-4, Tokai-Mura, Ibaraki 319-1195, Japan*
[5] *IUCF, Indiana University, Bloomington, Indiana 47408-0768 USA*
[6] *University of Groningen, Groningen, The Netherlands*

Abstract. The Chao matrix formalism allows analytic calculations of a beam's polarization behavior inside a spin resonance. We recently tested its prediction of polarization oscillations occurring in a stored beam of polarized particles near a spin resonance. Using a 1.85 GeV/*c* polarized deuteron beam stored in COSY, we swept a new rf solenoid's frequency rather rapidly through 400 Hz during 100 ms, while varying the distance between the sweep's end frequency and the central frequency of an rf-induced spin resonance. Our measurements of the deuteron's polarization for sweeps ending near and inside the resonance agree with the Chao formalism's predicted oscillations.

Keywords: Spin resonance, Spin dynamics, Chao matrix formalism.
PACS: 29.27.Bd, 29.27.Hj, 41.75.Ak

INTRODUCTION

Polarized stored hadron and lepton beams often provide the best technique for studying the spin dependence of hadronic interactions in the 1 GeV/*c* to 1 TeV/*c* region. Polarized beam experiments at storage rings need the ability to precisely control the beam's polarization. A stored beam's polarization can be manipulated in a well-controlled way by ramping an rf magnet's frequency through an rf-induced spin resonance.

The Froissart-Stora (F-S) formula [1] has been widely used to calculate a beam's polarization after crossing a spin resonance. However, it is valid only for a constant-rate linear crossing from far below to far above the resonance. Chao's new matrix formalism was proposed [2] to deal with conditions where the F-S formula is not valid. The Chao formalism can be used to calculate the spin dynamics anywhere inside

† also at SLAC, 2575 Sand Hill Rd., Menlo Park, CA 94025.
‡ also at Erlangen-Nürnberg Univ., D-91058, Germany.

CP1149, *Spin Physics, 18th International Spin Physics Symposium*
edited by D. G. Crabb, D. B. Day, S. Liuti, X. Zheng, M. Poelker, and Y. Prok

a piecewise linear resonance crossing. It allows one to analytically solve the spin equation of motion near an isolated spin resonance, if its crossing can be expressed as a series of linear segments. Each segment must have a fixed or linearly changing distance between the spin tune $\nu_s = G\gamma$ and the resonance tune $\nu_r \equiv k \pm f_r / f_c$. After obtaining, for each segment, the time-dependent matrix describing a spinor's evolution in the segment, one multiplies these matrices sequentially to obtain the final polarization P_f. We tested experimentally at the COoler SYnchrotron (COSY) the Chao formalism's striking prediction [3,4] of large polarization oscillations near a resonance by sweeping an rf solenoid's frequency near or through an rf-induced spin resonance.

EXPERIMENTAL PROCEDURE AND RESULTS

We tested Chao's matrix formalism with the technique shown in Fig. 1 using our new rf solenoid; it is a 25-turn air-core water-cooled copper coil, of length 57.5 cm and average diameter 21 cm. Its inductance was 41 ± 3 μH, and its longitudinal rf magnetic field was about 1.25 mT at its center. It was part of an RLC resonant circuit, which operated near 917 kHz, typically at an rf voltage of 5.7 kV rms producing a longitudinal rf $\int B_{\rm rms}\, dl$ of 0.69 ± 0.05 T mm.

The other apparatus for this experiment included the COSY storage ring [5], the EDDA polarimeter [6], the electron cooler [7], the low energy polarimeter [8], the injector cyclotron, and the polarized ion source [9]. The beam from the polarized D^- ion source was accelerated by the cyclotron to 75.7 MeV and then strip-injected into COSY. The Low Energy Polarimeter measured the D^- beam's polarization before injection into COSY to monitor the cyclotron's and the ion source's stability. The EDDA polarimeter [6] measured the beam's polarization in COSY.

We measured the rf solenoid's strength $\mathcal{E}$ by measuring the polarization after ramping its frequency trough the resonance with various ramp times Δt with its

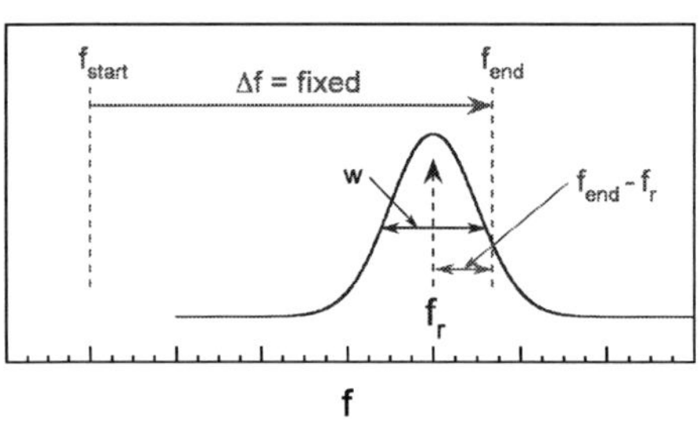

FIGURE 1. Schematic of the Chao formalism test. The rf solenoid's frequency $f_{\rm rf}$ was ramped, by a fixed range Δf, for different distances between the ramp's end frequency $f_{\rm end}$ and the resonance's center $f_{\rm r}$. The curve shows the resonance with a total (FWHM) width w.

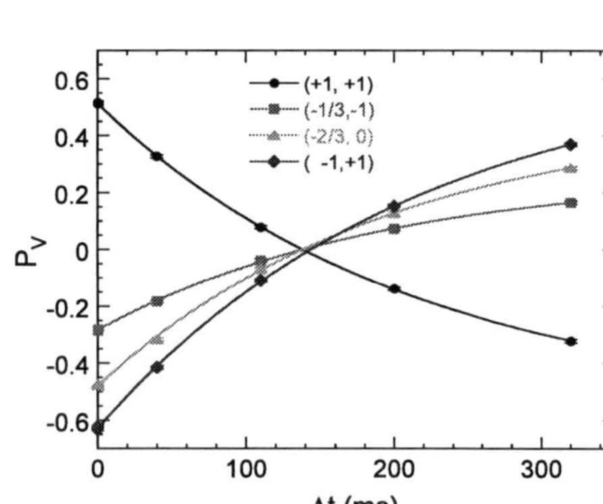

FIGURE 2. Measured 1.85 GeV/c deuteron vector polarizations plotted vs rf-solenoid ramp time Δt with 15 s electron cooling for the four indicated spin states. Their averaged fit to the Froissart-Stora equation [1] gave $\mathcal{E}$ of (1.05 ± 0.01) × 10^{-5}. The above Δt curve's frequency range Δf was 300 Hz. Averaging 13 similar Δt curves gave $\mathcal{E}$ of (1.060 ± 0.005) × 10^{-5}.

frequency range Δf and voltage fixed, as shown in Fig. 2. We then fit these data to the Froissart-Stora equation [1] to obtain the measured value of ε.

To study the Chao formalism's predicted dependence on the beam's momentum spread $\Delta p/p$, we varied the 20.6 keV electron Cooler's on-time at injection. It cooled the deuterons' emittances both longitudinally and transversely for 15 and 25 s. The deuterons were then accelerated to 1.85 GeV/c. The rf acceleration cavity was off and shorted during COSY's flat top; thus, there were no synchrotron oscillations.

We tested the Chao formalism by ramping the rf solenoid's frequency over a range Δf, which started at $f_{\rm start}$ (well outside the rf spin resonance centered at $f_{\rm r}$) and ended at $f_{\rm end}$ near or inside the resonance, as shown in Fig. 1. For each $f_{\rm end}$ data point, both Δf and the ramp time Δt were held fixed, at 400 Hz and 100 ms, respectively, while $f_{\rm end}$ was set to the values shown in Figs. 3 and 4. After $f_{\rm rf}$ reached $f_{\rm end}$, the rf solenoid was turned off abruptly (in a few μs) to preserve the vertical polarization at that instant. We then measured the deuterons' vector asymmetry in all polarization states. The resulting final vector polarization P_f for each nonzero spin state is plotted vs $f_{\rm end}$ in Figs. 3 and 4.

We first calculated [2-4] the Chao formalism's prediction for $\Delta p/p = 0$ using our measured $\varepsilon = 1.06 \times 10^{-5}$ from Fig. 2, our Δf of 400 Hz, and our Δt of 100 ms. To take into account the beam's momentum spread $\Delta p/p$, we next folded this result together with Gaussians representing different values of the beam's $f_{\rm r}$ spread $\delta f_{\Delta p}$ due to $\Delta p/p$. We then fit the data in Figs. 3 and 4 with $f_{\rm r}$ and $\delta f_{\Delta p}$ as the two free parameters. The Chao formalism fits are shown as solid lines for each nonzero spin state in each figure.

We calculated $\chi^2/(N\text{-}2)$ for each fit to compare its agreement with the data for each of the four nonzero spin states. Each χ^2 analysis included only the data's statistical errors and ignored systematic errors; nevertheless, all $\chi^2/(N\text{-}2)$ were near 1 despite the curves' complex shapes. The oscillations' positions and magnitudes are very sensitive to the values of $f_{\rm r}$ and $\delta f_{\Delta p}$, respectively.

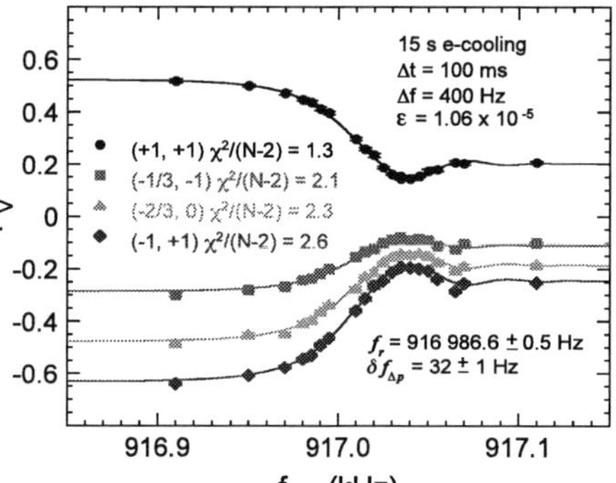

FIGURE 3. Measured 1.85 GeV/c deuteron vector polarizations plotted vs rf-solenoid end frequency $f_{\rm end}$. Its ramp time Δt was 100 ms; its frequency range Δf was 400 Hz, and its ε was 1.06×10^{-5}. The Chao formalism fits, shown by lines, gave a resonance frequency $f_{\rm r}$ of 916 986.6 ± 0.5 Hz and a Gaussian $\delta f_{\Delta p}$ of 32 ± 1 Hz FWHM. Only the statistical errors were used to obtain $\chi^2/(N\text{-}2)$; e cooling was on for 15 s.

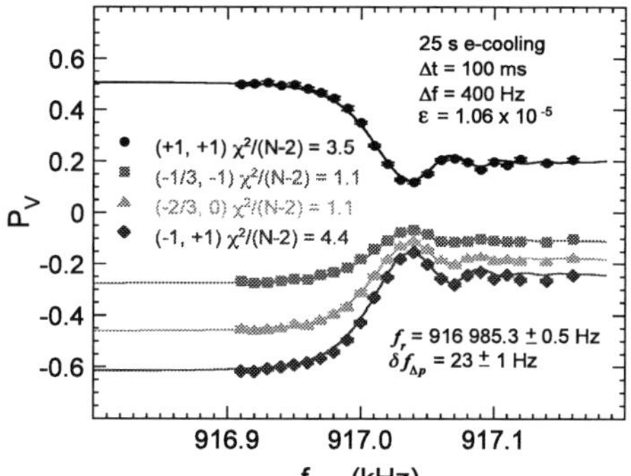

FIGURE 4. Measured deuteron vector polarizations plotted vs $f_{\rm end}$ as in Fig. 3. With 25 s e cooling, the fit curves gave $f_{\rm r}$ of 916 985.3 ± 0.5 Hz and $\delta f_{\Delta p}$ of 23 ± 1 Hz FWHM.

Minimizing the $\chi^2/(N\text{-}2)$ values in Fig. 3 gave f_r of 916 986.6 ± 0.5 Hz and $\delta f_{\Delta p}$ of 32 ± 1 Hz for 15 s electron cooling. Minimizing $\chi^2/(N\text{-}2)$ for the 25 s electron cooling data in Fig. 4 gave an f_r of 916 985.3 ± 0.5 Hz and $\delta f_{\Delta p}$ of 23 ± 1 Hz. The $\chi^2/(N\text{-}2)$ analyses in Figs. 3 and 4 both support the Chao formalism [2-4]. As predicted, the oscillation amplitude increased as $\delta f_{\Delta p}$ decreased.

SUMMARY

In summary, we tested the Chao formalism's prediction of polarization oscillations when crossing an isolated spin resonance, in a region where the Froissart-Stora formula is not valid. Using 1.85 GeV/*c* vertically polarized deuterons stored in COSY, we ramped an rf solenoid's frequency through a range Δf ending near a spin resonance; the magnitudes of both Δf and the ramp time Δt were fixed, while we varied the ramp's distance from the resonance. The good fits to the precise data from our new type of experiment, shown in Figs. 3 and 4, clearly demonstrate the Chao formalism's [2-4] ability to obtain the values of f_r, $\delta f_{\Delta p}$, and ε from the measured polarization's behavior inside a spin resonance.

ACKNOWLEDGMENTS

We thank the COSY staff for the successful operation of COSY with its injector cyclotron and polarized ion source. This research was supported by grants from the German BMBF Science Ministry, its JCHP-FFE program at COSY, the Netherlands NWO, the U.S. NSF, and the Helmholtz Association through funds provided to the virtual institute "Spin and strong QCD" (VH-VI-231).

REFERENCES

1. M. Froissart and R. Stora, Nucl. Instrum. Methods **7**, 297 (1960).
2. A.W. Chao, Phys. Rev. ST Accel. Beams **8**, 104001 (2005).
3. V.S. Morozov *et al.*, Phys. Rev. ST Accel. Beams **10**, 041001 (2007).
4. V.S. Morozov *et al.*, Phys. Rev. Lett. **100**, 054801 (2008).
5. R. Maier, Nucl. Instrum. Methods Phys. Res., Sect. A **390**, 1 (1997).
6. M. Altmeier *et al.* (EDDA Collaboration), Phys. Rev. Lett. **85**, 1819 (2000).
7. H. Stein *et al.*, At. Energ. **94**, 24 (2003).
8. D. Chiladze *et al.*, Phys. Rev. ST Accel. Beams **9**, 050101 (2006).
9. R. Weidmann *et al.*, Rev. Sci. Instrum. **67**, 1357 (1996).

Experimental Test of New Technique to Overcome Spin Depolarizing Resonances

R.S. Raymond[a], A.W. Chao[a*], A.D. Krisch[a], M.A. Leonova[a], V.S. Morozov[a], D.W. Sivers[a], V.K. Wong[a], A Garishvili[b], R. Gebel[b], A. Lehrach[b], B. Lorentz[b], R. Maier[b], D. Prasuhn[b], H. Stockhorst[b], D. Welsch[b], F. Hinterberger[c], A.M. Kondratenko[d]

[a]Spin Physics Center, University of Michigan, Ann Arbor, MI 48109-1040, USA
[b]Forschungszentrum Jülich, Institut für Kernphysik, Postfach 1913, D52425 Jülich, Germany
[c]Helmholtz-Institut für Stahlen- und Kernphysik, Universität Bonn, D-53115 Bonn, Germany
[d]GOO Zaryad, Russkaya St. 41, Novosibirsk, 630058 Russia

Abstract. We recently tested a new spin resonance crossing technique, Kondratenko Crossing (KC) by sweeping an rf solenoid's frequency through an rf-induced spin resonance with both the KC an traditional Fast Crossing (FC) patterns. Using both rf bunched and unbunched 1.85 GeV/c polarized deuterons stored in COSY, we varied the parameters of both crossing patterns. Compared to FC with the same crossing speed, KC reduced the depolarization by measured factors of 4.7 ± 0.3 and $19 \pm^{12}_{5}$ for unbunched and bunched beams, respectively. This clearly showed the large potential benefit of Kondratenko Crossing over Fast Crossing.

Keywords: Spin depolarizing resonance, Kondratenko Crossing, Preserving polarization.
PACS: 29.27.Bd, 29.27.Hj, 41.75.Ak

Kondratenko [1] proposed a technique for overcoming medium-strength depolarizing resonances, as an alternative to simple Fast Crossing (FC). This method, Kondratenko Crossing (KC), uses a more complicated crossing pattern, illustrated in Fig. 1, in which depolarizing phases before the resonance point are canceled by phases after the resonance. For a given maximum crossing rate, this pattern should result in less depolarization than FC. Recent tests of this method used an rf solenoid to produce the resonance; thus, Fig. 1 is presented as frequency vs. time; in other uses of KC the vertical axis might be the betatron tune or spin tune.

We recently tested KC using a 1.85 GeV/c polarized deuteron beam at COSY [2] by ramping an rf solenoid's frequency f through the Δf ranges during times Δt, as shown in Fig. 1. The rf solenoid [3] was a 25-turn, air-core, water-cooled copper coil, of length 57.5 cm and average diameter 21 cm. Its inductance was 41 ± 3 μH. It was part of an RLC circuit operating near 917 kHz, typically at a voltage of 5.7 kV_{rms}. Its on-axis longitudinal rf magnetic field was about 1.17 mT_{rms}, giving an rf $\int Bdl$ of 0.67 ± 0.03 $T{\cdot}mm_{rms}$. In COSY, the deuterons' average circulation frequency f_c was 1.14743 MHz at 1.85 GeV/c, where the Lorentz energy factor was $\gamma = 1.4046$. The deuteron's gyromagnetic anomaly G_d is -0.142987; thus, the spin tune ($\nu_s = G\gamma$) was -0.20084. We used the $(1 + G\gamma)$ rf spin resonance at a frequency $f_r = (1 + G\gamma)f_c \cong$ 917.0 kHz. The deuteron beam was electron cooled [4] at injection to minimize its

CP1149, *Spin Physics, 18th International Spin Physics Symposium*
edited by D. G. Crabb, D. B. Day, S. Liuti, X. Zheng, M. Poelker, and Y. Prok

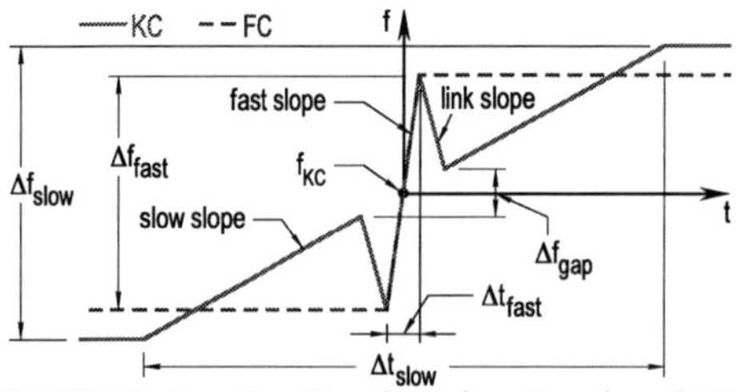

FIGURE 1. Our Kondratenko Crossing (KC) [solid line] and Fast Crossing (FC) [dashed line] patterns defining the parameters Δf_{fast}, Δt_{fast}, Δf_{slow}, and Δt_{slow}. The KC and FC patterns are both centered at f_{KC}.

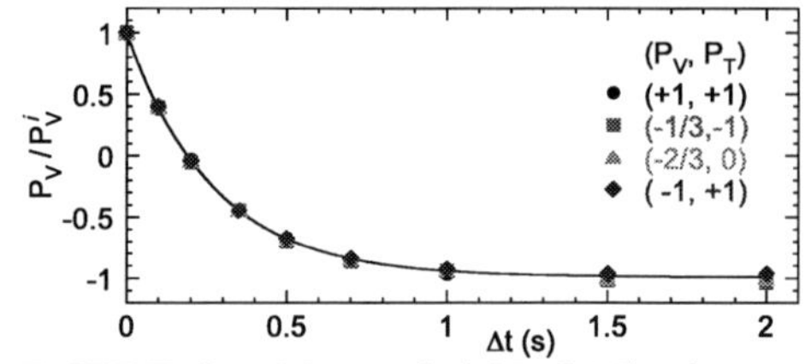

FIGURE 2. Measured 1.85 GeV/c deuteron vertical vector polarization ratio plotted vs. frequency ramp time Δt for the four indicated spin states. P_V^i is the initial polarization. The rf solenoid's voltage was 5.7 kV rms; the Δf sweep was 400 Hz and it was centered at 917 kHz; the beam was unbunched.

momentum spread Δp/p and then accelerated to 1.85 GeV/c. The EDDA detector [5] measured the beam's polarization.

We first obtained the resonance's strength ε by measuring the vertical vector polarization P_V after ramping the rf frequency through the resonance by a range Δf with various ramp times Δt; these data are shown in Fig. 2. We fit these data with the Froissart-Stora formula [6] to obtain the strength ε of $(1.067 \pm 0.003) \times 10^{-5}$.

We then used Kondratenko's optimization procedure [7] along with this measured resonance strength and the previously-measured [3] resonance frequency spread δf of 23 ± 1 Hz to calculate the optimal values of the KC pattern's parameters defined in Fig. 1. For our case, where $[df/dt]_{link} = [df/dt]_{fast}$, the rate $[df/dt]_{fast}$ must satisfy:

$$[df/dt]_{fast} \geq (\varepsilon f_c)^2 \quad \text{and} \quad [df/dt]_{fast} \geq 156(\delta f)^2. \qquad (1)$$

The crossing parameters were then chosen to satisfy:

$$\Delta f_{slow} \geq [df/dt]_{fast}^{1/2}; \qquad \Delta t_{slow} = 6.03\ \Delta f_{slow}/[df/dt]_{fast};$$
$$\Delta f_{fast} = 1.43([df/dt]_{fast})^{1/2}; \qquad \Delta t_{fast} = 1.43/([df/dt]_{fast})^{1/2}. \qquad (2)$$

The parameters chosen were:

$$\Delta f_{slow} = 400\text{Hz}; \quad \Delta t_{slow} = 160\text{ ms}; \quad \Delta f_{fast} = 185\text{ Hz}; \quad \text{and} \quad \Delta t_{fast} = 12\text{ ms}. \qquad (3)$$

We used Chao's matrix formalism [3,8] with these parameters to predict the polarization's behavior for unbunched beam. The rf solenoid's frequency was then programmed to form the KC pattern. To test the predicted behavior experimentally, we varied each parameter around its predicted optimal value; the resulting data are shown in Figs. 3-6.

We first checked that the KC pattern's central frequency f_{KC} was centered on the resonance frequency f_r by varying f_{KC}. The resulting data are plotted in Fig. 3 for both bunched and unbunched beam and for both KC and Fast Crossing (FC). For KC, bunching shifted the peak's central frequency by 5 Hz relative to unbunched KC; moreover, the bunched KC data had a broader flat-top. We fit the KC unbunched data to obtain f_r of 916 999.1 ± 0.1 Hz and δf of 24.4 ± 0.2 Hz. These values were used to predict the unbunched behavior as the parameters Δf_{fast}, Δt_{fast}, Δt_{slow}, and Δf_{slow} were individually varied; the resulting predictions and data are shown in Figs. 4-6. The KC and FC polarization ratios (P_V/P_V^i) at the KC peaks in Figs. 3-6 are listed in Table 1

FIGURE 3. (LEFT) Measured 1.85 GeV/c deuteron P_V/P_V^i, averaged for all nonzero spin states, plotted vs. the KC pattern's center frequency f_{KC}. The resonance strength $\mathcal{E}$ was 1.067×10^{-5}; Δf_{fast} was 185 Hz; Δt_{fast} was 12 ms; Δf_{slow} was 400 Hz; and Δt_{slow} was 160 ms. The KC unbunched solid curve is the Chao formalism prediction for these parameters; the long-dashed line through these unbunched points is a Chao formalism fit with parameters f_r and δf. The KC bunched solid curve is an empirical 2nd-order Lorentzian fit; the horizontal dashed line fit to the five highest points gives the peak KC bunched value of 0.990 ± 0.002. The errors are smaller than the data symbols in all figures. In all later figures, f_{KC} was set at 917 000 Hz and 916 995 Hz for unbunched and bunched beams, respectively.

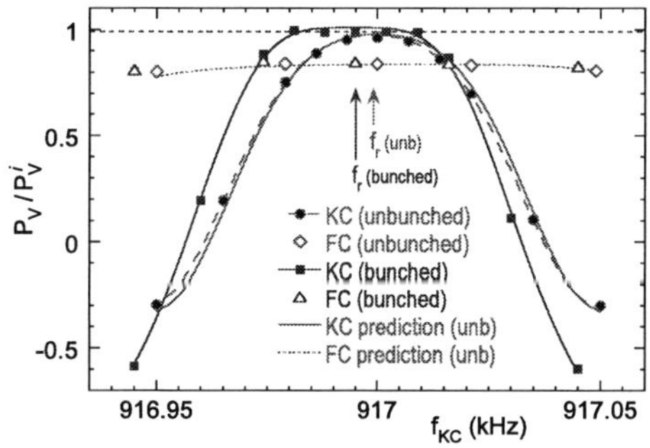

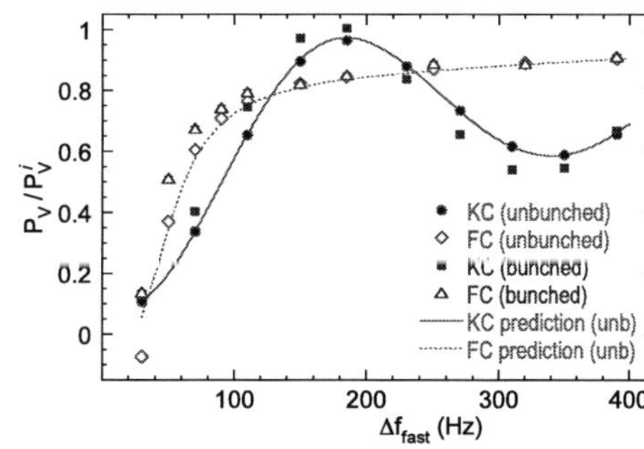

FIGURE 4. (RIGHT) Measured 1.85 GeV/c deuteron P_V/P_V^i, averaged for all nonzero spin states, plotted vs. fast frequency ramp range Δf_{fast}; Fig. 3 caption lists the parameters.

for both bunched and unbunched beams; the depolarizations ($1 - P_V/P_V^i$) are summarized in Fig. 7. With the optimized KC parameters, the average measured depolarizations were 3.3 ± 0.3% and 0.8 ± 0.3% for unbunched and bunched beams, respectively; the average measured FC depolarizations were 15.6 ± 0.2% and 15.0 ± 0.3%, respectively. Thus, KC reduced the depolarization far more than FC: by factors of 4.7 ± 0.3 and $19 \pm^{12}_{5}$ for unbunched and bunched beams, respectively.

While the Chao formalism cannot yet calculate the KC behavior for bunched beams, the measured 19-fold reduction in depolarization for KC over FC at the same crossing rate shows that Kondratenko Crossing may be quite valuable for the bunched beams used in accelerators.

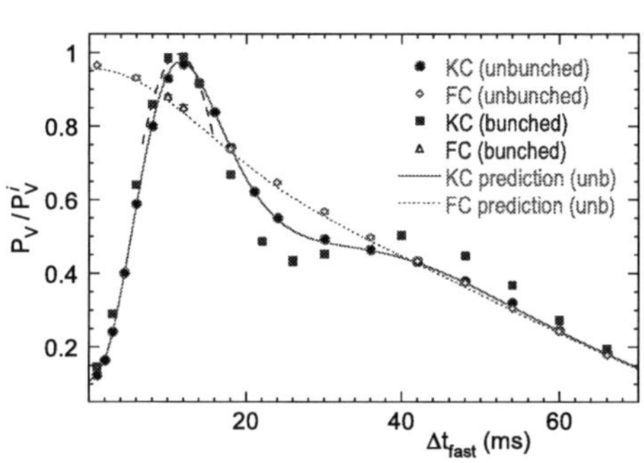

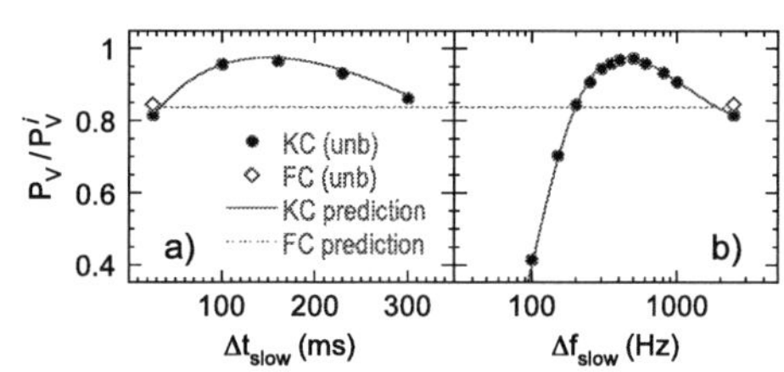

FIGURE 6. **(a)** Measured 1.85 GeV/c deuteron P_V/P_V^i plotted vs. slow frequency ramp time Δt_{slow}. **(b)** Measured 1.85 GeV/c deuteron P_V/P_V^i plotted vs. slow frequency ramp range Δf_{slow}. See Fig. 3 caption for parameters; the beam was unbunched.

FIGURE 5. Measured 1.85 GeV/c deuteron P_V/P_V^i, averaged for all nonzero spin states, plotted vs. fast frequency ramp time Δt_{fast}. The dashed line is an emperical 1st-order Lorentzian fit to the 4 highest bunched KC points; Fig. 3 caption lists the parameters.

TABLE 1. Summary of the KC and FC polarization and depolarization ratios at the KC peaks in Figs. 3-6 for the indicated parameters.

Fig. No.	3	4	5	6a	6b	P_V/P_V^i	$(1 - P_V/P_V^i)$	
Parameter	f_{KC}	Δf_{fast}	Δt_{fast}	Δf_{slow}	Δt_{slow}	**Average**	**Average**	**Pred.**
KC (unb.)	96.6±0.4%	96.5±0.4%	96.7±0.4%	96.8±0.3%	96.6±0.4%	96.7±0.2%	3.3±0.2%	2.6%
FC (unb.)	83.9±0.5%	84.4±0.3%	84.7±0.4%	84.5±0.5%	84.4±0.4%	84.4±0.2%	15.6±0.2%	16.4%
KC (bun.)	99.0±0.2%	100.5±0.7%	99.6±0.6%	N/A	N/A	99.2±0.3%	0.8±0.3%	N/A
FC (bun.)	84.8±0.3%	85.2±0.6%	86.0±0.7%	N/A	N/A	85.0±0.3%	15.0±0.3%	N/A

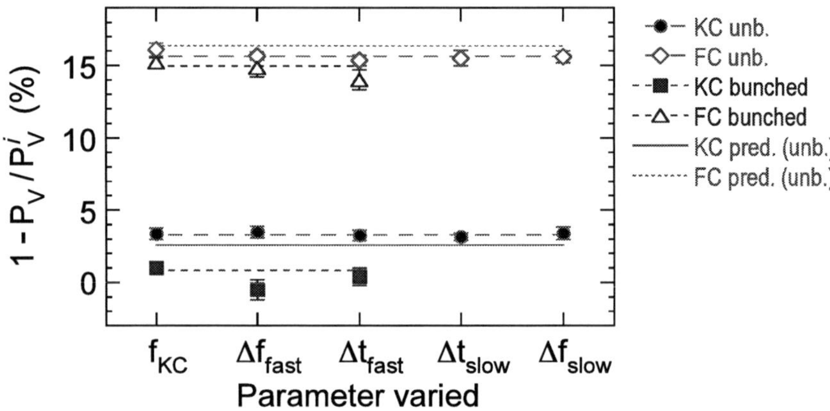

FIGURE 7. Summary of depolarization at each KC peak for both KC and FC, with both bunched and unbunched beam. The horizontal axis denotes the parameters varied to obtain the data listed in Table 1. The solid and dotted lines are the Chao formalism predictions (only for unbunched beam). The dashed lines are fits to the measured data that give the average depolarization values and errors.

ACKNOWLEDGMENTS

We thank the COSY staff for the successful operation of COSY, and E.D. Courant, Ya.S. Derbenev, Yu.F. Orlov, T. Roser, E.J. Stephenson and K. Ulbrich for their help and advice. This research was supported by grants from the German BMBF Science Ministry, its JCHP-FFE program at COSY and the Helmholtz Association through funds provided to the virtual institute "Spin and strong QCD" (VH-VI-231).

REFERENCES

* Also SLAC, 2575 Sand Hill Rd., Menlo Park, CA 94025

1. A.M. Kondratenko et al, Physics of Particles and Nuclei Letters **1**, 255 (2004).
2. R. Maier, Nucl. Instrum. Methods Phys. Res., Sect. A **390**, 1 (1997).
3. V.S. Morozov *et al.*, Phys. Rev. Lett. **100**, 054801 (2008).
4. H. Stein *et al.*, At. Energ. **94**, 24 (2003).
5. M. Altmeier *et al.* (EDDA Collaboration), Phys. Rev. Lett. **85**, 1819 (2000).
6. M. Froissart and R. Stora, Nucl. Instrum. Methods **7**, 297 (1960).
7. A.M. Kondratenko, "Choosing Crossing Parameters" SPIN@COSY Int. Report (November 2006). URL: http://spinbud.physics.lsa.umich.edu/OptKCpar.pdf
8. A.W. Chao, Phys. Rev. ST Accel. Beams **8**, 104001 (2005).
9. For additional references and details see V.S. Morozov *et al.*, "Experimental test of new technique to overcome spin-depolarizing resonances", submitted to Phys. Rev. Lett. (November 13, 2008).

Polarization Upgrade at ELSA

A. Balling[a], A. Dieckmann[a], F. Frommberger[a], W. Hillert[a]

[a]*Physikalisches Institut, Universitaet Bonn, Nussallee 12, D-53115 Bonn, Germany*

Abstract. The Electron Stretcher Accelerator ELSA at Bonn University is a fast ramping stretcher ring supplying a polarized electron beam of up to 3.5 GeV to hadron physics experiments. During acceleration (4.5 GeV/sec), sophisticated schemes for the compensation of depolarizing resonances are mandatory, such as betatron tune jumps for intrinsic resonances and harmonic closed orbit corrections for imperfection resonances. This contribution will present recently implemented techniques of dynamic closed orbit correction as well as the newly developed orbit correction system which reduces the time needed for an orbit bump application down to well below 20 milliseconds.

Keywords: Polarization; Depolarizing Resonances, Resonance Compensation; Orbit Correction
PACS: 29.27.Hj; 41.85.Lc

INTRODUCTION

The ELSA accelerator facility at the University of Bonn [1] consists of a 50 keV source of polarized electrons [2] and two thermionic electron guns, two injector LINACs, a booster synchrotron and the 3.5 GeV stretcher ring (see Figure 1). In order to supply a nearly continuous electron beam to the hadron physics experiments, several injections from the synchrotron, running at a repetition rate of 50 Hz, are accumulated in the stretcher ring (at 1.2 GeV typ.), accelerated to the extraction energy (3.5 GeV max.) and then extracted slowly by means of resonance extraction.

The hadron physics program performed at ELSA is mainly focused on double polarization experiments utilizing tagged photons (linearly and circularly polarized) and a polarized frozen spin target [3].

SPIN DYNAMICS IN CIRCULAR ACCELERATORS & DEPOLARIZING RESONANCES

Within a magnetic field $\vec{B}$, the spin $\vec{S}$ of a charged particle at relativistic speed precesses around the magnetic field vector according to the Thomas-BMT equation:

$$\vec{\Omega}_{BMT} = \frac{e}{m_o\gamma}\left(\underbrace{(1+\gamma a)\vec{B}_\perp}_{\text{Thomas term}} + \underbrace{(1+a)\vec{B}_\parallel}_{\text{Larmor term}} - \underbrace{(a+\frac{1}{\gamma+1})\gamma\vec{\beta}\times\frac{\vec{E}}{c}}_{\text{electric contribution}} \right) \quad (1)$$

where m_o and e denote rest mass and charge of the particle (in this case the electron), a its gyromagnetic anomaly, $\vec{\beta} = \vec{v}/c$ and $\gamma = E / E_o$ its relativistic parameters

CP1149, *Spin Physics, 18th International Spin Physics Symposium*
edited by D. G. Crabb, D. B. Day, S. Liuti, X. Zheng, M. Poelker, and Y. Prok

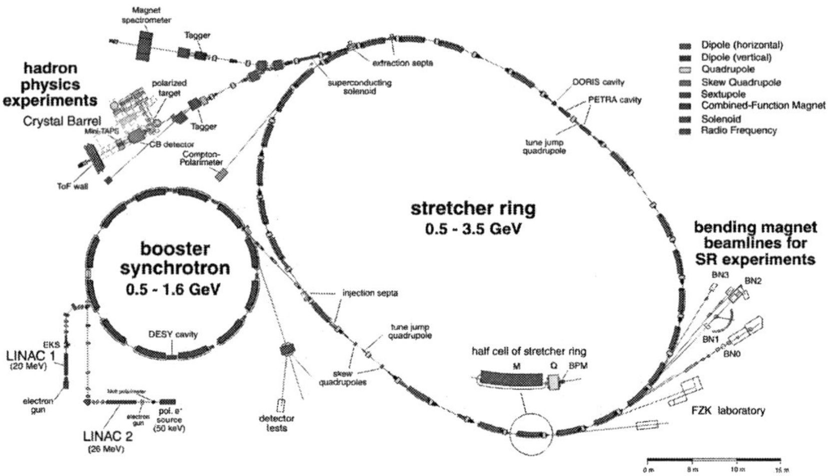

FIGURE 1. The ELSA accelerator facility at Bonn University.

($E_o = m_o c^2$), c the speed of light, $\vec{E}$ the electric field and $\vec{B}_\perp$, $\vec{B}_\parallel$ the components of the magnetic field perpendicular and parallel to the particle's momentum. For electron energies in the GeV regime and due to the lack of solenoid magnets within the stretcher ring the precession is dominated by the Thomas term. Subtraction of the cyclotron frequency $\vec{\Omega}_{cyc} = (e/m_o\gamma)\cdot\vec{B}_\perp$ leads to the so-called spin tune $Q_{sp} = \gamma \cdot a$ which is the energy-dependent number of spin precessions per particle revolution.

Due to the small but non-zero energy spread of the electrons within the beam, only polarization components parallel to the vertical magnetic field of the bending magnets are conserved during acceleration whereas horizontal components are lost within milliseconds. During acceleration, every $E_o/a = 440.63$ MeV the spin tune becomes an integer number and the spin precesses phase-synchronously with the particle's revolution. Under this condition of a so-called ***imperfection resonance***, horizontal field components (e.g. resulting from vertical orbit displacements in the focussing quadrupole magnets) cause the spin vector to tilt out of the vertical, thus leading to depolarization. ***Intrinsic resonances***, on the other hand, occur due to coupling between spin precession and vertical betatron oscillation.

RESONANCE COMPENSATION

Major key to polarization conservation during acceleration is to reduce horizontal magnetic fields of any kind, thus keeping the orbit flat. That requires a precise geodetic alignment of the accelerator components in the first place as well as a good orbit control during operation, especially on the fast energy ramp (up to $dE/dt = 7.5$ GeV/sec $\Leftrightarrow$ $dB/dt = 2$ T/sec). At ELSA, the closed orbit is stabilized with high precision using a system of 32 beam position monitors (one located at each quadrupole magnet) and 40 steerer magnets distributed along the ring. During the acceleration phase, the beam positions are read out from the BPMs at a rate of 1 kHz and energy-

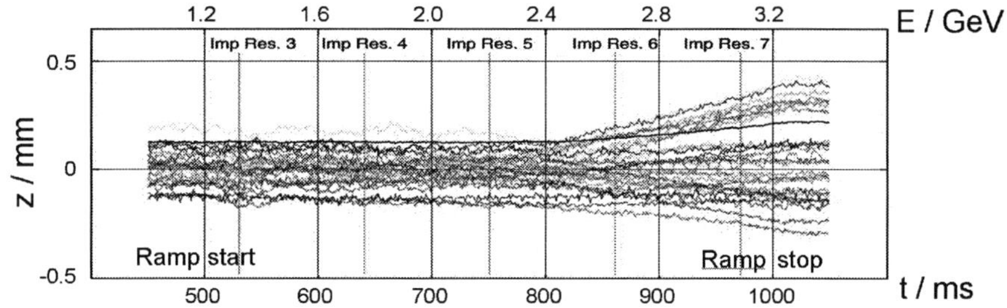

FIGURE 2. Vertical beam positions measured by the 32 BPMs during the fast energy ramp (bold line: rms) indicate that dynamic closed orbit correction works very well up to 2.4 GeV. For reference, the imperfection resonances occuring every 440.63 MeV are included into the figure.

dependent orbit corrections are applied accordingly by means of offline feed-forward techniques using the steerers. The system is thus sensitive to dynamic effects and a vertical orbit stabilization of 140 μm rms is achieved routinely up to an energy of 2.4 GeV as shown in Fig. 2, achieving a significant decrease in resonance strength.

In order to further reduce the strength of the imperfection resonances, additional harmonic corrections are applied: For each resonance the relevant obit harmonic is compensated individually by empirically optimized (parameters a and b) manipulations of the vertical closed orbit according to

$$B_{harmcorr}(\theta) = a \cdot \sin(Q_{sp} \cdot \theta) + b \cdot \cos(Q_{sp} \cdot \theta) \qquad (2).$$

For the compensation of intrinsic resonances fast betatron tune jumps are applied by dedicated pulsed quadrupole magnets, reducing the time span during which the resonance condition is fulfilled [4].

THE NEW ORBIT CORRECTION SYSTEM AT ELSA

The existing orbit correction system is suitable for orbit control and resonance compensation up to a beam energy of 2.4 GeV. Above however, it is limited (see Fig. 2) in two ways. On the one hand, the vertical steerer magnets begin to encounter field saturation and can therefore not provide sufficient field strength for every necessary orbit bump. On the other hand, the requirement on the rise times of the associated power supplies increase: at typical ramping speed of 4.5 GeV/sec the period between two consecutive imperfection resonances amounts to less than 100 msec, but in order not to interfere with the betatron tune jump for the compensation of the interjacent intrinsic resonance, the available time for application and withdrawal of a harmonic correction is on the order of only 40 msec. In a worst case scenario, the power supply must therefore be able to perform a jump from minimum current up to maximum current and back down within that period.

In order to avoid a reduced degree of polarization at higher energies a new orbit correction system has been developed: as horizontal steerers the so far unused trim windings of the 24 bending magnets will be utilized, while for vertical orbit corrections 29 newly designed steerer magnets will be installed together with new beam pipe segments optimized for the suppression of eddy currents. For both steerer

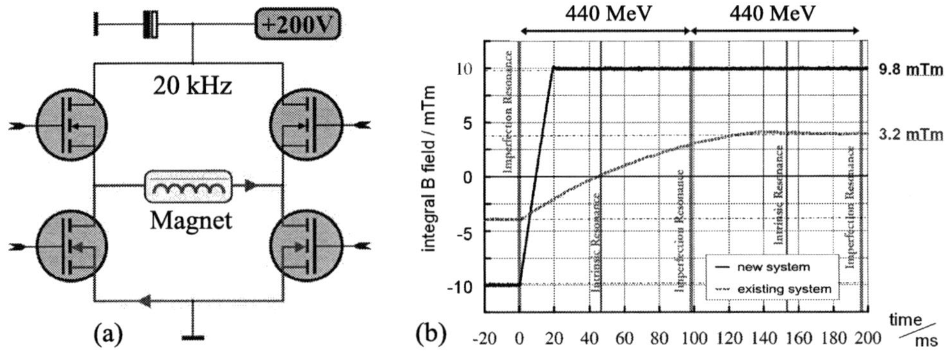

FIGURE 3. (a) The pulsed H bridge of the new power supplies. The inductance-driven current loop in between two consecutive loading pulses is indicated in the lower section (L ≈ 170 mH). (b) Comparison between the existing setup and the new system regarding integral field strength and achievable ramping times with respect to the occuring intrinsic and imperfection

types new programmable four-quadrant power supplies have been developed in-house. Their power sections are based on a 20 kHz H-bridge (see Fig. 3 (a)) and are operated by microcontrollers of the Microchip PIC18 series. The in total 53 power supplies are connected via four fast (1Mbaud) CAN bus segments to a Linux-PC which communicates with the top level accelerator control system via LAN.

Figure 3 (b) shows the results achieved for the vertical orbit correction subsystem: the available integral field strength has been increased by a factor of three compared to the existing setup and the time required for an orbit bump application has been reduced down to well below 20 milliseconds as required.

SUMMARY AND OUTLOOK

One major key to polarization conservation during acceleration within a circular accelerator is a precisely controlled closed orbit. In order to improve orbit control at higher energies, a new correction system has been deveolped and is being set up which will not only allow for the dynamic correction of the overall closed orbit up to the maximum beam energy but also for the harmonic compensation of all imperfection resonances on the fast energy ramp without interfering with the correction of the interjacent intrinsic resonances.

REFERENCES

1. W. Hillert, *Eur. Phys. J.* A **28** s01, 139-148 (2006).
2. W. Hillert et al., "The 50 kV inverted source of polarized electrons at ELSA" in Proceedings of the 14th international Spin Physics Symposium SPIN2000, edited by K. Hatanaka et al., AIP Conference Proceedings 570, American Institute of Physics, Melville, NY, 2001, pp. 961-964.
3. H. Dutz, *Nucl. Instrum. Methods* A **526**, 117-125 (2004).
4. M. Hoffmann et al, "The polarized electron beam at ELSA" in Proceedings of the 14th international Spin Physics Symposium SPIN2000, edited by K. Hatanaka et al., AIP Conference Proceedings 570, American Institute of Physics, Melville, NY, 2001, pp. 756-760 (2001).

Analysis of Data for Stored Polarized Beams Using a Spin Flipper

Yu. M. Shatunov*, S. R. Mane† and V. I. Ptitsyn**

*Budker Institute of Nuclear Physics, Novosibirsk, 630090, Russia
†Convergent Computing Inc., P.O. Box 561, Shoreham, NY 11786, USA
**Brookhaven National Lab., Upton, NY 11973, USA

Abstract. We employ the ASPIRRIN program to fit recent measurements of spin flip resonance widths, made using a radial field rf dipole spin flipper, for stored beams of polarized protons and deuterons at the COSY storage ring. We obtain a good fit to the data.

Keywords: polarized beams, spin flippers, spin resonances, circular accelerators
PACS: 29.27.Hj, 29.20.D-, 02.60.Lj

In a recent paper [1], we used the program ASPIRRIN [2, 3] ("Analysis of SPIn Resonances in RINgs") to fit recent experimental measurements of spin-flip resonance widths for stored polarized beams at the COSY storage ring, for protons [4] and deuterons [5]. The measurements in [4] and [5] were made using a radial field rf dipole spin flipper. The program ASPIRRIN implements the "spin response formalism" which is linear response theory applied to spin dynamics in storage rings. ASPIRRIN contains the full functionality to calculate spin-flip resonance widths for spin flippers, in rings of arbitrary structure (including full and/or partial Siberian Snakes).

The above data have received attention because of claims in [5] that the reported values of the spin-flip resonance widths, for polarized deuterons, did not agree with the theory of spin dynamics in storage rings. However, in addition to its direct spin coupling, an rf dipole spin flipper also induces coherent vertical betatron oscillations, and these can in turn couple to the spin, thereby modifying the overall spin-flip resonance strength. This latter contribution was omitted in the analyses in [4] and [5], but is included in ASPIRRIN.

Fig. 1 displays the theoretical calculation from [1]. The experimental data is taken from [6], which gives the actual data values for the graphs in [4] and [5]. Where there are multiple data points for the same value of the vertical betatron tune, we have plotted an equally weighted average data value. Note the data for protons in [4] had a factor of 2 normalization error. This was pointed out by Bai, MacKay and Roser [7]. The erroneous factor of 2 has been corrected in [6] and hence is not required in Fig. 1.

All the symbols below have their usual meanings. The independent variable is the ring azimuth θ and the ring circumference is $2\pi R$. The coordinate system is (x,y,z) where x is radially outward, y is along the reference orbit and z is vertical. A prime denotes differentiation with respect to θ. The spin tune is ν and the vertical betatron is ν_z. The particle's magnetic moment anomaly is $G = \frac{1}{2}(g-2)$ and its Lorentz factor is γ, and we define $\nu_0 = G\gamma$. We also define the dimensionless scaled magnetic fields on the reference

CP1149, *Spin Physics, 18th International Spin Physics Symposium*
edited by D. G. Crabb, D. B. Day, S. Liuti, X. Zheng, M. Poelker, and Y. Prok

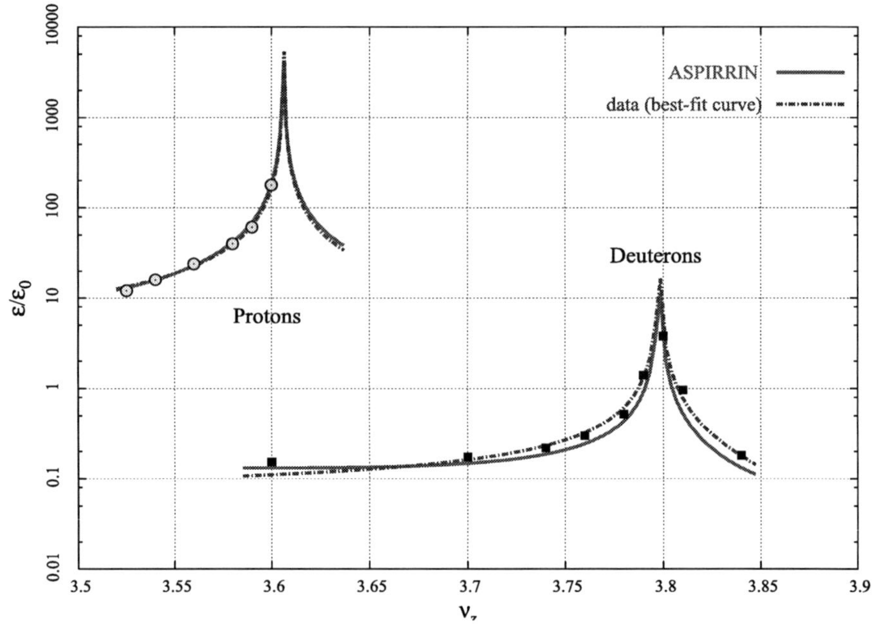

FIGURE 1. Graph of the ratio of the total spin-flip resonance strength ε to the direct spin kick ε_0 for a radial field rf dipole, for stored proton and deuteron beams at the COSY storage ring. The mathematical definitions of ε and ε_0 are presented in the main body of the text. The horizontal axis shows the vertical betatron tune ν_z.

orbit $K_{x,y,z} = B_{x,y,z}/B_0$, where B_0 is a reference value given by the arc dipoles

$$B_0 = \frac{1}{2\pi}\int_{\text{arcs}} B_z\, d\theta\,. \tag{1}$$

We also define the phase Θ via

$$\Theta = \int_0^\theta K_z(\tilde\theta)\, d\tilde\theta = \int_0^\theta \frac{R}{\rho(\tilde\theta)}\, d\tilde\theta\,, \tag{2}$$

where ρ is the local bend radius in the horizontal plane. Let the rf dipole spin flipper be located at the azimuth θ_0 (mathematically, a δ-function), and let its integrated field be

$$\int B_{\text{rf}}\, d\ell = B_{\text{osc}} L \cos(\nu_{rf}\theta + \chi)\,. \tag{3}$$

Here ν_{rf} is the tune of the spin flipper, χ is an initial phase and $B_{\text{osc}}L$ denotes the peak integrated field. The spin response formalism [2] gives the resonance strength as

$$\varepsilon = \frac{1}{4\pi}\frac{B_{\text{osc}}L}{B\rho}|F_3(\theta_0)|\,. \tag{4}$$

Here $B\rho$ is the magnetic rigidity. For a planar ring such as COSY,

$$F_3(\theta_0) = \frac{1}{2}\left\{\frac{f_z(\theta_0)}{e^{i2\pi(\nu-\nu_z)}-1}\int_{\theta_0}^{\theta_0+2\pi}[(\nu_0^2+G)K_z f_z^{*\prime}+(1+G)K_z' f_z^*]e^{i\nu_0\Theta}\,d\theta \right.$$
$$\left. -\frac{f_z^*(\theta_0)}{e^{i2\pi(\nu+\nu_z)}-1}\int_{\theta_0}^{\theta_0+2\pi}[(\nu_0^2+G)\,K_z f_z'+(1+G)K_z' f_z]e^{i\nu_0\Theta}\,d\theta\right\}. \quad (5)$$

Here $f_z = \sqrt{\beta_z/R}\,e^{i\phi_z}$, where β_z is the vertical beta function and the phase is given by $d\phi_z/d\theta = R/\beta_z$. We evaluate the integral at $\nu_0 = \nu_{\text{res}}$, the center of the resonance. We define the "direct" spin kick of an rf dipole via

$$\varepsilon_0 = \frac{G\gamma+1}{4\pi}\frac{B_{\text{osc}}L}{B\rho}. \quad (6)$$

The quantity plotted on the vertical axis of Fig. 1 is

$$\frac{\varepsilon}{\varepsilon_0} = \left|\frac{F_3(\theta_{rf})}{G\gamma+1}\right|. \quad (7)$$

In Fig. 2, we display the value of $|F_3|$ around the circumference of COSY, for protons and deuterons. For deuterons, the rf dipole is located at almost a minimum value of $|F_3|$. In Fig. 3, we display the value of $|F_{3a}|$ and $|F_{3b}|$, which are the contributions to F_3 from the terms in $(\nu_0^2+G)K_z$ and $(1+G)K_z'$, respectively. The former term is from the magnet body and the latter is from the dipole fringe fields. For protons, the magnet body term dominates, whereas for deuterons the fringe field term dominates the contribution to $|F_3|$.

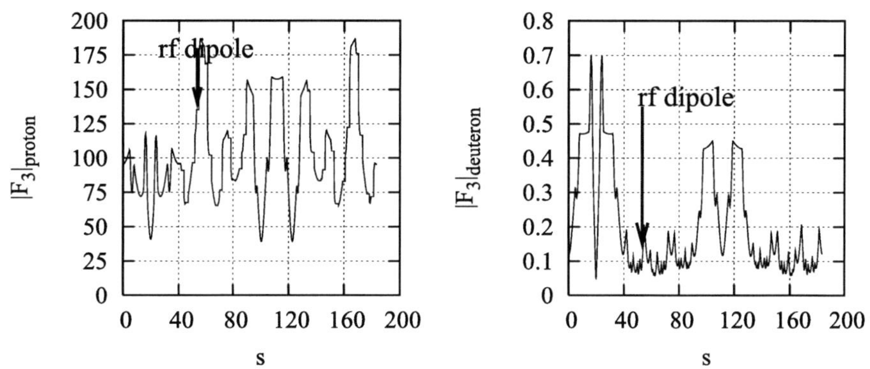

FIGURE 2. Graph of $|F_3|$, for protons (left) and deuterons (right), around the circumference of the COSY ring. The proton momentum is $p = 2.1$ GeV/c, and the vertical betatron tune is 3.525. The deuteron momentum is $p = 1.85$ GeV/c and the vertical betatron tune is 3.60. The location of the rf dipole is indicated.

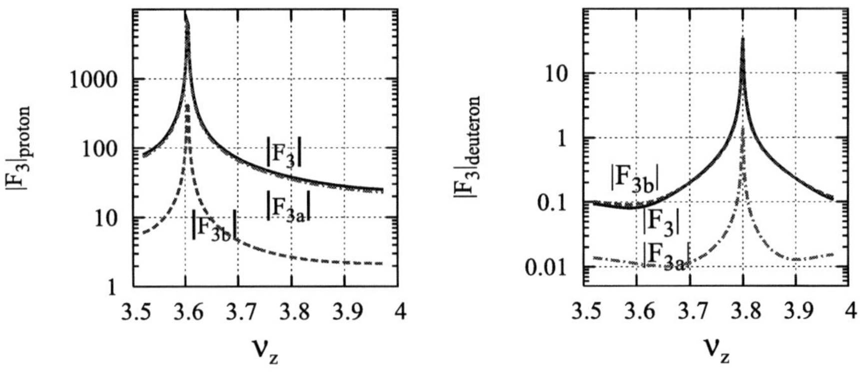

FIGURE 3. Graph of $|F_3|$ at the location of the rf dipole as a function of the vertical betatron tune ν_z, for polarized protons (left) at $p = 2.1$ GeV/c and polarized deuterons (right) at $p = 1.85$ GeV/c. The curves for F_{3a} and F_{3b} are the individual contributions to F_3 from the terms in $(\nu_0^2 + G)K_z$ and $(1+G)K_z'$, respectively, as shown in eq. (5).

ACKNOWLEDGMENTS

V. Ptitsyn is an employee of Brookhaven Science Associates, LLC and his work was funded under Contract No. DE-AC02-98CH10886 with the U. S. Department of Energy.

REFERENCES

1. Yu. M. Shatunov and S. R. Mane, *Phys. Rev. ST Accel. Beams*, **11** 094002 (2008).
2. V. Ptitsyn, Ph.D. thesis, Budker Institute of Nuclear Physics, Novosibirsk (1997) (in Russian).
3. E. A. Perevedentsev, Yu. M. Shatunov and V. Ptitsin, *Proceedings of SPIN02*, edited by Y. I. Makdisi, A. U. Luccio and W. W. MacKay, AIP Conference Proceedings 675, American Institute of Physics, New York, 2003, p. 761.
4. M. A. Leonova *et al.*, *Phys. Rev. ST Accel. Beams* **9** 051001 (2006).
5. A. D. Krisch *et al.*, *Phys. Rev. ST Accel. Beams* **10** 071001 (2007).
6. M. A. Leonova, Ph. D. Thesis, University of Michigan (2008).
7. M. Bai, W. W. MacKay and T. Roser, *Phys. Rev. ST Accel. Beams* **8** 099001 (2005).

Driven Spin Coherent Precession at RHIC

M. Bai[a], T. Roser[a], G. Bunce[a] and R. Gill[a]

[a]*Brookhaven National Laboratory, Upton, NY 11973*

Abstract. In a circular accelerator, exciting a coherent spin precession can be used to measure the spin precession tune, which is critical for understanding spin dynamics of the polarized beam. This can be achieved by driving the polarized beam with an rf dipole at an oscillating frequency f_{res} in the vicinity of the beam spin precession frequency f_s, namely, $f_{res} \approx f_s$. The driven coherent spin precession was measured with the RHIC Coulomb Nuclear Interference polarimeter using a carbon target. This paper presents the spin coherent precession observed with polarized proton beam at the BNL RHIC. Both experimental data and its application for spin tune measurement are discussed.

Keywords: spin coherent precession, single isolated spin resonance.
PACS: 29.20.db

INTRODUCTION

At the neighborhood of an isolated spin resonance, the stable spin direction[1,2] in the frame that rotates at the same frequency as the spin resonance is given by

$$S_y = \frac{Q_s - Q_{res}}{\sqrt{\left|\frac{\phi_{res}}{2\pi}\right|^2 + |Q_s - Q_{res}|^2}} \quad S_x = \frac{\frac{\phi_{res}}{2\pi}\cos\chi}{\sqrt{\left|\frac{\phi_{res}}{2\pi}\right|^2 + |Q_s - Q_{res}|^2}} \quad S_z = \frac{\frac{\phi_{res}}{2\pi}\sin\chi}{\sqrt{\left|\frac{\phi_{res}}{2\pi}\right|^2 + |Q_s - Q_{res}|^2}} \tag{1}$$

where S_x, S_y and S_z are the radial, vertical and longitudinal component of the stable spin direction respectively. Q_s is spin tune, i.e. number of spin precessions in one orbital revolution, Q_{res} is the spin resonance frequency in the unit of orbital revolution frequency, or in other words, spin resonance tune. ϕ_{res} is the amount of the spin rotation by the isolated resonance in the rotating frame, $\hat{x}\cos\chi + \hat{z}\sin\chi$ is the axis of spin rotation in the rotating frame. This shows that a coherent spin precession can be excited by adiabatically exciting an isolated spin resonance, and the ratio of the vertical projection of the coherent spin precession to the amplitude of the coherent spin precession projection in the horizontal plane is

$$\frac{S_y}{\sqrt{S_x^2 + S_z^2}} = (Q_s - Q_{res})\frac{2\pi}{\phi_{res}}. \tag{2}$$

Hence, a coherent spin precession can be very useful for spin manipulation as well as spin diagnostics such as measuring spin tune. The fact that this manipulation preserves the polarization makes it possible to measure the spin tune at store energy in a high energy accelerator.

CP1149, *Spin Physics, 18th International Spin Physics Symposium*
edited by D. G. Crabb, D. B. Day, S. Liuti, X. Zheng, M. Poelker, and Y. Prok

An isolated spin resonance can be excited by a localized rf spin rotator that rotates the spin vector by a fixed angle ϕ_{res} around an axis rotating in the horizontal plane at a tune of Q_{res}[3]. The one-turn transfer map of the of the two component spinor $\vec{\psi}$ for spin ½ particle[1] defined as $S_{x,y,z} =< \vec{\psi} \,|\, \sigma_{1,3,2} \,|\, \vec{\psi} >$ is given by

$$\vec{\psi}_{n+1} = e^{-\frac{i}{2}Q_s 2\pi\sigma_3} e^{-\frac{i}{2}\phi_{res}[\cos(Q_{res}2\pi+\chi)\sigma_1+\sin(Q_{res}2\pi+\chi)\sigma_2]}\vec{\psi}_n \qquad (3)$$

where $\sigma_{1,3,2}$ are Pauli matrices, n is the orbital revolution number. In the neighborhood of the resonance, i.e. $|Q_s - Q_{res}| << 1$ and for small ϕ_{res}, Eq. 3 in the resonance rotating frame becomes.

$$\vec{\psi}_{res,n+1} = e^{-\frac{i}{2}[(Q_s-Q_{res})\sigma_3+\phi_{res}(\cos\chi\sigma_1+\sin\chi\sigma_2]}\vec{\psi}_{res,n} \qquad (4)$$

with $\vec{\psi}_{res,n} = e^{\frac{i}{2}Q_{res}2\pi n\sigma_3}\vec{\psi}_n$.

Eq. 4 yields the same spin motion as in Eq. 1. The two left plots of Fig. 1 are turn-by-turn radial and vertical components of the spin vector of a numerical simulation of spin motion nearby an isolated spin resonance, which was adiabatically turned on and off.

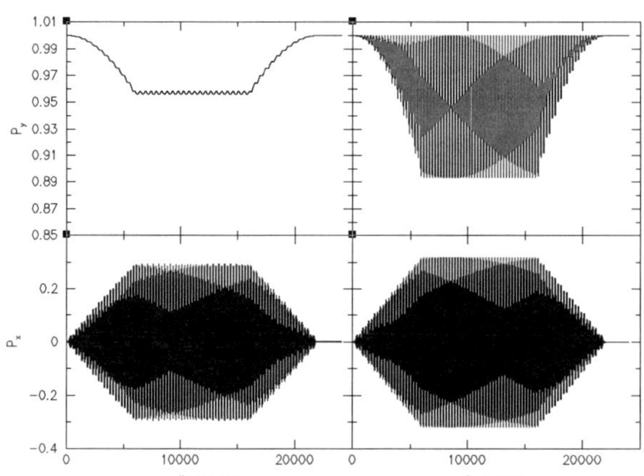

FIGURE 1. Simulation results of spin motion of an isolated spin resonance and spin motion excited by a single ac dipole. The left two plots are the vertical component(top plot) and radial component (bottom plot) of the spin vector in driven by an rf spin rotator. The right two plots are the vertical component(top plot) and radial component (bottom plot) of the spin vector driven by a single rf dipole.

Instead of the ideal rf spin rotator, which provides a rotating field, a single rf dipole is often used to manipulate spin. In this case, two resonances at $Q_s = Q_{res} + k$ and $Q_s = -Q_{res} + k$ are excited [3]. Here, *k* is an integer. Because the spin tune in RHIC is ½, these two resonances are very close and the spin motion becomes more complicated than the motion expressed in Eq. 1. The two right plots in Fig. 1 shows the turn-by-turn spin motion in the presence of a single rf dipole. Instead of a constant projection on the vertical direction, the vertical projection of the spin vector also oscillates.

Experimental Setup and Result

In RHIC, a dual Siberian snake configuration is employed for two accelerators to preserve polarization[4]. This yields a spin tune of ½. The experiment was carried out at RHIC injection energy during the RHIC polarized proton operation in 2008. A single rf dipole located next to one of the six interaction points of RHIC was used, and the polarization was measured by the RHIC CNI polarimeter[5] which consists of 6 detectors with two pairs located at the ±45 degrees above and below the beam direction and one pair located in the horizontal plane. The recoiled carbon events at each detector were recorded for each orbital revolution. The radial asymmetry A_{x,res,j_bin} in the resonance rotating frame was calculated in the following way:

$$A_{x,res,j\_bin} = \frac{N_{top,j\_bin} - N_{bottom,j\_bin}}{N_{top,j\_bin} + N_{bottom,j\_bin}} \quad (7)$$

with $$N_{top,botton,j\_bin} = \sum_{i}^{nturn} N_{top,bottom,i} \qquad \text{for } each\ i, (j\_bin - 1)\frac{2\pi}{N_{bin}} < \sin(2\pi Q_{res} i + \chi) < j\_bin \frac{2\pi}{N_{bin}}$$

where i is the ith orbital revolution, N_{bin} is the total number of bins of the resonance phase ranging from 0 to 2π. $N_{top,bottom,j\_bin}$ is the number of events of the two top detectors or two bottom detectors in the j_bin of the resonance phase. Vertical asymmetry in the resonance rotating frame was calculated similarly by just replacing the top and bottom events with left and right events.

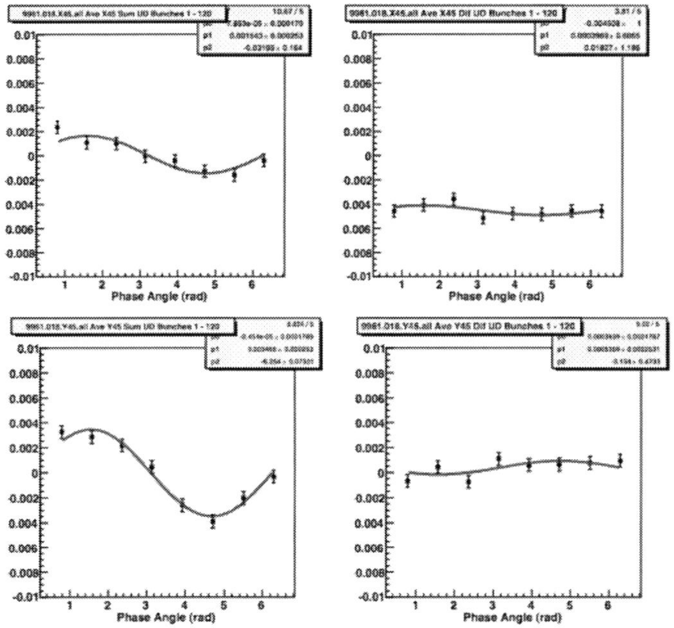

FIGURE 2. The right two plots are the measured vertical(top plot) and radial true asymmetry(bottom plot) in the resonance rotating frame. The left two plots are the vertical and radial false asymmetry. In all the plots, the horizontal axis is the resonance phase from 0 to 2π.

Since the rf dipole also induces coherent vertical orbital oscillation[6], the experiment was taken with not only spin up bunches but also spin down bunches to cancel the false asymmetry. The true asymmetry due to the spin coherent precession is the difference of the measured asymmetry for the spin up bunches and spin down bunches and the false asymmetry due to the coherent orbital oscillation is the average of the asymmetry of the bunches with opposite spin directions. During each data taking, the rf dipole was energized for more than 10,000 orbital revolutions at a tune close to ½. It was then left on for as long as 10 minutes for the RHIC CNI polarimeter to record the recoil carbon events at each orbital revolution.

Fig. 2 is a typical measured radial and vertical asymmetry in the rotating frame. The two left plots are the false asymmetries and the two right plots are the asymmetries due to the coherent spin precession. Both show that no significant spin coherent precession was observed. This seems to indicate that the spin tune is too far away from the rf dipole tune. For this particular case, the rf dipole was operated at $\phi_{res} = 0.0003 * 2\pi$ with its tune set at 0.491. For other cases when the rf dipole tune was placed at 0.493 or even closer to 0.5, 75% and more depolarization was observed. This could be due to the fact that when the rf dipole tune gets too close to 0.5, the two resonances inevitably interfere and cause polarization loss.

Conclusion

The experiments of using a single rf dipole to excite a coherent spin precession in RHIC where spin tune is ½ show it is very difficult to excite a coherent spin precession due to the interference between the two resonances excited by the rf dipole. We expect this situation will be avoided with a newly designed RHIC spin flipper using rotating fields[3].

ACKNOWLEDGMENTS

The authors would like to thank A. Zelenski, H. Huang and other polarimeter experts for their support with the CNI polarimeter. The authors also would like to thank the RHIC operations crew who helped in data taking and machine setup. The authors also would like to thank E. Stephenson from Indiana University for his advice on analyzing the systematic errors in the CNI polarimeter.

REFERENCES

1. S. Y. Lee *The New Physique*, Publisher: World Scientific, 2005, pp. 25-30.
2. A. Chao, Phys. Rev. ST Accel. Beams **8**, 104001 (2005)
3. M. Bai and T. Roser, Phys. Rev. ST Accel. Beams **11**, 091001 (2008)
4. K. Brown and et al, Polarized Proton Collider at RHIC Design Manual, 1998
5. J. Tojo et al., Phys. Rev. Lett. **89**, 052302 (2002).
6. E. Stephenson (private communication).

POLARIZED SOURCES AND TARGETS

CONTRIBUTIONS

Recent Advances in Polarized He-3 Targets

Jaideep Singh*, Peter Dolph*, Karen Mooney*, Vladimir Nelyubin*, Al Tobias*, Aidan Kelleher†, Todd Averett† and Gordon Cates*

*Physics Department, University of Virginia, P.O.Box 400714, Charlottesville, VA 22904-4714
†Dept. of Physics, The College of William & Mary, P.O. Box 8795, Williamsburg, VA 23187-8795

Abstract. Recently there have been many important advances made in spin-exchange optical-pumping (SEOP) polarized He-3 targets. Use of hybrid K-Rb SEOP achieved He-3 polarizations near 50% in-beam during the recent JLab G_E^n experiment (E02013). Combining well-optimized alkali-hybrid SEOP with high-power spectrally-narrowed laser diode arrays increases He-3 polarizations to 70%. We describe how these technologies work and why they improve target polarization.

Keywords: laser-polarized helium-3 targets for electron scattering, alkali-hybrid spin-exchange optical pumping, spectrally-narrowed laser diode arrays, hybrid potassium-rubidium (K-Rb) SEOP
PACS: 29.25.Pj,67.30.ep,32.80.Xx,25.30.c

INTRODUCTION

The focus of this talk is high-density two-chamber spin-exchange optical-pumping (SEOP) polarized ^{3}He target cells used in experiments conducted at Stanford Linear Accelerator Laboratory (SLAC) and Jefferson Laboratory (JLab). Over the last 20 years, they have been used to make precise measurements of the neutron electromagnetic form factors G_E^n & G_M^n, spin asymmetry A_1^n, and spin structure functions g_1^n & g_2^n. The first SLAC-type cells were expected to achieve 50% [1], whereas the average polarization in-beam during the SLAC experiments (E142 & E154) was more typically 35% [2, 3]. Until recently, the performance of the JLab cells had been marginally better (≈40%) and the ^{3}He polarization rarely exceeded 50%. The use of alkali-hybrid SEOP alone has consistently increased the ^{3}He polarization over 50%. In combination with high-power spectrally-narrowed (narrowband) laser diode arrays, the ^{3}He polarization now regularly approaches 70%. The purpose of this talk is to demonstrate how to optimize ^{3}He polarization using alkali-hybrid SEOP & narrowband lasers.

TRADITIONAL SPIN-EXCHANGE OPTICAL PUMPING

In traditional SEOP, a single alkali species, in our case rubidium (Rb), (1) is polarized by optical pumping and then (2) transfers its electronic polarization to the ^{3}He nuclei by spin-exchange collisions. Optical pumping occurs by exciting the D1 transition of the Rb vapor (795 nm) by circularly polarized laser light. If the light propagates along the axis of the applied magnetic field, then (ignoring the Rb nuclear spin) only one of the two Zeeman levels ($m_J = \pm 1/2$) of the $5S_{1/2}$ ground state is selectively excited to the $5P_{1/2}$ state. Because the levels of the excited states are mixed due to collisions, the Rb atom can decay back to either Zeeman level of the ground state with equal

CP1149, *Spin Physics, 18th International Spin Physics Symposium*
edited by D. G. Crabb, D. B. Day, S. Liuti, X. Zheng, M. Poelker, and Y. Prok

probability. Decays dominated by radiation would limit the optical-pumping efficiency; therefore, just enough N_2 is added to nonradiatively quench the $5P_{1/2} \rightarrow 5S_{1/2}$ transition. On a timescale of milliseconds, the Rb polarization reaches an equilibrium value of $P_{Rb} = 1/(1 + \Gamma_{Rb}/R)$, where the optical-pumping rate is given by:

$$R = \int_0 \phi(\nu) \cdot \sigma(\nu) \, d\nu \approx 115 \text{ kHz per } 75 \text{ W}/23 \text{ cm}^2/2 \text{ nm (broadband limit)} \qquad (1)$$

where $\phi(\nu)$ is the photon flux and $\sigma(\nu)$ is the Rb D1 photon-absorption cross section (which is a δ-function in the broadband limit). The Rb spin-relaxation rate is given by:

$$\Gamma_{Rb} = k_{Rb}[\text{Rb}] + k_{He}[^3\text{He}]_{pc} + k_{N2}[\text{N}_2]_{pc} \approx 1 \text{ kHz} \qquad (200^\circ\text{C} \ \& \ [^3\text{He}]_{pc} = 7 \text{ amg}) \ (2)$$

where k_X is the Rb spin-relaxation rate constant for Rb-X collisions and [X] refers to the X number density. The main source of this relaxation is a spin-orbit coupling between the Rb valence electron and the orbital angular momentum of the Rb-X pair. This is analogous to the spin-orbit coupling that occurs within an alkali atom A; therefore, the strength of the A-X spin-orbit coupling scales with the A spin-orbit splitting [4].

A hyperfine-like Fermi-contact interaction couples the valence electron spin of the Rb atom to the nucleus of the ^{3}He atom. The small but nonzero probability that spin exchange occurs between the Rb electron and the ^{3}He nuclear spin is quantified by the Rb-^{3}He spin-exchange rate constant ($k_{se} = 10^{-15}$ cm^3/4.08 hrs [5]). In the SLAC/JLab design, optical pumping and spin exchange occur only in the spherical pumping chamber (6–9 cm diameter, $[^3\text{He}]_{pc} = 7$ amg). The polarized ^{3}He nuclei subsequently diffuse through a transfer tube to the cylindrical target chamber (15–40 cm long, $[^3\text{He}]_{tc} =$ 10 amg), where the electron scattering occurs. The pumping chamber is heated to insure an adequate Rb vapor pressure. On a timescale of several hours, the ^{3}He polarization (in the fast diffusion limit) reaches an equilibrium value of $P_{He} = P_{Rb}/(1 + \langle\Gamma_{He}\rangle/\langle\gamma_{se}\rangle)$ where $\langle\Gamma_{He}\rangle = f_{pc}\Gamma_{pc} + f_{tc}\Gamma_{tc}$ is the cell-averaged ^{3}He spin-relaxation rate, $\langle\gamma_{se}\rangle = f_{pc}k_{se}[\text{Rb}]$ is the cell-averaged spin-exchange rate, $f_{pc(tc)}$ (≈ 0.5 for our cells) is the fraction of nuclei in the pumping (target) chamber, and $\Gamma_{pc(tc)}$ the ^{3}He spin-relaxation rate in the pumping (target) chamber.

Using nothing more than Eqns. (1) & (2), we find $P_{Rb} \geq 99\%$. Combining this with typical values for $\langle\Gamma_{He}\rangle = 0.02$/hrs & $\langle\gamma_{se}\rangle = 0.1$/hrs, we obtain $P_{He} = 83\%$. There are two reasons why this naive estimate is double what we regularly obtain with traditional SEOP using broadband lasers. First, it has been reported that there is a ^{3}He spin-relaxation mechanism, the X-factor, that appears to scale with the alkali density [6]. Second, the estimate for P_{Rb} is woefully inadequate because it doesn't account for the spatial variation of the photon flux due to the light absorption by Rb:

$$\partial\phi(\vec{r},\nu)/\partial z = -\phi(\vec{r},\nu)\sigma(\nu)[\text{Rb}](1 - P_{Rb}(\vec{r})) \qquad (3)$$

where z is the depth into the cell and $[\text{Rb}] \approx 10^{15}/\text{cm}^3$. Because the Rb polarization rate (kHz) is much faster than its diffusion rate (mHz), the *local* Rb polarization is given by the *local* optical-pumping rate, which in turn depends on the *local* photon flux. On the other hand, because the ^{3}He diffusion rate (mHz) is much faster than its polarization rate (μHz), the ^{3}He polarization actually depends on the average Rb polarization in the

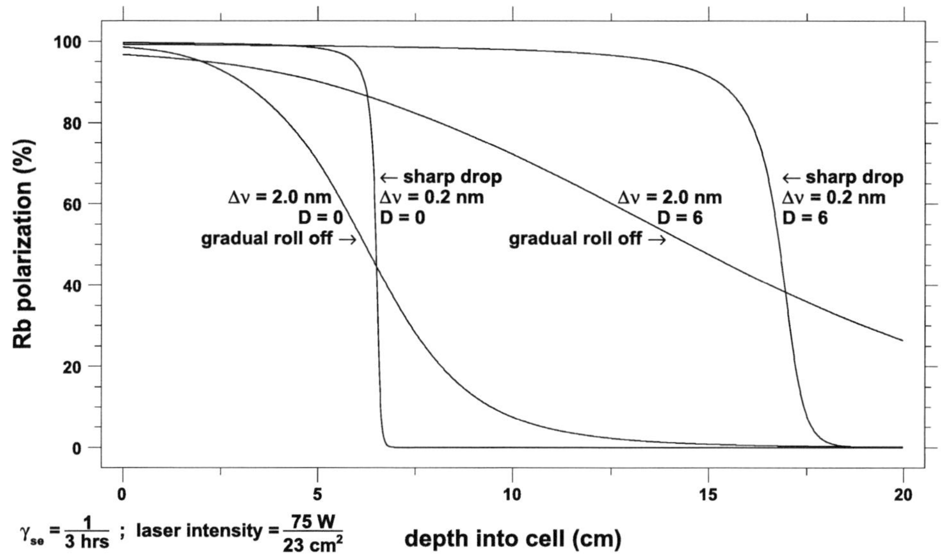

FIGURE 1. (NUMERICAL SIMULATION) Rb polarization gradient for fixed γ_{se} & laser intensity. Traditional (hybrid K-Rb) SEOP is depicted by the two curves labeled $D = 0$ ($D = 6$) on the left (right). Note the difference in shape between broadband ($\Delta\nu = 2$ nm) & narrowband ($\Delta\nu = 0.2$ nm) lasers.

pumping chamber ($\langle P_{Rb}\rangle_{pc}$). Thus the ^{3}He equilibrium polarization is more accurately given by $P_{He} = \langle P_{Rb}\rangle_{pc} / (1 + X + \langle\Gamma_{He}\rangle / \langle\gamma_{se}\rangle)$. We'll now focus on the shape of the Rb polarization gradient, which depends crucially on the spectral width of the laser.

SPECTRALLY-NARROWED LASER DIODE ARRAYS

The spectral profile of the type of laser diode arrays used for optical pumping can be represented by a gaussian line shape with a characteristic width given by the full width at half maximum $\Delta\nu$. The relevant scale with which to compare $\Delta\nu$ is the pressure-broadened Rb D1 absorption line width ($\Gamma_1 \approx [^3\text{He}\,]_{pc} \cdot 0.04\,\text{nm/amg} = 0.3\,\text{nm}$). For the last 10 yrs, we've used up to five 30 Watt Coherent FAP systems [7] with spectral widths of $\Delta\nu \approx 2$ nm. This is relatively broad and is a direct consequence of the naturally broad gain profile of the laser diode bars. Recently, we've performed tests with up to three 25 Watt Spectra-Physics Comets [8]. These lasers have volume Bragg gratings (VBG) coupled directly to the output of the laser diode bars. A narrow slice of the gain profile is amplified by reflecting wavelength-selected light by the VBG back into the laser diode bar. This "narrows" the spectral width to $\Delta\nu \approx 0.2$ nm for $\approx 90\%$ of the power.

The first advantage of these narrowband lasers is the dramatic factor of ≈ 5 increase in the optical-pumping rate for the same laser power [9]. The second advantage is the shape of the Rb polarization gradient that the light creates along its propagation path, see Fig. 1. At the front of the pumping chamber, the resonant part of the laser spectrum dominates the optical pumping and produces near unity Rb polarization. However, even at very high Rb polarization, a small amount of light is absorbed, mainly from the resonant part of the laser spectrum. After a sufficiently large distance, the resonant part of the laser spectrum is completely absorbed. At this depth, for a broadband laser, the off-resonant part of the laser spectrum continues to polarize the Rb, but at a much lower level. Consequently, the light penetrates further and the Rb polarization gradually rolls off to zero. On the other hand, for a narrowband laser, the *entire* spectrum is resonant. Therefore, once the Rb polarization starts to drop, the light is even more strongly absorbed and the Rb polarization drops sharply to zero.

If this sharp drop occurs well before the back end of the cell, then the Rb polarization averaged over the path is roughly the same for broadband and narrowband lasers with the same power. Therefore, to fully take advantage of narrowband lasers, the depth at which this sharp drop occurs must be located past the back end of the pumping chamber. Under this condition, the Rb polarization is near unity along the entire light path. We can accomplish this by lowering [Rb] and/or increasing the laser power.

ALKALI-HYBRID SPIN-EXCHANGE OPTICAL PUMPING

In alkali-hybrid SEOP [10, 11], a mix of two alkali species, in our case potassium (K) & rubidium (Rb), are used. As before, Rb is polarized by optical pumping. When the K to Rb vapor density ratio ($D = [\mathrm{K}]/[\mathrm{Rb}]$) is small ($D \ll 1000$), extremely efficient K-Rb spin-exchange collisions insure that $P_{\mathrm{K}} = P_{\mathrm{Rb}} = P_{\mathrm{A}}$. Now both K & Rb polarize ^{3}He by spin-exchange collisions. The equations for the equilibrium polarizations have the same form provided we relabel the subscript Rb $\rightarrow$ A and account for the spin relaxation & exchange due to K:

$$\Gamma_{\mathrm{A}} = \Gamma_{\mathrm{Rb}} + D(\Gamma_{\mathrm{K}} + 2k_{\mathrm{A}}[\mathrm{Rb}]) \qquad \& \qquad \langle\gamma_{\mathrm{se}}\rangle = f_{\mathrm{pc}} k_{\mathrm{se}}[\mathrm{Rb}]\left(1 + D\cdot k'_{\mathrm{se}}/k_{\mathrm{se}}\right) \quad (4)$$

where $\Gamma_{\mathrm{K}} \approx 0.2$ kHz (235°C & $[^3\mathrm{He}]_{\mathrm{pc}} = 7$ amg) is the K spin-relaxation rate, k_{A} is the mean K-Rb spin-relaxation rate constant, and $k'_{\mathrm{se}} = 10^{-15}\ \mathrm{cm}^3/5.05$ hrs [12] is the K-^{3}He spin-exchange rate constant. Since K has a much smaller spin-orbit splitting than Rb (3.5 nm vs. 15 nm), K also has a much smaller spin-relaxation rate than Rb.

Following the closing argument of the previous section, let's lower [Rb] while keeping the laser power fixed. In traditional SEOP ($D = 0$), lowering [Rb] has the undesirable effect of also lowering $\langle\gamma_{\mathrm{se}}\rangle$. On the other hand, in hybrid K-Rb SEOP ($D > 0$), we can lower [Rb] while keeping $\langle\gamma_{\mathrm{se}}\rangle$ fixed, provided we increase D by the correct amount. *If the increase in Γ_{A} (due to the increase in D) is not too large*, then the light would penetrate farther into the pumping chamber for the hybrid case (because K absorbs very little light near the Rb D1 wavelength). Consequently, for a narrowband laser, the sharp drop in Rb polarization would occur even farther away from the front of the pumping chamber, see Fig. 1. To achieve the best balance between Γ_{A} and $\langle\gamma_{\mathrm{se}}\rangle$, the optimal range for D must be found.

EXPERIMENTAL RESULTS & CONCLUSIONS

To make a cell that will have a prescribed ratio (D_{goal}) when the pumping chamber is hot, we must prepare a mix of solid alkali metal with the correct mole fractions of Rb & K (f_{Rb} & f_K). According to Raoult's law, the partial vapor pressure due to one constituent of an ideal solution is given by the product of that constituent's mole fraction and pure vapor pressure. A solution of K & Rb is nearly ideal because chemical interactions between K & K, Rb & Rb, and K & Rb are all similar. Consequently Raoult's law is satisfied very well and, combining it with the pure vapor pressure curves for K & Rb [13], gives the ratio of mole fractions

$$\frac{f_{Rb}}{f_K} \approx \frac{1}{6.31 \cdot D_{goal}} \times \left(\frac{[K]_{goal}}{10^{14}\ \text{cm}^{-3}} \right)^{0.097} \qquad \text{(valid for } D_{goal} \gg 0.2) \tag{5}$$

necessary to obtain D_{goal} at the oven temperature where the desired K density is $[K]_{goal}$. Using a glove box, we've found that we can prepare a break-seal ampule with the desired f_{Rb}/f_K to better than 15%. However, transferring the alkali mix from the ampule into the pumping chamber introduces variations. For every cell that we've measured the ratio under operating conditions (D_{meas}), our preliminary analysis has shown $D_{meas} \leq 1.2 \cdot D_{goal}$. In addition, we've found for cells with $D_{goal} = 20$ that $D_{meas} = 9.2 \pm 6.4$ and for cells with $D_{goal} = 5$ that $D_{meas} = 3.0 \pm 1.6$. Fig. 2 depicts the highest ^{3}He polarization achieved as a function of D with both broadband and narrowband lasers. Because these data were taken over several years with different cells, the laser conditions and the intrinsic ^{3}He spin-relaxation rates vary as well. With that caveat, we still feel that we can make the following preliminary claims: (1) the ideal vapor density ratio under our typical operating conditions is $2 \leq D \leq 6$, (2) even replacing a single broadband laser with a narrowband laser produces significant gains, (3) to achieve near 70% ^{3}He polarization in the pumping chamber, one needs to use both hybrid K-Rb SEOP and narrowband lasers and (4) because P_{He} is so high now, a magnetic field gradient (≈ 10 mG/cm) must be applied to prevent masing [3].

The variation in D_{meas} from D_{goal} almost perfectly overlaps the observed optimal range for D provided $D_{goal} = 5$. In other words, the level of control that we have over D for a given cell is sufficient to fall within the optimal range for D. Since we don't have arbitrary control over the spectral width of the laser, we've run several numerical simulations. Generally speaking, we've found that a spectral width that is roughly equal to the absorption line width is sufficiently narrow to obtain the significant benefits of optical pumping with narrowband lasers. Analysis is underway to calculate the X-factors for our cells by combining measurements of the alkali & ^{3}He polarizations, the alkali densities, and the ^{3}He polarization build-up & decay rates. The extraction of X is complicated by the polarization diffusion between the pumping & target chambers.

In conclusion, spectrally-narrowed lasers are most effective when the depth at which the sharp drop in Rb polarization occurs is beyond the back end of the pumping chamber. Because both the depth of this sharp drop & the back end of the pumping chamber vary with the radial distance from the propagation axis, care must be taken to match the laser beam size to the pumping chamber. In that case, the entire pumping chamber is illuminated and the average Rb polarization is nearly unity. Alkali-hybrid SEOP is

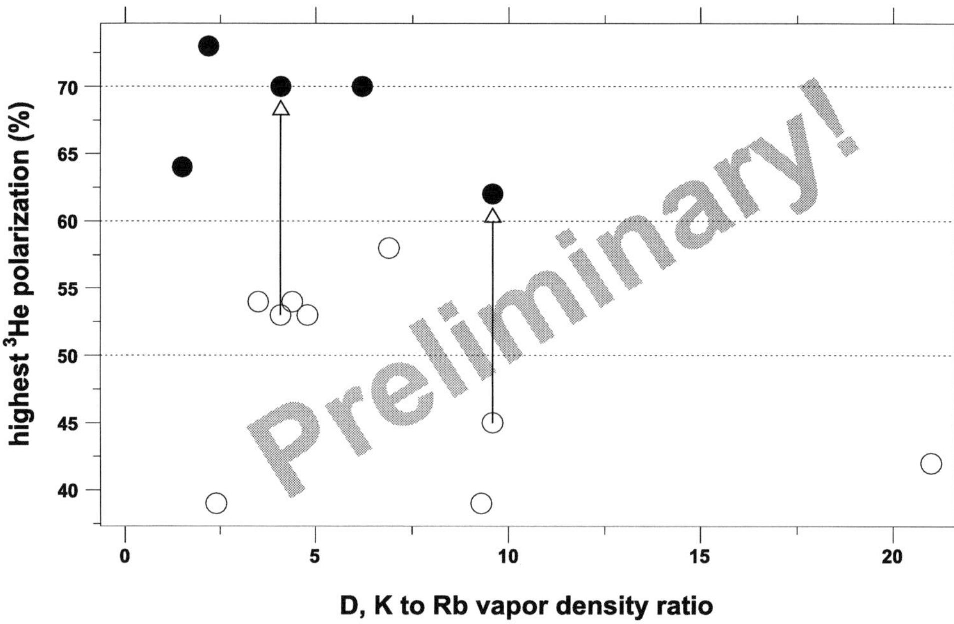

FIGURE 2. (PRELIMINARY DATA) Highest ^{3}He polarization achieved as a function of D using broadband (open circles) & narrowband (closed circles) lasers. The arrows indicate the increase in polarization for the same cell when the lasers are switched from broadband to narrowband.

effective because it partially decouples the photon cost from a high spin-exchange rate. Finally, we infer from measurements of the alkali polarization & the analysis presented in this talk that traditional SEOP with broadband lasers failed to achieve our naive expectations, at least in part, due to unexpectedly low *average* Rb polarization ($\langle P_{\mathrm{Rb}} \rangle_{\mathrm{pc}}$).

REFERENCES

1. E.W. Hughes, et al., *SLAC E-142 Proposal*, 1989.
2. H. L. Middleton, Ph.D. thesis, Princeton University (1994).
3. M. V. Romalis, Ph.D. thesis, Princeton University (1997).
4. T. G. Walker, J. H. Thywissen, and W. Happer, *Phys. Rev. A* **56**, 2090–2094 (1997).
5. A. Ben-Amar Baranga, et al., *Phys. Rev. Lett.* **80**, 2801–2804 (1998).
6. E. Babcock, et al., *Phys. Rev. Lett.* **96**, 083003 (2006).
7. FAPSys-Rubidium-30W-795to798-Single, Coherent, Inc., 5100 Patrick Henry Drive, Santa Clara, CA USA 95054, 800-527-3786.
8. VBG locked Comet, 30W, 794.7 nm, 400 μm core fiber, Spectra-Physics (a division of Newport Corporation), 3321 E. Global Loop, Tucson, AZ USA 85706, 520-746-1234.
9. B. Chann, et al., *J. Appl. Phys.* **94**, 6908–6914 (2003).
10. W. Happer, G. D. Cates, Jr., M. V. Romalis, and C. J. Erickson, *U.S. Patent No. 6,318,092* (2001).
11. E. Babcock, et al., *Phys. Rev. Lett.* **91**, 123003 (2003).
12. E. V. Babcock, Ph.D. thesis, University of Wisconsin-Madison (2005).
13. *CRC Handbook of Chemistry and Physics*, CRC Press, Boca Raton, FL, 1994, 75th student edn.

Polarized ^{3}He Targets at MAMI

J. Krimmer, W. Heil, S. Karpuk and Z. Salhi

Institut für Physik, Johannes Gutenberg-Universität Mainz
Staudinger-Weg 7,
55128 Mainz, Germany

Abstract. Polarized ^{3}He obtained from metastability exchange optical pumping is used for various double polarized experiments at the Mainz Microtron (MAMI). At the experimental area more than 70% initial target polarization is available. Besides a description of the target setup and its performance under electron beam conditions, the status of the polarized ^{3}He target for the tagged photon beam at MAMI will be given.

PACS: 29.25.Pj

INTRODUCTION

Due to the large S-state probability [1] polarized ^{3}He nuclei are a suitable substitute for a polarized neutron target. Experiments at the electron accelerator MAMI involve, for example, the extraction of the electric form factor of the neutron $G_{e,n}$ via electron scattering [2, 3], or measurements of the Gerasimov-Drell-Hearn (GDH) sum rule at the tagged photon beam. Furthermore, the possible use of polarized ^{3}He as an effective polarized proton target is studied via two-body breakup in the reaction $^3\overrightarrow{\mathrm{He}}(\vec{\mathrm{e}},\mathrm{e}'\vec{\mathrm{p}})\mathrm{d}$ [4].

^{3}He gas is polarized according to the method of metastability exchange optical pumping (MEOP) [5, 6]. A polarization of more than 70% can be obtained at a flux of 2 bar·l/h [7]. After filling at the polarizer in the Institute of Physics, the target cells are brought to the experiments at MAMI. This remote type of operation requires long relaxation times of the target cells and a minimal loss of polarization during the transport.

The following sections deal with the target setup and the performance under electron beam conditions for the $G_{e,n}$ measurement at Q^2=1.5(GeV/c)^2 [8]. Finally, the status of the polarized ^{3}He target for the tagged photon beam will be given.

SETUP FOR THE $G_{e,n}$ EXPERIMENT

A sketch of the experimental setup for the $G_{e,n}$ measurement is given in the right panel of Fig. 1. The target cell with the polarized ^{3}He is located in the middle of a box [9] providing a homogeneous holding field. The spin of the ^{3}He nuclei can be rotated in all directions via rotation of the magnetic field. This allows measurements with the target spin parallel or perpendicular to the direction of the momentum transfer which is a prerequisite for the extraction of $G_{e,n}$. Provided, that the relative field gradient $\frac{dB/dr}{B_0}$ is smaller than $5\cdot10^{-4}$ cm^{-1}, the partial relaxation time T_1^{grad} is larger than 1000 hours at a pressure of 5 bar [10]. Holes in the box for the primary electron beam (e) and

CP1149, *Spin Physics, 18th International Spin Physics Symposium*
edited by D. G. Crabb, D. B. Day, S. Liuti, X. Zheng, M. Poelker, and Y. Prok

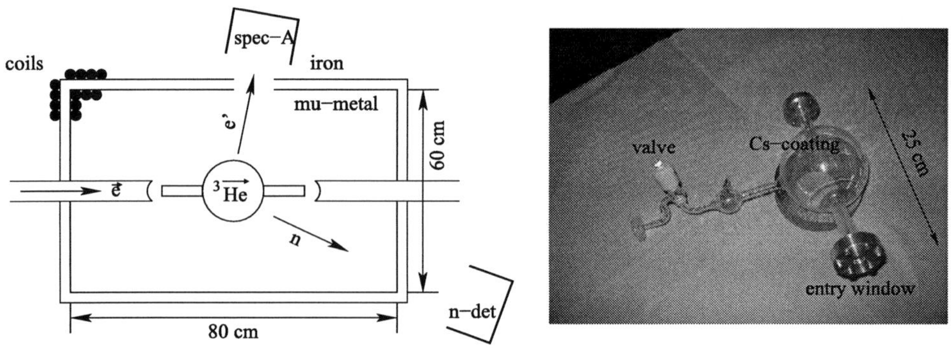

FIGURE 1. Left: Sketch of the target setup for the $G_{e,n}$ experiment. Right: Target cell for use in the electron beam.

for the scattered electrons (e′) spoil the perfect homogeneity of the magnetic field. The influence of the holes for the beam pipes on both ends of the box can be compensated for by means of correction coils, resulting in a relative gradient much smaller than the demanded $5 \cdot 10^{-4}$ cm^{-1}. The target cells for use in the electron beam (Fig. 1 right) consist of quarz glass with a total length of 25 cm and a diameter of the spherical part of 10 cm. They comprise 50 μm thick beryllium windows as entry windows for the electron beam. After coating with cesium the cells exhibit wall relaxation times T_1^{wall} between 100 and 200 hours. Polarization monitoring and measurements are accomplished by two methods. During the experiment the polarization is monitored by the nuclear magnetic resonance (NMR) techniques. Applying a static magnetic pulse perpendicular to the holding field rotates the magnetization out of the scattering plane by an angle $\alpha \approx 2°$. The signal amplitude (free induction decay monitored via pick-up coils) yields a relative measure of the polarization only, but allows to monitor the decay of the polarization (relaxation) with high precision. The relative loss of polarization due to this technique is 0.02% per measurement. The measurement of the magnetic field produced by a dense sample of spin polarized ^{3}He is used to determine the absolute polarization P_{He} [11]. The field B_{cell} produced by the target cell filled with polarized gas is on the order of 1mG. As this field is more than 1000 times smaller than the holding field B_0, B_{cell} is obtained from the difference $\Delta B = 2 \cdot B_{cell}$ of the total field measured before and after a 180° spin flip, where B_0 cancels out. The 180° flip of the magnetization is induced via Adiabatic Fast Passage (AFP) where a HF magnetic field pulse (the frequency is swept across the Larmor frequency) is applied via B_1 coils. The total field before and after the 180° flip is measured by a field probe (Fluxgate magnetometer Bartington MAG-03 MS). The polarization P_{He} can then be obtained via

$$P_{He} = \frac{r^3}{2\mu_{^3He}} \cdot \frac{1}{N} \cdot \Delta B \qquad (1)$$

where $\mu_{^3He} = 1.0746 \cdot 10^{-26}\text{Am}^2$ denotes the magnetic moment of the ^{3}He nucleus. The number N of atoms in a target cell with volume V at a temperature T and a pressure p is given by $\frac{p \cdot V}{k_B \cdot T}$ with k_B the Boltzmann constant. Due to the cubic dependence, a precise

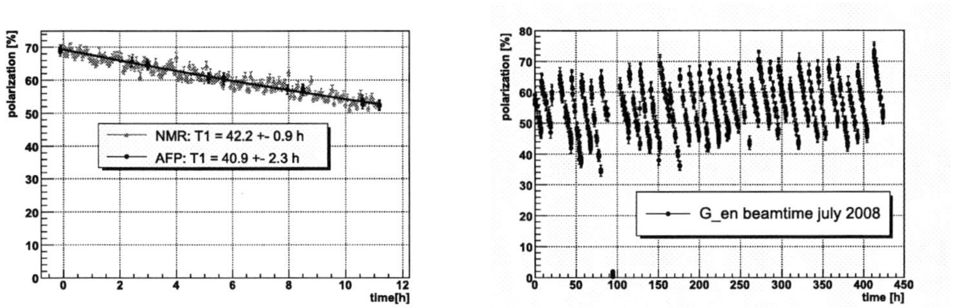

FIGURE 2. Target polarization during the $G_{e,n}$ run in July 2008.

knowledge of the distance r between the middle of the target cell and the field sensor is important. For calibration, a phantom with the shape of the target cell is used, providing a well-known dipole field. More details about this calibration and the complete target setup will be given in a forthcoming paper.

PERFORMANCE IN THE ELECTRON BEAM

In July 2008 a measurement of the electric form factor of the neutron has been performed at Q^2=1.5(GeV/c)^2. Figure 2 shows the target polarization during this beamtime. The polarization obtained via NMR and AFP, respectively, is given in the left part. The polarization values from NMR are scaled by a constant factor to match the AFP values. The T_1 times obtained with both methods agree among themselves. Under real experimental conditions additional relaxation terms come into play. At higher pressures, relaxation due to the dipole-dipole interaction (T_1^{dipole}) [12] becomes relevant, and the production of $^3He^+$ ions in a beam of charged particles adds an additional relaxation term (T_1^{beam}) [13]. All terms add up to T_1^{total} of about 30-40 hours: $\frac{1}{T_1^{total}} = \frac{1}{T_1^{grad}} + \frac{1}{T_1^{dipole}} + \frac{1}{T_1^{beam}} + \frac{1}{T_1^{wall}}$.
The right part of Fig. 2 shows the target polarization during the complete beamtime. The initial polarization steadily increased during the run reaching 72%, due to an improved target cell exchange procedure. With an exchange of the target cells twice per day, the mean polarization resulted in 55-60%.

STATUS OF THE TARGET FOR THE TAGGED PHOTON BEAM

A measurement of the GDH sum rule is planned for 2009 in the tagged photon hall of MAMI. The Crystal Ball detector sets the geometrical boundary conditions for the target setup (Fig. 3, left panel). The holding field along the beam axis will be provided by a solenoid with a length of 80 cm and a diameter of 8 cm. The solenoid has already been produced and the mean relative field gradient $\frac{dB/dr}{B_0}$ in the target volume is smaller than $1 \cdot 10^{-3}\text{cm}^{-1}$ [14].
The target cells will also obey the cylindrical symmetry. A prototype cell is shown in the

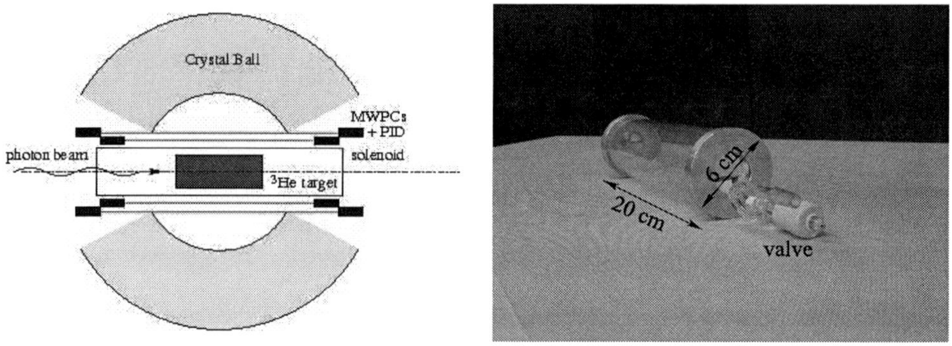

FIGURE 3. Left: target setup in the tagged photon hall. Right: prototype target cell for the photon beam.

right panel of Fig. 3. It is made from quarz glass with a length of 20 cm and a diameter of 6 cm. The entry windows consist of 50 μm Mylar metallized with aluminium. A T_1 measurement with the uncoated cell resulted in 4.5 h. From our experience with the cells for the electron beam (T_1^{wall} increases by a factor 7-10 after coating), we expect 30-40 h with Cs coating for the here described cell.

ACKNOWLEDGMENTS

This work is supported by the Deutsche Forschungsgemeinschaft (SFB 443).

REFERENCES

1. B. Blankleider, and R. M. Woloshyn, *Phys. Rev.* **C 29**, 538 (1984).
2. J. Bermuth *et al.*, *Phys. Lett.* **B 564**, 199 (2003).
3. A. Ott, Ph.D. thesis, University of Tübingen (in preparation).
4. M. Weinriefer, Ph.D. thesis, University of Mainz (in preparation).
5. F. Colegrove, L. Schearer, and G. Walters, *Phys. Rev.* **132**, 2561 (1963).
6. P. Nacher, and M. Leduc, *J. Physique* **46**, 2057 (1985).
7. E. Otten, *Europhysics News* **35**, 16 (2004).
8. S. Schlimme, Ph.D. thesis, University of Mainz (in preparation).
9. D. Rohe, Ph.D. thesis, University of Mainz (1998).
10. L. Schearer, and G. Walters, *Phys. Rev.* **A 139**, 1398 (1965).
11. E. Wilms, M. Ebert, W. Heil, and R. Surkau, *Nucl. Instr. Methods* **A 401**, 491 (1996).
12. N. Newbury, A. Barton, G. Cates, W. Happer, and H. Middleton, *Phys. Rev.* **A 48**, 4411 (1993).
13. T. Chupp, R. Loveman, A. Thomson, A. Bernstein, and D. Tieger, *Phys. Rev.* **C 45**, 915–932 (1992).
14. P. Aguar-Bartolome, Ph.D. thesis, University of Mainz (in preparation).

Optical Pumping Spin Exchange ^{3}He Gas Cells for Magnetic Resonance Imaging.

W.Kim[a,b], S.S.Stepanyan[a], A.Kim[a], Y.Jung[a], S.Woo[a], M.Yurov[a] and J.Jang[a]

[a]*Department of Physics, Kyungpook National University Daegu, 702-701, Republic of Korea*
[b]*Korea Basic Science Institute*

Abstract: We present a device for spin–exchange optical pumping system to produce large quantities of polarized noble gases for Magnetic Resonance Imaging (MRI). A method and design of apparatus for pumping the polarization of noble gases is described. The method and apparatus enable production, storage and usage of hyperpolarized noble gases for different purposes, including Magnetic Resonance Imaging of human and animal subjects. Magnetic imaging agents breathed into lungs can be observed by the radio waves of the MRI scanner and report back physical and functional information about lung's health and desease. The technique known as spin exchange optical pumping is used. Nuclear magnetic resonance is implemented to measure the polarization of hyperpolarized gas. The cells prepared and sealed under high vacuum after handling Alkali metals into the cell and filling with the ^{3}He-N_2 mixture. The cells could be refilled. The ^{3}He reaches around 50% polarization in 5-15 hours.

Keywords: ^{3}He, spin exchange, optical pumping, neutron, MRI, spin filter
PACS:29.25.Pj

INTRODUCTION

Polarized ^{3}He target plays an important role in studying form factors and spin structure functions of neutron with spin-dependent electron scattering. The use of polarized ^{3}He extends from nuclear physics experiments into other fields including medical diagnostics. It is widely used in application, for example, as Magnetic Resonance Imaging (MRI) and spin filter, etc. Presently we are developing polarized ^{3}He gas by using spin exchange optical pumping method. The vacuum and gas handling system was built for the glass cell filling. The construction and filling procedure of the target cell is important in determining the relaxation rates due to impurities in both glass and gases. Oven is constructed of Torlon material and can be heated to 220^{0}C using a forced hot air system to vaporize the Rubidium for optical pumping.

The nuclear magnetic resonance (NMR) system has been constructed for polarization measurements of the polarized ^{3}He target. The investigation has been carried out to determine the polarization in polarized ^{3}He target by comparing the NMR signal height of ^{3}He, with the 2,2-Diphenyl-1-Picrylhydrazyl (DPPH) sample and from the protons signal in water.

CELL FILLING SETUP

We employ the optical pumping spin-exchange method because it enables us to make a high density polarized ^{3}He rather easily and obtain high polarization. The cell filling is very important factor in determination of the final relaxation rate. Figure 1 shows

CP1149, *Spin Physics, 18th International Spin Physics Symposium*
edited by D. G. Crabb, D. B. Day, S. Liuti, X. Zheng, M. Poelker, and Y. Prok

schematically the hyperpolarized gas setup with gas handling system, pumping system and glass string. The cell filling includes ultra high vacuum system consisting of pumps, purifiers and gauges which allow ^{3}He to pass from the ^{3}He gas cylinder to the cell. The following factors [1-6] contribute to shortening of depolarization rate:

1) ^{3}He collisions with cell wall paramagnetic impurities. Cell wall collisions are inherently related to cell design and construction.

2) Magnetic dipole-dipole interactions between ^{3}He nuclei.

3) Ionization from experimental electron beam.

4) Magnetic holding field gradients.

5) Interactions with gas impurities.

High vacuum is produced by a turbo-molecular pump with backing before by a roughing pump. These two main pumps are capable of achieving a vacuum better than 10^{-8} Torr. The high vacuum is an indication of the cleanness of the system, which can be indicated by gas mass spectrometer RGA100.

The gases are supplied in stainless lecture bottles with stainless steel bellowed valves. Each bottle contains 25atm L gas. The manifold is wrapped by heater for periodic baking of the system at 100^0 C. In order to remove any impurities before filling system the glass cell is baked out in the solution oven under a vacuum of ~10^{-8} Torr at 400^0 C for 4 days, followed by 2 days at 200^0 C.

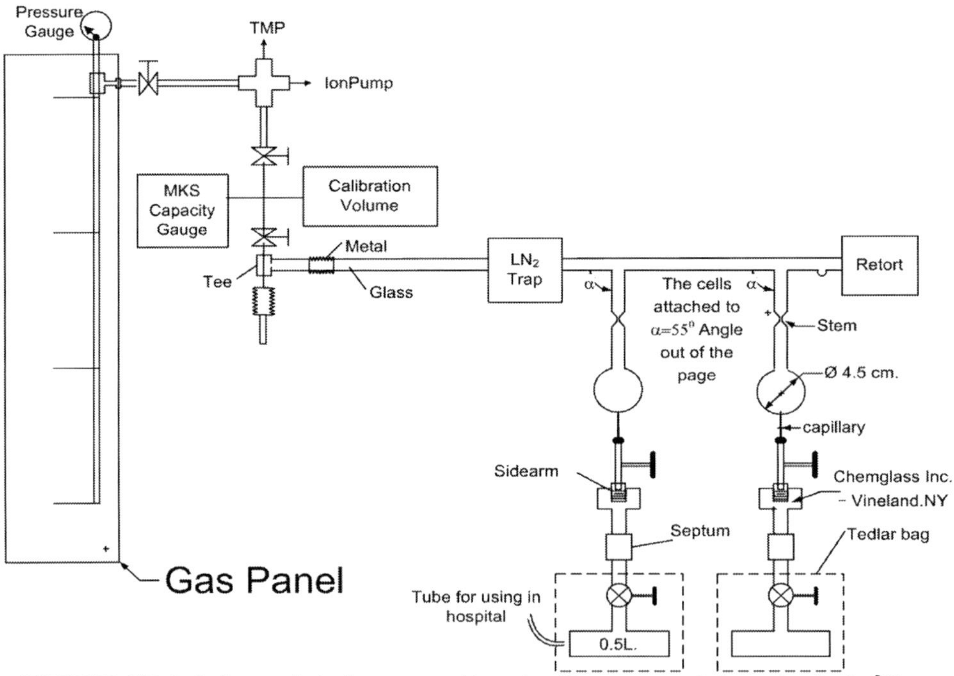

FIGURE1. This is the hyperpolarized gas setup with gas handling system and glass string for the ^{3}He gas filling.

Design with stopcock valves and capillary with using Jensen Inert tedlar packs with 0.5L volume was implemented. The cell body is spherical (5cm i.d.) and made from

Borosilicate glass by glass blower. Pyrex glass was selected for resistance to the corrosive effects of alkali metals at high temperatures and for low permeability to helium. Also the liquid nitrogen trap for vapor of Rb was implemented on glass string. During the glass cell filling all the valves are closed and two diode thermocouples are fitted for monitoring the temperature surrounding the cell. Liquid nitrogen is needed to cryogenically compress the gases in the cell to a pressure below 1 atm. during the cell sealing process.

THE POLARIZER AND POLARIZATION MEASUREMENT

The ^{3}He gas was polarized during spin-exchange collisions with polarized Rb atoms through the circularly polarized 794.7 nm laser light. A homogenous static magnetic field was produced by a Helmholtz pair. Apparatus for polarization and measurement of ^{3}He includes also force air oven, RF coils and the pickup coils. The polarization of the ^{3}He proceeds by optically pumping the 794.7nm 5s-5p transition for the Rb valence electron which transfers its spin to the ^{3}He nucleus via hyperfine interaction during the collisions [1-6]. With the present setup the optimum Rb density occurs at about 220°C. The laser and Rb absorption optical spectrum is continuously monitored with a high resolution Ocean Optics spectrometer. The resultant noble gas polarization saturates at value

$$P_N(t) = \langle P_A \rangle \frac{\gamma_{SE}}{\gamma_{SE} + \Gamma}\left[1 - e^{-(\gamma_{SE}+\Gamma)t}\right]$$

where $\langle P_A \rangle$ is the volume and time averaged alkali metal polarization produced by laser optical pumping and γ_{SE} is the spin exchange rate from the alkali metal to the noble gas nucleus, Γ is the spin relaxation rate due to paramagnetic impurities on the wall and in the gas [3]. Due to the slow rate of the spin-exchange polarization process relaxation rate $1/\Gamma \sim$ 50-100h required. Polarization is measured using NMR (adiabatic fast passage) technique. The absolute polarization is defined by comparing with the protons signal in water.

EXPERIMENTAL RESULTS

The polarization reached more than 50 %. It is shown on Fig. 2. The cells could be refilled several times and exposed to the laser at the temperatures of 180 s- 220^0 C. All cells contain rubidium and potassium. The ^{3}He cells polarization is limited by a temperature-dependent relaxation mechanism that is proportional to alkali-metal density and correlation with shape of cell [18]. The total estimated uncertainties for measurement of the polarization were around 7.5%. The measurement of the proton polarization had 3% uncertainties coming from the thermal equilibrium signal calibration. Also the significant uncertainties factors are from magnetic gradient of the tuned pickup coils, temperature and field measurements. Improvement has been made to fabricate optical pumping ^{3}He spin exchange cells. It increased the average polarization at high density alkali metal vapor mixture at oven temperature of 220°C. ^{3}He spin has got polarization over 50% during 5-15 hours. The system was left to run overnight to achieve maximum polarization.

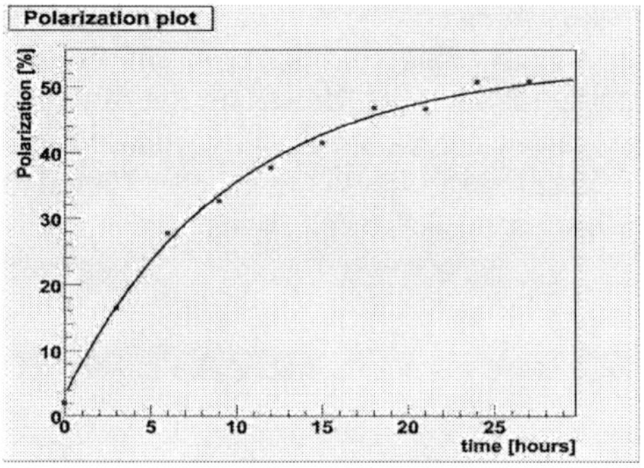

FIGURE 2. The absolute polarization with ^{3}He.

ACKNOWLEDGMENTS

The authors had a lot of useful discussions with Prof. T. D. Averett at College of William and Merry. This work was supported by a grant from the Korea Research Foundation.

REFERENCES:

1. M.A. Bouchiat, T.R. Carver, CM. Varnum, *Phys. Rev. Lett.* **5**, 373 (1960).
2. L. Rodger *et al., Phys. Rev.* **138**, A946 (1965).
3. T. E. Chupp *et al., Phys. Rev. C* **45**, 915 (1992).
4. M. Gatzke *et al., Phys. Rev. Lett.* **70**, 690 (1993).
5. C.H.Tseng *et al., Phys. Rev. Lett.* **81**, 3785 (1998).
6. J.Becker *et al., Nucl. Instr. and Meth. A* **402**, 327-336 (1998).
7. J.N.Zerger *et al., Appl. Phys. Lett.* **76**, 1798 (2000).
8. T. R. Gentile *et al., Journal of Research of the National Institute of Standards and Technology* **106**, 709-729 (2001).
9. R. E. Jacob, S. W. Morgan and B. Saam, *J. Appl. Phys.* **92**, 1588-1597 (2002).
10. T.W. Kornack and M.V. Romalis, *Phys. Rev. Lett.* **89**, 253002 (2002).
11. D.R.Rich *et al., Nucl. Instr. and Meth. A* **481**, 431 (2002).
12. R.E.Jacob *et al., Phys. Rev. A* **69**, 021401 (2004).
13. I.C. Ruset, *et al., Phys. Rev. Lett.* **96**, 053002 (2006).
14. M.S. Rosen, *et al., Phys. Rev. Sci. Instrum* **70**, 2, pp. 1546-1552 (1999).
15. G.D. Cates, *et al., Phys. Rev. Lett.* **65**, 2591 (1990).
16. B.Driehuys, *et al., Appl . Phys. Lett.* **69**, 12, 1668-1670 (1996).
17. Y.Y.Jau, et al., *Phys. Rev. A* **67**, 022720 (2003).
18. E. Babcock, et al., *Phys. Rev. Lett.* **96**, 083003 (2006).
19. W.Kim, S.S.Stepanyan, et al., Proceedings of the 17th International Spin Physics Symposium. AIP915 Kyoto, Japan, (2006).
20. J.F.Mata, *et al., J. Appl. Phys.* **102,** 1273–1280 (2007).

Analysis of the Ammonia Target Polarization

Kangkang L. Kovacs

Physics Dept. University of Virginia, Charlottesville, VA. Email: kangkangl@gmail.com

Abstract. The EG4 experiment (The proton/deuteron g_1 structure function and Gerasimov-Drell-Hearn (GDH) sum at low Q^2) ran in JLab Hall B in early 2006. The goal was to extend our knowledge of the deuteron and proton spin structure to the lowest possible momentum transfer ($0.015<Q^2<0.2$ GeV^2) in the resonance region (W < 2 GeV). In particular, the experiment will allow for the first time to test precisely the Chiral Perturbation Theory Calculations in a domain where it is expected to be valid. Polarized deuterons in ND_3 ammonia were used as the scattering targets in the EG4 experiment. There are three different ways to determine the ND_3 target polarization: the "area method" and the "peak ratio method" that extract the target polarization from the nuclear magnetic resonance (NMR) signals, and the scattering asymmetry method. Polarization results obtained from these methods will be compared and the error related to each method will be discussed.

Keywords: peak ratio method, area method, polarization analysis, NMR signals.
PACS: 29.25.Pj: 82.56.-b

INTRODUCTION

The EG4 experiment at the Jefferson Lab (E12-06-109) ran in Jlab Hall B in early 2006. The goal was to calculate the Gerasimov-Drell-Hearn (GDH) Sum at the lowest possible momentum transfer ($0.015 < Q^2 < 0.2$ GeV^2) in the resonance region (W < 2 GeV)

The experiment used a longitudinally polarized electron beam and longitudinally polarized solid ammonia targets (ND_3 and NH_3).

The experiment setup includes:

- Superconducting magnet
- Helium evaporation refrigerator
- Microwaves
- NMR system
- Insert housing the target material

The helium evaporation refrigerator can cool the target down to about 1K with the superconducting magnet creating a magnetic field around 5T to keep the target sufficiently polarized.

The particles scattered from the target are detected by the drift chambers, Cherenkov counter and the electro magnetic calorimeter.

CP1149, *Spin Physics, 18th International Spin Physics Symposium*
edited by D. G. Crabb, D. B. Day, S. Liuti, X. Zheng, M. Poelker, and Y. Prok

Target Preparation and Polarization Analysis

The ND3/NH3 ammonia target was prepared by slowly freezing the ammonia gas at 77K and then crushing it into small pieces, approximately 1-2mm in diameter. The target was polarized using the DNP (Dynamic Nuclear Polarization) method. The NH3 target achieved a polarization of about 80-90% while the ND3 target achieved a polarization of about 20-40%.

The target polarization during the experiment was measured by the nuclear magnetic resonance system (NMR). The method lies in inducing a detecting nuclear magnetic transitions, where the rate of the transitions is proportional to the population difference between the energy levels, therefore indicating the original polarization value.

There are two methods used to determine the target polarization from the NMR signals. One is the traditional area method. The target polarization is always proportional to the area under the NMR signal. Therefore, by calculating the target polarization at the thermal equilibrium temperature and measuring the area covered by the TE signal, we can calibrate the target polarization using the area of the signal at any other temperature:

$$P_{TE} = \frac{4\tanh(\frac{\hbar\omega_0}{2kT})}{3+\tanh^2(\frac{\hbar\omega_0}{2kT})}$$

$$P = P_{TE} * (A_{enhanced} / A_{TE})$$

In the EG4 experiment, the TE signal for the deuteron target was very small and noisy, which made it impossible to get an accurate measurement of the signal area. Therefore we used the ratio method to calculate the target polarization. The approximate relation between the peak ratio *r* and the target polarization can be described as follows: [2]

$$P(r) = \frac{r^2-1}{r^2+r+1}$$

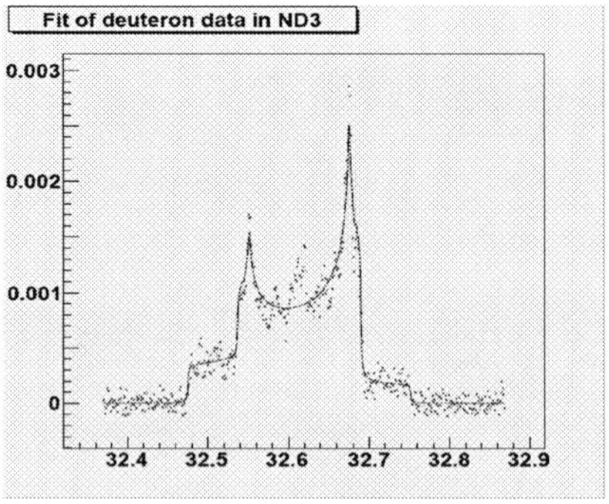

FIGURE 1: Fitting of the ND_3 NMR signal

Using a program that fits the NMR signal from the theoretical deuteron target lineshape function, we can extract the peak ratio and calculate the polarization.
Figure 1 shows the fitting to the NMR signal. Figure 2 is the polarization timeline in April, 2006 calculated using the ratio method. In April 18- 21 the target was taken out because of a helium liquifier outage.

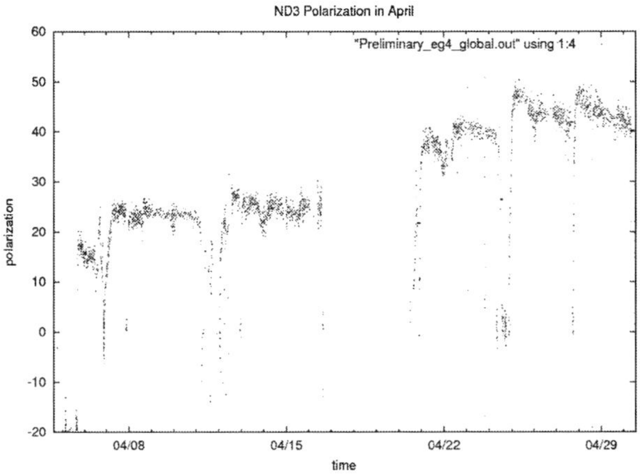

FIGURE 2: Polarization timeline in April 2006 It is known from past experiences that the target polarization increases with the amount of radiation dose and annealing it undergoes. The big sharp drop in the plot correspond to the annealing process while the small drops are due to the adjustments of the microwave frequency. We exposed the target under a strong electron beam to create the irradiation damage and then raised the temperature to around 100K for the annealing process.

The statistical error of the polarization measured by the ratio method is ~4.9%. In order to test the stability of the ratio method, we used the polarization data from the GEN and RSS experiments that took place earlier in Hall C at the Jefferson Lab and compared our ratio results with the results obtained by the area method during those two experiments. Figure 3 shows the comparison between the polarization measured by the ratio method and the polarization measured by the area method.

As we can see, the ratio method has its limitation at the low polarization range. Polarization values calculated by the ratio method are always too high compared to that of the area method. This is due to the fact that in the polarizing process there is a difference in the transition rates driving shoulders of the two peaks, therefore resulting in a false reading from the peak ratio method. But when the polarization gets above 25%, the two methods start to agree well with each other.

CONCLUSIONS

We were able to apply the peak ratio method on the ND_3 NMR signals and obtain stable polarization results. While understanding the limits of this method at low polarizations, we compared the results obtained via the ratio method to the results obtained via the are method on GEN and RSS data and showed that the peak ratio method was reliable for the polarization analysis for the EG4 experiment. We are also calculating the target polarization from the scattering asymmetry method as another source to check the reliability of the peak ratio method.

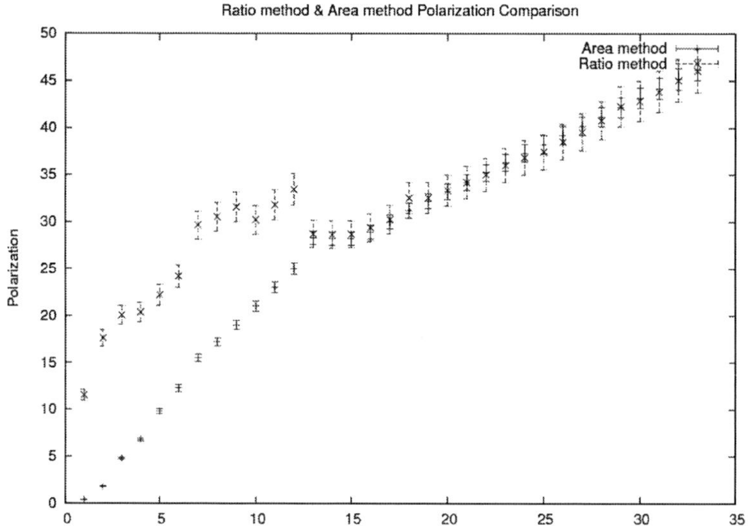

FIGURE 3. Results comparison between the area method and the peak ratio method using the GEN experiment data

ACKNOWLEDGMENTS

I wanted to thank my advisor, Donald Crabb for the instructions he gave me on the target polarization analysis. I also wanted to thank Karl Slifer, Donal Day and everybody in the EG4 experiment group at the Jefferson Lab for their great help. The code I used to calculate the peak ratio method was originally written by Chris Dulya.

This work was supported by a grant from the US Department of Energy.

REFERENCES

1. St. Goertz, W. Meyer, G.Reicherz, *Prog.Part.Nucl.Phys.* **49**, 403-489 (2002)
2. C. Dulya et Al. A Line-shape Analysis for Spin-1 NMR Signals, *Nuclear Instruments and Methods in Physics Research A* **398** (1997)
3. T.W. Donnelly and A.S. Raskin, *Annals of Physics* **169**, 247-351 (1986)

The E166 Experiment: Undulator Based Production of Polarized Positrons

Hermann Kolanoski (for the E166 Collaboration)

Humboldt-Universität zu Berlin

Abstract. The E166 experiment at the Stanford Linear Accelerator Center (SLAC) has demonstrated a scheme for the production of polarized positrons which is suitable for implementation in a future Linear Collider. A multi-GeV electron beam passed through a helical undulator to generate multi-MeV, circularly polarized photons which were then converted in a thin target to produce positrons (and electrons) with longitudinal polarization above 80% at 6 MeV. The results are in agreement with GEANT4 simulations that include the dominant polarization-dependent interactions of electrons, positrons and photons in matter.

Keywords: helical undulator, positron, polarization
PACS: 07.77.Ka, 13.88.+e, 29.27.Hj, 41.75.Fr

1. INTRODUCTION

The conceptual design of a future electron-positron linear collider, the International Linear Collider (ILC), is currently pursued through an international effort [1]. Scientifically, the ILC experiments will cover the energy range where the Higgs boson, Supersymmetry and other 'physics beyond the Standard Model' are expected. The ILC capabilities of identifying new particles by determining their quantum numbers and the Lorentz structure of their couplings are largely enhanced if the electron and positron beams are both polarized [2]. While the generation of polarized electrons is standard procedure, polarized positron sources with sufficient intensity are not existing. However, even in those cases where in principle the polarization of only one beam is sufficient, the polarization of both beams has large advantages: the higher effective polarization allows a better background reduction and at the same time a selective enhancement of the channel of interest, thereby providing clearer signatures and improved statistical accuracy. There are also cases which are only measurable if both beams are polarized, an example of this would be the separation of the selectrons which are partners to either the left-handed or right-handed electrons/positrons [2].

Polarized positrons can be generated via the pair production process by circularly polarized photons [3]. Different methods have been proposed to generate polarized photons. In the scheme proposed by Balakin and Mikhailichenko [4] the photons are produced by passing a multi-GeV electron beam through a helical undulator producing multi-MeV photons with circular polarization [5]. Alternatively, circularly polarized photons can also be produced by backscattering circularly polarized laser photons off an electron beam [6].

The E166 experiment [7, 8], described in this paper, has demonstrated that a polarized positron source based on the helical undulator scheme is suited for use at the ILC.

CP1149, *Spin Physics, 18th International Spin Physics Symposium*
edited by D. G. Crabb, D. B. Day, S. Liuti, X. Zheng, M. Poelker, and Y. Prok

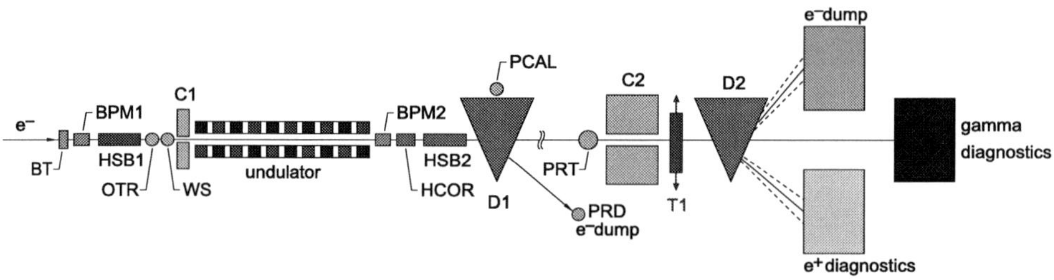

FIGURE 1. Conceptual layout (not to scale) of the E166 experiment in the SLAC FFTB. The electrons enter from the left and, after traversing the undulator, are dumped using magnet D1. The positron production target T1 and the positron and photon diagnostics are located 35 m downstream of the undulator. The spectrometer D2 allows to measure the energy dependence of the positron polarization. The labels BPM, HSB, OTR, BT, WS, HCOR, PCA, PRD, PRT indicate beam monitoring and steering devices and C1, C2 are aperture limiting collimators.

2. EXPERIMENTAL SETUP

The E166 experiment was carried out at the Final Focus Test Beam (FFTB) facility [9] at SLAC. Figure 1 shows a schematic layout of the experiment. A 46.6 GeV electron beam from the linac was passed through a 1 meter long pulsed helical undulator to generate polarized photons, the photon beam was then converted in a thin target to electron-positron pairs and the photons and positrons were analyzed for their polarization.

Undulator. A helical undulator consists of two wires wound adjacent to each other helically around a tube (Fig. 2 left). Electrical currents are sent in opposite directions through the wires thereby creating a magnetic field spiraling along the longitudinal helix axis. An electron beam propagating on axis through the undulator radiates circularly polarized photons due to interaction with the spiraling field. Parameters describing the behavior of an undulator are the energy of the electron beam, the helical period of the undulator and the so-called K-parameter which determines the photon intensity and spectrum (Fig. 2 right).

The E166 undulator was operated at a pulsed current of 2.3 kA, a pulse duration of about 12 μs and a repetition rate of 10 Hz. The positioning and alignment of the high-energy electron beam through the nominal 0.87 mm diameter aperture of the 1 meter long undulator presented special challenges.

Positron generation, transport and diagnostics. The undulator photons impinged on the conversion target T1 (0.2 r.l. tungsten) to produce electron-positron pairs (Fig. 1). Figure 3 depicts the diagnostics for photons and positrons beyond T1.

Photon polarimetry. The polarization of the MeV-photons from the undulator was determined using Compton transmission polarimetry which exploits the spin dependence of photon transmission through magnetized iron. The transmission asymmetry for opposite magnetizations, $\delta = (T^+ - T^-)(T^+ + T^-) \sim P_\gamma$, was measured behind the analyzer magnet TP2 (Fig. 3) using different detectors.

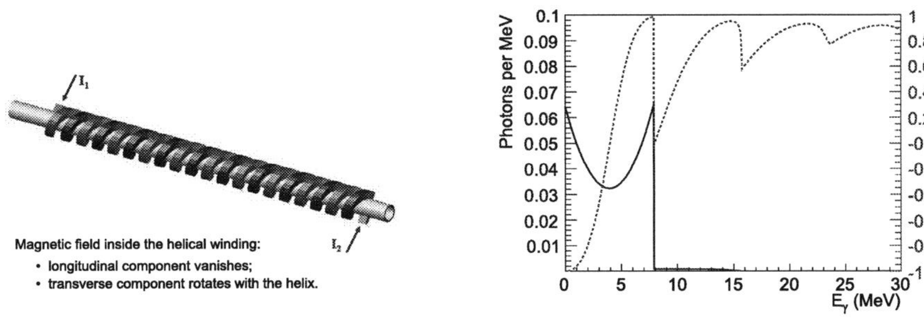

FIGURE 2. Left: Sketch of the left-handed, bifilar windings of the undulator. Right: The solid line shows the calculated photon number spectrum per beam electron of undulator radiation integrated over angle, plotted as a function of photon energy E_γ for electron beam energy 46.6 GeV, undulator period 2.54 mm and undulator strength parameter $K = 0.17$. The peak energy of the first harmonic (dipole) radiation was 7.9 MeV. The dashed line shows the longitudinal polarization P_γ of the undulator radiation as a function of energy.

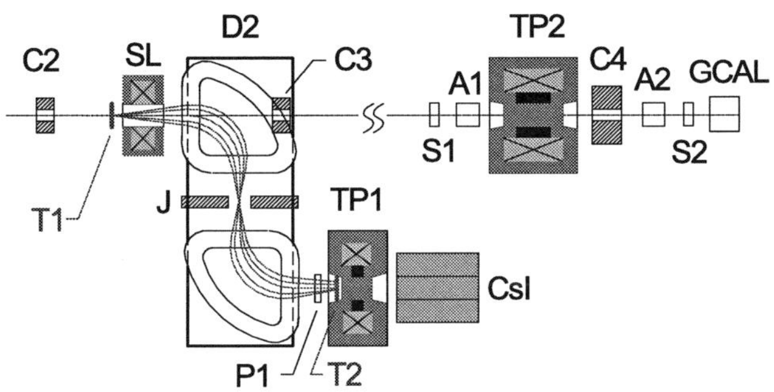

FIGURE 3. Schematic of the photon and positron diagnostics. SL = solenoid lens; D2 = dipole spectrometer magnet; J = movable jaws; C2-C4 = collimators; T1 = positron production target; T2 = reconversion target; TP1 = positron transmission polarimeter solenoid; TP2 = photon transmission polarimeter solenoid; CsI = 3 × 3 array of CsI crystals; GCAL = Si-W calorimeter; A1, A2 = aerogel Cherenkov detectors; P1, S1, S2 = Si-diode detectors.

Positron polarimetry. Compton transmission polarimetry was also employed to determine the positron polarization. This was possible by first transferring the positron polarization to photons via Bremsstrahlung and annihilation processes in a "reconversion" target (T2 in Fig. 3).

After generating polarized positrons in the conversion target (T1) the positrons passed through the spectrometer (SL+D2) which selected energy and focused the positrons onto the 'reconversion' target T2 (0.5 r.l. tungsten). The number of positrons intercepted by the target was counted in the silicon counter P1. The photons generated in the reconversion target partly inherited the positron polarization. The energy of the photons which traversed the magnetized iron core of the analyzer magnet TP1 was measured in the CsI calorimeter behind the magnet. The asymmetry δ of the energy deposition for opposite magnetization directions is related to the analyzing power A, the electron polarization in the iron, P^{Fe}, and the positron polarization P_{e^+} by

$$A = \frac{\delta}{P^{\mathrm{Fe}} P_{e^+}}. \tag{1}$$

The analyzing power was determined for different energy settings employing detailed simulations of polarization dependent cross sections in the reconversion target and the analyzer magnet. The simulation package GEANT4 has been extended to include spin dependence of the dominant electromagnetic processes [10].

3. DATA ANALYSIS AND RESULTS

Photon asymmetries. For the undulator photons the transmission asymmetries were measured behind the analyzer magnet TP2 with the three different detectors shown in Fig. 3. Since the transmission was obtained for the full undulator spectrum it was not possible to derive a polarization value for specific photon energies. Thus the measured asymmetries have just been compared to predictions from detailed simulations (the detectors have different energy thresholds and spectral sensitivities):

detector	asymmetry δ (%)	predicted δ (%)
S2 (silicon)	$3.88 \pm 0.12 \pm 0.63$	3.1
A2 (aerogel)	$3.31 \pm 0.06 \pm 0.16$	3.6
GCAL (Si/W)	$3.67 \pm 0.07 \pm 0.40$	3.4

Reasonable agreement with expectations was found for all detectors thus confirming the expected functionality of the undulator.

Positron asymmetries. A relatively large background, which was not associated with undulator photons, had to be taken into account to determine the positron induced energy deposition in the CsI calorimeter. This was done by pulsing the undulator only every other linac pulse (technically the undulator current pulse was just shifted in time). Figure 4 shows the distribution of undulator-on and undulator-off energy depositions in the central CsI crystal.

The data were taken in 'cycles' with the polarity of the analyzer magnet reversed for every other cycle. A cycle contained about 3000 events, alternating between undulator-on and undulator-off. For each cycle the corrected mean energy deposition was determined by plotting for all possible 'on-off' combinations (ij) the differences:

$$E_{ij} = E_{ij}^{\mathrm{on}} - E_{ij}^{\mathrm{off}} = \frac{E_i^{\mathrm{on}} - E_j^{\mathrm{off}} \frac{I_i^{\mathrm{on}}}{I_j^{\mathrm{off}}}}{\mathrm{P1}_i^{\mathrm{on}} - \mathrm{P1}_j^{\mathrm{off}} \frac{I_i^{\mathrm{on}}}{I_j^{\mathrm{off}}}}. \tag{2}$$

That means, the measured energy depositions were corrected for the actual currents $I_{i,j}^{\mathrm{on,off}}$ in each event and normalized to the number of positrons measured in the counter

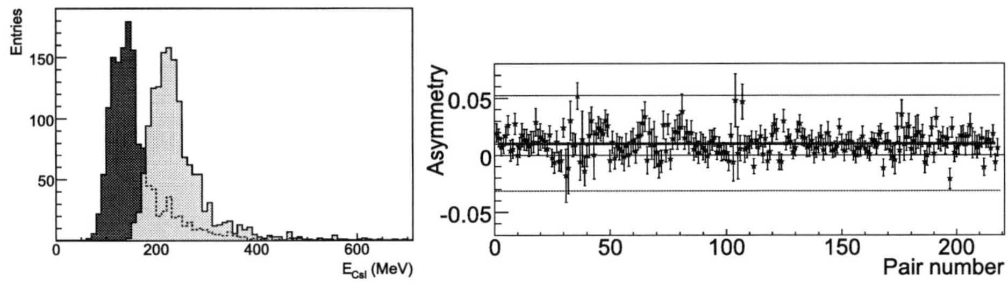

FIGURE 4. Left: Examples of energy distributions for events in the central CsI crystal (red/dark: undulator off, yellow/light: undulator on). Right: Positron-induced asymmetries δ_{e^+} in the central CsI crystal of the positron polarimeter for 220 pairs of cycles with opposite magnetization in the polarimeter. The central positron energy was 6.1 MeV. The average asymmetry was 0.0113, as indicated by the horizontal line.

TABLE 1. Positron (electron) asymmetries, analyzing powers and polarizations for different energies.

$E_{e^\pm}$ [MeV]	$\delta \pm \sigma_\delta$(stat) [%]	A	$P \pm \sigma_P$(stat) $\pm \sigma_P$(sys) [%]
4.6 (e^+)	0.69 ± 0.17	0.150	66 ± 16 ± 8
5.4 (e^+)	0.96 ± 0.08	0.156	89 ± 8 ± 9
6.1 (e^+)	1.13 ± 0.06	0.162	96 ± 6 ± 10
6.7 (e^+)	0.92 ± 0.08	0.165	80 ± 7 ± 9
6.7 (e^-)	0.94 ± 0.05	0.153	88 ± 5 ± 15
7.4 (e^+)	0.89 ± 0.20	0.169	76 ± 17 ± 12

P1 in front of the reconversion target (Fig. 3). The distribution of the E_{ij} (for each cycle) was fitted with two Gaussian functions (to account for the non-Gaussian tails). The average corrected signal energy S_{CsI} has been determined from the mean of all entries in the distribution within $\pm 2\sigma$ of the narrower Gaussian. This 'truncated mean' of all on-off combinations was found to be most efficient in eliminating the non-Gaussian fluctuations.

Using the corrected mean energy depositions the asymmetries were determined for each cycle pair with opposite magnetization (associating pairs in chronological order): $\delta = (S^-_{\mathrm{CsI}} - S^+_{\mathrm{CsI}})(S^-_{\mathrm{CsI}} + S^+_{\mathrm{CsI}})$. The asymmetries are plotted in Fig. 4 (right) for all analyzed cycle pairs of a particular spectrometer setting. The average asymmetries for each spectrometer setting are recorded in Table 1.

Positron polarization. Inverting the formula for the analyzing power (1) the positron polarization was determined from

$$P_{e^+} = \frac{\delta}{A\, P^{\mathrm{Fe}}}. \tag{3}$$

The results for different energies are shown in Fig. 5 and are recorded together with the asymmetries and analyzing powers in Table 1. In addition to five positron energy points the electron polarization was measured for one energy as a check of systematic uncertainties (the stray field of the analyzer magnet affects electrons and positrons differently). The measurements are compared in Fig. 5 with the expectations obtained

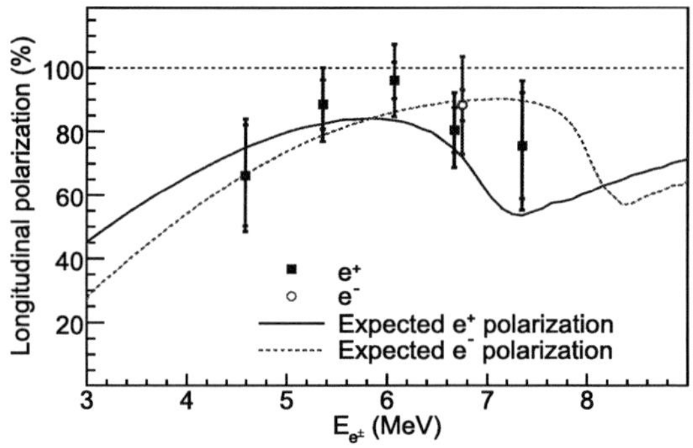

FIGURE 5. Longitudinal polarization $P_{e^\pm}$ as a function of energy $E_{e^\pm}$ of positrons and electrons as determined from the asymmetries observed in the central CsI crystal. The smaller error bars show the statistical uncertainty, and the larger bars indicate the statistical and systematic uncertainties combined in quadrature. Also shown are predictions by a GEANT4 simulation of the experiment.

from simulations. Within the given statistical and systematic uncertainties the agreement is very good.

4. SUMMARY

The E166 experiment has been carried out in the Final Focus Test Beam at SLAC to demonstrate production of polarized positrons suitable for implementation at the ILC. A 1 meter long helical undulator of 2.54 mm period produced circularly polarized photons (1st harmonic endpoint 7.9 MeV for a 46.6 GeV beam). The polarized photons were converted to polarized positrons in a thin target and analyzed by transmission polarimetry of photons obtained on "reconversion" of the positrons. The measured polarization reached 80% for positrons near 6 MeV and 90% for electrons near 7 MeV. The measurements agree well with simulations made with an upgraded GEANT4 version including the dominant polarization-dependent electromagnetic interactions.

REFERENCES

1. *International Linear Collider Reference Design Report* (Aug. 2007), http://www.linearcollider.org/rdr
2. G. A. Moortgat-Pick *et al.*, *Phys. Rept.* **460**, 131 (2008).
3. H. Olsen and L.C. Maximon, *Phys. Rev.* **114**, 887 (1959).
4. V.E. Balakin and A.A. Mikhailichenko, Budker Institute of Nuclear Physics, BINP 79-85 (1979).
5. R.C. Wingerson, *Phys. Rev. Lett.* **6**, 446 (1961).
6. E.G. Bessonov and A.A. Mikhailichenko, Budker Institute of Nuclear Physics, BINP 92-43 (1992); T. Omori *et al.*, *Phys. Rev. Lett.* **96**, 114801 (2006).
7. G. Alexander *et al.*, *Phys. Rev. Lett.* **100**, 210801 (2008).
8. G. Alexander et al., to be submitted to Nucl. Instr. and Meth. A.
9. M. Berndt *et al.*, SLAC-376 (Mar. 1991).
10. R. Dollan, K. Laihem and A. Schälicke, *Nucl. Instr. and Meth. A* **559**, 185 (2006); A. Schälicke, K. Laihem and P. Starovoitov, DESY 07-202 (Nov. 2007).

The RHIC Polarized H^- Source

A. Zelenski[a], A. Kponou[a], J. Ritter[a], A. Belov[b], and V. Zubets[b]

[a]*Brookhaven National Laboratory, Upton, New York 11973, U.S.A.*
[b]*INR RAS, Moscow 117312 Russia.*

Abstract. The depolarization factors in the multi-step spin-transfer polarization technique and basic limitations on maximum polarization in the Optically-Pumped Polarized H- Ion Source (OPPIS) are discussed. Results of simulations and experimental studies of the Sona-transition polarization transfer process are presented. The experimentally observed polarization oscillations and polarization maximum peak were reproduced in simulations. Detailed studies of polarization losses in the RHIC OPPIS and the source parameters optimization resulted in the OPPIS polarization increase to 86-90%. This contributed to increasing polarization in the AGS and RHIC to 65 ~ 70%.

Keywords: Polarized Ion Sources.
PACS: 29.25.Lg

INTRODUCTION

The RHIC complex is the first where the "Siberian snake" technique was successfully implemented to avoid resonance depolarization during beam acceleration in the AGS and RHIC[1]. A luminosity of $1.6 \cdot 10^{32}$ cm^{-2} sec^{-1} for polarized proton collisions in RHIC will be produced by colliding 110 bunches in each ring at $2 \cdot 10^{11}$ protons/bunch intensity. The RHIC polarized source routinely produces 0.5-1.0 mA (maximum 1.6 mA) current in 400 μs long pulses (which corresponds to about 10^{12} H^- ions/pulse) at 85% polarization. Since one source pulse is accelerated and transformed to one RHIC bunch, there is sufficient spare intensity out of the source to charge RHIC to its full capacity (limited by the beam-beam interaction) in spite of losses along the accelerator chain. In addition, an intentional beam loss of about 70% occurs during dynamic beam collimation after the Booster, to reduce beam emittance in AGS and RHIC. The polarization direction is switched in the source according to the required bunch spin direction pattern in the RHIC.

The RHIC OPPIS

In the OPPIS (Fig. 1), an ECR source produces a primary proton beam of 2.8 ~ 3.0 keV energy, which is converted to electron-spin polarized H atoms by electron pick-up in an optically pumped Rb vapor cell. Electrostatic deflection plates downstream of the polarized alkali remove any residual H^+ or other charged species. The electron-polarized H beam then passes through a magnetic field reversal region, where the polarization is transferred to the nucleus, via the hyperfine interaction (Sona-transition). The polarized H atoms are then negatively ionized in a Na-jet vapor cell to

CP1149, *Spin Physics, 18th International Spin Physics Symposium*
edited by D. G. Crabb, D. B. Day, S. Liuti, X. Zheng, M. Poelker, and Y. Prok

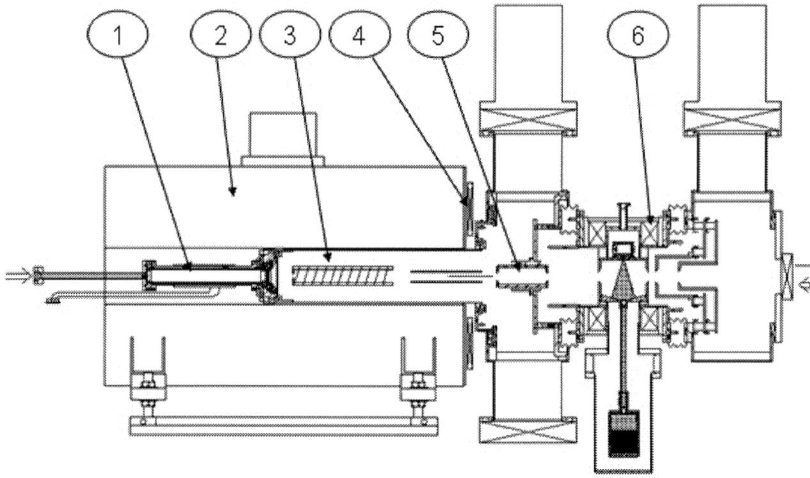

FIGURE 1. Layout of the RHIC OPPIS: 1) Electron Cyclotron Resonance primary proton source: 2). Super-conducting Solenoid; 3) Optically-Pumped Rubidium Cell; 4). Correction coil; 5). Sona-shield; 6). Na-jet ionizer.

form nuclear polarized H^- ions. Alternatively, the H atoms can be ionized in a He gaseous cell to form polarized protons[2].

The AGS cycle for polarized beam operation is 3 seconds, while OPPIS usually operates at 1 Hz repetition rate. Pulses not sent to the Booster are directed by a pulsed bending magnet to a 200 MeV p-Carbon polarimeter, for continuous polarization monitoring.

Sona-transition studies. The OPPIS technique is a multi-step polarization-transfer process. At each step there is some loss of polarization. The detailed studies and optimization of each of these factors resulted in 86% polarization in Run 2006 and further increase to 88~90% in Run 2007[3]. The recent increase is a result of the Sona-transition studies. The electron polarization is transferred to protons by the Sona-transition technique as the electron-spin polarized atomic H beam passes through a magnetic field reversal region between a high field superconducting solenoid and ionizer solenoid. Very strict restrictions are applied to the transverse magnetic field in the zero-crossing region (where the longitudinal field reverses direction) to avoid depolarization. The longitudinal field gradient generates a transverse field: $B_r = -r/2(dB/dz)$, and to fulfill Sona-transition conditions (for a Ø 2.0 cm atomic hydrogen beam) the condition $dB/dz \ll 0.2$ G/cm is required at the $B_z=0$ crossing point. A “soft” steel cylindrical Sona-shield and Correction Coil (CC) were designed and built to reduce the field gradient in the limited space between the superconducting solenoid 25 kG field and ionizer solenoid -1.5 kG field. As a result of Sona-shield and correction coil optimization, the gradient was reduced to less than 0.09 G/cm. For the Sona-transition efficiency studies and optimization the polarization dependence on correction coil current was measured for a wide range of CC field. Polarization was measured in the Lamb-shift polarimeter at 35 keV beam energy and at 200 MeV in the p-Carbon polarimeter.

Typical results of 200 MeV correction coil scan measurements are presented in Fig. 2. Since the CC field direction is opposite to superconducting solenoid field, the zero-crossing point is moving upstream inside the Sona-shield, when CC current is increased. This reduces the gradient and slightly increases polarization. At some current value the zero-crossing point is pushed out of the shield and the gradient is greatly increased, which causes a steep polarization drop.

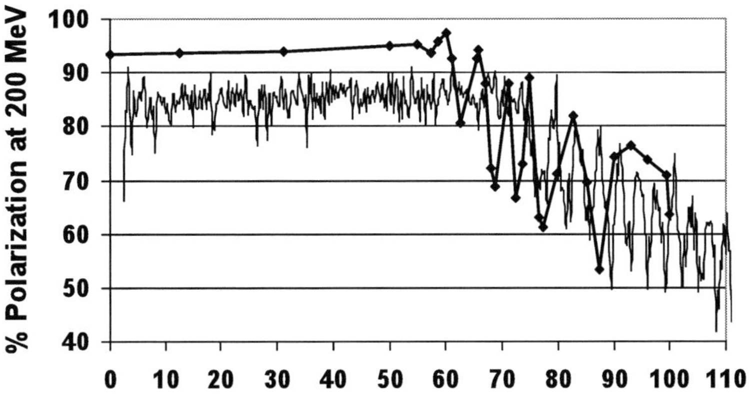

FIGURE 2. Measured and simulated (diamond) polarization (%) vs. correction coil current (A).

A new and stronger correction coil (and additional magnetic shielding at the superconducting solenoid flange) allowed significant expansion of the field variation range, which revealed unexpected periodic polarization oscillations, as seen in Fig 2. The polarization almost completely recovered at higher CC field, and the oscillations have a period of about 5 A correction coil current increment. The amplitude of the oscillations slowly decreases with higher CC field. To further explore this effect, the CC field was reversed. This pushed the zero-crossing downstream outside of the shield, and polarization oscillations were also observed.

The numerical simulations of the Sona–transition efficiency were made using code which was developed at INR, Moscow[4]. Results of numerical integration of Schrodinger equations with time-dependent magnetic fields for hydrogen atoms moving along z-axis with velocity of $7.6 \cdot 10^5$ m/sec (2.8 keV beam energy) are shown in Fig. 2. The resulting beam polarization is calculated by averaging over effective beam diameter of a 10 mm. The calculations correctly reproduce experimentally observed polarization dependence on the magnetic field distribution in the Sona-transition region. The minimal field gradient and maximum polarization is obtained just before the field dips negative (providing that all the other transverse field sources are eliminated by the Sona shield). An additional μ-metal shield layer was inserted inside the "soft" steel cylinder to suppress the small residual field of the steel itself. The polarizations for different beam diameters – 1.2 cm, 1.6 cm and 2.0 cm (beam diameter is defined by the collimator at the entrance of Sona-shield) were measured after this optimization and, within errors, the maximum polarization values are the same. This means that Sona-transition efficiency is close to 100% (the best value for further estimations is: E_{Sona}= 0.98±0.01.

The origin of polarization oscillations can be explained in the following way: Since the oscillations appear only when the longitudinal field crosses zero (as shown in Fig.3) there are three distinct regions for the polarization evolution. The **first** stage is the SONA transition itself at zero field crossing. The efficiency of the SONA transition will be less than 100% for off-axis trajectories. This is equivalent to rotation of spin direction for off- axis atoms (in spin state with F=1) for some angle relative to

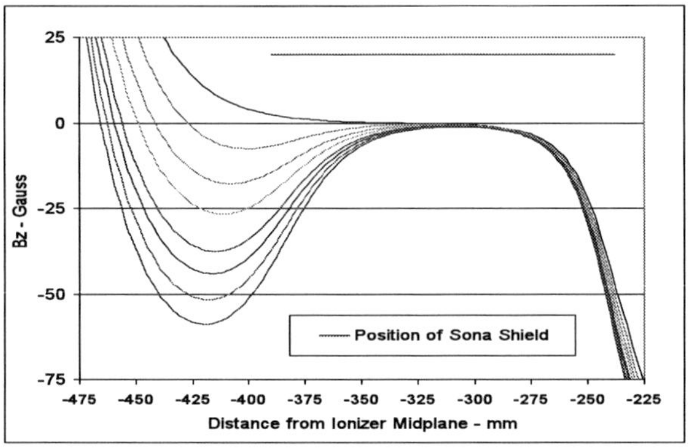

FIGURE 3. Magnetic field distribution in the SONA transition region for different currents in the correction coil.

the Z-axis, and the value of the angle is proportional to the R distance. Spin state IV (F=0, see Fig.4) is scalar and does not have any rotation in weak field. The **second** stage is spin rotation around the longitudinal magnetic field, which has the same direction as in the ionizer region. The average value of the magnetic field in this region is approximately proportional to the correction coil current and field strength is a few tens of G (see Fig.4). For hydrogen atoms in spin states with F=1, the precession rate is 1.4 MHz/G. The rotation angle for all hydrogen atoms is approximately the same, independent of radius, and the precession angle can be as large as π, 2π, 3π, … ,$n\pi$ spin flips about a transverse direction. Finally, in the **third** stage, the atomic beam passes (non-adiabatically) into the weak magnetic field region inside the SONA shield. During this passage the spin rotates again for some angle relative to the Z- axis. Again the rotation angle is proportional to the distance from axis. The rotation angle in the third stage can be added to, or subtracted from, the rotation angle obtained after passage of SONA transition region depending of rotation angle in the second stage. Generally, there will be vector summation of two rotation angles and spin projection on the Z-axis will oscillate after passage through the weak field region and then to the ionizer region. Remarkably, the precession angle phases are highly coherent and polarization amplitude is well preserved for many oscillations in the measurements, with the gradual reduction caused by the losses at the very first SONA transition stage in the increased gradient field (see Fig. 3).

The calculations correctly reproduce the observed oscillations. A somewhat surprising and very useful result of these studies is the polarization maximum just before the polarization drop, which was observed in earlier experiments. This maximum occurs at the CC value, which pushes the magnetic field zero-crossing outside the Sona-shield, where the gradient is not minimal. This non-trivial behavior was reproduced in calculations and is a result of fine optimization of spin precession evolution for different hyperfine sublevels (see example in Fig.4). The polarization gain of about 5-10 % was observed at this point for different Sona-shield configurations.

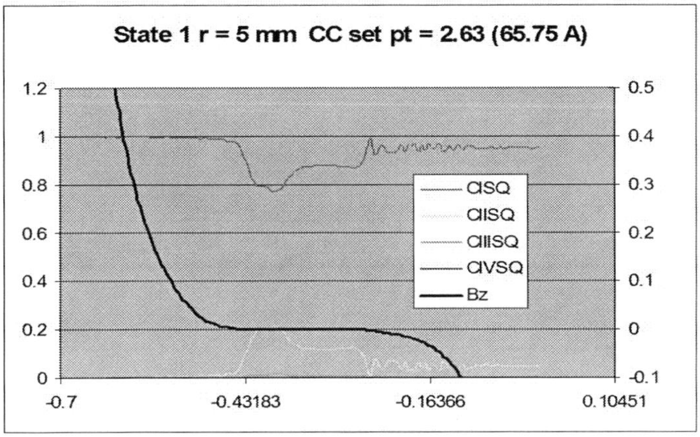

FIGURE 4. Hyperfine states population evolution during passage through the Sona–transition for initial hyperfine state 1. Left ordinate is probability amplitude, right is magnetic field in Tesla.

Polarization losses in Na-jet ionizer cell. A magnetic field of about 1.5 kG is required in the sodium-jet ionizer cell to break the electron-proton hyperfine interaction. Since the critical field for hydrogen atoms is 507.6 G, the theoretical polarization loss at 1.5 kG field is about 3%. The ionizer field increase over 1.5 kG is not very effective for polarization increase (the theoretical polarization gain between 1.5 kG and 2.0 kG is only ~1%) but it will cause beam emittance growth beyond the RFQ acceptance and result in beam intensity losses.

Significantly larger polarization increase at higher ionizer fields was observed in experiments (see Fig.5). Similar results were also observed in earlier measurements at the TRIUMF OPPIS, but it was explained by sodium vapor penetration in the low field region (in the oven-type sodium cell) and possible ionizer field influence on the Sona-transition region. In the present experiments, the sodium vapor is very well confined in the jet-cell within the center flattop field region. For every ionizer field strength value the correction coil scan was done to optimize the Sona-transition. The measurements were repeated with a smaller 1.2 cm diameter beam, which would reduce the possible additional contribution of the spatial polarization distribution (since the central fraction of the beam is better transported and accelerated in the linac to the 200 MeV polarimeter). Still, the polarization increase at higher field is not accounted for by the hyperfine interaction breaking alone. This means that there are other effects, which also depends on the ionizer field strength.

One possible explanation might be an observed effect of polarization dependence on the Na-jet vapor thickness. Usually the jet cell is operated at sodium reservoir temperature of ~500° C, which is required for the H^- ion yield saturation (the estimated total jet thickness is $\sim 2\cdot 10^{15}$ atoms/cm^2). At 470° C temperature, where the H^- current drops to about 80% of maximum and thickness is reduced to ~ $1.0\cdot 10^{15}$ atoms/cm^2 the polarization is increased for ~ 2-3%. Multiple charge-exchange processes are more likely at high sodium vapor thickness, which might cause some depolarization at each collision. The stronger ionizer field would suppress this depolarization process and reduce polarization losses.

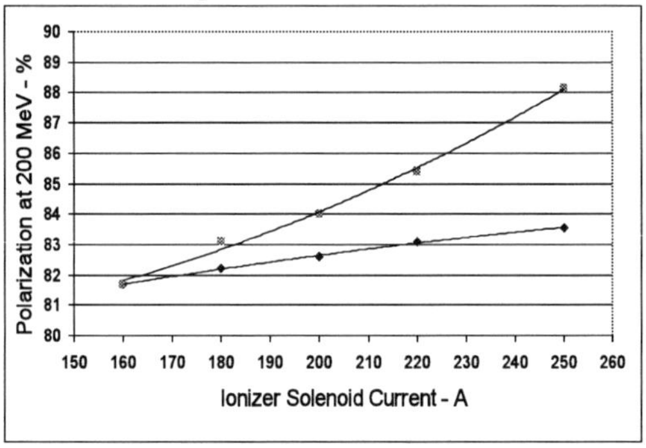

FIGURE 5. Measured (▒) and calculated (♦) polarization vs. ionizer solenoid field strength. Ionizer field at 250 A current is equal to 1.98 kG.

The benefit of the higher ionizer field for increasing polarization is partially offset by beam intensity reduction due to emittance growth, which is proportional to ~ $\mathbf{BR^2}$. For the typical RFQ acceptance of about 2.0 π mm·mrad, this limits the beam radius to less than ~ 0.8 cm, and ionizer magnetic field to less than 2.0 kG. A higher brightness polarized atomic H beam is required to keep beam emittance in the RFQ acceptance range at high field by using smaller diameter collimated beam.

CONCLUSIONS

Detailed studies of polarization losses in the RHIC OPPIS and the source parameters optimization resulted in the OPPIS polarization increase to 86-90% and to 65 ~ 70% in the AGS and RHIC. The authors thank Prof. T. Clegg, UNC and TUNL, for very helpful discussions on Sona transition features and the observed polarization oscillations.

REFERENCES

1. T. Roser, AIP Conf. Proc. **980**, p.15, (2008).
2. A. Zelenski et al., Rev. Sci. Instr., **73**, p.888, (2002).
3. A. Zelenski et al., AIP Conf.Proc., **980**, p.221, (2008).
4. E.P. Antishev and A.S. Belov, AIP Conf.Proc.,**980**, p.263, (2008).

Sources of Helicity-correlated Electron Beam Asymmetries

K.D. Paschke

University of Virginia, Charlottesville, VA 22904

Abstract. The availability of high current, high polarization electron beams from laser-driven GaAs photocathodes has enabled a broad program studying parity violation in electron scattering. Precision measurements of the tiny (<1 part per million) parity-violating beam-helicity asymmetry are used to study the structure of nuclei or to test the Standard Model of electroweak interactions. As these experiments grow ever more precise, asymmetric beam properties between the two beam helicity states threaten to become the leading source of experimental uncertainty.

These helicity-correlated beam asymmetries are predominantly created in the conversion of circularly polarized laser light to a polarized electron beam. In recent experiments at Jefferson Lab, improved techniques for configuring the laser optics of the polarized source have been used to control beam asymmetries to such a level as to be a negligible source of systematic uncertainty. This successful result serves as a promising start for the next generation of parity-violating electron-nuclear scattering measurements, which strive to improve precision by a factor of twenty.

Keywords: polarization, electron scattering, parity violation
PACS: 29.25.Bx, 29.27.Hj, 42.25.Ja, 11.30.Er

INTRODUCTION

The study of parity violation in electron scattering provides a way for isolating the weak neutral-current interaction between an electron and nucleus. It is also employed as a search for interactions beyond the electroweak Standard Model, which would be indicated by measurements which deviated from precisely predicted values based on the known quark and electron electroweak charges. In these experiments, longitudinally polarized electrons are scattered from unpolarized targets. For elastic scattering, or for fully inclusive measurements detecting only the electron, a cross-section asymmetry under change in the electron helicity violates parity, and in the Standard Model this difference in cross-section can only arise through the weak neutral current. This asymmetry, $A_{PV} = (\sigma_R - \sigma_L)/(\sigma_R + \sigma_L)$, ranges between 10^{-3}-10^{-7} for published experiments.

One challenge of these experiments is that changes in the beam properties (intensity, position, profile) will change the detected scattered flux. If the changes in the beam are correlated with the electron helicity, the result can mimic the tiny parity-violating asymmetry. Corrections for these changes are typically made with precision of around 10%. Helicity-correlated beam asymmetries (HCBAs) are therefore a potential systematic error in the measurement of small asymmetries.

The PREX experiment, to run at Jefferson Lab in Hall A in early 2010, and the QWeak experiment, which will start in mid-2010, have precision goals which are approximately an order of magnitude better than what has been previously achieved in scattering from nuclei. Of the two experiments, the PREX experiment, which aims for a precision of

CP1149, *Spin Physics, 18th International Spin Physics Symposium*
edited by D. G. Crabb, D. B. Day, S. Liuti, X. Zheng, M. Poelker, and Y. Prok

15×10^{-9}, is more demanding in terms of helicity-correlated beam changes due to the form-factor of the heavy lead nucleus which enhances the sensitivity of cross-section to scattering angle. The PREX experiment aims to keep helicity correlated beam motion to < 1 nm (as was accomplished for HAPPEX-II) and the bound the beam spot difference to $< 10^{-4}$ of the intrinsic beam size.

POLARIZED ELECTRON BEAM ASYMMETRIES

The polarized electron beam is created using a technology first developed at SLAC to enable the original measurement of parity violation in electron scattering [1]. (For a recent review of the development of the polarized electron photoemission source, see [2].) Laser light illuminates a semiconducting photocathode, the surface of which has been chemically treated to produce a negative work function. Laser light promotes electrons from the valence band to the conduction band of the semiconductor, where they are drawn out by an electrostatic potential for injection to an RF accelerator. Circularly polarized laser light will preferentially promote a specific electron spin-state, and splitting the valence band (with doping or substrate lattice mismatch) can enhance this selectivity. Electron beam polarizations of 85% are now routinely used at Jefferson Lab.

Since the electron polarization is fully determined by the circular polarization of the incident laser light, it is possible to rapidly flip the helicity of the electron beam by changing the laser polarization. This is accomplished using an electro-optic Pockels cell, acting as a quarter-wave plate to produce circularly polarized light from an initial linear polarization. This rapid reversal is critical for the precision measurement of small asymmetries, as it randomizes any slow-variations in the accelerator and detection apparatus across both helicity states.

There is significant operational experience in using the polarized electron source for parity-violation experiments at Jefferson Lab [3]. A sophisticated understanding of the sources of HCBA at JLab has been achieved. The dominant effects stem from imperfections in the laser beam polarization. Although the laser is highly circularly polarized it is very natural for there to be a small component of linear polarization which switches sign under helicity reversal, as described below. The strain induced in the photocathode, which is key to achieving high polarization, also introduces a "preferred direction" for the crystal, creating a variation in quantum efficiency which depends on the linear polarization of the incident light. This quantum-efficiency anisotropy of the photocathode (or *analyzing power*) interacts with the changing linear polarization of the laser light to produce a helicity-correlated intensity change in the electron beam.

Figure 1 illustrates relevant components of the laser optics of a polarized electron source. Initially linearly polarized light is converted to circular polarization, of either handedness, with a $\pm\pi/2$ phase shift in the electro-optic Pockels cell. If the applied voltage is too small, the result will be elliptic polarization, with the linear component in the same direction as the original laser polarization. If the applied voltage is too large, the residual linear component will be orthogonal to the original polarization. A small contribution from linear polarization which doesn't change when the beam helicity changes is not problematic, however, differences in the residual linear polarization

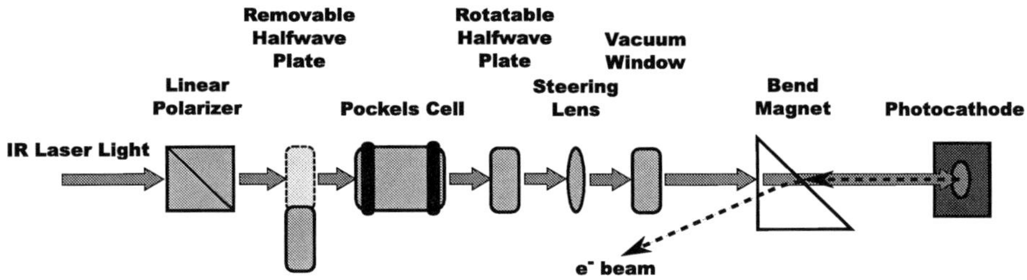

FIGURE 1. Schematic layout of polarized electron source laser optics.

between the states can create significant differences in beam intensity, position, or profile.

Consider the introduction of a small, constant birefringence, with axes aligned to the Pockels cell. This increases the net phase-shift magnitude for one handedness state and reduces it for the other, implying equal degree of ellipticity but orthogonal orientation of the major ellipse axes in the resulting polarization state. This sort of constant birefringence offset is referred to as a "asymmetric" phase shift, since it produces opposite residual linear polarization between the handedness states. One source of such an asymmetric phase shift is the intrinsic, stress-induced birefringence of the Pockels cell. Each element along the laser optics path may also contribute a constant birefringence offset. In particular, the vacuum window, being under mechanical stress, is significantly birefringent.

If, for one of the handedness states, the residual linear polarization is aligned to the preferred photocathode axis, then this state will have a maximal net quantum efficiency, the corresponding state of opposing handedness will have a minimal quantum efficiency, and the result will be that the electron beam will change intensity between the two polarization states. If the residual linear polarization components for each state are at $\pm 45°$ to the cathode analyzing power, then there is no difference in the effect between the two electron helicity states. The orientation of the polarization ellipses, relative to the photocathode analyzing power, is illustrated in Fig. 2. The rotatable halfwave plate (RHWP) (illustrated in Fig. 1) is used to tune the orientation of the residual linear polarization. Note that linear polarization from birefringence of downstream elements is not changed by rotation of this waveplate. A common technique is to use the rotating waveplate to keep the sensitivity to asymmetric phase shift in the Pockels cell small but not zero, and then use an asymmetric voltage shift on the Pockels cell setpoints to counteract the effects from downstream elements.

The residual linear polarization component is also not constant over the laser beam spot. This creates a varying intensity asymmetry over the beam spot. It is useful to picture the spatial moments of this asymmetry distribution; the zeroth-moment simply averages the intensity asymmetry over the entire spot. First spatial moments describe the movement of the centroid of the beam under helicity reversal (referred to as position differences), while second moments describe the change in root-mean-square size of the beam (referred to as spot-size differences). Variations in the linear polarization component, or variations in the quantum efficiency anisotropy, are the most important

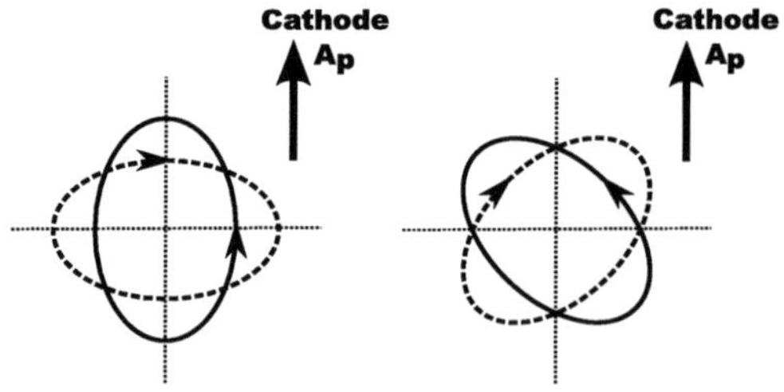

FIGURE 2. Polarization ellipse representation for right (solid) and left (dashed) handedness laser states. The figure on the left shows an orientation of the asymmetric residual linear light which is maximally sensitive to the cathode analyzing power, the figure on the right shows the configuration with minimal sensitivity.

contributors to helicity-correlated position and size differences.

It has long been appreciated that this non-uniform linear polarization component arises due to imperfections in the Pockels cell or in other optical components on the laser path. Recent work has demonstrated the critical role played by the alignment of the optic axis of the Pockels cell and the divergence of the laser spot. In a simple ray-trace view, the diverging optic rays across the beam spot can be seen to be subject to varying birefringence. This ultimately creates a position-dependent intensity asymmetry in the electron beam. If the beam is perfectly aligned to the optic axis of the Pockels cell, the net effect is a shape distortion between the two helicity states along orthogonal axes. The average zeroth, first, and second moments do not change between the helicity states. If the beam is off-axis of the Pockels cell, there are significant zeroth, first, and second moments of the intensity asymmetry. This effect is expected to scale with the beam divergence. For misalignments of around 1 milliradian or larger, this is expected to be the dominant mechanism in the laser optics for creating spot size and position differences. Data from a study of this effect is shown in Fig. 3.

A careful alignment algorithm was implemented for the HAPPEX-II experiment, which ran in Hall A in 2005, to minimize this and other similar effects [4]. The resulting helicity-correlated differences in beam trajectory, measured in the experimental hall, averaged over the month-long run, were less than 2 nm and 0.3 nanoradian in angle. The position and angle differences are plotted in Fig. 4. Similar results have since been obtained for subsequent parity-violation experiments at Jefferson Lab.

SLOW REVERSALS

A valuable technique for these experiments is the introduction of a beam helicity reversal, relative to some potential source of HCBA. The statistical consistency of data sets taken in different states of the reversal can be used to demonstrate the absence of large, unknown systematic errors, and the combination of data sets (appropriately sign-corrected) provides a method for further canceling possible unmeasured or poorly-corrected HCBA effects.

At present, only one slow-reversal is commonly employed at CEBAF. As shown in

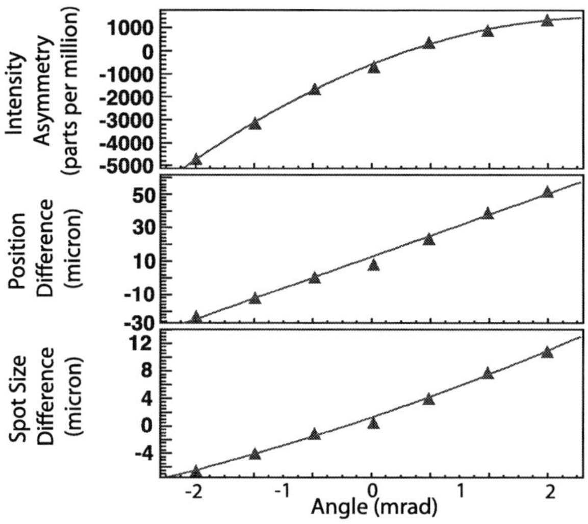

FIGURE 3. Measurement of laser beam intensity asymmetry, position differences and spot size differences during a scan of Pockels cell orientation, made using a 1-dimensional multi-element photodiode array. A linear polarizer provided the analyzing power. The spot size is approximately 1 mm, and the divergence around 1mrad.

Fig. 1, a half-wave plate can be inserted into the laser path to reverse the sign of laser polarization, relative to the voltage applied to the Pockels cell. This slow-reversal is particularly effective for canceling effects related to electrical signals, either from the logic or Pockels cell high voltage, which correlate to helicity. This was evident in the 2005 run of the HAPPEX-Helium experiment, when insufficiently grounded elements in the polarized source induced large beam motion coincident with the helicity logic signal. These beam asymmetries cancelled very closely under halfwave-plate insertion. However, the effects of residual linear laser polarization, which are typically the dominant mechanism for HCBA, change sign along with the beam helicity under waveplate insertion. For this reason, this insertable waveplate is not generally an ideal method for control of the largest sources of HCBA.

Another common slow helicity reversal would be the introduction of an additional half-cycle $g-2$ rotation, which would reverse the electron beam helicity at target with respect to the helicity of the beam created in the polarized source. This can be accomplished with a small energy change, if a high-energy beam is bent significantly in a magnetic field. Such a reversal was used for the E158 experiment at SLAC [5], where a 22° bend brought the beam into the end station. However, this technique is not feasible at the lower beam energies used at Jefferson Lab, and is anyway inconvenient at a multi-user facility.

A similarly effective slow reversal can be created using spin manipulation in the injector. Spin manipulation is necessary at most accelerators to align the in-plane launch angle for optimized longitudinal polarization at the experimental target. Typically, in the low energy portion of an injector, solenoids are used to rotate transverse beam

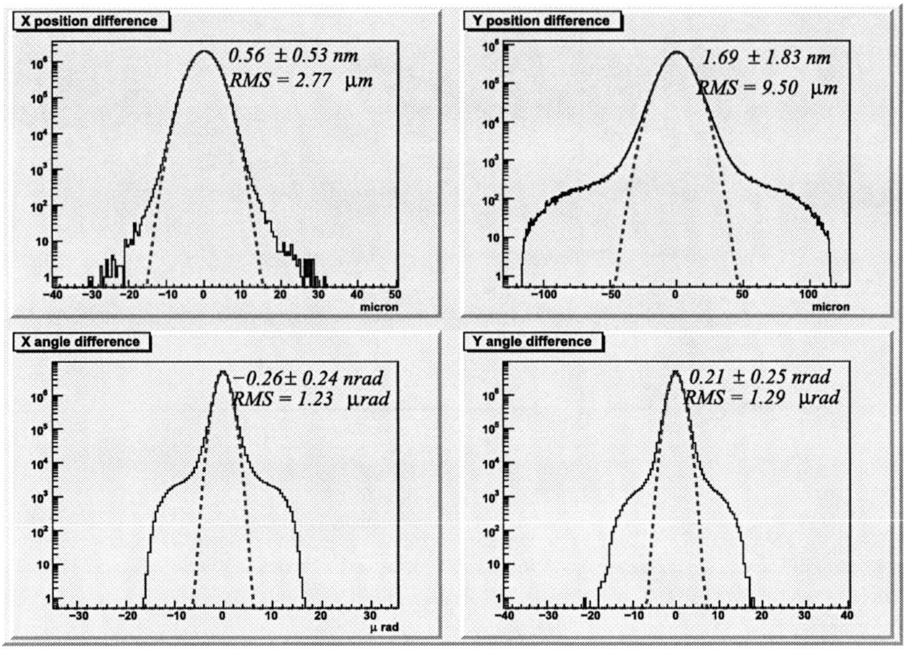

FIGURE 4. Differences in electron beam position and angle on target during the HAPPEX-II experiment at Jefferson Lab, measured for each helicity pair. The average over the course of the run is noted for each plot.

polarization, and a Wien spin rotator, with crossed E and B fields, is used to rotate transverse into longitudinal. These fields also focus the beam. In particular, the effect of the Wien fields on the beam are quite severe, and accomplishing a full π spin rotation would badly perturb the beam parameters, ruining the value of this rotation as a slow reversal. However, the focussing property of the solenoids go as the square of the B-field, implying that a pair of solenoids can be used to for a π spin rotation on a transversely polarized beam without changing the beam optics. The subsequent Wien rotation to optimize the launch angle would be unchanged for both states.

An upgrade of the 100 keV injector at Jefferson Lab is being designed to enable this slow reversal for future experiments. This is expected to provide a powerful new tool at JLab for the control of HCBA.

REFERENCES

1. C. Y. Prescott, et al., *Phys. Lett.* **B77**, 347–352 (1978).
2. C. K. Sinclair, *AIP Conf. Proc.* **980**, 25–36 (2008).
3. K. D. Paschke, *Eur. Phys. J.* **A32**, 549–553 (2007).
4. A. Acha, et al., *Phys. Rev. Lett.* **98**, 032301 (2007), nucl-ex/0609002.
5. P. L. Anthony, et al., *Phys. Rev. Lett.* **95**, 081601 (2005), hep-ex/0504049.

RCNP Project on Polarized ^{3}He Ion Sources - From Optical Pumping to Cryogenic Method

M. Tanaka*, T. Inomata†, Y. Takahashi†, M. Fujiwara†, T. Shimoda**, M. Yosoi†, K. Takahisa†, H. Kohri†, T. Ohta†, C. Morisaki†, T. Kunimatsu†, Yu. A. Plis‡, G. Frossati§ and J. P. Didelez¶

*Department of Clinical Technology, Kobe Tokiwa University, Ohtani-cho 2-6-2, Nagata-ku, Kobe 653-0838, Japan
†Research Center for Nuclear Physics, Osaka University, Mihogaoka 10-1, Ibaraki, Osaka 567-0047, Japan
**Department of Physics, Graduate School of Science, Osaka University, 1-1, Machikaneyama, Toyonaka, Osaka 560-0043, Japan
‡Joint Institute for Nuclear Research, Dubna, RU-141980, Moscow region, Russia
§Kamerlingh Onnes Institute, Leiden University, Leiden, The Netherlands
¶IN2P3, Institut de Physique Nucléaire, Orsay, France

Abstract. A polarized ^{3}He ion source has been developed at RCNP for intermediate and high energy spin physics. Though we started with an OPPIS (Optical Pumping Polarized Ion Source), it could not provide highly polarized ^{3}He beam because of fundamental difficulties. Subsequently to this unhappy result, we examined novel types of the polarized ^{3}He ion source, i.e., EPPIS (Electron Pumping Polarized Ion Source), and ECRPIS (ECR Polarized Ion Source) experimentally or theoretically, respectively. However, attainable ^{3}He polarization degrees and beam intensities were still insufficient for practical use. A few years later, we proposed a new idea for the polarized ^{3}He ion source, SEPIS (Spin Exchange Polarized Ion Source) which is based on enhanced spin-exchange cross sections at low incident energies for $^3He^+$ + Rb, and its feasibility was experimentally examined.

Recently, we started a project on polarized ^{3}He gas generated by the brute force method with low temperature ($\sim$4mK) and strong magnetic field ($\sim$17 T), and rapid melting of highly polarized solid ^{3}He followed by gasification. When this project will be successful, highly polarized ^{3}He gas will hopefully be used for a new type of the polarized ^{3}He ion source.

Keywords: polarized ^{3}He ion source, optical pumping, Lamb shift, metastability exchange, cryogenic, atomic beam
PACS: 29.25.Lg, 31.15.Ct, 52.20.Hv

HISTORY OF OUR DEVELOPMENT

We have been developing a polarized ^{3}He ion source at RCNP and Dept. of Physics, Osaka University, aiming at intermediate and high energy spin physics. We started with an OPPIS (Optical Pumping Polarized Ion Source) which uses an electron capture process between an alkali atom polarized by means of the laser optical pumping and an incident $^3He^{2+}$ ion. Our project was motivated by the success of the proton OPPIS [1]. However, the OPPIS applied to ^{3}He was not successful in spite of all our efforts [2, 3]. After that, we proposed a polarized ^{3}He ion source, EPPIS (Electron Pumping Polarized Ion Source) based on the principle of electron pumping[4] as a new type of polarized ^{3}He ion source to overcome the difficulties encountered in the OPPIS. Though the

CP1149, *Spin Physics, 18th International Spin Physics Symposium*
edited by D. G. Crabb, D. B. Day, S. Liuti, X. Zheng, M. Poelker, and Y. Prok

principle of electron pumping was experimentally verified, the realized ^{3}He polarization was still insufficient for practical use [5, 6, 7, 8, 9]. We made, then, a feasibility study of the polarized ^{3}He ion source based on the ionization of polarized ^{3}He gas with an ECR (Electron Cyclotron Resonance) ionizer as a more powerful polarized ^{3}He ion source than the EPPIS. This type of ion source was named ECRPIS (ECR Polarized Ion Source). However, as a result of the Monte Carlo simulation, it was found that a serious depolarization of ^{3}He could not be avoided in the ECR plasma, which is in a striking construct with the successful proton ECRPIS[10].

Finally, we arrived at a new idea, "SEPIS" (Spin Exchange Polarized Ion Source) as a novel polarized ^{3}He ion source to solve many difficulties encountered in OPPIS, EPPIS, and ECRPIS [11, 12, 13]. A hint of this idea came from the experimental results on the EPPIS. Extremely large spin-exchange cross sections, σ_{se} exceeding 10^{-14} cm^2, were theoretically predicted for a $^3He^+$ + Rb system at low $^3He^+$ incident energies (~1 keV/A) [8]. In addition, it was suggested that electron capture cross sections, σ_{ec} for a $^4He^+$ ion incident on an alkali atom, Cs decreases as decreasing a $^4He^+$ kinetic energy [14]. Since most of the SEPIS development was already presented somewhere, the detailed description is omitted in this paper. We are still on the way to the development of SEPIS. However, the energy dependence of ^{3}He polarization generated by the spin-exchange process clearly shows an increasing behavior as expected by the theory for the data down to the incident $^3He^+$ energy of 4 keV. To conclude the validity of SEPIS, we need the data at lower energies, say, at around 1 keV.

PRODUCTION OF POLARIZED ^{3}HE GAS BY CRYOGENIC METHOD

In the following, we will briefly present a new project on the production of polarized ^{3}He gas for the future medical application to the ^{3}He MRI (Magnetic Resonance Imaging). When this project will be successful, highly polarized ^{3}He gas may be used also for the polarized ^{3}He ion source as a more efficient method than the ^{3}He SEPIS itself.

The principle of polarization is based on, so called, brute force method, i.e., the production of nuclear polarized ^{3}He under the low temperature (~4 mK) and the high magnetic field (~17 T) according to the Curie's law. An attainable ^{3}He polarization degree is simply estimated to be larger than 95% under this condition. However, this is not true as far as liquid ^{3}He is concerned. Since liquid ^{3}He is a Fermi liquid, only a minor part of the ^{3}He nuclei around the magnetic Fermi temperature (T_F^{**} = 180 mK) can be polarized even when the brute-force condition is satisfied. As a result, the ^{3}He polarization for liquid ^{3}He could never exceed a % order. However, approximately 30 years ago, Castaing and Nozières proposed [15] an idea to produce highly polarized ^{3}He liquid, and their idea was later proved experimentally [16, 17]. Their idea is to produce polarized solid ^{3}He at first, and, then, to liquefy it within a time shorter than the ^{3}He spin relaxation time. Here, it is noted that solid ^{3}He can be highly polarized since solid ^{3}He does not obey the Fermi statistics, and exhibits paramagnetism under the presence of external magnetic field at $T > 1$ mK. If this principle can be extended to production of polarized ^{3}He gas, production of a large amount, say, ~1 m^3/day of highly polarized ($\geq$ 80%) ^{3}He may be available, which will hopefully open up the new application fields.

Toward this goal, we are developing an equipment enabling rapid melting of solid ^{3}He by employing a ^{3}He -^{4}He dilution refrigerator (DSR3000: Leiden Cryogenics), and a 17 T superconducting magnet (17T71:Jastec) currently used for the polarized HD target developed at RCNP.

^{3}He gas in a cooling cell connected to a mixing chamber of the dilution refrigerator is cooled down and liquefied. When liquid ^{3}He temperature is lower than 320 mK, it is, adiabatically, compressed by a ^{3}He gas, and further compressed by a ^{4}He gas through a mechanical piston of the cooling cell as shown in Fig. 1. Then, the cooling cell is thermally disconnected from the dilution refrigerator by a thermal switch[18], and liquid ^{3}He is solidified and cooled down to a few mK by the principle of Pomeranchuk cooling [19]. When the magnetic field of 17-T is applied to the cooling cell, the ^{3}He nuclei are highly polarized.

Finally, the polarized solid ^{3}He must be liquefied and further gasified for practical use by increasing the cooling cell temperature without depolarization. Since the ^{3}He relaxation time in the liquid phase is relatively short ($\leq$ 200 s) around 300 mK, solid ^{3}He should be rapidly (hopefully less than a few tens of seconds) melted and gasified [20]. A few methods of rapid melting are offered. One of them is to use a reverse process of the Pomeranchuk cooling, i.e., increasing the temperature by decompression and subsequent gasification, for example, by injecting hot water into the sample container [21]. Not only the temperature of the cooling cell but also the NMR (nuclear magnetic resonance) signals associated with the ^{3}He polarization will be monitored. A

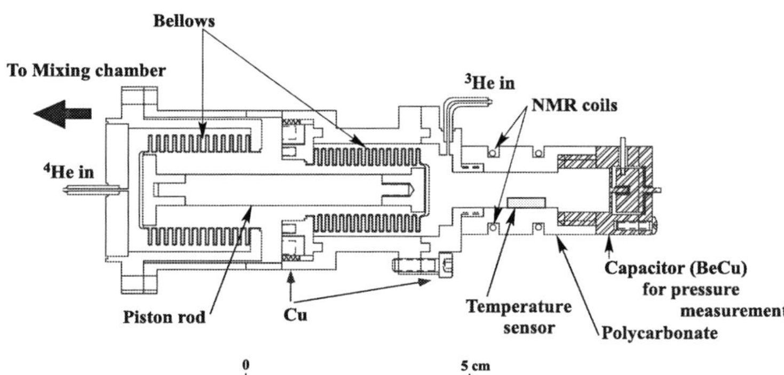

FIGURE 1. A prototype of Pomeranchuk cell

prototype of cooling cell named "Pomeranchuk cell" was constructed (see Fig. 1), and ready to be attached to another cryogenic system which has a cooling power down to 250 mK and has a 1-T superconducting Helmholtz coil for the test experiment. Basic data needed for design of practical device will be taken with this test system.

CONCLUSION

After long journey of the development, we have arrived at the SEPIS. To evaluate the SEPIS performance in the present work, we show a table of performances for the polarized ^{3}He ion sources tested or put to practical use obtained so far in the world institutes. Except for ours, most of the polarized ^{3}He ion sources sound rather old, say, older than 20 years. Unfortunately, our present results sound unexciting. However, our record will be improved remarkably by lowering the incident $^3He^+$ energy down to 1 keV as the theory predicts and by increasing the Rb polarization. Lastly, it should be emphasized that use of highly polarized ^{3}He gas produced by the brute force method would hopefully provide a highly polarized ^{3}He ion source in combination with an EBIS (Electron Beam Ion Source) ionizer[11].

REFERENCES

1. Y. Mori, A. Takagi, K. Ikegami, S. Fukumoto, and A. Ueno, *J. Phys. Soc. Jpn.* **55**, 453 (1986).
2. M. Tanaka, T. Ohshima, K. Katori, M. Fujiwara, H. Ogata, and M. Kondo, *Phys. Lett.* **B274**, 163 (1992).
3. M. Tanaka, N. Shimakura, T. Oshima, K. Katori, M. Fujiwara, H. Ogata, and M. Kondo, *Phys. Rev.* **A50**, 1184 (1994).
4. M. Tanaka, M. Fujiwara, S. Nakayama, and L.W. Anderson, *Phys. Rev.* **A52**, 392 (1995).
5. T. Yamagata,M. Tanaka, K. Yonehara, Y. Arimoto, T. Takeuchi, M. Fujiwara, Yu. A. Plis, L. W. Anderson, and R. Morgenstern, *Nucl. Instr. and Meth.* **A402**, 199 (1998).
6. M. Tanaka, T. Yamagata, K. Yonehara, T. Takeuchi, Y. Arimoto, M. Fujiwara, Yu. A. Plis, L.W. Anderson, and R. Morgenstern, *Phys. Rev.* **A60**, R3354 (1999).
7. Y. Arimoto, N. Shimakura, K. Yonehara, T. Yamagata, and M. Tanaka, *Eur. J.* **D8**, 305 (2000).
8. Y. Arimoto, N. Shimakura, T. Yamagata, K. Yonehara, and M. Tanaka, *Phys. Rev.* **A64**, 062714-1 (2001).
9. K. Yonehara, T. Yamagata, Y. Arimoto, T. Takeuchi, and M. Tanaka, *Nucl. Instr. and Meth.* **B184**, 391 (2001).
10. M. Tanaka, N. Shimakura, Yu. A. Plis, *Nucl. Instr. Meth.* **A524**, 46 (2004).
11. M. Tanaka, Yu. A. Plis, E.D. Donets, N. Shimakura, Y. Arimoto, T. Yamagata, and K. Yonehara, *Nucl. Instr. and Meth.* **A537**, 501 (2005).
12. M. Tanaka, Y. Takahashi, T. Shimoda, T. Furukawa, M. Yosoi, K. Takahisa, N. Shimakura, and S. Yasui, *Nucl. Instr. and Meth.* **A568**, 543 (2006).
13. M. Tanaka, Y. Takahashi, T. Shimoda, M. Yosoi, K. Takahisa, Yu. A. Plis, *Rev. Sci. Instr.* **79**, 02B308 (2008).
14. A. S. Schlachter, D. H. Loyd, P. J. Bjorkholm, L. W. Anderson, and W. Haeberli, *Phys. Rev.* **174**, 202 (1968).
15. B. Castaing, and P. Nozières, *J. Phys. (Paris)* **40**, 257 (1979).
16. M. Chapellier, G. Frossati, and F. Rasussen, *Phys. Rev. Lett.* **42**, 904 (1979).
17. G. Schumacher, D. Thoulouze, B. Castaing, Y. Chabre, P. Segransan, and J. Joffrin, *J. Phys. Lett. (Paris)* **40**, 143 (1979).
18. P. R. Roach, J. B. Ketterson, B. M. Abraham, P. D. Roach, and J. Monson, *Rev. Sci. Instr.* **46**, 207 (1975).
19. R. C. Richardson, *Rev. Modern Phys.*, **69**, 683 (1997).
20. G. Frossati, *Nucl. Instr. and Meth.* **A402**, 479 (1998).
21. J. H. Ardenkjaer-Larsen, B. Fridlund, A. Gram, G. Hansson, L. Hansson, M. H. Lercher, R. Servin, M. Thaning, and K. Golman, PNAS, **100** 10159.

Studies on Beam Formation in an Atomic Beam Source

A. Nass*, M. Stancari† and E. Steffens*

*Phys. Institut der Universität Erlangen Nürnberg, E.-Rommel-Str. 1, 91058 Erlangen, Germany
†Istituto Nazionale di Fisica Nucleare, 44100 Ferrara, Italy

Abstract. Atomic beam sources (ABS) are widely used workhorses producing polarized atomic beams for polarized gas targets and polarized ion sources. Although they have been used for decades the understanding of the beam formation processes is crude. Models were used more or less successfully to describe the measured intensity and beam parameters. ABS's are also foreseen for future experiments, such as PAX [1]. An increase of intensity at a high polarization would be beneficial. A direct simulation Monte-Carlo method (DSMC) [2] was used to describe the beam formation of a hydrogen or deuterium beam in an ABS. For the first time a simulation of a supersonic gas expansion on a molecular level for this application was performed. Beam profile and Time-of-Flight measurements confirmed the simulation results. Furthermore a new method of beam formation was tested, the Carrier Jet method [3], based on an expanded beam surrounded by an over-expanded carrier jet.

Keywords: Polarized Targets
PACS: 29.25.Pj

MOTIVATION

Polarized atomic beam sources (ABS's) are utilized to provide nuclear-polarized atomic hydrogen (H) and deuterium (D) beams. H_2 or D_2 gas is dissociated and the atomic gas expands through a cooled nozzle into the vacuum. A beam of high brightness is then formed by a skimmer and a collimator to adapt the beam to the acceptance of the subsequent system of sextupole magnets. Based on the Stern-Gerlach principle, these magnets focus (defocus) atoms with electron-spin projection +1/2 (-1/2) along the magnetic field within the magnet bores. The electron-spin polarized beam then enters a rf (radio frequency) transition unit, which changes the nuclear polarization by inducing transitions between the hyperfine states. To achieve a high output intensity, the atomic beam has to fulfill several requirements like high flow rate, low transversal and longitudinal temperature, and a high degree of dissociation α. The pressure within and at the exit of the nozzle corresponds to the transition region between laminar and molecular flow. There, the use of continuum-flow models is of restricted validity. Thus, a direct simulation Monte-Carlo (DSMC) method [2] was used to describe the processes during gas expansion.

CP1149, *Spin Physics, 18th International Spin Physics Symposium*
edited by D. G. Crabb, D. B. Day, S. Liuti, X. Zheng, M. Poelker, and Y. Prok

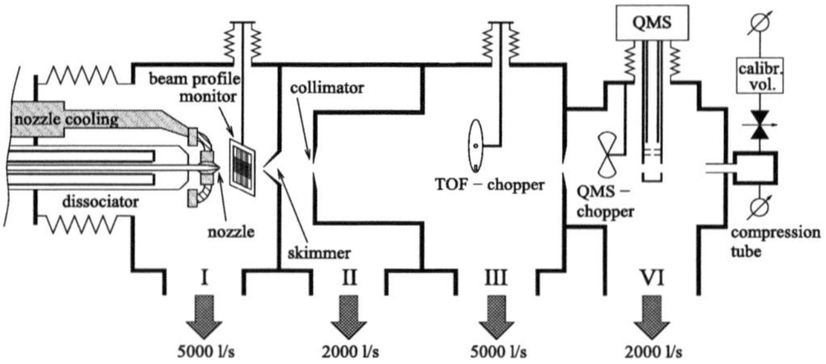

FIGURE 1. The experimental setup.

SUPERSONIC GAS EXPANSIONS AND DSMC

A free-jet atomic or molecular beam can be produced by a supersonic gas expansion from a high-pressure gas source (temperature T_0) into a low pressure background. As a result of the pressure difference the gas is accelerated. From energy considerations the maximum or terminal velocity can be derived:

$$v_\infty = \sqrt{\frac{2k_B}{m}\left(\frac{\gamma}{\gamma-1}\right)T_0}, \tag{1}$$

with the particle mass m, $\gamma = c_p/c_V$ and the Boltzmann constant k_B. For a pure atomic (molecular) hydrogen expansion $v_\infty = 2031\,(1436)$ m/s.

The DSMC method [2] is a technique for the computer modeling of a real gas by some thousands or millions of simulated particle trajectories. The velocity components and position coordinates of these particles evolve in time as the particles are concurrently followed through representative collisions and boundary interactions. The decoupling of the motion and collisions of the particles over small time steps and the division of the flow field into small cells are the key computational assumptions associated with the DSMC method. The time step should be much smaller than the mean collision time and a typical cell dimension should be much smaller than the local mean free path. For our purpose the geometry of the beam forming elements nozzle, skimmer, and collimator are implemented as boundary walls with temperature T (see Fig. 2).

THE EXPERIMENTAL SETUP

An atomic beam test stand (ABT) had been set up and equipped with several diagnostic devices (Fig. 1). A microwave dissociator [4] mounted on the first chamber produces H or D. α is determined with the quadrupole mass spectrometer (QMS). The velocity distribution of the particles in the atomic beam is determined with the time-of-flight (TOF) method. A fast chopper cuts a small bunch of particles out of the beam, and their arrival time at the QMS is measured. Since the particles have different velocities, a TOF distribution $F(t)$ is measured. By deconvolution the effects of the opening function

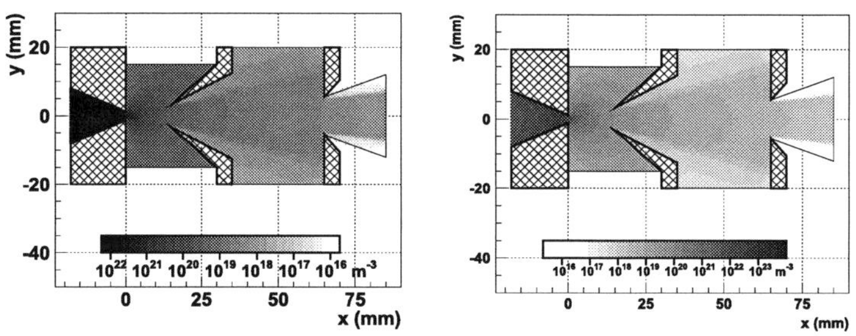

FIGURE 2. Calculated density distribution for a H_2 expansion (left) and the atomic part of a H/H_2 expansion (right, $\alpha = 0.63$) with a nozzle throughput of 1 mbar$l$/s and a nozzle temperature of 100 K.

TABLE 1. Comparison of the calculated and measured parameters of (i) a pure molecular and (ii) a partly dissociated hydrogen beam. For the partly dissociated beam $\alpha = 0.63$.

		v_x (m/s)	$T_{tr,x}$ (K)
molecular beam	measurement	$1274 \pm 8 \pm 13$	$19.0 \pm 1.1 \pm 0.9$
	simulation (original)	$1334 \pm 12\,^{+5}_{-15}$	$33.3 \pm 1.6\,^{+0}_{-3}$
	simulation (Ref. [6])	$1371 \pm 2\,^{+5}_{-15}$	$19.0 \pm 0.2\,^{+0}_{-3}$
	atoms		
partly dissociated beam	measurement	$1750 \pm 47 \pm 24$	$25.7 \pm 4.9 \pm 1.5$
	simulation	$1760 \pm 20\,^{+6}_{-19}$	$41.0 \pm 3.4\,^{+0}_{-3}$
	molecules		
	measurement	$1579 \pm 51 \pm 17$	$23.7 \pm 7.1 \pm 1.0$
	simulation	$1590 \pm 33\,^{+6}_{-17}$	$44.0 \pm 4.3\,^{+0}_{-3}$

of the chopper wheel and the electronics could be removed and the translational beam temperature $T_{tr,x}$ and the mean velocity v_x in beam direction could be determined. A beam-profile monitor [5] is used to measure intensity profiles of the atomic hydrogen or deuterium beam.

SIMULATION RESULTS AND EXPERIMENTAL VERIFICATION

DSMC simulations of expansions of H_2 and partially dissociated hydrogen were performed. The calculated particle-density distribution of a H_2 and a H expansion is shown in Fig. 2. The density near the nozzle follows a $\cos(\theta)$ distribution. Skimmer and collimator form a low-diverging and sharp-bound molecular beam. In Tab. 1, v_x and $T_{tr,x}$ are compared as they result from measurement and DSMC calculation. The calculated v_x is close to the measured one, but $T_{tr,x}$ is appreciably too high. The problem was studied [6] and it was found that $T_{tr,x}$ contrary to v_x, strongly depends on the parameters of the collision processes used in the DSMC. It seems that certain parameters, chosen by the editors of the used simulation code, are not fully correct for the low-T region.

Energy considerations showed a discrepancy between expected and measured beam

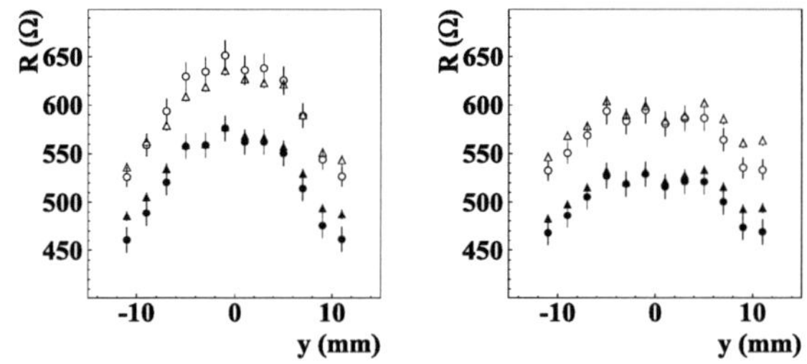

FIGURE 3. Comparison of the calculated (circles) and measured (triangles) distributions of the wire resistances at distances between wire plane and nozzle exit of 10 mm (left-hand side) and 20 mm (right-hand side). The nozzle throughput was 1 mbar*l*/s (full symbols) and 2 mbar*l*/s (open symbols).

parameters of partially dissociated hydrogen beams. This finding would be explained by the assumption that T_0 is higher than T_{nozzle}, i.e., that the gas does not reach the thermal equilibrium. This explanation was confirmed by DSMC calculations.

The data, obtained by the DSMC calculations for a partly dissociated hydrogen beam, could be further compared with the results of beam-profile measurements, made with the beam-profile monitor. Figure 3 shows the calculated distribution of the resistances compared with the measured one. Both distributions are not smooth due to the variations in the response functions of the wires but in good agreement.

The carrier jet method was proposed [3] to increase the phase-space density of the atomic beam and thus to reach a higher intensity through the collimator of the ABS. An over-expanded carrier jet, surrounding the inner atomic beam, was predicted to cool and to confine the inner beam. The mixing of the two gases has to be small and the carrier gas has to be removed by the skimmer and pumped away. The predicted carrier-jet effect could not be observed for H and D at the operational parameters of the ABS. It was, however, observed for an Ar beam surrounded by a N_2-carrier jet. Both experimental findings are consistent with the results of DSMC calculations.

SUMMARY AND OUTLOOK

It has been shown that the DSMC method is an excellent tool to describe the processes occurring in the expansion of light and also heavier gases in the transition region between laminar and molecular flow [7]. The results of the calculations were confirmed by the measurements, performed at an ABT. The origin of the discrepancies between simulated and measured temperatures was found, the problem could be explained by the influence of the input parameters of the DSMC code. The method could be used to design a new generation of ABS's by implementation of sextupole magnets into the DSMC code to finally understand the behavior of the output intensity of an ABS.

REFERENCES

1. *Antiproton-Proton Scattering Experiments with Polarization*, Technical Proposal for the HESR at FAIR, Jülich (2005), e-Print Archive: hep-ex/0505054.
2. G.A.Bird, *Molecular Gas Dynamics and the direct Simulation of Gas Flows*, Oxford (1998).
3. V.L.Varentsov et al., *PST1997 Urbana, AIP Conf. Proc. No.* **421**, 381 (1997).
4. N. Koch and E. Steffens, *Rev. Sci. Instr.* **70** 1631 (1999).
5. A. Vassiliev et al., *Proc. Int. Workshop on Polarized Sources and Targets (PST99), Erlangen 1999, A. Gute, S. Lorenz, E. Steffens (Eds.)*, (Friedrich-Alexander-Universität Erlangen-Nürnberg) 200 (1999).
6. M. Stancari, *Internal Note INFN Ferrara* **INFN/TC-06/11**, (2006) http://www.lnf.infn.it/sis/preprint.
7. A. Nass, *Nucl. Instr. Meth* **A 598**, 653 (2009).

Test Bench Studies and Simulations of Atomic Beam Sources

M. Stancari, L. Barion, G. Ciullo, M. Contalbrigo, P. F. Dalpiaz, P. Lenisa and M. Statera

Istituto Nazionale di Fisica Nucleare and Università di Ferrara, 44100 Ferrara Italy

Abstract. Two new test bench studies have deepened the understanding of polarized gas targets and the atomic beam sources (ABS) that fill them. The attenuation coefficient for beam loss due to rest gas scattering has been measured over a range of beam velocities. The total scattering cross sections can be extracted from these measurements for the first time.

Low conductance injection tubes have the potential to increase the thickness of storage cell targets. Injection tubes with internal fins were characterized, and the beam loss at the tube exit was larger than expected. Beam simulations agree with the measured intensity loss only when the atoms' trajectories have a non-zero azimuthal velocity component.

Keywords: Polarized gas targets; gas flow simulations
PACS: 07.77.Gx;29.25.Pj;37.20.+j

INTRODUCTION

The combination of internal polarized gaseous targets and high energy storage rings is a powerful tool in the study of the spin structure of the nucleon. Examples include the HERMES experiment[1], the beam polarimeter of the RHIC spin physics program[2], and the proposed antiproton polarizer for the PAX experiment[3]. However, the low target thickness is one of the obstacles that future experiments must overcome. The SpinLab laboratory at the Università di Ferrara focuses on the combination of test bench measurements and beam simulations to both understand the limits of existing targets and test new ideas for overcoming these limits. Two of the latest results are presented here. In the spirit of understanding of existing targets, a study of beam attenuation due to collisions with rest gas molecules was undertaken. In the spirit of testing new ideas, low conductance injection tubes were characterized, and their potential to increase the thickness of the storage cell target for the proposed PAX experiment was evaluated.

BEAM ATTENUATION STUDIES

Rest gas attenuation (RGA) inside the atomic beam sources (ABS) may be a limiting factor in existing and/or future targets. The relevant physical quantity is the total scattering cross section σ_{tot} for collisions of beam particles with rest gas particles, yet previous studies of RGA losses reported only the attenuation coefficient of the beam, referred to as A herein, for different nozzle temperatures. A depends not only on σ_{tot}, but also on the beam's velocity distribution, which varies from one ABS to another due to dif-

CP1149, *Spin Physics, 18th International Spin Physics Symposium*
edited by D. G. Crabb, D. B. Day, S. Liuti, X. Zheng, M. Poelker, and Y. Prok

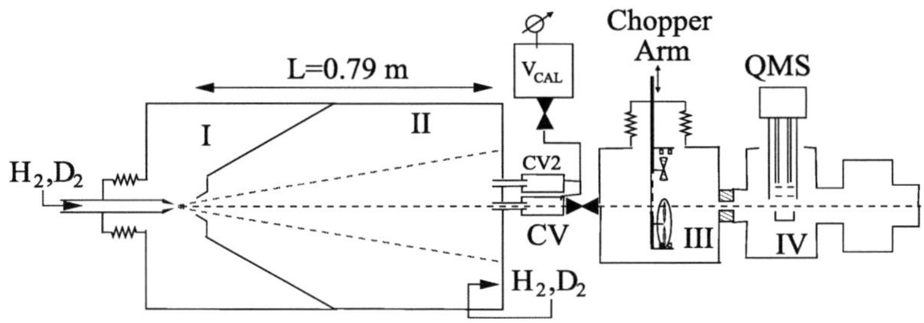

FIGURE 1. The experimental setup for the RGA measurements.

ferent nozzle temperatures and geometries. At SpinLab, the attenuation coefficient A of hydrogen and deuterium beams, both atomic and molecular, was measured for several different beam velocity distributions. For each measurement of A, the beam's velocity distribution was directly measured with a time of flight apparatus.

Figure 1 shows the experimental setup for the RGA measurements, using one of the two ABSs available at SpinLab. After dissociation, the beam is collimated by the skimmer that separates chambers I and II and continues to the Quadrupole Mass Spectrometer (QMS), where only 0.005% of the initial flux is collected. For these measurements, the rest gas pressure in chamber II was changed by adding gas through a side inlet. Chamber III contains two beam choppers. The first alternatively blocks and passes the beam, allowing background subtraction of the QMS signal offline. The second is a solid disk with a 2 mm wide radial slit. The sliver of beam that passes the slit expands spatially along the beam axis due to the different velocities of the atoms: the faster atoms arrive at the QMS before the slower ones. A time analysis of the resulting QMS signal reveals the beam's velocity distribution.

The dependence of the background-subtracted QMS signal S on the pressure P in chamber II is given by Beer's law

$$\ln(S) = \ln(S_0) - A \cdot P \cdot L / k_B T \tag{1}$$

where S_0 is the beam signal in the QMS for null pressure (no attenuation) in chamber II, $L = 0.79$ m is the length of chamber II, T is the temperature of the rest gas and k_B is Boltzmann's constant. The constant A was measured for a particular nozzle geometry and temperature by varying the pressure in chamber II in small steps until the QMS signal was 67% or less of its original value.

The collection of measurements for a molecular hydrogen beam is shown on the left in Fig. 2 for different operating conditions (nozzle diameter, discharge ON or OFF). The mean beam velocity was varied by changing the nozzle temperature. Similar measurements were performed for atomic hydrogen beams, atomic deuterium beams and molecular deuterium beams[5].

The attenuation coefficient A is a complicated combination of the beam's velocity distribution and the total scattering cross section σ_{tot}, which depends on the relative velocity of the colliding particles (The mathematical expression for A and its derivation can be found in references [4] and [5]). Assuming a functional form for σ_{tot}, the

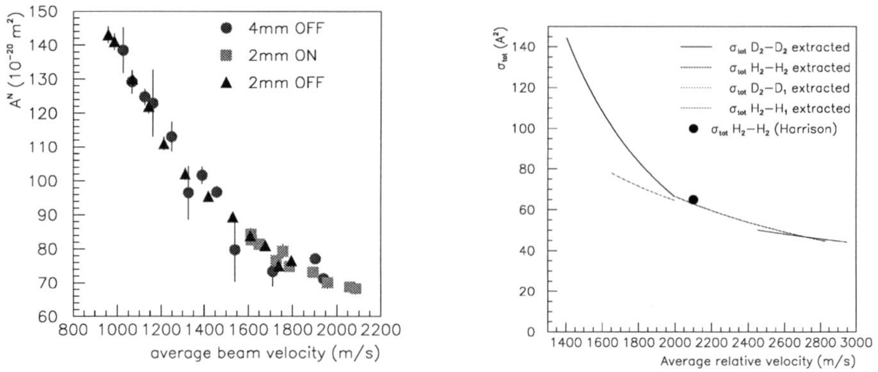

FIGURE 2. The set of A measurements for a molecular hydrogen beam (left) and preliminary determinations of σ_{tot} for the four processes studied (right).

values of the function's parameters which best agree with the measured values of A can be obtained from a simple χ^2 minimization. The right panel of Fig. 2 shows very preliminary results of these fits, illustrating the coverage of our data.

LOW CONDUCTANCE INJECTION TUBES

The geometry of a typical storage cell target is shown in the left panel of Fig. 3. Atoms from the ABS enter the center of the cell through the injection tube, then either diffuse out one end of the beam tube or diffuse back out the injection tube. The target thickness t is given by

$$t = IL/C_{tot} \quad \text{where} \quad C_{tot} = C_{beam} + C_{inj} \tag{2}$$

where I is the intensity of the incoming ABS beam, L is the half length of the beam tube and C_{tot} is the total cell conductance. Adding radial fins to a traditional cylindrical injection tube ideally does not obstruct the incoming ABS beam but does impede the diffusion of atoms back out the injection tube, increasing the target thickness.

If the incoming atoms from the ABS have a non-zero azimuthal velocity, they may collide with the radial fins, reducing the intensity I in Eqn. 2. Our test bench measurements quantified both I and C_{inj} for tubes with 0,5,10, and 12 fins[6]. A general parameterization of both I and C_{inj} in terms of the tube geometry was obtained and the calculated change in target thickness for the proposed PAX experiment is shown in the right panel of Fig. 3. No improvement is foreseen for the PAX target despite the 50% reduction in the conductance of the injection tube, which could not compensate the large intensity loss of the ABS beam. Beam simulations reproduce the measured beam intensity losses only when the beam atoms have a non-zero azimuthal velocity component.

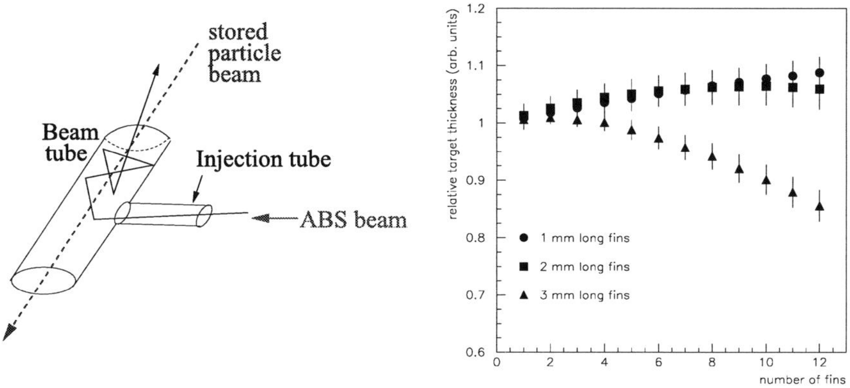

FIGURE 3. A typical storage cell (left) and the predicted thickness for the PAX experiment as a function of the injection tube geometry (right).

CONCLUSIONS

A parameterization of the total scattering cross section can be extracted from SpinLab's RGA data for the first time. This cross section is needed to calculate RGA losses for other sources and is a fundamental input for simulations of the beam formation process using the Direct Simulation Monte Carlo (DSMC) method. Low conductance injection tubes cannot increase the target thickness for the PAX storage cell target. This result is very sensitive to both the ABS beam characteristics and the storage cell geometry, therefore finned injection tubes may be useful in different experimental setups. The measurement of the beam's azimuthal velocity improves ABS sextupole tracking calculations because it discriminates among the various starting generators used to model the beam.

REFERENCES

1. A. Airapetian *et. al.*, *Nucl. Instr. and Methods* **A540**, 68-101 (2005).
2. A. Zelenski *et. al.*, *Nucl. Instr. and Methods* **A536**, 248-254 (2005).
3. V. Barone *et. al.*, arXiv:hep-ex/0505054 (PAX Technical Proposal).
4. Hans Pauly, Atom, Molecule, and Cluster Beams 1, Springer, 2000.
5. G. Pupillo, Laurea Thesis, Università degli Studi di Ferrara, 2008;
   L. Barion, Doctoral Thesis, Università degli Studi di Ferrara, 2009.
6. M. Stancari, *et. al.*, *Nucl. Instr. and Methods* **A594**, 126-131 (2008).

Dynamically Polarized Sample for Neutron Scattering At the Spallation Neutron Source

Josh Pierce[a], Don Crabb[b], and J.K. Zhao[a(*)]

[a]*Neutron Scattering Sciences Division, Spallation Neutron Source, Oak Ridge National Laboratory, Oak Ridge, TN 37831, USA*
[b]*Department of Physics, University of Virginia, Charlottesville, VA 22904*
[(*)]*zhaoj@ornl.gov*

Abstract. The recently constructed Spallation Neutron Source at the Oak Ridge National Laboratory is quickly becoming the world's leader in neutron scattering sciences. In addition to the world's most intense pulsed neutron source, we are continuously constructing state of the art neutron scattering instruments as well as sample environments to address today and tomorrow's challenges in materials research. The Dynamically Polarized Sample project at the SNS is aimed at taking maximum advantage of polarized neutron scattering from polarized samples, especially biological samples that are abundant in hydrogen. Polarized neutron scattering will allow us drastically increase the signal to noise ratio in experiments such as neutron protein crystallography. The DPS project is near completion and all key components have been tested. Here we report the current status of the project.

Keywords: Polarized Target, Neutron Scattering. Neutron Diffraction.
PACS: 61.05.F,29.25.pj.

INTRODUCTION

Hydrogen has a huge incoherent neutron scattering cross section ($\sigma_{\rm coh}$ =79.8 barns). It is one of the most limiting factors for neutron scattering studies on weak scattering samples, such as proteins in solution and protein crystallography. When polarized, this incoherent scattering cross section is drastically decreased. Simultaneously, the coherent scattering cross section is increased [1]. The spin dependent scattering amplitude (b) of a nucleus with spin **I** can be expressed as:

$$b = b_0 + b_1(\mathbf{I}\cdot\mathbf{s}) \tag{1}$$

where **s** is neutron's spin and b_0 and b_1 are the spin independent and dependent parts of the amplitude. For hydrogen, $\mathbf{I} = 1/2$, $b_0 = -3.74\,fm$, and $b_1 = 58.2\,fm$. The coherent ($\sigma_{\rm coh}$) and incoherent ($\sigma_{\rm inc}$) scattering cross sections can be expressed as:

$$\sigma_{coh} = 4\pi\left(b_0^2 + P_n P_I \cdot I b_0 b_1 + \tfrac{1}{4} P_I^2 I^2 b_1^2\right) \tag{2}$$

$$\sigma_{inc} = \pi\left[I(I+1)b_1^2 - P_n P_I \cdot I b_1^2 - P_I^2 I^2 b_1^2\right] \tag{3}$$

where P_I and P_n are the polarizations for the nuclei and neutrons, respectively. Figure. 1 illustrates polarization dependency of these cross sections. When both the neutrons

CP1149, *Spin Physics, 18th International Spin Physics Symposium*
edited by D. G. Crabb, D. B. Day, S. Liuti, X. Zheng, M. Poelker, and Y. Prok

and the hydrogen nuclei are polarized in the same direction, σ_{coh} increases from 1.8 barns for unpolarized neutron scattering to 14.7 barns for fully polarized sample. At the same time, σ_{inc} drops from 79.8 to 0 barns. Since about half of all atoms in biomolecules are hydrogen, the overall scattering changes for biomaterials will be just as dramatic.

The potential of polarized neutron scattering from polarized nuclei is well demonstrated by the study of L3 and L4 proteins in the 50S Ribosome. Ribosome is the molecular machine that synthesizes protein within cells. Before its structure was solved by protein crystallography, localizing ribosomal proteins and RNAs had provided crucial insights into Ribosome's function. However, because the 50S is very large (molecular weight: 1.5 $\times 10^6$ Da) and its protein subunits are relatively small (~20kDa or less), neutron scattering with conventional H/D substitution was resolution limited. In this case, the scattering signal from ribosomal proteins is too weak compared that of the ribosome. On the other hand, polarization dependent neutron scattering can significantly enhance the scattering signal from these small ribosomal proteins. In fact, using polarized neutron scattering from polarized samples, it was possible to determine not only the location of the ribosomal proteins L3 and L4 (figure 2), but also the asymmetries of these proteins [2].

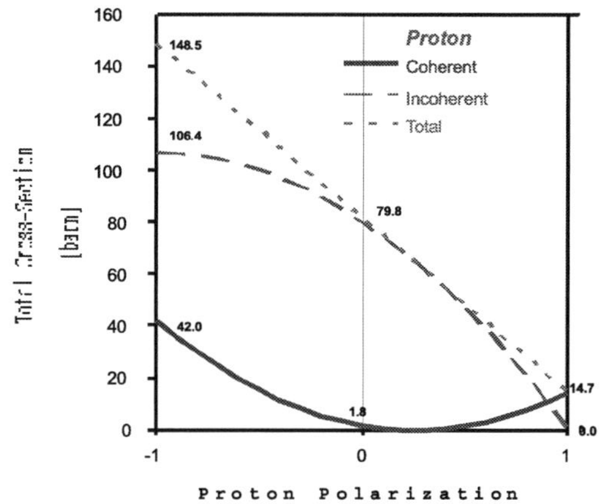

Figure 1. Total cross-sections of hydrogen as functions of the proton polarization. The polarization of the neutrons is assumed to be 100% [2].

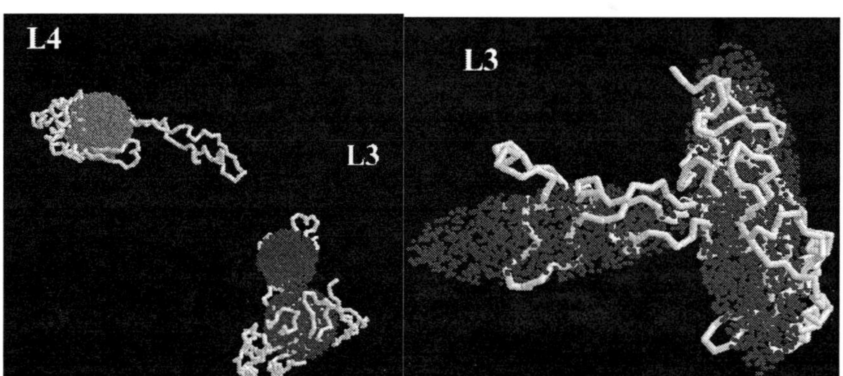

Figure 2. In situ L3 and L4 protein structures [2]. Left: Relative L3 and L4 locations with respect to each other from low resolution solution scattering with polarized neutrons (dots) [3,4], compared with their high resolution structure (coil) from protein crystallography [5]. Right: The asymmetry of the L3 protein from low resolution polarized neutron scattering (dots) [3,4], compared with its high resolution structure (coil) [5].

The Dynamically Polarized Sample Program at the SNS

The DPS project, at the SNS, is a collaboration between the SNS and the University of Virginia. Currently, all of the project's major components have been installed and tested. The system consists of a helium-4 cryostat, a 5T magnet, a 140 GHz microwave system and a 212 MHz NMR spectrometer (Figure 3).

The Magnet and The Helium-4 Cryostat

The magnet is a warm bore horizontal solenoid, capable of fields up to 7 Tesla (Figure 3). It will be run at 5 T to match resonance conditions of the microwave and NMR systems. The cryostat is a 1K fridge modified to fit into the magnet bore. It has been tested at 1K at the University of Virginia. The pumping station for the fridge consists of three stages, two roots pump stages and one rotary vane pump stage. The total system has a pumping speed of >10 [mmole/s] at the helium pressure of 0.2mbar, which corresponds to a cooling power of just below ~1W at the temperature of ~1K.

Figure 3. The 5 T magnet and the 1K refrigerator.

The Microwave System

The central piece of the microwave system is an Extended Interaction Oscillator provided by CPI Canada (www.cpii.com). (Figure 4). It has a center frequency of 140 GHz and a 2.1 GHz mechanical tuning range. The electronic tuning range is 250MHz. The power output within these frequency ranges varies between 10W and 23W. We use F-band (WR-08) microwave guide. A Phase Matrix microwave counter is used to monitor the frequency.

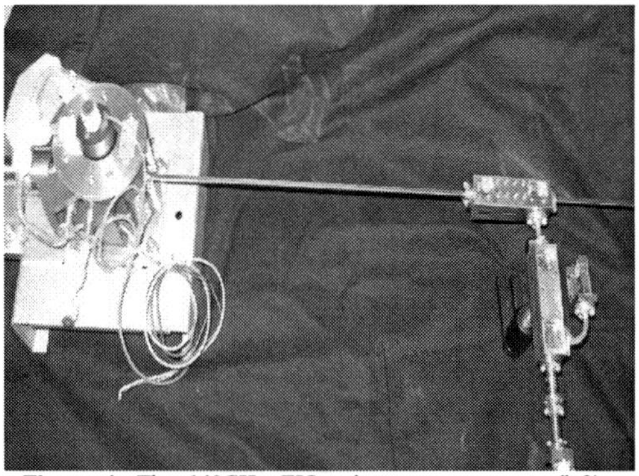

Figure 4. The 140GHz EIO microwave generator (left), and miscellaneous waveguide components.

The NMR System

The NMR system is based on the Liverpool Q-meter (Figure 5). Due to limited availability, efforts are underway to replace the Liverpool module [6]. with modern rf-components. On the front end, we use a frequency generator and a function generator from Agilent Technologies (www.agilent.com).

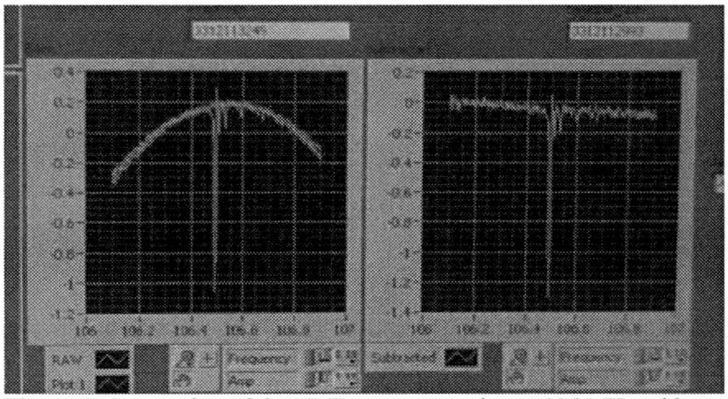

Figure 5. Screen shot of the NMR system running at 106 MHz with a quartz crystal.

Future Development

Parallel to these initial project setups, customized designs around specific instruments are being planned for the future. A candidate instrument is MANDI [7], the SNS neutron protein crystallography diffractometer. Protein crystals are typically very small (« 1mm) and MANDI will require a very compact setup. We will then likely be using a 2.5T magnet and a dilution refrigerator. The dilution refrigerator also offers the advantage of operating the polarized sample in the frozen spin mode.

ACKNOWLEDGMENTS

SNS is managed by UT-Battelle, LLC, under contract DE-AC05-00OR22725 for the U.S. Department of Energy.

REFERENCES

1. H. B. Stuhrmann, et al, *Eur. Biophys. J* 14 (1986) 1-6.
2. J.K. *Zhao, Physica B356, (*2005) *168*
3. J. Zhao "Bestimmung der in situ Struktur der Proteine L3 und L4 in der ribosomalen 50S Untereinheit mit Hilfe der Kernspinkontrastvariation", University of Hamburg, 1996.
4. J. Zhao and H.B. Stuhrmann, J. de Physique IV, **3**, (1993) 233-236
5. N. Ban et al, Science, 289 (2000), 905-920.
6. G. R. Court, D. W. Gifford, P. Harrison, W. G. Heyes, and M. A. Houlden, *Nucl. Instrum. Meth. A*, 324, (1993) 433-440.
7. J. Schultz, P. Thiyagarajan, J. P. Hodges, C. Rehm, D. A. A. Myles, P. Langan and A. D. Mesecar J. Appl. Cryst. (2005). 38, 964-974.

First Step Toward Ultrafast Nuclear-Spin Polarization: All-optical Control and Direct Detection of Ultrafast Electron-Spin Polarization Using Femtosecond Laser Pulses

Takashi Nakajima*,†, Yukari Matsuo† and Tohru Kobayashi†

*Institute of Advanced Energy, Kyoto University, Gokasho, Uji, Kyoto 611-0011, Japan
†RIKEN (The Institute of Physical and Chemical Research), 2-1 Hirosawa, Wako, Saitama 351-0198, Japan

Abstract. We demonstrate the first experimental observation of ultrafast electron-spin polarization upon photoionization using femtosecond and nanosecond lasers. For the optical detection of electron-spin polarization we measure the laser-induced fluorescence of photoions. The experimental results agree well with our theoretical results obtained by solving the time-dependent Schrödinger equations.

Keywords: electron-spin, nuclear-spin, polarization
PACS: 32.80.Rm, 32.80.Qk, 29.27.Hj

INTRODUCTION

In nuclear physics there is a great demand to produce spin-polarized nuclei [1], and there are a few well-known methods such as the use of the angle- and mass-resolved secondary beam after nuclear fragmentation at a target, optical pumping [2], and the combination of optical pumping and spin-exchange collisions in a buffer gas [3, 4].

In recent years we have been theoretically as well as experimentally working on new schemes to realize spin-polarization with pulsed (fs, ps, or ns) lasers [5, 6, 7, 8, 9], the advantage of which is their potential to achieve a high degree of spin-polarization in a very short time. In particular the ultrafast spin-polarization scheme with fs pulses [8, 9] is very appealing if a similar idea can be applied to polarize nuclei. Since the ultrafast spin-polarization scheme we discuss in this paper relies on the transient spin dynamics induced by spin-orbit interaction (to polarize electron-spin) or hyperfine interaction (to polarize nuclei) upon photoexcitation by short laser pulses, the underlying physics is very similar, and the only difference is the time scale needed to achieve spin polarization: Typically, it is hundreds of fs to tens of ps for the former, and sub-ns to tens of μs for the latter due to the entirely different interaction strengths. Nevertheless, we have theoretically shown that ultrafast nuclear-spin polarization within a few ns is indeed possible if appropriate states are chosen.

In this paper we report on the first experimental observation of ultrafast electron-spin polarization through photoionization using femtosecond and nanosecond lasers. Clearly this is a first step toward ultrafast nuclear-spin polarization.

CP1149, *Spin Physics, 18th International Spin Physics Symposium*
edited by D. G. Crabb, D. B. Day, S. Liuti, X. Zheng, M. Poelker, and Y. Prok

EXPERIMENT

The level scheme we employ is shown in Fig. 1(a). The essence of our scheme described above is to use a fs pump pulse to create a coherent superposition of fine structure manifold and ionize atoms by the fs probe pulse after the variable time delay [8] and to use an optical method to detect spin of photoions. That is, since our target is alkaline-earth atoms, we can directly detect spin-polarization of photoions through laser-induced fluorescence (LIF) by using the right-/left-circularly polarized ion detection laser [6, 7]. The choice of a multi-valence-electron system such as Sr also offers a possibility to simultaneously produce spin-polarized photoions and photoelectrons [5]. The experimental setup is shown in Fig. 1(b). It consists of a vacuum chamber, three ns lasers for ablation, excitation, and ion detection, two fs lasers for the pump and probe, and an optical detection system. A solid Sr disk on a rotating mount is placed in a vacuum chamber maintained at the pressure of $\sim 1 \times 10^{-4}$ Pa by two turbo-molecular pumps. The second-harmonic pulse of a Nd:YAG laser operated at 10 Hz is loosely focused onto the Sr disk with a f =25 cm lens to produce an ablation plume which consists of neutral Sr atoms as well as Sr ions. At 1 μs after the ablation pulse, we apply a pulsed electric field of 1 kV with 25 μs duration to an ion deflector electrode to prevent ablation ions from flying into the interaction area. About 50 μs after the ablation pulse, a right-circularly polarized excitation pulse (689 nm) with $\sim$15 ns duration is turned on to excite $5s5p\ ^3P_1$ $(m_J = +1)$, which serves as an initial state for the following pump-probe ionization. The linearly polarized pump pulse (397 nm, 150 fs), delivered from the amplified fs laser system (Hurricane, Spectral Physics) with a β-BBO crystal of 0.2 mm thickness, coherently excites the fine structure manifold, $5s6d\ ^3D_1$ and $^3D_2$. The time-delayed probe pulse (794 nm, 100 fs) ionizes the atoms in the $5s6d\ ^3D_1$ and $^3D_2$ states. After the fs pump-probe pulses we introduce the ns ion detection laser (421 nm) with right/left circular polarization to detect spin-polarized Sr^+ $5s_{1/2}$ $(m_J = \pm 1/2)$ ions through the LIF measurement with a time-gated photomultiplier [6, 9].

Now we show the experimental results. In Figs. 2(a) and 2(b) the LIF signals from photoions by the left-/right-circularly polarized ion detection laser are shown by open circles with error bars. Obviously they are proportional to the ion yields in Sr^+ $5s_{1/2}$ with

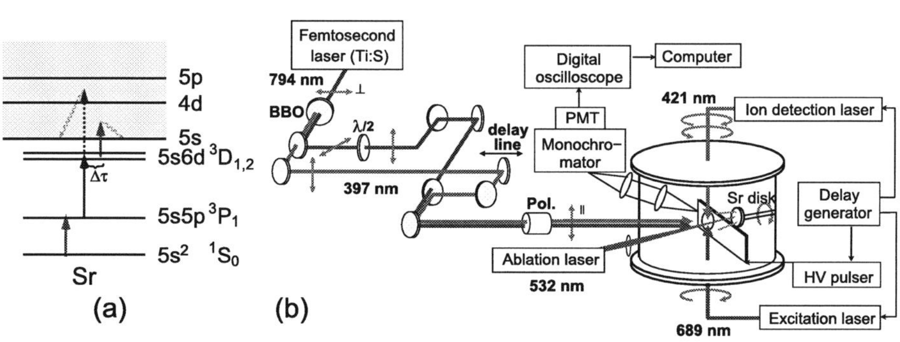

FIGURE 1. (a) Level scheme. (b) Experimental setup.

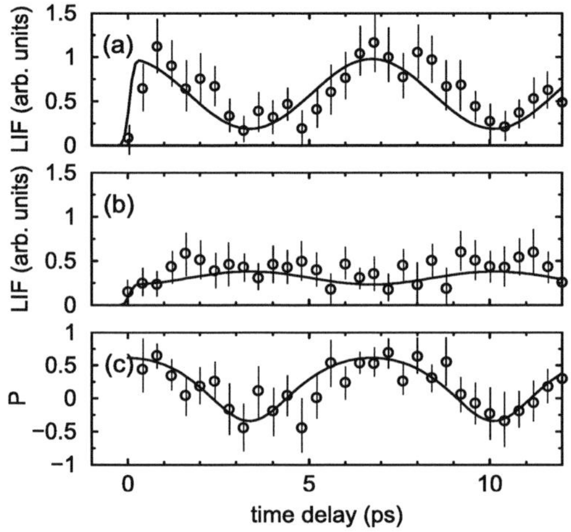

FIGURE 2. LIF signals from the (a) Sr^+ $5s_{1/2}(m_J = +1/2)$ and (b) Sr^+ $5s_{1/2}(m_J = -1/2)$ states by the left-/right-circularly polarized ion detection laser and (c) spin-polarization as a function of time delay between the pump and probe pulses. Ab-initio theoretical results are shown by solid lines in each graph.

up-/down-spin. As we can see in Fig. 2(a), Sr^+ ions with up-spin exhibits a modulation period of about 6.7 ps, which correspond to the spin-orbit coupling time of $5s6d$ $^3D_1$ and $^3D_2$. This is a quantum beat but in the spin-resolved signal. In contrast, as we see in Fig. 2(b), the signal modulation from the Sr^+ ions with down-spin is not clear due to the limited signal-to-noise ratio. Using data in Figs. 2(a) and 2(b), we plot the variation of spin-polarization in Fig. 2(c) where spin-polarization, P, is defined as $P = [I_{LCP} - I_{RCP}] / [I_{LCP} + I_{RCP}]$ where I_{RCP} (I_{LCP}) represents the LIF intensity by the right-circularly (left-circularly) polarized ion detection laser. It is clear that spin-polarization also modulates in time with the period of spin-orbit coupling time.

COMPARISON WITH THEORY

To compare with the experimental results shown in Fig. 2, we have carried out the theoretical analysis using a set of time-dependent Schrödinger equations for the essential bound states shown in Fig. 1(a) and also the spin-resolved continuum states. The equations read [8],

$$\dot{u}_0 = -i\Omega_1 u_1 - i\Omega_2 u_2 \,, \tag{1}$$

$$\dot{u}_1 = \left[i\delta_1 - \frac{1}{2}(\gamma_1 + \gamma_1^{sp}) \right] u_1 - i\Omega_1 u_0 - i\Omega_{12}^{(2)} u_2 \,, \tag{2}$$

$$\dot{u}_2 = \left[i\delta_2 - \frac{1}{2}(\gamma_2 + \gamma_2^{sp}) \right] u_2 - i\Omega_2 u_0 - i\Omega_{12}^{(2)} u_1 \,, \tag{3}$$

where u_j ($j=0,1,2$) are the probability amplitudes of states $|j\rangle$ ($|0\rangle = |5s5p\,^3P_1\rangle, |1\rangle = |5s6d\,^3D_1\rangle$, and $|2\rangle = |5s6d\,^3D_2\rangle$), and Ω_k ($k=1,2$) are the Rabi frequencies between $|0\rangle$ and $|k\rangle$. δ_k are the detunings defined as $\delta_k = E_0 + \hbar\omega_{pump} - E_k$ with E_k being the energy of $|j\rangle$. $\Omega_{12}^{(2)}$ is the *complex* two-photon Rabi frequency between $|1\rangle$ and $|2\rangle$ by the probe pulse. γ_j ($j=1,2$) are the ionization widths of state $|j\rangle$ by the probe laser, while γ_j^{sp} represent the natural widths, which are negligible for the time scale we consider in this work. Once the above equations have been solved, we can calculate the ion yields into each spin state, $R^{(\uparrow)}$ and $R^{(\downarrow)}$, by the formula we have derived before [8]. The temporal profiles of the pump and probe pulses are assumed to be Gaussian, i.e., $I_{pump} \propto exp(-((t/\tau_{pump})^2)$ and $I_{probe} \propto exp(-(t-t_{probe})/\tau_{probe})^2)$ where t_{probe} is the delay of the probe pulse with respect to the pump pulse, and we assume the pulse durations to be $\tau_{pump} = 150$ fs and $\tau_{probe} = 100$ fs, respectively. Finally we can calculate the spin polarization, $P = (R^{(\uparrow)} - R^{(\downarrow)})/(R^{(\uparrow)} + R^{(\downarrow)})$. The variation of $R^{(\uparrow)}$, $R^{(\downarrow)}$, and P as a function of time delay between the pump and probe pulses are presented in Fig. 2 by solid lines, which are in good agreement with the experimental data. Thus we have confirmed that we have successfully realized ultrafast electron-spin polarization for the first time [9].

SUMMARY

In summary we have shown the first experimental observation of ultrafast electron-spin polarization of photoions upon photoionization by femtosecond and nanosecond lasers, which would serve as a first step toward ultrafast nuclear spin polarization using a similar scheme. The advantage of utilizing the transient scheme is that the method can be applicable to even unstable nuclei which would be of great interest in nuclear physics. We have also carried out *ab-initio* theoretical calculations by solving the time-dependent Schrödinger equation, and obtained good agreement with the experimental results.

REFERENCES

1. D. Fick, G. Grawert, I.M. Turkiewicz, *Physics Reports* **214**, 1 (1992).
2. H. Reich and H.J. Jänsch, *Nucl. Instr. and Meth. A* **288**, 349 (1990).
3. G.D. Cates, D.R. Benton, M. Gatzke, W. Happer, K.C. Hasson, and N.R. Newbury, *Phys. Rev. Lett.* **65**, 2591 (1990).
4. A.N. Zelenskii, K. Jayamanna, C.D.P. Levy, M. McDonald, R. Ruegg, and P.W. Schmor, *Nucl. Instr. and Meth. A* **334**, 285 (1993).
5. T. Nakajima and N. Yonekura, *J. Chem. Phys.* **117**, 2112 (2002).
6. T. Nakajima, N. Yonekura, Y. Matsuo, T. Kobayashi, and Y. Fukuyama, *Appl. Phys. Lett.* **83**, 2103 (2003).
7. N. Yonekura, T. Nakajima, Y. Matsuo, T. Kobayashi, and Y. Fukuyama, *J. Chem. Phys.* **120**, 1806 (2004).
8. T. Nakajima, *Appl. Phys. Lett.* **88**, 111105 (2006).
9. T. Nakajima, Y. Matsuo, and T. Kobayashi, *Phys. Rev. A* **77**, 063404 (2008).

Precision Polarimetry of Solid HD Photonuclear Targets at LEGS

Craig Thorn

Brookhaven National Laboratory, Upton, NY 11973

Abstract. We have constructed and operated a NMR polarimeter for high-accuracy measurements of hydrogen and deuterium polarization in a frozen-spin solid HD photonuclear physics target. The heart of this polarimeter is a pair of crossed (orthogonal) RF coils that form a transformer that responds only to precessing spins in the sample. We have performed a complete electronic circuit analysis of this crossed coil polarimeter (CC-Meter), and have determined the optimum conditions for its operation. From this analysis, which is confirmed by NMR measurements on hydrogen and deuterium in solid HD, we conclude that the CC-meter can be operated to give a very high signal-to-noise ratio (SNR) for thermal equilibrium polarization while also producing a highly linear response for fully polarized targets. In general, we find that a well-designed and properly constructed CC-Meter can provide a larger SNR at similar non-linearity, compared to the more commonly used Q-meter. We have used this polarimeter to produce H polarization measurements for a new, high precision determination of the GDH integral for the proton.

Keywords: Polarized nuclear target, NMR, HD nuclear polarization.
PACS: 29.25.Pj, 25.20.Lj.

INTRODUCTION

Accurate target polarimetry is essential to the measurement of single and double polarization asymmetries in nuclear reactions. Since asymmetries are a ratio systematic errors tend to cancel and since counting statistics errors in the yields can be very small, the error in asymmetries is often just the error in determining the target polarization. Our goal is to achieve better than 1% uncertainty in the measurement of H and D polarizations in highly polarized, frozen spin HD targets (SPHICE targets) [1,2]. The technique that most commonly has been used to measure nuclear target polarizations is to observe NMR with the Q-Meter circuit [3-7]. The statistical accuracy that can be achieved with a Q-Meter is typically limited by the fact that in the absence of a polarized sample the instrument has a large output (the background). Any noise in the source of RF used for the measurement appears proportionately in the background, and hence in the net signal which is obtained by subtracting the background from the background plus signal measurement. This difficulty has been understood since the beginning of NMR, and to avoid it high-sensitivity measurements have used a "crossed-coil" arrangement with one coil to excite the Larmor precession and a second, magnetically orthogonal, coil to observe the precession [8]. Ideally, in such arrangement there is no response of the circuit without a polarized sample, and hence no noise from the RF source is added to the signal by background subtraction.

CP1149, *Spin Physics, 18th International Spin Physics Symposium*
edited by D. G. Crabb, D. B. Day, S. Liuti, X. Zheng, M. Poelker, and Y. Prok

Determining polarizations with an NMR instrument requires that the system be calibrated with a sample whose geometry is identical to the "unknown" polarized sample but of precisely known susceptibility. This "calibration" sample is the "unknown" sample itself, but in a state in which the nuclear spins are in thermal equilibrium with the lattice at an accurately known temperature. For hydrogen, this calibrates the instrument at a polarization of 6.7 10^{-5} at 12 MHz and 4.2 K. The fully polarized sample will have a polarization approaching 1, which produces a signal 14,925 times the thermal equilibrium signal. Clearly, an accurate NMR circuit must provide *both* low noise and a highly linear response to polarization in order to accommodate this large dynamic range.

To understand the conditions for achieving both low noise and a high linearity, we have performed a complete electronic circuit analysis of the crossed coil NMR polarimeter (CC-Meter), and from this analysis we have determined the optimal conditions for its operation for the LEGS SPHICE target system. In any practical realization, NMR polarimeters suffer from a significantly non-linear response to susceptibility for highly polarized samples. One important purpose of the circuit analysis is to specify conditions under which these non-linearities become insignificant, or to allow the computation of a correction to the polarization deduced from the NMR signal obtained in the presence of significant non-linearities.

THE ANALYSIS OF THE CC-METER NMR CIRCUIT

The CC-Meter consists of an electronic circuit made of an RF source, transmission lines, a transformer and other lumped-constant elements, and a phase sensitive detector to synchronously rectify the output of the circuit. A schematic diagram of the NMR circuit and a detail of the transformer equivalent circuit [9] are shown in Fig. 1.

The sensitivity of the circuit to the sample polarization enters through the dependence of the inductance of a coil on the susceptibility (χ) of the enclosed sample:

$$L = L_0[1+\phi(\chi)] = L_0[1+\eta(\chi)\chi] \quad (1)$$

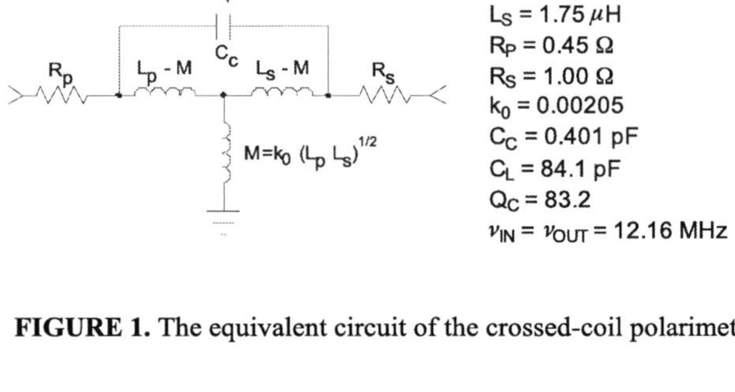

FIGURE 1. The equivalent circuit of the crossed-coil polarimeter.

The constant relating susceptibility to change in inductance, η, is called the filling factor. It can be computed from the RF magnetic field produced by the coil in the presence of the susceptible sample. For samples with high susceptibility (like highly polarized H in HD) the filling factor will be a function of susceptibility. In the case of two coupled coils we have three inductances (the primary and secondary self inductance and the mutual inductance)

$$\mathrm{M}_{jk} = \kappa\sqrt{\mathrm{L}_j \mathrm{L}_k} \qquad \mathrm{M}_{jk} = \mathrm{M}_{0,jk} + \Delta\mathrm{M}_{jk}(\chi) \tag{2}$$

and three filling factors

$$\eta_{jk} = \frac{1}{\chi}\frac{\Delta M_{jk}}{M_{0,jk}} = \frac{\int_{TGT}\overline{(\vec{H}_{0,j}\times\hat{z})}\cdot\overline{(\vec{H}_k\times\hat{z})}dv}{2\int_{\infty}\overline{H}_{0,j}\cdot\overline{H}_{0,k}dv} \tag{3}$$

The indices j and k denote the field produced by the primary (transmitter) or secondary (receiver) coils, $\hat{z}$ is a unit vector in the direction of the holding field, and $-1 \le \kappa \le 1$ is the coupling factor of the coils. The subscript 0 indicates that the quantity is computed in the absence of the sample (zero susceptibility). The integration in the numerator is over the target volume, and in the denominator it is over all space. The factor of one half accounts for the fact that a single coil produces a linearly polarized field, which can be resolved into two counter-rotating circularly polarized fields, only one of which rotates in the direction of the Larmor precession. For a single coil in the limit of zero susceptibility, the filling factor is just one half the ratio of the energy of the transverse field contained in the sample to the total field energy. For our coil geometry, using Maxwell3D we compute ηT=ηR=1.13 ηTR and ηTR=0.0211. For all targets the mean value measured for η_{TR} is 0.0238±0.0020.

The NMR circuit voltage transfer function can be expressed as a rational polynomial of frequency with coefficients containing the lumped circuit element values, the susceptibility and the filling factor:

$$\frac{V_{OUT}}{V_{In}} = T(\omega, \chi; \{R, L, C, \eta\}) \tag{4}$$

The response of the polarimeter to polarization can be displayed formally by expanding the transfer function in a power series in susceptibility.

$$T(\omega) = G_0(\omega) + G_1(\omega)\chi + G_2(\omega)\chi^2 + O(\chi^3) \qquad with \quad G_n(\omega) = \frac{1}{n!}\frac{\partial^n T}{\partial\chi^n}\bigg|_{\chi=0} \tag{5}$$

The response to the resonant susceptibility is determined by $G_1(\omega)$, which we call the *transducer gain* because of the analogy to the voltage response of, for example, pressure transducers to a physical parameter. The coefficient $G_0(\omega)$ is the voltage gain of the circuit in the absence of a polarized target: it is an undesirable *background* on which the signal due to the resonant susceptibility appears. $G_2(\omega)$ is the lowest order *non-linearity* in the susceptibility response of the circuit, and ideally it, and all higher order terms, would be zero. An optimal circuit has $G_0=0$, G_2 very small and G_1 large.

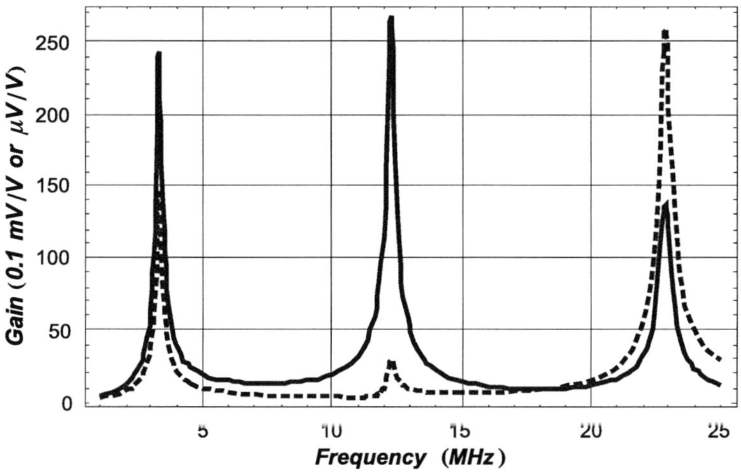

FIGURE 2. The computed magnitude of the background (dashed) and transducer gains for a 100% polarization H signal (solid) for the LEGS CC-Meter.

The component values for the circuit are measured or estimated values, which are then adjusted by fitting the measured background gain frequency dependence. The calculated gain curve is then virtually identical to that measured (chi-square per degree of freedom is 1.6). In Figure 2 the resulting computed background and transducer gains are shown for the circuit values in Figure 1. The background is generally about 100 times larger than the transducer gain, except in the vicinity of the second ($\lambda/2$) resonance, where it is much smaller. This is caused by the deliberate cancelation of the inductive (k_0) and capacitive (C_C) coupling of the primary and secondary coils at this frequency. The cancelation appears only in the background gain, not the transducer gain, so its effect on the circuit is to increase the signal-to-noise ratio (to 44 for a single 30 second NMR sweep) leaving the transducer gain unchanged. The sensitivity

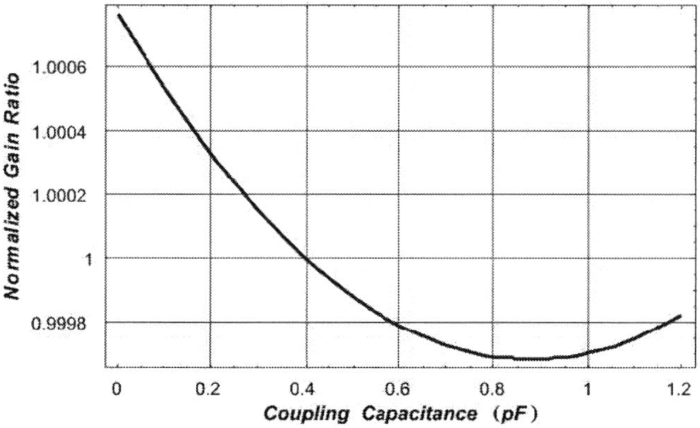

FIGURE 3. The change in the transducer gain with changes in the cross-coil coupling capacitance (C_C) for the LEGS CC-Meter, computed from the circuit model of Figure 1.

of the transducer gain to the value of the coupling capacitance, computed from the circuit model, is shown in Figure 3. Since the value of C_C is not easily controlled, it is important that this sensitivity be small to ensure the stability of the transducer gain.

Since the nonlinear gain terms in Equation 5 involve powers of the complex susceptibility function, they mix powers of the dispersive lineshape into the absorptive lineshape. The dispersive lineshape has very long "tails", so this mixing adds small, but very long, tails into the observed absorptive lineshape. A sensitive way to quantify these tails, and hence the nonlinearity, is to plot the integral of the observed absorption signal as a function of the interval over which the signal is integrated in magnetic field. This is shown in Figure 4 for the H NMR signal for both positive and negative polarizations. The slow change of the integral for large integration interval seen in the figure is a measure of the non-linear gain. Figure 4 also displays the calculated polarization signal, using the measured circuit parameters and adjusting the for a best fit. The best fit to the data is obtained for a filling factor of (0.0231±0.0045), and the uncertainty in this value corresponds to an uncertainty in the polarization, after correction for non-linearities, of 0.34% at 51% or a fractional error of 0.66% attributable to the non-linearity correction. The non-linear correction for H is approximately quadratic and has the form

$$P_{TRUE} = P_{NMR}[1+(0.48\pm 0.19)\,\Delta B^{-1}\,P_{NMR}] \tag{6}$$

with ΔB the integration half interval in gauss (typically ΔB=40 gauss).

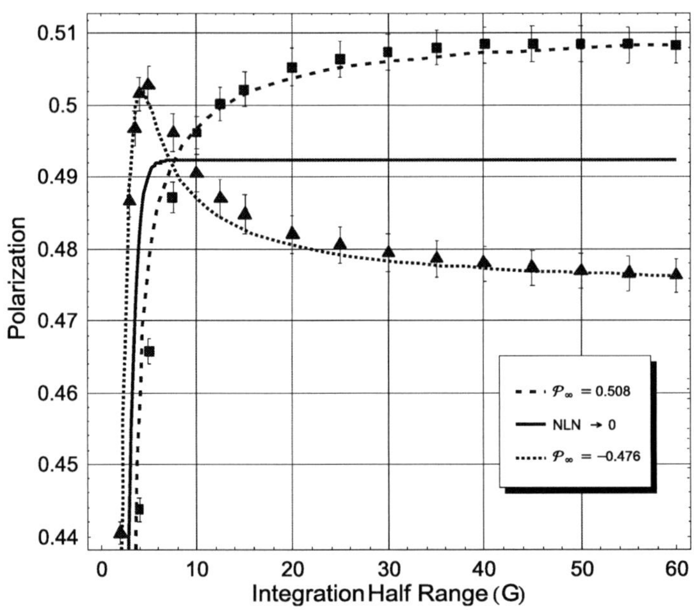

FIGURE 4. The deduced magnitude of H polarization as a function of the interval of integration of the NMR absorption signal for positive (squares – field parallel to polarization) and negative (triangles – field anti-parallel to polarization) polarization. Dashed and dotted lines are the calculations from the circuit model. The solid line is the calculation including only the linear term in Equation 5.

CONCLUSIONS

The circuit noise can be grouped into two classes, distinguished by their frequency dependence: frequency independent, Johnson noise which comes primarily from the output coil and cable resistances and from the lock-in amplifier input, and low-frequency divergent noise, primarily from mechanical motion of the transformer coils, temperature fluctuations in the cables, and the RF signal source. The white noise is typically 12 $nV/Hz^{1/2}$, which contributes an error of 0.2% to the thermal equilibrium calibration measurement. The low-frequency divergent noise is a more serious problem, since both the time between polarization calibration and use of the target (up to 3 months) and the useful polarized lifetime (T_1 up to 1 year) are so long. We have observed the measurement error in the system for periods up to 3 weeks, by which time the measured variance appears to increase with measurement time at a rate between 1/f and $1/f^2$. From the observed slope with time, we estimate an error due to this "drift" of the calibration of 1.8% after 6 months. The uncertainty in the H polarization due to uncorrected circuit non-linearities is estimated to be 0.7% as discussed above. The thermal equilibrium measurement also has uncertainties from the temperature measurement, possible H contamination in the materials around the target and coils, signal background subtraction, and incomplete polarization relaxation of the HD during the measurement. These uncertainties total to 0.8%. The total error from all calibration and polarized target measurement sources is then 2.1%.

There are, unfortunately, much larger errors that result from the fact that we were not able to calibrate the target in the In-Beam Cryostat (IBC). Instead, we calibrated and measured the polarization in a Production Dewar, and then transferred the target, and the polarization calibration, into the IBC. Various uncertainties in this process contributed an error of 5.4% to the polarization. These errors would not be present if the calibration was done in the IBC. In addition, uncertainties about the uniformity of the spatial distribution of polarization, and the different sampling of polarization by the NMR and the photon beam (the differential homogeneity) leads to an additional polarization uncertainty of 2.3%. The total error in polarization from all of these sources is 6.2%. We are presently working to eliminate the calibration transfer errors and to reduce the differential homogeneity error which will allow us to achieve 3% polarization accuracy.

REFERENCES

1. X. Wei et al. *NIM* **A526** (2004) 157.
2. G. Rouille, et al., NIM A464 (2001) 428.
3. G.R. Court, D.W. Gifford, P. Harrison, W.G. Heyes, and M.A. Houlden, *NIM* **A324** (1993) 433.
4. D.G. Crabb and W. Meyer, *Annu. Rev. Nucl. Part. Sci.* **47** (1997) 67.
5. Yu.F. Kisslev, C.M. Dulya, T.O. Niinikoski, NIM A354 (1995) 249.
6. T.O. Niinikoski, *NIM* **A356** (1995) 62.
7. Y.K.Semertzidis, *NIM* **A356** (1995) 83
8. A. Abragham, *The Principles of Nuclear Magnetism*, Oxford Univ. Press, Oxford, 1999, p76
9. Aram Budak, *Circuit Theory Fundamentals and Applications*, Prentice-Hall, Inc, Englewood Ckiffs, NJ (1978).

The JLab Frozen Spin Target

C.D. Keith

On behalf of the Jefferson Lab Target Group and Hall B Engineering Staff
Thomas Jefferson National Accelerator Facility, Newport News, VA 23606

Abstract. A polarized, frozen spin target has been designed and constructed at Jefferson Lab for use inside the CEBAF Large Acceptance Spectrometer. Protons in TEMPO-doped butanol are polarized via dynamic nuclear polarization (DNP) to approximately 90% using microwaves and an external, 5 T solenoid magnet. The target sample is then cooled to approximately 30 mK while an internal 0.56 T superconducting magnet is used to maintain the polarization. Relaxation times in excess of 3500 hours have been observed.

Keywords: Dynamic Nuclear Polarization, Polarized Target, Frozen Spin Target
PACS: 29.25.Pj

INTRODUCTION

The experimental program in Hall B at Jefferson Lab is largely focused upon the study of nucleon and nuclear structure using using polarized and unpolarized electron and photon beams. The centerpiece of the Hall B instrumentation package is the CEBAF Large Acceptance Spectrometer, or CLAS [1]. A dynamically polarized 1 K target suitable for use inside CLAS with electron or photon beams has been previously described [2]. However, the detection of particles scattered from this target is restricted to a half-angle cone of approximately 55° in the forward direction due to the 5 T Helmholtz coils used in the polarization process. Moreover, the geometry of the target cryostat permits only longitudinal target polarization within the CLAS detector. To overcome both of these limitations, a frozen spin target intended for use with tagged photon beams has been designed and constructed by the Jefferson Lab Target Group and Hall B Engineering staff. The design of this target allows detection of particles scattered at angles up to 135°, and the polarization can be oriented either longitudinally or transversely to the photon beam by judicious choice of the magnetic holding field. It has thus far been utilized in a series of meson and hyperon photoproduction experiments at JLab using both circularly and linearly polarized photons incident on longitudinally polarized protons. These experiments are scheduled to be repeated in 2010 using a transversely polarized proton sample.

GENERAL DESCRIPTION

The target sample consists of 5 g of TEMPO-doped butanol beads (1–2 mm diameter) cooled inside the mixing chamber of a high cooling-power, horizontal dilution refrigerator specifically designed to operate inside the CLAS detector. The sample is dynamically polarized outside CLAS by 140 GHz microwaves and a 5 T room-temperature

CP1149, *Spin Physics, 18th International Spin Physics Symposium*
edited by D. G. Crabb, D. B. Day, S. Liuti, X. Zheng, M. Poelker, and Y. Prok

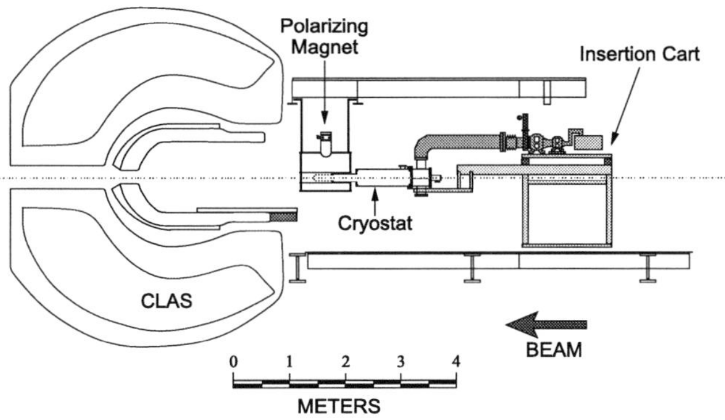

FIGURE 1. Side view of target in polarizing position.

bore solenoid. Once the sample is polarized (typically at 300 mK), the microwaves are switched off, and the refrigerator cools below 60 mK in about 30 minutes. Next the polarizing magnet is deenergized while a second magnet, located inside the target cryostat, is simultaneously ramped to a "holding" field of 0.56 T. The refrigerator, now below 50 mK, is then retracted from the polarizing magnet and moved to the center of CLAS where it continues to cool to about 30 mK.

The polarizing magnet is suspended from a rail system outside CLAS and can be moved away from the beam line approximately 1 m in the beam-left direction. Its field axis is oriented parallel to the direction of the photon beam, thus the target is polarized longitudinally during dynamic nuclear polarization (DNP). The target cryostat and all equipment essential to its operation (pumps, microwaves, electronics, gas systems, etc.) are mounted on an insertion cart that rides along a second rail system parallel to the beam. Thus the entire target system moves as a single unit and travels a total distance of about 2.5 m from the polarizing position to the data acquisition position at the center of CLAS. A side view of the system in the polarizing position is shown in Fig. 1.

REFRIGERATOR

The refrigerator, as shown in Fig. 2, is composed of two modules: a ^{4}He precooler and a ^{3}He-^{4}He dilution refrigerator, each constructed around a thin-walled stainless steel tube (50 mm and 40 mm diameters, respectively). The two tubes bolt together using commercial knife-edged flanges, thus forming a single tube that extends from room temperature to the mixing chamber. This tube serves both as a beam pipe for incoming photons and as a load lock tube for inserting the target beads into the mixing chamber.

The precooling module slides inside the 18 cm diameter ^{3}He pumping tube and contains two 1 liter volumes (or "pots") of liquid ^{4}He at 4 K and 1 K, respectively. Cold gas extracted from both pots is used to precool the circulating ^{3}He to approximately 2 K using a counterflow heat exchanger that consists of a series of nested, thin-wall tubes

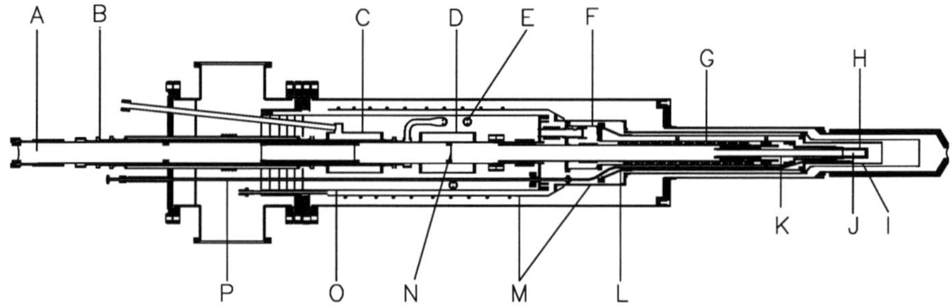

FIGURE 2. Sectional view of the target cryostat: A–beam pipe, B–precooling heat exchanger, C–4 K pot, D–1 K pot, E–1 K heat exchanger, F–still, G–sintered heat exchanger, H–mixing chamber, I–holding coil, J–target ampule, K–target insert, L–1 K heat shield, M–20 K heat shield, N–radiation baffle (one of three), O–^{3}He pump tube, P–waveguide. The overall length of the cryostat is approximately 2 m.

welded around the central beam pipe. The ^{3}He is condensed using a sintered copper heat exchanger located inside the 1 K pot. Cold gas from the 4 K pot also cools a heat shield surrounding the dilution refrigerator to about 20 K.

The dilution refrigerator consists of the customary still, heat exchanger, and mixing chamber. A second heat shield, cooled by the still to about 1 K, surrounds the latter two items. The heat exchanger is modeled after a design of Niinikoski [3] and is comprised of a 1.5 m length of stainless steel/copper fin tubing and five, 25 cm long segments of copper C122 tubing with copper powder (325 mesh and 5 micron) sintered inside and out. The heat exchanger is wound in a spiral groove that has been machined in a fiberglass mandrel. The mandrel slides tightly over the central beam pipe and is covered by a second tightly-fitting tube that attaches to the still at one end and the mixing chamber at the other.

The butanol beads are loaded under liquid nitrogen into a 15 mm diameter $\times$ 50 mm long polychlorotrifluoroethylene (PCTFE) ampule that attaches to the end of a short length of stainless steel tube, or target insert. The tube is sealed by a thin aluminum vacuum window at the downstream (target) end. Using a special wrench, the tube (and target ampule) is inserted through the central beam pipe and into the mixing chamber (at $\sim$10 K) where it makes a vacuum tight seal using a polyimide gasket. The wrench is then removed and replaced by a series of three aluminum foil radiation baffles heat sunk at 50, 4 and 1 K. A thin polyimide vacuum window is installed at the room temperature end of the beam pipe which is then evacuated. Two nuclear magnetic resonance (NMR) coils are located outside the PCTFE mixing chamber. The first is used at 213 MHz to measure the target polarization during DNP, and the second monitors the polarization during frozen spin mode at 24 MHz.

The dilution refrigerator operates with ^{3}He circulation rates that range from about 0.5 to 20 mmol s^{-1}. The base temperature is 26–28 mK with no beam on target, and 30–32 mK under a photon flux of about 10^7 s^{-1}. The cooling power is 1 mW at 52 mK, 10 mW at 110 mK, and 80 mW at 300 mK.

HOLDING COILS

A thin superconducting solenoid is used to maintain longitudinal target polarization while in frozen spin mode. The solenoid consists of three layers of 0.1 mm diameter copper-clad, multifilament NbTi wire that is wound around a 1 mm thick, 50 mm diameter aluminum extension of the 1 K heat shield. Additional windings at either end are used to improve the field homogeneity. A holding field of 0.56 T is produced by a current of 22.0 A.

The next series of experiments require transverse target polarization. In this case the holding solenoid will be replaced by a "racetrack" style dipole magnet which will rotate the spins 90° following the DNP procedure. The four-layer thick, 20 cm long dipole is fashioned from 0.15 mm NbTi wire. Both the dipole magnet and the spin rotation were successfully tested at 0.51 T following the longitudinal series of experiments.

RESULTS

The frozen spin target was used to acquire approximately 1100 h of photoproduction data over a three month run cycle in 2007 and 2008. Typical proton polarizations of 80–90% were achieved for both the positive and negative spin states. With beam on target, the spin-lattice relaxation time was typically 2700 h for the positive spin state and 1400 h for the negative, corresponding to a relative loss of polarization of 1–2% per day. Without beam, relaxation times up to 3600 h were observed for the positive polarization state. The target was repolarized about once a week, in nearly all instances to reverse the polarization. Three unscheduled repolarizations were required during the experiment due to failures of the holding coil power supply. Otherwise the target operated in a completely reliable manner throughout the three month run cycle.

ACKNOWLEDGMENTS

Authored by Jefferson Science Associates, LLC under U.S.DOE Contract No. DE-AC05-06OR23177. The U.S. Government retains a non-exclusive, paid-up, irrevocable, world-wide license to publish or reproduce this manuscript for U.S. Government purposes.

REFERENCES

1. B. A. Mecking, et. al, *Nucl. Instr. and Meth. A* **503**, 513–553 (2003).
2. C. D. Keith, et. al, *Nucl. Instr. and Meth. A* **501**, 327–339 (2003).
3. T. O. Niinikoski, *Proc. 6th Int. Conf. on Cryogenic Engineering, Grenoble* IPC Science and Technology Press, Guildford, UK, 1976, pp. 102–111.

First Experiments with the Polarized Internal Gas Target (PIT) at ANKE/COSY

R. Engels*, D. Chiladze†, S. Dymov**, K. Grigoryev*,‡, A. Kacharava§,¶, F. Klehr‖, B. Lorentz*, M. Mikirtychyants*,‡, S. Mikirtychyants‡, D. Prasuhn*, F. Rathmann*, J. Sarkadi*, H. Paetz gen. Schieck††, R. Schleichert*, H. Seyfarth*, E. Steffens§, H. Ströher* and A. Vasilyev‡

*Institut für Kernphysik, Forschungszentrum Jülich, Leo-Brandt-Str. 1, 52425 Jülich, Germany
†High Energy Physics Institute, Tbilisi State University, 380662 Tbilisi, Georgia
**Joint Institute for Nuclear Research, 141980 Dubna, Russia
‡Petersburg Nuclear Physics Institute, Orlova Rosha 2, 188300 Gatchina, Russia
§Physikalisches Institut II, Universität Erlangen-Nürnberg, 91058 Erlangen, Germany
¶High Energy Physics Institute, Tbilisi State University, 0186 Tbilisi, Georgia
‖Zentralabt. Technologie, Forschungszentrum Jülich, Leo-Brandt-Str. 1, 52425 Jülich, Germany
††Institut für Kernphysik, Universität zu Köln, Zülpicher Str. 77, 50937 Köln, Germany

Abstract. For future few-nucleon interaction studies with polarized beams and targets at COSY-Jülich, a polarized internal storage-cell gas target was implemented at the magnet spectrometer ANKE. First commissioning of the polarized Atomic Beam Source (ABS) at ANKE was carried out and some improvements of the system have been done. Storage-cell tests to determine the COSY beam dimensions have been performed. Electron cooling combined with stacking and stochastic cooling have been studied. Experiments with N_2 gas in the storage cell to simulate the background produced by beam interaction with the aluminum cell walls were performed to investigate the beam heating by the target gas. The analysis of the $\vec{d}\vec{p} \rightarrow dp$ and $\vec{d}\vec{p} \rightarrow (dp_{sp})\pi^0$ reactions showed that events from different positions of the extended target can be clearly identified in the ANKE detector system.

The polarization of the atomic beam of the ABS, positioned close to the strong dipole magnet $D2$ of ANKE, was tuned with a Lamb-shift polarimeter (LSP) beneath the target chamber. With use of the known analyzing powers of the quasi-free $np \rightarrow d\pi^0$ reaction, the polarization in the storage cell was measured to be $Q_y = 0.79 \pm 0.07$ in the vertical stray field of the $D2$ magnet acting as a holding field. The target thickness achieved was 2×10^{13} atoms/cm^2 for one hyperfine state populated in the ABS beam only. With a COSY beam intensity of 6×10^9 stored polarized deuterons in the ring, the luminosity for double polarized experiments was 1×10^{29} cm^{-2} s^{-1}.

Keywords: Atomic beam source, ANKE, Polarized target, Internal target, Polarimeter, Detection of atomic beams; Spin polarized hydrogen, Deuterium.
PACS: 29.25.-t; 29.25.Pj; 29.27.Fh; 07.30.-t; 07.77; 67.65

INTRODUCTION

In 2004 the Atomic Beam Source (ABS) [1] and the Lamb-shift polarimeter (LSP) [2] were transferred from the laboratory to the COSY building. After all necessary tests, where the parameters in Table 1 were determined, in the summer of 2005 the source was ready for installation at the spectrometer ANKE and further commissioning. Measurements to determine the COSY-beam dimensions at the ANKE-target position

CP1149, *Spin Physics, 18th International Spin Physics Symposium*
edited by D. G. Crabb, D. B. Day, S. Liuti, X. Zheng, M. Poelker, and Y. Prok

and first tests with storage-cell prototypes were carried out parallel to these studies.

Gas Type	Intensity, at/s	P_z	P_{zz}
Hydrogen	$(7.8 \pm 0.2) \times 10^{16}$	$+0.89 \pm 0.01$	-
		-0.96 ± 0.01	-
Deuterium	$(3.9 \pm 0.1) \times 10^{16}$	$+0.73$	$+0.77$
		-0.82	-1.17

Table 1: Main parameters of the polarized beam source of ANKE/COSY.

PIT AT ANKE

In order to achieve the maximum luminosity in the experiments with the internal gas target, one task was to determine the minimum possible size of the storage-cell tube. During the first test in February 2004 the diameter of the COSY beam at injection and after acceleration at the ANKE target position had to be measured. For this, a frame carrying various diaphragms was constructed. The diaphragm, which was mainly used, had dimensions of $50_{hor.}$x$25_{vert.}$ mm^2, i.e. larger than the expected beam size. During the tests, the supporting frame was moved across the beam by stepper motors. First, the center of the diaphragm was placed at the expected center of the COSY beam. By moving the diaphragm, the COSY beam is gradually destroyed and its full size can be measured. This measurements were done with 2.1 GeV proton beam without using any cooling procedures (electron cooling at injection, stochastic cooling after acceleration) and stacking procedure at injection. At injection, the beam had elliptical shape and its full size was $38_{hor.}$x$17_{vert.}$ mm^2. The accelerated beam without target had a size of $9_{hor.}$x$14_{vert.}$ mm^2. With the cluster-target beam (density: 10^{12} atoms/cm^2) it increased to $17_{hor.}$x$17_{vert.}$ mm^2 due to beam heating by the target.
To operate the ABS in the strong magnetic stray field of the *D*2 magnet requires shielding for several components. For example, most of the turbo pumps and all of the cryo-pumps must be shielded by soft iron to avoid their destruction. In addition, the weak-field transition (WFT) unit, following the last sextupole magnet of the ABS, could only be used with a strong shielding.

POLARIZED INTERNAL TARGET COMMISSIONING

Based on the measured results, two storage-cell prototypes were built from a 25 μm thick aluminum foil (99.95 Al). With acceleration of an unpolarized deuteron beam through the large cell ($30_{hor.}$x$20_{vert.}$ mm^2) to an energy of about 2.1 GeV, it was possible to store and accelerate more than 2/3 of the injected deuterons ($\sim 9 \times 10^9$) in the COSY ring. Using beam scrapers in the opposite section of the accelerator ring, the dimensions of the stored beam in the cell were decreased to $13_{hor.}$x$11_{vert.}$ mm^2. With a small cell of $15_{hor.} \times 15_{vert.}$ mm^2 1.7×10^9 deuterons, i.e. about 15% of the injected deuterons, were successfully stored in the COSY ring. The length of both cells was 220 mm.
For the first beamtime an aluminum foil, covered with PTFE to minimize depolarization on the surface, was used for the new prototype of the storage cell. The beam tube of this

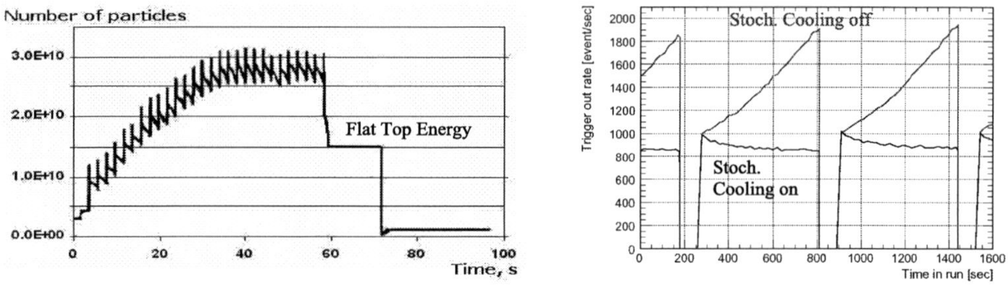

FIGURE 1. Left hand: The beam-current transformer signal and the number of the stored protons in the COSY ring during stacking (28 stacks) through a storage cell followed by 2 s of electron cooling and accelerating to flat top energy. **Right hand:** The trigger rate during data taking with stochastic cooling switched on and off during a set of cycles. The strong increase of the trigger rate at the flat top energy after acceleration with stochastic cooling off occurs due to the increase of beam-cell walls interactions, caused by beam heating.

prototype was 400 mm long and had a cross section of $20_{hor.} \times 20_{vert.}$ mm^2. During the run, stacking injection [3] and electron cooling were employed to increase the number of stored and accelerated protons with the storage cell in place (see left hand of Fig. 1). As a last step, the ANKE spectrometer magnet $D2$ was moved to the position which corresponds to a deflecting angle of 9.2° for the first beam bending magnet $D1$. At this configuration, 6.4×10^9 protons could be stored and accelerated in the ring. This is about 50% of the number of particles which can be accelerated without cell and stacking at injection. This number yields an appreciable luminosity of about 10^{29} cm^{-2}s^{-1} for double polarization experiments.

For beam energies higher than 831 MeV stochastic cooling can be used at COSY. This will compensate for the beam heating by the target. At the right hand side of Fig. 1 the full trigger rate during data acquisition is shown as a function of time during different beam cycles. When stochastic cooling is switched off and the beam is heated, the rate of beam hits on the cell walls is increased and the background grows significantly. With stochastic cooling, the beam heating could be compensated and the trigger rate is not increase during the cycles.

In addition to the storage cell tests, the ABS beam was used as a jet target. In a first experiment the target position along the COSY beam direction could not be identified by the ANKE detector system. The background gas pressure was too large. For a second experiment an ABS beam cryo catcher was built which was installed below the interaction point with the COSY beam. With this additional pumping power, the pressure in the target chamber decreased by one order of magnitude to 3.7×10^{-8} mbar. Now, with vertex reconstruction, the jet-target position could be clearly identified. The integral jet-target thickness of 1.5×10^{11} cm^{-2} perfectly fits the predicted value.

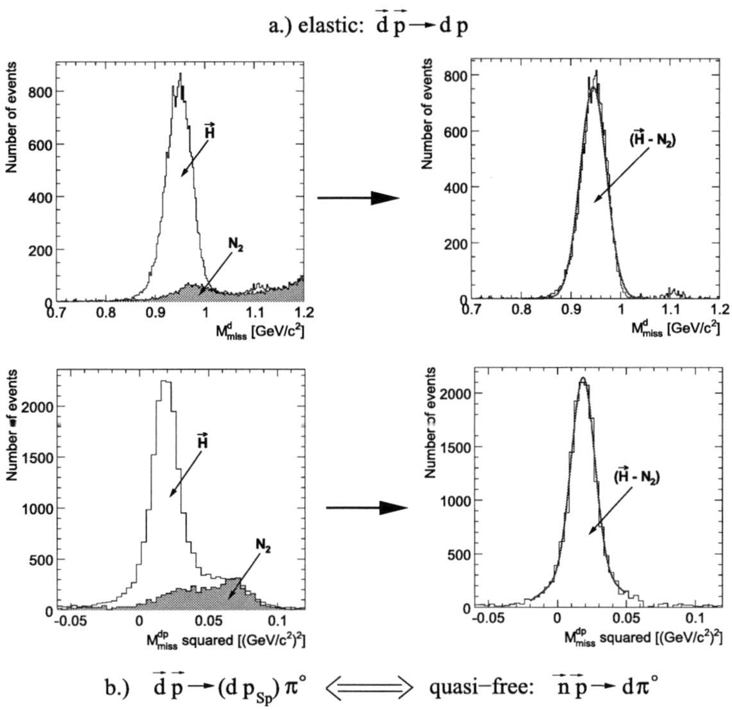

FIGURE 2. Missing-mass spectra for the proton from elastic $\vec{d}\vec{p}$ scattering (a.) and the π^0 from the quasi-free $np \rightarrow d\pi^0$ reaction (b.) before (figures at the left hand side) and after background subtraction (figures at the right hand side). The background subtraction is based on the additional measurement with N_2 gas in the target cell.

RESULTS OF THE COMMISSIONING

In early 2007, the LSP was used to tune and to control the polarization of the ABS beam. However, the magnetic stray field of the spectrometer magnet $D2$ caused a number of problems:

- The slow protons behind the Glavish-type ionizer of the LSP were partially deflected and, therefore, the sensitivity of the LSP was decreased.
- The quantization axis of the polarization, defined in the longitudinal solenoid field of the ionizer, was deflected, too. The LSP measures the projection of the polarization on the horizontal beam line only. The measured polarization was about 22% of the expected value only and, furthermore, it had had the wrong sign.

Nevertheless, the transition units could be tuned and the polarization, which was measured once per day, could be controlled to be stable within 5% during one week of operation.
During this beam time a storage cell ($15_{ver.} \times 20_{hor.} \times 380$ mm) was used. In addition

H_2 and N_2 could be injected into the cell by two separate gas feeding tubes. A first silicon tracking telescope (STT) was mounted around the cell. Polarized or unpolarized deuterons were accelerated to the flat-top energy of T_d = 1.2 GeV through the storage-cell tube filled with polarized hydrogen from the ABS or with unpolarized H_2 or N_2 gas from the calibrated gas supply system.

Fig. 2 shows sample spectra from the elastic scattering $\vec{d}\vec{p}$ (upper part) and the $\vec{d}\vec{p} \to (dp_{sp})\pi^0$ reaction (lower part). Both missing-mass spectra could be corrected for events, which are produced at the cell walls. For this reason data were taken with N_2 in the storage cell, of a pressure chosen to simulate the heating of the COSY beam by the hydrogen target gas [4]. The event distributions of these runs were subtracted from the original measured spectra with hydrogen in the cell and the results are shown on the right hand side of Fig. 2.

The analyzing powers for the reaction $d\vec{p} \to (dp_{sp})\pi^0$ for different scattering angles are known with a good precision as given in ref. [5]. Therefore, the polarization of the hydrogen atoms in the cell was determined with an unpolarized deuteron beam at COSY as $\sim 0.79 \pm 0.07$ based on the measured asymmetries. Vice versa, the polarization of the COSY beam can be observed with unpolarized hydrogen gas in the storage cell as well.

OUTLOOK

In the fall of 2009 a long beam time at ANKE on double-polarized $\vec{p}\vec{d}$ breakup is planned at flat top energies $T_d = 1.2$ or $T_d = 2.23$ GeV [6]. At this time a modified LSP with a rotatable Wien filter will be available to compensate for the deflection of the quantization axis by the magnetic stray fields of the ANKE spectrometer magnet *D*2.

ACKNOWLEDGMENTS

The authors want to thank the members of the ANKE collaboration and the COSY crew, who helped us during the installation and the commissioning studies of the polarized internal target.

REFERENCES

1. F. Rathmann *et al.*, Proc. 15th Int. Spin Physics Symposium, Upton, NY, USA,2002. AIP Conf. Proc. 675 553 (2003).
2. R. Engels *et al.*, *Rev. Sci. Instrum.* **74**, 4607 (2003).
3. H.J. Stein et al., Proc. 18th Conf. on Charged Particle Accelerators (RUPAC 2002), Obninsk, Russia, 2002. I.N. Meshkov (Ed. in chief), NRCRF Obninsk, 220 (2004).
4. F. Rathmann *et al.*, *Phys. Rev. C* **58**, 658 (1998).
5. D. Chiladze, *Phys. Rev. ST-AB* **9**, 050101 (2006).
6. COSY Proposal No. 172, Spokespesons: A. Kacharava, F. Rathmann and C. Wilkin, www.fz-juelich.de/ikp/anke/en/proposals.shtml

Considerations For The Optimal Polarization of ^{3}He Targets

Brielin Brown[a]

[a]*Department of Physics, University of Virginia, Charlottesville, Va 22904*

Abstract. The main characteristics of a polarized ^{3}He are maximum polarization, spin-up time, and cell lifetime. New hybrid K-Rb cells have higher polarization and shorter spin-up times. However, operating conditions also have a large impact on the operational characteristics of a cell. Here, hybrid K-Rb cells are polarized at varying temperatures and laser powers in order to measure the effects on the aforementioned operational characteristics, using nuclear magnetic resonance (NMR) and electron paramagnetic resonance (EPR) polarimetry to measure polarization. Results show manipulating temperature and laser power can have a large impact on polarization and spin-up time, yielding faster spin-up time and higher polarization in many cases.

Keywords: Polarized ^{3}He Target, Neutron Target, Spin-Exchange Optical Pumping,
PACS: 29.25.Pj, 32.10.dk

BACKGROUND

The ideal target for probing the fundamental quark-gluon structure of the nucleon is a free neutron target, but free neutrons are unstable and decay with a lifetime of under 15 minutes [1]. However, polarized ^{3}He can be used as an effective neutron target since the ground state wave function is predominantly an S state in which the spins of the protons are paired anti-parallel with each other. In this state, the spins of the protons cancel out so that the nucleus is left mainly with the spin of the unpaired neutron [2].

The adiabatic fast passage technique used to flip the spins during polarimetry results in a .3-.5% spin loss with each flip. Clearly, this loss of spin must be recuperated between spin flips. Short spin up times allow for more spin to be recovered between spin flips thus allowing a higher operational polarization. The higher polarization itself allows for more statistics to be collected in a shorter amount of time when using the ^{3}He targets in experiments [3].

MOTIVATION

Hybrid K-Rb cells provide faster spin-up times and polarization than Rb cells, however faster spin-up times and higher polarizations are still desirable. Research conducted by the Princeton group showed polarization and spin-up time have a strong dependence on temperature in some aspects and a weak dependence in others, in addition higher temperatures increase the frequency of atom-atom interactions [4].

CP1149, *Spin Physics, 18th International Spin Physics Symposium*
edited by D. G. Crabb, D. B. Day, S. Liuti, X. Zheng, M. Poelker, and Y. Prok

This implies that significant gains in polarization and spin-up time can be achieved by operating at a higher temperature.

However, it has been shown that the gas induced relaxation due to the interaction of the gasses in K-Rb cells increases with $T^{4.25}$, which implies lower levels of sustainable polarization at higher temperatures [4].

EXPERIMENT

Overview

We sought to find an optimal temperature and laser power for use in experiments involving a polarized ^{3}He target. To do so, we used Spin-Exchange Optical Pumping (SEOP) to polarize hybrid K-Rb cells at various temperatures. At the highest temperature we used SEOP to polarize the cell at various laser powers. During polarization, we used a combination of Nuclear Magnetic Resonance (NMR) and Electron Paramagnetic Resonance (EPR) to monitor spin-up time and polarization.

Setup

The exact experimental setup used for SEOP can be seen in Figure 1. As shown, the pumping chamber of the cell was held in an oven designed to hold temperatures of up to 300°C. The optical system itself contains 5 30 watt lasers and uses the system of quarter and half wave plates and mirrors detailed in Figure 2 to provide the pumping chamber of the cell with circularly polarized light of variable intensity.

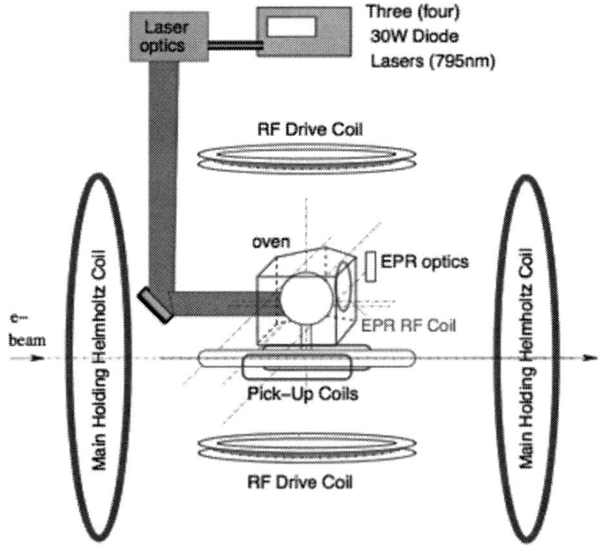

Figure 1: SEOP setup

Execution

The first hybrid cell was polarized at temperatures from 190°C to 240°C in 10°C increments. At 240°C a second hybrid cell was polarized at laser powers 90, 100, 125, and 150 watts. During the spinup procedure, NMR measurements were taken every 3 hours for a total of 7 measurements at each temperature and laser power. The program swept the holding field between 25G and 32G and recorded the signal induced in the pick-up coils. Afterwards, we fit the peak signal heights to an approximately logarithmic curve. This fit gave the expected maximum NMR and spin-up time of the cell.

Figure 2: Optical Setup for SEOP

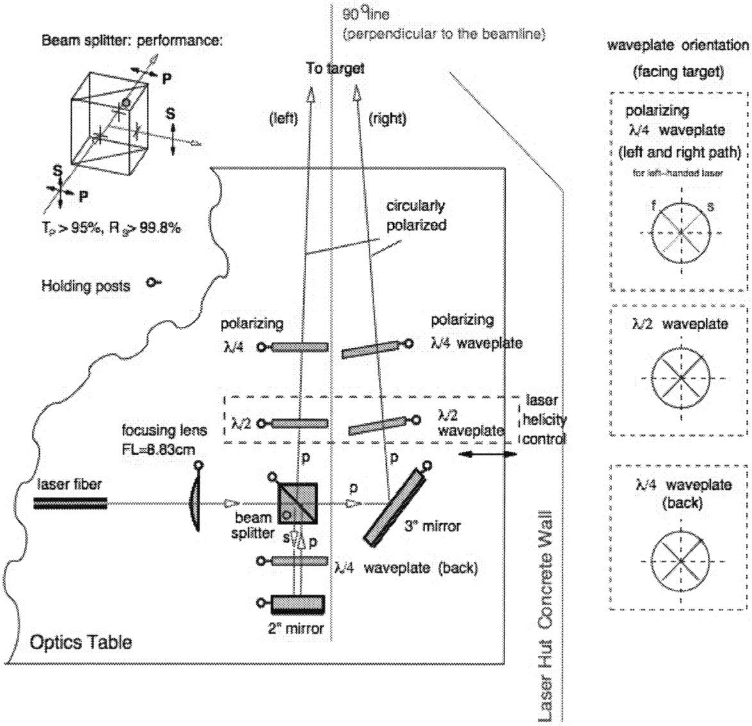

NMR Normalization

NMR measurements were used to obtain the maximum polarization and spin-up times during spin-up, however since NMR only provides a relative measurement, EPR was used for the normalization of NMR data. For each temperature and laser power a sequence of measurements (NMR, EPR, NMR) are taken in quick succession to minimize depolarization. This data is then used to normalize the NMR, and is extrapolated to the maximum NMR signal in the spin-up data.

RESULTS AND CONCLUSIONS

The data from the experiments is highlighted in tables 1 and 2. The results imply that theoretical increases in polarization and decreases in spin-up time gained by

Table 1: Temperature Manipulation Results

Temperature (°C)	Polarization (%)	Spin-Up Time (Hours)
200	51.7	8.9
210	54.0	7.9
220	51.3	6.6
230	58.5	5.3
240	49.4	4.1

Table 2: Laser Power Manipulation Results

Laser Power (Watts)	Polarization (%)	Spin-Up Times (Hours)
90	35.8	3.7
100	40.1	3.8
125	44.6	3.6
150	42.5	3.5

operating at higher temperature and power are physically realizable. These increases are extremely beneficial for the use of polarized ^{3}He targets. However, increasing the temperature or laser power too much has an adverse effect on polarization while continuing to lower spin-up time. This is most notable at the highest temperature and laser power, where the decrease in polarization is great enough that the increase in spin-up time gained would not be advantageous.

A major drawback of operating at such a high temperature and laser power is the decrease in cell lifetime, as the longevity of cells is compromised at high temperature. This would lead to more frequent cell changes and would increase the cost of experiments.

Our temperature data did turn out slightly different than expected: the trend in polarization in the temperature manipulation was dramatically different than that of the laser manipulation. This difference is dramatic enough to conjecture some sort of instrumentation error. While the results here are significant, this study should be continued with more cells to eliminate these possible errors.

ACKNOWLEDGMENTS

The author would like to thank his mentor JP Chen and the graduate students he worked with: Yi Qiang and Chiranjib Dutta, as well as the SULI program.

REFERENCES

1. W.-M. Yao et al, *Journal of Physics G* **33**, no. 1 (2006). (URL: http://pdg.lbl.gov)
2. T.W.Donnelly and A.S.Raskin, "Considerations of polarization in inclusive electron scattering from nuclei" *Annals of Physics* **169**, 247 (1986).
3. Jian-Ping Chen direct communication
4. S. Applet et al, "Theory of spin-exchange optical pumping of ^{3}He and ^{129}Xe," *Physics Review A*, **58**, no 2, 1412-1439 (1998).

A New ^{3}He Target for Jefferson Lab

Joseph M. Katich

College of William and Mary

Abstract. A polarized ^{3}He target has been successfully used in a number of experiments at Jefferson Lab to study the neutron spin structure as well as the ^{3}He structure. It has been upgraded for an upcoming set of experiments. The recent improvements include a rapid spin-flip process and polarization in the vertical (that is, normal to scattering plane) direction, allowing for polarization in any of three orthogonal directions. Nuclear Magnetic Resonance (NMR) and Electron Paramagnetic Resonance (EPR) polarimetry systems have also been improved.

Keywords: Polarized ^{3}He Target
PACS: 29.25.-t

AN OVERVIEW OF THE ^{3}HE TARGET

For nearly the past two decades a polarized ^{3}He target has been successfully used at Jefferson Lab as an effective neutron target, with early experiments achieving polarizations around 35%. Years of research have brought this number up to greater than 60%, as demonstrated by experiments E06-010 and E06-011, which concluded their run in early February 2009. The majority of this jump in polarization can be attributed to two major system upgrades including an advancement in Spin Exchange Optical Pumping (SEOP) via a hybrid alkali mixture cell and the use of line narrowed lasers to pump the cell.

The increase in maximum attainable polarization is not the only aspect of the ^{3}He target to experience significant advances. Measures have also been taken to improve both methods of target polarimetry, Electron Paramagnetic Resonance (EPR) and Nuclear Magnetic Resonance (NMR). Furthermore, a rapid spin-flip process can now be implemented to reduce systematic errors and false asymmetries. Finally, a radically new oven design as well as an extra sets of Helmholtz coils have been added to the set-up to allow for polarization in all the coordinate directions.

SEOP and Hybrid Cells

The first generation of ^{3}He target cells used at Jefferson Lab contained a small amount of Rb along with the ^{3}He. It is this Rb that, when vaporized, can easily be forced to the desired polarization via optical pumping. The Rb then transfers it's polarization to the ^{3}He via hyperfine-like coupling between the ^{3}He nucleus and valence Rb electron. Unfortunately, the spin-exchange efficiency between these two atoms is quite small. The introduction of K to the cell drastically increases the efficiency of the SEOP due to the

CP1149, *Spin Physics, 18th International Spin Physics Symposium*
edited by D. G. Crabb, D. B. Day, S. Liuti, X. Zheng, M. Poelker, and Y. Prok

much larger spin-exchange rate between Rb and K. These hybrid cells were first used in a precision measurement of G_E^n at JLab in 2006 with excellent results. Most recently, they have been successfully used in experiments E06-010 and E06-011, achieving beam polarizations greater than 60%.
The SEOP process is further improved by replacing the traditional broadband diode array pumping lasers with narrow band lasers. Much of the power output by the older broadband lasers goes to waste because the laser bandwidth is significantly wider than the linewidth of the Rb atom. The closely matched linewidths of the laser and Rb atom allows for much more efficient pumping.

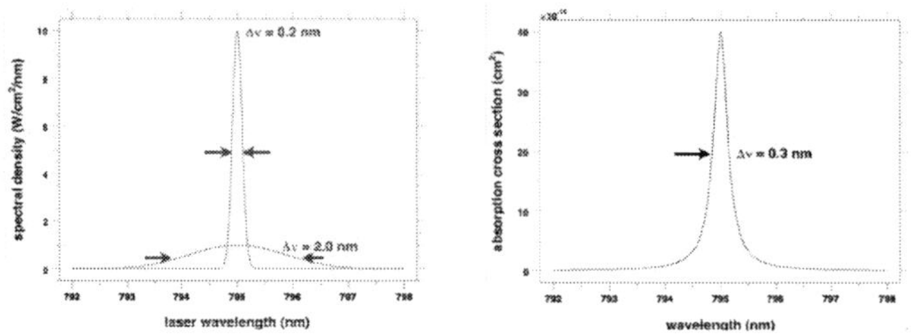

FIGURE 1. LEFT: Spectral density for old broadband laser compared to new narrow band lasers RIGHT: Absorption cross section of Rb vs wavelength

Improvements in Polarimetry

Polarization of the ^{3}He cell is measured using two independent techniques - EPR and NMR. NMR relies on an induced signal in a pair of 'pickup coils' as a result of a single spin flip. While this method produces a sizable signal, it has a major drawback in that it requires calibration against a known substance (in our case, water). A reliable calibration signal is notoriously difficult to obtain. The second method, EPR, is a direct measure of the polarization of the cell, needing no calibration. EPR takes advantage of the fact that there is a known Zeeman energy level splitting due to the target holding field of the system. Measurement of a deviation from this known energy level splitting is a result of an additional magnetic field created by the collective aligned spins of the ^{3}He atoms. Thus the deviation is a direct measure of the ^{3}He polarization.
One complication of EPR is getting a large enough radio frequency (RF) signal to the ^{3}He cell in order to drive the atomic transition of interest. In the new target setup the EPR RF coil has been placed *inside* the ^{3}He oven. This moves the RF coil drastically closer to the pumping chamber of the cell, leading to a much higher effective RF power, and thus an EPR signal that is easier to measure. Improvements to the NMR system include the addition of a set of pickup coils in each of the three coordinate directions surrounding the pumping chamber. In the past, NMR measurements could only be made on the scattering chamber of the cell in one direction. This new addition allows for easy polarimetry in all three directions.

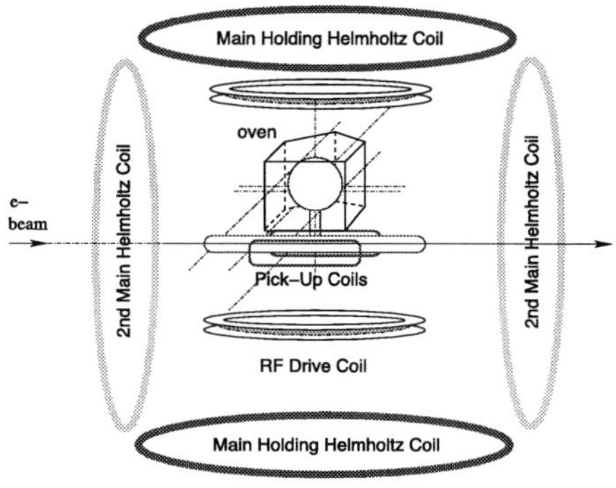

FIGURE 2. A schematic of a typical target setup - the Main Holding Helmoltz Coils hold the polarized ^{3}He in the desired direction. The RF Coils provide the signal to drive particular transitions of interest.

Modifications to the Target Set-Up

For years, the target polarization direction has been limited to in-plane scattering. This was mainly the result of difficulty in transporting laser light to pump in and out of plane direction. Furthermore, several extra coils are required, including a new set of vertical holding coils, an extra RF coil and NMR pickup coils. The newly designed ^{3}He setup has incorporated all of these extra pieces. Access for the vertical laser light is granted by aligning the cell in the oven at a 42° angle and shifting the cell pulloff by the same angle. Laser light can then safely pass through the pumping chamber in the vertical direction without interference from the rest of the cell. A drastically new oven was designed to accommodate for the new cell orientation.

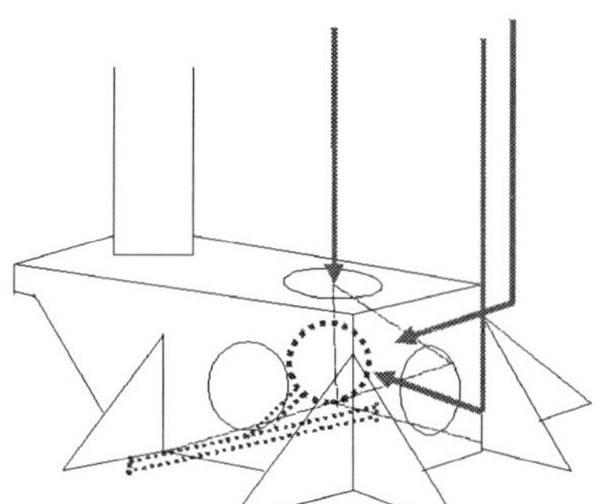

FIGURE 3. A schematic of the new 3D ^{3}He oven design

Rapid Polarization Spin Flip

A new layer of false asymmetries and systematic errors can be introduced when comparing cross-sections and scattering rates of a particular target state to it's corresponding anti-polarization state. For this reason, it is desirable to flip the spin of the target as rapidly as possible so as to cancel out any of these systematics. Unfortunately, there exists an intrinsic polarization loss with each spin flip. One must find a balance such that the time lag between spin flips is sufficient enough to recover any loss in polarization, but short enough to significantly reduce systematic error bars. Several spin-flip tests were carried out at JLab to determine the optimal time duration between spin flips. It was found that a 20 minute wait period between flips allows the target polarization to come to equilibrium at around 90% relative to the maximum polarization with no flips. Experiments E06-010 and E06-011 are the first experiments to use this rapid spin flip. Preliminary analysis shows excellent target performance while using this process.

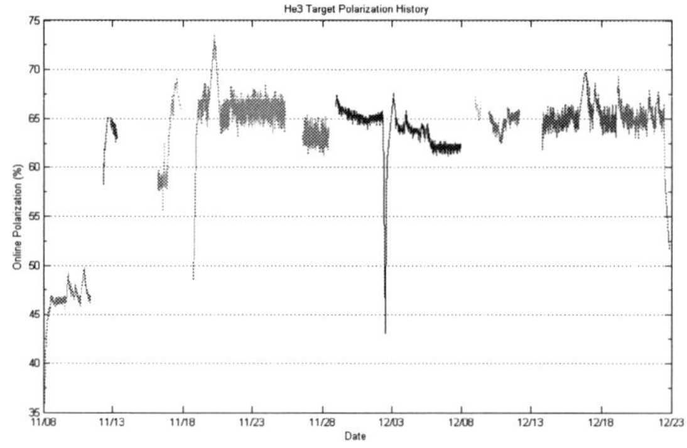

FIGURE 4. Target polarization performance during experiment E06-010 from early November 08 to late December 08. The preliminary results show a world record average in beam polarization of about 65% NOTE: Preliminary results! Probably not accurate to better than 10% (relative) until further systematic studies are done

ACKNOWLEDGMENTS

The author would like to thank all of the graduate students and postdocs who worked on making these improvements. They are (in alphabetical order): Chiranjib Dutta, Jin Huang, Vince Sulkosky, Yi Qiang and Yi Zhang. The work was done under the advisement of Jian-Ping Chen of Jefferson Lab and Todd Averett of the College of William and Mary.

REFERENCES

1. E. Babcock, I. Nelson, S. Kadlecek, B. Driehuys, L.W. Anderson, F.W. Hersman, T.G. Walker, *Hybrid Spin-Exchange Optical pumping of* ^{3}*He*, 2003.

APPLICATIONS OF SPIN, ASSOCIATED TECHNOLOGIES, AND FUTURE FACILITIES

Medical Imaging of Hyperpolarized Gases

G. Wilson Miller

Radiology Department, University of Virginia, Box 801339, Charlottesville, VA 22908, USA

Abstract. Since the introduction of hyperpolarized ^{3}He and ^{129}Xe as gaseous MRI contrast agents more than a decade ago, a rich variety of imaging techniques and medical applications have been developed. Magnetic resonance imaging of the inhaled gas depicts ventilated lung airspaces with unprecedented detail, and allows one to track airflow and pulmonary mechanics during respiration. Information about lung structure and function can also be obtained using the physical properties of the gas, including spin relaxation in the presence of oxygen, restricted diffusion inside the alveolar airspaces, and the NMR frequency shift of xenon dissolved in blood and tissue.

Keywords: hyperpolarized gas, lung imaging, magnetic resonance, MRI, helium-3, xenon-129
PACS: 87.61.Tg, 67.30.ep, 87.64.kj, 87.85.Pq

INTRODUCTION

Conventional magnetic resonance imaging (MRI) is based on nuclear magnetic resonance (NMR) of the hydrogen atoms in the human body. Clinical MR scanners provide a very large and uniform magnetic field (typically 1.5 Tesla) that generates a net alignment of the hydrogen proton spins. The size of this thermal equilibrium polarization is given by the Boltzmann distribution, and is only about 5 ppm at 1.5 T and body temperature. Although the net spin alignment is modest, it gives rise to a large and easily detectable NMR signal since nearly two-thirds of the atoms in the body are hydrogen. Conventional MRI is rarely used for lung imaging. The NMR signal from the lung is inherently weak due to the low tissue density (Fig. 1a), and the signal decays rapidly in the presence of magnetic field distortions caused by susceptibility mismatch at the numerous air-tissue interfaces in the lung. Thus specialized mechanisms of signal enhancement must generally be used to obtain useful information from lung MRI.

In 1994, an important advance toward improved lung imaging was made with the introduction of hyperpolarized noble-gas MRI [1]. With this method, a very large nuclear polarization (up to 5 orders of magnitude greater than the thermal equilibrium value at 1.5 T) is generated in either ^{3}He or ^{129}Xe gas, yielding a strong MRI signal source that can be inhaled by the subject. Both ^{3}He and ^{129}Xe are stable isotopes so no ionizing radiation is involved. Spin-exchange optical pumping is the most commonly used method for polarizing noble gases for medical imaging, yielding polarizations approaching or exceeding 50%. This large nuclear polarization allows the inhaled gas to be imaged with high signal-to-noise ratio despite its low physical density. This presentation will give an overview of the techniques and applications of hyperpolarized-gas lung MRI being developed at the University of Virginia.

CP1149, *Spin Physics, 18th International Spin Physics Symposium*
edited by D. G. Crabb, D. B. Day, S. Liuti, X. Zheng, M. Poelker, and Y. Prok

MRI OF HYPERPOLARIZED GASES

MRI of hyperpolarized gases can be performed using an ordinary clinical scanner with relatively minor modifications. The most important requirement is that the RF amplifier and receiver channels must be able to operate at non-hydrogen MR frequencies. This broadband capability is usually an option available from the manufacturer, although users must generally supply their own RF coils tuned to the specific frequency of interest. The resonance frequency of ^{3}He is 49 MHz at 1.5 T, whereas the resonance frequency of ^{129}Xe is 18 MHz at this field strength.

Static Ventilation Imaging

The most straightforward way to make use of the hyperpolarized MR signal is to simply image the inhaled gas at breath hold. We refer to this technique as static ventilation imaging. The resulting images show the ventilated regions of the lung airspaces with unprecedented detail. Figure 1b shows hyperpolarized ^{3}He images of a normal subject and a COPD patient. The regions of missing signal in the COPD image are referred to as ventilation defects. In patients with asthma, hyperpolarized gas MRI allows us to directly observe regional patterns of reversible airflow obstruction that are characteristic of this disease [2]. When used in conjunction with methacholine or exercise challenge, we can see that airway hyper-responsiveness in asthma is a heterogeneous process, leading to the closure of entire segments of the airway tree, and is not simply a diffuse, homogeneous narrowing of the airways [3,4].

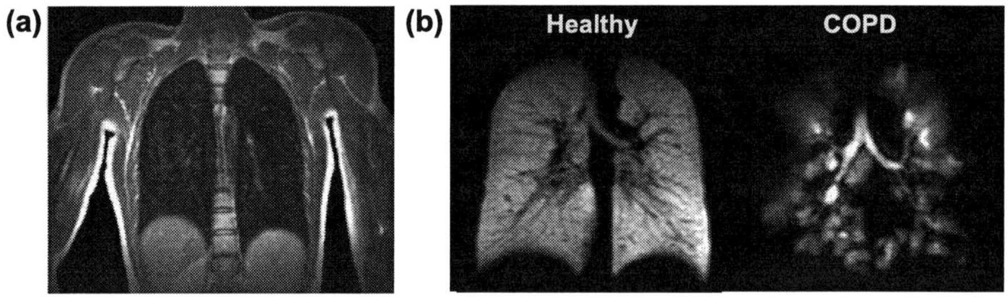

FIGURE 1. (a) Conventional ^{1}H MRI of a human torso showing the lack of signal from the lung. (b) Static ^{3}He ventilation images of a healthy subject (left) and a patient with chronic obstructive pulmonary disease (COPD).

To enable better characterization of the size and shape of ventilation defects, it is often desirable to acquire 3D images of the inhaled gas. This allows one to retrospectively determine the optimum view(s) for examining a particular ventilation defect, and reduces the chances that small defects will be overlooked or misidentified. Figure 2 shows 3D static ventilation images of a subject with a pneumatocele, which is a benign, air-filled cyst. The pneumatocele appears on a chest x-ray as a faint circle, indicated by the arrow in Fig. 2a, that is recognizable to a trained radiologist. The pneumatocele is more easily recognized by the untrained eye from the ^{3}He images, where the cyst appears as a spherical ventilation defect.

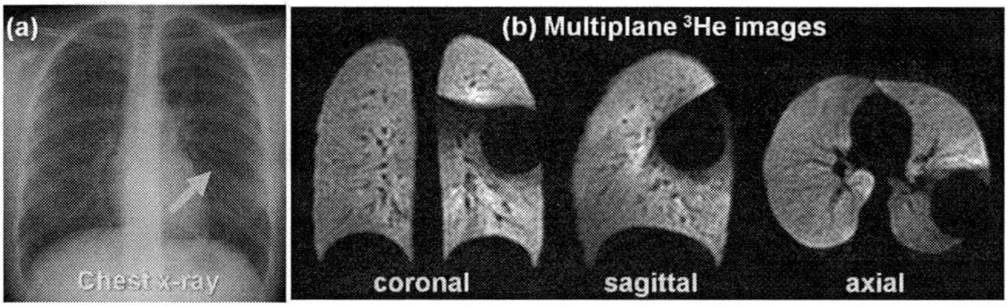

FIGURE 2. (a) Chest x-ray and (b) ^{3}He images of a pneumatocele, that show the level of detail obtainable from 3D acquisition. The ^{3}He imaging data were acquired on a 3D matrix with isotropic 3.3 mm resolution, allowing 2D images to be reconstructed in any of the three principal planes.

Grid Tagging of Lung Motion

Pulmonary ventilation is a dynamic process by which deoxygenated alveolar gases are exchanged for fully oxygenated air. Static ventilation imaging provides a snapshot of the ventilated lung volume at end-inspiration, but it does not provide insight into the dynamics of pulmonary motion during breathing. By using a technique called grid tagging, we can observe dynamic aspects of ventilation in a direct yet non-invasive way, and derive quantitative measures of pulmonary biomechanics [5].

After the subject inhales the hyperpolarized gas, specially designed RF pulses are applied that destroy the hyperpolarized magnetization along perpendicular sets of parallel lines. MR images of this "tagged" magnetization reveal a grid of uniformly spaced islands of signal (Fig. 3a, left image). Then the subject is instructed to exhale, and additional images of the tagged magnetization are acquired during the exhalation phase. Remarkably, the tag pattern remains largely intact at end-exhalation (Fig. 3a, right image), which demonstrates that not all the alveolar gas is replaced during each breathing cycle. By tracking the location of the individual grid elements, we can generate displacement maps that show the spatial and temporal variations of lung motion during breathing [6].

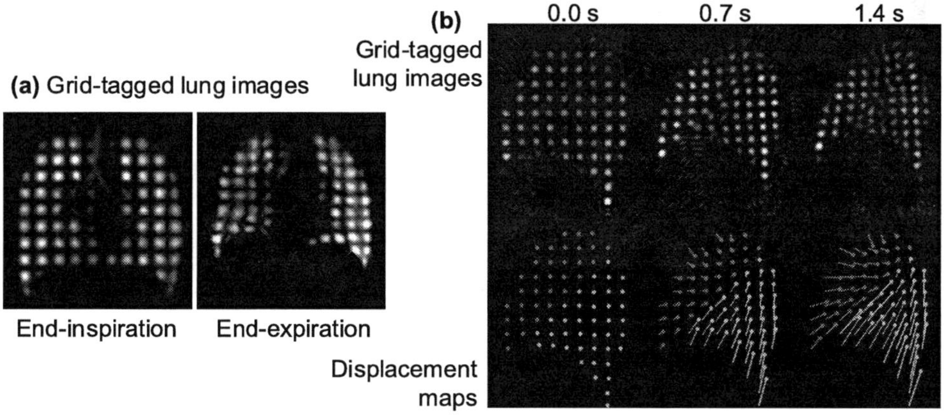

FIGURE 3. (a) Coronal images of grid-tagged ^{3}He at end-inspiration and end-expiration. (b) Sagittal images of grid-tagged ^{3}He, obtained at multiple time points during the exhalation phase, and corresponding displacement maps.

Figure 3b shows sagittal images of grid-tagged magnetization acquired at multiple time points during the exhalation phase together with the corresponding displacement maps. This particular slice spans two lobes of the right lung. To aid the eye in distinguishing the two lobes, the grid elements from upper and lower lobe are shown in different colors on the displacement map. We can see that the two lobes move largely independently. Not only do they contract in different directions, but the upper lobe begins emptying earlier than the lower lobe.

The grid-tagging technique has attracted interest from the field of radiation oncology, where it can be used to develop and validate models that predict regional lung motion during free breathing [7]. Such models are a crucial element of treatment planning for radiation therapy of lung cancer, as precise tracking of tumor motion during breathing can be used to minimize the radiation exposure of healthy tissue.

OTHER FORMS OF IMAGE CONTRAST

A crucial element of any medical imaging technique is its ability to generate contrast between different types of tissue, in order to identify particular anatomical features or to differentiate normal from abnormal tissue. Conventional MRI provides excellent soft tissue contrast, based not only on differences in tissue density but also on differences in the spin-lattice relaxation time T_1 and the spin-spin relaxation time T_2 of the hydrogen nuclei in different tissues. Among these forms of contrast, T_1 weighting is particularly useful in hyperpolarized ^{3}He lung MRI.

Oxygen-Weighted Imaging

The basic function of the lung is to exchange oxygen and carbon dioxide between the air and bloodstream. Efficient gas exchange requires local matching of air flow (ventilation) with blood flow (perfusion), and the effectiveness of this process determines the equilibrium concentration of oxygen in the alveoli. Under normal breathing of room air (21% oxygen) at atmospheric pressure, the steady-state alveolar oxygen concentration is about 14%, corresponding to an oxygen partial pressure (pO_2) of about 100 mm Hg.

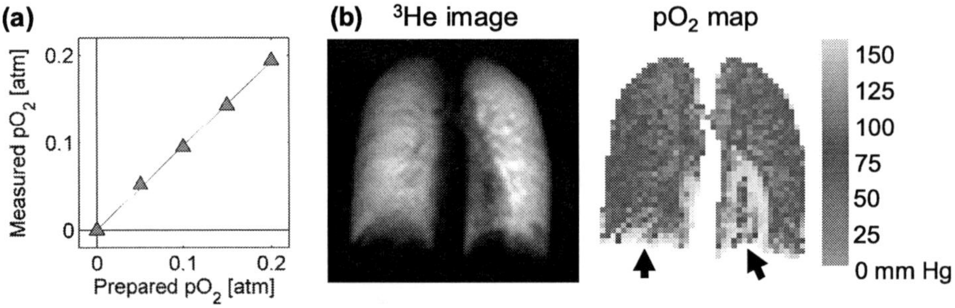

FIGURE 4. (a) Results of proof-of-principle imaging experiment, showing mean measured pO_2 versus prepared oxygen concentration. (b) Static ventilation image and pO_2 map from a healthy volunteer.

The longitudinal relaxation rate ($1/T_1$) of ^{3}He in the lung depends primarily on the alveolar pO_2. Thus T_1-weighted imaging can be used to generate contrast based on intrapulmonary pO_2 differences, and regional maps of absolute pO_2 can be acquired using carefully applied techniques [8,9]. Figure 4a shows the mean pO_2 values measured in phantoms containing a known mixture of ^{3}He and oxygen, which demonstrates that excellent quantitative accuracy is achievable. Figure 4b shows a static ventilation image and pO_2 map from a healthy subject. Anomalous pO_2 values are often observed near the heart and diaphragm (arrows), presumably due to motion that disrupts the measurement in these regions. The measured pO_2 values are uniform throughout the rest of the lung, with a mean value of 102 mm Hg.

Diffusion-Weighted Imaging

The typical spatial resolution of hyperpolarized-gas lung images is on the order of a millimeter, which is not fine enough to resolve individual alveoli. Instead, a technique called diffusion-weighted imaging can be used to probe the size and structure of the alveolar airspaces. This technique is based on consecutive application of equal but opposite magnetic field gradients (Fig. 5a) that attenuate the MR signal in the presence of diffusion. By comparing the signal obtained with and without diffusion-sensitizing gradients, we can extract a measurement the diffusion coefficient of the inhaled hyperpolarized gas. For restricted diffusion, the diffusion coefficient $D(t)$ is a deceasing function of the diffusion time t. The diffusion time of MR measurements is determined by the time between opposite gradient lobes.

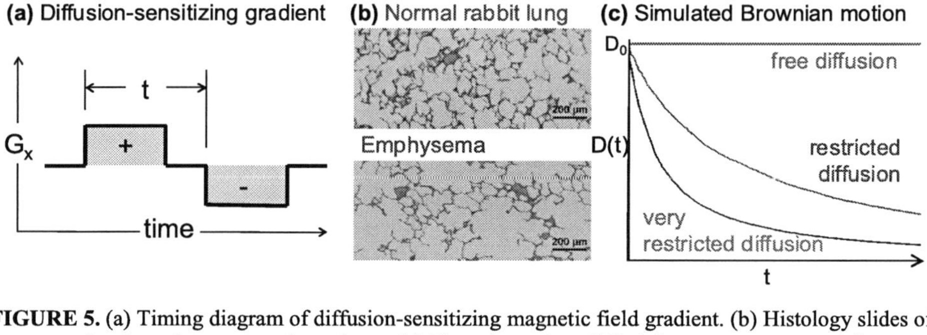

FIGURE 5. (a) Timing diagram of diffusion-sensitizing magnetic field gradient. (b) Histology slides of normal and emphysematous lung tissue. (c) Time-dependent diffusivity curves, generated from simulations of restricted gas diffusion within the airspaces shown in (b).

Monte-Carlo simulation of Brownian motion is a useful tool for understanding the quantitative dependence of the diffusion coefficient on the size and shape of the alveolar airspaces [10]. Figure 5b shows histological sections of lung tissue obtained from a normal rabbit and a rabbit with emphysema. These images were used to generate binary bitmaps delineating airspace from alveolar walls. Figure 5c shows simulated diffusivity curves, generated using Monte-Carlo simulations of restricted diffusion within these airspaces. At any given diffusion time, the diffusion coefficient is significantly higher in the emphysematous lung than in the normal lung, since the enlarged airspaces provide less restriction to the diffusive motion.

Xenon Transfer Contrast

Unlike helium, xenon is readily absorbed into tissue and blood and thus provides unique possibilities for additional forms of image contrast. The relatively large frequency difference (~200 ppm) between ^{129}Xe in the gas and dissolved phases makes it possible to selectively image either phase, and to directly observe gas exchange between the air and bloodstream [11]. There are also small frequency differences within the dissolved phase that make it possible to distinguish ^{129}Xe in tissue from ^{129}Xe in blood [12].

THE FUTURE

Hyperpolarized ^{3}He has been used much more extensively than ^{129}Xe, particularly for human studies, because it is easier to polarize, and because for the same level of polarization ^{3}He generates substantially greater MR signal due to its nearly three-fold larger gyromagnetic ratio. Since ^{3}He has negligible natural abundance, however, it is generally recognized that widespread use of hyperpolarized gases will only be possible using ^{129}Xe. The hyperpolarized gas images shown in this presentation were acquired using ^{3}He, but all of the techniques described could also be performed using ^{129}Xe. Recent advances in polarization technology have dramatically improved the polarization values and production rates of both hyperpolarized ^{129}Xe and ^{3}He [13,14].

ACKNOWLEDGMENTS

All MR images were acquired at the Center for In-Vivo Hyperpolarized-Gas MRI at the University of Virginia. Thanks to John Mugler, Michael Carl, Jing Cai, and Jaime Mata for contributing images to this presentation. Research supported by Siemens Medical Solutions and NIH grants R01-HL079077 and R21-HL089525 from the National Heart, Lung, and Blood Institute. The content is solely the responsibility of the author and does not necessarily represent the official views of NIH or NHLBI.

REFERENCES

1. M. S. Albert et al., *Nature* **370**, 199-201 (1994).
2. T. A. Altes et al., *J. Magn. Reson. Imaging* **13**, 378-384 (2001).
3. S. Samee et al., *J. Allergy Clin. Immunol.* **111**, 1205-1211 (2003).
4. E. E. de Lange et al., *J. Allergy Clin. Immunol.* **119**, 1072-1078 (2007).
5. J. Cai et al., *Magn. Reson. Med.* **58**, 373-380 (2007).
6. G. W. Miller et al., Proceedings of the 15th Annual Meeting of the ISMRM, Berlin, 2007, p. 945.
7. J. Cai et al., *Int. J. Radiation Oncology Biol. Phys.* **68**, 650-653 (2007).
8. A. J. Deninger et al., *J. Magn. Reson.* **141**, 207-216 (1999).
9. G. W. Miller et al., Proceedings of the 93rd Annual Meeting of the RSNA, Chicago, 2007.
10. G. W. Miller et al., *IEEE Trans. Med. Imaging* **26**, 1456-1463 (2007).
11. K. Ruppert et al., *Magn. Reson. Med.* **51**, 676-687 (2004).
12. B. Driehuys et al., *Proc. Natl. Acad. Sci. USA* **103**, 18278-83 (2006).
13. I. C. Ruset et al., *Phys. Rev. Lett.* **96**, 053002-5 (2006).
14. K. E. Mooney et al., Proceedings of the 17th Annual Meeting of the ISMRM, Honolulu, 2009.

Magnetism in $Ge_{(1-x)}Mn_{(x)}$ Thin Films and Quantum Dots Synthesized by Ion Implantation

Wenjing Yin[a], Li He[a], Melissa Commisso Dolph[b], Jiwei Lu[a], Robert Hull[c] and Stuart A. Wolf[a,b]

[a] *Department of Materials Science and Engineering, University of Virginia, 395 McCormick Rd, Charlottesville 22904, USA*
[b] *Department of Physics, University of Virginia, 382 McCormick Rd, Charlottesville 22904, USA*
[c] *Department of Materials Science and Engineering, Rensselaer Polytechnic Institute, 110 8th Street, Troy, NY 12180-3590, USA*

Abstract. Spin switching devices constructed using self-assembled magnetic Ge quantum dots are the ultimate goal of this effort. These devices are based on the manipulation of magnetic exchange between magnetic quantum dots and/or coherent spin states in the quantum dots. In this study, we introduced 1.1 atomic % of Mn ions into Ge thin films and 3 at.% Mn into Ge Quantum Dots by ion implantation in order to explore the ferromagnetism in Mn implanted Ge. In order to recrystallize the Ge and enhance the incorporation of Mn ions into the Ge lattice, rapid thermal annealing was applied following the implantation. A maximum saturation moment of 0.7 μ_B/Mn at 5 K was reached when the sample was annealed at 300°C and the moment decreased with higher annealing temperatures. The ferromagnetism in Mn:Ge thin films disappeared at ~ 15 K. However, superparamagnetism has been observed at 1 T up to 200 K for the 300°C annealed film and persisted to room temperature for the 400°C annealed film. Mn implanted Ge Quantum Dots were also found to be ferromagnetic at low temperatures and showed a superparamagnet transition around 200 K.

Keywords: diluted magnetic semiconductor, quantum dots, ferromagnetism, superparamagnetism, antiferromagnetism, GeMn, magnetic moment, ion implantation, rapid thermal annealing.
PACS: 75.50.Pp

INTRODUCTION

It is expected that the Moore's law progression of conventional CMOS technology will end in less than two decades and there is a worldwide search for new devices that can either extend Moore's law or provide a totally new pathway for information processing [1]. This research is focused on the use of spin as the state variable to develop spin exchange switches based upon the manipulation of magnetism in Ge quantum dots. The advantage of such a device is the ability to transfer information between adjacent circuit elements with energies far less than what is needed in conventional CMOS charge-based circuits. The material we use to create such a device is the diluted magnetic semiconductor (DMS) GeMn. Mn-based DMSs have hole-mediated ferromagnetic properties [2], which makes it possible to control the magnetism by electrically modulating the carriers in the semiconductor. It has been reported that Mn-doped Group IV semiconductors have room temperature ferromagnetism [3] and are compatible with

CP1149, *Spin Physics, 18th International Spin Physics Symposium*
edited by D. G. Crabb, D. B. Day, S. Liuti, X. Zheng, M. Poelker, and Y. Prok
2009 American Institute of Physics 978-0-7354-0686-5/09/$25.00

current Si-based CMOS technology [2]. In order to create a spin exchange switch, the initial step is to introduce the diluted ferromagnetism into Ge by incorporating Mn into the Ge lattice. We choose ion implantation because it is a well established and widely used technology in the semiconductor industry.

The penultimate method of fabrication is to selectively implant Mn ions into Ge Quantum dots using a focused ion beam. The top and side gates will be introduced to control the magnetism in the Quantum dots (QDs) and to control the coupling between the QDs, respectively. Figure 1 depicts two magnetic Ge Quantum dots side by side. In this study, we will characterize ferromagnetism in Mn doped Ge thin films and Ge QDs and demonstrate the modulation of ferromagnetism either by using electrical field or changing the initial carrier type/concentration.

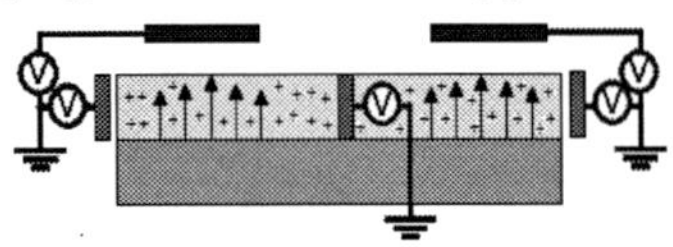

FIGURE 1. A schematic of a magnetic exchange switch.

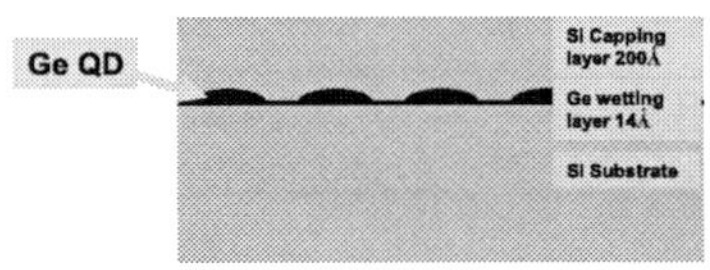

FIGURE 2. Single layer Ge Quantum dots.

EXPERIMENT

The samples were prepared by Mn ion implantation into Ge on insulator (GeOI) wafers and single layer Ge Quantum dots capped by Si layer [Fig. 2]. 200nm Ge (001) crystalline films were originally very lightly doped with Boron. A dual implantation process formed two concentration peaks in the film with doses of 3×10^{15} ions/cm^2 and 7×10^{15} ions/cm^2. The average Mn concentration is approximately 1.1 atomic % for Ge thin film and 3 at.% for Ge QDs. The substrate was maintained at 75°C during the implantation. Samples were processed by rapid thermal annealing (RTA) at 300°C, 400°C, 500°C and 600°C for 1 minute in the forming gas (95% N_2, 5% H_2) to avoid oxidation. The morphology of the films was studied by cross-sectional transmission electron microscopy (TEM). All magnetic and transport data were measured by a physical property measurement system (PPMS). Temperature-dependent magnetization (with an external field of 1 T) and remanence were carried out after cooling down the sample in 1 T field.

RESULTS AND DISCUSSIONS

Figure 3(a) is a TEM cross-sectional image of a 1.1 at.% Mn implanted GeOI sample annealed at 300°C for 1 minute. The surface roughness seen as indented structures was caused by ion implantation. The crystalline Ge layer after the implantation mostly turned into an amorphous phase, even after the 300°C annealing, and started to form polycrystals above 400°C. Both the bright field image [Fig. 3(b)] and the elemental mapping of Mn [Fig. 3(c)] clearly show a uniform distribution of Mn ions throughout the whole thickness of the Ge film. There were some Mn rich regions with non-regular shapes and diameters, which had also been found by Li [7] and Bougeard et al [4]. However, no secondary phases such as Mn_5Ge_3 [5] or $Mn_{11}Ge_8$ [6] intermetallic precipitates were observed.

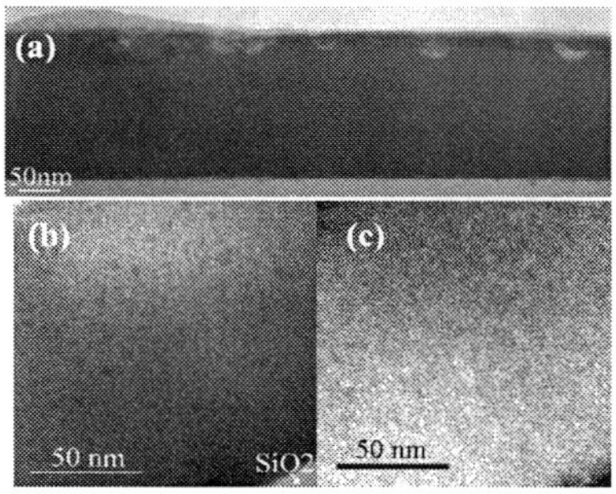

FIGURE 3. **(a)** Cross-sectional TEM image of a 1.1 at.% Mn doped Ge thin film undergoing RTA at 300°C for 1 minute; **(b)** Bright field image (zero-loss); **(c)** Mn M edge mapping

In order to study the process of Mn redistribution and its influence on magnetic and transport properties, we annealed the thin films at 300°C, 400°C, 500°C, and 600°C for 1 minute. Figures 4(a) and 4(b) show the magnetic moment versus applied field for the samples annealed at 300°C and 400°C. Comparison of magnetic hysteresis loops at 5 K shows that the saturation moment reached a maximum for the 300°C annealed sample, which was almost 0.7 μ_B/Mn and then the hysteresis loop started to shrink, and completely disappeared when the sample was annealed at 600°C. The coercive field increased with increasing annealing temperature. All the samples had soft hysteresis loops at low temperatures and as the temperature increased, the loop turned into an S-shape, which was characteristic of superparamagnetism. The related transition temperatures for 300°C and 400°C annealed samples were 15 K and 125 K respectively. This transition was also consistent with the point where the remanence disappeared [Fig. 5(b)]. We believe the strong ferromagnetism at the low temperature was mostly due to Mn rich clusters in which the holes were not itinerant but were localized. The superparamagnetism disappeared around 200 K for 300°C annealing and remained up to room temperature for the sample annealed at 400°C. For the 400°C annealed film, the M-T curve had a valley-like shape at low temperatures, which indicated the competition between ferromagnetism and antiferromagnetism. Since we did not observe any secondary phases from TEM images, this kind of antiferromagnetism might have come from Mn rich clusters, in which many Mn were nearest neighbors. It has been predicted that nearest neighbor Mn in Ge would be coupled antiferromagnetically [2] whereas Mn ions that are further apart can couple ferromagnetically through a double exchange with holes [8, 10].

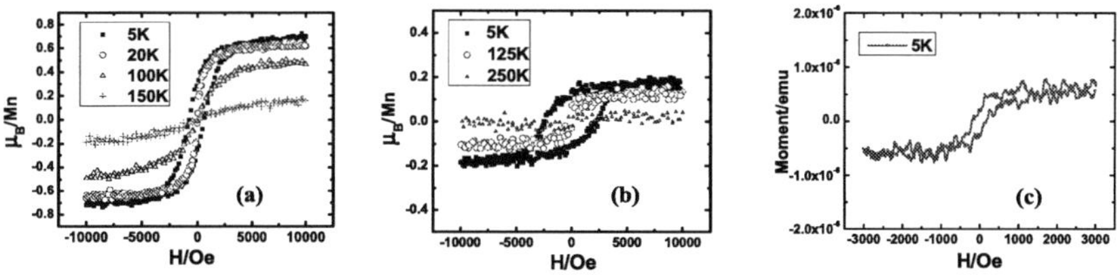

FIGURE 4. **(a)** Magnetization loops for (a) 300°C; **(b)** 400°C annealed thin film; **(c)** QDs taken at different temperatures.

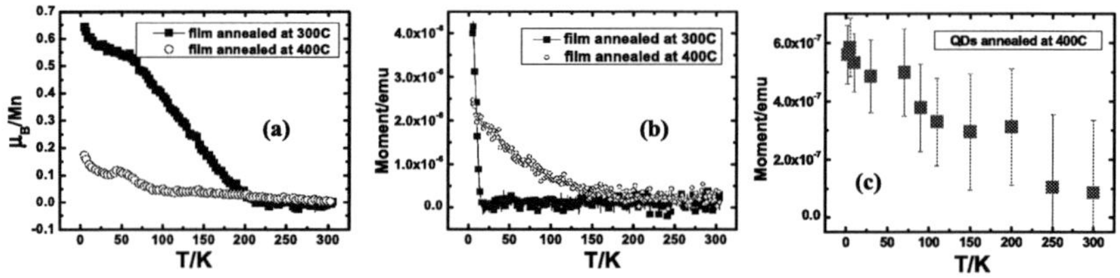

FIGURE 5. **(a)** Temperature-dependent magnetization M-T of Mn doped Ge thin film, measured under 1 Tesla; **(b)** Temperature-dependent remanence; **(c)** M-T for Mn doped Ge quantum dots.

Preliminary results in Fig. 4(c) and 5(c) show that 3 at.% Mn doped single layer Ge QDs are ferromagnetic at low temperatures and superparamagnetic up to 200K. The temperature-dependent magnetization also exhibited a similar transition trend as the 300°C annealed thin film, showing two ordering temperatures.

In summary, we successfully synthesized precipitate-free Mn doped Ge thin film with an average percentage of 1.1 at.% by ion implantation. Rapid thermal annealing was found to improve the diffusion and redistribution of Mn ions. As a consequence, the annealed GeMn samples exhibited superparamagnetism up to room temperature. The TEM study revealed that Mn was distributed throughout the film more or less uniformly, but there were definitely Mn rich clusters which contributed to both ferromagnetism and antiferromagnetism. In addition, we found ferromagnetic Mn doped Ge quantum dots with Tc $\approx$ 200 K. We are now in the process of modifying the magnetism by changing the carrier concentration and applying an electrical gate.

ACKNOWLEDGEMENTS

The authors gratefully acknowledge the support form NRI and NSF.

REFERENCES

1. S. A. Wolf, D. D. Awschalom, R. A. Buhrman et al., *Science* **294**, 1488 (2001).
2. Y. D. Park, A. T. Hanbicki, S. C. Erwin et al., *Science* **295**, 651 (2002).
3. W. Zhu, Z. Zhang and E. Kaxiras, *Phys. Rev. Lett.* **100**, 027205 (2008).
4. D. Bougeard, S. Ahlers, A. Trampert, N. Sircar and G. Abstreiter, *Phys. Rev. Lett.* **97**, 237202 (2006).
5. C. Bihler, C. Jaeger, T. Vallaitis et al., *Appl. Phys. Lett.* **88**, 112506 (2006)
6. E. Biegger, L. Stäheli, M. Fonin, U. Rüdiger and Yu. S. Dedkov, *J. Appl. Phys.* **101**, 103912 (2007).
7. A. P. Li, J. F. Wendelken, J. Shen, L. C. Feldman, J. R. Thompson, and H. H. Weitering, *Phys. Rev. B* **72**, 195205 (2005).
8. K. Sato and H. Katayama-Yoshida, *Semiconductor Science and Technology* **17**, 367 (2002).
9. A. Stroppa, S. Picozzi, A. Continenza and A. J. Freeman, *Phys. Rev. B* **68**, 155203 (2003).
10. H. Akai, *Phys. Rev. Lett.* **81**, 3002 (1998).

Photoinduced Spin Fluctuation in a Ferromagnetic Manganite Film

Xiaojun Liu

Department of Electronic Science and Engineering, Nanjing University, Nanjing 210093, China

Abstract. The transmission and resistance in a $La_{0.6}Sr_{0.4}MnO_3$ film (Tc~325 K) have been investigated under photo-irradiation at various temperatures by means of time-resolved pump-probe method. Below Tc, we have observed a negative transient transmission change $\Delta T/T$ at 2.54 eV as well as a positive resistance change $\Delta R/R$ under photo-irradiation at excited energy E_{exc} = 3.20 eV. With increasing the temperature to Tc, the $\Delta T/T$ shows a significantly reduction. We have further found that the photoinduced $\Delta R/R$ is suppressed with increase of the applied dc current I at 295 K. We have ascribed the photoinduced $\Delta T/T$ and $\Delta R/R$ of the $La_{0.6}Sr_{0.4}MnO_3$ film to the spin fluctuation in the ferromagnetic phase.

Keywords: Spin fluctuation; Photoinduced effect; Manganite film.
PACS: 75.47.Lx; 75.90.+w; 75.47.-m

Manganese oxides with distorted perovskite structures, $A_{1-x}B_xMnO_3$ (A is a trivalent rare earth ion and B is a divalent alkali earth ion), have been attracting extensive attention due to their colossal magnetoresistivity (CMR) effect [1]. Due to the strong coupling between the localized spins and the carriers, photo-irradiation or applied electric field is expected to influence not only the charge system, but also the spin system in doped manganites [2-3]. Matsuda *et al.* [2] reported the photoinduced demagnetization in a $(Nd_{0.5}Sm_{0.5})_{0.6}Sr_{0.4}MnO_3$ film at 5 K in the ferromagnetic metallic (FM) phase. They monitored the spin correlation by the characteristic absorption band at ~3 eV, which is ascribed to the interband transition between the exchange-split e_g bands (J-split bands) [4]. In their study, the photo-excited down-spin e_g carriers relax within ~ 1 ps with creating small-q spin-waves, and cause the disorder of the localized spins. The disorder of the spin system gradually disappears in a longer lifetime of order of 100 ns. On the other hand, Debnath *et al.* [5] and Gao *et al.* [6] found a current-induced depression of resistance in $La_{0.7}Ca_{0.3}MnO_3$ films, which is perhaps due to current-induced ordering of localized spins. If so, the photoinduced spin disorder can be controlled by applied electric current, which shows a great technological potential for CMR films.

In this paper, we have investigated the photoinduced effects on the transmission T and resistance R of a wide-W $La_{0.6}Sr_{0.4}MnO_3$ film (Tc~325 K) by means of the time-resolved pump-probe method. Below Tc, the transient changes of transmission $\Delta T/T$ and resistance $\Delta R/R$ have been observed, which are interpreted in terms of the photoinduced spin fluctuation in the ferromagnetic phase. We further have observed a reduction of the $\Delta R/R$-value with increasing the applied DC current I, suggesting the

CP1149, *Spin Physics, 18th International Spin Physics Symposium*
edited by D. G. Crabb, D. B. Day, S. Liuti, X. Zheng, M. Poelker, and Y. Prok

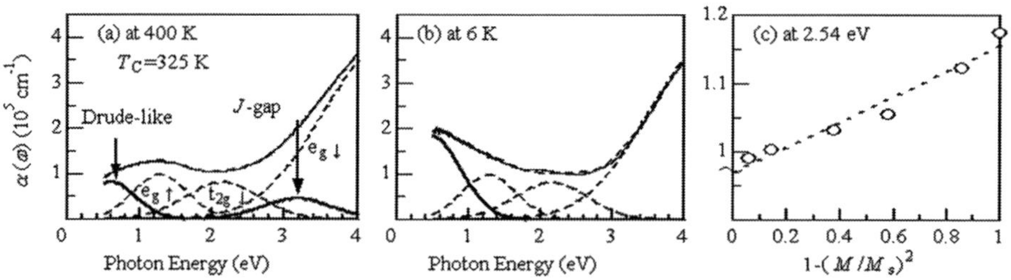

FIGURE 1. Absorption spectra of $La_{0.6}Sr_{0.4}MnO_3$ film at (a) 400 K and (b) 6 K. The spectrum consists of five components, that is, three charge-transfer ones (CT: broken curves), one "Drude-like" and one *J*-gap transition. The three CT components correspond to the transitions from O2p band to Mn$e_g\uparrow$-, Mn$t_{2g}\downarrow$- and Mn$e_g\downarrow$-levels, respectively. (c) Magnetization dependence of the absorption coefficient measured at 2.54 eV, where *M* and *Ms* are temperature-induced magnetization and saturation magnetization, respectively.

current-induced ordering of the localized spins even under photo-irradiation.

A patterned $La_{0.6}Sr_{0.4}MnO_3$ film with thickness of 200 nm was prepared by rf magnetron sputtering [7]. Curie temperature *Tc* (= 325 K) was determined from the inflection point of the magnetization-temperature curve measured at 0.05 T after cooling down to 5 K in the zero field (ZFC). Photoinduced transmission change $\Delta T/T$ was measured by means of a time-resolved pump-probe method by using a pulsed dye laser system as a pump source [8]. The pulse duration and repetition rate of the dye laser were 8 ns and 10 Hz, respectively. The photon energy of the pump pulse was fixed at E_{exc}=3.20 eV, which corresponds to the tail of the charge transfer transition from O2p level to Mn$e_g\downarrow$ state in $La_{0.6}Sr_{0.4}MnO_3$ [9]. We used a continuous wave (cw) Ar^+ laser (E_{probe}=2.54 eV) as probe light source. The transmission light was detected with a PIN photodiode, and analyzed with a digital oscilloscope. The experimental setup to measure the transient resistance change was described in Ref. 7.

Before describing details of the photoinduced effects on the doped manganite film, let us survey temperature variation of the absorption spectra. Figure 1 (a) and (b) show the absorption spectra of the $La_{0.6}Sr_{0.4}MnO_3$ film at 400 K (> *Tc*) and 6 K (<< *Tc*), respectively. At 400 K, the observed spectrum is composed of five components: three temperature-independent components due to charge-transfer transitions (broken curves) and two temperature-dependent components associated with e_g electrons (solid curves), i.e., one "Drude-like" component and one *J*-gap transition. The charge transfer transitions have been assigned to the transitions from O2p band to Mn$e_g\uparrow$, Mn$t_{2g}\downarrow$, and Mn$e_g\downarrow$ levels, respectively [3, 9]. The "Drude-like" component is ascribed to the intraband transition within the exchange-split band, and the *J*-gap transition is due to the interband transition between the exchange-split bands [4]. Figure 1 (c) shows that the absorption coefficient measured at 2.54 eV is proportional to $1-(M/Ms)^2$, where *M* and *Ms* are the temperature-induced magnetization and the saturation magnetization, respectively. Therefore, we can monitor the spin correlation by the transmission change $\Delta T/T$ at E_{probe}=2.54 eV.

We have observed a significant photoinduced transmission change $\Delta T/T$ at 2.54 eV in the $La_{0.6}Sr_{0.4}MnO_3$ film. Squares and circles in Fig. 2 (a) show the temporal

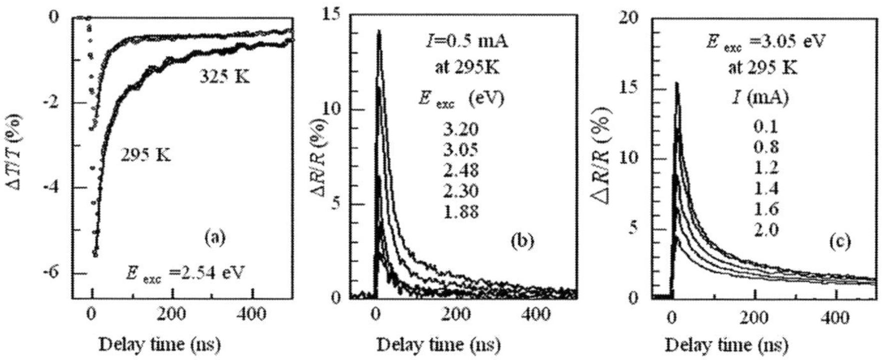

FIGURE 2. Temporal evolution of the transmission changes for (a) $\Delta T/T$ at 2.54 eV at 295 K (< *Tc*) and 325 K (~ *Tc*), (b) $\Delta R/R$ under various E_{exc} at 295 K, and (c) $\Delta R/R$ under various I at 295 K.

evolution for the $\Delta T/T$ at 2.54 eV at 295 K (< *Tc*) and 325 K (~ *Tc*) after the pump pulse excitation, respectively. The pulse energy density P is 3.3 mJ/cm^2. The negative $\Delta T/T$ value indicates the suppression of the spin correlation due to the photo-excited down-spin $e_g\downarrow$electrons via the strong on-site exchange coupling J. We have fitted the relaxation process from the photoinduced spin-disordered state by an exponential function. Solid curves in Fig. 2 (a) are the best-fitted curves. With increasing temperature, the magnitude of the $\Delta T/T$ is steeply reduced, while the relaxation time shows a negligible temperature dependence (~30 ns).

Figure 2 (b) shows the decay curves of the photoinduced $\Delta R/R$ under various E_{exc} at 295 K. The P-value is 1.5 mJ/cm^2 and the applied current I is 0.5 mA. At E_{exc} ~3.20 eV, the $\Delta R/R$ soon reaches maximum (~14%), and then decreases exponentially with time. With decreasing E_{exc}, the peak intensity ($\Delta R_{max}/R$) significantly decreases: ~14% at E_{exc} ~3.20 eV and 3% at E_{exc} ~1.88 eV, while the relaxation time (~30 ns) is nearly independent of E_{exc}. We have found that the $\Delta R_{max}/R$-value steeply increases when E_{exc} exceeds ~ 2.0 eV. Such threshold behavior is originated in the E_{exc} dependence of the spin-direction of the excited e_g electrons. Only the up-spin electrons are photo-excited, when E_{exc} is lower than ~2.0 eV [3], as shown in Fig. 1 (a). Then, the photo-irradiation scenery affects the spin system, since the up-spin e_g electrons cannot create extra spin-waves. With increasing E_{exc} beyond ~2.0 eV, however, the down-spin electrons start to be photo-excited. The down-spin electrons should cause the photoinduced spin-disorder due to the strong Hund coupling and hence the photoinduced enhancement of the resistance, because the effective transfer integral t [= $t_0\cos(\theta/2)$] is reduced in the spin-disordered state.

We further have found that the $\Delta R_{max}/R$-value strongly depends on the applied electric current I. Figure 2 (c) shows the typical $\Delta R/R$ variation with I at 295 K. The P and E_{exc} are 2.5 mJ/cm^2 and 3.05 eV, respectively. With increasing I, the $\Delta R_{max}/R$-value significantly decreases: it decreases from 18% at I = 0.1 mA to 3% at I = 2 mA. We also measured variation of the resistivity ρ for the film with I in absence of photo-excitation, and found that the ρ-value decreases with increase of I. The reduction of the ρ-value with increasing I cannot be ascribed to the conventional Joule heating effect, because the Joule heating effect should increase the ρ-value. As a result, the current-induced decrease of the $\Delta R_{max}/R$-value cannot be due to the heating effect.

Finally, let us discuss the current-induced reduction of photoinduced resistance change $\Delta R_{max}/R$. In the double-exchange doped manganites, the magnetic coupling between Mn ions is induced by the e_g electrons which prefer to hold their spin orientation when hopping among Mn ions in the FM phase. The coupling strength is proportional to the hopping rate of the e_g electron as predicted in the DE mechanism. Therefore, the strong Hund coupling J (~3 eV) forces the spins of e_g electrons to be parallel to the localized spins on the same site. If the nearest-neighboring localized spins are parallel to each other, the e_g electrons could easily move to the neighboring site. Otherwise, a higher energy is need to move the e_g electrons due to the Hund coupling. If the down-spin e_g electrons are photo-excited, the down-spin e_g electrons should cause the disorder of the localized spins due to the strong Hund coupling and hence the increased resistance, as observed. On the other hand, if strong current is applied, the e_g electrons are forced to move forward whether or not the neighboring localized spins are parallel to each other. As a result, the movement of the e_g electrons will produce a tendency to force the localized spins to be parallel and further decreases the resistance. Therefore, the photoinduced resistance change $\Delta R_{max}/R$ shows a decrease with applied current. This scenario also explains the decrease of the ρ-value with current in absent of the photo-irradiation [6]. Because ordering of localized spins may reduce the spin scattering of conduction electrons, a direct consequence of the increase of applied current may be reduction of the spin scattering for the e_g electrons and hence the decrease of ρ-value.

ACKNOWLEDGMENTS

This work was supported by the National Natural Science Foundation of China under grant Nos.10874088, 10874087 and the Key Project of Chinese Ministry of Education under grant No. 107051.

REFERENCES

1. S. Jin, T. H. Tiefel, M. McCormack, R. A. Fastnacht, R. Ramesh and L. H. Chen, *Science* **264**, 413 (1994).
2. K. Matsuda, A. Machida, Y. Moritomo and A. Nakamura, *Phys. Rev. B* **58**, R4203 (1998).
3. X. J. Liu, A. Machida, Y. Moritomo, M. Ichida and A. Nakamura, *Jpn. J. Appl. Phys.* **39**, L670 (2000).
4. Y. Moritomo, A. Machida, K. Matsuda, M. Ichida and A. Nakamura, *Phys. Rev. B* **56**, 5088 (1997).
5. A. K. Debnath and J. G. Lin, *Phys. Rev. B* **67**, 064412 (2003).
6. J. Gao, S. Q. Shen, T. K. Li and J. R. Sun, *Appl. Phys. Lett.* **82**, 4732 (2003).
7. X. J. Liu, Y. Moritomo, A. Nakamura and H. Asano, *Jpn. J. Appl. Phys.* **39**, L1233 (2000).
8. X. J. Liu, Y. Moritomo, A. Machida, A. Nakamura, H. Tanaka and T. Kawai, *Phys. Rev. B* **63**, 115105 (2001).
9. T. Arima, Y. Tokura and J. B. Torrance, *Phys. Rev. B* **48**, 17006 (1993).

Polarized Electrons for Experiments at Low Momentum Transfer SPIN @ S-DALINAC

C. Eckardt*, R. Barday*, J. Enders*, A. Göök*, C. Heßler*, W.F.O. Müller†, Y. Poltoratska*, B. Steiner†, M. Wagner* and T. Weiland†

*Institut für Kernphysik, Technische Universität Darmstadt, Germany
†Institut für Theorie Elektromagnetischer Felder, Technische Universität Darmstadt, Germany

Abstract. The superconducting Darmstadt linear electron accelerator S-DALINAC will be expanded by a newly developed polarized injector concept called SPIN. The polarized beam program includes experiments to search for parity violation in photo induced fission, the investigation of the 5th structure function at low momentum transfer and break up reactions of few body systems. These experiments and an overview of the current experiemental program will be covered in this paper.

Keywords: S-DALINAC, Polarized Electrons, Experiments
PACS: 29.20.Ej, 29.25.Bx, 29.27.Hj, 24.70.+s, 21.10.-k, 21.45.Ff, 25.85.Jg

The S-DALINAC [1] is a recirculating superconducting linear electron accelerator working at an energy range from a few MeV up to 130 MeV. The electron beam is presently produced from a thermionic gun. A complementary laser-driven GaAs electron source will expand the experimental possibilities by a polarized electron beam. As the S-DALINAC is the only accelerator in nuclear physics at this energy range, a polarized beam program might lead to a terra incognita.

CURRENT EXPERIMENTAL PROGRAM AT THE S-DALINAC

Around the S-DALINAC, a multifaceted nuclear-physics program is tailored. The research topics – along with the continuous upgrade of the accelerator – are the core of a center of excellence ("Sonderforschungsbereich") that was installed by the German Research Foundation (Deutsche Forschungsgemeinschaft) about five years ago. It focuses on the properties of nuclei and its constituents at very low momentum transfer using mainly electron- and photon-scattering experiments. For the layout of the S-DALINAC and its experimental sites, we refer to the contribution by Y. Poltoratska [2].

Since the S-DALINAC's commissioning around 1990, nuclear resonance fluorescence experiments are regularly performed behind the superconducting injector at energies between 3.5 and 10 MeV with average beam currents of up to 60 μA. The experimental setup is described in Ref. [3]. Recent emphasis of these photon-scattering experiments was on studying electric dipole excitations below the neutron separation energy where, in the region between 5 and 9 MeV, a concentration of dipole strength is found that is often referred to as "pygmy dipole resonance" (for an overview see, e.g. [4]). The same experimental site is used for photoactivation experiments of type (γ,n). Such measurements provide information on the photodisintegration near the separation energy

CP1149, *Spin Physics, 18th International Spin Physics Symposium*
edited by D. G. Crabb, D. B. Day, S. Liuti, X. Zheng, M. Poelker, and Y. Prok

that is relevant for nucleosynthesis in the astrophysical p (or γ) process [5].

The main linac section of the S-DALINAC may increase the beam energy up to 40 MeV. By recirculating the beam two times, three passes of the linac are possible. The electron beam behind the main linac is presently used for electron-scattering experiments. For that purpose, two electron spectrometers – a high-resolution energy-loss system [6] and a large-acceptance spectrometer of QClam type – are available. While at the former mainly form-factor measurements (e.g. Ref. [7]) are carried out, the latter is used for coincidence experiments such as described in Ref. [8] or single-arm scattering at 180°, see Ref. [9]. At the QClam spectrometer, experiments were recently performed also on very light nuclei. This includes the longitudinal-transverse interference in the deuteron breakup [8] and the precise analysis of the magnetic dipole strength just above the disintegration threshold of the deuteron in the astrophysically relevant energy window by 180° scattering [9].

Two setups provide photons behind the main linac section: (i) a bremsstrahlung site for about 50 – 100 MeV electron beams which is prepared for an experiment on the proton polarizability and (ii) a high-resolution photon tagger for astrophysically relevant photodisintegration and photon scattering studies between 10 and 20 MeV.

EXPERIMENTS WITH POLARIZED BEAMS

The development of a source of polarized electrons [10] and ultimately its installation at the S-DALINAC will lead to polarized beams between 3 and 130 MeV for electron and photon-scattering experiments. While polarized electrons and photons are used at other laboratories world-wide at higher energies, low-energy electron beams at energies below about 100 MeV have – to our knowledge – not been used before.

At the 10 MeV injector linac an experimental investigation of parity violation in nuclear fission is in preparation. Parity violation effects of a few times 10^{-4} have been observed in fission induced by thermal polarized neutrons [11, 12]. Like in investigations on the helicity-dependent transmission of polarized neutrons from various targets (see, e.g. Ref. [13] for an overview), the high level density is used to explain the enhancement with respect to the $\sim 10^{-8}$ level that would be expected from the ratio of the strong and the weak interaction coupling constants. However, open questions on the enhancement mechanism remain, and it would be highly desirable to obtain new data using a different probe. In order to study the feasibility of a photofission experiment with the required accuracy, first test experiments with unpolarized beam using a double ionization chamber with Frisch grids [14] are carried out. Parity violation would result in a fore-aft asymmetry of the light and heavy fission fragments, depending on whether the photon beam has left or right-circular polarization. While the test experiments demonstrated that such an ion chamber may be used in the environment of the S-DALINAC's high-intensity bremsstrahlung beam, the yields indicate that the required count rates cannot be achieved with the present technology. The quality of the data are shown in Fig. 1, where the two coincident anode signals, triggered by the two fission fragments leaving the cathode back to back, are displayed. We are currently considering the development of an active gas target containing ^{238}U. As first steps towards such an experiment, we are collecting precision data on photo-induced fission of ^{234,238}U, and

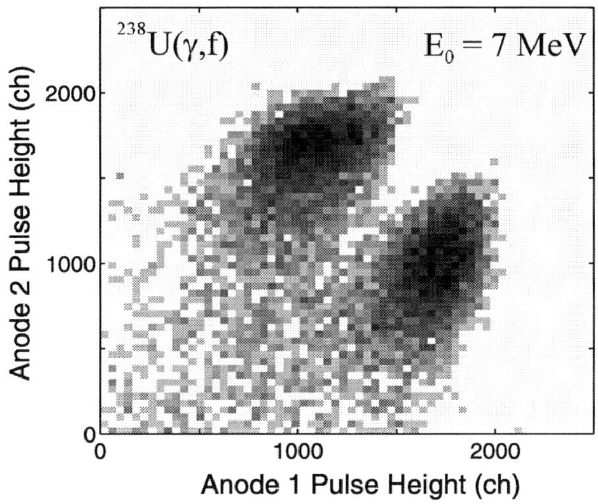

FIGURE 1. Event distribution (anode pulse heights) of ^{238}U fragments in a double ionization chamber for photo-induced fission at a bremsstrahlung endpoint energy of 7 MeV.

tests with UF_6 admixtures to counting gases will start soon.

Reactions on nuclear few-body systems at very low momentum transfers have recently attracted considerable interest due to the advent of an effective field-theoretical (EFT) approach in the form of chiral perturbation theory [15]. Up to 100 MeV nucleon-nucleon (NN) phase shifts are generally well reproduced with few parameters only, and bound-state energies as well as scattering observables are calculable with an accuracy comparable to NN potential models [16]. The energy range provided by the S-DALINAC is ideal for probing chiral perturbation theory. Prototype reactions with the polarized beam comprise the determination of the 5th structure function $f_{LT'}$ (see, e.g. [17]) especially on the deuteron. It is sensitive to the final state interaction and also allows one to test the recently proposed discrepancy of the LT interference term in the deuteron breakup at low q from potential models reported by von Neumann-Cosel and co-workers [8]. The LT' response function has been calculated within potential models for various kinematic regimes [18] and including polarization [19]. Calculations within a Faddeev scheme by Golak [20] along the lines of Ref. [21] indicate that the 5th structure function in the breakup of ^{3}He may be sensitive to the choice of the three-nucleon force. With the advent of polarized electron beams, we expect that the few-body physics program at the S-DALINAC will be strengthened.

In order to successfully analyze the polarization data quantitatively, the determination of the beam polarization close to the experimental sites is mandatory. A summary of the ongoing research and development projects concerning polarimeters are summarized in another contribution [22].

ACKNOWLEDGMENT

The S-DALINAC and this work are supported by the Deutsche Forschungsgemeinschaft through SFB 634. The photofission experiments are carried out in collaboration with S. and A. Oberstedt, M. Chernykh, P. von Neumann-Cosel, and A. Richter, whose support is gratefully appreciated. We thank J. Golak for providing us with preliminary calculations on the ^{3}He 5th structure function. The author likes to thank the SPIN 2008 organizing committee for financial support.

REFERENCES

1. A. Richter, "Operational Experience at the S-DALINAC," in *Proc. of the* 5th *EPAC*, 1996, p. 110.
2. Y. Poltoratska, et al., "'Polarized eletrons in Darmstadt: recent developments"', these proceedings (2009).
3. P. Mohr, et al., *Nucl. Instr. Meth. in Phys. Res. A* **423**, 480 (1999).
4. J. Enders, et al., *Acta Phys. Pol. B* **36**, 1077 (2005).
5. P. Mohr, et al., *Phys. Lett. B* **488**, 127 (2000).
6. T. Walcher, et al., *Nucl. Instr. Meth.* **153**, 17 (1978).
7. O. Burda, et al., *Phys. Rev. Lett.* **99**, 092503 (2007).
8. P. von Neumann-Cosel, et al., *Phys. Rev. Lett.* **88**, 202304 (2002).
9. N. Ryezayeva, et al., *Phys. Rev. Lett.* **100**, 172501 (2008).
10. C. Heßler, *Konzeption, Aufbau und Test einer Quelle spinpolarisierter Elektronen für den supraleitenden Darmstädter Elektronenlinearbeschleuniger S-DALINAC*, Ph.D. thesis, TU Darmstadt (2008).
11. G. Danilyan, et al., *JETP Lett.* **26**, 186 (1977).
12. V. Alfimenkov, et al., *Nucl. Phys. A* **645**, 31 (1999).
13. G. Mitchell, et al., *Phys. Rev.* **354**, 157 (2001).
14. C. Budtz-Jørgensen, et al., *Nucl. Instrum. Methods in Phys. Res. A* **258**, 209 (1987).
15. S. Beane, et al., *arXiv:nucl-th/0008064* (2000).
16. E. Epelbaum, et al., *Phys. Rev. Lett.* **86**, 4787 (2001).
17. T. Donnelly, et al., *Prog. Part. Nucl. Phys.* **13**, 183 (1984).
18. H. Arenhövel, et al., *Phys. Rev. C* **46**, 455 (1992).
19. H. Arenhövel, et al., *Phys. Rev. C* **53**, 1232 (1995).
20. J. Golak, private communication (2006).
21. J. Golak, et al., *Phys. Rep.* **415**, 89 (2005).
22. C. Eckardt, et al., Polarimetry at the Superconducting Darmstadt Electron Linac S-DALINAC, these proceedings (2009).

Measurement of Drell-Yan Process with Polarized Proton Beams at J-PARC

Yuji Goto

RIKEN, Wako, Saitama 351-0198, Japan
RIKEN BNL Research Center, Brookhaven National Laboratory, Upton, New York 11973-5000, USA

Abstract. We have proposed the construction of a spectrometer at J-PARC designed for the detection of high-mass dimuons produced in the interactions of 50 GeV primary proton beams. We have also proposed the acceleration of polarized proton beams at the J-PARC facility. Through the detections of the Drell-Yan events, a broad physics program addressing various issues in QCD is pursued. In the Drell-Yan transverse single-spin asymmetry measurement, we can explore the Sivers effect which is closely related to the orbital angular motion of partons inside the proton. Many other rich programs of spin-dependent structure function physics are carried out through the measurements of unpolarized, singly polarized, and doubly polarized Drell-Yan processes.

Keywords: J-PARC, polarized proton beam acceleration, Drell-Yan process
PACS: 13.88.+e, 14.20.Dh

Drell-Yan process is one of the simplest reactions and it has no QCD reaction in the final-state interaction. One famous experimental result obtained by using this process is a measurement of the flavor asymmetry of antiquark distributions in the nucleon by the FNAL-E866 experiment [1]. A possible origin of the flavor asymmetry of antiquark distributions is a π meson which virtually appear in the nucleon in the meson-cloud model or the chiral-quark model [2]. The π meson is a pseudo-scaler meson and it has a *P*-wave orbital angular momentum in the nucleon. This is another interesting aspect of this result because it is also related to the origin of the proton spin. By measuring the Drell-Yan process with polarized proton beams, we will be able to investigate more spin-structure of the proton.

A transverse-structure measurement with transversely-polarized beams is an important method to understand the proton multidimensionally. It extends our understanding of the collinear structure of the proton, and give us information of orbital angular momentum in the proton and shape of the proton. Frameworks to understand the structure of the proton are transverse-momentum dependent (TMD) distribution functions and generalized parton distribution (GPD) functions. A single-spin asymmetry (SSA) measurement of the Drell-Yan process will give a clear and important measurement of one of the TMD distribution functions, Sivers function [3]. The Sivers function represents a spin-correlated transverse momentum distribution of a parton in the proton, and it is closely related to the orbital angular momentum in the proton. In order to measure the Sivers function from the SSA, the measurement process needs to have an initial-state or final-state interaction with remnant partons in the proton. The deep-inelastic scattering process in the lepton scattering has a final-state interaction and the Drell-Yan process in the hadron reaction has an initial-state interaction with remnant partons in the proton.

CP1149, *Spin Physics, 18th International Spin Physics Symposium*
edited by D. G. Crabb, D. B. Day, S. Liuti, X. Zheng, M. Poelker, and Y. Prok

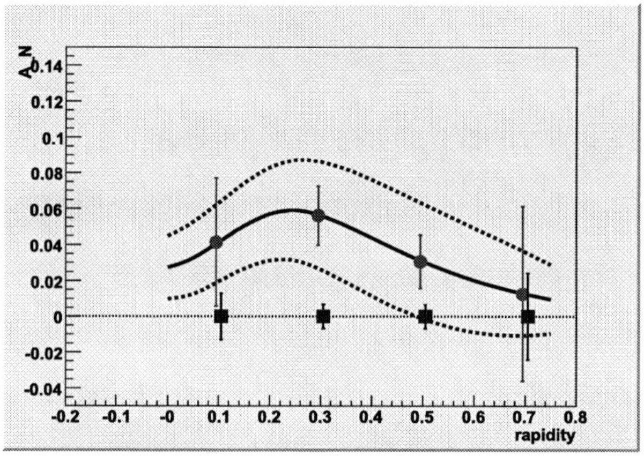

FIGURE 1. Drell-Yan SSA (A_N) integrated over the transverse momentum of the virtual photon and dimuon mass range $4 < M_{\mu^+\mu^-} < 5$ GeV at the 50-GeV J-PARC calculated with Sivers function from HERMES data fit. Expected statistical sensitivities are shown for a 120-day measurement, assuming 75% polarization for a 5×10^{12} ppp polarized proton beam with unpolarized liquid H_2 target (red circles) or nuclear target (blue squares).

Both processes give the Sivers function measurement. The TMD framework predicts the Sivers function measured by both processes should have the opposite sign each other [4]. It is an important and sensitive test of the TMD framework to investigate this property of the opposite sign with use of both the lepton scattering and the hadron reaction.

J-PARC (Japan Proton Accelerator Research Complex) is a high-intensity proton accelerator facility at Tokai in Japan as a joint project of KEK (High Energy Accelerator Research Organization) and JAEA (Japan Atomic Energy Agency). The facility consists of a linac, 3-GeV Rapid-Cycling Synchrotron, and 50-GeV Main Ring. At the Main Ring, the proton beam was successfully accelerated to 30 GeV in December, 2008, and the beam was extracted to the hadron experimental hall in January, 2009. Main scientific goals of the facility are nuclear and particle physics, material and life sciences, and R&D for nuclear transmutation. For the nuclear and particle physics, there are hadron experimental facility and neutrino beam facility to Kamiokande. Phase-1 of the project includes major accelerator components and a part of experimental facilities. At the initial stage of the operation of the Main Ring, 30-GeV proton beams are provided for the slow-extraction beam to the hadron experimental hall and the fast-extracted beam to the neutrino facility. The maximum beam intensity is 3.3×10^{14} ppp (particles per pulse), or 15 μA, while in phase-1, 2×10^{14} ppp, or 9 μA, will be achieved. The maximum beam power is 750 kW, while in phase-1, 270 kW will be achieved.

We have proposed the construction of a spectrometer at J-PARC designed for the detection of high-mass dimuons produced in the interactions of 50 GeV primary proton beams with various targets. Through the detections of the Drell-Yan events, a broad physics program addressing various issues in QCD is pursued. Unique measurements of the flavor asymmetry of antiquark distributions in the nucleon and the antiquark contents in nuclei can be performed over the large Bjorken-x region of $0.25 < x < 0.6$. We have also proposed the acceleration of polarized proton beams to 30-50 GeV at the J-PARC facility. Polarized proton beam acceleration with two partial helical Siberian snakes is shown to be feasible in discussion with J-PARC and BNL accelerator physicists.

By using a transversely polarized beam, the SSA measurement gives the Sivers function. We have proposed to use 5×10^{12} ppp beam with unpolarized liquid H_2 target, or nuclear target. Figure 1 shows the SSA integrated over the transverse momentum of the virtual photon and dimuon mass range $4 < M_{\mu^+\mu^-} < 5$ GeV calculated with Sivers function [5, 6] from HERMES data fit [7]. Expected statistical sensitivities are calculated by PYTHIA simulation for a 120-day measurement assuming 50% duty factor and 75% polarization. Red points show the sensitivity with 51-cm-long (5% interaction length) unpolarized liquid H_2 target located at 127-cm upstream from the entrance of the analyzing magnet. Blue points show that with 20%-interaction length nuclear target located at just downstream of the H_2 target. By these measurements, we can study not only the sign of the Sivers function, but also the shape of the function.

By using both the transversely polarized beam and target, the transversity distribution are measured in the double transverse-spin asymmetry (A_{TT}) measurement of the Drell-Yan process [8]. By using either of the transversely polarized beam or target, the SSA measurement gives the transversity distribution combined with the Boer-Mulders function. This measurement and the Sivers function measurement are distinguished by different angular distribution. The transversity distribution shows a transverse-spin distribution of a parton inside the transversely polarized proton. The Boer-Mulders function shows a correlation between the transverse-spin and the transverse momentum direction of a parton inside the unpolarized proton. The Boer-Mulders function is also measured by the angular distribution of the unpolarized Drell-Yan measurement.

Polarized proton beam at the 50-GeV J-PARC offers an opportunity for probing antiquark polarizations. Despite extensive work on polarized DIS, the helicity distributions of $\bar{u}$ and $\bar{d}$ quarks are still poorly known. Both the SMC [9] and the HERMES [10] experiments attempted to extract the antiquark polarizations via semi-inclusive polarized DIS measurements, and the results indicate small antiquark polarization consistent with zero. Different theoretical models make different predictions. In particular, the meson-cloud models, which successfully describe the unpolarized $\bar{d}/\bar{u}$ asymmetry, predict negligible amount of antiquark polarization [11]. The chiral-quark soliton model, on the other hand, predicts substantial antiquark polarization. A prediction of the chiral-quark model is the large flavor asymmetry of polarized antiquark distribution. This model predicts a significantly larger values for $\Delta\bar{u} - \Delta\bar{d}$ than for $\bar{d} - \bar{u}$ [12]. This prediction and the GRV94 LO parameterization [13] of $x(\bar{d} - \bar{u})$ are shown in the left panel of Fig. 2. The right panel of Fig. 2 shows the x_2 dependence of A_{LL}^{DY}, integrated over the spectrometer acceptance, with calculations using polarized PDF parameterization of G-S (set A and C) [14] and GRSV [15]. The statistical accuracy of such a measurement can test the predictions of various model. Note that the chiral-quark soliton model predicts a large positive A_{LL}^{DY} not shown in this figure.

A rich program of spin physics can be carried out through the measurements of unpolarized, singly polarized, and doubly polarized Drell-Yan processes. Measurement of the orbital angular momentum component in the nucleon is one of the most important goal of the spin physics program of the J-PARC dimuon experiment.

The author would like to thank people on the proposals, especially spokespersons

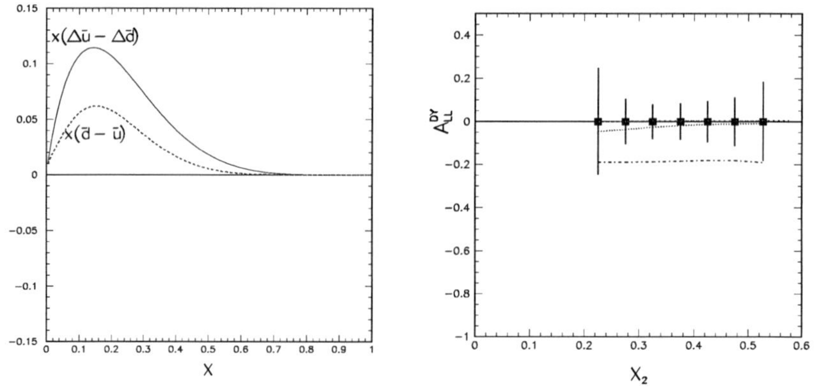

FIGURE 2. Left: $x(\Delta\bar{u} - \Delta\bar{d})$ predicted by the chiral-quark soliton model is shown as the solid curve. The GRV94 LO parameterization of $x(\bar{d} - \bar{u})$ is shown as the dashed curve. Right: Expected statistical accuracy for measuring A_{LL}^{DY} in polarized $p + p$ Drell-Yan at the 50-GeV J-PARC for a 120-day measurement, assuming 75% polarization for a 5×10^{11} ppp polarized proton beam and a polarized solid NH_3 target of 75% with a dilution factor of 0.15. The dashed, dotted, and dash-dotted curves correspond to calculations using polarized PDF parameterization of G-S (set A and C) and GRSV, respectively. The chiral-quark soliton model gives large positive A_{LL}^{DY}.

J. C. Peng, S. Sawada, and H. Sato, and is grateful to F. Yuan, and W. Vogelsang for discussions and theory calculations.

REFERENCES

1. E. A. Hawker et al., *Phys. Rev. Lett.* **80**, 3715 (1998).
2. S. Kumano, *Phys. Rept.* **303**, 183 (1998); G. T. Garvey and J. C. Peng, *Prog. Part. Nucl. Phys.* **47**, 203 (2001).
3. D. W. Sivers, *Phys. Rev.* **41**, 83 (1990).
4. S. J. Brodsky, D. S. Hwang and I. Schmidt, *Phys. Lett.* **530**, 99 (2002); *Nucl. Phys. B* **642**, 344 (2002); J. C. Collins, *Phys. Lett. B* **536**, 43 (2002); X. Ji and F. Yuan, *Phys. Lett. B* **543**, 66 (2002); A. V. Belitsky, X. Ji and F. Yuan, *Nucl. Phys. B* **656**, 165 (2003); D. Boer, P. J. Mulders and F. Pijlman, *Nucl. Phys. B* **667**, 201 (2003).
5. X. Ji, J. Qiu, W. Vogelsang and F. Yuan, *Phys. Rev. Lett.* **97**, 082002 (2006): *Phys. Rev. D* **73**, 094017 (2006).
6. F. Yuan and W. Vogelsang, private communications.
7. W. Vogelsang and F. Yuan, *Phys. Rev. D* **72**, 054028 (2005).
8. V. Barone, A. Drago and P. G. Ratcliffe, *Phys. Rept.* **359**, 1 (2002).
9. B. Adeva et al., *Phys. Lett. B* **420**, 180 (1998).
10. K. Ackerstaff et al., *Phys. Lett. B* **464**, 123 (1999).
11. K. G. Boreskov and A. B. Kaidalov, *Eur. Phys. J. C* **10**, 143 (1999): R. J. Fries and A. Schafer, *Phys. Lett. B* **443**, 40 (1998).
12. K. Goeke, P. V. Pobylitsa, M. V. Polyakov and D. Urbano, *Nucl. Phys. A* **680**, 308 (2001).
13. M. Gluck, E. Reya and A. Vogt, *Z. Phys. C* **67**, 433 (1995).
14. T. Gehrmann and W. J. Stirling, *Phys. Rev. D* **53**, 6100 (1996).
15. M. Glück, E. Reya, M. Stratmann and W. Vogelsang, *Phys. Rev. D* **53**, 4775 (1996).

Precision Spectroscopy of Hydrogen with a Lamb-Shift Polarimeter

R. Engels*, K. Grigoryev*,†, M. Mikirtytchyants*,†, H. Paetz gen. Schieck**, G. Schug*, H. Ströher*, A. Vasilyev† and M. Westig‡

*Institut für Kernphysik, FZ Jülich, Leo-Brandt-Str. 1, 52425 Jülich, Germany
†St. Petersburg Nuclear Physics Institute, Orlova Rosha, 188350 Gatchina, Russia
**Institut für Kernphysik, Universität zu Köln, Zülpicher Str. 77, 50937 Köln, Germany
‡I. Physikalisches Institut, Universität zu Köln, Zülpicher Str. 77, 50937 Köln, Germany

Abstract. A spinfilter, the most important component of a Lamb-shift polarimeter, is used to produce a beam of metastable hydrogen (deuterium) atoms in one hyperfine state (HFS)($\alpha 1$, $\alpha 2$ and together with the Sona transition $\beta 3$). As function of a magnetic field separated transitions between the $2S_{1/2}$ metastable Zeeman states seem to be observable as well as single transitions into the short-lived $2P_{1/2}$ and $2P_{3/2}$ states. The Breit-Rabi diagrams for these states and, therefore, the g factors can be measured with good precision. Furthermore, the hyperfine splittings and the Lamb shift can be observed as well. Application of this method to anti-hydrogen atoms is suggested.

Keywords: Lamb shift, Hyperfine Splitting, Spinfilter, Breit-Rabi Diagram, g-factor
PACS: 37.20.+j, 37.30.+i, 31.15.aj, 31.30.Gs, 31.30.J-, 32.10.Fn- 32.60.+i, 33.15.Pw

INTRODUCTION

In 1952 Rabi [1] suggested that the *Atomic Beam Resonance Method* [2] could be used to study also the excited states of atoms. This was realized in 1955 by Heberle et al. [3] with a modified setup for the metastable hydrogen state $2S_{1/2}$ in vanishing magnetic fields and in 1956 by Senitzky and Rabi [4] for the alkaline atoms. When Lamb and Retherford [5] measured the Lamb shift, they were not able to separate the different hyperfine levels of the $2S_{1/2}$, $2P_{1/2}$, and the $2P_{3/2}$ fine structure states completely. Since then, different methods have been applied to measure the hyperfine splittings and the Lamb shift of the excited states ($n = 2$) of the hydrogen atom near $B \sim 0$ G [6], [7], [8], [9]. However, the complete Breit-Rabi diagrams [10], i.e. the binding energies of the single Zeeman states with $n = 2$ as a function of the magnetic field, have not yet been measured. On the other hand, the binding energies and, therefore, the g-factors of the single Zeeman states, the Lamb shift and the hyperfine splittings, can be calculated as a function of the magnetic field with bound-state QED [12]. However, the accuracy of these calculations is limited by the uncertainties of a number of parameters. For example, the charge radius of the proton must be taken into account. With a fit to precision data these values could be determined.

CP1149, *Spin Physics, 18th International Spin Physics Symposium*
edited by D. G. Crabb, D. B. Day, S. Liuti, X. Zheng, M. Poelker, and Y. Prok

MEASUREMENT OF THE $2S_{1/2}$ HYPERFINE TRANSITIONS

With a spinfilter [13], the most important component of a Lamb-shift polarimeter [14], it is possible to produce an intense beam of hydrogen atoms in the ground state $1S_{1/2}$ and in single metastable HFS $\alpha 1$, $\alpha 2$ or $\beta 3$ of the $2S_{1/2}$ state only. This can be used to establish a setup similar to that of the *Atomic Beam Resonance Method* (Fig. 1). In our setup, the *analyzing magnets* of the Rabi apparatus are replaced by spinfilters. In an electron-impact ionizer H_2 molecules are dissociated and ionized. The energy of the produced protons is between 300 and 2000 eV, and a beam intensity up to 10 μA can be achieved. After deflection to a horizontal beam direction, a Wien filter is used to separate the protons from the other ions, which originate from the residual gas. The width of the proton-velocity distribution is minimized by two diaphragms at both ends of the Wien filter. By charge exchange with cesium vapour [15], metastable hydrogen atoms in the state $2S_{1/2}$ are produced from about 15% of the protons. In the first spinfilter all metastable atoms except those in one HFS are quenched into the ground state and only metastable atoms in the HFS $\alpha 1$ or $\alpha 2$ remain in the beam. In a homogeneous magnetic field, radio frequency (rf) transitions are induced with the electric field vector parallel to the velocity direction of the hydrogen atoms in order to avoid longitudinal Doppler shift and broadening. The direction of the magnetic field, produced by two Helmholtz-like coils ($B \leq 100$ G), can be aligned parallel or perpendicular to the velocity direction of the atoms. Thus, for a transverse magnetic field, the transitions $\alpha 1 \rightarrow \alpha 2$, $\alpha 1 \rightarrow \beta 4$, $\alpha 2 \rightarrow \alpha 1$ and $\alpha 2 \rightarrow \beta 3$ can be measured. Together with the Sona transition [16] behind the first spinfilter it is possible to exchange the population numbers of the HFS $\alpha 1$ and $\beta 3$. Therefore, the transition $\beta 3 \rightarrow \beta 4$ can be measured, too. In the setup, shown in figure 1, the first spinfilter is used to depopulate the HFS $\alpha 2$, $\beta 3$ and $\beta 4$, so that only metastable atoms in the HFS $\alpha 1$ are found in the beam. With the magnetic field in the transition unit perpendicular to the beam direction the rf is swept across the resonance to induce transitions from the HFS $\alpha 1$ to the HFS $\alpha 2$. A subsequent spinfilter is tuned to transmit only atoms in the state $\alpha 2$. After quenching the metastables into the ground state by the Stark effect, the intensity change is detected by counting the Lyman-α photons with a photomultiplier in the quenching region as a function of the rf to find the resonance frequency.

Estimated Results and Errors

Due to the long lifetime (0.14 s) of the metastable $2S_{1/2}$ state [17], the natural half-width of the resonance is about 1.1 Hz (4.6×10^{-15} eV). Because of the short time of the fast metastable atoms in the small interaction region and the Heisenberg uncertainty principle, a broadening of the resonance peak up to 40 MHz is expected to be the dominant source of the experimental error, which will be around $\Delta f = 10$ kHz for a count rate of more than 10^5 photons/s in the photomultiplier. The longitudinal Doppler effect is suppressed in first order, because the beam direction of the metastable atoms is perpendicular to the rf wave vector. Second order effects of the Doppler shift (caused, e.g., by misalignment of the beam axis) can be measured by rotating the rf

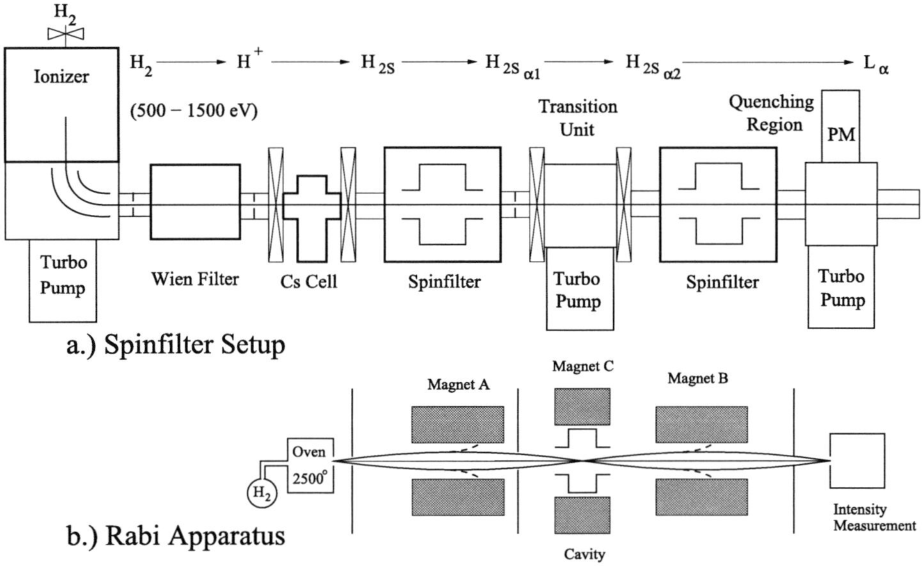

FIGURE 1. **a.)** The setup with two spinfilters for the measurement of the $2S_{1/2}$ Breit-Rabi diagram. **b.)** The analogous setup of the Rabi apparatus for the $1S_{1/2}$ HFS.

wave vector. The Doppler broadening of the resonance peak will be small due to the velocity distribution of the incoming metastable atoms of about 1%. The transverse Doppler shift will be around 200 Hz for a beam energy of 1 keV and around 50 Hz for 500 eV and can be measured by varying the beam energy. With measured frequencies at 100 different magnetic fields up to 100 G a relative error of 10^{-6} of the slope of the energy levels and, therefore, of the g-factor seems to be feasible. With an upgrade of the magnetic field up to 1000 G and decreasing of the frequency errors down to $\Delta f = 1$ kHz a relative error of 10^{-8} may be possible. A strong advantage of this method as compared to other measurements is the possibility to measure the hyperfine splitting of the $2S_{1/2}$ state independent of the magnetic field. The combinations of the transition frequencies $(\alpha 1 \rightarrow \beta 4) - (\alpha 2 \rightarrow \beta 3)$ and $(\alpha 1 \rightarrow \alpha 2) + (\beta 3 \rightarrow \beta 4)$ result in the values of the $2S_{1/2}$ hyperfine splitting for every magnetic field.

MEASUREMENT OF THE $2P_{1/2}$ HYPERFINE TRANSITIONS

Lamb and Retherford [5] proposed in their famous paper to induce the transitions from the metastable HFS into the HFS of the short-lived $2P_{1/2}$ state ($\tau = 10^{-9}$ s) with variable rf and a stable power level at a constant magnetic field. But at that time this was not possible because of technical constraints. Today, however, rf generators for the 1 GHz range with a precision better than 1 Hz and stable amplitude are available. Inducing transitions at various frequencies and constant rf power can be achieved by the use of a Lecher (TEM) transmission line. The TEM waveguide of a constant characteristic impedance transports the rf through the vacuum and at the opposite end the amplitude

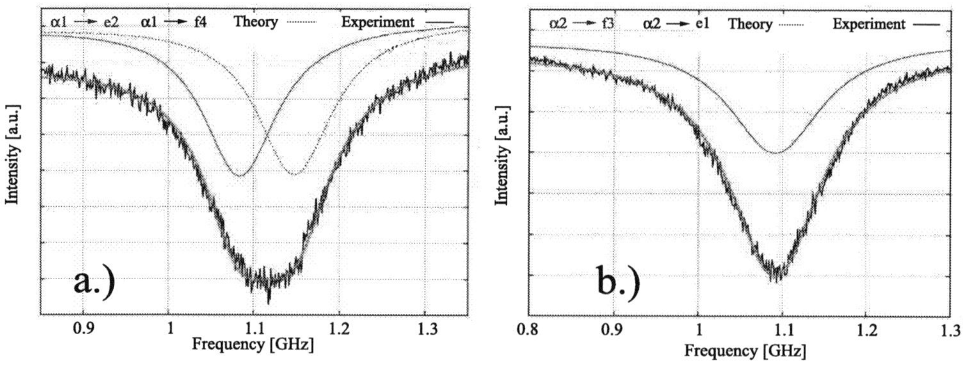

FIGURE 2. The observed transitions $\alpha 1 \rightarrow f4$ and $\alpha 1 \rightarrow e2$ (a.) or $\alpha 2 \rightarrow f3$ and $\alpha 2 \rightarrow e1$ (b.) at a small vertical magnetic field ($B \sim 0.5$ G) in the transition region.

can be controlled. Through holes in the conductors around the beam line the metastable atoms can reach the volume between the inner and the outer conductor of the TEM waveguide where the transitions are induced. With a proper design it is guaranteed that the electric rf field is homogeneous and its direction is parallel to the velocity vector of the hydrogen beam. In this case, the rf wave vector must be perpendicular to the atomic beam and the Doppler effect can be suppressed in first order. Through a hole in the outer conductor the Lyman-α photons produced by the decay of the $2P_{1/2}$ states are detected in a photomultiplier.

First Results

In a first proof-of-principle measurement [18] metastable hydrogen atoms in the HFS $\alpha 1$ or $\alpha 2$ are selected in the spinfilter and reach the TEM waveguide (Fig. 2). For a small transverse magnetic field about 0.5 G the transitions $\alpha 1 \rightarrow f4$ and $\alpha 1 \rightarrow e2$ (a.) or $\alpha 2 \rightarrow f3$ and $\alpha 2 \rightarrow e1$ (b.) are observed. The frequency difference of the two resonances of the transitions $\alpha 1 \rightarrow f4$ and $\alpha 1 \rightarrow e2$ in Fig. 2 a.) corresponds to the hyperfine splitting energy of the $2P_{1/2}$ state. The preliminary result from a fit ist 59.98 (2.03) MHz, which agrees with the prior result of 59.22 (14) MHz [8]. The transition frequencies from the $2S_{1/2}$ into the $2P_{1/2}$ state together with the HFS of these states allow to calculate a value of the classical Lamb shift. As a first result a value of 1057.34(1.11) MHz could be determined.

DISCUSSION

The dominant error in the first measurement was the inhomogeneity of the magnetic field and the uncertainty in the definition of the field direction. For example, an inhomogeneity of $\Delta B = 0.5$ G corresponds to an error of $\Delta f = 1$ MHz for the resonance frequency. A new setup with a set of Helmholtz coils to produce an inhomogeneity of $\Delta B/B = 10^{-4}$ and a much better shielding against external stray fields is in preparation. In addition, an error of $\Delta f = 100$ kHz is produced by the instability of the rf power of 3%. With available hardware this instability can be reduced to 0.002% or less. Because of the Heisenberg uncertainty relation $\Delta t \times \Delta W \sim h$ and the short transit time of the metastable atoms through the interaction region of $\Delta t = 4.4 \times 10^{-7}$ s (at 1 keV beam energy) the resonance half width will be increased from 100 to 109 MHz for the measurements of the transitions into the $2P_{1/2}$ states. For the transitions within the $2S_{1/2}$ states this effect will dominate the broadening to about 40 MHz. In addition, other effects like the velocity distribution of the incoming metastable hydrogen beam and the inhomogeneity of the magnetic field will influence the shape of the resonance peak. A large count rate of 10^7 photons/s in the multiplier allows very precise measurements of this shape. Several upgrades like a more efficient Lyman-α detection system and a larger proton beam after the ionizer will increase this number to 10^9 photons/s. Nevertheless, this deformation of the peak will result in an error of about $\Delta f = 100$ kHz. Instead of directly induced transitions it is possible to use the *separated-oscillatory-field* technique used by Ramsey and others [8] to overcome the large half-width of the resonance peak and to decrease the error. Until now no influence of the different Doppler shifts (see section 2.1) is observed. Other effects like the Stark shift caused by electric fields and the motional Stark effect can be neglected so far. An error of $\Delta f = 100$ kHz for a single measurement will lead to a relative error of 10^{-5} and after an upgrade 10^{-7} for the g-factors of the $2P_{1/2}$ states. Like the hyperfine splitting of the $2S_{1/2}$ state the hyperfine splitting of the $2P_{1/2}$ state can be measured by a combination of different transitions independent of the magnetic field ($f_{HFS_{(2P_{1/2})}} = (\alpha 1 \to f4) - (\alpha 2 \to f3) - (\alpha 1 \to e1) + (\alpha 2 \to e2)$). Therefore, it is possible to perform a large number of independent measurements at different magnetic fields and to average these results to decrease the error from 140 kHz [8] down zu 10 kHz. With a modified TEM waveguide, transitions from the $2S_{1/2}$ HFS into the $2P_{3/2}$ HFS can be induced and the hyperfine splitting of the $2P_{3/2}$ state can be measured. Those transitions are around 10 GHz. For deuterium the spinfilter allows to separate three metastable states $\alpha 1$, $\alpha 2$ and $\alpha 3$ which correspond to the three different nuclear spin substates with I=+1,0,-1. Therefore, a higher number of transitions can be measured and the results can be compared with hydrogen within the D_{21} theory. A spinfilter can be designed for other atoms, too. For example, a spinfilter for ^{3}He is in preparation. Before cold antihydrogen cannot be produced for 2-photon laser-spectroscopy experiments this method can also be applied to antihydrogen beams. During the recombination of the antiproton and the positron 30% of the produced antihydrogen atoms end up in the $2S_{1/2}$ metastable state. With a much better Lyman-α detection system 10 antihydrogen atoms per s will produce a count rate of 1 photon/s in the photomultplier.

REFERENCES

1. I.I. Rabi, *Phys. Rev.*, **87**, 379 (1952).
2. I.I. Rabi et al., *Phys. Rev.*, **55**, 526 (1939).
3. J.W. Heberle et al., *Phys. Rev.*, **101**, 612 (1955).
4. B. Senitzky and I.I. Rabi, *Phys. Rev.*, **103**, 315 (1956).
5. W.E. Lamb and R.C. Retherford, *Phys. Rev.*, **79**, 549 (1950).
6. T.W. Hänsch, *Rev. Mod. Phys.*, **78**, 1297 (2006).
7. N. Kolachevsky et al., *Phys. Rev. Lett.*, **92**, 033003 (2004).
8. S.R. Lundeen et al., *Phys. Rev. Lett.*, **34**, 377 (1975).

9. C. Schwob et al., *Phys. Rev. Lett.*, **82,** 4960 (1999).
10. G. Breit and I.I. Rabi, *Phys. Rev.*, **38**, 2082 (1931).
11. W.E. Lamb and R.C. Retherford, *Phys. Rev.,* **81**, 222 (1951).
12. D.L. Moskovskin and V.M. Shabaev, *Phys. Rev., A* **73**, 1 (2006).
13. J.L. McKibben et al., *Phys. Lett.*, **28B**, 594 (1969).
14. R. Engels et al., *Rev. Sci. Instr.*, **74**, 11 4607 (2003).
15. P. Pradel et al., *Phys. Rev. A*, **10**, 797 (1974).
16. P.G. Sona, *Energia Nucleare*, **14**, 295 (1968).
17. G. Breit and E. Teller, *Astrophys. J.*, **91**, 215 (1949).
18. M. Westig, Dipl. Thesis, University of Cologne/FZ Jülich (2008).

PHENIX Detector Upgrade for Triggering Fast Muons from W-Boson Decays Using RPC Technology

Xiaochun He (for the PHENIX Collaboration)

Georgia State University, 29 Peachtree Center Avenue, Suite 400, Atlanta, GA 30303, USA

Abstract. The BNL PHENIX experiment is designed for studying matter properties at ultra high temperatures and densities with heavy ion collisions, and the spin structure of the proton via polarized proton+proton collisions. A fast trigger system, based on the resistive plate chamber (RPC), is being developed for the muon arms that allows the measurement of the flavor structure of quark polarization in the proton through observation of W-bosons in proton+proton collisions at $\sqrt{s} = 500$ GeV. This talk will focus on RPC R&D, production of the RPC components, and assembly of the RPC modules at BNL. Challenges of RPC integration into PHENIX will also be presented.

Keywords: W-boson, Proton spin structure
PACS: 14.20.Dh, 14.60.Ef, 14.65.Bt, 14.70.Fm

INTRODUCTION

Parity violating W-boson production in polarized proton+proton collisions at the Relativistic Heavy Ion Collider (RHIC), at Brookhaven National Laboratory (BNL), probes quark and anti-quark helicity distributions in the proton. The measurements of single longitudinal spin asymmetry $A_L(p_T)$ for leptons in W production will provide a new elegant tool to directly measure the quark helicity distributions, $\Delta u(x,Q^2)$, $\Delta d(x,Q^2)$, $\Delta \bar{u}(x,Q^2)$, and $\Delta \bar{d}(x,Q^2)$ at a high scale, $Q^2 = 6400$ GeV^2/c^2, which is independent of quark fragmentation function information in comparison with other spin-related measurements at RHIC [1].

The PHENIX experiment was initially designed for studying the matter properties at extremely high density and temperature created from heavy ion collisions at RHIC. The PHENIX detector system was optimized for heavy ion physics and proton+proton comparison running at luminosity up to $L = 1 \times 10^{31}$ cm^{-2}s^{-1}. The existing online trigger does not have sufficient rejection power to sample the rare leptons from W decays at the high luminosity.

The experimental challenges for W measurement arise from extending the existing muon triggering capability to a proton+proton collision rate of 9.6 MHz at $\sqrt{s} =$ 500 GeV, and from using the PHENIX muon spectrometer for tracking muons above $p_T = 20$ GeV/c, where the W signals become dominant, as shown in Fig. 1. Additionally, the trigger detector must be able to reject dominant muons from hadronic decays and be able to fit into the PHENIX experimental aperture.

The muon trigger upgrade has two components: (1) RPC stations to be inserted in the PHENIX muon spectrometers; and (2) new trigger front-end electronics which are

CP1149, *Spin Physics, 18th International Spin Physics Symposium*
edited by D. G. Crabb, D. B. Day, S. Liuti, X. Zheng, M. Poelker, and Y. Prok

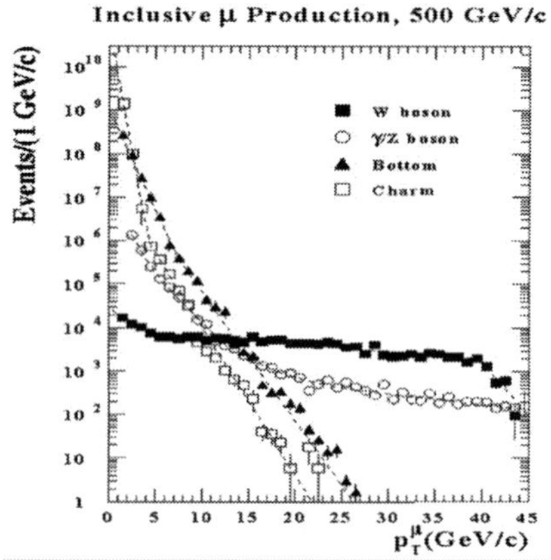

FIGURE 1. Transverse momentum distributions of muons from W, γ/Z, bottom and charm decays.

added to the existing muon tracker electronics. The production and installation of the muon tracker trigger electronics are presented in this proceeding volume by Tsutomu Mibe for the PHENIX collaboration.

FAST MUON TRIGGER DETECTOR UPGRADE - RPC

The PHENIX Forward Muon Trigger RPC has a standard double gas gap structure, which is similar to the RPC's of the Compact Muon Solenoid (CMS) experiment at the Large Hadron Collider at CERN. The main design requirements are the following:

- Fast signal and good timing resolution (less than 3 ns at > 95 % efficiency).
- Rate capability above 500 Hz/cm^2.
- Radiation hardness needed for the RHIC environment.
- Limited space for installing the detector.
- Low cost and relatively easy to construct.

RPC Technology

The cross-sectional view of an RPC gas gap is shown in Fig. 2. The gap is built with two parallel plates (of 2 mm thickness) of high resistivity material, like phenolic polymer (bakelite) or glass. The plates have a resistivity on the order of 10^{10} Ω·cm and serve as electrodes. The gap is typically 2 mm wide between the two plates which are glued together using side strips and spacers. The side strips and spacers are made of polycarbonate material. Two thin gas tubes are inserted into the gap from the opposite corners of the gap as the inlet and outlet.

The outside surfaces of the RPC gap are coated with graphite (via spraying or silk-painting) for distributing high voltage on one side and for electric grounding connection

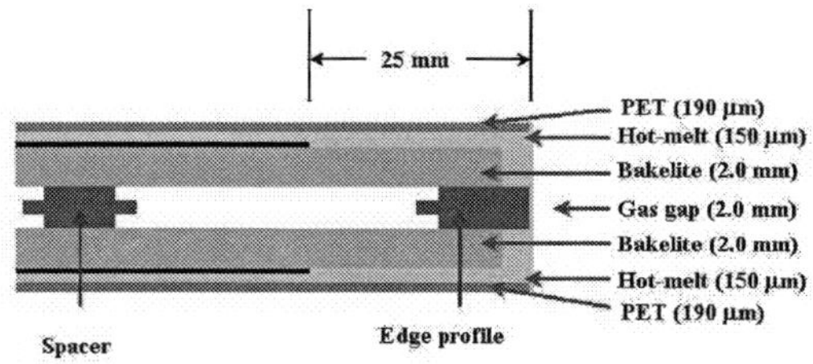

FIGURE 2. Cross-sectional view of an RPC gas gap.

on the other. The graphite covered area is indented inward from the sides of the gap for about 2 cm in order to prevent surface charge flowing from one side of the surface to the other. The surface resistivity of the graphite area is about couple of hundred kΩ per unit square.

The readout plane is typically made of copper strips or pads. This is one of the very attractive features of using an RPC where the readout is completely isolated from the gas environment. The gas mixture flows through 2 mm gas gap where a strong electric field is present. An avalanche is created by an ionizing particle traversing the gas volume. The drift of the charge in the avalanche within the gas volume then induces a pick-up current on the readout strips (or pads), which is read out by the RPC readout electronics. The signal charge has a very large dynamic range from 20 fC to 20 pC.

Production of the RPC Detector

The PHENIX RPC construction is a truly international operation. The bakelite sheets were made in Italy at the same factory where the CMS bakelite sheets were produced. The sheets were then inspected both by the factory technicians and representatives from PHENIX who traveled to Italy. The sheets which passed the inspection (mainly checking the uniformness of the volume resistivity of the bakelite) were shipped to another Italian factory for cleaning. The cleaned sheets were then cut to the proper sizes with the PHENIX specific design, and finally shipped to Korea University in Seoul, Korea.

The Nuclear Physics group at Korea University has been making RPC gaps for CMS for several years with a well established procedure and a sophisticated, expensive machine. The technicians at the university have extensive expertise for making RPC gaps. Once the bakelite sheets arrived at Korea University, they were inspected for any potential damages over the course of the shipment. The major steps of making a gas gap were: graphite coating, gluing bakelite sheets with side strips and spacers, and coating the gap inner surfaces with linseed oil. The completed gaps were further tested in Korea before shipping to BNL for assembly.

Another important component of an RPC module is the frame box which holds the gas gaps in place for sturdiness. The frame box also plays the role as a Faraday cage to shield external E&M noises. The frame side bars are made of aluminum. The top and bottom plates are made of aluminum honeycomb panels to reduce the overall weight of the

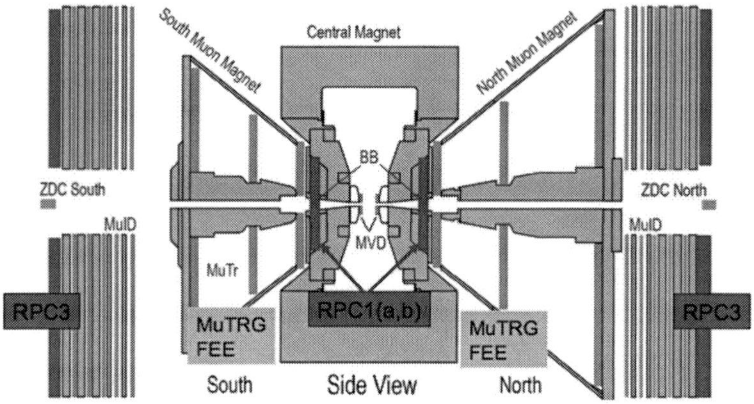

FIGURE 3. Integration RPC stations to the existing PHENIX aperture.

detector module. The parts of the frame boxes and the readout planes were manufactured in China, supervised by the PHENIX collaborators at CIAE in China.

The assembling of RPC detector modules is ongoing at the RPC factory at BNL. A special space has been allocated for quality assurance tests which includes gas leakage, dark current, and popped spacers for all gas gaps received. After the module is assembled, it is then put on a cosmic ray test stand for efficiency measurement.

Integration and Schedule

There will be two RPC stations (RPC1 and RPC3) in each of the PHENIX muon arms, as shown in Fig. 3, covering the same angular acceptance as the existing muon tracking stations. RPC1 is divided into octants. In each octant, there are three RPC gas gaps. Two small gaps are overlaid on top of a bigger gap at the bottom. The readout planes are sandwiched between the top smaller gaps and the bottom gap, which have sufficient angular segmentation for triggering purposes. For the RPC3 station, there will be 16 half-octants. Each half-octant consists of three RPC modules.

Currently, two half octants of prototype RPC's have been installed in the PHENIX experiment, and are ready for data taking during RHIC Run-9 period. The mass production started in December of 2008. The north arm will be ready for W-physics in the fall of 2009 for Run-10. The south arm will be ready in the fall of 2010 for Run-11.

REFERENCES

1. PHENIX Forward Trigger Upgrade CDR: http://www.phenix.bnl.gov/WWW/trigger/forwardupgrade/

Calorimeter Upgrade to the PHENIX Forward Spectrometers

Kenneth N. Barish for the PHENIX collaboration

University of California at Riverside, Department of Physics and Astronomy, Riverside CA 92521, USA

Abstract. The forward calorimeter upgrade to the PHENIX forward spectrometers aims to add capabilities at high rapidity ($0.9<\eta<3.0$) in order to significantly extend the kinematic reach of PHENIX's ΔG measurement for the prompt photon and photon-jet channels, study transverse spin physics through forward hadron asymmetries, study nucleon structure in nuclei at high parton densities in p+A collisions through the measurement of photons and neutral pions in the forward region, and enhance our quark and anti-quark spin measurements via W-production by providing an isolation cut and possibly providing an electron measurement at forward rapidities.

Keywords: Proton structure, Proton spin, RHIC, PHENIX, Forward Calorimeter
PACS: 14.20.Dh, 13.85.Ni, 13.88.+e

INTRODUCTION

The PHENIX forward calorimeter (FOCAL) upgrade expands PHENIX's kinematical coverage for jets, inclusive neutral pions, electrons, and photons (Fig.1) to forward rapidity. The detector will extend PHENIX's spin structure program to low-x with measurements of A_{LL} and A_N with γ, jets, h, and π^0, γ-jet. The feasibility of an $A_L(W\rightarrow e)$ measurement is also being evaluated. The detector requirements for these measurements include a γ/π^0 separation to p_t~15GeV/c, an angular resolution < .1 rad, and a resolution of ~150μm in ϕ.

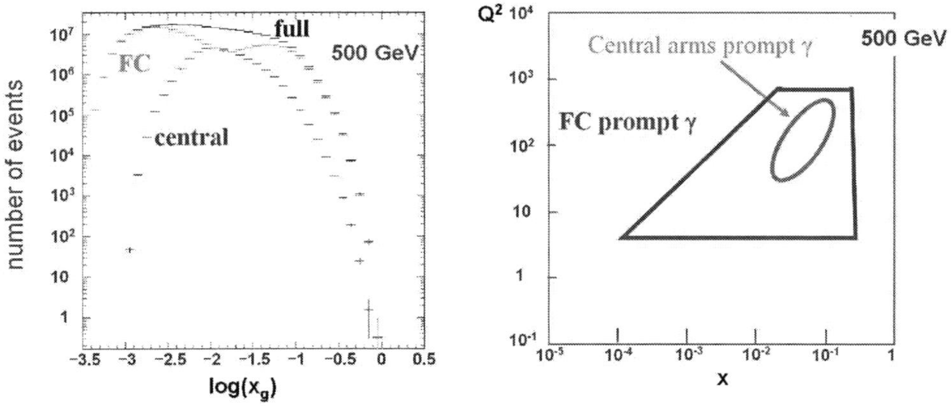

FIGURE 1. Kinematical coverage of the PHENIX FOCAL (FC) and central detectors for photons in 500 GeV p+p collisions. Statistics are for one year of running at RHIC II luminosity.

CP1149, *Spin Physics, 18th International Spin Physics Symposium*
edited by D. G. Crabb, D. B. Day, S. Liuti, X. Zheng, M. Poelker, and Y. Prok

PHENIX has limited space for a forward calorimeter. The FOCAL needs to be located 40 cm for the collision point and can be only 20 cm long, covering $0.9 < |\eta| < 3.0$. These constraints dictate a silicon pixel design interleaved with tungsten. The tungsten gives a short radiation length and large absorption/radiation length ratio limiting electromagnetic shower development to the first few cm of absorber. Several 500 μm silicon strip layers aid in the separation of photon showers from π^0s and add points for tracking muons. The detector is required to make good photon measurements, reasonably good jet measurements, and have triggering capability. The calorimeter will have a Molière radius of 1.4 cm, be ~42 radiation lengths and 1.6 absorption lengths deep, and have a lateral segmentation of 1.5cm x 1.5cm.

ΔG MEASUREMENTS AT LOW X

Prompt photon production provides the most direct access to the gluon polarization, as it is dominated by the gluon-Compton process. Present measurements indicate that ΔG is small in the kinematic range of the central region [1]. However, these measurements are not sensitive to low-x and do not rule out even very large values of ΔG. For example, GS-C [2], a parameterization of the gluon structure that is consistent with all experimental data, has a very large value of ΔG (Fig.2).

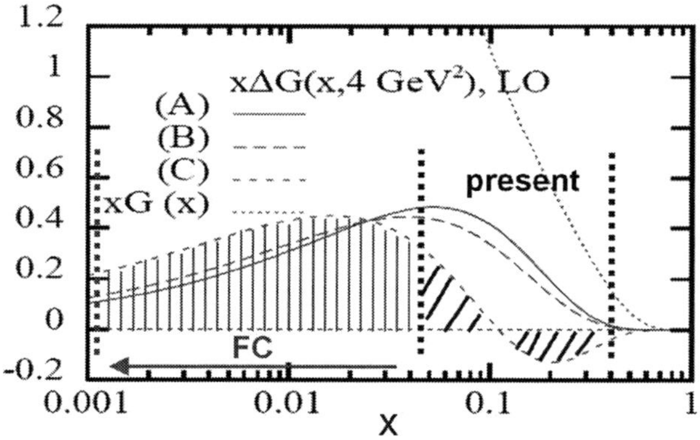

FIGURE 2. Parameterizations of the gluon structure function with large contributions at low-x cannot be distinguished without forward measurements.

The estimated sensitivity to A_{LL} is shown in Fig.3. We find that GS-C is clearly above 0, in contrast to the central arm sensitivity. This figure is at the Pythia level, but we have also performed a simulation where we have fully reconstructed the photons including an isolation cut, and the error bars only degrade slightly (Fig.4a). However, this plot also does not take into account a considerable background from π^0's. A factor of three improvements in the π^0/γ separation compared to our initial studies are needed. The π^0/γ separation is being improved: the isolation cut is being tuned for hadron rejection without a spectrometer in front. And the strip layer reconstruction algorithm, which is a highly correlated multi-dimensional problem, is being optimized.

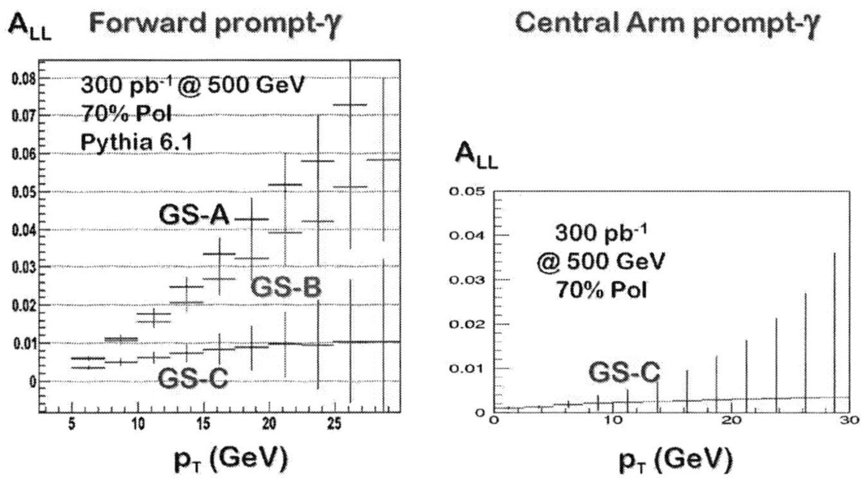

FIGURE 3. Analyzing power of prompt photons in the FOCAL compared to the central arms for one year of running at 500 GeV and 70% polarization.

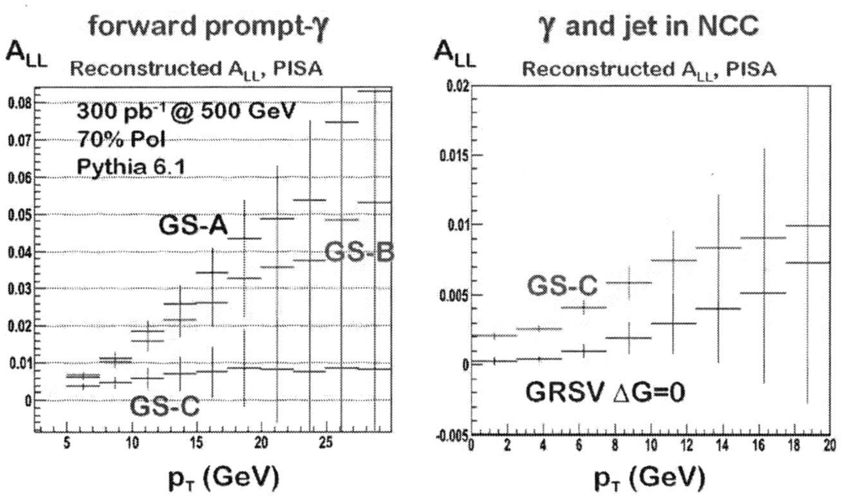

FIGURE 4. (a) Reconstructed version of Fig.3(a); (b) analyzing power of photon-jet A_{LL} in the FOCAL (labeled NCC).

The FOCAL will also allow for the determination of x_G of the gluon on an event-by-event basis by facilitating the detection of both hadron jet and final state photon (in conjunction with new central arm silicon vertex tracking detectors). We have evaluated the jet reconstruction in Pythia using the cone algorithm, and initial studies have found dE/E~100%/√E. The jet angular resolution is critical to the reconstruction of parton momentum fraction x in γ-jet events. Based on our studies we expect ~10% resolution on x and ~70% jet finding efficiency. γ-jet correlations with a photon and jet in the FOCAL is particularly sensitive to the parameterization of the gluon structure function at low x (Fig.4b). Other leading particles correlations, such as π^0-π^0 are also sensitive probes and being studied.

POSSILITY OF FLAVOR SEPARATION WITH THE FOCAL

The measurement of inclusive single spin lepton asymmetries from W's is the least biased way to probe $\Delta\bar{q}/\bar{q}$ and $\Delta q/q$. There is a complete theoretical treatment from first principles using Q_T re-summation and NLO techniques [3], and this novel technique does not suffer from scale uncertainties. Further, this is an experimentally clean measurement as A_L is parity violating so there are no false physics asymmetries and the measurement does not rely on the knowledge of fragmentation functions. The measurement takes advantage of the highest lumonisity p+p running at 500 GeV.

We are investigating the feasibility of using the FOCAL to make a measurement of $A_L(W\rightarrow e)$ [y: 1-1.5], which would complement the $A_L(W\rightarrow\mu)$ [y: 1-3] and $A_L(W\rightarrow e)$ [$|y|<0.35$] PHENIX measurements. There is enough $\int B \bullet dl$ to get a deflection which is of the order of the sizes of the elements of the PHENIX silicon vertex detector (VTX). The FOCAL can provide the electron momentum through the energy measurement, effectively giving a factor of two in measuring the sagita. However, the measurement requires excellent position resolution in the FOCAL (~150μm) and alignment. Feasibility studies are in progress.

OUTLOOK

The PHENIX Forward Calorimeter Upgrade facilitates ΔG measurements down to $x\sim10^{-3}$ with A_{LL} measurements of γ, γ-jet, and π^{0}-π^0 and transverse asymmetry measurements (jet-γ Sivers and π^0's in jets Collins) in the forward direction. The design is being optimized to minimize the π^0 background for the A_{LL} measurements. For A_N measurements, the jet axis can be identified, azimuthal distributions measured, and correlations can be made with the central arm or other particles in the forward calorimeter. The jet energy measurements also allow us to measure this vs. z. Simulations are underway to quantify these capabilities. The possibility of $A_L(W\rightarrow e)$ measurements in the forward direction is also being investigated. The measurement requires excellent position resolution and alignment in the PHENIX VTX and FOCAL detectors. The position resolution goal of the FOCAL is 150μm. The performance and backgrounds are being studied.

FOCAL R&D is ongoing, with a beam test of a W/Si prototype scheduled for early in 2009 at FNAL. Construction plans are under development.

REFERENCES

1. K. Boyle, these proceedings.
2. T. Gehrmann and W. Stirling, *Phys. Rev. D* **53**, 6100 (1996).
3. P.M. Nadolsky and C.P. Yuan, *Nucl. Phys. B* **666**, 31 (2003).

Spin Physics Capability with Silicon Vertex Tracker for PHENIX

M. Kurosawa (for the PHENIX vertex collaboration)

RIKEN Nishina Center for Accelerator Based Science, 2-1, Hirosawa, Wako, Saitama 351-0198, Japan

Abstract. The PHENIX detector is being upgraded with a silicon vertex tracker (VTX) to enhance physics capability both in spin physics and heavy ion physics. In order to check the performance of the VTX, PYTHIA simulation was performed with γ-jet production events under the condition of center-of-mass energy of 500 GeV and the integrated luminosity of 300 pb^{-1}. We will report the simulation results of a longitudinal double-spin asymmetry A_{LL} with the VTX.

Keywords: silicon vertex tracker, PYTHIA, longitudinal double-spin asymmetry
PACS: 14.20.Dh, 25.40.Ep, 13.85.Hd, 13.88+e

INTRODUCTION

PHENIX is an experiment aiming to study the spin structure of nucleons and hot and dense matter at Brookhaven National Laboratory's Relativistic Heavy Ion Collider (RHIC). The PHENIX detector [1] is being upgraded with the introduction of a VTX to extend its physics capability in both spin and heavy ion program [2]. The role of the VTX in spin program are to measure $\Delta G/G$ with charm and beauty, and to determine x-dependence of $\Delta G/G$ with γ-jet correlation in wide x range.

In the future, integrated luminosity is expected to reach 300 pb^{-1} at $\sqrt{s} = 500$ GeV. This report describes simulation results of a longitudinal double-spin asymmetry A_{LL} for γ-jet events with the VTX detector at the expected luminosity.

VTX DETECTOR

Heavy quarks are measured by the semi-leptonic decay of D and B meson. In order to separate charm and beauty, a detector with good secondary vertex resolution of 50 μm is required. Furthermore, large acceptance detector is necessary to reconstruct jet-axis in γ-jet measurement. The VTX was developed to satisfy these requirements.

The VTX comprises a four layer barrel detector, built from two internal layers of pixel detector and two external layers of strip detectors. Each layer was designed to have low material to reduce the background signal from photon conversion. Figure 1 shows beam view(left) and side view(right) of the VTX detector. The acceptance has $|\eta| < 1.2$ and almost 2π for ϕ direction. The vertex resolution was estimated to be less than 50 μm.

CP1149, *Spin Physics, 18th International Spin Physics Symposium*
edited by D. G. Crabb, D. B. Day, S. Liuti, X. Zheng, M. Poelker, and Y. Prok

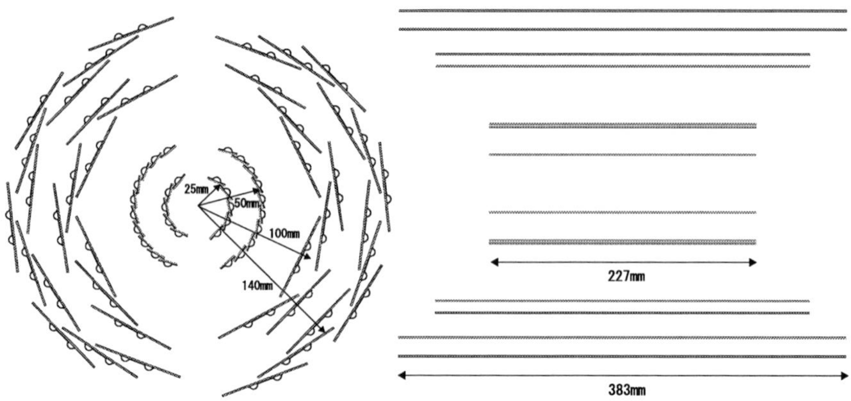

FIGURE 1. Cross sections of the VTX detector from beam view (left) and side view (right).

PYTHIA SIMULATION

The γ-jet production events were generated by using PYTHIA Monte Carlo program [3]. In order to reconstruct jet axis, we used a simple cone algorithm [4]. As the first step, the charged particles which has transverse momentum larger than 1 GeV/c at the opposite side of photon direction are selected. Then, first jet axis is determined by weighting the transverse momentum of these particles within the region of the VTX acceptance. After that, jet axis is calculated through below step.

1. Cone around the first jet axis was made and charged tracks inside the cone radius defined as below equation were selected:

$$R = \sqrt{(\Delta\eta)^2 + (\Delta\phi)^2}\,, \tag{1}$$

   where $\Delta\eta$ and $\Delta\phi$ are the difference in pseudorapidity and in azimuthal angle between the first jet axis and a track inside the cone, respectively.

2. Second jet axis was determined by weighting the transverse momentum inside the cone radius less than 0.5.

3. The difference between the first and second jet axis as defined below was calculated:

$$diff = \sqrt{(\eta_1 - \eta_2)^2 + (\phi_1 - \phi_2)^2}\,. \tag{2}$$

4. The second jet axis was redefined as first jet axis, and the steps from (1) to (3) were iterated until the *diff* minimize.

On the other hand, direct photons are reconstructed with the finely segmented electromagnetic calorimeters in PHENIX detector.

With current PHENIX detector, x_g is evaluated only by using transverse momentum of the direct photon as written by below equation:

$$x_g = \frac{2P_T}{\sqrt{s}} \tag{3}$$

With VTX detector, x_g can be calculated by using also jet axis information. The x_g is smaller one between x_1 and x_2 in below equation:

$$x_1 = \frac{P_T}{\sqrt{s}}(e^{+\eta_{jet}} + e^{+\eta_\gamma}), \ x_1 = \frac{P_T}{\sqrt{s}}(e^{-\eta_{jet}} + e^{-\eta_\gamma}), \ x_g = min\{x_1, x_2\} \tag{4}$$

Figure 2 (a) and (b) show the scatter plot of true x_g and reconstructed x_g with VTX and without VTX, respectively. The reconstructed x_g was improved with the VTX.

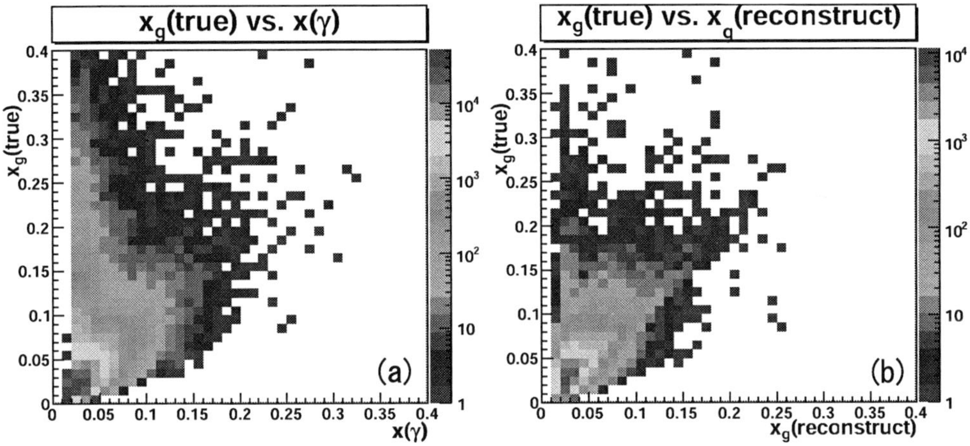

FIGURE 2. (a) and (b) show PYTHIA study to compare true x_g and reconstructed x_g with VTX and without VTX, respectively. x_g(true) is true x_g. $x_g(\gamma)$ and x_g(reconstruct) are obtained from Eq.(3) and Eq.(4), respectively.

Figure 3 shows the simulated A_{LL} of gluon as a function of x_g for assumption of $\Delta g = g$, $\Delta g = -g$ and GRSV standard model [5]. The polarization of proton was assumed to be 70%. Vertical error bars in the plot represent statistical and background subtraction errors. Horizontal error bars indicate reconstruct uncertainty of x_g.

CONCLUSIONS

PYTHIA simulation was performed under the condition of $\sqrt{s} = 500$ GeV and $L = 300\ \text{pb}^{-1}$. There was improvement for x reconstruction with γ-jet production with VTX. The A_{LL} was presented as a function of x_g.

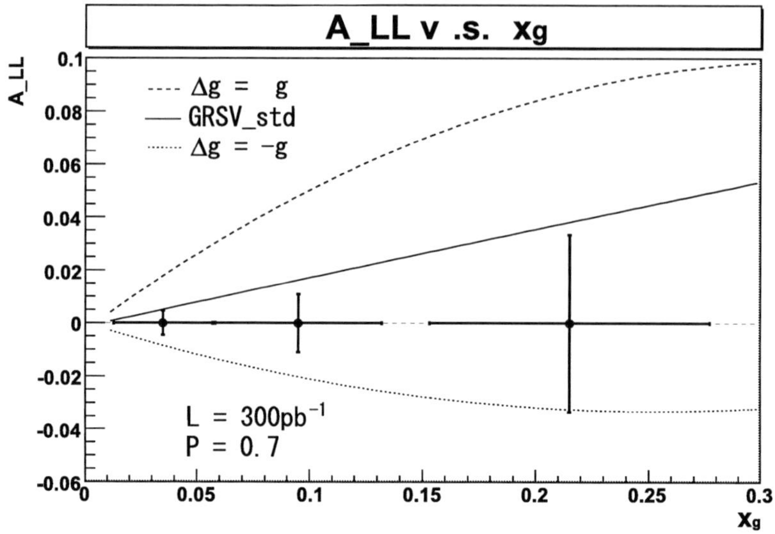

FIGURE 3. Simulated A_{LL} of gluon as a function of x_g. Dashed, dot-dashed and solid line correspond to A_{LL} assuming $\Delta g = g$, $\Delta g = -g$ and GRSV standard model, respectively.

ACKNOWLEDGMENTS

A simulation was performed using RIKEN Super Combined Cluster (RSCC) System and RIKEN Computing Center in Japan (CCJ). We thank the staff members of RSCC and CCJ.

REFERENCES

1. K. Adcox et al., *Nucl. Instr. and Meth.* A **499**, 469 (2003).
2. Y. Akiba, et al., Proposal for a Silicon Vertex Tracker (VTX) for the PHENIX Experiment (Brookhaven National Laboratory, Upton, NY, 2005).
3. PYTHIA: Event Generator, http://www.thep.lu.se/torbjorn/Pythia.html
4. F. Abe et al., *Phys. Rev.* D **45**, 2249 (1992).
5. M. Gluck, E. Reya, M. Stratmann and W. Vogelsang, *Phys. Rev. D* **63**, 094005 (2001).

Study of Polarized Sea Quark Distributions in Polarized Proton-Proton Collisions at $\sqrt{s}$ = 500GeV with PHENIX

Tsutomu Mibe for the PHENIX collaboration

High Energy Accelerator Research Organization, 1-1 Oho, Tsukuba, Ibaraki 305-0801 Japan

Abstract. The PHENIX spin program studies the flavor structure of the polarized sea quark distributions in polarized proton-proton collisions. Starting from 2009 run, the quark and anti-quark polarization, sorted by flavor, will be investigated with the parity-violating single-spin asymmetry of *W*-boson production at the collision energy of $\sqrt{s}$ =500GeV. High momentum muons from *W*-boson decay are detected in the PHENIX muon arms. The muon trigger is being upgraded to allow one to select high momentum muons.

Keywords: proton spin structure, proton-proton collisions, *W* bosons, trigger electronics
PACS: 14.20.Dh, 13.85.Hd, 13.88.+e. 14.70.Fm

INTRODUCTION

The spin structure of the nucleon is of fundamental importance in investigation of the system of strongly-interacting quarks and gluons. Especially sea quarks in the nucleon carry central information on the dynamical properties of the system. It is known that the average polarization of nucleon's sea quarks is small. Underlying mechanism behind the small sea quark polarization is further studied by its flavor structure. Weak interaction processes have been powerful tools to experimentally access the quark flavor in the nucleon as well as its polarization. The *W* boson production in polarized proton-proton collisions is of particular interest since the anti-up quark and anti-down quark polarization can be extracted from its parity violating single spin asymmetry [1].

PHENIX MUON TRIGGER UPGRADE

The parity violating single-spin asymmetry A_L of muons from *W* decay in the 500 GeV polarized proton-proton collisions will be measured at PHENIX in the future running. Muons are detected by the PHENIX muon arms which currently consist of three stations of cathode strip chambers (MuTr) in the muon magnet, followed by 80-cm-long steel absorber with Iarocci tubes (MuID). With the current PHENIX muon triggers, muons above 2 GeV/*c* are triggered by the MuID.

CP1149, *Spin Physics, 18th International Spin Physics Symposium*
edited by D. G. Crabb, D. B. Day, S. Liuti, X. Zheng, M. Poelker, and Y. Prok

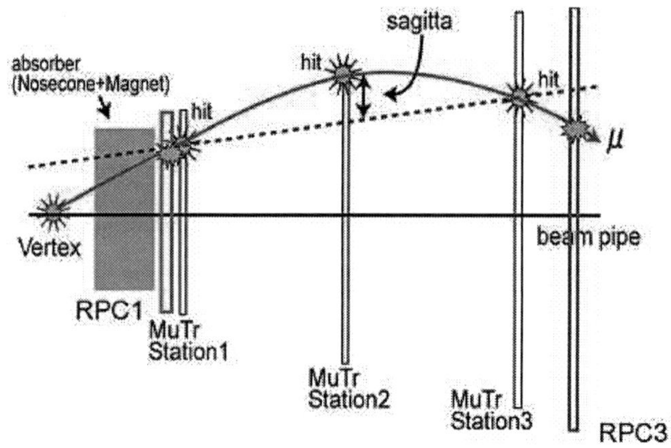

FIGURE 1. Schematic diagram of the Muon Trigger upgrade.

At the highest luminosity of 500 GeV collisions, trigger rate of the current muon system is expected to exceed the available DAQ bandwidth. The PHENIX muon trigger upgrade [2] plans to significantly increase the rejection power of the muon trigger by introducing momentum sensitivity in the trigger using MuTr signals, and a stringent timing cut from new Resistive Plate Chambers (RPC) [3]). The configuration of the upgraded muon trigger is shown in Fig. 1.

The main part of background can be rejected if the trigger selects high momentum muons. This is because that the transverse momentum (p_t) spectrum of the muons from W decay is almost flat over the wide range of p_t, while background muons from hadron decay sharply decrease with p_t. With the upgrade, a sagitta of the tracks from MuTr and RPC hit information is used to trigger on the high momentum tracks in the muon arms.

The cathode strips on the MuTr are placed in radial directions at a 1 cm pitch. Induced charge at each strip is read out with a Cathode PreAmplifier (CPA) followed by AMU ADC for the offline track position measurement. New trigger electronics uses a strip position as a coarse estimate of the track positions for the fast online trigger. Five percent of induced charge is skimmed before the CPA with a capacitor-based-splitter (Fig. 2). The skimmed signal passes through the preamplifier-discriminator board (ADTX board), and then transmitted to the counting house where the data merge/reformattor boards (MRG board) are located. The MRG board collects strip data from individual ADTX boards and reformats them in terms of strip location. The strip hit data from the MRG boards and RPC are processed by the Local Level 1 trigger board (LL1 board) [4] which makes a trigger decision for the muon detectors.

The development of the ADTX board has been completed in 2007. The series of tests at PHENIX have proven that the board meets requirements for the upgraded trigger system [5].

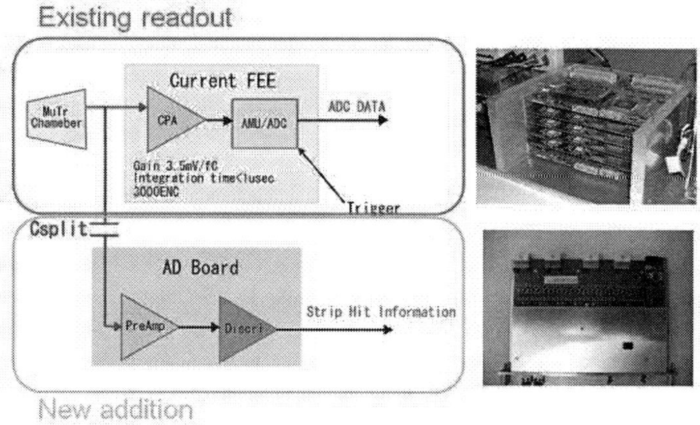

FIGURE 2. Diagram of the MuTr cathode signal readout

INSTALLATION TO PHENIX MUON ARMS

The mass production of the ADTX boards for the North muon arm has been carried out in Japan. The ADTX boards went through the burn-in processes at KEK, followed by the visual inspection and test of functions using the test read-out system at RIKEN [6]. After series of quality assurance tests of the boards and other installation equipments, ADTX boards were installed to the North muon arm. In total 192 boards were installed to cover the whole North muon arm by the end of the RHIC shutdown period (Fig. 3). ADTX boards were also installed to the part of South muon arm covering 1/16th of its acceptance where the RPC prototypes and additional absorber were installed as well. This way, although the acceptance is limited, the trigger performance of the final configuration of the muon trigger upgrade can be studied. The gain and noise level of the MuTr have been evaluated with the MuTr calibration system. It was confirmed that deterioration of the MuTr performance due to the addition of the new trigger electronics is minimal as designed.

The MRG board has been developed at KEK, Japan, and tested at PHENIX [7]. Integration to the PHENIX data acquisition system was tested with the prototype boards. The full chain of the signal processing from the MuTr wire to the PHENIX data collection module was succeeded. The mass production of the board will be completed in early February, followed by installation to the PHENIX.

FIGURE 3. Inside of muon magnet and MuTr in the PHENIX North muon arm after the installation of the ADTX boards (left). Magnified view of the ADTX board installed on the existing read out(right)

SUMMARY

The PHENIX muon trigger upgrade is essential to determine the spin contributions of the sea and valence quarks to the spin of the proton through measurements of single spin asymmetries A_L in the W boson production in polarized proton-proton collisions at 500 GeV. The momentum sensitivity is introduced in the trigger by measuring saggita of the tracks in the PHENIX MuTR in order to reinforce the rejection performance of the muon detection system. The amplifier-discriminator boards have been fully installed to the PHENIX North muon arm with the minimum deterioration of the MuTr performance. Integration to the PHENIX DAQ system is in progress. Together with the new RPCs, the upgraded muon trigger will make it possible to measure high momentum muons from W decay at the maximum luminosity of 500 GeV *pp* collisions.

REFERENCES

1. G. Bunce, N. Saito et al, *Ann. Rev. Nucl. Sci.*, **50**, 525 (2000).
2. Conceptual Design Report for a fast muon trigger at PHENIX (2007) http://www.phenix.bnl.gov/WWW /trigger/forwardupgrade/.
3. X. He, these proceedings.
4. J. Lajoie (private communication).
5. Technical Design Report on Amplifier-Discriminator board and Data Transfer board for MuTr FEE upgrade (2007).
6. Y. Fukao et al., RIKEN accelerator report (2009) (in preparation).
7. T. Mibe et al., RIKEN accelerator report (2009) (in preparation).

Future Investigations of the Flavor Dependence of Sea Quark Helicities at STAR

J. Sowinski *for the STAR Collaboration*

Indiana University Cyclotron Facility, 2401 Milo B. Sampson Ln., Bloomington, IN 47408, USA
sowinski@indiana.edu

Abstract. The flavor dependence of polarized and unpolarized quark distributions in the nucleon can lead to insights into the formation of the sea. Drell-Yan measurements have pointed to flavor asymmetries in the unpolarized distributions. Collisions at $\sqrt{s}$ = 500 GeV with polarized protons at RHIC will soon allow investigations of the flavor separated polarized quark distributions via W production to complement measurements from semi-inclusive DIS. We report on STAR's current plans, tracking upgrade, and expected sensitivities.

Keywords: Quark Polarization, Sea Quarks, W Boson Production
PACS: 13.88.+e, 14.20.Dh, 14.70.Fm, 13.38.Be, 11.30.Hv

INTRODUCTION

The identity of the nucleon comes from the valence quarks, two up and one down for the proton, two down and one up for the neutron. But high energy scattering processes such as deep inelastic scattering (DIS) and hadronic reactions, e.g. Drell-Yan (DY) and jet production, reveal that, in addition, there is a sea of quarks, anti-quarks and gluons residing in the nucleon. Moreover the distributions of these change with the scale or momentum transfer of the scattering. Much of this conference is devoted to reports on understanding how the spin of the nucleon is assembled from the intrinsic spin and orbital motion of these partons. One important aspect of this quest is separating how the various flavors of the quarks and anti-quarks differ in their contributions and exploring whether something more might be learned beyond the distributions themselves.

Simplified pictures of the origin of anti-quarks in the nucleon give dramatically different expectations for the ratios of $\bar{u}$ to $\bar{d}$ quarks in a proton. In perturbative QCD they arise in quark/anti-quark pair creation and, since the $\bar{u}$ and $\bar{d}$ masses are small, one would expect roughly equal numbers. On the other hand, if one considers the long range nuclear force mediated by pions, the proton has a component that is made of a neutron and a π^+ resulting in an expected excess of $\bar{d}$ quarks. (For a general overview see Ref. [1].) The E886 DY experiment by the NuSea collaboration has generated significant interest by demonstrating clear asymmetries in the momentum fraction (x) dependent distributions of $\bar{u}$ and $\bar{d}$ [2]. Corrections to pQCD models for generating the sea accounting for Pauli blocking effects have not been able to explain the data. A number of models, some related to quark bag models, and some chiral symmetry motivated are consistent with the unpolarized data. However they tend to make different predictions for the helicity dependence of the flavor asymmetry, with chiral quark soliton models predicting larger asymmetries for the polarized distributions than the bag models [3].

CP1149, *Spin Physics, 18th International Spin Physics Symposium*
edited by D. G. Crabb, D. B. Day, S. Liuti, X. Zheng, M. Poelker, and Y. Prok

In a recently published global analysis [4], the first to include both DIS and RHIC data, the most favored distributions demonstrate significant differences between the $\bar{u}$ and $\bar{d}$ helicity dependent distributions denoted by $\Delta\bar{u}$ and $\Delta\bar{d}$. However the uncertainties overlap and symmetric distributions are still allowed. In the following I will describe how measurements of W production in polarized proton-proton collisions at RHIC can provide further constraints on these distributions focusing in particular on the proposed measurements and techniques in the STAR detector.

W PRODUCTION AT STAR

While sensitivity to flavor separated distributions in DIS is primarily through coincident detection of hadrons in the exit channel [5, 6], the sensitivity in polarized proton-proton collisions will be attained via W production carried out at $\sqrt{s}$ = 500 GeV in RHIC. The production mechanism is primarily through the channel where a u quark combines with a $\bar{d}$ anti-quark to produce a W^+ and likewise a d with a $\bar{u}$ to produce a W^-. In addition, the V-A coupling of the weak interaction allows only left-handed quarks to combine with right-handed anti-quarks providing a maximal parity violating signal and thus the necessary spin determination. Finally, the handedness of the neutrinos in the W decay results in a partial transfer of the initial state kinematics, q and $\bar{q}$ relative momentum fractions, into the final state decay products. The required measurements consist of determining A_L, the change in cross section with a single beam spin helicity flip, for W^+ and W^- for flip of each beam independently. The two W^+ measurements provide sensitivity to two linear combinations of $\Delta u(x)$ and $\Delta\bar{d}(x)$ and the two W^- measurements to $\Delta d(x)$ and $\Delta\bar{u}(x)$. At forward angles for W^- the measurements separate almost completely into the two pure distributions.

At STAR we will focus on the W decay mode producing electrons or positrons, $W^+ \rightarrow e^+ + \nu$ or $W^- \rightarrow e^- + \bar{\nu}$. The momentum of the high p_T leptons will be measured with electromagnetic calorimeters and the crucial charge sign with tracking in a magnetic field. The STAR detector [7] covers a full 2π in azimuth and is embedded in a solenoidal magnet with field strength of 0.5 T. A major portion of the internal volume is occupied by a large time projection chamber (TPC) to track charged particles over a radius from 0.6 m to 1.9 m. The existing tracking should be sufficient for pseudo-rapidity (scattering angle) $|\eta| < 1$ ($\theta > 37°$). Outside of the TPC reside two Pb/scintillator sampling electromagnetic calorimeters with the barrel covering $-1 < \eta < 1$ and the endcap covering $1.1 < \eta < 2$ ($38° > \theta > 15°$). The calorimeters each have longitudinal segmentation to help in providing background rejection as will be discussed below. These include separate read out of the first few layers of the calorimeters to provide a pre-shower signal and fine position resolution shower maximum detectors to measure the transverse shower profile. In addition, the endcap calorimeter has a separately read out final layer to provide a post-shower signal.

To increase the reliable tracking range to $\eta \sim 2$ a Forward GEM Tracking upgrade [8] has been proposed and funded. Six GEM chambers transverse to the beam will be placed between the TPC and the beam in line with the endcap calorimeter. The $< 80\mu$m resolution of these planes along with a transverse beam constraint, points from the TPC where available, and the hit in the calorimeter shower max detector should provide

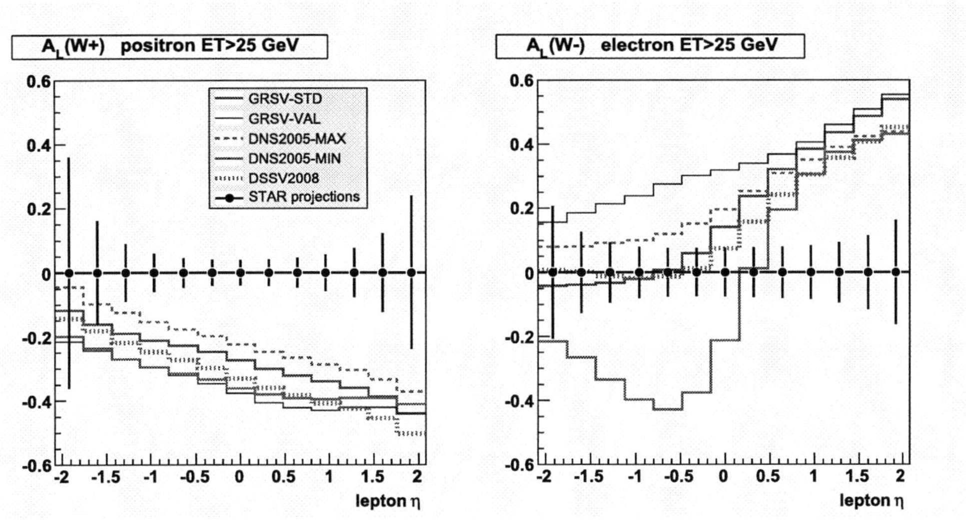

FIGURE 1. Projected sensitivity for 300 pb^{-1}. At left is shown the parity violating asymmetry for W^+ production as measured by positron detection. The asymmetry is plotted vs. the positron pseudo-rapidity. The curves represent predictions at NLO based on a range of PDFs that fit current data. At right are the same quantities for W^- production detected as electrons.

sufficient charge sign separation for the 20 GeV/$c < p_T <$ 40 GeV/c electrons and positrons of interest.

SIMULATIONS AND PROJECTED SENSITIVITY

Extensive simulations have been carried out to verify that the W signal can be separated from the voluminous hard pQCD hadronic backgrounds present and then estimate realistic expected statistical errors on the measurements including remaining necessary background subtraction and projected integrated luminosities. Pythia, which agrees well with NLO calculations of W yields [9], was used with GEANT models of our detector for these studies.

The signal for the channel of interest is an isolated electron with $p_T > 20$ GeV/c and a neutrino escaping undetected. The primary backgrounds are QCD jets which happen to have a particle that deposits a similarly high p_T in an electromagnetic calorimeter. The STAR detector is not sufficiently hermetic to reconstruct the neutrino with missing transverse energy. Thus we employ three classes of cuts to reject background while preserving the signal. The first is isolation around the electron candidate. Tracks and hits in the calorimeter within $r < 0.45$ in $\eta - \phi$ space are used to reject backgrounds based on the accompanying jet particles. The second class rejects events with tracks and calorimeter energy opposite in ϕ from the candidate, as most jets will have its partner of the dijet within the acceptance. Finally details of the candidate interaction in the detector are used, such as the transverse profile at the shower maximum detector and energy in the pre- and post-shower detectors. The calorimeters are only one hadronic interaction length long so most hadrons do not deposit a large fraction of their energy in the calorimeter. Mesons, such as π^0s, which decay electromagnetically need to be rejected by a lack of hits in the tracking system. Up to 30% conversions are allowed while still giving the background rejection considered here and sets a requirement on material in front of the tracking system. Together these cuts reduce the original background by up to a factor of 1000 while preserving over 80% of the signal and we find that for $p_T > 28$ GeV/c the signal to background is better than 1:1.

The projected sensitivity of our measurements for both electrons and positrons with 300 pb^{-1} of integrated luminosity is shown in Fig. 1. The asymmetries from the two beams have been combined into one plot by assigning negative pseudo-rapidity to scattering into the hemisphere from which the respective beam came. Also shown are projections for a range of recent parton distribution functions consistent with current data. For the W^- projections one sees significant sensitivity to the current range of uncertainty in $\Delta\bar{u}(x)$ for $\eta < 0$. For $\eta > 0$ the narrow range of projections reflects the relatively well known valence $\Delta d(x)$ distribution. For W^+ there is no clean separation between $\Delta\bar{d}(x)$ and $\Delta u(x)$ but still significant sensitivity for the central η range which can be used in constraining global fits to the parton distribution functions.

A run with polarized proton collisions at 500 GeV is planned to begin early in 2009. In this first run at that energy we project we could collect as much as 10 pb^{-1}. With polarization above 50% this should be sufficient to measure a statistically non-zero A_L for the W^+ channel and put us on track for the full statistics measurement.

ACKNOWLEDGMENTS

This work was funded in part by grant NSF-0758018.

REFERENCES

1. G. T. Garvey, and J.-C. Peng, *Prog. Part. Nucl. Phys.* **47**, 203–243 (2001), nucl-ex/0109010.
2. R. S. Towell, et al., *Phys. Rev.* **D64**, 052002 (2001), hep-ex/0103030.
3. V. Barone, T. Calarco, A. Drago, and M. C. Simani, *Phys. Lett.* **B571**, 50–54 (2003), hep-ph/0306225.
4. D. de Florian, R. Sassot, M. Stratmann, and W. Vogelsang, *Phys. Rev. Lett.* **101**, 072001 (2008), 0804.0422.
5. A. Airapetian, et al., *Phys. Rev.* **D71**, 012003 (2005), hep-ex/0407032.
6. M. Alekseev, et al., *Phys. Lett.* **B660**, 458–465 (2008), 0707.4077.
7. K. H. Ackermann, et al., *Nucl. Instrum. Meth.* **A499**, 624–632 (2003).
8. F. Simon, et al. (2008), 0811.2432.
9. P. M. Nadolsky, and C. P. Yuan, *Nucl. Phys.* **B666**, 31–55 (2003), hep-ph/0304002.

Registration and Reception		
Main Lounge and Art Gallery		
Sunday, October 5, 2008	06:00 PM - 09:00 PM	
Welcome		
Ball Room		
Monday, October 6, 2008		
Chair	D.G. Crabb	
Speaker	**Title**	**Time**
T. Skalak (University of Virginia)	Welcome on behalf of the University of Virginia	08:30 AM
W. Van Oers (Manitoba)	Welcome on behalf of IUPAP	08:45 AM
K. Imai (Kyoto)	Welcome on behalf of the International Spin Committee	09:00 AM
Plenary I		
Ball Room		
Monday, October 6, 2008		
Chair	K. Imai	
Speaker	**Title**	**Time**
E. Leader (Imperial College)	The Power of Spin: a scalpel-like probe of theoretical ideas	09:05 AM
K. Rith (Erlangen)	Spin Experiments: Technological Challenges and Selected Physics Highlights	09:50 AM
COFFEE	10:35 AM - 11:05 AM	
Plenary II		
Ball Room		
Monday, October 6, 2008	11:05 AM - 12:35 PM	
Chair	E. Steffens	
Speaker	**Title**	**Time**
J. Clarke (Cockcroft Inst./ Daresbury and Liverpool)	High Intensity Polarized Positron Sources for Future Linear Colliders	11:05 AM
R. Milner (MIT)	Future Electron-Ion/proton Collider	11:50 AM
Plenary III		
Ball Room		
Tuesday, 7 October	09:00 AM - 10:30 AM	
Chair	E. Stephenson	
Speaker	**Title**	**Time**
W. Marciano (BNL)	Fundamental Symmetries and Spin	09:00 AM
Y. Semertzidis (BNL)	The EDM Experiment with Polarized Deuteron Beams in a Storage Ring	09:45 AM
COFFEE	10:30AM - 11:00 AM	

Plenary IV		
Ball Room		
Tuesday, October 7, 2008	11:00 AM - 12:30 PM	
Chair	R. Milner	
Speaker	**Title**	**Time**
M. Burkardt (New Mexico State)	Parton Distributions to Impact Parameter Space	11:00 AM
M. Anselmino (Turin)	Transverse Spin Structure of the Nucleon	11:45 AM
Plenary V		
Ball Room		
Wednesday, October 8, 2008	09:00 AM - 10:30 AM	
Chair	H. Gao	
Speaker	**Title**	**Time**
K. Kumar (UMass)	Spin Experiments Beyond the Standard Model	09:00 AM
J. Erler (UNAM-Mexico)	Spin Beyond Standard Model: Theory	09:30 AM
E. Steffens (Erlangen)	Polarized Anti-protons - the Quest for a Missing Tool	10:00 AM
COFFEE	10:30 AM - 11:00 AM	
Plenary VI		
Ball Room		
Wednesday, October 8, 2008	11:00 AM - 12:30 PM	
Chair	G. Fidecaro	
Speaker	**Title**	**Time**
K. Asahi (Tokyo Tech.)	Nuclear Structure Studies with Polarized Radioactive Beams	11:00 AM
N. Makins (Illinois)	Transverse Spin and Momentum: Experimental Overview	11:45 AM
Plenary VII		
Ball Room		
Thursday, October 9, 2008	09:00 AM - 11:00 AM	
Chair	F. Rathman	
Speaker	**Title**	**Time**
W. Meyer (Bochum)	Polarized Solid and Gas Targets	09:00 AM
N. Saito (KEK/J-PARC)	Outlook for Spin at J-PARC	09:30 AM
A. Zelenski (Brookhaven)	Present Status of Polarized Ion Sources	10:00 AM
COFFEE	10:30 AM - 11:00 AM	
Plenary VIII		
Ball Room		

Thursday, October 9, 2008	11:00 AM – 12:30 PM	
Chair	N. Tyurin	
Speaker	**Title**	**Time**
S. Troshin (IHEP)	Polarization in Heavy Ion Physics	11:00 AM
C. Aidala (UMASS)	Spin in Hadron Reactions	11:30 AM
K. Slifer (Virginia)	Spin Content of the Nucleon: What We've Learned from Polarized Lepton Scattering	12:00 PM
Symposium Dinner		
Omni Hotel		
Thursday, October 9, 2008	07:30 PM – 08:00 PM Reception 08:00 PM – Dinner	
Plenary IX		
Ball Room		
Friday, October 10, 2008	09:00 AM – 10:30 AM	
Chair	G. Mallot	
Speaker	**Title**	**Time**
E. Rondio (SINS, Warsaw)	Gluons' Polarization in the Nucleon	09:00 AM
C. Weiss (JLab)	Generalized Parton Distributions: Status and Perspectives	09:45 AM
COFFEE	10:30 AM – 11:00 AM	
Plenary X		
Ball Room		
Friday, October 10, 2008	11:00 AM – 12:30 PM	
Chair	T. Roser	
Speaker	**Title**	**Time**
E. Voutier (Grenoble)	General Parton Distributions: the hunt for quark orbital momenta	11:00 AM
M. Leonova (Michigan)	Spin at COSY (p, d acceleration)	11:30 AM
M. Poelker (JLab)	Summary: PESP2008	12:00 PM
Special Director's Session		
Ball Room		
Friday, October 10, 2008	TBA	
Chair	A. Deshpande	
Speaker	**Title**	**Time**
F. Maas (for H. Stoecker)	Spin Physics at GSI, now and in future	TBA
S. Vigdor (for S. Aronson)	Spin Physics at BNL, now and in future	TBA
H. Montgomery	Spin Physics at JLab, now and in future	TBA
N. Tyurin	Spin Physics at IHEP, now and in future	TBA
R. Lednicky (for A. Sissakian)	Spin Physics at JINR, now and in future	TBA

N. Saito (for S. Nagamiya)	Spin Physics at J-PARC, now and in future	TBA
Plenary XI		
Ball Room		
Saturday, October 11, 2008	09:00 AM – 10:30 AM	
Chair	A.D. Krisch	
Speaker	**Title**	**Time**
T. Roser (BNL)	Future of High Energy Polarized Proton Beams	09:00 AM
T. Wakasa (Kyushyu U.)	Nuclear Spin Physics	09:45 AM
COFFEE	10:30 AM – 11:00 AM	
Plenary XI		
Ball Room		
Saturday, October 11, 2008	11:00 AM – 01:30 PM	
Chair	F. Bradamante and D.G. Crabb	
Speaker	**Title**	**Time**
R. Lednicky (JINR, Dubna)	Project of the Nuclotron-based Ion Collider fAcility (NICA) at JINR, Dubna: Perspectives of heavy ion and spin physics.	11:00 AM
J. Ralston (Kansas)	The Colorful Future of Spin	11:45 AM
S. Wolf (UVA)	Spintronics	12:30 PM
K. Imai (Kyoto)	Close Out	01:15 PM

	Fundamental Symmetry and Spin	**/Spin Beyond the Standard Model I**		
	South Meeting Room			
	Tuesday, October 7, 2008	*02:00 PM - 04:10 PM*		
	Chair	Xiaochao Zheng		
	Speaker	**Title**	**Talk**	***Time***
x	*Paver, Nello*	*Spin-2 identification of the Randall-Sundrum graviton at the LHC (abstract)*	*download*	*02:00 PM*
x	*Quayle, William*	*Measuring the properties of the Higgs with the ATLAS detector (abstract)*	*download*	*02:30 PM*
	King, Paul	*Parity-violating electron scattering asymmetries from hydrogen and deuterium at Q^2 of 0.63 and 0.23 GeV^2 (abstract)*	*download*	*02:50 PM*
x	*Fomin, Nadia*	*The NPDGamma experiment - A measurement of parity violation in polarized cold neutron capture (abstract)*	*download*	*03:10 PM*
x	*Kohl, Michael*	*The Time Reversal Experiment with Kaons (TREK) at J-PARC (abstract)*	*download*	*03:30 PM*

	Fundamental Symmetry and Spin	**/Spin Beyond the Standard Model III**		
	Commonwealth Room			
	Thursday, October 9, 2008	*02:00 PM - 04:00 PM*		
	Chair	David Armstrong		
	Speaker	**Title**	**Talk**	***Time***
	Gericke, Michael	*The Qweak Experiment: Searching for New Physics at the TeV Scale (abstract)*	*download*	*02:00 PM*
	Paschke, Kent	*PVDIS at JLab 12 GeV (Solenoid large acceptance device) (abstract)*	download	*02:20 PM*
x	*Subedi, Ramesh*	*{\bf{$e^{\hspace{-2mm}\rightarrow}$-^{2}H Parity Violating Deep Inelastic Scattering at Jefferson Laboratory at 6 GeV}} (abstract)*	*download*	*02:40 PM*
	Zhu, Xiaofeng	*The 3He injection test for the search of neutron electric dipole moment (nEDM) (abstract)*	*download*	*03:00 PM*
	Yoshimi, Akihiro	*Nuclear spin maser at highly stabilized low magnetic field and search for an atomic EDM (abstract)*	*download*	*03:20 PM*
x	*Rivas, Martin*	*On the kinematics of the center of charge of a spinning particle (abstract)*	*download*	*03:40 PM*

Longitudinal Spin Session	**I**	
Theatre		
Monday, October 6, 2008	02:00 PM - 04:00 PM	
Chair	*Roland Windmolders*	
Speaker	***Title***	***Time***
Ellinghaus, Frank	*Review of ΔG programs in DIS and pp*	*02:00 PM*
Bennett, Robert	*Longitudinal Double Spin Asymmetry and Cross Section for Direct Photon Production at Mid-rapidity in Polarized $\sqrt{s} = 200$ GeV pp Collisions at PHENIX*	*02:30 PM*
Datta, Amaresh	*Double Longitudinal Spin Asymmetry of Inclusive Charged Hadron Production in Polarized p+p Collisions at $sqrt(s) = 62.4$ GeV at PHENIX*	*02:50 PM*
Hoffman, Alan	*Recent STAR Results from Neutral Pion Production in Polarized p+p Collisions at Sqrt(s)=200GeV at RHIC*	*03:10 PM*
Kocoloski, Adam	*Recent STAR Results from Charged Pion Production in Polarized pp Collisions at $\sqrt{s}$ = 200~GeV at RHIC*	*03:30 PM*

Longitudinal Spin Session	**II**	
Theatre		
Monday, October 6, 2008	04:15 PM - 06:15 PM	
Chair	*Dimiter Stamenov*	
Speaker	***Title***	***Time***
Deur, Alexandre	*The strong coupling constant at low Q2*	*04:15 PM*
De Nardo, Lara	*HERMES results from Inclusive DIS on a transversely polarized H target*	*04:45 PM*
Honkanen, Heli	*New avenue to the Parton Distribution Functions: Self-Organizing Map*	*05:05 PM*
Kao, Chung-Wen	*Spin polarizabilities in Heavy baryon Chiral perturbation Theory*	*05:25 PM*
Koretune, Susumu	*Sum rules for the spin-dependent structure functions from the current algebra on the null-plane*	*05:45 PM*

Longitudinal Spin Session	**III**	
Theatre		
Tuesday, October 7, 2008	02:00 PM - 04:00 PM	
Chair	*Gail Dodge*	
Speaker	***Title***	***Time***
Bueltmann, Stephen	*Experimental Overview of Longitudinal Spin Structure Measurements*	*02:00 PM*
El Alaoui, Ahmed	*Measurement of the contribution of strange quarks to the proton spin*	*02:30 PM*
Guler, Nevzat	*Spin Structure of the Deuteron and the Neutron - New Results from CLAS*	*02:50 PM*
Liyanage, Nilanga	*Neutron Spin Structure in the Resonance Region and Quark-Hadron Duality*	*03:10 PM*
Sulkosky, Vincent	*Low Q^{2} Dependence of the Neutron and ^{3}He Spin Structure Functions*	*03:30 PM*

Longitudinal Spin Session	IV		
Theatre			
Tuesday, October 7, 2008	04:15 PM – 06:30 PM		
Chair	Werner Vogelsang		
Speaker	**Title**	**Time**	
Strattmann, Marco	Global QCD Analysis of Polarized Parton Densities	04:15 PM	
Myhrer, Fred	The Spin of the Proton	04:45 PM	
Kunne, Fabienne	Gluon Polarisation in the Nucleon from Open Charm at COMPASS	05:10 PM	
Kurek, Krzysztof	Gluon Polarisation in the Nucleon from high transverse momentum hadron pairs at COMPASS	05:30 PM	
Morreale, Astrid	fAccessing the Sign and Magnitude of Δg by Measuring the Double Longitudinal Spin Asymmetry in High pT Charged Pion Production in Polarized p+p Collisions at $\sqrt s=200$ GeV	05:50 PM	
Nakano, Kenichi	Determination of Delta g from inclusive pi0 measurement at RHIC-PHENIX	06:10 PM	

Longitudinal Spin Session	V		
South Meeting Room			
Thursday, October 9, 2008	04:15 PM – 06:25 PM		
Chair	N. Saito		
Speaker	**Title**	**Time**	
Stamenov, Dimiter	Status of Longitudinal Polarized Parton Densities and Higher Twist	04:15 PM	
Accardi, Alberto	Target mass corrections in collinear factorization	04:45 PM	
~~Musulmanbekov, Zhenis~~	~~Quarks as Vortical Solitons: Towards Understanding the Nucleon Spin~~	05:05 PM	no show
Ramsey, Gordon	Phenomenological determination of the orbital angular momentum	05:25 PM	
Rubin, Joshua	Newest Results on PDFs at HERMES	05:45 PM	
Liuti, Simonetta (for Fantoni, A.	The transition between non-perturbative and perturbative QCD	06:05 PM	

Longitudinal Spin Session	VI	
Theatre		
Friday, October 10, 2008	02:00 PM – 04:00 PM	
Chair	Alexander Bazilevski	
Speaker	**Title**	**Time**
Saito, Naohito (for the AAC)	Determination of polarized parton distribution functions with recent data on spin asymmetries	02:00 PM
Dahiya, Harleen	Semi-leptonic Octet Baryon Weak Axial-Vector Form Factors in the Chiral Constitutent Quark Model	02:30 PM
Miyachi, Yoshiyuki	Global QCD Analysis of Fragmentation and Parton Distribution Functions	02:50 PM
Pierce, Joshua	The Double Spin Asymmetry in Exclusive pi+ Electroproduction	03:10 PM

Riedl, Johann	*Longitudinally polarized hadroproduction of heavy quarks*	*03:30 PM*
Longitudinal Spin Session	***VII***	
Theatre		
Friday, October 10, 2008	04:15 PM - 06:15 PM	
Chair	*Bernd Surrow*	
Speaker	***Title***	***Time***
Windmolders, Roland	*The Strange Quark Polarisation from Charged Kaon Production on Deuterons*	*04:15 PM*
Christova, Ekaterina	*A model independent approach to fragmenation functions*	*04:35 PM*
Sakuma, Tai	*STAR Measurements of the Di-jet Production Cross Section and Double Longitudinal Spin Asymmetry in sqrt(s) = 200 GeV Proton+Proton Collisions at RHIC*	*04:55 PM*
Sarsour, Murad	*Constraints on DeltaG through Longitudinal Double Spin Asymmetry Measurements of Inclusive Jet Production in Polarized p+p Collisions*	*05:15 PM*
Wang, Xiaorong	*Measurement of Double Longitudinal Spin Asymmetry in Heavy Flavor Production at $\sqrt(s)=200GeV$ at PHENIX*	*05:35 PM*
Wissink, Scott	*Longitudinal Double-spin Asymmetries for Neutral Pion Production at Forward Rapidiites in pp Collisions at sqrt{s} = 200 GeV at STAR*	*05:55 PM*

	Transverse Spin Session	**I**	
	Ball Room		
	Tuesday, October 7, 2008	02:00 PM – 04:00 PM	
	Chair	Jian-ping Chen	
	Speaker	**Title**	**Time**
x	Feng Yuan	Theory Overview on Transverse Spin Physics	02:00 PM
x	Rainer Joosten	Measurement of Transverse Momentum Dependent Parton Distributions and Fragmentation Functions in Semi-Inclusive Deep Inelastic Scattering	02:30 PM
x	Contalbrigo Marco	Transverse Spin Physics at HERMES	03:00 PM
x	Heiner Wollny	Transverse Spin Effects at COMPASS	03:20 PM
x	Francesca Giordano	Measurement of Azimuthal Asymmetries of the Unpolarized Cross-section at HERMES	03:40 PM
	Transverse Spin Session	**II**	
	Ball Room		
	Tuesday, October 7, 2008	04:15 PM – 06:15 PM	
	Chair	Harut Avagyan	
	Speaker	**Title**	**Time**
	Leonard Gamberg	Gluon Pole Matrix Elements Universality and Transverse Single Spin Asymmetries	04:15 PM
x	Anselm Vossen	Fragmentation Function Measurement at Belle	04:45 PM
	Feng Wei	Measuring the Sivers Effect via Back-to-Back Di-Hadron Correlations in the PHENIX at RHIC	05:15 PM
x	Han Liu	Transverse Single Spin Asymmetries in Heavy Flavor Production in Polarized p+p Collisions at PHENIX	05:35 PM
x	Zhongbo Kang	Tri-gluon correlation and asymmetry in open charm production	05:55 PM
	Transverse Spin Session	**III**	
	Ball Room		
	Thursday, October 9, 2008	02:00 PM – 04:00 PM	
	Chair	Leonard Gamberg	
	Speaker	**Title**	**Time**
x	Alessandro Bacchetta	What Can We Learn from TMDs Measurements?	02:00 PM
	Xiaodong Jiang	Semi-Inclusive DIS Experiments at Jefferson Lab with a 6 GeV and 11 GeV Electron Beam	02:30 PM
	Yi Qiang	Measurement of Single Target-Spin Asymmetry in Semi-Inclusive Pion Electroproduction on a Tranversely Polairzed 3He Target	03:00 PM
	Xin Qian	The Study of Semi-Inclusive 3He(e,e'h) Reaction at 11 GeV in JLab Hall A	03:20 PM
x	Wes Gohn	Single-Spin Asymmetries from Semi-Inclusive Pion Electro-production in Deep Inelastic Scattering	03:40 PM
	Transverse Spin Session	**IV**	
	Ball Room		
	Thursday, October 9, 2008	04:15 PM – 06:15 PM	

	Chair	Zein-Eddine Meziani	
	Speaker	**Title**	**Time**
x	Alexei Prokudin	Transversity and Collins Fragmentation Functions: Towards a New Global Analysis	04:15 PM
-	Peter Bosted	Double Spin Asymmetries from Semi-Inclusive DIS from CLAS at JLab	04:45 PM
x	Barbara Pasquini	TMDs and Azimuthal Asymmetries in a Light-cone Quark Model	05:15 PM
x	Yuji Koike	Twist-3 Single-Spin Asymmetries for SIDIS and Its Azimuthal Structure	05:35 PM
	Mher Aghasyan	Beam-Spin Asymmetry Measurement at CLAS	05:55 PM

	Transverse Spin Session	**V**	
	Ball Room		
	Friday, October 10, 2008	02:00 PM - 04:00 PM	
	Chair	Alessandro Bacchetta	
	Speaker	**Title**	**Time**
x	Jen-Chieh Peng	Drell-Yan Process and Nucleon Spin	02:00 PM
x	Andreas Metz	Exploring TMDs through Drell-Yan Processes	02:30 PM
x	Kazuhiro Tanaka	Double Transverse-Spin Asymmetries for Small-Q_T Drell-Yan Pair Prodction in pp-bar Collisions	03:00 PM
x	Kazuhiro Tanaka	Double-Spin Asymmetry A_LT for Polarized Drell-Yan Process in p pbar Collisions: Wandzura-Wilczek Contribution	03:20 PM
x	Jian Zhou	Double Initial State Interactions in Drell-Yan Lepton Pair Production	03:40 PM

	Transverse Spin Session	**VI**	
	Ball Room		
	Friday, October 10, 2008	04:15 PM - 06:15 PM	
	Chair	Andreas Metz	
	Speaker	**Title**	**Time**
x	Werner Vogelsang	QCD Corrections in Transversely Polarized Scattering	04:15 PM
	Mickey Chiu	Transverse Single Spin Asymmetries at High xF in PHENIX	04:45 PM
x	Zhun Lu	Angular Dependence of Dijet Production in Unpolarized Hadron Scattering	05:15 PM
x	Nikola Poljak	Spin-dependent Forward Particle Correlations in p+p Collisions at sqrt(s) =200 GeV	05:35 PM
x	James Drachenberg	Forward Pion Detection in p+p Collisions at Sqrt(s)=200 GeV	05:55 PM

	Nucleon Spin Joint Session I		
	Ball Room		
	Monday, October 6, 2008	*2:00 PM - 4:00 PM*	
	Chair	*Christian Weiss*	
	Speaker	***Title***	***Time***
x	*Gary Goldstein*	*Exclusive Meson Electroproduction: A Mean to Determine and Relate GPDs, TMDs and Regge Theory*	*02:00 PM*
x	*Asmita Mukherrjee*	*Chiral-odd GPDs in Impact Parameter Space*	*02:30 PM*
x	Stephan Meissner	Properties and Applications of GTMDs	03:00 PM
x	Marc Schlegel	Two-Photon Exchange in Inclusive and Semi-Inclusive DIS	03:20 PM
x	*Petr Zavada*	*Polarized Distributions and Intrinsic Motion of the Quarks*	*03:40 PM*
	Nucleon Spin Joint Session II		
	Ball Room		
	Monday, October 6, 2008	4:15 PM - 6:15 PM	
	Chair	Matthias Burkardt	
	Speaker	**Title**	***Time***
x	Huey-Wen Lin	A Review of Nucleon Spin Calculations in Lattice QCD	*04:15 PM*
x	*Kieran Boyle*	*Recent Results and Future Prospects for the PHENIX Spin Program*	*04:45 PM*
x	*Kawtar Hafidi*	*Study of Partonic Distributions and Spin-Orbit Correlations Using Semi-Inclusive Production of Kaons at JLab*	*05:15 PM*
x	*Joe Seele*	*What Can We Learn from eta Production in p-p Collisons at PHENIX*	*05:35 PM*
x	Doug Fields	Asymmetry in Longitudinally Polarized pp Collisions at PHENIX	*05:55 PM*

GPD/Form Factors		
South Meeting Room		
Thursday, October 9, 2008	02:00 - 04:00PM	
Chair	M. Jones	
Speaker	**Title**	**Time**
Brash, Edward	The Gep-III Experiment at JLab	02:00 PM
Ralf Kaiser	Overview of future GPD measurements	02:30 PM
Kohl, Michael	Nucleon Form Factors from BLAST	3:00 PM
Pentchev, Lubomir	Proton Form Factor Measurements Using Polarization Method: Beyond Born Approximation	03:20 PM
Aparajita Bhattacharya	Magnetic Moments of Baryons and Exotics in Quasi-Particle Diquark Model	03:40 PM

GPD	**I**	
South Meeting Room		
Friday, October 10, 2008	02:00 - 04:00 PM	
Chair	G. Goldstein	
Speaker	**Title**	**Time**
Beni Zihlmann	Experimental overview of GPD results from HERMES	02:00 PM
Francois-Xavier Girod	Experimental overview of GPD results from CLAS	02:30 PM
Mussgiller, Andreas	Status and prospects of the HERMES Recoil Detector	03:00 PM
Egiyan, Hovanes	DVCS with longitudinally polarized target using CLAS12	03:20 PM
Saeed Ahmad	The Hybrid Generalized Parton Distributions Parametrization	03:40 PM

GPD	**II**	
South Meeting Room		
Friday, October 10, 2008	04:15 - 06:15 PM	
Chair	S. Liuti	
Speaker	**Title**	**Time**
Riordan, Seamus	Measurements of the Electric Form Factor of the Neutron at High Momentum Transfer	04:15 PM
Marukyan, Hrachya	HERMES results from a combined beam charge and spin analysis of DVCS on unpolarized hydrogen and deuterium targets	04:45 PM
Kawtar Hafidi	DVCS on 4He	05:05 PM
Biselli, Angela	DVCS with longitudinally polarized target using CLAS at 6 GeV	05:25 PM

Spin in Hadronic Reactions I	**Special Session on Lambdas**		
South Meeting Room			
Monday, October 6, 2008	04:15 PM – 06:25 PM		
Chair	*Haiyan Gao*		
Speaker	**Title**	**Talk**	**Time**
Homer Neal	Spin Studies at the LHC	download	04:15 PM
Xin Dong	Polarization Measurements of Hyperons and Vector Mesons in Heavy Ion Collisions at STAR	download	04:45 PM
Ernst Sichtermann	Longitudinal spin transfer of Lambda and anti-Lambda in polarized proton-proton collisions at 200 GeV	download	05:05 PM
Teresa Negrini	Lambda polarization with a transversely polarized proton target at COMPASS	download	05:25 PM
Stansilav Belostotski	Polarization effects in Lambda production at HERMES.	download	05:45 PM
Alexander Goritschnig	Proton - Antiproton into LambdaC - AntilambdaC within the Handbag Picture	download	06:05 PM

Spin in Hadronic Reactions II			
Commenwealth Room			
Thursday, October 9, 2008	04:15 PM – 06:15 PM		
Chair	*Sebastian Kuhn*		
Speaker	**Title**	**Talk**	**Time**
Vasily Mochalov	Spin physics program at IHEP-Protvino	download	04:15 PM
Yuji Koike	Soft-fermion-pole mechanism to the single spin asymmetry in hadronic pion production	download	04:45 PM
Donika Plyku	Polarized Proton-Proton Elastic Scattering at RHIC	download	05:05 PM
Andro Kacharava	Physics Program at COSY-Juelich with Polarized Hadronic Probes	download	05:25 PM
Miroslav Finger	Search for the NN to 6q phase transition in the np-polarized measurements at Tkin = 1 - 6 GeV		05:55 PM

Spin in Nuclei	**I**		
Room 389			
Monday, October 6, 2008	02:00 PM - 04:00 PM		
Chair	R. Lindgren		
Speaker	**Title**	**Talk**	**Time**
Kalantar, Nasser	What have we learned about three-nucleon systems at intermediate energies? abstract	download	02:00 PM
Dozono, Masanori	Study of 12C(pol-p,pol-n)12N(g.s.;1+) at 296MeV and nuclear correlation effects abstract	download	02:30 PM
Elbakyan, Garegin	Hadron Formation in Semi-Inclusive DIS on Nuclei at HERMES abstract	download	02:55 PM
Furukawa, Takeshi	A new laser spectroscopic method "OROCHI" (Optical RI-atom Observation in Condensed Helium as Ion-B4) abstract	download	03:20 PM

Spin in Nuclei	**II**		
Room 389			
Tuesday, October 7, 2008	02:00 PM - 04:00 PM		
Chair	D. Day		
Speaker	**Title**	**Talk**	**Time**
Silenko, Alexander	Potential for measurement of the tensor electric and magnetic polarizabilities of the deuteron and other nuclei in experiments with polarized beams abstract	download	02:00 PM
Stibunov, Victor	The tensor analysing power component $T_ {21}$ of exclusive π^{-} abstract	download	02:30 PM
Zetta, Luisa	117In and 118Sn Homologous State Identification via the 120Sn(ppol, ±)117In abstract	download	02:55 PM
Uzikov, Yury	Search for a quasi-bound $^3He-\eta$ state via spin-observables of the reaction $pd\to ^3He\eta$ abstract	download	03:20 PM

Accelerator, Storage and Polarimetry of Polarized Beams	**I**		
Commonwealth Room			
Monday, October 6, 2008	02:00 PM - 04:00 PM		
Chair	*Ed Stephenson*		
Speaker	**Title**	**Talk**	**Time**
Lorenzon, Wolfgang	Precision Electron Beam Polarimetry	download	02:00 PM
Poltoratska, Yuliya	Polarized electrons in Darmstadt: recent developments and polarimetry	download	02:30 PM
Silva, Marlene	Towards High-Precision Deuteron Polarimetry	download	02:45 PM
Bordovitsyn, Vladimir	Spin Light Identification against a Background of Powerful Synchrotron Radiation		03:00 PM
Bazilevsky, Alexander	Precision Measurements of the Absolute Proton Beam Polarization with RHIC Polarimeters	download	03:15 PM
Makdisi, Yousef	Towards Improved Polarimetry at RHIC	download	03:30 PM
Zelenski, Anatoli	THE RHIC P-CARBON CNI POLARIMETER UPGRADE FOR THE BEAM POLARIZATION AND INTENSITY PROFILE MEASUREMENTS	download	03:45 PM

Accelerator, Storage and Polarimetry of Polarized Beams	**II**		
Commonwealth Room			
Tuesday, October 7, 2008	02:00 PM - 04:45 PM		
Chair	*Y. S. Derbenev*		
Speaker	**Title**	**Talk**	**Time**
Ptitsyn, Vadim	Polarized beam st EIC	download	02:00 PM
Chevtsov, Pavel	Electron Spin Matching for ELIC, a High-Luminosity Ring-Ring Electron-Ion Collider	download	02:30 PM
Barber, Desmond	Spin motion and the invariant polarisation-tensor field for spin-1 particles in storage rings	download	02:50 PM
Higinbotham, Douglas	Electron Spin Precesion at CEBAF	download	03:10 PM
Montag, Christoph	RHIC polarized proton performance in RUN 08	download	03:25 PM
Luccio, Alfredo U	SPINK - a spin tracking code	download	03:45 PM
MacKay, William	Effect of various errors on the spin tune and stable spin axis	download	04:00 PM
Huang, Haixin	Fighting the Residual Polarization Loss in the AGS	download	04:30PM

Accelerator, Storage and Polarimetry of Polarized Beams	**III**		
Commonwealth Room			
Tuesday, October 7, 2008	04:15 PM - 06:30 PM		
Chair	*V. Ptitsyn*		
Speaker	**Title**	**Talk**	**Time**
Rathmann, Frank	Do unpolarized electrons affect the polarization of a stored proton beam?	download	04:15PM
Onderwater, Cornelis	Storage Rings for EDM Searches	download	04:45 PM
Lin, Fanglei	Study By Spin Tracking Of A Storage Ring For Deuteron Electric Dipole Moment	download	05:15PM
Luccio, Alfredo U	Simulation of deuteron polarization at COSY	download	05:35PM
Nass, Alexander	Spin Filtering Studies at COSY and AD	download	05:55PM

Bai, Mei	Simulation of spin interference and echo effects	download	06:15PM
Accelerator, Storage and Polarimetry of Polarized Beams	**IV**		
Commonwealth Room			
Friday, October 10, 2008	02:00 PM – 05:15 PM		
Chair	*A. Chao*		
Speaker	**Title**	**Talk**	**Time**
Kondratenko, Anatoliy	Multiple spin resonance crossing in accelerators	download	02:00 PM
Courant, Ernest	resonance strength calculation	download	02:30 PM
Morozov, Vasily	Experimental Verification of Predicted Oscillations near a Spin Resonance	download	02:50 PM
Raymond, Richard	Experimental Test of New Technique to Overcome Spin-Depolarizing Resonances	download	03:10 PM
Balling, Andreas	Polarization upgrade at ELSA	download	03:30 PM
Shatunov, Yuri, presented by V. Ptitsyn	Analysis of Spin-Flip Data using a Spin Flipper	download	04:15 PM
Bai, Mei	Observing Spin Coherent Precession Driven by an AC Dipole in RHIC	download	04:35 PM
Ranjbar, Vahid	Possible Hybrid Snake Spin Resonance in RHIC		04:55 PM

Polarized Sources and Targets Session I	**Polarized Gas Targets**			
South Meeting Room				
Monday, October 6, 2008	2:00 PM - 4:00 PM			
Chair	F. Piegsa			
Speaker	**Title**	**Talk**		**Time**
Singh, Jaideep	Recent Advances in SEOP Polarized He-3 Targets	download		2:00 PM
Krimmer, Jochem	Polarized 3He targets at MAMI	downlaod		2:30 PM
Kim, Wooyoung	Optical Pumping Spin Exchange 3He Gas Cells for Magnetic Resonance Imaging			2:50 PM
Kovacs, Kangkang	Deuteron Target Polarimetry Analysis for the EG4 Experiment	download		3:10 PM
Kolanoski, Hermann	The E166 Experiment: Undulator Based Production of Polarized Positrons	download		3:30 PM
Polarized Sources and Targets Session II	**Polarized Sources**			
Room 389				
Monday, October 6, 2008	4:15 PM - 6:15 PM			
Chair	M. Poelker			
Speaker	**Title**	**Talk**		**Time**
Zelenski, Anatoli	The RHIC polarized ion source.	download		4:15 PM
Paschke, Kent	Sources of Helicity-Correlated Electron Beam Asymmetries	download		4:45 PM
Masayoshi, Tanaka	RCNP Project on Polarized 3He Ion source--From Optical Pumping to Crogenic Method	download		5:15 PM
Nass, Alexander	Studies on Beam Formation in an Atomic Beam Source	download		5:35 PM
Stancari, Michelle	Test Bench Studies and Simulations of Atomic Beam Sources	download		5:55 PM
Polarized Sources and Targets Session III	**Solid Polarized Targets**			
Room 389				
Thursday, October 9, 2008	2:00 PM - 4:00 PM			
Chair	Thorn			
Speaker	**Title**	**Talk**		**Time**
Wakui, Takashi	Polarized Proton Target for Radioactive Beam Experiments			2:00 PM
Piegsa, Florian	Polarized Solid Targets at PSI: Recent Developments and Novel Applications in DNP	download	avi file	2:30 PM
Zhao, J.K.	Dynamically polarized solid target for neutron scattering	download		3:00 PM
Nakajima, Takashi	First step toward ultrafast nuclear-spin polarization: All-optical control and direct detection of ultrafast electron-spin polarization using fs laser pulses	download		3:20 PM
Urrego-Blanco, Juan Pablo	Development of Polarized Targets for Reactions with Radioactive Ion Beams	download		3:40 PM
Polarized Sources and Targets Session IV	**Polarized Targets**			
Room 389				
Friday, October 10, 2008	2:00 PM - 4:00 PM			
Chair	A. Zelenski			
Speaker	**Title**	**Talk**		**Time**
Thorn, Craig	Precision Polarimetry of Solid HD and the Convergence of the GHD Integral	download		2:00 PM
Keith, Chris	The Jefferson Lab Frozen Spin Target	download		2:30 PM

Engels, Ralf	First experiments with the Polarized Internal Target at ANKE/COSY	download	2:50 PM
Brown, Brielin	Considerations for the Optimal Polarization of 3He Target Cells	download	3:10 PM
Katich, Joe	A Polarized 3He Target System at Jefferson Lab	download	3:30 PM

	Applications and Future Facilities	I		
	South Meeting Room			
	Tuesday, October 7, 2008	04:15 PM - 06:15 PM		
	Chair	*Gordon Cates*		
	Speaker	**Title**	**Talk**	**Time**
	Mike Romalis	Recent advances in atomic spin magnetometry	download	04:15 PM
	Wilson Miller	Medical imaging of hyperpolairzed gases	download	04:45 PM
	Kevin West	Exploration and Development of Ferromagnetic Half-Metallic Materials for Spin Torque Transfer Random Access Memory (STT RAM)		05:15 PM
	Wenjing Yin	Diluted Magnetic Ge1-xMnx Thin Films and Quantum Dots Synthesized by Multiple Ion Implantation		05:35 PM
	XiaoJun Liu	PHOTOINDUCED SPIN-DISORDER IN La0.7Sr0.3MnO3 FILM		05:55 PM

	Applications and Future Facilities	II		
	Room 389			
	Thursday, October 9, 2008	04:15 PM - 06:15 PM		
	Chair	*TBA*		
	Speaker	**Title**	**Talk**	**Time**
	Pavel Chevtsov.	Electron Spin Matching for ELIC, a High-Luminosity Ring-Ring Electron-Ion Collider		04:15 PM
x	Christian Eckardt.	Polarized Electrons for Experiments at Low Momentum Transfer SPIN @ S-DALINAC*		04:35 PM
	Yuji Goto.	Measurement of Drell-Yan process with polarized proton beams at J-PARC		04:55 PM
	Ralf Engels.	Precision Spectroscopy of Metastable Hydrogen with a Lamb-Shift Polarimeter		05:15 PM
	Mesude Saglam	Spin-Flip Investigation of 1s-2p and 2p-3d Excitonic Transitions in Terms of Dirac Hydrogen Atom and Flux Quantization Argument		05:35 PM
x	Xiaochun He.	PHENIX Detector Upgrade for Triggering Fast Muons from W-Boson Decays Using RPC Technology	download	05:55 PM

	Applications and Future Facilities	III	
	Room 389		
	Friday, October 10, 2008	04:15 PM - 05:55 PM	
	Chair	*Sebastian Kuhn*	
	Speaker	**Title**	**Time**
	Nilanga Liyanage.	Measurement of neutron spin asymmetry A1n in the high x valence quark region	04:15 PM
x	Ken Barish.	Nose-Cone Calorimeter Upgrade to the PHENIX Forward Spectrometers	04:35 PM
	Maki Kurosawa.	Spin Physics Capability with the PHENIX Silicon Vertex Tracker	04:55 PM
	Mibe Tsutomu.	Study of polarized sea quark distributions in polarized proton-proton collisions at sqrt(s) =500GeV with PHENIX	05:15 PM
x	James Sowinski	Future Investigations of the Flavor Dependence of Sea Quark Helicities at STAR	05:35 PM

SPIN2008 Participants List

Name	Affiliation	Email Address
Abell, Dan	Tech-X Corp	dabell@txcorp.com
Accardi, Alberto	Hampton U. / Jefferson Lab	accardi@jlab.org
Aghasyan, Mher	LNF INFN	mher@jlab.org
Ahmad, Saeed	University of Wisconson, Whitewater	sa8y@virginia.edu
Aidala, Christine	Brookhaven National Laboratory	caidala@bnl.gov
Anselmino, Mauro	Torino University & INFN	mauro.anselmino@to.infn.it
Asahi, Koichiro	Tokyo Institute of Technology	asahi@phys.titech.ac.jp
Atashbar Tehrani, Shahin	Semnan University & IPM	atashbar@ipm.ir
Aulenbacher, Kurt	Institut f|r Kernphysik, Universitdt Mai	aulenbac@kph.uni-mainz.de
Avagyan, Harut	Jefferson Lab	avakian@jlab.org
Averett, Todd	College of William and Mary	tdaver@wm.edu
Bacchetta, Alessandro	Jefferson Lab	pampa@phys.jdvu.ac.in
Bacher, Andrew	Indiana University Bloomington	bacher@indiana.edu
Bai, Mei	Brookhaven National Lab	mbai@bnl.gov
Balewski, Jan	MIT	balewski@mit.edu
Balling, Andreas	University of Bonn, ELSA	balling@physik.uni-bonn.de
Barber, Desmond	DESY, University of Liverpool	mpybar@mail.desy.de
Barish, Kenneth	UC Riverside	Kenneth.Barish@ucr.edu
Bazilevsky, Alexander	Brookhaven National Laboratory	shura@bnl.gov
Belostotski, Stanislav	PNPI, St. Petersburg	belostot@mail.desy.de
Belov, Aleksandr	Institute for Nuclear Research of RAS	belov@inr.ru
Bennett, Robert	Stony Brook University	rbennett@grad.physics.sunysb.edu
Bhattacharya, Aparajita	Jadavpur University	pampa@phys.jdvu.ac.in
Biselli, Angela	Fairfield University	biselli@jlab.org
Bordovitsyn, Vladimir	Tomsk State Unniversity	vabord@sibmail.com
Bosted, Peter	Jefferson Lab	bosted@jlab.org
Boyle, Kieran	RIKEN BNL Research Center	kboyle@bnl.gov
Bradamante, Franco	Trieste University and INFN	franco.bradamante@ts.infn.it
Brown, Brielin	Jefferson Lab	bcb5q@virginia.edu
Bueltmann, Stephen	Old Dominion University	sbueltma@odu.edu
Burkardt, Matthias	NMSU/Jefferson Lab	burkardt@nmsu.edu
Careccia, Sharon	Old Dominion University	careccia@jlab.org
Cates, Gordon	University of Virginia	cates@virginia.edu
Chao, Alex	Stanford Linear Accelerator Center	achao@slac.stanford.edu
Chen, Jian-ping	Jefferson Lab	jpchen@jlab.org
Chernitskiy, Alexander	St. Univ. of Engineering and Economics	AAChernitskii@engec.ru

Chevtsov, Pavel	Jefferson Lab	chevtsov@jlab.org
Chiladze, Badri	Inst. of High Energy Physics and Information	badri@hepi.edu.ge
Chiu, Mickey	Brookhaven National Lab	chiu@bnl.gov
Christova, Ekaterina	Inst. for Nucl. Res. & Nucl. Energy	echristo@inrne.bas.bg
Clarke, Jim	Daresbury Laboratory	j.a.clarke@dl.ac.uk
Clendinin, James	SLAC	clen@slac.stanford.edu
Contalbrigo, Marco	INFN Ferrara	contalbrigo@fe.infn.it
Conti, Francesco	University of Pavia-Dip. Nucl. Fis. Teo.	francesco.conti@pv.infn.it
Courant, Ernest	Brookhaven National Laboratory	ecourant@msn.com
Crabb, Donald	University of Virginia	dgc3q@galileo.phys.virginia.edu
da Silva e Silva, Marlene	University of Groningen and KVI	silva@kvi.nl
Dahiya, Harleen	National Institute of Technology	harleen_dahiya@yahoo.com
Datta, Amaresh	University of Massachusetts, Amherst	amaresh@physics.umass.edu
Day, Donal	University of Virginia	dbd@virginia.edu
De Nardo, Lara	DESY	ldenardo@mail.desy.de
Derbenev, Yaroslav	Jefferson Lab	derbenev@jlab.org
Deshpande, Abhay	SUNY @ Stony Brook	abhay.deshpande@stonybrook.edu
Deur, Alexandre	Jefferson Lab	deurpam@jlab.org
Dodge, Gail	Old Dominion University	gdodge@odu.edu
Dong, Xin	Lawrence Berkeley National Lab	XDong@lbl.gov
Dozono, Masanori	Kyushu University	dozono@phys.kyushu-u.ac.jp
Drachenberg, James	Texas A&M University	jdrachenberg@physics.tamu.edu
Dzhordzhadze, Vasily	Brookhaven National Lab	wasiko@bnl.gov
Eckardt, Christian	TU Darmstadt	eckardt@ikp.tu-darmstadt.de
El Aloaoui, Ahmed	LPSC	ahmed.el-alaoui@lpsc.in2p3.fr
Elbakyan,	Yerevan Physics Institute	elbakian@mail.desy.de
Ellinghaus, Frank	University of Mainz	ellingha@uni-mainz.de
Engels, Ralf	FZ Juelich	r.w.engels@fz-juelich.de
Erler, Jens	Universidad Nacional Autsnoma de Mixico	erler@fisica.unam.mx
Fidecaro, Giuseppe	CERN	giuseppe.fidecaro@cern.ch
Fields, Douglas	University of New Mexico	fields@unm.edu
Finger, Miroslav	Charles University in Prague	miroslav.finger@cern.ch
Fomin, Nadia	University of Tennessee	nfomin@utk.edu
Furukawa, Takeshi	Nishina-Center, RIKEN	takeshi@riken.jp
Galindo-Uribarri, Alfredo	Oak Ridge National Laboratory	uribarri@ornl.gov
Gamberg, Leonard	Penn State Berks	lpg10@psu.edu
Gao, Haiyan	Duke Univ./TUNL	gao@phy.duke.edu
Gardner, Susan	University of Kentucky	gardner@pa.uky.edu
Gericke, Michael	University of Manitoba	mgericke@physics.umanitoba.ca
Giordano, Francesca	Universita` di Ferrara	giordano@fe.infn.it
Gohn, Wes	University of Connecticut	gohn@jlab.org

Goldstein, Gary	Tufts University	gary.goldstein@tufts.edu
Goritschnig, Alexander	Karl-Franzens-University Graz	alexander.goritschnig@uni-graz.at
Goto, Yuri	Brookhaven National Lab	goto@bnl.gov
Grosse Perdekamp, Matthias	University of Illinois	mgp@illinois.edu
Guler, Nevzat	Old Dominion University	nguler@jlab.org
Gupta, Rajiv	Guru Nanak Dev University	rajiv_gupta_mirage@yahoo.com
He, Xiaochun	Georgia State University	xhe@gsu.edu
Hermann, Kolanoski	Humboldt University Berlin	kolanosk@ifh.de
Higinbotham, Douglas	Jefferson Lab	doug@jlab.org
Hillert, Wolfgang	University of Bonn, Physics Institute	hillert@physik.uni-bonn.de
Hoffman, Alan	MIT	ahoff@mit.edu
Honkanen, Heli	Iowa State University	heli@iastate.edu
Huang, Haixin	Brookhaven National Lab	huanghai@bnl.gov
Illarionov, Alexey	SISSA	illario@sissa.it
Imai, Kenichi	Kyoto University	imai@scphys.kyoto-u.ac.jp
Jiang, Xiaodong	Los Alamos National Lab	jiang@jlab.org
Jin, Ge	University of Virginia	gj7f@virginia.edu
Jones, Mark	Jefferson Lab	jones@jlab.org
Joosten, Rainer	HISKP University Bonn	rainer.joosten@cern.ch
Kacharava, Andro	IKP, Forschungszentrum Juelich	a.kacharava@fz-juelich.de
Kageya, Tsuneo	Jefferson Lab	kageya@jlab.org
Kaiser, Ralf	University of Glasgow	r.kaiser@physics.gla.ac.uk
Kalantarians, Narbe	University of Virginia	Narbe@jlab.org
Kalantar-Nayestanaki, Nasser	KVI/University of Groningen	nasser@kvi.nl
Kang, Zhongbo	Iowa State University	kangzb@iastate.edu
Kao, Chung-Wen	Chung-Yuan Christian university	cwkao@phys.cycu.edu.tw
Keith, Christopher	Jefferson Lab	ckeith@jlab.org
King, Paul	Ohio University	pking@jlab.org
Kleinjan, David	Brookhaven Nat'l Laboratory	david.kleinjan@email.ucr.edu
Kocoloski, Adam	MIT	kocolosk@mit.edu
Kohl, Michael	Hampton University and Jefferson Lab	kohlm@jlab.org
Koike, Yuji	Niigata University	koike@nt.sc.niigata-u.ac.jp
Kondratenko, Anatoliy	GOO Zaryad	kondratenkom@mail.ru
Koretune, Susumu	Shimane University	koretune@riko.shimane-u.ac.jp
Korolija, Milorad	Ruder Boskovic Institute	korolija@irb.hr
Kovacs, Kangkang	University of Virginia	kangkangl@gmail.com
Krimmer, Jochen	University of Mainz	krimmer@uni-mainz.de
KRISCH, A. D.	University of Michigan	krisch@umich.edu
Kucuker, Serpil	University of Virginia	sk4yc@virginia.edu
Kuhn, Sebastian	Old Dominion University	skuhn@odu.edu
Kumar, Krishna	University of Massachusetts, Amherst	kkumar@physics.umass.edu

Kunne, Fabienne	CEA Saclay	fabienne.kunne@cea.fr
Kurek, Krzysztof	So3tan Institute for Nuclear Studies	kurek@fuw.edu.pl
Kurosawa, Maki	RIKEN	kurosawa@riken.jp
Kyle, Gary	New Mexico State University	kyle@nmsu.edu
Lajoie, John	Iowa State University	lajoie@iastate.edu
Leader, Elliot	Imperial College London	e.leader@imperial.ac.uk
Lednicky, Richard	JINR	lednicky@fzu.cz
Lenisa, Paolo	Universita di Ferrara and INFN	lenisa@mail.desy.de
Leonova, Maria	University of Michigan	leonova@umich.edu
Lin, Chuanzhe	University of Virginia	cl2jd@virginia.edu
Lin, Fanglei	Brookhaven National Lab	flin@bnl.gov
Lin, Huey-Wen	Jefferson Lab	hwlin@jlab.org
Lindgren, Richard	University of Virginia	ral5q@virginia.edu
Liu, Chen-Yu	Indiana University	CL21@indiana.edu
Liu, Han	Los Alamos National Lab	hliu@lanl.gov
Liu, Ming	Los Alamos National Lab	mliu@lanl.gov
Liuti, Simonetta	University of Virginia	sl4y@virginia.edu
Liyanage, Nilanga	University of Virginia	nilanga@virginia.edu
Long, Elena	Kent State University	ellie@jlab.org
Lorenzon, Wolfgang	University of Michigan	lorenzon@umich.edu
Lu, Zhun	Universidad Tecnica Federico Santa Maria	zhun.lu@usm.cl
Luccio, Alfredo U	Brookhaven National Laboratory	luccio@bnl.gov
Ma, Bo-Qiang	Peking University	mabq@phy.pku.edu.cn
Maas, Frank	GSI & Mainz University	f.maas@gsi.de
Mackay, William	Brookhaven National Laboratory	waldo@bnl.gov
Makdisi, Yousef	BNL C-AD	makdisi@bnl.gov
Makins, Naomi	University of Illinois Urbana-Champaign	makins@uiuc.edu
Mallot, Gerhard	CERN	Gerhard.Mallot@cern.ch
Martin, Anna	Trieste University and INFN	anna.martin@ts.infn.it
Marukyan, Hrachya	Yerevan Physics Institute	marukyan@mail.desy.de
Maxwell, James	University of Virginia	elmohoo@gmail.com
Meissner, Stephan	Institut fuer Theoretische Physik II	stephan.meissner@tp2.rub.de
Mekfhi, Mustapha	Universiti Oran	mekhfi@gmail.com
Metz, Andreas	Temple University	metza@temple.edu
Meziani, Zein-Eddine	Temple University	meziani@temple.edu
Mibe, Tsutomu	KEK	mibe@post.kek.jp
Mikheil, Nioradze	High Energy Physics Institute	niorad@hepi.edu.ge
Milner, Richard	MIT	milner@mit.edu
Minehart, Ralph	University of Virginia	rcm4y@virginia.edu
Miyachi, Yoshiyuki	Tokyo Tech	miyachi@nucl.phys.titech.ac.jp
Mochalov, Vasily	Institute for High Energy Physics	mochalov@ihep.ru

Montag, Christoph	Brookhaven National Laboratory	montag@bnl.gov
Morozov, Vasily	University of Michigan	morozov@umich.edu
Morreale, Astrid	University of California	astrid@bnl.gov
Mukherjee, Asmita	IIT Bombay	asmita@phy.iitb.ac.in
Mulholland, Jonathan	University of Virginia	jrm2zq@virginia.edu
Mussgiller, Andreas	Univertity of Erlangen	andreas.mussgiller@desy.de
Musulmanbekov, Zhenis	Joint Institute for Nuclear Research	genis@jinr.ru
Myhrer, Fred	University of South Carolina	myhrer@physics.sc.edu
Nakajima, Takashi	Inst. of Adv. Energy, Kyoto University	t-nakajima@iae.kyoto-u.ac.jp
Nakano, Kenichi	RIKEN	knakano@bnl.gov
Nasr, Mustafa Abdusalam	7th October University	bayio@yahoo.com
Nass, Alexander	University of Erlangen-N|rnberg	Alexander.Nass@physik.uni-erlangen.de
Neal, Homer	University of Michigan	haneal@umich.edu
Negrini, Teresa	HISKP - University of Bonn	Teresa.Negrini@gmail.com
Nguyen, Huong	University of Virginia	htn3r@virginia.edu
Nioradze, Mikheil	High Energy Physics Institute of TSU	niorad@hepi.edu.ge
Norum, Blaine	University of Virginia	ben@virginia.edu
Oakley, Chrostopher	Georgia State University	coakley2@student.gsu.edu
O'Brien, Edward	Brookhaven National Laboratory	eobrien@bnl.gov
Onderwater, Cornelis	KVI / Univ. of Groningen	onderwater@kvi.nl
Paschke, Kent	University of Virginia	paschke@virginia.edu
Pasquini, Barbara	University of Pavia	pasquini@pv.infn.it
Pate, Stephen	New Mexico State University	pate@nmsu.edu
Paver, Nello	University of Trieste and INFN	nello.paver@ts.infn.it
Pentchev, Lubomir	College of William and mary	pentchev@jlab.org
Penttila, Seppo	Oak Ridge National Laboratory	penttilasi@ornl.gov
Piegsa, Florian	Paul Scherrer Institute	florian.piegsa@psi.ch
Pierce, Josh	Oak Ridge National Laboratory	piercejj@ornl.gov
Plyku, Donika	Old Dominion University	dplyku@odu.edu
Poelker, Matt	Jefferson Lab	poelker@jlab.org
Poljak, Nikola	University of Zagreb	npoljak@phy.hr
Poltoratska, Yuliya	Darmstadt University of Technology (TUD)	yuliya@ikp.tu-darmstadt.de
Procureur, Sibastien	CEA-Saclay	Sebastien.Procureur@cea.fr
Ptitsyn, Vadim	Brookhaven National Laboratory	vadimp@bnl.gov
Qian, Xin	Duke University	xqian@jlab.org
Qiang, Yi	Duke University	yqiang@jlab.org
Quayle, William	University of Wisconsin-Madison	wbquayle@gmail.com
Radici, Marco	INFN - Sezione di Pavia	radici@pv.infn.it
Ralston, John	University of Kansas	ralston@ku.edu
Ramsey, Gordon	Loyola University Chicago	gpr@hep.anl.gov
Rathmann, Frank	Institut fuer Kernphysik, FZ Juelich	f.rathmann@fz-juelich.de

Raymond, Richard	University of Michigan	rraymond@umich.edu
Razzague, Abdur	Oak Trading	weimikus@yahoo.com
Riedl, Johann	University of Regensburg, Germany	Johann.Riedl@physik.uni-r.de
Riordan, Seamus	University of Viriginia	riordan@jlab.org
Rith, Klaus	University of Erlangen	klaus.rith@desy.de
Rivas, Martin	The University of the Basque Country	martin.rivas@ehu.es
Romalis. Michael,	Princeton University	romalis@princeton.edu
Rondio, Ewa	Soltan Institute for Nuclear Studies	ewar@fuw.edu.pl
Rondon, Oscar	University of Virginia	or@virginia.edu
Roser, Thomas	Brookhaven National Laboratory	roser@bnl.gov
Saito, Naohito	KEK	nsaito@post.kek.jp
Sakuma, Tai	MIT	sakuma@mit.edu
Sarsour, Murad	Georgia State University	msar@gsu.edu
Schlegel, Marc	Jefferson Lab, Theory Center	schlegel@jlab.org
Seele, Joseph	MIT	seelej@mit.edu
Semertzidis, Yannis	Brookhaven National Lab	yannis@bnl.gov
She, Jun	Peking University	10604805@pku.edu.cn
Sichtermann, Ernst	Lawrence Berkeley National Laboratory	EPSichtermann@lbl.gov
Silenko, Alexander	Research Institute for Nuclear Problems	silenko@inp.minsk.by
Silwal, Rupesh	University of Virginia	rs3te@virginia.edu
Singh, Jaideep	University of Virginia	js7uq@virginia.edu
Slifer, Karl	University of New Hampshire	karl.slifer@unh.edu
Song, Xiaotong	University of Virginia	songx3452@yahoo.com
Sowinski, James	Indiana University Cyclotron Facility	sowinski@indiana.edu
Stamenov, Dimiter	Institute for Nuclear Research	stamenov@inrne.bas.bg
Stancari, Michelle	Universita di Ferrara	michelle@fe.infn.it
Steffens, Erhard	University of Erlangen-N|rnberg	steffens@physik.uni-erlangen.de
Stepanyan, Samvel	Kyungpook National University	samuel@jlab.org
Stephenson, Edward	Indiana University Cyclotron Facility	stephene@indiana.edu
Stibunov, Victor	Nuclear Physics Institute of Tomsk PU	stib@npi.tpu.ru
Stratmann, Marco	University of W|rzburg	marco@ribf.riken.jp
Strvher, Hans	Forschungszentrum J|lich GmbH	h.stroeher@fz-juelich.de
Subedi, Ramesh	University of Virginia	rsubedi@jlab.org
Sukhoruchkin, Sergey	Petersburg Nuclear Physics Institute	sergeis@pnpi.spb.ru
Sulkosky, Vincent	Jefferson Lab	vasulk@jlab.org
Surrow, Bernd	MIT, Laboratory for Nuclear Science	surrow@MIT.EDU
Tanaka, Kazuhiro	Juntendo University	tanakak@sakura.juntendo.ac.jp
Tanaka, Masayoshi	Kobe Tokiwa University	tanaka@rcnp.osaka-u.ac.jp
Taneja, Swadhin	SUNY STONY BROOK	taneja@skipper.physics.sunysb.edu
Thorn, Craig	Brookhaven National Laboratory	thorn@bnl.gov
Tobias, Al	University of Virginia	wat4y@virginia.edu

Troshin, Sergey	IHEP, Protvino	Sergey.Troshin@ihep.ru
Tyukin, Valery	Inst. of Nuclear Physics, Uni Mainz	tioukine@kph.uni-mainz.de
Tyurin, Nikolay	Institute for High Energy Physics	tyurin@ihep.ru
Urrego-Blanco, Juan P	Paul Scherrer Institute	jurregob@utk.edu
Uzikov, Yury	Joint Institute for Nuclear Researches	uzikov@nusun.jinr.ru
Van Oers, Willem	University of Manitoba / TRIUMF	vanoers@triumf.ca
Vigdor, Steven	Brookhaven National Laboratory	vigdor@bnl.gov
Vladimir, Ladygin	Joint Institute for Nuclear Research	vladygin@jinr.ru
Vogelsang, Werner	Brookhaven National Laboratory	wvogelsang@bnl.gov
Vossen, Ansel	University of Illinois	vossen@illinois.edu
Voutier, Eric	LPSC/IN2P3?CNRS - UJF - INP	voutier@lpsc.in2p3.fr
Wakasa, Tomotsugu	Kyushu University	wakasa@phys.kyushu-u.ac.jp
Wakui, Takashi	CYRIC, Tohoku University	wakui@cyric.tohoku.ac.jp
Wang, xiaorong	New Mexico State University	xrwang@bnl.gov
Wei, Feng	Iowa State University	weifengisu@gmail.com
Weiss, Christian	Jefferson lab	weiss@jlab.org
Werner, Meyer	Ruhr-University Bochum	meyer@ep1.rub.de
Windmolders, Roland	University of Bonn	roland.windmolders@cern.ch
Wissink, Scott	Indiana University	wissink@indiana.edu
Wollny, Heiner	University of Freiburg	Heiner.Wollny@cern.ch
Yoshimi, Akihiro	RIKEN Nishina Center	yoshimi@ribf.riken.jp
Yuan, Feng	Lawrence Berkeley National Lab/RBRC	fyuan@lbl.gov
Zavada, Petr	Institute of Physics, Prague	zavada@fzu.cz
Zelenski, Anatoli	Brookhaven National Laboratory	zelenski@bnl.gov
Zetta, Luisa	INFN and University of Milano	luisa.zetta@mi.infn.it
Zhang, Bing	Peking University	bingzhang@pku.edu.cn
Zheng, Xiaochao	University of Virginia	xz5y@virginia.edu
Zhou, Jian	Lawrence Berkeley National Lab	Jzhou@lbl.gov
Zihlmann, Benedikt	Jefferson Lab	zihlmann@jlab.org

PART TWO

Workshop on Polarized Electron Sources and Polarimeters

October 1 -3, 2008
Jefferson Lab, Newport News, VA 23606 USA
Organizers: J. Grames, M. Stutzman, R. Suleiman, M. Poelker

Status Report of the New Darmstadt Polarized Electron Injector

Y. Poltoratska*, R. Barday*, U. Bonnes*, M. Brunken*, C. Eckardt*, R. Eichhorn*, J. Enders*, C. Heßler*, C. Ingenhaag*, W.F.O. Müller†, M. Platz*, M. Roth*, B. Steiner†, M. Wagner* and T. Weiland†

**Institut für Kernphysik, Technische Universität Darmstadt, Germany*
†Institut für Theorie Elektromagnetischer Felder, Technische Universität Darmstadt, Germany

Abstract. We present the status of the polarized injector for the superconducting Darmstadt electron linear accelerator S-DALINAC. An offline teststand of the source was built to test all components of the S-DALINAC polarized injector SPIN. The new electron source will deliver a 100 keV polarized beam and complement the present thermionic source operating at 250 kV. Results of the beam properties measurements will be introduced and an outlook on the upcoming installation of the new injector at the S-DALINAC will be given.

Keywords: S-DALINAC, strained superlattice GaAs, Mott polarimeter, Wien filter
PACS: 85.60.Ha, 79.60.-i, 29.27.Hj

INTRODUCTION

The superconducting Darmstadt electron linear accelerator S-DALINAC [1] is one of very few electron accelerators devoted to nuclear structure physics, including electron-scattering experiments. To complement the present scientific programm with the experiments that are focused on the investigation of violation of parity in the nuclei and breakup reactions of light nuclei [2] the new S-DALINAC Polarized injector (SPIN) was designed. It will provide - complementary to the existing 250 kV thermionic electron gun 100 keV polarized beams with polarization of 80 %, a mean current of 60 μA and 3 GHz cw or subharmonic time structure. The concept of the MAMI Gun [3] was used as a starting point for SPIN. To test the new developed polarized electron source independently from the operation of the S-DALINAC, an standalone teststand has been developed. In this contribution we describe the current status of the teststand and present the results of the measurements of the different beam properties.

TEST STAND OF THE POLARIZED INJECTOR

Figure 1 shows an schematic overview over the teststand of the polarized injector. The electrons are produced at a GaAs cathode by photoemission. As cathode material the strained superlattice crystals [4] are foreseen to achieve high degrees of polarization of more than 80%. Due to the limited space in the accelerator hall a very compact source had to be constructed, and major modifications from the Mainz design were necessary. One spin direction is selected by using circularly polarized light from diode laser system

CP1149, *Spin Physics, 18th International Spin Physics Symposium*
edited by D. G. Crabb, D. B. Day, S. Liuti, X. Zheng, M. Poelker, and Y. Prok

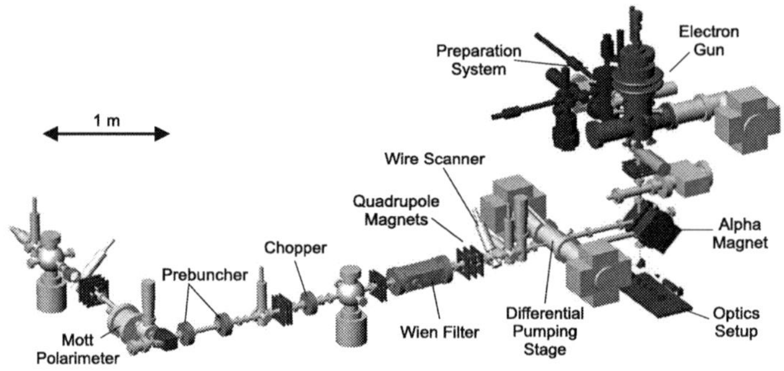

FIGURE 1. Schematic drawing of the teststand.

below the electron source focussed onto the photocathode surface. The photocathode crystal is installed at the front end of a highly polished stainless-steel electrode. The form of the electrode was designed following numerical simulations [5] and deviates from usual Pierce geometry [6]. To extract the electrons from the conduction band of the cathode crystal to the vacuum negative electron affinity (NEA) is needed. Therefore prior to installation into the cathode chamber the cathode is cleaned and activated by depositing thin CsO layer in a separate preparation chamber. With the aid of a load-lock system the photocathode activation can be handled quickly.

Afterwards the emitted electrons are accelerated electrostatically to an energy of 100 keV and injected by an alpha magnet into a horizontal beamline, where the beam properties are measured with various diagnostic elements. It is indispensable to create an ultra high vacuum conditions in the gun chamber to prevent a fast degradation of the NEA surface. After bake-out during 12 days at 220° C an end pressure of $< 2 \cdot 10^{-11}$ mbar has been achieved using non-evaporable getter (NEG) and ion-getter pumps.

Transverse beam quality

The transverse beam properties can be investigated qualitatively by fluorescent screens and quantitatively by a wire scanner [7]. The transverse emittance is determined by measuring of the beam radius for different focussing strengths of a solenoid preceding the wire scanner. The normalized transverse emittance has been determined to $\varepsilon_{n,x} = (0.146 \pm 0.037)$ mm mrad and $\varepsilon_{n,y} = (0.197 \pm 0.089)$ mm mrad, respectively.

Time Structure

The produced polarized electron beam needs to be modulated with the frequency of the S-DALINAS's superconducting cavities of 3 GHz for cw operation. This is achieved by modulating a laser diode [8]. The time structure was verified for the laser beam with a fast sensor and for the electron beam with an ultra-fast coaxial Faraday cup. For the capturing process of the beam into the superconducting accelerating structures the electron-bunch length need to be of the order of 5 ps. Therefore a new chopper and prebuncher system has been developed, based on a similar setup at MAMI [9].

Polarization manipulation and measurement

The degree of polarization of 100 keV electrons for the teststand is measured with a Mott polarimeter detecting count-rate asymmetries in electron scattering on nuclei due to the spin-orbit interaction. The asymmetry occurs only when the electron spin direction is not parallel to the scattering plane. Thus the initially longitudinal oriented spins must be rotated by 90° into the transverse direction. This is done by a Wien filter [10]. This device is also necessary tool for adjusting the spin orientation at different experimental areas of the S-DALINAC.

The ideal Wien filter acts as a drift space in the vertical plane, but its horizontal electron-optical properties are similar to those of a quadrupole: It focusses the electron beam. This effect was minimized: the focus has been set inside the filter with help of newly designed compact quadrupole triplet in front of the Wien filter. An additional quadrupole doublet installed after the device compensates its influence on the beam. The installed spin rotator can provide 100 degree spin rotation.

The 100 keV Mott polarimeter [11] houses 4 silicon-surface-barrier detectors which are oriented in two perpendicular scattering planes to measure both transverse polarization components simultaneously. As targets self-supporting gold foils of the thickness in the range from 42.5 nm to 500 nm are utilized. The polarimeter contains also a BeO viewer to control the beam position and one empty target for background measurements. The scattering angle of $(120 \pm 1.5)^\circ$ is determined by an aluminum collimator with 2 mm diameter. This angle matches the broad maximum of the analyzing power or so-called Sherman function at this angle. The typical energy resolution of the detector is

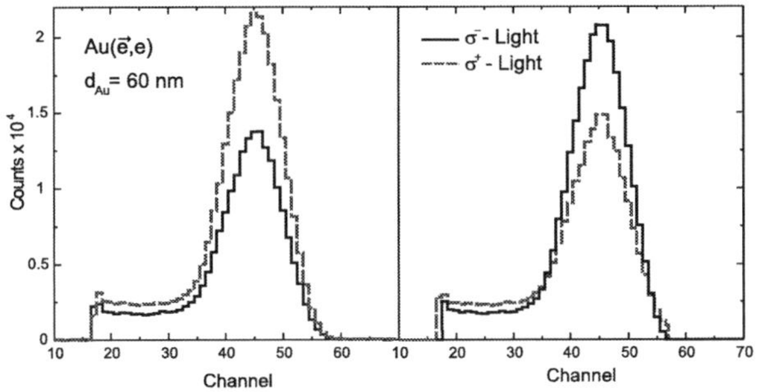

FIGURE 2. Measurement of Mott spectra with the two detectors of a scattering plane for reversing electron helicities. The helicity can be switched by reversing the polarization of the laser light.

about 20 keV and the peak-to-background ratio is found to be 14 : 1 for the 122 nm gold foil. The polarization of the electrons can be switched from parallel to antiparallel to the electron momentum by changing the incident light helicity.

To be independent from possible instrumental asymmetries two different measurements with opposite beam polarization must be performed.Typically measured spectra are shown in Fig. 2. From the measured data a degree of polarization was determined to be $(33.4 \pm 1.6)\%$ for bulk GaAs cathode and $(86 \pm 3)\%$ for strained superlattice cathode. Thereby the finite target thickness was taken into account by performing an extrapolation procedure to zero target thickness. A detailed explanation is given in reference [12].

SUMMARY AND OUTLOOK

In this paper we showed that the teststand for the S-DALINAC polarized injector has successfully been set up and put into operation. The measured transverse beam properties is comparable with others electron sources and the measured degrees of polarization for different cathode types is in good agreement with the literature. The next steps are the test of the chopper and prebuncher systems and implementation of the source at the S-DALINAC [13].

ACKNOWLEDGMENTS

We would like to thank K. Aulenbacher for useful discussions about Mott Polarimeter. Our thanks also due to SPIN 2008 organizing committee for financial support. This work is supported by the Deutsche Forschungsgemeinschaft through SFB 634.

REFERENCES

1. A. Richter, "Operational experience at the S-DALINAC", EPAC'96, Sitges, 1996, p. 110.
2. C. Eckardt et al., these proceedings.
3. K. Aulenbacher et al., *Nucl. Instrum. Methods in Phys. Research A* **391**, 1997, p. 498.
4. Yu. A. Mamaev et al.,"Spin polarized electron transport and emission from strained semiconductor heterostructures", SLAC-PUB-8391, 2000.
5. B. Steiner et al., "Recent Simulation Results of the Polarized Electron Injector (SPIN) of the S-DALINAC", EPAC'06, Edinburgh, 2006 p. 2188.
6. D. Pierce and F. Meier, *Phys. Rev. B*, **13**, 1976, pp. 5484.
7. C. Eckardt, "Emittanzmessung an der Quelle spinpolarisierter Elektronen am S-DALINAC", Diploma Thesis, TU Darmstadt, 2007, unpublished.
8. C. Heßler, "Commisioning of the offline-teststand for the S-DALINAC polarized injector SPIN", EPAC'08, Genoa, 2008, p. 1476.
9. V. I. Shvedunov et al., "Design of a Prebuncher for Increased Longitudinal Efficiency of MAMI", EPAC'96, Sitges, 1996, p. 1556.
10. B. Steiner et al.,"Wien filter as a spin rotator at low energy", PAC'07, Albuquerqe, 2007, p.170.
11. Y. Poltoratska, " Design and setup of a compact Mott polarimeter for the future S-DALINAC polarized injector", Diploma thesis, Kharkiv Karazin National University and TU Darmstadt, 2005, unpublished.
12. R. Barday et al., "Electron beam polarimetry at S-DALINAC", EPAC'08, Genoa, 2008 p. 1476.
13. Y. Poltoratska et al. "Polarized electrons in Darmstadt: recent developments", these proceedings.

Status of 200keV Beam Operations at Nagoya University

M. Yamamoto[a], T. Konomi[a], S. Okumi[a], Y. Nakagawa[a], N. Yamamoto[b], M. Tanioku[b], X. Jin[b], T. Ujihara[b], Y. Takeda[b], F. Furuta[c], H. Matsumoto[c], M. Yoshioka[c], M. Kuriki[d], C. Shonaka[d], and T. Nakanishi[a]

[a]*Graduate School of Science, Nagoya University, Furo-cho, Chikusa-ku, Nagoya, Aichi, Japan 464-8602*
[b]*Graduate School of Engineering, Nagoya University, Furo-cho, Chikusa-ku, Nagoya, Aichi, Japan 464-8603*
[c]*High Energy Accelerator Research Organization (KEK), 1-1 Oho, Tsukuba, Ibaraki, Japan 305-0801*
[d]*Graduate School of Advanced Sciences of Matter, Hiroshima University, 1-3-1 Kagamiyama, Higashi-Hiroshima, Hiroshima, Japan 739-8530*

Abstract. We have developed a 200keV polarized electron source for generating high bunch charge of >3.2nC within ~1ns bunch duration for the International Linear Collider (ILC) and >7.7pC within 30ps bunch duration for a basic source study of an Energy Recovery Linac (ERL). In order to improve gun operation voltage and an operational lifetime, we have employed new electrode materials of molybdenum and titanium to eliminate field emission dark current, and restructured a beam transport vacuum systems. As a beam generation for the ERL, bunch charge of up to 28pC was generated by irradiating with sliced laser pulse trains. The details are described in this report.

Keywords: Electron source; Polarized beams; Vacuum systems
PACS: 29.25.Bx; 29.27.Hj; 07.30.Kf

200KEV GUN AND BEAM TRANSPORT SYSTEM

A 200-keV polarized electron source has been developed for generating highly polarized electron beams with high bunch charge for requirements of ILC [1]. Remaining problems of this source are photocathode QE lifetime in the condition with actual beam current of more than 50μA and stable gun high voltage operation.

An ultra high vacuum ($\leq 10^{-9}$ Pa) environment of a beam transport system as well as of the gun is important to maintain NEA surface during beam operation. In order to avoid vacuum increase due to outgassing from a beam dump, we have applied four NEG modules (WP950) and a 100 L/s ion pump to the beam dump system, and two NEG pumps (GP100MK5) to a differential pumping system (DPS). The vacuum pressure of the beam transport line is important information for a photocathode QE lifetime measurement. The DPS vacuum pressure was monitored by an extractor-type

CP1149, *Spin Physics, 18th International Spin Physics Symposium*
edited by D. G. Crabb, D. B. Day, S. Liuti, X. Zheng, M. Poelker, and Y. Prok

vacuum gauge (AxTRAN X-11,ULVAC). By comparison of the DPS base pressure with and without NEG modules in a condition with 20μA beam dumping for ~10 minutes, the DPS base pressure of $1.2x10^{-9}$ Pa was increased to $1.7x10^{-9}$ Pa and $1.4x10^{-8}$ Pa, respectively.

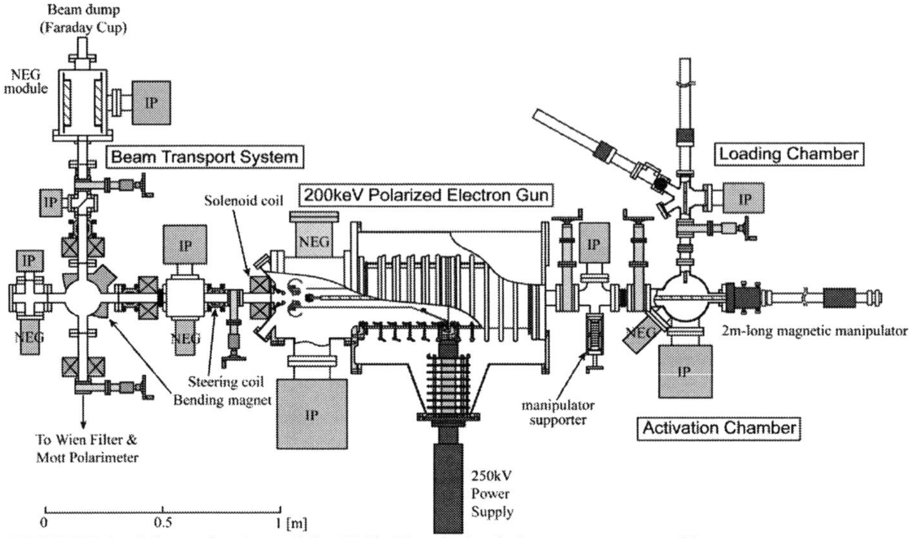

FIGURE 1. Schematic view of the 200keV polarized electron source and beam transport system.

To reduce the dark current in a high field gradient condition, we have employed a molybdenum cathode and titanium anode (Mo-Ti) [2]. A result of Mo-Ti electrode conditioning is shown in Fig. 2a.

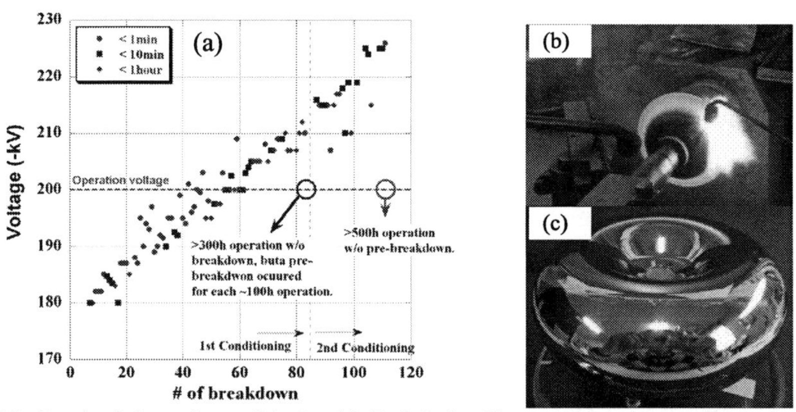

(b)

FIGURE 2. Result of electrode conditioning (a). Red circles, blue squares and green diamonds show high voltage duration less than 1 minute, 10 minutes and 1 hour respectively. The pictures show molybdenum cathode electrode fabrication processes of hot squeezing (b) and finished electrode (c).

The gradual increase of dark current observed before a breakdown in the SUS electrode was hardly detectable at a resolution of 1 nA, and breakdowns occurred suddenly. Thus, HV-conditioning was performed repeatedly by increasing the bias voltage up to the previous breakdown voltage. The breakdown voltage grew by ~0.4 kV per discharge. No significant damage was observed on either the Mo or Ti surface after more than 100 discharges. This feature is a great advantage of the new electrode.

After the 80 conditionings, the breakdown voltage had grown to 215 kV, though the dark current remained below 1 nA for this surface state under 200-kV operation. However, a pre-breakdown phenomenon was observed for this surface at a rare rate of once per 100 hours, causing fatal damage to the PC surface. In order to avoid this problem, the conditioning voltage was increased to 225 kV, for which the pre-breakdown phenomenon completely disappeared for more than 500 hours at 200-kV operation.

LIFETIME MEASUREMENT AND 50μA BEAM OPERATION

A dark-lifetime for bulk-GaAs PC with a -200 kV bias voltage applied to the cathode was measured under a vacuum of 2.3×10^{-9} Pa in the gun chamber (Fig. 3a). The dark current of the 200 kV electrode was well suppressed to be less than 1nA level during the measurement. The QE was once decreased for about 10 hours after the NEA surface activated, and it gradually increased again in several tens hours until the measurement was stopped. The initial QE was maintained for several hundred hours, showing that the effects of dark current and residual molecules adsorption are sufficiently suppressed in our 200-keV gun.

An operational-lifetime was measured under this vacuum condition with an extraction current of 50 μA. The laser diameter on the photocathode was adjusted to be 1 mm to maintain high beam transportation efficiency to the beam dump, and then the transportation efficiency of 96% was obtained at this experiment. The result of the operational lifetime was shown in Fig. 3b. The initial DPS vacuum pressure $1.2\text{x}10^{-9}$ Pa was slightly increased to $2.8\text{x}10^{-9}$ Pa and the initial QE of 4.3% was decreased to 1/e value for ~120 hours.

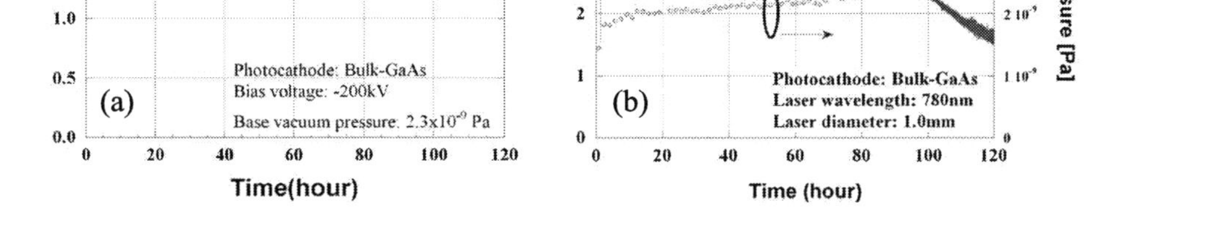

FIGURE 3. Results of dark-lifetime measurement with connecting beam transport line (a), and operational-lifetime measurement in the condition of 50μA constant beam current (b).

In order to research QE degradation behavior, a QE scan was done at the activation chamber. A laser light for scanning QE was transported by a single mode optical fiber and focused on ϕ 0.1 mm at the photocathode. The optical fiber system was fixed on the XY micro-motion stage and scanned for every 1 mm-steps for each direction. The Fig. 4a shows a flat QE distribution of 5~6% after NEA activation process, and Fig. 4b shows a degraded QE distribution after 50 μA beam operation. Comparing the data, it is clear that the damaged NEA surface region of ~ϕ10 mm is too large, while the electron emission area is considered to be as small as ϕ1 mm. It is presumed that the NEA damage is caused by back-bombardment not only by positive ions produced along the beam path between the HV electrodes but also by ions or outgases produced by the beam losses in the beam transport chambers [3,4].

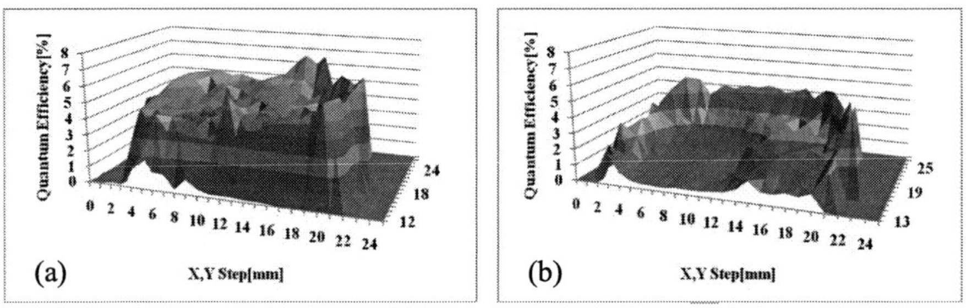

FIGURE 4. Result of QE scans after NEA activation (a) and 50μA beam operation (b).

PICO-SECOND BUNCH GENERATION

A high voltage DC-gun with an NEA-GaAs photocathode is considered to be major candidate to generate ultra-low emittance ($\varepsilon_{n.rms}$ < 1 mm.mrad) beams for the ERL. A pepper-pot (PP) method has been applied to measure the initial beam emittance of bulk-GaAs and GaAs-GaAsP superlattice photocathode by using the 200-keV gun system [5]. A duty factor controlled Ti:Sapphire mode-lock laser was employed because a thin film scintillator of the PP system is easy to charge up and to be damaged by a discharging under the beam average current of the order of more than micro-amperes. The Ti:Sapphire mode-lock (frequency of 81.25 MHz) laser output was sliced out 6.3 μs width with 20 Hz repetition. Since the rise and fall time of the pockels cell is less than 30 ns, less than 5 imperfect pulses contained in one sliced pulse-train which consists of ~500 pulses. By irradiating these sliced pulse trains with pulse length of ~30 ps and wavelength of 807 nm, the bunch charge of 28 pC could be generated from the gun (Fig. 5).

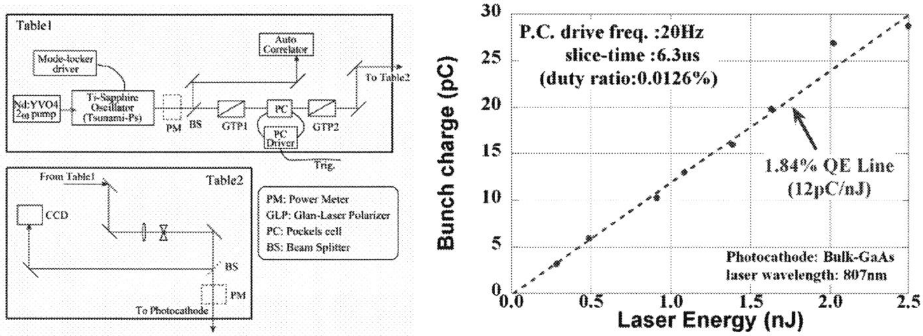

FIGURE 5. A schematic view of the Ti:Sapphire mode-lock laser system and the pulse slice system (left). A result of bunch generation from bulk-GaAs by illuminating sliced laser pulse trains (right).

CONCLUSION

The 200keV load-lock gun have been developed and operated more than 500 hours under condition with a low dark current of <1nA. The operational-lifetime measurement of 50 μA beam current was done and obtained ~120 hours that is available operation level for the ILC electron source. For the pico-second bunch generation, the bunch charge of 28 pC from bulk-GaAs could be emitted by irradiating sliced laser pulse trains. The emittance measurement with the bunch charge of more than 7.7 pC is underway.

ACKNOWLEDGMENTS

We wish to thank Mrs. K.Yamauchi, T.Katoh (Yorodu-Shibori Co., Ltd), T.Sashida, H.Uemori (Tosei Electrobeam Co., Ltd.), K.Satoh (Ultra Finish Technology Co., Ltd.), N.Takayama (SAKURAI Co., Ltd), T.Gotoh, M.Katsuki and K.Itoh (Technical Center of Nagoya University) for kindness support for fabricating the molybdenum and titanium electrode.

This work was has been supported partially by Grant-in-Aid for Scientific Research (No. 15204019, No. 20740129) and a research fund of KEK for cooperative development (No.2006-022, No.2007-023).

REFERENCES

1. T. Nakanishi et al., KEK Report 97-01 (1997) p.36-48
2. F. Furuta et al., Nucl. Instru. Meth. A538 (2005) p.33-44.
3. K. Aulenbacher et al., Proceedings of SPIN2004 p975-979
4 C.K. Sinclair et.al., Phys. Rev. ST Accel. Beams 10, 023501 (2007).
5 N. Yamamoto et.al., J. Appl. Phys. 102, 024904 (2007)

Polarized Photocathode R&D for Future Linear Colliders

F. Zhou, A. Brachmann, T. Maruyama, and J. C. Sheppard

SLAC National Accelerator Laboratory, 2575 Sand Hill Road, Menlo Park, CA 94025, USA.

Abstract. It is a challenge to generate full charge electrons from the electron sources without compromising polarization for the proposed ILC and CLIC. It is essential to advance polarized photocathodes to meet the requirements. SLAC has worldwide unique dedicated test facilities, Cathode Test System and dc-Gun Test Laboratory, to fully characterize polarized photocathodes. Recent systematic measurements on a strained-well InAlGaAs/AlGaAs cathode at the facilities show that 87% polarization and 0.3% QE are achieved. The QE can be increased to ~1.0% with atomic hydrogen cleaning. The surface charge limit at a very low current intensity and the clear dependence of the polarization on the surface charge limit are observed for the first time. On-going programs to develop photocathodes for the ILC and CLIC are briefly introduced.

Keywords: photocathode, photoemission, polarized beam.
PACS: 85.60.Ha, 79.60.-i, 29.27.Hj.

CHALLENGES OF THE ILC AND CLIC: HIGH-CURRENT HIGH-POLARIZATION ELECTRON SOURCES

Extensive developments on high-current high-polarization electron sources for the proposed International Linear Collider (ILC) and CERN Compact Linear Collider (CLIC) are being carried out at the SLAC and JLAB in the United States. Table 1 summarizes the major parameters of the polarized electron sources for the linear colliders. The major challenge for the ILC electron source is to generate full charge 1-ms-long macropulse without compromising polarization while the one for the CLIC is to generate a train of only-100-ps-long full-charge micropulses. In principle, the full charge production could be limited by space charge and/or surface charge effects in the linear colliders. A high quality polarized photocathode is required to overcome the surface charge limit, while a high-voltage dc-gun is needed to suppress the space charge effect. In addition, the high average beam current from the dc-gun sources for the linear colliders may cause serious ion-back bombardment onto the cathode, which may badly affect the QE lifetime due to the deterioration of the cathode. It requires us to explore solutions to reduce ions onto the cathode, such as cryo-pumping – condensing the gases onto the wall by cooling down the gun to a very low temperature, typically at 20-30 K. The major goals of the photocathode developments at the SLAC for the ILC and CLIC: 1) demonstration of full charge production without space charge and surface charge limit; 2) >85% polarization; 3) ~1% QE and long QE lifetime.

CP1149, *Spin Physics, 18th International Spin Physics Symposium*
edited by D. G. Crabb, D. B. Day, S. Liuti, X. Zheng, M. Poelker, and Y. Prok

TABLE 1. Major parameters of the ILC and CLIC high-current high-polarization electron sources.

Parameters	ILC	CLIC
Electrons Per Microbunch	$3x10^{10}$	$6x10^9$
Number Of Microbunch	2625	312
Width Of Microbunch	1 ns	100 ps
Time Between Microbunches	360 ns	500.2 ps
Width Of Macropulse	1 ms	156 ns
Macropulse Repetition Rate	5 Hz	50 Hz
Charge Per Macropulse	12600 nC	300 nC
Average Current From Dc-Gun	63 μA	15 μA
Peak Current Of Microbunch	4.8 A	9.6 A
Current Intensity (1cm Radius)	$1.5 A/cm^2$	$3.0 A/cm^2$
Polarization	>80%	>80%

COMPLETE EXPERIMENTAL CHARACTERIZATIONS OF A STRAINED-WELL InAlGaAs/AlGaAs

Strained-well InAlGaAs/AlGaAs structures designed and manufactured by St. Petersburg in Russia have the following advantages [1]: 1) larger valence band splitting due to the combination of deformation and quantum confinement effects in the quantum well; 2) a sharp band-bending-region; and 3) a reasonably thick working layer without strain relaxation. The measurements in Russia show that the structures have excellent performance – 1% QE and 90% polarization. To cross-check the Russian data and also provide all key performance of the cathode including charge limitation, polarization, QE and QE lifetime, the InAlGaAs/AlGaAs structures have been characterized at the SLAC cathode and gun test facilities.

Polarization and quantum efficiency were firstly measured at the SLAC Cathode Test System (CTS) [2]. The CTS is an ultra-high vacuum system equipped with a load-lock chamber through which samples can be introduced without venting the system vacuum. Polarization measurements are made using an electron transport column, an electrostatic spin-rotator and a 20 kV Mott polarimeter. Prior to the cathode installation, the sample is processed via standard chemical and heat cleaning methods. The CTS measurements for the InAlGaAs/AlGaAs sample demonstrate 0.3% QE and 87% polarization. The sample used at the SLAC is not a fresh one and may have some degradation of near-surface region and As-cap. The degradation could not be removed by the standard chemical and heating cleaning techniques. However, the QE of the InAlGaAs/AlGaAs cathode was recovered to 1% with atomic hydrogen cleaning [3].

The SLAC Gun Test Laboratory (GTL) [4] is capable of characterizing all parameters including the surface charge limit, QE and QE lifetime, and polarization. The GTL beamline consists of an ultra-high vacuum, high-voltage electrostatic dc-gun, a load-lock chamber for cathode transfer and activation, and a beamline with magnetic components for electron beam transport. With the cathode biased at 120 kV, the gun is capable of producing a space-charge-limited current of 15 A from the 20 mm diameter cathode. An electrically isolated, optically coupled nanoammeter is used to measure the average photoemission current. A fast Faraday Cup (FARC) is employed to measure the temporal profile of the electron beam from which the surface charge limitation can be measured. The electron beam is transported to the Mott chamber for polarization measurements. Fig. 1 shows the comparison of polarization spectrum of the InAlGaAs/AlGaAs cathode measured at St. Petersburg and at SLAC. It is shown that the polarization spectrum measured at the SLAC CTS and GTL is very close although there is a constant wavelength shift between them. The shift is probably caused by a wavelength calibration error in the CTS. The maximum polarization measured at the SLAC CTS/GTL is ~87% versus ~91% measured in Russia. The polarization difference is attributed to systematic differences in the measurement systems.

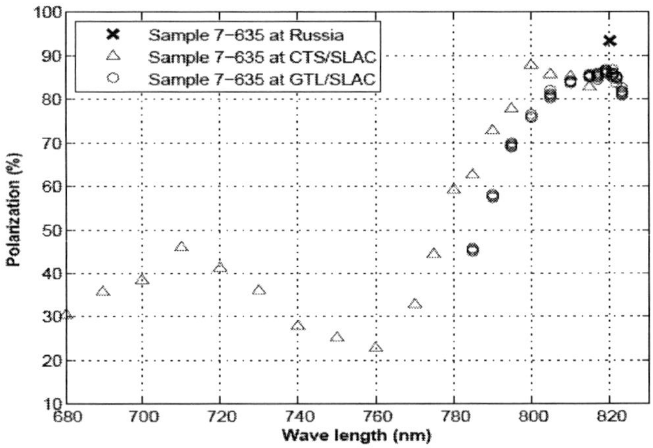

FIGURE 1. The comparison of the polarization spectrum measured at St. Petersburg in Russia, CTS/SLAC and GTL/SLAC.

The surface charge limit on the cathode is seen in Fig. 2 as the drop off of intensity during the current pulse. The limit is caused by the lower doping level in the surface layer, only $7\text{x}10^{18}/\text{cm}^3$ in the cathode. Further studies show that the onset of surface charge limitation occurs at a very low current intensity, only 0.06 A/cm^2. Increasing the doping level to $2\text{-}5\text{x}10^{19}/\text{cm}^3$ in the surface layer can effectively suppress surface charge limit [5].

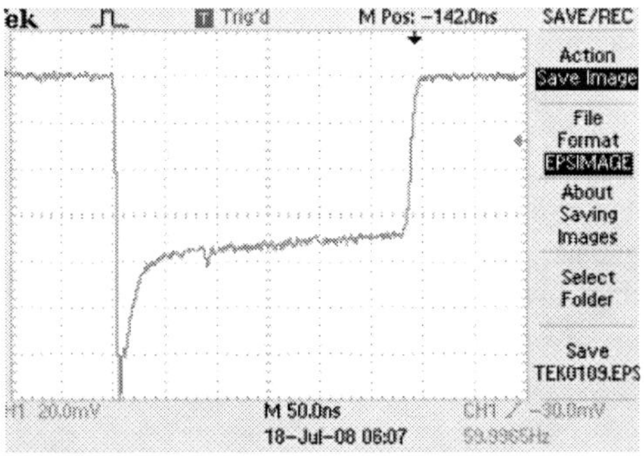

FIGURE 2. Observation of surface charge limit at $2.5\text{x}10^{10}$ electrons at 10 mm laser full size on the cathode. The abscissa is time, and the ordinate is the beam signal from Faraday Cup.

The dependence of the polarization on the surface charge limit is clearly observed for the first time. During the observations, only the laser position on the cathode is changed. At different laser positions A and B on the cathode, different levels of the surface charge limit are observed, as shown in Fig. 3(a) and 3(b), respectively. The surface charge limit at the laser position A is more severe than at position B. The beam generated at laser position B has a higher polarization, 2-3% more than at position A, as shown in Fig. 4. At the laser position A, the cathode is driven into saturation, and the electrons photoexcited into conduction band still can escape if they diffuse to a non-saturated region. Thus, these electrons spend a longer time inside the structure so it is likely that they suffer spin greater relaxation than at position B, which results in reduced polarization.

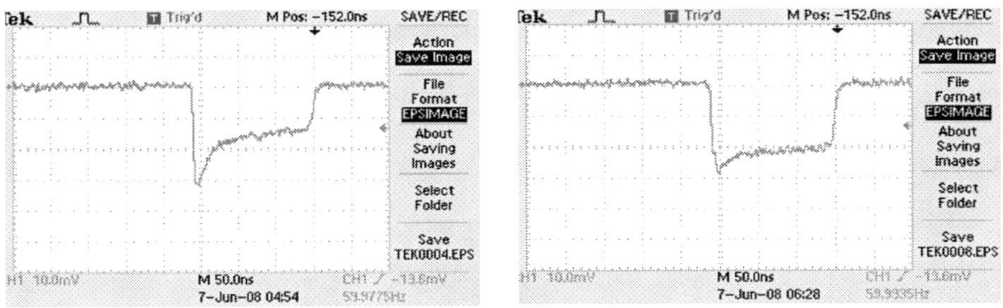

(a) (b)

FIGURE 3. Different degree of surface charge limit of the electron beam at different laser position on the cathode; (a) and (b) correspond to the laser position on the cathode at A and B, respectively. The abscissa is time, and the ordinate is the beam signal from Faraday Cup.

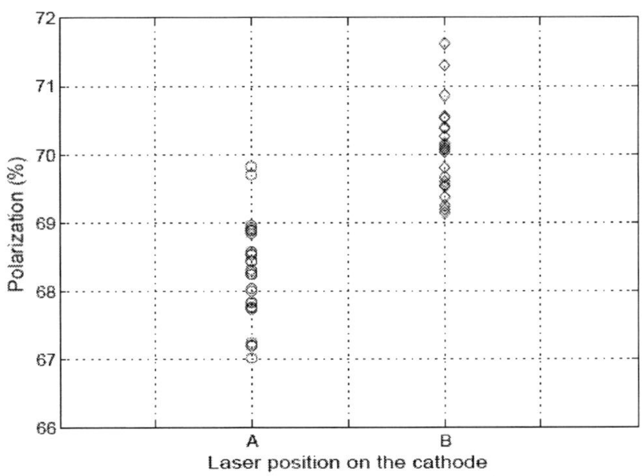

FIGURE 4. Polarizations corresponding to different laser position; data is taken at 800 nm wavelength.

SUMMARY AND PLANS

It is challenging to generate the ILC and CLIC full charge without compromising polarization. Advancing polarized photocathode technology to increase polarization, QE and QE lifetime along with reduced space charge and surface charge limit is one of the major factors to guarantee the success. Recent measurements at the SLAC test facilities show that the InAlGaAs/AlGaAs can produce 87% polarization, 0.3% QE. The QE can be recovered to ~1% with atomic hydrogen cleaning. Serious surface charge limit is observed at a very low current intensity. The clear dependence of the polarization on the surface charge limit is observed for the first time.

SLAC continues to be actively engaged in the development of improved polarized photocathodes through studies of the doping level in the structure of superlattice and of gradient doping in the active layer of the superlattice. Both QE and polarization are expected to be improved using these two techniques. The ILC laser prototype is expected to come online at the SLAC in 2009. This allows us demonstrate the ILC full charge production. In addition, the studies on ion effects on the QE lifetime are also being pursued.

ACKNOWLEDGMENTS

We would like to thank Jym Clendenin for his contributions and Y. Mamaev for the fruitful collaborations. The work is supported by U.S. DOE contract DE-AC03-76SF00515.

REFERENCES

1. Yu. A. Mamaev, et al., "InAlGaAs/AlGaAs superlattices for polarized electron photocathodes", SLAC-PUB-11403, August 2005.
2. G. A. Mulhollan, "Low energy polarized electron source test facilities", 1993 (unpublished).
3. V. Tioukine, "Atomic hydrogen cleaning of superlattice cathodes", Workshop on sources of polarized electrons and high brightness electron beams, Jefferson Laboratory, Newport News, VA, October 1-3, 2008.
4. P. J. Saez, "Polarization and charge limit studies of strained GaAs photocathodes", Ph. D thesis, SLAC-R-501, March 1997.
5. T. Maruyama, et al., "A very high charge, high polarization gradient-doped strained GaAs photocathode", SLAC-PUB-9133, March 2002.

High Intensity Polarized Electron Gun Studies at MIT-Bates

E. Tsentalovich

MIT-Bates Linear Accelerator Center, Middleton, MA

Abstract. A proposed new high-luminosity electron-ion collider requires a polarized electron source of extremely high intensity. An ion back bombardment of the cathode is the main factor limiting the intensity of modern polarized electron guns. This study examines the possibility of building a very high intensity polarized electron gun.

Keywords: Polarized electron source, ion bombardment
PACS: 29.25.Bx, 29.27.Hj

INTRODUCTION

Modern polarized electron sources [1-6] routinely produce average current of hundreds of μA with a polarization approaching 90%. This intensity satisfies the requirements of the existing accelerator facilities. New advances in nuclear physics are expected with the development of a high luminosity electron-ion collider (EIC). The concept of such a collider has been discussed in the nuclear physics communities around the world for the past decade. Several conceptual designs have been proposed. Brookhaven National Laboratory (BNL) develops two different versions of eRHIC [7], based on the existing Relativistic Heavy Ion Collider (RHIC) complex. ELIC project [8] is developed at Thomas Jefferson National Accelerator Facility (TJNAF). These EIC projects require a significantly higher intensity from a polarized electron source. The linac-ring version of eRHIC produces the most challenging requirements with an average current of up to 250 mA. MIT-Bates, in collaboration with BNL, is investigating the possibility of building a very high intensity polarized electron gun.

MAJOR CHALLENGES

Two main problems must be solved in order to build a polarized electron gun with a very high average current. The first challenge is the heat load on the cathode. Tens or even hundreds of Watts of laser power must be deposited on a high polarization (strained or superlattice) GaAs cathode in order to extract a current of hundreds of mA. Cathode holders in the existing guns can absorb less than 1W without temperature rise. Active cooling of the cathode is necessary to avoid overheating of the crystal. The cooling system must be compatible with the high voltage applied to the cathode and with UHV conditions of the polarized electron gun, and presents a formidable engineering challenge.

CP1149, *Spin Physics, 18th International Spin Physics Symposium*
edited by D. G. Crabb, D. B. Day, S. Liuti, X. Zheng, M. Poelker, and Y. Prok

The most difficult task is to achieve the high average current with a reasonable lifetime. The main problem is ion back bombardment. An electron beam ionizes molecules of the residual gases in the vacuum chamber; positive ions accelerate in the cathode-anode gap and strike the cathode, damaging the surface of the crystal. The effect is proportional to the product of the average electron current and the pressure in the gun chamber. It is difficult to expect a significant improvement of the vacuum conditions over present state-of-the-art installations, and the effect limits the average current in existing guns to several hundreds of μA.

Both problems could be addressed with increased cathode area. Heat transfer can be done more efficiently, and the ion damage will be distributed over a larger area, resulting in a higher lifetime of the cathode.

Measurements conducted at TJNAF [9] with a green laser indicated that the lifetime increased by a factor of ~10 when the laser spot diameter was increased by factor of 4. In a simplified picture, the average current that could be extracted from the cathode, with the same lifetime, is proportional to the emitting area. In reality, the picture is more complicated. Negatively charged electrons and positive ions follow a somewhat different trajectory, and ions tend to damage mostly the central area of the cathode. Existing electron guns are using the Gaussian shaped laser beams, with the maximum beam intensity near the cathode center, where the ion damage is most significant. For a high-intensity gun, it could be beneficial to form a ring-shaped laser beam, so the majority of the laser power is applied to the peripheral area of the cathode. The major difficulty in this approach is to avoid beam losses in the gun and its vicinity. Very detailed simulations including space charge effects must be conducted to design a large area cathode with a ring-shaped emission.

ION INDUCED DAMAGE STUDIES

The measurements have been conducted with the existing test beam line described in [6]. This beam line was designed for cathode tests and certification at very low current. It includes Mott polarimeter with silicon detectors and Mylar-backed golden foils. This section of the beam line cannot be baked at high temperature, and thus the vacuum in the whole beam line is rather inferior. The beam line has narrow apertures, and beam losses of 5-10% are typical. These losses produce out-gassing at high current, and degrade vacuum conditions even farther, resulting in a relatively short lifetime of the cathode. However, the poor performance of the test beam line at high current was beneficial for our studies. The ion damage accumulated quickly, and the measurements could be conducted in a relatively short time.

A fiber-coupled diode laser with a wavelength of λ=808nm was used for photoemission. The laser beam coming out of the fiber (0.4 mm diameter) has very large divergence (full open angle ≈25.4°). The first focusing lens L1 produced an almost parallel laser beam. An axicon (conical lens) in conjunction with the second focusing lens L2 formed a ring-shaped laser beam in the location of the cathode. With the axicon removed from the system, lenses L1 and L2 formed a point-to-point focusing between the output of the laser fiber and the cathode, resulting in a very small laser spot. A longitudinal translation of L2 allowed varying the beam spot size. The L2 lens was mounted on a translation stage for a precise translation in both transverse directions. With the axicon removed and the laser beam focused, one could scan the surface of the cathode using these translations and map the QE of the crystal.

The following procedure was used during the measurements. The crystal was heat-cleaned and activated. With the axicon removed, a very small diameter laser beam was scanned across the surface of the crystal, mapping the QE. The cathode was biased to only -64V during the mapping. Then the optical system was tuned to a certain configuration (small or large laser beam, ring-shaped beam etc.). The cathode was biased to -60kV, and the laser intensity was adjusted to produce a current of ≈120 μA. Each run continued for several days, and an integrated charge of 12-18 C was accumulated during each run. As the QE of the cathode decreased, the laser intensity was adjusted to keep the average current at about 100 μA. After the run the QE of the crystal was mapped again. The ratio of those two maps represents the change in the QE during the run.

Figure 1 shows the results of the runs in four different configurations: a) small Gaussian laser beam (σ≈0.15 mm) in the center of the cathode, b) large Gaussian beam (σ≈2 mm), c) ring-shaped laser beam and d) small Gaussian beam parked in the corner of the cathode.

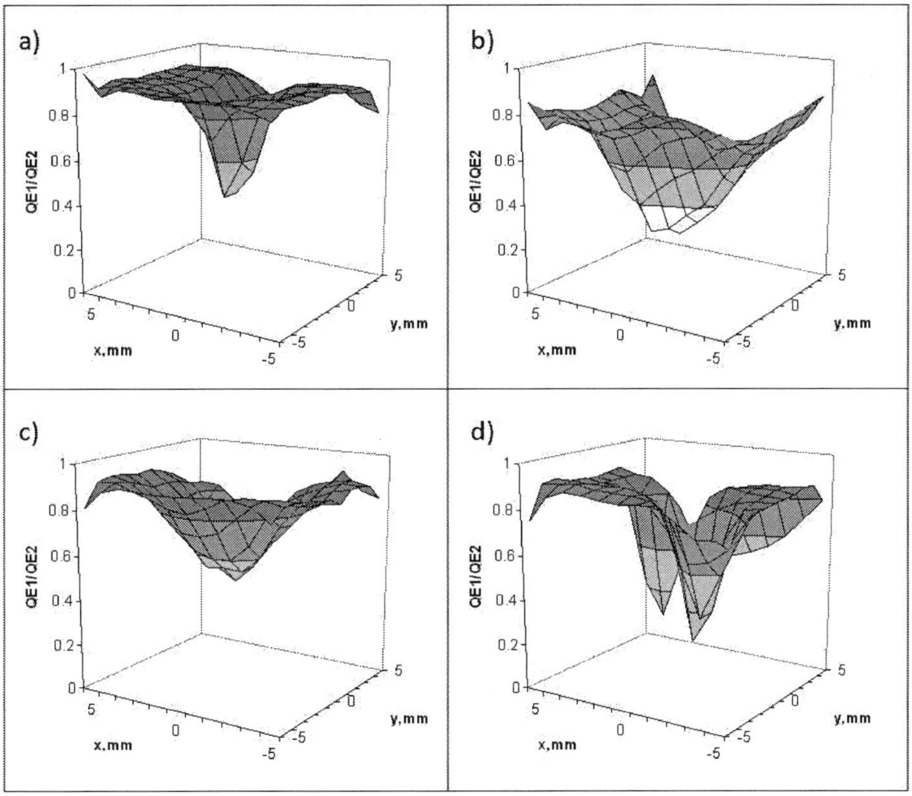

FIGURE 1. Map of the QE degradation after the run with a) – small centered beam, b) – large beam, c) – ring-shaped beam and d) – small beam parked at the corner of the cathode.

The run with a small laser spot centered on the cathode produced a deep "hole" in the QE distribution. Naturally, the "`hole" was located in the center of the crystal; the ions damaged mostly the area where the electrons have been emitted from. However, in the case of the ring-shaped beam, most of the damage occurred in the center of the crystal again, although the laser intensity was negligibly low there. The ring-shaped area where the electrons were emitted from actually received less damage than the central part.

The damage produced by a large Gaussian beam corresponds roughly to the beam shape, but the total amount of the damage is much higher. The Gaussian beam has tails, and significant amount of the electrons are emitted from the very edge of the cathode, where the transverse focusing forces are very large. Those electrons follow the extreme trajectories and get lost in the vicinity of the gun, causing out-gassing and increasing the ion-related damage. In contrast, an axicon-formed ring-shaped laser beam has very sharp edges with low tails and produces very few electrons with extreme trajectories.

A small laser beam parked in the corner of the cathode produced a “hole” in the emitting spot, a groove of damage extending toward the center of the crystal, and the second “hole” located on this groove slightly over the cathode center.

All these features can be understood by ion trajectory calculations. It should be noted that the anode hole acts as a defocusing element for the electrons, and a focusing element for ions. The ions produced in the vicinity of the cathode experience basically the same transverse forces as the electrons and follow approximately electron trajectories toward the cathode. The ions produced closer to the anode are affected by focusing fields of the anode at the part of the trajectory where their energy is still very low and the transverse forces are more effective. As a result, those ions tend to be focused toward the center of the cathode. The ions produced in the anode area are actually overfocused, but still damage mostly the central area of the crystal. These high energy ions are most likely responsible for the second “hole” on Fig.1d, giving a strong indication that although most ions are produced near cathode (due to high ionization cross section at low energy of the electrons), high energy ions produced near anode give a very significant contribution in ion-related damage.

The measurements have been conducted with CW beam, and ion trapping must be considered. The ions produced in the beam line after the anode could be trapped in the electron beam, travel toward the gun, get accelerated in the cathode-anode gap and strike the cathode. The preliminary measurements at TJNAF [10] indicated that this process could be important and that it can be suppressed by biasing the anode.

An additional run was conducted with the anode biased to +1kV and ring-shaped laser beam. A significant reduction in ion-related damage was measured, especially in the center area of the cathode, where most of the trapped ions are expected to be directed.

SIMULATIONS AND CATHODE COOLER

The simulations of the new gun and the beam line are under way. The simulations assume the current of up to 100 mA and gun voltage of 120 kV. The cathode diameter is 25 mm. The beam line includes two 90° dipole magnets, solenoidal lenses and a beam dump. The main purpose of the simulations is to minimize beam losses, especially in the gun vicinity. The losses are required to be less than 10^{-6} in the first straight section and less than 10^{-5} in the rest of the beam line with different emission pattern (flat-shaped beam, Gaussian beam, ring-shaped beam).

Additionally, the ion trajectories are calculated and the expected damage pattern estimated. The gun geometry and the emission pattern will be optimized to reduce the overlap of the emission pattern and the damage pattern.

A special test chamber is designed to study the cathode cooling. The cooling agent (fluid with very low electrical conductivity) will be delivered to the cathode via ceramic tubes. High Voltage and Ultra High Vacuum compatibilities will be tested, as well as a cooling efficiency.

ACKNOWLEDGMENTS

The work is partially supported by DOE grant DE-FG02-94ER40818. The author would like to thank the staff of Bates laboratory for a constant help and support. The author is grateful to C. Sinclair for his comments.

REFERENCES

1. R. Alley et al, *Nucl Instr. and Meth.* **A365**, p.1 (1995).
2. K. Aulenbacher et al, *Nucl. Instr. and Meth.* **A391**, p.498 (1997).
3. M. J. J. van den Putte et al, *Nucl. Instr. and Meth.* **A406**, p.50 (1998).
4. J. Grames et al, *AIP Conf. Proc.* **675,** p. 1047 (2002).
5. Wolther von Drachenfels et al, *AIP Conf. Proc.* **675,** p. 1053 (2002).
6. E.Tsentalovich et al , *Nucl. Instr. and Meth.* **A582**, p.413 (2007).
7. V.Ptitsyn, Proc. of the Particle Accelerator Conference. 2007, p.1927.
8. Y. Zhang et al., Proc. of the Particle Accelerator Conference, 2007, p.1935.
9. M.Poelker and J.Grames, Proc. of 11th Intern. PST Workshop, Tokyo, 2005, p.127.
10. J. Grames et al, Proc. of 12th Intern. PST Workshop, Upton, 2007, p.110.

Hydrogen Cleaning of Superlattice Photocathodes

V. Tioukine, K. Aulenbacher, E. Riehn

Institut für Kernphysik der Universität Mainz,
J.J. Becherweg 45, D-55099 Mainz, Germany, F. R.

Abstract. Surface cleaning with a beam of thermally dissociated hydrogen atoms was applied to superlattice cathodes which resulted in a considerable improvement of quantum efficiency. Compared to untreated samples no significant reduction of polarization of the emitted electrons was observed.

Keywords: Superlattices, Polarized beams, Surface Cleaning
PACS: 78.67.Pt 29.27.Hj 81.65.Cf

INTRODUCTION: AGING PROBLEM

Activation of photocathodes towards the state of negative electron affinity requires a clean surface before a deposition of cesium and oxygen takes place. Typically this condition is realized by the application of a passivation layer, e.g. elemental arsen, directly after the growth of the wafer. The sample may then be exposed to air and stored for long time. The As-layer can be removed by heating the sample up to 400 °C in Ultra High Vacuum (UHV) prior to an activation. So far, this strategy proved successful for our strained layer- and also for some superlattice (SL-) cathodes. However, a different behavior was observed for $InGaAlAs/GaAlAs$ samples which were designed at St. Petersburg State Technical University and grown by the Joffe Institute (see fig. 2). If measured directly after the growth, these samples offer attractive parameters [1]. However, we initially achieved only about 1/10 of the quantum efficiency (q.e.) values reported in [1]. Moreover, saturation effects were observed which increased the discrepancy at the high laser intensities necessary at our accelerator. We assumed an aging problem due to the fact that the samples were stored for about three months. This was supported by our collaborators in St. Petersburg who also observed similar effects after long term storage. In order to explain this we suspect insufficient passivation of the cathode surface. This will lead to formation of Arsenoxide (As_2O_3) on the surface immediately after the venting of the growth apparatus. As_2O_3 can effectively be removed at relatively low temperatures ($\approx 500\,°C$) thus explaining the good results achieved directly after growth. However, it was demonstrated [2] that As_2O_3 transforms into gallium oxides (Ga_2O_3), this reaction happening within months at room temperature. It cannot be avoided by storing the samples under inert gases or vacuum since the oxides are already present at the surface. Ga_2O_3 can not be effectively removed by heat treatment alone which can explain the problems observed. Alternative ways of surface cleaning must cope with the requirement to remove exclusively the oxides, since the functional structures close to the surface of the SL are extremely thin (fig. 2). Andreev et al. [3] studied several methods

CP1149, *Spin Physics, 18th International Spin Physics Symposium*
edited by D. G. Crabb, D. B. Day, S. Liuti, X. Zheng, M. Poelker, and Y. Prok

of surface cleaning for NEA photoemitters. Good results were achieved if a thermal hydrogen cracker was used, which may be due to the absence of energetic ions and neutrals in the atomic beam of this source.

HYDROGEN CLEANING APPARATUS

We have set up a dedicated apparatus, called HABS - Hydrogen Atomic Beam Source - which allows for the following operations

- Hydrogen cleaning by a beam of thermally cracked hydrogen atoms.
- Activation to NEA state and q.e. measurement in order to test cleaning effect.
- Transfer of samples under UHV to other apparatuses for further investigations.

The apparatus consists of a bakeable all metal UHV system (fig. 1). The hydrogen source is a commercial one, similar to the design in [4]. It consists of a heated tungsten capillary which reaches a temperature of about 2150 C at a heating power of 300 watts. H_2 flowing through the hot capillary is dissociated. A beam of atomic hydrogen then hits the sample surface which is located at a distance of about 10 cm from the end of the capillary. The dissociator is enclosed in a water cooled copper shielding which reduces the radiative heating of the inner surfaces and of the sample itself. Clean hydrogen is provided by using pure H_2 gas and further purification is achieved by letting it stream through a getter. H_2 is then stored in an all metal enclosure and introduced by a leak valve. Hydrogen pressure during treatment was $2 \cdot 10^{-6}$ mbar stabilized by a turbo pump (fig. 1). Typical exposure times were about 5 minutes during which the samples are heated to 550 °C. Then, the activation procedure is started, consisting of a 30 minute long heat treatment at 550 °C and co-deposition of cesium and oxygen after cooling down the sample. The q.e. achieved may be compared with the one from untreated samples (fig. 3). Once a promising result is achieved, the samples are cleaned again without subsequent activation. Afterwards they are transported to one of the 100 keV sources in our labs for further investigations which are not possible within HABS (e.g. polarimetry). For this purpose the samples are transferred into the transport vessel by a manipulator. The vessel is equipped with an ion pump. An intermediate loading chamber is pumped by the turbo already mentioned above. After shutting off the transfer vessel by a valve the loading chamber is vented and the vessel may be removed. Meanwhile, another (closed) valve allows to keep the vacuum in the HABS chamber. The vessel is then transported to one of the electron sources where similar loading chambers exist. All stages of the transport are performed under UHV.

HABS RESULTS

Photocathodes: The structure of the superlattice (SL-)cathodes labelled 7-395 and 7-396 is presented in fig. 2. In both cases the working layer consists of 12 periods of double layers with a composition $In_{.2}Al_{.19}Ga_{.61}As/Al_{.4}Ga_{.6}As$, yielding a total thickness of 91 nm for the working layer of the cathode. Due to the fact that all layers contain Al it is necessary to inhibit interaction with oxygen by placing a very thin layer of GaAs on top

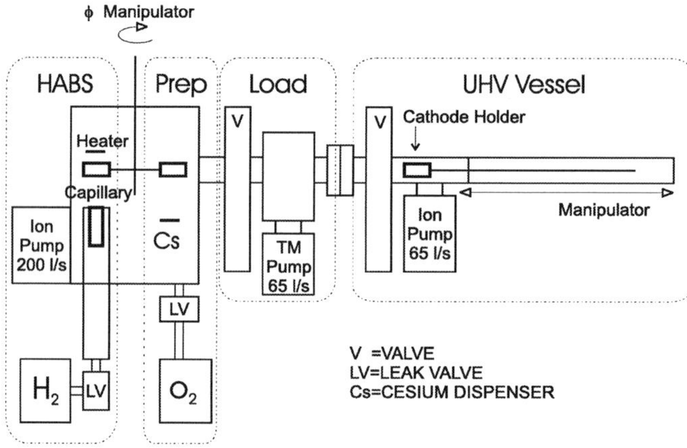

FIGURE 1. Schematic of hydrogen cleaning apparatus.

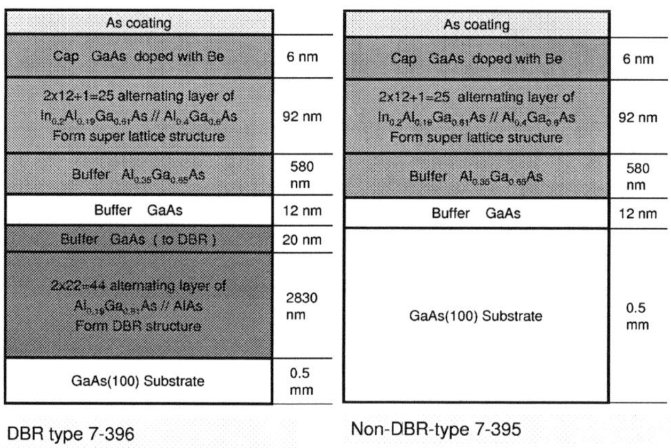

FIGURE 2. InAlGaAs/AlGaAs Superlattices. The As capping may not be fully present in our case.

of the SL. In contrast to the working layer, the GaAs is heavily p-doped, thus providing a thin band bending region which is essential for efficient emission in conjunction with low polarization losses. The SL 7-396 has an additional feature since it is grown on top of another superlattice (DBR) which reflects the incoming light. Together with the semiconductor/vacuum interface (Reflectivity ≈ 0.3) the DBR forms an optical cavity which allows to increase the internal absorbtion if the incoming wavelength corresponds to a cavity resonance. Both types suffered from the aging effect, which limited their q.e. to $\leq 1\%$ for light of $\lambda = 650\,\text{nm}$ which was used for the activation tests in HABS.

Activation results in HABS apparatus: The vertical bars in fig. 3 indicate when a cathode was hydrogen cleaned in addition to heat cleaning before the activation. The first two samples tested were bulk GaAs (activations 1-16). In both cases hydrogen cleaning resulted in an increase of q.e. by more than a factor of five. Following this,

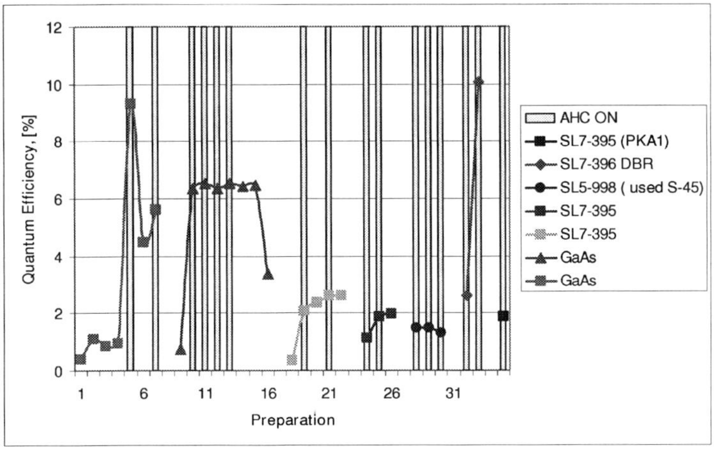

FIGURE 3. Sequence of activations in the HABS apparatus

maximum q.e. seems to remain constant for a few activations, even without further treatment, before a reduction is observed, see activations 12-16. The behavior of the SL's concerning hydrogen treatment is similar to bulk GaAs. The DBR cathode 7-396 has achieved the highest q.e., yielding more than 10% at 650 nm. 7-395/7-396 cathodes were transferred to our test source, a 7-395 went to the source at the MAMI accelerator.

Results in Production source: These tests were performed with a high intensity laser system operated at wavelengths of 785 and 802 nm, close to the maximum polarization response of the cathode. We observed a ten times increased q.e. at low intensities for the HABS treated sample if compared with an untreated one. Moreover, the saturation effects at high intensities had disappeared resulting in a factor 50 gain of maximum current. Tests were performed at a q.e. of 0.6% and 200 μA of average current (density of $\approx 200 mA/cm^2$) without showing nonlinear effects. Beam polarization was measured with a Mott polarimeter that offers a reproducibility of about 1%, whereas the absolute accuracy is $\Delta P/P = \pm 4\%$. After one week of operation at the accelerator the q.e. had dropped by a relative 15% which resulted in a relative increase of polarization by 6%, yielding an absolute value of $P = 85.5 \pm 3.2\%$. This value is within the highest ever observed at our production source.

Results in test source: Comparisons of maximum polarization for treated/untreated samples were done at the test source. We found no difference (at the 3% (relative) level) between the different samples once the initial q.e. of the hydrogen cleaned one was slightly reduced (see above). This provides evidence that hydrogen treatment has no detrimental effect on the bulk properties of the superlattice structure. A tunable laser was used for spectrally resolved measurements. The lower curve in figure 4 shows the results for an untreated sample of 7-395, the upper one those from a fresh one measured at St. Petersburg. The results for a hydrogen treated sample are close to the St. Petersburg measurements. Again, saturation effects were present for the non treated sample which would make the discrepancy in figure 4 even larger at high laser intensities. The DBR

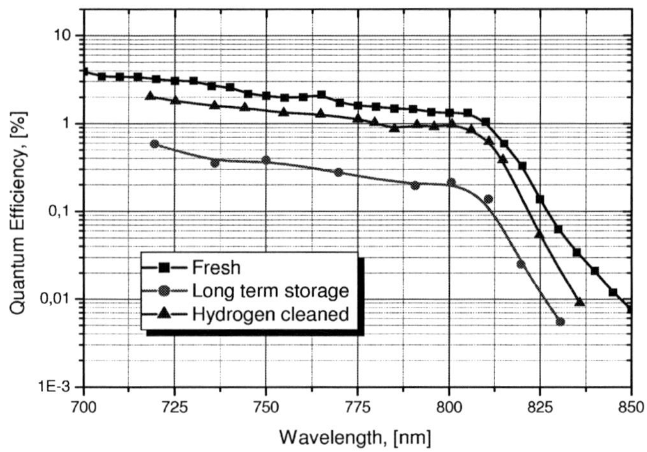

FIGURE 4. Comparison of results achieved under different conditions

sample 7-396 (not shown in fig. 4) achieved almost twice the q.e. of the non DBR sample 7-395 in figure 4, yielding q.e.=1.2% at the wavelength of maximum polarization which is even larger than the result obtained directly after growth.

CONCLUSION

We have demonstrated the usefulness of hydrogen cleaning for superlattice cathodes with very thin functional structures. We find no evidence for a depolarization with respect to untreated samples. The success of the method may be promoted by the specific technology of atomic hydrogen generation, based on thermal dissociation, which eliminates fast ions or neutral particles which may damage the surface.

ACKNOWLEDGMENTS

This work was supported by the SFB 443 of the Deutsche Forschungsgemeinschaft. We especially want to thank Yuri Mamaev and Yuri Yashin for stimulating discussions on the various aspects of surface treatment and semiconductor fabrication.

REFERENCES

1. Y. A. Mamaev et al., *Appl. Phys. Lett* **93**, 081114 (2008).
2. D. A. Allwood et al., *Thin Solid Films* **412**, 76 (2002).
3. V.E. Andreev et al., "Atomic and electronic engineering of p-GaAs-(Cs,O)-vacuum interface," in *SPIN 2000, AIP Proceedings Vol. 570*, edited by K. Hatanaka et al. , AIP, Melville, N.Y., 2001, p. 901.
4. U. Bischler and E. Bertel, *JVST A* **11**, 458 (1992).

Ultra Cold Photoelectron Beams for Ion Storage Rings

D. A. Orlov[a], C. Krantz[a], A. Shornikov[a], M. Lestinsky[a], J. Hoffmann[a], A.S. Jaroshevich[b], S.N. Kosolobov[b], A.S. Terekhov[b] and A. Wolf[a]

[a] *Max-Planck-Institut für Kernphysik, 69117, Heidelberg, Germany*
[b] *Institute of Semiconductor Physics, 630090, Novosibirsk, Russia*

Abstract. An ultra cold electron target with a cryogenic GaAs photocathode source, developed for the Heidelberg TSR, delivers electron currents up to a few mA with typical kinetic energies of few keV and provides unprecedented energy resolution below 1 meV for electron-ion recombination merged-beam experiments. For the new generation of low-energy electrostatic storage rings, cold electron beams from a photocathode source can bring additional benefits, improving the cooling efficiency of stored ions and making it possible to cool even heavy, slow molecules by electron beams of energies of only a few eV or even below.

Keywords: photocathode, electron beams, electron cooling, electron target, storage ring.
PACS: 52.25.Xz , 29.25.Bx, 29.20.D, 85.60.Ha, 34.80.Lx, 37.10-x

INTRODUCTION

Semiconductor photoemitters with negative electron affinity (NEA) are widely used in various scientific experiments for the generation of spin-polarized, short bunched high-density electron beams. Photoelectron sources also offer the possibility to employ extremely cold electrons with emission energy spread of about the cathode temperature, which can be as low as 7 meV at 100 K [1]. One of the main difficulties is to keep the beam cold as a space charge causes beam divergence at high currents. To avoid this heating, high magnetic guiding field is used in the electron coolers of storage rings, which decouples transverse and longitudinal electron motions. Electron-ion merged beams at storage rings have also opened a new domain for electron-ion recombination studies where ultra cold photoelectron sources can bring great benefits in high-resolution measurements. A cryogenic photocathode target was developed [2,3,4] for recombination measurements at the Heidelberg Test Storage Ring (TSR) and is in operation since 2002 delivering *dc* electron beams of mA currents and with ultra-cold temperatures below 1 meV [4,5]. For the upcoming electrostatic Cryogenic Storage Ring (CSR), where the cooling of stored slow ions has to be performed at typical electron energies of 10 eV down to the sub-eV range, the impact of the cold photocathode source can be of high additional importance. Indeed, at low energies of 1-100 eV the electron current drops down to the μA range due to a limited gun perveance, which strongly decreases the cooling efficiency. To make cooling possible, electron beams of low temperatures have to be used [6]. In this paper, we present the

CP1149, *Spin Physics, 18th International Spin Physics Symposium*
edited by D. G. Crabb, D. B. Day, S. Liuti, X. Zheng, M. Poelker, and Y. Prok

Heidelberg photoelectron target and demonstrate the performance of the cryogenic GaAs-photocathode source.

MAGNETIZED ELECTRON BEAMS FOR STORAGE RINGS

A cold magnetically guided electron beam, overlapped in a straight section of the ring with the stored hot ions moving at the same average velocity, provides efficient cooling of the ions [7]. Merged electron and ion beams can also be used for collision experiments if the electron energy is for a short time period detuned from the cooling velocity and recombination products (neutral or ions) are detected downstream. In the case of the Heidelberg TSR (see Fig.1) with a maximum magnetic rigidity of about 1.4 Tm, ions are injected at the ring with typical energies of 0.1-8 MeV/u, which corresponds to electron cooling energies (for singly charged ions) from about 50 eV up to 4.4 keV.

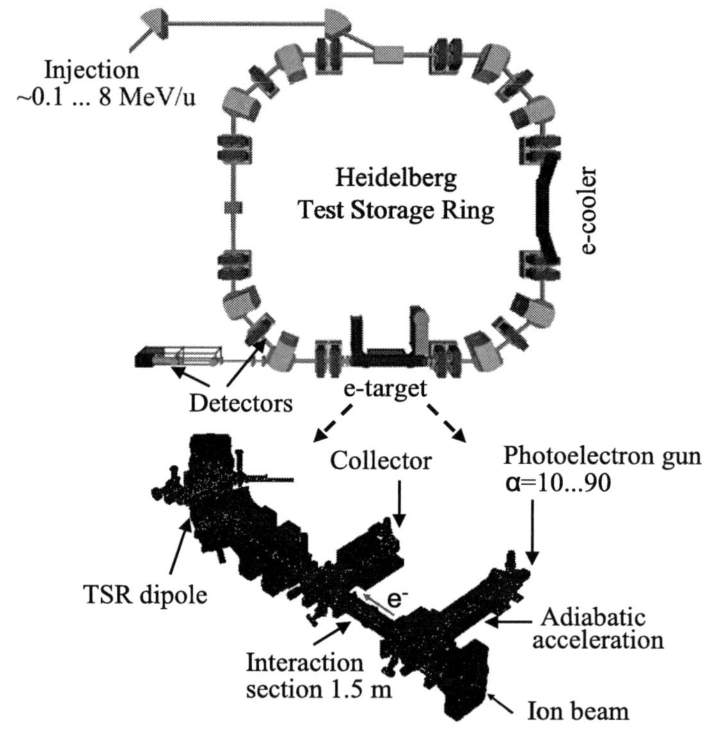

FIGURE 1. Schematic view of the Heidelberg TSR (top) and close up view of the e-target (bottom).

The TSR is the only storage ring equipped with two independent electron beam facilities, which allowed us to perform electron collision experiments at variable relative energies avoiding the resolution loss caused by the interruption of the electron cooling and by dragging of ions to the detuned electron energy [8]. The TSR electron cooler is equipped with a thermocathode source capable to deliver electron currents up to 0.3 A, while the electron target is dedicated to high resolution measurements with typical currents below one mA. A cryogenic GaAs photocathode source, developed for the electron target [2,3,4], delivers electron beams up to few mA with an emission energy spread of about 10 meV. By subsequent adiabatic magnetic expansion and acceleration electron beam temperatures below 1 meV are obtained [5,6]. A schematic view of the TSR electron target is also shown in the Fig.1. The photocathode vacuum setup (not shown here) is attached to the electron gun. Electrons are extracted from the cathode in a strong magnetic field up to 4 T, which is reduced in field strength by a factor of 20-40 (typically) to decrease the transverse temperature. Guided by the resulting final magnetic field of 0.02-0.1 T, the drifting electron beam is bent by a toroid field to overlap collinearly with the stored ion beam for a distance of about 1.5 m. After the interaction section, electrons are demerged from the ion beam by the second toroid and then follow the guiding field which brings them to a collector, where the beam profiles and energy distributions of electrons can be measured with the help of a retarding field energy analyzer.

Electron Beam Temperatures

Beam densities as well as electron temperatures are the key parameters of the electron beams used for electron cooling and for merged electron-ion beam experiments. The guiding magnetic field of a strength given above helps to avoid space-charge divergence of the beam, to provide adiabatic transport of electrons from the gun to the interaction area, and to suppress transverse longitudinal relaxation in the electron plasma. The decoupling of transverse and longitudinal electron motion results in a strongly anisotropic energy distribution of the electron beam with a longitudinal temperature $kT_{\|}$ of about 0.05-0.1 meV, which is much smaller than the transverse one $kT_{\perp}$ of about 1-10 meV.

Initially $kT_{\perp}$ is given by the cathode temperature T_C and is reduced by an adiabatic magnetic expansion to $kT_{\perp}=kT_C/\alpha$, where α is the expansion coefficient [9,10]. The longitudinal temperature $kT_{\|}$ of electrons in the co-moving frame (we have electron beams as the reference frame) is given by:

$$kT_{\|} = \frac{(kT_C)^2}{E_{kin}} + C\frac{e^2}{4\pi\varepsilon_0}n_e^{1/3}. \tag{1}$$

It is reduced due to kinematic transformation, when the beam is accelerated, to the first term in Eq. (1), where E_{kin} is the electron energy in laboratory frame. In this expression it is also taken into consideration that for $\alpha >> 1$ a large part of the transverse energy is transferred to the longitudinal one during adiabatic magnetic expansion, which increases the first term by a factor of about 2. The second term, with n_e representing electron densities, appears due to the relaxation of the internal energy of the moving plasma forming the electron beam, where the parameter C describes the quality of the acceleration [11]. It amounts to about 1.9 for fast acceleration and of about 0.7-0.8 for the TSR target using an adiabatic acceleration scheme. For electron energies *above* 100 eV, the kinematic term is small and the longitudinal temperature is defined mainly by the electron density and independent on emission energy spread.

However, it is emphasized here that for lower energies the situation reverses with the longitudinal temperature becoming strongly dependent on the cathode temperature.

Collision Energy Resolution, Cooling Efficiency

In recombination experiments, the electron temperatures lead to an asymmetric broadening δE of the observed cross-section resonances. The low-energy side of the resonance is broadened by $kT_\perp$, while $kT_\parallel$ effects in broadening the resonance symmetrically by $4(E_r\, kT_\parallel \ln 2)^{1/2}$, with E_r being the resonance energy [12]. Thus, the highest resolution (of about $kT_\perp$) is obtained at lowest energies typically below 0.1 eV.

The cooling time of the stored ions strongly depends on the electron density as well as on the transverse and the longitudinal temperatures of the electron beam, and on the charge Z_e and mass M_{ion} of the ions. The longitudinal momentum spread of the ions and the transverse emittance of the ion source, as well as the space-charge induced momentum spread of the electrons are also important to calculate cooling times. Assuming that ion beams are injected into the ring with a momentum spread lower than the electron momentum spread (as is typical for TSR), a first approximation of cooling times τ can be done in the model of a non-magnetized electron beam [13] with an electron temperature kT_e identical to the transverse temperature, yielding

$$\tau = \frac{3mc^3(4\pi\varepsilon_0)^2}{8\sqrt{2\pi}e^4}\,\frac{M_{ion}}{n_e\cdot Z^2\cdot L_C}\cdot\left(\frac{kT_e}{mc^2}\right)^{3/2}, \qquad (2)$$

where L_c is the Coulomb logarithm. The cooling time can be reduced further by using a magnetized electron beam with small longitudinal temperatures [14], which also defines the size of the cooled ion beam.

The cooling force of the TSR cooler is typically much stronger than for the target due to its higher electron currents. However, at low cooling energies below 100 eV, when the electron current is limited by gun perveance to values below 1 mA, the electron target cools the slow ions much faster due to the high quality (low temperatures) of its electron beam [15].

PHOTOCATHODE SETUP

The photocathode vacuum setup consists of three chambers (see Fig. 2): the photoelectron gun chamber (basic pressure $1.2\cdot10^{-11}$ mbar), the photocathode preparation chamber ($4\cdot10^{-12}$ mbar) and combined loading and atomic hydrogen chamber ($2\cdot10^{-9}$ mbar). The chambers are connected via all metal gate valves with photocathodes being transferred between the chambers by magnetically-driven manipulators. In the electron gun, the photocathode is cooled by continuously flowing liquid nitrogen, regulated by a mass-flow-control unit, which allows the cathode temperature to be changed in the range from 80 K up to 300 K.

We are using transmission–mode GaAs–photocathodes, consisting of a two–layer ($GaAs/Al_{0.5}Ga_{0.5}As$) semiconductor heterostructure, covered with an antireflection coating and bonded to a thick sapphire base plate via a thin (10 - 20 μm) glass layer [16]. Before loading, photocathodes are cleaned in a HCl – isopropanol solution in a glove-box filled by pure nitrogen and are then transferred to the loading chamber inside of a nitrogen–filled transfer vessel without any contact to the laboratory atmosphere. After attachment of the transfer vessel to the loading chamber and after pumping it out, photocathodes are moved into the preparation chamber and mounted on a carousel capable of keeping four cathodes. The carousel can be rotated into different positions for heat cleaning and (Cs,O) activation to NEA – state. This scheme of the photocathode preparation reproducibly provides photocathodes with high quantum efficiencies (QE) of ~ 20%. The main drawback of the scheme is that only ~ 3-4 restorations of degraded photocathodes by heat-cleaning and reactivation are possible. During the following cycles of heating and activation the QE drops strongly. In addition, the effect of current limitation due to surface photovoltage is observed for multiply recleaned cathodes. Formerly it was required to remove cathodes from the vacuum chamber for the chemical cleaning in the glove-box to restore their emission properties. With the low-energy atomic hydrogen treatment, multiple refreshings of the photocathodes within the UHV set up became possible.

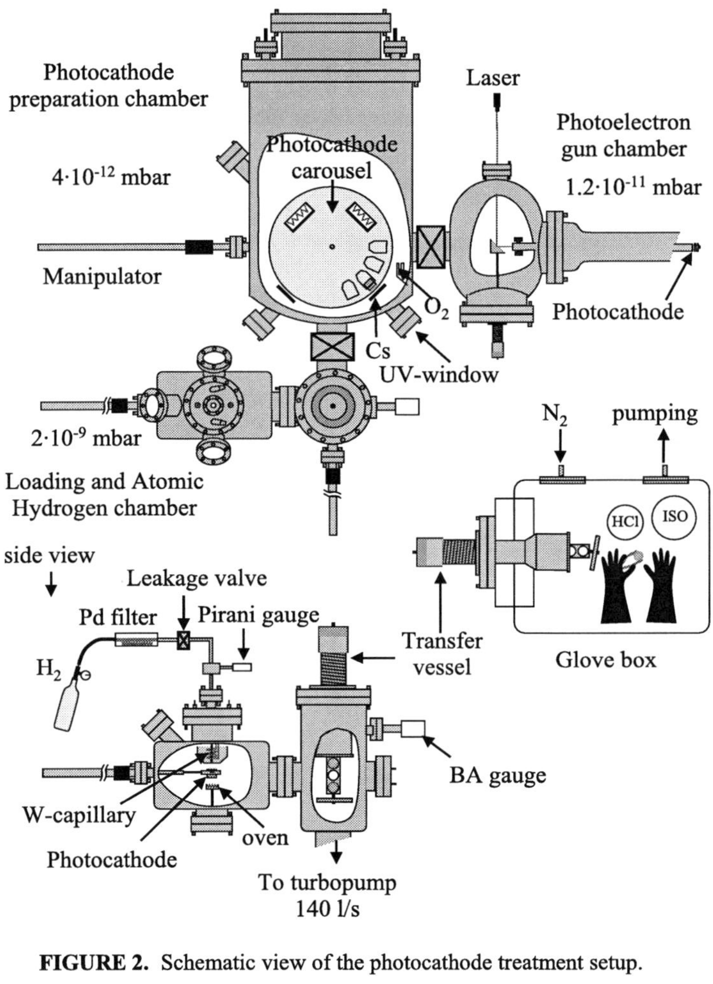

FIGURE 2. Schematic view of the photocathode treatment setup.

Atomic Hydrogen Treatment

Atomic hydrogen (AH), produced by thermal cracking of H_2-molecules inside a hot tungsten capillary, was implemented for the cleaning of transmission-mode GaAs–photocathodes with high quantum efficiency in a UHV multi–chamber photoelectron gun of the electron target. In contrast to previous measurements, where AH from RF sources was used [17,18], the low-energetic hydrogen from the capillary source was proved to be an efficient tool for *multiple* in-situ recleanings of the high QE cathodes without generating a "fatal" concentration of recombination centers within the near-surface region.

The design of our source (see Fig. 2) follows ideas presented in the paper [19] where a long tungsten capillary served as a high efficiency thermal cracker. Molecular hydrogen is brought into the hydrogen chamber through a heated palladium filter attached to the inlet of the capillary. We used a 60-mm-long capillary of 0.6 mm inner diameter. A 15-mm-part of the capillary near its outlet was heated by 21 W electron bombardment resulting in a capillary temperature of (1610 ± 25) ^{0}C. This temperature was high enough to crack 3-5% of the hydrogen molecules passing through the capillary, and was still sufficiently low to avoid a pollution of the GaAs surface by tungsten vapor. The pressure of molecular hydrogen at the inlet of the capillary was varied with a leakage valve in the range of 0.005-0.75 mbar, while the corresponding change of the pressure in the loading chamber, pumped by a 140 l/s TPU 180 H Balzers turbopump, was 10^{-7}-10^{-5} mbar. The exposure time was varied in the range of 2-15 minutes. For the AH treatment the cathode was heated to 450 ^{0}C. Afterwards the valve at the inlet of palladium filter was closed, and the heating filaments of the oven and of the capillary were turned off. The following activation of the cathode by (Cs,O) in the preparation chamber typically resulted in quantum efficiencies of 10% -15% at $\hbar\omega$ = 1.86 eV. It was also found experimentally that additional thermal annealing of the AH–treated photocathodes at about 450 ^{0}C during 30 min within the preparation chamber, increases the achievable QE to 15% - 20%, and up to 20-25% using two-step-activation (with a maximum value of 35%). After degradation during the cathode operation in the electron gun, the photocathodes were heat-cleaned and activated 3-4 times with the maximum QE typically being obtained for the second activation. After that, the AH cleaning was required to recover the emission properties. UV photoelectron spectroscopy was used to optimize and to control the efficiency of the atomic hydrogen and heat treatments. It was found that an atomic hydrogen exposure of 50-200 L is sufficient to remove the degraded activation layer from the GaAs surface.

Since 2002 AH–cleaning is routinely used for the refreshing of GaAs photocathodes at the TSR electron target. Ten photocathodes from two groups were used, two of which were grown by liquid phase epitaxy (LPE) and the others by metal–organic chemical vapor deposition (MOCVD). Photocathodes from the same group behaved in a similar way with respect of multiple AH-cleaning, but there was a difference between the groups. The LPE grown photocathodes, after cleaning with HCl–isopropanol, heat-cleaning and subsequent activation showed a maximum QE of about 15 – 25% achieved with two-step activation. A serious of AH treatments, each with a typical dose of about 400 L, was used for the LPE grown photocathode with a total exposure as high as 5800 L. Within some uncertainties, no degradation of the QE was observed. Photocathodes from the MOCVD group showed a different evolution of the quantum efficiency. With these photocathodes higher initial QE of about 25 - 30% were routinely obtained. Moreover, the first AH cleaning typically resulted in an increase of the QE with an obtained highest value of 35 %. However, further AH cleanings (21 with total exposure of 5100 L for the sample with longest history) and reactivations of these photocathodes lead to a slow, steady decay of the QE with a saturation value at about 15 %. We varied hydrogen exposures within 100 - 440 L but did not see strong improvements. The refreshing of the surface with HCL treatment inside glove box did not rise QE value proving that the surface cleanness is not responsible for the steady QE degradation along with the number of AH treatments. The possible reason of the degradation is believed to be due the insertion of a dislocation network inside the semiconductor structure during multiple heat-cleaning cycles of up to 450 ^{0}C. Indeed, transmission – mode GaAs photocathodes on glass or sapphire substrates are mechanically strained and potentially have a risk to develop a dislocation net along with heat-treatment cycles even at moderate temperature treatment due to the different thermal expansion coefficients of the semiconductor heterostructure and the sapphire base plate. Study of the surface morphology of degraded photocathodes with an atomic force microscope indeed revealed the existence of a mesh – like structure resulting from the dislocation network which is inserted into the heterostructure during multiple high temperature treatments under the influence of mechanical strain. More studies are needed to minimize the dislocation density in strained photocathodes.

TSR ELECTRON TARGET PERFORMANCE

Using the cryogenic photocathode source, ultra cold electron beams were obtained, with transverse and longitudinal temperatures of <1 meV and 0.03 meV, respectively [4,5]. By controlling and systematically reducing different degradation effects like cryosorption, back stream of ionised rest gas, and pressure rise due to electron-induced gas desorption from the chamber walls, the lifetime of the photocathodes was strongly improved. Presently, the lifetime at high currents is limited by back-stream of ions and amounts to about 24 hours at 1 mA current. A large number of high resolution recombination measurements on atomic and molecular beams were performed using the cryogenic photoelectron source [5,20,21]. Below we demonstrate the performance of the photoelectron beam device in the TSR source as a high resolution electron target as well as an efficient cooler for low-energy ion beams.

Dielectronic Recombination of $^{45}Sc^{18+}$: QED Test Bench for Li-like Ions

In dielectronic recombination (DR) between atomic ions and electrons a free electron is resonantly captured into a Rydberg state, accompanied by the excitation of electrons already bound by the target ion (core excitations). Radiative decay of this intermediate autoionizing state completes the recombination. DR takes place when the electron collision energy equals the difference of the core excitation energy and the

Rydberg binding energy. In many cases of lithium-like and beryllium-like systems, binding energies of high Rydberg states can be calculated with high accuracy. In these cases, measurements of the DR rate coefficient at high energy resolution are tools to precisely derive the transition energies of the core valence electrons. The highest precision is reached when studying resonances of low energies (ideally below ~ 0.1 eV). In this case the experimental resolution reaches an optimal value close to the transverse electron temperature $kT_{\perp}$, which is better than 1 meV for the TSR's ultra cold electron target.

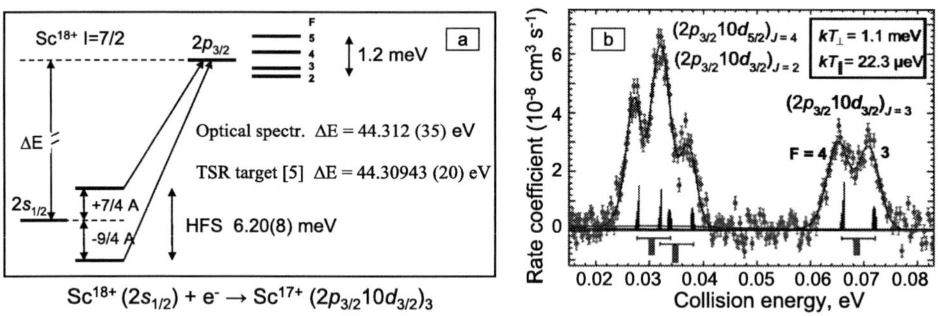

FIGURE 3. a) Energy level scheme for the $2s$-$2p_{3/2}$ transition of lithiumlike Sc^{18+}; b) high-resolution experimental DR rate coefficient spectrum of Sc^{18+} forming Sc^{17+} [5].

Measurements of low-energy DR resonances of lithium-like Sc^{18+} (see Fig.3), performed at the Heidelberg Test Storage Ring (TSR) with a cold photoelectron target, allowed us to determine the $2s_{1/2}$-$2p_{3/2}$ transition energy with an accuracy of 4.6 ppm, less than 1% of the few-body effects on radiative corrections [5]. The high resolution for the first time allows hyperfine components of DR resonances to be resolved in an electron collision experiment, revealing the hyperfine doublet of the $2s_{1/2}$ ground state due to the I=7/2 nuclear spin of ^{45}Sc. The analysis of the more complex hyperfine structure for the Rydberg DR resonances allows us to determine the resonance term energies from the hyperfine-split experimental peaks. The $2s_{1/2}$-$2p_{3/2}$ transition energy is found from these terms using relativistic many-body perturbation theory (RMBPT) and yielding the Rydberg binding energies to ~0.0001 eV.

CF^+ Cooling by 53 eV Electrons

Electron cooling experiments at low energies were performed at the TSR target on a CF^+ beam of energy 3 MeV (about 97 keV/u) [15]. The corresponding electron cooling energy is of about 53 eV. The diameter of the magnetically expanded electron beam was 13 mm (α=20), the electron current was limited by gun perveance to about 0.34 mA (electron density 3×10^6 cm^{-3}). In the case of the slow CF^+ ion beam the electron target was found to be by far stronger in cooling force than the TSR's e-cooler, providing much shorter cooling times. The CF^+ current in the storage ring after injection was estimated to be of a few 100 pA and the lifetime of the ion beam was about 4 s.

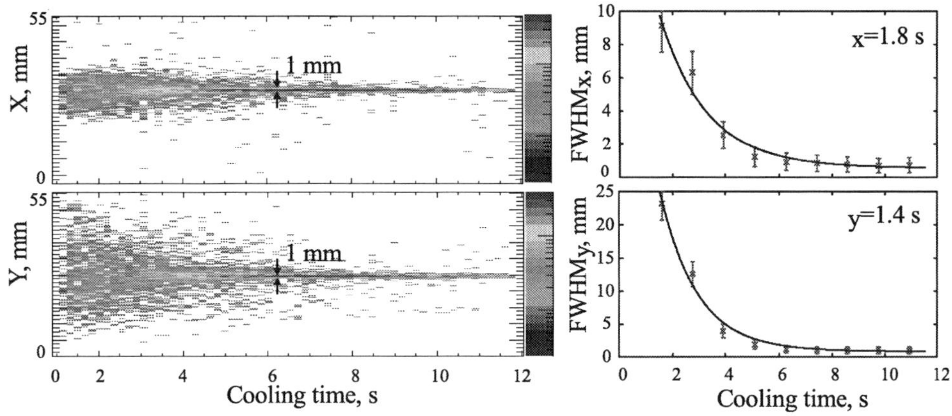

FIGURE 4. Transverse electron cooling of 97 keV/u CF^+ [15]. Centre-of-mass positions of the C and F neutral fragments vs time, recorded 12 m downstream of the electron target by the imaging detector and indicating rms divergence angles of about $3 \cdot 10^{-5}$.

Imaging detectors 12 m downstream of the electron target were used to measure neutral recombination fragments and to analyze the dissociation dynamics. The transverse momentum of the dissociation fragments is recorded event by event using a spatially resolving multi-hit detector (two-dimensional fragment imaging) [22].

For atomic ion beams the spatial profile of the neutral atoms produced by recombination in the electron target directly reflects the angular divergence of the ion beams [23]. For molecules, the imaging profile can not be used directly to monitor transverse ion beam properties during the phase space cooling due to the significant kinetic energy release in the dissociation process. However, this can be achieved by monitoring the centre-of-mass of all fragment hit positions for a given recombination event, which reflects the direction of the molecular ion before it captured an electron in the target [24].

Figure 4 (left) shows the distribution spread of the centre-of-mass as a function of the cooling time derived from correlated two-hit events for both transverse directions. The FWHM of the centre-of-mass distribution vs time is also shown in Fig. 4 (right). A transverse cooling time below 2 s was achieved, demonstrating a high cooling efficiency of the photocathode electron beam. An estimation of the cooling time from Eq. (2), assuming an isotropic electron beam with 1 meV temperature, L_c=3.3 and the length of the target to be about 2.2% of the ring circumference, gives a value of about 1.5 s for a cold ion beam and 6 s for a hot ion beam (with a diameter of the injected ion beam about 2 times larger than that of the electron beam). After 6 s a FWHM of the centre-of-mass distribution of 1 mm (horizontally) and 0.7 mm (vertically) was measured by the imaging detector. Assuming a longitudinal ion spread of about $5 \cdot 10^{-5}$ and using the horizontal TSR dispersion of 2 m and the β functions of 3.9 m (horizontal) and 1.5 m (vertical) at the target section, the divergence and size of the ion beam can be derived. The divergence was found to be of about $3 \cdot 10^{-5}$ in both transverse directions and the 1σ size was about 0.04 mm (vertically) and 0.2 mm (horizontally). The larger size of the ion beam in the horizontal direction arises from the ring dispersion and from fluctuations of the power supplies for dipole magnets and cathode voltage.

Thus, the short cooling times and the very small equilibrium beam size obtained in these experiments demonstrate the high potential of cryogenic photocathode electron beams for electron cooling of slow ions, in particular for electrostatic Cryogenic Storage Ring under construction at the Heidelberg MPI-K [25,26].

CONCLUSIONS

Ultra-cold photoelectron beams from a cryogenic GaAs electron gun, developed for the Heidelberg TSR target, was found to be excellent tool for high-resolution recombination measurements at storage rings as well as for eV-cooling of low-energy ions. The source can deliver electron currents up to few mA with a typical lifetime of about 24 hours and provides unprecedented energy resolution below 1 meV. Atomic hydrogen treatment, developed for in-situ refreshing of the high quantum efficiency GaAs photocathodes inside of the gun vacuum setup, provides conditions necessary for non-stop operation of the electron target.

REFERENCES

1. D.A. Orlov et al., *Appl. Phys. Letters* **78**, 2721-2724 (2001).
2. D.A. Orlov et al., *NIM A*, **532**, 418-421 (2004).
3. U. Weigel et al., *NIM A*, **536**, 323-328 (2005).
4. D.A. Orlov et al., *J.Phys.: Conf. Ser.* **4**, 290-295 (2005).
5. M. Lestinsky et al., *Phys. Rev. Lett.* **100**, 033001 (2008).
6. D.A. Orlov et al., "Ultra-cold electron beams for the Heidelberg CSR and TSR," in *Beam cooling and related topics*, edited by S. Nagaitsev and R.J. Pasquinelli, AIP Conference Proceedings 821, American Institute of Physics, Melville, NY, 2006, pp. 478-487.
7. N.Meshkov, *Phys. Part. Nucl.* 25, 631-661 (1994).
8. F. Sprenger et al., *NIM A*, **532**, 298-302 (2004).
9. T.M. O'Neil and P.G. Hjorth, *Phys Fluids*, **28**, 3241-3252 (1985)
10. S. Pastuszka et al., *NIM A*, **369**, 11-22 (1996).
11. N.S. Dikansky, Preprint **88-61**, Institute of Nuclear Physics, Novosibirsk, 1988.
12. Al-Khalili et.al., *Phys. Rev. A*, **68**, 042702 (2003).
13. L. Spitzer, "Physics of fully ionized gases", Inter-science, New York, 1956.
14. Ya.S. Derbenev, A.N. Skrinskii, *Sov. Phys. Rev.* **1**, 165-237 (1981).
15. D.A. Orlov et al., "*Electron cooling with photocathode electron beams applied to slow ions at TSR and CSR*", in *Proceedings of COOL 2007*, edited by R.W. Hasse and V. RW Schaa, Bad Kreuznach, Germany, 2007, pp.230-233.
16. Developed and manufactured in the group of Prof. A.S. Terekhov.
17. T. Maruyama et al., *Appl. Phys. Letters* **82**, 4184-4186 (2003).
18. M. Baylac et al., *Phys. Rev. ST AB* **8**, 123501(2005)
19. U. Bischler, E. Bertel, *J. Vac. Sci. Technol. A* **11**, 458-460 (1993)
20. H. Kreckel et al., *Phys. Rev. Lett.* **95**, 263201 (2005).
21. S. Novotny et al., *Phys. Rev. Lett.* **100**, 193201 (2008).
22. Z. Amitay, *Phys. Rev. A* **54**, 4032-4050 (1996)
23. G.I. Budker et al., Part. Accel. **7** 197-211 (1976).
24. A. Wolf et al., NIM A **532**, 69-78 (2004).
25. D. Zajfman et al., *J.Phys.: Conf. Ser.* **4**, 296-299 (2005).
26. A. Wolf et al., "The Heidelberg CSR: Stored Ion Beams in a Cryogenic Environment," in *Beam cooling and related topics*, edited by S. Nagaitsev and R.J. Pasquinelli, AIP Conference Proceedings 821, American Institute of Physics, Melville, NY, 2006, pp. 473-477.

Optimization of Semiconductor Superlattice for Spin Polarized Electron Source

L. G. Gerchikov[a], Yu. A. Mamaev[a], V. V. Kuz'michev[a], Yu. P.Yashin[a], V. S. Mikhrin[b], A. P. Vasiliev[b], A. E. Zhukov[b]

[a]*St. Petersburg State Polytechnic University, Russia*
[b]*A. F. Ioffe Physicotechnical Institute RAS, Russia*

Abstract. Optimized AlInGaAs/AlGaAs superlattice with strained quantum wells has been developed, fabricated and studied. The choice of hetero-layer composition and thicknesses is based on calculations of Sl's band energy spectrum, photoabsorption spectrum and transport properties. Electron emission from the developed photocathodes demonstrates maximal polarization above 90%.

Keywords: Polarization, Photoemission, Superlattice.
PACS: 72.25.Fe,73.21.Cd,79.60.-i

INTRODUCTION

The use of short-period superlattices (SL) opens up wide possibilities for band structure engineering of photoemitter working layer. The optimal choice of SL's layers composition and thickness providing the high energy splitting of the valence band together with good transport properties for photoelectrons have made it possible to design photocathodes with electron polarization above 90% [1-3]. Photocathodes based on GaAs/GaAsP [1] and AlInGaAs/GaAlAs [2] SL structures with strained quantum wells (QW), GaAs/AlInGaAs-based SLs [3] with strained barriers, and so-called compensated SLs based on AlInGaAs/GaAsP SL structures with opposite strains in the wells and barriers [4] have been developed.

These studies have also revealed several fundamental difficulties for achieving further progress. High electronic polarization, *P*, is achieved at the expense of quantum efficiency, *QE*. Indeed, the maximal spin orientation of photoelectrons takes place at the photoabsorption threshold where the light absorption coefficient is rather small. Maximal polarization of photoelectrons is limited by the mixture of light and heavy hole states which takes place even at the absorption edge due to the broadening of the valence band energy spectrum and smearing of the valence band edge. The mixture of light and heavy hole states can be reduced by increasing the energy splitting $\Delta E_{hh\text{-}lh}$ between the *hh* and *lh* subbands. However the splitting of SL valence band due to the deformation is limited by some critical value. Deformation beyond this level results in structural defects, smaller residual strain and lower polarization.

CP1149, *Spin Physics, 18th International Spin Physics Symposium*
edited by D. G. Crabb, D. B. Day, S. Liuti, X. Zheng, M. Poelker, and Y. Prok

The valence band splitting ΔE_{hh-lh} in SL cannot exceed the valence band offsets. Therefore, to create a significant splitting $\Delta E_{hh-lh} \approx 100$ meV, SL with large band offsets at the heterointerfaces must be used. In this case, high barriers of hundreds of meV usually appear in the conduction band that complicates the transport of photoelectrons to the photocathode surface. Low electron mobility along the SL axis reduces the *QE*. This leads to a partial loss of spin orientation by electrons during their transport to the surface due to D'yakonov–Perel' mechanism [5] for spin relaxation.

Thus, choosing the optimal structure of a SL to provide large ΔE_{hh-lh} splitting in the absence of strain relaxation and in combination with good transport properties is a difficult problem. The solution requires a reliable method to calculate the energy spectrum of the SLs on the basis of ternary and quaternary III–V semiconductor compounds. In addition, the maximum value of photoelectron polarization depends crucially on the structural quality of the SLs, which imposes heightened requirements on the technology of their growth.

VALENCE BAND SPLITTING

The initial polarization of the photoelectrons can be increased by choosing structures with a high valence band splitting. The SL structures with strained QW layers where heavy and light hole minibands, in addition to the strain splitting, are moved aside due to different light and heavy-hole confinement energies in the QW, are the best for this purpose. However strain induced splitting and splitting caused by quantum confinement do not combine additively and the resulting energy splitting is a complicated function of SL composition and layer thickness. Deformation splitting of the QW's valence band edge increases the QW depth for heavy holes and decreases it for the light holes. Thus the increase of deformation splitting reduces the quantum confinement contribution to ΔE_{hh-lh}. Fig. 1 illustrates the choice of optimal QW thickness for $Al_{0.19}In_{20}Ga_{0.61}As/Al_{0.4}Ga_{0.6}$ As SL with strained QW and 2.1nm thick barriers. Here we show the positions of the highest hole minibands together with the energy splitting ΔE_{hh-lh}. calculated within multiband Kane model [6] as a function of the QW width *a*. For our structure we have chosen the QW width $a = 5.4$nm because the increase of ΔE_{hh-lh} at $a > 5.4$ nm saturates while the risk of strain relaxation increases.

The deformation of QW layer and hence the deformation energy splitting of the valence band edge linearly varies with In concentration. The chosen for our structure In concentration 0.2 corresponds to 76meV of deformation splitting that together with quantum confinement contribution results in ΔE_{hh-lh} = 87eV. This valence band splitting is sufficient to provide the initial polarization of photoelectrons of up to 97% [2].

Aluminum concentration in the QW layer does not affect the deformation but changes the energy gap value. We have chosen it equal to 0.19 to adjust the photoabsorption threshold of SL and hence the position of the main polarization maximum to $\lambda = 820$nm. The Al concentration in the barrier layer 0.4 was chosen to achieve sufficiently high barriers in the SL's valence band in order to provide large ΔE_{hh-lh}. On the other hand the conduction band barriers should be transparent enough for effective electron transport. This was achieved by the choice of the barrier width *b*

= 2.1nm. For the chosen SL's composition the barriers for electrons, heavy and light holes are equal to 339 meV, 149 meV and 77 meV, respectively.

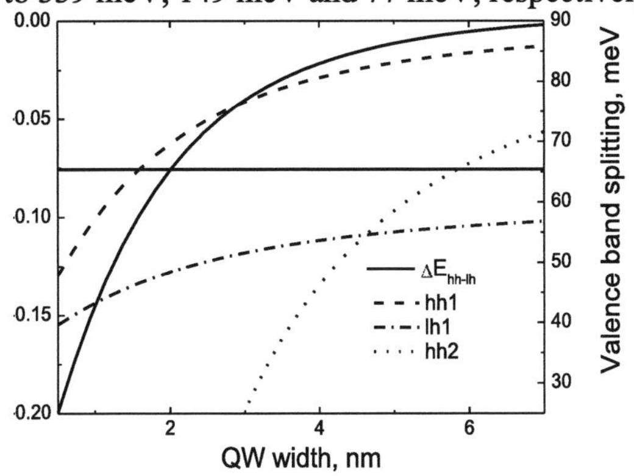

FIGURE 1. Edges of the first two heavy hole minibands, hh1 and hh2, and the first light hole miniband, lh1, are shown together with the energy splitting between hh1 and lh1 minibands ΔE_{hh-lh} as a functions of the QW width for $Al_{0.19}In_{20}Ga_{0.61}As/Al_{0.4}Ga_{0.6}As$ SL. The miniband energies are counted from the position of the heavy hole valence band edge of the QW layer. The position of the light hole valence band edge spitted by QW deformation is shown by horizontal line.

ELECTRON TRANSPORT

Optimization of electron transport along the SL axis is crucially important for SLs with strained QW because of the presence of relatively high barriers in the conduction band. To determine the transport properties of the SL we calculate the electron tunneling probability along the SL axes neglecting any electron collisions with phonons and impurities. In Fig. 2 we show the tunneling probability together with the tunneling time, τ, calculated as a function of electron energy within the range of the first electron miniband. We consider the stationary electronic current incidents on the SL and determine the tunneling time as the ratio of number of electrons inside the SL to the electron current penetrated through the SL. Each resonance peak of tunneling probability curve corresponds to electronic state localized within the SL. The resonance width Γ is proportional to the inverse tunneling time, $\Gamma = \hbar/\tau$. The energy distance between the resonances ΔE is proportional to the inverse period of electronic motion within the SL, $\Delta E = \hbar/\tau_{SL}$. Since $\Gamma << \Delta E$ the total tunneling time is much larger than the time of electron motion within the SL, $\tau >> \tau_{SL}$, i.e. τ is determined mainly by the tunneling through the last barrier. It is quite natural because the tunneling between the neighboring QWs is a resonant process and its probability is proportional to the first order of the tunneling exponent exp($-\kappa b$), where κ is the electronic wave vector under the barrier and b is the barrier width. The tunneling through the last barrier is a non resonant process and its probability is much smaller since it is proportional to the second order of the tunneling exponent exp($-2\kappa b$). It is clear that the optimal design of the SL providing equal probability for each tunneling will be achieved by taking the last barrier of half width *b*/2 [7]. In Fig. 2 we show by dash-dotted and dotted lines tunneling probability and tunneling time calculated for such optimized SL. In this case the tunneling resonances overlap, at least at the middle of the miniband, and the total tunneling time is decreased.

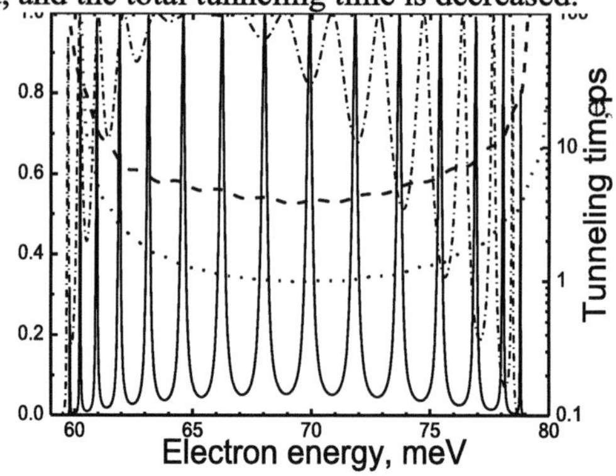

FIGURE 2. Tunneling probability (solid line) and tunneling time (dashed line) calculated for 15 periods $Al_{0.19}In_{20}Ga_{0.61}As$(5.4nm)/$Al_{0.4}Ga_{0.6}$ As(2.1nm) SL. Electron energy is counted from the bottom of electronic QW. Dash-dotted and dotted lines show the results obtained for the same SL with the last barrier of 1nm width.

The calculated tunneling time even for the optimized SL is larger by the order of magnitude than the typical collision time for electrons within the QW $\tau_p \approx 10^{-13}$s. Thus, to evaluate the real transport time one should take into account electron collisions with phonons and impurities. We do it by solving the kinetic equation within the constant relaxation time approximation. In the case of stationary optical pumping we derive the following expression for the transport time

$$\tau = \frac{(N-1/2)(N-1)}{6\tau_p}\tau_b^2 + N\tau_f, \qquad (1)$$

where N is the number of SL's periods, τ_b and τ_f are the tunneling times for the barriers within the SL and the last barrier, respectively. According to Eq.(1), the transport time is a sum of diffusion time within the SL given by the first term in r.h.s. of Eq.(1) and the time $N\tau_f$, taken to penetrate through the last barrier to the band bending region (BBR). Note, that obtained Eq.(1) is similar to the corresponding expression of the bulk case [8]. The first term in Eq.(1) corresponds to the diffusion term with diffusion coefficient $D = 2d^2\tau_p/\tau_b^2$ and the second term corresponds to the surface recombination term with surface recombination velocity $S = d/\tau_f$, $d = a + b$ is the SL period. For the developed $Al_{0.19}In_{20}Ga_{0.61}As$(5.4nm)/$Al_{0.4}Ga_{0.6}$ As(2.1nm) SL their values, D = 12 cm^2/s and $S = 3*10^6$ cm/s, are three times smaller than in the case of bulk GaAs. For 12 periods SL Eq. (1) gives the transport time τ = 5ps. The main contribution, 3ps, comes from the last tunneling. The use of last barrier with half width 1nm could make the time of the last tunneling five times smaller, $N\tau_f = 0.6$ps, and hence considerably decrease the total transport time, $\tau = 2.6$ps.

SUMMARY AND OUTLOOK

We have developed an optimized InAlGaAs/AlGaAs SL with strained quantum wells for photocathode working layer. The optimal SL design provides maximal valence band splitting $\Delta E_{hh-lh} = 87$eV with minimal risk of strain relaxation and good transport properties. Electron emission from the best photocathode based on the developed SL demonstrates the highest level of polarization of the photoelectrons P = 92% in combination with the largest quantum efficiency at polarization maximum QE = 0.85%. Further progress in development of polarized electron sources can be based on new structures with better structural quality and optimized transport properties. However maximal values of P and QE are limited by unavoidable losses of electrons and their polarization in BBR. Also, QE is strongly limited by the fact that the most part of incoming light is absorbed in the substrate. The use of Distributed Bragg Reflector can resolve this problem and considerably increase the QE of photocathode.

ACKNOWLEDGMENTS

This work was supported by Russian Ministry of Education and Science under grant N.P. 2.1.1.2215 and by Swiss National Science Foundation under grant SNSF IB7420-11111.

REFERENCES

1. T. Nishitani, T. Nakanishi, M. Yamamoto, S. Okumi, F. Furuta, M. Miyamoto, M. Kuwahara, N. Yamamoto, K. Naniwa, O. Watanabe, Y. Takeda, H. Kobayakawa, Y. Takashima, H. Horinaka, T. Matsuyama, K. Togawa, T. Saka, M. Tawada, T. Omori, Y. Kurihara, M. Yoshioka, K. Kato, T. Baba, *J. Appl. Phys* **97**, 094907 (2005).
2. Yu. A. Mamaev, L. G. Gerchikov, Yu. P. Yashin, D. A. Vasiliev, V. V. Kuzmichev ,V. M. Ustinov, A. E. Zhukov, and V. S. Mikhrin, *Appl.Phys.Lett.* **93**, 81114 (2008).
3. L. G. Gerchikov, Yu. A. Mamaev, A. V. Subashiev, Yu. P. Yashin, D. A. Vasil'ev, V. V. Kuz'michev, A. E. Zhukov, E. S. Semenova, A. P. Vasil'ev and V. M. Ustinov, *Semiconductors*, **40**, 1326 (2006).
4. A. V. Subashiev, L. G. Gerchikov, Yu. A. Mamaev, Yu. P. Yashin, J. S. Roberts, D.-A. Luh, T. Maruyama, and J. E. Clendenin, *Appl. Phys. Lett.* **86,** 171911 (2005).
5. *Optical Orientation*, edited by F. Meier and B. P. Zakharchenya, North-Holland, Amsterdam, 1984.
6. A.V. Subashiev, L.G. Gerchikov, and A.I. Ipatov. *J. Appl. Phys.,* **96**, 1511 (2004).
7. V. A. Tkachencko, V. E. Andreev, O. A. Tkachenko, V. L. Alperovich, A. S. Proc. *10th Intern. Symp. on High Energy Spin Physics*, Nagoya, Japan, 1992, p.857-860.
8. K. Aulenbacher, J. Schuler, D. v. Harrach, E. Reichert, J. Roethgen, A. Subashev, V. Tioukine, Y. Yashin, *J. Appl. Phys.*, 92, 7536 (2002).

A Study of the Activated GaAs Surface for Application as an Electron Source in Particle Accelerators

N. Chanlek[a,b], J.D. Herbert[b] , L.B. Jones[b] , R.M. Jones[a] and K.J. Middleman[b]

[a]*University of Manchester, Manchester, M13 9PL, United Kingdom*
[b]*STFC Daresbury Laboratory, Warrington, WA4 4AD, United Kingdom*

Abstract. The use of type III-V semiconductor materials as photocathodes has in recent years become a focus for the High Energy Physics community. Once activated to a negative electron affinity (NEA) state and illuminated by a laser, these materials can be used as a high-brightness source of both polarised and un-polarised electrons in some modern accelerators, for example, ALICE (Accelerators and Lasers in Combined Experiments) at Daresbury Laboratory. This paper will focus on the use of gallium arsenide (GaAs) as a photocathode, and detail the re-configuration and re-commissioning of two vacuum systems that support standard surface science techniques such as ultraviolet/X-ray photoelectron spectroscopy (UPS/XPS), low energy electron diffraction (LEED) and auger electron spectroscopy (AES). The paper will present details of cleaning GaAs in order to maximise quantum efficiency and will provide evidence from XPS and LEED to demonstrate what is happening at the atomic level.

Keywords: negative electron affinity (NEA), gallium arsenide (GaAs), photocathode, x-ray photoelectron spectroscopy (XPS) and low energy electron diffraction (LEED)
PACS: 29.20.-c, 29.25.Bx, 61.05.cj, 61.05.jh

INTRODUCTION

Gallium arsenide (GaAs) photocathodes are being used as electron sources for various modern particle accelerators and next generation light sources. For example, the ALICE ERL prototype, a 35 MeV energy recovery test facility, is currently under commissioning at Daresbury Laboratory [1]. The current design of this machine requires an electron source that can generate a continuous wave (CW) current of 6.5mA (electron beam) as a pre-cursor to the 100 mA injector with a normalised emittance of $< 2\,\pi$ mm mrad [2].

It is well known that once activated to a negative electron affinity (NEA) state, the GaAs photocathode can be used as a high-brightness, low emmitance electron source [3,4]. This state is prepared by depositing caesium and an oxidant (either O_2 or NF_3) onto its atomically clean surface. The preparation process involving handling, preparation and cleaning processes is an essential step in the performance of the NEA GaAs photocathode. However, the fundamental nature and mechanism of these processes is not well understood, and clarification has been the subject of research for many years.

CP1149, *Spin Physics, 18th International Spin Physics Symposium*
edited by D. G. Crabb, D. B. Day, S. Liuti, X. Zheng, M. Poelker, and Y. Prok

In this article we report on the progress of the re-configuration and re-commissioning of two vacuum systems which will allow the study and investigation of the photocathode activation process using the surface science techniques available within the same vacuum system. We also present the preliminary results of an X-ray photoelectron spectroscopy (XPS) and low energy electron diffraction (LEED) study on the effects of heating temperature on the GaAs surface in the heat cleaning process.

EXPERIMENT SET UP

Figure 1 shows our present experimental chamber configured at the Cockcroft Institute, STFC Daresbury Laboratory. The chamber consists of three sections, a loading chamber, cathode preparation chamber and surface analysis chamber which are isolated by gate valves. The photocathode sample is introduced to the system through the loading chamber which is pumped down via a dry turbomolecular pumping system. Once a suitable pressure has been reached the sample can be transferred between the sections using a transfer arm. The cathode preparation chamber includes a manipulator with sample holder and all of the components needed to clean and activate a cathode to the NEA state. The sample holder is isolated from the chamber wall by a ceramic ring and connected to a UHV electrical feedthrough so that a bias voltage can be applied to the sample. The chamber is also equipped with a caesium dispenser and a hydrogen gas cracker. The chamber is kept at a base pressure of 1×10^{-10} mbar by a dry turbomolecular pumping system, a DI ion pump and additional NEG pumping modules. The surface analysis chamber is equipped with a twin anode X-ray source, a hemispherical analyser for ultraviolet/X-ray photoelectron spectroscopy (UPS/XPS) technique, a LEED optics for low energy electron

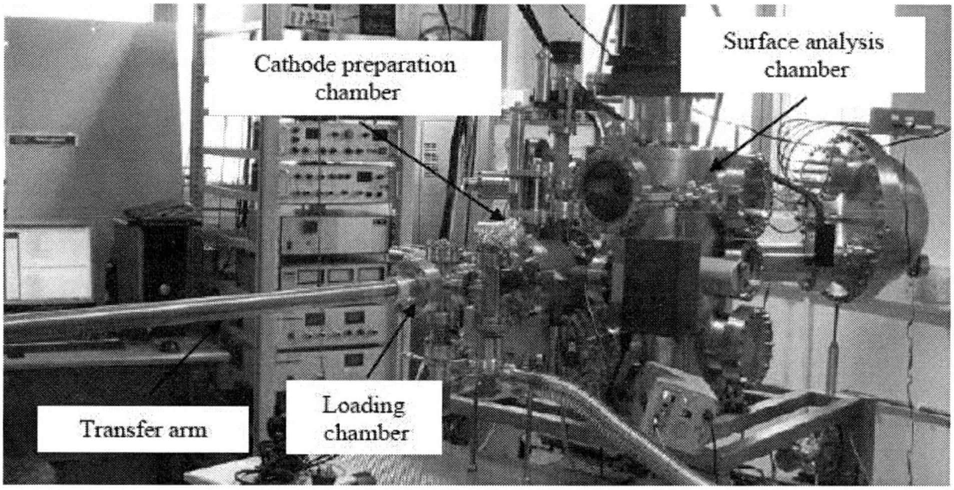

FIGURE 1. Picture of Photocathode experimental chamber which has been set up at the Cockcroft Institute, STFC Daresbury Laboratory.

diffraction (LEED) and auger electron spectroscopy (AES) technique, ion gun for sample sputter cleaning and RGA for residual gas analysis. This chamber is maintained at a base pressure of 1×10^{-10} mbar via a turbopump connected directly and supported by a turbomolecular drag dry pump system and two NEG pumps.

RESULTS AND DISCUSSION

For preliminary results, the heat cleaning procedures for a GaAs photocathode were studied. Bulk VGF (vertical gradient freeze) GaAs samples of p-type doping (Zn) without any chemical cleaning were loaded in order to leave native contaminants on the surface. The samples were heated to five different temperatures; 450, 500, 550, 600 and 625 °C for 60 minutes in the surface analysis chamber. During heating, the temperature of the sample surface was monitored via an infrared pyrometer which has the resolution of ±15 °C. Note that this error comes from the rough calibration because, presently, no accurate emissivity of the GaAs photocathode could be estimated. The XPS spectra of the GaAs surface excited by Al $K\alpha$ (1486.6 eV) radiation were taken before and after heat cleaning process. The typical contaminants for GaAs such as oxygen, carbon and chlorine were observed on the dirty sample surface [5]. The removal by heat cleaning of oxides which are the main coverage contaminants on the GaAs surface was studied, with representative results as shown in Fig. 2.

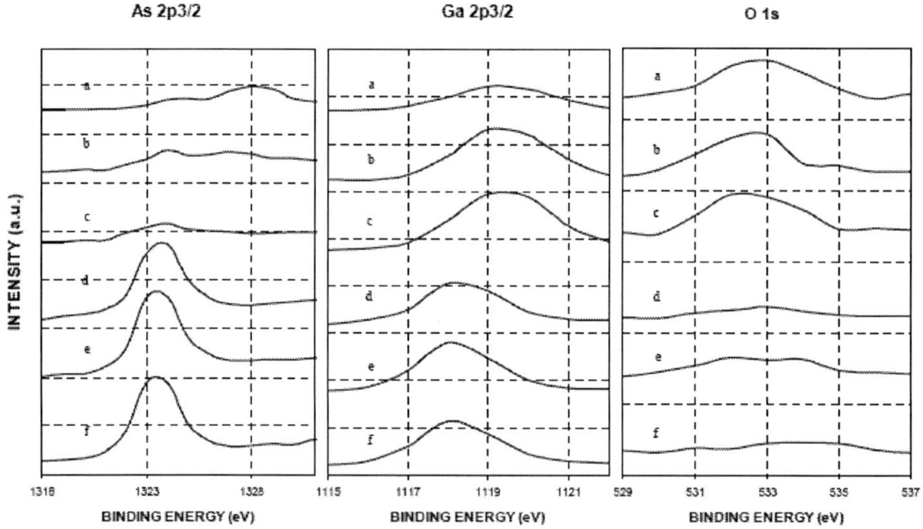

FIGURE 2. XPS spectra of the As $2p_{3/2}$, Ga $2p_{3/2}$ and O 1s for (a) the initial dirty sample, (b) after heating to 450°C for 60 min, (c) 500°C for 60 min, (d) 550°C for 60 min, (e) 600°C for 60 min and (f) 625°C for 60 min.

The core-level energy of the As $2p_{3/2}$ (1324 eV), Ga $2p_{3/2}$ (1117 eV) were observed and the presence of As_2O_3 and Ga_2O_3 were determined by the core-level energy shifts of 3 and 1 eV toward higher energies, respectively [6, 7]. The XPS spectra show that the energy shifts of the As $2p_{3/2}$, Ga $2p_{3/2}$ peaks decreased when the heating temperature was increased, and at the same time the intensity of As core-level energies increased. These results indicate that the surface oxides are desorbed from the GaAs surface under heat cleaning, and are completely removed after heating to a temperature higher than 550 °C for 60 min. This is also consistent with the reduction in the O 1s intensity.

The activation to NEA surface and measurement of the quantum efficiency (QE) was made after heating the sample to each of the different temperatures. Caesium and oxygen were deposited onto the sample by using the standard "yo-yo" method [3]. The samples were illuminated with a HeNe laser at 632.8 nm during the activation process to drive the photocurrent and calculate the highest QE of the sample. The QE of the GaAs samples are shown in Fig. 3. As is clearly seen, the QE of the GaAs samples increases when the cleaning temperature is increased, agreeing with the XPS results that the oxide overlayer is removed under heat cleaning. A QE of 4.12% was obtained as a result of heating the sample at 600°C, however, the QE dropped to 3.81% after heating to 625°C. This result shows that heating to excessive temperatures might damage the properties of the GaAs, resulting in a detrimental effect on the QE.

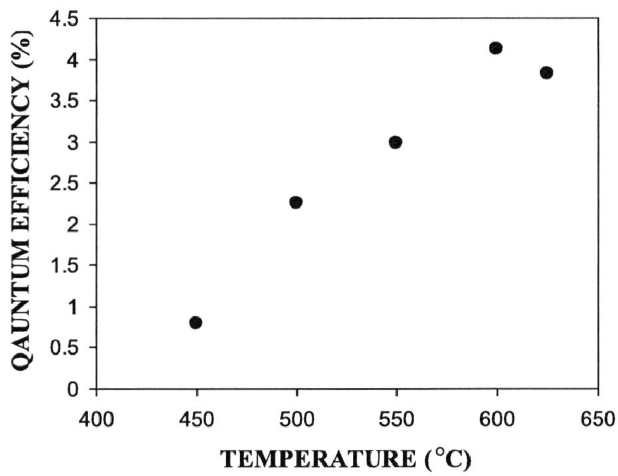

FIGURE 3. Quantum efficiency as a function of heat temperature in heat cleaning procedure.

To further our studies, the LEED pattern of the GaAs sample was also taken after heating to the different temperatures. It was found that a clear LEED pattern can be observed after heating the sample to a temperature of 550°C for 60 min, as shown in Fig. 4, indicating that the sample has a clean and well-ordered surface.

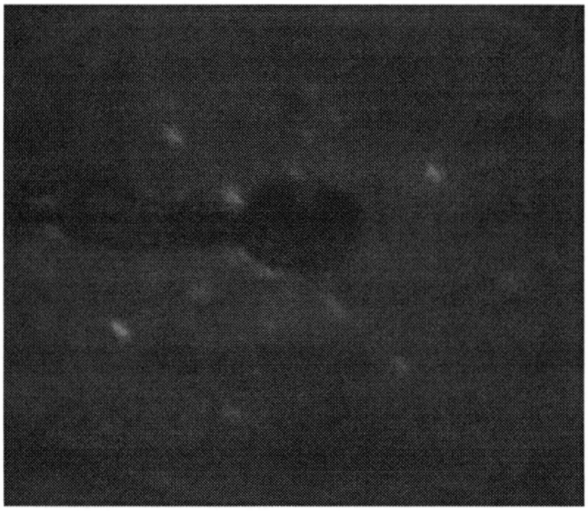

FIGURE 4. LEED pattern of the GaAs surface after heat the sample to 550°C for 60 min.

CONCLUSIONS

The amalgamation of two vacuum systems has been completed at the Cockcroft Institute, Daresbury Laboratory in order to study and investigate the photocathode material in detail. This experimental set up allows us to prepare and study the properties of the photocathode material using the application of surface science techniques such as ultraviolet/X-ray photoelectron spectroscopy (UPS/XPS), low energy electron diffraction (LEED) and auger electron spectroscopy (AES) within the same vacuum system.

The heat cleaning procedure for the NEA GaAs photocathode was studied. It was found that oxides from the GaAs surface can be removed after heating to a temperature higher than 550°C for 60 min. However, the heating temperature need to be optimized, as excessive temperatures can result in decreasing in the QE.

ACKNOWLEDGMENTS

The authors are grateful for support from ASTeC and STFC Daresbury Laboratory.

REFERENCES

1. Daresbury Laboratory, 4^{th}Generation Light Source, ERLP Construction, accesses September 2008, http://www.4gls.ac.uk/ERLP.htm
2. B.L. Militsyn (private communication).
3. D.T. Pierce and F. Meier, *Phys. Rev. B* **13**, 5484(1980).
4. T. Maruyama, D.-A. Luh, A. Brachmann, J.E. Clendenin, E.L. Garwin, S. Harvey, R.E. Kirby and C.Y. Prescott, *Appl. Phys. Lett.* **82**, 4184 (2003).
5. J. Uebbing, *J. Appl. Phys*. **41** 802(1969).
6. E. J. Petit and F. Houzay, *J. Vac. Sci. Technol. B*. **12**, 547(1994).
7. Z. Lu, M.T. Schmidt, D. Chen, R.M. Osgood, W.M. Holber, D.V. Podlesnik and J. Forster, *Appl. Phys. Lett.* **58**, 1143(1991).

Recent Progress toward Robust Photocathodes

G. A. Mulhollan and J. C. Bierman

Saxet Surface Science, Austin, TX 78744, USA

Abstract. RF photoinjectors for next generation spin-polarized electron accelerators require photocathodes capable of surviving RF gun operation. Free electron laser photoinjectors can benefit from more robust visible light excited photoemitters. A negative electron affinity gallium arsenide activation recipe has been found that diminishes its background gas susceptibility without any loss of near bandgap photoyield. The highest degree of immunity to carbon dioxide exposure was achieved with a combination of cesium and lithium. Activated amorphous silicon photocathodes evince advantageous properties for high current photoinjectors including low cost, substrate flexibility, visible light excitation and greatly reduced gas reactivity compared to gallium arsenide

Keywords: Photocathode, NEA, GaAs, Immunization, Photoyield decay, Amorphous silicon, FEL
PACS: 85.60.Ha, 29.25.Bx

INTRODUCTION

The motivations for robust photocathodes are manifold. Ideal photoemitters would perform well under very poor vacuum conditions. In practice, improved performance in operable vacuum is always desirable. It is well known that the environment in RF guns is still tough on photoemitters, though designs incorporating better pumping represent good progress. High electron spin-polarization photocathodes continue to demand ultra high vacuum (UHV) conditions for reliable operation. Better environmental immunity for these photocathodes is required for successful operation in RF guns, such as could be employed for the ILC. Photocathodes operated in high average current injectors manifest cumulative damage which limits operational lifetimes.

Photocathode environmental interactions may be viewed as stemming from either chemical reactions or charged particle fluence. The principal source for chemical reaction participants is the background gas. Electron beam induced desorption can increase the quantity of these reactants. Several gas species, such as CO_2, H_2O and O_2 are known to be harmful to the lifetime of negative electron affinity (NEA) photocathodes[1]. Detrimental charged particle fluences at photocathodes are primarily composed of ions. Low energy ions, most likely in RF guns, generate atomic displacements near the surface and can affect the activation layer as well as the near surface crystal structure. This damage can be annealed out and the photocathode activated again to an NEA state. High energy ions, such as found in DC guns can cause many displacements per particle if they impact the photocathode. The damage to the photocathode crystalline structure can be widespread and if the energy is large enough, the damage may not anneal out. Both high and low energy ion interactions can affect yield and the emitted electron spin-polarization.

CP1149, *Spin Physics, 18th International Spin Physics Symposium*
edited by D. G. Crabb, D. B. Day, S. Liuti, X. Zheng, M. Poelker, and Y. Prok

HIGH POLARIZATION PHOTOEMITTER IMMUNIZATION

Generation of high spin-polarization electrons from III-V and related semiconductors mandates the emitter be crystalline. Both single strained-layers and superlattices (strained and unstrained) have been employed in this role. The use of an activation layer to achieve NEA is also a requirement in common. The first goal of this work was to achieve a level of immunity against background gas chemical reactivity for such emitters. Several groups have studied the decay process of NEA photocathodes. While good progress toward understanding the chemical makeup of diminished yield photocathode surfaces has been made[2], both in process and susceptibility, this information has not provided insight into the prevention of the decay process, aside from straightforward improvements in vacuum base pressure leading to increased lifetimes. Our approach to the problem was to attempt to de-activate potential reaction sites. The overlayers on activated GaAs(100) are amorphous. This can be partly attributed to the large covalent radius of cesium, $\sim$2.3 Å, compared to the GaAs nearest neighbor distance. In a billiard ball model, large gaps are present between the cesium atoms. In fact, they are large enough for a much smaller covalent radius atom, e.g., oxygen, $\sim$0.73 Å, to fit through them and bond to the semiconductor surface and/or cesium. This argument suggests that if access to the underlying surface could be blocked during or following the activation process, then absorbed gas induced decay could be inhibited. The true picture must be formulated in terms of electron charge density with the gaps actually being regions of low charge density. An obvious choice for a blocking atom is another, smaller covalent radius alkali. While all alkali atoms lower the GaAs(100) electron affinity, the best performance is from cesium. This indicates that blocking the surface reactivity is likely to incurr a cost in the yield.

An alkali source flange containing Cs, Rb, K, Na and Li getter channels was constructed to allow testing of each alkali's effects on yield and sensitivity to reactive gas exposure. This flange was installed into an existing photocathode test system, described elsewhere[3]. Activations were first performed with each alkali individually to establish the proper operating curents. Standard activations were performed by depositing alkali until a peak in photocurrent was achieved. Following the peak, the oxidizer gas was introduced. The oxidizer partial pressure was adjusted for maximim photocurent increase (co-deposition) until the final peak value was achieved. In all cases, the activation oxidizer gas was NF_3. Carbon dioxide was chosen as the archtype decay gas. It is a known strong reactant with NEA surfaces and is readily distinguishable in residual gas spectra. Dosing data were acquired under computer control. The cathode bias was disconnected between data points and the light source was shuttered when not in use. First, dark current values were acquired for each point, then the light source was unshuttered for the emission measurement. Data were recorded once a minute. Dosing consisted of three segments. In the first 30 minutes, the photocathode was exposed only to the normal background gas, base pressure $4\text{x}10^{-11}$ Torr, primarily H_2. In the second segment, the pressure, as measured by a nude ion gauge, was increased to $1.5\text{x}10^{-10}$ Torr by introduction of the dosing gas, and in the final segment the pressure was increased to $5.0\text{x}10^{-10}$ Torr.

In keeping with the above argument, activations using two alkali types were initially attemped by introducing the second, non-Cs alkali after the first photocurrent peak was

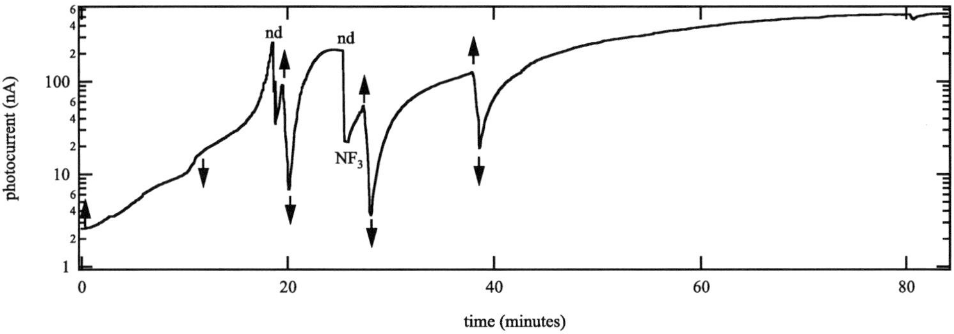

FIGURE 1. Activation curve for Cs + Li activated bulk GaAs. Up arrows indicate when the Li is on and down arrows when off. Neutral density filter insertion is indicated by 'nd'.

achieved. The second alkali was deposited at the rate required for the single alkali activations to complete in the same time as those using Cs alone. Both alkalis were used to complete the activation. These activations resulted in lower than normal photoyields and little resilience to the dosing gas. However, the decay curves for the Na and Li dual alkali activated photocathodes did evince a lower decay rate when dosed than the other combinations. By limiting the second alkali's exposure during most of the activation, reasonable yields were achieved and relatively better immunity to the dosing gas was achieved. An example of the activation with Li as the second alkali is shown in Fig. 1.

Figure 2 shows the yield decay upon dosing for bulk GaAs at 633 nm. The dual alkali activated photocathode maintained a yield ~2x that of the standard activated GaAs after the low pressure dosing and ~6x that of the standard activated GaA at the end of the high pressure dosing. Comparable, but not as dramatic results were observed for Cs and Na activated GaAs. Number and timing of deposition of the second alkali was investigated.

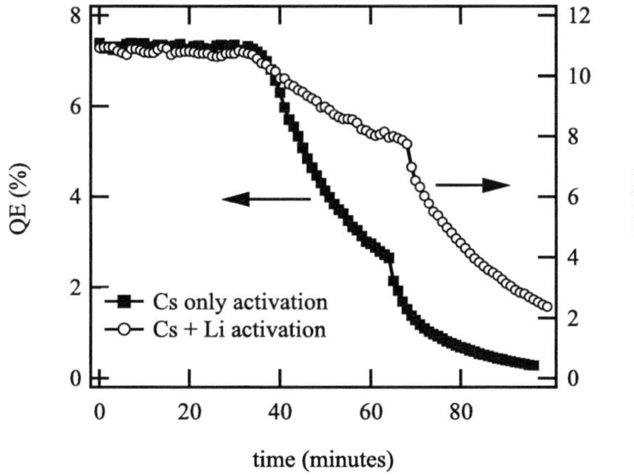

FIGURE 2. Comparison of yield decay at 633 nm for Cs only and Cs + Li activated bulk GaAs.

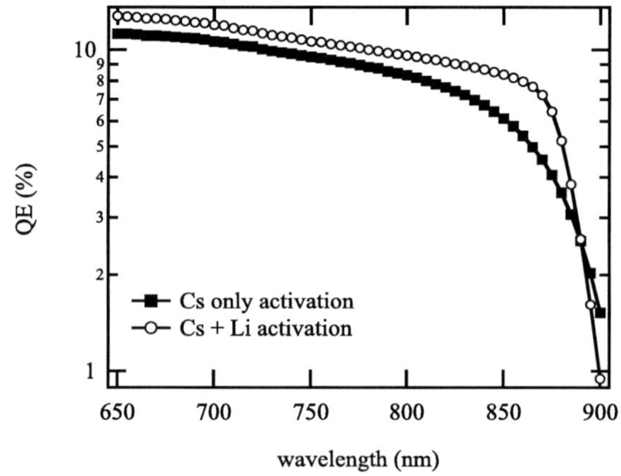

FIGURE 3. Yield curves for bulk GaAs activated with Cs only and Cs + Li.

Critical is the presence of the second alkali at the start of the activation. Subsequent depositions, both in timing and number, are less critical to achieve the best immunity. Yield data for standard Cs only activated and Cs + Li activated bulk GaAs are shown in Fig. 3. Overall the yield does not suffer from the addition of Li. Taking into account sample-to-sample variation, the numerous Cs + Li activated photocathodes, using the activation shedule of Fig. 1, all gave excellent photoyield. The yield curve, however, for the Cs + Li activated sample exhibits a much less rounded shape near the bandgap. This shape change is very repeatable and does not occur for the Cs + Na activated photocathodes. The addition of the Li changed not only the reactivity of the surface, but the shape of the surface barrier as well.

AMORPHOUS SiGe PHOTOEMITTERS FOR FEL SOURCES

Amorphous Si photocathodes have been grown in our laboratory both by sputter deposition and RF PECVD. Advantages of a-SiGe for use in an FEL injector gun include *ex situ* growth so that the growth apparatus is separated from the injector load system. Amorphous SiGe can be grown on many substrates including tantalum, glass, stainless steel and copper. The latter requires a diffusion barrier. Either transmission or reflection modes can be used. RF PECVD is a mature technology and a-SiGe can be grown to accommodate nearly any photocathode electrode shape. The chemical clean prior to vacuum introduction is easy, low cost and rapid. These photocathodes activate in a manner familiar to operators used to III-V photocathodes. The low temperature heat clean reduces stress on vacuum components and makes introduction systems simpler and less expensive. An a-SiGe photocathode can be reactivated without an additional heat clean if the yield is allowed to decay. Under background gas and ion exposure, a-SiGe photoemitters have shown considerably better performance than the III-V based photocathodes. As-grown photocathodes stored for 9 months activated as readily as freshly-grown samples. We have even demonstrated its use as a photoelectron source in stand-alone

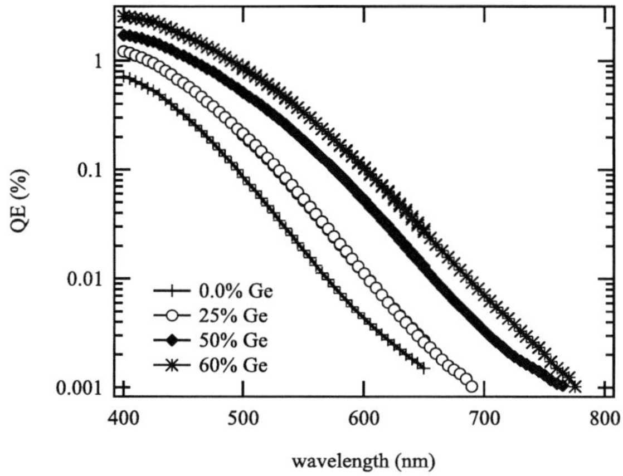

FIGURE 4. Yield curves from a-$Si_{(1-x)}Ge_x$ for various values of x.

vacuum tubes.

SUMMARY AND OUTLOOK

The addition of a second alkali to the activation of bulk GaAs has been shown to reduce its reactivity with carbon dioxide, a gas which shortens the lifetime of NEA GaAs. The addition of Li also provides an interesting modification to the surface barrier. Polarization measurements on dual activated bulk, thin unstrained and high polarization III-V photocathodes will take place soon. The surface chemistry will be studied with core level XPS. Measurements of the angular emission and charge density properties of a-Si are underway. Optimization of the material through growth parameter exploration continues. Amorphous SiGe holds great promise for use as a photoemitter in harsh environments and it may be feasible to employ it as the photoemitting layer for III-V photocathodes to provide them even greater environmental protection.

ACKNOWLEDGMENTS

This work was supported by the United States Department of Energy SBIR under Grants No. DE-FG02-06ER84475 and DE-FG02-07ER84832 and the United States Navy SBIR under Grant No. N65538-08-M-0126.

REFERENCES

1. T. Wada, T. Nitta, T. Nomura, M. Miyao and M. Hagino, *Jpn. J. Appl. Phys.* **29**, 2087–2091 (1990).
2. Z. Liu, Y. Sun, S. Peterson and Piero Pianetta *Appl. Phys. Lett.* **92**, 241107–241109 (2008).
3. G. A. Mulhollan and J. C. Bierman, *J. Vac. Sci. Technol. A* **26**, 1195-1197 (2008).

Superlattice Photocathode Damage Analysis

Marcy L. Stutzman and Joseph Grames

Center for Injectors and Sources
Thomas Jefferson National Accelerator Facility
12000 Jefferson Avenue, Newport News, VA 23606

Abstract. Jefferson Lab uses 100kV DC high voltage photoguns with high polarization strained superlattice GaAs photocathodes to achieve electron beam polarization over 80%. The photocathode is subjected to back-bombardment from ionized residual gasses during operation. We present surface analyses using transmission electron microscopy (TEM) and secondary ion mass spectrometry (SIMS) to characterize photocathode damage.

Keywords: Polarized electron source, superlattice photocathode, surface charge limit.
PACS: 29.25.Bx, 79.60.-i

PHOTOCATHODE DAMAGE: INTRODUCTION AND MOTIVATION

High polarization strained superlattice GaAs photocathode material is used inside 100kV DC high voltage photoguns at Jefferson Lab's CEBAF nuclear physics accelerator to generate electron beams with polarization over 80%. We use a GaAs/GaAsP superlattice structure developed in a partnership between SVT and SLAC [1]. This material is very reliable, with initial quantum efficiency (QE) near 1%, and minimal variation between batches [2]. Further, no degradation of the electron polarization has been observed over the course of experiments lasting many months and delivering hundreds of Coulombs at QE over 0.1% (Fig.1). (During operation with QE<0.1%, some degradation of polarization has been observed and is being investigated.)

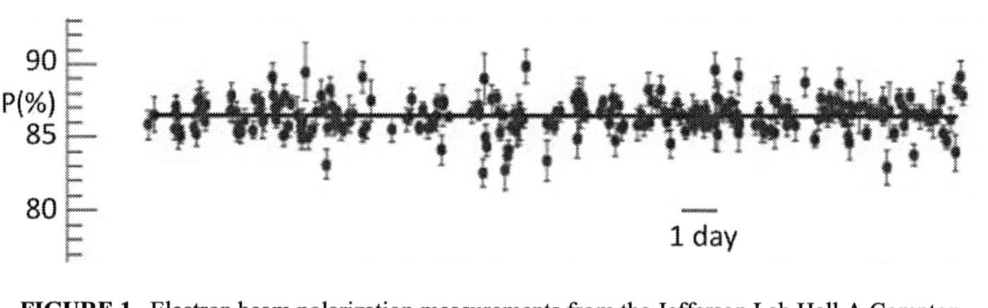

FIGURE 1. Electron beam polarization measurements from the Jefferson Lab Hall A Compton polarimeter [3] during a one month experimental run where 90 Coulombs were delivered. There is no significant variation in polarization. Note the QE is above 0.1% during this time.

CP1149, *Spin Physics, 18th International Spin Physics Symposium*
edited by D. G. Crabb, D. B. Day, S. Liuti, X. Zheng, M. Poelker, and Y. Prok

In contrast with the electron polarization, we do observe a degraded "surface charge limit" (SCL) [4], where quantum efficiency of the photocathode decreases as a function of laser intensity (Fig. 2). Although the SCL effect has been observed at pulsed machines operating with bunch charge over 1 nC, it was not expected at the sub pC bunch charge used at CEBAF.

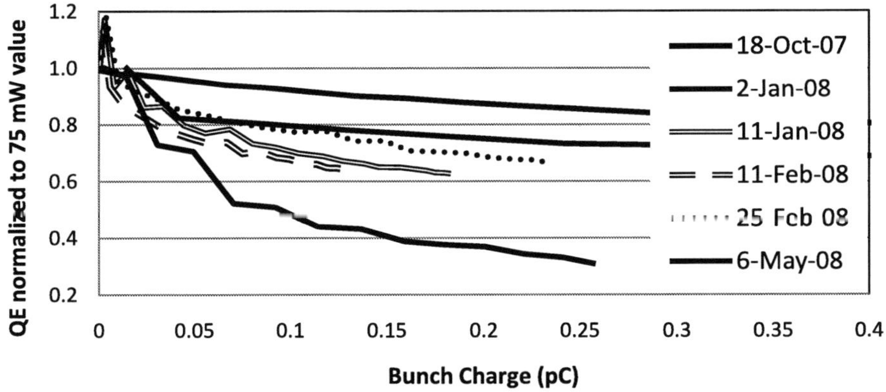

FIGURE 2 Surface charge limit measurements for a strained superlattice GaAs photocathode during its operational lifetime delivering beam in the CEBAF machine at Jefferson Lab. The total charge and charge since last heat/activation cycle are tabulated in Table 1.

TABLE 1. Summary of charge delivered from photocathode for data in Figure 1.

SCL Measurement	Total charge (C)	Charge since heat cycle (C)
October 18, 2007	2	2
January 2, 2008	256	0
January 11, 2008	305	49
February 11, 2008	422	166
February 25, 2008	444	0
May 6, 2008	618	167

Degradation of GaAs photocathodes has long been associated with the ionization of residual gasses in the photogun vacuum chamber and its acceleration into the photocathode material [5] where ions can affect both the surface chemistry and implant into the GaAs structure. The surface chemistry damage can be rejuvenated by simply adding cesium or by heating and reactivating the photocathode with cycles of Cs and an oxidant. For ions implanted into the photocathode material, thermal annealing is effective in repairing some of the damage. Measurements at Jefferson Lab using bulk GaAs indicate that extended heating (12 hours rather than a standard 2 hours at 525°C) are significantly more effective in rejuvenating the maximum yield from a photocathode, particularly in heavily damaged areas such as the electrostatic center where ions are focused.

Similarly there is evidence that heating the photocathode may be effective in reducing the effect of SCL, as shown in Fig. 2, where the photocathode was heated and reactivated between February 11 and February 25 and the SCL effect decreased. Understanding and mitigating the SCL effect is critical for future high current experiments planned at Jefferson Lab; at high currents, the photocathode is damaged more quickly by the higher rate of ion back-bombardment, and the SCL in a damaged photocathode will prevent adequate photocurrent delivery with laser powers that would otherwise be sufficient, requiring unacceptably frequent photocathode changes.

Presented here are the initial results from surface analyses of damaged strained superlattice GaAs photocathode that was removed from the CEBAF photoinjector after more than a year of beam delivery, using two complementary techniques: transmission electron microscopy (TEM) of a focused ion beam (FIB) prepared cross section of the material, and secondary ion mass spectrometry (SIMS) analysis of the composition of ions implanted in the used photocathode material.

SURFACE ANALYSIS TECHNIQUES AND RESULTS

Transmission Electron Microscopy

The first technique used to study the damage to the strained superlattice photocathode material is transmission electron microscopy (TEM). Sample preparation for TEM analysis of the strained superlattice structure was performed at North Carolina State University using a focused ion beam (FIB) milling and lift-out technique to prepare the samples without damaging the superlattice structure [6]. The GaAs is first protected through coating with an evaporative layer of gold and palladium, next a thicker layer of platinum is sputter coated using the FIB. The FIB is then used to mill a sample for TEM analysis. The FIB milled sample is a depth cross section of the superlattice photocathode and the substrate, where the layers of the superlattice are seen in cross section. Lattice dislocation are evident (see Fig. 3a) and present in both new and used material. Areas of significant darkening, typical of ion implantation damage, can be seen in the used sample, but are not evident in the unused control sample (Fig 3b).

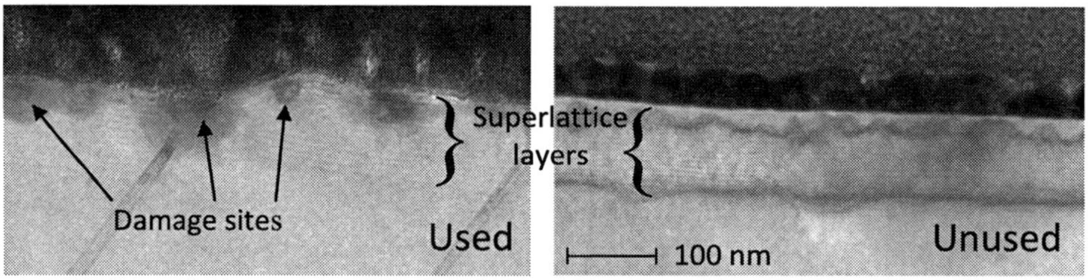

FIGURE 3: TEM images of used (left) and unused (right) superlattice photocathode material. The dark areas at the top are the Au/Pd and Pt protective layers. The 100 nm thick superlattice of 3 and 4 nm layers of GaAs/GaAsP can be seen. The used cathode shows deep "stains", which are not evident in the unused sample on the right. Blurring at the top of the unused cathode is due to a slight tilt during TEM imaging which images the gold overlayer as well as the cathode.

The damage is non-uniform across the surface of the material at this length scale. The laser spot is ~500μm diameter, far larger than the distance scales observed on this TEM image. The nature of this non-uniformity is not understood at this point.

Secondary Ion Mass Spectrometry

Secondary ion mass spectrometry (SIMS) was used to analyze composition of the ions implanted in the material. A cesium ion beam was used at several positions across the surface of the photocathode, and the secondary ions liberated were measured. As expected, the main components found in the sample are the residual gasses in the chamber: hydrogen, oxygen, carbon (from CO and CO_2) and fluorine which comes from the NF_3 used to activate the photocathode in this system.

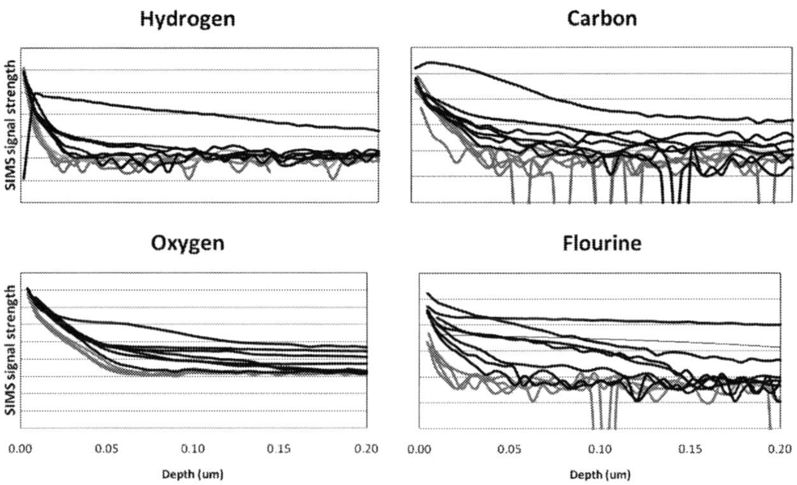

FIGURE 4. SIMS data for various species as a function of depth through the photocathode. Traces in black are from various spots across the damaged photocathode, and gray traces are from spots across the unused control sample. Signal strength is in arbitrary units, depths into the photocathode material are identical for all graphs, with the GaAs/GaAsP superlattice structure 0.1 μm thick.

FIGURE 4 shows an increased signal strength in the traces from the used samples (black lines) over the control unused samples (gray lines). Direct measurement of the common residual gasses implanted in the used photocathode is further evidence that the residual gasses are being implanted in the photocathode. The anomalously high data trace in the hydrogen, carbon and oxygen graphs is from the same position on the cathode, most likely corresponding to the electrostatic center of the photocathode, where ions are focused by the fields of the cathode/anode electrodes.

The damage analysis studies were performed on photocathode material that was used for more than a year in the single chamber horizontal gun where the photocathode is activated in the high voltage chamber, making fluorine a significant component of the residual gas. The SCL data presented above is from a new load-locked system where photocathode preparation is separate from the high voltage chamber and fluorine is no longer expected to be a significant residual gas.

Discussion of SIMS data and Comparison to Theory

Programs such as SRIM [7] can be used to calculate the depth of implantation for various residual gas species as a function of voltage. Comparison between the SRIM calculation and the SIMS data can provide an estimate of the energy of the implantation. In the case of hydrogen, energies as low as 10 kV are sufficient to implant to depths greater than the 100 nm superlattice structure from which electrons are photoemitted. The hydrogen SIMS data shows an increased hydrogen signal at depths less than 100 nm, indicating that low energy hydrogen ions implant in the photocathode. Oxygen shows increased signal strength from near the surface through to depths greater than 100 nm, while most of the carbon data shows no increase in signal until over 50 nm. SRIM simulations show a mean implantation depth of 50 nm for both carbon and oxygen at energies near 30 kV. Consequently, increasing gun voltage or decreasing cathode-anode gap, to reduce the volume of gas that ionizes at these intermediate range energies, may decrease photocathode damage.

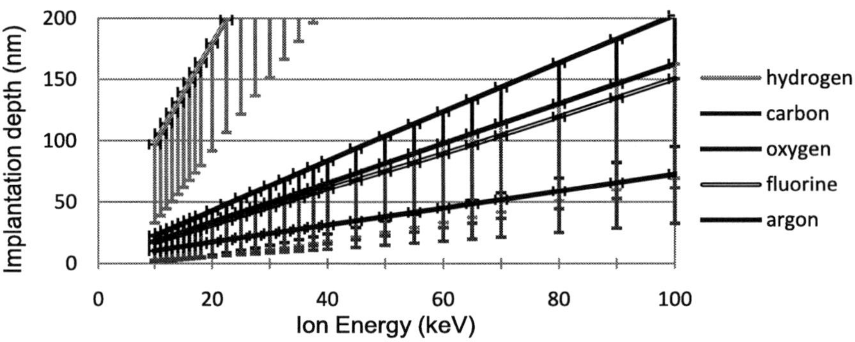

FIGURE 5. SRIM calculations showing mean stopping depths anticipated for various ion species implanted in GaAs.

CONCLUSIONS

Surface analyses of strained superlattice photocathode material used in a 100kV DC photogun show that the predominant residual gas species are implanted in the material, indicative of the ion back-bombardment model, and also provide information on the energy of the implanted ions. The non-uniform nature of the implantation sites is puzzling. We hope to perform studies of the diffusion of the beryllium dopant and any broadening of the interface between the GaAs/GaAsP superlattice layers using SIMS with an oxygen beam, which is more sensitive to the depth scale and species and aim to better understand and mitigate the observed surface charge limit effect.

ACKNOWLEDGMENTS

We would like to acknowledge the help of collaborators at North Carolina State University, Dr. Fred Stevie and Dr. Dale Batchelor, for the surface analysis techniques and interpretation of the results. Authored by Jefferson Science Associates, LLC under U.S. DOE Contract No. DE-AC05-06OR23177.

REFERENCES

1. T. Maruyama et al., Appl. Phys. Lett **85**, 2640 (2004)
2. P. Adderley et al., In the Proceedings of Particle Accelerator Conference (PAC 07), Albuquerque, New Mexico, 25-29 Jun 2007, pp 3130.
3. K.A. Aniol, et al., Phys. Lett. B 635 (2006) 275
4. T. Maruyama et al., PESP 2002 AIP Conf. Proc **675** 1083(2003).
5. K. Aulenbacher, SLAC 432-Rev., SLAC, Stanford, CA, 1994, p. 1.
6. F.A. Stevie et al., Surf. Interface Anal. 2001 **31** 345.
7. SRIM calculation program: the Stopping and Range of Ions in Matter. http://www.srim.org/

MBE Growth of Graded Structures for Polarized Electron Emitters

Aaron Moy,[a] T. Maruyama,[b] F. Zhou[b] and A. Brachmann[b]

[a]*SVT Associates, Eden Prairie, MN.*
[b]*SLAC National Accelerator Laboratory, Menlo Park, CA*

Abstract. SVT Associates, in collaboration with SLAC, has investigated two novel photocathode design concepts in an effort to increase polarization and quantum efficiency. AlGaAsSb/GaAs superlattice photocathodes were fabricated to explore the effect of antimony on device operation. In the second approach, an internal electrical field was created within the superlattice active layer by varying the aluminum composition in AlGaAs/GaAs. A 25% increase in quantum efficiency as a result of the gradient was observed.

Keywords: Polarized electrons, high polarization photocathode
PACS: 29.25.Bx, 29.27.Hj

INTRODUCTION

Polarized electrons have been essential for high-energy parity-violating experiments and measurements of nucleon spin structure, and polarized electron beams will be required for all future linear colliders. Polarized electrons are readily produced by GaAs photocathode sources. When a circularly polarized laser beam tuned to the bandgap minimum is directed to the negative-electron-affinity (NEA) surface of a GaAs crystal, longitudinally polarized electrons are emitted into vacuum. The electron polarization is easily reversed by reversing the laser polarization. The theoretical maximum polarization of 50% for natural GaAs was first exceeded in 1991 using the lattice mismatch of a thin InGaAs layer epitaxially grown over a GaAs substrate to generate a strain in the former that broke the natural degeneracy between the heavy- and light-hole valence bands [1].

In a previous collaboration between SVT Associates and SLAC, photocathodes were developed based on $GaAs_{0.64}P_{0.36}$/GaAs superlattice. These devices demonstrated polarizations as high as 86%, with quantum efficiencies exceeding 1% [2]. Two known factors limit the polarization of these cathodes: 1) a limited band splitting; and 2) a relaxation of the strain in the surface epilayer since the 10-nm critical thickness for the 1% lattice-mismatch is exceeded.

In an effort to advance photocathode operating characteristics, two different device structures were examined. The first design employed $Al_xGa_{1-x}As_ySb_{1-y}$ material to reduce depolarization. A second design employed an $Al_xGa_{1-x}As$/GaAs superlattice where the aluminum content was varied during the growth. This graded $Al_xGa_{1-x}As$ thus created an internal electrical field that would serve to accelerate electrons to the surface, increasing quantum efficiency.

CP1149, *Spin Physics, 18th International Spin Physics Symposium*
edited by D. G. Crabb, D. B. Day, S. Liuti, X. Zheng, M. Poelker, and Y. Prok

ANTIMONIDE-BASED PHOTOCATHODE STRUCTURES

To address material-specific spin-depolarization and increase the photocathode robustness which is essential for an RF gun, a strained superlattice structure based on GaAs and $Al_xGa_{1-x}As_ySb_{1-y}$ was explored. In this structure, the $Al_xGa_{1-x}As_ySb_{1-y}$ layers serve as energy barriers. The GaAs layers are wells and are strained. The addition of antimony (Sb) to the material was thought to have two benefits. Firstly, antimony has 3 orders of magnitude lower diffusivity than gallium and 2 orders of magnitude lower diffusivity than arsenic [3]. Diffusion is one of the main reasons for structural and surface decomposition of cathodes, especially during heat cleaning and surface bombardment.

Secondly, antimony has higher spin orbit coupling than As, so the effect of spin orbit coupling on depolarization can be studied. The increased spin orbit coupling allows structures with larger HH-LH separation, which can increase the polarization of the photocathode effectively. Thus it is believed that antimony-based photocathodes could play an important role in spin polarized RF guns because they can combine high polarization and high robustness in poor vacuum environment

Figure 1 displays a diagram of the AlGaAsSb/GaAs superlattice structure explored. Polarized emission continues from the GaAs layers in this structure, with the strain and barrier energies being defined by the $Al_xGa_{1-x}As_ySb_{1-y}$ alloy composition. The addition of antimony (Sb) atoms in the material growth was a novel approach to the design of photocathodes with polarized electron emission.

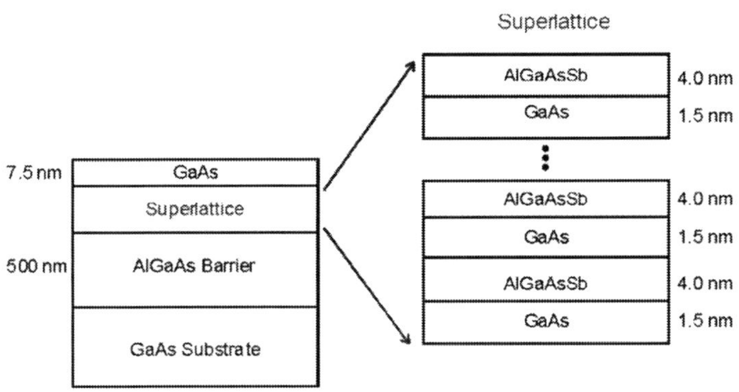

FIGURE 1. Diagram of the basic photocathode structure studied utilizing a superlattice consisting of GaAs wells and AlGaAsSb barriers.

The spin orbit (SO) coupling is more pronounced in superlattice structures due to heterointerface asymmetries. A 10-15nm GaAs layer of high p- doping (5x19 cm^{-3}) at the cathode surface is desirable in order to achieve NEA reliably. The SL structure should have low p-doping (~1x17 cm^{-3}) in order to increase mobility and avoid depolarization due to scattering. The increased doping at the surface results in an elevated CB at the surface that can be compensated by the SL miniband as long as the conduction miniband is higher that the conduction band level at the surface. The entire structure was grown on a GaAs substrate with an AlGaAs energy barrier layer.

In principle, the $Al_xGa_{1-x}As_ySb_{1-y}$ quaternary alloy provides design flexibility. The percentage of Sb can be used to experiment with the effect of spin orbit coupling and band lineup and the Al percentage can be used to control the band offsets and LH, HH valence band splitting.

Three device structures were grown on 2" diameter GaAs substrates by molecular beam epitaxy (MBE). The structures all utilized $Al_xGa_{1-x}As_{0.82}Sb_{0.18}$ alloy in the superlattice, where the aluminum composition was changed between the samples being x = 5, 10 or 15%.

X-ray diffraction was performed on the device wafer to confirm and measure the superlattice (Fig. 2). A simulation of the structure was also calculated assuming 100% strain in the structure. The superlattice peaks are observed in the measured data, as expected from the superlattice layers. However, the location of the sideband peaks is slightly shifted, indicating that some relaxation in the material may have occurred.

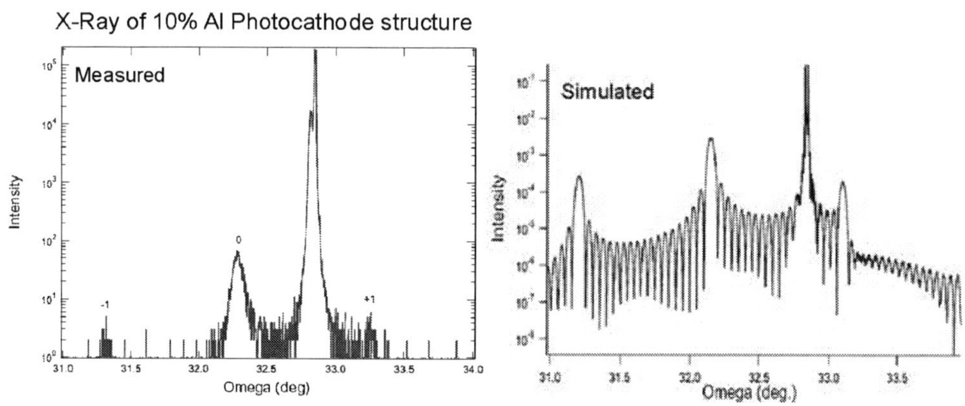

FIGURE 2. X-ray diffraction measurement and simulation of AlGaAsSb/GaAs superlattice photocathode structure. The simulation assumes 100% strain.

The wafers were sent to SLAC where they were cut into discs and cesium-activated. Prior to installation in the test system, the sample is degreased in a boiling solution of trichloroethane. After the protective oxide layer is removed in ammonium hydroxide, the sample is rinsed in distilled water and methanol. The cathode activation method used to obtain a negative-electron-affinity (NEA) surface consists of heat cleaning to 600° C for 1 hour, cool-down for an hour, followed by application of cesium until the photo-yield peaks, and then cesium and nitrogen-trifluoride co-deposition until the photo-yield is again maximized. The heat-cleaning temperature is monitored via an infrared pyrometer. The cathode is activated while monitoring the photo-yield with a white light and a 670 nm diode laser. Once a red response is observed from the diode laser, the white light is turned off and the diode laser is used to complete the activation. The absolute QE is measured using the diode laser at a photon wavelength of 670 nm. A tungsten lamp and a monochrometer are used to measure the relative QE as a function of photon wavelength, and these measurements are then normalized to the diode laser measurement at 670 nm.

Measurement of the quantum efficiency (QE) from the 10% aluminum sample was very low, ~0.01% at 650 nm excitation wavelength. Such emission was too low to measure polarization. The 5 and 15 % aluminum composition samples did yield higher currents, with QE of 0.2% and 0.1%, respectively. Polarization was measured from the 5% aluminum sample, but demonstrated polarization less than 50% (Fig. 3).

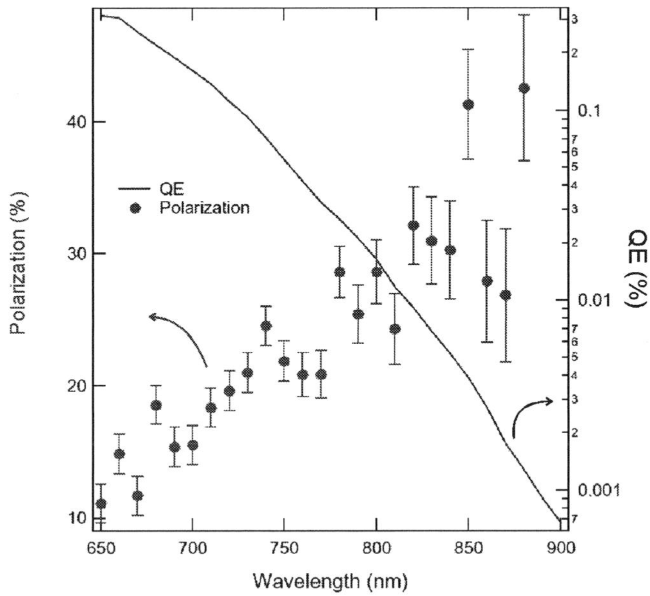

FIGURE 3. Quantum efficiency and polarization measurements from the AlGaAsSb/GaAs superlattice photocathodes with 5 % aluminum composition.

Unfortunately, the device results for antimony-based $Al_xGa_{1-x}As_{0.82}Sb_{0.18}/GaAs$ photocathode were not as anticipated. Both quantum efficiency and polarization values were very low. This could be a result of this superlattice design and not wafer quality. The hypothesis is that the confinement energy around the quantum wells was too high, trapping electrons inside and hampering their contribution to the photoemission. Another possibility is that the compressively-strained AlGaAsSb layers, which are normally grown at temperatures below 500 °C, could have been damaged during thermal removal of surface native oxide at 600 °C before surface activation process.

GRADED ALUMINUM GALLIUM ARSENIDE SUPERLATTICE PHOTOCATHODES

Another approach to increasing quantum efficiency is to create an electrical field internal to the photocathode active layer to accelerate electrons toward the negative electron affinity (NEA) surface. Figure 4 displays the concept, where the conduction band of the superlattice has a sloped gradient field to influence electron flow, as opposed to a typical design where the barrier heights exist at a constant level. To achieve such a gradient, $Al_xGa_{1-x}As$ alloys were used as barriers. This $Al_xGa_{1-x}As$ material permitted wide latitude in changing the bandgap energy by varying the composition x during growth, without creating undo changes to the overall lattice constant.

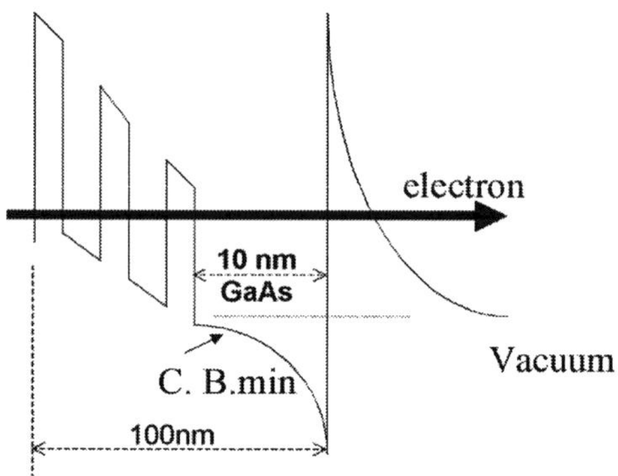

FIGURE 4. Diagram of the concept to create an accelerating field internal to the photocathode structure in order to drive electrons to the surface for increased quantum efficiency.

In the presence of such accelerating field, the electron transport time decreases by an order of magnitude to less than 1ps as it has been estimated by Monte Carlo simulations and verified experimentally [4, 5, 6]. The decrease of the transport time by an order of magnitude not only increases the current density that is required for an RF gun by the same amount but it also increases the polarization. In the case of thin films (<100nm) the polarization can increase by decreasing the transport time of the electrons and decreasing the total number of scattering events (Eq. 1) [7].

$$\frac{P_2}{P_1} = \frac{1+\frac{\tau_{em1}}{\tau_{s1}}}{1+\frac{\tau_{em1}}{\tau_{s1}}\left(\frac{s_2}{s_1}\frac{\tau_{em2}}{\tau_{em1}}\right)} \tag{1}$$

where τ_{em} is time of electron emission in vacuum, P is electron polarization upon excitation with circularly polarized light and s is the total number of scattering events. Based on the simple model of Eq. (1), an increase of 5-10% in polarization can be achieved with accelerating the electrons.

The main objective of these experiments was to demonstrate an internal bias using an $Al_xGa_{1-x}As$/GaAs superlattice structure. Figure 5 shows the reference sample layout, where the active layer consisted of 19 periods of superlattice layers, with 11 monolayers of $Al_xGa_{1-x}As$ and 7 monolayers of GaAs in each pair. For the reference sample, the aluminum composition x was held constant at 0.35. An internal bias was introduced into subsequent photocathodes by varying the aluminum fraction, x, throughout the active layer. By slightly varying the aluminum fraction between each pair through the superlattice, a graded field was created. This type of design, however, adds complexity to the epitaxial growth, as atomic mole fraction must be quickly and precisely varied.

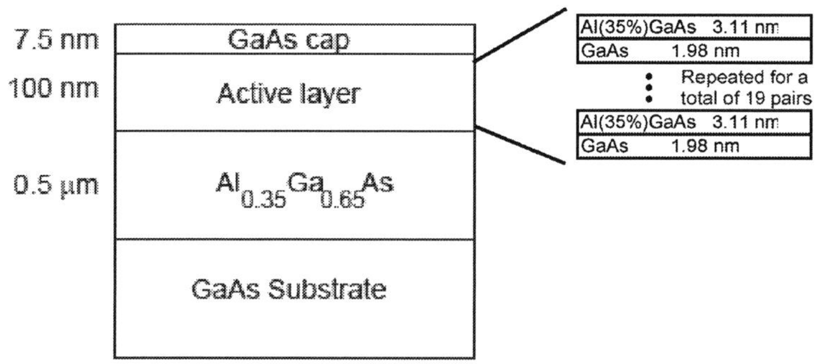

FIGURE 5. Schematic structure of the reference AlGaAs/GaAs superlattice photocathode.

The superlattice band structure was studied using the **k·p** envelope function method. Figure 6(a) shows the mini-band energies as a function of the superlattice well width for a fixed barrier width (7 monolayers). The two highest energy valence bands correspond to the heavy- and light-hole mini-bands, and the maximum energy splitting was expected to be about 28 meV. Figure 6(b) shows the expected polarization and QE as a function of the excitation wavelength. The band gap wavelength was expected at 760 nm.

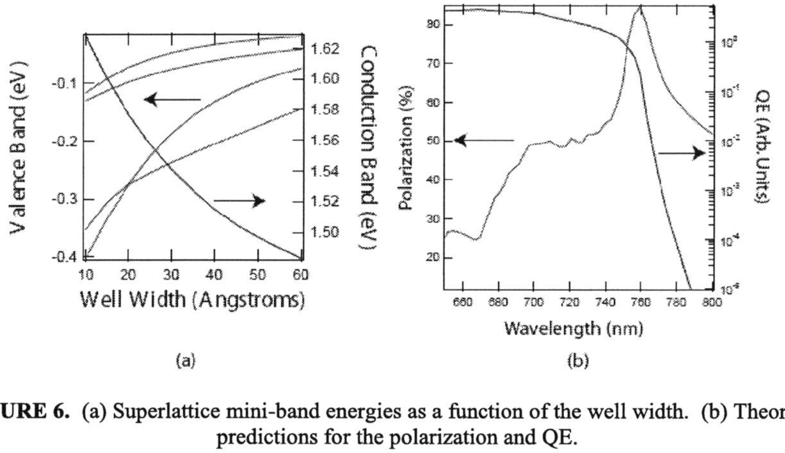

FIGURE 6. (a) Superlattice mini-band energies as a function of the well width. (b) Theoretical predictions for the polarization and QE.

Three wafers were grown with different aluminum grading values in the active layer: 1) no grading with x=0.35, 2) 10% grading from x=0.35 to 0.25, and 3) 20% grading from x=0.35 to 0.15. Because of the thin amount of the active layer and the extremely low lattice mismatch between the Al and Ga atoms, the x-ray diffraction simulation predicted two weak, widely spaced satellite peaks (Fig. 7) for the superlattice. Actual measured x-ray data of the grown wafer were in close agreement with the simulation (Fig. 8).

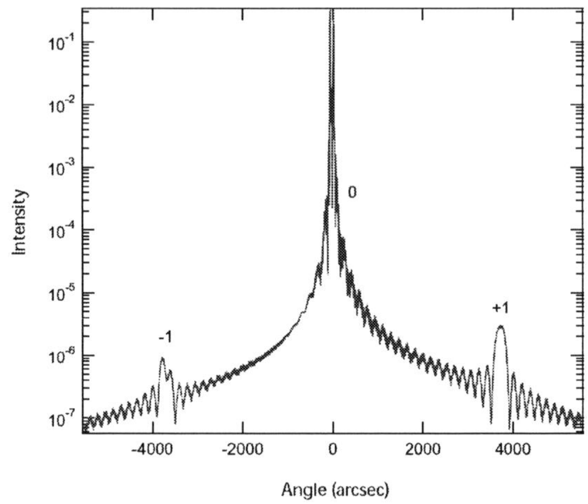

FIGURE 7. Simulation of anticipated x-ray diffraction curve based on the graded AlGaAs/GaAs superlattice structure.

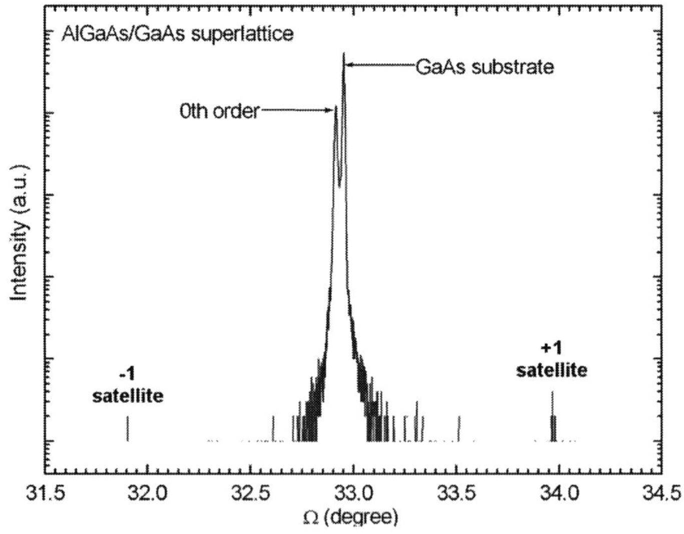

FIGURE 8. Measured x-ray diffraction curve of the grown AlGaAs/GaAs superlattice structure.

The photocathode wafers were cleaned and activated as described in the previous section. Figure 9 shows the QE at 650 nm for the three samples. As the internal bias is increased by increasing the aluminum grading, the QE increases as much as 25%. This is the first demonstration that an electric field can be incorporated in the conduction band and a QE enhancement is possible.

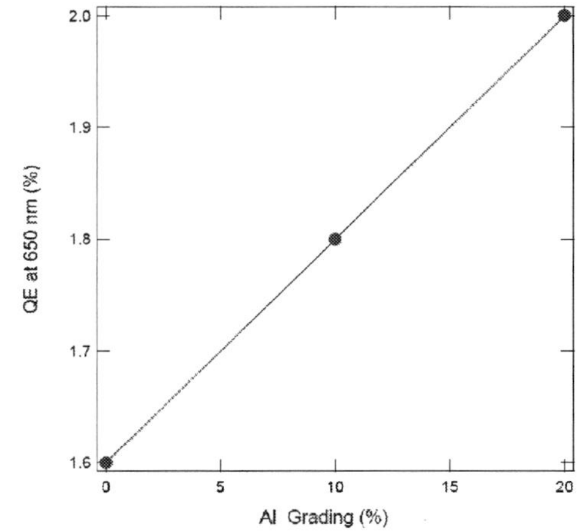

FIGURE 9. QE at 650 nm as a function of aluminum grading in AlGaAs/GaAs superlattice photocathodes.

Figure 10 shows QE and polarization as a function of wavelength for the three samples tested. As the aluminum grading was increased, the peak polarization decreased. Furthermore, the wavelength dependence of QE became steeper, eliminating the QE gain in the short wavelength region. This is a consequence of the structural change in the superlattice active layer. As the aluminum fraction is decreased, the effective bandgap energy gets smaller and the heavy- and light-hole splitting is reduced.

CONCLUSION

Two novel device structures were studied to improve the operating characteristics of photocathodes with polarized electron emission. $Al_xGa_{1-x}As_ySb_{1-y}/GaAs$ superlattices were created to take advantage of the lower diffusivity and higher spin orbit coupling of antimony atoms. These photocathodes in this instance yielded low quantum efficiency, attributed to the extremely deep well energy. Further investigation can continue in the future with a modified design.

The second approach was to create a superlattice with a self-generated internal electric field to accelerate electrons toward the photocathode surface. This was accomplished by varying the aluminum composition in a $Al_xGa_{1-x}As/GaAs$ superlattice. Quantum efficiency was observed to increase by up to 25% compared to a control sample that lacked an internally graded field. The low strain in $Al_xGa_{1-x}As/GaAs$ materials resulted in mild polarization values, but the internal gradient concept can be applied to other high polarization photocathodes.

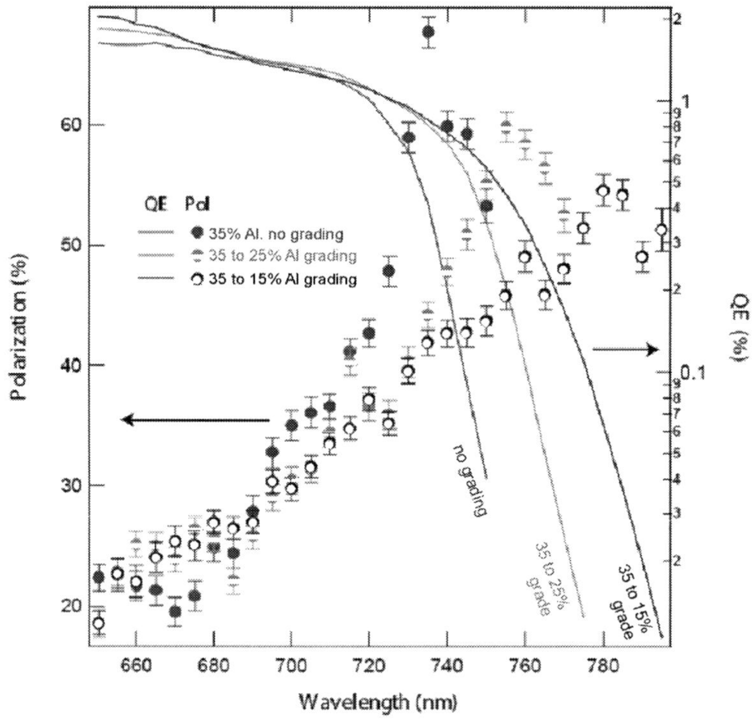

FIGURE 10. Polarization and QE as a function of the excitation wavelength for three AlGaAs/GaAs superlattice photocathodes with internal gradient fields.

ACKNOWLEDGMENTS

This research was supported by Small Business Innovation Research (SBIR) grants from the U.S. DOE under contracts DE-FG02-07ER86329 and DE-FG02-07ER86330.

REFERENCES

1. T. Maruyama, E.L. Garwin, R. Prepost, G.H. Zapalac, J.S. Smith, and J.D. Walker, Phys. Rev. Lett. **66**, 2376 (1991).
2. T. Maruyama, D.-A. Luh, A. Brachmann, J.E. Clendenin, E.L. Garwin, S. Harvey, J. Jian, R.E. Kirby, C.Y. Prescott, R. Prepost, and A.M. Moy, Appl. Phys. Lett. **85**, 2640 (2004).
3. H. Bracht, S.P. Nicols, W. Walukiewitz, J.P. Silveira, F. Briones, E.E. Haller, "Large disparity between gallium and antimony self diffusion in gallium antimonide", Nature, **408**, 69 (2000).
4. Schuler,J., et al., Polarized sources conf., Nagoya 2000.
5. K. Ioakeimidi, et al., "Photoelectronic analog-to-digital conversion: sampling and quantizing at 100 Gs/s", *IEEE Trans. Microwave Theory and Techniques*, Vol. 53, Issue 1, Jan. 2005.
6. K. Ioakeimidi et. al., "Sub-picosecond jitter in picosecond electron bunches", *J. Vac. Sci. Technol. B*, Vol. **23**, (2005).
7. K. Ioakeimidi et al., "Transport Mechanisms in Spin Polarized Semiconductor Superlattices", *SPIN 2006,* 1-7 October 2006, Kyoto, Japan.

Superlattice Photocathode with High Brightness and Long NEA-Surface Lifetime

Tomohiro Nishitani[a], Masao Tabuchi[b], Yoshikazu Takeda[c], Yuji Suzuki[c], Kazuya Motoki[c], Takashi Meguro[a]

[a]*Beam Application Team, RIKEN, Wako, Saitama, 351-0198, Japan*
[b]*Venture Business Laboratory, Nagoya University, Nagoya, 464-8601, Japan*
[c]*Department of Crystalline Materials Science, Nagoya University, Nagoya, 464-8601, Japan*

Abstract. We have suggested that a small momentum spread of photoelectrons and high quantum efficiency can be obtained concurrently by a photocathode using a semiconductor with a superlattice instead of a bulk. We have begun to search for the suitable semiconductor material of a superlattice photocathode for maintaining the surface with a negative electron affinity state for a long time. We measured quantum efficiency degradation of photocathodes using GaAs and AlGaAs semiconductors with various electron affinities. The AlGaAs semiconductor had a quantum efficiency lifetime of 10 times long compared with the GaAs semiconductor. We found that the AlGaAs semiconductor was the suitable material for the superlattice photocathode with the surface maintaining a negative electron affinity state for a long time.

Keywords: GaAs semiconductor, photocathode, electron affinity, superlattice, long lifetime.
PACS: 85.60.Ha, 29.25.Bx

INTRODUCTION

GaAs photocathodes are well known as photoemitters with a surface of a negative electron affinity state (NEA-surface) using semiconductors related with a GaAs semiconductor. GaAs photocathodes have played very important roles as spin-polarized electron sources in several fields of fundamental science[1,2,3]. Recently, a GaAs photocathode was found to be useful a high brightness electron source for a high energy accelerator based on linac[4], because GaAs photocathode has the advantages of a large emission current (several mA[5]) and a small momentum spread (energy of ~0.2eV[6]) of extracted electrons. The key technology of GaAs photocathode is that an NEA surface enables excited electrons in the conduction band minimum to escape to a vacuum. However, the NEA surface of the conventional GaAs photocathode has the short lifetime problem for a large emission current of several mA[7]. To realize the electron source with beam performance for the next generation synchrotron light source based on a linac[4], we have developed photocathodes which simultaneously realizes a high brightness and an NEA surface with a long lifetime.

CP1149, *Spin Physics, 18th International Spin Physics Symposium*
edited by D. G. Crabb, D. B. Day, S. Liuti, X. Zheng, M. Poelker, and Y. Prok

PHOTOCATHODE WITH NEA SURFACE USING SEMICONDUCTORS

An NEA surface of the conventional GaAs photocathode is realized by the band bending and an electric field by electric dipoles composed of cesium and gallium atoms on the surface as shown in FIGURE 1. When the surface of a GaAs semiconductor with a p-type doping is activated to be an NEA state by cesium, the electron affinity is given by the next formula:

$$\chi^{NEA} = \chi - \phi^{BB} - \phi^{D} < 0,$$

where χ is the electron affinity of the semiconductor, ϕ_{BB} is the surface band bending, ϕ_D is the electric field by the electric dipoles on the surface.

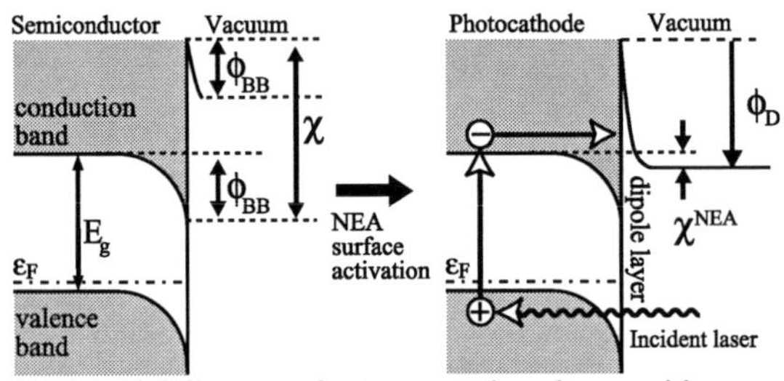

FIGURE 1. (a) The potential diagram of a GaAs semiconductor with a p-type doping. (b) The mechanism of photoemission illustrated on the potential diagram of a GaAs photocathode.

The χ^{NEA} of the conventional GaAs photocathode with a p-type doping GaAs semiconductor is -0.17eV[8] at room temperature. A GaAs semiconductor has electron affinity of 4.07eV and band gap energy of 1.42eV at a room temperature[9]. The ϕ_D can be estimated to be 3.5eV at a room temperature where it is assumed that the ϕ_{BB} of a GaAs semiconductor is a half of band gap energy. Therefore, a GaAs photocathode is possible to generate an electron beam with a small momentum spread of ~0.2eV by excitation photon energy tuned to band-gap energy of a semiconductor. However, quantum efficiency (QE) of a GaAs photocathode with a bulk structure at excitation photon energy tuned to its band-gap is low (3~4%) due to small electron energy density of state in the conduction band.

ALGAAS PHOTOCATHODE FOR MAINTAINING AN NEA STATE FOR A LONG TIME

An effective electron affinity (χ^{NEA}-ϕ^{BB}) of a semiconductor should be small for maintaining an NEA state (χ^{NEA}<0) for a long time. We considered that a semiconductor with a smaller electron affinity and larger band gap energy was a more suitable material of a photocathode for maintaining an NEA state for a long time. We fabricated test samples of GaAs and AlGaAs semiconductors with various Al fractions because the electron affinity of an AlAs semiconductor (3.50eV) is smaller than that of a GaAs semiconductor (4.07eV) and band gap energy of an AlAs semiconductor (2.95eV) is larger than that of a GaAs semiconductor (1.42eV)[10]. These samples have the same surface structure of a thin GaAs layer with a thickness of 5nm. TABLE 1 shows a electron affinity and band gap energy for each sample.

TABLE 1. Electron affinity and band gap energy for each sample.

Sample	Al fraction	Electron affinity (eV)	Band gap energy (eV)
GaAs	0	4.07	1.42
$Al_{0.168}GaAs$	0.168	3.89	1.64
$Al_{0.283}GaAs$	0.283	3.76	1.79

We measured the QE degradation of these samples using the photocathode electron source facility at Japan Atomic Energy Agency[11]. In the measurements, the excitation laser photon energy was chosen at ~0.1 eV above band gap energy for each sample, and the applied voltage for the extraction of a photocurrent was as high as 200 V to degrade the NEA surface by ion back bombardment[12]. A QE lifetime of AlGaAs samples was 10 times long compared with that of the GaAs sample. The QE lifetime of $Al_{0.283}GaAs$ sample was longer than that of the $Al_{0.168}GaAs$ sample.

SUPERLATTICE PHOTOCATHODE

The advantage of superlattice structure

We have suggested a photocathode with a superlattice structure for a high brightness electron source[13]. We consider a superlattice structure has intrinsic advantages in realization of higher QE, a smaller momentum spread of photoelectrons and a longer QE lifetime than those of the conventional GaAs photocathode due to the following two reasons.

(1) By selecting an appropriate semiconductor material for each layer in the superlattice structure, a superlattice semiconductor is possible to have large band gap energy, a small electron affinity for a long QE lifetime.

(2) Electron energy density of state (DOS) for a superlattice structure is a step function of excitation photon energy due to a quantum confinement effect shown in FIGURE 2. QE of a GaAs photocathode is proportional to its DOS integrated from band gap cnergy to the excitation photon energy[14]. A GaAs semiconductor with a bulk structure, which has a monotonically increasing DOS function, requires rather higher energy photons to achieve high QE operation at the expense of a large momentum spread of electrons. A superlattice structure, on the other hand, enable one to achieve high QE and a small momentum spread (ΔE_{SL}) simultaneously by choosing photon energy (E_g^{SL}) just above step energy of DOS.

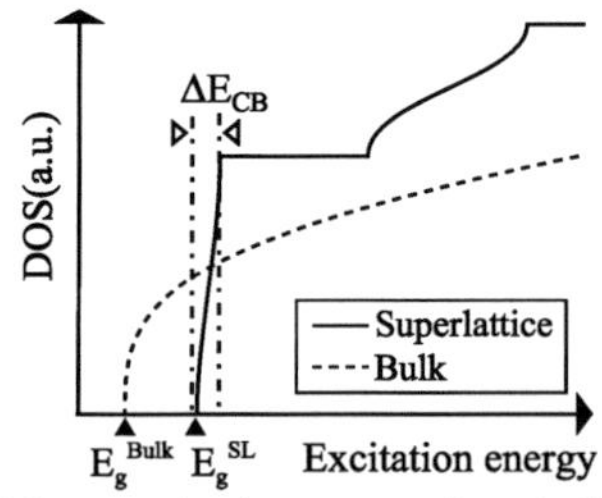

FIGURE 2. Dependence of DOS on excitation energy for a bulk and a superlattice structure.

AlGaAs-GaAs superlattice photocathode

We have decided to use AlGaAs and GaAs semiconductors as superlattice materials from our results of a QE lifetime using AlGaAs photocathodes. To optimize a superlattice structure for a small momentum spread of photoelectrons and a long QE lifetime, we estimated energy width of the conduction mini-band (Wcb) and the effective electron affinity for various superlattice structures. We calculated the dependence of these values on the well- (Lw) and barrier-layer (Lb) thickness of a GaAs-AlGaAs superlattice using Klonig-Penny-Bastard model[15]. FIGURE 3 shows the dependence of Wcb, the electron affinity and the band gap energy on Lw and Lb. Those layer thicknesses should be between 2 and 10nm for an effectively quantum confinement effect by a quality superlattice structure. We can see that superlattice structures with Lw = 2nm have the smallest electron affinity and the largest band gap energy. Superlattice structures with Lb more than 6nm have smaller Wcb than the thermal energy (26meV) at a room temperature. Superlattice structures with Lw = 2nm and Lb of 6~10nm have the almost same effective electron affinity of the AlGaAs photocathode with Al fraction of 0.168. From these results, we found that the superlattice photocathode with Lw = 2nm and Lb=6~10nm is a suitable photocathode for a smaller momentum spread of photoelectrons and longer QE lifetime than that of the conventional GaAs photocathode.

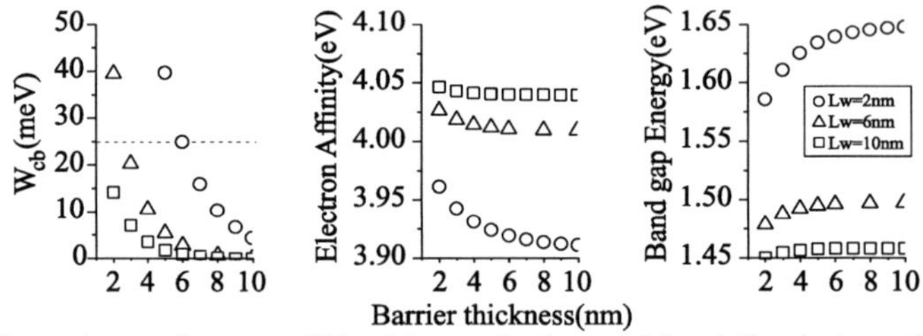

FIGURE 3. Dependence of energy width of the conduction mini-band, the electron affinity and band gap energy on the barrier and well layer thickness of the GaAs-AlGaAs superlattice with an Al fraction of 0.29.

For confirmation of improving a momentum spread of photoelectrons and a QE lifetime performance using superlattice photocathodes designed by our theoretical calculation, we fabricated test samples of the GaAs-AlGaAs superlattice with a well-layer thickness of 2nm and a barrier thickness of 6~10nm by a molecular beam epitaxial facility at Nagoya university. We will measure the QE degradation under a large photocurrent and dependence of QE on excitation energy of these test samples by the 30keV photocathode electron gun at RIKEN.

SUMMARY

To conclude, we measured the QE degradation using GaAs and AlGaAs photocathodes and designed a superlattice structure using theoretical calculation of an energy band model. We found AlGaAs photocathodes has 10 times longer QE lifetime than that of a GaAs photocathode. The results of the QE degradation using GaAs and AlGaAs photocathodes agreed that a photocathode with a smaller electron affinity is more suitable for maintaining an NEA state for a long time. We found that the AlGaAs-GaAs superlattice with the well layer thickness of 2nm and the barrier thickness of 6~10nm is an eminently suitable photocathode for a small momentum spread of photoelectrons and a long QE lifetime.

ACKNOWLEDGMENTS

The authors thank Dr. Eisuke J Minehara, Dr. Ryoichi Hajima and the energy recovery linac development group member at Japan Atomic Energy Agency for the useful discuss and the use of the photocathode electron source facility. This work was supported in part by the Kazato Research Encouragement Prize from the Kazato Research Foundation and Grant-in-Aid for Scientific Research (18740253) from the Ministry of Education, Culture, Sports, Science and Technology.

REFERENCES

1. SLD Collaboration Phys. Rev. Lett. 70 17, pp.23-31 (1993)
2. J. Kirschner, D. Rebenstorff, and H. Ibach, Phys. Rev. Lett. 53, 698- 701 (1984)
3. J. Unguris, A. Seiler, R. J. Celotta, D. T. Pierce, P. D. Johnson, and N. V. Smith, Phys. Rev. Lett. 49, 1047- 1050 (1982).
4. Sol M. Gruner, et al., Review of Scientific Instruments, Vol. 73 Issue3 pp.1402-1406 (2002)
5. G. R. Neil, et al., Phys. Rev. Lett. 84, pp.662-665 (2000)
6. D.A. Orlov, et al., Nucl. Instrum. Methods A 532, 418 (2004)
7. J. Grames et al., Proceedings of Particle Accelerator Conference, pp. 2875-2877 (2005) http://accelconf.web.cern.ch/AccelConf/p05/
8. D.A. Orlov et al., Nuclear Instruments and Methods in Physics Research A 532 (2004) 418–421
9. Sadao Adachi, J. Appl. Phys. 58 (3), (1985) pp. R1-R12
10. S. Adachi, "GaAs and Related Materials: Bulk Semiconducting and Superlattice Properties", World Scientific, (1994) pp. 145, 307
11. T. Nishitani, et al., Proceedings of International Free Electron Laser Conference, (2006) , pp.319-322, http://www.bessy.de/fel2006/proceedings/
12. J. Grames, P. Adderley, J. Brittian, J. Clark, J. Hansknecht, D. Machie, M. Poelker, E. Pozdeyev, M. Stutzman, and K. Surles-Law , AIP Conf. Proc. 980 (2008) pp. 110-117
13. T. Rao, A. Burrill, X.Y. Chang, J. Smedley, T. Nishitani, C. Hernandez Garcia, M. Poelker, E. Seddon, F.E. Hannon, C.K. Sinclair, J. Lewellen, D. Feldman, Nuclear Instruments and Methods in Physics Research A 557, pp.124–130, (2006)
14. T. Nishitani et al. J. Appl. Phys. 97 (2005) 094907
15. G. Bastard, Phys. Rev. B 24 5693 (1981)

High Brightness and high polarization electron source using transmission photocathode

Naoto Yamamoto*, Xiuguang Jin*, Atsushi Mano†, Yasuhide Nakagawa†, Tsutomu Nakanishi†, Toru Ujihara*, Shoji Okumi†, Masahiro Yamamoto†, Taro Konomi†, Yoshikazu Takeda*, Takashi Ohshima**, Takashi Saka‡, Toshihiro Kato§, Hiromichi Horinaka¶, Tsuneo Yasue|| and Takanori Koshikawa||

**Graduate School of Engineering, Nagoya University, Nagoya, Aichi,464-8602 Japan*
†Graduate School of Science, Nagoya University, Nagoya 464-8602, Japan
***Central Research Laboratory, Hitachi Ltd., Tokyo 185-8601, Japan*
‡Daido Institute of Technology, Nagoya 457-8531, Japan
§Daido steel Co. Ltd., Nagoya 457-8531, Japan
¶Faculty of Engineering, Osaka Prefecture University, Osaka 599-8531, Japan
||Fundamental Electronics Research Institute, Osaka Electro-Communication University, Osaka 572-8530, Japan

Abstract. A transmission photocathode was fabricated based on GaAs-GaAsP strained superlattice layers on a GaP substrate and a 20kV-gun was built to generate the polarized electron beams with the diameter of a few micro-meter. As the results, the reduced brightness of 1.3×10^7 A/cm^2/sr and the polarization of 90 % were achieved.

Keywords: polarized electron source, high brightness, transmission photocathode
PACS: 21.10.Hw,29.25.Bx,87.64.Ee

INTRODUCTION

The GaAs-type semiconductor photocathode (PC) with a negative electron affinity (NEA) surface is currently used as a conventionally polarized electron source (PES) to produce a highly polarized electron beam. Polarized electron beams have been widely used in high energy electron accelerators for spin physics studies of fundamental particles or hadrons [1, 2, 3, 4, 5, 6].

To answer the needs of future high energy accelerators, several groups, including our own, have made long-term efforts to improve PES (PC+gun) performances. For examples, electron spin polarization (ESP) higher than 85% has been achieved by strained GaAs [7] and GaAs-GaAsP strained super-lattice (SL) layers [8] grown on a GaAsP buffer substrate. A quantum efficiency (QE) higher than 0.3% has been simultaneously obtained for the same PCs. We have recently demonstrated experimentally that thermal emittance can be minimized to as low as 0.15 π mm mrad for a beam radius of 1.0 mm, by using a GaAs-GaAsP strained SL layer PC installed in a 200-kV DC-gun [9].

High brightness is one of the most important PES performances for the application to electron microscopes. Several types of point-like sources have already been developed to produce a needle-like beam, although it can not produce a polarized beam. Field

CP1149, *Spin Physics, 18th International Spin Physics Symposium*
edited by D. G. Crabb, D. B. Day, S. Liuti, X. Zheng, M. Poelker, and Y. Prok

emission sources can realize the highest brightness ($10^8 \sim 10^9$ A cm^{-2} sr^{-1}) and have been used for high space-resolution electron microscopes, such as scanning electron microscopes (SEM) where the radius of the W-tip is typically shaped to be as small as 100 nm.

For SEM applications, another source using an NEA-GaAs PC was developed to produce small energy spread beams by one of the co-authors (T. Ohshima) [10]. The electron generation from NEA-PC is based on two fundamental physical processes: 1) optical excitation to the conduction band and 2) electron emission through the surface potential barrier due to the NEA surface. In a conventional PC scheme, laser light is injected from the NEA surface side and the electrons are extracted to the same side. In contrast,to form a small diameter electron beam, they injected laser light from the back surface of a GaAs film grown on a glass sheet, with the excited electrons emitted into vacuum from the front NEA surface of the PC. Using this laser injection scheme, it is possible to decrease the distance between a focusing lens and the PC crystal and to minimize the laser spot size. Hereafter, we distinguish this PC from conventional PCs, by naming them "transmission PC (T-PC)" and "reflection PC (R-PC)", respectively.

In order to achieve high brightness and high polarization performances for polarized electron beams, a 20–kV gun system (named JPES-1) was designed and built to install the newly designed T-PC at Nagoya University.

EXPERIMENTAL SETUP

20-kV Electron guns

The T-PC PES system (JPES-1) was developed at Nagoya University as a prototype SPLEEM gun to demonstrate its high brightness performance [11]. JPES-1 consists of five chambers: an NEA surface preparation chamber, a gun chamber, a brightness measuring chamber, a spherical condenser chamber and a 100-keV Mott polarimeter chamber [12].

Two important apparatus performances must be met for a high voltage PES gun: an extreamely high vacuum ($\leq 10^{-10}$ Pa) environment around the NEA-PC and an extremely low dark current ($\leq$ 10 nA) between the high voltage electrodes. These performances are required to protect the NEA surface and reduce its degradation speed to maintain a long PC lifetime ($\geq$ several tens hours).

The NEA surface is known to be easily degraded by the desorption of residual gas molecules and the back bombardment of positive ion produced by the beam itself. Therefore the vacuum pressure near the photocathode was maintained less than 1.0×10^9 Pa by using an 100 L/s ion pump and a 2000 L/s non-evaporable getter pump.

The accelerating electrodes was designed by using the simulation cord [13], which was developed to calculate envelops and profiles of charged particles precisely. The field gradient on a photocathode surface was estimated to be 4 MV/m at 20–kV bias. Since titanium electrodes were found to suppress field emission currents by our previous study [14], the titanium electrodes were installed to the gun chamber.

In the experiments, the beam lifetime performance was monitored during JPES-1 keeping extremely low vacuum pressure ($\leq 2 \times 10^{-9}$ Pa) and extremely low dark

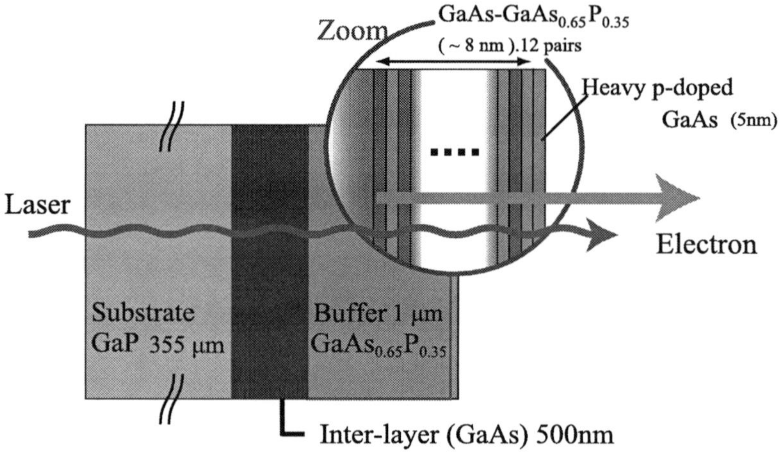

FIGURE 1. Details of the T-PC structure.

current ($\leq$ 1 nA) conditions. As typical data, an initial current of 3 μA extracted by 3.2-μm radius laser irradiation descreased exponentially with a lifetime of 36 hours, which corresponds to a charge lifetime density of 1.8×10^8 C/cm^2.

The ESP is measured using a standard 100-kV Mott polarimeter [12]. The Mott polarimeter has four different Au-foil targets of different thickness for self-calibration of the effective Sherman functions. The systematic error for the ESP measurement was estimated to be $\pm$ 6% (absolute value) at maximum.

Photocathode Design

The T-PCs were fabricated by using a low-pressure organometallic vapor phase epitaxy (OMVPE) system with a verticalcold-wall quartz reactor. The PC structure is shown in Fig 1, A commercially available GaP wafer with a Zn dopant of 1.4×10^{17} cm^{-3} was used as a substrate. To control the strain in the buffer and SL layers [15] [16], the GaAs inter-layer was grown on the GaP substrate. Then, 1 μm of the GaAsP buffer layer and 12 pairs of SL layers were grown with a Zn dopant consentration of 1.5×10^{18} cm^{-3}. Subsequently the SL structure was coated with a heavily doped 5nm layer of GaAs (Zn 6×10^{19} cm^3). This design effectively overcomes the surface charge effect that appears in the generation of a high current density beam [17],[18] and exhibits ESP and QE performances, as high as 92% and 0.5%, respectively [19].

RESULT & DISCUSSION

The beam performance of the T-PC was measured by JPES-1 with the laser focusing conditions. The full-width at half-maximum of the smallest laser spot was 1.3 μm. The highest brightness obtained under an extraction voltage of 20–kV was $(1.3 \pm 0.5) \times 10^7$

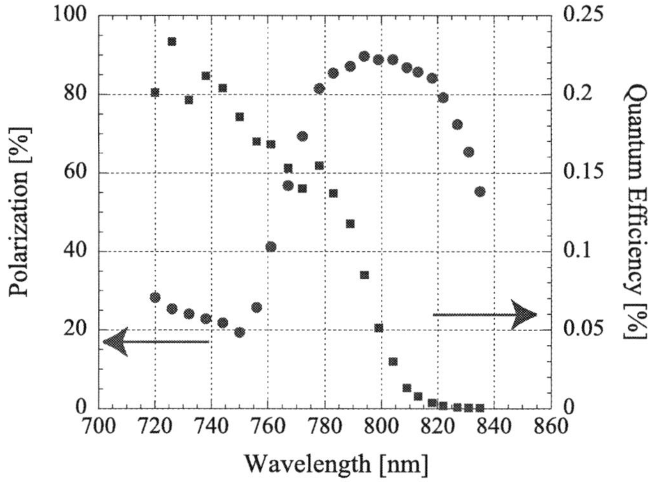

FIGURE 2. Observed ESP as a function of excitation laser wavelength.

A cm^{-2} sr^{-1} and the angular current density was 161 mA/sr. This brightness is large compared with those of un-polarized electron emitters. It is smaller than those of W-tip emitters by one or two orders, but one order higher than those of LaB_6 emitters.

ESP and QE spectra taken for the T-PC are shown in Fig. 2, where the closed circle and square points indicate ESP and QE, respectively. A maximum ESP of 90% with QE of 0.09% has been achieved at a wavelength of 794 nm. This ESP value is in the range of the values of our previous studies of R-PC. On the other hand, the QE value is smaller because the absorption of the pump laser light by a 500 nm-thick GaAs layer was about 50% and the QE performance was not optimized on this PC. Further efforts have been made in designing new T-PC samples with a higher QE.

SUMMARY

In order to achieve high brightness and high polarization performances for polarized electron beams, a 20–kV gun and T-PCs were developed. As the results, the brightness of $(1.3 \pm 0.5) \times 10^7$ A cm^{-2} sr^{-1} and the ESP of (90 ± 6)% were achieved. These performances are enough high for the applications to not only high energy electron accelerators but also electron microscopes.

The interference between the laser and electron beam lines in the conventional R-PC scheme is a serious problem for gun design and imposes a serious limit on beam performance. This T-PC scheme appears promising for applications of polarized beam in various types of high energy electron accelerators and microscopes, since there is sufficient free spaces along both the laser and electron beam lines to enable the optimization of design figures and parameters individually.

Furthermore, this T-PC scheme has an additional advantage for laser heating problem by high duty operation, such as Energy Recovery Linacs. In the electron generation process of NEA-PC with SL layers, only a several % of the pump laser light is absorbed in the SL layers. In the case of R-PCs, the other of that are absorbed in its substrate and the absorbed power causes the heating problem for the PCs. On the other hand, in the T-PCs case, since the substate is transparent for the pump laser light, it can be possible to suppress the effect of laser heating.

ACKNOWLEDGMENTS

This work has been financially supported by Japan Science and Technology (JST) as a technology development program for advanced measurement and analysis (program leader: T. Nakanishi).

REFERENCES

1. D. T. Pierce, F. Meier, and P. Zürcher, Appl. Phys. Lett. **26**, 670 (1975).
2. C. Y. Prescott, W. B. Atwood, R. L. A. Cottrell, H. DeStaebler, E. L. Garwin, A. Gonidec, R. H. Miller, L. S. Rochester, T. Sato, D. J. Sherden, et al., Phys. Lett. B **77**, 347 (1978).
3. M. Meyerhoff, D. Eyl, A. Frey, H. G. Andresen, J. R. M. Annand, K. Aulenbacher, J. Becker, et al., Physics Letters, Section B: Nuclear, Elementary Particle and High-Energy Physics **327**, 201 (1994).
4. SLD colaboration, I. Abt, C. J. Ahn, T. Akagi, W. W. Ash, D. Aston, N. Bacchetta, K. G. Baird, C. Baltay, H. R. Band, et al., Phys. Rev. Lett. **74**, 2880 (1995).
5. S. Mayer and J. Kessler, Phys. Rev. Lett. **74**, 4803 (1995).
6. U. Kolac, M. Donath, K. Ertl, H. Liebl, and V. Dose, Review of Scientific Instruments **59**, 1933 (1988).
7. T. Nakanishi, H. Aoyagi, H. Horinaka, Y. Kamiya, T. Kato, S. Nakamura, T. Saka, and M. Tsubata, Physics Letters A **158**, 345 (1991).
8. T. Nakanishi, K. Togawa, T. Baba, F. Furuta, H. Horinaka, T. Kato, Y. Kurihara, H. Matsumoto, T. Matsuyama, T. Nishitani, et al., Nuclear Inst. and Methods in Physics Research, A **455** (2000).
9. N. Yamamoto, M. Yamamoto, M. Kuwahara, R. Sakai, T. Morino, K. Tamagaki, A. Mano, A. Utsu, S. Okumi, T. Nakanishi, et al., Journal of Applied Physics **102**, 024904 (2007).
10. T. Ohshima and M. Kudo, Japanese Journal of Applied Physics **43**, 8335 (2004).
11. N. Yamamoto, T. Nakanishi, A. Mano, Y. Nakagawa, S. Okumi, M. Yamamoto, T. Konomi, X. Jin, T. Ujihara, Y. Takeda, et al., Journal of Applied Physics **103**, 064905 (2008).
12. T. Nakanishi, K. Dohmae, S. Fukui, Y. Hayashi, I. Hirose, N. Horinaka, T. Ikoma, Y. Kamiya, M. Kurashina, and S. Okumi, Jpn. J. Appl. Phys **25**, 766 (1986).
13. N. Yamamoto, A. Mizuno, K. Masuda, K. Kan, and M. Yamamoto, this Proceedings (2009).
14. F. Furuta, T. Nakanishi, S. Okumi, T. Gotou, M. Yamamoto, M. Miyamoto, M. Kuwahara, N. Yamamoto, et al., Nuclear Inst. and Methods in Physics Research, A **538**, 33 (2005).
15. X. Jin, N. Yamamoto, Y. Nakagawa, A. Mano, T. Kato, M. Tanioku, T. Ujihara, Y. Takeda, S. Okumi, M. Yamamoto, et al., Applied Physics Express **1**, 45002 (2008).
16. X. Jin, T. Saka, M. Tanioku, S. Fuchi, T. Ujihara, Y. Takeda, N. Yamamoto, Y. Nakagawa, A. Mano, S. Okumi, et al., Journal of Crystal Growth **310**, 5039 (2008).
17. K. Togawa, T. Nakanishi, T. Baba, F. Furuta, H. Horinaka, T. Ida, Y. Kurihara, H. Matsumoto, T. Matsuyama, M. Mizuta, et al., Nucl. Instr. and Meth. A **414**, 431 (1998).
18. T. Maruyama, A. Brachmann, J. Clendenin, T. Desikan, E. Garwin, R. Kirby, D. Luh, J. Turner, and R. Prepost, Nuclear Instruments and Methods in Physics Research, A **492** (2002).
19. T. Nishitani, T. Nakanishi, M. Yamamoto, S. Okumi, F. Furuta, M. Miyamoto, M. Kuwahara, N. Yamamoto, K. Naniwa, O. Watanabe, et al., Journal of Applied Physics **97**, 94907 (2005).

Cooled Transmission–Mode NEA–Photocathode with a Band–Graded Active Layer for High Brightness Electron Source

L.B. Jones*, S.A. Rozhkov†, V.V. Bakin**, S.N. Kosolobov**, B.L. Militsyn*, H.E. Scheibler**, S.L. Smith* and A.S. Terekhov†,**

**STFC Daresbury Laboratory, Warrington, WA4 4AD, United Kingdom*
†Novosibirsk State University, Novosibirsk, 630090, Russia
***Institute of Semiconductor Physics, Novosibirsk, 630090, Russia*

Abstract. A Free–Electron Laser (FEL) places many exacting demands on a Negative Electron Affinity (NEA) photocathode, such as the need for an ultra–fast response time, low energy spread for emitted electrons, high quantum efficiency (*Q.E.*) and a high average photocurrent. However, these key requirements are conflicting, and cannot be fulfilled by conventional photocathode design. For example, to achieve ~ 10 ps response time, the photocathode active layer should be thinned to $\sim 100-150$ nm, but this thickness is insufficient to provide near–complete absorption of light with $h\nu \approx \varepsilon_g$ so high *Q.E.* cannot be achieved. Complete optical absorption and high *Q.E.* can be obtained using a thin active layer at higher photon energies, but this generates photoelectrons with excess kinetic energy within the semiconductor. These photoelectrons do not thermalise in a thin active layer, so yield a broad energy distribution in the emitted electrons. Moreover, cooling of the conventional semiconductor photocathode structure is ineffective due to its fragility, so it cannot be pressed firmly to a heat sink to attain good thermal contact. Consequently, the maximum CW photocurrent is limited to a few milliamps. The goal of our work is to develop a new design of NEA–photocathode which is optimised for FEL applications.

Keywords: GaAs, photocathode, photoelectron, FEL, NEA, transmission mode, reflection mode
PACS: 29.25.Bx, 41.60.Cr, 81.05.Ea, 85.60.Ha

METHODOLOGY & EXPERIMENTAL DETAILS

Our approach is to design a transmission–mode GaAs (Cs,O) photocathode with a *composite band–graded* (BG) *active layer* on a buffered sapphire substrate. The energy diagram for this photocathode is shown in fig. 1 (a). During operation, the photocathode is back–illuminated through the sapphire and buffer layer with photons of ~ 2.3 eV. The BG layer is grown from $Al_xGa_{1-x}As$ with x varied over the range $0 < x < 0.1$, which absorbs most of the incident photons. The thickness of the BG layer is ~ 300 nm, and is sufficient for effective thermalisation of photoelectrons within this layer over a period of ~ 0.1 ps. The variation of the Al content across the BG layer creates a variation in the energy of the conduction band minima, and consequently, creates a 'built–in' electric field whose strength is $\sim 4 \times 10^4$ V/cm.

It is known from the literature [1] that this electric field accelerates thermalised photoelectrons to the saturation velocity, typically around 10^7 cm/sec. Thus photoelectrons are 'ejected' from the BG layer into the GaAs 'emitting' layer over an estimated time of approximately 3 ps. It then takes ~ 5 ps for photoelectrons to diffuse across the GaAs

CP1149, *Spin Physics, 18th International Spin Physics Symposium*
edited by D. G. Crabb, D. B. Day, S. Liuti, X. Zheng, M. Poelker, and Y. Prok

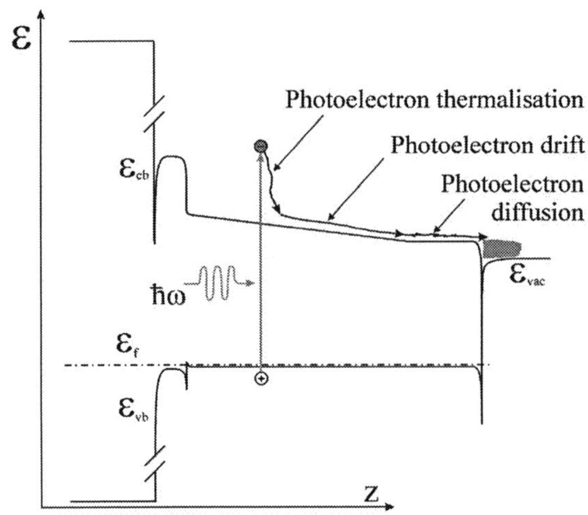

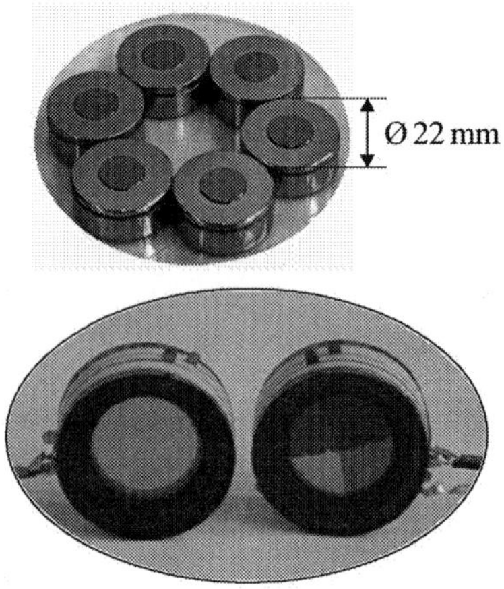

FIGURE 1. **(a)** Band–graded photocathode energy diagram. **(b)** Test photodiodes & photocathodes.

layer whose thickness is $\sim 100\,\text{nm}$, and finally escape into the vacuum [2]. Therefore, the estimated time response of a BG–photocathode does not exceed $\sim 10\,\text{ps}$.

To permit a comparative study of photoemission characteristics, a batch of test photocathodes were manufactured with both BG and homogenous (HM) active layers, with fig. 1 (b) showing the BG–photocathodes. The thickness of HM active layer was equal to 130 nm, so slightly exceeds the 100 nm GaAs layer present in the BG–photocathode. Both types of photocathode were hermetically sealed within test parallel plate photodiodes. The anodes of these photodiodes were made of glass, covered with transparent and conductive In_2O_3 layers, permitting photoemission to be studied in both transmission– and reflection–mode illumination geometries. Longitudinal energy distribution curves (LEDC) for the emitted photoelectrons were measured by applying a retarding potential between photocathode and anode [3]. *Q.E.*s were measured at low electric fields so that the influence of the Shottky effect could be neglected. Experiments were performed at both room and liquid nitrogen temperatures, specifically 300 K and 77 K respectively.

RESULTS

Q.E. spectra for two photocathodes with both HM and BG active layers, measured in both transmission– (TM) and reflection–mode (RM) illumination geometries are shown in fig. 2. One can clearly see that the *Q.E.*s for both photocathodes are low for photon energies close to the GaAs energy gap, but exceed 15 – 20% for $h\nu = 2.32\,\text{eV}$ ($\lambda = 532\,\text{nm}$), when most of exciting photons are absorbed within the active layers. LEDC measured at room temperature in RM illumination geometry for photocathodes with HM and BG active layers are shown in fig. 3 (a) & (b) respectively. It is clear that for this illumination geometry, an increase in photon energy is accompanied by an increase in both amplitude and width of the high–energy tail of LEDC. This observation

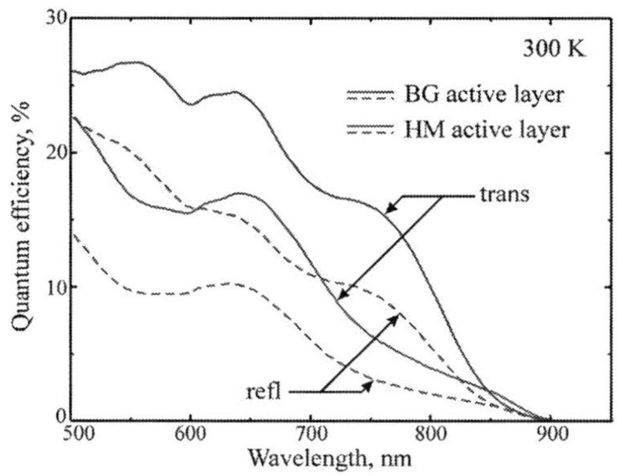

FIGURE 2. *QE* Spectra for photocathodes with both band–graded and homogeneous active layers in transmission–mode (*trans*) and reflection–mode (*refl*) illumination geometry.

is expected because at 'high' photon energies, the majority of photoelectrons with excess kinetic energy are generated in close proximity to the emitting surface. Consequently, a significant number of non–thermalised electrons are emitted before thermalisation, and their contribution to the photocurrent increases in tandem with the increase in photon energy. The BG layer cannot play any role in the RM illumination geometry, however the role of BG layer is changed significantly if the photocathode is illuminated in TM geometry.

Firstly, let us consider the photocathode with a HM active layer. If this photocathode is back–illuminated at 'high' photon energy, most of the photoelectrons with 'excess' kinetic energy are generated near the interface between the active and buffer layers. To be emitted into the vacuum, these photoelectrons must cross the active layer by diffusion. If the thickness of the active layer is approximately 100 nm, the diffusion process takes typically ~ 5 ps [2]. This time exceeds the photoelectron thermalisation time for p-doped GaAs [4], so a large portion of 'hot' photoelectrons generated near the back side of the HM active layer have to be thermalised during their diffusion to the emitting surface. The validity of this statement can be checked through analysis of the data presented on the fig. 4 (a) which shows the LEDC measured for the HM active layer in TM illumination geometry. One can see in fig. 4 that both the amplitude and width of the high–energy tail of the LEDC are approximately two times smaller than those shown in fig. 3 (a) which were measured from the same photocathode in RM illumination geometry. Nevertheless, it is shown that the thickness of the active layer (130 nm) is not sufficient for complete suppression of the 'hot' electron contribution to the photocurrent.

Analysis of data presented in fig. 3 (a) has shown that the thickness of the active layer of this photocathode is close to the optical absorption length for $h\nu = 2.32$ eV, and to the thermalisation length of 'hot' photoelectrons. This is the basis for the remarkable contribution from 'hot' electrons which is detected in the LEDC. To suppress photoemission of 'hot' photoelectrons from a conventional TM photocathode, one has to increase the thickness of HM active layer. However, if this is done, the response time of the photocathode increases rapidly because it is proportional to the squared thickness of the active

Left (a): Homogeneous (HM) layer **Right (b):** Band–graded (BG) layer

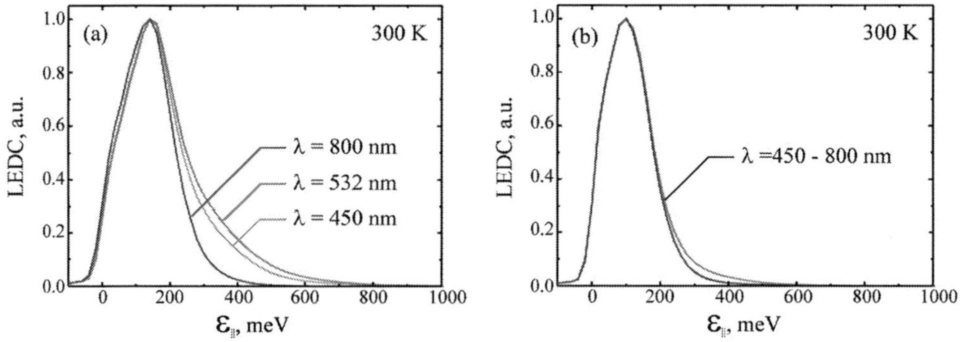

FIGURE 3. LEDC measured for the reflection–mode photocathode at room temperature.

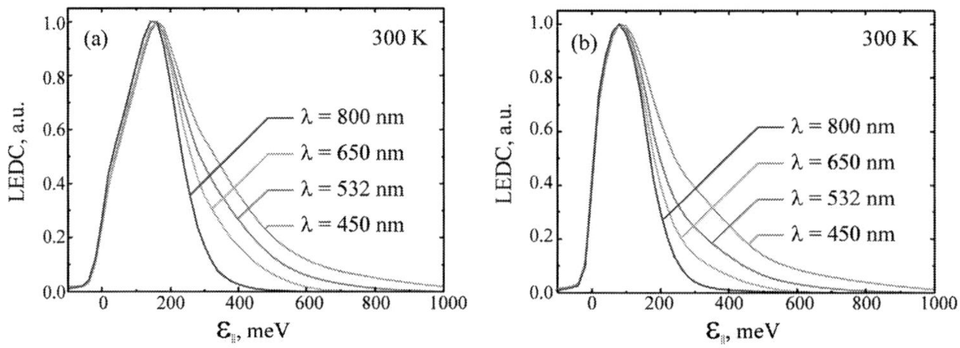

FIGURE 4. LEDC measured for the transmission–mode photocathode at room temperature.

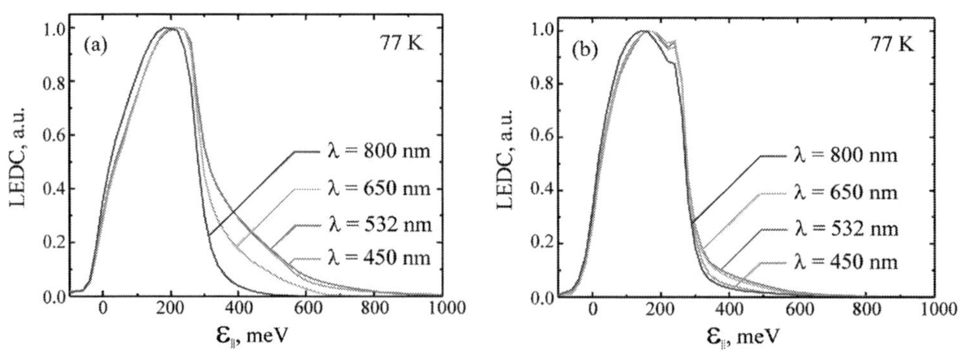

FIGURE 5. LEDC measured for the transmission–mode photocathode at liquid nitrogen temperature.

layer.

To overcome this problem, we have increased the total thickness of the active layer using the BG layer, thus creating the strong built–in electric field. This field is sufficient to accelerate photoelectrons to saturation velocity [1], boosting the photoelectron's 'slow' diffusion to a 'rapid' drift. Our estimate shows that it takes $\sim 3\,\text{ps}$ to extract photoelectrons from the BG layer and to inject them into the GaAs emitting layer. Therefore, use of a BG layer confers a low response time to the photocathode.

To check the effectiveness of the BG layer to suppress the 'hot' electron contribution, we have measured the LEDC for the photocathode with a composite active layer, as shown in fig. 4 (b). One can see from this figure that the 'hot' electron contribution in this photocathode is much less than those measured for the photocathode with a HM active layer. The similar data, shown in fig. 5 (a) & (b), were obtained for these two photocathodes at liquid nitrogen temperature. One can see, also, that the cooling of NEA–photocathode increases the width of the photoelectron energy distribution due to the increase of NEA–value.

SUMMARY

A new design for a cooled transmission–mode (TM) NEA photocathode on a sapphire base plate with a composite active layer, appropriate for FEL applications, is proposed and realised. To suppress the contribution of 'hot' photoelectrons in the broadening of the photoelectron energy distribution without loss of photocathode time response, a 'thick' band–graded (BG) layer with a strong built–in electric field was incorporated in the photocathode semiconductor structure. The estimated response time does not exceed $10\,\text{ps}$ and $Q.E.$ at $h\nu = 2.32\,\text{eV}$ exceeds 20%. Future plans include optimisation of the band–graded active layer parameters and measurement of response time.

ACKNOWLEDGMENTS

This work has been funded by the Science and Technology Facilities Council (STFC) as R&D in support of a possible fourth–generation light source facility in the United Kingdom. The work has been carried out by the Novosibirsk Institute of Semiconductor Physics and Novosibirsk State University, in collaboration with the Accelerator Science and Technology Centre at STFC Daresbury Laboratory.

REFERENCES

1. H. Arabshahi, M.R. Khalvati & M. Rezaee Rokn–Abadi, *Brazilian J. of Physics*, **38**, page 293 (2008).
2. K. Aulenbacher, J. Schuler, D.V. Harrach, E. Reichert, J. Rothgen, A. Subashiev, V. Tiokine & Y. Yashin, *J. Appl. Phys.*, **92**, page 7536 (2002).
3. A.S. Terekhov & D.A. Orlov, *Pis'ma Zh. Exp. Teor. Fiz.*, **59**, page 864 (1994).
4. *Optical orientation* edited by F. Meier & B.P. Zakharchenya, North–Holland publishing (1984).

K_2CsSb Cathode Development

John Smedley[a], Triveni Rao[a] and Erdong Wang[b]

[a]*Brookhaven National Laboratory, Upton, NY 11973*
[b] *Institute of Heavy Ion Physics, Peking University, Beijing, 100871, China*

Abstract. K_2CsSb is an attractive photocathode for high current applications. With a quantum efficiency of >4% at 532nm and >10% at 355nm, it is the only cathode to have demonstrated an average current of 35mA in an accelerator environment. We describe ongoing cathode development work for the energy recovery linac being constructed at BNL. Several cathodes have been created on both copper and stainless steel substrates, and their spatial uniformity and spectral response have been characterized. Preliminary lifetime measurements have been performed at high average current densities (>1mA/mm^2).

Keywords: Photoemission, Photocathode
PACS: 85.60.Ha

INTRODUCTION

Alkali antimonide cathodes have been used in photoinjectors for many years [1-2]. They have high quantum efficiency for visible light, but are challenging to implement in an accelerator; the lifetime of a K_2CsSb cathode in a normal-conducting (NC) RF cavity is measured in hours [1] due to the relatively poor vacuum in the injector. With the advent of superconducting (SRF) injectors, it may be possible to lengthen this lifetime considerably, as the cavity vacuum can be orders of magnitude better (~0.01nTorr in an SRF cavity, compared to ~1nTorr in an NCRF cavity).

One project based on such an SRF injector is the energy recovery linac (ERL) project at Brookhaven National Laboratory [3]. This machine has two modes of operation, with average currents of 50mA and 0.5A. The cathode being developed for 50mA operation is cesium potassium antimonide (K_2CsSb). The cathode will be mounted on the stalk of a ¼ wave choke filter, and will be kept at 77K to reduce the radiative heat load on the cavity. The laser spot on the cathode will be ~7mm in diameter, resulting in a design current density of 1.3mA/mm^2. For 0.5A operation, a diamond amplifier [4] will be used; in this case the average current required from the K_2CsSb photocathode is lower, but the cathode will be operated in transmission mode, with the laser incident from the back of the cathode through a transparent conductor.

This paper describes the current status of the cathode development for the ERL. Four cathodes have been created, and all have reached the design quantum efficiency (QE) of 10% at 355nm in reflection mode. The spectral response of each cathode has been measured, and preliminary measurements of lifetime at the design current density have been performed.

CP1149, *Spin Physics, 18th International Spin Physics Symposium*
edited by D. G. Crabb, D. B. Day, S. Liuti, X. Zheng, M. Poelker, and Y. Prok

METHODS

The cathodes are created in a vacuum system with a base pressure of ~0.02nTorr. This vacuum is achieved using a 400l/s ion pump in conjunction with a titanium sublimation pump. The deposition geometry is vertical – the sources evaporate up onto the substrate, which is moved via a linear manipulator into position above each source sequentially. The antimony source is a tantalum boat packed with antimony pellets. The potassium and cesium sources are "V" sources from alvatech. The source to substrate distance is 4cm. A crystal monitor can be used to monitor the deposition rate of antimony and potassium. The cesium deposition rate is not measured; instead the QE is monitored during deposition by illuminating the cathode with a green laser (532nm), and the deposition ended when the QE stops increasing. In addition to the laser port used for deposition, the system has two horizontally aligned optical ports. After deposition, the cathode can be rotated to face one of these ports, with the back of the cathode facing the other. This arrangement allows the optical transmission of a transparent portion of the cathode substrate to be measured to check the deposition thickness, and it allows the QE to be measured in both transmission and reflection mode. Spectral response measurements are made through the reflection port by using a lamp (DH2000 from Ocean Optics) with a fiber-coupled monochromator (Edmond Optics) to illuminate the cathode. This source has a 2nm bandwidth, and a 5x5mm^2 spot on the cathode. The measurements with the lamp source require all external light sources to be off, as even the light from the filament of an ion gauge produces more current than the lamp. A 532nm CW laser can be used for lifetime and high current measurements.

For high current tests, the cathode can be moved in front of a wire mesh anode, with an electrode separation of 2.5cm. The anode can be biased to 5kV, allowing higher currents to be extracted from the cathode without space charge limitation. The parallel geometry of the cathode and anode in this case allows the electric field on the cathode to be estimated. The wire mesh is chosen to allow 100% transmission of the 532nm laser while maintaining a nearly uniform electric field between electrodes.

The copper cathode substrate is 2.5cm in diameter, polished with 1μm diamond polishing compound to a mirror finish. Portions of the copper substrate can be covered with shims made of other materials, allowing comparison of the QE for cathodes deposited on different substrates (to date, only copper and stainless steel have been measured). A 6mm diameter hole in the substrate exposes a glass slide covered with a transparent conductor. For the current measurements, 20-40nm of sputtered copper was used; indium-tin oxide (ITO) coated slides will be tested in the future. This 6mm region can be illuminated from either side, allowing the QE to be compared in reflection and transmission modes, and allowing the optical absorption of the cathode layer to be measured. The cathode is electrically isolated from the vacuum translation arm, enabling the cathode to be biased and the charge leaving the cathode to be measured. The arm includes a vacuum heater capable of raising the substrate to 150C, and liquid nitrogen (LN_2) cooling lines capable of cooling the cathode to -80C.

The cathode is deposited sequentially, following the general method of D. Dowell [1]. The substrate is at 150C for the antimony deposition, and the rate is set to ~0.5nm/s. 20nm of antimony is deposited, after which the substrate is cooled to ~140C for the potassium deposition. 40nm of potassium is deposited, again with a rate of ~0.5nm/s. The substrate is cooled to 135C, and the substrate is biased to -20V. The cathode is illuminated with ~0.5mW of 532nm light, and the emitted current is monitored. Cesium is deposited until the emitted current stops rising, at which point the cathode is cooled to room temperature by flowing LN_2 through the arm.

RESULTS

Figure 1 shows the spectral response of cathode 2 deposited on both copper and stainless steel substrates. The spatial uniformity of this cathode is shown in figure 2. The QE was also measured on the unpolished copper fork which holds the substrate puck. To test the linearity of the response with flux, a 532nm CW laser was used to deliver 47.7mW to the cathode, resulting in 0.526mA of emitted current. The QE was the same as that measured with the lamp source, which provides 20nW and 226pA at 532nm. Cathode 2 was also cooled to -80C, with no change in the spectral response. Figure 3 shows the spectral response of 4 cathode depositions. Cathode 4 had a second cathode deposited on top of the first after the lifetime measurements, restoring its QE. Cathode 3 was deposited with the substrate at 120C (rather than 150C-135C); this exhibited somewhat inferior QE.

The performance of the cathodes in transmission mode depends strongly on the cathode thickness. Cathode 3 was the best, with a QE in transmission mode half that of reflection mode after accounting for the 20% transmission of the sputtered copper layer used to provide conduction.

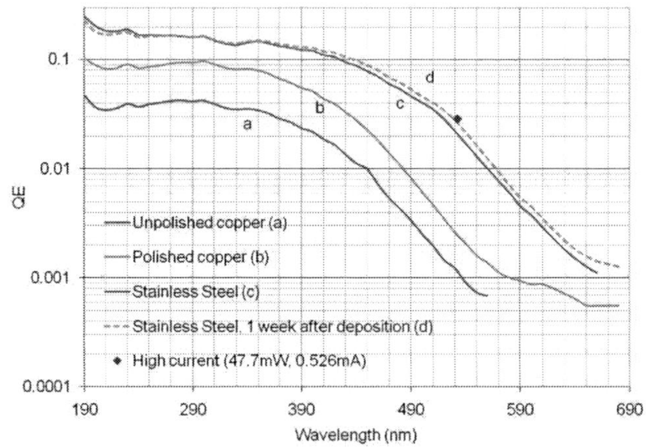

FIGURE 1. Spectral Response of a K_2CsSb cathode on copper and stainless steel substrates. The high current point was taken 8 days after deposition, immediately after response curve (d).

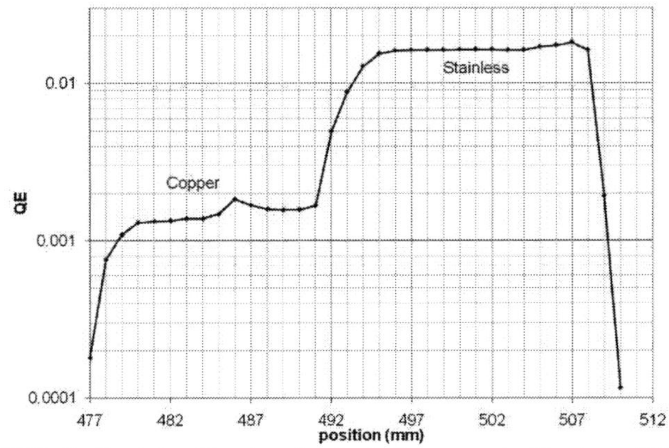

FIGURE 2. Spatial uniformity of cathode 2 measured with a laser (532mn).

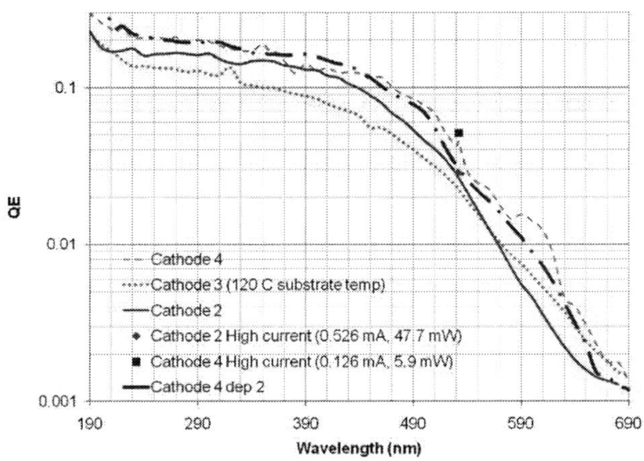

FIGURE 3. Spectral Response of a K_2CsSb cathodes on stainless steel substrates.

Lifetime

Cathode 4 was used to test the performance of K_2CsSb under continuous operation at the ERL current density. Tests on earlier cathodes have shown that the QE is unchanged after 40 days of storage in the chamber vacuum. The measurement station with the anode was used for these tests to achieve a field necessary to overcome space charge. The 532nm laser was focused to 80μm FWHM on the cathode to achieve a current density of up to 2.2mA/mm^2 with a bias of 3kV (1.45mW optical power, 10.9μA emitted current). The optical power was modulated to keep the current density near the ERL goal of 1.3mA/mm^2; for a 1kV bias on the anode, the cathode emission was space charge limited to 1.1mA/mm^2. Figure 4 shows the result of this test. For 2kV and 3kV anode bias, the QE decay is consistent with a 1/e lifetime of 100hrs; as the emission was space charge limited at 1kV, no estimate of lifetime can be made. A similar measurement was made with a 3mm diameter laser spot on the cathode and a bias of 500V, with 60μA of emitted current. In this case the QE decayed from 2.3% to 2.0% in 40 hrs, yielding a 1/e lifetime of 360hrs. Two possible causes of this decay are ion bombardment of the cathode and local vacuum pressure rise near the cathode due to electron stimulated desorption from the anode (as the 1-3keV electrons strike the anode).

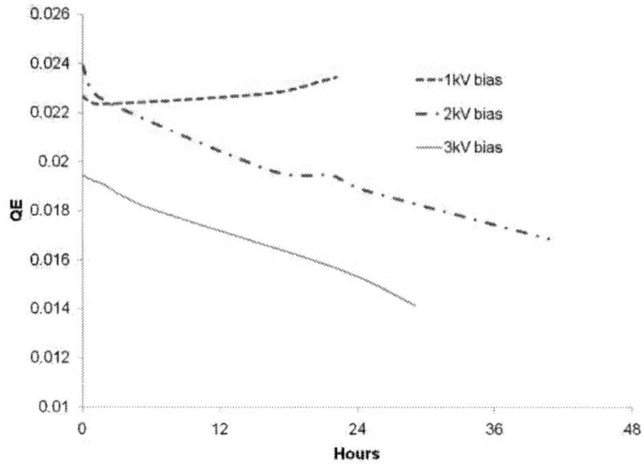

FIGURE 4. Decay of QE under continuous extraction of 1.3mA/mm^2.

CONCLUSIONS

K_2CsSb is an attractive cathode for high average current injectors. It has less stringent vacuum requirements than Cs:GaAs and has high QE for visible light. It has been shown to have a limited lifetime in a NCRF injector, but in the better vacuum of a SRF or DC injector, lifetimes of days to weeks should be possible, even at high current density. Cathodes have been fabricated that meet the QE and current density requirements of the BNL ERL.

ACKNOWLEDGMENTS

The authors would like to thank John Walsh for his expert technical assistance, Andrew Burrill for his advice and Veljko Radeka for his support. This work funded by the U.S DOE under Contract No. DE-AC02-98CH10886.

REFERENCES

1. D.H. Dowell, S.Z. Bethel and K.D. Friddell, *Nucl. Inst. and Meth. in Phys. Res.*, **A356** 167 (1995)
2. P. Michelato et al, "R & D Activity on High QE Alkali Photocathodes for RF Guns", Proceedings of PAC '95, p 1049 (1995). accelconf.web.cern.ch/AccelConf/p95/ARTICLES/WPC/WPC23.PDF.
3. V. Litvinenko et al, "R & D Energy Recovery Linac at Brookhaven National Laboratory", Proc. of EPAC '08, p 193 (2008). accelconf.web.cern.ch/AccelConf/e08/papers/mopc057.pdf.
4. J. Smedley, et al., "Diamond Amplified Photocathodes", in Diamond Electronics—Fundamentals to Applications II, Mater. Res. Soc. Symp. Proc. 1039, Warrendale, PA, (2007), 1039-P09-02.

Photoemission Spectroscopic Study of Cesium Telluride Thin Film Photocathode

Harue Sugiyama, Koji Ogawa, Junpei Azuma, Kazutoshi Takahashi, and Masao Kamada

Synchrotron Light Application Center, Saga University, 1 Honjo, Saga 840-8502, Japan

Abstract. The photoemission spectroscopy using synchrotron radiation has been carried out to study the high quantum efficiency and long working lifetime of cesium telluride (Cs_xTe_y) thin film photocathode. The electron affinity derived from the observed energy-distribution curves provides an important hint for long persistency of the photocathode.

Keywords: Cesium-Telluride, Photocathode
PACS: 29.25.Bx, 73.61.–r

INTRODUCTION

It has been known that Cs_xTe_y (cesium telluride) thin film photocathodes for RF guns have high quantum efficiency and long working lifetime. In fact, Cs_2Te thin film photocathodes have been used as RF-gun photocathodes for over 10 years [1]. Why do Cs_xTe_y thin films maintain such high quantum efficiency with long lifetime against the damage by grinding ion-bombardment under high RF-voltages? In order to understand this issue, we take up the first step to study Cs_xTe_y characteristics.

In the present study, Cs_xTe_y thin films were formed with vacuum evaporation. The photoemission spectroscopy was applied to measure the energy distribution of the electrons emitted from the films. The threshold energy or work function was estimated from the observed spectra of secondary electrons and valence-band photoelectrons. Finally, the electron affinity of Cs_xTe_y is discussed, giving an important hint for long persistency of the photocathode.

EXPERIMENT

Saga University has constructed a new beam line at BL13 of Saga Synchrotron Light Source (SAGA-LS), which has 1.4 GeV beam energy [2]. The BL13 consists of a VLS-type monochromator with a planar undulator and a PGM-type monochromator with bending radiation. The PGM monochromator provides photons in the energy range from few eV to 150 eV. The present photoemission study has been carried out at the PGM station.

The PGM station consists of three chambers and the transfer system. An air-lock chamber makes it possible to install samples quickly. A preparation chamber and an

CP1149, *Spin Physics, 18th International Spin Physics Symposium*
edited by D. G. Crabb, D. B. Day, S. Liuti, X. Zheng, M. Poelker, and Y. Prok

analyzing chamber are used for sample preparation and analyses, respectively. Samples can be transferred in the ultra high vacuum (UHV) between them. In the preparation chamber, cesium and tellurium sources, a leak valve for oxygen gas, a thickness monitor, and an ion-sputtering gun are installed. The sample annealing and the total-yield measurement with a He-Ne laser are also conducted in the chamber. The analyzing chamber involves an OMICROM HR-125 hemi-spherical electron-analyzer and a PHI 10-155A/CMA Auger spectrometer, which can be used to measure the photoemission spectra and to check the surface contaminations, respectively. The drain current from the samples can be observed with a Keithley 616 pico-ammeter.

Cs_xTe_y thin films were formed with a vacuum evaporation method in the preparation chamber having a base pressure of 3×10^{-8} Pa. Molybdenum plates were used as sample substrates. The substrates were heated up to 550 °C in the preparation chamber before deposition. Cs_xTe_y thin films were formed at room temperature by first deposition of tellurium on the substrates and subsequent deposition of cesium. Tellurium bits of 99.9999 % in purity were purchased from Nilaco and installed in the deposition device, while a commercially available cesium dispenser was obtained from SAES getters. The details of the sample preparation have been already reported elsewhere [3]. The Cs_xTe_y thin films were transferred in UHV to the analyzing chamber. All measurements were carried out in the analyzing chamber having a base pressure of 3×10^{-8} Pa.

PHOTOEMISSION SPECTRA

The photoemission spectroscopy was applied to measure the energy distribution of the electrons emitted from the films. The observed spectra consist of secondary electrons and photoelectrons. The former is the majority of electrons emitted from the films, and then firstly we focus our attention on the secondary electrons. The cut-off energy of the secondary electrons corresponds to the vacuum level of the samples, while the threshold energy or the work function can be derived from the observed spectra of secondary electrons and valence-band maximum (VBM).

Figure 1 (a) shows the energy distribution curves of secondary electrons from the cesium telluride thin film. The distribution curves were measured in biased condition in order to observe the electrons of less energy than the analyzer work function with avoiding stray electrons in the chamber. The ordinate is normalized at the peak intensity and the abscissa is given relative to the vacuum level of the analyzer. Figure 1 (a) shows the spectra of the film (about 50 nm as total thickness) from fresh off to 6 days after the fabrication. It is found that the low energy cut-off is shifted from -3.3 eV (1 hour) to -2.5 eV (6 days). The vacuum levels from the Fermi level correspond to the value of 1.3 eV (1 hour) and 2.1 eV (6 days).

Figure 1 (b) shows valence-band spectra of the film. The arrows indicate the Fermi level that was calibrated with photoemission of gold. As seen in Fig. 1 (b) the VBMs of the film are observed at 2.5 eV lower than the Fermi level. It is noticed that the VBMs are not shifted within several days.

The present results of the photoemission spectroscopy indicate that the threshold energy is changed from 3.8 eV to 4.6 eV. A cesium telluride thin film as a photocathode is irradiated with the laser, which has the energy of 4.7 eV in general. Therefore, the change in the electron affinity does not make any difference.

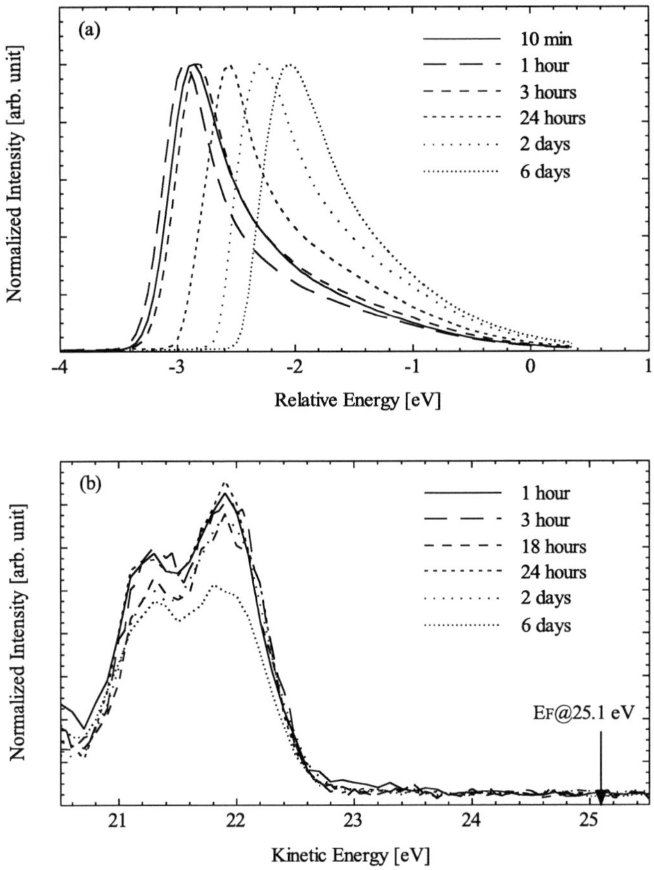

FIGURE 1. (a) Secondary electron spectra and, (b) valence-band spectra.

The VBMs are not shifted within several days. It is inferred from this result that the energy gap is not changed. It is possible to consider that the change of the low energy cut-off is attributed to the temporal changes in the electron affinity. The values of the electron affinity, that can be derived from the results of the photoemission spectroscopy and reported value of the energy gap [5] are 0.5 eV (1 hour) and 1.3 eV (6 days). The temporal change in the electron affinity and in the shape of valence-band are interesting observations, but need further analysis.

AUGER ELECTRON SPECTRA

Figure 2 shows the Auger-electron spectra of the same film as those shown in Fig. 1, and the substrate before a heat treatment by way of comparison. The Auger-electron spectrum of the film was observed after the photoemission measurement. It is stressed that there were no peaks except for Te MNN and Cs MNN Auger peaks. The peaks originated from carbon or oxygen on the spectrum of the film were out of the detection limit. Therefore, a contamination on the surface is not a main cause of the temporal changes as shown in Fig. 1.

The elemental ratio between cesium and tellurium (Cs:Te) has been derived from the Auger spectra. The cross-section ratio of Cs MNN to Te MNN Auger peaks is about 1:5 in the literature [4]. As seen in Fig. 2, the peaks a and b, which are assigned as Te MNN and Cs MNN Auger peaks, respectively, show the intensity ratio of 7:5. Therefore, the elemental ratio is roughly designated as Cs_7Te.

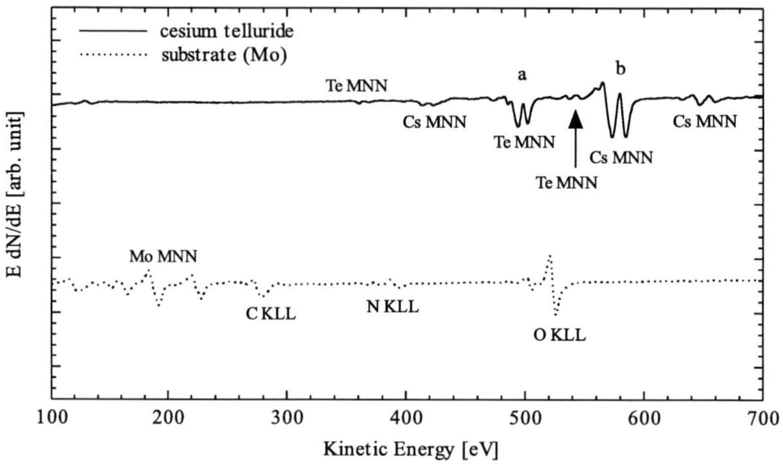

FIGURE 2. Auger-electron spectra.

ACKNOWLEDGMENTS

I express my greatest appreciation to everyone at Jefferson Lab for hosting the workshop and to all other people for giving me such a wonderful opportunity. This research was partially supported by the Ministry of Education, Science, Sports and Culture, Grant-in-Aid for Young Scientists (B).

REFERENCES

1. S. H. Kong, J. Kinross-Wright, D. C. Nguyen, and R. L. Sheffield, J. Appl. Phys. **77** (11), 6031 (1995).
2. K. Takahashi, Y. Kondo, J. Azuma, and M. Kamada, J. Elect. Spect. and Rel. Phenom. 144/147, 1093 (2005).
3. H. Sugiyama, "Characteristics of Cesium Telluride Thin Film Photocathode", Ph.D. Thesis, Nagoya University, 2005.
4. "Handbook of Auger Electron Spectroscopy" 3rd ed. (PHYSICAL ELECTRONICS, ISBN: 0-964824-0-1).
5. R. A. Powell, W. E. Spicer, G. B. Fisher, and P. Gregory, Phys. Rev. B **8**, 3987 (1973).

DC High Voltage Conditioning of Photoemission Guns at Jefferson Lab FEL

C. Hernandez-Garcia, S. V. Benson, G. Biallas, D. Bullard, P. Evtushenko, K. Jordan, M. Klopf, D. Sexton, C. Tennant, R. Walker, and G. Williams

Thomas Jefferson National Accelerator Facility, Newport News, VA 23606

Abstract. DC high voltage photoemission electron guns with GaAs photocathodes have been used to produce polarized electron beams for nuclear physics experiments for about 3 decades with great success. In the late 1990s, Jefferson Lab adopted this gun technology for a free electron laser (FEL), but to assist with high bunch charge operation, considerably higher bias voltage is required compared to the photoguns used at the Jefferson Lab Continuous Electron Beam Accelerator Facility. The FEL gun has been conditioned above 400kV several times, albeit encountering non-trivial challenges with ceramic insulators and field emission from electrodes. Recently, high voltage processing with krypton gas was employed to process very stubborn field emitters. This work presents a summary of the high voltage techniques used to high voltage condition the Jefferson Lab FEL photoemission gun.

Keywords: DC high voltage, photoemission electron guns, krypton processing.
PACS: 29.25.Bx, 81.65.Ps, 84.70.+p, 07.30.-t, 79.60.Bm

INTRODUCTION

The surface chemistry of the Cs:GaAs photocathode imposes extreme requirements on the vacuum environment in photoemission electron guns, while the need to extract electrons from the photocathode and create electrostatic focusing demands bias voltage at least ~ 100kV, with gradients as high as 5 MV/m. Besides the challenges to obtain and maintain extreme high vacuum conditions, another non-trivial challenge is presented by the ceramic insulator, which electrically isolates the electrode from ground. The insulator must be vacuum compatible and able to stand-off the voltage. Extreme care is taken in polishing and cleaning the electrodes to minimize field emission, which can puncture the insulator by charge build-up, as well as severely degrade vacuum conditions in the gun chamber. Surface preparation of the electrode structure is not sufficient to eliminate field emission. Once the gun is assembled in a clean environment and the electrodes are under ultra high vacuum conditions, the voltage needs to be increased slowly and systematically to prevent electrical breakdown. This is a time consuming but critical process before the gun can be operated at the desired voltage. In the process, field emitters are encountered and processed out. At higher voltages, gas desorption takes place with every kV increase, contributing to the complexity of the process. Once the field emission is extinguished, and the gas desorption no longer present, the gun is fully conditioned. Typically, the gun is conditioned to higher voltage than desired for nominal operations.

In the early 1990s, the DC photoemission gun technology used by the Continuous Electron Beam Accelerator Facility at Jefferson Lab (JLab) was adopted by the IR-

CP1149, *Spin Physics, 18th International Spin Physics Symposium*
edited by D. G. Crabb, D. B. Day, S. Liuti, X. Zheng, M. Poelker, and Y. Prok

Demo Free Electron Laser (FEL) [1] to produce un-polarized electron beam, with charge per bunch ~100pC, which required voltages in excess of 300kV to mitigate space charge effects [2]. The JLab FEL has successfully conditioned two generations of electron guns to 425kV for operation at 350kV, albeit with a multitude of problems with punctured insulators and damaged electrodes [3,4].

The JLab FEL DC Photoemission Gun

In the JLab FEL gun [3,4], the high voltage cathode electrode is supported by a long tube, also at high voltage, that is coaxial to the cylindrical insulator as shown in Fig. 1. Field emission from the support tube can produce charge build-up in the bulk of the insulator. When the insulator discharges it creates micro-channels that eventually lead to vacuum failure. The ceramic cannot be repaired at that point. It is very expensive and requires highly skilled shops for the bracing of large diameter kovar rings. This is why high voltage processing must be done very carefully. Early in the JLab IR Demo FEL, the inside of the ceramic insulators were platinum implanted to mitigate the charge build-up by adding a conductive path to ground, although at very high resistivity (in the order of tens of Giga-Ohms across the length of the ceramic).

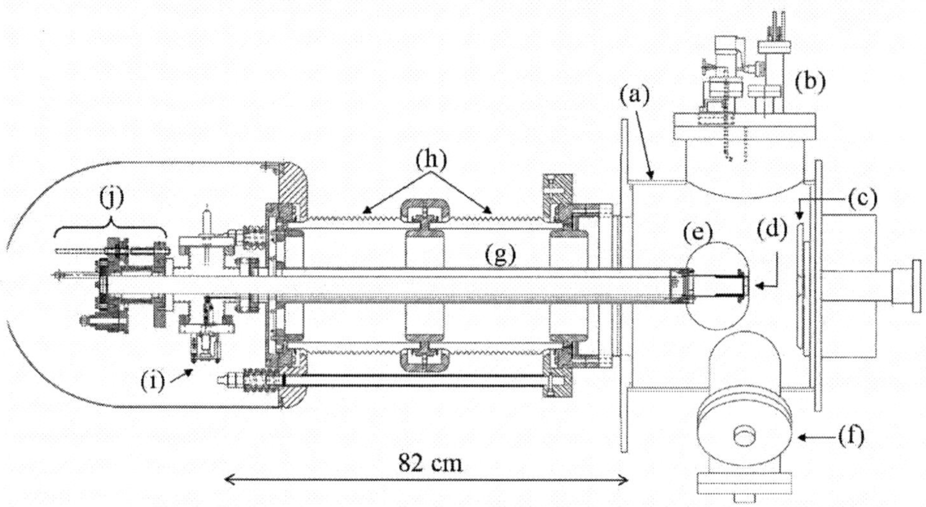

FIGURE 1. Schematic showing the JLab FEL DC photoemission gun. (a) vacuum chamber, (b) instrumentation, (c) anode, (d) 25 mm diameter GaAs photocathode, (e) ball cathode, (f) NEG pumps, (g) ball support tube, (h) ceramics, (i) shield door actuator mechanism, and (j) stalk retracting mechanism.

HIGH VOLTAGE CONDITIONING

High voltage conditioning in DC guns is nominally performed under vacuum conditions. Usually, the field emitters can be processed out by slowly increasing the voltage to control the field emission current. There are commonly three cases of field emission: a) erratic current, b) current increasing rapidly with time, and c) steady current. In case a), the usual procedure is to sustain the voltage for a few minutes until the field emitter 'burns off', causing the field emission current to decrease or extinguish almost completely. For case b), while the voltage is held constant, the current reaches the trip limit turning off the high voltage power supply. In some occasions, enough energy is released to burn off the emitter when the voltage is restored; in other occasions, the field emitter can sharpen (by surface migration) leading to much larger current at the same voltage when originally generated. A sharp emitter is somewhat easy to burn off, by going to lower voltage limiting the field emission current and slowly increasing the voltage as the current decreases. Case c) is probably the most difficult to process. Since the field emission current can be sustained for hours at levels beyond 100 micro-amperes, at hundreds of kV there is enough power to heat the copper gaskets in the large diameter conflat flanges opening leaks, or worse, puncturing the ceramic insulator.

The diagnostics include: total current from the high voltage power supply, the ion pump current to monitor pressure and a residual gas analyzer (RGA) to look for potential leaks. Another diagnostic that proved essential in the recent FEL gun conditioning is radiation monitors. Two ionization chambers were installed next to the gun, one on the stainless steel vacuum chamber (closest to the ball cathode) and one more on the SF_6 aluminum tank to monitor field emission from the support tube electrode, which is coaxial to the ceramic insulator (See Fig. 1).

In the FEL gun, field emitters are typically encountered between 100kV and 250kV. Once the emitters are processed up to that voltage, a new phase in the processing begins; something we have called "voltage-induced gas desorption". With every 1 kV increment, the pressure in the gun vacuum chamber increases from low 1×10^{-10} Torr to mid 1×10^{-8} Torr. The time it takes for the vacuum to recover is voltage dependent: below 300 kV, it takes about 5 to 10 minutes but at 420 kV it can take up to one hour for each 1 kV step.

The RGA reports increased H_2, CH_4, CO and CO_2 partial pressure with each step up in voltage. This gas desorption phase is not fully understood yet, but it has been observed every time the gun has been conditioned, and it has a very sharp onset. For example, if the gun is fully conditioned to 350kV, it can operate for years at that voltage, but if the voltage is increased by 1 kV, then gas desorption is present again.

Despite these complications, the FEL gun has been successfully conditioned in several occasions to 450 kV, taking about 100-man hours in the process. However, during September 2008, high-voltage conditioning of the FEL gun took much longer. The first significant field emitter was encountered at 270 kV. After ~50 hours of processing under good vacuum conditions, the emitter could not be processed out. At that point it was decided to try gas processing.

High Voltage Processing with Helium Gas

When field emitters cannot be processed under the nominal conditioning (i.e. the excess current is above 100 uA at voltages below 150kV and is self-sustained for hours, or if the emitter returns at the same voltage it first appeared after several hours of processing), an inert gas can be used in combination with the voltage.

Helium is commonly used and has shown good results for the Cornell gun [5]. However, there is always a risk to develop a leak especially in the ceramic insulator due to the lack of vacuum diagnostics at the $1x10^{-5}$ Torr level where the inert gas processing takes place. Note that this process is not a DC glow discharge, the pressure is too low for igniting a plasma. Most ion gauges do not operate at that pressure and the ion pumps need to be turned off. The NEG pumps keep pumping though, and do not pump inert gasses. Actually, at the $1x10^{-11}$ Torr range most of the pumping is provided by the NEGs and the ion pumps are mostly used as gauges. In the FEL gun, an RGA capable of operating up to the $1x10^{-4}$ Torr range was used to monitor the overall pressure with its integrated ion gauge, and its mass analyzer was used to monitor argon for potential leaks during the high voltage processing.

In the $1x10^{-5}$ Torr range, helium gas molecules are ionized by the field emission current and are accelerated towards the negatively biased electrode, effectively "back-ion" bombarding the field emitter until the geometry or the work function is altered. In the FEL gun, helium processing worked very effectively in the sense of processing field emitter at voltages below 270kV. The high voltage power supply current was at baseline and the voltage could then be increased up to 375kV. When the helium was pumped out and the voltage was ramped up back to where the field emission first started at 270kV, the gas desorption phase was observed, but at 274 kV the field emission started again and the emitter could not be processed out.

High Voltage Processing with Krypton Gas

Since processing with helium showed some improvement, but could not process the emitter at 270kV, it was proposed to use a heavier gas. Ideally one would start with neon. Argon could work better, but the problem is that the Argon peak is used to monitor for leaks during the processing. The only gas available at the time was krypton, which is about 20 times heavier than helium. For a detailed procedure on krypton processing, see Reference 6.

Starting with $5x10^{-6}$ Torr, krypton processing quickly burned off some emitters at voltages below 250kV. After one hour of processing between 290kV and 315kV, the current was at baseline and very stable, but the radiation was high. At that point a pattern was observed in the current and radiation with every 1 kV increase. The pattern resembled the gas desorption phase observed under nominal vacuum high voltage conditioning, only this time the diagnostic was the signal from the radiation monitors instead of the current from the ion pumps. The radiation tracked the current perfectly, showing a sharp rise and an exponential decay with every 1-kV step. The krypton processing was then successful in eliminating emitters and at the same time the processing had evolved into the gas desorption phase, which continued up to 375kV for tens of hours, but the field emitter at that voltage was not fully processed yet. The krypton was then pumped down and with the gun in nominal vacuum conditions, the voltage was ramped up effortlessly to 364 kV, where the first indication of gas desorption was observed. Krypton processing continued for tens of hours up to 415kV followed by soaking at 365kV under vacuum for several hours, concluding successfully the gun high voltage conditioning.

GUN OPERATIONS AFTER KRYPTON PROCESSING

The GaAs wafer suffered surface damage after ~200 hours of high voltage conditioning. The cesiated photocathode exhibited large amounts of field emission seen on the phosphor screen downstream of the gun. This effect has been observed previously in the IR-Demo FEL, but heat cleaning the wafer for 5 hours proved back then to improve the cathode QE lifetime and to minimize field emission from the cesiated surface [6]. This time, the field emission from the wafer limited the gun operating voltage to 290kV, which proved insufficient to drive the FEL. It was decided to vent the gun and change the wafer, which showed the anodization area almost gone. The non-anodized area showed a myriad of small pits, but a thorough surface analysis needs to be done. No evidence of krypton in the wafer bulk was found with XPS.

Since the gun had been conditioned up to 365kV under vacuum conditions and up to 415kV with krypton, it was expected for the gun to remain conditioned after the wafer change-out and to exhibit mainly gas desorption, as it had been previously observed. However, during the initial conditioning after replacing the wafer, field emission in excess of 100 microamperes was present again at 270kV, as if no processing had occurred at all. It appears as the gun bake cycle at 250C had re-activated the field emitter.

The gun had to be re-conditioned again up to 375kV with krypton. Then under vacuum, the gun remained conditioned up to 360kV, until the wafer was heat cleaned at 550C as part of the photocathode activation process. We did not expect the heat clean, which is made in-situ by retracting the wafer inside the ball electrode, to affect the electrode conditioning, but it did. The field emission went from zero to 100 micro-amperes at 350kV, but this time the field emission was generated at the ball cathode, not the wafer. After a few weeks of operating the gun at 350kV, the field emission current slowly decreased and settled down to 10 microamperes. The gun has been operating with this field emission since November 2008, decreasing significantly the photocathode lifetime. The field emission is inducing some charge build-up in the ceramic insulator that discharges every 30 to 45 minutes. This has no effect on FEL operations, other than a pressure surge in the gun from $5x10^{-10}$ Torr to $1x10^{-9}$ Torr.

CONCLUSIONS

Kyrpton processing was very useful, and provided a means to achieve bias voltage unattainable using conventional HV processing techniques. Helium processing was less effective, in particular it did not contribute to raising the on-set voltage for field emission neither to the gas desorption phase. One speculation is that if the emitter is not processed, most of the available 'energy' is going through the field emission current, limiting the available energy for desorbing gas. Once the emitter was processed with krypton, the gas desorption phase continued as usual. It is not clear if neon or argon would have helped processing the emitter in the same was as krypton did. At the pressures used for gas processing ($1x10^{-5}$ Torr), there is little chance for plasma formation and therefore sputtering on the electrodes surfaces. Besides, the electrodes are biased negatively. In plasma sputtering systems, the pressure is in the $1x10^{-3}$ Torr range and the electrodes are biased positively. Radiation monitors proved to be an essential diagnostics during the gun high voltage conditioning. High voltage processing with krypton gas works.

ACKNOWLEDGMENTS

This work supported by the Office of Naval Research, the Commonwealth of Virginia, and by DOE Contract DE-AC05-060R23177. Many thanks to Michael Kelley for doing the krypton analysis on the wafer, and to CEBAF's polarized source group for quickly making available for us a bottle of krypton gas.

REFERENCES

1. G. R. Neil, et al., *Physical Review Letters*, **84**, 4, 662-665 (2000).
2. D. Engwall, et al., 1997 Particle Accelerator Conference, Vancouver, B. C., Canada, IEEE, Piscataway, NJ, 1998.
3. T. Siggins, et al., *Nucl. Insttrum. Methods Phys Res., Sect. A* **475**, 549 (2001).
4. C. Hernandez-Garcia, et al., Proceedings of the 2005 Particle Accelerator Conference, Knoxville, TN, pp. 3117-3119.
5. B. Dunham (private communication).
6. C. Hernandez-Garcia, JLAB-TN-08-065.

Design and Performance of the Cornell ERL DC Photoemission Gun

K. Smolenski, I. Bazarov, B. Dunham, H. Li, Y. Li, X. Liu, D. Ouzounov, C. Sinclair

CLASSE, Cornell University, Ithaca, NY 14853, U.S.A.

Abstract. Cornell University is planning to build an Energy Recovery Linac (ERL) X-ray facility. For an ERL, it is well known that the x-ray beam brightness for the users is mainly determined by the initial electron beam emittance provided by the injector. To address technical challenges of producing very low emittance beams at high average current as required for an ERL, Cornell University has proposed a prototype injector with 5-15 MeV beam energy, 100 mA maximum average current and 77 pC/bunch. In this article, we describe the design, construction and initial results for a DC photoemission gun now under operation.

Keywords: DC high voltage GaAs photogun, Energy recovery linac.
PACS: 29.25.Bx, 81.65.Ps, 84.70.+p, 07.30.-t, 79.60.Bm

INTRODUCTION

An electron injector for an ERL-based light source has many challenges. To provide the x-ray beam quality that users demand for the future, the injector needs to meet the requirements show in Table 1 [1].

TABLE 1. Injector Requirements (values in ( ) are goals for the prototype system of this paper)

Beam Energy	10-15 (5-15) MeV
Charge per bunch	77 (77) pC
Average Current	100 (100) mA
Bunch Length	2-3 (2-3) ps
Transverse Emittance	0.3 (2) μm
Operating Frequency	1.3 (1.3) GHz

At Cornell University, we have undertaken a program to develop and test such an injector (see Fig. 1) towards realization of this difficult set of requirements. The design was based on detailed simulations using genetic algorithms to find the optimum solutions to the multi-parameter design space [2]. The electron source is a DC photoemission gun using a GaAs cathode, providing a 1.3 GHz bunch train of 77 pC/bunch with a 20-40 ps 'beer can' distribution. This is followed by a short section

CP1149, *Spin Physics, 18th International Spin Physics Symposium*
edited by D. G. Crabb, D. B. Day, S. Liuti, X. Zheng, M. Poelker, and Y. Prok

for emittance compensation solenoids and a normal conducting buncher cavity [3]. The beam is then accelerated through a cryomodule containing five 2-cell niobium superconducting RF (SRF) cavities, each with individual control of phase and gradient. The cavities have two opposing 50 kW input couplers to feed in 100 kW per cavity. The available RF power allows for either 100 mA at 5 MeV or 33 mA at 15 MeV. After the SRF cavities, an extensive suite of diagnostics allows for a complete characterization of the transverse and longitudinal phase space of the beam. The beam is terminated in an aluminum dump with a capacity for disposing 600 kW of average power. The current status of the complete injector commissioning and beam performance has been reported elsewhere. [4]

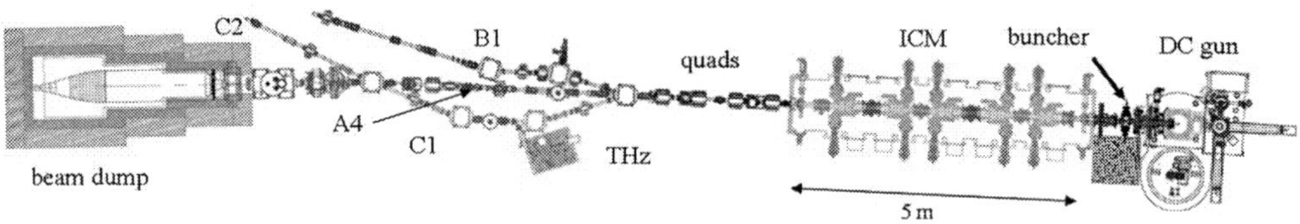

FIGURE 1. The layout of the Cornell prototype ERL injector.

DC PHOTOEMISSION GUN

Based on the experience of other labs [5] and of the authors, DC photoemission guns provide the best chance of producing the low emittance, high average power beam to meet the needs of an ERL. The present record for average current belongs to the Boeing normal conducting RF (NCRF) gun [6] at 32 mA (25% duty factor), and while other projects continue to push for higher current with NCRF guns, no improvements have been realized. Work on SRF guns has made excellent progress recently [7], but the prospect of obtaining 100 mA average current is still many years away. The DC gun used for the Jefferson Lab FEL project [5] has reliably provided 135 pC bunches at an average current of ~9 mA, and an extension of that technology is the most likely path to meet the needs of an ERL injector in the near future.

High Voltage Gun

Common sense dictates that high initial beam energy and high electric field at the cathode are necessary to overcome the space charge forces in bunched beams and obtain the best possible emittance. Simulations [2] show that higher gun voltage is important for obtaining low emittance up to a certain point, after which the improvement is relatively small. Thus, a DC gun operating in the range of 500-600 kV should meet the emittance goals with the appropriate cathode. To minimize dark current at the operating voltage, the gun must be processed to roughly 25% above the operating value; consequently the Cornell gun has been designed to withstand a 750 kV maximum voltage.

A schematic cutaway of the Cornell DC photoemission gun is shown in Fig. 2, and was designed to meet the requirements in Table 2. The gun was operated for over a year using a test beamline to measure the performance of the gun and cathode before mating it to the rest of the injector. Details of these measurements have been published elsewhere [8, 9], with the main result that the emittance measurements at 77 pC/bunch and 250 kV beam energy match the simulations very closely, giving confidence that the simulations for the entire system are valid. Two difficulties observed involved the laser profile and laser stability. Tails in the phase space distribution were traced (through simulation) to non-uniformities and laser stability across the 'flat-top' laser distribution and variations in the distribution over time. The pointing stability of the laser at the cathode was also an issue, causing jumps in the phase space measurements. The laser itself is stable to 20-30 μm rms, but this increases upon demagnification to the cathode. We have since purchased an active position stabilization device (MRC Systems GmbH) which will reduce the position jitter at the cathode to < 10 μm rms.

TABLE 2. DC Gun and Laser Requirements

Operating voltage	500-600 kV
Maximum voltage	750 kV
Average Current	100 mA
Vacuum during operation	$< 1\times10^{-11}$ Torr
External SF_6 pressure	4 atm
Laser wavelength	520 nm
Laser pulse shape	'beer can'
Laser pulse length	20-40 ps
Phase jitter	< 1 ps

All guns of this type suffer from the problem of controlling field emitted electrons from the high voltage surfaces. These electrons can land on the insulator, and if the charge builds up punch-through can occur, causing a vacuum leak. We purchased an insulator with an internal resistive coating (CPI, Inc.) to bleed off these electrons, but it has only been successful up to 450 kV during processing, above which punch-through occurs. In addition, the coating has not adhered well, leaving a layer of dust on the electrodes, certain demise for reaching 750 kV. Two such insulators have been built and tested with similar damage occurring to both. Correlation between the location of edges of the segmented electrode support stalk and the vacuum breaches caused by punch-through have led to the adoption of a single piece stalk to replace the one shown in fig. 2.

Colleagues at Daresbury Lab [10] have built an insulator using a new material from Morgan Advanced Ceramics (AL-970CD) which is more resistant to field emitted electrons, at least up to 500 kV. We are in the process of obtaining a new insulator using this material, and are also investigating the use of segmented insulators [11] which completely block the line of sight between the electrodes and the insulator, but have a much more complicated mechanical structure. For now, we are limiting the gun voltage to 300 kV to reduce the chance of damage, until a spare is obtained.

If one can make electrodes that do not field emit, the insulator design is not so important – this is the Holy Grail for gun design (and for SRF cavities). We have built the present gun emulating the best techniques known for cavity cleaning, namely electro-polishing followed by high pressure rinsing. In a test chamber, gradients of 30 MV/m have been routinely reached on large area electrodes, but these results have not been reproduced in the real gun, most likely due to dust contamination from the insulator coating.

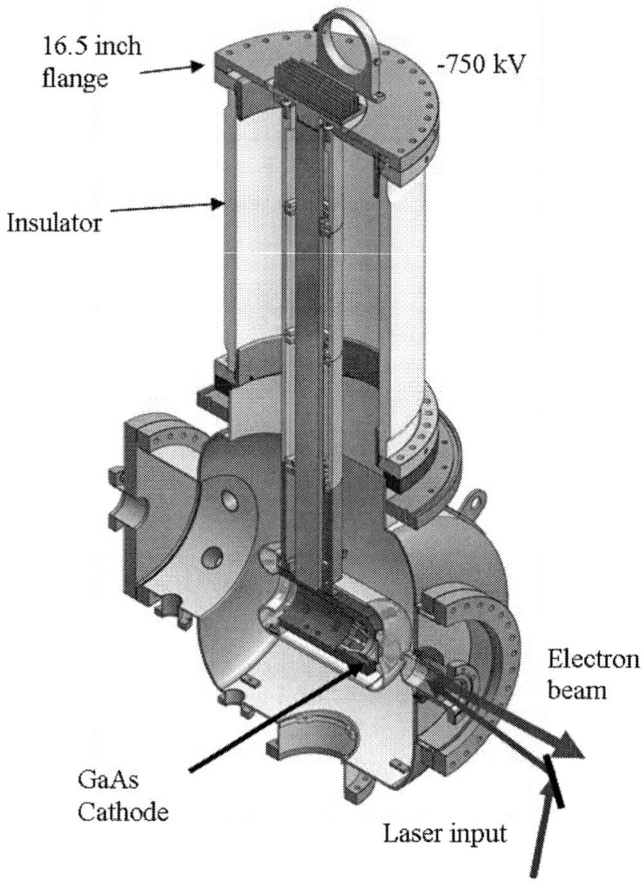

FIGURE 2. The Cornell Photoemission Gun.

Laser System

To produce 100 mA from a GaAs cathode with a 1% QE, over 20 Watts of laser power at 520 nm is required at the cathode (2 W for 10% QE). Accounting for possible losses in the laser transport and beam shaping, a laser with 40 W of average power is desired. The system must provide pulses at a repetition rate of 1.3 GHz synchronized to the RF master clock (with a timing jitter < 1 ps), and a configurable pulse shape in time and space for minimizing the electron beam emittance.

We have chosen to use a Yb-doped fiber laser system to meet these requirements [12]. Initially, the oscillator was made in-house, but difficulties were encountered in accurately synchronizing it to the RF system. Subsequently, a commercial fiber laser 'clock' was purchased from PriTel Inc. They modified a standard product to work at our pulse repetition rate. The laser is triggered by the RF master clock signal, and the jitter between the output pulse the clock signal is less than 500 fs. The pulses are fed to a single mode fiber amplifier where the pulse energy is boosted to 150 nJ, low enough to prevent any nonlinear distortion. The pulse energy is further increased through amplification in a double clad large mode area fiber amplifier built to work in nearly single mode regime. Currently we have implemented only one such stage and achieved average power of 35 watts at 1040 nm (27 nJ pulse energy). The IR pulses are frequency doubled in a LBO crystal to produce pulses centered at 520 nm and energy of 9 nJ (12 W average power). Additional thermal management is needed to reach higher powers. Laser shaping of the pulses has been described in detail elsewhere in these proceedings [13].

Photocathode Materials

A perfect photocathode for an accelerator electron source would have high efficiency at a convenient laser wavelength, fast response time, long lifetime and a low thermal emittance. Unfortunately, no such cathode exists today, although the search continues. A number of different photocathodes meet some of these criteria, so tradeoffs have to be made depending on the requirements of the particular system. For an ERL, obtaining 100 mA average current means high quantum efficiency (QE) photocathodes are a necessity.

Semiconductor photocathodes are currently the best choice for high QE and low emittance. Examples are GaAs, Cs_2Te, GaN, and K_2CsSb. Both Cs_2Te and GaN show promise but require UV light but there are no laser systems available to produce enough average power in the UV. Both GaAs and K_2CsSb have 5-10% QE for ~520 nm light, where high average power lasers are readily available. At Cornell, we have chosen to use GaAs, but are still considering other cathodes depending on the application.

As mentioned earlier, it is possible to recover the intrinsic thermal emittance from the cathode using a carefully designed emittance compensation scheme for the bunch charges of interest to an ERL [2]. Of all the cathodes available, GaAs has the lowest thermal emittance [7], so is the cathode of choice for an ERL. Unfortunately, the QE is near a minimum when the thermal emittance is smallest (close to the band gap), thus unsuitable for high average current. In addition, it is well known that at ~780 nm a fast laser pulse will generate an electron beam with a 20-40 ps tail, which is not acceptable for low emittance operation. For shorter wavelengths (~520nm) recent measurements [10] show that the response time is quite fast (~1 ps) as long as the QE is not too high (< 10%). This points out some of the tradeoffs one has to make even when using GaAs, and we have chosen an operating wavelength of 520 nm as the best compromise between QE, thermal emittance and response time.

The last important parameter is cathode lifetime. The cathodes described above are all sensitive to chemical poisoning to some extent, and require ultra-high vacuum. GaAs is the most sensitive, unfortunately, requiring vacuum levels < 10^{-11} Torr for successful operation. An additional lifetime limiter is ion back-bombardment which further reduces QE. The electron beam can ionize residual gas molecules anywhere along their path, which can then be accelerated back towards the cathode surface. Jefferson Lab [14] has carried out extensive measurements at 10 mA average current, and has measured cathode lifetimes as high as 10^6 C/cm^2 (the amount of charge extracted per cm^2 when the QE has fallen by 1/e). One can use this data to estimate that a 10 W maximum power laser system should be able to provide 100 mA over 100 hours with a 1.8 mm diameter laser spot [15]. Such performance has not been demonstrated yet, and it is certainly an optimistic estimate.

Vacuum System

In order to meet the extreme vacuum levels required for sufficient cathode lifetime both reductions in outgassing and massive pumping have been implemented. The stainless steel vacuum vessel and all stainless internal components including the electrodes have been air-fired for 100 hours at 400°C to reduce their hydrogen outgassing rate, the primary residual gas constituent [16]. A typical 160°C vacuum bake follows assembly of the gun. The pumping scheme consists of a 400 l/sec Perkin-Elmer ion pump and 20 NEG modules lining the walls of the gun vessel. The NEG modules are of the WP1650-ST 707 type supplied by SAES Getters and provide approximately 15,000 l/sec of pumping for hydrogen. Typical static pressure in the gun is below 6×10^{-12} mbar.

Small vacuum leaks have opened at the Conflat flange joints between the insulator and the gun vessel during processing, perhaps caused by heating of the gasket by field emission. New triple point protection rings that cover the vacuum joint were installed and no further leaks have occurred.

Cathode Preparation System

A modular cathode preparation system is attached to rear of the gun vessel to load, clean, prepare, and store the semiconductor cathodes. Vacuum isolation between each chamber (load lock, hydrogen cleaning and heating, cathode surface preparation) allows for multiple cathodes to be accommodated at once. For example a backup cathode could be heat cleaned at the same time the primary cathode is being cesiated. Its design allows for flexibility in the cathode material and is extensible for future cathode material development. 1m long orthogonal manipulator arms transfer the cathode pucks between processing chambers and also move the prepared cathode into the electrode structure.

COMMISSIONING RESULTS AND PLANS

Processing of the electrode structure at a rate of 1kV/Hr. has been successful only to 450KV where punch through occurred (as described earlier). The gun typically operates at 250 kV and we are not pushing the voltage higher until a backup insulator is acquired. Now in operation for two years, the gun routinely provides pulsed beams of varying bunch charge for diagnostic measurements. Average current as high as 20mA at 250kV has been achieved for short periods. The near term plans include extensive sets of phase space measurements along with the push for higher currents.

ACKNOWLEDGEMENTS

This work was supported by NSF grant PHY 0131508.

REFERENCES

1. S. M. Gruner and M. Tigner, Eds., "Study for a proposed Phase I Energy Recovery Linac (ERL) synchrotron light source at Cornell University", CHESS Technical Memo 02-003, Jefferson Laboratory Report No. JLABACT-01-04 (2001).
2. I.V. Bazarov, *et al.*, "Multivariate optimization of a High Brightness DC Gun Photoinjector", PRST AB, **8**, 034202 (2005).
3. S. Belomestnykh, *et al.*, “Commissioning of the Cornell ERL Injector RF Systems”, 2008 European Particle Acc. Conf., Genoa, Italy, 832-834 (2008).
4. B. Dunham, *et al.*, “First Tests of the Cornell University ERL Injector”, Linac 08 Proceedings, In Press (2008).
5. C. Hernandez-Garcia, *et al.*, “A High Average Current DC GaAs Photocathode Gun for ERLs and FELs”, 2005 Particle Acc. Conf., Knoxville, TN, 3117-3119 (2005).
6. D. Dowell, *et al.*, Nucl. Instr. Meth. A **356**, 167 (1995).
7. A. Arnold, *et al.*, “Development of a Superconducting RF Photoinjector”, Nucl. Instr. Meth., A **577**, 440 (2007).
8. I. Bazarov *et al.*, “Thermal emittance and response time measurements of negative electron affinity photocathodes”, J. Appl. Phys., **103** (2008) 054901.
9. I. Bazarov *et al.*, “Efficient temporal shaping of electron distributions for high brightness photoemission guns”, PRST AB 11 (2008) 040702.
10. R.J. Smith, Daresbury Lab, personal communication.
11. J. Haimson, IEEE Trans. Nucl. Sci. **NS 22**, 1354 (1975.
12. D. Ouzounov, *et al.*, “The Laser System for the ERL Electron Source at Cornell”, 2007 Particle Acc. Conf., Albuquerque, NM, 530-532 (2007).
13. D. Ouzounov, *et al.*, These Proceedings.

Status of the ALICE Energy Recovery Linac

L.B. Jones, S.P. Jamison, Y.M. Saveliev, K.J. Middleman and S.L. Smith

ASTeC – Accelerator Science and Technology Centre, Science and Technology Facilities Council, Daresbury Laboratory, Warrington WA4 4AD, United Kingdom

Abstract. The ERLP energy recovery linac prototype at Daresbury Laboratory has been re-branded, and is now called *ALICE* (**A**ccelerators and **L**asers **I**n **C**ombined **E**xperiments). This paper gives an overview of the project and its status, outlines some of the challenges experienced during the commissioning of the photoinjector and superconducting accelerating systems, and presents our photoinjector gun commissioning results. An outline is given of the planned photon science program for ALICE, and its under-pinning role in the EMMA NS–FFAG[1] project.

Keywords: superconducting, linac, photoinjector, GaAs, electron, FEL, CBS, THz
PACS: 29.20.Ej, 29.25.Bx, 29.27.Eg, 29.27.Fh, 41.75.Ht, 42.72.-g

INTRODUCTION

The ALICE ERL prototype was conceived as a test–bed for the key concepts and technologies expected to feature in the 4th Generation Light Source (4GLS). Despite the cancellation of 4GLS, it now has a broader role as an accelerator physics and technology test facility, and for developing fourth generation light source science. The facility is undergoing commissioning at present, and the past year has been a frustrating combination of progress and setbacks. A short productive period of gun commissioning has been book–ended with several gun vacuum problems. However in parallel to the beam commissioning work, progress has been made with the cryogenic and RF systems.

ALICE ACCELERATOR SYSTEMS OVERVIEW

The photoinjector is a DC photocathode gun operating at 350 kV, based on the Jefferson Laboratory IR-FEL gun [1]. It combines a caesiated gallium arsenide (Cs:GaAs) reflection–mode cathode (activated in–situ) with a laser system specified to deliver 80 pC per electron bunch. The gun uses a one–piece bulk–doped ceramic insulator, which has demonstrated extremely good high voltage performance during conditioning to 450 kV.

The drive laser is a *High Q* IC10000 Picotrain model, delivering >10 W at 1064 nm wavelength. The lasing medium is diode–pumped neodymium yttrium vanadate ($Nd:YVO_4$), mode–locked to generate 7 ps FWHM pulses at a repetition rate of 81.25 MHz, this being the 16th sub–harmonic of the 1.3 GHz RF frequency. The drive laser optical system [2] defines and delivers *macrobunches* to the cathode at a repetition rate up to 20 Hz, the length of which can vary from a single laser pulse up to a 100 μs

[1] Electron Model of Many Applications non–scaling Fixed–Field Alternating Gradient accelerator

CP1149, *Spin Physics, 18th International Spin Physics Symposium*
edited by D. G. Crabb, D. B. Day, S. Liuti, X. Zheng, M. Poelker, and Y. Prok

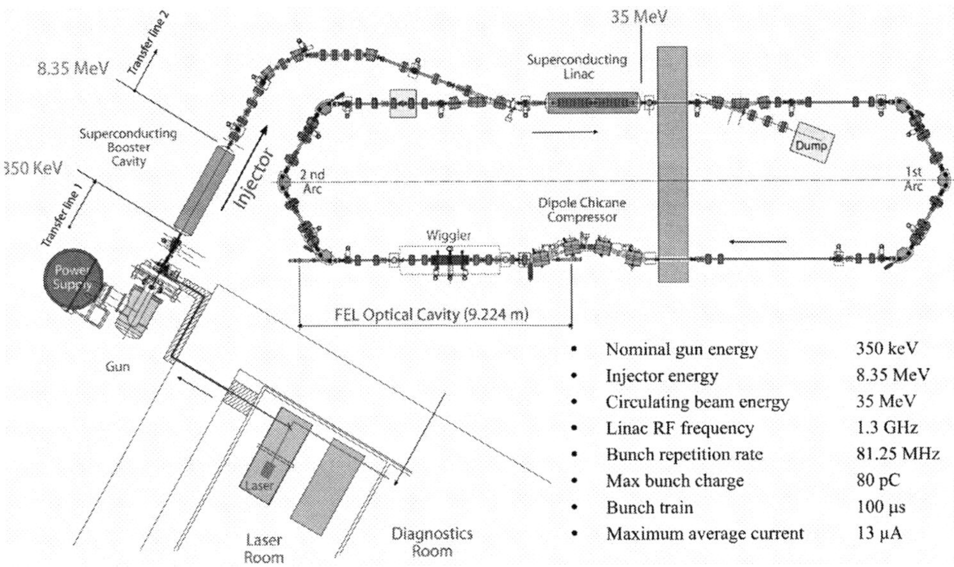

FIGURE 1. Schematic showing the layout of the ALICE machine and its main parameters.

pulse train. This constitutes a maximum duty cycle of 0.2% and represents a maximum average electron beam current of 13 μA.

Cryogenic system and accelerating modules

The cryogenic system [3] was installed during 2006 as a collaborative effort between Daresbury Laboratory, Linde, ACCEL and DeMaco, with both accelerating modules delivered in the first half of that year. Both modules were cooled to 2 K in December of 2006, and were tested with a simulated dynamic heat load of 112 W. Pressure stability of $\pm$ 0.03 mbar at 2 K and $\pm$ 0.10 mbar at 1.8 K has also been demonstrated, and the cryogenic system is now performing beyond its design specification.

ALICE uses two Stanford/Rossendorf cryomodules assembled by ACCEL, each containing 2 $\times$ 9-cell TESLA cavities with a total cryogenic load of $\sim$ 180 W at 2 K. One module is configured as the *booster* and the other as the *main linac*, both using a JLab HOM coupler. The booster has up to 52 kW of RF power to provide a 4 MV/m gradient to accelerate the beam by 8 MeV. Due to energy recovery, the linac has only 16 kW of RF power to provide a 13.5 MV/m gradient accelerating the beam to 35 MeV.

Extreme levels of field emission were observed during RF commissioning at Daresbury, cavity performance being significantly worse than during vertical tests at DESY, as shown in table 1. This necessitated the re–location of sensitive equipment close to the booster linac, and installation of a lead shield around the main linac, with a long–term plan of aggressive cavity conditioning fitted around the beam commissioning program.

TABLE 1. Booster & main linac gradient test results.

	Booster		Linac	
	Cavity 1	Cavity 2	Cavity 1	Cavity 2
	Vertical tests at DESY, July – December 2005			
E_{acc} (MV/m)	18.9	20.8	17.1	20.4
Q_o	5×10^9	5×10^9	5×10^9	5×10^9
	Acceptance tests at Daresbury, May – September 2007			
Max E_{acc} (MV/m)	10.8	13.5	16.4	12.8
Measured Q_o	3.5×10^9 @ 8.2 MV/m	1.3×10^9 @ 11 MV/m	1.9×10^9 @ 14.8 MV/m	7.0×10^9 @ 9.8 MV/m
Limitation	FE Quench	FE Quench	RF Power	FE Quench

COMMISSIONING STATUS

The gun delivered first beam at 01:08 on Wednesday 16th August 2006, has been operated since then for several commissioning periods into a dedicated diagnostics beamline [4]. Problems were experienced with poor quantum efficiency (*Q.E.*) and cathode lifetime, but steady improvements have been achieved through better cathode handling and activation procedures, and improved vacuum. Additionally, beam halo, field emission and high voltage breakdown have been encountered during gun commissioning to–date, giving the ALICE team useful experience in dealing with these challenges. We have also suffered repeated failure of the one–piece ceramic at its flange welds, and are now using a smaller spare loaned to us by Stanford which limits our operating voltage to $\sim$ 250 kV.

Commissioning goals achieved so far include: beam energy of 350 keV; bunch charge of $>$ 80 pC; *Q.E.* of 3.7% with a $\frac{1}{e}$ lifetime of $\sim$100 hours; macrobunch length from a single 7 ps pulse to 100 μs train and train repetition rate up to 20 Hz.

ALICE EXPLOITATION AND UPGRADE PROGRAM

ALICE will drive a mid–IR FEL using a wiggler device loaned from Jefferson Laboratory. The gap will be tuneable (12 – 20 mm), which combined with a variable electron beam energy (24 – 35 MeV), will yield radiation tuneable over a range of $\lambda = 4 - 12\,\mu$m. Additionally, a multi–10 TW laser system will provide beams to generate X–rays through Compton backscattering (CBS), and for a longitudinal electro–optic (EO) electron bunch length diagnostic [5] which will be critical to optimise FEL performance. The CBS source will use head–on and side–on geometries in collisions with the electron beam, yielding $h\nu$ to 30 keV [6]. The bunch compressor will be an *intense* pulsed THz source, and will be used to study exposure in live biological cells [7].

In addition to the light derived directly from ALICE (mid–IR FEL, CBS X–rays & THz), an exciting research programme will also use *combinations* of these with a free–standing tunable femtosecond laser and the multi–10 TW laser, mainly for pump–probe experiments that will use a combination of light sources. ALICE exploitation capabilities will be extended into bioscience after completion of a tissue culture laboratory.

The photoinjector will be up–graded by adding a load–lock chamber and re–designed cathode ball, permitting external activation of photocathodes, so improving its HV performance and reducing cathode–change downtime from several weeks to hours [8]. There are also plans to re–install a gun diagnostic beamline as a permanent feature. The ultimate upgrade comes in the form of EMMA: a prototype non–scaling FFAG electron accelerator which will use ALICE as its injector. It is hoped that NS–FFAG technology will provide a cost–effective means to build hadron therapy centres in the UK [9].

PHOTOINJECTOR GUN COMMISSIONING RESULTS

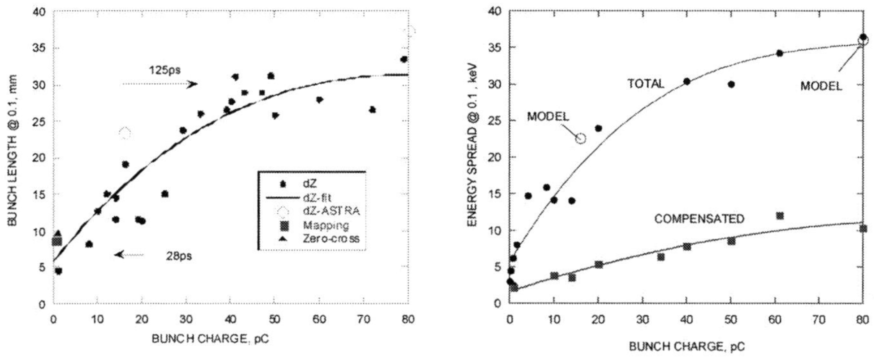

FIGURE 2. **(Left)** Bunch length at 10% of peak **(Right)** Total and tilt–compensated energy spread

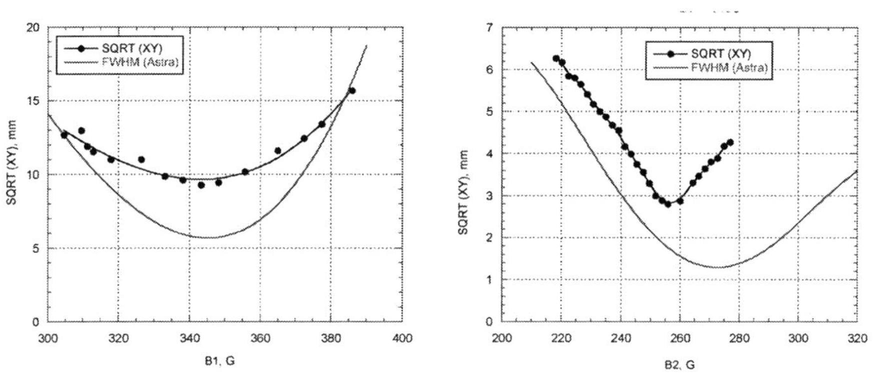

FIGURE 3. **(Left)** Beam size at the 1st solenoid **(Right)** Beam size at the 2nd solenoid

Having overcome the problems described earlier, an improved cathode activation process produced a *QE* of 3.5%, and a bunch charge of > 100 pC was achieved. This allowed injector performance to be measured at high bunch charge, and several highly–productive weeks of measurements were made to characterise the gun until field emission from the cathode again became a problem [10]. This had been managed to that point by setting the solenoid parameters to sub–optimum values. In summary:

- The gun has been repeatedly HV conditioned to 450 kV;

- The beam was fully–characterised operating at 350 keV (emittance, bunch length and energy spectra) in a wide range of bunch charges from 1 to 80 pC [10];
- Despite sub–optimal experimental conditions (e.g., field emission and non–uniform cathode *QE*), good agreement between ASTRA simulations and experimental data was found for the bunch length and energy spread (figure 2) and beam size (figure 3);
- Figure 4 compares some of the bunch characteristics measured using 7 ps and 28 ps drive laser pulses, at a bunch charge of 16 pC. This indicated little difference in bunch quality, though appreciable improvement in emittance is expected with longer laser pulses at higher bunch charges, in the ideal case of a near flat–top laser pulse [11];
- A complete review of the commissioning results is contained in references [10, 11].

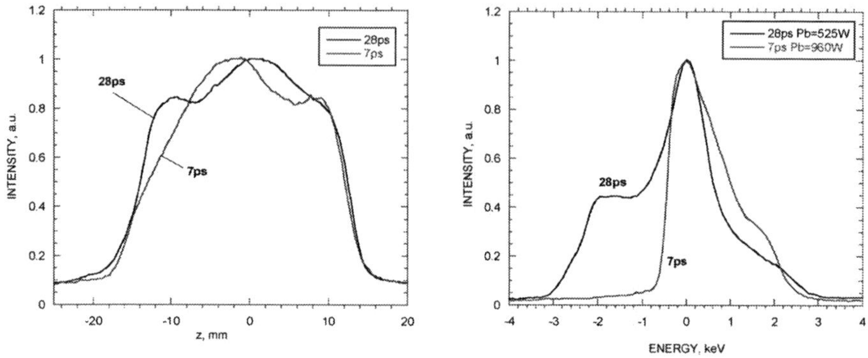

FIGURE 4. **(Left)** Bunch length with 7 and 28 ps laser pulses **(Right)** Compensated energy spread

REFERENCES

1. T. Siggins, C.K. Sinclair, C. Bohn, D. Bullard, D. Douglas, A. Grippo, J. Gubeli, G.A. Krafft & B. Yunn, *Performance of a DC GaAs photocathode gun for the Jefferson lab FEL*, Nuclear Instruments & Methods A; 475 (2001), pages 549–553.
2. L.B. Jones, *Status of the ERLP Drive Laser*, ERL '07 Proceedings, pages 212–213.
3. A.R. Goulden *et. al*, *Installation and Commissioning of the Superconducting RF Linac Cryomodules for the ERLP*, Advances in Cryogenic Engineering, 53B (2007), page 1573.
4. D.J. Holder *et. al*, *ERLP Gun Commissioning Beamline Design*, Proc. LINAC '04, page 93.
5. P.J. Phillips, W.A. Gillespie, S.P. Jamison, A.M. MacLeod, *Electro–optic diagnostics on the daresbury energy recovery linac*, TUPCH041, Proc. EPAC '06
6. G. Priebe, D. Laundy, M.A. MacDonald, G.P. Diakun, S.P. Jamison, L.B. Jones, D.J. Holder, S.L. Smith, P.J. Phillips, B.D. Fell, B. Sheehy, N. Naumova, I.V. Sokolov, S. Ter-Avetisyan, K. Spohr, G.A. Krafft, J.B. Rosenzweig, U. Schramm, F. Gruner, G.J. Hirst, J. Collier, S. Chattopadhyay & E.A. Seddon, *Inverse Compton backscattering source driven by the multi-10 TW laser installed at Daresbury*, Laser and Particle Beams 26 (4), doi:10.1017/S0263034608000700 (2008)
7. M.A. Bowler, M. Surman, F.M. Quinn, S.P. Higgins & A.D. Smith, *Radiation sources on the ERL Prototype at Daresbury Laboratory*, 9th Conf. on SR Inst., AIP Conf. Proc. 879 (2007), page 627.
8. K.J. Middleman, *Load–lock upgrade for the ALICE photoinjector*, these proceedings.
9. R. Edgecock *et. al*, *EMMA – The World's First Non–Scaling FFAG*, Proc. EPAC'08.
10. Y.M. Saveliev, D.J. Holder, B.D. Muratori & S.L. Smith, *Results from ALICE (ERLP) DC photoinjector gun commissioning*, MOPC062, Proc. EPAC '08.
11. Y.M. Saveliev, S.P. Jamison, L.B. Jones & B.D. Muratori, *Characterisation of electron bunches from ALICE DC photoinjector gun at two different laser pulse lengths*, MOPC063, Proc. EPAC '08.

Photocathode Preparation System for the ALICE Photoinjector

Keith J Middleman[a], I. Burrows[a], R. Cash[a], B. Fell[a], L.B. Jones[a], J. McKenzie[a], B. Militsyn[a] and A.S. Terekhov[b]

[a]STFC Daresbury Laboratory, Keckwick Lane, Warrington, Cheshire, UK, WA4 4AD.
[b]Institute of Semiconductor Physics, Novosibirsk, 630090, Russian Federation

Abstract. ALICE - Accelerators and Lasers in Combined Experiments - is a relatively new accelerator built at Daresbury Laboratory that will demonstrate the process of energy recovery by the end of 2008. The project is a research facility to develop the technology required to build a New Light Source (NLS) in the UK. This paper details the current ALICE photoinjector design and highlights the limitations before focusing on a photoinjector upgrade. The key component of the upgrade is a three-stage extreme high vacuum load-lock system that will be incorporated into the ALICE photoinjector in 2010. The load-lock system has *de facto* become a standard component of a type III-V semiconductor photocathode injector and comprises: 1) loading chamber to allow new photocathodes to be introduced, 2) cleaning chamber for atomic hydrogen cleaning of the photocathodes and, 3) a preparation and activation chamber where the photocathodes will be activated to the NEA state ready for use on the ALICE accelerator. Once commissioned the load-lock system will allow rapid transfer of photocathodes between the load-lock system and the ALICE photoinjector whilst maintaining the integrity of the vacuum system and providing many other benefits. The new load-lock system will not only remove the problems with the existing set-up, it will also permit a new vacuum chamber to be designed for the gun itself. This new design will also aim to improve vacuum performance by addressing some of the major vacuum associated problems ALICE has encountered in the past 2 years.

Keywords: photoinjector, load-lock, ALICE, photocathode, vacuum, GaAs, lifetime.
PACS: 29.20.Ej; 29.25.Bx; 81.05.Ea; 85.60.Ha.

INTRODUCTION

The current ALICE photoinjector is based on the Jefferson Laboratory IR-FEL injector design [1] which is a 350 kV DC injector with a single vacuum chamber design as shown in figure 1. The semiconductor GaAs is used as the electron source, although it requires activating to the negative electron affinity (NEA) state to maximise the electron emission (quantum efficiency: QE). The activation of the GaAs photocathode is done by depositing alternate layers of Cs/O_2 or NF_3 all of which has to be done inside the photoinjector. For successful operation the integrity of the vacuum system is critical to cathode performance [2], the vacuum requirements for the ALICE photoinjector were to achieve 5 x 10^{-11} mbar whilst ensuring the residual gas species are not detrimental to the GaAs photocathode.

The single-chamber design has a number of limitations some of which have been problematic during ALICE operations. The main limitation is the fact it is a single vacuum chamber design so only has the ability to accommodate one GaAs

CP1149, *Spin Physics, 18th International Spin Physics Symposium*
edited by D. G. Crabb, D. B. Day, S. Liuti, X. Zheng, M. Poelker, and Y. Prok

photocathode. This limitation means that whenever a new GaAs photocathode is required the photoinjector must be vented to atmospheric pressure with dry nitrogen. Given that the vacuum integrity is critical then a minimum period of one month is required to condition the vacuum system back to its original state prior to venting. This conditioning requires an extensive bakeout to 250°C and this itself can create further vacuum problems. The activation of the GaAs photocathode also impacts on vacuum performance and high voltage, operating Cs dispensers at high temperature whilst injecting O_2 or NF_3 through a fine leak valve meaning the vacuum recovery time can be a number of hours.

Given these basic design problems it was decided that a new multi-chamber design was required that would allow the introduction of photocathodes without compromising the ALICE photoinjector vacuum system. This new design would also allow the activation of the photocathode to be performed external to the ALICE photoinjector.

FIGURE 1. Current ALICE Photoinjector Based On JLab IR-FEL Design

New Multi Chamber Load-Lock Design

Figure 2(a) shows the new load-lock design with 2(b) showing a plan view of how it will be incorporated on to the ALICE photoinjector. It is a 3-stage load-lock design with each chamber having a purpose-designed function. There is a loading chamber where photocathodes will be introduced to the system, a hydrogen-cleaning chamber to remove contaminants from the photocathode surface and a preparation chamber where photocathodes will be activated to the NEA state before being transferred into the ALICE photoinjector. The system is primarily pumped using Sputter Ion Pumps (SIP) and Non Evaporable Getter (NEG) strips, and the vacuum specification was to achieve a minimum of $5e^{-12}$ mbar in the preparation chamber. To help achieve this, the system was vacuum fired at a temperature of 950°C for 5 hours to deplete the 316LN Stainless Steel of hydrogen.

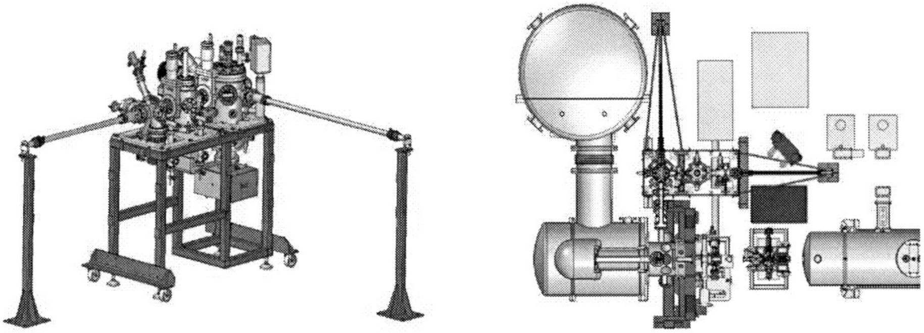

FIGURE 2. (a) New load-lock system, (b) plan view of ALICE photoinjector with new system

Loading Chamber Overview

Figure 3(a) shows a section through of the loading chamber, whilst 3(b) shows the z-translation stage with a magazine holder that can accommodate up to four photocathodes at any one time. To load new photocathodes the z-stage is removed from the loading chamber and transported under dry nitrogen to a nitrogen-purged glove box. The newly prepared photocathodes are inserted into the magazine holder, and the z-stage is closed such that the 'O' ring seals (Fig. 3(b)). The photocathodes are thus stored in a leak tight nitrogen environment. The z-stage is then re-inserted in to the loading chamber, the 'O' ring seal opened and the chamber evacuated to ensure the photocathodes are not exposed to any contaminants from the atmosphere.

(a) (b)

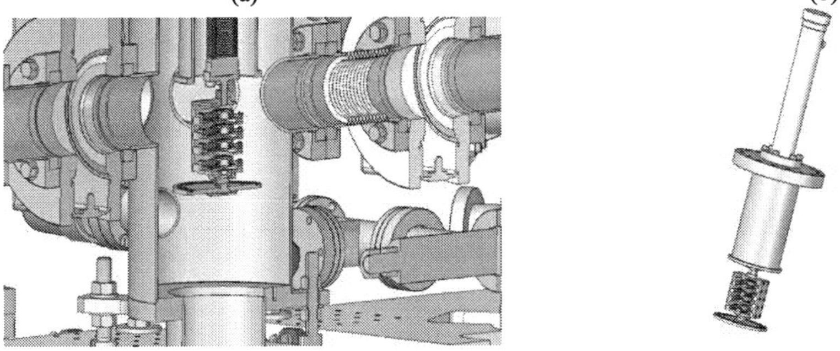

FIGURE 3. (a) Section-through of loading chamber, (b) Z-translation stage with magazine holder

Hydrogen Chamber Overview

Figure 4(a) shows the hydrogen cleaning chamber, whilst 4(b) shows a section through the chamber. Here the cathode is heated to an appropriate temperature ready for hydrogen cleaning (~300ºC). This heating process is done via the use of a 250W halogen bulb, shielded by tantalum to avoid any radiative heating of other components whilst focusing the heat onto the photocathode, thus minimising the power requirements. The hydrogen cleaning process makes use of a thermal gas cracker that uses electron bombardment of a tungsten capillary to thermally dissociate the gas passing through it. Given the right conditions the thermal cracking efficiency of hydrogen is very high and this is important in order to minimise the number of H^+/H^- ions in the cleaning process. Once hydrogen cleaned the photocathode is then transferred through to the preparation chamber ready for activation.

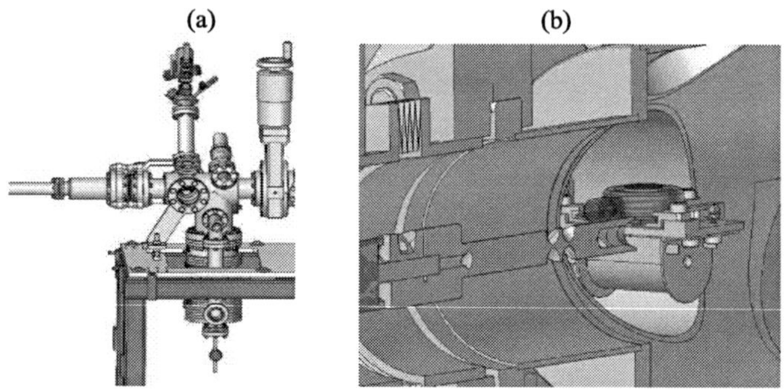

FIGURE 4. (a) Hydrogen cleaning chamber, (b) section-through hydrogen chamber

Preparation Chamber Overview

Figures 5 (a) and (b) show the preparation chamber, (a) showing the complete preparation chamber, with (b) showing a section through of the preparation chamber.

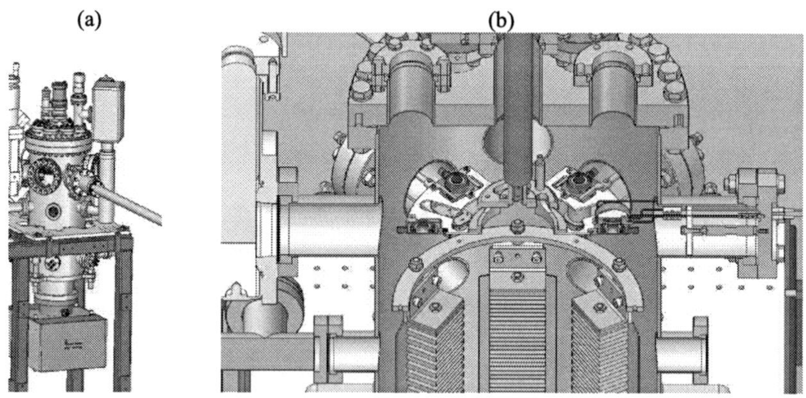

FIGURE 5. (a) Preparation chamber, (b) section-through view of preparation chamber

Once a photocathode has been loaded into the system and hydrogen cleaned, it is ready for a full activation in the preparation chamber using the standard 'yo-yo' method. As shown in Fig.5 the preparation chamber has a carousel with the capability to hold six photocathodes. There are two possible heating positions again using halogen bulb heaters, they are in the 10 and 2 o'clock positions in Fig.5. After heating to ~560-580°C for upto 3 hours [3] and cooling to room temperature the carousel is rotated such that the photocathode to be activated is in the 3 o'clock position in Fig.5. The Cs dispensers are positioned within 10mm of the photocathode surface, as is the charge collector used to measure the photocurrent. The O_2/NF_3 is injected into the system via a piezo-electric fine leak valve which is positioned on the 2.75 inch conflat flange that sits directly above the 3 o'clock position in Fig.5. Once activated to the NEA state the photocathode is ready for transport in to the ALICE photoinjector and is ready for use.

Insertion Of Photocathode Into ALICE Photoinjector

The activated photocathode will be transferred into the ALICE photoinjector via a side loading mechanism into the cathode ball. Figure 6(a) shows the slot in the cathode ball were the photocathode will be inserted [4] whilst Fig.6(b) shows the mechanism inside the cathode ball from the opposite side to Fig.6(a). A separate transfer arm will locate onto the small joystick shown in Fig.6(b) and through rotation of the joystick in a clockwise/anti-clockwise direction the photocathode can be moved forward or backward until it is located in its correct position. Once in position the activated photocathode is ready for use as an electron source.

FIGURE 6. (a) Slot in cathode ball for insertion of photocathode, (b) internal mechanism for movement of photocathode

REFERENCES

1. T. Siggins, C.K. Sinclair, C. Bohn, D. Bullard, D. Douglas, A. Grippo, J. Gubeli, G.A. Krafft & B. Yunn, Performance of a DC GaAs photocathode gun for the Jefferson lab FEL, Nuclear Instruments & Methods A; 475 (2001), pages 549–553.
2. T. Reichelt *et al,* Applied Surface Science. 143 (1999) P319–322
3. N. Chanlek, Study of Activated GaAs Surface for Application as an Electron Source in Particle Accelerators, these proceedings.
4. J. McKenzie, 3D Modeling of the ALICE Photoinjector Upgrade, these proceedings.

Development of an electron gun for an ERL based light source in Japan

N. Nishimori*, R. Nagai*, H. Iijima*, Y. Honda†, T. Muto†, M. Kuriki**, M. Yamamoto‡, S. Okumi‡, T. Nakanishi‡ and R. Hajima*

*Japan Atomic Energy Agency (JAEA), Tokai, Naka, Ibaraki 319-1195, Japan
†High Energy Accelerator Research Organization (KEK), Oho, Tsukuba, Ibaraki 305-0801, Japan
**Hiroshima University, Kagamiyama, Higashihiroshima, Hiroshima 739-8530, Japan
‡Graduate School of Science, Nagoya University, Nagoya 464-8602, Japan

Abstract. We have developed a prototype DC photoemission electron gun for a future energy recovery linac (ERL) based light source in Japan. The prototype gun is operated at up to 250 kV and is designed to deliver up to 50 mA beam current. A diagnostic beam line for emittance and bunch length measurements has been constructed. We have also designed a new 500 kV DC gun which is capable of producing up to 10 mA electron current with emittance lower than 1 mm-mrad. The high voltage terminal will be isolated by ten ceramics stacked in series. The new gun will satisfy performance required as an injector for the compact ERL, which will be constructed at KEK site as a prototype of the future ERL light source.

Keywords: ERL, High brightness electron beam, Photoinjector
PACS: 29.25.Bx, 41.75.Hz, 07.30.Kf, 79.60.-i

INTRODUCTION

An electron gun capable of producing a high brightness electron beam with emittance lower than 1 mm-mrad and current up to 100 mA needs to be developed for next generation Energy Recovery Linac (ERL) light sources (LS). A DC gun equipped with a GaAs photocathode is considered to be the most promising candidate, since high current operation has been successfully demonstrated at Jefferson Laboratory. The DC gun for the JLab 10 kW IR Upgrade FEL provides a high current electron beam of 9.1 mA CW with normalized emittance lower than 8 mm-mrad [1]. The JLab 100 kV load locked electron gun delivers 10 mA CW electron beam with lifetime exceeding 2000 C [2].

A Japanese ERL light source project has been jointly launched by High Energy Accelerator Research Organization (KEK), Japan Atomic Energy Agency (JAEA), the Institute for Solid State Physics of the University of Tokyo and other laboratories [3, 4]. We are planning to construct together the Compact ERL as an ERL test facility at KEK site [5]. The tentative parameters for the Compact ERL are the beam current 100 mA, the injection energy 5 MeV, normalized emittance 1 mm-mrad, beam energy at main linac 60 MeV, and the bunch length 100 fs. We started to develop a photocathode DC gun for the ERL LS [6].

Hajima et al. have proposed to use this gun technology to develop a nondestructive assay system of nuclear waste [7]. The ERL technology will be employed to produce a high-flux Compton gamma ray and identify the nuclide in the nuclear waste using nuclear resonance fluorescence based detection system. Many other different applications

CP1149, *Spin Physics, 18th International Spin Physics Symposium*
edited by D. G. Crabb, D. B. Day, S. Liuti, X. Zheng, M. Poelker, and Y. Prok

will be inspired by the gun development for ERLs.

In the present paper, we describe our current status of development of two photocathode DC guns. One is a 250 kV prototype gun, which will be used for demonstration purposes such as high current operation up to 50 mA, emittance and bunch length measurements. We have generated 1 μA beam from the prototype gun and constructed a diagnostic beam line downstream of the gun. The other is a 500 kV gun, which will be used as the gun for the Compact ERL. We have designed a 500 kV Cockcroft Walton power supply for generation of 10 mA beam, a segmented insulator tube where ten ceramics are stacked in series, and SF6 tank where the power supply and the insulator are placed.

DEVELOPMENT OF 250 KV PROTORYPE GUN

We have developed a 250 kV 50 mA DC gun equipped with an NEA photocahode shown in Fig. 1 [6]. A Cockcroft Walton power supply is used to apply high voltage to an insulator tube. A photocathode sample is introduced into a loading chamber and heat cleaned by an IR heater. The cathode sample is then transferred into the preparation chamber where cesium and oxygen are applied on the cathode. The activated cathode is finally transferred to the main chamber and the photoemitted electron beam is produced upon laser illumination.

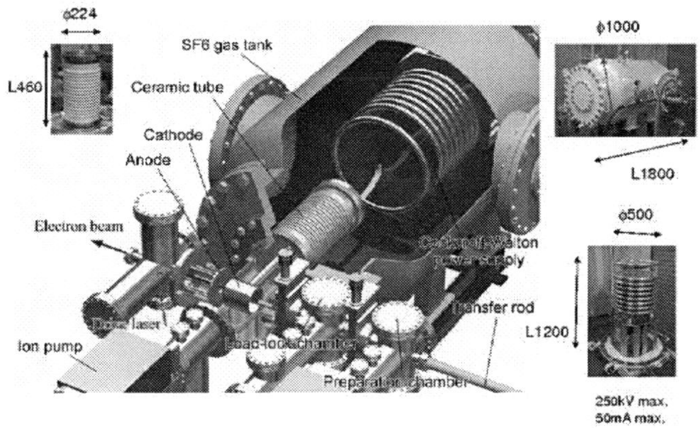

FIGURE 1. JAEA photocathode electron gun.

All the gun chambers are made of titanium which has outgassing rate significantly lower than stainless steel [8]. The flanges are made of a titanium alloy (KS100, Kobe Steel, Ltd.) with hardness of 230 Hv, which is higher than stainless steel (SUS304) of 190 Hv. A 300 l/s turbo pump and a 1700 l/s cryo pump are installed in the loading chamber for rough evacuation, a 500 l/s ion pump and a 2000 l/s NEG pump are used in the preparation chamber, and two pairs of an ion pump and a 2000 l/s NEG pump have been installed upstream and downstream the cathode holder (see Fig. 2). The vacuum pressures at the loading, preparation, and high voltage chambers are 7×10^{-9}

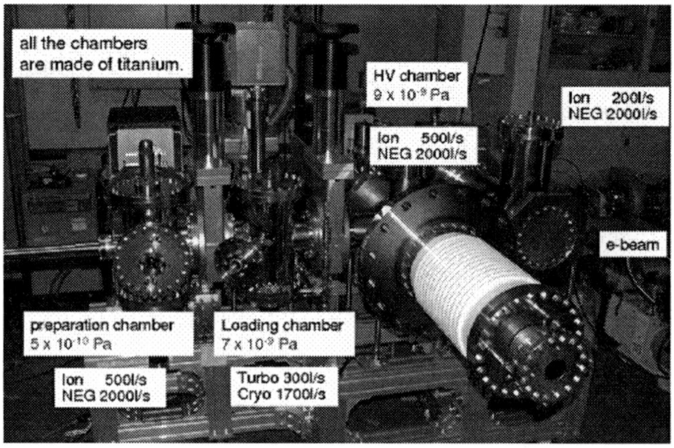

FIGURE 2. Preparation, loading, and HV chambers from left to right.

Pa, 5×10^{-10} Pa, and 9×10^{-9} Pa, respectively. A hot cathode type ionization vacuum gauge (AxTRAN [9]) whose measuring range is from 1×10^{-1} to 5×10^{-11} Pa has been installed at each chamber.

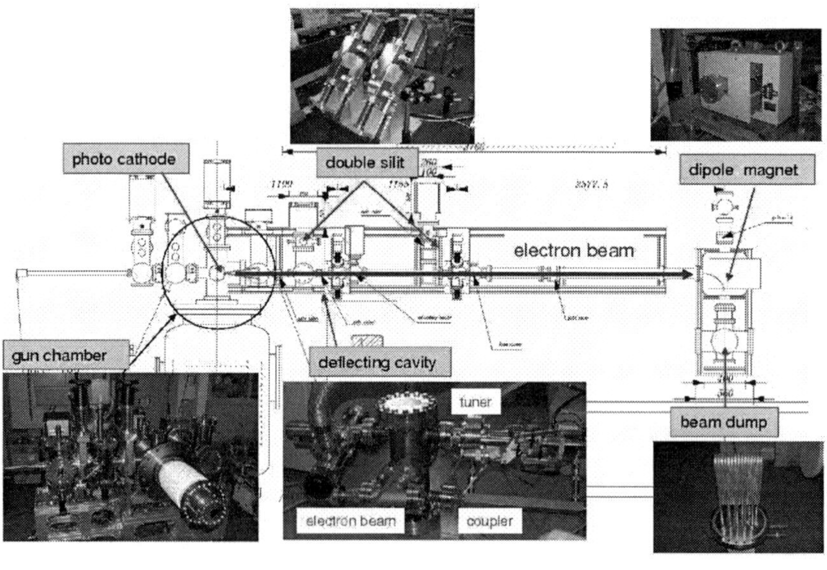

FIGURE 3. A diagnostic beam line for JAEA photocathode electron gun.

The typical quantum efficiency (QE) of GaAs cathode right after Cs activation is 4 % with a laser diode at wavelength of 670 nm. We have generated 1 μA beam at 150 kV using HeNe laser of 2.4 mW. This corresponds to QE of 0.08 %. The small value of QE may be attributed to short life time of the cathode and poor alignment of the cathode center along the downstream beam axis.

We have constructed a diagnostic beam line downstream of the gun chamber consisting of a double silt system for emittance measurement, a deflecting cavity for bunch length measurement, a dipole magnet to deflect the beam to a water cooled beam dump for beam current measurement. The diagnostic beam line is shown in Fig. 3. A 83.3 MHz Ti:sapphire laser with 500 mW average power has been installed for further experiments. The laser pulse length will be varied 5-20 ps (FWHM) by temporal shaping technique with birefringent crystals [10].

DESIGN STUDY OF 500 KV GUN

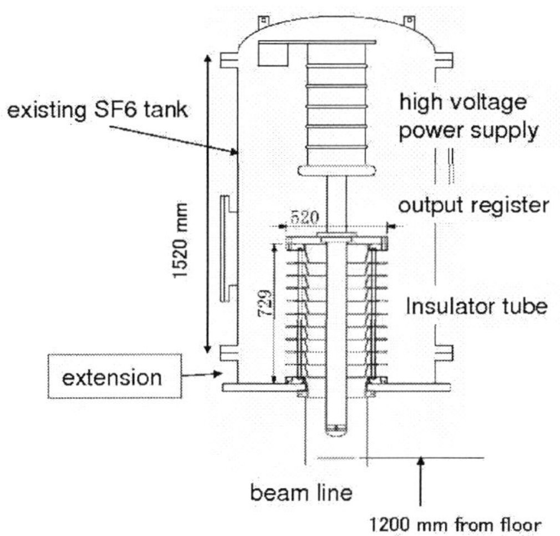

SF6 tank for JAEA
thermionic cathode gun

FIGURE 4. A schematic of SF6 tank for 500 kV gun.

We have started design of 500 kV 10 mA gun for the Compact ERL. A Cockcroft Walton power supply which can deliver beam current up to 10 mA will be constructed by March 2009. The maximum voltage is 550 kV, and the voltage ripple is less than 10^{-4}. The size is 400 mm in diameter and 700 mm in height. The design is similar to the existing high power supply for 250 kV prototype gun.

We designed a segmented insulator tube where ten ceramics are stacked in series. Kovar ring electrode is sandwiched between two ceramics and blazed. The diameters of the ceramics, and flange are 400 mm and 520 mm, respectively. The height of the tube is 730 mm. The insulator tube is similar to that for JAEA thermionic cathode gun used for FEL experiments [11] and that for polarized electron gun at Nagoya university [12].

The diameter of a rod supporting cathode electrode has been determined to be 106 mm for the field on the rod to be minimum. The guard ring attached to each electrode is optimized to suppress secondary electron emission from the ceramics by preventing field emitted electrons from reaching ceramics. The voltage between two consecutive electrodes is limited to 50 kV over 65 mm space. This segmented electrode configuration may help to avoid electric field concentration on some particular point of the ceramics.

The design of SF6 tank is shown in Fig. 4 where the high voltage circuit, output register and insulator tube are placed in a straight line. This configuration will help one get an axially symmetrical field around the tube and power supply. The existing SF6 tank for JAEA thermionic cathode gun shown in the right of Fig. 4 will be reused as the SF6 tank for 500 kV gun for cost reduction, although an extension tank needs to be added. The top height is estimated to be 3.5 to 3.8 m from the floor when the beam line is 1.2 m high.

We also need to design cathode preparation system consisting of loading and NEA preparation chambers. We will fabricate these two chambers by March 2009.

SUMMARY

We have developed a 250 kV photo-emission DC gun and generated 150 keV, 1μA electron beam. We have designed 500 kV segmented insulator tube with optimized guard rings as well as a SF6 tank with symmetrical configuration. We will design and begin construction of cathode preparation system for 500 kV gun soon.

ACKNOWLEDGMENTS

This work was partially supported by Grants-in-Aid for Scientific Research (Grants No. 20360424) from the Ministry of Education, Culture, Sports, Science and Technology of Japan.

REFERENCES

1. C. Hernandez-Garcia *et al.*, "A High Average Current DC GaAs Photocathode Gun for ERLs and FELs", Proc. of PAC2005, 3117 (2005).
2. J. Grames *et al.*, "Measurements of Photocathode Operational Lifetime at Beam Currents up to 10 mA using an Improved DC High Voltage GaAs Photogun", Proc. of the 17th International Spin Physics Symposium and Workshop, Kyoto, Japan, 2–7 October 2006, 1037.
3. T. Kasuga *et al.*, "Status of R&D Efforts toward the ERL-based Future Light Source in Japan", Proc. of PAC2007, 1016 (2007).
4. S. Sakanaka *et al.*, "Progress in R&D Efforts on the Energy Recovery Linac in Japan", Proc. of EPAC2008, (2008) 205.
5. R. Hajima *et al.* (ed.), KEK Report 2007-7/JAEA-Research 2008-032 (2008) [in Japanese].
6. N. Nishimori *et al.*, "Development of an Electron Gun for the ERL Light Source in Japan", Proc. of ERL07 (2007) (to be published).
7. R. Hajima *et al.*, "Proposal of Nondestructive Radionuclide Assay Using a High-Flux Gamma-Ray Source and Nuclear Resonance Fluorescence", J. Nucl. Sci. and Tech. 45, 441 (2008).
8. H. Kurisu *et al.*, J. Vac. Sci Technol. A21, L10 (2003).
9. AxTRAN, ULVAC, Inc., `http://www.ulvac.com/vacuum/vacgauge_axtran.asp`.
10. S. Zhou *et al.*, Applied Optics 46, 8488 (2007).
11. N. Nishimori *et al.*, "A Thermionic Electron Gun System for the JAERI Superconducting FEL", Proc of EPAC2000, 1672 (2000).
12. M. Yamamoto *et al.*, "200 keV Polarized Electron Source for Linear Collider", Proc of LINAC2002, 680 (2002).

KSI's Cross Insulated Core Transformer Technology

Uwe Uhmeyer

Kaiser Systems, Inc, 126 Sohier Road, Beverly, MA 01915

Abstract. Cross Insulated Core Transformer (CCT) technology improves on Insulated Core Transformer (ICT) implementations. ICT systems are widely used in very high voltage, high power, power supply systems. In an ICT transformer ferrite core sections are insulated from their neighboring ferrite cores. Flux leakage is present at each of these insulated gaps. The flux loss is raised to the power of stages in the ICT design causing output voltage efficiency to taper off with increasing stages. KSI's CCT technology utilizes a patented technique to compensate the flux loss at each stage of an ICT system. Design equations to calculate the flux compensation capacitor value are presented. CCT provides corona free operation of the HV stack. KSI's CCT based High Voltage power supply systems offer high efficiency operation, high frequency switching, low stored energy and smaller size over comparable ICT systems.

Keywords: High Voltage Power Supply, HVPS, Insulated Core Transformer, ICT, Cross Core Transformer, CCT, Flux Compensation, corona free
PACS: 29.25.Bx

INTRODUCTON

Dr. James Cross developed Cross Transformer Technology (CCT) for high voltage (HV) generation improves upon existing techniques for HV generation at higher power levels. CCT builds upon Insulated Core Transformer (ICT) techniques adding several significant innovations. Dr. Cross was awarded US patents #5,631,815 and 6,026,004 for his inventions. Kaiser Systems is the exclusive worldwide licensee of the CCT patents. CCT power supplies have been demonstrated from 25 kV to 1000 kV at power levels up to 200kW and are suitable for SF_6, oil or solid insulation systems.

A design issue for conventional HV transformers is the creepage distance in the insulation system between the HV secondaries and the transformer core. Ferrites are conductive (~ 10^6 Ω/cm) and will draw corona. A 500 kV HV transformer with a conventional core would require several inches of creepage between secondary windings and the core in addition to the size of the windings and core itself, resulting in large size, high weight and high cost. Electrically, the drawbacks of large magnetic structures are increased leakage inductance and decreased efficiency.

CCT and ICT have similar basic features. The HV DC output is produced by multiple sections wired in series. Sections are stacked one above the other. Each section has its own secondary winding(s) and rectifiers associated with its own piece of magnetic core material, electrically connected to its rectified output, but insulated

CP1149, *Spin Physics, 18th International Spin Physics Symposium*
edited by D. G. Crabb, D. B. Day, S. Liuti, X. Zheng, M. Poelker, and Y. Prok

from its neighbors. The core-to-winding insulation requirement for each section is never more than the localized Vdc output of that section.

Conventional ICT implementations use 3 phase line frequency excitation. Each 20kV (typical stage voltage) stage contains three series output arc limiting resistors, six rectifiers/capacitors and three magnetic core pieces and secondary windings. Stage sections vary as higher turns ratio are needed on higher stages to keep similar output voltage per stage. The most common regulation technique is a motor driven variable transformer to vary primary voltage and as such, control BW is limited to tens of Hz.

A conceptual diagram of an ICT system is shown in Fig 1. Take note that the secondary windings are in close proximity to their respective core sections. Each section has its own ferrite sections that are insulated from the decks above and below. At the fringes of the insulated gaps in the magnetic circuit, magnetic flux leaks out.

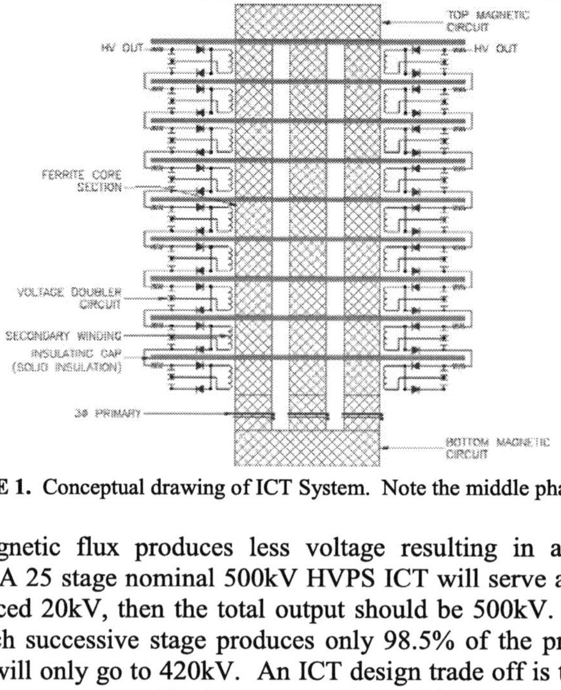

FIGURE 1. Conceptual drawing of ICT System. Note the middle phase circuitry is not shown

Less magnetic flux produces less voltage resulting in a reduction of voltage efficiency. A 25 stage nominal 500kV HVPS ICT will serve as an example. If each stage produced 20kV, then the total output should be 500kV. However, due to flux leakage, each successive stage produces only 98.5% of the previous stage; the total HV output will only go to 420kV. An ICT design trade off is to decrease the number of stages at the expense of higher stage-to-stage voltage. Often the deck transformer turns ratio is increased with each higher stage to keep stage-to-stage voltage the same.

CROSS TRANFORMER TECHNOLOGY (CCT):

CTT has many advantages over ICT. The most significant is a uniform stage-to-stage voltage due to compensation of flux leakage. Additional benefits include extremely low stored energy, fast transient response time, greater efficiency and an inherently corona-free design. This translates into higher reliability.

Figure 2 shows a 500 kV CCT stack. Each of the 40 stack cards are identical, producing up to 12.5 kV. The magnetic circuit is formed by the ferrite bars at top and bottom and joined by the ferrite tiles associated with each section. At the outside of the assembly, grading rings control the electrical field towards the enclosure vessel that houses the stack. The dome at the top of the stack is the output of the power supply. A wire or cable can be attached at that point.

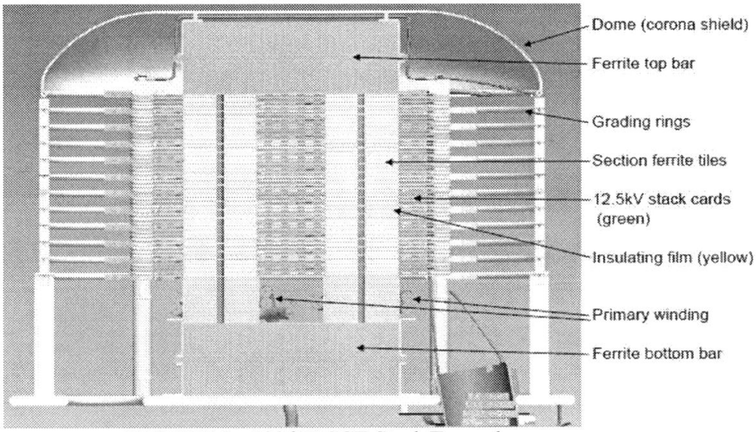

FIGURE 2. 500 kV CCT Stack Example

FIGURE 3. CCT HV Card

Figure 3 shows a CCT HV stack card. The card contains thirty-two identical circuits connected in series. Sixteen circuits at the top are fed from transformer secondaries from the right side ferrite tile, while the other 16 circuits from the left side ferrite tile. With 32 circuits connected in series, the card produces up to 12,500 Vdc.

Each circuit, as shown in Fig. 4, is composed of a fuse, voltage doubler, a flux compensating capacitor and an output arc current limiting resistor and is capable of producing up to 400 Vdc. Using low voltage components in a HV environment, there is minimal voltage stress across the stack board. The components in the circuit see less than 400V. The local electrical field is no more than 1 kV/inch. The corona inception voltage is never exceeded. [1] There is no gradual degradation of insulating materials by corona. Fault tolerance is provided by the fuse which will blow should a stage be damaged. The stage will generally fail in a shorted mode, thereby maintaining its series connection and the output current path. The rest of the circuit elements continue to operate and the power supply functions but with a small loss of voltage efficiency.

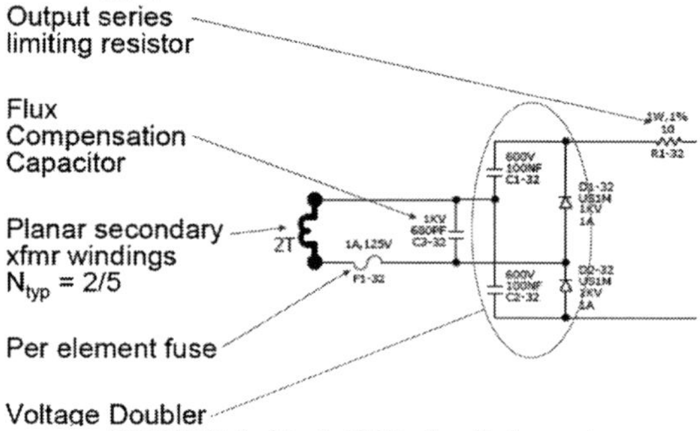

FIGURE 4. Basic CCT circuit element

An important benefit of the topology is very low stored energy. With only 50 nF (0.05uF) of capacitance per stage stored energies of only 0.98J, 4.88J and 7.32J for 100kV, 500kV and 750 kV systems, respectively, are present. These low stored output energies minimize arc currents and their duration should arcs occur.

The construction technique employed in CCT design lowers costs. The stack cards use inexpensive 2-layer Printed Wiring Board technology. Planar transformer design (PCB traces to form transformer windings) removes the need for individually wound secondary windings. Surface Mount Technology (SMT) components are low cost and can be assembled with automated SMT equipment. Lastly, all stack cards in HV stacks are identical to each other. Only one type of spare HV stack card is needed.

FLUX COMPENSATION:

From Patent 6,026,004: “The segmentation of the magnetic core in the transformer introduces gaps in the magnetic structure with permeability essentially that of air. This greatly increases the reluctance of the magnetic structure and produces leakage of magnetic flux. As a result, the upper sections of the magnetic core carry less flux than the lower sections of the core, which results in lower generated voltage per turn on the secondary windings” [2]. Flux compensation restores the lost magnetomotive force (MMF) per gap. The energy associated with the ‘leaking’ fields is associated with the value of the ‘leakage inductance” parameter of transformers. Compensating for the leakage flux, in effect, cancels out the leakage inductance.

$$C = \frac{R}{N^2 \omega^2} \tag{1}$$

$$C = \frac{l}{\mu_0 A N^2 (2\pi f)^2} \tag{2}$$

Dr. Cross ends patent 6,026,004 with Eq. 1. KSI derived Eq. 2 based on Eq. 1 as a more useful form. Examination of the equation reveals that the value of the flux compensation capacitor C is a function of only the transformer physical properties and the operating frequency. It is independent of the output voltage or current. A fixed frequency sinusoidal voltage waveform is required at the primary of the transformer.

SUMMARY

TABLE 1. Selected Characteristics For A Typical 750 kV, 100 mA HV System.

Description	Value
Output Voltage and Operating Range	Meets all the efficiency, stability and regulation specifications over its normal operating voltage range of 100 kV to 750 kV
High Voltage Section Insulation	Pressurized SF6 gas, maximum pressure of 5 atm, absolute (59 psig)
HV Driver	Separate cabinet with control module and inverter module system. Inverters require water cooling
Power Supply Input Voltages	
Inverter supply	480Vac ±10%, 3-phase, 50-60 Hz
Controls and interlocks power	120 Vac ±15%
Efficiency	> 80% overall. Typically 92% at full voltage
Line and Load Regulation	< ±0.5% for a change of ±10% (lower values available on request)
Stability & Ripple	±0.5% total variation for fixed output voltage, current and temperature. (lower values available on request)
Temperature Coefficient	< 200ppm/°C (lower values available on request)
Reproducibility	-up (lower values available on request)
Operating Temperature	15°C to 40°C

Typical specifications for a CCT system are shown above in Table 1. KSI has delivered practical implementations using Phase Shift Modulation (PSM) topology operating at 90 kHz. PSM allows for Zero Voltage Switching (ZVS), which greatly reduces switching losses at mid-level to full power operation.

CCT builds upon ICT but has several advantages. The flux compensation technique minimizes stage-to-stage voltage differences yielding a higher efficiency design. Uniform stage-to-stage voltages allow all deck cards to be identical without compromising the voltage rating of the components on the higher stages. The uniform stage-to-stage voltage ensures inherent scalability of the topology over a very wide range of output voltages. The use of multiple low voltage circuits connected in series minimizes electric field stress, creating a corona free design, providing longer life and higher reliability. The lower cost SMT components and automated assembly make for an easier-to-assemble, and therefore lower cost, implementation. Low stored energy minimizes arc energy and also allows for excellent transient response.

ACKNOWLEDGMENTS

Contributions to this presentation include the work and input from David Johns, Yuri Botnar, Ken Kaiser and Steve Swech, all employees of Kaiser Systems, Inc.

REFERENCES

1. US Patent #5,631,815 "High Voltage Power Supply", Inventor: Cross; James D. (41 Allen Street East, Waterloo, Ontario, CA)
2. US Patent #6,026,004 "Modular high voltage power supply with integral flux leakage compensation", Inventor: Cross; James D. (Waterloo, CA)

Low Emittance Electron Gun for XFEL Application

C.Gough, M.Paraliev and S.Ivkovic

Paul Scherrer Institute, Villigen, Switzerland

Abstract. The Paul Scherrer Institute (PSI) in Switzerland is planning to build a cost-effective X-ray Free Electron Laser (XFEL) facility for 0.1-10 nm wavelength and 10-100 fsec pulse length, requiring only 6 GeV electron energy. The facility will consist of a Low Emittance electron Gun (LEG) with high gradient acceleration and advanced accelerator technology for preserving the emittance during acceleration and bunch compression. To demonstrate feasibility of the project, a 4 MeV test stand followed by a new 250 MeV test stand will be used at PSI. An emittance of <0.1 mm-mrad is desired, and this extreme value has prompted the development of several novel features: gated field emitting array, a pulsed high gradient gun, combined photo-field emission, pulsed solenoid focusing and a two-frequency cavity. The LEG should give stable emission of >200 pC, with >700 keV energy and >125 MV/m gradient.

Keywords: relativistic electron beams, electron sources, electron guns.
PACS: 29.25.Bx, 79.70 +q.

TEST STAND OVERVIEW

Figure 1 shows the present 4 MeV test stand. The initial acceleration of electrons is across a ~4 mm anode-cathode gap. After initial solenoid focusing, the electrons pass through a 2-cell, 1.5 GHz cavity, two further DC solenoids before passing into an emittance monitor. Details of the LEG development are given [1-3].

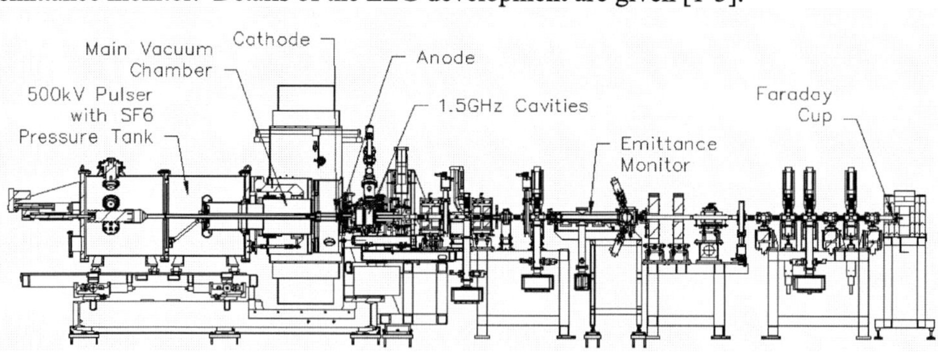

FIGURE 1. Cross-section of 4 MeV test stand.

The pulsed cathode voltage is generated by an optimized Tesla coil [4]. The shortest possible pulse was wanted to minimize the risk of breakdown, but this was balanced against the need for reasonable lifetime and low jitter for a long development program

CP1149, *Spin Physics, 18th International Spin Physics Symposium*
edited by D. G. Crabb, D. B. Day, S. Liuti, X. Zheng, M. Poelker, and Y. Prok

of field emitting arrays and photoemission studies. Finally, thyratron switches with a Tesla coil were found to be our best compromise. There is a practical limit to the pulse width possible from a Tesla coil; in our case, ~17 secondary turns mean over 20 meters of secondary conductor, and due to speed of light constraints, this limits the FWHM pulse length to the range of ~250 ns.

ELECTRODE MEASUREMENTS

Figure 2a shows a rather primitive cathode electrode that was used for earlier tests. It was expected that the unpolished sharp transition from flat to elliptical geometry would give such high field enhancement that breakdown must occur on this circular line. Surprisingly, no breakdowns occurred there, but rather in random locations over the newly machined flat surface. Figure 2b shows one of several surprising features of this surface were found with an electron microscope.

FIGURE 2a. Elliptical copper electrode, 80 mm in diameter, but with the front surface machined flat and not polished.

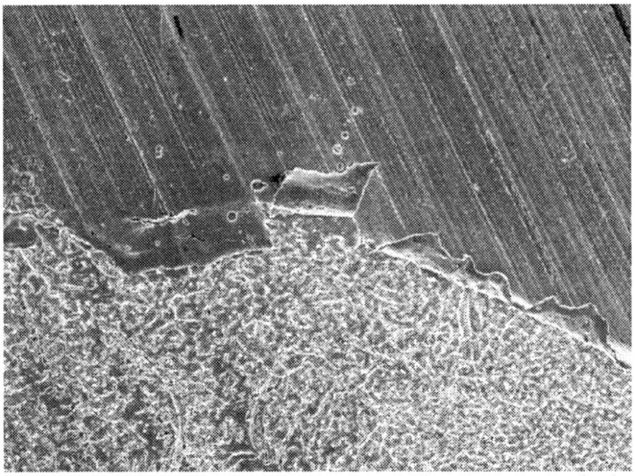

FIGURE 2b. Enlargement of one breakdown site, image width ~100 μm. In the lower part, the copper has been melted. The upper part shows a hardened surface of ~1 μm thickness which has peeled back. Surprisingly, these sharp edges with high β were not sites for later breakdowns.

FIGURE 3. Copper cathode and corresponding copper anode plate, after multiple breakdowns.

Figure 3 shows the results from another rather primitive test, where the anode was flat, but perforated by holes from 1 mm to 6 mm in diameter, and the cathode was a polished copper elliptical electrode. The expectation was that the rather sharp geometry changes around the holes would give preferential breakdown at these sites. Fig.3 shows that most breakdowns (in the 30-40 MV/m range) were not related to the anode holes with right angle edges.

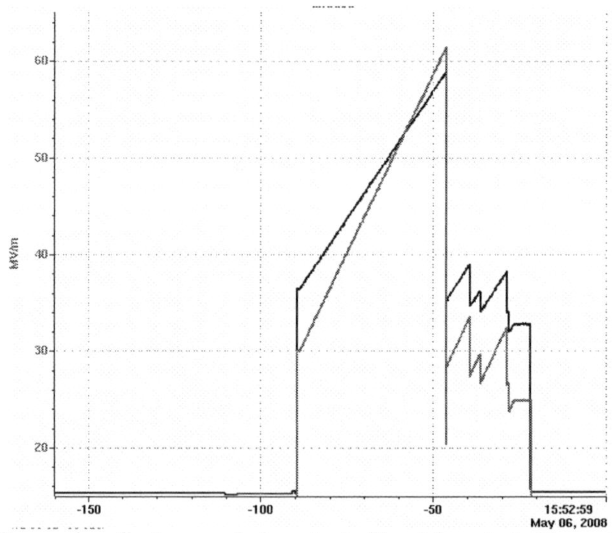

FIGURE 4. Voltage and gradient curves during a typical breakdown test. The gradient started at 30 MV/m and reached 61 MV/m about 45 minutes later before the first breakdown occurred. This event destroys the electrode surface and thereafter the breakdown gradient is around 35 MV/m.

From these results, we have a general knowledge of what is important for controlling breakdown, and that is solely the surface polishing. The following are <u>not</u> dominant in determining breakdown up to 130 MV/m:

- electrode geometry
- material type and purity
- surface scratches
- anode thickness
- argon discharge cleaning
- vacuum pressure
- surface cleanliness
- time in vacuum
- time in service and voltage ramp-up rate

Improving skills in hand polishing gave ever higher breakdown gradients, but also practical difficulties: X-ray emission was completely absent before breakdown so there is no warning of impending breakdown; the first breakdown completely destroys the surface; and to permit uniform polishing, the electrode surfaces should not have a transition from flat to curved geometry on the front surface.

PULSED SOLENOID

A novel feature of the gun is the use of a pulsed focusing solenoid after the initial acceleration, as shown in Fig.5. In this limited space, a DC solenoid would have a limited number of turns, making field asymmetry a problem. To solve this, a pulsed primary solenoid together with a precision stainless steel cylinder as a secondary "winding" was developed.

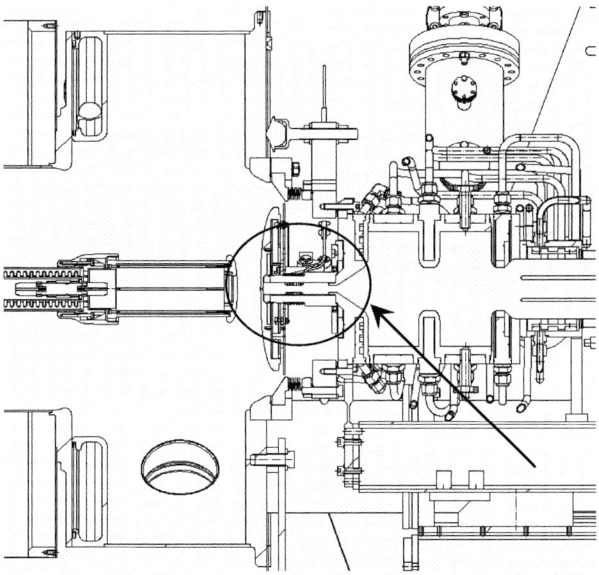

FIGURE 5a. Cross-section of the LEG, showing location of the pulsed solenoid between AK gap and RF cavity.

The primary and the stainless steel cylinder "secondary" are mounted in a precise ceramic flange, with tolerances in the +/-20 μm range. The operation is shown in Fig.5b. A pulsed current is applied to the primary for about 150 μsec; initially, there is a modest circulating current in the secondary which opposes the change of flux. The primary current is then turned off rapidly; the opposing secondary current is then in the kiloamp current range, and decays with a time constant of ~30 μsec. During this time, the magnetic field is from the secondary only, and has near-perfect centering and symmetry. Additional advantages of the pulsed solenoid are the zero leakage field at the cathode and low thermal dissipation.

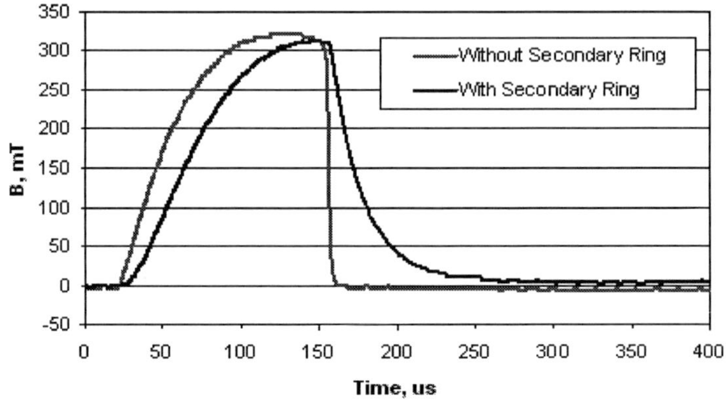

FIGURE 5b. Magnetic field waveforms. Without the secondary, the field is switched rapidly to zero. With the secondary, the field decays with a time constant ~30 μs.

SUMMARY AND PRESENT STATUS

Over 45 electrode pairs have been tested in the LEG pulser and considerable experience was gained with different materials and surface conditions. Another 40+ pairs are scheduled for testing in the first half of 2009.

Breakdown is presently limited by surface polishing. A gradient of 60 MV/m is standard for most materials, and well polished electrodes could reach 130 MV/m. Stainless steel has become our standard material because of low cost and ease of polishing.

Improvement in surface treatment by electron beam melting and surface deposition is being investigated. Gradients of >200 MV/m at 400 kV have been reached but further tests are needed to confirm this result.

REFERENCES

1. R.Ganter et al., *Phys Rev.Lett.* **100**,064801 (2008)
2. F.Le Pimpec et al., FEL 2008 Conference, Gyeongju, Korea (2008).
3. Project overview and publications are available from http://fel.web.psi.ch
4. M.Paraliev et al., IEEE Pulsed Power Conference, Monterey, USA (2005).

Ion bombardment in RF guns

E. Pozdeyev, D. Kayran and V.N. Litvinenko

Brookhaven National Laboratory, Upton, NY 11973-5000

Abstract. We use the ponderomotive potential of the RF field to study the motion of ions in an RF gun. We apply this method to the BNL 1/2-cell SRF photogun and demonstrate that a significant portion of the ions produced in the gun can reach the cathode if no special precautions are taken. Also, we propose simple mitigation techniques that can reduce the rate of ion bombardment.

Keywords: electron source, ion bombardment, cathode, RF gun, ponderomotive potential
PACS: 29.25.Bx,85.60.Ha

INTRODUCTION

Operational experience with DC guns has demonstrated that ion backbombardment is the major cause of degradation of the quantum efficiency of GaAs cathodes. Numerical simulations described in publications [1],[2] revealed that ion backbombardment also can occur in RF guns. In this paper, we use the method of a rapidly oscillating field [3] to analyze the motion of ions in an RF gun. We briefly describe the method and examine the ion motion in the BNL 1/2-cell SRF gun. We would like to note that electron backbombardment and multipacting also can affect the cathode lifetime in an RF gun. However, we do not discuss this phenomenon herein.

MOTION OF IONS IN A RAPIDLY OSCILLATING RF FIELD

The motion of an ion with the mass m and charge q in a rapidly oscillating electromagnetic field can be described as a superposition of a fast oscillating term, $\mathbf{a}$, and a term describing the ion motion averaged over the fast oscillations, $\mathbf{x}(t) = \overline{\mathbf{r}(t)}$: $\mathbf{r} = \mathbf{x} + \mathbf{a}$. If the RF electric field has the following form $\mathbf{E} = \vec{\mathscr{E}}(\mathbf{r})\cos(\omega t + \psi)$ where $\omega t + \psi$ is the RF phase, the fast oscillating term is given by

$$\mathbf{a} = -\frac{\lambda^2}{4\pi^2}\frac{q\vec{\mathscr{E}}\cos(\omega t + \psi)}{mc^2} \tag{1}$$

and the equation for the averaged motion is given by

$$\ddot{\mathbf{x}} = -\frac{\lambda^2 c^2}{16\pi^2}\left(\frac{q}{mc^2}\right)^2 \nabla\vec{\mathscr{E}}^2, \tag{2}$$

where λ is the RF wavelength. The method is applicable provided that the amplitude of the fast oscillations, $|\mathbf{a}|$, is small compared to the characteristic size of the inhomogeneity of the RF field, L: $|\mathbf{a}|/L \ll 1$.

CP1149, *Spin Physics, 18th International Spin Physics Symposium*
edited by D. G. Crabb, D. B. Day, S. Liuti, X. Zheng, M. Poelker, and Y. Prok

Equation (2) shows that the effect of the RF field averaged over the fast oscillations can be described by the effective potential energy

$$U_e = \frac{mc^2}{16\pi^2}\left(\frac{\lambda q\vec{\mathscr{E}}}{mc^2}\right)^2 \tag{3}$$

The potential energy (3) frequently is referred to as the "ponderomotive" potential energy. If an external static electromagnetic field is present, the Lagrange function of the averaged motion is

$$L = T_e - U_e - U + \frac{e}{c}\mathbf{A}\cdot\dot{\mathbf{x}}, \tag{4}$$

where U and $\mathbf{A}$ are the scalar and vector potentials of the external field, respectively, and T_e is the effective kinetic energy $m\dot{\mathbf{x}}/2$.

In treating the ion motion, we assume that the ions are generated only in collisions of the electron beam with the residual gas. The energy transferred in the ionization process is mostly absorbed by knocked out electrons. Therefore, the cross section of collisions with a large energy transfer to an ion is much smaller than the ionization cross section. Thus, we can assume that ions originate at rest and find the initial effective velocity $\dot{\mathbf{x}}_0$:

$$\dot{\mathbf{x}}_0 = -\frac{\lambda c}{2\pi}\frac{q\vec{\mathscr{E}}\sin(\phi_0)}{mc^2}, \tag{5}$$

where ϕ_0 is the ionization RF phase. The associated initial effective kinetic energy is given by:

$$T_{e0} = m\frac{\dot{\mathbf{x}}_0^2}{2} = \frac{mc^2}{8\pi^2}\left(\frac{\lambda q\vec{\mathscr{E}}}{mc^2}\right)^2 \sin^2(\phi_0) = 2U_e\sin^2(\phi_0). \tag{6}$$

BNL 1/2-CELL SRF GUN

Brookhaven National Laboratory and Advanced Energy Systems, Inc. are jointly developing a 1/2-cell superconducting radio-frequency (SRF) photogun [4]. Figure 1 shows the SuperFish [5] model of the gun. The gun will operate with the following nominal parameters: a beam energy of 2 MeV, an accelerating gradient of 29 MeV/m, a fundamental RF frequency of 703.75 MHz, a bunch repetition rate of 9.38 MHz, a charge per bunch up to 5 nC, and an average beam current up to 50 mA.

The motion of ions on the gun axis is one-dimensional because of the axial symmetry of the gun. The electric field on the gun axis was calculated by SuperFish. The program Parmela [6] was used to calculate the RF phase at which electron bunches pass a given coordinate, yielding the ionization phase ϕ_0 as a function of z. In this simulation, the initial beam phase was chosen to minimize the beam emittance. Figure 2 shows the effective potential energy (in red) of H_2^+ ions on the gun axis. Also, Figure 2 shows the total effective energy of H_2^+ ions as a function of the ionization coordinate. The total energy curve is divided into two branches: Ions whose velocity $\dot{\mathbf{x}}_0$ points towards the cathode (in green) and those whose velocity points out of the gun (in blue). The total

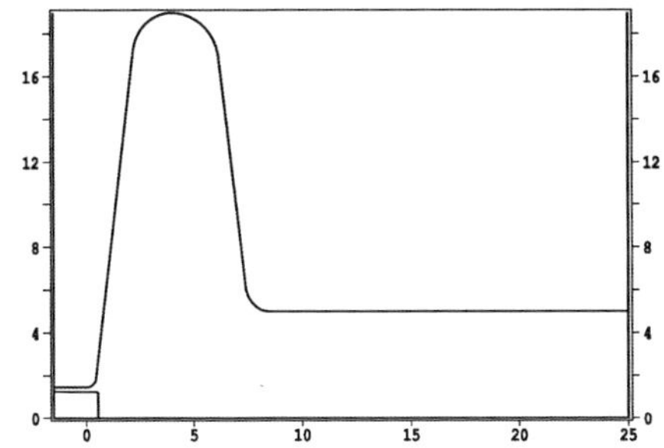

FIGURE 1. SuperFish model of the BNL 1/2-cell SRF gun. The horizontal axis corresponds to the gun axis.

effective energy of ions originating at $z < 3.4$ cm is larger than the effective potential energy at the cathode. Thus, those ions originating at $z < 3.4$ cm will reach the cathode. All other ions produced at $z \geq 3.4$ cm will be expelled from the gun.

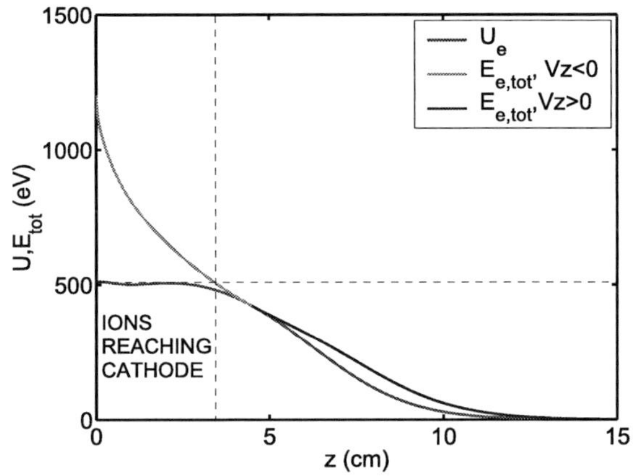

FIGURE 2. The effective potential energy (red) and the total effective energy (green and blue curves) of H_2^+ ions in the BNL 1/2-cell SRF Gun. The green and blue curves, respectively, show the total energy of ions with the effective velocity $\dot{\mathbf{x}}_0$ pointing towards the cathode and away from the cathode.

Off-axis, ions experience a deflecting force. The effective potential energy off the gun axis is given to the second order in r by

$$U_e = \frac{mc^2}{16\pi^2}\left(\frac{\lambda q}{mc^2}\right)^2\left(\mathscr{E}_a^2 - \frac{\mathscr{E}_a}{2}\left(\mathscr{E}_a'' + \frac{\mathscr{E}_a}{\lambda^2}\right)r^2 + \frac{(\mathscr{E}_a')^2}{4}r^2\right), \tag{7}$$

where $\mathscr{E}_a(z)$ is the electric field on the gun axis, and $'$ denotes a derivative with respect to z. Because we assumed that ions originate at rest, the equation for the average ion radius can be written as $m\ddot{r} = -\partial U_e/\partial r$. This equation has to be solved simultaneously with the Lagrange equation for z. However, because we are interested only in ion trajectories with a small deviation from the gun axis, we can assume that the axial motion does not depend on r and use the value of $\dot{x}_z$ on the gun axis.

Generally, the equation for the radial motion has to be solved numerically. Its solution also can be found by iterations. If the trajectory radius changes little, we can limit the solution to the first iteration and solve the problem analytically. Figure 3 shows a deviation of ion trajectories from the initial radius calculated at the cathode and normalized to the initial radius vs. the ionization coordinate. The blue curve shows a result of numerical solution of the equation of the radial motion. The green curve depicts the first iteration obtained analytically. The points show results of particle tracking in a SuperFish field map conducted to validate the results obtained by using the ponderomotive potential. There is a good agreement among all the results for $z < 3$ cm. The first iteration fails to predict the motion of ions correctly for ions originating at $3 < z < 3.4$ cm.

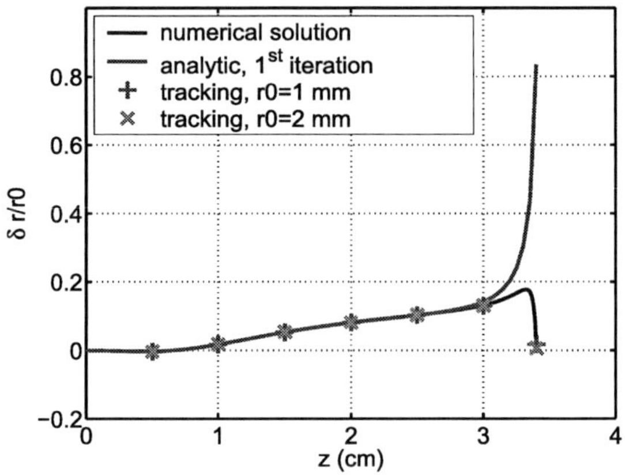

FIGURE 3. Normalized deviation of ion trajectories from the initial radius at the cathode as a function of the ionization coordinate. Tracking results for $r_0 = 1$ mm and $r_0 = 2$ mm practically coincide with each other, confirming the linearity of ion motion close to the gun axis.

The described results were obtained without including the field of the beam. Simple estimates of the beam field that are not presented here show that we can disregard the beam field up to a beam current of a few hundred milliamperes.

Rate of ion bombardment. Comparison to a DC gun.

The number of ions bombarding the cathode normalized to the extracted charge is given by

$$\frac{\mathrm{d}N}{\mathrm{d}Q} = \frac{n_i}{e} \int_0^D \sigma(E(z))\mathrm{d}z, \tag{8}$$

where n_i is the particle density of the residual gas, σ is the ionization cross section, and D is the distance from which ions can reach the cathode. For a residual hydrogen pressure of $5 \cdot 10^{-12}$ Torr and a distance D of 3.4 cm, equation (8) yields $(\mathrm{d}N/\mathrm{d}Q)_{\mathrm{BNL}} = 1.7 \cdot 10^6$ ions/C. We can compare this number to the number of ions produced in a high voltage DC gun. For this example, we choose the following parameters: a beam energy of 650 keV, an accelerating gap of 5 cm, and the residual gas pressure the same as that in the example of the BNL gun, viz., $5 \cdot 10^{-12}$ Torr. With these parameters, equation (8) yields $(\mathrm{d}N/\mathrm{d}Q)_{\mathrm{HVDC}} = 2.4 \cdot 10^6$ ions/C.

This example indicates that rates of cathode ion bombardment can be comparable in RF and DC guns. However, in reality, a cross-comparisson of the cathode lifetime between these two types of guns based on the over-all rate of ion bombardment can be inaccurate. First, the ion energy spectra in the two differ substantially. Second, DC guns are frequently operated with the laser spot shifted from the cathode center. Therefore, a detailed knowledge of the ion spectra and the efficiency of QE damage as a function of the ion energy are required to predict the cathode lifetime with a reasonable accuracy.

MITIGATION TECHNIQUES

The dependence of the initial drift velocity $\dot{\mathbf{x}}_0$ on the accelerating phase can be employed to suppress ion bombardment in a single-cell gun. Assuming the RF field is accelerating in the phase range from $-\pi/2$ to $\pi/2$, the initial drift velocity $\dot{\mathbf{x}}_0$ of all ions will point away from the cathode if the RF phase changes between 0 and $\pi/2$ during the acceleration of electron bunches. Ions with the total effective energy greater than the effective potential energy will exit the gun. Only a small portion of ions originating close to the cathode still will be able to strike the cathode because positively charged ions are accelerated towards the cathode immediately after ionization.

The phasing method described above is not applicable to a multi-cell gun because the accelerating phase cannot be limited to the range between 0 and $\pi/2$. Cathode biasing can be used in this case. Also, cathode biasing can be effective in a single-cell gun. For the BNL gun, a bias voltage of 800 V might suffice to significantly reduce the rate of ion bombardment.

ACKNOWLEDGMENTS

We would like to thank Ilan Ben-Zvi for bringing the subject of ion bombardment in RF guns to our attention. This work was performed under the auspices of the U.S. Department of Energy.

REFERENCES

1. R. Fliller, T. Anderson, H. Edwards, H. Bluem, T. Schultheiss, C. Sinclair, and M. Huening, "Progress on Using NEA Cathodes in an RF Gun," in *Proceedings of PAC 2005, Knoxville, Tennessee, 2005*, IEEE, Piscataway, New Jersey, 2005, p. 2708.
2. J. W. Lewellen, *Phys. Rev. ST Accel. Beams* **5, 020101** (2002).
3. L. D. Landau, and E. M. Lifshitz, *Mechanics (Course of Theoretical Physics)*, Butterworth-Heinemann; 3rd edition, 1982.
4. R. Calaga, I. Ben-Zvi, M. Blaskiewicz, X. Chang, D. Kayran, and V. Litvinenko, *Physica C* **441**, 159 (2006).
5. K. Halbach, and R. Holsinger, *Particle Accelerators* **7**, 21 (1976).
6. L. Young, Parmela, Tech. Rep. LA-UR-96-1835, LANL (1996), documentation by J.H. Billen, rev. 2004.

High-Fidelity RF Gun Simulations with the Parallel 3D Finite Element Particle-In-Cell Code Pic3P

A. Candel, A. Kabel, L. Lee, Z. Li, C. Limborg, C. Ng, G. Schussman and K. Ko

SLAC National Accelerator Laboratory, Menlo Park, CA 94025, USA

Abstract. SLAC's Advanced Computations Department (ACD) has developed the first parallel Finite Element 3D Particle-In-Cell (PIC) code, Pic3P, for simulations of RF guns and other space-charge dominated beam-cavity interactions. Pic3P solves the complete set of Maxwell-Lorentz equations and thus includes space charge, retardation and wakefield effects from first principles. Pic3P uses higher-order Finite Element methods on unstructured conformal meshes. A novel scheme for causal adaptive refinement and dynamic load balancing enable unprecedented simulation accuracy, aiding the design and operation of the next generation of accelerator facilities. Application to the Linac Coherent Light Source (LCLS) RF gun is presented.

Keywords: higher-order finite element time domain (FETD), electromagnetic particle-in-cell (PIC), space charge, electron dynamics, moving window, parallel computing
PACS: 41.20.Jb, 14.60.Cd, 41.75.Fr, 03.65.Pm, 02.60.Cb, 02.70.Dh

INTRODUCTION

A detailed understanding of emission, acceleration and transport phenomena is necessary for advancing the development of high-quality electron guns for next-generation light sources and colliders to be operated at ever more demanding parameter regimes. Joining in this effort, SLAC has developed the first high-performance higher-order Finite Element (FE) 3D Particle-In-Cell (PIC) code Pic3P, for realistic modeling of low energy, space-charge dominated electron dynamics in injectors. Running on large-scale supercomputing facilities, Pic3P provides advanced capabilities that enable RF gun simulations with unprecedented accuracy and speed.

THE PARALLEL CODE PIC3P

In Pic3P, the full set of Maxwell's equations is solved numerically in time domain using parallel higher-order FE methods. Electron macro-particles are pushed self-consistently in space charge, wake- and external drive fields.

CP1149, *Spin Physics, 18th International Spin Physics Symposium*
edited by D. G. Crabb, D. B. Day, S. Liuti, X. Zheng, M. Poelker, and Y. Prok

Finite Element Time-Domain Field Solver

Ampère's and Faraday's laws are combined and integrated over time to yield the inhomogeneous vector wave equation for the time integral of the electric field $\mathbf{E}$:

$$\left(\varepsilon\frac{\partial^2}{\partial t^2}+\sigma_{\text{eff}}\frac{\partial}{\partial t}+\nabla\times\mu^{-1}\nabla\times\right)\int^t \mathbf{E}(\mathbf{x},\tau)\,\mathrm{d}\tau=-\mathbf{J}(\mathbf{x},t), \tag{1}$$

with permittivity ε and permeability μ. The effective conductivity σ_{eff} provides a simple model for damping in lossy materials.

The computational domain is discretized into curved tetrahedral elements and $\int^t \mathbf{E}\,\mathrm{d}\tau$ in Equation (1) is expanded into a set of hierarchical Whitney vector basis functions $\mathbf{N}_i(\mathbf{x})$ up to order p within each element:

$$\int^t \mathbf{E}(\mathbf{x},\tau)\,\mathrm{d}\tau=\sum_{i=1}^{N_p} e_i(t)\cdot\mathbf{N}_i(\mathbf{x}). \tag{2}$$

Substituting Equation (2) into Equation (1), multiplying by a test function and integrating over the computational domain results in a system of linear equations for the coefficients e_i, second-order in time. Numerical integration is performed with the unconditionally stable implicit Newmark-Beta scheme [1]. More detailed information about the employed methods has been published earlier [2].

Higher-Order Particle-Field Coupling

Electron macro particles are specified by position $\mathbf{x}$, momentum $\mathbf{p}$, rest mass m and charge q. The total current density $\mathbf{J}$ in Equation (1) is then approximated as

$$\mathbf{J}(\mathbf{x},t)=\sum_i q_i\cdot\delta(\mathbf{x}-\mathbf{x}_i(t))\cdot\mathbf{v}_i(t), \tag{3}$$

with $\mathbf{v}=\frac{\mathbf{p}}{\gamma m}$, $\gamma^2=1+|\frac{\mathbf{p}}{mc}|^2$. The classical relativistic collision-less Newton-Lorentz equations of motion are integrated using the standard Boris pusher, an explicit method splitting the momentum update into one magnetic and two electric contributions [3].

Maintaining numerical charge conservation during the self-consistent simulation of charged particles and electromagnetic fields is achieved by starting with proper initial conditions and then fulfilling the discrete versions of Equation (1) and the continuity equation

$$\frac{\partial\rho}{\partial t}+\nabla\cdot\mathbf{J}=\mathbf{0} \tag{4}$$

simultaneously during time integration.

The use of higher-order finite elements not only significantly improves field accuracy and dispersive properties [4], but also leads to intrinsic higher-order accurate particle-field coupling equivalent to, but much less laborious than, complicated higher-order interpolation schemes commonly found in finite-difference methods.

Causal Moving Window

The methods used in Pic3P are based on unstructured meshes of up to second-order curved tetrahedral elements – this allows modeling of small geometric features and scattering effects with high accuracy. In combination with higher-order field representation, highly efficient use of computational resources and unprecedented simulation accuracy are obtained.

Furthermore, orders of magnitude in computational resources can be saved, without sacrificing modeling accuracy, by restricting field computations to the causal field region around the particle bunch. This 'moving window' technique can be efficiently implemented using adaptive p-refinement – every element in the overall domain uses an individual basis order $p \geq 0$.

Figure 1a) shows a Minkowski spacetime diagram indicating the causal domain for a bunch transiting through a compact structure. Figure 1b) shows the corresponding parallel dynamic load balancing of the causal field domain during a typical RF gun PIC simulation with Pic3P, where both the field calculation and communication efforts are balanced.

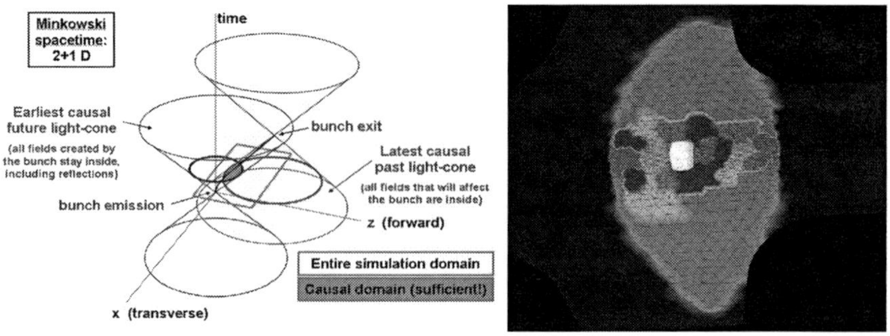

(a) (b)

FIGURE 1. Snapshot in time during bunch transit through compact structure with causal moving window technique. (a) Minkowski spacetime diagram indicating the causal domain.
(b) Pic3P simulation of a quarter model of an RF gun with particle bunch in transit. Elements in non-causal domain are omitted in the field computations ($p = 0$). Elements and fields in the causal domain ($p > 0$) around the particle bunch are partitioned onto all CPUs with dynamic parallel load balancing.

RESULTS

PIC simulations of the 1.6-cell S-band LCLS RF gun are presented [5]. In the simulations, the gun is driven by the π-mode with a peak accelerating field gradient of 120 MV/m at the cathode (the cavity wall). A cold, uniform, 10 ps flat-top, cylindrically symmetric electron bunch of 1 mm radius is emitted, centered around a phase of -58° with respect to the crest. Solenoidal focusing fields are neglected for simplicity. These parameters allow comparisons between the 3D results from simulations with Pic3P and PARMELA and the 2D results from simulations with Pic2P and MAFIA.

For Pic3P simulations, a conformal, unstructured 3D (1/4) mesh model with 305k tetrahedral elements is used, with mesh refinement along the center of the beam pipe. High fidelity cavity mode fields are obtained with the parallel FE frequency domain code Omega3P and directly loaded into Pic3P as drive fields. Figure 2a) shows a comparison of transverse emittance results by the different codes and Figure 2b) shows the parallel scalability of Pic3P.

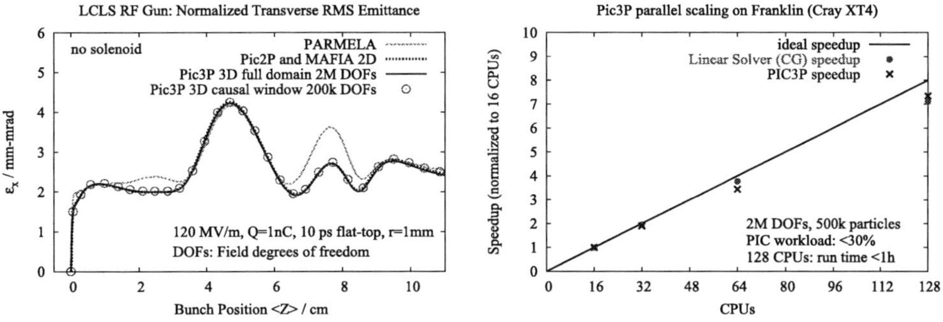

(a) (b)

FIGURE 2. (a) Comparison of normalized transverse RMS emittance as a function of beam position in the LCLS RF gun as calculated with PARMELA, Pic2P and MAFIA 2D (both agree), and Pic3P, where the causal moving window technique reduces the problem size by one order of magnitude.
(b) Parallel scalability of Pic3P for the full domain simulation with 2M DOFs.

Excellent agreement between 3D results from Pic3P and the 2D results from Pic2P and MAFIA is found, as expected from the high cylindrical symmetry in the fields and the convergence behavior of the codes. PARMELA results differ as space-charge effects are significant, presumably because wakefield and retardation effects are ignored, as detailed in a previous study [2].

Pic3P allows large-scale 3D PIC simulations of RF guns with realistic particle distributions, including intra-bunch effects. Starting with an initial particle distribution obtained from measurements, another LCLS RF gun simulation was performed. Figure 3a) shows a snapshot of the scattered fields during bunch transit. Figure 3b) shows the transverse and longitudinal phase phase of the particle bunch near the exit of the gun.

CONCLUSIONS

The first successful implementation of a parallel Finite Element 3D electromagnetic PIC code is presented. The new code Pic3P was used to model space-charge effects in the LCLS RF gun from first principles, including wakefield and retardation effects. It employs state-of-the-art parallel Finite Element methods on conformal, unstructured meshes with unconditionally stable time integration and self-consistent higher-order particle-field coupling. In combination with novel causal moving window techniques and dynamic load balancing, Pic3P allows unprecedented accurate and efficient simulations of low-emittance electron injectors, aiding the design and operation of the next generation of light sources and colliders.

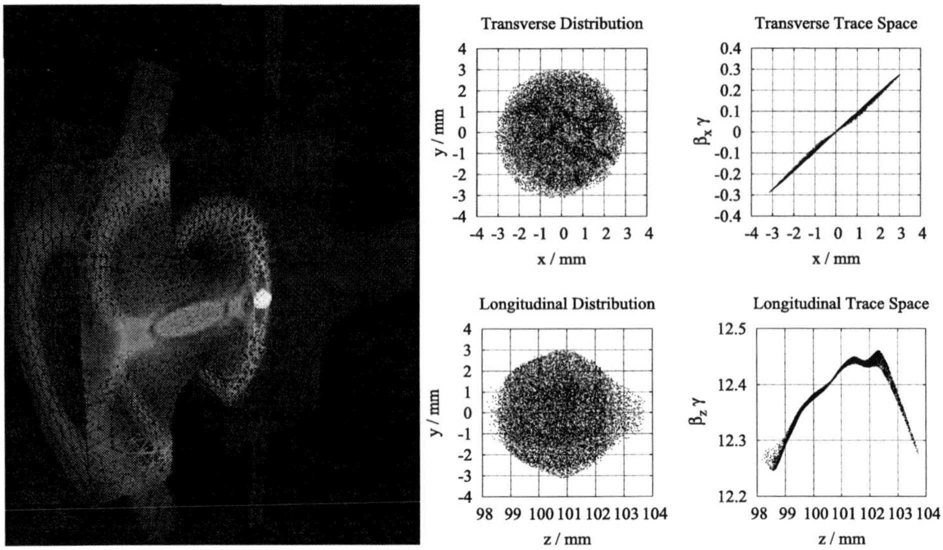

(a) (b)

FIGURE 3. Snapshot during LCLS RF gun simulation with Pic3P, starting with a non-uniform electron bunch distribution obtained from measurements. (a) Scattered electric fields (b) Phase space

ACKNOWLEDGMENTS

This work was supported by the US DOE ASCR, BES, and HEP Divisions under contract No. DE-AC002-76SF00515. This research used resources of the National Energy Research Scientific Computing Center, and of the National Center for Computational Sciences at Oak Ridge National Laboratory, which are supported by the Office of Science of the U. S. Department of Energy under Contract No. DE-AC02-05CH11231 and No. DE-AC05-00OR22725. – We also acknowledge the contributions from our SciDAC collaborators in numerous areas of computational science.

REFERENCES

1. N. M. Newmark, "A method of computation for structural dynamics", Journal of Eng. Mech. Div., ASCE, vol. 85, pp. 67-94, July 1959.
2. A. Candel et al., "Parallel Higher-order Finite Element Method for Accurate Field Computations in Wakefield and PIC Simulations", Proc. ICAP 2006, Chamonix Mont-Blanc, France, Oct. 2-6, 2006.
3. J. P. Boris, "Relativistic plasma simulation-optimization of a hybrid code", Proc. Fourth Conf. Num. Sim. Plasmas, Naval Res. Lab, Wash. D.C.,pp. 3-67, Nov. 2-3, 1970.
4. M. Ainsworth, "Dispersive properties of high-order Nedelec/edge element approximation of the time-harmonic Maxwell equations", Philos. trans.-Royal Soc., Math. phys. eng. sci., vol. 362, no. 1816, pp. 471-492, 2004.
5. L. Xiao et al., "Dual Feed RF Gun Design for the LCLS", Proc. PAC 2005, Knoxville, Tennessee, May 15-20, 2005.

The Superconducting RF Photoinjector at ELBE – First Operational Experience

J. Teichert[a], A. Arnold[a], H. Büttig[a], D. Janssen[a], M. Justus[a], U. Lehnert[a], P. Michel[a], P. Murcek[a], A. Schamlott[a], C. Schneider[a], R. Schurig[a], F. Staufenbiel[a], R. Xiang[a], T. Kamps[b], J. Rudolph[b], M. Schenk[b], G. Klemz[c], and I. Will[c]

[a]*Forschungszentrum Dresden-Rossendorf, P.O. Box510119, 01314 Dresden, Germany.*
[b]*BESSY, Albert-Einstein-Str. 15, 12489 Berlin, Germany.*
[c]*Max-Born-Institut, Max-Born-Str.2A ,12489 Berlin, Germany.*

Abstract. An RF photoinjector with a superconducting cavity (SRF gun) for installation at the Radiation Source ELBE was developed within a collaboration of BESSY, DESY, FZD, and MBI. This new and promising injector type allows CW operation and has the potential for the production of high-brightness electron beams. The gun cryostat, the electron diagnostic beamline, and the driver laser with optical beamline were installed. In November 2007 the first beam was produced. Results of the beam parameter measurements with Cs_2Te photo cathodes are presented.

Keywords: electron source, photo electron injector, superconducting cavity, SRF gun, radio frequency acceleration, cesium telluride photo cathode.
PACS: 07.77.Ka, 29.25.Bx, 29.27.Ac, 29.27.Fh.

INTRODUCTION

One of the main concerns in the development of electron guns for future accelerator based light sources is the combination of high beam brightness and high average current. State-of the-art normal conducting RF photo injectors have demonstrated very high brightness at large bunch charges. The success based on direct generation of short electron bunches by laser pulses and the immediate strong acceleration of the electrons from the surface of the photo cathode in the RF field. By exploiting superconducting technology the RF power losses can be minimized and operation in continuous wave (CW) mode with high average current is possible. In the Research Center Dresden-Rossendorf (FZD) a superconducting RF photo injector (SRF gun) with a 3½-cell niobium cavity has been developed. The SRF gun is installed at the superconducting electron accelerator ELBE, a user facility for the production of electro-magnetic radiation (infrared, x-rays, gamma radiation) and secondary beams (neutrons, positrons).

The SRF gun project at FZD, carried out in collaboration with BESSY, DESY, and MBI, has two main goals: the installation of a high-brightness photo-injector for ELBE, and contributing to the SRF gun development as a promising future technology. Now the new SRF gun has opened an unique opportunity for experimental

CP1149, *Spin Physics, 18th International Spin Physics Symposium*
edited by D. G. Crabb, D. B. Day, S. Liuti, X. Zheng, M. Poelker, and Y. Prok

studies like long term studies of photo cathodes operation and lifetime, cavity quality degradation, or emittance compensation methods.

On November 12, 2007, the first accelerated electron beam was generated which was extracted from a Cu photo cathode. In March 2008 the cathode transfer system was mounted and the first set of Cs_2Te photo cathodes was produced. Since May 2008 the gun has been operated with these cathodes. This paper reports on the operational experience and on beam parameter measurements within the first operation run until October 2008. The RF properties of the SRF gun cavity are presented in a separate paper [1]. An overview of the SRF gun design is published in ref. [2].

PHOTO CATHODES AND DRIVE LASER

During the commissioning and the first measurement period two types of photo cathodes were used: a Cu cathode and later semiconductor cesium telluride (Cs_2Te) cathodes. The Cu cathode was mounted into the cavity during the assembly of the cryomodule. It was used during the commissioning until the cathode transfer system was installed. The quantum efficiency (QE) was about 1×10^{-6}, which was rather low. However, an electron beam could be produced and observed on YAG screens.

After installation of the cathode transfer system, the first set of Cs_2Te photo cathodes for the SRF gun was produced in May 2008 in the preparation lab with QE of 4-5 %. After transfer to the gun the QE dropped down to 0.05 %. The first Cs_2Te cathode was in operation for 44 h. This cathode and all the others of the first set were damaged due to an accidental venting during the summer shut down.

A second set of Cs_2Te photo cathodes was prepared and transported to the SRF gun in July 2008. One was installed in the gun and was in operation for about 350 h until the end of the measurement period. Its QE after preparation was 3 %. Again the QE measurement in the SRF gun after insertion delivers a much lower value of about 0.1 %. Although the vacuum of 10^{-9} mbar in the transportation chamber was now good enough the worse vacuum during cathode exchange in the transfer chamber causes the drop-down. Nevertheless the QE was high enough to produce bunch charges up to 200 pC and to conduct the beam parameter measurements. The QE measurement was regularly repeated without any further significant changes during the whole measurement period.

The cathode cooling, support and transfer system worked properly. The cooling of the cathode with liquid N_2 was efficient. As desired the cathodes can be exchanged without warming-up.

In order to fulfill the requirements of the SRF gun specifications, a UV laser system with two channels has been developed. The two channels have different oscillators, but jointly used amplifier and UV conversion. At present, the 500 kHz channel is used for the experiments. The frequency-quadrupled Nd:YLF system for CW operation consists of a mode-locked oscillator with 26 MHz, a regenerative amplifier, and a two-stage frequency conversion. The Pockels cells in the amplifier allow variable repetition rates up to 500 kHz. The laser has a Gaussian temporal beam shape with a width of 15 ps FWHM. The laser spot is enlarged with a telescope on the laser table. Then the beam is cut with an aperture to obtain a circular flat top profile with 2.7 mm diameter.

BEAM PARAMETER MEASUREMENTS

For the electron beam characterization a diagnostic beam line has been installed [3]. The beam properties were measured using the following parameters: cavity acceleration gradient of 5 MV/m, photo cathode about 2.5 mm retracted, 5 kV DC voltage at the cathode, and the gun was operated in CW.

In a laser phase scan the arrival time of the laser pulse onto the cathode is varied with respect to the RF phase while measuring the bunch charge in a Faraday cup at constant laser pulse energy. In Fig. 1 such a measurement is shown for the whole phase range from -180° to 180°. The RF field is given by $B_z(z) = B_{z0}(z)\sin\varphi_{RF}$ and the z-axis corresponds to the beam direction. Due to its negative charge, electrons are emitted in the phase range -180° to 0°, whereas emission is suppressed between 0° and 180°. This behavior is clearly seen in the measured curve. The curve delivers important information on synchronization, phase jitter, and proper operation of the laser system. But the main purpose of the measurement is the determination of the optimum laser launch phase.

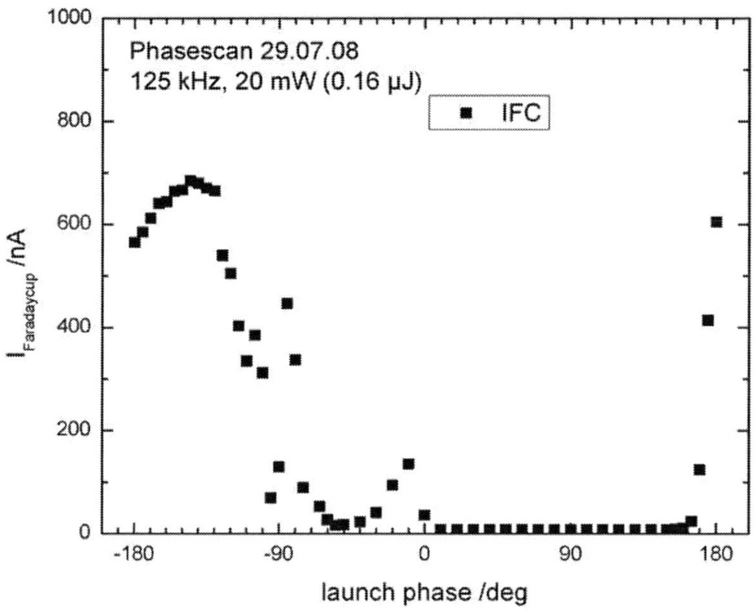

FIGURE 1. Laser phase scan with low bunch charge measured for a full RF period from -180° to 180° (T = 1/f = 769.2 ps). The dark current measured in the phase range between about 0° and 180° is 8 nA.

The phase scan has been carried out for different laser pulse repetition rates from 125 kHz down to 1 kHz with nearly the same laser power of 20 mW for each scan. Whereas the average current is about the same, the bunch charge is increasing with lower laser pulse rate. With increasing laser pulse energy and correspondingly higher electron bunch charge, the space charge effect in front of the cathode becomes more and more important. As visible in Fig. 2, this effect smoothes the curve structures and causes the saturation in the bunch charge. Except these space charge effects, the behavior of the curves is independent of the laser repetition rate which confirms the proper laser synchronization.

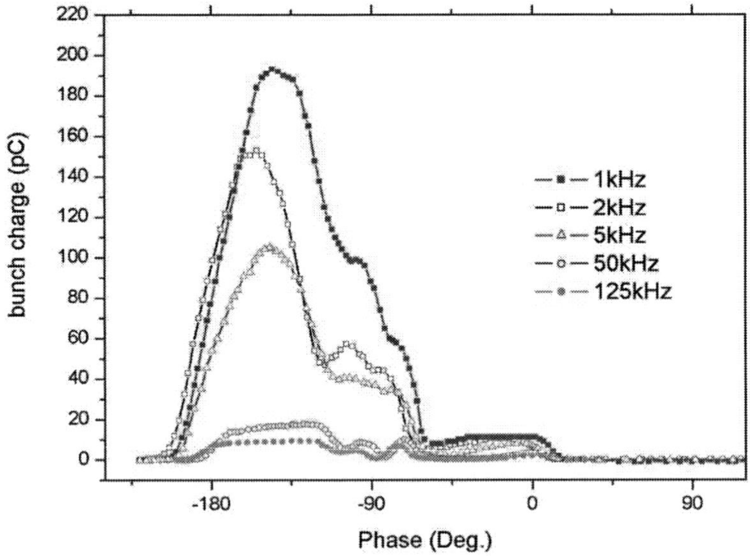

FIGURE 2. Bunch charge versus laser launch phase measured for different laser repetition rates.

The beam energy as a function of launch phase is presented in Fig. 3 with a maximum beam energy of 2.06 MeV. The comparison with the laser phase scan in Fig. 2 shows that the energy drops down at phases > -160° whereas the emission current is still constant until -120°.

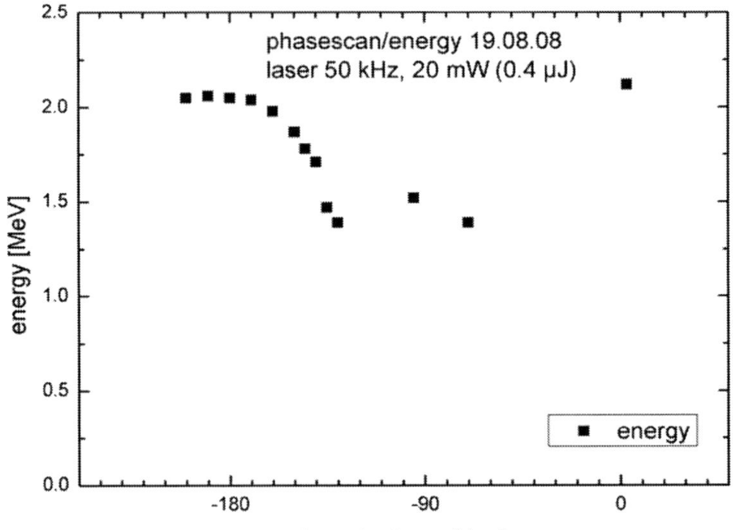

FIGURE 3. Electron beam energy versus launch phase measured by means of the 180° bending magnet in the diagnostic beam line.

The transverse emittance have been measured with the solenoid scan method. For these measurements the gun solenoid and two following screens have been used. Between the solenoid and the screens there is only a drift space of 38 cm and 170 cm, respectively. The magnetic axial field distribution of the solenoid was precisely measured. Thus the focal strength in dependence on the excitation current I can be calculated. The measured beam size $\sigma^2_{x,y}$ on the screen is a parabolic function in I^2 containing the phase space ellipse parameters. From a quadratic fit these parameters and the emittance can be determined. In order to establish the operation phase of the gun, the tranverse emittance was first measured as a function of laser phase. Considering also the other results the laser operational phase was set to -160°. For this laser phase the transverse emittance as a function of bunch charge has been measured with the results in Fig. 4. Measurements were carried out till 70 pC. For higher bunch charges the method could not be applied. The strong space charge effect causes a solenoid current dependence of the beam size which could not be fitted with the theoretical curve. Furthermore the beam shows an increasing halo. To measure the beam size CCD camera images with 8 bit and 660 x 495 pixels were produced, the camera gain was adapted, back ground subtracted and a region of interest (RoI) selected. In the RoI the beam spot size was obtained by a direct rms analysis or by a Gaussian fit to the projected profiles. All theses measurement points are shown in the figure.

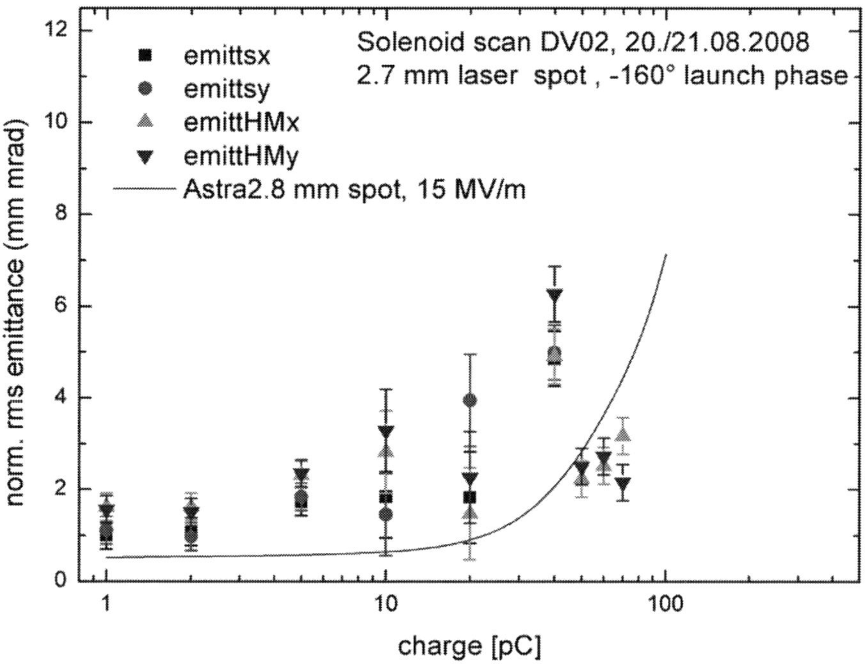

FIGURE 4. Normalized transverse emittance as a function of bunch charge at -160° laser launch phase measured with the solenoid scan method using beam spot on screen 2. Measurement in horizontal (x) and vertical (y) direction, direct rms beam spot size (emittHM), and Gaussian fit (emits).

In Fig. 4 the curve shows an ASTRA simulation result with parameters corresponding to the present experimental situation. The peak value of the RF field was chosen to 15 MV/m, the cathode 2 mm retraced, for the launch phase the optimum value was taken for each bunch charge. In the simulation bunches could be accelerated up to a bunch charge of about 100 pC. For higher bunch charges particle loss occurs due the space charge in front of the cathode. The transverse emittance starts with about 0.5 mm mrad and increases to about 7 mm mrad at 100 pC. This simulation agrees rather well with the experimental data. Especially there is a good agreement in the predicted space charge limit.

For the next runs of the SRF gun in 2009 it is planned to deliver beam to the ELBE accelerator for user operation. Thereby bunch charges above 100 pC are important. It is obvious that the present low acceleration gradient of the gun is the crucial point. As reported in [1] the acceleration gradient could be increased to about 18 MV/m peak field by means of RF high power processing. Furthermore the space charge effect can be reduced by a larger laser spot size. Both changes can improve the electron beam parameters. For a prediction ASTRA simulations have been carried out with 18 MV/m peak field. The cathode position, laser launch phase and laser spot size have been optimized. It turned out that for the higher field and enlarged laser spot of 5.2 mm diameter the space charge limit at the cathode is about 750 pC. The limitation is the transverse emittance now. If 10 mm mrad are acceptable, bunch charge up to about 500 pC can be delivered.

ACKNOWLEDGMENTS

We acknowledge the support of the European Community-Research Infrastructure Activity under the FP6 "Structuring the European Research Area" programme (CARE, contract number RII3-CT-2003-506395) and the support of the German Federal Ministry of Education and Research grant 05 ES4BR1/8.

REFERENCES

1. Arnold, et al., "Rossendrof SRF-Gun Cavity Characteristics," these proceedings.
2. Arnold, et al., "Development of a Superconducting Radio frequency Photo Electron Injector," Nucl. Instr. and Meth. A **577**, 440-454 (2007).
3. T. Kamps, et al., "Electron beam diagnostics for a superconducting radio frequency photoelectron injector," Rev. Sci. Instrum. **79**, 093301 (2008).

Rossendorf SRF-Gun Cavity Characteristics

A. Arnold[a], H. Büttig[a], D. Janssen[a], M. Justus[a], T. Kamps[b], G. Klemz[d], U. Lehnert[a], P. Michel[a], K. Möller[a], P. Murcek[a], J. Rudolph[b], A. Schamlott[a], M. Schenk[b], Ch. Schneider[a] R. Schurig[a], F. Staufenbiel[a], J. Teichert[a], V. Volkov[c], I. Will[d], R. Xiang[a]

[a]*Forschungszentrum Dresden-Rossendorf, P.O. Box 510119, 01314 Dresden, Germany.*
[b]*BESSY, Albert-Einstein-Str. 15, 12489 Berlin, Germany.*
[c]*BINP, 11, akademika Lavrentieva prospect, Novosibirsk, 630090, Russia.*
[d]*Max Born Institute, Max-Born-Str. 2A , 12489 Berlin, Germany.*

Abstract. At the Forschungszentrum Dresden-Rossendorf the development and the setup of the 2nd superconducting radio frequency photo electron injector (SRF-Photo-Gun) is finished. This new injector is placed next to the existing thermionic gun of the superconducting linear accelerator ELBE. A connection between the accelerator and the SRF-Gun will provide improved beam parameters for the users at the second half of 2009. At the moment the commissioning is fully under way. We will report on important results concerning cavity commissioning like measurements of: Q vs. E, microphonics, Lorentz detuning, tuner parameters, pressure sensibility and in-situ fundamental mode field distribution calculated from measured pass band.

Keywords: superconducting cavity, radio frequency acceleration, photo electron injector, SRF-Gun.
PACS: 07.77.Ka, 29.25.Bx, 29.25.Pj, 29.27.Ac.

INTRODUCTION

For future FEL light sources and high energy linear accelerators a high current electron gun with high brilliance is absolutely essential. Thus, an innovative superconducting RF photo injector test bench (SRF-Gun) is in use at the Forschungszentrum Dresden Rossendorf (ELBE) since November 2007. This electron source was developed by a collaboration of BESSY, DESY, MBI and FZD and supported by the European Community.

The use of a superconducting cavity allows CW mode operation with an average beam current of 1 mA and electron energy up to 9.4 MeV. Moreover, the source can deliver short pulses and high-brightness electron beams, as known from the conventional normal conducting RF photo injectors.

Table 1 shows the important cavity design parameters. A detailed description of the cavity and cryomodule is published elsewhere [1]. In this paper we will report on important RF cavity properties measured during the first period of operation. Further results concerning the first beam parameter measurements are presented in [2].

CP1149, *Spin Physics, 18th International Spin Physics Symposium*
edited by D. G. Crabb, D. B. Day, S. Liuti, X. Zheng, M. Poelker, and Y. Prok

TABLE 1. RF cavity parameters calculated with MWS© and normalized to 50MV/m peak axis field.

Parameter	Value
stored energy U	32.5 J
quality factor Q_0	$1x10^{10}$
dissipated power P_C	25.8 W
geometry factor G	241.9 Ω
Acceleration voltage V_{acc}	9.4 MV
Acceleration gradient E_{acc}	18.8 MV/m
shunt impedance $R_a = V_{acc}^2 / 2Pc$	$1.72x10^{12}$ Ω
R_a/Q_0	166.6 Ω
E_{peak}/E_{acc}	2.66
B_{peak}/E_{acc}	6.1 mT/(MV/m)

1ST COOL DOWN

The first cool down of the SRF-Gun took place in August, 2007. For cooling and filling with liquid nitrogen two days were needed. Then the cavity was cooled down with He gas (10 K) for about 24 h. After a break in which the two ELBE modules were cooled with He gas, the tank was filled with liquid He. Finally, the tank was pumped to 30 mbar in order to get the working temperature of 2 K. Pressure stabilization is performed with cold compressors for all three cryostats of ELBE together using a pressure sensor near the first ELBE cryomodule. During the cool-down, the frequency of the π-mode (acceleration mode) was monitored (Fig. 1).

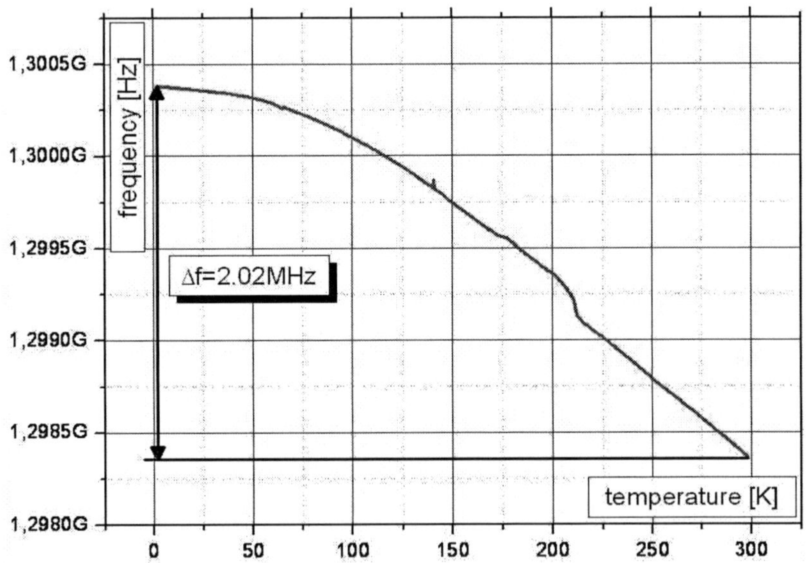

FIGURE 1: Cool down curve of the SRF-Gun cryomodule: cavity resonance frequency versus temperature. The final temperature was 2.0 K.

For the frequency shift from room temperature (RT) to 2 K a value of 2.02 MHz was found which is equal to the shift of TESLA resonators in the ELBE cryomodules. Due to the uniformly shrinkage of the cavity, the coefficient of thermal expansion can be calculated to:

$$\alpha = \frac{f_0}{f_0 + \Delta f} - 1 \rightarrow \alpha \;\; 300K - 2K \;\; = \Delta l / l = -1.55 \cdot 10^{-3} . \tag{1}$$

Unfortunately, the pre-stress adjustment of the SRF-Gun tuners was wrong. Thus, the final frequency obtained is about 400 kHz higher than expected. For the next operation period in 2009, this was corrected by another adjustment.

PRESSURE SENSITIVITY

To estimate the pressure sensitivity of the cavity, the resonant frequency and the pressure of the helium liquefier were measured at the same time. The result is shown in Fig. 2. To get the pressure dependence of the resonant frequency, the time domain signal versus the pressure variation is plotted for the same time. The linear fit function of this plot gives a slew rate (SR) of approximately SR=230 Hz/mbar. In comparison to the measured value of SR=35 Hz/mbar for the ELBE modules, the value is seven times higher. This could be a problem for operations at high dissipated power which causes slow oscillations of the helium pressure and thus the phase controller could exceed its regulation limit. To improve the frequency behavior the design of a new cavity was modified. Further a control loop using the phase signal and the forward power as process variable and the cavity tuner as actuator can be used at the existing cavity.

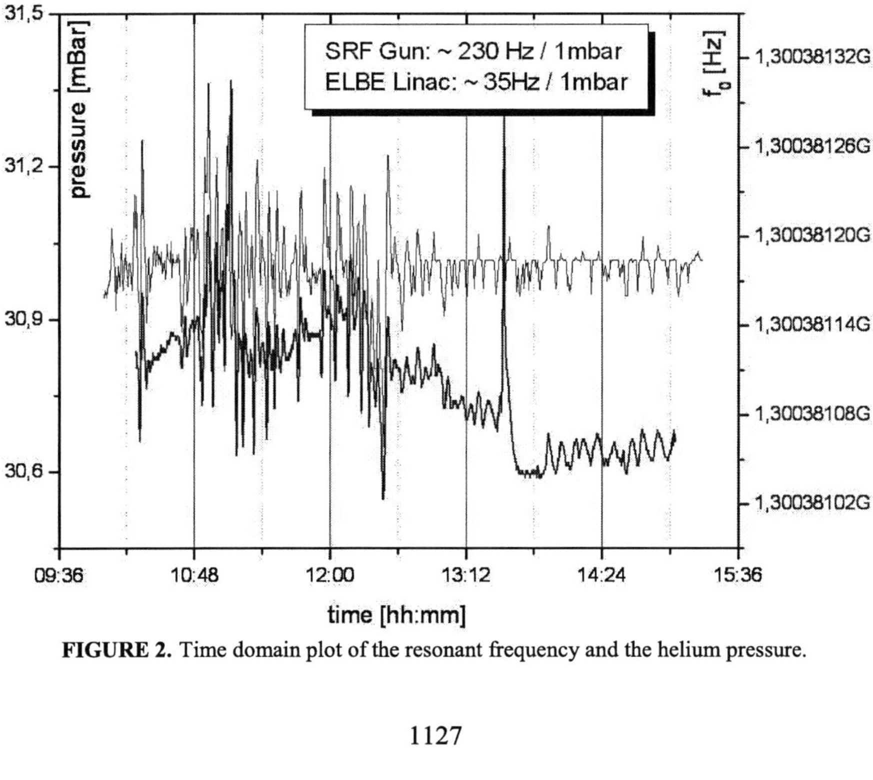

FIGURE 2. Time domain plot of the resonant frequency and the helium pressure.

PASSBAND AND ON-AXIS FIELD DISTRIBUTION

In a mathematical model the RF properties of the 3½-cell cavity can be described by an eigenwert problem. The eigen values are the passband resonance frequencies and the eigen vectors are the field amplitudes in the four cells. Then, the field distribution of the cavity in operation can be calculated from its fundamental mode passband measured at a temperature of 2K. The initial point for that calculation is the last measured cavity state right before the final BCP treatment. The change of the passband frequencies caused by cool down shrinking, change of permittivity and the final BCP cleaning must be taken into account. A detailed explanation of the used algorithm can be found elsewhere [3]. The measured frequencies are shown in Table 2. The obtained field distribution is in good agreement with the design values of 60% in the half cell and 100% in the Tesla cells.

TABLE 2. Measured pass band frequencies at 2 K and calculated acceleration field distribution.

¼ π [MHz]	**½ π [MHz]**	**¾ π [MHz]**	**π [MHz]**
1267.346959	1282.669318	1294.920031	1300.381168
↓	↓	↓	↓
half cell	**1st TESLA cell**	**2nd TESLA cell**	**3rd TESLA cells**
63.1%	100%	96.6%	97.0%

CAVITY TUNER

Due to the different mechanical stiffness, the cavity has two tuners, one for the three TESLA cells, and one for the half-cell. The 3D model and a simplified principle of the mechanism are shown in Fig. 3. Detailed information can be found in [1].

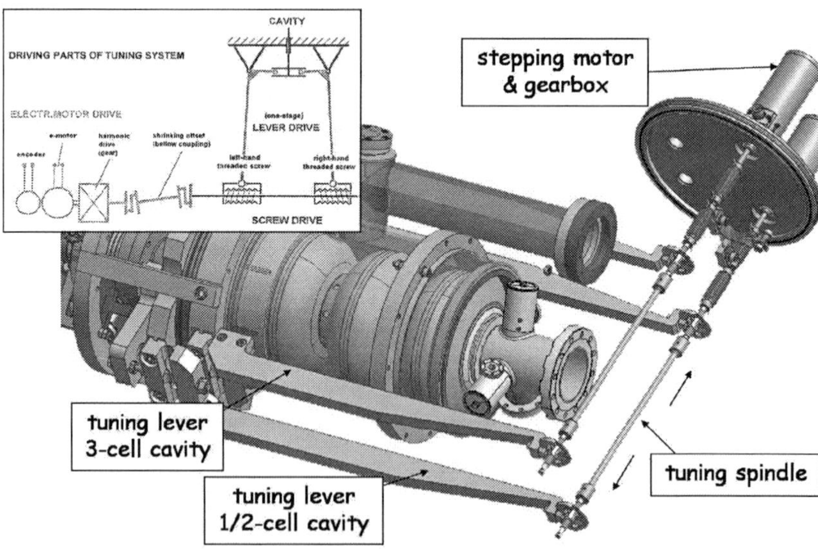

FIGURE 3. 3D model and simplified principle of both tuner systems.

In order to quantify the resolution, the frequency shifts of both tuners were measured versus motor steps while they were moved ±900 steps around the point of origin. To find the overall tuning range the frequency was measured at the mechanical end positions. The results are arranged in Table 3.

TABLE 3. Summary of the measured tuner parameters.

Parameter	Unit	½ cell	3 cells
tuning span	kHz	± 78	± 225
mechanical resolution	nm / step	1.3	2.1
frequency resolution	Hz / step	0.3	0.9

A strong hysteresis, found during the first measurement, could be corrected. The achieved tuning span and the resolution are sufficient for CW operation and in good agreement with the design values.

LORENTZ FORCE DETUNING

In order to measure the Lorentz force detuning of the SRF-Gun cavity, a network analyzer (NWA) was used as a driver for the klystron amplifier. Due to the fact that an increasing RF field detunes the cavity frequency downward, the NWA has to sweep the klystron from higher frequencies over the cavity resonance to lower values. Otherwise the maximum field could not be reached. To increase the cavity field, the output power of the NWA was raised in steps of 1dB while the gradient was calculated from the calibrated pickup voltage. Using this method, the achieved resonance curves for different forward power levels are shown in Fig. 4.

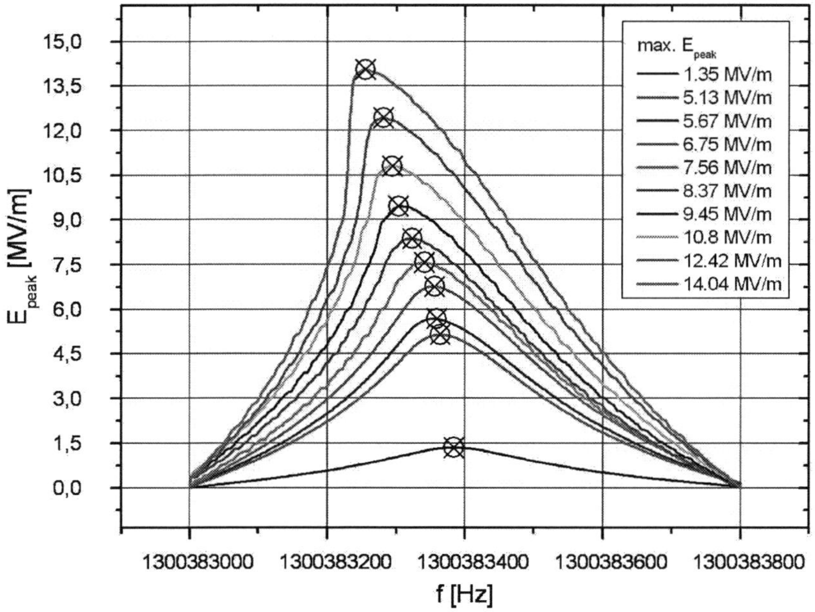

FIGURE 4. Measurement of the peak axis field in the TESLA cells for increasing gradients.

If the frequency shift on crest is plotted versus the applied peak field in the cavity, one can find a quadratic dependence. The determined constant of proportionality is:

$$\Delta f = -k \cdot E_{peak}^2 \quad ; \quad k = 0.7\,Hz/\,(MV/m)^2 . \tag{2}$$

In comparison to TESLA cavities (with k=-0.25 $(MV/m)^2$), the detuning of the SRF-Gun is about three times higher. The reason is the mechanical weakness of the half cell. At present, for gradients of E_{acc} = 5-6 MV/m the frequency shift is in the order of one bandwidth and causes no operational problems.

MICROPHONICS

Microphonics is known as an excitation of the mechanical eigen modes of the cavity. This results in a permanent detuning of the acceleration mode frequency. Due to the low bandwidth and the weak niobium walls this is typically a problem for superconducting cavities. The mechanical vibrations are induced by pumps, compressors and other heavy machinery next to the SRF-Gun module as well as from helium pressure fluctuation.

The cavity detuning results in a phase shift and amplitude drop of the acceleration mode. In order to stabilize both, a low level RF system is used [4]. The residual detuning (error signal) is measured via the demodulated low level phase controller signal in time domain. A following fast Fourier transformation (FFT) of the stored time signal shows the corresponding discrete power spectral density (PSD). The integrated PSD gives the variance of the measured phase deviation which can be seen as a weighted spectral distribution (Fig. 5).

The overall phase noise up to 1000 Hz of the closed loop operation is $\sigma = 0.055\,°$ (RMS). This is comparable to the known value of $\sigma \leq 0.05\,°$ (RMS) measured at the four ELBE cavities and ensures an energy spread of the accelerated electrons below 10^{-3} [4].

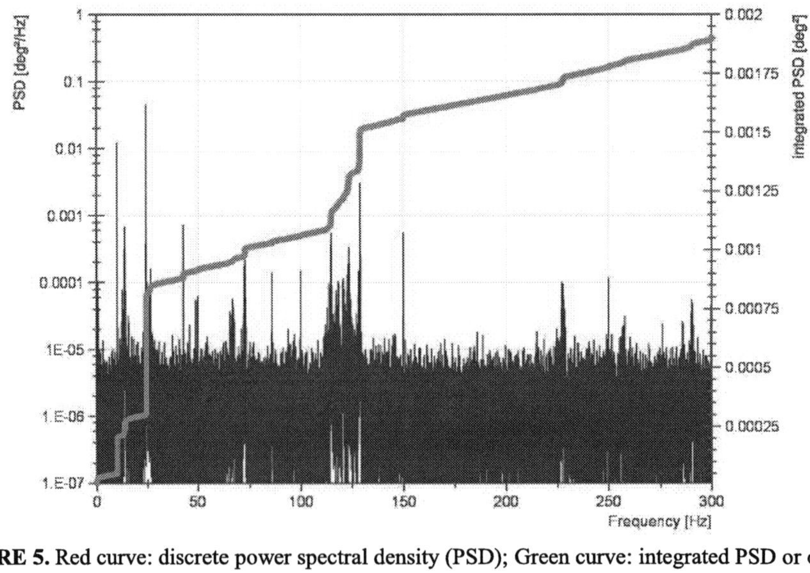

FIGURE 5. Red curve: discrete power spectral density (PSD); Green curve: integrated PSD or $\sigma(f)^2$

CAVITY PERFORMANCE & HIGH POWER PROCESSING

The SRF-Gun cryostat has an electrical heater in the helium tank. For regular operating conditions, the liquid He flow into the cryostat is constant. The different consumption due to variations of the acceleration gradient is compensated by adequate changes of the heater power. It is controlled by a level sensor in the He tank. Therefore the change of the heater power between the status RF-off and a given gradient is equal to the RF loss of the cavity. These values were measured as function of the acceleration gradient E_{acc}. The on-axis peak field E_{peak} in the TESLA like cells is related to the acceleration gradient by a factor of 2.7, i.e. $E_{peak} = 2.7 \times E_{acc}$.

In Figure 6 the second measurement on September 19, 2007 and the eighth measurement on August 28th, 2008 are compared. For the second measurement the red curve shows the corresponding radiation level caused by field emission in the cavity. The maximum achievable field is only one third of the designed value (E_{peak}=50MV/m) and the unloaded Q at low field levels is 10 times lower than measured in the vertical test before assembling. Possible reasons are the pollution of the inner surface during clean room preparation or frozen magnetic flux produced by RF cables, screws and other ferrous materials inside the μ-metal shielding. Especially the last assumption was strengthened by a magnetized RF feed through found during the last maintenance. Nevertheless, it is obvious that the drop down of Q_0 is caused by field emission. Fortunately, after one year of operation the use Cu and Cs_2Te cathodes did not changed the cavity performance.

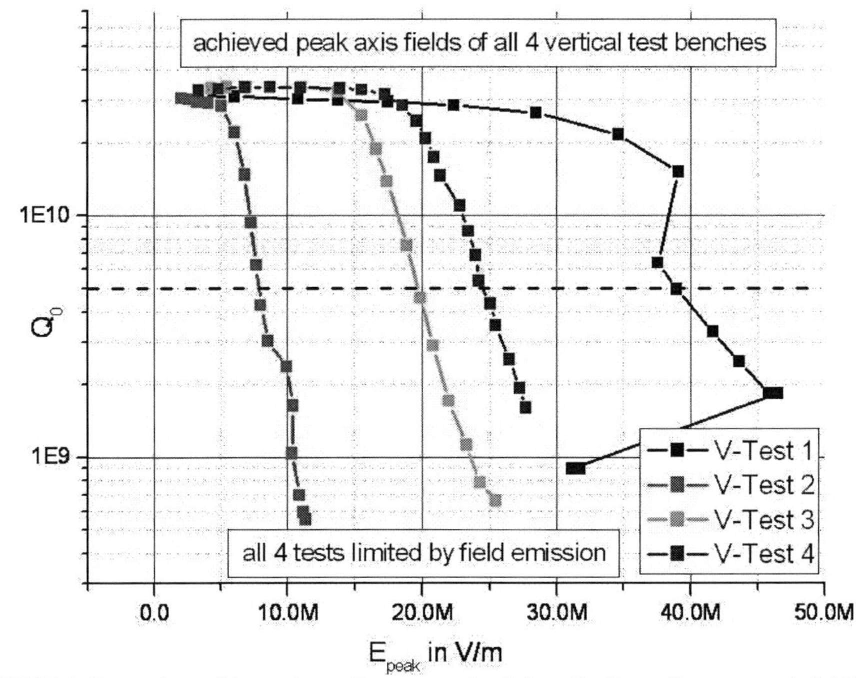

FIGURE 6. Comparison of the cavity performance, unloaded quality factor Q_0 versus peak field E_{peak}, before and after high power RF processing of the cavity in September 2008.

In order to increase the quality factor at a certain RF field level, high power RF processing (HPP) of the SRF-Gun cavity was done at the end of the first measurement period in September 2008. Therefore the RF klystron was pulsed with a low duty cycle (30ms/1000ms) to reduce the dissipated power. Within the short pulses a maximum field of E_{peak} = 25 MV/m was reached. The 9th measurement in Fig. 6 shows, that this results in a stable CW operation up to E_{peak}=17.6 MV/m or E_{acc} = 6.5 MV/m which is a reduction of the dissipated power by a factor of two at the same field level. Above this level the cavity starts to quench. The higher gradient allows maximal particle energy of 3 MeV at the exit of the electron source.

ACKNOWLEDGMENTS

We acknowledge the support of the European Community-Research Infrastructure Activity under the FP6 "Structuring the European Research Area" programme (CARE, contract number RII3-CT-2003-506395) and the support of the German Federal Ministry of Education and Research grant 05 ES4BR1/8.

REFERENCES

1. A. Arnold, et al., "Development of a Superconducting Radio Frequency Photo Electron Injektor", Nucl. Instr. and Meth. A 577, 440 (2007).
2. J. Teichert, et al., "The Superconducting RF Photoinjector at ELBE – First Operational Experience", these proceedings.
3. A. Arnold, "Field Reconstruction by Band Pass Frequency Measurement at the Rossendorf SRF-Gun Cavity", Proc. SRF 2007, Peking, China.
4. A. Büchner, F. Gabriel, H. Langenhagen, "Noise Measurements at the RF System of the ELBE Superconducting Accelerator" EPAC'2002, Paris, June 2002.

The Polarized SRF Gun Experiment

Jőrg Kewisch[a], Ilan Ben-Zvi[a], Triveni Rao[a], Andrew Burrill[a], David Pate[a], Erdong Wang[a], Rob Todd[a], Hans Bluem[b], Doug Holmes[b], Tom Schultheiss[b]

[a]*Brookhaven National Laboratory, Upton, NY 11973*
[b]*Advanced Energy Systems, Medford, NY*

Abstract. An experiment is under way to prove the feasibility of a super-conducting RF gun for the production of polarized electrons. We report on the progress of the experiment and on simulations predicting the possibility of success.

Keywords: Polarization, electron sources, quantum efficiency
PACS: 29.25.Bx, 29.27.Hj

INTRODUCTION

Photo-cathode RF electron guns are superior to DC guns in producing high brightness electron beams, because the electric field on the cathode can be much higher without the technical difficulty of voltages higher than 100 kV. However, all successful existing guns for polarized electrons are DC guns because the environment inside an RF gun is usually hostile to the GaAs cathode material necessary for polarization. Ions created by scattering of electrons on residual gas can be accelerated back into the cathode. The number of back-scattered ions is proportional to the vacuum pressure inside the gun and while the typical vacuum pressure in a DC gun is better than 10^{-11} Torr the vacuum in an RF gun is in the order of 10^{-9} Torr. The cathode can also suffer damage from electrons which are emitted from the cavity walls or the cathode itself and are accelerated during the negative RF phase to impact the cathode with high energy.

Experiments at BINP Novosibirsk in the late 1990s demonstrated the following effects [1, 2]: the quantum efficiency in a pulsed normal-conducting 2.8 GHz RF gun was destroyed in as little as 10 pulses. We are now working to repeat this experiment with modifications that we hope will give a positive result: First, instead of a normal-conductive gun we use a superconductive gun. Since the RF fields will not heat up the cavity walls we avoid out-gassing. In fact, the cold walls will act as a cryogenic pump and we expect the vacuum pressure to be better that 10^{-11} Torr. Secondly, we will use a lower frequency half cell gun (1.3 GHz) which will help to avoid electron multipacting involving the cathode.

CP1149, *Spin Physics, 18th International Spin Physics Symposium*
edited by D. G. Crabb, D. B. Day, S. Liuti, X. Zheng, M. Poelker, and Y. Prok
2009 American Institute of Physics 978-0-7354-0686-5/09/$25.00

THE EXPERIMENT

The experimental setup has two parts. The first is the electron accelerator, where the gun is placed in a 100 liter cryostat. The layout is shown in Fig. 1. The gun points upwards, so that the beam exits the cryostat on the top and is bend by a 90 degree dipole (shown in Fig. 2) into a Faraday cup. A window on top of the dipole is used for the laser. The polarization of the beam is not measured, since we are only interested in the decay of the quantum efficiency. In order to measure the beam current the gun is insulated from the rest of the apparatus with a ceramic break, which is protected from mechanical stress by a G10 sleeve.

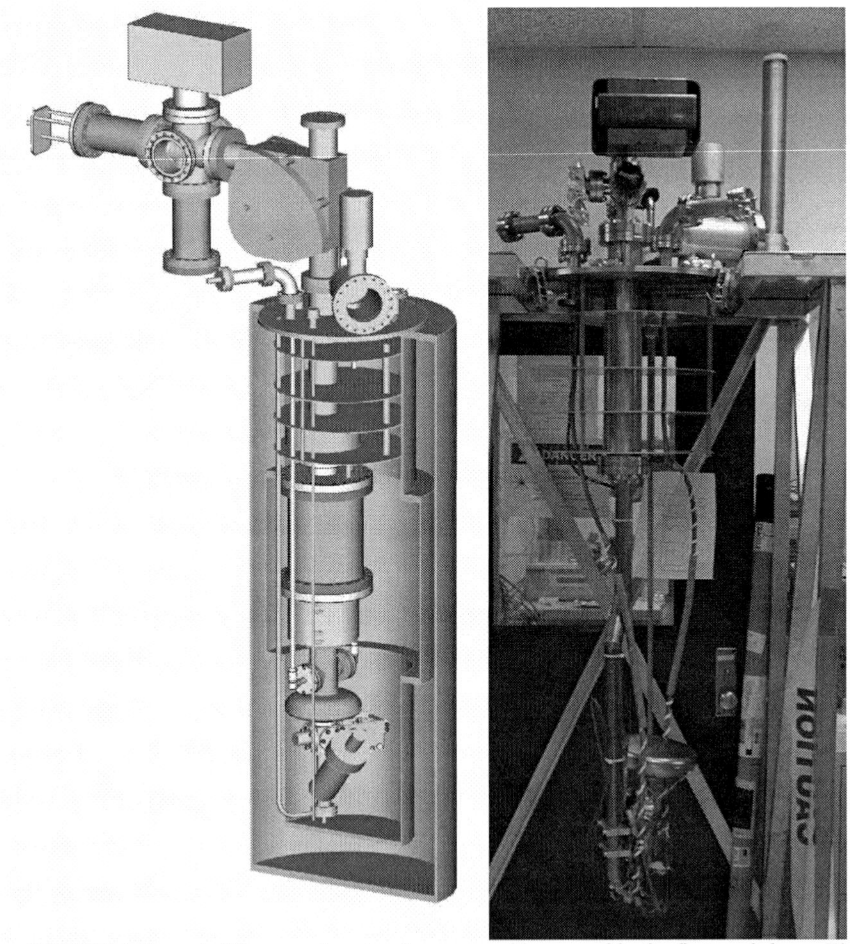

FIGURE 1. Layout of the experiment and gun assembly as used in the first cool-down. The gun is mounted upside-down in this test to prevent particulate from dropping into the gun cavity. The ion pump on to of the assembly was used during the cool-down to measure the vacuum pressure.

A tank filled with NEG pumps close to the gun is maintaining the vacuum when the gun is warm and is not cryo-pumping. Another NEG pump and an ion pump is placed next to the Faraday cup, where the impact of the beam will cause degradation of the vacuum. A set of four aluminum baffles insulates the liquid helium from the top plate and a high temperature superconducting focusing solenoid is placed between the lowest baffles, where a temperature of 17 K was measured

FIGURE 2. The 90° bending dipole and its 3-way vacuum chamber and the high temperature superconducting focusing solenoid.

The other part is the cathode preparation system, which allows heating the GaAs cathode to 600° C for cleaning and the deposition of a mono-layer of Cesium. In normal-conducting guns the preparation system is integrated into the gun, in our case it must be separate with a transporter which allows transfer of the cathode without breaking the vacuum.

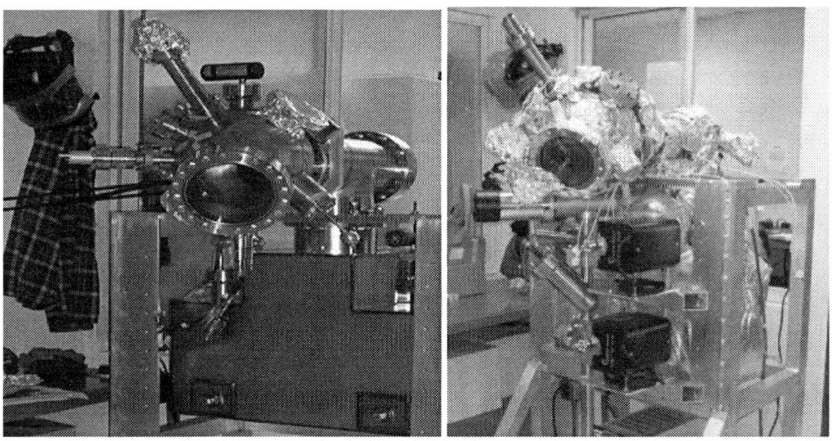

FIGURE 3. The preparation system on the left with attached TSP and on the right wrapped for baking with attached cathode transporter. The two ion pumps are part of the transporter.

FIRST COOL-DOWN

In the first cool-down the Q of the cavity was verified and the thermal properties and vacuum pressure of the system was measured. After the cryostat was filled the ion pump was turned off and the pressure fell by cryo-pumping alone below the range of the pressure gauge. The heat load of the system was approximately 5 Watts.

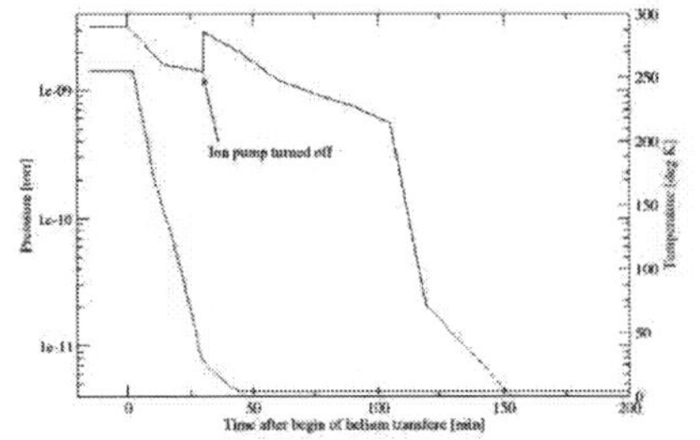

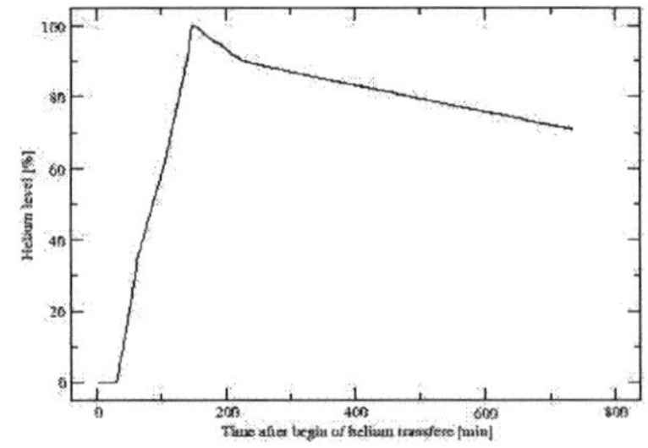

FIGURE 4. The preparation system on the left with attached TSP and on the right wrapped for baking with attached cathode transporter. The two ion pumps are part of the transporter.

ION AND ELECTRON BOMBARDMENT

The above measurement gives us confidence that the vacuum pressure will be comparable or better than found in DC guns. Also, only a fraction of the generated ions will return to the cathode and the impact energy will be significantly lower than in a DC gun [3]. But as suggested by Aleksandrov [2], electron bombardment may be the cause of the short cathode lifetime in the BINP experiment.

Electrons emitted from the cathode at the RF phase between approximately 100 and 180 degrees in the RF phase will not leave the gun cavity, but will be accelerated backwards and impact the cathode, in our case with energies up to 350 MeV.

The effect is twofold: one is the damage to the cathode; the other is to lift thousands of secondary electrons from the valence band into the conduction band. These electrons will drift under the influence of the RF field to the surface and leave the cathode due to the negative affinity.

A tracking program was developed to investigate the possibility of multipacting involving the negative affinity cathode. The program calculates the maximum phase advance from an emission of an electron to the emission of its secondary electrons including the motion inside the GaAs, were the penetration depth of the electric field and the dependence of the drift velocity on the field must be considered.

The results are shown in Fig. 5, where the maximum phase advance of the electron impact and the emission of the last secondary electron are plotted vs. the gradient at the cathode. The condition for multipacting is that the phase advance is 360 degrees. Therefore, if the range overlaps the 360 degree line multipacting must be expected.

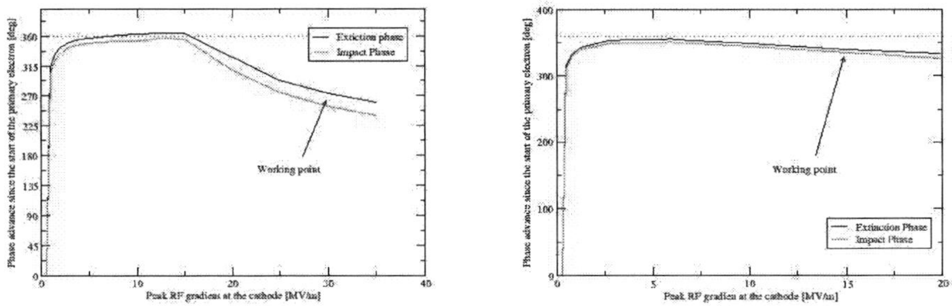

FIGURE 5. Maximum phase advance between the emissions of an electron to the emission of its secondary electrons, on the left for a pill box cavity similar to the BINP gun, on the right for our 1.3GHz gun. Since the exact dimensions of the BINP gun were not available, the gun shape was approximated by a pill box cavity, matching the data given in [1]. The green curve shows the impact, the black curve the last emitted secondary electron.

We see that this is the case for the BINP gun while the in the electric gradient range between 10 and 15 MV/m. This agrees with the observation that dark current was observed above 10 MV/m. In order to improve the vacuum in the gun the RF was pulsed. Therefore the multipacting condition was fulfilled every pulse. The calculations for our experiment show that we escape this disaster, but with little room to spare.

These calculations were done assuming bulk GaAs with p-doping of $10^{18}/cm^3$. The skin depth is in our case 45μ. In strained GaAs, where only a 5 nm thick surface layer is doped, it will be important to make the cathode crystal thin enough to prevent multipacting.

ACKNOWLEDGMENTS

Work performed under contract No. DE-AC02-98CH10886 with the Department of Energy of the United States.

REFERENCES

1. A. V. Aleksandrov et.al.: "A Prototype of RF photogun with GaAs photocathode for injector of VEPP-5", 1996 European Particle Accelerator Conference.
2 A. V. Aleksandrov et.al.: "Experimental Study of GaAs Photocathode Performance in RF Gun", 1999 Particle Accelerator Conference
3. E. Pozdeyev: these proceedings.

Non Evaporable Getter (NEG) Pumps: a Route to UHV-XHV

Paolo Manini

SAES Getters SpA, Viale Italia 77, 20010 Lainate (MI), Italy

Abstract. Non Evaporable Getter (NEG) technology has been developed in the 1970's and since then adopted by industry, R&D labs, research centres and in large physics projects like accelerators, synchrotrons and fusion reactors. NEG pumps are very compact and vibration-free devices able to deliver very high pumping with minimal power requirement and electromagnetic interference. In the present paper, main features and performances of getter pumps are reviewed and discussed with a special focus to photocathode gun application, where UHV or XHV conditions are mandatory to ensure adequate gun life. NEG coating and future challenges for NEG technology are also discussed.

Keywords: Non evaporable getter, NEG pumps, photocathode gun, UHV-XHV.
PACS: 07.30.Cy, 81.65.Tx, 81.20.Ev.

NEG PUMP GENERAL FEATURES

NEG pumps are capture pumps which remove, by a chemical reaction, molecules impinging on their surface. They use high surface area reactive alloys, generally made of Zr or Ti. NEG pumping speed and sorption capacity is optimized by selecting getter alloy composition, its geometrical and physical configuration and the operating temperature [1]. NEG pumps, are generally combined with sputter ion [2], turbo-molecular [3] or cryo- pumps [4], to improve the final vacuum, reduce the evacuation time and increase bake-out effectiveness in vacuum systems.

They generally operate in the 10^{-5} $^{-}10^{-11}$ Pa range and present some distinctive features:

- High pumping speed for H_2, the main residual gas in UHV-XHV systems
- Irreversible sorption of all active gases, like O_2, CO, CO_2, H_2O, N_2.
- Constant pumping speed in UHV-XHV regime
- Minimal power requirements.
- Oil and vibration-free.
- Negligible interference with electro-magnetic fields.
- Extremely compact and light, can be mounted inside the vacuum system.

NEG do not pump noble gases and CH_4, a relatively stable molecule difficult to dissociate at room temperature. Pumping of CH_4 is however possible at higher temperatures (250-350°C), while sorption of larger and less stable hydrocarbons and organics, like toluene, decane and methyl-metacrylate, occurs at room temperature [5].

CP1149, *Spin Physics, 18th International Spin Physics Symposium*
edited by D. G. Crabb, D. B. Day, S. Liuti, X. Zheng, M. Poelker, and Y. Prok

Principles of Operation

To sorb gases, NEG pumps need first to be "activated". The activation process is carried out under vacuum (or noble gas atmosphere) heating the getter at moderate temperature (350-500°C) for a short time (typically ~ 1 hour). This treatment allows surface oxides and carbides covering the surface to decompose and diffuse inside the getter material bulk structure, leaving a clean and reactive metal surface, ready to chemi-sorb impinging molecules [6, 7].

As the getter surface progressively gets saturated by chemisorbed molecules, pumping efficiency slows down. When the pumping speed drops below a specified value (application dependent), the getter needs to be activated again ("reactivation"). Getters can be re-activated many times providing a long lasting and low total cost of ownership pumping solution.

Range of Configurations

The getter material can be configured in many different ways and geometries [1, 8]. They can be laminated onto a metal strip which can be mounted inside a chamber to provide distributed pumping.

The strip can also be folded into concertina-like or wafer panel (WP) modules having higher trapping efficiency for gases [9]. This allows achieving very large speed in compact volumes. Getter powder can also be sintered as high porosity bodies like disks or fins. These are arranged into high efficiency cartridges which are then assembled in a range of NEG pumps [2-4, 8]. Depending on the pump model and the flange size, speeds from 20 to 2000 l/s (H_2) are available. Larger, custom made pumps have been also realized with speed exceeding 10,000 l/s for H_2 [10].

In most of these configurations, the getter material is activated by radiation (e.g. emitted by a heating element integrated in the cartridge) or, as in the case of the WP, by direct passage of current across the strip. Passive activation during the chamber bake-out is also a possibility.

Another possible configuration is obtained depositing ~50-100 micron getter powder onto a thin metal sheet which undergoes a vacuum sintering process [11]. The outcome is a well adhered getter layer onto a flexible support which can be also pre-cut or folded to fit specific shapes (HPTF or High Porosity Thick Film).

Sputter deposited thin getter films (~ 1micron thick) have been also proposed. They find applications in MEMS [12] and other micro-devices as well as in the vacuum pipes of accelerators and synchrotrons [13].

In particular, CERN has investigated a range getter alloys [14] (composition centered around $Ti_{40}V_{30}Zr_{30}$) activatable at low temperature (<200°C), compatible with most of the accelerator bake-out procedures. The getter film is here directly sputtered onto the internal surface of the accelerator beam pipe [13]. After activation, it displays large in-situ pumping which is particularly beneficial for long and narrow gap pipes, like insertion devices, which can be hardly otherwise pumped due to conductance issues. In the past 5 years, NEG coating has been successfully adopted by an increasing number of accelerators, boosting their vacuum performances [15-17].

Examples of Applications

NEG pumps cover a broad range of applications, encompassing surface science, vacuum equipments for material preparation and characterization, semiconductor processing, inspection and review tools or metrological equipments. Being very compact and with minimal power requirements they are ideal in portable and miniaturized equipments (like hand-held mass spectrometers, radioactivity counters or UHV suitcases). The absence of electromagnetic interference allows to mount NEG pumps very close to ion/electron sources, like in the case of scanning and transmission electron microscopes and lithographic equipments, where vacuum is most critical.

NEGs are also used in the main rings and beam lines of colliders and synchrotron machines as well as in a number of other experimental facilities [8].

Application to Photoelectron Guns

Photoelecron guns are one of the most demanding applications, vacuum wise. Most of the photo-cathodes require very good vacuum conditions to deliver stable currents for a long time. Gases present in the gun chamber can poison the cathode surface or damage it by ion back bombardment, both effects reducing the cathode useful life.

To avoid cathode deterioration, pressures in the range of 10^{-8} -10^{-10} Pa are required. This poses serious challenges in term of gun chamber design, material selection and processing as well as the choice of the pumping group. The use of arrays of NEG pumps mounted inside the gun chamber is one of the preferred choices to achieve very low pressure [18-21]. A typical configuration is based on the use of WP modules which, thanks to their design, can be mounted side by side inside the gun to cover a large area and provide extremely large pumping speed for H_2, the main residual gas. An example is showed in Fig.1, where pumping speed exceeding 20,000 l/s has been reported [18].

FIGURE 1. An array of WP 1650-St 707 modules is placed inside the photocathode gun chamber , A few modules are also placed inside the ion pump pumping port (Courtesy of Cornell University).

The use of cartridge pumps mounted in the cathode chamber as well as in the preparation and load-locked chambers [21, 22] is also quite popular (Fig.2).

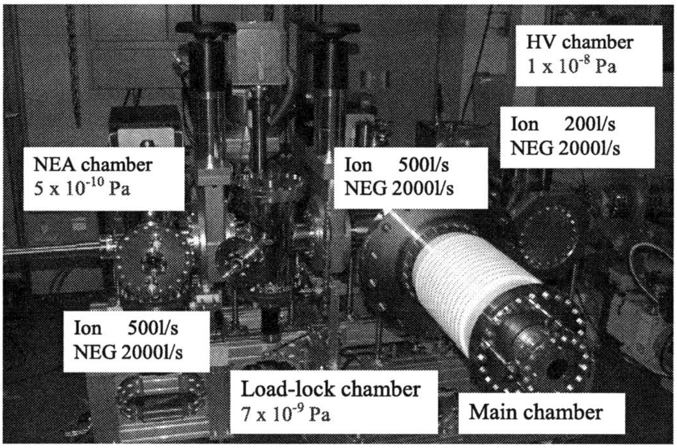

FIGURE 2. NEG cartridge pumps (Capacitorr D 2000 model) mounted in the NEA, preparation and main chambers of the DC gun developed for the ERL project at JAEA (Courtesy of JAEA).

NEG coating has also been proposed to reduce H_2 out-gassing and improve the vacuum [19, 23]. To deposit films with large pumping speed, careful control and optimization of the sputter deposition parameters is key. This is not always easy in chambers with irregular geometries but has definitively an impact on the quality and reproducibility of the deposited getter film [24]. If this aspect is not taken into account, lower than expected speed and modest vacuum improvement may be obtained [23].

LIMITATIONS AND ULTIMATE PERFORMANCES OF NEGS

The use of NEG pumps, coupled with a small sputter or turbo pump for inert gases, is one of the best routes to achieve XHV. The ability of NEG to reach ~$3x10^{-12}$ Pa was measured at CERN [25] in a series of experiments where the NEG was the main pump, backed by small sputter ion pumps for the pumping of noble gases and CH_4.

The two main mechanisms which can intrinsically limit NEG pumps in achieving extremely low pressures are the generation of methane and the H_2 emission due to the equilibrium always existing between H_2 in the gaseous phase and hydrogen atoms in solid solution in the getter lattice.

CH_4 can be released from a getter surface due to the reaction of surface carbon atoms and hydrogen present in the getter volume or coming from the vacuum system. However, due to the difficulty of the measurement (CH_4 can be in fact generated also by hot filaments and gauges and emitted by other sources), the identification of a clear and unambiguous mechanism for CH_4 generation is still controversial.

At low H_2 concentration and moderate temperature, NEG alloys used in pump applications, follow Sievert's law [1]. This has been confirmed also in the case of Ti-V-Zr based NEG coatings in the 100-250°C range and down to 10^{-9} Pa [26]. However, no data are available in XHV regime, due to the complexity of measuring equilibrium isotherms in a controlled and reproducible way below 10^{-10} Pa. For St 707 ($Zr_{57}V_{36}Fe_7$), Sievert's law extrapolated at 25°C and 100 l x $Pa(H_2)$/g(alloy) gives an equilibrium value of ~2 $x10^{-14}$ Pa. This might be considered as an indication of the pressure limit for this alloy. Other getter alloys have been developed which have lower equilibrium pressure than St 707, so in principle, lower pressure could be even reached using different getters. Specific investigation is ongoing at SAES Getters on this point.

These considerations suggest that NEG pumps can be definitively effective also in new, more demanding, electron sources at present under study or being built. From this point of view it must be stressed however, that achieving extreme performances in such sophisticated guns is possible only when the integration of the NEG pumps in the vacuum system is optimized and carefully designed to avoid interferences between sub-systems. Being a reactive surface, the NEG can in fact release and/or adsorb gases during the various different phases of evacuation, conditioning and baking procedures. This has to be taken in due account to avoid, for example, that gases released by the getter during the activation, contaminate the system or partially load the other installed pumps or, vice-versa, that a fully active getter surface is saturated during bake-outs or degassing procedures, thus preventing the achievement of the best performances.

REFERENCES

1. B. Ferrario "Getters and Getter Pumps" in *Foundation of Vacuum Science and Technology,* edited by J.M.Lafferty, John Wiley & Sons, 1998, pp. 261-315.
2. R. Caloi et al, *Proceedings 14th Italian Vacuum Association Conference*, Vicenza, Italy, 5-7 May, 1998 (Ed. Compositori, Bologna, Italy)
3. R. Giannantonio et al., *Vacuum*, **55**, 1999, pp 27-37
4. R. Giannantonio et al. , *J. Vac.Sci. Technol.*, A **15**(1), Jan/Feb. 1997, pp 187-191
5. A. Conte, *Proceedings 55th AVS Conference*, Boston, 2008
6. K. Ichimura et al, , *J. Vac.Sci. Technol.*, A **5**(2), 1987, pp 220.
7. M. Sancrotti et al, *J. Vac.Sci. Technol.*, A **9**(2), Mar/Apr 1991, pp 182-189
8. F. Mazza , *Non Evaporable Getters for UHV application*, SAES literature , 2002
9. B. Ferrario and L. Rosai, *Proceedings 7th Int.Vac. Conference*, Wien (Austria), 1977, pp. 359-362
10. R. Giannantonio et al., *J. Vac.Sci. Technol.*, A **17**(4), Jul/Aug 1999, pp 2093-2098
11. E. Giorgi, *Journal of Vac. Sci. & Technol.* A **7** (2), Mar/Apr. 1989, pp. 218-222
12. M. Moraja et al. *Proceeding of SPIE Photonic West 2004*, 2003, Vol.4980.
13. A. Conte et al *Proceedings Particle Accelerator Conference*, 2007, Albuquerque, USA.
14. C. Benvenuti et at., Vacuum, **53** , 1999, pp 219-225
15. P. Chiggiato, 41st IUVSTA workshop, Brdo ori Kranju, Slovenia, June 2004
16. C. Herbeaux et al., *Proceedings European Particle Accelerator Conference*, 2008, Genova, Italy
17. H. Huang et al. *Proceedings European Particle Accelerator Conference,* 2006, Edinburgh, UK
18. B.M. Dunham et al *Proceedings Particle Accelerator Conference*, 2007, Albuquerque, USA
19. C. K. Sinclair, *Physical Review Special Topics- Accelerators and Beams*, **10**, 2007, 023501
20. S. Pastuszka et al. *Journal of Applied Physics*, **88**(11), December 2000
21. M. Yamamoto et al, *Proceedings LINAC 2004*, 2004, Lubeck Germany
22. N. Nishimori et al, *Proceeding ERL07*, 2007, Daresbury Laboratory, UK
23. M.Poelker et al. *Proceedings Particle Accelerator Conference*, 2007, Albuquerque, USA
24. A. Bonucci et al *Proceedings Particle Accelerator Conference*, 2007, Albuquerque, USA
25. C. Benvenuti, *Proceedings Particle Accelerator Conference*, 2001, Chicago, USA, pp 602-606
26. A. Rossi, *Proceedings European Particle Accelerator Conference*, 2006, Edinburgh, UK

Sensitive Ion Pump Current Monitoring Using an In-House Built Ion Pump Power Supply

J. Hansknecht, P. Adderley, M. L. Stutzman and M. Poelker

Thomas Jefferson national Accelerator Facility
12000 Jefferson Ave., Newport News, VA 23606

Abstract. Ion pumps are common vacuum pumps on DC high voltage photoguns and baked-accelerator beamlines. Commercial ion pump power supplies provide a measure of the electrical current drawn by the pump, but typically have resolution to only ~ 0.1uA, which corresponds to pressure ~ 10^{-9} Torr, a value considerably higher than the minimum pressure required by photoguns and nearby beamline. This submission describes a very sensitive in-house-built ion pump power supply with current monitoring capability good to less than 1nA, and corresponding pressure in the low-10^{-11} Torr range. Besides providing "free" pressure monitoring on a scale equivalent to the best available commercial pressure gauges, the ion pump power supply also serves as a sensitive diagnostic for detecting field emission from the photogun cathode electrode and bad electron beam orbits that could diminish photogun operating lifetime. Since its inception, this ion pump power supply has become an invaluable tool for operating the CEBAF polarized electron source. It is also a very useful low-cost diagnostic for ultrahigh vacuum studies in the laboratory.

Keywords: Ion pump; Polarized electron source
PACS: 29.25.Bx; 29.27.Hj; 07.30.Cy; 07.30.Dz

INTRODUCTION

Ion pumps have been used for decades at particle accelerators [1] and are common components on DC high voltage polarized electron sources [2], where they serve to pump gas species not effectively pumped by non-evaporable getter (NEG) pumps, for example, argon and methane. At the heart of the ion pump is the Penning cell [3], which is used to ionize and capture residual gas. There are different types of ion pumps, composed of different electrode materials, with different shapes and electrical configurations. The "diode" ion pump is widely used and well suited for UHV systems. The term "diode" describes a Penning cell composed of a positively biased tubular anode situated between grounded cathode plates, immersed in a magnetic field created by permanent magnets affixed to the outside of the ion pump vacuum chamber. Energetic free electrons provided by the cathode travel in long helical trajectories, trapped by the applied magnetic field, striking and ionizing residual gas along the way. Ions generated by these collisions are only weakly affected by the magnetic field (because of their relative mass compared to electrons). They are repelled by the biased anode and accelerated toward the cathode end plates. Gas is removed from the vacuum chamber when these ions are embedded within the cathode end plates. If the cathode end plates are made of titanium, there is the added benefit of

CP1149, *Spin Physics, 18th International Spin Physics Symposium*
edited by D. G. Crabb, D. B. Day, S. Liuti, X. Zheng, M. Poelker, and Y. Prok

chemical gettering that results when titanium is sputtered away, which boosts pump speed, especially for hydrogen gas.

It has long been known that the electrical current associated with the ion-pump process is an indication of the vacuum level, or pressure inside the chamber. This is illustrated in Fig. 1, which shows ion pump current measured using a sensitive in-house built ion pump power supply [4], as a function of pressure inside one of the polarized electron sources at CEBAF/Jefferson Lab. Pressure was measured using a Leybold extractor gauge [5] while gradually venting the chamber to atmospheric pressure with hydrogen gas, prior to photocathode replacement. The pressure values in Fig. 1 are "raw", i.e., uncorrected for gas species. Ion pump current increases with pressure, following a typical and expected power rule relationship to produce a straight line on the log/log plot. The slope of the line depends on specific features of the Penning cell(s) used inside the pump, like anode geometry, anode bias voltage, and magnetic field. The remarkable feature of Fig. 1 is that data extend to less than 1 nA, while commercial ion pump power supplies are limited to ~ 0.1uA, more than a factor of 100 larger. This submission describes the inexpensive in-house built ion pump power supply with sub-nA current monitoring capability that provides accurate and "free" pressure monitoring into the low-10^{-11} Torr vacuum range. Besides providing a measure of static vacuum inside UHV systems, the device has become a useful diagnostic for photogun set-up, characterization and troubleshooting. Some examples are presented below.

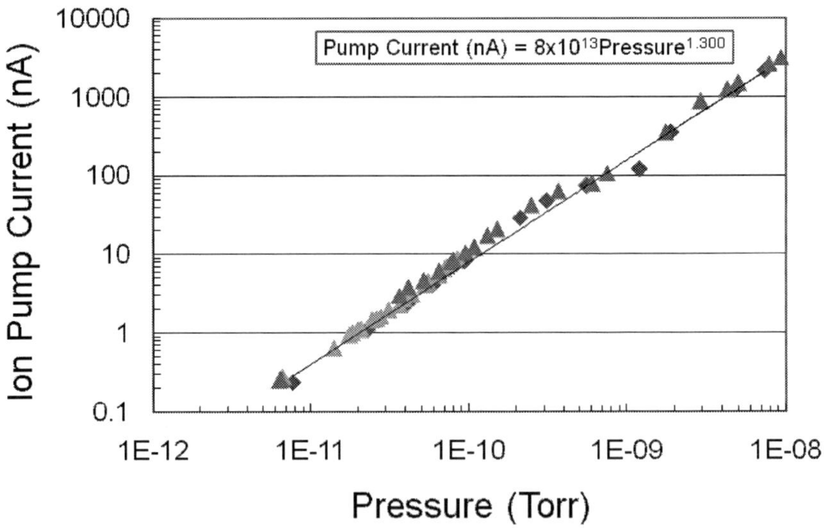

FIGURE 1. Ion pump current v. pressure inside a CEBAF/Jefferson Lab DC high voltage GaAs photogun. The data were obtained while venting the photogun to atmospheric pressure with hydrogen gas prior to photocathode replacement. The three data sets correspond to three different photocathode installations. The expected functional form $I_p=kP^x$ provides a good fit to the data.

ELECTRICAL DESIGN

At the heart of the ion-pump power supply is a high voltage transformer with "floating" secondary winding circuit attached to a pico-ammeter (Fig. 2). Good electrical isolation between the two windings ensures that any current flowing through the ion pump also flows through the pico-ammeter, and this current is not "contaminated" by uninteresting current from other parts of the circuit. The design relies on a high frequency transformer (~ 400kHz) that provides a number of benefits, including; 1) it shrinks the size and weight of the power supply while increasing electrical efficiency, 2) higher frequencies are easier to rectify and filter and, 3) it permits the use of a Cockcroft-Walton multiplier to raise voltage to the ultimate desired level. Since the transformer runs at lower voltage, the potential for leakage to ground is greatly reduced. Pump current as low as ~ 100pA can be accurately measured while the total cost for all components is modest, < \$500. The only disadvantage of this design compared to commercial power supplies is that pump current must be < 0.5mA, which limits operation of the ion pump to 10^{-7} Torr.

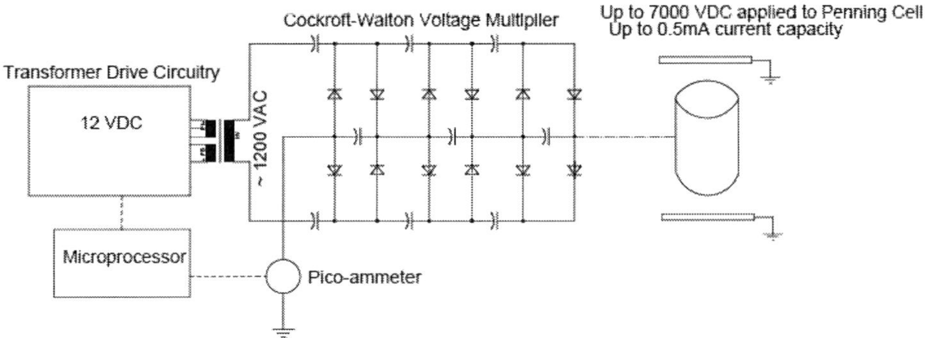

FIGURE 2: Schematic of ion pump power supply with sensitive current monitoring capability.

APPLICATIONS

Field emission from the cathode electrode inside a DC high voltage GaAs photogun is problematic. Large amounts of field emission lead to dramatic reduction of photocathode quantum efficiency, and can be easily identified. Small amounts of field emission however, also spell trouble for the photogun, contributing to shorter than expected operating lifetime, but can be more difficult to diagnose. The ion pump power supply with sensitive current monitoring capability has proven to be a very useful tool for detecting small amounts of field emission inside CEBAF/Jefferson Lab GaAs polarized photoguns. Figure 3 shows ion pump current versus applied photogun bias voltage, indicating vacuum degradation (due to electron stimulated desorption of gas associated with field emission striking the photogun anode electrode and/or vacuum chamber wall) when bias voltage exceeds 100kV. From Fig. 3, it would appear field emission is an obvious problem, however it must be pointed out pressure inside the photogun remains very good, with ion pump current < 1nA, even at bias voltages that produce field emission. This information cannot be obtained using a commercial ion pump power supply because small changes in ion pump current (tens to hundreds of pA) cannot be resolved. And it would be difficult to obtain similar data using a conventional vacuum gauge, since hot filament gauges produce light that generate electron beam from the activated GaAs photocathode, thereby degrading vacuum in a similar (or worse) manner, masking the problem associated with field emission.

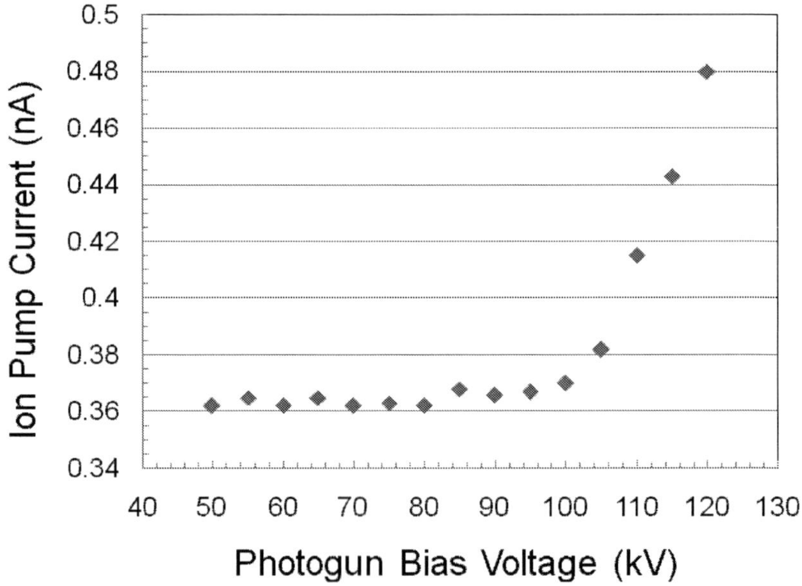

FIGURE 3. Ion pump current versus applied photogun bias voltage. Low-level field emission degrades vacuum inside the photogun for bias voltage > 100kV that can be easily detected using the ion pump power supply with sensitive current monitoring capability.

Another example illustrating the utility of sensitive ion pump current monitoring can be seen in Fig. 4, which shows a time-dependent "chart-recorder" trace of ion pump current for five ion pumps downstream of the CEBAF/Jefferson Lab GaAs polarized photoguns. These data were obtained while inadvertently operating one of the photoguns with a bad beam orbit – a very small portion of the electron beam was striking a retracted phosphorescent view screen. Electrons would accumulate on the view screen, causing an electrostatic deflection of the electron beam. Charge on the view screen would build and eventually arc to ground, thereby restoring the normal beam orbit. This "back and forth" beam motion was disrupting the physics program at one the CEBAF endstations. Pinpointing the origin of the problem was difficult using conventional beam position monitors but easily tracked to a specific location by analyzing the relative intensity of the vacuum bursts associated with each event. All of the beamline ion pumps near the CEBAF photoguns operate with less than 70nA current – these vacuum bursts would not have been visible using commercial ion pump power supplies. Accelerator operators now monitor ion pump current to prevent these types of beam orbit problems.

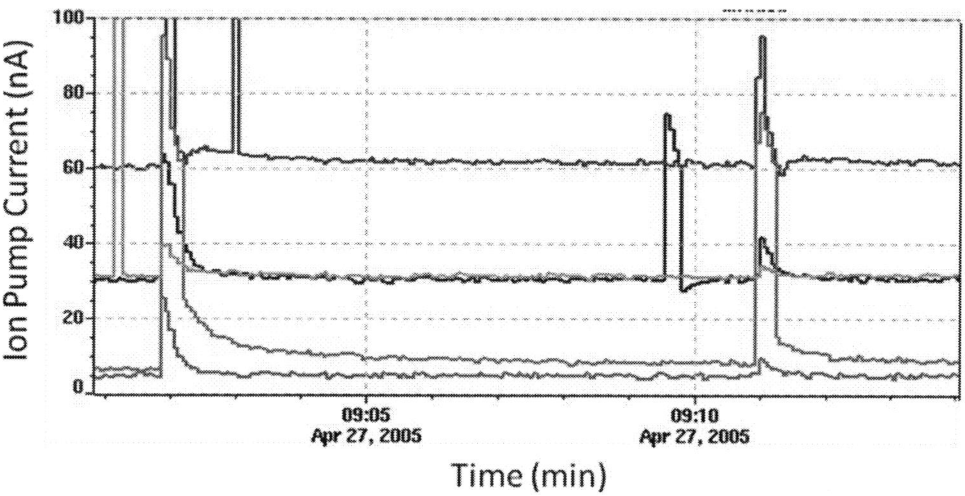

FIGURE 4. Ion pump current versus time, for five ion pumps attached to a baked beamline near the GaAs photoguns at CEBAF/Jefferson Lab. Vacuum "bursts" correspond to charging-discharging of a retracted phosphorescent view screen, due to a bad beam orbit.

CONCLUSION

An ion pump power supply with extremely sensitive current monitoring capability has been constructed that exploits the relationship between pump current and pressure to provide "free" pressure monitoring of the CEBAF/Jefferson Lab DC high voltage GaAs photoguns and injector beamline, at a level equivalent to the best available commercial vacuum gauges. The device has also proven useful for characterizing and troubleshooting photogun performance.

ACKNOWLEDGEMENTS

Authored by Jefferson Science Associates, LLC under U.S. DOE Contract No. DE-AC05-06OR23177.

REFERENCES

1. K. Welch Capture Pumping Technology (Elsevier, Amsterdam, 2001), p33
2. C. K. Sinclair, P. A. Adderley, B. M. Dunham, J. C. Hansknecht, P. Hartmann, M. Poelker, J. S. Price, P. M. Rutt, W. J. Schneider, and M. Steigerwald, Phys. Rev. ST Accel. Beams 10, 023501 (2007).
3. F.M. Penning, Philips Technical Review, 2 (1937) 201.
4. Ultra-high vacuum ion pump power supply with pump current resolution < 1 nA. Model EOS-902, from Eye On Science LLC, 527 Lakeshead Dr., Williamsburg, Va 23185.
5. Leybold extractor gauge, model IE 514 with IONIVAC IM 520 controller.

High Brightness and High Polarization Electron Source for Spin-LEEM

T. Nakanishi[1], N. Yamamoto[1], A. Mano[1], Y. Nakagawa[1], T. Konomi[1], M. Yamamoto[1], S. Okumi[1], X. Jin[2], T. Ujihara[2], Y. Takeda[2], T. Ohshima[3], T. Kohashi[3], T. Yasue[4], T. Koshikawa[4], T. Saka[5], and T. Kato[6]

[1]Graduate School of Science, Nagoya University, Nagoya 464-8602, Japan
[2] Graduate School of Engineering, Nagoya University, Nagoya 464-8601, Japan
[3] Central Research Laboratory, Hitachi Ltd. Tokyo, 185-8601, Japan
[4]Fundamental Electronics Research Center, Osaka Electro-Communication University, Osaka 572-8530, Japan
5Daido Institute of Technology, Nagoya 457-8531, Japan
[6]Daido Steel Co. Ltd., Nagoya 457-8531, Japan

Abstract. A point-like emission mechanism is required for a GaAs polarized electron source to produce an electron beam with high brightness. This is realized by changing the direction of injection laser from a front-side to a back-side of the photocathode. Based on this concept, a 20keV gun (JPES-1) was constructed and a transmission photocathode including an active layer of a GaAs-GaAsP superlattice was installed. It produced a laser spot diameter as small as 1.3μm for the 760~810nm laser wavelength, and the brightness of $\sim 2\times10^7 A\cdot cm^{-2}\cdot sr^{-1}$ corresponding to a reduced brightness of $\sim 1.0\times10^7\ A\cdot cm^{-2}\cdot\ sr^{-1}\cdot V^{-1}$ was obtained for an extracted current of 5.3μA. This brightness is still smaller than those of W-field-emitters, but one order of magnitude higher than those of LaB_6 emitters. The peak polarization of ~90% was achieved at the same time by the photocathode which has an inserted GaAs thin layer between the GaAsP substrate and the GaAaP buffer layer. It demonstrated that the strain-control of the GaAsP buffer layer is indispensable to achieve the highest polarization. The charge density lifetime of 1.8×10^8 C/cm^2 was observed for an extracted current of ~3μA. The second gun system (JPES-2) was already constructed and mounted to a LEEM apparatus operated at OECU..

Keywords: Polarized electrons, LEEM, SPLEEM, transmission photocathode
PACS: 29.25.Bx, 79.60.-i

MOTIVATION

Low Energy Electron Microscopy (LEEM) is a powerful tool to observe the structure and morphology of ultra-thin metal film grown on the metal substrate [1]. Using polarized electron beam for LEEM (SPLEEM) the magnetic domain structure can be observed, since the magnetic sensitivity is based on the spin-dependent exchange scattering between beam and specimen. Images of the spin-up and spin-down asymmetry can pick up the exclusive features that purely originate in the magnetism of the sample [2].

In order to enable a real time observation of the film growth, one image should be taken within 0.1 seconds and it has been realized already by LEEM with the LaB_6

CP1149, *Spin Physics, 18th International Spin Physics Symposium*
edited by D. G. Crabb, D. B. Day, S. Liuti, X. Zheng, M. Poelker, and Y. Prok

emitter but it takes more than 10 seconds by the available SPLEEM, as shown in Table 1. The motivation of this work is to overcome such a dilemma by improving the brightness of the PES by two order magnitude or more.

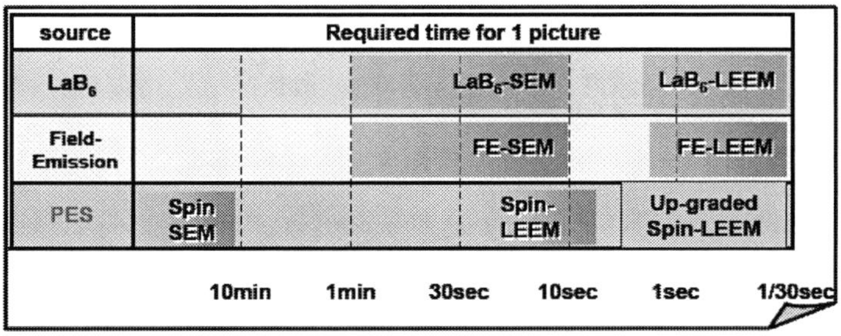

TABLE 1. Required time for taking one picture by various microscopes.

KEY TECHNOLOGY

In order to minimize the beam spot, a new geometry is proposed as shown in Figure1, where the laser lights transmit through the photocathode and the excited electrons are emitted in the forward direction. Using this transmission type photocathode, the distance between a focus lens and the photocathode can be shortened to a few mm and the laser spot size less than 10μm will be easily obtained.

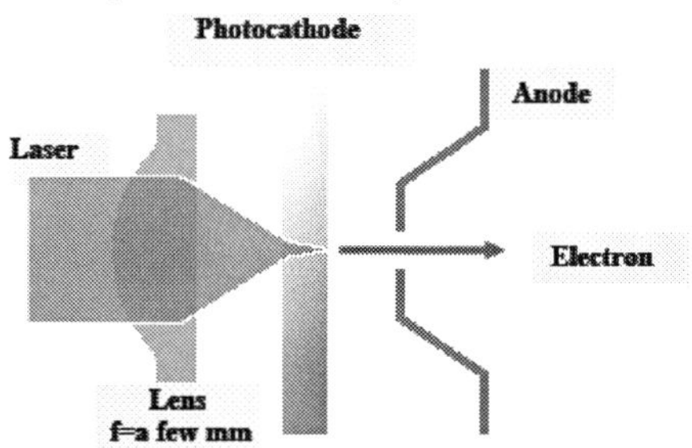

FIGURE 1. Conceptual view of transmission type photocathode to minimize electron source size.

EXPERIMENTAL EQUIPMENT

Laser optical system

The design of our laser optical system is based on that developed by one of the authors (T.O.) for a 1keV-SEM source with high brightness and mono-chromaticity [3]. The components of our laser system and the measured laser spot profile are shown in Figure2 [5]. The input laser light is guided to the gun by a single mode optical fiber. Its output light is collimated and converted to circularly polarized light by combination of a polarizing beam splitter and a quarter wave-plate, and finally focused on the active layers of the photocathode by the focusing lens (f=4.2mm, NA=0.5). The focusing lens is mounted on a translation XYZ stage to optimize the laser illumination conditions.

The laser focusing spot size was monitored by using reflected laser light from the surface of photocathode. The 180° reflected light is separated by a polarizing beam splitter and focused on a CCD by an imaging lens (f=100.0mm). Combining an image magnification factor of 23.8 on the CCD and 1pixel size of 4.65μm, the resolution of this CCD monitor is expected to be 0.20μm/pixel, and the obtained laser spot size was 1.3~1.4μm (FWHM) for 780nm laser wavelength.

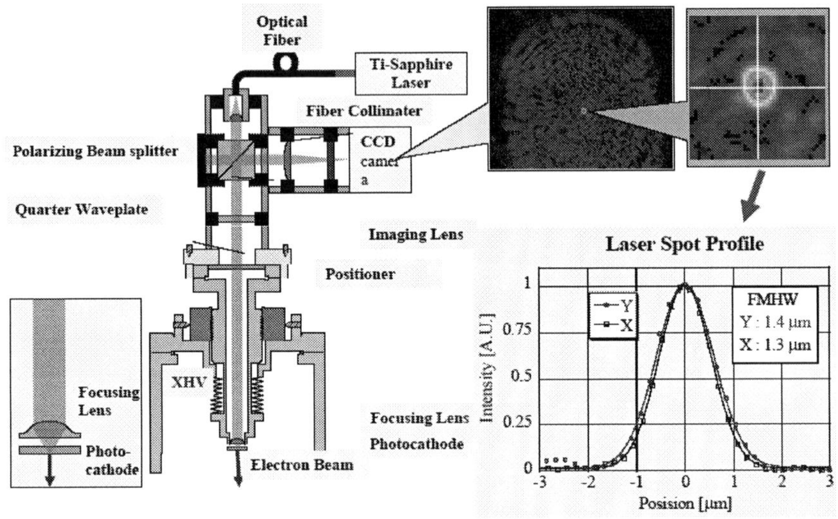

FIGURE 2. Components of laser optics equipment and the laser spot profiles on the photocathode

20keV gun

A schematic illustration of our PES system (named JPES-1) is shown in Figure3 [5]. The JPES-1 consists of four vacuum chambers including the activation chamber, the gun chamber, and other chambers for measurements of beam brightness and spin polarization. The activation chamber has the vacuum level of ~10^{-9}Pa, and the gun chamber maintains the ultrahigh vacuum environment of an order of 10^{-10}Pa near the photocathode surface.

The heat cleaning and NEA activation are done in the activation chamber and the photocathode pack is installed in the electrode using load lock mechanism. The beam brightness is measured at the downstream of the gun. The beam is deflected by 90° angle using a spherical condenser to change the polarization vector from longitudinal to transverse, and the beam polarization is measured by a 100kV Mott chamber analyzer.

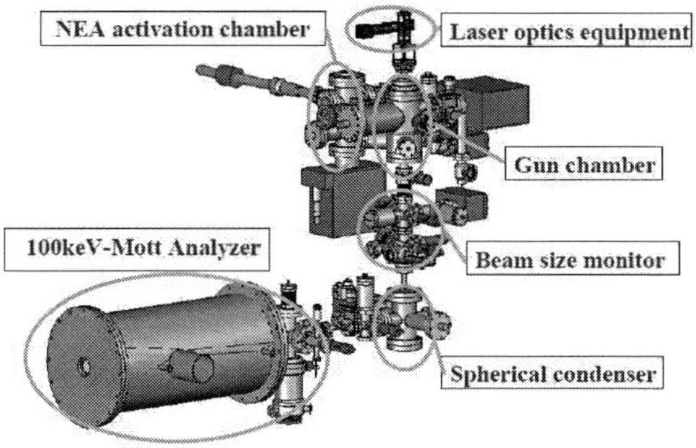

FIGURE 3. Transmission-type polarized electron source for high brightness performance

As the beam energy of the LEEM operated at OEC Univ. is 20keV, the same beam energy is employed for the gun. A gap distance of electrode is set to be 4mm, and the electric field gradient to be 5MV/m. In order to minimize the dark current, Mo and Ti are used for the cathode and the anode materials, respectively [4]. The electrode surface was finished by the electro-buff-compound polishing, and the dark current was well suppressed below 6nA level at 25kV bias voltage.

Transmission photocathode and spin polarization

The GaAs/GaAsP strained superlattice layer is used to produce the highly polarized electrons (≥85% @780nm) [5, 8]. As a substrate material GaP is chosen, since it is transparent for 780nm laser light due to the wide band-gap corresponding to 550nm.

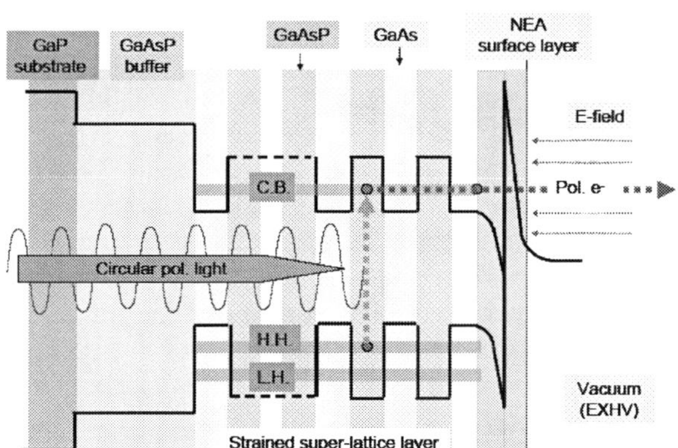

FIGURE 4. Schematic view of energy band structure for the transmission photocathode

We fabricated two types of photocathode to achieve 90% polarization, as shown in Figure5. Maximum polarization was limited to below ~60% using the first photocathode. Polarization improved to ~90% by using a thin GaAs interlayer between the GaP substrate and the GaAsP buffer layer [6, 7]. It is the same magnitude of electron polarization obtained by the conventional photocathode with the same superlattice structure but grown on the GaAs substrate [8].

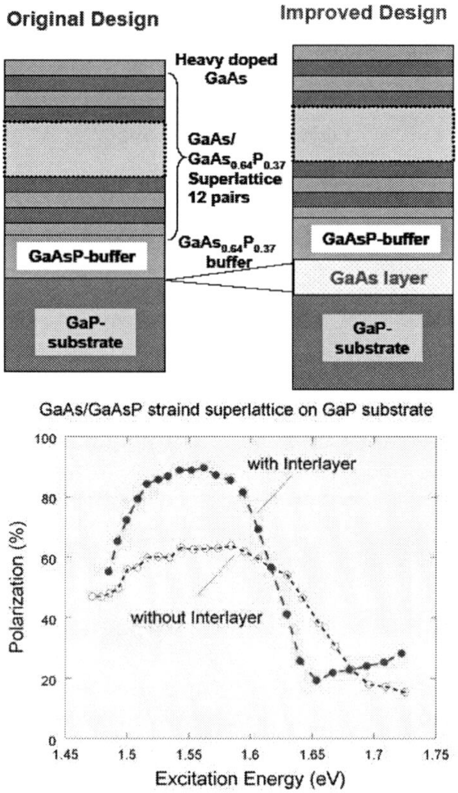

FIGURE 5. Two designs for photocathodes and the corresponding polarization spectra

The large difference between the maximum polarizations clearly shows the importance of strain control in the buffer layer. With the various studies on strain property of our crystals it was concluded that the spin-polarization of electrons produced from the photocathode was strongly related to the types of defects induced in the buffer layer. The tensile strain in the buffer layer was relieved by cracks, while the compressive strain was relieved by dislocations, and those cracks and dislocations propagated into the active superlattice layer. From our data, it was demonstrated that the cracks parallel to the polarized electron escape direction slightly influenced to the polarization, while the dislocations intersecting with escape direction caused the sizable spin-flip depolarization [7]. The quantum efficiency for the 90% polarization photocathode was ~0.1% at 790nm laser wavelength.

Brightness measurements

The procedure of brightness measurement is shown in Figure 6 [5]. The beam profile was measured using a pair of knife-edge slits located 53.1cm downstream from the photocathode, and the beam radius (FWHM) was determined by Gaussian fit of this profile. The reduced brightness can be derived using parameters of this beam radius, the estimated electron source radius and the emission current by using the formula in Figure 6, where the obtained value is also shown.

Conditions: Beam energy (U) = 20keV, Beam current (I)=5.3 μA

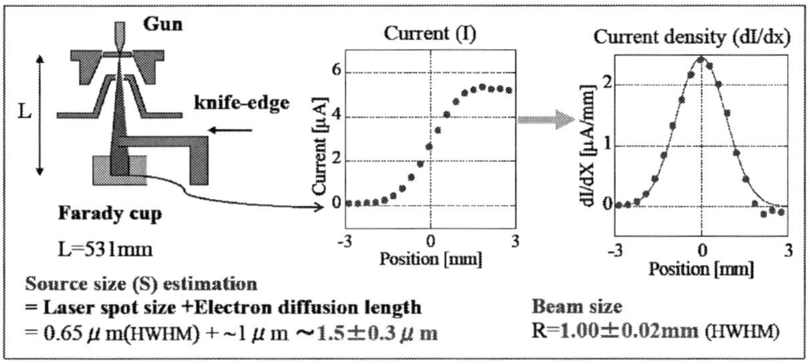

Reduced Brightness $B_r = I \dfrac{1}{\pi S^2} \dfrac{L^2}{\pi (R-S)^2} \dfrac{1}{U}$ =1.0±0.4 × 107 A m-2 sr-1 V-1

FIGURE 6. Procedure of beam brightness measurement and the obtained brightness results

The dependence of reduced brightness on the extracted voltage was measured and the result is shown in Figure 7 (left), where the saturation behavior is obviously observed above 10kV bias voltage. The current dependence of the reduced brightness was also studied at 20kV with the beam source radius kept at 0.95~1.00mm and the result is also shown in Figure 7 (right), where the reduced brightness increases linearly up to the maximum current of 5.3μA. For this current, the angular current density and the current density were estimated to be 250mA/sr, ~5.1A/mm^2 respectively, and the reduced brightness to be (1.0±0.4)×10^{7}A/m^2/sr/V. This brightness is three orders of magnitude higher than that of the available polarized electron source. Moreover, it still smaller than those of the W-field-emitters, but one order of magnitude higher than those of LaB_6 emitters. The charge density lifetime of 1.8×10^8 C/cm^2 was observed for an extracted current of ~3μA [5].

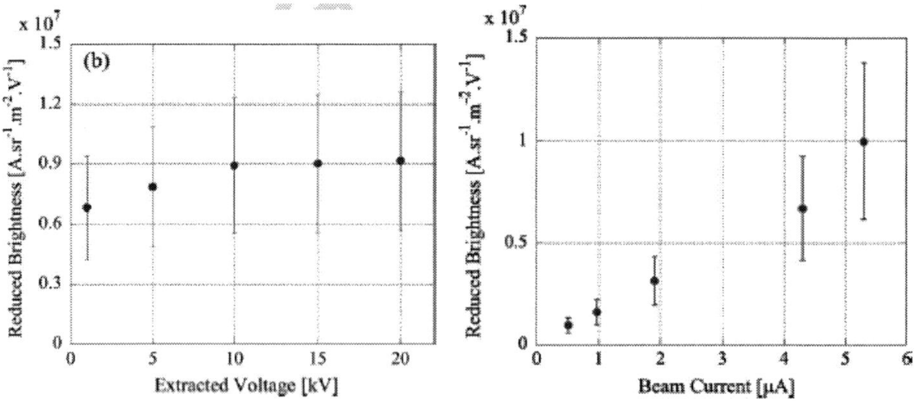

FIGURE 7. Reduced brightness dependence on the extracted voltage and beam current

By the work with JPES-1, we could demonstrate that the polarized electron source with transmission photocathode is highly effective to achieve high brightness and high polarization simultaneously. Based on this work, the second gun system (JPES-2) was already constructed at Nagoya Univ. and now it was mounted for the LEEM apparatus operated at OECU.

ACKNOWLEDGEMENTS

This work has been financially supported by the Japan Science and Technology (JST) Agency as a technology development program for advanced measurements and analysis (program leader: T. Nakanishi).

REFERENCES

1. E. Bauer, T. Duden and R. Zdyb: "Spin-polarized low energy electron microscopy of ferromagnetic thin films", J. of Phys. D: Appl. Phys., 35 (2002) pp. 2327-2331.
2. K. L. Man et al.: "Growth morphology, structure, and magnetism of ultrathin Co films on W (111)", Phys. Rev. B 67, p. 184402 (2003).
3. T. Ohshima and M. Kudo: "Electron Beam Brightness from Negative-Electron-Affinity Photocathodes for Scanning Electron microscopy Application", Jpn. J. Appl. Phys., Vol.43, No.12 (2004) pp. 8335-8340.
4. F. Furuta, T. Nakanishi, et al.: "Reduction of field emission dark current for high-field gradient electron gun by using molybdenum cathode and titanium anode", Nucl. Intrum. Methods A538 (2005) pp. 33-44.
5. N. Yamamoto, T. Nakanishi, A Mano, X. Jin, T. Ujihara et al., "High brightness and high polarization electron source using transmission photocathode with GaAs-GaAsP superlattice layers" J. Appl. Phys., Vol.103, No. 6 (2008), 064905.
6. X. Jin, N. Yamamoto, T. Ujihara, T. Nakanishi et al.,"Super-high brightness and high spin-polarization photocathode" Appl. Phys. Express Vol. 1 (2008), No. 4, 045002.
7. X. Jin, N. Yamamoto et al., "Highly spin-polarized electron photocathode based on GaAs-GaAsP superlattice grown on mosaic-structure buffer layer" J. of Crystal Growth 310 (2008) pp. 5039-5043
8. T. Nishitani, T. Nakanishi, et al.: "Highly polarized electrons from GaAs-GaAsP and InGaAs-AlGaAs strained-layer superlattice photocathodes", J. Appl. Phys., Vol.97, No.9, 094907 (2005).

Ab Initio Calculation of Effective Sherman Functions in MeV Mott Scattering

K. Aulenbacher, V. Tioukine

Institut f. Kernphysik der Universität Mainz,
J.J. Becherweg 45, D-55099 Mainz, Germany, F. R.

Abstract. A Mott polarimeter operating at energies from 1 to 3.5 MeV and at backward scattering angle of 164 degree has been installed at the MAMI accelerator. Its detection system is based on magnetic spectrometers allowing for efficient background suppression. The useful range of primary currents is between 10 nA and 100 μA, hence adapted to the typical beam currents at MAMI. The reduction of analyzing power due to multiple scattering can be calculated directly from the cross sections with the help of Monte Carlo simulation.

Keywords: Pol. beams, multiple scattering, Polarization in interactions and scattering
PACS: 29.27.Hj -11.80.La -13.88+e

INTRODUCTION

Mott detectors determine the beam polarization P_{beam} by measuring the experimental asymmetry $A_{exp} = P_{beam} \cdot S_{eff}$ in elastic scattering of polarized electrons off targets with high nuclear charge, typically gold. P_{beam} can then be determined once that the effective analyzing power S_{eff} is known. The statistical efficiency $Q = S_{eff}^2 \cdot (I_{scat}/I_0)$ (I_{scat} =scattered current, I_0 = primary current on target) reaches a maximum at about 25 keV beam energy and then decreases since I_{scat} is proportional to the differential cross section being roughly proportional to $1/E^2$. It seems therefore not attractive to use higher energies since the size of the apparatus increases, whereas the statistical efficiency decreases. However, both effects are only of minor concern at the MAMI accelerator, which evidently provides high energies and where the low efficiency is compensated for by high beam currents. Hence, in our case, the following advantages of 'high' energies - in the range of several MeV - become effective:

- MeV beam energies could allow for better reproducibility, e.g. because of lower sensitivity with respect to varying magnetic fields, high pointing stability and low geometric emittance.
- The reduction of the effective analyzing power with target thickness is much less pronounced thus allowing to use much thicker -i.e. less delicate- targets.
- The energy loss dE/dx is lower; leading to less heat dissipation in a thin target.
- In a single scattering experiment, the effective analyzing power S_{eff} can only be determined if the analyzing power S_0 which corresponds to single elastic scattering has been calculated theoretically. As an input for such calculations details of the electronic and nuclear charge distribution are required. However, the momentum transfer at 3.5 MeV and backward scattering angle is sufficiently high that the

CP1149, *Spin Physics, 18th International Spin Physics Symposium*
edited by D. G. Crabb, D. B. Day, S. Liuti, X. Zheng, M. Poelker, and Y. Prok

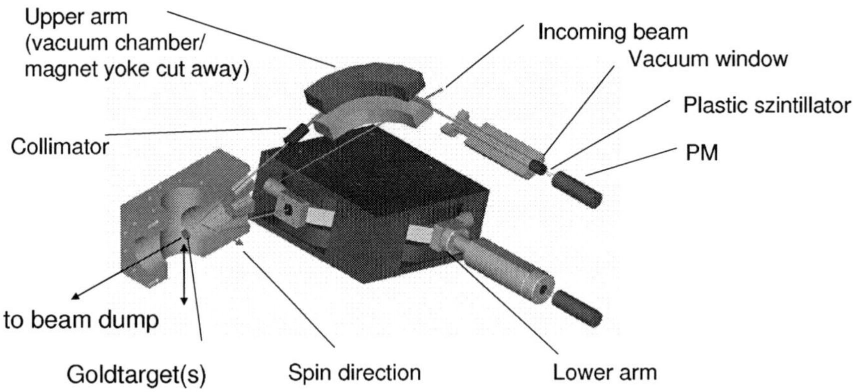

FIGURE 1. Mott polarimeter with two magnetic spectrometers for backward scattered electrons

atomic electrons become rather unimportant for scattering whereas the momentum transfer is still small enough to make the nucleus appear as a point charge. The variation of S_0 for a point nucleus in comparison to an exact calculation with realistic nuclear and atomic charge distributions is only 1% (relative). The uncertainty introduced by the model dependence of charge distributions is therefore negligible.

APPARATUS AT MAMI

After passing the 3.5 MeV injector linac, the electron beam is deflected from the main beam line towards the Mott polarimeter (fig. 1). The energy may be varied by changing the r.f.- phase of the third acceleration section of the linac, which allows to achieve energies as low as 1 MeV. Typically the spin direction is rotated into the horizontal transverse direction with a Wien filter before entering the linac [1]. The Mott asymmetry is then observed for up/down scattered electrons, the scattering angle being 164 degree. The solid angle is defined by cylindrical diaphragms with 4 mm aperture located at 240 mm distance from the scattering center. Double focusing 90 degree spectrometers are used. Because the detectors lie out of the line of sight to the target the background is effectively suppressed. Electrons are sufficiently energetic to pass a 0.3 mm thick aluminium vacuum window and reach the air mounted plastic szintillators. Gold targets of different thicknesses ($d = 0.1, 0.25, 0.5, 1, 15 \mu m$) can be moved into the beam, alternatively a szintillation screen can be used for inspection of beam shape and position.

RESULTS

Statistical efficiency: Typically, a scattering rate of 1.7 kHz is observed per μA of beam current hitting a 1 μm thick gold target, which is in agreement with the expected rate from the elastic cross section. Figure 2 also shows that for the thickest target (15 μm), the rate is increased over-proportionally, which results from the multiple scattering

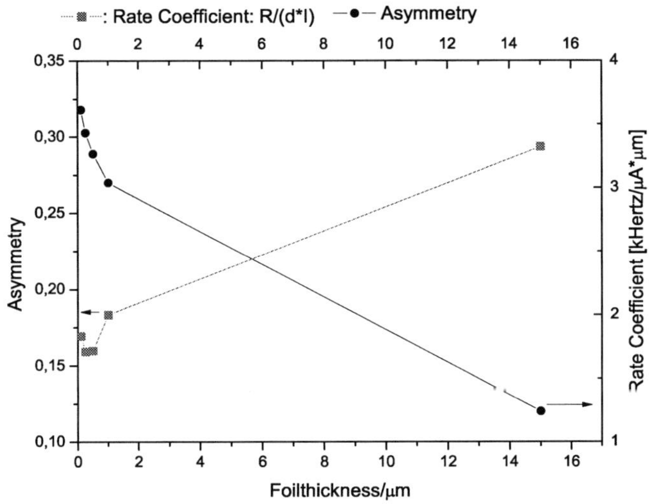

FIGURE 2. Rate coefficient and asymmetry vs. target thickness. Beam energy 3.5 MeV.

processes to be discussed below. Multiple scattering is also the main reason for the dramatic reduction of effective analyzing power $S_{eff} = A_{exp}/P_{beam}$ for the 15μm target. However, if statistical efficiency is considered, this is more than compensated by the higher rate. Therefore asymmetry measurements have the highest efficiency if performed with the 15 μm target, leading to a measurement time of 1 hour for 1% accuracy at a primary beam current of 10 nA. Measurement times at the higher beam currents which are characteristic for electron scattering experiments are of the order of a few seconds. One may worry that ionization losses in the target will lead to eventual destruction due to heating. Such problems may be alleviated by using the thinner targets, since radiative surface cooling should remain the same but the dissipated power is $\propto d$. We have already tested the $0.1\mu m$ target at the maximum beam current possible at MAMI, $100\mu A$.

Reproducibility: Several beam parameters (position of the foil, beam current, beam movement in position and angle) were varied, the asymmetry being insensitive to this manipulations at a level of $\Delta A/A < 1\%$. Long term observations showed a relative drift of asymmetry by 0.3% in 8 hours, which may be due to a change of beam polarization because of cathode aging. An independent simultaneous monitor is presently under construction in order to disentangle changes of polarization from apparative drifts.

AB INITIO CALCULATION OF S_{eff}

In a real experiment the single elastic scattering analyzing power S_0 is modified, which leads to an 'effective' analyzing power S_{eff}. For example, a determination of S_{eff} must take into account the reduction of S_0 due to multiple scattering. Analytic approaches [2] in order to extrapolate to zero target thickness introduce errors: we estimate presently

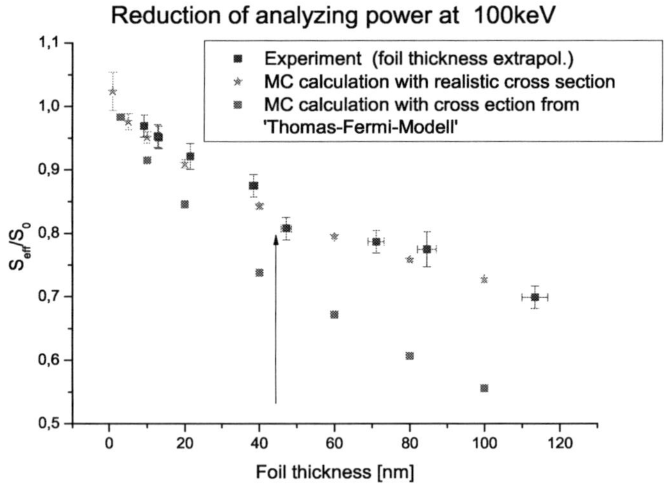

FIGURE 3. Comparison of MC simulation and data based on extrapolation (100keV).

the relative error caused by the extrapolation for our apparatus to be 4%. However these ambiguities can be reduced if the reduction is determined by a Monte Carlo (MC-) simulation, which was successfully demonstrated by M. Khakoo et al. [3] for energies up to 120 keV. Such 'ab initio' calculations are limited in accuracy only by the input accuracy of the cross sections and the corresponding spin rotation parameters. The MC results then predict the reduction of analyzing power for a target of finite thickness, hence eliminating the need for an extrapolation. We have checked this method for previous results obtained by us with a 100 keV Mott and found good agreement (fig. 3). Besides the agreement, figure 3 reveals that good knowledge of the atomic charge distribution is important, a simple atomic model with a Thomas Fermi charge distribution which reproduces cross section and Sherman function at large angles does not reproduce the data. This probably happens because the forward cross sections at angles below 5 deg are not correct in this model. At thicknesses larger than $d \approx 40\,nm$ an increasing number of electrons is subject to energy losses larger than the energy resolution of the experimental set up and does not contribute to the signal. This results in a change of slope, which is indicated by the arrow in fig. 3. For larger d the MC results therefore depend on the value assumed for the detector resolution. Albeit a reasonable value for the energy resolution was chosen, it should be noted that a highly accurate determination for large thicknesses requires a corresponding knowledge of the properties of the detection system.

Preliminary results for 1 and 2 MeV are shown in figure 4 which are the result of 100 hours runtime each on a conventional PC, the statistical accuracy can be increased easily by using a small PC farm. The experimental data points are normalized to the extrapolated asymmetry at d=0, which leads to the above mentioned 4% error bar. The MC data are normalized to S_0 and carry statistical error bars. Even for the limited

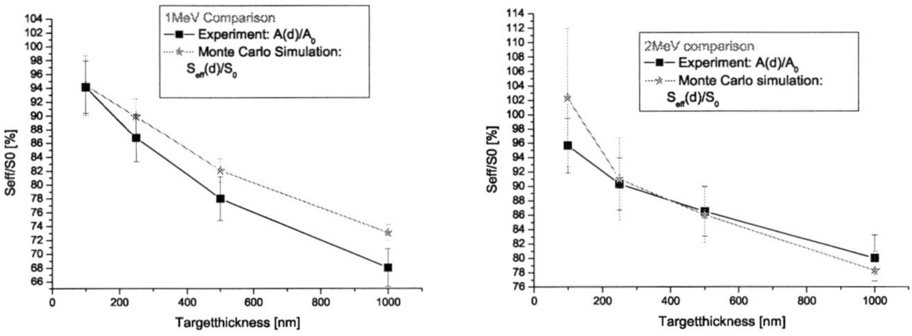

FIGURE 4. Decrease of analyzing power with target thickness for 1 and 2 MeV.

statistical accuracy some discrepancy seems to be present at 1 MeV. This is probably due to the fact that, since at these energies high accuracy cross section calculations as provided by [4],[5] are not available to us, a Thomas-Fermi model was used again. However, the extension of high accuracy cross section calculations to higher energies seems to be no principal problem.

CONCLUSION

A 3.5 MeV Mott polarimeter is in operation at the accelerator MAMI in Mainz. It offers sufficient measurement speed and good reproducibility for primary currents in the range 0.01-100 μA which coincides with the current requirements of most MAMI experiments. The effective analyzing power of the targets for elastic scattering can be simulated by a MC calculation which takes into account the microscopic cross sections ('ab initio' calculation). It seems possible to increase the accuracy of the Mott polarimeter by supporting the foil thickness extrapolation method by MC data. We believe that a reduction of the total systematic error below 2% is possible.

ACKNOWLEDGMENTS

This work was supported by the SFB 443 of the Deutsche Forschungsgemeinschaft.

REFERENCES

1. V. Tioukine and K. Aulenbacher, *NIM* **A 568** *537* (2006).
2. T.J. Gay et al. , *Rev. Sci. Instrum.* **63 (1)** *114* (1992).
3. M. A. Khakoo et al. , *Phys. Rev.* **A 64** *052713* (2001).
4. G. Holzwarth and H.J. Meister, *Nucl. Phys.* **59** *56* (1964).
5. A. W. Ross and M. Fink, *Phys. Rev.* **A 38,(12)***6055* (1988).

Compton Polarimetry at ELSA

Wolfgang Hillert, Bastian Aurand, and Jürgen Wittschen

Physikalisches Institut, Universität Bonn,
Nussallee 12, D-53115 Bonn, Germany

Abstract. Part of the future polarization program performed at the Bonn accelerator facility ELSA will rely on precision Compton polarimetry of the stored transversely polarized electron beam. Precise and fast polarimetry poses high demands on the light source and the detector which were studied in detail performing numerical simulations of the Compton scattering process. In order to experimentally verify these calculations, first measurements were carried out using an argon ion laser as light source and a prototype version of a counting silicon microstrip detector. Calculated and measured intensity profiles of backscattered photons are presented and compared, showing excellent agreement. Background originating from beam gas radiation turned out to be the major limitation of the polarimeter performance. In order to improve the situation, a new polarimeter was constructed and is currently being set up. Design and expected performance of this polarimeter upgrade are presented.

Keywords: Compton scattering; Microstrip detector; Polarimetry; Polarization
PACS: 13.60.Fz; 29.40.Wk; 95.75.Hi; 29.27.Hj

INTRODUCTION

The experimental hadron physics program at the **EL**ectron **S**tretcher **A**ccelerator ELSA ([1], see Fig. 1) is based on the availability of polarized beams and polarized solid state targets. A beam of polarized electrons is generated with a dedicated source [2] and post accelerated in a linear accelerator, the booster synchrotron and the stretcher ring. During post acceleration in the stretcher ring, crossing of depolarizing resonances will cause significant beam depolarization unless these effects are being compensated by application of sophisticated correction schemes [3]. Therefore, precise determination and monitoring of the beam polarization is mandatory and is performed in the low energy beam-line (by means of Mott polarimetry) and the external high energy beam-line to the experiment (by means of Møller polarimetry). In order to measure the polarization of the stored electron beam in the stretcher ring directly on a time scale of minutes, a Compton polarimeter has been developed [4] and installed at ELSA, and is currently being upgraded. It is based on Compton backscattering of circularly polarized photons off the stored beam of transversely polarized electrons. The electron beam polarization is extracted from measuring the shift of the center of the backscattered photon spatial distribution when switching the polarization of the incoming photons from right-handed to left-handed. In order to record the intensity profile of the backscattered photons, a counting silicon microstrip detector [5] is utilized.

CP1149, *Spin Physics, 18th International Spin Physics Symposium*
edited by D. G. Crabb, D. B. Day, S. Liuti, X. Zheng, M. Poelker, and Y. Prok

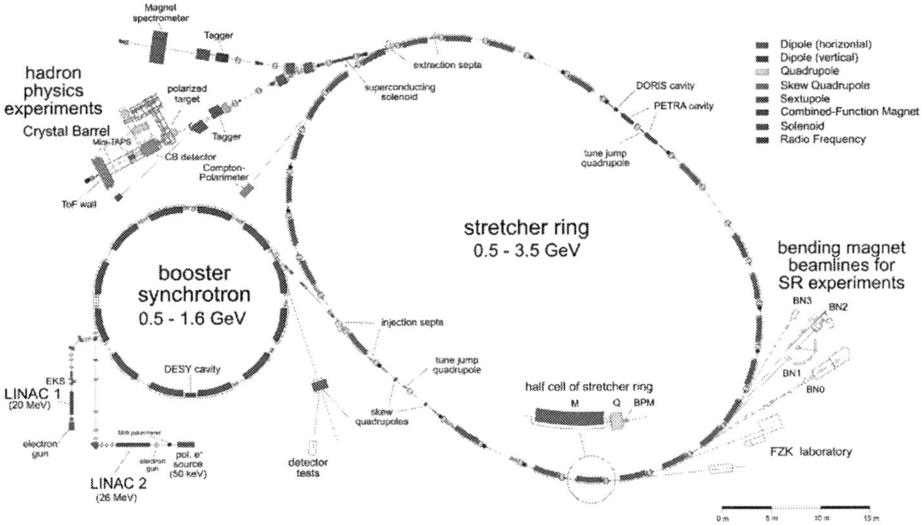

FIGURE 1. The Electron Stretcher Accelerator ELSA at Bonn University.

POLARIMETER SET-UP

In order to achieve a minimum electron beam divergence, the crossing region of photon and electron beams is placed in a defocusing quadrupole magnet of the stretcher ring. A crossing angle of about 3 mrad was chosen to protect the final dielectric deflection mirror, which is mounted off axis in vacuum, from synchrotron radiation emitted by the electron beam (see Fig. 2 (b)). Compton-scattered photons are detected 15 m away from the interaction region. Assuming typical vertical beam widths of the electron and the photon beam of about 2 mm, the interaction length is below 1.4 m. In order to suppress background produced by beam gas radiation, the interaction region was placed in one of the shortest straight sections of the stretcher ring, where a defocusing quadrupole magnet is directly surrounded by dipole magnets, as shown in Fig. 2 (a).

The photon counting detector is based on a silicon microstrip sensor of type BABAR 1 (size 38.4 mm x 41.3 mm), which contains 768 horizontally oriented strips with a pitch of 50 microns. In the first version of the detector electronics, every second strip was read out using dedicated front-end chips (see [5]), partially developed by the ATLAS pixel group of Bonn University [6].

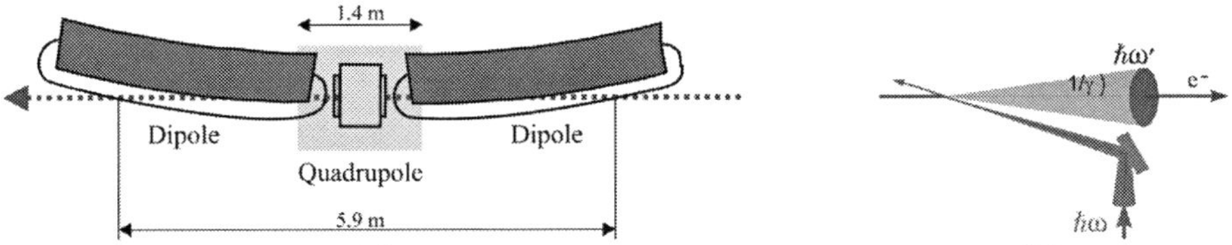

FIGURE 2. The interaction region in ELSA. Photon and electron beams are crossed in a defocusing quadrupole magnet, directly surrounded by dipole magnets (a). The crossing angle is 3 mrad in order to protect the deflection mirror from synchrotron radiation (b).

In order to be detected, the Compton-backscattered photons (with energies in the MeV-range) are pair converted in a lead converter of two radiation lengths thickness, which is placed directly in front of the sensor.

NUMERICAL SIMULATIONS

The photon counting rate was calculated for each of the 384 sensitive strips of the sensor by numerical integration over the 5-dim phase space of a 100% transversely polarized electron beam of fixed energy (chosen in the range between 1 and 3 GeV) and a 100% circularly polarized photon beam (wavelength $\lambda = 514.5$ nm). Each strip was divided into 41 sections and the number of photons $\dot{N}$ impinging on a section centered at (x_d, z_d) was derived by solving

$$\dot{N}(x_d, z_d) = \int_{S_0}^{S_1} ds \int_{Z_{\min}}^{Z_{\max}} dz \int_{X_{\min}}^{X_{\max}} dx \int_{Z'_{\min}}^{Z'_{\max}} dz' \int_{X'_{\min}}^{X'_{\max}} dx' \; \frac{\partial(\vartheta^*, \varphi^*)}{\partial(x,z)} \cdot \rho_e \cdot \rho_\gamma \cdot \frac{d\sigma}{d\Omega^*} \cdot \eta(E_\gamma) \quad (1)$$

where electron and photon beams are represented by their phase space distribution functions ρ_e and ρ_γ. The Compton cross-section $d\sigma/d\Omega^*$ was taken from [7], and the pair conversion efficiency $\eta(E_\gamma)$ for a 1.7 radiation lengths converter was derived from [8] using Compton kinematics to determine E_γ.

The shift of the center of the profile of backscattered photons was determined from subtraction of the weighted sums

$$\Delta\bar{z} = \sum_{i,j} z_j \dot{N}^+(x_i, z_j) - \sum_{i,j} z_j \dot{N}^-(x_i, z_j) \quad (2)$$

over the calculated 2-dim profiles in case of right-handed ($\dot{N}^+$) and left-handed ($\dot{N}^-$) circularly polarized photons and is presented in Figure 3.

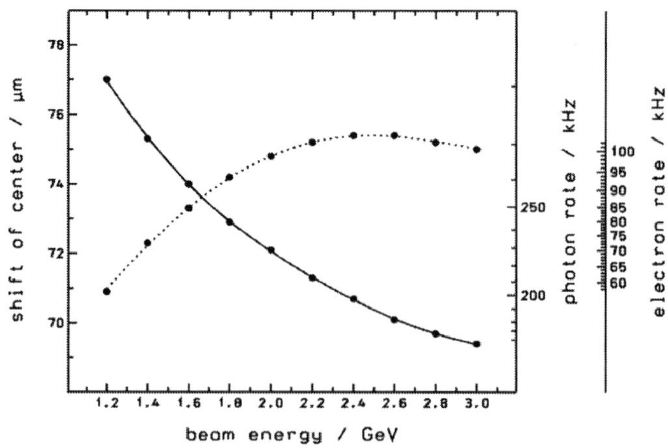

FIGURE 3. The shift of the center of the profile of backscattered photons (solid line) and pair converted electrons / positrons (dashed line) in case of switching from right-handed to left-handed circular polarization of the laser beam. Corresponding overall counting rates are indicated at the right hand side of the figure for a laser power of 10 W and an electron beam current of 100 mA.

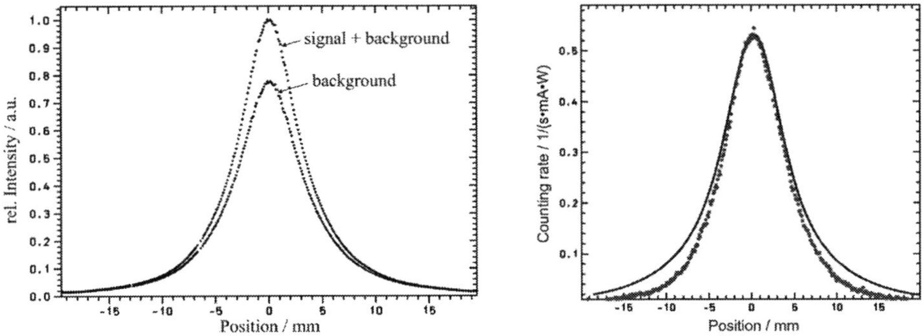

(a) **(b)**

FIGURE 4. Measured and calculated intensity profiles of pair converted backscattered photons. The left figure (a) shows profiles derived with the argon ion laser switched on (signal + background) and off (background). The right figure (b) presents a comparison of the measured signal (dots) and a simulated profile (solid line), normalized to 1 mA beam current and 1 W laser power.

COMPARISON WITH MEASURED DATA

First measurements were carried out using a 10 W cw argon ion laser (green line) and the prototype version of a counting microstrip detector in combination with a two radiation lengths lead converter. Background produced from beam gas radiation was derived from an additional measurement with the laser beam switched off.

Figure 4 (a) shows the measured intensity profiles in case of a 2.3 GeV, 20 mA stored electron beam. Due to the limited laser power and moderate vacuum conditions in the interaction region (pressure in the 10^{-7} mbar regime) the signal to background ratio is less than 1:3. Due to this considerably low value, a measuring time of about one hour is expected in order to reach a polarization error of 2 %.

Figure 4 (b) presents a comparison of measured and calculated profiles of backscattered photons, both normalized to 1 mA electron beam current and 1 W laser power. The profiles agreement is excellent for the center part. Slightly lower counting rates observed in the wings of the measured profile may be attributed to the difference in radiation lengths of the lead converters used in the measurement (2 X_0) and the numerical simulation (1.7 X_0).

POLARIMETER UPGRADE

In order to significantly enhance the signal to background ratio, a new 2x20 W two-beam solid state laser system was purchased and is currently being set up at the accelerator. In addition, the vacuum chamber of the interaction region was replaced by a newly developed one including electron beam position monitoring, water cooling, and an additional ion getter pump. With these improvements, a signal to background ratio of at least 3:1 is expected.

Figure 5 shows the set-up of the laser system and the beam-line in the accelerator tunnel. The two laser beams become circularly polarized by means of rotatable quarter wave plates. Two beam telescopes focus the beams to rms diameters of less than 3 mm in the interaction region, where the beams are crossed. Each laser beam polarization is monitored after the interaction by means of a second quarter wave plate, a polarizing beam splitter and two photo-diodes. The correct positioning of the laser beams can be controlled online using two four-element photo-diodes, detecting the small fraction of laser light being transmitted through the dielectric mirrors which deflect the laser beams to the polarization analyzer box.

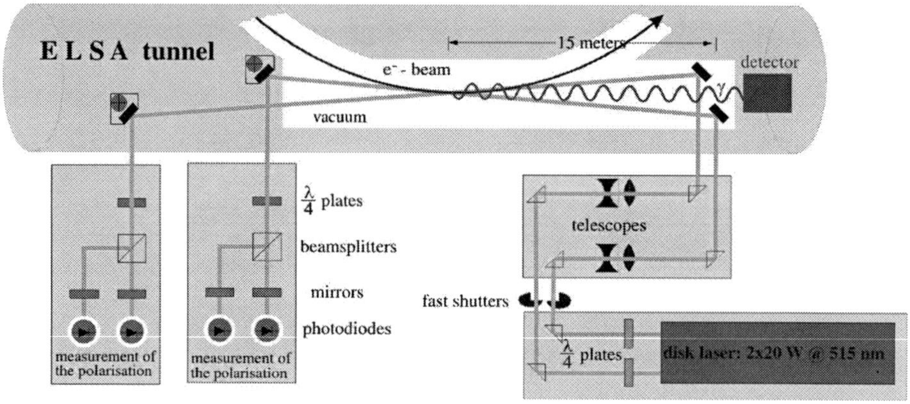

FIGURE 5. Set-up of the upgraded Compton polarimeter based on a 2x20 W two-beam solid state laser system and a counting silicon microstrip detector with improved LVDS micro-electronics.

The improved electronics of the silicon microstrip detector consists of six front end chips, each having 128 channels in order to read out all 768 strips of the sensor. Each channel consists of a low noise charge sensitive amplifier, a CR-RC shaper, a comparator, and a 13 bit asynchronous binary counter. The digital part of the front end chips is built in low voltage differential signal (LVDS) technology which offers a high rate acceptance of up to 150 MHz. With a spatial resolution of 14 microns, the position error turns out to be below 0.7 microns, which results in an achievable absolute error of polarimetry of less than 1 %. Due to the high signal to background ratio and laser power, a polarization measurement error of 2 % should be obtained within minutes of measuring time.

REFERENCES

1. W. Hillert, *Eur. Phys. J.* A **28** s01, 139-148 (2006).
2. W. Hillert, M. Gowin, and B. Neff, "The 50 kV inverted source of polarized electrons at ELSA", in Proceedings of the 14th international Spin Physics Symposium SPIN2000, edited by K. Hatanaka et al., AIP Conference Proceedings 570, American Institute of Physics, Melville, NY, 2001, pp. 961-964.
3. M. Hoffmann et al, "The polarized electron beam at ELSA", in Proceedings of the 14th international Spin Physics Symposium SPIN2000, edited by K. Hatanaka et al., AIP Conference Proceedings 570, American Institute of Physics, Melville, NY, 2001, pp. 756-760 (2001).
4. W. Hillert et al., "The Compton Polarimeter at ELSA", in Proceedings of the 12th international Symposium on High Energy Spin Physics SPIN96, edited by C.W. de Jager et al., World Scientific, Singapore, 1997, pp. 785-787.
5. D. Doll et al., *Nucl. Instrum. Methods* A **492**, 356-364 (2002).
6. P. Fischer, *Nucl. Instrum. Methods* A **378**, 297-300 (1996).
7. F.W. Lipps, and H.A. Tolhoek, *Physica* **20**, 395-405 (1954).
8. D.H. Fitzgerald et al., Nucl. Instrum. Methods **192**, 193-203 (1982).

A Green Fabry-Pérot Cavity for Jefferson Lab Hall A Compton Polarimetry

Abdurahim Rakhman[†], Sirish Nanda[*], Paul Souder[†]

[*]*Thomas Jefferson National Accelerator Facility, Newport News, VA 23606, USA*
[†]*Department of Physics, Syracuse University, Syracuse, NY 13244, USA*

Abstract. A green laser (CW, 532 nm) based Fabry-Pérot cavity for high precision Compton Polarimetry is under development in Hall A of the Jefferson Laboratory. In this paper, we present the principle and the preliminary studies for our test cavity.

Keywords: Fabry-Pérot Cavity, Compton Polarimetry
PACS: 01.30.Cc

INTRODUCTION

Many experiments at Jefferson Lab require accurate knowledge of the polarization of the electron beam. In Hall A, beam polarization measurement is achieved by using both Møller and Compton Polarimeters. The Møller Polarimeter will be suitable only at low current (typically 10 μA) and the measurement cannot be done simultaneously with the experimental data taking. A Compton Polarimeter gives a non-intrusive measurement for higher current beam in which the electrons are scattered from the photons trapped in a high-finesse Fabry-Pérot cavity [1]. The presently operational Polarimeter uses an infra-red (1064 nm, 1.16 eV) laser as its photon source which is not capable of giving good accuracy at beam energies below 2 GeV. To reach 1% accuracy for high precision parity violation experiments like PREx [2] in Hall A, a green (532 nm, 2.33 eV) Fabry-Pérot cavity was proposed [3]. In this paper, we explain the principle of power buildup inside such a cavity and present some preliminary results from our test setup. As a conclusion, we also discuss how to achieve our design goal before installing the new system in the beam line.

THE COMPTON POLARIMETER

In a Compton Polarimeter, the longitudinal polarization P_e of the electron beam is calculated from the measurement of the experimental asymmetry A_{exp} in elastic scattering of circularly polarized (P_γ) photons off the electrons.

$$A_{\text{exp}} \equiv \frac{n^{+} - n^{-}}{n^{+} + n^{-}} = P_e P_\gamma \langle A_l \rangle \tag{1}$$

CP1149, *Spin Physics, 18th International Spin Physics Symposium*
edited by D. G. Crabb, D. B. Day, S. Liuti, X. Zheng, M. Poelker, and Y. Prok

where n^+ and n^- are the counting rates of scattering events before and after the electron spin flip ($P_e \rightarrow -P_e$). The asymmetry $< A_l >$ on the Compton cross section with longitudinally polarized electrons is calculated in the framework of QED. The longitudinal asymmetry A_l is plotted on Figure 1 as a function of the normalized energy of the scattered photons.

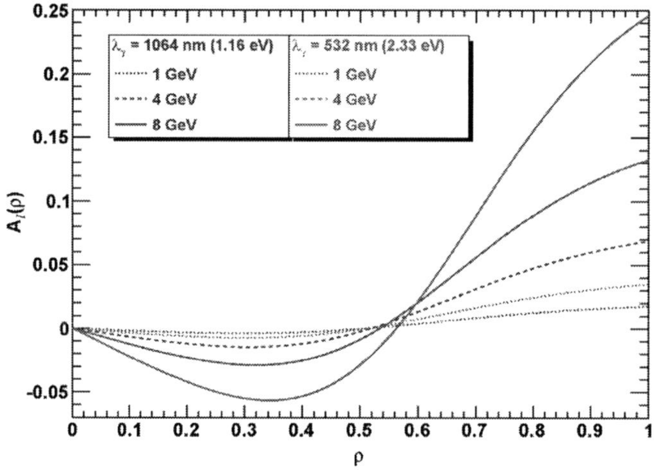

FIGURE 1. Longitudinal asymmetry for a 1 GeV, 4 GeV, 8 GeV electron beam and a green (2.33 eV) and an IR (1.16 eV) laser beam versus the scattered photon energy normalized to the maximum scattered photon energy.

The time needed to get the electron beam polarization with a statistical precision ($\Delta P_e/P_e$) is given by [4],

$$t \approx \frac{1}{L\left(\frac{\Delta P_e}{P_e}\right)^2_{stat} P_e^2 P_\gamma^2 \sigma_t \langle A_l^2 \rangle} \qquad (2)$$

where σ_t and $< A^2{}_l >$ are the total cross section and the squared theoretical asymmetry. For 1 GeV electrons, if the IR laser was used, the maximum value of A_l is 1.8 %, whereas with the green laser, its value is 3.5 %. The luminosity L can be calculated using the relation [4],

$$L \approx \sqrt{\frac{2}{\pi}} \frac{I_e}{e} \frac{P_L \lambda}{hc^2} \frac{1}{\sqrt{\sigma_e^2 + \sigma_\gamma^2}} \frac{1}{\sin \alpha} \qquad (3)$$

where I_e is the electron beam current, P_L is the power of the laser beam, σ_e and σ_γ are the Gaussian width of the electron and laser beams at the Compton Interaction Point (CIP), α is the crossing angle between the beams. Using Eq. 2 and Eq. 3, one can evaluate the time of measurement to get a 1% level statistical error. The proposed alternative [3] will use a tunable CW green laser (532 nm) coupled with optical cavity to increase the L and $< A^2_l >$ respectively.

The schematic of the Compton Polarimeter at Jefferson Lab is shown in Figure 2. A Fabry-Pérot cavity provides a high-density circularly polarized photon target at the CIP. A magnetic chicane is used for vertical displacement of the electron beam and optimizing the crossing of the electron and photon beams. The back-scattered photons are collected by the photon calorimeter. The silicon micro-strips in the electron detector tag the momentum of the scattered electrons at the end of the third dipole and provide a complimentary polarization measurement.

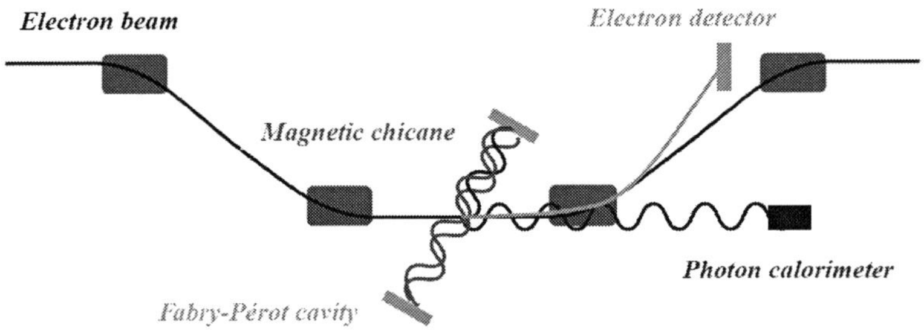

FIGURE 2. A schematic of Compton Polarimeter in Jefferson Lab hall A

THE FABRY-PÉROT CAVITY

The Fabry-Pérot cavity is made of two high reflective (R > 99 %) mirrors separated by integral multiple of the half wavelength of the input light. When the focusing of the injected Gaussian laser light correctly matches the fundamental TEM_{00} mode of the cavity, the circulated light inside the cavity is amplified by a large factor with respect to the incident beam power. The amplification factor or cavity gain is defined by the properties of the mirrors and the optical coupling of the light to the cavity. In other words, the cavity mirrors should be high reflective and low loss type. Maximum coupling is achieved by good alignment of the laser beam to the optical axes of the cavity and proper focusing of it at the cavity center. However, a small variance in the cavity length brings a large change in the cavity resonance frequency and we always have to keep the laser frequency adjusted to the change in cavity resonance frequency. The critical part of Fabry-Pérot cavity system is sustaining the TEM_{00} resonance mode of the cavity in order to have continuous high power inside the cavity. One of the most widely used techniques is the Pound-Drever-Hall (PDH) locking scheme [5] which controls the laser frequency based on the analysis of the reflected light by the cavity.

Our laser (Prometheus, Innolight Gmbh [6], Hannover, Germany) is a diode pumped solid-state NPRO laser which has a primary wavelength of 1064 nm. After single pass frequency doubling with a PPKTP crystal, it delivers up to 100 mW of CW green laser (532 nm) with longitudinal mode TEM_{00} and a narrow linewidth (1 kHz over 100 ms). Frequency tunability is achieved changing the frequency of the pump laser by means of a temperature (SLOW) control of the NPRO crystal with a tuning range of 60 GHz or of a voltage (FAST) applied to a piezoelectric bonded to the crystal with a dynamic range of 200 MHz.

Our test cavity (Figure 3.) consists of two identical high reflective (manufacturer specification R = 99.9 %) concave mirrors (from CVI Laser, ROC = 0.5 m, 0.5 in. diameter) separated by 0.85 m. They are mounted on gimbal mount on two ends of the cavity and can be controlled remotely by picomotors. Our home-made locking system uses the PDH scheme. Based on an Analog Devices 32-bit microprocessor, the system automatically scans for a TEM_{00} cavity mode and engages the feedback servo to acquire lock. The shape of the resonance mode is monitored by a Spiricon CCD camera. From the transmitted and the reflected power measured by two pre-calibrated photodiodes (PDT and PDR), one can estimate the intra cavity power and the optical coupling as well. The purpose of this test cavity is to check the locking electronics and understand the basics of optical setup.

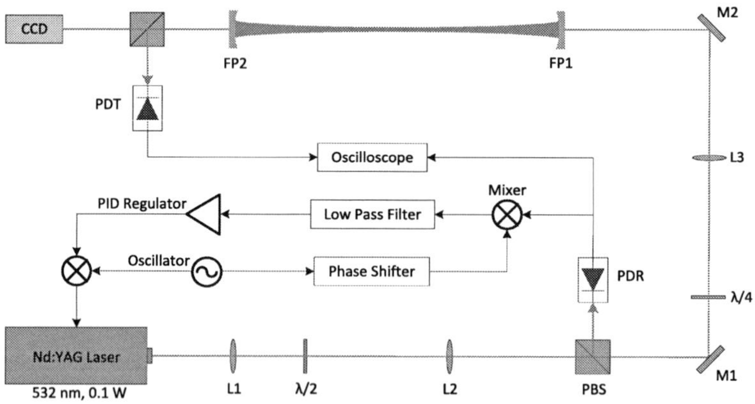

FIGURE 3. A schematic of the optical setup for the test cavity

RESULTS AND DISCUSSION

Our PDH feedback servo system was able to lock this test cavity successfully. During cavity locking tests, no careful optimization of the input optics was performed resulting in poor optical coupling. Furthermore, substantial loss (absorption, scattering) in these stock mirrors limited the power gain in the cavity. With 55 mW injected laser power, the intra-cavity power was about 6 Watts. We measured the finesse of the mirrors from cavity decay time as well as cavity bandwidth, to be 2695 corresponding to a reflectivity of 99.896 %. Losses in the mirrors were estimated to be 500 ppm.

In order to improve optical coupling efficiency, the laser beam was analyzed with the Spiricon Laser Beam Analyzer. Lens positions were determined by simulation of the transport optics with OptoCad [7] to achieve better mode-matching. The simulated and measured beam profiles agree reasonably well (Figure 4). To control losses in mirrors, new sets of super-polished (surface RMS roughness = 0.5 Å) low-loss mirrors (from General Optics, Moorpark, CA) were ordered. We expect to achieve less than 10 ppm losses in the new mirrors, thus substantially improving cavity gain. The feedback servo system needs to be re-optimized for higher gain cavities. Our goal for operational green cavity for high precision Compton Polarimetry is to get 1500 W intra cavity power.

The future goal also includes the polarization transfer function of laser beam in the system. The accurate determination of the circular polarization of the beam at the CIP is crucial to electron beam polarization measurement.

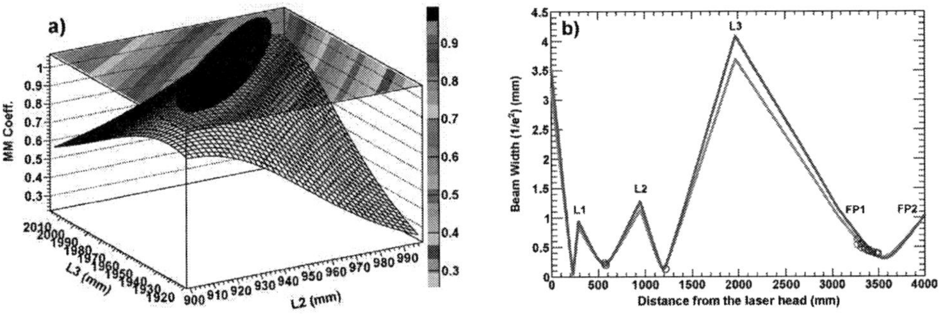

FIGURE 4. a) The calculated mode matching coefficients for L2 and L3 lens configurations, while L1 is fixed, b) calculated and measured beam diameters along the beam path from the laser head. The green (magenta) curve corresponds to the calculated value of tangential (sagittal) component. The red (blue) circles show the measured value of tangential (sagittal) component with Spiricon Laser Beam Analyzer.

CONCLUSIONS AND OUTLOOK

A new green Fabry-Pérot cavity for high precision Compton Polarimetry is under development. We have developed our own feedback control system based on PDH scheme. Test cavity has been assembled and tested for mechanical and ultra-high vacuum integrity. Low power gain locking has accomplished. More work is necessary to improve the light coupling to the cavity. We are working on developing a proper mirror cleaning and handling method to minimize losses. We are also working on feedback electronics, in order to get 1500 W power with a higher gain cavity. Alternative ways, such as injecting more green power to lower gain cavity is also an option, provided we have tunable high power green laser available.

REFERENCES

1. N. Falletto et al., *Nucl. Instrum. Methods A*, **459** (2001), 412-425.
2. K. Kumar et al., CEBAF experiment E06002, http://hallaweb.jlab.org/parity/prex/
3. S. Nanda and D. Lhuillier, "Conceptual Design Report of a Hall A Compton Polarimeter Upgrade", Jefferson Lab, 2004.
4. G. Bardin et al., "Conceptual Design Report of a Compton Polarimeter for CEBAF Hall A", CEA-Saclay, 1996.
5. R. V. Pound, *Rev. Sci. Instrum.* **17**, 490 (1946); R. W. P. Drever, J. L. Hall, *Appl. Phys. B: Photophys. Laser Chem.* **31**, 97 (1983); R. W. P. Drever et al., *Appl. Phys. B* **31**, 145 (1983)
6. I. Freitag, InnoLight GmbH, http://www.innolight.de/
7. R. Schilling, The Rechenzentrum Garching (RZG), http://www.rzg.mpg.de/~ros/OptoCad/

Polarimetry at the Superconducting Darmstadt Electron Linac S-DALINAC

C. Eckardt*, R. Barday*, U. Bonnes*, M. Brunken*, R. Eichhorn*, J. Enders*, C. Heßler*, W.F.O. Müller†, M. Platz*, Y. Poltoratska*, B. Steiner†, M. Wagner* and T. Weiland†

**Institut für Kernphysik, Technische Universität Darmstadt, Germany*
†*Institut für Theorie Elektromagnetischer Felder, Technische Universität Darmstadt, Germany*

Abstract. A spin polarized electron source will be installed at the S-DALINAC to extend the existing experimental capabilities polarization. In order to manipulate and monitor the beam polarization a Wien filter and different polarimeters will be installed. The already installed Wien filter and 100 keV Mott polarimeter and the planned 5 MeV Mott, 30-130 MeV Møller and Compton transmission polarimeters are described.

Keywords: S-DALINAC, Polarimetry
PACS: 29.20.Ej, 29.25.Bx, 29.27.Hj, 29.27.Fh, 24.70.+s

The superconducting Darmstadt electron linac S-DALINAC [1] will soon start its polarized beam program. As a first step a laser-driven GaAs electron source [2] at 100 keV has been set up at a separate test stand [3, 4] for development purposes. The source will be moved to the S-DALINAC in 2009. The beam line set up of the test stand includes a Wien filter to switch from longitudinally polarized beam for experiments to transversally polarized beam for polarimetry and to adjust for spin precession through the bending magnets. A 100 keV Mott polarimeter [5] is installed to measure the polarization of the electron beam produced by the source. As the S-DALINAC has an adjustable energy range of about 4 up to 130 MeV and many bending magnets because of its two recirculations it is necessary to determine the polarization at different positions close to the experimental sites as indicated in Fig. 1. For this purpose a 5-10 MeV Mott polarimeter will be installed at the low-energy experimental area, a 30-100 MeV Møller polarimeter in the extraction section and Compton transmission polarimeters at the end of the beam line at different experimental sites.

100 KEV MOTT POLARIMETER

The 100 keV Mott polarimeter has been installed and tested in the low energy beam line section at the test stand of the polarized source. A description of the test stand status can be found in these proceedings in ref. [6]. The Wien filter is placed on the beam axis to switch the polarization into the transverse direction for Mott polarimetry and to account for the bending magnets of the S-DALINAC. Inside the Mott polarimeter four silicon surface-barrier detectors are placed, one in each quadrant at 120° with respect to the beam. This allows one to measure both transverse beam polarizations.

CP1149, *Spin Physics, 18th International Spin Physics Symposium*
edited by D. G. Crabb, D. B. Day, S. Liuti, X. Zheng, M. Poelker, and Y. Prok

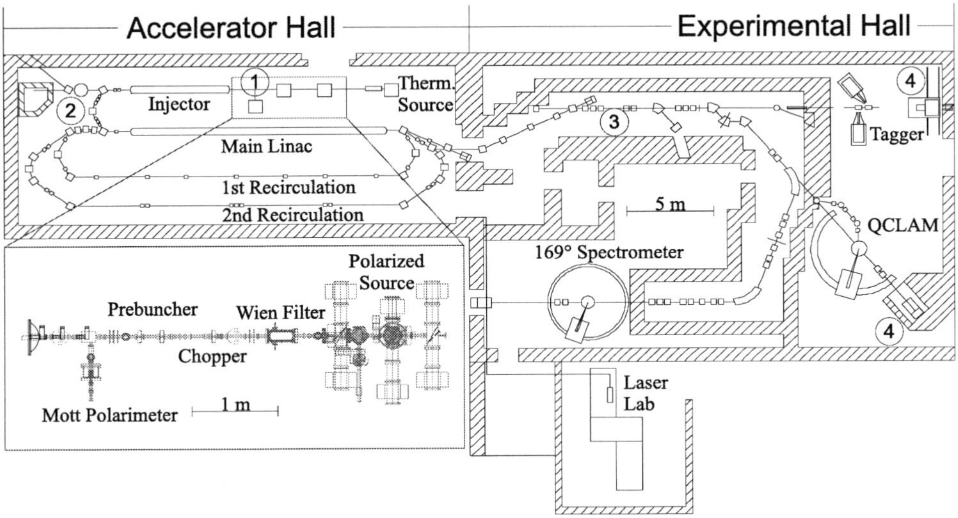

FIGURE 1. Layout of the S-DALINAC. The polarized source seen in the lower left will be installed between the thermionic source and the injector part of the linac. The laser beam runs through a fiber or an evacuated laser beam transport line originating in the laser lab. The position of the different polarimeters is as follows: 1. 100 keV Mott polarimeter 2. 5-10 MeV Mott polarimeter 3. 30-130 MeV Møller polarimeter 4. Compton transmission polarimeter

A target wheel with an empty slot for the beam, a BeO target to monitor the beam position and several self-supporting gold foils ranging from 42.5 to 500 nm is positioned in front of the detectors. Targets with different thicknesses are needed to extrapolate to a target thickness of zero, as the analyzing strength is only known for this target thickness. With this set up an asymmetry was measured at the test stand and it has been possible to determine the beam polarization for bulk GaAs to 35.5$\pm$1.4% and to 86$\pm$3% respectively for strained superlattice GaAs at a laser wavelength of 830 nm.

5-10 MeV Mott Polarimeter

A 5-10 MeV Mott polarimeter will be installed behind the injector section of the S-DALINAC, in front of the low energy experimental area. Here an energy up to 10 MeV and currents up to 60 μA are possible. This polarimeter has to be installed to monitor the beam polarization after the acceleration to the MeV range. Again, self-supporting gold foils of various thickness will be used for the extrapolation of the asymmetry to thickness zero. The uncertainty due to the foil thickness is smaller at higher energies. Due to limited space of only 0.5 m at the beam line the optimal angle for the Sherman function of 173° for 5 MeV or 176.5° for 10 MeV cannot be selected, instead an angle of 165° is chosen. The polarimeter will be installed with two scintillators for the detection of the scattered electrons.

30-130 MEV MØLLER POLARIMETER

A 30-130 MeV Møller Polarimeter will be installed at the beam extraction section behind the main accelerator. Energies up to 130 MeV with a maximum current of 20 μA are possible while typical values are 30-80 MeV with beam currents of a few μA only. (This prohibits, e.g. the use of a laser-backscattering Compton polarimeter). At this energy range Møller scattering is the method of choice. The polarized beam hits a polarized target, in this case a 20 μm vacoflux foil rotated by 20° with respect to the beam polarized by the magnetic field of two Helmholtz coils. As space is limited to 1 m and the scattering angle shows a large variation over this energy range, a large-acceptance compact magnet set up has to be developed to separate the scattered Møller electrons from the beam and to focus these electrons on the focal plane of a plastic scintillator detector array for coincident detection of the scattered and the recoiling electrons near a center of mass scattering angle of 90°. Present design considerations include a quadrupole magnet with collimators to separate the scattered electrons from the direct beam that has undergone small-angle scattering and a dipole magnet to bend the Møller electrons for momentum analysis. The main beam has to be safely dumped. The schematic layout of the Møller polarimeter and the energy dependence of the scattering angle are shown in Fig. 2. Note the logarithmic scale. Further information can be found in ref. [7].

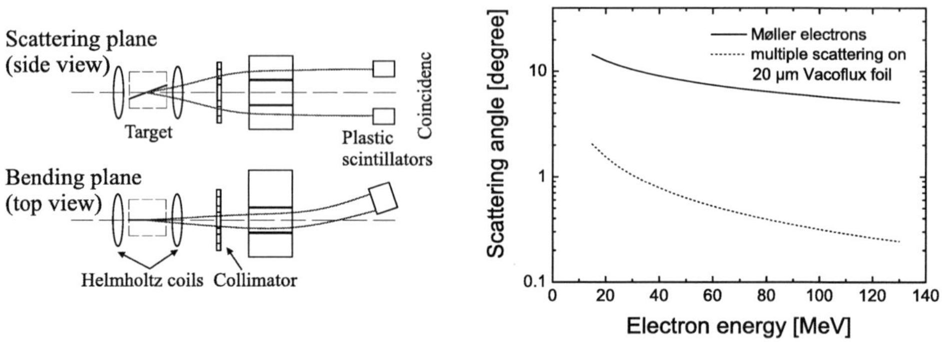

FIGURE 2. Schematic layout of the Møller polarimeter. The magnetic field of the two Helmholtz coils polarizes a Vacoflux foil. Scattered electrons are separated by a quadrupole and collimator and bent by a dipole magnet out of the beam path. On the right the energy dependence of the scattering angle for Møller electrons and multiple scattered electrons is shown.

COMPTON TRANSMISSION POLARIMETER

In the experimental hall three experimental sites are in use, the NEPTUN photon tagging system, the QCLAM spectrometer with 180° chicane and the 169° spectrometer. The NEPTUN tagger and the QCLAM spectrometer have experiments planned with polarized electrons. Therefore it is necessary to monitor the polarization of the electron beam for both experimental sites. The tagger is placed behind the Møller polarimeter.

The beam line to the QCLAM spectrometer is bent by 40°. To monitor the polarization at the spectrometer during long experimental runs a Compton transmission spectrometer is foreseen. A similar device has recently been developed for the A4 experimental setup at Mainz [8]. The use of Compton backscattering polarimeters is not possible because of the limited current at these experimental sites.

SUMMARY

The polarized beam program at the S-DALINAC is soon going to start. First planned experiments are sketched in ref. [9]. The test stand set up for the polarized source has been completed and the installation at the S-DALINAC will soon proceed. Necessary beam diagnostic elements to monitor the polarization of the beam at different areas of the beam line are completed or planned. First experiments with polarized beams will start in 2010 and collaboraters for further experiments are welcome.

ACKNOWLEDGMENTS

The S-DALINAC and this work are supported by the Deutsche Forschungsgemeinschaft through SFB 634. We would like to thank K. Aulenbacher and V. B. Asgekar for helpful discussions in the development of the 100 keV Mott polarimeter and H. Schmieden and co-workers for support with the polarization measurement of the vacoflux foil. The author likes to thank the SPIN 2008 organizing committee for financial support.

REFERENCES

1. A. Richter, "Operational Experience at the S-DALINAC," in *Proc. of the 5th EPAC*, 1996, p. 110.
2. D. Pierce, et al., *Rev. Sci. Instrum.* **51**, 478 (1980).
3. C. Heßler, *Konzeption, Aufbau und Test einer Quelle spinpolarisierter Elektronen für den supraleitenden Darmstädter Elektronenlinearbeschleuniger S-DALINAC*, Doctoral thesis D17, TU Darmstadt (2008).
4. C. Heßler, et al., "Comissioning of the Offline-Teststand for the S-DALINAC Polarized Injector SPIN," in *Proc. of the 11th EPAC*, 2008, p. 1482.
5. Y. Poltoratska, *Setup and Design of a 100 keV Mott-Polarimeter for the future S-DALINAC Polarized Injector*, Diploma thesis, Kharkiv Karazin National University and TU Darmstadt (2005).
6. Y. Poltoratska, et al., Polarized eletrons in Darmstadt: recent developments, these proceedings (2008).
7. R. Barday, et al., "Electron beam polarimetry at S-DALINAC," in *Proc. of the 11th EPAC*, 2008, p. 1476.
8. C. Weinrich, *Eur. Phys. J. A* **24**, 129 (2005).
9. C. Eckardt, et al., Polarized Eletrons for Experiments at Low Momentum Transfer SPIN @ S-DALINAC, these proceedings (2008).

Benchmark studies among own and commercial beam tracking codes on fundamental conditions

Naoto Yamamoto*, Akihiko Mizuno†, Kai Masuda**, Koichi Kan‡, Shuji Matsumoto§ and Masashi Yamamoto¶

*Graduate School of Engineering, Nagoya University, Furo-cho, Chikusa-ku, Nagoya, Aichi,464-8602 Japan
†JASRI/SPring-8, Mikazuki, Sayo, Hyogo 679-5198, Japan
**Institute of Advanced Energy, Kyoto University, Uji, Kyoto 611-0011, Japan
‡The Institute of Scientific and Industrial Research (ISIR) Osaka University, Mihogaoka 8-1, Ibaraki, Osaka 567-0047. JAPAN
§KEK. 1-1 Oho. Tsukuba-shi. Ibaraki-ken. 305- 0801. Japan
¶Accuthera Inc.,2-7-6 Kurigi, Asaoku, Kawasaki-city, Kanagawa, Japan

Abstract. In order to design accelerators or to analyze experimental data, commercially or non-commercially programs for tracking particles have been developed and used. However the reliabilities of those programs are not clear, especially if the beam quality is as high as that required for future applications. Therefore, for the purpose to study the reliabilities and to improve their accuracy, we planed to benchmark those programs. As a first step, the calculations were carried out in the both condition of free space and homogeneously accelerating-field by using three own codes and three commercial codes (MAGIC, PARMELA, and GPT)

Keywords: beam tracking, electron source, simulation cord
PACS: 29.25.Bx, 41.75.-i

INTRODUCTION

The generation and transportation of low emittance and high brightness electron beams are important technologies for future generation accelerators. For example, an energy recovery linac requires a beam emittance of 0.1 π mm mrad with a peak current higher than 4 A at the source [1].

To design accelerators satisfied with such requirements, particle tracing become a very powerful tool. For this purpose, some simulation cords such as PARMELA [2], GPT [3] and MAGIC [4], have been provided commercially and some researchers have developed their own cords.

One of authors (N. Y) has also developed his own cord to design polarized electron sources (PESs). The cord is possible to trace thousands of macro particles through complex electromagnetic fields taking all 3D effects, space-charge forces and image-charge forces near the source material into account by using 4th-order Runge-Kutta method as a function of time. In fact, this tracing cord was successfully used to design a emittance measurement system for the 200–kV PES [5] and a high brightness gun for the 20–kV PES [6] [7] at Nagoya University.

Though the tracing cords described above have been used in various reseach groups, the reliabilities of those cords are not clear in the especially case that the beam perfor-

CP1149, *Spin Physics, 18th International Spin Physics Symposium*
edited by D. G. Crabb, D. B. Day, S. Liuti, X. Zheng, M. Poelker, and Y. Prok

mances are as high as that required for future applications. Therefore, for the purpose to study the reliabilities and to improve those accuracies, we planed to benchmark those tracing cords. We have to note that the purpose of our bench mark is not to decide the superiority or inferiority of the codes.

In this report, the collaborator and their cords, the assignment for benchmarking are described. A example of benchmarking result is also reported, however, the data analysis and detail discussions will be given in a forth coming paper.

SIMULATION CORDS

The benchmark studies have been carried out with 6 researchers in Japan and using 6 codes. The list of tracing cords is shown in Table 1. There are three own cords and two (own cord 1 & 2) of them can trace macro particles by using 3D point to point method. One of them (own cord 3) uses the method of particle in cell instead of 3D point to point. Detail descriptions of own cord 1 and 3 were given in ref [8] and [9] respectively. In addition to three own cords, three commercially cords are also employed. Those are MAGIC, PARMELA and GPT.

BENCHMARK ASSIGNMENT

For the benchmarking test, we decided the definition of beam parameters (beam size, bunch length, energy spread and emittance) and four conditions of benchmarking test written below have been assigned.

- Comparison with numerical analysis (Envelop Equation)
  1. Beam envelop in free space
  2. Beam envelop in DC / RF fields
- Comparison with experimental results
  3. RF-gun system in Osaka Univ. (ISIR) [10]
  4. SPring-8 Compact SASE Source (SCSS) [11]

The conditions of 1 and 2 were assigned for the purpose to find the basic character of each cord, such as dependences of accuracy on beam energy, charge densities in bunch, number of macro particles. In addition, in these conditions, the result can be compared with analytical result calculated by the envelop equation [12], as follows:

TABLE 1. the list of simulation cords.

index name in this paper	calculation method	cord name & collaborator	
own cord 1	3D point to point	own cord	A. M
own cord 2	3D point to point	own cord	N. Y
own cord 3	2D particle in cell	own cord	K. M
MAGIC	2D particle in cell	MAGIC	S. M
PARMELA	2D	PARMELA	K. K
GPT	3D particle in cell & point to point	General Particle Tracer	M. Y

$$\frac{d^2x}{ds^2} = \frac{eQ}{4\pi\varepsilon_0\, m_0c^2\, \gamma^3\beta^2\, x} \frac{1}{\sqrt{\frac{L^2}{4} + \frac{x^2}{\gamma^2}}} \tag{1}$$

$$\frac{d^2z}{ds^2} = \frac{eQ}{4\pi\varepsilon_0\, m_0c^2\, \gamma^3\beta^2\, R^2\, z} \left(\sqrt{4z^2 + \frac{R^2}{\gamma^2}} - \sqrt{\frac{R^2}{\gamma^2}} - 2z \right) \tag{2}$$

where, x and z are a transverse distance and a longitudinal distance from bunch center, z is a longitudinal distance of bunch center from the origin, R and L are a radial beam size and a bunch length, and β and γ are relativistic factor of the electron.

Simultaneously, we planed to be compared our calculation to experimental results obtained by various groups. At present, RF-gun system in Osaka Univ (ISIR) and SPring-8 Compact SASE Source (SCSS) was scheduled.

PRELIMINARY RESULT

As examples, comparison of calculation results with each cord and the envelop equation are shown in Fig 1. In this calculation, a electron beam in free space was traced for the traveling distance of 1 m and the beam paremeter at the traveling distance of 0 m was set with initial conditions of charge per bunch of 1 nC, a beam diameter of 1 mm flat-top, a beam bunch length of 200 mm and a beam energy of 500–kV. The alphabets (a, b, c and d) of each plot indicate the comparison results of beam radius, pulse length, normalized emittance and energy spread, respectively.

In Fig 1, we obtained good agreement for the envelop of beam radius and pulse length between each cord and the envelop equation. On the other hand, the disagreements were apparent on the envelop of normalized emittance and energy spread. Though other data was not shown in this report, similar behaviors were obtained with the different initial conditions, which is a beam bunch length of 3 mm and beam energies of 1, 3 and 6 MeV.

As results of this series of calculations, however this is only one example of our achievements, it is found that some cords do not have enough accuracy in the range of low energy, which mean the conditions that the energy of particles is less than or around equal to a few MeV. Futhermore, we considered that differences in calculation methods and inital distribution of particles between tracing cords caused these disagreements. Then additional calculations have been made with more optimized conditions.

SUMMARY

The benchmark studies have been carried out by using three own cords and three commercial cords (MAGIC, PARMELA, and GPT). As a first step, the calculations in the both conditions of free space and homogeneously accelerating-field were carried out and some discrepancies were apparent on the envelop of normalized emittance and energy spread. Additional benchmark studies have been made to understand the reason of those discrepancies and to improve our own cords. The detail descriptions will be given in a forth coming paper.

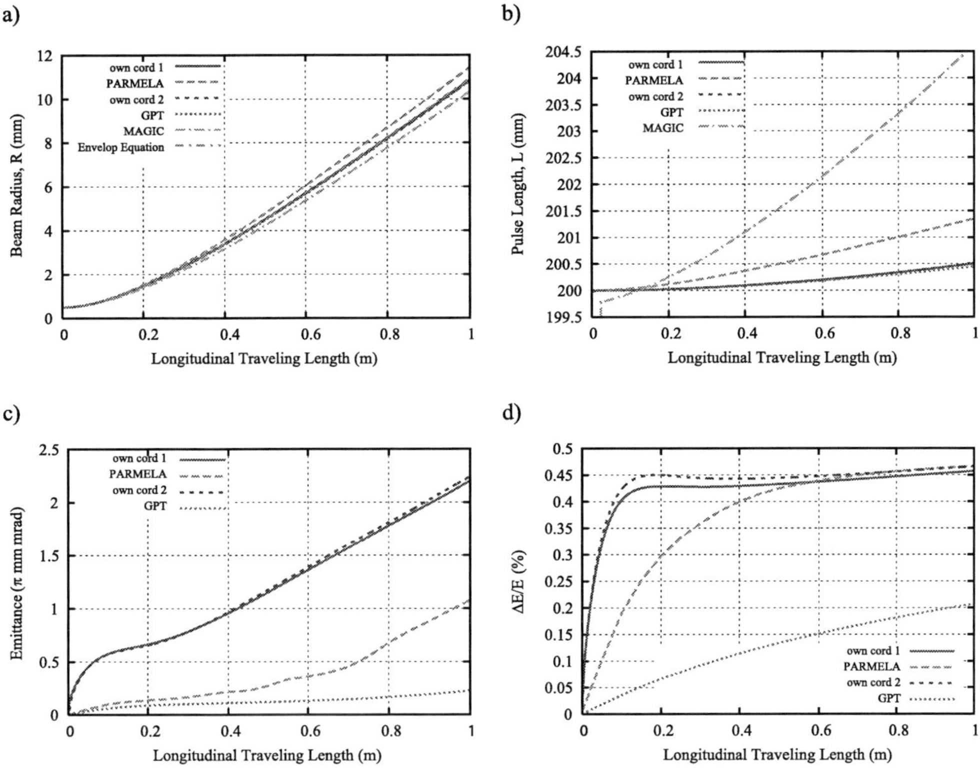

FIGURE 1. Comparison of calculation results in free space by each cord as a function of the longitudinal traveling distance. The alphabets of each plot indicate a) beam size, b) pulse length, c) normalized emittance and d) energy spread. The initial beam parameters are a energy of 500–kV, charge per bunch of 1 nC, a beam diameter of 1 mm flat-top and a pulse length of 200mm

Simultaneously, the calculation in the accelerating field and more realistic conditions, such as RF Gun system in Osaka University and SPring-8 compact SASE Source have been started.

ACKNOWLEDGMENTS

They wish to thank Dr. J. Yang and T. Shintake for providing the detail information of experimental setup and experimental results at each facility. One of authors (N. Yamamoto) extends his gratitude to Professor T. Nakanishi for the continuous encouragements of his tracing cord developing research. He has been financially supported by the Hori Information Science Promotion Foundation.

REFERENCES

1. S. M. Gurner, D. Bilderback, I. Bazarov, K. Finkelstein, G. Krafft, L. Merminga, H. Padamsee, Q. Shen, C. Sinclair, and M. Tigner, Review of Scientific Instruments **73**, 1402 (2002).
2. L. Young and J. Billen, Tech. Rep., LA-UR-96-1835 (1998).
3. M. de Loos and S. van der Geer, in *Proc. 5th Eur. Particles Accelator Conf* (1998).
4. B. Goplen, L. Ludeking, D. Smith, and G. Warren, Computer Physics Communications **87**, 54 (1995), ISSN 0010-4655.
5. N. Yamamoto, M. Yamamoto, M. Kuwahara, R. Sakai, T. Morino, K. Tamagaki, A. Mano, A. Utsu, S. Okumi, T. Nakanishi, et al., Journal of Applied Physics **102**, 024904 (2007).
6. N. Yamamoto, T. Nakanishi, A. Mano, Y. Nakagawa, S. Okumi, M. Yamamoto, T. Konomi, X. Jin, T. Ujihara, Y. Takeda, et al., Journal of Applied Physics **103**, 064905 (2008).
7. N. Yamamoto, X. Jin, A. Mano, Y. Nakagawa, and Nakanishi, this Proceedings (2009).
8. A. Mizuno, T. Asaka, H. Dewa, T. Kobayashi, S. Suzuki, T. Taniuchi, H. Tomizawa, K. Yanagida, H. Hanaki, and M. Uesaka, Nuclear Instruments and Methods in Physics Research Section A: Accelerators, Spectrometers, Detectors and Associated Equipment **528**, 387 (2004), ISSN 0168-9002, proceedings of the 25th International Free Electron Laser Conference, and the 10th FEL Users Workshop.
9. K. Masuda, K. Yoshikawa, M. Ohnishi, Y. Yamamoto, H. Toku, M. Sobajima, and J. Kitagaki, Microwave Theory and Techniques, IEEE Transactions on **46**, 1180 (1998), ISSN 0018-9480.
10. J. Yang, T. Kondoh, K. Kan, T. Kozawa, Y. Yoshida, and S. Tagawa, Nuclear Instruments and Methods in Physics Research Section A: Accelerators, Spectrometers, Detectors and Associated Equipment **556**, 52 (2006), ISSN 0168-9002.
11. T. Shintake, T. Tanaka, T. Hara, K. Togawa, T. Inagaki, Y. Kim, T. Ishikawa, H. Kitamura, H. Baba, H. Matsumoto, et al., Nuclear Inst. and Methods in Physics Research, A **507**, 382 (2003).
12. A. Mizuno (2006), unpublished electronic article, URL `http://acc-web.spring8.or.jp/~workshop/e-gun/benchmark/envelop.pdf`.

3D Modeling of the ALICE Photoinjector Upgrade

J.W. McKenzie, B.L. Militsyn, Y.M. Saveliev

STFC Daresbury Laboratory, Warrington, WA4 4AD, United Kingdom

Abstract. The injector for the ALICE machine (Accelerators and Lasers In Combined Experiments) at Daresbury Laboratory is based around a 350 kV DC photocathode electron gun. An upgrade is proposed to introduce a load-lock GaAs photocathode preparation facility to allow rapid transfer of photocathodes to the gun without breaking the vacuum system. In the current design this requires side-loading of the photocathodes into the cathode ball. An alternative is to relocate the ceramic insulator vertically which will allow back-loading and also back-illumination of the photocathodes. 3D electrostatic simulations of the gun chamber are presented for both options along with 3D beam dynamic simulations for an off-axis photocathode, introduced to increase photocathode lifetime by reducing damage by ion back-bombardment. Beam dynamic simulations are also presented for the entire injector beamline as well as for a proposed extension to the injector beamline to include a diagnostic section.

Keywords: Electron sources; linear accelerators; electron and positron beams
PACS: 29.25.Bx; 29.20.Ej; 41.75.Fr

INTRODUCTION

ALICE is a 35 MeV energy recovery linac undergoing commissioning at Daresbury Laboratory [1]. The injector is based around a 350 kV DC photocathode gun which is a modified version of the gun developed for the Jefferson Laboratory Infra-Red FEL [2]. Electron bunches with charge 80 pC are produced from a GaAs photocathode illuminated by laser pulses of wavelength 532 nm. Currently the photocathodes are activated *in situ* in the gun chamber by means of a Cs and NF_3 or O_2 "yo-yo" procedure. Replacing the photocathode takes weeks due to the requirement to break the vacuum, replace the photocathode, and then restore the vacuum to the operational level of 10^{-11} mbar. An upgrade is planned to introduce an external load-lock photocathode preparation facility to reduce changeover downtime to a matter of hours [3]. Details of this three-chamber design can be found in [4]. This facility will use new GaAs photocathodes currently under development [5], with a reduction in diameter from 25 mm, as currently used in ALICE, to 10 mm.

SIDE-LOADING GUN OPTION

It is proposed that the photocathodes are loaded from the preparation chamber into the side of the gun, in order to avoid disruption to the existing ALICE infrastructure. This involves a re-design of the cathode ball from the present rear-loading mechanism. The

CP1149, *Spin Physics, 18th International Spin Physics Symposium*
edited by D. G. Crabb, D. B. Day, S. Liuti, X. Zheng, M. Poelker, and Y. Prok

ball requires a slot in the side for loading of the photocathode. This has been positioned on the cylindrical part of the ball surface to keep the field distortion low. The photocathode then has to be moved forward into position, requiring a second slot further back in the cathode ball for insertion of a magnetic screwdriver to drive the winding mechanism. A third slot, perpendicular to the loading slot, might be required as a viewport to ensure the photocathode is loaded properly. Figure 1 shows the electric fields at 350 kV on the cathode ball surface, modelled in CST Studio [6]. The electric field has been kept lower than 10 MV/ on the curved surface of the ball and also around the edges of the slots. A focussing electrode has been added and optimised by performing beam dynamic simulations in ASTRA [7]. Figure 2 shows that the transverse beam properties for the new gun design compared to the existing gun which lacks the focussing electrode.

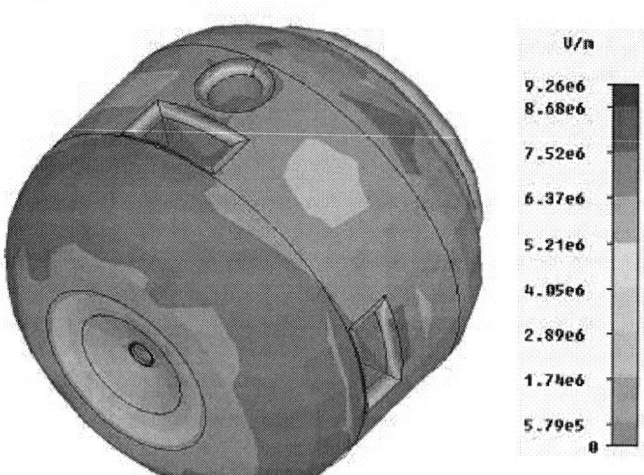

FIGURE 1. The cathode ball with slots and focusing electrode showing electric fields.

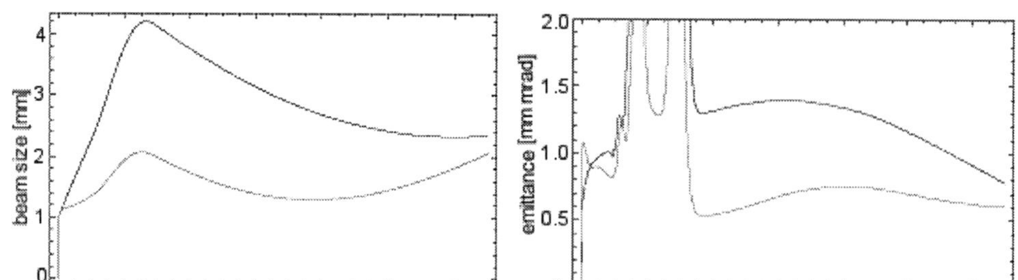

FIGURE 2. RMS beam size (left) and transverse emittance (right) for the new gun design (green) compared to the current gun (red) including a 330 G solenoid at 0.25 m.

VERTICAL CERAMIC GUN OPTION

As an alternative to the side-loading design, the ceramic insulator can be relocated vertically to allow back-loading of the photocathodes. This would also allow back-illumination of transmission-mode photocathodes and should be more mechanically stable. However, this requires relocation of the gun power supply and ancillary equipment. Electrostatic simulations have been performed using CST Studio, shown in figure 3. The cathode stem introduces a vertical component of electric field along the electron beam trajectory, peaking at 1.5 kV/m. This is small compared to the peak of the longitudinal electric field of 5 MV/m. 3D fieldmaps were extracted and used in GPT [8] to perform particle tracking simulations with space charge. These show that the vertical field displaces the beam by an almost negligible amount.

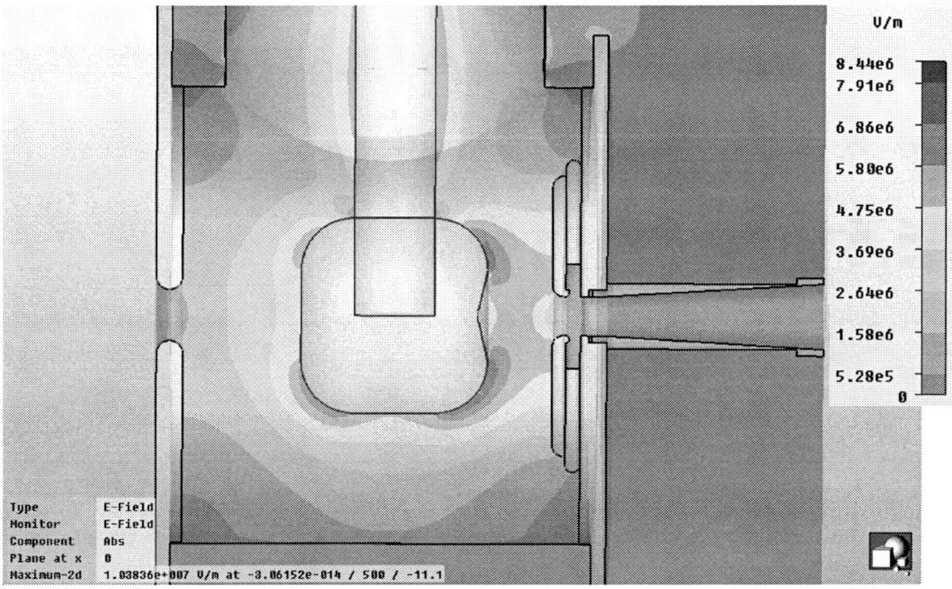

FIGURE 3. Electric fields in the gun chamber for the vertical ceramic design at 350 kV.

OFF-AXIS BEAM TRANSPORT

The lifetime of photocathodes is a major issue for regular operation of photoelectron guns. The principle cause of photocathode degradation is ion back-bombardment. When the gun is in operation, ions are produced by collision of the electron beam with residual gases and accelerated directly back towards the cathode. These ions acquire a large acceleration but little transverse displacement. Therefore, citing the photocathode off-centre should result in fewer ions striking the photocathode surface, thus increasing the photocathode lifetime. One method currently in use is to direct the drive laser off-centre on the photocathode [9]. When designing a new gun, two further options present themselves - situating the photocathode off-centre on the cathode ball, or situating the whole cathode ball off-axis. Previous GPT simulations [10] show that for the offset ball, the electron beam still moves parallel to the optical axis, thus ion damage should remain the same, whereas the off-centre photocathode produces a beam which moves directly away from the photocathode. Both horizontal and vertical steering coils are placed around the beampipe at the same position as the solenoid and simulations in GPT show that these can be used to steer the beam parallel to the beampipe whilst retaining an emittance of 1 π mm mrad, as can be seen in figure 4.

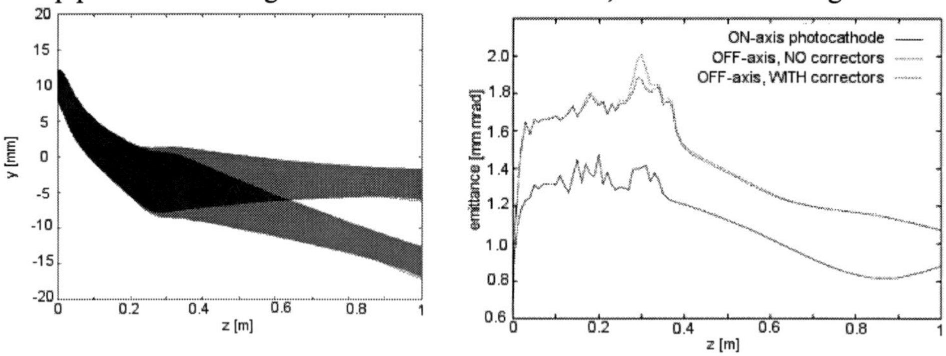

FIGURE 4. **(a)** Particle trajectories for a 10 mm off-centre photocathode including a 300 G solenoid both with (blue) and without (red) steering coils. **(b)** Transverse emittance for the off-centre photocathode compared to the on-axis version.

GUN BEAMLINE UPGRADE

During the ALICE gun commissioning phase a dedicated diagnostic beamline was used to characterize the beam. For the commissioning and operation of the full ALICE machine, this was removed, leaving just two focusing solenoids, three sets of steering coils, a buncher cavity, and a single YAG screen before the superconducting booster cavity. An extension to the current gun beamline is proposed to reinstate some diagnostics. The layout is shown in figure 5.

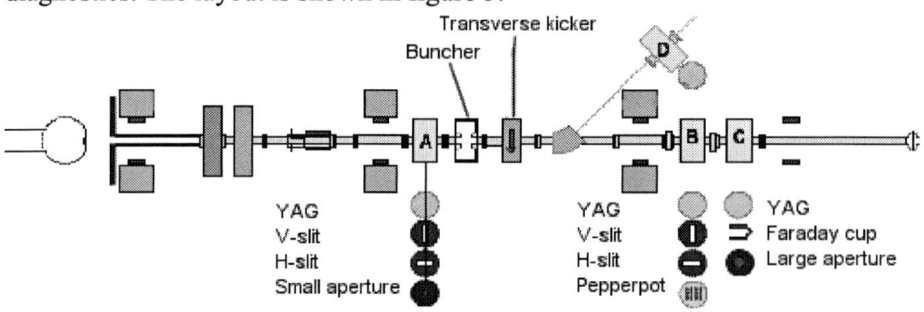

FIGURE 5. Schematic of the gun beamline upgrade showing components in each diagnostic unit.

A RF transverse kicker and dipole leading to a YAG screen orientated above the gun beamline will allow for longitudinal profile, energy, and bunch length measurements. A pair of horizontal and vertical slits plus a pepperpot provide the ability for emittance measurements and an insertable Faraday cup will be used for measuring the bunch charge. Two apertures are included: one small (~1 mm) to give the option of operating ALICE at low bunch charges without having to reset the setup of the photocathode gun; and one larger aperture to cut off any halo that may be produced along with the core beam. The positions of the first diagnostic unit, buncher and second solenoid have been reversed from the initial gun beamline in order to perform measurements with the solenoid switched on. An additional solenoid and set of steering coils are needed to provide good beam transport into the booster module for this extended beamline. The total additional length of the beamline is 1.2 m which involves moving the entire gun and power supply systems backwards. The extra length of this beamline means that there is a smaller range of parameters for good beam transport. Figure 6 shows the results of beam dynamics simulations performed using ASTRA, compared to the existing injector design. They show that although there is an increase in transverse emittance, there is a decrease in both bunch length and longitudinal emittance which combined with the benefit of the added diagnostics, will improve the operation of the ALICE facility.

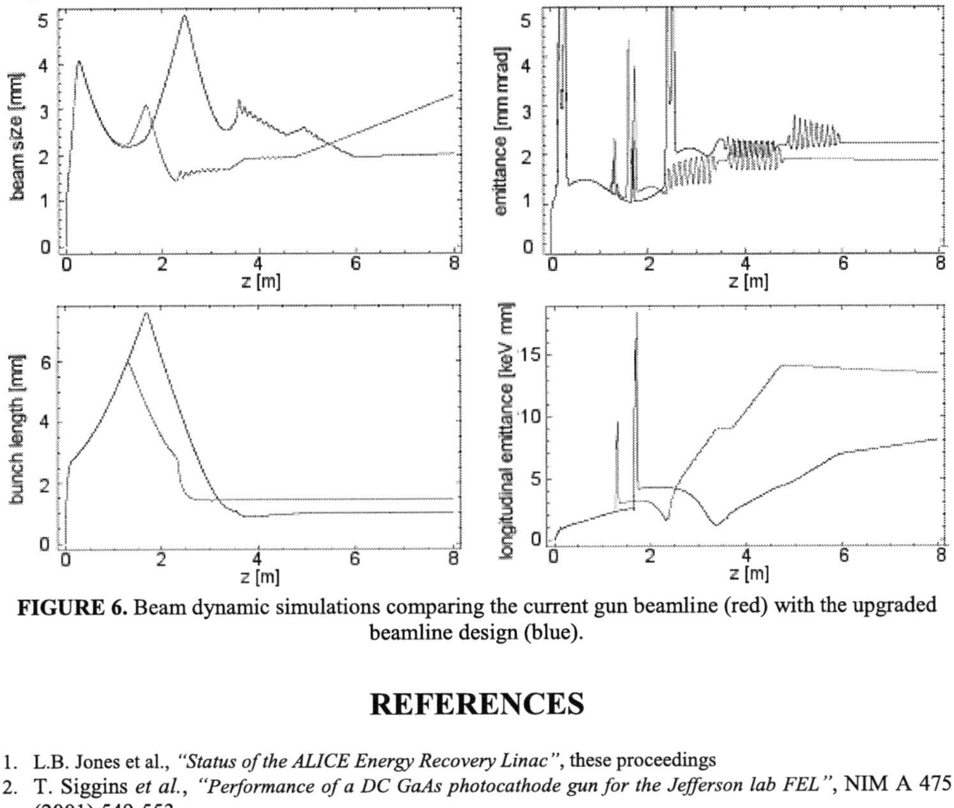

FIGURE 6. Beam dynamic simulations comparing the current gun beamline (red) with the upgraded beamline design (blue).

REFERENCES

1. L.B. Jones et al., *"Status of the ALICE Energy Recovery Linac"*, these proceedings
2. T. Siggins *et al.*, *"Performance of a DC GaAs photocathode gun for the Jefferson lab FEL"*, NIM A 475 (2001) 549-553
3. B.L Militsyn et al., *"Design of an upgrade to the ALICE photocathode electron gun"*, Proceedings of EPAC 2008
4. K.J. Middleman et al., *"Photocathode preparation system for the ALICE photoinjector"*, these proceedings
5. L.B. Jones et al., *"Cooled transmission–mode NEA–photocathode with a band–graded active layer for high brightness electron source"*, these proceedings
6. CST Studio Suite, http://www.cst.de
7. K. Flöttmann, ASTRA, http://www.desy.de/~mpyflo
8. M.J. de Loos and S.B. van der Geer, General Particle Tracer, http://www.pulsar.nl/gpt
9. C. Hernandez-Garcia et al., *"A high average current DC GaAs photocathode gun for ERLs and FELs"*, Proceedings of PAC 2005
10. J.W. McKenzie and B.L. Militsyn, *"3D simulations of a non-axisymmetric high average current DC photocathode electron gun"*, Proceedings of EPAC 2008

Polarized Positrons at Jefferson Lab

J. Dumas[a,b], J. Grames[b], E. Voutier[a]

[a]*Laboratoire de Physique Subatomique et de Cosmologie*
IN2P3/CNRS – Université Joseph Fourier - INP
53, rue des Martyrs, 38026 Grenoble Cedex, France
[b]*Thomas Jefferson National Accelerator Facility*
12000 Jefferson Avenue, Newport News, VA 23606, USA

Abstract. A novel concept for producing polarized positrons at Jefferson Lab using the CEBAF polarized electron photo-injector is presented. This approach relies on the polarization transfer from low energy highly polarized electrons to positrons via bremsstrahlung followed by pair production in a conversion target. An experiment to test this concept is discussed on the basis of GEANT4 simulations. It is shown that this low energy approach, which benefits from recent advances in high current high polarization electron sources, can yield positron longitudinal polarization up to 40%.

Keywords: Positron source, polarization transfer, electron sources.
PACS: 29.25.Bx, 41.75.Ht, 23.20.Ra, 07.77.Ka

INTRODUCTION

The high spin polarization of the electron beam at the Continuous Electron Beam Accelerator Facility (CEBAF) at Jefferson Laboratory (JLab) is a significant feature of this facility. More generally, the polarization of a beam is a key element for the completion of a physics program and the achievement of precision physics. In addition, the sensitivity of electromagnetic processes to the charge of a lepton beam is a fundamental property, revealing singular features of a system when comparing negative and positive lepton responses. In the JLab context, a *polarized positron beam* would offer a unique opportunity for the study of Generalized Parton Distributions [1] or the two-photon exchange mechanisms, and globally the deep scattering processes.

An efficient scheme for positron production, widely used in particle accelerators, relies on the creation of electron-positron pairs from high energy photons. A significant aspect of the process is the dependence on the polarization, in particular, the circular polarization of the photon transfers to the longitudinal polarization of the positron [2]. A concept to generate polarized positrons from undulator-generated polarized photons [3] has recently been demonstrated by the E-166 experiment [4], in view of its application to the International Linear Collider (ILC) project. There, circularly polarized photons (~10 MeV) are created via synchrotron radiation using very high energy electrons (~50 GeV) travelling through a helical undulator, and are then converted into positrons via pair-creation. In contrast with this scheme, a new concept using a low energy (10 MeV) highly polarized electron beam to produce polarized positrons for the JLab physics program is hereafter discussed.

CP1149, *Spin Physics, 18th International Spin Physics Symposium*
edited by D. G. Crabb, D. B. Day, S. Liuti, X. Zheng, M. Poelker, and Y. Prok

JLAB POLARIZED POSITRON SOURCE CONCEPT

Similarly to pair creation, the bremsstrahlung process is a polarization sensitive mechanism. This property has been widely used at un-polarized electron accelerators to produce linearly polarized photon beams. In addition to the intrinsic linear polarization (L), the photons have a circular component (C) when the incoming electron beam is polarized, such that the bremsstrahlung of polarized electrons most generally lead to elliptically polarized photons [2,5]. The polarization component ratio C/L depends on the electron and photon energies, and is proportional to the electron beam polarization.

The proposed JLab polarized positron source concept takes advantage of recent advances of high-current high-polarization electron sources using GaAs-based photocathodes that simultaneously have both high electron polarization (>80%) and high quantum efficiency (~1%). Recent advances in electron gun performance have resulted in improved photocathode lifetime and sustainable high average current (1 mA) in a configuration suitable for an accelerator has been demonstrated [6]. At the CEBAF photo-injector a high intensity and highly polarized continuous electron beam would interact in a thin tungsten target foil to produce polarized positrons via bremsstrahlung followed by pair-creation. A thin target is chosen to limit the emittance of the positrons exiting the target as well as limit the extent of radiation. In this respect, a low electron beam energy (~10 MeV) helps limit photo-neutron production. The emitted positrons would then be selected in a given energy range, suitable for subsequent acceleration by the CEBAF system [7]. An experiment at the CEBAF photo-injector is presently considered to test and characterize this concept. Some key aspects are summarized in Table 1 along with those of the E-166 experiment. Practical ingredients for testing this concept are the pre-existing electron photo-injector and relatively small foot-print and cost for a conversion target and diagnostics beam line. An actual source for the CEBAF accelerator would be more sophisticated and costly, yet continue to be based upon the pre-existing photo-injector.

TABLE 1. Key aspects of the JLab and E-166 polarized positron generation schemes.

	JLab	**E-166 [4]**
Electron beam energy	~10 MeV	~50 GeV
Electron beam polarization	85 %	Un-polarized
Photon production	Bremsstrahlung	Synchrotron
Converter target	Tungsten foil	Tungsten foil
Positron polarization	40% (simulated)	80% (measured)

SIMULATIONS

To study the proposed concept, particle interactions in a material are simulated with GEANT4 [8]. In particular, we expect the production of particles to follow the bremsstrahlung and pair creation cross-sections which are respectively given in [9] and [10]. One consideration is the incident electron energy. The electromagnetic shower is larger for high energy electrons in a high Z (atomic number) medium and photons created with higher energy are more likely to produce positrons, however, the energy distribution of bremsstrahlung photons decrease as $1/E_\gamma$. Another consideration is the transfer of polarization in an electromagnetic shower, also implemented in

GEANT4 [11]. Using Stokes parameters [5], each particle is followed, step-by-step, including the polarization transfer for bremsstrahlung and pair creation calculated by Olsen and Maximon [2], as well as depolarization due to bremsstrahlung, ionization, Compton and Moller/Bhabha scattering. For these processes, the polarization transfer is determined [3] as a function of the particle kinematics. These kinematic quantities are also combined into two parameters that determine the Coulomb and screening corrections of the cross-section and polarization transfer coefficients. These corrections to the Born approximation have been evaluated for the ultra-relativistic and small angle limits, which are known to be inaccurate at small energy [12]. In particular for the JLab concept, where energy distribution of the photon is near the pair-creation threshold, it is legitimate to question the effects of the mass-less electron and Lorentz focusing approximations. This problem is presently under study in view of an improved implementation in GEANT4. In the context of these simulations, calculations are limited to an ultra-relativistic Coulomb correction [13] and to the complete screening approximation, which presently appears in GEANT4 as the prescription in the considered low energy range.

Finally, GEANT4 was used to simulate the energy, yield and polarization distributions for a CEBAF electron beam with total energy of E = 5 MeV, longitudinal polarization P = 85% and current I = 1 mA. The target used in the simulation is a 250 μm thick tungsten foil. A contour plot of the distribution is shown in Fig. 1 where the positron energy is normalized to incident electron beam energy. In this simulation, first moments of the distribution yield a positron polarization of 43% and energy of 1.5 MeV.

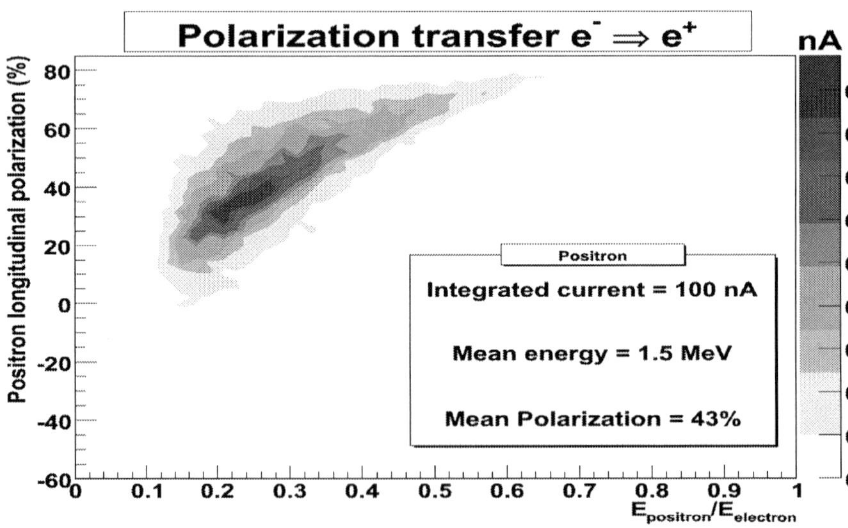

FIGURE 1. GEANT4 positron distribution of an electron beam E = 5 MeV, P = 85% and I = 1 mA incident on a 250 μm thick tungsten target foil.

TEST OF CONCEPT

An experiment is planned to test the proposed concept, making use of the existing CEBAF electron photo-injector. A highly polarized electron beam (>80%) is produced using a strained superlattice GaAs/GaAsP photocathode inside a 100 kV DC high voltage electron gun. Electron bunches are emitted from the photocathode using a high-power fiber-based laser pulsing synchronously (1497 MHz) with the accelerating frequency. The low energy (100 keV) electrons are bunched, captured and accelerated to intermediate energy (<10 MeV) with small emittance ($\varepsilon_{n,rms} < 1$ mm-mrad) and low energy spread ($\delta E/E < 2 \times 10^{-5}$). At this point the electron beam may be deflected to a Mott polarimeter to measure the electron polarization, a diagnostics beam line to measure the electron beam momentum (energy) or transverse profile, or toward the main accelerator. To test the polarized positron concept a new beam line is required containing the conversion target and means to characterize the positron distribution, limit the background and dump the primary electron beam.

The converter target is an important part of the positron source. The target should have a high atomic number (Z) to be more efficient for the production of an electromagnetic shower, but have sufficiently high thermal conductivity and melting point to dissipate deposited beam energy loss. Tungsten (Z = 74) is a good choice with a high melting point (3695 K) and can be made in a thin foil. A thinner target reduces positron yield, however, collection becomes easier because the electromagnetic shower is limited. As an example, the total integrated positron yield and deposited power versus target thickness resulting from an incident 5 MeV electron beam with current of 1 mA is shown in Fig. 2. The optimum positron yield ($Ne^+/Ne^- \sim 10^{-4}$) occurs at a target thickness of 0.5 mm, corresponding to considerable power deposited in the target. To initially avoid high power issues a low duty factor pulsed electron beam may be used, however, optimization of target thickness, target cooling, positron yield and polarization are ultimately to be considered in any practical source operating a high average current.

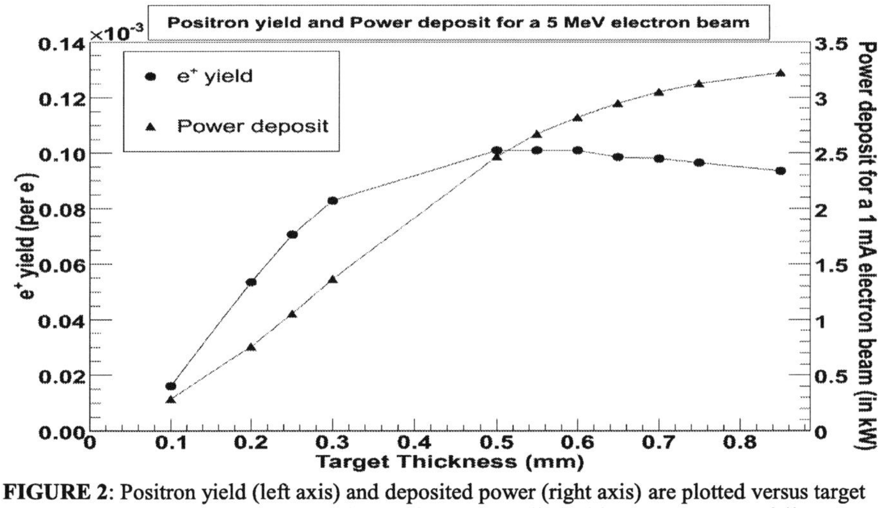

FIGURE 2: Positron yield (left axis) and deposited power (right axis) are plotted versus target thickness for a 5 MeV, 1 mA electron beam normally incident on a tungsten foil.

Ultimately, the shower of particles exiting the conversion target must be separated and isolated using a combination of spectrometer magnets and collimators to be analyzed and counted. Of interest are testing the positron and photon polarization when compared to the theory and simulation routines used by GEANT4, particularly at low energy and with respect to the aforementioned Coulomb and screening corrections. The intent is to measure the positron and photon polarization using a Compton transmission polarimeter [14].

SUMMARY & OUTLOOK

A novel concept for producing polarized positrons at Jefferson Lab using the CEBAF polarized electron photo-injector is presented. The concept relies on transferring the electron beam spin polarization, via bremsstrahlung and pair production in a conversion target, to the positrons. Initial GEANT4 simulations indicate a positron polarization of 40% should be accessible with the present CEBAF photo-injector. An experiment to test this concept, benchmark modeling tools and collect valuable technical information for the design of an actual source is planned.

REFERENCES

1. E. Voutier, *Proc. of the XVIIIth Symposium on Spin Physics*, Charlottesville (Virginia, USA), October 6-11, 2008.
2. H.A. Olsen, L.C. Maximon, *Phys. Rev.* **114** (1959) 887.
3. V.E Balakin and A. A. Mikhailichenko, in *Budker Institute of Nuclear Physics Report No. BINP 79-85* (1979).
4. P. Schuler et al, in *Phys. Rev. Lett.* **100** (2008) 210801.
5. W.H. McMaster, *Rev. Mod. Phys.* **33** (1961) 8.
6. J. Grames et al., in *Proc. of the 2007 Particle Accelerator Conference, THPMS064,* 3130.
7. S. Golge, in *Proc. of the 2007 Particle Accelerator Conference, THPMS067,* 3133.
8. S. Agnostinelli et al., Nucl. Inst. Meth. **A 506** (2003) 250.
9. H.W. Koch, J.W. Motz, *Rev. Mod. Phys.* **31** (1959) 920.
10. J.W. Motz, H.A. Olsen, H.W. Koch, *Rev. Mod. Phys.* **41** (1969) 581.
11. R. Dollan et al., *Nucl. Inst. Meth.* **A 559** (2006) 185.
12. E. Borie, *Nuo. Cim.* **A 11** (1972) 969.
13. H. Davies, H.A. Bethe, L.C. Maximom, *Phys. Rev.* **93** (1954) 788.
14. H. Schopper, *Nucl. Instr.* **3** (1958) 158.

Workshop on Sources of Polarized Electrons and High Brightness Electron Beams	
7:30-8:30	Continental breakfast and Registration
Introduction (chair: M. Poelker)	
8:30-8:45 8:45-9:15	Welcome by Larry Cardman, Dir. Physics Division Polarized electron beams: opening new doors to physics (Gordon Cates, Univ. of Virginia)
Polarized Guns/Programs (chair: W. Hillert)	
9:15-9:45 9:45-10:15	Status report of the new Darmstadt polarized electron injector (Y. Poltoratska, Darmstadt Univ.) Status of 200keV beam operations at Nagoya University (Masahiro Yamamoto, Nagoya Univ.)
10:15-10:30	Coffee break
10:30-11:00 11:00-11:30 11:30-12:00	Recent polarized photocathode R&D at SLAC and future plans (Feng Zhou, SLAC) High Intensity Polarized Electron Gun Studies at MIT-Bates (E. Tsentalovich) Polarized Electron Beams from a Rubidium Spin Filter (Tim Gay, Univ. Nebraska)
12:00-13:30	Lunch on own, Option: JLab cafeteria adjacent to auditorium
Photocathodes (chair: J. Grames)	
13:30-14:00 14:00-14:30 14:30-15:00 15:00-15:30	Atomic hydrogen cleaning of superlattice cathodes (V. Tioukine, MAMI) Atomic Hydrogen Cleaning at TSR Ring Heidelberg (A Terekhov and D. Orlov) Ultra cold photoelectron beams for ion storage rings (D. Orlov, MPI Heidelberg) Optimization of Semiconductor Superlattice for Spin-Polarized Electron Source (L. Gerchikov, St Petersburg State Polytechnic University)
15:30-15:45	Coffee Break
15:45-16:15 16:15-16:45 16:45-17:15 17:15-17:45 17:45-18:00	Study of the Activated GaAs Surface for Application as an Electron Source in Particle Accelerators (N. Chanlek, STFC Daresbury Laboratory) Super-high brightness and high-spin-polarization electron source based on a novel transmission-type GaAs/GaAsP strained superlattice - Defects and polarization (Toru Ujihara, Nagoya Univ.) Recent Progress toward Robust Photocathodes (Gregory Mulhollan, Saxet Surface Science) Polarized Photocathode Development at SVT Associates (Aaron Moy, SVTA) K2CsSb Photocathodes (J. Smedley, BNL)
18:00-20:00	Reception at CEBAF Center
20:00	Socialize at City Center, Location to be announced

Workshop on Sources of Polarized Electrons and High Brightness Electron Beams	
8:00-8:30	Continental breakfast and Registration
DC High Voltage Guns - Unpolarized (chair: B. Dunham)	
8:30-9:00	Status Jefferson Lab FEL High Voltage Guns (Carlos Hernandez-Garcia, JLab)
9:00-9:30	Status Cornell High Voltage Gun (Karl Smolenski, Cornell Univ.)
9:30-10:00	Status of the ALICE Energy-Recovery Linac (L. Jones, STFC Daresbury Laboratory)
10:00-10:15	Coffee Break
10:15-10:45	Photocathode Preparation System for the ALICE Photoinjector Gun, (K.J. Middleman, STFC Daresbury Laboratory)
10:45-11:15	Development of an electron gun for an ERL based light source in Japan (Nobuyuki Nishimori, ERL Development Group, Japan Atomic Energy Agency
High Voltage	
11:15-11:45	KSI's Cross Insulated Core Transformer Technology (Uwe Uhmeyer, Kaiser Systems, Inc.)
11:45-13:30	Lunch Provided, CEBAF Center Room F113 Option1 – for interested participants: Lunch discussion concerning high voltage, insulators, field emission, etc. Option2 – tour of CEBAF injector and FEL Gun Test Stand. *Sign-up required.*
RF Guns (chair: K. Aulenbacher)	
13:30-14:00	Low Emittance Gun for XFEL Application (Christopher Gough, Paul Scherrer Institute)
14:00-14:30	Ion Bombardment in RF Guns: Analytical Approach (E. Pozdeyev, BNL)
14:30-15:00	High-Fidelity RF Gun Simulations with the Parallel 3D Finite Element Particle-In-Cell Code Pic3P (A. Candel, SLAC)
15:00-15:15	Coffee Break
15:15-15:45	Polarized Electron PWT Photoinjectors (David Yu, DULY)
15:45-16:15	15:45-16:15 The Superconducting RF Photoinjector at Elbe – First Operational Experience (J. Teichert, Forschungszentrum Dresden-Rossendorf, FZD)
16:15-16:45	16:15-16:45 The BNL Polarized SRF gun (J. Kewisch, BNL)
17:00-18:00	Poster Session at CEBAF Center Atrium
18:00-20:00	Conference Dinner at CEBAF Center Atrium
20:00-	Socialize at City Center, Location to be announced

Workshop on Sources of Polarized Electrons and High Brightness Electron Beams	
8:00-8:30	Continental breakfast
RF Guns (cont.)	
8:30-9:00	A Polarized SRF Gun for the ILC (D. Holmes, Advanced Energy Systems)
9:00-9:30	Status of the 1.3 GHz Photoinjector for GaAs testing (A. Burrill, BNL)
Lasers (chair: M. Poelker)	
9:30-10:15	Status of the Cornell High Power Fiber-based Drive Laser (Dimitre Gueorguiev Ouzounov and Heng Li, Cornell Univ.)
10:15-10:30	Coffee Break
Vacuum	
10:30-11:00	Vacuum Considerations for GaAs Photoelectron Guns (C. Sinclair, Cornell Univ.)
11:00-11:30	NEG Pump Technology (Paolo Manini, SAES Getters SpA)
11:30-13:30	Lunch Provided, CEBAF Center Room F113 **Option1** – for interested participants: Lunch discussion concerning photocathode issues **Option2** – tour of CEBAF injector and FEL Gun Test Stand. *Sign-up required.*
Other Applications of Polarized Electrons (chair: R. Suleiman)	
13:30-14:00	High brightness and high polarization PES for SPLEEM and/or High brightness PES with transmission photocathode (T. Nakanishi, Nagoya Univ.)
Parity Violation Experiments	
14:00-14:30	Sources of Helicity Correlated Beam Asymmetries (K. Paschke, UVa)
Polarimetry	
14:30-15:00	Ab initio calculation of effective Sherman function in MeV Mott scattering (K. Aulenbacher, MAMI)
15:00-15:30	Compton Polarimeter at Bonn (W. Hillert, Univ. Bonn)
15:30-15:45	Coffee Break
Beam Modeling	
15:45-16:15	Benchmark studies among own and commercial beam tracking codes on fundamental conditions (Naoto Yamamoto, Nagoya Univ.)
16:15-16:45	3D Modeling of the ALICE Photoinjector Upgrade (J.W. McKenzie, STFC Daresbury Laboratory)
Closeout	

PESP Posters

- Ultrahigh Vacuum Ion Pump Power Supply with Sensitive Current Monitoring Capability (J. Hansknecht and P. Adderley, JLab)

- Fast and Inexpensive Pockels Cell Flipping (J. Hansknecht, JLab)

- Efficiency Measurements of the Nebraska Retarding Potential Mott Polarimeter (J. McCarter, JLab/UVa)

- Electron Spin Precession at CEBAF (Douglas Higinbotham, Marie-Isabelle Holdrinet, JLab)

- High Brightness Electron Source Using a Superlattice Photocathode with Long Life NEA-Surface (Tomohiro Nishitani, RIKEN)

- Photoemission Spectroscopic Study of Cesium Telluride Thin Film Photocathode (Harue Sugiyama, Synchrotron Light Application Center, Saga Univ., Japan)

- High Brightness and High Polarization Electron Source Using Transmission Photocathode (Naoto Yamamoto, Nagoya Univ.)

- New Transmission-Type Photocathode Structure Based on Strain-Compensated Superlattice (Xiuguang Jin, Nagoya Univ.)

- Cooled Transmission-Mode NEA- Photocathode with Band-Graded Active Layer for High – Brightness Electron Source (L. Jones, STFC Daresbury Laboratory, with A. Terekhov, Institute of Semiconductor Physics, Novosibirsk)

- Measurements of the Rossendorf SRF-Gun Cavity Characteristics (Andre Arnold, Forschungszentrum Dresden Rossendorf, FZD)

- Polarimetry and Planned Experiments at the Superconducting Darmstadt Electron Linac S-DALINAC (Christian Eckardt, Darmstadt Univ.)

- Hall C Compton Polarimeter with RF Pulsed Green Light (Gaskell/Chuyu, JLab)

- Surface Analysis of Damaged Superlattice Photocathode (M. Stutzman, JLab)

- Field Emission Measurements from Cathodes Electrodes Prepared Using Electro and Chemical Polishing (Ken Surles-Law, JLab)

- Polarized Positron Source at Jefferson Lab (J. Dumas, JLab)

- Benchmarking PARMELA against Beam-Based Measurements at the CEBAF Photoinjector (Ashwini Jayaprakash, K. Surles-Law, Y. Zhang, Old Dominion University and Jefferson Lab)

- Hall A Compton Polarimeter – DC Green Light and Optical Storage Cavity (Abdurahim Rakhman, JLab)

PESP 2008
LIST OF PARTICIPANTS

Adderley , Philip
Jefferson Lab
12050 Jefferson Avenue
Suite 500
Newport News, VA 23606
adderley@jlab.org

Arnold , Andre
Forschungszentrum Dresden
Rossendorf
Bautzner Landstraße 128
Dresden, Saxony 01314
Germany
a.arnold@fzd.de

Aulenbacher , Kurt
Institut für Kernphysik, Universität
Mainz
J.J. Becherweg 45
Mainz, RLP 55099
Federal Republik of Germany
aulenbac@kph.uni-mainz.de

Balling , Andreas
University of Bonn
Nussallee 12
Bonn, NRW 53115
Germany
balling@physik.uni-bonn.de

Bazarov , Ivan
Cornell University
373 Wilson Lab
Ithaca, NY 14853
ib38@cornell.edu

Benesch , Jay
Jefferson Lab
12000 Jefferson Avenue
Suite 19
Newport News, VA 23606
benesch@jlab.org

Benson , Stephen
Jefferson Lab
12000 Jefferson Avenue
Suite 19
Newport News, VA 23606
felman@jlab.org

Biallas, George
Jefferson Lab
12050 Jefferson Avenue
Suite 402
Newport News, VA 23606
biallas@jlab.org

Bullard, Don
Jefferson Lab
12000 Jefferson Avenue
Suite 17
Newport News, VA 23606
bullard@jlab.org

Burrill, Andrew
Brookhaven National Laboratory
MS 911A
Upton, NY 11973
aburrill@bnl.gov

Burrows, Ian
STFC Daresbury Laboratory
Keckwick Lane
Daresbury, Warrington WA4 4AD
United Kingdom
i.burrows@dl.ac.uk

Candel, Arno
SLAC National Accelerator Laboratory
2575 Sand Hill Road
Menlo Park, CA 94025
candel@slac.stanford.edu

Chanlek, Narong
STFC Daresbury Laboratory
Keckwick Lane
Daresbury, Warrington WA4 4AD
United Kingdom
n.chanlek@dl.ac.uk

Clendenin, James
SLAC National Accelerator Laboratory
2575 Sand Hill Rd.
Menlo Park, CA 94025
clen@slac.stanford.edu

Comer, Amy
Jefferson Lab
12000 Jefferson Avenue
Suite 19
Newport News, VA 23606
acomer@jlab.org

Dabiran, Amir
SVT Associates, Inc.
7620 Executive Drive
Eden Prarie, MN 55344
dabiran@svta.com

Dreiling, Joan
University of Nebraska
115 Ferguson Hall
Lincoln, NE 68588
jmdreiling2@gmail.com

Dumas, Jonathan
Jefferson Lab
12050 Jefferson Avenue
Suite 500
Newport News, VA 23606
dumas@jlab.org

Dunham, Bruce
Cornell University
373 Wilson Lab
Ithica, NY 14853
bmd29@cornell.edu

Eckardt, Christian
TU Darmstadt
Schlossgartenstrasse 9
Darmsradt 64289
Germany
eckardt@ikp.tu-darmstadt.de

Farkhondeh, Manouchehr
DOE Office of Nuclear Physics
SC-26.2, Germantown Building
Washington, DC 208585
manouchehr.farkhondeh@science.doe.gov

Gay, Timothy
University of Nebraska
116 Brace Lab
Lincoln, NE 68588
tgay1@unl.edu

Gerchikov, Leonid
St. Petersburg State Polytechnic University
Polytechnicheskaya 29
St. Petersburg 197373
Russia
lgerchikov@rambler.ru

Gough, Christopher
Paul Scherrer Institute
WSLA-015
Paul Scherrer Institut
Villigen AG 5232
Switzerland
christopher.gough@psi.ch

Grames, Joseph
Jefferson Lab
12050 Jefferson Avenue
Suite 500
Newport News, VA 23606
grames@jlab.org

Gulliford, Colwyn
Cornell University
610 Mitchell St
Ithaca, NY 14850
cg248@cornell.edu

Hernandez-Garcia, Carlos
Jefferson Lab
12000 Jefferson Avenue
Suite 19
Newport News, VA 23606
chgarcia@jlab.org

Hannon, Fay
Jefferson Lab
12000 Jefferson Ave.
Newport News, VA 23606
fhannon@jlab.org

Hansknecht, John
Jefferson Lab
12050 Jefferson Avenue
Suite 500
Newport News, VA 23606
hansknec@jlab.org

Higinbotham, Douglas
Jefferson Lab
12000 Jefferson Avenue
Suite 4
Newport News, VA 23606
doug@jlab.org

Hillert, Wolfgang
University of Bonn
Physics Institute
Nussallee 12
Bonn, NRW 53115
Germany
hillert@physik.uni-bonn.de

Hofler, Alicia
Jefferson Lab
12050 Jefferson Avenue
Suite 704
Newport News, VA 23606
hofler@jlab.org

Holmes, Douglas
Advanced Energy Systems
27 Industrial Blvd
Unit E
Medford, NY 11763
doug_holmes@mail.aesys.net

Jayaprakash, Ashwini
Old Dominion University
12000 Jefferson Avenue
Newport News, VA 23606
ashwini@jlab.org

Jones, Lee
STFC Daresbury Laboratory
The Cockcroft Institute, CCS 07
Keckwick Lane, Daresbury
Warrington, Cheshire WA4 4AD
United Kingdom
l.b.jones@dl.ac.uk

Kewisch, Jorg
Brookhaven National Lab
Building 911
Upton, NY 11973
jorg@bnl.gov

Klopf, Michael
Jefferson Lab
12000 Jefferson Avenue
Suite 19
Newport News, VA 23606
klopf@jlab.org

Li, Heng
Cornell University
214 Wilson Lab
Ithica, NY 14853
hl344@cornell.edu

Manini, Paolo
Saes Getters Spa
Viale Italia 77
Lainate 20020
Italy
paolo_manini@saes-group.com

McCarter, James
Jefferson Lab
12000 Jefferson Ave.
Newport News, VA 23606
mccarter@jlab.org

McKenzie, Julian
STFC Daresbury Laboratory
Keckwick Lane
Warrington, Cheshire WA4 4AD
United Kingdom
j.w.mckenzie@dl.ac.uk

Middleman, Keith
STFC Daresbury Laboratory
Keckwick Lane
Warrington, Cheshire WA4 4AD
United Kingdom
k.j.middleman@dl.ac.uk

Moy, Aaron
SVT Associates
7620 Executive Drive
Eden Prairie, MN 55344
moy@svta.com

Mulhollan, Gregory
Saxet Surface Science
3913 Todd Lane
Suite 303
Austin, TX 78744
mulhollan@saxetsurfacescience.com

Murcek, Petr
Forschungszentrum Dresden-Rossendorf
P.O. Box 510119
Dresden 01314
Germany
p.murcek@fzd.de

Nakanishi, Tsutomu
Nagoya University
Department of Physics, Graduate School of Science
Nagoya, Aichi 464-8602
Japan
nakanisi@spin.phys.nagoya-u.ac.jp

Neil, George
Jefferson Lab
12000 Jefferson Ave.
Newport News, VA 23606
neil@jlab.org

Nishimori, Nobuyuki
JAEA
2-4 Shirakata Shirane, Tokai-mura, Naka
Ibaraki 319-1195
Japan
nishimori.nobuyuki@jaea.go.jp

Nishitani, Tomohiro
RIKEN
Hirosawa 2-14
Wako, Saitama 351-0198
JAPAN
tnishitani@riken.jp

Orlov, Dmitry
MPI-K
Saupfercheckweg 1
69117 Heidelberg
Germany
orlov@mpi-hd.mpg.de

Ouzounov, Dimitre
LEPP, Cornell University
226A Wilson Lab
Ithaca, NY 14853
dgo4@cornell.edu

Paff, John
Spectra-Mat, Inc
100 Westgate Dr
Watsonville, CA 95076
john_paff@saes-group.com

Paschke, Kent
University of Virginia
Department of Physics
PO Box 400714
Charlottesville, VA 22904
paschke@virginia.edu

Poelker, Matt
12050 Jefferson Avenue
Suite 500
Newport News, VA 23606
poelker@jlab.org

Poltoratska, Yuliya
Darmstadt University of Technology
Institute for Nuclear Physics
Schlossgartenstrasse 9
64289 Darmstadt
Germany
yuliya@ikp.tu-darmstadt.de

Pozdeyev, Eduard
Brookhaven National Lab
Bldg. 911
Upton, NY 11973
pozdeyev@bnl.gov

Rao, Triveni
Brookhaven National Lab
Bldg. 911
Upton, NY 11973
triveni@bnl.gov

Silwal, Rupesh
University of Virginia
Department of Physics
PO Box 400714
Charlottesville, VA 22904
rs3te@virginia.edu

Sinclair, Charles
Cornell University (retired)
230 Wilson Lab
Ithaca, NY 14853
cks26@cornell.edu

Smedley, John
Brookhaven National Laboratory
Bldg. 911
Upton, NY 11973
smedley@bnl.gov

Smolenski, Karl
Cornell University
230 Wilson Lab
Ithaca, NY 14853
karl.smolenski@cornell.edu

Stoltz, Peter
Tech-X Corporation
5621 Arapahoe Ave
Boulder, CO 80303
pstoltz@txcorp.com

Stutzman, Marcy
Jefferson Lab
12050 Jefferson Avenue
Suite 500
Newport News, VA 23606
marcy@jlab.org

Sugiyama, Harue
Saga University
1 Honjo
Saga 840-8502
Japan
harues@hep.phys.saga-u.ac.jp

Suleiman, Riad
Jefferson Lab
12050 Jefferson Avenue
Suite 500
Newport News, VA 23606
suleiman@jlab.org

Surles-Law, Ken
Jefferson Lab
12050 Jefferson Avenue
Suite 500
Newport News, VA 23606
law@jlab.org

Teichert, Jochen
Forschungszentrum Dresden-
Rossendorf
P.O. Box 510119
Dresden 01314
Germany
j.teichert@fzd.de

Tennant, Chris
Jefferson Lab
12000 Jefferson Avenue
Suite 19
Newport News, VA 23606
tennant@jlab.org

Terekhov, Alexander
Institute of Semiconductor Physics,
Russian Academy of Sciences,
Siberian Branch
Lavrentev ave., 13,
Novosibirsk 630090
Russian Federation
terek@thermo.isp.nsc.ru

Trantham, Kenneth
Fort Hays State University
600 Park Street
Physics Department
Hays, KS 67601
ktrantha@fhsu.edu

Tsentalovich, Evgeni
MIT-Bates
21 Manning Rd
Middleton, MA 01949
evgeni@mit.eduj

Tyukin, Valery
Inst. of Nuclear Physics
J.-J. Becherweg 45
Mainz 55099
Germany
tioukine@kph.uni-mainz.de

Uhmeyer, Uwe
Kaiser Systems, Inc
126 Sohier Road
Beverly, MA 01915
uuhmeyer@kaisersystems.com

Ujihara, Toru
Nagoya University
Furo-cho, Chikusaku
Nagoya, Aichi 4648603
Japan
ujihara@numse.nagoya-u.ac.jp

Voutier, Eric
LPSC/IN2P3/CNRS
Laboratoire de Physique Subatomique
et de Cosmologie
53 rue des Martyrs
Grenoble 38026
France
voutier@lpsc.in2p3.fr

Walker, Richard
Jefferson Lab
12000 Jeffereson Avenue
Suite 19
Newport News, VA 23606
richard@jlab.org

Williams, Gwyn
Jefferson Lab
12000 Jefferson Ave.
Newport News, VA 23606
gwyn@jlab.org

Xiuguang, Jin
Nagoya University
Furo-cho Chikusa-ku
Nagoya 464-8603
Japan
kim@mercury.numse.nagoya-u.ac.jp

Yamamoto, Masahiro
Nagoya University
Furo-cho Chikusa-ku
Nagoya 464-8603
Japan
yamamoto@spin.phys.nagoya-u.ac.jp

Yamamoto, Naoto
Nagoya University
Furo-cho Chikusa-ku
Nagoya 464-8603
Japan
naoto@nagoya-u.jp

Yu, David
DULY Research Inc.
1912 MacArthur Street
Rancho Palos Verdes, CA 90275
davidyu@pacbell.net

Zhang, Shukui
Jefferson Lab
12000 Jefferson Avenue
Suite 19
Newport News, VA 23606
shukui@jlab.org

Zhang, Yuhong
Jefferson Lab
12050 Jefferson Avenue
Suite 704
Newport News, VA 23606
yzhang@jlab.org

Zhou, Feng
SLAC National Accelerator Laboratory
2575 Sand Hill road
Menlo Park, CA 94025
zhoufeng@slac.stanford.edu

AUTHOR INDEX

A

B

C

D

E

F

G

H

I

J

K

L

M

N

O

P

T

U

V

W

X

Y

Z